D1074167

The Classification of Quasithin Groups

II. Main Theorems: The Classification of Simple QTKE-groups

Mathematical
Surveys
and
Monographs

Volume 112

The Classification of Quasithin Groups

II. Main Theorems: The Classification of Simple QTKE-groups

Michael Aschbacher
Stephen D. Smith

American Mathematical Society

EDITORIAL COMMITTEE

2000 *Mathematics Subject Classification.* Primary 20D05; Secondary 20C20.

For additional information and updates on this book, visit
www.ams.org/bookpages/surv-112

Library of Congress Cataloging-in-Publication Data

Aschbacher, Michael, 1944–
    The classification of quasithin groups / Michael Aschbacher, Stephen D. Smith.
        p. cm. — (Mathematical surveys and monographs, ISSN 0076-5376 ; v. 111–112)
        Contents: 1. Structure of strongly quasithin $\mathcal{K}$-groups — 2. Main theorems : the classification of simple QTKE-groups.
        Includes bibliographical references and index.
        ISBN 0-8218-3410-X (Volume 111); ISBN 0-8218-3411-8 (Volume 112)
        1. Finite simple groups—Classification.   I. Smith, Stephen D., 1948–   II. Title.   III. Mathematical surveys and monographs ; 111–112.

QA177.A8   2004
512′.2—dc22

2004054548

To Pam and Judy

Contents of Volumes I and II

Preface   xiii

Volume I: Structure of strongly quasithin $\mathcal{K}$-groups   1

Introduction to Volume I   3
- 0.1.   Statement of Main Results   3
- 0.2.   An overview of Volume I   5
- 0.3.   Basic results on finite groups   7
- 0.4.   Semisimple quasithin and strongly quasithin $\mathcal{K}$-groups   7
- 0.5.   The structure of SQTK-groups   7
- 0.6.   Thompson factorization and related notions   8
- 0.7.   Minimal parabolics   10
- 0.8.   Pushing up   10
- 0.9.   Weak closure   11
- 0.10.   The amalgam method   11
- 0.11.   Properties of $\mathcal{K}$-groups   12
- 0.12.   Recognition theorems   13
- 0.13.   Background References   15

Chapter A.   Elementary group theory and the known quasithin groups   19
- A.1.   Some standard elementary results   19
- A.2.   The list of quasithin $\mathcal{K}$-groups: Theorems A, B, and C   32
- A.3.   A structure theory for Strongly Quasithin $\mathcal{K}$-groups   41
- A.4.   Signalizers for groups with $\mathbf{X} = \mathbf{O}^2(\mathbf{X})$ 56
- A.5.   An ordering on $\mathcal{M}(T)$ 61
- A.6.   A group-order estimate   64

Chapter B.   Basic results related to Failure of Factorization   67
- B.1.   Representations and FF-modules   67
- B.2.   Basic Failure of Factorization   74
- B.3.   The permutation module for A_n and its FF*-offenders   83
- B.4. $\mathbf{F}_2$-representations with small values of q or $\hat{q}$ 85
- B.5.   FF-modules for SQTK-groups   98
- B.6.   Minimal parabolics   112
- B.7.   Chapter appendix: Some details from the literature   118

Chapter C.   Pushing-up in SQTK-groups   121
- C.1.   Blocks and the most basic results on pushing-up   121
- C.2.   More general pushing up in SQTK-groups   143
- C.3.   Pushing up in nonconstrained 2-locals   148

C.4.   Pushing up in constrained 2-locals   151
C.5.   Finding a common normal subgroup   154
C.6.   Some further pushing up theorems   164

Chapter D.   The qrc-lemma and modules with $\hat{q} \leq 2$ 171
D.1.   Stellmacher's qrc-Lemma   171
D.2.   Properties of q and $\hat{q}$: $\mathcal{R}(G,V)$ and $\mathcal{Q}(G,V)$ 177
D.3.   Modules with $\hat{\mathbf{q}} \leq \mathbf{2}$ 192

Chapter E.   Generation and weak closure   209
E.1. $\mathcal{E}$-generation and the parameter $\mathbf{n(G)}$ 209
E.2.   Minimal parabolics under the SQTK-hypothesis   215
E.3.   Weak Closure   230
E.4.   Values of $\mathbf{a}$ for $\mathbf{F_2}$-representations of SQTK-groups.   240
E.5.   Weak closure and higher Thompson subgroups   242
E.6.   Lower bounds on $\mathbf{r(G,V)}$ 244

Chapter F.   Weak BN-pairs and amalgams   259
F.1.   Weak BN-pairs of rank 2   259
F.2.   Amalgams, equivalences, and automorphisms   264
F.3.   Paths in rank-2 amalgams   269
F.4.   Controlling completions of Lie amalgams   273
F.5.   Identifying $\mathbf{L_4(3)}$ via its $\mathbf{U_4(2)}$-amalgam   299
F.6.   Goldschmidt triples   304
F.7.   Coset geometries and amalgam methodology   310
F.8.   Coset geometries with $\mathbf{b > 2}$ 315
F.9.   Coset geometries with $\mathbf{b > 2}$ and $\mathbf{m(V_1) = 1}$ 317

Chapter G.   Various representation-theoretic lemmas   327
G.1.   Characterizing direct sums of natural $\mathbf{SL_n(F_{2^e})}$-modules   327
G.2.   Almost-special groups   332
G.3.   Some groups generated by transvections   337
G.4.   Some subgroups of $\mathbf{Sp_4(2^n)}$ 338
G.5. $\mathbf{F_2}$-modules for $\mathbf{A_6}$ 342
G.6.   Modules with $\mathbf{m(G,V) \leq 2}$ 345
G.7.   Small-degree representations for some SQTK-groups   346
G.8.   An extension of Thompson's dihedral lemma   349
G.9.   Small-degree representations for more general SQTK-groups   351
G.10. Small-degree representations on extraspecial groups   357
G.11. Representations on extraspecial groups for SQTK-groups   364
G.12. Subgroups of $\mathbf{Sp(V)}$ containing transvections on hyperplanes   370

Chapter H.   Parameters for some modules   377
H.1. $\mathbf{\Omega_4^\epsilon(2^n)}$ on an orthogonal module of dimension $\mathbf{4n}$ $(\mathbf{n > 1})$ 378
H.2. $\mathbf{SU_3(2^n)}$ on a natural $\mathbf{6n}$-dimensional module   378
H.3. $\mathbf{Sz(2^n)}$ on a natural $\mathbf{4n}$-dimensional module   379
H.4. $\mathbf{(S)L_3(2^n)}$ on modules of dimension $\mathbf{6}$ and $\mathbf{9}$ 379
H.5.   7-dimensional permutation modules for $\mathbf{L_3(2)}$ 385
H.6.   The 21-dimensional permutation module for $\mathbf{L_3(2)}$ 386
H.7. $\mathbf{Sp_4(2^n)}$ on natural $\mathbf{4n}$ plus the conjugate $\mathbf{4n^t}$.   388

H.8. $\mathbf{A_7}$ on $\mathbf{4} \oplus \overline{\mathbf{4}}$ 389
H.9. $\mathbf{Aut(L_n(2))}$ on the natural $\mathbf{n}$ plus the dual $\mathbf{n}^*$ 389
H.10. A foreword on Mathieu groups 392
H.11. $\mathbf{M_{12}}$ on its 10-dimensional module 392
H.12. $\mathbf{3M_{22}}$ on its 12-dimensional modules 393
H.13. Preliminaries on the binary code and cocode modules 395
H.14. Some stabilizers in Mathieu groups 396
H.15. The cocode modules for the Mathieu groups 398
H.16. The code modules for the Mathieu groups 402

Chapter I. Statements of some quoted results 407
I.1. Elementary results on cohomology 407
I.2. Results on structure of nonsplit extensions 409
I.3. Balance and 2-components 414
I.4. Recognition Theorems 415
I.5. Characterizations of $L_4(2)$ and $Sp_6(2)$ 418
I.6. Some results on TI-sets 424
I.7. Tightly embedded subgroups 425
I.8. Discussion of certain results from the Bibliography 428

Chapter J. A characterization of the Rudvalis group 431
J.1. Groups of type Ru 431
J.2. Basic properties of groups of type Ru 432
J.3. The order of a group of type Ru 438
J.4. A $^2\mathbf{F_4(2)}$-subgroup 440
J.5. Identifying G as Ru 445

Chapter K. Modules for SQTK-groups with $\hat{q}(G, V) \leq 2$. 451
Notation and overview of the approach 451
K.1. Alternating groups 452
K.2. Groups of Lie type and odd characteristic 453
K.3. Groups of Lie type and characteristic 2 453
K.4. Sporadic groups 457

Bibliography and Index 461

Background References Quoted
(Part 1: also used by GLS) 463

Background References Quoted
(Part 2: used by us but not by GLS) 465

Expository References Mentioned 467

Index 471

Volume II: Main Theorems; the classification of simple QTKE-groups 479

Introduction to Volume II 481
0.1. Statement of Main Results 481

0.2.   Context and History   483
0.3.   An Outline of the Proof of the Main Theorem   487
0.4.   An Outline of the Proof of the Even Type Theorem   495

Part 1.   Structure of QTKE-Groups and the Main Case Division   497

Chapter 1.   Structure and intersection properties of 2-locals   499
1.1.   The collection $\mathcal{H}^e$ 499
1.2.   The set $\mathcal{L}^*(G,T)$ of nonsolvable uniqueness subgroups 503
1.3.   The set $\Xi^*(G,T)$ of solvable uniqueness subgroups of G 508
1.4.   Properties of some uniqueness subgroups   514

Chapter 2.   Classifying the groups with $|\mathcal{M}(T)| = 1$ 517
2.1.   Statement of main result   518
2.2.   Bender groups   518
2.3.   Preliminary analysis of the set Γ_0 521
2.4.   The case where Γ_0^e is nonempty   527
2.5.   Eliminating the shadows with Γ_0^e empty   550

Chapter 3.   Determining the cases for $L \in \mathcal{L}_f^*(G,T)$ 571
3.1.   Common normal subgroups, and the qrc-lemma for QTKE-groups   571
3.2.   The Fundamental Setup, and the case division for $\mathcal{L}_f^*(G,T)$ 578
3.3.   Normalizers of uniqueness groups contain $N_G(T)$ 585

Chapter 4.   Pushing up in QTKE-groups   605
4.1.   Some general machinery for pushing up   605
4.2.   Pushing up in the Fundamental Setup   608
4.3.   Pushing up $L_2(2^n)$ 613
4.4.   Controlling suitable odd locals   619

Part 2.   The treatment of the Generic Case   627

Chapter 5.   The Generic Case: $L_2(2^n)$ in $\mathcal{L}_f$ and $n(H) > 1$ 629
5.1.   Preliminary analysis of the $L_2(2^n)$ case   630
5.2.   Using weak BN-pairs and the Green Book   646
5.3.   Identifying rank 2 Lie-type groups   658

Chapter 6.   Reducing $\mathbf{L_2(2^n)}$ to $\mathbf{n = 2}$ and V orthogonal   663
6.1.   Reducing $\mathbf{L_2(2^n)}$ to $\mathbf{L_2(4)}$ 663
6.2.   Identifying M_{22} via $L_2(4)$ on the natural module   679

Part 3.   Modules which are not FF-modules   693

Chapter 7.   Eliminating cases corresponding to no shadow   695
7.1.   The cases which must be treated in this part   696
7.2.   Parameters for the representations   697
7.3.   Bounds on w 698
7.4.   Improved lower bounds for r 699
7.5.   Eliminating most cases other than shadows   700
7.6.   Final elimination of $\mathbf{L_3(2)}$ on $\mathbf{3 \oplus \bar{3}}$ 701
7.7.   mini-Appendix: $\mathbf{r > 2}$ for $\mathbf{L_3(2).2}$ on $\mathbf{3 \oplus \bar{3}}$ 703

Chapter 8.   Eliminating shadows and characterizing the J_4 example   711
  8.1.   Eliminating shadows of the Fischer groups   711
  8.2.   Determining local subgroups, and identifying J_4 714
  8.3.   Eliminating $L_3(2) \wr 2$ on 9   723

Chapter 9.   Eliminating $\Omega_4^+(2^n)$ on its orthogonal module   729
  9.1.   Preliminaries   729
  9.2.   Reducing to $n = 2$ 730
  9.3.   Reducing to $n(H) = 1$ 732
  9.4.   Eliminating $n(H) = 1$ 735

Part 4.   Pairs in the FSU over F_{2^n} for $n > 1$.   739

Chapter 10.   The case $L \in \mathcal{L}_f^*(G,T)$ not normal in M.   741
  10.1.   Preliminaries   741
  10.2.   Weak closure parameters and control of centralizers   742
  10.3.   The final contradiction   755

Chapter 11.   Elimination of $L_3(2^n)$, $Sp_4(2^n)$, and $G_2(2^n)$ for $n > 1$ 759
  11.1.   The subgroups $N_G(V_i)$ for T-invariant subspaces V_i of V   760
  11.2.   Weak-closure parameter values, and $\langle V^{N_G(V_1)} \rangle$ 766
  11.3.   Eliminating the shadow of $L_4(q)$ 770
  11.4.   Eliminating the remaining shadows   775
  11.5.   The final contradiction   778

Part 5.   Groups over F_2   785

Chapter 12.   Larger groups over F_2 in $\mathcal{L}_f^*(G,T)$ 787
  12.1.   A preliminary case: Eliminating $L_n(2)$ on $n \oplus n^*$ 787
  12.2.   Groups over F_2, and the case V a TI-set in G 794
  12.3.   Eliminating A_7 807
  12.4.   Some further reductions   812
  12.5.   Eliminating $L_5(2)$ on the 10-dimensional module   816
  12.6.   Eliminating A_8 on the permutation module   822
  12.7.   The treatment of $\hat{A}_6$ on a 6-dimensional module   838
  12.8.   General techniques for $L_n(2)$ on the natural module   849
  12.9.   The final treatment of $L_n(2)$, $n = 4, 5$, on the natural module   857

Chapter 13.   Mid-size groups over F_2 865
  13.1.   Eliminating $L \in \mathcal{L}_f^*(G,T)$ with $L/O_2(L)$ not quasisimple   865
  13.2.   Some preliminary results on A_5 and A_6 876
  13.3.   Starting mid-sized groups over F_2, and eliminating $U_3(3)$ 884
  13.4.   The treatment of the 5-dimensional module for A_6 896
  13.5.   The treatment of A_5 and A_6 when $\langle V_3^{G_1} \rangle$ is nonabelian   915
  13.6.   Finishing the treatment of A_5 926
  13.7.   Finishing the treatment of A_6 when $\langle V^{G_1} \rangle$ is nonabelian   935
  13.8.   Finishing the treatment of A_6 946
  13.9.   Chapter appendix: Eliminating the A_{10}-configuration   969

Chapter 14. $L_3(2)$ in the FSU, and $L_2(2)$ when $\mathcal{L}_f(G,T)$ is empty   975

14.1.   Preliminary results for the case $\mathcal{L}_{\mathbf{f}}(\mathbf{G}, \mathbf{T})$ empty   975
14.2.   Starting the $\mathbf{L}_2(\mathbf{2})$ case of $\mathcal{L}_{\mathbf{f}}$ empty   981
14.3.   First steps; reducing $\langle \mathbf{V}^{\mathbf{G}_1} \rangle$ nonabelian to extraspecial   989
14.4.   Finishing the treatment of $\langle \mathbf{V}^{\mathbf{G}_1} \rangle$ nonabelian   1005
14.5.   Starting the case $\langle \mathbf{V}^{\mathbf{G}_1} \rangle$ abelian for $\mathbf{L}_3(\mathbf{2})$ and $\mathbf{L}_2(\mathbf{2})$ 1013
14.6.   Eliminating $\mathbf{L}_2(\mathbf{2})$ when $\langle \mathbf{V}^{\mathbf{G}_1} \rangle$ is abelian   1020
14.7.   Finishing $\mathbf{L}_3(\mathbf{2})$ with $\langle \mathbf{V}^{\mathbf{G}_1} \rangle$ abelian   1042
14.8.   The QTKE-groups with $\mathcal{L}_{\mathbf{f}}(\mathbf{G}, \mathbf{T}) \neq \emptyset$ 1078

Part 6.   The case $\mathcal{L}_f(G, T)$ empty   1081

Chapter 15.   The case $\mathcal{L}_{\mathbf{f}}(\mathbf{G}, \mathbf{T}) = \emptyset$ 1083
15.1.   Initial reductions when $\mathcal{L}_{\mathbf{f}}(\mathbf{G}, \mathbf{T})$ is empty   1083
15.2.   Finishing the reduction to $\mathbf{M}_{\mathbf{f}}/\mathbf{C}_{\mathbf{M}_{\mathbf{f}}}(\mathbf{V}(\mathbf{M}_{\mathbf{f}})) \simeq \mathbf{O}_4^+(\mathbf{2})$ 1104
15.3.   The elimination of $\mathbf{M}_{\mathbf{f}}/\mathbf{C}_{\mathbf{M}_{\mathbf{f}}}(\mathbf{V}(\mathbf{M}_{\mathbf{f}})) = \mathbf{S}_3 \text{ wr } \mathbf{Z}_2$ 1120
15.4.   Completing the proof of the Main Theorem   1155

Part 7.   The Even Type Theorem   1167

Chapter 16.   Quasithin groups of even type but not even characteristic   1169
16.1.   Even type groups, and components in centralizers   1169
16.2.   Normality and other properties of components   1173
16.3.   Showing L is standard in G 1177
16.4.   Intersections of $\mathbf{N}_{\mathbf{G}}(\mathbf{L})$ with conjugates of $\mathbf{C}_{\mathbf{G}}(\mathbf{L})$ 1182
16.5.   Identifying $\mathbf{J}_1$, and obtaining the final contradiction   1194

Bibliography and Index   1205

Background References Quoted
            (Part 1: also used by GLS)   1207

Background References Quoted
            (Part 2: used by us but not by GLS)   1209

Expository References Mentioned   1211

Index   1215

Volume II: Main Theorems; the classification of simple QTKE-groups

In Volume II we establish our Main Theorem classifying the simple QTKE-groups. The proof uses machinery from Volume I. Also in chapter 16 we establish the Even Type Theorem, which uses our Main Theorem to provide a classification of the quasithin group satisfying the "even type" hypothesis of the Gorenstein-Lyons-Solomon project [**GLS94**].

Introduction to Volume II

The treatment of the "quasithin groups of even characteristic" is one of the major steps in the Classification of the Finite Simple Groups (for short, the Classification). Geoff Mason announced a classification of a subclass of the quasithin groups in about 1980, but he never published his work, and the preprint he distributed [**Mas**] is incomplete in various ways. In two lengthy volumes, we treat the quasithin groups of even characteristic; in particular we close that gap in the proof of the Classification.

Each volume contains an Introduction discussing its contents. For further background, the reader may also wish to consult the Introduction to Volume I; that volume records and develops the machinery needed to prove our Main Theorem, which classifies the simple quasithin $\mathcal{K}$-groups of even characteristic. Volume II implements the proof of the Main Theorem.

Section 0.1 of this Introduction to Volume II gives the statement of the two main results of the paper, together with a few definitions necessary to state those results. Section 0.2 discusses the role of quasithin groups in the larger context of the Classification; we also compare the hypotheses of the original quasithin problem with those of more recent alternatives, and give some history of the problem. In sections 0.3 and 0.4, we introduce further fundamental concepts and notation, and give an outline of the proofs of our two main theorems.

The Introduction to Volume I describes the references we appeal to during the course of the proof; see section 0.12 on recognition theorems and section 0.13 on Background References.

0.1. Statement of Main Results

We begin by defining the class of groups considered in our Main Theorem, and since the definitions are somewhat technical, we also supply some motivation. For definitions of more basic group-theoretic notation and terminology, the reader is directed to Aschbacher's text [**Asc86a**].

The quasithin groups are the "small" groups in that part of the Classification where the actual examples are primarily the groups of Lie type defined over a field of characteristic 2. We first translate the notion of the "characteristic" of a linear group into the setting of abstract groups: Let G be a finite group, $T \in Syl_2(G)$, and let $\mathcal{M}$ denote the set of maximal 2-local subgroups of G. [1] We define G to be of *even characteristic* if

$$C_M(O_2(M)) \leq O_2(M) \text{ for all } M \in \mathcal{M}(T),$$

---

[1]A *2-local subgroup* is the normalizer of a nonidentity 2-subgroup.

where $\mathcal{M}(T)$ denotes those members of $\mathcal{M}$ containing T. The class of simple groups of even characteristic contains some families in addition to the groups of Lie type in characteristic 2. In particular it is larger than the class of simple groups of characteristic 2-type (discussed in the next section), which played the analogous role in the original proof of the Classification.

The Classification proceeds by induction on the group order. Thus if G is a minimal counterexample to the Classification, then each proper subgroup H of G is a $\mathcal{K}$-*group*; that is, all composition factors of each subgroup of H lie in the set $\mathcal{K}$ of known finite simple groups.

Finally quasithin groups are "small" by a measure of size introduced by Thompson in the N-group paper [**Tho68**]. Define

$$e(G) := \max\{m_p(M) : M \in \mathcal{M} \text{ and } p \text{ is an odd prime}\}$$

where $m_p(M)$ is the p-*rank* of M (namely the maximum rank of an elementary abelian p-subgroup of M). When G is of Lie type in characteristic 2, $e(G)$ is a good abstract approximation of the Lie rank of G. Janko called the groups with $e(G) \leq 1$ "thin groups", leading Gorenstein to define G to be *quasithin* if $e(G) \leq 2$. The groups of Lie type of characteristic 2 and Lie rank 1 or 2 are the "generic" simple quasithin groups of even characteristic.

Define a finite group H to be *strongly quasithin* if $m_p(H) \leq 2$ for all odd primes p. Thus the 2-locals of quasithin groups are strongly quasithin.

We combine the three principal conditions into a single hypothesis:

Main Hypothesis. Define G to be a *QTKE-group* if

(QT) G is quasithin,

(K) all proper subgroups of G are $\mathcal{K}$-groups, and

(E) G is of even characteristic.

We prove:

THEOREM 0.1.1 (Main Theorem). *The finite simple QTKE-groups consist of:*

(1) (Generic case) Groups of Lie type of characteristic 2 *and Lie rank at most* 2, *but* $U_5(q)$ *only for* $q = 4$.

(2) (Certain groups of rank 3 or 4) $L_4(2)$, $L_5(2)$, $Sp_6(2)$.

(3) (Alternating groups:) A_5, A_6, A_8, A_9.

(4) (Lie type of odd characteristic) $L_2(p)$, p *a Mersenne or Fermat prime;* $L_3^\epsilon(3)$, $L_4^\epsilon(3)$, $G_2(3)$.

(5) (sporadic) M_{11}, M_{12}, M_{22}, M_{23}, M_{24}, J_2, J_3, J_4, HS, He, Ru.

We recall that there is an "original" or "first generation" proof of the Classification, made up by and large of work done before 1980; and a "second generation" program in progress, whose aim is to produce a somewhat different and simpler proof of the Classification. The two programs take the same general approach, but often differ in detail. Our work is a part of both efforts.

In particular Gorenstein, Lyons, and Solomon (GLS) are in the midst of a major program to revise and simplify the proof of part of the Classification. We also prove a corollary to our Main Theorem, which supplies a bridge between that result and the GLS program. We now discuss that corollary:

There is yet another way to approach the characterization of the groups of Lie type of characteristic 2. The GLS program requires a classification of quasithin groups which again satisfy (QT) and (K), but instead of condition (E) they impose a more technical condition (see p. 55 of [**GLS94**], and 16.1.1 in this work):

(E') G is of *even type*.

The condition (E') allows certain components [2] in the centralizers of involutions t (including involutions in $Z(T)$, which are not allowed under our hypothesis of even characteristic); but these components can only come from a restricted list. To be precise, a quasithin group G is of even type if:

$(E'1)$ $O(C_G(t)) = 1$ for each involution $t \in G$, and

$(E'2)$ If L is a component of $C_G(t)$ for some involution $t \in G$, then one of the following holds:

(i) $L/O_2(L)$ is of Lie type and in characteristic 2, but L is not $SL_2(q)$, $q = 5, 7, 9$ or $A_8/\mathbf{Z}_2$, and if $L/O_2(L) \cong L_3(4)$ then $O_2(L)$ is elementary abelian.

(ii) $L \cong L_3(3)$ or $L_2(p)$, p a Fermat or Mersenne prime.

(iii) $L/O_2(L)$ is a Mathieu group, J_2, J_4, HS, or Ru.

In order to supply a bridge between our Main Theorem and the GLS program, we also establish (as Theorem 16.5.14):

THEOREM 0.1.2 (Even Type Theorem). *The Janko group J_1 is the only simple group of even type satisfying (QT) and (K) but which is not of even characteristic.*

Since the groups appearing as conclusions to our Main Theorem are in fact of even type, the quasithin simple groups of even type consist of J_1 together with that list of groups.

0.2. Context and History

In this section we discuss the role of quasithin groups in the Classification, focusing on motivation for our basic hypotheses. We also recall some of the history of the quasithin problem. Occasionally we abbreviate 'Classification of the Finite Simple Groups" by CFSG.

0.2.1. Case division according to notions of even or odd "characteristic". The Classification of the Finite Simple Groups proceeds by analyzing the p-local subgroups of an abstract finite simple group G for various primes p. Further for various reasons, which we touch upon later, the 2-local subgroups are preferred.

On the other hand the generic example of a simple group is a group G of Lie type over a field of some prime characteristic p, which is the *characteristic* of that group of Lie type. Such a group G can be realized as a linear group acting on some space V over a finite field of characteristic p, and the local structure of G is visible from this representation. For example if $g \in G$ is a p'-element (i.e., $(|g|, p) = 1$) then g is semisimple (i.e., diagonalizable over some extension field), so its centralizer $C_G(g)$ is well-behaved in that it is essentially the direct product of quasisimple groups of Lie type in characteristic p corresponding to the eigenspaces of g. There are standard methods for exploiting the structure of these *components*. On the other hand, if g is a p-element, then g is unipotent (i.e., all its eigenvalues are 1), so $C_G(g)$ has no components; instead its structure is dominated by the unipotent subgroup

$$F^*(C_G(g)) = O_p(C_G(g))$$

---

[2]See section 31 of [**Asc86a**] for the definition of a *component* of a finite group (namely quasisimple subnormal subgroup), and corresponding properties.

and in particular is more complex, so that this centralizer is more difficult to deal with.

We seek to translate these properties of linear groups, and in particular the notion of "characteristic", into analogous notions for abstract groups. If G is a finite group and p is a prime, G is defined to be of *characteristic p-type* if each p-local subgroup H of G satisfies

$$F^*(H) = O_p(H),$$

or equivalently $C_H(O_p(H)) \leq O_p(H)$. Every group of Lie type in characteristic p is of characteristic p-type; indeed for large p, they are the only examples of p-rank at least 2—though for small primes, there are groups of characteristic p-type which are not of Lie type in characteristic p.

If a simple group G of p-rank at least 3 is "connected" at the prime p (as discussed in the next section) but is not of characteristic p-type, then the centralizer of some element of order p will behave like the centralizer of a semisimple element in a group of Lie type—that is, it will have components, making it easier to analyze. Thus the aim is to find a prime p such that G has a reasonably rich p-local structure, but G is not of characteristic p-type. Recall also that one chooses p to be 2 whenever possible. The original proof of the Classification partitioned the simple groups into two classes: those of characteristic 2-type, and those not of characteristic 2-type; furthermore different techniques were used to analyze the two classes.

In the remainder of this subsection, we'll try to give some insight into how more recent work (done since the original proof of the Classification) has suggested that it is useful and natural to change the boundary of this even/odd partition. We mentioned earlier that in the GLS program, the notion of even type replaces the notion of characteristic 2-type. However for the purpose of dealing with quasithin groups, our notion of even characteristic seems to be more natural than that of even type. Notice that a group of characteristic 2-type *is* of even characteristic, since the former hypothesis requires all 2-locals to be of characteristic 2, while the latter imposes this constraint only on locals containing the Sylow group T. Thus the class of groups of even characteristic is larger than the class of groups of characteristic 2-type, since the 2-locals in the former class are more varied.

In a moment, we will discuss two classes of groups where this extra flexibility is useful. But before doing so, we'll say a word about the influence of these groups and others on our work. In December 1996, Helmut Bender gave a talk at the conference in honor of Bernd Fischer's 60th birthday, in which he suggested approaching classification problems like ours with a list of groups in mind, to serve as a guide to where difficulties are likely to occur. However, that list should include not only the "examples"—the groups which appear in the conclusion of the theorem; it should also include "shadows"—groups not in the conclusion, but whose local structure is very close to that of actual examples, since these configurations of local subgroups will also arise in the analysis, and typically they can be eliminated only with real effort. Thus in our exposition, we try to emphasize not only how the examples arise, but also where the shadows are finally eliminated. Our Index lists occurrences in the proof of examples and shadows.

In particular we must deal with shadows of the following two classes which are QTKE-groups but not simple—since it is hard to recognize *locally* that the groups are not simple.

Two non-simple configurations. Let L be a simple group of Lie type in characteristic 2, and assume either

(a) $G = L\langle t\rangle$ is L extended by an involutory outer automorphism t, or
(b) $G = (L \times L^t)\langle t\rangle$, for some involution t; i.e., G is the wreath product of L by $\mathbf{Z}_2$.

Then G is in fact of even characteristic, but rarely of characteristic 2-type, since $C_G(t)$ usually has a component. However the components of $C_G(t)$ are of Lie type in characteristic 2, so G is also usually of even type. During the proof of the CFSG, groups with the 2-local structure of those in (a) and (b) often arise. Under the original approach, lengthy and difficult computations were required, to reduce to a situation where transfer could be applied to show the group was not simple. In the opinion of GLS (and we agree), the proof should be restructured to avoid these difficulties.

This is achieved in GLS by replacing the old partition into characteristic 2-type/not characteristic 2-type by the partition into even type/odd type, while we achieve it for quasithin groups with the partition into even characteristic/not even characteristic. Locals like those in the two classes of nonsimple groups above are allowed under both the even characteristic hypothesis and the even type hypothesis, but were not allowed under the older characteristic 2-type hypothesis. Thus under the old approach, such groups would be treated in the "odd" case by focusing on the "semisimple" element t—rather artificially, as its order is *not* coprime to the characteristic of its components—and usually at great expense in effort. Under the new approach, such groups arise in the "even" case, where the focus is not on $C_G(t)$.

In the generic situation when G is "large" (see the next subsection for a discussion of size), GLS are able to avoid considering such centralizers by passing to centralizers of elements of odd prime order, which can therefore be naturally regarded as semisimple. However, quasithin groups G are "small", and in particular the p-rank of G is too small to pass to p-locals for odd p; so we avoid difficulties when G is of even characteristic by using unipotent methods applied to overgroups of T, rather than semisimple methods applied to $C_G(t)$. The case where G is of even type but not of even characteristic is discussed later in section 0.4 of this Introduction. There we will again encounter local subgroups resembling those in our two classes, when they appear as shadows in the proof of the Even Type Theorem.

0.2.2. Case division according to size. After the case division into characteristic 2-type/not characteristic 2-type or even type/odd type described above, both generations of the CFSG proceed by also partitioning the simple groups according to notions of size. Here the underlying idea is that above some critical size, there should be standard "generic" (i.e., size-independent) methods of analysis; but that "small" groups will probably have to be treated separately.

In the even/odd division of the previous subsection, we indicated that the generic examples for the even part of the partition should be the groups of Lie type in characteristic 2. For these groups the appropriate measure of size is the Lie rank of the group, and as we mentioned in section 0.1, $e(G)$ is a good approximation of the Lie rank for G of Lie type and characteristic 2. From this point of view, the quasithin groups are the small groups of even characteristic, so our critical value defining the partition into large and small groups occurs at $e(G) = 2$.

This leaves the question of *why* the boundary of the partition according to size occurs when $e(G) = 2$, rather than $e(G) = 1$ or 3 or something else. The answer is that when one passes to p-locals for odd primes p, $e(G) \geq 3$ is needed in order to use signalizer functors. (See e.g. chapter 15 of [**Asc86a**]). Namely such methods can only be applied to subgroups E which are elementary abelian p-groups of rank at least 3, and E needs to be in a 2-local because of connectedness theorems for the prime 2 (which will be discussed briefly in the next section). Using both signalizer functors and connectedness theorems for the prime 2, one can show that the centralizer of some element of E looks like the centralizer of a semisimple element in a group of Lie type and characteristic 2. Then this information is used to recognize G as a group of Lie type. [3]

Thus, in both programs, the two partitions of the simple groups indicated above, into groups of "even" and "odd" characteristic, and into large and small groups, give rise to a partition of the proof of the Classification into four parts. Since groups of even characteristic include those of characteristic 2-type, our Main Theorem determines the groups in one of the four parts—the small even part—in the first generation program.

To integrate our result into the GLS second-generation proof, we need to reconcile our notion of "even characteristic" with the GLS notion of "even type". The former notion is more natural in the context of the unipotent methods of this work, but the latter fits better with the GLS semisimple methods. Our Even Type Theorem provides the transition between the two notions, and is relatively easy to prove. We will say a little more about that result in section 0.4 of this introduction. The Main Theorem, together with the Even Type Theorem, determine the groups in the small even part of the second generation program.

0.2.3. Some history of the quasithin problem.

We close this section with a few historical remarks about quasithin groups, and more generally small groups of even characteristic.

The methods used in attacking the problem go back to Thompson in the N-group paper [**Tho68**]; in an N-group, all local subgroups are assumed to be solvable. In particular, Thompson introduced the parameter $e(G)$, and used weak closure arguments, uniqueness theorems, and work of Tutte [**Tut47**] and Sims [**Sim67**]. We discuss some of these techniques in the next section; a more extended discussion appears in the Introduction to Volume I.

Groups G of characteristic 2-type with $e(G)$ small were subsequently studied by various authors. Note that $e(G) = 0$ means that all 2-locals are 2-groups, which is impossible in a nonabelian simple group of even order by an elementary argument going back to Frobenius; cf. the Frobenius Normal p-Complement Theorem 39.4 in [**Asc86a**]. In [**Jan72**], Janko defined G to be *thin* if $e(G) = 1$, and used Thompson's methods to determine all thin groups of characteristic 2-type in which all 2-locals are solvable. His student Fred Smith extended that classification from thin to quasithin groups in [**Smi75**]. The general thin group problem was solved by Aschbacher in [**Asc78b**]. Mason went a long way toward a complete treatment of the general quasithin case in [**Mas**], which unfortunately has never been published. See however his discussion of that work in [**Mas80**].

---

[3]In both the original proof of CFSG and in the GLS project, the case $e(G) = 3$ requires special treatment.

There have since been new treatments of portions of the N-group problem due to Stellmacher [**Ste97**] and to Gomi and his collaborators [**GT85**], using an extension of the Tutte-Sims theory which has come to be known as the *amalgam method*. The Thin Group Paper [**Asc78b**] used some early versions of such extensions due to Glauberman, which eventually were incorporated in the proof of the Glauberman-Niles/Campbell Theorem [**GN83**]. Goldschmidt initiated the "modern" amalgam method in [**Gol80**], and this was extended and the amalgam method modified in [**DGS85**] by Goldschmidt, Delgado, and Stellmacher, and in [**Ste92**] by Stellmacher. Those techniques and more recent developments are used in places in this work; our approach is a bit different from the standard approach, and is described briefly in section 0.10 of the Introduction to Volume I.

0.3. An Outline of the Proof of the Main Theorem

In this section we introduce some fundamental concepts and notation, and give a rough outline of the proof of the Main Theorem. Throughout the section, assume G is a simple QTKE-group and $T \in Syl_2(G)$. Recall that $\mathcal{M}$ is the set of maximal 2-local subgroups of G, and $\mathcal{M}(T)$ is the collection of maximal 2-locals containing T.

0.3.1. Setting up the Thompson amalgam strategy. An overall strategy for studying groups of even characteristic originated in Thompson's N-group paper [**Tho68**]; generically it involves exploiting the interaction of distinct maximal 2-locals $M, N \in \mathcal{M}(T)$. (We sometimes refer to this as the "Thompson amalgam strategy").

Of course prior to this generic case, we must first deal with the "disconnected" case where T lies in a unique maximal 2-local. To indicate that $|\mathcal{M}(T)| = 1$, we will usually write $\exists!\mathcal{M}(T)$, to emphasize the existence of the unique maximal 2-local overgroup of T. Recall that in the generic conclusion of the Main Theorem, where G is of Lie type of Lie rank at least 2, there are distinct maximal parabolics above T. So for us, the disconnected case will have as its generic conclusion the groups of Lie type of characteristic 2 and Lie rank 1. We handle this in Theorem 2.1.1, which says:

Theorem 2.1.1 If G is a simple QTKE-group such that $\exists!\mathcal{M}(T)$, then G is a rank 1 group of Lie type and characteristic 2, $L_2(p)$ with $p > 7$ a Mersenne or Fermat prime, $L_3(3)$, or M_{11}.

A finite group G is *disconnected* at the prime 2 if the *commuting graph* on vertices given by the set of nonidentity 2-elements of G (whose edges are pairs of vertices which commute as subgroups) is disconnected. The groups of Lie type and characteristic 2 of Lie rank 1 are the simple groups of 2-rank at least 2 which are disconnected at the prime 2. The classification of these groups is due to Bender [**Ben71**] and Suzuki [**Suz64**]; indeed the groups (namely $L_2(2^n)$, $Sz(2^n)$, $U_3(2^n)$) are often referred to as *Bender groups*. However when working with groups of even characteristic, a weaker notion of disconnected group is also important: namely a group G of even characteristic should be regarded as disconnected if $\exists!\mathcal{M}(T)$ for $T \in Syl_2(G)$.

In view of Theorem 2.1.1, henceforth we will assume that $|\mathcal{M}(T)| \geq 2$. Thompson's strategy now fixes a particular maximal 2-local $M \in \mathcal{M}(T)$. Then instead of working with another maximal 2-local, it will be more advantageous (for reasons which will emerge below) to work with a subgroup H which is *minimal* subject to $T \leq H$, $H \not\leq M$, and $O_2(H) \neq 1$. For example if G is a group of Lie type and characteristic 2, then M is a maximal parabolic over T, and $H = O^{2'}(P)$, where P is the unique parabolic of Lie rank 1 over T not contained in M. Similar remarks hold for other simple groups G with diagram geometries.

We introduce some further definitions to formalize this approach in our abstract setting. We will need to work not only with 2-local subgroups, but also with various subgroups of 2-locals, so we define

$$\mathcal{H} = \mathcal{H}_G := \{H \leq G : O_2(H) \neq 1\},$$

and for $X \subseteq G$, define $\mathcal{H}(X) = \mathcal{H}_G(X) := \{H \in \mathcal{H} : X \subseteq H\}$. Note that any $H \in \mathcal{H}$ lies in the 2-local $N_G(O_2(H))$, and hence is contained in some member of $\mathcal{M}$. Thus as G is quasithin, each $H \in \mathcal{H}$ is in fact *strongly quasithin*; that is H satisfies:

(SQT) $m_p(H) \leq 2$ for each odd prime p.

In addition each $H \in \mathcal{H}$ must also be a $\mathcal{K}$-group by our hypothesis (K), so H in fact satisfies

(SQTK) H is a $\mathcal{K}$-group satisfying (SQT).

The possible simple composition factors for SQTK-groups are determined in Theorem C (A.2.3) in Volume I. The proof of the Main Theorem depends on general properties of $\mathcal{K}$-groups, but also on numerous special properties of the groups in Theorem C, so we refer to the list of groups in that Theorem frequently throughout our proof. We must also occasionally deal with proper subgroups which are not contained in 2-locals. Such groups are quasithin $\mathcal{K}$-groups but not necessarily SQTK-groups; thus we also require Theorem B (A.2.2), which determines all simple composition factors of such groups.

In view of Theorem 2.1.1, the set

$$\mathcal{H}(T, M) := \{H \in \mathcal{H}(T) : H \not\leq M\}$$

is nonempty. Write $\mathcal{H}_*(T, M)$ for the minimal members of $\mathcal{H}(T, M)$, partially ordered by inclusion. Note that for $H \in \mathcal{H}_*(T, M)$, $H \cap M$ is the unique maximal subgroup of H containing T by the minimality of H. Further if $N_G(T) \leq M$ (and we will show in Theorem 3.3.1 that this is usually the case), then T is not normal in H. These conditions give the definition of an abstract *minimal parabolic*, originating in work of McBride; see our definition B.6.1. The condition strongly restricts the structure of H. In particular, the possibilities for H are described in sections B.6 and E.2. In the most interesting case, $O^2(H/O_2(H))$ is a Bender group, so H does resemble a *minimal parabolic* in the Lie theoretic sense for a group of Lie type: namely $O^{2'}(P)$ where P is a parabolic of Lie rank 1.

Thus for each $M \in \mathcal{M}(T)$, we can choose some $H \in \mathcal{H}_*(T, M)$. By the maximality of M, $\langle M, H \rangle$ is not contained in a 2-local subgroup, so that $O_2(\langle M, H \rangle) = 1$. Thompson's weak closure methods and the later amalgam method depend on the latter condition, rather than on the maximality of M, so often we will be able to replace M by a smaller subgroup. We say U is a *uniqueness subgroup* of G if $\exists! \mathcal{M}(U)$. Furthermore we usually write $M = !\mathcal{M}(U)$ to indicate that M is the unique overgroup of U in $\mathcal{M}$. Notice that if $M = !\mathcal{M}(U)$, then from the definition of uniqueness subgroup, $O_2(\langle U, H \rangle) = 1$, so again we can apply weak closure arguments or the amalgam method to the pair U, H.

In the next subsection 0.3.2, we describe how to obtain a uniqueness subgroup U with useful properties, while subsection 0.3.3 discusses how to determine a list of possiblities for U. Here is a brief summary: No nontrivial subgroup T_0 of T can be normal in both U and H; in particular, $Z := \Omega_1(Z(T))$ is not in the center of Y for some $Y \in \{U, H\}$. This places strong restrictions on the $\mathbf{F}_2$-module $\langle Z^Y \rangle$, and on the action of Y on this module. Our approach concentrates on the situation where Y is the uniqueness group U. Roughly speaking, we can classify the possibilities for U and $\langle Z^U \rangle$, resulting in a list of cases to be analyzed when $Y = U$. The bulk of the proof of the Main Theorem then involves the treatment of these cases, a process which is outlined in the final subsection 0.3.4.

0.3.2. Finding a uniqueness subgroup. We put aside for a while the groups M and H from the previous subsection, to see how the hypothesis that G is a QTKE-group gives strong restrictions on the structure of 2-local subgroups of G.

We begin with the definition of objects similar to components: For $H \leq G$, let $\mathcal{C}(H)$ be the set of subgroups L of H minimal subject to

$$1 \neq L = L^\infty \trianglelefteq \trianglelefteq H.$$

We call the members of $\mathcal{C}(H)$ the $\mathcal{C}$-*components* of H. To illustrate and motivate this definition, consider the following

Example. Suppose G is a group of Lie type over a field $\mathbf{F}_{2^n}$ with $n > 1$, and H is a maximal parabolic. If H corresponds to an end node of the Dynkin diagram Δ of G, then H^∞ will be the unique member of $\mathcal{C}(H)$. But suppose instead that G is of Lie rank at least 3 and H corresponds to an interior node δ of Δ. Then the minimality of a $\mathcal{C}$-component L of H says that L covers only that part of the Levi complement corresponding to just one connected component of $\Delta - \{\delta\}$. Furthermore H^∞ is then the product of the $\mathcal{C}$-components of H, and distinct $\mathcal{C}$-components commute modulo $O_2(H)$.

We list some facts about $\mathcal{C}$-components and indicate where these facts can be found; see also section 0.5 of the Introduction to Volume I. In section A.3 we develop a theory of $\mathcal{C}$-components in SQTK-groups. Then in 1.2.1 we use this theory to show that two of the properties in the Example in fact hold for each $H \in \mathcal{H}$ in a QTKE-group G: namely $\langle \mathcal{C}(H) \rangle = H^\infty$, and for distinct $L_1, L_2 \in \mathcal{C}(H)$, $[L_1, L_2] \leq O_2(L_1) \cap O_2(L_2) \leq O_2(H)$. The quasithin hypothesis further restricts the number of factors and the structure of the factors in such commuting products: If $L \in \mathcal{C}(H)$, then either $L \trianglelefteq H$, or $|L^H| = 2$ and $L/O_2(L) \cong L_2(2^n)$, $Sz(2^n)$, $L_2(p)$ with p an odd prime, or J_1. In particular for $S \in Syl_2(H)$, $\langle L^S \rangle \trianglelefteq H$, and $\langle L^S \rangle$ is L or LL^s for some $s \in S$. Moreover 1.2.1.4 shows that almost always $L/O_2(L)$ is quasisimple. Since the cases where $L/O_2(L)$ is not quasisimple cause little difficulty, it is probably best for the expository purposes of this Introduction to ignore the non-quasisimple cases.

To get some control over how 2-locals intersect, and in particular to produce uniqueness subgroups, we also wish to see how $\mathcal{C}$-components of $H \in \mathcal{H}$ embed in other members of $\mathcal{H}$. To do so, we keep appropriate 2-subgroups S of H in the picture, and define $\mathcal{L}(H, S)$ to be the set of subgroups L of H with

$$L \in \mathcal{C}(\langle L, S \rangle), \ S \in Syl_2(\langle L, S \rangle), \ \text{and} \ O_2(\langle L, S \rangle) \neq 1.$$

Again to movitate this definition, consider the case where G is the shadow obtained by extending $G_0 := L_4(2^n)$ for $n > 1$ by an involutory graph automorphism of G_0, with P the middle node maximal parabolic over $T \cap G_0$, and $H := PT$. Then $H \geq \langle L, T \rangle$ for an $L \in \mathcal{L}(G, T)$ with $|L^T| = 2$.

We partially order $\mathcal{L}(G, T)$ by inclusion and let $\mathcal{L}^*(G, T)$ denote the maximal members of this poset. In our earlier example where H is a parabolic of a group of Lie type, notice that any $L \in \mathcal{C}(H)$ is contained in a maximal parabolic determined by some end node. Thus the $\mathcal{C}$-components of such parabolics are the members of $\mathcal{L}^*(G, T)$.

In an abstract QTKE-group G, the members of $\mathcal{L}^*(G, T)$ can be used to produce uniqueness subgroups: For by 1.2.4, when $S \in Syl_2(H)$, any $L \in \mathcal{L}(H, S)$ is contained in some $K \in \mathcal{C}(H)$. Then a short argument in 1.2.7 shows that whenever $L \in \mathcal{L}^*(G, T)$,

$$N_G(\langle L^T \rangle) = !\mathcal{M}(\langle L, T \rangle).$$

Thus $\langle L, T \rangle$ is a uniqueness subgroup in our language, achieving the goal of this subsection.

But it could also happen (for example in a group of Lie type over the field $\mathbf{F}_2$) that the visible 2-locals are solvable, so that $\mathcal{L}(G, T)$ is empty. To deal with such situations, and with the case where $L/O_2(L)$ is not quasisimple for some $L \in \mathcal{L}^*(G, T)$, we also show that certain solvable minimal T-invariant subgroups are uniqueness subgroups. The quasithin hypothesis allows us to focus on p-groups of small rank: Define $\Xi(G, T)$ to consist of those T-invariant subgroups $X = O^2(X)$ of G such that

$XT \in \mathcal{H}$, $X/O_2(X) \cong E_{p^2}$ or p^{1+2} for some odd prime p, and T is irreducible on the Frattini quotient of $X/O_2(X)$.

For example, in the extension of $L_4(2^n)$ discussed above, if we take $n = 1$ instead of $n > 1$, then $H = PT \in \Xi(G, T)$ for $p = 3$.

If X is not contained in certain nonsolvable subgroups, then XT will be a uniqueness subgroup. Thus we are led to define $\Xi^*(G, T)$ to consist of those $X \in \Xi(G, T)$ such that XT is not contained in $\langle L, T \rangle$ for any $L \in \mathcal{L}(G, T)$ with $L/O_2(L)$ quasisimple. We find in 1.3.7 that if $X \in \Xi^*(G, T)$, then

$$N_G(X) = !\mathcal{M}(XT),$$

so that XT is a uniqueness subgroup.

0.3.3. Classifying the uniqueness groups and modules. We now return to our pair M, H with $M \in \mathcal{M}(T)$ and $H \in \mathcal{H}_*(T, M)$ from subsection 0.3.1. The structure of H is restricted since H is a minimal parabolic, but *a priori* M could be a fairly arbitrary quasithin group, subject to the constraint $F^*(M) = O_2(M)$; in particular, the composition factors of M could include arbitrary simple SQTK-groups acting on arbitrary "internal modules" (elementary abelian M-sections) involved in $O_2(M)$

To obtain a more tractable set of possibilities, we exploit a uniqueness subgroup U produced by one of the two methods in the previous subsection 0.3.2; that is, we take U of the form $\langle L, T \rangle$ with $L \in \mathcal{L}^*(G, T)$, or XT with $X \in \Xi^*(G, T)$, and take $M := N_G(O^2(U)) = !\mathcal{M}(U)$. Recall that $Z := \Omega_1(Z(T))$ cannot be central in both U and H. The case where $Z \leq Z(U)$ for all choices of U is essentially a "small" case, treated in Part 6, so most of the analysis deals with the case $[Z, U] \neq 1$.

We introduce notation to cover both the situations discussed in subsection 0.3.2: Define $\mathcal{X}$ to consist of those subgroups $X = O^2(X)$ of G such that $F^*(X) = O_2(X)$. For example $\mathcal{L}(G, T)$ and $\Xi(G, T)$ are contained in $\mathcal{X}$. To describe the members with a "faithful action", write $\mathcal{X}_f$ for those $X \in \mathcal{X}$ such that $[\Omega_1(Z(O_2(X))), X] \neq 1$, with a similar use of the subscript to define subsets $\mathcal{L}_f(G, T)$ and $\Xi_f(G, T)$. Our analysis focuses on the faithful uniqueness groups U in $\mathcal{L}_f^*(G, T)$ and $\Xi_f^*(G, T)$.

If $Y \in \mathcal{H}(T)$, so that $F^*(Y) = O_2(Y)$ by 1.1.4.6, then by a standard lemma B.2.14, $V := \langle Z^Y \rangle$ is elementary abelian and *2-reduced*: that is, $O_2(Y/C_Y(V)) = 1$. Following Thompson, define $\mathcal{R}_2(Y)$ to be the set of 2-reduced elementary abelian normal 2-subgroups of Y. By B.2.12 (26.2 in [**GLS96**]), the product of members of $\mathcal{R}_2(Y)$ is again in $\mathcal{R}_2(Y)$, so $\mathcal{R}_2(Y)$ has a unique maximal member $R_2(Y)$. We regard $R_2(Y)$ as an $\mathbf{F}_2 Y$-module.

Observe that if $L \in \mathcal{L}_f^*(G, T)$ with $L/O_2(L)$ quasisimple, or $X \in \Xi_f^*(G, T)$, then $C_U(R_2(U)) \leq O_{2,\Phi}(U)$. [4] Then the representation of $U/C_U(R_2(U))$ on $R_2(U)$ (or indeed on any $V \in \mathcal{R}_2(U)$ with $V \not\leq Z(U)$) is particularly effective, since for any weakly closed subgroup W of $C_T(V)$, W is normal in the uniqueness subgroup U, so that $N_G(W) \leq M$. That is $M = !\mathcal{M}(U)$ contains the normalizers of various weakly closed subgroups W of T.

For $M := N_G(O^2(U))$ and U a uniqueness subgroup of the form $\langle L, T \rangle$ with $L \in \mathcal{L}^*(G, T)$, or XT with $X \in \Xi^*(G, T)$, we prove in Theorem 3.3.1 that $N_G(T) \leq M$. It follows that T is *not* normal in H in those cases, so that H is a minimal parabolic in the sense of Definition B.6.1, and hence we can use the explicit description of $H/O_2(H)$ from section E.2 mentioned earlier.

We next turn to Theorem 3.1.1, which is used in a variety of ways; it says:

Theorem 3.1.1 *If* $M_0, H \in \mathcal{H}(T)$, *such that* T *is in a unique maximal subgroup of* H, *and* $R \leq T$ *with* $R \in Syl_2(O^2(H)R)$ *and* $R \trianglelefteq M_0$, *then* $O_2(\langle M_0, H \rangle) \neq 1$.

For example in our standard setup we can take M_0 to be the uniqueness group U and $R := C_T(V)$—and conclude that $R \notin Syl_2(O^2(H)R)$, since $H \not\leq M =!\mathcal{M}(U)$; hence $O_2(\langle U, H \rangle) = 1$. In particular we use Theorem 3.1.1 to rule out the first case which occurs in Stellmacher's *qrc*-lemma D.1.5 (see below), and in the remaining cases the *qrc*-lemma gives us strong information on a module V for the action of U. That information is given in terms of small values of certain parameters, which we now introduce. For X a finite group, let $\mathcal{A}^2(X)$ denote the set of nontrivial elementary abelian 2-subgroups of X. Given a faithful $\mathbf{F}_2 X$-module V, define

$$q(X, V) := \min\{\frac{m(V/C_V(A))}{m(A)} : 1 \neq A \in \mathcal{A}^2(X) \text{ such that } 0 = [V, A, A]\}$$

and the analogous parameter correponding to cubic rather than quadratic action:

$$\hat{q}(X, V) := \min\{\frac{m(V/C_V(A))}{m(A)} : 1 \neq A \in \mathcal{A}^2(X) \text{ such that } 0 = [V, A, A, A]\}.$$

For example V is a *failure of factorization module* (FF-module—see section B.1) for X precisely when $q(X, V) \leq 1$.

Using Theorem 3.1.1 and Stellmacher's *qrc*-Lemma (see Theorem D.1.5), we obtain:

Theorem 3.1.6 *Let* $T \leq M_0 \leq M \in \mathcal{M}(T)$ *and* $H \in \mathcal{H}_*(T, M)$ *Assume* $V \in \mathcal{R}_2(M_0)$ *with* $C_T(V) = O_2(M_0)$, *and* $H \cap M$ *normalizes* $O^2(M_0)$ *or* V. *Then one of the following holds:*

---

[4]Here $O_{2,\Phi}(U)$ denotes the preimage of the Frattini subgroup $\Phi(U/O_2(U))$; elsewhere we use similar notation such as $O_{2,F}(U)$, $O_{2,E}(U)$, etc.

(1)$O_2(\langle M_0, H \rangle) \neq 1$, so M_0 is not a uniqueness subgroup of G.
(2) $V \nleq O_2(H)$ and $\hat{q}(M_0/C_{M_0}(V), V) \leq 2$.
(3) $q(M_0/C_{M_0}(V), V) \leq 2$.

When we apply this result with M_0 our uniqueness subgroup U from subsection 0.3.1, case (1) does not arise, so the module V satisfies $\hat{q} \leq 2$.

In section D.3, we determine the groups and modules satisfying this strong restriction (and a suitable minimality assumption) under the SQTK-hypothesis. Since the most general SQTK-group H of characteristic 2 could have arbitrary internal modules as sections of $O_2(H)$, Theorem 3.1.6 leads to a solution in section 3.2 of the First Main Problem for QTKE-groups:

First Main Problem. Show that a simple QTKE-group G does not have the local structure of the general nonsimple strongly quasithin $\mathcal{K}$-group Q with $F^*(Q) = O_2(Q)$, but instead has a more restrictive structure resembling that of the examples in the conclusion of the Main Theorem, or the shadows of groups with similar local structure.

A solution of the First Main Problem amounts to showing that there are relatively few choices for $L/O_2(L)$ and its action on V, where $L \in \mathcal{L}_f^*(G, T)$, $V \in \mathcal{R}_2(\langle L, T \rangle)$, and $[V, L] \neq 1$. Indeed in most cases, $L/O_2(L)$ is a group of Lie type in characteristic 2 and V is a "natural module" for $L/O_2(L)$. This leads us in section 3.2 to define the *Fundamental Setup* FSU (3.2.1), and to the possibilities for $L/O_2(L)$ and V listed in 3.2.5–3.2.9. The proof can be roughly summarized as follows: Apply Theorem 3.1.6 to $M_0 := U = \langle L, T \rangle$. As M_0 is a uniqueness subgroup, conclusion (1) of 3.1.6 cannot hold. Then from section D.3, the restrictions on q and $\hat{q}$ in conclusions (2) and (3) of 3.1.6 allow us to determine a short list of possibilities for $M_0/C_{M_0}(V)$ and its action on V.

0.3.4. Handling the resulting list of cases. We continue to restrict attention to the most important case where $L \in \mathcal{L}^*(G, T)$ with $L/O_2(L)$ quasisimple, and let $L_0 := \langle L^T \rangle$ and $M := N_G(L_0)$. Then in the FSU, there is $1 \neq V = [V, L_0] \in \mathcal{R}_2(L_0 T)$ with $V/C_V(L_0)$ an irreducible $L_0 T$-module. Set $V_M := \langle V^M \rangle$ and $\bar{M} := M/C_M(V_M)$. By 3.2.2, $V_M \in \mathcal{R}_2(M)$, and by Theorems 3.2.5 and 3.2.6, we may choose V so that one of the following holds:

(1) $V = V_M \trianglelefteq M$.
(2) $C_V(L) = 1$, $V \trianglelefteq T$, and V is a TI-set under M. [5]
(3) $\bar{L} \cong L_3(2)$, $L < L_0$, and subcase 3.c.iii of Theorem 3.2.6 holds.

Further the choices for L and V are highly restricted, and are listed in Theorems 3.2.5 and 3.2.6, with further information given in 3.2.8 and 3.2.9.

The bulk of the proof of our Main Theorem consists of a treatment of the resulting list of possibilities for L and V. The analysis falls into several broad categories: The cases with $|L^T| = 2$ are handled comparatively easily in chapter 10; so from now on assume that $L \trianglelefteq M$. The Generic Case where $\bar{L} \cong L_2(2^n)$ (leading to the generic conclusion in our Main Theorem of a group of Lie type and characteristic 2 of Lie rank 2) is handled in Part 2. Most cases where V is not an FF-module for $LT/O_2(LT)$ are eliminated in Part 3. The remaining cases where V is an FF-module are handled in Parts 4 and 5.

───────────

[5]Recall a TI-set is a set intersecting trivially with its distinct conjugates.

In order to discuss these cases in more detail, we need more concepts and notation.

First, another consequence of Theorem 3.1.1 (established as part (3) of Theorem 3.1.8) is that either

  (i) $L = [L, J(T)]$, or
  (ii) $\mathcal{H}_*(T, M) \subseteq C_G(Z)$, where $Z = \Omega_1(Z(T))$.

Here $J(T)$ is the Thompson subgroup of T (cf. section B.2). In case (i), V is an FF-module; so when V is not an FF-module, we know $[Z, H] = 1$ for all $H \in \mathcal{H}_*(T, M)$. In particular $C_V(L) = 1$ since H is not contained in the uniqueness group M for LT, whereas if $C_V(L)$ were nontrivial then $C_Z(L)$ would be nontrivial and centralized by H as well as LT.

Second, in section E.1, we introduce a parameter $n(H)$ for $H \in \mathcal{H}$. The parameter involves the generation of H by minimal parabolics, but the definition of $n(H)$ is somewhat more complicated; for expository purposes one can oversimplify somewhat to say that roughly $n(H) = 1$ unless H has a composition factor which is of Lie type over $\mathbf{F}_{2^n}$—in which case $n(H)$ is the maximum of such n. Thus for example in a twisted group H of Lie type, $n(H)$ is usually the exponent n of the larger of the orders of the fields of definitions of the Levi factors of the parabolics of Lie rank 1 of H. In particular if $H \in \mathcal{H}_*(T, M)$, then either $n(H) = 1$, or (using section B.6) $O^2(H/O_2(H))$ is a group of Lie type over $\mathbf{F}_{2^n}$ of Lie rank at most 2, $O^2(H) \cap M$ is a Borel subgroup of $O^2(H)$, and $n(H) = n$. In that event, we call the Hall $2'$-subgroups of $H \cap M$ *Cartan subgroups* of H. Our object is to show that $n(H)$ is roughly bounded above by $n(L)$, and to play off Cartan subgroups of H against those of L when $L/O_2(L)$ is of Lie type. It is easy to see that if $n(H) > 1$ and B is a Cartan subgroup of $H \cap M$, then $H = \langle H \cap M, N_H(B) \rangle$, so that $N_G(B) \not\leq M$. On the other hand, if $n(H)$ is small relative to $n(L)$ (e.g. if $n(H) = 1$), then weak closure arguments can be effective.

Third, except in certain cases where V is a small FF-module, we obtain the following important result, which produces still more uniqueness subgroups:

Theorem 4.2.13 With small exceptions, if $I \leq LT$ with $L \leq O_2(LT)I$ and $I \in \mathcal{H}$, then I is also a uniqueness subgroup.

Theorem 4.2.13 has a variety of consequences, but perhaps its most important application is in Theorem 4.4.3, to show that (except when V is a small FF-module) if $1 \neq B$ is of odd order in $C_M(V)$, then $N_G(B) \leq M$. In particular from the previous paragraph, if $H \in \mathcal{H}_*(T, M)$ with $n(H) > 1$ and B is a Cartan subgroup of $H \cap M$, then $[V, B] \neq 1$. If $[Z, H] = 1$, this forces B to be faithful on L, so that it is possible to compare $n(H)$ to $n(L)$ and show that $n(H)$ is not large relative to $n(L)$.

0.3.4.1. *Weak Closure methods.* Thompson introduced weak closure methods in the N-group paper [**Tho68**]. When $n(H)$ is small relative to $n(L)$ and (roughly speaking) $q(LT/O_2(LT), V)$ is not too small, weak closure arguments become effective. We will not discuss weak closure in any detail here, but instead direct the reader to the discussion in section 0.9 of the Introduction to Volume I, and to section E.3 of Volume I, particularly the exposition introducing that section and the introductions to subsections E.3.1 and E.3.3. However we will at least say here that weak closure, together with the constellation of concepts and techniques introduced earlier in this subsection, plays the largest role in analyzing those cases in the

FSU where V is not an FF-module. The only quasithin example which arises from those cases is J_4, but shadows of groups like the Fischer groups and Conway groups complicate the analysis, and are only eliminated rather indirectly because they are not quasithin. When V is not an FF-module, the pair $L/O_2(L)$, V is usually sufficiently far from pairs in examples or shadows, that the pair can be eliminated by comparing various paramters from the theory of weak closure.

0.3.4.2. *The Generic Case.* In the Generic Case, $\bar{L} \cong L_2(2^n)$ and $n(H) > 1$ for some $H \in \mathcal{H}_*(T, M)$. We prove in Theorem 5.2.3 that the Generic Case leads to the bulk of the groups of Lie type and characteristic 2 in the conclusion of our Main Theorem; to be precise, one of the following holds:

(1) V is the A_5-module for $L/O_2(L) \cong L_2(4)$.

(2) $G \cong M_{23}$.

(3) G is Lie type of Lie rank 2 and characteristic 2.

To prove Theorem 5.2.3, we proceed by showing that if neither (1) nor (2) holds, and D is a Cartan subgroup of L, then the amalgam

$$\alpha := (LTB, BDT, HD)$$

is a *weak BN-pair* of rank 2 in the sense of the "Green Book" [**DGS85**]; then by Theorem A of the Green Book and results of Goldschmidt [**Gol80**] and Fan [**Fan86**], the amalgam α is determined up to isomorphism. At this point there is still work to be done, as this determines G only up to "local isomorphism". Fortunately there is a reasonably elegant argument to complete the final identification of G as a group of Lie type and characteristic 2; this argument is discussed in the Introduction to Volume I, in section 0.12 on recognition theorems. It also requires the extension 4.3.2 of Theorem 4.2.13 to show that $G = \langle L, H \rangle$.

After dealing with the Generic Case, we still have to consider the situation where $L/O_2(L) \cong L_2(2^n)$ and $n(H) = 1$ for all $H \in \mathcal{H}_*(T, M)$; in Theorem 6.2.20, we show that then either V is the A_5-module for $L/O_2(L)$, or $G \cong M_{22}$. Thus from now on, if $L/O_2(L) \cong L_2(2^n)$, we may assume $n = 2$ and V is the A_5-module.

0.3.4.3. *Other FF-modules.* Next in Theorem 11.0.1, we eliminate the cases where $\bar{L}$ is $SL_3(2^n)$, $Sp_4(2^n)$, or $G_2(2^n)$ for $n > 1$. From the list in section 3.2, this leaves the cases where $\bar{L}$ is essentially a group of Lie type defined over $\mathbf{F}_2$; that is, $\bar{L}$ is $L_n(2)$, $n = 3, 4, 5$; A_n, $n = 5, 6$; or $U_3(3) = G_2(2)'$—and V is an FF-module. Roughly speaking, these cases, together with certain cases where $\mathcal{L}_f(G, T)$ is empty, are the cases left untreated in Mason's unpublished preprint. They are also the most difficult cases to eliminate.

We first show either that there is $z \in Z \cap V^\#$ with $G_z := C_G(z) \nleq M$, or G is A_8, A_9, M_{22}, M_{23}, M_{24}, or $L_5(2)$. In the latter case the groups appear as conclusions in our Main Theorem, so we may now assume the former.

Let $\tilde{G}_z := G_z/\langle z \rangle$, $L_z := O^2(C_L(z))$, and $V_z := \langle V_2^{L_z} \rangle$, where V_2 is the preimage of $C_{\tilde{V}}(T)$, and $U := \langle V_z^{G_z} \rangle$. Then $\tilde{U} \leq Z(O_2(\tilde{G}_z))$ by B.2.14, and our next task is to reduce to the case where U is elementary abelian. If not, then $U = Z(U)Q_U$, where Q_U is an extraspecial 2-group, and then to analyze $\tilde{G}_z$, we can use some of the ideas from the the theory of groups with a large extraspecial 2-group (cf. [**Smi80**]) in the original CFSG: We first show that if $Z(U) \neq \langle z \rangle$, then $G \cong Sp_6(2)$ or HS. Hence we may assume in the remainder of this case that U is extraspecial. Then we repeat some of the elementary steps in Timmesfeld's analysis in [**Tim78**], followed by appeals to results on $\mathbf{F}_2$-modules in section G.11, to pin down the structure of

G_z. At this point our recognition theorems show that G is $G_2(3)$, $L_4^\epsilon(3)$, or $U_4(2)$. The shadow of the Harada-Norton group F_5 also arises to cause complications.

We have reduced to the case where U is abelian. In this difficult case, we show that only $G \cong Ru$ arises. Our approach is to use a modified version of the amalgam method on a pair of groups (LT, H), where $H \in \mathcal{H}(L_z T)$ with $H \not\leq M$. Using the fact that U is abelian, we can show that $\langle V^{G_z} \rangle$ is abelian, and hence conclude that $[V, V^g] = 1$ if $V \cap V^g \neq 1$. In the context of the amalgam method, this shows that the graph parameter b is odd and greater than 1. Then we show that $q(H/C_H(\tilde{U}), \tilde{U}) \leq 2$, which eventually leads to the elimination of all choices for $L/O_2(L)$, V, $H/C_H(\tilde{U})$, and U other than the 4-tuple leading to the Ru example.

We have completed the outline of our treatment of quasithin groups in the main case, when there is $L \in \mathcal{L}_f^*(G, T)$ with $L/O_2(L)$ quasisimple. The case where $L/O_2(L)$ is not quasisimple is handled fairly easily in section 13.1. That leaves:

0.3.5. The case $\mathcal{L}_f(G, T)$ empty. In Part 6 we handle the case $\mathcal{L}_f(G, T) = \emptyset$. Part of the analysis here has some similarities to the $\mathbf{F}_2$-case just discussed, and leads to the groups J_2, J_3, $^3D_4(2)$, the Tits group $^2F_4(2)'$, $U_3(3)$, M_{12}, $L_3(2)$, and A_6.

To replace the uniqueness subgroup $\langle L, T \rangle$, we introduce the partial order $\stackrel{<}{\sim}$ on $\mathcal{M}(T)$ defined in section A.5, choose $M \in \mathcal{M}(T)$ maximal with respect to $\stackrel{<}{\sim}$, and set $Z := \Omega_1(Z(T))$ and $V := \langle Z^M \rangle$. Then by A.5.7, for each overgroup X of T in M with $M = C_M(V)X$, we obtain $M = !\mathcal{M}(X)$. The case where $C_G(Z)$ is not a uniqueness subgroup is relatively easy, and handled in the last section of Part 6; in this case $G \cong L_3(2)$ or A_6. The case where $C_G(Z)$ is a uniqueness subgroup is harder; the subcase where $m(V) = 2$ and $Aut_M(V) \cong L_2(2)$ presents the greatest difficulties, and is handled in Part 5—in tandem with the cases where V is the natural module for $L/O_2(L) \cong L_n(2)$ for $n = 4$ and 5. The elimination of these cases completes the proof of our Main Theorem.

0.4. An Outline of the Proof of the Even Type Theorem

Assume in this section that G is a simple QTK-group of even type, but G is not of even characteristic. We outline our approach for showing G is isomorphic to the smallest Janko group J_1.

As G is of even type, there is an involution $z \in Z(T)$ and a component L of $C_G(z)$. As G is quasithin of even type, the possibilities for L are few. Our object is to show that L is a *standard* subgroup of G: That is, we must show that L commutes with none of its conjugates, $N_G(L) = N_G(C_G(L))$, and $C_G(L)$ is *tightly embedded* in G. This last means that $C_G(L)$ is of even order, but if $g \in G - N_G(L)$ then $C_G(L) \cap C_G(L^g)$ is of odd order. Once this is achieved, the facts that $z \in Z(T)$ and that L is highly restricted will eventually eliminate all configurations except $L \cong L_2(4)$ and $C_G(z) = \langle z \rangle \times L$, where $G \cong J_1$ via a suitable recognition theorem.

Here are some details of the proof. We first observe that if i is an involution in $C_T(L)$ and $|S : C_T(i)| \leq 2$ for some $S \in Syl_2(C_G(i))$, then L is a component of $C_G(i)$: For L is a component at least of $C_{C_G(i)}(z)$, and hence contained in KK^z for some component K of $C_G(i)$ by "L-balance" (see I.3.1). Now the hypothesis that $|S : C_T(i)| \leq 2$, together with the restricted choices for K, leads to $L = K$ as desired.

This fundamental lemma can be used to show first that $L \trianglelefteq C_G(z)$—which is very close to showing that L commutes with none of its conjugates. Then the fundamental lemma also shows that $L \trianglelefteq C_G(i)$ for each involution $i \in C_G(L)$, after which it is a short step to showing that $C_G(L)$ is tightly embedded in G, and L is standard in G.

At this point, we could quote some of the theory of standard subgroups and tightly embedded subgroups (developed in [**Asc75**] and [**Asc76**]) to simplify the remainder of the proof. But since GLS do not use this machinery, we content ourselves with using only elementary lemmas from that theory which are easy to prove; the lemmas and their proofs are reproduced in sections I.7 and I.8. In particular, we use I.8.2 to see that our hypothesis that G is not of even characteristic shows that for some L^g distinct from L, an involution of $C_G(L^g)$ normalizes L; this provides the starting point for our analysis. Then, making heavy use of the fact that z is 2-central, and that the component L is highly restricted by the even type hypothesis, we eliminate all configurations except $N_G(L) \cong \mathbf{Z}_2 \times L_2(4)$. Then we identify $G = J_1$ via the structure of $C_G(z)$ as noted above. Along the way, we encounter various shadows coming from groups which are not perfect, like the groups in the examples in subsection 0.2.1. In most such cases it is possible to apply transfer to contradict $G = O^2(G)$, given the fact that the Sylow 2-group T of G normalizes L.

This shows the advantages of introducing the notion of a group of "even characteristic", and hence of the the partition of the quasithin groups of even type into those of even characteristic, and those of even type which are not of even characteristic: The first subclass we studied via unipotent methods, and the latter by semisimple methods at the prime 2. If instead we had used unipotent methods to treat only the more restricted subclass of groups of characteristic 2-type, then our semisimple analysis at the prime 2 would have had to deal with the shadows of the nonsimple configurations in subsection 0.2.1, in which involution centralizers $C_G(z)$ with components do not contain a Sylow 2-group T of G. When z is not 2-central the road to obtaining T, so that one can show G is not simple via transfer, is much longer and very bumpy.

As a final remark, we recall that for the generic groups of even type, GLS are able switch to semisimple analysis of elements of *odd* prime order, and so are able to avoid dealing with shadows of the nonsimple examples of subsection 0.2.1. Thus they do not need the concept of groups of "even characteristic" in their generic analysis.

Part 1

Structure of QTKE-Groups and the Main Case Division

See the Introductions to Volumes I and II for terminology used in this overview.

In this first Part, we obtain a solution to the First Main Problem: that is, we show that a simple QTKE-group G (with Sylow 2-subgroup T) does not have the local structure of the arbitrary nonsolvable SQTK-group Q with $F^*(Q) = O_2(Q)$, but instead has more restricted 2-locals resembling those in examples and shadows. More precisely, we establish the existence of a "large" member of $\mathcal{H}(T)$ (i.e., a uniqueness subgroup of G) resembling a maximal 2-local in an example or shadow. Then the cases corresponding to the possible uniqueness subgroups will be treated in subsequent Parts of this Volume.

Here is an outline of Part 1:

In chapter 1 we use the results in sections A.2 and A.3 of Volume I to establish tools for working in 2-local subgroups H of G, using the fact that our 2-locals are strongly quasithin. In particular we obtain a good description of the last term H^∞ of the derived series for H, primarily in terms of the $\mathcal{C}$-components of H, and some information about $F(H/O_2(H))$. We then go on to show that certain subgroups of G are "uniqueness subgroups" contained in a unique maximal 2-local M. In particular, we show that members of the sets $\mathcal{L}^*(G,T)$ and $\Xi^*(G,T)$ are uniqueness subgroups.

The "disconnected" case where T itself is a uniqueness subgroup and so contained in a unique maximal 2-local, is treated in chapter 2, which characterizes certain small groups via this property. Consequently after Theorem 2.1.1 is proved, we are able to assume during the remainder of the proof of the Main Theorem that T is contained in at least two maximal 2-locals of G. Hence there exist 2-locals H with $T \leq H \not\leq M$.

Next in chapter 3, we begin by proving two important preliminary results: Theorem 3.3.1 which says that $N_G(T) \leq M$ when $M = !\mathcal{M}(L)$ with L in $\mathcal{L}^*(G,T)$ or $\Xi^*(G,T)$; and Theorem 3.1.1, which among other things is needed to apply Stellmacher's qrc-lemma D.1.5 to the amalgam defined by M and H. The qrc-lemma gives strong restrictions on certain internal modules U for M via the bound $\hat{q}(Aut_M(U), U) \leq 2$. Section 3.2 then uses those restrictions to determine the list of possibilities for $L/O_2(L)$ with $L \in \mathcal{L}_f^*(G,T)$, and for the internal modules $V \in \mathcal{R}_2(\langle L, T \rangle)$. This provides the Main Case Division for the proof of the Main Theorem. One consequence of Theorem 3.3.1 is that members of $\mathcal{H}_*(T, M)$ are minimal parabolics, in the sense of the Introduction to Volume II.

The first Part concludes with chapter 4, which uses the methods of pushing up from chapter C of Volume I to establish some important technical results: In particular, we show in Theorem 4.2.13 that unless V is an FF-module and L is "small", then for each $I \leq L$ with $O_2(I) \neq 1$ and $L = O_2(L)I$, we have $M = !\mathcal{M}(I)$. This large family of uniqueness subgroups then allows us (in Theorems 4.4.3 and 4.4.14) to control the normalizers of nontrivial subgroups of odd order centralizing V. This control is in turn important later, particularly in Part 2 and in chapter 11, when we deal with cases where $L/O_2(L)$ (or $H/O_2(H)$ for $H \in \mathcal{H}_*(T, M)$) is of Lie type over $\mathbf{F}_{2^n}$ for some $n > 1$, allowing us to exploit the existence of nontrivial Cartan subgroups.

CHAPTER 1

Structure and intersection properties of 2-locals

In this chapter we show how the structure theory for SQTK-groups from section A.3 of Volume I translates into a description of the 2-local subgroups of a QTKE-group G. We then use this description to establish the existence of certain uniqueness subgroups, which are crucial to our analysis. We will concentrate on C-components of 2-locals, and the two families $\mathcal{L}(G,T)$ and $\Xi(G,T)$ of subgroups of G discussed in the Introduction to Volume II.

In this chapter, and indeed unless otherwise specified throughout the proof of the Main Theorem, we adopt the following convention:

NOTATION 1.0.1 (Standard Notation). G is a simple QTKE-group, and $T \in Syl_2(G)$.

Recall from the Introduction to Volume I that a finite group G is a *QTKE-group* if

(QT) G is quasithin,

(K) every proper subgroup of G is a $\mathcal{K}$-group, and

(E) $F^*(M) = O_2(M)$ for each maximal 2-local subgroup M of G of odd index.

Also as in the Introductions to Volumes I and II, let $\mathcal{M}$ denote the set of maximal 2-local subgroups of G, for $X \subseteq G$ define

$$\mathcal{M}(X) := \{N \in \mathcal{M} : X \subseteq N\},$$

and recall that a subgroup $U \leq M \in \mathcal{M}$ is a *uniqueness subgroup* if $M = !\mathcal{M}(U)$. (Which means $\mathcal{M}(U) = \{M\}$ in the notation more common in the earlier literature). The members of $\mathcal{M}$ are of course uniqueness subgroups, but for our purposes it is preferable to work with smaller uniqueness subgroups, which have better properties in various arguments involving amalgams, pushing up, etc. We summarize some useful properties of uniqueness subgroups in the final section of the chapter.

1.1. The collection $\mathcal{H}^e$

DEFINITION 1.1.1. Define $\mathcal{H}^e = \mathcal{H}_G^e$ to be the set of subgroups H of G such that $F^*(H) = O_2(H)$; equivalently $C_H(O_2(H)) \leq O_2(H)$ or $O^2(F^*(H)) = 1$.

Using this notation, Hypothesis (E)—namely that G is of *even characteristic*— just says

$$\mathcal{M}(T) \subseteq \mathcal{H}^e.$$

The property that $H \in \mathcal{H}^e$ has many important consequences which we can exploit later—notably the existence of 2-reduced internal modules for H, such as in lemma B.2.14. Thus we want $\mathcal{H}^e$ to be as large as possible, so in this section we establish several sufficient conditions to ensure that a subgroup is in $\mathcal{H}^e$.

We begin by defining some notation.

DEFINITION 1.1.2. Set

$$\mathcal{H} = \mathcal{H}_G := \{H \leq G : O_2(H) \neq 1\};$$

and for $X \subseteq G$, set

$$\mathcal{H}(X) = \mathcal{H}_G(X) := \{H \in \mathcal{H} : X \subseteq H\}.$$

For $X \subseteq Y \subseteq G$, set

$$\mathcal{H}(X, Y) = \mathcal{H}_G(X, Y) := \{H \in \mathcal{H}(X) : H \not\subseteq Y\}.$$

Define $\mathcal{H}^e(X)$ (resp. $\mathcal{H}^e(X, Y)$) as the intersection of $\mathcal{H}^e$ with $\mathcal{H}(X)$ (resp. $\mathcal{H}(X, Y)$).

The subgroups in $\mathcal{H}$ are the primary focus of our proof, so we record here the following elementary (but important) observations: Notice that by (QT), H is an SQT-group. As G is simple and $O_2(H) \neq 1$, certainly H is proper in G; hence by (K), simple sections of subgroups of H are in $\mathcal{K}$, so that H is an SQTK-group. Then by (2) of Theorem A (A.2.1), all simple sections of H are also SQTK-groups.

We are interested in conditions on members H of $\mathcal{H}$ which will ensure that $H \in \mathcal{H}^e$. For example, in 1.1.4.6 below, we show that each member of the collection $\mathcal{H}(T)$ is in $\mathcal{H}^e$. We begin with some well known results in that spirit, which we use frequently:

LEMMA 1.1.3. *Let* $M \in \mathcal{H}^e$. *Then*

(1) If $1 \neq N \trianglelefteq \trianglelefteq M$, *then* $N \in \mathcal{H}^e$.
(2) If X *is a 2-subgroup of* M, *and* $XC_M(X) \leq H \leq N_M(X)$, *then* $H \in \mathcal{H}^e$ *and* $C_M(X) \in \mathcal{H}^e$.
(3) If $H \leq M$ *and* $B_1, \ldots, B_n$ *are 2-subgroups of* H *with* $B_j \leq N_H(B_i)$ *for all* $i \leq j$ *and* $H = \bigcap_{i=1}^n N_M(B_i)$, *then* $H \in \mathcal{H}^e$.

PROOF. As $N \trianglelefteq \trianglelefteq M$, $O^2(F^*(N)) \leq O^2(F^*(M)) = 1$. Thus (1) holds. If X is a 2-subgroup of M, then $N_M(X) \in \mathcal{H}^e$ by 31.16 in [**Asc86a**], so $C_M(X) \in \mathcal{H}^e$ by (1). If $XC_M(X) \leq H \leq N_M(X)$, then $X \leq O_2(H)$, so $O^2(F^*(H))$ centralizes X, and hence $O^2(F^*(H)) \leq O^2(F^*(C_M(X))) = 1$, so that $H \in \mathcal{H}^e$. Thus (2) holds, and (3) follows from (2) by induction on n. □

For $X \leq G$ let $\mathcal{S}_2(X)$ be the set of nontrivial 2-subgroups of X, and let $\mathcal{S}_2^e(G)$ consist of those $S \in \mathcal{S}_2(G)$ such that $N_G(S) \in \mathcal{H}^e$. Here is a collection of conditions sufficient to ensure that various overgroups and subgroups are in $\mathcal{H}^e$:

LEMMA 1.1.4. *(1) If* $U \in \mathcal{S}_2^e(G)$ *and* $U \leq V \in \mathcal{S}_2(G)$, *then* $V \in \mathcal{S}_2^e(G)$.
(2) If $1 \neq U \trianglelefteq T$, *then* $U \in \mathcal{S}_2^e(G)$. *In particular 2-locals containing* T *are in* $\mathcal{H}^e$.
(3) If $U \in \mathcal{S}_2(G)$ *and* $1 \neq Z(T) \cap U$, *then* $U \in \mathcal{S}_2^e(G)$.
(4) If $1 \neq N \leq M \leq G$ *with* $M \in \mathcal{H}^e$ *and* $C_{O_2(M)}(O_2(N)) \leq N$, *then* $N \in \mathcal{H}^e$.
(5) If $1 \neq N \leq M \in \mathcal{M}(T)$ *with* $C_{O_2(M)}(O_2(N)) \leq N$, *then* $N \in \mathcal{H}^e$.
(6) $\mathcal{H}(T) \subseteq \mathcal{H}^e$.
(7) If $M \in \mathcal{H}^e$, $S \in Syl_2(M)$, *and* $1 \neq M_1 \leq M$ *with* $|S : S \cap M_1| \leq 2$, *then* $M_1 \in \mathcal{H}^e$.

PROOF. Assume the hypotheses of (1) and set $N := N_G(U)$. Then by hypothesis $N \in \mathcal{H}^e$. Now if $U \trianglelefteq V$ then $V \leq N$, so $N_N(V) \in \mathcal{H}^e$ by 1.1.3.2. But

$$O^2(F^*(N_G(V))) \leq C_G(V) \leq C_G(U) \leq N,$$

so $O^2(F^*(N_G(V))) \leq O^2(F^*(N_N(V))) = 1$ as $N_N(V) \in \mathcal{H}^e$. Therefore $N_G(V) \in \mathcal{H}^e$ as desired. This shows that (1) holds when $U \trianglelefteq V$. Then as $U \trianglelefteq \trianglelefteq V$, (1) holds by induction on $|V : U|$.

Under the hypotheses of (2), $N_G(U)$ is contained in some $X \in \mathcal{M}(T)$, and, as we remarked earlier, $X \in \mathcal{H}^e$ by Hypothesis (E). Then as $N_G(U) = N_X(U)$, $N_G(U) \in \mathcal{H}^e$ by 1.1.3.2, proving (2).

For (3), observe $Z(T) \cap U \in \mathcal{S}_2^e(G)$ by (2), and then $U \in \mathcal{S}_2^e(G)$ by (1).

Now assume the hypotheses of (4) and set $R := C_{O_2(M)}(O_2(N))$. As $R \leq N \leq M$ by hypothesis, we conclude $R \leq O_2(N)$; and then $O_2(N)$ and R are centralized by $O^2(F^*(N)) =: L$. Then as $L = O^2(L)$, the Thompson $A \times B$-lemma A.1.18 says L centralizes $O_2(M)$. But $O_2(M) = F^*(M)$ as $M \in \mathcal{H}^e$, so that $L \leq Z(O_2(M))$, and then $L = O^2(L)$ forces $L = 1$. Thus (4) is established.

As G is of even characteristic, $\mathcal{M}(T) \subseteq \mathcal{H}^e$, so (4) implies (5).

If $N \in \mathcal{H}(T)$, then $O_2(N) \neq 1$, so there is M such that

$$T \leq N \leq N_G(O_2(N)) \leq M \in \mathcal{M}(N_G(O_2(N))).$$

Then as $T \in Syl_2(M)$, $M \in \mathcal{H}^e$ by (E), and also $O_2(M) \leq N$ by A.1.6. Therefore $N \in \mathcal{H}^e$ by (5), proving (6).

Finally assume the hypotheses of (7) and set $M_2 := M_1 O_2(M)$. By (4), $M_2 \in \mathcal{H}^e$. But as $|S : S \cap M_1| \leq 2$, $|M_2 : M_1| \leq 2$, and so $M_1 \trianglelefteq M_2$. Then $M_1 \in \mathcal{H}^e$ by 1.1.3.1 establishing (7).

This completes the proof of 1.1.4. $\square$

We also need to control members of $\mathcal{H}$ which are not in $\mathcal{H}^e$. The following result gives some control in an important special case. For example, the subsequent result 1.1.6 shows that the hypotheses are achieved in any sufficiently large subgroup of a 2-local subgroup.

Recall our convention in Notation A.3.5 that $\hat{A}_6$, $\hat{A}_7$, and $\hat{M}_{22}$ denote the nonsplit 3-fold covers of A_6, A_7, and M_{22}.

LEMMA 1.1.5. *Let $H \in \mathcal{H}$, $S \in Syl_2(H)$, and $M \in \mathcal{H}^e(S)$. Assume that*

$$C_{O_2(M)}(O_2(H \cap M)) \leq H,$$

and $M \in \mathcal{H}(C_G(z))$ for some $1 \neq z \in \Omega_1(Z(S))$. Then:

(1) $F^*(H \cap M) = O_2(H \cap M)$.

(2) z inverts $O(H)$.

(3) If L is a component of H, then $L = [L, z] \not\leq M$ and one of the following holds:

 (a) L is simple of Lie type and characteristic 2, described in conclusion (3) or (4) of Theorem C (A.2.3), and z induces an inner automorphism on L.

 (b) $1 \neq Z(L) = O_2(L)$ and $L/O_2(L)$ is $L_3(4)$ or $G_2(4)$, with z inducing an inner automorphism on L.

 (c) $L \cong A_6$ or $\hat{A}_6$, and z induces a transposition on L.

 (d) $L \cong A_7$ or $\hat{A}_7$, and z acts on L with cycle structure 2^3.

(e) $L \cong L_3(3)$ or $L_2(p)$, p a Fermat or Mersenne prime, and z induces an inner automorphism on L.

(f) $L/O_2(L)$ is a Mathieu group, J_2, J_4, HS, He, or Ru; and z induces a 2-central inner automorphism on L.

PROOF. Part (1) follows from 1.1.4.4 applied with $H \cap M$ in the role of "N", in view of our hypothesis.

Next $C_G(z) \leq M$ by hypothesis, so

$$C_{O(H)}(z) \leq O(H) \cap M \leq O(H \cap M) = 1$$

by (1), giving (2).

Now assume L is a component of H. If $L \leq M$ then $L \leq E(H \cap M)$, contrary to (1). Thus $L \not\leq M$ so in particular $L \not\leq C_G(z)$.

As $z \in Z(S)$ and $S \in Syl_2(H)$, z normalizes each component of H; so as $L \not\leq C_G(z)$, $L = [L, z]$.

Set $R := N_S(L)$ and $(RL)^* := RL/O_2(RL)$. Then $R \in Syl_2(RL)$ so $R^* \in Syl_2(R^*L^*)$ and z^* is an involution in the center of R^*. By hypothesis, $C_G(z) \leq M$, so $C_H(z) = C_{H \cap M}(z)$. Now $H \cap M \in \mathcal{H}^e$ by (1), so by 1.1.3.2, $C_{H \cap M}(z) \in \mathcal{H}^e$. Since $L \trianglelefteq \trianglelefteq H$ we have

$$C_L(z) \trianglelefteq \trianglelefteq C_H(z) = C_{H \cap M}(z),$$

so $C_L(z) \in \mathcal{H}^e$ by 1.1.3.1. Also $O^2(C_{L^*}(z^*)) = O^2(C_L(z))^*$ by Coprime Action, while $O_2(RL) \cap L \leq O_2(L) \leq Z(L)$, so we conclude $F^*(C_{L^*}(z^*)) = O_2(C_{L^*}(z^*))$ from A.1.8.

If z induces an inner automorphism on L then z centralizes $Z(L)$, so $O(L) = 1$ by (2), and hence $Z(L) = O_2(L)$. Put another way (recalling L is quasisimple), if $O(L) \neq 1$ then z induces an outer automorphism on L.

As H is an SQTK-group, we may examine the possibilities for $L/Z(L)$ appearing on the list of Theorem C.

Suppose first that $L/Z(L)$ is of Lie type and characteristic 2; then L^* appears in conclusion (3) or (4) of Theorem C. Now $z^* \in Z(R^*)$, so from the structure of $Aut(L^*)$, either $z^* \in L^*$, or L^* is A_6 or $\hat{A}_6$ with z^* inducing a transposition on L^*. However in the latter case as $O_2(L) = 1$, or else $L/O(L) \cong SL_2(9)$ by I.2.2.1, so that the transposition z does not centralize a Sylow 2-subgroup of L, contrary to $z \in Z(R)$; hence (c) holds. Thus we may assume $z^* \in L^*$, so by an earlier remark, $O(L) = 1$. Thus either $Z(L) = 1$, so L is simple and (a) holds; or from the list of Schur multipliers in I.1.3, L^* is $L_2(4)$, A_6, $Sz(8)$, $L_3(4)$, $G_2(4)$, or $L_4(2)$. Then as z centralizes a Sylow 2-group of L, when $L^* \cong L_2(4) \cong A_5$, A_6, or $Sz(8)$, we obtain a contradiction from the structure of the covering group L in (1) or (4) of I.2.2, or in 33.15 of [**Asc86a**]. This leaves covers of $L_3(4)$ and $G_2(4)$, which are allowed in (b).

We have shown that the lemma holds if $L/Z(L)$ is of Lie type and characteristic 2. But $A_5 \cong \Omega_4^-(2)$, $A_6 \cong Sp_4(2)'$, and $A_8 \cong L_4(2)$, so if L^* is an alternating group, then from conclusion (1) of Theorem C and I.1.3, $L^* \cong A_7$ or $\hat{A}_7$. As $F^*(C_{L^*}(z^*)) = O_2(C_{L^*}(z^*))$, $z^* \notin L^*$ and z^* is not a transvection, so we conclude z^* has cycle structure 2^3. As z centralizes a Sylow 2-group of L, we conclude that $O_2(L) = 1$, from the structure of the double cover of A_7 in 33.15 of [**Asc86a**]. So (d) holds.

Next assume $L/Z(L)$ is of Lie type and odd characteristic; then L^* appears in conclusion (2) of Theorem C. If $L^* \cong L_2(p^2)$ then as $L_2(9) \cong Sp_4(2)'$, we may assume $p > 3$. Therefore as $z^* \in Z(R^*)$, either $z^* \in L^*$ and $O(C_{L^*}(z^*)) \neq 1$ or z^* induces a field automorphism on L^* so that $C_{L^*}(z^*)$ has a component; in either case, this is contrary to $F^*(C_{L^*}(z^*)) = O_2(C_{L^*}(z^*))$. The same argument eliminates $L_2(p)$ unless p is a Fermat or Mersenne prime, which is allowed in (e); as before, the fact that $z \in Z(R)$ rules out the double covers $SL_2(p)$, the only possibilities with $Z(L) \neq 1$ by I.1.3. Similarly if $L^* \cong L_3^\epsilon(p)$ then as $z^* \in Z(R^*)$, $z^* \in L^*$; then unless $p = 3$, $C_{L^*}(z^*)$ has an $SL_2(p)$-component, for our usual contradiction. Finally $U_3(3) \cong G_2(2)'$ was covered earlier, while if $L^* \cong L_3(3)$ then $Z(L) = 1$ by I.1.3, so conclusion (e) holds.

This leaves the case L^* sporadic, so L^* appears in conclusion (5) of Theorem C. First J_1 is ruled out by the existence of a component in $C_{L^*}(z^*)$. Then as usual $z^* \in L^*$ since $z \in Z(R)$, so that (f) holds.

This completes the proof of 1.1.5.   □

LEMMA 1.1.6. *Let B be a nontrivial 2-subgroup of G, $H \leq G$ with $BC_G(B) \leq H \leq N_G(B)$, $S \in Syl_2(H)$, T a Sylow 2-subgroup of G containing S, z an involution in $Z(T)$, and $M \in \mathcal{M}(C_G(z))$. Then the hypotheses of 1.1.5 are satisfied.*

PROOF. As $z \in Z(T)$, $M \in \mathcal{M}(T)$, so $M \in \mathcal{H}^e$ since G is of even characteristic. Thus as $S \leq T$, $M \in \mathcal{H}^e(S)$. Next $B \leq O_2(H) \leq S \leq M$ so that $B \leq O_2(H \cap M)$, and hence

$$C_{O_2(M)}(O_2(H \cap M)) \leq C_G(B) \leq H.$$

Also $z \in C_T(B) \leq T \cap H = S$ as $S \in Syl_2(H)$, so $z \in Z(S)$; hence the hypotheses of 1.1.5 are satisfied. The proof is complete.   □

1.2. The set $\mathcal{L}^*(G,T)$ of nonsolvable uniqueness subgroups

In this section we use our results on the structure of SQTK-groups in section A.3 to establish tools for working in 2-local subgroups of G; such appeals are possible since our 2-locals are strongly quasithin. In particular we obtain a description of H^∞ for $H \in \mathcal{H}$, and also properties of the poset of perfect members of $\mathcal{H}$, partially ordered by inclusion. Such results then lead to the existence of uniqueness subgroups of G.

We begin by recalling Definition A.3.1 which defines $\mathcal{C}$-components: For $H \leq G$, let $\mathcal{C}(H)$ be the set of subgroups $L \leq H$ minimal subject to

$$1 \neq L = L^\infty \trianglelefteq \trianglelefteq H.$$

The members of $\mathcal{C}(H)$ are the *$\mathcal{C}$-components* of H. As we will see, usually we can expect there will be $H \in \mathcal{H}$ with $\mathcal{C}(H)$ nonempty.

We recall also that the elementary results in A.3.3 hold for arbitrary finite groups. By contrast, the later results in section A.3 requiring Hypothesis A.3.4 apply only to an SQTK-group X with $O_2(X) = 1$. We apply those results to $H/O_2(H)$ for $H \in \mathcal{H}$, and then pull them back to obtain results about H.

Recall that $\pi(X)$ denotes the set of primes dividing the order of a group X.

PROPOSITION 1.2.1. *Let $H \in \mathcal{H}$. Then*

(1) $\langle \mathcal{C}(H) \rangle = H^\infty$.

(2) If L_1, L_2 are distinct members of $\mathcal{C}(H)$, then $[L_1, L_2] \leq O_2(L_1) \cap O_2(L_2) \leq O_2(H)$.

(3) If $L \in \mathcal{C}(H)$, then either $L \trianglelefteq H$; or $|L^H| = 2$ and $L/O_2(L) \cong L_2(2^n)$, $Sz(2^n)$, $L_2(p)$, p an odd prime, or J_1.

(4) Let $L \in \mathcal{C}(H)$ and $\bar{H} = H/O_2(H)$. Then one of the following holds:

(a) $\bar{L}$ is a simple component of $\bar{H}$ on the list of Theorem C (A.2.3).

(b) $\bar{L}$ is a quasisimple component of $\bar{H}$, $Z(\bar{L}) \cong \mathbf{Z}_3$, and $\bar{L}$ is $SL_3^\epsilon(q)$, $q = 2^e$ or q an odd prime, $\hat{A}_6$, $\hat{A}_7$, or $\hat{M}_{22}$.

(c) $F^*(\bar{L}) \cong E_{p^2}$ for some prime $p > 3$, and $F^*(\bar{L})$ affords the natural module for $\bar{L}/F^*(\bar{L}) \cong SL_2(p)$.

(d) $F^*(\bar{L})$ is nilpotent with $Z(\bar{L}) = \Phi(F^*(\bar{L}))$, $\bar{L}/F^*(\bar{L}) \cong SL_2(5)$, and for each $p \in \pi(F^*(\bar{L}))$:

(i) either $p^2 \equiv 1 \mod 5$ or $p = 5$; and
(ii) either $O_p(\bar{L}) \cong p^{1+2}$, or $O_p(\bar{L})$ is homocyclic of rank 2.

(5) If $L \in \mathcal{C}(H)$ satisfies $O_2(L) \leq Z(L)$ and $m_2(L) > 1$, then L is quasisimple.

PROOF. As we observed at the start of the section, since $H \in \mathcal{H}$, H is an SQTK-group, and hence so is $\bar{H} := H/O_2(H)$. Certainly $O_2(\bar{H}) = 1$—so we may apply the results of section A.3. to $\bar{H}$. Further by A.3.3.4:

(*) The map $L \mapsto \bar{L}$ is an H-equivariant bijection of $\mathcal{C}(H)$ with $\mathcal{C}(\bar{H})$—with inverse $\bar{K} \mapsto K^\infty$, where K is the full preimage of $\bar{K}$ in H.

Thus for $L \in \mathcal{C}(H)$, we have $\bar{L} \in \mathcal{C}(\bar{H})$ and the possibilities in (4) are just those from A.3.6. Similarly the existence of the equivariant bijection in (*), together with A.3.7, A.3.9, and (1) and (3) of A.3.8, implies (2), (1), and (3), respectively.

Assume the hypotheses of (5). If $L/O_2(L)$ is quasisimple, then as $O_2(L) \leq Z(L)$ and L is perfect, L is quasisimple. Thus we may assume that case (4c) or (4d) holds. Then as $O_2(L) \leq Z(L)$, $O_{2,F}(L) = O_2(L) \times O(L)$. Thus $L/O(L)$ is the central extension of the 2-group $O_{2,F}(L)/O(L)$ by $L/O_{2,F}(L) \cong SL_2(p)$. But the multiplier of $SL_2(p)$ is trivial (I.1.3), so we conclude $O_2(L) = 1$. Now $m_2(L) = m_2(L/O(L))$ and $L/O(L) \cong SL_2(p)$ has 2-rank 1, contrary to the hypothesis that $m_2(L) > 1$. This establishes (5), and completes the proof of 1.2.1. $\qquad\square$

As we mentioned in the Introduction to Volume II, in the bulk of the proof, there will be $H \in \mathcal{H}$ with H nonsolvable; and in that case by 1.2.1.1, $\mathcal{C}(H)$ is nonempty.

LEMMA 1.2.2. Let $H \in \mathcal{H}$, $\bar{H} := H/O_2(H)$, $L \in \mathcal{C}(H)$, and p an odd prime.

(a) If $|L^H| = 2$ and $p \in \pi(\bar{L})$, then $O^{p'}(H) = \langle L^H \rangle$.

(b) If $m_p(L) = 2$ then $L \trianglelefteq H$.

PROOF. Part (b) follows as $m_p(\bar{L}) = 1$ for each of the groups $\bar{L}$ listed in 1.2.1.3.

Assume the hypotheses of (a), and set $L_0 := \langle L^H \rangle$. Recall L_0 is normal in H by 1.2.1.3. Then $m_p(\bar{L}_0) = 2$, so $C_{\bar{H}}(\bar{L}_0)$ is a p'-group as $m_p(H) \leq 2$. As $|L^H| = 2$, $O^{p'}(H)$ normalizes L. Recall from the Introduction to Volume I that we refer to [GLS98] for the structure of the outer automorphism groups of the groups listed in Theorem C. For those $\bar{L}$ listed in 1.2.1.3, $O^2(Out(\bar{L}))$ is a group of field automorphisms (or trivial), and $O^2(Aut(\bar{L}))$ splits over $Inn(\bar{L}) \cong \bar{L}$. Therefore if $O^{p'}(H) \not\leq L_0$, there is x of order p in $N_H(L) - L_0$. Then x centralizes nontrivial elements of order p in each factor of $P \in Syl_p(L_0)$, contradicting $m_p(H) \leq 2$. This contradiction gives $O^{p'}(H) \leq L_0$, while $L_0 = O^{p'}(L_0)$ as $\bar{L}$ is simple and $p \in \pi(\bar{L})$. This proves (a). $\qquad\square$

Next we extend the notation of $\mathcal{L}(X,Y)$ in Definition A.3.10 to our QTKE-group G. This will help us keep track of the possible embeddings of $\mathcal{C}$-components of a subgroup $H_1 \in \mathcal{H}$ in some other $H_2 \in \mathcal{H}$, as long as H_1 and H_2 share a common Sylow 2-subgroup.

DEFINITION 1.2.3. For H a finite group, and S a 2-subgroup of H, let $\mathcal{L}(H,S)$ be the set of subgroups L of H such that

(1) $L \in \mathcal{C}(\langle L,S \rangle)$,
(2) $S \in Syl_2(\langle L,S \rangle)$, and
(3) $O_2(\langle L,S \rangle) \neq 1$; that is, $\langle L,S \rangle \in \mathcal{H}_H$.

Assume for the moment that $H \in \mathcal{H}$, $S \in Syl_2(H)$, $\bar{H} := H/O_2(H)$, and $L \in \mathcal{L}(H,S)$. Then by Hypotheses (QT) and (K), $\bar{H}$ satisfies Hypothesis A.3.4, with $\bar{S} \in Syl_2(\bar{H})$; so from condition (1) of the definition of $\mathcal{L}(H,S)$ and A.3.3, $\bar{L} \in \mathcal{L}(\bar{H}, \bar{S})$, defined only for $\bar{H}$ in section A.3. Also applying 1.2.1.3 to $\langle L,S \rangle$, either $L^S = L$ and $\langle L,S \rangle = LS$, or $L^S = \{L, L^s\}$ and $\langle L,S \rangle = LL^sS$. Further as in A.3.11, $\mathcal{C}(H) \subseteq \mathcal{L}(H,S)$, so when $\mathcal{C}(H)$ is nonempty, $\mathcal{L}(H,S)$ is nonempty.

Now just as in section A.3, we wish to see how members of $\mathcal{L}(H,S)$ embed in H.

LEMMA 1.2.4. Let $H \in \mathcal{H}$, with $S \in Syl_2(H)$; set $\bar{H} := H/O_2(H)$, and assume $B \in \mathcal{L}(H,S)$. Then $B \leq L$ for a unique $L \in \mathcal{C}(H)$, and the pair $(\bar{B}, \bar{L})$ is on the list of lemma A.3.12. In particular

(+) If S normalizes B, then $L \trianglelefteq H$.

PROOF. We apply A.3.12 to conclude $\bar{B}$ is contained in a unique $\bar{L} \in \mathcal{C}(\bar{H})$, with the pair $(\bar{B}, \bar{L})$ on the list of A.3.12. Then using the one-to-one correspondence from A.3.3.4, $\bar{L}$ is the image of a unique $L \in \mathcal{C}(H)$; and as $B \leq O_2(H)L$ we see $B = B^\infty \leq (O_2(H)L)^\infty = L$. This completes the proof, as (+) follows from the uniqueness of L. $\square$

LEMMA 1.2.5. Let $H \in \mathcal{H}$, $S \in Syl_2(H)$, $R \leq S$ with $|S:R| = 2$, and suppose $L \in \mathcal{L}(H,R)$. Then there exists a unique $K \in \mathcal{C}(H)$ with $L \leq K$.

PROOF. The proof is much like that of A.3.12. Let $H^* := H/O_\infty(H)$. By 1.2.1.1, $H^\infty = K_1 \cdots K_r$ where $K_i \in \mathcal{C}(H)$, and by 1.2.1.2, $H^{\infty*} = K_1^* \times \cdots \times K_r^*$. Now $L = L^\infty \leq H^\infty$, so for some i (which we now fix), the projection P^* of L^* on $K^* := K_i^*$ is nontrivial. As P^* is a homomorphic image of $L^* \in \mathcal{C}(L^*)$, $P^* \in \mathcal{C}(P^*)$ by A.3.3.4.

As $S \in Syl_2(H)$ and K is subnormal in H, $S \cap K \in Syl_2(K)$, and similarly $R \cap L \in Syl_2(L)$ using our hypothesis that $L \in \mathcal{L}(H,R)$. Then as $R \leq S$, $S \cap L = R \cap L \in Syl_2(L)$, so $S \cap P \in Syl_2(P)$, for P the preimage of P^*. Then $|S \cap P : R \cap P| \leq |S : R| \leq 2$; so $(R \cap P)^* \not\leq O_\infty(P^*)$, as otherwise $P^*/O_\infty(P^*)$ has Sylow 2-groups of order at most 2, and so is solvable using Cyclic Sylow 2-Subgroups A.1.38, contrary to $P^* \in \mathcal{C}(P^*)$ nonsolvable. Hence $[L, R \cap P] \not\leq O_\infty(L)$. However as $(R \cap P)^*$ acts on P^* and permutes the $\mathcal{C}$-components L^R of $\langle L,R \rangle$, $R \cap P$ acts on L; so by A.3.3.7, $L = [L, R \cap P] \leq [L, K] \leq K$. Finally K is unique since $K_i \cap K_j \leq O_\infty(H)$ for any $j \neq i$. This completes the proof of 1.2.5. $\square$

Lemma 1.2.4 gives information about $\mathcal{L}(H,S)$ considered as a set partially ordered by inclusion. This leads us to define $\mathcal{L}^*(H,S)$ to be the maximal members of this poset.

We will focus primarily on the case where the role of S is played by $T \in Syl_2(G)$. In this case when $H \in \mathcal{H}(T)$, then T is also Sylow in H, so an earlier remark now specializes to:

LEMMA 1.2.6. $\mathcal{C}(H) \subseteq \mathcal{L}(H,T) \subseteq \mathcal{L}(G,T)$ for each $H \in \mathcal{H}(T)$.

THEOREM 1.2.7 (Nonsolvable Uniqueness Groups). If $L \in \mathcal{L}^*(G,T)$ then

(1) $L \in \mathcal{C}(H)$ for each $H \in \mathcal{H}(\langle L, T \rangle)$.
(2) $F^*(L) = O_2(L)$.
(3) $N_G(\langle L^T \rangle) = !\mathcal{M}(\langle L, T \rangle)$.
(4) Set $L_0 := \langle L^T \rangle$ and $Z := \Omega_1(Z(T))$ Then $C_Z(L_0) \cap C_Z(L_0)^g = 1$ for $g \in G - N_G(\langle L^T \rangle)$.

PROOF. Let $H \in \mathcal{H}(\langle L, T \rangle)$. As $T \in Syl_2(G)$, $T \in Syl_2(H)$, so also $L \in \mathcal{L}(H,T)$. Then by 1.2.4, $L \leq K \in \mathcal{C}(H)$ for some K. But by 1.2.6, $\mathcal{C}(H) \subseteq \mathcal{L}(G,T)$; so $L = K$ from the maximal choice of L. Hence (1) holds.

Next by 1.1.4.6, $F^*(H) = O_2(H)$; so as L is subnormal in H, (2) holds by 1.1.3.1.

Set $L_0 := \langle L^T \rangle$. As $L \in \mathcal{C}(H)$, $L_0 \trianglelefteq H$ by 1.2.1.3. Hence $H \leq M := N_G(L_0)$, and as $O_2(L) \neq 1$ by (2), $O_2(M) \neq 1$. In particular if $H \in \mathcal{M}(T)$, we conclude $H = M$. Thus (3) holds.

To prove (4), assume $Z_0 := C_Z(L_0) \cap C_Z(L_0)^g \neq 1$. Then $L_0 T, L_0^g T^g \leq C_G(Z_0)$, so using (3), $M = !\mathcal{M}(C_G(Z_0)) = M^g$; but then $g \in N_G(M) = M$ as $M \in \mathcal{M}$, contrary to $g \notin M$. $\qquad\square$

Part (3) of 1.2.7 says that if $L \in \mathcal{L}^*(G,T)$ then $\langle L, T \rangle$ is a uniqueness subgroup of G. This fact plays a crucial role through most of our work.

Next we obtain some further restrictions on chains in the poset $\mathcal{L}(G,T)$. For example we see in part (4) of 1.2.8 that for many choices of $L/O_2(L)$, $L \in \mathcal{L}(G,T)$ is already maximal. In parts (2) and (3) of 1.2.8 we see that if L is not T-invariant, then usually L is maximal.

LEMMA 1.2.8. Let S be a 2-subgroup of G, and $L, K \in \mathcal{L}(G,S)$ with $L \leq K$. Then

(1) $N_S(L) = N_S(K)$. So if $L \neq L^s$ then $LL^s \leq KK^s$ for $K \neq K^s$.
(2) If $L < \langle L^S \rangle$, then either

(a) $L = K$, or
(b) $L/O_2(L) \cong A_5$, and $K/O_2(K)$ is either J_1 or $L_2(p)$ for some prime p with $p^2 \equiv 1 \mod 5$.

(3) If $L < \langle L^S \rangle$, then either $L \in \mathcal{L}^*(G,S)$, or $L/O_2(L) \cong A_5$.
(4) We have $L \in \mathcal{L}^*(G,S)$ if $L/O_2(L)$ is any of the following: $\hat{A}_7$; $L_2(r^2)$, $r > 3$ an odd prime; $(S)L_3^\epsilon(p)$, p an odd prime; M_{11}, M_{12}, M_{23}, J_1, J_2, J_4, HS, He, Ru, $L_5(2)$, or $(S)U_3(2^n)$; a group of Lie type of characteristic 2 and Lie rank 2, other than $L_3(2)$ or $L_3(4)$.

PROOF. Let $H := \langle K, S \rangle$, and recall $\mathcal{C}(H) = \{K\}$ or $\{K, K^s\}$. By 1.2.4, K is the unique $\mathcal{C}$-component of H containing L, so that $N_S(L) \leq N_S(K)$. The opposite inclusion follows from A.3.12, as we check that in each of the embeddings listed there, K does not contain a product of two copies of L, so that L is $N_S(K)$-invariant. Hence (1) holds.

Assume as in (2) that $L \neq L^s$, and that $L < K$; then $K^s \neq K$ by (1). Then by 1.2.1.3, $K/O_2(K)$ is $L_2(2^n)$, $Sz(2^n)$, $L_2(p)$, or J_1, and $L/O_2(L)$ is also in this list. Consulting A.3.12, we see the only possible proper embeddings of L in K are those given in (2). This establishes (2) and (3).

Finally (4) is established similarly: from the list of groups in Theorem C, we extract the sublist *not* occurring as an initial possibility in A.3.12. $\square$

We next wish to study the action of members of $\mathcal{L}(G,T)$ on their internal modules. To do so, we use some of the results from section A.4 of Volume I. Recall from Definition A.4.5 that $\mathcal{X}$ consists of the nontrivial subgroups Y of G satisfying $Y = O^2(Y)$ and $F^*(Y) = O_2(Y)$. Notice the second condition says that $\mathcal{X} \subseteq \mathcal{H}^e$.

Now for $L \in \mathcal{L}(G,T)$, $L = L^\infty$ by the definition of $\mathcal{C}$-component, while $L \in \mathcal{H}^e$ by 1.2.7.2, so that $\mathcal{L}(G,T) \subseteq \mathcal{X}$. Next recall that for $Y \in \mathcal{X}$ and $R \in \mathcal{M}_{N_G(Y)}(Y,2)$, from Definition A.4.6

$$V(Y,R) := [\Omega_1(Z(R)), Y] \quad \text{and} \quad V(Y) := V(Y, O_2(Y)).$$

There we also defined $\mathcal{X}_f$ to consist of those $Y \in \mathcal{X}$ with $V(Y) \neq 1$. The subscript "f" stands for "faithful"; for example, if $X \in \mathcal{X}_f$ with $X/O_2(X)$ simple, then $X/O_2(X)$ is faithful on the module $V(X)$. Define

$$\mathcal{L}_f(G,T) := \mathcal{L}(G,T) \cap \mathcal{X}_f,$$

and also define

$$\mathcal{L}_f^*(G,T) := \mathcal{L}^*(G,T) \cap \mathcal{X}_f,$$

which of course coincides with $\mathcal{L}_f(G,T) \cap \mathcal{L}^*(G,T)$. Now by definition, elements of $\mathcal{L}_f^*(G,T)$ are maximal in the subposet $\mathcal{L}_f(G,T)$; in the next lemma we see that the converse holds.

LEMMA 1.2.9. *Let* $L \in \mathcal{L}_f(G,T)$. *Then*

(1) If $L \leq K \in \mathcal{L}(G,T)$, *then* $V(L, O_2(N_T(L)L)) \leq V(K, O_2(N_T(K)K))$, *and so* $K \in \mathcal{L}_f(G,T)$.

(2) If L *is maximal in* $\mathcal{L}_f(G,T)$ *with respect to inclusion, then* $L \in \mathcal{L}^*(G,T)$, *and hence* $L \in \mathcal{L}_f^*(G,T)$.

PROOF. Let $L \leq K \in \mathcal{L}(G,T)$ and $R := N_T(L)$. By 1.2.8.1, $R = N_T(K)$. Thus $R \in Syl_2(N_{KR}(L))$, so $O_2(KR) \leq R$ by A.1.6, and $O_2(RL) = C_R(L/O_2(L))$. Hence we may apply parts (2) and (3) of A.4.10 to obtain (1). Then (1) implies (2). $\square$

LEMMA 1.2.10. *Let* $T \in Syl_2(G)$, $H \in \mathcal{H}(T)$, *and* $L \in \mathcal{C}(H)$. *Then the following are equivalent:*

(1) $L \in \mathcal{L}_f(G,T)$.
(2) There is $V \in \mathcal{R}_2(H)$ *with* $[V, L] \neq 1$.
(3) $[R_2(H), L] \neq 1$.
In particular the result applies to $L \in \mathcal{L}^*(G,T)$ *and* $H \in \mathcal{H}(\langle L, T \rangle)$.

PROOF. We have $F^*(H) = O_2(H)$ by 1.1.4.6, and from 1.2.1.4 we see that all non-central 2-chief factors of L lie in $O_2(L)$. These are the hypotheses for A.4.11, whose conclusions are exactly the assertions of 1.2.10. $\square$

LEMMA 1.2.11. *Let $H \in \mathcal{H}$ with $T \cap H =: T_H \in Syl_2(H)$, and $K \in \mathcal{C}(H)$. Assume $z \in Z := \Omega_1(Z(T))$ lies in $K \cap T_H$ and:*

(a) $z \in [U, O^2(N_H(U))]$ for some elementary abelian 2-subgroup U of H;
(b) $C_T(O_2(H)) \leq T_H$.

Then either K is quasisimple or $H \in \mathcal{H}^e$.

PROOF. Assume K is not quasisimple; we must show $H \in \mathcal{H}^e$. By A.3.3.1, $E(K) = 1$, so $F^*(K) = F(K)$, and hence $C_K(F(K)) \leq F(K)$. We claim first that z centralizes $O(K)$: As H is an SQTK-group, $m_r(O(H)) \leq 2$ for all primes r. Then hypothesis (a) allows us to apply A.1.26.2 to $O^2(N_H(U))$, $[U, O^2(N_H(U))]$ in the roles of "X, V", to conclude $z \in [U, O^2(N_H(U))] \leq C_H(O(H)) \leq C_H(O(K))$. Next our assumption that $z \in K \cap T_H$ gives $z \in C_K(O(K)O_2(K)) \leq C_K(F(K)) \leq F(K)$, and hence $z \in O_2(K) \leq O_2(H)$.

Let $G_z := C_G(z)$ and $H_z := C_H(z)$. As $z \in O_2(H)$, $O^2(F^*(H)) \leq O^2(F^*(H_z))$, so it suffices to show $H_z \in \mathcal{H}^e$. As T_H is Sylow in H and $T_H \leq H_z$, $O_2(H) \leq O_2(H_z)$ by A.1.6. Therefore using (b),

$$C_{O_2(G_z)}(O_2(H_z)) \leq C_T(O_2(H)) \leq T_H \cap G_z \leq H_z.$$

Then as $G_z \in \mathcal{H}^e$ by 1.1.4.3, we get $H_z \in \mathcal{H}^e$ by 1.1.4.4. □

1.3. The set $\Xi^*(G, T)$ of solvable uniqueness subgroups of G

As noted in the Introduction to Volume II, it might happen that there are no nonsolvable locals $H \in \mathcal{H}(T)$, so that $\mathcal{L}(G, T)$ is empty; in this case we will need to produce some solvable uniqueness groups. Notice also that any L occurring in cases (c) or (d) of 1.2.1.4 involves interesting (and potentially tractable) solvable subgroups in $O_{2,F}(L)$.

Motivated particularly by the latter example:

DEFINITION 1.3.1. Define $\Xi(G, T)$ to consist of the subgroups $X \leq G$ such that:

(1) $X = O^2(X)$ is T-invariant with $XT \in \mathcal{H}$,
(2) $X/O_2(X) \cong E_{p^2}$ or p^{1+2} for some odd prime p, and
(3) T is irreducible on the Frattini quotient of $X/O_2(X)$.

Notice that each $X \in \Xi(G, T)$ is in $\mathcal{H}^e$ by 1.1.4.6 and 1.1.3.1, so as $X = O^2(X)$ we see $\Xi(G, T) \subseteq \mathcal{X}$.

Subsets $\Xi_-(G, T)$ and $\Xi_+(G, T)$ of $\Xi(G, T)$ appear in Definition 3.2.12.

We first collect some useful elementary properties of the members of $\Xi(G, T)$:

LEMMA 1.3.2. *Let $X \in \Xi(G, T)$. Then*

(1) X is a $\{2, p\}$-group for some odd prime p and $X = O_2(X)P$ for some $P \in Syl_p(X)$.
(2) $X = \langle P^X \rangle = \langle P^{O_2(X)} \rangle$ and $O_2(X) = [O_2(X), P]$.
(3) $T = O_2(X)N_T(P)$ and $N_T(P)$ is irreducible on $P/\Phi(P)$.
(4) $P = [P, \Phi(N_T(P))]$.
(5) If $H \in \mathcal{H}(XT)$, then $X = O^2(O_2(H)X)$.

PROOF. Part (1) is immediate from condition (2) in the definition of $\Xi(G, T)$ and Sylow's Theorem. As $X = O^2(X)$ in condition (1) of the definition of $\Xi(G, T)$, conclusion (1) now implies

$$X = \langle Syl_p(X) \rangle = \langle P^X \rangle = \langle P^{O_2(X)} \rangle$$

and $O_2(X) = [O_2(X), P]$, giving conclusion (2). Notice $XT = PT$, so the Dedekind Modular Law gives $N_{XT}(P) = PN_T(P)$; then a Frattini Argument on $X = O_2(X)P$ gives $T = O_2(X)N_T(P)$. Now $N_T(P)$ is irreducible on $P/\Phi(P)$ by condition (3) of the definition of $\Xi(G,T)$, so conclusion (3) is proved.

Let $S := N_T(P)$ and $S^* := S/C_T(P)$. Now S^* is irreducible on $P/\Phi(P)$ by (3), so each involution $i^* \in Z(S^*)$ inverts $P/\Phi(P)$. Thus for each $I \leq S$ with $i^* \in I^*$, $P = [P, I]$. In particular if $\Phi(S^*) \neq 1$, we can choose $I = \Phi(S)$, so that (4) holds in this case. Otherwise $\Phi(S^*) = 1$, and then S^* is reducible on $P/\Phi(P)$ by A.1.5. This contradiction completes the proof of (4).

Under the hypotheses of (5), $O_2(H) \leq T$, while by condition (1) of the definition, $T \leq N_G(X)$ and $X = O^2(X)$, so $X = O^2(O_2(H)X)$, as required. $\qquad \square$

Assume for the moment that $L \in \mathcal{L}(G,T)$ with $L/O_2(L)$ not quasisimple, as in cases (c) and (d) of 1.2.1.4. Then L is T-invariant by 1.2.1.3. Given an odd prime p, define

$$\Xi_p(L) := O^2(X_p), \quad \text{where} \quad X_p/O_2(L) := \Omega_1(O_p(L/O_2(L)));$$

then define $\Xi_{rad}(G,T)$ to be the collection of subgroups $\Xi_p(L)$, for $L \in \mathcal{L}(G,T)$ with $L/O_2(L)$ not quasisimple, and $p \in \pi(F(L/O_2(L)))$.

We observe that $X \in \Xi_{rad}(G,T)$ satisfies conditions (2) and (3) in the definition of $\Xi(G,T)$, using the action of $L/O_{2,F}(L) \cong SL_2(r)$ ($r = p$ or 5) in cases (c) and (d) of 1.2.1.4. By construction, $X = O^2(X)$, while X is T-invariant as X char $L \trianglelefteq LT$. Finally $LT \in \mathcal{H}^e$ by 1.1.4.6, so that $1 \neq O_2(LT) \leq O_2(XT)$ by A.1.6, the last requirement of condition (1) of the definition. So we see:

LEMMA 1.3.3. $\Xi_{rad}(G,T) \subseteq \Xi(G,T)$.

Define $\Xi^*(G,T)$ to consist of those $X \in \Xi(G,T)$ such that XT is not contained in $\langle L, T \rangle$ for any $L \in \mathcal{L}(G,T)$ with $L/O_2(L)$ quasisimple. So for Ξ (in contrast to $\mathcal{L}$), the superscript * will not denote maximality under inclusion in the poset $\Xi(G,T)$. However the following result will be used in 1.3.7 (which is the analogue of 1.2.7.1) to prove that XT is a uniqueness subgroup for each member X of $\Xi^*(G,T)$. Furthermore the list of possible embeddings of members of $\Xi(G,T)$ in nonsolvable groups appearing in the lemma will also be very useful.

PROPOSITION 1.3.4. Let $X \in \Xi(G,T)$, $P \in Syl_p(X)$ a complement to $O_2(X)$ in X, and $H \in \mathcal{H}(XT)$. Then either $X \trianglelefteq H$, or $X \leq \langle L^T \rangle$ for some $L \in \mathcal{C}(H)$ with $L/O_2(L)$ quasisimple, and in the latter case one of the following holds:

(1) L is not T-invariant and $P = (P \cap L) \times (P \cap L)^t \cong E_{p^2}$ for $t \in N_T(P) - N_T(L)$. Either $L/O_2(L) \cong L_2(2^n)$ with n even and $2^n \equiv 1 \mod p$, or $L/O_2(L) \cong L_2(q)$ for some odd prime q.

In the remaining cases, L is T-invariant and satisfies one of:

(2) $P \cong E_{p^2}$ and $L/O_2(L) \cong (S)L_3(p)$.

(3) $P \cong E_{p^2}$, $L/O_2(L) \cong Sp_4(2^n)$ with n even and $2^n \equiv 1 \mod p$, and $Aut_T(P)$ is cyclic.

(4) $p = 3$, $P \cong E_9$, and $L/O_2(L) \cong M_{11}$, $L_4(2)$, or $L_5(2)$.

PROOF. Set $\bar{H} := H/O_2(H)$. We first consider $F(\bar{H})$. So let r be an odd prime, and $\bar{R}$ a supercritical subgroup of $O_r(\bar{H})$. (Cf. A.1.21). As usual $m_r(\bar{R}) \leq 2$ since $m_r(H) \leq 2$. Therefore by A.1.32, $[\bar{R}, \bar{P}] = 1$ if $p \neq r$; while if $p = r$, then either

$\bar{R} = \bar{P}$—or $\bar{R} \cong \mathbf{Z}_p$, $\bar{P} \cong p^{1+2}$, and $\bar{R} = Z(\bar{P})$. In particular by A.1.21, $\bar{P}$ centralizes $O^p(F(\bar{H}))$.

Suppose for the moment that $O_p(\bar{H}) \neq 1$, and choose $r = p$. If $m_p(\bar{R}) = 2$ then $\bar{P} = \bar{R} \trianglelefteq \bar{H}$, so $X = O^2(R) \trianglelefteq H$ by 1.3.2.5, and the lemma holds. Therefore we may assume $m_p(\bar{R}) = 1$. Then as $\bar{R}$ is supercritical, it contains all elements of order r in $C_{O_r(\bar{H})}(\bar{R})$, so $O_p(\bar{H})$ is cyclic.

Thus in any case, we may assume that $O_p(\bar{H})$ is cyclic. In particular $\bar{P} \not\leq F(\bar{H})$ as $\bar{P}$ is noncyclic. Hence as $Aut(O_p(\bar{H}))$ is cyclic and $P = [P, N_T(P)]$, $\bar{P}$ centralizes $O_p(\bar{H})$; therefore as $\bar{P}$ centralizes $O^p(F(\bar{H}))$, $\bar{P}$ centralizes $F(\bar{H})$.

By 1.3.2.3, $N_T(P)$ is irreducible on $\bar{P}/\Phi(\bar{P})$, so as $O_p(\bar{H})$ is cyclic, $\bar{P} \cap O_p(\bar{H}) \leq \Phi(\bar{P})$; therefore as $\bar{P}$ centralizes $F(\bar{H})$, we conclude $C_{\bar{P}}(E(\bar{H})) \leq \Phi(\bar{P})$. Thus there is a component $\bar{L}_1$ of $\bar{H}$ with $[\bar{L}_1, \bar{P}] \neq 1$. By A.3.3.4, there is $L \in \mathcal{C}(H)$ with $\bar{L} = \bar{L}_1$. Set $K := \langle L^T \rangle$, so that $K \trianglelefteq H$ by 1.2.1.3. As $1 \neq [\bar{L}, \bar{P}]$, $[L, P] \not\leq O_2(L)$, so $L \leq [L, P] \leq [K, P]$ by A.3.3.7. Then as T acts on P, $K = \langle L^T \rangle = [K, P]$.

We claim $P \leq K$. Suppose first that $L < K = LL^t$. Then $\Phi(N_T(P)) \leq N_T(L)$ as $|L^T| = 2$. Notice that the groups listed in 1.2.1.3 have $Out(\bar{L})$ abelian. But by 1.3.2.4,
$$P = [P, \Phi(N_T(P))] = [P, N_T(P) \cap N_T(L)],$$
so P induces inner automorphisms on L and then also on K. Then by 1.2.2.a, $P \leq O^{p'}(H) = K$, establishing the claim in this case.

Next suppose that $L = K$. This time we examine $Out(\bar{L})$ for the groups $\bar{L}$ appearing in Theorem C, to see in each case there are no noncyclic p-subgroups U whose normalizer is irreducible on $U/\Phi(U)$—as would be the case for the image of P in $Out(\bar{L})$, if P did not induce inner automorphisms on $\bar{L}$. Thus $\bar{P} \leq \bar{L}C_{\bar{H}}(\bar{L})$. Then as $N_T(P)$ is irreducible on $P/\Phi(P)$, either $P \leq L = K$ as claimed, or $P \cap L \leq \Phi(P)$. However as $C_{\bar{P}}(\bar{L}) \leq \Phi(\bar{P})$, $m_p(\bar{L}) = 2$; so in the case where $P \cap L \leq \Phi(P)$, there exists x of order p in $C_{PL}(\bar{L}) - L$, and hence $m_p(L\langle x \rangle) > 2$, contradicting H an SQTK-group. This completes the proof of the claim.

Thus $P \leq K$ by the claim. Then by 1.3.2.2, $X = \langle P^{O_2(X)} \rangle \leq K$.

We next establish the lemma in the case $L < K = LL^t$. Here $m_p(L) = 1$ by 1.2.1.3, so
$$P = (P \cap L) \times (P \cap L)^t \cong E_{p^2},$$
and by 1.2.1.3, $\bar{L}$ is $L_2(2^n)$, $Sz(2^n)$, $L_2(q)$ for some odd prime q, or J_1. If $\bar{L}$ is $L_2(q)$ for some odd prime q, then conclusion (1) of the lemma holds, so we may assume we are in one of the other cases. Then as $PT = TP$, $\bar{P}$ lies in the Borel subgroup $N_{\bar{K}}(\bar{T} \cap \bar{K})$ of $\bar{K}$ if $\bar{L}$ is a Bender group, and similarly $P \leq N_K(T \cap K)$ when $\bar{L}$ is J_1. In the first case $\bar{P}$ lies in a Cartan subgroup, so $2^n \equiv 1 \mod p$, and in the second, $p = 3$ or 7. Further as P acts on $T \cap K$, $[N_{T \cap K}(P), P] \leq (T \cap K) \cap P = 1$. Therefore $Aut_T(P)$ is isomorphic to a subgroup of $Out(K)$, so as $N_{Aut_T(K)}(Aut_P(K))$ is irreducible on $Aut_P(K)/\Phi(Aut_P(K))$ by 1.3.2.4, $|Out(K)|_2 > 2$. Then as $|Out(\bar{K})| = 2 |Out(\bar{L})|^2$, $Out(\bar{L})$ is of even order, which reduces us to $\bar{L} \cong L_2(2^n)$, n even—so that conclusion (1) of the lemma holds.

It remains to treat the case $K = L \trianglelefteq H$, where we must show that one of conclusions (2)–(4) holds. Thus $P \leq L$ by the claim.

Suppose first that $p > 3$. Then the possibilities for $\bar{L}$ and $\bar{P}$ with $PT = TP$ are determined in A.3.15. Suppose case A.3.15.3 holds. Then p plays the role of "r" in that result, and it follows that the signs δ and ϵ there conincide. Thus $\bar{L} \cong (S)L_3^\delta(q)$ with $q \equiv \delta \mod 4$; further $C_{\bar{L}}(Z(\bar{T}))^\infty \cong SL_2(p)$ plays the role of

"K" in that result, so that $\bar{P} \cap C_{\bar{L}}(Z(\bar{T}))^\infty$ is cyclic of order p dividing $q - \delta$. This contradicts the irreducible action of $N_T(P)$ on $P/\Phi(P)$. Suppose case A.3.15.2 holds. Then $\bar{L} \cong (S)L_3(p)$ and $\bar{P} \cong E_{p^2}$; here the parabolic $N_{\bar{L}}(\bar{P})$ induces $SL_2(p)$ on $\bar{P}$, and in particular the action of $\bar{T}$ on $\bar{P}$ is irreducible. This case appears as our conclusion (2)—using I.1.3 to see that the only cover of $L_3(p)$ is $SL_3(p)$. In cases (1), (4), (6), and (7) of A.3.15 $\bar{P}$ is cyclic, whereas P is noncyclic, so those cases do not arise here. Thus it remains to consider the cases in A.3.15.5, with $\bar{L}$ of Lie type over $\mathbf{F}_{2^n}$ with $n > 1$, As $P \le L$, $\bar{P}$ lies in a Cartan subgroup of $\bar{L}$ by that result, so $\bar{L}$ is of Lie rank at least 2, and hence $\bar{L}$ is of one of the following Lie types: A_2, B_2, G_2, ${}^3D_4$, or ${}^2F_4$. As in an earlier case, $Aut_T(P)$ is isomorphic to a subgroup of $Out(\bar{L})$. In the last three types, $Out_T(\bar{L})$ consists only of field automorphisms; so as P is a p-group, $N_T(P)$ normalizes each subgroup of P, contradicting the irreducible action of $N_T(P)$ on $P/\Phi(P)$. If $\bar{L}$ is $Sp_4(2^n)$ then $Out(\bar{L})$ is cyclic as $n > 1$; cf. 16.1.4 and its underlying reference. So as $N_T(P)$ is irreducible, n is even and hence conclusion (3) holds. Finally if $\bar{L}$ is $(S)L_3(2^n)$ then $Out_T(L)$ is the product of groups generated by a field automorphism and a graph automorphism of order 2. However the field automorphism acts on each subgroup of P as above, and any automorphism of P of order 2 is not irreducible on P, so $N_T(P)$ is not irreducible on P. This eliminates $(S)L_3(2^n)$, completing the proof for $p > 3$.

We have reduced to the case $p = 3$. Here a priori $\bar{L}/Z(\bar{L})$ can be any group appearing in the conclusion of Theorem C. To eliminate the various possible cases, ordinarily we first apply the restriction $m_3(L) = 2$ (as P is noncyclic), and then the restriction $PT = TP$; a final sieve is provided by the irreducibility of $N_T(P)$ on $P/\Phi(P)$.

Thus from the cases in conclusion (2) of Theorem C: We do not have $\bar{L} \cong L_2(q^e)$ for $q > 3$, as $m_3(L) = 2$, and $\bar{L}$ is not $L_2(3^2) \cong A_6$ as $PT = TP$. The latter argument eliminates $U_3(3)$; while $\bar{L} \cong L_3(3)$ appears in conclusion (2). The groups $L_3^\delta(q)$ for $q > 3$ are eliminated when $q \equiv -\delta \mod 3$ since $m_3(L) = 2$; and when $q \equiv \delta \mod 3$, since $PT = TP$ and $N_T(P)$ is irreducible on $P/\Phi(P)$.

We next turn to conclusion (1) of Theorem C: A_5 is eliminated as $m_3(L) = 2$, and A_6 is impossible since $PT = TP$ as just noted. In A_7 there is indeed a subgroup $PT = TP \cong \mathbf{Z}_2/(A_4 \times A_3)$; but even in $Aut(A_7) = S_7$, we see that $S_4 \times S_3$ fails the requirement $N_T(P)$ irreducible on $P/\Phi(P)$. Finally $A_8 \cong L_4(2)$ appears in conclusion (4) of our proposition, as do the groups $L_4(2)$ and $L_5(2)$ arising in conclusion (4) of Theorem C.

In conclusion (3) of Theorem C, $\bar{L}/Z(\bar{L})$ is of Lie type in characteristic 2. Then as $P \le L$ and $PT = TP$, $\bar{P}$ is contained in a proper parabolic of $\bar{L}$, and unless possibly $\bar{L}$ is defined over $\mathbf{F}_2$, we have $\bar{P}$ in the Borel subgroup $N_{\bar{L}}(\bar{T} \cap \bar{L})$. The case where $\bar{P}$ is contained in a Borel subgroup was treated above among the embeddings in A.3.15. In the case where $\bar{L}$ is defined over $\mathbf{F}_2$, proper parabolics have 3-rank at most 1, contradicting P noncyclic.

This leaves only conclusion (5) of Theorem C, where $\bar{L}/Z(\bar{L})$ is sporadic. Notice the case $\bar{L} \cong M_{11}$ appears in conclusion (4) of our lemma, while $\bar{L}$ is not J_1 as $m_3(L) = 2$. In the other cases, we use [**Asc86b**] to see that $PT = TP$ rules out all but M_{23}, M_{24}, J_2, J_4—which contain 2-groups extended by $GL_2(4)$, $S_3 \times L_3(2)$, $\mathbf{Z}_2/(S_3 \times \mathbf{Z}_3)$, $S_5 \times L_3(2)$, respectively. In these cases (even in $Aut(J_2)$) $N_T(P)$ is not irreducible on $P/\Phi(P)$.

This completes the proof of 1.3.4. $\square$

We have the following corollaries to Proposition 1.3.4:

PROPOSITION 1.3.5. *If $X \in \Xi^*(G,T)$ and $H \in \mathcal{H}(XT)$, then $X \trianglelefteq H$.*

PROOF. Notice that the proposition follows from 1.3.4, since by 1.2.6, $\mathcal{C}(H) \subseteq \mathcal{L}(G,T)$ for $H \in \mathcal{H}(T)$. $\qquad\square$

LEMMA 1.3.6. *If $X \in \Xi(G,T)$ with $X/O_2(X) \cong p^{1+2}$, then $X \in \Xi^*(G,T)$.*

PROOF. This is immediate from 1.3.4, which says that if $X \leq \langle L^T \rangle$ with $L/O_2(L)$ quasisimple, then $P \cong E_{p^2}$. $\qquad\square$

Now, as promised, we see that if $X \in \Xi^*(G,T)$, then XT is a uniqueness subgroup of G:

THEOREM 1.3.7 (Solvable Uniqueness Groups). *If $X \in \Xi^*(G,T)$ then $N_G(X) = !\mathcal{M}(XT)$.*

PROOF. Let $M \in \mathcal{M}(XT)$. By 1.3.5, $X \trianglelefteq M$, so maximality of M gives $M = N_G(X)$. $\qquad\square$

Recall that $\Xi_{rad}(G,T)$ consists of the subgroups $\Xi_p(L)$, for $L \in \mathcal{L}(G,T)$ such that $L/O_2(L)$ is not quasisimple; and by 1.3.3, $\Xi_{rad}(G,T) \subseteq \Xi(G,T)$. Define $\Xi_{rad}^*(G,T)$ to consist of those $X \in \Xi_{rad}(G,T)$ such that $X \trianglelefteq L \in \mathcal{L}^*(G,T)$. We see next that XT is a uniqueness subgroup for each $X \in \Xi_{rad}^*(G,T)$. This fact will allow us to avoid most of the difficulties caused by those $L \in \mathcal{L}^*(G,T)$ for which $L/O_2(L)$ is not quasisimple, by replacing the uniqueness group LT with the smaller uniqueness subgroup $\Xi_p(L)T$.

PROPOSITION 1.3.8. $\Xi_{rad}^*(G,T) \subseteq \Xi^*(G,T)$.

PROOF. Let $X \in \Xi_{rad}^*(G,T)$. Then $X \trianglelefteq L \in \mathcal{L}^*(G,T)$ by definition. By 1.3.3, $X \in \Xi(G,T)$, so there is an odd prime p such that $X = O_2(X)P$ for $P \in Syl_p(G)$. Indeed by 1.2.1.4, $p > 3$ and either $L/X \cong SL_2(p)$ or $L/O_{2,F}(L) \cong SL_2(5)$. Thus in any case there is a prime r with $L/O_{2,F}(L) \cong SL_2(r)$; r has this meaning throughout the remainder of the proof of the proposition.

By 1.2.1.3, T normalizes L. Then by 1.2.7.3, $M := N_G(L) = !\mathcal{M}(LT)$. We will see shortly how this uniqueness property can be exploited. As X is characteristic in L, $X \trianglelefteq M$, so we also get $M = N_G(X)$ using the maximality of $M \in \mathcal{M}$.

We will next establish a condition used to apply the methods of pushing up. Set $R := O_2(XT)$. Recall the definition of $C(G,R)$ from Definition C.1.5. We claim that

$$C(G,R) \leq M. \qquad\qquad (*)$$

The proof of the claim will require a number of reductions.

We begin by introducing a useful subgroup Y of $N_G(R)$: Recall $X \trianglelefteq LT$, and R is Sylow in $C_{LT}(X/O_2(X))$ by A.4.2.7; so by a Frattini Argument,

$$LT = C_{LT}(X/O_2(X))N_{LT}(R). \qquad\qquad (**)$$

Thus if we set $Y := N_L(R)^\infty$, then Y contains X, and also by the factorization $(**)$, $N_Y(P)$ has a section $SL_2(r)$, where $r = p$ or 5. So our construction gives $1 \neq Y \in \mathcal{C}(N_{LT}(R))$, such that $Y/O_2(Y)$ is not quasisimple. Further T normalizes R, so in fact using 1.2.6, $Y \in \mathcal{C}(N_{LT}(R)) \subseteq \mathcal{L}(G,T)$.

Next we obtain some restrictions on L. If $R \trianglelefteq LT$ then any $1 \neq C$ char R is normal in LT, so as $M = !\mathcal{M}(LT)$ by 1.2.7.3, we conclude $N_G(C) \leq M$, establishing our claim (*). Thus we may assume R is not normal in LT.

Suppose next that p is the only odd prime in $\pi(O_{2,F}(L))$; in particular this holds in case (c) of 1.2.1.4. Then $X = O^2(O_{2,F}(L))$, so as R centralizes $X/O_2(X)$,

$$[R,L] \leq C_L(X/O_2(X)) \leq O_{2,F}(L) \leq XR$$

and hence $RX \trianglelefteq RL$. But then $R = O_2(RX) \trianglelefteq LT$, contrary to our assumption. Thus we may assume L is in case (d) of 1.2.1.4, and hence $r = 5$ with either $p = 5$ or $p \equiv \pm 1 \mod 5$. Further we have shown there is an odd prime $q \in \pi(O_{2,F}(L))$ with $p \neq q$. Notice that $q \geq 5$.

For $1 \neq C$ char R, let $L_C := N_L(C)^\infty$, and set $X_q := \Xi_q(L)$. Notice $Y \leq L_C$ and $O_2(X_q) \trianglelefteq XT$, so $O_2(X_q) \leq R$. Therefore $R \in Syl_2(X_qR)$. Then as $q \geq 5$, by Solvable Thompson Factorization B.2.16,

$$X_qR = N_{X_qR}(J(R))C_{X_qR}(\Omega_1(Z(R))).$$

So for $C_0 := J(R)$ or $\Omega_1(Z(R))$, $N_{X_qR}(C_0) \not\leq O_{2,\Phi}(X_q)$. Therefore as Y is irreducible on $X_q/O_{2,\Phi}(X_q)$, we conclude $X_q \leq N_G(C_0)$, so $X_q = [X_q, Y] \leq L_{C_0}$. Hence $\pi(O_{2,F}(L_{C_0}))$ contains at least two odd primes p and q.

We are now in a position to complete the proof of the claim. Assume (*) fails. Then there is $1 \neq C$ char R with $N := N_G(C) \not\leq M$. As $YT \leq N_G(R)$, $N \in \mathcal{H}(YT)$; in particular $Y \in \mathcal{L}(N,T)$ as we saw $Y \in \mathcal{L}(G,T)$. So we may apply 1.2.4 to embed $Y \leq Y_C \in \mathcal{C}(N)$, with the inclusion described in A.3.12. Notice in particular that $Y_C \trianglelefteq N$, by 1.2.2.b, since Y contains X of p-rank 2. Also $Y \leq L_C$ by the previous paragraph, so $L_C = [L_C, Y] \leq Y_C$ as $L_C \in \mathcal{L}(G,T)$.

Now if $X \trianglelefteq Y_C$, then X char Y_C using 1.2.1.4, and hence $N \leq N_G(X) = M$, contrary to our choice of $N \not\leq M$. Thus we may assume X is not normal in Y_C. As $X \trianglelefteq Y$, it follows that $Y < Y_C$. In addition the fact that X is not normal in Y_C means $X \not\leq O_\infty(Y_C)$, which rules out cases (21) and (22) of A.3.12, leaving only case (10) of A.3.12 with $Y_C/O_2(Y_C) \cong L_3(p)$. In particular, $p = r = 5$, $Y = L_C$, and $\pi(O_{2,F}(L_C)) = \{2, p\}$. But we saw earlier that $\pi(O_{2,F}(L_{C_0}))$ contains at least two odd primes, so we conclude that our counterexample C cannot be the subgroup C_0 constructed earlier. That is, $N_G(R) \leq N_G(C_0) \leq M$.

Let $(Y_CR)^* := Y_CR/O_2(Y_CR)$. Then P^* is the unipotent radical of a maximal parabolic of $Y_C^* \cong L_3(5)$, so $\mathcal{U}_{Y_C^*R^*}(P^*, 2) = 1$, giving $R^* = 1$ and hence $R \leq O_2(Y_CR)$. On the other hand $O_2(Y_CR) \leq O_2(XT) = R$, so $R = O_2(Y_CR)$. But then $Y_C \leq N_G(R) \leq M$, impossible as X is normal in M, but not in Y_C. This establishes (*); namely $C(G,R) \leq M$.

We now use (*) and results on pushing up from section C.2 to complete the proof of 1.3.8: Assume $X \notin \Xi^*(G,T)$. Then $XT < \langle K, T\rangle =: H$ for some $K \in \mathcal{L}(G,T)$, with $K/O_2(K)$ quasisimple, and H is described in 1.3.4. As $p > 3$, H does not satisfy 1.3.4.4. Now $Aut_{T\cap L}(P)$ is quaternion in cases (c) and (d) of 1.2.1.4, so $Aut_T(P)$ is not cyclic and hence 1.3.4.3 does not hold. Thus we have reduced to cases (1) and (2) of 1.3.4. As X is not normal in H but $X \trianglelefteq M$, $K \not\leq M$. Further from 1.3.4, R acts on K in each case.

We observe next that the property $R \in \mathcal{B}_2(H)$ from Definition C.1.1 and Hypothesis C.2.3 of Volume I hold for $M_H := H \cap M$: Namely $C(H,R) \leq M_H$ using (*); and then by A.4.2.7, $R = O_2(N_H(R))$ and $R \in Syl_2(\langle R^{M_H}\rangle)$.

Furthermore $H \in \mathcal{H}^e$ by 1.1.4.6. So as $K \not\leq M$ and R acts on K, we have the hypotheses of C.2.7, and we conclude K appears on the list of C.2.7.3. In 1.3.4.2, $K/O_2(K) \cong (S)L_3(p)$, whereas no such K appears in C.2.7.3. So we have reduced to 1.3.4.1 where $K/O_2(K) \cong L_2(2^n)$ or $L_2(q)$ for q an odd prime. Then by C.2.7.3, either K is an $L_2(2^n)$-block or an A_5-block, or else $K/O_2(K) \cong L_2(7) \cong L_3(2)$. But the latter two cases are eliminated as $PT = TP$ with $p > 3$. Therefore K is an $L_2(2^n)$-block and 1.3.4.1 holds, so $H = KK^tT$, where $t \in N_T(P) - N_T(K)$. In particular $[K, K^t] = 1$ as distinct blocks commute by C.1.9, so $X = WW^t$ with $W := O^2(X \cap K)$ and $[W, W^t] = 1$. Thus

$$X/Z(X) = WZ(X)/Z(X) \times W^tZ(X)/Z(X)$$

and then $N_G(X)$ permutes $\{WZ(X), W^tZ(X)\}$ by the Krull-Schmidt Theorem A.1.15. In particular, $L = O^2(L)$ acts on $WZ(X)$. This is impossible, as $N_L(P)$ is irreducible on $P/\Phi(P)$ from the structure of L in cases (c) and (d) of 1.2.1.4.

The proof of 1.3.8 is complete. $\qquad\square$

As in the previous section, we want to study the action of members of our new class of solvable uniqueness subgroups on their internal modules. So let $\Xi_f(G, T)$ consist of those $X \in \Xi(G, T)$ with $X \in \mathcal{X}_f$, and let $\Xi_f^*(G, T) := \Xi_f(G, T) \cap \Xi^*(G, T)$.

LEMMA 1.3.9. *Let* $X \in \Xi(G, T)$, $L \in \mathcal{L}(G, T)$, *and* $X \leq K := \langle L^T \rangle$. *Then*

(1) If $L/O_2(L)$ *is quasisimple, then* $L \in \mathcal{L}^*(G, T)$.

(2) If $X \in \Xi_f(G, T)$, *then* $V(X, C_{T \cap K}(X/O_2(X))) \leq V(K)$ *and* $L \in \mathcal{L}_f(G, T)$.

PROOF. Part (2) follows from A.4.10, just as in the proof of 1.2.9. Thus it remains to establish (1).

Assume $L/O_2(L)$ is quasisimple. Then L is described in 1.3.4. In cases (2)–(4) of 1.3.4, $L \in \mathcal{L}^*(G, T)$ by 1.2.8.4—unless possibly $L/O_2(L) \cong L_4(2)$. But in the latter case, if (1) fails, then $L < Y \in \mathcal{L}(G, T)$, and from 1.2.4 and A.3.12, $Y/O_2(Y) \cong L_5(2)$, M_{24}, or J_4. Now $X = O_2(X)P$ with $P \cong E_9$ and $N_T(P)$ is irreducible on P, so T acts nontrivially on the Dynkin diagram of $L/O_2(L) \cong L_4(2)$. This is impossible, as no such outer automorphism is induced in $Aut(Y/O_2(Y))$.

Therefore 1.3.4.1 must hold. Then by 1.2.8.2, $P \cong E_9$, $L/O_2(L) \cong A_5$, and $Y/O_2(Y) \cong J_1$ or $L_2(p)$. But again as $N_T(P)$ is irreducible on P, some element of $N_T(L)$ induces an outer automorphism on $L/O_2(L) \cong A_5$, whereas no such automorphism is induced in $N_T(Y)$.

Thus 1.3.9 is established. $\qquad\square$

1.4. Properties of some uniqueness subgroups

In this section we summarize some basic properties of the families $\mathcal{L}^*(G, T)$ and $\Xi^*(G, T)$ of uniqueness subgroups, which will be used heavily later.

So we consider some L contained either in $\mathcal{L}^*(G, T)$ or in $\Xi^*(G, T)$. Note that the assertion in 1.4.1.1 below is the starting point (as we just saw in the proof of 1.3.8) for arguments using pushing up (sections C.2 etc.).

LEMMA 1.4.1. *Let* $L \in \mathcal{L}^*(G, T) \cup \Xi^*(G, T)$ *and set* $L_0 := \langle L^T \rangle$ *and* $Q := O_2(L_0T)$. *Then* $M := N_G(L_0) =!\mathcal{M}(L_0T)$, *so* L_0T *and* $N_G(Q)$ *are both uniqueness subgroups, and*

(1) $C(G, Q) \leq M$.

(2) $Q \in \mathcal{U}_G^*(L_0, 2) = Syl_2(C_M(L_0/O_2(L_0)))$.

(3) $C_G(Q) \leq O_2(M) \leq Q$.

(4) If $L \in \mathcal{X}_f$, then there is $V \in \mathcal{R}_2(L_0 T)$ with $[V, L_0] \neq 1$.

(5) If $L \in \mathcal{L}_f^*(G, T)$, assume that $L/O_2(L)$ is quasisimple. Let $V \in \mathcal{R}_2(L_0, T)$ with $[V, L_0] \neq 1$. Then $C_{L_0 T}(V) \leq O_{2,\Phi}(L_0 T)$, $C_T(V) = Q$, and $\Omega_1(Z(Q)) = R_2(L_0 T)$.

PROOF. First $M := N_G(L_0) = !\mathcal{M}(L_0 T)$ by 1.2.7.3 or 1.3.7. Then since $L_0 T \leq N_G(Q)$ by definition of Q, also $M = !\mathcal{M}(N_G(Q))$.

Next if $1 \neq R$ char Q, then embedding $N_G(R) \leq N \in \mathcal{M}$, we have $N_G(Q) \leq N_G(R) \leq N$, forcing $N = M$ as $M = !\mathcal{M}(L_0 T)$. So (1) holds.

Now (2) follows from A.4.2.7. By (1), $C_G(Q) \leq M$, and by (2), $O_2(M) \leq Q$. Also $M \in \mathcal{H}(T) \subseteq \mathcal{H}^e$ by 1.1.4.6, so

$$C_G(Q) \leq C_M(Q) \leq C_M(O_2(M)) \leq O_2(M),$$

giving (3).

Next when $L \in \mathcal{L}_f(G, T)$, there exists $V \in \mathcal{R}_2(L_0 T)$ with $[V, L] \neq 1$ by 1.2.10.3, while this follows from A.4.11 when $L \in \Xi_f(G, T)$, since there all 2-chief factors lie in $O_2(L)$. Thus (4) holds. Finally assume that either $L \in \Xi_f^*(G, T)$ or $L \in \mathcal{L}_f^*(G, T)$ with $L/O_2(L)$ quasisimple. Therefore $L_0 Q/O_{2,\Phi}(L_0 T) = F^*(L_0 T/O_{2,\Phi}(L_0 T))$ is a chief factor for $L_0 T$, so as $[V, L_0] \neq 1$, $C_{L_0 T}(V) \leq O_{2,\Phi}(L_0 T)$. But Q is Sylow in $O_{2,\Phi}(L_0 T)$ so $C_T(V) \leq Q$, while as V is 2-reduced, $Q = O_2(L_0 T) \leq C_T(V)$. This completes the proof of (5). $\square$

CHAPTER 2

Classifying the groups with $|\mathcal{M}(T)| = 1$

Recall from the outline in the Introduction to Volume II that the bulk of the proof of the Main Theorem proceeds under the Thompson amalgam strategy, which is based on the interaction of a pair of distinct maximal 2-local subgroups containing a Sylow 2-subgroup T of G. Clearly before we can implement that strategy, we must treat the case where T is contained in a unique maximal 2-local subgroup.

In Theorem 2.1.1 of this chapter, we determine the simple QTKE-groups G in which a Sylow 2-subgroup T is contained in a unique maximal 2-local subgroup.

This condition is similar to the hypothesis defining an abstract minimal parabolic B.6.1, where T lies in a unique maximal subgroup of G, so we can expect many of the examples arising in E.2.2 to appear as conclusions in Theorem 2.1.1.

The generic examples of simple QTKE-groups with $|\mathcal{M}(T)| = 1$ are the Bender groups. Recall a *Bender group* is a simple group of Lie type and characteristic 2 of Lie rank 1; namely $L_2(2^n)$, $Sz(2^n)$, or $U_3(2^n)$. The Bender groups also appear in case (a) of E.2.2.2. In addition, some groups from cases (c) and (d) of E.2.2.2 also satisfy the hypotheses of Theorem 2.1.1, as does M_{11} which is not a minimal parabolic.

However, shadows of various groups which are not simple also intrude, and eliminating them is fairly difficult. We mention in particular the shadows of certain groups of Lie type and Lie rank 2 of characteristic 2, extended by an outer automorphism nontrivial on the Dynkin diagram: namely as in cases (1a) and (2b) of E.2.2, extensions of the groups $L_2(2^n) \times L_2(2^n)$, $Sz(2^n) \times Sz(2^n)$, $L_3(2^n)$, and $Sp_4(2^n)$. These groups are not simple, but they *are* QTKE-groups with the property that the normalizer of a Borel subgroup is the unique maximal 2-local containing a Sylow 2-subgroup. We will eliminate the first two families of shadows in 2.2.5 by first using the Alperin-Goldschmidt Fusion Theorem to produce a strongly closed abelian subgroup, and then arguing that G is a Bender group to derive a contradiction. However it is difficult to see that the shadows of the latter two families are not simple, until we have reconstructed in Theorem 2.4.7 most of their local structure, and are then able to transfer off the graph automorphisms and so obtain a contradiction.

Also certain groups of Lie type and odd characteristic are troublesome: The groups $L_2(p) \times L_2(p)$, p a Fermat or Mersenne prime, extended by a 2-group interchanging the components (a subcase of case (b) of E.2.2.1); and the group $L_4(3) \cong P\Omega_6^+(3)$ extended by a group of automorphisms not contained in $PO_6^+(2)$. These groups are also minimal parabolics but not strongly quasithin. Shadows related to the last group appear in many places in the proof.

2.1. Statement of main result

Our main theorem in this chapter is:

THEOREM 2.1.1. *Assume G is a simple QTKE-group, $T \in Syl_2(G)$, and $M = !\mathcal{M}(T)$. Then G is a Bender group, $L_2(p)$ for $p > 7$ a Fermat or Mersenne prime, $L_3(3)$, or M_{11}.*

Of course the groups appearing in the conclusion of Theorem 2.1.1 also appear in the conclusion of our Main Theorem. Thus after Theorem 2.1.1 is proved, we will be able to assume that $|\mathcal{M}(T)| \geq 2$ in the remainder of our work.

Throughout chapter 2, we assume that G, M, T satisfy the hypotheses of Theorem 2.1.1. Thus $M = !\mathcal{M}(T)$, and hence by Sylow's Theorem, also $M = !\mathcal{M}(T')$ for each Sylow 2-subgroup T' of M, so we are free to let T vary over $Syl_2(M)$.

2.2. Bender groups

As we mentioned, the generic examples in Theorem 2.1.1 are Bender groups. These groups were originally characterized by Bender as the simple groups G with the property that the Sylow 2-normalizer M is *strongly embedded* in G; that is (cf. I.8.1), $N_G(D) \leq M$ for all nontrivial 2-subgroups D of M.

If we assume that G is not a Bender group, then there is $1 < D \leq T$ with $N_G(D) \not\leq M$, so that $N_G(D) \in \mathcal{H}(D, M)$ in our notation. If we pick D so that $U := N_T(D)$ is of maximal order subject to this constraint, then since $M = !\mathcal{M}(T)$ by hypothesis, U is a proper subgroup of T Sylow in $N_G(D)$ with $N_G(U) \leq M$. Our proof will focus on pairs (U, H_U) such that $U \leq T$, $U \in Syl_2(H_U)$, and $H_U \in \mathcal{H}^e(U, M)$. While the pair $(U, N_G(D))$ satisfies the first two conditions, and $N_G(D) \in \mathcal{H}(U, M)$, it may not be the case that $N_G(D) \in \mathcal{H}^e$. Thus to ensure that such pairs exist, we use an approach due to GLS (cf. p. 97 in [**GLS94**]) to produce a nontrivial strongly closed abelian 2-subgroup in the absence of such pairs. Then we argue as in the GLS proof [1] of Goldschmidt's Fusion Theorem, to show that G is a Bender group. Our extra hypotheses makes the proof here much easier. We identify G using a special case of Shult's Fusion Theorem, which appears in Volume I as Theorem I.8.3, and is deduced in Volume I from Theorem ZD in [**GLS99**].

We now begin to implement the GLS approach. Instead of considering arbitrary subgroups D of T, we focus on the members of the Alperin-Goldschmidt conjugation family: Using the language of Theorem 16.1 in [**GLS96**] (a form of the Alperin-Goldschmidt Fusion Theorem):

DEFINITION 2.2.1. Given a finite group G and $T \in Syl_2(G)$, define $\mathcal{D}$ to be the set of all nontrivial subgroups D of T such that

(a) $N_T(D) \in Syl_2(N_G(D))$,
(b) $C_G(D) \leq O_{2',2}(N_G(D))$, and
(c) $O_{2',2}(N_G(D)) = O(N_G(D)) \times D$.

The set $\mathcal{D}$ is called the *Alperin-Goldschmidt conjugation family* for T in G.

Next recall that a subgroup X of T is *strongly closed* in T with respect to G if for each $g \in G$, $X^g \cap T \subseteq X$.

---

[1]See the proof of Theorem SA in section 24 of [**GLS99**]—but recall that we will not make use of their hypothesis of even type.

PROPOSITION 2.2.2. *Assume for each D in the Alperin-Goldschmidt conjugation family that $N_G(D) \le M$ for T in G. Then*

(1) Each normal 2-subgroup of M is strongly closed in T with respect to G.

(2) G is a Bender group.

PROOF. We first prove (1). Let U be a normal 2-subgroup of M, $u \in U$ and $g \in G$ with $u^g \in T$. We must show $u^g \in U$, so assume otherwise. By the Alperin-Goldschmidt Fusion Theorem (the elementary result 16.1 in [**GLS96**], proved as X.4.8 and X.4.12 in [**HB85**]), there exist $u =: u_1, \ldots, u_n := u^g$ in T, $D_i \in \mathcal{D}$, and $x_i \in N_G(D_i)$, $1 \le i < n$, such that $u^g = u^{x_1 \cdots x_{n-1}}$, $\langle u_i, u_{i+1} \rangle \le D_i$, and $u_i^{x_i} = u_{i+1}$. As $u = u_1 \in U$ but $u_n = u^g \notin U$, there exists a least i such that $u_{i+1} \notin U$. Thus $u_i \in U$, and by hypothesis $x_i \in N_G(D_i) \le M$; therefore as $U \trianglelefteq M$, also $u_{i+1} = u_i^{x_i} \in U$, contrary to the choice of i. Thus (1) holds.

We could now appeal to Goldschmidt's Fusion Theorem [**Gol74**] to establish (2). However the version of this theorem in our list of Background References (cf. Theorem SA in [**GLS99**]) assumes that G is of even type, whereas in the Main Theorem we assume G is of even characteristic. Fortunately the even type hypothesis is unnecessary, and we now extract an easier version of the proof from section 24 of [**GLS99**] under our own hypotheses:

Let U be a minimal normal 2-subgroup of M. Then U is elementary abelian, M is irreducible on U, $M = N_G(U)$, and U is strongly closed in G by (1). Thus for $u \in U^\#$, $u^G \cap M \subseteq U$ and M controls fusion in U by Burnside's Fusion Lemma, so $u^G \cap M = u^M$. Set $G_u := C_G(u)$.

As $U \trianglelefteq T$, we may choose $z \in Z(T) \cap U^\#$. Hence $G_z \le M$ as $M = !\mathcal{M}(T)$, so as $z^G \cap M = z^M$, M is the unique point fixed by z in the representation of G by right multiplication on the coset space G/M (cf. 46.1 in [**Asc86a**]). We use this fact to show:

(*) For each 2-subgroup S of G containing z, $N_G(S) \le M$.

For $C_S(z)$ fixes the unique fixed point M of z on G/M, and hence M is the unique fixed point of $C_S(z)$ on G/M. Then as each subgroup of S is subnormal in S, we conclude by induction on $|S|$ that M is the unique fixed point of S of G/M. Hence $N_G(S) \le M$.

First assume $G_u \le M$ for every $u \in U^\#$. Then as U is not normal in G, Remark I.8.4 and Theorem I.8.3 tell us that G is a Bender group.

Thus we may assume that $\mathcal{J} := \{u \in U^\# : G_u \not\le M\}$ is nonempty, and it remains to derive a contradiction. In particular $U > \langle z \rangle$, so the elementary abelian group U is noncyclic. Let $u \in \mathcal{J}$, set $H := G_u$, $M_H := M \cap H$, and let $U \le S \in Syl_2(H)$. By (*), $S \le M$, so conjugating in M we may assume that $S \le T$. By 1.1.6 applied to the 2-local $G_u = H$, the hypotheses of 1.1.5 are satisfied, so $M_H \in \mathcal{H}^e$ by 1.1.5.1.

Suppose $H \in \mathcal{H}^e$. Then as $S \le T$ and $S \in Syl_2(H)$, $z \in Z(S) \le O_2(H)$, so $H \le N_G(O_2(H)) \le M$ by (*), contradicting $H \not\le M$. Thus $H \notin \mathcal{H}^e$.

Let W be any hyperplane of U. Then $|z^M \cap U| > 1$ as U is noncyclic, so $z^M \cap W \ne \emptyset$ by A.1.43. Now as $G_z \le M$, $C_G(W) \le C_G(z^M \cap W) \le M$. Hence using Generation by Centralizers of Hyperplanes A.1.17, $O(H) = \langle C_{O(H)}(W) : m(U/W) = 1 \rangle \le M$, so $O(H) = 1$ since $M_H \in \mathcal{H}^e$.

Thus as $H \notin \mathcal{H}^e$, there is a component L of H, and by 1.1.5, $L = [L, z] \not\le M$ and L is described in 1.1.5.3. Set $L_0 := \langle L^H \rangle$. As U is strongly closed in S with respect to H, $Aut_U(L_0)$ is strongly closed in $Aut_H(L_0)$, so by inspection of the groups in 1.1.5.3, L is a Bender group with $Aut_U(L) = \Omega_1(S \cap L)$. In particular, U acts on each component of H.

Let U_L and U_C be the projections of U on L and $C_H(L)$, respectively. As $z \in U \leq U_L U_C$, $N_G(U_L U_C) \leq M$ by (*). As $L = [L, z]$, the projection of z on L is nontrivial, while as L is a Bender group, $N_L(U_L)$ is irreducible on $\Omega_1(S \cap L)$. Therefore $U_L = [z, N_L(U_L)] \leq U$ and hence $U = U_C \times U_L$. In particular $m(U) = m(U_L) + m(U_C)$.

Now pick $u \in \mathcal{J}$ so that L is maximal among components of G_j for $j \in \mathcal{J}$. Let $v \in U_C^\#$. Since G_v contains $L \not\leq M$, $v \in \mathcal{J}$, so by earlier remarks, U acts on each component of G_v and $O(G_v) = 1$. Then as $u \in U$, u acts on each component of G_v, so L is contained in a component L_v of G_v by I.3.2. Hence $L = L_v$ by maximality of L.

Suppose $g \in M$ with $U_C^g \cap U_C \neq 1$; we claim that $L = L^g$, so that $U_C = U_C^g$ as $M = N_G(U)$. Assume the claim fails and let $1 \neq v \in U_C \cap U_C^g =: V$. By the previous paragraph, L and L^g are components of G_v, and we may assume $L \neq L^g$, so that $[L, L^g] = 1$. It will suffice to show that M acts on $\{L, L^g\}$, since then M permutes $\{U_C, U_C^g\}$, and hence M acts on $1 \neq V = U_C \cap U_C^g \leq U_C$, contradicting the irreducible action of M on U. Now

$$m(U_L) + m(U_C) = m(U) = 2m(U_L) + m(V),$$

so $m(U_C) = m(U_L) + m(V) > m(U)/2$ since $m(V) > 0$. Then for each $x \in M$, $1 \neq U_C \cap U_C^x$. Hence if $L^x \notin \{L, L^g\}$, by symmetry between x and g, also $[L, L^x] = 1$. Then $U_L \leq U_C^g \cap U_C^x$, so also $[L^x, L^g] = 1$. But now for p an odd prime divisor of $|N_L(U_L)|$, $m_{2,p}(LL^g L^x) > 2$, contradicting G quasithin. This completes the proof of the claim.

The claim shows that U_C is a TI-set under M. Further $Aut_L(U)$ is cyclic and regular on $U_L^\#$, and is invariant under $N_{Aut_M(U)}(U_C)$. Hence $(Aut_M(U), U)$ is a Goldschmidt-O'Nan pair in the sense of Definition 14.1 of [**GLS96**]. So by O'Nan's lemma, Proposition 14.2 in [**GLS96**], one of the four conclusions of that result holds. Neither conclusion (i) nor (iii) holds, as M is irreducible on U. As $G_z \leq M$ but $\mathcal{J} \neq \emptyset$, M is not transitive on $U^\#$, so conclusion (iv) does not hold. Thus conclusion (ii) holds, so that $N_G(U_C)$ is of index 2 in M. However as M is the unique point fixed by z in G/M, by 7.4 in [**Asc94**], M controls G-fusion of 2-elements of M. Therefore by Generalized Thompson Transfer A.1.37.2, $O^2(G) \cap M \leq N_M(U_C)$, contrary to the simplicity of G. This contradiction completes the proof of Proposition 2.2.2. $\quad\square$

Recall that $\mathcal{S}_2(G)$ is the set of nonidentity 2-subgroups of G, and (cf. chapter 1) that $\mathcal{S}_2^e(G)$ consists of those $S \in \mathcal{S}_2(G)$ such that $N_G(S) \in \mathcal{H}^e$. We next verify:

LEMMA 2.2.3. *The Alperin-Goldschmidt conjugation family lies in $\mathcal{S}_2^e(G)$.*

PROOF. By (b) and (c) of the definition of the Alperin-Goldschmidt conjugation family $\mathcal{D}$ for T in G, $O^{2'}(C_G(D)) \leq D$ for each $D \in \mathcal{D}$. Thus as $D \leq T$, $Z(T) \leq D$. Therefore $D \in \mathcal{S}_2^e(G)$ using 1.1.4.3. $\quad\square$

NOTATION 2.2.4. Define $\delta = \delta_M$ to consist of those $D \in \mathcal{S}_2^e(G)$ such that $D \leq M$, but $N_G(D) \not\leq M$. Let $\delta^* = \delta_M^*$ denote the maximal members of δ under inclusion.

THEOREM 2.2.5. *If $\delta = \emptyset$, then G is a Bender group.*

PROOF. Let $\mathcal{D}$ be the Alperin-Goldschmidt conjugation family for T in G. By 2.2.3, $\mathcal{D} \subseteq \mathcal{S}_2^e(G)$. Therefore if $\delta = \emptyset$, then $N_G(D) \leq M$ for each $D \in \mathcal{D}$. Hence by Proposition 2.2.2.2, G is a Bender group. $\qquad\square$

REMARK 2.2.6. The idea of using the Alperin-Goldschmidt Fusion Theorem and Goldschmidt's Fusion Theorem in this way is due to GLS. This approach allows us to avoid considering the case where the centralizer of some involution i has a component which is a Bender group: For if i is such an involution then $\mathcal{U}(C_G(i)) = \emptyset$ (in the language of Notation 2.3.4 established later), whereas Theorem 2.2.5 allows us to assume $\delta \neq \emptyset$, which supplies us with 2-locals H such that $\mathcal{U}(H) \neq \emptyset$. It is these 2-locals which we will exploit during the remainder of this chapter. In particular, as mentioned in the introduction to the chapter, this allows us to avoid difficulties with the shadows of Bender groups extended by involutory outer automorphisms, and also with the shadows of the wreathed products $L_2(2^n)$ wr $\mathbf{Z}_2$ and $Sz(2^n)$ wr $\mathbf{Z}_2$.

2.3. Preliminary analysis of the set Γ_0

Since the Bender groups appear in the conclusion of Theorem 2.1.1, by Theorem 2.2.5, we may assume for the remainder of this chapter that

$$\delta \neq \emptyset, \text{ so that also } \delta^* \neq \emptyset.$$

Recall from the second paragraph of the previous section that there exist pairs (U, H_U) such that $U \in Syl_2(H_U)$, $N_G(U) \leq M$, and $H_U \in \mathcal{H}(U, M)$. Using the fact that δ is nonempty, we will produce such pairs with H_U in $\mathcal{H}^e(U, M)$. Moreover we will see that we can choose U to have a number of useful properties which we list in the next definition:

NOTATION 2.3.1. Let $\beta = \beta_M$ consist of those $U \in \mathcal{S}_2(G)$ such that

(β_0) $U \leq M$, so in fact $U \in \mathcal{S}_2(M)$;
(β_1) For all $U \leq V \in \mathcal{S}_2(M)$, $N_G(V) \leq M$; and
(β_2) $C_{O_2(M)}(U) \leq U$.

Notice that (β_0)–(β_2) are inherited by any overgroup of U in $\mathcal{S}_2(M)$, so all such overgroups are also in β. Some other elementary consequences of this definition include:

LEMMA 2.3.2. Assume $U \in \beta$, and $U \leq H \leq G$. Then

(1) If $U \leq V \in \mathcal{S}_2(G)$, then $V \in \beta$. In particular all 2-overgroups of U in G lie in M.

(2) $|H|_2 = |H \cap M|_2$.

(3) If $H \in \mathcal{H}^e$, then $O_2(H) \in \mathcal{S}_2^e(G)$. In particular $\beta \subseteq \mathcal{S}_2^e(G)$.

PROOF. To prove (1), assume $U \leq V \in \mathcal{S}_2(G)$. Recall that each 2-overgroup V of U in M is in β, so it only remains to show that $V \leq M$. If $U \trianglelefteq V$, then $V \leq N_G(U) \leq M$ by (β_1). So as $U \trianglelefteq \trianglelefteq V$, $V \leq M$ by induction on $|V : U|$, completing the proof of (1).

Next let $U \leq S \in Syl_2(H)$. Then $S \in \beta$ by (1), so $S \leq M$ by (β_0), giving (2).

Finally set $Q := O_2(H)$, so that $Q \leq S$ since $S \in Syl_2(H)$. As $S \leq M$, we may assume that $S \leq T$. Then $O_2(M) \leq T$ as $T \in Syl_2(M)$, so $Z(T) \leq C_{O_2(M)}(S) \leq Z(S)$ by (β_2). Under the hypothesis of (3), $Q = F^*(H)$, so $Z(T) \leq Z(S) \leq$

$C_H(Q) \leq Q$, so $Q \in \mathcal{S}_2^e(G)$ by 1.1.4.3. In particular applying this observation to U in the role of "H", $U = O_2(U) \in \mathcal{S}_2^e(G)$, completing the proof of (3). $\quad\square$

We now use our assumption that $\delta^* \neq \emptyset$ to verify that $\beta \neq \emptyset$:

LEMMA 2.3.3. *Let $D \in \delta^*$ and $S \in Syl_2(N_M(D))$. Then:*

(1) $U \in \beta$ for each U in $\mathcal{S}_2(M)$ with $D < U$.

(2) $D < S$, $|S| < |M|_2$, $S \in \beta$, and $S \in Syl_2(N_G(D))$.

PROOF. To prove (1), assume $D < U \in \mathcal{S}_2(M)$. Then U satisfies (β_0) in Notation 2.3.1. By definition of $D \in \delta$ in Notation 2.2.4, $D \in \mathcal{S}_2^e(G)$, so also $U \in \mathcal{S}_2^e(G)$ by 1.1.4.1; hence by maximality of D, $U \notin \delta$, so that $N_G(U) \leq M$. Applying this observation to any $W \in \mathcal{S}_2(M)$ containing U, we obtain (β_1) for U. Next set $E := O_2(N_G(D))$. If $D < E$, then $N_G(D) \leq N_G(E) \leq M$ by the observation, contradicting $D \in \delta$. Thus $D = O_2(N_G(D))$. We saw $D \in \mathcal{S}_2^e(G)$, so that $N_G(D) \in \mathcal{H}^e$, and hence $C_{O_2(M)}(D) \leq C_{N_G(D)}(D) \leq D$. Thus (β_2) holds for D, and hence also for the 2-overgroup U. This completes the proof that $U \in \beta$, giving (1).

Next let $S \in Syl_2(N_M(D))$; we may assume $S \leq T$, and hence $S = N_T(D)$. As $D \in \delta$, $S \leq N_G(D) \not\leq M = !\mathcal{M}(T)$, so $S < T$. In particular $D < T$, so $D < N_T(D) = S$. Then $S \in \beta$ by (1), and hence $S \in Syl_2(N_G(D))$ by 2.3.2.2, completing the proof of (2). $\quad\square$

We now introduce further notation suggested by the GLS proof of the Global C(G,T)-Theorem, in as yet unpublished notes slated to appear in the GLS series; an outline of their proof appears in Sec 2.10 of [**GLS94**].

NOTATION 2.3.4. Let $\mathcal{U}(G) = \mathcal{U}_M(G)$ denote the set of pairs (U, H_U) such that $U \in \beta$ and $H_U \in \mathcal{H}^e(U, M)$. Write $\mathcal{U} = \mathcal{U}_M$ for the set of $U \in \beta$ such that $\mathcal{H}^e(U, M) \neq \emptyset$. For $H \in \mathcal{H}$, let $\mathcal{U}(H) = \mathcal{U}_M(H)$ consist of those $(U, H_U) \in \mathcal{U}(G)$ such that $H_U \leq H$.

Recall that there exists $D \in \delta^*$. By 2.3.3.2, a Sylow 2-group S of $N_M(D)$ is in β, so $N_G(D) \in \mathcal{H}^e(S, M)$ by the definition of δ in Notation 2.2.4. Thus $(S, N_G(D)) \in \mathcal{U}(G)$ and $S \in \mathcal{U}$, so that

$$\mathcal{U}(G) \text{ and } \mathcal{U} \text{ are nonempty,}$$

and by 2.3.3, $S \in Syl_2(N_G(D))$ and $N_G(S) \leq M$. Observe that if $H, H_1 \in \mathcal{H}$ with $H \leq H_1$ then $\mathcal{U}(H) \subseteq \mathcal{U}(H_1)$.

NOTATION 2.3.5. Let $\Gamma = \Gamma_M$ be the set of all $H \in \mathcal{H}$ such that $\mathcal{U}(H) \neq \emptyset$. Let $\Gamma^* = \Gamma_M^*$ consist of those $H \in \Gamma$ such that $\mathcal{U}(H)$ contains some member (U, H_U) with U of maximal order among members of $\mathcal{U}$, and subject to that constraint, with $|H|_2$ maximal. Let $\Gamma_* = \Gamma_{*,M}$ consist of those $H \in \Gamma$ such that $|H|_2$ is maximal among members of Γ. Finally let $\Gamma_0 = \Gamma_{0,M} := \Gamma^* \cup \Gamma_*$.

If $D \in \delta^*$ and $S \in Syl_2(N_M(D))$, then we saw a moment ago that $(S, N_G(D)) \in \mathcal{U}(N_G(D))$, so that $N_G(D) \in \Gamma$ and hence $\Gamma \neq \emptyset$. As Γ is nonempty, also Γ^* and Γ_* are nonempty.

Observe that by that by 2.3.2.2, $|H|_2 = |H \cap M|_2$ for each $H \in \Gamma$, so the constraints on the maximality of $|H|_2$ amount to constraints on $|H \cap M|_2$.

LEMMA 2.3.6. *If $H \in \Gamma_0$, then $|H|_2 \geq |V|$ for any $V \in \mathcal{U}$.*

PROOF. Let $U \in \mathcal{U}$ be of maximal order and $H_U \in \mathcal{H}^e(U, M)$. Then $|V| \leq |U| \leq |H|_2$ for $H \in \Gamma_0$. $\qquad \square$

The remainder of the proof of Theorem 2.1.1 focuses on the members of Γ_0. We need to consider members of Γ maximal in the two different senses of Notation 2.3.5 because: On one hand, at a number of points in the proof we produce members of Γ_* (for example in 2.3.7.1), so we need results on the structure of such subgroups. On the other hand, near the end of the proof, particularly in 2.5.10, we need to work with those $H \in \Gamma$ such that $\mathcal{U}(H)$ contains a member (U, H_U) with $|U|$ maximal in $\mathcal{U}$. Thus at that point we choose $H \in \Gamma^*$.

We often use the following observations to produce members of Γ_0:

LEMMA 2.3.7. *Assume $H \in \Gamma$, and let $(U, H_U) \in \mathcal{U}(H)$ and $U \leq S \in Syl_2(H)$.*

(1) Assume $|T : S| = 2$. Then $H \in \Gamma_*$. If $H_1 \in \Gamma$ with $|H_1|_2 \geq |S|$, then $|H_1|_2 = |S|$, and $H_1 \in \Gamma_*$.

(2) Assume $H \in \Gamma_0$ and $H_1 \in \mathcal{H}(H)$. Then $H_1 \in \Gamma_0$, and S is Sylow in H_1 and $H_1 \cap M$.

(3) Assume $H \in \Gamma_0$ and $S \leq H_1 \in \Gamma$; when $H \in \Gamma^*$, assume in addition that $|U|$ is maximal among members of $\mathcal{U}$ and that $\mathcal{H}^e(U, M) \cap H_1 \neq \emptyset$. Then $H_1 \in \Gamma_0$, and S is Sylow in H_1 and $H_1 \cap M$.

(4) Under the hypotheses of (2) and (3), if $H \in \Gamma^*$, Γ_*, then $H_1 \in \Gamma^*$, Γ_*, respectively.

PROOF. Since $U \leq S$ by hypothesis, $S \leq M$ by 2.3.2.1.

Assume $|T : S| = 2$ and $H_1 \in \Gamma$. As $M = !\mathcal{M}(T)$, $|H_1|_2 \leq |T|/2 = |S| = |H|_2$, so $H \in \Gamma_*$, and if $|H_1|_2 \geq |S|$, then $H_1 \in \Gamma_*$, establishing (1).

Now assume the hypotheses of (2); then $H_1 \in \mathcal{H}(H) \subseteq \Gamma$. When $H \in \Gamma_*$, maximality of $|S|$ forces $H_1 \in \Gamma_*$, with S Sylow in H_1, and hence in $H_1 \cap M$. Thus (2) and the corresponding part of (4) hold in this case. When $H \in \Gamma^*$ there is some $(U, H_U) \in \mathcal{U}(H) \subseteq \mathcal{U}(H_1)$, with U of maximal order in $\mathcal{U}$, so by the maximality of $|S|$ subject to this constraint, $H_1 \in \Gamma^*$ and S is Sylow in H_1 and in $H_1 \cap M$. This completes the proof of (2), along with the corresponding part of (4).

Assume the hypotheses of (3); the proof is very similar to that of (2): Again if $H \in \Gamma_*$, then as $S \leq H_1$, $H_1 \in \Gamma_*$ by maximality of $|S|$. Thus we may assume that $H \in \Gamma^*$. Then by hypothesis $|U|$ is maximal in $\mathcal{U}$ and there is $H_2 \in \mathcal{H}^e(U, M) \cap H_1$. Thus $(U, H_2) \in \mathcal{H}(H_1)$, and hence $H_1 \in \Gamma$. Then by maximality of $|U|$ and maximality of $|S|$ subject to that constraint, $H_1 \in \Gamma^*$. $\qquad \square$

The next result 2.3.8 lists various properties of members of Γ. In particular part (4) of that lemma is the basis for our analysis of the case where Γ_0 contains a member of $\mathcal{H}^e$ in the next section.

LEMMA 2.3.8. *Let $H \in \Gamma$, $(U, H_U) \in \mathcal{U}(H)$, and $U \leq S \in Syl_2(H)$. Then*

(1) $|S| < |T|$ and $S \in \beta$. In particular, $S \leq M$, so $S \in Syl_2(H \cap M)$.

(2) $O_2(H_U) \in \mathcal{S}_2^e(G)$.

(3) $(U, H_U) \in \mathcal{U}(N_G(O_2(H)))$, and $N_G(O_2(H)) \in \Gamma$.

(4) If $H \in \Gamma_0 \cap \mathcal{H}^e$, then $C(H, S) \leq H \cap M$, so $H = (H \cap M)L_1 \cdots L_s$ with $s \leq 2$ and L_i an $L_2(2^n)$-block, A_3-block, or A_5-block such that $L_i \not\leq M$ and $L_i = [L_i, J(S)]$.

(5) Assume $H \in \Gamma_0$. Then

    (a) $N_G(J(S)) \leq M$.

(b) If $J(S) \leq R \leq S$ with $|S : R| = 2$ and $C_{O_2(M)}(R) \leq R$, then $R \in \beta$.

(c) If $H \in \mathcal{H}^e$ then $C_{O_2(M)}(R_0) \leq R_0$ for each overgroup R_0 of $O_2(H)$ in S.

(6) If $H \in \Gamma_0$, then the hypotheses of 1.1.5 are satisfied for each involution $z \in Z(S)$ which is 2-central in M.

PROOF. As $U \in \beta$, $S \in \beta$ by 2.3.2.1, so $S \in Syl_2(H \cap M)$. As $S \leq M$, we may assume $S \leq T$. As $M = !\mathcal{M}(T)$, $|S| < |T|$ completing the proof of (1). Part (2) follows from 2.3.2.3. Next $N_G(O_2(H)) \in \mathcal{H}(H) \subseteq \Gamma$ and $(U, H_U) \in \mathcal{U}(H) \subseteq \mathcal{U}(N_G(O_2(H)))$, so (3) holds.

Assume $H \in \mathcal{H}^e \cap \Gamma_0$. Then $S \in \beta$ by (1), and $H \in \mathcal{H}^e(S, M)$ so that $(S, H) \in \mathcal{U}(H)$ and $S \in \mathcal{U}$. Assume that $C(H, S) \nleq M$. Then there is a nontrivial characteristic subgroup R of S such that $N_H(R) \nleq M$. Now $N_H(R) \in \mathcal{H}^e$ using 1.1.3.2, so $(S, N_H(R)) \in \mathcal{U}(N_G(R))$ and thus $N_G(R) \in \Gamma$. Then we may apply 2.3.7.3 with $N_G(R)$ in the role of "H_1" to conclude that $S \in Syl_2(N_G(R))$. But $S < T$ by (1), so $S < N_T(S) \leq N_T(R)$, contradicting $S \in Syl_2(N_G(R))$. This contradiction shows that $C(H, S) \leq H \cap M$. Then as $S \in Syl_2(H)$, we may apply the Local $C(G, T)$-Theorem C.1.29 to complete the proof of (4).

We next turn to (5), so we assume $H \in \Gamma_0$, and set $J := J(S)$. By (1), $S \in \beta$, so $Z(T) \leq C_{O_2(M)}(S) \leq S$ using (β_2) from the definition in 2.3.1. Then $\Omega_1(Z(T)) \leq \Omega_1(Z(S)) \leq J$ using B.2.3.7, so that $J \in \mathcal{S}_2^e(G)$ by 1.1.4.3. Suppose $N_G(J) \nleq M$. Then $(S, N_G(J)) \in \mathcal{U}(N_G(J))$ so $N_G(J) \in \Gamma$ and $S \in Syl_2(N_G(J))$ by 2.3.7.3. This is impossible as $S < N_T(S) \leq N_T(J)$. Therefore $N_G(J) \leq M$, proving part (a) of (5).

Next assume that $H \in \mathcal{H}^e$, and consider any R_0 with $Q := O_2(H) \leq R_0 \leq S$. By 2.3.7.2, $N_G(Q) \in \mathcal{H}(H) \subseteq \Gamma_0$, with S Sylow in $N_G(Q)$ and $N_M(Q)$. Therefore

$$E := C_{O_2(M)}(Q) \leq O_2(N_M(Q)) \leq S \leq H.$$

Also $F^*(H) = O_2(H) = Q$ as $H \in \mathcal{H}^e$, so $E \leq C_H(Q) \leq Q$. Then as $Q \leq R_0$,

$$C_{O_2(M)}(R_0) \leq C_{O_2(M)}(Q) = E \leq Q \leq R_0,$$

establishing part (c) of (5).

So to complete the proof of (5), it remains to establish part (b). Thus we assume that $J \leq R \leq S$ with $|S : R| = 2$, and $C_{O_2(M)}(R) \leq R$. We must show that $R \in \beta$; as R satisfies (β_0) since $S \leq M$, and R satisfies (β_2) by hypothesis, we may assume that (β_1) fails for R, and it remains to derive a contradiction. Then for some $R \leq V \in \mathcal{S}_2(M)$, $N_G(V) \nleq M$, and we may choose V maximal subject to this constraint. As usual, we may assume that $V \leq T$. By hypothesis $J(S) \leq R$, so $J(S) = J(R)$ by B.2.3.3, and hence $N_G(R) \leq N_G(J(S)) \leq M$ by part (a) of (5). Therefore $R < V$. Further $N_G(V) \nleq M = !\mathcal{M}(T)$, so that $V < T$ and hence $V < N_T(V) := W$. Then W satisfies (β_0), and also (β_2), since this condition is inherited by overgroups of R. Further by maximality of V, $N_G(X) \leq M$ for each X satisfying $W \leq X \in \mathcal{S}_2(M)$, establishing ($\beta_1$) for W. Hence $W \in \beta$. We saw earlier that $J(S) = J \in \mathcal{S}_2^e(G)$, and by hypothesis $J \leq R \leq V$, so $V \in \mathcal{S}_2^e(G)$ by 1.1.4.1, and hence $N_G(V) \in \mathcal{H}^e$. Then $(W, N_G(V)) \in \mathcal{U}(N_G(V))$ so $N_G(V) \in \Gamma$. However by hypothesis $|S : R| = 2$, while $R < V < W$, so that $|W| > |S|$. This contradicts the maximality of $|H|_2$ in Notation 2.3.5 when $H \in \Gamma_*$, and the maximality of $|U|$ when $H \in \Gamma^*$. This contradiction completes the proof of (5).

It remains to prove (6). By (1), $S \leq M$ and $S \in Syl_2(H \cap M)$. Assume that $H \in \Gamma_0$. Then by 2.3.7.2, $N_G(O_2(H)) \in \Gamma$, and S is Sylow in $N_G(O_2(H))$ and $N_M(O_2(H))$. Thus $C_{O_2(M)}(O_2(H)) \leq O_2(N_M(O_2(H))) \leq S$. Now $O_2(H) \leq O_2(H \cap M)$ by A.1.6, so

$$C_{O_2(M)}(O_2(H \cap M)) \leq C_{O_2(M)}(O_2(H)) \leq S \leq H,$$

establishing one of the hypotheses of 1.1.5. Finally if z is an involution central in $T' \in Syl_2(M)$, then $C_G(z) \leq M = !\mathcal{M}(T')$, establishing the remaining hypothesis for that result. This establishes (6), and so completes the proof of 2.3.8. $\square$

The final section of this chapter will focus on components of a member of Γ_0. Using part (6) of 2.3.8, the next result describes these components.

LEMMA 2.3.9. Let $H \in \Gamma_0$, $Q := O_2(H)$, $(U, H_U) \in \mathcal{U}(H)$, and $U \leq S \in Syl_2(H)$. Then

(1) S is Sylow in $N_G(Q)$ and $N_M(Q)$, and $N_G(Q) \in \Gamma_0$. If $H \in \Gamma^*$, then $N_G(Q) \in \Gamma^*$.

(2) $C_{O_2(M)}(Q) \leq S$.

(3) $Z(T) \leq S < T$ for some $T \in Syl_2(M)$ depending on H. In particular, $Z(T) \leq Z(S)$.

(4) $F^*(H \cap M) = O_2(H \cap M)$.

(5) Let z be an involution in $Z(T)$ for T as in (3). Then $C_G(z) \leq M$ and z inverts $O(H)$.

(6) If L is a component of H, then $L = [L, z] \not\leq M$, and L is contained in a component L_Q of $N_G(Q)$.

(7) If L is a component of H then z induces an inner automorphism on L unless possibly $L/Z(L) \cong A_6$ or A_7. Moreover one of the following holds:

(a) L is a Bender group.

(b) $L \cong Sp_4(2^n)'$ or $L_3(2^n)$, or $L/O_2(L) \cong L_3(4)$ or $L \cong \hat{A}_6$.

(c) $L \cong A_7$ or $\hat{A}_7$, and $L \cap M$ is the stabilizer in L of a partition of type $2^3, 1$.

(d) $L \cong L_3(3)$ or M_{11}, and $L \cap M = C_L(z_L)$ where z_L is the projection of z on L.

(e) $L \cong L_2(p)$, p a Fermat or Mersenne prime, and $L \cap M = S \cap L$.

(f) $L \cong M_{22}$ or M_{23}, and $L \cap M \cong A_6/E_{16}$ or A_7/E_{16}, respectively.

(g) $L \cong L_4(2)$, S is nontrivial on the Dynkin diagram of L, and $L \cap M = C_L(z_L)$, where z_L is the projection of z on L.

(8) Assume $|S : R| = 2$, with R containing $J(S)$, $O_2(H)$, and $C_S(R)$. Then $R \in \beta$.

PROOF. By 2.3.8.3, $N_G(Q) \in \Gamma$; then (1) follows from parts (2) and (4) of 2.3.7.

By (1), S is Sylow in $N_M(Q)$, so $C_{O_2(M)}(Q) \leq O_2(N_M(Q)) \leq S$, proving (2). By 2.3.8.1, $S \in \beta$, so in particular $S \leq M$ and $S < T$ for some $T \in Syl_2(M)$, As $F^*(M) = O_2(M)$, $Z(T) \leq O_2(M)$, so as $Q \leq S \leq T$, $Z(T) \leq S$ by (2), completing the proof of (3).

By (3), $Z(T) \leq Z(S)$, so by 2.3.8.6 the hypotheses of 1.1.5 are satisfied for each involution $z \in Z(T)$, and in particular $C_G(z) \leq M$. Therefore 1.1.5.1 implies (4), while 1.1.5.2 says z inverts $O(H)$, completing the proof of (5).

Similarly if L is a component of H, then by 1.1.5.3, $L = [L, z] \not\leq M$, and the possibilities for L are listed in 1.1.5.3. Notice that L is a component of $\langle L, S \rangle$ and S is Sylow in $N := N_G(Q)$ by (1), so by 1.2.4, $L \leq L_Q \in \mathcal{C}(N)$. Since $L = [L, z]$, also $L_Q = [L_Q, z]$. As $N \in \Gamma_0$ by (1), z inverts $O(N)$ by (5); so as $L_Q = [L_Q, z]$, L_Q centralizes $O(N)$. Similarly z centralizes $O_2(N)$ as $z \in Z(S)$ and $S \in Syl_2(N)$, so $L_Q = [L_Q, z]$ centralizes $O_2(N)$. Thus L_Q centralizes $F(L_Q)$, so L_Q is quasisimple by A.3.3.1, and hence L_Q is a component of N. This completes the proof of (6).

To prove (7), we must refine the possibilities listed in 1.1.5.3. If $L/Z(L)$ is a Bender group, then $Z(L) = 1$ by 1.1.5.3, so conclusion (a) of (7) holds in this case. Hence we may assume $L/Z(L)$ is not a Bender group.

In this paragraph, we make a slight digression, to construct some machinery to deal with groups of Lie rank at least 2. Assume $L \leq \langle H_1, H_2 \rangle$ with $H_i \in \mathcal{H}^e(S)$. Suppose $H_i \not\leq M$ for some i. Then from the definitions in Notation 2.3.4, $(S, H_i) \in \mathcal{U}(H_i)$, so $H_i \in \Gamma_0$ by 2.3.7.3. Consequently H_i is described in 2.3.8.4.

Now suppose $L/Z(L)$ appears in one of cases (a)–(c) of 1.1.5.3; then as $L/Z(L)$ is not a Bender group, $L/Z(L)$ is a group of Lie type and characteristic 2 of rank at least 2 in Theorem C (A.2.3). If there do not exist two distinct maximal $N_S(L)$-invariant parabolics K_1 and K_2, then (cf. E.2.2.2) $L/Z(L) \cong L_3(2^n)$ or $Sp_4(2^n)'$ with S nontrivial on the Dynkin diagram of $L/Z(L)$, and then conclusion (b) of (7) holds. Thus we may assume K_1 and K_2 exist, take $H_i := \langle K_i, S \rangle$, and apply the observations in the previous paragraph. By (6), $L \not\leq M$, and hence $H_i \not\leq M$ for some i, so K_i is a block described in 2.3.8.4. Then we check that the only groups in (a)–(c) of 1.1.5.3 with such a block are those in conclusions (b) and (g) of (7), keeping in mind that $Z(L) = O_2(L)$ in case (b) of 1.1.5.3. Similar arguments, using generation by a pair of members of $\mathcal{H}^e(S)$ in LS, eliminate those cases where $L/Z(L)$ is M_{12}, M_{24}, J_2, J_4, HS, He, or Ru; thus in case (f) of 1.1.5, $L/Z(L)$ is M_{11}, M_{22} or M_{23}.

If case (d) of 1.1.5.3 holds, then z has cycle structure 2^3 and as $C_G(z) \leq M$, $L \cap M$ contains the stabilizer K in $L\langle z \rangle$ of a partition of type $2^3, 1$ determined by z. So as K is a maximal subgroup of $L\langle z \rangle$ and $L \not\leq M$, $K = M \cap L\langle z \rangle$; thus conclusion (c) of (7) holds.

In the cases $L_3(3)$, $L_2(p)$, M_{11}, M_{22}, M_{23} remaining from (e) and (f) of 1.1.5.3, the description of z determines the maximal subgroup of $L\langle z \rangle$ described in conclusions (d), (e), (d), (f), and (f) of (7), respectively. Finally by 1.1.5.3, z induces an inner automorphism on L, except possibly when $L/Z(L)$ is A_6 or A_7, completing the proof of (7).

Assume the hypotheses of (8). Because we are assuming that $Q \leq R \geq C_S(R)$, $C_{O_2(M)}(R) \leq C_S(R) \leq R$ by (2). Then since $|S : R| = 2$ and $J(S) \leq R$ by hypothesis, we have the hypotheses of 2.3.8.5b, and that lemma completes the proof of (8), and hence of 2.3.9. $\qquad\qquad\square$

LEMMA 2.3.10. *If S is of index 2 in T and $\mathcal{H}(S) \not\subseteq M$, then $S \in \beta$.*

PROOF. As $S \leq T \leq M$, condition (β_0) from the definition in Notation 2.3.1 holds. As $|T : S| = 2$, $N_G(S) \leq M = !\mathcal{M}(T)$, and then the only proper 2-overgroups of S are Sylow groups T' of M, so (β_1) holds as $M = !\mathcal{M}(T')$. Finally by hypothesis, there is $H \in \mathcal{H}(S)$ with $H \not\leq M$; enlarging H if necessary, we may assume $H = N_G(O_2(H))$. As $M = !\mathcal{M}(T')$ for $T' \in Syl_2(M)$, $S \in Syl_2(H \cap M)$.

Thus $O_2(H \cap M) \leq S$ and $C_G(O_2(H)) \leq H$, so

$$C_{O_2(M)}(S) \leq C_{O_2(M)}(O_2(H)) \leq O_2(M) \cap H \leq O_2(H \cap M) \leq S,$$

establishing (β_2). $\qquad\square$

NOTATION 2.3.11. Set $\Gamma^e = \Gamma_M^e := \Gamma \cap \mathcal{H}^e$ and $\Gamma_0^e = \Gamma_{0,M}^e := \Gamma_0 \cap \mathcal{H}^e$

The proof of Theorem 2.1.1 now divides into two cases: Either Γ_0^e is nonempty or Γ_0^e is empty. In the first case we focus on a member of Γ_0^e; the structure of such groups is described in 2.3.8.4. In the second case 2.3.9 gives us information about the members of Γ_0, particularly about their components. The two cases are treated in the remaining two sections of this chapter.

2.4. The case where Γ_0^e is nonempty

In this section, we treat the case where Γ_0^e is nonempty. Here by 2.3.8.4, H has a very restricted structure dominated by χ_0-blocks. We will use this fact to identify the groups in the conclusion of Theorem 2.1.1 which are not Bender groups, and eliminate some difficult shadows. The main result of this section is:

THEOREM 2.4.1. If there exists $H \in \Gamma_0$ with $F^*(H) = O_2(H)$, then G is $L_2(p)$, $p > 7$ a Mersenne or Fermat prime, $L_3(3)$, or M_{11}.

In the remainder of this section, we assume that

$H \in \Gamma_0^e$, and the pair G, H afford a counterexample to Theorem 2.4.1.

The groups appearing in the conclusion of Theorem 2.4.1 will emerge during the proof of 2.4.26.

Choose $S \in Syl_2(H)$ as in 2.3.8, and choose $T \in Syl_2(M)$ as in 2.3.9.3; then

S is Sylow in H and in $H \cap M$, $T \in Syl_2(M)$, and $Z(T) \leq S < T$.

LEMMA 2.4.2. If $S_1 \in Syl_2(H_1 \cap M)$ for $H_1 \in \Gamma^e$, then $S_1 \in Syl_2(H_1)$, $(S_1, H_1) \in \mathcal{U}(H_1)$, and $S_1 \in \mathcal{U}$.

PROOF. From the definition of Γ in Notation 2.3.5, $\mathcal{U}(H_1)$ contains a member (U, H_U) with $U \leq S_1$. Then $S_1 \in Syl_2(H_1)$ and $S_1 \in \beta$ by 2.3.8.1, so as $H_1 \in \mathcal{H}^e$ it follows that $(S_1, H_1) \in \mathcal{U}(H_1)$ and $S_1 \in \mathcal{U}$. $\qquad\square$

In particular

$$S \in \mathcal{U} \text{ and } (S, H) \in \mathcal{U}(H)$$

by 2.4.2. For the remainder of this section, we set

$$Q := O_2(H) \text{ and } G_Q := N_G(Q).$$

LEMMA 2.4.3. (1) $S \in Syl_2(G_Q)$ and $G_Q \in \Gamma_0^e$. In particular, $F^*(G_Q) = O_2(G_Q) = Q$.

(2) Assume $H_1 \in \Gamma^e$ and $|H_1|_2 \geq |H|_2$. Then $|H_1|_2 = |H|_2$ and $H_1 \in \Gamma_0^e$.

PROOF. First $S \in Syl_2(G_Q)$ and $G_Q \in \Gamma_0$ by 2.3.9.1. Then using A.1.6, $Q \leq O_2(G_Q) \leq O_2(H) = Q$, so $Q = O_2(G_Q)$. As $(S, H) \in \mathcal{U}(H)$, $Q = O_2(H) \in \mathcal{S}_2^e(G)$ by 2.3.8.2, so $G_Q \in \Gamma_0^e$, completing the proof of (1).

Next assume the hypotheses of (2). Recall $\Gamma_0 = \Gamma^* \cup \Gamma_*$ from the definitions in Notation 2.3.5. If $H \in \Gamma_*$, then $|H|_2 \geq |H_1|_2$ by maximality of $|H|_2$, so as $|H_1|_2 \geq |H|_2$ by hypothesis, $H_1 \in \Gamma_*$, so that $H_1 \in \Gamma_0^e$ in this case. Thus we may assume $H \in \Gamma^*$, so S is a member of $\mathcal{U}$ of maximal order. Choose $S_1 \in Syl_2(H_1 \cap M)$; by

2.4.2, $S_1 \in Syl_2(H_1)$ and $S_1 \in \mathcal{U}$. Then as $|H_1|_2 \geq |H|_2 = |S|$ by hypothesis, we conclude from maximality of $|S|$ over $\mathcal{U}$ that $|S_1| = |S|$. Then by maximality of $|H|_2$ over members of Γ containing a pair with a member of $\mathcal{U}$ of maximal order, we conclude $H_1 \in \Gamma^*$, so that $H_1 \in \Gamma_0^e$ in this case as well, completing the proof.   □

Since $H \in \Gamma_0^e$, 2.3.8.4 says $H = (H \cap M)L_1 \cdots L_s$, where L_i is an $L_2(2^n)$-block with $n > 1$, an A_3-block, or an A_5-block; further $L_i \not\leq M$, and $s \leq 2$. Since S, H play the roles of "U, H_U" in the previous section, in the remainder of this section U will instead denote the module $U(L_1) = [O_2(L_1), L_1]$ in the notation of Definition C.1.7. Furthermore we set:

$$L := L_1, \ L_0 := \langle L^S \rangle, \text{ and } U_0 := \langle U^S \rangle.$$

Then $L_0 \trianglelefteq H$ by 1.2.1.3, so $L_0 \in \mathcal{H}^e$ by 1.1.3.1, and hence $L_0 S \in \mathcal{H}^e$. Further $L_0 S \not\leq M$, so $(S, L_0 S) \in \mathcal{U}(L_0 S)$ and hence $L_0 S \in \Gamma^e$. Then $L_0 S \in \Gamma_0^e$ by 2.4.3.2, so replacing H by $L_0 S$, we may assume $H = L_0 S$. Then from section B.6:

$H = L_0 S$ is a minimal parabolic and L is a χ_0-block.

LEMMA 2.4.4. *If* $1 \neq S_0 \leq S$ *with* $S_0 \trianglelefteq H$, *then* $N_T(S_0) = S$.

PROOF. By 2.3.7.2, $N_G(S_0) \in \Gamma_0$ and $S \in Syl_2(N_G(S_0))$. In particular $S = N_T(S_0)$.   □

LEMMA 2.4.5. *(1) Hypotheses C.5.1 and C.5.2 are satisfied with S in the roles of both "T_H, R" for any subgroup M_0 of T with S a proper normal subgroup of M_0.*
(2) Assume $S \leq M_0 \leq T$ with $|M_0 : S| = 2$ and set $D := C_S(L_0)$. Then

    (a) $Q = U_0 D \in \mathcal{A}(S)$.
    (b) For each $x \in M_0 - S$, $1 = D \cap D^x$ and $U_0^x \not\leq Q$.

(3) Assume either that L is an A_3-block, or that $L = L_0$ is an $L_2(2^n)$-block or an A_5-block. Then the hypotheses of Theorem C.6.1 are satisfied with T, S in the roles of "Λ, T_H".
(4) $U_0 = O_2(L_0)$.

PROOF. We saw that $H = L_0 S$ is a minimal parabolic, and the rest of Hypothesis C.5.1 is straightforward. As S is proper in M_0, Hypothesis C.5.2 follows from 2.4.4. Thus (1) holds.

Choose M_0 as in (2) and set $D_0 := C_{\text{Baum}(S)}(L_0)$. This is the additional hypothesis for C.5.6.7; and that result implies (4); and also says that $Q = U_0 D_0$, $Q \in \mathcal{A}(S)$, and $D_0 \cap D_0^x = 1$ for each $x \in M_0 - S$. As $Q \in \mathcal{A}(S)$, $D_0 = C_S(L_0) = D$. By C.5.5, there exists $y \in M_0$ with $U_0^y \not\leq Q$. Then as $U_0 \trianglelefteq S$ and $|M_0 : S| = 2$, $M_0 - S = \{x_0 \in M_0 : U_0^{x_0} \not\leq Q\}$, so the proof of (2) is complete.

Finally assume the hypotheses of (3). The first three conditions in the hypothesis of Theorem C.6.1 are immediate, while condition (iv) follows from 2.4.4, establishing (3).   □

LEMMA 2.4.6. $L_0 \trianglelefteq G_Q$.

PROOF. By 2.4.3.1, $S \in Syl_2(G_Q)$ with $G_Q \in \Gamma_0^e$. Hence we may apply 2.3.8.4 to G_Q, to conclude that G_Q is the product of $N_M(Q)$ with a product of χ_0-blocks. But using 1.2.4 and A.3.12, no larger χ_0-block contains an S-invariant product L_0 of χ_0-blocks, so we conclude $L_0 \trianglelefteq G_Q$.   □

2.4.1. Shadows of groups of rank 2 with $L_2(2^n)$-blocks. In this subsection we continue the proof of Theorem 2.4.1 by eliminating the shadows of $L_3(2^n)$ and $Sp_4(2^n)$ extended by an outer automophism nontrivial on the Dynkin diagram. To be more precise, we will show that if L is an $L_2(2^n)$-block, then H essentially has the structure of a maximal parabolic of $L_3(2^n)$ or $Sp_4(2^n)$. Then we will show that $O^2(G) < G$ via transfer.

The main result of this subsection is:

THEOREM 2.4.7. *L is not an $L_2(2^n)$-block for $n > 1$.*

Throughout this subsection, G and H continue to be a counterexample to Theorem 2.4.1, with $H = L_0 S$ and $L_0 = \langle L^S \rangle$. Moreover we also assume that H is a counterexample to Theorem 2.4.7, so that L is an $L_2(2^n)$-block with $n > 1$. Set $q := 2^n$. Fix a Hall $2'$-subgroup D of $L_0 \cap M$ normalizing $L_0 \cap S$; thus $L_0 \cap M = (L_0 \cap S)D$ is a Borel subgroup of L_0. Of course $D \neq 1$ as $n > 1$.

The proof divides into two cases: $s = 1$ and $s = 2$. Further the case where $s = 1$ is by far the more difficult, as that is where the shadows of $L_3(q)$ and $Sp_4(q)$ extended by outer automorphisms arise. Thus the treatment of that case involves a long series of lemmas.

In the remainder of this subsection, set

$$R := J(S).$$

The Case $s = 1$.

Until this case is complete, we assume that $s = 1$, so that $L_0 = L_1 = L$, and $H = LS$.

LEMMA 2.4.8. *(1) $|T : S| = 2$. Hence T normalizes S and R.*
(2) $O_2(L) = U = Q = O_2(H)$.
(3) $R = Baum(S) = UU^x \in Syl_2(L)$, $U \cap U^x = Z(R)$, and $\mathcal{A}(S) = \mathcal{A}(T) = \{U, U^x\}$ for each $x \in T - S$.
(4) D^x acts on L, and either

(a) $Z(L) = 1$ and $L \cong P^\infty$ for P a maximal parabolic in $L_3(q)$, or
(b) $Z(L) \cong E_{2^n}$, D^x is regular on $Z(L)^\#$, and $L \cong P^\infty$ for P a maximal parabolic in $Sp_4(q)$.

(5) $\langle T, D \rangle = TB$, where B is an abelian Hall $2'$-subgroup of $\langle T, D \rangle$ containing D, S normalizes RD, $R \trianglelefteq RB \trianglelefteq BT$, $C_R(B) = 1$, and T is the split extension of R by $N_T(B)$. If $x \in N_T(B) - S$, then $B = DD^x$.
(6) If $Z(L) \neq 1$, then $B = D \times D^x$; while if $Z(L) = 1$, then $Aut_B(Z(R)) = Aut_D(Z(R)) \cong D$ is regular on $Z(R)^\#$.
(7) U and U^x are the maximal elementary abelian subgroups of R.

PROOF. By 2.4.5.3, we have the hypotheses of Theorem C.6.1, with T, S in the roles of "Λ, T_H", so we may appeal to Theorem C.6.1. In particular, conclusion (a) of C.6.1.6 holds, since L is of type $L_2(2^n)$ for $n > 1$. Thus (1) holds and $R = J(S) = J(T)$. By C.6.1.1, $Baum(S) = R = QQ^x$ for each $x \in T - S$ and by C.6.1.3, $\{Q, Q^x\} = \mathcal{A}(S)$. As $R = QQ^x$ with $Q \in \mathcal{A}(S)$, $Q \cap Q^x = Z(R)$ using B.2.3.7. Since $Q^x \not\leq Q$, Q^x is an FF-offender on U by Thompson Factorization B.2.15, so as $H = LS$ with L an $L_2(2^n)$-block, $R/Q = QQ^x/Q$ is Sylow in LQ/Q

by B.4.2.1, and hence $R \in Syl_2(LQ)$. Then (3) will follow once we prove (2). However, we will first establish (5) and the assertion in (4) that D^x normalizes L.

As $L = O^2(H)$, $C_R(L) \leq Q$, so as Q is abelian and $R \leq LQ$, $C_R(L) \leq Z(R)$. Further by (1), we may apply 2.4.5.2 with T in the role of "M_0", to conclude that $Q = UC_S(L)$. Then as $U \leq L$ and $R \leq LQ$, $R = (R \cap L)C_R(L)$. As R is Sylow in LQ, $S \cap L = R \cap L$; then $M \cap L = (R \cap L)D$ is a Borel subgroup of L, and $R = (R \cap L)C_R(L)$ is D-invariant.

Now S normalizes the Borel subgroup $(R \cap L)D$ over $R \cap L$, and hence normalizes RD. Thus S also normalizes $(RD)^x = RD^x$. Also T permutes Q and Q^x, and so acts on $G_Q \cap G_Q^x =: Y$. We saw D normalizes R, so as $D = O^2(D)$, D normalizes the two members Q and Q^x of $\mathcal{A}(R)$; that is, $D \leq Y$, and hence also $D^x \leq Y$. By 2.4.6, G_Q normalizes $L_0 = L$; in particular D^x normalizes L, giving the first assertion in (4). Now $M \cap G_Q$ normalizes $M \cap L = (R \cap L)D$ as well as $Q = UC_R(L)$, and hence normalizes their product RD. Then as $D^x \leq M \cap Y$, D^x also normalizes RD. As $x^2 \in S$ normalizes RD and $(RD)^x = RD^x$,

$$RDD^x \trianglelefteq \langle RDD^x, S, x \rangle = RDD^x T = DD^x T.$$

Hence by a Frattini Argument, we may take x to act on a Hall $2'$-subgroup B of RDD^x containing D.

We can now obtain the conclusions of (5), except possibly for $C_R(B) = 1$: First D^x normalizes $RD \cap B = D$, so DD^x is a subgroup of B, and hence $DD^x = B$. Then T normalizes $RDD^x = RB$, while R is normalized by D and hence also by D^x, and further $\langle T, D \rangle = BT$. We saw $D \trianglelefteq DD^x = B$, so also $D^x \trianglelefteq B$, and then $D^x[D, D^x] \leq D^x$. As D^x is abelian this shows that no element of D^x induces an outer automorphisms on $L/U \cong L_2(2^n)$, so that $B = DD^x = D \times C_B(L/U)$. Then since $B = DD^x$ and D is abelian, it follows that B is abelian. By a Frattini Argument on $RB \trianglelefteq TB$, $T = RN_T(B)$. This extension splits once we show $C_R(B) = 1$.

Thus to complete the proof of (5), it remains to show that $C_R(B) = 1$. We saw that $R = (R \cap L)C_R(L)$, so $[R, D] = [R \cap L, D] = R \cap L$ since L is an $L_2(2^n)$-block; thus also $[R, D^x] = R \cap L^x$. Further we saw $Q \cap Q^x = Z(R) \geq C_R(L)$, so $R/Z(R) = [R/Z(R), D]$. Then also $R/Z(R) = [R/Z(R), D^x]$, so

$$R \cap L = [R \cap L, D^x] \leq [R, D^x] = R \cap L^x;$$

so as $(R \cap L)^x = R \cap L^x$, $R \cap L = R \cap L^x$. Therefore $R \cap L = [R, D^x]$, so $[R, B] = [R, DD^x] = R \cap L$. As $C_{R/C_R(L)}(D) = 1$, $C_R(B) \leq C_R(L) \leq Z(R)$, so $C_R(B) \trianglelefteq LRN_S(B) = LS = H$. Also x normalizes R and B, and hence also $C_R(B)$, so that $C_R(B) = 1$ by 2.4.4, completing the proof of (5).

Now by Coprime Action, $R = [R, B] = R \cap L$, so that $R \leq L$. As $Q \leq R$, $Q = O_2(L) = U$ by 2.4.5.1, so that (2) holds. This also completes the proof of (3) as mentioned earlier.

So it remains to complete the proof of (4) and establish (6) and (7). As L is an $L_2(q)$-block, L is indecomposable on U with $U/Z(L)$ the natural module for L/U. From the cohomology of that module in I.1.6, $m(Z(L)) \leq n$. Further $Z(L) = C_R(D)$ with D semiregular on $R/Z(L)$, so D^x is semiregular on $R/C_R(D^x)$. Thus as $C_R(B) = 1$, D^x is semiregular on $Z(L)^\#$, so as $m(Z(L)) \leq n$, either $Z(L) = 1$ or $m(Z(L)) = n$. In each case (using I.1.6 in the latter) the representation of L/U on U is determined up to equivalence, and as the Sylow group $R = UU^x$ of L splits over U, L also splits over U by Gaschütz's Theorem A.1.39. Therefore L is determined up to isomorphism in each case. The parabolics P in cases (a) and (b) of (4) exhibit such extensions, so this completes the proof of (4). Part (4) implies (7).

When $Z(L) \neq 1$ the fact that D^x is semiregular on $Z(L) = C_R(D)$ shows that $B = D \times D^x$. When $Z(L) = 1$, D is regular on $Z(R)^{\#}$, so $D \cong Aut_D(Z(R))$ is self-centralizing in $GL(Z(R))$; thus as B is abelian, $Aut_B(Z(R)) = Aut_D(Z(R))$. This completes the proof of (6), and hence also of 2.4.8. $\qquad\square$

REMARK 2.4.9. The cases (a) and (b) of 2.4.8.4 were treated separately in sections 3 and 4 of [**Asc78a**]. However many of the arguments for the two cases are parallel, so we give a common treatment here where possible.

NOTATION 2.4.10. During the remainder of the treatment of the case $s = 1$, x denotes an element of $T - S$. By 2.4.8.1, $|T : S| = 2$, so as S acts on L, H, U, the conjugates L^x, H^x, U^x are independent of the choice of x.

LEMMA 2.4.11. *(1)* $G_Q = N_G(L)$.
(2) $G_Q = !\mathcal{M}(L)$.
(3) If $Z(L) \neq 1$ then $Z(L)$ is a TI-subgroup of G with $G_Q = N_G(Z(L))$.

PROOF. Recall $G_Q \le N_G(L)$ by 2.4.6; so as $Q = U = O_2(L)$ by 2.4.8.2, $N_G(L) \le G_Q$, so (1) holds.

As $Q = O_2(LS)$ while $S \in Syl_2(G_Q)$ by 2.4.3.1, $\mathcal{W}_{G_Q}^*(L, 2) = \{Q\}$. Then as L is irreducible on $Q/Z(L)$ and indecomposable on Q, if $1 \neq V \in \mathcal{W}_{G_Q}(L, 2)$ then either $V = Q$ or $V \le Z(L)$.

Let $X \in \mathcal{H}(L)$; to prove (2), we must show that $L \trianglelefteq X$, so assume otherwise. Let $P := O_2(X)$. Then $1 \neq P_0 := N_P(Q) \le G_Q$, so $P_0 \in \mathcal{W}_{G_Q}(L, 2)$. Thus by the previous paragraph, either $P_0 = Q$ or $P_0 \le Z(L)$. In either case $P_0 \le Q$, so that $N_{PQ}(Q) = P_0 Q = Q$, and then $PQ = Q$ so that $P = P_0$. If $P = Q$, then $X \le N_G(Q) = G_Q$, contrary to assumption; hence $P \le Z(L)$. This shows that (2) holds when $Z(L) = 1$. Thus for the rest of the proof, we may assume $Z(L) \neq 1$, since this is also the hypothesis of (3). In particular, case (b) of 2.4.8.4 holds.

Next we claim that $C_G(v) \le G_Q$ for each $v \in Z(L)^{\#}$. Assume otherwise; then we may choose $C_G(v)$ in the role of "X" in the previous paragraph. As B is transitive on $Z(L)^{\#}$ by 2.4.8.4, we may assume $S \le X$. Thus $H = LS \le X$, so by 2.3.7.2, $X \in \Gamma_0$ and $S \in Syl_2(X)$. Then by 1.2.4, $L \le K \in \mathcal{C}(X)$. As the Sylow 2-subgroup S of X normalizes L, but we are assuming L is not normal in X, $L < K$. Now $O_2(K) \le O_2(X) = P \le Z(L)$ by the previous paragraph, so $K = [K, L]$ centralizes $O_2(K)$. Also $m_2(K) \ge m_2(L) > 1$, so we conclude from 1.2.1.5 that K is quasisimple, and hence K is a component of X. Thus K is described in 2.3.9.7. Since $1 \neq v \in L \cap Z(X) \le Z(K)$, $Z(K)$ is of even order, so $K/O_2(K)$ is $L_3(4)$ or A_6 and $Z(K) = O_2(K)$. If $K/Z(K)$ is A_6, then $K \cong SL_2(9)$ by I.2.2.1, a contradiction as $L \le K$ with $m_2(L) \ge 4$. Thus $K/O_2(K) \cong L_3(4)$, $L/Q \cong L_2(4)$, and $Z(K) = Z(L) = P$ as L is irreducible on $Q/Z(K)$ and we saw $Z(K) \le P \le Z(L)$. In particular, $P \trianglelefteq H$. Further $Z(L) \cong E_4$ since $n = 2$ and case (b) of 2.4.8.4 holds. Observe since $K/Z(K) \cong L_3(4)$ that $L = N_K(Q)$, so as R is Sylow in L by 2.4.8.3, R is Sylow in K. Now consider $x \in T - S$ as in Notation 2.4.10. By parts (2) and (3) of 2.4.8, $\mathcal{A}(R) = \{Q, Q^x\}$, so $N_K(Q^x)$ is the maximal parabolic of K over R distinct from L. Therefore $N_K(Q^x)/Q^x \cong L_2(4)$ and $P = Z(K) = Z(N_K(Q^x))$. Hence $L^x = \langle R^{N_G(Q^x)} \rangle \ge N_K(Q^x)$, so we conclude

$L^x = N_K(Q^x)$ since $N_K(Q^x) \cong N_K(Q) = L$. Thus $Z(L) = P = Z(L^x)$, contrary to 2.4.4. This contradiction completes the proof that $C_G(v) \leq G_Q$.

In the remainder of the proof, X again denotes an arbitrary member of $\mathcal{H}(L)$ not normalizing L; thus $1 \neq O_2(X) = P \leq Z(L)$ by earlier remarks. Now $C_G(Z(L)) \leq C_G(v) \leq G_Q = N_G(L)$ by the previous paragraph, so $L \in \mathcal{C}(N_G(Z(L)))$. As $H \leq N_G(Z(L))$, $S \in Syl_2(N_G(Z(L)))$ by 2.3.7.2, so $L \trianglelefteq N_G(Z(L))$ by 1.2.1.3. Therefore $N_G(Z(L)) = N_G(L) = G_Q$ using (1). Also B is transitive on $Z(L)^\#$ and $C_G(v) \leq G_Q = N_G(Z(L))$, so $Z(L)$ is a TI-subgroup of G by I.6.1.1, completing the proof of (3). Then as $1 \neq O_2(X) \leq Z(L)$, $X \leq N_G(Z(L)) = G_Q$ by (3), contrary to assumption. This contradiction completes the proof of the lemma. □

We next repeat some arguments from sections 3 and 4 of [**Asc78a**], which force the 2-local structure of G to be essentially that of an extension of $L_3(2^n)$ or $Sp_4(2^n)$; this information is used later in transfer arguments to eliminate these shadows.

In fact, by 2.4.8 and 2.4.11, the hypotheses of section 3 or 4 in [**Asc78a**] are satisfied, in cases (a) or (b) of 2.4.8.4, respectively. Thus we could now appeal to Theorems 2 and 3 of [**Asc78a**]. However those results are not quite strong enough for our present purposes, and in any event we wish to keep our treatment as self-contained as possible, as discussed in the Introduction to Volume I under Background References. Thus we reproduce those arguments from [**Asc78a**] necessary to complete our proof.

LEMMA 2.4.12. *(1) H is the split extension of L by a cyclic subgroup F of S inducing field automorphisms on L/Q. Thus S is the split extension of R by F.*

(2) If f is an involution in F, then all involutions in fR are fused to f under R, $C_L(f)$ is an $L_2(q^{1/2})$-block (or S_4 or $\mathbf{Z}_2 \times S_4$ if $q^{1/2} = 2$), and either

(a) $Z(L) = 1$, and $C_R(f)$ is special of order $q^{3/2}$; in this case we say $C_R(f)$ is of type $L_3(q^{1/2})$.

(b) $Z(L) \cong E_q$, with $|C_{Z(L)}(f)| = q^{1/2}$ and $|C_R(f)| = q^2$; in this case we say $C_R(f)$ is of type $Sp_4(q^{1/2})$.

PROOF. Recall $H = LS$, while by parts (3) and (5) of 2.4.8, S is the split extension of $R \in Syl_2(L)$ by $N_S(B) =: F$. Thus $F \cap L = F \cap R = 1$, so that (1) holds.

Suppose f is an involution in F. As f induces a field automorphism on $\bar{L} := L/Q$, $q = r^2$, $C_{\bar{L}}(f) \cong L_2(r)$, and $C_{Q/Z(L)}(f)$ is the natural module for $C_{\bar{L}}(f)$. Indeed if $Z(L) \neq 1$, then $Z(L) \cong E_q$ by 2.4.8.4, so from I.1.6, Q is the largest indecomposable extension of a submodule centralized by $\bar{L}$ by a natural $\bar{L}$-module; hence $m(Z(L)) = 2m(C_{Z(L)}(f))$. Thus in any event $m(Q) = 2m(C_Q(f))$, so Q is transitive on the involutions in fQ. Then by a Frattini Argument, $C_{\bar{L}}(f) = \overline{C_L(f)}$, so $C_L(f)$ is as claimed in (2). Further by Exercise 2.8 in [**Asc94**], R is transitive on involutions in fR, completing the proof of (2). □

DEFINITION 2.4.13. Relaxing somewhat the usual definition in the literature, we define a *Suzuki 2-group* to be a 2-group I admitting a cyclic group of automorphisms transitive on its involutions, with $[I, I] = Z(I)$.

LEMMA 2.4.14. *Assume $t \in T - S$ with $t^2 \in R$. Then $\langle t \rangle R$ splits over R, R is transitive on the involutions in tR, and choosing t to be an involution, one of the following holds:*

(1) $Z(L) = 1$, $Z(R) = C_R(t)$, R is transitive on $t[R, t]$, and $[R, t]$ is transitive on $tZ(R)$; in this case we say $C_R(t)$ is of type $L_2(q)$.

(2) $Z(L) = 1$, n is even, and $C_R(t)$ is a Suzuki 2-group of order $q^{3/2}$ with $|\Omega_1(C_R(t))| = q^{1/2}$; in this case we say $C_R(t)$ is of type $U_3(q^{1/2})$.

(3) $Z(L) \cong E_q$, and $C_R(t)$ is a Suzuki 2-group of order q^2 with $|\Omega_1(C_R(t))| = q$; in this case we say $C_R(t)$ is of type $Sz(q)$.

PROOF. Since $t \in T - S$, t serves in the role of the element "x" in Notation 2.4.10; in particular, we may apply 2.4.8. As $RB \trianglelefteq TB$ by 2.4.8.5, by a Frattini Argument we may choose t to normalize B. Also by 2.4.8.5, $R \trianglelefteq RB$ and $C_R(B) = 1$, so as $t^2 \in R$, $[B, t^2] \leq B \cap R = 1$ and hence t is an involution. In particular $R\langle t \rangle$ splits over R.

We recall from Notation 2.4.10 that $Q^t = Q^x$ is independent of the choice of $x \in T - S$, so by 2.4.8.3, $m(R/Z(R)) = 2m(C_{R/Z(R)}(t))$ and $C_{R/Z(R)}(t) = [R/Z(R), t]$. Let R_t denote the preimage of $C_{R/Z(R)}(t)$, so that R_t contains $C_R(t)$. By 2.4.8.7, Q and Q^t are the maximal elementary abelian subgroups of R, so $Z(R) = \Omega_1(R_t)$, and hence $C_{Z(R)}(t) = \Omega_1(C_R(t))$.

Assume that $Z(L) \neq 1$. Then $Z(L) \cong E_q$ is a TI-subgroup of G by 2.4.11.3, while $|Z(R)| = 2^{2n} = |Z(L)|^2$ by 2.4.8.4, so $Z(R) = Z(L) \times Z(L)^t$. Thus by Exercise 2.8 in [**Asc94**], R is transitive on the involutions in tR, and $R_t = Z(R)C_R(t)$. As $B = D \times D^t$ by 2.4.8.6, $C_B(t)$ is a full diagonal subgroup of B, and so $C_B(t)$ is regular on $C_{Z(R)}(t)^{\#} = Z(C_R(t))^{\#}$. Further $C_R(t)$ is nonabelian, so that $[C_R(t), C_R(t)] = Z(C_R(t))$; thus $C_R(t)$ is a Suzuki 2-group of order q^2, so that conclusion (3) holds.

Now assume instead that $Z(L) = 1$. Set $(TB)^* := TB/C_{TB}(Z(R))$. As t normalizes B and $B^* = D^*$ is regular on $Z(R)^{\#}$ by 2.4.8.6, either $t^* = 1$ or $m(Z(R)) = 2m(C_{Z(R)}(t))$, and in either case $C_{B^*}(t^*) = C_B(t)^*$ is regular on $C_{Z(R)}(t)^{\#}$. Assume first that $t^* = 1$. Then as $R_t/Z(R) = [R/Z(R), t]$, with $\Omega_1(R_t) = Z(R)$, t inverts an element r of order 4 in each coset of $Z(R)$ in R_t. So as r is of order 4, $C_R(t) = C_{R_t}(t) = Z(R)$, and conclusion (1) holds. Further $|R : C_R(t)| = |R_t|$, so R is transitive on tR_t and hence on the involutions in tR. Now assume instead that $t^* \neq 1$, so that $m(Z(R)) = 2m(C_{Z(R)}(t))$. By Exercise 2.8 in [**Asc94**], $|C_R(t)| = q^{3/2}$ and R is transitive on the involutions in tR. As $C_B(t)$ is transitive on $C_{Z(R)}(t)^{\#} = Z(C_R(t))^{\#}$, and $C_R(t)$ is nonabelian so that $[C_R(t), C_R(t)] = Z(C_R(t))$, $C_R(t)$ is a Suzuki 2-group of order $q^{3/2}$. Thus conclusion (2) holds. $\square$

NOTATION 2.4.15. In the remainder of our treatment of the case $s = 1$, we define Z as follows: If $Z(L) = 1$, set $Z := Z(R)$, while if $Z(L) \neq 1$ set $Z := Z(L)$.

LEMMA 2.4.16. *(1) $Z \cong E_{2^n}$ and $Z \trianglelefteq S$.*
(2) For $x \in T - S$, either
    (a) $Z(L) = 1$ and $Z^x = Z = Z(R)$, or
    (b) $Z(L) \neq 1$ and $Z(R) = Z \times Z^x$.
(3) $U = \langle (Z^x)^L \rangle = \langle Z^G \cap U \rangle$.

PROOF. Part (1) follows from 2.4.8.4. Next x normalizes $J(S) = R$, so conclusion (a) of (2) holds when $Z(L) = 1$ as $Z = Z(R)$ in that case. If $Z(L) \neq 1$ then $Z = Z(L)$ is a TI-subgroup of G by 2.4.11.3, and $Z \neq Z^x$ by 2.4.4, so conclusion (b) of (2) holds as $|Z(R)| = |Z|^2$.

If $Z(L) = 1$ then $Z = Z^x$ by (2); hence $(Z^x)^L = Z^L$ gives the partition of the natural module U by its 1-dimensional $\mathbf{F}_q$-subspaces, so (3) holds in this case. If $Z(L) \neq 1$ then $Z(R) = ZZ^x$ and $(Z(R)/Z)^L$ is the corresponding partition of U/Z, so again (3) holds as L is indecomposable on U. $\qquad\square$

LEMMA 2.4.17. *(1) Either $R = C_T(Z)$; or case (1) of 2.4.14 holds, so that $Z(L) = 1$ and $C_T(Z) = R\langle t\rangle$ for some involution t in $T-S$ with $Z = Z(R) = C_R(t)$. (2) If $Z(L) = 1$ then $T/C_T(Z)$ is cyclic.*

PROOF. By 2.4.12.1, $C_S(Z) = R$, since the field automorphisms in F do not centralize Z. Assume $C_T(Z) > R$. Then $C_T(Z) = R\langle u\rangle$ for some $u \in C_T(Z) - S$, so $u^2 \in R$ and hence 2.4.14 completes the proof of (1).

Assume $Z(L) = 1$, so that $Z = Z(R)$. By 2.4.8.6, $Aut_B(Z)$ is cyclic and regular on $Z^\#$, so $Aut_{GL(Z)}(Aut_B(Z))$ is the multiplicative group of $\mathbf{F}_q$ extended by $Aut(\mathbf{F}_q)$. Since $Aut_B(Z)$ is normal in $Aut_{BT}(Z)$ by 2.4.8.5, we conclude $Aut_T(Z)$ is cyclic, so that (2) holds. $\qquad\square$

LEMMA 2.4.18. *(1) Z is a TI-subgroup of G. (2) If $Z(L) = 1$ then $N_G(Z) = M$. (3) if $Z(L) \neq 1$ then $N_G(Z) = G_Q$.*

PROOF. If $Z(L) \neq 1$ then (1) and (3) hold by 2.4.11.3. Thus we may assume $Z(L) = 1$, so $Z = Z(R)$ from Notation 2.4.15. Set $P := O_2(M)$. As T normalizes R by 2.4.8.1, there is an involution z in $Z \cap Z(T)$. As $F^*(M) = O_2(M)$, $z \in C_M(P) = Z(P)$. Then as $D \leq M$ and D is irreducible on Z, $Z \leq Z(P)$. It suffices to show that $Z \trianglelefteq M$: For then $M = N_G(Z)$ since $M \in \mathcal{M}$, so that (2) holds. Further as $M = !\mathcal{M}(T)$, $C_G(z) \leq M$, and hence as as D is transitive on $Z^\#$, Z is a TI-set in G by I.6.1.1, so that (1) also holds.

Thus it remains to show that $Z \trianglelefteq M$. If $R \leq P$, then as $R = J(T)$, also $R = J(P)$ by B.2.3, so that $Z = Z(J(P)) \trianglelefteq M$. Thus we assume that $R \not\leq P$. Now for $x \in T - S$, $R = UU^x$ by 2.4.8.3, so $U \not\leq P$. Then as $Z \leq P$ and D is irreducible on U/Z, $Z = U \cap P$, and then also $Z = U^x \cap P$. So since U and U^x are the maximal elementary subgroups of R by 2.4.8.7, $Z = \Omega_1(R \cap P)$. We now assume Z is not normal in M, and it remains to derive a contradiction. We saw $Z \leq Z(P)$, so that $Z < Z_P := \Omega_1(Z(P))$ and hence there is an involution $t \in Z_P - Z$. As $Z = \Omega_1(R \cap P)$, $t \notin R$, so as t centralizes Z, the second case of 2.4.17.1 holds. Therefore $Z = C_R(t)$ and t is described in case (1) of 2.4.14. But $[R,t] \leq [R, Z_P] \leq R \cap Z_P \leq C_R(t) = Z$, impossible as $[R,t] > Z$ in case (1) of 2.4.14. $\qquad\square$

LEMMA 2.4.19. *R is the weak closure of Z in T.*

PROOF. By 2.4.16.3, $Q = U = \langle(Z^x)^L\rangle$, and $R = QQ^x$ by 2.4.8.3. Hence R is contained in the weak closure of Z in T. Thus we may assume that there is $g \in G$ with $Z^g \leq T$ but $Z^g \not\leq R$, and it remains to derive a contradiction. By 2.4.16.1, $|Z| = 2^n > 2 = |T : S| \geq |T : N_T(Z)|$, so that $N_{Z^g}(Z) \neq 1$. Then as Z is a TI-subgroup of G by 2.4.18.1, and $\langle Z, Z^g\rangle$ is a 2-group, $Z^g \leq C_T(Z)$ by I.6.2.1. As $Z^g \not\leq R$, there is an involution $t \in Z^g - S$ with $C_T(Z) = R\langle t\rangle$ by 2.4.17.1. Then t satisfies conclusion (1) of 2.4.14 with $C_R(t) = Z$. Hence as $|Z^g| > 2 = |C_T(Z) : R|$, $R \cap Z^g \neq 1$. But $R \cap Z^g \leq C_R(t) = Z$, so as Z is a TI-subgroup of G, $Z^g = Z \leq R$, contrary to $Z^g \not\leq R$. $\qquad\square$

LEMMA 2.4.20. *Assume $Z(L) \neq 1$. Then for $x \in T - S$:*

(1) $B = D \times D^x$ *is regular on* $\Delta := Z(R) - (Z \cup Z^x)$.

(2) *For $u \in \Delta$, u is 2-central in M and hence 2-central in G, $C_G(u) \leq M$, and $u^G \cap Z = \emptyset$.*

(3) *All involutions in R are fused to $u \in \Delta$ or $z \in Z^\#$.*

(4) $R \trianglelefteq M$, *so* $R \trianglelefteq C_G(u)$ *for $u \in \Delta$.*

(5) *For $z \in Z^\#$, Sylow 2-subgroups of $C_G(z)$ are in S^G.*

(6) *If $u \in \Delta$ and $X = \langle Z^G \cap X \rangle$ is a 2-subgroup of $C_G(u)$, then $X \leq R$.*

PROOF. As $Z(L) \neq 1$, $E_{2^n} \cong Z(L) = Z$. By 2.4.11.3, Z is a TI-subgroup of G with $N_G(Z) = G_Q$, so for $z \in Z^\#$, $C_G(z) \leq G_Q$. Further $S \in Syl_2(G_Q)$ by 2.4.3, so (5) holds and z is not 2-central in G.

By 2.4.8.6, $B = D \times D^x$, while $Z(R) = Z \times Z^x$ by 2.4.16.2. By 2.4.8.4, D^x is regular on $Z^\#$, so as x interchanges D and D^x and Z and Z^x, D is regular on $(Z^x)^\#$. Thus $Z = C_{Z(R)}(D)$, completing the proof of (1). Next Q and Q^x are the maximal elementary abelian subgroups of R by 2.4.8.7, while all elements of Q are fused into $Z(R)$ under L, so (3) holds. Then as z is not 2-central in G, but $Z \times Z^x = Z(R) \trianglelefteq T$ since T normalizes R by 2.4.8.1, $u \in Z(T)$ for some $u \in \Delta$. So as $M = !\mathcal{M}(T)$, $G_u := C_G(u) \leq M$, and then (2) follows from the transitivity of D on Δ in (1).

Next we prove (4). Set $P := O_2(M)$. As $R = J(T)$ by 2.4.8.3, it suffices to show that $R \leq P$, since then $R = J(P)$ by B.2.3.3. As $F^*(M) = P$, $u \in C_M(P) = Z(P)$, so by (1), $Z(R) = \langle u^{BT} \rangle \leq Z(P)$. Let $W := \langle Z^G \cap P \rangle$. By 2.4.19, $W \leq R$, so as B is irreducible on $Q/Z(R)$, either $W = Z(R)$ or $W = R$. Since $W \trianglelefteq M$, (4) holds if $W = R$. If $W \neq R$ then $Z(R) = W \trianglelefteq M$ so that $M = N_G(Z(R))$ since $M \in \mathcal{M}$. But then as Z is a TI-subgroup of G, it follows from (1) and (2) that $M = N_M(Z)\langle x \rangle$. Now $N_G(Z) = G_Q = N_G(L)$ by 2.4.11, so $N_M(Z)$ normalizes $O_2(N_{M \cap L}(Z)) = R$. As x also normalizes R, we conclude (4) holds in this case also.

Finally assume the hypotheses of (6). Then $X \leq C_G(u) \leq M$ by (2), and as X is a 2-group, $X \leq T^m$ for some $m \in M$. Then $X \leq \langle Z^G \cap T^m \rangle = R^m$ by 2.4.19, so that $X \leq R$ by (4). $\square$

In the remainder of the treatment of the case $s = 1$, we let z denote an involution of $Z^\#$. If $Z(L) \neq 1$, let u denote an element of the set Δ defined in 2.4.20.1.

LEMMA 2.4.21. (1) R *is the strong closure of Q in T.*

(2) $i^G \cap T \subseteq R$ *for each involution i in R.*

PROOF. By parts (2) and (7) of 2.4.8, all involutions in R are fused into Q, so (1) implies (2).

By 2.4.19, R is contained in the strong closure of Q in T. Hence we may assume that a is an involution in $T - R$ fused into Q, and it remains to derive a contradiction. If $Z(L) = 1$ then L is transitive on $Q^\#$, so $a = z^g$ for some $g \in G$. If $Z(L) \neq 1$ then by 2.4.20.3, either $a = z^g$, or $a = u^g$ for $u \in \Delta$. Set $I := C_R(a)$ and let $I \leq T^* \in Syl_2(C_G(a))$ and set $R^* := J(T^*)$.

We claim that if $Z(L) \neq 1$ then $a \in S$. Thus we assume $Z(L) \neq 1$ and $a \in T - S$, and it remains to derive a contradiction. By 2.4.14, I is of type $Sz(q)$, so the involutions of I lie in Δ rather than in Z or $Z^x = Z^a$, since $a \in T - S$. Assume first that $a = z^g$. By 2.4.20.5, $T^* \in S^G$, and by 2.4.12.1, T^*/R^* is cyclic, so $Z(I) = [I, I] \le R^*$. Now we saw that involutions of $Z(I)$ lie in Δ, so we may assume that $u \in Z(I)$. Thus $Z(R^*) \le C_G(u) \le M$ by 2.4.20.2. By 2.4.16.2, $Z(R^*)$ is generated by a pair of conjugates of Z, so $Z(R^*) \le R \le S$ by 2.4.20.6. As $a \in Z(R^*)$, this contradicts our assumption that $a \notin S$. Therefore $a = u^g$, and so $T^* \in T^G$. Let $Q^* \in \mathcal{A}(T^*)$ and $S^* = N_{T^*}(Q^*)$. Then $|T^* : S^*| = 2$, so arguing much as before, $a \in Z(T^*) \le Z(R^*)$ and $Z(I) = [I, I] \le S^*$. Then as S^*/R^* is cyclic by 2.4.12.1, either $Z(I)$ is noncyclic so that $Z(I) \cap R^* \neq 1$, or $Z(I)$ is of order 2 so that $q = 4$. In the former case we obtain a contradiction as before, and in the latter T^*/R^* is of order at most 4 and hence abelian, so again $[I, I] \le R^*$, for the same contradiction. This completes the proof of the claim.

We now summarize the remaining possibilities: If $Z(L) \neq 1$ then $a \in S = N_T(Z)$ by the claim, so that I is of type $Sp_4(q^{1/2})$ by 2.4.12.2. So assume that $Z(L) = 1$. Then $a = z^g$ and $T = N_T(Z)$, so again a normalizes Z. If $a \notin S$, then by 2.4.14, either $I = Z$ is of type $L_2(q)$, or I is of type $U_3(q^{1/2})$. Finally if $a \in S$, then I is of type $L_3(q^{1/2})$ by 2.4.12.2.

Assume that a centralizes Z. Then by the previous paragraph, $Z(L) = 1$, $a = z^g \in T - S$, and $I = Z$ is of type $L_2(q)$. Since Z is a TI-subgroup by 2.4.18.1 and $a = z^g$ centralizes Z, $[Z, Z^g] = 1$ by I.6.2.1. Thus $aZ \subseteq V := ZZ^g \cong E_{2^{2n}}$. However $[R, a]$ is transitive on aZ by 2.4.14. Thus for $r \in [R, a]$, $a^r \in aZ \subseteq V \le C_G(V)$, so again by I.6.2.1, $Z^{gr} \le C_G(V)$. Then as $m(V) = 2n = m_2(T)$, $Z^{gr} \le V$, so $[R, a]$ normalizes $\langle Z^{g[R,a]} \rangle = V$. Notice $V \in \mathcal{A}(G) = Q^G$ in view of 2.4.8.3, and of course $Z \in Z^G \cap V$. Now $|[R, a]V| = q^3 = |R|$ and by 2.4.17.1, R is Sylow in $G_Q \cap C_G(Z)$, so that $[R, a]V = R^h$ for some $h \in G$. By 2.4.14, R is transitive on $a[R, a]$, so for $s \in R$, $a^s \in R^h$. Thus a^s is contained in some conjugate of Z contained in R^h, so as Z is a TI-subgroup of G, $Z^{gs} \le R^h$. Then $V = \langle Z^{g[R,a]} \rangle \le \langle Z^{gR} \rangle =: X$ is a subgroup of R^h normalized by R. It follows that R normalizes R^h: for if $X < R^h$, then $V = J(X)$, so that R normalizes $[R, a]V = R^h$. So as $R^h = J(T^h)$ is weakly closed in T^h, $R = R^h$. But then $a \in V \le R^h = R$, contradicting our observation that $a \notin S$.

Therefore $[a, Z] \neq 1$, so from our earlier summary, I is of type $Sp_4(q^{1/2})$, $U_3(q^{1/2})$, or $L_3(q^{1/2})$. In each case $[I, I] = Z(I)$. Furthermore setting $Z_a := C_Z(a)$, either $Z_a \le [I, I]$, or $q = 4$ and $I \cong \mathbf{Z}_2 \times D_8$ is of type $Sp_4(2)$.

Suppose first that $a = z^g$. Assume $Z(L) = 1$. Then by 2.4.17.2, $T^*/C_{T^*}(Z^g)$ is cyclic, and using the previous paragraph, $Z_a \le [I, I] \le C_{T^*}(Z^g)$. Thus as $1 \neq Z_a \le Z$, $[Z^g, Z] = 1$ by I.6.2.1, contradicting $[a, Z] \neq 1$. Thus $Z(L) \neq 1$ so $T^* \in S^G$ by 2.4.20.5, and T^*/R^* is cyclic by 2.4.12.1. Hence $[I, I] \le R^* = C_{T^*}(Z^g)$. Thus if $Z_a \le [I, I]$, we get the same contradiction as above, so from the previous paragraph, $q = 4$ and $[I, I] =: \langle u \rangle \le R^* = C_{T^*}(Z^g)$. Then $a \in Z^g \le \langle Z^G \cap C_G(u) \rangle$, as $C_G(u) \le M$ by 2.4.20.2, so $a \in R$ using 2.4.20.6. Again this contradicts $[a, Z] \neq 1$, so $a \notin z^G$.

Therefore $a = u^g$, so that $Z(L) \neq 1$ by our previous summary; and it also now follows from our remarks at the start of the proof that R is the weak closure of z in T. From our summary, $a \in S$ and I is of type $Sp_4(q^{1/2})$. We may assume $z \in I$. Then $I = \langle z^G \cap I \rangle \le \langle z^G \cap C_G(a) \rangle =: Y$. Since $C_G(a) \le M^g$ by 2.4.20.2, and R is the weak closure of z in T, we conclude from 2.4.20.4 that $z \in Y \le R^g$. But then z is contained in a conjugate of Z in $R^g = R^*$, so as Z is a TI-subgroup of G, $Z \le R^g \le C_G(a)$, again contradicting $[a, Z] \neq 1$. This finally completes the proof of (1), and hence of 2.4.21. $\qquad\square$

At this point, we have obtained strong control over the 2-local structure and 2-fusion of G, which we can use to obtain contradictions via transfer arguments.

LEMMA 2.4.22. *(1) T/R is not cyclic.*
(2) $R < S$.

PROOF. If $R = S$ then $|T : R| = 2$ by 2.4.8.1, so T splits over R by 2.4.14, and hence there is an involution $t \in T - R$. On the other hand if $R < S$, there is an involution t in $S - R$ by 2.4.12.1. Thus in any case there is an involution $t \in T - R$.

As T/R is cyclic if $S = R$, it remains to assume T/R is cyclic and derive a contradiction. By 2.4.21, $t^G \cap R = \emptyset$. Then by Generalized Thompson Transfer A.1.36.2, $t \notin O^2(G)$, contrary to the simplicity of G.   □

By 2.4.22.2, $R < S$; so since S splits over R by 2.4.12.1, there is an involution $S - R$. It is convenient to use the notation s for this involution; there should be no confusion with the earlier numerical parameter "s", as in the branch of the argument for several pages before and after this point, that parameter has the value 1. Let $G_s := C_G(s)$, $L_s := C_L(s)$, etc.

We use the standard notation that for x an integer, x_2 denotes the 2-primary part of x.

LEMMA 2.4.23. *(1) Either L_s is an $L_2(2^{n/2})$-block with $U_s = U(L_s)$, or $q = 4$ and $L_s \cong S_4$ or $S_4 \times \mathbf{Z}_2$.*
(2) R_s is the strong closure of Q in T_s.
(3) $U_s = O_2(L_s)$ and $N_G(U_s) \leq G_Q$.
(4) $T = RT_s$, there exists $x \in T_s - S$, and $T_s \in Syl_2(G_s)$.
(5) Assume $Z(L) \neq 1$ and $q = 2^n > 4$. Set $K_s := \langle L_s, L_s^x \rangle$. Then $K_s \cong Sp_4(2^{n/2})$, $C_{T_s}(K_s) = \langle s \rangle$, and $T_s/\langle s \rangle R_s$ is cyclic of order $n_2 = |Out(K_s)|_2$.

PROOF. Part (1) follows from 2.4.12.2. By (1), $U_s = O_2(L_s)$. Part (2) follows from 2.4.21.1.

From (1) and the proof of 2.4.16.3, $U_s = \langle (Z_s^x)^G \cap U_s \rangle$. But $N_G(U_s)$ permutes $(Z_s^x)^G \cap U_s$ and Z is a TI-subgroup of G, so $N_G(U_s)$ permutes $(Z^x)^G \cap U$ and hence $N_G(U_s) \leq N_G(U) = G_Q$ by 2.4.16.3, and as $Q = U$ by 2.4.8.2. This completes the proof of (3).

By 2.4.12.1, S/R is cyclic, so $\langle s \rangle R \trianglelefteq T$. By 2.4.12.2, R is transitive on the involutions in sR, so by a Frattini Argument $T = RT_s$, and as $S \in Syl_2(G_Q)$ by 2.4.3.1, S_s is Sylow in $N_{G_s}(U_s)$ by (3). As $S < T$, there is $x \in T_s - S$ and by 2.4.8.7, U and U^x are the maximal elementary abelian subgroups of R, so $\mathcal{A}(R_s) = \{U_s, U_s^x\}$. Therefore $N_G(R_s) = N_G(U_s)\langle x \rangle$. So using (2), $N_G(T_s) \leq N_G(U_s)\langle x \rangle$. Thus as S_s is Sylow in $N_{G_s}(U_s)$ and $S_s\langle x \rangle = T_s$, $T_s \in Syl_2(G_s)$, so that (4) holds.

Assume the hypotheses of (5), and set $K_s := \langle L_x, L_x^s \rangle$. Let Θ be the set of subgroups of S_s invariant under L_s. From the action of S and L, $U_s\langle x \rangle$ is the unique maximal member of Θ, and if $Y \in \Theta$ with $U_s \not\leq Y$, then $Y \leq \langle s \rangle C_U(L)$. Therefore as $R_s = U_s U_s^x$ and $C_U(L) \cap C_U(L)^x = 1$, $\langle s \rangle$ is the largest subgroup of T_s invariant under L_s and L_s^x, and hence $\langle s \rangle = O_2(K_s T_s)$. As $q > 4$ by hypothesis, $L_s \in \mathcal{L}(G_s, S_s)$, so since $|T_s : S_s| = 2$ with $T_s \in Syl_2(G_s)$ by (4), $L_s \leq K \in \mathcal{C}(G_s)$ by 1.2.5. As x acts on $R_s \leq K$, x acts on K, so $K_s \leq K$. Then using (4) and A.1.6, $O_2(K) \leq O_2(KT_s) \leq O_2(K_s T_s) = \langle s \rangle \leq C_G(K)$. As $m_2(K) \geq m_2(L_s) > 1$, K is quasisimple by 1.2.1.5. By (1), L_s is an $L_2(q^{1/2})$-block with $Z_s = Z(L_s) \neq 1$, so as $R_s = U_s U_s^x$ is a Sylow 2-subgroup of L_s, we conclude by examination of the possibilities in Theorem C (A.2.3) that $K_s = K \cong Sp_4(q^{1/2})$, and x induces an outer automorphism on K nontrivial on the Dynkin diagram. Then $C_{T_s}(K_s) = O_2(K_s T_s) = \langle s \rangle$. Finally $Out(K_s)$ is cyclic and $R_s \in Syl_2(K_s)$, so $T_s/R_s\langle s \rangle$ is cyclic. Further (cf. 16.1.4 and its underlying reference) a Sylow 2-subgroup of $Out(K_s)$ is generated by the image of any 2-element nontrivial on the Dynkin diagram of K, so $|T_s : R_s\langle s \rangle| = n_2 = |Out(K)|_2$, completing the proof of (5). $\qquad\square$

LEMMA 2.4.24. *Let* $T_B := N_T(B)$. *Then*

(1) T *is the split extension of* R *by* T_B.

(2) $Z(L) = 1$.

(3) $T_B = \langle x \rangle \times F$, *where* x *is an involution such that* $C_R(x) = Z$, $R\langle x \rangle = C_T(Z)$, *and* F *is cyclic and induces field automorphisms on* L/Q.

PROOF. Part (1) is one of the conclusions of 2.4.8.5. By 2.4.12.1, $T_B = F\langle x \rangle$, where $F := N_S(B)$ is cyclic and induces field automorphisms on L/Q, and $x \in T_B - S$. By 2.4.22.1, T_B is noncyclic. Choose $s \in F$.

Suppose first that $Z(L) = 1$. By 2.4.17.1, either $C_T(Z) = R$, or there exists some involution $x \in T_B - S$ with $Z = C_R(x)$ such that $C_T(Z) = R\langle x \rangle$. In the former case, T_B is cyclic by 2.4.17.2, contrary to the previous paragraph, so the latter must hold. Then $[x, F] \leq C_F(Z) = 1$, so $T_B = \langle x \rangle \times F$, establishing (3). Since (2) holds by assumption, the lemma holds in this case. Thus we may assume that $Z(L) \neq 1$ and it remains to derive a contradiction.

Suppose first that $n/2$ is odd. Then $|S : R| = 2$ since $R < S$ by 2.4.22, so $|T : R| = 4$ using 2.4.8.1. Hence $T_B \cong T/R \cong E_4$, since T_B is noncyclic, so there is an involution x in $T - S$ and by 2.4.14, $C_R(x)$ is of type $Sz(q)$, so $V := \Omega_1(C_R(x)) \cong E_q$ and $V^\# \subseteq \Delta$. It will suffice to show that V is the strong closure of u in a Sylow 2-subgroup T_x of $C_G(x)$ containing $C_T(x)$: For by 2.4.23.2, R_s is the strong closure of u in a Sylow 2-group of $C_G(s)$, and hence is nonabelian by 2.4.12.2. So as V is the strong closure of u in T_x, it follows that $s \notin x^G$. Further $x^G \cap R = \emptyset$ by 2.4.21.2, so as all involutions in $S - R$ are fused to s by 2.4.12.2, we conclude that $x^G \cap S = \emptyset$. Then $x \notin O^2(G)$ by Thompson Transfer, for the usual contradiction to the simplicity of G.

So it remains to show that V is strongly closed in T_x. Now conjugates of u generate R by 2.4.20; so by 2.4.21 and 2.4.20.4, R is the strong closure of u in $C_G(u)$. Therefore as $V = \Omega_1(C_R(x))$ and $V^\# \subseteq u^G$, V is strongly closed in T_x. As we mentioned, this completes the elimination of the case $n/2$ odd.

Therefore $n/2$ is even, so $q > 4$. Thus by 2.4.23.5, $T_s/R_s\langle s \rangle$ is cyclic of order $n_2 \geq 4$, and $n_2 = |Out(K_s)|_2$. Let $tR_s\langle s \rangle$ denote the involution of $T_s/R_s\langle s \rangle$; then this involution lies in the cyclic subgroup of index 2 in $T_s/R_s\langle s \rangle$ inducing field automorphisms, so any preimage t of $tR_s\langle s \rangle$ induces an involutory field automorphism on L_s/U_s. Thus t induces a field automorphism of order 4 on L/Q, so t is not an involution. Since $s \in T_B$, $T_B/\langle s \rangle \cong T/R\langle s \rangle \cong T_s/R_s\langle s \rangle$ using 2.4.23, so s is the unique involution in T_B. Also T_B is not quaternion since $T_B/\langle s \rangle$ is cyclic. Therefore T_B is cyclic, contrary to our earlier reduction. This contradiction completes the proof. $\qquad\square$

We can now finally eliminate the case where the numerical parameter we denoted earlier by "s" has the value 1: Let $T_C := C_T(Z)$. By 2.4.24.2, $Z(L) = 1$. Then $Z = Z(R) \trianglelefteq T$, so $T_C \trianglelefteq T$. By 2.4.24.3, there is an involution $x \in T - S$ such that $Z = C_R(x)$, $T_C = R\langle x \rangle$, and $T/T_C \cong T_B/\langle x \rangle \cong F$ is cyclic. It will suffice to show that $s^G \cap T \subseteq S$, for the involution we have been denoting by s: For then

$$s^G \cap T_C \subseteq s^G \cap T_C \cap S = s^G \cap R = \emptyset$$

using 2.4.21.2. Then as T/T_C is cyclic, $s \notin O^2(G)$ by Generalized Thompson Transfer A.1.37.2, as usual contrary to the simplicity of G.

Thus it remains to show that $s^G \cap T \subseteq S$. By 2.4.23, R_s is the strong closure of Q in $T_s \in Syl_2(G_s)$. As $Z(L) = 1$, R_s is of type $L_3(2^{n/2})$ by 2.4.12. Finally by 2.4.14, for each involution $i \in T - S$, $C_R(i)$ is of type $L_2(2^n)$ or $U_3(2^{n/2})$, and in either case, $\Omega_1(C_R(i)) \leq Z$. To show that $s^G \cap T \subseteq S$, we must show that $i \notin s^G$ for each such i; so we assume that that $i \in s^G$, and it remains to derive a contradiction.

Assume first that $C_R(i)$ is of type $L_2(2^n)$. Then i centralizes Z of order 2^n, whereas for each $g \in G$ with $Z^g \cap R_s \neq 1$, $|C_{Z^g}(s)| = 2^{n/2}$, contrary to $i \in s^G$ and 2.4.21.1.

Therefore $R_i := C_R(i)$ is of type $U_3(2^{n/2})$. Set $Z_i := C_Z(i) = Z(R_i)$. Then $i^g = s$ for some $g \in G$, and for suitable $c \in G_s$, $R_i^{gc} \leq T_s$ as $T_s \in Syl_2(G_s)$ by 2.4.23.4. Then $Z_i^{gc} \leq R_s$ by 2.4.23.2. Interchanging U and U^x if necessary, we may assume that $Z_i^{gc} \leq U_s$. Indeed we claim $Z_i^{gc} = Z_s$: For assume otherwise. By 2.4.18.1, Z_i^{gc} and Z_s are TI-subgroups of G_s of order $q^{1/2}$, so $U_s = Z_s \times Z_i^{gc}$, and hence $R_i^{gc} \leq C_{T_s}(U_s/Z_s) = R_s\langle s \rangle$. Then $Z_i^{gc} = \Phi(R_i^{gc}) \leq \Phi(R_s\langle s \rangle) = Z_s$, a contradiction establishing the claim that $Z_i^{gc} = Z_s$.

By the claim, $R_i^{gc} \leq C_{T_s}(Z_s)$. But $R\langle x \rangle = C_T(Z)$ with $T/R\langle x \rangle$ cyclic, so $R\langle x, s \rangle = C_T(Z_s)$ as Z is a TI-subgroup. Thus $|C_{T_s}(Z_s) : R_s\langle s \rangle| \leq 2$, so as $|R_i| = q^{3/2} = |R_s|$, also $|C_{T_s}(Z_s) : R_i^{gc}\langle s \rangle| \leq 2$, and hence $|U_s\langle s \rangle : U_s\langle s \rangle \cap R_i^{gc}\langle s \rangle| \leq 2$. Now $U_s\langle s \rangle$ is elementary abelian of order $2q$, while $\Omega_1(R_i^{gc}\langle s \rangle) = Z_i^{gc}\langle s \rangle$ is elementary of order $2q^{1/2}$, so $2q \leq 4q^{1/2}$, and hence we conclude $q = 4$. Therefore $T_s = \langle s \rangle \times R_s\langle x \rangle$, with x an involution by 2.4.24.3, and $R_s = U_s U_s^x \cong D_8$, so $R_s\langle x \rangle \cong D_{16}$. This is impossible, as the group R_i of type $U_3(2)$ is isomorphic to Q_8, and $\mathbf{Z}_2 \times D_{16}$ contains no such subgroup.

This contradiction finally completes the treatment of the case $s = 1$ of Theorem 2.4.7.

The case $s = 2$.

So we turn to the case $s = 2$. Here we will produce members of Γ_0 other than $H = L_0 S$, which we use to obtain a contradiction.

As $s = 2$, $L_0 = L_1 L_2$ with $L = L_1$, and we set $U_i := U(L_i)$, so that $U_0 = \langle U^S \rangle = U_1 U_2$. By 2.4.5.1, Hypotheses C.5.1 and C.5.2 are satisfied with S in the roles of both "T_H" and "R", for any subgroup M_0 of T with $|M_0 : S| = 2$. Observe U_0, Baum(S) play the roles that "U, S" play in section C.5. Further as $|M_0 : S| = 2$, the hypotheses of C.5.6.7 are satisfied by 2.4.5.2.

Recall from the beginning of this subsection 2.4.1 that $R = J(S)$, and also that D is defined there; and from the opening few pages of this section 2.4 that $Q = O_2(H) = O_2(L_0 S)$. By 2.4.5.2, $Q = U_0 C \in \mathcal{A}(S)$, where $C := C_S(L_0)$, and $U_0^x \not\leq Q$. As $s = 2$, case (iii) of C.5.6.7 holds; hence there are two S-invariant members $\{Q, Q^x\}$ of $\mathcal{A}(S)$, and $QQ^x = R = $ Baum(S) since Baum(S) contains R, and RQ is Sylow in $L_0 Q$ by B.4.2.1.

We can now argue much as in the proof of 2.4.8.5, but using M_0, L_0 in the roles of "T, L", to show that $B := DD^x$ is abelian of odd order, omitting details except to point out where the argument differs slightly: Notice this time that D normalizes Q and the unique member Q^x of $\mathcal{A}(S)$ with $R = QQ^x$. Further $D^x = O^2(D^x)$, so D^x normalizes each of the two conjugates L_1 and L_2 of L in $L_0 S$.

Now G_Q is an SQTK-group, so $m_p(G_Q) \leq 2$ for p a prime divisor of $|D|$; then as $m_p(D) = 2$ it follows that $B = D = D^x$.

Next $x \in T - S$ acts on B and hence on $C_R(B)$. As $B = D$ and L is an $L_2(2^n)$-block, $C_R(B) = C_R(L_0)$, and as Q is abelian, $C_R(L_0) \leq Z(R)$ so that $C_R(B) \trianglelefteq L_0 N_S(B) = L_0 S = H$. Hence $1 = C_R(B) = C_R(D)$ by 2.4.4. Then $C_{U_1}(L) = 1$ and $Q = CU_0 = U_0 \leq L_0$, so that $G_Q = N_G(L_0)$, just as in the proof of 2.4.11.1. Also as RQ is Sylow in $L_0 Q$, $R = QQ^x$ is Sylow in L_0. Then $L_0 = L_1 \times L_2$ so $R = R_1 \times R_2$, where $R_i := R \cap L_i \in Syl_2(L_i)$ is of order q^3, and $D = D_1 \times D_2$, where $D_i := D \cap L_i$, and D_1 and D_2 are the subgroups of D maximal subject to $C_R(D_i) \neq 1$. Therefore as $N_S(D)$ interchanges D_1 and D_2, we may choose x in $M_0 - S$ so that x normalizes D_1 and D_2, and hence x acts on $C_R(D_{3-i}) = R_i$. As $L_2 = [L_2, Q^x]$, $x \notin N_G(L_2)$. Thus $L_2 < K := \langle L_2, L_2^x \rangle \leq C_G(R_1 D_1)$, and $S_1 := \langle x \rangle N_S(R_1) = N_{M_0}(R_1)$ normalizes K. Observe that $|S : N_S(R_1)| = 2$ with $R = J(S) \leq N_S(R_1)$, $Q = O_2(H) \leq N_S(R_1)$, and $H \in \Gamma_0^e$. Thus $N_S(R_1) \in \beta$ by 2.3.8.5b. Then as $L_2 N_S(R_1) \in \mathcal{H}^e$ and $L_2 \not\leq M$, from the definitions in Notation 2.3.4 and Notation 2.3.5, $(N_S(R_1), L_2 N_S(R_1)) \in \mathcal{U}(KS_1)$, so that $KS_1 \in \Gamma$.

We claim next that $R = J(M_0)$: For suppose $A \in \mathcal{A}(M_0)$ with $A \not\leq R$. By 2.4.3, $S = N_T(Q)$, so as $R = J(S)$, there is an involution $a \in A - S$; hence $Q^a = Q^x$, since $M_0 = S\langle x \rangle = S\langle a \rangle$ and S acts on Q. If $R_1^a = R_2$ then $C_R(a) \cong R_1$ is of rank $2n$, while if $R_1^a = R_1$, then as $Q^a = Q^x$, $\Omega_1(C_{R_i}(a)) \leq Z(R_i)$, and so again $m_2(C_R(a)) \leq 2n$. Now S/R is contained in the wreath product of a cyclic group of field automorphisms of $L_2(2^n)$ by $\mathbf{Z}_2$, so that $m_2(S/R) \leq 2$; hence

$$4n \leq m(A) \leq m(M_0/S) + m(S/R) + m(A \cap R) \leq 1 + 2 + m(C_R(a)) \leq 3 + 2n < 4n$$

since $n \geq 2$. This contradiction establishes the claim that $R = J(M_0)$.

Next from the proof of C.5.6.7, $|\mathcal{A}(R)| = 4$, and $M_0 - N_T(R)$ induces a 4-subgroup on $\mathcal{A}(R)$ generated by a pair of commuting transpositions. Thus either $M_0 = N_T(R)$ and $Q^{N_T(R)} = \{Q, Q^x\}$ is of order 2, or $M_0 < N_T(R)$ with $Q^{N_T(R)} = \mathcal{A}(R)$ and $N_T(R)$ inducing D_8 on $\mathcal{A}(R)$.

Assume that the latter case holds. Now D acts on each member of $\mathcal{A}(R)$, so for each $y \in N_T(R)$, $D \leq G_Q^y = N_G(L_0^y)$, and by 1.2.2, $D \leq L_0^y$. It follows that $N_T(R)$ normalizes the intersection RD of the groups L_0 and L_0^y; hence $RD \trianglelefteq N_T(R)D$, so $N_T(R) = R(N_T(R) \cap N_T(D))$ by a Frattini Argument. Then arguing as above, $N_T(R)$ permutes the subgroups D_i maximal subject to $C_R(D_i) \neq 1$, and so permutes their fixed spaces $\{R_1, R_2\}$. Therefore $N_S(R_1)$ is of index 2 in a subgroup $S_2 \leq N_T(R)$ such that S_2 acts on R_1 and U_2. We have seen that $N_S(R_1) \in \beta$, so $S_2 \in \beta$ by 2.3.2.1. Next $R_1 U_2 = QQ^s$ for $s \in S_2 - S$ with $\mathcal{A}(R_1 U_2) = \{Q, Q^s\}$, so $N := N_G(R_1 U_2) = (N_G(Q) \cap N_G(Q^s))S_2$. By 2.4.3.1, $Q \in \mathcal{S}_2^e(G)$, so by 1.1.4.1, $N \in \mathcal{H}^e$. Then as $L_2 \leq N$ with $L_2 \not\leq M$, $(S_2, N) \in \mathcal{U}(N)$, so $N \in \Gamma$. But $|S_2| = |S|$, so by 2.4.3.2, $N \in \Gamma_0^e$ and $S_2 \in Syl_2(N)$. Now $H_1 := \langle S_2, L_2 \rangle \leq N$ and as $S_2 \in Syl_2(N)$ and $N \in \mathcal{H}^e$, $H_1 \in \mathcal{H}^e$ by 1.1.4.4. Thus $(S_2, H_1) \in \mathcal{U}(H_1)$, so $H_1 \in \Gamma$; then $H_1 \in \Gamma_0^e$ by 2.4.3.2. Therefore as $H_1 \leq N_G(R_1)$, $S_2 \in Syl_2(N_G(R_1))$ by 2.3.7.2. This is impossible as $|N_T(R) : N_T(R_1)| = 2$ since $N_T(R)$ permutes

$\{R_1, R_2\}$ transitively, so that $|N_T(R_1)| \geq 2|S| = 2|S_2| > |S_2|$. This contradiction eliminates the case $M_0 < N_T(R)$.

Therefore $M_0 = N_T(R)$. Then as $N_T(M_0) \leq N_T(J(M_0)) = N_T(R) = M_0$, we conclude $M_0 = T$, and hence $|T| = 2|S|$. Recall $S_1 = \langle x \rangle N_S(R_1) = N_{M_0}(R_1)$; thus $|S_1| = |S| = |T|/2$. Then by 2.3.7.1, $H \in \Gamma_*$, and as we saw $KS_1 \in \Gamma$, similarly $KS_1 \in \Gamma_*$ with $S_1 \in Syl_2(KS_1)$.

As $L_2 \in \mathcal{L}(KS_1, N_S(R_1))$ and $|S_1 : N_S(R_1)| = 2$, $L_2 \leq K_2 \in \mathcal{C}(KS_1)$ by 1.2.5. By construction S_1 normalizes R_1, and K centralizes $R_1 D_1$; indeed much as in the proof of 2.4.23.5, R_1 is the largest subgroup of S_1 invariant under L_2 and x, so that $R_1 = O_2(KS_1) \geq O_2(K_2)$. As K centralizes R_1, we conclude that $O_2(K_2) \leq Z(K_2)$. Then as $m_2(K_2) \geq m_2(L_2) > 1$, we conclude from 1.2.1.5 that K_2 is quasisimple, and hence is a component of KS_1. Thus K_2 is described in 2.3.9.7; so as $K_2 \cap M$ contains the $L_2(q)$-block L_2 and $C_{U_2}(L_2) = 1$, we conclude that $K = K_2$ and $K/O_2(K) \cong L_3(q)$. But now $m_p(KD_1) > 2$, for p a prime divisor of $q - 1$, contradicting KD_1 an SQTK-group.

This contradiction shows that the case $s = 2$ cannot occur in Theorem 2.4.7. Hence the proof of Theorem 2.4.7 eliminating $L_2(2^n)$ blocks for $n > 1$ is at last complete.

2.4.2. The small examples and shadows of extensions of $L_4(3)$. In this subsection, we complete the proof of Theorem 2.4.1. Thus we continue the hypotheses and notation from the beginning of this section. By Theorem 2.4.7, the block L is of type A_3 or A_5, and in the latter case $L_0 \cap M$ is a Borel subgroup of L_0 as $L_0 S$ is a minimal parabolic. Therefore $Z(L) = 1$ by C.1.13.c.

Recall from the beginning of this section 2.4 that $Q := O_2(H)$. However in this new subsection, $J(S)$ is no longer denoted by R, but instead

$$R := \mathrm{Baum}(S).$$

Recall also from 2.3.8.4 that $L_i = [L_i, J(S)]$ for each i, so that R normalizes L_i by C.1.16.

LEMMA 2.4.25. *If L is an A_5-block, then $s = 1$.*

PROOF. Assume otherwise, so that $s = 2$. Recall we defined $R = \mathrm{Baum}(S)$ just above, and set $Q_i := O_2(L_i R)$, $I := C_R(L)$, $S_I := N_S(L)$, and $T_0 := N_T(S)$. By 2.4.5.1, Hypotheses C.5.1 and C.5.2 are satisfied with S in the role of both "T_H" and "R", for each subgroup M_0 of T_0 with S a proper normal subgroup of M_0. As R denotes $\mathrm{Baum}(S)$, U_0, R play the roles played by "U, S" in section C.5, while I plays the role of "D_1".

By C.5.4.3, $Q_2 = U_2 \times D_2$ where $D_2 := C_R(L_2)$ and $U_2 := O_2(L_2)$, and $R/Q_2 \cong E_4$ is generated by two transpositions in $L_2 R/Q_2 \cong S_5$. Also from the proof of C.5.4.3, $RQ_2 = J(S)Q_2$ and for $A \in \mathcal{A}(S)$ with $A \not\leq Q_2$, $|U_2 : C_{U_2}(A)| = |A : (A \cap Q_2)|$. It follows that $[A, U_1] = 1$ so $[A, L] \leq C_L(U_1) = U_1$, and hence $A = U_1 \times (A \cap I)$. Thus I/Q_I is generated by two transpositions in $L_2 I/Q_I \cong S_5$, where $Q_I := O_2(L_2 I) = U_2 \times D_0$, and $D_0 := C_R(L_0)$. Thus $[U_2, I] \leq \Phi(I) \leq Q_I$, and as U_2 is the A_5-module for L_2, it follows that $\Phi := C_{\Phi(I)}(S_I) = \Phi_2 \times D_\Phi$ centralizes $O^{3'}(M \cap L_2)$, where $\Phi_2 := C_{U_2}(S_I) \cong \mathbf{Z}_2$ and $D_\Phi := C_{\Phi(I) \cap D_0}(S_I)$. Therefore $O^{3'}(M \cap L_0) = O^{3'}(M \cap L)O^{3'}(M \cap L_2)$ centralizes Φ. Observe also that $S_I = N_S(\Phi(I)) = N_S(\Phi)$.

By C.5.5, we may choose $x \in M_0$ with $U^x \not\leq Q$, and and so as $Q_1^t = Q_2$ with $Q = Q_1 \cap Q_2$, $U^x \not\leq Q_1$. By C.5.6.4, $\Phi(I)^x = \Phi(I)$. Then as x also acts on S, x acts on S_I and hence also on Φ. Let $G_I := N_G(\Phi)$, and set $S_0 := \langle S_I, x \rangle$. Then $S_0 \leq N_T(\Phi)$. As $J(S) \leq R$, $O_2(L_0 S) J(S) \leq N_S(L) = S_I$, while $N_S(L)$ is of index 2 in S, so $S_I \in \beta$ by 2.3.8.5b. As $N_H(\Phi) \in \mathcal{H}^e$ by 1.1.3.2, and $N_H(\Phi)$ contains $L \not\leq M$, from the definitions in Notations 2.3.4 and 2.3.5, $(S_I, N_H(\Phi)) \in \mathcal{U}(G_I)$, and hence $G_I \in \Gamma$. By 1.1.6, the 2-local G_I satisfies the hypotheses of 1.1.5 in the role of "H".

As $U^x \not\leq Q_1 = O_2(LR)$, $U^x \not\leq O_2(LS_I)$. Therefore as $U^x \leq R \leq S_I$, while L^x is irreducible on U^x, $U^x \cap O_2(G_I) = 1$. Notice $LS_I \in \mathcal{H}^e$, so that $(S_I, LS_I) \in \mathcal{U}(LS_I)$, and hence $LS_I \in \Gamma$. Define $\mathcal{H}_1$ to consist of the subgroups H_1 satisfying:

$H_1 \in \mathcal{H}^e(LS_I) \cap G_I$, and

$H_1 = \langle L, S_1 \rangle$ for some $S_1 \in Syl_2(H_1)$ containing S_I.

Then $\mathcal{H}_1$ is nonempty, since $LS_I \in \mathcal{H}_1$.

We next claim

(*) $L \in \mathcal{C}(H_1)$ for any $H_1 \in \mathcal{H}_1$.

It is clear that (*) holds if $S_1 = S_I$, so assume instead that $S_1 > S_I$. Then as $|S : S_I| = 2$, $|S_1| \geq |S|$. Since $H_1 \in \mathcal{H}^e$, $(S_I, H_1) \in \mathcal{U}(H_1)$ and $H_1 \in \Gamma$. Then by 2.4.3.2, $|S_1| = |S|$ and $H_1 \in \Gamma_0^e$. As $|S_1 : S_I| = 2$ and $L \in \mathcal{L}(H_1, S_I)$, $L \leq K \in \mathcal{C}(H_1)$ by 1.2.5. Then as $H_1 \in \Gamma_0^e$, K is a χ_0-block of H_1 by 2.3.8.4. Since no χ_0-block has a proper A_5-block, $K = L$, completing the verification of (*).

Now $L \leq G_I^\infty$, and by 1.2.1.1, G_I^∞ is a product of $\mathcal{C}$-components $K_1, \ldots, K_r$, with L inducing inner automorphisms on $K_i/O_2(K_i)$ for each i. However using 1.2.1.1, $C_{G_I^\infty}(G_I^\infty/O_2(G_I^\infty)) = O_2(G_I^\infty)$, so as $U \cap O_2(G_I) = 1$, $L \cap O_2(G_I^\infty) = 1$. Hence $Aut_U(K_i/O_2(K_i)) \neq 1$ for some $K_i \in \mathcal{C}(G_I)$. Choose notation so that the projection L_{K_i} of L on $K_i/O_2(K_i)$ is nontrivial iff $1 \leq i \leq t$. Since $L \cap O_2(G_I^\infty) = 1$, it follows that $L \leq L_{K_1} \cdots L_{K_t}$. Observe for $i \leq t$ that L_{K_i} has a quotient A_5, so that $m_3(L_{K_i}) \geq 1$.

We claim that $t = 1$: For $t \leq m_3(G_I) \leq 2$ since G_I is an SQTK-group, so that $t = 2$ if $t > 1$, and then the proof of 1.2.2 (which does not depend on conjugacy of the $\mathcal{C}$-components in the lemma) shows that $O^{3'}(G_I) = K_1 K_2$. Since $O^{3'}(M \cap L_0)$ centralizes Φ, $O^{3'}(M \cap L_0) \leq O^{3'}(G_I) \leq K_1 K_2$. Therefore for $i = 1$ or 2, there exists y of order 3 in $L_0 \cap K_i$ with $L = [L, y]$. Then $L = [L, y] \leq K_i$, so that $L \leq L_{K_i}$, and hence $L = L_{K_i}$ with $L_{K_j} = 1$ for $j \neq 1$, contrary to our assumption that $t = 2$. This contradiction establishes the claim that $t = 1$. Hence $L = L_{K_1} \leq K_1 =: K$. Since $U \cap O_2(G_I) = 1$, $m_2(K/O_2(K)) \geq m(U) = 4$, ruling out cases (c) and (d) of 1.2.1.4, and hence showing that $K/O_2(K)$ is quasisimple.

Suppose first that $F^*(K) = O_2(K)$. Now $S_I = N_S(L)$ normalizes L and hence normalizes K. Then $KS_I \in \mathcal{H}_1$ so $L \in \mathcal{C}(KS_I)$ by (*). Thus $L \in \mathcal{C}(K)$ and hence $L = K$, contrary to $U \cap O_2(G_I) = 1$.

Thus $F^*(K) > O_2(K)$, so as $K/O_2(K)$ is quasisimple, we conclude that K is quasisimple, and hence K is a component of G_I. Thus K is on the list of 1.1.5.3. Indeed as K contains the A_5-block L, we conclude from that list that K is either of Lie type and characteristic 2 of Lie rank at least 2, but not $L_3(2)$, or one of M_{22}, M_{23}, M_{24}, J_4, HS, He, or Ru. Let $S \leq T_I \in Syl_2(G_I)$; then T_I normalizes K by 1.2.1.3. Let $X \in \mathcal{H}^e \cap KT_I$, $S_I \leq S_1 \in Syl_2(X)$, and $Y := \langle L, S_1 \rangle$. Then $S_1 \in Syl_2(Y)$, and $Y \in \mathcal{H}^e$ by 1.1.4.4, so $Y \in \mathcal{H}_1$. Then by (*), L is subnormal in Y, so $L \in \mathcal{L}(X, S_1)$. Thus we have shown that for each $X \in \mathcal{H}^e \cap KT_I$ and

$S_1 \in Syl_2(X)$ with $T_I \leq S_1$, we have $L \in \mathcal{L}(X, S_1)$. This is a contradiction, since from the 2-local structure of the groups K on our list, none contains an A_5-block L, such that for each overgroup X of LS_+ in K with $F^*(X) = O_2(X)$ and $S_+ \in Syl_2(N_K(L))$, $L \in \mathcal{L}(X, S_1)$ for $S_+ \leq S_1 \in Syl_2(X)$. This completes the proof of 2.4.25. □

By 2.4.7 and 2.4.25, either L is an A_5-block with $s = 1$, or L is an A_3-block with $s = 1$ or 2. So by 2.4.5.3, the hypotheses of Theorem C.6.1 are satisfied with T, S in the roles of "Λ, T_H". Similarly by 2.4.5.1, we can appeal to results from section C.5, with S, S, L_0, U_0, Baum(S) in the roles of "T_H, R, K, U, S".

We will first show that when $s = 1$ and L is an A_3-block, then G is a group in the conclusion of Theorem 2.4.1. Since G is a counterexample to Theorem 2.4.1, this will establish the following reduction:

LEMMA 2.4.26. *If L is an A_3-block, then $s = 2$.*

PROOF. Assume L is an A_3-block with $s = 1$. By Theorem C.6.1, $H \cong S_4$ or $\mathbf{Z}_2 \times S_4$.

Suppose first that $H \cong S_4$, so that case (b) of Theorem C.6.1.6 holds, and in particular T is dihedral or semidihedral. Then by I.4.3, G is $L_2(p)$, p a Fermat or Mersenne prime, A_6, $L_3(3)$, or M_{11}. As $M = !\mathcal{M}(T)$, G is not $L_2(7)$ or A_6. As $\delta \neq \emptyset$, G is not $L_2(5)$. This leaves the groups in Theorem 2.4.1, contradicting the choice of H, G as a counterexample.

Therefore $H \cong \mathbf{Z}_2 \times S_4$, so case (a) of Theorem C.6.1.6 holds. Then $|T : S| = 2$ and $J(S) = S = J(T)$. By C.6.1.1, $S = QQ^x$ for $x \in T - S$. Define y and z by $\langle y \rangle = Z(H)$ and $\langle z \rangle = \Phi(S)$; by 2.4.4, $S = C_T(y)$. Since $S = J(T)$ is weakly closed in T, by Burnside's Fusion Lemma A.1.35, $N_G(S)$ controls fusion in $Z(S)$, so $y \notin z^G$. Thus $y^x = yz$, and H is transitive on $yU - \{y\}$, so all involutions in yUU^x are in y^G, and all involutions in UU^x are in z^G.

Suppose first that $y^G \cap T \subseteq S$. Then $y^G \cap T \subseteq yUU^x$. Now T/UU^x is of order 4 and hence abelian, so by Generalized Thompson Tranfer A.1.37.2, $y \notin O^2(G)$, contradicting the simplicity of G.

Thus we may take $x \in y^G$; in particular, x is now an involution. Let $u \in U - \langle z \rangle$. Then $\langle u, x \rangle \cong D_{16}$, and we saw $[x, y] = z$, so $S_1 := \langle xy, u \rangle \cong SD_{16}$, with xy of order 4. Hence all involutions in S_1 are in UU^x and therefore lie in z^G. Therefore $y^G \cap S_1 = \emptyset$, so Thompson Transfer produces our usual contradiction to the simplicity of G, completing the proof. □

By 2.4.25 and 2.4.26, the structure of S is similar in the two remaining cases where L_0 is either an A_5-block or the product of two A_3-blocks; we summarize some of these common features in the next lemma:

LEMMA 2.4.27. *(1) $|T : S| = 2$ and $R = Baum(S) = J(S) = J(T)$.*
(2) $\mathcal{A}(T) = \{Q, Q^x, A_1, A_1^r\}$ for $x \in T - S$, $r \in S - R$, and $|A_1 : A_1 \cap Q| = 2$.
(3) Let $T_C := C_T(L_0)$. Then $\Phi(T_C) = 1$, $Q = T_C \times U_0$, and $T_C \cap T_C^x = 1$ for each $x \in T - S$.
(4) $R = T_C \times U_0 U_0^x$, with $L_0 = [L_0, U_0^x]$.

PROOF. Let $M_0 := N_T(S)$; by Theorem C.6.1, $|M_0 : S| = 2$. Thus by 2.4.5.2, the hypotheses of C.5.6.7 are satisfied. Further by C.6.1.1, $QQ^x = R = Baum(S) = J(S)$. By 2.4.25 and 2.4.26, L_0 is an A_5-block or the product of two A_3-blocks, so by C.6.1.4, $\mathcal{A}(R) = \mathcal{A}(S)$ is described in (2). Thus to complete the proof of (2), it remains to show that $\mathcal{A}(S) = \mathcal{A}(T)$, or equivalently to establish the assertion $J(S) = J(T)$ in (1).

As $T_C = C_T(L_0) \leq Q \in \mathcal{A}(S)$, $\Phi(T_C) = 1$. Further if L_0 is an A_5-block, then $Q = T_C \times U$ by C.5.4.3, and this holds when L_0 is a product of A_3-blocks as $S_4 = Aut(A_4)$. Also C.5.6.7 says $T_C \cap T_C^x = 1$ for $x \in M_0 - S$; hence (3) will also follow, once we have established the equality $|T : S| = 2$ in (1). Thus to prove (1)–(3), it remains to establish (1).

Suppose (1) fails. Since we saw that $R = J(S)$, either $|T : S| \neq 2$ or $J(S) \neq J(T)$; thus conclusion (a) of C.6.1.6 does not hold. By 2.4.26, L_0 is not an A_3-block, so conclusion (b) of C.6.1.6 does not hold. Hence conclusion (c) of C.6.1.6 holds. Define $A_1 \in \mathcal{A}(S)$ as in C.6.1.4 and set $W := A_1 Q$ and $S_W := N_T(W)$. By conclusion (c) of C.6.1.6, $|S_W| \geq |S|$, and $Q^y = A_1$ for some $y \in S_W$, since $N_T(N_T(S))$ induces D_8 on $\mathcal{A}(S) = \mathcal{A}(N_T(S))$. Then $\mathcal{A}(W) = \{Q, Q^y\}$, so $G_W := N_G(W) = (G_Q \cap G_Q^y)S_W$. By 2.4.3.1, $G_Q \in \mathcal{H}^e$. As $G_Q \cap G_Q^y = N_{G_Q}(W)$, $G_Q \cap G_Q^y \in \mathcal{H}^e$ by 1.1.3.2; therefore $G_W = (G_Q \cap G_Q^y)S_W \in \mathcal{H}^e$. From the structure of L_0, $Q \leq J(S) = R = N_S(W)$, $|S : R| = 2$, and $N_H(W) \not\leq M$, so $G_W \not\leq M$. As $H \in \Gamma_0^e$, 2.3.8.5c says $C_{O_2(M)}(R) \leq R$. Then we conclude from 2.3.8.5b that $R \in \beta$. Then as usual $(R, G_W) \in \mathcal{U}(G_W)$, so $G_W \in \Gamma$. Hence as $|S_W| \geq |S|$, $G_W \in \Gamma_0$ by 2.4.3.2, so that $G_W \in \Gamma_0^e$. Thus G_W satisfies the hypotheses for H in this section. In particular as we showed that $Q = O_2(H)$ is abelian, by symmetry between H and G_W, $O_2(G_W)$ is abelian. This is a contradiction, as $W \leq O_2(G_W)$ and $W = A_1 Q$ is nonabelian since $Q \in \mathcal{A}(S)$. This contradiction establishes (1), and completes the proof of (1)–(3).

By C.5.6.2, for each $x \in T - S$, $R = U_0^x Q$ and $[U_0, U_0^x] = U_0 \cap U_0^x$. Thus $L_0 = [L_0, U_0^x]$ and $U_0^x \cap Q \leq U_0$, so as $U_0 U_0^x$ and T_C are normal in R, $R = T_C \times U_0 U_0^x$. That is, (4) holds. $\qquad\square$

REMARK 2.4.28. In the next lemma, we deal with the shadows of extensions of $L_4(3) \cong P\Omega_6^+(3)$ which are not contained in $PO_6^+(3)$. In this case, L is an A_5-block. The subcase where $C_T(L) \neq 1$ is quickly eliminated using 2.3.9.7: that subcase is the shadow of $Aut(L_4(3))$, which is not quasithin since an involution in $C_T(L)$ has centralizer $\mathbf{Z}_2 \times PO_5(3)$. The remaining cases we must treat correspond to the two extensions of $L_4(3)$ of degree 2 distinct from $PO_6^+(3)$, which are in fact quasithin. These subcases are eventually eliminated by using transfer to show G is not simple, but only after building much of the 2-local structure of such a shadow.

Shadows of extensions of $L_4(3)$ will also appear several more times in later reductions.

LEMMA 2.4.29. L is an A_3-block. Hence $H = L_0 S$ where L_0 is a product of two S-conjugates of L.

PROOF. The second statement follows from the first in view of 2.4.26. We assume L is not an A_3-block, and derive a contradiction. Then L is an A_5-block, and $s = 1$ by 2.4.25. Set $T_C := C_T(L)$. By 2.4.27, $Q \leq J(T) = J(S) = \text{Baum}(S) = R$, $\Phi(T_C) = 1$, $Q = T_C \times U$, and $R = T_C \times UU^x$. By C.5.4.3, $R/Q \cong E_4$ and $LR/Q \cong S_5 = Aut(A_5)$, so that $LS = LR$. Recall $L \cap M$ is a Borel subgroup of L.

Let $K := O^2(M \cap L)$ and $P := O_2(K)$. Then $P \cong Q_8^2$, and $S = PR = PUU^x T_C$ centralizes T_C, so $\Phi(S) = \Phi(UU^x)\Phi(P)[UU^x, P] \leq P$. Therefore $Z := Z(P) = \langle z \rangle = \Phi(S) \cap Z(S)$. Since S is of index 2 in T by 2.4.27.1, $z \in Z(T)$.

Set $G_Z := C_G(Z)$ and $\tilde{G}_Z := G_Z/Z$. Then $\tilde{P}\tilde{T}_C = J(\tilde{S})$ is x-invariant, so $PT_C \unlhd \langle x, KS \rangle =: M_1$. Observe $T \leq M_1 \leq G_Z$ and $T_C Z = Z(PT_C) \unlhd M_1$. By 2.4.27.3, $T_C \cap T_C^x = 1$, so as x normalizes $Z(PT_C) = ZT_C$, and T_C is of index 2 in $T_C Z$, $|T_C| \leq 2$ with $[x, T_C] = Z$ in case of equality. As $|ZT_C| \leq 4$ and $M_1 \leq N_{G_Z}(ZT_C)$, $O^2(M_1)$ centralizes ZT_C. As S centralizes T_C, $H = LS$ centralizes T_C.

Next $PT_C \cong Q_8^2$ or $Q_8^2 \times \mathbf{Z}_2$, so $Aut_{Aut(PT_C)}(PT_C/ZT_C) \cong O_4^+(2)$. Let $M_1^+ := M_1/C_{M_1}(PT_C/ZT_C)$. Then $M_1^+ \leq O_4^+(2)$ and $K^+R^+ \cong S_3 \times \mathbf{Z}_2$ with $U^+ = O_2(K^+R^+)$. As $U^x \not\leq O_2(KR)$ and $x \in M_1$, $U \not\leq O_2(M_1)$; then as $M_1^+ \leq O_4^+(2)$, $M_1^+ \cong O_4^+(2)$. In particular, M_1 is irreducible on PT_C/ZT_C.

Suppose first that $T_C \neq 1$. As $H = LS \leq C := C_G(T_C)$, we conclude from 2.3.7.2 that $C \in \Gamma_0$ and $S \in Syl_2(C)$. By 1.2.4, $L \leq L_C \in \mathcal{C}(C)$, and the embedding of L in L_C is described in A.3.14. From the previous paragraph, $O^2(M_1) \leq C$ but $U \not\leq O_2(M_1)$, so $U \not\leq O_2(C)$. Hence as L is irreducible on U, $O_2(C) = T_C \leq Z(C)$. Therefore as $m_2(L_C) \geq m_2(L) > 1$, L_C is quasisimple by 1.2.1.5, and so L_C is a component of C. But the list of A.3.14 contains no embedding $L_C > L$ with L an A_5-block.

Therefore $T_C = 1$, so $Q = T_C \times U = U$, and hence $U = O_2(N_G(U)) = F^*(N_G(U))$ by 2.4.3.1. In particular, $C_G(U) = U$, so $C_G(L) = C_U(L) = 1$. By 2.4.6, $L \unlhd N_G(U)$, so as $LS = Aut(L)$, $H = LS = N_G(U)$.

As $T \leq M_1 \leq G_Z$ and $M = !\mathcal{M}(T)$, $M_1 \leq G_Z \leq M$. As G is of even type, $M \in \mathcal{H}^e$, so $Z \leq C_M(O_2(M)) \leq O_2(M) =: P_M$. Also $Q_8^2 \cong P = C_T(\tilde{P})$, so as $M_1^+ \cong O_4^+(2)$, $P = O_2(M_1)$. Therefore as $T \leq M_1 \leq M$, $P_M \leq P$ by A.1.6. Then as M_1 is irreducible on P/Z, P_M is either P or Z. As $M \in \mathcal{H}^e$, the latter is impossible, so $P = P_M$. Then as $Z = Z(P)$, $M \leq G_Z$, so that $M = G_Z$ as $M \in \mathcal{M}$. Since $M_1/P \cong O_4^+(2) \cong Out(P)$, $M = M_1 = G_Z = C_G(z)$. In particular, M is solvable.

Let $u \in Z(R) - Z$. As U is the S_5-module for H/U, we can adopt the notation of section B.3 to describe U, and choose $u = e_{1,2}$. Then $z = e_{1,2}e_{3,4} = uu^s$ for a suitable $s \in S - R$. Set $G_u := C_G(u)$, $H_u := C_H(u)$, etc. Then as $H/U \cong S_5$, $H_u \cong D_8 \times S_4$, so $R = S_u$ is of index 2 in S. Further $C_U(O^2(H_u))^\# = \{u, e_{1,3,4,5}, e_{2,3,4,5}\}$ with $e_{1,3,4,5}$ and $e_{2,3,4,5}$ in z^L. As $\langle u, z \rangle = Z(R) \unlhd T$ and $u^s = uz$, there is $x \in T_u - S$, and $T_u = \langle x \rangle R$ is of index 2 in T with $T_u = M_u$. As $T_u \unlhd T$, $N_G(T_u) \leq M = !\mathcal{M}(T)$. Then as $T_u = M_u$, $T_u \in Syl_2(G_u)$, and in particular $u \notin z^G$. Also $H_u \not\leq M$ with $|T_u| = |S| = |T|/2$, so by 2.3.7.1, $G_u \in \Gamma_*$.

Suppose first that $F^*(G_u) = O_2(G_u)$, so that $G_u \in \Gamma_0^e$. Then we may apply the results of this section to G_u in the role of "H". By 2.3.8.4, $G_u = M_u K_1 \cdots K_t$ is a product of blocks K_i, where K_i is an A_5-block or A_3-block, since 2.4.7 eliminated the case where some K_i is an $L_2(2^n)$-block. Indeed as we saw $M_u = T_u$ is a 2-group, and $K_i \cap M$ is a Borel subgroup of K_i in 2.3.8.4, each K_i is in fact an A_3-block. Then as T_u is of order 2^7 and 2-rank 4 with $1 \neq u \in Z(G_u)$, $t = 1$. But now $K_1 = O^2(G_u) \cong A_4$ contains $O^2(H_u) \cong A_4$, so that $O^2(G_u) = O^2(H_u)$. As S_u is of index 2 in T_u, H_u is of index 2 in G_u, and hence is normal in G_u. Then as $U \leq O_2(H_u)$ and $x \in T_u$, $U^x \leq O_2(H_u)$, which is not the case.

Thus $F^*(G_u) \neq O_2(G_u)$. As $O^2(H_u) \cong A_4$, $O_2(O^2(H_u))$ centralizes $O(G_u)$ by A.1.26, so $z \in \langle u \rangle O_2(O^2(H_u)) \leq C_{G_u}(O(G_u))$; hence $O(G_u) = 1$, as z inverts $O(G_u)$ by 2.3.9.5. Therefore G_u has a component K, which must appear in 2.3.9.7. We further restrict the list of 2.3.9.7 using the facts that $M_u = C_{G_u}(z)$ is a 2-group of order 2^7 and rank 4, and $H_u = N_{G_u}(U) \cong D_8 \times S_4$, to conclude that $K \cong A_6$, $L_2(7)$, or $L_2(17)$. Next $O^2(N_{G_u}(U)) = O^2(H_u) \cong A_4$ and $O^2(N_K(U)) \cong A_4$ in each of the possibilities for K, so $O^2(H_u) \leq K$. Now $z \in \langle u \rangle O^2(H_u) \leq \langle u \rangle K$, but $z \neq u$, so T_u normalizes the component K, and hence $K \trianglelefteq G_u$ by 1.2.1.3. As $J(T_u) = R = UU^x$, K is not $L_2(17)$, and in the remaining two cases, x induces an outer automorphism on K interchanging the two 4-groups in $R \cap K \in Syl_2(K)$, so that $K = \langle O^2(H_u), O^2(H_u)^x \rangle$. Also $z = uu^s$ for $s \in S - R$, and $u^s \in O^2(H_u) \leq K$; so as K has one class of involutions, by a Frattini Argument,

$$G_u = KC_{G_u}(u^s) = KC_{G_u}(z) = KM_u = KT_u.$$

Let $D := C_R(O^2(H_u))$ and $U_D := D \cap U$. Then $D \cong D_8$, U_D is a 4-group, and $U_D - \langle u \rangle \subseteq z^L$ from an earlier remark. Hence as $C_G(z)$ is solvable, $C_{U_D}(K) = \langle u \rangle$. But if K is $L_2(7)$, then $C_{Aut(K)}(O^2(H_u)) = 1$, so we conclude that $K \cong A_6$ and $v \in U_D - \langle u \rangle$ induces a transposition on K. As $G_u = KT_u$ and K is simple, $B := C_{G_u}(K) = C_{T_u}(K) \leq C_{T_u}(O^2(H_u)) = D$, so $B = C_D(K)$ is of order 4, with $G_u/B \cong Aut(A_6)$, since x interchanges the two 4-groups in $R \cap K$. As $R = UU^x$, x also interchanges the two 4-groups in $R/(R \cap K) = D(R \cap K)/(R \cap K) \cong D$, and hence $B \cong \mathbf{Z}_4$, since $U_D \not\leq B$.

Let $I := O^2(M)$. Then $I = I_1 I_2$ with $I_i \cong SL_2(3)$ and $[I_1, I_2] = 1$. Further there exists $y \in T - S$ centralizing I_1 with $y^2 \in Z$: namely any y inducing an orthogonal transvection on $\tilde{P}$ centralizing I_1. Moreover each $t \in T - S$ with $t^2 \in Z$ is conjugate under M to y or ya, where $a \in I_1 \cap P$ is of order 4, and exactly one of y and ya is an involution. Thus M is transitive on the set $\mathcal{I}$ of involutions in $M - IS$, and either y or ya is a representative i for $\mathcal{I}$. Let $j := ia$; then $j^2 = z$. Observe $j^G \cap S = \emptyset$: For if $j^g \in S$ then $z^g = (j^g)^2 \in \Phi(S) \leq P$. But as $u \in P$ and $u \notin z^G$ while M is transitive on the involutions in $P - Z$, Z is weakly closed in P with respect to G; so $z = z^g$ and hence $g \in G_Z = M$, contradicting $IS \trianglelefteq M$.

As $j^G \cap S$ is empty but $G = O^2(G)$ with $|T : S| = 2$, we can apply Generalized Thompson Transfer A.1.37 to j in the role of "g", to see that $j^2 = z$ must have a G-conjugate in $T - S$; so $i = z^g$ for some $g \in G$. Now if $y = i$ then $SL_2(3) \cong I_1 \leq C_G(i) = M^g$, so $z \in O_2(I_1) \leq O_2(M^g) = P^g$. However we saw in the previous paragraph that $z^G \cap P = \{z\}$, so $z = z^g = i$, contradicting $i \notin S$. Therefore y is of order 4 and $i = ya$ centralizes bc, where $b \in I_2$ is of order 4 and inverted by y, and $O_2(I_1) = \langle a, c \rangle$. As $bc \in u^M$, we may assume $bc = u$, so that u centralizes i. Then $i \in T_u - S$ acts on $K \cong A_6$. As $S \geq U_D$ and $v \in U_D - \langle u \rangle$ induces a transposition on K, KS induces the S_6-subgroup of $Aut(K)$ on K, so as $i \notin S$, i does not induce an automorphism in S_6. Then as i is an involution, i induces an automorphism in $PGL_2(9)$ rather than M_{10}, and hence $C_K(i) \cong D_{10}$. This is impossible as $i \in z^G$ and $M = C_G(z)$ is a $\{2, 3\}$-group. The proof of 2.4.29 is complete. $\qquad \square$

By 2.4.29, we have reduced to the case where L_0 is the product of two A_3-blocks. Henceforth we let s denote an element of $S - N_S(L)$. Thus $H = L_0 S$ and $L_0 = L \times L^s$. Let $U_1 := U$ and $U_2 := U^s$.

LEMMA 2.4.30. *(1)* $QQ^x = R = Baum(S) = J(S)$ for $x \in T - S$.
(2) $H \in \Gamma_*$.
(3) $\{Q, Q^x\}$ are the S-invariant members of $\mathcal{A}(R)$.
(4) $RL_0 = C_S(L_0) \times L_0 U_0^x$ with $\Phi(C_S(L_0)) = 1$ and $L_0 U_0^x \cong S_4 \times S_4$.
(5) R is of index 2 in $S = R\langle s \rangle$, so $|T : R| = 4$.

PROOF. By 2.4.27.1, $|T : S| = 2$, so (2) holds by 2.3.7.1. By 2.4.27, $R = J(S) = T_C \times U_0 U_0^x$, where $T_C := C_S(L_0)$, and $Q = T_C \times U_0$, so $R = QQ^x$. Thus (1) holds, and (3) follows from 2.4.27.2. By 2.4.27.4, $R = T_C \times U_0 U_0^x$ and $L_0 = [L_0, U_0^x]$, so (4) holds. Further as $L_1^s = L_2$, $H/Q \cong S_3$ wr $\mathbf{Z}_2$, and as $R = J(S)$ acts on L_1 and $Q \leq R$, R is Sylow in $RL_0 = N_H(L_1)$ of index 2 in S, so (5) holds. $\square$

REMARK 2.4.31. In the remainder of the proof of Theorem 2.4.1, we are again faced with a shadow of an extension of $L_4(3)$, but now approached from the point of view of a 2-local with two A_3-blocks. We will construct the centralizer of the involution z_2 defined below, as a tool for eventually obtaining a contradiction to the absence of an A_5-block in any member of Γ_0. In $P\Omega_6^+(3)$, z_2 is an involution whose commutator space on the orthogonal module is of dimension 2 and Witt index 0, and whose centralizer has a component $\Omega_4^-(3) \cong L_2(9) \cong A_6$.

Now let $\langle z_i \rangle = C_{U_i}(R)$. Then by 2.4.30, $\langle z_1, z_2 \rangle = \Phi(R) \trianglelefteq T$ and $L_{3-i} \leq C_G(z_i) =: G_i$, so $G_i \not\leq M$. Since $Q \leq R$, we conclude by 2.3.8.5c that $C_{O_2(M)}(R) \leq R$. Then since $|S : R| = 2$ and $R = J(S)$ by 2.4.30.1, the first sentence of 2.3.8.5b says $R \in \beta$. So since $L_i \not\leq M$, we conclude as usual from the definitions in Notation 2.3.4 and Notation 2.3.5 that $(R, L_{3-i}R) \in \mathcal{U}(G_i)$ and $G_i \in \Gamma$. Next $z_1^s = z_2$, so $z := z_1 z_2$ generates $Z(T) \cap \Phi(R)$, and replacing x by xs if necessary, we may assume $x \in G_i$, for $i = 1$ and 2. Let $S_1 := R\langle x \rangle$. Then $|T : S_1| = 2 = |T : S|$, so by 2.3.7.1, $G_i \in \Gamma_*$ and $S_1 \in Syl_2(G_i)$.

Observe that $F^*(G_2) \neq O_2(G_2)$: For otherwise by 2.3.8.4 and 2.4.29, $G_2 = C_M(z_2)K_0$, where K_0 is the product of two A_3-blocks. But $R = J(S_1)$, so applying 2.4.30.4 to $K_0 S_1$, $C_{\Phi(R)}(K_0) = 1$, contradicting $z_2 \in C_{\Phi(R)}(K_0)$.

Next $O_2(L) \cong E_4$ centralizes $O(G_2)$ by A.1.26, so $z \in \langle z_2 \rangle O_2(L) \leq C_{G_2}(O(G_2))$, and hence $O(G_2) = 1$ since z inverts $O(G_2)$ by 2.3.9.5. Thus as $F^*(G_2) \neq O_2(G_2)$, there exists a component K of G_2, and K is described in 2.3.9.7. By 2.3.9.6, $K = [K, z]$, so L is faithful on K since $z \in \langle z_2 \rangle L$.

Recall $S_1 \in Syl_2(G_2)$ and $|S_1 : R| = 2$ with $R = N_{S_1}(L)$; therefore $Aut_{S_1}(K) \in Syl_2(Aut_{G_2}(K))$ with $|Aut_{S_1}(K) : N_{Aut_{S_1}(K)}(Aut_L(K))| \leq 2$. Further we saw L is faithful on K, so $Aut_L(K) \cong A_4$. Inspecting the 2-locals of the automorphism groups of the groups K listed in 2.3.9.7 for such a subgroup, and recalling $O(G_2) = 1$, we conclude that K is one of A_5, A_6, A_7, A_8, $L_2(7)$, $L_2(17)$, $L_3(3)$, or M_{11}. Moreover if L_K is the projection of L on K, then as $|S_1 \cap K : N_{S_1 \cap K}(L)| \leq 2$ (since L is irreducible on $O_2(L)$ of rank 2), $O_2(L_K) \leq N_K(L)$, and then $O_2(L_K) = [O_2(L_K), L] = O_2(L) = U$. As S_1 centralizes z and z_2, S_1 centralizes $z_1 = zz_2 \in U \leq K$, so S_1 acts on K and hence $K \trianglelefteq G_2$ by 1.2.1.3. If $K \cong A_5$ or A_7, then $U \trianglelefteq S_1$, contradicting $x \notin N_T(U)$. If K is A_8, then L is an A_4-subgroup moving 4 of the 8 points permuted by K, so z_1 is not 2-central in K, a contradiction. If K is $L_3(3)$, M_{11}, or $L_2(17)$, there is $x_K \in S_1 \cap K$ with $Q^{x_K} \neq Q$, so we may take $x = x_K$; but now $|Q^x : C_{Q^x}(Q)| = 2$, contradicting 2.4.30.1 which shows this index is 4. Therefore:

LEMMA 2.4.32. $G_2 \in \Gamma_*$ and $L \leq K \cong A_6$ or $L_2(7)$.

Next $z_1^{G_2} = z_1^K$ since A_6 and $L_3(2)$ have one class of involutions; so by a Frattini Argument, $G_2 = KC_{G_2}(z_1) = KC_{G_2}(z) = KM_2$, where $M_2 := M \cap G_2$. As $G_2 \in \Gamma_*$, $F^*(M_2) = O_2(M_2)$ by 2.3.9.4. Then as $C_{G_2}(K) \leq M_2$, $F^*(G_2) = KO_2(G_2)$. In particular:

LEMMA 2.4.33. $K = E(G_2)$ and $F^*(G_2) = KO_2(G_2)$.

Now suppose that $U_2 \leq C_G(K)$. For $g \in L_2$, $z_2^g \in U_2 \leq C_G(z_2) = G_2$, so K is a component of $C_{G_2}(z_2^g)$ by 2.4.33. By I.3.2 and 2.4.33, $K \leq O_{2',E}(C_G(z_2^g)) = K^g$. We conclude $K = K^g$, and hence $K = E(C_G(u))$ for each $u \in U_2^{\#}$. Therefore $K^s = E(C_G(u))$ for each $u \in U_1^{\#} = U^{\#}$. Also x centralizes z_1 and hence normalizes $K^s = E(C_G(z_1))$, so $K^s = E(C_G(u^x))$ for each $u^x \in (U^x)^{\#}$. Further $L = [L, U_0^x]$ by 2.4.30.4, so as $U_2^x \leq C_G(K)$, $L = [L, U^x]$. Thus using the structure of K in 2.4.32,
$$K = \langle C_K(u), C_K(u^x) : u \in U^{\#} \rangle \leq N_G(K^s).$$
As z_2 centralizes K, z_1 centralizes K^s, so $K = [K, z_1] \leq C_G(K^s)$, and hence $T = S_1 \langle s \rangle$ normalizes $KK^s = K \times K^s$. Let $I := KK^sT$. Since I contains $L \not\leq M = !\mathcal{M}(T)$, $O_2(I) = 1$. As G is quasithin, $m_{2,3}(KK^s) \leq 2$, so $K \cong L_3(2)$ rather than A_6. As $O_2(I) = 1$, $m_2(T) \leq m_2(Aut(KK^s)) = 4$, so $Q = U_0$ and $R \cong D_8 \times D_8$. It follows that $R \in Syl_2(KK^s)$ and $T = R\langle x, s \rangle$, with x an involution inducing an outer automorphism on K and K^s, and s an involution centralizing x. Then I has 5 classes of involutions, with representatives z, z_2, x, s, and sx. Now $O_2(G_2) \leq C_{S_1}(K) \cong D_8$, so $O^2(G_2)$ centralizes $O_2(G_2)/\langle z_2 \rangle$ and z_2, and hence by Coprime Action also centralizes $O_2(G_2)$. Therefore as $F^*(C_{G_2}(K)) = O_2(G_2)$ using 2.4.33, we conclude that $C_{G_2}(K)$ is a 2-group, and hence $C_{G_2}(K) = C_{S_1}(K) = O_2(G_2)$. Thus $G_2/KO_2(G_2) \leq Out(K)$ which is a 2-group, so G_2/K is a 2-group, and hence $K = O^2(G_2)$, so $m_3(G_2) = 1$.

Now $C_I(s) = \langle s \rangle \times K_s \langle x \rangle$ with $K_s \cong L_3(2)$, and the involutions in the subgroup K_s diagonally embedded in $K \times K^x$ are in z^G as $z = z_1 z_2$; thus $s \notin z_2^G$, since the involutions in $K = G_2^{\infty}$ are in z_2^G. Similarly $sx \notin z_2^G$. Next $C_I(x) = \langle x, s \rangle (I_1 \times I_2)$ with $I_1 := C_K(x) \cong S_3$ and $I_1^s = I_2$. In particular as $m_3(G_2) = 1$, $x \notin z_2^G$. As $O(C_{G_2}(x)) \neq 1$, $F^*(C_G(x)) \neq O_2(C_G(x))$ by 1.1.3.2, so $x \notin z^G$.

But as $G = O^2(G)$, by Thompson Transfer, $x^G \cap S \neq \emptyset$. Therefore as we saw x is not conjugate to z or z_2, it must be conjugate to s. Arguing similarly with S replaced by $\langle sx \rangle UU^x$, we conclude $sx \in s^G$. So $x^G = s^G = (sx)^G$, and hence by the previous two paragraphs, s, z, and z_2 are representatives for the conjugacy classes of involutions of G. Thus s is in fact *extremal* in T: that is, $T_s := C_T(s) \in Syl_2(C_G(s))$. But each involution in $C_I(s)$ is fused in I to s, x, sx, or z, so $z_2^G \cap T_s = \emptyset$. This is impossible as $z_2 \in C_G(x)$ with x conjugate to s. This contradiction shows $U_2 \not\leq C_G(K)$, and hence:

LEMMA 2.4.34. $K = [K, U_2]$.

Now $U_2 \leq C_{G_2}(L)$. But if $K \cong L_3(2)$, then $C_{G_2}(L) = C_{G_2}(K)$ from the structure of $Aut(K)$, so U_2 centralizes K, contrary to 2.4.34. Therefore part (1) of the following lemma holds:

LEMMA 2.4.35. (1) $K \cong A_6$, and some $u \in U_2 - \langle z_2 \rangle$ induces a transposition on K centralizing L.

(2) The automorphism induced by x on K is not in S_6.

For if part (2) of 2.4.35 fails, then setting $(KS_1)^+ := KS_1/C_{KS_1}(K)$, $x^+ \in K^+ R^+$, so $U_0^{x+} \in U_0^{+K}$. Then as $U_0^+ = O_2(L^+ R^+)$ is weakly closed in R^+ with respect to K^+ from the structure of A_6, $U_0^+ = U_0^{x+}$, contrary to 2.4.30.4.

LEMMA 2.4.36. (1) $R = R_K \times R_K^s$ with $R_K := R \cap K \in Syl_2(K) \cong D_8$.

(2) $C_R(K) = C_{G_2}(K)$ is cyclic of order 4, and $G_2/C_R(K) \cong \mathrm{Aut}(A_6)$.
(3) $C_T(L_0) = 1$ and $|T| = 2^8$.

PROOF. We claim that z_2 is the unique involution in $C_R(K)$. Assume the claim fails, and let $z_2 \neq r \in C_R(K)$ be an involution. Recall $R \leq G_2$.

Under this assumption, we establish a second claim: namely that $K \trianglelefteq G_r := C_G(r)$. First K is a component of $C_{G_r}(z_2)$ using 2.4.33, so by I.3.2, there is a 2-component K_r of G_r such that either $K \leq K_r$, or $K \leq K_r K_r^{z_2}$ with $K_r \neq K_r^{z_2}$—and in the latter case, $K_r/O_\infty(K_r) \cong K$. As $K \cong A_6$ by 2.4.35, the former case holds by 1.2.1.3. As K_r is a 2-component of G_r, $K_r \in \mathcal{C}(G_r)$ and $O_2(K_r) \leq Z(K_r)$. As $m_2(K_r) \geq m_2(K) > 1$ and $O_2(K_r) \leq Z(K_r)$, K_r is quasisimple by 1.2.1.5.

Now as $m_3(K_r) \geq m_3(K) = 2$, $K_r \trianglelefteq G_r$ using 1.2.1.3, so our second claim holds if $K = K_r$. Thus we may assume that $K < K_r$, and it remains to derive a contradiction. We verify the hypotheses of 1.1.5 for G_r in the role of "H": Let $C_R(r) \leq T_r \in Syl_2(G_r)$, and $T_r \leq T^g$, so that $z^g \in Z(T^g) \leq T_r$, and hence $z^g \in Z(T_r)$; thus z^g, T_r, M^g play the roles of "z, S, M". As $r \in O_2(G_r \cap M^g)$, trivially $C_{O_2(M^g)}(O_2(G_r \cap M^g)) \leq G_r$. This completes the verification of the hypotheses of 1.1.5. As $K \cong A_6$ is a component of $C_{K_r}(z_2)$, we conclude from inspection of the list of 1.1.5.3 that one of the following holds:

(i) z_2 induces a field automorphism on $K_r \cong Sp_4(4)$.
(ii) z_2 induces an outer automorphism on $K_r \cong L_4(2)$ or $L_5(2)$.
(iii) z_2 induces an inner automorphism on $K_r \cong HS$.

Recall that $|T : R| = 4$, while $|R : C_R(r)| \leq 2$ by 2.4.30, and $z_2 \in Z(R)$. Thus

$$|T_r : C_{T_r}(z_2)| \leq |T_r : C_R(r)| < |T : C_R(r)| \leq 8,$$

where the strict inequality holds since r is not 2-central in G, as $G_r \notin \mathcal{H}^e$. Since z_2 centralizes K but not K_r, we conclude (ii) holds, with $K_r \cong L_4(2) \cong A_8$. Now L is an A_4-subgroup of K_r fixing 4 of the 8 points permuted by K_r, so it centralizes an A_4-subgroup L_r of K_r. Then using A.3.18 and the fact that $z_1 = z_2^s \in O_2(L)$,

$$K_0 := \langle L_r, L_2 \rangle \leq O^{3'}(C_G(L)) \leq O^{3'}(G_1) = K^s.$$

Now $K^s \cong A_6$ with $z_2 \in L_2 \leq K^s$ and z_2 induces an outer automorphism on L_r. Thus $\langle z_2 \rangle L_r \cong S_4$, so $\langle z_2 \rangle L_r$ is a maximal subgroup of K^s. It follows that $K^s = K_0 \leq C_G(L)$, so $m_{2,3}(LK_0) = 3$, contradicting G quasithin. This contradiction establishes the second claim, namely that $K = K_r$ is a normal component of G_r for each involution $r \in C_R(K)$.

Set $E_r := \langle z_2, r \rangle$. Using 2.4.35.2, $C_{K^s S_1}(z_2)$ is a maximal subgroup of $K^s S_1$, which does not contain $C_{K^s S_1}(a)$ for any involution $a \notin z_2 C_G(K^s)$. Thus in the notation of Definition F.4.41, $K^s S_1 = \Gamma_{1,E_r}(K^s S_1)$, so $K^s \leq N_G(K)$ using the second claim. Then as $m_{2,3}(N_G(K)) \leq 2$ since G is quasithin, $K = K^s$. This is impossible as $z_1 \in K$ but $z_2 = z_1^s$ centralizes K. This contradiction completes the proof of the first claim that z_2 is the unique involution in $C_R(K)$.

By 2.4.35, $C_R(K)$ is of index 2 in $C_R(L) \cong C_Q(L_0) \times D_8$, so by the uniqueness of z_2, $C_R(K)$ is cyclic of order 4 and $C_Q(L_0) = 1$. Then $C_T(L_0) = C_Q(L_0) = 1$. Therefore $R \cong D_8 \times D_8$ by 2.4.30.4, so $|T| = 4|R| = 2^8$ by 2.4.30.5, completing the proof of (3).

As $R_K \cong D_8$ and $R_K \cap R_K^s \trianglelefteq S$ but $Z(R_K) = \langle z_1 \rangle \not\leq Z(S)$, we conclude $R_K \cap R_K^s = 1$. Thus $R \geq R_K R_K^s = R_K \times R_K^s$; so as $|R| = |R_K|^2$, $R = R_K \times R_K^s$, and (1) holds.

Let $\bar{G}_2 := G_2/C_{G_2}(K)$. By 2.4.35, $\bar{S}_1\bar{K} \cong Aut(A_6)$ and hence $\bar{G}_2 \cong Aut(A_6)$. In particular $|\bar{S}_1| = 2^5$, so as $C_R(K) \cong \mathbf{Z}_4$ and $|S_1| = |T|/2 = 2^7$, it follows that $C_R(K) \in Syl_2(C_{G_2}(K))$. Then by Cyclic Sylow 2-Subgroups A.1.38, $C_{G_2}(K) = O(G_2)C_R(K)$. Recall that $z = z_1z_2$ with $z_1 \in K$, so that $C_{G_2}(K) \le C_G(z)$. However by 2.3.9.5, z inverts $O(G_2)$, so $O(G_2) = 1$, completing the proof of (2). $\square$

LEMMA 2.4.37. $z_2^G \cap R = (z_2^G \cap R_K) \cup (z_2^G \cap R_K^s)$ with $|z_2^G \cap R_K| = 5$.

PROOF. Recall $A_1 \in \mathcal{A}(T)$ is defined in 2.4.27.2. Further by 2.4.27.2, T induces the 4-group
$$\langle(Q, Q^x), (A_1, A_1^r)\rangle$$
of permutations on $\mathcal{A}(T)$. Thus $y := x$ or xr acts on A_1, so $S_A := R\langle y\rangle$ is of index 2 in T and S_A normalizes A_1. As $z \in A_1$, $H_1 := N_G(A_1) \in \mathcal{H}^e$ by 1.1.4.3. Now $N_H(A_1)$ contains $L_2 \not\le M$. Also $Q \le R = J(S)$, so by 2.3.8.5c, $C_{O_2(M)}(R) \le R$. Then $R \in \beta$ by 2.3.8.5b, so as usual $H_1 \in \Gamma$. Then as $|S_A| = |S|$, $H_1 \in \Gamma_*$ by 2.3.7.1, so we may apply the results of this section to H_1 in the role of "H". In particular we conclude from 2.4.29 [2] that H_1 induces $O_4^+(2)$ on A_1. Therefore for each $A \in \mathcal{A}(T)$, $A = A^1 \times A^2$ with $A^i \cong E_4$ and $A^{1\#} \cup A^{2\#} = z_2^G \cap A$. By 2.4.36.1, $R = R_K \times R_K^s$, so $A = (A \cap R_K) \times (A \cap R_K^s)$ with $A \cap R_K \cong A \cap R_K^s \cong E_4$. Thus as all involutions in R_K are in z_2^G, $A^1 = A \cap R_K$ and $A^2 = A \cap R_K^s$. Therefore as each involution in R is in a member of $\mathcal{A}(T)$, the lemma holds. $\square$

We are now in a position to obtain a contradiction, and hence complete the proof of Theorem 2.4.1. By 2.4.36, $B := C_{G_2}(K) = C_R(K) \cong \mathbf{Z}_4$ and $R = R_K \times R_K^s$. Let $\bar{G}_2 := G_2/B$; then $\bar{R} = \bar{R}_K\langle\bar{u}\rangle$, where $u \in U - \langle z_2\rangle$. By 2.4.35, $\bar{u}$ induces a transposition on $\bar{K}$, so $\bar{R} = \langle\bar{u}\rangle \times \bar{R}_K \cong \mathbf{Z}_2 \times D_8$ is Sylow in $\bar{R}\bar{K} \cong S_6$.

Next each involution in $\bar{R} - \bar{K}$ is either a transposition or of cycle type 2^3, and there are a total of 6 involutions in $\bar{R} - \bar{K}$. Further $u \in z_2^G$ and $\bar{u}$ is a transposition, so as x induces an outer automorphism on $\bar{R}\bar{K}$, $\bar{u}^x$ is of type 2^3. Thus $\Delta := z_2^G \cap (R - K)$ is of order $6m$, where $m := |z_2^G \cap uB|$. However by 2.4.37, Δ is s-conjugate to $z_2^G \cap R_K$ of order 5.

This contradiction finally completes the proof of Theorem 2.4.1.

2.5. Eliminating the shadows with Γ_0^e empty

The groups occurring in the conclusion of Theorem 2.1.1 have already appeared in Theorems 2.2.5 and 2.4.1, so from now on we are working toward a contradiction. We have also dealt with the most troublesome shadows, although a number of other shadows are still to appear.

By Theorem 2.4.1, we may assume Γ_0^e is empty: that is no member of Γ_0 is contained in $\mathcal{H}^e$. In 2.5.3, we will produce a component K of H, consider the various possibilities for K listed in 2.3.9.7, and analyze the structure of $C_S(\langle K^S\rangle)$, where $S \in Syl_2(H)$. Eventually we eliminate all configurations, completing the proof of Theorem 2.1.1.

We continue to assume that G is a counterexample to Theorem 2.1.1. Therefore as the groups in Theorem 2.4.1 are conclusions of Theorem 2.1.1, in the remainder of the section we assume that
$$\Gamma_0^e = \emptyset.$$

---

[2]As mentioned earlier, our use of 2.4.29 here to exclude A_5-blocks is essentially eliminating the shadow configuration.

In addition we define $\mathcal{T}$ to consist of the 4-tuples (H, S, T, z) such that $H \in \Gamma_0$, $S \in Syl_2(H \cap M)$, $T \in Syl_2(M)$ with $Z(T) \le S < T$, and z is an involution in $Z(T)$. For each $H \in \Gamma_0$, there exists a tuple in $\mathcal{T}$ whose first entry is H, using 2.3.9.3. Throughout this section (H, S, T, z) denotes a member of $\mathcal{T}$.

LEMMA 2.5.1. *(1) $\mathcal{H}^e(S) \subseteq M$.*
(2) $H \cap M$ is the unique maximal member of $\mathcal{H}^e(S) \cap H$.
(3) $S \in Syl_2(H)$.

PROOF. By 2.3.8.1, $S \in \beta$ and $S \in Syl_2(H)$, so that (3) holds. Suppose there is $X \in \mathcal{H}^e(S)$ with $X \not\le M$. Then from the definitions in Notation 2.3.4 and Notation 2.3.5, $(S, X) \in \mathcal{U}(X)$, so $X \in \Gamma$. Then by 2.3.7.4, $X \in \Gamma_0$, contrary to our assumption in this section that $\Gamma_0^e = \emptyset$. Thus (1) holds. By 2.3.9.4, $H \cap M \in \mathcal{H}^e$, so that (1) implies (2). $\square$

From now on we use without comment the fact from 2.5.1.3, that S is Sylow in H.

LEMMA 2.5.2. *Suppose L is a component of H and set $M_L := M \cap L$. Then z induces an inner automorphism on L, $L = [L, z] \not\le M$, and one of the following holds:*

(1) L is a Bender group and M_L is a Borel subgroup of L.
(2) $L \cong Sp_4(2^n)'$ or $L_3(2^n)$ or $L/O_2(L) \cong L_3(4)$. Further $N_S(L)$ is nontrivial on the Dynkin diagram of $L/Z(L)$, and M_L is a Borel subgroup of L.
(3) $L \cong L_3(3)$ or M_{11} and $M_L = C_L(z_L)$, where z_L is the projection on L of z.
(4) $L \cong L_2(p)$, $p > 7$ a Mersenne or Fermat prime, and $M_L = S \cap L$.

PROOF. Observe L is described in 2.3.9.7, and $L = [L, z] \not\le M$ by 2.3.9.6. If $L \cong L_4(2)$, M_{22}, M_{23}, A_7, or $\hat{A}_7$, then from the description of M_L in 2.3.9.7, there is $H_1 \in \mathcal{H}^e(S) \cap H$ with $H_1 \cap L \not\le M_L$, contradicting 2.5.1.2. Similarly if conclusion (b) of 2.3.9.7 holds, then by 2.5.1.2, S is nontrivial on the Dynkin diagram of $L/Z(L)$, and M_L is as described in (2)—in particular, observe we cannot have $L \cong A_6$ or $\hat{A}_6$ with z inducing a transposition, since S is nontrivial on the Dynkin diagram, while $z \in Z(S)$ as $(H, S, T, z) \in \mathcal{T}$. So when L is A_6 or $\hat{A}_6$, z induces an inner automorphism of L. Indeed as $z \in LC_S(L)$, and z inverts $O(H)$ by 2.3.9.5, L is not $\hat{A}_6$ for any action of z on L. If conclusion (a) of 2.3.9.7 holds, then by 2.5.1.2, M_L is a Borel subgroup of L, so that (1) holds. The remaining cases (d) and (e) of 2.3.9.7 appear as (3) and (4). Since we have eliminated the case where z induces an outer automorphism on $L/Z(L) \cong A_6$ or A_7, in each case z induces an inner automorphism on L by 2.3.9.7. $\square$

Part (4) of the next result produces the component of H on which the remainder of the analysis in this section is based. Furthermore it eliminates case (1) of 2.5.2 where the component is a Bender group.

LEMMA 2.5.3. *Assume*

$$H = \bigcap_{i=1}^{k} N_G(B_i) \text{ for some 2-subgroups } B_1, \ldots, B_k \text{ of } H,$$

and let $(U, H_U) \in \mathcal{U}(H)$. Set $Q_U := O_2(H_U)$. Then
(1) If $O_2(H) \le Q_U$, then $N_H(Q_U) \in \mathcal{H}^e$.

(2) If $Q_1 \in \mathcal{N}_H(H_U, 2)$, then $(U, H_U Q_1) \in \mathcal{U}(H)$.

(3) If L is a component of H which is a Bender group and $\mathcal{N}_H(H_U, 2) \subseteq Q_U$, then $Q_U \cap L \in Syl_2(L)$.

(4) There exists a component K of H such that K is not a Bender group, and if $H \in \Gamma^*, \Gamma_*$, then $\langle K, S \rangle \in \Gamma^*, \Gamma_*$, respectively.

PROOF. By 2.3.8.2, $Q_U \in \mathcal{S}_2^e(G)$, so $H_0 := N_G(Q_U) \in \mathcal{H}^e$. Assume $O_2(H) \leq Q_U$. By hypothesis, $B_i \leq H = \bigcap_{j=1}^k N_G(B_j)$, so $B_i \leq O_2(H) \leq Q_U$. Thus

$$N_H(Q_U) = \bigcap_{i=1}^k N_{H_0}(B_i) \in \mathcal{H}^e$$

by 1.1.3.3. Hence (1) holds.

Assume the hypotheses of (2), and let $Q_2 := Q_U Q_1$ and $H_2 := H_U Q_1$. As $F^*(H_U) = O_2(H_U) = Q_U$ since $H_U \in \mathcal{H}^e$, also $F^*(H_2) = Q_2 = O_2(H_2)$; so as $U \leq H_U \leq H_2$ with $U \in \beta$, $(U, H_2) \in \mathcal{U}(H)$, and hence (2) holds.

Assume the hypotheses of (3), and let $L_0 := \langle L^H \rangle$. First, $O_2(H) \in \mathcal{N}_H(H_U, 2) \subseteq Q_U$ by hypothesis; so by (1), $N_H(Q_U) \in \mathcal{H}^e$, and then by 1.1.3.1,

$$F^*(N_{L_0}(Q_U)) = O_2(N_{L_0}(Q_U)). \tag{$*$}$$

Set $P_U := L_0 C_H(L_0) \cap Q_U$, and let P_L, P_1 denote the projections of P_U on L, L_0, respectively. If $L < L_0 = LL^s$, let P_{L^s} be the projection of P_U on L^s. If $P_1 = 1$ then $Aut_{Q_U}(L_0) \cap Inn(L_0) = 1$, so as L is a Bender group, from the structure of $Aut(L_0)$, $O^2(F^*(C_{L_0}(Q_U))) \neq 1$, contrary to $(*)$. Thus $P_1 \neq 1$ and $P_1 \in \mathcal{N}_H(H_U, 2) \subseteq Q_U$. Similarly if $L < L_0$, $P_1 \leq P_L P_{L^s} \in \mathcal{N}_H(H_U, 2) \subseteq Q_U$, and as $P_1 \neq 1$, either $P_L \neq 1$ or $P_{L^s} \neq 1$. Further if $P_L = 1$, then Q_U acts on $P_1 = P_{L^s}$ and hence on L, and $Aut_{Q_U}(L) \cap Inn(L) = 1$ so again $O^2(F^*(C_L(Q_U)) \neq 1$, contrary to $(*)$. Thus $P_L \neq 1$, and if $L < L_0$ also $P_{L^s} \neq 1$. Therefore as L is a Bender group, there is a unique Sylow 2-group P_0 of L_0 containing P_1, so $P_0 \in \mathcal{N}_H(H_U, 2) \subseteq Q_U$ and hence $P_0 = Q_U \cap L_0$, establishing (3).

It remains to prove (4). Let L_+ be the product of all Bender-group components of H, with $L_+ := 1$ if no such components exist. Partially order $\mathcal{U}(H)$ by $(U_1, H_1) \leq (U_2, H_2)$ if $U_1 \leq U_2$ and $H_1 \leq H_2$, and choose (U, H_U) maximal with respect to this order. Then by (2) and maximality of (U, H_U), $\mathcal{N}_H(H_U, 2) \subseteq H_U$, and hence

$$\mathcal{N}_H(H_U, 2) \subseteq Q_U \text{ and in particular } O_2(H) \leq Q_U. \tag{!}$$

Observe by (!) that we may apply (1) and (3).

Replacing (U, H_U) by a suitable conjugate under $H \cap M$, we may assume $S \cap H_U \in Syl_2(H_U \cap M)$. Set $Q_+ := S \cap L_+ \in Syl_2(L_+)$. Then $Q_+ = Q_U \cap L_+$ by (3), and so $H_U \leq X := N_H(Q_+)$. When $L_+ \neq 1$, $M_+ := M \cap L_+ = N_{L_+}(Q_+)$ by 2.5.2.1. In any case by a Frattini Argument, $H = L_+ X$. Further $S \in Syl_2(X)$ since $S \in Syl_2(H)$ by 2.5.1.3. Also $(U, H_U) \in \mathcal{U}(X)$, so $X \in \Gamma$. As $(U, H_U) \in \mathcal{U}(X)$ is maximal with respect to our ordering and $S \leq X$, it follows from parts (3) and (4) of 2.3.7 that $X \in \Gamma^*, \Gamma_*$, when $H \in \Gamma^*, \Gamma_*$, respectively. Moreover the components of X are the components of H not in L_+, so by definition of L_+, X has no Bender components. Thus replacing (H, S, T, z) by $(X, S, T, z) \in \mathcal{T}$, and adjoining Q_+ to $B_1, \ldots, B_k$, we may assume $L_+ = 1$; that is, that H has no Bender components.

Let $H \in \Gamma_*, \Gamma^*$; it remains to show that there is a component K of H with $\langle K, S \rangle \in \Gamma_*, \Gamma^*$, respectively.

We first consider the case where $E(H) \neq 1$; thus there is a component K of H. As $L_+ = 1$, K is not a Bender group, and so K is described in one of cases (2)–(4) of 2.5.2. Set $K_0 := \langle K^S \rangle$ and $R_U := K_0 C_H(K_0) \cap Q_U$, and let R_0 denote the projection of R_U on K_0.

We now argue as in the proof of (3) using (!) to conclude that $N_{K_0}(Q_U) \in \mathcal{H}^e$ and $R_0 \leq Q_U$. Further $z \in Q_U$, so by the initial statement in 2.5.2, we conclude that $R_U \not\leq C_H(K_0)$. Therefore $R_0 \neq 1$. Indeed since $O_2(N_{K_0}(R_0)) \in \mathcal{M}_H(H_U, 2)$, $O_2(N_{K_0}(R_0)) \leq Q_U \cap K_0 \leq R_0$, so that $R_0 = O_2(N_{K_0}(R_0))$. From the description of K in cases (2)–(4) of 2.5.2, $N_{K_0}(R_0) \in \mathcal{H}^e$. Thus if $N_{K_0}(R_0) \not\leq M$, we can argue as in case (ii) that (4) holds.

Therefore we may assume that $N_{K_0}(R_0) \leq M$. It follows that $O_2(M \cap K_0) \leq O_2(N_{K_0}(R_0)) = R_0$. Now from the description of K and $M \cap K$ in cases (2)–(4) of 2.5.2, either $O_2(M \cap K_0) \in Syl_2(K_0)$, or case (3) holds with $K = K_0 \cong M_{11}$ or $L_3(3)$ and $O_2(M \cap K_0) = C_K(z)$ is of index 2 in a Sylow 2-group of K. Hence $R_0 = O_2(M \cap K_0)$, and either $R_0 = S \cap K_0 \in Syl_2(K_0)$, or case (3) holds and $R_0 = O_2(M \cap K_0) = O_2(C_{K_0}(z))$. In any case $R_0 \trianglelefteq S$ and $H_U \leq N_H(Q_U) \leq N_H(R_0)$. Further $R_0^H = R_0^{K_0}$, either by Sylow's Theorem or as M_{11} and $L_3(3)$ have one class of involutions. Therefore by a Frattini Argument, $H = K_0 X_0$, where $X_0 := N_H(R_0)$. Now $(U, H_U) \in \mathcal{U}(X_0)$, so that $X_0 \in \Gamma$, and as usual $X_0 \in \Gamma^*$, Γ_*, when $H \in \Gamma^*$, Γ_*, respectively. Now $(X_0, S, T, z) \in \mathcal{T}$ and adjoining R_0 to $B_1, \ldots, B_k$, X_0 satisfies the hypotheses for H, so we conclude (4) holds by induction on the number of components of H.

We have reduced to the case where $E(H) = 1$, where to complete the proof we derive a contradiction.

As $F^*(H) \neq O_2(H)$ and $E(H) = 1$, $Y := O(H) \neq 1$. By 2.3.9.5, z inverts Y, so Y is abelian. By (!) and (1), $O_2(N_H(Q_U)) = Q_U$ and $N_H(Q_U) \in \mathcal{H}^e$. Then by our maximal choice of (U, H_U), $N_H(Q_U) = H_U$ and $U \in Syl_2(H_U)$ so $Q_U \leq U$. Then as $U \leq S$, $z \in Z(S) \leq C_H(Q_U) \leq C_{N_H(Q_U)}(Q_U) = Z(Q_U)$.

As $E(H) = 1$, $F^*(H) = F(H) = O_2(H)Y$. Further $O_2(H) \leq S \leq C_H(z)$, so $[z, H] \leq C_H(O_2(H))$, while as z inverts Y, $[z, H] \leq C_H(Y)$, and Y is abelian, so

$$[z, H] \leq C_H(F^*(H)) = Z(F^*(H)) = Z(O_2(H))Y.$$

Hence setting $O_2(H)\langle z \rangle =: D$, $DY \trianglelefteq H$, so by a Frattini Argument, $H = YN_H(D)$. As $z \in D$, $D \in \mathcal{S}_2^e(G)$ by 1.1.4.3, so $N_G(D) \leq M$ by 2.5.1.1. Now $O_2(H) \leq Q_U$ by (!), and $z \in Z(Q_U)$ by the previous paragraph, so $D \leq Q_U$. Hence $D = Q_U \cap DY \trianglelefteq H_U$, so that $H_U \leq N_G(D) \leq M$, contradicting $H_U \not\leq M$. Therefore (4) is finally established, completing the proof of 2.5.3. $\qquad\square$

In view of 2.5.3.4, we are led to define Γ^+ to consist of those $H \in \Gamma_0$ such that $H = \langle K, S \rangle$, for some component K of H and $S \in Syl_2(H \cap M)$, such that K is not a Bender group.

We verify that Γ^+ is nonempty: For given any $(H_0, S, T, z) \in \mathcal{T}$, we conclude from 2.3.9.1 that $H_1 := N_G(O_2(H_0)) \in \Gamma_0$, $S \in Syl_2(H_1)$, and if $H_0 \in \Gamma^*$, then also $H_1 \in \Gamma^*$. Now applying 2.5.3.4 to the 2-local H_1, we obtain a component K of H_1 such that K is not a Bender group, $H_2 := \langle K, S \rangle \in \Gamma_0$, and $H_2 \in \Gamma^*$ if $H_0 \in \Gamma^*$. Thus $H_2 \in \Gamma^+$, so Γ^+ is nonempty, and since we saw in section 1 that Γ^* is nonempty, also $\Gamma^+ \cap \Gamma^*$ is nonempty.

NOTATION 2.5.4. Let $\mathcal{T}^+$ consist of the tuples (H, S, T, z) in $\mathcal{T}$ such that $H \in \Gamma^+$. In the remainder of the section we pick $(H, S, T, z) \in \mathcal{T}^+$ and let $K \in \mathcal{C}(H)$ and

$K_0 := \langle K^S \rangle$. Set $S_K := S \cap K$, $S_{K_0} := S \cap K_0$, $S_C := C_S(K_0)$, and $\bar{H} := H/S_C$. Let $x \in N_T(S) - S$ with $x^2 \in S$.

As $H \in \Gamma^+$, K_0 is the product of at most two conjugates of the component K of H, and $H = K_0 S$. Further K is not a Bender group, and $S \in Syl_2(H)$, so $S_K \in Syl_2(K)$, $S_{K_0} \in Syl_2(K_0)$, and $S_C = O_2(H) \in Syl_2(C_H(K_0))$. As $H \in \Gamma \subseteq \mathcal{H}$, $1 \neq S_C$. By 2.5.2, z induces an inner automorphism on K with $K = [K, z]$. Thus $z \in K_0 S_C - S_C$, so z has nontrivial projection in $Z(S_K)$ and in $Z(S_{K_0})$.

We begin to generate information about S_C:

LEMMA 2.5.5. *(1) $S_C \cap S_C^x = 1$, so $S_C^x \cong S_C$ is isomorphic to a subgroup of $\bar{S}$.*
(2) $S_C S_C^x = S_C \times S_C^x$, so in particular $S_C^x \leq C_S(S_C)$.

PROOF. Recall $S_C = O_2(H) \trianglelefteq S$. Then as x normalizes S, S_C^x is also normal in S. As $x^2 \in S$, $S_0 := S_C \cap S_C^x \trianglelefteq S_1 := S\langle x \rangle$, and $S_0 \leq S_C$, so $S_0 \trianglelefteq K_0 S = H$. Thus if $S_0 \neq 1$, then by 2.3.7.2, $N_G(S_0) \in \Gamma_0$ and $S \in Syl_2(N_G(S_0))$. This is a contradiction since $S < S_1 \leq N_G(S_0)$. So $S_0 = 1$, and hence (1) holds. Then as both S_C and S_C^x are normal in S, (1) implies (2). $\qquad\square$

LEMMA 2.5.6. *If $1 \neq E \leq S_C$ with $E \trianglelefteq S$, then $G_E := N_G(E) \in \Gamma_0$, $S \in Syl_2(G_E)$, and $G_E \in \Gamma^*$ if $H \in \Gamma^*$. Further either*
(1) K is a component of G_E, or
(2) $K = K_0 \cong A_6$, $H/S_C \cong M_{10}$, and $K_E := \langle K^{G_E} \rangle \cong M_{11}$.

PROOF. As $E \leq S_C$ and $E \trianglelefteq S$, $H = K_0 S \leq G_E$. Thus by parts (2) and (4) of 2.3.7, $G_E \in \mathcal{H}(H) \subseteq \Gamma_0$, $S \in Syl_2(G_E)$, and $G_E \in \Gamma^*$ if $H \in \Gamma^*$. Next by 1.2.4, $K \leq K_E \in \mathcal{C}(G_E)$. Then by 2.3.7.2, $\langle K_E, S \rangle \in \Gamma_0$, and $\langle K_E, S \rangle \notin \mathcal{H}^e$ by our assumption in this section that $\Gamma_0^e = \emptyset$. As $m_2(K_E/O_2(K_E)) \geq m_2(K) > 1$, $K_E/O_2(K_E)$ is quasisimple by 1.2.1.4. So as $\langle K_E, S \rangle \notin \mathcal{H}^e$, K_E is a component of G_E. Then K_E is described in 2.5.2, K is described in one of cases (2)–(4) of 2.5.2, and if $K < K_E$, then the embedding of K in K_E is described in A.3.12. We conclude that the lemma holds. $\qquad\square$

We next show that K is essentially defined over $\mathbf{F}_2$:

LEMMA 2.5.7. *If $K/O_2(K) \cong L_3(2^n)$ or $Sp_4(2^n)$, then $n = 1$.*

PROOF. Assume that $n > 1$ and set $B := K \cap M$. By 1.2.1.3, $K_0 = K$, so that $H = KS$. By 2.5.2, some element s in S is nontrivial on the Dynkin diagram of $K/O_2(K)$ and B is a Borel subgroup of K. Let K_1 be a maximal parabolic of K over B, set $L_1 := K_1^\infty$ and $V := O_2(L_1)$.

We first observe that as case (2) of 2.5.2 holds, either $Z(K) = 1$, or $Z(K) = O_2(K)$ with $K/Z(K) \cong L_3(4)$. In the latter case, $\Phi(Z(K)) = 1$: for otherwise from the structure of the covering group in I.2.2.3a, $Z(S) \leq C_S(K) = S_C$; and as $x \in N_T(S)$, this is contrary to 2.5.5.1. By this observation and the structure of the covering group in I.2.2.3b when $Z(K) \neq 1$, in each case $\Phi(V) = 1$ and $V/C_V(L_1)$ is the natural module for $L_1/V \cong L_2(2^n)$.

Recall from Notation 2.5.4 that $S_K = S \cap K$ and $S_K \in Syl_2(K)$. Set $R := J(S)$ and $R_C := S_C \cap R = C_R(K)$. Observe since s is nontrivial on the Dynkin diagram of $K/O_2(K)$ that $S_K = VV^s$ and $\mathcal{A}(S_K) = \{V, V^s\}$ are the maximal elementary abelian subgroups of S_K.

We claim that $R = S_K R_C$: For let $A \in \mathcal{A}(S)$. Suppose first that $A \leq N_S(L_1)$. As $V/C_V(L_1)$ is the natural module for $L_1/V \cong L_2(2^n)$, either A centralizes V, or by B.2.7 and B.4.2.1, $Aut_A(V)$ is Sylow in $Aut_{AL_1}(V)$. In the former case $V \leq A$ since $A \in \mathcal{A}(S)$, so as V is self-centralizing in $Aut(K)$, $A = VC_A(K)$, where $C_A(K) \leq R_C$. In the latter case A induces an elementary abelian group of inner automorphisms on K not centralizing V, and hence A centralizes V^s, so by symmetry between V and V^s, $A = V^sC_A(K)$. Thus the claim holds if $R \leq N_S(L_1)$, so we may assume there is $a \in A - N_S(L_1)$. Then $m_2(C_{K/Z(K)}(a)) = n$, so $m(C_A(K)) \geq m(A) - (n+1)$. Hence as $A \in \mathcal{A}(S)$, and $n > 1$ by hypothesis, we conclude that

$$m(A) \geq m(VC_A(K)) \geq 2n + m(C_A(K)) \geq m(A) + n - 1 > m(A),$$

since we are assuming that $n > 1$. This contradiction completes the proof of the claim.

Next suppose that $\Phi(R_C) = 1$. Set $Q := O_2(L_1S_C) = VS_C$. By the claim, $Q_R := Q \cap R = V(S_C \cap R) = VR_C$. Then $Q_RS_C = Q$ and $N_S(Q_R) = N_S(Q)$. Since $\mathcal{A}(S_K) = \{V, V^s\}$, and we are assuming that R_C is elementary abelian, $Q_R = VR_C \in \mathcal{A}(S)$, and $\mathcal{A}(S) = \{Q_R, Q_R^s\}$ is of order 2. Hence $|S : N_S(Q_R)| = 2$, and for $T_Q := N_T(S) \cap N_T(Q_R)$, $N_T(S) = T_Q\langle s \rangle$. Also $|T_Q| \geq |S|$, since $S < N_T(S)$ because $S < T$. As $RS_C = S_KS_C \leq L_1S_C$ normalizes $O_2(L_1S_C) = Q$, $RS_C \leq N_S(Q) = N_S(Q_R)$. Thus we have shown that $|S : N_S(Q_R)| = 2$ and both $J(S) = R$ and $S_C = O_2(H)$ lie in $N_S(Q_R)$. Also $C_S(N_S(Q_R)) \leq C_S(Q_R) \leq N_S(Q_R)$. Therefore applying 2.3.9.8 to $N_S(Q_R)$ in the role of "R", we conclude that $N_S(Q_R) \in \beta$. So as $N_S(Q_R) \leq T_Q$, $T_Q \in \beta$ by 2.3.2.1. We saw earlier that $N_S(Q_R) = N_S(Q)$. Further $N_H(Q) = L_1N_S(Q)$, so $Q = O_2(N_H(Q))$ and $N_H(Q) \in \mathcal{H}^e$. Also $N_H(Q) \not\leq M$ since $K_0 \cap M$ is a Borel subgroup of K_0. Therefore $(N_S(Q), N_H(Q)) \in \mathcal{U}(N_G(Q))$, and hence $N_G(Q) \in \Gamma$. Then by 2.3.8.2, $Q = O_2(N_H(Q)) \in \mathcal{S}_2^e(G)$, so $N_G(Q) \in \mathcal{H}^e$. Since we saw above that $T_Q \in \beta$, $(T_Q, N_G(Q)) \in \mathcal{U}(N_G(Q))$. However $|T_Q| \geq |S| \geq |U_1|$ for each $U_1 \in \mathcal{U}$ by 2.3.6. Hence by the maximality of $|U|$ and/or $|S|$ in the definitions of $H \in \Gamma^*$ or Γ_* in Notation 2.3.5, $N_G(Q) \in \Gamma_0$, and therefore $N_G(Q) \in \Gamma_0^e$, contrary to our assumption in this section that $\Gamma_0^e = \emptyset$. This contradiction shows that $\Phi(R_C) \neq 1$.

By 2.5.5.1, $R_C^x \cap S_C = 1$, while $R_C^x \leq R^x = R$; so as $R = S_KR_C$, R_C^x is isomorphic to a subgroup of $S_K/Z(K)$. Indeed we further claim that the members of $\mathcal{A}(S)$ are of the form $A_C \times A_K$ with $A_X \in \mathcal{A}(R_X)$ for each $X \in \{C, K\}$: If $Z(K) = 1$, then $R = R_C \times S_K$, so the second claim is clear in this case. Otherwise $K/O_2(K) \cong L_3(4)$, and as $\Phi(Z(K)) = 1$, from the structure of the covering group K in I.2.2.3b, each elementary subgroup of $S_K/Z(K)$ lifts to an elementary subgroup of S_K, completing the proof of the second claim.

Hence as $\Phi(R_C) \neq 1$ and R_C is isomorphic to a subgroup of $S_K/Z(K)$, which has exactly two maximal elementary subgroups $V/Z(K)$ and $V^s/Z(K)$, we conclude that $\mathcal{A}(S_C) = \{A_1, A_2\}$, where A_1 and A_2 are the two maximal elementary abelian subgroups of R_C.

Now suppose that $[V, V^x] \neq 1$. Then as $V \leq A \in \mathcal{A}(S)$, $m(V^x/C_{V^x}(V)) = n = m(R/C_R(V))$. Similarly $m(V/C_V(V^x)) = m(R/C_R(V^x))$, so $R = VV^xC_R(VV^x)$ with $\Phi(C_R(VV^x)) \leq R_C$, and as V and V^x are normal in R, $[V, V^x] = V \cap V^x = C_{V^x}(V) = V^x \cap Z(R)$. By symmetry, $\Phi(C_R(VV^x)) \leq R_C^x$, so $\Phi(C_R(VV^x)) = 1$ by 2.5.5.1. Further for $v \in V - Z(R)$, $m([v, R]) = n$ and $[v, R] \cap R_C = 1$; so for $u \in V^s - Z(R)$, $m([u, R]) = n$ and $[u, R] \cap R_C = 1$. Now for $w \in V^x - Z(R)$, since $R = S_KR_C$, $w = uc$ for some $u \in V^s - Z(R)$ and $c \in R_C$, so $[V, w] = [V, u]$

is of rank n, and hence $[R, w] = [V, u]$ and $[R, w] \cap R_C = [V, u] \cap R_C = 1$. Thus $[R_C, w] = 1$, so $[R_C, V^x] = 1$, and hence $\Phi(R_C) \le \Phi(C_R(VV^x)) = 1$, contrary to an earlier reduction. This contradiction shows that $V^x \le C_R(V) = VR_C$, and hence $V^x \le VA_i$ for $i = 1$ or 2 using the second claim.

Next suppose that x normalizes $N_S(V)$. Set $I := \Omega_1(Z(T))VV^x$. Then $I \trianglelefteq N_S(V) = N_S(V)^x$ using our assumption. Further as $J(S) = R = S_K R_C$, $\Omega_1(Z(T)) \le VR_C$. Therefore as $V^x \le VR_C$, $I \le VR_C$ with $[VR_C, L_1] = V \le I$, and hence
$$I \trianglelefteq L_1 N_S(V).$$
Also arguing as above using 2.3.9.8, $N_S(V) \in \beta$. As $\Omega_1(Z(T)) \le I$, $I \in \mathcal{S}_2^e(G)$ by 1.1.4.3. Hence as $N_G(I)$ contains $L_1 \not\le M$, $(N_S(V), N_G(I)) \in \mathcal{U}(N_G(I))$ and thus $N_G(I) \in \Gamma^e$. However $S_1 := \langle N_S(V), x \rangle \le N_T(I)$ with $|S_1| = |S|$, so again by 2.3.6, $|S_1| \ge |U_1|$ for each $U_1 \in \mathcal{U}$. Hence again from the maximality of $|U|$ and/or $|S|$ in the definitions of $H \in \Gamma^*$ or Γ_* in Notation 2.3.5, $N_G(I) \in \Gamma_0$. Then $N_G(I) \in \Gamma_0^e$, contrary to our assumption in this section that $\Gamma_0^e = \emptyset$.

Therefore x does not normalize $N_S(V)$. Set $W := N_S(V) \cap N_S(V)^x$ and $T_W := S\langle x \rangle$. As $|S : N_S(V)| = 2$ and $N_S(V) \ne N_S(V^x)$, $S/W \cong E_4$, $T_W/W \cong D_8$, and we can choose x with $s := x^2 \in S - N_S(V)$. Thus $(V^x, V^{x^{-1}}) = (V, V^s)^x$. Hence setting $D := [V, V^s]$, $D^x = [V^x, V^{x^{-1}}]$. We showed $[V, V^x] = 1$, and by symmetry between x and x^{-1}, $V^{x^{-1}}$ also centralizes V, so $\langle V^x, V^{x^{-1}} \rangle$ centralizes V. Thus conjugating by s,
$$\langle V^x, V^{x^{-1}} \rangle \le C_S(VV^s) = R_C D.$$
Therefore $D^x \le \Phi(R_C D) \le R_C$. Also $D^x \trianglelefteq S$, so as K is not A_6 since $n > 1$, $K \trianglelefteq N_G(D^x)$ by 2.5.6.

Let p be a prime divisor of $2^n - 1$, and for $J \le G$, let $\theta(J) := O^{p'}(J)$ if $p > 3$, and $\theta(J) := \langle j \in J : |j| = 3 \rangle$ if $p = 3$. By A.3.18, either $K = \theta(N_G(D^x))$, or $p = 3$ and $\theta(N_G(D^x)) / O_{3'}(\theta(N_G(D^x))) \cong PGL_3(2^n)$. Thus, except possibly in the exceptional case, as $x^2 \in N_S(D)$ and $D^x \le R_C$, we have $\theta(N_K(D)) \le K^x \le C_G(D)$, impossible as $[D, \theta(N_K(D))] \ne 1$. Thus $K/Z(K) \cong L_3(2^n)$; 3 is the only prime divisor of $2^n - 1$, so that $n = 2$; and $K/Z(K) \cong L_3(4)$ and a subgroup X of order 3 in $N_K(D)$ induces outer automorphisms on K^x. Now $X \le Y \in Syl_3(N_G(D) \cap N_G(D^x) \cap N_G(R))$ with $Y = X(Y \cap K^x) \cong E_9$. By a Frattini Argument, we may assume x acts on Y. Now $R_C = C_R(X)$ from the structure of K, so as $R_C \cap R_C^x = 1$ by 2.5.5.1, $R_C^x = [R_C^x, X] \le K$. Now Y normalizes R and K^x, so Y normalizes R_C^x; then as R_C^x is not elementary abelian, $R_C^x = S_K$. This is impossible, as X^x centralizes R_C^x, but is faithful on S_K. This contradiction completes the proof of 2.5.7. $\square$

As a consequence of 2.5.7, the groups remaining in cases (2)–(4) of 2.5.2 have the following common features:

LEMMA 2.5.8. *(1)* $Out(K)$ *is a 2-group.*

(2) K *is simple so* $K_0 S_C = K_0 \times S_C$.

(3) Either S_K *is dihedral of order at least 8 or* S_K *semidihedral of order 16.*

(4) $Z(S) = (Z(S) \cap S_{K_0}) \times (Z(S) \cap S_C)$ *and* $Z(S) \cap S_{K_0} = \langle z_K \rangle$ *is of order 2, where* z_K *is the projection on* K_0 *of* z.

(5) For each 4-subgroup F *of* K, $N_K(F) \cong S_4$; *and furthermore if* $F \le S_K$, *then* $C_{Aut_H(K)}(F) \le Aut_S(K)$.

PROOF. First either K appears in case (3) or (4) of 2.5.2, or by 2.5.7, K appears in case (2) with $n = 1$. Now (1)–(3) and (5) follow by examination of those groups. Then $Z(S_K)$ is of order 2 by (3), so $Z(S) \cap K_0$ is of order 2. By 2.5.2, z induces a nontrivial inner automorphism on K_0, so $Z(S) \cap K_0 = \langle z_K \rangle$. Further $Z(\bar{S}) = Z(\bar{S}_{K_0})$, since S is nontrivial on the Dynkin diagram when $K = K_0 \cong A_6$ by 2.5.7. Then (2) completes the proof of (4). $\qquad\square$

Just before establishing Notation 2.5.4, we verified that $\Gamma^* \cap \Gamma^+ \neq \emptyset$, and hence there is a member of $\mathcal{T}^+$ with first entry in this set. We now take advantage of this flexibility:

NOTATION 2.5.9. In the remainder of the section, we choose $(H, S, T, z) \in \mathcal{T}^+$ with $H \in \Gamma^*$. Let $\mathcal{U}^*(H)$ denote the pairs $(U, H_U) \in \mathcal{U}(H)$ with U of maximal order in $\mathcal{U}$. By definition of Γ^*, $\mathcal{U}^*(H) \neq \emptyset$.

LEMMA 2.5.10. *(1) If $(U, H_U) \in \mathcal{U}(H)$, then $N_G(O_2(H_U)) \in \mathcal{H}^e$.*
(2) If $(U, H_U) \in \mathcal{U}^*(H)$, then $U \in Syl_2(N_G(O_2(H_U)))$, so $U \in Syl_2(H_U)$. If also $U \leq S$ then $z \in Z(S) \leq Z(U)$.

PROOF. By 2.3.8.2, $N := N_G(O_2(H_U)) \in \mathcal{H}^e$, establishing (1) and showing $(U, N) \in \mathcal{U}(N)$. Then if $(U, H_U) \in \mathcal{U}^*(H)$, U is Sylow in H_U and N by 2.3.2.2 and maximality of $|U|$, so the first statement in (2) holds. Finally if $U \leq S$, then as $U \in Syl_2(N)$, $O_2(N) \leq U = S \cap N$, and so using (1) we conclude

$$z \in Z(S) \leq C_H(U) \leq C_H(O_2(N)) \leq O_2(N) \leq U,$$

completing the proof of (2). $\qquad\square$

LEMMA 2.5.11. *(1) $Z(T) = \langle z \rangle$ is of order 2 and $Z(S) = \langle t, z \rangle = \langle t, t^x \rangle = \langle t, z_K \rangle \cong E_4$, where t is an involution in S_C and z_K is the projection of z on K_0.*
(2) $H = K_0 S \leq C_G(t) \in \Gamma^*$, with $S \in Syl_2(C_G(t))$. In particular, $t \notin z^G$.

PROOF. By 2.5.8.4, $Z(S) = \langle z_K \rangle \times Z_{S,C}$, where $Z_{S,C} := Z(S) \cap S_C$, and z_K is the projection on S_{K_0} of z. In the discussion following Notation 2.5.4 we observed $1 \neq O_2(H) = S_C$, so $Z_{S,C} \neq 1$. Then as $Z_{S,C}$ is of index 2 in $Z(S)$ while $Z_{S,C} \cap Z_{S,C}^x = 1$, we conclude from 2.5.5.1 that $\langle t \rangle := Z_{S,C}$ is of order 2 and $Z(S) = \langle t, t^x \rangle$. Now (2) follows from 2.5.6.1. Finally as $1 \neq z \in Z(T) \leq Z(S)$ from the definition of $\mathcal{T}$, $Z(T) = \langle z \rangle$ is of order 2, completing the proof of (1). $\qquad\square$

For the remainder of the section, let t be defined as in 2.5.11, and set $G_t := C_G(t)$.

LEMMA 2.5.12. *Assume $K \trianglelefteq H$, and let $(U, H_U) \in \mathcal{U}^*(H)$ with $U \leq S$. Then*
(1) $H_U = N_H(E)$ and $U = N_S(E)$ for some 4-subgroup E of S_K.
(2) $O^2(H_U) \cong A_4$ and $E = O_2(O^2(H_U)) = C_K(E)$.
(3) The map $E \mapsto (N_S(E), N_H(E))$ is a bijection of the set of 4-subgroups of S_K with

$$\{(U', H_{U'}) \in \mathcal{U}^*(H) : U' \leq S\}.$$

In particular, $N_S(E) \in Syl_2(N_H(E))$.
(4) If Q_E is a 2-group with $z \in Q_E \trianglelefteq H_U$, then $N_G(Q_E) \in \Gamma$ and $U \in Syl_2(N_G(Q_E))$.

PROOF. By Notation 2.5.4, $H \in \Gamma^+$ so that $H = K_0 S$ with K a component of H, $K_0 = \langle K^H \rangle$, and $K/O_2(K)$ is not a Bender group. Thus as $K \trianglelefteq H$ by hypothesis, $H = KS$ and $K = O^2(H)$. Further by 2.5.8.5, for each 4-subgroup F of K, $N_K(F) \cong S_4$, and if $F \leq S_K$ then $C_{Aut_H(K)}(F) \leq Aut_S(K)$. It follows as $H = KS$ with $S \in Syl_2(H)$ that if $F \leq S_K$ then $N_H(F) = N_K(F)C_S(F)$, and in particular $N_S(F) \in Syl_2(N_H(F))$.

Next as $(U, H_U) \in \mathcal{U}^*(H)$ by hypothesis, $U \in Syl_2(H_U)$ by 2.5.10.2. Hence $H_U = O^2(H_U)U \in \mathcal{H}^e$ with $O^2(H_U) \leq O^2(H) = K$. Set $E := \langle z_K^{H_U} \rangle$. Now $z_K \in Z(S) \leq Z(U)$ by 2.5.10.3, so by B.2.14, $E \leq O_2(H_U)$ and E is elementary abelian. In particular, $E \leq U$ as $U \in Syl_2(H_U)$. As $H_U \leq G_t$ by 2.5.11.2 and $H_U \not\leq M$ but $C_G(z) \leq M = !\mathcal{M}(T)$, we conclude $m(E) > 1$. Then as $O^2(H) \leq K$ and $m_2(K) = 2$, $E \cong E_4$. Now $H_U \leq N_H(E)$, and we saw in the previous paragraph that $N_H(E) = N_K(E)C_S(E)$, with $N_K(E) \cong S_4$ and $N_S(E) \in Syl_2(N_H(E))$. Since $H_U \not\leq M$, $A_4 \cong O^2(N_K(E)) = O^2(H_U)$ and $E = O_2(O^2(H_U))$, so that (2) holds. Further $N_H(E) \in \mathcal{H}^e$ and $U \leq N_S(E)$ so that $N_S(E) \in \beta$ by 2.3.2.1. Therefore $(N_S(E), N_H(E)) \in \mathcal{U}(H)$ and $N_S(E) \in \mathcal{U}$, so as $(U, H_U) \in \mathcal{U}^*(H)$, we conclude $N_S(E) = U \in Syl_2(H_U)$, and hence $N_H(E) = O^2(H_U)N_S(E) = H_U$. This completes the proof of (1). Further (3) follows from (1) since we saw that $N_S(E) \in Syl_2(N_H(E))$.

Now assume that $z \in Q_E \trianglelefteq H_U$ with Q_E a 2-group. Then as $z \in Q_E$, $N_G(Q_E) \in \mathcal{H}^e$ by 1.1.4.3. So as $U \in \mathcal{U}$, and $H_U \leq N_G(Q_E)$ with $H_U \not\leq M$, $(U, N_G(Q_E)) \in \mathcal{U}(N_G(Q_E))$ and $N_G(Q_E) \in \Gamma$. Then since $(U, H_U) \in \mathcal{U}^*(H)$ by hypothesis, we conclude $U \in Syl_2(N_G(Q_E))$ using 2.3.2.1. This completes the proof of (4), and hence of 2.5.12. $\qquad\square$

LEMMA 2.5.13. *(1)* $|N_T(S) : S| = 2$, *and* $t^x = tz$ *for each* $x \in N_T(S) - S$.
(2) If $\langle z_K \rangle$ *char* S, *or more generally if* $[x, z_K] = 1$, *then* $z = z_K$ *and* $t^x = tz_K$.
(3) If $tz_K \in t^G$, *then* $z = z_K$ *and* $t^x = tz_K$.

PROOF. By 2.5.11.1, $Z(S) = \langle z, t \rangle \cong E_4$ with $\langle z \rangle = Z(T)$. By 2.5.11.2, $S \in Syl_2(G_t)$ and hence $S = C_T(t)$, so (1) follows. Then (1) implies (2). Further $z \notin t^G$ by 2.5.11.2, so (1) also implies (3). $\qquad\square$

REMARK 2.5.14. There are extensions of $L_4(3) \cong P\Omega_6^+(3)$ by a 2-group, with involution centralizer $\mathbf{Z}_2 \times L_3(3)$ or $\mathbf{Z}_2 \times Aut(L_3(3))$, which are of even characteristic, and in which a Sylow 2-group is contained in a unique maximal subgroup. The first extension is even quasithin. The next lemma eliminates the shadows of such extensions.

LEMMA 2.5.15. K *is not* M_{11} *or* $L_3(3)$.

PROOF. Assume otherwise. Then case (3) of 2.5.2 holds, and $K = K_0 \trianglelefteq H$ by 1.2.1.3. As z induces inner automorphisms on K, $K_z := O^2(C_K(z)) \cong SL_2(3)$ from the structure of K.

By 2.5.11.2, $H = KS \leq G_t$, so by 2.5.6, $K \trianglelefteq G_t$. Then $K = O^{3'}(G_t)$ by A.3.18. By 2.5.11.1, $Z_S := Z(S) = \langle z, t \rangle \cong E_4$. Then as $K = O^{3'}(G_t)$, $K_z = O^{3'}(C_G(Z_S))$, so x acts on $Z(K_z) = \langle z_K \rangle$. Hence by 2.5.13.2, $z = z_K \in K$ and $t^x = tz$.

Next as $Aut(K_z)$ is induced in $K_z S$, we may choose $x \in C_T(K_z)$. Furthermore as $\langle z \rangle = C_K(K_z)$, $M_{11} = Aut(M_{11})$, and $|Aut(L_3(3)) : L_3(3)| = 2$ with $C_{Aut(K)}(K_z) \cong \mathbf{Z}_4$ if $K \cong L_3(3)$, either:

(i) S induces inner automorphisms on K, and $C_S(K_z) = S_C \times \langle z \rangle$, or

(ii) $\bar{H} \cong Aut(L_3(3))$ and $C_S(K_z) = S_C\langle y \rangle$, where y induces an outer automorphism on K with $\bar{y}^2 = \bar{z}$.

Recall from Notation 2.5.9 that we may choose $(U, H_U) \in \mathcal{U}^*(H)$ with $U \leq S$. By 2.5.12.3, $H_U = N_H(E)$ for some 4-subgroup E of S_K and $U = N_S(E) \in Syl_2(H_U)$. Then as $O^2(H_U) \cong A_4$ by 2.5.12.2, $Q_E := O_2(H_U) = C_S(E)$. In case (i) S induces inner automorphisms on K, so $S = S_C \times S_K$, and hence as $E = C_K(E)$ by 2.5.12.2, $Q_E = S_C \times E$. On the other hand in case (ii), we compute that $e \in E - \langle z \rangle$ inverts y, so $Q_E = (S_C \times E)\langle f \rangle$, where $f = yk$ and k is one of the two elements of $O_2(K_z)$ of order 4 inverted by e.

Recall $x \in N_T(S) \cap C_T(K_z)$, so x normalizes $C_S(K_z)$, and hence

$$[e, x] \in S \cap C_T(K_z) = C_S(K_z). \qquad (*)$$

But if case (i) holds then $C_S(K_z) = S_C\langle z \rangle \leq Q_E$, and by the previous paragraph $S_C E = Q_E$, so $x \in N_G(Q_E)$. Then $U < N_S(Q_E)\langle x \rangle \leq N_G(Q_E)$, contradicting 2.5.12.4.

Therefore case (ii) holds. Here x normalizes $C_S(K_z) = S_C\langle y \rangle$, while $S_C \cap S_C^x = 1$ by 2.5.5.1, so as $t \in S_C$, S_C is cyclic of order 2 or 4.

Assume $S_C \cong \mathbf{Z}_4$. Then by 2.5.5.2, $S_C S_C^x = S_C \times S_C^x$, so as $\bar{y}$ and S_C are of order 4, $C_S(K_z) = S_C \times S_C^x$ is abelian. In particular y centralizes S_C, so since $S = S_C S_K \langle y \rangle$, $Z(S)$ contains $S_C \cong \mathbf{Z}_4$, contrary to 2.5.11.1.

Therefore $S_C = \langle t \rangle$, so $C_S(K_z) = \langle t, y \rangle$, and as $\bar{y}^2 = \bar{z}$, $y^2 = z$ or tz. Hence as we saw $t^x = tz$, while x normalizes $\Phi(S_C\langle y \rangle) = \langle y^2 \rangle$, $y^2 = z$. Therefore as $H = KS$,

$$H = \langle t \rangle \times A,$$

where $A := K\langle y \rangle \cong Aut(L_3(3))$. Observe that $S_C\langle z \rangle = \langle t, z \rangle = Z(S)$ using 2.5.11.1.

Assume that $[e, x] \in \langle t, z \rangle$. Then as x acts on $Z(S) = \langle z \rangle$, x acts on $S_C E \trianglelefteq H_U$, so that $N_S(E) < N_S(E)\langle x \rangle \leq N_G(S_C E)$, again contrary to 2.5.12.4. Therefore $[e, x] \notin \langle t, z \rangle$.

Next A is transitive on involutions in $A - K$, and on E_8-subgroups of A, with representatives f and $F := \langle f, E \rangle$, respectively. Further we may choose notation so that $C_A(f) = \langle f \rangle \times C_K(f)$ with $C_K(f) = N_K(E) \cong S_4$. Now x acts on $C_S(K_z) = \langle t, y \rangle$, and we've seen that $[e, x] \in C_S(K_z) - \langle t, z \rangle$, so replacing y by a suitable element of $y\langle t, z \rangle$, we may take $e^x = ey$. Thus $ey \in A - K$ is an involution in $e^G = z^G$, so all involutions in $F^\#$ are in z^G. On the other hand, we saw that $tz = t^x \in t^G$, so all involutions in tK are in t^G, and in particular $te \in t^G$. Further

$$(te)^x = t^x e^x = tzey = tey^{-1},$$

with ey^{-1} an involution in $A - K$; so all involutions in $H - A$ are in t^G.

As F^A is the set of E_8-subgroups of A, and $Q_E = O_2(N_H(E)) = \langle t \rangle \times F$, Q_E^H is the set of E_{16}-subgroups of H. By 2.5.11.2, $G_t \in \Gamma^*$ and $S \in Syl_2(G_t)$. So $\langle t \rangle$ is Sylow in $C_{G_t}(K)$, and hence using Cyclic Sylow-2 Subgroups A.1.38 we conclude that $C_{G_t}(K) = O(G_t)\langle t \rangle$. We saw that $K \trianglelefteq G_t$ so $z \in K \leq C(O(G_t))$. Thus $O(G_t) = 1$ since z inverts $O(G_t)$ by 2.3.9.5. Hence $G_t = KS = H$. Therefore $C_G(t) = H$ is transitive on its E_{16}-subgroups with representative Q_E, so by A.1.7.1, $N_G(Q_E)$ is transitive on $t^G \cap Q_E = Q_E - F$ of order 8. Then $|N_G(Q_E) : N_{G_t}(Q_E)| = 8$, whereas $N_S(E) \in Syl_2(N_G(Q_E))$ by 2.5.12.4, and $N_S(E) \leq G_t$. Hence the proof of 2.5.15 is at last complete. $\qquad \square$

Observe that by 2.5.7 and 2.5.15, we have reduced the list of possibilities for K in 2.5.2 to:

LEMMA 2.5.16. *One of the following holds:*

(1) $K \cong L_2(p)$, $p > 7$ a Mersenne or Fermat prime.
(2) $K \cong L_3(2)$ and $N_H(K)/C_S(K) \cong Aut(L_3(2))$.
(3) $K \cong A_6$ and $N_H(K)/C_S(K) \cong M_{10}$, $PGL_2(9)$, or $Aut(A_6)$.

REMARK 2.5.17. All of these configurations appear in some shadow which is of even characteristic, and in which a Sylow 2-group is in a unique maximal 2-local. Usually the shadow is even quasithin. The group is not simple, but it takes some effort to demonstrate this and hence produce a contradiction.

The groups $L_2(p) \times L_2(p)$ extended by a 2-group interchanging the components are shadows realizing the configurations in (1) and (2), while $L_4(3) \cong P\Omega_6^+(2)$ extended by a suitable group of outer automorphisms realize the configurations in (3). The last case causes the most difficulties, and consequently is not eliminated until the final reduction.

LEMMA 2.5.18. *(1) K is a component of G_t.*
(2) $G_t = K_0 S C_{G_t}(K_0)$ with $C_G(K_0)S \leq M$.
(3) $C_{G_t}(K_0) \in \mathcal{H}^e$, so $O(G_t) = 1$.

PROOF. By 2.5.11.3, $H \leq G_t \in \Gamma^*$ and $S \in Syl_2(G_t)$. Thus if K is not a component of G_t, we may apply 2.5.6 with $\langle t \rangle$ in the role of "E", to conclude that $K = K_0 \cong A_6$ and $K_t := \langle K^{G_t} \rangle \cong M_{11}$. Since $H \in \Gamma^+ \cap \Gamma^*$, we conclude from parts (2) and (4) of 2.3.7 that $K_t S \in \Gamma^+ \cap \Gamma^*$, contrary to 2.5.15.

Thus (1) holds, so as $S \in Syl_2(G_t)$, $K_0 \trianglelefteq G_t$, and by 2.5.8.1, $G_t = K_0 S C_{G_t}(K_0)$. Then $C_{G_t}(K_0) \leq C_{G_t}(z_K) \leq C_{G_t}(z) \leq M$, proving (2). By 2.3.9.4, $G_t \cap M \in \mathcal{H}^e$, so (2) implies (3). □

LEMMA 2.5.19. *Assume i is an involution in $C_S(K)$ such that K is not a component of $C_G(i)$. Then*

(1) $K = K_0$.
(2) $C_S(i) \cap C_S(K) = \langle t, i \rangle$.
(3) There exists a component K_i of $C_G(i)$ such that either:

(I) $K_i \neq K_i^t$, $K = C_{K_i K_i^t}(t)^\infty$, and $K_i \cong K \cong L_2(p)$, $p \geq 7$, or
(II) $K = C_{K_i}(t)^\infty$, and one of the following holds:

(a) $K \cong L_3(2)$, and t induces a field automorphism on $K_i \cong L_3(4)$ or $L_3(4)/\mathbf{Z}_2$.
(b) $K \cong L_3(2)$, and t induces an outer automorphism on $K_i \cong J_2$.
(c) $K \cong A_6$ and $K_i \cong Sp_4(4)$, $L_5(2)$, HS, or A_8.

(4) Either $z = z_K \in K$ and $tz \in t^G$, or $K_i \cong A_8$ and t induces a transposition on K_i.

PROOF. Let $G_i := C_G(i)$ and $R := G_i \cap C_S(K)$. As $t \in Z(S) \cap S_C$, $\langle t, i \rangle \leq R$ by our hypothesis on i. As K is not a component of G_i, $i \neq t$ by 2.5.18. Therefore $i \notin Z(S)$, or otherwise i centralizes $\langle K^S \rangle = K_0$, whereas $Z(S) \cap S_C = \langle t \rangle$ by 2.5.11.1. By 2.5.18, $C_{G_t}(K_0) \leq M$ and $S \in Syl_2(G_t)$, so conjugating in $C_{G_t}(K_0)$ we may assume $C_S(\langle i, K_0 \rangle) \in Syl_2(C_G(\langle t, i, K_0 \rangle))$.

Next K is a component of $C_{G_i}(t)$ in view of 2.5.18, so by I.3.2 there is $K_i \in \mathcal{C}(G_i)$ with $K_i/O(K_i)$ quasisimple, such that for $K_+ := \langle K^{O_{2'}, E(G_i)} \rangle$, either

(i) $K_+ = K_i K_i^t$, $K_i \neq K_i^t$, $K_i / O_{2',2}(K_i) \cong K$, and $K = C_{K_+}(t)^\infty$, or

(ii) $K_+ = K_i = [K_i, t]$ and K is a component of $C_{K_i}(t)$.

Set $R_0 := C_R(K_+)$. In case (ii) as $K_i / O(K_i)$ is quasisimple, $O_2(K_i) \leq Z(K_i)$, so as $m_2(K_i) \geq m_2(K) > 1$, K_i is quasisimple by 1.2.1.5. Similarly if (i) holds, then $O(K_i) = 1$ by 1.2.1.3, so that K_i is quasisimple. Thus in any case K_i is a component of G_i.

Let $g \in G$ with $T_i := C_{T^g}(i) \in Syl_2(G_i)$; then applying 1.1.6 to the 2-local G_i, the hypotheses of 1.1.5 are satisfied with G_i, M^g, z^g in the roles of "H, M, z". Therefore K_i is described in 1.1.5.3.

Suppose for the moment that case (i) holds. Then by 1.2.1.3 applied to K_i, K is not A_6, so by 2.5.16, K is $L_2(p)$ for $p \geq 7$ a Fermat or Mersenne prime. Then as $K_i / Z(K_i) \cong K$ in (i), $K_i \cong K$ by 1.1.5.3. Therefore all involutions in tK_+ are conjugate, and hence $tz_K \in t^G$, so $z = z_K$ by 2.5.13.3 and hence $tz \in t^G$. Therefore conclusion (I) of (3) and the first alternative in (4) hold in case (i). Thus in case (i), it remains only to verify (1) and (2). Observe also in this case that $N_R(K_i)$ centralizes the full diagonal subgroup K of K_+, so $R_0 = N_R(K_i)$ and $R = \langle t \rangle \times R_0$.

Next suppose for the moment that case (ii) holds. Comparing the groups in 2.5.16 to the components of centralizers of involutions in $Aut(K_i / Z(K_i))$ for groups K_i on the list of 1.1.5.3, we conclude that one of the following holds:

(α) $K \cong L_3(2)$, and t induces a field automorphism on $K_i / Z(K_i) \cong L_3(4)$.

(β) $K \cong L_3(2)$, and t induces an outer automorphism on $K_i / Z(K_i) \cong J_2$.

(γ) $K \cong A_6$, and t induces one of: an inner automorphism on $K_i / Z(K_i) \cong HS$, an outer automorphism on $K_i \cong L_4(2)$ or $L_5(2)$, or a field automorphism on $K_i \cong Sp_4(4)$.

Thus to prove that conclusion (II) of (3) holds in case (ii), it remains to show that $|Z(K_i)| \leq 2$ if (α) holds, and to show that $Z(K_i) = 1$ when (β) holds, or when $K_i / Z(K_i) \cong HS$ and (γ) holds.

Notice also when (ii) holds that from the structure of $C_{Aut(K_i / Z(K_i))}(t)$ for the groups in (α)–(γ), either $R_0 = C_R(K_i)$ is of index 2 in R, or else $K_i / Z(K_i) \cong HS$— and in the latter case some $r \in R$ induces an outer automorphism on K_i, with $|R : R_0| = 4$, and $C_{K_i}(R_1)^\infty \cong A_8$ for some subgroup R_1 of index 2 in R.

In the next few paragraphs, we will reduce the proof of 2.5.19 to the proof of (2). So until that reduction is complete, suppose that (2) holds; that is that $R = \langle i, t \rangle \cong E_4$.

We first deduce (1) from (2), so suppose that (1) fails. Thus $K_0 = KK^u$ for some $u \in S - N_S(K)$. Therefore i also acts on K^u, and hence also on $S \cap K^u$, so that $|C_{\langle i \rangle (S \cap K^u)}(i)| > 2$. Since $S \cap K^u \leq C_S(K)$ and $t \notin \langle i \rangle (S \cap K^u)$ because t centralizes K_0, $|R| > 4$, contrary to assumption. This contradiction shows that (2) implies (1).

As remarked earlier, (1) and (2) suffice to prove the entire result when case (i) holds. Thus to complete the proof of the sufficiency of (2), we may now assume that case (ii) holds, and it remains to establish (3) and (4). Recall that at the start of the proof we chose $C_S(\langle i, K_0 \rangle) \in Syl_2(C_G(\langle i, t, K_0 \rangle))$, so as $K = K_0$ by (1), $R \in Syl_2(C_G(\langle t, i, K \rangle))$.

As $K \trianglelefteq H$, $N_S(C_S(i))$ acts on $C_S(i) \cap C_S(K) = R$. We saw $i \notin Z(S)$, so $C_S(i) < N_S(C_S(i))$. Then as $N_S(C_S(i))$ acts on $R = \langle i, t \rangle$, $it \in i^{N_S(C_S(i))}$. But by A.3.18, $K_i = O^{3'}(E(G_i))$, so $i \notin t^G$ by 2.5.18, and hence as $it \in i^G$, also $it \notin t^G$. As $K \leq K_i = [K_i, t]$ and $R \in Syl_2(C_G(\langle t, i, K \rangle))$, $\langle i \rangle = C_{O_2(K_i C_S(i))}(t)$.

Therefore if $\langle i \rangle < O_2(K_i C_S(i))$, then $it \in t^{O_2(K_i C_S(i))}$, contradicting $it \notin t^G$. Thus $\langle i \rangle = O_2(K_i C_S(i))$.

Suppose case (α) or (β) holds. If $O_2(K_i) = 1$ then (3) holds, and from the structure of $Aut(K_i)$, K_i is transitive on involutions in tK_i, so $tz_K \in t^G$, and hence $z = z_K$ by 2.5.13.3, establishing (4). Thus we may assume that $O_2(K_i) \neq 1$, so $\langle i \rangle = O_2(K_i)$ from the previous paragraph. If (β) holds, then from the embedding of $K_i/\langle i \rangle$ in $G_2(4)$, t acts faithfully on some root subgroup $Q/\langle i \rangle$, with $Q \cong Q_8$, so that $ti \in t^Q$, contrary to a remark in the previous paragraph. Thus (α) holds, with $K_i/\langle i \rangle \cong L_3(4)$, so (3) holds in this case. Further the field automorphism t normalizes each maximal parabolic P of K_i over $C_S(i) \cap K_i$. From the structure of the covering group in I.2.2.3b, $V := O_2(P)$ is an indecomposable P-module such that $V/\langle i \rangle$ is the natural module for $P/V \cong L_2(4)$. Now t centralizes X of order 3 in P, and $V = [V, X] \times \langle i \rangle$ with

$$C_{[V,X]}(t) = [V, X, t] \leq O^2(C_{K_i}(t)) = K.$$

It follows that $tz_K \in t^G$, so $z = z_K$ by 2.5.13.3, and hence (4) holds.

Thus it remains to consider the case where (γ) holds. If $K_i \cong A_8$, then the lemma holds, since there we do not assert that $tz_K \in t^G$. If $K_i \cong L_5(2)$ or $Sp_4(4)$, then K_i is transitive on involutions in tK_i, so that $tz_K \in t^G$, and hence $z = z_K$ by 2.5.13.3, so the lemma holds. Thus we have reduced to the case $K_i/O_2(K_i) \cong HS$. Assume first that $Z(K_i) \neq 1$. Then as before, $\langle i \rangle = Z(K_i) = O_2(K_iC_S(i))$, so as we are assuming $\langle t, i \rangle = R$ and t is inner on K_i in (γ), $t \in K_i C_{K_iC_S(i)}(K_i) = K_i$. Thus $t \in C_{K_i}(K)$ so t is not 2-central in K_i. However, an element of the covering group K_i projecting on a non-2-central involution of HS is of order 4 by I.2.2.5b. This contradiction shows that K_i is HS, so that (3) holds. Furthermore if u is the projection on K_i of t, then $uz_K \in u^{K_i}$ and $iuz_K \in (iu)^{K_i}$. Therefore as $t = u$ or iu, $tz_K \in t^{K_i}$, and again (4) follows from 2.5.13.3. This completes the proof of the reduction of the proof of the lemma to the proof of (2).

We have shown that it suffices to prove that $R = \langle i, t \rangle$. Thus we assume that $\langle i, t \rangle < R$, and derive a contradiction. Choose i so that $R = C_S(i) \cap C_S(K)$ is maximal subject to K not being a component of G_i. Further if $i \in Z(C_S(K))$ then $R = C_S(K)$, and we choose i so that $C_S(i)$ is maximal subject to the constraint that $R = C_S(K)$.

Recall we showed soon after stating (i) that that assumption implies $|R : R_0| = 2$. Inspecting the groups in cases (α)–(γ) of (ii), we check that either $|R : R_0| = 2$, or $K_i/Z(K_i) \cong HS$ and $|R : R_0| = 4$. When $|R : R_0| = 2$ we set $R_2 := R_0$, and when $|R : R_0| = 4$ we let R_2 be the subgroup R_1 of index 2 in R with $C_{K_i}(R_1)^\infty \cong A_8$ discussed earlier. Thus in either case, $i \in R_0 \leq R_2$ and $|R : R_2| = 2$.

We next claim that $K < K_0$ and $i \in Z(N_S(K))$. Thus we assume that at least one of the two assertions of the claim fails, and derive a contradiction. As $i \notin Z(S)$ there is $s \in N_S(C_S(i)) - C_S(i)$ with $s^2 \in C_S(i)$. Furthermore we observe when $K < K_0$ that $C_S(i)$ normalizes K: For otherwise i centralizes some $u \in C_S(i) - N_S(K)$ and $K_+ \neq K_+^u$. But in all cases appearing in (i) and (ii), $m_3(K_+) = 2$; therefore as K_+ and K_+^u are products of components of G_i, $m_3(K_+ K_+^u) > 2$, impossible as G_i is an SQTK-group. Thus in any case, $C_S(i)$ normalizes K, and hence $C_S(i)$ normalizes $C_S(K)$ and $N_S(K)$.

During the remainder of the proof of the claim, we choose the element $s \in N_S(C_S(i)) - C_S(i)$ with $s^2 \in C_S(i)$ as follows:

(A) If $R < C_S(K)$ choose $s \in C_S(K)$.

(B) If $R = C_S(K)$, choose $s \in N_S(K)$; we check this choice is possible: When $K = K_0$ this is trivial, while when $K < K_0$, by assumption $i \notin Z(N_S(K))$, so again the choice is possible.

In either (A) or (B), $s \in N_S(K)$. Hence as $s \in N_S(C_S(i))$, s normalizes $C_S(i) \cap C_S(K) = R$.

In case (A) set $W := R\langle s \rangle$, and in case (B) set $W := C_S(i)\langle s \rangle$. In either case, $W = C_W(i)\langle s \rangle$. Furthermore $s^2 \in C_W(i)$: As $s^2 \in C_S(i)$, this is immediate from the definition of W in case (B), while in case (A) we chose $s \in C_S(K)$, so that $s^2 \in C_S(i) \cap C_S(K) = R = C_W(i)$.

We now show that $R_2 \trianglelefteq C_W(i)$: In case (A), this holds as R_2 is of index 2 in $R = C_W(i)$, so assume case (B) holds. Then $C_W(i) = C_S(i)$ normalizes $C_S(i) \cap C_S(K_+) = R_0$, so the claim holds when $|R : R_0| = 2$, since in that case $R_2 = R_0$. Thus we may assume $|R : R_0| = 4$ and $K_i/Z(K_i) \cong HS$, so that R_2 is the subgroup R_1 of R_0 with a component A_8 in its centralizer. But $C_S(i)$ acts on the 4-group R/R_0, and hence also on the unique subgroup R_1/R_0 of order 2 with $K < E(C_{K_i}(R_1))$. So indeed $R_2 \trianglelefteq C_W(i)$.

As $R_2 \trianglelefteq C_W(i)$ and $s^2 \in C_W(i)$, $W = C_W(i)\langle s \rangle$ normalizes $R_2 \cap R_2^s$. Assume $R_2 \cap R_2^s \neq 1$; then $C_{R_2 \cap R_2^s}(W) \neq 1$. Let r be an involution in $C_{R_2 \cap R_2^s}(W)$; from the definition of R_2, K is not a component of $C_G(r)$. In case (A), $R < W \leq C_S(r) \cap C_S(K)$, contrary to the maximality of R. In (B), $R = C_S(K) \leq C_S(i) < W \leq C_S(r)$, contrary to the maximality of $C_S(i)$ in our choice of i, R under the constraint that $R = C_S(K)$. Therefore $R_2 \cap R_2^s = 1$, so as $|R : R_2| = 2$, $|R| = 4$, contrary to our assumption that $R \neq \langle i, t \rangle$. This finally completes the proof of the claim.

By the claim, $K_0 = KK^u$ for $u \in S - N_S(K)$ and $i \in Z(N_S(K))$. Therefore by 1.2.1.3, K is described in case (1) or (2) of 2.5.16, so $K \cong L_2(p)$ for $p \geq 7$ a Fermat or Mersenne prime. In case (i) we showed that $K_i \cong K$, so K_+ is the direct product of two t-conjugates of a copy of K. In case (ii), $K \cong L_3(2)$, so (α) or (β) holds.

Let j be an involution in $R_0 = C_R(K_+)$, $G_j := C_G(j)$, $L_0 := \langle K_i^{O_{2'},E(G_j)} \rangle$, and $L_+ := \langle K_+^{O_{2'},E(G_j)} \rangle$. Then $K < K_+ \leq G_j$, so as K is not subnormal in K_+, K is not a component of G_j. Indeed we claim that $K_+ \trianglelefteq G_j$. As K_i is a component of $C_{G_j}(i)$, we may apply the initial arguments of the proof of 2.5.19 to j, i, K_i in the roles of "i, t, K". We conclude that there is a component L of G_j such that either $L = L_0$ is i-invariant, or $L < L_0 = LL^i$ with $C_{L_0}(i)^\infty$ a component of $C_{G_j}(i)$ isomorphic to $K_i \cong L_2(p)$ for suitable p. It follows that $L_+ = L_0 L_0^t$. Similarly in case (i) where $K \cong K_i$, if $L = L_0$ we may apply 1.1.5 to conclude that L is $L_3(4)$ or J_2 of 3-rank 2.

If $K_+ = L_+$, then we conclude from A.3.18 in case (ii) or from 1.2.2 in case (i) that $L_+ = O^{3'}(E(G_j)) \trianglelefteq G_j$. Thus to establish the claim that $K_+ \trianglelefteq G_j$, it will suffice to show that $K_+ = L_+$.

Suppose that case (ii) holds. Then K_i is described in (α) or (β), so that 1.2.1.3 rules out the case $L < L_0$. Thus $L_0 = L$, and $L = [L, t]$ as t acts on K_i. Then our earlier argument applied to t, j, K in the roles of "t, i, K" shows that L is $L_3(4)$ or J_2. But then as K_i is a component of $C_L(i)$, $L = K_i$. Then as $L = L_0 = K_i$, $L_+ = LL^t = K_i K_i^t = K_+$, as desired.

So assume that case (i) holds. Suppose first that $L < L_0$. We saw that $L \cong K_i \cong L_2(p)$ for a suitable prime p, so $L_0 = L \times L^i$ with $K_i = C_{L_0}(i)$ a full diagonal subgroup of L_0. By 1.2.2, $L_0 = O^{3'}(G_j)$, so t acts on L_0 and then on $C_{L_0}(i) = K_i$, contrary to our assumption that case (i) holds. Thus $L = L_0$, and by an earlier remark, $L = [L, i]$ is $L_3(4)$ or J_2. But then t acts on L by 1.2.1.3, so $K_+ = K_i K_i^t \leq L$, a contradiction as $L_3(4)$ and J_2 contain no such subgroup. This completes our proof that $K_+ \trianglelefteq G_j$.

We showed that in case (ii), that $K_i/Z(K_i)$ is not HS; hence in either case (i) or (ii), $|R : R_0| = 2$, so $R = R_0 \times \langle t \rangle$. As $i \in Z(N_S(K))$ and $K^u \langle t \rangle$ centralizes K, $S \cap K^u \langle t \rangle \leq C_S(\langle i, K \rangle) = R$. Therefore $S \cap K^u \langle t \rangle = S_0 \times \langle t \rangle$, where $S_0 := R_0 \cap (S \cap K^u) \langle t \rangle$. Thus $S_0 \langle t \rangle$ is Sylow in $K^u \langle t \rangle$, so from the structure of $Aut(K) \cong PGL_2(p)$ for $p \geq 7$ a Fermat or Mersenne prime, and using the second claim,

$$K^u = \langle C_{K^u}(j) : j \text{ an involution of } S_0 \rangle \leq N_G(K_+).$$

Therefore $K^u \leq (N_G(K_+) \cap C_G(K))^\infty \leq C_G(K_+)$ from the structure of $C_{Aut(K_+)}(K)$. But now as $m_3(K_+) = 2$ in cases (i) and (ii), $m_{2,3}(K_+ K^u) > 2$, contradicting G quasithin. This contradiction completes the proof of (2), which we saw suffices to establish 2.5.19. $\qquad\square$

LEMMA 2.5.20. $K = K_0$.

PROOF. Assume $K < K_0$. By 2.5.16 and 1.2.1.3, $K \cong L_2(p)$ with $p \geq 7$ a Fermat or Mersenne prime, and $K_0 = KK^u$ for $u \in S - N_S(K)$. By 2.5.19.1, K is a component of $C_G(i)$ for each $i \in C_S(K)$.

We claim that $K_0 = O^{3'}(N_G(K^u))$. For let i be an involution in $K^u \cap S = S_K^u$. Then as $K^u \cong L_2(p)$ has one class of involutions, by a Frattini Argument, $N_G(K^u) = K^u I_i$ where $I_i := C_G(i) \cap N_G(K^u)$. Further we just saw that K is a component of $C_G(i)$, and hence K is a component of I_i. As $K \cong L_2(p)$ has no outer automorphism of order 3, $O^{3'}(N_{I_i}(K)) = KO^{3'}(C_{I_i}(K)) = KO^{3'}(C_{I_i}(K_0))$. As G is quasithin and $m_{2,3}(K_0) = 2$, $O^{3'}(C_{I_i}(K_0)) = 1$, so $O^{3'}(N_{I_i}(K)) = K$ and hence $K = O^{3'}(I_i)$ as K is subnormal in I_i. Thus $O^{3'}(N_G(K^u)) = K^u O^{3'}(I_i) = K^u K$, establishing the claim.

Then as u interchanges K and K^u, also $K_0 = O^{3'}(N_G(K))$, so that $K^u = O^{3'}(C_G(K))$ and hence $N_G(K) = N_G(K^u)$. Thus $C_G(i) \cap N_G(K) = C_G(i) \cap N_G(K^u) = I_i$, so that $O^{3'}(C_G(i)) \cap N_G(K)) = O^{3'}(I_i) = K$. We saw K is subnormal in $C_G(i)$, so

$$O^{3'}(C_G(i)) = K,$$

and hence $C_G(i) \leq N_G(K) = N_G(K^u)$. Thus if there is an involution $i \in K^u \cap K^{ug}$, then $K = O^{3'}(C_G(i)) = K^g$, so $g \in N_G(K) = N_G(K^u)$; that is, K^u is *tightly embedded* in G. Then as S_K is nonabelian, I.7.5 says that distinct conjugates of S_K in T commute. Suppose $S_K^g \leq T$ with $S_K \neq S_K^g \neq S_K^u$. Then $S_K^g \leq C_G(S_K S_K^u) \leq N_G(K^u) = N_G(K_0)$ since K^u is tightly embedded. Then since the center of a Sylow 2-subgroup of $Aut_S(K)$ is elementary abelian, $\Phi(S_K^g) \leq \Phi(C_T(S_K S_K^u) \cap N_G(K_0)) \leq C_T(K_0)$, and then $KK^u = K_0 \leq O^{3'}(C_G(\Phi(S_K^g))) = K^{ug}$, a contradiction. Therefore $\{S_K, S_K^u\} = S_K^G \cap T$, so T permutes the set Δ of groups $O^{3'}(C_G(j))$ for j an involution in $S_K \cup S_K^u$. We showed $\Delta = \{K, K^u\}$, so T acts on K_0. Therefore $H = K_0 S \leq K_0 T \leq M = !\mathcal{M}(T)$, contradicting $H \not\leq M$. This completes the proof of 2.5.20. $\qquad\square$

We now eliminate all possibilities for K remaining in 2.5.16 except for the one corresponding to the most stubborn remaining shadow discussed earlier:

LEMMA 2.5.21. $\bar{H} \cong Aut(A_6)$.

PROOF. First $K_0 = K$ by 2.5.20, so $H = KS$. Assume $\bar{H}$ is not $Aut(A_6)$. Then by 2.5.16, either $K \cong L_2(p)$ for $p \geq 7$ a Fermat or Mersenne prime, or $\bar{H} \cong PGL_2(9)$ or M_{10}. Therefore either $\bar{S}$ is dihedral, or $\bar{H} \cong M_{10}$ and $\bar{S} \cong SD_{16}$. Hence by 2.5.5.1, S_C is cyclic or dihedral, unless possibly $S_C \cong Q_8$ or SD_{16} when $\bar{H} \cong M_{10}$. In each case $|\bar{H} : \bar{K}| \leq 2$.

Assume that S_C is of order 2, so that $S_C = \langle t \rangle$. As $|\bar{H} : \bar{K}| \leq 2$, $|S : S_K| \leq 4$, so S/S_K is abelian and hence $[S, S] \leq S_K$. Also $\bar{S}$ is dihedral or semidihedral of order at least 8, so $\Omega_1([\bar{S}, \bar{S}]) = \langle \bar{z}_K \rangle$. Therefore $\Omega_1([S, S]) = \langle z_K \rangle$. Then x centralizes z_K, so by 2.5.13.2, $z = z_K$ and $t^x = tz$. Thus all involutions in K are in z^G, and all involutions in tK are in t^G. Choose $(U, H_U) \in \mathcal{U}^*(H)$. Then by 2.5.12, there is a 4-subgroup E of S_K such that $U = N_S(E)$ and $H_U = N_H(E)$. Since $S_C = \langle t \rangle$ is of order 2, $N_H(E) = N_H(F) \cong \mathbf{Z}_2 \times S_4$, where $F := ES_C = O_2(N_H(E)) = O_2(H_U) \cong E_8$. In particular $N_S(F) = U \in Syl_2(N_G(F))$ by 2.5.10.2. If $F^x \in F^S$, then by a Frattini Argument, we may take $x \in N_T(F)$, contradicting $N_S(F) \in Syl_2(N_G(F))$. Thus $F^x \notin F^S$.

Assume first that $S \not\leq KS_C$. $H = KS$ is transitive on E_8-subgroups of KS_C, so $F^x \not\leq KS_C$. But all involutions in M_{10} are in $E(M_{10})$, so if $K \cong A_6$ then $\bar{H} \cong PGL_2(9)$. Thus $\bar{H} \cong PGL_2(q)$ for q a Fermat or Mersenne prime or 9. But x acts on $Z(S) = \langle z, t \rangle \leq KS_C$, so as $F^x \not\leq KS_C$ by the previous paragraph, $e^x \notin KS_C$ for $e \in E - \langle z \rangle$. As $e^x \notin KS_C$ and $\bar{H} \cong PGL_2(q)$, $O(C_K(e^x)) \neq 1$, so since K is a component of G_t by 2.5.18, $1 \neq O(C_K(e^x)) \leq O(C_G(\langle e^x, t \rangle))$. Hence $C_G(e^x) \notin \mathcal{H}^e$ by 1.1.3.2, contradicting $e^x \in z^G$.

Therefore $S \leq KS_C$, so $H = K \times S_C$, and hence $S = S_K \times S_C$. This rules out cases (2) and (3) of 2.5.16 in which S is nontrivial on the Dynkin diagram of K, so $K \cong L_2(p)$ for $p > 7$ a Fermat or Mersenne prime. We saw earlier that $tE \subseteq t^G$, so there is $g \in G$ with $t^g \in F - \langle t, z \rangle$. As $S_C = \langle t \rangle$ is of order 2, $C_{G_t}(K) = O(C_{G_t}(K))S_C$ by Cyclic Sylow 2-Subgroups A.1.38. By 2.5.18, $G_t = KSC_{G_t}(K)$ and $O(C_{G_t}(K)) = O(G_t) = 1$, so $G_t = KS_C = H$. Thus $F \leq G_t^g = H^g = K^g S_C^g$, so $Aut_{K^g}(F) \cong S_3$, and hence $\langle Aut_K(F), Aut_{K^g}(F) \rangle$ is the parabolic in $GL(F)$ stabilizing $\langle z^G \cap F \rangle = K \cap F = E$. As this group is transitive on $F - E$ of order 4 and $S \leq G_t$, we conclude $|N_G(F) : N_S(F)|_2 \geq 4$, contradicting our earlier remark that $N_S(F) \in Syl_2(N_G(F))$. Therefore $|S_C| > 2$.

Suppose next that S_C is abelian. From remarks at the start of the proof, either S_C is cyclic, or possibly $S_C \cong E_4$ when $\bar{H} \cong M_{10}$. By 2.5.11.1, $Z(S) \cap S_C = \langle t \rangle$, so $S_C \not\leq Z(S)$, and hence $S \not\leq KS_C$. Indeed as $|\bar{S} : \bar{S}_K| \leq 2$, $|S : S_K S_C| = 2$ and $C_S(S_C) = S_K S_C$. Thus conjugating by x, also $|S : C_S(S_C^x)| = 2$, so $|\bar{S} : C_{\bar{S}}(\bar{S}_C^x)| \leq 2$. Hence as $\bar{S}$ is dihedral or semidihedral of order at least 16, while $S_C^x \cong \bar{S}_C^x$ is abelian of order at least 4 by 2.5.5.1, we conclude $\bar{S}_C$ is cyclic and $\bar{S}_C^x \leq \bar{K}$. Since $S_C S_C^x = S_C \times S_C^x$ by 2.5.5.1.2 we conclude $S_C \times S_C^x \leq S_C \times Y$, where Y is the cyclic subgroup of index 2 in S_K, and $C_S(S_C^x) = S_C \times Y$. This is impossible, as

$$C_S(S_C^x) = C_S(S_C)^x \cong C_S(S_C) = S_C \times S_K,$$

and S_K is nonabelian.

This contradiction shows that S_C is nonabelian. So again by our initial remarks, either S_C is dihedral of order at least 8, or $H/S_C \cong M_{10}$ and $S_C \cong Q_8$ or SD_{16}.

Set $S_0 = C_S(S_C)$. In any case, $\langle t \rangle = Z(S_C)$ and $S_K \leq S_0$, so as $|\bar{S} : \bar{S}_K| \leq 2$, $|S_0 : S_K| \leq 4$ and hence $z_K \in [S_K, S_K] \leq [S_0, S_0] \leq S_K$. Let Y be the cyclic subgroup of index 2 in S_K. Then $\Omega_1(Y) = \langle z_K \rangle$ and $[\bar{S}, \bar{S}] \leq \bar{Y}$, so $[S_0, S_0] \leq Y$ and hence $\Omega_1([S_0, S_0]) = \langle z_K \rangle$. However $S_C^x \leq C_S(S_C)$ by 2.5.5.2, and hence $[S_C^x, S_C^x] \leq [S_0, S_0]$, so $t^x = z_K$ and $z = t z_K \neq z_K$. Therefore $t z_K \notin t^G$ in view of 2.5.11.2.

We next show that K is a component of $C_G(i)$ for each involution $i \in S_C$. We assume i is a counterexample and derive a contradiction: As $z \neq z_K$, 2.5.19.4 says $K \cong A_6$ and $K < K_i \trianglelefteq C_G(i)$ with $K_i \cong A_8$ and t induces a transposition on K_i. But then $C_{K_i}(t) \cong S_6$, whereas $S \in Syl_2(G_t)$ by 2.5.11.2, and no element of S induces an outer automorphism in S_6 on K since $\bar{H} \cong PGL_2(9)$ or M_{10}. This contradiction shows K is a component of $C_G(i)$.

Next we claim that $K \trianglelefteq C_G(i)$ for each involution i of $C_{G_t}(K)$: For assume $u \in C_G(i)$ with $K \neq K^u$. Then $\langle K, K^u \rangle = K \times K^u$ as K is a component of $C_G(i)$, and $i \neq t$ by 2.5.20. Now $\langle i, t \rangle$ is not Sylow in $K^u \langle i, t \rangle \cap G_t$, so $\langle i, t \rangle$ is not Sylow in $C_G(\langle i, t \rangle K)$. On the other hand as S is Sylow in G_t, we may assume $C_{S_C}(i) \in Syl_2(C_{G_t}(K \langle t \rangle))$, a contradiction as S_C is dihedral, semidihedral or quaternion. This contradiction establishes the claim that K is normal in $C_G(i)$ for each involution i of S_C.

Now assume S_C is not Q_8; in this part of the proof we eliminate the shadows of subgroups of $PSL_2(p) \text{ wr } \mathbf{Z}_2$. By our earlier remarks, either S_C is dihedral of order at least 8, or $H/S_C \cong M_{10}$ and $S_C \cong SD_{16}$. Recall from 2.5.5.1 that $S_C \cap S_C^x = 1$, so $K \neq K^x$ and $S_C^x \cong \bar{S}_C^x$. Since $K \cong L_2(q)$ for q a Fermat or Mersenne prime or 9, we compute from the possibilities for $\bar{H} \leq Aut(K)$ that

$$K = \langle C_K(j) : j \text{ an involution of } S_C^x \rangle,$$

so that $K \leq N_G(K^x)$ by the claim in the previous paragraph. By symmetry K^x acts on K, so $[K, K^x] = 1$, so K is not A_6 since G is quasithin. Thus $K \cong L_2(p)$ for $p \geq 7$ a Fermat or Mersenne prime. Let $K_+ := KK^x$, $M_+ := N_G(K_+)$, and $S_+ := S\langle x \rangle$.

Next $S_+ \leq MK_+$, and as we saw that $t^x = z_K \in K$, $t \in K^x$. Then as $K \cong L_2(p)$ has one class of involutions, by a Frattini Argument, $M_+ = K_+ N_{M_+}(\langle t, t^x \rangle)$. Then as $S \in Syl_2(G_t)$, $S_+ \in Syl_2(M_+)$. Also $C_S(K_+) = S_C \cap S_C^x = 1$, and hence $C_{S_+}(K_+) = 1$ so $C_G(K_+)) = O(C_G(K_+))$. As $K = K_0$, $G_t = KSC_{G_t}(K)$ and $O(G_t) = 1$ by 2.5.18. Then as $K^x \leq C_G(K) \leq G_{t^x}$ since $t^x \in K$, K^x is normal in $C_G(K)$. Thus $C_G(K) = K^x C_G(K_+) = K^x O(C_G(K_+)) = K^x O(N_G(K))$. Then $C_{G_t}(K) = C_{K^x}(K)O(N_G(K)) = S_K^x O(N_G(K))$, so

$$G_t = KSO(N_G(K)) = KSO(G_t) = KS = H \leq M_+.$$

In particular $K = O^2(G_t)$.

We claim that $t^G \cap M_+$ is the set $\mathcal{I}$ of involutions in $K \cup K^x$. We saw earlier that $z_K = t^x$ and K has one class of involutions, so $\mathcal{I} \subseteq t^G \cap M_+$. Furthermore we saw that $z = t z_K = t t^x$, so that the diagonal involutions in K_+ are in z^G, and hence these involutions are not in t^G by 2.5.11.3. Thus if the claim fails, there is $i := t^g \in S_+ - \mathcal{I}$, such that either i induces an outer automorphism on K or K^x, or $K^x = K^i$. In the latter case, $C_{K_+}(i) =: K_i \cong K$, so $K_i = K^g$ since $K = O^2(G_t)$; this is impossible as the involutions in K_i are in z^G, while those in K are in t^G. Thus we may assume that i induces an outer automorphism on K.

Suppose first that i either centralizes K^x or induces an outer automorphism on K^x. If i induces an outer automorphism on K^x, then $K^x\langle i \rangle \cong PGL_2(p)$, so in either situation, i centralizes an E_{q^2}-subgroup of K_+, where q is an odd prime divisor of $p + \epsilon$, $p \equiv \epsilon \mod 4$, and $\epsilon = \pm 1$. This is impossible, as $K = O^2(G_t) \cong L_2(p)$ is of q-rank 1. Therefore i induces a nontrivial inner automorphism on K^x. Then $t \in C_{K^x}(i) \cong D_{p-\epsilon}$ centralizes $C_K(i) \cong D_{p+\epsilon}$, so

$$t \in \Phi(\, C_{K^x}(i) \cap C_G(C_K(i)) \,) \leq C_{G_i}(K^g),$$

since the centralizer in $Aut(K^g)$ of a $D_{p+\epsilon}$-subgroup of K^g is of order 2. Then as $K = O^2(G_t)$, $K^g = K$, a contradiction as i centralizes K^g but not K. This establishes the claim that $\mathcal{I} = t^G \cap M_+$.

We've shown that $t^G \cap M_+ = \mathcal{I}$, so $t^G \cap M_+ = t^{M_+}$. We also showed that $G_t \leq M_+$, so by 7.3 in [**Asc94**], t fixes a unique point in the representation of G by right multiplication on G/M_+. Therefore as T is nilpotent (cf. the proof of 2.2.2), $T \leq M_+$. Further M_+ is the unique fixed point of each member of $S_K^\#$, so $S_K \cup S_K^x$ is strongly closed in T with respect to G. Thus the hypotheses of 3.4 in [**Asc75**] are satisfied with S_K, S_K^x, M_+ in the roles of "A_1, A_2, H", so that result says $G = M_+$, contradicting G simple.

We have reduced to the case where $S_C \cong Q_8$. In particular $\bar{H} \cong M_{10}$. Now $S_C S_C^x = S_C \times S_C^x$ by 2.5.5.2. Since $S_K \cong D_8$, $\bar{S}_C^x \not\leq \bar{K}$, so $\bar{H} = \bar{K}\bar{S}_C^x$. Thus $[\bar{S}_C^x, \bar{S}_K]$ is the image of the cyclic subgroup Y of index 2 in S_K. Then as $S_C^x \trianglelefteq S$, $Y = [S_C^x, S_K] \leq S_C^x$, so $S_C^x = Y\langle v \rangle$ for $v \in S_C^x - K$. Then as $[S_C, S_C^x] = 1$ and v induces an outer automorphism on K with $v^2 = z_K \in Y \leq K$, it follows that $H = S_C \times S_C^x K$, so $S = S_C \times S_C^x S_K$ with $S_C^x S_K$ a Sylow 2-subgroup of M_{10}. Since $z_K = t^x$, $C_S(S_C^x) = S_C \times Z(S_C^x) = S_C\langle t^x \rangle$, and hence

$$|C_S(S_C^x)| = 16 < 32 = |C_S(S_C)|,$$

a contradiction as x acts on S. This finally completes the proof of 2.5.21. $\qquad\square$

In view of 2.5.21, it only remains to eliminate the case $\bar{H} \cong Aut(A_6)$. In particular $K \cong A_6$ and $S_K \cong D_8$.

LEMMA 2.5.22. *(1) If $z^g \in S$ for some $g \in G$, then $K = [K, z^g]$, and z^g induces an automorphism in S_6 on K.*
(2) $H = G_t$ and $C_H(z) = S$.

PROOF. Assume $z^g \in S$ for some $g \in G$. Then as $C_{G_t}(z^g) \in \mathcal{H}^e$ by 1.1.3.2, $C_K(z^g) \in \mathcal{H}^e$ using 1.1.3.1. Further $K = K_0 \trianglelefteq G_t$ by 2.5.20, so since $\bar{H} \cong Aut(A_6)$ by 2.5.21, (1) follows.

Let $C := C_{G_t}(K)$. By 2.5.18, $G_t = KSC$ and $C \in \mathcal{H}^e$. Thus $R := O_2(G_t) \leq S_C$ and $R \trianglelefteq S$. By 2.5.5.1, $S_C \cap S_C^x = 1$, so $R \cong \bar{R}^x \trianglelefteq \bar{S}$. As $\bar{S}$ is Sylow in $\bar{H} \cong Aut(A_6)$, it follows that either
  (i) $\bar{R}$ is abelian and $m(\bar{R}) \leq 2$, or
  (ii) $[\bar{R}, \bar{R}] =: \bar{Y}$ is the cyclic subgroup of index 2 in $\bar{S}_K$, and either $m(\bar{R}/\bar{Y}) \leq 2$ or $\bar{R} = \bar{S}$.
We conclude that $Aut(R)$ is a $\{2,3\}$-group, and hence C is a $\{2,3\}$-group. However as H is an SQTK-group, C is a $3'$-group, so as $F^*(C) = O_2(C)$, C is a 2-group. Thus $G_t = KSC = KS = H$, so $C_H(z) = S$ as $K \cong A_6$. $\qquad\square$

LEMMA 2.5.23. $\mathcal{U}^*(H) = \{(N_S(E_i), N_H(E_i)) : i = 1, 2\}$, where E_1 and E_2 are the 4-subgroups of S_K, and $N_S(E_i) \in Syl_2(N_H(E_i))$.

PROOF. This follows from 2.5.12.3.   □

LEMMA 2.5.24. $\Phi(S_C) \neq 1$.

PROOF. Assume $\Phi(S_C) = 1$, define E_1 and E_2 as in 2.5.23, and set $Q_i :=$
$O_2(N_H(E_i))$. Now $\bar{S}/\bar{S}_K \cong E_4$ since $\bar{H} \cong Aut(A_6)$ by 2.5.21, so that $\Phi(S) \leq$
$S_K S_C$. Let Y denote the cyclic subgroup of S_K of index 2. Then $Y \leq [S_K, S] \leq$
$[S, S] \leq \Phi(S)$. Since $\bar{Y} = \Phi(\bar{S}) \geq \overline{\Phi(S)}$, $\Phi(S) \leq Y \times S_C$. Then using the Dedekind
Modular Law, $\Phi(S) = Y \times \Phi_C$, where $\Phi_C := \Phi(S) \cap S_C$. In particular as $\Phi(S_C) = 1$,
$\Phi(\Phi(S)) = \Phi(Y) = \langle z_K \rangle$, so by 2.5.13.2, $z = z_K \in K$ and $t^x = tz_K = tz$.

Next $C_S(Y) = S_1$, where $\bar{S}_1$ is the modular subgroup M_{16} (see p. 107 in
[**Asc86a**]) of $\bar{S}$. Thus

$$S_+ := \Omega_1(C_S(\Phi(S))) = \Omega_1(C_{S_1}(\Phi_C)) \text{ is either } S_C\langle z \rangle \text{ or } S_0\langle z \rangle$$

where S_0 is the preimage in S of the subgroup generated by the transposition in
$\bar{H} \cong Aut(A_6)$ centralizing $\bar{Y}$. Thus as $S_C \cap S_C^x = 1$ by 2.5.5.1 while x acts on
S_+, we conclude as usual that $m(S_C) \leq 2$, with $S_+ = S_0\langle z \rangle = S_C \times S_C^x$ in case of
equality.

Suppose the latter case holds. Then $m(S_C) = 2$, and S_C^x contains an element
inducing a transposition on K. Thus $\mathcal{A}(S) = \{Q_1, Q_2\}$, and $Q_i = S_C S_C^x E_i \cong$
E_{32}. Further S is transitive on $\mathcal{A}(S)$, so by a Frattini Argument, we may take
$x \in N_T(S) \cap N_T(Q_i)$ for each i, and hence $N_S(E_i) < N_S(E_i)\langle x \rangle$, so $N_S(E_i) \notin$
$Syl_2(N_G(Q_i))$. But by 2.5.23, $(N_S(E_i), N_H(E_i)) \in \mathcal{U}^*(H)$, whereas 2.5.10.2 says
$N_S(E_i)$ is Sylow in $N_G(Q_i)$. This contradiction eliminates the case $m(S_C) = 2$.

Therefore $m(S_C) = 1$, so as we are assuming S_C is elementary abelian, in
fact $S_C = \langle t \rangle$ is of order 2. Suppose first that $E_1^x \leq KS_C$. Then x acts on
$S_- := E_1 S_C (E_1 S_C)^x$. If x does not normalize $E_1 S_C$ then $\mathcal{A}(S_-) = \{E_1 S_C, E_2 S_C\}$,
so S is transitive on $\mathcal{A}(S_-)$, and again by a Frattini Argument we may replace x by
$x' \in N_T(S) \cap N_T(S_C E_1)$, and assume x acts on $E_1 S_C$. Thus x acts on $E_1 S_C$ and
hence on $C_S(E_1 S_C) = Q_1$, allowing us to obtain a contradiction as in the previous
paragraph.

Thus $E_1^x \not\leq S_C S_K$. We showed $z = z_K$, so $E_1^\# \subseteq z^G$. Therefore by 2.5.22.1,
e^x induces a transposition on $K \cong A_6$ for some $e \in E_1 - \langle z \rangle$. Now some conjugate
v of e^x in $S_K e^x$ centralizes S_K, so $Q_i = S_C \times E_i\langle v \rangle \cong E_{16}$, and S is transitive on
$\mathcal{A}(S) = \{Q_1, Q_2\}$, so by a Frattini Argument we may choose $x \in N_T(S) \cap N_T(Q_i)$,
leading to the same contradiction as in the two previous paragraphs.   □

LEMMA 2.5.25. $t^x = z_K$ and $z = tz_K$.

PROOF. Assume otherwise. Then by 2.5.11.1, $z = z_K$, $t^x = tz_K$, and $\langle t \rangle =$
$Z(S) \cap S_C$, so $\langle t^x \rangle = Z(S) \cap S_C^x$. But $S_K \cap S_C^x$ is normal in S, so if $1 \neq S_K \cap S_C^x$
then $1 \neq Z(S) \cap S_K \cap S_C^x$, contradicting $t^x = tz_K$. Hence $S_K \cap S_C^x = 1$. Thus
$[S_K, S_C^x] \leq S_K \cap S_C^x = 1$, so $S_C^x \leq C_S(S_K) =: S_0$, and hence S_C^x is isomorphic
by 2.5.5.1 to a subgroup of $\bar{S}_0 \cong E_4$, whereas S_C is not elementary abelian by
2.5.24.   □

LEMMA 2.5.26. $m_2(S_C) = 1$.

PROOF. Assume $m_2(S_C) > 1$. In the first few paragraphs of the proof, we will
establish the claim that K is a component of $C_G(i)$ for each involution $i \in S_C$.
Assume otherwise; by 2.5.18, $i \neq t$, and by 2.5.19.2, $C_{S_C}(i) = \langle i, t \rangle$. Further
$z \neq z_K$ by 2.5.25, so by 2.5.19.4, $K \leq K_i \trianglelefteq C_G(i)$ where $K_i \cong A_8$, and t induces a

transposition on K_i. As $C_{S_C}(i) = \langle i, t \rangle$, S_C is dihedral or semidihedral by a lemma of Suzuki (cf. Exercise 8.6 in [**Asc86a**]), so as S_C is not elementary abelian by 2.5.24, $|S_C| \geq 8$. Using 2.5.5.2, $S_C^x \leq C_S(S_C) \leq C_S(i)$. However, as $C_{S_C}(i) = \langle i, t \rangle$, $K_i C_S(i) \cong \langle i \rangle \times S_8$. Therefore a Sylow 2-subgroup of $K_i C_S(i) \cap C_G(t)$ is isomorphic to $E_8 \times D_8$, which contains no D_{16} or SD_{16}-subgroup, so $S_C \cong D_8$. Hence $|S| = 2^8$ since $\bar{H} \cong Aut(A_6)$ by 2.5.21.

Let V denote the cyclic subgroup order 4 in S_C. By 2.5.22.2 $G_t = KS$, so $V \trianglelefteq G_t$, and thus V is a TI-set in G. Hence as V is not elementary abelian, $\langle V^G \cap T \rangle$ is abelian by I.7.5.

Assume $V^g \leq T$ for some $g \in G$. Then by the previous paragraph, $V^g \leq C_T(V) = C_S(V)$ and hence $\Phi(V^g) \leq \Phi(S) \leq S_K S_C$ since $\bar{S}/\bar{S}_K \cong E_4$. Now no involution in $\bar{S}_K - \langle \bar{z} \rangle$ is a square in $\bar{S}$, so no involution in $S_K S_C - \langle z \rangle S_C$ is a square in S. Hence

$$\langle t^g \rangle = \Phi(V^g) \leq \Omega_1(C_{\langle z \rangle S_C}(V)) = \Omega_1(V \langle z \rangle) = \langle t, z \rangle.$$

Therefore $t^g \in t^G \cap \langle t, z \rangle$, so that t^g is t or t^x by 2.5.11. Hence V^g is either V or V^x.

Since $V^G \cap T = \{V, V^x\}$, $V V^x \trianglelefteq T$, so $\Omega_1(V V^x) = \langle t, t^x \rangle \trianglelefteq T$. Then as $S \in Syl_2(G_t)$ by 2.5.11.2, $|T| = 2|S| = 2^9$.

Let $H_0 := K_i \langle i, t \rangle$, $T_i \in Syl_2(H_0)$, and $T_i \leq T^g$ for suitable $g \in G$. As K_i is a component of $C_G(i)$, $H_0 \not\leq M^g$ by 1.1.3.2. As $H_0 \cong \mathbf{Z}_2 \times S_8$, $H_0 = \langle H_1, H_2 \rangle$, where H_1 and H_2 are the maximal 2-locals of H_0 over T_i; thus we may assume $H_1 \not\leq M^g$. As $|T_i| = 2^8 = |T|/2$, $T_i^{g^{-1}} \in \beta$ by 2.3.10, so $(T_i^{g^{-1}}, H_1^{g^{-1}}) \in \mathcal{U}(H_1^{g^{-1}})$ and $H_1^{g^{-1}} \in \Gamma$ from the definitions in Notation 2.3.4 and Notation 2.3.5. Then by 2.3.7.1, $H_1^{g^{-1}} \in \Gamma_0^e$, contrary to the hypothesis of this section. This contradiction finally completes the proof of the claim.

By the claim, K is a component of $C_G(i)$ for each involution $i \in S_C$. Further $K \trianglelefteq C_G(i)$ by 1.2.1.3. Recall $t^x = z_K$ and $S_C S_C^x = S_C \times S_C^x$, so for any $i \in S_C$ distinct from t, $i^x \notin t^x S_C = z_K S_C$. Therefore from the 2-local structure of $Aut(A_6)$, $C_K(i^x) \not\leq S_K$. Hence as $S = C_{KS}(t^x)$ is a maximal subgroup of KS,

$$KS = \langle C_{KS}(t^x), C_{KS}(i^x) \rangle \leq N_G(K^x)$$

using the claim. By symmetry, K^x acts on K, and $K \neq K^x$ as t^x centralizes K^x but not K. Therefore $[K, K^x] = 1$, a contradiction as $m_{2,3}(KK^x) \leq 2$ since G is quasithin. This contradiction completes the proof of 2.5.26. $\qquad \square$

LEMMA 2.5.27. $S_C \cong \mathbf{Z}_4$, $\mathbf{Z}_8$, or Q_8.

PROOF. By 2.5.26, $m_2(S_C) = 1$; by 2.5.24, S_C is not elementary abelian; and by 2.5.5.1, $S_C \cong S_C^x$ is isomorphic to a subgroup of $\bar{S}$. Thus the lemma holds as the three groups listed in the lemma are the only subgroups X of $\bar{S} \in Syl_2(Aut(A_6))$ of 2-rank 1 with $\Phi(X) \neq 1$. $\qquad \square$

We are now ready to complete the proof of Theorem 2.1.1.

By 2.5.27 there is a cyclic subgroup V of S_C of order 4 normal in S. Let Y be cyclic of order 4 in S_K, and S_0 the preimage in S of the subgroup generated by the transposition in $C_{\bar{S}}(\bar{S}_K)$. As $V \trianglelefteq S$, $V^x \trianglelefteq S$, so $\bar{V}^x \trianglelefteq \bar{S}$ and hence $\bar{V}^x/\langle \bar{z} \rangle \leq Z(\bar{S}/\langle \bar{z} \rangle) = \bar{Y} \bar{S}_0/\langle \bar{z} \rangle$. Therefore $\bar{V}^x \bar{S}_0 = \bar{Y} \bar{S}_0$. Let E be a 4-subgroup of S_K and $e \in E - \langle z_K \rangle$. As $\bar{V}^x \bar{S}_0 = \bar{Y} \bar{S}_0$ and $\bar{e}$ inverts $\bar{Y}$, $\bar{e}$ inverts $\bar{V}^x$, and hence e inverts V^x and centralizes S_C. Therefore e^x inverts V and centralizes S_C^x, so $e^x \notin S_K$ as S_K centralizes V.

As $S_K \trianglelefteq S$ and x acts on S, $S_K \cap S_K^x \trianglelefteq S$. However $t^x = z_K$ by 2.5.25, so

$$Z(S) \cap S_K \cap S_K^x \leq Z(S_K) \cap Z(S_K^x) = \langle t^x \rangle \cap \langle t \rangle = 1,$$

and hence $S_K \cap S_K^x = 1$. Thus $[S_K, e^x] \leq [S_K, S_K^x] \leq S_K \cap S_K^x = 1$, so $\bar{e}^x \in \Omega_1(C_{\bar{S}}(\bar{S}_K)) = \bar{S}_0\langle \bar{z} \rangle$. Hence as $e \in K$ centralizes S_C,

$$\bar{S}_C^x \leq C_{\bar{S}}(\bar{e}^x) = C_{\bar{S}}(\bar{S}_0) = \bar{S}_0 \times \bar{S}_K \cong \mathbf{Z}_2 \times D_8.$$

Thus as $S_C^x \cong \bar{S}_C^x$ by 2.5.5.1, and $\mathbf{Z}_2 \times D_8$ contains no Q_8 or $\mathbf{Z}_8$ subgroups, we conclude from 2.5.27 that $S_C = V \cong \mathbf{Z}_4$, and hence $|S| = 2^7$.

Next $A := E \times E^x = \langle t, t^x, e, e^x \rangle \cong E_{16}$, and

$$N_H(A) = \langle e^x, V \rangle \times N_K(E) \cong D_8 \times S_4,$$

as e^x inverts V and centralizes $N_K(E)$. It follows that $N_{H^x}(A) \cong D_8 \times S_4$ and $I := \langle N_H(A), x \rangle$ acts on A. Now $N_S(A) = N_S(E) \in \mathcal{U}^*(H)$ by 2.5.23, so $(N_S(A), N_H(A)) \in \mathcal{U}(I) \subseteq \mathcal{U}(N_G(A))$ from the definitions in Notation 2.3.4. As $T \cap I$ contains $\langle N_S(A), x \rangle$ of order $2^7 = |S|$ where $S \in Syl_2(H)$ for $H \in \Gamma^*$, and U has maximal order in $\mathcal{U}$, from the maximality of these groups in the definition of Γ^* in Notation 2.3.5, also $N_G(A) \in \Gamma^* \subseteq \Gamma_0$. This is impossible: for $z \in A$, so that $A \in \mathcal{S}_2^e(G)$ by 1.1.4.2; hence $N_G(A) \in \mathcal{H}^e$, so that $N_G(A) \in \Gamma_0^e$, contradicting our hypothesis in this section that $\Gamma_0^e = \emptyset$.

This contradiction completes the proof of Theorem 2.1.1.

CHAPTER 3

Determining the cases for $L \in \mathcal{L}_f^*(G, T)$

By Theorem 2.1.1, we may assume in the remainder of the proof of our Main Theorem that the Sylow 2-subgroup T of our QTKE-group G is contained in at least two distinct maximal 2-local subgroups. Thus we may implement the Thompson amalgam strategy described in the outline in the Introduction to Volume II: We choose $M \in \mathcal{M}(T)$ to contain a uniqueness subgroup of the sort considered in 1.4.1, and choose a 2-local subgroup H not contained in M. Indeed we may choose H minimal subject to this constraint:

DEFINITION 3.0.1. $\mathcal{H}_*(T, M)$ denotes the members of $\mathcal{H}(T)$ which are minimal subject to not being contained in M.

In this chapter, we establish two important technical results, and define and begin to analyze the Fundamental Setup, which will occupy us for most of the proof of the Main Theorem.

We begin in section 3.1 by proving Theorem 3.1.1 and various corollaries of that result. Theorem 3.1.1 ensures that suitable pairs of subgroups are contained in a common 2-local subgroup of G. We appeal to this theorem and its corollaries many times during the proof of the Main Theorem, but most particularly in applying Stellmacher's qrc-lemma D.1.5, and in proving the main result of section 3.3.

In section 3.2 we define the Fundamental Setup and use the qrc-lemma to determine the cases that can arise there. A discussion of this important part of the proof can be found in the introduction to section 3.2.

Finally in section 3.3, we prove that if L is in $\mathcal{L}^*(G, T)$ or $\Xi^*(G, T)$ with $M :=$ $!\mathcal{M}(\langle L, T \rangle)$ as in 1.4.1, then $N_G(T) \leq M$. We use this result often, most frequently via its important consequence that each $H \in \mathcal{H}_*(T, M)$ is a minimal parabolic in the sense of Definition B.6.1.

3.1. Common normal subgroups, and the qrc-lemma for QTKE-groups

In this section we assume G is a simple QTKE-group, $T \in Syl_2(G)$, $Z :=$ $\Omega_1(Z(T))$, and $M \in \mathcal{M}(T)$. We derive various consequences for QTKE-groups from Theorem C.5.8 of Volume I, in one case by applying the result in conjunction with the qrc-lemma D.1.5. We begin with a restatement of Theorem C.5.8.

THEOREM 3.1.1. *Assume that* $M_0, H \in \mathcal{H}(T)$, T *is in a unique maximal subgroup of* H, *and* $1 \neq R \leq T$ *with* $R \in Syl_2(O^2(H)R)$ *and* $R \trianglelefteq M_0$. *Then there is* $1 \neq R_0 \leq R$ *with* $R_0 \trianglelefteq \langle M_0, H \rangle$.

PROOF. We verify the hypotheses of Theorem C.5.8, most particularly Hypothesis C.5.1: As $H \in \mathcal{H}(T)$, $F^*(H) = O_2(H)$ by 1.1.4.6, and as G is a $QTKE$-group, $m_3(H) \leq 2$. By the hypotheses of Theorem 3.1.1, T is in a unique maximal subgroup of H—completing the verification of C.5.1.1. Again by those hypotheses,

$R \trianglelefteq M_0$ and $R \in Syl_2(O^2(H)R)$, so C.5.1.2 holds. Thus Hypothesis C.5.1 is indeed satisfied, while by the hypotheses of this section, $T \in Syl_2(G)$ and G is a simple QTKE-group, supplying the remaining hypotheses of Theorem C.5.8. Of course the conclusion of C.5.8 is the existence of a nontrivial normal subgroup of $\langle M_0, H \rangle$ contained in R, so Theorem 3.1.1 is established. $\square$

We sometimes use the following easy observation:

LEMMA 3.1.2. *If* $T \leq Y \leq H \in \mathcal{H}(T)$, *then also* $Y \in \mathcal{H}(T) \subseteq \mathcal{H}^e$.

PROOF. As $H \in \mathcal{H}$, $O_2(H) \neq 1$. Further $T \in Syl_2(Y)$, so $1 \neq O_2(H) \leq O_2(Y)$ by A.1.6, and hence also $Y \in \mathcal{H}$. Finally $Y \in \mathcal{H}^e$ by 1.1.4.6. $\square$

In view of Theorem 2.1.1, we may assume that our fixed $M \in \mathcal{M}(T)$ is not the unique maximal 2-local subgroup of G containing T, so that $\mathcal{H}_*(T, M)$ is nonempty. During the remainder of our proof of our Main Theorem, we typically implement the Thompson amalgam strategy exploiting the interaction of M with some member of $\mathcal{H}_*(T, M)$.

Recall also from Definition B.6.2 that a subgroup X of G is in $\mathcal{U}_G(T)$ if T is contained in a unique maximal subgroup of X; and X is in $\hat{\mathcal{U}}_G(T)$ if $X \in \mathcal{U}_G(T)$ and T is not normal in X. In the terminology of Definition B.6.1, the members of $\hat{\mathcal{U}}_G(T)$ are called *minimal parabolics*.

As mentioned in the Introduction to Volume II and at the start of this chapter, once we have established Theorem 3.3.1 in the final section of this chapter, part (2) of the next lemma will ensure that members of $\mathcal{H}_*(T, M)$ are minimal parabolics for suitable choices of M.

LEMMA 3.1.3. *Assume* $H \in \mathcal{H}_*(T, M)$. *Then*

(1) $H \cap M$ *is the unique maximal subgroup of* H *containing* T. *That is,* $\mathcal{H}_*(T, M) \subseteq \mathcal{U}_G(T)$.

(2) If $N_G(T) \leq M$ *or* H *is not 2-closed, then* $H \in \hat{\mathcal{U}}_G(T)$. *Thus* H *is a minimal parabolic, and so is described in B.6.8, and in E.2.2 if* H *is nonsolvable.*

PROOF. Since $H \not\leq M$, $T \leq H \cap M < H$. If $T \leq Y < H$, then by 3.1.2, $Y \in \mathcal{H}(T)$; thus $Y \leq H \cap M$ by the minimality of H in the definition of $\mathcal{H}_*(T, M)$, so that (1) holds. If $N_G(T) \leq M$ or H is not 2-closed, then T is not normal in H, so (2) holds. $\square$

LEMMA 3.1.4. *Assume that* $H \leq G$ *and* V *is an elementary abelian 2-subgroup of* $H \cap M$ *such that* V *is a TI-set under* M *with* $N_G(V) \leq M$ *and* $H \leq N_G(U)$ *for some* $1 < U \leq V$. *Then*

(1) $H \cap M = N_H(V)$.

(2) $H \not\leq M$ *iff* $H \not\leq N_G(V)$, *in which case* $H \cap M = N_H(V) < H$.

PROOF. As we assume $N_G(V) \leq M$, $N_H(V) \leq H \cap M$. Conversely as V is a TI-set in M, $N_M(U) \leq N_M(V)$. Then as $H \leq N_G(U)$ by hypothesis, $H \cap M = H \cap N_M(U) \leq N_H(V)$, so that (1) holds. Then (2) follows. $\square$

Usually we will apply Theorem 3.1.1 under one of the hypotheses in Hypothesis 3.1.5—which will hold in the Fundamental Setup (3.2.1).

Recall from Definition B.2.11 the set $\mathcal{R}_2(M_0)$ of 2-reduced modules for M_0 from the Introduction to Volume II, and see the discussion in chapter B of Volume I.

HYPOTHESIS 3.1.5. $T \leq M_0 \leq M$, $H \in \mathcal{H}_*(T, M)$, and $V \in \mathcal{R}_2(M_0)$ such that $R := O_2(M_0) = C_T(V)$. Further either

(I) $H \cap M \leq N_G(O^2(M_0))$, or
(II) $H \cap M \leq N_G(V)$.

Observe that Hypothesis 3.1.5 includes the hypotheses of Theorem 3.1.1, other than the condition that $R \in Syl_2(O^2(H)R)$: For example, T is in a unique maximal subgroup of H by 3.1.3.1.

The next result is a corollary to Stellmacher's qrc-lemma D.1.5 using Theorem 3.1.1.

THEOREM 3.1.6. *Assume Hypothesis 3.1.5. Then one of the following holds:*

(1) There exists $1 \neq R_0 \leq R$ such that $R_0 \unlhd \langle M_0, H \rangle$.
(2) $V \not\leq O_2(H)$ and $\hat{q}(M_0/C_{M_0}(V), V) \leq 2$. If in addition V is a TI-set under M, then in fact $\hat{q}(M_0/C_{M_0}(V), V) < 2$.
(3) $q(M_0/C_{M_0}(V), V) \leq 2$.

PROOF. Assume that conclusion (1) does not hold. We verify Hypothesis D.1.1, with M_0, H in the roles of "G_1, G_2": By Hypothesis 3.1.5, T lies in both M_0 and H—so it is Sylow in both, since it is Sylow in G. By 3.1.5, $V \in \mathcal{R}_2(M_0)$ and $H \in \mathcal{H}_*(T, M)$, so that $H \cap M$ is the unique maximal overgroup of T in H by 3.1.3.1, giving (1) of D.1.1. By 3.1.5, $R = O_2(M_0) = C_T(V)$, which is (2) of D.1.1. Finally, our assumption that (1) fails is (3) of D.1.1. Thus we may apply the qrc-Lemma D.1.5, to see (on combining its conclusions (2) and (4) in conclusion (ii) below) that one of the following holds:

(i) $V \not\leq O_2(H)$.
(ii) $q(M_0/C_{M_0}(V), V) \leq 2$.
(iii) V is a dual FF-module.
(iv) $R \cap O_2(H) \unlhd H$, and $U := \langle V^H \rangle$ is elementary abelian.

Observe in case (ii) that conclusion (3) of Theorem 3.1.6 holds, so we may assume that (ii) fails, and it remains to treat cases (i), (iii), and (iv).

Suppose case (iii) holds and let V^* be the dual of V as an M_0-module. Then V^* is a faithful $\mathbf{F}_2$-module for $Aut_{M_0}(V^*) \cong Aut_{M_0}(V)$, so $O_2(Aut_{M_0}(V^*)) = 1$ since $V \in \mathcal{R}_2(M_0)$. As (iii) holds, $J^* := J(Aut_{M_0}(V^*), V^*) \neq 1$. Also M_0 is an SQTK-group using our QTKE-hypothesis, and hence so is the preimage in M_0 of J^*. Therefore Hypothesis B.5.3 is satisfied with J^*, V^* in the role of "G, V", so we may apply B.5.13 to see that conclusion (3) again holds, completing the treatment of case (iii).

As we are assuming that (ii) fails, $q(M_0/C_{M_0}(V), V) > 1$, so we may apply D.1.2. By (2) and (3) of D.1.2,

$$J(T) = J(R) \not\leq O_2(H).$$

By (4) of D.1.2, H is a minimal parabolic in the sense of Definition B.6.1, and is described in B.6.8.

In case (i), we argue that conclusion (2) holds: We will apply E.2.13, so we need to verify that Hypothesis E.2.8 is satisfied with $H \cap M$ in the role of "M", and that $F^*(H) = O_2(H)$. We just saw that H is a minimal parabolic in the sense of Definition B.6.1, and is described in B.6.8. As $H \in \mathcal{H}(T)$, using our QTKE-hypothesis and 1.1.4.6, H is an SQTK-group with $F^*(H) = O_2(H)$. By

Hypothesis 3.1.5, $V \in \mathcal{R}_2(M_0)$, so V is elementary abelian, normal in T, and contained in $\Omega_1(Z(O_2(M_0)))$. Further $T \leq M_0 \leq M$ so that $O_2(M) \leq O_2(M_0)$ by A.1.6; and $M \in \mathcal{M}(T) \subseteq \mathcal{H}^e$ since G is of even characteristic. Therefore $V \leq C_M(O_2(M)) \leq O_2(M)$, and hence $V \leq O_2(H \cap M)$. Finally $V \not\leq O_2(H)$ in case (i), and $O_2(H) = \ker_{H \cap M}(H)$ by B.6.8.5. This completes the verification of the hypothesis of E.2.13. Hence we conclude from E.2.13.3, that $\hat{q}(Aut_H(V), V) \leq 2$. Therefore since T is Sylow in both H and M_0, $\hat{q}(M_0/C_{M_0}(V), V) \leq 2$. Further if V is a TI-set under M, then we have the hypotheses for E.2.15, so that result shows that $\hat{q}(Aut_H(V), V) < 2$, and hence $\hat{q}(M_0/C_{M_0}(V), V) < 2$. Thus (2) holds, as claimed.

Thus we may assume that cases (i)–(iii) do not hold. In particular, case (iv) holds; and as (i) fails, now $V \leq O_2(H)$. By our observation following Hypothesis 3.1.5, it suffices to prove that $R \in Syl_2(O^2(H)R)$, since then Theorem 3.1.1 shows that conclusion (1) of Theorem 3.1.6 holds.

Set $Q_H := O_2(H)$, $K := O^2(H)$, and $H^* := H/Q_H$. As case (iv) holds, $Q := R \cap Q_H \unlhd H$, so as $C_T(V) = R$ and $V \leq Q_H$ by the previous paragraph, $V \leq Z(Q)$. Therefore $U \leq Z(Q)$.

We saw earlier that $J(T) = J(R) \not\leq Q_H$, and H is a minimal parabolic described in B.6.8. Now by Hypothesis 3.1.5, $Q_H \leq T \leq M_0 \leq N_G(R)$, so $[Q_H, J(R)] \leq Q_H \cap R = Q$, and hence $[K, J(R)]J(R)$ centralizes Q_H/Q. Next $[K, J(R)]J(R)$ is normal in $KT = H$, but $J(R) \not\leq Q_H$, so $K \leq [K, J(R)]J(R)$ by B.6.8.4, and then K centralizes Q_H/Q. Therefore $[O_2(K), K] \leq Q$.

If K centralizes U then K centralizes V, so $C_T(V) = R$ is Sylow in $C_G(V)$ and hence R is Sylow in KR, which as we observed earlier suffices to complete the proof. Thus we may assume that K does not centralize U. Then $C_H(U) \leq \ker_{H \cap M}(H)$ and $C_T(U) = C_{Q_H}(U)$ by B.6.8.6.

As $J(R) \not\leq Q_H$, there is some $A \in \mathcal{A}(R)$ with $A^* \neq 1$. As $A \leq R$ and $U \leq Z(Q)$, $A \cap Q_H = A \cap Q \leq C_A(U)$, so $A \cap Q_H = C_A(U)$ by the previous paragraph. Then as $A \in \mathcal{A}(R)$, $r_{A^*, U} \leq 1$ by B.2.4.1. Now U might not be in $\mathcal{R}_2(H)$, but each nontrivial H-chief section W on U is an irreducible for $H/C_H(W)$, so that $O_2(H/C_H(W)) = 1$. Furthermore $C_H(W) \leq \ker_{H \cap M}(H)$ and $C_T(W) = C_{Q_H}(W)$ by B.6.8.6, so $m(A^*) = m(Aut_A(W))$ and hence $r_{Aut_A(W), W} \leq r_{A^*, U} \leq 1$. Therefore W is an FF-module for $Aut_H(W)$. Hence by B.6.9 and E.2.3, $m(W/C_W(A^*)) = m(A^*)$, $K = K_1$ or $K_1 K_2$, and $[W, K_i]$ is the natural module for $K_i^* \cong L_2(2^n)$, A_3, or A_5. Furthermore as $m(U/C_U(A)) \leq m(A^*) = m(W/C_W(A^*))$, we conclude K_i has a unique noncentral chief factor $\tilde{U}_i$ on U, where $\tilde{U}_i = U_i/C_{U_i}(K_i)$ is the natural module for K_i^*, and $[U, K_i] = U_i$.

Set $B := H \cap M$ and observe that B is solvable: This is clear if H is solvable, while if H is not solvable then by E.2.2 and the previous paragraph, $B^* \cap K^*$ is a Borel subgroup of K^*, and in particular B is solvable. By Hypothesis 3.1.5, either (I) holds and B normalizes $L := O^2(M_0)$, or (II) holds and B normalizes V. In case (I), let $D := C_B(L/O_2(L))$, and in case (II), let $D := C_B(V)$. Then B normalizes D in either case.

We claim that R is Sylow in D, and $D \unlhd B$: In case (II), $R = C_T(V)$ is Sylow in $C_G(V)$, and hence also in $C_B(V) = D$. As B normalizes V in (II), $D \unlhd B$. In case (I), we apply parts (4) and (5) of A.4.2 with L, M_0 in the roles of "X, M", to see that $R = O_2(M_0)$ is Sylow in $C_{M_0}(L/O_2(L))$. Hence R is also Sylow in $C_B(L/O_2(L)) = D$. As B normalizes L in (I), $D \unlhd B$.

Let Y denote a Hall $2'$-subgroup of B. As $D \trianglelefteq B$ by the previous paragraph, $Y_D := Y \cap D$ is also Hall in D, so $D = Y_D R = R Y_D$. Further $Y \le B \le N_G(D)$, so

$$YR = YY_DR = YD = DY = RY_DY = RY.$$

Then R is Sylow in the group YR, and Y normalizes $O_2(YR) \le R$.

We claim that $T \cap K \le R$; this is the crucial step in showing that R is Sylow in RK, and hence in completing the proof. Since T is transitive on the groups K_i, it suffices to show that $T_i := T \cap K_i \le R$ for some i. Let $Q_i := O_2(K_i)$, $T_0 := N_T(K_i)$, $Y_i := Y \cap K_i$, and $\overline{K_iT_0} := K_iT_0/O_2(K_iT_0)$. Then $A \le T_0$ by B.1.5.4, and as $A \not\le Q_H$, while K_i^* is quasisimple or of order 3, we may choose i so that $K_i = [K_i, A]$. Next

$$P_i := [Q_i, K_i] \le Q_i \le Q \le R. \tag{$*$}$$

But if $K_i^* \cong A_3$ then $P_i = Q_i \in Syl_2(K_i)$ since $K_i = O^2(K_i)$, so that $T_i = P_i \le R$ by $(*)$, as claimed.

Suppose next that $\tilde{U}_i$ is the natural module for $K_i^* \cong L_2(2^n)$ with $n > 1$. Then by B.4.2.1, the FF^*-offender $\bar{A}$ is Sylow in $\bar{K}_i$, so that $T_i \le J(R)Q_i$ with $J(R) \le O_2(Y_iT_0)$. Thus $J(R) \le O_2(YT_0)$, so

$$J(R) \le O_2(YT_0) \cap YR \le O_2(YR) \le R,$$

so Y acts on $J(O_2(YR)) = J(R)$ using B.2.3.3, and hence again using $(*)$,

$$T_i = [J(R), Y_i]P_i \le RP_i \le R.$$

Finally if U_i is the natural module for $K_i \cong A_5$, then by B.3.2.4, the FF^*-offender $\bar{A}$ is generated by one or two transpositions. Thus $[A, T_i] \le R \cap K_i =: R_i$, so as $[A, T_i] \not\le Q_i$, $(*)$ says

$$T_i = \langle R_i^{Y_i} \rangle P_i = R_iP_i \le R.$$

We have established the claim that $T \cap K \le R$. Since T is Sylow in H and $K \trianglelefteq H$, $T \cap K$ is Sylow in K, so R is Sylow in RK, completing the proof of Theorem 3.1.6. $\qquad\square$

The next result is another corollary of Theorem 3.1.1, in the same spirit as Theorem 3.1.6. Recall that Z is $\Omega_1(Z(T))$, and the Baumann subgroup of T from Definition B.2.2 is $\mathrm{Baum}(T) = C_T(\Omega_1(Z(J(T))))$.

LEMMA 3.1.7. *Assume Hypothesis 3.1.5, with $J(T) \le R$. Then either*

(1) $Z \le Z(H)$ and $Z(M_0) = 1$, or

(2) There is $1 \ne R_0 \le R$ with $R_0 \trianglelefteq \langle M_0, H \rangle$.

PROOF. By hypothesis $J(T) \le R = C_T(V)$. Then $J(T) = J(R)$ and $S := \mathrm{Baum}(T) = \mathrm{Baum}(R)$ by B.2.3.5 with V in the role of "U". Therefore if $J(T) \trianglelefteq H$, then (2) holds with $J(T)$ in the role of "R_0". Thus we may assume $J(T)$ is not normal in H, so H is not 2-closed. Hence $H \in \hat{\mathcal{U}}_G(T)$ and H is described in B.6.8 by 3.1.3.

Suppose $Z \le Z(H)$. If $Z(M_0) = 1$ then conclusion (1) holds, so we may assume $Z(M_0) \ne 1$. By Hypothesis 3.1.5, $M_0 \in \mathcal{H}(T)$, and hence $M_0 \in \mathcal{H}^e$ by 1.1.4.6. Therefore $Z(M_0)$ is a 2-group, so $\Omega_1(Z(M_0)) \le Z \le Z(H)$, and hence conclusion (2) holds with $\Omega_1(Z(M_0))$ in the role of "R_0".

Thus we may assume that $Z \not\le Z(H)$. Let $U_H := \langle Z^H \rangle$ and $K := O^2(H)$. As $H = KT$, $K \not\le C_H(Z)$, so $K \not\le C_H(U_H)$. We saw in the previous paragraph that

$H \in \mathcal{H}^e$, so $U_H \in \mathcal{R}_2(H)$ by B.2.14. As $K \not\le C_H(U_H)$, $C_H(U_H) \le \ker_{M \cap H}(H)$ by B.6.8.6, and $C_H(U_H)$ is 2-closed by B.6.8.5. So as $J(T)$ is not normal in H, $J(T) \not\le C_H(U_H)$. Hence by E.2.3, $K = K_1 \cdots K_s$, with $s = 1$ or 2, T permutes the K_i transitively, $K_1/C_{K_1}(U_H) \cong L_2(2^n)$, A_3, or A_5, $S = \text{Baum}(T) = \text{Baum}(R)$ acts on K_i, and either S is Sylow in $K_i S$, or $[U_H, K_i]$ is the A_5-module for $K_i/O_2(K_i)$. In the latter case, by E.2.3.3, S is of index 2 in a Sylow 2-group S_i of SK_i and $S_i \le \langle S^{K_i \cap M} \rangle$. Then by an argument near the end of the proof of 3.1.6, $S_i \le R$. So in either case, $R \cap K \in Syl_2(K)$, and hence $R \in Syl_2(KR)$. As we observed after Hypotheses 3.1.5, this is sufficient to establish the hypotheses of Theorem 3.1.1. Hence conclusion (2) holds by that result, completing the proof. $\qquad \square$

Finally we extend Theorems 3.1.6 and 3.1.7, by bringing uniqueness subgroups into the picture:

THEOREM 3.1.8. *Assume* $L_0 = O^2(L_0) \trianglelefteq M$ *with* $M = !\mathcal{M}(L_0 T)$, *and* $V \in \mathcal{R}_2(L_0 T)$ *such that* $O_2(L_0 T) = C_T(V)$. *Then*

 (1) $\hat{q}(L_0 T/C_{L_0 T}(V), V) \le 2$.
 (2) Either

  (i) $q(L_0 T/C_{L_0 T}(V), V) \le 2$, *or*
  (ii) For each $H \in \mathcal{H}_*(T, M)$, $V \not\le O_2(H)$. *If in addition* V *is a TI-set under* M, *then* $\hat{q}(L_0 T/C_{L_0 T}(V), V) < 2$.

 (3) Either:

  (i) $J(T) \not\le C_T(V)$, *so* V *is an FF-module for* $L_0 T/C_{L_0 T}(V)$, *or*
  (ii) $J(T) \le C_T(V)$, $Z \le Z(H)$ *for each* $H \in \mathcal{H}_*(T, M)$, *and* $Z(L_0 T) = 1$.

PROOF. Set $M_0 := L_0 T$, and consider any $H \in \mathcal{H}_*(T, M)$. Observe that case (I) of Hypothesis 3.1.5 holds. Further as $M = !\mathcal{M}(M_0)$ and $H \not\le M$, $O_2(\langle M_0, H \rangle) = 1$. In particular, neither conclusion (1) of Theorem 3.1.6, nor conclusion (2) of 3.1.7 holds. Therefore since $\hat{q}(Aut_{L_0 T}(V), V) \le q(Aut_{L_0 T}(V), V)$ from the definitions B.1.1 and B.4.1, we conclude from Theorem 3.1.6 that conclusions (1) and (2) of Theorem 3.1.8 hold.

If $J(T) \not\le C_T(V)$, then conclusion (i) of (3) holds by B.2.7. On the other hand, if $J(T) \le C_T(V)$, then by the previous paragraph, conclusion (1) of 3.1.7 holds, so conclusion (ii) of (3) is satisfied. $\qquad \square$

In certain situations we will require a refinement of the qrc-Lemma making use of information in D.1.3 and definition D.2.1.

LEMMA 3.1.9. *Assume case (II) of Hypothesis 3.1.5 holds, with* $H \in \mathcal{H}_*(T, M)$. *Further assume:*

 (a) $q(M_0/C_{M_0}(V), V) = 2$.
 (b) $M = !\mathcal{M}(M_0)$.
 (c) $V \le O_2(H)$.
 (d) V *is not a dual FF-module for* M_0.
Set $U_H := \langle V^H \rangle$ *and* $Z := \Omega_1(Z(T))$. *Then* U_H *is elementary abelian, and*

 (1) H *has exactly two noncentral chief factors* U_1 *and* U_2 *on* U_H.
 (2) There exists $A \in \mathcal{A}(T) = \mathcal{A}(C_T(V))$ *with* $A \not\le O_2(H)$, *and for each such* A *chosen with* $AO_2(H)/O_2(H)$ *minimal,* A *is quadratic on* U_H.
 (3) For A *as in (2), set* $B := A \cap O_2(H)$. *Then* $B = C_A(U_i)$,

$$2m(A/B) = m(U_H/C_{U_H}(A)) = 2m(B/C_B(U_H)),$$

$$2m(B/C_B(V^h)) = m(V^h/C_{V^h}(B))$$

for each $h \in H$ with $[V^h, B] \neq 1$, $m(A/B) = m(U_i/C_{U_i}(A))$, and $C_{U_H}(A) = C_{U_H}(B)$.

(4) Define

$$m := \min\{m(D) : D \in \mathcal{Q}(Aut_M(V), V)\}.$$

Then $m(A/B) \geq m$.

(5) Assume $O^2(C_M(Z)) \leq C_M(V)$. Then $H/C_H(U_i) \cong S_3$, S_3 wr $\mathbf{Z}_2$, S_5, or S_5 wr $\mathbf{Z}_2$, with U_i the direct sum of the natural modules $[U_i, F]$, as F varies over the S_3-factors or S_5-factors of $H/C_H(U_i)$. Further $J(H)C_H(U_i)/C_H(U_i) \cong S_3$, $S_3 \times S_3$, S_5, or $S_5 \times S_5$, respectively.

(6) Assume that each $\{2,3\}'$-subgroup of $C_M(Z)$ permuting with T centralizes V, $m \geq 2$, and each subgroup of order 3 in $C_M(Z)$ has at least three noncentral chief factors on V. Then $H/C_H(U_i) \cong S_3$ wr $\mathbf{Z}_2$.

PROOF. Observe that hypothesis (a) implies:

(a') V is not an FF-module for M_0.

We will first show that (a') and (b)–(d) lead to the hypotheses of the qrc-lemma D.1.5.

Set $R := C_T(V)$. By (a'), $J(T) \leq C_T(V) = R$. Thus the hypothesis of Theorem 3.1.7 holds, and by B.2.3.3, $J(T) = J(R)$.

Next by (b), there is no $1 \neq R_0 \leq R$ with $R_0 \trianglelefteq \langle M_0, H \rangle$. Thus conclusion (1) of Theorem 3.1.7 holds, so that $Z \leq Z(H)$, and in particular $H \cap M \leq C_G(Z)$. Further $J(T)$ is not normal in H, so we conclude from 3.1.3.2 that H is a minimal parabolic in the sense of Definition B.6.1. Also (as at the start of the proof of Theorem 3.1.6) Hypothesis D.1.1 holds with M_0, H in the roles of "G_1, G_2". Thus we can appeal to results in section D.1, and in particular to the qrc-lemma D.1.5.

Observe that (c) rules out conclusion (1) of D.1.5, and (a') and (d) rule out conclusions (2) and (3), respectively. We rule out conclusion (5) of D.1.5 just as in the proof of 3.1.6, using (c) to eliminate case (i) in that proof. Thus conclusion (4) of D.1.5 holds, so U_H is abelian, and H has more than one noncentral chief factor on U_H. This last condition together with (c) and (a') are the hypotheses of D.1.3. Furthermore (a') gives the hypothesis of D.1.2, so by part (4) of that result, H is a minimal parabolic in the sense of Definition B.6.1, and is described in B.6.8.

Next (a) supplies the hypothesis of part (3) of D.1.3. Then (1) follows from D.1.3.3. We saw earlier that $J(T) = J(R)$, so by D.1.3.2 there is $A \in \mathcal{A}(T)$ with $A \not\leq O_2(H)$ and A quadratic on U_H. Indeed from the proof of D.1.3.2, our choice of $A \in \mathcal{A}(T) - \mathcal{A}(O_2(H))$ with $AO_2(H)/O_2(H)$ minimal guarantees that A is quadratic on U_H, and that $B := A \cap O_2(H) = C_A(U_i)$ for $i = 1, 2$. Thus (2) holds, and D.1.3 establishes the remaining assertions of (3).

By (3), $m(V^h/C_{V^h}(B)) = 2m(B/C_B(V^h))$, and B is quadratic on V^h by (2), so $Aut_B(V^h) \in \mathcal{Q}(Aut_{M^h}(V^h), V^h)$ by (a). Thus

$$m \leq m(B/C_B(V^h)) \leq m(B/C_B(U_H)) = m(A/B),$$

establishing (4).

Set $H^* := H/C_H(U_i)$. As H is irreducible on U_i, $O_2(H^*) = 1$, so $U_i \in \mathcal{R}_2(H)$. As $B = C_A(U_i)$ and $m(A/B) = m(U_i/C_{U_i}(A))$ by (3), $A^* \cong A/B$ is an FF*-offender on U_i. Therefore by B.6.9, $H = YT$ where $Y := J(H, V)$, $Y^* = Y_1^* \times \cdots \times Y_s^*$, and $U_i = U_{i,1} \oplus \cdots \oplus U_{i,s}$ with $U_{i,j}$ the natural module for $Y_j^* \cong L_2(2^n)$ or S_{2^k+1}. By

A.1.31.1, $s \leq 2$; by E.2.3.2, if Y_j^* is a symmetric group, then Y_j^* is S_3 or S_5; and in any case $H \cap M$ is the product of T with the preimages of the Borel subgroups over $T^* \cap Y_j^*$ in Y_j^*. Further if $s = 2$, then as U_i is irreducible under H, $\{U_{i,1}, U_{i,2}\}$ is permuted transitively by T.

Assume for the moment that $Y_j^* \cong L_2(2^n)$ with $n > 1$ and some $U_{i,j}$ the natural module. Then by B.4.2.1, A^* is Sylow either in Y^* or in some Y_j^*. Now A^* is also an FF*-offender on U_{3-i}, and B.4.2 says that the only other possible FF*-module for Y_j^* is the A_5-module when $n = 2$, whereas the FF*-offenders on that module are not Sylow in Y_j^*. Thus in any case U_1 is Y-isomorphic to U_2.

Let $K := O^2(H)$, and W an H-submodule of U_H maximal subject to $U_0 := [U_H, K] \not\leq W$. Set $U_H^+ := U_H/W$. Thus $U_0^+ \neq 0$, H is irreducible on U_0^+, and $C_{U_H^+}(K) = 0$. As $U_H = \langle V^H \rangle$, $U_H^+ = \langle V^{+H} \rangle$, so $V_0^+ := C_{V^+}(T) \neq 0$. As $C_{U_H^+}(K) = 0$, $V_0^+ \leq U_0^+$ using Gaschütz's Theorem A.1.39. As H is irreducible on U_0^+, we may take $U_1 = U_0^+$. Further

$$0 \neq V_0^+ \leq C_{U_1}(J(R)^*), \tag{$*$}$$

and as case (II) of Hypothesis 3.1.5 holds,

$$H \cap M \text{ acts on } V^+. \tag{$**$}$$

Let X denote a Cartan subgroup of $Y_j \cap M$.

Suppose that $Y_j^* \cong L_2(2^n)$ with $n > 1$ and $U_{1,j}$ the natural module. Then as $J(R)^* \in Syl_2(Y^*)$, we conclude from ($*$) and ($**$) that

$$V_j^+ := V^+ \cap U_{1,j} = C_{U_{1,j}}(J(R)^*) \tag{!}$$

is the $J(R)^*$-invariant 1-dimensional $\mathbf{F}_{2^n}$-subspace of $U_{1,j}$. In particular X acts faithfully on V. This is a contradiction to the hypotheses of (5), and under the hypotheses of (6), $O^3(X) = 1$ so $n = 2$. But now V_j^+ is the only noncentral chief factor for X on V^+, and the image of $[V \cap W, X]$ in $U_{2,j}$ is contained in $C_{U_{2,j}}(J(R)^*)$, so X has a single noncentral chief factor on $V \cap W$. Thus X has just two noncentral chief factors on V, contrary to the hypotheses of (6).

We have completed the proof of (5), so we may assume the hypotheses of (6) with $U_{i,j}$ the natural module for $Y_j^* \cong S_3$ or S_5. By (4) and the hypothesis of (6), $m(A^*) \geq m \geq 2$, so H^* is not S_3. Thus we may assume $Y_j^* \cong S_5$. Then from the description of FF^*-offenders in B.3.2.4, $O^2((H \cap M)^*) = [O^2(H \cap M)^*, J(R)^*]$, so as $H \cap M$ acts on V and $J(R)$ centralizes V, X centralizes V, contrary to the hypotheses of (6). This completes the proof of (6). $\qquad\square$

3.2. The Fundamental Setup, and the case division for $\mathcal{L}_f^*(G, T)$

The bulk of the proof of the Main Theorem involves the analysis of various possibilities for $L \in \mathcal{L}_f^*(G, T)$. In this section we establish a formal setting for treating these subgroups, and provide the list of groups L and internal modules V which can arise in that setting. In the language of the Introduction to Volume II, this gives a solution to the First Main Problem—reducing from an arbitrary choice for L, V to the much shorter list arising in what we call below our Fundamental Setup (FSU).

In this section we assume G is a simple QTKE-group, $T \in Syl_2(G)$, $Z := \Omega_1(Z(T))$, and $M \in \mathcal{M}(T)$. The notation $Irr_+(X, V)$ and $Irr_+(X, V, Y)$ appears in Definition A.1.40. We will be primarily interested in

HYPOTHESIS 3.2.1 (Fundamental Setup (FSU)). *G is a simple QTKE-group,* $T \in Syl_2(G)$, $L \in \mathcal{L}_f^*(G, T)$ *with* $L/O_2(L)$ *quasisimple,* $L_0 := \langle L^T \rangle$, $M := N_G(L_0)$, *and* $V_\circ \in Irr_+(L_0, R_2(L_0T), T)$. *Set* $V := \langle V_\circ^T \rangle$, $V_M := \langle V^M \rangle$, $M_V := N_M(V)$, $\bar{M}_V := M_V/C_{M_V}(V)$, *and* $\tilde{V}_M := V_M/C_{V_M}(L_0)$.

In our first lemma we apply results from section D.3 to subgroups $M \in \mathcal{M}(T)$ such that M is the normalizer of one of the uniqueness subgroups constructed in chapter 1. We will also see in 3.2.3 that case (i) of 3.2.2 includes the Fundamental Setup, as the similar notation in the lemmas suggests.

LEMMA 3.2.2. *Assume there is* $M_+ = O^2(M_+) \trianglelefteq M$ *such that either*

(i) $M_+ = \langle L^T \rangle$ *for some* $L \in \mathcal{L}_f(G, T)$ *with* $L/O_2(L)$ *quasisimple, or*

(ii) $M_+ = O_{2,p}(M_+)$ *for some odd prime* p, *with* T *irreducible on* $M_+/O_{2,\Phi}(M_+)$.

Let $V_\circ \in Irr_+\big(M_+, R_2(M_+T), T\big)$ *and set* $V_M := \langle V_\circ^M \rangle$, $V := \langle V_\circ^T \rangle$, *and* $\tilde{V}_M := V_M/C_{V_M}(M_+)$. *Then*

(1) $C_{M_+}(V_M) \leq O_{2,\Phi}(M_+)$.

(2) $V_M \in \mathcal{R}_2(M)$.

(3) $V_M = [V_M, M_+]$, $\tilde{V}_M$ *is a semisimple* M_+-*module, and* M *is transitive on the* M_+-*homogeneous components of* $\tilde{V}_M$.

(4) $C_{V_M}(M_+) = \langle C_{V_\circ}(M_+)^M \rangle = \langle C_V(M_+)^M \rangle$.

(5) *If* $C_{V_\circ}(M_+) = 0$, *then* $V_\circ$ *is a TI-set under* M.

(6) *If* $C_{V_M}(M_+) \neq 0$ *and* $M = !\mathcal{M}(M_+T)$, *then* $M_+ = [M_+, J(T)]$ *and* V *is an FF-module for* M_+T.

(7) *Hypothesis D.3.1 is satisfied with* $Aut_M(V_M)$, $Aut_{M_+}(V_M)$, $V_\circ$ *in the roles of* "M, M_+, V".

(8) $V \in \mathcal{R}_2(M_+T)$ *and* $O_2(M_+T) = C_T(V)$.

(9) *Assume* $M = !\mathcal{M}(M_+T)$. *Then the hypothesis of Theorem 3.1.8 is satisfied with* M_+ *in the role of* "L_0", *and D.3.10 applies.*

PROOF. By A.1.11, $R_2(M_+T) \leq R_2(M)$. Now it is straightforward to verify that Hypothesis D.3.2 is satisfied with M, T, M_+, $R_2(M)$, 1, $V_\circ$ in the roles of "$\dot{M}$, $\dot{T}$, $\dot{M}_+$, Q_+, Q_-, V". Notice that V, V_M play the roles of "V_T, V_M" in Hypothesis D.3.2 and lemma D.3.4. Now (1) and (7) follow from parts (2) and (1) of D.3.3.

By (7), we may apply D.3.4 to $Aut_M(V_M)$; then conclusions (1)–(4) and (6) of D.3.4 imply conclusions (2)–(5) of 3.2.2.

Set $M_0 := M_+T$ and $R := O_2(M_0)$. By D.3.4.1, $O_2(M_0/C_{M_0}(V)) = 1$, so $V \in \mathcal{R}_2(M_0)$ and hence $R \leq C_T(V)$. By D.3.4.2, $C_{M_+}(V) \leq O_{2,\Phi}(M_+)$, so as $M_+ = O^2(M_0)$, $C_{M_0}(V) \leq RO_{2,\Phi}(M_+)$ and hence $R = C_T(V)$, completing the proof of (8).

Now assume that $M = !\mathcal{M}(M_+T)$. Then (9) follows from (8), so it remains to prove (6); thus we assume that $C_{V_M}(M_+) \neq 0$. Then $Z_0 := C_Z(M_+T) \neq 0$ and $Z_0 \leq Z(M_0)$. By (9) we may apply Theorem 3.1.8.3 to conclude that $J(T) \not\leq C_T(V)$. From the structure of M_+ in cases (i) and (ii) of the lemma, $\Phi(M_+/O_2(M_+))$ is the largest M_0-invariant proper subgroup of $M_+/O_2(M_+)$, so we conclude that $M_+ = [M_+, J(T)]O_2(M_+)$. Then as $M_+ = O^2(M_+)$, also $M_+ = [M_+, J(T)]$, completing the proof of (6), and hence of 3.2.2. $\qquad\square$

LEMMA 3.2.3. *Assume* $L \in \mathcal{L}_f^*(G, T)$ *with* $L/O_2(L)$ *quasisimple, and let* $L_0 := \langle L^T \rangle$. *Then* $M := N_G(L_0) \in \mathcal{M}(T)$, $M = !\mathcal{M}(L_0T)$, *and for each member* I *of*

$Irr_+(L_0, R_2(L_0T))$ *there exists* $V_\circ \in Irr_+(L_0, R_2(L_0T), T)$ *with* $V_\circ/C_{V_\circ}(L_0)$ L_0-*isomorphic to* $I/C_I(L_0)$. *In particular* L *and* $V := \langle V_\circ^T \rangle$ *satisfy the Fundamental Setup (3.2.1).*

PROOF. By 1.2.7.3, $M = !\mathcal{M}(L_0T)$. By A.1.42.2, there exists a member $V_\circ$ of $Irr_+(L_0, R_2(L_0T), T)$ with $V_\circ/C_{V_\circ}(L_0)$ isomorphic as L_0-module to $I/C_I(L_0)$. Hence the lemma holds.   □

REMARK 3.2.4. Given $L \in \mathcal{L}_f^*(G,T)$ with $L/O_2(L)$ quasisimple, lemma 3.2.3 shows that we can choose V so that L and V satisfy the Fundamental Setup. Then by 3.2.2.7, we may apply the results of section D.3 to analyze V, V_M, and $Aut_{L_0}(V_M)$. By 3.2.2, we may also appeal to Theorem 3.1.8, and in view of 3.2.2.4, 3.2.2.6 supplies extra information when $C_V(L) \neq 0$.

In the next few lemmas, we determine the list of modules V and V_M that can arise in the Fundamental Setup for the various possible $L \in \mathcal{L}_f^*(G,T)$. The first result 3.2.5 below gives us a qualitative description of what goes on in the case $L = L_0$, including a fairly complete description of the case where $V_\circ < V$. Then 3.2.8 gives more detailed information when $L = L_0$ but $V_\circ = V$.

Recall that $V_M := \langle V_\circ^M \rangle$ and that V plays the role of "V_T" played in lemma D.3.4. Also recall that in the FSU, $\bar{M}_V$ denotes $N_M(V)/C_M(V)$.

THEOREM 3.2.5. *Assume the Fundamental Setup (3.2.1), with* $L = L_0$. *Then* $\hat{q}(\bar{L}\bar{T}, V) \leq 2 \geq \hat{q}(Aut_M(V_M), V_M)$, *and one of the following holds:*

  (1) $V_\circ = V = V_M$; *that is,* $V_\circ \trianglelefteq M$.
  (2) $V_\circ = V \trianglelefteq T$, $C_{V_\circ}(L) = 0$, *and* V *is a TI-set under* M.
  (3) $\bar{L} \cong SL_3(2^n)$ *or* $Sp_4(2^n)$ *for some* n, A_6, $L_4(2)$, *or* $L_5(2)$; $C_{V_\circ}(L) = 0$ *and either* $V_\circ$ *is a natural module for* $\bar{L}$ *or* $V_\circ$ *is a 4-dimensional module for* $\bar{L} \cong A_7$; *and* $V_M = V = V_\circ \oplus V_\circ^t$ *with* $t \in T - N_T(V_\circ)$, *and* $V_\circ^t$ *not* $\mathbf{F}_2 L$-*isomorphic to* $V_\circ$.

PROOF. As discussed in Remark 3.2.4, we may apply 3.2.2, Theorem 3.1.8, and results in section D.3. Recall that in our setup, $V_\circ$ and V play the roles of "V" and "V_T" in Hypothesis D.3.2 and lemma D.3.4.

Set $\hat{q} := \hat{q}(\bar{L}\bar{T}, V)$ and $q := \hat{q}(Aut_M(V_M), V_M)$. As $L = L_0$ by hypothesis, conclusion (1) of Theorem 3.1.8 gives $\hat{q} \leq 2$.

Next we will show that $q \leq 2$ by an appeal to Theorem 3.1.6. Set $R := C_T(V_M)$, so that $R \in Syl_2(C_M(V_M))$. We first verify that for any $H \in \mathcal{H}_*(T, M)$, Hypothesis 3.1.5 is satisfied with $M_0 := N_M(R)$ and V_M in the role of "V": First as $V_M \trianglelefteq M$, hypothesis (II) of 3.1.5 is satisfied. By a Frattini Argument, $M = C_M(V_M)M_0$, so $Aut_M(V_M) \cong Aut_{M_0}(V_M)$, and hence as $V_M \in \mathcal{R}_2(M)$ by 3.2.2.2, also $V_M \in \mathcal{R}_2(M_0)$. As $R \trianglelefteq M_0$, $R \leq O_2(M_0)$. As $V_M \in \mathcal{R}_2(M_0)$, $O_2(M_0) \leq C_M(V_M)$, so as R is Sylow in $C_M(V_M)$, $R = O_2(M_0)$. This completes the verification of Hypothesis 3.1.5.

Next $V \leq V_M$, so $R \leq C_T(V)$, while $C_T(V) \trianglelefteq LT$ by 3.2.2.8. Thus $R = C_T(V) \cap C_M(V_M) \trianglelefteq LT$, so as $M = !\mathcal{M}(LT)$, $M = !\mathcal{M}(M_0)$. Therefore conclusion (1) of Theorem 3.1.6 is not satisfied, so one of conclusions (2) or (3) holds, and in either case, $q \leq 2$ as desired.

We have shown that $\hat{q} \leq 2 \geq q$, so it remains to show that one of conclusions (1)–(3) holds. Suppose first that $C_{V_\circ}(L) \neq 0$. Then by 3.2.2.6, $L = [L, J(T)]$, so that (in the language of Definition B.1.3) $Aut_L(V_M) \leq J(Aut_M(V_M), V_M)$ by B.2.7. Thus we have the hypotheses for D.3.20, which gives conclusion (1). Therefore we may assume that $C_{V_\circ}(L) = 0$. Then by 3.2.2.5, $V_\circ$ is a TI-set under M. If $V_\circ = V$ then conclusion (2) holds, so we may assume that $V_\circ < V$. As $\hat{q} \leq 2 \geq q$, the hypotheses of Theorem D.3.10 are satisfied; therefore as we have reduced to the case where $V_\circ < V$, conclusion (2) of Theorem D.3.10 holds. But this is precisely conclusion (3) of Theorem 3.2.5, so the proof is complete. $\qquad\square$

The notation $\hat{\mathcal{Q}}(X, W)$ appears in Definition D.2.1.

THEOREM 3.2.6. *Assume the Fundamental Setup (3.2.1) with $L < L_0$. Set $M^* := M/C_M(V_M)$, $U := [V_M, L]$, and let $t \in T - N_T(L)$. Then $\hat{q}(\bar{L}_0\bar{T}, V) \leq 2 \geq \hat{q}(M^*, V_M)$, and one of the following holds:*

(1) $L^* \cong L_2(2^n)$ and $V_\circ = V = V_M$ is the $\Omega_4^+(2^n)$-module for L_0^*.

(2) $L^* \cong L_3(2)$ and $V_\circ = V = V_M$ is the tensor product of natural modules for L^* and L^{*t}.

(3) Each of the following holds:

(a) $\tilde{V}_M = \tilde{U} \oplus \tilde{U}^t$, where $U = [V_M, L] \leq C_{V_M}(L^t)$.

(b) Each $A \in \hat{\mathcal{Q}}_*(M^*, V_M)$ acts on U, so $\hat{q}(Aut_{L_0T}(U), U) \leq 2$.

(c) One of the following holds:

(i) $U = V_\circ$ and $V = V_M$.

(ii) $Aut_M(L^*) \cong Aut(L_3(2))$, $V = V_M$, $U = V_\circ \oplus V_\circ^s$ for $s \in N_T(L) - LO_2(LN_T(L))$, and $m(V_\circ) = 3$.

(iii) $L^* \cong L_3(2)$, U is the sum of four isomorphic natural modules for L^*, and $O^2(C_{M^*}(L_0^*)) \cong \mathbf{Z}_5$ or E_{25}.

PROOF. Proceeding as in the proof of Theorem 3.2.5, and recalling the discussion in Remark 3.2.4, we verify Hypothesis 3.1.5 for $M_0 := N_M(R)$ where $R := C_T(V_M)$, and apply Theorems 3.1.6 and 3.1.8 as before to conclude

$$\hat{q}(\bar{L}_0\bar{T}, V) \leq 2 \geq \hat{q}(M^*, V_M).$$

Recall from the remark before that result that we may reduce case (3) to case (1) by a new choice of V. If $V < V_M$, then conclusion (2) of D.3.21 holds, so that conclusion (3) of 3.2.6 holds, with case (iii) of part (c) of (3) satisfied.

So we may suppose instead that $V = V_M$, as in conclusion (1) of D.3.21. Assume first that $V_\circ < V$. In particular we have the hypotheses of D.3.6, and conclusions (1) and (2) of that result give parts (a) and (b) of conclusion (3) of 3.2.6, while the two alternatives in part (3) of D.3.6 are cases (i) and (ii) of part (c) of conclusion (3) of 3.2.6.

Thus the Theorem holds when $V_\circ < V$, so assume instead that $V_\circ = V$. Then we have the hypotheses of D.3.7, and its conclusions (1) and (2) give the corresponding conclusions of 3.2.6. The proof is complete. $\qquad\square$

We often need to know that V is a TI-set under M. The previous two results say that this is almost always the case:

LEMMA 3.2.7. *Assume the Fundamental Setup (3.2.1). Then either*

(1) V is a TI-set under M, or

(2) $\bar{L} \cong L_3(2)$, $L < L_0$, and subcase (3.c.iii) of Theorem 3.2.6 holds.

PROOF. Suppose V is not a TI-set under M. Then in particular V is not normal in M, so that $V < V_M$. Therefore $L < L_0$, since if $L = L_0$ then either

$V = V_M$ or V is a TI-set under M, by 3.2.5. Thus L_0 and V are described in 3.2.6, where $V < V_M$ occurs only in subcase (3.c.iii). $\qquad\square$

With 3.2.6 in hand, we return to the case in the Fundamental Setup where $L = L_0$, and we obtain more information in the subcase where $V = V_\circ$. As in the proof of the Main Theorem, we divide our analysis into the case where V is an FF-module and the case where V is not an FF-module.

LEMMA 3.2.8. *Assume the Fundamental Setup (3.2.1) with $L = L_0$ and $V_\circ = V$. Assume further that V is an FF-module for $Aut_{GL(V)}(\bar{L})$. Then one of the following holds:*

(1) $\bar{L} \cong L_2(2^n)$ and $\tilde{V}$ is the natural module.

(2) $\bar{L} \cong SL_3(2^n)$, and either V is a natural module or V is a 4-dimensional module for $L_3(2)$.

(3) $\bar{L} \cong Sp_4(2^n)$ and $\tilde{V}$ is a natural module.

(4) $\bar{L} \cong G_2(2^n)'$ and $\tilde{V}$ is the natural module.

(5) $\bar{L} \cong A_5$ or A_7, and V is the natural module.

(6) $\bar{L} \cong A_6$ and $\tilde{V}$ is a natural module.

(7) $\bar{L}\bar{T} \cong A_7$ and $m(V) = 4$.

(8) $\bar{L} \cong \hat{A}_6$ and $m(V) = 6$.

(9) $\bar{L}\bar{T} \cong L_n(2)$, $n = 4$ or 5, and V is a natural module.

(10) $\bar{L} \cong L_4(2)$ and $\tilde{V}$ is the 6-dimensional orthogonal module.

(11) $\bar{L}\bar{T} \cong L_5(2)$ and $m(V) = 10$.

PROOF. This is a consequence of Theorem B.4.2, using the 1-cohomology of those modules listed in I.1.6. $\qquad\square$

PROPOSITION 3.2.9. *Assume the Fundamental Setup FSU (3.2.1), with $L = L_0$ and $V_\circ = V$. Further assume V is not an FF-module for $Aut_{GL(V)}(\bar{L})$. Set $q := q(\bar{L}\bar{T}, V)$ and $\hat{q} := \hat{q}(\bar{L}\bar{T}, V)$. Then one of the following holds:*

(1) $\bar{L} \cong L_2(2^{2n})$, $n > 1$, V is the $\Omega_4^-(2^n)$-module, and $q = \hat{q} \geq 3/2$, or $q \geq 4/3$ if $n = 2$.

(2) $\bar{L} \cong U_3(2^n)$, V is a natural module, and $q = \hat{q} = 2$.

(3) $\bar{L} \cong Sz(2^n)$, V is a natural module, and $q = \hat{q} = 2$.

(4) $\bar{L} \cong (S)L_3(2^{2n})$, $m(V) = 9n$, $q > 2$, and $\hat{q} = 5/4$. Further $\bar{T}$ is trivial on the Dynkin diagram of $\bar{L}$.

(5) $\bar{L}\bar{T} \cong Aut(M_{12})$, $m(V) = 10$, $q > 2$, and $\hat{q} > 1$.

(6) $\bar{L} \cong \hat{M}_{22}$, $m(V) = 12$, and $\hat{q} > 1$.

(7) $\bar{L} \cong M_{22}$, $m(V) = 10$, $q \geq 2$, $\hat{q} > 1$, and $q > 2$ if V is the cocode module.

(8) $\bar{L} \cong M_{23}$, $m(V) = 11$, $q > 2$, and $\hat{q} > 1$.

(9) $\bar{L} \cong M_{24}$, $m(V) = 11$, $q > 2$, and $\hat{q} > 1$.

PROOF. By hypothesis, V is not an FF-module for $\bar{L}\bar{T}$, so $J(T) \leq C_T(V)$ by B.2.7; hence we conclude $C_V(L) = 0$ from 3.2.2.6. Then as $V \in Irr_+(L, R_2(LT))$, L is irreducible on V. By 3.2.5, $\hat{q} \leq 2$. Then the result follows from the list in B.4.5, plus the following remarks: The cases in B.4.5 where $\bar{L}$ is A_7 or $G_2(2)'$ do not arise here because of our hypothesis that V is not an FF-module for $Aut_{GL(V)}(\bar{L})$. If $\bar{L} \cong (S)L_3(2^{2n})$ and $m(V) = 9n$, then V may be regarded as an $\mathbf{F}_{2^n}$-module, and $\mathbf{F}_{2^{2n}} \otimes_{\mathbf{F}_{2^n}} V = N \otimes N^\sigma$, where N is the natural $\mathbf{F}_{2^{2n}}$-module for $SL_3(2^{2n})$ and σ is the involutory field automorphism of $\mathbf{F}_{2^{2n}}$. Hence V is not invariant under an automorphism nontrivial on the Dynkin diagram. Finally we eliminate the cases in part (iii) of B.4.5, via an appeal to Theorem 3.1.8.2: For in these cases, $q > 2 = \hat{q}$ in the notation of B.4.5. As $q > 2$, case (i) of 3.1.8.2 does not hold. But V is a TI-set under M by 3.2.7, so as $\hat{q} = 2$, case (ii) of 3.1.8.2 does not hold either, a contradiction. $\qquad\square$

In our final result on the Fundamental Setup, we collect some useful properties that hold when $J(T) \leq C_T(V)$—and hence in particular under the hypotheses of 3.2.9 where V is not an FF-module.

Recall that $J_1(T)$ appears in Definition B.2.2, Further $n(X)$ appears in E.1.6, $r(G, V_+)$ in E.3.3, and $W_0(T, V_+)$ in E.3.13.

PROPOSITION 3.2.10. *Assume the Fundamental Setup (3.2.1). Set* $V_+ := V$, *except in case (3.c.iii) of 3.2.6, where we take* $V_+ := V_M$. *Assume* $J(T) \leq C_T(V_+)$. *Then*

(1) $N_G(J(T)) \leq M$.

(2) $N_M(V_+)$ *controls fusion in* V_+.

(3) *For each* $U \leq V_+$, $N_G(U)$ *is transitive on* $\{V_+^g : U \leq V_+^g\}$.

(4) *For each* $U \leq V_+$, $|N_G(U) : N_M(U)|$ *is odd.*

(5) *If* $U \leq V_+$ *with* $\langle V_+^{N_G(U)}\rangle$ *abelian, then* $[V_+, V_+^g] = 1$ *for all* $g \in G$ *with* $U \leq V_+^g$.

(6) *Suppose* $U \leq V_+$ *with* $V_+ \leq O_2(N_G(U))$, *and either*

   (a) $[V_+, W_0(T, V_+)] = 1$, *or*

   (b) V_+ *is not an FF-module for* $\operatorname{Aut}_{L_0 T}(V_+)$.

Then $[V_+, V_+^g] = 1$ *for each* $g \in G$ *with* $U \leq V_+^g$.

(7) *If* $J_1(T) \leq C_T(V_+)$ *and* $r(G, V_+) > 1$, *then* $n(H) > 1$ *for each* $H \in \mathcal{H}_*(T, M)$.

(8) *If* $J(T) \leq S \in \mathcal{S}_2(G)$, *then* $J(T) = J(S)$ *and so* $N_G(S) \leq M$.

(9) $C_Z(L_0) = 1 = C_{V_+}(L_0)$.

PROOF. By 3.2.3, $M = !\mathcal{M}(L_0 T)$. We have $C_T(V_+) \leq C_T(V) = O_2(L_0 T)$ by 3.2.2.8. Hence as $J(T) \leq C_T(V_+)$ by hypothesis, using B.2.3.3,

$$J(T) = J(C_T(V_+)) = J(O_2(L_0 T)) \trianglelefteq L_0 T,$$

so that (1) holds. Notice the same argument establishes (8). Further $Z(L_0 T) = 1$ by Theorem 3.1.8.3, so (9) follows.

Observe that V_+ is a TI-set under M: This holds in case (3.c.iii) of 3.2.6 as $V_+ = V_M$ is normal in M in that case, and in the remaining case $V_+ = V$ is a TI-set under M by 3.2.7.

Also $V_+ \leq E := \Omega_1(Z(J(T)))$. As $J(T)$ is weakly closed in T, by Burnside's Fusion Lemma A.1.35, $N_G(J(T))$ controls fusion in E and hence in V_+. Thus as V_+ is a TI-subgroup under M, (1) implies (2). Then (2) implies (3) using A.1.7.1.

Let $U \leq V_+$ and $S \in \operatorname{Syl}_2(N_M(U))$. As $J(T) \leq C_G(V_+)$ by hypothesis, we may assume $J(T) \leq S$. Then $N_G(S) \leq M$ by (8), so $S \in \operatorname{Syl}_2(N_G(U))$, establishing (4). Assume the hypotheses of (5), and let $U \leq V_+^g$. By (3), we may take $g \in N_G(U)$; then as $\langle V_+^{N_G(U)}\rangle$ is abelian by hypothesis, $[V_+, V_+^g] = 1$—so that (5) is established.

Assume the hypotheses of (6). Then $V_+ \leq O_2(N_G(U))$, so $\langle V_+^{N_G(U)}\rangle \leq W_0(T, V_+)$. Hence if $[V_+, W_0(T, V_+)] = 1$ as in (6a), then $\langle V_+^{N_G(U)}\rangle \leq C_T(V_+)$, so $\langle V_+^{N_G(U)}\rangle$ is abelian, and thus (5) implies (6) in this case. Now assume the hypothesis of (6b). We may take $g \in N_G(U)$ by (3), so

$$\langle V_+, V_+^g \rangle \leq O_2(N_G(U)) \leq S \cap S^g \leq N_M(U) \cap N_M(U^g) \leq N_M(V_+) \cap N_M(V_+^g),$$

where the last inclusion holds since V_+ is a TI-set under M. Reversing the roles of V_+ and V_+^g if necessary, we may assume that $m(V_+^g/C_{V_+^g}(V_+)) \geq m(V_+/C_{V_+}(V_+^g))$. Thus as $Aut_{L_0T}(V_+)$ is not an FF-module by hypothesis, $[V_+, V_+^g] = 1$. This completes the proof of (6).

As L_0T normalizes $O_2(L_0T) \cap C_M(V_+) = C_T(V_+)$, $M = !\mathcal{M}(N_{N_M(V_+)}(C_T(V_+)))$. Thus Hypothesis E.6.1 is satisfied with V_+ in the role of "V", so part (7) follows from E.6.26 with 1 in the role of "j". □

Sometimes in arguments where we can pin down the structure of a pair in the FSU (especially when we can show L is a block), we encounter the following situation:

LEMMA 3.2.11. *Assume the Fundamental Setup (3.2.1). Assume further that $V = O_2(L_0T)$. Then $O_2(M) = V = C_G(V)$ and $M = M_V$. If further $\bar{M}_V = \bar{L}_0\bar{T}$, then $M_V = M = L_0T$.*

PROOF. By A.1.6, $O_2(M) \leq O_2(L_0T) = V \leq O_2(L_0) \leq O_2(M)$, so that $O_2(M) = V$, and in particular $M = M_V$ as $M \in \mathcal{M}$. Now as $F^*(M) = O_2(M)$, $C_G(V) \leq Z(O_2(M)) \leq V$, so that $C_G(V) = V$. The result follows. □

Our last two results of the section involve the collection $\Xi(G,T)$ of Definition 1.3.1, and appearing in case (ii) of the hypothesis of 3.2.2.

DEFINITION 3.2.12. Define $\Xi_-(G,T)$ to consist of those $X \in \Xi(G,T)$ such that either

(a) X is a $\{2,3\}$-group, or
(b) $X/O_2(X)$ is a 5-group and $Aut_G(X/O_2(X))$ a $\{2,5\}$-group.

Set $\Xi_+(G,T) := \Xi(G,T) - \Xi_-(G,T)$.

LEMMA 3.2.13. $\Xi_f^*(G,T) \subseteq \Xi_-(G,T)$.

PROOF. Assume $X \in \Xi_f^*(G,T)$. Then $X/O_2(X) \cong E_{p^2}$ or p^{1+2} for some odd prime p, and T is irreducible on $X/O_{2,\Phi}(X)$. By 1.3.7, $M = !\mathcal{M}(XT)$, where $M := N_G(X)$. Let $(XT)^* := XT/C_{XT}(R_2(XT))$. By A.4.11, $V := [R_2(XT), X] \neq 1$, so as T is irreducible on $X/O_{2,\Phi}(X)$, $C_X(V) \leq O_{2,\Phi}(X)$. Thus as $R := O_2(XT)$ centralizes $R_2(XT)$, $X^* = F^*(X^*T^*)$, so as X^* is faithful on V, also X^*T^* is faithful on V. Hence $C_T(V) = R$ and $V \in \mathcal{R}_2(XT)$. Therefore the hypotheses of Theorem 3.1.8 are satisfied with X in the role of "L_0", so $\hat{q} := \hat{q}(X^*T^*, V) \leq 2$ by 3.1.8.1. As $\hat{q} \leq 2$, D.2.13 says $p = 3$ or 5. We may assume by way of contradiction that $X \notin \Xi_-(G,T)$, so $p = 5$ and $Aut_G(X/O_2(X))$ is not a $\{2,5\}$-group. By D.2.17 and D.2.12, $X^* = X_1^* \times \cdots \times X_s^*$ and $V = V_1 \oplus \cdots \oplus V_s$, where $X_i^* \cong \mathbf{Z}_5$, $V_i := [V, X_i]$ is of rank 4, and $s \leq 2$. As $m_5(X/O_{2,\Phi}(X)) = 2$, $s = 2$. As $T \in Syl_2(N_G(X))$, $R \in Syl_2(C_G(X/O_2(X)))$ by A.4.2.5; so by a Frattini Argument, $Aut_G(X/O_2(X)) = Aut_H(X/O_2(X))$, where $H := N_G(X) \cap N_G(R)$. Thus $Aut_H(X/O_2(X))$ is not a $\{2,5\}$-group, so $Aut_H(X^*)$ is not a $\{2,5\}$-group. As R centralizes $R_2(XT)$, $R_2(XT) \leq \Omega_1(Z(R))$. Then as $V \leq R_2(XT)$,

$$C_{XT}(\Omega_1(Z(R))) \leq C_{XT}(V) \leq RO_{2,\Phi}(X),$$

so $\Omega_1(Z(R))$ is 2-reduced. Therefore $R_2(XT) = \Omega_1(Z(R))$, so H acts on

$$[\Omega_1(Z(R)), X] = [R_2(XT), X] = V,$$

and hence $O^2(H)$ acts on V_i and X_i^*. This is a contradiction as $Aut_H(X^*)$ is not a $\{2,5\}$-group, but $Aut(\mathbf{Z}_5)$ is a 2-group. $\qquad\square$

Lemma 3.2.13 allows us to establish a result about those $L \in \mathcal{L}(G,T)$ such that $L/O_2(L)$ is not quasisimple. Recall from chapter 1 that $\Xi_p(L)$ is $O^2(X_p)$ where X_p is the preimage of $\Omega_1(O_p(L/O_2(L)))$.

LEMMA 3.2.14. *If $L \in \mathcal{L}(G,T)$ and $L/O_2(L)$ is not quasisimple, then $O_\infty(L)$ centralizes $R_2(LT)$.*

PROOF. We assume L is a counterexample, and it remains to derive a contradiction.

By 1.2.1.4, $L/O_{2,F}(L) \cong SL_2(q)$ for some prime $q > 3$, and T normalizes L by 1.2.1.3. Set $V := R_2(LT)$; by hypothesis $[V, L] \neq 1$ so $L \in \mathcal{L}_f(G,T)$.

Let $L \leq K \in \mathcal{L}^*(G,T)$; then $K \in \mathcal{L}_f^*(G,T)$ by 1.2.9. In the cases in A.3.12 where "$B/O_2(B)$" is not quasisimple, either $O_\infty(L) \leq O_\infty(K)$ in case (21) or (22), or $K/O_2(K) \cong (S)L_3(r)$ for some prime $r > 3$ in case (9). In the latter case by 3.2.3, K is listed in one of 3.2.5, 3.2.8, or 3.2.9, but of course $(S)L_3(r)$ for a prime $r > 3$ does not appear on any of those lists. Thus $O_\infty(L) \leq O_\infty(K)$, so replacing L by K, we may assume $L \in \mathcal{L}_f^*(G,T)$.

Let $\pi := \pi(O_{2,F}(L)/O_2(L))$, $p \in \pi$, and $X := \Xi_p(L)$. Since $L \in \mathcal{L}_f^*(G,T)$, $X \in \Xi_{rad}^*(G,T)$ by the definition in chapter 1, so $X \in \Xi^*(G,T)$ by 1.3.8. As $Aut_L(X/O_2(X))$ contains $SL_2(q)$ for $q > 3$, $X \notin \Xi_-(G,T)$, so X centralizes V by 3.2.13. Hence

$$Y := \prod_{p \in \pi} \Xi_p(L) \leq C_L(V).$$

Let $I_p := O^{p'}(O_\infty(L))$. If I_p centralizes V for each $p \in \pi$, then $O_{2,F}(L) \leq O_2(L)Y \leq C_L(V)$, so $O_\infty(L)$ centralizes V as $L/O_{2,F}(L) \cong SL_2(q)$ and V is 2-reduced. Thus as L is a counterexample, there is $p \in \pi$ such that $I := I_p$ does not centralize V, so $I \neq X_p$ and hence case (d) of 1.2.1.4 holds and $I/O_2(I) \cong \mathbf{Z}_{p^e}^2$ for some $e > 1$. As case (d) of 1.2.1.4 holds, $L/O_{2,F}(L) \cong SL_2(5)$. Since $e > 1$, we conclude from A.1.30 that $p > 5$.

Set $R := C_T(V)$. As $V = R_2(LT)$, $O_2(LT) \leq R$. As $L/O_{2,F}(L) \cong SL_2(5)$, $O_2(LT) = O_2(IT)$, and then as $[I, V] \neq 1$, $R = O_2(IT)$ and $V \in \mathcal{R}_2(IT)$. As $X \in \Xi^*(G,T)$, $M := N_G(X) = !\mathcal{M}(XT)$ by 1.3.7. As $L \in \mathcal{L}^*(G,T)$, $L \trianglelefteq M$, so as I char L, $I \trianglelefteq M$. Thus for each $H \in \mathcal{H}_*(T,M)$, $H \cap M$ normalizes I, so case (I) of Hypothesis 3.1.5 is satisfied with IT in the role of "M_0". As $M = !\mathcal{M}(XT)$, $O_2(\langle IT, H \rangle) = 1$, so conclusion (2) or (3) of Theorem 3.1.6 holds. In either case $\hat{q}(IT/C_{IT}, V) \leq 2$. As $p > 5$ and $[V, I] \neq 1$, this contradicts D.2.13. $\qquad\square$

3.3. Normalizers of uniqueness groups contain $N_G(T)$

The bulk of the proof of the Main Theorem analyzes the situation where $\mathcal{L}_f(G,T)$ is nonempty, leading (as we saw in 3.2.3) to the Fundamental Setup (3.2.1) and the extended analysis of the cases arising there. The very restricted situation where $\mathcal{L}_f(G,T)$ is empty will be treated only at the end of the proof after that analysis.

In this section, in Theorem 3.3.1 we establish an important property of maximal 2-locals containing T and suitable uniqueness subgroups. Theorem 3.3.1 will be used repeatedly in our analysis of the cases arising from the Fundamental Setup.

It turns out that case (2) of Theorem 3.3.1 is not actually required to prove the Main Theorem, contary to what we expected when we proved the result. However as the proof for this case is short, we have retained its statement and proof here.

THEOREM 3.3.1. *Assume G is a simple QTKE-group, $T \in Syl_2(G)$, $M \in \mathcal{M}(T)$, and either*

(1) *$L \in \mathcal{L}^*(G, T)$ with $L/O_2(L)$ quasisimple and $L \leq M$, or*
(2) *$X \in \Xi^*(G, T)$ with $X \leq M$.*

Then $N_G(T) \leq M$.

We first record an elementary but important consequence of Theorems 2.1.1 and 3.3.1, that we will use repeatedly in the remainder of the paper: In the Fundamental Setup, the members of $\mathcal{H}_*(T, M)$ are minimal parabolics.

COROLLARY 3.3.2. *Assume G is a simple QTKE-group, $T \in Syl_2(G)$, and $L \in \mathcal{L}^*(G, T)$ with $L/O_2(L)$ quasisimple. Set $M := N_G(\langle L^T \rangle)$. Then*

(1) *$M = !\mathcal{M}(\langle L, T \rangle)$.*
(2) *$|\mathcal{M}(T)| > 1$, so $\mathcal{H}_*(T, M) \neq \emptyset$.*
(3) *$N_G(T) \leq M$.*
(4) *For each $H \in \mathcal{H}_*(T, M)$, $H \cap M$ is the unique maximal subgroup of H containing T, and $H \in \hat{\mathcal{U}}_G(T)$ so that H is a minimal parabolic described in B.6.8, and in E.2.2 when H is nonsolvable.*

PROOF. Part (1) follows from 1.2.7. Part (2) holds since 2-locals of odd index in the groups G in the conclusion of Theorem 2.1.1 are solvable, so that $\mathcal{L}(G, T)$ is empty. Part (3) follows from Theorem 3.3.1. Finally (4) follows from (3) and 3.1.3. $\qquad\square$

REMARK 3.3.3. In the simple QTKE-groups G, $N_G(T) \leq M$ under the hypotheses of Theorem 3.3.1. However there is an almost simple shadow where this assertion fails: In the extension G of $\Omega_8^+(2)$ by a graph automorphism of order 3, there is a maximal parabolic L of $E(G)$ which is an A_8-block and is a member of $\mathcal{L}^*(G, T)$, but which is not invariant under an element of order 3 in $N_G(T)$ inducing the triality outer automorphism on $E(G)$. This extension is of even characteristic, but it is neither simple nor quasithin. However it is difficult to verify these global properties just from the point of view of the 2-local L, so that the shadow of this group causes difficulties in the proof of 3.3.21.f. Also the proof of 3.3.24 is complicated by the shadow of the non-maximal parabolic $L_3(2)/2^{3+6}$ in this same extension G.

Case (2) of the hypothesis of Theorem 3.3.1 will be eliminated fairly early in the argument in 3.3.10.3. Thus the bulk of the proof is devoted to case (1) of the hypothesis.

NOTATION 3.3.4. In case (1) of the hypothesis of Theorem 3.3.1, where $L \in \mathcal{L}^*(G, T)$ with $L/O_2(L)$ quasisimple, set $M_+ := L_0 := \langle L^T \rangle$. In case (2) of that hypothesis, where $X \in \Xi^*(G, T)$, set $M_+ := X$. As $N_G(T)$ is 2-closed and hence solvable, $N_G(T) = TD$, where D is a Hall $2'$-subgroup of $N_G(T)$.

We recall that M_+T is a uniqueness subgroup in the language of chapter 1:

LEMMA 3.3.5. *(1)* $M = N_G(M_+)$.
(2) $M = !\mathcal{M}(M_+T)$.
(3) $F^*(M_+T) = O_2(M_+T)$.

PROOF. Parts (1) and (2) are a consequence of 1.2.7.3 and 1.3.7. By definition $M_+T \in \mathcal{H}(T)$, so (3) follows from 1.1.4.6. $\square$

Throughout this section, we assume we are working in a counterexample to Theorem 3.3.1, so that $N_G(T) \not\leq M$. Our arguments typically derive a contradiction by violating one of the consequences of 3.3.5.2 in the following lemma:

LEMMA 3.3.6. *(a)* $D \not\leq M$.
(b) $O_2(\langle M_+T, D\rangle) = 1$. *Thus if* $1 \neq X \trianglelefteq M_+T$, *then* $D \not\leq N_G(X)$.
(c) No nontrivial characteristic subgroup of T *is normal in* M_+T.
(d) Assume case (1) of Theorem 3.3.1 holds with $L/O_{2,Z}(L)$ *of Lie type and Lie rank 2 in characteristic 2. Then* T *acts on* L *unless possibly* $L/O_2(L) \cong L_3(2)$; *and if* T *acts on* L, *then* (LT, T) *is an MS-pair in the sense of Definition C.1.31.*

PROOF. Part (a) holds as $T \leq M$, but $TD = N_G(T) \not\leq M$. Then (b) follows from (a) and 3.3.5.2, and (c) follows from (b).
Assume the hypothesis of (d). Then unless $L/O_2(L) \cong L_3(2)$, T acts on L by 1.2.1.3. Assume T acts on L. Then (LT, T) satisfies hypothesis (MS1) in Definition C.1.31 by 3.3.5.3, hypothesis (MS2) is satisfied as T is Sylow in LT, and hypothesis (MS3) holds by (c). $\square$

Set $Z := \Omega_1(Z(T))$, $V := \langle Z^{M_+}\rangle = \langle Z^{M_+T}\rangle$, $\overline{M_+T} := M_+T/C_{M_+T}(V)$, and $\tilde{V} := V/C_V(M_+)$.

LEMMA 3.3.7. *(1)* $C_{M_+T}(V) \leq O_{2,\Phi}(M_+T)$ *and* $C_T(V) = O_2(M_+T)$.
(2) $J(T) \not\leq C_T(V)$, *so* V *is a failure of factorization module for* $\overline{M_+}\bar{T}$.
(3) $V \in \mathcal{R}_2(M_+T)$, *so* $O_2(\bar{M}_+\bar{T}) = 1$.
(4) $[V, M_+] = [Z, M_+]$ *and* $V = [V, M_+]C_Z(M_+)$.

PROOF. Since $F^*(M_+T) = O_2(M_+T)$ by 3.3.5.3, part (3) is a consequence of B.2.14. As $V = \langle Z^{M_+}\rangle$, $V = [V, M_+]Z$, so that $V = [V, M_+]C_Z(M_+)$ using Gaschütz's Theorem A.1.39. If $\bar{M}_+ = 1$, then $V = Z$ and $M_+T \leq C_G(Z)$, contrary to 3.3.6.c. Thus $\bar{M}_+ \neq 1$, so (1) follows from (3) and 1.4.1.5 with M_+ in the role of "L_0". If $J(T) \leq C_T(V)$, then by B.2.3.3, $J(T) = J(C_T(V)) = J(O_2(M_+T)) \trianglelefteq M_+T$, contrary to 3.3.6.c. Thus $J(T) \not\leq C_T(V)$, so V is an FF-module for $\bar{M}_+\bar{T}$ by B.2.7. $\square$

We now use 3.3.7 to determine a list of possibilities for $\bar{M}_+$ and V, which we will eliminate during the remainder of the proof. Notice if case (2) of the hypothesis of Theorem 3.3.1 holds, then conclusion (1) of the next lemma holds with $\bar{L}_i \cong \mathbf{Z}_3$.

LEMMA 3.3.8. *One of the following holds:*
(1) $\bar{M}_+ = \bar{L}_1 \times \bar{L}_2$ *with* $\bar{L}_i \cong L_2(2^n)$, $L_3(2)$, *or* $\mathbf{Z}_3$, *and* $\bar{L}_1^t = \bar{L}_2$ *for some* $t \in T - N_T(L_1)$. *Further* $[\tilde{V}, M_+] = \tilde{V}_1 \oplus \tilde{V}_2$, *where* $\tilde{V}_i := [\tilde{V}, L_i]$, *and either* $\tilde{V}_i$ *is the natural module for* $\bar{L}_i$, *the* A_5-*module for* $\bar{L}_i \cong A_5$, *or the sum of two isomorphic natural modules for* $\bar{L}_i \cong L_3(2)$.
(2) $\bar{M}_+ \cong L_2(2^n)$ *with* $n > 1$, *and* $[\tilde{V}, M_+]$ *is the natural module for* $\bar{M}_+$.

(3) $\bar{M}_+ \cong A_5$ or A_7, and $[V, M_+]$ is the natural module for $\bar{M}_+$.

(4) $\bar{M}_+ \cong SL_3(2^n)$, $Sp_4(2^n)'$, or $G_2(2^n)'$, and $[\tilde{V}, M_+]$ is either the natural module for $\bar{M}_+$ or the sum of two isomorphic natural modules for $\bar{M}_+ \cong SL_3(2^n)$.

(5) $\bar{M}_+ \cong A_7$, and $[V, M_+]$ is of rank 4.

(6) $\bar{M}_+ \cong \hat{A}_6$, and $[V, M_+]$ is of rank 6.

(7) $\bar{M}_+ \cong L_4(2)$ or $L_5(2)$, and the possiblities for $[V, M_+]$ are listed in Theorem B.5.1.1.

PROOF. By 3.3.7.2, V is an FF-module for $\bar{M}_+\bar{T}$, and by 3.3.7.3, $O_2(\bar{M}_+\bar{T}) = 1$. Hence the action of $\bar{J} := J(\bar{M}_+\bar{T}, V)$ on $[V, \bar{J}]$ is described in Theorem B.5.6.

In case (2) of Theorem 3.3.1, M_+T is a minimal parabolic, and using 3.3.7.1, $\bar{M}_+$ is noncyclic, so conclusion (1) of the lemma holds by B.6.9. Thus we may assume case (1) holds. Therefore $F^*(\bar{M}_+\bar{T}) = \bar{M}_+ = \bar{L}$ or $\bar{L}\bar{L}^t$ for $t \in T - N_T(L)$. Therefore as $1 \neq \bar{J} \trianglelefteq \bar{M}_+\bar{T}$, $\bar{M}_+ = F^*(\bar{J})$. Further if $L < M_+$, then $\bar{L} \cong L_2(2^n)$, $Sz(2^n)$, $L_2(p)$ or J_1 by 1.2.1.3. Therefore conclusion (1) of the lemma holds by B.5.6.

Thus we may assume that $L = M_+$, so that $\bar{L} = F^*(\bar{J}) = F^*(\bar{M}_+\bar{T})$ is quasisimple. Hence the action of L on V is described in Theorem B.5.1. The conclusions of the lemma include cases (ii), (iii), and (iv) of B.5.1.1 in which $[\tilde{V}, L]$ is reducible, so we may assume $[\tilde{V}, L]$ is irreducible. Hence by B.5.1 the possibilities for the action of $\bar{L}\bar{T}$ on $[\tilde{V}, L]$ are listed in Theorem B.4.2, and again our conclusions contain all those cases. $\square$

LEMMA 3.3.9. $C_{M_+}(Z) = C_{M_+}(Z \cap [V, M_+])$.

PROOF. Since $Z = (Z \cap [V, M_+])C_Z(M_+)$ by 3.3.7.4, the lemma follows. $\square$

We now begin to eliminate cases from 3.3.8:

LEMMA 3.3.10. (1) If $H \in \mathcal{H}(T)$ and T is contained in a unique maximal subgroup of H, then $O_2(\langle H, D \rangle) \neq 1$.

(2) $\bar{M}_+$ is not $L_2(2^n)$, eliminating case (2) of 3.3.8 and the A_5-subcase of case (3) of 3.3.8.

(3) If case (1) of 3.3.8 holds, then $\bar{L}_i \cong L_3(2)$.

(4) Case (1) of the hypothesis of Theorem 3.3.1 holds.

PROOF. Part (1) follows from Theorem 3.1.1, with TD, T in the roles of "M_0, R". In particular if T lies in a unique maximal subgroup of M_+T, then (1) contradicts 3.3.6.b. Parts (2) and (3) follow from this observation. Finally, as we observed earlier, if case (2) of the hypothesis of Theorem 3.3.1 holds, then conclusion (1) of 3.3.8 holds with $\bar{L}_i \cong \mathbf{Z}_3$. Thus (3) implies (4). $\square$

REMARK 3.3.11. By 3.3.10.4, case (1) of Notation 3.3.4 holds. Therefore $M_+ = \langle L^T \rangle$, where $L \in \mathcal{L}^*(G, T)$ with $L/O_2(L)$ quasisimple. Thus L has this meaning from now on.

LEMMA 3.3.12. Suppose $Y \in \mathcal{L}(L, T)$ and $O_2(H) \neq 1$ where $H := \langle Y, TD \rangle$. Then

(1) $Y \leq K \in \mathcal{C}(H)$.

(2) $K \trianglelefteq H$.

(3) One of the following holds:

    (a) $D \leq N_G(Y)$, or

(b) $Y/O_2(Y) \cong L_2(4)$, $K/O_2(K) \cong J_1$, $D = (K \cap D)N_D(Y)$, and $|D : N_D(Y)| = 7$. Further T induces inner automorphisms on $Y/O_2(Y)$.

(c) $Y/O_2(Y) \cong A_6$, $K/O_2(K) \cong U_3(5)$, and D of order 3 induces an outer automorphism on $K/O_2(K)$ centralizing a subgroup isomorphic to the double covering of S_5 which is not $GL_2(5)$.

PROOF. Part (1) follows from 1.2.4 applied with Y, H in the roles of "B, H".

By 3.3.6.b, $Y < L$. Applying 1.2.4 with Y, L in the roles of "B, H", and comparing the embeddings described in A.3.12 to the list of possibilities for L in 3.3.8, we conclude that $Y/O_2(Y)$ is $L_2(2^n)$, $L_3(2)$, A_6, or $L_4(2)$. Furthermore L is not $L_3(2)$, so we conclude from 3.3.10.3 and 3.3.8 that $M_+ = L$. Now by 1.2.8.1, T normalizes Y, and then T also normalizes K. Thus (2) follows from 1.2.1.3.

Assume that conclusion (a) of (3) fails; we must show that conclusion (b) or (c) of (3) holds. By (2), $Y < K$. Then $Y/O_2(Y)$ is described in the previous paragraph, and the possible proper overgroups K of Y are described in A.3.12.

Set $H^* := H/C_H(K/O_2(K))$, and let Y_H be the preimage of Y^* in H. We claim that $Y \trianglelefteq N_H(Y^*)$: By hypothesis, $Y \in \mathcal{L}(L, T)$, so Y is the unique member of $\mathcal{C}(O_2(K)Y)$. Then as $YO_2(K) \trianglelefteq Y_H$, $Y \in \mathcal{C}(Y_H)$ by A.3.3.2. Therefore as T acts on Y, $Y \trianglelefteq Y_H$ by 1.2.1.3, establishing the claim.

By assumption, $D \not\leq N_H(Y)$, so by the claim:
$$N_{D^*}(Y^*) = N_D(Y)^*, \quad \text{so} \quad D^* \not\leq N_{H^*}(Y^*). \tag{$*$}$$

In particular, $D^* \neq 1$. Similarly $C_D(K^*) \leq C_D(Y^*) < D$. Next $T_K := T \cap K \in Syl_2(K)$ and $1 \neq D^* \leq N_{H^*}(T_K^*)$, so
$$N_{H^*}(T_K^*) \geq T_K^* D^* > T_K^*. \tag{$**$}$$

Assume that K^* is sporadic; that is, K appears in one of cases (11)–(20) of A.3.12. Then $Out(K^*)$ is a 2-group, so $D^* \leq K^*$, and we conclude from ($**$) that $K^* \cong J_1$ or J_2. In the latter case, $Y^* \cong A_5/2^{1+4}$ is uniquely determined by A.3.12, and $D^* \leq Y^*$, contrary to ($*$). In the former case, $N_{H^*}(T_K^*) = N_H(T)^*$ is a Frobenius group of order 21, and T^* induces inner automorphisms on $Y^* \cong A_5$, so that $|D^* : N_{D^*}(Y^*)| = 7$. Thus $D = (D \cap K)N_D(Y)$ and $|D : N_D(Y)| = 7$ by ($*$). Then since the multiplier of J_1 is trivial by I.1.3, $K/O_2(K) \cong J_1$, so case (b) of conclusion (3) holds.

Thus we may assume K^* satisfies one of cases (2), (4)–(9), (21), or (22) of A.3.12. In cases (4)–(7), $Out(K^*)$ is a 2-group, so that $D^* \leq K^*$, and ($**$) supplies a contradiction. In case (2), K^* is of Lie type and Lie rank 2 in characteristic 2, with $Y^* = P^{*\infty}$ for some T-invariant maximal parabolic P^* of K^*. Thus as there are exactly two such parabolics,
$$D^* \leq O^2(N_{H^*}(T_K^*)) \leq N_{H^*}(P^*) \leq N_{H^*}(Y^*),$$
again contrary to ($*$).

In cases (21) and (22), T_K^* is contained in a unique complement K_1^* to $O(K^*)$ in K^*, with $K_1^* \cong SL_2(p)$ for an odd prime $p > 3$. By the uniqueness of K_1^*, $Y^* \leq K_1^*$ and D^* acts on K_1^*, so that $Y^* < K_1^*$ by ($*$). So replacing K by the $\mathcal{C}$-component K_1 of the preimage of K_1^*, we reduce the treatment of these cases to the elimination of the subcase of case (8) where $H^* \cong L_2(p)$ for some prime $p \equiv \pm 3 \mod 8$ and $Y^* \cong L_2(5)$. Then as $D^* \neq 1$ normalizes T^*, $A_4 \cong N_{H^*}(T^*) = T^* D^* \leq Y^* T^* \leq N_{H^*}(Y^*)$, again contrary to ($*$). In the remaining subcase of (8), $K^* \cong L_2(p^2)$ for an odd prime p. Here T_K^* is dihedral of order greater than 4 and self-centralizing in $Aut(K^*)$, so that $N_{H^*}(T_K^*)$ is a 2-group, and then $D^* = 1$, contrary to (*).

Thus case (9) of A.3.12 holds, with $K^* \cong L_3^\epsilon(p)$. If $Y^* \cong SL_2(p)$, $Y^* = C_{K^*}(Z(T_K^*))^\infty$ is D^*-invariant, again contrary to (*).

In the remaining subcase of (9), $K^* \cong U_3(5)$, with $Y^* \cong A_6$. Here $X^* = O^2(C_{Aut(K^*)}(T_K^*))$ is of order 3 and induces outer automorphisms on K^* with $C_{K^*}(X^*)$ the double covering of S_5 which is not $GL_2(5)$. We conclude $D^* = X^*$. Finally $K/O_2(K)$ is not $SU_3(5)$ by A.3.18. Therefore $K/O_2(K) \cong U_3(5)$, so case (c) of conclusion (3) holds.

This completes the treatment of the cases appearing in A.3.12, and hence completes the proof of the lemma. $\square$

LEMMA 3.3.13. *If $H \in \mathcal{H}(T)$ with $H/O_2(H) \cong S_3$ wr $\mathbf{Z}_2$, then $D \leq N_G(H)$.*

PROOF. Let $H_0 := \langle H, D \rangle$; by 3.3.10.1, $O_2(H_0) \neq 1$. Set $Y := O^2(H)$ and notice $Y \in \Xi(G,T)$. If D normalizes Y, then D normalizes $YT = H$ and the lemma holds, so we assume that D does not act on Y. Therefore Y is not normal in H_0, so by 1.3.4, $Y < K_0 := \langle K^T \rangle$ for some $K \in \mathcal{C}(H_0)$, and K_0 is a normal subgroup of H_0 described in cases (1)–(4) of 1.3.4 with 3 in the role of "p". Let $(K_0TD)^* := K_0TD/C_{K_0TD}(K_0/O_2(K_0))$. Notice that $O_2(K_0) \leq O_2(H_0) \leq O_2(H)$ using A.1.6, so that $N_{D^*}(Y^*) = N_D(Y^*)$. Hence D^* does not act on Y^* and in particular $D^* \neq 1$, so that

(*) T^* is not self-normalizing in $K^*T^*D^*$.
Further $H^*/O_2(H^*) \cong H/O_2(H) \cong O_4^+(2)$, so

(**) $T^*/O_2(Y^*T^*) \cong D_8$.

Inspecting the list in 1.3.4 for cases in which (*) and (**) are satisfied, we conclude that case (1) of 1.3.4 holds, with $K^* \cong L_2(2^n)$ for $2^n \equiv 1$ mod 3, and H^* is contained in the T-invariant Borel subgroup B^* of K_0^*. As D^* acts on T^*, D^* acts on B^* and hence also on the characteristic subgroup Y^* of B^*, contrary to an earlier remark. This completes the proof. $\square$

LEMMA 3.3.14. $L = M_+ \trianglelefteq M$, *eliminating case (1) of 3.3.8.*

PROOF. Assume otherwise. Then by 3.3.10.3, $\bar{L}_i \cong L_3(2)$. Therefore $M_+T = \langle H_1, H_2 \rangle$, where $H_i := \langle H_{i,1}, T \rangle$ and $\bar{H}_{i,1}$, $i = 1, 2$, are the maximal parabolics of $\bar{L}_1$ over $\bar{T} \cap \bar{L}_1$. Notice that $H_i/O_2(H_i) \cong S_3$ wr $\mathbf{Z}_2$, so by 3.3.13, D normalizes H_i. But then D normalizes $M_+T = \langle H_1, H_2 \rangle$, contrary to 3.3.6.b. $\square$

Our next lemma puts us in a position to exploit an argument much like that in the proof of 3.3.14, to eliminate many cases where L is generated by a pair of members of $\mathcal{L}(L,T)$.

LEMMA 3.3.15. *Suppose $LT = \langle Y_1, Y_2, T \rangle$ with $Y_j \in \mathcal{L}(L,T)$. Set $H_j := \langle Y_j, TD \rangle$, and assume $O_2(H_j) \neq 1$ for $j = 1$ and 2. Then for $i = 1$ or 2: D does not normalize Y_i, $Y_i/O_2(Y_i) \cong L_2(4)$ or A_6, $Y_i < K \in \mathcal{C}(H_i)$ such that $K/O_2(K) \cong J_1$ or $U_3(5)$, respectively, $K \trianglelefteq H_i$, and $D \not\leq M$. When $K/O_2(K) \cong J_1$, T induces inner automorphisms on $Y_i/O_2(Y_i)$ and $K \cap D \not\leq M$.*

PROOF. Notice Y_j, H_j satisfy the hypotheses of 3.3.12 in the roles of "Y, H", so we can appeal to that lemma. Suppose D normalizes both Y_1 and Y_2. Then D normalizes $\langle Y_1, Y_2, T \rangle = LT$, contradicting 3.3.6.b. Thus D does not normalize some Y_i, so the pair Y_i, H_i is described in case (b) or (c) of 3.3.12.3. Further $D \not\leq M$ by 3.3.6.a, and when $K/O_2(K) \cong J_1$, $K \cap D \not\leq M$ by 3.3.12. $\square$

LEMMA 3.3.16. $\bar{L}$ is not $SL_3(2^n)$, $Sp_4(2^n)$, or $G_2(2^n)$ with $n > 1$.

PROOF. Assume otherwise. Let $T_L := T \cap L$ and $\bar{M}_i$, $i = 1, 2$, be the maximal parabolics of $\bar{L}$ containing $\bar{T}_L$. Set $Y_i := M_i^\infty$; then $Y_i \in \mathcal{L}(L, T)$ with $Y_i/O_2(Y_i) \cong L_2(2^n)$ and $LT = \langle Y_1, Y_2, T \rangle$. By 3.3.10.1, $H_i := \langle Y_i, TD \rangle \in \mathcal{H}(T)$. Thus by 3.3.15, we may assume that D does not normalize $Y_1 =: Y$, $n = 2$, and $Y < K \in \mathcal{C}(H_1)$ with $K \trianglelefteq H_1$, $K/O_2(K) \cong J_1$, $K \cap D \not\leq M$, and T induces inner automorphisms on $Y/O_2(Y)$. Set $H_1^* := H_1/C_{H_1}(K/O_2(K))$. Then $Y^* \cong L_2(4)$, so $O_2(Y^*) = 1$.

By 3.3.6.d, (LT, T) is an MS-pair, and so we may apply the Meierfrankenfeld-Stellmacher result Theorem C.1.32. Since $n = 2$, $L/O_2(L)$ is $SL_3(4)$ or $Sp_4(4)$ or $G_2(4)$. By Theorem C.1.32, $L/O_2(L)$ is not $G_2(4)$, and if $L/O_2(L) \cong Sp_4(4)$, then L is an $Sp_4(4)$-block.

As T induces inner automorphisms on $Y/O_2(Y)$, T induces inner automorphisms on $L/O_2(L)$ from the structure of $Aut(L/O_2(L))$. From the structure of $L/O_2(L)$, $X := C_{D \cap L}(Y/O_2(Y))$ is of order 3, and as X normalizes T, $Q := [X, T]$ is a 2-group. Now $X \leq D \leq H_1$, and we saw that $K \trianglelefteq H_1$. As $[X^*, Y^*] \leq O_2(Y^*) = 1$ and $C_{Aut(K^*)}(Y^*)$ is of order 2 since $K^* \cong J_1$, we conclude $X^* = 1$. Therefore $Q^* = [X^*, T^*] = 1$, so $Q = [X, O_2(KT)] = O_2(O^2(XO_2(KT))) \trianglelefteq KT$. But if $L/O_2(L) \cong SL_3(4)$, then $O_2(L)X = O_{2,Z}(L) \trianglelefteq LT$, so that $Q = [O_2(L), X]$ is also normal in LT, and hence $K \leq N_G(Q) \leq M = !\mathcal{M}(LT)$, contradicting $K \cap D \not\leq M$.

Therefore L is an $Sp_4(4)$-block. Now $O_2(L)$ is of order at most 2^{10} using the value for 1-cohomology of the natural module in I.1.6; thus Q is of order at most

$$|O_2(Y) : O_2(L)||O_2(L)| \leq 2^6 \cdot 2^{10} = 2^{16}.$$

Therefore as 19 divides the order of J_1 but not of $L_{16}(2)$, K centralizes Q. This is impossible as $Y \leq K$ and Y is nontrivial $QO_2(L)/O_2(L)$. This contradiction completes the proof. $\square$

LEMMA 3.3.17. If $\bar{L} \cong A_7$, then $m([V, L]) = 6$, eliminating case (5) of 3.3.8.

PROOF. Assume the lemma fails. Then by 3.3.8, $m([V, L]) = 4$. We work with two of the three proper subgroups in $\mathcal{L}(L, T)$. First, let $M_1 := C_L(Z)^\infty$. By 3.3.9, $C_L(Z) = C_L(Z \cap [V, L])$, so $\bar{M}_1 = C_{\bar{L}}(Z) \cong L_3(2)$. Then $1 \neq Z \leq O_2(\langle TD, M_1 \rangle)$. Second, there is $M_2 \in \mathcal{L}(L, T)$ with $\bar{M}_2\bar{T} \cong S_5$, so by 3.3.10.1, $O_2(\langle M_2, TD \rangle) \neq 1$. As $LT = \langle M_1, M_2, T \rangle$ and $M_iT/O_2(M_iT)$ is not isomorphic to $L_2(4)$ or A_6, 3.3.15 supplies a contradiction. $\square$

LEMMA 3.3.18. If $\bar{L} \cong L_n(2)$ with $n = 4$ or 5, then $[V, L]$ is not the direct sum of isomorphic natural modules.

PROOF. Assume otherwise; then $[V, L] = V_1 \oplus \cdots \oplus V_r$, where the V_i are isomorphic natural modules for $\bar{L}$. Therefore T induces inner automorphisms on $L/O_2(L)$, and in particular normalizes each parabolic of L containing $T \cap L$.

Let $Y_1 := C_L(Z)^\infty$, and recall $C_L(Z) = C_L(Z \cap [V, L])$ by 3.3.9. As the natural submodules V_i are isomorphic, $C_L(Z \cap [V, L])$ is the parabolic stabilizing a vector in each V_i, so that $\bar{Y}_1 \cong L_{n-1}(2)/E_{2^{n-1}}$, and hence $Y_1 \in \mathcal{L}(L, T)$.

Let W_1 be the T-invariant 3-subspace of V_1, and set $Y_2 := N_L(W_1)^\infty$. Then $\bar{Y}_2 \cong L_3(2)/E_8$ or $L_3(2)/E_{64}$ for $n = 4$ or 5, respectively, so $Y_2 \in \mathcal{L}(L, T)$. If some nontrivial characteristic subgroup of T were normal in Y_2T, then $O_2(\langle Y_2T, D \rangle) \neq 1$; so as $L = \langle Y_1, Y_2, T \rangle$, and $Y_2/O_2(Y_2) \cong L_3(2)$ rather than $L_2(4)$ or A_6, we have a contradiction to 3.3.15. It follows that (Y_2T, T) is an MS-pair in the sense of

Definition C.1.31. As $Y_2/O_2(Y_2) \cong L_3(2)$, case (5) of Theorem C.1.32 holds, so that Y_2T is described in C.1.34. Since T is Sylow in Y_2T, case (5) of C.1.34 does not hold, so that one of cases (1)–(4) of C.1.34 holds.

Let $Q := [O_2(Y_2T), Y_2]$ and $U := Z(Q)$. By B.2.14, $Z \leq \Omega := \Omega_1(Z(O_2(Y_2T)))$, so $[Y_2, Z] \leq Q \cap \Omega = U$ and hence $W_1 \leq [Z, Y_2] \leq U$ and Y_2 acts on UZ. Then by 12.8 in [**Asc86a**], $UZ = UZ_0$, where $Z_0 := C_Z(Y_2)$, so $Z = Z_0(Z \cap U)$. On the other hand $C_{V_i}(Y_2) = 1$ for each i, so $C_V(Y_2) = 1$ and hence $Z_0 = C_Z(L)$ by 3.3.7.4. Then as $M = !\mathcal{M}(LT)$, $M = !\mathcal{M}(C_G(z_0))$ for each $z_0 \in Z_0^\#$.

Assume that case (4) of C.1.34 holds. Then $U = U_0 \oplus U_1$, where $U_0 := C_U(Y_2T)$ is of rank 2 and U_1 is a natural module for $Y_2T/O_2(Y_2T) \cong L_3(2)$. Thus $U \cap Z = U_0 \oplus Z_1$, where $Z_1 := U_1 \cap Z$ is of order 2, so as $Z = Z_0(U \cap Z)$, $|Z : Z_0| = 2$. Further $m(Z) \geq m(U \cap Z) = 3$, so for each $d \in D$, $Z_0 \cap Z_0^d \neq 1$. Finally by an earlier remark,

$$M^d = !\mathcal{M}(C_G(z^d)) = M \text{ for some } z \in Z_0^\# \text{ with } z^d \in Z_0.$$

Thus $D \leq N_G(M) = M$ as $M \in \mathcal{M}$, contradicting 3.3.6.a. Hence case (4) of C.1.34 is eliminated.

Next $\bar{Y}_2$ has $m := 1$ or 2 noncentral 2-chief factors in $O_2(\bar{Y}_2)$, for $n = 4$ or 5, respectively, and Y_2 has $r \geq 1$ noncentral 2-chief factors in $[V, L] \leq O_2(L)$. Therefore Y_2 is not an $L_3(2)$-block, eliminating case (1) of C.1.34. Next the chief factor(s) for Y_2 in $O_2(\bar{Y}_2)$ are isomorphic to $W_1 \leq U$, so case (3) of C.1.34 is also eliminated, since there the noncentral 2-chief factors of Y_2 other than U lie in Q/U and are dual to U. Thus case (2) of C.1.34 holds, so $Q = U = U_1 \oplus U_2$ is the sum of two isomorphic natural modules U_i, and in particular Y_2 has exactly two noncentral 2-chief factors. Thus $m + r \leq 2$, so as $m \geq 1 \leq r$, it follows that $m = r = 1$, and therefore $n = 4$ and $V = V_1 = [O_2(L), L]$. We may choose notation so that $W_1 \leq U_1$. As $V = [O_2(L), L]$, L is an $L_4(2)$-block, so $P := O_2(Y_1) \cong D_8^3$, $P/Z(P) = P_1/Z(P) \oplus P_2/Z(P)$ is the sum of two nonisomorphic natural modules $P_i/Z(P)$ for Y_1/P, and we may choose notation so that $V_1 = P_1$. Thus as we saw that D normalizes Y_1, D normalizes $O_2(Y_1) = P$, and hence $D = O^2(D)$ also normalizes P_1. Then as $P_1 = V_1 \trianglelefteq LT$, $D \leq N_G(P_1) \leq M = !\mathcal{M}(LT)$, contradicting 3.3.6.b. This completes the proof. $\square$

LEMMA 3.3.19. $\bar{L}$ is not $L_5(2)$.

PROOF. Assume otherwise, and let $Y := C_L(Z)^\infty$. As $V = \langle Z^L \rangle$, part (4) of Theorem B.5.1 shows that $V = [V, L] \oplus C_Z(L)$. Since 3.3.18 eliminates case (iv) of B.5.1.1, either case (iii) of that result holds with $[V, L]$ the sum of the natural module and its dual, or case (i) there holds, with $[V, L]$ irreducible. In the latter case by Theorem B.4.2 and 3.3.18, $[V, L]$ is a 10-dimensional irreducible.

Assume first that $[V, L]$ is the sum of the natural module and its dual. Then by B.5.1.6, $\bar{Y} \cong L_3(2)/2^{1+6}$, so $Y \in \mathcal{L}(L, T)$. By 3.3.12.3, D acts on Y, and then also on $J(O_2(YT))$. But again by B.5.1.6 (notice we can apply B.2.10 with $O_2(YT)$ in the role of "R"), we see that $J(O_2(YT)) \leq C_T(V) = O_2(LT)$, so $J(O_2(YT)) = J(O_2(LT))$ by B.2.3.3. Hence $D \leq N_G(J(O_2(LT))) \leq M = !\mathcal{M}(LT)$, contradicting 3.3.6.b.

Therefore $[V, L]$ is irreducible of dimension 10, and in particular is the exterior square of a natural module. So this time (see e.g. K.3.2.3) Y is the parabolic determined by the stabilizer of a 2-space in that natural module; again $Y/O_2(Y) \cong L_3(2)$ so $Y \in \mathcal{L}(L, T)$ and as before D normalizes Y by 3.3.12.3. Now $O_2(\bar{Y}T)$ does not contain the unipotent radical of the maximal parabolic determined by the end node stabilizing a 4-space in the natural module. Thus by B.4.2.11 (again for more detail see K.3.2.3), $J(O_2(YT)) \leq C_T(V)$, so again $J(O_2(YT)) = J(O_2(LT))$, for the same contradiction. The proof is complete. $\qquad\square$

The next technical result has the same flavor as 3.3.12, and will be used in a similar way. In particular it will help to eliminate the shadows discussed earlier.

LEMMA 3.3.20. *Assume* $X = O^2(X)$ *is* T-*invariant with* $XT/O_2(XT) \cong S_3$, *and* D *does not normalize* $R := O_2(XT)$. *Let* $Y := \langle X^D \rangle$, *and let* γ *denote the number of noncentral 2-chief factors for* X. *Then*

(1) $\langle XT, D \rangle \in \mathcal{H}(T)$ *and* $Y \trianglelefteq \langle XT, D \rangle = YTD$.

(2) $YT/O_2(YT) \cong L_2(p)$ *for a prime* $p \equiv \pm 11 \mod 24$.

(3) $O_2(X) \leq O_2(Y)$, $XT/O_2(YT) \cong D_{12}$, *and* $|D : N_D(X)| = 3$.

(4) $\gamma \geq 3$.

(5) *If* $\gamma \leq 4$, *then:*

> (a) Y *has a unique noncentral 2-chief factor* W, $m(W) \geq 10$, *and* $|T| \geq 2^{12}$.

> (b) $\Phi(O_2(X)) \leq Z(Y)$.

> (c) *If* $Z(YT) \neq 1$, *then* $|T| > 2^{12}$.

PROOF. Let $B := \langle XT, D \rangle$. As D does not act on R, $R \neq 1$. Thus $XT \in \mathcal{H}(T)$ and T is maximal in XT as $XT/R \cong S_3$. Therefore $B \in \mathcal{H}(T)$ by 3.3.10.1. Also $X^{TD} = X^D$ so $Y \trianglelefteq B$, establishing (1).

Notice using A.1.6 that $O_2(B) \leq O_2(XT) = R$. Let B_0 be maximal subject to $B_0 \trianglelefteq B$ and $XT \cap B_0 \leq R$. Then $XT \cap B_0 = R \cap B_0 = T \cap B_0 =: T_0$ contains $O_2(B)$ and is invariant under XT and D, so $T_0 \trianglelefteq B$. Thus $T_0 = O_2(B)$. As D does not act on R by hypothesis, $T_0 < R$.

Set $B^* := B/B_0$. As $T_0 = XT \cap B_0 < R$, $R^* \neq 1 \neq X^*$. Then as $XT/R \cong S_3$, $|T^*| = 2|R^*| > 2$.

Let B_1^* be a minimal normal subgroup of B^*. By maximality of B_0, $XT \cap B_1 \not\leq R = O_2(XT)$, so $X^* \cap B_1^*$ is not a 2-group. So as $|X : O_2(X)| = 3$ and $X = O^2(X)$, $X^* \leq B_1^*$. Then by minimality of B_1^*, $B_1^* = \langle X^{*D} \rangle = Y^*$. In particular Y^* is the unique minimal normal subgroup of B^*, so $Y^* = F^*(B^*)$; hence T^* is faithful on Y^*.

Suppose Y^* is solvable. Then $Y^* \cong E_{3^n}$ as Y^* is a minimal normal subgroup of B^*. As $Y^* = \langle X^{*D} \rangle$, and D acts on T with X^* a simple T-submodule of Y^*, Y^* is a semisimple T-module. Therefore as T^* is faithful on Y^*, $\Phi(T^*) = 1$, and as $|T^*| > 2$, $m(T^*) > 1$. Then by (1) and (2) of A.1.31, $m(T^*) = 2$ and $m(C_{Y^*}(t^*)) \leq 1$ for each $t^* \in T^{*\#}$, so that $n = 2$ or 3. Further if $n = 3$, then as $B = YTD$ by (1), $T^*D^* \cong A_4$ is irreducible on Y^*, contrary to A.1.31.3. Thus $n = 2$, so $T^*D^* \leq GL_2(3)$. Then as $\Phi(T^*) = 1$ and D^* is a subgroup of $GL_2(3)$ of odd order normalizing T^*, $D^* = 1$. Hence $Y^* = \langle X^{*D^*} \rangle = X^*$, contradicting $n = 2$.

So Y^* is not solvable, and hence $Y^* = F^*(B^*) = Y_1^* \times \cdots \times Y_s^*$ is the direct product of isomorphic simple groups Y_i^* permuted transitively by TD. Then (1.a) of Theorem A (A.2.1) holds, so $m_q(Y^*) \leq m_q(B) \leq 2$ for each odd prime q, so that $s \leq 2$ and Y^* is an SQTK-group. Thus as $D = O^2(D)$, D normalizes each Y_i^*, so T is transitive on the Y_i^*. Therefore if T acts on Y_1^*, then $s = 1$ and Y^* is simple. As $Y^* = \langle X^{*D^*} \rangle$ and Y^* is not solvable, $D^* \neq 1$.

As D does not act on X, there is $g \in D - N_G(X)$. Set $G_1 := XT$, $G_2 := X^gT$, and $G_0 := \langle G_1, G_2 \rangle$. As $XT/R \cong S_3$ and D acts on T, (G_0, G_1, G_2) is a Goldschmidt triple as in Definition F.6.1. Thus if g does not act on $R = O_2(XT)$, $O_2(XT) \neq O_2(X^gT)$, so $G_0^+ := G_0/O_{3'}(G_0)$ is described in Theorem F.6.18 by F.6.11.2.

Suppose for each $g \in D - N_G(X)$ that the group G_0^+ defined by g satisfies case (1) or (2) of F.6.18. Then $O_2(G_0) = R \cap R^g$ is normalized by XT for all $g \in D$, so

$$R_D := \bigcap_{d \in D} R^d = \bigcap_{d \in D} (R \cap R^d)$$

is invariant under XT and D, and hence $R_D \leq O_2(B) = T_0$. Therefore as $T_0 \leq R$, $R_D = T_0$. Also $\Phi(T) \leq R \cap R^d$ since $T/(R \cap R_d) \cong \mathbf{Z}_2$ or E_4 in cases (1) and (2) of F.6.18, so $\Phi(T) \leq T_0$ and hence $\Phi(T^*) = 1$. Thus T^* acts on each Y_i^* as $T^* \cap Y_i^* \neq 1$, so $s = 1$ and $Y^* = F^*(B^*)$ is a simple SQTK-group by earlier remarks. As $\Phi(T^*) = 1$, we conclude from Theorem C (A.2.3) that $Y^* \cong L_2(2^n)$, J_1, or $L_2(p)$ for a prime $p \equiv \pm 3 \mod 8$. As $X^*T^*/R^* \cong S_3$, the first two cases are eliminated. In the third case $B^* = Y^*$ as $Y^* = F^*(B^*)$ and $\Phi(T^*) = 1$. Thus $X^*T^* \cong D_{12}$, and $N_{B^*}(T^*) \cong A_4$. Then from the list of maximal subgroups of B^* in Dickson's Theorem A.1.3, $B^* = Y^*T^* = \langle X^*T^*, X^{*g}T^* \rangle$, contrary to our assumption that each $g \in D - N_G(X)$ defines a group G_0^+ satisfying case (1) of (2) of F.6.18.

Therefore we may choose $g \in D - N_G(X)$ so that G_0^+ satisfies one of the remaining cases (3)–(13) of F.6.18. In particular inspecting those cases, $1 \neq G_0^{+\infty} = E(G_0^+)$ is quasisimple. Then as $O_{3'}(G_0)$ is solvable by F.6.11.1, we conclude from 1.2.1.1 that $K_0 := G_0^\infty$ is the unique member of $\mathcal{C}(G_0)$, and $K_0^+ = E(G_0^+)$. Hence $K_0 \in \mathcal{L}(G,T)$. By 1.2.4, $K_0 \leq K \in \mathcal{C}(B)$, and $K \trianglelefteq B$ as T acts on K_0. As $T \cap B_0 = O_2(B_0)$, $K^* \neq 1$, so as Y^* is the unique minimal normal subgroup of B^*, $K^* = Y^* = F^*(B^*)$ is simple.

Assume for the moment that $K_0^* < K^*$. Set $T_K := T \cap K \in Syl_2(K)$. We compare the possiblities for K_0^+ described in F.6.18 to the embeddings described in A.3.12, to obtain a list of possiblities for K^*. Cases (2), (3), (15), (16), and (22) of A.3.12 do not arise, since there the candidate "$B/O_{3'}(B)$" for K_0^+ does not appear in F.6.18; this also eliminates the subcase of (8) with $K^* \cong L_2(p)$ for $p \equiv \pm 3$ mod 8 and $K_0^* \cong A_5$. In cases (4)–(7), (11)–(14), and (17)–(21), and also in the remaining subcase of (8) where $K^* \cong L_2(p^2)$, $Aut(K^*)$ is a 2-group, so $B^* = K^*T^*$ since $K^* = Y^* = F^*(B^*)$. Furthermore in each case T_K^* is self-normalizing in $Aut(K^*)$, so $N_{B^*}(T^*) = T^*$ in these cases.

Next assume we are in the subcase of (9) where $K^* \cong U_3(5)$ and $K_0^* \cong A_6$. As in the proof of 3.3.12, D^* induces a group of outer automorphisms of order 3 on K^* centralizing T_K^*, and as D^* normalizes T^*, T^* induces inner automorphisms on K^* so that $B^* = K^*D^*$ and $T_K^* = T^*$. Now as D centralizes $T_K^* = T^* \in Syl_2(B^*)$, D centralizes $O_2(X^*T^*)$, so D normalizes the preimage S in B of $O_2(X^*T^*)$, and hence as $O_{2,Z}(K)$ is 2-closed, D normalizes $O^{2'}(S) = O_2(XT) = R$, contrary to the hypothesis of the lemma.

Finally in the remaining subcase of (9) and in (10), $K^* \cong L_3^\epsilon(p)$ with $K_0^* \cong SL_2(p)$ or $SL_2(p)/E_{p^2}$ for an odd prime p, since $K^* = Y^*$ is simple.

Thus we have shown that one of the following holds:

(a) $K_0^* = K^*$.
(b) $N_{B^*}(T^*) = T^*$.

(c) $K_0^* < K^* \cong L_3^\epsilon(p)$ for some odd prime p.

In case (b), $D^* = 1$, contrary to an earlier remark. Suppose case (c) holds. Then from the structure of $N_{Aut(K^*)}(T^*)$, $D^* \le D_0^*$, where D_0^* is a cyclic subgroup of K^* of order dividing $p - \epsilon$ centralizing T^*. Further we saw that $K_0^* \cong SL_2(p)$ or $SL_2(p)/E_{p^2}$. But now $[K_0^*, D^*] \le [K_0^*, D_0^*] \le O(K_0^*)$, contradicting $G_0^* = \langle X^*T^*, X^{*g}T^* \rangle$.

Therefore case (a) holds, with $K_0^* = K^* = Y^* = F^*(B^*)$, and D^* acts on $Y^*T^* = G_0^*$, so $G_0^* \trianglelefteq B^* = Y^*T^*D^* \le Aut(K^*)$. Recall G_0^+ satisfies one of cases (3)–(13) of F.6.18, but does not satisfy (b). As $F^*(G^*) = K^*$ is simple and K_0^+ is quasisimple, $K^* \cong K_0^+/Z(K_0^+)$. Examining F.6.18 for groups with $T^* < N_{G^*}(T^*)$, we conclude case (4) or (10) of F.6.18 holds. However $G_2 = G_1^g$, so $G_2^* \cong G_1^*$, ruling out case (10) of F.6.18, since $G_1^+Z(K_0^+)/Z(K_0^+) \cong G_1^* \cong G_2^* \cong G_2^+Z(K_0^+)/Z(K_0^+)$. This leaves case (4) of F.6.18, so we conclude that $G_0^+ = K_0^+ \cong L_2(p)$, $p \equiv \pm 11$ mod 24, and $X^+T^+ \cong D_{12}$. As G_0^+ is simple, $G_0^+ \cong G_0^* = K^*$. Further $Aut(K^*)$ is a 2-group, so $B^* = G_0^*D^* = K^* \cong G_0^+$.

Next there is $t \in T \cap K$ with $X^* = [X^*, t^*]$, so $X = [X, t] \le K$, and hence $Y = \langle X^D \rangle \le K$ as $K \trianglelefteq B$. By (1), $Y \trianglelefteq B = YTD$, so since $K \in \mathcal{C}(B)$ with $K^* = B^* \cong G_0^+ \cong L_2(p)$, we conclude from 1.2.1.4 that either (2) holds, or $Y/O_2(Y) \cong SL_2(p)/E_{p^2}$. However in the latter case, by a Frattini Argument, $Y = O_p(Y)Y_0$, where $Y_0 := N_Y(T_1)$ and $T_1 := T \cap O_\infty(Y)$. But then XT and D act on T_1, so $T_1 \le O_2(B)$, whereas $T_1 \not\le O_2(Y)$. Thus (2) is established.

We saw that $X^*T^* \cong D_{12}$, and from (2), $N_{B^*}(T^*) \cong A_4$, so (3) follows. Further we observed earlier that $B^* \cong G_0^+$, so $B^* = \langle X^*T^*, X^{*g}T^* \rangle$ for $g \in D$ with $g^* \ne 1$.

Let W be a noncentral 2-chief factor of Y, $n := m(W)$ and $\alpha := m([W, X^*])/2$. Then α is the number of noncentral chief factors for X^* on W, so $\alpha \le \gamma$. As $B^* = \langle X^*T^*, X^{*g}T^* \rangle$, $C_W(X) \cap C_W(X^g) = 0$, so $n \le 2m([W, X^*]) = 4\alpha$. On the other hand, a Borel subgroup of B^* is a Frobenius group of order $p(p-1)/2$, so $n \ge (p-1)/2$ and hence $p \le 2n + 1 \le 8\alpha + 1$. Thus either $\alpha > 4$ or $p \le 33$, and in the latter case as $p \equiv \pm 11$ mod 24, $p = 11$ or 13. As neither 11 nor 13 divides the order of $GL_9(2)$, we conclude that $n \ge 10$ and hence $\alpha \ge n/4 > 2$. Thus as $\gamma \ge \alpha$, (4) holds.

It remains to prove (5), so assume that $\gamma \le 4$. Then $\alpha \le 4$, so by the previous paragraph, W is the unique noncentral 2-chief factor for Y, and $m(W) \ge 10$. Then as $|T^*| = 4$, $|T| \ge 2^{12}$, with equality only if $p = 11$ and $W = O_2(YT)$, so that $Z(YT) = 1$. Therefore parts (a) and (c) of (5) hold. Finally $W = U/U_0$ where $U := [O_2(Y), Y]$ and $U_0 := C_U(Y)$, and as $O_2(X) \le O_2(Y)$ by (3), $O_2(X) = [O_2(X), X] \le U$. Then as U/U_0 is elementary abelian and $X \le Y$, $\Phi(O_2(X)) \le U_0 \le Z(Y)$, establishing part (b) of (5). This completes the proof. $\qquad\square$

In the next lemma, we eliminate the first occurrence of the shadow of $\Omega_8^+(2)$ extended by triality.

PROPOSITION 3.3.21. $\bar{L}$ is not $L_4(2)$, eliminating case (7) of 3.3.8.

PROOF. Assume otherwise. Arguing as in the proof of 3.3.19 via appeals to Theorems B.5.1, B.4.2, and 3.3.18, we conclude:

(a) Either

(1) $[V, L] = U_1 \oplus U_2$, where U_1 is a natural submodule of V and U_2 is the dual of U_1, or

(2) $[\tilde{V}, L]$ is the 6-dimensional orthogonal module for $\bar{L}$.

Next by 3.3.9, and appealing to B.5.1.6 in case (a1):

(b) In case (a1), $C_{\bar{L}}(Z) \cong S_3/2^{1+4}$.

(c) In case (a2), $C_{\bar{L}}(Z) \cong (S_3 \times S_3)/E_{16}$.

Let $R := O_2(C_L(Z)T)$. Then $\bar{R}$ is the unipotent radical of the parabolic $C_{\bar{L}}(Z)$ of $\bar{L}$, so $N_M(R) \leq N_M(O^2(C_L(Z)))$. By B.5.1.6 and B.4.2.10, $J(R) \leq C_T(V) = O_2(LT)$, so that $J(R) = J(O_2(LT))$ by B.2.3.3, and hence $N_G(R) \leq N_G(J(R)) \leq M$ as $M = !\mathcal{M}(LT)$. Therefore as we just showed that $O^2(C_L(Z))$ is normal in $N_M(R)$:

(d) $J(R) = J(O_2(LT))$, and $O^2(C_L(Z)) \trianglelefteq N_G(R) \leq M$. Thus D does not act on R, and hence does not act on $O^2(C_L(Z))$.

Next we show:

(e) $C_Z(L) = 1$, so $Z \leq [V, L] = V$. Further when (a2) holds, L is irreducible on V. For if $C_Z(L) \neq 1$, then $C_G(Z) \leq C_G(C_Z(L)) \leq M = !\mathcal{M}(LT)$, so $O^{3'}(C_G(Z)) = O^{3'}(C_M(Z)) = O^2(C_L(Z))$ is D-invariant, contrary to (d). Then since $V = [V, L]C_Z(L)$ by 3.3.7.4, $V = [V, L]$.

Our final technical result requires a lengthier proof:

(f) T is nontrivial on the Dynkin diagram of $\bar{L}$.

Assume that T is trivial on the Dynkin diagram of $\bar{L}$. Then $\bar{T} \leq \bar{P}_i \leq \bar{L}$, for $i = 1, 2$, with $\bar{P}_i \cong L_3(2)/E_8$. Let $Y_i := P_i^\infty$, so that $Y_i \in \mathcal{L}(L, T)$. Then $LT = \langle Y_1, Y_2, T \rangle$.

We now repeat some of the proof of 3.3.18: By 3.3.15 we may assume there is no nontrivial characteristic subgroup of T normal in YT for $Y := Y_1$, so the MS-pair (YT, T) is described in C.1.34. As T is Sylow in G, case (5) of C.1.34 does not hold. By (a) and (e), $m(C_Z(Y)) \leq 1$, so case (4) does not hold. Next Y has a nontrivial 2-chief factor on $O_2(\bar{Y})$ and two on $[V, L]$ from (a1) and (a2), eliminating cases (1) and (2) of C.1.34 where there are at most two such factors. Therefore case (3) of C.1.34 holds. Set $Q := [O_2(YT), Y]$ and $U := Z(Q)$; then U is a natural module for $Y/O_2(Y)$ and Q/U the sum of two copies of the dual of U. In particular, Y has exactly three noncentral 2-chief factors. Then $C_Q(Y) = 1$, eliminating case (a1) where $C_{[V,Y]}(Y) \neq 1$ and $[V, Y] \leq Q$. Thus case (a2) holds and L is an A_8-block.

As $\bar{T} \leq \bar{L}$, $LT = O_2(LT)L$. By (e), $C_T(L) = 1$, so by C.1.13.b and B.3.3, either $V = O_2(LT)$ or $O_2(LT)$ is the 7-dimensional quotient of the permutation module for $\bar{L}$. But in the latter case, as $T = O_2(LT)(L \cap T)$, $J(T) \leq C_T(V)$ by B.3.2.4, contradicting 3.3.7.2.

Thus $O_2(LT) = V$, so $T \leq L$ and $|T| = 2^{12}$. Let L_i, $i = 1, 2$, be the rank-1 parabolics of $C_L(Z)$ over T, and set $X_i := O^2(L_i)$, and $R_i := O_2(L_i)$. By (d), D does not act on R, so as $R = R_1 \cap R_2$, D does not act on R_i for some i, say $i = 1$. We now apply 3.3.20 to X_1 in the role of "X": Let $Y := \langle X_1^D \rangle$, and observe that the number γ of noncentral 2-chief factors of X_1 is four, and $Z \leq Z(YT)$. Thus as $|T| = 2^{12}$, part (c) of 3.3.20.5 supplies a contradiction, which establishes (f).

We now complete the proof of lemma 3.3.21.

Let P be the parabolic of L with $P/O_2(P) \cong S_3 \times S_3$, and set $H := PT$. Then by (f), $H/O_2(H) \cong S_3 \text{ wr } \mathbf{Z}_2$, so by 3.3.13, $D \leq N_G(H)$. However in case (a2), $J(O_2(H)) \leq C_T(V)$ by B.3.2, so that $J(O_2(H)) = J(O_2(LT))$ by B.2.3.3; hence D normalizes $J(O_2(LT))$, contradicting 3.3.6.b. Therefore case (a1) must hold.

We have $Z \leq [V, L] = V$ by (e), and $T \nleq LO_2(LT)$ by (f), so $V = W \oplus W^t$ for $t \in T - LO_2(LT)$ with $W := U_1$ the natural module for $\bar{L}$ and W^t dual to W. In particular $Z \cong \mathbf{Z}_2$ is D-invariant, and we saw $D \leq N_G(H)$, so D normalizes

$$U := \langle Z^H \rangle = (U \cap W) \oplus (U \cap W)^t,$$

with $U \cap W \cong E_4$. Now H acts as $O_4^+(2)$ on U, so $Aut_H(U)$ is self normalizing in $GL(U)$ and $Aut_T(U)$ is self normalizing in $Aut_H(U)$; thus we conclude $[U, D] = 1$. Hence $[H, D] \leq C_H(U) = O_2(H)$; in particular D centralizes $T/O_2(H)$, so D acts on $S := T \cap LO_2(LT)$, and hence on $Z_W := C_W(S)$, since $Z_W \leq U$ and D centralizes U.

Let $L_W := C_L(Z_W)^\infty$. Then $L_W/O_2(L_W) \cong L_3(2)$, and L_W has noncentral chief factors on each of W/Z_W, W^t, and $O_2(\bar{L}_W)$. We will now apply earlier arguments to see that $(L_W S, S)$ cannot be an MS-pair; then since (MS1) and (MS2) hold, we can conclude (MS3) does not hold. So suppose (MS3) does hold: then we may apply C.1.32, and as before one of cases (1)–(4) of C.1.34 holds. Since we saw there are at least three noncentral 2-chief factors, cases (1) and (2) of C.1.34 are eliminated. As $Z_W \leq W = [W, L_W]$ is a nonsplit extension of a natural quotient over a trivial submodule, case (3) of C.1.34 does not hold. We've seen $m(Z) = 1$, so as $|T : S| = 2$, $m(Z(S)) \leq 2$, and hence case (4) of C.1.34 does not hold. This contradiction shows that (MS3) fails, so there is $1 \neq C$ char S with $C \trianglelefteq L_W S$. But then $C \trianglelefteq \langle L_W, T \rangle = LT$, while D normalizes S and hence also C, contradicting 3.3.6.b. $\qquad\square$

LEMMA 3.3.22. $\bar{L}$ is not A_7, eliminating case (3) of 3.3.8.

PROOF. If $\bar{L} \cong A_7$ then by 3.3.8 and 3.3.17, $[V, L]$ is the natural module for $\bar{L}$. We adopt the notational conventions of section B.3; that is we regard $\bar{L}\bar{T} \cong S_7$ as the group of permutations on $\Omega := \{1, \dots, 7\}$, $[V, L]$ as the set of even subsets of Ω, and take $\bar{T}$ to have orbits $\{1, 2, 3, 4\}$, $\{5, 6\}$, $\{7\}$ on Ω. Set $\theta := \Omega - \{7\}$; then

$$Z_V := Z \cap [V, L] = \langle e_{5,6}, e_\theta \rangle.$$

Let $L_\theta := C_L(e_\theta)^\infty$. Observe $\bar{L}_\theta \cong A_6$ and $R := O_2(LT) = O_2(L_\theta T)$, with $C(G, R) \leq M$ by 1.4.1.1.

Consider any $z \in C_Z(L)e_\theta$, and set $G_z := C_G(z)$ and $M_z := C_M(z)$. Then $z \in Z$, so that $G_z \in \mathcal{H}^e$ by 1.1.4.6. As $L_\theta \trianglelefteq M_z$, $R \in \mathcal{B}_2(G_z)$ and $R \in Syl_2(\langle R^{M_z} \rangle)$ by A.4.2.7, so as $C(G, R) \leq M$, it follows that Hypothesis C.2.3 is satisfied with G_z, M_z in the roles of "H, M_H". Further by 1.2.4, $L_\theta \leq K \in \mathcal{C}(G_z)$. Now $F^*(K) = O_2(K)$ by 1.1.3.1, and $m_3(L_\theta) = 2$, so $K/O_2(K)$ is quasisimple by 1.2.1.4 and T acts on K by 1.2.1.3. Assume $L_\theta < K$, so that $K \nleq N_G(L) = M$. Then C.2.7 supplies a contradiction, as in none of the cases listed there does there exist a T-invariant $L_\theta \in \mathcal{C}(M \cap K)$ with $L_\theta/O_2(L_\theta) \cong A_6$. Hence $L_\theta = K \trianglelefteq G_z$. Thus by A.3.18

$$L_\theta = O^{3'}(C_G(z)) \quad \text{for each} \quad z \in C_Z(L)e_\theta. \tag{$*$}$$

Similarly for $z \in C_Z(L)$, $C_G(z) \leq M$ as $M = !\mathcal{M}(LT)$, so by A.3.18:

$$L = O^{3'}(C_G(z)) \quad \text{for each} \quad z \in C_Z(L). \tag{$**$}$$

Now as $C_{[V,L]}(L) = 0$, 3.3.7.4 says that $V = [V, L] \oplus C_Z(L)$ and $Z = Z_V \oplus C_Z(L)$. We claim that $C_Z(L) = 1$, so that $V = [V, L]$ and $Z = Z_V$. Assume otherwise. Then $m(Z) > 2$, and $Z_\theta := \langle C_Z(L), e_\theta \rangle$ is a hyperplane of Z, so for each $d \in D$, $1 \neq Z_\theta \cap Z_\theta^d$. Hence we may choose $z \in Z_\theta^\#$ with $z^d \in Z_\theta$. First suppose $z \in$

$C_Z(L)$. By (**), $O^{3'}(C_G(z^d)) = L^d$ and $L^d \neq L_\theta$ since $|L| > |L_\theta|$. Therefore by (*), $z^d \notin C_Z(L)e_\theta = Z_\theta - C_Z(L)$, and hence $z^d \in C_Z(L)$, so again using (**), $L = O^{3'}(C_G(z^d)) = L^d$. Thus $d \in N_G(L) = M$ by 1.4.1 in this case. In the remaining case, $z \in Z_\theta - C_Z(L) = C_Z(L)e_\theta$, where by (*), $O^{3'}(C_G(z)) = L_\theta$, and hence $O^{3'}(C_G(z^d)) = L_\theta^d \neq L$. Therefore by (**), $z^d \in Z_\theta - C_Z(L) = C_Z(L)e_\theta$, and then $L_\theta = O^{3'}(C_G(z^d))$ by (*), and hence $L_\theta = L_\theta^d$. Thus d normalizes $L_\theta T$ and hence also $O_2(L_\theta T) = O_2(LT)$, so again $d \in M$. Therefore $D \leq M$, contrary to 3.3.6.a, establishing the claim.

Next $C_D(Z) \leq C_D(e_\theta)$, and $C_D(e_\theta)$ normalizes $O^{3'}(C_G(e_\theta)) = L_\theta$ using (*), and hence also normalizes $O_2(L_\theta T) = R$. Therefore $C_D(Z) \leq N_G(R) \leq M$ as $C(G,R) \leq M$, so $C_D(Z) < D$ as $D \not\leq M$. As Z is of rank 2, we conclude $|D : D \cap M| = 3$, with D transitive on $Z^\#$. In particular there is $d \in D$ with $e_{5,6}^d = e_\theta$. Let $L_{5,6} := C_L(e_{5,6})^\infty$. Thus $\bar{L}_{5,6}\bar{T} \cong \mathbf{Z}_2 \times S_5$, and $L_{5,6}^d \leq O^{3'}(C_G(e_\theta)) = L_\theta$. This is impossible, as $T = T^d$ acts on $L_{5,6}^d$ and L_θ, whereas there is no T-invariant subgroup of $L_\theta/O_{2,Z}(L_\theta) \cong A_6$ isomorphic to A_5.

We have shown that $\bar{L}$ is not A_7. Thus case (3) of 3.3.8 does not hold by 3.3.10.2. This completes the proof of 3.3.22. □

Notice that at this point, cases (1), (2), (3), (5), and (7) of 3.3.8 have been eliminated by 3.3.14, 3.3.10.2, 3.3.22, 3.3.17, and 3.3.21. Thus leaves case (6) of 3.3.8, where $\bar{L} \cong \hat{A}_6$, and case (4) of 3.3.8, where $\bar{L} \cong L_3(2)$, A_6, or $U_3(3)$ by 3.3.16. In each of these cases, $L/O_{2,Z}(L)$ is of Lie type and Lie rank 2 in characteristic 2, and T normalizes L. Threfore by 3.3.6.d, (LT, T) is an MS-pair in the sense of Definition C.1.31. Thus we may apply C.1.32 to LT to conclude that either L is a block, or $\bar{L} \cong L_3(2)$ is described in C.1.34. We first investigate the latter possibility in more detail:

LEMMA 3.3.23. *If $\bar{L}$ is $L_3(2)$, then either*

(1) L is an $L_3(2)$-block, and D acts on the preimage T_0 in T of $Z(\bar{T})$, or

(2) L has two or three noncentral 2-chief factors, and D does not act on $O_2(C_L(Z)T)$.

PROOF. As in earlier arguments we conclude that one of cases (1)–(4) of C.1.34 holds. In particular $[V, L]$ is a sum of $r \leq 2$ isomorphic natural modules, so by 3.3.7.4, $V = [V, L] \oplus Z_L$ and $Z = (Z \cap [V, L]) \oplus Z_L$, where $Z \cap [V, L]$ has rank r.

Suppose case (4) of C.1.34 holds; we argue as in the proof of 3.3.18, although many details are now easier: As $M = !\mathcal{M}(LT)$, $M = !\mathcal{M}(C_G(z))$ for each $z \in Z_L^\#$, and in case (4) of C.1.34, $m(Z_L) \geq 2$ and $r = 1$ so Z_L is a hyperplane of Z, leading to the same contradiction as in the proof of 3.3.18.

Thus we are in case (m) of C.1.34 for some $1 \leq m \leq 3$, where L has m noncentral 2-chief factors. This gives the first statements of (1) and (2). Next in each case of C.1.34, $T \leq LO_2(LT)$. Set $X := O^2(C_L(Z))$ and $R := O_2(XT)$. Now LR, R also satisfy (MS1) and (MS2), but if $m = 2$ or 3, then (LR, R) is not an MS-pair as the corresponding cases of C.1.34 exclude this choice of R. Therefore (MS3) must fail for R, so there is a nontrivial characteristic subgroup C of R normal in LR, and hence normal in LT as $R \triangleleft T$. Thus $N_G(R) \leq N_G(C) \leq M = !\mathcal{M}(LT)$, so D does not act on R as $D \not\leq M$ by 3.3.6.a, proving the second statement in (2).

Finally if $m = 1$, let P_i, $i = 1, 2$, denote the maximal parabolics of LT over T. Then P_i has just two noncentral 2-chief factors, so D acts on $O_2(P_i)$ by 3.3.20.4. Thus D acts on $T_0 := O_2(P_1) \cap O_2(P_2)$, completing the proof of the lemma. $\qquad \square$

In the proof of the next lemma, we encounter the shadow of the non-maximal parabolic in $\mathbf{Z}_3/\Omega_8^+(2)$, and we eliminate this shadow using 3.3.20.

LEMMA 3.3.24. L is a block of type A_6, $\hat{A}_6$, $G_2(2)$, or $L_3(2)$.

PROOF. We observed earlier that either L is a block of type A_6, $\hat{A}_6$, or $G_2(2)$, or $\bar{L} \cong L_3(2)$. Thus appealing to 3.3.23, we only need to eliminate the cases arising in 3.3.23.2, where L has $k := 2$ or 3 noncentral 2-chief factors.

Let $Q := [O_2(LT), L]$. When $k = 2$, C.1.34.2 says that Q is the direct sum of two isomorphic natural modules for $L/O_2(L)$; then LT acts on at least one of the three natural submodules V_0 of Q, and we set $Z_0 := Z \cap V_0$. When $k = 3$, $V_0 := Z(Q)$ is a natural $L/O_2(L)$ module, and Q/V_0 is the direct sum of two copies of the dual of V_0. In this case we again set $Z_0 := Z \cap V_0$. Thus in either case Z_0 is of rank 1 and $V_0 = \langle Z_0^L \rangle = [Z_0, L]$ is an LT-invariant natural $L/O_2(L)$-module.

Set $R := O_2(C_L(Z)T)$, $X := O^2(C_L(Z))$, and $Y := \langle X^D \rangle$. Then X has $k + 1 \le 4$ noncentral 2-chief factors. By 3.3.23.2, D does not act on R, so we can apply 3.3.20.5 to conclude that Y has a unique noncentral 2-chief factor W, and that $Z_0 \le \Phi(O_2(X)) \le Z(Y)$. Set $\widetilde{YT} := YT/Z_0$, $R_Y := O_2(YT)$ and $U := \langle V_0^Y \rangle$. As X is irreducible on $\tilde{V}_0$, we may apply G.2.2.1 with V_0, Z_0, YT in the roles of "V, V_1, H", to conclude that $\tilde{U} \le \Omega_1(Z(\tilde{R}_Y))$, so $\Phi(U) \le Z_0$. As $V_0 = [V_0, X]$, $U = [U, Y]$, so by uniqueness of W, $W = U/U_0$ where $U_0 := C_U(Y)$. By 3.3.20.3, $O_2(X) \le R_Y$, so as $X = O^2(X)$, $O_2(X) = [O_2(X), X] \le [R_Y, Y] = U$. Then $\Phi(O_2(X)) \le Z_0$, eliminating the case $k = 2$, for there $\Phi(O_2(X)) = C_Q(L)$ is of rank 2. Thus $k = 3$, and here we compute that $Q/(O_2(X) \cap Q) \cong E_4$ and $[O_2(X), a] \not\le Z_0$ for each $a \in Q - O_2(X)$. Therefore setting $(YT)^* := YT/R_Y$, $Q^* \cong E_4$. This is impossible, since by 3.3.20.3, $X^*T^* \cong D_{12}$, whereas $Q^* \trianglelefteq X^*T^*$. $\qquad \square$

LEMMA 3.3.25. (1) L is a block of type A_6, $G_2(2)$, or $L_3(2)$.

(2) Assume $C_T(L) \ne 1$ and $\bar{L}$ is not $L_3(2)$, and let $X := O^2(C_L(Z))$ and $R := O_2(XT)$. Then D acts on X and R, but does not act on any nontrivial D-invariant subgroup of R normal in LT.

(3) If $C_T(L) = 1$, then either $V = O_2(LT)$, or L is an A_6-block.

PROOF. Let $X := O^2(C_L(Z))$ and $R := O_2(XT)$. Inspecting the cases listed in 3.3.24, $XT/R \cong S_3$.

We first prove (2), so suppose $C_T(L) \ne 1$ and $\bar{L}$ is not $L_3(2)$. Then $C_Z(L) \ne 1$, so as usual $C_G(Z) \le C_G(C_Z(L)) \le M = !\mathcal{M}(LT)$, and then $C_G(Z) = C_M(Z)$. As $\bar{L}$ is not $L_3(2)$, $m_3(L) = 2$ and so by A.3.18, L is the subgroup $\theta(M)$ generated by all elements of M of order 3. Therefore $X = \theta(C_G(Z))$, so D acts on X and hence also on R. Then the final statement of (2) follows from 3.3.6.b.

In view of 3.3.24, to prove (1) we may assume $\bar{L} \cong \hat{A}_6$, and it remains to derive a contradiction. By B.4.2, $J(R) \le C_T(V) = O_2(LT)$, so that $J(R) = J(O_2(LT))$ by B.2.3.3. Therefore $C_T(L) = 1$ by (2). Then as the $\hat{A}_6$-module has trivial 1-cohomology by I.1.6, $V = O_2(LT)$ by C.1.13.b. But again using B.4.2, there is a unique member $\bar{A}$ of $\mathcal{P}(\bar{T}, V)$, $m(\bar{A}) = 2$, and $C_V(\bar{A}) = C_V(\bar{a})$ for each $\bar{a} \in \bar{A}^\#$. Therefore by B.2.21, there is a unique member $A \in \mathcal{A}(T)$ with $[A, V] \ne 1$, and hence $\mathcal{A}(T) = \{A, V\}$ is of order 2. Therefore as D is of odd order, D acts on V, contrary to 3.3.6.b. So (1) is estabished.

Finally we prove (3), so we assume that $C_T(L) = 1$, and $V < O_2(LT)$. By (1), we may assume $\bar{L}$ is $L_3(2)$ or $U_3(3)$, and it remains to derive a contradiction. As $C_T(L) = 1$, $Q := O_2(LT)$ is elementary abelian by C.1.13.a. Further by C.1.13.b, B.4.8, and B.4.6, Q is the indecomposable module with natural irreducible submodule V and trivial quotient, of rank 4 or 7, respectively. By 3.3.7.2, V is an FF-module, so by B.4.6.13, $\bar{L}$ is not $U_3(3)$. Thus $\bar{L} \cong L_3(2)$, and by B.4.8.3, there is a unique member $\bar{A}$ of $\mathcal{P}(\bar{T}, Q)$. As $C_Q(\bar{A}) = C_Q(\bar{a})$ for each $\bar{a} \in \bar{A}^{\#}$ and $Q = C_{LT}(Q)$, we may apply B.2.21 to obtain the same contradiction as earlier. This completes the proof of (3). $\qquad \square$

Observe now that as L is a block by 3.3.25.1, Hypothesis C.6.2 is satisfied with L, T, T, TD in the roles of "L, R, T_H, Λ", For example, if $1 \neq R_0 \leq T$ with $R_0 \trianglelefteq LT$, then $D \not\leq N_G(R_0)$ by 3.3.6.b, which verifies part (3) of Hypothesis C.6.2. As Hypothesis C.6.2 is satisfied, we can apply C.6.3 to conclude:

LEMMA 3.3.26. *There exists $d \in D - M$ with $V^d \not\leq O_2(LT)$.*

In the remainder of the section, let d be defined as in 3.3.26. Set $Q_L := O_2(LT)$ and $T_C := C_T(L)$.

LEMMA 3.3.27. *Assume $T_C = C_T(L) \neq 1$. Then*

(1) $T_C \cap T_C^d = 1$.
(2) $\Phi(T_C) = 1$.
(3) Either $T_C^d \leq Q_L$ or $T_C \leq Q_L^d$.

PROOF. As L centralizes $T_C \trianglelefteq T$ and D acts on T, also $T_C^d \trianglelefteq T$, and then $T_C \cap T_C^d$ is normal in LT and in $L^d T$. Thus if $T_C \cap T_C^d \neq 1$, then

$$M^d = !\mathcal{M}(L^d T) = !\mathcal{M}(N_G(T_C \cap T_C^d)) = !\mathcal{M}(LT) = M,$$

contradicting our choice of $d \in D - M$ in 3.3.26, and so establishing (1). Then applying (1) to d^2 in the role of "d", $T_C \cap T_C^{d^2} = 1$, so also $T_C^{d^{-1}} \cap T_C^d = 1$.

Now as L is a block, $\Phi(Q_L) \leq T_C$ by C.1.13.a. Suppose (2) fails, so that $\Phi(T_C) \neq 1$. If $T_C^d \leq Q_L$, then $1 \neq \Phi(T_C^d) \leq \Phi(Q_L) \leq T_C$, contradicting (1); therefore $T_C^d \not\leq Q_L$, so by symmetry $T_C \not\leq Q_L^d$, and thus (3) fails. Hence (3) implies (2), so it remains to assume that (3) fails, and to derive a contradiction. Thus $T_C^d \not\leq Q_L$ and $T_C \not\leq Q_L^d$, so also $T_C^{d^{-1}} \not\leq Q_L$.

Suppose for the moment that $\bar{L}$ is $L_3(2)$. Then by 3.3.23.1, D acts on the preimage T_0 in T of $Z(\bar{T})$. Therefore as $\bar{T}_0$ is of order 2 and $T_C^d \not\leq Q_L$, $\bar{T}_C^d = \bar{T}_0$.

Now suppose that $\bar{L}$ is not $L_3(2)$. Then by 3.3.25.2, D acts on $X := O^2(C_L(Z))$ and on $R := O_2(XT)$. Therefore as $T_C \trianglelefteq XT$, $T_C^d \trianglelefteq XT$, and as T_C centralizes X, $1 \neq \bar{T}_C^d$ centralizes $\bar{X}$. Now $\bar{L} \cong A_6$ or $G_2(2)'$ by 3.3.25.1, and $\bar{T}$ is trivial on the Dynkin diagram of $\bar{L}$ if $\bar{L} \cong A_6$ since $\bar{L}$ is an A_6-block. Inspecting $Aut(\bar{L})$, we find that $C_{Aut(\bar{L})}(\bar{X}) = 1$ unless $\bar{L}\bar{T} \cong S_6$, whereas we saw $\bar{T}_C^d \neq 1$ centralizes $\bar{X}$. Therefore $\bar{L}\bar{T} \cong S_6$ and $\bar{T}_C^d = Z(\bar{X}\bar{T}) = \bar{T}_0$ is of order 2.

Thus $\bar{L}\bar{T} \cong S_6$ or $L_3(2)$ and $\bar{T}_C^d = \bar{T}_0$ is of order 2. As $T_C^d \trianglelefteq T$, $1 \neq [V, T_0] = [V, T_C^d] \leq T_C^d$. Similarly $[V, T_0] = [V, T_C^{d^{-1}}] \leq T_C^{d^{-1}}$, so $1 \neq [V, T_0] \leq T_C^d \cap T_C^{d^{-1}}$, contrary to the final remark in paragraph one. $\qquad \square$

LEMMA 3.3.28. *If $\bar{L}$ is $L_3(2)$ or $U_3(3)$, then $Q_L = V \times T_C$ and $\Phi(Q_L) = 1$. Indeed if $\bar{L}$ is $U_3(3)$, then $T_C = 1$ and $Q_L = V$.*

PROOF. Assume that $\bar{L}$ is $L_3(2)$ or $U_3(3)$ and set $T_C := C_T(L)$. By 3.3.26, there is $d \in D - M$ with $V^d \nleq Q_L$.

Suppose first that $\bar{L} \cong L_3(2)$. As case (1) of 3.3.23 holds, D acts on the preimage T_0 in T of $Z(\bar{T})$. Then as $|\bar{T}_0| = 2$, $T_0 = V^d Q_L$ and $m(T_0/C_{T_0}(V)) = 1$, so $m(Q_L/C_{Q_L}(V^d)) = 1 = m(V/C_V(V^d))$, and hence $Q_L = V C_{Q_L}(V^d)$. Now if $Q_L/C_T(L)$ is the unique nonsplit extension of V with a 1-dimensional submodule described in B.4.8, then the fixed points of $\bar{T}_0$ are contained in $V C_T(L)$, contrary to $Q_L = V C_{Q_L}(V^d)$ with $\bar{T}_0 = \bar{V}^d$. Therefore $Q_L = V \times T_C$, so as $\Phi(T_C) = 1$ by 3.3.27.2, the lemma holds in this case.

Thus we may assume $\bar{L} \cong U_3(3)$. Notice that if $T_C = 1$, then $V = O_2(LT)$ by 3.3.25.3, so that the lemma holds. Therefore we may assume that $T_C \neq 1$, and it remains to derive a contradiction.

Set $X := O^2(C_L(Z))$ and $R := O_2(XT)$. By 3.3.25.2, D acts on R and X. Then V^d is elementary abelian and normal in the parabolic subgroup XT, so using B.4.6, $m(\bar{V}^d) = 2$ or 3, and hence $m(V/C_V(V^d)) = 3$. Then by symmetry between V and V^d, $m(V^d/C_{V^d}(V)) = 3$. Thus $m(\bar{V}^d) = 3$ so as $\bar{V}^d \trianglelefteq \bar{X}$, $\bar{V}^d = C_{\bar{R}}(\bar{V}^d)$ is the unique FF-offender on V in $\bar{R}$ by B.4.6.13. Therefore $C_R(V^d) \leq V^d Q_L$, so $C_R(V^d) = V^d C_{Q_L}(V^d)$. Also $|C_R(V^d)| = |C_R(V)| = |Q_L|$, so $|Q_L : C_{Q_L}(V^d)| = |\bar{V}^d|$. Then as $|\bar{V}^d| = |V : C_V(V^d)|$, $Q_L = C_{Q_L}(V^d)V$. However in the unique nonsplit extension of $V/C_V(L)$ over a 1-dimensional submodule described in B.4.6, the fixed points of $\bar{V}^d$ are contained in $V/C_V(L)$. Thus as $Q_L = V C_{Q_L}(V^d)$, $Q_L = V T_C$. Then since $\Phi(T_C) = 1$ by 3.3.27.2, $\Phi(Q_L) = 1$.

Again by B.4.6.13, $\bar{V}^d$ is the unique member of $\mathcal{P}(\bar{R}, V)$, and $C_V(\bar{V}^d) = C_V(\bar{a})$ for each $\bar{a} \in \bar{V}^d - \bar{L}$. Therefore as $Q_L = V C_T(V^d)$ and $m(\bar{V}^d) = m(V/C_V(V^d))$, B.2.21 applied with Q_L in the role of "V" says Q_L^d is the unique member of $\mathcal{A}(R)$ with $[Q_L, Q_L^d] \neq 1$, so $\mathcal{A}(R)$ is of order 2. Then as D of odd order acts on R, D normalizes Q_L, contrary to 3.3.6.b. This completes the proof. $\qquad\square$

LEMMA 3.3.29. *$\bar{L}$ is not $L_3(2)$.*

PROOF. Assume $\bar{L}$ is $L_3(2)$. By 3.3.23.1, D acts on the preimage T_0 in T of $Z(\bar{T})$. Thus as $D \nleq M$ by 3.3.6.a and $M = !\mathcal{M}(LT)$, no D-invariant subgroup of T_0 is normal in LT. Hence $J(T_0) \nleq Q_L$ by B.2.3.3, so there is $A \in \mathcal{A}(T_0)$ with $A \nleq Q_L$. Then as $|\bar{T}_0| = 2$, $T_0 = \langle a \rangle Q_L$ for $a \in A - Q_L$. Now $\Phi(Q_L) = 1$ by 3.3.28, so $C_{Q_L}(A) = C_{Q_L}(a)$. Therefore by B.2.21, $\mathcal{A}(T_0) = \{A, Q_L\}$ is of order 2. Thus as D is of odd order, D acts on Q_L, so that $D \leq M = !\mathcal{M}(LT)$, contrary to $D \nleq M$. $\qquad\square$

LEMMA 3.3.30. *L is an A_6-block.*

PROOF. Assume otherwise. Then by 3.3.25.1 and 3.3.29, L is a $G_2(2)$-block, and it remains to derive a contradiction. By 3.3.28, $T_C = 1$ and $V = Q_L$, while by 3.3.7.2, V is an FF-module for $\bar{L}\bar{T}$, so $V \cong E_{64}$ is the natural module for $LT/V \cong G_2(2)$.

Define $\bar{A}_1$ as in B.4.6. Then by B.4.6, $m(\bar{A}_1) = 3$, $\mathcal{P}(\bar{L}\bar{T}, V) = \bar{A}_1^{\bar{L}}$, and $C_V(\bar{A}_1) = C_V(\bar{a})$ is of rank 3 for each $\bar{a} \in \bar{A}_1 - \bar{L}$. Let A_0 be the preimage in M of A_1; by B.2.21 there is a unique member A of $\mathcal{A}(A_0)$ with image $\bar{A}_1$. Hence $\mathcal{A}(A_0) = \{V, A\}$. By Burnside's Fusion Lemma A.1.35, $N_{\bar{L}\bar{T}}(\bar{T}) = \bar{T}$ is transitive on the members of $\bar{A}_1^{\bar{L}}$ normal in $\bar{T}$, so that $\bar{A}_1$ is the only such member. Thus $\{A,V\} = \{B \in \mathcal{A}(T) : B \trianglelefteq T\}$ is D-invariant, so as usual D acts on V. Then $D \le N_G(V) \le M$, contrary to 3.3.6.a. $\qquad\square$

By 3.3.30, $\bar{L}\bar{T} \cong A_6$ or S_6, so we can represent $\bar{L}\bar{T}$ on $\Omega := \{1,\dots,6\}$ so that $\bar{T}$ has orbits $\{1,2,3,4\}$ and $\{5,6\}$, and permutes the set of pairs $\{\{1,2\},\{3,4\}\}$. Further we adopt the notation of section B.3.

LEMMA 3.3.31. $T_C = 1$.

PROOF. Assume otherwise; then in particular, $C_Z(L) \ne 1$. By 3.3.25.2, D acts on $Y := O^2(C_L(Z))$ and on $R := O_2(YT)$. Then by 3.3.26, there is $d \in D - M$ with $V^d \not\le Q_L$. As $V^d \trianglelefteq YT$, either $\bar{V}^d = \langle(5,6)\rangle$, or $\bar{V}^d$ contains $\langle(1,2)(3,4),(1,3)(2,4)\rangle$. The latter is impossible, since as $V^d \trianglelefteq T$, V^d acts quadratically on V. Thus $\bar{V}^d = \langle(5,6)\rangle$, and in particular $LT/Q_L \cong S_6$ rather than A_6.

By Sylow's Theorem, D acts on some B of order 3 in Y, and so D acts on $C_R(B)$. Now for $v \in C_V(B) - Z$, $V = \langle v^T \rangle$, so $v \notin Q_L^d$ since $V \not\le Q_L^d$. Therefore by symmetry, $v^d \notin Q_L$, and thus $\bar{v}^d = (5,6)$.

Next $|C_R(B) : C_{Q_L}(B)| = 2$ and $C_R(B) = \langle v^d \rangle C_{Q_L}(B)$ since $YT/Q_L \cong S_4 \times \mathbf{Z}_2$. As $Q_L = C_R(V)$, $C_{Q_L}(B) = C_{Q_L}(VB) = C_R(VB)$. Conjugating by d, $|C_R(B) : C_R(V^d B)| = 2$, so as $C_R(B) = \langle v^d \rangle C_{Q_L}(V^d)$, $|C_{Q_L}(B) : C_{Q_L}(V^d)| = 2$. Then as $[C_V(B)/C_V(L), v^d] \ne 1$, $C_{Q_L}(B) = C_V(B)C_{Q_L}(BV^d)$, so as $T_C \le C_{Q_L}(B)$ $T_C \le C_{Q_L}(BV^d)$ and hence V^d centralizes T_C. Thus $C_R(B) = \langle v^d \rangle C_V(B)C_{Q_L}(BV^d)$. Finally by Coprime Action, $Q_L = VC_{Q_L}(B)$, so $Q_L = VC_{Q_L}(BV^d)$.

Set $S := Q_L V^d$. As $C_{\bar{T}}(\bar{B}) = \bar{V}^d = \bar{S}$, $C_T(B) = C_S(B) = C_R(B)$ and $[V^d, B] \le [Q_L, B] = [V, B]$. So by symmetry, $[V, B] \le [V^d, B] = [V, B]^d$, and hence $[V, B] = [V^d, B] = [V, B]^d$ as these groups have the same order. Thus d acts on $C_T(B)[V,B] = \langle v^d \rangle C_V(B)C_{Q_L}(BV^d)[V,B] = \langle v^d \rangle VC_{Q_L}(BV^d) = V^d Q_L = S$.

By 3.3.7.4, $V = [V,L]C_Z(L)$, so that $Z = ([V,L] \cap Z)C_Z(L)$. Therefore $|Z : C_Z(L)| = |(Z \cap [V,L]) : C_{[V,L]}(L)| = 2$. We saw $C_Z(L) \ne 1$, so as $T_C \cap T_C^d = 1$ by 3.3.27.1, $Z \cong E_4$ and $C_Z(L) \cong \mathbf{Z}_2$.

Suppose $\Phi(Q_L) = 1$. Then as d normalizes S and $\bar{S} = \bar{V}^d$ is of order 2, $\mathcal{A}(S) = \{Q_L, Q_L^d\}$, so as d is of odd order, $d \in N_G(Q_L) \le M = !\mathcal{M}(LT)$, contrary to our choice of $d \in D - M$. Thus $\Phi(Q_L) \ne 1$. So as $\Phi(T_C) = 1$ by 3.3.27.2, $T_C V < Q_L$. As we saw $Q_L = VC_{Q_L}(V^d B)$, we may choose $u \in C_{Q_L}(V^d B) - T_C V$.

Now $|Q_L : T_C V| \le 2$ by C.1.13.b and B.3.1, so $Q_L = \langle u \rangle T_C V$ and $T = \langle u \rangle (T \cap L)T_C V^d$. Also $\Phi(T_C) = 1$, T_C commutes with L by definition, and we saw V^d centralizes T_C. Therefore as $T = \langle u \rangle (T \cap L)T_C V^d$, $1 \ne C_{T_C}(u) = Z \cap T_C \le C_Z(L)$, so as $C_Z(L)$ is of order 2, $C_{T_C}(u)$ is of order 2. As $u^2 \in VT_C \le C_G(T_C)$ and T_C is elementary abelian, it follows that $m(T_C) \le 2$.

Assume first that $T_C \cong \mathbf{Z}_2$. Then as $\Phi(Q_L) \ne 1$ while $\Phi(Q_L) \le T_C$ by C.1.13.a, u^2 generates T_C. Recall we chose u to centralize V^d and V^d centralizes T_C. Therefore $Z(S) = C_V(V^d)T_C\langle u \rangle$, with $C_V(V^d)T_C$ elementary, so that $\Phi(Z(S)) = T_C$ is d-invariant, contradicting 3.3.27.1.

Thus $T_C \cong E_4$, so $\langle u \rangle T_C \cong D_8$. Hence $S = S_1 \times S_2 \times E$, where $S_i \cong D_8$ and $E \cong E_4$. But then as d is of odd order, the Krull-Schmidt Theorem A.1.15, says d acts on $Z(S)S_i$ for $i = 1$ and 2, so d centralizes $\Phi(Z(S)S_i)$ of order 2, and hence also centralizes $\Phi(S)$. This contradicts 3.3.27.1, since $T_C \cap \Phi(S) \ne 1$. $\qquad\square$

LEMMA 3.3.32. *(1) Either $Q_L = V$ is irreducible, or $Q_L \cong E_{32}$ is the quotient of the permutation module on Ω modulo $\langle e_\Omega \rangle$, denoted by "$\tilde{U}$" in section B.3.*

(2) $\bar{L}\bar{T} \cong S_6$.

(3) D acts on the preimage T_0 in T of $\bar{A}_2 := \langle (1,2)(3,4), (5,6) \rangle$.

PROOF. As $T_C = 1$ by 3.3.31, (1) follows from C.1.13 and B.3.1. Let P_1 be the stabilizer in LT of $\{5,6\}$, and P_2 the stabilizer of the partition $\{\{1,2\},\{3,4\},\{5,6\}\}$; set $R_i := O_2(P_i)$, and $X_i := O^2(P_i)$. Then P_1 and P_2 are the maximal parabolics of LT over T, P_1 has two noncentral 2-chief factors, P_2 has three noncentral 2-chief factors, and $O_2(X_2)$ is nonabelian with $Z(X_2) = 1$. Then P_1 does not satisfy conclusion (4) of 3.3.20 and P_2 does not satisfy conclusion (5c) of 3.3.20, so D acts on R_1 and R_2. Thus D acts on $T_1 := R_1 \cap R_2$.

If $\bar{L}\bar{T} \cong S_6$, then $T_0 = T_1$ and the lemma holds, so we may assume $\bar{L}\bar{T} \cong A_6$. Thus $\bar{T}_1 = \langle (3,4)(5,6) \rangle$. But then $\mathcal{P}(\bar{T}_1, Q_L)$ is empty by B.3.4.1, so $J(T_1) \leq C_{LT}(Q_L) = Q_L$. Then as Q_L is elementary abelian by (1), $J(T_1) = Q_L \trianglelefteq LT$, and hence $D \leq N_G(Q_L) \leq M$, contrary to 3.3.6.a. Thus the lemma is established. $\square$

We can now obtain a contradiction, and complete the proof of Theorem 3.3.1.

In view of 3.3.32.1, Q_L is either the natural module for $\bar{L}$ denoted by "$\tilde{U}_0$" in B.3.2, or the quotient denoted "$\tilde{U}$" of the permutation module. Define $\bar{A}_1 := \langle (5,6) \rangle$, and $\bar{A}_2$ as in 3.3.32.2. By 3.3.32.3, D acts on the preimage T_0 of $\bar{A}_2$ in T, and as $D \not\leq M$ by 3.3.6.a, D acts on no nontrivial subgroup of T_0 normal in LT. In particular $J(T_0) \not\leq Q_L$ by B.2.3.3, so there is $A \in \mathcal{A}(T_0)$ with $A \not\leq Q_L$. By B.3.2, $\bar{A} = \bar{A}_i$ for $i = 1$ or 2. By inspection, $C_{Q_L}(\bar{A}) = C_{Q_L}(\bar{a})$ for some $\bar{a} \in \bar{A}$, so by B.2.21 there is at most one member of $\mathcal{A}(T_0)$ projecting on $\bar{A}_i$; if such a member exists, we denote it by A_i. Thus $\mathcal{A}(T_0) \subseteq \{Q_L, A_1, A_2\}$. Therefore as D acts on $\mathcal{A}(T_0)$ but not on Q_L, and D is of odd order, D_L is transitive on $\mathcal{A}(T_0)$ of order 3. Further D is transitive on the 2-subsets of $\mathcal{A}(T_0)$. This is impossible as $|A_1 Q_L| < |A_2 Q_L|$.

This contradiction completes the proof of Theorem 3.3.1.

CHAPTER 4

Pushing up in QTKE-groups

Recall that in chapter C of Volume I, we proved "local" pushing up theorems in SQTK-groups. In this Chapter we use those local theorems to prove "global" pushing up theorems in QTKE-groups. Let L, V be a pair in the Fundamental Setup (3.2.1), $L_0 := \langle L^T \rangle$, and $M := N_G(L_0)$. We use L_0T and our pushing up theorems to show that large classes of subgroups must be contained in M.

For example, in Theorem 4.2.13 we use the fact that L_0T is a uniqueness subgroup to prove roughly that if the pair L, V in the FSU is not too "small", then each subgroup I of L_0 which covers L_0 modulo $O_2(L_0T)$ with $O_2(I) \neq 1$ is also a uniqueness subgroup. Then we use Theorem 4.2.13 to prove Theorem 4.4.3, which shows that for suitable subgroups B of odd order centralizing V, $N_G(B) \leq M$. As a corollary, we see in Theorem 4.4.14 that for $H \in \mathcal{H}_*(T, M)$ with $n(H) > 1$, a Hall $2'$-subgroup of $H \cap M$ must act faithfully on V. This gives the inequality $n(H) \leq n'(N_M(V)/C_M(V))$, (cf. E.3.38) which is used crucially in many places in this work.

4.1. Some general machinery for pushing up

Our eventual goal is to show roughly in most cases of the FSU that if $\mathcal{I}$ is the set of subgroups I of L_0T covering L_0 modulo $O_2(L_0T)$ with $O_2(I) \neq 1$, then each member of $\mathcal{I}$ is also a uniqueness subgroup. If some member of $\mathcal{I}$ fails to be a uniqueness subgroup, then we study a maximal counterexample I using the theory of pushing up from chapter C of Volume I. Our starting point is 1.2.7.3, which says that L_0T is a uniqueness subgroup. We develop some fairly general machinery to implement this approach. So in this section we assume the following hypothesis (which we will see in 4.2.2 holds in the FSU):

HYPOTHESIS 4.1.1. *Assume G is a simple QTKE-group, $T \in Syl_2(G)$, $M \in \mathcal{M}(T)$, and $M_+ = O^2(M_+) \trianglelefteq M$. Further assume that $M = !\mathcal{M}(I)$ for each subgroup I of M such that*

$$M_+ C_T(M_+/O_2(M_+)) \leq I \quad and \quad M = C_M(M_+/O_2(M_+))I.$$

Let $\Sigma(M_+)$ consist of those subgroups M_- of M containing $M_+ C_M(M_+/O_2(M_+))$.

LEMMA 4.1.2. *Let $R_+ \in Syl_2(C_M(M_+/O_2(M_+)))$. Then $M = !\mathcal{M}(N_M(R_+))$.*

PROOF. By hypothesis T is Sylow in M, so as $M_+ \trianglelefteq M$, we may assume $R_+ = C_T(M_+/O_2(M_+))$. Also $M_+ = O^2(M_+)$, so by A.4.2, $M_+R_+ \leq N_G(R_+)$. Now $M = C_M(M_+/O_2(M_+))N_M(R_+)$ by a Frattini Argument. So by Hypothesis 4.1.1 with $N_M(R_+)$ in the role of "I", $M = !\mathcal{M}(N_M(R_+))$. $\qquad\square$

Next we define some more technical notation. We will study overgroups of M_+ which (in contrast to the subgroups I in 4.1.1) need not cover *all* of M modulo

$C_M(M_+/O_2(M_+))$, but just cover M_- modulo $C_M(M_+/O_2(M_+))$ for some $M_- \in \Sigma(M_+)$. For example in the FSU, take $M_+ := L_0$, and $M_- := L_0 C_M(L_0/O_2(L_0))T$, or more generally $M_- \in \Sigma(M)$ with $M_- \leq L_0 C_M(L_0/O_2(L_0))T$ and $L^T = L^{M_-}$.

In the remainder of the section pick $M_- \in \Sigma(M_+)$ and define $\eta = \eta(M_+, M_-)$ to be the set of all subgroups I of M_- such that $IC_M(M_+/O_2(M_+)) = M_-$ and $M_+ \leq IO_2(M_+)$ with $O_2(I) \neq 1$. We wish to show that each $I \in \eta$ is a uniqueness subgroup; thus we consider the set of counterexamples to this conclusion, and define $\mu = \mu(M_+, M_-)$ to consist of those $I \in \eta$ such that $\mathcal{H}(I, M) \neq \emptyset$, where

$$\mathcal{H}(I, M) := \{H \in \mathcal{H}(I) : H \not\leq M\}.$$

Finally define a relation $\stackrel{<}{\sim}$ on η by $I_1 \stackrel{<}{\sim} I_2$ if $O_2(I_1) \leq O_2(I_2)$ and $I_1 \cap M_+ \leq I_2 \cap M_+$. Let $\mu^* = \mu^*(M_+, M_-)$ consist of those $I \in \mu$ such that $O_2(I)$ is not properly contained in $O_2(I_1)$ for any $I_1 \in \mu$ such that $I \stackrel{<}{\sim} I_1$.

We begin to study this set μ^* of "maximal" members of μ.

LEMMA 4.1.3. *Let $I \in \eta$, $I \leq I_0 \leq M_-$, and $I_1 \leq I_0$ with $1 \neq O_2(I_1)$. Assume $I_0 = I_1 C_{I_0}(M_+/O_2(M_+))$ and $M_+ \cap I_0 \leq I_1 O_2(M_+)$. Then*

(1) $I_1 \in \eta$.

(2) If $I \in \mu^*$, $I \stackrel{<}{\sim} I_1$, and $O_2(I) < O_2(I_1)$, then $M = !\mathcal{M}(I_1)$.

PROOF. By hypothesis $I \in \eta$ and $I \leq I_0 \leq M_-$, so from the definition of η,

$$M_- = IC_M(M_+/O_2(M_+)) \leq I_0 C_M(M_+/O_2(M_+)) \leq M_-, \qquad (*)$$

and hence all inequalities in (*) are equalities. Again from the definition of η, $M_+ \leq IO_2(M_+) \leq I_0 O_2(M_+)$.

Next as $I_0 = I_1 C_{I_0}(M_+/O_2(M_+))$ by hypothesis, and (*) is an equality,

$$M_- = I_0 C_M(M_+/O_2(M_+)) = I_1 C_M(M_+/O_2(M_+)) \leq M_-,$$

and again this inequality is an equality. As $M_+ \leq I_0 O_2(M_+)$ and $M_+ \cap I_0 \leq I_1 O_2(M_+)$, $M_+ = (I_0 \cap M_+)O_2(M_+) \leq I_1 O_2(M_+)$. Then as $O_2(I_1) \neq 1$ by hypothesis, $I_1 \in \eta$, and hence (1) holds.

Assume the hypothesis of (2). If $M \neq !\mathcal{M}(I_1)$, then $\mathcal{H}(I_1, M) \neq \emptyset$, so that $I_1 \in \mu$. As $I \stackrel{<}{\sim} I_1$ and $O_2(I) < O_2(I_1)$, this contradicts $I \in \mu^*$, establishing (2). $\square$

The next two results are used to establish Hypothesis C.2.8 in various situations; see 4.2.4 for one such application. Hypothesis C.2.8 allows us to apply the pushing up results in chapter C of Volume I.

LEMMA 4.1.4. *Suppose $I \in \mu^*$, and let $R := O_2(I)$ and $H \in \mathcal{H}(I, M)$. Set $H_+ := O^2(M_+ \cap H)$. Then*

(1) $R \leq C_M(M_+/O_2(M_+))$.

(2) $C(G, R) \leq M$.

(3) $M_+ = H_+ O_2(M_+)$ and $H_+ \trianglelefteq H \cap M$.

(4) $R \in Syl_2(C_H(H_+/O_2(H_+))) \cap Syl_2(C_{H \cap M}(H_+/O_2(H_+)))$.

(5) $R = O_2(N_H(R))$ so that $R \in \mathcal{B}_2(H)$, and $O_2(H) \leq O_2(H \cap M) \leq R$.

(6) $F^*(H \cap M) = O_2(H \cap M)$.

PROOF. Let $I_+ := O^2(M_+ \cap I)$. As $M_+ \leq IO_2(M_+)$ by definition of η, while $M_+ = O^2(M_+)$ by Hypothesis 4.1.1, $M_+ = I_+ O_2(M_+)$. Therefore (1) follows from A.4.3.1, with M_+, I_+ in the roles of "X, Y". Also (3) follows as $I_+ \leq H_+$.

Set $M_1 := N_{M_-}(R)$, and pick $R_+ \in Syl_2(C_M(M_+/O_2(M_+))$ so that $R_1 := N_{R_+}(R) \in Syl_2(C_{M_1}(M_+/O_2(M_+)))$. If $R = R_+$, then (2) holds by 4.1.2, so we may assume that $R < R_+$, and hence $R < R_1$. We will verify the hypotheses of 4.1.3, with M_1, $N_{M_1}(R_1)$ in the roles of "I_0, I_1". First $I \leq M_1$, and $O_2(M_1) \neq 1 \neq O_2(N_{M_1}(R_1))$, since $1 \neq O_2(I) = R \leq O_2(M_1) \cap O_2(N_{M_1}(R_1))$. By a Frattini Argument,

$$M_1 = N_{M_1}(R_1)C_{M_1}(M_+/O_2(M_+)).$$

Finally M_+ acts on R_+ by A.4.2.4, and hence $M_+ \cap M_1 = N_{M_+}(R) \leq N_{M_1}(R_1)$, completing the verification of the hypotheses of 4.1.3. Thus $N_{M_1}(R_1) \in \eta$ by 4.1.3. Also $[N_{M_+}(R), R_1] \leq O_2(M_+) \cap M_1 \leq R_1$ as $R_1 \in Syl_2(C_{M_1}(M_+/O_2(M_+)))$, so $I \cap M_+ \leq N_{M_+}(R) \leq N_{M_+}(R_1)$. By construction $O_2(I) = R < R_1 \leq O_2(N_{M_1}(R_1))$, so $I \overset{<}{\sim} N_{M_1}(R_1)$. Therefore as $I \in \mu^*$ by hypothesis, $M = !\mathcal{M}(N_{M_1}(R_1))$ by 4.1.3.2. Then as $M_1 \leq N_G(R)$, (2) follows.

A similar argument shows $R \in Syl_2(C_{H \cap M}(H_+/O_2(H_+)))$: Assume that

$$R < R_H \in Syl_2(C_{H \cap M}(H_+/O_2(H_+))).$$

As $C_M(M_+/O_2(M_+)) \leq M_-$, R_H is also Sylow in $C_{H \cap M_-}(M_+/O_2(M_+))$. Set $H_1 := N_{H \cap M_-}(R)$ and choose R_H so that $R_1 := N_{R_H}(R) \in Syl_2(C_{H_1}(M_+/O_2(M_+))$. By a Frattini Argument, $H_1 = N_{H_1}(R_1)C_{H_1}(M_+/O_2(M_+))$. By (3),

$$M_+ \cap H = H_+O_2(M_+ \cap H) \leq H_+R_H,$$

and by A.4.2.4, H_+ acts on R_H, so

$$M_+ \cap H_1 = N_{M_+ \cap H_1}(R) \leq N_{H_1}(R_1).$$

Hence applying 4.1.3.1 to H_1, $N_{H_1}(R_1)$ in the roles of "I_0, I_1", we conclude $N_{H_1}(R_1) \in \eta$. By construction, $H_1 \leq H \not\leq M$, so $\mathcal{H}(N_{H_1}(R_1), M) \neq \emptyset$, and hence $N_{H_1}(R_1) \in \mu$. Also by construction, $O_2(I) = R < R_1 \leq O_2(N_{H_1}(R_1))$ and arguing as above, $I \overset{<}{\sim} N_{H_1}(R_1)$. This contradicts our hypothesis that $I \in \mu^*$, completing the proof that $R \in Syl_2(C_{H \cap M}(H_+/O_2(H_+)))$. Then (4) follows using (2).

As $H_+ \trianglelefteq H \cap M$, $R \in \mathcal{B}_2(H \cap M)$ by C.1.2.4. By (2), $N_H(R) \leq H \cap M$, so $R \in \mathcal{B}_2(H)$ by C.1.2.3. By C.2.1.2, both $O_2(H)$ and $O_2(H \cap M)$ lie in $R \leq H \cap M$, so in fact $O_2(H) \leq O_2(H \cap M) \leq R$, completing the proof of (5).

Let $H \leq H_1 \in \mathcal{M}$. Then $H_1 \in \mathcal{H}(I, M)$, so all results proved for H also apply to H_1. In particular by (5), $O_2(H_1 \cap M) \leq R \leq H \cap M$, and hence $O_2(H_1 \cap M) \leq O_2(H \cap M)$. Now if $F^*(H_1 \cap M) = O_2(H_1 \cap M)$, then

$$C_{H \cap M}(O_2(H \cap M)) \leq C_{H_1 \cap M}(O_2(H_1 \cap M)) \leq O_2(H_1 \cap M) \leq O_2(H \cap M),$$

so (6) holds. That is, if (6) holds for H_1, then it also holds for H, so we may assume $H = H_1 \in \mathcal{M}$. Now $C_G(O_2(H)) \leq N_G(O_2(H)) = H$, while $O_2(H) \leq O_2(H \cap M)$ by (5). Thus $C_{O_2(M)}(O_2(H \cap M)) \leq C_M(O_2(H)) \leq H \cap M$, so $H \cap M \in \mathcal{H}^e$ by 1.1.4.5, proving (6). This completes the proof of 4.1.4. $\qquad\square$

LEMMA 4.1.5. *Let* $R_+ \in Syl_2(C_M(M_+/O_2(M_+))$, *and assume*

$$1 \neq V = [V, M_+] \leq \Omega_1(Z(R_+)).$$

Suppose $I \in \mu^*$ *and* $R := O_2(I) \leq R_+$. *Then*

(1) $V \leq Z(R)$.

(2) *If* $V = [\Omega_1(Z(R_+)), M_+]$, *then* $N_G(V) \leq M$.

(3) *Let* $H \in \mathcal{H}(I, M)$, *and set* $H_+ := O^2(M_+ \cap H)$. *Then* $V = [V, H_+]$.

PROOF. Notice that the pair I,R satisfies the hypotheses of 4.1.4 for any $H \in \mathcal{H}(I,M)$. Since $I \in \mu$, there is $H_1 \in \mathcal{M}(I) - \{M\}$. By 4.1.4.5, $O_2(H_1) \le O_2(H_1 \cap M) \le R$, while $R \le R_+ \le C_G(V)$. Then $V \le C_G(O_2(H_1)) \le H_1$ as $H_1 \in \mathcal{M}$, so as $F^*(H_1 \cap M) = O_2(H_1 \cap M)$ by 4.1.4.6, $V \le C_{H_1 \cap M}(O_2(H_1 \cap M)) \le O_2(H_1 \cap M) \le R$. Hence $V \le Z(R)$, proving (1).

Next $N_M(R_+)$ acts on R_+ and M_+, and hence also on $[\Omega_1(Z(R_+)), M_+]$, so (2) follows from 4.1.2. Let $H \in \mathcal{H}(I,M)$. By 4.1.4.3, $M_+ = H_+ O_2(M_+)$, so as $O_2(M_+) \le R_+ \le C_M(V)$, $V = [V, M_+] = [V, H_+ O_2(M_+)] = [V, H_+]$, establishing (3). □

4.2. Pushing up in the Fundamental Setup

In this section, we apply the machinery of the previous section in the context of our Fundamental Setup (3.2.1). Recall from the discussion in Remark 3.2.4 that under the following assumption, the FSU holds for some $V \in \mathcal{R}_2(\langle L, T \rangle)$:

HYPOTHESIS 4.2.1. G is a simple QTKE-group, $T \in Syl_2(G)$, $M \in \mathcal{M}(T)$; and $L \in \mathcal{L}_f^*(G,T) \cap M$ with $L/O_2(L)$ quasisimple.

LEMMA 4.2.2. Hypothesis 4.1.1 holds with $M_+ := \langle L^T \rangle$.

PROOF. By 1.2.1.3, $M_+ \trianglelefteq M$, and by 1.2.7.3, $M = !\mathcal{M}(M_+ T)$. Further by 1.4.1.2 $O_2(M_+ T) = C_T(M_+/O_2(M_+))$ is Sylow in $C_M(M_+/O_2(M_+))$, so any subgroup satisfying the hypotheses on "I" in Hypothesis 4.1.1 contains a Sylow 2-group of M, and hence conjugating in M we may assume $T \le I$. But then $M_+ T \le I$, so that $M = !\mathcal{M}(I)$, and so Hypothesis 4.1.1 is satisfied. □

HYPOTHESIS 4.2.3. Assume Hypothesis 4.2.1, and set

$$M_+ := \langle L^T \rangle \quad and \quad R_+ := C_T(M_+/O_2(M_+)).$$

Further assume $M_- \le M$ with $M_+ C_M(M_+/O_2(M_+)) \le M_-$ and $L^T = L^{M_-}$, $I \in \mu^*(M_+, M_-)$, and $R := O_2(I) \le R_+$.

LEMMA 4.2.4. Assume Hypothesis 4.2.3 and $H \in \mathcal{H}(I,M)$. Set $M_H := H \cap M$, $L_H := (L \cap H)^\infty$, $M_0 := \langle L_H^{M_H} \rangle$, and $V := [\Omega_1(Z(R_+)), M_+]$. Then
(1) The hypotheses of 4.1.4 and 4.1.5 are satisfied, with $M_0 = O^2(M_+ \cap H)$ in the role of "H_+".
(2) Hypothesis C.2.8 is satisfied.
(3) $R_+ = O_2(M_+ T) = C_T(V)$.

PROOF. By construction $V \le Z(R_+)$, so that $R_+ \le C_T(V)$. As $L/O_2(L)$ is quasisimple and $[L, V] \ne 1$, $C_{M_+}(V) \le O_{2,Z}(M_+)$, so $C_T(V) \le R_+$, establishing (3).

By hypothesis, $H \in \mathcal{H}$, so $O_2(H) \ne 1$ and H is an SQTK-group. Of course $R \le H \cap M = M_H$. By 4.2.2 and Hypothesis 4.2.3, the hypotheses of 4.1.4 are satisfied, so $F^*(M_H) = O_2(M_H)$ by 4.1.4.6. Thus part (1) of Hypothesis C.2.8 is established.

By 4.1.4.3, $L = L_H O_2(L)$, so $L_H \in \mathcal{C}(M_H)$. Using Hypothesis 4.2.3, $L^M = L^{M_-} = L^I \subseteq L^{M_H} \subseteq L^M$, so that $O^2(M_+ \cap H) = \langle L_H^{M_H} \rangle = M_0$. Hence M_0 plays the role of "H_+" in 4.1.4. Now part (2) of Hypothesis C.2.8 holds by 4.1.4.

Since $R_2(M_+ T) \le \Omega_1(Z(O_2(M_+ T))) = \Omega_1(Z(R_+))$, $V \ne 1$ by 1.2.10. Since $R \le R_+$ by Hypothesis 4.2.3, the hypotheses of 4.1.5 are satisfied. In particular,

(1) holds, and $N_H(V) \leq M_H$ and $V = [V, M_0]$ by 4.1.5. As $V \leq Z(R_+)$ and $O_2(M_0R) \leq R_+$, $V \leq Z(O_2(M_0R))$. Thus part (3) of Hypothesis C.2.8 holds, completing the verification of that Hypothesis, and establishing (2). $\square$

THEOREM 4.2.5. *Assume Hypothesis 4.2.3 and $H \in \mathcal{M}(I) - \{M\}$. Then*

$$O_{2,F^*}(H) \not\leq M.$$

The proof of Theorem 4.2.5 involves a short series of reductions, culminating in 4.2.10. Until it is complete, assume I, H afford a counterexample; that is, assume $O_{2,F^*}(H) \leq M$.

By 4.2.4, the quintuple H, $M_H := H \cap M$, $L_H := (L \cap H)^\infty$, R, $V := [\Omega_1(Z(R_+)), M_+]$ satisfies Hypothesis C.2.8, so we can apply results in the latter part of section C.2 to this quintuple.

LEMMA 4.2.6. *(1) $M_+ = L$.*
(2) $H_+ := L_H \in \mathcal{C}(H)$, and $M_H = H \cap M = N_H(L_H)$ is of index 2 in H.

PROOF. As we are assuming $O_{2,F^*}(H) \leq M$, we may apply C.2.13. Since $M \neq H \in \mathcal{M}$ we have $M_H < H$, so case (1) of C.2.13 does not hold. Thus case (2) of C.2.13 holds, so that (2) holds. By Hypothesis 4.2.3, $L^I = L^M$, while $L^I \subseteq L^{M_H} = \{L\}$ by (2), so (1) holds. $\square$

We now reverse the roles of H, M—applying suitable results on pushing up to M instead of H.

Set $Q := O_2(M_H)$. By assumption $O_{2,F^*}(H) \leq M$, so $Q = O_2(H)$ by A.4.4.1. Now as $H \in \mathcal{M}$, $H = N_G(O_2(H))$, and $C(M, Q) = M_H$ by A.4.4.2. Thus $Q \in \mathcal{B}_2(M)$ and Q is Sylow in $\langle Q^{M_H} \rangle = Q$, so the triple Q, M_H, M satisfies Hypothesis C.2.3 in the roles of "R, M_H, H". Therefore we can apply the results from Section C.2 based on Hypothesis C.2.3 to this triple. Further as $Q \in \mathcal{B}_2(M)$,

$$O_2(M) \leq Q$$

by C.2.1.2.

LEMMA 4.2.7. *(1) $L = L_H \in \mathcal{C}(H)$.*
(2) $L^H = \{L, L^h\}$ for each $h \in H - M$.

PROOF. By 4.2.6.1, $M_+ = L \in \mathcal{C}(M)$. By 4.2.4, we may apply 4.1.4. Then by 4.1.4.3, $L = L_H O_2(L)$, so as $O_2(L) \leq O_2(M) \leq Q \leq H$, $L = L_H$. Thus (1) holds, and then (2) follows from 4.2.6.2. $\square$

In the remainder of the proof of Theorem 4.2.5, let h denote an element of $H - M$. Set $H_0 := \langle L^H \rangle$. Then $H_0 \leq N_H(L) = M_H \leq M$ using 4.2.6.2 and 4.2.7.1. As $H_0 \trianglelefteq H$ and $H \in \mathcal{M}$, we have:

LEMMA 4.2.8. $H = N_G(H_0)$.

LEMMA 4.2.9. $O_{2,F}(M) \leq H$.

PROOF. Recall that Q, M satisfy Hypothesis C.2.3 in the roles of "R, H". We may assume that $O_{2,F}(M) \not\leq H$, so by C.2.6, there is a subnormal A_4-block Y of M with $Y \not\leq H$. As $m_3(M) \leq 2$, $H_0 \leq O^2(M) \leq N_M(Y)$, so as $Aut(Y/O_2(Y))$ is a 2-group, $[Y, H_0] \leq O_2(Y) \leq O_2(M) \leq Q$. But then Y acts on $O^2(H_0Q) = H_0$, so $Y \leq H$ by 4.2.8. This contradicts $Y \not\leq H$, completing the proof. $\square$

By A.4.4.3, $O_{2,F^*}(M) \not\leq H$, so in view of 4.2.9, there is $K \in \mathcal{C}(M)$ with $K/O_2(K)$ quasisimple and $K \not\leq H$.

LEMMA 4.2.10. *(1)* $L^h \leq K \cap H < K$.
(2) $m_p(K) = 1$ for each odd prime $p \in \pi(L)$, and $K \trianglelefteq M$.

PROOF. First by 4.2.6.1, $M = N_G(L)$ since $M \in \mathcal{M}$. Then $L^h \leq C_H(L/O_2(L))$ by 4.2.7 and 1.2.1.2, and hence $L^h \leq C_M(L/O_2(L))$. Similarly $L \neq K$ as $K \not\leq H$, so $K \leq C_M(L/O_2(L))$ by 1.2.1.2. Hence by 1.2.1.1, $KL^h \leq \langle \mathcal{C}(C_M(L/O_2(L))) \rangle =: K_0 \trianglelefteq M$.
Let $p \in \pi(L)$ be an odd prime. As M is an SQTK-group, $m_p(M) \leq 2$, so as $K_0 \leq C_M(L/O_2(L))$, $m_p(K_0) \leq 1$. Thus $L^h \leq O^{p'}(K_0) =: K_1$, and $K_1 \in \mathcal{C}(K_0)$. If $K \neq K_1$ then K acts on $LL^h = H_0$, so that $K \leq N_G(H_0) = H$ by 4.2.8, contradicting $K \not\leq H$. Therefore $L^h \leq K_1 = K$, and then (1) holds as $K \not\leq H$. Further as $K = K_1$, $m_p(K) = 1$ and $K = O^{p'}(K_0)$ by earlier observations, so (2) holds as $K_0 \trianglelefteq M$.   □

We are now in a position to complete the proof of Theorem 4.2.5. First $K \trianglelefteq M$ by 4.2.10.2, so Q acts on K. Set $(KQ)^* := KQ/C_{KQ}(K/O_2(K))$ and $J := L^h$. Then K^* and the action of Q^* on K^* are described in C.2.7. Now $J \leq K \cap M_H$ by 4.2.10.1, while by 4.2.6.2 and 4.2.7, $M_H = N_H(L) = N_H(J)$. Hence $J^* \trianglelefteq (K \cap M_H)^*$. As J^* is not solvable, inspecting the list of possibilities in C.2.7.3, cases (a)–(d) and (f) are eliminated, as are the cases in (h) where the parabolic is solvable. The condition in 4.2.10.2 that $m_p(K) = 1$ for each odd prime $p \in \pi(J^*)$ then eliminates the remaining cases. This contradiction completes the proof of Theorem 4.2.5.

NOTATION 4.2.11. Assume Hypothesis 4.2.1, set $M_+ := \langle L^T \rangle$, and let $\mathcal{I}$ be the set of subgroups I of M such that

$$L \leq IO_2(\langle L, T \rangle), \quad L^T = L^I, \quad \text{and} \quad O_2(I) \neq 1.$$

LEMMA 4.2.12. *Assume Hypothesis 4.2.1, $I \in \mathcal{I}$, and $H \in \mathcal{M}(I) - \{M\}$. Let $O_2(I) \leq R_+ \in Syl_2(C_M(M_+/O_2(M_+)))$. Then*

(1) $M_- := M_+C_M(M_+/O_2(M_+))I \in \Sigma(M_+)$ *and* $I \in \mu(M_+, M_-)$.
(2) *Assume* $I \in \mu^*$ *and set* $L_H := (L \cap I)^\infty$. *Then* $M_+ = L$, $L_H \in \mathcal{C}(H \cap M)$ *is normal in* $H \cap M$, $[\Omega_1(Z(R_+)), L_H] = [\Omega_1(Z(R_+)), L] = [R_2(LT), L]$, *and* $L_H \leq K \in \mathcal{C}(H)$ *with* $K \not\leq M$, $K/O_2(K)$ *quasisimple, and* K *is described in one of cases (1)–(9) of Theorem C.4.8.*

PROOF. Set $R := O_2(I)$. Since $T \in Syl_2(G)$, we may assume that $R \leq T \cap I \in Syl_2(I)$. By 4.2.2, Hypothesis 4.1.1 is satisfied. By construction, $M_- \in \Sigma(M_+)$. By definition of $I \in \mathcal{I}$ in Notation 4.2.11, $L^I = L^T$, $1 \neq R$, and $L \leq IR_+$, where $R_+ := O_2(\langle L, T \rangle)$. By A.4.2.4, $R_+ = C_T(M_+/O_2(M_+))$. As $L^T = L^I$, $M_+ \leq IR_+$, and hence $M_+ \leq IO_2(M_+)$, so $R = O_2(I) \leq C_T(M_+/O_2(M_+)) \leq R_+$ and $M_- = C_M(M_+/O_2(M_+))I$. Thus $I \in \eta$, and as $H \in \mathcal{M}(I) - \{M\}$, $I \in \mu$. That is, (1) is established.
Assume $I \in \mu^*$ and set $V_+ = [\Omega_1(Z(R_+)), M_+]$, $M_H := M \cap H$, $L_H := (L \cap H)^\infty$, and $M_0 := O^2(M_+ \cap H)$. As Hypothesis 4.2.3 holds, by 4.2.4 we may apply 4.1.4 and 4.1.5. By 4.1.5.3, $V_+ = [V_+, M_0]$. Also by 4.2.4, $M_0 = \langle L_H^{M_H} \rangle$ and the quintuple H, L_H, M_H, R, V_+ satisfies Hypothesis C.2.8.

We now appeal to Theorem C.4.8. By Theorem C.4.8, $L_H \trianglelefteq M_H$, so $L = L_0 \trianglelefteq M$ since $L^T = L^I$. As $O_{2,F^*}(H) \not\leq M$ by 4.2.5, one of cases (1)–(9) of Theorem C.4.8 holds. By Theorem C.4.8, $L_H \leq K \in \mathcal{C}(H)$ with $K \not\leq M$ and $K/O_2(K)$ quasisimple. As $L/O_2(L)$ is quasisimple, $\Omega_1(Z(R_+)) = R_2(LT)$, so $V_+ = [R_2(LT), L]$. This completes the proof of (2). $\qquad\square$

Now we come to a fundamental result, showing that many subgroups of LT covering $L/O_2(L)$ are uniqueness subgroups, whenever V is not on a short list of FF-modules.

THEOREM 4.2.13. *Assume Hypothesis 4.2.1 and let $I \in \mathcal{I}$. Then either $M = !\mathcal{M}(I)$; or $L \trianglelefteq M$, $V := [R_2(LT), L]$ is an FF-module for $LT/O_2(LT)$, and one of the following holds:*

(1) $L/O_2(L) \cong L_2(2^n)$.

(2) $L/O_2(L) \cong L_3(2)$ or $L_4(2)$, and $V/C_V(L)$ is either the sum of isomorphic natural modules, or the 6-dimensional orthogonal module for $L_4(2)$.

(3) $O^2(I \cap L)$ is an A_6-block or an exceptional A_7-block.

(4) $O^2(I \cap L)$ is a block of type $\hat{A}_6$, and for each $z \in C_V(T)^\#$, $V \not\leq O_2(C_G(z))$.

(5) $O^2(I \cap L)$ is a block of type $G_2(2)$, and if $m(V) = 6$ and V_3 is the $(T \cap I)$-invariant subspace of V of rank 3, then $C_G(V_3) \not\leq M$.

PROOF. Assume $I \in \mathcal{I}$, $H \in \mathcal{M}(I) - \{M\}$, and set $R := O_2(I)$. Since $T \in Syl_2(G)$, we may assume that $R \leq T \cap I \in Syl_2(I)$. Define M_- as in 4.2.12; by 4.2.12.1, $I \in \mu$.

Let $I \stackrel{<}{\sim} I_1 \in \mu$. Then $I_1 \in \mathcal{I}$, and if I_1 satisfies one of the conclusions (1)–(5) of the Theorem, then so does I since $I \cap M_+ \leq I_1 \cap M_+$. Thus we may assume $I \in \mu^*$. Hence Hypothesis 4.2.3 is satisfied. Similarly let $I_2 := (T \cap I)(M_+ \cap I)$. Then $I = I_2 C_I(M_+/O_2(M_+))$, so the hypotheses of 4.1.3 are satisfied with I, I_2 in the roles of "I_0, I_1", and hence $I_2 \in \eta$ by that lemma. Then by construction, $I_2 \in \mu^*$, so replacing I by I_2, we may assume $I \leq M_+T$.

Set $M_H := M \cap H$ and $L_H := (L \cap H)^\infty$. As $I \in \mu^*$, 4.2.12.2 says $M_+ = L \trianglelefteq M$, $V = [\Omega_1(Z(R_+)), L_H] \leq L_H$, $L_H \leq K \in \mathcal{C}(H)$ with $K \not\leq M$ and $K/O_2(K)$ quasisimple, and one of cases (1)–(9) of Theorem C.4.8 holds. We first eliminate case (9): for in that case, K is the double cover of A_8 with $Z(K) = Z(L_H)$; but then $1 \neq Z(L_H) = C_V(L_H) = C_V(L)$ is LT-invariant, so that $K \leq M = !\mathcal{M}(LT)$, contrary to $K \not\leq M$. Among the remaining cases, only case (6) is not included among the conclusions of Theorem 4.2.13—although in cases (5) and (7) of C.4.8, we still need to show that the extra constraints in conclusions (4) and (5) of Theorem 4.2.13 hold. We will eliminate case (6) of C.4.8 later.

In case (5) of C.4.8, L_H is a block of type $\hat{A}_6$ with $m(V) = 6$ and $K \cong M_{24}$ or He. Therefore for each $z \in C_V(T \cap L)^\#$, $V \not\leq C_K(z)$, so that conclusion (4) of Theorem 4.2.13 holds.

Assume that case (7) of C.4.8 holds, so that L_H is a $G_2(2)$-block and $K \cong Ru$. We may assume that $m(V) = 6$, and it remains to show that $C_K(V_3) \not\leq M \cap K$. To see this, we will use facts about the 2-locals of $K \cong Ru$ found in chapter J of Volume I. Observe that $M \cap K = N_K(L_H)$ with $(M \cap K)/V \cong G_2(2)$. Let V_1 be the $(T \cap L_H)$-invariant subspace of V of rank 1; then $M_1 := C_{M \cap K}(V_1)$ is of order $3 \cdot 2^{12}$, so $3 \in \pi(C_K(V_1))$ and hence V_1 is 2-central in K by J.2.7.4 and J.2.9.1. Let $K_1 := C_K(V_1)$, $Q_1 := O_2(K_1)$, and $X_1 \in Syl_3(M_1)$. From (Ru2) in the definition of groups of type Ru in chapter J of Volume I, $K_1^* := K_1/Q_1 \cong S_5$, and from J.2.3, $C_{Q_1}(X_1) \cong Q_8$. Let $v \in C_V(X_1) - V_1$; it follows that v^* is of order 2 in $C_{M_1^*}(X_1^*)$, so $M_1^* \cong D_{12}$. Hence $P_1 := Q_1 \cap M_1$ is of order 2^{10} with $[O_2(M_1), X_1] \leq P_1$ and $|C_{P_1}(X_1)| = 4$. Then $V_3 \leq \Phi([O_2(M_1), X_1]) \leq \Omega_1(Q_1)$, and $\Omega_1(Q_1)$ is the group denoted by "U" in (Ru2). Thus by J.2.2.3, $C_{Q_1}(X_1) \leq C_{Q_1}(U) \leq C_{Q_1}(V_3)$. Hence as $|C_{Q_1}(X_1)| = 8 > |C_{P_1}(X_1)|$, $C_K(V_3) \not\leq M$, as claimed.

Thus to complete the proof of Theorem 4.2.13, we may assume that case (6) of Theorem C.4.8 holds, and it remains to derive a contradiction. Then L_H is a block of type M_{24} or $L_5(2)$, and $K \cong J_4$. In particular, K is a component of the maximal 2-local H, and so centralizes $O_2(H) \neq 1$. As $Out(J_4) = 1$, $H = K \times C_H(K)$, with $O_2(H) \leq C_H(K)$. Hence $I = L_H N_{T \cap K}(L_H) \times C_I(K)$, and setting $R_C := C_R(K)$, $R = O_2(I) = V \times R_C$. As V is self-centralizing in K, $R_C = C_R(K) = C_R(L_H)$. By 4.1.4.5, $O_2(H) \leq R$, so $O_2(H) \leq R_C$.

Recall we reduced in the first two paragraphs of the proof to the case where $I \leq LT$. Thus as $I = L_H N_{T \cap K}(L_H) \times C_I(K)$, $I = L_H(N_{T \cap K}(L_H)) \times R_C$. Let $S := N_{T \cap K}(L_H) \times R_C$ and r an involution in $Z(R_C)$; thus $S \in Syl_2(I)$ and $r \in Z(S)$. Next $O^2(I) = L_H \leq K \leq C_H(r)$ as $r \in R_C$, and hence r centralizes $O^2(I)S = I$, so without loss $H \in \mathcal{M}(C_G(r))$. Then in particular K is a component of $C_G(r)$.

From the structure of L_H in case (6) of Theorem C.4.8, there is X of order 3 in L_H with $C_V(X) \neq 1$. Let $K_X := C_K(X)^\infty$ and $G_X := C_G(X)$. Then K_X is quasisimple with $Z(K_X) \cong \mathbf{Z}_6$ and $K_X/Z(K_X) \cong M_{22}$. Thus K_X is also a component of $C_{G_X}(r)$, and hence by I.3.2, $K_X \leq L_X \in \mathcal{C}(G_X)$ with $\bar{L}_X := L_X/O(L_X)$ quasisimple. We claim $K_X = L_X$, so assume that $K_X < L_X$. Then as $K_X \in \mathcal{C}(C_{G_X}(r))$, r is faithful on L_X, and in particular on the quasisimple quotient $\bar{L}_X$. Now case (1.a) of Theorem A (A.2.1) holds since $\bar{L}_X$ is quasisimple, so $\bar{L}_X$ is quasithin. Then inspecting the list of groups in Theorem B (A.2.2), we find that none possesses an involutory automorphism r whose centralizer has a component $\bar{K}_X$ which is a covering of M_{22}. This contradiction establishes the claim that $K_X = L_X \in \mathcal{C}(G_X)$.

Recall from Hypothesis 4.2.3 that $R \leq R_+ = C_T(M_+/O_2(M_+)) = O_2(\langle L, T \rangle)$, and set $R_1 := N_{R_+}(R)$ and $R_1^* := R_1/R$. Recall also from our application of 4.2.12.2 early in the proof that $L \trianglelefteq M$, $V \leq L_H$, and $V = [R_2(LT), L]$, so V is T-invariant. If L_H is an $L_5(2)$-block, then by Theorem C.4.8, V is one of the 10-dimensional modules for L_H/V, so as V is T-invariant, T induces inner automorphisms on $L/O_2(L)$. Of course T induces inner automorphisms on $L/O_2(L)$ if L_H is an M_{24}-block as $Out(M_{24}) = 1$. Thus $LT = LR_+$, so as $I < LT$ (since $M = !\mathcal{M}(LT)$), $R = O_2(I) < R_+$ and hence $R < R_1$. By 4.1.4.4, $R = R_+ \cap H$, so as $C_G(r) \leq H$ we have $R = C_{R_1}(r)$. As $R = V \times R_C$, we can choose $r \in R_C$ so that $rV \in C_{R/V}(R_1)$. Hence the map $\chi : x^* \mapsto [r, x]$ is an L_H-isomorphism of R_1^* with V: Since $V \leq \Omega_1(Z(R_+))$, the map is a homomorphism by a standard commutator formula 8.5.4 in [**Asc86a**]; then injectivity follows from $R = C_{R_1}(r)$, and surjectivity as L_H is irreducible on V. Now there is $v \in C_V(X) - C_V(K_X)$, and for $s \in \chi^{-1}(v) \cap G_X$, $r^s = rv$. As M_{22} is not involved in the groups in A.3.8.2, $K_X \trianglelefteq G_X$, so as $[r, K_X] = 1$, also $[r^s, K_X] = 1$ and hence $[v, K_X] = 1$, contrary to the choice of v. This contradiction completes the proof of Theorem 4.2.13. $\qquad\square$

4.3. Pushing up $L_2(2^n)$

In the first exceptional case of Theorem 4.2.13 where $L \trianglelefteq M$ and $L/O_2(L) \cong L_2(2^n)$ for $n > 1$, it is possible to obtain a result weaker than Theorem 4.2.13, but still stronger than $M = !\mathcal{M}(LT)$: Namely in Theorem 4.3.2, we show in this case that at least L is also a uniqueness subgroup. Theorem 4.3.2 will be used in the Generic Case of the proof of the Main Theorem. Therefore:

Throughout this section we assume Hypothesis 4.2.1, with $L/O_2(L) \cong L_2(2^n)$, and $L \trianglelefteq M$.

LEMMA 4.3.1. *Let S be a 2-subgroup of M, $T_H \in Syl_2(N_M(S))$, and assume that $S \cap L \in Syl_2(L)$ and $M = !\mathcal{M}(LT_H)$. Then $N_G(S) \leq M$.*

PROOF. Assume otherwise, and pick S to be a counterexample to 4.3.1 such that T_H is of maximal order subject to this constraint. We may assume $T_H \leq T$. We claim that $T_H \in Syl_2(N_G(S))$. If $T_H = T$ this is clear, so we may assume that $T_H < T$, and hence $T_H < N_T(T_H)$. As $S \leq T_H$, $T_H \cap L = S \cap L \in Syl_2(L)$ and by hypothesis $M = !\mathcal{M}(LT_H)$, so by maximality of $|T_H|$, $N_G(T_H) \leq M$. Hence if $T_H \leq T_S \in Syl_2(N_G(S))$, then $N_{T_S}(T_H) \leq T_S \cap M \leq N_M(S)$, so $T_H = T_S$ as claimed.

Observe next that Hypothesis C.5.1 of chapter C of Volume I is satisfied with LT_H, $N_G(S)$, S in the roles of "H, M_0, R". Further we may assume that Hypothesis C.5.2 is satisfied, or otherwise $O_2(\langle LT_H, N_G(S)\rangle) \neq 1$, so that $N_G(S) \leq M = !\mathcal{M}(LT_H)$, as desired. Thus we may apply C.5.6.6, and obtain a contradiction to $L \trianglelefteq M$. This completes the proof. $\qquad\square$

THEOREM 4.3.2. $M = !\mathcal{M}(L)$.

The proof of Theorem 4.3.2 involves a series of reductions, culminating in 4.3.16.

Assume the Theorem fails, and pick I so that $L \leq I \leq LO_2(LT)$ and I is maximal subject to $\mathcal{M}(I) \neq \{M\}$. Set $R := O_2(I)$ and $R_+ := O_2(LT)$, so that

$$I = LR \text{ and } R = I \cap R_+.$$

Set $V := [\Omega_1(Z(R_+)), L]$. Choose $H \in \mathcal{M}(I) - \{M\}$, and set $M_H := H \cap M$.

Define $\mathcal{I}$ as in Notation 4.2.11 and observe $I \in \mathcal{I}$. Set $M_- := LC_M(L/O_2(L))$; by 4.2.12.1, $M_- \in \Sigma(L)$ and $I \in \mu$. Then by maximality of I, $I \in \mu^*$, so Hypothesis 4.2.3 is satisfied and hence by 4.2.4, the quintuple H, L, M_H, R, V satisfies Hypothesis C.2.8. By 4.2.12.2, $L \leq K \in \mathcal{C}(H)$, with $K \not\leq M$, $K/O_2(K)$ quasisimple, and K appears in one of cases (1)–(9) of Theorem C.4.8. As $L/O_2(L) \cong L_2(2^n)$, case (1) of Theorem C.4.8 holds, so that either $V/C_V(L)$ is the natural module for $L/O_2(L)$, or $n = 2$ and V is the A_5-module. Furthermore M_H acts on K by Theorem C.4.8. By 1.2.1.5, either $F^*(K) = O_2(K)$, or K is quasisimple and hence a component of H. Therefore K is described in either Theorem C.4.1 or Theorem C.3.1, respectively. Set $M_K := M \cap K$.

Recall from 4.2.4.3 that $R_+ = O_2(LT)$, and $R_+ = C_T(L/O_2(L))$ by 1.4.1.2. Without loss $S := T \cap H \in Syl_2(M_H)$, and we choose $H \in \mathcal{M}(I) - \{M\}$ so that S is maximal. As $L \leq H$ and $M = !\mathcal{M}(LT)$, $T \not\leq H$, so $S < T$, and hence also $S < N_T(S)$.

LEMMA 4.3.3. *(a) If $S < X \leq T$, then $M = !\mathcal{M}(LX)$.*
(b) $N_G(S) \leq M$, so $S \in Syl_2(H)$ and $H = N_G(K)$.

PROOF. As $I = LR \leq LS$, maximality of S implies (a). Then as $S < N_T(S)$, (a) and 4.3.1 imply $N_G(S) \leq M$. Therefore as $S \in Syl_2(M_H)$, $S \in Syl_2(H)$. As we saw earlier that K is M_H-invariant, $K \trianglelefteq H$ by 1.2.1.3, so $H = N_G(K)$ as $H \in \mathcal{M}$. $\square$

LEMMA 4.3.4. $R = S \cap R_+$. In particular, S normalizes R and $R = O_2(IS)$.

PROOF. As $I \leq L(S \cap R_+)$, this follows from maximality of I. $\square$

We next choose an element $t \in N_T(S) - S$ with $t^2 \in S$. If $R < R_+$, then $R_+ \not\leq S$ by 4.3.4, so in this case we may choose t so that also $t \in R_+$ and $t^2 \in S \cap R_+ = R$. By convention, t will denote such an element throughout the proof.

As $t \in N_T(S)$, t normalizes $S \cap R_+ = R$. Further $t \notin S$, so by 4.3.3.a:

LEMMA 4.3.5. $M = !\mathcal{M}(LS\langle t \rangle)$.

LEMMA 4.3.6. $F^*(K) = O_2(K)$.

PROOF. Assume otherwise. Then from our remarks following the statement of Theorem 4.3.2, K is a component of H described in Theorem C.3.1. As $L/O_2(L) \cong L_2(2^n)$, we conclude that either

(i) $K/Z(K)$ is of Lie type and Lie rank 2 over $\mathbf{F}_{2^n}$, and M_K is a maximal parabolic of K, or

(ii) $K/Z(K) \cong M_{22}$ or M_{23}, and L is an $L_2(4)$-block.

Let $\overline{KS} := KS/C_{KS}(K)$ and $S_K := S \cap K$. Now $L \leq K \leq C_H(O_2(H))$ as K is a component of H, and $1 \neq O_2(H) \leq R$ by 4.1.4.5, so $1 \neq R_0 := C_R(L)$. Recall from 4.3.4 and our choice of t that $S\langle t \rangle$ acts on R and L and hence also on R_0, so $N_G(R_0) \leq M = !\mathcal{M}(LS\langle t \rangle)$ by 4.3.5. Then $[K, R_0] \neq 1$ as $K \not\leq M$, so $1 \neq \bar{R}_0 \leq C_{\bar{R}}(\bar{L})$. Inspecting the automorphism groups of the groups in (i) and (ii) (e.g., 16.1.4 and 16.1.5) for such a 2-local subgroup, we conclude $K/Z(K) \cong Sp_4(2^n)$. Indeed $Z(K) = 1$ since the multiplier of $Sp_4(2^n)$ for $n > 1$ is trivial by I.1.3. Furthermore $V = O_2(L)$ is the maximal nonsplit extension of the natural module for $L/O_2(L)$ over a trivial module by I.1.6, and $C_V(L)$ is a root subgroup of K. Since $Aut(K)$ fuses the two K-classes of root subgroups, we may regard $C_V(L)$ as a short root subgroup of K, and take $Z \leq C_V(S_K)$ to be a long root subgroup of K.

Set $G_Z := N_G(Z)$. As $Z = [C_V(T \cap L), N_L(T \cap L)]$ and T acts on L and V, T acts on Z; hence $F^*(G_Z) = O_2(G_Z) =: Q_Z$ by 1.1.4. Let $K_2 := N_K(Z)^\infty$ where $N_K(Z)$ is the maximal parabolic of K containing S_K and distinct from $N_K(C_V(L))$. As $L \leq M$ but $K = \langle L, K_2 \rangle \not\leq M$, $K_2 \not\leq M$. Further $T \not\leq N_G(K_2)$, or otherwise T normalizes $\langle L, K_2 \rangle = K$, and hence $T \leq H$ by 4.3.3.b, contrary to our observation just before 4.3.3. We will now analyze G_Z, and eventually obtain a contradiction by showing that $T \leq N_G(K_2)$.

First, a Cartan subgroup Y of the Borel group $M_K \cap N_K(Z)$ of K decomposes as $Y = Y_1 \times Y_2$, where $Y_1 := C_Y(K_2/O_2(K_2))$ and $Y_2 := Y \cap K_2^\infty$ are cyclic of order $2^n - 1$, Y_1 is regular on $Z^\#$, and $N_K(Z) = Y_1 K_2^\infty$.

Next by 1.2.1.1, K_2 is contained in the product $L_1 \cdots L_r$ of those members L_i of $\mathcal{C}(C_G(Z))$ with $L_i = [L_i, K_2]$. If $r > 1$, then for a prime divisor p of $2^n - 1$, $m_p(L_1 \cdots L_r Y_1) > 2$, contradicting G_Z an SQTK-group. Therefore $K_2 \leq L_1 =: K_Z \in \mathcal{C}(C_G(Z))$. Recall from the remarks after (i) and (ii) above that $K \cong Sp_4(2^n)$ is simple, so that K_2 contains a Levi complement isomorphic to $L_2(2^n)$, and in particular $K_Z/O_{2,F}(K_Z)$ is not $SL_2(p)$ for any odd prime p. This rules out cases (c) and (d) in 1.2.1.4, so that $K_Z/O_2(K_Z)$ is quasisimple. Furthermore as $Y_1 \leq G_Z$ is faithful on Z, $K_Z \trianglelefteq G_Z$ by 1.2.2. Similarly as $m_p(K_Z Y_1) \leq 2$, we conclude from A.3.18 that $m_p(K_Z) = 1$ for each prime divisor p of $2^n - 1$—unless possibly $p = 3$ (so that n is even), and a subgroup of Y_1 of order 3 induces a diagonal automorphism on $K_Z/O_2(K_Z) \cong L_3^\epsilon(q)$, with $q \equiv \epsilon \mod 3$. (If case (3b) of A.3.18 were to hold, then $m_3(Y_1 K_2 O_{2,3}(K_Z)) = 3$.)

Set $U := \langle C_V(L)^{G_Z} \rangle$. Now T acts on V and L, and hence on $C_V(L)$, so as $C_V(L) \neq 1$, $C_V(LT) \neq 1$. Then as $G_Z \in \mathcal{H}^e$, $C_V(LT) \leq \Omega_1(Z(Q_Z))$, so as Y is irreducible on $C_V(L)$ and $O_2(K_2) = \langle C_V(L)^{K_2} \rangle$,

$$O_2(K_2) \leq \langle C_V(L)^{G_Z} \rangle = U \leq \Omega_1(Z(Q_Z)). \tag{$*$}$$

In particular U is generated by G_Z-conjugates of elements of $Z(T)$, so $U \in \mathcal{R}_2(G_Z)$ by B.2.14.

Let $G_Z^* := G_Z/C_{G_Z}(U)$. As $K_Z/O_2(K_Z)$ is quasisimple, so is K_Z^*. As $V/C_V(L)$ is the natural module for $L/O_2(L) \cong L_2(2^n)$, $C_T(C_V(L)Z) = C_T(V)(T \cap L)$ with $C_T(V)(T \cap L)/C_T(V) \cong E_{2^n}$, and in fact $C_T(V)(T \cap L) = C_T(V)O_2(K_2)$. Further $O_2(K_2) \leq Q_Z$ by $(*)$; and also $[Q_Z, V] \leq Q_Z \cap V = O_2(K_2) \cap V \leq C_V(T \cap L)$, so that $Q_Z \leq C_T(V)(T \cap L)$. Hence

$$m(O_2(K_2)/C_{O_2(K_2)}(V)) = n = m(Q_Z/C_{Q_Z}(V)) \text{ and } Q_Z = O_2(K_2)C_{Q_Z}(V). \tag{$**$}$$

By $(*)$, $O_2(K_2) \leq U$, so as $m(V/V \cap O_2(K_2)) = n$ with $C_V(U) \leq C_V(O_2(K_2)) = V \cap O_2(K_2)$, $m(V^*) = n$. By $(*)$ and $(**)$, $m(U/C_U(V)) \leq m(Q_Z/C_{Q_Z}(V)) = n$. Therefore U is a failure of factorization module for K_Z^* with FF*-offender V^*. In particular $K_Z/O_2(K_Z)$ is not $L_3^\epsilon(q)$ with $q \equiv \epsilon \mod 3$, since in that event as U is an FF-module, Theorem B.5.6.1 says $K_Z^* \cong SL_3(q)$, whereas $SL_3(q)$ is not isomorphic to $L_3(q)$ when $q \equiv 1 \mod 3$. This eliminates the exceptional case in our discussion above, so we conclude that $m_p(K_Z) = 1$ for each p dividing $2^n - 1$. Therefore by inspection of the lists in Theorems B.5.1 and B.4.2, $K_Z^* = K_2^*$, and U/Z is the natural module for $K_2^* \cong L_2(2^n)$ or the orthogonal module for $L_2(4)$. Thus as $O_2(K_2)/Z$ is the natural module for K_2^*, $U = O_2(K_2)$ by $(*)$, and as $Q_Z = O_2(K_2)C_{Q_Z}(V)$ by $(**)$, we conclude $[V, Q_Z] = [V, U] \leq U$. Then as $K_2 = \langle V^{K_2} \rangle$, $[K_2, Q_Z] = U \leq K_2$,

$$K_Z = \langle K_2^{K_Z} \rangle \leq \langle K_2^{K_2 Q_Z} \rangle = K_2,$$

and hence $K_2 = K_Z$ is normalized by T, contrary to our earlier observation that $T \not\leq N_G(K_2)$. This contradiction completes the proof of 4.3.6. $\qquad\square$

By 4.3.6, $F^*(K) = O_2(K)$; so as we observed following the statement of Theorem 4.3.2, K is described in Theorem C.4.1, and as $L/O_2(L) \cong L_2(2^n)$, one of cases (1)–(3) of Theorem C.4.1 holds.

LEMMA 4.3.7. K is not a block.

PROOF. Assume otherwise. Inspecting cases (1)–(3) of Theorem C.4.1, we conclude that either K is an $SL_3(2^n)$-block, or $n = 2$ and K is an A_7-block or an $Sp_4(4)$-block. Set $U := U(K)$ in the notation of Definition C.1.7. Now S normalizes K by 4.3.3.b, so as t normalizes S, S also normalizes U^t. Therefore if $U^t \leq O_2(KS)$, then as $[O_2(KS), K] \leq U \leq UU^t$, $UU^t \trianglelefteq KS\langle t \rangle$, forcing $K \leq M$ by 4.3.5, contrary to $K \not\leq M$. Hence $K = [K, U^t]$. Recall also that $V = [\Omega_1(Z(R_+)), L]$ is T-invariant, so $V = V^t$. As $R = O_2(LS)$ by 4.3.4, while $S \in Syl_2(H)$ by 4.3.3.b, $O_2(KR) \leq R$.

Suppose first that K is an $Sp_4(4)$-block. Then $Z_K := C_U(K) \leq V$ using I.2.3.3, and U/V is the natural $L_2(4)$-module for $L/O_2(L)$. So as $V = V^t$, U^t/V is also the natural module, with $C_U(K) \leq V \leq U \cap U^t < U$, impossible as $O_2(L)O_2(K)/O_2(K)$ is a non-split extension of a trivial submodule by a natural module, so that there is no natural L-submodule.

Suppose next that K is an A_7-block. Then by C.4.1.1, S induces a transposition on $L/O_2(L)$, so that $LS/O_2(LS) \cong S_5 = Aut(L/O_2(L))$, and hence $T = SR_+$. Hence as $R \leq S < T$, $R < R_+$ and so $R < N_{R_+}(R)$. We claim that K, R, S, R_+, KS satisfy Hypothesis C.6.2, and the hypotheses of C.6.4, in the roles of "L, R, T_H, Λ, H". Most requirements are either immediate or have been established earlier—except possibly for C.6.2 and C.6.4.II (recall the latter result uses C.6.3 and in particular verifies its hypotheses), which we now verify: If $1 \neq R_0 \leq R$ satisfies $R_0 \trianglelefteq KS$, then by 4.3.3.a, $N_T(R_0) = S$ as $K \not\leq M$; so by 4.3.4 $N_{R_+}(R_0) = R < N_{R_+}(R)$, completing the verification of those hypotheses. As $T = SR_+$, we conclude from C.6.4.10 that $e_{1,2} \in Z(T)$. Then as $e_{1,2}$ centralizes L, $C_G(e_{1,2}) \leq M = !\mathcal{M}(LT)$. Now $v := e_{3,4}$ is in V, and there is $k \in K$ with $e_{1,2}^k = v$. Then $R_+ \leq C_G(V) \leq C_G(v) \leq M^k$, so R_+ acts on L^k. But then R_+ acts on $K = \langle L, L^k \rangle$, so $T = SR_+ \leq N_G(K) = H$, which we saw earlier is not the case.

Therefore K is an $SL_3(2^n)$-block. Thus case (3) of C.4.1 holds, so L is the stabilizer of the line V of U, so that $[U, L] = V$. Therefore as t acts on V and L, also $[U^t, L] = V \leq U$. This is impossible as we saw $K = [K, U^t]$, whereas $K/O_2(K)$ admits no involutory automorphism centralizing $LO_2(K)/O_2(K)$. This contradiction completes the proof of 4.3.7. $\square$

LEMMA 4.3.8. $K/O_2(K) \cong SL_3(2^n)$, (KR, R) is an MS-pair described in one of cases (2)–(4) of Theorem C.1.34, and $S \in Syl_2(H)$.

PROOF. Recall that K is described in one of cases (1)–(3) of Theorem C.4.1. As $L/O_2(L) \cong L_2(2^n)$ and K is not a block by 4.3.7, conclusion (3) of C.4.1 holds, so that $K/O_2(K) \cong SL_3(2^n)$, and one of cases (1)–(4) of C.1.34 holds. Further 4.3.7 rules out case (1) where K is an $SL_3(2^n)$-block. By 4.3.3, $S \in Syl_2(H)$. $\square$

LEMMA 4.3.9. $C_S(K) = 1$.

PROOF. Let $U := \Omega_1(Z(O_2(KS)))$; as $K \trianglelefteq H$, $[U, K] \leq O_2(K)$. By 4.3.8, K is described in one of cases (2)–(4) of C.1.34, so that $[U, K]$ is the sum of one or two isomorphic natural modules for $K/O_2(K)$. So as the natural module has trivial 1-cohomology by I.1.6 since $n > 1$, we conclude that $U = C_U(K) \oplus [U, K]$. Further L stabilizes an $\mathbf{F}_{2^n}$-line in the natural summands of $[U, L]$ by C.4.1, so $C_{[U,K]}(L) = 0$. Thus $C_U(K) = C_U(L)$, so $C_Z(L) = C_Z(K)$, where $Z := \Omega_1(Z(S))$. But $N_T(S)$ normalizes $C_Z(L)$, so if $C_Z(L) \neq 1$ then $N_G(C_Z(L)) \leq M$ by 4.3.5. Therefore as $C_Z(K) = C_Z(L)$ and $K \not\leq M$, $C_Z(K) = 1$, establishing the lemma. $\square$

LEMMA 4.3.10. K satisfies conclusion (3) of Theorem C.1.34.

PROOF. By 4.3.8, one of conclusions (2)–(4) of Theorem C.1.34 holds, and as $C_S(K) = 1$ by 4.3.9, conclusion (4) does not hold. Thus we may assume conclusion (2) holds, and it remains to derive a contradiction. Then $U = O_2(K)$ is the sum of two isomorphic natural modules. As $C_S(K) = 1$, we may apply C.1.36, to conclude that $\mathcal{A}(S) = \{U, A\}$ is of order 2 with $V = U \cap A$ of rank $4n$. We now obtain a contradiction similar to that in the $L_3(2^n)$-case of 4.3.7: Again $U^t \not\leq O_2(KS)$ using 4.3.5 and $V = [U, L]$ by C.4.1. As $U^t \not\leq O_2(KS)$ and $\mathcal{A}(S) = \{U, A\}$, $U^t = A$, while as $[U, L] = V$ is t-invariant, also $V = [L, U^t]$. This is a contradiction as $[A/U, L] = A/U \neq 1$. $\qquad\square$

Set $Q := [O_2(K), K]$ and $U := Z(Q)$. By 4.3.10, conclusion (3) of C.1.34 holds; that is, U is the natural module for $K/O_2(K)$ and Q/U is the direct sum of two copies of the dual of U. In particular, S is trivial on the Dynkin diagram of $K/O_2(K)$, and hence normalizes both maximal parabolics over $S \cap K$.

Set $S_L := S \cap L$ and $Z_S := C_V(S_L)$. Set $G_Z := N_G(Z_S)$. By C.1.34, V is an $\mathbf{F}_{2^n}$-line in U, so Z_S an $\mathbf{F}_{2^n}$-point. As $S_L = T \cap L$ and V are T-invariant, Z_S is T-invariant.

Set $K_2 := C_K(Z_S)^\infty$, $R_2 := O_2(K_2 S)$, and let Y be a Hall $2'$-subgroup of $O_{2,2'}(N_K(Z_S))$. Thus Y is cyclic of order $2^n - 1$, with $[K_2, Y] \leq O_2(K_2)$, and Y faithful on Z_S. Further Y is fixed point free on the natural module U for $K/O_2(K)$, so as we saw above just after 4.3.10 that the composition factors of Q are natural and dual, $Q = [Q, Y]$. Appealing to 4.3.9, we conclude from C.1.35.3 that:

LEMMA 4.3.11. $Q = O_2(KS)$ so $O_2(KS) = [O_2(KS), Y]$.

Next by 1.2.1.1, K_2 is contained in the product $L_1 \cdots L_s$ of those members L_i of $\mathcal{C}(G_Z)$ such that K_2 projects nontrivially on $L_i/O_2(L_i)$. Therefore for each prime divisor p of $2^n - 1$, p divides the order of L_i. But if $s > 1$, then as Y is faithful on Z_S, and $Y = O^2(Y)$ acts on each L_i by 1.2.1.3, $m_p(YL_1L_2) > 2$, contradicting $G_Z Y$ an SQTK-group. Thus $s = 1$. Set $K_Z := L_1$. A similar argument shows K_Z is the unique member of $\mathcal{C}(G_Z)$ of order divisible by p, so that $K_Z \trianglelefteq G_Z$. If $p = 3$ and K_Z appears in case (3b) of A.3.18, then $m_3(YK_2O_{2,Z}(K_Z)) = 3$, contradicting G_Z an SQTK-group. Therefore we may appeal to A.3.18 to obtain:

LEMMA 4.3.12. (1) $K_2 \leq K_Z \in \mathcal{C}(G_Z)$ and $K_Z \trianglelefteq G_Z$.

(2) For p a prime divisor of $2^n - 1$, either $m_p(K_Z) = 1$, or $p = 3$ and a subgroup of order 3 in Y induces a diagonal automorphism on $K_Z/O_2(K_Z) \cong L_3^\epsilon(q)$ for $q \equiv \epsilon$ mod 3.

If T normalizes K_2, then T acts on $\langle L, K_2 \rangle = K$, contradicting $M = !\mathcal{M}(LT)$. This shows:

LEMMA 4.3.13. $K_2 < K_Z$.

LEMMA 4.3.14. (1) $N_G(R_2) \leq N_H(K_2)$.
(2) $R_2 = O_2(N_{L_1 T}(R_2))$.
(3) $O_2(K_Z T) \leq R_2$ and $K_2 < O_2(K_Z)K_2$.

PROOF. Suppose $H_1 \in \mathcal{M}(KS)$. Then as $I \leq KS$ and $K \not\leq M$, $H_1 \in \mathcal{M}(I) - \{M\}$, so the reductions of this section also apply to H_1. In particular by 4.3.3, $H_1 = N_G(K) = H$; that is, $H = !\mathcal{M}(KS)$.

Next K_2 is the maximal parabolic over $S \cap K$ stabilizing the point Z_S of the natural module U. Now (KR_2, R_2) satisfies (MS1) and (MS2) of Definition C.1.31. If (KR_2, R_2) satisfies (MS3), C.1.34 would apply to R_2, whereas here $R_2 = O_2(C_{KS}(Z_S))$ which is explicitly excluded in case (3) of C.1.34, which holds by 4.3.10. Thus (MS3) fails, so there is a nontrivial characteristic subgroup C of R_2 normal in KS, and hence $N_G(R_2) \leq N_G(C) \leq H = !\mathcal{M}(KS)$. Then

$N_G(R_2) = N_H(R_2)$ acts on the parabolic K_2 of K, since we saw after 4.3.6 that $K \trianglelefteq H$, so (1) holds.

Next using A.4.2.4, R_2 is Sylow in $Syl_2(C_H(K_2/O_2(K_2)))$, Now $K_2 \trianglelefteq G_Z \cap H$, so by C.1.2.4, $R_2 \in \mathcal{B}_2(N_{K_Z T \cap H}(R_2))$. Therefore (2) follows from C.1.2.3. By (2) and C.2.1, $O_2(K_Z T) \leq R_2$, so by (1) $K_2 = O^2(K_2 O_2(K_Z T))$. Then 4.3.13 completes the proof of (3). $\square$

Set $G_0 := L_1 R_2 Y$ and $G_0^* := G_0/C_{G_0}(L_1/O_2(L_1))$. By 4.3.14.3, $O_2(K_Z R_2) \leq R_2$. As Y acts on R_2, $O_2(K_Z R_2) \in Syl_2(C_{G_0}(K_Z/O_2(K_Z)))$, so $N_{G_0}(R_2)^* = N_{G_0^*}(R_2^*)$ by a Frattini Argument. Thus $K_2^* \trianglelefteq N_{G_0^*}(R_2^*)$; so in view of 4.3.13 and 4.3.14:

LEMMA 4.3.15. $R_2^* \neq 1$.

Now $K_Z^*/Z(K_Z^*)$ is a group appearing in Theorem C (A.2.3), satisfying the restrictions on prime divisors of $2^n - 1$ in 4.3.12.2.

Inspecting the automorphism groups of those groups for a proper 2-local subgroup $N_{K_Z^*}(R_2^*)$ with a normal subgroup K_2^* such that $K_2^*/O_2(K_2^*) \cong L_2(2^n)$, we conclude:

LEMMA 4.3.16. *One of the following holds:*

(1) $K_Z/O_2(K_Z) \cong L_2(2^{2^i n})$ *for some* $i \geq 1$.
(2) $K_Z/O_2(K_Z) \cong (S)U_3(2^n)$.
(3) $n = 2$ *and* $K_Z/O_2(K_Z) \cong L_3(5)$ *or* J_1.
(4) $n = 2$, $K_Z/O_2(K_Z) \cong L_3(4)$ *or* $U_3(5)$, *and* Y *induces outer automorphisms on* $K_Z/O_2(K_Z)$.

We are now in a position to complete the proof of Theorem 4.3.2.

Assume that one of cases (1)–(3) of 4.3.16 holds and let p be a prime divisor of $2^n - 1$. As Y^* centralizes $K_2^*/O_2(K_2^*)$ and hence K_2^*, but the groups in those cases do not admit an automorphism of order p centralizing K_2^*, we conclude that $Y^* = 1$. By 4.3.11, $O_2(KS) = [O_2(KS), Y]$, so as $R_2/O_2(KS) = [R_2/O_2(KS), Y]$, also $R_2 = [R_2, Y]$. Then since $Y^* = 1$, $R_2^* = 1$, contrary to 4.3.15.

Thus case (4) of 4.3.16 holds. Choose X of order 5 in K_2. Recall that K has three noncentral 2-chief factors, U and two copies of the dual of U on Q/U. Thus K_2 has four noncentral 2-chief factors, and each is a natural module for $K_2 R_2/R_2$. Therefore X has four nontrivial chief factors on R_2. As $G_Z \in \mathcal{H}(T)$ and $K_Z \trianglelefteq G_Z$, $F^*(K_Z) = O_2(K_Z)$, so at least one of those chief factors is in $O_2(K_Z)$.

Suppose that $K_Z/O_2(K_Z) \cong U_3(5)$. Then $X = Z(P)$ for some $P \in Syl_5(K_Z)$, and $P \cong 5^{1+2}$. Thus from the representation theory of extraspecial groups, X has five nontrivial chief factors on any faithful P-chief factor in $O_2(K_Z)$. But $O_2(K_Z) \leq R_2$ by 4.3.14.3, and we saw that X has just four nontrivial chief factors on R_2, with at least one in $O_2(K_Z)$.

Therefore $K_Z/O_2(K_Z) \cong L_3(4)$. Therefore $K_Z/O_2(K_Z) \cong L_3(4)$. Let X be a subgroup of order 3 in $O_{2,Z}(K)$. Then X is faithful on Z_S, so $X \leq G_Z$ but $X \not\leq K_Z$, and hence $X K_Z/O_2(K_Z) \cong PGL_3(4)$ by A.3.18. Further X centralizes $K_2/O_2(K_2)$, and from the structure of $[O_2(K), K]$ in C.1.34.3, there are four nontrivial K_2-chief factors in $O_2(K)$, all natural modules for $K_2/O_2(K_2) \cong L_2(4)$, and $C_{R_2}(X)/C_{R_2}(K_2 X)$ is a natural module for $K_2/O_2(K_2)$. It follows from B.4.14 that each nontrivial $K_Z X$-chief factor W in $O_2(K_Z)$ is the adjoint module for $K_Z/O_2(K_Z)$, and $C_W(X)/C_W(X K_2)$ is an indecomposable of $\mathbf{F}_4$-dimension 4 for

$K_2/O_2(K_2)$, contrary to $C_{R_2}(X)/C_{R_2}(K_2X)$ the natural module for $K_2/O_2(K_2)$. This contradiction completes the proof of Theorem 4.3.2.

THEOREM 4.3.17. *If $S \le T$ with $S \cap L \in Syl_2(L)$, then $N_G(S) \le M$.*

PROOF. By Theorem 4.3.2, $M = !\mathcal{M}(L)$, so the assertion follows from 4.3.1. $\square$

4.4. Controlling suitable odd locals

In this section, we apply Theorem 4.2.13 to force the normalizers of suitable subgroups of odd order to lie in M. The main results are Theorem 4.4.3 and its corollary Theorem 4.4.14.

During most of this section, we assume:

HYPOTHESIS 4.4.1. *(1) Hypothesis 4.2.1 holds. Set $M_+ := \langle L^T \rangle$ and $R_+ := O_2(M_+T) = C_T(M_+/O_2(M_+))$.*
(2) $1 \ne B \le C_M(M_+/O_2(M_+))$, with B abelian of odd order and $BT_+ = T_+B$ for some $T_+ \le T$ with $L^T = L^{T_+}$.
(3) $1 \ne V_B = [V_B, M_+] \le C_M(B)$ with V_B an M_+T-submodule of $\Omega_1(Z(R_+))$.

REMARK 4.4.2. Observe that if $L \trianglelefteq M$, then it is unnecessary to assume the existence of T_+. For example, we could then take $T_+ = 1$. Thus if Hypothesis 4.2.1 holds with $L \trianglelefteq M$ and $V \in \mathcal{R}_2(LT)$ with $[V, L] \ne 1$, then appealing to 1.4.1.4, Hypothesis 4.4.1 is satisfied for each nontrivial abelian subgroup B of $C_M(V)$ of odd order with V in the role of "V_B".

In this section we prove:

THEOREM 4.4.3. *Assume Hypothesis 4.4.1. Then either*
(1) $N_G(B) \le M$; or
(2) $L \trianglelefteq M$, $L/O_2(L)$ is isomorphic to $L_2(2^n)$, $L_3(2)$, $L_4(2)$, A_6, A_7, $\hat{A}_6$, or $U_3(3)$, and one of the following holds:
 (i) V_B is an FF-module for $LT/C_{LT}(V_B)$. Further:
  (a) If $L/O_2(L) \cong L_n(2)$, then either V_B is the sum of one or more isomorphic natural modules for $L/O_2(L)$, or V_B is the 6-dimensional orthogonal module for $L/O_2(L) \cong L_4(2)$.
  (b) If $L/O_2(L) \cong \hat{A}_6$, then for each $z \in C_{V_B}(T \cap L)^\#$, $V_B \not\le O_2(C_G(z))$.
  (c) If $L/O_2(L) \cong U_3(3)$ and $m(V_B) = 6$, then $C_G(V_3) \not\le M$, for V_3 the $(T \cap L)$-invariant subspace of V_B of rank 3.
 (ii) $L/O_2(L) \cong L_2(2^{2n})$, and V_B is the $\Omega_4^-(2^n)$-module.
 (iii) $L/O_2(L) \cong L_3(2)$, and V_B is the core of a 7-dimensional permutation module for $L/O_2(L)$.

Set $G_B := N_G(B)$, $M_B := N_M(B)$, $L_B := C_{M_+}(B)^\infty$, and $T_B := N_{T_+}(B)$. Making a new choice of T_+ if necessary, we may assume $T_B \in Syl_2(M_B)$. As G is simple, $G_B < G$, so G_B is a quasithin $\mathcal{K}$-group.

Before working with a counterexample to Theorem 4.4.3, we first prove two preliminary lemmas which assume only parts (1) and (2) of Hypothesis 4.4.1.

LEMMA 4.4.4. *Assume parts (1) and (2) of Hypothesis 4.4.1. Then $T_+ = [O_2(T_+B), B]T_B$.*

PROOF. Let $X := T_+B$, $Q := O_2(X)$ and $X^* := X/Q$. Then $F(X^*)$ is of odd order, so as B^* is an abelian Hall $2'$-subgroup of X, $B^* \leq C_{X^*}(F(X^*)) \leq F(X^*)$, so $B^* = F(X^*)$. Thus $BQ \trianglelefteq X$, so by a Frattini Argument (using the transitivity of a solvable group on its Hall subgroups in P. Hall's Theorem, 18.5 in [**Asc86a**]), $X = QN_X(B) = QT_BB$, so that $T_+ = QT_B$. Also $Q = C_Q(B)[Q, B]$ by Coprime Action, with $C_Q(B) \leq T_B$, so $T_+ = [Q, B]T_B$. $\qquad\square$

LEMMA 4.4.5. *Assume parts (1) and (2) of Hypothesis 4.4.1. Then* $M_+ = L_BO_2(M_+)$.

PROOF. By 4.4.1.2, $[M_+, B] \leq O_2(M_+)$, so M_+ acts on $BO_2(M_+)$; hence by a Frattini Argument, $M_+ = O_2(M_+)C_{M_+}(B)$. Now M_+ is perfect by Hypothesis 4.2.1 in 4.4.1.1, so $M_+ = O_2(M_+)C_{M_+}(B)^\infty = O_2(M_+)L_B$. $\qquad\square$

In the remainder of this section, we assume we are in a counterexample to Theorem 4.4.3; in particular, $G_B \not\leq M$.

LEMMA 4.4.6. *(1)* $M = !\mathcal{M}(L_BT_B)$.
(2) If $L \trianglelefteq M$ *then* $M = !\mathcal{M}(L_B)$.
(3) $N_G(V_B) \leq M$.

PROOF. Set $I := L_BT_B$ and $V_L := [R_2(LT), L]$. Observe that (cf. Notation 4.2.11) $I \in \mathcal{I}$: By 4.4.5, $L \leq IR_+$; $L^T = L^{T_+} = L^{T_B}$ by 4.4.1.2 and 4.4.4 (since $[O_2(T_+B), B] \leq R_+$); and $1 \neq V_B \leq O_2(I)$ by 4.4.1.3. Thus if (1) fails, then by Theorem 4.2.13, $L \trianglelefteq M$, and $L_B/O_2(L_B) \cong L/O_2(L)$ appears on the list of Theorem 4.2.13. Further 4.2.13 says that V_L is an FF-module for $Aut_{LT}(V_L)$, so the LT-submodule V_B is an FF-module for $Aut_{LT}(V_B)$ by B.1.5. Suppose $L/O_2(L) \cong L_n(2)$ for $n = 3$ or 4. Then case (2) of 4.2.13 holds, so either V_L is the sum of one or more isomorphic natural modules, or V_L is the 6-dimensional orthogonal module for $L_4(2)$. Therefore the submodule V_B satisfies the same constraints, so conclusion (i.a) of case (2) of Theorem 4.4.3 holds. Similarly if conclusion (4) or (5) of 4.2.13 holds, then $V_B = V_L$ and conclusion (i.b) or (i.c) of part (2) of Theorem 4.4.3 holds. In the remaining cases in Theorem 4.2.13, subcase (i) of case (2) of Theorem 4.4.3 imposes no further restriction on V_B; hence subcase (i) of case (2) in 4.4.3 holds. This contradicts our assumption that we are in a counterexample to Theorem 4.4.3, so we conclude that (1) holds. Under the hypothesis of (2), $L^T = L$, so by Remark 4.4.2, we may take $T_+ = 1$ and $I := L_B$; thus (2) follows from (1). Finally (1) implies (3), completing the proof of 4.4.6. $\qquad\square$

LEMMA 4.4.7. *(1)* $O_2(G_B) = 1$.
(2) M_B *is a maximal 2-local subgroup of* G_B.

PROOF. By 4.4.6.1, $M = !\mathcal{M}(M_B)$. Hence (2) holds, and as $G_B \not\leq M$, (2) implies (1). $\qquad\square$

LEMMA 4.4.8. $O(G_B) \leq M_B$.

PROOF. By Hypothesis 4.4.1 and 4.4.5, $1 \neq V_B = [V_B, L_B]$. As L_B is perfect, $m(V_B) \geq 3$, and in case of equality, L_B acts irreducibly as $L_3(2)$ on V_B, so $V_B \cap Z^*(G_B) = 1$. Therefore applying A.1.28 with G_B in the role of "H", we conclude that $m_p(O_p(G_B)) \leq 2$ for each odd prime p. Thus by A.1.26, $V_B = [V_B, L_B] \leq C_G(O_p(G_B))$. Hence $V_B \leq C_{V_BO(G_B)}(F(V_BO(G_B))) \leq F(V_BO(G_B))$, so $V_B = O_2(V_BO(G_B))$ and thus $O(G_B) \leq N_G(V_B) \leq M$ by 4.4.6.3. $\qquad\square$

LEMMA 4.4.9. *If K is a component of G_B, then $|K^{G_B}| \leq 2$, and in case of equality, $K \cong L_2(2^n)$, $Sz(2^n)$, $L_2(p^e)$, for some prime $p > 3$ and $e \leq 2$, J_1, or $SU_3(8)$.*

PROOF. Since we saw that G_B is a QTK-group, this follows from (1) and (2) of A.3.8; notice we use 4.4.7.1 to guarantee $O_2(K) = 1$, and I.1.3 to see that the Schur multiplier of $SU_3(8)$ is trivial, and in the remaining cases the multiplier of $K/Z(K)$ is a 2-group, so that K is simple. $\square$

By 4.4.8, V_B centralizes $O(G_B)$, and by 4.4.7.1, $O_2(G_B) = 1$, so V_B is faithful on $E(G_B)$. Thus there is a component K of G_B with $[K, V_B] \neq 1$. Set $K_0 := \langle K^{M_B} \rangle$ and $M_K := M \cap K$. Recall that G_B is a quasithin $\mathcal{K}$-group, and hence so is K by (a) or (b) of (1) in Theorem A (A.2.1), so that $K/Z(K)$ is described in Theorem B (A.2.2).

LEMMA 4.4.10. *(1) $K \not\leq M_B$.*
(2) $V_B \leq K_0$.
(3) $C_{G_B}(K_0) = O(G_B)$.

PROOF. As $[K, V_B] \neq 1$ and $V_B \leq O_2(M_B)$, (1) holds. As $L_B = O^2(L_B)$, L_B acts on K by 4.4.9, so $1 \neq V_B = [V_B, L_B]$ acts on K. Indeed as $Out(K)$ is 2-nilpotent for each K in Theorem B, V_B induces inner automorphisms on K_0, so that $V_B \leq K_0 H$ where $H := C_{G_B}(K_0)$. Then the projection of V_B on H is an M_B-invariant 2-group Q. If $Q \neq 1$, then by 4.4.7.2, $M_B = N_{G_B}(Q)$; but then $K \leq C_{G_B}(Q) \leq M_B$ contrary to (1). Thus $Q = 1$, giving (2). Now $H \leq C_{G_B}(V_B) \leq M_B$ by 4.4.6.3. Set $S := T_B \cap H$. As t_b IS Sylow in M_B, and $H \trianglelefteq M_B$, S is Sylow in H, $S \trianglelefteq T_B$, and

$$[S, L_B] \leq C_{L_B}(V_B) \cap H \leq O_2(L_B) \cap H \leq O_2(H) \leq O_2(G_B) = 1,$$

in view of 4.4.7.1. Thus $L_B T_B \leq N_G(S)$, so if $S \neq 1$ then $N_G(S) \leq M$ by 4.4.6.1; as S centralizes K, this contradicts (1). Thus the Sylow 2-group S of H is trivial, so (3) holds. $\square$

LEMMA 4.4.11. *(1) $K = K_0 \trianglelefteq G_B$.*
(2) $L_B \leq M_K$.

PROOF. Observe $Out(K_0)$ is solvable, since $|K^{G_B}| \leq 2$ by 4.4.9 and the Schreier property is satisfied for the groups in Theorem B. Also $C_{G_B}(K_0)$ is solvable by 4.4.10.3. Hence $L_B = L_B^\infty \leq K_0$. Thus (2) will follow from (1).

Assume K is not normal in G_B. By 4.4.9, $K_0 = K_1 K_2$ where $K_1 := K$ and $K_2 := K^s$ for $s \in G_B - N_{G_B}(K)$, and K is a simple Bender group, $L_2(p^e)$, J_1, or $SU_3(8)$. But then K has no nonsolvable 2-local M_K with $O_2(M_K)$ not in the center of M_K, contradicting $L_B \leq M \cap K_0$. This establishes (1). $\square$

LEMMA 4.4.12. *$K/Z(K)$ is not of Lie type and characteristic 2.*

PROOF. Assume otherwise. By 4.4.11.1 and 4.4.10.3, $O(G_B) = C_G(K)$, so T_B is faithful on K. By 4.4.10.2, $V_B \leq K$, so $Q_B := O_2(M_B) \cap K \not\leq Z(K)$. Therefore as $K/Z(K)$ is of Lie type and characteristic 2 by hypothesis, M_B acts on some proper parabolic of K (e.g. using the Borel-Tits Theorem 3.1.3 in [**GLS98**]). Hence by 4.4.7.2, M_K is a maximal M_B-invariant parabolic of K. Furthermore from Theorem B, $K/Z(K)$ either has Lie rank at most 2, or is $L_4(2)$ or $L_5(2)$ or $Sp_6(2)$, so as we chose $T_B \in Syl_2(M_B)$, T_B is transitive on each orbit of M_B on parabolics of K containing $T_B \cap K$, and hence M_K is a maximal T_B-invariant parabolic.

As L_B is a nonsolvable subgroup of M_K, K is of Lie rank at least 2, and M_K is of Lie rank at least 1. Assume that K is of Lie rank exactly 2. Then as M_K is a proper parabolic of rank at least 1, it must be of rank exactly 1, and hence is a maximal parabolic. Also $L_B = M_K^\infty$ as $M_K^\infty / O_2(M_K)^\infty$ is quasisimple. Then as $V_B \le Z(O_2(L_B))$ and $V_B = [V_B, L_B]$ we conclude by inspection of the parabolics of the rank 2 groups that $M_+/O_2(M_+) \cong L_B/O_2(L_B) \cong L_2(2^n)$, and either V_B is an FF-module, or (when K is unitary) V_B is the $\Omega_4^-(2^{n/2})$-module for $L_B/O_2(L_B)$. These are cases (i) and (ii) of conclusion (2) in Theorem 4.4.3, and in case (i) there are no further restrictions on V_B since $L/O_2(L) \cong L_2(2^n)$. This contradicts the choice of B as a counterexample to Theorem 4.4.3.

Therefore K is of Lie rank at least 3, so as we saw from Theorem B, $K \cong L_4(2)$, $L_5(2)$, or $Sp_6(2)$. Thus $M_+/O_2(M_+) \cong L_B/O_2(L_B) \cong L_3(2)$, $L_4(2)$, or A_6, and either V_B is an FF-module, which is a natural module in the first two cases, or $K \cong Sp_6(2)$, $L_B/O_2(L_B) \cong L_3(2)$, and $V_B = O_2(L_B)$ is the core of a 7-dimensional permutation module for $L_B/O_2(L_B)$. But then case (i) or (iii) of Theorem 4.4.3.2 holds, contrary to the choice of B as a counterexample, and completing the proof of 4.4.12. $\qquad\square$

We are now in a position to complete the proof of Theorem 4.4.3.

By 4.4.12, $K/Z(K)$ is not of Lie type and characteristic 2. By 4.4.10.2, $V_B \le K$.

Assume first that $m(V_B) \le 4$. Then inspecting the list of quasisimple subgroups of $GL_4(2)$, $L_B/O_2(L_B)$ is one of $L_2(4)$, $L_3(2)$, $L_4(2)$, A_6, or A_7, with V_B an FF-module, or an A_5-module for $L_2(4)$. Further if $L_B/O_2(L_B) \cong L_3(2)$ or $L_4(2)$, then either V_B is a natural module for $L_B/O_2(L_B)$, so condition in (a) of subcase (i) of case (2) of Theorem 4.4.3 is satisfied, or $m(V_B) = 4$ and $L_B/O_2(L_B) \cong L_3(2)$. The former case contradicts our assumption that B is a counterexample, so we may assume the latter holds. Then as $V_B = [V_B, L_B]$, $Z_B := C_{V_B}(L_B)$ is of rank 1 and V_B/Z_B is a natural module. By 4.4.6.1, $M_K T_B = C_{K T_B}(Z_B)$, so $L_B \trianglelefteq C_K(Z_B)$. Examining involution centralizers in the groups appearing in Theorem B for such a normal subgroup, we conclude $K \cong M_{23}$; but there L_B is not normal in $N_K(V_B) \cong A_7/E_{16}$.

Thus we may assume that $m(V_B) > 4$, and hence $m_2(K) > 4$. Then from the list of Theorem B, $K/Z(K)$ is not $L_2(p^e)$, $L_3^\epsilon(p)$, $PSp_4(p)$, $L_4^\epsilon(p)$, $G_2(p)$, A_7, A_9, a Mathieu group other than M_{24}, a Janko group other than J_4, HS, or Mc.

Since $K/Z(K)$ is not of Lie type and characteristic 2 by 4.4.12, we conclude from Theorem B that $K/Z(K)$ is M_{24}, J_4, He, and Ru. Since the multipliers of these groups are 2-groups by I.1.3, while $O_2(K) = 1$ by 4.4.7.1, it follows that K is simple. Again by 4.4.6.1, $M_K T_B$ is the unique maximal 2-local subgroup of KT_B containing $L_B T_B$. Inspecting the maximal 2-locals of $Aut(K)$ for a nonsolvable 2-local $M_K T_B$ such that $L_B \trianglelefteq M_K T_B$ and $1 \ne V_B = [V_B, L_B] \le Z(O_2(L_B))$, we conclude one of the following holds:

    (a) $K \cong J_4$ and L_B is a block of type M_{24} or $L_5(2)$.

    (b) K is M_{24} or He, and L_B is a block of type $\hat{A}_6$.

    (c) K is Ru and L_B is a block of type $G_2(2)$.

    (d) $K \cong Ru$ and $L_B/O_2(L_B) \cong L_3(2)$.

    (e) $K \cong M_{24}$, and $L_B/O_2(L_B) \cong L_4(2)$ or $L_3(2)$.

    (f) $K \cong J_4$ and $L_B/O_2(L_B) \cong L_3(2)$.

In cases (d)–(f), V_B is a natural module for $L_B/O_2(L_B)$, so that subcase (i) of case (2) of Theorem 4.4.3 holds, contrary to our assumption that B affords a counterexample to Theorem 4.4.3. Hence it only remains to dispose of cases (a)–(c).

Assume first that case (b) holds. Then from the structure of $K \cong M_{24}$ or He, $V_B \not\leq O_2(C_K(z))$ for each $z \in C_{V_B}(T \cap L)^{\#}$. Hence $V_B \not\leq O_2(C_G(z))$, so condition in (b) of subcase (i) of case (2) in Theorem 4.4.3 holds, again contrary to our choice of a counterexample. Similarly if case (c) holds then from the structure of Ru (cf. the case corresponding to Ru in the proof of Theorem 4.2.13, using facts from chapter J) of Volume I, $C_K(V_3) \not\leq M_K$. Thus condition (c) of subcase (i) of case (2) in Theorem 4.4.3.2 holds, for the same contradiction.

Therefore we may assume case (a) holds. Set $Z_B := C_{V_B}(T_B)$ and $G_Z := C_G(Z_B)$. Observe that Z_B is of order 2 and $K_Z := C_K(Z_B)^{\infty} \cong \hat{M}_{22}/2^{1+12}$. Arguing as in the last paragraph of the proof of Theorem 4.2.13, T induces inner automorphisms on $L/O_2(L)$, and hence $LT = LR_+$; therefore as $V_B \leq Z(R_+)$, $Z_B \leq Z(T)$, so $T \leq G_Z$. By 1.2.1.1, K_Z is contained in the product of the members of $\mathcal{C}(G_Z)$ on which it has nontrivial projection. Since $m_3(K_Z) = 2$ and G_Z is an SQTK-group, there is just one such member, so that $K_Z \leq L_Z \in \mathcal{C}(G_Z)$, and from 1.2.1.4, $L_Z/O_2(L_Z)$ is a quasisimple group described in Theorem C. Set

$$(L_Z B)^* := L_Z B/C_{L_Z B}(L_Z/O_2(L_Z)).$$

Then $K_Z^* \in \mathcal{C}(C_{L_Z^*}(B^*))$ with $K_Z^*/O_2(K_Z^*) \cong \hat{M}_{22}$ or M_{22}. Inspecting the p-locals (for odd primes p) of the groups in Theorem C, we conclude that either $K_Z^* = L_Z^*$ or $L_Z^* \cong J_4$ and $B^* = Z(K_Z^*)$ is of order 3. In the latter case, $K_Z \leq I_Z \leq L_Z$ with $I_Z \in \mathcal{L}(G,T)$ and $I_Z^* \cong \hat{M}_{22}/2^{1+12}$. Thus replacing L_Z by I_Z in this case, and replacing the condition that $L_Z \in \mathcal{C}(G_Z)$ by $L_Z \in \mathcal{L}(G,T)$, we may assume $L_Z = K_Z O_2(L_Z)$.

Thus in either case, $L_Z \in \mathcal{L}(G,T)$ with $L_Z = K_Z O_2(L_Z)$ and $[L_Z, B] \leq O_2(L_Z)$. Let $X := \langle B^T \rangle$; then $X = O^2(X) = O^2(XT)$. As $[L, B] \leq O_2(L)$, $[L, X] \leq O_2(L) \leq T \leq N_G(X)$, so that $X = O^2(XO_2(L)) \trianglelefteq LTX$, and hence $N_G(X) \leq M = !\mathcal{M}(LT)$. Similarly as $[L_Z, B] \leq O_2(L_Z)$, $L_Z \leq N_G(X)$, and hence $K_Z \leq L_Z T \leq N_G(X)$. Now $K = \langle L_B, K_Z \rangle \leq N_G(X) \leq M$, contradicting 4.4.10.1.

This final contradiction completes the proof of Theorem 4.4.3.

We interject a lemma which is often used in applying Theorem 4.4.3. Recall the notation $n(H)$ in Definition E.1.6.

LEMMA 4.4.13. *Assume that G is a simple QTKE-group, $H \in \mathcal{H}$ with $n(H) > 1$, $S \in Syl_2(H)$, and S is contained in a unique maximal subgroup M_H of H. Then $M_H \cap O^2(H)$ is 2-closed, and if we let B denote a Hall $2'$-subgroup of M_H, then:*

(1) If A is an elementary abelian p-subgroup of B with $AS = SA$, then $H = \langle M_H, N_H(A) \rangle$. In particular $N_H(A) \not\leq M_H$.

(2) Assume that $M \in \mathcal{M}(S)$, $M_H = M \cap H$, and $M_+ = O^2(M_+) \trianglelefteq M$. Then $C_B(M_+/O_2(M_+))S = SC_B(M_+/O_2(M_+))$.

PROOF. As $n(H) > 0$, S is not normal in H, so as M_H is the unique maximal subgroup of H over S, H is a minimal parabolic in the sense of Definition B.6.1. As $n := n(H) > 1$, E.2.2 then says that $K_0 := O^2(H) = \langle K^S \rangle$ for some $K \in \mathcal{C}(H)$ with $K/O_2(K)$ a Bender group over $\mathbf{F}_{2^n}$, $(S)L_3(2^n)$, or $Sp_4(2^n)$, and in the latter two cases S is nontrivial on the Dynkin diagram of $K/O_2(K)$. Set $H^* := H/O_2(H)$ and $M_0 := M_H \cap K_0$. By E.2.2, M_0 is the Borel subgroup of K_0 over $S \cap K_0$. In particular, M_0 is 2-closed, and a Hall $2'$-subgroup B of M_0 is abelian of p-rank at most 2 for each odd prime p.

In proving (1), we may take $A \neq 1$. Then $1 \leq m_p(A) \leq m_p(B) \leq 2$ for each $p \in \pi(A)$. It will suffice to show $N_{H^*}(A^*) \nleq M_H^*$, since then as M_H is a maximal subgroup of H, $H = \langle M_H, N_H(A) \rangle$, so that (1) holds.

Suppose first that $m_p(A) = m_p(B)$ for some p. Then $A = \Omega_1(O_p(B))$ and so $N_H(B) \leq N_H(A)$. But as B^* is a Cartan subgroup of K_0^*, $N_{K_0^*}(B^*) \nleq M_0^*$, and this suffices as we just observed.

So assume $m_p(B) = 2$ and $m_p(A) = 1$. Then by E.2.2, one of the following holds:

(i) $K < K_0$ and $K^* \cong L_2(2^n)$ or $Sz(2^n)$.
(ii) $K^* \cong Sp_4(2^n)$.
(iii) $K^* \cong (S)L_3(2^n)$.

In cases (i) and (ii), there is an element in $K_0^* - M_0^*$ inverting B^*, so $N_{K_0^*}(A^*) \nleq M_0^*$, which suffices to establish (1) in this case as we indicated. Thus we may assume case (iii) holds, so some $t \in S$ acts nontrivially on the Dynkin diagram of K^*, and by a Frattini Argument we may take $t \in N_S(B)$. Then as $AS = SA$, A is t-invariant. Let $U^* := N_{H^*}(B^*)$, $\tilde{U} := U^*/B^*$, and $\tilde{W}$ the image of $N_{K^*}(B^*)$ in $\tilde{U}$. Then $\tilde{W} \cong S_3$ is the Weyl group of K^* and $\tilde{t} = \tilde{s}\tilde{w}$, where $\tilde{w}$ is an involution in $\tilde{W}$, and $\tilde{s} \in C_{\tilde{U}}(\tilde{W})$. Pick preimages w^* and s^* of $\tilde{w}$ and $\tilde{s}$. As $\tilde{W}$ acts indecomposably on $\Omega_1(O_p(B))$, $\tilde{s}$ inverts or centralizes B^*, so s^* and t^* act on A^*, and hence $w \in N_H(A) - M_H$ completing the proof of (1).

So we may assume the hypotheses of (2). Let $D := C_B(M_+/O_2(M_+))$ and $Q := O_2(BS)$. Then, as in the proof of 4.4.4, a Frattini Argument gives $S = QN_S(B)$. Now as $M_+ \trianglelefteq M$, $N_S(B)$ acts on M_+ and hence also on $D = C_B(M_+/O_2(M_+))$. Therefore $DN_S(B)$ is a subgroup of G acting on Q, and hence $DN_S(B)Q = DS$ is a subgroup of G, completing the proof of (2). $\qquad \square$

Usually we use Theorem 4.4.3 via an appeal to the following corollary:

THEOREM 4.4.14. *Assume Hypothesis 4.2.1, and let* $M_+ := \langle L^T \rangle$, $V_0 \in \mathcal{R}_2(M_+T)$, *and* $H \in \mathcal{H}_*(T, M)$. *Assume*

(a) $V := [V_0, M_+] \neq 1$, $V_0 = \langle C_{V_0}(T)^{M_+} \rangle$, *and* V *is not an FF-module for* $M_+T/C_{M_+T}(V)$.

(b) $n(H) > 1$.

Then one of the following holds:

(1) $O^2(H) \cap M$ *is 2-closed, and a Hall $2'$-subgroup of* $H \cap M$ *is faithful on* $M_+/O_2(M_+)$.

(2) $M_+/O_2(M_+) \cong L_2(2^{2n})$, *and* V *is the* $\Omega_4^-(2^n)$-module.

(3) $M_+/O_2(M_+) \cong L_3(2)$, *and* V *is the core of a* 7-*dimensional permutation module for* $M_+/O_2(M_+)$.

PROOF. Let $Z := \Omega_1(Z(T))$ and $K := O^2(H)$. We observed in Remark 3.2.4 that Hypothesis 4.2.1 allows us to apply Theorem 3.1.8. As V is not an FF-module, $J(T) \leq C_T(V)$ by B.2.7, so $H \leq C_G(Z)$, by 3.1.8.3. Similarly by 3.3.2.4, H is a minimal parabolic described in E.2.2. Since $n(H) > 1$ by hypothesis, E.2.2 shows that $K/O_2(K)$ is of Lie type in characteristic 2 and of Lie rank at most 2, and $K \cap M$ is a Borel subgroup of K, so in particular $K \cap M$ is 2-closed. Let B_H be a Hall $2'$-subgroup of $H \cap M$; thus B_H is abelian of odd order.

Assume (1) fails. Then $B := C_{B_H}(M_+/O_2(M_+)) \neq 1$. Observe that we have the hypotheses of 4.4.13 with T, B_H, B in the roles of "S, B, A", so $BT = TB$ by 4.4.13.2. Hence parts (1) and (2) of Hypothesis 4.4.1 are satisfied, with T in the role of "T_+". Thus by 4.4.5, $M_+ = L_B O_2(M_+)$, where $L_B := C_{M_+}(B)^\infty$.

Next since $H \leq C_G(Z)$, $C_{V_0}(T) = Z \cap V_0 \leq C_G(B)$, so $V_0 = \langle (Z \cap V_0)^{M_+} \rangle = \langle (Z \cap V_0)^{L_B} \rangle \leq C_G(B)$ by (a). Therefore part (3) of Hypothesis 4.4.1 is also satisfied, with V in the role of "V_B", so that we may apply Theorem 4.4.3. By (a), V is not an FF-module for $L_B/O_2(L_B)$, which rules out subcase (i) of case (2) of Theorem 4.4.3. By 4.4.13.1, $N_H(B) \not\leq M$, ruling out case (1) of Theorem 4.4.3. Thus subcase (ii) or (iii) of case (2) of Theorem 4.4.3 must hold, and these are conclusions (2) and (3) of Theorem 4.4.14. $\qquad\square$

Part 2

The treatment of the Generic Case

Part 1 has set the stage for the proof of the Main Theorem by supplying information about the structure of 2-locals, establishing the Fundamental Setup (3.2.1), and proving that in the FSU, the members of $\mathcal{H}_*(T, M)$ are minimal parabolics. We now begin the analysis of the various possibilites for $L \in \mathcal{L}_f^*(G, T)$ and $V \in \mathcal{R}_2(L_0 T)$ arising in the FSU. Recall the FSU includes the hypotheses that G is a simple QTKE-group, $T \in Syl_2(G)$, and $L \in \mathcal{L}_f^*(G, T)$ with $L/O_2(L)$ quasisimple and V a suitable member of $\mathcal{R}_2(LT)$.

In Part 2, we consider the Generic Case of our Main Theorem. This is the case where $L/O_2(L) \cong L_2(2^n)$ with $L \trianglelefteq M$ and $n(H) > 1$ for some $H \in \mathcal{H}_*(T, M)$. We show in Theorem 5.2.3 of chapter 5 that in the Generic Case, (modulo the sporadic exception M_{23} and the "$\mathbf{F}_2$-case") G is one of the generic conclusions in our Main Theorem: namely G is of Lie type of Lie rank 2 and characteristic 2. In chapter 6 we consider the remaining case where $n(H) = 1$ for each $H \in \mathcal{H}_*(T, M)$, and show in that case that $n = 2$ and V is the A_5-module. The case where V is the A_5-module is treated in Part 5 on groups over $\mathbf{F}_2$, since the A_5-module is the module for $\Omega_4^-(2)$.

Thus once we have dealt with the groups $L_2(p)$ and the Bender groups in Theorem 2.1.1, and the groups of Lie type in characteristic 2 of Lie rank 2 in Theorem 5.2.3, we will have handled all the infinite families of groups appearing as conclusions in the Main Theorem.

CHAPTER 5

The Generic Case: $L_2(2^n)$ in $\mathcal{L}_f$ and $n(H) > 1$

In this chapter we assume the following hypothesis:

HYPOTHESIS 5.0.1. G is a simple QTKE-group, $T \in Syl_2(G)$, $L \in \mathcal{L}_f^*(G, T)$ with $L/O_2(L) \cong L_2(2^n)$ and $L \trianglelefteq M \in \mathcal{M}(T)$.

As L is nonsolvable, $n \geq 2$. Further $M = !\mathcal{M}(LT)$ by 1.2.7.3 and $M = N_G(L)$. Set

$$Z := \Omega_1(Z(T)).$$

From the results of section 1.2, there exists $V \in \mathcal{R}_2(LT)$ with $[V, L] \neq 1$; choose such a V and set $\overline{LT} := LT/C_{LT}(V)$. By 3.2.3 it is possible to choose V so that the pair L, V satisfies the hypotheses of the Fundamental Setup (3.2.1). However occasionally we need information about other members of $\mathcal{R}_2(LT)$, so usually in this chapter we do not assume V satisfies the hypotheses of the FSU. Later, when appropriate, we sometimes specialize to that case.

By Theorem 2.1.1, $\mathcal{H}_*(T, M)$ is nonempty.

In the initial section 5.1, we determine the possibilities for V and provide restrictions on members of $\mathcal{H}_*(T, M)$. The following section begins the proof of Theorem 5.2.3, which supplies very strong information when $n(H) > 1$ for some $H \in \mathcal{H}_*(T, M)$. Indeed in the FSU, if V is not the A_5-module, then either G is of Lie type and Lie rank 2 over a field of characteristic 2, or G is M_{23}; hence we refer to this situation as the Generic Case . The final section 5.3 completes the proof of Theorem 5.2.3.

Our primary tool for proving Theorem 5.2.3 is the main theorem of the "Green Book" of Delgado-Goldschmidt-Stellmacher [**DGS85**], which gives a local description of weak BN-pairs of rank 2. To apply the Green Book, we must achieve the setup of Hypothesis F.1.1. There are two major obstacles to verifying this hypothesis: Let D be a Hall $2'$-subgroup of $N_L(T \cap L)$, and $K := O^2(H)$. We must first show that D acts on K, unless the exceptional case in part (1) of Theorem 5.2.3 holds. Second, we must construct a normal subgroup S of T such that S is Sylow in SL and SK, and so that there exists an S-invariant subgroup K_1 of K such that $K_1/O_2(K_1)$ a Bender group. Now $K/O_2(K)$ is of Lie type in characteristic 2 of Lie rank 1 or 2. If K is of Lie rank 1, we take $K_1 := K$; if K is of Lie rank 2, we choose K_1 to be a rank one parabolic of K. In either case, we take S to be $O_2(H \cap M)$, unless $K/O_2(K) \cong L_3(4)$, which provides a final obstruction that we deal with in Theorem 5.1.14.

After producing our weak BN-pair and identifying it up to isomorphism of amalgams using the Green Book, we still need to identify G. To do so we appeal to Theorem F.4.31 as a recognition theorem; ultimately Theorem F.4.31 depends upon the Tits-Weiss classification of Moufang generalized polygons, although the Fong-Seitz classification of split BN-pairs of rank 2 would also suffice. There is also an obstacle to applying this recognition theorem: the case where $K \notin \mathcal{L}^*(G,T)$, leading to M_{23}. This case is dealt with in Theorem 5.2.10.

5.1. Preliminary analysis of the $L_2(2^n)$ case

5.1.1. General analysis of V and H. Since this is the first case in the FSU which we analyze, we begin with a lemma summarizing some of the basic tools (developed in Volume I and earlier chapters of Volume II) to deal with the FSU. We thank Ulrich Meierfrankenfeld for several improvements to the proofs in this section.

LEMMA 5.1.1. *(1) $C_T(V) = O_2(LT)$.*
(2) Each $H \in \mathcal{H}_*(T, M)$ is a minimal parabolic described in B.6.8, and in E.2.2 if $n(H) > 1$.
(3) For each $H \in \mathcal{H}_*(T, M)$, case (I) of Hypothesis 3.1.5 is satisfied with LT in the role of "M_0".
(4) LT is a minimal parabolic.

PROOF. Part (1) follows from 1.4.1.4, (2) follows from 3.3.2.4, (3) follows from (1) and the fact that $L \trianglelefteq M$, and (4) is well known and easy.   □

We begin by discussing the possibilities for V:

LEMMA 5.1.2. *One of the following holds:*
(1) $J(T) \le C_M(V)$, so $J(T)$ and $Baum(T)$ are normal in LT and $M = !\mathcal{M}(N_G(J(T))) = !\mathcal{M}(N_G(Baum(T)))$.
(2) $[V, J(T)] \ne 1$ and $V/C_V(L)$ is the natural module for $\bar{L}$.
(3) $[V, J(T)] \ne 1$, $n = 2$, and $V = C_V(LT) \oplus [V, L]$ with $[V, L]$ the S_5-module for $\bar{L}\bar{T} \cong S_5$.

PROOF. By 5.1.1.1, $C_T(V) = O_2(LT)$. Thus if $J(T) \le C_M(V)$, then $J(T) = J(O_2(LT))$ and $Baum(T)) = Baum(O_2(LT))$ by B.2.3, so LT acts on $J(T)$ and $Baum(T)$. However by 1.2.7.3, $M = !\mathcal{M}(LT)$, so (1) holds in this case. So assume $[V, J(T)] \ne 1$. Then V is an FF-module for $\bar{L}\bar{T}$ by B.2.7, so by B.5.1.1, $I := [V, L] \in Irr_+(L, V)$, and by B.5.1.5, $V = I + C_V(L)$. By B.4.2, either $I/C_I(L)$ is the natural module, or $n = 2$ and $I/C_I(L)$ is the A_5-module. In the former case (2) holds as $V = I + C_V(L)$, and in the latter (3) holds by B.5.1.4.   □

LEMMA 5.1.3. *One of the following holds:*
(1) V is the direct sum of two natural modules for $\bar{L}$.
(2) $n = 2$ and V is the direct sum of two S_5-modules for $\bar{L}\bar{T} \cong S_5$.
(3) $[V, L]/C_{[V,L]}(L)$ is the natural module for $\bar{L}$.
(4) n is even and V is the $O_4^-(2^{n/2})$-module for $\bar{L}$.
(5) $V = [V, L] \oplus C_V(LT)$, and $[V, L]$ is the S_5-module for $\bar{L}\bar{T} \cong S_5$.

REMARK 5.1.4. Recall that the A_5-module and the $O_4^-(2)$-module are the same. Notice however that in case (4) we may have $\bar{L}\bar{T} \cong A_5$, which is not allowed in (5). On the other hand in case (5) we may have $C_V(L) \ne 1$, which is not allowed in (4).

PROOF. If $[V, J(T)] \ne 1$ then (3) or (5) holds by 5.1.2. Thus we may assume $[V, J(T)] = 1$, so that $C_V(L) = 1$ by 3.1.8.3.

Next $\hat{q}(\bar{L}\bar{T}, V) \le 2$ by 3.1.8.1. Hence in the language of Definition D.2.1, there is $\bar{A} \in \hat{\mathcal{Q}}(\bar{T}, V)$. Recall that we are not yet assuming the FSU, so we will work with the results of section D.3 rather than those of section 3.2 based on the FSU. By A.1.42.2, there is $I \in Irr_+(L, V, T)$. Now Hypothesis D.3.1 is satisfied with $\bar{L}\bar{T}$, $\bar{L}$, I, $V_M := \langle I^T \rangle$ in the roles of "M, M_+, V, V_M". Hence we may apply D.3.10 to conclude that $I \trianglelefteq LT$.

Suppose first that $I < [V, L]$, and choose an LT-submodule V_1 of V with $[V, L] \not\leq V_1 \geq I$. As $\bar{L} = F^*(\bar{L}\bar{T})$ is simple, $\bar{L}$—and hence also $\bar{A}$—is faithful on V_1 and on $\tilde{V} := V/V_1$. Thus

$$2 \geq r_{\bar{A},V} \geq r_{\bar{A},V_1} + r_{\bar{A},\tilde{V}}$$

in the language of Definition B.1.1. On the other hand, by B.6.9.1, $r_{\bar{A},W} \geq 1$ for each faithful $\bar{L}\bar{A}$-module W, so $r_{\bar{A},V_1} = r_{\bar{A},\tilde{V}} = 1$. Then by another application of B.6.9, V_1 and $\tilde{V}$ have unique noncentral chief factors, and either both factors are natural, or $n = 2$ and at least one is an A_5-module. Now if a factor is natural, then $\bar{A} \in Syl_2(\bar{L})$, while if a factor is an A_5-module, then $\bar{A} \not\leq \bar{L}$. So if one factor is an A_5-module, then both are A_5-modules; then as A_5-modules have trivial 1-cohomology by I.1.6, and we saw $C_V(L) = 1$, (2) holds. This leaves the case where both factors are natural modules. Here we choose V_1 maximal subject to $[\tilde{V}, L] \neq 1$, so as $\tilde{V}$ is an FF-module, $\tilde{V}$ is natural by B.5.1.5. Also V_1 is an FF-module, so $V_1/C_{V_1}(L)$ is natural by B.5.1.5; hence as $C_V(L) = 1$, both $V_1 = I$ and V/I are natural. Further as $r_{\bar{A},V} = 2$ with $m(V/C_V(\bar{i})) = 2n = 2m(\bar{L})$ for each involution $\bar{i} \in \bar{L}$, $\bar{A} \in Syl_2(\bar{L})$ with $C_V(\bar{A})) = C_V(\bar{a}) = [V, \bar{a}]$ for each $\bar{a} \in \bar{A}^\#$. Therefore V is semisimple by Theorem G.1.3, and hence (1) holds.

Thus we may assume that $I = [V, L]$, and therefore that LT is irreducible on $W := [V, L]/C_{[V, L]}(L)$. Then as $\hat{q}(\bar{L}\bar{T}, V) \leq 2$, it follows from B.4.2 and B.4.5 that either W is the natural module, or n is even and W is the orthogonal module. In the first case (3) holds, so assume the second holds. Then $H^1(\bar{L}, W) = 0$ by I.1.6, so as $C_V(L) = 1$, V is irreducible and hence (4) holds. This completes the proof. $\square$

Recall that by Theorem 2.1.1, there is $H \in \mathcal{H}_*(T, M)$.

LEMMA 5.1.5. *Let $H \in \mathcal{H}_*(T, M)$ and D_L a Hall $2'$-subgroup of $N_L(T \cap L)$. Then*

(1) $H \cap M$ acts on $T \cap L$ and on $O^2(D_LT)$, and

(2) if $n(H) > 1$, then $H \cap M$ is solvable, and some Hall $2'$-subgroup of $H \cap M$ acts on D_L.

PROOF. Let $T_L := T \cap L$ and $B := N_L(T_L)$. Since $L/O_2(L) \cong L_2(2^n)$, B is the unique maximal subgroup of L containing T_L. But as $M = !\mathcal{M}(LT)$ and $H \not\leq M$, $L \not\leq H$, so $H \cap L \leq B$; hence $H \cap M$ acts on $O_2(H \cap L) = T_L$ and on $N_L(T_L) = B$. Thus (1) holds.

Assume $n(H) > 1$. Then $H \cap M$ is solvable by E.2.2, so as $H \cap M$ acts on B and B is solvable, $(H \cap M)B$ is solvable. Therefore by Hall's Theorem, a Hall $2'$-subgroup D_H of $H \cap M$ is contained in a Hall $2'$-subgroup D of $(H \cap M)B$, and $D \cap B$ is a Hall $2'$-subgroup of B. By Hall's Theorem there is $t \in T_L$ with $(D \cap B)^t = D_L$, so as $T_L \leq H$, the Hall $2'$-subgroup D_H^t of $H \cap M$ acts on D_L, completing the proof of (2). $\square$

LEMMA 5.1.6. *Let $H \in \mathcal{H}_*(T, M)$, D_L a Hall $2'$-subgroup of $N_L(T \cap L)$, and assume $O_2(\langle D_L, H \rangle) = 1$. Then n is even and one of the following holds:*

(1) $n = 2$, V is the direct sum of two natural modules for $\bar{L}$, and $[Z, H] = 1$.

(2) $n = 2$ or 4, $[V, L]$ is the natural module for $\bar{L}$, and $[Z, H] = 1$.

(3) $n = 2$, $[V, L]$ is the S_5-module for $\bar{L}\bar{T} \cong S_5$, and $Z(H) = 1$.

(4) $n \equiv 0 \mod 4$, V is the $\Omega_4^-(2^{n/2})$-module for $\bar{L}$, and $[Z, H] = 1$. Furthermore if we take D_ϵ to be the subgroup of D_L of order $2^{n/2} - \epsilon$, $\epsilon = \pm 1$, and $X_\epsilon := \langle D_\epsilon, H \rangle$, then $Z \le Z(X_-)$ and either $O_2(X_+) \ne 1$, or $n = 4$ or 8.

PROOF. Let $X := \langle D_L, H \rangle$. Then by hypothesis, $O_2(X) = 1$. Recall from the start of the chapter that $Z = \Omega_1(Z(T))$, and set $V_D := \langle Z^{D_L} \rangle$ and $V_Z := \langle Z^L \rangle$. Observe that $V_Z \in \mathcal{R}_2(LT)$ and $V_D \in \mathcal{R}_2(TD_L)$ by B.2.14. In each case of 5.1.3,

$$V = \langle (Z \cap V)^L \rangle \le V_Z.$$

Suppose first that $T \trianglelefteq TD_L$. Then applying Theorem 3.1.1 with TD_L, T in the roles of "M_0, R", we contradict $O_2(X) = 1$. Therefore $T \ntrianglelefteq TD_L$.

Since $\bar{L} \cong L_2(2^n)$, it follows that n is even, and also that $\bar{L}\bar{T} = \bar{L}\bar{S}$ where $S \le T$, $\bar{S} \ne 1$, $\bar{L} \cap \bar{S} = 1$, and $\bar{S}$ acts faithfully as field automorphisms of $\bar{L}$.

As $V_Z \in \mathcal{R}_2(LT)$, we can apply 5.1.2 and 5.1.3 to V_Z in the role of "V". For example by 5.1.2 and 3.1.8.3, either

(i) $[Z, H] = 1 = C_{V_Z}(L)$, or

(ii) $[V_Z, J(T)] \ne 1$, and either $V_Z / C_{V_Z}(L)$ is the natural module for $\bar{L}$, or $[V_Z, L]$ is the S_5-module for $\bar{L}\bar{T} \cong S_5$.

To complete the proof, we consider each of the possibilities for V arising in 5.1.3.

Suppose first that V is described in case (1) of 5.1.3. As the overgroup V_Z of V is also described in one of the cases in 5.1.3, we conclude that $V = V_Z$. By the previous paragraph, $[Z, H] = 1$. From the structure of V, $V_D \le C_V(T \cap L)$ which is of rank $2n$ in V of rank $4n$, D_L is faithful on V_D so that $m(V_D) \ge n$, with

$$(T \cap L)C_T(V) = O_2(TD_L) = C_T(V_D) = C_{TD_L}(V_D),$$

and $T / C_T(V_D)$ is cyclic. Thus as $H \cap M$ normalizes TD_L by 5.1.5.1, Hypothesis 3.1.5 is satisfied by TD_L, V_D in the roles of "M_0, V". As $O_2(X) = 1$, we conclude from 3.1.6 that $\hat{q}(TD_L / O_2(TD_L), V_D) \le 2$. Hence as $T / C_T(V_D)$ is cyclic and $m(V_D) \ge n$, we conclude that $n = 2$, so that conclusion (1) holds.

Similarly if V appears in case (3) of 5.1.3, we conclude as in the previous paragraph that V_Z appears in case (1) or (3) of 5.1.3, that Hypothesis 3.1.5 is satisfied with TD_L, V_D in the roles of "M_0, V", and that $\hat{q}(TD_L / O_2(TD_L), V_D) \le 2$. Hence either $n = 2$, or possibly $n = 4$ in case V_Z satisfies conclusion (3) of 5.1.3—since $m(V_D / C_{V_D}(t)) = n/2$ for $t \in T - C_T(V_D)$ with $t^2 \in C_T(V_D)$ when V_Z satisfies that conclusion. Further $J(T) \le C_T(V_D)$ by B.4.2.1, so $[H, Z] = 1 = C_Z(L)$ by Theorem 3.1.7, which completes the proof that conclusion (2) holds in this case.

Suppose next that V appears in case (2) or (5) of 5.1.3, or in case (4) with $n = 2$. These are the cases where $n = 2$ and L has an A_5-submodule on V, and hence also on V_Z, so that V_Z must also satisfy one of these three conclusions. Therefore $D_L \le C_G(Z)$. Recall $H \in \mathcal{H}(T) \subseteq \mathcal{H}^e$ by 1.1.4.6, so if $Z(H) \ne 1$ then $Z \cap Z(H) \ne 1$. Thus as $O_2(X) = 1$, $Z(H) = 1$, so that case (ii) holds; therefore V_Z satisfies conclusion (3), and hence so does V.

This leaves the case where V satisfies case (4) of 5.1.3 with $n > 2$. Thus $V = V_Z$ as before, and hence (ii) does not hold, leaving case (i) where $[Z, H] = 1 = C_Z(L)$. Now V is a 4-dimensional FL-module, where $F := \mathbf{F}_{2^{n/2}}$, and $Z = C_U(T)$ where U is the 1-dimensional F-subspace of V stabilized by $\bar{S} := \bar{T} \cap \bar{L}$. Further setting $A := N_{GL(V)}(\bar{L})$, A is the split extension of $\bar{L}$ by $\langle \sigma \rangle$ where σ is a field automorphism.

Also if s is the involution in $\langle \sigma \rangle$, then $C_A(U) = \bar{S}\langle s \rangle D_-$ and $U = \langle Z^{D_+} \rangle$, so $U = V_D$. In particular $[D_-, Z] = 1$, so $Z \leq Z(X_-)$. If $n \equiv 2 \mod 4$, then $\bar{T} \leq \bar{S}\langle s \rangle$, so $Z = U$ is D_+-invariant; hence $X = \langle H, D_L \rangle \leq N_G(Z)$, contrary to $O_2(X) = 1$. Thus $n \equiv 0 \mod 4$. Finally D_+ is faithful on V_D, so applying 3.1.6 with TD_+, V_D in the roles of "M_0, V" as before, either $O_2(X_+) \neq 1$ or $\hat{q}(D_+T/O_2(D_+T), V_D) \leq 2$. In the latter case, as $T/C_T(V_D)$ is cyclic and $m(V_D/C_{V_D}(t)) \geq n/4$ for $t \in T - C_T(V_D)$, $n = 4$ or 8. Thus (4) holds. $\square$

LEMMA 5.1.7. *(1)* $N_G(Baum(T)) \leq M$.
  (2) Let $H \in \mathcal{H}_*(T, M)$ and set $K := O^2(H)$. Assume $[Z, H] \neq 1$. Then

  (i) $L = [L, J(T)]$.
  (ii) $K = [K, J(T)]$.
  (iii) Either $O_2(\langle N_L(T \cap L), H \rangle) \neq 1$, or $[V, L]$ is the S_5-module for $\bar{L}\bar{T} \cong S_5$, and $Z(H) = 1$.

PROOF. We first prove (1). Let $S := Baum(T)$. If $J(T) \leq C_T(V)$, then (1) follows from 5.1.2. Thus we may assume $J(T) \not\leq C_T(V)$, so by 5.1.2, either $V/C_V(L)$ is the natural module for $\bar{L}$ or $[V, L]$ is the A_5-module. In the former case, $S \cap L \in Syl_2(L)$ by E.2.3.2, so (1) follows from 4.3.17.

Therefore we may assume that $[V, L]$ is the A_5-module. As $[V, J(T)] \neq 1$, we conclude from E.2.3 that $\bar{L}\bar{T} \cong S_5$, $\bar{S} = \overline{J(T)} \cong E_4$ is generated by the two transvections in $\bar{T}$, and $\langle Z^L \rangle = [V, L] \oplus C_Z(L)$. We may assume $V = [V, L]$.

Assume that $N_G(S) \not\leq M$; then no nontrivial characteristic subgroup of S is normal in LT as $M = !\mathcal{M}(LT)$. Hence by E.2.3.3, L is an A_5-block, so $V = O_2(L) \trianglelefteq M$. Let $Q := O_2(LS)$. It follows using C.1.13.b that $Q = V \times Q_C$, where $Q_C := C_S(L)$.

For any $1 \neq S_0 \leq S$ normalized by LT, we have $N_G(S_0) \leq M = !\mathcal{M}(LT)$, so $N_G(S) \not\leq N_G(S_0)$ by our assumption. Thus Hypothesis C.6.2 is satisfied with L, S, T, $N_G(S)$ in the roles of "L, R, T_H, Λ". Therefore by C.6.3.1 there is $g \in N_G(S)$ with $V^g \not\leq Q$. As $V \trianglelefteq M$, $g \notin M$.

Suppose that $Q_C \not\leq Q^g$. Since $[Q_C, V^g] \cap [V, V^g] \leq Q_C \cap V = 1$, from the action of S on V and hence on V^g, we conclude that Q_C and V induce distinct transvections on V^g. Thus as $|S : Q^g| = 4$, $S = Q_C V Q^g$. Let $x \in [Q_C, V^g]^{\#}$; then $x \in Q_C \leq C_G(L)$, so as $M = !\mathcal{M}(L)$ by Theorem 4.3.2, $C_G(x) \leq M$, so $V \leq O_2(C_G(x))$. Since Q_C induces a transvection on the A_5-module V^g for L^g, $C_{L^g S}(x)Q_C Q^g/Q_C Q^g \cong S_3$, so $V \leq O_2(C_{L^g S}(x)Q_C Q^g) = Q_C Q^g$, contrary to V and Q_C inducing distinct transvections on V^g.

Therefore $Q_C \leq Q^g$. Hence

$$\Phi(Q_C) \leq \Phi(Q^g) = \Phi(Q_C^g V^g) = \Phi(Q_C^g),$$

so $\Phi(Q_C) = \Phi(Q_C^g)$. Thus as $\Phi(Q_C) \trianglelefteq LT$ and $g \notin M = !\mathcal{M}(LT)$, $\Phi(Q_C) = 1 = \Phi(Q)$.

Next we claim we can choose g so that $S = QQ^g$. If not then $Q \cap Q^g$ is a hyperplane of Q and Q^g centralized by Q^g, so Q^g induces a transvection on Q and hence $S = Q^g Q^{gt}$ for $t \in T - SO_2(LT)$. Thus as $g \in N_G(S)$, $S = QQ^{gtg^{-1}}$, establishing the claim.

As $S = QQ^g$ with $\Phi(Q) = 1$ and $Q_C \leq Q^g$, $S = Q_C \times D_1 \times D_2$, where $D_1 \cong D_2$ is dihedral of order 8. By the Krull-Schmidt Theorem A.1.15, $N_G(S)$

permutes $\{D_1 Z(S), D_2 Z(S)\}$. Then $O^2(N_G(S))$ acts on $D_i Z(S)$, and indeed centralizes $D_i Z(S)/Z(S)$ as $D_i Z(S)/Z(S)$ is of order 4 and contains a unique coset of $Z(S)$ containing elements of order 4. Thus $O^2(N_G(S))$ acts on Q, and hence $O^2(N_G(S)) \leq M = !\mathcal{M}(N_G(Q))$. But then $N_G(S) = O^2(N_G(S))T \leq M$, contrary to assumption. This contradiction completes the proof of (1).

As (1) is established, we may assume the hypotheses of (2). Thus $[Z, H] \neq 1$, so $J(T) \not\leq C_T(V)$ by 3.1.8.3, and then part (i) of (2) holds by B.6.8.6.d. Therefore by 5.1.2, either $[V, L]$ is the S_5-module for $\bar{L}\bar{T} \cong S_5$, or $V/C_V(L)$ is the natural module for $\bar{L}$. Set $U := \langle Z^H \rangle$, so that $U \in \mathcal{R}_2(H)$ by B.2.14. By (1), $S \neq \mathrm{Baum}(O_2(H))$. Then as $[Z, H] \neq 1$, $J(T) \not\leq C_T(U)$ by B.6.8.3.d, and (ii) follows. Finally if $O_2(\langle N_L(T \cap L), H \rangle) = 1$, we may apply 5.1.6; as $[Z, H] \neq 1$, conclusion (3) of 5.1.6 holds, completing the proof of (iii). $\square$

5.1.2. Further analysis when $n(H) > 1$. Recall that in this Part we focus on the "generic" situation, where $n(H) > 1$ for some $H \in \mathcal{H}_*(T, M)$. Later in Theorem 6.2.20, we will reduce the case where $n(H) = 1$ for each $H \in \mathcal{H}_*(T, M)$ to $n = 2$ with $\bar{L} = L_2(4) \cong A_5$ acting on $[Z, L]$ as the sum of at most two A_5-modules. That situation is treated later in those Parts dedicated to groups defined over $\mathbf{F}_2$.

So in the remainder of this section we assume the following hypothesis:

HYPOTHESIS 5.1.8. *Hypothesis 5.0.1 holds, and there is $H \in \mathcal{H}_*(T, M)$ with $n(H) > 1$. Set $K := O^2(H)$, $M_H := M \cap H$, and $M_K := M \cap K$.*

NOTATION 5.1.9. By 5.1.5.2, we may choose a Hall $2'$-subgroup B of M_H, and a B-invariant Hall $2'$-subgroup D_L of $N_L(T \cap L)$. This notation is fixed throughout the remainder of the section.

Observe $M_H = BT = TB$ since $T \in Syl_2(M_H)$. Further B and T normalize $N_L(T \cap L) = D_L(T \cap L)$ by 5.1.5.1, so $D_L BT$ is a subgroup of G.

Our goal (oversimplifying somewhat) is to show in the following section that $(LTB, D_L TB, D_L H)$ forms a weak BN-pair of rank 2 in the sense of [**DGS85**], as in our Definition F.1.7. Indeed we already encounter such rank 2 amalgams in this section.

The next few results study the structure of K and the embedding of K in members X of $\mathcal{H}(H)$, and show that usually $D_L \cap X$ acts on K. This last type of result is important, since to achieve Hypothesis F.1.1 and show $(LTB, TD_L B, HD_L)$ is a weak BN-pair of rank 2, we need to show D_L acts on K.

LEMMA 5.1.10. *Let $k := n(H)$ and $H^* := H/O_2(H)$. Then K^* is a group of Lie type over $\mathbf{F}_{2^k}$ of Lie rank 1 or 2, M_K^* is a Borel subgroup of K^*, and B^* is a Cartan subgroup of K^*. More specifically, $K = \langle K_1^T \rangle$ for some $K_1 \in \mathcal{C}(H)$, and one of the following holds:*

(1) $K_1 < K$ and $K_1^* \cong L_2(2^k)$ or $Sz(2^k)$.

(2) $K_1 = K$ and K^* is a Bender group over $\mathbf{F}_{2^k}$.

(3) $K_1 = K$, $K^* \cong (S)L_3(2^k)$ or $Sp_4(2^k)$, and T is nontrivial on the Dynkin diagram of K^*.

PROOF. As $n(H) > 1$, this follows from E.2.2. $\square$

From now on, whenever we assume Hypothesis 5.1.8, we also take $K_1 \in \mathcal{C}(H)$.

LEMMA 5.1.11. *Let $S := O_2(M_H)$ and $H^* := H/O_2(H)$. Then*

(1) $S \cap K \in Syl_2(K)$.
(2) $S \cap L \in Syl_2(L)$.
(3) If K^* is of Lie rank 2, then either

    (i) S acts on both rank one parabolics of K^*, or
    (ii) K^*S^* is $L_3(4)$ extended by a graph automorphism.

PROOF. Note that $O_2(H) \leq S$ by A.1.6. By 5.1.10, M_K^* is 2-closed and $O_2(M_K^*) \in Syl_2(K^*)$, so (1) follows. By 5.1.5, B acts on $T \cap L$, and hence $T \cap L \leq O_2(BT) = O_2(M_H) = S$, so $S \cap L \in Syl_2(L)$, proving (2).

Note by 5.1.10 that B^* is a Cartan subgroup of K^*. Thus by inspection of the groups $L_2(2^k) \times L_2(2^k)$, $Sz(2^k) \times Sz(2^k)$, $(S)L_3(2^k)$, and $Sp_4(2^k)$ of Lie rank 2 listed in 5.1.10, either $C_{T^*}(B^*) = 1$—so that (i) holds; or $K^* \cong L_3(4)$, and $C_{T^*}(B^*)$ is of order 2 and induces a graph automorphism on K^*, giving (ii). Hence (3) holds. $\square$

LEMMA 5.1.12. *For each $X \in \mathcal{H}(H)$, K_1 lies in a unique $\hat{K}_1(X) \in \mathcal{C}(X)$, $K \leq \hat{K}(X) := \langle \hat{K}_1(X)^T \rangle$, and one of the following holds:*

(1) $K = \hat{K}(X)$.
(2) $K_1 < K$, $K_1/O_2(K_1) \cong L_2(4)$, and $\hat{K}_1(X)/O_2(\hat{K}_1(X)) \cong J_1$ or $L_2(p)$, p prime with $p^2 \equiv 1 \mod 5$ and $p \equiv \pm 3 \mod 8$.
(3) $K/O_2(K) \cong Sz(2^k)$ and $\hat{K}(X)/O_2(\hat{K}(X)) \cong {}^2F_4(2^k)$.
(4) $K/O_2(K) \cong L_2(2^k)$ and $\hat{K}(X)/O_2(\hat{K}(X))$ is of Lie type and characteristic 2 and Lie rank 2.
(5) $K/O_2(K) \cong L_2(4)$ and $K < \hat{K}(X)$ with $\hat{K}(X)/O_2(\hat{K}(X))$ not of Lie type and characteristic 2. The possible embeddings are listed in A.3.14.

PROOF. By 1.2.4, K_1 lies in a unique $\hat{K}_1(X) \in \mathcal{C}(X)$, and the embedding is described in A.3.12. If $K_1 < K$, then (1) or (2) holds by 1.2.8.2, so we may assume $K_1 = K$, and hence $\hat{K}_1(X) = \hat{K}(X)$ by 1.2.8.1. We may assume (1) does not hold, so that $K < \hat{K}(X)$.

As $K_1 = K$, $K/O_2(K)$ satisfies conclusion (2) or (3) of 5.1.10. In conclusion (3) of 5.1.10 as $k \geq 2$, $K/O_2(K) \cong L_3(4)$ by 1.2.8.4, and then $\hat{K}(X)/O_2(\hat{K}(X)) \cong M_{23}$ by A.3.12. However this case is impossible as T is nontrivial on the Dynkin diagram of $K/O_2(K)$, whereas this is not the case for the embedding in M_{23}.

Thus we may assume conclusion (2) of 5.1.10 holds. By 1.2.8.4, $K/O_2(K)$ is not unitary, while if $K/O_2(K)$ is a Suzuki group, then (3) holds by A.3.12. Thus we may assume $K/O_2(K) \cong L_2(2^k)$. Then by A.3.12 and A.3.14, (4) or (5) holds. $\square$

LEMMA 5.1.13. *Let $X \in \mathcal{H}(H)$, define $\hat{K} := \hat{K}(X)$ as in 5.1.12, and set $D := D_L \cap X$. Then either $D \leq N_G(K)$, or the following hold:*

(1) $K/O_2(K) \cong L_2(4)$.
(2) $L/O_2(L) \cong L_2(4)$.
(3) V is the sum of at most two copies of the A_5-module.
(4) $\hat{K} \leq C_G(Z)$.
(5) $\hat{K}/O_2(\hat{K}) \cong A_7$, J_2, or M_{23}.
(6) $\hat{K}(C_G(Z)) = O^{3'}(C_G(Z))$, and either $\hat{K} = \hat{K}(C_G(Z))$ or

$$\hat{K}/O_2(\hat{K}) \cong A_7 \text{ with } \hat{K}(C_G(Z))/O_2(\hat{K}(C_G(Z))) \cong M_{23}.$$

PROOF. We may assume D does not act on K, so in particular, $D \neq 1$. As $\hat{K} \trianglelefteq X$ by 1.2.1, D acts on $\hat{K}$ but not on K, so $K < \hat{K}$ and the possibilities for the embedding of K in $\hat{K}$ are described in 5.1.12.

If $\hat{K}/O_2(\hat{K})$ is of Lie type of characteristic 2 and Lie rank 2, then $K = P^\infty$, where $P/O_2(P)$ is one of the two maximal parabolics of $\hat{K}/O_2(\hat{K})$ containing $(T \cap \hat{K})/O_2(\hat{K})$. Then as D permutes with T, and T acts on P, also D acts on P, and hence also on K, contrary to assumption.

Therefore we may assume that case (2) or (5) of 5.1.12 holds. Let $D_c := C_D(\hat{K}/O_2(\hat{K}))$. Then $[D_c, K] \leq [D_c, \hat{K}] \leq O_2(\hat{K}) \leq O_2(KT)$, so D_c acts on $O^2(KO_2(KT)) = K$. Thus $D_c < D$.

Set $(\hat{K}TD)^* := \hat{K}TD/C_{\hat{K}TD}(\hat{K}/O_2(\hat{K}))$; then $1 \neq D^* \leq (\hat{K}TD)^* \leq Aut(\hat{K}^*)$. If D^* acts on K^* with preimage K_+, then D acts on $K = K_+^\infty$, contrary to our assumption; thus we may also assume that D^* does not act on K^*, and in particular that $D^* \not\leq B^*$ and so $D^* \neq 1$.

Suppose that case (2) of 5.1.12 holds. The case $\hat{K}_1^* \cong L_2(p)$ can be handled as in the case $\hat{K}^* \cong L_2(p)$ below, so take $\hat{K}_1^* \cong J_1$. As $K_1 < K$, $B \cong E_9$ is a Sylow 3-subgroup of $N_{\hat{K}}(T \cap \hat{K})$. Recall B normalizes D, so we may embed B^*D^* in a Hall $2'$-subgroup $E^* \cong (Frob_{21})^2$ of $N_{\hat{K}^*}(T^* \cap \hat{K}^*)$. Now D^* is cyclic as $D \leq D_L$. Also D permutes with T, so D^* is invariant under $N_{T^*}(E^*)$. But $N_{T^*}(E^*) = \langle t^* \rangle$ is of order 2, where t^* interchanges the two components of $\hat{K}^*$, so D^* is diagonally embedded in $\hat{K}^*$. Then as D^* is cyclic and B^*-invariant, $O_7(D^*) = 1$. So $D^* \leq B^*$, contradicting an earlier reduction. Therefore case (5) of 5.1.12 holds, establishing (1).

By (1), $B \cong B^*$ is of order 3. It remains to consider the corresponding possibilities for $\hat{K}^*$ in A.3.14. Furthermore the possibilities of Lie type in characteristic 2 in case (1) of A.3.14 were eliminated earlier.

Suppose first that $\hat{K}^*$ is not quasisimple. Then by 1.2.1.4, $\hat{K}^*/O(\hat{K}^*) \cong SL_2(p)$ for some odd prime p. Let R be the preimage in T of $O_{2',2}(\hat{K}^*)$. As $DT = TD$, D^* centralizes R^*, and so acts on $C_{\hat{K}}(R^*)^\infty =: K_R$; notice $K_R < \hat{K}$ as $K_R/O_2(\hat{K})) \cong SL_2(p)$. Similarly $K \leq K_R$ and T acts on K_R; so as $K_R/O_2(K_R)$ is quasisimple, D acts on K by induction on the order of $\hat{K}$, contrary to assumption. Thus we may assume $\hat{K}^*$ is quasisimple.

Suppose $\hat{K}^* \cong L_2(p)$ for some odd prime p. Recall in this case that $p \equiv \pm 3$ mod 8, so that $B^*T^* \cong A_4$; so as B^* acts on $1 \neq D^* \leq Aut(\hat{K})^*$ and $D^*T^* = T^*D^*$, we conclude $D^* = B^*$, contrary to an earlier reduction. As mentioned earlier, this argument suffices also when $K_1 < K$, where $B^*T^* \cong A_4$ wr $\mathbf{Z}_2$.

Suppose $\hat{K}^* \cong (S)L_3^\epsilon(5)$. Then $K^* = E(C_{\hat{K}^*}(Z(T^*)))$, and as D^* is cyclic and permutes with T^*, we conclude from the structure of $Aut(\hat{K}^*)$ that either $D^* \leq C_{\hat{K}^*}(Z(T^*)) \leq N_{\hat{K}^*}(K^*)$, or $\hat{K}^* \cong L_3(5)$ and D^*T^* is the normalizer in $\hat{K}^*T^*$ of the normal 4-subgroup E^* of $T^* \cap \hat{K}^*$. In the former case we contradict our assumption that D^* does not act on K^*; in the latter, $B^* \leq N_{K^*}(D^*T^*) = T^*$, contradicting B^* of order 3. Similarly if $\hat{K}^* \cong L_2(25)$ then as D^* permutes with T^*B^*, from the structure of $Aut(\hat{K}^*)$, $D^*T^* = B^*T^* \leq K^*T^*$, a contradiction.

Next suppose that $|D^*| = |D : D_c|$ is not a power of 3. Then as $DT = TD$, and $\hat{K}^*$ is not of Lie type and characteristic 2, A.3.15 says that $\hat{K}^* \cong J_1$, $L_2(q^e)$, $L_3^\delta(q)$, for q a suitable odd prime and $e \leq 2$. Then comparing these groups to our list of embeddings of A_5 in A.3.14, we conclude $\hat{K}^* \cong J_1$. As $D \not\leq N_G(K)$ is cyclic, we conclude that $D^* = [D^*, B^*]$ is of order 7; hence as $D \leq D_L = N_L(T \cap L)$, $n = 3m$ for some m. In particular as B does not centralize D, B induces a group of field automorphisms of order 3 on $L/O_2(L)$. Further $D \cap \hat{K} =: D_7$ is the subgroup of D_L of order 7. If all noncentral 2-chief factors of L on V are natural, then $C_D(Z) = 1$. If not, then by 5.1.3, m is even so that $m = 2s$ for some s, and the unique noncentral chief factor is orthogonal; so as 7 divides $2^{3s} - 1 = 2^{n/2} - 1$, $[Z, D_7] \neq 1$. Hence in any case $[Z, D_7] \neq 1$, so as $D_7 \leq \hat{K}$, $[Z, \hat{K}] \neq 1$. Thus $\langle Z^{\hat{K}} \rangle \in \mathcal{R}_2(\hat{K})$ by B.2.14, so that $\hat{K} \in \mathcal{L}_f(G, T)$. Then $\hat{K} \in \mathcal{L}_f^*(G, T)$ by 1.2.8.4. Now by 3.2.3, a suitable module for $\hat{K}$ satisfies the FSU. As J_1 does not appear among the possibilities for "$\bar{L}$" given in 3.2.6–3.2.9, this is a contradiction.

Thus D^* is a 3-group, and we have seen $D^* \not\leq B^*$, so $B^* D^*$ is a 3-group of order at least 9 permuting with T^*. Inspecting the possibilities for $\hat{K}$ in the remaining cases of A.3.14, we conclude that $\hat{K}/O_2(\hat{K}) \cong A_7, \hat{A}_7, J_2$, or M_{23}, and D^* is of order 3 and inverted by some $t \in \hat{K} \cap T$. (There are groups of order 9 in J_4 containing B^* and permuting with T^*, but each such group acts on K^*). Since $D^* \not\leq B^*$ and B acts on the cyclic group D, $\hat{K}/O_2(\hat{K})$ is not $\hat{A}_7$, establishing (5).

Next $\hat{K} = O^{3'}(X)$ by A.3.18, so $BO_3(D) \leq \hat{K}$. Hence as D^* is a 3-group, $D = O_3(D) \times D_c$, with $O_3(D) =: D_3$ of order 3 and $D_c = O^3(D)$.

Now $\hat{K} \leq \tilde{K} \in \mathcal{L}^*(G, T)$ and $D_3 \leq \hat{K} \leq \tilde{K}$ with $D_3 \not\leq N_{\hat{K}}(K)$. Therefore $\tilde{K}$ satisfies the hypotheses of $\hat{K}$, and hence replacing $\hat{K}$ by $\tilde{K}$ if necessary, we may assume $\tilde{K} = \hat{K} \in \mathcal{L}^*(G, T)$.

We next prove (4) by contradiction, so we assume that $\hat{K} \not\leq C_G(Z)$ and choose V so that $Z \leq V$; this argument will require several paragraphs. By 5.1.7.1, Baum(T) is not normal in $\hat{K}T$, so $\hat{K} = [\hat{K}, J(T)]$ using B.6.8.6.d. Set $U := [\langle Z^{\hat{K}} \rangle, \hat{K}]$, so that $U \in \mathcal{R}_2(\hat{K}T)$ by B.2.14 and U is an FF-module for $\hat{K}T$ by B.2.7. Then as M_{23} and J_2 do not have FF-modules by B.4.2, $\hat{K}/O_2(\hat{K}) \cong A_7$. Hence as $B^* D^*$ is of order 9, $B^* D^* T^*$ is the stabilizer of a partition of type $3, 4$ in the 7-set permuted by $\hat{K}^* T^*$, and $K^* T^*$ is the stabilizer of a partition of type $2, 5$. By B.5.1 and B.4.2, U is irreducible of dimension 4 or 6, with $\langle Z^{\hat{K}} \rangle = UZ = U \times C_Z(\hat{K})$. Then from the action of $\hat{K}$ on U, $[Z \cap U, K] \neq 1$, so by 3.1.8.3, $L = [L, J(T)]$. Therefore by 5.1.2, $V/C_V(L)$ is the natural module or the A_5-module for $\bar{L}$.

Suppose first that $V/C_V(L)$ is the A_5-module. Then $D_L = D = D_3 \leq C_G(Z)$. But if $m(U) = 6$, then $B = C_{BD}(Z)$, contradicting $B^* \not\leq D^*$. Hence $m(U) = 4$. However from the description of FF^*-offenders in B.4.2.7, $N_{\hat{K}^*}(J(T))$ is the stabilizer in $\hat{K}^*$ of a partition of type $3, 4$, so $J(T) \trianglelefteq BDT$; while as $[V, L]$ is the S_5-module, $J(T)$ is not normal in DT.

Therefore $V/C_V(L)$ is the natural module. Then $J(T) \leq (T \cap L)O_2(LT)$ by B.4.2.1, so that $J(T) \trianglelefteq DT$. If $m(U) = 6$, then $J(T)$ is not normal in D_3T using the discussion of FF^*-offenders in B.3.2.4; hence $m(U) = 4$.

As $V/C_V(L)$ is the natural module, $[Z, D_3] \neq 1$ and $C_Z(D_3) = C_Z(L)$. Then as $m(U) = 4$ and $[Z, D_3] \neq 1$, with $UZ = U \times C_Z(\hat{K})$, $C_Z(D_3) = C_U(\hat{K}) = C_Z(\hat{K})$. Therefore $C_Z(L) = C_Z(\hat{K})$, so $C_Z(L) = C_Z(\hat{K}) = 1$ as $H = KT \not\leq M = !\mathcal{M}(LT)$. Then $Z \leq U$, so $C_Z(K) \leq C_U(K) = 1$. Next by C.1.28, either there is a nontrivial characteristic subgroup C of Baum(T) normal in both LT and KT, or one of L or K is a block. As $M = !\mathcal{M}(LT)$ but $K \not\leq M$, L or K is a block.

Suppose first that K is a block. Then so is $\hat{K}$, and of the four subgroups of BD_3 of order 3, B has three noncentral chief factors on $O_2(BD_3T)$ and all others have two such factors. Thus D_3 has at most three noncentral chief factors on $O_2(BD_3T)$, so L is a $L_2(4)$-block. But then $D_L = D_3$ has exactly three noncentral chief factors, so $D = B$, contrary to $D^* \not\leq B^*$.

Consequently L is a block. But if $n = 2$, then as $C_Z(L) = 1$, T is of order at most 2^7, so K is also a block, the case we just eliminated. Hence $n > 2$. Further as K is not a block, we saw that there is a $C \trianglelefteq KT$; then as $C \trianglelefteq D_L T$, $\hat{K}T = \langle KT, D_3 \rangle \leq N_G(C)$—so that $D_L \leq N_G(C) \leq N_G(\hat{K}) = !\mathcal{M}(\hat{K}T)$ by 1.2.7.3, since we chose $\hat{K} \in \mathcal{L}^*(G, T)$. Now D_3 is inverted by $t \in T \cap \hat{K}$, so t induces a nontrivial field automorphism on $L/O_2(L)$, and hence n is even. Then the subgroup D_- of D_L of order $2^{n/2} + 1$ satisfies $D_- = [D_-, t] \leq D_L \cap \hat{K}$ as $t \in \hat{K}$. As $\hat{K}/O_2(\hat{K}) \cong A_7$, this forces $D_- = D_3$. But then $n = 2$, a case we eliminated at the start of the paragraph. This contradiction shows that $\hat{K} \leq C_G(Z)$, establishing (4).

We have established (1), (4), and (5) and also showed $\hat{K} = O^{3'}(X)$. As we could take $X = C_G(Z)$, it follows that (6) holds: for $A_7/E_{2^4} < M_{23}$ is the only proper inclusion in A.3.12 among the groups in (5).

As $K \leq \hat{K} \leq C_G(Z)$ by (4), as usual $C_Z(L) = 1$ using 1.2.7.3. Hence 5.1.3 says either V is the $O_4^-(2^{n/2})$-module and indeed $n/2$ must be odd, or V is the sum of two S_5-modules. In the latter case, (2) and (3) hold. In the former case, the subgroup D_- of D_L of order $2^{n/2} + 1$ centralizes Z. Now in each of the possibilities for $\hat{K}$ in (5), D_3 is inverted by $t \in T \cap \hat{K}$. Then the final few sentences in the proof of (4) show that $n = 2$. This completes the proof of (2) and (3) and hence of the lemma. $\qquad\square$

5.1.3. More detailed analysis of the case $\mathbf{K/O_2(K) = L_3(4)}$. The remainder of the section is devoted to an analysis of the subcase of 5.1.10.3 where $K/O_2(K) \cong L_3(4)$. This case is the remaining major obstruction to applying the Green Book [**DGS85**] and beginning the identification of G as a rank 2 group of Lie type and characteristic 2 in Theorem 5.2.3 of the next section.

THEOREM 5.1.14. *Let $H^* := H/O_2(H)$ and assume $K^* \cong L_3(4)$. Then*

(1) $K \in \mathcal{L}^*(G, T)$, so $N_G(K) = !\mathcal{M}(H)$ but $K \notin \mathcal{L}_f^*(G, T)$.

(2) $[Z, H] = 1$ and $C_G(z) \leq N_G(K)$ for each $z \in Z^\#$.

(3) $C_Z(L) = 1$.

(4) $n = 2$, V is the sum of one or two copies of the S_5-module for $\bar{L}\bar{T} \cong S_5$, and $D_L = B$.

(5) $C_G(K/O_2(K))$ is a solvable $3'$-group.

In the remainder of this section assume the hypotheses of Theorem 5.1.14, and set $H^* := H/O_2(H)$. We will prove Theorem 5.1.14 by a series of reductions.

Note that B has order 3, since $K^* \cong L_3(4)$, and B^* is a Cartan subgroup of K^*. By 5.1.12, $K \in \mathcal{L}^*(G, T)$. In particular $N_G(K) = !\mathcal{M}(H)$ by 1.2.7.3. On the other hand, $K \notin \mathcal{L}_f^*(G, T)$: For if $K \in \mathcal{L}_f^*(G, T)$ then by 3.2.3, there exist $V_K \in \mathcal{R}_2(KT)$ such that the pair K, V satisfies the FSU. By 5.1.10.3, T is nontrivial on the Dynkin diagram of K^*, so case (4) of 3.2.9 in the FSU is excluded, while $L_3(4)$ (as opposed to $SL_3(4)$) does not arise anywhere else in 3.2.8 or 3.2.9. This contradiction establishes conclusion (1) of Theorem 5.1.14.

Now as $K \notin \mathcal{L}_f^*(G, T)$, K centralizes $R_2(KT)$ by 1.2.10, so that $H = KT$ centralizes Z. Then the remaining statement in conclusion (2) follows as $N_G(K) = !\mathcal{M}(H)$; and conclusion (2) implies conclusion (3) as $H \nleq M = \mathcal{M}(LT)$.

Thus it only remains to prove parts (4) and (5) of Theorem 5.1.14. Moreover throughout the remainder of the proof we can and will appeal to the first three parts of Theorem 5.1.14.

Set $M_+ := N_G(K)$; by 5.1.14.1, $M_+ \in \mathcal{M}(T)$. If n is even, define D_ϵ for $\epsilon = \pm 1$ as in Lemma 5.1.6.

LEMMA 5.1.15. *One of the following holds:*

(1) $D_L \leq M_+$.
(2) $n = 2$ and V is the direct sum of two natural modules for $\bar{L}$.
(3) $n = 2$ or 4 and $[V, L]$ is the natural module for $\bar{L}$.
(4) $n = 4$ or 8, V is the $\Omega_4^-(2^{n/2})$-module for $\bar{L}$, and $D_- \leq M_+$.

PROOF. First if $D \leq D_L$ and $O_2(\langle D, H \rangle) \neq 1$, then by 5.1.14.1, $D \leq M_+$. However we may assume conclusion (1) does not hold, so $D_L \nleq M_+$ and hence $O_2(\langle D_L, H \rangle) = 1$. Cases (1) and (2) of 5.1.6 appear as cases (2) and (3) of 5.1.15. Case (3) of 5.1.6 cannot occur since there $Z(H) = 1$, contrary to 5.1.14.2. Finally in case (4) of 5.1.6, $O_2(\langle D_-, H \rangle) \neq 1$, so $D_- \leq M_+$. Thus as $D_+ D_- = D_L \nleq M_+$, $O_2(\langle D_+, H \rangle) = 1$, so $n = 4$ or 8 by 5.1.6.4. Hence 5.1.15.4 holds. $\square$

We now begin to make use of the local classification of weak BN-pairs of rank 2 in the Green Book [**DGS85**]. We recognize weak BN-pairs of rank 2 by verifying Hypothesis F.1.1.

LEMMA 5.1.16. *Let $C_K := C_G(K/O_2(K))$. Then*

(1) C_K is a $3'$-group.
(2) If C_K is not solvable, then $C_K^\infty / O_2(C_K^\infty) \cong Sz(2^k)$ for some odd $k \geq 3$, $C_K^\infty \nleq M$, and $D_L \nleq M_+$.

PROOF. Part (1) follows as H is an SQTK-group. Thus it remains to prove (2), so we assume $C_K^\infty \neq 1$. Hence by 1.2.1, there exists $K_+ \in \mathcal{C}(C_K)$. Then any such K_+ satisfies $K_+ / O_2(K_+) \cong Sz(2^k)$ for some odd $k \geq 3$. Further $m_5(K_+) = 1 = m_5(K)$, while $m_5(M_+) \leq 2$ as M_+ is an SQTK-group, so $K_+ = C_K^\infty$ by 1.2.1.1, establishing the first assertion of (2). Further $M_+ = N_G(K_+)$ since we saw $M_+ \in \mathcal{M}$. Let B_+ be a Borel subgroup of K_+; then $B_+ \leq N_G(T) \leq M = N_G(L)$ using Theorem 3.3.1. Now if $K_+ \leq M$, then $[K_+, L] \leq O_2(L)$, so that L normalizes $O^2(K_+ O_2(L)) = K$ and hence $L \leq N_G(K_+) = M_+$ contradicting $M = !\mathcal{M}(LT)$. Thus $K_+ \nleq M$, proving the second statement of (2).

To complete the proof of (2), we suppose by way of contradiction that $D_L \leq M_+ = N_G(K_+)$. We claim that under this assumption, Hypothesis F.1.1 is satisfied with L, K_+, T in the roles of "L_1, L_2, S". Let $G_+ := \langle LT, H \rangle$. As $K_+ \nleq M = !\mathcal{M}(LT)$, $O_2(G_+) = 1$, establishing hypothesis (e) of F.1.1. We have seen that $B_+ \leq M = N_G(L)$, and we are assuming $D_L \leq N_G(K_+)$, so hypothesis (d) of F.1.1 holds. The remaining conditions in F.1.1 are easy to verify, in particular since we take S to be the Sylow 2-subgroup T of G; therefore Hypothesis F.1.1 is satisfied as claimed. We conclude from F.1.9 that $\alpha := (LTB_+, D_L TB_+, D_L TK_+)$ is a weak BN-pair of rank 2. Indeed $T \trianglelefteq B_+ T$, so by F.1.12.I, α is of type $^2F_4(2^k)$, with $n = k$—as this is the only type where a parabolic possesses an $Sz(2^k)$

composition factor. By F.1.12.II, $T \leq K_+$. But then $T \leq K_+ \leq C_K$, contradicting $T \cap K \nleq C_K$. □

Notice now that to complete the proof of Theorem 5.1.14, it suffices to prove part (4) of 5.1.14: Namely we have already established the first three parts of Theorem 5.1.14. Further if part (4) holds then $D_L = B \leq M_+$, which by 5.1.16 forces C_K to be a solvable $3'$-group, establishing part (5) of 5.1.14.

LEMMA 5.1.17. n is even.

PROOF. Assume n is odd, so in fact $n \geq 3$ as $n > 1$. Let $F := \mathbf{F}_{2^n}$. Then T induces inner automorphisms on $\bar{L}$, so $\bar{T} \leq \bar{L}$. By 5.1.15, $D_L \leq M_+$; then as D_L is a $\{2, 3\}'$-group acting on T and $K/O_2(K) \cong L_3(4)$, we conclude that $D_L \leq C_K := C_G(K/O_2(K))$.

We now specialize our choice of V to be the module "V" in the Fundamental Setup (3.2.1) for L, as we may by 3.2.3. As $L/O_2(L) \cong L_2(2^n)$, case (1) or (2) of Theorem 3.2.5 holds, so L is irreducible on $V/C_V(L)$ and V is a TI-set under M. Since n is odd, $V/C_V(L)$ is the the natural module for $\bar{L}$ by 5.1.3; then as $C_Z(L) = 1$ by 5.1.14.3, V is a natural module. Let $Z_1 := Z \cap V$. Notice as $\bar{T} \leq \bar{L}$, Z_1 is the 1-dimensional F-subspace of V stabilized by T. In particular Z_1 is a TI-set under $N_M(V)$, so as V is a TI-set under M, Z_1 is a TI-set under M.

Observe also that L is not a block: For if it were, then as $C_Z(L) = 1$, $C_T(D_L) = 1$, contradicting $D_L \leq C_K$. Also C_K is a solvable $3'$-group by 5.1.16, since we saw $D_L \leq M_+$.

Let $S := \mathrm{Baum}(T)$, and recall from 5.1.7.1 that $N_G(S) \leq M$.

We claim Z_1 is a TI-set in G. For let $Z_0 := \langle Z^{C_K} \rangle$; then $Z_0 \in \mathcal{R}_2(C_K T)$ by B.2.14. As C_K is a solvable $3'$-group, by Solvable Thompson Factorization B.2.16, $[Z_0, J(T)] = 1$, so that $S = \mathrm{Baum}(C_T(Z_0))$ using B.2.3. Now by a Frattini Argument, $C_K = C_{C_K}(Z_0) N_{C_K}(S)$. Then as $Z_1 \leq Z_0$ while $N_{C_K}(S) \leq M$ and Z_1 is a TI-set under M, Z_1 is a TI-set under C_K. Now $n \neq 6$ since n is odd, so by Zsigmondy's Theorem [**Zsi92**], there is a *Zsigmondy prime divisor* p of $2^n - 1$, namely such that a suitable element of order p is irreducible on Z_1. Let $P \in Syl_p(C_K)$. As $D_L \leq C_K = C_{C_K}(Z_0) N_{C_K}(S)$ with $N_{C_K}(S) \leq M$, we may choose P so that $P = C_P(Z_0)(P \cap M)$ and $P_L := P \cap D_L \in Syl_p(D_L)$. By the choice of p, $P \cap M = P_L \times C_{P \cap M}(Z_1)$, so $P = P_L C_P(Z_1)$, and P is irreducible on Z_1. Therefore Z_1 is a TI-set under $N_{M_+}(P)$. Further by a Frattini Argument, $M_+ = C_K N_{M_+}(P)$, so as Z_1 is a TI-set under C_K, Z_1 is a TI-set under M_+. Finally by 5.1.14.2, $C_G(z) \leq M_+$ for each $z \in Z_1^\#$, so as $D_L \leq M_+$ is transitive on $Z_1^\#$, Z_1 is a TI-set under G by I.6.1.1, and hence the claim holds.

Let $G_1 := N_G(Z_1)$ and $\tilde{G}_1 := G_1/Z_1$. Recall by 5.1.14.2 that $H \leq C_G(Z_1)$, so $G_1 \leq M_+$ by 5.1.14.1.

Consider any H_1 with $HD_L \leq H_1 \leq G_1$, and set $Q_1 := O_2(H_1)$ and $U := \langle V^{H_1} \rangle$. Observe that Hypothesis G.2.1 is satisfied with Z_1 and H_1 in the roles of "V_1" and "H". Therefore $\tilde{U} \leq Z(\tilde{Q}_1)$ and $\Phi(U) \leq Z_1$ by G.2.2.

Suppose by way of contradiction that $\Phi(U) \neq 1$. Then $U = \langle V^{H_1} \rangle$ is not elementary abelian, so $U \nleq C_T(V)$. Thus $\bar{U} \neq 1$, and hence the hypotheses of G.2.3 are satisfied. Therefore $\bar{U} \in Syl_2(\bar{L})$ by G.2.3.1. Set $I := \langle U^L \rangle$ and $W := O_2(I)$. By G.2.3.4, there exists an I-series

$$1 = W_0 \leq W_1 \leq W_2 \leq W_3 = W,$$

where $W_1 = V$, $W_2 = U \cap U^l$, for some $l \in L - G_1$, and W/W_2 is the sum of r natural modules for $L/O_2(L)$ and some $0 \leq r$, with $(U \cap W)/W_2 = C_{W/W_2}(\bar{U})$. In particular $W = [W, D_L]W_2$. But $D_L \leq C_K$ and C_K is a solvable 3'-group, so by A.1.26.2, $[W, D_L] \leq O_2(C_K) \leq O_2(M_+) \leq O_2(H_1) = Q_1$ using A.1.6. Thus as $W_2 \leq U \leq Q_1$,

$$W \leq Q_1 \leq C_G(\tilde{U}).$$

Therefore as $Z_1 \leq W_2 \leq U \cap W$ and $(U \cap W)/W_2 = C_{W/W_2}(\bar{U})$, it follows that $W \leq U$. But in G.2.3.6, $(U \cap W)/W_2$ is a proper direct summand of W/W_2 if $r > 0$, so we conclude $W = W_2$ and thus $[O_2(I), I] \leq W_2$. Then as $L \leq I$ and $[W_2, L] = V$, we conclude $V = [O_2(L), L]$, so that L is an $L_2(2^n)$-block, contrary to an earlier observation.

This contradiction shows that U is elementary abelian. Applying this result to G_1 in the role of "H_1", we conclude that $\langle V^{G_1} \rangle$ is abelian. But L is transitive on $V^\#$ and Z_1 is a TI-set in G, so (cf. A.1.7.1) G_1 is transitive on $\{V^g : Z_1 \cap V^g \neq 1\}$, and hence as $\langle V^{G_1} \rangle$ is abelian, $[V, V^g] = 1$ whenever $Z_1 \cap V^g \neq 1$. This verifies part (a) of Hypothesis F.8.1 with Z_1, HD_L in the roles of "V_1, H".

During the remainder of the proof take $H_1 := HD_L$. Then part (b) of Hypothesis F.8.1 is part of Hypothesis G.2.1 verified earlier. Next using 3.1.4.1, $C_{H_1}(\tilde{V}) \leq N_{H_1}(V) = H_1 \cap M = TBD_L$. As V is the natural module for $\bar{L}$, $C_{N_{GL(V)}(\bar{L})}(\tilde{V}) \cong Z_{2^n-1}$, so as D_L is a Hall subgroup of TBD_L and D_L is faithful on $\tilde{V}$, we conclude $C_{H_1}(\tilde{V}) = C_{TB}(V)$. Therefore $\ker_{C_{H_1}(\tilde{V})}(H_1) \leq \ker_{TB}(H_1) = Q_1$, so part (c) of F.8.1 holds. Finally part (d) holds as $H \not\leq M = !\mathcal{M}(LT)$. Thus we have verified Hypothesis F.8.1, so we can apply the results of section F.8.

Define b, γ, etc. as in section F.8. By F.8.5.1, $b \geq 3$ is odd, so G_γ is a conjugate of H_1 and hence as $D_L \leq C_K$,

$$\hat{G}_\gamma := G_\gamma/O_2(G_\gamma) \cong H_1^+ := H_1/Q_1 = KT/Q_1 \times D_LQ_1/Q_1$$

with KT/Q_1 an extension of $L_3(4)$ and $D_LQ_1/Q_1 \cong D_L \cong \mathbf{Z}_{2^n-1}$.

As $D_L^+ \trianglelefteq H_1^+$ and $\tilde{V} = [\tilde{V}, D_L]$, $\tilde{U} = [\tilde{U}, D_L]$. Thus each KD_L-irreducible is the sum of n K-irreducibles $\tilde{I}$, as $\mathbf{F}_4$ is a splitting field for K^* and n is odd. We claim $m(H_1^+, \tilde{U}) \geq 9$: For if y is an involution in H^+ with $m([\tilde{U}, y]) < 9$, then as $m(\tilde{I}) \geq 9$, y^+ acts on $\tilde{I}$. Then by H.4.7, either $m([\tilde{I}, y]) \geq 4$, or $m(\tilde{I}) = 9$ and $m([\tilde{I}, y]) = 3$. So $\tilde{I}_D := \langle \tilde{I}^{D_L} \rangle$ is the sum of $n \geq 3$ conjugates of $\tilde{I}$, so $m([\tilde{I}_D, y]) = m([\tilde{I}, y])n \geq 9$, proving the claim. In particular $\tilde{U}$ is not an FF-module for H_1^+ by B.4.2.

Recall from section F.8 that $Q_1 = C_{H_1}(\tilde{U})$, there is $g_b \in G$ with $\gamma = \gamma_1 g_b$, $A_1 := Z^{g_b}$, $D_\gamma := C_{U_\gamma}(\tilde{U})$, and $D_{H_1} := C_U(U_\gamma/A_1)$.

Suppose U_γ centralizes $\tilde{U}$, so that $U_\gamma = D_\gamma$. By F.8.7.7, $[D_{H_1}, U_\gamma] = 1$. By F.8.7.5, $[V, U_\gamma] \neq 1$, so $[Z_1^l, U_\gamma] \neq 1$ for some $l \in L$. If $1 \neq Z_1^l \cap D_{H_1}$, then

$$U_\gamma \leq O^{2'}(C_G(Z_1^l \cap D_{H_1})) \leq C_G(Z_1^l)$$

as Z_1 is a TI-set in G in the center of $T \in Syl_2(G)$. Of course this contradicts the choice of Z_1^l, so we conclude that $1 = Z_1^l \cap D_{H_1}$, and hence Z_1^l is isomorphic to a subgroup of $\hat{G}_\gamma$. Therefore

$$4 = m_2(\hat{G}_\gamma) \geq m(Z_1) = n,$$

so as n is odd, $n = 3$. As we are assuming $D_\gamma = U_\gamma$, $[U_\gamma, V] \leq Z_1$ by F.8.7.6; so for $1 \neq y \in Z_1^l$, $m([U_\gamma, y]) \leq m(Z_1) = 3$, contradicting $m(H_1^+, \tilde{U}) \geq 9$.

This contradiction shows that $D_\gamma < U_\gamma$. Therefore there is $\beta \in \Gamma(\gamma)$ with $V_\beta \not\leq Q_1$, and $d(\beta, \gamma_1) = b$ by minimality of b. Thus we have symmetry between γ_0, γ_1, γ and β, γ, γ_1; so reversing the roles of these triples if necessary, we may assume that $m(U_\gamma^+) = m(U_\gamma/D_\gamma) \geq m(U/D_{H_1})$. Thus if $\tilde{D}_{H_1} \leq C_{\tilde{U}}(U_\gamma)$, then $\tilde{U}$ is an FF-module for H_1^+, contrary to an earlier observation. Therefore $[D_{H_1}, U_\gamma] \neq 1$, so there is $g \in G$ with $Z_1^g = Z_\gamma$ (so that $V^g \leq U_\gamma$) and $[D_{H_1}, V^g] \neq 1$. By F.8.7.6, $[D_{H_1}, U_\gamma] \leq A_1$, so D_{H_1} acts on V^g; then since V is the natural module for $\bar{L}$ and n is odd,

$$m(D_{H_1}/C_{D_{H_1}}(V^g)) \leq m_2(\bar{L}\bar{T}) = n = m(V^g/C_{V^g}(D_{H_1})).$$

Also $V^g \cap Q_1 \leq D_\gamma \leq C_G(D_{H_1})$ by F.8.7.7. Thus

$$4 = m_2(H_1^+) \geq m(V^{g+}) \geq m(V^g/C_{V^g}(D_{H_1})) = n,$$

so $n = 3$ and

$$m(\tilde{U}/C_{\tilde{U}}(V^g)) \leq m(U/D_{H_1}) + m(D_{H_1}/C_{D_{H_1}}(V^g)) \leq m(U_\gamma^+) + 3 \leq 7,$$

contradicting $m(H_1^+, \tilde{U}) \geq 9$. This contradiction completes the proof of 5.1.17. □

As n is even by 5.1.17, there is a unique subgroup D_3 of order 3 in D_L.

LEMMA 5.1.18. *If $D_3 \leq M_+$, then $D_3 = B$, so that $[Z, D_3] = 1$.*

PROOF. Notice the final statement follows from the first, as $B \leq H \leq C_G(Z)$ by 5.1.14.2.

Assume $D_3 \leq M_+$. It suffices to assume $D_3 \neq B$ and establish a contradiction. If D_3 induces inner automorphisms on K^* then $D_3 \leq K$ by 5.1.16.1. Then as BT is the largest solvable subgroup of KT containing T, $D_3 \leq BT$ and hence $D_3 = B$, contrary to assumption. Therefore D_3 induces outer automorphisms on K^*, and $K^*D_3^* \cong PGL_3(4)$. Set $D := D_L \cap M_+$ and $S := O_2(DBT)$. Arguing as in 5.1.11, $T \cap L \leq S$ and hence $S \cap L \in Syl_2(L)$; similarly $S \cap K \in Syl_2(K)$. From the structure of $Aut(L_3(4))$, $C_{T^*}(B^*D_3^*) = 1$, so $S = (T \cap K)C_S(K^*) = (S \cap K)O_2(KS)$. Let P_2 be a rank-1 parabolic of K over $S \cap K$, and set $K_2 := O^2(P_2)$. Then SD acts on K_2, and as $K \not\leq M$ with T nontrivial on the Dynkin diagram of $K/O_2(K)$, $K_2 \not\leq M$. Thus $O_2(G_0) = 1$, where $G_0 := \langle LS, K_2 \rangle$, since $M = !\mathcal{M}(L)$ by Theorem 4.3.2.

Suppose that $D_L \leq M_+ = N_G(K)$. Then $D_L = D$ acts on K_2, so that $N_L(S \cap L) = D(S \cap L)$ acts on K_2. Now it is easy to verify the remainder of Hypothesis F.1.1 with K_2, L in the roles of "L_1, L_2": For example as $O_2(M) \leq S \geq O_2(M_+)$, $L_iSBD \in \mathcal{H}^e$ by 1.1.4.5. So by F.1.9, $\alpha := (K_2SD, BSD, BSL)$ is a weak BN-pair of rank 2. Further by construction $S \trianglelefteq SBD$, so α is described in F.1.12. Since $K_2/O_2(K_2) \cong L_2(4)$ and $L/O_2(L) \cong L_2(2^n)$ with n even, it follows from F.1.12 that α is the amalgam of a (possibly twisted) group of Lie type over $\mathbf{F}_4$. Then as K_2 centralizes Z, α is the amalgam of $G_2(4)$ or $U_4(4)$. But now K_2 has only two noncentral chief factors, which is incompatible with the embedding of K_2 in K with $F^*(K) = O_2(K)$.

Therefore $D_L \not\leq M_+$, so one of the last three cases of 5.1.15 must hold. However by hypothesis, $D_3 \leq M_+$, so $D_L > D_3$ and hence $n > 2$. Thus either case (3) of 5.1.15 holds with $n = 4$, or case (4) holds with $n = 8$—since in that case $D_- \leq M_+$, so that $D_L = D_3D_- \leq M_+$ if $n = 4$. Similarly in either case, $D_5 \not\leq M_+$, where D_5 is the subgroup of D_L of order 5, since in case (3), $D_L = D_3D_5$, while in case (4), $D_L = D_3D_5D_-$ with $D_- \leq M_+$.

Recall $S \cap L \in Syl_2(L)$, so $SD_5 \trianglelefteq TD_5$, and hence $S_0 D_5$ is a subgroup of G for each subgroup S_0 of T containing S. As B acts on D_5 and S, it acts on SD_5. As $S \cap K \in Syl_2(K)$ and $S \trianglelefteq T$, $1 \neq O_2(\langle N_G(S), H\rangle)$ by Theorem 3.1.1. Then $N_G(S) \leq M_+$ by 5.1.14.1, and hence $D_5 \not\leq N_G(S)$.

Let $X := \langle SBD_5, K_2\rangle$. Suppose first that $O_2(X) = 1$. We just saw D_5 does not act on S, so $SD_5/O_2(SD_5) \cong D_{10}$ or $Sz(2)$. Therefore Hypothesis F.1.1 is satisfied with K_2, SD_5 in the roles of "L_1, L_2". Thus $\beta := (K_2 S, BS, BSD_5)$ is a weak BN-pair of rank 2 by F.1.9, and as S is self-normalizing in SD_5, β is on the list of F.1.12. But D_{10} or $Sz(2)$ occur as factors of $L_i/O_2(L_i)$ only in the amalgams of $^2F_4(2)'$ and $^2F_4(2)$, where the rank-1 parabolic over S other than K_2 in those amalgams is solvable, a contradiction as K_2 is not solvable.

This contradiction shows that $O_2(X) \neq 1$. Set $T_0 := N_T(K_2)$. We saw earlier that T acts on SD_5 and similarly T acts on SB. Thus T acts on SBD_5, so T_0 acts on X, and hence on $O_2(X)$. Embed $T_0 \leq T_1 \in Syl_2(XT_0)$; as $|T : T_0| = 2$, $|T_1 : T_0| \leq 2$. As $O_2(KT_0) \leq T_0$ and $K \notin \mathcal{L}_f(G, T)$ by 5.1.14.1, $[Z(T_0), K] = 1$ using B.2.14; hence $N_G(T_0) \leq M_+$ by 5.1.14.1. Also by 4.3.17, $N_G(T_0) \leq M$, so $T_1 \leq M \cap M_+$. Thus if $T_0 < T_1$ we may take $T_1 = T$. However if $T_1 = T$, then $KT = \langle K_2, T\rangle \leq XT_0 \in \mathcal{H}$, so that $D_5 \leq X \leq M_+$ using 5.1.14.1, contrary to an earlier reduction. Hence $T_0 \in Syl_2(X)$.

We claim D_5 acts on K_2; assume otherwise. As $K_2 \in \mathcal{L}(X, T_0)$ and $T_0 \in Syl_2(X)$, $K_2 < K_X \in \mathcal{C}(X)$ by 1.2.4, with the embedding described in A.3.14. Let $Y := K_X T_0 D_5$ and $Y^* := Y/C_Y(K_X/O_2(K_X))$. Arguing as in the beginning of the proof of 5.1.13, $C_{D_5}(K_X/O_2(K_X))$ normalizes K_2; so as we are assuming $D_5 \not\leq N_G(K_2)$, $D_5^* \neq 1$. As $S \leq T_0$, D_5 permutes with T_0 and so $D_5 T_0$ is a subgroup of G by an earlier remark; therefore K_X^* appears on the list of A.3.15. Comparing that list to the list of A.3.14, we conclude that case (3) of A.3.14 holds with $K_X^* \cong L_2(p)$, $p^2 \equiv 1 \mod 5$ and $p \equiv \pm 3 \mod 8$. But B acts on D_L, so that $[B, D_5] = 1$. Then D_5^* permutes with the subgroup $(T_0 \cap K_X)^* B^* \cong A_4$ of K_X^*, which is not the case in $Aut(L_2(p))$.

This contradiction establishes our claim that D_5 acts on K_2. By symmetry, D_5 also acts on $K_3 := O^2(P_3)$, where P_3 is the second rank one parabolic of K over $T \cap K$. Therefore D_5 acts on $K = \langle K_2, K_3\rangle$, a contradiction as we showed $D_5 \not\leq M_+$. This completes the proof of 5.1.18. $\qquad\square$

From this point on, we assume H is a counterexample to Theorem 5.1.14. Under this assumption we show:

LEMMA 5.1.19. *One of the following holds:*

(1) $D_3 \not\leq M_+$, *and either*

    (i) $n = 2$, *and V is the direct sum of two natural modules for $\bar{L}$, or*

    (ii) $n = 2$ *or* 4, *and $[V, L]$ is a natural module for $\bar{L}$.*

(2) $n = 4$ *or* 8, V *is the $\Omega_4^-(2^{n/2})$-module for $\bar{L}$, and $D_3 \not\leq M_+$.*

(3) $n \equiv 2 \mod 4$, $n > 2$, 3 *does not divide n, $D_3 = B \leq M_+$, and V is the $\Omega_4^-(2^{n/2})$-module for $\bar{L}$.*

PROOF. First suppose $D_3 \leq M_+$. Then by 5.1.18, $[Z, D_3] = 1$ and $D_3 = B$. This forces one of cases (2), (4), or (5) of 5.1.3 to hold, with $n \equiv 2 \mod 4$ in (4).

Assume first that $n = 2$, so that $D_L = D_3 = B$. As $C_Z(L) = 1$ by part (3) of Theorem 5.1.14, V is the sum of at most two copies of the S_5-module, so part (4) of Theorem 5.1.14 holds. Hence by our remark after 5.1.16, Theorem 5.1.14 holds, contrary to our assumption that H is a counterexample to that Theorem.

So $n > 2$, and then case (4) of 5.1.3 holds, with $n \equiv 2 \mod 4$. Thus $D_L \leq M_+$ by 5.1.15. Further 3 does not divide n, or otherwise D_L contains a cyclic subgroup of order 9, which must be faithful on K^* as $D_3 = B$ is faithful. However this is impossible as $Aut(K^*)$ has no cyclic subgroup of order 9 permuting with T^*. So conclusion (3) holds when $D_3 \leq M_+$.

Therefore we may assume $D_3 \not\leq M_+$. Then one of the last three cases of 5.1.15 must hold. Cases (2) and (3) give conclusion (1), and case (4) gives conclusion (2). $\qquad\qquad\square$

LEMMA 5.1.20. $D_3 \not\leq M_+$, so $O_2(\langle H, D_3 \rangle) = 1$.

PROOF. If $D_3 \not\leq M_+$, then $O_2(\langle H, D_3 \rangle) = 1$ by 5.1.13.1. Thus it suffices to assume $D_3 \leq M_+$, and derive a contradiction. As $D_3 \leq M_+$, case (3) of 5.1.19 holds; thus $n \equiv 2 \mod 4$, $n > 2$, and 3 does not divide n, so $n \geq 10$. Set $S := (T \cap L)O_2(LT)$; then $S \in Syl_2(LS)$. Also $S = O_2(D_3T)$, so as $D_3 \leq M_+$, $S \in Syl_2(KS)$.

Next as case (3) of 5.1.19 holds, $J(T) \trianglelefteq LT$ by 5.1.2, so $J(T) \leq O_2(LT) \leq S$ and hence $J(T) = J(S)$ by B.2.3.3. As $K \not\leq M = !\mathcal{M}(LT)$, $J(S)$ is not normal in KS. By B.5.1 and B.4.2, K^*S^* has no FF-modules, so as $m_2(K^*S^*) = 4$, E.5.4 says $E := \Omega_1(Z(J_4(S))) \trianglelefteq KS$. Therefore as $K \not\leq M$ and $M = !\mathcal{M}(LS)$, $J_4(S) \not\leq O_2(LS) = C_S(V)$. By E.5.5, there is $\bar{A} \in \mathcal{A}^2(\bar{S})$ with $m(V/C_V(\bar{A})) - m(\bar{A}) \leq 4$. But by construction $\bar{S} \leq \bar{L}$, so by H.1.1.3 applied with $n/2$ in the role of "n",

$$n/2 \leq m(V/C_V(\bar{A})) - m(\bar{A}) \leq 4.$$

Thus $n \leq 8$, whereas we saw earlier that $n \geq 10$. This contradiction completes the proof. $\qquad\qquad\square$

By 5.1.20, $D_3 \not\leq M_+$. So by 5.1.19, case (1) or (2) of 5.1.19 holds. In particular, $n = 2$, 4, or 8. However by 5.1.14.1 we may apply Theorem 3.3.1 to K, to conclude $N_G(T) \leq M_+$; hence $D_3 \not\leq N_G(T)$. Therefore $\bar{L}\bar{T} \cong Aut(L_2(2^n))$.

By B.2.14, $V_Z := \langle Z^L \rangle \in \mathcal{R}_2(LT)$, so we can apply the results of this section to V_Z in the role of "V". In particular as $\bar{L}\bar{T} = Aut(\bar{L})$, from the structure of the modules in case (1) or (2) of 5.1.19, either Z is of order 2, in which case we set $Z_1 := Z$; or V_Z is the sum of two natural modules for $\bar{L} \cong L_2(4)$, where we take $Z_1 := Z \cap V_1$ for some $V_1 \in Irr_+(L, V_Z)$. Thus in any case Z_1 is of order 2, and $V_2 := \langle Z_1^{D_3} \rangle \cong E_4$. Set $G_1 := C_G(Z_1)$, $G_2 := N_G(V_2)$, and consider any H_1 with $H \leq H_1 \leq G_1$. Set $U := \langle V_2^{H_1} \rangle$, $Q_1 := O_2(H_1)$, $\tilde{G}_1 := G_1/Z_1$, and $L_2 := \langle D_3^T \rangle = D_3[O_2(D_3T), D_3]$. Observe Hypothesis G.2.1 is satisfied with L_2, V_2, Z_1, H_1 in the roles of "L, V, V_1, H", so by G.2.2 we have:

LEMMA 5.1.21. $\tilde{U} \leq Z(\tilde{Q}_1)$ and $\Phi(U) \leq Z_1$.

LEMMA 5.1.22. (1) $C_G(V_2) = C_T(V_2)B \leq M$.
(2) $n = 2$ or 4, and $[V, L]$ is the sum of at most two natural modules for $\bar{L}$.
(3) $[V_2, O_2(K)] = Z_1$ and $D_3 O_2(C_G(V_2)) \trianglelefteq G_2$.

PROOF. Notice (1) implies (2), since if case (2) of 5.1.19 holds, then $1 \neq C_{D_L}(V_2)$ is a 3'-group.

If K normalizes V_2, then by 5.1.14.1, $D_3 \leq G_2 \leq M_+$, contradicting 5.1.20. Thus $[K, V_2] \neq 1$. Set $Q_K := O_2(K)$. Then $V_2 \not\leq Z(Q_K)$, for otherwise $K \in$

$\mathcal{L}_f(G, T)$ using 1.2.10, contrary to 5.1.14.1. Thus 5.1.21 says $[V_2, Q_K] = Z_1$, proving the first assertion of (3). Hence as $V_2 = [V_2, L_2]$, $L_2 = [L_2, Q_K]$. Now $K \trianglelefteq G_1$ by 5.1.14.2, so $C_G(V_2) \leq G_1 \leq N_G(Q_K)$, and hence $C_{Q_K}(V_2) \leq O_2(G_2)$. Therefore $P := \langle C_{Q_K}(V_2)^{G_2} \rangle \leq O_2(G_2)$, and $[C_G(V_2), Q_K] \leq C_{Q_K}(V_2) \leq P$. Then $L_2 = [L_2, Q_K] \leq C_G(C_G(V_2)/P)$, so as $G_2 = L_2 T C_G(V_2)$, $L_2 P \trianglelefteq G_2$. Then as $P \leq O_2(G_2) \leq T \leq N_G(L_2)$, $L_2 = O^2(L_2 P) \trianglelefteq G_2$. Now since $L_2 = D_3 O_2(L_2)$ with $O_2(L_2) = C_{L_2}(V_2)$, $D_3 O_2(C_G(V_2)) \trianglelefteq G_2$. Therefore (3) holds, and it remains to establish (1).

Now B acts on D_3 and $B \leq K \leq C_G(Z_1)$, so B centralizes $\langle Z_1^{D_3} \rangle = V_2$. On the other hand as G_2 is an SQTK-group, $m_3(G_2) \leq 2$, so by (3), $m_3(C_G(V_2)) = 1$. Further $C_G(V_2) = C_{G_1}(V_2)$, with $G_1 \leq M_+$. As C_K is a $3'$-group by 5.1.16.1, either $O^{3'}(M_+) = K$, or $O^{3'}(M_+)/O_{3'}(O^{3'}(M_+)) \cong PGL_3(4)$. In particular as Sylow 3-groups of $PGL_3(4)$ are of exponent 3 and $m_3(C_G(V_2)) = 1$, $B \in Syl_3(C_G(V_2))$. Therefore as $B \leq K$ and $C_G(V_2) \leq G_1 \leq N_G(K)$, $Y := O^{3'}(C_G(V_2)) \leq K$. Then as BT is the unique maximal subgroup of KT containing BT, and $[K, V_2] \neq 1$, we conclude $Y = O^{3'}(TB)$. Thus to complete the proof of (1) and hence of the lemma, it remains to show $X := O^{\{2,3\}}(C_G(V_2)) = 1$. As X is BT-invariant and $Aut_{BT}(K/O_2(K))$ is maximal in $Aut_{KT}(K/O_2(K))$, $X \leq C_K$. Therefore $\langle H, D_3 \rangle \leq N_G(X)$, so if $X \neq 1$, then by 5.1.14.1, $D_3 \leq N_G(X) \leq M_+$, contradicting 5.1.20. This establishes (1), and completes the proof of 5.1.22. $\qquad\square$

LEMMA 5.1.23. $\langle V_2^{G_1} \rangle$ is abelian.

PROOF. We specialize to the case $H_1 = G_1$, and recall Hypothesis G.2.1 is satisfied with L_2, V_2, Z_1, G_1 in the roles of "L, V, V_1, H". Our proof is by contradiction, so we assume that U is nonabelian. Then $[V_2, U] = Z_1$ using 5.1.21, so $L_2 = [L_2, U]$, and hence the hypotheses of G.2.3 are also satisfied. So setting $I := \langle U^{G_2} \rangle$, G.2.3 gives us an I-series

$$1 = S_0 \leq S_1 \leq S_2 \leq S_3 = S := O_2(I)$$

such that $S_1 = V_2$, $S_2 = U \cap U^g$ for $g \in D_3 - G_1$, $[S_2, I] \leq S_1 = V_2$, and S/S_2 is the sum of natural modules for $I/S \cong L_2(2)$ with $(U \cap S)/S_2 = C_{S/S_2}(U)$. As L_2 has at least two noncentral chief factors on V and one on $(S \cap L)/C_{S \cap L}(V)$, $m := m((U \cap S)/S_2) > 1$.

Let $G_1^* := G_1/C_{G_1}(\tilde{U})$, $W := U \cap S$, and $A := U^g \cap S$. Observe

$$\tilde{S}_2 = \widetilde{A \cap U} \leq C_{\tilde{U}}(A)$$

and $[U, a] \not\leq S_2$ for each $a \in A - S_2$. Thus as $Z_1 \leq S_2$, $S_2 = C_A(\tilde{U})$. Therefore as $m(U/(U \cap S)) = 1$ since $I/S \cong L_2(2)$,

$$m(A^*) = m(A/S_2) = m((U \cap S)/S_2) = m = m(\tilde{U}/\tilde{S}_2) - 1 \geq m(\tilde{U}/C_{\tilde{U}}(A^*)) - 1,$$

so $A^* \in \hat{\mathcal{Q}}_r(G_1^*, \tilde{U})$, where $r := (m+1)/m < 2$ as $m > 1$. Let $C_1 := C_{G_1}(K/O_2(K))$; we apply D.2.13 to G_1^* in the role of "G". By 5.1.16.1, C_1 is a $3'$-group, so as $r_{A^*, \tilde{U}} \leq r < 2$, D.2.13 says that $[F(C_1^*), A^*] = 1$. But as $G_1 \leq N_G(K)$, $F^*(G_1^*) = K^* F^*(C_1^*)$, so either A^* is faithful on K^*, or by 5.1.16.2, A^* acts nontrivially on a component $X^* \cong Sz(2^k)$ of C_1^*. Let $Y := K$ in the first case, and $Y := X$ in the second. By A.1.42.2 there is $\tilde{W} \in Irr_+(\tilde{U}, Y^*, T^*)$; set $\tilde{U}_T := \langle \tilde{W}^T \rangle$. As $Y^* = [Y^*, A^*]$, $C_A(U_T) < A$. Then by D.2.7,

$$\hat{q} := \hat{q}(Aut_{YT}(\tilde{U}_T), \tilde{U}_T) \leq r < 2.$$

Observe that Hypothesis D.3.1 is satisfied, with Y^*T^*, Y^*, $\tilde{U}_T$, $\tilde{W}$ in the roles of "M, M_+, V_M, V". So as $\hat{q} < 2$, we conclude from D.3.8 that $Y^* \not\cong Sz(2^k)$; hence $Y = K$. By construction $\tilde{U}_T$ plays the role of both "V_T" and "V_M" in Hypothesis D.3.2 and lemma D.3.4, so the hypotheses of D.3.10 are satisfied. Thus we conclude from D.3.10 that $\tilde{W} = \tilde{U}_T$. Then B.4.2 and B.4.5 show that $\hat{q} > 2$, keeping in mind that K^* is $L_3(4)$ rather than $SL_3(4)$, and $\dim(\tilde{W}) \neq 9$ as T is nontrivial on the Dynkin diagram of K^*. This contradiction completes the proof of 5.1.23. $\square$

We are now in a position to obtain a contradiction which will establish Theorem 5.1.14. We specialize to the case $H_1 = H$. As L_2 is transitive on $V_2^\#$ and Z_1 is of order 2, G_1 is transitive on $\{V_2^g : Z_1 \leq V_2^g\}$ by A.1.7.1. So by 5.1.23, $[V_2, V_2^g] = 1$ whenever $Z_1 \leq V_2^g$. Also $C_H(\tilde{U}) = O_2(H)$, since otherwise by Coprime Action, K centralizes V_2, contrary to 5.1.22.1 as $K \not\leq M$. Further as $D_3 \leq L_2$, $O_2(\langle L_2T, H \rangle) = 1$ by 5.1.20. Hence Hypothesis F.8.1 is satisfied with Z_1, V_2, L_2 in the roles of "V_1, V, L". As Z_1 is of order 2, Hypothesis F.9.8 is satisfied with V_2 in the role of "V_+" by Remark F.9.9). Therefore by F.9.16.3 $q(H^*, \tilde{U}) \leq 2$. However we observe that the argument at the end of the proof of 5.1.23, with H^*, $\tilde{U}$ in the roles of "G_1^*, $\tilde{U}_T$", shows that $q(H^*, \tilde{U}) > 2$.

The proof of Theorem 5.1.14 is complete.

5.2. Using weak BN-pairs and the Green Book

In this section, we continue to assume Hypothesis 5.1.8—in particular, $n(H) > 1$.

We work toward the goal of constructing a weak BN-pair of rank 2. This will be accomplished by establishing Hypothesis F.1.1. In our construction, L plays the role of "L_1" in Hypothesis F.1.1, and we choose L_2 to be a suitable subgroup of K. To be precise, if $K_1/O_2(K_1)$ is a Bender group in 5.1.10, we let $L_2 := K_1$. Otherwise $K/O_2(K) \cong (S)L_3(2^n)$ or $Sp_4(2^n)$, in which case we let P_+ be a maximal parabolic of K over $T \cap K$, and take $L_2 \in \mathcal{C}(P_+)$. Notice in either case that $T \cap L_2 \in Syl_2(L_2)$. Further $K = \langle L_2^T \rangle$ and $H \not\leq M$, so that $L_2 \not\leq M$.

In any case, $L_2/O_2(L_2)$ is a group of Lie type of Lie rank 1, and of course $L/O_2(L) \cong L_2(2^n)$ in this chapter. Next set $S := O_2(M_H) = O_2(BT)$. By 5.1.11, $S \cap K \in Syl_2(K)$, and $S \cap L \in Syl_2(L)$. Then as $S \cap K = T \cap K$, $S \cap L_2 \in Syl_2(L_2)$ by a remark in the previous paragraph. Further by 5.1.11.3:

LEMMA 5.2.1. *If $K/O_2(K)$ is not $L_3(4)$ then S acts on L_2.*

Next the Cartan group B of K lies in M, and so normalizes L; therefore to achieve condition (d) of F.1.1, we need to show that D_L acts on L_2. To show D_L acts on L_2, we first show that—modulo an exceptional case where we view L as defined over $\mathbf{F}_2$—D_L acts on K. Then we deduce that D_L acts on L_2. Eventually it turns out that $L_2 = K$.

LEMMA 5.2.2. *Either*

(1) $D_L \leq N_G(K)$, or

(2) $K/O_2(K) \cong L/O_2(L) \cong L_2(4)$, V is the sum of at most two copies of the A_5-module, and $K \leq K_Z := O^{3'}(C_G(Z))$, with $K_Z/O_2(K_Z) \cong A_7$, J_2, or M_{23}.

PROOF. Assume that neither (1) nor (2) holds. In particular $D_L \not\leq B$ as (1) fails. For $D \leq D_L$ let $X_D := \langle D, H \rangle$. Let $\mathcal{D}$ consist of those $D \leq D_L$ such that $O_2(X_D) = 1$. If $D \in \mathcal{D}$ then $D \not\leq N_G(K)$ as $O_2(K) \neq 1$ and $K \trianglelefteq H$. If $O_2(X_D) \neq 1$, then by 5.1.13, either $D \leq N_G(K)$ or the various conclusions of 5.1.13 hold, and the latter contradicts our assumption that (2) fails. Thus $O_2(X_D) \neq 1$ iff $D \leq N_G(K)$ iff $D \notin \mathcal{D}$. Finally if Δ is a collection of subgroups generating D_L, then as (1) fails, $D \not\leq N_G(K)$ for some $D \in \Delta$, so that $\Delta \cap \mathcal{D} \neq \emptyset$.

In particular $D_L \in \mathcal{D}$. We conclude from 5.1.6 that one of the four cases of 5.1.6 holds. Now in the first three cases of 5.1.6, $n = 2$ or 4. If case (4) holds, then we may take Δ to consist of D_- and D_+. However $1 \neq Z \leq O_2(X_{D_-})$ in that case by 5.1.6, so that $D_+ \in \mathcal{D}$. Therefore $n = 4$ or 8 by 5.1.6.4.

So in any case, we have $n = 2$, 4, or 8. Next let D_p denote the subgroup of D_L of order p. When $n = 2$, $D_3 = D_L$ so $D_3 \in \mathcal{D}$. When $n = 4$, $D_L = \langle D_3, D_5 \rangle$, so $D_p \in \mathcal{D}$ for $p = 3$ or 5. Finally when $n = 8$, $D_+ = \langle D_3, D_5 \rangle$, and we saw D_- acts on K, so again $D_p \in \mathcal{D}$ for $p = 3$ or 5. Thus in each case, $D_p \in \mathcal{D}$ for $p = 3$ or 5; choose p with this property during the remainder of the proof.

As $D_L \not\leq B$, $K/O_2(K)$ is not $L_3(4)$ by part (4) of Theorem 5.1.14. Hence S acts on L_2 by 5.2.1, and, as we observed at the beginning of this section, $S \cap K \in Syl_2(K)$ and $S \cap L \in Syl_2(L)$. Recall B normalizes $O_2(BT) = S$ and L_2. Set $G_0 := \langle D_p, L_2 S \rangle$.

We first suppose that $O_2(G_0) = 1$. This gives part (e) of Hypothesis F.1.1, with $D_p S$ and L_2 in the roles of "L_1" and "L_2". Part (f) follows from 1.1.4.5, as M and H are in $\mathcal{H}^e$ and S contains $O_2(H)$ and $O_2(M)$. To check part (c), we only need to prove that S is not normal in $D_p S$, since then $D_p S / O_2(D_p S) \cong L_2(2)$, D_{10}, or $Sz(2)$. But if $S \trianglelefteq SD_p$, then as $S \trianglelefteq T$ and $S \cap K \in Syl_2(K)$, Theorem 3.1.1 says $1 \neq O_2(\langle D_p T, H \rangle) \leq O_2(H) \leq O_2(BT) = S \leq G_0$ using A.1.6, contrary to our assumption that $O_2(G_0) = 1$. The remaining parts of Hypothesis F.1.1 are easily verified.

Now by F.1.9, $\alpha := (D_p S B_2, S B_2, S L_2)$ is a weak BN-pair of rank 2, where $B_2 := B \cap L_2$. Indeed since S is self-normalizing in SD_p, α is described in F.1.12. As we saw in 5.1.18, when $p = 5$ the amalgams in F.1.12 have solvable parabolics, and so are ruled out as L_2 is not solvable. So $p = 3$ and $D_3 S / O_2(D_3 S) \cong L_2(2)$; then as L_2 is not solvable, we conclude that α is of type J_2, $Aut(J_2)$, $^3D_4(2)$, or $U_4(2)$. In each case, $Z(S)$ is of order 2, and is centralized by one of the parabolics in the amalgam.

Suppose first that α has type $U_4(2)$. Then $D_3 S$ is the solvable parabolic centralizing $Z(S)$, with $[O_2(SD_3), D_3] \cong Q_8^2$, and L_2 is an A_5-block with $O_2(L_2) = F^*(L_2 S)$. Thus $O_2(L_2)$ is the unique 2-chief factor for $L_2 S$, so $K = L_2$. Also $C_S(L_2) = 1$, so $Z(H) = 1$. As $Z(H) = 1$, from the discussion above we are in case (3) of 5.1.6, so that $[V, L]$ is the A_5-module for $L/O_2(L)$; in particular $n = 2$ and $D_L = D_3 \not\leq B$. As $[D_3, O_2(D_3 S)] \cong Q_8^2$, L also is an A_5-block. But then as $D_3 < D_3 B$ and $S = O_2(BT)$, $1 \neq C_{BD_3}(L) \leq O(LTB)$, contradicting $F^*(LTB) = O_2(LTB)$).

Thus we may suppose α is of type J_2, $Aut(J_2)$, or $^3D_4(2)$. In each case $L_2 S$ is the parabolic centralizing $Z(S)$, so as $Z = \Omega_1(Z(T)) \leq Z(S)$ and $Z(S)$ is of order 2, we conclude $Z(S) = Z$ centralizes $\langle L_2, T \rangle = H$. Again in each case $Q := O_2(L_2 S)$ is extraspecial and L_2 is irreducible on Q/Z; so as $H \in \mathcal{H}^e$, $Q = O_2(H)$ using A.1.6. Arguing as above, as Q/Z is the unique noncentral factor for L_2 and $Z \leq \Phi(Q)$, again $K = L_2$. Then $B \leq K = L_2$, so as $S = O_2(BT)$, α is not the $Aut(J_2)$-amalgam. Now either α is of type $^3D_4(2)$ and $O^{2'}(Aut(K)) = Inn(K)$, or α is of type J_2 and $O^{2'}(Aut(K)) = Aut(K) \cong S_5/E_{16}$. So either $T = S \leq K$; or α is the J_2-amalgam, $|T : S| = 2$, and (D_3TB, TB, TK) is a weak BN-pair extending α, and hence is the $Aut(J_2)$-amalgam. Therefore if α is of type J_2, then T is a Sylow 2-subgroup of either J_2 or $Aut(J_2)$, so $m_2(T) = 4$ and T has no normal E_{16}-subgroup. This is impossible as $m(V) \geq 4$ from 5.1.6. Thus α is of type $^3D_4(2)$ with $S = T$, so $K/O_2(K) \cong L_2(8)$, B is of order 7, and T is a Sylow 2-subgroup of $^3D_4(2)$. We are free to choose V to be $\langle Z^L \rangle$; thus $Z \leq V$, so $V_2 := \langle Z^{D_3} \rangle \leq V$. From the structure of α, $V_2 \leq C_T(B) \cong D_8$. As B acts on L and Z, B acts on $\langle Z^L \rangle = V$. Therefore $V_2 = C_V(B)$ and in particular $[B, V] \neq 1$, so B is faithful on $L/O_2(L)$. This is impossible as $n = 2$, 4, or 8 and B acts on $S \cap L = T \cap L$ with $|B| = 7$.

This contradiction shows that $O_2(G_0) \neq 1$. Let $T_0 := N_T(L_2)$. As TB acts on D_pS and S, and T_0B acts on L_2, T_0B acts on G_0; hence $O_2(G_0T_0B) \neq 1$. Thus as $O_2(X_{D_p}) = 1$ and $D_p \leq G_0$, $H \not\leq G_0T_0$; hence $T_0 < T$ and $L_2 < K$. Therefore either case (1) of 5.1.10 holds with $L_2 = K_1 < K$, or case (3) holds with $L_2 < K_1 = K$. In either case $L_2 < K$, $T_0 < T$, and $K = \langle L_2, L_2^t \rangle$ for $t \in T - T_0$. Furthermore as T acts on D_pT_0, $(L_2^t)^{D_pT_0} = (L_2^{D_pT_0})^t$, so as $D_p \not\leq N_G(K)$ it follows that $D_p \not\leq N_G(L_2)$.

Embed T_0 in $T_1 \in Syl_2(G_0T_0B)$. As $|T : T_0| = 2$, $|T_1 : T_0| \leq 2$. As $S \leq T_0$, $N_G(T_0) \leq M$ by 4.3.17; hence T_1 acts on $T_0 \cap L$, and then as $D_p(T_0 \cap L) \trianglelefteq N_M(T_0 \cap L)$, T_1 acts on D_pT_0.

By Theorem 3.1.1, applied with T_0, $N_G(T_0)$, H in the roles of "R, M_0, H", we conclude $O_2(X) \neq 1$, where $X := \langle N_G(T_0), H \rangle$. Now $K_1 \in \mathcal{L}(X, T)$ and $T \in Syl_2(X)$, so by 1.2.4, $K_1 \leq K_X \in \mathcal{C}(X)$, and we set $K_+ := \langle K_X^T \rangle$. Recalling that $L_2 < K$, we conclude from A.3.12 and 1.2.8 that either $K = K_+ \trianglelefteq X$, or $K_1/O_2(K_1) \cong L_2(4)$ and $K_1 < K_X$, with $K_X \neq K_X^t$ for $t \in T - T_0$ and $K_X/O_2(K_X) \cong J_1$ or $L_2(p)$. In any case, $T \cap K_+ = S \cap K_+ = T_1 \cap K_+$.

Suppose that $T_0 < T_1$. Set $H_0 := \langle L_2, T_1 \rangle$ and $K_0 := \langle L_2^{T_1} \rangle$. As $T, T_1 \in Syl_2(X)$, and $T \cap K_+ = T_1 \cap K_+$, $K = \langle L_2^T \rangle = \langle L_2^{T_1} \rangle$ from the structure of K_+T. Thus $K \in \mathcal{L}(H_0, T_1)$, $K = \langle L_2^{T_1} \rangle \leq G_0T_0B$, and applying 5.1.13 with G_0T_0B, T_1, H_0 in the roles of "X, T, H", we conclude that either $K/O_2(K) \cong L_2(4)$, or D_p acts on K. The first case is impossible as we saw $L_2 < K$, and the second is impossible as we chose p so that D_p does not act on K.

This contradiction shows that $T_1 = T_0 \in Syl_2(G_0T_0)$. Now we can repeat parts of the proof of 5.1.13 with G_0T_0B, L_2T_0, D_p in the roles of "X, H, D" to obtain a contradiction: We know $G_0T_0B \in \mathcal{H}(L_2T_0)$ and $D_p \not\leq N_G(L_2)$ from earlier reductions. Then $L_2 < \hat{L}_2 \in \mathcal{C}(G_0T_0)$ using 1.2.4, and arguing as in 5.1.12 with L_2 in the role of "K_1", one of conclusions (2)–(5) of that result must hold. Indeed as L_2 is normalized by the Sylow group T_0, conclusion (2) of that result cannot arise. Then the argument in the second paragraph of the proof of 5.1.13 shows $\hat{L}_2/O_2(\hat{L})$ is not of Lie type in characteristic 2 of Lie rank 2, so that conclusions (3) and (4) of 5.1.12 are ruled out. Hence we are reduced to case (5) of 5.1.12, and in particular, $L_2/O_2(L_2) \cong L_2(4)$, with the embedding $L_2 < \hat{L}_2$ described in A.3.14. We saw $K/O_2(K) \not\cong L_3(4)$, so by 5.1.10, $K/O_2(K)$ is $Sp_4(4)$ or $L_2(4) \times L_2(4)$, and in either case $B \cong E_9$. Next proceeding as in the proof of 5.1.13 with D_p in the role of "D", we obtain $p = 3$; notice that here $\hat{L}_2/O_2(\hat{L}_2)$ is not J_1, since here $p = 3$ or 5 rather than 7. Since B acts on D_3 by 5.1.5.2, B centralizes D_3. But also $B \leq N_G(L_2)$ so $D_3 \nleq B$; hence $M \geq D_3 B \cong E_{27}$, a contradiction as M is an SQTK-group. This completes the proof of 5.2.2. $\qquad\square$

We now state the main result of this chapter:

THEOREM 5.2.3. *Assume G is a simple QTKE-group, $T \in Syl_2(G)$, and $L \in \mathcal{L}_f^*(G,T)$ with $L/O_2(L) \cong L_2(2^n)$ and $L \trianglelefteq M \in \mathcal{M}(T)$. In addition assume $H \in \mathcal{H}_*(T,M)$ with $n(H) > 1$, let $K := O^2(H)$, $Z := \Omega_1(Z(T)$, and $V \in \mathcal{R}_2(LT)$ with $[V,L] \neq 1$. Then one of the following holds:*

(1) $n = 2$, V is the sum of at most two copies of the A_5-module for $L/O_2(L) \cong A_5$, and $K \leq K_Z \in \mathcal{C}(C_G(Z))$. Further either $K/O_2(K) \cong L_2(4)$ with $K_Z/O_2(K_Z) \cong A_7$, J_2, or M_{23}, or $K = K_Z$ and $K/O_2(K) \cong L_3(4)$.

(2) $G \cong M_{23}$.

(3) G is a group of Lie type of characteristic 2 and Lie rank 2, and if G is $U_5(q)$ then $q = 4$.

Note that conclusions (2) and (3) of Theorem 5.2.3 are also conclusions in our Main Theorem. Thus once Theorem 5.2.3 is proved, whenever $L \in \mathcal{L}_f^*(G,T)$ is T-invariant with $L/O_2(L) \cong L_2(2^n)$, we will be able to assume that either conclusion (1) of Theorem 5.2.3 holds, or $n(H) = 1$ for each $H \in \mathcal{H}_*(T, N_G(L))$. The treatment of these two remaining cases is begun in the following chapter 6, and eventually completed in Part 5, devoted to those $L \in \mathcal{L}_f^*(G,T)$ with $L/O_2(L)$ defined over $\mathbf{F}_2$.

5.2.1. Determining the possible amalgams. The proof of Theorem 5.2.3 will not be completed until the final section 5.3 of this chapter. In this subsection, we will produce a weak BN-pair α, and use the Green Book [**DGS85**] to identify α up to isomorphism of amalgams. This leaves two problems: First, show that the subgroup G_0 generated by the parabolics of α is indeed a group of Lie type. Second, show that $G_0 = G$. In one exceptional case, G_0 is proper in G; the second subsection will give a complete treatment of that branch of the argument, culminating in the identification of G as M_{23}.

Assume the hypotheses of Theorem 5.2.3. Notice that Hypothesis 5.1.8 holds, since in Theorem 5.2.3 we assume $n(H) > 1$. During the proof of Theorem 5.2.3, write D for D_L.

Notice that if $K/O_2(K) \cong L_3(4)$, then conclusion (1) of Theorem 5.2.3 holds by Theorem 5.1.14. Thus we may assume during the remainder of the proof of Theorem 5.2.3 that $K/O_2(K)$ is not $L_3(4)$. Therefore by 5.1.11.3, S acts on the rank one parabolics of K, and hence on the group L_2 defined at the start of the section.

Next if $D \nleq N_G(K)$, then conclusion (2) of 5.2.2 is satisfied, so again conclusion (1) of Theorem 5.2.3 holds. Thus we may also assume during the remainder of the proof that D acts on K; we will show under this assumption that conclusion (2) or (3) of Theorem 5.2.3 holds. The following consequences of these observations are important in producing our weak BN-pair:

LEMMA 5.2.4. *(1) $D \leq N_G(K)$.*

(2) $D \leq N_G(B)$ and $B \leq N_G(D)$.

(3) $B \leq N_G(S)$, $D \leq N_G(S \cap L_2)$, and $DS = SD$.

(4) DSB acts on L_2.

PROOF. By construction in Notation 5.1.9, $B \leq N_G(D)$. Part (1) holds by assumption, and says D acts on $M_K := M \cap K = (S \cap K)B$. Thus D acts on $DB \cap (S \cap K)B = B$, completing the proof of (2). Further as the Borel subgroup M_K is 2-closed by 5.1.10, D acts on $S \cap K$. As D acts on $S \cap K$ and there are at most two rank one parabolics of K over $S \cap K$, D acts on each such parabolic. So as $L_2 = P^\infty$ for one of these parabolics, D acts on L_2 and hence also on $S \cap L_2$.

By definition of S, $S = O_2(BT)$, so B acts on S. As $N_L(S \cap L) = (S \cap L)D$, $DS = SD$, completing the proof of (3). As B acts on SD, DSB is a group. By 5.2.1, S acts on L_2, while by definition B is a Cartan subgroup acting on L_2. This completes the proof of (4). □

We now verify that Hypothesis F.1.1 is satisfied with L, L_2, S in the roles of "L_1", "L_2", "S". Set $B_2 := B \cap L_2$, $G_1 := LSB_2$, $G_2 := DSL_2$, and $G_{1,2} := G_1 \cap G_2$. As $L \trianglelefteq M$ and B_2 normalizes S by 5.2.4.3, G_1 is a subgroup of G with $L \trianglelefteq G_1$. Again using 5.2.4, G_2 is a subgroup of G with $L_2 \trianglelefteq G_2$. Thus $L_i = G_i^\infty$ as DSB is solvable. Notice conditions (a), (b), and (c) of F.1.1 follow from remarks at the beginning of the section, together with the fact that S acts on L_2. Further condition (d) of F.1.1 holds as $N_{L_j}(S \cap L_j) \leq DSB \leq G_i$, and we saw $L_i \trianglelefteq G_i$. Condition (f) follows from 1.1.4.5, since $G_1 \leq M$, $G_2 \leq N_G(K)$, and S contains $O_2(M)$ and $O_2(H)$, and hence contains $O_2(N_G(K))$ using A.1.6. Finally we establish (e) of F.1.1 in the following lemma:

LEMMA 5.2.5. $O_2(\langle G_1, G_2 \rangle) = 1$.

PROOF. Let $G_0 := \langle G_1, G_2 \rangle$. By 4.3.2, $M = !\mathcal{M}(L)$, so as $L_2 \not\leq M$, $O_2(G_0) = 1$. □

We now use the Green Book [**DGS85**] (via an appeal to F.1.12) to determine the possible amalgams that can arise; these will subsequently lead us to the "generic" quasithin groups in conclusion (3) of Theorem 5.2.3, and to M_{23} in conclusion (2) of 5.2.3.

PROPOSITION 5.2.6. $\alpha := (G_1, G_{1,2}, G_2)$ is a weak BN-pair of rank 2. Further $L_2 = K = G_2^\infty$, with $O_2(G_i) = O_2(L_i)$ for $i = 1$ and 2, and one of the following holds:

(1) α is the $L_3(2^n)$-amalgam and L and K are $L_2(2^n)$-blocks.

(2) α is the $Sp_4(2^n)$-amalgam and L and K are $L_2(2^n)$-blocks.

(3) α is the $G_2(q)$-amalgam for $q = 2^n$, $L/O_2(L) \cong K/O_2(K) \cong L_2(q)$, $O_2(K) \cong q^{1+4}$, and $|O_2(L)| = q^5$.

(4) α is the $^3D_4(q)$-amalgam for $q = 2^n$, $L/O_2(L) \cong L_2(q)$, $|O_2(L)| = q^{11}$, $K/O_2(K) \cong L_2(q^3)$, and $O_2(K) \cong q^{1+8}$.

(5) α is the $^2F_4(q)$-amalgam for $q = 2^n$, $L/O_2(L) \cong L_2(q)$, $|O_2(L)| = q^{11}$, $K/O_2(K) \cong Sz(q)$, and $|O_2(K)| = q^{10}$.

(6) $n > 2$ is even, α is the $U_4(q)$-amalgam for $q = 2^{n/2}$ or its extension of degree 2, L is an $O_4^-(q)$-block, $K/O_2(K) \cong L_2(q)$, and $O_2(K) \cong q^{1+4}$.

(7) $n = 4$, α is the $U_5(4)$-amalgam, $L/O_2(L) \cong L_2(16)$, $|O_2(L)| = 2^{16}$, $K/O_2(K) \cong SU_3(4)$, and $O_2(K) \cong 4^{1+6}$.

Moreover $O_2(KT) = O_2(KS)$, and either

(a) $S \leq L_i$ and $O_2(L_iS) = O_2(L_i)$ for $i = 1$ and 2, or

(b) α is an extension of the $U_4(q)$ amalgam of degree 2 and $O_2(KS)$ is the extension of $O_2(K)$ by an involution t such that $C_K(t) \cong P^\infty$ for P a maximal parabolic of $Sp_4(q)$.

PROOF. We have already verified Hypothesis F.1.1, so by F.1.9, α is a weak BN-pair of rank 2. By 5.2.4.2, $B_2 \le N_G(S)$, so that we may apply F.1.12 to determine α. As L_1 and L_2 are not solvable, cases (8)–(13) of F.1.12.I are ruled out. Together with F.1.12.II, this shows that $S \le L_i$ and hence also $O_2(L_i) = O_2(G_i)$ for $i = 1$ and 2, unless possibly α is the extension of the $U_4(q)$ amalgam of degree 2. In the latter case by F.4.29.5, (II.i) fails only weakly, in the sense that $O_2(L) = O_2(LS)$ and $|S : S \cap L| = |S : S \cap L_2| = |O_2(L_2S) : O_2(L_2)| = 2$. Further by F.4.29.4, $O_2(G_i) = O_2(L_i)$. Now the remaining amalgams in cases (1)–(7) of F.1.12.I are those given in 5.2.6; notice that the numbering convention for L_1 and L_2 in F.1.12 differs in some cases from that used here in 5.2.6. We are using the facts that $L/O_2(L) \cong L_2(2^n)$ and $1 \ne [Z, L]$.

We next show that $L_2 = K$; that is, we eliminate cases (1) and (3) of 5.1.10. First suppose $L_2 = K_1 < K$. Then for $t \in T - N_T(K_1)$, $O_2(L_2S)$ contains $S \cap L_2^t$ with $O_2(L_2) \cap L_2^t \le O_2(L_2^t)$ and $|S \cap L_2^t : O_2(L_2^t)| > 2$; therefore $|O_2(L_2S) : O_2(L_2)| > 2$, contrary to an earlier observation. So we may suppose instead that $K/O_2(K)$ is $(S)L_3(2^k)$ or $Sp_4(2^k)$. We recall in this case that $L_2 = P_+^\infty$ for a maximal parabolic P_+ of K. Thus $L_2/O_2(L_2) \cong L_2(2^k)$, $O_2(L_2)O_2(K)/O_2(K)$ has a natural chief factor, and there is at least one more noncentral 2-chief factor for L_2 in $O_2(K)$. Thus L_2 has at least two noncentral 2-chief factors, so that α is not the $L_3(q)$ or $Sp_4(q)$-amalgam. As $L_2/O_2(L_2) \cong L_2(2^k)$, rather than $Sz(q)$ or $SU_3(q)$, α is not the $^2F_4(q)$ or $U_5(4)$-amalgam. If α is the amalgam for $G_2(q)$ or $^3D_4(q)$, then L_2D has just one noncentral 2-chief factor, and that factor is *not* natural. This leaves the $U_4(q)$-amalgam, where L_2 has two natural 2-chief factors on the Frattini quotient of $O_2(L_2)$, but L_2D is irreducible on the Frattini quotient. However D acts on $O_2(K)$ and hence on the 2-chief factor for $O_2(L_2)$ in $O_2(K)$. This contradiction shows that $L_2 = K$, completing the proof of the claim.

Recall $S = O_2(M_H)$, so that $O_2(KT) \le S$ by A.1.6, and hence $O_2(KT) = O_2(KS)$. By F.1.12.II, $O_2(L) = O_2(LS)$, and either $O_2(K) = O_2(KS)$ or α is the extension of the $U_4(q)$-amalgam of degree 2. In the latter case by F.4.29.5, $O_2(KS) = O_2(K)\langle t \rangle$, where t induces a graph-field automorphism on $U_4(q)$, and hence $C_{U_4(q)}(t) \cong Sp_4(q)$, so that $C_K(t)$ is as claimed. This completes the proof of 5.2.6. $\qquad\square$

In the remainder of this section, if α is the extension of degree 2 of the $U_4(q)$-amalgam, we replace α by its subamalgam of index 2. Thus in effect, we are replacing $S = O_2(BT)$ by $S \cap L$ of index 2 in S. Subject to this convention:

LEMMA 5.2.7. (1) $\alpha := (G_1, G_{1,2}, G_2)$ is the amalgam of $L_3(q)$, $Sp_4(q)$, $G_2(q)$, $^3D_4(q)$, $^2F_4(q)$, $U_4(q)$, with $q > 2$, or $U_5(4)$.

(2) $G_i = L_iBD$ and $G_{1,2} = SBD$, where $L_1 = L$, $L_2 = K$; and $S = T \cap L_1 = T \cap L_2 = O_2(G_{1,2})$.

(3) $O_2(L_i) = O_2(G_iT)$.

PROOF. Parts (1) and (2) are immediate from 5.2.6 and the convention for $U_4(q)$. Let $S_0 := O_2(BT)$. By 5.2.6, $O_2(G_iS_0) = O_2(L_i)$ for $i = 1, 2$. Further as $B \le G_i$, $O_2(G_iT) \le O_2(BT) = S_0$ using A.1.6, so $O_2(G_iT) \le O_2(G_iS_0) = O_2(L_i)$ and hence $O_2(G_iT) = O_2(L_i)$, establishing (3). $\qquad\square$

Recall the notion of a completion of an amalgam from Definition F.1.6. Let $G(\alpha)$ denote the simple group of Lie type for which there is a completion $\xi : \alpha \to G(\alpha)$; that is, α is an amalgam of type $G(\alpha)$. To establish conclusion (3) of Theorem 5.2.3, we must show that $G \cong G(\alpha)$. Let $2m(\alpha)$ be the order of the Weyl group of $G(\alpha)$.

LEMMA 5.2.8. *Either*

(1) $K \in \mathcal{L}^*(G, T)$, *or*

(2) α *is the* $L_3(4)$-*amalgam, and* $K < \hat{K} \in \mathcal{L}^*(G, T)$ *with* $\hat{K}$ *an exceptional* A_7-*block.*

PROOF. Assume $K < \hat{K} \in \mathcal{L}^*(G, T)$ and let $Q := O_2(K)$ and $\hat{K}^* := \hat{K}/O_2(\hat{K})$. Then K/Q is not $SU_3(4)$, since in that event $K \in \mathcal{L}^*(G, T)$ by 1.2.8.4. Thus we may assume α is not of type $U_5(4)$.

Recall $\hat{K} \in \mathcal{H}^e$ by 1.1.3.1, so $1 \neq [O_2(\hat{K}), K] \leq K \cap O_2(\hat{K})$. If $K/Q \cong Sz(q)$, then α appears in case (5) of 5.2.6, and $\hat{K}^* \cong {}^2F_4(q)$ by 1.2.4 and A.3.12. But then K is isomorphic to its image K^* in $\hat{K}^*$, so $K \cap O_2(\hat{K}) = 1$, contrary to our earlier observation. Thus we have eliminated the case where α is the ${}^2F_4(q)$-amalgam.

If α is the $L_3(q)$ or $Sp_4(q)$ amalgam, then K is an $L_2(q)$-block, so it has a unique noncentral 2-chief factor, and hence the same holds for $\hat{K}$, with $Q \leq O_2(\hat{K})$. By 5.2.6, $Q = O_2(KT)$, so $Q = O_2(\hat{K})$. Therefore $K^* \cong L_2(q)$ is a T-invariant quasisimple subgroup of $\hat{K}^*$, so by A.3.12, $q = 4$; and then by A.3.14, $\hat{K}^*$ is A_7, $\hat{A}_7$, J_1, $L_2(25)$, or $L_2(p)$, $p \equiv \pm 3 \mod 8$ and $p^2 \equiv 1 \mod 5$. As α is of type $L_3(4)$ or $Sp_4(4)$, Q is an extension of a natural module for $K/Q \cong L_2(4)$ and $m(Q) = 4$ or 6. As $\hat{K} \in \mathcal{H}^e$ and $\hat{K}^*$ is quasisimple, $\hat{K}^*$ is faithful on Q, so that $\hat{K}^* \leq GL(Q)$. Comparing the possiblities for $\hat{K}^*$ listed above to those in G.7.3, we conclude from G.7.3 that $\hat{K}^* \cong A_7$, and then as $m(Q) = 4$ or 6, $\hat{K}$ is an A_7-block or an exceptional A_7-block. In the former case, the noncentral chief factor for K on Q is not the $L_2(4)$-module, so the latter case holds, forcing α to be the $L_3(4)$-amalgam. Thus (2) holds in this case.

Suppose α is the $U_4(q)$-amalgam. From 5.2.6, $K/Q \cong L_2(q)$ for $q = 2^{n/2} > 2$ and Q is special of order q^{1+4} with K trivial on $Z(Q)$. Further by 5.2.6, either $Q = O_2(KT)$, or $O_2(KT) = Q\langle t \rangle$ where t is an involution with $C_Q(t) \cong E_{q^3}$.

We claim $Q \trianglelefteq \hat{K}$, so assume otherwise. Suppose first that $Q \leq R := O_2(\hat{K})$. Then as $R \leq O_2(KT)$, and $Q < R$ by assumption, $R = O_2(KT) = Q\langle t \rangle$. But now $Z(Q) = Z(R) \trianglelefteq \hat{K}$, and $Q/Z(Q) = J(R/Z(Q))$, so $Q \trianglelefteq \hat{K}$, contrary to assumption. Thus $Q \not\leq R$, so as K has two natural chief factors on $Q/Z(Q)$ and $[R, K] \neq 1$, we conclude $(Q \cap R)Z(Q)/Z(Q)$ is one of these chief factors. Thus $Z(Q) = [Q \cap R, Q] \leq R$ and $Q \cap R \cong E_{q^3}$. Again as $R \leq O_2(KT)$, either $R = Q \cap R$ or $|R : Q \cap R| = 2$. In the latter case $Q \cap R = [Q, t] = C_Q(t)$, so $R = (Q \cap R)\langle t \rangle$.

In any case K^* is an $L_2(q)$-block with $|O_2(K^*)| = q^2$. The only possibilities for such an embedding in A.3.12 are that $\hat{K}^* \cong (S)L_3(q)$, or $q = 4$ and $\hat{K}^* \cong M_{22}$, $\hat{M}_{22}$, or M_{23}. The last three cases are impossible, as those groups are of order divisible by 11, a prime not dividing the order of $GL_7(2)$. Thus $\hat{K}^* \cong SL_3(q)$ and $[R, \hat{K}]$ is the natural module for $\hat{K}^*$, so $[R, \hat{K}] = [R, K] = Q \cap K$. However as α is the $U_4(q)$-amalgam, $J(T) = O_2(L)$ is normal in LT, so $N_G(J(T)) \leq M = !\mathcal{M}(LT)$. From the action of $\hat{K}$ on R, $K_1 := N_{\hat{K}}(J(T))$ is the scond maximal parabolic of $\hat{K}$ over $\hat{K} \cap T$. Thus as $T \cap L = T \cap K$ by 5.2.7.2, $K_1^\infty = [K_1^\infty, K \cap T] \leq L$, and then as $|L| = |K_1^\infty|$, $L \leq \hat{K}$, contradicting $M = !\mathcal{M}(LT)$. This contradiction completes the proof of the claim.

Finally we treat the case where α is the $U_4(q)$-amalgam and $Q \trianglelefteq \hat{K}$, along with the remaining two cases where α is the amalgam of $G_2(q)$ or $^3D_4(q)$. In these last two cases Q is special and K is irreducible on $Q/Z(Q)$, so as in the earlier cases of the $L_3(q)$ and $Sp_4(q)$ amalgams, there is a unique noncentral 2-chief factor under the extension of K by a Cartan subgroup, and again we get $O_2(\hat{K}) = Q$. Thus in each of our three cases, $Q \trianglelefteq \hat{K}$, so $\hat{K} \in \mathcal{C}(N_G(Q))$ by 1.2.7 as $\hat{K} \in \mathcal{L}^*(G, T)$. Further $K^* \cong L_2(q)$ when α is the amalgam for $U_4(q)$ or $G_2(q)$, and $K^* \cong L_2(q^3)$ when α is the $^3D_4(q)$-amalgam. As above, A.3.12 gives a proper extension with "$O_2(B) = 1$" only when "B" is $L_2(4)$. This eliminates the $^3D_4(q)$ amalgam, and forces α to be the amalgam of $U_4(4)$ or $G_2(4)$. Therefore $Q \cong 4^{1+4}$ and there is X of order 3 in $C_{D_L B}(K/Q)$ with $Q/\Phi(Q) = [Q/\Phi(Q), X]$, so X acts on $\hat{K} \in \mathcal{C}(N_G(Q))$ by 1.2.1.3. But as in our application of A.3.12 above, $\hat{K}^* \cong A_7$, $\hat{A}_7$, J_1, $L_2(25)$, or $L_2(p)$ for $p \equiv \pm 1 \mod 5$ and $p \equiv \pm 3 \mod 8$, and X centralizes $A_5 \cong K^* \leq \hat{K}^*$, so we conclude from the structure of $Aut(\hat{K}^*)$ that X centralizes $\hat{K}^*$. Thus $\hat{K}^*$ is not A_7, for otherwise $m_3(\hat{K}X) = 3$, contradicting $N_G(\hat{K})$ an SQTK-group. Further as $Q/\Phi(Q) = [Q/\Phi(Q), X]$ is of rank 8, and 8 is not divisible by 3, $\hat{K}^*$ is not $\hat{A}_7$. Finally G.7.2 eliminates the remaining possiblities for $\hat{K}^*$. This completes the proof of 5.2.8. $\qquad\square$

Conclusions (1) and (2) of 5.2.8 will lead to conclusions (3) and (2) of Theorem 5.2.3, respectively, so we adopt notation reflecting the groups arising in those conclusions. Namely we define G to be of *type* $X_r(q)$ if α is the $X_r(q)$-amalgam and $K \in \mathcal{L}^*(G, T)$. Define G to be of *type* M_{23} if α is the $L_3(4)$-amalgam and $K \notin \mathcal{L}^*(G, T)$. Thus in this language, we can summarize what we have accomplished in 5.2.6 and 5.2.8:

THEOREM 5.2.9. *One of the following holds:*
(1) G is of type $L_3(q)$, $Sp_4(q)$, $G_2(q)$, $^3D_4(q)$, or $^2F_4(q)$, for some even $q > 2$.

(2) $n > 2$ is even and G is of type $U_4(2^{n/2})$.
(3) G is of type $U_5(4)$.
(4) G is of type M_{23}.

5.2.2. Characterizing M_{23}. The remainder of this section is devoted to a proof that:

THEOREM 5.2.10. *If G is of type M_{23} then G is isomorphic to M_{23}.*

The proof of Theorem 5.2.10 involves a short series of reductions. Assume G is of type M_{23}. Then by 5.2.8, α is the $L_3(4)$-amalgam and $K < \hat{K} \in \mathcal{L}^*(G, T)$ with $\hat{K}$ an exceptional A_7-block. Let $M_2 := \hat{K}$, $M_1 := M$, and $M_{1,2} := M_1 \cap M_2$. Set $V_i := O_2(M_i)$, $V := V_1$, and $U := V_2$. Then $V \cong U \cong E_{16}$ with $M_2/U \cong A_7$. Hence we can represent M_2/U on $\Omega = \{1, \ldots, 7\}$ so that T has orbits $\{1, 2, 3, 4\}$, $\{6, 7\}$, and $\{5\}$ on Ω. Indeed:

LEMMA 5.2.11. *(1) H is the global stabilizer in M_2 of $\{6, 7\}$.*
(2) $M_{1,2}$ is the global stabilizer in M_2 of $\{5, 6, 7\}$.
(3) $M/V \cong \Gamma L_2(4)$.
(4) $M_2 \in \mathcal{M}(T)$.

(5) $|T : S| = 2$.

PROOF. Let $M_2^* := M_2/U$. There is a unique T^*-invariant subgroup $K_T^* \cong A_5$ of M_2^*, and $K_T^* T^*$ is the global stabilizer in M_2^* of $\{6, 7\}$, so (1) holds. Then VU/U is the 4-group with fixed-point set $\{5, 6, 7\}$ and $N_{M_2}(VU) = N_{M_2}(V) = M_{1,2}$ as $M \in \mathcal{M}(T)$, so (2) holds.

Let $M_2 \le M_0 \in \mathcal{M}(T)$. By 5.2.8, $M_2 \in \mathcal{L}^*(G, T)$, so $M_2 \trianglelefteq M_0$ by 1.2.7.3. Then $U = O_2(M_2) \le O_2(M_0)$, so as $T \le M_2$, $U = O_2(M_0)$ by A.1.6. As $M_0 \in \mathcal{H}^e$, $M_0/U \le GL(U)$, so as M_2/U is self-normalizing in $GL(U)$, $M_0 = M_2$, proving (4).

As $V = O_2(LT)$, $O_2(M) = V = C_G(V)$ by 3.2.11, so $M/V \le GL(V)$. Next $UV \in Syl_2(L)$, so by a Frattini Argument, $M = L N_M(UV) \ge L N_M(U) = L M_{1,2}$ using (4). From the structure of M_2, $M_{1,2}/V$ is isomorphic to a Borel group of $\Gamma L_2(4)$, so $L M_{1,2}/V = N_{GL(V)}(L/V)$ as $N_{GL(V)}(L/V) \cong \Gamma L_2(4)$. Then as $L \trianglelefteq M$, (3) holds, and (3) implies (5). $\qquad\square$

LEMMA 5.2.12. *(1)* $Z(T) = \langle z \rangle$ *is of order 2.*
(2) $C_G(z) = C_{M_2}(z)$ *is an* $L_3(2)$*-block.*
(3) M_2 *is transitive on* $U^\#$.
(4) U *is a TI-set in* G.

PROOF. Parts (1) and (3) are easy consequences of the fact that M_2 is an exceptional A_7-block containing T. As another consequence, $Y := C_{M_2}(z)$ is an $L_3(2)$-block. Let $G_z := C_G(z)$ and $G_z^* := G_z/\langle z \rangle$. As $T \le G_z$, $F^*(G_z) = O_2(G_z)$ by 1.1.4.6, so $F^*(G_z^*) = O_2(G_z)^*$ by A.1.8. Thus as $U = O_2(Y) \ge O_2(G_z)$ by A.1.6, and Y is irreducible on U^*, $U = O_2(G_z)$. Thus $G_z \le N_G(U) = M_2$ using 5.2.11.4. Therefore (2) holds. Then (2), (3), and I.6.1.1 imply (4). $\qquad\square$

LEMMA 5.2.13. G *has one conjugacy class of involutions.*

PROOF. All involutions of V are conjugate under M and hence fused into $U \cap V$. Similarly all involutions in U are conjugate under M_2, so as U and V are the maximal elementary abelian subgroups of UV, all involutions in UV are fused in G. From the structure of M_2, each involution in M_2 is fused into UV in M_2. So the lemma holds, as M_2 contains a Sylow 2-group T of G. $\qquad\square$

LEMMA 5.2.14. *(1)* G *is transitive on its elements of order 3 which centralize involutions.*
(2) *All elements of order 3 in* $M_1 \cup M_2$ *are conjugate in* G.

PROOF. By 5.2.12.2, $C_G(z)$ has one class of elements of order 3, so 5.2.13 implies (1). Next M_2 has two classes of elements of order 3, those with either 1 or 2 cycles of length 3 on Ω. The first class centralizes an involution in M_2/U and hence has centralizer of even order. The second class centralizes an involution in U. Thus (1) implies all elements of order 3 in M_2 are conjugate in G. Then as $M_{1,2}$ contains a Sylow 3-group of M_1 and M_2, (2) holds. $\qquad\square$

LEMMA 5.2.15. *Let* $X \in Syl_3(C_M(L/O_2(L)))$. *Then* $N_G(X) = N_M(X) \cong \Gamma L_2(4)$.

PROOF. First $N_M(X) \cong \Gamma L_2(4)$. On the other hand by 5.2.14, X is conjugate to $Y \le C_G(z)$ and $C_G(Y\langle z \rangle) = Y \times C_U(Y) \cong \mathbf{Z}_3 \times E_4$. Let $G_Y := C_G(Y)$ and $G_Y^* := G_Y/Y$. Then $C_{G_Y^*}(z^*) \cong E_4$, and as $C_M(X)$ is not 2-closed, neither is G_Y^*. Thus by Exercise 16.6.8 in [**Asc86a**], $G_Y^* \cong A_5$. Therefore $|C_M(X)| = |G_Y|$, so $C_M(X) = C_G(X)$. Then as $|N_M(X) : C_M(X)| = 2 = |Aut(X)|$, the lemma follows. $\square$

Recall the definition of the subgroups G_1 and G_2 in our amalgam α from the previous subsection, and let $G_0 := \langle G_1, G_2 \rangle$.

LEMMA 5.2.16. (1) $G_0 \cong L_3(4)$.
(2) $G_0 T$ is G_0 extended by a field automorphism.

PROOF. Notice in the $L_3(4)$-amalgam that we have $B = D = BD$. Thus to prove (1), it suffices by F.4.26 to show that there exist involutions $s_i \in N_{L_i}(B)$, such that $|s_1 s_2| \leq 3$. Then (2) follows from (1), since T acts on G_i, with $|T : S| = 2$ by 5.2.11.5, and $O_2(L_i T) = O_2(L_i)$ by 5.2.7.3. Thus it remains to exhibit involutions $s_i \in N_{L_i}(B)$, with $|s_1 s_2| \leq 3$.

Notice that each involution $s_i \in N_{L_i}(B)$ inverts B. Now $B \leq M_1$, so by 5.2.14.2, B is conjugate to the subgroup X defined in 5.2.15. Therefore as s_1 inverts B, $N_G(B) = (B \times L_B)\langle s_1 \rangle$, where $L_B \cong A_5$, s_1 inverts B, and s_1 induces a transposition on L_B. But s_2 also inverts B, so replacing s_1 by a suitable member of $B s_1$, we may assume $s_1 s_2 \in L_B$. Thus s_1 and s_2 are distinct transpositions in $L_B \langle s_1 \rangle \cong S_5$, so $|s_1 s_2| = 2$ or 3, completing the proof. $\square$

We now define some notation to use in our identification of G with M_{23}. Let $\bar{G} := M_{23}$ act on $\Theta := \{1, \ldots, 23\}$. Then (cf. chapter 6 in [Asc94]) we may take our 7-set Ω to be a block in the Steiner system $(\Theta, \mathcal{C})$ on Θ preserved by $\bar{G}$, so that $N_{\bar{G}}(\Omega) = \bar{M}_2$ is the split extension of $\bar{U} = \bar{G}_\Omega \cong E_{16}$ by A_7, and $\bar{M}_2$ is an exceptional A_7-block.

LEMMA 5.2.17. There is a permutation equivalence $\zeta : M_2 \to \bar{M}_2$ of M_2 and $\bar{M}_2$ on Ω.

PROOF. As B is of order 3 in $K \cap M_{1,2}$, it follows from parts (1) and (2) of 5.2.11 that we may choose B to act on Ω as $\langle (1, 2, 3) \rangle$. Then as $C_U(B) = 1$, $N_{M_2}(B) \cong \mathbf{Z}_2/(\mathbf{Z}_3 \times A_4)$ has Sylow 2-groups of order 8. Thus T splits over U, so M_2 splits over U by Gaschütz's Theorem A.1.39. Thus there is an isomorphism $\zeta : M_2 \to \bar{M}_2$, and adjusting by a suitable inner automorphism, this map is a permutation equivalence. $\square$

For the remainder of this section, define ζ as in 5.2.17.

Let $\Gamma := \Theta^2$ be the set of unordered pairs of elements from Θ and fix $\bar{x} := \{6, 7\}$ and $\bar{y} := \{5, 6\}$ in Γ. From chapter 6 of [Asc94]:

LEMMA 5.2.18. (1) $\bar{G}_{\bar{x}}$ is the extension of $L_3(4)$ by a field automorphism.
(2) $\Theta - \{6, 7\}$ is a projective plane over $\mathbf{F}_4$ with lines $\{C - \{6, 7\} : \{6, 7\} \subseteq C \in \mathcal{C}\}$, and $\bar{G}_{\bar{x}}$ preserves this structure.
(3) The global stabilizer $\bar{I}$ of $\{4, 5, 6, 7\}$ in $\bar{G}$ is the global stabilizer in $\bar{M}_2$ of $\{4, 5, 6, 7\}$.

PROOF. In [Asc94], the Steiner system $(\Theta, \mathcal{C})$ is constructed so that (1) and (2) hold. As each 4-point subset of Θ is contained in a unique block of the Steiner system, (3) holds. $\square$

Regard Γ as a a graph by decreeing that $a, b \in \Gamma$ are adjacent if $|a \cap b| = 1$. We wish to show $G \cong \bar{G}$. To do so, we write G_x for $G_0 T$ and essentially show there is a graph structure on $\Gamma_G := G/G_x$ isomorphic to the graph Γ, such that the representations of $\bar{G}$ on Γ (which is in turn $\bar{G}$-isomorphic to the analogous graph on $\Gamma_{\bar{G}} := \bar{G}/\bar{G}_x$) and G on Γ_G are equivalent. This leads us to write x for G_x regarded as a point of Γ_G—namely the coset of G_x containing the identity. Thus G_x is indeed the stabilizer of the point $x \in \Gamma_G$.

Let I denote the global stabilizer in M_2 of $\{4, 5, 6, 7\}$. Notice that the representation of M_2 on $\Omega \subseteq \Theta$ induces a representation of M_2 on $\Omega^2 \subseteq \Gamma$; this is the representation implicit in the next lemma.

LEMMA 5.2.19. *(1) $G_x \cap M_2 = H$ is the stabilizer in M_2 of $\bar{x} = \{6, 7\} \in \Gamma$, and the stabilizer in M_2 of $x = G_x \in \Gamma_G$.*

(2) The representation of M_2 on $xM_2 \subseteq \Gamma_G$ is equivalent to its representation on $\Omega^2 \subseteq \Gamma$.

(3) $\zeta : M_2 \to \bar{M}_2$ restricts to an isomorphism $\zeta : I \to \bar{I}$.

(4) $\zeta(I_{\bar{x}}) = \bar{I}_{\bar{x}}$ and $\zeta(I_{\bar{x},\bar{y}}) = \bar{I}_{\bar{x},\bar{y}}$.

PROOF. By 5.2.11.1, H is the stabilizer of $\bar{x} = \{6, 7\} \in \Gamma$, and hence is a maximal subgroup of M_2. Therefore $H = G_x \cap M_2$, and thus H is also the stabilizer in M_2 of the coset $G_x \in \Gamma_G$, which we are denoting by x. Therefore (1) holds. Then (1) implies (2), while 5.2.17 and the definition of I and $\bar{I}$ imply (3) and (4).   □

Using the equivalence of 5.2.19.2, the point $\bar{y} = \{5, 6\} \in \Gamma$ corresponds to a point $y \in \Gamma_G$; namely the coset $y = G_x t$, where $t \in I$ has cycle $(\bar{x}, \bar{y})$ on Γ. Such a t exists, as I is the global stabilizer of $\{4, 5, 6, 7\}$ in M_2, and hence induces the full symmetric group on that subset. The coset y is independent of t by 5.2.19.1.

Recall as in [**Asc94**] that $I(\{x, y\})$ denotes the global stabilizer in I of $\{x, y\}$.

LEMMA 5.2.20. *(1) $G_{x,y} = L$.*

(2) I_x is the stabilizer in M_2 of the partition $\{1, 2, 3\}$, $\{4, 5\}$, $\{6, 7\}$ of Ω, and $I_x/U \cong S_3 \times \mathbf{Z}_2$.

(3) $I_{x,y} = UB$ and $\zeta(I(\{x, y\})) = \bar{I}(\{\bar{x}, \bar{y}\})$.

(4) $G_x = \langle G_{x,y}, I_x \rangle$.

(5) There is an isomorphism $\beta : G_x \to \bar{G}_{\bar{x}}$ agreeing with ζ on I_x, such that $\beta(G_{x,y}) = \bar{G}_{\bar{x},\bar{y}}$.

PROOF. By 5.2.11.2, $M_{1,2}$ is the global stabilizer in M_2 of $\{5, 6, 7\}$, so there is $t \in (M_{1,2})_4 \le I \cap M$ with cycle $(\bar{x}, \bar{y})$. Then by a remark preceding this lemma, t has cycle (x, y). As $L \le G_0 T = G_x$, L fixes x, so that $L = L^t$ fixes $xt = y$, and then $L \le G_{x,y}$. But LT and G_0 are the only maximal subgroups of G_x containing L, and by 5.2.19.2, $T \nleq G_{x,y} \ngeq K$. So (1) holds.

Parts (2) and (3) are easy calculations given 5.2.19. As observed earlier, LT and G_0 are the only maximal subgroups of G_x containing L and $G_{x,y} = L$ by (1). So as $I_x \nleq G_0 \cap M_2 = K$ and $I_x \nleq M_{1,2}$, (4) holds.

By 5.2.16 and 5.2.18.1, there is an isomorphism $\beta : G_x \to \bar{G}_{\bar{x}}$, which we may take to map T to $\bar{T} := \zeta(T)$, B to $\bar{B} := \zeta(B)$, and K and L to the parabolics $\bar{K} := \zeta(K)$ and $\bar{L} := \bar{G}_{\bar{x},\bar{y}}$ of $O^2(\bar{G}_{\bar{x}})$. Now by (2) and (3), $I_x = UB\langle t, r \rangle$, where $t := (1, 2)(6, 7)$ and $r := (4, 5)(6, 7)$ on Ω. In particular $I_x = UN_{KT}(B)$, so

$$\beta(I_x) = \beta(U)N_{\beta(K)\beta(T)}(\beta(B)) = \bar{U}N_{\bar{K}\bar{T}}(\bar{B}) = \bar{I}_x.$$

Finally let $\gamma := \zeta^{-1} \circ \beta$, regarded as an automorphism of I_x, so that $\gamma \in Aut(I_x)$. Notice $|N_{GL(U)}(Aut_{I_x}(U)) : Aut_{I_x}(U)| = 2$ and $U = C_{Aut(I_x)}(U)$, so $|Aut_I(I_x) : Inn(I_x)| = 2$. Then as $|N_I(I_x) : I_x| = 2$, $Aut(I_x) = Aut_I(I_x)$. Indeed as $\beta(T) = \bar{T} = \zeta(T)$, $\gamma(t) \in O^2(I_x)t$, so $\gamma \in Inn(I_x)$. Thus adjusting β by the inner automorphism of G_x which acts on I_x as γ^{-1}, we may choose $\beta = \zeta$ on I_x, proving (5). $\square$

LEMMA 5.2.21. $G = \langle M, M_2 \rangle = \langle G_x, I \rangle$.

PROOF. Let $Y := \langle M, M_2 \rangle$. If $Y < G$, then by induction on the order of G, $Y \cong M_{23}$. In particular, Y has one class of involutions; while by (1) and (2) of 5.2.12, $N_G(T) \le C_G(z) \le Y$. Thus Y is a strongly embedded subgroup of G (see I.8.1), so by 7.6 in [**Asc94**], Y has a subgroup of odd order transitive on the involutions in Y. Now Y has

$$i := 3 \cdot 5 \cdot 11 \cdot 23$$

involutions, but no subgroup of odd order divisible by i. This contradiction shows $G = \langle M, M_2 \rangle$. But $M = LM_{1,2}$ and $M_2 = \langle K, I \rangle$, so

$$G = \langle M, M_2 \rangle = \langle LT, K, I \rangle = \langle G_x, I \rangle,$$

completing the proof. $\square$

LEMMA 5.2.22. $I = \langle I(\{x,y\}), I_x \rangle$.

PROOF. Notice I_x contains the kernel UB of the action of I on $\Lambda := \{4,5,6,7\}$. Further I_x contains elements inducing $(4,5)$ and $(6,7)$ on Λ, while $I(\{x,y\})$ contains an element inducing $(5,6)$. So as the symmetric group on Λ is generated by these three transpositions, the lemma holds. $\square$

We are now in a position to complete the proof of Theorem 5.2.10, by appealing to the theory of uniqueness systems in section 37 of [**Asc94**]. Namely write Γ_G for the graph on $\Gamma_G = G/G_x$ with edge set $(x,y)G = (G_x, G_x t)G$, and let Γ_I be the subgraph with vertex set xI and edge set $(x,y)I$. By 5.2.19.2, Γ_I is isomorphic to the subgraph $\Gamma_{\bar{I}} := \{4,5,6,7\}^2$ of Γ.

Observe that $\mathcal{U} := (G, I, \Gamma_G, \Gamma_I)$ is a uniqueness system in the sense of [**Asc94**]. Namely by 5.2.21, $G = \langle G_x, I \rangle$; by 5.2.20.4, $G_x = \langle G_{x,y}, I_x \rangle$; and by 5.2.22, $I = \langle I(\{x,y\}), I_x \rangle$. This verifies the defining conditions for uniqueness systems (see (U) on page 198 of [**Asc94**]). Similarly $\bar{\mathcal{U}} := (\bar{G}, \bar{I}, \Gamma, \Gamma_{\bar{I}})$ is a uniqueness system.

Now by 5.2.19 and 5.2.20, $\beta : G_x \to \bar{G}_{\bar{x}}$ and $\zeta : I \to \bar{I}$ define a similarity of uniqueness systems, as defined on page 199 of [**Asc94**]. Next we will apply Theorem 37.10 in [**Asc94**], to prove this similarity is an equivalence.

In applying Theorem 37.10, we take L in the role of the group "K" in the Theorem, and take $t, h \in I$ to be elements acting on Ω as

$$t := (1,2)(5,7), \ h := (1,2)(5,6).$$

Then $t, h \in M_{1,2} \le N_G(L)$, and by construction t has cycle (x,y), $t^h = (1,2)(6,7) \in K \le G_x$, and $\zeta(h) \in \bar{M}_{1,2} \le N_{\bar{G}}(\bar{L})$, so that hypothesis (2) of Theorem 37.10 holds. Next $L = G_{x,y}$ by 5.2.20.1, so trivially $G_{x,y} = \langle L_y, I_{x,y} \rangle$, which is hypothesis (3) of 37.10. Finally $L \cap I = BU$, and from the structure of the $L_2(4)$-block L, we check that $C_{Aut(L)}(BU) = 1$; this verifies hypothesis (1) of 37.10.

Therefore $\mathcal{U}$ is equivalent to $\bar{\mathcal{U}}$. It remains to check that $\Gamma_{\bar{I}}$ is a *base* for $\bar{\mathcal{U}}$ in the sense of p.200 of [**Asc94**]: for then as $\bar{G} = M_{23}$ is simple, Exercise 13.1 in [**Asc94**] says $G \cong \bar{G}$, completing the proof of Theorem 5.2.10.

Recall from page 200 of [**Asc94**] that $\Gamma_{\bar{I}}$ is a base for $\bar{\mathcal{U}}$ if each cycle in the graph Γ is in the closure of the conjugates of cycles of $\Gamma_{\bar{I}}$. But each triangle in Γ is conjugate to one of:

$$\{6,7\}, \{5,6\}, \{5,7\} \text{ or } \{6,7\}, \{5,7\}, \{4,7\},$$

which are triangles of $\Gamma_{\bar{I}}$. So it remains to show Γ is *triangulable* in the sense of section 34 of [**Asc94**]; that is, that each cycle of Γ is in the closure of the triangles, or equivalently the graph Γ is simply connected. This is the crucial advantage of working with Γ as opposed to Γ_G; one can calculate in Γ to check it is triangulable.

As Γ is of diameter 2, by Lemma 34.5 in [**Asc94**], it suffices to show each r-gon is in the closure of the triangles, for $r \leq 5$. For $r = 2, 3$ this holds trivially, and we now check the cases with $r = 4$ and 5, using the 4-transitivity of M_{23} on Θ.

It follows from 34.6 in [**Asc94**] that 4-gons are in the closure of triangles: Namely a pair of points at distance 2 are conjugate to $\{1,2\}$ and $\{3,4\}$, whose common neighbors are $\{1,3\}, \{1,4\}, \{2,4\}, \{2,3\}$—forming a square in Γ, which is in particular connected. Finally it follows from Lemma 34.8 in [**Asc94**] that 5-gons are in the closure of the triangles: for if x_0, x_1, x_2, x_3 is a path in Γ with $d(x_0, x_2) = d(x_0, x_3) = d(x_1, x_3) = 2$, then up to conjugation under $\bar{G}$, $x_0 = \{1,2\}$, $x_1 = \{2,3\}$, $x_2 = \{3,4\}$, and $x_3 = \{4,a\}$ for some $a \in \Theta - \{1,2,3,4\}$. Then as x_0, x_2, and x_3 are all connected to $\{2,4\}$, it follows that $x_0^\perp \cap x_2^\perp \cap x_3^\perp \neq \emptyset$, in the language of [**Asc94**].

Thus the proof of Theorem 5.2.10 is complete.

5.3. Identifying rank 2 Lie-type groups

In this section, we complete the proof of Theorem 5.2.3. Recall the definition of groups of type $X_r(q)$ and type M_{23} appearing before the statement of Theorem 5.2.9. If G is of type M_{23}, then conclusion (2) of Theorem 5.2.3 holds by Theorem 5.2.10. Therefore by Theorem 5.2.9, we may assume that one of conclusions (1)–(3) of Theorem 5.2.9 holds. Thus G is of *type* $X_r(q)$ for some even $q > 2$ and some X_r of Lie rank 2. Recall from 5.2.7 that $\alpha = (G_1, G_{1,2}, G_2)$ is an $X_r(q)$-amalgam, where $G_i = L_i BD$, $G_{1,2} = SBD$, and $S = T \cap L_1 = T \cap L_2 = O_2(G_{1,2})$. We write $G(\alpha)$ for the corresponding group $X_r(q)$ of Lie type defining the amalgam. To establish Theorem 5.2.3, we must show $G \cong G(\alpha)$.

Set $M_i := N_G(L_i)$ and $M_{1,2} := M_1 \cap M_2$, and let $\gamma := (M_1, M_{1,2}, M_2)$ be the corresponding amalgam.

LEMMA 5.3.1. *(1)* $L_i \in \mathcal{L}(G,T)$ *and* $M_i = !\mathcal{M}(L_i T)$ *with* $M_1 \neq M_2$.
(2) $F^*(M_i) = O_2(M_i) = O_2(L_i)$.
(3) $N_G(S) = M_{1,2} = N_{M_i}(S)$.
(4) $M_i = L_i M_{1,2}$.

PROOF. By the hypothesis of Theorem 5.2.3, $L_1 = L \in \mathcal{L}^*(G,T)$, and by 5.2.7.2, $L_2 = O^2(H)$ so that $L_2 \not\leq M_1$. By 5.2.8 and our assumption that G is not of type M_{23}, $L_2 = K \in \mathcal{L}^*(G,T)$. Thus (1) holds by 1.2.7. Hence $F^*(M_i) = O_2(M_i)$ by 1.1.4.6. By 5.2.7.3, $O_2(G_i T) = O_2(L_i)$, so $O_2(M_i) = O_2(L_i)$ using A.1.6, completing the proof of (2).

To prove (3), it will suffice to show $N_G(S) \leq M_i$ for $i = 1$ and 2. For then $N_G(S) \leq M_{1,2}$. On the other hand $N_{M_i}(S)$ is maximal in M_i, and M_1 and M_2 are distinct maximal 2-locals by (1), so $M_{1,2} = N_{M_i}(S)$ and hence (3) holds.

Thus it remains to show $N_G(S) \leq M_i$. But as $S \in Syl_2(L_i)$ and T is in a unique maximal subgroup of L_iT, we conclude from Theorem 3.1.1 that $O_2(\langle N_G(S), L_i\rangle) \neq 1$. Therefore $N_G(S) \leq M_i = !\mathcal{M}(L_iT)$ by (1). Thus (3) is established. Then (4) follows from (3) via a Frattini Argument. $\qquad\square$

LEMMA 5.3.2. γ is an $M(\alpha)$-extension of α (in the sense of Definition F.4.3), for some extension $M(\alpha)$ of $G(\alpha)$.

PROOF. Let $M_0 := \langle M_1, M_2\rangle$. We first verify that γ satisfies Hypothesis A of the Green Book [**DGS85**], with L_i in the role of "P_i^*".

By 5.3.1.1, $O_2(M_0) = 1$. By 5.3.1.2, $F^*(M_i) = O_2(M_i) = O_2(L_i)$, so condition (ii) of Hypothesis A holds. Then as $O_2(M_i) = O_2(L_i)$, condition (i) holds by 5.3.1.4. Condition (iii) follows from 5.3.1.3, and the list of possibilities for L_i in 5.2.6. This completes the verification of Hypothesis A.

As Hypothesis A holds, and $q > 2$ by 5.2.7.1, case (a) of Theorem A in the Green Book [**DGS85**] holds, so that γ is an extension of the Lie amalgam α of $G(\alpha)$. That is, γ determines subgroups $M_i(\alpha) \cong M_i$ of $Aut(G(\alpha))$, with corresponding completion $M(\alpha) := \langle M_1(\alpha), M_2(\alpha)\rangle \leq Aut(G(\alpha))$. So the lemma holds. $\qquad\square$

Let $Z_S := Z(S)$ and $Z_i := Z(L_i)$.

LEMMA 5.3.3. *Either*

(1) The hypotheses of Theorem F.4.31 are satisfied, with G in the role of "M",

or

(2) $G(\alpha) \cong L_3(q)$, and $C_G(z) \not\leq M_{1,2}$ for each involution $z \in Z_S$.

PROOF. By 5.3.2, γ is an extension of the Lie amalgam α, so that $M(\alpha)$ plays the role of "$\bar{M}$" in Theorem F.4.31. Hypothesis (d) of F.4.31 holds for G in the role of "M", as $T \leq M_{1,2}$ and $T \in Syl_2(G)$. Hypothesis (e) holds as G is simple. Hypothesis (a) follows from the fact that L_iT is a uniqueness subgroup by 5.3.1.1. Further BD is transitive on $Z_i^{\#}$. Thus if $Z_i \neq 1$, each involution in Z_i is conjugate under M_i to some $z \in Z(L_iT)$, and therefore $C_G(z) \leq M_i$ using 5.3.1.1. Similarly if $G(\alpha) \cong L_3(q)$, then BD is transitive on $Z_S^{\#}$, so if $C_G(z_0) \leq M_1$ for some $z_0 \in Z_S^{\#}$, then $C_G(z) \leq M_1$ for all $z \in Z_S^{\#}$. Hence the first statement in Hypothesis (c) holds, and either Hypothesis (b) holds, or conclusion (2) of 5.3.3 holds. If $G(\alpha)$ is $Sp_4(q)$, then each involution z in Z_S is fused into $Z(T)$ under BD, and hence $C_G(z) \in \mathcal{H}^e$ by 1.1.4.6. This completes the verification of Hypothesis (c). Therefore either the hypotheses of F.4.31 are satisfied, so that conclusion (1) of 5.3.3 holds, or conclusion (2) of 5.3.3 holds. $\qquad\square$

THEOREM 5.3.4. *Either*

(1) $G \cong G(\alpha)$, *or*
(2) $G(\alpha) \cong L_3(q)$, and $C_G(z) \not\leq M_{1,2}$ for each involution $z \in Z_S$.

PROOF. If 5.3.3.1 holds, we may apply Theorem F.4.31 to conclude $G \cong M(\alpha)$. Since G is simple, we must in fact have $M(\alpha) \cong G(\alpha)$. $\qquad\square$

By Theorems 5.2.9, 5.2.10, and 5.3.4, Theorem 5.2.3 holds unless possibly G is of type $L_3(q)$ and conclusion (2) of 5.3.4 holds. We will finish by showing (in 5.3.7 below) that the latter case leads to a contradiction.

Thus in the remainder of this section, we assume G is of type $L_3(q)$ and conclusion (2) of 5.3.4 holds.

Pick $z \in Z^\#$ and set $G_z := C_G(z)$. Set $V_i := O_2(L_i)$ and observe $S = V_1 V_2 = J(T)$ using 5.3.2 and F.4.29.6. Similarly by F.4.29.2, if $t \in T - S$, then t induces a field automorphism on L_i, so $[Z_S, t] \neq 1$; that is, $S = C_T(Z_S)$.

LEMMA 5.3.5. $N_G(Z_S) = M_{1,2}$.

PROOF. Set $G_Z := C_G(Z_S)$. Then $S = V_1 V_2 \in Syl_2(G_Z)$, as we just observed. As $T \leq N_G(Z_S)$, $F^*(G_Z) = O_2(G_Z)$ by 1.1.4.6, and therefore also $F^*(G_Z/Z_S) = O_2(G_Z/Z_S)$ by A.1.8. Hence as S/Z_S is abelian, $S/Z_S = O_2(G_Z/Z_S)$, so $S = O_2(G_Z)$. Then as $\mathcal{A}(S) = \{V_1, V_2\}$ with $V_i \trianglelefteq T$, $N_G(Z_S) \leq N_G(V_i) = M_i$ as $M_i \in \mathcal{M}$ by 5.3.1.1. On the other hand by 5.3.1.3, $M_{1,2} = N_G(S) \leq N_G(Z_S)$. $\square$

LEMMA 5.3.6. (1) V_i is weakly closed in T with respect to G.
(2) $Z_S^G \cap V_i = Z_S^{L_i}$ is of order $q + 1$.

PROOF. We saw $\mathcal{A}(T) = \{V_1, V_2\}$ and $V_i \trianglelefteq T$; in particular, $V_1^G \cap T \subseteq \mathcal{A}(T)$, so to establish (1) we only need to show $V_2 \notin V_1^G$. But if $V_2 \in V_1^G$ then as $V_i \trianglelefteq T$ and $N_G(T)$ controls fusion of normal subgroups of T by Burnside's Fusion Lemma, V_2 is in fact conjugate to V_1 in $N_G(T)$, and hence in $O^2(N_G(T))$ as T normalizes V_1 and V_2. This is impossible as $|\mathcal{A}(S)| = 2$, establishing (1). By (1) and Burnside's Fusion Lemma, M_i controls fusion in V_i, so (2) follows. $\square$

LEMMA 5.3.7. $G_z \leq M_{1,2}$.

PROOF. As $Z_S \trianglelefteq T$ and $M_{1,2}$ is transitive on $Z_S^\#$, we may take $z \in Z(T)$. Therefore $F^*(G_z) = O_2(G_z) =: R$ by 1.1.4.6. Next unless $q = 4$ and $M_i/V_i \cong S_5$ for $i = 1$ and 2, $S = V_1 V_2 = O_2(C_{M_i}(z))$ for each i. Assume for the moment that the exceptional case does not hold. Then as $S \in Syl_2(C_G(Z_S))$ by 5.3.5, $R \leq S$ by A.1.6, so $Z_S = Z(S) \leq \Omega_1(C_{G_z}(R)) = \Omega_1(Z(R)) =: Z_R$. If $Z_S = Z_R$ then $R \leq N_G(Z_S) \leq M_{1,2}$ by 5.3.5, and the lemma holds; so assume instead that $Z_S < Z_R$.

Let $\hat{G}$ denote our target group $G(\alpha) \cong L_3(q)$ and $\hat{M}_i$ the subgroups $M_i(\alpha)$ in 5.3.2. Recall we have a corresponding isomorphism of amalgams $\beta : \hat{\gamma} := (\hat{M}_1, \hat{M}_{1,2}\hat{M}_2) \to \gamma$. Thus $S \cong \hat{S}$ is isomorphic to a Sylow 2-group of $L_3(q)$, so V_1 and V_2 are the maximal elementary abelian subgroups of S. Therefore $V_i \cap Z_R > Z_S$ for $i = 1$ or 2, so that $R \leq C_S(V_i \cap Z_R) = V_i$. Then $V_i \leq C_{G_z}(R) \leq R$, so $V_i = R$. But then by 5.3.6.1, $G_z \leq N_G(V_i) = M_i$, so that $G_z = G_z \cap M_i \leq M_{1,2}$ using β, and the lemma holds.

It remains to treat the exceptional case where $q = 4$ and $M_i/V_i \cong S_5$ for $i = 1$ and 2. Let $\bar{G}_z := G_z/\langle z \rangle$, so that $F^*(\bar{G}_z) = O_2(\bar{G}_z)$ by A.1.8. Now M_i is determined up to isomorphism, so in particular T is isomorphic to a Sylow 2-subgroup of M_{22}. Therefore $J(\bar{T}) = \bar{Q} \cong E_{16}$ with $Q \cong Q_8^2$ and $C_T(Q) \leq Q$. Let $V_z := \langle Z_S^{G_z} \rangle$. As $\bar{Z}_S \leq Z(\bar{T})$, $\bar{V}_z$ is elementary abelian by B.2.14, so $\Phi(V_z) \leq \langle z \rangle$.

Suppose first that V_z is abelian, and therefore elementary abelian. Then $V_z \leq C_T(Z_S) = S$ using an earlier observation. As V_1 and V_2 are the maximal elementary abelian subgroups of S, $V_z \leq V_i$ for $i = 1$ or 2. If $V_z = Z_S$, then the lemma holds by 5.3.5, so we may assume $Z_S < V_z \leq V_i$. But $V_i = C_G(A)$ for each hyperplane A of V_i through Z_S, so $V_z = A$ or V_i, and in any case $V_i \trianglelefteq G_z$. Hence the lemma holds, again since $G_z \cap M_i \leq M_{1,2}$.

Thus we may suppose instead that V_z is not abelian. Now if $V_z \leq S$, then $Z_S = Z(V_z) \trianglelefteq G_z$, and the lemma holds by 5.3.5. Hence there is $v \in V_z - S$; we will see this leads to a contradiction. Now from the structure of M_1, $E_4 \cong [v, S/Z_S] \leq (V_z \cap S)/Z_S$, so $m(\bar{V}_z) \geq 4$. Therefore as $E_{16} \cong \bar{Q} = J(\bar{T})$, we must have $V_z = Q$. Next as $C_T(Q) \leq Q$, $G_z/Q \leq Out(Q) \cong O_4^+(2)$, so $|G_z : T| = 3$ or 9. As

$$|G_z : T| \geq |\bar{Z}_S^{G_z}| \geq m(\bar{V}_z) = 4,$$

$|G_z : T| = 9$. As $m_2(Q) = 3$ and $m(V_i) = 4$, $V_i \not\leq Q$; indeed $[V_i, v]Z_S \leq Q$ and $[V_i, Q] \leq V_i$, so that V_iQ/Q has order 2 and induces an involution of type a_2 on $\bar{Q}$, so it centralizes a nontrivial element in $O^2(G_z/Q) \cong E_9$. Therefore $O^2(N_{G_z}(V_iQ)) \neq 1$. However by 5.3.6.1, V_i is weakly closed in V_iQ; so $O^2(N_{G_z}(V_iQ)) \leq O^2(G_z \cap M_i) = 1$, contradicting the previous remark. This contradiction completes the proof of 5.3.7.   □

Observe that 5.3.7 contradicts our assumption that 5.3.4.2 holds. So the proof of Theorem 5.2.3 is complete.

CHAPTER 6

Reducing $L_2(2^n)$ to n = 2 and V orthogonal

In this chapter, we continue our analysis of simple QTKE-groups G for which there exists a T-invariant $L \in \mathcal{L}_f^*(G, T)$ with $L/O_2(L) \cong L_2(2^n)$. Recall that we began this analysis in chapter 5. In particular in Theorem 5.2.3 we showed under these hypotheses, and the hypothesis that $n(H) > 1$ for some $H \in \mathcal{H}_*(T, M)$, that either

(I) G is M_{23} or a group of Lie type of characteristic 2 and Lie rank 2, or

(II) the conclusion of 5.2.3.1 holds; in particular $n = 2$ and $[R_2(LT), L]$ is the sum of at most two A_5-modules.

In Theorem 6.2.20 of this chapter, we complete the reduction to the situation where $n = 2$ and $[R_2(LT), L]$ is the sum of A_5-modules by considering the remaining case where $n(H) = 1$ for each $H \in \mathcal{H}_*(T, M)$, and $[R_2(LT), L]$ is not the sum of A_5-modules when $n = 2$. Section 6.1 carries out the reduction to the subcase $n = 2$. Then section 6.2 shows that the only quasithin example to arise in this subcase is M_{22}.

This reduction allows us thereafter to regard $L/O_2(L) \cong L_2(4)$ as $\Omega_4^-(2)$. We treat that case in Part 5, which deals with the situation where there exists $L \in \mathcal{L}_f^*(G, T)$ with $L/O_2(L)$ a group of Lie type group defined over $\mathbf{F}_2$.

6.1. Reducing $L_2(2^n)$ to $L_2(4)$

As mentioned above, we wish to complete the reduction to the situation where $n = 2$ and $[R_2(LT), L]$ is the sum of A_5-modules, under the hypothesis of chapter 5. By Theorem 5.2.3, we may assume Hypothesis 5.1.8 fails. Thus in this section, we assume the following hypothesis:

HYPOTHESIS 6.1.1. G is a simple QTKE-group, $T \in Syl_2(G)$, and $L \in \mathcal{L}_f^*(G, T)$ with $L/O_2(L) \cong L_2(2^n)$, $L \trianglelefteq M \in \mathcal{M}(T)$, and $V \in \mathcal{R}_2(LT)$ with $[V, L] \neq 1$. In addition, assume

(1) $[V, L]$ is not the sum of one or two copies of the A_5-module for $L/O_2(L) \cong A_5$.

(2) $n(H) = 1$ for each $H \in \mathcal{H}_*(T, M)$.

REMARK 6.1.2. Notice Hypothesis 6.1.1.1 has the effect of excluding cases (2) and (5) of 5.1.3 plus case (4) of 5.1.3 when $n = 2$. Thus either case (1) or (3) of 5.1.3 holds, or $n > 2$ and case (4) of 5.1.3 holds. Similarly 6.1.1.1 excludes case (3) of 5.1.2; therefore by 5.1.2, either case (3) of 5.1.3 holds, or $J(T) \leq C_T(V) = O_2(LT)$, so that $J(T) \trianglelefteq LT$ and hence $M = !\mathcal{M}(N_G(J(T)))$.

Throughout this section, define $Z := \Omega_1(Z(T))$, $V_L := [V, L]$, and $T_L := T \cap L$. Set $M_V := N_M(V)$, and $\bar{M}_V := M_V/C_M(V)$.

In contrast to the previous chapter, we find now when $n(H) = 1$ for each $H \in \mathcal{H}_*(T, M)$ that weak-closure methods are frequently effective.

LEMMA 6.1.3. *(1) If V is a TI-set under M, then Hypothesis E.6.1 holds.*
(2) Either

> *(I) $r(G, V) = 1$, or*
> *(II) $J_1(T) \not\le C_T(V)$.*

PROOF. Part (1) of Hypothesis E.6.1 follows from Hypothesis 6.1.1. We saw $C_T(V) = O_2(LT)$, so that $M = !\mathcal{M}(LT) = !\mathcal{M}(N_{M_V}(C_T(V)))$, giving part (3) of Hypothesis E.6.1. This establishes (1). Further $n(H) = 1$ for all $H \in \mathcal{H}_*(T, M)$ by Hypothesis 6.1.1.2, so the hypotheses of E.6.26 are satisfied with "j" equal to 1. Therefore (2) follows from E.6.26. $\square$

6.1.1. Initial reductions. In this subsection, we establish various reductions culminating in the two cases of Proposition 6.1.15; eliminating the first of those cases is then the goal of the second subsection.

LEMMA 6.1.4. $V_L/C_{V_L}(L)$ *is the natural module for $\bar{L}$.*

PROOF. Assume that the lemma fails. This assumption excludes case (3) of 5.1.3, so by Remark 6.1.2 and 5.1.3, either

(A) $n > 2$ is even and V is the $O_4^-(2^{n/2})$-module, or
(B) V is the sum of two copies of the natural module.

Similarly by Remark 6.1.2 and 5.1.2, $J(T) \trianglelefteq LT$ and $M = !\mathcal{M}(N_{LT}(J(T)))$. Thus $[Z, H] = 1$ for each $H \in \mathcal{H}_*(T, M)$ by 5.1.7. Further by Hypothesis 6.1.1.2, $n(H) = 1$. Enlarging V if necessary, we may take $V = R_2(LT)$.

Assume that there is $A \in \mathcal{A}_1(T)$ with $\bar{A} \ne 1$. Then by B.2.4.1,

$$m(V/C_V(A)) \le m(\bar{A}) + 1 \le m_2(\bar{L}\bar{T}) + 1 = n + 1. \qquad (*)$$

But in case (B), $m(V/C_V(A)) \ge 2n > n+1$ since $n > 1$, contrary to (*), so case (A) holds. Hence by H.1.1.2 with $n/2$ in the role of "n", $n = 4$, and $\bar{A}$ is of rank 1 and generated by an orthogonal transvection. Further for $t \in T - C_T(V)$, $m(V/C_V(t)) \ge 2n$ in case (A), and $m(V/C_V(t)) \ge n$ in case (B) by H.1.1.1. Therefore we have shown that either:

(i) $n = 4$, V is the $O_4^-(4)$-module, and if $\overline{J_1(T)} \ne 1$ then $\overline{J_1(T)}$ is generated by an orthogonal transvection, or
(ii) $m(\bar{L}\bar{T}, V) > 2$ and $J_1(T) \le C_T(V) = O_2(LT)$, so that $J_1(T) \trianglelefteq LT$.

Suppose first that case (ii) holds. Then $r(G, V) = 1$ by 6.1.3.2. Now if Hypothesis E.6.1 is satisfied, then since $m(\bar{L}\bar{T}, V) > 2$ in case (ii), $r(G, V) > 1$ by Theorem E.6.3, a contradiction. Thus V is not a TI-set in M by 6.1.3.1. Therefore as $L \trianglelefteq M$, L is not irreducible on V, so case (B) holds where $V = V_1 \oplus V_2$ is the sum of two natural modules V_1 and V_2. Further we may choose V_1 to be T-invariant (cf. the proof of A.1.42.1). As L is irreducible on V_i, V_i is a TI-set under M. As $r(G, V) = 1$, there is a hyperplane W of V with $C_G(W) \not\le N_G(V)$. Set $U_i := W \cap V_i$. Then $C_G(W) \le C_G(U_i)$ and $m(V_i/U_i) \le m(V/W) = 1$, so $U_i \ne 1$ as $m(V_i) \ge 4$. Thus if $C_G(W) \le M$, then as V_1 and V_2 are TI-sets in M, $C_G(W)$ normalizes $V_1 \oplus V_2 = V$, contrary to our choice of W. Therefore $C_G(W) \not\le M$, so $C_G(U_1) \not\le M$. But as $V_1 \trianglelefteq LT$, $N_G(V_1) \le M$, so $r(G, V_1) = 1$. As V_1 is a TI-set in

M, Hypothesis E.6.1 holds by 6.1.3.1 applied to V_1 in the role of "V". However as L is transitive on the hyperplanes of V_1, and the stabilizer in LT of a hyperplane contains a Sylow 2-subgroup of LT, we may take $T \leq N_G(U_1)$. Thus $C_G(U_1) \leq M$ by E.6.13, contrary to an earlier observation.

This contradiction shows that case (i) holds. The elimination of case (i) will be lengthier. As L is irreducible on V, V is a TI-set in M, so that by 6.1.3.1, Hypothesis E.6.1 is satisfied, and we may appeal to results in section E.6.

We first claim $r(G, V) > 1$. If not, there is a hyperplane U of V with $C_G(U) \not\leq M$, and by E.6.13, U is not T-invariant. Thus U contains the subspace U_0 orthogonal to a nonsingular $\mathbf{F}_4$-point of the orthogonal space V. Therefore U contains a 2-central involution. As $V = R_2(LT)$, $V = \Omega_1(Z(Q))$, where $Q := O_2(LT)$. Finally $C_V(N_T(U)) \leq U$, so as $C_{\bar{L}\bar{T}}(U) = 1$,

$$\Omega_1(Z(N_T(U))) = C_{\Omega_1(Z(Q))}(N_T(U)) = C_V(N_T(U)) \leq U,$$

contrary to E.6.10.4, establishing the claim that $r(G, V) > 1$.

Let $M_1 \in \mathcal{M}(C_G(Z))$, and set $Q_1 := O_2(M_1)$, so that $M_1 = N_G(Q_1)$. As $H \leq C_G(Z)$ for $H \in \mathcal{H}_*(T, M)$, $M \neq M_1$. Suppose that $O_{2,F^*}(M_1) \leq M$. Then

$$Q_1 = O_2(M \cap M_1) = O_2(N_M(Q_1))$$

by A.4.4.1, so that $Q_1 \in \mathcal{B}_2(M)$. By A.4.4.2, $C(M, Q_1) = M \cap M_1$, so Hypothesis C.2.3 is satisfied, with M, $M \cap M_1$, Q_1 in the roles of "H, M_H, R". Now since V is the orthogonal module and $n > 2$, L is not a χ-block; so for L in the role of "K", the conclusions of C.2.7 do not hold, and hence $L \leq M \cap M_1$. But then $M_1 = !\mathcal{M}(LT) = M$, contradicting $M_1 \neq M$. This contradiction shows that $O_{2,F^*}(M_1) \not\leq M$.

Next $Z \leq R_2(LT) = V$ by B.2.14, so $Z = C_V(T)$. Let $X := O^{5'}(N_L(T_L))$. Then $X/O_2(X) \cong \mathbf{Z}_5$ and $XT \leq C_G(Z) \leq M_1$, from the structure of $O_4^-(4) \cong L_2(16)$ and its action on V. Let $S := O_2(XT)$, so that $S = T_L Q$. Then $J_1(S) \leq C_S(V)$ since case (i) holds. Define

$$\mathcal{H}_S := \{M_S \leq M_1 : S \in Syl_2(M_S) \text{ and } T \leq N_G(M_S)\}.$$

As $r(G, V) > 1$, E.6.26 says $M_S \leq M$ for each $M_S \in \mathcal{H}_S$ with $n(M_S) = 1$.

Now $O_2(M_1) = Q_1 \leq O_2(XT) = S$ by A.1.6. Then S is Sylow in $SO_{2,F}(M_1)$, so that $SO_{2,F}(M_1) \in \mathcal{H}_S$—and since $n(O_{2,F}(M_1)) = 1$ by E.1.13, $O_{2,F}(M_1) \leq M$ by the previous paragraph. We saw $O_{2,F^*}(M_1) \not\leq M$, so there is $K_1 \in \mathcal{C}(M_1)$ with $K_1 \not\leq M$, and $K_1/O_2(K_1)$ quasisimple. Let $K_0 := \langle K_1^T \rangle$ and observe that $X = O^2(X)$, so X normalizes K_1 by 1.2.1.3.

Next as $K_1 \not\leq M$, there is $H_S \in \mathcal{H}_*(T, M) \cap K_0 T$. Now $n(H_S) = 1$ by Hypothesis 6.1.1.2. Thus if $S \in Syl_2(O^2(H_S)S)$, then $O^2(H_S)S \in \mathcal{H}_S$, and hence $H_S \leq M$ by an earlier remark. Therefore S is not Sylow in $O^2(H_S)S$, and hence S is not Sylow in $K_0 S$. But if X normalizes $T \cap K_0 \in Syl_2(K_0)$, then $T \cap K_0 \leq O_2(XT) = S$; thus we conclude $X \not\leq N_G(T \cap K_0)$. In particular, $[X, K_1] \not\leq O_2(K_1)$, so a Sylow 5-subgroup X_5 of X acts faithfully on $K_1/O_2(K_1)$. Then as $X_5 T = T X_5$, this quotient is described in A.3.15. In cases (5)–(7) of A.3.15, X normalizes $T \cap K_0$, contrary to an earlier observation. As $X/O_2(X)$ is of order 5, cases (2) and (4) are ruled out. So it follows from A.3.15 that either

(a) $K_1/O_2(K_1) \cong L_2(p^e)$ and $(M \cap K_1)/O_2(K_1) \cong D_{p^e - \epsilon}$, or

(b) $K_1/O_2(K_1) \cong L_3^\delta(p)$, and there is an X-invariant $K_2 \in \mathcal{L}(G,T) \cap K_1$ with $K_2 O_2(K_1)/O_2(K_1) \cong SL_2(p)$.

In case (b), if the projection of X_5 on K_1 centralizes $K_2/O_2(K_2)$, then from the structure of $L_3^\delta(p)$, X_5 centralizes a Sylow 2-group of $K_1/O_2(K_1)$, which is not the case as X does not normalize $T \cap K_0$. Thus the projection is inverted in $T \cap K_2$, so as $X \trianglelefteq XT$, $X \leq K_2$. Similarly in case (a) the projection is inverted in $T \cap K_1$, so $X \leq K_1$. Now $L \cap M_1$ contains T_L and so is contained in a Borel subgroup of L, and hence $X \trianglelefteq M_1 \cap M$. Thus in case (b), $K_2 \not\leq M$ as $X < K_2$. In this case, we replace K_1 by K_2, reducing to the case where $K_1 \in \mathcal{L}(G,T)$, $K_1 \not\leq M$, and $K_1/O_2(K_1) \cong L_2(p^e)$ as in case (a). (We no longer require $K_1 \in \mathcal{C}(M_1)$.) As $X \leq K_1$ is normalized by T, $K_0 = \langle K_1^T \rangle = K_1$.

Let $K_1^* T^* := K_1 T/O_2(K_1 T)$. Recall by Remark 6.1.2 that $M = !\mathcal{M}(N_G(J(T)))$ and $J(T) = J(O_2(LT)) \trianglelefteq XT$. Thus $J(T)$ is not normal in $K_1 T$ as $K_1 \not\leq M$, so there is $A \in \mathcal{A}(T)$ with $A^* \neq 1$. As $J(T) \trianglelefteq XT$, $A^* \leq J(T)^* \leq O_2(X^* T^*)$. But from the structure of $Aut(L_2(p^e))$, each nontrivial elementary abelian 2-subgroup of $O_2(X^* T^*)$ is fused under K_1^* to a subgroup of T^* not in $O_2(X^* T^*)$, contrary to $J(T)^* \leq O_2(X^* T^*)$. This contradiction finally completes the proof of 6.1.4. $\qquad\square$

LEMMA 6.1.5. $\mathcal{H}_*(T,M) \subseteq C_G(Z)$.

PROOF. Assume that $H \in \mathcal{H}_*(T,M)$ with $[H,Z] \neq 1$, and let $K := O^2(H)$. Let D_L be a Hall $2'$-subgroup of $N_L(T_L)$. Enlarging V if necessary, we may take $V = R_2(LT)$, so $Z \leq V$. By 5.1.7.2, $K = [K,J(T)]$ and $L = [L,J(T)]$.

Let $\tilde{V}_L := V_L/C_{V_L}(L)$ and $Z_L := Z \cap V_L$. As $\tilde{V}_L$ is the natural module for $\bar{L}$ by 6.1.4, $\overline{J(T)} = \bar{T}_L$ by B.4.2.1. Hence $J(T) \leq T_L Q$ where $Q := O_2(LT)$, so D_L normalizes $J(T_L Q) = J(T)$. Also $V_L = [Z_L, L]$ and $C_{LT}(\tilde{V}_L) = C_{LT}(V_L) = Q$. Let $S := \text{Baum}(T)$. As $L = [L, J(T)]$, and $\tilde{V}_L$ is the natural module, E.2.3.2 says $S \in Syl_2(LS)$ and hence $S \cap L \in Syl_2(L)$. As $\overline{J(T)} = \bar{T}_L$ and $T_L Q = C_T(C_V(T_L Q))$, also $S = \text{Baum}(T_L Q)$, so that D_L normalizes S.

As $\tilde{V}_L$ is the natural module for $\bar{L}$, the normalizer N of $\bar{L} \cong SL_2(2^n)$ in $GL(\tilde{V}_L)$ is $\Gamma L_2(2^n)$, with $C_N(\bar{L}) \cong \mathbf{Z}_{2^n-1}$, and $O^2(C_N(\tilde{Z}_L))$ is the product of $\bar{T}$ with a diagonal subgroup of $C_N(\bar{L}) \times \bar{L}$ isomorphic to $\mathbf{Z}_{2^n-1}$. Therefore $C_Z := C_M(Z_L)$ acts on T_L and on $[Z_L, L] = V_L$, and $O^2(\bar{C}_Z/\bar{T}_L)$ is a subgroup of $\mathbf{Z}_{2^n-1}$.

Let $U_H := \langle Z^H \rangle$ and set $\hat{H} := H/C_H(U_H)$. Observe $U_H \in \mathcal{R}_2(H)$ by B.2.14. By Hypothesis 6.1.1, $n(H) = 1$. Recall by 3.3.2.4 that we may apply results of section B.6 to H. So as $K = [K, J(T)]$ and $[H,Z] \neq 1$, H appears in case (2) of E.2.3, with $\hat{H} \cong S_3$ or S_3 wr $\mathbf{Z}_2$ and $S \in Syl_2(KS)$. By parts (a) and (b) of B.6.8.6, $C_T(U_H) \trianglelefteq H$.

We claim $C_H(U_H) = O_2(H)$, so assume otherwise. By B.6.8.6.a, $C_H(U_H) \leq O_{2,\Phi}(H)$, so by B.6.8.2, $H/O_2(H) \cong D_8/3^{1+2}$. Thus there is a T-invariant subgroup $Y = O^2(Y)$ of $O_{2,\Phi}(K)$ with $Y = [J(T),Y]$ and $|Y : O_2(Y)| = 3$, and Y centralizes U_H by assumption. Then by B.6.8.2, $Y \leq O_{2,\Phi}(K) \leq M$, so as Y centralizes U_H and $Z \leq U_H$, Y centralizes Z and normalizes $[Z, L] = V_L$. If Y centralizes V_L then $[Y, L] \leq C_L(V_L) = O_2(L)$, so that LT normalizes $O^2(Y O_2(L)) = Y$, and hence $N_G(Y) \leq M = !\mathcal{M}(LT)$. As $K \leq N_G(Y)$, this contradicts $K \not\leq M$. Hence $\bar{Y} \neq 1$, and as $Y \leq C_Z$, we conclude from paragraph three that $\overline{J(T)} = \bar{T}_L \trianglelefteq \bar{T}\bar{Y}$. This contradicts $Y = [Y, J(T)]$, and so completes the proof that $C_H(U_H) = O_2(H)$. It follows that $H = J(H)T$ with $H/O_2(H) \cong S_3$ or S_3 wr $\mathbf{Z}_2$, and in particular that $H \cap M = T$.

Let $X := \langle D_L, H \rangle$. Then $X \in \mathcal{H}(T)$ by 5.1.7.2.iii, as V_L is not the S_5-module. Set $U := \langle Z^X \rangle$, $Q_X := O_2(X)$ and $X^* := X/C_X(U)$. As $\tilde{V}_L$ is the natural module and $Z \leq V$, for $d \in D_L^{\#}$ we have $C_Z(d) = C_Z(L) < Z$, so that D_L is faithful on U. Thus $C_{D_L T}(U) = C_T(U)$. Also $C_H(U) \leq C_H(U_H) = O_2(H)$ from an earlier reduction. Thus $C_T(U) = C_H(U)$, so $C_T(U)$ is normal in $X = \langle D_L, H \rangle$. Finally $Q_X \leq C_T(U)$ as $U \in \mathcal{R}_2(X)$, so $Q_X = C_T(U)$ is Sylow in $C_X(U)$.

We next show that D_L^* does not act on K^*, so we assume that $D_L^* \leq N_{X^*}(K^*)$, and derive a contradiction during the next few paragraphs. First D_L acts on the preimage $KC_X(U)$ of K^*. Recall D_L acts on S, so that D_L normalizes $[C_U(S), K^*] = [C_U(S), K] =: U_K$. We saw that $S \in Syl_2(SK)$, so that $U_K \in \mathcal{R}_2(SK)$ by B.2.14. As $K = [K, J(T)]$, we may apply E.2.3.2 to U_K to conclude $K^*S^* = H_1^* \times \cdots \times H_s^*$ and $U_K = U_1 \oplus \cdots \oplus U_s$ with $s \leq 2$, $H_i^* \cong S_3$, and $U_i := [U_K, H_i] \cong E_4$. As $s \leq 2$, D_L normalizes H_i^* and U_i. Therefore D_L acts on $C_{U_i}(S) \cong \mathbf{Z}_2$, so D_L centralizes K^*S^* and U_K. Then as T normalizes K and $C_Z(D_L) = C_Z(L)$,

$$1 < Z \cap U_K \leq C_Z(D_L) = C_Z(L),$$

so that $C_X(U) \leq C_X(Z \cap U_K) \leq M = !\mathcal{M}(LT)$. Thus $C_X(U) \leq C_Z \leq N_G(T_L) \cap N_G(V_L)$ using paragraph three. Set $X_0 := O^2(C_X(U))$ and $C := C_{X_0}(\tilde{V}_L)$.

Suppose for the moment that there exists an odd prime divisor p of $|X_0|$ coprime to $2^n - 1$. Then as $O^2(\bar{C}_Z/\bar{T}_L)$ is a subgroup of $\mathbf{Z}_{2^n-1}$ by paragraph three, $O^{p'}(X_0) \leq C$. In this case set $X_1 := O^{p'}(X_0)$; then X_1 char $X_0 \trianglelefteq X$, so that $X_1 \trianglelefteq X$. Now suppose instead that q is any prime divisor of $2^n - 1$. Then $m_q(M) \leq 2$ as M is an SQTK-group, so as D_L is faithful on U, $m_q(X_0) \leq 1$. Thus if all odd prime divisors of $|X_0|$ divide $2^n - 1$, and C is not a 2-group, then for some odd prime p, $X_1 := O^{p'}(O_{2,p}(C)) \neq 1$, and X_0 has cyclic Sylow p-groups, so again X_1 char X_0, and $X_1 \trianglelefteq X$.

We have shown that if C is not a 2-group, then there is $1 \neq X_1 = O^2(X_1) \leq C$ with $X_1 \trianglelefteq X$. Thus $[L, X_1] \leq C_L([\tilde{V}, L]) = O_2(L)$, so that LT normalizes $O^2(O_2(L)X_1) = X_1$. But then $X \leq N_G(X_1) \leq M = !\mathcal{M}(LT)$, contradicting $H \nleq M$. We conclude that C is a 2-group, and so $C_{X_0 T}(\tilde{V}_L) = C_T(\tilde{V}_L)C = Q$ from paragraph two. Then as we saw that $C_X(U)$ normalizes V_L and T_L, X_0 normalizes $\mathrm{Baum}(T_L Q) = S$. Therefore as D_L acts on S and KX_0, D_L acts on $\langle S^{KX_0} \rangle = \langle S^K \rangle$, and hence on $O^2(\langle S^K \rangle) = K$.

Let $K_1 := O^2(K \cap H_1)$. We saw that H appears in case (2) of E.2.3, so S acts on K_1 with S Sylow in SK_1 and $SK_1/O_2(SK_1) \cong S_3$. As D_L normalizes H_1, D_L normalizes $K_1 S$. Thus parts (a)–(d) of Hypothesis F.1.1 hold with LS, $K_1 S$ in the roles of "L_1", "L_2". By Theorem 4.3.2, $M = !\mathcal{M}(LS)$, so $O_2(\langle LS, K_1 \rangle) = 1$, giving part (e). Finally as $LS \trianglelefteq LT$, $LS \in \mathcal{H}^e$ by 1.1.3.1, and similarly $K_1 S \in \mathcal{H}^e$, giving part (f). Thus $\alpha := (LS, SD_L, K_1 D_L S)$ is a weak BN-pair of rank 2 by F.1.9. Indeed as $N_{L_2}(S) \leq S$, α is described in F.1.12. Then α is not of type $L_3(q)$ since $n(K_1) = 1 < n(L)$. In all other cases of F.1.12, one of LS or $K_1 S$ centralizes $Z(S) \geq Z$, which is not the case. This contradiction shows that D_L^* does not act on K^*.

Recall that $H = J(H)T$, and U_H is an FF-module for $H/O_2(H) \cong S_3$ or $S_3 \text{ wr } \mathbf{Z}_2$. Thus U is also an FF-module for X^*. By Theorem B.5.6, $J(X)^* = L_1^* \times \cdots \times L_s^*$ is a direct product of $s \leq 2$ subgroups L_i^* permuted by H, with either $L_i^* \cong L_2(2)$ or $F^*(L_i^*)$ quasisimple. In particular as $s \leq 2$, $O^2(X)$ normalizes each L_i^*. Choose numbering so that $L_0^* := L_1^* \cdots L_r^*$ is the product of those factors L_i^* upon which some X-conjugate of K projects nontrivially; in particular $K^* = [K^*, J(T)^*] \leq L_0^*$, $1 \leq r \leq 2$, and by construction $L_0^* \trianglelefteq X^*$. Thus $X^* = \langle K^*, D_L^* T^* \rangle = L_0^* D_L^* T^*$ and D_L acts on each L_i^*.

Now for $1 \leq i \leq r$, $[U, L_i^*]$ is an FF-module for L_i^*, and we claim L_i^* is on the following list: $L_k(2)$, $k = 2, 3, 4, 5$; S_k, $k = 5, 6, 7, 8$; A_k, $k = 6, 7, 8$; $\hat{A}_6$, or $G_2(2)$. For no L_i^* can be isomorphic to $L_2(2^m)$, $SL_3(2^m)$, $Sp_4(2^m)$, or $G_2(2^m)$ with $m > 1$, acting on the natural module, since in those cases $J(T)^*$ induces inner automorphisms on L_i^*, whereas T acts on the solvable group K and $K = [K, J(T)]$. Thus the claim follows from B.5.6 and B.4.2. Furthermore L_i^* is not isomorphic to $L_2(2)$ for all $i \leq r$, since D_L does not normalize K^* by a previous reduction.

As $D_L T = T D_L$ and the groups L_i^* do not appear in A.3.15, we conclude $O^3(D_L^*)$ centralizes L_i^*. So as D_L^* does not normalize $K^* \leq L_0^*$, $O^3(D_L^*) < D_L^*$. As $L/O_2(L) \cong L_2(2^n)$, it follows that 3 divides $2^n - 1$, so that n is even. As $Out(L_i^*)$ is a 2-group for each L_i^*, D_L induces inner automorphisms on L_0^*. Then as D_L is cyclic and L_i^* has no element of order 9, $D_L^*/C_{D_L^*}(L_1^* \cdots L_r^*)$ is of order 3.

Set $D_0 := O^2(D_L T)$ and let A_i^* be the projection of D_0^* on L_i^*. By the previous paragraph, $1 \neq A_i^*$ for some i, and $A_i^* = O_2(A_i^*) B^*$ for B^* of order 3. As D_0 is invariant under the Sylow group T, we conclude by inspection of the possibilities for L_i^* listed above that $A_i^* = O^2(P^*)$, where P^* is either a rank one parabolic over $T^* \cap L_i^*$, or a subgroup isomorphic to S_3 or S_4 containing $T^* \cap L_i^*$ in case $O_2(L_i^*) \cong A_7$. Let L_i denote the preimage of L_i^*. In each case $A_i^* = [T \cap L_i, A_i^*]$, so $O^{3'}(D_0) = [O^{3'}(D_0), T \cap L_i] \leq L_i$. It follows as D_L is cyclic that $A_i^* \neq 1$ for a unique i, and $T \cap L_i$ centralizes a subgroup of index 3 in $D_0/O_2(D_0)$. We conclude from the structure of $Aut(L/O_2(L))$ that $n = 2$; hence $D_L = O_3(D_L) \leq L_i$ and $D_0 T/O_2(D_0 T) \cong S_3$. We may choose notation so that $i = 1$.

As T acts on D_0, T acts on L_1, so as $O^2(X)$ normalizes each L_i, $L_1 \trianglelefteq X$. Recall by definition that the projection A^* of K^* on L_1^* is nontrivial. As A^* is T-invariant with $A^*/O_2(A^*) \cong \mathbf{Z}_3$ or E_9, arguing as in the previous paragraph, we conclude that $A^* = [A^*, T \cap L_1]$. Then as T acts on K, $A^* \cap K^* \neq 1$, so as T is irreducible on $K/O_2(K)$, $K^* = A^* \leq L_1^*$. Now as X acts on L_1, and D_L and K are contained in L_1, $X = \langle D_L, KT \rangle = L_1 T$.

Assume L_1^* is $L_2(2)$ or S_5. Then there is a unique T^*-invariant nontrivial solvable subgroup $Y^* = O^2(Y^*)$ of L_1^*. Hence $K^* = Y^* = D_0^*$, impossible as D_L^* does not act on K^*. Therefore L_1^* is $L_k(2)$, $3 \leq k \leq 5$, S_k or A_k, $6 \leq k \leq 8$, $\hat{A}_6$, or $G_2(2)$.

Suppose that $H/O_2(H) \cong S_3 \text{ wr } \mathbf{Z}_2$. Then as $K^* \leq L_1^*$ and $X = L_1 T$, $X^* \cong Aut(L_k(2))$, $k = 4$ or 5, and K^* a rank-2 parabolic determined by a pair of non-adjacent nodes. As T normalizes D_0^*, with $D_0^*/O_2(D_0^*)$ of order 3, $k = 4$. Then as $[K_j, Z] \neq 1$ for $j = 1$ and 2, Theorems B.5.1 and B.4.2 show that $[U, L_1]$ is the sum of the natural module and its dual. But then $J(T)^* = O_2(K^*)$, contrary to $K = [K, J(T)]$.

This contradiction shows that $H/O_2(H) \cong S_3$. Recall also $D_0 T/O_2(D_0 T) \cong S_3$. Now $X = \langle H, D_0 T \rangle$, so that $O^2(X^*)$ is generated by K^* and D_0^*. We conclude $O^2(L_1^*)$ is $L_3(2)$, $U_3(3)$, A_6, A_7, or $\hat{A}_6$. Further neither D_0 nor K centralizes Z, so we conclude $X^* \cong S_7$ and $[U, L_1^*]$ is the natural module for X^*. From the description of offenders in B.3.2.4, $J(T)^*$ is generated by the three transpositions in T^*, so as $J(T) \trianglelefteq D_0 T$, it follows that D_0^* permutes these transpositions transitively, and hence $C_Z(D_0T) \cap [U, L_1]$ is a vector of weight 6, so that $C_{X^*}(C_Z(D_0T)) \cong S_6$. Now $C_Z(D_0T) = C_Z(D_L) = C_Z(L)$, so $C_X(C_Z(D_0T)) \le M = !\mathcal{M}(LT)$. But this is impossible as $D_0 \unlhd X \cap M$, completing the proof of 6.1.5. $\square$

LEMMA 6.1.6. *(1)* $C_Z(L) = 1$, *and hence* $C_T(L) = 1$.
(2) V_L *is the natural module for* $\bar{L}$, *and* $V = V_L$ *if* $L = [L, J(T)]$.
(3) $V_L = [R_2(LT), L]$.

PROOF. If $C_Z(L) \ne 1$, then $C_G(Z) \le C_G(C_Z(L)) \le M = !\mathcal{M}(LT)$. But then for $H \in \mathcal{H}_*(T, M)$, $H \le M$ by 6.1.5, contrary to $H \not\le M$. This contradiction establishes (1). Then 6.1.4 and (1) imply V_L is the natural module for $\bar{L}$. The final statement of (2) follows as $V = C_V(L)[V, L]$ by E.2.3.2. Finally $V \le R_2(LT)$, so $V_L \le [R_2(LT), L]$. On the other hand, applying (2) to $R_2(LT)$ in the role of "V", L is irreducible on $[R_2(LT), L]$, so (3) holds. $\square$

Now replacing V by V_L if necessary, we assume throughout the rest of this section that

$$V = V_L.$$

Thus by 6.1.6.2, V is the natural module for $\bar{L} \cong L_2(2^n)$. Since $L \unlhd M$, and L is irreducible on V, using 6.1.3.1 we have:

LEMMA 6.1.7. *(1)* V *is a TI-set under* M. *Thus if* $1 \ne U \le V$, *then* $N_M(U) \le N_M(V) = M_V$.
(2) *Hypothesis E.6.1 holds, so we may apply results from section E.6.*

Using 3.1.4.1, 6.1.7, and 6.1.5 we have:

LEMMA 6.1.8. *If* $H \le N_G(U)$ *for* $1 \ne U \le V$, *then* $H \cap M = N_H(V)$. *In particular* $H \cap M = N_H(V)$ *for each* $H \in \mathcal{H}_*(T, M)$.

Let $Z_S := C_V(T_L)$, so that Z_S is a 1-dimensional $\mathbf{F}_{2^n}$-subspace of the natural module V. Let $S := C_T(Z_S)$.

LEMMA 6.1.9. *(1)* $S = T_L O_2(LT)$ *and* $S \in Syl_2(C_G(Z_S))$.
(2) $N_G(S) \le M$.
(3) $F^*(N_G(Z_S)) = O_2(N_G(Z_S))$.
(4) $V \le O_2(C_G(Z_S))$ *and* $V/Z_S \le Z(S/Z_S)$.
(5) $N_G(Z_S) = C_G(Z_S)N_M(Z_S) = C_G(Z_S)N_{M_V}(Z_S)$.
(6) $J(T) = J(S)$ *and* $Baum(T) = Baum(S)$.

PROOF. As $T \le N_G(Z_S)$, (3) holds by 1.1.4.6, and also $C_T(Z_S) = S \in Syl_2(C_G(Z_S))$. As V is the natural module for $\bar{L}$, the remaining assertion of (1) holds, and also $V/Z_S \le Z(S/Z_S)$. Then an application of G.2.2.1, with $N_G(Z_S)$ in the role of "H", establishes the remaining assertion of (4).

Now using (1), we may apply a Frattini Argument to conclude that $N_G(Z_S) = C_G(Z_S)(N_G(Z_S) \cap N_G(S))$. Thus (5) will follow from (2) since V is a TI-set in M by 6.1.7; so it remains to prove (2) and (6).

If $J(T) \le C_T(V)$, then in particular $J(T) \le S$. On the other hand, if $J(T)$ does not centralize V, then as V is the natural module for $\bar{L}$, $J(T) \le S$ by B.4.2.1. Therefore as $S = C_T(Z_S)$, (6) follows from B.2.3.5. Finally Theorem 4.3.17 implies (2). $\square$

LEMMA 6.1.10. *(1)* $r(G, V) \geq n$.

(2) $s(G, V) = m(Aut_M(V), V) = n$.

(3) Suppose that V^g normalizes but does not centralize V for some $g \in G$. Then $m(V^g/C_{V^g}(V)) = n$.

PROOF. As V is the natural module for $\bar{L}$, $m(Aut_M(V), V) = n$. By 6.1.7.2, V satisfies Hypothesis E.6.1. Thus if $n > 2$, (1) and (2) hold by Theorem E.6.3. So assume $n = 2$, and let $U \leq V$ with $m(V/U) = 1$. As V is the natural module, L is transitive on $\mathbf{F}_2$-hyperplanes of V, so we may choose $U \trianglelefteq T$. Then E.6.13 says $C_G(U) \leq M$. Thus in any case, $r(G, V) \geq n = m(Aut_M(V), V)$, so that (1) and (2) are established.

Assume the hypotheses of (3), and set $U := C_{V^g}(V)$. As $V^g \leq N_G(V)$,

$$m(V^g/U) \leq m_2(LT/C_{LT}(V)) = n.$$

On the other hand as $V \not\leq C_G(V^g)$, $m(V^g/U) \geq s(G, V) = n$ by E.3.7 and (2), establishing (3). $\square$

LEMMA 6.1.11. *Suppose $V^g \leq T$ with $1 \neq [V, V^g] \leq V \cap V^g$. Then $Z_S = [V, V^g] = V \cap V^g$ and $V^g \in V^{C_G(Z_S)}$.*

PROOF. Let $A := V^g$. By 6.1.10.3, $m(A/C_A(V)) = m(V/C_V(A)) = n$, so that $\bar{A}$ is an FF^*-offender on V. Therefore by B.4.2.1, $\bar{A} \in Syl_2(\bar{L})$ and $Z_S = [A, V] = C_V(A)$. As V normalizes A by hypothesis, we have symmetry between A and V, so $Z_S = C_A(V)$. Therefore $Z_S^{g^{-1}} = C_V(V^{g^{-1}})$ is a 1-dimensional $\mathbf{F}_{2^n}$ subspace of V, and hence $Z_S^{g^{-1}} = Z_S^h$ for some $h \in L$ by transitivity of L on such subspaces. Thus $V^g = V^{hg}$ with $hg \in N_G(Z_S)$, so $V^g \in V^{C_G(Z_S)}$ by 6.1.9.5. $\square$

LEMMA 6.1.12. *(1) Either $N_G(W_0(T, V)) \not\leq M$ or $C_G(C_1(T, V)) \not\leq M$.*

(2) If $n > 2$, then $Z_S \leq C_1(T, V)$.

(3) $W_0(T, V) \leq S$.

PROOF. By Hypotheses 6.1.1, $n(H) = 1$ for each $H \in \mathcal{H}_*(T, M)$. Hence as $H \not\leq M$, part (1) follows from 6.1.10.2 and E.3.19. Assume $A \leq V^g \cap T$, with $w := m(V^g/A)$ satisfying $n - w \geq 2$. By 6.1.10, $n = s(G, V)$, so by E.3.10, either $\bar{A} = 1$ or $\bar{A} \in \mathcal{A}_2(\bar{T}, V)$. In either case, $\bar{A} \leq \bar{T}_L$, so that $A \leq S$ by 6.1.9.1. Since $n \geq 2$, (3) follows from this observation in the case $w = 0$. If $n > 2$, (2) follows from the observation in the case $w = 1$. $\square$

LEMMA 6.1.13. *Let $U \leq V$ with $m(V/U) = n$. Then one of the following holds:*

(1) $C_G(U) \leq N_G(V)$.

(2) $U \in Z_S^L$.

(3) n is even, and $U = C_V(t)$ for some $t \in M$ inducing an involutory field automorphism on $\bar{L}$.

PROOF. If U does not satisfy either (2) or (3), then $C_M(U) = C_M(V)$. Then as $r(G, V) \geq n > 1$ by 6.1.10.1, (1) holds by E.6.12. $\square$

LEMMA 6.1.14. *Assume n is even and $U = C_V(t)$ for some $t \in T$ inducing an involutory field automorphism on $\bar{L}$. Choose notation so that $T_U := N_T(U) \in Syl_2(N_M(U))$. Then*

(1) $R := Q\langle t \rangle \in Syl_2(C_G(U))$, $N_G(J(R)) \leq M$, and $T_U \in Syl_2(N_G(U))$.

(2) $N_G(U)$ and $C_G(U)$ are in $\mathcal{H}^e$.

(3) $W_0(R,V) \leq C_T(V)$, and if $n > 2$ then $W_1(R,V) \leq C_T(V)$.

(4) $V_U := \langle V^{N_G(U)} \rangle$ is elementary abelian, and $V_U/U \in \mathcal{R}_2(N_G(U)/U)$; further $[O_2(N_G(U)), V_U] \leq U$.

(5) Assume further that $n > 2$, and $V^g \leq C_G(U)$ is V-invariant with $[V, V^g] \neq 1$. Then $C_G(Z_S) \not\leq M$.

PROOF. Observe that $R := C_T(U) = Q\langle t \rangle$, where $Q := C_T(V)$, and U and V/U are the natural module for $N_L(U)/O_2(N_L(U)) \cong L_2(2^{n/2})$. Now $\mathcal{A}(R) = \mathcal{A}(Q)$, so $J(R) = J(Q) \trianglelefteq LT$, and hence $N_G(R) \leq N_G(J(R)) \leq M = !\mathcal{M}(LT)$, so $R \in Syl_2(C_G(U))$. Similarly $T_U := N_T(U) \in Syl_2(N_G(U))$ since $J(T_U) = J(Q)$. Thus (1) holds. As $U \cap Z \neq 1$, $F^*(N_G(U)) = O_2(N_G(U))$ by 1.1.4.3. Then $C_G(U) \in \mathcal{H}^e$ by 1.1.3.1, so (2) holds.

Next by 6.1.12, $W_i := W_i(R,V) \leq C_R(Z_S) \leq Q$ for $i = 0$ when $n \geq 2$, and for $i = 1$ when $n > 2$. Thus (3) holds.

Let $V_U := \langle V^{N_G(U)} \rangle$ and $N_G(U)^* := N_G(U)/C_G(V_U)$. We may apply G.2.2 with U, V, $O^2(C_L(U))$, T_U, $N_G(U)$ in the roles of "V_1, V, L, T, H". By G.2.2.4, $V_U/U \in \mathcal{R}_2(N_G(U)/U)$. By G.2.2.1, $V_U \leq O_2(C_G(U))$ and $[O_2(N_G(U)), V_U] \leq U$. Then $V_U \leq O_2(C_G(U)) \leq R$ using (1), so that $V_U \leq W_0(R,V) = W_0$. Therefore as $W_0 \leq C_T(V)$ by (3), $V_U = \langle V^{N_G(U)} \rangle$ is elementary abelian. This establishes (4).

Now assume the hypotheses of (5). First $m(V/C_V(V^g)) = n$, by applying 6.1.10.3 with the roles of V, V^g reversed. Then as $U \leq C_V(V^g)$ with $m(U) = m(V/U) = n$, we conclude $U = C_V(V^g)$. As $n > 2$, $L_U := O^2(N_L(U)) \in \mathcal{L}(N_G(U), T_U)$. As $T_U \in Syl_2(N_G(U))$ by (1), $L_U \leq K \in \mathcal{C}(N_G(U))$ by 1.2.4. As $[U, L_U] = U$, $C_K(U) \leq O_\infty(K)$. By (1) and a Frattini Argument, $KR = C_{KR}(U)N_{KR}(J(R)) = C_K(U)(K \cap M)R$. Now $L_U = L_U^\infty \trianglelefteq K \cap M$, and $K/O_\infty(K)$ is simple by A.3.3.1, so $K = L_U C_K(U)$. Thus if $C_K(U) \leq M$, then $K \leq M$, so that $K = K^\infty = L_U$. On the other hand, if $C_K(U) \not\leq M$, then also $O_\infty(K) \not\leq M$.

By (3) and E.3.16, $N_G(W_0) \leq M \geq C_G(C_1(R,V))$. Each solvable subgroup X of $C_G(U)$ containing R satisfies $n(X) = 1$ by E.1.13, and so is contained in M by E.3.19. This eliminates the exceptional case $O_\infty(K) \not\leq M$ of the previous paragraph, so that $L_U = K$. Since T_U normalizes $L_U \in \mathcal{C}(N_G(U))$, and is Sylow in $N_G(U)$ by (1), $N_G(U)$ normalizes L_U by 1.2.1.3. Then as $O_2(L_U) \leq Q \leq C_G(V)$, $O_2(L_U) \leq C_G(V_U)$.

Recall V_U is elementary abelian by (4). As V is the direct sum of two copies of the natural module U for $L_U/O_2(L_U)$, and $L_U \trianglelefteq N_G(U)$, V_U is the sum and hence the direct sum of copies of the natural module for $L_U/O_2(L_U)$. Next as $V^g \leq C_G(U)$, $V^g \leq R^h$ for some $h \in C_G(U)$, so by (3)

$$V^g \leq W_0(R^h, V) \leq Q^h \leq O_2(T_U^h L_U).$$

Thus $[V^g, L_U] \leq [O_2(T_U^h L_U), L_U] \leq O_2(L_U) \leq C_G(V_U)$. Thus L_U normalizes $Z_1 := [V^g C_G(V_U), V] = [V^g, V]$. We saw earlier that $U = C_V(V^g)$ with $m(V/U) = n$. Then as $n > 2$, $V \leq S^g$, so that in fact $S^g = VO_2(L^g S^g)$. Hence $Z_1 = [V, V^g] = [S^g, V^g] = Z_S^g$.

We finally assume that $C_G(Z_S) \leq M$. Then $N_G(Z_S) \leq M$ by 6.1.9.5, so

$$L_U \leq N_G(Z_1) = N_G(Z_S^g) = N_{M^g}(Z_S^g) \leq N_{M^g}(V^g),$$

since V is a TI-set in M by 6.1.7. This is impossible, as the L_U^*-submodule $V \cap V^g = Z_1 = Z_S^g$ of rank n in V_U is natural by an earlier remark, whereas $Aut_M(Z_S)$ is solvable. This contradiction establishes (5), and so completes the proof of the lemma. $\square$

PROPOSITION 6.1.15. *Either*

(1) $C_G(Z_S) \not\leq M$, *or*
(2) $n = 2$, *and either* $N_G(W_0(T, V)) \not\leq M$ *or* $W_1(T, V) \not\leq S$.

PROOF. Set $W_0 := W_0(T, V)$. Suppose first that $N_G(W_0) \not\leq M$. Then as $M = !\mathcal{M}(LT)$, $W_0 \not\leq O_2(LT) = C_T(V)$ by E.3.16.1, so there is $V^g \leq T \leq N_G(V)$ which does not centralize V. Set $U := C_{V^g}(V)$; then $m(V^g/U) = n$ by 6.1.10.3, so that 6.1.13 applies to U with V^g is in the role of "V". If V acts on V^g, then by 6.1.11, $V^g \in V^{C_G(Z_S)}$, while $V^M \leq O_2(L) \leq C_G(V)$, so (1) holds. Therefore we may assume $V \not\leq N_G(V^g)$. In particular $C_G(U) \not\leq N_G(V^g)$, so that case (1) of 6.1.13 does not hold. If case (2) of 6.1.13 holds, then again (1) holds. If case (3) of 6.1.13 holds with $n > 2$, then $v \in V - C_V(V^g)$ induces a field automorphism on V^g with $U = C_{V^g}(v)$ and V is V^g-invariant with $1 \neq [V, V^g]$, so by 6.1.14.5, (1) holds yet again. Finally if $n = 2$, then (2) holds as we are assuming that $N_G(W_0) \not\leq M$.

Thus we may instead assume that $N_G(W_0) \leq M$. Therefore by 6.1.12.1, $C_G(C_1(T, V)) \not\leq M$. Thus if $Z_S \leq C_1(T, V)$, then (1) holds. On the other hand if $Z_S \not\leq C_1(T, V)$ then $n = 2$ by 6.1.12.2, and also $W_1(T, V) \not\leq S$, so (2) holds. $\square$

6.1.2. Reducing to $C_G(Z_S) \leq M$ and $n = 2$. In this subsection, we consider the first case of 6.1.15, where $C_G(Z_S) \not\leq M$. Our object is to establish a contradiction and so eliminate that case; this is accomplished in Theorem 6.1.27. In the following chapter, we show that in the second case, G is isomorphic to M_{22}.

Hence in this subsection, we assume:

HYPOTHESIS 6.1.16. $C_G(Z_S) \not\leq M$, *where* $Z_S := C_V(T_L)$.

Let $I := C_G(Z_S)$ and

$$\mathcal{H}_S := \{H \in \mathcal{H}(T) : H \not\leq M \text{ and } O^2(H) \leq I\}.$$

In particular $IT \in \mathcal{H}_S$, so that $\mathcal{H}_S$ is nonempty.

Let H denote some arbitrary member of $\mathcal{H}_S$. As $O^2(H) \leq I$, $H = O^2(H)T \leq IT \leq N_G(Z_S)$. Set $U_H := \langle V^H \rangle$, $H_S := C_H(Z_S)$, $Q_H := O_2(H_S)$, and $\tilde{H} := H/Z_S$.

Notice that $U_{IT} = \langle V^I \rangle$, $(IT)_S = I$, and $Q_{IT} = O_2(I)$. Also a Hall 2'-subgroup D_L of $N_L(T_L)$ normalizes Z_S and hence I, but $D_L \cap I = 1$. Then as $N_G(Z_S)$ is an SQTK-group,

$$m_p(D_L I) \leq 2 \text{ for each odd prime } p.$$

LEMMA 6.1.17. *(1)* $V \leq Q_H$, $S \in Syl_2(H_S)$, $F^*(H_S) = O_2(H_S) = Q_H$, *and* $H_S \trianglelefteq H = H_S T$.
    (2) $\tilde{U}_H \in \mathcal{R}_2(\tilde{H}_S)$, *so* $\tilde{U}_H \leq Z(\tilde{Q}_H)$.
    (3) $Q_H = C_{H_S}(\tilde{U}_H)$.
    (4) For $s \in S - C_S(V)$ and $Z_S \leq Y \leq V$, $[V, s] = Z_S$ and $m([Y, s]) = m(Y/Z_S)$.
    (5) If $Z_S \leq Y \leq V$ with $|V : Y| = 2$, and $\bar{S}_0$ is a noncyclic subgroup of $\bar{S}$, then $Z_S = [Y, S_0]$.

PROOF. As $H \in \mathcal{H}(T)$, $F^*(H) = O_2(H)$ by 1.1.4.6. We saw $H \leq N_G(Z_S)$, so that $H_S = C_H(Z_S) \trianglelefteq H$; then $F^*(H_S) = O_2(H_S) = Q_H$ by 1.1.3.1, and $S = C_T(Z_S) \in Syl_2(H_S)$. Recall also $T \leq H \leq IT$, so that $H = T(H \cap I) = TH_S$. As

V is the natural module for $\bar{L}$, $[S, V] = Z_S$; therefore $\tilde{V}$ is central in $\tilde{S} \in Syl_2(\tilde{H}_S)$, and hence $\tilde{U}_H = \langle \tilde{V}^H \rangle \in \mathcal{R}_2(\tilde{H}_S)$ by B.2.14. This establishes (2).

Next $C_H(\tilde{U}_H) \leq N_H(V) \leq M$, and further $X := O^2(C_{H_S}(\tilde{U}_H)) \leq C_M(V) \leq C_M(L/O_2(L))$, so that LT normalizes $O^2(XO_2(L)) = X$. Hence if $X \neq 1$, then $H \leq N_G(X) \leq M = !\mathcal{M}(LT)$, contradicting $H \not\leq M$. Therefore $X = 1$, so $C_{H_S}(\tilde{U}_H) \leq O_2(H_S) = Q_H$; then (3) follows from (2). Parts (4) and (5) follow from the fact that V is the natural module for $\bar{L}$. $\qquad \square$

Let $G_1 := LT$, $G_2 := H$, and $G_0 := \langle G_1, G_2 \rangle$. Notice Hypothesis F.7.6 is satisfied: in particular $O_2(G_0) = 1$ as $G_2 \not\leq M = !\mathcal{M}(G_1)$. Form the coset geometry $\Gamma := \Gamma(G_0; G_1, G_2)$ as in Definition F.7.2, and adopt the notation in section F.7. In particular for $i = 1, 2$ write γ_{i-1} for G_i regarded as a vertex of Γ, let $b := b(\Gamma, V)$, and pick $\gamma \in \Gamma$ with $d(\gamma_0, \gamma) = b$ and $V \not\leq G_\gamma^{(1)}$. Without loss, γ_1 is on the geodesic

$$\gamma_0, \gamma_1, \cdots, \gamma_b := \gamma.$$

Observe in particular that U_H plays the role played by "V_{γ_1}" in section F.7. For $\alpha := \gamma_0 x \in \Gamma_0$ let $V_\alpha := V^x$. For $\beta := \gamma_1 y \in \Gamma_1$ let $Z_\beta := Z_S^y$ and $U_\beta = U_H^y$.

Notice that by 6.1.17.1 and F.7.7.2, $V \leq Q_H \leq G_{\gamma_1}^{(1)}$, so that by F.7.9.3:

LEMMA 6.1.18. $b > 1$.

LEMMA 6.1.19. *Suppose there exists $H \in \mathcal{H}_S \cap \mathcal{H}_*(T, M)$ with b odd. Then $n = 2$ and $\langle V^I \rangle$ is nonabelian.*

PROOF. Assume b is odd. By 6.1.18, $b > 1$, so $b \geq 3$. Then U_H is elementary abelian by F.7.11.4.

Further by F.7.11.5, $U_H \leq G_\gamma$ and $U_\gamma \leq H$, so applying 6.1.12.3 to suitable Sylow 2-subgroups of G_γ and H, we obtain:

$$U_H \leq C_{G_\gamma}(Z_\gamma), \text{ and } U_\gamma \leq C_{G_{\gamma_1}}(Z_S) = H_S. \qquad (!)$$

Observe that the hypotheses of F.7.13 are satisfied: We just verified hypothesis (a) of F.7.13, and hypothesis (c) holds by 6.1.1.2 as $H \in \mathcal{H}_*(T, M)$. Also as $H \in \mathcal{H}_*(T, M)$, $H \cap M$ is the unique maximal subgroup of H containing T by 3.3.2.4. Finally $H \cap M = N_H(V)$ by 6.1.8, so hypothesis (b) of F.7.13 holds. Applying F.7.13 to $A := U_H$, we conclude there is $\alpha \in \Gamma(\gamma)$ with $B := N_A(V_\alpha)$ of index 2 in A. Write $E := V_\alpha$. If $[E, B] = 1$, then as $s(G, V) > 1$ by 6.1.10.2, for each $h \in H$

$$E \leq C_G(B) \leq C_G(B \cap V^h) \leq C_G(V^h).$$

But then $[E, A] = 1$, contrary to $B < A$. Therefore $[E, B] \neq 1$. So as $A \leq C_{G_\gamma}(Z_\gamma)$ by (!), $[E, B] = Z_\gamma$ by 6.1.17.4.

Suppose that $E \leq Q_H$. Then $[A, E] \leq Z_S$ by 6.1.17.2, so that $Z_\gamma = [B, E] \leq [A, E] \leq Z_S$. Hence $Z_S = Z_\gamma$, as these groups are conjugate and so have the same order. This is impossible, as $V \leq O_2(C_G(Z_S))$ by 6.1.17.2, while $V \not\leq O_2(G_\gamma)$ by choice of γ, and $G_\gamma \leq N_G(Z_\gamma)$.

Therefore $E \not\leq Q_H$, so since $U_\gamma \leq H_S$ by (!), also $E \not\leq O_2(H)$. But as $H \in \mathcal{H}_*(T, M)$, by 3.3.2.4 we may apply B.6.8.5 to conclude that $O_2(H) = O^{2'}(G_1^{(1)})$, so that $E \not\leq G_{\gamma_1}^{(1)}$. Thus $d(\alpha, \gamma_1) = b$ with $\alpha, \gamma, \cdots, \gamma_1$ a geodesic, so we have symmetry between γ and γ_1. Using this symmetry, and applying F.7.13 to E in the role of "A", we conclude there is $\delta \in \Gamma(\gamma_1)$ such that $F := N_E(V_\delta)$ is a hyperplane of E.

Then applying the subsequent arguments with F in the role of "B", $[F, V_\delta] = Z_S$ and $V_\delta \not\leq Q_\gamma$, so replacing γ_0 by δ, we may assume that $\delta = \gamma_0$ and $V_\delta = V$.

Let $V_B := V \cap B = N_V(E)$. Notice V_B is of index at most 2 in V, as B is of index 2 in U_H. Then $[V_B, F] \leq Z_S \cap Z_\gamma$, with V_B, F of index 2 in V, E. Therefore $[V_B, F]$ is of index at most 2 in Z_S and Z_γ by (!) and 6.1.17.4. Further if $[V_B, F] = Z_S$, then $Z_S = [V_B, F] = Z_\gamma$, which we saw earlier is not the case. Hence $[V_B, F]$ is of index 2 in both Z_S and Z_γ, so $|V : V_B| = 2$ by 6.1.17. Therefore by 6.1.17.5, $n = 2$, and $\langle z \rangle := [V_B, F] = Z_S \cap Z_\gamma$ is of order 2. Thus we have established the first assertion of 6.1.19.

As D_L is transitive on $Z_S^\#$, z is 2-central in LT, so we may assume $T \leq G_z$. Thus $H \leq G_z$. As D_L is transitive on $Z_S^\#$, and L is transitive on $V^\#$, we conclude from A.1.7.1 that $G_z := C_G(z)$ is transitive on the G-conjugates of Z_S and V containing z. Then $Z_\gamma = Z_S^g$ for $g \in G_z$. Similarly if $V \leq O_2(G_z)$, then $E \leq O_2(G_z)$ as $E \in V^{G_z}$; but then $E \leq O_2(G_z) \leq O_2(H)$, contrary to an earlier reduction. We conclude $V \not\leq O_2(G_z)$.

Let $W_0 := \langle V^I \rangle$; to complete the proof, we assume W_0 is abelian and it remains to derive a contradiction. Let $Q_z := \langle Z_S^{G_z} \rangle$. By 1.1.4.6, $F^*(G_z) = O_2(G_z)$. As $n = 2$, $[Z_S, T] \leq \langle z \rangle$, so $Q_z \leq O_2(G_z)$ by B.2.14 applied in $\hat{G}_z := G_z/\langle z \rangle$, and hence $Q_z \leq T$. Let $W := W_0 \cap Q_z$; as $I \leq G_z$, $I \leq N_G(W)$. Set $I^* := I/C_I(\hat{W})$. Now $Q_z \leq T \leq N_G(V)$, so $Q_z \leq \ker_{G_z}(N_{G_z}(V))$. Then as $E \in V^{G_z}$, Q_z acts on E, and in particular W acts on E. We have seen that $E \leq I \leq N_G(W)$, so that $[W, E] \leq W \cap E$. Next as $V_B \leq W_0$, $[V_B, E] \leq W_0$. But $Z_\gamma \leq Q_z$ as $Z_\gamma \in Z_S^{G_z}$, and $Z_\gamma = [V_B, E]$ by (!) and 6.1.17.4, so $Z_\gamma \leq W \cap E$. Finally if $Z_\gamma < E \cap W$, then $m(E/(E \cap W)) \leq 1$ since $n = 2$. Then as $V \leq W_0$ and W_0 is abelian by assumption, $V \leq C_G(E \cap W) \leq C_G(E)$ by 6.1.10.2, contrary to $[V_B, F] = \langle z \rangle$. Thus $[E, W] \leq E \cap W = Z_\gamma$, so $[E^*, \hat{W}] \leq \hat{Z}_\gamma$ of order 2, and hence E^* is trivial or induces a group of transvections on $\hat{W}$ with center $\hat{Z}_\gamma = \hat{Z}_S^g$.

Note that $C_I(\hat{W}) \leq N_G(Z_S^g) \leq N_G(O_2(I^g))$, so that

$$O_2(I^g) \cap C_I(\hat{W}) \leq O_2(C_I(\hat{W})) \leq O_2(I). \qquad (*)$$

Then as $E \leq U_\gamma \leq O_2(I^g)$, but we saw $E \not\leq O_2(H)$, we conclude from (*) that E does not centralize $\hat{W}$, so that $E^* \neq 1$. As W_0 is abelian, $Z_\gamma \leq C_I(\tilde{W})$, so we conclude $1 \leq m(E^*) \leq m(E/Z_\gamma) = n = 2$.

Let $P := \langle E^I \rangle$. As E centralizes Z_S but $N_E(V) = F < E$, $P \not\leq M$ by 6.1.7.1. As $E \leq O_2(I^g)$ and we saw $C_I(\hat{W})$ acts on $O_2(I^g)$, it follows from (*) that

$$[E, C_I(\hat{W})] \leq O_2(I^g) \cap C_I(\hat{W}) \leq O_2(I),$$

so we conclude that $C_P(\hat{W}) \leq O_{2,Z}(P)$. Let P_0 denote the preimage in P of $O_2(P^*)$. Then $P_0 \leq O_{2,Z,2}(P) = O_{2,Z}(P)$, so that $P_0 = O_2(P)C_P(\hat{W})$, and hence $O_2(P^*) = O_2(P)^*$. On the other hand, by 6.1.17.2, $O_2(P) \leq O_2(I) \leq C_I(\hat{W}_0) \leq C_I(\hat{W})$, so $O_2(P^*) = O_2(P)^* = 1$, and then $\hat{W} \in \mathcal{R}_2(P^*)$. Thus as E^* induces a group of transvections on $\hat{W}$ with center $\hat{Z}_\gamma$ of order 2, we see from G.6.4 that P^* is the direct product of subgroups X_i^* isomorphic to S_m or $L_k(2)$ for suitable m and k. So either $X_i^* \cong L_2(2) \cong S_3$, or X_i^* is nonsolvable, in which case as the preimage X_i is normal in P and P is subnormal in $N_G(Z_S)$, $X_i^\infty \in \mathcal{C}(N_G(Z_S))$. In that case, as $D_L \cong \mathbf{Z}_3$ and $D_L \cap I = 1$, we conclude from A.3.18 that $m_3(X_i^*) = 1$. Therefore

$X_i^* \cong S_3$, S_5 or $L_3(2)$. In particular now $O_{2,Z}(P) = O_2(P)$ as the multiplier of these groups is a 2-group. Thus $P^* = P/O_2(P)$.

Note that $O^2(I) \leq N_I(X_i^*)$ by G.6.4.3. Next if X_i^* is not S_3, then D_L normalizes $O^2(X_i) = X_i^\infty$ by 1.2.1.3. On the other hand, if $X_i^* \cong S_3$, then for $d \in D_L$, $O^2(X_i)^d \leq O^2(I) \leq N_I(X_i^*)$. Then recalling that $m_3(I) \leq 2$, either $O^2(X_i) = O^2(X_i)^d$, or else $X_iO^2(X_i^d)/O_2(X_iO^2(X_i)^d) \cong S_3 \times \mathbf{Z}_3$ and $O^2(X_i)O^2(X_i)^d = O^{3'}(I) =: J$. In the latter case, $I/C_I(J/O_2(J)) \cong S_3 \times S_3$ or S_3 wr $\mathbf{Z}_2$, whose outer automorphism groups are 2-groups, so the former must hold. Thus in any case, D_L and $O^2(I)$ act on each X_i. So as $m_3(ID_L) \leq 2$, $P = X_1$ and $O^2(P) = O^{3'}(I)$. If $P^* \cong S_5$, then the T-invariant Borel subgroup of P is not contained in M—for otherwise, $TP \in \mathcal{H}_*(T, M)$ with $n(PT) > 1$, contrary to 6.1.1.2. If P^* is $L_3(2)$ then T induces inner automorphisms on P^* by G.6.4.2a. Thus in each case there exists a TD_L-invariant parabolic subgroup P_1 of P, with $P_1 \not\leq M$ and $TP_1/O_2(P_1) \cong S_3$. Then $\theta := (LT, D_LT, P_1D_LT)$ satisfies Hypothesis F.1.1, and so by F.1.9 defines a weak BN-pair. Moreover the hypotheses of F.1.12 are satisfied by P_1D_LT, so that θ is described in one of the cases of F.1.12.I. Since $L/O_2(L) \cong L_2(4)$ and $P_1/O_2(P_1) \cong L_2(2)$, the only possibility there is the $U_4(2)$-amalgam, which cannot occur here, since in that amalgam V is the A_5-module for $L/O_2(L)$. This contradiction completes the proof of 6.1.19. $\qquad \square$

Let $U := \langle V^I \rangle$ and recall $\tilde{H} = H/Z_S$.

LEMMA 6.1.20. *(1)* $U \leq O_2(I)$ and $\tilde{U} \leq Z(O_2(\tilde{I}))$.
(2) U is nonabelian.
(3) For $x \in U - Z(U)$, $[U, x] = Z_S$.
(4) $U/C_U(V) \cong E_{2^n}$. Further for $g \in I$ with $[V, V^g] \neq 1$, $U = VV^gC_U(VV^g)$, and $\{V, V^g\}$ is the set of maximal elementary abelian subgroups of VV^g.

PROOF. Pick $H \in \mathcal{H}_*(T, M)$. If b is odd, then (2) holds by 6.1.19. On the other hand, if b is even, then $1 \neq [V, V_\gamma] \leq V \cap V_\gamma$ by F.7.11.2, so that $V_\gamma \leq N_G(V) \leq M$, and we may take $V_\gamma \leq T$. Then by 6.1.11, $Z_S = [V, V_\gamma]$ and $V_\gamma \in V^I$. So (2) is established in this case also.

Part (1) folows from 6.1.17.2 applied to IT in the role of "H". For $x \in U - Z(U)$, x does not centralize all I-conjugates of V; so replacing x by a suitable I-conjugate, we may assume $[x, V] \neq 1$. Then as $x \in O_2(I) \leq S$, $[x, V] = Z_S$ by 6.1.17.4, so (3) holds. By (2) we may choose $g \in I$ with $[V, V^g] \neq 1$; by (1), $V^g \leq N_S(V)$. Then by 6.1.10, $m(V^g/C_{V^g}(V)) = n = m(S/C_S(V))$, so $S = V^gC_S(V)$, and hence also $U = V^gC_U(V)$. Then we conclude that (4) holds from the symmetry between V and V^g. $\qquad \square$

For the remainder of the section, we choose $H := N_G(Z_S)$; in contrast to our earlier convention, this "H" is not in $\mathcal{H}_S$. We also pick $g \in I$ with $[V, V^g] \neq 1$; such a g exists by 6.1.20.2. As $N_L(Z_S)$ is irreducible on V/Z_S, Hypothesis G.2.1 is satisfied with Z_S in the role of "V_1". Recall from section G.2 that the condition U nonabelian in 6.1.20.2 is equivalent to $\tilde{U} \neq 1$. Thus we have the hypotheses of G.2.3, so we can appeal to that lemma.

LEMMA 6.1.21. *Let* $l \in L - H$, *and set* $L_1 := \langle U, U^l \rangle$, $R := O_2(L_1)$, *and* $E := U \cap U^l$. *Then*
(1) $L_1 = \langle U^{M_V} \rangle \trianglelefteq M_V$ *and* $L_1 = LU$.
(2) $R = C_U(V)C_{U^l}(V)$ *and* $UR \in Syl_2(L_1)$.

(3) $\Phi(E) = 1$, $E/V \le Z(L_1/V)$, *and* $E = \ker_U(M_V) \trianglelefteq M_V$.

(4) $\Phi(R) \le E$, *and* $R/E = C_U(V)/E \times C_{U^l}(V)/E$ *is the sum of natural modules for* L_1/R *with* $C_U(V)/E = C_{R/E}(U)$.

(5) $M_V \le N_G(R)$; *in particular*, $R \le O_2(M_V)$.

PROOF. As we just observed, we may apply G.2.3 with Z_S in the role of "V_1"; in that application, L_1, R, E play the roles of "I, S, S_2".

Now $L_1 = LU$ by G.2.3.2 and $LU = LO_2(LU)$ by G.2.3.1, so $O_2(L_1) = L_1 \cap O_2(LT)$. Hence $U \cap O_2(L_1) = C_U(V)$, so that $C_U(V)$ plays the role of "W". Then (2) follows from parts (3) and (1) of G.2.3, while (4) follows from G.2.3.6. By G.2.3.5, $E/V \le Z(L_1/V)$ and $\Phi(E) = 1$. Thus it remains to establish the first statement of (1), the last statement of (3), and (5).

Now $U = \langle V^{C_G(Z_S)} \rangle$, so as $N_G(Z_S) = N_{M_V}(Z_S)C_G(Z_S)$ by 6.1.9.4, $N_G(Z_S)$ acts on U. Next $Z_S^{M_V} = Z_S^L$, so that $M_V = N_{M_V}(Z_S)L \le N_{M_V}(U)L$. Then as $L_1 = LU$, $L_1 \trianglelefteq M_V$, completing the proof of (1). Similarly $\ker_U(M_V) = \ker_U(L_1) \le U \cap U^l = E$ and $E \trianglelefteq L_1$ by G.2.3.4, so $E = \ker_U(L_1)$, completing the proof of (3). Finally (5) holds as $R = O_2(L_1)$ and $L_1 \trianglelefteq M_V$ by (1). $\quad\square$

During the remainder of the section, R and E are as defined in 6.1.21.

LEMMA 6.1.22. $E < R$.

PROOF. Assume that $R = E$. In particular $R \le U$, and hence $R = C_U(V)$ by 6.1.21.2. By 6.1.20.2, we may choose $g \in I$ with $[V, V^g] \ne 1$; then $U = VV^gC_U(VV^g)$ by 6.1.20.4. Also $C_U(VV^g) = C_R(V^g) = C_E(V^g)$. By 6.1.21.3, $\Phi(E) = 1$, while by 6.1.20.4, the maximal elementary abelian subgroups of VV^g are V and V^g, so the maximal elementary abelian subgroups of U are $R = C_U(V)$ and $R^g = C_U(V^g)$. By 6.1.21.5, LT acts on R, so T normalizes both members of $\mathcal{A}(U)$, and hence both R and R^g are normal in $O^2(I)C_T(Z_S) = I$. But then $I \le N_G(R) \le M = !\mathcal{M}(LT)$, contradicting Hypothesis 6.1.16. This completes the proof. $\quad\square$

LEMMA 6.1.23. *If* $S_0 \le S$ *with* $RU \le S_0$, *then* $N_G(S_0) \le M$.

PROOF. By 6.1.21, RU is Sylow in $L_1 = LU$, so that $S_0 \cap L$ is Sylow in L. Thus the assertion follows from Theorem 4.3.17. $\quad\square$

Recall $H = N_G(Z_S)$. Let $H^* := H/C_H(\tilde{U})$ and set $q := 2^n$. By 6.1.21.4 and 6.1.22, R/E is the sum of $s \ge 1$ natural modules for $L_1/R \cong L_2(q)$.

LEMMA 6.1.24. *(1)* $C_U(V) = C_R(\tilde{U})$.

(2) $R^* \cong E_{q^s}$, *and* $R^* = [R^*, D]$ *for each* $1 \ne D \le D_L$.

(3) $[R^*, F(I^*)] = 1$.

(4) $O_2(I^*) = 1$.

PROOF. By 6.1.20.4, $U = V^gC_U(V)$. Also by 6.1.21.4, $C_{R/E}(U) = C_U(V)/E$, so that $[U, r] \not\le E$ for $r \in R - C_U(V)$; as $\tilde{U}$ is abelian by 6.1.17.2, we conclude (1) holds. By 6.1.21.4, $R/E \cong E_{q^{2s}}$ is the sum of s natural modules for L_1/R with $C_U(V)/E$ the centralizer in R/E of U, so

$$R^* = C_{U^l}(V)^* \cong C_{U^l}(V)/E = [R^*, D] \cong E_{q^s}$$

for each $1 \ne D \le D_L$. That is, (2) holds.

By 6.1.17.2, $\tilde{U} \in \mathcal{R}_2(I)$. Hence $O_2(I^*) = 1$, which proves (4), and also shows that $F(I^*) \le O(I^*)$. Then as $R^* = [R^*, D_L]$ by (2), (3) follows from A.1.26. $\quad\square$

By 6.1.24.2, $R^* \neq 1$ as $s \geq 1$. By 6.1.24.4, R^* is faithful on $F^*(I^*)$. Thus by 6.1.24.3, R^* is faithful on $E(I^*)$, so there is $K \in \mathcal{C}(I)$ with $K/O_2(K)$ quasisimple and $[K^*, R^*] \neq 1$. As $|K^H| \leq 2$ by 1.2.1.3, D_L acts on K; further $D_L \cap I = 1$. So as $R^* = [R^*, D_L]$ by 6.1.24.2, R also acts on each member of K^H, and hence $[K^*, R^*] = K^*$. Let $M_K := M \cap K$, and $S_K := S \cap K$; then $S_K \in Syl_2(K)$ as $S \in Syl_2(I)$.

We claim that $K \not\leq M$, so that $M_K^* < K^*$ as $C_H(\tilde{U}) \leq N_G(V) \leq M$: For otherwise $K \leq C_M(Z_S) \leq M_V \leq N_G(R)$ using 6.1.7.1 and 6.1.21.5, contradicting $[K^*, R^*] = K^*$.

LEMMA 6.1.25. (1) $n = 2$.
(2) $K^* \cong L_2(p)$, $p \equiv \pm 3 \mod 8$, $p \geq 11$.
(3) $s = 1$, so that R/E is the natural module for L_1/R.

PROOF. First D_L normalizes $S \in Syl_2(I)$, and hence also normalizes $S_K^* \in Syl_2(K^*)$. If $D_K := C_{D_L}(K^*) \neq 1$, then as we saw R^* acts on K^*, $R^* = [R^*, D_K] \leq C_{I^*}(K^*)$ by 6.1.24.2, contrary to the choice of K. Thus D_L is faithful on K^*. Therefore either

(A) D_L is a 3-group, and hence of order 3 with $n = 2$, or
(B) $K^*/Z(K^*)$ is described in A.3.15 with $Z(K^*)$ of odd order by 6.1.24.4.

Assume for the moment that (B) holds. As D_L acts on S_K^*, it follows from A.3.15 that one of the following holds:

(a) K^* is of Lie type and characteristic 2.
(b) K^* is J_1 and $n = 3$ as D_L has order 7.
(c) K^* is $(S)L_3^\epsilon(p)$ and $D_L^* \cap K^* = 1$.

However in case (c), using the description in A.3.15.3, D_L centralizes S_K^*. As $R^* = [R^*, D_L]$ and $Out(K^*) \cong S_3$, R^* induces inner automorphisms on K^*, impossible as $1 \neq R^* = [R^*, D_L]$ and D_L centralizes S_K^*. This eliminates case (c).

Now assume for the moment that (A) holds. We check the list of Theorem C (A.2.3) for groups $K^*/Z(K^*)$ in which the normalizer of S_K^* in $Aut(K^*/Z(K^*))$ contains a subgroup of order 3, and conclude that either K^* is of Lie type and characteristic 2, or K^* is $L_2(p)$ with $p \equiv \pm 3 \mod 8$ or J_2. The case where $K^* \cong J_2$ is ruled out by A.3.18 as $D_L \cap I = 1$.

Next suppose (A) or (B) holds and K^* is of Lie type over $\mathbf{F}_{2^k}$. Then as D_L acts on S_K^*, either $k > 1$, or K^* is $^3D_4(2)$ and D_L is of order 7—so that $n = 3$. In any case, D_L acts on a Borel subgroup B^* of K^* containing S_K^*. Further either K^* is of Lie rank 1, in which case we set $K_1 := K$, or K^* is of Lie rank 2. In the latter case, as $K \not\leq M$, either

(i) $D_L T$ acts on a maximal parabolic P^* of K with preimage P satisfying $K_1 := O^{2'}(P) \not\leq M$, or
(ii) K^* is $Sp_4(2^k)$ or $(S)L_3(2^k)$ and T is nontrivial on the Dynkin diagram of K^*, and we set $K_1 := K$.

In any case, $K_1 \not\leq M$.

Suppose first that $B \leq M$. Then $H_2 := \langle K_1, T \rangle \in \mathcal{H}_*(T, M)$ with $n(H_2) > 1$—unless possibly $K^* \cong {}^3D_4(2)$ with $n = 3$, and K_1 is solvable. In the former case, Hypothesis 6.1.1 is contradicted. In the latter case, our usual argument with the Green Book [DGS85] supplies a contradiction: That is, just as in the proofs of 6.1.5 and 6.1.19, $\alpha := (LT, D_L T, D_L H_2)$ satisfies Hypothesis F.1.1, so that α is a weak BN-pair by F.1.9. Also $D_L H_2$ satisfies the hypothesis of F.1.12, so α must be in the list of F.1.12. As $n = 3$ and $k = 1$, the only possibility is the $^3D_4(2)$ amalgam of F.1.12.I.4. However, in that case Z is central in the parabolic L_1 with $L_1/O_2(L_1) \cong L_2(8)$, contradicting V the natural module for $L/O_2(L) \cong L_2(8)$.

This contradiction shows that $B \not\leq M$. In particular K^* is not $^3D_4(2)$, so $K_1 \in \mathcal{L}(G,T)$. Next as $R^* = [R^*, D_L]$, and $Out(K^*)$ is 2-nilpotent for each K^*, R^* induces inner automorphisms on K^*, so that $R^* \leq O_2(B^*R^*) := C^*$. Then $RU \leq S_0 := S \cap C \in Syl_2(C)$, and as $K_1/O_2(K_1)$ is quasisimple, $S_0 = O_2(C)$. However $N_G(S_0) \leq M$ by 6.1.23, contradicting $B \not\leq M$.

This contradiction shows K^* is not of Lie type and characteristic 2. Thus by our earlier discussion, either $n = 2$ and $K^* \cong L_2(p)$ for $p \equiv \pm 3 \mod 8$ or J_1, or $n = 3$ and $K^* \cong J_1$. In each case as $R^* = [R^*, D_L]$, $R^* \leq O_2(N_{K^*}(S_K^*)R^*) := C^*$; then the argument of the previous paragraph shows $N_{K^*}(S_K^*) \leq M_K^*$.

Suppose $K^* \cong J_1$. Then $N_{K^*}(S_K^*) \cong Frob_{21}/E_8$ is maximal in K^*, so $M_K^* = N_{K^*}(S_K^*)$. Now $D_L T_L \trianglelefteq M_K$, so we conclude D_L is of order 7 rather than 3, and $D_L \leq [D_L, M_K] \leq K \leq C_G(Z_S)$—impossible, as $[Z_S, D_L] = Z_S$.

Therefore $K^* \cong L_2(p)$ with $p \equiv \pm 3 \mod 8$ and $n = 2$. As K^* is not $L_2(4)$ by an earlier reduction, $p \geq 11$. Therefore (1) and (2) are established.

As $n = 2$, D_L is of order 3, so as $m_3(D_L I) \leq 2$, $m_3(I) = 1$ and hence $K = O^{3'}(I)$. As D_L is not inverted in $D_L S$ and D_L is faithful on K^*, S induces inner automorphisms on K^*. As $K = O^{3'}(I)$, if $K_0 \in \mathcal{C}(I)$ with $K_0 \neq K$, then $K_0/O_2(K_0) \cong Sz(2^k)$. As $D_L = O^2(D_L)$, D_L acts on each member of K_0^I by 1.2.1.3, and hence so does $R^* = [R^*, D_L]$. The case $[R^*, K_0^*] \neq 1$ was eliminated in our earlier treatment of groups of Lie type in characteristic 2. Therefore R^* centralizes K_0^{*I}, so R^* centralizes $C_{F^*(I^*)}(K^*)$ in view of 6.1.24.3. Recall S^* induces inner automorphisms on K^*, so as $O_2(I^*) = 1$ by 6.1.24.4, we conclude $R^* \leq K^*$. Thus $R^* \leq S_K^*$, so as $R^* = [R^*, D_L]$, we conclude $R^* = S_K^*$. In particular R^* is of order 4, so by 6.1.24.2, $s = 1$ and hence (3) holds.   □

LEMMA 6.1.26. *If there exists $e \in E - V$, then:*

(1) R *is transitive on* eV.

(2) $|E : V| \leq 4$.

PROOF. Set $L_0 := \langle V^g, V^{gl} \rangle$, where $l \in L$ is as in 6.1.21. Then $\bar{V}^g = \bar{U}$ by 6.1.20.4, and so $\bar{V}^{gl} = \bar{U}^l$. Therefore by 6.1.21.1, $\bar{L} = \bar{L}_1 = \bar{L}_0$ and $L \leq L_1 = L_0 R$. By 6.1.20.4, $m(U/C_U(V^g)) = 2$, so $m(E/C_E(V^g)) \leq 2 = m(Z_S)$. Then as $C_{Z_S}(V^{gl}) = 1 = C_{Z_S^l}(V^g)$ and L acts on E by 6.1.21.3,

$$E = Z_S C_E(V^{gl}) = Z_S^l C_E(V^g),$$

so that $E = V C_E(L_0)$.

If $E = V$ then the lemma is trivial, so assume $e \in E - V$. As $E = V C_E(L_0)$ there is $f \in eV \cap C_E(L_0)$. If $[R, f] = 1$, then f is centralized by R and L_0, so $L \leq L_0 R \leq C_G(f)$, a contradiction as $C_T(L) = 1$ by 6.1.6.1. This contradiction shows $[R, f] \neq 1$. But by 6.1.21.3, $[R, f] \leq V$, so as L_0 is irreducible on V, $[R, f] = V$. Therefore (1) holds, and we may take $e = f \in C_E(L_0) =: F$. Now $V^g E \leq C_U(F)$ and $\bar{V}^g = \bar{U}$, so $|U : C_U(F)| \leq |U : V^g E| \leq |U \cap R : E|$. But $n = 2$ by 6.1.25.1, and R/E is the natural module for L_1/R by 6.1.25.3, so we conclude $|U : C_U(F)| \leq 4$. We saw R does not centralize f, so as L_0 centralizes F and acts irreducibly on R/E, $[U \cap R, F] \neq 1$. Thus there is $u \in (U \cap R) - C_U(F)$, and for each such u, $[F, u] \leq Z_S$ by 6.1.17.2. Then $|F/C_F(u)| \leq |Z_S| = 4$ by Exercise 4.2.2

in [**Asc86a**]. Therefore to prove (2), it remains to show $F_u := C_F(u) = 1$—since this shows $|F| \leq 4$, and we saw earlier that $E = VF$.

As $\bar{L} = \bar{L}_0$, we may take $D_L \leq L_0 \leq C_G(F)$, so $D_L \leq C_L(F_u)$. Thus as D_L is irreducible on $(U \cap R)/E$ and $u \in (U \cap R) - E$, $U \cap R$ centralizes F_u. Then $R \leq \langle U^{L_0} \rangle \leq C_G(F_u)$, so $L \leq L_0 R \leq C_G(F_u)$, and hence $F_u = 1$ by 6.1.6.1, as desired. $\square$

We now complete this section by eliminating case (1) of 6.1.15—hence reducing Hypothesis 6.1.1 to the case leading to M_{22} in the following chapter:

THEOREM 6.1.27. *Assume Hypothesis 6.1.1 and set* $V_L := [V, L]$. *Then*

(1) $n = 2$.
(2) V_L *is the natural module for* $L/O_2(L) \cong L_2(4)$ *and* $C_T(L) = 1$.
(3) Let $Z_S := C_{V_L}(T_L)$. *Then* $C_G(Z_S) \leq M$.
(4) Either $N_G(W_0(T, V_L)) \not\leq M$ *or* $W_1(T, V_L) \not\leq C_T(Z_S)$.

PROOF. By 6.1.6.2, V_L is the natural module for $L/O_2(L) \cong L_2(2^n)$, and $C_T(L) = 1$ by 6.1.6.1. Thus to complete the proof of (2), it suffices to prove (1).

As the statements in Theorem 6.1.27 concerning V are about V_L, we may as well assume $V = V_L$, so that we may apply the results following 6.1.6, which depend upon that assumption.

Suppose first that $C_G(Z_S) \leq M$. Then (3) holds and we are in case (2) of 6.1.15, so (1) and (4) also hold. Therefore as (1) implies (2), Theorem 6.1.27 holds in this case.

Therefore we may assume that $C_G(Z_S) \not\leq M$, so that Hypothesis 6.1.16 is satisfied. Thus we can apply the lemmas in this subsection, which assume Hypothesis 6.1.16. We will derive a contradiction to complete the proof of the Theorem.

First $n = 2$ by 6.1.25.1, so $|U : C_U(V)| = 4$ by 6.1.20.4. Then by 6.1.21.4 and 6.1.25.3, $|C_U(V)/E| = 4$. Finally V is of order 16, and $|E : V| \leq 4$ by 6.1.26.2, so we conclude $|U| \leq 4^5$. Hence $m(\tilde{U}) \leq 8$.

Let W be a noncentral chief factor for K on $\tilde{U}$. By 6.1.25.2, for each extension field F of $\mathbf{F}_2$, the minimal dimension of a faithful FK^*-module is $(p-1)/2$. Hence as $m(\tilde{U}) \leq 8$, $p \leq 17$, so $p = 11$ or 13 by 6.1.25.2. But then $p - 1$ is the minimal dimension of a nontrivial $\mathbf{F}_2\mathbf{Z}_p$-module, so we have a contradiction to $m(\tilde{U}) \leq 8$. This contradiction completes the proof of Theorem 6.1.27. $\square$

6.2. Identifying M_{22} via $L_2(4)$ on the natural module

In this section, we complete the treatment of groups satisfying Hypothesis 6.1.1, by showing in Theorem 6.2.19 that M_{22} is the only group satisfying the conditions established in Theorem 6.1.27. Then applying results in chapter 5, the treatment of those groups containing a T-invariant $L \in \mathcal{L}_f^*(G, T)$ with $L/O_2(L) \cong L_2(2^n)$ is reduced in Theorem 6.2.20 to the case where $n = 2$ and V is the sum of at most two orthogonal modules for $L/O_2(L)$ regarded as $\Omega_4^-(2)$. We treat that final case in Part F2, which is devoted to the groups containing $L \in \mathcal{L}_f^*(G, T)$ with $L/O_2(L)$ a group over $\mathbf{F}_2$.

So in this section, we continue to assume Hypothesis 6.1.1, and as in section 6.1, we let $Z_S := C_V(T \cap L)$, $V_L := [V, L]$, and $S := C_T(Z_S)$. As usual, Z denotes $\Omega_1(Z(T))$. By Theorem 6.1.27, $n = 2$, and by 6.1.6, $C_Z(L) = 1$ and V_L is the natural module for $L/O_2(L) \cong L_2(4)$. Applying these observations to $R_2(LT)$ in the role of V, $Z \leq V_L$. Further replacing V by V_L if necessary, we may assume V *is* the natural module.

By Theorem 6.1.27, $C_G(Z_S) \leq M$, so by 6.1.7.1, $C_G(Z_S) \leq M_V := N_M(V)$; hence by 6.1.9.5:

LEMMA 6.2.1. $N_G(Z_S) \leq N_G(V) \leq M$.

Observe that Z_S is the T-invariant 1-dimensional $\mathbf{F}_4$-subspace of V regarded as a 2-dimensional $\mathbf{F}_4$-space. Let $\bar{M}_V := M_V/C_M(V)$.

LEMMA 6.2.2. *(1)* $\bar{L}\bar{T} \cong S_5$.
(2) Z is of order 2.
(3) $C_G(Z) \not\leq M$.

PROOF. Part (3) follows from 6.1.5. Recall $Z \leq V$, so if $\bar{L}\bar{T} \cong A_5$, then $Z_S = C_V(T) = Z$, and (3) contradicts 6.2.1. Hence (1) holds and $Z = C_V(T)$ is of order 2 by (1), establishing (2). $\qquad\square$

LEMMA 6.2.3. *If* $g \in G$ *with* $V \leq N_G(V^g)$ *and* $V^g \leq N_G(V)$, *then* $[V, V^g] = 1$.

PROOF. If $[V, V^g] \neq 1$, then 6.1.11 says $V^g \in V^{C_G(Z_S)}$. But $C_G(Z_S) \leq N_G(V)$ by 6.2.1, contradicting our assumption that $1 \neq [V, V^g]$. $\qquad\square$

LEMMA 6.2.4. *Assume* $U \leq V$ *with* $m(V/U) = 2$ *and* $H := C_G(U) \not\leq N_G(V)$. *Choose notation so that* $T_U := N_T(U) \in Syl_2(N_M(U))$, *and let* $Q := C_T(V)$, $L_U := O^2(N_L(U))$, $U_H := \langle V^H \rangle$, $\tilde{H} := H/U$, *and* $H^* := H/C_H(\tilde{U}_H)$. *Then*

(1) $U = C_V(t)$ *for some* $t \in T$ *inducing a field automorphism of order 2 on* $\bar{L}$.
(2) $F^*(H) = O_2(H)$, $R := Q\langle t \rangle \in Syl_2(H)$, $N_G(R) \leq N_G(J(R)) \leq M$, $T_U \in Syl_2(N_G(U))$, *and* $|T : T_U| = 2$.
(3) $W_0(R, V) \leq Q$.
(4) U_H *is elementary abelian,* $\tilde{U}_H \leq Z(O_2(\tilde{H}))$, *and* $C_H(\tilde{U}_H) = O_2(H)$, *so* $\tilde{U}_H \in \mathcal{R}_2(\tilde{H})$.
(5) $L_U/O_2(L_U) \cong \mathbf{Z}_3$ *with* $O_2(L_U) = L_U \cap H$.
(6) There is at most one $K \in \mathcal{C}(H)$ *of order divisible by 3, and if such a* K *exists then either*

(i) $K = O^{3'}(H)$ *and* $m_3(K) = 1$, *or*
(ii) $K/O_2(K) \cong (S)L_3^\epsilon(q)$, *and a subgroup of order 3 in* L_U *induces a diagonal automorphism on* $K/O_2(K)$.

PROOF. Observe by 6.1.8 that as $H = C_G(U)$, $H \cap M = N_H(V)$, so that our hypothesis $H \not\leq N_G(V)$ is equivalent to $H \not\leq M$. As $C_G(Z_S) \leq M$, case (3) of 6.1.13 must hold, proving (1). Next by (1), $|T : T_U| = 2$, and the remaining statements of (2)–(4) follow from 6.1.14, except for the inclusion $O_2(H) \geq C_H(\tilde{U}_H)$ in part (4). Part (5) follows from (1), and (6) follows from A.3.18 in view of (5).

Finally $C_H(\tilde{U}_H) \leq N_G(V) \leq M$, and by Coprime Action, $Y := O^2(C_H(\tilde{U}_H)) \leq C_M(V) \leq C_M(L/O_2(L))$. Thus LT normalizes $O^2(YO_2(L)) = Y$. Therefore if $Y \neq 1$ then $H \leq N_G(Y) \leq M = !\mathcal{M}(LT)$, contradicting our initial observation that $H \not\leq M$. Thus $C_H(\tilde{U}_H)$ is a 2-group, completing the proof of (4), and hence also the proof of 6.2.4. $\qquad\square$

Define a 4-subgroup F of V^g to be of *central type* if F is centralized by a Sylow 2-subgroup of L^g; of *field type* if F is centralized by an element of M_V^g inducing a field automorphism on $L^g/O_2(L^g)$ and V^g; and of *type 3* if F is of neither of the first two types. By 6.2.2, there exist 4-subgroups of V of field type, and of course Z_S is of central type.

LEMMA 6.2.5. *(1) Let ξ denote the number of orbits of M_V on the 4-subgroups of V of type 3. Then $\xi = 0$ or 1 for $|\bar{M}_V : \bar{L}| = 6$ or 2, respectively. The orbits are of length 0 or 20, respectively.*

(2) M_V is transitive on 4-subgroups of V of each type.

(3) For each 4-subgroup U of V, $U^G \cap V = U^{LT}$.

(4) V is the unique member of V^G containing Z_S.

(5) If $g \in G$ with $V \cap V^g$ noncyclic, then $[V, V^g] = 1$.

(6) V is the unique member of V^G containing any hyperplane of V.

PROOF. As V is the natural module for $\bar{L}$, $\bar{L}$ preserves an $\mathbf{F}_4$-space structure $V_{\mathbf{F}_4}$ on V, in which the central 4-subgroups are the five 1-dimensional subspaces of $V_{\mathbf{F}_4}$, and $\bar{M}_V \leq Aut_{GL(V)}(\bar{L}) = \Gamma L(V_{\mathbf{F}_4})$. In particular, L is transitive on 4-subgroups of central type, and there are 30 4-subgroups not of central type, which form an orbit under $Aut_{GL(V)}(\bar{L})$. This orbit splits into three orbits of length 10 under $\bar{L}$, and $Aut_{GL(V)}(\bar{L})$ induces S_3 on this set of orbits. By 6.2.2, $\bar{M}_V \cong S_5$ or $\Gamma L_2(4)$, so it follows that (1) and (2) hold.

By 6.2.4, we can choose a representative U for each orbit so that $N_T(U) \in Syl_2(N_G(U))$. Now $T = N_T(U)$ iff U is of central type, so groups of central type are not fused to groups of field type or type 3. Similarly if $N_T(U) < T$, then $|T : N_T(U)| = 2$ or 4 for U of field type, or type 3, respectively, so distinct M-orbits are not fused in G. Thus (3) holds.

By (3) and A.1.7.1, $N_G(Z_S)$ is transitive on G-conjugates of V containing Z_S; then as $N_G(Z_S) \leq N_G(V)$ by 6.2.1, (4) holds. As V is a self-dual $\mathbf{F}_2 L$-module and L is transitive on $V^\#$, L is transitive on hyperplanes of V, so (4) implies (6).

Assume the hypotheses of (5), and let U be a 4-subgroup of $V \cap V^g$; then by (3) and A.1.7.1, $N_G(U)$ is transitive on G-conjugates of V containing U. Furthermore for U of each type, $Aut_G(U) \cong S_3 \cong Aut_{M_V}(U)$, so that $N_G(U) = C_G(U)N_{M_V}(U)$; we conclude that $C_G(U)$ is transitive on the G-conjugates of V containing U. Thus if $C_G(U) \leq N_G(V)$, then $V = V^g$ and (5) is trivial. If $C_G(U) \not\leq N_G(V)$, then U is of field type by 6.2.4.1, so $\langle V, V^g \rangle$ is abelian by 6.2.4.2, completing the proof of (5). $\square$

LEMMA 6.2.6. *Assume $A := V^g \cap N_G(V)$ and $U := V \cap N_G(V^g)$ are of index 2 in V^g and V, respectively. Then either*

(1) $\bar{A}$ and $U/C_U(V^g)$ are of order 2, $C_A(V)$ and $C_U(V^g)$ are of field type, and $\langle V, V^g \rangle$ is a 2-group, or

(2) $\bar{A} \cong E_4$, $\bar{A} \not\leq \bar{L}$, $Y := \langle V, V^g \rangle \cong S_3/Q_8^2$, $V \cap V^g = [A, U]$ is of order 2, and $O_2(Y) \leq O^2(Y)$.

PROOF. Without loss, we may assume $A \leq T$. First $B := [A, U] \leq A \cap U$, so $B \neq Z_S$ by 6.2.5.4, and hence $\bar{A} \not\leq Syl_2(\bar{L})$. Also $\bar{A} \neq 1$, as otherwise $V \leq C_G(A) \leq N_G(V^g)$ by 6.2.5.6, contrary to hypothesis.

Suppose first that $\bar{A}$ is of order 2. Then $A_0 := C_A(V)$ is of codimension 2 in V^g, so as $V \leq C_G(A_0)$ but $V \not\leq N_G(V^g)$, we conclude from 6.2.4.1 that A_0 is of field type. Then as U centralizes A_0, $U_0 := C_U(V^g)$ is of index 2 in U since $|C_G(A_0) : C_G(V^g)|_2 = 2$ by 6.2.4.2. Thus we have symmetry between V and V^g, so

U_0 is also of field type. Also $\langle V, V^g \rangle \leq C_G(U_0)$, and by 6.2.4.4, $V \leq O_2(C_G(U_0))$, so $\langle V, V^g \rangle$ is a 2-group and (1) holds.

Thus we may assume that $\bar{A}$ is of order 4, so $\bar{A} \not\leq \bar{L}$ as we saw $\bar{A} \notin Syl_2(\bar{L})$. From (1), our hypotheses are symmetric in V and V^g, so also $Aut_U(V^g)$ is a 4-group not contained in $Aut_{L^g}(V^g)$. Let $Q := UA$ and $\tilde{Q} := Q/B$. From the action of $\bar{A}$ on V, $B = C_V(\bar{A})$ is of order 2 and $C_A(V)$ is the centralizer in A of each hyperplane of V. Also $|V^g : B| = 8$, so as $|V^g| = 16$, it follows that $B = C_A(V) = C_A(U) = V \cap V^g$. Then we conclude $Q \cong Q_8^2$ with $B = Z(Q)$. Further $[[V, A], A] \leq C_V(A) = B \leq A$, so $[V, A] \leq N_V(A) \leq N_V(V^g) = U$ by 6.2.5.6, and thus we conclude $[V, A] = U$ as both groups have rank 3. Thus $[V, A] \leq Q$, so V acts on Q, and then by symmetry, V^g acts on Q. Hence $Y := \langle V, V^g \rangle$ acts on Q. Set $Y^* := Y/Q$, so that Y^* is dihedral, as V^* and V^{g*} are of order 2. We have seen that $[\tilde{A}, V^*] = \tilde{U}$, so we conclude $[\tilde{Q}, V^*] = \tilde{U} = C_{\tilde{Q}}(V^*)$. Therefore V^* is generated by an involution of type a_2 in $Out(Q) \cong O_4^+(2)$, $Y^*/C_{Y^*}(Q) \cong S_3$ with $Q \leq O^2(Y)$, and the images of V^* and V^{*g} are conjugate in this quotient. Thus $\tilde{U} = C_{\tilde{Q}}(V^*)$ is conjugate to $\tilde{A} = C_{\tilde{Q}}(V^{g*})$ in Y, and hence U is conjugate to A in Y. Therefore V^g is conjugate to V in Y by 6.2.5.6. Thus V^* is conjugate to V^{*g} in Y^*, so that $|Y^*| \equiv 2 \mod 4$. Again by 6.2.5.6, $C_Y(Q) \leq N_G(V)$, so as V^* inverts $O(Y^*)$, $C_Y(Q)^* = C_{Y^*}(Q) = 1$. Thus $C_Y(Q) = Z(Q) = B$, so $Y \cong S_3/Q_8^2$, completing the proof of (2). $\square$

LEMMA 6.2.7. $W_0(T, V) \leq C_T(V) = O_2(LT)$, so that $N_G(W_0(T, V)) \leq M$.

PROOF. By E.3.34.2, it suffices the prove the first assertion. So assume by way of contradiction that $W_0(T, V) \not\leq C_T(V)$. Then there is $g \in G$ such that $V \leq T^g$ but $[V, V^g] \neq 1$. By 6.2.3, $V^g \not\leq N_G(V)$. Let $U := C_V(V^g)$. Then $m(V/U) = 2$ by 6.1.10.3, and as $V^g \not\leq N_G(V)$, $C_G(U) \not\leq N_G(V)$, so the hypotheses of 6.2.4 are satisfied. Adopt the notation of that lemma (e.g., $H = C_G(U)$, $\tilde{H} = H/U$, $U_H = \langle V^H \rangle$, etc.) and let $A := V^g$, $B := Z_S^g$, and D_U of order 3 in L_U. Then $V = [V, D_U]$. By 6.1.10.2 and E.3.10, $VC_G(A)/C_G(A) \in \mathcal{A}_2(N_G(A)/C_G(A), A)$, so $S^g = VC_{S^g}(V^g)$ and $[A, V] = B$.

We claim next that if K^* is a subgroup of $C_{H^*}(D_U^*)$ with $A^* \leq K^*$, then $[\tilde{U}_H, K^*, D_U^*] \neq 1$: For otherwise using the Three-Subgroup Lemma, $A^* \leq K^* \leq C_{H^*}([\tilde{U}_H, D_U^*]) \leq C_{H^*}(\tilde{V})$, contrary to the fact that A does not act on V.

Now $A \leq H := C_G(U)$ and $V \leq U_H$, so $B = [A, V] \leq U_H$, which is abelian by 6.2.4.4. Thus $U_H \leq C_G(B) \leq N_G(A)$ by 6.2.1, so we may take $U_H \leq T^g$. Indeed as U_H centralizes B, we have $V \leq U_H \leq C_{T^g}(Z_S^g) = S^g$. Then $U_H = VC_{U_H}(A)$ by the first paragraph of the proof, so $[U_H, A] = [V, A] = B$, $m(U_H/C_{U_H}(A)) = 2$, and $B = C_A(U_0)$ for $C_{U_H}(A) < U_0 \leq U_H$.

We saw $V \leq C_G(B)$, so $B \leq N_A(V)$. If $B < N_A(V)$, then as $S^g = VC_{S^g}(A)$, $B = [V, N_A(V)] \leq V$; but now $Z_S^g = B \leq V \neq V^g$, contrary to 6.2.5.4. Hence $B = N_A(V)$. We saw $B \leq U_H$, so in particular $B = C_A(\tilde{U}_H)$ as $C_G(\tilde{U}_H) \leq N_G(V)$, and hence $A^* \cong E_4$.

Let $B < A_1 \leq A$. Suppose that $\tilde{U}_1 := C_{\tilde{U}_H}(A_1) > \widetilde{C_{U_H}(A)}$. We saw $B = C_A(U_0)$ for $C_{U_H}(A) < U_0 \leq U_H$. Thus $B = C_A(U_1)$, so as $[U_H, A] = B$, $1 \neq [U_1, A_1] =: B_1 \leq U \cap B$. We will show that $1 \neq U \cap B$ leads to a contradiction. For B is of rank 2, so $m(\tilde{B}) \leq 1$. Then since $[A, U_H] = B$, A^* induces a 4-group of transvections on $\tilde{U}_H$ with center $\tilde{B}$. Thus by G.3.1, there is $K \in \mathcal{C}(H)$

such that $K = [A, K]$, $A^* \leq K^*$, K^* induces $GL(\tilde{U}_1)$ on $\tilde{U}_1 := \langle \tilde{B}^K \rangle$ of rank at least 3, and the kernel of the action lies in $O_2(K^*)$. But $O_2(H^*) = 1$ by 6.2.4.4, so $K^* \cong GL(\tilde{U}_1)$. Then by 6.2.4.6, $K^* \cong L_3(2)$ (so that $m(\tilde{U}_1) = 3$) and $K = O^{3'}(H)$. As $[\tilde{U}_H, A] = \tilde{B} \leq \tilde{U}_1$ and $K = [K, A]$, $\tilde{U}_1 = [\tilde{U}_H, K]$. We saw $m(U_H/C_{U_H}(A)) = 2$, so $\tilde{U}_H = \tilde{U}_1 \oplus C_{\tilde{U}_H}(K)$. (cf. B.4.8.3). Now D_U of order 3 in L_U acts on the subgroup R of 6.2.4.2, and then on $R_K := R \cap K$ in view of 1.2.1.3. But $R_K^* \in Syl_2(K^*)$, so R_K^* is self-normalizing in K^* and hence $[D_U^*, K^*] = 1$. Then D_U centralizes $\tilde{U}_1$ since $K^* = Aut(\tilde{U}_1)$. As $A^* \leq K^*$, this contradicts our claim in paragraph two.

This contradiction shows that $B \cap U = 1$ and that

$$C_{\tilde{U}_H}(A_1) = \widetilde{C_{U_H}(A)} \text{ for each } 1 \neq A_1^* \leq A^*. \tag{$*$}$$

Since $A^* \cong E_4$, (*) says

$$A^* \in \mathcal{A}_2(H^*, \tilde{U}_H); \tag{$**$}$$

and since $B \cap U = 1$ we have

$$\tilde{B} = [\tilde{U}_H, A^*] \cong E_4. \tag{!}$$

Further applying (*) when $A_1 = A$ and recalling $m(U_H/C_{U_H}(A)) = 2$, we conclude

$$m(\tilde{U}_H/C_{\tilde{U}_H}(A)) = 2. \tag{!!}$$

Thus A^* is an offender on the FF-module $\tilde{U}_H$. Recall $\tilde{U}_H \in \mathcal{R}_2(\tilde{H})$ by 6.2.4.4, and let $K_A^* := \langle A^{*H} \rangle$. By (**) and E.4.1, A^* centralizes $O(H^*)$, so that $F(K_A^*) \leq Z(K_A^*)$. Next (!!) restricts the possible components K^* of K_A^* in the list of Theorem B.5.6 to alternating groups or groups defined over $\mathbf{F}_2$ or $\mathbf{F}_4$. Now K^* is the image of $K \in \mathcal{C}(H)$, and by 6.2.4.6 and inspection of our restriced list from B.5.6, either

(i) $m_3(K^*) = 1$, so that $K^* \cong L_2(4)$ or $L_3(2)$, or

(ii) $K^* \cong SL_3(4)$ and D_U induces outer automorphisms on K^*.

In particular $K^* = J(K_A^*)^\infty$ is described by Theorem B.5.1. As in the previous paragraph, $R_K := R \cap K \in Syl_2(K)$.

Suppose first that case (ii) holds. By Theorem B.5.1.1, either $\tilde{V}_K := [\tilde{U}_H, K^*] \in Irr_+(K^*, \tilde{U}_K)$, or $\tilde{V}_K$ is the sum of two isomorphic natural modules for K^*. In the former case, $\tilde{V}_K$ is a natural module by B.4.2. In either case, A.3.19 contradicts the fact that $D_U \not\leq K$.

Thus case (i) holds. By Theorem B.5.1.1, either $\tilde{V}_K := [\tilde{U}_H, K^*] \in Irr_+(K^*, \tilde{U}_H)$, or $K^* \cong L_3(2)$ and $\tilde{V}_K$ is the sum of two isomorphic natural modules.

Assume first that $K^* \cong L_3(2)$. If $\tilde{V}_K$ is the sum of two isomorphic natural modules, then by (*), A^* induces the group of transvections with a fixed axis on each of the natural summands, contrary to (!). Thus $\tilde{V}_K \in Irr_+(K^*, \tilde{U}_H)$. Then by B.4.8.4, $\tilde{V}_K = [\tilde{U}_H, K^*]$ is either the natural module or the extension in B.4.8.2. Now as D_U acts on $R_K^* \in Syl_2(K^*)$, D_U centralizes K^* and $\tilde{V}_K/C_{\tilde{V}_K}(K^*)$, and hence D_U centralizes $\tilde{V}_K$ by Coprime Action. As $A^* \leq K^*$, this contradicts our claim in paragraph two.

This contradiction shows $K^* \cong L_2(4)$, so $\tilde{V}_K \in Irr_+(\tilde{U}_H, K)$. Then by B.4.2, either $\tilde{V}_K$ is the A_5-module, or $\tilde{V}_K/C_{\tilde{V}_K}(K)$ is the natural module. The first case is impossible by (*). Thus the second case holds, and $A^* \in Syl_2(K^*)$ by B.4.2.1. Further $C_{\tilde{V}_K}(K) = 1$ by (!), so $\tilde{V}_K$ is the natural module, and $\tilde{U}_H = \tilde{V}_K \oplus C_{\tilde{U}_H}(K)$ by B.5.1.4.

Set $L_K := O^2(N_K(R_K))$, so that $L_K/O_2(L_K) \cong \mathbf{Z}_3$. First suppose $L_K \le M$. As $K \le H = C_G(U)$, by 6.1.8 we obtain $L_K \le K \cap M = N_K(V)$. Then as $[L_K, U] = 1$ and U is of field type, $[L_K, V] = 1$. But $C_{\tilde{U}_H}(L_K) = C_{\tilde{U}_H}(K)$, so $\tilde{V} \le C_{\tilde{U}_H}(K^*) \le C_{\tilde{U}_H}(A^*)$, and then $A \le N_G(V)$, contrary to paragraph one. Therefore $L_K \not\le M$. By 6.2.4.2, $N_G(R) \le M$. If $[D_U^*, K^*] \ne 1$, then

$$R = O_2(D_U R) = O_2(KR)(K \cap R) \trianglelefteq L_K R,$$

so $L_K \le N_G(R) \le M$, contradicting the reduction just obtained; hence $[D_U^*, K^*] = 1$. Thus as $A^* \le K^*$, $[\tilde{V}_K, D_U^*] \ne 1$ by our claim in paragraph two. Thus $D_U^* K^*$ acts on $\tilde{V}_K$ as $GL_2(4)$ with $D_U^* = Z(D_U^* K^*)$. As $N_G(R) \le M$ but $L_K \not\le M$, $R^* \ne R_K^*$, so there is $r \in R$ inducing an involutory field automorphism on K^*. This is impossible, as the field automorphism r^* inverts the center D_U^* of $GL_2(4)$, whereas $R \trianglelefteq RD_U$. This contradiction completes the proof of 6.2.7. $\qquad \square$

For the remainder of the section, let z denote the generator of Z, set $G_z := C_G(z)$, and $\tilde{G}_z := G_Z/Z$. By 6.2.2.3, $G_z \not\le M$, so $\mathcal{H}_1 \ne \emptyset$, where

$$\mathcal{H}_1 := \{H \le \mathcal{H}(T) : H \le G_z \text{ and } H \not\le M\}.$$

Consider any $H \in \mathcal{H}_1$, and observe that Hypothesis F.7.6 is satisfied with LT, H in the roles of "G_1, G_2". Form the coset graph Γ as in section F.7, and more generally adopt the notational conventions of section F.7. By 6.2.3 and F.7.11.2, $b := b(\Gamma, V)$ is odd.

LEMMA 6.2.8. $V \not\le O_2(G_z)$.

PROOF. Choose H minimal in $\mathcal{H}_1$; then $H \in \mathcal{H}_*(T, M) \cap G_z$. Thus $n(H) = 1$ by Hypothesis 6.1.1.2. We assume that $V \le O_2(G_z)$ and derive a contradiction. Then $V \le O_2(H)$ so $V \le G_{\gamma_1}^{(1)}$ by F.7.7.2, and hence $b > 1$; thus $b \ge 3$ as we saw b is odd. Let $U_H := \langle V^H \rangle \le O_2(H)$. As $b \ge 3$, U_H is abelian by F.7.11.4. As usual, let $\gamma \in \Gamma$ with $d(\gamma_0, \gamma) = b$, and γ_i at distance i from γ_0 on a fixed geodesic from γ_0 to γ. By F.7.11.6, $[U_H, U_\gamma] \le U_H \cap U_\gamma$, where U_γ is the conjugate of U_H defined in section F.7.

As $H \le G_z$, $H \cap M = N_H(V)$ by 6.1.8. By 3.3.2.4, $H \cap M$ is the unique maximal subgroup of H containing T. Hence we may apply F.7.13 to U_H in the role of "A" to conclude there exists $\alpha \in \Gamma(\gamma)$ such that $m(U_H/N_{U_H}(V_\alpha)) = 1$.

As U_H does not act on V_α, there exists $\beta \in \Gamma(\gamma_1)$ such that V_β does not act on V_α; we consider any β satisfying these two conditions. Notice that as $m(U_H/N_{U_H}(V_\alpha)) = 1$, also $m(V_\beta/N_{V_\beta}(V_\alpha)) = 1$. Let $U_\beta := C_{V_\beta}(V_\alpha)$, so that $U_\beta \le N_{V_\beta}(V_\alpha) < V_\beta$. Then $V_\alpha \not\le N_G(V_\beta)$, since otherwise $[V_\alpha, V_\beta] = 1$ by 6.2.7, contradicting $U_\beta < V_\beta$. As $m(V_\beta/N_{V_\beta}(V_\alpha)) = 1$, $C_G(N_{V_\beta}(V_\alpha)) \le N_G(V_\beta)$ by 6.1.10.1. So as V_α centralizes U_β but does not normalize V_β, $U_\beta < N_{V_\beta}(V_\alpha)$; hence $U_\alpha := [N_{V_\beta}(V_\alpha), V_\alpha]$ is a noncyclic subgroup of V_α. But $U_\alpha \le [U_H, V_\alpha] \le U_H$, so as $V_\beta \le U_H$ which is abelian, $U_\alpha \le C_{V_\alpha}(V_\beta)$. Now as $V_\beta \not\le N_G(V_\alpha)$, $C_G(C_{V_\alpha}(V_\beta)) \not\le N_G(V_\alpha)$, so that $m(C_{V_\alpha}(V_\beta)) \le 2$ by 6.1.10.1; as U_α is noncyclic, we conclude $U_\alpha = C_{V_\alpha}(V_\beta)$ is a 4-group. Then U_α is of field type by 6.2.4.1. So as V_β centralizes U_α, $m(N_{V_\beta}(V_\alpha)/U_\beta) = 1$ by 6.2.4.2, with $N_{V_\beta}(V_\alpha)$ inducing a field automorphism on V_α. Then $m(V_\beta/N_{V_\beta}(V_\alpha)) = 1 = m(N_{V_\beta}(V_\alpha)/U_\beta)$, so U_β is also a 4-group. Therefore as V_α centralizes U_β but does not normalize V_β, $C_G(U_\beta) \not\le N_G(V_\beta)$, and then U_β is also of field type by 6.2.4.1.

As $V_\alpha \not\leq N_G(V_\beta)$, $V_\alpha \not\leq G_{\gamma_1}^{(1)}$, so $d(\alpha) = b$ with $\alpha, \gamma, \cdots, \gamma_1$ a geodesic, and we have symmetry between γ and γ_1. By this symmetry (as in the proof of 6.1.19) we can apply F.7.13 to V_α in the role of "A", to conclude that there exists $\beta' \in \Gamma(\gamma_1)$ such that $m(V_\alpha/N_{V_\alpha}(V_{\beta'})) = 1$, and also that there exists $h \in H$ such that V_α^h fixes β' and $I := \langle V_\alpha, V_\alpha^h \rangle$ is not a 2-group.

Observe next that if $\mu, \nu \in \Gamma_0$, and V_μ acts on V_ν, then $[V_\mu, V_\nu] = 1$ by 6.2.7. But as V_α^h fixes β', and $\beta' = \gamma_0^g$ for some $g \in G$, $V_\alpha^h \leq G_1^g \leq N_G(V_{\beta'})$, so V_α^h centralizes $V_{\beta'}$. Similarly as V_α does not centralize $V_{\beta'}$, $V_{\beta'}$ does not act on V_α. Thus β' satisfies the two conditions for "β" in our earlier argument, so we may take $\beta' = \beta$. Then $m(V_\alpha/N_{V_\alpha}(V_\beta)) = 1$, so that we have symmetry between α and β. Thus as we showed that $[N_{V_\beta}(V_\alpha), V_\alpha] = C_{V_\alpha}(V_\beta) = U_\alpha$, by symmetry between α and β, $[N_{V_\alpha}(V_\beta), V_\beta] = C_{V_\beta}(V_\alpha)$. In particular as $U_\beta = C_{V_\beta}(V_\alpha)$, we also have symmetry between U_α and U_β. Further $N_{V_\beta}(V_\alpha)$ and $N_{V_\alpha}(V_\beta)$ are each of rank 3, and induce a field automorphism on V_α and V_β, respectively. Hence

$$1 \neq U_{\alpha,\beta} := [N_{V_\beta}(V_\alpha), N_{V_\alpha}(V_\beta)] \leq U_\alpha \cap U_\beta.$$

Now $U_{\alpha,\beta} \leq U_\beta$ centralizes I as V_α centralizes U_α and V_α^h centralizes $V_{\beta'} = V_\beta$. Thus for $z_0 \in U_{\alpha,\beta}^\#$, $z_0 \in V_\alpha$, but $V_\alpha \not\leq O_2(G_{z_0})$—since $I \leq G_{z_0}$, and $V_\alpha \not\leq O_2(I)$ as $I = \langle V_\alpha, V_\alpha^h \rangle$ is not a 2-group. As the pair (V, z) is conjugate to (V_α, z_0), 6.2.8 is established.   □

In the remainder of this section, choose

$$H := G_z,$$

and let $M_z := C_M(z)$, $U := \langle Z_S^H \rangle$, $K := \langle V^H \rangle$, $M_K := K \cap M$, and $H^* := H/C_H(\tilde{U})$. By 6.2.8, $V \not\leq O_2(K)$, so $K \not\leq N_G(V)$. By 6.2.1, $N_G(Z_S) \leq N_G(V) = M_V$, and as V is the natural module for $\bar{L}$, $C_{M_V}(z) \leq N_M(Z_S)$. As $H = G_z$, by 6.1.8 we conclude:

LEMMA 6.2.9. $H \cap M = N_H(V) = N_H(Z_S)$ and $M_K = N_K(V) = N_K(Z_S)$.

LEMMA 6.2.10. (1) $F^*(H) = O_2(H) =: Q_H$ and $\tilde{U} \leq Z(\tilde{Q}_H)$.
(2) $C_H(\tilde{U}) \leq N_G(V) \leq M$, so $C_V(\tilde{U}) \leq Q_H$.
(3) $O_2(H^*) = 1$.
(4) $V^* \neq 1$.
(5) $[V, U] \leq V \cap U$.

PROOF. The first assertion in (1) holds by 1.1.4.6. Hypothesis G.2.1 is satisfied with Z, Z_S in the roles of "V_1", "V", so G.2.2 completes the proof of (1) and establishes (3). By 6.2.9, $C_H(\tilde{U}) \leq N_H(Z_S) = N_H(V) \leq M$, so $C_V(\tilde{U}) \leq O_2(C_H(\tilde{U})) \leq Q_H$, proving (2). By (1), $\tilde{U}$ is abelian, so by (2), U acts on V. Also $V \leq H \leq N_G(U)$, so (5) holds. As $V \not\leq Q_H$ by 6.2.8, (4) follows from (2).   □

LEMMA 6.2.11. V^* is of order 2.

PROOF. Assume the lemma fails; then as $V^* \neq 1$ by 6.2.10.4, $m(V^*) \geq 2$. By 6.2.10.1, $Z_S \leq C_V(\tilde{U})$, so that $m(V^*) \leq m(V/Z_S) = 2$. Thus $m(V^*) = 2$, and $Z_S = V \cap Q_H = V \cap U = C_V(\tilde{U})$. Next by (4) and (5) of 6.2.10, $1 \neq [V^*, \tilde{U}] \leq \widetilde{V \cap U} = \tilde{Z}_S$ of order 2. Thus V^* induces a 4-group of transvections on $\tilde{U}$ with center $\tilde{Z}_S$. Also $O_2(H^*) = 1$ by 6.2.10.3. Thus we may apply G.3.1 and the results of section G.6 to H^*. In particular, since $\tilde{U} = \langle Z_S^H \rangle$, we conclude from G.3.1 that

K^* is the direct product of copies of $GL_m(2)$ for some $m \geq 3$. Next as $V \trianglelefteq T$, $1 \neq V^* \cap Z(T^*)$, so by G.6.4.4, $K^* = GL(\tilde{U})$. By 6.2.9,

$$C_{H^*}(\tilde{Z}_S) = M_z^* = N_M(V)^*.$$

Thus as V^* is a 4-group we conclude $m(\tilde{U}) = 3$ and $H^* \cong L_3(2)$.

As $\Phi(Z_S) = 1$ and H^* is transitive on $\tilde{U}^\#$, $\Phi(U) = 1$, so $U \cong E_{16}$. As V^* is the group of transvections with center $\tilde{Z}_S$, $\tilde{Z}_S = C_{\tilde{U}}(V^*)$, so $Z_S = C_U(V)$. Further $U \leq C_T(Z_S) = T_L C_T(V)$, where $T_L := T \cap L$; thus $|\bar{U}| = |U/C_U(V)| = |U : Z_S| = 4 = |\bar{T}_L|$, so $\bar{U} = \bar{T}_L \in Syl_2(\bar{L})$.

Now $[C_H(\tilde{U}), V] \leq C_V(\tilde{U}) \leq V \cap Q_H$ by 6.2.10.2, and we saw that $V \cap Q_H \leq U$. Hence $K = \langle V^H \rangle$ centralizes $C_H(\tilde{U})/C_H(U)$. Next $C_H(\tilde{U})/C_H(U)$ is a subgroup of the group X of all transvections on U with center Z, and $\tilde{U}$ is the dual of X as a module for $C_{GL(U)}(Z)$. Thus as $\tilde{U}$ is the natural module for K^* and K centralizes $C_H(\tilde{U})/C_H(U)$, we conclude $C_H(\tilde{U}) = C_H(U)$.

Next $L = [L, U]$ with $[U, O_2(LT)] \leq C_U(V) = Z_S \leq V$, so L is an $L_2(4)$-block. Also $C_{T^*}(V^*) = V^*$ as V^* is a 4-subgroup of $H^* \cong L_3(2)$; thus $C_T(V) \leq V C_T(\tilde{U})$. Therefore as $C_T(\tilde{U}) = C_T(U)$ by the previous paragraph, we conclude $C_T(V) = V C_T(UV)$. Then as $\bar{U} \in Syl_2(\bar{L})$, it follows from Gaschütz's theorem A.1.39 and C.1.13.a that $LO_2(LT) = LC_T(L)$. On the other hand, $C_T(L) = 1$ by 6.1.6.1. Therefore $V = O_2(LT) = O_2(M)$ using A.1.6. Then $T_L = J(T)$ with $\mathcal{A}(T) = \{A_1, A_2\}$ and $A_1 = V$, so as $m(U) = 4$, $U = A_2$. Thus as $N_L(T_L)$ acts on V, it also acts on U, so that $L_0 := \langle N_L(T_L), H \rangle$ acts on U, and hence $\hat{L}_0 := L_0/C_{L_0}(U) \leq GL(U) \cong A_8$. As $N_L(T_L)$ is transitive on $Z_S^\#$ and H is transitive on $U - Z$, L_0 is transitive on $U^\#$. Further $C_{\hat{L}_0}(z) = \hat{H} \cong L_3(2)$, so we conclude $\hat{L}_0 \cong A_7$. Moreover setting $M_0 := M \cap L_0$, $N_{L_0}(Z_S) \leq M_0 < L_0$ by 6.2.1. The stabilizer of any 4-subgroup of U in $\hat{L}_0$ is the global stabilizer in $\hat{L}_0$ of 3 of the 7 points permuted by $\hat{L}_0$ in its natural representation, which is a maximal subgroup of $\hat{L}_0$. Thus $\hat{M}_0 = N_{L_0}(Z_S)$. Now we can also embed $T \leq Y \leq L_0$ with $\hat{Y} \cong S_5$ and $|Y : Y \cap M_0| = 5$. Thus $Y \in \mathcal{H}_*(T, M)$ with $n(Y) = 2$ by E.2.2, contradicting Hypothesis 6.1.1.2. $\square$

LEMMA 6.2.12. *(1) $O^2(H \cap M) \leq C_M(V) \leq C_M(L/O_2(L))$.*
(2) $O^2(C_H(\tilde{U})) = 1$, so $C_H(\tilde{U}) = Q_H$.

PROOF. As V^* has order 2 by 6.2.11, we conclude from 6.2.9 and 6.2.2 that $H \cap M$ acts on the series $V > C_V(\tilde{U}) > Z_S > Z$, and all factors in the series are of rank 1. Therefore $O^2(H \cap M)$ centralizes V by Coprime Action. Then $O^2(H \cap M)$ centralizes $L/O_2(L)$, proving (1).

Next using 6.2.10.2 and (1), $X := O^2(C_H(\tilde{U})) \leq O^2(H \cap M)$. Thus X centralizes $L/O_2(L)$, so that L normalizes $O^2(XO_2(L)) = X$. Now if $X \neq 1$, then $O_2(X) \neq 1$ by 1.1.3.1, since $H \in \mathcal{H}^e$ by 1.1.4.6. But then $H \leq N_G(O_2(X)) \leq M = !\mathcal{M}(LT)$, contradicting $H \not\leq M$. This shows that $C_H(\tilde{U})$ is a 2-group, and then 6.2.10.1 completes the proof of (2). $\square$

We can now isolate the case leading to M_{22}, which we identify via a recent characterization of Chao Ku. Recall that $U = \langle Z_S^H \rangle$, so that $Z_S \leq V \cap U$.

PROPOSITION 6.2.13. *If $Z_S = V \cap U$, then $G \cong M_{22}$.*

PROOF. Assume $Z_S = V \cap U$. We begin by arguing much as at the start of the proof of 6.2.11, except this time V^* has order 2 by 6.2.11. By 6.2.10.5, $1 \neq [V^*, \tilde{U}] \leq \widetilde{V \cap U} = \tilde{Z}_S$ of order 2, so that V^* is generated by a transvection on $\tilde{U}$ with center $\tilde{Z}_S$. As $\tilde{U} = \langle \tilde{Z}_S^H \rangle$ and $\tilde{Z}_S = [\tilde{U}, V^*]$, $\tilde{U} = [\tilde{U}, K^*]$ by G.6.2. As $V \trianglelefteq T$, $V^* \leq Z(T^*)$, so G.6.4.4 shows that $K^* \cong L_n(2)$, $2 \leq n \leq 5$, S_6, or S_7; and by G.6.4.2, $\tilde{U}$ is the natural module or the core of the permutation module for S_6. In each case $K^* = N_{GL(\tilde{U})}(K^*)$, so $H^* = K^*$. Next by 6.2.9:

$$C_{H^*}(\tilde{Z}_S) = N_H(Z_S)^* = M_z^* = N_H(V)^* = C_{H^*}(V^*).$$

But if H^* is $L_n(2)$ with $3 \leq n \leq 5$, then V^* is not normal in $C_{H^*}(\tilde{Z}_S)$. Thus $H^* = K^* \cong L_2(2)$, S_6, or S_7. In each case, $V^* \not\leq O^2(H^*)$, so in particular $V \not\leq O^2(X)$, where $X := O^2(M_z)$, and hence $V > V \cap X$. By 6.2.12.1, L acts on $O^2(O^2(H \cap M)O_2(L)) = X$, so L acts on $V \cap X$. Therefore as L is irreducible on V, $V \cap X = 1$.

Suppose first that H^* is S_6 or S_7. Then there are $x, y \in H$ such that $I := \langle V^x, V^y \rangle \leq M_z$ and $I^* \cong S_3$. Then $V^x \not\leq N_G(V^y)$, but $C_{V^x}(\tilde{U}) \leq N_G(Z_S^y) \leq N_G(V^y)$ by 6.2.1; so as V^{*x} has order 2, $N_{V^x}(V^y) = C_{V^x}(\tilde{U})$ is of index 2 in V^x. Similarly $|V^y : N_{V^y}(V^x)| = 2$, so as I is not a 2-group, $O_2(I) \leq O^2(I) \leq X$ and $|Z(O_2(I))| = 2$ by 6.2.6. But as $x, y \in G_z$, $Z \leq V^x \cap V^y = Z(O_2(I))$, so $Z \leq V \cap X$, contrary to the previous paragraph.

This contradiction shows that $H^* \cong S_3$, so $H^* = \langle V^*, V^{g*} \rangle$ for $g \in H - M$ and $|V^H| = |H : M_z| = 3$. Thus $V^H \leq \langle V, V^g \rangle$, so that $K = \langle V, V^g \rangle$. Therefore case (2) of 6.2.6 holds with $K \cong S_3/Q_8^2$, and $Z = V \cap V^g = Z(P)$, where $P := O_2(K)$.

Notice as $Z_S \leq P \trianglelefteq H$ that $U = \langle Z_S^H \rangle \leq P$. Then $R := C_H(\tilde{P}) \leq C_H(\tilde{U}) = Q_H$ by 6.2.12.2. Also as case (2) of 6.2.6 holds, $\overline{N_{V^g}(V)} = \bar{P} \cong E_4$, $C_P(V) = P \cap V$, and $\bar{P} \not\leq \bar{L}$. Therefore $\bar{T} = \bar{P}\langle \bar{t} \rangle$, where $t \in T \cap L$ acts nontrivially on $\bar{P}$. Thus t is nontrivial on $P/(P \cap V)$, so that $t^* \notin V^*$ since $[P, V] \leq P \cap V$. Therefore as $N_{Out(P)}(K^*) \cong S_3 \times S_3$ and $H = KT$, we conclude $H/R \cong S_3 \times \mathbf{Z}_2$ and $C_T(V) = VC_R(V)$. Now $R = PC_R(P)$ as $Inn(P) = C_{Aut(P)}(\tilde{P})$ by A.1.23. But $C_R(P) \leq C_R(V \cap P) = C_R(V)$ by 6.1.10.2, so as $C_R(P) \trianglelefteq H$, $C_R(P)$ centralizes $\langle V^H \rangle = K$. Therefore $C_R(P) = C_R(K)$, so $R = PC_R(K)$ and $C_R(K) \leq C_R(V)$. Thus $C_R(V) = C_R(K)C_P(V) = C_R(K)(P \cap V)$, and hence $C_T(V) = VC_R(V) = VC_R(K) = VC_R(P)$. Then $[P, C_T(V)] = [P, V] \leq V$, so as $L = [L, P]$, $[L, O_2(LT)] \leq V$, and hence L is an $L_2(4)$-block. Now $\Phi(C_T(V)) \leq C_T(L) = 1$ by C.1.13.a and 6.1.6.1. Then since $C_T(V) = VC_R(K)$, $C_R(K)$ is also elementary abelian. Also we chose $t \in T \cap L$ with $\overline{T \cap L} \leq \langle \bar{t} \rangle \bar{P}$; so as $C_T(L) = 1$, by Gaschütz's Theorem A.1.39 $C_T(V) \cap C_G(P\langle t \rangle) = C_V(P\langle t \rangle) = Z$. Thus as $C_R(K)$ centralizes P, $C_R(K) \cap C_G(t) = Z$. But $[t, C_R(K)] \leq C_{[T \cap L, C_T(V)]}(K) = C_V(K) = Z$, so we conclude $m(C_R(K)) \leq 2$, and in case of equality, $[t, C_R(K)] = Z$.

In any case, V is of index at most 2 in $Q := O_2(LT)$. By 1.1.4.6, $F^*(M) = O_2(M)$. Then as Q contains $O_2(M)$ by A.1.6 and Q is abelian, $Q \leq C_M(O_2(M)) \leq O_2(M)$, so $O_2(M) = Q$. Next by 6.2.12.1, $O^2(H \cap M)$ centralizes V, so by Coprime Action, $O^2(H \cap M) \leq C_M(Q) \leq Q$, so $O^2(H \cap M) = 1$. In particular, $C_M(V) = Q$, so that $\bar{M} = M/Q$. An involution in V^g induces a nontrivial inner automorphism on $\bar{L}$, so L/V is not $SL_2(5)$ and hence $V = O_2(L)$.

Now $V = O_2(L) \trianglelefteq M$, so $S_5 \cong \bar{L}\bar{T} \leq \bar{M} \leq N_{GL(V)}(\bar{L}) \cong \Gamma L_2(4)$. Further if $\bar{M} \cong \Gamma L_2(4)$, then an element of order 3 whose image is diagonally embedded in

$\bar{L} \times C_{\bar{M}}(\bar{L})$ centralizes z and hence lies in H, contrary to $O^2(H \cap M) = 1$. Thus $S_5 \cong \bar{M} = \bar{L}\bar{T}$, so that $M = LT$.

Assume first that $C_R(K) = Z$ is of order 2. Thus $M \cong S_5/E_{16}$, with $H = K\langle t \rangle \cong (S_3 \times \mathbf{Z}_2)/Q_8^2$. Then as $C_R(K) = Z$, $C_H(P) \leq P$, and $Z = C_P(X)$ for $X \in Syl_3(H)$; thus G satisfies the Hypothesis on page 295 of C. Ku in [**Ku97**]. (Note that the term Z_z there is unnecessary, and also that $\mathbf{Z}_1$ in H/Q should read $\mathbf{Z}_2$). We next verify that G is of *type* M_{22} as defined on p. 295 of that paper— namely we show there exists $z \neq z^d \in P$ with $m(P \cap P^d) = 2$: Let D_L be of order 3 in $N_L(T \cap L)$, and pick $d \in D_L^\#$. Then $Z_S = \langle z, z^d \rangle$ and $P \cap P^d = Z_S \cong E_4$, as $\bar{P} \cap \bar{P}^d = 1$ and $Z_S = P \cap P^d \cap V$ from the structure of $\bar{L}$ and its action on V. Thus G is of type M_{22}, so we may apply the Main Theorem of that paper to conclude that $G \cong M_{22}$.

So now we assume that $C_R(K) \cong E_4$, and it remains to derive a contradiction. Then $M \cong S_5/E_{32}$, with $Q \cong E_{32}$. As $C_T(L) = 1$ by 6.1.6.1, Q does not split over V as an L-module. Thus $Q = J(T)$.

Next all involutions in P are fused into V in K, and all involutions in V are fused in L, as are all involutions in $L - V$. Thus all involutions in L are conjugate in G, and are fused to some $j \in P - L$. Next j induces a field automorphism on L/V, so all involutions in jL are conjugate in L. Let $T_0 := P(T \cap L) = \langle j \rangle (T \cap L)$, so that all involutions in T_0 are in z^G. Let $r \in C_R(K) - Z$. Then $r \in Q - V$, and as $Q = J(T)$, $M = N_G(Q)$ controls fusion in Q by Burnside's Fusion Lemma A.1.35. Hence $r \notin z^G$. Therefore $r^G \cap T_0 = \emptyset$, so by Thompson Transfer, $O^2(G) < G$, contradicting simplicity of G. This completes the proof of 6.2.13. $\qquad\square$

By 6.2.13, we may assume during the remainder of the section that $Z_S < V \cap U =: V_U$; in Theorem 6.2.19, we will obtain a contradiction under this assumption. Let $Z_U := Z(U)$.

As V^* has order 2 by 6.2.10.4, $m(V_U) \leq m(V \cap Q_H) = 3$, so as $Z_S < V_U$:

LEMMA 6.2.14. $V_U = V \cap Q_H$ *is of rank* 3.

LEMMA 6.2.15. *(1)* $U = Z_U * U_0$ *is a central product, where* U_0 *is extraspecial of width at least* 2 *and rank at least* 3.

(2) For $v \in V - U$ *there exists* $g \in H$ *with* $v^* v^{*g}$ *not a 2-element, and for each such* g, $|v^* v^{*g}| = 3$ *and* $\langle V, V^g \rangle \cong S_3/Q_8^2$ *with* $V_U V_U^g = O_2(\langle V, V^g \rangle) \leq U$.

(3) $Z_U \leq Z(K)$ *and* K^* *is faithful on* U/Z_U.

PROOF. By 6.2.10.3, $O_2(H^*) = 1$, so by the Baer-Suzuki Theorem A.1.2, there is $g \in H$ with $v^* v^{*g}$ not a 2-element. Then $V \not\leq N_G(V^g)$, and so $V_U \leq N_V(V^g) < V$, so by 6.2.14, $V_U = N_V(V^g)$ is of index 2 in V. Similarly $V_U^g = N_{V^g}(V)$ is of index 2 in V^g, so part (2) follows from 6.2.6. As Z_U centralizes V_U, it centralizes V by 6.1.10.2, so Z_U centralizes $K = \langle V^H \rangle$. Thus $C_{K^*}(U/Z_U) \leq O_2(K^*) \leq O_2(H^*) = 1$ using 6.2.10.3, so that K^* is faithful on U/Z_U, completing the proof of (3). As $\Phi(U) \leq Z$ of order 2 by 6.2.10.1, and U is nonabelian by (2), $\Phi(U) = Z$. We conclude (1) holds, using (2) to see that U_0 is of width at least 2 and rank at least 3. $\qquad\square$

Let $\hat{H} := H/Z_U$ and $\dot{H} := H/C_H(\hat{U})$, and identify Z with $\mathbf{F}_2$. Thus by 6.2.15.1, $\hat{U} = \hat{U}_0$ is an $\mathbf{F}_2\dot{H}$-module, and $\dot{H}$ preserves the symplectic form $(\hat{u}_1, \hat{u}_2) := [u_1, u_2]$ on $\hat{U}$, so $\dot{H} \leq Sp(\hat{U})$.

LEMMA 6.2.16. *(1)* $V \cap Z_U = Z$, *so* $\dim(\hat{V}_U) = 2$.
(2) $\hat{U} = \langle \hat{Z}_S^H \rangle$ *and* $O_2(\hat{H}) = 1$.
(3) $K^* \cong \dot{K}$.
(4) $\hat{V}_U = [\dot{V}, \hat{U}]$, *and* $\dot{V}$ *is generated by an involution in* $Sp(\hat{U})$ *of type* a_2.
(5) $C_H(\dot{V}) = N_H(\hat{V}_U) = H \cap M$.
(6) $N_{\dot{H}}(\hat{V}_U)$ *is not transitive on* $\hat{V}_U^{\#}$.

PROOF. Part (1) follows from 6.2.15.2, and part (3) from 6.2.15.3. As $U = \langle Z_S^H \rangle$, $\hat{U} = \langle \hat{Z}_S^H \rangle$, so as $\hat{Z}_S \leq Z(\hat{T})$, $\hat{U} \in \mathcal{R}_2(\hat{H})$ by B.2.13, establishing (2). By 6.2.10.2, $[U, V] \leq V_U$, so by 6.2.15.2, $\hat{V}_U = [\hat{U}, \dot{V}]$ is of rank 2 and $U = V^g C_U(V)$ for some $g \in H$. Thus $\dot{V}$ is generated by an involution of type a_2 or c_2 in $Sp(\hat{U})$ in the sense of Definition E.2.6. Indeed for $y \in V^g - Z$ and $v \in V - U$, $[y, v] \in C_{V_U}(y)$ as y induces an involution on V, so $(\hat{y}, \hat{y}^v) = 0$ and hence $\dot{v}$ is of type a_2, establishing (4). As there is a unique involution $i \in Sp(\hat{U})$ of type a_2 with $[\hat{U}, i] = \hat{V}_U$, it follows that $N_{\dot{H}}(\hat{V}_U) = C_{\dot{H}}(\dot{V})$.

Let $h \in C_H(\dot{V})$; then $V^{*h} = V^*$ by (3), so that h acts on $[\tilde{U}, V^*] = \tilde{V}_U$. Thus $C_H(\dot{V}) \leq N_H(V_U)$. But by the previous paragraph, $N_{\dot{H}}(\hat{V}_U) = C_{\dot{H}}(\dot{V})$, so $N_H(V_U) = N_H(\hat{V}_U) = C_H(\dot{V})$. Finally $N_H(V_U) \leq H \cap M$ by 6.2.5.6, while $H \cap M = N_H(V)$ by 6.2.9, and $N_H(V)$ acts on $V \cap U = V_U$, so (5) holds.

By 6.2.9, $H \cap M$ acts on Z_S, so (5) implies (6). $\qquad\square$

Let $L_S := O^2(N_L(Z_S))$, $l \in L_S - H$, $E := U \cap U^l$, $W := C_U(Z_S)$, and $X := C_{U^l}(Z_S)$. Observe as $Z_S \leq U$ that $Z_S \leq U^l$, and hence

$$Z \leq Z_S \leq E.$$

LEMMA 6.2.17. *(1)* $Z_U \cap Z_U^l = 1$.
(2) $Z_U \cap U^l = (Z_U \cap Z_U^l)Z$.
(3) $\hat{W} = \hat{Z}_S^{\perp}$ *and* $[\dot{X}, \hat{W}] \leq \hat{E}$.
(4) $\hat{E}$ *is totally singular.*
(5) For $\dot{x} \in \dot{X} - \dot{Z}_U^l$, $C_{\hat{U}}(\dot{x}) \leq \hat{W}$.
(6) $C_X(\hat{U}) = EC_{Z^l}(\hat{U})$.
(7) $\dot{X}$ *induces the full group of transvections on* $\hat{E}$ *with center* $\hat{Z}_S$.
(8) $C_{\hat{E}}(\dot{X}) = \hat{Z}_S$.
(9) $\dot{V} \leq \dot{X}$.
(10) $m(\hat{E}) + m(\dot{X}/\dot{Z}_U^l) = m(\hat{U}) - 1$.

PROOF. Part (1) follows as $V \cap Z_U = Z$ by 6.2.16.1.

Next $\Phi(U^l) = Z^l$ and X acts on Z_U, so $[Z_U \cap U^l, X] \leq Z_U \cap Z^l = 1$ by (1). Thus $Z_U \cap U^l \leq Z(X)$. By 6.2.15.1, $U = U_0 Z_U$ with U_0 extraspecial, so $Z_U^l Z = Z(C_{U^l}(Z)) = Z(X)$. Therefore $Z_U \cap U^l \leq Z_U^l Z$, so as $Z \leq Z_U \cap U^l$, (2) holds.

Observe Hypothesis G.2.1 is satisfied with Z, Z_S, L_S, H in the roles of "V_1, V, L, H", and set $I := \langle U, U^l \rangle$ and $P := O_2(I)$. As U is nonabelian by 6.2.15.1, while $L_S/O_2(L_S) \cong L_2(2)'$, the hypotheses of G.2.3 are also satisfied. So by that lemma, $I = L_S U$, $P = WX$, $1 < Z_S \leq E \leq P$ is an I-series such that $[I, E] \leq Z_S$, and for some nonnegative integer s, and $P/E = W/E \oplus X/E$ is the sum of s natural modules for $I/P \cong L_2(2)$ with $W/E = C_{P/E}(U)$. Now $V = [V, L_S] \leq L_S \leq I$, so $V \leq P$ and hence $\dot{V} \leq \dot{P} = \dot{W}\dot{X} = \dot{X}$, establishing (9).

By definition of the bilinear form on $\hat{U}$, $\hat{Z}_S^{\perp}$ is the image of $C_U(Z_S) = W$ in $\hat{U}$, and the image of a subgroup Y of U in $\hat{U}$ is totally singular iff Y is abelian. As P/E is abelian, $[X, W] \leq E$, completing the proof of (3). As $\Phi(E) \leq \Phi(U) \cap \Phi(U^l) = Z \cap Z^l = 1$, (4) holds.

Pick $u \in U - W$; from the action of I on P/E, the map $\varphi : X \to W/E$ defined by $\varphi(x) := [x, u]E$ is a surjective linear map with kernel E. In particular as $Z \leq E$, $C_{\tilde{U}}(x) \leq \tilde{W}$ for each $x \in X - E$. Further setting $D := \varphi^{-1}(Z_U E/E)$, $D = C_X(U/Z_U E)$. As P/E is a sum of natural modules for I/P, $DZ_U = \langle Z_U^I \rangle E = Z_U Z_U^l E$, so $D = Z_U^l E$. Thus $C_X(\hat{U}) \leq C_X(U/Z_U E) = D = Z_U^l E$. In particular for $u \in U - W$, $C_X(\hat{u}) \leq Z^l E$, and hence (5) follows.

Let $R := C_T(\hat{U})$, and $\tilde{U}_R := C_{\tilde{U}}(R)$ with preimage U_R. By a Frattini Argument, $H = C_H(\hat{U})N_H(R)$, so as $Z_S \leq U_R$ and $U = \langle Z_S^H \rangle$, $U = U_R Z_U$. Therefore as $Z_U \leq W < U$, R centralizes $\tilde{u} \in \tilde{U} - \tilde{W}$. In particular $C_X(\hat{U}) \leq C_X(\hat{u}) \leq Z^l E$, so (6) holds.

Let $E_0 := EZ_U^l \cap U_0^l$ and $Z_0 := ZZ_U^l \cap U_0^l$. Then $EZ_U^l \leq U^l = U_0^l Z_U^l$, so $EZ_U^l = E_0 Z_U^l$, and similarly $Z_S Z_U^l = ZZ_U^l = Z_0 Z_U^l$. Thus $X = C_{U^l}(Z_S) = C_{U^l}(Z_0)$. As EZ_U^l is abelian, so is E_0. Therefore as U_0 is extraspecial, we conclude that:

X induces the full group of transvections on E_0 with center Z^l centralizing Z_0.
$$(!)$$

Let $\hat{e} \in \hat{E} - \hat{Z}_S$. As $EZ_U^l = E_0 Z_U^l$, $eZ_U^l = e_0 Z_U^l$ for some $e_0 \in E_0$. By (2),

$$E \cap Z_S Z_U = Z_S(E \cap Z_U) = Z_S(U^l \cap Z_U) = Z_S(Z_U \cap Z_U^l) = E \cap Z_S Z_U^l.$$

Thus as $\hat{e} \notin \hat{Z}_S$, $e \notin Z_S Z_U^l$, so as $Z_S Z_U^l = Z_0 Z_U^l$, $e_0 \notin Z_0 Z_U^l$. Thus $[e, X] = [e_0, X] = Z^l$ by (!). Hence (7) holds and of course (7) implies (8). Finally

$$m(\hat{U}) = m(\hat{E}) + m(\hat{W}/\hat{E}) + 1 = m(\hat{E}) + m(X/EZ_U^l) + 1 = m(\hat{E}) + m(\dot{X}/\dot{Z}_U^l) + 1,$$

where the last equality follows from (6). Thus (10) holds. $\quad\square$

LEMMA 6.2.18. *(1) $\dot{X}$ and $\dot{Z}_U^l$ are normal in $C_{\dot{H}}(\dot{V})$.*
(2) $\dot{H}$ and its action on $\hat{U}$ satisfy one of the conclusions of Theorem G.11.2.

PROOF. We first verify that $\hat{U}$, $\dot{H}$, $\hat{Z}_S$, $\hat{E}$, $\dot{X}$, $\dot{Z}_U^l$ satisfy Hypothesis G.10.1 in the roles of "V, G, V_1, W, X, X_0". As $\Phi(X) \leq Z^l \leq U$, $\dot{X}$ is elementary abelian, and $\hat{E}$ is totally singular by 6.2.17.4. By construction condition (a) of part (2) of Hypothesis G.10.1 holds. Conditions (b), (c), (d), and (e) are parts (10), (3), (5), and (7) of 6.2.17, respectively. So Hypothesis G.10.1 is indeed satisfied.

Let $M_H := H \cap M$. By 6.2.16.5, $\dot{M}_H = C_{\dot{H}}(\dot{V})$, and by 6.2.9, $M_H = N_H(Z_S)$, so since $[Z_S, U] = Z$, we conclude $M_H = UC_H(Z_S)$. Then as X and Z_U^l are normal in $C_H(Z_S)$, (1) holds. Next we verify Hypothesis G.11.1. Case (ii) of condition (3) of that Hypothesis holds by 6.2.17.9 and 6.2.16.4. As $\dot{M}_H$ contains the Sylow 2-subgroup $\dot{T}$ of $\dot{H}$, condition (4) of Hypothesis G.11.1 follows from part (1) of this lemma. So Hypothesis G.11.1 is verified. Then part (2) of the lemma follows from Theorem G.11.2. $\quad\square$

We can now complete the elimination of the case remaining after 6.2.13.

THEOREM 6.2.19. *If G satisfies Hypothesis 6.1.1, then $G \cong M_{22}$.*

PROOF. By 6.2.13, we may assume $Z_S < U \cap V$, so the subsequent lemmas in this section are applicable. In particular by 6.2.18.2, $\dot{H}$ and its action on $\hat{U}$ are described in Theorem G.11.2.

By 6.2.16.4, $\dot{V}$ is generated by an involution $\dot{v}$ of type a_2 in $Sp(\hat{U})$ and by 6.2.17.9, $\dot{v} \in \dot{X}$. However in cases (8) and (10)–(13) of G.11.2, $\dot{X}$ contains no involution i with $m([\hat{U}, i]) = 2$, so none of these cases holds. Similarly in case (9), we must have $\dot{H} = \dot{H}_1 \times \dot{H}_2$ with $\dot{H} \cong S_5$, $\dot{H}_2 \cong L_2(2)$, $\hat{U}$ is the tensor product of the natural modules for $\dot{H}_1$ and $\dot{H}_2$, and $\dot{v}$ is a transposition in $\dot{H}_1$. But then $\dot{H}_2$ is transitive on $[\hat{U}, \dot{v}]^{\#}$, contrary to parts (4) and (6) of 6.2.16. The same argument eliminates case (3) of G.11.2, as there $\dot{v}$ centralizes $Z(O(\dot{H}))$ which is transitive on $[\hat{U}, \dot{v}]^{\#}$.

Let $d := \dim(\hat{U})$. By 6.2.15.1, $d \geq 4$, so case (1) of Theorem G.11.2 does not hold.

In case (2) of G.11.2, $d = 4$ so $Sp(\hat{U}) \cong S_6$ acts naturally on $\hat{U}$. Thus as $\dot{v}$ is of type a_2, $\dot{v}$ is of cycle type 2^3 in S_6 and $3 \in \pi(\dot{H})$, so 15 or 18 divides $|\dot{H}|$ by G.11.2. Therefore $\dot{H}$ is S_6, S_5 with $\hat{U}$ the $L_2(4)$-module, or a subgroup of $O_4^+(2)$ of order divisible by 9. In each case $N_H(\hat{F})$ is transitive on $\hat{F}^{\#}$ for each totally singular line $\hat{F}$ in $\hat{U}$, contrary to 6.2.16.6.

As $\dot{v}$ is of type a_2 in $Sp_d(2)$, $|\dot{v}\dot{v}^h| \leq 4$ for each $h \in H$. Thus in case (4) of Theorem G.11.2, $\dot{v}$ is a transposition; in case (5), $\dot{v}$ is a transposition or of type 2^4; in case (6), $\dot{v}$ is a long root involution; and case (7) is eliminated. As $m([\hat{U}, \dot{v}]) = \hat{V}_U$ is of rank 2, while transpositions in cases (4) and (5) act as transvections on $\hat{U}$, we conclude that case (4) does not hold, and in case (5), that $\dot{v}$ is of type 2^4. But now $N_{\dot{H}}(\hat{V}_U)$ is transitive on $\hat{V}_U^{\#}$, contrary to 6.2.16.6. This contradiction completes the proof of the Theorem. $\square$

We summarize the work of the previous two chapters in:

THEOREM 6.2.20. *Assume G is a simple QTKE-group, $T \in Syl_2(G)$, $L \in \mathcal{L}_f^*(G, T)$ with $L/O_2(L) \cong L_2(2^n)$ and $L \trianglelefteq M \in \mathcal{M}(T)$, and $V \in \mathcal{R}_2(LT)$ with $[V, L] \neq 1$. Then one of the following holds:*

(1) $L/O_2(L) \cong A_5$, and $[V, L]$ is the sum of at most two A_5-modules for $L/O_2(L)$. Further $n(H) = 1$ for all $H \in \mathcal{H}_*(T, M)$.

(2) G is a rank-2 group of Lie type and characteristic 2, but G is $U_5(q)$ only if $q = 4$.

(3) $G \cong M_{22}$ or M_{23}.

PROOF. Suppose first that Hypothesis 5.1.8 holds. Then we may apply Theorem 5.2.3, whose conclusions are among those of (2) and (3) in Theorem 6.2.20. Thus we may suppose that Hypothesis 5.1.8 fails, and hence $n(H) = 1$ for all $H \in \mathcal{H}_*(T, M)$. Then we are done if the first statement in conclusion (1) of 6.2.20 holds; so we may assume it fails, and then we have Hypothesis 6.1.1. Then Theorem 6.2.19 says $G \cong M_{22}$, so that (3) holds. $\square$

In particular, since the groups in conclusions (2) and (3) appear in the list of our Main Theorem, the treatment of QTKE-groups G containing some T-invariant $L \in \mathcal{L}_f^*(G, T)$ with $L/O_2(L) \cong L_2(2^n)$ is reduced the case where conclusion (1) is satisfied. As mentioned at the outset, we treat this case later in Part F2, which deals with the situation where there exists $L \in \mathcal{L}_f^*(G, T)$ with $L/O_2(L)$ defined over $\mathbf{F}_2$.

Part 3

Modules which are not FF-modules

In Part 3, we consider most cases where the Fundamental Setup (3.2.1) holds for a pair L, V such that V is not a failure of factorization module for $N_{GL(V)}(Aut_{L_0}(V))$ where $L_0 := \langle L^T \rangle$. The object of Part 3 is to eliminate all but one of the pairs considered here: we will show that $G \cong J_4$ when V is the cocode module for $M/V \cong M_{24}$, and that none of the other pairs lead to examples. However we will also have to deal with a number of shadows whose local subgroups possess the pairs considered in this chapter.

THEOREM Assume the Fundamental Setup (3.2.1). Then one of the following holds:

(1) V is an FF-module for $N_{GL(V)}(Aut_{L_0}(V))$.

(2) V is the cocode module for $M/V \cong M_{24}$ and $G \cong J_4$.

(3) V is the orthogonal module for $Aut_{L_0}(V) \cong L_2(2^{2n}) \cong \Omega_4^-(2^n)$, with $n > 1$.

(4) Conclusion (3) of 3.2.6 is satisfied. In particular $L < L_0$ and $L/O_2(L) \cong L_2(2^n)$, $Sz(2^n)$, or $L_3(2)$.

Note that case (3) and a part of case (1) were handled earlier in Part 2; while case (4) and the remainder of case (1) will be handled later in Part 4 and Part 5.

In the initial chapter of Part 3, we begin to implement the outline for weak closure arguments described in subsection E.3.3. The cases not corresponding to shadows or J_4 will then be quickly eliminated by comparing various parameters associated to the representation of L_0T on V. The remaining two chapters in Part 3 will pursue the deeper analysis required when the configurations do correspond to shadows or J_4.

CHAPTER 7

Eliminating cases corresponding to no shadow

Recall we wish to prove:

THEOREM 7.0.1. *Assume the Fundamental Setup (3.2.1). Then one of the following holds:*

(1) V is an FF-module for $N_{GL(V)}(Aut_{L_0}(V))$.

(2) V is the cocode module for $M/V \cong M_{24}$ and $G \cong J_4$.

(3) V is the $\Omega_4^-(2^n)$-module for $Aut_{L_0}(V) \cong L_2(2^{2n})$.

(4) Conclusion (3) of 3.2.6 is satisfied. In particular $L < L_0$ and $L/O_2(L) \cong L_2(2^n)$, $Sz(2^n)$, or $L_3(2)$.

Recall also that in Part 3, we concentrate on the cases in the FSU not appearing in cases (1), (3), or (4) of Theorem 7.0.1; so we assume the following hypothesis:

HYPOTHESIS 7.0.2. *(1) The Fundamental Setup (3.2.1) holds. In particular $L \in \mathcal{L}_f^*(G, T)$ with $L/O_2(L)$ quasisimple, $L_0 := \langle L_0^T \rangle$, and $M := N_G(L_0)$.*

(2) V is not an FF-module for $N_{GL(V)}(Aut_{L_0}(V))$.

(3) Case (3) of 3.2.6 does not hold.

(4) V is not the orthogonal module for $Aut_{L_0}(V) \cong \Omega_4^-(2^n)$.

Part (1) of Hypothesis 7.0.2 has various consequences including the following: As $L \in \mathcal{L}^*(G, T)$, by 1.2.7.3 $L_0 T$ is a uniqueness subgroup with $M =! \mathcal{M}(L_0 T)$. Furthermore by 3.2.2.8, our module V for $L_0 T$ is 2-reduced, and we have various other properties including $Q := O_2(L_0 T) = C_T(V)$, $V \trianglelefteq T$, and $M =! \mathcal{M}(N_G(Q))$, so that $C(G, Q) \leq M$, as in 1.4.1.

By part (2) of Hypothesis 7.0.2 and Remark B.2.8, $J(T) \leq C_G(V)$, so Q contains $J(T)$. By 3.2.10, a number of useful properties follow from this fact; for example, $N_G(J(T)) \leq M$, so that $J(T) \leq S \leq T$ implies $N_G(S) \leq M$. Further there are restrictions on the subgroups $H \in \mathcal{H}_*(T, M)$: By 3.1.8.3, H centralizes $Z := \Omega_1(Z(T))$ and $C_V(L_0) = 1$.

Finally by part (3) of Hypothesis 7.0.2 and 3.2.7, V is a TI-set under M. It follows that $H \cap M \leq C_M(Z) \leq N_G(V) = M_V$.

In this chapter we begin the anaysis of groups satisfying Hypothesis 7.0.2. In the first section, we list the cases that can arise. The last of these cases seems difficult to treat using only the methods of this chapter, so in the third section we also add Hypothesis 7.3.1, which excludes that case; the case is treated in the final chapter of part 3. Also the penultimate case and the case where $L_0/O_2(L_0) \cong L_3(2)$ and $m(V) = 6$ cause difficulties, requiring extra analysis; these cases are treated in the last sections of this chapter and the next chapter.

7.1. The cases which must be treated in this part

Recall we are assuming Hypothesis 7.0.2 and the notation established in the discussion following that Hypothesis in the introduction to this chapter.

Section 3.2 determines the list of possibilities for $\bar{L}_0$ and V. We first extract the sublist consisting of those cases where V is not an FF-module for $N_{GL(V)}(Aut_{L_0}(V))$. We begin that deduction, later summarizing the final results in the Table of Proposition 7.1.1.

Recall in the Fundamental Setup that $V = \langle V_\circ^T \rangle$ for some member $V_\circ$ of $Irr_+(L_0, R_2(L_0), T)$, while $V_M := \langle V^M \rangle$, $M_V := N_M(V)$, and $\bar{M}_V := M_V/C_{M_V}(V)$ $= Aut_G(V)$. We wish to determine the cases where V is not an FF-module for $N_{GL(V)}(Aut_{L_0}(V))$.

We first consider the case where $T \not\leq N_G(L)$. Here 3.2.6 applies, and we see that in cases (1) and (2) of 3.2.6, V is not an FF-module and $V_M = V = V_\circ$; these examples appear as the last two cases (below the second horizontal line) in the Table of Proposition 7.1.1. By part (3) of Hypothesis 7.0.2, case (3) of 3.2.6 does not hold. These are the modules where $V \neq V_\circ$; they are treated later in chapter 10 of part 4 in a uniform manner, although some of these examples are FF-modules and some are not.

Therefore we may assume that $T \leq N_G(L)$, so $L_0 = L$ and $\langle L, T \rangle = LT$. We first consider the case where $T \not\leq N_G(V_\circ)$, so that case (3) of 3.2.5 holds. These modules satisfy $V_M = V = V_\circ \oplus V_\circ^t$ for $t \in T - N_T(V_\circ)$; the examples with $\bar{L} \cong L_4(2)$ or $L_5(2)$ are FF-modules, but the others are not, and so the latter appear as the second group in the Table (between the horizontal lines).

Thus we are reduced to the case $T \leq N_G(V_\circ)$, so that $V = V_\circ$. Furthermore $C_V(L) = 1$ as remarked in the introduction to this chapter, so V is an irreducible L-module. These cases are listed in 3.2.9, and form the first group in the Table— except for the first case 3.2.9.1, which is excluded by part (4) of Hypothesis 7.0.2. This case was handled in part 2 in the "Generic Case", since the unitary groups $U_4(2^n)$ arise in that case.

This completes the deduction of Proposition 7.1.1.

We also indicate, in the last two columns of the Table of that result, first the "shadows"(that is, groups having such a local configuration but which are not quasithin or simple), and then the single simple quasithin example given by J_4.

Three of the cases seem to require treatment different from the fairly uniform approach used to treat the remaining cases. In the final case where V is the orthogonal module for $\bar{L}_0 = \Omega_4^+(2^n)$, we have $m = 2$ when $n = 2$—and worse, $a = m = n$ for any n, and as L is not normal in M, we can't appeal to Remark 4.4.2. Because of these difficulties, this case will be treated by more direct methods in the third and final chapter of this part. The penultimate case poses similar difficulties, and is treated in the last section of the second chapter 8 of this part. Finally the case where $\bar{L}_0\bar{T} \cong Aut(L_3(2))$ and V is the sum $3 \oplus \bar{3}$ of the natural and dual module requires special treatment, particularly as $m = 2$ makes it difficult to establish lemma 7.3.2. This case is dealt with at the end of chapter 7.

We have established the list of cases to be treated under Hypothesis 7.0.2:

PROPOSITION 7.1.1. *The cases where V is not an FF-module, and which appear in neither conclusion (3) nor (4) of Theorem 7.0.1, are:*

$\bar{M}_V \geq$	restr. on n	$\dim V$	descr. V	shadows	example
$U_3(2^n)$	$n \geq 2$	$6n$	natural		
$Sz(2^n)$	odd $n \geq 3$	$4n$	natural		
$L_3(2^{2n})$	$n \geq 2$	$9n$	$3 \otimes 3^\sigma$	$U_6(2^n), U_7(2^n)$	
$Aut(M_{12})$		10	irred.perm.		
$\hat{M}_{22}$		12	unitary		
M_{22}		10	code	Co_2	
		$\overline{10}$	cocode	F_{22}	
M_{23}		11	code		
		$\overline{11}$	cocode	F_{23}	
M_{24}		11	code	Co_1	
		$\overline{11}$	cocode	F_{24}	J_4
$SL_3(2^n).2$		$6n$	$3n \oplus \overline{3n}$		
$Sp_4(2^n)'.2$		$8n$	$4n \oplus 4n^{\bar{t}}$		
S_7		8	$4 \oplus \bar{4}$		
$L_3(2) \wr 2$		9	$3 \otimes 3^t$	$L_6(2).2, L_7(2).2$	
$L_2(2^n) \wr 2$	$n \geq 2$	$4n$	$2n \otimes 2n^{\bar{t}}$	$L_4(2^n).2, L_5(2^n).2$	

7.2. Parameters for the representations

Our main task in chapter 7 will be to eliminate the cases not corresponding to a shadow or example. We use the weak closure methods of section E.3. These methods are "numerical", in the sense that they compare parameters—such as a, m, n', α, β determined only by the representation of M on V, and on other parameters r, s, w determined by suitable subspaces U of V with $C_G(U) \leq M$. We will obtain a numerical contradiction from the Fundamental Weak Closure Inequality involving these parameters, established in E.3.29. [1]

Because the initial steps in the weak closure argument involve primarily the parameters m_2 of $\bar{M}_V$ and m, a of the module V, estimates on these values are included in the early columns of the Table in Proposition 7.2.1 below.

Proofs that the parameters are indeed as indicated in the Table appear in corresponding sections of chapter H of Volume I—with the exception of the parameter n', which is determined in 7.3.4. Certain values in the table are given in parentheses; these are values which seem to be well known, but which we do not require in our argument, and hence are not verified in chapter H. The last two columns of the table list parameters α and β primarily relevant to an application of E.6.27 later in this chapter; the derivation of these parameters also appears in chapter H, except in some cases like the last case where they are not used.

We now describe the Table in more detail: Column 1, labeled "case", indicates the pair $\bar{L}_0, V$ discussed in the corresponding row. Column 2, labeled "$a \leq$", gives an upper bound on $a := a(\bar{M}_V, V)$. Column 3, labeled "$m \geq$", gives a lower bound on $m := m(\bar{M}_V, V)$. The definitions of these parameters appear as E.3.9 and E.3.1. Column 4, labeled "$w \geq$", gives the resulting lower bound on the difference $m - a$, which is in turn a lower bound on the parameter w of Definition E.3.23 by 7.3.3. Column 5, labeled "n'", is the parameter $n' := n'(Aut_G(V))$ given in Definition

---

[1] Of course, local configurations $\bar{L}, V$ that actually exist in shadows are not eliminated numerically. So in the following chapter 8, we instead show that those configurations provide the *unique* solution to the FWCI; and then eliminate the cases by showing those configurations violate our SQTK hypothesis.

E.3.37; by 7.3.4 this column will give an upper bound on w. Column 6, labeled "$m_2 \leq$", gives an upper bound on $m_2 := m_2(\bar{M}_V)$. Columns 7 and 8, labeled "$\beta \geq$" and "$\alpha \geq$", give the minimum codimension of a subspace U of V such that $O^2(C_M(U)) \not\leq C_M(V)$, or such that $C_{\bar{M}_V}(U)$ contains an $(F-1)$ offender, respectively. If there are no $(F-1)$-offenders, then $J_1(T)$ centralizes V and column 8 contains ∞. We remark that the minimum of α and β by 7.4.1 gives a lower bound for the parameter r of Definition E.3.3 in the cases where $L \trianglelefteq M$.

PROPOSITION 7.2.1. *The values of various parameters for our modules are:*

case	$a \leq$	$m \geq$	$w \geq$	n'	$m_2 \leq$	$\beta \geq$	$\alpha \geq$
$SU_3(2^n)/6n$	n	$2n$	n	n	$n+1$	$4n$	∞
$Sz(2^n)/4n$	n	$2n$	n	n	n	$\frac{8}{3}n$	∞
$(S)L_3(2^{2n})/9n$	$3n$	$3n$	0	$2n$	$4n$	$4n$	$\infty; 5 \ if \ n=1$
$M_{12}/10$	2	4	2	2	4	6	∞
$3M_{22}/12$	3	4	1	2	5	8	∞
$M_{22}/10$	3	3	0	2	5	6	6
$M_{22}/\overline{10}$	3	3	0	2	5	6	5
$M_{23}/11$	3	4	1	2	4	6	∞
$M_{23}/\overline{11}$	3	4	1	2	4	6	5
$M_{24}/11$	3	4	1	2	6	6	7
$M_{24}/\overline{11}$	3	4	1	2	6	6	5
$SL_3(2^n).2/3n \oplus 3n$	n	$2n$	n	n	$2n$	$4n$	$\infty; 2 \ if \ n=1$
$Sp_4(2^n)'.2/4n \oplus 4n^{\bar{t}}$	$< 2n$	$3n$	$> n$	n	$3n$	$4n$	∞
$S_7/4 \oplus \bar{4}$	2	4	2	2	3	4	∞
$L_3(2) \wr 2/3 \otimes 3^t$	2	3	1	2	4	6	3
$L_2(2^n) \wr 2/2n \otimes 2n^{\bar{t}}$	(n)	n	0	n	$2n$	$(2n)$	$\infty; 2 \ if \ n=1$

7.3. Bounds on w

We now implement the outline discussed in subsection E.3.3.

As remarked earlier, in chapter 7 and the next chapter 8, we exclude the final case in the Tables of Propositions 7.1.1 and 7.2.1:

HYPOTHESIS 7.3.1. *V is not the orthogonal module for $\bar{L}_0 \cong \Omega_4^+(2^n)$.*

Recall that the case excluded by Hypothesis 7.3.1 will be treated by other methods in the third chapter 9 of this part 3. Thus in this chapter and the next, discussion of "all" cases in the Tables refers to the remaining cases, with the final row of the Tables excluded.

We first discuss the parameters r and s. See Definitions E.3.3, E.3.5, E.3.1, and E.3.9 for the parameters r, s, m, and a.

PROPOSITION 7.3.2. *$r \geq m$, so that $s = m$.*

PROOF. This follows from Theorem E.6.3 when $m > 2$, which we see from Table 7.2.1 holds in all cases except for $L_3(2)$ on $3 \oplus \bar{3}$. In that case we make a direct argument, but as the methods are of a different flavor from the uniform treatment in this chapter, we banish those details to a mini-Appendix at the end of the chapter; see 7.7.1 for the proof. $\square$

In view of 7.3.2, the column headed $m \geq$ in Table 7.2.1 also provides a lower bound for the parameter s. Then comparison with a gives us information on w. Recall from Definition E.3.23 that

$$w := \min\{m(V^g/V^g \cap T) : g \in G \text{ and } [V, V^g \cap T] \neq 1\}.$$

LEMMA 7.3.3. *The column "$w \geq$" of Table 7.2.1 gives a lower bound for w.*

PROOF. Recall from E.3.34.1 that $w \geq s - a$. As $s = m$ by 7.3.2, we subtract the column for a from the column for m in the Table, and obtain the result. $\square$

Having established a lower bound on w, we now apply E.3.35 in order to obtain an upper bound for w.

Let H denote an arbitrary member of $\mathcal{H}_*(T, M)$, although from time to time we may temporarily impose further constraints on H.

PROPOSITION 7.3.4. $w \leq n(H) \leq n'(\bar{M}_V) = n' < s$, *where* n' *is listed in the column headed "n'" in Table 7.2.1.*

PROOF. Let k denote the value of n' given in Table 7.2.1; we first assume $n' = k$. Recall that $s = m$ by 7.3.2, and observe further that $m > n'$ in all cases in the Table, so that $s > n'$. Next we check that Hypothesis E.3.36 is satisfied: We observed in the introduction to this chapter that $V \trianglelefteq T$, $M = !\mathcal{M}(N_G(Q))$, and V is a TI-set under M, with $H \leq C_G(Z)$, and $H \cap M \leq C_M(Z) \leq N_M(V)$. Further by Hypothesis 7.0.2, V is neither an FF-module nor the orthogonal module for $L_2(2^{2n})$, so whenever $n(H) > 1$ we can apply Theorem 4.4.14 to conclude that a Hall $2'$-subgroup B of $H \cap M$ is faithful on $\bar{L}_0$, and hence also on V. It follows that $C_{H \cap M}(V) \leq O_2(H \cap M)$, completing the verification of Hypothesis E.3.36. Now since $n' < s \leq r$, the lemma holds by E.3.39.1.

Thus it remains to verify that $k = n'$. If $\bar{L}$ is $L_3(2)$ on $3 \oplus \bar{3}$ or $Sp_4(2)' \cong A_6$ on $4 \oplus \bar{4}$, then T is nontrivial on the Dynkin diagram of $\bar{L}$, and hence $\bar{T}$ permutes with no nontrival subgroup of $\bar{M}_V$ of odd order, so that $n' = 1 = k$. In all other cases where $\bar{L}$ is of Lie type, $\bar{T}$ permutes with a Cartan subgroup of $\bar{L}$, which contains a cyclic subgroup of order $2^k - 1$, so that $n' \geq k$ in these cases. Similarly when $\bar{L}$ is sporadic, $\bar{T}$ permutes with a subgroup of order 3 and $k = 2$, so $n' \geq k$. Finally if $n' > k$ then $n' > 2$ and we may apply A.3.15 to some prime $p > 3$ which does not divide $k(2^k - 1)$ and obtain a contradiction which completes the proof. $\square$

We can already see that when $\bar{L}$ is $Sp_4(2^n)$, the value in the column $w \geq$ strictly exceeds the value in the column n', so that 7.3.3 and 7.3.4 provide our first example of a numerical contradiction, eliminating one of our cases from Table 7.1.1:

COROLLARY 7.3.5. $\bar{L}$ *is not* $Sp_4(2^n)'$. [2]

7.4. Improved lower bounds for r

We saw earlier in 7.3.2 that $r \geq m \geq 2$. In many cases, we can improve this bound on r using E.6.28: First $r > 1$, giving hypothesis (1) of E.6.28. As V is not an FF-module, hypothesis (2) of E.6.28 holds. Finally if $L \trianglelefteq M$, and X is an abelian subgroup of $C_M(V)$ of odd order, then $N_G(X) \leq M$ by Theorem 4.4.3.

---

[2]It would also be possible to eliminate case (iii) of 3.2.6.3.c at this point (adjusting for the fact that V might not be a TI-set under M). However, it seems more natural to treat all cases of 3.2.6.3.c uniformly in chapter 10.

Note that when $L \trianglelefteq M$, Hypothesis 4.4.1 is satisfied by any abelian subgroup X of $C_M(V)$ of odd order, in view of Remark 4.4.2. Thus the hypotheses of E.6.28 are satisfied, so $r \geq \min\{\alpha, \beta\}$ by that result, while column 7 and 8 in Table 7.2.1 give lower bounds on α and β, so:

PROPOSITION 7.4.1. *If* $L \trianglelefteq M$ *then* $r \geq \min\{\alpha, \beta\}$, *the bound appearing in the final column of Table 7.2.1.*

7.5. Eliminating most cases other than shadows

We begin with the cases which are simplest to eliminate. Recall the Fundamental Weak Closure Inequality E.3.29:

LEMMA 7.5.1. (**FWCI**) $m_2 + w \geq r$.

We add the adjacent columns for $w \leq$ and $m_2 \leq$ in Table 7.2.1, and compare this sum S with the bound R given by the final column $\min\{\alpha, \beta\}$ of the Table. We find in the following cases that we get the contradiction $S < R$ to the FWCI, in view of 7.4.1:

LEMMA 7.5.2. *(1)* $\bar{L}$ *is not* $U_3(2^n)$, $Sz(2^n)$, *or* $\hat{M}_{22}$.
(2) If $\bar{L}_0$ *is* $L_3(2^n)$ *on* $3 \oplus \bar{3}$, *then* $n = 1$.

Certain other cases are not immediately ruled out, but require only a slight extension of this argument.

For the rest of the section, adopt the notation of the latter part of section E.3: Let $A := N_{V^g}(V)$, be a "w-offender" on V; that is $m(V^g/A) = w$ with $A \not\leq C_G(V)$, so that $\bar{A} \neq 1$.

LEMMA 7.5.3. *(1) Assume the inequality in 7.5.1 is an equality, and let*
$$\mathcal{B} := \{B \leq A : |B : C_A(V)| = 2\}, \quad \text{and} \quad W := \langle C_V(B) : B \in \mathcal{B} \rangle.$$
Then $m(\bar{A}) = m_2$, $r = m(V^g/C_A(V))$, *and* $W \leq N_V(V^g)$. *Further* $m(V/W) \geq w$, *and in case of equality,* $W = N_V(V^g)$ *is a* w-offender on V^g and $m(W/C_V(A)) = m_2$.
(2) $m(\bar{A}) \geq r - w$.
(3) $C_V(A) = C_V(V^g)$.

PROOF. By 7.3.4, $w < s$, so (3) follows from E.3.6. By part (2) of Hypothesis 7.0.2, Hypothesis E.3.24 is satisfied. Thus (1) follows from E.3.31 and (3), and (2) from E.3.28.3.   □

In certain cases when the FWCI has a unique solution, the embedding of $\bar{A}$ in $\bar{M}_V$ is determined, which leads to a contradiction:

LEMMA 7.5.4. $\bar{L}$ *is neither* M_{12}, *nor* M_{23} *on the code module 11.*

PROOF. Assume otherwise. From Table 7.2.1 and 7.4.1, the FWCI is an equality with $w = 2$. Therefore by 7.5.3.1, $m(\bar{A}) = m_2 = 4$ and $r = 6 = m(V^g/C_A(V))$. Define W as in 7.5.3.1, and observe that $W \leq N_V(V^g)$ and $m(V/W) \geq w = 2$ by that result. But if $\bar{M}_V = M_{23}$, then as $m(\bar{A}) = 4$, H.16.8 says $m(V/W) < 2$, a contradiction.

Therefore $\bar{M}_V = M_{12}$. Here as $m(\bar{A}) = 4$, $U = C_V(A)$ is of dimension at most 3 and $m(W) \geq 8$ by H.11.1.4. But then $m(W/U) \geq 5 > 4 = m_2$, contrary to 7.5.3.1. This contradiction completes the proof.   □

In the case of A_7, we can dig a little deeper to increase r:

LEMMA 7.5.5. $\bar{L}$ is not A_7.

PROOF. Assume $\bar{L}$ is A_7. First $r \geq 4$ by 7.4.1, and by the FWCI 7.5.1, it suffices to show that $r > 5$. We appeal to E.6.27 with $j = 1$: As ∞ is in the column for α in Table 7.2.1, V is not an $(F - 1)$-module for $Aut_{\bar{M}}(\bar{L})$, hence $J_1(M) \leq C_M(V)$. From the proof of 7.4.1, $C_G(X) \leq M$ for any $1 \neq X \leq C_M(V)$ of odd order. Thus for $U \leq V$ with $O^2(C_M(U)) \leq C_M(V)$, E.6.27 says $C_G(U) \leq M$. Let $\mathcal{U}$ consists of those $U_1 \leq V$ with $O^2(C_M(U_1)) \not\leq C_M(V)$; it suffices to show $C_G(U_1) \leq M$, for each $U_1 \in \mathcal{U}$ with $m(V/U_1) < 6$. But if $U_1 \in \mathcal{U}$ with $m(V/U_1) < 6$, then $U_1 < U_s := C_V(\bar{s})$ where $\bar{s}$ is a 3-element of cycle type 3^2 in A_7. Thus it will suffice to show that $C_G(U_1) \leq M$, for each U_1 of codimension at most 1 in U_s. Choose a counterexample U_1, and let $U_1 \leq U_2 \leq V$ be maximal subject to $C_G(U_2) \not\leq M$. Note that $C_{\bar{M}}(U_1) = \langle \bar{s} \rangle$, and in particular $O^{2'}(C_M(U_1)) \leq C_M(V)$: For $V = V_1 \oplus V_2$ where $\{V_1, V_2\} = Irr_+(L, V)$, so that $U_s = (U_s \cap V_1) \oplus (U_s \cap V_2)$ and $U_s \cap V_j$ is a 2-subspace of V_j. If $\bar{i}$ is an involution in $\bar{M}$ centralizing U_1, then i must act on $U_1 \cap V_j \neq 0$ and hence on V_j. Thus i centralizes the projection $U_{1,j}$ of U_1 on V_j, and so for $j = 1$ or 2, $U_{1,j} = U_s \cap V_j$. This is impossible as $C_{\bar{L}}(U_s \cap V_j) = \langle \bar{s} \rangle$. So U_1, and hence also U_2, lies in the set Γ of Definition E.6.4. Then U_2 satisfies the hypotheses of E.6.11, so as $m(V/U_2) < 6$ and $m(V/U_2) \geq r \geq 4$, we conclude from E.6.11 that $Aut_{C_M(U_2)}(V)$ contains an element of order 15 or 31, whereas A_7 has no such element. This contradiction shows that $C_G(U_1) \leq M$, completing the proof of the lemma. $\qquad \square$

Finally our weak closure methods provide some numerical information which will be useful in the next chapter in treating two cases arising in certain shadows:

LEMMA 7.5.6. (1) If $\bar{L}$ is M_{22} on the code module then $w > 0$.
(2) If $\bar{L}$ is $(S)L_3(2^{2n})$ on $9n$ then $w \geq n$.

PROOF. Assume that the lemma fails. From Table 7.2.1 and 7.4.1, $r \geq 4n$ if $\bar{L}$ is $(S)L_3(2^{2n})$, while $r \geq 6$ if $\bar{L}$ is M_{22} on the code module. From Table 7.2.1, $m_2 \leq 5$ when $\bar{L}$ is M_{22}, so 7.5.1 supplies a contradiction to our assumption that $w = 0$ in that case. Thus $\bar{L}$ is $(S)L_3(2^{2n})$.

By E.3.10, $\bar{A} \in \mathcal{A}_{s-w}(\bar{M}_V, V)$, while by 7.3.2, $s = m$. Thus $s \geq 3n$ by Table 7.2.1, so as $w < n$, $\bar{A} \in \mathcal{A}_{2n+1}(\bar{M}_V, V)$. By 7.5.3.2, $m(\bar{A}) \geq r - w > 3n$. Thus we have verified the hypotheses of lemma H.4.5.

Next if $\bar{B}_1 \leq \bar{A}$ with $m(\bar{A}/\bar{B}_1) \leq 3n$ and B is the preimage in A of $\bar{B}_1$, then $m(V^g/B) \leq 3n + w < 4n \leq r$, so $C_V(\bar{B}_1) = C_V(B) \leq N_G(V^g)$ by E.3.4. Thus

$$W_A = \langle C_V(\bar{B}_1) : m(\bar{A}/\bar{B}_1) \leq 3n \rangle \leq N_V(V^g).$$

Therefore $[W_A, A] \leq W_A \cap V^g \leq C_{W_A}(A)$, so A is quadratic on W_A, contrary to H.4.5.2. This contradiction completes the proof of (2) and establishes the lemma. $\qquad \square$

7.6. Final elimination of $\mathbf{L_3(2)}$ on $\mathbf{3 \oplus \bar{3}}$

In this section, we eliminate the case left open in 7.5.2.2. This "small" case of $L_3(2)$ on $3 \oplus \bar{3}$ seems to require special treatment: For example, we've already seen in 7.3.2 that the fact that $m = 2$ requires arguments of a different flavor to prove that $r \geq m$; indeed recall that we are postponing that proof that $r \geq m$ until Theorem 7.7.1 in the final section of the chapter.

A second difficulty is that we cannot improve our lower bound on r using E.6.28: since when V is the $3 \oplus \bar{3}$-module for $L_3(2)$, the elementary groups of rank 1 or 2 in $\bar{L}$ centralize subspaces of codimensions 2 or 3 in V, respectively, and hence are $(F-1)$-offenders. In the next lemma, we use *ad hoc* methods to complete the treatment of the case of $L_3(2)$ on $3 \oplus \bar{3}$.

LEMMA 7.6.1. $\bar{L}_0$ *is not* $L_3(2)$ *on* $3 \oplus \bar{3}$.

PROOF. From Table 7.2.1, $n' = 1$, so $n(H) = 1$ for each $H \in \mathcal{H}_*(T, M)$ by 7.3.4. Also $w > 0$ by 7.3.3 and Table 7.2.1, while $w \leq n(H) = 1$ by 7.3.4; so in fact $w = 1$.

Next $r \geq 3$ as we will show in 7.7.1 in the final section, so as $m_2 = 2$, 7.5.1 is an equality; hence $m(\bar{A}) = 2$ and $m(V^g/C_A(V)) = 3 = r$ by 7.5.3.1.

Suppose first that $\bar{A} \not\leq \bar{L}$. Then by H.4.3.1, $U := C_V(A)$ has dimension 2, and (for $\mathcal{B}$ as in 7.5.3.1) $A_1 := \langle C_V(B) : B \in \mathcal{B} \rangle$ is of dimension 5, while $A_1 \leq N_V(V^g)$ by 7.5.3.1. Also $U = C_V(V^g)$ by 7.5.3.3, so that $m(Aut_{A_1}(V^g)) = 3$, contradicting $m_2(\bar{M}) = 2$.

Thus $\bar{A} \leq L$. In the notation of 7.7.1 and subsection H.4.1 of chapter H of Volume I, $V = V_1 \oplus V_2$ with $V_i \in Irr_+(L, V)$, $V_2 = V_1^t$ for $t \in T - N_T(V_1)$, and V_1 has basis denoted by $1, 2, 3$. By H.4.3.2, we may take $\bar{A}$ to be the unipotent radical of the centralizer of the vector $1 \in V_1$; then $U := C_V(A) = \langle 1 \rangle \oplus \langle 2^t, 3^t \rangle$ is of rank 3, and

$$A_1 = \langle C_V(\bar{a}) : \bar{a} \in \bar{A}^{\#} \rangle = V_1 \oplus \langle 2^t, 3^t \rangle$$

is of rank 5. So by 7.5.3.1, $A_1 = N_V(V^g)$; thus we have symmetry between V and V^g, in that A_1 is also a w-offender on V^g. Set $(M^g)^* := M^g/C_G(V^g)$. Then $A_1^* \leq L^{g*}$ by the previous paragraph, so $U_1 := C_{V^g}(A_1^*)$ is 3-dimensional and $U_1 = C_A(V)$.

In particular $Z_1 := [A, A_1] \leq V \cap V^g$, and by H.4.3.2, Z_1 is generated by the vector $1 \in V_1$. Thus

$$X := \langle V^g, V \rangle \leq G_1 := N_G(Z_1) = C_G(Z_1).$$

Now A centralizes U and V/U, and by symmetry, A_1 centralizes U_1 and V^g/U_1. It follows that X centralizes the quotients in the series

$$1 < UU_1 < AA_1.$$

Set $\tilde{X} := X/AA_1$. As $\tilde{V}$ and $\tilde{V}^g$ have order 2, $\tilde{X}$ is dihedral; set $\tilde{Y} := O(\tilde{X})$. A Hall $2'$-subgroup Y_0 of the preimage of $\tilde{Y}$ centralizes AA_1 by Coprime Action, and then as $r = 3$ while $m(V/A_1) = 1 = m(V^g/A)$, Y_0 centralizes $\langle V^g, V \rangle = X$. As $\tilde{Y}$ is dihedral, $\tilde{Y}_0 = 1$, so X is a 2-group.

We can now finish the proof of the lemma using later Proposition 7.7.2, which says that $G_1 \in \mathcal{H}^e$; we postpone the statement and proof of Proposition 7.7.2 until the next section, as it is proved in parallel with lemma 7.7.6.

Set $\tilde{G}_1 := G_1/Z_1$; then as $G_1 \in \mathcal{H}^e$, $F^*(\tilde{G}_1) = O_2(\tilde{G}_1)$ by A.1.8. Recall $T_1 := C_T(Z_1)$ is Sylow in G_1 by 3.2.10.4. Now $T_1 \leq LO_2(LT)$, so $C_{\tilde{V}_1}(T_1)$ and $C_{\tilde{V}_1^t}(T_1)$ are nontrivial, and by B.2.14 both lie in $O_2(\tilde{G}_1)$. Then as $C_L(Z_1)$ is irreducible on $\tilde{V}_1 = \widetilde{A_1 \cap V_1}$ and $\langle \tilde{2}^t, \tilde{3}^t \rangle = \widetilde{A_1 \cap V_2}$, it follows that $A_1 \leq O_2(G_1)$.

Since $Z_1 \leq V \cap V^g$ with $[V, V^g] \neq 1$, 3.2.10.6 says that $V \not\leq O_2(G_1)$. So as $|V : A_1| = 2$, $A_1 = V \cap O_2(G_1)$, and hence $m(V/V \cap O_2(G_1)) = 1$. Then for any $h \in G_1$, we have $m(V^h/V^h \cap O_2(G_1)) = 1$, with $V^h \cap O_2(G_1) \leq T_1 \leq N_G(V)$. If V^h centralizes V, then $\langle V, V^h \rangle = VV^h$ is a 2-group, while if V^h does not centralize V then $V^h \cap O_2(G_1)$ is a w-offender on V, so our argument above for V^g applies to V^h to show $\langle V, V^h \rangle$ is again a 2-group. Therefore the Baer-Suzuki Theorem forces $V \leq O_2(G_1)$, which we saw is not the case. This completes the proof. $\qquad \square$

7.7. mini-Appendix: $r > 2$ for $\mathbf{L_3(2).2}$ on $3 \oplus \bar{3}$

Our goal in this section is to prove the following two results:

THEOREM 7.7.1. *If $\bar{L}_0$ is $L_3(2)$ on $3 \oplus \bar{3}$, then $r > 2$. In particular, $s = m = 2$.*

PROPOSITION 7.7.2. *Assume $\bar{L}_0$ is $L_3(2)$ on $3 \oplus \bar{3}$, and $r > 2$. Then $F^*(C_G(v_1)) = O_2(C_G(v_1))$ for each $V_1 \in Irr_+(L, V)$ and $v_1 \in V_1^\#$.*

So throughout this section, assume we are in the case where $\bar{L}_0$ is $L_3(2)$ on $3 \oplus \bar{3}$. Recall $L \in \mathcal{L}^*(G, T)$, $L \trianglelefteq M \in \mathcal{M}(T)$, $V \in \mathcal{R}_2(LT)$ is normal in M, $\bar{M} := M_V/C_M(V) \cong Aut(L_3(2))$, and $V = V_1 \oplus V_2$, where $V_2 := V_1^t$ for $t \in T - N_T(V_1)$ and V_2 is the dual of the natural module V_1. Recall $Q := O_2(LT)$.

The module V is discussed in subsection H.4.1 of chapter H of Volume I, where we find that we can view $\bar{L}$ as the group of invertible 3×3 matrices over $\mathbf{F}_2$, with respect to some basis of V_1 denoted by $\{1, 2, 3\}$, with $\bar{t}$ the inverse-transpose automorphism.

7.7.1. Reduction to $\mathbf{C_G(V_0) \leq M}$ for $\mathbf{V_0 := \langle 1, 1^t \rangle}$.

Our goal in Theorem 7.7.1 is to show that $r(G, V) > 2$, so we need to prove that $C_G(U) \leq M$ for each $U \leq V$ with $m(V/U) \leq 2$. It turns out this can be accomplished by controlling the centralizer of the single subspace $V_0 := \langle 1, 1^t \rangle$, by showing:

PROPOSITION 7.7.3. $G_0 := C_G(V_0) \leq M$.

In this short subsection, we prove that Theorem 7.7.1 can be deduced from Proposition 7.7.3.

So assume Proposition 7.7.3, and suppose that for some $U \leq V$ with $m(V/U) \leq 2$, we have $C_G(U) \not\leq M$.

We first consider the case where $m(V/U) = 1$. Since V admits an orthogonal form, $U = v^\perp$ for some $v \in V$. Now replacing the orbit representatives in H.4.2 by conjugates $v = 2, 2 + 3^t, 2 + 2^t$, we see using the form in H.4.1 that $V_0 \leq v^\perp = U$, so that $C_G(U) \leq C_G(V_0) \leq M$ by Proposition 7.7.3.

Thus we have established that $r > 1$, so it remains to treat the case $m(V/U) = 2$.

First assume U is centralized by no involution of $\bar{M}$. Then Q is Sylow in $C_M(U)$, and no nontrivial element of odd order in $\bar{M}$ centralizes a subspace of V of codimension 2, so that $C_M(U) = C_M(V)$. Hence as $r > 1$, we get $C_G(U) \leq M$ from E.6.12.

This leaves the case where U is centralized by some involution $\bar{i} \in \bar{M}$. Since $m(V/U) = 2$, we must have $\bar{i} \in \bar{L}$, and conjugating in $\bar{L}$, we may take $\bar{i}$ to be given by the matrix for the permutation $(2, 3)$ (and hence also $(2^t, 3^t)$). So again $V_0 \leq U$, and Proposition 7.7.3 gives $C_G(U) \leq M$.

This completes the proof of Theorem 7.7.1 modulo Proposition 7.7.3. So the remainder of this section is devoted to the proof of Propositions 7.7.3 and 7.7.2.

7.7.2. More detailed properties of V_0 and its centralizer. Observe $C_{\bar{M}}(V_0)$ is the subgroup of $\bar{L}$ fixing 1 and acting on the subspace $\langle 2, 3 \rangle$, so $C_{\bar{M}}(V_0) \cong L_2(2) \cong S_3$.

Set $L^0 := O^2(C_L(V_0))$, so that $\bar{L}^0/O_2(\bar{L}^0)$ is of order 3. Let $\theta \in L^0$ be of order 3. Observe

$$[V, L^0] =: V_- = \langle 2, 3 \rangle \oplus \langle 2^t, 3^t \rangle = V_0^\perp.$$

and

$$V = V_0 \oplus V_-.$$

Set $T_0 := C_T(V_0)$ and $M_0 := C_M(V_0)$. Then $C_{LT}(V_0) = L^0 T_0$, $T_0 \in Syl_2(C_M(V_0))$, and $\overline{T_0}$ of order 2 is generated by the involution $\bar{i}$ defined in the previous subsection.

Let $Z_1 := \langle 1 \rangle$, $G_1 := C_G(Z_1)$, and $L_1 := O^2(C_L(Z_1))$. Thus $Z_1 \leq V_0$, so $G_0 \leq G_1$ and $L^0 \leq L_1$. Again $L_1/O_2(L_1)$ is of order 3, but $L_1/Q \cong A_4$ while $L^0/Q \cong \mathbf{Z}_3$.

Let V_+ denote either V_0 or Z_1, and define $G_+ := C_G(V_+)$, $L_+ := O^2(C_L(V_+))$, $M_+ := C_M(Z_+)$, and $T_+ := C_T(V_+)$. Then

$$M_+ = C_M(V)L_+T_+,$$

and by 3.2.10.4, T_+ is Sylow in G_+.

We emphasize that

$$Q = O_2(L^0 T_0),$$

and that this property is crucial to our proof that $G_0 \leq M$.

LEMMA 7.7.4. *If Y is an abelian subgroup of $C_M(V_+)$ of odd order, then*

(1) $Y_C := C_Y(V)$ *is of index at most 3 in Y, and*

(2) *if $Y_C \neq 1$, then $N_G(Y_C) \leq M$.*

PROOF. As Y is of odd order in $O^2(C_M(V_+)) = O^2(C_M(V))L_+$ and $|\bar{L}_+ : O_2(\bar{L}_+)| = 3$, $|Y : Y_C| \leq 3$. By Theorem 4.4.3 and Remark 4.4.2, $N_G(Y_C) \leq M$. $\square$

LEMMA 7.7.5. *If $w \in V^\#$ is 2-central in G, and $L_+T_+ \leq H \leq G_+$, then*

$$F^*(C_G(w)) = O_2(C_H(w)).$$

PROOF. We show that the hypotheses of 1.1.4.4 are satisfied with $G_w := C_G(w)$ in the role of "M", and $H \cap G_w$ in the role of "N". First $G_w \in \mathcal{H}^e$ by 1.1.4.3 and our hypothesis that w is 2-central. Set $G_{+,w} := C_G(V_+\langle w \rangle)$, and embed $Q \leq T_w \in Syl_2(G_{+,w})$. Then $J(T) \leq Q \leq T_w$ so $T_w \leq N_G(T_w) \leq M$ by 3.2.10.8. Consequently $T_w \leq M_+$, which we saw above is $C_M(V)L_+T_+$. Then by Sylow's Theorem, $T_w^c \leq L_+T_+$ for some $c \in C_M(V) \leq G_{+,w}$, so without loss $T_w \leq L_+T_+ \leq H$. Hence $V_+ \leq H \cap O_2(G_+) \cap G_w \leq O_2(H \cap G_w)$. So

$$C_{O_2(G_w)}(O_2(H \cap G_w)) \leq C_{O_2(G_w)}(V_+) \leq O_2(G_w) \cap G_{+,w} \leq O_2(G_{+,w})$$

$$\leq T_w \leq H \cap G_w.$$

Thus we finally have the hypothesis for 1.1.4.4, and we conclude from 1.1.4.4 that $H \cap G_w \in \mathcal{H}^e$. $\square$

7.7.3. Proof of Proposition 7.7.2. In the remaining two subsections of the section, we assume that either

(H0) $V_+ = V_0$ and $G_0 \not\leq M$, or

(H1) $V_+ = Z_1$, $r > 2$, and $G_1 \notin \mathcal{H}^e$.

In each case, we work toward a contradiction. In this subsection, we assume (H1) and obtain a contradiction establishing Proposition 7.7.2, and hence also completing the proof of lemma 7.6.1, which depended upon that Proposition. At the same time, we will prove a lemma 7.7.6, necessary for the proof of Proposition 7.7.3. Then in the final subsection we assume (H0) and complete the proof of Proposition 7.7.3, on which various earlier results depended.

Under (H0), choose $H \in \mathcal{H}_*(L^0 T_0, M)$ with $H \leq G_0$. Under (H1), choose $H \in \mathcal{H}(L_1 T_1, M)$ with $H \leq G_1$, and H minimal subject to $H \notin \mathcal{H}^e$.

In either case set $M_H := H \cap M$. As $H \in \mathcal{H}$, H is an SQTK-group. Set $A := V_+ V_-$; and observe that $A = V$ under (H0), while A is a hyperplane of V under (H1). Therefore $C_G(A) \leq M$ under either hypothesis, since $r > 1$ in Hypothesis (H1).

Under Hypothesis (H0) we will prove:

LEMMA 7.7.6. *Assume Hypothesis (H0). Then*

(1) $T_0 \in Syl_2(H)$.

(2) $H = J(H) L^0 T_0$.

(3) $F^*(H) = O_2(H)$.

We prove lemma 7.7.6 and Proposition 7.7.2 together.

First assume just Hypothesis (H0). Since T_0 is Sylow in G_0, part (1) of 7.7.6 holds. As $O_2(L^0 T_0) = Q$, with T_0 Sylow in both $L^0 T_0$ and H, we conclude from A.1.6 that $O_2(H) \leq Q$. By a Frattini Argument, $H = J(H) N_H(R)$, where $R := T_0 \cap J(H) \in Syl_2(J(H))$, and $J(T) = J(R)$. Then $N_H(R) \leq M$ by 3.2.10.8, so as $H \not\leq M$, also $J(H) \not\leq M$—and hence part (2) of 7.7.6 follows from minimality of H.

It now remains to prove part (3) of 7.7.6, as well as Proposition 7.7.2. Thus we assume either (H0) or (H1), and it remains to show that $F^*(H) = O_2(H)$. As a first step, A.1.6 says $O_2(M) \leq Q \leq T_+ \leq H$, so by 1.1.4.5, $F^*(M_H) = O_2(M_H)$.

Next applying A.1.26.1 to L^0 on $V_- = [V_-, L^0]$, V_- centralizes $O(H)$. Therefore

$$O(H) \leq C_H(V_-) = C_H(V_+ V_-) = C_H(A).$$

Thus given our earlier observation that $C_G(A) \leq M$, $O(H) \leq O(M_H)$, so $O(H) = 1$ since $M_H \in \mathcal{H}^e$.

It remains to show that $E(H) = 1$. Thus we may assume that there is a component K of H. If $K \leq M$, then $K \leq E(M_H)$, contradicting $M_H \in \mathcal{H}^e$; thus $K \not\leq M$. By 1.2.1.3, $L_+ = O^2(L_+) \leq N_H(K)$, so $L_+ T_+$ acts on $K_0 := \langle K^{T_+} \rangle$. Therefore by minimality of H, $H = K_0 L_+ T_+$.

Next as L^0 acts on K, so does $V_- = [V_-, L^0]$. We claim V_- acts faithfully on K, so assume otherwise; the proof will require several paragraphs. First $V_+ < W := C_A(K)$, so as L^0 acts on W, W contains at least one of the five nontrivial orbits $\mathcal{O}$ of $\langle \theta \rangle$ on $V^\#$. Now $\mathcal{O} = W_-^\#$ for some 2-subspace W_- of W. Observe W contains no involution w 2-central in G: For if w is such an involution, then $K \leq E(H) \cap G_w \leq E(H \cap G_w)$, while $E(H \cap G_w) = 1$ by 7.7.5.

Suppose first that (H0) holds. Then W contains the orthogonal sum of the hyperbolic 2-space V_0 with W_-, and either W_- lies in V_1 or V_2 and hence is totally singular, or W_- is diagonal and definite. Set $w := v_0 + w_-$ for $0 \neq w_- \in W_-$, where we choose v_0 to be singular in $V_0 \cap V_{3-i}$ in case $W_- \leq V_i$, or the non-singular vector in V_0 in case W_- is definite. Then by construction w is singular and diagonal, so by H.4.2, w is 2-central, contrary to the previous paragraph. This establishes the claim when (H0) holds.

So suppose instead that (H1) holds. As W contains no 2-central involution, W is not $C_V(O_2(L_1))$, so $\mathcal{O}$ does not contain 2^t. Therefore W is not centralized by an involution of $\bar{M}$—so that $W \in \Gamma$ in the language of Definition E.6.4. By (H1), $r > 2$, so as $m(W) = 3$, W is maximal subject to $C_G(W) \not\leq M$. But then E.6.11.2 says there is a subgroup of order 7 normal in $N_{\bar{M}}(W)$, which cannot happen— since $N_{\bar{M}}(W)$ is a $7'$-group unless $W = V_1$, where $N_{\bar{M}}(W) \cong L_3(2)$ has no normal subgroup of order 7. This completes the proof of the claim that V_- is faithful on K.

Next observe that V_- induces inner automorphisms on K: We check that the groups listed in Theorem C (A.2.3) have no A_4-group of outer automorphisms, whereas $V_- = [V_-, \theta]$. Thus the projection V_K of V_- on K is faithful of rank 4.

Let $Z_+ := 1$ under (H0) and $Z_+ := Z_1$ under (H1). In either case, set $\tilde{H} := H/Z_+$. Now $O_2(L_+)Q$ is of index 2 in T_+, and centralizes $\tilde{A} = \tilde{V}_+ \tilde{V}_-$. Thus $\tilde{A}$ centralizes a subgroup of $\tilde{T}_+$ of index 2. Therefore $\tilde{V}_K$ is centralized by $Q_K := O_2(L_+)Q \cap K$ of index at most 2 in $T_K := T_+ \cap K$, so $\tilde{Z}_K := C_{\tilde{V}_K}(T_K)$ is noncyclic and contained in $Z(\tilde{T}_K)$. Therefore $m_2(K/Z(K)) \geq 4$ and $Z(\tilde{T}_K)$ is noncyclic. We check the groups on the list of Theorem C for groups with these properties: $m_2(K/Z(K)) \geq 4$ eliminates the groups in cases (1) and (2) of Theorem C (other than A_8 which also appears in case (4)), while $Z(\tilde{T}_K)$ noncyclic eliminates those in cases (4) and (5) as well as those in case (3) over the field $\mathbf{F}_2$. Therefore $K/Z(K)$ is of Lie type over $\mathbf{F}_{2^n}$ for some $n > 1$. Now if $\tilde{R}$ is a root group of $\tilde{K}$ in $\tilde{T}_K$, then $1 \neq \tilde{R} \cap \tilde{Q}_K$, so $\tilde{V}_K \leq C_{\tilde{T}_K}(\widetilde{R \cap Q_K}) \leq C_{\tilde{T}_K}(\tilde{R})$, and hence $\tilde{V}_K \leq Z(\tilde{T}_K)$, so $\tilde{A}$ centralizes $\tilde{T}_K$. In particular $m_2(Z(\tilde{T}_K)) \geq 2$, so either $n \geq 4$ or $K/Z(K)$ is $Sp_4(4)$. Thus by I.1.3, the multiplier of $K/Z(K)$ is of odd order, so as $[\tilde{A}, \tilde{T}_K] = 1$, $[A, T_K] \leq K \cap Z_+ \leq O_2(K) = 1$. Therefore $T_K \leq C_T(A) = Q$, so Q is Sylow in QK_0. However $C(G, Q) \leq M$, so $C(K_0, Q) \leq K_0 \cap M < K_0$. Thus we may apply the local $C(G, T)$-theorem C.1.29 to the maximal parabolics of K_0. Now if K is of Lie type G_2, $^3D_4$, or $^2F_4$, neither of the two maximal parabolics of K are blocks, so by C.1.29, each is contained in M. Thus $K \leq M$ as K is generated by these maximal parabolics, a contradiction. This reduces us to the case where $K/Z(K)$ is a Bender group over F_{2^n}, $L_3(2^n)$, or $Sp_4(2^n)$, and $M \cap K_0$ is either a Borel subgroup of K_0 or a maximal parabolic K_1 of $K \cong L_3(2^n)$ or $Sp_4(2^n)$. In any case $M \cap K_0$ contains a Borel subgroup B of K_0 normalizing T_K. By an earlier remark, either $n \geq 4$ or $K \cong Sp_4(4)$.

Now let Y be a Cartan subgroup of B. By 7.7.4, $Y_C := C_Y(V)$ is of index at most 3 in Y. But when $n \geq 4$, certainly $|Y : Y_C| > 3$, since Y_C centralizes V and hence centralizes $V_K \leq Z(T_K)$, while some subgroup of Y isomorphic to $\mathbf{Z}_{2^n-1}$ is semiregular on $Z(T_K)$. Therefore K_0 is $Sp_4(4)$, with Y_C of order 3—again centralizing V and hence V_K. This is impossible, as the Cartan group of B is faithful on $Z(T_K)$ in $Sp_4(2^n)$.

This completes the proof of Lemma 7.7.6 and Proposition 7.7.2.

7.7.4. Proof of Proposition 7.7.3. Now that Proposition 7.7.2 is established, we work under Hypothesis (H0), and it remains to obtain a contradiction, establishing Proposition 7.7.3.

We are in a position to exploit Thompson factorization: First, lemma B.2.14 tells us that

$$U := \langle \Omega_1(Z(T_0))^H \rangle \in \mathcal{R}_2(H),$$

so setting $H^* := H/C_H(U)$, we have $O_2(H^*) = 1$. Further

$$V = \langle C_V(T_0)^{L^0} \rangle \leq U,$$

so

$$C_H(U) \leq C_H(V) \leq M_H.$$

We saw early in the proof of 7.7.6 that $J(H) \not\leq M$, so $J(H)^* \neq 1$.

Next $J(H)^*$ is described in Theorem B.5.6. In particular as $J(H) \not\leq M$, either $O_3(J(H)^*) \not\leq M_H^*$ or some component K^* of $J(H)^*$ is not contained in M_H^*.

Assume the first case holds. Then

$$X^* := O_3(J(H)^*) = X_1^* \times \cdots \times X_d^*$$

with $X_i^* \cong \mathbf{Z}_3$ and $[U, X] = U_1 \oplus \cdots \oplus U_d$ where $U_i := [U, X_i]$ is of rank 2. Further $d \leq 2$ so that $L^0 = O^2(L^0)$ acts on each U_i. As $J(T) \trianglelefteq L^0 T_0$ and L^0 acts on U_i, L^0 acts on $C_{U_i}(J(T)) \cong \mathbf{Z}_2$, so $[U_i, L^0] = 1$. Then $1 = [U, X, L^0]$, and $[X^*, L^{0*}] = 1$ which says $[X, L^0, U] = 1$. So by the Three-Subgroup Lemma we have $[L^0, U, X] = 1$. But recall $V_- = [L^0, V] \leq [L^0, U]$. Thus X centralizes $V_0 V_- = V$, contradicting $X \not\leq M$.

Therefore some component K_+^* of $J(H)^*$ is not contained in M_H^*, so taking $K \in \mathcal{C}(H)$ with $K_+^* = K^*$ and setting $K_0 := \langle K^{T_0} \rangle$, $H = K_0 L^0 T_0$ by minimality of H. Similarly by a Frattini Argument, $H = C_H(U) N_H(C_{T_0}(U))$, so that $K/O_2(K)$ is quasisimple by 1.2.1.4 and minimality of H.

LEMMA 7.7.7. *Hypothesis C.2.3 is satisfied with Q in the role of "R".*

PROOF. Recall $C(G, Q) \leq M$, so $C(H, Q) \leq M_H < H$. By A.4.2.4, $Q \in Syl_2(C_0)$, where $C_0 := C_{M_H}(L^0/O_2(L^0)) \trianglelefteq M_H$; then $C_0 \geq \langle Q^{M_H} \rangle$, so Q is also Sylow in the latter group. Also $Q \in \mathcal{B}_2(M_H)$ by C.1.2.4, so that $Q \in \mathcal{B}_2(H)$ by C.1.2.3. Thus we have verified Hypothesis C.2.3. $\square$

LEMMA 7.7.8. $Q \leq N_H(K)$.

PROOF. Assume otherwise. Then by C.2.4, $Q \cap K \in Syl_2(K)$, and as $K \not\leq M$, K is a χ_0-block. Further as K^* is quasisimple and $K < K_0$, we conclude from the list in A.3.8.3 that $K^* \cong L_2(2^n)$ with $n \geq 2$. Then by C.2.4, $K_0 \cap M$ is the Borel subgroup B normalizing $Q \cap K_0$. Let Y be a Cartan subgroup of B. By 7.7.4, $|Y : Y_C| \leq 3$ and $N_G(Y_C) \leq M$ because $Y_C \neq 1$ since K_0 is the product of two conjugates of K. On the other hand, $Y T_0 = T_0 Y$ and T_0 acts on L, so also $Y_C T_0 = T_0 Y_C$. Then as $H \not\leq M$, $N_H(Y_C) \not\leq M$ by 4.4.13.1. This contradiction completes the proof. $\square$

Now that $Q \leq N_H(K)$ by 7.7.8 and $K/O_2(K)$ is quasisimple, we may apply C.2.7 to conclude that K is desribed in C.2.7.3.

LEMMA 7.7.9. *(1) If case (a) of C.2.7.3 holds, then K is an A_7-block.*
(2) $[L^{0*}, T_0^* \cap K^*] \not\leq O_2(L^{0*})$.

PROOF. Suppose case (a) of C.2.7.3 holds, where K is a χ-block. Suppose first that K is an $L_2(2^n)$-block, and either $n > 2$ or $K < K_0$. Let Y be a Cartan subgroup of $K_0 \cap M$. An argument in the proof of the previous lemma shows that $Y_C \neq 1$, and supplies a contradiction. Thus $K = K_0$ is a block of type $L_2(4)$, A_5, or A_7.

Suppose next that K is a block of type A_5 or $L_2(4)$, and let $Y \in Syl_3(M \cap KL^0)$, $Y_C := C_Y(V)$, $Y_L := Y \cap L^0$, and $Y_K := Y \cap K$. By 7.7.4.1, $|Y : Y_C| \leq 3$. As $N_K(Y_K) \not\leq M$, 7.7.4.2 says that $Y_K \neq Y_C$, and hence $Y = Y_K Y_C = Y_L Y_C$ as $|Y : Y_C| \leq 3$. Then

$$V_- = [V, L^0] = [V, Y_L] = [V, Y_K] \leq U \cap K \leq O_2(K).$$

But as K is of type A_5 or $L_2(4)$, $m(O_2(K)/Z(K)) = 4 = m(V_-)$, so $V_- Z(K) = O_2(K)$. This is impossible, as $Q \cap K \in Syl_2(K)$, and Q centralizes V but not $O_2(K)$. This establishes (1); in particular K is not a χ_0-block.

Assume that $[L^{0*}, T_0^* \cap K^*] \leq O_2(L^{0*})$. Then $T_0 \cap K \leq C_{T_0}(L^0/O_2(L^0)) = Q$, so $Q \in Syl_2(K_0 Q)$. Therefore K is a χ_0-block by C.2.5, contrary to the previous paragraph. Thus (2) holds. $\square$

LEMMA 7.7.10. $L^0 \leq K$, *and hence* $T_0 \leq N_H(K)$, *so that* $K = K_0$.

PROOF. We may assume $L^0 \not\leq K$, and it suffices to derive a contradiction. Since $1 \neq [V, L^0] \leq [U, L^0]$, we have $L^{0*} \neq 1$. We will appeal frequently to the fact that L^{0*} is normal in M_H^*, and hence is normalized by $M_K := M \cap K$, with $L^{0*}/O_2(L^{0*})$ of order 3.

Inspecting the groups listed in C.2.7.3 and appealing to 7.7.9.1, either $m_3(K) = 2$ or $K^* \cong SL_3(2^n)$ with n odd. In the former case we apply A.3.18, and A.3.19 when $K^* \cong SL_3(2^n)$ with n even; we conclude that K is the subgroup of H generated by all elements of order 3 so that $L^0 \leq K$, and the lemma holds in this case.

Therefore we are reduced to the case where $K^* \cong SL_3(2^n)$ with n odd, and M_K^* is a maximal parabolic. Assume $L^0 \not\leq K_0$. Then $[L^{0*}, M_K^*] \leq L^{0*} \cap M_K^* \leq O_2(M_K^*)$, so as $C_{Aut(K^*)}(M_K^*/O_2(M_K^*))$ is a $3'$-group, $[L^{0*}, K^*] = 1$, contrary to 7.7.9.2. This contradiction shows $L^0 \leq K_0$. As we are assuming $L^0 \not\leq K$, we must have $K < K_0 = KK^s$ for $s \in T_0 - N_{T_0}(K)$. Hence $K^* \cong L_3(2)$ by A.3.8.3. As $L^0 \not\leq K$ and T_0 acts on L^0, L^{0*} is diagonally embedded in K_0^*. But the Sylow group T_0^* acts on no such diagonally embedded subgroup with Sylow 3-subgroup of order 3, completing the proof of the lemma. $\square$

As $L^0 \leq K$, $L^{0*} \trianglelefteq M_K^*$. Hence as $L^{0*}/O_2(L^{0*})$ is of order 3, K^* is not $L_3(2^n)$ with $n > 1$ odd. Similarly if $K^* \cong SL_3(2^n)$ with n even, then $L^{0*} = Z(K^*)$, so that $[L^{0*}, K^*] = 1$, contrary to 7.7.9.2. Thus $n = 1$ in case (g) of C.2.7.3.

Assume we are in the subcase of case (e) of C.2.7.3 where $K^* \cong Sp_4(4)$ and M_K^* is a maximal parabolic. Then as $L^{0*} \trianglelefteq M_K^*$, $L^{0*} = O_{2,3}(M_K^*))$. But then $[L^{0*}, T_K^*] \leq O_2(L^{0*})$, contrary to 7.7.9.2.

Thus we are left with the subcase of case (a) of C.2.7.3 where K is an A_7-block, or one of cases (b)–(d), case (e) with $K^* \cong A_6$, case (f), case (g) with $n = 1$, or case (h).

We now eliminate the cases (a)–(d), (e) with $K^* \cong A_6$, and (f); in all these cases, K is a block. We have $V_- = [V_-, L^0] \leq K$ using 7.7.10. Recalling that $V \leq U \leq O_2(H)$, we see that $V_- \leq O_2(K)$. Let W be the unique noncentral 2-chief factor of the block K, and W_- the image of V_- in W. As $C_{V_-}(L^0) = 1$, $W_- \cong V_-$. Further Q centralizes W_- and Q is of index 2 in the Sylow group T_0. However in each case, W is of dimension 4 or 6, and no subgroup of index 2 in a Sylow group centralizes a 4-subspace of W.

We are left with case (h), and with the subcase of case (g) where $n = 1$. Thus $K^* \cong L_m(2)$ with $m := 3, 4, 5$. As L^{0*} is normal in the parabolic M_K^* and T_0-invariant, $L^{0*}T_K^*$ is a rank one parabolic determined by a node δ in the Dynkin diagram adjacent to no node in M_K^*. So when m is 4 or 5, unless $K^*T_0^* \cong S_8$ and δ is the middle node, there is an $L^0 T_0$-invariant proper parabolic which does not lie in M, contrary to the minimality of H. When $K^*T_0^* \cong S_8$, Theorems B.5.1 and B.4.2 say $I := [U, K]/C_{[U,K]}(K)$ is either the orthogonal module or the sum of the natural module and its dual. But in either case, $m(C_I(T_0)) = 1$, impossible as V_- is isomorphic to an $L^0 T_0$-submodule of I and $m(C_{V_-}(T_0)) = 2$.

Therefore $K^* \cong L_3(2)$, and C.2.7.3 says that K is described in Theorem C.1.34. As $m(C_{V_-}(T_0)) = 2$, there are at least two composition factors on $U \leq Z(O_2(K))$, ruling out all but case (2) of C.1.34. Hence $O_2(K) = U = U_1 \oplus U_2$ is the sum of two isomorphic natural modules for $K^* = K/U$, with $V_- = W_1 \oplus W_2$ where $W_i = C_{U_i}(Q)$. Then an element θ of L^0 of order 3 has a unique nontrivial composition factor on $O_2(L^{0*})$, (which is realized on Q/U) plus two nontrivial composition factors W_1 and W_2 in U (realized in V). Thus L^0 has just one nontrivial composition factor on Q/V, which is impossible since the outer automorphism $\bar{t}$ of $\bar{L} \cong L_3(2)$ must interchange any natural module and its dual, and these are the only irreducibles with a unique nontrivial L^0-composition factor. This contradiction finally completes the proof of Proposition 7.7.3 and hence also of Theorem 7.7.1.

CHAPTER 8

Eliminating shadows and characterizing the J_4 example

We begin by reviewing the cases remaining after the work of the previous chapter, which eliminated those cases which do not lead to examples or shadows.

We continue to assume Hypotheses 7.0.2 and 7.3.1 from the previous chapter. The latter hypothesis excludes the case where $\bar{L}_0$ is $\Omega_4^+(2^n)$ on its orthogonal module; that case will be treated in chapter 9 of this part, because the methods used to attack that case are different from those in the remaining cases.

The cases $\bar{L}_0/V$ remaining from Table 7.1.1 that were not eliminated in the previous chapter 7, and are not among the cases to be treated in later chapters, are: $L_3(2^{2n})/9n$, $M_{22}/10$ or $M_{22}/\bar{10}$, $M_{23}/\bar{11}$, $M_{24}/11$ or $M_{24}/\bar{11}$, and $(L_3(2) \wr 2)/9$.

In the case of $(L_3(2) \wr 2)/9$, technical complications also arise, primarily because the existence of small $(F-1)$-offenders on V only gives $r \geq 3$. As a result, different methods are required to treat this case; thus we will defer its treatment to 8.3.1 in the final section of this chapter.

As indicated in Table 7.1.1 in the previous chapter, the subgroups M we study in this chapter do arise as maximal 2-locals in various shadows, and in the case of M_{24} on its cocode module $\overline{11}$, in the quasithin example J_4. Thus we should not expect the methods of the previous chapter to eliminate these configurations on simple numerical grounds. Instead we seek to show that our bounds determine a unique solution for the various parameters: namely, the solution corresponding to the shadow or example. Then to eliminate the shadows, we go on to show that this unique solution leads (via study of w-offenders and subgroups $H \in \mathcal{H}_*(T, M)$) to a local subgroup other than M which is not an SQTK-group. In the $M_{24}/\overline{11}$ case, we construct the centralizer of a 2-central involution, which allows us to identify G as J_4.

8.1. Eliminating shadows of the Fischer groups

In this section, we assume $\bar{L}$ is M_{22}, M_{23}, or M_{24} and V is the cocode module for $\bar{L}$. In these cases we take a shortcut bypassing the uniform route we just outlined. This is because the initial bound on r given by the columns in Table 7.2.1 is a little too weak to pin down the structure of appropriate 2-locals, without a much more detailed analysis of elementary subgroups of $\bar{M}$ and their fixed points on V, and we wish to avoid that analysis.

In fact we will be able to eliminate these configurations, which correspond to the shadows of the Fischer groups, not by directly constructing a local subgroup that is not strongly quasithin, but instead by the use of techniques of pushing up from sections C.2, C.3, and C.4. These results implicitly rule out a number of locals which are not SQTK-groups; as a consequence we obtain an improved bound on r, and this slight improvement makes the remaining weak closure analysis much easier. Since this improved bound on r now exceeds the value occurring in the shadows, our calculations will in effect eliminate the Fischer groups—and in the case of M_{24}, will produce the centralizer of a 2-central involution resembling that in J_4.

In brief, we will use methods of pushing up to show for certain $x \in V$ that $C_G(x) \leq M$. Consequently any $U \leq V$ with $C_G(U) \not\leq M$ must contain only elements in conjugacy classes other than that of x. This restriction, added to those from Table 7.2.1, produces the improved bound on r. Then the remaining weak closure analysis proceeds rapidly.

In this section, we will by convention order the cases so that the case $\bar{L} \cong M_{22}$ is first, the case $\bar{L} \cong M_{23}$ is second, and the case $\bar{L} \cong M_{24}$ is third. When we make an argument simultaneously for all cases, we will list values of parameters for the cases in that order, without explicitly writing "respectively". Thus for example, the module V is the cocode module, which we are denoting by $\overline{10}, \overline{11}, \overline{11}$.

We take the standard point of view (cf. section H.13 of Volume I) that the cocode modules are sections of the space spanned by the 24 letters permuted by M_{24}, modulo the 12-dimensional subspace given by the Golay code. For M_{24}, the 11-dimensional cocode module V is the image of the subspace of all subsets of even size. The orbits of M_{24} on V consist of the set $\mathcal{O}_2$ of images of 2-sets and the set $\mathcal{O}_4$ of images of 4-sets, with the latter determined only modulo the code—that is, $\mathcal{O}_4$ is in 1-1 correspondence with the sextets in the terminology of Conway [**Con71**] and Todd [**Tod66**]. For M_{23} and M_{22} we can consider 2-sets containing just one of the letters fixed by this subgroup, and denote the corresponding vector orbit by $\mathcal{O}_2$.

Our subgroup M corresponds to a local subgroup $\dot{M}$ in the shadow group $\dot{G} := F_{22}, F_{23}, F_{24}$. Notice in these shadows that for $\dot{x} \in \dot{\mathcal{O}}_2$, $C_{\dot{G}}(\dot{x}) \not\leq \dot{M}$; in fact $C_{\dot{G}}(\dot{x})$ has a component, which is not strongly quasithin. We will see that the results on pushing up in section C.2 apply, and in fact rule out these components which arise in the shadows, forcing $C_G(x) \leq M$.

PROPOSITION 8.1.1. $C_G(x) \leq M$ for $x \in \mathcal{O}_2$.

PROOF. By H.15.1.1,

$$C_{\bar{L}}(x) \cong M_{21}, \ M_{22}, \ Aut(M_{22}),$$

where $M_{21} \cong L_3(4)$. Let $H := C_G(x)$, $M_H := H \cap M$, and $L_H := C_L(x)^\infty$. Replacing x by a suitable M-conjugate if necessary, we may assume $T_H := C_T(x) \in Syl_2(C_M(x))$. As $F^*(C_{\bar{L}}(x))$ is simple, $O_2(C_L(x)T_H) = Q = O_2(LT)$.

Next we show that Hypothesis C.2.8 is satisfied with Q, L_H in the roles of "R, M_0". Recall first that as part of the general setup in the introduction to chapter 7, $C(G, Q) \leq M$. By A.4.2.7, Q is Sylow in $C_{M_H}(L_H/O_2(L_H))$, so that the second hypothesis of C.2.8 is satisfied. By H.15.1.2, we have $V = [V, L_H]$ for the cocode modules. By construction $Q = O_2(L_H Q)$ centralizes V, with $N_G(V) \leq M$, so that the third hypothesis of C.2.8 is satisfied. Finally $O_2(M) \leq Q \leq H$ using A.1.6, so that $M_H \in \mathcal{H}^e$ by 1.1.4.4, establishing the first hypothesis of C.2.8.

Thus Hypothesis C.2.8 holds, and we may apply Theorem C.4.8. If $C_G(x) \not\leq M$, then $M_H < H$. But L_H is not listed among the possibilities in C.4.8. This contradiction show that $C_G(x) \leq M$. $\qquad\square$

COROLLARY 8.1.2. $r \geq 6, 7, 8$.

PROOF. Suppose that $U \leq V$ with $C_G(U) \not\leq M$. We must show that $m(U) \leq 4$, 4, 3. By Proposition 8.1.1, $U \cap \mathcal{O}_2 = \emptyset$. During the proof of 7.4.1, we verified the hypotheses of E.6.28; and hence (as observed in the proof of that result), also hypotheses (1) and (4) of E.6.27 with $j = 1$. So since the conclusion $C_G(U) \leq M$ of that latter result fails, hypothesis (2) or (3) of that result must fail; hence U centralizes either some $(F - 1)$-offender on V, or some nontrivial element of $\bar{M}_V$ of odd order.

First we consider the case where $U \leq C_V(\bar{A})$ for some $(F - 1)$-offender $\bar{A}$. By H.15.2.3, if $U \leq C_V(\bar{A})$ with $U \cap \mathcal{O}_2 = \emptyset$, then $m(U) \leq 4, 4, 3$, completing the proof in this case.

So it remains to consider the case where $U \leq W := C_V(\bar{y})$ for some nontrivial element $\bar{y}$ of $\bar{L}$ of odd order. In the case of M_{22}, $m(W) \leq 4$ as $\beta = 6$ in Table 7.2.1. When $\bar{L}$ is M_{23} or M_{24}, then as $U \leq W$ with $U \cap \mathcal{O}_2 = \emptyset$, $m(U) \leq 4$ or 2 by H.15.7.3, completing the proof. $\qquad\square$

Using this improved bound on r, it is not hard to eliminate the shadows of the Fischer groups, and isolate the configuration leading to J_4:

THEOREM 8.1.3. *If V is the cocode module for $\bar{L} \cong M_{22}$, M_{23}, or M_{24}, then $\bar{L} \cong M_{24}$, and there is a unique solution of the Fundamental Weak Closure Inequality 7.5.1. Indeed that solution satisfies $r = 8$, $m(C_A(V)) = 3$, $w = n(H) = 2$, and $\bar{A} = K_T$ of rank 6, for A a w-offender on V and $H \in \mathcal{H}_*(T, M)$.*

PROOF. Let A be a w-offender, with $A \leq V^g$ for suitable $g \in G$. By 8.1.2, $r \geq 6, 7, 8$, while by Table 7.2.1, $w \leq 2$ and $m_2 \leq 5, 4, 6$. Thus the FWCI is violated when $\bar{L} \cong M_{23}$. When $\bar{L} \cong M_{24}$, the FWCI is an equality, so all inequalities are equalities, and hence $w = 2$ and $r = 8$. Finally when $\bar{L} \cong M_{22}$, $w \geq 1$ by the FWCI. Further $m(\bar{A}) \geq r - w \geq 4$ by 7.5.3.2, and when these inequalities are equalities, we must have $w = 2$ and $r = 6$—since we saw $w \leq 2$ and $r \geq 6$.

In particular, we have eliminated M_{23}. Suppose next that $\bar{L} \cong M_{24}$, where we have shown the FWCI is an equality with $r = 8$ and $w = 2$. Let W be the subspace of V defined in 7.5.3.1, and note $W = \xi_V(\bar{A})$ in the language of H.10.1. As $w = 2 = n'$, $n(H) = 2$ by 7.3.4. By 7.5.3.1, $m(\bar{A}) = m_2 = 6$. Therefore by H.14.1.1, $\bar{A}$ is K_T or K_S. If $\bar{A} = K_S$, then $W = V$ by H.15.3.3, contrary to 7.5.3.1. Thus $\bar{A} = K_T$, so that the Theorem holds in this case.

We have reduced to the case where $\bar{L} \cong M_{22}$. This case is a little harder. Recall $m_2 \leq 5$, $w \leq 2$, and $m(\bar{A}) \geq 4$, with $w = 2$ in case $m(\bar{A}) = 4$. Let $\mathcal{B}$ be the set of $B \leq A$ with $C_A(V) \leq B$ and $m(V^g/B) = 5$. Then for $B \in \mathcal{B}$, $m(V^g/B) < 6 \leq r$, so $C_G(B) \leq N_G(V^g)$ and hence

$$W := \langle C_V(B) : B \in \mathcal{B} \rangle \leq N_V(V^g).$$

Further $m(\bar{B}) = m(\bar{A}) - 5 + w$, so $m(\bar{B}) = 1$ if $m(\bar{A}) = 4$ (since in that case we showed $w = 2$); while if $m(\bar{A}) = 5$, then $m(\bar{B}) = w$ is either 1 or 2. As $W \leq N_V(V^g)$, $m(V/W) \geq w \geq 1$ by definition of w, so in particular $W < V$.

If $m(\bar{A}) = 5$, then by H.14.3.1, $\bar{A} = K_Q$. Then $W = V$ by H.15.4.4, contrary to the previous paragraph. Thus $m(\bar{A}) = 4$, so as $W < V$, H.15.5 says $m(V/W) \leq 1$. But by earlier remarks, $w = 2$ and $m(V/W) \geq w$. This contradiction completes the proof of the Theorem. $\qquad \square$

8.2. Determining local subgroups, and identifying J_4

In this section we treat the remaining cases other than $L_3(2)$ wr $\mathbf{Z}_2$, which we consider in the final section of the chapter; thus in addition to Hypotheses 7.0.2 and 7.3.1, we assume:

HYPOTHESIS 8.2.1. $\bar{L}_0$ *is not* $L_3(2) \times L_3(2)$ *on the tensor module 9.*

As a result of the previous section, we have eliminated M_{22} and M_{23} on their cocode modules, and in the case of M_{24} on its cocode module, we showed there is a unique solution for the weak closure parameters of a w-offender A on V. Indeed in that case we showed that $\bar{A} = K_T$ and $C_V(A) = C_V(K_T)$ is of dimension 3.

Because of Hypothesis 8.2.1, the other cases to be treated in this section are:

$$\bar{L} \cong (S)L_3(2^{2n})/9n, \ M_{22}/10, \ M_{24}/11.$$

As before we will use this ordering in common arguments, and we adjoin $M_{24}/\overline{11}$ as the fourth case on our list. In the first three cases we will show (as we did in case four) that there is a canonical choice for our w-offender A, and for each such canonical A, $C_V(A)$ is determined. Then in all four cases, we construct a sizable part of the local subgroup $N := N_G(C_V(A))$. In some cases N will not be strongly quasithin, so those cases are eliminated. In the surviving cases we study $C := C_G(z)$, where z is a 2-central involution in V; from $C_M(z)$ and $C_N(z)$ we can construct enough of C to see that either C is not strongly quasithin, or that $M \cong M_{24}/\overline{11}$ and C has the structure of the centralizer of an involution in J_4. Then we identify G as J_4 in the final subsection of this section.

8.2.1. Isolating a w-offender. As usual let $H \in \mathcal{H}_*(T, M)$. Recall H is a minimal parabolic by 3.3.2.4, with $H \cap M$ the unique maximal overgroup of T in H. We see in the next lemma that $V \not\leq O_2(H)$, so from lemma E.2.9, the set $\mathcal{I}(H, T, V)$ of Definition E.2.4 is nonempty.

PROPOSITION 8.2.2. *(1)* $V \not\leq O_2(H)$.
(2) There exists $h \in H$ *such that* $I := \langle V, V^h \rangle$ *is in the set* $\mathcal{I}(H, T, V)$ *and* $h \in I$.
(3) $1 \neq Z_I := V \cap V^h \leq Z(I)$.
(4) $T_I := T \cap I \in Syl_2(I)$ *and* $M_I := M \cap I = N_I(V)$.
(5) $\ker_{M_I}(I) = O_2(I)$, *and* $I^* := I/O_2(I) \cong D_{2m}$, m *odd (in which case we set* $k := 1$*),* $L_2(2^k)$, *or* $Sz(2^k)$, *for some suitable* k *dividing* $n(H)$.
(6) $V^* = Z(T_I^*)$ *and* $M_I^* = N_{I^*}(T_I^*)$.
(7) $A := V^h \cap O_2(I) = N_{V^h}(V)$, $C_A(V) = Z_I$, A *is cubic on* V, $r_{Aut_A(V),V} < 2$, $m(\bar{A}) = m(V/Z_I) - k$, *and* $C_V(A) \leq B := V \cap O_2(I)$.
(8) If $k > 1$, *then* $C_V(\bar{X}) = Z_I$ *for* $\bar{X}$ *of order* $2^k - 1$ *in* $\bar{M}_I$.

PROOF. From Table B.4.5, either $\bar{M}_V = Aut(M_{22})$ and V is the code module; or $q(\bar{M}_V, V) > 2$, so that $V \not\leq O_2(H)$ by 3.1.8.2, and (1) holds.

Therefore we may assume that $V \leq O_2(H)$ with V the code module for $\bar{M}_V = Aut(M_{22})$, and it remains to derive a contradiction. We first verify that the hypotheses of 3.1.9 hold with LT in the role of "M_0": Recall we saw after Hypothesis 7.0.2 that $H \cap M \leq M_V$; thus case (II) of Hypothesis 3.1.5 holds. Now part (c) in the hypothesis of 3.1.9 holds by hypothesis, part (a) is a consequence of Table B.4.5, and (d) follows as the dual V^* of V satisfies $q(\bar{L}\bar{T}, V^*) > 2$. Finally $M = !\mathcal{M}(LT)$ by 1.2.7.3, so (b) holds.

By A.3.18, $L = O^{3'}(M)$. Then we observe that each element of order 3 in M_{22} has six 3-cycles on 22 points, so it has three noncentral chief factors on each of V and V^*. Set $L_z := O^2(C_L(z))$; then $\bar{L}_z = O^2(C_{\bar{M}}(z))$ and $L_z/O_2(L_z) \cong A_6$. Thus each $\{2,3\}'$-subgroup of $C_M(Z)$ permuting with T centralizes V. As $q(\bar{M}_V, V) = 2$, but $m(\bar{M}_V, V) = 3$ by H.14.4, each member of $\mathcal{Q}(M_V, V)$ has rank at least 2. Thus 3.1.9.6 says that $H/O_2(H) \cong S_3 \text{ wr } \mathbf{Z}_2$ or $D_8/3^{1+2}$; in particular $X := O^2(H) \in \Xi(G,T)$. Next by 1.2.4, $L_z \leq K_z \in \mathcal{C}(C_G(z))$; and by A.3.12, either $K_z = L_z$ or $K_z/O_2(K_z) \cong A_7$, M_{11}, M_{22}, M_{23}, or $U_3(5)$. By A.3.18, $K_z = O^{3'}(C_G(z))$, so $X \leq K_z$. Thus $K_z/O_2(K_z) \cong M_{11}$ by 1.3.4. But then $H/O_2(H) \cong SD_{16}/E_9$, impossible as $H/O_{2,3}(H) \cong D_8$. This completes the proof of (1).

Then as H is a minimal parabolic, $V \not\leq \ker_{M \cap H}(H)$ by B.6.8.5, so that Hypothesis E.2.8 holds. Then (2) follows from E.2.9. By E.2.11.5, $O_2(I) = \ker_{M_I}(I)$. By 7.3.3 and 7.5.6, $w > 0$, so $W_0(T, V)$ centralizes V. Therefore I^* is not $Sp_4(2^k)$ by E.2.13.5; in particular the remainder of (5) holds by definition of $\mathcal{I}(H, T, V)$. As $q(\bar{M}, V) > 1$, E.2.13.4 says that (3) holds. We recall from the introduction to the previous chapter that V is a TI-set under M, so that with (3), we have the hypotheses of E.2.14. Now (4) follows from the definition of $\mathcal{I}(H, T, V)$ and E.2.14.1, while (6) follows from E.2.14.2. The first few statements in (7) follow from E.2.13.1 and E.2.15. Then we compute $m(\bar{A})$ using (5), (6), and the fact that $C_A(V) = Z_I$. By E.2.10.1, $AB \trianglelefteq I$, while by parts (3), (4), or (10) of E.2.14, $C_I(AB) \leq \ker_{M_I}(I) = O_2(I)$; thus $C_V(A) \leq C_V(AB) \leq O_2(I) \cap V = B$, completing the proof of (7). Finally, (8) follows from (5) using E.2.14.9. $\qquad\square$

As in 8.2.2, pick $I = \langle V, V^h \rangle \in \mathcal{I}(H, T, V))$, and adopt the rest of the notation established in the lemma; e.g., $T_I := T \cap I \in Syl_2(I)$, $M_I := M \cap I$, $I^* := I/O_2(I) = I/\ker_{M_I}(I)$, $k := n(I)$, etc.

PROPOSITION 8.2.3. $k = n(I) = w = n, 1, 1, 2$, and A is a w-offender on V.

PROOF. By 8.2.2.5, k divides $n(H)$, so $k \leq n(H)$. By definition $w \leq m(V^*)$, while $m(V^*) = k$ using 8.2.2.6. Then we can extend the inequality in 7.3.4 to

$$w \leq m(V^*) = n(I) = k \leq n(H) \leq n' = 2n, 2, 2, 2 \qquad (*)$$

using the values in Table 7.2.1.

In the fourth case $M_{24}/\overline{11}$, $w = 2$ by 8.1.3, so the lemma follows from (*).

Thus we may assume $\bar{L}$ is not M_{24} on $\overline{11}$. If $w = k$, then A is a w-offender. By Table 7.2.1 and 7.5.6, $w \geq n, 1, 1$. Thus if $k \leq n, 1, 1$, then $w = k$ by (*) and the lemma holds. Therefore by (*), we may assume that $k = 2$ if $\bar{L}$ is M_{22} or M_{24}, while $n < k \leq 2n$ if $\bar{L} \cong (S)L_3(2^{2n})$, and it remains to derive a contradiction.

Assume first that $\bar{L} \cong (S)L_3(2^{2n})$. Then $k > n \geq 1$, so $I^* \cong L_2(2^k)$ or $Sz(2^k)$ and hence $Aut_I(V)$ contains a cyclic subgroup $\bar{X}$ of order $2^k - 1 \geq 3$ acting nontrivially on $\bar{A}$. Therefore as $Out(\bar{L})$ is 2-nilpotent, $1 \neq [\bar{A}, \bar{X}] \leq \bar{L}$ is an X-invariant 2-group. Hence $\bar{X}$ acts on some parabolic of $\bar{L}$, and indeed on a maximal parabolic as $\bar{X}$ has odd order. Therefore $2^k - 1$ divides $(2^{4n} - 1)n$, so as $n < k \leq 2n$, it follows that $k = 2n$. Thus $m(\bar{A}) \leq m_2 = 4n = 2k$, so by E.2.14.7,

$m(V/Z_I) = 3k = 6n$ and $m(\bar{A}) = 4n$. Therefore $m(Z_I) = m(V) - 6n = 3n$. By 8.2.2.8, $Z_I = C_V(\bar{X})$. This contradicts H.4.4.4, which says if $m(\bar{A}) = 4n$, no subgroup of $\bar{M}_V$ of order $2^{2n} - 1$ centralizes a subspace of $C_V(\bar{A})$ of rank exactly $3n$.

Therefore $\bar{L}$ is M_{22} or M_{24}, with $k = 2$. This time $m(\bar{A}) \leq m_2 \leq 6$, so by E.2.14.8, $I^* \cong L_2(4)$, $m(\bar{A}) = 2s$ for $s := 2$ or 3, and $m(V/Z_I) = 2(s+1)$. Again $Z_I = C_V(\bar{X})$, contradicting H.16.7, which says there is no subgroup $\bar{X}$ of order 3 centralizing a subspace of $C_V(\bar{A})$ of corank $2(s+1)$ in V. So the lemma is established. $\qquad\square$

We can now eliminate the shadows of the groups $U_6(2^n)$ or $U_7(2^n)$, when $\bar{L} \cong (S)L_3(2^{2n})$ and $n > 1$. Recall that $U_6(2)$ can be regarded as a Fischer group F_{21}.

LEMMA 8.2.4. *If $\bar{L} \cong (S)L_3(2^{2n})$ then $n = 1$, $\bar{L} \cong L_3(4)$, $r = 5$, $k = w = 1$, $m(\bar{A}) = 4$, and $C_A(V) = Z_I$ is of rank 4.*

PROOF. By 8.2.3, $k = w = n$. By 7.4.1 and Table 7.2.1, $r \geq 4n$ with equality only if:

(*) $C_G(U) \not\leq M$ for some U of rank $5n$ where U is the centralizer of an element $\bar{y} \neq 1$ of odd order in $\bar{M}_V$.

So by E.3.28.3, $m(\bar{A}) \geq r - w \geq 3n$, and hence by H.4.4.3, $m(V/C_V(\bar{A})) \geq 5n$. But by 8.2.2.7,

$$m(\bar{A}) = m(V/Z_I) - k \geq m(V/C_V(\bar{A})) - n \geq 4n,$$

so as $m(\bar{A}) \leq m_2 = 4n$, we conclude that all inequalities are equalities, so that $m(\bar{A}) = 4n$ and $Z_I = C_V(A)$ is of rank $4n$. Then by the FWCI, $r \leq m(\bar{A}) + w = 5n$.

Assume $n = 1$. Then from H.4.4.7, $\bar{L} \cong L_3(4)$, and we saw earlier that $k = w = n = 1$, $m(\bar{A}) = 4n = 4$, and $Z_I = C_V(A)$ is of rank $4n = 4$. The lemma holds when $r = 5$, so as $4 \leq r \leq 5$, we may assume $r = 4$, and it remains to derive a contradiction. Thus (*) holds. By H.4.6.1, $\langle \bar{y} \rangle = C_{\bar{M}_V}(U)$ is of order 3, so U is in the set Γ of Definition E.6.4. But now E.6.11.2 contradicts the fact that U is not centralized by an element of $\bar{M}_V$ of order 15.

Thus we may take $n > 1$, and it remains to derive a contradiction. As $n = k$, there is $\bar{X}$ of order $2^n - 1$ in $\bar{M}_V$ with $C_V(\bar{X}) = Z_I$ by 8.2.2.8. However this contradicts H.4.4.5, completing the proof. $\qquad\square$

If $\bar{L} \cong (S)L_3(2^{2n})$ then $\bar{L} \cong L_3(4)$ by 8.2.4 and H.4.4.7. In that event, let U_L denote the unipotent radical of the stabilizer of a line in the natural module for $L_3(4)$. We now obtain the analogue of lemma 8.1.3 in our remaining cases:

PROPOSITION 8.2.5. *Let $U := C_V(A)$. Then $w = k = n(I)$ and:*

$\bar{L}/V$	w	r	$\bar{A}$	$m(\bar{A})$	$m(U)$
$L_3(4)/9$	1	5	U_L	4	4
$M_{22}/10$	1	6	K_Q	5	4
$M_{24}/11$	1	7	K_S	6	4
$M_{24}/\overline{11}$	2	8	K_T	6	3

In each case, $U \trianglelefteq T$, so $N_G(U) \in \mathcal{H}(T)$. Further $U = C_A(V) = Z_I \leq Z(I)$ and so $I \leq C_G(U)$.

PROOF. Recall $Z_I = V \cap V^h \leq Z(I)$ by 8.2.2.3, and $w = k = n(I)$ by 8.2.3. By 8.2.2.7, $Z_I = C_A(V)$, so $m(Z_I) = m(V) - k - m(\bar{A})$. Thus if $m(U)$ and $m(\bar{A})$

are as described in the Table, then $m(U) = m(Z_I)$, so $Z_I = U$. Further the Table says $\bar{A} \trianglelefteq \bar{T}$, so $U = C_V(A) \trianglelefteq T$. Hence it remains to verify the Table.

When $\bar{L}$ is M_{24} on the cocode module, we verified the Table in 8.1.3. If $\bar{L}$ is $L_3(4)$, the Proposition follows from 8.2.4 modulo the following remark: As both $U = C_V(A)$ and $\bar{A}$ have rank 4, H.4.4.2 says that $\bar{A} = U_L$.

Thus we may assume $\bar{L}$ is M_{22} or M_{24} on the code module. By 8.2.3, $k = w = 1$. By 7.4.1 and the values in Table 7.2.1, $r \geq 6$.

Suppose $\bar{L} \cong M_{24}$ and $r = 6$. Then arguing as in the proof of (*) in the previous lemma, $C_G(U_0) \not\leq M$ for some subspace U_0 of V of rank 5 which is the centralizer of an element $\bar{y}$ of order 3 in $\bar{M}_V$. By H.16.6, $\langle \bar{y} \rangle = C_{\bar{M}_V}(U_0)$ so that $U_0 \in \Gamma$. Then by E.6.11.2, there is an element of order 63 centralizing U_0, contradicting H.16.6. Thus $r \geq 7$ when $\bar{L} \cong M_{24}$ on the code module.

Now by E.3.28.3,

$$m(\bar{A}) \geq r - w = r - 1,$$

so as $r - 1 \geq 5, 6 = m_2$, we conclude $m(\bar{A}) = m_2 = r - 1 = 5, 6$. Then by 8.2.2.6,

$$m(\bar{A}) = m(V/Z_I) - k \geq m(V/C_V(\bar{A})) - 1, \qquad (*)$$

so $m(V/C_V(\bar{A})) \leq m(\bar{A}) + 1 = 6, 7$. Since V is not an FF-module, this inequality is an equality, so the inequality in (*) is also an equality. Thus $U = Z_I$ is of rank 4. Further it follows from H.16.5 that $\bar{A} = K_Q, K_S$, so the proof is complete. $\square$

8.2.2. Constructing $N_G(U)$. We now use the results from the previous subsection to study the subgroup $N := N_G(U)$, where U is defined in 8.2.5. Let $\tilde{N} := N/U$ and $L_U := N_L(U)^\infty$. Recall from 8.2.5 that $T \leq N$, so $N \in \mathcal{H}^e$ by 1.1.4.6.

As $k \leq 2$ by 8.2.5, 8.2.2.5 says $I^* \cong D_{2m}$ or $L_2(4)$. Thus case (i) of E.2.14.2 holds, with $P := O_2(I) = AB$ and $A = B^h$. By 8.2.5, $U = Z_I$; it follows from E.2.14 that $P = [P, O^2(I)]U$.

We first observe:

LEMMA 8.2.6. *(1) $L_U \in \mathcal{C}(N_M(U))$.*

(2) L_U acts naturally on U as A_5, A_5, A_6, $L_3(2)$.

(3) Either $O_2(L_U) = C_{L_U}(U)$, or $\bar{L}$ is M_{24} on the code module, $L_U/O_2(L_U) \cong \hat{A}_6$, and $C_{L_U}(U) = O_{2,Z}(L_U)$.

PROOF. Part (1) follows from the definitions. Parts (2) and (3) follow from H.4.6.2, H.16.3.2, H.16.1.2, and H.15.6.2. $\square$

As $T \leq N_M(U)$, T acts on L_U, so by 8.2.6.1 and 1.2.4, $L_U \leq K_U \in \mathcal{C}(N)$ with $T \leq N_N(K_U)$.

LEMMA 8.2.7. *$K_U/O_2(K_U)$ is quasisimple.*

PROOF. Assume not. Then by 1.2.1.4, $K_U/O_{2,F}(K_U) \cong SL_2(q)$ for some odd prime q. Then as $L_U \leq K_U$, A.3.12 says that either $K_U = L_U O_{2,F}(K_U)$ or $L_U/O_2(L_U) \cong L_2(4)$ and $q \equiv \pm 1 \mod 5$. In any case (in the notation of chapter 1) $X := \Xi_p(K_U) \neq 1$ for some prime $p > 3$, and by 1.3.3, $X \in \Xi(G, T)$. By 1.2.1.4 either $p = q$ and $X = O^2(O_{2,F}(K_U))$, or $K_U = L_U O_{2,F}(K_U)$ and $L_U/O_2(L_U) \cong L_2(4)$. In particular V is not the code module for $\bar{L} \cong M_{24}$, since $\hat{A}_6$ is not isomorphic to $L_2(p)$ for any odd prime p.

Now $X = [X, L_U]$, so as $L_U \trianglelefteq N_M(U)$, $X \not\leq M$; hence $XT \in \mathcal{H}_*(T, M)$, so replacing H by XT if necessary, we may take $H = XT \leq N$. Then H and the subgroup I of H are solvable, so that $1 = n(I) = k$ by E.1.13; hence by 8.2.3, V is not the cocode module for M_{24}. Thus $\bar{L}$ is $L_3(4)$ or M_{22}.

Let $Y \in Syl_p(X)$, so that also $Y \not\leq M$. Then $Y \cong E_{p^2}$ or p^{1+2} by definition of $X \in \Xi_p(G, T)$. Suppose $Y \cong p^{1+2}$. Then $\Phi(Y) \leq M$ by B.6.8.2, so as $p > 3$, Y centralizes U from the action of $Aut_M(U)$ on U in 8.2.6. Then as $p > 3$, $[V, \Phi(Y)] = 1$ by H.4.6.3 and H.16.3.4. Thus $Y \leq N_G(\Phi(Y)) \leq M$ by 4.4.3 and Remark 4.4.2, contradicting our observation that $Y \not\leq M$. We conclude $Y \cong E_{p^2}$.

Let $\hat{H} := H/O_2(H)$. As $k = 1$, $H = O_{2,p,2}(H)$ by B.6.8.2. Thus as we saw $O_2(I) = P = [P, O^2(I)]U$ and $H \leq N$, $P \leq O_2(H)$. Then as $U = Z_I \leq Z(I)$ by 8.2.2.3, and $I \leq O^2(H) = X$, there is a chief factor W for H on $O_2(X)U/U$ with $W = [W, Y]$. As $V \not\leq O_2(H)$ by 8.2.2.1, $V \not\leq O_2(X)$; and V/B is of rank $k = 1$, $B = V \cap P = V \cap O_2(H) = V \cap O_2(X)$, so that $\hat{V}$ is of rank 1. Therefore as $\hat{T}$ is irreducible on $\hat{Y}$, $\hat{V}$ inverts $\hat{Y}$, so $m(W) = 2m([W, V])$. But $[O_2(X)U, V] \leq O_2(X) \cap V = B$, so $[W, V] \leq W_B$, where W_B is the image of B in W. Thus

$$m(W) = 2m([W, V]) \leq 2m(W_B) \leq 2m(B/U) \leq 10$$

using 8.2.5. But this is impossible, as $SL_2(q)/E_{p^2}$ for $p > 3$ has no faithful module of dimension less than $5^2 - 1 = 24$. $\square$

PROPOSITION 8.2.8. *(1)* $L_U = K_U \trianglelefteq N$.
(2) $[L_U, C_G(U)] \leq O_2(L_U)$.
(3) Either

(a) $I/P \cong L_2(2^k)$, or
(b) $\bar{L}$ is $L_3(4)$ or M_{24} on the code module, and $I/P \cong D_{10}$.

(4) L_U acts on I and P with $O_2(L_U I) = PO_2(L_U) = C_{L_U I}(\tilde{P})$.
(5) Let $\tilde{J} \in Irr_+(I, \tilde{P})$ and set $F := \mathbf{F}_2$ in case (a) of (3), and $F := \mathbf{F}_4$ in case (b). Then $\tilde{P}$, $\tilde{J}$, and $\tilde{B}$ can be regarded as F-modules $\tilde{P}_F$, $\tilde{J}_F$ and $\tilde{B}_F$, for $L_U I$, I, and L_U, respectively, and $\tilde{P}_F = \tilde{J}_F \otimes \tilde{B}_F$ as an $FL_U I$-module.
(6) If V is the code module for $\bar{L} \cong M_{24}$, then case (b) of (3) holds and T does not act on $O^2(I)$.

PROOF. By 8.2.7, $K_U/O_2(K_U)$ is quasisimple, while $L_U \leq K_U$ and $C_{L_U}(U) = O_{2,Z}(L_U)$ by 8.2.6.3. Therefore $C_{K_U}(U) \leq O_{2,Z}(K_U)$. But $[K_U, C_G(U)] \leq C_{K_U}(U)$, so $[K_U, C_G(U)] \leq O_2(K_U)$. Hence (2) follows.

Choose h as in 8.2.2.2. By 8.2.5 and (2), $h \in I \leq C_G(U) \leq N_G(L_U O_2(K_U))$. Therefore as $L_U O_2(K_U)$ acts on V, $L_U O_2(K_U)$ also acts on V^h, and hence on $\langle V, V^h \rangle = I$ and on $O_2(I) = P$.

Set $Y := IL_U$ and $\dot{Y} := Y/C_Y(\tilde{P})$. Since $\tilde{B}$ is an L_U-submodule of rank $m(\tilde{A})$ given in 8.2.5, in the various cases the $L_U/O_2(L_U)$-module $\tilde{B}$ is identified as: the natural module for $L_2(4)$ by H.4.6.2; the 5-dimensional indecomposable (with trivial quotient) for $L_2(4)$ by H.16.3.3; a natural module for $\hat{A}_6$ by H.16.1.3; the sum of two isomorphic natural modules for $L_3(2)$ by H.15.6.3. Furthermore in each case $C_{L_U}(\tilde{B}) = O_2(L_U)$. In particular, the number of $\dot{L}_U$-constituents on $\tilde{B}$ is 1, 1, 1, 2, and hence is equal to k by 8.2.5.

Now by E.2.10.2, $\tilde{P} = \tilde{B} \oplus \tilde{A}$ is the sum of two I-conjugates of $\tilde{B}$, and $P = C_I(\tilde{P})$ by E.2.14. Therefore as $[L_U, I] \leq O_2(L_U) = C_{L_U}(\tilde{B})$ by (2), $O_2(L_U) = C_{L_U}(\tilde{P})$ and $\dot{L}_U \cong L_U/O_2(L_U)$ is quasisimple and centralized by $\dot{I} \cong I/P$, so $\dot{Y} = \dot{I} \times \dot{L}_U$ and (4) holds.

If $\dot{I}$ is $L_2(2)$, then conclusion (a) of (3) holds for $k = 1$, and $\tilde{P}$ is the sum of copies of the natural module $\tilde{J}$ with $End_{\mathbf{F}_2 I}(\tilde{J}) = \mathbf{F}_2$, so (5) follows from 27.14 in [**Asc86a**] in this case.

Suppose $\bar{L}$ is M_{22}. Then $\tilde{P}/[\tilde{P}, L_U]$ is of rank 2, so as $\tilde{P}$ is the sum of copies of $\tilde{J}$, it follows that $\dot{I}$ is $L_2(2)$, so that (3a) and (5) hold in this case by the previous paragraph. In the remaining cases for $\bar{L}$, $\tilde{B}$ is the sum of k copies of the natural irreducible module Λ for $\dot{L}_U$, so $\tilde{P}$ is the sum of $2k$ copies of Λ. Further $\Delta := End_{\mathbf{F}_2 \dot{L}_U}(\Lambda)$ is $\mathbf{F}_4$, $\mathbf{F}_4$, $\mathbf{F}_2$, respectively; and by 27.14 in [**Asc86a**], $\tilde{P}$ has the structure $\tilde{P}_\Delta$ of a Δ-module for $\dot{L}_U \Sigma$, where $\Sigma := C_{GL(\tilde{P})}(\dot{L}_U) = GL(\Theta)$ for some $2k$-dimensional Δ-module Θ, and $\tilde{P}_\Delta = \Lambda \otimes \Theta$ as a $\dot{L}_U \Sigma$-module. Then $\dot{I} \le \Sigma$, and among the possibilities for $\dot{I}$ listed in 8.2.2.5, the only ones which are subgroups of $GL_{2k}(\Delta)$ are $\dot{I} \cong L_2(2^k)$, or D_{10} in the case $k = 1$ and $\Delta = \mathbf{F}_4$. Further $\tilde{J}$ is $\mathbf{F}_2\dot{I}$-isomorphic to Θ by parts (3) and (10) of E.2.14. This completes the proof of (3) and (5).

Suppose V is the code module for M_{24}. Then by (3), $\dot{L}_U \dot{I} \cong \hat{A}_6 \times D_{2m}$ for $m := 3$ or 5. Therefore as $m_3(N) \le 2$ since N is an SQTK-group, $m = 5$. [1] Next $\bar{T}\bar{L}_U/O_2(\bar{T}\bar{L}_U) \cong \hat{S}_6/E_{64}$ with $\bar{A} = O_2(\bar{L}_U)$, where each involution in $\bar{T}$ is fused into $\bar{A}$ under $\bar{M}$, and there is an involution in $\bar{T} - \bar{L}_U$. Therefore there is an involution $t \in T - L_U O_2(L_U T)$. Assume T acts on $O^2(I)$. Then as $I = \langle V, V^h \rangle = O^2(I)(T \cap I)$ since $T \cap I \in Syl_2(I)$, while $V \trianglelefteq T$, $I = O^2(I)V$, so that T acts on I. Extend the earlier "dot notation" to $Y_T := YT$ by defining $\dot{Y}_T := Y_T/O_2(Y_T)$, and let $v \in V - B$. Then $\dot{s} := \dot{t}$ or $\dot{t}\dot{v}$ centralizes $\dot{I}$. Thus $\dot{I}$ acts on $C_{\tilde{P}}(\dot{s})$, whereas by (5), $C_{\tilde{P}}(\dot{s})$ is of 2-rank 6, while all irreducibles for $\dot{I}$ on $\tilde{P}$ are of rank 4. This contradiction completes the proof of (6).

It remains to establish (1). As $K_U \trianglelefteq N$, we must show that $K_U = L_U$. First $Aut_{L_U}(U) \le Aut_{K_U}(U)$, and by 8.2.7 and 8.2.6.3, either $C_{K_U}(U) = O_2(K_U)$, or V is the code module for M_{24} and $C_{K_U}(U) = O_2(K_U)O^2(O_{2,3}(L_U))$. If $\bar{L}$ is M_{24} on the cocode module then $Aut_{L_U}(U) = GL(U)$, so $K_U = L_U C_{K_U}(U) = L_U O_2(K_U)$, and hence $L_U = K_U$ in this case. Thus we may assume one of the first three cases holds, so $m(U) = 4$ by 8.2.5.

Suppose case (a) of (3) holds. Then by (6), one of the first two cases holds. Now $m_3(I) = 1 = m_3(L_U)$, $L_U \le K_U$ with $[K_U, I] \le O_2(K_U)$, and N is an SQTK-group, so $m_3(K_U) = 1$. Also $Aut_{L_U}(U) \cong A_5$, and $Aut_T(U)$ acts on $Aut_{L_U}(U)$. The proper overgroups of $Aut_{TL_U}(U)$ in $GL(U)$ have 3-rank at least 2, so as $m_3(K_U) = 1$, we conclude again that $Aut_{K_U}(U) = Aut_{L_U}(U)$ and $K_U = L_U$.

Finally assume case (b) of (3) holds. As $O_2(K_U)$ acts on I, $X := O^2(I) = O^2(IO_2(K_U))$. Thus as $[K_U, I] \le O_2(K_U)$, K_U acts on X, and hence also on $O_2(X)U = P$. Now $\mathbf{F}_{16} = End_{\mathbf{F}_2 X}(\tilde{J})$, and $\tilde{P}$ is the sum of $e := m(\tilde{P})/4 = 2$ or 3 copies of $\tilde{J}$, so by 27.14 in [**Asc86a**], $K_U/C_{K_U}(\tilde{P}) \le GL(\Omega)$, where Ω is an e-dimensional space over $\mathbf{F}_{16}$. Arguing as in the previous paragraph, $m_5(I) = 1 = m_5(L_U)$ so that $m_5(K_U) = 1$. Then inspecting the overgroups of $Aut_{TL_U}(\Omega)$, we conclude as before that $K_U = L_U$. This completes the proof of the lemma. $\quad\square$

LEMMA 8.2.9. *(1) T acts on $O^2(I)$, and $H = IT$.*

(2) T normalizes VA.

(3) V is not the code module for $\bar{L} \cong M_{24}$.

---

[1] We just eliminated the shadow of Co_1, where $m = 3$ in the 2-local N.

PROOF. We begin with the proof of (1), although we will obtain (3) along the way. Set $H^+ := H/O_2(H)$ and $H_0 := \langle I, T \rangle$. Then $T \leq H_0 \leq H$ but $H_0 \not\leq M$, so $H_0 = H$ by minimality of $H \in \mathcal{H}_*(T, M)$. By 7.3.4 and Table 7.2.1, $n(H) \leq 2$. Next $H = \langle I^T \rangle T$, so $O^2(H) \leq \langle I^T \rangle \leq C_G(U)$ since $I \leq C_G(U)$ and $U \trianglelefteq T$ by 8.2.5. Now I acts on L_U by 8.2.8.2, and hence so does $H = \langle I, T \rangle$; therefore $m_3(HL_U) \leq 2$ as HL_U is an SQTK-group. We conclude from 8.2.8.2 and the description of L_U in 8.2.6 that $O^2(H)$ centralizes $L_U/O_2(L_U)$, and $m_3(H) \leq 1$.

Suppose that V is the code module for $\bar{L} \cong M_{24}$. Then $L_U/O_2(L_U) \cong \hat{A}_6$, so the argument of the previous paragraph shows that H is a $3'$-group. Therefore as $n(H) \leq 2$, and $5 \in \pi(H)$ by 8.2.8.6, we conclude from E.2.2 and B.6.8.2 that H is a $\{2, 5\}$-group. Then as $m_5(L_U H) \leq 2$, it follows that $O^2(I) = O^2(H)$, whereas T does not act on $O^2(I)$ by 8.2.8.6. This establishes (3).

Suppose that $\bar{L} \cong M_{24}$, so that V is the cocode module by the previous paragraph. Then $n(H) = 2 = k = n(I)$ by 8.1.3 and 8.2.5, and $I/O_2(I) \cong L_2(4)$ by 8.2.8.3. As $m_3(H) \leq 1$ by the first paragraph, inspecting the possibilities in E.2.2, we conclude that $O^2(H^+) \cong L_2(4)$ or $U_3(4)$. In the former case, $H = IT$ and $O^2(H) = O^2(I)$ so that (1) holds; so we may assume the latter. Then $I^+ \cong L_2(4)$ is generated by the centers of a pair of Sylow 2-groups of $O^2(H^+)$ and hence I^+ is centralized by a subgroup X of $H \cap M$ of order 5. Recall $H \cap M$ acts on V since V is a TI-set under M, so X acts on $\langle V^{O_2(H)I} \rangle = \langle V^I \rangle = I$. Thus X acts on $O_2(I) = P$. As $m(U) = 3$ by 8.2.5, $GL(U)$ is a $5'$-group, as is $C_{GL(\tilde{P})}(Aut_{L_U I}(\tilde{P}))$ by 8.2.8.5. Thus X centralizes P by Coprime Action, and then as $m(V/V \cap P) = k = 2$, X centralizes V. Then as $I = O_2(I)C_I(X)$, X centralizes $\langle V^{C_I(X)} \rangle = I$. Therefore $I \leq N_H(X) \leq H \cap M$ by Remark 4.4.2 and 4.4.3, impossible as we saw that V is normal in $H \cap M$ but not in I.

Thus we may assume that $\bar{L}$ is $L_3(4)$ or M_{22}. Hence $k = n(I) = 1$ by 8.2.5, and by 8.2.8, either

(i) $\bar{L} \cong L_3(4)$, $I/O_2(I) \cong D_{2m}$ for $m := 3$ or 5, and $\tilde{B} = [\tilde{B}, L_U]$, or
(ii) $\bar{L} \cong M_{22}$, $I/O_2(I) \cong L_2(2)$, and $|\tilde{B} : [\tilde{B}, L_U]| = 2$.

Recall H acts on L_U and U, so that $B \leq O_2(L_U)U \leq O_2(H)$ in case (i), and similarly $|B : B \cap O_2(H)| \leq 2$ in case (ii). As $m(V/B) = 1$ and $V \not\leq O_2(H)$, either

(I) $B = V \cap O_2(H)$, so that $V^+ = \langle v^+ \rangle$ is of order 2, or
(II) case (ii) holds and $m(V^+) = 2$.

Suppose case (II) holds. As $n(H) \leq 2$, $V \trianglelefteq H \cap M$, and $m_3(H) = 1$, we conclude from E.2.2 that $O^2(H^+) \cong L_2(4)$ or $U_3(4)$ and V^+ is the center of $T^+ \cap O^2(H^+)$. This contradicts $I \in \mathcal{I}(H, T, V)$ with $n(I) = 1$. The argument also shows that $n(H) = 1$.

Therefore case (I) holds and $n(H) = 1$. Thus for any $g \in H$ with $1 \neq |v^+ v^{+g}|$ an odd prime power, $I_1 := \langle V, V^g \rangle \in \mathcal{I}(H, T, V)$. Therefore by 8.2.8.4, $|v^+ v^{+g}| \in \pi$, where $\pi := \{3, 5\}$ or $\{3\}$, in case (i) or (ii), respectively. Also we saw earlier that $m_3(H) \leq 1$, and as $m_5(L_U H) \leq 2$ while $m_5(L_U) = 1$, $m_5(H) \leq 1$. We conclude by inspection of the list of possibilities for H with $n(H) = 1$ in B.6.8.2 and E.2.2 that either H^+ is $L_2(2)$ or $Aut(L_3(2))$, or case (i) holds and $O^2(H^+)$ is $\mathbf{Z}_5$ or $L_2(31)$. [2]

---

[2]In particular, we cannot have $H^+ \cong U_3(2)$; thus in the first case we are eliminating the shadow of $U_7(2)$, where N is not an SQTK-group—though the shadow of $U_6(2)$ still survives in that first case.

If $O^2(H^+)$ is of order 3 or 5, then $H = IT$, so that (1) holds. Thus we may assume $O^2(H^+) \cong L_3(2)$ or $L_2(31)$.

Let W be a chief section for $L_U H$ on $O_2(\tilde{L}_U \tilde{H})$ with $[W, O^2(H)] \neq 1$ and set $(L_U H)^! := L_U H / C_{L_U H}(W)$. As $L_U H$ is irreducible on W, $O_2(H) = C_H(W)$ and $O_2(L_U) \leq C_{L_U}(W)$. Then as $O^2(H)$ centralizes $L_U/O_2(L_U)$, $H^+ \cong H^!$ centralizes $L_U^!$, and W is the sum of isomorphic irreducibles for $H^!$ and for $L_U^!$ by Clifford's Theorem. Recall $\tilde{P} = \tilde{A} \oplus \tilde{B}$, with $\tilde{B}$ either natural or a 5-dimensional indecomposable for $\dot{L}_U \cong SL_2(4)$. Thus we may choose W so that W is the sum of $d \geq 2$ copies of the natural module for $L_U^!$, and W is the tensor product of the natural module for $L_U^!$ with a d-dimensional $O^2(H)$-submodule D of W. As case (I) holds, $[O_2(\tilde{L}_U \tilde{H}), V] \leq V \cap \widetilde{O_2(H)} = \tilde{B}$, so $[W, V]$ is the image of $\tilde{B}$ in W. Therefore L_U is irreducible on $[W, v^+]$, so it follows that v^+ induces a transvection on D. Therefore D is a natural module for $O^2(H^!) \cong L_3(2)$, which is impossible as $H^+ \cong Aut(L_3(2))$ and W is a homogeneous $L_U^!$-module. Therefore (1) is established.

Finally V is T-invariant, and by (1) so is $O_2(I) = AB$, establishing (2). $\qquad\square$

LEMMA 8.2.10. *(1) L is a block of type $L_3(4)/9$, $M_{22}/10$, or $M_{24}/\overline{11}$.*
(2) $C_T(L) = 1$.
(3) $V = O_2(L)$.
(4) $Z = C_V(T)$ is of order 2.

PROOF. By 8.2.9.3, V is not the code module for $\bar{L} \cong M_{24}$. By 8.2.9.2, T normalizes VA, so $[O_2(L), A] \leq O_2(L) \cap VA \leq V C_A(V) = VU = V$. Then $L = [L, A]$ centralizes $O_2(L)/V$, so that (1) holds. By 3.2.10.9, $C_Z(L) = 1$, so (2) follows. By (1), $[Z, L] \leq V$. Then as the Sylow group T centralizes Z, we conclude from (2) and Gaschütz's Theorem A.1.39 that $VZ = VC_Z(L) = V$. Therefore $Z = C_V(T)$, so Z is of order 2, completing the proof of (4). By (1), L/V is quasisimple, and as $F^*(L) = O_2(L)$, $Z(L/V)$ is a 2-group. Thus as the multiplier of M_{24} is trivial, (3) holds when $\bar{L} \cong M_{24}$; and similarly (3) holds when $Z(L/V) = 1$, so we may assume that $Z(L/V) \neq 1$. If $\bar{L} \cong L_3(4)$, we may consider a quotient of L/V with center of order 2; then from the structure of the covering group in (3b) of I.2.2, $O_2(L_U)V/V$ is an indecomposable extension of a natural $L_2(4)$ module over a nonzero trivial submodule, which is not isomorphic to $\tilde{B}$ as an L_U-module, contrary to 8.2.8.5. Since an extension of M_{22} over a center of order 2 restricts to such an extension of $L_3(4)$, this argument also eliminates extensions of M_{22}. This completes the proof of (3). $\qquad\square$

8.2.3. Constructing $\mathbf{C_G(z)}$. At this stage, in view of 8.2.10.1, the cases remaining are

$$L_3(4)/9, \quad M_{22}/10, \quad \text{and} \quad M_{24}/\overline{11}.$$

By 8.2.10.4, $Z = C_V(T)$ is of order 2. In this section we let z denote a generator of Z, and set $C := C_G(z)$.

Using the subgroup of C generated by $C_M(z)$ and H (appearing essentially as $K_z T$ in the proof of 8.2.13), we will show that $O_2(C)$ is extraspecial with center Z. Then using the fact that C is an SQTK-group, we eliminate the $L_3(4)/9$ and $M_{22}/10$ cases, where $C/O_2(C)$ is $U_4(2)$ or $Sp_6(2)$ in the shadows $U_6(2)$ or Co_2. This reduces us to the case where $L/V \cong M_{24}$ and V is the cocode module. There we show C has the structure of the centralizer of a 2-central involution in J_4, which allows us to identify G as J_4.

Let $L_z := C_L(z)^\infty$ and $\tilde{C} := C/Z$.

LEMMA 8.2.11. (1) $L_z \in \mathcal{C}(C_M(z))$.

(2) There exists an $L_z T$-series $1 < Z < V_z < V$ with $V_z := [V, O_2(L_z)]$, and $\tilde{V}_z$ is the natural module for $L_z/O_2(L_z) \cong L_2(4)$, A_6, $\hat{A}_6$.

(3) V/V_z is the A_5-module, the core of the 6-dimensional permutation module, the 4-dimensional irreducible, respectively.

(4) $O_2(L_z)/V$ induces the group of transvections on V_z with center Z, so $O_2(\bar{L}_z\bar{T}) = O_2(\bar{L}_z)$ is $L_z T$-isomorphic to the dual of $\tilde{V}_z$.

PROOF. Parts (2)–(4) follow from H.4.6.4, H.16.4, and H.15.3. Then (2) implies (1). □

Set $E := \langle V_z^C \rangle$.

LEMMA 8.2.12. (1) $E \cong D_8^e$ is extraspecial, for $e := 4, 4, 6$.

(2) $E = O_2(C)$.

(3) $O_2(L_z) = EV$ and $V_z = E \cap V$.

PROOF. By 1.1.4.6, $F^*(C) = O_2(C) =: Q_C$, so $F^*(\tilde{C}) = \tilde{Q}_C$ by A.1.8. Therefore as $V_z \trianglelefteq T$, $1 \neq C_{\tilde{V}_z}(T) \leq Z(\tilde{Q}_C)$. Then as L_z is irreducible on $\tilde{V}_z$ by 8.2.11.2, $\tilde{V}_z \leq Z(\tilde{Q}_C)$, so $\tilde{E} = \langle \tilde{V}_z^C \rangle \leq Z(\tilde{Q}_C)$.

Let $Q_M := O_2(LT)$. By parts (2) and (5) of 8.2.8, $[V, L_U] = B$. Then by H.4.6.5, H.16.4.4, and H.15.8, $V_z \leq B$ but $V_z \not\leq U$; therefore $V_z^h \leq A$ but $V_z^h \not\leq U$. Thus as $U = A \cap Q_M$ and $V_z^h \leq E$, $E \not\leq Q_M$. But by 8.2.11.4, L_z is irreducible on $O_2(\bar{L}_z\bar{T}) = O_2(\bar{L}_z)$, so $\bar{E} = O_2(\bar{L}_z)$. Thus as $V = O_2(L)$ by 8.2.10.3, $EV = O_2(L_z)$, establishing the first statement in (3).

Now $Z \leq V = O_2(L)$ with L irreducible on V, so if $\Phi(Q_M) \neq 1$ then $\Phi(Q_M) \geq V$. But $C_{LT}(Q_M) \leq Q_M$, so each x of odd order in L is faithful on $Q_M/\Phi(Q_M)$, whereas $[Q_M, x] \leq V$ by 8.2.10.1. Thus $\Phi(Q_M) = 1$. Similarly as $Z \leq V$, L is indecomposable on Q_M. But by earlier remarks, $\widetilde{Q_M \cap Q_C} \leq C_{\tilde{Q}_M}(\tilde{E}) \leq C_{\tilde{Q}_M}(O_2(\bar{L}_z))$. Next from the structure of indecomposable extensions of V by a trivial quotient (obtained from the duals of modules described in I.1.6), $C_{\tilde{Q}_M}(O_2(\bar{L}_z)) \leq C_{\tilde{V}}(O_2(\bar{L}_z))$, while $C_{\tilde{V}}(O_2(\bar{L}_z)) = \tilde{V}_z$ by H.4.6.6, H.16.4.5, and H.15.3.4. Hence $V_z = Q_M \cap Q_C$. Thus $V_z = V \cap E$, completing the proof of (3). Now using (3) we have

$$|E| = |V_z||E : V_z| = |V_z||EV : V| = |V_z| \cdot |O_2(\bar{L}_z)| = 2^{1+2e}$$

where $e := 4, 4, 6$. By 8.2.11.4, $Z = C_{V_z}(E)$, so (1) holds. (As $e + 1 = m(V_z)$, $E \cong D_8^e$).

As $E \leq Q_C \leq O_2(C_{LT}(z)) = EQ_M$, and $Q_C \cap Q_M = V_z \leq E$, (2) holds. □

PROPOSITION 8.2.13. (1) V is the cocode module for $L/V \cong M_{24}$.

(2) $L = M$ and $C/E \cong \hat{M}_{22}.2$

PROOF. By 8.2.11.1, $L_z \in \mathcal{C}(C_M(z))$, and of course L_z is T-invariant. Then by 1.2.4, $L_z \leq K_z \in \mathcal{C}(C)$, and the possibilities for $K_z/O_2(K_z)$ are described in A.3.12.

By 8.2.12.2, $E = O_2(C)$. Let $C^* := C/E$. As $K_z \trianglelefteq C$ and $E = O_2(C)$, $O_2(K_z^*) = 1$; in particular $L_z < K_z$ by 8.2.12.3. Indeed using that result, $O_2(L_z^*) \cong V/V_z$ is described in 8.2.11.3. We inspect the lists in A.3.12 and A.3.14 for such subgroups, and conclude that $\bar{L}$ is M_{24} and $K_z^* \cong \hat{M}_{22}$; in particular, notice when

$L_z^*/O_2(L_z^*) \cong A_5$ that the A_5-module V/V_z does not arise in A.3.14. [3] That is, (1) holds.

By 8.2.12.1, $\tilde{E}$ is of rank 12. Therefore by H.12.1, $\tilde{E}$ is irreducible and $End_{K_z^*}(\tilde{E}) \cong \mathbf{F}_4$, so $Z(K_z^*) = C_{C^*}(K_z^*)$. Finally there is $t \in T \cap L$ inducing an outer automorphism on $L_z/O_2(L_z)$ and hence also on K_z^*, so as $|Aut(K_z^*) : K_z^*| = 2$, $C = TK_zC_C(K_z) = TK_z$ with K_z of index 2 in C. Therefore $C_M(z) = L_zT$ with $|T| = 2^{21} = |L \cap T|$. Then as $M = LC_M(z)$, $L = M$, so (2) holds. $\square$

As a corollary we get:

THEOREM 8.2.14. $G \cong J_4$.

PROOF. By 8.2.12, $E = F^*(C) \cong D_8^6$, and by 8.2.13, $C/E \cong Aut(\hat{M}_{22})$. Also $z^L \cap V_z \neq \{z\}$, so z is not weakly closed in E with respect to G. These are the hypotheses of Aschbacher-Segev [AS91], so we conclude from the main theorem of that paper that $G \cong J_4$. We mention that their work uses the graph-theoretic methods used elsewhere in this work to establish recognition theorems. $\square$

8.3. Eliminating $L_3(2) \wr 2$ on 9

In this final section of chapter 8, we treat the exceptional case of $L_3(2) \wr 2$ on its tensor product module, which we have been postponing since the previous chapter. We prove:

THEOREM 8.3.1. The case $\bar{L}_0 \cong L_3(2) \times L_3(2)$ on its 9-dimensional tensor product module cannot arise.

We begin by defining notation: Let $L_1 := L$, $L_2 := L^t$, $L_0 := L_1L_2$, $V_1 \in Irr_+(L_1, V)$ with V_1 $N_T(L)$-invariant, and $V_2 := V_1^t$, so that V is the tensor product of V_1 and V_2 as an $\bar{L}_0$-module. Thus we can appeal to subsection H.4.4 of chapter H of Volume I.

Let $V_{i,m}$ be the $N_T(L)$-invariant m-dimensional subspace of V_i, and adopt the following notation for the unipotent radicals of the corresponding parabolic subgroups: $R_i := C_{T \cap L_i O_2(L_0T)}(V_{i,2})$, and $S_i := C_{T \cap L_i O_2(L_0T)}(V_i/V_{i,1})$. Let $R := R_1R_2$, $S := S_1S_2$, and as usual set $Q := O_2(L_0T)$. Notice $T_0 := RS$ is Sylow in L_0Q, and of index 2 in T. Let $W_j := W_j(T, V)$ for $j = 0, 1$.

From 3.2.6.2, we have $V = V_M$, so $M = M_V$.

LEMMA 8.3.2. $s(G, V) = 3$.

PROOF. This follows from 7.3.2 and Table 7.2.1. $\square$

Recall from 7.3.3 and Table 7.2.1 that $w \geq 1$; indeed we can show:

LEMMA 8.3.3. Either

(1) W_1 centralizes V, so that $w > 1$; or
(2) $\bar{W}_1 = \bar{R}$ and $W_1(S, V) = W_1(Q, V)$.

PROOF. Suppose $A \leq V^g \cap T$ with $m(V^g/A) \leq 1$, but $\bar{A} \neq 1$. By 8.3.2, $s = 3$, so that $\bar{A} \in \mathcal{A}_2(\bar{M}, V)$ by E.3.10. Then by H.4.11.2, $\bar{A} \leq \bar{R}$. So if W_1 does not centralize V, $\bar{W}_1 = \bar{R}$ since $N_M(R)$ is irreducible on $\bar{R}$. Similarly $\bar{R} \cap \bar{S}$ contains no members of $\mathcal{A}_2(\bar{M}, V)$, so $W_1(S, V) = W_1(Q, V)$. $\square$

---

[3] We just eliminated the shadows of $U_6(2)$ and Co_2, where $C/E \cong U_4(2)$, $Sp_6(2)$ are not SQTK-groups.

REMARK 8.3.4. The second case of lemma 8.3.3 in fact arises in the shadows of $G = Aut(L_n(2))$, $n = 6$ and 7. In those shadows, H is the parabolic determined by the node(s) complementary to those determining the maximal T-invariant parabolic M. Further $w = 1$, and $U = C_V(R)$ is the centralizer of a w-offender. In most earlier cases in this chapter, we were able to use elementary weak closure arguments to show that the configuration correponding to a shadow is the unique solution of the Fundamental Weak Closure Inequality FWCI, and then obtain a contradiction to the fact that $N_G(U)$ is an SQTK-group. But here, as in our treatment of the cases corresponding to the Fischer groups, we instead use the fact that G is quasithin to show that $C_G(U) \leq M$ for suitable subgroups U of V, and then use weak closure to obtain a contradiction.

LEMMA 8.3.5. $N_G(W_0(S,V)) \leq M \geq C_G(C_1(S,V))$.

PROOF. By 8.3.3, $W_1(S,V) = W_1(Q,V)$. As $W_0(S,V) \leq W_1(S,V)$ and $M = !\mathcal{M}(N_G(Q))$, the lemma follows from E.3.16.   □

LEMMA 8.3.6. If $H \in \mathcal{H}^e$ with $S \in Syl_2(H)$ and $n(H) = 1$, then $H \leq M$.

PROOF. Since $s(G,V) = 3$ by 8.3.2, this follows from 8.3.5 using E.3.19 with 0, 1 in the roles of "i, j".   □

As usual we wish to show that $C_G(U) \leq M$ for various subspaces U of V. Usually these subspaces will contain a 2-central involution, so it will be useful to establish some restrictions on the centralizers of such involutions.

Let z be a generator for $C_V(T)$; in the notation of subsection H.4.4, we may take z to the involution $x_{1,1}$ generating $V_{1,1} \otimes V_{2,1}$. Set $G_z := C_G(z)$, $M_z := C_M(z)$, $X := O^2(C_{L_0}(z))$, and $K_z := \langle X^{G_z} \rangle$. Note that $O_2(XT) = S$.

LEMMA 8.3.7. $G_z = K_z M_z$, where either

(i) $K_z = KK^s$ for some $K \in \mathcal{C}(G_z)$ and $s \in T - N_T(K)$ with $K/O_2(K) \cong L_2(p)$, p prime, or

(ii) $K_z/O_2(K_z) \cong L_4(2)$ or $L_5(2)$.

PROOF. Let $P \in Syl_3(X)$; then $X \in \Xi(G,T)$, $P \cong E_9$, and $Aut_T(P) \cong D_8$. We apply 1.3.4 to $G_z \in \mathcal{H}(XT)$ in the role of "H". If $X \triangleleft G_z$, define $K_z := X$; otherwise 1.3.4 gives $X \leq K_z := \langle K^T \rangle$, where $K \in \mathcal{C}(G_z)$ is described in one of the cases of 1.3.4. Notice case (3) of 1.3.4 is ruled out, as there $Aut_T(P)$ is cyclic. Similarly case (2) of 1.3.4 and case (4) with $K_z/O_2(K_z) \cong M_{11}$ are eliminated, as in those cases $Aut_T(P)$ contains a quaternion subgroup. We may assume the lemma fails. Thus neither of the remaining possiblities in case (4) of 1.3.4 holds, so case (1) of 1.3.4 holds and we may take $K_z = KK^s$ with $K/O_2(K) \cong L_2(2^n)$ and $n \geq 4$ even, as $p = 3$ and $L_2(4) \cong L_2(5)$.

Note in either case that $K_z \trianglelefteq G_z$. Set $Y_z := C_{G_z}(X/O_2(X))$. As $T \in Syl_2(G)$ acts on X, $T \cap Y_z \in Syl_2(Y_z)$, so by A.4.2.4, $S = T \cap Y_z$. If $K/O_2(K) \cong L_2(2^n)$, X is characteristic in $N_{K_z}(T \cap K_z)$ and $T \cap K_z = O_2(K_z)O_2(X))$, so by Sylow's Theorem $X^{N_G(K_z)} = X^{K_z}$. This holds trivially if $K_z = X$. Hence by a Frattini Argument, $G_z = K_z N_{G_z}(X) = K_z N_{G_z}(Y_z)$. Then as $S \in Syl_2(Y_z)$, $G_z = K_z Y_z N_{G_z}(S)$ by another Frattini Argument. As $J(T) \leq Q \leq S$, $N_G(S) \leq M$ by 3.2.10.8, so $G_z = K_z Y_z M_z$. Next $Y_z = XY$, where $Y := O^3(Y_z)$ is a $3'$-group as $m_3(G_z) = 2$. As $S \in Syl_2(Y_z)$, $S \in Syl_2(YS)$.

We claim $Y \leq M$. If Y is solvable, then $n(Y) = 1$ by E.1.13, so $Y \leq M$ by 8.3.6. So suppose Y is not solvable. Then there is $Y_1 \in \mathcal{C}(Y)$ with $Y_1/O_2(Y_1) \cong Sz(2^k)$. Now a Borel subgroup B of Y_1 is solvable, so as before $B \leq M$ using 8.3.6. Set $H := \langle Y_1, T \rangle$; then $n(H) = k$ is odd and $k \geq 3$. If $H \not\leq M$ then as $B \leq M$, we get $H \in \mathcal{H}_*(T, M)$, contradicting 7.3.4, which says $n(H) \leq 2$. So $H \leq M$, and in particular $Y_1 \leq M$. These arguments apply to each minimal parabolic H of YS over S, so as this set of parabolics generates $O^{2'}(YS)$ by B.6.5, $O^{2'}(YS) \leq M$. Finally as $S \in Syl_2(YS)$, by a Frattini Argument $YS = O^{2'}(YS)N_{YS}(S) \leq M$, since we saw $N_G(S) \leq M$. This completes the proof of the claim.

As $G_z = K_z Y_z M_z$, and $Y_z = XY$ with $X \leq K_z$, we conclude $G_z = K_z M_z$, establishing the first assertion of the lemma.

If $K_z = X$, then $G_z = X M_z = M_z \leq M$, contradicting 3.1.8.3.ii, which shows $H \leq G_z$ for each $H \in \mathcal{H}_*(T, M)$. Thus $X < K_z$, so $K/O_2(K) \cong L_2(2^n)$ with $n > 2$. But now we replace Y_1 by K in the argument above, and again obtain a contradiction to $n(H) \leq 2$ in 7.3.4. This completes the proof. $\square$

We can now essentially eliminate the shadows of the linear groups:

LEMMA 8.3.8. $C_G(C_V(R)) \leq M$.

PROOF. Set $U := C_V(R)$; our proof relies on the following properties:

(a) $z \in U$.

(b) $N_{L_0}(R) \leq N_{L_0}(U)$, and there is a subgroup $P \cong E_{3^2}$ of $N_{L_0}(R)$ faithful on U.

(c) $T \leq N_G(U)$.

Since $C_G(U) \not\leq M$, using (c) we may choose $H \in \mathcal{H}_*(T, M)$ with $I := O^2(H) \leq C_G(U)$. By (a), $I \leq G_z$, and by (b) and A.1.27, $C_G(U)$ is a $3'$-group.

Next $G_z = K_z M_z$ by 8.3.7. As $I \not\leq M_z$, the projection K_I^* of I on $(K_z T)^* := K_z T/O_2(K_z T)$ is non-trivial. Furthermore $C_G(U)$, and hence also K_I^*, is a T-invariant $3'$-group. In case (ii) of 8.3.7, $K_I^*(T^* \cap K_I^*)$ contains a Sylow 2-subgroup of K_z^* and hence is a parabolic subgroup of K^*; as this parabolic is a $3'$-group, $I \leq T M_z$, contradicting $I \not\leq M$. So instead case (i) of 8.3.7 holds, and $K_z = K K^s$ with $K \cong L_2(p)$. Now $m_2(L_2(p)) = 2$, so if P^* is a $3'$-subgroup of K^*, then $O^2(P^*) = O(P^*)$. Thus as $I = O^2(I)$, the $3'$-group K_I^* is of odd order, so $O_2(I) \leq O_2(K_z T)$, and $O_2(I)$ is Sylow in I. Then since $X \leq K_z$, by A.1.6 we have $O_2(I) \leq O_2(K_z T) \leq O_2(XT) = S$. It follows that $S \in Syl_2(IS)$. But $n(I) = 1$ as I is solvable, so $I \leq M$ by 8.3.6, a contradiction which establishes the lemma. $\square$

Now we achieve our initial goal:

PROPOSITION 8.3.9. $n(H) = 2$ for each $H \in \mathcal{H}_*(T, M)$.

PROOF. Recall $n(H) \leq 2$ by 7.3.4. As $w > 0$, $N_G(W_0) \leq M$ by E.3.16.1. Also $s = 3$ by 8.3.2. Thus if $C_G(C_1(T, V)) \leq M$, then $n(H) = 2$ by E.3.19, so the lemma holds. However if $W_1 \leq C_G(V)$, then $C_G(C_1(T, V)) \leq M$ by E.3.16.1.3, so we may assume that $W_1 \not\leq C_G(V)$. Then by 8.3.3, $\overline{W_1} = \overline{R}$. Therefore $C_V(R) \leq C_T(W_1) = C_1(T, V)$, so $C_G(C_1(T, V)) \leq M$ by 8.3.8, completing the proof. $\square$

LEMMA 8.3.10. (1) $K_z = K K^s$ with $K/O_2(K) \cong L_2(5) \cong L_2(4)$, $K_z T \in \mathcal{H}_*(T, M)$, and $X(T \cap K_z) = K_z \cap M$ is a Borel subgroup of K_z.

(2) $G_z = K_z T$ and $M = L_0 T$.

PROOF. We first prove (1). Assume the first statement in (1) fails. We claim then that $n(H) = 1$ for each $H \in \mathcal{H}_*(T, M)$ with $H \leq K_z T$. We examine the groups listed in 8.3.7. The claim follows in case (i) of 8.3.7 from E.1.14.6 when $p \geq 7$, and in case (ii) from E.1.14.1. Thus the claim is established, and of course it contradicts 8.3.9. Thus the first part of (1) holds, and as the Borel subgroup $X(K_z \cap T)$ of K_z is the unique T-invariant maximal subgroup of K_z, the remaining statements of (1) hold.

Next by 8.3.7, $G_z = K_z M_z$. Assume that $G_z > K_z T$. Then $Y := O^2(C_{G_z}(K_z/O_2(K_z))) \neq 1$, and $Y \leq M_z$. But Y is a $3'$-subgroup of M_z by 1.2.2.a, so as $\bar{M}_z$ is a $\{2,3\}$-group, Y centralizes V. Then $[L_0, Y] \leq C_{L_0}(V) = O_2(L_0)$, so that $L_0 T$ normalizes $O^2(Y L_0) = Y$, and hence $G_z \leq N_G(Y) \leq M = !\mathcal{M}(L_0 T)$, contradicting $K_z \not\leq M$. Thus $G_z = K_z T$, so $M_z = XT$, and hence $C_M(V)$ is a 2-group. Therefore $M = L_0 T$, completing the proof of (2). $\qquad\square$

LEMMA 8.3.11. $r(G, V) > 3$.

PROOF. Recall $r(G, V) \geq 3$ by 7.3.2. Assume $r(G, V) = 3$. Then there is $U \leq V$ with $m(V/U) = 3$ and $C_G(U) \not\leq M$. By E.6.12, $Q < C_M(U)$, and $C_M(U) = C_{L_0 T}(U)$ by 8.3.10.2. Therefore by H.4.12.3 and H.4.10, $U = C_V(\bar{i})$ for some involution $\bar{i} \in \bar{L}_0 \bar{T}$. By H.4.12.3, $C_{\bar{M}}(U)$ is a 2-group, so by E.6.27, U is centralized by an $(F-1)$-offender. Thus $\bar{i} \in \bar{L}_0$ by H.4.10.3. Consequently as $m(V/C_V(\bar{i})) = 3$, we may assume $\bar{i} \in \bar{R}_1$, so that $U = C_V(R_1)$. But of course $R_1 \leq R$ and $C_G(C_V(R)) \leq M$ by 8.3.8. This contradiction establishes 8.3.11. $\qquad\square$

LEMMA 8.3.12. $V \leq O_2(G_z)$.

PROOF. Let $Q_z := O_2(K_z T)$. If $V \leq Q_z$, then the lemma holds, since $G_z = K_z M_z$ by 8.3.7 and $V \trianglelefteq M$. So we assume $V \not\leq Q_z$. Let $\tilde{G}_z := G_z/\langle z \rangle$ and $K_z^* T^* := K_z T/Q_z$. By H.4.9.2, XT is irreducible on $\tilde{V}_5$ and V/V_5, where V_5 denotes the 5-dimensional space in H.4.9.2.

We claim that $V_5 = V \cap Q_z$: to see this, we apply G.2.2, which is designed for such situations. Note that Hypothesis G.2.1 is satisfied with $\langle z \rangle$, V_5, L_0, X, $K_z T$ in the roles of "V_1, V, L, L_1, H". We conclude from G.2.2 that

$$\tilde{U} := \langle \tilde{V}_5^{K_z} \rangle \leq Z(\tilde{Q}_z),$$

and $\tilde{U}$ is a 2-reduced module for K_z^*. Further as $V \not\leq Q_z$ and XT is irreducible on V/V_5, $V_5 = V \cap Q_z$ as claimed.

Notice as $U \leq Q_z \leq T$ that $[U, V] \leq V \cap Q_z = V_5$; so as $m(\tilde{V}_5) = 4 = m(V/V_5)$, $\tilde{U}$ is a dual FF-module for $K_z^* T^*$, with dual FF^*-offender V^*. Now V^* is a normal E_{16}-subgroup of the Borel subgroup $(M \cap K_z T)^*$ in 8.3.10, so $V^* \in Syl_2(K_z^*)$. In particular there is $h \in K_z$ with $K_z^* = \langle V^*, V^{*h} \rangle$. Observe $\tilde{U} = [\tilde{U}, K_z^*]$, since $\tilde{V}_5 = [\tilde{V}_5, X]$ and $\tilde{U} = \langle \tilde{V}_5^{K_z} \rangle$. Then as $[\tilde{V}_5, V^*] \leq \tilde{V}_5$ and $K_z^* = \langle V^*, V^{*h} \rangle$, we conclude

$$\tilde{U} = [\tilde{U}, K_z^*] = \tilde{V}_5 + \tilde{V}_5^h,$$

so that $\tilde{U}$ has dimension at most 8, and hence is itself an FF-module, with quadratic FF^*-offender V^*. By Theorems B.5.6 and B.5.1, the only possibility is $\tilde{U} = \tilde{U}_K \oplus \tilde{U}_K^s$ for $\tilde{U}_K$ a natural module for K^*. But now P of order 3 in X diagonally embedded in KK^s is fixed-point-free on $\tilde{U}$, and hence on $\tilde{V}_5$ of rank 4. Also X is fixed-point-free on V^*, so $C_V(X) = \langle z \rangle$, contradicting H.4.12.1. This completes the proof of 8.3.12. $\qquad\square$

LEMMA 8.3.13. *If $V^g \cap V \cap z^G \neq \emptyset$, then $V^g \leq C_G(V)$.*

PROOF. This is a consequence of 8.3.12 and 3.2.10.6.   □

LEMMA 8.3.14. $W_2(S, V) \leq C_G(V)$.

PROOF. If not, there is V^g with $m(V^g/A) = 2$ and $A := V^g \cap M$ satisfies $1 \neq \bar{A} \leq \bar{S}$. As $A \not\leq C_G(V)$ we may assume without loss that $A \not\leq C_G(V_1)$—so $\bar{A}$ has non-trivial projection $\bar{A}_1$ on $\bar{S}_1$. If $\bar{A}_1 = \bar{S}_1$, then for any hyperplane $\bar{B}$ of $\bar{A}_1$, $\bar{A}$ is non-trivial on the proper subspace $C_{V_1}(B)$ of V_1. On the other hand, if $\bar{A}_1$ is of rank 1, the same is true for the hyperplane $B := S_2 \cap A$ of A with $C_{V_1}(B) = V_1$. Since $V_1^\# \subseteq z^G$, without loss we may assume $z \in [C_{V_1}(B), A]$. By construction, $m(V^g/B) = 3$, so as $r > 3$ by 8.3.11, $C_{V_1}(B) \leq N_G(V^g)$. Therefore $z \in [C_{V_1}(B), A] \leq V \cap V^g$. But now 8.3.13 says $V^g \leq C(V)$, contrary to our choice of V^g. This establishes the lemma.   □

Now we are in a position to complete the proof of Theorem 8.3.1. Recall $S = O_2(XT)$, so from the embedding of X in K_z in 8.3.10, S is Sylow in $K_z S$ and $n(K_z) = 2$. From 8.3.14 and E.3.16.3, $C_G(C_2(S, V)) \leq M$; and from 8.3.5, $N_G(W_0(S, V)) \leq M$. Therefore as $s = 3$ by 8.3.2, E.3.19 says $K_z \leq M$, contradicting 8.3.10.

CHAPTER 9

Eliminating $\Omega_4^+(2^n)$ on its orthogonal module

The results in chapters 7 and 8 almost suffice to establish Theorem 7.0.1, our main result on pairs L, V in the Fundamental Setup (3.2.1) where V is not an FF-module. The only case left to treat is the case where $L_0/O_2(L_0) \cong L_2(2^n) \times L_2(2^n) \cong \Omega_4^+(2^n)$ with $n > 1$, and V is the orthogonal module for $L_0/O_2(L_0)$.

The standard weak closure arguments that handle most of the pairs in chapters 7 and 8 are not so effective in this case. Difficulties are already apparent from the parameters in Table 7.2.1: For example if T contains an orthogonal transvection σ, then $m(\bar{M}, V) = n$, so that if $n = 2$ we cannot immediately apply Theorem E.6.3 to obtain $r(G, V) \geq m(\bar{M}, V)$ as in 7.3.2. We are able to circumvent this difficulty in Lemma 9.2.3 below. There are more serious problems, however: First, $a(\bar{M}, V) = n = s(G, V)$, so 7.3.3 is ineffective. Second, L is not normal in M, so we can't appeal to 7.4.1 to get an effective lower bound on r. Thus we will instead use the fact that G is a QTKE-group to restrict various 2-locals, in order to show that r is large and $n(H)$ is small for each $H \in \mathcal{H}_*(T, M)$. Then weak closure will become effective.

9.1. Preliminaries

We begin by establishing some notation and a few properties of M.

Let $F := \mathbf{F}_{2^n}$ and regard V as a 4-dimensional orthogonal space V_F over F. As usual, let $Q := O_2(L_0T)$. Notice that we are in case (1) of 3.2.6, and in that case $V = V_M \unlhd M$, so $M_V = M$.

LEMMA 9.1.1. $L_0 = O^{p'}(M)$ for each prime divisor p of $2^{2n} - 1$.

PROOF. This follows from 1.2.2.a. $\qquad\square$

LEMMA 9.1.2. (1) $\bar{M} := M/C_G(V)$ is a subgroup of $N_{GL(V)}(\bar{L}_0) = N_{\Gamma L(V_F)}(\bar{L}_0)$, which is the product of $\bar{L}_0$ with the F-scalar maps, extended by $\langle f, \sigma \rangle \cong \mathbf{Z}_2 \times \mathbf{Z}_n$, where σ induces an F-transvection on V_F normalizing $\bar{T}$, with $\bar{L}^\sigma = \bar{L}^t$, and f generates the group of field automorphisms (simultaneously) on $\bar{L}$ and $\bar{L}^t$.

(2) There are elements in $\bar{T} - N_{\bar{T}}(\bar{L})$ of the form σf_0 with $f_0 \in O_2(\langle f \rangle)$.

(3) L_0 has two orbits on F-points of V, consisting of the singular and nonsingular F-points.

(4) $V_N := [V, \sigma]$ is a nonsingular F-point, and setting $L_N := O^2(N_{L_0}(V_N))$, $N_{L_0Q}(V_N) = L_N Q$ with $\bar{L}_N \cong L_2(2^n)$ and $[V, L_N] = C_V(\sigma) = V_N^\perp$ an indecomposable $3n$-dimensional $\bar{L}_N$-module, with $C_V(\sigma)/V_N$ the natural $\bar{L}_N$-module.

(5) Let V_1 denote the singular F-point stabilized by T. Then $N_{L_0T}(V_1)$ is a Borel subgroup of L_0T, and is transitive on $V_1^\#$.

PROOF. This is straightforward. $\qquad\square$

9.2. Reducing to n = 2

Our first goal is to show that $n \leq 2$. We cannot use the uniform approach of chapters 7 and 8, but we can still use some of the underlying techniques. For example we will not be able to bound r as in 7.4.1 using E.6.28 (which relies on E.6.27), but we can instead use extended Thompson factorization to achieve the hypotheses of E.6.26, which we use in place of E.6.27:

LEMMA 9.2.1. *(1)* $[V, J_{n-2}(T)] = 1$.
(2) Either $[V, J_1(T)] = 1$, *or* $n = 2$ *and* $\sigma \in \bar{T}$.

PROOF. This follows from H.1.1.2 and B.2.4.1. $\qquad\square$

Recall $Z := \Omega_1(Z(T))$.

LEMMA 9.2.2. *(1)* $V = \Omega_1(Z(Q))$.
(2) If $Q \leq S \leq T$, *then* $\Omega_1(Z(S)) \leq C_V(S)$.
(3) $Z \leq V_1 = C_V(T \cap L)$.

PROOF. By 3.2.10.9, $C_Z(L_0) = 1$. Assume (1) fails. Now $H^1(\bar{L}_0, V) = 0$ (e.g., using Exercise 6.4 in [**Asc86a**]). So we obtain $[\Omega_1(Z(Q)), L_0] \not\leq V$. But $\hat{q}(\bar{L}_0\bar{T}, V) > 1$ since V is not an FF-module, and $\hat{q}(Aut_{L_0T}(W), W) \geq 1$ for any nontrivial L_0T-chief factor W on $\Omega_1(Z(Q))$ by B.6.9.1, so $\hat{q}(Aut_{L_0T}(\Omega_1(Z(Q)), \Omega_1(Z(Q))) > 2$, contrary to 3.1.8.1. Thus (1) is established. Further for $Q \leq S \leq T$, $Z(S) \leq Q$ since $L_0T \in \mathcal{H}^e$, so (1) implies (2) and (3). $\qquad\square$

We can now prove the analogue of 7.3.2 in the case $\bar{L}_0 \cong \Omega_4^+(2^n)$, using 9.2.2 as an alternative to E.6.3 when $n = 2$:

LEMMA 9.2.3. $r(G, V) \geq n$.

PROOF. As $m(\bar{M}, V) = n$, this follows from Theorem E.6.3 when $n > 2$. Thus we may assume that $n = 2$ and $r = 1$; that is $C_G(U) \not\leq M$ for some U of corank 1 in V—and without loss, $N_T(U) \in Syl_2(N_M(U))$. Now U contains a unique F-hyperplane U_0, and from 9.1.2.3, there are two M-orbits on F-hyperplanes, each of the form $W^\perp$ for an F-point W of V. Next $T_0 := N_{T \cap L}(U_0) \leq N_T(U)$, so that

$$C_V(N_T(U)) \leq C_V(T_0) \leq U_0 \leq U. \qquad (*)$$

But $U \cap Z \neq 1$, so by E.6.10.4, $\Omega_1(Z(N_T(U))) \not\leq U$. On the other hand by 9.2.2.2, $\Omega_1(Z(N_T(U))) \leq C_V(N_T(U))$, so $C_V(N_T(U)) \not\leq U$, contradicting $(*)$. $\qquad\square$

From now on, let $H \in \mathcal{H}_*(T, M)$. Recall that H is a minimal parabolic in the sense of Definition B.6.1 by 3.3.2.4. Further by 3.1.8, H centralizes Z. Set $K := O^2(H)$. If $n(H) > 1$, let B be a Cartan subgroup of $H \cap M$.

LEMMA 9.2.4. *(1)* $n(H) \geq n - 1$.
(2) If $n(H) = 1$, *then* $[V, J_1(T)] \neq 1$ *and* $n = 2$.

PROOF. To prove (1), we may assume $n \geq 3$; we will apply E.6.26 with $j := n - 2 \geq 1$. By 9.2.3, $r > j$, and by 9.2.1.1, $J_j(T) \leq C_T(V)$; therefore (1) follows from E.6.26. Similarly (2) follows from E.6.26, this time using $j := n - 1$ and 9.2.1.2. $\qquad\square$

LEMMA 9.2.5. *Either* $n(H) = n$, *or* $n = 2$ *and* $n(H) = 1$.

PROOF. Recall that H centralizes Z. By 9.2.4.1, $k := n(H) \geq n - 1$, so either $k > 1$ or $n = 2$. Thus we may assume $k > 1$, and it remains to show that $k = n$. As $k > 1$, $K/O_2(K)$ is of Lie type over $\mathbf{F}_{2^k}$ by E.2.2.

If $k \neq 6$, let p be a Zsigmondy prime divisor of $2^k - 1$; recall by Zsigmondy's Theorem [**Zsi92**] that this means that a suitable element of order p in $GL_k(2)$ acts irreducibly. If $k = 6$, let $p = 3$. Set $B_p := O_p(B)$. By Theorem 4.4.14, B is faithful on $\bar{L}_0$, so as $BT = TB$, either some $b \in B_p^{\#}$ induces an inner automorphism on $\bar{L}_0$, or $|B_p|$ divides n and B_p induces field automorphisms on $\bar{L}_0$. Assume the former. If p is a Zsigmondy prime divisor of $2^k - 1$, then k divides n; while if $k = 6$, then $p = 3$ so that n is even. Hence as $k \geq n - 1$, either $n = k$ and the lemma holds, or $k = 6$ and $n = 2$ or 4, impossible as then B_3 of order 9 is faithful on $\bar{L}_0$. Therefore we may assume that B_p induces field automorphisms on $\bar{L}$ and $\bar{L}^t$, and $|B_p|$ divides n. Then as $k \geq n - 1$, $k \neq 6$. Thus p is a Zsigmondy prime divisor of $2^k - 1$, so k divides $p - 1$. Hence as p divides n and $k \geq n - 1$, we conclude $p = n = k + 1$. Then n is odd, and so $V_1 = Z \leq Z(H)$ by 9.2.2.3, a contradiction as $[V_1, B_p] \neq 1$ since B_p induces field automorphisms on $\bar{L}_0$. This establishes the lemma. $\square$

LEMMA 9.2.6. *If $n(H) > 1$, then B is contained in a Cartan subgroup D of L_0 acting on $T \cap L_0$.*

PROOF. This is a consequence of 9.1.1 and 9.2.5. $\square$

Lemma 9.2.6 has essentially eliminated the shadows of $Aut(L_m(2^n))$ for $m := 4$ or 5, since in those groups $B \not\leq D$: our argument above that $B \leq D$ assumes G quasithin, whereas in those groups the parabolic $M = N_G(V)$ has 3-rank 3. So the remainder of the proof (or more precisely, the reduction to $n(H) = 1$ in the next section) can be viewed as showing that any embedding of B in D leads to a contradiction. In the previous chapter 8, the road after eliminating the configurations corresponding to shadows was typically fairly short; unfortunately in this case the only route after that we know is fairly long and hard.

We can at least immediately eliminate all cases where $n > 2$:

PROPOSITION 9.2.7. *(1) $n = 2$.*
(2) $n(H) = 1$ or 2.
(3) If $n(H) = 2$, then $K/O_2(K) \cong L_2(4)$, B is cyclic of order 3, and $B = C_D(V_1)$ with $\bar{B} = [\bar{D}, \sigma]$.

PROOF. If $n = 2$ then (2) holds by 9.2.5, so it only remains to prove that (3) holds; thus in this case we may assume $n(H) = 2 = n$. On the other hand if $n > 2$, then $n(H) = n$ by 9.2.5. So in any event we may assume that $n(H) = n > 1$.

By 9.2.6, $B \leq D$, so as $B \leq K \leq C_G(Z)$ and $C_D(Z) = C_D(V_1)$ is cyclic of order $2^n - 1$, B is cyclic of order at most $2^n - 1$. Therefore as $n(H) = n > 1$, E.2.2 says $K/O_2(K) \cong L_2(2^n)$ or $Sz(2^n)$ and $|B| = 2^n - 1$, so $B = C_D(V_1)$. By 9.1.2.2, $\bar{T}$ contains $\bar{t} = \sigma f_0$ with f_0 a field automorphism of order a power of 2. Observe σ inverts $\bar{B} = C_{\bar{D}}(V_1) = [\bar{D}, \sigma]$. Pick a preimage $t \in N_T(D)$. Then either t acts nontrivially on B, or $n = 2$, $f_0 \neq 1$, and t centralizes B. In the latter case the lemma holds, so we may assume the former.

As B is not inverted by an inner automorphism of $K/O_2(K)$ in T, t induces an outer automorphism on $K/O_2(K)$. Therefore n is even, and hence $K/O_2(K) \cong L_2(2^n)$ and t induces a field automorphism of some order 2^i on $K/O_2(K)$. Therefore $n = 2^i m$ and $|C_B(t)| = 2^m - 1$. If $\bar{t} = \sigma$, then t inverts B so $m = 1 = i$, and hence

$n = 2$, so the lemma holds. Finally if $\bar{t} \neq \sigma$, then t induces an automorphism on B of order $|f_0|$, so that $|f_0| = 2^i$. Then since $B = C_D(V_1)$, we calculate in L_0 that $|C_B(t)| = 2^m + 1$. This is impossible as $2^m - 1 \neq 2^m + 1$. Thus the Proposition is established. $\square$

9.3. Reducing to n(H) = 1

In this subsection, we assume $n(H) = 2$, and eventually arrive at a contradiction.

Set $G_1 := N_G(V_1)$. By 9.2.7.3, $K/O_2(K) \cong L_2(4)$, $\bar{B} = [\bar{D}, \sigma]$, and $B = C_D(V_1)$.

PROPOSITION 9.3.1. D acts on K and $[K, V_1] = 1$.

PROOF. Define $\bar{D}_\sigma := C_{\bar{D}}(\sigma)$. Then $\bar{D} = [\bar{D}, \sigma]\bar{D}_\sigma$, and hence $D = BD_\sigma$ for a suitable preimage D_σ in D of $\bar{D}_\sigma$. Thus D_σ is of order 3 and faithful on V_1. The proof begins with a series of three reductions:

First, notice if $D_\sigma \leq N_G(K)$, then $D \leq N_G(K)$, and hence $V_1 \leq \langle Z^{D_\sigma} \rangle \leq C_G(K)$, so that we are done. Thus we may assume $D_\sigma \not\leq N(K)$; in particular, K is not normal in G_1.

Second, suppose that $K \leq G_1$. Then $K \in \mathcal{L}(G_1, T)$, so by 1.2.4, $K \leq K_1 \in \mathcal{C}(G_1)$, and indeed $K < K_1$ by the previous paragraph, so K_1 is described in A.3.14. Suppose $m_3(K_1) = 2$. Then $K_1 \trianglelefteq G_1$ by 1.2.2.b. As $D_\sigma \not\leq K$, comparing the list in A.3.18 to that of A.3.14, we conclude D_σ induces diagonal automorphisms on $K_1/O_2(K_1) \cong L_3(4)$ or $U_3(5)$, and so D normalizes K from the embedding described in A.3.14. Thus in this case we are done by our first reduction, so we may assume that $m_3(K_1) = 1$. Then by A.3.14, $K_1/O_2(K_1)$ is J_1, $L_2(25)$, or $L_2(p)$, or $K_1/O_{2,2'}(K_1) \cong SL_2(p)$ for suitable p. We can reduce the fourth case to the third case by noting that $K_0 := N_{K_1}(T \cap O_{2,2',2}(K_1))^\infty$ is D-invariant. But in the first three cases, $D = C_D(K_1/O_2(K_1))B$ acts on K, contrary to the first reduction. Therefore we may assume that $K \not\leq G_1$. In particular, $[K, V_1] \neq 1$.

Third, we recall that K centralizes Z, so $K \leq G_1$ if $\bar{T} \leq \bar{L}_0\langle\sigma\rangle$ by 9.2.2.3, contrary to the second reduction.

In view of our three reductions, we may assume D does not act on K, $K \not\leq G_1$, and $\bar{T} \not\leq \bar{L}_0\langle\sigma\rangle$. To complete the proof, we construct an overgroup X of K, and obtain a contradiction in X.

By the third reduction and 9.1.2.2, there is $t \in T$ with $\bar{t} = \sigma f$, where f is an involution inducing a field automorphism on $\bar{L}_0$. As σ and f invert $\bar{B}$, t centralizes $\bar{B}$, so $T_2 := \langle t \rangle O_2(DT)$ is B-invariant and $D_\sigma T_2/O_2(D_\sigma T_2) \cong S_3$. Set $X := \langle D, H \rangle$. Suppose $O_2(X) = 1$. Then K, $D_\sigma T_2$, T satisfies Hypothesis F.1.1 in the roles of "L_1, L_2, S", so the amalgam $\alpha := (KT, BT, DT)$ is a weak BN-pair of rank 2 by F.1.9. Further T_2 is maximal in $D_\sigma T_2$, so the hypotheses of F.1.12 are satisfied, and hence α is one of the amalgams listed in that lemma. As $D_\sigma T_2/O_2(D_\sigma T_2) \cong L_2(2)$ and $K/O_2(K) \cong L_2(4)$, α is of type $U_4(2)$, J_2, or $Aut(J_2)$, so that $|T| \leq 2^8$. This contradicts $|V| = 2^8$ with $V < T$.

Thus $O_2(X) \neq 1$, so $X \in \mathcal{H}(T) \subseteq \mathcal{H}^e$ by 1.1.4.6. By 1.2.4, $K \leq K_X \in \mathcal{C}(X)$, and $K_X \trianglelefteq X$ by (+) in 1.2.4, so $X = K_X TD$. As $D \not\leq N_G(K)$, $K < K_X$. Next $V_1 \leq V_X := \langle Z^X \rangle \in \mathcal{R}_2(X)$ by B.2.14. As $[K, V_1] \neq 1$, $[K_X, V_X] \neq 1$. Set $X^* := X/C_X(V_X)$. Then $K^* \neq 1$. Also $K = [K, J(T)]$, or else $K \leq N_G(J(T)O_2(K)) \leq M$ using 3.2.10.8. Thus $J(T)^* \neq 1$, so V_X is an FF-module for $K_X^* T^*$. Comparing the list in A.3.14 with the list of FF-modules in B.5.6, we conclude $K_X^* \cong SL_3(4)$, $Sp_4(4)$, $G_2(4)$, or A_7. In the first three cases, D induces inner-diagonal automorphisms on K_X^* in a Cartan group stabilizing the parabolic of K_X^* normalizing K^* and hence K, contrary to an earlier reduction. In the last case as $[Z, K] = 1$ we have a contradiction since $C_{K_X^*}(C_{V_X}(T))$ contains no A_5-subgroup when V_X is either of the FF-modules of dimension 4 and 6 for $K_X^* \cong A_7$ listed in B.4.2. This finally establishes 9.3.1. $\qquad\square$

Define $T_K := T \cap K$ and $T_L := T \cap L_0 \in Syl_2(L_0)$.

LEMMA 9.3.2. $T_L Q = O_2(TD) = T_K O_2(HD)$.

PROOF. First $T_L Q = O_2(TD)$ and $TD = DT$ from the structure of $\bar{L}_0 \bar{T}$. Also $H = KT$ with $D \le N_G(K)$ by 9.3.1. Then as $K/O_2(K) \cong L_2(4)$, we conclude $HD/O_2(HD)$ is a subgroup of $S_3 \times S_5$, containing $GL_2(4)$. Then from the structure of this group, $O_2(TD) = T_K O_2(HD)$. $\qquad\square$

Our strategy for the remainder of the section, much as in the proof of 9.3.1, is to construct an overgroup M_0 of K and L, and use this overgroup to obtain a contradiction.

Set $T_1 := N_T(L)$. Then T_1 is Sylow in $N_M(L)$ of index 2 in M, so $|T : T_1| = 2$. In particular T_1 contains $T_L Q$, so $T_1 \in Syl_2(LDT_1)$ by 9.3.2. Similarly as $T_L Q = T_K O_2(HD)$, T_1 is Sylow in KDT_1 as well.

Define $M_0 := \langle LDT_1, K \rangle$, and $V_2 := \langle V_1^L \rangle$. Of course $M_0 \not\le M$ as $K \not\le M$. Observe V_2 is a natural module for $L/O_2(L) \cong L_2(4)$.

LEMMA 9.3.3. $O_2(M_0) \ne 1$.

PROOF. Assume otherwise and let $S := T_L Q$. Then Hypothesis F.1.1 is satisfied by K, L, S in the roles of "L_1, L_2, S", and $S \trianglelefteq DS$ so $\alpha := (KDS, DS, LDS)$ is a weak BN-pair of rank 2 described in F.1.12. As $K/O_2(K) \cong L/O_2(L) \cong L_2(4)$, the amalgam is one of the untwisted types A_2, B_2, G_2 over $\mathbf{F}_4$. As $[K, V_1] = 1$ by 9.3.1, while V_2 is the natural module for $L/O_2(L)$, we conclude α is of type $G_2(4)$. But then $O_2(LS) = [O_2(LS), L] \le L$, which is not the case since $T_L \cap L^t \not\le L$. $\qquad\square$

LEMMA 9.3.4. $T_1 \in Syl_2(M_0)$.

PROOF. Recall $J(T) \le T_1$, so if $T_1 \le T_0 \in Syl_2(M_0)$, then $T_0 \le M$ by 3.2.10.8. If $T_1 < T_0$, then $T_0 \in Syl_2(G)$ as $|T : T_1| = 2$, and then $L_0 T_0 \le M_0 \not\le M$, contradicting $M = !\mathcal{M}(L_0 T_0)$. $\qquad\square$

LEMMA 9.3.5. (1) $[V_2, K] \ne 1$.
(2) $[L, K] \not\le O_2(L)$.

PROOF. First $B \le K$; and B is faithful on V_2 as V_2 is the natural module for $L/O_2(L) \cong L_2(4)$ while $B = C_D(V_1)$. Thus (1) holds. If (2) fails, then as $[V_1, K] = 1$ by 9.3.1, K centralizes $V_2 = \langle V_1^L \rangle$, contrary to (1). $\qquad\square$

Now $M_0 \in \mathcal{H}$ by 9.3.3. As $L \in \mathcal{L}(G, T_1)$, and $T_1 \in Syl_2(M_0)$ by 9.3.4, $L \le K_L \in \mathcal{C}(M_0)$ by 1.2.4. Similarly $K \le K_K \in \mathcal{C}(M_0)$. If $K_L \ne K_K$, then by 1.2.1.2 $[K, L] \le [K_K, K_L] \le O_2(M_0)$, contrary to 9.3.5.2; therefore $K_K = K_L =: K_0 \in \mathcal{C}(M_0)$ and $\langle L, K \rangle \le K_0$.

LEMMA 9.3.6. $M_0 = K_0 T_1 \in \mathcal{H}^e$, $K_0 = O^2(M_0)$, and $Z(M_0) = 1$.

PROOF. By 9.2.7, $B = C_D(V_1) \leq K$ is diagonally embedded in LL^t, so $D = B(D \cap L) \leq \langle K, L \rangle \leq K_0$. As $M_0 = \langle LDT_1, K \rangle$, $O^2(M_0) = \langle L, K \rangle \leq K_0$, so $M_0 = K_0 T_1$ and $K_0 = O^2(M_0)$. Next using parts (2) and (3) of 9.2.2, $\Omega_1(Z(T_1)) \leq C_V(T_1) \leq V_1$. Hence as $O_2(M_0) \cap \Omega_1(Z(T_1)) \neq 1$, and as D is irreducible on V_1, $V_1 \leq O_2(M_0)$. Therefore $N_G(O_2(M_0)) \in \mathcal{H}^e$ by 1.1.4.1. Next $V_2 = \langle V_1^L \rangle \leq O_2(M_0)$, and then

$$C_T(O_2(M_0)) \leq C_T(V_2) \leq N_T(L) \leq T_1 \leq M_0,$$

so $M_0 \in \mathcal{H}^e$ by 1.1.4.4 with $N_G(O_2(M_0))$ in the role of "M". Also $Z(M_0) = 1$ as $\Omega_1(Z(T_1)) \leq V_1$ and $C_{V_1}(D) = 1$. $\qquad\square$

We now proceed as in the last paragraph of the proof of 9.3.1. Let $U := \langle V_1^{M_0} \rangle$. As $V_1 = \langle Z^D \rangle$, $U = \langle Z^{M_0} \rangle$, so by B.2.14, $U \in \mathcal{R}_2(M_0)$. Set $M_0^* := M_0/C_{M_0}(U)$. By 9.3.5.1, $K^* \neq 1$. Now as in the proof of 9.3.1, $K = [K, J(T)]$ and hence $[U, J(T)] \neq 1$, so U is an FF-module for K_0^*. Then we obtain the same four possiblities for K_0^* as in the proof of 9.3.1, and eliminate the fourth case $K_0^* \cong A_7$ as in that proof, to conclude:

LEMMA 9.3.7. $K_0^* \cong SL_3(4)$, $Sp_4(4)$, or $G_2(4)$, and U is an FF-module for M_0^*.

LEMMA 9.3.8. K_0^* is not $SL_3(4)$.

PROOF. Otherwise $Z(K_0^*) = C_D(L^*) = (D \cap L^t)^*$, as each is of order 3. But then $K/O_2(K) \cong L_2(4)$ is centralized by $\langle (D \cap L^t)^{N_T(D)} \rangle = D$, a contradiction since $B \leq D \cap K$. $\qquad\square$

LEMMA 9.3.9. $K_0^* \cong Sp_4(2^n)$.

PROOF. If not, by 9.3.7 and 9.3.8, $K_0^* \cong G_2(4)$. Now L and K are normalized by T, so $L = P_1^\infty$ and $K = P_2^\infty$, where P_1^* and P_2^* are the maximal parabolics of K_0^* containing $(T \cap K_0)^*$. By 9.3.7, U is an FF-module for K_0^*, and by 9.3.6, $Z(M_0) = 1$—so U is the natural $G_2(4)$-module by Theorems B.5.1 and B.4.2.4. Therefore by B.4.6.14, $D \cap L$ centralizes $K/O_2(K)$. We again use the action of $N_T(D)$ to obtain the same contradiction obtained at the end of the proof of 9.3.8. $\qquad\square$

LEMMA 9.3.10. K_0 is an $Sp_4(4)$-block.

PROOF. Recall T_1 is of index 2 in T. If $1 \neq C$ char T_1 with $C \trianglelefteq M_0$, then $N_G(C)$ contains $M_0 \not\leq M$ and $\langle L, T \rangle = L_0 T$, contradicting $M = !\mathcal{M}(L_0 T)$. Thus no such characteristic subgroup exists, giving the condition (MS3) of Definition C.1.31. We obtain (MS1) and (MS2) using 9.3.9. Then the lemma follows from C.1.32.3. $\qquad\square$

We are now in a position to obtain a contradiction, eliminating the case $n(H) = 2$. For by 9.3.6, $Z(M_0) = 1$, so U is the natural module for the $Sp_4(4)$-block K_0 by 9.3.10. Now V/V_2 is the natural module for $L/O_2(L)$. However $L = P^\infty$ for some maximal parabolic P^* of K_0^*, so $O_2(L)/U$ is an indecomposable of F-dimension 3 with no natural submodule. Therefore $V \leq U$, so $V = U$ as both are of order 2^8. Then $K_0 \leq N_G(U) = N_G(V) = M$, a contradiction.

9.4. Eliminating n(H) = 1

As we just showed $n(H) \neq 2$, $n(H) = 1$ for all $H \in \mathcal{H}_*(T, M)$ by 9.2.5. This makes weak closure arguments effective, once we obtain restrictions on the weak closure parameters r and w.

Define V_N and L_N as in 9.1.2.4, and let $U_N := V_N^\perp$. By 9.1.2.4, $U_N = [V, L_N]$ and U_N/V_N is the natural module for $L_N/O_2(L_N) \cong L_2(4)$. For $v \in V_N^\#$, set $G_v := C_G(v)$.

PROPOSITION 9.4.1. $L_N \lhd G_v$.

PROOF. Assume the lemma fails. Then $G_v \not\leq M$. We can assume $T_v := C_T(v) \in Syl_2(C_M(v))$, and then by 3.2.10.4, $T_v \in Syl_2(G_v)$. By 1.2.4, $L_N \leq L_v \in \mathcal{C}(G_v)$ with L_v described in A.3.14, and $L_v \trianglelefteq G_v$ by (+) in 1.2.4 applied to T_v. We are done if $L_N = L_v$, so assume $L_N < L_v$; thus $L_v \not\leq M$.

We claim that $L_v T_v \in \mathcal{H}^e$. Suppose first that L_v is quasisimple. As $v \in [V, L_N] \leq L_N \leq L_v$, $v \in Z(L_v)$, so the multiplier of $L_v/Z(L_v)$ is of even order. Also $C_V(L_v) \leq C_V(L_N) = V_N$, so $m_2(Aut(L_v)) \geq m(V/V_N) = 6$. Inspecting the lists of A.3.14 and I.1.3 for groups with an automorphism group of 2-rank at least 6, we conclude $L_v/Z(L_v) \cong G_2(4)$. But then by I.1.3, $Z(L_v)$ is of order 2, so $\langle v \rangle = C_{V_N}(L_v)$ and hence $m_2(Aut_V(L_v)) = 7 > m_2(Aut(G_2(4)))$, a contradiction. Thus L_v is not quasisimple. As $z \in U_N = [U_N, L_N]$ and $C_T(O_2(L_v T_v)) \leq C_T(v) = T_v$, we conclude using 1.2.11 that $L_v T_v \in \mathcal{H}^e$.

As $L_v T_v \in \mathcal{H}^e$, it follows from B.2.14 that $U := \langle Z^{L_v} \rangle \in \mathcal{R}_2(L_v T_v)$. Notice using 9.1.2.4 that U contains U_N and V_1. Set $(L_v T_v)^* := L_v T_v/C_{L_v T_v}(U)$.

We next claim that $L_v^* = L_N^*$, so assume otherwise.

Suppose first that $J(T) \not\leq C_G(U)$. Then $[L_v^*, J(T)^*] \neq 1$ and U is an FF-module for $L_v^* T_v^*$. If L_v appears in case (c) or (d) of 1.2.1.4 then $O_\infty(L_v)^*$ is a $3'$-group, so by B.5.6, $[O_\infty(L_v^*), J(T)^*] = 1$. Therefore as $[L_v^*, J(T)^*] \neq 1$ and $L_v^*/O_\infty(L_v^*)$ is quasisimple, $L_v^* = [L_v^*, J(T)^*]$. On the other hand, if $L_v/O_2(L_v)$ is quasisimple, then so is $L_v^* = [L_v^*, J(T)^*]$. Thus in any case, $L_v^* = [L_v^*, J(T)^*]$ is quasisimple. Now L_v^* appears in A.3.14 and B.5.1, and hence as in a previous argument is $SL_3(4)$, $Sp_4(4)$, $G_2(4)$, or A_7. Further by B.5.1 and B.4.2, $[U, L_v]/C_{[U, L_v]}(L_v)$ is either the natural module or the sum of two natural modules for $L_3(4)$. As $v \in [U_N, L_N]$, $v \in C_{[U, L_v]}(L_v)$. Hence the 1-cohomology of the natural module is nontrivial, so that by I.1.6, $L_v^* \cong Sp_4(4)$ or $G_2(4)$, and $[U, L_v]$ is a quotient of a 5-dimensional orthogonal space or the 7-dimensional Cayley algebra over $\mathbf{F}_4$, respectively. Further $L_N^* = P^{*\infty}$ for some maximal parabolic P^* of L_v^*. Then $C_U(O_2(L_N^*)) = C_U(O_2(P^{*\infty}))$ contains U_N, which does not split over V_N, and $v \in C_{V_N}(L_v^*)$. This is impossible, since from the structure of these two modules, $C_U(O_2(P^*)) = C_U(L_v^*) \oplus [C_U(O_2(P^*)), P^*]$.

Therefore $J(T) \leq C_G(U)$. By a Frattini Argument, $L_v^* T_v^* = N_{L_v T_v}(J(T))^*$, so as $N_G(J(T)) \leq M$ by 3.2.10.1, $L_v^* = L_N^*$, completing the proof of our second claim.

In particular as $L_N/O_2(L_N)$ is simple and U is 2-reduced, the second claim says $L_v^* = L_N^* \cong L_2(4)$; hence $O_\infty(L_v) \leq C_{L_v}(U)$. Therefore as $L_v \not\leq M$, $C_{L_v}(U) \not\leq M$, so case (c) or (d) of 1.2.1.4 holds. In the notation of chapter 1, there is at least one prime $p > 3$ with $1 \neq X := \Xi_p(L_v)$. Then X is characteristic in L_v and hence normal in G_v. Further X centralizes U, and hence centralizes $V_1 V_N$. By 1.3.3,

$X \in \Xi(G_v, T_v)$. Now T acts on $V_1 V_N$ and there is $g \in N_{L_0}(V_1 V_N)$ with $v^g \notin V_N$ and $v^g \in Z(T_v)$. Then $V_1 V_N \leq U^g$, so $X^g \leq C_G(v) = G_v$, and hence X^g acts on X. Further $T_v \leq G_v^g \leq N_G(X^g)$, and X centralizes v^g as $v^g \in V_1 V_N$, so X acts on X^g. Recall from the definition of $\Xi(G_v, T_v)$ that $X = PO_2(X)$ with $P \cong E_{p^2}$ or p^{1+2}. Set $(XX^g T_v)^+ := XX^g T_v / O_2(XX^g T_v)$. Then T_v is irreducible on $P^+/\Phi(P^+)$ and $P^{g+}/\Phi(P^{g+})$, so $P^+ \cap P^{g+}$ is 1, $\Phi(P^+)$, or P^+. As $m_p(XX^g) \leq 2$, the last case holds, so $X = X^g$. Therefore X is normal in G_v and G_{v^g}, so $L_0 = \langle L_N, L_N^g \rangle$ acts on X. Then as $Aut(X/O_{2,\Phi}(X))^\infty \cong SL_2(p)$, either L or L^t centralizes $X/O_2(X)$, and thus $L_0 = \langle L^{T_v} \rangle$ centralizes $X/O_2(X)$, contradicting $X = [X, L_N]$. This finally establishes 9.4.1. $\square$

LEMMA 9.4.2. *(1) If $v \in V_N^\#$, $g \in L_0 - N_G(V_N)$, and $u \in V_N^{g\#}$, then $C_G(\langle u, v \rangle) \leq M$.*

(2) V is the unique member of V^G containing $V_1 V_N$.

PROOF. Part (1) follows as $C_G(\langle u, v \rangle)$ acts on $\langle L_N, L_N^g \rangle = L_0$ by 9.4.1. As $V_1 V_N = V_N V_N^l$ for suitable $l \in L$, $C_G(V_1 V_N) \leq M = N_G(V)$ by (1). By 3.2.10.2, M controls fusion in V, so we conclude that $N_G(V_1 V_N) \leq M$, and that (2) follows from the proof of A.1.7.2. $\square$

We can finally begin to implement our standard weak closure strategy.

LEMMA 9.4.3. *$r(G, V) > 3$.*

PROOF. Suppose $U \leq V$ with $m(V/U) \leq 3$ and $C_G(U) \not\leq M$. As $m(V/U) \leq 3$, $C_{\bar{M}}(U)$ is a 2-group by 9.1.2. Recall from 9.2.3 that $r > 1$, so by E.6.12, $C_{\bar{M}}(U)$ is a nontrivial 2-group. As $m(V/U) < 4$, we may take $U \leq C_V(t) = U_N$. Now for each $V_N^g \leq U_N$, $1 \neq V_N^g \cap U$ as $m(U_N/U) \leq 1$, so the lemma follows from 9.4.2. $\square$

LEMMA 9.4.4. *$W_0 := W_0(T, V)$ centralizes V, so $w > 0$.*

PROOF. Suppose $A := V^g \leq T$ with $\bar{A} \neq 1$. If $m(C_A(V)) \geq 5$, then $V \leq N_G(V^g)$ by 9.4.3, contrary to E.3.11. Hence $m(\bar{A}) \geq 4$, so as $m_2(\bar{M}) = 4$ and $\bar{T}_L = J(\bar{T})$, $\bar{A} = \bar{T}_L$. Then $C_V(\bar{A}) = V_1$, so if U_1 is the L-irreducible containing V_1, then $C_A(\bar{L})$ centralizes $\langle V_1^L \rangle = U_1$. Now $m(A/C_A(\bar{L})) = 2$, so as $r > 3$, $U_1 \leq M^g$. Similarly $U_1^t \leq M^g$, so $U_1 U_1^t = V_1^\perp \leq M^g$, and $[U_1 U_1^t, A] = V_1$, so $U_1 U_1^t$ induces F-transvections on A with center V_1. This is impossible since M controls fusion in V by 3.2.10.2, while the center of an F-transvection on V is nonsingular by 9.1.2.4, and V_1 is singular by 9.1.2.5. $\square$

LEMMA 9.4.5. *$W_1(T, V)$ centralizes V, so $w > 1$.*

PROOF. If not, then arguing as in the proof of the previous lemma, there is a hyperplane $A := V^g \cap T$ of V^g with $\bar{A} \neq 1$, and this time $m(\bar{A}) \geq 3$. Suppose first $\bar{A} \not\leq \bar{L}_0$. Then $\bar{A}$ has maximal rank (namely 3) subject to $\bar{A} \not\leq \bar{L}_0$, so $\bar{A} \in \mathcal{A}(C_{\bar{M}}(\bar{a}))$ for each $\bar{a} \in \bar{A} - \bar{L}_0$. Observe $m(V^g/C_A(V_1)) \leq 2$, so $V_1 \leq M^g$ since $r > 3$ by 9.4.3. Thus if A does not centralize V_1, then $Z = [V_1, A] \leq M \cap V^g = A$. As $C_A(V_1)$ is of codimension at most 2 in V^g, V_1 induces an F-transvection on V^g with Z contained in the center $[V^g, V_1]$, a contradiction as in the proof of the previous lemma. Therefore $[A, V_1] = 1$, so as $\bar{A} \not\leq L_0$, there is $t \in A$ with $\bar{t} = \sigma$ and $\bar{A} = \langle \bar{t} \rangle (\bar{A} \cap \bar{L}_N)$. But then $m(V^g/C_A(U_N)) \leq 3$, so $U_N \leq M^g$ since $r > 3$. Therefore $V_N V_1 = [A, U_N] \leq A \leq V^g$, contrary to 9.4.2.2.

Thus $\bar{A} \leq \bar{T}_L$, so $\bar{A}$ has rank 3 or 4. We now argue as in the proof of 9.4.4: First $C_V(\bar{A}) = C_V(\bar{T}_L) = V_1$, and $C_A(V^g)$ has rank 4,3, with $C_{\bar{A}}(\bar{L})$ of rank 1,2, respectively. So in any case $m(V^g/C_A(\bar{L})) = 3 < r$, and hence we can continue the argument in the proof of 9.4.4 to get $U_1 U_1^t \leq M^g$, and obtain the same contradiction. $\qquad\square$

Observe that by 9.4.4, 9.4.5, and E.3.16, $N_G(W_0) \leq M \geq C_G(C_1(T, V))$. As $m(\bar{M}, V) \geq 2$, $s(G, V) \geq 2$ by 9.4.3. Then as $n(H) = 1$, E.3.19 forces $H \leq M$, a contradiction. This contradiction finally shows that case (1) of 3.2.6 cannot occur, and hence completes the proof of Theorem 7.0.1 begun in chapter 7.

Part 4

Pairs in the FSU over $\mathbf{F}_{2^n}$ for $n > 1$.

In part 4, we prove two theorems about pairs L, V in the Fundamental Setup (3.2.1): In chapter 10, we show that $L = L_0$. Then in chapter 11, we show that L is not of Lie type of Lie rank 2 over $\mathbf{F}_{2^n}$ for $n > 1$.

A counter example in chapter 10 is of the form $L_0 = LL^t$ with $t \in T - N_T(L)$ and $L/O_2(L)$ isomorphic to $L_2(2^n)$ or $Sz(2^n)$ with $n > 1$, or to $L_3(2)$. In the first two cases, we can view $L_0/O_2(L_0)$ as of Lie type of Lie rank 2 over $\mathbf{F}_{2^n}$. Thus the majority of the effort in part 4 is devoted to the elimination of cases in the FSU where $\bar{L}_0$ is of Lie type and Lie rank 2 over $\mathbf{F}_{2^n}$ for some $n > 1$.

One of the main tools for treating such groups is the study of Cartan subgroups, both of L_0 and of $H \in \mathcal{H}_*(T, M)$: a Cartan subgroup of $X := L_0$ or H is defined to be a Hall $2'$-subgroup of $N_X(T \cap X)$.

The most difficult cases are those where the Cartan subgroup is small or trivial: that is, when $n = 2$, or in chapter 10 when $\bar{L} \cong L_3(2)$ is defined over $\mathbf{F}_2$.

CHAPTER 10

The case $L \in \mathcal{L}_f^*(G, T)$ not normal in M.

In this chapter we prove:

THEOREM 10.0.1. *Assume G is a simple QTKE-group, $T \in Syl_2(G)$, and $L \in \mathcal{L}_f^*(G, T)$ with $L/O_2(L)$ quasisimple. Then $T \leq N_G(L)$.*

10.1. Preliminaries

Assume Theorem 10.0.1 is false, and pick a counterexample L. Let $L_0 := \langle L^T \rangle$ and $M := N_G(L_0)$. By 3.2.3, there is $V_o \in Irr_+(L_0, R_2(L_0T))$ such that L and $V_T := \langle V_o^T \rangle$ are in the Fundamental Setup 3.2.1. Set $V := \langle V_T^M \rangle$, and note that this differs from the notation in the FSU where V_T, V are denoted by "V, V_M". Note in particular that by construction $V \trianglelefteq M$, so that $M = N_G(V)$.

As $L < L_0$, we can appeal to Theorem 3.2.6. In the first two cases of Theorem 3.2.6, V_T is not an FF-module, and those cases were eliminated in Theorem 7.0.1. Thus we are left with case (3) of Theorem 3.2.6. We recall from that result that $V = V_1 V_1^t$ for $t \in T - N_T(L)$, with $V_1 := [V, L] \leq C_V(L^t)$.

Recall that in the FSU with $V \trianglelefteq M$, we set $\bar{M} := M/C_M(V)$ and $\tilde{V} = V/C_V(L_0)$. Also set $L_1 := L$, $L_2 := L^t$ for $t \in T - N_T(L)$, and $V_i := [V, L_i]$.

The cases to be treated are listed in the following lemma. Subcases (ii) and (iii) of 3.2.6.3 appear as cases (5) and (6) in 10.1.1. In subcase (i) $V_1 \in Irr_+(L, V)$, and by 3.2.6.3b, $\hat{q}(Aut_{L_0T}(V_1), V_1) \leq 2$, so $\tilde{V}$ appears in B.4.2 or B.4.5. As $L < L_0$, $\bar{L}$ appears in 1.2.1.3. Intersecting those lists leads to the remaining cases in 10.1.1.

LEMMA 10.1.1. $V = V_1 V_2 \in \mathcal{R}_2(M)$ with $V_i := [V, L_i] \leq C_V(L_{3-i})$, $\tilde{V} = \tilde{V}_1 \oplus \tilde{V}_2$, and one of the following holds:

(1) $\tilde{V}_1$ is the natural module for $\bar{L} \cong L_2(2^n)$, with $n > 1$.

(2) V_1 is the A_5-module for $\bar{L} \cong A_5$.

(3) $\tilde{V}_1$ is the natural module for $\bar{L} \cong L_3(2)$.

(4) V_1 is the orthogonal module for $\bar{L} \cong \Omega_4^-(2^n)$, with $n > 1$.

(5) V_1 is the sum of a natural module for $\bar{L} \cong L_3(2)$ and its dual, with the summands interchanged by an element of $N_T(L)$.

(6) V_1 is the sum of four isomorphic natural modules for $\bar{L} \cong L_3(2)$, and $O^2(C_{\bar{M}}(\bar{L})) \cong \mathbf{Z}_5$ or E_{25}.

(7) V_1 is the natural module for $\bar{L} \cong Sz(2^n)$.

Let $Z := \Omega_1(Z(T))$, $t_0 := T \cap L_0$, $T_1 := N_T(L)$; and $B_0 := O^2(N_{L_0}(T_0))$. Note that $B_0 T = T B_0$ and (except when $\bar{L} \cong L_3(2)$ where $B_0 = 1$) $\bar{B}_0$ is a Borel subgroup of $\bar{L}_0$. Set $S := \text{Baum}(T)$.

LEMMA 10.1.2. (1) *Except possibly in the first three cases of 10.1.1, V is not an FF-module for $Aut_{L_0T}(V)$, so $J(T) \leq C_T(V)$.*

(2) $J(T) \leq N_T(L) = T_1$.

(3) $C_T(V) = O_2(L_0 T)$ except in case (6) of 10.1.1, where at least $C_T(V) \trianglelefteq L_0 T$. In any case, $M = !\mathcal{M}(N_G(C_T(V))) = !\mathcal{M}(N_G(J(C_T(V))))$.

(4) Except possibly in cases (1) and (3) of 10.1.1, $C_V(L_0) = 1$.

(5) Assume $\bar{L}$ is not $L_3(2)$ and let D be a Hall $2'$-subgroup of B_0. Then either:

   (a) $C_D(Z) = 1$, or

   (b) V_1 is the orthogonal module for $\bar{L} \cong \Omega_4^-(2^n)$ and $C_D(Z) \cong \mathbf{Z}_{2^n+1}^2$.

(6) In cases (1) and (2) of 10.1.1, $L_0 T$ is a minimal parabolic in the sense of Definition B.6.1, with $N_{L_0 T}(T_0)$ the unique maximal overgroup of T in $L_0 T$. Thus if $J(T) \not\leq C_T(V)$ then $L_0 T$ is described in E.2.3 and $S \leq N_T(L) = T_1$.

PROOF. Part (2) is clear if $J(T) \leq C_T(V)$, while if $J(T) \not\leq C_T(V)$, it follows from B.1.5.4. Except in cases (5) and and (6) of 10.1.1, $\tilde{V}_1$ is an irreducible for L, and (1) follows from B.4.2. In cases (5) and (6), V is not an FF-module for $Aut_{L_0 T}(V)$ by Theorem B.5.6, so (1) is established. Next in all cases of 10.1.1 except case (6), $V = V_T$, so that $C_T(V) = O_2(L_0 T)$ by 1.4.1.4. In case (6), $C_{L_0 T}(V) \leq O_2(L_0 T)$, so as $V \trianglelefteq L_0 T$, $C_T(V) = C_{L_0 T}(V) \trianglelefteq L_0 T$, and hence (3) holds in that case too. Part (4) follows in the final four cases of 10.1.1 from (1) and 3.2.10.9; in the second case it follows from I.1.6. Part (5) follows easily from (4) and the structure of the modules in 10.1.1. Finally the first two remarks in (6) are elementary observations, and then if $J(T) \not\leq C_T(V)$, the remaining remarks are a consequence of E.2.3. $\square$

LEMMA 10.1.3. $L_0 = O^{p'}(M)$ for each prime divisor p of $|\bar{L}|$.

PROOF. This follows from 1.2.2. $\square$

10.2. Weak closure parameters and control of centralizers

We will make use of weak closure, together with control of centralizers of elements of $V_1^\#$. In 10.2.3, we will use the fact that G is a QTKE-group to show $n(H) \leq 2$ for $H \in \mathcal{H}_*(T,M)$; subsequent results provide lower bounds on the weak-closure parameters $r(G,V)$ and $w(G,V)$. In 10.2.13, we will eliminate most cases using the relation $n(H) \geq w(G,V)$ in E.3.39.

LEMMA 10.2.1. Except possibly in case (3) of 10.1.1, $N_G(S) \leq M$.

PROOF. We may assume case (3) of 10.1.1 does not hold. If $J(T) \leq C_T(V)$, then as $J(T) \leq S$, $N_G(S) \leq M$ by 3.2.10.8. Thus we may assume $J(T) \not\leq C_T(V)$, so by 10.1.2.1, we are reduced to cases (1) and (2) of 10.1.1. In those cases, $L_0 T$ is a minimal parabolic, and is described in E.2.3 by 10.1.2.6.

In case (1) of 10.1.1, E.2.3.2 says $S \in Syl_2(L_0 S)$, so we can apply Theorem 3.1.1 with $L_0 T$, $N_G(S)$, S in the roles of "H, M_0, R", to conclude that $O_2(\langle N_G(S), L_0 T \rangle) \neq 1$. Thus $N_G(S) \leq M = !\mathcal{M}(L_0 T)$, as desired.

Therefore we may assume case (2) of 10.1.1 holds; the proof for this case will be longer. Moreover for each $S_+ \trianglelefteq T$ with $T_0 = T \cap L \leq S_+$, $S_+ \in Syl_2(L_0 S_+)$; hence applying 3.1.1 as in the previous paragraph, we conclude that $N_G(S_+) \leq M$. In particular $N_G(T_1) \leq M$. We may also assume that $N_G(S) \not\leq M$, so as $M = !\mathcal{M}(L_0 T)$, no nontrivial characteristic subgroup of S is normal in $L_0 T$. Then E.2.3.3 says that L_1 is an A_5-block.

Suppose first that $C_Z(L_0) = 1$. Then $O_2(L_0 T) = V$ by C.1.13.c, so that $V = O_2(M)$ using A.1.6. Further using E.2.3, $S = S_1 \times S_2$, where $S_i := C_S(L_{3-i}) = $

$S_{i,1} \times S_{i,2}$ with $S_{i,j} \cong D_8$, and T acts transitively as D_8 on the four members of $\Delta := \{S_{i,j} : i, j\}$. As S is the direct product of the subgroups in Δ, by the Krull-Schmidt Theorem A.1.15, $N_G(S)$ permutes $\Gamma := \{DZ(S) : D \in \Delta\}$. Let K be the kernel of $N_G(S)$ on Γ and $N_G(S)^\Gamma := N_G(S)/K$. Then $D_8 \cong T^\Gamma \le N_G(S)^\Gamma \le S_4$. Observe that for $F \in \Gamma$, $\mathcal{A}(F) = \{V_F, A_F\}$ is of order 2, where $V_F := V \cap F$. Thus $O^2(K)$ acts on each V_F. Then as $V \trianglelefteq T$, $K = O^2(K)(K \cap T)$ acts on $\langle V_F : F \in \Gamma \rangle = V$. Hence $K \le N_G(V) = M$, so as we are assuming $N_G(S) \not\le M$, there is $x \in N_G(S)$ with x inducing a 3-cycle on Γ. Therefore $N_G(S)^\Gamma \cong S_4$. Let K_R be the preimage in $N_G(S)$ of $O_2(N_G(S)^\Gamma)$ and $R := T \cap K_R$. By a Frattini Argument, $N_G(S) = K(N_G(S) \cap N_G(R))$, so we may take $x \in N_G(R)$. But R normalizes just two members V and $A := \langle A_F : F \in \Gamma \rangle$ of $\mathcal{A}(S) = \mathcal{A}(T)$, so x acts on V and A. Therefore $N_G(S) = KT\langle x \rangle \le N_G(V) = M$, contrary to our assumption.

Thus in the remainder of the proof, we assume that $C_Z(L_0) \ne 1$. We may choose $H \in \mathcal{H}_*(T, M)$ with $H \le N_G(S)$. Let $E := \Omega_1(Z(S))$, $V_H := \langle Z^H \rangle$, and $H^* := H/C_H(V_H)$. As usual $V_H \in \mathcal{R}_2(H)$ by B.2.14. Now $Z \le E$ and hence $V_H \le E$. As $C_Z(L_0) \ne 1$, $C_H(V_H) \le C_G(C_Z(L_0)) \le M = !\mathcal{M}(L_0 T)$, so $H^* \ne 1$. Observe applying E.2.3.3 to $L_0 T$ that for $t_i \in T \cap L_i - S$, t_i induces a transvection on E with center $v_i \in V_i$. If $t_1 \in C_H(V_H)$, then

$$S_0 := T_0 S = \langle t_1, t_2, S \rangle \le C_T(V_H).$$

But we saw earlier that $N_G(S_+) \le H$ for each $S_+ \trianglelefteq T$ with $T_0 \le S_+$, so by a Frattini Argument, $H = N_H(C_T(V_H))C_H(V_H) \le M$, contrary to our assumption.

Thus $t_i^* \ne 1$, so as $V_H \le E$, t_i^* induces a transvection on V_H with center v_i. Then comparing the possibilities in E.2.3 to the list of groups in G.6.4 containing $\mathbf{F}_2$-transvections, we conclude that either $H^* \cong O_4^+(2)$ with $m([V_H, H]) = 4$, or H^* is one of S_5 or S_5 wr $\mathbf{Z}_2$. The latter cases are out, as then $N_M(S)$ is not a $3'$-group, contrary to 10.1.3 and the fact that $N_{L_0}(S)$ is a $3'$-group. So $[V_H, H]$ is the orthogonal module for $H^* \cong O_4^+(2)$. Let $Y := O^2(C_H(v_2))$; then $Y^* \cong \mathbf{Z}_3$, and $Y \cap M \le O_2(H)$.

Let $X := C_G(v_2)$. Then $T_1 = C_T(v_2)$, $|T : T_1| = 2$, and $L \le X$. As $T \not\le X$ but $N_G(T_1) \le M$, $T_1 \in Syl_2(X)$. Thus by 1.2.4, $L \le I \in \mathcal{C}(X)$. Suppose first that $L = I$. Then $L \trianglelefteq X$ by 1.2.1.3, so X acts on $[O_2(L), L] = V_1$. As $Y = [Y, T_1]$ while $Y \cap M \le O_2(H)$, we conclude from the structure of $Aut(L/O_2(L))$ that $Y \le O^2(C_X(L/O_2(L)))$. Further $End_{\mathbf{F}_2(L/O_2(L))}(V_1) \cong \mathbf{F}_2$, so that Y must centralize V_1. However, Y does not centralize $v_1 \in V_1$. This contradiction shows that $L < I$.

Suppose that $V_1 \le O_2(I)$. Then since the A_5-block L has a unique nontrivial 2-chief factor V_1, and V_1 is projective, $W := \langle V_1^I \rangle = V_1 \oplus C_W(L) \le Z(O_2(I))$ and I has a unique nontrivial 2-chief factor. In particular $W \in \mathcal{R}_2(I)$ and setting $\hat{I} := I/C_I(W)$, $[W, \hat{a}] = [V_1, \hat{a}]$ for each involution $\hat{a} \in \hat{L}$, so $q(\hat{I}, W) \le 2$. Also we conclude from A.3.14 that $I/O_2(I) \cong A_7$, $\hat{A}_7$, J_1, $L_2(25)$, or $L_2(p)$ with $p \equiv \pm 1$ mod 5 and $p \equiv \pm 3$ mod 8. Then as $q(\hat{I}, W) \le 2$, we conclude from B.4.2 and B.4.5 that $I/O_2(I) \cong A_7$. Since the unique nontrivial L-chief factor V_1 is the A_5-module, we conclude that W is the A_7-module, so I is an A_7-block. However $I = O^{3'}(X)$ by A.3.18, so $1 \ne O^{3'}(C_{L_2}(v_2)) \le O^2(C_I(L))$, contradicting $O^2(C_I(L)) = 1$.

Therefore $V_1 \not\le O_2(I)$, so as L is irreducible on V_1, $V_1 \cap O_2(I) = 1$. Set $\dot{I} := I/O_2(I)$. Then $\dot{L}$ is a T_1-invariant A_5-block in $\dot{I}$, a situation that does not occur in A.3.14. This contradiction completes the proof. $\square$

LEMMA 10.2.2. *Assume $H \in \mathcal{H}_*(T,M)$ with $[Z,H] \neq 1$, and set $W := \langle Z^H \rangle$. Then*

(1) $L_0 = [L_0, J(T)]$ and one of the first three cases of 10.1.1 holds.

If in addition case (1) or (2) of 10.1.1 holds, then:

(2) $O^2(H) = [O^2(H), J(T)]$ and $J(T) \not\leq C_T(W)$, so W is an FF-module for $H/C_H(W)$.

(3) If case (1) of 10.1.1 holds, then $B_0 \leq N_G(S)$ and $S \in Syl_2(L_0 S)$.

(4) $O_2(\langle N_G(S), H \rangle) \neq 1$.

PROOF. As $[Z,H] \neq 1$, $[V, J(T)] \neq 1$ by 3.1.8.3. Thus $L_0 = [L_0, J(T)]$, and then 10.1.2.1 completes the proof of (1). In the remaining assertions we may assume case (1) or (2) of 10.1.1 holds. Then by 10.1.2.6, $L_0 T$ is a minimal parabolic described in E.2.3, and $S \leq T_1$.

In case (1) of 10.1.1, E.2.3.2 says that $S \in Syl_2(L_0 S)$ and $S \leq T_+ := T_0 O_2(L_0 T)$, so that $S = \text{Baum}(T_+)$. But B_0 normalizes T_+ so $B_0 \leq N_G(S)$, completing the proof of (3).

Assume $[O^2(H), J(T)] < O^2(H)$. Then as $[Z,H] \neq 1$ we conclude from B.6.8.3d that $S = \text{Baum}(O_2(H))$ and hence $H \leq N_G(S)$. However since we are excluding case (3) of 10.1.1, $N_G(S) \leq M$ by 10.2.1. This contradicts $H \not\leq M$, so (2) holds.

If $J(H)^*$ is the product of copies of $L_2(2^m)$ then by E.2.3.2, $S \in Syl_2(O^2(H)S)$. Then using Theorem 3.1.1 as in the proof of 10.2.1, (4) follows. Since we may assume (4) fails, we conclude from E.2.3.1 that $J(H^*)$ is a product of $s \leq 2$ copies of S_5, and that no nontrivial characteristic subgroup of S is normal in H. Therefore by E.2.3.3, $O^2(H) = K_1 \times \cdots \times K_s$ is the product of A_5-blocks K_i.

Next $O^2(H \cap M) \leq L_0$ by 10.1.3, so $O^2(H \cap M) \leq B_0$ and a Sylow 3-subgroup P of $O^2(H \cap M)$ is contained in $P_0 \in Syl_3(B_0)$. As $O^2(H)$ is a product of A_5-blocks, P centralizes Z, so case (2) of 10.1.1 holds since $C_{P_0}(Z) = 1$ in case (1) by 10.1.2.5. Then since $L_0 = [L_0, J(T)]$ by (1), $\bar{L}_0 \bar{T} \cong S_5$ wr $\mathbf{Z}_2$ in view of B.4.2.5, so $B_0 \in \Xi(G,T)$. Since $O^2(H \cap M)$ is T-invariant and lies in B_0, while T is irreducible on $B_0/O_2(B_0)$, we conclude $O^2(H \cap M) = B_0$. Therefore $P = P_0$ is of order 9, so $s = 2$ and $O^2(H) = K_1 \times K_2$ is the product of two blocks. Therefore $O^2(H \cap M) = X_1 \times X_2$ with $X_i := O^2(K_i \cap M) \cong \mathbf{Z}_3/Q_8^2$. Now as $O^2(H \cap M) = B_0$, while X_i has just two noncentral 2-chief factors, X_i cannot be diagonally embedded in L_0, so (interchanging L_1 and L_2 if necessary) $X_i = B_0 \cap L_i$. Then X_i is T_1-invariant, and L_i is an A_5-block as X_i has two noncentral 2-chief factors. Now K_i is T_1-invariant, $I := \langle L_1, K_1 \rangle \leq C_G(X_2)$, I is T_1-invariant, and $S = \text{Baum}(T_1)$ since we saw $S \leq T_1$. Hence $N_G(T_1) \leq M$ by 10.2.1. Therefore $N_T(X_2) = T_1 \in Syl_2(N_G(X_2))$, so $T_1 \in Syl_2(IT_1)$. Hence we can apply 1.2.4 to embed $L_1 \leq L_I \in \mathcal{C}(I)$, and then $K_1 = [K_1, X_1] \leq L_I$, so $L_1 < L_I$ since $K_1 \cap M = X_1$. Now $N_G(X_2)$ is an SQTK-group, so $m_3(N_G(X_2)) \leq 2$ and hence $m_3(L_I) = 1$. This rules out the possibility that $O_2(L_1) \leq O_2(L_I)$ and $L_I/O_2(L_I) \cong A_7$ in A.3.14. We now obtain a contradiction via the argument in the last two paragraphs of the proof of 10.2.1. This contradiction completes the proof of (4), and of the lemma. $\square$

PROPOSITION 10.2.3. *If $H \in \mathcal{H}_*(T,M)$ with $n(H) > 1$, then*

(1) $n(H) = 2$.

(2) A Hall $2'$-subgroup of $H \cap M$ is faithful on $\bar{L}_0$.

(3) If $\bar{L} \cong L_3(2)$, then $T_0 O^2(H \cap M)$ is a maximal parabolic in L_0 and $H/O_2(H) \cong S_5$ wr $\mathbf{Z}_2$.

(4) Case (1) of 10.1.1 does not hold; that is, $n(H) = 1$ for each H in that case.

PROOF. Let B_H be a Hall $2'$-subgroup of $H \cap M$. Notice B_H permutes with T, so that $B_+ := B_H \cap L_0$ permutes with T_0.

We first establish (2). If V is not an FF-module for $L_0T/C_{L_0T}(V)$, then (2) follows from Theorem 4.4.14; so we may assume that $B := C_{B_H}(\bar{L}_0) \neq 1$ and V is an FF-module for $\bar{L}_0\bar{T}$. We first verify Hypothesis 4.4.1 and then we apply Theorem 4.4.3: By 4.4.13.2 we have $BT = TB$, giving (1) and (2) of Hypothesis 4.4.1. As $BT = TB$, $N_H(B) \not\leq M$ by 4.4.13.1. As $V_i \trianglelefteq O^2(M)$, B acts on V_i. But by 10.1.3, $(|B|, |\bar{L}|) = 1$, so as $|End_{\bar{L}_i}(\tilde{V}_i)|$ divides $|\bar{L}|$, $[V, B] = 1$. Thus we also have 4.4.1.3, with V in the role of "V_B". Since $L < L_0$, case (1) of Theorem 4.4.3 must hold, contradicting our earlier observation that $N_H(B) \not\leq M$. So (2) is established.

Appealing to (2), 10.1.3, and the structure of $Aut(\bar{L}_i)$, we conclude that either

(i) $\bar{L}$ is not $L_3(2)$ and $B_H = B_+F$, with $B_+ \leq B_0$ (since B_+ permutes with T_0), and F induces field automorphisms on $\bar{L}_0$, or

(ii) $\bar{L} \cong L_3(2)$ and $B_H = B_+ \leq L_0$.

Assume first that (ii) holds; this case corresponds to cases (3), (5), and (6) of 10.1.1. Then as B_H permutes with T_0, B_H is a 3-group, and so $n(H) = 2$. Further $B_HO_2(B_HT)$ is T-invariant, so $\bar{B}_H\bar{T}$ contains a Sylow 3-group of $\bar{L}_0$, and hence $B_HT_0 = O^2(H \cap M)T_0$ is a maximal parabolic in L_0. In particular, $(H \cap M)/O_2(H \cap M) \cong S_3 \text{ wr } \mathbf{Z}_2$, and the only case in E.2.2 with $n(H) = 2$ satisfying this condition is $H/O_2(H) \cong S_5 \text{ wr } \mathbf{Z}_2$. For example case (2b) of E.2.2 is ruled out as here $(H \cap M)/O_{2,3}(H \cap M) \cong D_8$. Thus we have established (3), and also proved (1) in this case. So from now on, we may assume that (i) holds.

Suppose next we are in case (2) of 10.1.1, where $\bar{L} \cong A_5$. Then $F = 1$, so that $B_H = B_+ \leq B_0$. Now we may argue much as in the previous paragraph: As B_H permutes with T, it is a 3-group and so $n(H) = 2$, completing the proof of (1) and hence of the lemma in this case.

So at this point, we have reduced to one of cases (1), (4), or (7) of 10.1.1. Since $B_H = B_+F$ by (i), there is a B_H-invariant Hall $2'$-subgroup D of B_0, and $B_+ \leq D$. By 10.1.2.5, $C_D(Z) = 1$ in cases (1) and (7) of 10.1.1, while $C_D(Z) \cong \mathbf{Z}_{2^n+1}^2$ in case (4). Further in any case, $C_F(Z) = 1$.

Suppose first that $[Z, H] = 1$. Then $F = C_F(Z) = 1$, so $B_+ = B_H \leq C_D(Z)$, and hence $C_{B_0}(Z) \neq 1$ so that case (4) of 10.1.1 holds by the previous paragraph. Set $m := n(H) \geq 2$. From E.2.2, B_H has a cyclic subgroup B of order $2^m - 1$. As $B \leq C_D(Z)$, $2^m - 1$ divides $2^n + 1$, so m divides $2n$. If m divides n then $2^m - 1$ divides $2^n - 1$, impossible as $(2^n + 1, 2^n - 1) = 1$. Thus $m = 2d$ is even and d divides n, so as $(2^n + 1, 2^n - 1) = 1$, $2^d - 1 = 1$ and hence $m = 2$. Therefore the lemma holds in this case.

We may now assume that $[Z, H] \neq 1$. Then $L_0 = [L_0, J(T)]$ by 10.2.2.1, eliminating cases (4) and (7) of 10.1.1, leaving only case (1), where it remains to derive a contradiction in order to complete the proof of the lemma. Recall in this case that $C_{DF}(Z) = 1$.

By 10.2.2.2, $O^2(H) = [O^2(H), J(T)]$. By E.2.3.1, $O^2(H) = \langle K^T \rangle$ where $K \in \mathcal{C}(H)$ with $K/O_2(K) \cong L_2(2^m)$ or A_5, and setting $W := \langle Z^H \rangle$ and $V_K := [W, K]$, $V_K/C_{V_K}(K)$ is the natural module for $K/O_2(K)$. Observe V_K is not the A_5-module as $B_H \leq DF$ and $C_{DF}(Z) = 1$, whereas if V_K were the A_5-module then $[Z, B_H] = 1$.

We next claim that $B_0 \leq N_G(K)$: By 10.2.2.3, $B_0 \leq N_G(S)$, so by 10.2.2.4, $\langle B_0, H \rangle \leq M_1 \in \mathcal{M}(T)$. Hence by 1.2.4, $K \leq I \in \mathcal{C}(M_1)$, and as $K = [K, J(T)]$

and $[Z,K] \neq 1$, also $I = [I, J(T)]$ and $1 \neq U := [Z,I] \in \mathcal{R}_2(I)$ using B.2.14. Thus $J(T) \not\leq C_T(U)$ and U is an FF-module. We conclude from intersecting the lists of A.3.12 and B.4.2 that one of the following holds:

    (a) $K = I$.
    (b) $I/O_2(I) \cong SL_3(2^m)$, $Sp_4(2^m)$, or $G_2(2^m)$.
    (c) $m = 2$ and $I/O_2(I) \cong A_7$ or $\hat{A}_7$, with $I/C_I(U) \cong A_7$.
    (d) $I/O_2(I)$ is not quasisimple, $I = O_{2,F}(I)K$, and $O_{2,F}(I)$ centralizes U.

By 1.2.1.3, $B_0 = O^2(B_0)$ normalizes I, so we may assume that $K < I$, and so one of (b)–(d) holds. Hence T acts on I by 1.2.1.3, and then also T acts on K by 1.2.8. In case (d), as $C_D(Z) = 1$, $B_0 \cap O_{2,F}(I) \leq T$, so B_0 acts on the unique $(T \cap I)$-invariant supplement K to $O_{2,F}(I)$ in I. Suppose case (c) holds. By A.3.18, $I = O^{3'}(M_1)$, so $D = D_I \times D_C$, where $D_C := O^3(D) = C_D(I/O_2(I))$ and $D_I := O_3(D) = D \cap I$. As D_C acts on K, we may assume $D_I \not\leq K$. Then $D_I \in Syl_3(I)$. But from the structure of the FF-modules for A_7, $C_{D_I}(Z) \neq 1$, contradicting $C_D(Z) = 1$. Suppose case (b) holds. Then $K = P^\infty$ for some T-invariant parabolic P in $I/O_2(I)$, so as $B_0 = O^2(B_0)$ permutes with T, it must also act on K, completing the proof of the claim.

By the claim B_0 acts on K, and by symmetry B_0 also acts on K^t if there is $t \in T - N_T(K)$. Thus B_0 acts on $O^2(H)$. Recall that by construction B_H acts on D, so D acts on $O^2(H) \cap DB_H = B_H$. Therefore $[B_H, D] \leq B_H \cap D \leq C_D(B_H)$ since the Hall subgroup B_H of $O^2(H \cap M)$ is abelian. Now if $F \neq 1$, then F does not centralize $[F, D]$; thus $F = 1$, and hence $B_H = B_+ \leq D$. Since $B_H \cap K$ is cyclic of order $2^m - 1$, while $D \cong \mathbf{Z}_{2^n-1}^2$, m divides n.

In the remainder of the proof, we will show that $B_H = D$, and that $K \neq K^t$ for some $t \in T - T_1$. Then we will see that the embeddings of D in LL^t and KK^t are incompatible.

As $M = !\mathcal{M}(L_0T)$, $C_Z(\langle L_0, H \rangle) = 1$. As $\tilde{V}_K$ is the natural module for $K/O_2(K) \cong L_2(2^m)$, $C_Z(H) = C_Z(b)$ for each $b \in B_H^\#$. Similarly $C_Z(L_0) = C_Z(d)$ for each $d \in D^\#$, so as $1 \neq B_H \leq D$, we conclude $C_Z(L_0) = C_Z(H) = 1$. Thus V_1 and V_K are natural modules, with $C_{V_1}(L) = 1 = C_{V_K}(K)$.

Next C.1.26 says that there are nontrivial characteristic subgroups $C_1(T) \leq Z$ of T and $C_2(T)$ of S, such that one of the following holds: K is a block, $C_1(T) \leq Z(H)$, or $C_2(T) \trianglelefteq H$. As $C_Z(H) = 1$, $C_1(T) \not\leq Z(H)$, so either K is a block or H normalizes $C_2(T)$. Similarly either L is a block or $C_2(T) \trianglelefteq L_0T$. However $C_2(T)$ cannot be normal in both H and L_0T, since $M = !\mathcal{M}(L_0T)$; therefore either K or L is a block.

Next set $E := \Omega_1(Z(J(T)))$ and $E_0 := \langle E^{L_0} \rangle$. By 10.1.2.6 we may apply E.2.3.2 to L_0T, to conclude that $E_0 = C_{E_0}(L_0)V$. Therefore as $C_Z(L_0) = 1$, $E_0 = V = V_1 \times V_2$ is of rank $4n$. In particular, $E \leq V$ and $E = E_1 \times E_2$ with $E_i := E \cap V_i$ of rank n. Also $D = D_1 \times D_2$, where $D_i := D \cap L_i = C_D(E_{3-i})$. Notice that $C_E(d) = 1$ for $d \in D - (D_1 \cup D_2)$. Similarly applying E.2.3.2 to H and using $C_Z(H) = 1$, we conclude that $E = E_K \times C_E(K) = E_K \times C_E(B_K)$, where $E_K := E \cap V_K$ has rank $m = n(H)$, and $B_K := B \cap K$. We saw m divides n, so $m \leq n$, and hence

$$m(C_E(B_K)) = m(E) - m(V_K) = 2n - m \geq n. \qquad (*)$$

So as $C_E(d) = 1$ for each $d \in D - (D_1 \cup D_2)$, we conclude (interchanging the roles of L and L^t if necessary) that $B_K \leq D_1$, so $E_K = [E, B_K] = [E, D_1] = E_1$ and

$C_E(B_K) = C_E(D_1) = E_2$ are of rank n. Thus $m = n$ by (*), and $D_1 = B_K$ as $B_K \leq D_1$ and $|D_1| = 2^n - 1 = 2^m - 1 = |B_K|$.

Next recall from E.2.3.2 that S normalizes L and K. Therefore as $D_1 = B_K$, $S_1 := [S, D_1] \leq L \cap K$. Since either L or K is a block, and $C_Z(L_0) = C_Z(H) = 1$, we conclude S_1 is special of order 2^{3n}; then it follows that both L and K are blocks, with $O_2(L)$ and $O_2(K)$ of rank $2n$, and S_1 is Sylow in both L and K.

Next L_1 and L_2 commute by C.1.9, so $[S_1, D_2] \leq [L_1, D_2] = 1$. So as D_2 centralizes $S_1 \in Syl_2(K)$, D_2 centralizes K from the structure of $Aut(K)$. Similarly $S_2 := [S, D_2] \in Syl_2(L_2)$ and S_2 centralizes K. But $S_2 = S_1^t$ for $t \in T - T_1$, so S_2 is Sylow in the block K^t and $K^t \neq K$. Hence $O^2(H) = KK^t = K \times K^t$ since $C_{V_K}(K) = 1$. Setting $K_1 := K$ and $K_2 := K^t$, $S_i D_i$ is Borel in both L_i and K_i.

Set $M_1 := N_G(S_1 D_1)$. As L_2 centralizes L_1, $L_2 T_1 \leq M_1$, with $T_1 = T \cap M_1$. Similarly $K_2 T_1 \leq M_1$. Embed $T_1 \leq T_+ \in Syl_2(M_1)$. Recall that $S \leq T_1$, so $S = Baum(T_+)$, and hence $T_+ \leq N_G(S) \leq M$ by 10.2.1. If $T_1 < T_+$ then T_+ is also Sylow in M, so $M = !\mathcal{M}(L_0 T_+)$ by 1.2.7.3. However $L_0 T_+ = \langle L_2, T_+ \rangle \leq M_1$, so $K_2 \leq M_1 \leq M$, contradicting $K_2 \not\leq M$.

This contradiction shows that $T_1 = T_+$ is Sylow in M_1. Hence $L_2 \leq L_+ \in \mathcal{C}(M_1)$ by 1.2.4. Now $K_2 = O^2(K_2)$ normalizes L_+ by 1.2.1.3, so as $D_2 \leq L_2 \leq L_+$, also $K_2 = [K_2, D_2] \leq L_+$, and hence $L_2 < L_+$. As L_2 and K_2 are distinct members of $\mathcal{L}(L_+, T_1)$ and both are blocks of type $L_2(2^n)$ with trivial centers, we conclude from A.3.12 that $O_2(L_+) = 1$ and $L_+ \cong (S)L_3(2^n)$. Now L_+ normalizes $S_1 D_1$, and so in fact centralizes $S_1 D_1$ since S_1 is special of order 2^{3n}. Therefore for p a prime divisor of $2^n - 1$, $m_p(D_1 L_+) > 2$, contradicting M_1 an SQTK-group. This completes the elimination of case (1) of 10.1.1 when $n(H) > 1$, and hence establishes 10.2.3. $\square$

LEMMA 10.2.4. *Assume that* $\bar{L} \cong L_3(2)$, *but case (5) of 10.1.1 does not hold, so that* $\bar{L}\bar{T}_1 \cong L_3(2)$. *Let* P *be one of the two maximal subgroups of* $L_0 T$ *containing* T. *Set* $X := O^2(P)$, *assume* $H \in \mathcal{H}(XT)$, *and set* $K := \langle X^H \rangle$. *Then one of the following holds:*

(1) $K = X$.

(2) $K = K_1 K_1^s$ *with* $K_1 \in \mathcal{C}(H)$, $K_1/O_2(K_1) \cong L_2(2^m)$ *for some even* m *or* $L_2(p)$ *for some odd prime* p, *and* $s \in T - N_T(K_1)$.

(3) $K \in \mathcal{C}(H)$ *and* $KT/O_2(KT) \cong Aut(L_k(2))$, $k = 4$ *or* 5.

PROOF. As $X \in \Xi(G, T)$, K is described in 1.3.4 with $p = 3$. Further $XT/O_2(XT) \cong S_3$ wr $\mathbf{Z}_2$, which reduces the list to the cases appearing in the lemma. $\square$

LEMMA 10.2.5. $N_G(T_1) \leq M$.

PROOF. If $J(T) \leq C_T(V)$ then the lemma follows from 3.2.10.8, so we may assume $J(T) \not\leq C_T(V)$. Then one of the first three cases of 10.1.1 holds by 10.1.2.1. If case (1) or (2) of 10.1.1 holds then $S \leq T_1$ by 10.1.2.6, so $S = Baum(T_1)$ and then $N_G(T_1) \leq M$ by 10.2.1.

Thus we may assume case (3) of 10.1.1.3 holds, so $\bar{L} = \bar{L}\bar{T} \cong L_3(2)$. Let H_1 and H_2 be the two maximal subgroups of $L_0 T$ containing T. Thus $X_i := O^2(H_i) \in \Xi(G, T)$ and $H_i/O_2(H_i) \cong S_3$ wr $\mathbf{Z}_2$. Since $O_2(X_i) \leq T_1$, $T_1 \in Syl_2(X_i T_1)$. Further T is a maximal subgroup of H_i, so applying Theorem 3.1.1 with H_i, $N_G(T_1)$, T_1 in the roles of "H, M_0, R", we conclude $O_2(G_i) \neq 1$, where $G_i := \langle N_G(T_1), H_i \rangle$.

It will suffice to show $N_G(T_1)$ acts on X_i for $i = 1$ and 2, since then $N_G(T_1)$ acts on $\langle X_1, X_2 \rangle = L_0$, so $N_G(T_1) \leq N_G(L_0) = M$, as desired. Therefore we may assume $N_G(T_1) \not\leq N_G(X_i)$ for some i, and we now fix that value of i.

Set $K_j := \langle X_j^{G_j} \rangle$ and $K_j^* T^* := K_j T / O_2(K_j)$ for each j. Notice $O_2(K_j) \leq O_2(H_j) \leq T_1$ using A.1.6. Now $N_G(T_1) \leq G_j$ so $N_G(T_1)$ acts on K_j. Thus as $N_G(T_1) \not\leq N_G(X_i)$, $X_i < K_i$, and hence by 10.2.4 either $K_i = K_{i,1} K_{i,1}^s$ with $K_{i,1} \in \mathcal{C}(G_i)$ and $s \in T - N_T(K_{i,1})$, or $K_i^* T^* \cong Aut(L_k(2))$, $k := 4$ or 5. In either case, $N_G(T_1)$ acts on $R := T_1 \cap K_i$ and $O_2(X_i(T \cap K_i)) \leq R$.

Suppose first that either $K_i^* \cong L_k(2)$, or $K_{i,1}^* \cong L_2(2^m)$. As $H_i \cap K_i$ is T-invariant, $H_i \cap K_i \leq J_i$, where J_i^* is a T-invariant parabolic of K_i^* such that X_i is the characteristic subgroup generated by the elements of order 3 in J_i. Notice

$$O_2(J_i) \leq O_2(X_i(T \cap K_i)) \leq R \leq T \cap K_i. \qquad (*)$$

Now when $K_{i,1}^* \cong L_2(2^m)$, $T \cap K_i = O_2(J_i)$, so the inequalities in $(*)$ are equalities, and then $N_G(T_1) \leq N_{G_i}(R) \leq N_{G_i}(J_i) \leq N_G(X_i)$, contrary to our assumption. On the other hand if $K_i^* \cong L_k(2)$, then $O_2(J_i^*)$ is a unipotent radical, and so by I.2.5 is weakly closed in $(T \cap K_i)^*$ with respect to G_j; thus $N_G(T_1) \leq N_{G_i}(O_2(J_i))) \leq N_G(X_i)$, for the same contradiction.

This leaves the case where $K_{i,1}^* \cong L_2(p)$. If $p \equiv \pm 3 \mod 8$, then again $T \cap K_i = O_2(X_i(T \cap K_i)) \leq R$, so $R = T \cap K_i$; and $N_G(T_1)$ normalizes $N_{K_i}(T \cap K_i) = X_i(T \cap K_i)$ and hence also $O^2(X_i(T \cap K_i)) = X_i$, for our usual contradiction. Therefore $p \equiv \pm 1 \mod 8$, and $(T \cap K_{i,1})^*$ is a nonabelian dihedral 2-group. Since $(X_i \cap K_{i,1})^*$ is a $T \cap K_i$-invariant A_4-subgroup of $K_{i,1}^*$, $|(T \cap K_{i,1})^*| = 8$.

Next R is of index $r \leq |T : T_1| = 2$ in $T \cap K_i$. Further if $r = 2$, then $O_2(X_i(T \cap K_i))^* = J(R^*)$, so $N_G(T_1) \leq N_{G_i}(O_2(X_i(T \cap K_i)) \leq N_G(X_i)$, again contrary to assumption. Therefore $R = T \cap K_i$ and there are exactly two subgroups Y of $K_{i,1}$ with $R \cap K_{i,1} \leq Y$ and $Y^* \cong S_4$. So $O^2(N_G(T_1))$ acts on both such subgroups, and in particular on $X_i \cap K_{i,1}$. Similarly $O^2(N_G(T_1))$ acts on $X_i \cap K_{i,1}^s$, and hence on the product X_i of these two subgroups, so $N_G(T_1) = T O^2(N_G(T_1)) \leq N_G(X_i)$, for our final contradiction. $\qquad \square$

LEMMA 10.2.6. *(1)* $M = !\mathcal{M}(L_0 T_1)$.
(2) $N_G(V_i) \leq M \geq N_G(L)$.

PROOF. Notice (1) implies (2), so it suffices to prove (1). Suppose that there is $H \in \mathcal{M}(L_0 T_1) - \{M\}$. Then $|T : T_1| = 2$, and $N_G(T_1) \leq M$ by 10.2.5. By 1.2.7.3, $M = !\mathcal{M}(L_0 T_+)$ for each $T_+ \in Syl_2(M)$, so that $T_1 \in Syl_2(H)$. Thus by 1.2.4, $L_i \leq K_i \in \mathcal{C}(H)$, and $K_i \trianglelefteq H$ by $(+)$ in 1.2.4. Now from A.3.12, K_i does not contain $L_0 = L_1 L_2$, so $K_1 \neq K_2$. Thus as $m_p(H) \leq 2$ for each prime divisor p of $|\bar{L}|$, while $L_{3-i} \leq C_H(K_i/O_2(K_i))$, we conclude $m_p(K_i) = 1$ for each such prime. As $H \neq M = N_G(L_0)$, L_0 is not normal in H, so $L_i < K_i$ for $i := 1$ or 2; we fix this value of i.

Now if $\bar{L} \cong Sz(2^n)$, A.3.12 says L_i is properly contained in no K_i with $m_p(K_i) = 1$ for each prime p dividing $2^n - 1$, and similarly L_i is proper in no K_i with $m_7(K_i) = 1 = m_3(K_i)$ when $\bar{L} \cong L_3(2)$. Therefore $\bar{L} \cong L_2(2^n)$.

Assume $F^*(K_i) = O_2(K_i)$. Set $H_0 := K_i L_{3-i} T_1$ and $R := O_2(L_0 T)$. Then $L_0 T_1 \leq M_0 := M \cap H_0$. As $M = !\mathcal{M}(L_0 T)$, $C(H_0, R) \leq M_0$, and by A.4.2.7, $R \in \mathcal{B}_2(H_0)$ and $R \in Syl_2(\langle R^{M_0} \rangle)$. Thus Hypothesis C.2.3 is satisfied, so K_i is described in C.2.7.3. Comparing the list of possibilities for K_i appearing there such that $m_p(K_i) \leq 1$ for each $p \in \pi(|\bar{L}|)$ to the list of embeddings of $L_2(2^m)$ in A.3.12, we obtain a contradiction.

Therefore we may assume instead that $F^*(K_i) \neq O_2(K_i)$. By A.1.26, $V = [V, L_0]$ centralizes $O(K_i)$, so $O(K_i) \leq C_G(V) \leq M$. Then as $O_2(M) \leq T_1$, $[O_2(M), O(K_i)] \leq O_2(M) \cap O(K_i) = 1$, so $O(K_i) = 1$ as $M \in \mathcal{H}^e$. Thus as $K_i/O_\infty(K_i)$ is quasisimple, K_i is quasisimple. As L_i does not centralize V_i, $O_2(L_i) \not\leq Z(K_i)$. But now each possibile embedding of L_i in K_i in A.3.12 with $O_2(L_i) \not\leq Z(K_i)$ has $m_p(K_i) > 1$ for some odd prime p dividing $|\bar{L}|$, again contradicting our earlier observation. This completes the proof. $\square$

At this point, we eliminate the sixth case of 10.1.1; this will avoid complications in the proof of 10.2.9.

LEMMA 10.2.7. *Case (6) of 10.1.1 does not hold. In particular,* $C_T(V) = O_2(L_0 T)$.

PROOF. The second statement follows from the first by 10.1.2.3. Assume the first statement fails. Then $m := m(\bar{M}, V) = 4$ and $a := a(\bar{M}, V) = 2$. By Theorem E.6.3, $r := r(G, V) \geq m$, so $r \geq 4$ and $s := s(G, V) = 4$.

Indeed we show $r > 4$: For suppose $U \leq V$ with $m(V/U) = 4$ and $C_G(U) \not\leq M$. If $U \leq C_V(\bar{x})$ for some $\bar{x} \in \bar{M}^\#$, then $\bar{x}$ is an involution and $U = C_V(\bar{x}) \geq V_i$ for $i = 1$ or 2. But then $C_G(U) \leq C_G(V_i) \leq M$ by 10.2.6, a contradiction. Therefore $C_M(U) = C_M(V)$, and E.6.12 supplies a contradiction.

We observe next that 10.1.2.3 and 10.2.3.2 establish Hypothesis E.3.36. A maximal cyclic subgroup of odd order in $\bar{M}$ permuting with $\bar{T}$ is of order 15, so $n'(Aut_G(V)) = 4 < r$. Finally by 10.2.3.1, $n(H) \leq 2$ for each $H \in \mathcal{H}_*(T, M)$. Therefore by E.3.39.2,

$$2 = s - a \leq w \leq n(H) \leq 2$$

where $w := w(G, V)$ is the weak closure parameter defined in E.3.23. Thus $w = 2$. Let $A \leq V^g$ be a w-offender in the sense of Definition E.3.27. By E.3.33.4, $\bar{A} \in \mathcal{A}_2(\bar{M}, V)$. Thus $1 \neq C_{V_1}(N_A(V_1)) \leq C_V(A)$, so A acts on V_1. As $\bar{A} \in \mathcal{A}_2(\bar{M}, V)$, $\bar{A}$ centralizes $O(\bar{M})$ by E.3.40, so $m(A/C_A(V_1)) \leq m_2(Aut(\bar{L})) = 2$. Thus $m(V^g/C_A(V_1)) \leq w + 2 = 4 < r$, and hence $V_1 \leq C_G(C_A(V_1)) \leq M^g$. Similarly $V_2 \leq M^g$, so $V \leq M^g = N_G(V^g)$, contrary to E.3.25 since $w > 0$. $\square$

LEMMA 10.2.8. *Assume* $\bar{L} \cong L_3(2)$ *and* $C_{V_1}(L) \neq 1$. *Set* $Q := C_T(V)$. *Then*

(1) $[Z, L] = 1$.

(2) $Z_Q := \Omega_1(Z(Q)) = Z_{T_1} V$, *where* $Z_{T_1} := \Omega_1(Z(T_1)) = C_{Z_Q}(L_0)$.

(3) $L = [L, J(T)]$, *and* $[Z, H] \neq 1$ *for each* $H \in \mathcal{H}_*(T, M)$.

(4) Set $\tilde{U}_i := C_{\tilde{V}_i}(T_1)$, let R_i be the preimage in T of $O_2(C_{\bar{L}_i}(\tilde{U}_i))$, $R := R_1 R_2 Q$, and $v_2 \in U_2 - C_{V_2}(L_0)$. *Then* $C_{\bar{L}_0 \bar{T}}(v_2) \cong A_4 \times L_3(2)$, $R = J(T)Q$, $C_T(v_2) = (T \cap L)R_2 Q$, *and* $\Omega_1(Z(C_T(v_2))) = Z_{T_1}\langle v_2 \rangle$.

PROOF. As $Z_{T_1} \leq C_T(V) = Q \leq T_1$, $Z_{T_1} = C_{Z_Q}(T_1)$. As $Z_i := C_{V_i}(L_0) \neq 1$ and $\tilde{V}_1$ is a natural module for $\bar{L}$, $Z_i \cong \mathbf{Z}_2$ by B.4.8.1. In particular $C_Z(L) \neq 1$, so (3) follows from 3.1.8.3, since $H \not\leq M = !\mathcal{M}(L_0 T) = !\mathcal{M}(C_G(C_Z(L_0)))$.

By 1.4.1.5, $Z_Q = R_2(L_0 T)$ with $Q = C_{L_0 T}(Z_Q) = C_{L_0 T}(V)$ and $V \leq Z_Q$. By (3), Z_Q is an FF-module for $L_0 T$. As $V_1 \in Irr_+(Z_Q, L)$ with $C_{V_1}(L) \neq 1$, by part (1) of Theorem B.5.1, $V = [Z_Q, L_0]$, and that for any $A \in \mathcal{A}(T)$ with $L = [L, A]$ and $\bar{A}$ minimal subject to this constraint, $\bar{A} \leq \bar{L}$ and $Z_Q = V_1 C_{Z_Q}(A)$.

By B.4.8.2, $\bar{A} = \bar{R}_1$ and $r_{Z_Q,\bar{A}} = 1$, so by B.4.8.4, $Z_Q = V_1 C_{Z_Q}(L)$. This shows $Z_Q = V C_{Z_Q}(L_0)$, $R = J(T)Q$, and $C_T(v_2) = (T \cap L)R_2 Q$, establishing (4) except for its final assertion. Notice it also shows $Z \cap V \leq Z_{T_1} \cap V \leq C_V(L_0)$. But $T_1 = T_0 Q$, so $C_{Z_Q}(L_0) \leq Z_{T_1}$. Conversely, $Z_{T_1} \leq Z_Q$ and we saw $V \cap Z_{T_1} \leq C_V(L_0)$, so $Z_{T_1} \leq C_{Z_Q}(L_0)$, and hence (2) holds. Further $Z \leq Z_{T_1}$, so (2) implies (1). Finally $Q \leq C_T(v_2)$ and $Q = F^*(L_0 T)$, so $\Omega_1(Z(C_T(v_2))) \leq Z_Q = V Z_{T_1}$; therefore $\Omega_1(Z(C_T(v_2))) = Z_{T_1} C_V(C_T(v_2)) = Z_{T_1}\langle v_2 \rangle$, completing the proof of (4), and hence of the lemma. $\qquad\square$

We are now in a position to produce a crucial bound on the weak closure parameter r of Definition E.3.3:

PROPOSITION 10.2.9. *(1)* $C_G(v) \leq M$ *for each* $v \in V_i^\#$.
(2) $r(G,V) \geq m(V_i)$.
(3) If $v \in V_i - C_{V_i}(L_0)$, *then* $C_G(v) \leq N_M(V_i)$.

PROOF. Part (3) follows from (1) and the fact that M permutes $\{V_1, V_2\}$ and $V_1 \cap V_2 = C_V(L_0)$. Also (1) implies (2), so it remains to prove (1).

Let $v \in V_2^\#$, and suppose by way of contradiction that $H := C_G(v) \not\leq M$. Without loss $T_v := C_T(v) \in Syl_2(C_M(v))$. By 10.2.6.1, $v \notin C_{V_2}(L_0 T_1)$.

We claim first that $N_G(T_v) \leq M$. If $J(T) \leq C_T(V)$, this follows from 3.2.10.8; so by 10.1.2.1 we may assume that one of the first three cases of 10.1.1 holds. Suppose first that case (3) of 10.1.1 holds, and also $C_{V_1}(L) \neq 1$. Then by 10.2.8.2, $Z_{T_1} := \Omega_1(Z(T_1)) \geq C_V(L_0)$, so $v \notin C_V(L_0)$ using our observation in the previous paragraph. Therefore as L_2 is transitive on $\tilde{V}_2^\#$, we may assume $\langle \tilde{v} \rangle = C_{\tilde{V}_2}(T_1)$. Hence by 10.2.8.4, $T_v = (T \cap L)R_2 Q$, and $Z_v := \Omega_1(Z(T_v)) = Z_{T_1}\langle v \rangle$. By 10.2.8.1, L_0 centralizes Z, so $C_G(Z_v) \leq C_G(Z) \leq M = !\mathcal{M}(L_0 T)$, and hence by 10.1.3, L is the unique member of $\mathcal{C}(C_G(Z_v))$ of order divisible by 3. Therefore $N_G(T_v) \leq N_G(Z_v) \leq N_G(L) \leq M$ using 10.2.6.2. We now turn to the remaining subcase of case (3) of 10.1.1, where $C_{V_1}(L) = 1$. Then $T_v = T_1$, so $N_G(T_v) \leq M$ by 10.2.5. Finally in cases (1) and (2) of 10.1.1, $S \leq T_1$ by 10.1.2.6; and in case (2), S centralizes both singular and nonsingular vectors. So in either case, $S \leq T_v$. Therefore $S = \mathrm{Baum}(T_v)$ and $N_G(T_v) \leq N_G(S) \leq M$ by 10.2.1. This completes the proof of the claim.

As $N_G(T_v) \leq M$ by the claim, while we chose $T_v \in Syl_2(C_M(v))$, $T_v \in Syl_2(H)$. Also $L \leq H$, so by 1.2.4, $L \leq I \in \mathcal{C}(H)$, with $I \trianglelefteq H$ by (+) in 1.2.4. By 10.2.6, $N_G(L) \leq M$, so $L < I$ and hence $I \not\leq M$. Thus I is described in A.3.12.

Suppose first that I is quasisimple. Then $V_1 \cap Z(I) \leq C_{V_1}(L)$, so $\tilde{V}_1 \cong V_1/C_{V_1}(L)$ is a subquotient of $R_2(LZ(I)/Z(I))$. Inspecting the list in A.3.12 for embeddings with such a subquotient appearing in 10.1.1, we conclude that case (1) or (3) of 10.1.1 holds; and keeping in mind that $N_G(V_1) \leq M$ so that $L \trianglelefteq N_I(V_1)$, we conclude that either:

(i) $\bar{L} \cong L_2(2^n)$, and either $I/Z(I)$ is of Lie type and Lie rank 2 over $\mathbf{F}_{2^n}$, or $n = 2$ and $I/Z(I)$ is M_{22}, $\hat{M}_{22}$, or M_{23}; or

(ii) case (3) of 10.1.1 holds with $C_{V_1}(L) \leq Z(I)$, and $I/Z(I)$ is $L_4(2)$, $L_5(2)$, M_{24}, J_4, HS, or Ru.

In particular either $C_T(L) = C_T(I)$, or $I \cong Sp_4(2^n)$ in (i), using I.1.3 to conclude the Schur multiplier of $Sp_4(2^n)$ is trivial when $n > 1$. When $C_T(L) = C_T(I)$, $V_2 \leq C_T(L) = C_T(I)$, so $I \leq C_G(V_2) \leq M$ by 10.2.6, contradicting $I \not\leq M$. On the other hand if $I \cong Sp_4(2^n)$, then L is indecomposable on $O_2(L)$, so $V_1 = O_2(L)$. Then there is $X \leq N_I(L)$ of order $2^n - 1$ centralizing L/V_1 and faithful on V_1. Thus $X \leq N_G(L) \leq M$, so $X \leq L_0$ by 10.1.3, impossible as there is no such subgroup of L_0.

Thus I is not quasisimple. So $E(I) = 1$ by A.3.3.1. We claim $F^*(IT_v) = O_2(IT_v)$: If not, then $O(I) \neq 1$ as $E(I) = 1$. But by A.1.26.1, $V_1 = [V_1, L]$ centralizes $O(I)$, so $O(I) \leq M$ by 10.2.6.2, and hence $O(C_M(v)) \neq 1$, a contradiction as $C_M(v) \in \mathcal{H}^e$ by 1.1.3.2.

We have shown that $F^*(IT_v) = O_2(IT_v)$. So $V_I := \langle C_{V_1}(T_v)^I \rangle \in \mathcal{R}_2(IT_v)$ by B.2.14. Let $(IT_v)^* := IT_v/C_{IT_v}(V_I)$. Now $V_v := \langle C_{V_1}(T_v)^L \rangle \leq V_I$, and from the action of L_0 on V in 10.1.1, either $V_1 = V_v$ or case (3) of 10.1.1 holds with $C_{V_1}(T_v) = C_{V_1}(L_0) \neq 1$ and $V_v = C_{V_1}(L_0)$. Therefore either $C_V(L_0) \neq 1$, or $N_G(V_v) \leq M$ by 10.2.6.2. In the former case, $1 \neq C_Z(L_0) \leq V_I$, so $C_G(V_I) \leq C_G(C_Z(L_0)) \leq M = !\mathcal{M}(L_0T)$; in the latter, $C_G(V_I) \leq C_G(V_v) \leq M$. So in any case, $C_G(V_I) \leq M$, and hence $L^* < I^*$ as $I \not\leq M$, while $L^* \neq 1$ as $I = \langle L^I \rangle$.

Next observe that $J(T) \leq T_v$, so that $J(T) = J(T_v)$ and $S = \text{Baum}(T_v)$: If $J(T) \leq C_T(V)$ this is clear, so by 10.1.2.1 we may assume that one of the first three cases of 10.1.1 holds. But in each of these cases v centralizes some M-conjugate of $J(T)$, so again the remark holds.

We next claim that $I^* = [I^*, J(T_v)^*]$ is quasisimple. Suppose not, so that either $[V_I, J(T_v)] = 1$ or I^* is not quasisimple. Suppose first that $J(T_v)^* \neq 1$. Then I^* is not quasisimple, so I^* is described in case (c) or (d) of 1.2.1.4, and hence $[X^*, J(T_v)^*] \neq 1$ for $X := \Xi_p(I)$ and some prime $p > 3$, contradicting Solvable Thompson Factorization B.2.16. Thus we may take $J(T_v)^* = 1$. However $L^* \neq 1$, so $J(T) \leq O_2(LT_v)$ and hence $J(T) \trianglelefteq L_0T$, so that $N_G(J(T)) \leq M$. Then by a Frattini Argument, $I = C_I(V_I)N_I(J(T)) \leq M$, contradicting $I \not\leq M$. So the claim is established.

By the claim, V_I is an FF-module for $I^*T_v^*$. Now intersecting the list of possibilities for the embedding of L^* in I^* in A.3.12 with the list of B.4.2, we get the following cases:

(a) $\bar{L} \cong L_2(2^n)$, $I^* \cong SL_3(2^n)$, $Sp_4(2^n)$, or $G_2(2^n)$, and $O_2(L^*) \neq 1$.
(b) $\bar{L} \cong A_5$ or $L_3(2)$, and $I^* \cong A_7$ with $O_2(L^*) = 1$.
(c) $\bar{L} \cong L_3(2)$ and $I^* \cong L_4(2)$ or $L_5(2)$, with $O_2(L^*) \neq 1$.

Observe in particular that I does not appear in case (c) or (d) of 1.2.1.4, so $I/O_2(I)$ is quasisimple.

Assume case (a) holds. Recall we saw earlier that $V_1 = V_v \leq V_I$ and the FF-module V_I is described in Theorem B.5.1. Then $L = N_I(V_1)^\infty$ and $N_{I^*}(V_1)$ is a maximal parabolic of I^*, so $N_I(L)$ contains a subgroup X of order $2^n - 1$ centralizing $L/O_2(L)$ and nontrivial on V_1. We now get a contradiction much as in the earlier case of $Sp_4(2^n)$ where I was quasisimple: for $X \leq N_G(L) \leq M$, and hence $X \leq L_0$ by 10.1.3, whereas there is no such subgroup of L_0.

Thus we have shown that (b) or (c) holds, so $\bar{L} \cong A_5$ or $L_3(2)$. We next show: In case (b) either

(b1) I is an exceptional A_7-block, $I^*T_v^* \cong A_7$, and V_I is the natural module for $L^* \cong L_2(4)$, or an indecomposable of rank 3 or 4 for $L^* \cong L_3(2)$, or

(b2) I is an A_7-block, $I^*T_v^* \cong S_7$, and $[V_I, L]$ is the A_5-module for $L^* \cong A_5$.

For assume case (b) holds. We saw that $S = \text{Baum}(T_v)$, so applying C.1.24 with I, T_v, T_v in the roles of "L, T, R", either I is an A_7-block or an exceptional

A_7-block, or there is a nontrivial characteristic subgroup C of S normal in IT_v. However in the last case $G_0 := \langle I, T \rangle \leq N_G(C)$, so as $L \leq I$, $L_0 T \leq G_0$ and hence $I \leq G_0 \leq M = !\mathcal{M}(L_0 T)$. This contradicts $I \not\leq M$, so I is a block. Further if I is an A_7-block, then as $I = [I, J(T_v)]$, $I^* T_v^* \cong S_7$, so $L/O_2(L)$ is not $L_3(2)$ as $L \in \mathcal{L}(IT_v, T_v)$. If I is an A_7-block, then I^* is self-normalizing in $GL(V_I)$, so $I^* T_v^* = I^*$. Thus (b1) or (b2) holds.

In particular in case (b), $O_2(I) = C_I(V_I)$. In case (c) since $I/O_2(I)$ is quasisimple, the list of Schur multipliers in I.1.3 says $I/O_2(I) \cong I^*$, so again $O_2(I) = C_I(V_I)$.

Assume $\bar{L} \cong L_3(2)$; this argument will be fairly lengthy. By 10.2.7, case (3) or (5) of 10.1.1 holds. In case (b), subcase (b1) holds; so L^* is self-normalizing in $I^* T_v^* \cong A_7$, and hence T_v induces inner automorphisms on $\bar{L}$ so that case (3) of 10.1.1 holds. Similarly in case (c): if $I^* \cong L_4(2)$, then $L^* \cong L_3(2)/E_8$, and so T_v induces inner automorphisms on $\bar{L}$ and L^* is self-normalizing in I^*; while if $I^* \cong L_5(2)$, then either T_v induces inner automorphisms on $\bar{L}$, or $I^* T_v^* \cong Aut(L_5(2))$, L^* is the T_v-invariant nonsolvable rank-2 parabolic, and L^* is self-normalizing in I^*. Except in this last case, case (3) of 10.1.1 holds.

Set $Y := O^2(C_{L_2}(v))$. In case (3) of 10.1.1, $Y/O_2(Y) \cong \mathbf{Z}_3$. In case (5) of 10.1.1, either $Y/O_2(Y) \cong \mathbf{Z}_3$, or v is diagonally embedded in the two summands with $Y = 1$, and $T_v = T_1$ with $LT_v/O_2(LT_v) \cong Aut(L_3(2))$.

Suppose $Y \neq 1$. By A.3.18, $I = O^{3'}(H)$ so $Y \leq N_I(L)$. As we saw $C_I(V_I) = O_2(I)$, $1 \neq Y^* \leq N_{I^*}(L^*)$ and $Y^* \not\leq L^*$. Thus $L^* < O^2(N_{I^*}(L^*))$, so by the previous two paragraphs, $I^* T_v^* \cong L_5(2)$, $Y^* L^* T_v^* \cong S_3 \times L_3(2)$, and case (3) of 10.1.1 holds. On the other hand if $Y = 1$, then by the previous two paragraphs, case (5) of 10.1.1 holds, and $I^* T_v^* \cong Aut(L_5(2))$. Therefore in any case for Y, $I^* \cong L_5(2)$.

Suppose that $C_V(L_0) \neq 1$. Then case (3) of 10.1.1 holds by 10.1.2.4, so by the previous paragraph, $LYT_v/O_2(LYT_v) \cong L_2(2) \times L_3(2)$, contrary to 10.2.8.4, which says that $LYT_v/O_2(LYT_v) \cong \mathbf{Z}_3 \times L_3(2)$.

Therefore $C_V(L_0) = 1$. By B.4.2 and Theorem B.5.1 V_I is either an irreducible of rank either 5 or 10, the sum of the 5-dimensional module and its dual, or the sum of isomorphic 5-dimensional modules. If $Y = 1$, we saw that $I^* T_v^* \cong Aut(L_5(2))$ and L^* is the nonsolvable T_v^*-invariant rank 2 parabolic. Thus $V_I = V_{I,1} \oplus V_{I,2}$ with $V_{I,1}$ a natural I^*-submodule and $V_{I,2}$ its dual. But we also saw that case (5) of 10.1.1 holds, and in that case we saw that $V_v = V_1 \leq V_I$. However V_1 is the sum of a natural module for $\bar{L}$ and its dual, whereas the parabolic L^* has no such submodule on V_I.

Thus $Y \neq 1$, $I^* T_v^* \cong L_5(2)$, and $L^* Y^* T_v^* \cong S_3 \times L_3(2)$. In case (5) of 10.1.1, $V_1 \leq V_I$ and V_1 is the sum of a natural module for $\bar{L}$ and its dual. However examining the possibilities for V_I listed above, we see that the parabolic $L^* Y^* T_v^*$ has no such submodule.

Therefore case (3) of 10.1.1 holds. Since $C_V(L_0) = 1$, V_1 is the natural module for L. But from the our list of possibilities for V_I, each natural submodule for L is contained in an I-irreducible. Thus as $V_I = \langle V_1^I \rangle$, V_I is an I-irreducible, and hence $\dim(V_I) = 5$ or 10.

Again since $C_V(L_0) = 1$, $T_v = T_1$, so that T normalizes T_v. Let $t \in T - T_v$, $u := v^t$, and $E := \langle u, v \rangle$. Then $\langle u \rangle = C_{V_1}(T_v)$ and $C_{G_v}(E) = C_{G_v}(u)$. Since V_I is an irreducible of dimension 5 or 10, $C_{I^* T_v^*}(u)$ is a maximal parabolic of $I^* T_v^*$, and so from the structure of such parabolics,

$$C_{IT_v}(E) = O^{3'}(C_G(E))T_v \leq I^t T_v,$$

as $I^t = O^{3'}(C_G(u))$ since $I = O^{3'}(H)$. Then $C_{IT_v}(E) = C_{I^tT_v}(E)$, so that t acts on $C_{IT_v}(E)$.

Let P be the rank one parabolic of IT_v over T_v not contained in M, and let P_c and P_f be the rank one parabolics of L centralizing and not centralizing u, respectively. Observe that as $L^t = L_2$, t interchanges Y and P_c. If $m(V_I) = 10$, then $C_{I^*T_v^*}(u)$ is an $L_3(2) \times L_2(2)$ parabolic and $C_{IT_v}(u) = \langle Y, P \rangle P_c$. Therefore as t interchanges Y and P_c, and t acts on $C_{IT_v}(E) = C_{IT_v}(u)$ by the previous paragraph, $P = P^t$. This is impossible, as $\langle Y, P \rangle$ is of type $L_3(2)$, while PP_c is of type $L_2(2) \times L_2(2)$. Therefore $m(V_I) = 5$, and $C_{IT_v}(u) = \langle Y, P, P_c \rangle$ is of type $L_4(2)$; again $P^t = P$, and as P_f acts on $O^2(P)$, so does P_f^t. This is impossible, as P centralizes E, but $P_f P_f^t$ contains a E_9-subgroup D with $C_E(D) = 1$ so $m_3(DO^2(P)) = 3$, contradicting $DO^2(P)$ an SQTK-group. This concludes the treatment of the case $\bar{L} \cong L_3(2)$.

Therefore $\bar{L} \cong L_2(4)$ and case (b1) or (b2) holds. In (b1), $V_1 = V_I \trianglelefteq I$, so $I \leq N_G(V_1) \leq M$ by 10.2.6.2, contrary to $I \not\leq M$. In (b2), $[V_I, L]$ is the A_5-module, so case (2) of 10.1.1 holds with $V_1 = [V_I, L]$. Then $Y := O^{3'}(C_{L_2}(v)) \neq 1$, and $Y \leq I$ as $O^{3'}(H) = I$ by A.3.18. Hence $1 \neq Y^* \leq N_{I^*}(L^*)$ but $Y^* \not\leq L^*$, contradicting $L^* = O^2(N_{I^*}(L^*))$. This contradiction finally completes the proof of 10.2.9. $\qquad\square$

LEMMA 10.2.10. (1) For $g \in G - M$, $V_2 \cap V_2^g = 1$.
(2) If $C_V(L_0) = 1$, then V_i is a TI-set in G.

PROOF. As M permutes $\{V_1, V_2\}$ transitively and $V_1 \cap V_2 = C_V(L_0)$, (1) implies (2).

Suppose $g \in G$ with $1 \neq v \in V_2 \cap V_2^g$. By 10.2.9.1, $C_G(v) \leq M \cap M^g$. Let p be an odd prime divisor of $|\bar{L}|$, and for $X \leq G$ let $\theta(X) := O^{p'}(X^\infty)$. By 10.1.3, $L_0 = \theta(M)$, so $L^g \leq L_0$; and $L_0 \leq L_0^g$ if $v \in C_{V_2}(L_0)$. In the latter case $g \in N_G(L_0) = M$, so we may assume $v \notin C_{V_2}(L_0)$. Thus $L = \theta(C_{L_0}(v))$, so $L^g = L$. Then $g \in M$ by 10.2.6.2, establishing (1). $\qquad\square$

LEMMA 10.2.11. Assume case (3) of 10.1.1 holds with $C_V(L_0) = 1$. Let $1 \neq v_i \in C_{V_i}(T_1)$, set $E := \langle v_1, v_2 \rangle$, and $z := v_1 v_2$. Let $G_z := C_G(z)$, $X := O^2(C_{L_0}(z))$, $K_z := \langle X^{G_z} \rangle$, and $V_z := \langle E^{G_z} \rangle$. Then
(1) $V_z \leq Z(O_2(G_z))$ and $C_{G_z}(V_z) \leq N_M(V_1)$.
(2) If $X < K_z$ then $V_z \in \mathcal{R}_2(G_z)$.
(3) $V \leq O_2(G_z)$.

PROOF. By construction, $z \in Z(T)$, so $G_z \in \mathcal{H}^e$ by 1.1.4.6. As $XT \leq G_z$, $O_2(G_z) \leq O_2(XT)$ by A.1.6; then as $O_2(XT) \leq T_1 \leq C_{G_z}(E)$, $V_z \leq Z(O_2(G_z))$. Further

$$C_{G_z}(V_z) \leq C_{G_z}(v_1) \leq N_M(V_1)$$

by 10.2.10, since by hypothesis $C_V(L_0) = 1$, so (1) holds.

Set $G_z^* := G_z/C_{G_z}(V_z)$ and let R denote the preimage in T of $O_2(G_z^*)$. By a Frattini Argument, $G_z = C_{G_z}(V_z)N_G(R)$. Thus if $R \leq T_1$, then R centralizes E, and hence also $\langle E^{N_{G_z}(R)} \rangle = V_z$, so that $V_z \in \mathcal{R}_2(G_z)$. Thus to prove (2), we may assume $R \not\leq T_1$. In particular $[X, R] \not\leq O_2(X)$, so as T is irreducible on $X/O_2(X)$ and normalizes R, $X = [X, R]$. Thus $X^* = [X^*, R^*] \leq R^*$, so X^* is a 2-group and hence $X = O^2(X) \leq C_{G_z}(V_z)$. By 10.1.3, $X = O^{3'}(G_z \cap M)$, so by (1), $X = O^{3'}(C_{G_z}(V_z)) \trianglelefteq G_z$ and hence $X = K_z$, establishing (2).

Assume (3) fails. If V centralizes V_z, then as $C_{G_z}(V_z) \le M$ by (1), $V \le O_2(C_{G_z}(V_z)) \le O_2(G_z)$, contrary to assumption. Hence as XT is irreducible on V/E, $E = C_V(V_z)$. If $X \trianglelefteq G_z$, then as X centralizes E, it centralizes V_z; then $V = [V, X]E$ centralizes V_z, a contradiction. Thus $X < K_z$, and hence $K_z \not\le M$ so $V_z \in \mathcal{R}_2(G_z)$ by (2).

As $E = C_V(V_z)$, V_1^* is a 4-group. By (1), $V_z \le N_M(V_1)$, so $[V_z, V_1] \le V_z \cap V_1 = \langle v_1 \rangle$. That is V_1^* is a 4-group inducing transvections on V_z with center v_1. Further $K_z^* T^*$ is described in case (2) or (3) of 10.2.4. Appealing to G.3.1, the only group $K_z^* T^*$ listed there containing a 4-group of $\mathbf{F}_2$-transvections with a fixed center in some representation is $L_3(2) \text{ wr } \mathbf{Z}_2$ with $[V_z, K_z] = V_{z,1} \oplus V_{z,2}$, where $V_{z,i} := [V_z, K_{z,i}]$ is a natural module. However in that case, $V_i^* \le K_{z,i}^*$ with $v_i = [V_z, V_i^*] \le V_{z,i}$, so $z = v_1 v_2 \in [V_z, K_z]$, which is impossible as $z \in Z(G_z)$ but $C_{[V_z, K_z]}(K_z^*) = 1$. This contradiction completes the proof. $\qquad\square$

We can now prove our major weak closure result, which establishes an effective lower bound on the parameter $w(G, V)$.

PROPOSITION 10.2.12. *One of the following holds:*

(1) $w(G, V) > 2$.
(2) $w(G, V) = 2$, *and case (3) of 10.1.1 holds.*
(3) $w(G, V) = 2$, *and case (1) of 10.1.1 holds with $n = 2$.*

PROOF. In case (3) of 10.1.1, and in case (1) when $n = 2$, set $j := 1$. Otherwise set $j := 2$. We must prove $w(G, V) > j$, so we may assume $A := V^g \cap M$ with $k := m(V^g/A) \le j$ and $[V, V^g] \ne 1$, and it remains to derive a contradiction.

Let $m := m(\tilde{V}_1)$ and $a := a(Aut_M(V_1), V_1)$. Observe $m > j + 1$. Recall $a \le m_2(Aut_M(V_1))$ and in case (2) of 10.1.1, $a = 1$. Thus $k < m - a$ unless case (3) of 10.1.1 holds and $k = 1$.

For $i = 1, 2$, set $A_i := V_i^g \cap A$ and $B_i := N_{A_i}(V_1)$. Suppose $A_1 A_2$ centralizes V_1. Then by 10.2.9.1, $V_1 \le N_{M^g}(V_i^g)$, so $C_{V_1}(V_i^g) \ne 1$ since $m(V_i) < m_2(Aut_M(V_1))$ in each case. Then $A = V^g$ by another application of 10.2.9.1. But then $V^g = A_1 A_2 \le C_M(V_1) = C_M(V)$, contrary to our choice of V^g. Thus we may assume A_i does not centralize V_1 for some choice of $i := 1$ or 2.

Next $m(V_i^g/A_i) \le k$ with $m(A_i/B_i) \le 1$, so $m(V_i^g/B_i) \le k + 1 < m = m(V_i^g/C_{V_i^g}(L_0^g))$ by paragraph two. Thus $B_i \not\le C_{V^g}(L_0^g)$, so there exists $b \in B_i - C_{V^g}(L_0^g)$. For each such b and each $r = 1, 2$, we may apply 10.2.9.1 to get

$$C_{V_r}(b) \le N_{V_r}(V_i^g) =: U_r,$$

so $V_0 := [A_i, C_{V_1}(b)] \le V_i^g \cap V$ and $[B_i, C_{V_1}(b)] \le V_i^g \cap V_1 = 1$ by 10.2.10.1. Thus if $V_0 \ne 1$ then $A_i > B_i$ and $V \le C_G(V_0) \le M^g$ by 10.2.9.1. Thus $[A_i, V] \le V^g \cap V \le C_V(b)$, and as $A_i > B_i$, for any $a \in A_i - B_i$, $V = [a, V]V_2$, so b centralizes V/V_2. Thus $b \in C_T(V/V_2) = C_T(V_1)$, so $V_1 = C_{V_1}(b)$ and $V = V_0 V_2$. Then by 10.2.9.1,

$$L = \langle C_L(v_0) : v_0 \in V_0^\# \rangle \le M^g,$$

so $L_0 = \langle L^{A_i} \rangle \le M^g$ and hence $L_0 = L_0^g$ by 10.1.3, contradicting $g \notin M$. Therefore $V_0 = 1$, so

$$C_{V_1}(A_i) = C_{V_1}(b). \tag{$*$}$$

As $C_{V_1}(b) \not\le C_{V_1}(L)$ from the structure of the modules in 10.1.1, A_i acts on V_1 by $(*)$, so $A_i = B_i$. Then as A_i does not centralize V_1, $(*)$ says

$$Aut_{A_i}(V_1) \in \mathcal{A}_{m-k}(Aut_M(V_1), V_1).$$

Thus $k \geq m - a$, so by paragraph two, case (3) of 10.1.1 holds with $w(G, V) = k = 1$. Hence $V \not\leq M^g$ by E.3.25.

Assume first that $C_V(L_0) \neq 1$. Then $m(V_1) = 4$ by I.1.6, so $m(A_i) \geq 3$ as $k = 1$, and hence $C_{A_i}(V_1) \neq 1$ as $m_2(Aut_M(V_1)) = 2$. But then $V_1 \leq M^g$ by 10.2.9.1, and similarly $V_2 \leq M^g$, contradicting $V \not\leq M^g$.

Therefore $C_V(L_0) = 1$, so V_1 is a TI-set in G by 10.2.10.2. As $Aut_{A_i}(V_1) \in \mathcal{A}_2(Aut_M(V_1), V_1)$, $Aut_{A_i}(V_1)$ is a 4-group of transvections with a fixed axis U_1, so $A_i \cong E_4 \cong U_1$.

Set $I := \langle V_i^g, V_1 \rangle$. We've shown that

$$A_i = B_i = N_{V_i^g}(V_1) \neq 1 \neq U_1 = N_{V_1}(V_i^g).$$

By I.6.2.2a, $O_2(I) = A_i \times U_1$ is of rank 4 with $C_I(V_1) = U_1$, and as $|V_1 : U_1| = 2$, $I/O_2(I)$ is dihedral of order $2d$, with d odd. As $D_{2d} \leq GL_4(2)$, $d = 3$ or 5. Now A_i is of index 2 in V_i^g, so as $k = 1$, $A = A_1 A_2 \langle c \rangle$ with $c = c_1 c_2$, where $c_i \in V_i^g - A_i$. Further as $I/O_2(I) \cong D_{2d}$, there is an involution in I interchanging V_1 and V_i^g, and $U := V \cap M^g = U_1 U_2 \langle w \rangle$, where $w = w_1 w_2$ with $w_r \in V_r - U_r$. If w acts on V_i^g then $1 \neq [A_i, w] \leq V_i^g \cap V$, so that $V \leq C_G([w, A_i]) \leq M^g$ by 10.2.9.1, contrary to an earlier reduction. Thus w interchanges L_1^g and L_2^g, so by symmetry, $L^c = L_2$. Now $[c, U_1] \leq V^g$ is diagonally embedded in V, so we may take $z \in Z^\#$ to lie in $[c, U_1] \leq V^g$. Then $V, V^g \leq G_z$, so $I \leq G_z$. Hence as $V_r \not\leq O_2(I)$, $V \not\leq O_2(G_z)$, contradicting 10.2.11.3. This completes the proof. $\square$

COROLLARY 10.2.13. *Case (3) of 10.1.1 holds with* $w(G, V) = 2 = n(H)$ *for each* $H \in \mathcal{H}_*(T, M)$.

PROOF. Take $H \in \mathcal{H}_*(T, M)$. By 10.1.2.3 and 10.2.3.2, Hypothesis E.3.36 holds. By 10.2.9.2, $r(G, V) \geq m(V_1)$ and it is easy to check in each case of 10.1.1 that $n'(\bar{M}) < m(V_1)$. Thus the hypotheses of lemma E.3.39 are satisfied. By 10.2.3.1, $n(H) \leq 2$, with $n(H) = 1$ in case (1) of 10.1.1. Thus by E.3.39.1, $w(G, V) \leq n(H) \leq 2$, so 10.2.12 completes the proof of the corollary. $\square$

10.3. The final contradiction

LEMMA 10.3.1. *(1)* $C_V(L_0) = 1$.
(2) V_i *is a TI-set in* G.

PROOF. By 10.2.10.2, (1) implies (2). Thus we may assume $C_V(L_0) \neq 1$, and it remains to derive a contradiction. Let $H \in \mathcal{H}_*(T, M)$ and set $U := \langle Z^H \rangle$ and $H^* := H/C_H(U)$. By 10.2.13, case (3) of 10.1.1 holds, so by 10.2.8.1, $C_H(U) \leq C_G(Z) \leq M = !\mathcal{M}(L_0 T)$, and hence $H^* \neq 1$. By 10.2.13, $n(H) = 2$, so by 10.2.3.3, $H^* \cong S_5$ wr $\mathbf{Z}_2$ and $O^2(H \cap M) T_0$ is a maximal parabolic of L_0. In particular by 10.2.8.1,

$$[O^2(H \cap M), Z] \leq [L_0, Z] = 1.$$

Then as 3-elements are fixed-point-free on natural modules for $L_2(4)$, any $I \in Irr_+(H, U)$ satisfies either

(a) $I = I_1 \oplus I_2$, where $I_i := [I, K_i]$ is the A_5-module for $K_i \in \mathcal{C}(H)$, or

(b) $I = I_1 \otimes I_2$ is the tensor product of A_5-modules I_i for K_i.

In either case we compute directly that $a(H^*, I) = 1$. But by 10.2.9, $r(G, V) \geq m(V_1)$ and $m(V_1) = 4$ by I.1.6, so $s(G, V) = m(\bar{M}, V) = 2$ using B.4.8.2. Set $W_0 := W_0(T, V)$. By 10.2.13, $w(G, V) > 0$, so $N_G(W_0) \leq M$ by E.3.16. If $W_0^* = 1$, then

$W_0 \leq O_2(H)$ by B.6.8.3d, so $W_0 = W_0(O_2(H), V)$ and then $H \leq N_G(W_0) \leq M$, contradicting $H \not\leq M$. Thus $W^* \neq 1$; since $s(G,V) = 2$, we must have $a(H^*, I) \geq 2$ by E.3.18, contradicting $a(H^*, I) = 1$. $\square$

LEMMA 10.3.2. $C_G(z) \not\leq M$ for $z \in Z^\# \cap V$.

PROOF. Assume that $C_G(z) \leq M$. We first prove that V is a TI-subgroup of G: For as $C_V(L_0) = 1$ by 10.3.1.1, each diagonal involution in V is conjugate in L_0 to z, and hence has centralizer contained in M by hypothesis. By 10.2.9.1, centralizers of nondiagonal involutions are contained in M. Thus these involutions are not 2-central in G, so they are not fused in G to diagonal involutions, and hence M controls fusion of involutions in V. Therefore V is a TI-set in G by I.6.1.1.

As V is a TI-subgroup of G, $r(G,V) = m(V) = 6$. Let A be a w-offender on V. By 10.2.13, $w(G,V) = 2$, so as $m_2(Aut_G(V)) = 4$, $m(\bar{A}) = 4$ by E.3.28.2. But as V is a TI-subgroup of G, I.6.2.2a says that $C_V(a) = V \cap M^g$ for each $a \in A^\#$. This is impossible as no rank-4 subgroup of $\bar{M}$ satisfies $C_V(\bar{a}) = C_V(\bar{A})$ for each $\bar{a} \in \bar{A}^\#$. This contradiction completes the proof. $\square$

By 10.2.13 and 10.3.1.1, the hypotheses of 10.2.11 hold. So for the remainder of the section, we adopt the notation of that lemma; in particular, we study the group $K_z = \langle X^{G_z} \rangle$.

LEMMA 10.3.3. $K_z T/O_2(K_z T) \cong S_5$ wr $\mathbf{Z}_2$.

PROOF. We first observe that if $Y \in \mathcal{H}(T)$ is generated by $N_Y(T)$ and a set Δ of minimal parabolics D such that $n(D) = 1$ for each $D \in \Delta$, then $Y \leq M$ by Theorem 3.3.1 and 10.2.13. In particular each solvable member H of $\mathcal{H}(T)$ is contained in M by E.1.13 and B.6.5, since $H = O^{2'}(H)N_H(T)$ by a Frattini Argument.

Let $J := G_z^\infty$. By a Frattini Argument, $G_z = JN_{G_z}(T \cap J)$, and as G_z/J and $N_J(T \cap J)$ are solvable, $N_{G_z}(T \cap J)$ is a solvable member of $\mathcal{H}(T)$. Therefore $N_{G_z}(T \cap J) \leq M$ by the previous paragraph, so $J \not\leq M$ by 10.3.2. Hence by 1.2.1.1 there is $I \in \mathcal{C}(G_z)$ with $I \not\leq M$.

Suppose $I/O_2(I)$ is a Bender group. Then a Borel subgroup of $I_0 := \langle I^T \rangle$ lies in M by the first paragraph, so $I_0 T \in \mathcal{H}_*(T, M)$. Hence by 10.2.13, $n(I) = 2$. Then by 10.2.3.3, $I_0 T/O_2(I_0 T) \cong S_5$ wr $\mathbf{Z}_2$ and $X \leq I_0$, so $I_0 = K_z$, and the lemma holds.

Therefore we may assume $I/O_2(I)$ is not a Bender group. Suppose next that $X = K_z$. As G_z is an SQTK-group, $m_3(IX) \leq 2$, so I is a $3'$-group. Thus $I/O_2(I)$ is a Suzuki group and hence a Bender group, contrary to our assumption.

Thus $X < K_z$, so by 10.2.4, $K_z = \langle I^T \rangle$, for $I \in \mathcal{C}(G_z)$ with $I \not\leq M$, and as $I/O_2(I)$ is not a Bender group, $I/O_2(I) \cong L_2(p)$ for an odd prime $p > 5$, $L_4(2)$, or $L_5(2)$. But then $K_z T$ is generated by $N_{K_z T}(T)$ and minimal parabolics D with $n(D) = 1$, contrary to an earlier remark. $\square$

We are now in a position to obtain our final contradiction.

By 10.3.3, $K_z = \langle K^T \rangle$ with $K \in \mathcal{L}(G,T)$ and $K/O_2(K) \cong A_5$. In particular $X < K_z$, so by 10.2.11.2, $V_z \in \mathcal{R}_2(G_z)$. Thus $K \in \mathcal{L}_f(G,T)$.

Let $M_+ \in \mathcal{M}(K_z T)$ and set $J_z := \langle X^{M_+} \rangle$. As $X < K_z \leq J_z$, by 10.2.4, $J_z = \langle I_z^T \rangle$ for $I_z \in \mathcal{C}(M_+)$. Furthermore arguing as in the proof of 10.3.3, J_z is not generated by minimal parabolics D with $n(D) = 1$, so from 10.2.4, $I_z/O_2(I_z) \cong$

$L_2(2^m)$ with $m \geq 2$. However the embedding $K < I_z$ does not occur in the list of A.3.14, so we conclude that $K_z = J_z$. Therefore $K \in \mathcal{L}_f^*(G, T)$ with $M_+ = N_G(K_z)$. Thus the hypotheses of Theorem 10.0.1 are satisfied with K in the role of "L". As $K/O_2(K)$ is A_5 rather than $L_3(2)$, 10.2.13 applied to K in the role of "L" supplies a contradiction. This contradiction completes the proof of Theorem 10.0.1.

CHAPTER 11

Elimination of $L_3(2^n)$, $Sp_4(2^n)$, and $G_2(2^n)$ for n > 1

In this chapter, we complete the elimination of the groups possessing a pair L, V arising in the Fundamental Setup (3.2.1) such that $L/O_2(L)$ is of Lie type of Lie rank 2 over a field of order 2^n, $n > 1$.

Choose V so that L, V are in the FSU and $L/O_2(L)$ is of Lie type of Lie rank 2 over a field of order $q := 2^n$, $n > 1$. By Theorem 7.0.1, V is an FF-module. The weak closure parameters of FF-modules make it difficult to do weak closure without first doing some extra work. Furthermore corresponding local configurations do actually occur in suitable maximal parabolics in non-quasithin shadows given by certain groups G of Lie type and Lie rank 3: namely for $\bar{L} \cong SL_3(q)$, in $G \cong L_4(q)$, $Sp_6(q)$, $\Omega_8^+(q).2$, and $\Omega_8^-(q)$; and for $\bar{L} \cong Sp_4(q)$, in $G \cong Sp_6(q)$.

We restrict attention at this point to $q = 2^n$ for $n > 1$, largely because for such q, $\bar{L}$ has a Cartan subgroup X of p-rank 2 for primes p dividing $q - 1$. Using our quasithin hypothesis, G contains no member of $\mathcal{H}(X)$ of larger p-rank, whereas the the groups of Lie type in the previous paragraph do contain such subgroups. This leads to a contradiction, which does not arise in the shadows of groups over the small field $\mathbf{F}_2$; the more complicated treatment needed for the subcase of L of rank 2 over $\mathbf{F}_2$ is postponed to part 5.

Thus in this chapter we will prove:

THEOREM 11.0.1. *Assume G is a simple QTKE-group, $T \in Syl_2(G)$, and $L \in \mathcal{L}_f^*(G, T)$. Then $L/O_2(L)$ is not isomorphic to $(S)L_3(2^n)$, $Sp_4(2^n)$, or $G_2(2^n)$ with $n > 1$.*

Throughout this chapter we assume L is a counterexample to Theorem 11.0.1

By 1.2.1.3, L is T-invariant, so by 3.2.3, $M := N_G(L) \in \mathcal{M}(T)$, $M = !\mathcal{M}(LT)$, and we can choose V so that L and V are in the FSU. In particular let $V_M := \langle V^M \rangle$, $\tilde{V}_M := V_M/C_{V_M}(L)$, $M_V := N_M(V)$, and $\bar{M}_V := M_V/C_M(V)$. Let $T_L := T \cap LO_2(LT)$ and let X be a Hall $2'$-subgroup of $N_L(T_L)$; since $n > 1$, $m_p(X) = 2$ for each prime divisior p of $|X|$ (see 11.0.4). As mentioned earlier, the Cartan subgroup X will provide a main focus for our analysis. Set $Z := \Omega_1(Z(T))$ and abbreviate $q := 2^n$.

Lemmas 11.0.2, 11.0.3, and 11.0.4 collect observations from various earlier results, and provide a starting point for the analysis.

LEMMA 11.0.2. *(1) $V \in Irr_+(L, R_2(LT))$ and V is T-invariant. Moreover T is trivial on the Dynkin diagram of $L/O_2(L)$.*

(2) $V/C_V(L)$ is the natural module for $L/O_2(L) \cong \bar{L} \cong SL_3(q)$, $Sp_4(q)$, or $G_2(q)$.

PROOF. By Theorem 7.0.1, V is an FF-module for $Aut_{GL(V)}(\bar{L})$. By construction in the FSU, $V = \langle V_o^T \rangle$ for some $V_o \in Irr_+(L, R_2(LT), T)$, so V is T-invariant. If $V > V_o$, then V is described in case (3) of Theorem 3.2.5. However in that case by Theorem B.5.1, V is not an FF-module for $Aut_{GL(V)}(\bar{L})$. Therefore $V = V_o$, so as V is an FF-module, (2) follows since one of cases (2), (3), or (4) of 3.2.8 must hold. Then as V is T-invariant, T is trivial on the Dynkin diagram of $L/O_2(L)$, completing the proof of (1). $\qquad\square$

LEMMA 11.0.3. *(1)* $V_M \in \mathcal{R}_2(M)$.

(2) $\tilde{V}_M$ *is a homogeneous* $\mathbf{F}_2 L$*-module.*

(3) *Either* $C_V(L) = C_{V_M}(L) = 1$; *or* $\bar{L} \cong Sp_4(q)$ *or* $G_2(q)$, $V = V_M$, $m(C_V(L)) \le n$, *and* $L = [L, J(T)]$.

(4) V *is a TI-set under* M.

(5) *If* $\bar{L}$ *is* $Sp_4(q)$ *or* $G_2(q)$ *then* $H \cap M \le N_M(V)$ *for each* $H \in \mathcal{H}_*(T, M)$.

PROOF. Part (1) is 3.2.2.2; part (2) follows from 3.2.2.3; and as $n > 1$, part (4) is a consequence of 3.2.7. By 3.2.2.4, $C_{V_M}(L) = \langle C_V(L)^M \rangle$. If $L \cong SL_3(q)$, then as $n > 1$ we have $H^1(L, V/C_V(L)) = 0$ by I.1.6, so $C_V(L) = 1$. Hence $C_{V_M}(L) = 1$, so that (3) holds in this case. If $C_V(L) \ne 1$, then $L = [L, J(T)]$ by 3.2.2.6, and $V = V_M$ by Theorem 3.2.5, since now neither cases (2) nor (3) of that result hold. Further by I.1.6, $m(C_V(L)) \le m(H^1(L, V/C_V(L)) = n$, completing the proof of (3).

Finally assume the hypotheses of (5), and suppose $H \in \mathcal{H}_*(T, M)$ with $H \cap M \not\le N_M(V)$. In particular $V < V_M$ as $V_M \trianglelefteq M$. As V is a TI-set under M by (4), while $Z \cap V \ne 1$, $[Z, H \cap M] \ne 1$ and hence $[Z, H] \ne 1$. Thus $J(T) \not\le C_T(V)$ by 3.1.8.3, and so $L = [L, J(T)]$. So setting $M^* := M/C_M(V_M)$, by B.2.7 there is $A^* \in \mathcal{P}(M^*, V_M)$ with $L^* = [L^*, A^*]$. Then by Theorem B.5.6, $F^*(J(M^*, V_M)) = L^*$, and then Theorem B.5.1 supplies a contradiction to $V < V_M$. $\qquad\square$

LEMMA 11.0.4. $L = O^{p'}(M)$ *for each prime* p *such that*

(1) p *divides* $q^2 - 1$, *if* $\bar{L}$ *is* $Sp_4(q)$ *or* $G_2(q)$; *or*

(2) p *divides* $q - 1$ *and* $p > 3$, *if* $\bar{L}$ *is* $SL_3(q)$.

Moreover if $\bar{L} \cong SL_3(q)$ *with* n *even, then* L *contains each element of* M *of order* 3.

PROOF. The primes p are chosen so that $m_p(L) = 2$; hence the lemma follows from A.3.18, using A.3.19 for the final assertion. $\qquad\square$

11.1. The subgroups $\mathbf{N_G(V_i)}$ for $\mathbf{T}$-invariant subspaces $\mathbf{V_i}$ of $\mathbf{V}$

By 11.0.2.2, $\tilde{V}$ is the natural $\mathbf{F}_q \bar{L}$-module; thus the two classes of maximal parabolics of $\bar{L}$ preserve $\mathbf{F}_q$-subspaces of dimension 1 and 2. We will use our structure theory of QTKE-groups to restrict the normalizers of these subspaces. The results in this section roughly have the effect of forcing these normalizers to resemble those in the shadows mentioned earlier.

For $i = 1, 2, 3$, let $\mathcal{V}_i$ denote the set of $U \le V$ such that $C_V(L) \le U$ and $\tilde{U}$ is an i-dimensional $\mathbf{F}_q T'$-subspace of $\tilde{V}$ for some $T' \in Syl_2(M)$. Further for $i = 1, 2$, set $L(U) := N_L(U)^\infty$.

Denote by V_i the unique T-invariant member of $\mathcal{V}_i$. For $i = 1, 2$, let $L_i := L(V_i)$ and $R_i := O_2(L_i T)$. Then $L_i/O_2(L_i) \cong L_2(2^n)$. By construction $T \le N_G(V_i)$, so that $N_G(V_i) \in \mathcal{H}^e$ by 1.1.4.6. Notice when $\bar{L} \cong SL_3(q)$ that $V_3 = V$, while in the other cases, from the action of $N_{GL(\tilde{V})}(\bar{L})$ on $\tilde{V}$, $N_M(V_3) = N_M(V_1)$.

We begin by considering the embedding of L_i in a $\mathcal{C}$-component K_i of $N_G(V_i)$.[1]

LEMMA 11.1.1. *Assume either*

(i) *$i = 1$, $1 \neq V_0 \leq V_1$, $H := N_G(V_0)$, and $T_0 := N_T(V_0) \in Syl_2(H)$, or*
(ii) *$i = 2$ and $H := N_G(V_2)$.*

Then $L_i \leq K \in \mathcal{C}(H)$ with $K \trianglelefteq H$, and one of the following holds:

(1) *$L_i = K$.*
(2) *$K/O_2(K) \cong (S)L_3(q)$, $Sp_4(q)$, $G_2(q)$, $^2F_4(q)$, $^3D_4(q)$, or $^3D_4(q^{1/3})$.*
(3) *$n = 2$ and $K/O_2(K)$ is isomorphic to A_7, $\hat{A}_7$, $L_2(p)$ for a prime p with $p \equiv \pm 1 \mod 5$ and $p \equiv \pm 3 \mod 8$, $L_2(25)$, $(S)L_3^\epsilon(5)$, M_{22}, $\hat{M}_{22}$, M_{23}, J_1, J_2, J_4, HS, Ru, $SL_2(5)/P_0$ for a suitable nilpotent group P_0 of odd order, or $SL_2(p)/E_{p^2}$ for a prime p satisfying the congruences above.*

PROOF. If $i = 1$, V_0 and T_0 are defined in (i); if $i = 2$, set $V_0 := V_2$ and $T_0 := T$. Thus in either case $H = N_G(V_0)$, and $T_0 \in Syl_2(H)$ acts on L_i, so $L_i \in \mathcal{L}(H, T_0)$. Thus by 1.2.4, L_i is contained in a unique $K \in \mathcal{C}(H)$, and the embedding $L_i \leq K$ appears on the list of A.3.12. As T_0 acts on L_i, T_0 also acts on K, so $K \trianglelefteq H$ by 1.2.1.3. The possibilities for K are determined by restricting the list of A.3.12 to $L_i/O_2(L_i) \cong L_2(q)$. The groups in (2) are the groups of Lie type, characteristic 2, and Lie rank 2 in Theorem C (A.2.3). When $n = 2$, we use the list in A.3.14, and get the further examples in (3). $\square$

We next determine the possible embeddings of L_i in $N_G(V_i)$ for $i = 1$ and 2. Recall that X is a Hall $2'$-subgroup of $N_L(T_L)$, so $X \leq N_L(V_i)$.

PROPOSITION 11.1.2. *For $i = 1, 2$, $L_i \leq K_i \in \mathcal{C}(N_G(V_i))$ with $K_i \trianglelefteq N_G(V_i)$ and $K_i \in \mathcal{H}^e$. Furthermore for $K := K_i$ either $L_i = K$, or $i = 1$, $q = 4$, and one of the following holds:*

(1) *$K/O_{2,2'}(K) \cong SL_2(p)$ where $p = 5$, or $p \geq 11$ is prime.*
(2) *$K/O_2(K) \cong L_2(p)$ for a suitable prime $p \geq 11$, and $L/O_2(L)$ is not $SL_3(4)$.*
(3) *$KX/O_2(KX) \cong PGL_3(4)$. Further if K_0 denotes the member of $\mathcal{L}(G,T) \cap K$ distinct from K and L_1, and $I := \langle K_0, L_2 \rangle$, then $I \in \mathcal{L}_f^*(G,T)$, and interchanging the roles of L and I if necessary, $L/O_2(L) \cong G_2(4)$ and $I/O_2(I) \cong Sp_4(4)$.*

PROOF. By 11.1.1, $L_i \leq K_i \trianglelefteq N_G(V_i)$. Recall $N_G(V_i) \in \mathcal{H}^e$, so $K_i \in \mathcal{H}^e$ by 1.1.3.1. So we may assume $L_i < K_i =: K$, and K appears in case (2) or (3) of 11.1.1, but not among the conclusions of 11.1.2. In particular $K \not\leq M$.

Let $G_i := C_G(V_i)$; observe that $X \leq N_G(V_i)$, and set $(G_iX)^* := G_iX/O_2(K)$. As $N_G(V_i) \in \mathcal{H}^e$, $G_i \in \mathcal{H}^e$ by 1.1.3.1. Set $X_i := C_X(L_i/O_2(L_i))$. By 11.0.2.2, $|X_i| = q - 1$.

Suppose first that $i = 1$. By inspection of the possiblities for K, namely in (2) and (3) of 11.1.1 but not in 11.1.2, $K/O_2(K)$ is quasisimple and either

(i) $m_p(K) = 2$ for some prime p dividing $q - 1$, or
(ii) $q = 4$ and $K/O_2(K) \cong L_2(p)$ for a prime $p \geq 11$, $L_2(25)$, $L_3(5)$, or J_1.

---

[1]Notice that in the shadows we expect $L_i = K_i \trianglelefteq N_G(V_i)$.

Next $L_1 = L_1^\infty \le C_G(V_1) = G_1$, so $K = [K, L_1] \le G_1$. As X_1 is faithful on V_1, the product KX_1 is semidirect. Thus for each prime p dividing $q - 1$, K does not contain all elements of order p centralizing $L_1/O_2(L_1)$, so applying A.3.18 we conclude that in case (i):

(*) $q = 4$, $K^* \cong L_3(4)$, and $K^*X_1^* \cong PGL_3(4)$ with X_1 inducing outer automorphisms on K^*.

We will return to case (*), after treating case (ii). There $q = 4$ so that $|X_1^*| = 3$. If $K^* \cong J_1$ then $K = \langle L_1, N_K(T) \rangle \le M$ using Theorem 3.3.1, contradicting $K \not\le M$. Thus $K/O_2(K)$ is $L_2(p)$ or $L_2(25)$ or $L_3(5)$, so $Out(K^*)$ is a $3'$-group, and hence X_1^* centralizes K^*. If $K/O_2(K) \cong L_2(25)$ or $L_3(5)$, then some $t \in T \cap K$ induces an outer automorphism on $L_1/O_2(L_1)$, so t induces a field automorphism on $L/O_2(L)$, impossible as $[t, X_1] \le O_2(X_1T)$. Thus $K/O_2(K) \cong L_2(p)$, so conclusion (2) will hold in this case, once we show $L/O_2(L)$ is not $SL_3(4)$. But in that case, $X_1O_2(L) = O_{2,Z}(L) \trianglelefteq M$, so $Y := O^2(X_1T) \trianglelefteq LT$, and hence $N_G(Y) \le M = !\mathcal{M}(LT)$. Then as $[K, X_1] \le O_2(K) \le T$, K normalizes $O^2(YO_2(K)) = Y$ so that $K \le N_G(Y) \le M$, contrary to $K \not\le M$.

Thus to complete the treatment of the case $i = 1$, we assume (*) holds; as this is the first requirement of conclusion (3), it remains to establish the remaining assertions of (3). This argument will require several pages.

Our strategy will be to use K and L to construct a third group I, and obtain a triple $L = \langle L_1, L_2 \rangle$, $K = \langle L_1, K_0 \rangle$, and $I := \langle L_2, K_0 \rangle$—where K_0 is essentially the maximal parabolic of K over $T \cap K$ other than $L_1(T \cap K)$. We will be able to exploit some symmetry in this triangle of subgroups.

Let K_0 denote the member of $\mathcal{L}(G, T) \cap K$ distinct from L_1 and K—that is, $K_0/O_2(K)$ is normal in the maximal parabolic of $K/O_2(K)$ stabilized by XT which is distinct from $N_K(L_1)$. In particular, $K_0/O_2(K_0) \cong L_2(4)$, and $K_0 \in \mathcal{H}^e$. Set $S := O_2(XT)$, $H_1 := K_0SX$, $H_2 := L_2SX$, and $H_{1,2} = SX$.

Assume that there is no nontrivial normal subgroup of T normal in $H := \langle H_1, H_2 \rangle$. Then Hypothesis F.1.1 is satisfied with K_0, L_2, S in the roles of "L_1, L_2, S", so by F.1.9, $\alpha := (H_1, H_{1,2}, H_2)$ is a weak BN-pair of rank 2. As $S \trianglelefteq H_{1,2}$, α appears on the list of F.1.12. Indeed α must be one of the (untwisted) cases where the nonabelian chief factor of H_1 and H_2 is isomorphic to $L_2(4)$. As L_2 has at least two noncentral 2-chief factors, α is not the $PGL_3(4)$, $SL_3(4)$, or $Sp_4(4)$ amalgam, so α is the $G_2(4)$-amalgam. By construction V_1 is $H_{1,2}$-invariant, and hence plays the role of the long root group of $G_2(4)$ normal in a maximal parabolic. The parabolic H_1 stabilizing this long root group is irreducible on $O_2(H_1)/V_1$, and $H_1^\infty/O_2(H_1)$ has two A_5-modules on this section; but in our construction K_0 has a natural $L_2(4)$-chief factor on $O_2(K_0)$.

This contradiction shows that $O_2(H) \ne 1$, and hence $HT \in \mathcal{H}(T) \subseteq \mathcal{H}^e$ using 1.1.4.6. Now by 1.2.4, $L_2 \le I \in \mathcal{C}(HT)$, and $I \trianglelefteq HT$ by 1.2.1.3 as L_2 is T-invariant, so also $I \in \mathcal{H}^e$. Similarly $K_0 \le I_0 \in \mathcal{C}(HT)$. We conclude from (*) that

$$X_K := X \cap K = X \cap L_1 = X \cap K_0.$$

But as $[V_1, L_1] = 1$ while $V_1 = [V_1, X_2]$ from the action of L on V, $X \cap L_1 \ne X_2$; so X_K does not centralize $L_2/O_2(L_2)$, and hence $[I, X_K] \not\le O_2(I)$. Thus $[I, I_0] \not\le O_2(I)$, so $K_0 \le I_0 = I$. Therefore

$$X = (X \cap L_1)(X \cap L_2) = X_K(X \cap L_2) \le I,$$

so $m_3(I) = 2$. Also $\mathcal{L}(G,T) \cap I$ contains two members L_2, K_0 with $L_2 X/O_2(L_2) \cong K_0 X/O_2(K_0) \cong GL_2(4)$; inspecting the list of A.3.14, we conclude $I/O_2(I) \cong SL_3(4), Sp_4(4)$, or $G_2(4)$ and $I = \langle K_0, L_2 \rangle$. Furthermore $O^2(H) = \langle K_0, L_2, X \rangle \leq I$, so $I = O^2(HT)$ and $HT = IT$.

Suppose first that $\bar{L} \cong SL_3(4)$; we must eliminate this case as part of our proof that (3) holds. This subcase will require approximately a page of argument.

First

$$X_1 = X_2,$$

and X_1 is Sylow in $O_{2,Z}(L)$. Thus $I/O_2(I)$ is not $SL_3(4)$ or else $X_1 = X_2 = C_X(L_2/O_2(L_2)) = C_X(K_0/O_2(K_0))$, which is not the case in $K^* X^* \cong PGL_3(4)$. In the remaining cases the subgroup $X_1 = X_2$ of the Cartan group X of I is inverted by a 2-element projecting on the center of the Weyl group (D_8 or D_{12}) of $I/O_2(I)$, so this element is not in $L_2 X$. Thus $N_I(X_2) \not\leq L_2 X = I \cap M$. Therefore as $X_1 = X_2$, $G_{X_1} := N_G(X_1) \not\leq M$.

Let $L_{X_1} := N_L(X_1)^\infty$, so that $L = O_2(L_1)L_{X_1}$ and $X_1 \leq L_{X_1}$. We now show that it suffices to prove $Q_{X_1} := [O_2(L_{X_1}), L_{X_1}] \neq 1$: For then $O_2(L_{X_1}) \neq 1$, so that Theorem 4.2.13 says $M = !\mathcal{M}(L_{X_1})$. Then as $G_{X_1} \not\leq M$ by the previous paragraph, $O_2(G_{X_1}) = 1$. Next as $Q_{X_1} \neq 1$, L_{X_1} has a 2-chief section of rank at least 6, so $m_2(G_{X_1}) \geq m(Q_{X_1}) > 3$. Therefore $m_p(O_p(G_{X_1})) \leq 2$ for each odd p by A.1.28, so $L_{X_1} \leq C_{X_1} := C_{G_1}(O(F(G_{X_1})))$. As $O_2(G_{X_1}) = 1$, $F^*(C_{X_1}) = EZ(C_{X_1})$, where $E := E(G_{X_1})$). Further using (1) of Theorem A (A.2.1), $|J^{G_{X_1}}| \leq 3$ for each component J of G_{X_1}, so $G_{X_1}^\infty$ normalizes J. Hence $E = C_{X_1}^\infty$ as J satisfies the Schreier Conjecture. Then $L_{X_1} \leq E$, so that Q_{X_1} projects nontrivially on some component K_{X_1} of G_{X_1}. As G is quasithin, $m_{2,3}(E) = 2$, so K_{X_1} is the unique component not centralized by L_{X_1}, and hence $L_{X_1} \leq K_{X_1}$, so $X_1 \leq Z(K_{X_1})$. However $K_{X_1}/Z(K_{X_1})$ appears in Theorem B (A.2.2), and inspecting such groups for a 2-local containing a subgroup $\hat{L}$ with $\hat{L}/O_2(\hat{L}) \cong L_3(4)$ and $[O_2(\hat{L}), \hat{L}] \neq 1$, we conclude $K_{X_1}/Z(K_{X_1}) \cong J_4$. This is contradiction as $X_1 \leq Z(K_{X_1})$ but the multiplier of J_4 is trivial. This completes the proof that to eliminate $L/O_2(L) \cong SL_3(q)$, it is sufficient to show $Q_{X_1} \neq 1$.

So we assume $Q_{X_1} = 1$, and it remains to derive a contradiction. We set up the apparatus to apply lemma G.2.5. Set $U := \langle V^{G_1} \rangle$ and $\hat{G}_1 := G_1/V_1$. It is straightforward to check that Hypothesis G.2.1 is satisfied, with $N_G(V_1)$, $O^2(N_L(V_1))$, $G_1 X_1 T$, U, V in the roles of "G_1, L_1, H, U, V". Therefore $\hat{U} \leq Z(O_2(\hat{G}_1))$ by G.2.2.

Let P be a Sylow 3-subgroup of KX containing X with $X_K = Z(P)$, so that $P \cong 3^{1+2}$. As $Z(P) = X_K = X \cap L_1$ is nontrivial on V, P is faithful on U; so as $P \cong 3^{1+2}$, $1 \neq [C_U(X_1), X] = [C_U(X_1), X_K]$. If $Y := [C_U(X_1), X] \leq O_2(LT)$, then $L_{X_1} = \langle X_K^{L_{X_1}} \rangle$ is nontrivial on $O_2(L_{X_1})$, which we saw above suffices; so we may assume $Y \not\leq O_2(LT)$. In particular $U \not\leq O_2(LT)$ so the hypotheses of G.2.5 are also satisfied. Thus by G.2.5.1, $R_1 = UO_2(LT)$. Furthermore by G.2.5.2, $L \leq J \leq LT$, where J plays the role of "I" in G.2.5—with the structure of $O_2(J)$ described in detail in the remaining parts of G.2.5. Let $L_+ := N_{L_1}(X_1)^\infty$. As $1 = Q_{X_1}$, $C_{R_1}(X_1)/C_{O_2(LT)}(X_1)$ is the unique noncentral chief factor for L_+ on $C_{R_1}(X_1)$, so $C_{R_1}(X_1)/C_{R_1}(L_+ X_1)$ is the natural module for $L_+/O_2(L_+) \cong L_2(4)$. Let W be a KX-chief factor in $O_2(KT)$ with K nontrivial on W. By G.2.5, the nontrivial L-constituents on $O_2(J)$ are natural or dual, so the nontrivial L_+-constituents are all natural. Therefore by B.4.14, W is the adjoint module for $K/O_2(K)$ and $C_W(X_1)$ is indecomposable of $\mathbf{F}_4$-dimension 4 for $L_+/O_2(L_+)$. In particular $C_W(X_1)$ does not split over $[C_W(X_1), L_+]$, a contradiction as $C_{R_1}(X_1)/C_{R_1}(L_+X_1)$ is the natural module for $L_+/O_2(L_+)$. This contradiction finally completes the elimination of the case where (*) holds and $L/O_2(L) \cong \bar{L} \cong SL_3(4)$.

Thus in view of 11.0.2.2, we have shown that if (*) holds then $L/O_2(L) \cong \bar{L} \cong Sp_4(4)$ or $G_2(4)$, and to complete our treatment of the case $i = 1$ it remains to show that (*) implies the remaining statements in (3). Recall $I/O_2(I) \cong SL_3(4)$, $Sp_4(4)$, or $G_2(4)$, so $I \in \mathcal{L}^*(G,T)$ by 1.2.8.4, and then as $[Z, L_2] \neq 1$, even $I \in \mathcal{L}_f^*(G,T)$. This begins to establish some symmetry between L and I; in particular applying 11.0.2 to I, we conclude there is $V_I \in \mathcal{R}_2(IT)$ with $V_I/C_{V_I}(I)$ the natural module.

Assume $[Z, K] \neq 1$. Then $K \leq K_+ \in \mathcal{L}_f^*(G,T)$ by 1.2.9. Now since $K/O_2(K) \cong L_3(4)$, by A.3.12, either $K = K_+$ or $K_+/O_2(K_+) \cong M_{23}$. By B.4.2, neither $L_3(4)$ nor M_{23} has an FF-module, so Theorem 7.0.1 supplies a contradiction.

Therefore $[Z, K] = 1$, so if $C_{V_I}(I) \neq 1$ then $Z_I := C_Z(I) \neq 1$. But then by 1.2.7.3, $N_G(I) = !\mathcal{M}(IT) = !\mathcal{M}(C_G(Z_I))$, so $L_1 \leq K \leq N_G(I) \geq L_2$, and hence $K \leq M = !\mathcal{M}(LT)$, for our usual contradiction. Thus $C_{V_I}(I) = 1$. As $[Z, K] = 1$, K_0 stabilizes the 1-dimensional $\mathbf{F}_4$-subspace $V_{I,1}$ of V_I stabilized by T. Thus K_0 plays the same role in I that L_1 plays in L. As $XT = TX$ and we saw $X \leq I$, X is also a Cartan subgroup of I and $V_{I,1} = [Z \cap V_{I,1}, X_K] \leq C_G(K)$. Therefore K is the member of $\mathcal{C}(N_G(V_{I,1}))$ containing K_0, so K plays the role of "K" for I as well as for L. In particular, (*) is also satisfied by I. Therefore applying our previous reduction to I, $I/O_2(I)$ is not $SL_3(4)$. Notice also that L_2 plays the same role in both L and I: L_2 is the derived group of the stabilizer of a line of V and V_I.

Suppose $L/O_2(L) \cong G_2(4)$. Then $X_1 \leq L_2$ by B.4.6.14. From the previous paragraph, K_0 centralizes $V_{I,1}$, so if $I/O_2(I) \cong G_2(4)$, then by the same argument,

$$C_X(K_0/O_2(K_0)) = X \cap L_2 = X_1.$$

But from (*), $[K_0, X_1] \not\leq O_2(K_0)$, a contradiction. Therefore if $L/O_2(L) \cong G_2(4)$, then $I/O_2(I) \cong Sp_4(4)$ and so (3) holds.

This leaves the case $L/O_2(L) \cong Sp_4(4)$. Interchanging the roles of L and I if necessary, and appealing to the previous paragraph, we may assume that also $I/O_2(I) \cong Sp_4(4)$. As $L/O_2(L) \cong Sp_4(4)$, $X_K = X \cap L_1$ and X_1 are the two diagonally-embedded subgroups of order 3 with respect to the decomposition

$$X = X_2 \times (X \cap L_2).$$

Therefore as K_0 centralizes $V_{I,1}$, and $X_K = X \cap K_0$, from the structure of $I/O_2(I) \cong Sp_4(4)$, the second diagonal subgroup X_1 centralizes $K_0/O_2(K_0)$. But again this does not hold in (*), a contradiction completing the treatment of the case $i = 1$.

Now we turn to the easier case $i = 2$. We assume that $L_2 < K_2 = K$, and it remains to derive a contradiction. Here $V_2/C_{V_2}(L)$ is the natural module for $L_2/O_2(L_2) \cong L_2(q)$, and by 11.0.3.3, either $C_{V_2}(L) = 1$, or $\bar{L}$ is $Sp_4(q)$ or $G_2(q)$ with $m(C_{V_2}(L)) \leq n$. Thus $m(V_2) \leq 3n$. Examining the possibilities in (2) and (3) of 11.1.1 for cases where K possesses a nontrivial module of rank at most $3n$, we conclude that one of the following holds:

(a) $K/O_2(K) \cong SL_3(q)$ and V_2 is the natural module.
(b) $q = 4$, $K/O_2(K) \cong A_7$, and V_2 is the natural module.

(c) $q = 4$, $K/O_\infty(K) \cong L_2(5)$, and $O_2(K) < O_\infty(K)$ centralizes V_2 of rank at least 4.

In case (b), $K = O^{3'}(N_G(V_2))$ by A.3.18, so $X_2 \leq C_K(L_2/O_2(L_2))$, contrary to the structure of A_7. In case (a), $m(V_2) = 3n$ so $m(C_V(L)) = n$, and from the action of $\bar{L} \cong Sp_4(2^n)$ or $G_2(2^n)$ on V in I.2.3.1.ii.a, R_2 centralizes V_2, whereas this is not the case for the parabolic L_2^* in $K^* \cong SL_3(q)$ on V_I. Hence case (c) holds so $K/O_2(K)$ is not quasisimple and there is $Y \in \Xi(G,T)$ contained in $O_{2,F}(K)$ by 1.3.3; in particular Y is normalized by L_2T. Next $YT \leq C_G(V_2)T \leq N_G(V_1)$, so as $Y \in \Xi(G,T)$, we may apply 1.3.4 to conclude that either $Y \trianglelefteq N_G(V_1)$, or $Y \leq K_Y \in \mathcal{C}(N_G(V_1))$ with K_Y described in 1.3.4. Therefore either $K_1 \leq N_G(Y)$ or $K_1 = K_Y$. However comparing the list of possibilities for K_Y in 1.3.4 to the list of possiblities for K_1 in this lemma, we find no overlap. Thus $K_1 \leq N_G(Y)$, so

$$LT = \langle L_1, L_2T \rangle \leq N_G(Y).$$

Then $K \leq N_G(Y) \leq M = !\mathcal{M}(LT)$, for our usual contradiction. This completes the treatment of the case $i = 2$, and hence the proof of 11.1.2. $\qquad\square$

In the remainder of the section, we obtain several further technical restrictions on the normalizers of the subspaces V_i.

LEMMA 11.1.3. L_2 is the unique member of $\mathcal{C}(N_G(V_2))$ which does not centralize V_2.

PROOF. By 11.1.2, $L_2 \in \mathcal{C}(N_G(V_2))$. If there is $L_2 \neq K \in \mathcal{C}(N_G(V_2))$, then by 1.2.1.2, $[K, L_2] \leq O_2(L_2) \leq C_G(V_2)$, so as $V_2 \in Irr_+(L_2, V_2)$, K centralizes V_2 by A.1.41. $\qquad\square$

LEMMA 11.1.4. $C_G(V_3/V_1) \leq M \geq N_G(V_3) \cap N_G(L_1)$.

PROOF. If $\bar{L}$ is $SL_3(q)$ then $V_3 = V$, so $N_G(V_3) \leq M = !\mathcal{M}(LT)$. Hence we may take $\bar{L}$ to be $Sp_4(q)$ or $G_2(q)$. Let $\Delta := V_2^{L_1}$ and $H := N_G(V_3)$; note that V_1 is the intersection of the members of Δ, while V_3 is their span. Then by 11.1.3, $N_H(\Delta)$ acts on

$$\langle L(U) : U \in \Delta \rangle = L,$$

where we recall $L(U) = N_L(U)^\infty$. Therefore $N_H(\Delta) \leq N_G(L) = M$. In particular $C_G(V_3/V_1) \leq M$. Further Δ is the set of subspaces $C_{V_3}(S)$ for $S \in Syl_2(L_1)$, so $N_H(L_1) \leq M$. $\qquad\square$

LEMMA 11.1.5. Either

(1) $N_G(R_1) \leq M$, or

(2) $V = V_M$, L is an $SL_3(q)$-block or $Sp_4(4)$-block, $C_T(L) = 1$ and $V_1 = \Omega_1(Z(R_1))$.

PROOF. Assume $N_G(R_1) \not\leq M$. Then as $M = !\mathcal{M}(LT)$, there is no nontrivial characteristic subgroup of R_1 normal in LT. Therefore L, R_1 is an MS-pair in the language of Definition C.1.31. so L appears on the list of Theorem C.1.32. Therefore L is an $Sp_4(4)$-block or an $SL_3(q)$-block, since the remaining possibilities in C.1.34 explicitly exclude the case where R_1 is the unipotent radical of the point stabilizer. In particular $V = V_M$.

Set $Q := O_2(LT)$ and $Q_1 := VC_T(V)$. If $V = Q$ then $\Omega_1(Z(R_1)) = C_V(R_1) = V_1$ and the lemma holds, so we may assume $V < Q$. By C.1.13, $\Phi(Q) \leq C_T(L)$ and

$m(Q/Q_1) \leq m(H^1(\bar{L}, \tilde{V}))$, so $H^1(\bar{L}, \tilde{V}) \neq 0$. Therefore by I.1.6, L is an $Sp_4(4)$-block and $m(Q/Q_1) \leq 2$, and by I.2.3.1, $Q/C_T(L)$ is a submodule of the dual of the natural 5-dimensional module over $\mathbf{F}_4$ for $\Omega_5(4) \cong Sp_4(4)$. Here we compute (e.g. by restricting to the subgroup $Sp_4(2) \cong S_6$) that $C_Q(R_1) \leq Q_1$ and $C_V(R_1) = V_1$. Therefore $Z_1 := \Omega_1(Z(R_1)) = V_1 C_{Z_1}(L)$.

If $C_T(L) = 1$ then $Z_1 = V_1$ by the previous paragraph, and hence (2) holds. Thus we may assume that $C_T(L) \neq 1$. Now $[O_2(LT), X] \leq [O_2(LT), L] = V$, so

$$Z_1 := \Omega_1(Z(R_1)) = Z_X \times Z_C,$$

where $Z_X := [V_1, X]$, $Z_C := C_{Z_1}(X) = C_{Z_1}(L)$, and $Z_C \neq 1$ as $C_T(L) \neq 1$. For $D \leq G$, let $\theta(D)$ be the subgroup generated by all elements of D whose order lies in Δ, where Δ is the set of divisors of $2^n - 1$ if $\bar{L}$ is $SL_3(2^n)$ with n odd, and $\Delta := \{3\}$ otherwise. Thus $L = \theta(M)$ by 11.0.4. By Theorem 4.2.13, $M = !\mathcal{M}(L)$, so Z_C is a TI-set under the action of $Y := N_G(R_1)$, with $Y_M := Y \cap M = N_Y(Z_C)$: for if $y \in Y$ with $1 \neq Z_C \cap Z_C^y$, then $\langle L, L^y \rangle \leq C_G(Z_C \cap Z_C^y)$, so as $M = !\mathcal{M}(L)$, $L^y \leq \theta(C_G(Z_C \cap Z_C^y)) \leq \theta(M) = L$, and hence $y \in N_Y(L) = Y_M$. Notice $Y_M < Y$ as (1) fails. Set $Y^* := Y/C_Y(Z_1)$. Then X^* is regular on $Z_X^\#$, and normal in Y_M^* since $L_1 R_1$ centralizes V_1. Thus we have the hypotheses for a Goldschmidt-O'Nan pair in the sense of Definition 14.1 in [**GLS96**]; so we may apply O'Nan's Lemma 14.2 in [**GLS96**, 14.2], with Y^*, X^*, Z_1 in the roles of "X, Y, V". Observe conclusion (iv) of that result must hold—since in (i), Y normalizes Z_C giving $Y_M = Y$; while in (ii) and (iii), T does not normalize Z_C. In conclusion (iv) of 14.2 of [**GLS96**, 14.2], $q = 4$, $Z_1 \cong E_8$, and Y^* is a Frobenius group of order 21. Next $C_G(Z_1) \leq C_G(Z_C) \leq M = !\mathcal{M}(L)$, so $L_1 \in \mathcal{C}(N_G(Z_1))$ and hence Y acts on L_1. If L is an $SL_3(4)$-block, the noncentral 2-chief factors for L_1 are $VZ(L_1)/Z(L_1)$ and $O_2(L_1)/VZ(L_1)$, and both are natural modules. Therefore the induced action of $N_G(L_1)$ on $Irr_+(L_1, O_2(L_1)/Z(L_1))$ is contained in $\Gamma L_2(4)$, so $O^{7'}(Y)$ acts on $VZ(L_1)$ and then on $[VZ(L_1), L_1] = V$. But then $Y = O^{7'}(Y)Y_M \leq N_G(V) \leq M$, contradicting $Y_M < Y$. Similarly if L is an $Sp_4(4)$-block, then Y acts on $[V, L_1] = V_3$, so $Y \leq M$ by 11.1.4, for the same contradiction. This completes the proof. $\square$

11.2. Weak-closure parameter values, and $\langle \mathbf{V^{N_G(V_1)}} \rangle$

Since V is an FF-module, we do not have the ideal situation for weak closure described in subsection E.3.3; however, we will be able to establish at least some restrictions on the weak closure parameters $r(G, V)$, $w(G, V)$, and $n(H)$ discussed in Definitions E.3.3, E.3.23, and E.1.6. Recall that the paramter $n'(Aut_M(V))$ is defined in Definition E.3.37, and notice that $n'(Aut_M(V)) = n > 1$: for example this follows from A.3.15.

LEMMA 11.2.1. For $H \in \mathcal{H}_*(T, M)$, either

(1) $n(H) \leq n$, or
(2) $\bar{L} \cong SL_3(q)$, V_M is the sum of two isomorphic natural modules for $L/O_2(L)$, $C_V(H) = 1$, $L = [L, J(T)]$, and $n(H) \leq 2n$. [2]

PROOF. Assume (1) fails, so that $n(H) > n > 1$. Then by E.2.2, $O^2(H/O_2(H))$ is of Lie type over $\mathbf{F}_{2^m}$, for $m := n(H) > n$, and $H \cap M$ is a Borel subgroup of

---

[2]Notice this essentially eliminates the shadow of $\Omega_8^-(2^n)$, in which $n(H) = 2n$ but $V \trianglelefteq M$. Our use of the quasithin hypothesis is via reference to the pushing up result Theorem 4.4.3.

H. Let B be a Hall $2'$-subgroup of $H \cap M$. If $A := C_B(V) \neq 1$, then by 4.4.13.1, $N_G(A) \not\leq M$, contrary to Theorem 4.4.3 using Remark 4.4.2. Thus $C_B(V) = 1$.

Suppose that B normalizes V. Then B is faithful on V, giving Hypothesis E.3.36—so that by E.3.38 we have $n(H) \leq n'(Aut_M(V)) = n$, contrary to assumption.

Hence we may assume that B does not normalize V, so in particular $V < V_M$. By 11.0.3.5, $\bar{L} \cong SL_3(q)$, and then by 11.0.3.4, $C_V(B) = 1$, so $C_V(H) = 1$. In particular as $Z \cap V \neq 1$, $[Z, H] \neq 1$, so $L = [L, J(T)]$ by 3.1.8.3. Now the argument in the final paragraph of the proof of 11.0.3 and an appeal to B.5.1.1.ii shows V_M is the sum of two isomorphic natural modules for $\bar{L} \cong SL_3(q)$. Thus $C_{GL(V_M)}(L) \cong GL_2(q)$, so if $m > 2n$ then $C_B(V_M) \neq 1$, contrary to paragraph one. Thus $m \leq 2n$, completing the verification of (2) and hence the proof. $\qquad\square$

Recall $M_V = N_M(V) = N_G(V)$ and $T_L = T \cap LO_2(LT) = T \cap LC_T(V)$.

LEMMA 11.2.2. *Set* $m := 2n$ *if* $\bar{L} \cong G_2(q)$ *and* $m := n$ *otherwise. Let* $U \leq V$, *and set* $k := m(V/U)$. *Then*

(1) $m(\bar{M}_V, V) = m$.
(2) Assume that $O^{2'}(C_M(U)) \leq C_M(V)$ *and* $k < 2m$. *Then* $C_G(U) \leq M$, *and so* $O^{2'}(C_G(U)) \leq C_M(V)$.
(3) Either $r(G, V) > m$; *or* $\bar{L} \cong SL_3(q)$, $r(G, V) = m$, *and* $C_G(V_2) \not\leq M$. *In particular,* $s(G, V) = m$.
(4) $W_j(T, V) \leq T_L$ *for* $j < m - 1$, *so* $V_1 \leq C_V(W_j(T, V))$.
(5) If $\bar{L} \cong G_2(q)$ *then* $W_0(T, V) \leq C_T(V)$, *so* $N_G(W_0(T, V)) \leq M$; *that is,* $w(G, V) > 0$.
(6) If $\bar{L} \cong G_2(q)$ *then* $C_G(C_1(R_1, V)) \leq M$.

PROOF. Part (1) is a standard fact about the natural module and its nonsplit central extensions in I.2.3.1; cf. B.4.6 when $\bar{L} \cong G_2(q)$.

Next we claim $r(G, V) \geq m$. By (1), $m(\bar{M}_V, V) \geq m$; so if $m > 2$, the claim follows from Theorem E.6.3. Assume $m \leq 2$; then $m = 2 = n$, and $\bar{L}$ is $SL_3(4)$ or $Sp_4(4)$. If $\bar{L}$ is $Sp_4(4)$, assume further that $C_V(L) = 1$. Then L is transitive on non-zero vectors in the dual of V, and hence transitive on $\mathbf{F}_2$-hyperplanes U of V, so in particular each hyperplane is invariant under a Sylow 2-subgroup of M. Hence as $m(\bar{M}_V, V) = m \geq n > 1$ by (1), $r(G, V) > 1$ by E.6.13. Thus we may assume $L/O_2(L) \cong Sp_4(4)$, $C_V(L) \neq 1$, and U is a hyperplane of V with $C_G(U) \not\leq M$. By Theorem 4.2.13, $M = !\mathcal{M}(L)$, so $C_U(L) = 1$; hence U is an $\mathbf{F}_2$-space complement to $C_V(L)$, and so $m(C_V(L)) = 1$. Now V is a quotient of the full covering $\hat{V}$ of the natural module $\tilde{V}$ for $\bar{L}$, which has the structure of a 5-dimensional orthogonal space over $\mathbf{F}_4$. From this structure, L has two orbits on the $\mathbf{F}_4$-complements to $C_{\hat{V}}(L)$ in $\hat{V}$, with representatives $\hat{U}_\epsilon$, $\epsilon = \pm 1$, such that $Aut_{\bar{L}}(\hat{U}_\epsilon) \cong O_4^\epsilon(4)$. Moreover each $\mathbf{F}_2$-hyperplane of $\hat{V}$ supplementing $C_{\hat{V}}(L)$ contains such an $\mathbf{F}_4$-hyperplane, so the images U_ϵ of $\hat{U}_\epsilon$, for $\epsilon = \pm 1$, are representatives for the orbits of L on $\mathbf{F}_2$-complements to $C_V(L)$. In particular $N_{LT}(U)$ is maximal in LT but not of index 2, and there is a subgroup Y of order 3 in $N_L(U)$ faithful on U. Next $Z \cap V \not\leq C_V(L)$, so that $Z \cap U \neq 1$, and hence $N_G(U) \in \mathcal{H}^e$ by E.6.6.4. By E.6.7.1, $C_G(U)$ contains a χ-block invariant under $Y = O^2(Y)$. Then as Y is faithful on U, while $m_3(YC_G(U)) \leq 2$ as M is an SQTK-group, $m_3(C_G(U)) \leq 1$. Hence we have the hypotheses of E.6.14, and that lemma supplies a contradiction, completing the proof of the claim that $r(G, V) \geq m$.

Assume the hypotheses of (2). By 11.0.3.4, $C_M(U) \leq M_V$. Then $C_M(U) = C_M(V)$ since if Y is of odd prime order in $\bar{M}_V$, then $m(V/C_V(Y)) \geq 2m$; notice we use 11.0.4 and A.1.41 to conclude $C_{\bar{M}_V}(\bar{L}) = Z(\bar{L})$, and to exclude diagonal outer automorphisms. Now (2) follows from E.6.12 and the fact that $r(G, V) \geq m > 1$.

We have shown $r(G, V) \geq m$. Further in case of equality, we may pick U with $k = m$, $C_G(U) \not\leq M$, and $O^{2'}(C_{\bar{M}_V}(U)) \neq 1$ by (2). But then $U = C_V(i)$ for a suitable root involution $i \in \bar{L}$, and up to conjugacy either:

(i) $\bar{L} \cong Sp_4(q)$ or $G_2(q)$ and $V_3 \leq U$, or
(ii) $\bar{L} \cong SL_3(q)$ and $U = V_2$.

Case (i) contradicts 11.1.4, so that (3) holds.

Let $A \leq V^g \cap T$ with $m(V^g/A) =: j < m - 1$. To prove (4), we must show that $A \leq T_L$, so we may assume $\bar{A} \neq 1$. Then for $B \leq A$ with $m(V^g/B) < m = s(G, V)$, $\bar{A} \in \mathcal{A}_{m-j}(\bar{M}_V, V) \subseteq \mathcal{A}_2(\bar{M}_V, V)$, using E.3.10. Hence (using B.4.6.9 in case $\bar{L}$ is $G_2(q)$) $\bar{A} \leq \bar{L}$, so $W_j(T, V) \leq T_L$. Then as $V_1 = C_V(T_L)$, (4) holds.

Assume next that $\bar{L} \cong G_2(q)$ and $j := m(V^g/A) = 0$ or 1 with $\bar{A} \neq 1$. By B.4.6.3, there is $E_{q^3} \cong A_1 \trianglelefteq N_{\bar{L}}(V_1)$ with $C_V(A_1) \in \mathcal{V}_3$. As $\bar{L}$ is $G_2(q)$, $m = 2n$, so as $n \geq 2$, $\bar{A} \in \mathcal{A}_{m-j}(\bar{T}, V) \subseteq \mathcal{A}_{n+1}(\bar{T}, V)$ by the previous paragraph. Hence by B.4.6.9, $\bar{A} \leq A_1^h$ for some $h \in L$, and if $j = 0$, then $\bar{A} = A_1^h$. Further if $j = 1$ and $A \leq R_1$, then by B.4.6.12, $\bar{A} \leq A_1$. Thus $C_V(W_1(R_1, V)) \geq C_V(A_1) = V_3$, so $C_G(C_1(R_1, V)) \leq C_G(V_3) \leq M$ by 11.1.4. That is, (6) holds.

Now take $j = 0$. Thus $\bar{A} = A_1^h$, so without loss $\bar{A} = A_1$. Let $D \leq A$ with $\bar{D}$ a long root group of $\bar{L}$. Then $m(V^g/D) = m(A/D) = 2n = m < r(G, V)$ by (3), so $C_V(D) \leq N_G(A)$. Set

$$E := \langle C_V(D) : D \leq A, \ \bar{D} \text{ is a long root subgroup of } \bar{L} \rangle.$$

Then by B.4.6.3, $m(V/E) = n$ and $[E, A] = V_3$. We just saw $E_D := C_V(D) \leq N_G(A)$, so E acts on A, and hence $V_3 = [E, A] \leq A$. Thus $V_3 \leq V \cap V^g$, so as V is a TI-set under M by 11.0.3.4, $g \notin M$. Furthermore $D = C_A(E_D)$, so $m(A/C_A(E_D)) = m(A/D) = 2n$. Therefore by B.4.6, the image of E_D in $L^g/O_2(L^g)$ is contained in a long root group, and $[E_D, A] =: A_D \in \mathcal{V}_2^g =: \mathcal{V}_2(A)$. But as $\bar{A} = A_1$, also $[E_D, A] \in \mathcal{V}_2 =: \mathcal{V}_2(V)$. So if we define $\Delta(V_3, V) := \mathcal{V}_2(V) \cap V_3$, we see

$$\Delta(V_3, V) = \Delta(V_3, V)^g := \Delta(V_3, A). \tag{$*$}$$

Define

$$L(V_3, V) := \langle L(I) : I \in \Delta(V_3, V) \rangle.$$

By $(*)$ and 11.1.3, $L(V_3, V) = L(V_3, V)^g =: L(V_3, A)$. But we check that $L(V_3, V) = L$, so by Theorem 4.2.13,

$$M = !\mathcal{M}(L(V_3, V)) = !\mathcal{M}(L(V_3, A)) = M^g,$$

contradicting $g \notin M$. Together with E.3.16.1, this establishes (5). $\qquad\square$

LEMMA 11.2.3. *(1) If $C_V(L) = 1$, then $N_G(V_1) = C_G(V_1)N_M(V_1)$.*
(2) If $C_G(V_1) \leq M$, then $N_G(V_1) \leq M$.

PROOF. Set $Y := N_G(V_1)$ and $Y^* := Y/C_G(V_1)$. Now $C_T(V_1) = T_L$ and $T = \langle f \rangle T_L$ where f induces a field automorphism on $\bar{L}$, so T^* is cyclic. Hence by Cyclic Sylow 2-Subgroups A.1.38, $Y^* = O(Y^*)T^*$.

Assume $C_V(L) = 1$. Then X^* is regular on $V_1^{\#}$, so by A.1.12, X^* is normal in any overgroup of odd order in $GL(V_1)$. Hence $O(Y^*) \leq N_{GL(V_1)}(X^*)$, where the latter group consists of X^* extended by $\mathbf{Z}_n$, and contains T^*. Then $X^*T^* \trianglelefteq O(Y^*)T^* = Y^*$, so by a Frattini Argument, $Y^* = X^*N_Y(T)^* \leq N_M(V_1)^*$, since $N_G(T) \leq M$ by Theorem 3.3.1. Thus (1) holds.

Assume $G_1 := C_G(V_1) \leq M$. In view of (1), we may also assume that $C_V(L) \neq 1$. Hence $N_G(R_1) \leq M$ by 11.1.5. As $G_1 \leq M$, $L_1 \trianglelefteq G_1$, so as T acts on L_1, $L_1 \trianglelefteq N_G(V_1)$ by 1.2.1.3. Then as $R_1 \in Syl_2(C_{G_1}(L_1/O_2(L_1)))$, by a Frattini Argument $Y = C_{G_1}(L_1/O_2(L_1))N_Y(R_1) \leq M$, so that (2) holds. $\qquad\square$

LEMMA 11.2.4. *Assume* $\langle V^{N_G(V_1)} \rangle$ *is abelian, and* $[V, W_0(T, V)] \neq 1$. *If* $\bar{L} \cong SL_3(q)$, *assume further that* $C_G(V_2) \leq M$. *Then*

(1) $W_0(T, V) \leq R_2$.
(2) *If* $V^g \leq T$ *with* $[V, V^g] \neq 1$, *then* $V \not\leq N_G(V^g)$.
(3) $r(G, V) \leq 2n$.

PROOF. By hypothesis $w(G, V) = 0$, so by 11.2.2.5, $\bar{L}$ is not $G_2(q)$ and $s(G, V) = n$ by 11.2.2.3. Furthermore there is $A := V^g \leq T$ with $\bar{A} \neq 1$. As $s(G, V) = n$, $\bar{A} \in \mathcal{A}_n(\bar{T}, V)$ by E.3.10. Let $\hat{A} := A/C_A(L^g)$ and $\dot{L}^g := L^g/C_{L^g}(A)$.

Our hypothesis that $C_G(V_2) \leq M$ when $\bar{L} \cong SL_3(q)$, together with 11.2.2.3, says that $r(G, V) > n$. Thus if $m(A/B) \leq n$, then $C_G(B) \leq N_G(A)$. Also by hypothesis $\langle V^{N_G(V_1)} \rangle$ is abelian, so $g \notin N_G(V_1)$ as $[V, A] \neq 1$.

We next claim there is no $W \leq V$ with $[\tilde{W}, A] = \tilde{V}_1$ and $m(A/C_A(W)) = n = m(W/C_W(A))$. For if so, $W \leq C_G(C_A(W)) \leq N_G(A)$ by the previous paragraph, and then W induces transvections on the $\mathbf{F}_q$-space $\hat{A}$ with axis $\widehat{C_A(W)}$. If $\bar{L}$ is $SL_3(q)$ then $C_V(L) = 1$ and by hypothesis $V_1 = [A, W]$, so V_1 is a 1-dimensional $\mathbf{F}_q$-subspace of A. If $\bar{L}$ is $Sp_4(q)$ then as $m(W/C_W(A)) = n$, $Aut_W(A)$ is a root subgroup of $\dot{L}^g$ inducing transvections on A, so $[\hat{A}, W]$ is a 1-dimensional $\mathbf{F}_q$-subspace of $\hat{A}$, and $C_A(L^g) \leq [A, W]$ by I.2.3.1.ii.b. Thus as $[\tilde{W}, A] = \tilde{V}_1$, $[W, A] = V_1$. Now in either case L^g is transitive on 1-subspaces of $\hat{A}$ with representative $\hat{V}_1^g$, so conjugating in $N_G(A)$ we may assume $g \in N_G(V_1)$, contrary to the previous paragraph.

Next assume that $A \not\leq R_2$. Then $Aut_A(V_2) \in \mathcal{A}_n(Aut_T(V_2), V_2)$, so as R_2 centralizes V_2 and $\tilde{V}_2$ is the natural module for $L_2/O_2(L_2)$, $Aut_A(\tilde{V}_2) \in Syl_2(Aut_{L_2}(\tilde{V}_2))$. Hence $[\tilde{V}_2, A] = \tilde{V}_1$, and $m(A/C_A(V_2)) = n = m(V_2/C_{V_2}(A))$, contary to the claim applied to V_2 in the role of W. Thus (1) is established.

By (1), $A \leq R_2$, and hence $[V, A] \leq V_2$. Suppose that $[\tilde{V}, A] < \tilde{V}_2$. Then $m([A, \tilde{V}]) = n$ and $\bar{A}$ is contained in the root subgroup of a transvection in $\bar{R}_2$. In particular, $m(\bar{A}) = m(A/C_A(V)) = n$ and conjugating in L_2, we may assume $[\tilde{V}, A] = \tilde{V}_1$, contrary to the claim applied to V in the role of W. Therefore $[\tilde{V}, A] = \tilde{V}_2$, so $C_V(A) = V_2$ and $C_{\tilde{V}}(A) = \tilde{V}_2$ since $A \leq R_2$.

We next reduce (3) to (2). Namely as $A \leq R_2$,

$$V = \langle C_V(B) : m(A/B) \leq 2n \rangle,$$

so if $r(G, V) > 2n$ then $V \leq N_G(A)$, contrary to (2).

Thus it remains to prove (2), so we may assume that $V \leq N_G(A)$.

Assume first that $V_2 = [A, V]$. Then as $V \leq N_G(A)$, $V_2 = [A, V] \leq V \cap A$. By symmetry between A and V, since $\tilde{V}_2 = C_{\tilde{V}}(A)$, $\hat{V}_2 = \widehat{C_A(V)}$ is an L^g-conjugate of

$\hat{V}_2^g$. Thus V_2 is an L_2-conjugate of V_2^g, and hence we may assume $g \in N_G(V_2)$. In particular $V_1^g \leq V_2$. Now by 11.1.3, $g \in N_G(L_2)$, so g permutes the fixed points of Sylow 2-groups of L_2. Of course $V_1^{L_2}$ is the set of these subspaces, so conjugating in L_2, we may assume $g \in N_G(V_1)$, contradicting an earlier observation. Thus $[A, V] < V_2$, so as $\tilde{V}_2 = [A, \tilde{V}]$, $C_V(L) \neq 1$, and hence $\bar{L} \cong Sp_4(q)$.

Now $V_2 = C_V(A)$, so $m(Aut_V(A)) = 2n$, and by symmetry $m(\bar{A}) = 2n$. If some $\bar{a} \in \bar{A}$ induces an $\mathbf{F}_q$-transvection on $\tilde{V}$, then without loss $[V_3, a] = 1$. Hence $\dot{V}_3$ is the root group of a transvection on $\hat{A}$, and then by symmetry $\bar{A}$ contains the root group centralizing V_3 and $[\tilde{V}_3, A] = \tilde{V}_1$. Then $m(A/C_A(V_3)) = n$, contrary to the claim applied to V_3 in the role of W. Therefore $\bar{A}$ contains no transvections.

Let $\bar{R}_l$ and $\bar{R}_k$ be root groups of transvections contained in $\bar{R}_2$, $\bar{S} := \bar{R}_l \bar{R}_k$, and $\bar{A}_S = \bar{A} \cap \bar{S}$. Then $m(\bar{S}/\bar{A}_S) \leq m(\bar{R}_2/\bar{A}) = n$, so as $\bar{R}_l \cap \bar{A} = 1$, we conclude $\bar{S} = \bar{R}_l \times \bar{A}_S$ and $m(\bar{A}_S) = n$. Now $\bar{R}_k \leq \bar{Y} \leq C_{\bar{L}}(\bar{R}_l)$ with $\bar{Y} \cong L_2(q)$, and setting $V_Y := [V, Y]$, $\tilde{V}_Y$ is a nondegenerate 2-dimensional $\mathbf{F}_q$-subspace of $\tilde{V}$ with $V_Y \leq C_V(\bar{R}_l)$. Indeed taking $\bar{R}_k \unlhd \bar{T}$, $V_1 = [V_Y, \bar{R}_k]$, so $V_1 = [V_Y, \bar{S}] = [V_Y, A_S]$; hence as $\tilde{V}_2 = [V, A]$, $V_2 = [V, A]$. This contradicts an earlier reduction, and completes the proof. $\qquad\square$

LEMMA 11.2.5. *Let $H \in \mathcal{H}(T)$ with $H \not\leq M$. If $\bar{L} \cong SL_3(q)$, assume further $C_G(V_2) \leq M$. Then either*

(1) $W_0(T, V) \not\leq O_2(H)$, or

(2) $\langle V^{N_G(V_1)} \rangle$ is nonabelian.

PROOF. We observe that Hypothesis F.7.6 is satisfied with LT, H in the roles of "G_1, G_2". Adopt the notation of section F.7, and assume $W_0(T, V) \leq O_2(H)$. Then the parameter b of Definition F.7.8 is even by F.7.9.4. Thus by F.7.11.2, there exists $g \in G$ with $1 \neq [V, V^g] \leq V \cap V^g$, so $\langle V^{N_G(V_1)} \rangle$ is nonabelian in view of 11.2.4.2, and hence (2) holds. $\qquad\square$

11.3. Eliminating the shadow of $\mathbf{L_4(q)}$

Notice that when $\bar{L} \cong SL_3(q)$, the condition $C_G(V_1) \leq M$ distinguishes $L_4(q)$ from the other shadows. In this section, we eliminate that troublesome configuration, and also (when we show $C_V(L) = 1$) eliminate the shadow of $Sp_6(q)$ in the case $\bar{L} \cong Sp_4(q)$.

Throughout this section we assume:

HYPOTHESIS 11.3.1. *(1) There exists $H \in \mathcal{H}_*(T, M)$ with $[Z, H] \neq 1$.*
(2) If $\bar{L} \cong SL_3(q)$ then $C_G(V_1) \leq M$.

The object of this section is to prove:

PROPOSITION 11.3.2. *Assume Hypothesis 11.3.1. Then*

(1) $\bar{L} \cong Sp_4(q)$.
(2) $C_G(V_1) \not\leq M$.
(3) $C_V(L) = 1$.
(4) If $W_0(T, V) \leq O_2(H)$, then $\langle V^{C_G(V_1)} \rangle$ is nonabelian.

During this section we assume the pair G, L is a counterexample to Proposition 11.3.2. We begin a series of reductions.

LEMMA 11.3.3. $W_0(T, V) \not\leq O_2(H)$.

PROOF. Assume that $W_0(T, V) \leq O_2(H)$. We will show that 11.3.2 holds, contrary to our choice of G as a counterexample. When $\bar{L} \cong SL_3(q)$, we have $C_G(V_2) \leq C_G(V_1) \leq M$ by Hypothesis 11.3.1.2, so we may apply 11.2.5 to conclude that $\langle V^{N_G(V_1)} \rangle$ is nonabelian. As V is a TI-set under M by 11.0.3.4, this forces $N_G(V_1) \not\leq M$, so conclusion (2) of 11.3.2 follows from 11.2.3.2. Next $C_V(L) \leq V_1 \trianglelefteq T$, so if $C_V(L) \neq 1$ then $C_{V_1}(LT) \neq 1$, and hence $C_G(V_1) \leq C_G(C_{V_1}(LT)) \leq M = !\mathcal{M}(LT)$, whereas we just saw 11.3.2.2 holds; this contradiction establishes conclusion (3) of 11.3.2. Hence $N_G(V_1) = C_G(V_1)N_M(V_1)$ by 11.2.3.1, so $\langle V^{C_G(V_1)} \rangle$ is nonabelian as $N_M(V_1) \leq M_V$, establishing conclusion (4) of 11.3.2.

By Hypothesis 11.3.1.2 and as 11.3.2.2 holds, L is not $SL_3(q)$. Since $H \not\leq M = !\mathcal{M}(N_G(O_2(LT)))$, while H normalizes $W_0(T, V)$ by E.3.15 since $W_0(T, V) \leq O_2(H)$, $\bar{L}$ is not $G_2(q)$ by 11.2.2.5. Thus $\bar{L} \cong Sp_4(q)$, so that conclusion (1) of 11.3.2 holds. But now the choice of G as a counterexample is contradicted. $\square$

Set $W_0 := W(T, V)$. By 11.3.3, $W_0 \not\leq O_2(H)$, so part (4) of Proposition 11.3.2 is vacuously satisfied. Thus we only need to establish parts (1)–(3).

Let $V_H := \langle Z^H \rangle$, $H^* := H/C_H(V_H)$, $m := s(G, V)$, and $k := n(H)$. By B.2.14, $V_H \in \mathcal{R}_2(H)$. As $W_0 \not\leq O_2(H)$ while $C_H(V_H)$ is 2-closed by B.6.8, there exists $A := V^g \leq T$ with $A^* \neq 1$.

LEMMA 11.3.4. *(1)* $A^* \in \mathcal{A}_m(H^*, V_H)$.
(2) $\bar{L}$ *is* $SL_3(q)$ *or* $Sp_4(q)$.
(3) *Either* $k = n$, *or* $k > n$ *and conclusion (2) of 11.2.1 holds.*
(4) $N_G(V_1) \leq M_V$.

PROOF. Part (1) follows from E.3.6. By E.3.20, $k \geq m$. By 11.2.2, $m \geq n$, and $m = 2n$ if $\bar{L}$ is $G_2(2^n)$. Finally by 11.2.1, either $k \leq n$, or conclusion (2) of 11.2.1 holds. Thus (2) and (3) hold.

Suppose $C_G(V_1) \not\leq M$. Then conclusion (2) of Proposition 11.3.2 holds, and hence (as we saw during the proof of 11.3.3) also conclusion (3) of 11.3.2 holds. Then by Hypothesis 11.3.1.2, $\bar{L}$ is not $SL_3(q)$, so that conclusion (1) of Proposition 11.3.2 holds, contrary to our choice of G as a counterexample. Thus $C_G(V_1) \leq M$, and hence $N_G(V_1) \leq M_V$ by 11.2.3.2 and 11.0.3.4. This establishes conclusion (4), and completes the proof. $\square$

LEMMA 11.3.5. *(1)* $O^2(H) = \langle K^H \rangle$, *with* $K \in \mathcal{C}(H)$ *and* $K/O_2(K) \cong L_2(2^k)$.
(2) *If* $k > n$ *assume* $k = 2n$. *Then* $K = O^2(H)$.

PROOF. By 11.3.4.3, $k \geq n > 1$. Therefore by E.2.2, $O^2(H) = \langle K^H \rangle$, for $K \in \mathcal{C}(H)$ described in E.2.2; in particular $K/O_2(K)$ is of Lie type over $\mathbf{F}_{2^k}$. As $[Z, H] \neq 1$ by Hypothesis 11.3.1.1, $K \in \mathcal{L}_f(G, T)$, so $K \leq K_+ \in \mathcal{L}_f^*(G, T)$ by 1.2.9.2. Now the possibilities for the embedding of K in K_+ are described in the list of A.3.12. In particular if $K/O_2(K)$ is not $L_2(2^k)$, then we conclude by comparing that list with those of Theorems B.5.1 and B.5.6, that K_+T has no FF-module—contrary to Theorem 7.0.1.

Thus (1) is established, so we assume the hypotheses of (2) with $K < O^2(H)$. By 11.3.4.3 and the hypotheses of (2), $k = n$ or $2n$.

Let D be a Hall $2'$-subgroup of $H \cap M$, p a prime divisor of $q - 1$, and $D_p := \Omega_1(O_p(D))$. As $k = n$ or $2n$, $D_p \neq 1$. As $K < O^2(H)$, $D = D_1 \times D_1^t$ for $D_1 := D \cap K$ and $t \in N_T(D) - N_T(K)$. Thus $[D_p, t] \neq 1 \neq C_{D_p}(t)$.

By 11.0.4, $D_p \le L$; so as $D_p T = T D_p$, $\bar{D}_p$ is contained in a Cartan subgroup of $\bar{L}$. As $[D_p, t] \ne 1$, t induces a field automorphism on $\bar{L}$. But then either $C_{D_p}(t) = 1$ or t centralizes D_p, contrary to the previous paragraph. This completes the proof of (2). $\qquad\square$

LEMMA 11.3.6. $k = n$.

PROOF. Assume $k > n$ and let $\hat{M} := M/C_M(V_M)$ and $C_T(V_M) := Q_M$. By 11.3.4.3, conclusion (2) of 11.2.1 holds, so $\bar{L} \cong SL_3(q)$, V_M is the sum of two conjugates of V, $k \le 2n$, and $m(\hat{M}, V_M) = 2n$. We first observe that the weak closure hypothesis E.6.1 is satisfied with V_M in the role of "V"; in particular LT normalizes $C_T(V) \cap C_M(V_M) = Q_M$, so $M = !\mathcal{M}(N_M(Q_M))$. As $m(\hat{M}, V_M) = 2n > 2$, $m(\hat{M}, V_M) = 2n = s(G, V_M)$ by Theorem E.6.3.

Suppose first that $B := V_M^y \le T$ for some $y \in G$ with $\hat{B} \ne 1$. Then $m(\hat{B}) \le m_2(\hat{M}) = m_2(\hat{L}) = 2n = s(G, V_M)$. On the other hand, $\hat{B} \in \mathcal{A}_{2n}(\hat{T}, V_M)$ by E.3.10, so $m(\hat{B}) = 2n$. Now assume further that $V_M \le N_G(B)$. Then $[V_M, B] \le V_M \cap B$, so by symmetry between B and V_M, $m(V_M/C_{V_M}(B)) = 2n$. Thus from the structure of the natural module V for $\bar{L} \cong SL_3(q)$, $\hat{B} = \hat{R}_2$ and $V_2 = C_V(B)$. Now for $v \in V - V_2$, $[v, B] = [V, B]$, so by the symmetry between V_M and B, v induces a root element of $M^y/C_{M^y}(B)$, and V induces the corresponding root group. Thus $[V, V^y] = V_1$, and by symmetry, $[V, V^y] = V_1^y$; then $y \in N_G(V_1) \le M_V$ by 11.3.4.4, contrary to $[V, V^y] = V_1$. Thus we have shown that if $[V_M, V_M^y] \le V_M \cap V_M^y$, then $[V_M, V_M^y] = 1$.

We next reproduce the argument establishing 11.2.5: Namely we now have Hypothesis F.7.6 with $N_M(Q_M)$, H, V_M in the roles of "G_1, G_2, V"; for example $M = C_M(V_M) N_M(Q_M)$ by a Frattini Argument, so $V_M \in \mathcal{R}_2(N_M(Q_M))$ by 11.0.3.1. If $W_0(V_M, T) \le O_2(H)$, then as in 11.2.5, the parameter b_M for the graph as in Definition F.7.8 is even using F.7.9.4, so $1 \ne [V_M, V_M^g] \le V_M \cap V_M^g$ for some $g \in G$ by F.7.11.2, contrary to the previous paragraph.

Thus there is $B := V_M^g \le T$ with $B \not\le O_2(H)$. By E.3.20, $k \ge s(G, V_M) = 2n$, so as $k \le 2n$, $k = 2n$. Therefore $O^2(H) = K \in \mathcal{C}(H)$ with $K/O_2(K) \cong L_2(2^{2n})$ by 11.3.5.

Next $N_{GL(V_M)}(\bar{L})$ is an extension of $GL_3(q) \times GL_2(q)$ by field automorphisms. Using 11.0.4, we conclude that $N_{\bar{M}}(\bar{L})$ is an extension of $\bar{L}$ by field automorphisms. Therefore for each $U \le V_M$ with $m(V_M/U) = 2n$, $C_{Aut_M(V_M)}(U)$ is a 2-group. Also $m_2(Aut_M(V_M)) \le 3n$.

We claim $r(G, V_M) > 2n$. For assume $U \le V_M$ with $m(V_M/U) = 2n$ and $C_G(U) \not\le M$. Then $1 \ne V \cap U$, so U contains a 2-central involution. By the previous paragraph $O^2(C_M(U)) \le C_M(V_M)$. Thus $O^{2'}(C_M(U)) \not\le C_M(V_M)$ by E.6.12, so conjugating in L if necessary, $V_1 \le U$. But then $C_G(U) \le C_G(V_1) \le M$, so the claim is established.

As $r(G, V_M) > 2n$ while $m(\hat{B}) = 2n$, $V_H \le C_G(C_B(V_H)) \le N_G(B)$. By E.3.6, $B^* \in \mathcal{A}_k(H^*, V_H)$, so as $K^* \cong L_2(2^{2n})$, $B^* \in Syl_2(K^*)$ is of rank $2n$ and $C_{V_H}(B^*) = C_{V_H}(b^*)$ for all $b^* \in B^{*\#}$. Thus by G.1.6, $V_H/C_{V_H}(K)$ is a direct sum of $s \ge 1$ copies of the natural module for K^*. Since V_H normalizes B, $m(Aut_{V_H}(B)) = ks = 2ns$, so as $m_2(Aut_M(V_M)) \le 3n$ by an earlier remark, we conclude that $s = 1$ and $V_H/C_{V_H}(K)$ is the natural module for K^*. Now $m(Aut_{V_H}(B)) = k = m(B/C_B(V_H))$, so as B is the sum of two isomorphic natural modules for $SL_3(q)$, V_H induces R_2^g on B, $m([B, V_H]) = 4n$, and $V_2^g \le [B, V_H]$.

As $m([B, V_H]) = 4n$, V_H is the $6n$-dimensional maximal central extension of the natural module for K^* appearing in I.2.3.1. Then from the structure of that module as orthogonal 3-space over $\mathbf{F}_{2^n}$, $[V_H, a] \cap [V_H, b] = 1$ for $a^* \neq b^*$ in B^*; whereas from the action of R_2^g on V^g, for elements $a, b \in V^g$ in distinct cosets of V_2^g, we $[a, V_H] = [a, R_2^g] = V_2 = [b, R_2^g] = [b, V_H]$. This contradiction completes the proof of the lemma. $\qquad\square$

We now obtain successive restrictions forcing various 2-locals to closely resemble those in the shadow $L_4(q)$.

Set $K := O^2(H)$, let D be a Hall $2'$-subgroup of $H \cap M$, and further set $D_0 := O^3(D)\Omega_1(O_3(D))$. Recall X is a Cartan subgroup of L acting on T_L.

LEMMA 11.3.7. *(1)* $K \in \mathcal{C}(H)$ and $K^* \cong K/O_2(K) \cong L_2(2^n)$.
(2) $D_0 \leq L$.
(3) $C_Z(L) = C_V(L) = 1 = C_{V_H}(K)$.
(4) $V = A$ and $V^* \in Syl_2(K^*)$.
(5) $V_1 = [V_1, D_0]$ and we may take $D_0 \leq X$.

PROOF. Part (1) follows from 11.3.6 and 11.3.5.2, and then (2) follows from 11.0.4. Next $\bar{L} \cong SL_3(q)$ or $Sp_4(q)$ by 11.3.4.2, so that $m = s(G, V) = n$ by 11.2.2. Thus $A^* \in \mathcal{A}_n(H^*, V_H)$ by 11.3.4.1, so it follows that $A^* \in Syl_2(K^*)$ is of rank n and $C_{V_H}(A^*) = C_{V_H}(a^*)$ for all $a^* \in A^{*\#}$. Then by G.1.6, $V_H/C_{V_H}(K)$ is a direct sum of s copies of the natural module for K^*.

Let $V_L := [\langle Z^L \rangle, L] = [Z, L]$, so that $V \leq \langle Z^L \rangle = V_L C_Z(L)$ using B.2.14. Similarly $V_H = [V_H, K]C_Z(K)$. As $[Z, H] \neq 1$ by Hypothesis 11.3.1.1, $L = [L, J(T)]$ by Theorem 3.1.8.3; therefore by Theorem B.5.1.1, either $V_L = V$ and $\tilde{V}$ is the natural module for $\bar{L}$, or $\bar{L} \cong SL_3(q)$ and V_L is the sum of two isomorphic natural modules for $\bar{L}$. As $D_0 \leq L$ and $TD_0 = D_0T$, we may take $D_0 \leq X$, and either $C_Z(D_0) = C_Z(L)$, or conjugating in $N_L(V_2)$ if necessary, we may assume $[V_1, D_0] = 1 = [Z, D_0]$. On the other hand as $V_H/C_{V_H}(K)$ is a sum of natural modules for K^* and $V_H = [V_H, K]C_Z(K)$, $C_Z(D_0) = C_Z(K)$. In particular, $[Z, D_0] \neq 1$, so $C_Z(D_0) = C_Z(L)$. Then as $K \not\leq M = !\mathcal{M}(LT)$, $1 = C_Z(K) = C_Z(D_0) = C_Z(L)$, so (3) follows, [3] $[V_H, K] = V_H$, $V_1 = [V_1 \cap Z, D_0] \leq V_H$, and the proof of (5) is complete. By 11.2.2.4, $V_1 \leq C_{V_H}(W_0) \leq C_{V_H}(A)$.

Next $C_G(V_2) \leq C_G(V_1) \leq M$ using 11.3.4.4. Thus as $m(A/C_A(V_H)) = n$, 11.2.2.3 says $V_H \leq C_G(C_A(V_H)) \leq N_G(A)$. Now V_H centralizes $C_A(V_H)$ of corank n in A, so $V_H/C_{V_H}(A)$ is contained in the group Λ of all $\mathbf{F}_q$-transvections on A with axis $C_A(V_H)$. From the action of A^* on V_H, $C_{V_H}(A)$ is of rank sn, so as $m_2(\Lambda) = 2n, n$ for $\bar{L} \cong SL_3(q), Sp_4(q)$, conjugating in L^g if necessary, either:

(i) $s = 1$, $m(V_H/C_{V_H}(A)) = n$, and $[A, V_H] = V_1^g$, or
(ii) $s = 2$, $\bar{L} \cong SL_3(q)$, $Aut_{V_H}(A) = Aut_{R_2^g}(A)$, and $V_2^g = [A, V_H]$.

In case (i), $V_1^g = [A, V_H] = C_{V_H}(A)$, so $V_1^g = V_1$ using our earlier observation that $V_1 \leq C_{V_H}(A)$; hence $A = V$ by 11.3.4.4. Similarly in case (ii), from the action of R_2^g on A, for each $u \in V_H - V_2^g$, $[u, A] = V_1^{gx}$ for some $x \in L_2^g \leq N_G(A)$. Further V_H is the sum of two natural modules for K^* and $V_1 = [V_1, D_0] \leq C_{V_H}(A)$, so V_1 is a 1-dimensional $\mathbf{F}_q$-subspace of $C_{V_H}(A)$. Therefore $V_1 = [A, u]$ for some $u \in V_H - V_2^g$, so again $V_1 = V_1^{gx}$ for some $x \in N_G(A)$, and hence we may assume $V_1 = V_1^g$, so $A = V$ by 11.3.4.4. This completes the proof of (4). $\qquad\square$

---

[3] So we have eliminated the shadow of $Sp_6(q)$ where $\bar{L} \cong Sp_4(q)$ and $C_V(L) \neq 1$.

By 11.3.7.4, $V = A \not\le O_2(H)$. Therefore we have Hypothesis E.2.8 with H, T, $H \cap M$ in the roles of "H, T, M". Let $h \in H - M$ and set $I := \langle V, V^h \rangle$. By 11.3.7, $K^* \cong L_2(2^n)$ with $V^* \in Syl_2(K^*)$, so $I^* = K^*$. Thus I is in the set $\mathcal{I}(H, T, V)$ defined in Definition E.2.4. Therefore by E.2.9.1, $O_2(H)$ acts on I, so $K = O^2(I)$, and $I \trianglelefteq H$ as T acts on V and $H = KT$.

Next by E.2.11, the hypotheses of E.2.10 are satisfied with I, $M \cap I$, $T \cap I$, V in the roles of "H, M, T, V". Notice also that $B := C_V(V_H) = V \cap O_2(H)$ plays the role of "B" in E.2.10, and by that result $P := BB^h$ is a normal 2-subgroup of I. Since $V \cap V^h \le Z(I)$ and $K = O^2(I)$, $V \cap V^h = 1$ since $C_{V_H}(K) = 1$ by 11.3.7.3. Therefore $P = B \times B^h$ and $B = C_P(V^*)$ by E.2.10.2. By E.2.10.7, $P = O_2(I)$, and by G.1.7, P is a sum of j natural modules for $K/O_2(K) \cong L_2(q)$, so $P = [P, K]$ and hence $P = O_2(K)$. Therefore B^h acts faithfully on V, with $B = C_V(B^h)$ of corank n in V and $m(B^h) = m(B) = jn$. Thus B^h is a group of transvections with axis B, so $\bar{L} \cong SL_3(q)$, $j = 2$, and B is T-invariant of rank $2n$; hence $B = V_2$ and $\bar{P} = \bar{B}^h = \bar{R}_2$. Thus P is of rank $4n$ with $P \cap V = V_2$. As $V_2 \cap Z \le V_1$, and $V_1 = [V_1, D_0]$ by 11.3.7.5, $V_K := \langle (Z \cap P)^H \rangle = \langle V_1^H \rangle$. As P is a sum of two natural modules for $K^* \cong L_2(2^n)$, V_K is a natural submodule of P of rank $2n$ and

$$[O_2(LT), V_K] \le O_2(LT) \cap V_K = V_1.$$

Therefore $L = [L, V_K]$ centralizes $O_2(LT)/V$, so L is an $SL_3(q)$-block. Thus L/V is a covering group of $SL_3(q)$, so from the list of Schur multipliers in I.1.3, either $V = O_2(L) = C_L(V)$, or $q = 4$ and $O_2(L/V) \ne 1$. However in the latter case $(R_2 \cap L)/V$ does not split over $O_2(L)/V$ by I.2.2.3b, whereas $\bar{P} = \bar{R}_2$ and $PV/V \cong P/V_2 \cong E_{q^2}$ is T-invariant. Thus $V = C_L(V) = O_2(L)$, and then as $P = J(PV) \cong E_{q^4}$, $P = O_2(L_2)$.

Recall X is a Cartan subgroup of L acting on T_L and we may take $D_0 \le X$ by 11.3.7.5. Notice that in the shadow $L_4(q)$, the Cartan group D of $H \cap M$ is not contained in the derived subgroup L of M, and DX is a group of rank 3 for primes dividing $q - 1$. Indeed, $D_0 \not\le L$, so it remains to show that the unnatural inclusion $D_0 \le L$ leads to a contradiction. This is accomplished by studying the action of X on P and V.

Assume n is even. Then the subgroup D_3 of D_0 of order 3 is contained in X. However $C_P(D_3) = 1 = C_V(D_3)$ as P is the sum of natural modules for K^* and $V^* = [V^*, D_3]$, whereas $C_X(L/V) \in Syl_3(O_{2,Z}(L))$ is the only subgroup of X of order 3 having no fixed points on V, and it centralizes PV/V.

Thus n is odd, so $D = D_0 \le X$ and $T = T_L$. Now $C_T(L) = 1$ by 11.3.7, and $H^1(L/V, V) = 0$ by I.1.6, so that $V = O_2(LT)$ by C.1.13.b. Thus $T = T_L \in Syl_2(L)$. Set $G_P := N_G(P)$, $\hat{G}_P := G_P/P$, and $Y := \langle L_2, I \rangle$. We've seen that $P = O_2(L_2) = O_2(K)$ with $\hat{L}_2 \cong \hat{K} \cong L_2(q)$ and $\hat{T} \cong E_{q^2}$. As $K \in \mathcal{L}(G_P, T)$, $K \le K_+ \in \mathcal{C}(G_P)$ by 1.2.4, and if $K < K_+$, then the embedding of K in K_+ is described in A.3.12. As $\hat{K} \cong L_2(2^n)$ with n odd and $\hat{T}$ is abelian, we conclude $K = K_+ \in \mathcal{C}(G_P)$. Similarly $L_2 \in \mathcal{C}(G_P)$. Next $\hat{L}_2 \ne \hat{K}$ since $K \not\le M$, so $\hat{Y} = \hat{K} \times \hat{L}_2 \cong \Omega_4^+(q)$. As $|\hat{T}| = q^2 = |\hat{Y}|_2$, $Y = O^{2'}(G_P)$. As P is the sum of two copies of the natural module for $\hat{K}$, and V and P/V are natural modules for $\hat{L}_2$, P is the orthogonal module for $\hat{Y}$. As $X \le N_G(P)$ and $m_p(N_G(P)) \le 2$ for each prime divisor p of $q - 1$, X is a Cartan subgroup of Y.

We next show:

LEMMA 11.3.8. *There exist $\mathbf{F}_q$-structures on P and V, preserved by Y and L, respectively, which agree on $P \cap V = V_2$.*

PROOF. Let

$$E_V := End_{\mathbf{F}_q L}(V), \quad E_{P \cap V} := End_{\mathbf{F}_q L_2 X}(P \cap V), \quad \text{and } E_P := End_{\mathbf{F}_q Y}(P).$$

Then $E_W \cong \mathbf{F}_q$ for each $W \in \{P, V, P \cap V\}$. In particular we may regard $E_{P \cap V}$ as the restriction of E_W to $P \cap V$ for $W := P, V$, so the lemma holds. $\square$

Let $X_1 := C_X(L_1/O_2(L_1))$ and $X_2 := X \cap L_2$. Let W_1 be an X-complement to $V \cap P$ in V, and W_2 an X-complement to $P \cap R_1$ in P. Finally let $W_3 := [W_1, W_2]$. Then W_3 is X-invariant, and $\langle W_1, W_2 \rangle = W_1 W_2 W_3$ is a special group of order q^3 with center W_3. By 11.3.8, the $\mathbf{F}_q$-structures on P and V restrict to X-invariant $\mathbf{F}_q$-structures on W_i, which agree on W_3. Thus we may regard W_i as an $\mathbf{F}_q X$-module.

LEMMA 11.3.9. *The map $c : W_1 \times W_2 \to W_3$ defined by $c(w, w') := [w, w']$ is X-invariant and $\mathbf{F}_q$-bilinear.*

PROOF. Since X acts on W_i, c is X-invariant and $\mathbf{F}_2$-bilinear. Pick generators w_i for W_i as an $\mathbf{F}_q$-space with $[w_1, w_2] = w_3$. Using the $\mathbf{F}_q$-structure on P, we may write $X_2 = \{x(\lambda) : \lambda \in \mathbf{F}_q^{\#}\}$ so that $x(\lambda) w_2 = \lambda w_2$. Next $[V, X_2] \leq [V, L_2] \leq V_2 = P \cap V$ from the action of L on V, so as W_1 is X-invariant, $[W_1, X_2] = 1$. As W_1 centralizes X_2, it acts on the λ-eigenspace of $x(\lambda)$ on P; then as W_2 is contained in that eigenspace, so is $W_3 = [W_1, W_2]$—and hence $x(\lambda) w_3 = \lambda w_3$. Thus

$$\lambda w_3 = x(\lambda) w_3 = [x(\lambda) w_1, x(\lambda) w_2] = [w_1, \lambda w_2],$$

and hence c is linear in its second variable. Similarly X_1 centralizes W_2, since W_2 covers a Sylow 2-group of $L_1/O_2(L_1)$, so $W_1 W_3$ is an eigenspace for each member of $X_1^{\#}$ on V, and the same argument shows c is linear in its first variable. $\square$

We are now in a position to obtain a contradiction, and hence finally eliminate the shadow of $L_4(q)$. Let y be a generator for $X \cap K = D$. Then y has two eigenspaces on P: $P \cap V = V_2$ and $\langle W_2^{L_2} \rangle$. Let λ be the eigenvalue on the second space; then as y is of determinant 1 on P, y has eigenvalue λ^{-1} on $P \cap V$. Similarly y is of determinant 1 on V, so the eigenvalue for y on W_1 is λ^2. Then by 11.3.9, the eigenvalue for y on W_3 is the product $\lambda^2 \lambda = \lambda^3$ of its eigenvalues on W_1 and W_2. This is impossible, as $W_3 = [W_1, W_2] \leq [V, P] = P \cap V$ and the eigenvalue for y on $P \cap V$ is λ^{-1}. This contradiction completes the proof of Proposition 11.3.2.

11.4. Eliminating the remaining shadows

Recall from earlier discussion that the shadows other than $Sp_6(q)$ and $\Omega_8^+(q).2$ with $\bar{L} \cong SL_3(q)$, have been eliminated. In these remaining shadows, the centralizer of a 2-central involution is not quasithin, and we essentially eliminate those configurations in 11.4.4 in this section.

LEMMA 11.4.1. *(1) $C_V(L) = 1$. In particular, L is transitive on $V^{\#}$.*
(2) If $L_1 < K \in \mathcal{C}(C_G(V_1))$, then K is described in case (1) or (2) of 11.1.2.
(3) $R_1 \leq O_{\infty}(KT)$, and $[R_1, X] \leq O_2(KT)$.

PROOF. Let $H \in \mathcal{H}_*(T, M)$. If $C_V(L) \neq 1$, then $\bar{L}$ is not $SL_3(q)$ by 11.0.3.3, and $[Z, H] \neq 1$ as $H \nleq M = !\mathcal{M}(LT)$, contrary to 11.3.2.3. Then 11.0.2.2 completes the proof of (1).

Suppose 11.1.2.3 holds. Observe I satisfies the hypotheses for L, so applying 11.0.2.2 and (1) to $V_I \in Irr_+(I, R_2(IT))$ in the Fundamental Setup (3.2.1), we conclude V_I is the natural module for $I^* := I/C_I(V_I) = I/O_2(I) \cong Sp_4(4)$. From the proof of 11.1.2.3, L_2 stabilizes a line of V_I. In 11.1.2.3, we also have $L/O_2(L) \cong \bar{L} \cong G_2(4)$. By 11.2.2.5, $W_0 := W_0(T, V) \leq C_T(V)$, and hence $W_0 \leq C_{L_2T}(V_2) = O_2(L_2T)$; furthermore $N_G(W_0) \leq M$, so as $I \nleq M$, $W_0 \nleq O_2(IT) = C_{IT}(V_I)$. Thus $1 \neq W_0^* \leq O_2(L_2^*T^*) = O_2(L_2^*) = R_2^*$, and R_2^* is of rank 6. Let $A := V^g \leq T$ with $A^* \neq 1$. Let R^* be a root subgroup of $O_2(L_2^*)$, and A_R the preimage in A of $A^* \cap R^*$. Then $m(A^*/A_R^*) \leq m(O_2(L_2^*)/R^*) = 4$. By 11.2.2.3 $r(G, V) > 2n = 4$, so $C_G(A_R) \leq N_G(A)$. Hence from the action of R_2^* on V_I,

$$V_I = \langle C_{V_I}(A_R) : R^* \leq O_2(L_2^*) \rangle \leq N_G(A).$$

On the other hand by 11.2.2.3, $s(G, V) = 4$, so $m(A^*) \geq 4$ by E.3.10. It follows that $[V_I, A] = C_{V_I}(A)$ is of rank 4 so $m(V_I/C_{V_I}(A)) = 4$; thus as A is the natural module for $\bar{L}^g$, $Aut_{V_I}(A)$ is contained in a long root group of $Aut_{L^g}(A) \cong G_2(4)$ of rank 2 (e.g. see (6) and (13) of B.4.6), forcing $m(V_I/C_{V_I}(A)) \leq 2$. This contradiction establishes (2).

If $L_1 = K$, then (3) is immediate. Otherwise by (2), K is described in case (1) or (2) of 11.1.2. In those cases, observe that L_1 has no nontrivial 2-signalizers in $Aut(K/O_\infty(K))$, so that $R_1 \leq O_\infty(KT)$. Since $O_{2,F}(KT)$ is of index 1 or 2 in $O_\infty(KT)$, $[R_1, X] \leq O_2(KT)$, so that (3) holds in these cases also. $\qquad \square$

We can now return to our study of the embedding of L_1 in $C_G(V_0)$ for $1 \neq V_0 \leq V_1$, begun in the initial section of the chapter. For $1 \neq V_0 \leq V_1$ with $T \leq N_G(V_0)$, define $K(V, V_0) := \langle L_1^{N_G(V_0)} \rangle$; and for $z \in V_1^{\#}$, let $K(V, z) = K(V, \langle z \rangle)$.

LEMMA 11.4.2. Let $z \in C_V(T)^{\#}$. Then $K(V, z) = K(V, V_1)$.

PROOF. Let $K := K(V, z)$ and $K_1 := K(V, V_1)$. By 11.1.1, $K \in \mathcal{C}(N_G(\langle z \rangle))$ and $K_1 \in \mathcal{C}(N_G(V_1))$. Then $K_1 \leq C_G(V_1) \leq C_G(z)$, so $K_1 = \langle L_1^{K_1} \rangle \leq K$.

We assume that $K_1 < K$ and derive a contradiction. As $K_1 \trianglelefteq C_G(V_1)$, $K \nleq C_G(V_1)$. Further $L_1 < K$, so K is described in case (2) or (3) of 11.1.1. As $K \in \mathcal{L}(G, T)$, $K \leq K_+ \in \mathcal{L}^*(G, T)$.

Assume first that $K < K_+$. Then the embedding of K in K_+ is described in A.3.12. The pairs $K/O_2(K)$, $K_+/O_2(K_+)$ appearing there with K described in case (2) or (3) of 11.1.1 (cf. also A.3.13) are: $L_3(4)$, M_{23}; A_7, M_{23}; $L_2(p)$, $(S)L_3^\epsilon(p)$, for a prime $p \geq 11$; $SL_2(p)/E_{p^2}$, $(S)L_3(p)$ for a prime $p \geq 5$; M_{22}, M_{23}; $\hat{M}_{22}$, J_4; and $SL_2(5)/P_0$, $SL_2(5)/P_1$, where P_0 and P_1 are suitable nilpotent groups of odd order. Moreover as $K < K_+$, $[z, K_+] \neq 1$, so in the last case $[z, O_\infty(K_+)] \neq 1$, contrary to 3.2.14. Therefore $K_+/O_2(K_+)$ is quasisimple. Further as $[z, K_+] \neq 1$, $K_+ \in \mathcal{L}_f^*(G, T)$. But this contradicts Theorem 7.0.1, since K_+ has no FF-module by Theorem B.4.2.

Therefore $K = K_+ \in \mathcal{L}^*(G, T)$. As $K_1 < K$, $[V_1, K] \neq 1$. But by A.1.6, $O_2(KT) \leq R_1$, and $V_1 \leq Z(R_1)$, so

$$V_1 \leq \Omega_1(Z(O_2(KT))) =: V_K.$$

In particular $[V_K, K] \neq 1$, so $K \in \mathcal{L}_f^*(G, T)$ by A.4.9.

Suppose first that $K/O_2(K)$ is not quasisimple. Then $O_\infty(K)$ centralizes $R_2(KT)$ by 3.2.14, and we conclude from A.4.11 that $O_{2,F}(K)$ centralizes V_K and hence also V_1. By 11.1.1, either $K = O_{2,F}(K)L_1$; or $K/O_{2,F}(K) \cong SL_2(p)$ for a prime p with $p \equiv \pm 1 \mod 5$ and $p \equiv \pm 3 \mod 8$. However in the former case, K centralizes V_1, contrary to an earlier remark, so the latter case holds.

By 11.4.1.3, $R_1 \leq O_\infty(KR_1)$. Thus $R_1 \in Syl_2(O_\infty(KR_1))$, so by a Frattini Argument, $K = O_{2,F}(K)K_J$, where $K_J := N_K(R_1)^\infty$. Therefore $K_J/O_2(K_J) \cong K/O_\infty(K) \cong L_2(p)$. As $O_{2,F}(K)$ centralizes V_1 but K does not, $K_J \not\leq C_G(V_1)$.

First $XT \leq N_G(R_1) =: N$. We claim that XT normalizes K_J. By 1.2.4, $K_J \leq K_0 \in \mathcal{C}(N)$. The claim is immediate if $K_J = K_0$, so we may assume that $K_J < K_0$, and hence the embedding is described in A.3.12. As R_1 is Sylow in $O_\infty(KR_1)$ and normal in N, $K_0/O_2(K_0)$ is not $SL_2(p)/E_{p^2}$, so $K_0/O_2(K_0)$ is quasisimple. Hence as $p \geq 11$ since $p \equiv \pm 1 \mod 5$, we conclude that $K_0/O_2(K_0) \cong L_2(p^2)$. But then T does not act on K_1, establishing the claim. Therefore as $K_J \leq C_G(z)$ while $X \leq N_N(K_J)$ by the claim, $V_1 = [z, X] \leq C_G(K_J)$, contrary to the previous paragraph.

Therefore $K/O_2(K)$ is quasisimple, so $V_K \in \mathcal{R}_2(KT)$ and $O_2(KT) = C_T(V_K)$ by 1.4.1.4. Set $K^* := K/C_K(V_K)$. We saw $O_2(KT) \leq R_1$. Thus if $J(R_1) \leq O_2(KT)$ then $J(R_1) = J(O_2(KT))$. Hence $KT \leq N_G(J(R_1))$, so $N_G(J(R_1)) \leq N_G(K)$ by 1.2.7.3. Thus $X \leq N_G(K)$, so as $K \leq C_G(z)$ and $C_V(L) = 1$ by 11.4.1.1, $V_1 = [z, X] \leq C_G(K)$, for our usual contradiction.

Therefore $J(R_1)^* \neq 1$, so we may apply B.2.10.2 with K, T, R_1 in the roles of "L, T, R" to conclude that V_K is an FF-module for K^*T^*, with $K^* = [K^*, J(R_1)^*] = J(K^*T^*, V_K)$. Then Theorems B.5.1 and B.4.2 reduce the list in 11.1.1 to those cases where either K^* is $SL_3(q)$, $Sp_4(q)$, or $G_2(q)$, or $q = 4$ and K^* is A_7.

Assume K^* is not A_7, and let Y be a Hall $2'$-subgroup of $N_K(R_1)$. Then Y centralizes z, and induces a group of order $q - 1$ on $Z_1 := \Omega_1(Z(R_1))$ containing V_1, so as V_1 has order q, we conclude $V_1 < Z_1$. Hence $Y \leq N_G(R_1) \leq M$ by 11.1.5. Then by 11.0.4, $Y_1 := O^3(Y)\Omega_1(O_3(Y)) \leq L$. As Y^* has p-rank 2 for primes p dividing $q - 1$, so does $\bar{Y}$. Therefore as V is the natural module for $L/O_2(L)$. $z \in C_{V_1}(Y_1) = C_V(L)$, contrary to 11.4.1.1.

Thus K^* is A_7. As L_1^* centralizes $z \in V_K - C_{V_K}(K)$, $[V_K, K]$ is not a 4-dimensional irreducible for K^*. Therefore by B.4.2.5, $[V_K, K]$ is the natural 6-dimensional module, $J(R_1)^*$ is generated by a transposition, and $q(K^*T^*, V_K) = 1$, so $m_2(T) = m_2(R_1)$ by B.2.4.3. Thus conjugating in K, there is $A \in \mathcal{A}(T)$ such that A^* induces a field automorphism on L_1^*. However this is impossible since $J(T) \leq LC_T(\bar{L})$ for the natural modules in 11.0.2.2 by (2)–(4) of B.4.2. This contradiction finally shows that $K_1 = K$, completing the proof. $\square$

In the remainder of the section, fix $z \in C_V(T)^\#$.

Set $G_z := C_G(z)$, $M_z := C_M(z)$, and $K := K(V, V_1)$. Then $K = K(V, z) \trianglelefteq G_z$ by 11.4.2. By 11.4.1.2, either $K = L_1$, or K is described in case (1) or (2) of 11.1.2. Furthermore if $G_z \leq M$, then as $z \in V_1$, $C_G(V_1) \leq G_z \leq M$, so for $H \in \mathcal{H}_*(T, M)$, $[Z, H] \geq [z, H] \neq 1$; thus Hypothesis 11.3.1 holds, so $C_G(V_1) \not\leq M$ by 11.3.2.2, a contradiction. Therefore

$$G_z \not\leq M.$$

Define $G_1 := N_G(V_1)$ (as opposed to $C_G(V_1)$ in earlier sections), and $M_1 := N_M(V_1)$. Recall that $X \leq G_1$.

LEMMA 11.4.3. (1) $G_z = KC_{G_z}(K/O_2(K))M_z$. Therefore if $K = L_1$, then $C_{G_z}(K/O_2(K)) \not\leq M$.

(2) $G_1 = K(C_{G_1}(V_1) \cap C_{G_1}(K/O_2(K)))M_1$.

PROOF. Recall $K = K(V, V_1) = K(V, z)$ is normal in G_1 and G_z. Set $Y := C_{G_z}(K/O_2(K))$. From 11.1.2, $Out(K/O_2(K))$ is 2-closed, so $YKT \trianglelefteq G_z$, and hence by a Frattini Argument, $G_z = YKN_{G_z}(T)$. Now by Theorem 3.3.1, $N_G(T) \leq M$, proving the first assertion of (1). So if $K = L_1$, then as $G_z \not\leq M$, $Y \not\leq M$, giving the remaining assertion of (1).

Now instead set $Y := C_{G_1}(V_1) \cap C_{G_1}(K/O_2(K))$. The same proof shows that $C_{G_1}(V_1)T = YTKC_{M_1}(V_1)$. As $C_V(L) = 1$ by 11.4.1.1, $G_1 = C_{G_1}(V_1)M_1$ by 11.2.3.1, proving (2). $\qquad\square$

Since $G_z \not\leq M$, there is $H \in \mathcal{H}_*(T, M) \cap G_z$; H has this meaning for the remainder of the chapter.

Since $n(H) \geq n$ in the shadows, our next result eliminates those groups:

LEMMA 11.4.4. $n(H) = 1$.

PROOF. Assume $n(H) > 1$; then from E.2.2, $O^2(H) = \langle I^T \rangle$ for some $I \in \mathcal{L}(G, T)$ with $I \not\leq M$. By 1.2.4 $I \leq I_1 \in \mathcal{C}(G_z)$, and then by 1.2.1, either $[L_1, I] \leq [K, I_1] \leq O_2(K) \leq O_2(L_1T)$, so $[L_1, I] \leq O_2(L_1)$, or $I \leq I_1 \leq K$. But by 11.4.1.2, either $K = L_1$ or K is described in case (1) or (2) of 11.1.2; in either case L_1 is the unique minimal member of $\mathcal{L}(G, T) \cap K$. Therefore if $I \leq K$ then $I = L_1 \leq M$, contradicting $I \not\leq M$. Thus $[L_1, I] \leq O_2(L_1)$.

Let B be a Hall $2'$-subgroup of $I \cap M$. Then $B \leq C_M(L_1/O_2(L_1))$ and B centralizes z. For each prime divisor p of $q - 1$, L contains each subgroup B_p of B of order p by 11.0.4, so $B_p \leq C_L(z) \cap C_L(L_1/O_2(L_1)) = L \cap R_1$ from the action of L on the natural module V. Therefore $(|B|, q - 1) = 1$.

As B centralizes z, $B \leq M_V$ by 11.0.3.4. Then as $BT = TB$, $[L_1, B] \leq O_2(L_1)$, and $(q - 1, |B|) = 1$, it follows from the action of $N_{Aut(V)}(\bar{L})$ on V that $[V, B] = 1$. But then using Remark 4.4.2, $N_G(B) \leq M$ by Theorem 4.4.3; so $H = \langle H \cap M, N_H(B) \rangle \leq M$, contradicting $H \not\leq M$. $\qquad\square$

11.5. The final contradiction

We now work to obtain a contradiction, by analyzing the normal closure $\langle V^{G_1} \rangle$ of V in G_1. The analysis falls into two cases, depending on whether $\langle V^{G_1} \rangle$ is abelian or not. The strong restriction in 11.4.4 will make weak closure methods more effective.

LEMMA 11.5.1. $L/O_2(L)$ is $SL_3(q)$ or $Sp_4(q)$.

PROOF. In view of 11.0.2, we may assume $\bar{L} \cong G_2(q)$. Hence by parts (5) and (6) of 11.2.2,

$$C_G(C_1(R_1, V)) \leq M \geq N_G(W_0(R_1, V)).$$

Thus it will suffice to find $H_1 \leq H$ with $H_1 \not\leq M$, $n(H_1) = 1$, and $R_1 \in Syl_2(H_1)$: For since $n(H_1) = 1$ and $s(G, V) > 1$ by 11.2.2.3, we may apply E.3.19 to conclude that $H_1 \leq M$, contrary to our choice of H_1.

Suppose first that $K = L_1$. Then by 11.4.3.1, $C_{G_z}(L_1/O_2(L_1)) \not\leq M$, so we may choose H with $[L_1, O^2(H)] \leq O_2(L_1)$ and set $H_1 := R_1O^2(H)$; then $n(H_1) = 1$ using 11.4.4. On the other hand if $L_1 < K$, then by 11.4.1.2, K is described in case

(1) or (2) of 11.1.2. In case (1) of 11.1.2, $O_{2,2'}(K) \not\leq M$, so we may choose $R_1 \leq H_1 \leq R_1 O_{2,2'}(K)$, and then $n(H_1) = 1$ by E.1.13. As $R_1 \in Syl_2(C_G(L_1/O_2(L_1)))$, $R_1 \in Syl_2(H_1)$, completing the proof in these two cases by paragraph one.

In case (2) of 11.1.2, $K/O_2(K) \cong L_2(p)$ for a prime $p \geq 11$, and L_1 has no nontrivial 2-signalizer in $Aut(K/O_2(K))$, so $R_1 = O_2(KT)$. Thus $K \leq N_G(R_1) \leq N_G(W_0(R_1, V)) \leq M$, a contradiction. $\qquad \square$

As $G_z \not\leq M$, 11.4.3.1 says $C_{G_z}(K/O_2(K))$ or K is not contained in H. Thus the following choice is possible:

From now on we choose H so that either $[K, O^2(H)] \leq O_2(K)$, or $O^2(H) \leq K$.

In particular notice that if $K = L_1$, then as $H \not\leq M$, $[K, O^2(H)] \leq O_2(K)$. By 11.4.4, $n(H) = 1$. Recall $G_1 = N_G(V_1)$ and set $\tilde{G}_1 := G_1/V_1$.

LEMMA 11.5.2. *(1) If $z \in V \cap V^g$ then $V^g \in V^{G_z}$. That is G_z is transitive on $\{V^g : z \in V^g\}$.*

(2) $K = K(V^g, z)$ for each $g \in G_z$.

PROOF. By 11.4.1.1, L is transitive on $V^\#$, so (1) holds using A.1.7.1.

As $K \trianglelefteq G_z$, for $g \in G_z$

$$K = K^g = K(V, z)^g = K(V^g, z),$$

so (2) holds. $\qquad \square$

LEMMA 11.5.3. *Assume $K = L_1$ and set $m := 2n, 3n$, for $\bar{L} \cong SL_3(q), Sp_4(q)$, respectively. Then*

(1) For each $g \in G - N_G(V)$, $V \cap V^g \leq V_1^y$ for some $y \in L$.

(2) $r(G, V) \geq m$.

(3) $\langle V^{G_1} \rangle$ is nonabelian.

PROOF. Let $g \in G - N_G(V)$. Our first goal is to prove (1), so we may suppose $1 \neq U := V \cap V^g$. By transitivity of L on $V^\#$, we may assume $U \cap V_1 \neq 1$, and we may suppose that $U \not\leq V_1$. For $u \in U^\#$, $V^g = V^{g_u}$ for some $g_u \in C_G(u)$ by 11.5.2.1. Now $u \in V_1^x$ for some $x \in L$, and $K(u) := K(V, u) = K(V^{g_u}, u) = K(V^g, u)$ by 11.5.2.2, while $K(u) = L_1^x$ by our hypothesis that $K = L_1$. Since $U \not\leq V_1$, $L = \langle K(u) : u \in U^\# \rangle$; so as $U^{g^{-1}} \leq V$, $L^{g^{-1}} = L$, and hence $g \in M$; indeed $g \in M_V$ since V is a TI-set under M by 11.0.3.4. This contradicts the choice of g, so (1) is established.

If $U \leq V$ with $m(V/U) < m$, then $U \not\leq V_1^y$ for any $y \in L$, so $C_G(U) \leq M_V$ by (1). Thus (2) holds.

Suppose that $\langle V^{G_1} \rangle$ is abelian. By (2), $C_G(V_2) \leq M$. Hence $W_0 := W_0(T, V) \not\leq O_2(H)$ by 11.2.5. Then since H is a minimal parabolic with $H \cap M$ the unique maximal overgroup of T, $N_H(W_0) \leq H \cap M$. As $m(\bar{M}_V, V) > 1$, $s(G, V) > 1$ by (2). Hence as $n(H) = 1$ by 11.4.4, it will suffice to show $C_G(C_1(T, V)) \leq M$, since then E.3.19 supplies a contradiction. Indeed as $C_G(V_2) \leq M$, it suffices to show $V_2 \leq C_1(T, V)$.

So suppose $A := V^g \cap T$ is of corank at most 1 in V^g, but $[V_2, A] \neq 1$. Let $B := C_A(V_2)$. Then

$$m(V^g/B) \leq m_2(Aut_M(V_2)) + 1 = n + 1 < 2n \leq m,$$

so by (2), $V_2 \leq C_G(B) \leq N_G(V^g)$. Hence $D := C_{V^g}(V_2)$ is of corank at most $n + 1$ in V^g, so from the action of M_V^g on V^g either:

(i) D is of corank exactly n in V^g and V_2 induces transvections with axis D on V^g, or

(ii) $n = 2$, $\bar{L} = SL_3(4)$, and V_2 induces a group of field automorphisms on $Aur_{L^g}(V^g)$.

By (1), $[V_2, A] \leq V \cap V^g \leq V_1^y \cap V_1^{gw}$ for some $y \in L$ and $w \in L^g$, so $\bar{A} \leq \bar{T}_L$ and (i) holds; then since $[V_2, \bar{A}] \neq 1$, $[V_2, A] = V_1$. As V_2 induces transvections with axis D on V^g, $V_1 = [A, V_2] \in V_1^{gL^g}$, so we may take $g \in G_1$. But then as $[V, V^g] \neq 1$, we have a contradiction to our assumption that (3) fails. This shows $V_2 \leq C_1(T, V)$, and completes the proof. $\qquad \square$

LEMMA 11.5.4. $C_G(V_1) \not\leq M$, so we may choose $H \leq G_1$ with $O^2(H) \leq C_G(V_1)$.

PROOF. If $L_1 < K$, then $K \leq C_G(V_1)$ but $K \not\leq M$. On the other hand, if $L_1 = K$, then by 11.5.3.3 $\langle V^{G_1} \rangle$ is nonabelian, so $G_1 \not\leq M$ since $V \leq Z(O_2(M))$. Hence $C_G(V_1) \not\leq M$ by 11.2.3.2, so using 11.4.3.2 and the argument we made just before 11.5.2, we can choose $H \leq G_1$ with $O^2(H) \leq C_G(V_1)$, while aintaining the condition $O^2(H) \leq K$ or $[K, O^2(H)] \leq O_2(K)$. $\qquad \square$

Because of 11.5.4, the set $\mathcal{H}_1 := \mathcal{H}(L_1T, M) \cap G_1$ is nonempty. In the remainder of this section we choose $H_1 \in \mathcal{H}_1$ and set $U_H := \langle V_3^{H_1} \rangle$.

LEMMA 11.5.5. (1) $U_H \leq O_2(H_1)$.
(2) $\tilde{U}_H \leq Z(O_2(\tilde{H}_1))$, and $\Phi(U_H) \leq V_1$.
(3) If $K = L_1$, then $\tilde{U}_H$ is a direct sum of natural modules for $K/O_2(K)$.

PROOF. Observe that Hypothesis G.2.1 is satisfied with V_3, H_1 in the roles of "V, H"; hence (1) and (2) hold by G.2.2. Further $\tilde{V}_3$ is the natural module for $L_1/O_2(L_1)$, so if $L_1 = K$, then as $K \trianglelefteq G_1$, (3) holds. $\qquad \square$

LEMMA 11.5.6. Let $Y := C_G(V_1) \cap C_G(K/O_2(K))$. Then
(1) $(|Y|, q - 1) = 1$.
(2) $m_3(Y) \leq 1$, and if n is even, then Y is a $3'$-group.
(3) If $I \in \mathcal{C}(Y)$ then $[I, X] \leq O_2(Y)$.
(4) If $P = [P, X] \leq T$ and $\Phi(P) \leq O_2(G_1)$, then $[P, Y] \leq O_2(Y)$.
(5) $[O_2(LT), X] \leq O_2(KT)$.
(6) If $O^2(H_1) \leq K$, then $m(A/A \cap O_2(H_1)) \leq 1$ for each elementary subgroup A of R_1.

PROOF. For p a prime divisor of $q - 1$, $m_p(X) = 2$ and $C_X(V_1) = X \cap L_1 = X \cap K$, so $X \cap Y = 1$. Next $O_p(X)$ normalizes Y and hence a Sylow p-group Y_p of Y—so as G_1 is an SQTK-group, $Y_p = 1$, proving (1). Similarly as Y centralizes $L_1/O_2(L_1)$ of order divisible by 3, $m_3(Y) \leq 1$, the first requirement of (2).

Assume n is even. Then Y is a $3'$-group by (1), completing the proof of (2). Further if $I \in \mathcal{C}(Y)$, then $I/O_2(I) \cong Sz(2^k)$. If (3) fails then by (1), $X/C_X(I/O_2(I))$ is a nontrivial group of field automorphisms on $I/O_2(I)$. Let B be an XT-invariant Borel subgroup of $I_0 := \langle I^T \rangle$. Then using 1.2.1.3 as usual, either $B = N_B(T)$, or $I < I_0$ and $N_B(T)T/T \cong \mathbf{Z}_{2^k-1}$. In either case, X acts nontrivially on $N_B(T)T/T$. By 3.3.1, $N_B(T) \leq C_M(V_1) \cap C_G(L_1/O_2(L_1))$; thus a Hall $2'$-subgroup B_0 of $N_B(T)$

acts on R_1X, and hence by a Frattini Argument can be taken to normalize X. Then $[X, B_0] \leq X \cap B_0 T = 1$ using (1), contrary to an earlier observation. Thus (3) holds when n is even.

So assume instead n is odd. Then X is a $3'$-group of odd order coprime to $|I|$ by (1). Therefore as $m_3(Y) \leq 1$ by (2), (3) follows from an examination of the list of A.3.15, unless possibly $X/C_X(I/O_2(I))$ induces a nontrivial group of field automorphisms on $I/O_2(I) \cong L_2(2^k)$ or $L_3(2^k)$. In that event, $k = 2^a m$ for some odd m divisible by $|X : C_X(I/O_2(I))|$, and $X/O_2(X)$ induces a faithful group of field automorphisms on the subgroup I_1 of I with $I_1/O_2(I_1) \cong L_2(2^m)$ or $L_3(2^m)$. Further a Borel subgroup of I_1 acts on T, unless possibly some $t \in T$ induces a graph automorphism on $I_1/O_2(I_1) \cong L_3(2^m)$, in which case a subgroup of order $2^m - 1$ acts on T. Then arguing as in the previous paragraph, $[X, B_0] = 1$, contradicting the fact that $X/C_X(I/O_2(I))$ induces nontrivial field automorphisms on $I_1/O_2(I_1)$. This contradiction completes the proof of (3).

Assume the hypotheses of (4). Applying (3) and appealing to 1.2.1.1, we conclude X centralizes $Y^\infty/O_2(Y)$, and hence so does $[P, X] = P$. Thus P centralizes $E(Y/O_2(Y))$. As $\Phi(P) \leq O_2(G_1)$, $P = [P, X]$ centralizes $F(Y/O_2(Y))$ by A.1.26. Thus P centralizes $F^*(Y/O_2(Y))$, establishing (4).

If $L_1 = K$, then $O_2(LT) \leq O_2(KT)$, while if $L_1 < K$, then by 11.4.1.3, $[O_2(LT), X] \leq [O_2(L_1T), X] \leq O_2(KT)$. Thus (5) is established.

Finally if $O^2(H_1) \leq K$, then $L_1 < K$ so K is described case (1) or (2) of 11.1.2. In particular in each case, $m_2(R_1/C_{R_1}(K/O_2(K)) \leq 1$ as $R_1 = O_2(L_1T)$. But $C_{R_1}(K/O_2(K)) \leq O_2(H_1)$, since $O^2(H_1) \leq K$. Thus (6) holds. $\square$

PROPOSITION 11.5.7. $\langle V^{G_1} \rangle$ is abelian.

PROOF. Assume that $\langle V^{G_1} \rangle$ is nonabelian. Set $U := \langle V_3^{G_1} \rangle$. By 11.5.5 applied to G_1 in the role of "H_1", $\tilde{U} \leq Z(O_2(\tilde{G}_1))$ and $\Phi(U) \leq V_1$. Let $Y := C_G(V_1) \cap C_G(K/O_2(K))$.

We first treat the case $\bar{L} \cong SL_3(q)$. Then $V_3 = V$ so that U is nonabelian by assumption. Let $x, y \in N_L(X)$ with $V = V_1 \oplus V_1^x \oplus V_1^y$. As $U = \langle V^{G_1} \rangle$ is nonabelian, $V \not\leq Z(U)$, so $\bar{U} \neq 1$. From the proof of 11.5.5.1, Hypothesis G.2.1 is satisfied, so by G.2.5, $L \leq I := \langle U, U^x, U^y \rangle = LU$, $\bar{U} = O_2(\bar{L}_1) = \bar{R}_1$,

$$S := O_2(I) = C_U(V)C_{U^x}(V)C_{U^y}(V),$$

$US/S = O_2(L_1)S/S$, S has an L-series

$$1 =: S_0 \leq S_1 \leq S_2 \leq S_3 \leq S_4 := S$$

such that $S_1 := V$, $S_2 := U \cap U^x \cap U^y$, (and setting $W_i := S_i/S_{i-1}$) L centralizes W_2, W_3 is the direct sum of r copies of the dual V^* of V, and W_4 is the direct sum of s copies of V. As $I = \langle U^L \rangle$, M_1 acts on I and hence on S, as does L since $L \leq I$.

We claim that $S \leq O_2(G_1)$. Set $E := U^x \cap U^y$. By 11.5.5.2, $\Phi(E) \leq V_1^x \cap V_1^y = 1$. From the discussion above, $W_4 = [W_4, L_1]$ and for each irreducible J in W_3, the image of E in J is the X-invariant complement to $[J, L_1]$ in J. Hence $S = [S, L_1]E$. Further X acts on E and $S/S_2 = [S/S_2, X]$, so $E = [E, X]S_2$ and $S = [S, X](U \cap S)$. Now as $\Phi(E) = 1$, $[E, X]$ centralizes $Y/O_2(Y)$ by 11.5.6.4, so as $S_2 \leq U \leq O_2(G_1)$, $E = [E, X]S_2$ centralizes $Y/O_2(Y)$. Then as $[L_1, Y] \leq [K, Y] \leq O_2(Y)$, $S = [S, L_1]E$ centralizes $Y/O_2(Y)$. Also we saw that $S \leq O_2(LT)$, so $[S, X]$ centralizes $K/O_2(K)$ by 11.5.6.5, and hence so does $S = [S, X](S \cap U)$. Thus S centralizes

$KY/O_2(KY)$, so as $G_1 = KYM_1$ by 11.4.3.2 and M_1 acts on S, we conclude that $S \le O_2(G_1)$, completing the proof of the claim.

As $S \le O_2(G_1)$, $[\tilde{U}, S] = 1$ by 11.5.52. Consequently $r = 0 = s$, so that $S = S_2 \le U$ and L is an $SL_3(q)$-block. Since $H^1(\bar{L}, V) = 0$ by I.1.6, C.1.13.b says that $S = C_S(L)V$. As $S_2 \le U \cap E$ and $\Phi(E) = 1$, $C_S(L) = C_U(L)$ is abelian. As $\bar{U} = \bar{R}_1 = [\bar{R}_1, \bar{L}_1]$, $U = [U, L_1]C_U(L) = (U \cap L)C_U(L)$ and $(U \cap L)/C_{U \cap L}(L)$ is special of order q^5. Further $C_U(L_1) = C_{Z(U)}(L)V_1 = Z(U)$, so that $U/Z(U)$ is of rank $4n$. As L_1 centralizes $Z(U)$, also $[K, L_1] = K$ centralizes $Z(U)$.

Assume that $L_1 < K$. Then by 11.1.2, $n = 2$ and $q = 4$, so that $U/Z(U)$ is of rank $4n = 8$ by an earlier observation. As we are assuming that $\bar{L} \cong SL_3(4)$, case (2) of 11.1.2 does not arise, so by 11.4.1.2, K is described in case (1) of 11.1.2. As $L_1 \trianglelefteq M_1$ and $V \le U$, $C_K(\tilde{U})$ acts on L_1, so $C_K(\tilde{U}) \le O_{2,Z}(K)$. Then as K centralizes $Z(U)$, $C_K(U/Z(U)) \le O_{2,Z}(K)$, impossible as $K/O_{2,Z}(K)$ is not a section of $GL_8(2)$.

Therefore $L_1 = K$. As L is an $SL_3(q)$-block, L_1 has two noncentral 2-chief factors, so 11.5.5.3 says that $\tilde{U}$ is a sum of exactly two copies of the natural module for $K/O_2(K) \cong L_2(q)$. In particular $|U| = q^5$. Therefore as $U = C_{Z(U)}(L)(U \cap L)$ and $|(U \cap L)/C_{U \cap L}(L)| = q^5$, we conclude that $U = U \cap L = T \cap L = O_2(L_1) = O_2(K)$, and $C_{U \cap L}(L) = 1$ so that $V = O_2(L)$.

As $K = L_1 \le M$, $K_H := O^2(H) \le Y$ by our choice of H. Let $X_1 := C_X(K/O_2(K))$ and $C := \langle R_1, K_H, X_1 \rangle$. Further since $C \le C_{G_1}(K/O_2(K))$ and $R_1 = C_T(K/O_2(K))$, $R_1 \in Syl_2(C)$. Set $\hat{C} := C/C_C(\tilde{U})$. As $\tilde{U}$ is a sum of two absolutely irreducible modules for $K/O_2(K)$ over $\mathbf{F}_q$, $\hat{C} \le C_{GL(\tilde{U})}(K/O_2(K)) \cong GL_2(q)$. Since L is an $SL_3(q)$-block with $V = O_2(L)$, as before C.1.13.b says

$$T_L = (T \cap L)C_T(L). \qquad (*)$$

In particular as $U \le L$ and $\bar{U} = \bar{R}_1$, $R_1 = UC_{R_1}(L)$ by (*) and $C_{R_1}(L)$ centralizes U. Thus $\hat{C}$ is a subgroup of $GL_2(q)$ of odd order. Next $C_H(\tilde{U}) \le N_H(V) \le H \cap M$, so $C_H(\tilde{U}) \le \ker_{H \cap M}(H)$. Therefore by B.6.8, $K_H = O_2(K_H)D$ for some p-group D with $D \cap M = \Phi(D)$. By 11.5.6.1, $(p, q - 1) = 1$, so as $\hat{C}$ is a subgroup of $GL_2(q)$ of odd order, $\hat{D}$ is cyclic of order dividing $q + 1$ and $[\hat{D}, \hat{X}_1] = 1$. If n is even then $X_0 := C_X(L/V)$ is a subgroup of X_1 of order 3. Then as $\hat{D}$ centralizes $\hat{X}_1$, $H = DT$ acts on $[U, X_0] = V$, contradicting $K_H \not\le M$. Therefore n is odd so $T = T_L$. Then using (*) and our earlier observations that $O_2(K) = U = T \cap L$,

$$T = (T \cap L)C_T(L) = UC_T(L) = UC_T(U). \qquad (**)$$

Further $[C_T(U), K_H] \le C_{K_H}(U) \le O_{2, \Phi}(K_H)$, so H acts on $C_T(U)$ and hence by (**), $H \le N_G(T) \le M$ using Theorem 3.3.1, a contradiction. This completes the treatment of the case $\bar{L} \cong SL_3(q)$.

Therefore by 11.5.1 it remains to treat the case $\bar{L} \cong Sp_4(q)$. At several places we use the fact that:

(!) $\bar{L}_1\bar{X}$ is indecomposable on $O_2(\bar{L}_1)$ with chief series $1 < Z(\bar{L}_1) < O_2(\bar{L}_1)$, and $Z(\bar{L}_1) = C_{\bar{M}_V}(V_3)$.

We first observe that $V \le O_2(G_1)$: For $V = [V, X]$, so by parts (4) and (5) of 11.5.6, V centralizes $KY/O_2(KY)$. Then recalling that $G_1 = KYM_1$ and $V \trianglelefteq M_1$ since V is a TI-set under M by 11.0.3.3.4, the observation is established.

Suppose U is nonabelian. Then as $U = \langle V_3^{G_1} \rangle$, U does not centralize V_3, so $\bar{U} \neq 1$ and $1 \neq [V_3, U] \leq V_1$ by 11.5.5.2. As $U \trianglelefteq L_1 T$, $\bar{U} = O_2(\bar{L}_1)$ by (!). Therefore $[V, U] = V_3$, which is impossible since by the previous paragraph, $V \leq O_2(G_1)$, so $[V, U] \leq V_1$ by 11.5.5.2.

Thus U is abelian. If $[V, U] \neq 1$ then as $[V_3, U] = 1$, $\bar{U} = Z(\bar{L}_1)$ by (!). In this case set $W := U = U_H$. On the other hand if $[V, U] = 1$, set $W := W_H$, where $W_H := \langle V^{G_1} \rangle$. As $U \trianglelefteq G_1$, $[U, W] = 1$. As $V \leq O_2(G_1)$, $W \leq O_2(G_1)$, so as we are assuming that W is nonabelian, and as $[V_3, W] = 1$, again $\bar{W} = Z(\bar{L}_1)$. Therefore in either case, $\bar{W} = Z(\bar{L}_1)$, so $[V, W] = [V, Z(\bar{L}_1)] = V_1$, and hence $\Phi(W) = V_1$.

Choose an element $y \in L$ so that $\langle Z(\bar{L}_1), Z(\bar{L}_1)^y \rangle \cong L_2(q)$, and $I := \langle W, W^y \rangle$ contains $X_1 = C_X(L_1/O_2(L_1))$. Observe that $\bar{I} = \langle \bar{W}, \bar{w}^y \rangle$ for each $1 \neq \bar{w} \in \bar{W}$, and $V = V_3 \oplus V_1^y$. Then as $[W, O_2(I)] \leq W \cap O_2(I)$,

$$Q := (W \cap O_2(I))(W^y \cap O_2(I)) \trianglelefteq I,$$

with $[O_2(I), I] \leq Q$. Since $I = \langle W, W^y \rangle$ and $\Phi(W) = V_1$, $[W \cap W^y, I] \leq V_1 V_1^y \leq V$. Also $\Phi(W) = V_1 \leq V$, and

$$Q/(W \cap W^y)V = (WV \cap Q)/(W \cap W^y)V \times (W^y V \cap Q)/(W \cap W^y)V,$$

with $(WV \cap Q)/(W \cap W^y)V = C_{Q/(W \cap W^y)V}(w)$, as $\bar{I} = \langle \bar{W}, \bar{w}^y \rangle$.

We claim that $Q \leq O_2(G_1) =: Q_1$: It follows from G.1.6 that $Q/(W \cap W^y)V$ is a sum of natural modules for $\bar{I}$, so that $C_Q(X_1) \leq (W \cap W^y)V \leq Q_1$. Thus as $Q = [Q, X_1]C_Q(X_1)$, it remains to show $[Q, X_1] \leq Q_1$. Next $\Phi(Q) \leq (W \cap W^y)V \leq Q_1$ and $Q \leq O_2(I) \leq O_2(LT)$, so using parts (4) and (5) of 11.5.6, $[Q, X_1]$ centralizes $KY/O_2(KY)$. Next L_1 is transitive on $V_1^G \cap (V - V_3)$, so by a Frattini Argument, $M_1 = L_1 N_{M_1}(V_1^y)$. Thus as $N_{M_1}(V_1^y)$ acts on $[Q, X_1]$, $G_1 = KYM_1 = KYN_{M_1}([Q, X_1])$, so as $[Q, X_1]$ centralizes $KY/O_2(KY)$, the claim is established.

Next $\bar{L}$ is generated by three conjugates $\bar{I}^{l_i}$, $1 \leq i \leq 3$, of $\bar{I}$ under L_1. As $O_2(L)$ acts on W, it acts on W^{yl_i} for each i, so $[O_2(L), L]$ is the product of $[O_2(L), W] \leq Q \leq Q_1$ and $[O_2(L), W^{yl_i}] \leq Q_1$, so $[O_2(L), L] \leq Q_1$. Then as the Schur multiplier of $Sp_4(q)$ is trivial by I.1.3, $O_2(L) = [O_2(L), L] \leq Q_1$.

Now as $Z(\bar{L}_1) = \bar{W} \leq \bar{Q}_1 \trianglelefteq \bar{L}_1$, $\bar{Q}_1 = \bar{R}_1$ or $\bar{W}$. However in the latter case as $O_2(L) \leq Q_1$, $Q_1 \in Syl_2(IQ_1)$, and then by C.1.29, there is a nontrivial characteristic subgroup Q_0 of Q_1 normal in IQ_1. But then $Q_0 \trianglelefteq \langle I, L_1 T \rangle = LT$, so $G_1 \leq N_G(Q_1) \leq N_G(Q_0) \leq M = !\mathcal{M}(LT)$, contradicting 11.5.4. Therefore $\bar{Q}_1 = \bar{R}_1$, so $R_1 \cap LQ_1 = Q_1$. Then by C.1.32, either there is a nontrivial characteristic subgroup Q_0 of Q_1 normal in LQ_1, or L is an $Sp_4(4)$-block. The former case leads to the same contradiction as before.

Therefore L is an $Sp_4(4)$-block. Since $H^1(\bar{L}, V) = 0$ by I.1.6, and we have seen that $O_2(L) = [O_2(L), L]$, we conclude from C.1.13.b that $O_2(L) = V$.

We now treat the case that $[V, U] = 1$. Here we recall that $W = W_H$ and $\bar{W} = Z(\bar{L}_1)$, so $[W, L_1] \leq O_2(L) = V$, and hence $[W, L_1] = [V, L_1] = V_3 \leq U$. Therefore $K = \langle L_1^K \rangle$ centralizes W/U, so $W = UV$. But then as $[V, U] = 1$ and U and V are abelian, $W = W_H$ is abelian, contrary to our assumption.

Therefore $[V, U] \neq 1$. In this case we recall that $W = U$. As $1 \neq [V, U]$, $\bar{W}_H \not\leq C_{\bar{L}\bar{T}}(V_3) = Z(\bar{L}_1)$, so as W_H is $L_1 X$-invariant, $\bar{W}_H = \bar{R}_1$ and there exists $g \in G_1$ with $\bar{V}^g \leq \bar{R}_1$ but $\bar{V}^g \not\leq \bar{Z}(\bar{L}_1)$. Thus conjugating in L_1 if necessary, $V_1 \leq C_V(V^g) \leq V_2 \leq [V, V^g]$. But $W_H \leq Q_1 \leq N_G(V)$, so $V^g \trianglelefteq W_H$, and thus $[V, V^g] \leq V \cap V^g \leq C_V(V^g)$. Hence $V_2 = V \cap V^g$ is conjugate to V_2^g under L^g, so we may choose $g \in N_G(V_2)$. By 11.1.3, g acts on L_2, so as $V = [V, L_2]$, also $V^g = [V^g, L_2]$. But then $\bar{V}^g = [\bar{R}_2, L_2]$, contradicting $\bar{V}^g \leq \bar{R}_1$. This final contradiction completes the proof of 11.5.7. $\square$

With 11.5.7 in hand, we are finally in a position to obtain a contradiction to 11.5.3.3.

PROPOSITION 11.5.8. $K = L_1$.

PROOF. Assume that $L_1 < K$; then we may take $H_1 := KT$ to be our chosen member of $\mathcal{H}_1$. Observe that Hypothesis F.7.6 is satisfied with LT, H_1, L_1T in the roles of "G_1, G_2, $G_{1,2}$". Adopt the notation of section F.7, and in particular let $b := b(\Gamma, V)$. By 11.5.7, $U := \langle V^{H_1} \rangle$ is elementary abelian, so $b > 1$ by F.7.7.2.

Assume first that b is even. Then by F.7.11.2, there is $g \in G$ with $1 \neq [V, V^g] \leq V \cap V^g$, and by F.7.11.5 with the roles of γ_0 and γ reversed, we may choose $V^g \leq O_2(G_{1,2}) = R_1$. Then inspecting the subgroups of $\bar{R}_1$ acting quadratically on V, either

(i) $V_1 = [V, V^g]$ is a 1-dimensional $\mathbf{F}_q$-subspace of V and V^g, or

(ii) $\bar{L} \cong Sp_4(q)$ and (conjugating in L_1 if necessary) $[V, V^g] = V_2$, and $[V_3, V^g] = V_1$ is a 1-dimensional $\mathbf{F}_q$-subspace of V and V^g.

In either case, as L^g is transitive on 1-dimensional $\mathbf{F}_q$-subspaces, we may choose $g \in G_1$. Then 11.5.7 contradicts our choice of $1 \neq [V, V^g]$.

So b is odd. Pick γ at distance b as in F.7.11, choose a geodesic

$$\gamma_0, \gamma_1, \ldots, \gamma_b := \gamma$$

in Γ, and choose g so that $\gamma_1 g = \gamma$. Thus $V \nleq O_2(H_1^g)$ and as γ_1 is on the geodesic, $[U, U^g] \leq U \cap U^g$ by F.7.11.1. By 11.5.6.6, $U_1 := U \cap O_2(H_1^g)$ and $U_0 := U^g \cap O_2(H_1)$ are of index at most 2 in U and U^g, respectively. Further $V_1 \cap V_1^g = 1$—or else by 11.4.2, $K = K^g$, so $[U, O^2(H_1^g)] \leq [U, K^g] \leq [U, K] \leq U$, contradicting $V \nleq O_2(H_1^g)$. Thus $[U_1, U_0] \leq V_1 \cap V_1^g = 1$, so U_0 centralizes $U_1 \cap V$ of corank at most 1 in V. However by 11.2.2.3, $s(G, V) = m(\bar{M}, V) = n > 1$, so U_0 centralizes V by E.3.6. Then as $1 \neq [V, V^g] \leq [V, U^g]$, V induces a group of transvections on U^g with axis U_0. As $V \nleq G_\gamma^{(1)}$, by F.7.7.2, $V \nleq O_2(G_\gamma)$.

Since $L_1 < K$, K is described in case (1) or (2) of 11.1.2 by 11.4.1.2. As $C_{H_1}(U) \leq C_{H_1}(V) \leq M_1$ and $L_1 \trianglelefteq M_1$, we conclude from the structure of those groups that $C_K(U) \leq O_{2,Z}(K)$. Thus we may pick an H_1^g-chief section W of U^g such that $F := O_{2,F^*}(K)$ is nontrivial on W. Again from the structure of K, as $V \nleq O_2(H_1^g)$, V is nontrivial on $Aut_F(W)$, so V induces a transvection on W. But comparing the groups in 11.1.2 to those in G.6.4.2, $Aut_{H_1}(W)$ contains no transvection, completing the proof of the lemma. $\square$

By 11.5.8, $K = L_1$, so $\langle V^{G_1} \rangle$ is nonabelian by 11.5.3.3, contrary to 11.5.7. This contradiction completes the proof of Theorem 11.0.1.

Part 5

Groups over F_2

Results in the previous parts have reduced the choices for L, V in the Fundamental Setup (3.2.1) to the case where $L/O_{2,Z}(L)$ is essentially a group of Lie type defined over $\mathbf{F}_2$, and V is highly restricted. We adopt the convention that A_5 (regarded as $\Omega_4^-(2)$), and A_7 (a subgroup of $A_8 \cong L_4(2) \cong \Omega_6^+(2)$) are considered to be defined over $\mathbf{F}_2$. For a precise description of the pairs L, V which remain to be considered, see conclusion (3) of Theorem 12.2.2 early in this part.

The first chapter 12 of this part contains a number of useful reductions which smooth out the situation. For example, some reductions treat or eliminate certain larger possibilities for L or V. These reductions use special and comparatively elementary techniques, such as the weak-closure methods from section E.3, or control of centralizers of certain elements of V.

The cases that remain after these sections are then treated in chapters 13 and 14, using "generic" techniques for groups over $\mathbf{F}_2$, such as versions of the theory of large extraspecial 2-subgroups in the original classification literature, and variants on the amalgam method from section F.9

CHAPTER 12

Larger groups over $\mathbf{F}_2$ in $\mathcal{L}_f^*(G, T)$

In this chapter we consider the cases remaining in the Fundamental Setup (3.2.1) after the work of the previous parts. Then we reduce that list further, concentrating on cases which can be treated by methods such as weak closure and control of centralizers of certain elements of V.

After an initial reduction in the first section 12.1, the cases that remain are listed in part (3) of Theorem 12.2.2 in the second section. Then in Hypothesis 12.2.3, we add the assumption that G is not one of the groups already treated in earlier analysis; the latter groups are listed in conclusions (1) or (2) of Theorem 12.2.2. In the remaining cases $L/C_L(V)$ is essentially a group defined over $\mathbf{F}_2$. Then the main goal of this chapter is to treat, and in most cases eliminate, the largest of those groups over $\mathbf{F}_2$: namely $\hat{A}_6$, A_7, $L_4(2)$, and $L_5(2)$.

12.1. A preliminary case: Eliminating $\mathbf{L_n(2)}$ on $\mathbf{n \oplus n^*}$

In this section we complete our analysis of case 3.2.5.3 of the Fundamental Setup (3.2.1), where V is a sum of two T-conjugates of $V_\circ \in Irr_+(L, R_2(LT), T)$. Recall that most such cases were eliminated in Theorem 7.0.1. Thus it remains to consider the cases where $L/C_L(V) \cong L_4(2)$ or $L_5(2)$, and $V_\circ$ is a natural module for $L/C_L(V)$. We eliminate these cases using the weak-closure techniques of part 3, together with reductions from chapters E.6 and 11. We must work a little harder however, because $m(M/C_M(V), V) = 2$, so that Theorem E.6.3 is not available to give an initial lower bound on $r(G, V)$.

Once this case is eliminated, we will have completed the treatment of the cases in the FSU where L is T-invariant and L is not irreducible on $V/C_V(L)$; for recall chapter 10 completed the treatment of the case where L is not T-invariant, while Theorems 6.2.20 and 7.0.1 treated the cases where V is not an FF-module.

Thus at the end of this section, the treatment of the FSU will be reduced to the cases described in 3.2.8. The first four subcases of 3.2.8 include all cases where $L/C_L(V)$ is defined over $\mathbf{F}_{2^n}$ with $n > 1$, and those cases were handled in Theorems 6.2.20 and 11.0.1. Hence after this section it remains only to treat the cases where $L/C_L(V)$ is a group defined over $\mathbf{F}_2$; by convention we include $\hat{A}_6$ and A_7 among such groups.

While in this section $L/C_L(V)$ is also a group over $\mathbf{F}_2$, the fact that L is not irreducible on V makes the treatment of this case easier, and different from the treatment of the generic case of groups over $\mathbf{F}_2$.

So in this section we assume G is a simple QTKE-group, $T \in Syl_2(G)$, $L \in \mathcal{L}_f^*(G, T)$ with $L/O_2(L) \cong L_n(2)$, $n = 4$ or 5, $M := N_G(L)$, $V \in \mathcal{R}_2(M)$, $\bar{M} := M/C_M(V) \cong Aut(L_n(2))$, and $V = V_1 \oplus V_2$, with V_1 the natural module for $\bar{L}$ and $V_2 = V_1^t$ for $t \in T - LQ$, where $Q := O_2(LT) = C_T(V)$. Thus V_2 is the dual of V_1 as an $\mathbf{F}_2 L$-module. Let $T_1 := N_T(V_1)$, and for $v \in V$, let $M_v := C_M(v)$, $G_v := C_G(v)$, and $L_v := O^2(C_L(v))$.

By 3.2.5.3, $V \trianglelefteq M$; hence

$$M = N_G(V)$$

as $M \in \mathcal{M}$.

Recall that the module V is described in section H.9 of Volume I. We adopt the notation of that section, including the description of the orbits $\mathcal{O}_i$ $(1 \le i \le 3)$ of M on $V^{\#}$.

We now proceed to analyze our group G. Eventually we obtain a contradiction, and hence show no such group exists.

LEMMA 12.1.1. *Let $v \in \mathcal{O}_3$. Then $G_v = C_G(v) \le M$.*

PROOF. First from lemma H.9.1.4, $Q = O_2(L_v T_v)$, where $T_v := C_T(v) \in Syl_2(M_v)$. Now $C(G, Q) \le M$ by 1.4.1.1. Thus as $L_v \trianglelefteq M_v$, we conclude from A.4.2.7 that $Q \in \mathcal{B}_2(G_v)$ and Q is Sylow in $\langle Q^{M_v} \rangle$. Therefore Hypothesis C.2.3 is satisfied by G_v, M_v, Q in the roles of "H, M_H, R".

Let $W := [V, L_v]$. By lemma H.9.1.4, $L_v/O_2(L_v) \cong L_{n-1}(2)$ and $W = W_1 \oplus W_2$ with W_1 the natural module for $L_v/O_2(L_v)$ and W_2 its dual. Let z generate $C_W(T_v)$, and observe z is 2-central in G by H.9.1.2.

For each value of n we define a subgroup $K_v \in \mathcal{C}(G_v)$ with $L_v \le K_v \trianglelefteq G_v$: If $n = 4$, then W is not an FF-module for L_v by Theorem B.5.1.1, so $J(T_v) = J(Q)$ by B.2.7. Then as $C(G, Q) \le M$, $N_G(T_v) \le M_v$ and hence $T_v \in Syl_2(G_v)$. So by 1.2.4, $L_v \le K_v \in \mathcal{C}(G_v)$, and as T_v acts on L_v, $K_v \trianglelefteq G_v$ by 1.2.1.3. On the other hand, if $n = 5$, then by 1.2.1.1, L_v projects nontrivially on some $K_v \in \mathcal{C}(G_v)$, so K_v has a section isomorphic to $L_v/O_2(L_v) \cong L_4(2)$. Therefore $K_v = O^{3'}(G_v)$ by A.3.18, so that again $L_v \le K_v \trianglelefteq G_v$.

Suppose that there is a component K of G_v. Then $L_v = O^2(L_v)$ acts on K by 1.2.1.3. By A.1.6, $O_2(M) \le Q \le G_v$, so $M_v \in \mathcal{H}^e$ by 1.1.4.4; thus $K \not\le M_v$. Similarly $G_z \in \mathcal{H}^e$ by 1.1.4.6, so $G_{v,z} := G_v \cap G_z \in \mathcal{H}^e$ by 1.1.3.2; thus $K \not\le G_{v,z}$, so $K \not\le G_z$. But $z \in W = [W, L_v] \le L_v$, so $[K, L_v] \ne 1$. Therefore $[K, K_v] \ne 1$, and hence $K = K_v$ by 1.2.1.2. Then $C_W(K) \le C_W(L_v) = 1$. Set $G_v^* := G_v/C_{G_v}(K)$. Then $W^* \cong W$ as an L_v^*-module, and $L_v^* \trianglelefteq M_v^*$. But no group with such a 2-local M_v^* appears on the list of Theorem C (A.2.3).

This contradiction shows that $E(G_v) = 1$. Next $W = [W, L_v]$, so W centralizes $O(G_v)$ by A.1.26. Therefore $O(G_v) \le G_{v,z}$, and hence $O(G_v) = 1$ as $G_{v,z} \in \mathcal{H}^e$. Thus we have shown $O^2(F^*(G_v)) = 1$, so that $G_v \in \mathcal{H}^e$.

We now assume that $G_v \not\le M$, and derive a contradiction.

Suppose that $L_v \trianglelefteq G_v$ and set $Y := C_{G_v}(L_v/O_2(L_v))$. Then as $Aut(L_v/O_2(L_v))$ is induced in $L_v T_v$, $G_v = L_v T_v Y$, so $Y \not\le M$ as $G_v \not\le M$. Next embed $T_v \le X \in Syl_2(G_v)$; then $N_X(T_v)$ normalizes $C_{T_v}(L_v/O_2(L_v)) = Q$, and so lies in M_v—hence $T_v = X \in Syl_2(G_v)$. Thus $Q = T_v \cap Y$ is Sylow in Y, so we conclude from the $C(G, T)$-Theorem C.1.29 that there is a χ_0-block B of Y with $B \not\le M$. If B is an $L_2(2^n)$-block, then a Cartan subgroup D of B lying in $B \cap M$ centralizes $L_v/O_2(L_v)$; hence D centralizes V, as $\bar{M} = N_{GL(V)}(\bar{L})$ and $C_{\bar{M}}(\bar{L}_v) = 1$. Thus $V \le C_{T_v}(B \cap M) = C_{T_v}(B)$, and hence $B \le C_G(V) \le M$, contrary to the choice of B. If B is an A_5-block, then $O^2(B \cap M) \le O^{3'}(M_v) = L_v$ by A.3.18, whereas $Z(L_v/O_2(L_v)) = 1$. Thus B is an A_3-block. Notice since B centralizes $L_v/O_2(L_v)$ of order divisible by 3, and G_v is an SQTK-group, that $B \trianglelefteq G_v$. Set $H := BT_v$,

の位置

so that $T_v \in Syl_2(H)$, $B = O^2(H)$, and $Q \in Syl_2(QB)$. As $B \not\leq M = !\mathcal{M}(LT)$, there is no $1 \neq R_0 \leq Q$ with $R_0 \trianglelefteq \langle LT, H \rangle$. Thus Hypotheses C.5.1 and C.5.2 are satisfied with LT, Q in the roles of "M_0, R". Further $L_v T_v$ is maximal in LT, so $L_v T_v = N_{LT}(B)$. Then we have the hypotheses of C.5.7, and as $|LT : L_v T_v| \neq 2$, C.5.7 supplies a contradiction.

This contradiction shows that L_v is not normal in G_v. Thus $K := K_v > L_v$, so as $L_v \trianglelefteq M_v$, $K \not\leq M_v$. As $G_v \in \mathcal{H}^e$, $K \in \mathcal{H}^e$ by 1.1.3.1. By C.2.6.2, $O_{2,F}(K) \leq M_v \leq N_G(L_v)$, so $K/O_2(K)$ is quasisimple by 1.2.1.4. As $L_v \leq K$ and T_v is nontrivial on the Dynkin diagram of $L_v/O_2(L_v)$, K is not a χ_0-block, so Q normalizes K by C.2.4. Thus we have the hypotheses of C.2.7, so K is described in C.2.7.3. Thus as T_v is nontrival on the Dynkin diagram of $L_v/O_2(L_v) \cong L_3(2)$ or $L_4(2)$, and $L_v \trianglelefteq M_v \cap K$, we conclude that case (h) of C.2.7.3 holds with $KT_v/O_2(KT_v) \cong Aut(L_5(2))$ and $L_v T_v \cap K$ is the parabolic subgroup determined by the middle two nodes; in particular $n = 4$. Let $Z_v := \Omega_1(Z(O_2(KT_v)))$, $Y := \langle Z_v^K \rangle$, and $(KT_v)^+ := KT_v/O_2(KT_v)$. By C.2.7.2, Y is an FF-module for $K^+ T_V^+ \cong Aut(L_5(2))$, so we conclude from Theorem B.5.1.1 that $[Y, K] = U \oplus U^t$ for $t \in T_v - N_{T_v}(U)$. By B.2.14, $Y = [Y, K] \oplus C_{Z_v}(K)$. Thus the parabolic $L_v T_v \cap K$ determined by the middle nodes of K centralizes Z_v, whereas from the action of LT on V, $C_V(T)$ is not centralized by $L_v T_v$. This contradiction completes the proof of 12.1.1. $\qquad\square$

From Lemma H.9.1, V has the structure of an orthogonal space preserved by $\bar{M}$, and $\mathcal{O}_3$ is the set of nonsingular vectors in that space.

LEMMA 12.1.2. *(1) If $U \leq V$ with $C_G(U) \not\leq M$, then U is totally singular.*
(2) $r(G, V) \geq n$, so that $s(G, V) = m(Aut_M(V), V) = 2$.

PROOF. Part (1) follows from 12.1.1 and the fact that $\mathcal{O}_3$ is the set of nonsingular vectors in V. Then (1) implies (2). $\qquad\square$

Using the lower bound on the parameter $r(G, V)$ in 12.1.2.2, we can apply the weak-closure machinery in section E.3 (subsection E.3.3) to establish successively better lower bounds on the parameter $w(G, V)$. Often results are easier to establish in the case $n = 5$; for example, the analogue of 12.1.3 below is not established for $n = 4$ until 12.1.7.

LEMMA 12.1.3. *If $n = 5$ then $W_0 := W_0(T, V)$ centralizes V.*

PROOF. Suppose that $n = 5$ but $W_0 \not\leq C_T(V)$. Then there is $A := V^g \leq T$ with $\bar{A} \neq 1$. Recall $M = N_G(V)$ and $M^g = N_G(A)$.

We begin by showing we may choose A with $m(\bar{A}) \geq 5$. Suppose first that $V \not\leq N_G(A)$; then as $r(G, V) \geq 5$ by 12.1.2.2, $m(\bar{A}) \geq 5$ by E.3.4.2. So suppose instead that $V \leq N_G(A)$. Here, interchanging the roles of A and V if necessary, we may assume that $m(\bar{A}) = m(A/C_A(V)) \geq m(V/C_V(A))$; equivalently $m(C_V(A)) \geq m(C_A(V))$. Suppose that $m(\bar{A}) < 5$. Then by our assumption above, $m(C_V(A)) \geq m(C_A(V)) > 5$. Hence $C_V(A)$ is not totally singular and $1 \neq C_{V_1}(A)$, so $\bar{A} \leq \bar{T}_1 \leq \bar{L}$. Then as A centralizes a nonsingular vector $v \in V$, by lemma H.9.1.4, $\bar{A} \leq \bar{L}_v \cong L_4(2)$ and $V = C_V(L_v) \oplus W$, where W is the sum of a natural module and its dual. Now $m(\bar{A}) \geq m(W/C_W(A))$, so that $\bar{A}$ contains a member $\bar{B}$ of $\mathcal{P}(\bar{L}_v, W)$ by B.1.4.4. Then B.4.9.2iii determines $\bar{B}$ uniquely as $J(C_{\bar{T}}(v))$, so that $\bar{B} = J(C_{\bar{T}_1}(v)) = \bar{A}$. In particular $\bar{A}$ is the unipotent radical of the stabilizer in $\bar{L}_v \cong L_4(2)$ of a 2-subspace of the natural module W. Thus $m(\bar{A}) = m(V/C_V(A))$,

$\bar{A}$ is faithful on each V_i, and for $u \in V_1 - C_{V_1}(A)$, $[u,A] = [V_1,A]$ is of rank 2. Set $A_i := V_i^g$. Now $V \leq N_G(A)$ by our assumption, and we saw that $m(\bar{A}) = m(V/C_V(A)) = 4 < 5$, so we have symmetry between V and A. Reversing the roles of V and A, we conclude from that symmetry that $V/C_V(A)$ is faithful on A_i, so as $[A,V_1]$ is of rank 2, V_1 induces transvections on A_i with center $[A,V_1] \cap A_i$. This is impossible as $m(V_1/C_{V_1}(A)) = 2 > 1$ and A_2 is dual to A_1.

This contradiction establishes the claim that we may take $m(\bar{A}) \geq 5$. Hence by lemma H.9.2.3, we may take $\bar{A} \leq \bar{A}_0$, where $\bar{A}_0$ is the centralizer in $\bar{T}_1$ of a 3-subspace of V_1. Then by H.9.2.5,

$$V = \check{\Gamma}_{4,\bar{A}}(V) = \check{\Gamma}_{4,A}(V);$$

so as $r(G,V) \geq 5$, $V = \check{\Gamma}_{4,A}(V) \leq N_G(A)$ by E.3.32. Then $[V,A] \leq V \cap A \leq C_V(A)$, and applying lemma H.9.2.4 to the action of $\bar{A}$ on V, $C_V(A) = [V,A] = V \cap A$ is of rank 5. Thus $m(V/C_V(A)) = 5$, so we have symmetry between V and A, and by that symmetry $C_A(V) = V \cap A$ is of rank 5. Hence $m(\bar{A}) = 5$. We saw that $\bar{A} \leq \bar{A}_0$, so $\bar{A}$ acts faithfully on each V_i. In particular, $V_2 \not\leq A$, and for $v \in V_2 - A$, $[v,A] \leq A \cap V_2 = C_{V_2}(A)$, with $m(C_{V_2}(A)) = 2$ by H.9.2.4. By symmetry, V_2 normalizes but does not centralize V_i^g, so as $m([V_2,A]) = 2$ and $m(V_2/C_{V_2}(A)) > 1$, we have the same contradiction as in the previous paragraph. This completes the proof of 12.1.3. $\square$

LEMMA 12.1.4. *Assume $n = 4$ and let v generate $C_{V_1}(T_1)$. Then $T_1 \in Syl_2(G_v)$.*

PROOF. Let $T_1 \leq T_0 \in Syl_2(G_v)$. If the lemma fails, then as $|T : T_1| = 2$, $T_0 \in Syl_2(G)$. But $T_1 \in Syl_2(M_v)$, so $T_0 \not\leq M$, and hence $N_G(T_1) \not\leq M$. If C is a nontrivial characteristic subgroup of T_1 normal in $L_v T_1$, then $C \trianglelefteq \langle T, L_v \rangle = LT$, so $N_G(T_1) \leq N_G(C) \leq M = !\mathcal{M}(LT)$, contrary to the previous sentence. Thus no such C exists, so (L_v, T_1) is an MS-pair in the sense of Definition C.1.31. However L_v has at least three noncentral 2-chief factors, and as $v \in V_1 = [V_1, L_v] \leq L_v$ by H.9.1.3, $v \in Z(L_v)$. Hence L_v must satisfy case (4) of C.1.34. Therefore $Z_1 := \Omega_1(Z(T_1))$ is of rank at least 3, with $m(Z_1/C_{Z_1}(L_v)) = 1$. Now $L = \langle L_v, L_v^t \rangle$ for $t \in T - T_1$, so $1 \neq C_{Z_1}(L_v) \cap C_{Z_1}(L_v^t) \leq C_{Z_1}(L)$, and hence $C_Z(L) \neq 1$.

Next $J(T) \leq T_1$ by B.1.5.4. Thus $J(T_1) = J(T)$, so as $N_G(T_1) \not\leq M$, $N_G(J(T)) \not\leq M$. In particular $J(T) \not\leq Q$, and hence $R_2(LQ) = V \oplus C_{Z_1}(L)$ by B.5.1.4. Now an FF^*-offender in $\bar{T}$ lies in $\mathcal{P}(\bar{T},V)$ by B.2.7, and by B.4.9.2iii the unique member $\overline{J(T)}$ of $\mathcal{P}(\bar{T},V)$ is the unipotent radical of the stabilizer in L of a 2-subspace of V_1. Thus $N_{LT}(J(T)) = XT$, where $X \in \Xi(G,T)$ with $XT/O_2(XT) \cong S_3 \text{ wr } \mathbf{Z}_2$. As $R_2(LQ) = V \oplus C_{Z_1}(L)$ with $C_V(X) = 1$, $C_{Z_1}(X) = C_{Z_1}(L)$. As $J(T) \leq T_1$, $T_0 \leq N_G(T_1) \leq N_G(J(T)) =: G_J$, and of course $TX \leq G_J$. Thus $H := \langle T_0, TX \rangle \leq G_J$, so $H \in \mathcal{H}(XT)$. Suppose $X \trianglelefteq H$. Then T_0 and T act on T_1 and hence on $C_{Z_1}(X) = C_{Z_1}(L)$, so $T_0 \leq N_G(C_{Z_1}(L)) \leq M = !\mathcal{M}(LT)$, contrary to $T_0 \not\leq M$. Therefore X is not normal in H, so by 1.3.4, $X < K_0 := \langle K^T \rangle$ for some $K \in \mathcal{C}(H)$, and K_0 is described in 1.3.4. Now $K \in \mathcal{L}(G,T)$ and $[Z,X] \neq 1$, so $K \in \mathcal{L}_f(G,T)$. By 1.3.9, $K \in \mathcal{L}_f^*(G,T)$, so by 3.2.3 there is $V_K \in \mathcal{R}_2(K_0 T)$ such that the pair K, V_K satisfies the Fundamental Setup. Hence by Theorem 3.2.5, this pair is listed in 3.2.5.3, 3.2.8, or 3.2.9. By Theorem 10.0.1, $K = K_0$, so case (1) of 1.3.4 does not hold. Theorem 11.0.1 eliminates case (3) of 1.3.4. Case (2), and case (4) with $K/O_2(K) \cong M_{11}$, do not appear in the indicated lists for the FSU. Thus $KT/O_2(KT) \cong S_8$ or $Aut(L_5(2))$. Let $H^* := H/C_H(K/O_2(K))$, so that $H^* = K^*T^*$ since $K^*T^* = Aut(K^*)$. Then $X^* = O^2(P^*)$, where P^* is the parabolic of K^* determined by the end nodes of the Dynkin diagram. Therefore $R^* := O_2(X^*) \leq T_1^*$, and hence as it is the unipotent radical of P^*, R^* is weakly closed in T^* with respect to H^* by I.2.5. Since $C_T(K^*) \leq C_T(X^*) \leq T_1$, a Frattini Argument shows that $N_H(T_1)^* = N_{H^*}(T_1^*)$. Thus $T_0^* \leq N_H(T_1)^* = N_{H^*}(T_1^*) \leq N_{H^*}(R^*) = N_{H^*}(P^*) = N_{H^*}(X^*)$, so

$$K_0^* T^* = H^* = \langle T_0^*, X^* T^* \rangle \leq N_{H^*}(X^*) = X^* T^*,$$

a contradiction. This completes the proof of 12.1.4. □

LEMMA 12.1.5. *(1) Let v generate $C_{V_1}(T_1)$. Then $T_1 \in Syl_2(G_v)$.*
(2) M controls fusion of involutions in V.

PROOF. If $n = 4$, then (1) follows from 12.1.4. If $n = 5$, then by 12.1.3, $W_0 \leq Q \leq T_1$, so $N_G(T_1) \leq N_G(W_0)$ by E.3.15. As $M = !\mathcal{M}(N_G(Q))$ by 1.4.1, $N_G(W_0) \leq M$ by E.3.34.2. Thus $N_{G_v}(T_1) \leq N_{G_v}(W_0) \leq M_v$, so as T_1 is Sylow in M_v, (1) also holds in this case.

By (1), $|G_v|_2 = |T|/2$ for any $v \in \mathcal{O}_1$, while $|G_z|_2 = |T|$ for $z \in \mathcal{O}_2$. Finally by 12.1.1, $|G_v|_2 < |T|/2$ for $v \in \mathcal{O}_3$. Thus the distinct M-classes of involutions in V are in different G-classes, so (2) holds. □

LEMMA 12.1.6. *(1) For $v \in \mathcal{O}_1$, $\langle V^{G_v} \rangle$ is abelian.*
(2) If $V_1 \cap V^g \neq 1$, then $[V, V^g] = 1$.

PROOF. Let $v \in \mathcal{O}_1$. By 12.1.5.2, $M = N_G(V)$ is transitive on G-conjugates of v in V, so by A.1.7.1, G_v is transitive on G-conjugates of V containing v. Thus (2) follows from (1), so it suffices to establish (1).

We may as well choose v to generate $C_{V_1}(T_1)$. By 12.1.5.1, $T_1 \in Syl_2(G_v)$. By lemma H.9.1, $U := [V, L_v] = V_1 \oplus U_2$, where $U_2 := v^\perp \cap V_2$. Let v_2 generate $C_{V_2}(T_1)$. Then $z := vv_2$ generates $C_V(T)$, and $v_2 \in U_2$.

By 1.1.6, the hypotheses of 1.1.5 are satisfied with G_v, G_z in the roles of "H, M". But as $z \in U$, $[O(G_z), z] = 1$ by A.1.26, so $O(G_v) = 1$ as z inverts $O(G_z)$ by 1.1.5.2.

Suppose first that $G_v \notin \mathcal{H}^e$. Then as $O(G_v) = 1$, there is a component K of G_v, and by 1.1.5.3, $K = [K, z] \not\leq M$. As $T_1 \in Syl_2(G_v)$, $L_v \leq K_v \in \mathcal{C}(G_v)$ by 1.2.4. Now $z \in U = [U, L_v] \leq K_v$, so as $K = [K, z]$, we conclude $K = [K, K_v] = K_v$ from 1.2.1.2. Thus $v \in L_v \leq K$. Set $K^* := K/O_2(K)$. Then $U \cap Z(K) = \langle v \rangle$, so $U^* \unlhd L_v^*$, with $U^* = V_1^* \oplus U_2^*$ the sum of the natural module and its dual for $L_v/O_2(L_v) \cong L_{n-1}(2)$. As no group on the list of 1.1.5.3 has such a subgroup invariant under a Sylow 2-group T_1^*, we have a contradiction.

This contradiction shows that $G_v \in \mathcal{H}^e$. Let $Q_v := O_2(G_v)$, and $\tilde{G}_v := G_v/\langle v \rangle$. Now $T_1 \in Syl_2(G_v)$, and L_v is irreducible on $\tilde{V}_1$, so Hypothesis G.2.1 holds with $\langle v \rangle$, V_1, T_1, G_v in the roles of "V_1, V, T, H". Then by G.2.2.1, $\tilde{V}_1 \leq Z(O_2(\tilde{G}_v)) = Z(\tilde{Q}_v)$. Similarly as L_v is irreducible on U_2, $[Q_v, U_2] \leq \langle v \rangle \cap U_2 = 1$, so that $U_2 \leq Z(Q_v)$. In particular $U = V_1 U_2 \leq Q_v$, so for any $g \in G_v$, U_2 centralizes U^g. Hence by 12.1.2.2, $U_2 \leq C_G(U^g) = C_{M^g}(V^g)$.

Suppose first that $V \leq Q_v$. Then for all $g \in G_v$, $V^g \leq Q_v \leq M = N_G(V)$. By the previous paragraph, V^g centralizes U_2, so V^g acts on V_1 and V_2; hence by symmetry, V acts on V_1^g and V_2^g. Then as $v \notin V_2$,

$$[V_1, V_2^g] \leq [V_1, Q_v] \cap V_2^g \leq \langle v \rangle \cap V_2^g = 1,$$

so that $V_1 \leq C_{M^g}(V_2^g) = C_{M^g}(V^g)$. Then $V^g \leq C_M(V_1) = C_G(V)$, so (1) holds.

So assume instead that $V \not\leq Q_v$. Then by the Baer-Suzuki Theorem, there is $g \in G_v$ such that $I := \langle V, V^g \rangle$ is not a 2-group. We showed $U_2 \leq C_G(V^g)$, and by symmetry U_2^g centralizes V, so $U_2 U_2^g \langle v \rangle \leq Z(I)$ and $V_1^g \leq C_{T_1}(U_2)$. But $C_{\bar{T}_1}(U_2)$ is the group of transvections on V_2 with axis U_2, so as V_1 is dual to V_2, $C_{\bar{T}_1}(U_2)$ is the group of transvections on V_1 with center $\langle v \rangle$. Hence $[V, V_1^g] \leq U_2 \langle v \rangle = U_2 \langle z \rangle$, and then $[V_1 U_2 V_1^g U_2^g, I] \leq U_2 U_2^g \langle z \rangle$. Therefore $O^2(I) \leq C_G(V_1 U_2) = C_G(V)$ using 12.1.2.2, contradicting I not a 2-group. This completes the proof of 12.1.6. $\square$

LEMMA 12.1.7. $W_0 := W_0(T, V)$ centralizes V, so that $w := w(G, V) > 0$.

PROOF. Assume that $W_0 \not\leq C_T(V)$. Then $n = 4$ by 12.1.3, and there is $A := V^g \leq T$ with $\bar{A} \neq 1$.

Suppose first that $V \leq N_G(A)$. Then interchanging the roles of A and V if necessary, we may assume $m(A/C_A(V)) \geq m(V/C_V(A))$. Then by B.1.4.4, $\bar{A}$ contains a member of $\mathcal{P}(\bar{T}, V)$, which is $J(\bar{T})$ by B.4.9.2iii. Thus $\bar{A}$ is the unipotent radical of the stabilizer in $\bar{L}$ of a 2-subspace of V_1, so that $[V_1, A]$ is of rank 2. As $\bar{A}$ normalizes V_1, and $V_1 \leq V \leq N_G(A)$, $1 \neq [V_1, A] \leq V_1 \cap A$, contrary to 12.1.6.2.

Therefore we may assume that $V \not\leq N_G(A)$. As $r(G, V) \geq 4$ by 12.1.2.2, $m(\bar{A}) \geq 4$ by E.3.4, so that $m(\bar{A}) = 4 = m_2(\bar{L}\bar{T})$. Then by lemma H.9.3.3, we may take $\bar{A}$ to be one of the groups denoted there by $\bar{A}_i$ for $0 \leq i \leq 2$. As $r(G, V) \geq 4$, we conclude from E.3.32 that

$$\check{\Gamma}_{3,\bar{A}}(V) = \check{\Gamma}_{3,A}(V) \leq U := N_V(A). \qquad (*)$$

As we are assuming $U < V$, $i \neq 0$ by lemma H.9.3.4. Let $B := N_A(V_1)$. Then $\bar{B} = \bar{A} \cap \bar{L}$ has rank 3, as $\bar{A}$ is $\bar{A}_1$ or $\bar{A}_2$. Set $U_i := V_i \cap U$. By 12.1.6.2,

$$1 = V_i \cap A \geq [U_i, B].$$

But for any $\bar{b} \in \bar{B}^{\#}$, $C_{V_i}(b) \leq U_i$ by $(*)$, so $C_{V_i}(b) = U_i = C_V(B)$. However this is not the case as $\bar{B}$ contains at least one transvection on V_1, but not all elements of $\bar{B}^{\#}$ induce transvections on V_1. This contradiction completes the proof. $\square$

LEMMA 12.1.8. $W_1 := W_1(T, V)$ centralizes V, so that $w > 1$.

PROOF. Assume that $W_1 \not\leq C_T(V)$. As $w > 0$ by 12.1.7, $w = 1$. Thus there is a w-offender $A := N_{V^g}(V) \leq T$ with A a hyperplane of V^g and $\bar{A} \neq 1$. Now $V \not\leq N_G(V^g)$ by E.3.25. As $r(G, V) \geq n$ by 12.1.2.2, $m(\bar{A}) \geq n - 1$ by E.3.28.3. As $r(G, V) \geq n$, E.3.32 says that

$$\check{\Gamma}_{n-2,A}(V) = \check{\Gamma}_{n-2,\bar{A}}(V) \leq U := N_V(V^g). \qquad (*)$$

Let $U_i := V_i \cap U$ and $B := N_A(V_1)$. Then $m(A/B) \leq 1$, so $m(V^g/B) \leq 2$. Also $[U_i, B] \leq V_i \cap V^g = 1$ by 12.1.6.2, so that $U_i \leq C_{V_i}(\bar{B})$.

Suppose $m(\bar{A}) = n - 1$. Then by $(*)$, $C_{V_i}(\bar{b}) \leq U_i$ for each $\bar{b} \in \bar{B}^{\#}$, so that $U_i = C_{V_i}(\bar{b})$ for each such $\bar{b}$. However, if $\bar{b}$ is a transvection in $\bar{L}$, then U_i is a hyperplane of V_i, so that $\bar{B}$ must be the full group of transvections with axis U_i for $i = 1$ and 2. This is not the case as V_1 is dual to V_2. Thus $\bar{B}$ contains no transvections, and hence $\dim(U_i) = n - 2$ and $\bar{B}$ lies in the unipotent radical $\bar{R}_i$ centralizing U_i. However $m(\bar{R}_1 \cap \bar{R}_2) = 4$ and $\bar{R}_1 \cap \bar{R}_2$ contains a 4-group $\bar{R}$ with each member of $R^{\#}$ a transvection, so as $\bar{B}$ contains no transvections, $m(\bar{B}) \leq m((\bar{R}_1 \cap \bar{R}_2)/\bar{R}) = 2$. Thus as $m(\bar{B}) \geq n - 2$, we conclude $n = 4$ and $\bar{A} > \bar{B}$, so $C_{U_1}(A) = 1$ and hence U_1 is faithful on V^g. But U_1 centralizes the subspace B of codimension 2 in V^g; this forces U_1 to induces a group of transvections on

V_i^g with fixed axis $B \cap V_i^g$ for $i = 1$ and 2. Thus as V_1 is dual to V_2, $m(U_1) \leq 1$, contradicting $m(U_1) = n - 2 = 2$.

This contradiction shows that $m(\bar{A}) \geq n$. Suppose $n = 5$. Then by lemma H.9.2.3, we may take $\bar{A} \leq \bar{A}_0$, where $\bar{A}_0$ is the centralizer in $\bar{T}_1$ of a 3-subspace X of V_1. Let W be a hyperplane of V_1 containing X. Then $m(\bar{A}/C_{\bar{A}}(W)) \leq m(\bar{A}_0/C_{\bar{A}_0}(W)) = 3$, so $W \leq C_V(C_{\bar{A}}(W)) \leq U$ by (*). As this holds for each such hyperplane, we conclude $V_1 \leq U$. But then $1 \neq [V_1, A] \leq V_1 \cap V^g$, contrary to 12.1.6.2.

Therefore $n = 4$. Then by lemma H.9.3.3, we may assume $\bar{A}$ is one of the subgroups there denoted $\bar{A}_i$ for $0 \leq i \leq 2$. Now $U < V$ as $V \not\leq N_G(V^g)$, so $i \neq 0$ in view of (*) and lemma H.9.3.4. Therefore by parts (5) and (6) of lemma H.9.3, $m(U) \geq 6$, and $C_U(A)$ is of rank 1 or 2 for $i = 1$ or 2, respectively. Next as $s(G, V) = 2$ by 12.1.2.2, $C_U(A) = C_U(V^g)$ by E.3.6. Thus $m(U/C_U(V^g)) \geq 5$ or 4 in the respective cases; so as $m_2(\bar{M}) = 4$, we conclude that $\bar{A} = \bar{A}_2$ and $m(U/C_U(V^g)) = 4$. But as $r(G, V) > m(V/U)$, $C_G(U) \leq N_G(V)$; hence $C_A(V) = C_{V^g}(U)$ since $N_{\bar{A}}(U)$ is faithful on U by H.9.3.6. Therefore as $m(\bar{A}) = 4$, $C_A(V) = C_{V^g}(U)$ is of rank 3. This contradicts parts (4)–(6) of lemma H.9.3, which say that $m(C_{V^g}(U)) = 1$, 2, or 4, since $m(U/C_U(V^g)) = 4$. This completes the proof of 12.1.8. $\qquad\square$

Having shown that $w > 1$, we turn to the other weak closure parameters of section E.3; as usual we will obtain a contradiction from their interrelations.

Recall by 3.3.2 that we may apply the results of section B.6 to any $H \in \mathcal{H}_*(T, M)$.

LEMMA 12.1.9. (1) If $1 \neq X$ is of odd order in $C_M(V)$, then $N_G(X) \leq M$.
(2) If $H \in \mathcal{H}_*(T, M)$, then $n(H) = 2$.
(3) $r(G, V) \geq n + 2$.

PROOF. Assume X is as in (1); replacing X with $Z(F(X))$, we may assume that X is abelian. Then $[X, L] \leq O_2(L)$. By Remark 4.4.2, Hypothesis 4.4.1 is satisfied. As V is not the sum of isomorphic natural modules for $L/O_2(L) \cong L_n(2)$ or $\Omega_6^+(2)$, $N_G(X) \leq M$ by Theorem 4.4.3.

Next suppose $U \leq V$ with $G_U := C_G(U) \not\leq M$. By 12.1.2.1, U is totally singular. Conjugating in L, we may take $T_U := C_T(U)$ Sylow in $C_M(U)$. Let $H \in \mathcal{H}_*(T_U, M) \cap G_U$. By 12.1.8, $W_i = W_i(T, V) \leq Q \leq T_U$ for $i = 0, 1$, so by E.3.15, $W_i = W_i(T_U, V) = W_i(Q, V)$, and $N_G(T_U) \leq N_G(W_i)$. But $M = !\mathcal{M}(N_G(Q))$ by 1.4.1, so by E.3.34.2,

$$N_G(W_0(T_U, V)) \leq M \geq C_G(Z(W_1(T_U, V))) \geq C_G(C_1(T_U, V)).$$

In particular $N_{G_U}(T_U) \leq M_U$, and hence $T_U \in Syl_2(G_U)$. Then as $s(G, V) = 2$ by 12.1.2.2, $n(H) > 1$ by E.3.19, so we may apply E.2.2 to conclude that a Cartan subgroup B of the Borel subgroup $H \cap M$ is nontrivial. For each odd prime p and $1 \neq X \in Syl_p(B)$, $H = \langle H \cap M, N_H(X) \rangle$, so $N_G(X) \not\leq M$ as $H \not\leq M$. If $T = T_U$ and $p > 3$, then as $XT = TX$, $X \leq M = N_G(V)$, while $\bar{M} = \bar{L}\bar{T}$ has no nontrivial p-subgroup permuting with $\bar{T}$, we conclude $[X, V] = 1$. This contradicts (1); thus if $H \in \mathcal{H}_*(T, M)$ then $p = 3$ so that $n(H) = 2$, establishing (2).

Indeed this argument shows more generally that $\bar{X} \neq 1$. As X is of odd order, $C_V(X)$ is a nondegenerate subspace of V of codimension at least 4, so as U is a totally singular subspace of $C_V(X)$, $\dim(C_V(X)) \geq 2\dim(U)$. Thus

$$\dim(V/U) \geq \dim(V/C_V(X)) + \dim(C_V(X))/2$$

$$= 2n - \dim(C_V(X))/2 \geq 2n - (2n-4)/2 = n+2,$$

establishing (3). $\square$

LEMMA 12.1.10. $W_2(T,V)$ *centralizes* V, *so that* $w > 2$.

PROOF. Assume that $W_2(T,V) \not\leq C_T(V)$. Then $w = 2$ by 12.1.8, so there is a w-offender $A := V^g \cap M \leq T$ with $m(V^g/A) = 2$ and $\bar{A} \neq 1$. Let $U := N_V(V^g)$; then $m(V/U) \geq 2$ as $w = 2$. By 12.1.9.3, $m(\bar{A}) \geq n$. Then by E.3.32,

$$\check{\Gamma}_{n-1,A}(V) = \check{\Gamma}_{n-1,\bar{A}}(V) \leq U < V. \tag{$*$}$$

Suppose first that $n = 4$. Then $m(\bar{A}) = 4 = m_2(\bar{M})$, so by lemma H.9.3.3, we may take $\bar{A}$ to be one of the subgroups there denoted by $\bar{A}_i$ for $0 \leq i \leq 2$. Set $\bar{B}_i := \bar{A}_i \cap \bar{L}$. By ($*$) and H.9.3.4, $i \neq 0$. Then we conclude from the last two parts of H.9.3 that $\check{\Gamma}_{2,\bar{A}}(V)$ is of rank 6. As $m(V/U) \geq 2$, $U = \check{\Gamma}_{2,\bar{A}}(V) = \check{\Gamma}_{3,\bar{A}}(V)$, whereas $C_V(\bar{a}) \not\leq \check{\Gamma}_{2,\bar{A}}(V)$ for $\bar{a} \in \bar{A}_1 - \bar{B}_1$, or for $\bar{a}$ a transvection in $\bar{B}_2$.

This contradiction reduces us to the case $n = 5$. Then by lemma H.9.2.3, we may take $\bar{A} \leq \bar{A}_0$ in the notation of that result. Now lemma H.9.2.5 contradicts ($*$), completing the proof. $\square$

We are now in a position to establish a contradiction. Pick $H \in \mathcal{H}_*(T,M)$. By 12.1.9.2, $n(H) = 2$. However by 12.1.10, $w \geq 3$, while by 12.1.9.3, $r(G,V) \geq 6$. Thus $H \not\leq M$ with $n(H) < \min\{w, r(G,V)\}$, contrary to E.3.35.1.

This contradiction shows:

THEOREM 12.1.11. *Assume* G *is a simple QTKE-group,* $T \in Syl_2(G)$, *and* $L \in \mathcal{L}_f^*(G,T)$ *with* $L/O_2(L) \cong L_4(2)$ *or* $L_5(2)$. *Let* $M := N_G(L)$. *Then there is no* $V \in \mathcal{R}_2(M)$ *such that* $M/C_M(V)) \cong Aut(L/O_2(L))$ *and* V *is the sum of the natural module and its dual for* $L/O_2(L)$.

By Theorems 12.1.11, 3.2.5, 3.2.8, and 3.2.9, the subcase of the Fundamental Setup with $L/O_2(L) \cong L_4(2)$ or $L_5(2)$ is reduced to the cases (i.e. cases (9), (10), and (11) of 3.2.8) with $V/C_V(L)$ a natural module or its exterior square. These cases will be treated along with the other cases where $L/O_2(L)$ is defined over $\mathbf{F}_2$; in particular they are completed in section 12.6, and in the final three sections of this chapter.

12.2. Groups over $\mathbf{F}_2$, and the case V a TI-set in G

We now begin a fairly unified treatment of those simple QTKE-groups G for which there exists $L \in \mathcal{L}_f^*(G,T)$ such that the section $L/O_2(L)$ has not yet been eliminated from the list of cases in section 3.2. Thus in section 12.2, and indeed in many subsequent sections, we assume the following hypothesis:

HYPOTHESIS 12.2.1. G *is a simple QTKE-group,* $T \in Syl_2(G)$, *and* $L \in \mathcal{L}_f^*(G,T)$ *with* $L/O_2(L)$ *quasisimple.*

We begin with a Theorem which summarizes much of what we have accomplished up to this point:

THEOREM 12.2.2. *Assume Hypothesis 12.2.1. Then one of the following holds:*

(1) G is a group of Lie type of Lie rank 2 over $\mathbf{F}_{2^n}$, $n > 1$, but $G \cong U_5(2^n)$ only for $n = 2$.

(2) $G \cong M_{22}, M_{23}$, or J_4.

(3) $T \le M := N_G(L)$, and there exists $V \in Irr_+(L, R_2(LT), T)$. For each such V, $V \trianglelefteq T$, $V \in \mathcal{R}_2(LT)$, the pair L, V is in the Fundamental Setup (3.2.1), V is a TI-set under M, and either $V \trianglelefteq M$ or $C_V(L) = 1$. In addition, one of the following holds:

    (a) V is the natural module of rank n for $L/O_2(L) \cong L_n(2)$, with $n = 3$, 4, or 5.

    (b) $m(V) = 4$ and V is indecomposable under $L/O_2(L) \cong L_3(2)$.

    (c) $L/O_2(L) \cong L_5(2)$, and V is an irreducible of rank 10.

    (d) $V/C_V(L)$ is the natural module for $L/C_L(V) \cong A_n$, with $5 \le n \le 8$.

    (e) $m(V) = 4$, and $L/C_L(V) \cong A_7$.

    (f) $V/C_V(L)$ is the natural module of rank 6 for $L/O_2(L) \cong G_2(2)' \cong U_3(3)$.

    (g) V is a faithful irreducible of rank 6 for $L/O_2(L) \cong \hat{A}_6$.

PROOF. By Theorem 10.0.1, $T \le N_G(L)$. Hence $\langle L, T \rangle = LT$ and by 3.2.3, there exists $V_\circ \in Irr_+(L, R_2(LT), T)$ and for each such $V_\circ$, L and $V := \langle V_\circ^T \rangle$ satisfy the FSU. Therefore by Theorem 3.2.5, one of the following holds:

(i) $V = V_\circ \trianglelefteq M$.

(ii) $V = V_\circ \trianglelefteq T$, $C_V(L) = 1$, and V is a TI-set under M.

(iii) Case (3) of Theorem 3.2.5 holds.

Case (iii) was eliminated in Theorem 7.0.1 and Theorem 12.1.11. Thus case (i) or (ii) holds, so that $V = V_\circ \in Irr_+(L, R_2(LT))$ and $V \trianglelefteq T$. As $O_2(LT)$ centralizes $R_2(LT)$ and $L/O_2(L)$ is quasisimple, $O_2(LT) \le C_{LT}(V) \le O_2(LT)O_{2,Z}(L)$, so that $V \in \mathcal{R}_2(LT)$. In case (i), $V \trianglelefteq M$, and in case (ii), $C_V(L) = 1$, so in either case V is a TI-set under M. Thus it remains only to show either that G is described in (1) or (2), or that L and its action on V are as described in one of the cases (a)–(g) of part (3) of Theorem 12.2.2.

The possibilities for the pair (L, V) when V is not an FF-module under the action of $Aut_{GL(V)}(L/C_L(V))$ are listed in 3.2.9. If the first case of 3.2.9 holds, then V is the $\Omega_4^-(2^n)$-module for $L_2(2^{2n})$ with $n > 1$, so by Theorem 6.2.20, either $G \cong U_4(2^n)$, or $n = 2$ and $G \cong U_5(4)$. Thus conclusion (1) of Theorem 12.2.2 holds in this case. The remaining cases of 3.2.9 were treated in Theorem 7.0.1, where it was shown that G is isomorphic to J_4, so that conclusion (2) of Theorem 12.2.2 holds.

Thus we have reduced to the case where V is an FF-module for $Aut_{GL(V)}(\bar{L})$, where $\bar{L} := L/C_L(V)$. Therefore $\bar{L}$ and its action on $\tilde{V} := V/C_V(L)$ are listed in 3.2.8. In the first case of 3.2.8, $\bar{L} \cong L_2(2^n)$ and V is the natural module. Then by Theorem 6.2.20, the only groups G arising are: the groups of Lie rank 2 and characteristic 2 (arising in our Generic Case), so that conclusion (1) of 12.2.2 holds; and M_{22} and M_{23}, so that conclusion (2) of 12.2.2 holds. Indeed the only case of the FSU with $\bar{L} \cong L_2(2^n)$ left open by Theorem 6.2.20 is the case where $n = 2$ and V is the A_5-module; this case is one of the subcases of 3.2.8.5, and it appears as a subcase of case (d) of conclusion (3) of 12.2.2. The cases with $n > 1$ in (2), (3), and (4) of 3.2.8, were eliminated in Theorem 11.0.1. On the other hand when $n = 1$, one of the conclusions of Theorem 12.2.2 holds—namely (a), (b), the subcase of (d) with $Sp_4(2)' \cong A_6$, or (f). Thus Theorem 12.2.2 holds in the first four cases of 3.2.8. In the remaining cases of 3.2.8, one of the conclusions of part (3) of Theorem 12.2.2 holds; notice that 3.2.8.10 correponds to the subcase of case (d) of conclusion (3) of 12.2.2 where $\bar{L} \cong L_4(2) \cong A_8$. So the proof is complete. □

Thus in the remainder of this section, and in many subsequent sections, we will assume:

HYPOTHESIS 12.2.3. *Hypothesis 12.2.1 holds, and G is not one of the groups in conclusions (1) and (2) of Theorem 12.2.2. Thus conclusion (3) of Theorem 12.2.2 holds. Set $M := N_G(L)$, and let $V \in Irr_+(L, R_2(LT), T)$.*

Since Hypothesis 12.2.3 implies that conclusion (3) of Theorem 12.2.2 holds, the remainder of our treatment of the Fundamental Setup is devoted to the groups and modules listed there.

Observe also that Hypothesis 12.2.3 imposes constraints on all members of $\mathcal{L}_f^*(G, T)$:

REMARK 12.2.4. Assume Hypothesis 12.2.3. Then for any $K \in \mathcal{L}_f^*(G, T)$ with $K/O_2(K)$ quasisimple, Hypothesis 12.2.3 holds for K in the role of "L". Thus K is described in conclusion (3) of Theorem 12.2.2, T normalizes K, there exists $V_K \in Irr_+(K, R_2(KT), T)$, and any such V_K is described in conclusion (3) of Theorem 12.2.2.

Indeed observe that any KT-submodule of $R_2(KT)$ which is irreducible module K-fixed points must contain such a V_K, and hence must itself be of the form V_K. However rather than introducing further notation for KT, we will continue to use the existing notation of $Irr_+(K, R_2(KT), T)$.

Usually when we assume Hypothesis 12.2.3, we adopt the following notational conventions:

NOTATION 12.2.5. (1) $Z := \Omega_1(Z(T))$ and $Q := O_2(LT)$.
(2) $M_V := N_M(V) = N_G(V)$ and $\bar{M}_V := M_V/C_{M_V}(V)$.
(3) For $v \in V^\#$, $G_v := C_G(v)$, $M_v := C_M(v)$, $L_v := O^2(C_L(v))$, and $T_v := C_T(v)$. We have the properties:

    (a) $L_v \trianglelefteq M_v$.
    (b) Conjugating in L, we may choose v so that $T_v \in Syl_2(M_v)$.
    (c) $M_v \in \mathcal{H}^e$.
    (d) $C(G_v, Q) \leq M_v$.
    (e) $M_v \leq M_V$.

(4) For $z \in V \cap Z^\#$, set $\tilde{G}_z := G_z/\langle z \rangle$.

PROOF. We establish the properties claimed in part (3): Part (a) follows since $L \trianglelefteq M$; (b) follows since $T \in Syl_2(G)$, (c) follows from 1.1.3.2; (d) follows as $C(G, Q) \leq M$ by 1.4.1.1; (e) is a special case of 12.2.6, established in the next subsection. □

12.2.1. Preliminary results under Hypothesis 12.2.3. Recall that we are assuming Hypothesis 12.2.3, so that in particular Theorem 12.2.2.3 holds. Our first result follows from Theorem 12.2.2.3 and 3.1.4.1:

LEMMA 12.2.6. V is a TI-set in M, so if $1 \neq U \leq V$ and $H \leq N_G(U)$, then $H \cap M = N_H(V)$.

LEMMA 12.2.7. Assume $C_G(Z) \leq M$ and $H \in \mathcal{H}_*(T, M)$. Let $K := O^2(H)$ and $V_H := \langle Z^H \rangle$. Then

(1) $V_H \in \mathcal{R}_2(H)$ and $C_T(V) = O_2(H) \leq C_H(V_H) \leq \ker_{N_H(V)}(H)$.

Assume further that H is not solvable. Then

(2) $K/O_2(K) \cong L_2(4)$.

(3) $K \leq Y \in \mathcal{L}_f^*(G, T)$, and either

(i) $Y = K$ and $[V_H, K]$ is the sum of at most two A_5-modules for $K/O_2(K)$,

or

(ii) $Y/O_{2,Z}(Y) \cong A_7$, Hypothesis 12.2.3 is satisfied with Y in the role of "L", and for each $V_Y \in Irr_+(Y, R_2(YT), T)$, V_Y is T-invariant and $m(V_Y) = 4$ or 6.

PROOF. First $V_H \in \mathcal{R}_2(H)$ by B.2.14, so $O_2(H) \leq C_H(V_H)$. Then as $C_G(Z) \leq M$ but $H \not\leq M$, $K \not\leq C_H(V_H)$. The remaining statements in (1) follow from B.6.8.6 and 12.2.6.

Now assume H is not solvable. By E.2.2, $K = \langle X^T \rangle$ for a suitable $X \in \mathcal{C}(H)$ with $X/O_2(X)$ quasisimple and $X \not\leq M$. As $[V_H, X] \neq 1$ by (1), $X \in \mathcal{L}_f(G, T)$. We may embed $X \leq Y \in \mathcal{L}^*(G, T)$, and then by 1.2.9, $Y \in \mathcal{L}_f^*(G, T)$.

Suppose first that $Y/O_2(Y)$ is quasisimple. Then by Remark 12.2.4, Hypothesis 12.2.3 is satisfied with Y in the role of "L". In particular Y is T-invariant; and for each $V_Y \in Irr_+(Y, R_2(YT), T)$, V_Y is T-invariant, and Y, V_Y satisfies one of the conclusions of 12.2.2.3. Therefore T acts on X by 1.2.8.1, so that $X = K$ with KT described in E.2.2.2.

Assume first that $K = Y$. Comparing the lists of 12.2.2.3 and E.2.2.2, we conclude that $K/O_{2,Z}(K) \cong L_2(4)$, $L_3(2)$, or A_6. However if $K/O_{2,Z}(K)$ is $L_3(2)$ or A_6, then by E.2.2.2, T is nontrivial on the Dynkin diagram of $K/O_2(K)$, a contradiction as 12.2.2.3 says $V_Y/C_{V_Y}(K)$ is a natural module. Thus $K/O_2(K) \cong L_2(4)$. Hence by the exclusions in Hypothesis 12.2.3 and Theorem 6.2.20, $[V_H, K]$ is the sum of at most two A_5-modules. Therefore (2) and (3i) hold in this case.

So we may assume that $K < Y$. Therefore by 1.2.4, the embedding of K in Y is described in A.3.12. Searching for pairs K, Y with K appearing in E.2.2.2 and Y appearing in 12.2.2.3, we conclude that either $K/O_{2,Z}(K) \cong L_2(4)$, $L_3(2)$, or A_6, with $Y/O_{2,Z}(Y) \cong A_7$; or $K/O_2(K) \cong L_3(2)$, with $Y/O_2(Y) \cong L_4(2)$ or $L_5(2)$. But again when $K/O_2(K)$ is $L_3(2)$ or A_6, T is nontrivial on the Dynkin diagram of $K/O_2(K)$, whereas there is no such embedding of $KT/O_2(KT)$ in S_7 or $Aut(L_4(2))$, so $KT/O_2(KT) \cong Aut(L_3(2))$ and $YT/O_2(YT) \cong Aut(L_5(2))$. However this is also impossible as V_Y is a T-invariant natural module for $Y/O_2(Y)$ by 12.2.2.3, so $Y/O_2(Y)$ is self-normalizing in $GL(V_Y)$. Thus $K/O_2(K) \cong A_5$ and $Y/O_{2,Z}(Y) \cong A_7$, so that (2) holds. Also as Y, V_Y appears in 12.2.2.3, (3ii) holds.

Finally assume that $Y/O_2(Y)$ is not quasisimple. Then by 1.2.1.4, $Y/O_{2,2'}(Y) \cong SL_2(p)$ for some prime $p > 3$. Now T acts on Y by 1.2.1.3, so again $X = K$ is T-invariant by 1.2.8.1, and hence appears in E.2.2.2; in particular as $K/O_2(K)$ is quasisimple, $K < Y$. Again by 1.2.4, the embedding $X < Y$ is described in A.3.12, so by A.3.12, $K/O_2(K) \cong L_2(p)$ or $L_2(5)$. Now for some odd prime q, $X_0 := \Xi_q(Y) \in \Xi_{rad}(G, T)$, and as $Y \in \mathcal{L}^*(G, T)$, by definition $X_0 \in \Xi_{rad}^*(G, T)$. Then by 1.3.8, $X_0 \in \Xi^*(G, T)$. By 3.2.14 applied to Y, $[Z, X_0] = 1$, so $X_0 T \leq C_G(Z) \leq M$. Then

$M = N_G(X_0)$ since $N_G(X_0) = !\mathcal{M}(X_0 T)$ by 1.3.7, so $H \leq YT \leq N_G(X_0) \leq M$, contradicting $H \not\leq M$. $\square$

Given a group A, write $\theta(A)$ for the subgroup of A generated by all elements of order 3 in A.

LEMMA 12.2.8. *One of the following holds:*

(1) $O^{3'}(M) = L$.

(2) $L/O_2(L) \cong A_5$ or $L_3(2)$.

(3) $L/O_2(L) \cong \hat{A}_6$ or $\hat{A}_7$ and $L = \theta(M)$ is the subgroup of M generated by all elements of M of order 3.

PROOF. First L is described in 12.2.2.3, so if $m_3(L) = 1$, then (2) holds; thus we may assume $m_3(L) = 2$. Then (1) or (3) holds by 12.2.2 and A.3.18. $\square$

LEMMA 12.2.9. (1) *If $C_Z(L) \neq 1$, then $C_G(Z) \leq M$.*

(2) *If $C_G(Z) \leq M$, then $L = [L, J(T)]$.*

PROOF. As $M = !\mathcal{M}(LT)$, (1) holds. Theorem 3.1.8.3 implies (2). $\square$

LEMMA 12.2.10. (1) $C_{\bar{M}_V}(\bar{L}) = Z(\bar{L})$.

(2) $\bar{L} = O^2(\bar{M}_V)$ *and* $\bar{M}_V = \bar{L}\bar{T}$.

PROOF. In each case listed in conclusion (3) of Theorem 12.2.2, $Out(L/O_2(L))$ is a 2-group, so $O^2(\bar{M}_V) \leq \bar{L}C_{\bar{M}_V}(\bar{L})$. Further in cases (a)–(f), the irreducible module $I := V/C_V(L)$ satisfies $E := End_{\bar{L}}(I) \cong \mathbf{F}_2$, so that $C_{\bar{M}_V}(\bar{L}) = 1$. Hence (1) and (2) hold in these cases. In case (g), $I = V$ and $E \cong \mathbf{F}_4$, with $Z(\bar{L})$ inducing $E^\#$, so again (1) and (2) follow. $\square$

LEMMA 12.2.11. *Assume $H \in \mathcal{H}_*(T, M)$ with $H \leq N_G(U)$ for some $1 \neq U \leq V$. Assume also that one of the following holds:*

(a) $L/O_2(L) \cong L_5(2)$.

(b) $L/O_2(L) \cong \hat{A}_6$, *and* $V \leq O_2(C_G(v))$ *for* $v \in C_V(T)^\#$.

(c) $L/O_2(L) \cong G_2(2)'$ *and* $C_G(V_3) \leq M$, *where V_3 is the $(T \cap L)$-invariant subspace of V of rank 3.*

Then

(1) $n(H) \leq 2$, *and*

(2) *if $n(H) = 2$, then a Hall $2'$-subgroup of $H \cap M$ is a nontrivial 3-group.*

PROOF. The lemma is vacuously true if $n(H) \leq 1$, so we may assume that $n(H) \geq 2$. Then by E.2.2, $H \cap M$ is the preimage of the normalizer of a Borel subgroup of the group $O^2(H/O_2(H))$ of Lie type and characteristic 2. We take C to be a Hall $2'$-subgroup of $H \cap M$, so that C is abelian and $CT = TC$. We may assume that either:

(I) $n(H) = 2$, but there is a prime $p > 3$ such that $B := O_p(C) \neq 1$, or

(II) $n(H) > 2$, in which case p and B also must exist.

Then also $A := \Omega_1(B) \neq 1$, $BT = TB$, and $AT = TA$. As $n(H) > 1$ and $AT = TA$, $N_H(A) \not\leq M$ by 4.4.13.1.

Next as $B \leq H \cap M \leq N_M(U)$ by hypothesis, B normalizes V by 12.2.6. We claim in fact that B centralizes V: For otherwise $1 \neq \bar{B} \leq O^2(\bar{M}_V) = \bar{L}$ by 12.2.10.2. Thus $\bar{B}$ is an abelian p-subgroup of $\bar{L}$ with $p > 3$, and $\bar{B}\bar{T} = \bar{T}\bar{B}$, so we may apply A.3.15. However, the list of possibilities for $L/O_2(L)$ from A.3.15 does not intersect the list from Theorem 12.2.2.3.

Therefore $B \leq C_{M_V}(V) \leq C_{M_V}(\bar{L})$, and hence $B \leq C_{M_V}(L/O_2(L))$. Visibly Hypothesis 4.2.1 is satisfied, so Hypothesis 4.4.1 is satisfied by Remark 4.4.2. Thus we can apply Theorem 4.4.3 to obtain a contradiction: Namely we showed that $N_G(A) \not\leq M$, so that one of the conclusions of 4.4.3.2 must hold, which is not the case as we are assuming one of hypotheses (a)–(c). $\qquad\square$

The statement of the following lemma makes use of Notation 12.2.5.

LEMMA 12.2.12. *Assume* $v \in V^{\#}$ *with* $O_2(\bar{L}_v\bar{T}_v) = 1$, *and choose* $T_v \in Syl_2(M_v)$. *Then*

(1) $Q = O_2(L_vT_v)$, $Q \in Syl_2(C_{G_v}(L_v/O_2(L_v)))$, *and Hypothesis C.2.3 is satisfied with* G_v, M_v, Q *in the roles of "H, M_H, R".*

(2) Assume $\bar{L}_v = \bar{L}_v^{\infty}$ *and* $V = [V, \bar{L}_v]$. *Then Hypothesis C.2.8 is satisfied with* G_v, M_v, L_v^{∞}, Q *in the roles of "H, M_H, L_H, R".*

PROOF. Since $V \in \mathcal{R}_2(LT)$, $Q \leq T_v$, so as $O_2(\bar{L}_v\bar{T}_v) = 1$, $Q = O_2(L_vT_v)$ by 1.4.1.4. We chose $T_v \in Syl_2(M_v)$ while $L_v \trianglelefteq M_v$ and $C(G_v, Q) \leq M_v$ by 12.2.5.3. Therefore (1) follows from A.4.2.7.

Assume $\bar{L}_v = \bar{L}_v^{\infty}$ and $V = [V, L_v^{\infty}]$. Then $L_v^{\infty} \in \mathcal{C}(M_v)$, and as $\bar{L}_v = \bar{L}_v^{\infty}$, the argument in the previous paragraph shows that $Q \in Syl_2(C_{M_v}(L_v^{\infty}/O_2(L_v^{\infty})))$. Also $M_v \in \mathcal{H}^e$ by (3c) of 12.2.5, and the verification of the remainder of Hypothesis C.2.8 is straightforward. $\qquad\square$

12.2.2. The treatment of V a TI-set in G. We now come to the main result of this section, in which we treat the case where V is a TI-set in G.

THEOREM 12.2.13. *Assume Hypothesis 12.2.3. Then one of the following holds:*

(1) $C_G(v) \not\leq M$ *for some* $v \in V^{\#}$.

(2) L *is an* $L_n(2)$-block for $n = 3$ or 4, *and* $G \cong L_{n+1}(2)$.

(3) L *is an* $L_3(2)$-block, *and* $G \cong A_9$.

(4) L *is an* $L_4(2)$-block, *and* $G \cong M_{24}$.

REMARK 12.2.14. The groups M_{22} and M_{23} contain a pair (L, V) failing 12.2.13.1, with V of rank 4 and $L/O_2(L) \cong A_6$ or A_7, but these groups are explicitly excluded by Hypothesis 12.2.3. Their shadows are eliminated via an appeal to 12.2.7.3, which is violated in M_{22} and M_{23}.

REMARK 12.2.15. As the groups appearing in conclusions (2), (3), and (4) of Theorem 12.2.13 appear as conclusions in our Main Theorem, we will sometimes assume that G is not one of those groups. Then Theorem 12.2.13 tells us that $C_G(v) \not\leq M$ for some $v \in V^{\#}$.

Until the proof of Theorem 12.2.13 is complete, assume that $C_G(v) \leq M$ for each $v \in V^{\#}$. We must show that one of (2)–(4) holds. We begin a series of reductions. Recall we have adopted Notation 12.2.5.

LEMMA 12.2.16. V *is a TI-set in* G.

PROOF. By 12.2.6, V is a TI-set in M. Thus if the lemma fails, there is $g \in G - M$ and $v \in V^{\#}$ with $u := v^g \in V$. As we are assuming that conclusion (1) of 12.2.13 fails, $G_w = M_w$ for $w \in V^{\#}$, so

(1) $M_v = G_v \cong G_u = M_u$.

Next if $u^x = v$ for some $x \in M$, then $gx \in G_v \leq M$, so $g \in Mx^{-1} = M$, contrary to the choice of g. Hence

(2) $u \notin v^M$.

By (1) and (2) there are $u, v \in V^\#$ with $M_u \cong M_v$ but $v \notin u^M$. Inspecting the list of Theorem 12.2.2.3, we first eliminate the cases where L is irreducible on V. In the remaining cases, let z denote the generator of $C_V(L)$. We also eliminate case (b), since there $Z \cap V = C_V(L)$ by B.4.8.2, so that each $v \in V - C_V(L)$ is T-conjugate to vz, and hence all members of $V - C_V(L)$ are M-conjugate. This leaves the subcases of (d) where V is the core of the permutation module of degree n for $\bar{L} \cong A_n$, $n = 6$ or 8, and the subcase of (f) where V is the Weyl module of dimension 7 for $\bar{L} \cong G_2(2)'$. In the former, we may take v of weight 2, and u of weight $n - 2$; in the latter, we may take v singular and u nonsingular in $V - C_V(L)$. Now conjugating in L, we may assume $u = vz$. As $C_V(L) \neq 1$, $V \trianglelefteq M$ by 12.2.2.3, so $z \in Z(M)$ and hence $M = G_z$.

Without loss, $T_v := C_T(v) \in Syl_2(M_v)$, so as $G_v \leq M$, $T_v \in Syl_2(G_v)$. As $u = vz$ with z central in $M_u = G_u$, also $T_v \in Syl_2(G_u)$. Then replacing g by a suitable member of $gC_G(u)$, we may assume $g \in N_G(T_v)$. However if $\bar{L} \cong A_6$ or $G_2(2)'$, then v is 2-central in M so that $T = T_v$, so $g \in N_G(T) \leq M$ by Theorem 3.3.1, contrary to (2). Hence $\bar{L} \cong A_8$.

Let $Z_v := \Omega_1(Z(T_v))$ and $V_v := \langle Z_v^L \rangle$. As $Q \leq T_v$, Q centralizes Z_v and hence V_v. On the other hand, Z_v contains $Z \cap V \not\leq C_V(L)$ by I.2.3.1i; so $V \leq V_v$ as $V \in Irr_+(L, R_2(LT))$, and hence $C_{LT}(V_v) \leq C_{LT}(V) = Q$. Therefore $Q = C_{LT}(V_v)$, and hence $V_v \in \mathcal{R}_2(LT)$. If $[V_v, J(T)] = 1$, then as $V \leq V_v$, also $[V, J(T)] = 1$, and then 3.2.10.2 contradicts (2). Thus $[V_v, J(T)] \neq 1$, so by B.2.7, V_v is an FF-module for $LT/C_{LT}(V_v)$. Then as V is the core of the permutation module for A_8, by Theorem B.5.1.1, $V = [V_v, L]$, and hence $V_v = VZ_L$ by B.2.13, where $Z_L := C_{Z_v}(L)$. Thus $Z_v = \langle v \rangle Z_L$. Now $T = T_v(T \cap L) \leq C_G(Z_L)$, so $Z_L \cap Z_L^g = 1$ using (2), since $M = !\mathcal{M}(LT)$ and $M^g = !\mathcal{M}(L^g T^g)$. Hence as $g \in N_G(T_v) \leq N_G(Z_v)$ and Z_L is a hyperplane of Z_v, we conclude Z_L is of order 2. Therefore $Z_L = \langle z \rangle$ and $Z_v = \langle z, v \rangle$. Then as $v^g = u \notin z^G$ by (1), z is weakly closed in Z_v. Therefore $g \in G_z = M$, contrary to (2), completing the proof of 12.2.16. $\square$

LEMMA 12.2.17. $W_0(T, V) \leq C_T(V) = Q$, so that $N_G(W_0(T, V)) \leq M$.

PROOF. Let $V^g \leq T$. By 12.2.16, V is a TI-set in G, so as $V^g = N_{V^g}(V)$, we conclude from I.6.2.1 that $[V, V^g] = 1$. Now the final statement follows from E.3.34.2. $\square$

During the remainder of the proof of Theorem 12.2.13, pick $H \in \mathcal{H}_*(T, M)$, and set $K := O^2(H)$, $V_H := \langle Z^H \rangle = \langle Z^K \rangle$, and $H^* := H/C_H(V_H)$. Observe that $C_G(Z) \leq C_G(Z \cap V) \leq N_G(V) \leq M$ by 12.2.16; thus we may apply 12.2.7 during the course of the proof.

LEMMA 12.2.18. $V \not\leq O_2(H)$, so that $V^* \neq 1$.

PROOF. By 12.2.7.1, $O_2(H) = C_T(V_H)$. Thus $V \leq O_2(H)$ iff $V^* = 1$, so we may assume $V \leq O_2(H)$, and it remains to derive a contradiction.

Similarly if $W_0 := W_0(T, V) \leq O_2(H)$, then $H \leq N_G(W_0) \leq M$ by E.3.15 and 12.2.17, contrary to $H \not\leq M$. Thus there is $A := V^g \leq T$, with $A^* \neq 1$, and $K \leq \langle A^H \rangle$ by B.6.8.4.

Suppose that $A \cap O_2(H) = 1$. Then $A^* \cong A$, so $m(A^*) = m(V) \geq 3$ by 12.2.2.3. But if H is solvable, then $m_2(H^*) \leq 2$ as $H = O_{2,p,2}(H)$ for some odd prime p by B.6.8.2, so that $H/O_{2,p}(H)$ is a subgroup of $GL_2(p)$. On the other hand if H is nonsolvable, then by 12.2.7.2, $K^* \cong L_2(4)$, so that again $m_2(H^*) \leq 2$. Thus in either case, we have a contradiction to $m(A^*) \geq 3$.

This contradiction shows that $1 \neq A \cap O_2(H)$. Thus for each $h \in H$, $1 \neq A^h \cap O_2(H) \leq N_{A^h}(V)$ by 12.2.7.1. However as $V \leq O_2(H)$, $\langle V, A^h \rangle$ is a 2-group, so $[V, A^h] = 1$ by I.6.2.1. We saw $K \leq \langle A^H \rangle$, so $K \leq C_G(V) \leq M$, contradicting $H \nleq M$. This completes the proof of 12.2.18. $\qquad \square$

LEMMA 12.2.19. H is solvable.

PROOF. Assume that H is not solvable. Then by 12.2.7, $K/O_2(K) \cong K^* \cong L_2(4)$. By 12.2.18, $V^* \neq 1$. As $V \leq O_2(M)$, $V^* \leq O_2(M \cap H)^* = T^* \cap K^* \in Syl_2(K^*)$. Thus either $V^* = T^* \cap K^* \cong E_4$, or $V^* \leq K^*$ is of order 2. Pick $h \in K - M$, and let $U := V \cap O_2(H)$, $I := \langle V, V^h \rangle$, and $W_I := O_2(I)$. Then either $|V^*| = 4$ and $I^* = K^* \cong L_2(4)$, or V^* is of order 2 and $I^* \cong D_{2m}$, $m = 3$ or 5. As $m(V) \geq 3 > m(V^*)$, $U \neq 1$. Then as $U \leq O_2(H) \leq N_H(V^h)$ by 12.2.7.1, $N_V(V^h) \neq 1$. It follows from (a) and (c) of I.6.2.2 that $W_I := U \times U^h$ is a sum of natural modules for $I/W_I \cong I^*$; in particular if $I^* \cong D_{2m}$, an element of order m is fixed point free on W_I. If V^* is of order 2, pick $x \in H \cap M$ of order 3 and let $K_I := \langle V^x, I \rangle$ and $W := U^x W_I$; if $|V^*| = 4$ let $K_I := I$ and $W := W_I$. Thus in either case $K^* = K_I^*$.

We claim that K_I acts on W, and W is elementary abelian: Suppose first that $|V^*| = 2$. We saw that $U \neq 1$ normalizes V^x, and as $\langle V^*, V^{*x} \rangle$ is a 2-group by our choice of x, $\langle V, V^x \rangle$ is a 2-group. Therefore V centralizes V^x by I.6.2.1. Now by symmetry between $I = \langle V, V^h \rangle$ and $\langle V^x, V^h \rangle$, $\langle V^x, V^h \rangle$ acts on $U^x \times U^h$, so $K_I = \langle V, V^x, V^h \rangle$ acts on $W = UU^x U^h$ and W is elementary abelian. On the other hand if $|V^*| = 4$ then $K_I = I$ acts on $W = U \times U^h$, completing the proof of the claim.

Next by 12.2.7.1, $O_2(H)$ acts on V and V^h, and also on V^x when $|V^*| = 2$, so $O_2(H)$ acts on K_I and W. Thus $KO_2(H) = K_I O_2(H)$ acts on K_I and W. Therefore as $K_I O_2(H)/K_I \cong O_2(H)/(O_2(H) \cap K_I)$ is a 2-group, $K = O^2(H) \leq K_I$.

Now if $|V^*| = 4$, then $K_I = I$, so $W = O_2(I)$ is a sum of natural modules for $K^* \cong K_I/W$. Suppose on the other hand that V^* is of order 2. We saw earlier that V centralizes V^x; hence $W = U^x W_I = C_W(V)W_I$, so that $W = C_W(i) \times W_I$ for i of order m in I, which is fixed point free on W_I; and $C_W(i) = C_W(I) \leq C_W(V) = UU^x = C_W(V^x)$. Thus $C_W(I) = C_W(K_I)$. As $[W, V] = U$ and $[W, V^x] = U^x$ with VV^x abelian, $T^* \cap K^* = V^* V^{*x}$ is quadratic on W. Also i is fixed-point-free on W_I, so by G.1.5 and G.1.7, $W/C_W(I)$ is a sum of natural modules for K^*.

Now $Z \cap V \neq 1$, so $1 \neq V_Z := \langle (Z \cap V)^K \rangle \in \mathcal{R}_2(KT)$ by B.2.14. As V is a TI-set in G by 12.2.16 and $K \nleq M \geq N_G(V)$, $C_V(K) = 1$. As $Z \cap V \leq O_2(H) \cap V = U \leq W$, $V_Z \leq W$, so by the previous paragraph $1 \neq V_Z/C_{V_Z}(K)$ is a sum of natural modules for K^*.

By 12.2.7.3, $K \leq Y \in \mathcal{L}_f^*(G, T)$, with Y described in case (i) or (ii) of that result. Let $V_Y := \langle (Z \cap V)^Y \rangle$ and $\hat{Y} := Y/C_Y(V_Y)$; then $V_Z \leq V_Y$. As $V_Z/C_{V_Z}(K)$ is a sum of natural $L_2(4)$-modules, case (i) of 12.2.7.3 cannot arise, since there the noncentral chief factors for K on V_Z are A_5-modules. [1] Therefore case (ii) of 12.2.7.3 occurs, so $\hat{Y} \cong A_7$, and each $J \in Irr_+(Y, V_Y, T)$ is T-invariant and of rank 4 or 6. Again using the fact that the noncentral chief factors for K on V_Z are $L_2(4)$-modules, we conclude that J is of rank 4 and the natural module for $\hat{K}$. Therefore $[J, T \cap K] = [J, w] \cong E_4$ for each $w \in V - O_2(K)$. Now $[J, w] \le V$, and from the action of A_7 on J,

$$Y = \langle C_Y(v) : v \in V \cap J^\# \rangle \le M,$$

whereas $K \le Y$ with $K \not\le M$. This contradiction completes the proof of 12.2.19. $\qquad \square$

By 12.2.19, H is solvable, so we may apply B.6.8.2 to conclude that

$$H = PT,$$

where P is a p-group for some odd prime p, $P^* = F^*(H^*)$, and $\Phi(P) = P \cap M$. Thus $[P^*, V^*] \ne 1$ by 12.2.18, and hence $P^* = [P^*, V^*]$, since $V \trianglelefteq T$ and T^* is irreducible on $P^*/\Phi(P^*)$ by B.6.8.2. By Coprime Action, we may pick $h \in P - \Phi(P)$ so that $C_{V^*}(h^*)$ is a hyperplane of V^*. Let $I := \langle V, V^h \rangle$ and $U := O_2(I)$. By I.6.2.2, $U = (V \cap U) \times (V^h \cap U)$ is a sum of natural modules of $I/U \cong D_{2p}$. Thus $V^h \cap U$ is of rank $m(V) - 1 \ge 2$, and induces the full group of transvections in $GL(V)$ with axis $V \cap U$. Therefore we may apply the dual of G.3.1 to the action of LT on V, to restrict the cases in 12.2.2.3 to:

LEMMA 12.2.20. $\bar{L} = GL(V) \cong L_n(2)$ for $n = 3$, 4, or 5, V is the natural module for $\bar{L}$, $U \cap V$ is a hyperplane of V, and U induces the full group of transvections with axis $U \cap V$ on V. In particular as $T \le N_G(V)$, $T \le LC_T(V)$.

Observe that we are beginnning to show that G has a 2-local structure similar to that of one of the groups in conclusions (2)–(4) of Theorem 12.2.13.

LEMMA 12.2.21. $U \le O_2(H)$, so V^* is of order 2 and inverts $P^*/\Phi(P^*)$.

PROOF. We saw $U = [U, O^2(I)]$, so if $U \not\le O_2(H)$, then $U^* = [U^*, O^2(I^*)] \ne 1$, impossible as H^* is 2-nilpotent. We also saw that $P^* = [P^*, V^*]$, so as $V \cap U$ is a hyperplane of U by 12.2.20, V^* is of order 2, and V^* inverts $P^*/\Phi(P^*)$. $\qquad \square$

Next we obtain some restrictions on the structure of H and its action on $\langle (U \cap V)^H \rangle$.

We observed earlier for each $h \in P - \Phi(P) = P - M$ that $I^* \cong D_{2p}$, so that h has order p. Then by A.1.24, $P \cong \mathbf{Z}_p$, E_{p^2}, or p^{1+2}. Let $v \in V - U$. By the Baer-Suzuki Theorem, v inverts an element h' of order p in K; replacing P by a Sylow group containing h', we may take $h' \in P$. Then as V^* is of order 2 by 12.2.21, we may take $h' = h$ in the definition of I. Let $W := \langle (U \cap V)^H \rangle$. Then $W \le O_2(H)$ by 12.2.21, and $U = (V \cap U)(V^h \cap U) \le \langle (U \cap V)^P \rangle \le W$. Indeed as T acts on $U \cap V = O_2(H) \cap V$, $(U \cap V)^H = (U \cap V)^{TP} = (U \cap V)^P = U^P$, so $W = \langle U^P \rangle$. In particular as $U \le [W, O^2(I)] \le [W, P]$, $W = \langle U^P \rangle \le [W, P]$, and hence $W = [W, P]$. Now $1 \ne U \cap V = O_2(H) \cap V \trianglelefteq O_2(H)$, so $U \cap V$ commutes with its H-conjugates by I.6.2.2, and hence W is elementary abelian. From the action of I on U in I.6.2.2a, $[U, v] = [U \cap V^h, v] = U \cap V$. Also $[W, v] \le W \cap V = U \cap V$, so $W = (U \cap V^h) \oplus C_W(v)$. Similarly $W = (U \cap V) \oplus C_W(v^h)$, so $W = U \oplus C_W(I)$

---

[1]This is where Hypothesis 12.2.3 eliminates M_{22} and M_{23} as possible conclusions in Theorem 12.2.13.

with $U = [W, I] = [W, h]$. Further $[W, V] = [U, V] = V \cap U = V \cap W$ is of rank $n - 1$.

We next claim that we may choose P invariant under v; and when P is non-abelian, that $\Phi(P) \le N_{H \cap M}(V)$. If $P \cong \mathbf{Z}_p$, then v acts on P and we are done. Suppose $P \cong E_{p^2}$. Then $\langle P, v \rangle \le N := N_H(\langle h \rangle)$. Set $\dot{N} := N/\langle h \rangle$. As v^* inverts P^*, $\dot{v} \notin O_2(\dot{N})$, so we may apply the Baer-Suzuki Theorem again in $\dot{N}$, to conclude that v inverts y of order p in $C_H(h) - \langle h \rangle$. Thus $\langle h, y \rangle \in Syl_p(H)$ is v-invariant, and choosing $P := \langle h, y \rangle$, we are done in this case also. Finally, suppose $P \cong p^{1+2}$. Then $\Phi(P) = Z(P)$ centralizes h and so acts on $[W, h] = U$. Further $O_{2,\Phi}(H) = O_2(H)\Phi(P)$ and V centralizes $O_{2,\Phi}(H)/O_2(H)$, so that $[\Phi(P), V] \le O_2(H)$. Thus $\Phi(P)$ acts on $1 \ne C_U(O_2(H)V) \le U \cap V$, and so since V is a TI-set in M by 12.2.6, $\Phi(P)$ acts on V, establishing the final assertion of the claim. Next $\Phi(P)$ centralizes $V/(U \cap V)$ of rank 1, and hence centralizes some $v_0 \in V - U$, which we may take to be v. Set $P_1 := \Phi(P)\langle h \rangle$, $H_1 := N_H(P_1)$, and $\dot{H}_1 := H_1/P_1$. We apply the Baer-Suzuki Theorem one more time to $\dot{H}_1$: As v^* inverts $P^*/\Phi(P^*)$, $\dot{v} \notin O_2(\dot{H}_1)$, so $\dot{v}$ inverts an element $\dot{k}$ of order p in $\dot{H}_1$, and then the preimage of $\langle \dot{k} \rangle$ is a v-invariant Sylow p-group P of H. This completes the proof of the claim.

So in any event we may assume v acts on P. Thus $VPW = \langle v \rangle PW$ is a subgroup of H. Further $[O_2(H), v] \le V \cap O_2(H) = V \cap U \le W$, so that v centralizes $O_2(H)/W$. Then as $P = [P, v]$, $[O_2(H), P] \le W$, and hence $PW \trianglelefteq O_2(H)P$. Thus as $K = O^2(H) \le PO_2(H)$, $K \le PW$. We saw earlier that $W = [W, P]$, so $W \le O^2(PW) \le O^2(H) = K$, and hence $K = PW$. Summarizing:

LEMMA 12.2.22. $P \cong \mathbf{Z}_p$, E_{p^2}, or p^{1+2}, and we may choose P so that P is invariant under $v \in V - U$, $W = \langle U^P \rangle = [W, P]$ is elementary abelian, $[W, V] = V \cap W$ is of rank $n - 1$, $K = PW$, and $\Phi(P) \le N_{H \cap M}(V)$.

LEMMA 12.2.23. $P \cong \mathbf{Z}_p$.

PROOF. Assume P is not $\mathbf{Z}_p$, and let $H_P := KV$, $\Phi := \Phi(P)$, $W_P := C_W(\Phi)$, and $\hat{H}_P := H_P/C_{H_P}(W_P)$.

Suppose $W_P \ne 1$. By 12.2.22, $W = [W, P]$, so as T^* is irreducible on $P^*/\Phi(P^*)$, $\Phi = C_P(W_P)$ and $\hat{P} \cong E_{p^2}$. Then by Generation by Centralizers of Hyperplanes A.1.17, W_P is generated by nontrivial subgroups $W_i := C_{W_P}(P_i)$, where P_i runs over a nonempty collection of subgroups of index p in P generating P. As v inverts P/Φ, v acts on each subgroup P_i and hence on each W_i; further as $W = [W, P]$ so that $C_W(P) = 1$, also $W_i = [W_i, P]$, so that v is nontrivial on W_i. Thus $1 \ne [W_i, v] \le W_i \cap V$. Therefore as V is a TI-set in G by 12.2.16, $P_i \le C_G(W_i \cap V) \le N_G(V) \le M$, so $H = KT = PT \le M$, contrary to $H \not\le M$.

Therefore $W_P = 1$, so $P \cong p^{1+2}$ and $W = [W, \Phi]$. By 12.2.22, $\Phi \le N_{H \cap M}(V)$, so $W \cap V = [W \cap V, \Phi]$ is of rank $n - 1$ by 12.2.22, and then $m([V, \Phi]) = n - 1$. In particular, Φ is faithful on V. Now $\bar{\Phi} \le \bar{M}_V = \bar{L} = L/O_2(L) = GL(V) \cong L_n(2)$ by 12.2.20. Therefore as $\Phi T = T\Phi$, $p = 3$ and $\bar{\Phi}\bar{T}$ is a rank one parabolic of $\bar{L}$. However for any X of order 3 in a rank one parabolic, $[V, X]$ is of rank 2; so as $m([V, \Phi]) = n - 1$ we conclude $n = 3$.

As $n = 3$, $U \cap V$ is of rank 2. Now $KV = WPV = \langle V, V^x, V^y \rangle$ for x, y chosen so that $P := \langle x, y \rangle$. Thus

$$W = [W, KV] = [W, V][W, V^x][W, V^y] = (U \cap V)(U \cap V^x)(U \cap V^y).$$

Furthemore $m(W) \leq 3m(U \cap V) = 6$, so $m(W) = 6$, since this is the minimal dimension of a faithful module for $P \cong 3^{1+2}$.

Let $Q_L := O_2(L)$ and $T_L := T \cap L$. Then $T = QT_L$ by 12.2.20, while $Q = C_{LT}(V)$ and $\overline{LT} \cong L_3(2)$. As $\bar{\Phi}$ is inverted in $\bar{T}_L$ and Φ permutes with T, Φ is inverted in $N_{T_L}(X)$ so $\Phi \leq L$.

Now $K = PW$ by 12.2.22, and $W = [W, \Phi]$, so $Y := \Phi W = O^2(\Phi T)$. As ΦW char $PW \trianglelefteq H$, $Y \trianglelefteq H$. Now as $\bar{L} \cong L_3(2)$, $O_2(\bar{Y}) = \bar{W} \cong W/(W \cap Q)$ is of rank 2, as is $W \cap V$; so as W is of rank 6, $(W \cap Q)/(W \cap V) \cong (W \cap Q)V/V$ is of rank 2. Further $(W \cap Q)V/V = [Q/V, \Phi]$ since $W = [W, \Phi] = O_2(Y)$. Therefore L has a unique noncentral chief factor $[Q, L]/Q_0$ (for some suitable Q_0 containing V) on $[Q, L]/V$. Also since the unique noncentral chief factors for Φ on $[Q, L]/Q_0$ and V are in the centralizer of the unipotent radical $\bar{W}$, it follows from the representation theory of $L_3(2)$ that $[Q, L]/Q_0$ is isomorphic to V as an L-module. Further $W[Q, L]/[Q, L] \cong E_4$, so $L/[Q, L]$ is not $SL_2(7)$; and hence as $L = O^2(L)$, $[Q, L] = Q_L$. As W is abelian by 12.2.22, $\bar{W}$ centralizes $(W \cap Q)V/V = [Q/V, X]$, so Q_L/V is not the 4-dimensional indecomposable of B.4.8.2. Thus $V = Q_0$. Then as $V \leq Z(Q)$, while L is transitive on $(Q_L/V)^\#$, and $W \cap Q_L$ contains involutions not in V, it follows that $Q_L \cong E_{64}$.

Let $Q_H := O_2(H)$. As H is irreducible on W, $W \leq Z(Q_H)$, so $C_{Q_H}(Y) = C_{Q_H}(\Phi)$. Each involution in $C_T(\Phi)$ is in $Q_H V$, so from the action of L on Q_L, $C_{Q_L}(\Phi) = \langle q, v \rangle$ with $q \in C_{Q_H}(\Phi) = C_{Q_H}(Y)$ and $Q_L = \langle q^T \rangle V$. We saw $Y \trianglelefteq H$, so $C_{Q_H}(Y) \trianglelefteq H$, and hence $\langle q^T \rangle \leq C_T(Y)$, contradicting $Q_L = \langle q^T \rangle V$. This contradiction completes the proof of 12.2.23. $\qquad\square$

By 12.2.22 and 12.2.23,

$$K = PW = O^2(I), \text{ and } U = W \trianglelefteq H.$$

In particular as $P \nleq M$ by construction,

$$I \nleq M.$$

As $V \trianglelefteq T$, also

$$I = PWV = KV \trianglelefteq KT = H.$$

Furthermore as T acts on $U = O_2(I)$ and $Q = C_T(V)$,

$$[Q, U] \leq C_U(V) = U \cap V \leq V;$$

and hence as $\bar{L} = \langle \bar{U}^L \rangle$, we have:

LEMMA 12.2.24. L is an $L_n(2)$-block.

We remark that 12.2.24 establishes the first statement in each of conclusions (2)–(4) of Theorem 12.2.13, so it only remains to identify G. Our next result shows that $M = L$, and hence determines the structure of $C_G(z)$ as $C_G(z) \leq M$.

By 12.2.20, $U \cap V$ is a hyperplane of V, and U induces on V the full group of transvections with axis $U \cap V$ on V. Let $Y := O^2(N_L(U \cap V))$, so that $Y/O_2(Y) \cong L_{n-1}(2)$. Then $\bar{U} = O_2(\bar{Y}\bar{T}) = \overline{C_T(U \cap V)}$, so that $C_T(U \cap V) = UC_T(V) \trianglelefteq YT$.

LEMMA 12.2.25. (1) $M = L$ and $V = O_2(L)$.

(2) $n = 3$ or 4, $p = 3$, $U = C_G(U)$, $N_G(U) = YIT$, $YIT/U \cong L_{n-1}(2) \times L_2(2)$, and U is the tensor product of the natural modules for the factors.

(3) $|Z| = 2$.

PROOF. Let $R := C_T(U)$. By a Frattini Argument, $R = UN_R(P)$, and then $R = U \times C_T(PU)$. Also $[C_T(PU), v] \leq C_V(PU) = 1$, so $C_T(PU) = C_T(I)$ and $R = U \times C_T(I)$. Thus $C_T(UV) = C_R(V) = (U \cap V) \times C_T(I)$. Next from the structure of $Aut_{GL(U)}(I/U)$, $C_T(U \cap V) = VC_T(U)$, so $C_T(U \cap V) = UV \times C_T(I)$.

Next $Q \leq C_T(U \cap V)$, so by the previous paragraph, $Q = C_T(I) \times V = VC_Q(U)$. By 12.2.24, L is an $L_n(2)$-block for $3 \leq n \leq 5$, so by C.1.13, $m(Q/VC_T(L)) \leq m(H^1(L/O_2(L), V))$. If $n \neq 3$, then $H^1(L, V) = 0$, by (6) and (8) of I.1.6 so that $Q = V \times C_T(L)$ If $n = 3$, the same conclusion holds since $Q = VC_Q(U)$ and $\bar{U}$ is of elementary abelian of order 4 in $\bar{L}$, ruling out the indecomposable in B.4.8.2.

Now $[C_T(L), U] \leq C_U(L) = 1$, so $C_T(L) \leq R$. Thus $C_Q(U) = C_T(L) \times C_V(U) = C_T(L) \times (U \cap V)$, so as $C_T(U \cap V) = UQ$, we conclude $R = C_{UQ}(U) = UC_Q(U) = U \times C_T(L)$. We have shown:

$$R = C_T(U) = U \times C_T(I) = U \times C_T(L). \qquad (*)$$

Next LT and I act on $T_0 := C_T(L) \cap C_T(I)$, so if $T_0 \neq 1$ then $I \leq N_G(T_0) \leq M = !\mathcal{M}(LT)$, contrary to $I \not\leq M$. Therefore $T_0 = 1$. But by $(*)$, $\Phi(C_T(L)) = \Phi(R) = \Phi(C_T(I))$, so $\Phi(R) \leq T_0 = 1$, and hence R is elementary abelian.

From 12.2.20, $T = (T \cap L)Q$, and we saw $Q = V \times C_T(L)$ with $\Phi(C_T(L)) = 1$, so $T = (T \cap L) \times C_T(L)$ and $Z = C_V(T) \times C_T(L)$. So as $|C_V(T)| = 2$, $C_T(L)$ is a hyperplane of Z. Next as $C_T(L) \leq Z$, from $(*)$ we see that $[C_T(I), T] \leq [U, T] \cap C_T(I) \leq C_U(I) = 1$; thus $C_T(I) \leq Z$ by $(*)$ since R is elementary abelian. Then $C_T(I)$ is also a hyperplane of Z, since $|C_T(L)| = |C_T(I)|$ by $(*)$. Hence as $T_0 = 1$, $|C_T(I)| = |C_T(L)| \leq 2$. Thus $|R : U| \leq 2$ in view of $(*)$. Also we saw $C_T(U \cap V) = UV \times C_T(I)$, with $J(UV) = U$, so $R = J(C_T(U \cap V)) \trianglelefteq YT$, since YT acts on $C_T(U \cap V)$ by an observation just before 12.2.25; in particular,

$$Y \leq N_G(R).$$

We suppose for the moment that $C_T(L) = 1$. Then $Q = V$, so $O_2(M) \leq V$ by A.1.6. Therefore as $L \trianglelefteq M$, $V = O_2(M) = F^*(M)$. Thus $T \leq L$ and $M = LC_M(L/V)$ by 12.2.20. Then as $End_L(V) = \mathbf{F}_2$, $C_M(L/V) = C_M(V) = V$, so $M = L$. Thus (1) will hold once we show that $C_T(L) = 1$. Also $C_T(L) = 1$ implies $U = C_T(U)$ by $(*)$. But $N_G(U) \in \mathcal{H}^e$ by 1.1.4.6, so that $C_G(U) \in \mathcal{H}^e$ by 1.1.3.1, so that $C_G(U) = U$. Thus $U = C_G(U)$ also follows once we establish $C_T(L) = 1$.

Assume first that $n > 3$. Then $Y \in \mathcal{L}_f(G, T)$, so $Y \leq Y_R \in \mathcal{C}(N_G(R))$ by 1.2.4, with $Y_R \in \mathcal{L}(G, T)$. Then $Y_R \leq Y_0 \in \mathcal{L}_f^*(G, T)$ by 1.2.9. If $Y_0/O_2(Y_0)$ is quasisimple, then applying Theorem 12.2.2.3 to restrict the list of A.3.12, we conclude that either $Y_0/O_2(Y_0) \cong L_m(2)$ for some $n - 1 \leq m \leq 5$, or $n = 4$ and $Y_0/O_{2,3}(Y_0) \cong A_7$. If $Y_0/O_2(Y_0)$ is not quasisimple, then from A.3.12, $n = 4$ and $Y_0/O_2(Y_0) \cong SL_2(7)/E_{49}$. As $Y \leq Y_R \leq Y_0$, we conclude that either $Y_R/O_2(Y_R) \cong L_k(2)$ with $n - 1 \leq k \leq m \leq 5$; or $n = 4$ and either $Y_R/O_{2,3}(Y_R) \cong A_7$, or $Y_R/O_2(Y_R) \cong SL_2(7)/E_{49}$. However $Y_R \leq N_G(R)$, so that $YR/R \cong L_{n-1}(2)$ is a T-invariant subgroup of $Y_R R/R$; so we conclude that either $Y = Y_R \trianglelefteq N_G(R)$, or $n = 4$ and $Y_R R/R \cong A_7$. Assume this last case holds. We showed $|R : U| \leq 2$, so $m(R) \leq 7$. Then as Y has two isomorphic 3-dimensional composition factors on U, we conclude that $U = [R, Y_R]$ is the 6-dimensional permutation module for $Y_R R/R$. This is impossible, as in the A_7-module, U/V is dual to V as a Y-module.

This contradiction shows that $Y = Y_R \trianglelefteq N_G(R)$. Then using $(*)$, $C_T(L) = C_R(Y) \trianglelefteq N_G(R)$. Now I normalizes R as $R = C_{IT}(U)$, so if $C_T(L) \neq 1$ then

$I \leq N_G(R) \leq N_G(C_T(L)) \leq M = !\mathcal{M}(LT)$, contrary to $I \not\leq M$. Thus $C_T(L) = 1$, so that $R = U$ by (*). Recall that this completes the proof of (1), and shows that $U = C_G(U)$.

Set $G_U := N_G(U)$ and $\dot{G}_U := G_U/U = Aut_G(U)$. We showed that $Y \trianglelefteq N_G(R) = G_U$, so as $\dot{Y}$ centralizes $\dot{V}$, we conclude that $I = \langle \dot{V}^I \rangle$ centralizes $\dot{Y}$. Now $\dot{Y} \cong L_{n-1}(2)$ has two chief factors on U, both isomorphic to the natural module $U \cap V$, while $\dot{I} \cong D_{2p}$; it follows that IY is irreducible on U. Then as $End_{\dot{Y}}(U \cap V) = \mathbf{F}_2$, $\dot{G}_U = \dot{Y} \times C_{\dot{G}_U}(\dot{Y}) \cong L_{n-1}(2) \times L_2(2)$, so $\dot{I} = C_{\dot{G}_U}(\dot{Y}) \cong L_2(2)$, and U is the tensor product of the natural modules for the factors. In particular $p = 3$. Then as $m_3(YI) \leq 2$ as G_U is an SQTK-group, it also follows that $n < 5$. This completes the proof of (2), and hence of 12.2.25, under the assumption that $n > 3$.

We turn to the case $n = 3$. This time let $G_R := N_G(R)$, $L_R := N_L(R)$, $M_R := N_M(R)$, and $\dot{G}_R := G_R/R$. Since $R = C_T(U)$, R is Sylow in $C_G(R)$, while $G_R \in \mathcal{H}^e$ by 1.1.4.4.6; thus $R = C_G(R)$. As $n = 3$, U is of rank 4; so as $|R : U| \leq 2$, R is of rank $k := 4$ or 5. Thus $\dot{G}_R \leq GL(R) = GL_k(2)$. Further $\dot{I} \cong D_{2p}$ with $U = [R, P]$ of rank 4, so $p = 3$ or 5. As H acts on R and I, as $R = U \times C_R(I)$ by (*), and as $|C_R(I)| \leq 2$, H centralizes $C_R(I)$. Thus $\dot{H}$ is faithfully embedded in $GL(U) \cong GL_4(2)$, with $D_{2p} \cong \dot{I} \trianglelefteq \dot{H}$. We conclude that $\dot{H} \cong S_3$, $\mathbf{Z}_2 \times S_3$, D_{10}, or $Sz(2)$. Hence $\dot{T}$ is cyclic or a 4-group. On the other hand, $\dot{L}_R \cong \dot{V} \times S_3$, so that $\dot{T}$ is noncyclic. Hence $\dot{H} \cong \mathbf{Z}_2 \times S_3$, and in particular $p = 3$. Furthermore $L_R \trianglelefteq M_R$, so $\dot{M}_R$ centralizes $C_R(\dot{L}_R)$ as $|C_R(L_R)| \leq 2$; hence $\dot{M}_R$ is faithful on the complement $[R, L_R]$ to $C_R(\dot{L}_R)$ in R in (*). Next M_R normalizes $[R \cap O_2(L), L_R] = V \cap U$, and hence normalizes V as V is a TI-set in G by 12.2.16. Therefore $\dot{M}_R$ centralizes $\dot{V}$ as $|\dot{V}| = 2$. Thus $O^2(\dot{L}_R) = O^2(\dot{M}_R)$ from the structure of the normalizer of $O^2(\dot{L}_R)$ in $GL([R, L_R]) \cong GL_4(2)$, so that $\dot{M}_R = \dot{L}_R \dot{T} = \dot{L}_R$. Next $C_{\dot{G}_R}(\dot{V})$ acts on $[R, V] = U \cap V$, so again as V is a TI-set in G, $C_{\dot{G}_R}(\dot{V}) \leq \dot{M}_R$, and hence $C_{\dot{G}_R}(\dot{V}) = \dot{L}_R \cong \mathbf{Z}_2 \times S_3$.

Let $\dot{i} \in \dot{T} - \dot{V}$; then $1 \neq [U \cap V, \dot{i}] \leq U \cap V$. But if $\dot{i} = \dot{v}^g$ for some $g \in G_R$, then $[R, \dot{i}] = [R, \dot{v}]^g = (U \cap V)^g$, so as V a TI-set in G, we conclude $V = V^g$, contradicting $\dot{v} \neq \dot{i}$. Therefore $\dot{i}^G \cap \dot{V} = \emptyset$, so by Burnside's Transfer Theorem 37.7 in [**Asc86a**], $\dot{G}$ is 2-nilpotent. As $\dot{Y} = C_{O(\dot{G}_R)}(\dot{V}) \cong \mathbf{Z}_3$ and $\dot{P} \leq O(\dot{G}_R)$, we conclude from the structure of $GL_5(2)$ that $\dot{G}_R \cong S_3 \times S_3$ and $C_T(L) = C_R(Y) = C_R(P) = C_T(I)$. We saw earlier that $T_0 = C_T(L) \cap C_T(I) = 1$, so we conclude $C_T(L) = 1$ and $R = U$. As observed earlier, this establishes (1) and shows that $U = C_G(U)$. Along the way we established the other assertions of (2), and (1) implies (3). Thus the proof of 12.2.25 is complete. $\qquad \square$

By 12.2.25.2, $N_{YV}(P)$ is a complement to U in $N_L(U)$. Further U is a homogeneous $C_{YT}(P)$-module, so there is a $C_{YT}(P)$-complement U_0 to $V \cap U$ in U, and hence $U_0 C_{YT}(P)$ is a complement to V in $N_L(U)$. As $N_L(U)$ contains the Sylow 2-group T of L, we conclude from Gaschütz's theorem A.1.39:

LEMMA 12.2.26. *L splits over V, and $N_G(U) \cap N_G(P) \cong L_{n-1}(2) \times S_3$ is a complement to U in $N_G(U)$.*

By 12.2.26, the structure of L, and hence also of $C_G(z)$, are determined, so we can move toward the identification G using recognition theorems from our Background References.

PROPOSITION 12.2.27. *(1) If $n = 4$, then $G \cong M_{24}$ or $L_5(2)$.*
(2) If $n = 3$ then $G \cong L_4(2)$ or A_9.

PROOF. Let $z \in V \cap Z^\#$. By assumption, $C_G(z) \leq M$, so we conclude from 12.2.25.1 that $C_G(z) = C_M(z) = C_L(z)$. By 12.2.26, L is determined up to isomorphism, so as $L_{n+1}(2)$ satisfies the hypotheses on G, $C_G(z)$ is isomorphic to the centralizer of a transvection in $L_{n+1}(2)$. Hence if $n = 4$ then by Theorem 41.6 in [**Asc94**], $G \cong M_{24}$ or $L_5(2)$. Similarly if $n = 3$ then $G \cong L_4(2)$ or A_9 by I.4.6. $\square$

By 12.2.20, $L/O_2(L) \cong L_n(2)$, and by 12.2.25.2, $n = 3$ or 4. Thus one of the conclusions of Theorem 12.2.13 holds by 12.2.27. Therefore the proof of Theorem 12.2.13 is complete.

12.3. Eliminating A_7

In section 12.3 we eliminate the cases where $L \in \mathcal{L}_f^*(G, T)$ with $L/O_{2,Z}(L) \cong A_7$; namely we prove:

THEOREM 12.3.1. *Assume Hypothesis 12.2.3. Then $L/O_{2,Z}(L)$ is not A_7.*

We adopt the conventions of Notation 12.2.5, including $Z = \Omega_1(Z(T))$.
After Theorem 12.3.1 is established, case (ii) of 12.2.7.3 cannot arise, so we obtain:

COROLLARY 12.3.2. *Assume Hypothesis 12.2.3, and further assume $C_G(Z) \leq M$. Let $H \in \mathcal{H}_*(T, M)$ and set $K := O^2(H)$ and $V_K := \langle Z^K \rangle$. Then either*
(1) H is solvable, or
(2) $K/O_2(K) \cong L_2(4)$, $K \in \mathcal{L}_f^*(G, T)$, and $[V_K, K]$ is the sum of at most two A_5-modules for $K/O_2(K)$.

We mention some shadows which the analysis must at least implicitly handle: As we noted in Remark 12.2.14, in the QTKE-group $G = M_{23}$ there is $L \in \mathcal{L}_f^*(G, T)$ with $L \cong A_7/E_{16}$. The case $G = M_{23}$ is explicitly excluded by Hypothesis 12.2.3, and its shadow is eliminated early in this section by an appeal to Theorem 12.2.13.
The group $G = McL$ is quasithin but not of even characteristic, in view of the involution centralizer isomorphic to $\hat{A}_8$; this group has $L \in \mathcal{L}^*(G, T)$ with $L \cong A_7/E_{16}$. Further $G = \Omega_7(3)$ is not quasithin but has $L \in \mathcal{L}^*(G, T)$ with $L \cong A_7/E_{64}$. The shadows of these two groups are eliminated by control of the centralizer of a 2-central element of V whose centralizer in the shadow is not in $\mathcal{H}^e$.

In the remainder of this section we assume G, L, M afford a counterexample to Theorem 12.3.1. Choose a $V \in Irr_+(L, R_2(LT), T)$; then V is described in 12.2.2.3.
We now begin a series of reductions.

LEMMA 12.3.3. *V is the natural permutation module of rank 6 for $\bar{L} \cong A_7$.*

PROOF. By 12.2.2.3, either $V/C_V(L)$ is the natural module for $\bar{L}$ or $m(V) = 4$. In the first case since $V = [V, L]$ and the 1-cohomology of the natural module is trivial by I.1.6, the lemma holds.

Thus we may assume that V is a 4-dimensional irreducible for A_7, and it remains to derive a contradiction. Then L is transitive on $V^\#$. As V is not invariant under S_7, $\bar{L} = \bar{M}_V \cong A_7$. Since the groups in conclusions (2)–(4) of Theorem 12.2.13 do not have a member $L \in \mathcal{L}_f^*(G,T)$ of this form, conclusion (1) of Theorem 12.2.13 must hold: [2] that is, $G_v \not\le M$ for each $v \in V^\#$. Now $Z \cap V = \langle z \rangle$ is of order 2, so that $G_z \not\le M$. Recall from Notation 12.2.5 that $L_z = O^2(C_{M_v}(z))$; set $K_z := L_z^\infty$. From the structure of V as an L-module, $\bar{L}_z = \bar{K}_z \cong L_3(2)$ and $V = [V, \bar{L}_z]$ is the indecomposable L_z-module of B.4.8.2 with $V/\langle z \rangle$ a natural module. Thus $Q = O_2(K_zT) \in Syl_2(C_{G_z}(K_z/O_2(K_z))$ and Hypothesis C.2.8 is satisfied with G_z, M_z, K_z, Q in the roles of "H, M_H, L_H, R" by 12.2.12. Now $K_z \in \mathcal{L}_f(G,T)$, so by 1.2.4, $K_z \le K \in \mathcal{C}(G_z)$, and then $K \in \mathcal{L}_f(G,T)$ by 1.2.9.1.[3]

We claim that $K_z = K$. Suppose that $K_z < K$. Then $K \not\le M_z$ by 12.2.5.3a. If $K/O_2(K)$ is not quasisimple, then $K/O_2(K) \cong SL_2(7)/E_{49}$ by A.3.12. On the other hand if $K/O_2(K)$ is quasisimple, then $K/O_2(K) \cong L_4(2)$ or $L_5(2)$ by Theorem C.4.1. In either case $V \le V_K := [\Omega_1(Z(O_2(KT))), K]$ by 1.2.9.1. But if $K/O_2(K) \cong SL_2(7)/E_{49}$, then by 3.2.14, $\Xi_7(K) \le C_G(V_K) \le C_G(V) \le M$, so $K = \Xi_7(K)L_z \le M_z$, contrary to $K \not\le M_z$. Thus $K/O_2(K)$ is $L_n(2)$ for $n = 4$ or 5. As our tuple satisfies Hypothesis C.2.8, it also satisfies Hypothesis C.2.3. Hence by C.2.7.2, $J(Q) \not\le O_2(KQ)$ and V_K is an FF-module for $KT/O_2(KT)$. Then V_K is described in Theorem B.5.1. As $z \in V \le V_K$, $C_{V_K}(K) \ne 0$, so by Theorem B.5.1.2, $n = 4$, $V_K \in Irr_+(K, V_K)$, and $V_K/C_{V_K}(K)$ is the 6-dimensional irreducible for $K/O_2(K) \cong A_8$. Indeed as the 1-cohomology of that module is 1-dimensional by I.1.6.1, V_K is the 7-dimensional core of the permutation module for A_8. But then from the structure of that module, $O_2(N_K(V))$ induces the full group of transvections with center $\langle z \rangle$ on V, contrary to $O_2(N_K(V)) \le O_2(K_zT) = Q \le C_G(V)$.

Therefore $K_z = K \trianglelefteq G_z$, so $V = [V, K_z] \le K_z = K$. Let $Y := C_{G_z}(K/O_2(K))$ and recall that $Q \in Syl_2(Y)$. Then by a Frattini Argument, $G_z = YN_{G_z}(Q) = YM_z$, and hence $Y \not\le M$ as $G_z \not\le M$. Further $m_3(G_z) \le 2$ as G_z is an SQTK-group, and L_z contains a subgroup of order 3 intersecting Y trivially, so $m_3(Y) \le 1$. Notice $Y \in \mathcal{H}^e$ by 1.1.3.1. Then as $Q \in Syl_2(Y)$, while $C(G,Q) \le M$ by 1.4.1.1, Hypothesis C.2.3 is satisfied now with Y, $Y \cap M$, Q in the roles of "H, M_H, R". Therefore as $m_3(Y) \le 1$, we conclude from C.2.5 that $Y = (Y \cap M)X$, where $X \not\le M$ is a block of type A_3, A_5, or $L_2(2^n)$ for some n, and X is normal in G_z. As $[K, X] \le O_2(K)$, we conclude from C.1.10 that K centralizes X. Hence $X \le C_G(K) \le C_G(V) \le M$, a contradiction which completes the proof of 12.3.3. $\qquad\square$

By 12.3.3, V is the 6-dimensional irreducible module for $\bar{L} \cong A_7$, so we now adopt the notation of section B.3 in discussing the action of M_V on V. In particular:

Lemma 12.3.4. *(1) L has three orbits $\mathcal{O}_m$, $m = 2, 4, 6$, on $V^\#$, where $\mathcal{O}_m$ is the set of vectors in V of weight m.*

(2) $(Z \cap V)^\# = \{e(m) : m = 2, 4, 6\}$, with $e(m) := e_{\theta_m}$ of weight m, where $\theta_2 := \{1, 2\}$, $\theta_4 := \{3, 4, 5, 6\}$, and $\theta_6 := \Omega - \{7\}$.

---

[2]This application of 12.2.13 eliminates the "shadow" of M_{22} in Theorem 12.3.1.

[3]Notice this eliminates the shadow of $G = McL$, in which $K \cong \hat{A}_8$; thus $O_2(K) = \langle z \rangle$, and hence $K \notin \mathcal{L}_f(G,T)$.

(3) $\bar{M}_V \cong S_7$ *or* A_7*, and for* $m = 2, 4, 6$*,* $C_{\bar{M}_V}(e(m))$ *is isomorphic to* $\mathbf{Z}_2 \times S_5$ *or* S_5*;* $S_4 \times S_3$ *or a subgroup of index 2 in* $S_4 \times S_3$*;* S_6 *or* A_6*; respectively.*

LEMMA 12.3.5. *L controls fusion of involutions in V.*

PROOF. Recall $N_G(T) = N_M(T)$ by Theorem 3.3.1, and this subgroup controls fusion of involutions in Z by Burnside's Fusion Lemma A.1.35. We saw in 12.3.4.3 that the three involutions in $Z \cap V$ are not M_V-conjugate; hence they are not M-conjugate since V is a TI-set in M by 12.2.6. Further by 12.3.4, each member of V is fused into $Z \cap V$ under L, so the lemma holds. $\qquad\square$

LEMMA 12.3.6. *$G_{e(6)} \le M$.*

PROOF. Let $e := e(6)$. By 12.3.4, $O_2(\bar{L}_e\bar{T}) = 1$ and $\bar{L}_e \cong A_6$, so applying 12.2.12, Hypothesis C.2.3 is satisfied by G_v, M_v, Q. Also $L_e/O_2(L_e) \cong A_6$ or $\hat{A}_6$ for $L/O_2(L) \cong A_7$ or $\hat{A}_7$, respectively, so $L_e \in \mathcal{L}(G,T)$ and hence $L_e \le K \in \mathcal{C}(G_e) \subseteq \mathcal{L}(G,T)$. [4] As L_e involves A_6, $K/O_2(K)$ is quasisimple by 1.2.1.4; further $G_e \in \mathcal{H}^e$ by 1.1.4.2, so that $K \in \mathcal{H}^e$ by 1.1.3.1. Then if $L_e < K$, K and $K \cap M$ are described in the list of conclusion (3) of Theorem C.2.7; but we find no case where $K \cap M$ contains a T-invariant subgroup L_e with $L_e/O_{2,Z}(L_e) \cong A_6$.

Thus $L_e = K$. Now $\theta(G_e) = L_e$ by A.3.18, so $\theta(G_e) \le M$. Set $Y := C_{G_e}(K/O_2(K))$; then $Q \in Syl_2(Y)$ by 12.2.12.1. Thus $G_e = YN_{G_e}(Q) = YM_e$ by a Frattini Argument. Further Hypothesis C.2.3 is satisfied with Y, $Y \cap M$, Q in the roles of "H, M_H, R", so by C.2.5, Y is the product of $Y \cap M$ with χ_0-blocks. Hence as each χ_0-block is generated by elements of order 3, $G_e = \theta(Y)M_e \le M$, completing the proof. $\qquad\square$

LEMMA 12.3.7. *(1)* $L = [L, J(T)]$*, so* $\bar{M}_V \cong S_7$ *and* $\Omega_1(Z(O_2(LT))) = V \oplus C_Z(L)$*.*
(2) Let $K_e := L_{e(2)}^\infty$*. Then* $K_e = [K_e, J(T)]$ *and* $\Omega_1(Z(O_2(K_eT))) = [V, K_e] \oplus C_Z(K_e)$*.*

PROOF. By 12.3.6, $C_G(Z) \le M$, so $L = [L, J(T)]$ by 12.2.9.2. Hence by B.3.2.4, $\bar{M}_V \cong S_7$, and if $A \in \mathcal{A}(T)$ with $\bar{A} \ne 1$, then $\bar{A}$ is generated by transvections and $m(\bar{A}) = m(V/C_V(A))$. In particular $K_e = [K_e, A]$ for some such A. Let $Z_X := \Omega_1(Z(O_2(X))$ for $X := LT$ or K_eT. As $m(\bar{A}) = m(V/C_V(A))$, $Z_{LT} = VC_{Z_{LT}}(A)$, so $V = [Z_{LT}, L]$. Then as the 1-cohomology of V under $L/O_2(L) \cong A_7$ is trivial by I.1.6.1, $Z_{LT} = V \oplus C_{Z_{LT}}(LA)$. Hence as $LT = LAO_2(LT)$, $C_{Z_{LT}}(L) \le C_{Z_{LT}}(T) \le Z$, and (1) follows. Similarly $Z_{K_eT} = [V, K_e] \oplus C_Z(K_e)$, so that (2) holds. $\qquad\square$

LEMMA 12.3.8. *$G_{e(2)} \le M$.*

PROOF. Let $e := e(2)$ and $K_e := L_e^\infty$. Then $K_e \le K \in \mathcal{C}(G_e) \subseteq \mathcal{L}(G,T)$, and $K \le K_0 \in \mathcal{L}^*(G,T)$. As $K_e \in \mathcal{L}_f(G,T)$, K and K_0 are also in $\mathcal{L}_f(G,T)$ by 1.2.9.1, and $K_0 \in \mathcal{L}_f^*(G,T)$ by 1.2.9.2. Let $V_0 := \Omega_1(Z(O_2(K_0T)))$. Then $e, e(6) \in Z \le V_0$ since $F^*(K_0T) = O_2(K_0T)$ by 1.1.4.6, so $[V, K_e] = [e(6), K_e] \le V_0$, and hence $[V_0, K_0] \ne 1$. By 12.3.7.2, $K_e = [K_e, J(T)]$, so $K_0 = [K_0, J(T)]$.

Suppose $K_0/O_2(K_0)$ is not quasisimple. Then $K_e < K_0$, so as $K_e/O_2(K_e) \cong A_5$, the embedding of K_e in K_0 is described in cases (13) or (14) of A.3.14. As

---

[4]Just as for McL in 12.3.3, the shadow of $\Omega_7(3)$ is now eliminated by the application of 1.2.9.1, as in that group K would be $\Omega_6^-(3)$.

$K_0 = [K_0, J(T)]$ does not centralize V_0, but $O_\infty(K_0)$ centralizes V_0 by 3.2.14, we conclude that $K_0/C_{K_0}(V_0) \cong L_2(p)$ for $p = 5$ or $p \geq 11$. But $p \geq 11$ is ruled out by Theorem B.4.2, so $K_0 = C_{K_0}(V_0)K_e$. However as $e, e(6) \in V_0$, $C_{K_0}(V_0) \leq G_{e(6)} \cap G_e \leq M_e$ by 12.3.6, and then K_0 acts on K_e, a contradiction.

Thus $K_0/O_2(K_0)$ is quasisimple, so by Remark 12.2.4, Hypothesis 12.2.3 is satisfied with K_0 in the role of "L". Thus K_0 and its action on any $I \in Irr_+(K_0, V_0, T)$ are described in Theorem 12.2.2.3. Then comparing that list with the possible embeddings in A.3.14, we conclude that either $K_e = K_0$ or $K_0/C_{K_0}(V_0) \cong A_7$.

Suppose first that $K_e < K$. Then as usual $K \not\leq M$ and $K < K_0$ is eliminated, since by A.3.14, there is no $K \in \mathcal{L}(G,T)$ with $K_e < K < K_0$ when $K_0/C_{K_0}(V_0) \cong A_7$. Then $K = K_0 \in \mathcal{L}_f^*(G,T)$, so that K satisfies our hypothesis in this section that $K/O_{2,Z}(K) \cong A_7$. Hence we may apply the results in this section to K. In particular by 12.3.3 and 12.3.7, $I := [V_0, K_0]$ is the A_7-module, $V_0 = I \oplus C_Z(K)$, and

$$[V, K_e] = [\Omega_1(Z(O_2(K_eT))), K_e] = [V_0, K_e].$$

Pick $v \in [V, K_e]$ of weight 4. Then ev is of weight 6 in V, so $C_G(ev) \leq M$ by 12.3.6. But $C_K(v) = C_K(ev)$ as $K \leq G_e$, so $K = \langle K_e, C_K(v) \rangle \leq M$, contrary to $K \not\leq M$.

This contradiction shows that $K_e = K \trianglelefteq G_e$. Then as $Out(K/O_2(K))$ is a 2-group, $G_e = KTY$, where $Y := C_{G_e}(K/O_2(K))$, so it remains to show that $Y \leq M$. Set $U := \langle Z^{G_e} \rangle$ and $G_e^* := G_e/C_{G_e}(U)$. Then $U \in \mathcal{R}_2(G_e)$ by B.2.14. As $K = [K, J(T)]$, Theorems B.5.1 and B.5.6 imply $[U, K] = [V, K]$; so as $End_{K^*}([V, K]) = \mathbf{F}_2$, $Y \leq C_{G_e}([V, K]) \leq C_{G_e}(ev) \leq M$, completing the proof of 12.3.8. $\qquad\square$

Recall the weak closure parameters $r := r(G, V)$ and $w := w(G, V)$ from Definitions E.3.3 and E.3.23.

LEMMA 12.3.9. *(1) If $g \in G - N_G(V)$, then $V^\# \cap V^g \subseteq \mathcal{O}_4$.*
(2) $r(G, V) \geq 4$.

PROOF. By 12.2.6, V is a TI-set in M; so by 12.3.5 and A.1.7.3, if $u \in V^\#$ with $G_u \leq M$, then u is in a unique conjugate of V. Thus (1) follows from 12.3.6 and 12.3.8. Up to conjugation, $\langle e_{1,2,3,4}, e(4) \rangle$ is the unique maximal subspace U of V with $U^\# \subseteq \mathcal{O}_4$, so (1) implies (2) since $m(V) = 6$. $\qquad\square$

LEMMA 12.3.10. *$W_1(T, V) \leq C_T(V)$, so $w(G, V) > 1$.*

PROOF. Assume the lemma fails. Then we may choose $A := V^g \cap M \leq T \leq N_G(V)$ to be a w-offender in the sense of subsection E.3.3. Thus $\bar{A} \neq 1$ and $w := m(V^g/A) \leq 1$. Now from the action of S_7 on V, for each $\bar{a} \in \bar{A}^\#$, $[V, a]^\# \not\subseteq \mathcal{O}_4$. But if $V \leq N_G(V^g)$, then $[V, a] \leq V \cap V^g$, contrary to 12.3.9.1, so we conclude $V \not\leq N_G(V^g)$. Therefore $m(V^g/C_A(V)) \geq r(G, V) \geq 4$ by 12.3.9.2, so that $m(\bar{A}) \geq 3 = m_2(\bar{L}\bar{T})$. Thus these inequalities must be equalities, so $m(\bar{A}) = 3$, $w = 1$, and $r(G, V) = 4$. Hence $\bar{A}$ is fused under L to

$$\bar{A}_1 := \langle (1,2), (3,4), (5,6) \rangle \text{ or } \bar{A}_2 := \langle (1,2), (3,4)(5,6), (3,5)(4,6) \rangle.$$

Now the Fundamental Weak Closure Inequality of Remark E.3.29 is an equality, so by E.3.31.1:

$$V_A := \langle C_V(\bar{a}) : \bar{a} \in \bar{A}^\# \rangle \leq N_G(V^g).$$

Therefore $[A, V_A] \leq V \cap V^g$, and hence $[A, V_A]^\# \subseteq \mathcal{O}_4$ by 12.3.9.1. We compute that this does not hold if $\bar{A} = \bar{A}_1$. Similarly $[V_A, A] \leq V^g \leq C_G(A)$, so that $[V_A, A, A] =$

1, and we compute that this does not hold if $\bar{A} = \bar{A}_2$. This contradiction completes the proof of 12.3.10. $\qquad\square$

LEMMA 12.3.11. *If $H \in \mathcal{H}(T)$ with $n(H) = 1$, then $H \leq M$.*

PROOF. By 12.3.9.2 and 12.3.10, $\min\{r, w\} > 1$, so the lemma follows from E.3.35.1. $\qquad\square$

Let $e := e(4)$. Since L does not appear in conclusions (2)–(4) of Theorem 12.2.13, conclusion (1) of Theorem 12.2.13 holds: $G_v \not\leq M$ for some $v \in V^{\#}$. By 12.3.4, 12.3.6, and 12.3.8, we may take $v = e$. Thus there is $H \in \mathcal{H}_*(T, M)$ with $H \leq G_e$. Set $K := O^2(H)$; as usual, $K \not\leq M$.

LEMMA 12.3.12. $K \trianglelefteq G_e$, $K/O_2(K) \cong A_5$, $K = [K, J(T)]$, *and* $[Z, K]$ *is the A_5-module.*

PROOF. By 12.3.11 and E.1.13, H is not solvable. By 12.3.6, $C_G(Z) \leq M$, so we may apply 12.2.7.2 to conclude that $K/O_2(K) \cong A_5$. By 1.2.4, we may embed $K \leq K_e \in \mathcal{C}(G_e) \subseteq \mathcal{L}(G, T)$, and $K_e \leq K_0 \in \mathcal{L}^*(G, T)$. As $[V_H, K] \neq 1$ by 12.2.7.1, 1.2.9.1 says $K_0 \in \mathcal{L}_f^*(G, T)$. Then by 12.2.7.3, either $K = K_0$ or $K_0/O_{2,Z}(K_0) \cong A_7$.

Assume first that $K < K_0$. Then by 12.2.7.3, Hypothesis 12.2.3 is satisfied with K_0 in the role of "L". Hence as $K_0/O_{2,Z}(K_0) \cong A_7$, the hypotheses of this section hold with K_0 in the role of L, so we may apply the results obtained so far to K_0. Set $V_0 := \Omega_1(Z(O_2(K_0 T)))$. By 12.3.7, $V_0 = V_K \oplus C_Z(K_0)$, with $V_K = [Z, K_0]$, $[Z, K]$ is the A_5-module, and $K = [K, J(T)]$. Thus the lemma holds in this case if $K = K_e$. On the other hand if $K < K_e$, then $K_e = K_0$ by A.3.14. Further by 12.2.8, K_0 contains all elements of order 3 in G_e, so in particular $L_e \leq K_0$. But K is the unique member of $\mathcal{L}(K_0 T, T)$ with $K/O_2(K) \cong A_5$, so $K = L_e^\infty \leq M$, contrary to $K \not\leq M$.

Thus we may assume that $K = K_0 = K_e \in \mathcal{L}^*(G, T)$. Therefore $G_e \leq N_G(K) = !\mathcal{M}(KT)$ by 1.2.7.3. Then there is $H_1 \in \mathcal{H}_*(T, N_G(K))$, and in particular $H_1 \not\leq G_e$. Thus $[Z, H_1] \neq 1$, so $K = [K, J(T)]$ and $[Z, K]$ is an FF-module by Theorem 3.1.8.3. By 12.2.7.3, $[Z, K]$ the sum of A_5-modules, and then by Theorem B.5.1.1, $[Z, K]$ is an A_5-module, completing the proof of the lemma. $\qquad\square$

Next by 12.3.4 and 12.3.7.1, $C_{\bar{M}_V}(e) = \bar{M}_1 \times \bar{M}_2$, where $\bar{M}_1 \cong S_4$ is the pointwise stabilizer in $\bar{M}_V$ of $\{1, 2, 7\}$, and $\bar{M}_2 \cong S_3$ is the pointwise stabilizer of $\{3, 4, 5, 6\}$. Let $L_i := O^{3'}(M_i)$, so that $L_e = L_1 L_2$, and $L_1 = O^{3'}(C_L(Z \cap V)) = O^{3'}(C_L(Z))$ using 12.3.7.1. Let $P \in Syl_3(L_e)$. By 12.3.12, $K \trianglelefteq G_e$, so $P = (P \cap K) \times C_P(K/O_2(K))$, and hence $P \not\cong 3^{1+2}$. Therefore $O_{2,Z}(L) = O_2(L)$, and appealing to 12.2.8:

LEMMA 12.3.13. $L/O_2(L) \cong A_7$ *and* $L = O^{3'}(M)$.

We are now in a position to complete the proof of Theorem 12.3.1. As $L = O^{3'}(M)$ by 12.3.13 and $C_G(Z) \leq M$ by 12.3.6, $L_1 = O^{3'}(C_L(Z)) = O^{3'}(C_G(Z))$. By 12.3.12, $[Z, K]$ is the A_5-module, so $O^2(K \cap M)$ centralizes $Z \cap [Z, K]$, and hence $O^2(K \cap M)$ centralizes Z by B.2.14. Thus $L_1 = O^{3'}(K \cap M)$. Then as L_1 and L_2 are the T-invariant subgroups $X = O^2(X)$ of L_e with $|X : O_2(X)| = 3$, it follows that $L_2 = O^2(C_{L_e}(K/O_2(K)))$.

Let $Y := KL_2T$, $U := \langle Z^Y \rangle$, and $Y^* := Y/C_Y(U)$. As $L_e = L_1L_2$, $L_1 \leq C_K(Z)$, and $Z = C_Z(L)(Z \cap V)$ by 12.3.7.1,

$$[Z, L_e] = [Z \cap V, L_2] = \langle e_{1,7}, e(2) \rangle.$$

Then $C_{L_2}([Z, L_2]) = O_2(L_2)$, and $C_K([Z, K]) = O_2(K)$ by 12.3.12, so $C_Y(U) = O_2(Y)$. Thus $Y^* \cong S_5 \times S_3$ since $M_1M_2/O_2(M_1M_2) \cong S_3 \times S_3$, and $U \in \mathcal{R}_2(Y)$. Also $K = [K, J(T)]$ by 12.3.12, and $L_2 = [L_2, J(T)]$ using 12.3.7.1, so $Y^* = J(Y)^*$. Therefore by Theorem B.5.6,

$$[U, Y] = [U, K] \oplus [U, L_2] = [Z, K] \oplus [Z, L_2],$$

so in particular $K \leq C_G([Z, L_2]) \leq C_G(e(2)) \leq M$ by 12.3.8, contrary to $K \not\leq M$. This contradiction completes the proof of Theorem 12.3.1.

12.4. Some further reductions

We begin section 12.4 with a technical lemma 12.4.1, which we use in particular to prove the main result 12.4.2 of the section.

As we will be assuming Hypothesis 12.2.3, as usual we adopt the conventions of Notation 12.2.5, including $Z = \Omega_1(Z(T))$.

LEMMA 12.4.1. *Assume Hypothesis 12.2.3. In addition assume:*

(i) $C_G(Z) \leq M$, and
(ii) $s(G, V) > 1$.

Then there exists $g \in G$ with $1 \neq [V, V^g] \leq V \cap V^g$.

PROOF. Assume the lemma is false. Let $H \in \mathcal{H}_*(T, M)$, $K := O^2(H)$, $V_H := \langle Z^H \rangle$, and $H^* := H/C_H(V_H)$. As $C_G(Z) \leq M$ by (i), 12.3.2 says either H is solvable, or $[V_H, K]$ is the sum of at most two A_5-modules for $K^* \cong A_5$. Then $a(H^*, V_H) = 1$, by E.4.1 in the former case, or by an easy direct computation in the latter.

Observe that the triple $G_1 := LT$, $G_2 := H$, V satisfies Hypothesis F.7.6. Form the coset geometry Γ as in that section, with parameter $b := b(\Gamma, V)$. If $W_0(T, V) \leq O_2(H)$, then by F.7.14, b is even. Hence by F.7.11.2, there exists $g \in G$ with $1 \neq [V, V^g] \leq V \cap V^g$, contrary to our assumption that the lemma fails. Therefore $W_0(T, V) \not\leq O_2(H)$. So there is $A := V^g$ with $A^* \neq 1$. Now as $s(G, V) > 1$ by (ii), $A^* \in \mathcal{A}_2(H^*, V_H)$ by E.3.10, contradicting $a(H^*, V_H) = 1$. $\square$

The main result of this section is Theorem 12.4.2. It eliminates two of the four cases in 12.2.2.3 where $C_V(L) \neq 1$ (cases (b) and (f)), leaving only A_6 and A_8 in case (d). In particular when $\bar{L}$ is $L_3(2)$ or $G_2(2)'$, the result reduces V to the natural module. The analogous reduction will be carried out later for $L_4(2)$ and $L_5(2)$ in Theorems 12.6.34 and 12.5.1. Theorem 12.4.2 also moves in the direction (begun in 12.2.13) of showing that $C_G(V \cap Z) \not\leq M$.

THEOREM 12.4.2. *Assume Hypothesis 12.2.3. Then*

(1) If $L/O_2(L) \cong L_3(2)$ or $G_2(2)'$, then $C_V(L) = 1$.
(2) If $L/O_2(L) \cong L_5(2)$ and $\dim(V) = 10$, then $C_G(Z \cap V) \not\leq M$.

Until the proof of Theorem 12.4.2 is complete, assume G, L, V affords a counterexample. Let $Z_V := C_V(L)$.

When $L/O_2(L) \cong L_3(2)$ or $G_2(2)'$, $Z_V \neq 1$ as we are in a counterexample to Theorem 12.4.2. Hence by Theorem 12.2.2.3, V is an indecomposable for $L/O_2(L)$, and V/Z_V is a natural module for $L/O_2(L)$. Then as the 1-cohomology of the dual of V/Z_V in I.1.6 is 1-dimensional, $Z_V = \langle z \rangle$ is of order 2. As $M = !\mathcal{M}(LT)$, $G_z \leq M$.

On the other hand, when $L/O_2(L) \cong L_5(2)$, we have $m(V) = 10$ by hypothesis; and as we are in a counterexample to the theorem, $C_G(Z \cap V) \leq M$. As $\dim(V) = 10$, $Z \cap V$ is of order 2, and in this case we take z to be a generator for $Z \cap V$. Thus $G_z \leq M$ in this case also.

We begin a series of reductions.

LEMMA 12.4.3. (1) $C_G(Z) \leq M$.
(2) $L = [L, J(T)]$.

PROOF. As $G_z \leq M$ and $z \in Z$, (1) holds; then (2) follows from 12.2.9.2. $\qquad\square$

We are already in a position to complete the proof of part (2) of Theorem 12.4.2:

LEMMA 12.4.4. $L/O_2(L)$ is not $L_5(2)$.

PROOF. Assume $L/O_2(L)$ is $L_5(2)$. Then L has two orbits on $V^{\#}$, represented by z and some further involution t.

We claim that $t \notin z^G$. First $L = O^{3'}(M)$ by 12.2.8. Next $L_z/O_2(L_z) \cong L_3(2) \times \mathbf{Z}_3$, so as $G_z \leq M$, $m_3(G_z^\infty) = 1$. However $L_t/O_2(L_t) \cong A_6$ is of 3-rank 2, so $t \notin z^G$, establishing the claim.

It follows from the claim that L is transitive on $z^G \cap V$, so as $G_z \leq M$, while V is a TI-set under M by 12.2.6, V is the unique member of V^G containing z by A.1.7.3.

Next $m(\bar{M}_V) = 3$, so $s(G, V) \geq 3$ by Theorem E.6.3. By 12.4.3.1, $C_G(Z) \leq M$, so by 12.4.1, there exists $g \in G$ with $1 \neq [V, V^g] \leq V \cap V^g$. Conjugating in M_V if necessary, we may assume $V^g \leq T$. Let $A := V^g$. Interchanging the roles of A and V if necessary, we may assume $m(\bar{A}) \geq m(V/C_V(A))$. Then by B.1.4.4, $\bar{A}$ contains a member of $\mathcal{P}(\bar{M}_V, V)$. Therefore by B.4.2.11, $C_V(A) = [V, A]$ is a 6-dimensional subspace of A, and $\bar{A}$ of rank 4 is the unipotent radical of the maximal parabolic of $\bar{L}$ over $\bar{T}$ stabilizing $[V, A]$. In particular, $[V, A]$ is T-invariant, so the generator z of $Z \cap V$ is in $[V, A] \leq V \cap V^g$. This contradicts our earlier observation that z is in a unique conjugate of V, completing the proof. $\qquad\square$

By 12.4.4, $L/O_2(L) \cong L_3(2)$ or $G_2(2)'$, so as we are in a counterexample to the Theorem, $Z_V \neq 1$. Hence $V \trianglelefteq M$ by Theorem 12.2.2.3. Then M normalizes $C_V(L) = Z_V = \langle z \rangle$, so since $M \in \mathcal{M}$,

$$M = C_G(z) = N_G(V).$$

When $\bar{L} \cong L_3(2)$, let E be the T-invariant 4-subgroup of V, choose $v \in E - Z_V$, and let $L_1 := O^2(C_L(E))$ and $R_1 := C_T(E)$.

LEMMA 12.4.5. If $\bar{L} \cong L_3(2)$, then
(1) $[Z, L] = 1$.
(2) $Z_Q := \Omega_1(Z(Q)) = ZV$.
(3) $\bar{R}_1 := \bar{A}$ for each $A \in \mathcal{A}(R_1)$ with $A \nleq Q$.

PROOF. Observe that (3) holds by B.4.8.2. By 1.4.1.4, $Z_Q = R_2(LT)$, so $V = [Z_Q, L]$ by B.5.1.1. Then $Z_Q = C_Z(L)V$ by B.4.8.4, so (2) holds. Again By B.4.8.2, $Z \cap V = Z_V$, so (2) implies (1). $\qquad\square$

LEMMA 12.4.6. *If* $\bar{L} \cong L_3(2)$, *then* $R_1 \in Syl_2(G_v)$ *and* $|T : R_1| = 2$.

PROOF. First $R_1 \in Syl_2(M_v)$ and $|T : R_1| = 2$. Thus the result holds if $N_G(R_1) \leq M$. So we assume that $N_G(R_1) \nleq M$, and it remains to derive a contradiction.

Let $Z_1 := \Omega_1(Z(R_1))$. Then by 12.4.5.2,

$$Z_1 = C_{Z_Q}(R_1) = Z C_V(R_1) = Z\langle v \rangle.$$

Now $[Z, L] = 1$ by 12.4.5.1, so $Z \cap Z^g = 1$ for $g \in N_G(R_1) - M$ by 1.2.7.4. Hence as Z is a hyperplane in Z_1, we conclude Z is of order 2, so that $Z = Z_V$ and $E = Z_1 = \Omega_1(Z(R_1))$. In particular,

$$N_G(R_1) \leq N_G(E).$$

Furthermore $C_G(E) \leq C_G(Z) \leq G_z = M$, and $Aut_G(E) \cong S_3$ with $Aut_M(E)$ of order 2; thus $|N_G(R_1) : N_M(R_1)| = 3$.

Next as $N_G(R_1) \nleq M = !\mathcal{M}(LT)$, there is no nontrivial characteristic subgroup of R_1 normal in LT. Thus (LR_1, R_1) is an MS-pair as in Definition C.1.31, so that C.1.34 applies. As V is indecomposable, conclusion (5) of C.1.34 holds; hence L is a block with $Q = V C_T(L)$ and $C_{R_1}(L_1) = E C_T(L)$.

Suppose that $N_G(R_1) \leq N_G(L_1)$. Then $N_G(R_1)$ normalizes $\Phi(C_{R_1}(L_1)) = \Phi(E C_T(L)) = \Phi(C_T(L))$, so $\Phi(C_T(L)) = 1$ since $N_G(R_1) \nleq M = !\mathcal{M}(LT)$. Thus $C_T(L)$ is central in $V C_T(L) = Q$ and in $Q(T \cap L) = T$. We conclude $C_T(L) = Z = Z_V$, so that $Q = V C_T(L) = V$. Since the nontrivial characteristic subgroup $J(R_1)$ of R_1 is not normal in LT, $J(R_1) \nleq O_2(LT) = C_T(V)$, so there is $A \in \mathcal{A}(R_1)$ with $\bar{A} \neq 1$. By 12.4.5.3, $\bar{A} = \bar{R}_1$. Thus $\mathcal{A}(R_1) = \{A, V\}$ by B.2.21, since V is self-centralizing in G and $C_V(A) = C_V(a)$ for $\bar{a} \in \bar{A}^\#$ by B.4.8.2. Hence $O^2(N_G(R_1))$ acts on V, so $O^2(N_G(R_1)) \leq M$, contradicting $|N_G(R_1) : N_M(R_1)| = 3$.

Thus there is $g \in N_G(R_1) - N_G(L_1)$. We have seen that $N_G(R_1) \leq N_G(E)$ and $C_G(E) \leq M$; so as $L_1 \trianglelefteq C_M(E)$ while $m_3(C_M(E)) \leq 2$, $L_1 L_1^g =: X = \theta(C_G(E))$ and $X/O_2(X) \cong E_9$. Then $\bar{X}_0 := C_X(\bar{L})$ is of order 3, so by C.1.10, $X_1 := O^2(X_0)$ centralizes L and $X_1/O_2(X_1) \cong \mathbf{Z}_3$. Next X_1 is centralized by $t \in T \cap L - R_1$ inverting $L_1/O_2(L_1)$, so L_1 and X_1 are the two T-invariant members of the set $\mathcal{Y}$ of subgroups Y of X such that $Y = O^2(Y)$ and $|Y : O_2(Y)| = 3$. Now $N_G(R_1)$ normalizes X and hence permutes $\mathcal{Y}$. Since $N_G(R_1) \nleq N_G(L_1)$, while L_1 is stabilized by $N_M(R_1)$ of index 3 in $N_G(R_1)$, the $N_G(R_1)$-orbit of L_1 has length 3, and the fourth member of $\mathcal{Y}$ is fixed by $N_G(R_1)$. Since $T \leq N_G(R_1)$ and X_1 is the only T-invariant member of $\mathcal{Y}$ other than X_1, we conclude $X_1 \trianglelefteq N_G(R_1)$. However $X_1 \trianglelefteq XLT$, so

$$N_G(R_1) \leq N_G(X_1) \leq M = !\mathcal{M}(LT),$$

contrary to our earlier reduction. This completes the proof of 12.4.6. $\square$

LEMMA 12.4.7. *L controls fusion of involutions in V.*

PROOF. Suppose first that $\bar{L} \cong L_3(2)$. By 12.4.6, $v^G \cap Z_V = \emptyset$. Thus as L is transitive on $V - Z_V$, the lemma holds in this case.

Next take $\bar{L} \cong G_2(2)'$. Then $Z \cap V$ is a 4-group containing a representative of each of the three orbits of M on $V^\#$. But $N_G(T) \leq M$ by Theorem 3.3.1, and $N_G(T)$ controls fusion in Z by Burnside's Fusion Lemma A.1.35, so the lemma holds in this case also. $\square$

LEMMA 12.4.8. $r(G, V) > 1$.

PROOF. Assume that $r(G, V) = 1$. Then there is a hyperplane U of V with $C_G(U) \not\leq N_G(V)$. Let $G_U := C_G(U)$, $M_U := C_M(U)$, and $L_U := N_L(U)$. Then $Z_V \not\leq U$ as $C_G(Z_V) = C_G(z) = M$.

Now consider some hyperplane U_0 of U, and set $G_{U_0} := C_G(U_0)$ and $M_{U_0} := C_M(U_0)$; then $G_U \leq G_{U_0}$, so also $G_{U_0} \not\leq M$. As $Z_V \not\leq U$ and $m(V/U) = 1$, also $Z_V \not\leq U_0$ and $m(V/U_0 Z_V) = 1$. For any involution $\bar{t} \in \bar{M}$, $Z_V \leq C_V(\bar{t})$ and $m(V/C_V(\bar{t})) \geq 2$ (cf. B.4.8.2 and B.4.6), so $\bar{t}$ does not centralize U or U_0. Thus U and U_0 lie in the set Γ of Definition E.6.4, and we may apply appropriate results from that section. In particular by E.6.5.1, Q is Sylow in G_{U_0} and G_U. Also M_{U_0} centralizes the quotients of the series $V > U_0 Z_V > U_0 > 1$, so by Coprime Action, $\bar{M}_{U_0}$ is a 2-group. But we just observed that $\bar{M}_{U_0}$ does not contain involutions, so we conclude that $M_{U_0} = C_M(V)$, and hence also $M_U = C_M(V)$.

Now if $\bar{L} \cong L_3(2)$, then T is regular on hyperplanes not containing Z_V, so U is determined up to conjugation under T, and $\bar{L}_U \cong Frob_{21}$. On the other hand, if $\bar{L} \cong G_2(2)'$, then by Theorem 2 in [**Asc87**], L has two orbits on hyperplanes not containing Z_V, exhibited by conclusions (3) and (4) of Theorem 3 in [**Asc87**], and given by representatives U_1 and U_2, where $\bar{L}_{U_1} \cong PSL_2(7)$ and $\bar{L}_{U_2} \cong Q_8/3^{1+2}$. When $\bar{L}\bar{T} = G_2(2)$, the stabilizers in $\bar{L}\bar{T}$ are twice as large. In each case $N_{LT}(U)$ is maximal in LT, but not of index 2; further L_U contains X_U of order 3 faithful on U. Thus if $F^*(G_U) = O_2(G_U)$, then the hypotheses of lemma E.6.14 are satisfied with U and LT in the roles of "W" and "M_0", so by that lemma, $G_U = C_G(U) \leq M$, contrary to our assumption.

This contradiction shows that $G_U \notin \mathcal{H}^e$. Suppose next that there is a component K of G_U. Then K is described in E.6.8, and in particular $K \not\leq M$. Now $K \cap M \leq M_U = C_M(V)$, so that $[V, K \cap M] = 1$. If case (1) of E.6.8 occurs, this forces $n = 1$, so that $K \cong L_3(2)$; we regard this group as $L_2(7)$, and treat it with the groups $L_2(p)$ arising below. In particular K is not a Suzuki group. The existence of X_U of order 3 faithful on U and an appeal to A.3.18 eliminates all cases of 3-rank 2—namely (2)–(4) of E.6.8, and all cases of E.6.8.5 except $K \cong L_2(p)$, p a Fermat or Mersenne prime. Notice now that the case $U = U_2$ for $\bar{L} \cong G_2(2)'$ cannot arise: For in that case there is $Y_U \cong 3^{1+2}$ faithful on U, so as $m_3(N_G(U)) \leq 2$, G_U is a $3'$-group, whereas K is not a Suzuki group. In the remaining two cases choose $l \in N_L(X_U) - L_U$ with $l^2 \in L_U \cap L_U^l$, and choose the hyperlane U_0 of U to be $U_0 := U \cap U^l$. As l acts on X_U and X_U is faithful on U, X_U acts faithfully on U_0. We saw $Q \in Syl_2(G_{U_0})$, so $K \leq K_0 \in \mathcal{C}(G_{U_0})$ by 1.2.4. Since $K \cong L_2(q)$ rather than $SL_2(q)$, K centralizes $O(G_{U_0})$ by A.1.29. Now by I.3.1, K is contained in the product of a U-orbit of 2-components of G_{U_0}, so as K centralizes $O(G_{U_0})$, we conclude those 2-components are ordinary components. Hence $K \leq E(G_{U_0})$, so K_0 is a component of G_{U_0}. As X_U is faithful on U_0, we may argue as before that $K_0 \cong L_2(q)$ for q a Fermat or Mersenne prime. But no proper embedding $K < K_0$ of these groups appears in A.3.12, so we conclude $K_0 = K$ is also a component of $C_G(U_0)$. Indeed $K = O^{3'}(E(G_{U_0}))$ since $m_3(N_G(U_0)) \leq 2$ and X_U is faithful on U_0. But $l^2 \in L_U \cap L_U^l$ so that $U_0 = U_0^l$. Hence $K^l = O^{3'}(E(G_{U_0^l})) = O^{3'}(E(G_{U_0})) = K$. Then as $K \leq G_U$, K centralizes $UU^l = V$, contrary to $K \not\leq M$.

Therefore $F^*(G_U) = F(G_U)$. As $G_U \notin \mathcal{H}^e$, we conclude $O_U := O(G_U) \neq 1$. Again let U_0 denote a hyperplane of U. By 1.1.6, the hypotheses of 1.1.5 are satisfied with G_{U_0}, $M = C_G(z)$ in the roles of "H, M". In particular by 1.1.5.2,

z inverts $O_{U_0} := O(G_{U_0})$. Similarly z inverts O_U, as we may also apply 1.1.6 and 1.1.5 to U. Now O_U is a nontrivial Q-invariant subgroup of $O(C_{G_{U_0}}(U))$.

Suppose first that O_U acts nontrivially on K_0, for some component K_0 of G_{U_0}. Then $1 \neq Aut_{O_U}(K_0) \leq O(C_{Aut(K_0)}(U))$ is Q-invariant. Inspecting the list of 1.1.5.3 for such a centralizer, we conclude $K_0/Z(K_0) \cong A_7$, U induces a group of inner automorphisms of order 2 on K_0, and $Aut_{O_U}(K_0) \cong \mathbf{Z}_3$. But by 1.1.5.3d, z induces an involution of cycle type 2^3, so that $V = Z_V U$ is not normal in $C_{K_0 Z_V}(z)$, contradicting $G_z = N_G(V)$.

Therefore O_U centralizes $E(G_{U_0})$. As z inverts O_U and O_{U_0}, O_U centralizes O_{U_0}. By 31.14.1 in [**Asc86a**], O_U centralizes $O_2(G_{U_0})$. Thus $O_U \leq C_{G_{U_0}}(F^*(G_{U_0})) \leq F(G_{U_0})$, so in particular $O_U \leq O_{U_0}$. Further O_{U_0} abelian since it is inverted by z.

Now given any $l \in M - N_M(U)$, we may choose $U \cap U^l$ as our hyperplane U_0 of U. Then $\langle O_U, O_U^l \rangle$ is contained in the abelian group $O_{U \cap U^l}$, and in particular, O_U and O_U^l commute. Therefore $1 \neq P := \langle O_U^M \rangle$ is abelian of odd order. Thus $LT \leq N_G(P) < G$ as G is simple; and $N_G(P)$ is quasithin. As $m_2(N_G(P)) \geq m(V) \geq 4$, V cannot act faithfully on $O_p(P)$ for any odd prime p by A.1.5, so $m_p(P) \leq 2$ for each odd prime p. Therefore $V = [V, L]$ centralizes P by A.1.26, which is impossible as z inverts O_U and so acts nontrivially on P. This contradiction finally completes the proof of 12.4.8. $\qquad\square$

LEMMA 12.4.9. *If* $A := V^g \leq N_G(V)$ *with* $[A, V] \neq 1$, *then* $V \nleq N_G(A)$.

PROOF. Assume otherwise. Interchanging the roles of A and V if necessary, we may assume $m(\bar{A}) \geq m(V/C_V(A))$, so that $\bar{A}$ contains a member of $\mathcal{P}(\bar{M}, V)$ by B.1.4.4. Then by B.4.6.13 or B.4.8.2, $\bar{A}$ is determined up to conjugacy in $\bar{M}$, and $m(\bar{A}) = m(V/C_V(A))$. In particular, we have symmetry between A and V.

Suppose $\bar{L} \cong L_3(2)$. Then $E = [A, V]$, so by symmetry $E = E^g$. Then as Z_V is weakly closed in E by 12.4.7, $g \in C_G(Z_V) = M = N_G(V)$, contradicting $[V, V^g] \neq 1$.

Therefore $\bar{L} \cong G_2(2)'$ and $\bar{A}\bar{L} = \bar{M} \cong G_2(2)$. Thus by B.4.6.3, $[V, A] = C_V(A)$, and again $Z_V \leq C_V(A)$ and by symmetry $[V, A] = [V, A]^g$. Then again Z_V is weakly closed in $[V, A]$ by 12.4.7, and we obtain the same contradiction. This completes the proof of 12.4.9. $\qquad\square$

We are now in a position to complete the proof of Theorem 12.4.2. Recall $m(\bar{M}, V) = 2$, so by 12.4.8, $s(G, V) > 1$. Then by 12.4.3.1, we may apply 12.4.1 to conclude that there is $g \in G$ with $1 \neq [V, V^g] \leq V \cap V^g$. In particular, $V^g \leq N_G(V)$ and $V \leq N_G(V^g)$, contrary to 12.4.9. This contradiction completes the proof of Theorem 12.4.2.

12.5. Eliminating $\mathbf{L}_5(2)$ on the 10-dimensional module

In this section we eliminate the exterior-square module in case (3c) of Theorem 12.2.2, hence reducing the treatment of $L_5(2)$ to the natural module in case (3a). This is analogous to the reduction for $L_4(2)$ in Theorem 12.6.34 of the next section. Specifically we prove:

THEOREM 12.5.1. *Assume Hypothesis 12.2.3 with* $L/O_2(L) \cong L_5(2)$. *Then* V *is the natural module for* $L/O_2(L)$.

Assume G, L, V afford a counterexample to Theorem 12.5.1. Then case (3c) of Theorem 12.2.2 occurs, so V is one of the 10-dimensional irreducibles for $L/O_2(L)$.

We mention that there is $L \in \mathcal{L}_f^*(G,T)$ with $L \cong L_5(2)/E_{2^{10}}$ in the non-quasithin groups $Sp_{10}(2)$, $\Omega_{10}^+(2)$, $\Omega_{12}^-(2)$, and $O_{12}^+(2)$. These shadows cause little trouble, as they are essentially eliminated immediately in 12.5.3 below.

The proof involves a series of reductions. As usual we adopt the conventions of Notation 12.2.5. Observe that as T acts on V, T induces inner automorphisms on $\bar{L}$, so $\bar{M}_V = \bar{L}$.

We next discuss the parabolic subgroups of $\bar{L}$ over $\bar{T}$, and their action on the module V. Let Γ be the natural 5-dimensional module for $\bar{L}$ with a basis for Γ denoted by $\{1, \ldots, 5\}$, and let $\Gamma_k := \langle 1, \ldots, k \rangle$. Choose notation so that T acts on Γ_k for each k. We regard V as the exterior square $\Lambda^2(\Gamma)$, so that V has basis $i \wedge j$ for $1 \leq i < j \leq 5$. Then T acts on the subspaces V_k of dimension k defined by

$$V_1 := \Lambda^2(\Gamma_2) = \langle 1 \wedge 2 \rangle, \quad V_3 := \Lambda^2(\Gamma_3) = \langle 1 \wedge 2, 1 \wedge 3, 2 \wedge 3 \rangle,$$

$$V_4 := \Gamma_1 \wedge \Gamma = \langle 1 \wedge i : 1 < i \leq 5 \rangle,$$

$$V_6 := \Lambda^2(\Gamma_4) = \langle i \wedge j : 1 \leq i < j < 5 \rangle, \quad V_7 := \Gamma_2 \wedge \Gamma = \langle 1 \wedge i, 2 \wedge j : i > 1, 3 \leq j \leq 5 \rangle.$$

For $i = 1, 3, 4, 6$, set $G_i := N_G(V_i)$, $M_i := N_{LT}(V_i)$, $L_i := N_L(V_i)^\infty$, and $R_i := O_2(L_iT)$.

Notice that $L_i \in \mathcal{L}(G,T)$ for each i.

LEMMA 12.5.2. *(1)* $\bar{M}_1 = N_{\bar{L}}(\Gamma_2)$, $\bar{M}_1/O_2(\bar{M}_1) \cong L_3(2) \times L_2(2)$, $\bar{L}_1/\bar{R}_1 \cong L_3(2)$, *and* $0 < V_1 < V_7 < V$ *is a chief series for* M_1.
(2) $\bar{M}_3 = N_{\bar{L}}(\Gamma_3)$, $\bar{M}_3/O_2(\bar{M}_3) \cong L_2(2) \times L_3(2)$, $\bar{L}_3/\bar{R}_3 \cong L_3(2)$, *and* V_3 *is a natural module for* $\bar{L}_3/\bar{R}_3$.
(3) $\bar{M}_4 = \bar{L}_4 = N_{\bar{L}}(\Gamma_1)$, *and* V_4 *is a natural module for* $\bar{L}_4/\bar{R}_4 \cong L_4(2)$.
(4) $\bar{M}_6 = \bar{L}_6 = N_{\bar{L}}(\Gamma_4)$, V_6 *is the orthogonal module for* $\bar{L}_6/\bar{R}_6 \cong L_4(2) \cong \Omega_6^+(2)$, *and* V/V_6 *is a natural module isomorphic to* $\bar{R}_6$.

PROOF. These are easy calculations. $\qquad\qquad\qquad\qquad\qquad\qquad\qquad\square$

Observe that from 12.5.2.1, $M_1 = PL_1$, where P is the minimal parabolic of LT over T which is in M_1, but not in L_1T. Further $O^2(P) = O^{3'}(P) \trianglelefteq M_1$ with $[O^2(P), L_1] \leq O_2(M_1)$. Similarly $O^2(P) \leq L_3 \cap L_6$, P is a minimal parabolic of L_iT for $i = 3, 6$, and $M_1 \cap M_i$ is the product of P with the minimal parabolic $P_i := M_i \cap L_1T$.

Recall from chapter 1 the definition of $\Xi_p(X)$ for $X \in \mathcal{L}(G,T)$ with $X/O_2(X)$ not quasisimple.

LEMMA 12.5.3. *For each* $i = 1, 3, 4, 6$, $L_i \leq K_i \in \mathcal{C}(N_G(V_i))$ *with* $K_i \trianglelefteq N_G(V_i)$, *and one of the following holds:*
(1) $L_i = K_i$.
(2) $i = 1$, $K_1/O_2(K_1) \cong L_5(2)$, M_{24}, *or* J_4, *and* $O^2(P) \leq K_1$.
(3) $i = 1$, $K_1 = \Xi_7(K_1)L_1$ *and* $K_1/O_2(K_1) \cong SL_2(7)/E_{49}$.

PROOF. The existence and normality of K_i follows from 1.2.4 and the fact that T normalizes L_i. By 12.5.2, $L_i/O_2(L_i) \cong L_3(2)$ if $i = 1, 3$ and $L_4(2)$ if $i = 4, 6$.

We first treat the case $i = 1$. We may assume that $L_1 < K_1$, so that $K_1/O_2(K_1)$ is described in the sublist of A.3.12 where $B/O_2(B) \cong L_3(2)$. If $K_1/O_2(K_1) \cong L_5(2)$, M_{24}, or J_4, then $K_1 = O^{3'}(G_1)$ by A.3.18, so $O^2(P) \leq K_1$ and hence (2) holds. Thus we may assume that $K_1/O_2(K_1)$ is not one of these groups, nor $SL_2(7)/E_{49}$, so $K_1/O_2(K_1)$ is one of $L_4(2)$, A_7, $\hat{A}_7$, $L_2(49)$, $(S)L_3^\epsilon(7)$, M_{23}, HS,

He, or Ru. To rule out $L_2(49)$ or $(S)L_3^\epsilon(7)$, observe that in those groups, some element of T induces an outer automorphism on $L_1/O_2(L_1) \cong L_3(2)$, while as $\bar{T} \leq \bar{L}$, this is not the case in LT. In the remaining cases by A.3.18, K_1 is the characteristic subgroup $\theta(G_1)$ of G_1 generated by all elements of order 3. Hence K_1 contains $I := O^{3'}(M_1)$, since $I/O_2(I) \cong \mathbf{Z}_3 \times L_3(2)$ by 12.5.2.1. Further $M_1 = IT$ and $M_1/O_2(M_1) \cong S_3 \times L_3(2)$, whereas $Aut(K_1/O_2(K_1))$ contains no such overgroup of a Sylow 2-group. This completes the proof of the lemma when $i = 1$, although we need some more information about K_1 which we develop in the next paragraph.

Set $(K_1T)^* := K_1T/O_2(K_1T)$. When $K_1^* \cong M_{24}$, $L_5(2)$, or J_4, the overgroups of T^* are described by a 2-local diagram, cf. [**RS80**] or [**Asc86b**]; we now describe the embedding of $L_1^*T^*$ and M_1^* in K_1^* in terms of the minimal parabolics in the sense of Definition B.6.1 indexed by the nodes of of this diagram: As $L_1^*T^*/O_2(L_1^*T^*) \cong L_3(2)$ and $M_1^*/O_2(M_1^*) \cong L_2(2) \times L_3(2)$, it follows that if $K_1^* \cong L_5(2)$ then (up to a symmetry of the diagram) $L_1^*T^*$ is generated by the third and fourth minimal parabolics of K_1^*, and the remaining parabolic P^* of M_1^* is the first parabolic of K_1^*. If $K_1^* \cong M_{24}$ or J_4, then $L_1^*T^*$ is generated by the parabolics indexed by the "square node" and by the adjacent node in those diagrams. Further in M_{24}, P^* is the parabolic P_K^* indexed by the node which is adjacent to neither of these nodes, while in J_4, the corresponding parabolic P_K satisfies $P_K^*/O_2(P_K^*) \cong S_5$, and P^* is the Borel subgroup of that parabolic. Thus when $K_1^* \cong M_{24}$, M_1^* is the trio stabilizer in the language used in chapter H of Volume I, while when $K_1^* \cong J_4$, $P_K^*M_1^* \cong S_5 \times L_3(2)/2^{3+12}$ and $M_1^* \cong (S_4 \times L_3(2))/2^{3+12}$.

We next treat the cases $i = 3$ or 4. Here $L_i/C_{L_i}(V_i) = GL(V_i)$ by 12.5.2, so that $K_i = L_iC_{K_i}(V_i)$. Hence if $K_i/O_2(K_i)$ is quasisimple, then $L_i = K_i$, as required. Therefore we may assume that $K_i/O_2(K_i)$ is not quasisimple, and it remains to derive a contradiction. As $K_i/O_2(K_i)$ is not quasisimple and $L_i \leq K_i$, we conclude from A.3.12 that $i = 3$, $K_3/O_2(K_3) \cong SL_2(7)/E_{49}$, and $K_3 = XL_3$ for $X := \Xi_7(K_3)$. Set $Y := O^2(P)$. Recall X char $K_3 \triangleleft G_3$. and $O_2(X) \neq 1$ using 1.1.3.1. Also X centralizes $V_3 \geq V_1$, so $X \leq K_{1,3} := O^2(C_{K_3}(V_1)) \leq G_1 \cap G_3$. We saw earlier that $Y = O^2(P) \leq L_3$, so $Y \leq K_{1,3}$. Then $K_{1,3}T/O_2(K_{1,3}T) \cong GL_2(3)/E_{49}$ and $K_{1,3} = YX = \langle Y^{K_{1,3}} \rangle$.

Suppose first that K_1^* is $L_5(2)$, M_{24}, or J_4. We saw earlier that $Y \leq K_1$. Hence $K_{1,3} = \langle Y^{K_{1,3}} \rangle \leq K_1$, so $K_{1,3}^*T^*$ is a subgroup of $K_1^*T^*$ containing T^*. But from the description of overgroups of a Sylow 2-group in terms of the 2-local diagrams for $L_5(2)$, M_{24}, and J_4 mentioned earlier, no such group has a $GL_2(3)/E_{49}$-section.

So we may suppose instead that $K_1 = L_1$ or $K_1^* \cong SL_2(7)/E_{49}$. By an earlier observation $[L_1, Y] \leq O_2(L_1)$. Thus if $K_1 = L_1$, Y centralizes K_1^*, and we claim this also holds when $K_1^* \cong SL_2(7)/E_{49}$: For $K_1 = \Xi_7(K_1)L_1$ and there is $Y < Y_0 \leq P$ with $Y_0/O_2(Y) \cong L_2(2)$ and $[L_1, Y_0] \leq O_2(L_1)$. Then as $L_1^* \cong SL_2(7)$ is centralized in $Aut(\Xi_7(K_1)^*)$ by $Z(GL_2(7)) \cong \mathbf{Z}_6$ which is abelian, we conclude $Y = [Y_0, Y_0]$ centralizes $\Xi_7(K_1^*)$. Hence $Y = O^7(Y)$ centralizes $K_1^* = O^{7'}(K_1^*)$, establishing the claim.

By the claim, $\langle Y^{K_{1,3}} \rangle = K_{1,3}$ centralizes K_1^*, so as $X \leq K_{1,3}$, X centralizes K_1^*. By construction $X \in \Xi(G,T)$, so by 1.3.4, either $X \trianglelefteq G_1$, or $X < K_0 \in \mathcal{C}(G_1)$ with $m_3(K_0) = 2$. In the latter case $K_0 = \langle X^{K_0} \rangle$ centralizes $K_1/O_2(K_1)$, so that $m_3(K_0K_1) > 2$, contradicting G_1 an SQTK-group. In the former case

$LT \leq \langle G_1, G_3 \rangle \leq N_G(X)$, so as $M = !\mathcal{M}(LT)$ and $O_2(X) \neq 1$, $G_1 \leq M$, contrary to 12.4.2.2. This completes the proof that $K_i = L_i$ if $i = 3$ or 4.

Finally we treat the case $i = 6$. Then either $K_6 = L_6$ as required, or as $L_6/O_2(L_6) \cong L_4(2)$, we obtain $K_6/O_2(K_6) \cong L_5(2)$, M_{24}, or J_4 from A.3.12. The latter three cases are impossible, since L_6 acts as $\Omega_6^+(2)$ on V_6, and this action does not extend to any 6-dimensional module for $L_5(2)$, while M_{24} and J_4 have no nontrivial modules of dimension 6. This completes the proof of 12.5.3. $\square$

LEMMA 12.5.4. $G_3 \leq M \geq G_6$.

PROOF. Let $i := 3$ or 6. By 12.5.3, $L_i \trianglelefteq G_i$, so as $N_{GL(V_i)}(Aut_{L_i}(V_i)) = Aut_{M_i}(V_i)$, $G_i = M_i C_G(V_i)$. Thus to show $G_i \leq M$, it suffices to show $C_G(V_i) \leq M$. If $C_G(V_i)$ acts on L_1, it acts on $\langle L_1^{L_i} \rangle = L$, so that $C_G(V_i) \leq N_G(L) = M$. Thus we may assume $C_G(V_i) \not\leq N_G(L_1)$.

Set $G_1^* := G_1/O_2(G_1)$. By 12.5.3, $Out(K_1^*)$ is a 2-group, so $G_1 = K_1 T C_{G_1}(K_1^*)$. Thus as $C_G(V_i) \leq G_1$, and T and $C_{G_1}(K_1^*)$ act on $O^2(L_1 O_2(K_1)) = L_1$, we may assume the preimage Y in K_1 of the projection of $G_i \cap K_1$ with respect to the decomposition $K_1^* \times C_{G_1^*}(K_1^*)$ is not contained in M_1. Therefore $K_1 \neq L_1$. As $T \leq G_i$, $[T \cap K_1, Y] \leq [G_i \cap K_1, Y] \leq Y \cap G_i$.

Suppose that case (3) of 12.5.3 holds, and set $Y_i := O^{3'}(N_{L_1}(V_i))$. Then $Y_i T$ is a minimal parabolic of $L_i T$, so as $L_1 \not\leq G_i$, $Y_i T = M_i \cap L_1 T = P_i$. Then as $[G_i \cap K_1, Y] \leq Y \cap G_i$ and Y_i acts irreducibly as $SL_2(3)$ on $\Xi_7(K_1)^* \cong E_{49}$, $Y \leq G_i$ and $Y = P_i \cap K_1$ or $\Xi_7(K_1)(P_i \cap K_1)$; as $Y \not\leq M_1$, the latter case holds. Then $\Xi_7(K_1) \leq \langle Y_i^Y \rangle \leq L_i$ as Y normalizes L_i by 12.5.3, contrary to $m_7(L_i) = 1$.

Therefore $K_1^* \cong L_5(2)$, M_{24}, or J_4. Recall from the discussion before 12.5.3 that $M_1 \cap M_i = PP_i$ is the product of the two minimal parabolics P and P_i, and $O^2(P) \leq K_1$. Then Y^* is a proper overgroup of $O^2(P^*)O^2(P_i^*)(T^* \cap K^*)$ which does not contain L_1^*. Let $Y_0 := \langle O^2(P)^Y \rangle$ and recall that the discussion during the proof of 12.5.3 determined the embedding of M_1^* in K_1^*. If $K_1^* \cong M_{24}$, the conditions above on Y^* force $Y^*/O_2(Y^*) \cong \hat{S}_6$ and $Y = Y_0$. If $K^* \cong L_5(2)$, then $Y/O_2(Y) \cong L_4(2)$ or $L_3(2) \times L_2(2)$, and $Y_0/O_2(Y_0) \cong L_4(2)$ or $L_3(2)$, respectively. If $K^* \cong J_4$, then $Y/O_2(Y) \cong \hat{M}_{22}$ or $S_3 \times S_5$, and $Y_0/O_2(Y_0) \cong \hat{M}_{22}$ or S_5, respectively. In particular, in each case $Y_0 \not\leq M$, since $M_1 = PL_1$. Further as $[Y, G_i \cap Y] \leq G_i$, $Y_0 \leq G_i$. But as $O^2(P) \leq L_i \trianglelefteq G_i$ by 12.5.3, $Y_0 \leq L_i \leq M$, contrary to the previous remark. This contradiction completes the proof of 12.5.4. $\square$

LEMMA 12.5.5. (1) $\bar{L}$ has two classes of involutions with representatives j_1 and j_2, where $m([\Gamma, j_i]) = i$, $m([V, j_2]) = 4$, and $C_V(j_1) = V_4 + V_6$ is of codimension 3 in V.

(2) L has two orbits on the points of V with representatives V_1 and
$$V_1' := \langle 1 \wedge 2 + 3 \wedge 4 \rangle.$$

(3) $C_{\bar{L}}(V_1')$ is $\bar{R}_6$ extended by S_6.

PROOF. These are straightforward calculations. $\square$

In the remainder of the section, we let V_1' be defined as in 12.5.5.2

LEMMA 12.5.6. $r(G, V) > 3 = m(M_V, V) = s(G, V)$.

PROOF. By 12.5.5.1, $m(\bar{M}_V, V) = 3$, so $r(G, V) \geq 3$ by Theorem E.6.3. Further if $U \leq V$ with $m(V/U) = 3$ and $C_G(U) \not\leq M$, then U is conjugate to $C_V(j_1)$

by E.6.12; then as U is normal in some Sylow 2-subgroup of LT, E.6.13 supplies a contradiction.　　□

LEMMA 12.5.7. $W_0 := W_0(T,V)$ centralizes V, so $w := w(G,V) > 0$ and $N_G(W_0) \le M$.

PROOF. Suppose $A := V^g \le T$ with $[A,V] \ne 1$. By 12.5.6, $s(G,V) = 3$, so that $\bar{A} \in \mathcal{A}_3(\bar{T},V)$ by E.3.10. But $Aut_{\bar{M}}(V_3) \cong L_3(2)$ is of 2-rank 2, so we conclude A centralizes V_3. Next T acts on Γ_1 and Γ_4, and hence acts on $V_3' := \Gamma_1 \wedge \Gamma_4 = \langle 1 \wedge 2, 1 \wedge 3, 1 \wedge 4 \rangle$, with $Aut_{\bar{M}}(V_3) \cong L_3(2)$, so the same argument shows A also centralizes V_3'. Similarly A acts on V_6, so that $A \le C := C_{M_6}(V_3 + V_3')$. By 12.5.2.4, V_6 is the orthogonal module for L_6/R_6, so that $m(C/R_6) = 1$; again as $\bar{A} \in \mathcal{A}_3(\bar{T},V)$, we conclude $A \le C_{LT}(V_6) = R_6$. By 12.5.5.1, V_6 is a hyperplane of $C_V(\bar{r})$ for each $\bar{r} \in \bar{R}_6^\#$, while by 12.5.2.4, V/V_6 is a natural module for L_6/R_6 isomorphic to $\bar{R}_6$. It follows that $V = W$, where $W := \langle C_V(\bar{r}) : \bar{r} \in \bar{R}_6^\# \rangle$, and no hyperplane of $\bar{R}_6$ lies in $\mathcal{A}_3(\bar{T},V)$. We conclude that $m(\bar{A}) > 3$, so that $\bar{A} = \bar{R}_6$, and hence $W = \langle C_V(B) : m(A/B) \le 3 \rangle$. Now $r(G,V) > 3$ by 12.5.6, so $W \le N_G(A)$ by E.3.32. Hence as $V = W$, we have symmetry between A and V. As $\bar{A} = \bar{R}_6$, $V_6 = [V,A]$; then by symmetry between A and V, $[A,V]$ is conjugate to V_6^g in L^g. Thus we may take $g \in G_6$, so $g \in N_M(V_6)$ by 12.5.4, and hence $g \in M_V$ by 12.2.6, contrary to $[V,V^g] \ne 1$. This contradiction shows $W_0 \le C_T(V) = O_2(LT)$, and so $N_G(W_0) \le M$ by E.3.34.2.　　□

Let $U := \langle V^{G_1} \rangle$ and $\tilde{G}_1 := G_1/V_1$.

LEMMA 12.5.8. (1) $V \le O_2(G_1)$.
(2) U is elementary abelian.

PROOF. Let $Y := O^2(M_1)$ and $U_1 := \langle V_7^{G_1} \rangle$. By 12.5.2.1, Y has chief series $0 < V_1 < V_7 < V$. Thus Hypothesis G.2.1 is satisfied with Y, G_1, G_1, Y, V_7 in the roles of "L, G, H, L_1, V", so by G.2.2, $\tilde{U}_1 \in \mathcal{R}_2(\tilde{G}_1)$ and $\tilde{U}_1 \le \Omega_1(Z(O_2(\tilde{G}_1)))$. In particular, $V_7 \le O_2(G_1)$.

Next $V = [V,L_1] \le [O_2(L_1),L_1]$, so if $K_1 = L_1$ or $K_1/O_2(K_1) \cong SL_2(7)/E_{49}$, then $V \le O_2(K_1) \le O_2(G_1)$, and hence (1) hold in these cases. We assume that (1) fails, so $K_1^* := K_1/O_2(K_1) \cong L_5(2)$, M_{24}, or J_4 by 12.5.3. Also $V^* \cong V/V_7$ is the natural module for $L_1^*/O_2(L_1^*) \cong L_3(2)$. Then $[V,U_1] \le V \cap U_1 = V \cap O_2(G_1) = V_7$. Further V^* is invariant under M_1^* by 12.2.6, and from the discussion in the proof of 12.5.3, $M_1^*/O_2(M_1^*) \cong L_2(2) \times L_3(2)$, with the embedding of M_1^* in K_1^* determined. When $K_1^* \cong M_{24}$ or $L_5(2)$, this is contrary to the action of Y^* on $O_2(Y^*)$ as the tensor product module of rank 6. Finally suppose $K_1^* \cong J_4$. Our discussion of the embedding of M_1^* showed that $M_1^* < N^*$, with $O_2(N^*)$ special of order 2^{3+12} and $N^*/O_2(N^*) \cong S_5 \times L_3(2)$. It follows that $V^* = Z(O_2(N^*)) \trianglelefteq N^*$. Since $L_1^* \not\le C_{G_1^*}(\tilde{V}_7)$, $K_1^* = \langle L_1^{*K_1} \rangle \not\le C_{G_1^*}(\tilde{U}_1)$, and in particular $V^* \not\le C_{G_1^*}(\tilde{U}_1)$; as we saw $[V,U_1] \le V_7$ and Y^* is irreducible on $\tilde{V}_7$, we conclude $[U_1,V] = V_7$. Therefore N^* normalizes $[\tilde{U}_1,V^*] = \tilde{V}_7$. But this is impossible as $\tilde{V}_7$ is the tensor product module for $Y^*/O_2(Y^*) \cong L_2(2) \times L_3(2)$, and this action does not extend to $N^*/O_2(N^*) \cong S_5 \times L_3(2)$.

Thus (1) is established. By 12.5.7, $V \le \Omega_1(Z(W_0(O_2(G_1),V)))$, so (2) follows from (1).　　□

LEMMA 12.5.9. *Let* $T_1 := C_T(V_1')$, *and choose notation with* $T_1 \in Syl_2(C_M(V_1'))$. *Then*

(1) $|T : T_1| = 4$, *and* $T_1 \in Syl_2(C_G(V_1'))$.
(2) $V_1' \notin V_1^G$.
(3) $V_1^G \cap V = V_1^L$.
(4) *If* $g \in G - N_G(V)$ *with* $V_1 \leq V^g$, *then* $V^g \in V^{G_1}$ *and* $[V, V^g] = 1$.

PROOF. The first part of (1) follows from 12.5.5.3. By 12.5.7, $W_0 := W_0(T, V) = W_0(T_1, V)$ and $N_G(W_0) \leq M$, so $T_1 \in Syl_2(C_G(V_1'))$. Thus (1) holds, and (1) implies (2). Then (2) and 12.5.5.1 imply (3). Finally under the hypothesis of (4), (3) and A.1.7.1 imply $V^g \in V^{G_1}$, and then 12.5.8.2 implies $[V, V^g] = 1$. $\square$

LEMMA 12.5.10. (1) *Either* $W_1 := W_1(T, V)$ *centralizes* V, *or* $\bar{W}_1 = \bar{R}_6$.
(2) $C_G(C_1(T, V)) \leq M$.

PROOF. Suppose $A := V^g \cap M \leq T$ with $[A, V] \neq 1$, and $m(V^g/A) \leq 1$. By 12.5.7, $m(V^g/A) = 1$ and $V > I := N_V(V^g)$. We now argue much as in 12.5.7: This time $r(G, V) > 3 = s(G, V)$ by 12.5.6, so $\bar{A} \in \mathcal{A}_2(\bar{T}, V)$ by E.3.10. Therefore either A centralizes V_3, or $Aut_A(V_3) \in \mathcal{A}_2(Aut_T(V_3), V_3)$, so that $m(A/C_A(V_3)) = m_2(L_3/O_2(L_3)) = 2$, and hence $V_3 \leq I$ as $r(G, V) > 3$. But in the latter case, $V_1 \leq [V_3, A] \leq V^g$, contrary to 12.5.9.4. We conclude A centralizes V_3, and similarly that A centralizes the space V_3' of 12.5.7; so again $A \leq C := C_{M_6}(V_3 + V_3')$ with $m(C/R_6) = 1$, and as $\bar{A} \in \mathcal{A}_2(\bar{T}, V)$, $A \leq C_{LT}(V_6) = R_6$. Thus as L_6 is irreducible on $\bar{R}_6$, $\bar{W}_1 = \bar{R}_6$. Hence (1) holds.

By (1), $V_6 \leq C_1(T, V)$, so (2) follows from 12.5.4. $\square$

For the remainder of the section, let $H \in \mathcal{H}_*(T, M)$. Recall from 3.3.2.4 that H is described in B.6.8 and E.2.2.

LEMMA 12.5.11. (1) $n(H) > 1$.
(2) $K_1 = L_1$.

PROOF. By 12.5.7 and 12.5.10, $N_G(W_0) \leq M \geq C_G(C_1(T, V))$; so as $s(G, V) = 3$, (1) follows from E.3.19 with $i, j = 0, 1$. Suppose $L_1 < K_1$, so that in particular $K_1 \not\leq M$. Then using the description of the embedding of M_1^* in K_1^* in 12.5.3 and its proof, there is $H \in \mathcal{H}(T)$ with $H \leq K_1 T$, $H \not\leq M$, and either $H/O_2(H) \cong S_3$, or $K_1/O_2(K_1) \cong SL_2(7)/E_{49}$ and $H := \Xi_7(K_1)T$. Thus $H \in \mathcal{H}_*(T, M)$ with $n(H) = 1$, contrary to (1). Thus (2) is also established. $\square$

LEMMA 12.5.12. *If* $H \leq G_1$, *then* $n(H) = 2$, *and a Hall* $2'$-*subgroup of* $H \cap M$ *is a nontrivial* 3-*group.*

PROOF. By 12.5.11.1, $n(H) > 1$. Then applying 12.2.11 to V_1 in the role of "U", the lemma holds. $\square$

We are now in a position to complete the proof of Theorem 12.5.1.

By Theorem 12.4.2.2, $G_1 \not\leq M$, so we may choose $H \in \mathcal{H}_*(T, M) \cap G_1$. Hence by 12.5.12, $n(H) = 2$ and a Hall $2'$-subgroup of $H \cap M$ is a nontrivial 3-group. Set $K := O^2(H)$, so that $K \not\leq M$, and $X := C_{G_1}(L_1/O_2(L_1))$. Then as $L_1 = K_1 \trianglelefteq G_1$ by 12.5.11.2, and $\bar{T} \leq \bar{L}$, $G_1 = L_1 X$. In particular, $X \not\leq M$ as $G_1 \not\leq M$, so we may choose $H \leq XT$. As $m_3(L_1) = 1$ and $m_3(G_1) \leq 2$, $m_3(X) \leq 1$. Therefore $m_3(X) =$

$m_3(H) = 1$, so as $n(H) = 2$, and a Hall $2'$-subgroup of $H \cap M$ is a nontrivial 3-group, $K/O_2(K) \cong L_2(4)$. Also $O^{3'}(H \cap M) \leq O^{3'}(X \cap M) = O^{3'}(X \cap L)$ using 12.2.8, so $M_1 = L_1(H \cap M)$ by 12.5.2.1.

Now just as in the proof of 12.5.8, Hypothesis G.2.1 is satisfied with $H_1 := L_1 H$, $O^2(M_1)$, V_7 in the roles of "H, L_1, V". Let $\tilde{H}_1 := H_1/V_1$, $U_H := \langle V_7^H \rangle$, and $Q_H := O_2(H_1)$. Then $\tilde{U}_H \leq Z(\tilde{Q}_H)$ by G.2.2.1, and $U_H \leq U$, so that U_H is elementary abelian by 12.5.8.2. If $[U_H, K] = 1$ then as $V_3 \leq V_7$, $K \leq G_3 \leq M$ by 12.5.4, contrary to $K \not\leq M$. Thus as $H_1 = KTL_1$ with $KL_1Q_H/Q_H \cong A_5 \times L_3(2)$, we conclude $Q_H = C_H(\tilde{U}_H)$.

Let $H_1^* := H_1/Q_H$. Now $W_0 := W_0(T,V) \leq C_T(V)$ and $N_G(W_0) \leq M$ by 12.5.7. Hence as $H \not\leq M$, $W_0 \not\leq O_2(H)$ by E.3.15, so there is $A := V^g \leq T$ with $A \not\leq O_2(H)$. As $A \leq W_0(T,V) \leq C_T(V) \leq O_2(M_1) \leq Q_H(T \cap K)$, $A^* \leq K^*$. Let $B := A \cap Q_H = C_A(\tilde{U}_H)$. Then $m(A/B) \leq m_2(K^*) = 2$. Further $[U_H, B] \leq V_1$, so for $u \in U_H$, $m(B/C_B(u)) \leq m_2(V_1) = 1$, and hence $m(A/C_B(u)) \leq 3$. Now $r(G,V) > 3 = s(G,V)$ by 12.5.6, so $u \in N_G(A)$, and hence $U_H \leq N_G(A)$. Thus if $[U_H, B] \neq 1$ then $V_1 = [U_H, B] \leq A$. But then 12.5.9.4 and 12.5.8.1 show $A \in V^{G_1} \subseteq O_2(G_1)$, contrary to $A \not\leq O_2(H)$. Thus U_H centralizes B, so as $m(A/B) < s(G,V)$, A centralizes U_H by E.3.6. But then $[K,A] = K$ centralizes U_H, contrary to our earlier observation that $Q_H = C_H(\tilde{U}_H)$.

This final contradiction completes the proof of Theorem 12.5.1.

12.6. Eliminating A_8 on the permutation module

The main result of this section is Theorem 12.6.34, which eliminates the A_8-subcase in case (d) of Theorem 12.2.2.3, reducing the treatment of $L/O_2(L) \cong L_4(2)$ to case (a) where V is a 4-dimensional natural module. This leaves only one case of Theorem 12.2.2.3 where it is possible that $C_V(L) \neq 1$: case (d) with $\bar{L} \cong A_6$. That case will be treated in section 13.4 of the following chapter.

We mention that $L_4(2)/E_{64}$ arises as $L \in \mathcal{L}_f^*(G,T)$ in the non-quasithin shadows $G \cong \Omega_8^+(2)$, $O_{10}^+(2)$, $Sp_8(2)$, and $P\Omega_8^+(3)$. Also such a 6-dimensional internal module appears in a suitable non-maximal member of $\mathcal{L}_f(G,T)$ in other non-quasithin groups, such as larger orthogonal and symplectic groups, as well as the sporadic groups J_4 and F'_{24}. As a result, the analysis in this case is fairly long and difficult. In particular, these shadows are not eliminated until 12.6.26.

So in section 12.6 we assume Hypothesis 12.2.3, and adopt the conventions of Notation 12.2.5, including $Z = \Omega_1(Z(T))$. In addition set $Z_V := C_V(L)$ and $\hat{V} := V/Z_V$.

Throughout this section, we assume that $\bar{L} \cong A_8$ and that $\hat{V}$ is the orthogonal module for $\bar{L} \cong \Omega_6^+(2)$. In particular notice $O_2(L) = O_{2,Z}(L) = C_L(V)$ by 1.2.1.4, since the Schur multiplier of A_8 is of order 2 by I.1.3.1. Further $\bar{M}_V = \bar{L}\bar{T} \cong A_8$ or $S_8 = Aut(A_8)$.

We adopt the notational conventions of section B.3, and assume T preserves the partition $\{\{1,2\}, \{3,4\}, \{5,6\}, \{7,8\}\}$ of the set Ω of 8 points. In particular by B.3.3, if $Z_V \neq 1$, then V is the core of the permutation module for $\bar{L}$ on Ω, and Z_V is generated by e_Ω. In that case, $V \trianglelefteq M$ by Theorem 12.2.2.3; hence $Z_V = C_V(L) \trianglelefteq M$, and we conclude $M = C_G(Z_V)$ as $M \in \mathcal{M}$. In any case $\hat{V}$ is the quotient of the core of the permutation module, modulo $\langle e_\Omega \rangle$. We can also view $\hat{V}$ as a 6-dimensional orthogonal space for $\bar{L} \cong \Omega_6^+(2)$. Thus we can speak of singular vectors in $\hat{V}$, nondegenerate subspaces of $\hat{V}$, etc. For $i = 1, 2, 5$, let V_i denote the preimage in V of the i-dimensional subspace of $\hat{V}$ fixed by T. Let $G_i := N_G(V_i)$, $M_i := N_M(V_i)$, $L_i := O^2(N_L(V_i))$, and R_i the preimage in T of $O_2(\bar{L}_i\bar{T})$.

12.6.1. Preliminary results.

LEMMA 12.6.1. *(1) L has two orbits on $\hat{V}^\#$, consisting of the singular and nonsingular vectors of $\hat{V}$.*
(2) If $Z_V \neq 1$, then $\mathbf{Z}_2 \cong Z_V = Z(T) \cap V$.
(3) Either $J(T) = J(C_T(V))$, or $|R_2(LT) : VC_{R_2(LT)}(L)| \leq 2$.
(4) $J(R_1) = J(C_T(V))$. Hence $N_G(R_1) \leq M$.
(5) $O^{3'}(M) = L$.
(6) If $L = [L, J(T)]$ and $Z_V \neq 1$, then L centralizes Z.

PROOF. Part (5) follows from 12.2.8. Recall that either

(a) $Z_V \neq 1$, and V is the 7-dimensional core of the permutation module for $\bar{L}$,

or

(b) $Z_V = 1$, and V is the 6-dimensional quotient of that core, modulo $\langle e_\Omega \rangle$.

Hence (1) and (2) are well known, easy calculations. Also either case (5) or (6) of B.3.2 holds, so if $\bar{A} \in \mathcal{P}(\bar{T}, V)$, then one of the following holds:

(i) $\bar{A} = \langle D \rangle$, for some $D \subseteq \Delta = \{(1, 2), (3, 4), (5, 6), (7, 8)\}$.
(ii) $\bar{A} = \langle \Delta \rangle \cap \bar{L}$.
(iii) $\bar{A}$ is conjugate under $\bar{L}$ to $\bar{A}_0 := \langle (1, 2)(3, 4), (1, 3)(2, 4), (5, 6), (7, 8) \rangle$, or to a hyperplane in the group of (ii), given by either

$$\langle (1, 2)(3, 4), (1, 2)(5, 6), (7, 8) \rangle, \text{ or } \langle (1, 2)(3, 4), (5, 6), (7, 8) \rangle.$$

(iv) $Z_V = 1$ and $\bar{A} \cong E_8$ is the unipotent radical of an $L_3(2)/E_8$ parabolic of $\bar{L}$.

Now $\bar{R}_1 \cong E_{16}$ is the unipotent radical of the stabilizer of the partition

$$\{\{1, 2, 3, 4\}, \{5, 6, 7, 8\}\},$$

so $\bar{R}_1$ contains no transpositions and hence contains no subgroup of type (i) or (iii); nor does it contain a subgroup of type (ii) or (iv). Thus $\bar{R}_1$ contains no FF*-offenders, so that $J(R_1) = J(O_2(LT))$, and hence $N_G(R_1) \leq N_G(J(O_2(LT))) \leq M = !\mathcal{M}(LT)$, so that (4) holds.

Also if $J(T) \not\leq C_T(V)$, then $\hat{V}$ is the unique noncentral chief factor for L on $R_2(LT)$ by Theorem B.5.1.1. Then (3) follows as $H^1(L, \hat{V}) \cong \mathbf{Z}_2$ by I.1.6.1. If in addition $Z_V \neq 1$, then $Z_V = Z \cap V$ by (2), and we've just seen that $V = [R_2(LT), L]$, so (6) follows as $\langle Z^L \rangle = VC_Z(L)$ by B.2.14. $\square$

In the shadows mentioned earlier (such as $\Omega_8^+(2)$), $C_G(v) \leq M$ for each nonsingular $v \in V$, as in the first main reduction Theorem 12.6.2 below. However, this result does eliminate the sporadic configurations in J_4 and F_{24}', since in those groups $C_G(v) \not\leq M$.

THEOREM 12.6.2. *$C_G(v) \leq M$ for each $v \in V$ with $\hat{v}$ nonsingular.*

Until the proof of Theorem 12.6.2 is complete, let $v \in V$ with $\hat{v}$ nonsingular and set $R_v := O_2(C_{LT}(v))$. Conjugating in L, we may assume $\hat{v} = \hat{e}_{1,2}$. Thus $\bar{L}_v \cong A_6$, so as $O_2(L) = O_{2,Z}(L) = C_L(V)$, $L_v = C_L(v)^\infty$, $L_v/O_2(L_v) \cong A_6$, and either $R_v = Q$ or $\bar{R}_v = \langle (1,2) \rangle$. By choice of v, $T_v := C_T(v) \in Syl_2(M_v)$ and $|T : T_v| = 4$. As $C_{LT}(v) = L_v T_v$, $R_v = O_2(L_v T_v)$.

LEMMA 12.6.3. *There exists* $K_v \in \mathcal{C}(G_v)$ *with* $L_v \leq K_v = O^{3'}(G_v^\infty) \trianglelefteq G_v$, *and* $K_v/O_2(K_v)$ *quasisimple.*

PROOF. By 1.2.1.1, L_v is contained in the product of the $\mathcal{C}$-components of G_v, so L_v projects nontrivially on $K_v/O_\infty(K_v)$ for some $K_v \in \mathcal{C}(G_v)$. As $L_v/O_2(L_v) \cong A_6$, it follows from A.3.18 that $m_3(K_v) = 2$ and $K_v = O^{3'}(G_v^\infty)$. Thus $L_v \leq K_v \trianglelefteq G_v$, and as $L_v/O_2(L_v) \cong A_6$, $K_v/O_2(K_v)$ is quasisimple by 1.2.1.4. $\square$

Let $T_v \leq S \in Syl_2(G_v)$, set $(K_v S)^* := K_v S/O_2(K_v S)$, and choose S so that $N_S(L_v) \in Syl_2(N_{G_v}(L_v))$. Hence $R := C_S(L_v^*/O_2(L_v^*)) \in Syl_2(C_{K_v S}(L_v^*/O_2(L_v^*)))$ and $R_v = R \cap T_v$. Then:

LEMMA 12.6.4. $|S : T_v| \leq |T : T_v| = 4 \geq |R : R_v|$. *Further* $O_2(K_v S) \leq R$.

LEMMA 12.6.5. R *normalizes* L_v, *and therefore* $[L_v, R] \leq O_2(L_v)$ *and* $R = C_S(L_v/O_2(L_v))$.

PROOF. Let X be the preimage in $K_v S$ of $O_2(L_v^*)$. As $[R, L_v] \leq X$ while $|R : R_v| \leq 4$, $(L_v X)^\infty = L_v$, so the lemma holds. $\square$

Let $V_v := [V, L_v]$. Then V_v is the 5-dimensional core of the permutation module for $L_v/O_2(L_v) \cong A_6$. In particular V_v is generated by the L_v-conjugates of a vector of weight 4 in that module, which is central in $T_v \in Syl_2(L_v T_v)$, so that $V_v \in \mathcal{R}_2(L_v T_v)$ and $V_v \leq \Omega_1(Z(R_v))$ by B.2.14. Let v_0 denote the generator of $C_{V_v}(L_v)$; thus v_0 has weight 6 in V and V_v, even though v itself may have weight 2 rather than 6 in V.

LEMMA 12.6.6. *If* $J(T) \not\leq Q$ *then either*

(1) $L_v = [L_v, J(T_v)]$, *or*
(2) $J(T_v) = J(Q)$.

PROOF. By hypothesis there is some $A \in \mathcal{A}(T)$ not in Q. Assume first that A satisfies one of (i)–(iii) in the proof of 12.6.1. Then some L-conjugate of A centralizes v and is nontrivial on $L_v/O_2(L_v)$, so that $J(T_v) \not\leq C_{T_v}(V_v)$, $L_v = [L_v, J(T_v)]$, and $L_v T_v/O_2(L_v T_v) \cong S_6$, and hence conclusion (1) holds. Thus if $J(T_v) \leq C_{T_v}(V_v)$, then each $\bar{A}$ must satisfy (iv), so in particular $Z_V = 1$ and V_v is a hyperplane of V. But since $\bar{A}$ satisfies (iv), $\bar{A}$ centralizes no vector of weight 2, so $\bar{A} \not\leq \bar{T}_v$ and hence $J(T_v) \leq Q$. As $Q \leq T_v$, we conclude $J(T_v) = J(Q)$ using B.2.3.3, so that conclusion (2) holds. $\square$

LEMMA 12.6.7. (1) *If* $J(T) \leq Q$, *then* $J(T) = J(Q) = J(T_v)$.
(2) *If* $L_v = [L_v, J(T_v)]$, *then* $L = [L, J(T)]$.

PROOF. As $Q \leq T_v \leq T$, (1) holds. Then (1) implies (2). $\square$

LEMMA 12.6.8. *If* $J(T_v) \leq Q$, *then*

(1) $S = T_v$ *and* $R = R_v$.
(2) $J(T_v) = J(R_v) = J(Q)$.
(3) $N_G(J(S)) \leq M$.

PROOF. Assume $J(T_v) \leq Q$. Then $J(T_v) = J(Q)$, so $N_G(T_v) \leq N_G(J(T_v)) = N_G(J(Q)) \leq M = !\mathcal{M}(LT)$. Hence as $T_v \in Syl_2(M_v)$, (1) and (3) hold. Then as $Q \leq O_2(L_vT_v) = R_v$, (2) holds.   □

LEMMA 12.6.9. If $J(T_v) \not\leq Q$, then $J(T) \not\leq Q$ and $L_v = [L_v, J(T_v)]$.

PROOF. Assume $J(T_v) \not\leq Q$. Then by 12.6.7.1, $J(T) \not\leq Q$. So by 12.6.6, $L_v = [L_v, J(T_v)]$.   □

LEMMA 12.6.10. Let $\Delta(v)$ be the set of vectors of weight 2 in V_v. Then

$$L(v) := \langle L_u : u \in \Delta(v) \rangle = L.$$

PROOF. Straightforward.   □

During the remainder of the proof of Theorem 12.6.2, we assume that $G_v \not\leq M$. In addition when $Z_V \neq 1$ and $G_u \not\leq M$ for some u of weight 2 in V, we choose v to be of weight 2 rather than 6.

LEMMA 12.6.11. $L_v < K_v$, so $K_v \not\leq M$.

PROOF. Assume $L_v = K_v$. Then $L_v = O^{3'}(G_v)$ by A.3.18. Furthermore $C_{G_v}(V_v)$ permutes $\{L_u : u \in \Delta(v)\}$, and hence $C_{G_v}(V_v) \leq N_G(L(v))$, so $C_{G_v}(V_v) \leq N_G(L) = M$ by 12.6.10. We deduce several consequences of this fact: First, $V_v \leq O_2(L_v) \leq O_2(G_v)$, so $O^2(F^*(G_v)) \leq C_{G_v}(V_v) \leq M$; then $O^2(F^*(G_v)) \leq O^2(F^*(M_v)) = 1$ using 1.1.3.2—that is, $G_v \in \mathcal{H}^e$. Second, suppose that $V_v \trianglelefteq G_v$. Then as $L_v \trianglelefteq G_v$,

$$Aut_{G_v}(V_v) \leq N_{GL(V_v)}(Aut_{L_v}(V_v)) \cong S_6 \cong Aut_{L_vT_v}(V_v),$$

so $G_v = L_vT_vC_{G_v}(V_v) \leq M$, contrary to our choice of v with $G_v \not\leq M$. Therefore V_v is not normal in G_v.

Suppose first that $J(T_v) \leq Q$. Let $H_v := C_{G_v}(L_v/O_2(L_v))$. By 12.6.8, $S = T_v$ and $N_G(J(S)) \leq M$. As $Out(A_6)$ is a 2-group, $G_v = L_vSH_v$, so $H_v \not\leq M$; then as $G_v \not\leq N_G(V_v)$, also $H_v \not\leq N_{G_v}(V_v)$. Therefore $V_v < \langle V_v^{H_v} \rangle =: U$. Recall that the core V_v of the permutation module for A_6 is generated by L_v-conjugates of a vector of weight 4 in that module, which is central in $T_v = S \in Syl_2(G_v)$. Then as $G_v \in \mathcal{H}^e$, $U \leq \Omega_1(Z(O_2(G_v)))$ by B.2.14. As $L_v = O^{3'}(G_v)$ and $Z(L_v/O_2(L_v)) = 1$, H_v is a 3'-group. Then we conclude from Theorem B.5.6 that U is not a failure of factorization module for $H_v/C_{H_v}(U)$, and hence $J(S) \leq C_{G_v}(U)$ by B.2.7. Now by a Frattini Argument, $H_v = C_{H_v}(U)N_{H_v}(J(S)) \leq C_G(V_v)N_G(J(S)) \leq M$, contrary to our remark that $H_v \not\leq M$.

Therefore $J(T_v) \not\leq Q$. Then by 12.6.9, $L_v = [L_v, J(T_v)]$, so $[R_2(G_v), L_v] = V_v$ by Theorems B.5.6 and B.5.1. Then $V_v \trianglelefteq G_v$, contrary to an earlier reduction.   □

LEMMA 12.6.12. K_v is not quasisimple.

PROOF. Assume K_v is quasisimple. Then $m_2(K_v) \geq m(V_v) = 5$, so $K_v/Z(K_v)$ is not M_{23}; and if $K_v/Z(K_v) \cong M_{22}$, then $\langle v_0 \rangle = C_{V_v}(L_v) \leq Z(K_v)$ and L_v is an A_6-block. Next as a 2-local of $K_v/Z(K_v)$ contains a quotient of L_v, as $L_v/O_2(L_v) \cong A_6$, and as $[O_2(L_v), L_v] \neq 1$, we eliminate most possibilities for $K_v/Z(K_v)$ in the list of Theorem C (A.2.3), reducing to $K_v/Z(K_v) \cong L_5(2)$, M_{22}, M_{24}, or J_4. As $|S : T_v| \leq 4$ by 12.6.4 with $S \cap K_v \in Syl_2(K_v)$, we conclude that $K_v/Z(K_v) \cong M_{22}$. However $C_V(L_v)$ is of corank 5 in V, so $C_V(K_v) \leq C_V(L_v)$ is of corank at least 5 in V. Hence $V/C_V(K_v)$ is of rank at least 5 in $Aut_{G_v}(K_v)$ and centralizes $V_v/\langle v_0 \rangle$, whereas $\langle v_0 \rangle = C_{V_v}(L_v) = C_{V_v}(K_v)$ with $V_v/\langle v_0 \rangle$ of rank 4 and self-centralizing in $Aut(K_v)$. This contradiction completes the proof. $\quad\square$

Set $U := \Omega_1(Z(O_2(K_vS)))$. Recall $(K_vS)^* = K_vS/O_2(K_vS)$.

LEMMA 12.6.13. *(1)* $F^*(K_vS) = O_2(K_vS) = C_{K_vS}(U)$.
(2) K_v^* *is simple.*
(3) $V_v \leq [U, K_v]$.

PROOF. By 12.6.12, K_v is not quasisimple, while by 12.6.3, K_v^* is quasisimple. Since $L_v/O_2(L_v)$ has trivial center and contains an E_9-subgroup, if $O_2(K_v) < O_{2,3}(K_v)$ then $m_3(L_vO_{2,3}(K_v)) = 3$, contrary to G_v an SQTK-group. Therefore from the list of possibilities in 1.2.1.4b, K_v^* is simple, so $F^*(K_vS) = O_2(K_vS)$ as K_v is not quasisimple. We showed $V_v \in \mathcal{R}_2(L_v)$, so that $L_v \in \mathcal{X}_f$ by A.4.11. By 12.6.4 and 12.6.5, $O_2(K_vS) \leq R \leq N_S(L_v) \in Syl_2(N_{G_v}(L_v))$, so we may apply A.4.10.3 with L_v, K_v, S in the roles of "X, Y, T" to conclude that $K_v \in \mathcal{X}_f$; then $[R_2(K_vS), K_v] \neq 1$ by A.4.11. As K_v^* is simple, $U = R_2(K_vS)$ and $C_S(U) = O_2(K_vS)$. Similarly by A.4.10.2, $V_v \leq [U, K_v]$. $\quad\square$

LEMMA 12.6.14. $J(T_v) \not\leq Q$.

PROOF. Assume $J(T_v) \leq Q$. By 12.6.8, $S = T_v$, $R = R_v$, and $J(S) = J(R) = J(Q)$, so that $N_G(R) \leq N_G(J(R)) \leq M = !\mathcal{M}(LT)$. By 12.6.4, $O_2(K_vS) \leq R$. Thus $N_{K_v^*}(R^*) = N_{K_v}(R)^* \leq M_v^*$. Then as $K_v \not\leq M$ by 12.6.11, it follows that $R^* \neq 1$. Since $T_v \in Syl_2(G_v)$, by 1.2.4 the embedding $L_v < K_v$ is described in in A.3.12; so as $R^* \neq 1$, we conclude that $K_v^* \cong M_{22}$ or M_{23}. Then K_v^* has no FF-module by B.4.2, so that $J(S) \leq C_S(U)$ by B.2.7. But $C_S(U) = O_2(K_vS)$ by 12.6.13.1, so $K_v \leq N_G(J(S)) \leq M$, contrary to $K_v \not\leq M$. This contradiction completes the proof of 12.6.14. $\quad\square$

LEMMA 12.6.15. U *is an FF-module for* $K_v^*S^*$.

PROOF. By 12.6.14 and 12.6.9, $L_v = [L_v, J(T_v)]$. Thus $K_v = [K_v, J(T_v)]$, so $r_{V,A^*} \leq 1$ for some $A \in \mathcal{A}(T_v)$ by B.2.4.1, and hence U is an FF-module for $K_v^*S^*$. $\quad\square$

LEMMA 12.6.16. *Assume* $Z_V \neq 1$ *and* $S \in Syl_2(G)$. *Then*

(1) $N_G(R_v) \not\leq M$, *and*
(2) L *is not a block.*

PROOF. By 12.6.14 and 12.6.9, $L = [L, J(T)]$; so as $Z_V \neq 1$ by hypothesis, L centralizes Z by 12.6.1.6. Therefore $C_G(z) \leq M = !\mathcal{M}(LT)$ for each $z \in Z^\#$.

As $S \in Syl_2(G)$ by hypothesis, v is 2-central in G, so there is $g \in G$ with $S^g = T$ and hence $v^g \in Z$. Further $L_v^g < K_v^g$ by 12.6.11, so as $G_v^g \leq M$, $K_v^g \leq O^{3'}(M) = L$ by 12.6.1.5. Also $L \leq C_G(Z) \leq G_v^g$, and $K_v^g = O^{3'}(G_v^{g\infty})$ by 12.6.3, so $K_v^g = L$. Thus $L_v^g \leq L$, and L_v^g is normal in the preimage of $\bar{L}_v^g$ in L by 12.6.4. Hence as $\bar{L}$ is transitive on its subgroups isomorphic to A_6, $L_v^g \in L_v^L$; then as L centralizes Z, without loss $L_v^g = L_v$. Then $R_v^g \in R_v^{N_{LT}(L_v)}$, so we also take $R_v^g = R_v$.

Thus if $N_G(R_v) \leq M$, then $g \in N_G(R_v) \leq M = N_G(L)$, so $K_v^g = L = L^g$, and hence $K_v = L \leq M$, contrary to 12.6.11. Therefore (1) is established.

As $N_G(R_v) \not\leq M \geq N_G(Q)$, $Q < R_v$. Then as we saw at the start of the proof of Theorem 12.6.2, $\bar{R}_v = \langle (1,2) \rangle$, so that $LR_v = LT$. As $(K_vS)^g = LT$ and $L_v^g = L_v$, $R = O_2(N_{LT}(L_v)) = R_v$, and hence $R^g = R_v^g = R_v = R$. Since it only remains to establish (2), we may assume that L is a block. Let $a := g^{-1}$ Notice that $V \trianglelefteq R$, so that also $V^a \trianglelefteq R$.

Suppose first that $[V, V^a] = 1$. Then $V^a \leq C_R(V) = Q$, so as L is a block, $[VV^a, L] \leq [Q, L] \leq V \leq VV^a$. Similarly $[VV^a, K_v] \leq VV^a$. Therefore as $LR = LT$, $K_v \leq M = !\mathcal{M}(LT)$, contrary to 12.6.11.

Thus $[V^a, V] \neq 1$, so as $V^a \leq R = R_v$, $\bar{V}^a = \langle (1,2) \rangle$, and hence $[V, V^a] = \langle e_{1,2} \rangle$. Since $Z_V \neq 1$ by hypothesis, v is chosen to have weight 2, so $v = e_{1,2}$. By symmetry, $[V^a, V] = \langle v^a \rangle$, so $v = v^a = v^{g^{-1}}$, and hence $g \in G_v$, impossible as v^g centralizes L. $\qquad \square$

LEMMA 12.6.17. $N_G(R_v) \not\leq M$.

PROOF. Assume that $N_G(R_v) \leq M$. Then as $R_v \in Syl_2(C_M(L_v/O_2(L_v)))$, $R_v = R$. Hence $N_{K_v^*}(R^*) = N_{K_v}(R_v)^* \leq M^*$, so $R^* \neq 1$ as $K_v \not\leq M$ by 12.6.11. In view of 12.6.15, K_v^* appears in the list of Theorem B.4.2, so since $K_v^*S^*$ has a nontrivial 2-subgroup R^* such that $L_v^* \trianglelefteq N_{K_v^*}(R^*) \geq T_v^*$ with $L_v^*/O_2(L_v^*) \cong A_6$ and $|S^* : T_v^*| \leq 4$, we conclude that $K_v^*S^* \cong S_8$ and R^* induces a transposition on K_v^*. Then $|S^* : T_v^*| = 4 = |T : T_v|$, so $S \in Syl_2(G)$. By 12.6.13.3, $V_v \leq [U, K_v] =: U_v$. Then from Theorem B.5.1.1, $U_v/C_{U_v}(K_v)$ is the 6-dimensional quotient of the core of the permutation module for K_v^*. Further $[C_{V_v}(L_v), K_v] \neq 1$, as v_0 is of weight 6 in both U_v and V. But v is central in G_v, so that $v \notin C_{V_v}(L_v) = \langle v_0 \rangle$, and hence $Z_V \neq 1$. Now 12.6.16.1 supplies a contradiction, completing the proof of the lemma. $\qquad \square$

LEMMA 12.6.18. (1) L is an A_8-block.
(2) K_v is an A_7-block.
(3) $Z_V \neq 1$.
(4) $LT = LR_v$.

PROOF. By 12.6.17, $N_G(R_v) \not\leq M$. So as $N_G(Q) \leq M$, $Q < R_v$, and hence $\bar{R}_v = \langle (1,2) \rangle$, so (4) holds. Then $\bar{R}_v \leq \bar{X} \leq \bar{M}$ for some $\bar{X} \cong S_3$—so either there is $1 \neq C$ char R_v with $C \trianglelefteq X$, or we may apply C.1.29 to $R_v \in Syl_2(X)$, to conclude that $O^2(X)$ is an A_3-block. In the former case, $C \trianglelefteq \langle X, C_{LT}(v) \rangle = LT$, so $N_G(R_v) \leq N_G(C) \leq M = !\mathcal{M}(LT)$, contrary to 12.6.17. Thus X an A_3-block, so L that (1) holds and L_v is an A_6-block. Further as L_v is trivial on R/R_v, V_v is the unique non-central chief factor for L_v on R, so V_v is the unique noncentral chief factor for L_v on $O_2(K_vS)$. Thus K_v is also a block. By 12.6.15, U is an FF-module for $K_v^*S^*$, so by Theorem B.4.2, K_v^* is either of Lie type and characteristic 2, or A_7. (The case of $\hat{A}_6$ is ruled out as $L_v/O_2(L_v) \cong A_6$.) Indeed as $SL_3(2^n)$ and $G_2(2^n)$ have no subgroup X with $X/O_2(X) \cong A_6$, $K_v^* \cong Sp_4(2^n)$, A_7, A_8, or $L_5(2)$. As T_v^* acts on L_v^* and $|S^* : T_v^*| \leq 4 = |T : T_v|$, K_v^* is A_7, A_8, or $L_5(2)$. Furthermore in the latter two cases $|S : T_v| = 4 = |T : T_v|$, so that $S \in Syl_2(G)$, and we calculate that $R = R_v$, and $L_v^* \cong A_6$ or A_6/E_{16}, respectively. In particular K_v^* is not $L_5(2)$, since in that group $[O_2(L_v^*), R^*] \neq 1$, whereas L_v has a unique noncentral 2-chief factor. Then as $V_v \leq [U, K_v]$, Theorem B.5.1.1 says $U/C_U(K_v)$ is the natural module for $K_v^* \cong A_7$ or A_8. Now we argue as in the proof of 12.6.17: In either case $[v_0, K_v] \neq 1$, so as $v \in Z(G_v)$, $v \neq v_0$ and hence (3) holds. Finally if K_v is an A_8-block, we showed $S \in Syl_2(G)$; then (1) contradicts 12.6.16.2. Thus K_v is an A_7-block, so (2) holds. $\qquad \square$

We are now in a position to complete the proof of Theorem 12.6.2. By 12.6.18.2, K_v is an A_7-block. Therefore $S = T_v$, since A_6 is of odd index in A_7. Hence $R_v = R = C_S(L_v/O_2(L_v)) = O_2(K_vS)$. Since the natural module for A_7 has trivial 1-cohomology by I.1.6.1, $O_2(K_vS) = UC_S(K_v)$ by C.1.13.b. Then from the structure of the A_7-module, $C_S(L_v) = C_S(K_v) \times \langle v_0 \rangle$. By 12.6.18.4, $LT = LS$, so $T_0 := C_T(L) \cap C_S(K_v) \trianglelefteq LT = LS$, and hence $T_0 = 1$ as $K_v \not\le M$ by 12.6.11. Then as $C_T(L) \le C_S(L_v)$, $|C_T(L)| \le |C_S(L_v) : C_S(K_v)| = 2$, so $C_T(L) = Z_V$ as $Z_V \ne 1$ by 12.6.18.3. Therefore as the 1-cohomology of the natural module for A_8 is 1-dimensional by I.1.6.1, we conclude from C.1.13.b that either $Q = V$, or Q is isomorphic to the 8-dimensional permutation module P as an L/V-module. In either case $R_v = \langle r \rangle Q$ with $\bar{r} = (1, 2)$ and $[R_v, r] = [Q, r] = \langle v \rangle$. Also $|R_v| = 2|Q| = 2^8$ or 2^9. On the other hand as $R_v = C_S(K_v) \times U$ with U the 6-dimensional permutation module for $K_v/O_2(K_v) \cong A_7$, $r \in C_{R_v}(L_v) = \langle v_0 \rangle \times C_S(K_v)$. In particular r centralizes U, so as $[R_v, r] = \langle v \rangle$, $[C_S(K_v), r] = \langle v \rangle$. As R_v is nonabelian, but U is central in in R_v, we conclude $C_S(K_v)$ is nonabelian, so $|C_S(K_v)| \ge 8$ and hence $|R_v| \ge 2^9$. Then using our earlier bounds, $|R_v| = 2^9$, with $Q \cong P \cong E_{2^8}$ and $R_v \cong D_8 \times E_{64}$. As $R_v = O_2(K_vS)$ and $K_v = O^2(K_v)$, K_v acts on both E_{2^8}-subgroups of R_v, so that $K_v \le N_G(Q) \le M$, for our usual contradiction to 12.6.11.

This contradiction completes the proof of Theorem 12.6.2. $\blacksquare$

In the remainder of the subsection, we show $Z_V = 1$.

LEMMA 12.6.19. *(1) L controls G-fusion in V.*
(2) If $Z_V \ne 1$ and v_4 is of weight 4 in V, then $|C_G(v_4) : C_M(v_4)|$ is odd.

PROOF. Suppose $Z_V = 1$. Then (2) is vacuous, and L has two orbits on $V^\#$, consisting of the singular and nonsingular vectors, with the singular vectors 2-central. By Theorem 12.6.2, the nonsingular vectors are not 2-central, so (1) holds in this case.

Thus we may assume $Z_V \ne 1$. In this case, L has four orbits on $V^\#$, with representatives v_2, v_4, v_6, v_8, where v_m is of weight m. By Theorem 12.6.2, $|C_G(v_m)|_2 = |T|/4$ for $m = 2, 6$. Notice v_8 is 2-central, and $|C_M(v_4)|_2 = |T|/2$. Assume that (1) fails. Then it follows that $v_4^g = v_8$ for some $g \in G$. We may choose v_4 so that $V_1 = \langle v_4, v_8 \rangle$; thus $T_4 := C_T(v_4) \in Syl_2(C_M(v_4))$ and $O^2(C_L(v_4)) = O^2(N_L(V_1)) = L_1$. Then $L_1^g \le C_G(v_8) \le M$, so $L_1^g \le L$ by 12.6.1.5, and we may take $T_4^g \le T$. As $|T : T_4| = 2$ and L_1T_4/R_1 is of index at most 2 in $S_3 \times S_3$, we conclude $L_1^g \in L_1^L$; thus as L centralizes v_8, we may take $L_1^g = L_1$. But then $R_1^g = O_2(L_1T_4)^g = R_1$. Now using 12.6.1.4, $g \in N_G(R_1) \le M$, contrary to $v_4 \notin v_8^M$. Hence (1) is established.

Suppose finally that (2) fails. Then $v_4^g \in Z$ for some $g \in G$. If $[V, J(T)] = 1$, then $N_G(J(T)) \le M$ by 3.2.10.1; and by Burnside's Fusion Lemma, $N_G(J(T))$ controls fusion in $Z(J(T)) \ge VZ$, contrary to v_4 not 2-central in M. Thus we may take $L = [L, J(T)]$, so as $Z_V \ne 1$, L centralizes Z by 12.6.1.6. Hence $L_1^g \le C_G(v_4^g) \le M = !\mathcal{M}(LT)$. We now repeat the argument of the previous paragraph to obtain the same contradiction, completing the proof of (2). $\square$

LEMMA 12.6.20. *Let $\mathcal{S}$ be the set of vectors in V of weight 4. If $g \in G - N_G(V)$ with $V \cap V^g \ne 1$, then*

(1) $V \cap V^g \subseteq \mathcal{S}$, so $m(V \cap V^g) \le 3$.

(2) $V^g \in V^{C_G(v)}$ for each $v \in V \cap V^g$.

(3) $r(G,V) \geq 3$, and $r(G,V) \geq 4$ if $Z_V \neq 1$.

(4) If $1 \neq [V,V^g] \leq V \cap V^g$, then $Z_V = 1$, $V \cap V^g$ is a totally singular 3-subspace of V, and $\bar{V}^g$ is the unipotent radical of an $L_3(2)/E_8$ parabolic.

PROOF. Part (2) follows from A.1.7.1 in view of 12.6.19.1. If $v \in V$ is of weight 8 then $G_v = M$ as we saw at the start of the section, while if v has weight 2 or 6, then $G_v \leq M$ by Theorem 12.6.2. So by 12.6.19.1, we may apply A.1.7.2 to see that each element of weight 2, 6, or 8 lies in a unique conjugate of V. Then (1) follows, and (1) implies (3).

Assume the hypotheses of (4). Interchanging V and V^g if necessary, we may assume $m(\bar{V}^g) \geq m(V/C_V(V^g))$. Then by B.1.4.4, $\bar{V}^g$ contains a member of $\mathcal{P}(\bar{T},V)$, so the possibilities for $\bar{V}^g$ are described in the discussion near the beginning of the proof of 12.6.1. As $[V,V^g] \leq V \cap V^g$, (1) says $[V,V^g]^{\#} \subseteq \mathcal{S}$, and the only possibility satisfying this restriction is that given in (4). $\qquad \square$

LEMMA 12.6.21. $C_G(v) \not\leq M$ for each $v \in V^{\#}$ of weight 4.

PROOF. As the groups in conclusions (2)–(4) of Theorem 12.2.13 do not have a member $L \in \mathcal{L}_f^*(G,T)$ with $\bar{L} \cong A_8$ and $V/C_V(L)$ the permutation module, conclusion (1) of 12.2.13 holds: $G_v \not\leq M$ for some $v \in V^{\#}$. By 12.6.20, $G_v \leq M$ for v of weight 2, 6, or 8, so the lemma holds. $\qquad \square$

LEMMA 12.6.22. If $Z_V \neq 1$, then

(1) $W_0 := W_0(T,V)$ centralizes V.

(2) If $m(V^g/V^g \cap M) \leq 1$ for some $g \in G$, then $V^g \leq N_G(V)$.

(3) $w(G,V) > 1$.

PROOF. Notice (1) and (2) imply (3), so it remains to prove (1) and (2). Assume $Z_V \neq 1$. Then $M = N_G(V)$ by 12.2.2.3. Suppose $A := V^g \cap M$ with $\bar{A} \neq 1$. Assume $k := m(V^g/A) \leq 1$, and if (1) fails, choose $k = 0$. Thus $V \not\leq N_G(V^g)$ if $k = 1$: for otherwise by assumption (1) does not fail, so that $V \leq C_G(V^g)$, contradicting $\bar{A} \neq 1$. Now by 12.6.20.3 and E.3.4, $m(\bar{A}) \geq r(G,V) - k \geq 4 - k$. Similarly using 12.6.20.3 as in E.3.32,

$$U := \langle C_V(B) : B \leq A \text{ and } m(V^g/B) \leq 3 \rangle \leq N_G(V^g),$$

so $[A,U] \leq V \cap V^g$. Now $\bar{L}\bar{T}$ is A_8 or S_8, and the maximal elementary abelian 2-subgroups of S_8 are

(i) $D_1 \cong E_8$ regular on Ω.

(ii) $D_2 \cong E_{16}$ with two orbits of length 4.

(iii) $D_3 \cong E_{16}$ with one orbit of length 4, and two of length 2.

(iv) $D_4 \cong E_{16}$ with four orbits of length 2.

If $k = 0$, then $m(\bar{A}) = 4$, so $\bar{A} = D_i$ for $i = 2,3,4$; while if $k = 1$, then $m(\bar{A}) \geq 3$, so either $\bar{A} = D_i$ for $i = 1,2,3,4$ or $\bar{A}$ is of index 2 in D_j for $j = 2,3,4$. In each case we find that $[A,U]$ contains a vector of weight 2 or 8. This contradicts 12.6.20.1, as $[A,U] \leq V \cap V^g$, so the proof is complete. $\qquad \square$

LEMMA 12.6.23. $C_V(L) = 1$. Thus V is the 6-dimensional orthogonal module for $\bar{L}$.

PROOF. Assume $Z_V = C_V(L) \neq 1$. Let $H \in \mathcal{H}_*(T, M)$, $K := O^2(H)$, $V_H := \langle Z^H \rangle$, and $H^* := H/C_H(V_H)$. By 12.6.20.3 and 12.6.22.3, $\min\{r(G, V), w(G, V)\} > 1$, so each solvable member of $\mathcal{H}(T)$ is contained in M by E.3.35.1 and E.1.13. In particular H is not solvable. By 12.2.9.1, $C_G(Z) \leq M$, so by Corollary 12.3.2, $1 \neq V_K := [V_H, K]$ is the sum of at most two A_5-modules for $K^* \cong A_5$. As $H = KT$, $O_2(H) = C_H(V_H) = C_T(V_H)$. Let H_M^* be the Borel subgroup of H^* over T^*; then $H_M = H \cap M$ by 3.3.2.4. Now $O^2(H_M) = O^{3'}(H_M) \leq O^{3'}(M) = L$ by 12.6.1.5.

Next if $W_0 := W_0(T, V) \leq C_T(V_H)$, then $H \leq N_G(W_0) \leq M$ by 12.6.22.1 and E.3.34.2, contrary to $H \not\leq M$. Therefore $W_0 \not\leq C_T(V_H)$, so there is $A := V^g \leq T$ with $A^* \neq 1$. Now by 12.6.22.1, $A \leq W_0(T, V) \leq C_T(V) = Q$. Thus as $O^2(H_M) \leq L$, $H_M \leq LT$, so that $A \leq O_2(H_M)$.

Let $B := A \cap O_2(H)$; then $m(A/B) = m(A^*) =: k \leq 2 = m_2(H^*)$. As $k \leq 2$, $V_H \leq C_G(B) \leq N_G(A)$ by 12.6.20.3, so $[A, V_H] \leq A \cap V_H$. However $O_2(H_M^*)$ is not quadratic on V_H, so $A^* < O_2(H_M^*)$. Therefore $k = 1$, so V_H induces a group of transvections with axis B on A. Thus $|V_H : C_{V_H}(A)| = 2$ from the action of $\bar{L}\bar{T} \cong S_8$ on V. This is impossible, since as $A^* \leq O_2(H_M^*)$, $m(V_H/C_{V_H}(A)) > 1$ from the action of involutions in A_5 on its permutation module. This contradiction completes the proof of 12.6.23.   □

12.6.2. The amalgam setup, and the elimination of V_H nonabelian. With 12.6.23 in place, we can begin to use some of the techniques from section F.9, which are more representative of the arguments for the $\mathbf{F}_2$-case in the three chapters of this part.

By 12.6.23, $Z \cap V = V_1$ is of order 2. Let z be the generator of V_1. Recall from Notation 12.2.5 that $G_z = C_G(z)$ and $\tilde{G}_z := G_z/V_1$. Now z is of weight 4 in V, so $G_z \not\leq M$ by 12.6.21. Recall that $L_1 = O^2(N_L(V_1)) = O^2(C_L(z)) = L_z$, V_5 is the 5-subspace $V_1^{\perp}$ of V, $\bar{L}_1 \cong E_9/E_{16}$, and $\bar{L}_1\bar{T}/O_2(\bar{L}_1\bar{T})$ acts on $\tilde{V}_5$ as $\Omega_4^+(2)$ or $O_4^+(2)$ on its natural module.

We now make a definition which, will be repeated in many later sections of the chapters on the $\mathbf{F}_2$ case: let

$$\mathcal{H}_z := \{H \in \mathcal{H}(L_1T) : H \not\leq M \text{ and } H \leq G_z\}.$$

As $G_z \not\leq M$, $G_z \in \mathcal{H}_z$, and hence $\mathcal{H}_z \neq \emptyset$.

For the remainder of the section, H will denote an element of $\mathcal{H}_z$.

Set $Q_H := O_2(H)$, $U_H := \langle V_5^H \rangle$, $V_H := \langle V^H \rangle$, and $H^* := H/Q_H$.

LEMMA 12.6.24. *(1) L_1T is irreducible on $\tilde{V}_5$.*
(2) $U_H \leq O_2(H)$. In particular, $V_5 \leq O_2(G_1)$.
(3) $V \leq O_2(G_2)$.
(4) $G_5 \leq N_G(V) = M_V$.
(5) If $g \in G$ with $V_1 < V \cap V^g$, then $\langle V, V^g \rangle$ is a 2-group.
(6) $C_G(\tilde{V}_5) = C_G(V)R_1 = C_M(V)R_1$ and $C_G(V_5) = C_G(V) = C_M(V)$.
(7) Hypothesis F.9.1 is satisfied for each $H \in \mathcal{H}_z$, with V_5 in the role of "V_+".
(8) $Q_H = C_H(\tilde{U}_H)$, so $H^* = Aut_H(\tilde{U}_H)$.

PROOF. Part (1) is standard and an easy calculation. Since $m(V_5) = 5 > 3$, (4) follows from 12.6.20.1. By (4), $C_G(\tilde{V}_5) \leq M$; so as $\bar{R}_1 = C_{\bar{M}_V}(\tilde{V}_5)$ and $\bar{M}_V$ contains no transvection with axis V_5, (6) holds.

We next prove (7). Most parts of Hypothesis F.9.1 are easy to check: For example, (4) implies hypothesis (c) of F.9.1, (1) implies (b), and (d) holds as $M = !\mathcal{M}(LT)$ but $H \not\leq M$. Assume the hypothesis of (e); as the conclusion of (e) holds trivially if $[V, V^g] = 1$, we may assume $1 \neq [V, V^g]$. Then 12.6.20.4 says $\bar{V}^g$ is the unipotent radical $\bar{R}$ of the stabilizer of some 3-subspace of V, while (6) and the hypothesis of (e) that V^g centralizes $\tilde{V}_5$ forces $\bar{V}^g \leq \bar{R}_1$. This contradicts $\bar{R} \not\leq \bar{R}_1$, and completes the proof of (e). As $H \in \mathcal{H}(T)$, $H \in \mathcal{H}^e$ by 1.1.4.6. If X is a normal subgroup of H centralizing $\tilde{V}_5$, then by (6) and Coprime Action, $O^2(X)$ centralizes V, so $[L, O^2(X)] \leq O_2(L)$. Hence LT normalizes $O^2(XO_2(L)) = O^2(X)$. Then if X is not a 2-group, $H \leq N_G(O^2(X)) \leq M = !\mathcal{M}(LT)$, contrary to $H \not\leq M$. Therefore $X \leq O_2(H) = Q_H$, so that hypothesis (a) of F.9.1 holds. Thus we have established (7).

Next (7) and parts (1) and (3) of F.9.2 imply (2) and (8), respectively. By (2), $V_5 \leq O_2(G_1) \cap C_G(V_2)$, so as $C_G(V_2) \leq G_1$, $V_5 \leq O_2(C_G(V_2)) \leq O_2(G_2)$. Then $V = \langle V_5^{L_2} \rangle \leq O_2(G_2)$. Hence (3) holds. If $g \in G - M_V$ with $V_1 < V \cap V^g$, then by 12.6.20.1, $V \cap V^g$ is totally singular; so without loss $V_2 \leq V^g$, so that $V^g \leq G_2$. But then $\langle V, V^g \rangle$ is a 2-group by (3). Hence (5) holds. $\square$

LEMMA 12.6.25. *The following are equivalent:*

(1) U_H is abelian.

(2) $V \leq Q_H$.

(3) $V \leq C_G(U_H)$.

(4) V_H is abelian.

PROOF. Parts (2) and (3) are equivalent by F.9.3, which applies by 12.6.24.7. Similarly the hypotheses of F.9.4.3 are satisfied by 12.6.24.6, so (1), (2), and (4) are equivalent by F.9.4.3. $\square$

Observe since $H \leq G_z = G_1$, that if V_H is nonabelian, then V_{G_1} is also nonabelian.

In the non-quasithin shadows mentioned earlier (such as $\Omega_8^+(2)$), V_H is nonabelian. Hence these shadows are eliminated in the next result:

LEMMA 12.6.26. V_H *is abelian.*

PROOF. Assume that V_H is nonabelian. Then by 12.6.25, U_H also is nonabelian and $V^* \neq 1$. Notice the hypotheses of F.9.5 are now satisfied, so in particular V^* is of order 2 and $[\tilde{U}_H, V^*] = \tilde{V}_5$ by parts (1) and (2) of F.9.5. By 12.6.24.5, the hypothesis of part (5) of F.9.5 is satisfied. Also $m(V) = 6$, and by 12.6.24.6, $C_H(V_5) = C_H(V)$, so the hypotheses of LL.5.6F.9.5.6ii are satisfied, and hence we can appeal to that result also. For example if $g^* \in H^*$ such that $I^* := \langle V^*, V^{*g} \rangle$ is not a 2-group, then by F.9.5.5, I^* is faithful on $U_I := V_5 V_5^g$; and by F.9.5.6ii, $U_I \cong Q_8^4$ and $I^* \cong D_6$, D_{10}, or D_{12}. Therefore elements of odd order in H^* inverted by V^* are of order 3 or 5.

We show first that $V \leq O_2(C_H(L_1/O_2(L_1)))$: Assume otherwise. Then by the Baer-Suzuki Theorem, for some $g \in C_H(L_1/O_2(L_1))$, $I^* := \langle V^*, V^{*g} \rangle$ is not a 2-group. By the previous paragraph, I^* is faithful on $U_I \cong Q_8^4$ and I^* is dihedral of order $2m$, $m = 3$, 5, or 6. Also as $g \in C_H(L_1/O_2(L_1))$, as V^* centralizes L_1^*, and as $O_2(L_1^*) = C_{L_1^*}(\tilde{V}_5)$, we conclude that I^* centralizes L_1^* and L_1 acts on V_5^g with $O_2(L_1^*) = C_{L_1^*}(\tilde{V}_5^g)$. Let $\widehat{IL_1} := Out_{IL_1}(U_I)$. Then $\hat{L}_1 \cong E_9$ is faithful on $\tilde{V}_5$

and $\tilde{V}_5^g$, so from the structure of $\hat{O} := Out(U_I) \cong O_8^+(2)$, $\hat{I} \leq C_{\hat{O}}(\hat{L}_1) \cong S_3 \times S_3$. But now $m_3(H) > 2$, whereas H is an SQTK-group. This contradiction shows that $V \leq O_2(C_H(L_1/O_2(L_1)))$.

We next show that V^* centralizes $F(H^*)$. Assume otherwise. Let $P \in Syl_3(L_1)$ and set $C^* := C_{O_3(H^*)}(P^*)$. If $[O_3(H^*), V^*] \neq 1$, then as V^* centralizes P^*, V^* inverts a subgroup X^* of C^* of order 3 by the Thompson $A \times B$ Lemma; but then $m_3(P^*X^*) = 3$, contrary to H an SQTK-group. Therefore V^* centralizes $O_3(H^*)$, so we may assume that $[V^*, O_p(H^*)] \neq 1$ for some prime $p > 3$. We saw that elements of odd order in H^* inverted by V^* are of order 3 or 5, so $p = 5$. Let Y^* be a supercritical subgroup of $O_5(H^*)$. By the previous paragraph $V \leq O_2(C_H(L_1/O_2(L_1)))$, so $O_5(H^*) \not\leq C_{H^*}(L_1^*)$. Therefore we conclude from A.1.25 that Y^* is E_{25} or 5^{1+2} and P is irreducible on $Y^*/\Phi(Y^*)$. Thus as V^* centralizes P^*, V^* inverts $Y^*/\Phi(Y^*)$. If $Y^* \cong 5^{1+2}$, then a faithful chief section W for Y^*V^* on $\tilde{U}_H$ is of dimension 20 over $\mathbf{F}_2$, and $m([W, V^*]) \geq 8$. If $Y^* \cong E_{25}$, then a faithful Y^*V^*-chief section W for P^*Y^* is of dimension 12 over $\mathbf{F}_2$ and $m([W, V^*]) = 6$. So in any case $m([\tilde{U}_H, V^*]) \geq 6$, whereas we saw $[\tilde{U}_H, V^*] = \tilde{V}_5$ is of rank 4. This contradiction establishes the claim that V^* centralizes $F(H^*)$.

So as $O_2(H^*) = 1$, $[K^*, V^*] \neq 1$ for some $K \in \mathcal{C}(H)$ such that K^* is a component of H^*. Then $K \not\leq M$ as $V \leq O_2(M)$. As $V^* \leq Z(T^*)$, V^* normalizes K^*, so that $K^* = [K^*, V^*]$. Further $L_1 = O^2(L_1)$ normalizes K by 1.2.1.3. As $V \leq O_2(C_H(L_1/O_2(L_1)))$, $K^* \not\leq C_{H^*}(L_1^*)$, so that $K^* = [K^*, L_1^*]$. Set $X := KL_1V$ and $\hat{X} := X/C_X(K^*)$. By F.9.5.3, $C_{H^*}(V^*) = N_H(V)^*$. Further $N_H(V) \leq H \cap M \leq N_H(L_1)$ and $C_X(\hat{V}) = C_K(\hat{V})L_1V$. Now $C_K(\hat{V})$ centralizes $V^*Z(K^*)/Z(K^*)$, so as $O_2(K^*) = 1$, $C_{\hat{K}}(\hat{V}) = \widehat{C_K(V^*)} \leq \widehat{K \cap M}$, and then

$$\hat{L}_1 \leq O_{2,3}(C_{\hat{X}}(\hat{V})) \text{ with } \hat{V} \text{ of order 2 in } Z(\hat{T}) \text{ and } 3 \in \pi(\hat{L}_1). \qquad (*)$$

Since all elements of odd order in $\hat{K}$ inverted by $\hat{V}$ are of order 3 or 5, we conclude $\hat{K}$ is not $Sz(2^n)$. Hence $m_3(K) = 1$ or 2.

Suppose first that $m_3(K) = 2$. Then as $m_3(P) = 2 = m_3(PK)$, either P is faithful on K^*, or $1 \neq P \cap O_{2,Z}(K)$, so that by 1.2.1.4b, $K^* \cong SL_3^\epsilon(q)$, $\hat{A}_6$, $\hat{A}_7$, or $\hat{M}_{22}$. Further in the latter case, $K/O_2(K) \cong \hat{A}_7$ or $\hat{M}_{22}$ by $(*)$.

Suppose that $K^* \cong \hat{A}_7$. Then $\hat{R}_1 = O_2(\hat{L}_1\hat{T}) \cong E_4$, while $N_K(R_1) \leq M$ by 12.6.1.4, so $O^{3'}(N_K(R_1)) \leq O^{3'}(M) = L$ by 12.6.1.5. Thus $O^{3'}(N_K(R_1)) \leq O^2(C_L(z)) = L_1$. This is impossible, as $N_K(R_1)$ has Sylow 3-group 3^{1+2}, while $E_9 \cong P \in Syl_2(L_1)$.

Suppose next that $K^* \cong \hat{M}_{22}$. As $H \leq G_z$, $H \cap M = N_H(V)$ by 12.2.6, so $(M \cap K)^* = N_K(V)^* = C_K(V^*)$, and hence $(M \cap K)^* \cong (S_3 \times \mathbf{Z}_2)/(\mathbf{Z}_3 \times Q_8^2)$. We saw $\tilde{V}_5 = [\tilde{U}_H, V^*]$ is of rank 4, so we conclude from H.12.1.3 that $\tilde{U}_K := [\tilde{U}_H, K]$ is the 12-dimensional irreducible for K^*. Choose v_2 of weight 2 in V_5; by parts (5) and (7) of H.12.1, $C_{\hat{K}}(v_2) \cong S_5/E_{16}$ or A_5/E_{16}. In particular $C_K(v_2)$ involves A_5, so $C_K(v_2) \not\leq M$, as we saw earlier that $(M \cap K)^*$ is solvable. But this contradicts Theorem 12.6.2.

We have shown that if $m_3(K) = 2$ then P is faithful on K^*, and hence also on $\hat{K}$. Then $m_3(\hat{P}) = 2$, so that $m_3(O_{2,3}(C_{\hat{X}}(\hat{V}))) = 2$ by $(*)$. But no simple group $\hat{K}$ of 3-rank 2 in the list of Theorem C satisfies this restriction. This contradiction completes the treatment of the case $m_3(K) = 2$.

Therefore $m_3(K) = 1$. As $m(P) = 2 = m_3(PK)$, we conclude that $P_K := P \cap K$ is of rank 1. Again inspecting the list of Theorem C, and using (*), we conclude that $\hat{K}$ is $L_2(p^e)$ for some odd prime p with $p^e \equiv \pm 1 \mod 12$. Recalling that elements of odd order in $\hat{K}$ inverted by $\hat{V}$ are of order 3 or 5, we reduce to the case $\hat{K} \cong L_2(p^e)$, with $p \neq 23$ or 25, and $\hat{V}$ inverts no element of order p if $p > 5$, so in fact $p^e \equiv -1 \mod 12$.

Set $H_0 := \langle K, L_1 T \rangle$, so that $H_0 \in \mathcal{H}_z$ as $K \not\leq M$. As $K^* = [K^*, V^*]$, $V \not\leq O_2(H_0)$, so U_{H_0} is nonabelian by 12.6.25. Thus replacing H by H_0, we may assume $H = \langle K, L_1 T \rangle$.

Next as $K^* \cong L_2(p^e)$, K^*V^* is generated by 3 conjugates of V^*. Thus as $\tilde{V}_5 = [\tilde{U}_H, V^*]$ is of rank 4, for each nontrivial chief section W for K^* on $\tilde{U}_H$, $m([W, K^*]) \leq 3m([W, V^*]) \leq 12$. Thus p^e divides $|L_{12}(2)|$, so as $p^e \neq 23$ or 25 and $p^e \equiv -1 \mod 12$, it follows that $p^e = 11$. But as 11 does not divide $|L_9(2)|$, the smallest nontrivial irreducible for K^* is of rank at least 10, so we conclude $\tilde{U}_K := [\tilde{U}_H, K] \in Irr_+(\tilde{U}_H, K)$, $\tilde{V}_5 = [\tilde{U}_K, V^*]$, and $10 \leq m(\tilde{U}_K)$. Thus if $K \neq K^t$ for some $t \in T$, then K^t centralizes $\tilde{U}_K$. However as T acts on V_5 and K does not centralize $\tilde{V}_5$, neither does K^{*t}, a contradiction. Thus T normalizes K, so $K \trianglelefteq H$ by 1.2.1.3, and so $H = \langle K, L_1 T \rangle = KL_1 T = KPT$.

Next $P = P_C \times P_K$ where $P_C := C_P(K^*)$ and $P_K = P \cap K$ are of order 3. As $H = KPT$ and $O_2(H^*) = 1$, $H^* = K^* P_C^* T^*$ and $P_C^* = O^2(C_{H^*}(K^*) \trianglelefteq H^*$. In particular P_C^* is T-invariant, so $\tilde{V}_5 = [\tilde{V}_5, P_C]$ since $L_1 T$ is irreducible on $\tilde{V}_5$. Then as P_C^* and K^* are normal in H^*, $\tilde{V}_5 \leq \tilde{U}_K$, and $U_H = \langle V_5^H \rangle$, $\tilde{U}_K = \tilde{U}_H = [\tilde{U}_H, P_C]$. As $\mathcal{N}_{C_{H^*}(V^*)}(P^*, 2) \subseteq C_{H^*}(P^*)$ from the structure of $Aut(L_2(11))$, $O_2(L_1^*) = 1$, and so $O_2(L_1) \leq Q_H$. Next $[Q_H, V] \leq Q_H \cap V = V_5 \leq U_H$, so as $K = [K, V]$, $[Q_H, K] \leq U_H$. As P_K^* is inverted by a conjugate of V^*, it follows from the first paragraph of the proof that $[U_H, P_K] \cong Q_8^4$. Then as $O_2(L_1) \leq Q_H$ by the previous paragraph, $[O_2(L_1), P_K] \leq [Q_H, K] \leq U_H$, so that $[O_2(L_1), P_K] = [U_H, P_K] \cong Q_8^4$ is of order 2^9. Therefore as $[V, P_K] \cong E_{16} \cong [\bar{R}_1, \bar{P}_K]$, we conclude that L is an A_8-block and $O_2(L_1) = [O_2(L_1), P_K] \cong Q_8^4$. We saw $\tilde{U}_H = [\tilde{U}_H, P_C]$, so $U_H = [U_H, P_C] \leq O_2(L_1)$. Thus $2^{11} \leq |U_H| \leq |O_2(L_1)| = 2^9$. This contradiction completes the proof of 12.6.26. $\square$

12.6.3. Restrictions on H, and the final contradiction.

In this section, we use machinery from section F.9 to handle the case V_H abelian. The same approach will be used many times in the remainder of our treatment of groups over $\mathbf{F}_2$.

LEMMA 12.6.27. *If $g \in G$ with $V \cap V^g \neq 1$, then $[V, V^g] = 1$.*

PROOF. By 12.6.20.1, we may assume $z \in V \cap V^g$. Then by 12.6.20.2, we may take $g \in G_z$. Applying 12.6.26 to G_z in the role of "H", we conclude $V_H = \langle V^{G_z} \rangle$ is abelian, so the lemma holds. $\square$

LEMMA 12.6.28. *(1) $\mathcal{A}_3(\bar{M}_V, V) = \emptyset$.*

(2) If $\bar{B} \in \mathcal{A}_2(\bar{M}_V, V)$, then $m(\bar{B}) \geq 3$ and there exists $\bar{D}$ of index 8 in $\bar{B}$ with $|\mathcal{E}(\bar{B}, \bar{D})| > 2$, where

$$\mathcal{E}(\bar{B}, \bar{D}) := \{\bar{E} \leq \bar{B} : m(\bar{E}/\bar{D}) = 1 \text{ and } C_V(\bar{E}) > C_V(\bar{B})\}.$$

(3) $W_0 := W_0(T, V)$ centralizes V, so $N_G(W_0) \leq M$.

(4) $W_1(T, V)$ centralizes V, so $w(G, V) > 1$.

PROOF. Recall $\bar{M}_V$ acts as A_8 or S_8 on the set Ω of eight points. Thus if $\bar{i}$ is an involution in $\bar{M}_V$, the $\bar{M}_V$-conjugacy class of $\bar{i}$ is determined by the number $n(\bar{i})$ cycles of $\bar{i}$ of length 2 on Ω. Thus $n(\bar{i}) = 1, 2, 3, 4$, and we check in the respective cases that $\bar{i}$ is of Suzuki type (cf. Definition E.2.6) b_1, c_2, b_3, a_2 on the orthogonal 6-space V. In particular,

$$C_V(\bar{i}) \neq C_V(\bar{j}) \text{ for involutions } \bar{j} \neq \bar{i}. \tag{$*$}$$

We first prove (1) and (2), so we assume that either $\bar{A} \in \mathcal{A}_3(\bar{M}_V, V)$ or $\bar{B} \in \mathcal{A}_2(\bar{M}_V, V)$. Then $m(\bar{A}) > 3$ by $(*)$, so that $m(\bar{A}) = 4 = m_2(\bar{L}\bar{T})$. Similarly $m(\bar{B}) \geq 3$.

Now the possibilities for $\bar{A}$ of rank 4 are described in cases (ii)–(iv) in the proof of 12.6.22. If $\bar{A} \leq \bar{L}$, then $\bar{A}$ is in case (ii); thus $\bar{A}$ is conjugate to $\bar{R}_1 = J(\bar{L} \cap \bar{T})$, whereas $\bar{R}_1 \notin \mathcal{A}_3(\bar{M}_V, V)$. Therefore $\bar{A} \not\leq \bar{L}$, so we are in case (iii) or (iv), and hence we may take $\bar{i} = (1,2) \in \bar{A}$. Let $W := C_V(\bar{i})$ and $\bar{X} := C_{\bar{M}_V}(\bar{i})$. Then $\langle \bar{i} \rangle = C_{\bar{M}_V}(W)$ and $\text{Aut}_{\bar{A}}(W) \in \mathcal{A}_3(\text{Aut}_{\bar{X}}(W), W)$. However W is the core of the 6-dimensional permutation module for S_6, and we compute directly that $a(S_6, W) \leq 2$. This contradiction completes the proof of (1).

Next $Z(\bar{T})$ is generated by

$$\bar{t} := (1,2)(3,4)(5,6)(7,8);$$

and

$$J := C_V(\bar{t}) = I \oplus \langle v \rangle,$$

where $v := e_{1,3,5,7}$ and $I := \langle e_{1,2}, e_{3,4}, e_{5,6}, e_{7,8} \rangle$. Let $I_0 := [V, \bar{t}]$ and $\bar{Y} := C_{\bar{M}_V}(\bar{t})$. Then I is isomorphic to the 3-dimensional quotient of the permutation module for $\text{Aut}_{\bar{Y}}(I) \cong S_4$ or A_4 on $\{e_{1,2}, e_{3,4}, e_{5,6}, e_{7,8}\}$, $\text{Aut}_{\bar{Y}}(I) \leq N_{GL(I)}(I_0)$, and $C_{\bar{Y}}(I) = C_{\bar{M}_V}(I)$ is

$$\bar{X} := \langle (1,2), (3,4), (5,6), (7,8) \rangle,$$

when $\bar{T} \not\leq \bar{L}$, or $\bar{X} \cap \bar{L}$, when $\bar{T} \leq \bar{L}$. In either case, $C_{\bar{Y}}(I)/\langle \bar{t} \rangle$ acts faithfully as a group of transvections on J with axis I.

Since $\bar{t}$ is 2-central in $\bar{M}$, we may take $\bar{B} \leq C_{\bar{M}_V}(\bar{t}) = \bar{Y}$, and then either $\bar{B}$ centralizes I or $\text{Aut}_{\bar{B}}(I) \in \mathcal{A}_2(\text{Aut}_{\bar{Y}}(I), I)$.

Suppose first that $\bar{B}$ centralizes I. Then $\bar{B} \leq C_{\bar{Y}}(I) \leq \bar{X}$, so as $m(\bar{B}) \geq 3$, we check that $C_V(\bar{B}) = I$. If $m(\bar{B}) = 3$ then $m(\bar{B} \cap \bar{L}) \geq 2$, and for each $\bar{b} \in \bar{B}^\# \cap \bar{L}$, $m(C_V(\bar{b})) = 4 > m(I)$, so $\langle \bar{b} \rangle \in \mathcal{E}(\bar{B}, 1)$. Thus $|\mathcal{E}(\bar{B}, 1)| > 2$, and hence (2) holds with $\bar{D} = 1$. On the other hand if $m(\bar{B}) = 4$ then $\bar{B} = \bar{X}$ and $C_V(\bar{B}) = I$. In this case we take $\bar{D} := \langle \bar{d} \rangle$ for $\bar{d} := (1,2)$. Then for each of the three other transpositions $\bar{d}'$ in $\bar{B}$, $m(C_V(\langle d, d' \rangle)) = 4 > m(I)$, so that $\langle d, d' \rangle \in \mathcal{E}(\bar{B}, D)$, and again (2) holds.

So suppose instead that $\text{Aut}_{\bar{B}}(I) \in \mathcal{A}_2(\text{Aut}_{\bar{Y}}(I), I))$. We saw that $\text{Aut}_{\bar{Y}}(I)$ is a subgroup of $P := N_{GL(I)}(I_0) \cong S_4$ containing A_4, so from the action of $GL(I)$ on I, $\mathcal{A}_2(P, I) = \{O_2(P)\}$, and hence $\text{Aut}_{\bar{B}}(I) = O_2(P)$ is of rank 2. Hence it is easy to calculate that $\langle \bar{t} \rangle = C_{\bar{X}}(\bar{B})$, so as $m(\bar{B}) \geq 3$ and $C_{\bar{B}}(I) \leq C_{\bar{X}}(\bar{B})$, we conclude that $C_{\bar{B}}(I) = \langle \bar{t} \rangle$ and $m(\bar{B}) = 3$. In particular each member of $\bar{B}^\#$ is regular on Ω, of rank 3, and for each $\bar{b} \in \bar{B}^\#$, $C_V(\bar{b})$ is of rank 4, so that $\langle \bar{b} \rangle \in \mathcal{E}(\bar{B}, 1)$. Hence $|\mathcal{E}(\bar{B}, 1)| = 7$, so that (2) holds with $\bar{D} = 1$.

It remains to prove (3) and (4), so we may assume that for some $y \in G$, $A := V^y \cap T$ with $[A, V] \neq 1$ and $k := m(V^y/A) \leq 1$. Hence $[V^y, V] \neq 1$, so by 12.6.27, $V \cap V^y = 1$. Then $[V \cap N_G(V^y), A] = 1$, so in particular $V \not\leq N_G(V^y)$. On the other hand for each $A_0 \leq A$ with $m(V^y/A_0) \leq 2$, $C_G(A_0) \leq N_G(V^y)$ by 12.6.20.3. Thus for each $A_0 \leq A$ with $m(A/A_0) < 3 - k$, $C_V(A_0) \leq V \cap N_G(V^y) \leq C_V(A)$, so that $C_V(A_0) = V \cap N_G(V^y)$ for all such A_0. That is, $\bar{A} \in \mathcal{A}_{3-k}(\bar{M}_V, V)$. Therefore $k \neq 0$ by (1). The remaining statement in (3) holds in view of E.3.34.2.

Hence we have reduced to the case $k = 1$, so that $m(A) = 5$, and $\bar{A} \in \mathcal{A}_2(\bar{M}_V, V)$. Now by (2), there exists $\bar{D}$ of corank 3 in $\bar{A}$ satisfying $|\mathcal{E}(\bar{A}, \bar{D})| > 2$. Consider any $\bar{A}_1 \in \mathcal{E}(\bar{A}, \bar{D})$; thus the preimage A_1 in V^y has rank 3. Since $C_V(\bar{A}_1) > C_V(\bar{A}) = V \cap N_G(V^y)$ from the previous paragraph, $C_V(A_1) \not\leq N_G(V^y)$. We conclude from Theorem 12.6.2 that the 3-subspace A_1 is totally singular in V^y; in particular, D is a totally singular 2-subspace of V^y. But then D lies in just two totally singular 3-subspaces of V^y, whereas there are at least 3 choices for $\bar{A}_1 \in \mathcal{E}(\bar{A}, \bar{D})$. This contradiction shows that $k \neq 1$, and so completes the proof of (4). $\qquad\square$

LEMMA 12.6.29. *If $H_0 \in \mathcal{H}(T)$ with $n(H_0) = 1$ or H_0 solvable, then $H_0 \leq M$.*

PROOF. Assume that $n(H_0) = 1$. We may apply 12.6.20.3 and 12.6.28.4 to see that $\min\{r(G, V), w(G, V)\} > 1$, so $H_0 \leq M$ E.3.35.1. Recall also that if H_0 is solvable, then $n(H_0) = 1$ by E.1.13 $\qquad\square$

As $H \not\leq M$, $n(H) > 1$ and H is not solvable by 12.6.29. Thus $H^\infty \neq 1$. Suppose $H^\infty \leq M$. Then as $C_{\bar{L}}(z)$ is solvable, $H^\infty \leq C_M(V) \leq C_M(L/O_2(L))$, so L normalizes $(H^\infty O_2(L))^\infty = H^\infty$. But then $H \leq N_G(H^\infty) \leq M = !\mathcal{M}(LT)$, a contradiction. We conclude $H^\infty \not\leq M$, so that by 1.2.1.1, there exists $K \in \mathcal{C}(H)$ with $K \not\leq M$. As usual by 1.2.1.3, $L_1 = O^2(L_1)$ normalizes K. Let $K_0 := \langle K^T \rangle$, so that $K_0 \trianglelefteq H$ by 1.2.1.3.

Notice that $K_0 L_1 T \in \mathcal{H}_z$.

For the rest of the section, we assume $H = K_0 L_1 T$, where $K \in \mathcal{C}(H_1)$ for some $H_1 \in \mathcal{H}_z$ with $K \not\leq M$.

Let $M_H := M \cap H$. Notice that $L_1^* \leq O_{2,3}(M_H^*)$.

LEMMA 12.6.30. *(1) Hypothesis F.9.8 is satisfied for each $H_2 \in \mathcal{H}_z$ with V_5 in the role of "V_+". In particular it holds for $H = K_0 L_1 T$.*

(2) $q(H^*, \tilde{U}_H) \leq 2$.

(3) $K/O_2(K)$ is quasisimple, and K_0^* and its action on $\tilde{U}_H$ are described in (4) or (5) of F.9.18.

PROOF. By 12.6.24.7, Hypothesis F.9.1 is satisfied, while F.9.8.f holds by 12.6.27, and case (i) of F.9.8.g holds by 12.6.24.6. Thus (1) holds. Then (1) and F.9.16.3 imply (2).

Suppose $K/O_2(K)$ is not quasisimple. Then $K \trianglelefteq H$ by 1.2.1.3, and by 1.2.1.4, $X := \Xi_p(K) \neq 1$ for some prime $p > 3$. By 12.6.29, $X \leq M = N_G(L)$. By 1.3.3, $X \in \Xi(G, T)$; so $X \trianglelefteq LXT$ by 1.3.4 since L cannot play the role of "L" in that result. Thus $H \leq N_G(X) \leq M = !\mathcal{M}(LT)$, a contradiction. Therefore $K/O_2(K)$ is quasisimple, so as $H = K_0 L_1 T$, (3) follows from F.9.18. $\qquad\square$

LEMMA 12.6.31. *One of the following holds:*

(1) $H^* \cong \operatorname{Aut}(L_3(4))$ or $SL_3(4)$ extended by a 4-group. Further M_H^* is the product of T^* with a Borel subgroup of $O^2(H^*)$.

(2) H^* is of index at most 2 in $S_5 \operatorname{wr} \mathbf{Z}_2$, and M_H^* is the product of T^* with a Borel subgroup of K_0^*.

(3) $H^* \cong S_5 \times S_3$, $|H : M_H| = 5$, and $R_1^* = O_2(M_H^*) \cong E_4$.

PROOF. By 12.6.30.3 we may apply F.9.18 to conclude that $K^*/Z(K^*)$ is a Bender group, $L_3(2^n)$, $Sp_4(2^n)'$, $G_2(2^n)'$, $L_4(2)$, $L_5(2)$, or A_7, or $K_0^* = K^* \cong M_{22}$ or $\hat{M}_{22}$. By 12.6.29, $n(H) > 1$ as $H \not\le M$, so by E.1.14 either

(i) $K/O_{2,Z}(K)$ is a Bender group over $\mathbf{F}_{2^n}$, $L_3(2^n)$, $Sp_4(2^n)$, or $G_2(2^n)$, with $n := n(H) > 1$, or

(ii) $K_0^* \cong M_{22}$ or $\hat{M}_{22}$, and $n(H) = 2$.

Pick $\tilde{I} \in Irr_+(K_0^*, \tilde{U}_H, T^*)$, and adopt the notation of F.9.18; in particular $I_H := \langle I^H \rangle$. Let $L_C := O^2(C_{L_1}(K_0^*))$ and $L_K := O^2(L_1 \cap K_0)$. In case (i), a Borel subgroup of K_0 over $T \cap K_0$ is contained in M by 12.6.29.

Suppose first that $m_3(K_0) = 0$. Then $K^* \cong Sz(2^n)$. In particular, $L_K = 1$ as $L_1 = O^{3'}(L_1)$. By F.9.18.3, $q(Aut_{K_0 T}(\tilde{I}), \tilde{I}) \le 2$, so $\tilde{I}$ is described in B.4.2 or B.4.5. Hence $\tilde{I}$ is the natural module for K^*, and either F.9.18.4i holds with $K = K_0$ and $I = I_H$, or F.9.18.5iiia holds, with $K < K_0$ and $\tilde{I}_H = \tilde{I} \oplus \tilde{I}^t$ for $t \in T - N_T(K)$. Now $Sz(2^n)$ has no FF-modules by B.4.2, so by F.9.18.7, $I_H = [U_H, K_0]$. As $L_1 = [L_1, T]$, while $Out(K^*)$ is cyclic, either $L_1 = L_C$; or $K < K_0$, L_C^* is of order 3, and an element of order 3 in $L_1 - L_C$ acts as a nontrivial field automorphism on each component of K_0^*. As L_C^* is $L_1 T$-invariant and nontrivial, $\tilde{V}_5 = [\tilde{V}_5, L_C]$ since $L_1 T$ is irreducible on $\tilde{V}_5$. Then as $L_C \trianglelefteq H$ and $U_H = \langle V_5^H \rangle$, $\tilde{U}_H = [\tilde{U}_H, L_C]$ and L_C^* is a 3-group, so $C_{\tilde{U}_H}(L_C^*) = 1$ by Coprime Action. However L_C stabilizes K and I, and $End_K(\tilde{I}) \cong \mathbf{F}_{2^n}$ with n odd so that 3 does not divide $2^n - 1$. Therefore $[I_H, L_C] = 1$, contradicting $C_{\tilde{U}_H}(L_C^*) = 1$. Therefore $m_3(K_0) > 0$.

Suppose next that $m_3(K_0) > 1$. Then comparing A.3.18 to the list of groups in (i) and (ii), either $K_0 = \theta(K_0)$, so that $L_1 \le K_0$, or $L_1 K_0/O_{2Z}(K_0) \cong PGL_3^\epsilon(2^n)$ or $L_3^{\epsilon,\circ}(2^n)$.

Suppose first that $K^* \cong M_{22}$ or $\hat{M}_{22}$. Then there is $H_0 \in \mathcal{H}(T) \cap H$ with $O^2(H_0^*/O_{2,Z}(H_0^*)) \cong A_6$. Therefore $n(H_0) = 1$ by E.1.11, E.1.13, and E.1.14.1, so $H_0 \le M$ by 12.6.29. But then $O^{3'}(H_0) \le O^{3'}(M) = L$ by 12.6.1.5. impossible as L has no T-invariant A_6-section.

Therefore as $m_3(K_0) = 2$, K_0^* is one of the Lie-type groups $L_2(2^n) \times L_2(2^n)$, $(S)L_3^\epsilon(2^n)$, $Sp_4(2^n)$, or $G_2(2^n)$ determined earlier. As $L_1^* \le O_{2,3}(M_H^*)$, and $M_H^* T^*$ is the normalizer of a parabolic subgroup while $n > 1$, it follows that $M_{K_0} = M \cap K_0$ is a Borel subgroup of K_0, with n even; hence K^* is not $(S)U_3(2^n)$ as $m_3(K) = 2$. Recall $L_1 T/O_2(L_1 T) \cong S_3 \times S_3$ or S_3 wr $\mathbf{Z}_2$, so $T/O_2(L_1 T)$ is noncyclic. Thus as $T \cap K_0 \le O_2(L_1 T)$, $T/O_2(L_1 T)$ projects on a noncyclic 2-subgroup of $Out(K_0^*)$, so $K_0^* \cong (S)L_3(2^n)$ or $L_2(2^n) \times L_2(2^n)$. We return to these cases in a moment.

We now consider the case $m_3(K_0) = 1$. Here $K = K_0$, and $K^* \cong L_2(2^n)$, $L_3(2^m)$, m odd, or $U_3(2^k)$, k even. As $L_1 = [L_1, T]$, and $Out(K^*)$ is abelian, L_1 induces inner automorphisms on K^*, so that $L_1^* = L_C^* \times L_K^*$. Then $L_C^* \ne 1$ as $m_3(K) = 1$, while $L_C^* < L_1^*$ as L_1^* has 3-rank 2. Now T normalizes K and L_1, and hence normalizes L_K and L_C; hence for $X \in \{K, C\}$, $L_X T/O_2(L_X T) \cong S_3$. As $TL_K = L_K T$, K^* is not $L_3(2^m)$ since m is odd, and if $K^* \cong L_2(2^n)$, then n is even.

We now handle together the remaining cases: $K_0^* \cong (S)L_3(2^n)$, $L_2(2^n) \times L_2(2^n)$, $L_2(2^n)$, and $U_3(2^n)$, with n even. Let $T_L := T \cap L$; then $L_1 = [L_1, T_L]$ and T_L acts on L_K, so $L_K = [L_K, T_L]$. Moreover from the structure of $Aut(K_0^*)$, unless $K^* \cong (S)L_3(4)$ or $L_2(4)$, there exists a prime $p > 3$ and a nontrivial p-subgroup X of the Borel subgroup M_{K_0} with $XT = TX$ and $X = [X, T_L]$, so that

$$X = [X, T_L] \le [X, L] \le L.$$

This is a contradiction as T permutes with no nontrivial p-subgroups of L for $p > 3$ as $\bar{L} \cong L_4(2)$.

Therefore $K^* \cong (S)L_3(4)$ or $L_2(4)$. If $K^* \cong SL_3(4)$ and $L_1 \not\le K$, then as M_H^* contains a Borel subgroup of H^*, a Sylow 3-subgroup of M_H is of order 27, whereas by 12.6.1.5, a Sylow 3-subgroup of M is of order 9. Therefore if K^* is $(S)L_3(4)$, then as $T/O_2(L_1T)$ projects on a noncyclic subgroup of $Out(K^*)$, (1) holds. So suppose $K^* \cong L_2(4)$. If $K < K_0$ then as $T/O_2(L_1T)$ is noncyclic, (2) holds. If $K = K_0$ our earlier analysis shows $L_1^* = L_C^* \times L_K^*$ with $L_1^*T^* \cong S_3 \times S_4$, so that (3) holds. $\qquad \square$

LEMMA 12.6.32. $K^* \cong L_2(4)$.

PROOF. Assume otherwise. Then case (1) of 12.6.31.1 holds, so $K_0^* = K^* \cong (S)L_3(4)$, and $T^*K^*/K^* \cong E_4$. We pick $\tilde{I} \in Irr_+(K, \tilde{U}_H, T)$, and by 12.6.30.3, we may adopt the notation of F.9.18.4; in particular $I_H := \langle I^H \rangle$. As T is nontrivial on the Dynkin diagram of K^*, it follows from B.5.1 and B.4.2.2 that K^*T^* has no FF-modules. Thus by F.9.18.7, $[\tilde{U}_H, K] = \tilde{I}_H$. If $I_H = I$, then $q(H^*, \tilde{I}) \le 2$ by F.9.18.2; so as K^*T^* has no FF-modules, $\tilde{I}/C_{\tilde{I}}(K)$ must appear in B.4.5. As the tensor-product module for $L_3(4)$ in B.4.5 has $q > 2$, we have a contradiction. Thus $I < I_H$, so case (iii) of F.9.18.4 occurs; that is, $K^* \cong SL_3(4)$ and $\tilde{I}_H = \tilde{U}_1 \oplus \tilde{U}_2$, where $\tilde{U}_1 = \tilde{I}$ is a natural K^*-module, and $\tilde{U}_2 = \tilde{U}_1^t$, for $t \in T$ nontrivial on the Dynkin diagram of K^*. As $\tilde{V}_5 = [\tilde{V}_5, L_1] \le \tilde{I}_H$, $\tilde{U}_H = \langle \tilde{V}_5^H \rangle \le \tilde{I}_H$, so $\tilde{U}_H = \tilde{I}_H$.

Next as M_H^* is the product of T^* with a Borel subgroup of K^* by 12.6.31.1, $M_H = L_1T$, so $T^* \cap K^* = O_2(M_H^*) = R_1^*$ is Sylow in K^*. As $\tilde{V}_5$ is L_1T-invariant and centralized by R_1^*, we conclude $\tilde{V}_5 = \tilde{V}_{5,1} \oplus \tilde{V}_{5,2}$, with $\tilde{V}_{5,i} = C_{\tilde{U}_i}(T \cap K)$ an $\mathbf{F}_4$-point in $\tilde{U}_i$. In particular $\tilde{V}_{5,i} = C_{\tilde{V}_5}(X_i)$ for some X_i of order 3 in L_1; so $V_{5,i}$ contains a nonsingular vector u_i of V. Now $C_{K^*}(\tilde{u}_i)$ is a maximal parabolic of K^*, so in particular $C_K(u_i)^*$ does not lie in the Borel group $M_{K_0}^*$. This is a contradiction as $C_G(u_i) \le M$ by Theorem 12.6.2. This contradiction completes the proof of 12.6.32. $\qquad \square$

LEMMA 12.6.33. $H^* \cong S_5 \times S_3$.

PROOF. Assume otherwise. As 12.6.32 eliminates case (1) of 12.6.31, we must be in case (2), where H^* of index at most 2 in S_5 wr $\mathbf{Z}_2$. Thus there is $t \in T - N_T(K)$, and we let $K_1 := K$ and $K_2 := K^t$. For $X \in Syl_3(L_1)$, $X = X_1 \times X_2$ with $X_i \in Syl_3(K_i)$ and $V_5 = [V_5, X] \le [U_H, K_0] = U_1U_2$, where $U_i := [U_H, K_i]$. Thus $U_H = U_1U_2 = [U_H, K_0]$.

Pick $\tilde{I} \in Irr_+(K_0, \tilde{U}_H, T)$; by 12.6.30.3, we may adopt the notation of F.9.18.5. We claim that either

(a) $[U_H/I_H, K_1, K_2] \le I_H$, and $[I_H, K_1, K_2] = 1$, or

(b) $\tilde{U}_H$ is the $\Omega_4^+(4)$-module for K_0^*.

For notice by Theorems B.5.6 and B.4.2 that H^* has no strong FF-modules. Thus it follows from F.9.18.6 that either $\tilde{U}_H = \tilde{I}_H$, or both $\tilde{I}_H$ and U_H/I_H are FF-modules for H^*. Observe that only cases (i) and (iiia) of F.9.18.5 can arise. In case (i), $\tilde{I}_H$ is not an FF-module for H^*, so $\tilde{U}_H = \tilde{I}_H$ and (b) holds. Suppose case (iiia) holds. Then (a) holds if $\tilde{U}_H = \tilde{I}_H$, so assume otherwise. Thus U_H/I_H is an FF-module for H^*; then as $U_H = [U_H, K_0]$, it follows from B.5.6 that $[U_1, K_2] \le I_H$, so again (a) holds. This completes the proof of the claim.

Suppose now that $[\tilde{U}_1, K_2] = 1$. Then $\tilde{V}_5 = \tilde{V}_{5,1} \oplus \tilde{V}_{5,2}$, where $\tilde{V}_{5,i} := [\tilde{V}_5, X_i] \cong E_4$. Then, as in the proof of 12.6.32, $V_{5,i}$ contains a nonsingular point u_i, and $K_{3-i} \leq C_G(u_i) \leq M$ by Theorem 12.6.2, contrary to $K_0 \nleq M$.

This contradiction shows $[\tilde{U}_1, K_2] \neq 1$. Suppose now that case (a) holds. Then $[U_1, K_2] = [U_H, K_1, K_2] \leq [I_H, K_1] \leq C_{U_H}(K_2)$. So $[U_1, K_2] = [U_1, K_2, K_2] = 1$, contrary to $[\tilde{U}_1, K_2] \neq 1$.

Therefore case (b) holds. As in the proof of 12.6.32, $R_1^* = T^* \cap K_0^*$ and $\tilde{V}_5 \leq C_{\tilde{U}_H}(R_1^*)$. But as $\tilde{U}_H^*$ is the $\Omega_4^+(4)$-module, $C_{\tilde{U}_H}(R_1^*)$ is an $\mathbf{F}_4$-point of $\tilde{U}_H$, whereas $E_{16} \cong \tilde{V}_5 \leq C_{\tilde{U}_H}(R_1^*)$. This contradiction completes the proof of 12.6.33. $\qquad\square$

We are at last in a position to obtain a contradiction under the hypotheses of this section.

By 12.6.33, $H^* = H_1^* \times H_2^*$ with $H_1^* \cong S_3$ and $H_2^* \cong S_5$. In particular $L_1 T/O_2(L_1 T) \cong S_3 \times S_3$, so $\bar{T} \leq \bar{L}$ and $\bar{M}_V = \bar{L} \cong A_8$. Also $X \in Syl_3(L_1)$ is of the form $X = X_1 \times X_2$ with $X_i \in Syl_3(H_i)$. As $X_i T = T X_i$, we conclude each X_i moves 6 points of Ω; hence $C_V(X_i)$ is a nondegenerate space of dimension 2 and $\tilde{V}_5 = [\tilde{V}_5, X_i]$. In particular as $X_1^* \trianglelefteq H^*$ and $\tilde{U}_H = \langle \tilde{V}_5^H \rangle$, $\tilde{U}_H = [\tilde{U}_H, X_1^*]$, and then

$$\tilde{U}_H = \tilde{U}_1 \oplus \tilde{U}_2$$

is an H_2-decomposition of $\tilde{U}_H$, where $\tilde{U}_i := [\tilde{U}_H, t_i^*]$ for t_1^* and t_2^* distinct involutions in H_1^*. As the third involution t_3^* in H_1^* commutes with H_2^* while interchanging $\tilde{U}_1$ and $\tilde{U}_2$, and $F^*(H_2^*) = K^*$ is simple, H_2^* is faithful on $\tilde{U}_1$, and $\tilde{U}_2$ is isomorphic to $\tilde{U}_1$ as an H_2-module. Recalling that $H_2^* \cong S_5$ has no strong FF-modules, we must be in case (b) of F.9.18.6, so that U_i^* is an FF-module for H_2^*. Hence either $[\tilde{U}_i, H_2^*]$ is the S_5-module, or $[\hat{U}_i, H_2^*]$ is the $L_2(4)$-module, where $\hat{U}_H := \tilde{U}_H / C_{\tilde{U}_H}(K)$.

In particular, no member of H^* induces a transvection on $\tilde{U}_H$. Thus by F.9.16.1, $D_\gamma < U_\gamma$, in the notation of F.9.16. Hence by F.9.16.4, we can choose γ so that $0 < m := m(U_\gamma^*) \geq m(U_H/D_H)$. Further by F.9.13.2, $U_\gamma \leq O_2(G_{\gamma_1,\gamma_2}) = R_1^h$ for suitable $h \in H$, so $m \leq 2$ as $H^* \cong S_5 \times S_3$. Next for $b \in U_\gamma - D_\gamma$, $[D_H, b] \leq A_1$ by F.9.13.6, where A_1 is the conjugate of V_1 defined in section F.9; thus $m(A_1) = 1$ and

$$m([\tilde{U}_H, b^*]) \leq m(U_H/D_H) + m([\tilde{D}_H, b]) \leq m + 1 \leq 3,$$

impossible as $m([\tilde{U}_H, b]) = m([\tilde{U}_1, b]) + m([\tilde{U}_2, b]) = 4$, since $b^* \in U_\gamma^* \leq R_1^{h*} \leq K^*$ and $\hat{U}_i$ is the natural or A_5-module for K^*.

This contradiction finally eliminates the A_8-subcase of Theorem 12.2.2.3d, and hence establishes:

THEOREM 12.6.34. *If Hypothesis 12.2.3 holds with $\bar{L} \cong L_4(2)$, then V is the natural module for $\bar{L}$.*

12.7. The treatment of $\hat{A}_6$ on a 6-dimensional module

In this section we prove

THEOREM 12.7.1. *Assume Hypothesis 12.2.3 with $L/C_L(V) \cong \hat{A}_6$. Then G is isomorphic to M_{24} or He.*

We recall that M_{24} has already appeared in Theorem 12.2.13, in the case that V is a TI-set in G. However in this section, our argument does not require Theorem

12.2.13 until after both He and M_{24} have been independently identified; 12.2.13 is only used when we are working toward the final contradiction.

We mention that the groups He and M_{24} will be identified via Theorem 44.4 in [**Asc94**] in our Background References.

12.7.1. Preliminary results. The proof of Theorem 12.7.1 involves a series of reductions.

Assume L, V arise in a counterexample G. Then by Theorem 12.2.2, V is a 6-dimensional module for $L/C_L(V) \cong \hat{A}_6$ and $C_L(V) = O_2(L)$. We adopt the conventions of Notation 12.2.5. Let $T_L := T \cap L$, and P_1 and P_2 the two maximal subgroups of LQ containing T_LQ. Let $R_i := O_2(P_i)$ and $X := O^2(O_{2,Z}(L))$. We can regard V as a 3-dimensional vector space $_FV$ over $F := \mathbf{F}_4$, with $\bar{L} \leq SL(_FV)$ and $\bar{X}$ inducing F-scalars on $_FV$.

LEMMA 12.7.2. *(1) Either $\bar{M}_V = \bar{L} \cong \hat{A}_6$ or $\bar{M}_V = \bar{L}\bar{T} \cong \hat{S}_6$.*
(2) P_1T is the stabilizer in LT of an F-point V_1 of $_FV$, and $\bar{R}_1 \cong E_4$ is a group of F-transvections on $_FV$ with center V_1 and $C_V(R_1) = V_1$.
(3) P_1 is irreducible on V/V_1 and V_1.
(4) P_2T is the stabilizer of an F-line V_2 of $_FV$, and $\bar{R}_2 \cong E_4$ is a group of F-transvections on $_FV$ with axis V_2 and $[V, R_2] = V_2$.
(5) $O^2(P_i) = L_iX$, where $\{L_i, X\}$ are the unique T-invariant subgroups $I = O^2(I)$ of P_i with $|I : O_2(I)| = 3$.
(6) $L = \theta(M)$ is the characteristic subgroup of M generated by all elelements of order 3 in M.

PROOF. The calculations in (1)–(5) are well-known and easy. Notice in (1) that automorphisms of $A_6 = Sp_4(2)'$ nontrivial on the Dynkin diagram are ruled out, as they do not preserve V. Part (6) follows from 12.2.8. $\qquad \square$

In the remainder of the section, we adopt the notation of L_1 and L_2 as in 12.7.2.5.

LEMMA 12.7.3. $\mathcal{A}_2(\bar{T}, V) = \{\bar{R}_2\}$ *and* $a(\bar{M}_V, V) = 2$.

PROOF. Let $\bar{A} \in \mathcal{A}_2(\bar{T}, V)$. Then $C_V(\bar{A}) = C_V(\bar{B})$ for each hyperplane $\bar{B}$ of $\bar{A}$, so as $1 \neq \bar{A}$, $m(\bar{A}) > 1$. If $\bar{A} \not\leq \bar{L}$, then $\bar{B} := \bar{A} \cap \bar{L}$ is a hyperplane of $\bar{A}$, and $C_V(\bar{B})$ is an F-subspace of V, whereas $\bar{a} \in \bar{A} - \bar{L}$ is nontrival on each $\bar{a}$-invariant F-point since $\bar{a}$ inverts $\bar{X}$. We conclude that $\bar{A} \leq \bar{L}$, so as $\bar{R}_i$, $i = 1, 2$, are of rank 2 and are the maximal elementary abelian subgroups of $\bar{T}_L$, $\bar{A} = \bar{R}_i$ for some i. By 12.7.2.2, $i \neq 1$, and 12.7.2.4 shows that $\bar{R}_2 \in \mathcal{A}_2(\bar{T}, V)$, so $\bar{A} = \bar{R}_2$. Since $m(\bar{R}_2) = 2$, $a(\bar{M}_V, V) = 2$. $\qquad \square$

LEMMA 12.7.4. *(1) L has two orbits on $V^{\#}$ with representatives $z \in V_1$ and t.*
(2) Let V_t be the F-point of $_FV$ containing t. Then $N_{\bar{L}}(V_t) \cong GL_2(4)$, and V is an indecomposable module for $C_L(V_t)$ with V/V_t the natural module.
(3) $t \in V = [V, L_t] \leq L_t$ and $L_t = \theta(M_t)$.
(4) V_2 is partitioned by two conjugates of V_t and three conjugates of V_1.

PROOF. From 12.7.2, $|V_1^L| = |L : P_1| = 15$, leaving a set $\mathcal{O}$ of 6 F-points of $_FV$ not in V_1^L. As 6 is the minimal degree of a faithful permutation representation for $\bar{L}/\bar{X} \cong A_6$, it follows that L is transitive on $\mathcal{O}$ (so that (1) holds), and the stabilizer in $\bar{L}$ of $V_t \in \mathcal{O}$ is isomorphic to $GL_2(4)$. As $V = [V, X]$, V/V_t is the natural module for $N_{\bar{L}}(V_t) \cong GL_2(4)$, so a Sylow 2-subgroup $\bar{S}$ of $N_{\bar{L}}(V_t)$ centralizes an

F-hyperplane of $_F V$. Hence as $\bar{R}_i$, $i = 1, 2$, are representatives for the conjugacy classes of 4-subgroups of $\bar{L}$, we may take $\bar{S} = \bar{R}_2$ since $\bar{R}_1$ centralizes no hyperplane by 12.7.2.2. Then as $[V, R_2]$ is not a point by 12.7.2.4, V does not split over V_t as an $N_L(V_t)$-module. This completes the proof of (2), and (2) and 12.7.2.6 imply (3). Further V_2/V_t is the 1-dimensional F-subspace centralized by $\bar{S} = \bar{R}_2$, so $V_t \le V_2$ and P_2 has two orbits on F-points of V_2 of length 2 and 3, and then (4) follows as T acts on V_1. $\square$

For the rest of the section, t and V_t have the meaning given in 12.7.4. Observe that $L_t/O_2(L_t) \cong A_5$ using 12.7.4.2, so that $L_t = L_t^\infty$.

LEMMA 12.7.5. *Either*

(1) $G_t \le M$, or

(2) $K_t := \langle L_t^{G_t} \rangle$ is a component of G_t with $V_t = Z(K_t)$ and $K_t/V_t \cong L_3(4)$.

PROOF. Assume that (1) fails, and choose t so that $T_t = C_T(t) \in Syl_2(M_t)$. From 12.7.4, $O_2(\bar{L}_t \bar{T}_t) = 1$ and $V = [V, \bar{L}_t]$, and we saw $L_t = L_t^\infty$, so we may apply 12.2.12.2 to conclude that Hypothesis C.2.8 is satisfied with G_t, M_t, L_t, Q in the roles of "H, M_H, L_H, R". By C.2.10.1, $O(G_t) = 1$. By Theorem C.4.8, $L_t \le K \in \mathcal{C}(G_t)$ with $K/O_2(K)$ quasisimple and K described in one of the conclusions of that result. If conclusion (10) of C.4.8 holds, then for $g \in G_t - M_t$, $L_t \ne L_t^g \le M_t$, so $L_t^g \le \theta(M_t) = L_t$ by 12.7.4.2, a contradiction. Thus by C.4.8, $L_t < K$, $K/O_2(K)$ is quasisimple, and K is described in C.3.1 or C.4.1. By 12.7.4.3, $L_t = \theta(M_t)$ and $V = [V, L_t]$, so $L_t = \theta(K \cap M)$ and $t \in V_t \le V \le L_t \le K$, so $t \in Z(K)$.

Suppose first that $F^*(K) = O_2(K)$. Examining the list of C.4.1 for "M_0" with $M_0/O_2(M_0) \cong L_2(4)$ acting naturally on V/V_t, we see conclusion (2) of C.4.1 holds: K is an $Sp_4(4)$-block with $V_t \le Z(K)$, and $M \cap K$ is the parabolic stabilizing the 2-dimensional F-space V/V_t in $U(K)/V_t$. As $U(K)$ is a quotient of the orthogonal FK-module of dimension 5, V splits over V_t as an L_t-module—contrary to 12.7.4.2.

Thus as $K/O_2(K)$ is quasisimple, K is a component of G_t; and $Z(K)$ is a 2-group since $O(G_t) = 1$. This time examining the list of C.3.1 for "M_0" given by L_t acting as $L_2(4)$ on V/V_t, we see that one of cases (1), (3), or (4) must occur. Then as $V_t \le Z(L_t)$ and $t \in V_t \cap Z(K)$, we conclude that $V_t \le Z(K)$. Now by I.1.3, the only case with a multiplier of 2-rank 2 is $K/Z(K) \cong L_3(4)$. As V is L_t-invariant and elementary abelian, $V_t = O_2(K)$ from the structure of the covering group of $L_3(4)$ in I.2.2.3b. Thus as $Z(K)$ is a 2-group, (2) holds in this case, completing the proof of 12.7.5. $\square$

LEMMA 12.7.6. *Assume that L is a $\hat{A}_6$-block. Then*

(1) $Q = O_2(LT) = V \times C_T(L)$.

(2) If $C_T(L) = 1$ then $O_2(L) = V = O_2(M) = C_G(V)$ and $M = LT$.

PROOF. Since the 1-cohomology of V is trivial by I.1.6, (1) follows from C.1.13.b. Assume $C_T(L) = 1$. By (1), $O_2(LT) = V$. Now (2) follows from 3.2.11. $\square$

12.7.2. The identification of He. In this subsection we prove:

THEOREM 12.7.7. *If $G_t \not\le M$, then G is isomorphic to He.*

PROOF. Assume $G_t \not\le M$ and let $K_t := \langle L_t^{G_t} \rangle$. By 12.7.5, K_t is quasisimple with $V_t = Z(K_t)$ and $K_t/V_t \cong L_3(4)$. In particular by A.3.18, K_t is the unique component of G_t of order divisible by 3. Therefore as X normalizes V_t, for each

$x \in X$, $K_{t^x} = K_t^x \leq C_G(V_t^x) = C_G(V_t) \leq G_t$, so that $K_t = K_{t^x}$. Hence X acts on K_t and $XK_t/Z(K_t) \cong PGL_3(4)$.

From the structure of K_t, L_t is an $L_2(4)$-block, so L is an $\hat{A}_6$-block. Let $Y_X \in Syl_3(X)$. As $X = O^2(O_{2,Z}(L))$ and L is an $\hat{A}_6$-block, $X = VY_X$. As $K_t \trianglelefteq G_t$ and $\Omega_1(O_2(K_t)) = V_t$, $G_t \leq N_G(V_t)$. Then as X is transitive on $V_t^{\#}$, V_t is a TI-set in G by I.6.1.1.

Next as $V \leq L_t$ by 12.7.4.2, $C_G(L_t) \leq C_{G_t}(L_t) = C_{G_t}(K_t)$, so as $C_{Aut(K_t)}(L_t) = 1$, $C_G(L_t) = C_{G_t}(K_t)$. Similarly $C_G(L_t) \leq C_G(V) = C_M(V)$, and $[L, C_M(X)] \leq C_L(X) = Z(L)$, so as L is perfect, $C_G(L_tX) = C_G(L)$ is LT-invariant. Further $[C_G(L_t), X] \leq C_X(L_t) = V_t$, so by a Frattini Argument, $C_G(L_t) = V_tC_G(L_tY_X) = C_G(L_tX)$. On the other hand, we saw that $C_G(L_t) = C_{G_t}(K_t)$, so if $C_G(L_tX) \neq 1$, then $K_t \leq C_G(C_G(L_t)) \leq M = !\mathcal{M}(LT)$, contrary to $K_t \not\leq M_t$. Therefore $C_G(L_tX) = C_G(L) = 1$, and $C_{G_t}(K_t) = C_G(L_t) = V_tC_G(L_tX) = V_t$. Thus $V = O_2(M)$ and $M = LT$ by 12.7.6.2. Then by 12.7.2.1, either $M = L$, or $|M : L| = 2$ with $M/V \cong \hat{S}_6$.

Choose t so that $T_t := C_T(t) \in Syl_2(M_t)$. As $K_t \trianglelefteq G_t$, $G_t \notin \mathcal{H}^e$, so t is not 2-central in G by 1.1.4.6. hence $P = C_Y(\tilde{P})$ since $Inn(P)$ induces $C_{Aut(P)}(\tilde{P})$ by A.1.23. Therefore

$$Y/P \leq Aut(\tilde{P}) \cong O_6^+(2),$$

and $D_8 \cong T/P \in Syl_2(Y/P)$. Further $\alpha := (M_z/P, T/P, N_{G_z}(U)/P)$ is a Goldschmidt triple as in Definition Aa.t:defnGldtrpl. As $O_2(M_z/P) \neq O_2(N_{G_z}(U)/P)$, case (i) of F.6.11.2 holds, and so the image in $Y/O_{3'}(Y)$ of α is a Goldschmidt amalgam; therefore as Y is an SQTK-group, $Y/O_{3'}(Y)$ is described in Theorem F.6.18. In view of (*), $Y/O_{3'}(Y)$ appears in case (6) of Theorem F.6.18; that is, $Y/O_{3'}(Y) \cong L_2(q)$ for $q \equiv \pm 7 \mod 16$. Then as $Y/P \leq O_6^+(2)$, we conclude $Y/P \cong L_3(2)$ or A_6.

Next $\tilde{P}^+$ is the sum of the natural module and its dual for $Y^+/P^+ \cong L_3(2)$, so M_z^+ and $N_{G_z^+}(U^+)$ stabilize unique points of $\tilde{P}^+$. Indeed $\tilde{V}_1^+$ is the point stabilized by M_z^+, and we write $\tilde{U}_1^+$ for the point stabilized by $N_{G_z^+}(U^+)$. Applying φ, M_z stabilizes only $\tilde{V}_1$ and $N_{G_z}(U)$ stabilizes only $\tilde{U}_1$. As $\tilde{V}_1^+ \neq \tilde{U}_1^+$, $\tilde{V}_1 \neq \tilde{U}_1$. But if $Y/P \cong A_6$, then Y stabilizes a point of $\tilde{P}$, so $\tilde{V}_1 = C_{\tilde{P}}(Y) = \tilde{U}_1$, contrary to the previous remark. We conclude $Y/P \cong L_3(2)$.

Now $S_4 \cong M_z/P$ is the stabilizer in Y/P of $\tilde{V}_1$, so $\tilde{P}_1 := \langle \tilde{V}_1^Y \rangle$ is a nontrivial quotient of the 7-dimensional permutation module on Y/M_z, and similarly $\tilde{P}_2 := \langle \tilde{U}_1^Y \rangle$ is a nontrivial quotient of the permutation module on $Y/N_{G_z}(U)$. Hence by H.5.3, either $\tilde{P} = \tilde{P}_i$ is the 6-dimensional core of the permutation module for $i = 1$ or 2, or else $\tilde{P} = \tilde{P}_1 \oplus \tilde{P}_2$ with $\dim(\tilde{P}_i) = 3$ for $i = 1$ and 2. Next $\varphi : T^+ \to T$ is an isomorphism, and for each 3-dimensional indecomposable $\tilde{W}$ for a rank one parabolic Y_0^+ of Y_+ containing the fixed point of Y_0^+, $\tilde{P}^+$ splits over $\tilde{W}$ as a T^+-module. However this is not the case when $\tilde{P}$ is the core of the permutation module, and that module is indecomposable. Hence the former case is impossible, so the latter holds.

Now $Q_z \leq P$ is Y-invariant, so $Q_z = \langle z \rangle$, P_i, or P. As $F^*(G_z) = Q_z$, the first case is out. Next suppose $Q_z = P_i$. Now $P_i \cong E_{16}$, and as $T \in Syl_2(G)$ normalizes P_i, $N_G(P_i) \in \mathcal{H}^e$ by 1.1.4.6, so $C_G(P_i) \in \mathcal{H}^e$ by 1.1.3.1. Therefore as $C_T(P_i) = P_i$, we conclude $C_G(P_i) = P_i$. Now $GL(P_i) = Aut(P_i)$ with $Y/P_i = C_{GL(P_i)}(z)$, so $G_z = YC_G(P_i) = Y$ normalizes P, contrary to $O_2(G_z) = Q_z = P_i < P$. Thus

$Q_z = P \trianglelefteq G_z$, and then as $T/P \cong D_8$ is Sylow in $G_z/P \le O_6^+(2) \cong S_8$, we conclude from the list of maximal subgroups of S_8 that either $Y = G_z$ or $G_z/P \cong A_7$. From the structure of $G_t \cong P\Gamma L_3(4)/E_4$, we see that $G_{t,z}$ is of order $3 \cdot 2^9$, so that Y is transitive on $t^G \cap P$ of order 14. Thus if G_z/P is A_7, $G_{t,z}$ contains an A_6-section, contradicting $G_{z,t}$ a $\{2,3\}$-group.

Therefore $G_z = Y$. We have also seen that z is not weakly closed in P with respect to z, so that Theorem 44.4 of [**Asc94**] applies. Since $M_V/V \cong \hat{S}_6$, $G \not\cong L_5(2)$, and since $G_t \not\le M$, $G \not\cong M_{24}$. Therefore as G is simple, Theorem 44.4 in [**Asc94**] shows $G \cong He$. $\qquad\square$

12.7.3. The case $\mathbf{V} \not\le \mathbf{O_2(G_z)}$, including the identification of $\mathbf{M_{24}}$. Because of Theorem 12.7.7, we can assume in the remainder of this section that

LEMMA 12.7.8. $G_t \le M$.

LEMMA 12.7.9. *(1) M controls fusion of its involutions.*
(2) G_v is transitive on $\{V^g : v \in V^g\}$ for each $v \in V$.
(3) V is the unique conjugate of V containing t.

PROOF. By 12.7.8 and 12.7.4.2, t is not 2-central in G, so $t \notin z^G$. Thus (1) follows from 12.7.4.1. Then (1) and A.1.7.1 imply (2), and (2) and 12.7.8 imply (3) using A.1.7.2. $\qquad\square$

LEMMA 12.7.10. *(1) $m(\bar{M}_V, V) = 2$.*
(2) $r(G, V) > 2$. Hence $s(G, V) = 2$.
(3) If $\bar{L} < \bar{M}_V$ then there are two classes $\mathcal{O}_j$, $j = 1, 2$, of involutions in $\bar{M}_V$ not in $\bar{L}$. Further $\bar{i}_j \in \mathcal{O}_j$, where $\langle \bar{i}_j \rangle = Z(\bar{L}_j\bar{T})$, and $m(C_V(\bar{i}_j)) = 3$. Finally $\bar{i}_2$ acts on a conjugate of V_t, but $\bar{i}_1$ does not.
(4) If $U \le V$ with $m(V/U) = 3$, then one of the following holds:
(i) $C_M(U) = C_M(V)$.
(ii) Up to conjugation in L, U is a hyperplane of V_2 and $C_M(U) = C_M(V)R_2$.
(iii) $U = C_V(\bar{i})$ for some involution $\bar{i} \in \bar{M}_V - \bar{L}$, and $C_M(U) = \langle i \rangle C_M(V)$.
(5) If $U \le V$ with $m(V/U) = 3$ and $C_G(U) \not\le M$, then $U = C_V(\bar{i})$ for some $\bar{i} \in \mathcal{O}_1$.

PROOF. First L is transitive on the set $\mathcal{O}$ of involutions in $\bar{L}$, and by 12.7.2.4, $V_2 = C_V(\bar{i})$ for $\bar{i} \in \mathcal{O} \cap \bar{R}_2$. Assume $\bar{L} < \bar{M}_V$. Then $\bar{M}_V \cong \hat{S}_6$ by 12.7.2.1, so there are two classes $\mathcal{O}_j$, $j = 1, 2$, of involutions in $\bar{M}_V - \bar{L}$, and we can choose notation so that $\bar{i}_j \in \mathcal{O}_j$, where $\bar{i}_j$ is defined in (3). As $\bar{i}_j$ inverts $\bar{X}$, $m([V, \bar{i}_j]) = 3$, completing the proof of (1). If we represent $\bar{M}_V$ on the set Ω of 6 cosets of $\bar{H} := N_{\bar{M}_V}(V_t)$, then each involution $\bar{i} \in \bar{H} - \bar{L}$ induces a transposition on Ω. Consequently the members of the other class $\mathcal{O}_1$ have cycle type 2^3 on Ω. This completes the proof of (3), and part (4) also follows since $C_M(U)$ is a 2-group for each $U \le V$ with $m(V/U) < 4$.

Next let $U \le V$. If U is a hyperplane of V, then $1 \ne U \cap V_t$, so $C_G(U) \le M$ by 12.7.8. Thus $r(G, V) > 1$. Assume $U \le V$ with $C_G(U) \not\le M$ and $k := m(V/U) < 4$. By E.6.12, $C_M(U) > C_M(V)$. Hence U is centralized by some involution $\bar{i} \in \bar{M}_V$ by (4). Thus if $k = 2$, we can take $U = V_2$ by the previous paragraph; however $V_2 = V_1 V_t$, so $C_G(U) \le M$ by 12.7.8. We conclude $k = 3$, so $r(G, V) > 2$, and hence $s(G, V) = 2$ using (1), so (2) holds. Indeed this argument shows $U \not\le V_2$, as each hyperplane of V_2 intersects V_t nontrivially. Thus $U = C_V(\bar{i})$ and $\bar{i} \notin \bar{L}$

by (4). Finally if $\bar{i} \in \mathcal{O}_2$, we may take $\bar{i}$ to act on V_t by (3), so $1 \neq C_{V_t}(\bar{i}) \leq U$, contradicting $C_G(U) \not\leq M$ in view of 12.7.8. This completes the proof of (5), and hence of 12.7.10. $\square$

LEMMA 12.7.11. $W_0 := W_0(T, V) \leq C_T(V)$, so that $w(G, V) > 0$ and $N_G(W_0) \leq M$.

PROOF. Suppose $A := V^g \leq T$ with $\bar{A} \neq 1$. By 12.7.10.2, $s(G, V) > 1$, so by E.3.10, $\bar{A} \in \mathcal{A}_2(\bar{T}, V)$, and hence $\bar{A} = \bar{R}_2$ by 12.7.3. Now $m(A/C_A(V)) = 2$, so by 12.7.10.2, $V \leq C_G(C_A(V)) \leq N_G(A)$. Thus $V_2 = [V, A] \leq A$ by 12.7.2.4. This contradicts 12.7.9.3, as $t \in V_2$. Hence $W_0 \leq C_T(V) = O_2(LT)$, and $N_G(W_0) \leq M$ by E.3.34.2. $\square$

LEMMA 12.7.12. $C_T(L) = 1$.

PROOF. If $C_T(L) \neq 1$, also $C_Z(L) \neq 1$; so by 12.2.9.1, $C_G(Z) \leq M$. But by 12.7.10.2, $s(G, V) > 1$, so by 12.4.1 there is $g \in G$ with $V^g \leq T$ and $[V, V^g] \neq 1$, contrary to 12.7.11. $\square$

LEMMA 12.7.13. (1) Hypothesis G.2.1 is satisfied for each $H \in \mathcal{H}(P_1 T) \cap N_G(V_1)$, with P_1 in the role of "L_1".
(2) $V \leq O_2(N_G(V_1))$.

PROOF. By 12.7.2.3, P_1 is irreducible on V/V_1, so (1) holds. Then (2) follows from G.2.2.1. $\square$

Much of the rest of the section is devoted to the proof of the following result, which identifies the remaining group in the conclusion of Theorem 12.7.1.

THEOREM 12.7.14. If $V \not\leq O_2(G_z)$, then as $G_t \leq M$, G is isomorphic to M_{24}.

Until the proof of Theorem 12.7.14 is complete, assume $V \not\leq O_2(G_z)$. Recall the subgroup L_1 defined in 12.7.2.5. Set $K := \langle V^{G_z} \rangle$, $U := \langle V_1^K \rangle$, $\tilde{G}_z := G_z/\langle z \rangle$, $H := KL_1 T$, $Q_z := O_2(H)$, and $H^* := H/C_H(\tilde{U})$. As V_1 is $L_1 T$-invariant, $U \trianglelefteq H$. As $V \not\leq O_2(G_z)$, $O^2(K) \neq 1$ and $V \not\leq O_2(H)$, so $K \not\leq M$ by 12.2.6.

LEMMA 12.7.15. (1) $\Phi(U) \leq \langle z \rangle$, and $\tilde{U} \in \mathcal{R}_2(\tilde{H})$, so that $O_2(H^*) = 1$.
(2) Either
(a) $\bar{U} = \bar{R}_1$, or
(b) $\bar{M}_V \cong \hat{S}_6$ and $\bar{U}$ is either $Z(\bar{L}_1 \bar{T})$ of order 2 or $O_2(\bar{L}_1 \bar{T})$ of order 8.
(3) If U is abelian, then $\bar{U} = \bar{R}_1$.
(4) $m(V^*) = 2$ or 4 and $V^* = [V^*, L_1^*]$, so $L_1^*/O_2(L_1^*) \cong \mathbf{Z}_3$.

PROOF. Observe that Hypothesis G.2.1 is satisfied with $\langle z \rangle$, V_1, G_z, in the roles of "V_1, V, G_1"; hence (1) holds by G.2.2. As

$$C_H(\tilde{U}) \leq C_H(\tilde{V}_1) \leq N_G(V_1),$$

and $V \not\leq O_2(H)$, V does not centralize $\tilde{U}$ by 12.7.13.2. Thus $V^* \neq 1$, so $\bar{U} \neq 1$. However $U \trianglelefteq H$, so $\bar{U} \trianglelefteq \bar{L}_1 \bar{T}$, and hence (2) follows. Further if U is abelian then $\bar{U} \leq C_{\bar{M}}(V_1)$, so (3) holds. As $V/V_1 = [V/V_1, L_1]$, $V^* = [V^*, L_1]$, so as $V^* \neq 1$ and $L_1/O_2(L_1) \cong \mathbf{Z}_3$, (4) holds. $\square$

We now deal with the case leading to the remaining conclusion of Theorem 12.7.1:

LEMMA 12.7.16. *If $\bar{U} = \bar{R}_1$, then $G \cong M_{24}$.*

PROOF. Assume that $\bar{U} = \bar{R}_1$. By 12.7.2.2, $[\tilde{U}, V^*] = \tilde{V}_1$, so V^* induces a group of transvections on $\tilde{U}$ with center $\tilde{V}_1$. Also $m(V^*) = 2$ or 4 by 12.7.15.4. Thus if $\Delta_1, \ldots, \Delta_s$ are the orbits of K on $\tilde{V}_1^{G_z}$, then by G.3.1, $\tilde{U} = \tilde{U}_1 \oplus \cdots \oplus \tilde{U}_s$ and $K^* = K_1^* \times \cdots \times K_s^*$, where $\tilde{U}_i := \langle \Delta_i \rangle$ is of dimension $n \geq 3$, K_i^* is generated by the transvections in K^* with centers in Δ_i, $[K_i, \tilde{U}_j] = 0$ for $i \neq j$, and (as $O_2(K_i^*) = 1$ by 12.7.15.1) K_i^* acts faithfully as $GL(\tilde{U}_i)$ on $\tilde{U}_i$. Observe in particular that $L_1^* \leq K^*$. Now each preimage K_i contains a member of $\mathcal{C}(H)$ by 1.2.1.1, so by 1.2.1.3, $s \leq 2$; and in case of equality, $H^* \cong L_3(2)$ wr $\mathbf{Z}_2$. Therefore $s = 1$, as T^* acts on L_1^* and $L_1^*/O_2(L_1^*) \cong \mathbf{Z}_3$ by 12.7.15.4. Thus by Theorem C (A.2.3), $K^* = GL(\tilde{U}) \cong L_n(2)$, $n = 3$, 4, or 5. Then K is transitive on $\tilde{U}^\#$, so $\Phi(U) = 1$. By 12.7.15.4, $L_1^* T^*$ is a rank one parabolic of K^*.

If $n = 5$, then $C_K(V_1)^* \cong L_4(2)/E_{16}$, so as X is faithful on V_1, $m_3(N_G(V_1)) > 2$, contrary to $N_G(V_1)$ an SQTK-group.

Thus $n = 3$ or 4. As we saw $V^* \leq C_{K^*}(U/V_1) \cong E_{2^{n-1}}$ and $m(V^*) = 2$ or 4, we conclude $m(V^*) = 2$. Next

$$m(Q \cap U) = m(U) - m(\bar{U}) = n + 1 - 2 = n - 1.$$

As $V_1 \leq V \cap U \leq C_V(U) = C_V(\bar{R}_1) = V_1$, we conclude $V_1 = C_V(U) = V \cap U$. Thus

$$m((Q \cap U)V/V) = n - 1 - m(V_1) = n - 3 \leq 1.$$

Now $[Q, U] \leq Q \cap U$, so that $m([W, U]) \leq 1$ for W any noncentral chief factor for L on Q/V. However $\bar{U} = \bar{R}_1$ does not induce transvections on any nontrivial irreducible for $\hat{A}_6$; hence $[U, Q] \leq V$, and L is an $\hat{A}_6$-block. In particular, L_1 has exactly three noncentral 2-chief factors.

Suppose $n = 4$. Then as $L_1^* = O^2(P^*)$ for some rank one parabolic P^* of K^*, L_1 has one noncentral chief factor $\tilde{W}$ on the natural module $\tilde{U}$, and two such factors on $O_2(L_1^*)$. We conclude $[Q_z, L_1] \leq U$, so that $[Q_z, K] \leq U$. Thus by the Thompson $A \times B$-Lemma, $O^2(K)/U$ is faithful on $C_U(O_2(KT)/U)$, so as K is irreducible on $\tilde{U}$, $O_2(KT)$ centralizes U. Then as $H^1(K^*, \tilde{U}) = 0$ by I.1.6, $U = [U, K] \oplus \langle z \rangle$. This is impossible as $z \in [V, R_1]$ by 12.7.2.2, while $UC_T(V) = R_1$ by hypothesis, so that $z \in [V, U]$

Therefore $n = 3$, so $U \cong E_{16}$. Assume $\bar{M}_V = \bar{L}$. Then $V_1 \leq Z(T)$, so by B.2.14, $U \in \mathcal{R}_2(G_z)$, and hence $C_{G_z}(U) = C_{G_z}(\tilde{U})$. However we saw that $4 = |V^*| = |V : C_V(\tilde{U})|$ and $C_V(U) = V_1$ is of index 16 in V. Thus $\bar{M}_V \cong \hat{S}_6$ and $U \notin \mathcal{R}_2(G_z)$, so that $C_G(\tilde{U})/C_G(U) \neq 1$. Therefore from the action of $H^* = GL(\tilde{U})$, $C_G(\tilde{U})/C_G(U)$ is the full group of transvections on U with center z, and affords the K^*-module dual to $\tilde{U}$. Recall that L is a $\hat{A}_6$-block, while $C_T(L) = 1$ by 12.7.12. Then by 12.7.6.2, $V = O_2(M)$ and $M = LT$, so that $M/V \cong \hat{S}_6$. Therefore $|T| = 2^{10}$, so as $|T^*| = 8 = |C_T(\tilde{U})/C_T(U)|$ and $|U| = 16$, $C_T(U) = U$. As T normalizes U, $N_G(U) \in \mathcal{H}^e$ by 1.1.4.6, so $U = C_G(U)$. Hence as $Aut_K(U) = C_{Aut(U)}(z)$, $H = N_{G_z}(U) = K$ with $Q_z = O_2(K)$ of order 2^7. In particular, K has 2-chief series

$$1 < \langle z \rangle < U < Q_z.$$

As Q_z/U is dual to $\tilde{U}$, K is transitive on $(Q_z/U)^\#$ and $\tilde{U}^\#$. As $V \cap Q_z \not\leq U$, there are involutions in $Q_z - U$. It follows that $\Phi(\tilde{Q}_z) = 1$, so $Q_z \cong 2^{1+6} \cong D_8^3$. Now G_z normalizes K and hence normalizes $O_2(K) = Q_z$.

Let $G^+ := M_{24}$. Arguing as in the proof of Theorem 12.7.7, M is determined up to isomorphism, so as G^+ satifies the hypotheses of this Theorem, there is an isomorphism $\varphi : M^+ \to M$. As $\tilde{Q}_z^+ = J(\tilde{T}+)$, $\varphi(Q_z^+) = Q_z$. Now either $\tilde{U}$ is the socle of K on $\tilde{Q}_z$, or $\tilde{Q}_z$ is the sum of $\tilde{U}$ and its dual as a K-module. As in the final three paragraphs of the proof of 12.7.7, any $L_1^+ T^+$-submodule of $\tilde{Q}_z^+$ isomorphic to $\tilde{U}^+$ splits over $\tilde{U}^+$, so applying φ the same holds in K, and hence again $\tilde{U}$ is a semisimple K-module. Thus $Aut_{GL(\tilde{Q}_z)}(K^*) \cong Aut(L_3(2))$, so as $K \trianglelefteq G_z$ and $T^* \cong D_8$, $K^* = Aut_{G_z}(\tilde{Q}_z)$. Therefore $K = G_z$. As T^+ splits over Q_z^+, applying φ, T splits over Q_z; so K splits over Q_z, and hence $G_z = K$ is determined up to isomorphism. We have seen that z is not weakly closed in Q_z with respect to G, so that we may apply Theorem 44.4 in [**Asc94**]. This time as $G_t \leq M$ by 12.7.8, we conclude that $G \cong M_{24}$. $\qquad\square$

By 12.7.16, to complete the proof of Theorem 12.7.14, we may assume that $\tilde{U} \neq \tilde{R}_1$, and it remains to derive a contradiction. In particular U is not abelian by 12.7.15.3, and $\Phi(U) = \langle z \rangle$ by 12.7.15.1. Let $\bar{U}_1 := Z(\bar{L}_1 \bar{T})$. By 12.7.15.2, $\bar{M}_V \cong \hat{S}_6$, so $O_2(\bar{L}_1 \bar{T}) = \bar{U}_1 \times \bar{R}_1 \cong E_8$, and either $\bar{U} = \bar{U}_1$, or $\bar{U} = \bar{O}_2(\bar{L}_1 \bar{T})$ contains $\bar{U}_1$. In any case $E_8 \cong [V, U_1] \leq U \cap V$, and L_1 is irreducible on $V/V_1[V, U_1] \cong E_4$, so $V \cap U = V_1[V, U_1]$, and hence:

LEMMA 12.7.17. $V^* \cong E_4$.

LEMMA 12.7.18. $\bar{U} = O_2(\bar{L}_1 \bar{T}) \cong E_8$.

PROOF. If not, by the discussion before 12.7.17, $\bar{U} = \bar{U}_1$ is of order 2. Then V^* induces a group of transvections on $\tilde{U}$ with axis $\widetilde{U \cap Q}$, so using the dual of G.3.1 as in the proof of 12.7.16, $L_1^* \leq K^* \cong L_n(2)$ with $n = 3, 4$, or 5. This time since we are arguing in the dual of $\tilde{U}$, $[\tilde{U}, K^*]$ is the natural module for K^*. Then $\tilde{U} = [\tilde{U}, K^*] \oplus C_{\tilde{U}}(K^*)$ as K^* is genrated by $m([\tilde{U}, K^*])$ transvections. Next as $[V, U_1]$ is of rank 3 and contains z,

$$[\tilde{U}, V^*] = [U_1, V]/\langle z \rangle = [U_1, V, L_1]\langle z \rangle / \langle z \rangle = [\tilde{U}, V^*, L_1^*]$$

is of rank 2. Thus in its action on the natural module $[\tilde{U}, K^*]$, $L_1^* T^*$ is the rank one parabolic stabilizing the line $[\tilde{U}, V^*]$ and centralizing $[\tilde{U}, K^*]/[\tilde{U}, V^*]$. In particular $L_1^* T^*$ fixes no point in the natural module, so $\tilde{V}_1 \leq C_{\tilde{U}}(L_1^* T^*) = C_{\tilde{U}}(K^*)$, contradicting $\tilde{U} = \langle \tilde{V}_1^K \rangle$. $\qquad\square$

We may represent LT on $\Omega := \{1, \ldots, 6\}$ so that $P_2 T$ is the global stabilize of $\{1, 2\}$. Then by 12.7.18, $\bar{U} = \langle (1, 2), (3, 4), (5, 6) \rangle$. Pick $g \in L$ with $\bar{U}^g = \langle (1, 6), (2, 3), (4, 5) \rangle$. Then

$$\bar{L} = \langle \bar{U}, \bar{U}^g \rangle = \langle \bar{U}, \bar{x} \rangle$$

for each $1 \neq \bar{x} \in \bar{U}^g$ which is not a transposition. Let $I := \langle U, U^g \rangle$. Arguing as in the the proof of G.2.3, $L \leq I$ and

$$[O_2(I), I] =: P = (P \cap U)(P \cap U^g)$$

with $[I, U \cap U^g] \leq V$, and setting $I/(U \cap U^g) V =: I^+$,

$$P^+ = (U \cap P)^+ \oplus (U^g \cap P)^+$$

with $C_{P^+}(U) = (U \cap P)^+$. Indeed if $1 \neq \bar{x} \in \bar{U}^g$ such that $\bar{x}$ is not a transposition, then $I = \langle U, x \rangle$, so

$$C_{(P \cap U)^+}(x) \leq C_{(P \cap U)^+}(I) = (U \cap U^g)^+ = 1,$$

so $(P \cap U^g)^+ = C_{P^+}(x)$.

Recall $X = O^2(O_{2,Z}(L))$, so $X \leq L \leq I \leq N_G(P)$.

LEMMA 12.7.19. $[P^+, X] = 1$.

PROOF. Assume otherwise. Take $\bar{x} \in \bar{L} \cap \bar{U}^g$, and take $\bar{y} \in \bar{U}^g$ to be the product of 3 transpositions. Then $[C_{[P^+, \bar{X}]}(\bar{x}), \bar{y}] \neq 1$ as $\bar{y}$ inverts $\bar{X}$, so $C_{P^+}(\bar{x}) \neq C_{P^+}(\bar{y})$, whereas $C_{P^+}(\bar{x}) = (U^g \cap P)^+ = C_{P^+}(\bar{y})$ since neither $\bar{x}$ nor $\bar{y}$ is a transposition. This contradiction establishes the lemma. $\qquad \square$

LEMMA 12.7.20. L is a $\hat{A}_6$ block, $V = O_2(M)$, and $M = LT$.

PROOF. As $[P, X] \leq (U \cap U^g)V$ by 12.7.19 and I centralizes $(U \cap U^g)V/V$, it follows by Coprime Action that $[P, X] = V$. Thus $X = VY$ where Y has order 3, so $LT = VN_{LT}(Y)$ by a Frattini Argument.

If L is a $\hat{A}_6$-block then as $C_T(L) = 1$ by 12.7.12, $O_2(M) = V$ and $M = LT$ by 12.7.6.2. Thus we may assume that L is not a $\hat{A}_6$-block. Then $1 \neq Z_Y :=$ $\Omega_1(Z(N_T(Y))) \cap O_2(LT)$, and Z_Y is in the center of $VN_T(Y) = T$, so that $Z_Y \leq Z$. Let $V_Y := \langle Z_Y^L \rangle$; then $V_Y \in \mathcal{R}_2(LT)$ by B.2.14 and $V_Y \leq C_G(Y)$. As $C_T(L) = 1$, $C_{V_Y}(L) = 1$. Let $V_0 := V_Y V$, so that also $V_0 \in \mathcal{R}_2(LT)$ by B.2.12. By B.4.2.8, the unique FF*-offender in $\bar{L}\bar{T}$ on V is $\bar{R}_2$ and $m(V/C_V(\bar{R}_2)) = m(\bar{R}_2)$. Then it follows from B.4.2 and B.3.4 that $\hat{q} := \hat{q}(Aut_{LT}(V_0), V_0) \geq 2$, with equality only if $V_Y/C_{V_Y}(L)$ is the A_6-module (so that $m(V_Y) = 4$ since we saw $C_{V_Y}(L) = 1$) and either

(i) $\bar{R}_2$ is an FF*-offender on V_Y and hence L_1 centralizes Z_Y, or

(ii) There is a strong FF*-offender $\bar{A}$ in $\bar{T}$ on V_0 with $m(V/C_V(\bar{A})) = m(\bar{A}) + 1$, so that $\bar{A} = O_2(\bar{L}_2\bar{T})$ and again L_1 centralizes Z_Y.

However by 3.1.8.1, $\hat{q} \leq 2$, so indeed $\hat{q} = 2$. Therefore V_Y is the 4-dimensional A_6-module in which L_1T centralizes a point, so as $\bar{U} = O_2(\bar{L}_1\bar{T})$ by 12.7.18, $\bar{U}$ is not quadratic on V_Y, impossible as $[V_Y, U, U] \leq [V_Y \cap U, U] \leq V_Y \cap \langle z \rangle = 1$ using 12.7.15.1. $\qquad \square$

We are now ready to complete the proof of Theorem 12.7.14.

By 12.7.20, L is a $\hat{A}_6$-block, $V = O_2(M)$, and $M = LT$. By 12.7.18, $\bar{U} \cong E_8$, so $M/V \cong \hat{S}_6$. In particular M, and hence also T, are determined up to isomorphism, so T is isomorphic to a Sylow 2-group of He. Thus $J(\tilde{T}) \cong E_{64}$. But by our remark before 12.7.17, $V \cap U = V_1[V, U_1]$ is of rank 4, so as $\bar{U} \cong E_8$,

$$|U| = |\bar{U}||U \cap V| = 8 \cdot 16 = 2^7,$$

so $\tilde{U} \cong E_{64}$ and hence U is the preimage D_8^3 of $J(\tilde{T})$ in T and $L_1T/U \cong S_4$.

As T is Sylow in He, $C_T(U) \leq U$, so as U induces $C_{Aut(U)}(\tilde{U})$ by A.1.23, $U = C_H(\tilde{U})$. Thus $H^* \leq Out(U) \cong O_6^+(2)$. Recall $O_2(H^*) = 1$ by 12.7.15.1. As $V^* = [V^*, L_1]$, $[O(H^*), V^*] = 1$ by A.1.26. Then since $K = \langle V^{G_z} \rangle$, K^* centralizes $O(H^*)$. Further $H = KL_1T$ and $L_1^*T^* \cong S_4$ with $V^* = O_2(L_1^*T^*)$. Now examining $O_6^+(2)$ for subgroups satisfying these conditions, we conclude H^* is $L_3(2)$, A_6, A_7, S_5, or $\Gamma L_2(4)$. Next $\tilde{U}_T := C_{\tilde{U}}(T) = C_{\widetilde{U \cap V}}(T) \cong E_4$ and $C_{\tilde{U}}(L_1T) = \tilde{V}_1$. Therefore H^* is not A_6 or S_5, since those groups fix a point of of $\tilde{U}$, but K moves $\tilde{V}_1$. If H^* is $\Gamma L_2(4)$, then $[\tilde{U}, K]$ is the A_5-module for K^*, impossible as V^* is quadratic on $\tilde{U}$.

Next the preimage U_T is isomorphic to E_8 and contains V_1, so by 12.7.10.5, $C_H(U_T) \leq M_z = L_1 T$. Then $C_H(U_T) = C_{L_1 T}(U_T) \leq T$, and hence $C_{H^*}(\tilde{U}_T)$ is a 2-group by Coprime Action, so that H^* is not A_7. Therefore $H^* \cong L_3(2)$. Then arguing as in the proof of 12.7.7, $\tilde{U} = \tilde{U}_1 \oplus \tilde{U}_2$ is the sum of two nonisomorphic natural modules for H^*. Therefore as $\tilde{V}_1 = C_{\tilde{U}}(L_1^* T^*)$, $\tilde{V}_1 \leq \tilde{U}_i$ for some i, so $U = \langle V_1^H \rangle \leq U_i < U$. This contradiction finally completes the proof of Theorem 12.7.14.

12.7.4. The final contradiction. Because of Theorem 12.7.14, we can assume in the remainder of the section that:

LEMMA 12.7.21. $V \leq O_2(G_z)$.

LEMMA 12.7.22. (1) If $g \in G$ with $V \cap V^g \neq 1$, then $[V, V^g] = 1$.
(2) Either $W_1 := W_1(T, V)$ centralizes V, or $\bar{W}_1 = \bar{R}_2$ and $r(G, V) = 3$.
(3) $C_G(C_1(T, V)) \leq M$.
(4) If $r(G, V) > 3$, then $C_G(C_2(T, V)) \leq M$.
(5) If $C_V(V^g) \neq 1$, then $\langle V, V^g \rangle$ is a 2-group.

PROOF. Under the hypotheses of (1), we may take $z \in V^g$ by 12.7.9.3 and 12.7.4.1. Then by 12.7.9.2, we may take $g \in G_z$. Now by 12.7.21, $V^g \leq O_2(G_z) \leq T$, so by 12.7.11, $[V, V^g] = 1$. That is, (1) holds.

We next prove (2), (3), and (4). Let $A := V^g \cap M \leq T$ be a w-offender. Thus $\bar{A} \neq 1$ and $w := m(V^g/A)$. By 12.7.11, $w > 0$. If $w > 1$, then W_1 centralizes V by definition, so that (2) holds, and then $C_G(C_1(T, V)) \leq M$ by E.3.34.2, so that (3) holds. That result also shows that (4) holds if $w > 2$.

Next as $1 \neq [A, V] \leq [V^g, V]$, $V \cap V^g = 1$ by (1). If $B \leq A$ with $m(V^g/B) < r(G, V) =: r$, then $C_V(B) \leq N_G(V^g)$, so $[C_V(B), A] \leq V \cap V^g = 1$; thus $C_V(B) = C_V(A)$, so that $\bar{A} \in \mathcal{A}_{r-w}(\bar{T}, V)$ Then by 12.7.3, $r - w \leq 2$; and in case of equality, $\bar{A} = \bar{R}_2 = \overline{W_w(T, V)}$. Thus if $r - w = 2$, then $V_2 = C_V(R_2) \leq C_w(T, V)$, so that $C_G(C_w(T, V)) \leq G_t \leq M$ by 12.7.8.

By 12.7.10.2, $r \geq 3$. Assume first that $r > 3$. Then by the previous paragraph: first $w > 1$; and then either $w > 2$—or $w = 2$ and $r = 4$, so that (4) holds. Thus the lemma holds when $r > 3$ by paragraph two, so we may assume that $r = 3$. Then (4) is vacuous, and (2) and (3) hold by paragraph two when $w > 1$, so we may assume that $w = 1$. Then $r - w = 2$, so that (2) and (3) hold by paragraph three. This completes the proof of (2), (3) and (4).

Assume the hypotheses of (5). By 12.7.4.1, we may assume V^g centralizes $v := t$ or z. We observe $V \leq O_2(G_v)$: if $v = t$, this follows from 12.7.8, and if $v = z$ it follows from 12.7.21. Hence $\langle V, V^g \rangle$ is a 2-group, proving (5). $\qquad \square$

If $G_z \leq M$, then by 12.7.4.1 and 12.7.8, we may apply Theorem 12.2.13 to conclude that $G \cong M_{24}$; but then $V \not\leq O_2(G_z)$, contrary to 12.7.21. Therefore $G_z \not\leq M$, so we can choose $H \in \mathcal{H}_*(T, M)$ with $H \leq G_z$. By 3.3.2.4, we may apply the results of section B.6 to H.

LEMMA 12.7.23. (1) $n(H) = 2$.
(2) $O^2(H/O_2(H)) \cong L_2(4)$ or $L_3(4)$.
(3) $L_1 T = H \cap M$.

PROOF. Let $K_H := O^2(H)$. By 12.7.10.2, $s(G, V) > 1$, and by 12.7.11, $N_G(W_0) \leq M$. As $C_G(C_1(T, V)) \leq M$ by 12.7.22.3, E.3.19 says that $n(H) \geq$

2. Then H is not solvable by E.1.13, so H is described in E.2.2; in particular $(K_H \cap M)O_2(H)/O_2(H)$ is a Borel subgroup of $H/O_2(H)$. As $V \le O_2(G_z)$ by 12.7.21, 12.2.11.1 says $n(H) \le 2$, proving (1). By 12.2.11.2, $(K_H \cap M)/O_2(H)$ is a nontrivial 3-group. Next by 12.2.8,

$$\theta(H \cap M) \le \theta(M) = L,$$

where we recall that $\theta(Y)$ is the characteristic subgroup generated by all elements of order 3 in a group Y. Thus $\theta(H \cap M) \le O^2(C_L(z)) = L_1$, so as $|L_1|_3 = 3$, we conclude that $\theta(H \cap M) = L_1$. Then inspecting the list of groups in E.2.2 with $n(H) = 2$, $H \cap M$ a $\{2,3\}$-group, and $\theta(H \cap M)/O_2(\theta(H \cap M))$ of order 3, we conclude that (2) holds and $O^2(H \cap M) = \theta(H \cap M)$, so that (3) holds. $\square$

LEMMA 12.7.24. $r(G,V) > 3$.

PROOF. Assume otherwise. Then $r(G,V) = 3$ by 12.7.10.2, and then by 12.7.10.5, $C_G(U) \not\le M$ where $U := C_V(\bar{i}_1)$. Now T acts on U, so we may choose $H \le C_G(U)T$, so that $H = C_H(U)T$. Hence $L_1 \le O^2(H) \le C_H(U)$ by 12.7.23.3, which is impossible as $C_{LT}(U) = Q\langle i_1 \rangle$ by 12.7.10.4. $\square$

We are now in a position to obtain the final contradiction establishing Theorem 12.7.1.

By 12.7.11, $N_G(W_0) \le M$; hence as $H \not\le M$, $W_0 \not\le O_2(H)$ by E.3.16, so that there exists $A := V^g \le T$ with $A \not\le O_2(H)$. Let $K_H := O^2(H)$ and $H^+ := H/O_2(H)$. In case $K_H^+ \cong L_2(4)$, set $H_1 := H$, $K_1 := K_H$, and $T_1 := T$. Otherwise by 12.7.23.2, $K_H^+ \cong L_3(4)$. Here as $A \le Q \le O_2(L_1T)$ by 12.7.11 and $L_1T = H \cap M$ by 12.7.23.3, $A \le O_2(H \cap M)$. Therefore we have two subcases: either $A^+ \le K_H^+$; or $A^+ = \langle a^+ \rangle A_K^+$, where $A_K^+ := A^+ \cap K_H^+$, and a^+ induces a graph automorphism on K_H^+. In the former subcase, replacing A by a suitable conjugate if necessary, $A \not\le O_2(P)$ for one of the two maximal parabolics P of K_H. In this subcase, we let $H_1 := N_H(P)$, $K_1 := O^2(H_1)$, and $T_1 := T \cap H_1$, and observe that as $A \not\le O_2(P)$, $C_{T^+}(A^+) \le T_1^+$. Finally in the latter subcase, $C_{K_H^+}(a^+) \cong L_2(4)$. In this subcase, let $a^+ \ne b^+ \in a^+(T^+ \cap K^+ \cap Z(C_{T^+}(a^+)))$, H_1 the preimage in H of $C_{H^+}(b^+)$, $K_1 := O^2(H_1)$, and $T_1 := T \cap H_1$.

In each case, $K_1/O_2(K_1) \cong L_2(4)$, $T_1 \in Syl_2(H_1)$, $A \not\le O_2(H_1)$, and $C_{T^+}(A^+) \le T_1^+$. Also in each case, $K_1 \not\le M$ as $H \cap M = L_1T$. Let $Q_1 := O_2(H_1)$, $H_1^* := H_1/Q_1$, $B := A \cap Q_1$, and $D := C_2(Q_1, V)$. As $A \le O_2(H \cap M)$, $A^* \le O_2((H_1 \cap M)^*) = (T_1 \cap K_1)^* \in Syl_2(K_1^*)$. As $r(G,V) > 3$ by 12.7.24, and $K_1 \not\le M$, we conclude from 12.7.22.4 that $K_1 \not\le C_G(C_2(T,V))$. As $n(H) = 2$, $K_1 \in \mathcal{E}_2(H,T,A)$ in the sense of Definition E.1.5 by construction. So we apply E.3.17.1 with 0, 2, 2 in the roles of "i, j, k", to conclude $C_2(T,V) \le D$, so that $K_1 \not\le C_G(D)$, and $A \not\le C_G(D)$ by E.1.4. But $m(A/B) \le m_2(H_1^*) = 2$, so $D \le C_G(B) \le N_G(A)$ as $r(G,V) > 3$. Indeed as D centralizes B with $m(A/B) \le 2$, but does not centralize A, we conclude from 12.7.10 that $m(A/B) = m(A^*) = 2$, and we may take $B = V_2^g$ and $D \le R_2^g$. As $A^* \le (T_1 \cap K_1)^*$ and $m(A^*) = 2$, $A^* = (T_1 \cap K_1)^* \in Syl_2(K_1^*)$. Thus $m(D/C_D(A)) \ge 2$, as $m(W/C_W(A^*)) \ge 2$ for any nontrivial chief section W for K_1^* on D. So as $m(R_2/Q) = 2$, we conclude $R_2^g = DQ^g$ and $|D : C_D(A)| = 4$. Then by 12.7.2.4,

$$B = V_2^g = [R_2^g, A] = [D,A] \le D.$$

Let $k \in K_1 - M$; then $K_1^* = \langle A^*, A^{*k} \rangle$. Now $[B^k, A] \le [D,A] = B$, so A acts on BB^k, and by symmetry, so does A^k, so that $I := \langle A, A^k \rangle$ acts on $U := BB^k$, and

$I^* = K_1^* \cong L_2(4)$. Indeed $|D : C_D(A)| = 4$, so

$$|B : C_B(I)| = |B : C_B(A^k)| \leq |D : C_D(A^k)| = 4.$$

So as $m(B) = 4$, $C_B(I) \neq 1$. But this contradicts 12.7.22.5, since I is not a 2-group. This final contradiction completes the proof of Theorem 12.7.1.

12.8. General techniques for $L_n(2)$ on the natural module

When Hypothesis 12.2.3 holds and $\bar{L} \cong L_n(2)$ for $n = 3, 4, 5$, Theorems 12.4.2.1, 12.6.34, and 12.5.1 tell us that V is the natural module for $\bar{L}$. We will encounter a similar setup involving $L_2(2)$ after completing our treatment of the Fundamental Setup. Thus in this section we establish some general techniques for treating all four case simultaneously.

The hypotheses below reflect one difference between the treatments for $n = 2$ and $n > 2$: For $n > 2$, we have already analyzed the case where V is a TI-set in G in Theorem 12.2.13, so we simply exclude the groups appearing in conclusions (2)–(4) of 12.2.13 as part of the operating hypothesis 12.8.1 of this section; then by 12.2.13, $C_G(Z \cap V) \not\leq M$ as L is transitive on $V^{\#}$. However, the treatment of the case where $n = 2$ and V is a TI-set in G does not appear until the end of the analysis of that case, so for the moment we instead assume $Z \leq V$ and $C_G(Z) \not\leq M$ as part of our operating hypothesis when $n = 2$.

Thus in this section, we assume the following hypothesis:

HYPOTHESIS 12.8.1. *Either (1) or (2) holds:*

(1) Hypothesis 12.2.3 holds, with $L/O_2(L) \cong L_n(2)$, $n = 3, 4, 5$, and V the natural module for $L/O_2(L)$. Further G is not $L_{n+1}(2)$, A_9, or M_{24}.

(2) G is a simple QTKE-group, $T \in Syl_2(G)$, $Z := \Omega_1(Z(T))$, $M \in \mathcal{M}(T)$, $V := \langle Z^M \rangle$ is of rank 2, $L = O^2(L) \trianglelefteq M$ with $M = !\mathcal{M}(LT)$, $C_{LT}(V) = O_2(LT)$, and $LT/O_2(LT) \cong L_2(2) \cong S_3$. Furthermore assume $C_G(Z) \not\leq M$.

We adopt the following notation, which is consistent with that in Notation 12.2.5 when $n > 2$:

NOTATION 12.8.2. (1) $Z := \Omega_1(Z(T))$, $M := N_G(L)$, $M_V := N_M(V)$, and $\bar{M}_V := M_V/C_M(V)$.

(2) $n := m_2(V)$, and for $1 \leq i \leq n$, let V_i denote the i-dimensional subspace of V invariant under T, $G_i := N_G(V_i)$, and $M_i := N_M(V_i)$. Let $L_i := O^2(N_L(V_i))$, unless $n = 5$ and $i = 2$ or 3, where $L_i := N_L(V_i)^{\infty}$. Set $R_i := O_2(L_iT)$.

(3) Let z be the generator of V_1 and $\tilde{G}_1 := G_1/V_1$. Set

$$\mathcal{H}_z := \{H \in \mathcal{H}(L_1T) : H \leq G_1 \text{ and } H \not\leq M\}.$$

For $H \in \mathcal{H}_z$, set $U_H := \langle V^H \rangle$, $Q_H := O_2(H)$, and $H^* := H/Q_H$.

Note when $n = 2$ that $V \trianglelefteq M$, so that $M_i \leq M_V$, and $L_1 = 1$. When $n > 2$, V is a TI-subgroup in M and $M_i \leq M_V$ by 12.2.6.

12.8.1. General preliminary results.

LEMMA 12.8.3. *(1) $\bar{M}_V = GL(V)$, and either $\bar{M}_V = \bar{L}$, or $n = 2$ and $\bar{M}_V = \bar{L}\bar{T}$.*

(2) L is transitive on i-dimensional subspaces of V, for each i.

(3) G_i is transitive on $\{V^g : V_i \leq V^g\}$.

(4) $G_1 \not\leq M$.

PROOF. Part (1) is an immediate consequence of Hypothesis 12.8.1. Then (1) implies (2), and (2) and A.1.7.1 imply (3).

Assume case (1) of Hypothesis 12.8.1 holds. Then Hypothesis 12.2.3 holds, but conclusions (2)–(4) of Theorem 12.2.13 are excluded by that hypothesis. Thus conclusion (1) of Theorem 12.2.13 holds, so that $C_G(v) \not\leq M$, and then (4) follows from the transitivity of L on nonzero vectors of V in (2). Finally when case (2) of Hypothesis 12.8.1 holds, (4) is a consequence of the assumption in that hypothesis that $C_G(Z) \not\leq M$. $\qquad\square$

By 12.8.3.4, $G_1 \in \mathcal{H}_z$, so $\mathcal{H}_z \neq \emptyset$. Observe that $\mathcal{H}_z \subseteq \mathcal{H}^e$ by 1.1.4.6.

LEMMA 12.8.4. *Let $H \in \mathcal{H}_z$. Then*

(1) Hypothesis G.2.1 is satisfied.

(2) $\tilde{U}_H \leq \Omega_1(Z(\tilde{Q}_H))$ and $\tilde{U}_H \in \mathcal{R}_2(\tilde{H})$.

(3) $\Phi(U_H) \leq V_1$.

(4) $Q_H = C_H(\tilde{U}_H)$, so H^* is the image of H in $GL(\tilde{U}_H)$ under the representation of H on $\tilde{U}_H$ by conjugation.

PROOF. As L_1 is irreducible on $\tilde{V}$, (1) holds. Then G.2.2 implies (2) and (3). If (4) fails, then $Y := O^2(C_H(\tilde{U}_H)) \neq 1$. But by Coprime Action, $Y \leq C_G(V) \leq M_V$, so $[Y, L] \leq C_L(V) = O_2(L)$. Hence L normalizes $O^2(YO_2(L)) = Y$, so that $H \leq N_G(Y) \leq M = !\mathcal{M}(LT)$, contrary to the choice of $H \not\leq M$. $\qquad\square$

LEMMA 12.8.5. *Assume $n > 2$, so that $L_1 \neq 1$.*

(1) If $H \in \mathcal{H}_z$ with $L_1 \trianglelefteq H$, then $\tilde{U}_H$ is the direct sum of copies of the natural module $\tilde{V}$ for $L_1^* \cong L_{n-1}(2)'$.

(2) If $L_1 \trianglelefteq G_1$, then for $1 < i < n$, $G_i \leq M$ and V is the unique member of V^G containing V_i, so that $m(V \cap V^g) \leq 1$ for $g \in G - M_V$.

PROOF. Observe $\tilde{V}$ is the natural module for $L_1/O_2(L_1) \cong L_{n-1}(2)'$, so (1) holds as $U_H = \langle V^H \rangle$. Now assume $L_1 \trianglelefteq G_1$. Then for $1 < i < n$, $N_L(V_i)$ induces $GL(V_i)$ on V_i by 12.8.3.1, so that $G_i = C_G(V_i)N_L(V_i)$ and $C_G(V_i) \leq G_1 \leq N_G(L_1)$. Hence G_i acts on

$$\langle L_1^g : g \in G_i \rangle = \langle L_1^g : g \in N_L(V_i) \rangle = L.$$

So $G_i \leq N_G(L) = M$. Then $G_i = M_i \leq M_V$, so the remaining assertions of (2) follow from 12.8.3.3. $\qquad\square$

The next lemma 12.8.6 shows that the condition "U_H is abelian for all $H \in \mathcal{H}_z$" is equivalent to "$\langle V^{G_1} \rangle$ abelian". Much of our remaining work on the $\mathbf{F}_2$-Case is partitioned via the cases "U_H abelian for all $H \in \mathcal{H}_z$" versus "$\langle V^{G_1} \rangle$ nonabelian". We will discuss this distinction further after 12.8.6.

LEMMA 12.8.6. *The following are equivalent:*

(1) U_H is abelian for each $H \in \mathcal{H}_z$.

(2) $\langle V^{G_1} \rangle$ is abelian.

(3) If $g \in G$ with $V \cap V^g \neq 1$, then $[V, V^g] = 1$.

(4) Hypothesis F.8.1 is satisfied for each $H \in \mathcal{H}_z$.

(5) Hypothesis F.9.8 is satisfied for each $H \in \mathcal{H}_z$, with V in the role of "V_+".

PROOF. First (1) implies (2) trivially as $G_1 \in \mathcal{H}_z$. By 12.8.3.3 and the transitivity of L on $V^\#$, (2) implies (3). If (3) holds, then condition (a) of Hypothesis

F.8.1 is satisfied for each $H \in \mathcal{H}_z$, while the remaining conditions are easily verified; for example, 12.8.4.4 says $\ker_{C_H(\tilde{V})}(H) = Q_H$, giving (c). Thus (3) implies (4). Finally (4) implies (1) by F.8.5.2, and (4) and (5) are equivalent by Remark F.9.9. $\square$

REMARK 12.8.7. Notice that if one of the equivalent conditions in 12.8.6 holds, then from condition (5), $\langle V^H \rangle = \langle V_+^H \rangle$. Thus the subgroups denoted by "U_H" and "V_H" in section F.9 both coincide with the group denoted by U_H in this section.

When U_H is abelian for all $H \in \mathcal{H}_z$, by parts (4) and (5) of 12.8.6, we can apply lemmas from sections F.8 and F.9 to analyze the amalgam defined by LT and H. Notice in particular by F.8.5 and F.9.11 that in this case the amalgam parameter "b" of those sections is odd and at least 3. On the other hand when U_H is nonabelian, we normally specialize to the case $H = G_1$, and apply methods from the theory of large extraspecial 2-subgroups, which are developed further in the following subsection.

12.8.2. $\langle V^{G_1} \rangle$ nonabelian almost extraspecial subgroups. In this subsection, we consider the case where $\langle V^{G_1} \rangle$ is nonabelian. The analysis in the subsection continues to develop the theory of almost extraspecial 2-subgroups U (i.e., U is nonabelian and $|\Phi(U)| = 2$) begun in section G.2 of Volume I. The theory is a variant of the theory of large extraspecial 2-subgroups appearing in the original classification literature.

In the remainder of the section we take $H := G_1$ and assume that $U := U_H = \langle V^H \rangle$ is nonabelian.

As U is nonabelian, 12.8.4.3 says that:

$$\Phi(U) = V_1.$$

Set $\hat{H} := H/Z(U)$ and $\dot{H} := H/C_H(\hat{U})$.

LEMMA 12.8.8. *(1) $U = U_0 Z(U)$, with U_0 an extraspecial 2-group and $\Phi(U_0) = V_1$.*

(2) Regard V_1 as $\mathbf{F}_2$. Then the map

$$(\tilde{u}_1, \tilde{u}_2) := [u_1, u_2]$$

defines a symmetric bilinear form on $\tilde{U}$ with radical $\widetilde{Z(U)}$ preserved by H^*, which induces an H-invariant symplectic form on $\hat{U}$. If $\Phi(Z(U)) = 1$, then

$$q(\tilde{u}) := u^2$$

defines an H^*-invariant quadratic form on $\tilde{U}$ with bilinear form $(\ ,\ )$, which induces an H^*-invariant orthogonal space structure on $\hat{U}$.

(3) $V \cap Z(U) = V_1$.

(4) Assume $n \leq 3$, let $I := \langle U^L \rangle$, and $S := O_2(I)$. Then $L \leq I$ and S has the I-chief series

$$1 =: S_0 \leq S_1 \leq \cdots \leq S_{n+1} := S$$

described in G.2.3 or G.2.5, for $n = 2$ or 3, respectively.

(5) If $L_1 \trianglelefteq H$ then $n \leq 3$, and when $n = 3$ the chief series in (3) becomes

$$1 =: S_0 < S_1 < S_3 = S$$

with

$$S_1 := V = U \cap U^g \cap U^h, \ S = (U \cap U^g \cap S)(U \cap U^h \cap S)(U^g \cap U^h \cap S),$$

and S/V the sum of copies of the dual of $\tilde{V}$ as an L_1^*-module, for each $g, h \in L$ with $V = V_1 \oplus V_1^g \oplus V_1^h$.

(6) $U = \langle V_2^H \rangle$.

PROOF. Recall $\Phi(U) = V_1$; then (1) and (2) follow from standard arguments (cf. 23.10 in [**Asc86a**]). As L_1 is irreducible on $\tilde{V}$, either (3) holds or $V \leq Z(U)$, and the latter is impossible as $U = \langle V^H \rangle$. By 12.8.4, Hypothesis G.2.1 is satisfied, and we recall as in section G.2 that as U is nonabelian, the hypothesis in G.2.3 and G.2.5 that $U \not\leq C_T(V) = O_2(LT)$ is satisfied, so that (4) follows from those results.

Assume $L_1 \trianglelefteq H$. Then by 12.8.5.1, $\tilde{U}$ is the direct sum of copies of $\tilde{V}$ as a module for $L_1^* \cong L_{n-1}(2)'$. By (2) and (3), the bilinear form $(\ ,\ )$ induces an L_1-equivariant isomorphism between $U/C_U(V)$ and the dual space of $\tilde{V}$. But if $n > 3$, then $\tilde{V}$ is not isomorphic to its dual as an L_1^*-module; so we conclude $n \leq 3$. Assume $n = 3$. Then $L_1^* \cong \mathbf{Z}_3$, $\tilde{U} = [\tilde{U}, L_1^*]$, and all chief factors for L_1 on $(S \cap U)/V$ are 2-dimensional. Therefore by (4) and G.2.5,

$$V =: S_1 = S_2 = U \cap U^g \cap U^h$$

since $[I, S_2] \leq V$ by G.2.5.5. Similarly $S = S_3$, as if $S/S_3 \neq 1$, then from G.2.5.7, L_1 has a 1-dimensional chief factor on $(U \cap S)S_3/S_3$. This completes the proof of (5).

Next

$$U = \langle V^H \rangle = \langle V_2^{L_1 H} \rangle = \langle V_2^H \rangle,$$

giving (6). This completes the proof of 12.8.8. $\qquad\qquad\square$

We continue to establish analogues of results in the literature on large extraspecial subgroups. In Hypotheses G.10.1 and G.11.1 in Volume I, we axiomatized some of the properties that are satisfied by $C_{G_0}(z_0)/O_2(G_0)$ acting on $O_2(G_0)/\langle z_0 \rangle$, when $O_2(G_0)$ is a large almost extraspecial 2-subgroup of a group G_0. In 12.8.12, we verify these hypotheses in our setup, and after that we appeal to the results in sections G.10 and G.11, particularly Theorem G.11.2.

Notice for example that G.10.2 is an analogue of 3.8 in Timmesfeld [**Tim78**]. If G is of Lie type, with the involution centralizer G_1 a maximal parabolic, the subgroup I_2 below corresponds to the complementary minimal parabolic. In the theory of large extraspecial 2-subgroups, the inequality in G.10.2 typically produced a lower bound on the 2-rank of $H/C_H(\hat{U})$. But here H is an SQTK-group over which we have some control, so that G.10.2 serves as an upper bound on $m(\hat{U})$, which we then use in 12.8.12 (via an appeal to Theorem G.11.2) in order to strongly restrict the structure of $\dot{H}$ and its action on $\hat{U}$.

Let P be the minimal parabolic of LT acting nontrivially on V_2; notice under part (2) of Hypothesis 12.8.1 that $P = LT$. Set $I_2 := \langle U^P \rangle$, $W := C_U(V_2)$, and let $g \in P - H$. Set $E := W \cap W^g$, $X := W^g$, and $Z_U := Z(U)$. Observe $Z_U \leq W$ as $V_2 \leq U$.

LEMMA 12.8.9. *(1) $\langle U^H \rangle = O^2(P)U = I_2 = \langle U, U^g \rangle$, $C_{I_2}(V_2) = O_2(I_2)$, and $O^2(P)$ and I_2 are normal in G_2.*

(2) $O_2(I_2) = WX$, $[E, I_2] = V_2$, and $O_2(I_2)/E = W/E \oplus X/E$ *is the direct sum of natural modules for* $I_2/O_2(I_2) \cong L_2(2) \cong S_3$.

(3) $C_{O_2(I_2)/E}(u) = [O_2(I_2)/E \,,\, u] = [X/E \,,\, u] = W/E$ *for* $u \in U - W$.

(4) For $y \in X - W$, $C_{\tilde{U}}(y) \le \tilde{W}$.

(5) $C_X(\hat{U}) = C_X(\tilde{U}) = E$.

(6) For $u \in U - W$, $C_X(\hat{u}) \le Z_U^g E$.

(7) $V_1^g \cap Z_U = 1$.

PROOF. Observe (7) holds as $g \in N_L(V_2) - G_1$, and $V \cap Z_U = V_1$ by 12.8.8.3. As V is the natural module for $\bar{L}\bar{T}$ and $O_2(LT) = C_{LT}(V)$, $P = O^2(P)T$ with $C_P(V_2) = O_2(P)$ and $P/O_2(P) = GL(V_2) \cong L_2(2)$. As U is nonabelian, $[V_2, U] \ne 1$ by 12.8.8.6, so $O^2(P) = [O^2(P), U]$ and $P = \langle U, U^g \rangle O_2(P)$. Thus $I_2 = O^2(P)U$. As $Aut_P(V_2) = GL(V_2)$, $G_2 = C_G(V_2)P$, so as $C_G(V_2) \le G_1 \le N_G(U)$, we conclude $I_2 = \langle U, U^g \rangle = \langle U^{G_2} \rangle \trianglelefteq G_2$, so $O^2(P) = O^2(I_2) \trianglelefteq G_2$, completing the proof of (1).

By 12.8.8.6, Hypothesis G.2.1 is satisfied with $O^2(P)$, V_2, 1 in the roles of "L, V, L_1"; further $U = \langle V_2^H \rangle$ by 12.8.8.6, so (2) and (3) follow from G.2.3.

Pick $u \in U - W$; by (3), $[X, u] \le W$, so we can define $\varphi : X \to W/E$ by $\varphi(x) := [x, u]E$. Set $D := \varphi^{-1}(Z_U E/E)$. By (3), $C_X(\hat{U}) \le D$, and

$$m(X/D) = m(W/Z_U E).$$

As $O_2(I_2)/E$ is the sum of natural modules for $I_2/O_2(I_2) \cong S_3$ by (2),

$$DZ_U = \langle Z_U^{I_2} \rangle E = Z_U Z_U^g E,$$

so $D = Z_U^g E$. Thus if $y \notin Z_U^g E$, then $[y, u] \notin Z_U E$, and in particular $[y, u] \notin Z_U$, so (6) holds. Similarly for $y \in X - W$ and $u \in U - W$, $[y, u] \notin E$ by (2) and in particular $[y, u] \notin V_1$, so $[y, \tilde{u}] \ne 1$. Thus (4) holds.

Of course $E \le C_X(\tilde{U}) \le C_X(\hat{U})$. Let $R := C_T(\hat{U})$ and $\tilde{V}_0 := C_{\tilde{U}}(R)$. By a Frattini Argument, $H = C_H(\hat{U})N_H(R)$; so as $\tilde{V}_2 \le \tilde{V}_0$, as $\tilde{V}_0$ is normalized by $N_H(R)$, and as $\hat{U} = \langle \hat{V}_2^H \rangle$, we conclude that $\tilde{U} = \tilde{V}_0 \tilde{Z}_U$. In particular as $\tilde{Z}_U \le \tilde{W} < \tilde{U}$, R centralizes some $\tilde{u} \in \tilde{U} - \tilde{W}$, so by (4), $X \cap R \le X \cap W = E$, completing the proof of (5). $\square$

LEMMA 12.8.10. *(1)* $Z_U \cap U^g = (Z_U \cap Z_U^g)V_1$.

(2) $Z_U \cap Z_U^g = Z(I_2)$.

(3) If $Z_U \cap U^g > V_1$, *then* $Z \cap Z(I_2) \ne 1$.

(4) $[W, Z_U^g] \le Z_U V_2$, *so* $m([\hat{U}, x]) \le 2$ *for* $x \in Z_U^g$.

(5) If $Z_U^g \le U$, *then* $Z_U = Z(I_2) \times V_1$ *and* $[L, Z(I_2)] = 1$.

(6) $C_{Z_U^g}(\hat{U}) = C_{Z_U^g}(\tilde{U}) = Z_U^g \cap Q_H = Z_U^g \cap U = (Z_U^g \cap Z_U)V_1^g = Z(I_2)V_1^g \le U$.

PROOF. By 12.8.9.7, $V_1^g \cap Z_U = 1$, and by 12.8.8.1, $Z(W) = V_2 Z_U$. Thus by symmetry between U and U^g, $V_1 \cap Z_U^g = 1$ and $Z(X) = V_2 Z_U^g = V_1 Z_U^g$.

By 12.8.4.2, $[Z_U \cap U^g, X] \le V_1^g \cap Z_U = 1$, so $Z_U \cap U^g \le Z(X)$. Therefore $Z_U \cap U^g \le Z_U^g V_1$ by the previous paragraph, so as $V_1 \le Z_U \cap U^g$, (1) holds by the Dedekind Modular Law.

By 12.8.9, $I_2 = \langle U, U^g \rangle$, so $Z_U \cap Z_U^g \le Z(I_2)$. To prove the reverse inclusion, observe by 12.8.9.2 that $Z(I_2) \le W \cap X$, so $Z(I_2) = Z(I_2) \cap U \le Z(U)$, and similarly $Z(I_2) \le Z(U^g)$. Thus (2) holds. As T acts on V_2, T acts on I_2 by 12.8.9.1, and hence on $Z(I_2)$. Further if $Z_U \cap U^g > V_1$ then $Z(I_2) \ne 1$ by (1) and (2), so $C_{Z(I_2)}(T) \ne 1$ and hence (3) holds.

Next $[Z_U^g, W] \leq Z_U^g \cap U \leq Z_U V_2$ by (1) and symmetry between U and U^g. Thus any $x \in Z_U^g$ either centralizes the hyperplane $\hat{W}$ of $\hat{U}$, or induces a transvection on $\hat{W}$ with center $\hat{V}_2$, so (4) follows.

To prove (5), assume $Z_U^g \leq U$. Then by (1) and symmetry between U and U^g, $Z_U^g = (Z_U \cap Z_U^g)V_1^g$, so $Z_U^g = Z(I_2) \times V_1^g$ by (2). Then as U is conjugate to U^g in I_2, the first assertion of (5) holds.

Next let $P_1, \ldots, P_{n-1}$ denote the minimal parabolics of L with the usual ordering so that $N_L(V_i) = \langle P_j : j \neq i \rangle$. Define $H_i := \langle O^2(P_j) : j \leq i \rangle$. We argue by induction on j that each H_j centralizes $Z(I_2)$; and then in particular $H_{n-1} = L$ centralizes $Z(I_2)$, which will complete the proof of (5). First $H_1 = O^2(P) = O^2(I_2)$ from 12.8.9.1, and hence H_1 centralizes $Z(I_2)$. Now suppose that $[Z(I_2), H_j] = 1$ for some $1 \leq j < n-1$. Then $H_j T$ is a maximal parabolic subgroup of $H_{j+1} T$, and so there is $k \in P_{j+1} - H_j T$ such that $H_{j+1} = \langle H_j, H_j^k \rangle$ centralizes $F := Z(I_2) \cap Z(I_2)^k$. Now $k \in P_{j+1} \leq N_L(V_1) \leq H \leq N_G(U)$, so that $Z(I_2)$ and $Z(I_2)^k$ are hyperplanes of Z_U using the result of the previous paragraph. Hence FV_1 is of codimension at most 1 in Z_U and is centralized by H_{j+1}, so $H_{j+1} = O^2(H_{j+1})$ centralizes $Z_U \geq Z(I_2)$ by Coprime Action. This completes our inductive proof of the remaining assertion of (5).

Finally by 12.8.9.5, $C_{Z_U^g}(\hat{U}) = C_{Z_U^g}(\tilde{U}) \leq Z_U^g \cap U$, and the reverse inclusion is immediate. Further $C_{Z_U^g}(\tilde{U}) = Z_U^g \cap Q_H$ by 12.8.4.4, and the remaining equalities in (6) follow from (1) and (2). $\qquad\square$

LEMMA 12.8.11. *(1)* $[W, X] \leq E$.

(2) $\Phi(E) = 1$, so $\hat{E}$ is totally isotropic in the symplectic space $\hat{U}$.

(3) X induces the full group of transvections on $\hat{E}$ with center $\hat{V}_2$.

(4) $C_{\hat{E}}(X) = \hat{V}_2$.

(5) $m(\hat{E}) + m(\dot{X}/\dot{Z}_U^g) = m(\hat{U}) - 1$.

(6) If $C_{\hat{U}}(X) > \hat{V}_2$, then there exists $1 \neq \dot{x} \in \dot{X}$ such that $m([\hat{U}, \dot{x}]) \leq 2$ and $\hat{V}_2 \leq [\hat{U}, \dot{x}]$.

PROOF. By 12.8.9.2, (1) holds. As $E = W \cap X$, $[E, X] \leq V_1^g$ by 12.8.4.2. As $\Phi(E) \leq \Phi(U) \cap \Phi(U^g) = V_1 \cap V_1^g = 1$, (2) holds.

By 12.8.8.1, $U = U_0 Z_U$ with U_0 extraspecial. Let $E_0 := EZ_U^g \cap U_0^g$ and $V_0 := V_1 Z_U^g \cap U_0^g$. Then $EZ_U^g = E_0 Z_U^g$ and $V_2 Z_U^g = V_1 Z_U^g = V_0 Z_U^g$. As $V_1 Z_U^g = V_0 Z_U^g$, $X = W^g = C_{U^g}(V_1) = C_{U^g}(V_0)$. As E is abelian and centralizes Z_U^g, E_0 is also abelian. Therefore as U_0 is extraspecial, we conclude from these two remarks that:

(!) X induces the full group of transvections on E_0 which have center V_1^g, and centralize V_0.

Let $\hat{e} \in \hat{E} - \hat{V}_2$. As $EZ_U^g = E_0 Z_U^g$, $eZ_U^g = e_0 Z_U^g$ for some $e_0 \in E_0$. Now by 12.8.10.1, $Z_U \cap E \leq Z_U \cap U^g = (Z_U \cap Z_U^g)V_1 \leq Z_U \cap E$, so that all inequalities are equalities. Hence $E \cap V_2 Z_U = V_2(Z_U \cap E) = V_2(Z_U \cap Z_U^g)$, and so by symmetry between U and U^g, $E \cap V_2 Z_U = E \cap V_2 Z_U^g$. Thus as $\hat{e} \notin \hat{V}_2$, $e \notin V_2 Z_U^g$, so as we saw that $V_2 Z_U^g = V_0 Z_U^g$, $e_0 \notin V_0 Z_U^g$. Thus $[e, X] = [e_0, X] = V_1$ by (!). Hence (3) holds, and of course (3) implies (4).

Next

$$m(\hat{U}) = m(\hat{E}) + m(\hat{W}/\hat{E}) + 1 = m(\hat{E}) + m(X/EZ_U^g) + 1 = m(\hat{E}) + m(\dot{X}/\dot{Z}_U^g) + 1,$$

as $E = C_X(\hat{U})$ by 12.8.9.5. That is, (5) holds.

Let $\hat{F} := C_{\hat{U}}(X)$ and suppose $\hat{F} > \hat{V}_2$. Then by (4), $\hat{F} \not\le \hat{E}$, while by 12.8.9.5, $C_U(U^g/Z_U^g) = E$, so $F \not\le C_U(U^g/Z_U^g)$. Now by 12.8.9.2, $F \le O_2(I_2) \le N_G(X)$, so $[X, F] \le Z_U \cap W^g \le Z_U^g V_2$ by 12.8.10.1. Hence conjugating in I_2, $\dot{X}_0 := C_{\dot{X}}(\hat{W}/\hat{V}_2) \ne 1$. If $1 \ne \dot{x} \in \dot{X}_0$ centralizes $\hat{W}$, then $\dot{x}$ is a transvection on $\hat{U}$ with axis $\hat{W}$ and center $\hat{V}_2$, so (6) holds. If $\dot{x}$ does not centralize $\hat{W}$, then $\hat{V}_2 = [\hat{W}, x] \le [\hat{U}, x]$ so as $\hat{W}$ is a hyerplane of $\hat{U}$, $m([\hat{U}, x]) = 2$ and again (6) holds. Thus (6) is established. $\square$

We are in a position to appeal to results in Volume I on centralizers with a large almost extraspecial subgroup:

LEMMA 12.8.12. *(1) Hypothesis G.10.1 is satisfied with $\dot{H}$, $\hat{U}$, $\hat{V}_2$, $\hat{E}$, $\dot{X}$, $\dot{Z}_U^g$ in the roles of "G, V, V_1, W, X, X_0".*

(2) Let $H_2 := H \cap G_2$. Then $\dot{X}$ and $\dot{Z}_U^g$ are normal in $\dot{H}_2$, so in particular $\dot{X} \trianglelefteq \dot{T}$.

(3) Hypothesis G.11.1 is satisfied.

(4) $\dot{H}$ and its action on $\hat{U}$ satisfy one of the conclusions of Theorem G.11.2.

PROOF. By 12.8.8.2, $\hat{U}$ is a symplectic space and $\dot{H} \le Sp(\hat{U})$, so Hypothesis G.10.1.1 holds. By 12.8.11.2, $\hat{E}$ is totally isotropic. By 12.8.8.6, $\hat{U} = \langle \hat{V}_2^H \rangle$. As T acts on V_2, $\dot{T}$ fixes the point $\hat{V}_2$ of $\hat{U}$, so part (a) of Hypothesis G.10.1.2 holds. By 12.8.9.5, E is the kernel of the action of X on $\hat{U}$. Observe also that $\hat{W} = \hat{V}_2^{\perp}$; thus if $\dot{x} \in \dot{X} - \dot{Z}_U^g$, then $x \notin Z_U^g E$, so 12.8.9.6 shows that $C_{\hat{U}}(x) \le \hat{W} = \hat{V}_2^{\perp}$, establishing hypothesis (d). Hypothesis (b) follows from 12.8.11.5, hypothesis (c) from 12.8.11.1, and hypothesis (e) from 12.8.11.3. This completes the proof of (1).

Next as $[V_2, U] = V_1$, $H_2 = C_G(V_2)U$; so to prove (2), it suffices to show that X and Z_U^g are normal in $C_G(V_2)$. But this follows as $C_G(V_2)$ acts on U^g and V_1.

Observe that part (4) of Hypothesis G.11.1 follows from (2), and hypothesis (3) follows from 12.8.11.6. Thus (3) holds. Finally $\dot{H}$ is a quotient of the SQTK-group H, so (3) and Theorem G.11.2 imply (4). $\square$

Using Hypothesis 12.8.1, we can refine some of the results from sections G.10 and G.11:

LEMMA 12.8.13. *(1) $V \le E$.*

(2) Z_U^g centralizes V.

(3) If $n = 2$, then $Z_U \cap Z_U^g = Z(I_2) = 1$, so $\dot{Z}_U^g \cong \tilde{Z}_U$ and $[Z_U, Z_U^g] = 1$.

(4) If $Z_U > V_1$ then $\dot{Z}_U^g \ne 1$.

(5) If $\hat{U}$ is the 6-dimensional orthogonal module for $F^*(\dot{H}) \cong A_8$, then $O^{3'}(H) =: K \in \mathcal{C}(H)$ with $K/O_2(K) \cong A_8$, $Z_D := Z \cap Z(I_2) \ne 1$, $V_D := \langle Z_D^K \rangle \le Z_U$, $V_D \in \mathcal{R}_2(KT)$, $1 \ne [Z_D, K]$, and $K = [K, Z_U^g] \in \mathcal{L}_f(G, T)$.

(6) Conclusion (4) of G.11.2 does not hold; that is, $\hat{U}$ is not the natural module for $F^*(\dot{H}) \cong A_7$.

(7) Conclusion (12) of G.11.2 does not hold.

(8) $m_3(C_H(\tilde{V}_2)) \le 1$.

PROOF. As $V \le U$ and $g \in N_G(V)$, $V \le U^g$, so (1) and (2) hold.

If $n = 2$, then by Hypothesis 12.8.1, $\mathbf{Z}_2 \cong Z \le V$, so $Z = V_1 \not\le Z(I_2)$, and hence $Z(I_2) = 1$. Therefore $[Z_U, Z_U^g] \le Z_U \cap Z_U^g = Z(I_2) = 1$. It follows from

12.8.10.6 that $C_{Z_U^g}(\hat{U}) = V_1^g$, so that $\dot{Z}_U^g \cong Z_U/V_1 = \tilde{Z}_U$, completing the proof of (3).

Suppose (4) fails; then $\tilde{Z}_U \neq 1$ but Z_U^g centralizes $\hat{U}$. First $n > 2$ as there $\dot{Z}_U^g \cong \tilde{Z}_U$ by (3). Next by 12.8.10.6, $Z_U^g = C_{Z_U^g}(\hat{U}) \leq U$; so by 12.8.10.5, $Z_U = V_1 \times Z(I_2)$ with $[L, Z(I_2)] = 1$. Then $N_G(Z(I_2)) \leq M = !\mathcal{M}(LT)$. Let $J := \langle L_1^H \rangle$ and suppose $L_1 < J$. Now $L_1 \leq C_L(V_1) = C_L(V_1 Z(I_2)) = C_L(Z_U)$, so $J \leq C_H(Z_U) \leq N_H(Z(I_2)) \leq M_1$. If $n > 3$ then by 12.2.8, $J \leq O^{3'}(H \cap M) = L_1$ contrary to assumption. Hence $n = 3$, so $L_1/O_2(L_1) \cong \mathbf{Z}_3$. Then since M is an SQTK-group and $J = \langle L_1^H \rangle$, $J/O_2(J) \cong E_9$ and $J = L_1 J_C$, where $J_C := O^2(C_J(\bar{L}))$. As $L_1 = [L_1, T \cap L]$, L_1 and J_C are the only T-invariant subgroups Y_1 of J with $|Y_1 : O_2(Y_1)| = 3$. Thus H is not transitive on the four subgroups Y of J with $|Y : O_2(Y)| = 3$, and we conclude $|H : N_H(L_1)| = 3$ and $J_C \trianglelefteq H$. But $J_C \trianglelefteq LT$, so $H \leq N_G(J_C) \leq M = !\mathcal{M}(LT)$, a contradiction. Therefore $L_1 = J \trianglelefteq H$. Thus by 12.8.5.1, $C_{\tilde{U}}(L_1) = 1$. However by hypothesis $\tilde{Z}_U \neq 1$, and we had seen that L_1 centralizes Z_U. This contradiction establishes (4).

By 12.8.9.1, $I_2 \trianglelefteq G_2$, and a Sylow 3-subgroup of I_2 is faithful on V_2. Thus if (8) fails, then $C_H(V_2)$ contains $Y \cong E_9$ and $m_3(G_2) \geq m_3(I_2 Y) > 2$, contradicting G_2 an SQTK-group. So (8) is established.

Define $H_2 := O^{3'}(H \cap G_2)$ and observe that $H_2 = O^{3'}(C_H(\tilde{V}_2))$. We see that if $m_3(C_{\dot{H}}(\hat{V}_2)) > 1$, then by (8), $O^{3'}(C_H(\hat{V}_2)) < H_2$, so H_2 does not centralize Z_U by Coprime Action. Hence $Z_U > V_1$, so $\dot{Z}_U^g \neq 1$ by (4) under this assumption.

Asume the hypotheses of (5). Then by 1.2.1.1, there is $K \in \mathcal{C}(H)$ with $\dot{K} = F^*(\dot{H})$, so by 1.2.1.4, $K/O_2(K) \cong A_8$. Then $K = O^{3'}(H)$ by A.3.18. Next $O^{3'}(C_{\dot{H}}(\hat{V}_2)) \cong E_9/E_{16}$, so by the previous paragraph, $[Z_U, K] \neq 1$ and $\dot{Z}_U^g \neq 1$. As $\dot{Z}_U^g \neq 1$, $K = [K, Z_U^g]$; so as $[Z_U, K] \neq 1$, $Z_U^g \not\leq C_H(Z_U)$. Then $1 \neq [Z_U, Z_U^g] \leq Z_U \cap Z_U^g = Z(I_2)$ by 12.8.10.2. Thus $Z_D := Z \cap Z(I_2) \neq 1$ by 12.8.10.3. Therefore $n > 2$ by (3), so $L_1 \neq 1$ and $L_1 \leq O^{3'}(H) = K$. Thus if $[Z_D, K] = 1$, then $LT = \langle I_2, L_1 T \rangle$ centralizes Z_D, so $K \leq C_G(Z_D) \leq M = !\mathcal{M}(LT)$. This is impossible as $L_1 \trianglelefteq M_1$, but L_1 is not normal in K as $K/O_2(K) \cong A_8$. Thus $[Z_D, K] \neq 1$. As $V_D \in \mathcal{R}_2(KT)$ by B.2.13, $K \in \mathcal{L}_f(G, T)$, so (5) holds.

Assume that (7) fails; thus $F^*(\dot{H}) = \dot{K} \times \dot{K}^x$ for $x \in H - N_H(K)$, $\hat{U} = [\hat{U}, \dot{K}] \oplus [\hat{U}, \dot{K}^x]$ with $[\hat{U}, \dot{K}]$ the A_5-module for $\dot{K} \cong L_2(4)$, and $\dot{X} = \langle \dot{x} \rangle (\dot{X} \cap \dot{K} \dot{K}^x) \cong E_8$. Then $O^2(C_{\dot{H}}(\hat{V}_2))$ is a Borel subgroup of $E(\dot{H})$, and hence of 3-rank 2, so $\dot{Z}_U^g \neq 1$ by an earlier observation. On the other hand, in this case $m(\dot{X}) = 3$ and $m(\hat{U}) = 8$ so that $m(\hat{E}) \leq 4$ as $\hat{E}$ is totally isotropic by 12.8.11.2. Therefore by 12.8.11.5, $m(\hat{E}) = 4$ and $m(\dot{W}^g/\dot{Z}_U^g) = 3$, contradicting $\dot{Z}_U^g \neq 1$. So (7) is established.

Finally assume that (6) fails; that is, $F^*(\dot{H}) \cong A_7$ and $\hat{U}$ the 6-dimensional permutation module. Then $\hat{U}$ is described in section B.3, and we adopt the notation of that section. By 12.8.11.5:

$$m(\hat{E}) = m(\hat{U}) - m(\dot{X}/\dot{Z}_U^g) - 1 \geq 5 - m_2(\dot{H}) \geq 2. \qquad (*)$$

Thus $\hat{E} > \hat{V}_2$ as $m(\hat{V}_2) = 1$, so we conclude from 12.8.11.3 that $\hat{V}_2 = [\hat{E}, X] \leq [\hat{V}_2^\perp, T] \leq \hat{V}_2^\perp$; it follows that a generator $\hat{v}$ for $\hat{V}_2$ is not of weight 2 or 6, so that $\hat{v}$ is of weight 4. Hence, in the notation of section B.3, we may choose $\hat{v} = e_{1,2,3,4}$, so

$$\hat{V}_2^\perp = \{e_J : |J| \text{ and } |J \cap \{1,2,3,4\}| \text{ are even}\}.$$

In particular $\hat{U}_+ := \{0, e_{5,6}, e_{5,7}, e_{6,7}\} \leq \hat{V}_2^\perp$ is T-invariant but does not contain $\hat{V}_2$, so by 12.8.11.4, $\hat{U}_+ \cap \hat{E} = 1$. Then by 12.8.11.1, $[\hat{U}_+, X] \leq \hat{U}_+ \cap \hat{E} = 1$.

Next $O^2(C_{\hat{H}}(\hat{V}_2)) \cong A_4 \times \mathbf{Z}_3$, so by an earlier observation, $\dot{Z}_U^g \neq 1$. Thus (*) implies $m(\hat{E}) \geq 3$, with equality only if $m(\dot{X}) = 3$. By 12.8.11.2, $\hat{E}$ is totally isotropic so that $m(\hat{E}) \leq 3 = m(\dot{X})$. However we showed in the previous paragraph that $\dot{X} \leq C_{\dot{T}}(\hat{U}_+)$, which is a contradiction as $C_{\dot{T}}(\hat{U}_+) \cong D_8$.

This completes the proof of 12.8.13. $\qquad\qquad\square$

12.9. The final treatment of $\mathbf{L_n(2)}$, $n = 4, 5$, on the natural module

In this section we prove:

THEOREM 12.9.1. *Assume Hypothesis 12.2.1 with $L/O_2(L) \cong L_n(2)$, $n = 4$ or 5. Then $n = 4$, and G is isomorphic to $L_5(2)$ or M_{24}.*

We recall that the QTKE-groups $G \cong L_5(2)$ and M_{24} appear as conclusions in Theorem 12.2.13. In proving Theorem 12.9.1 we verify that Hypothesis 12.8.1 holds and apply Theorem 12.2.13 to establish 12.8.3.4. Two groups appear as shadows: The sporadic group Co_3 has a 2-local $L \in \mathcal{L}_f^*(G, T)$ with $L \cong L_4(2)/E_{2^4}$; but Co_3 is neither quasithin nor of even characteristic, in view of the involution centralizer $Sp_6(2)/\mathbf{Z}_2$, and is essentially eliminated in 12.9.3 below. Similarly the sporadic Thompson group F_3 contains $L \in \mathcal{L}_f^*(G, T)$ with $L \cong L_5(2)/E_{2^5}$, but F_3 is not quasithin in view of the involution centralizer $A_9/2^{1+8}$, and is eliminated in 12.9.4.

Furthermore in many groups of large rank there is $L \in \mathcal{L}_f(G, T)$ which is not maximal, but satisfies the rest of the hypothesis of Theorem 12.9.1: namely in many groups of Lie type over $\mathbf{F}_2$, as well as in the sporadic groups F_3, the Baby Monster, and the Monster. In addition the Conway group Co_2 has a 2-local L not containing a Sylow group with structure $L_4(2)/(E_{2^4} \times 2^{1+6})$. These groups are of course not quasithin, and the configurations are also eliminated in 12.9.3 and 12.9.4.

The proof of Theorem 12.9.1 involves a series of reductions. Assume G, L afford a counterexample to Theorem 12.9.1, and choose the counterexample so that $n = 5$ if that choice is possible. Neither A_9 nor the groups appearing in conclusions (1) and (2) of Theorem 12.2.2 contain $L \in \mathcal{L}_f^*(G, T)$ with $L/O_2(L) \cong L_n(2)$ for $n = 4$ or 5. Thus Hypothesis 12.2.3 is satisfied, we can pick V as in Theorem 12.2.2.3, and $N_G(L) =: M \in \mathcal{M}(T)$. Then Theorems 12.5.1 and 12.6.34 eliminate cases (c) and (d) of Theorem 12.2.2.3, so we conclude that V is the natural module for $L/O_2(L)$. As G, L affords a counterexample to Theorem 12.9.1, G is neither $L_5(2)$ nor M_{24}. Also G is not $L_6(2)$ as G is quasithin, and G is not A_9 as we observed earlier. Thus Hypothesis 12.8.1 is satisfied, so we can appeal to the results of section 12.8, and adopt the conventions of Notation 12.8.2 of that section. Recall that $G_1 \not\leq M$ by 12.8.3.4, so that $G_1 \in \mathcal{H}_z$.

LEMMA 12.9.2. *If $n = 4$, then there is no $K \in \mathcal{L}_f^*(G, T)$ with $K/O_2(K) \cong L_5(2)$, M_{24}, or J_4.*

PROOF. Assume such a K exists. By Remark 12.2.4, Hypothesis 12.2.1 is satisfied with K in the role of "L" and conclusion (3) of Theorem 12.2.2 holds, so $K/O_2(K) \cong L_5(2)$. This is a contradiction as $n = 4$ by hypothesis, contrary to our choice of $n = 5$ if such a choice is possible. $\qquad\square$

LEMMA 12.9.3. Let $1 \leq i < 5$ when $n = 5$,, and $i = 1$ or 3 when $n = 4$. Then $L_i \leq K_i \in \mathcal{C}(N_G(V_i))$ with $K_i \trianglelefteq N_G(V_i)$, and one of the following holds:

(1) $L_i = K_i$.

(2) $i = 1$, and $K_1/O_2(K_1) \cong L_5(2)$, M_{24}, or J_4.

(3) $i = 1$, $n = 4$, and $K_i/O_2(K_i) \cong L_4(2)$, A_7, $\hat{A}_7$, M_{23}, HS, He, Ru, or $SL_2(7)/E_{49}$.

PROOF. The proof is of course similar to that of 12.5.3: First $L_i \in \mathcal{L}(G, T)$, so the existence and normality of K_i follow from 1.2.4. If $K_i > L_i$, the possibilities for $K_i/O_2(K_i)$ are given by the sublist of A.3.12 for $L_i/O_2(L_i) \cong L_k(2)$ for a suitable choice of $k := 3$ or 4. When $k = 4$ we obtain the groups in conclusion (2). When $k = 3$ we obtain the groups in conclusions (2) and (3), along with $L_2(49)$ and $(S)L_3^\epsilon(7)$—but these last cases are out, since there T acts nontrivially on the Dynkin diagram of $L_1/O_2(L_1)$, which is not the case by 12.8.3.1.

Thus when $i = 1$ the lemma is established, so we may assume $i > 1$ and $L_i < K_i$, and it remains to derive a contradiction. Set $K_1^* T^* := K_1 T/O_2(K_1 T)$.

Assume first that $i = 3$ or 4. Then $L_i/C_{L_i}(V_i) = GL(V_i)$, so $K_i = L_i C_{K_i}(V_i)$. Hence $K_i = L_i$ if $K_i/O_2(K_i)$ is quasisimple, contrary to our assumption, so that $K_i/O_2(K_i)$ is not quasisimple. Then from the first paragraph, $K_i/O_2(K_i) \cong SL_2(7)/E_{49}$, $K_i = XL_i$, where $X := \Xi_7(K_i)$, and $i = 3$ since $L_4/O_2(L_4) \cong L_4(2)$ is not involved in K_i. We argue much as in the proof of 12.5.3: Set $K_{1,3} := O^2(C_{K_3}(V_1))$. Then $K_{1,3} T/O_2(K_{1,3} T) \cong SL_2(3)/E_{49}$, since $X \leq C_{K_3}(V_3) \leq C_G(V_1)$. Further $K_{1,3} = YX$ where $Y := O^{3'}(N_{L_1 \cap L_3}(V_1)) \leq K_1$, so that $K_{1,3} = \langle Y^X \rangle \leq K_1$ since $K_1 \trianglelefteq N_G(V_1)$. Now $K_{1,3}^* T^*$ is a subgroup of $K_1^* T^*$ containing T^*. But from the structure of the overgroups of T^* in the groups listed in (2) and (3), no subgroup of these groups containing a Sylow 2-subgroup has a $GL_2(3)/E_{49}$-section, except when K_1^* is also $SL_2(7)/E_{49}$. In this last case, $X = O^{7'}(K_{1,3}) = \Xi_7(K_1)$ is normal in K_3 and K_1, so that $L = \langle L_1, L_3 \rangle \leq N_G(X)$. Hence $X \leq N_G(X) \leq M = !\mathcal{M}(LT)$, so that $X = [X, L_3] \leq L$, contrary to $m_7(L) = 1$. This contradiction completes the proof that $K_i = L_i$ if $i = 3$ or 4.

Finally take $i = 2$. Thus $n = 5$ by our choice of i in the hypothesis, so $L_2/O_2(L_2) \cong L_3(2)$, and $L_2 \leq L_1$ with $L_1/O_2(L_1) \cong L_4(2)$. In particular $m_3(L_1) = 2$, and $L_1 = O^{3'}(N_G(L_1))$ by A.3.18. We conclude $G_2 \not\leq N_G(L_1)$, since G_2 contains a subgroup X of order 3 faithful on V_2, whereas if $G_2 \leq N_G(L_1)$, then $X \leq O^{3'}(N_G(L_1)) = L_1 \leq G_1$. Similarly when $L_1 < K_1$ we conclude that $G_2 \not\leq N_G(K_1)$.

We now claim

$$L_2 T < K_2 T < K_1 T \text{ and } K_2 T \neq L_1 T.$$

First as $\dim(V_2) = 2$, $K_2 = K_2^\infty \leq C_G(V_2) \leq C_G(V_1)$. Then as $L_2 < L_1 \leq K_1 \trianglelefteq N_G(V_1)$, $K_2 = [K_2, L_2] \leq K_1$. As G_2 does not act on L_1 or K_1, $K_2 < K_1$ and $K_2 \neq L_1$. Finally by assumption, $L_2 < K_2$, so the claim holds.

Now if $L_1 = K_1$, then $L_2 T$ is maximal in $L_1 T = K_1 T$, contrary to $L_2 T < K_2 T < K_1 T$. Thus $L_1 < K_1$, so that K_1^* is in the list of (2). Observe in each of those three groups that $L_1^* T^*$ is determined (up to outer automorphism when K_1^* is $L_5(2)$) as the unique overgroup of T^* in K_1^* with $L_1^* T^*/O_2(L_1^* T^*) \cong L_4(2)$. Suppose first that $K_1^* \cong M_{24}$. Then from the list of overgroups of T^*, L_2^* is normal in each overgroup of $L_2^* T^*$ other than $L_1^* T^*$, contradicting $L_2 T < K_2 T < K_1 T$ with L_2 not normal in $K_2 T \neq L_1 T$. Therefore $K_1^* \cong L_5(2)$ or J_4, and a similar argument shows that $K_2/O_2(K_2)$ is isomorphic to $L_4(2)$ in the former case, and

M_{24} in the latter. Now we can repeat our argument in the first paragraph of the proof for the case $i = 2$: In each case $K_2 = O^{3'}(N_G(K_2))$ by A.3.18. Also we saw $K_2 \leq C_G(V_2)$, while G_2 contains a subgroup X of order 3 fixed-point-free on V_2, so $X \leq O^{3'}(N_G(K_2)) = K_2 \leq G_1$, a contradiction. This completes the proof of 12.9.3. $\square$

For the remainder of the section, let K_1 be defined as in 12.9.3.

LEMMA 12.9.4. $\langle V^{G_1} \rangle$ is abelian.

PROOF. Assume otherwise; then we have the hypotheses of the latter part of section 12.8, so we can appeal to the results there. Adopt the notation of the second subsection of section 12.8; in particular take $H := G_1$, $U := U_H = \langle V^H \rangle$, and $H^* := H/C_H(\tilde{U})$.

As $n \geq 4$, we conclude from 12.8.8.5 that L_1 is not normal in H, so that $L_1 < K_1$ and in particular $K_1 \not\leq M$. Hence $K_1/O_2(K_1)$ is described in (2) or (3) of 12.9.3. By 12.8.12.4, $\dot{H}$ and its action on $\hat{U}$ are described in Theorem G.11.2. As $[\hat{V}, L_1] \neq 1$, $\dot{L}_1 \neq 1$, so $\dot{K}_1$ is a nontrivial normal subgroup of $\dot{H}$, and is also a quotient of K_1^*.

If $K_1^* \cong SL_2(7)/E_{49}$ then either $\dot{K}_1 = \dot{L}_1 \cong L_3(2)$ or $\dot{K}_1 \cong K_1^*$. However by inspection of the list in Theorem G.11.2, $\dot{H}$ has no such normal subgroup. Thus one of the remaining cases holds, where K_1^* is quasisimple, and hence $\dot{K}_1/Z(\dot{K}_1) \cong K_1^*/Z(K_1^*)$. Comparing the list in (2) and (3) of 12.9.3 to the normal subgroups of groups listed in Theorem G.11.2, we conclude $n = 4$ and one of conclusions (4), (5), or (8) of Theorem G.11.2 holds. Conclusion (8) does not occur, as there $\dot{H} \cong S_7$, so that there is no $\dot{T}$-invariant subgroup $\dot{L}_1$ with $\dot{L}_1/O_2(\dot{L}_1) \cong L_3(2)$. Conclusion (4) does not hold by 12.8.13.6.

Thus conclusion (5) of Theorem G.11.2 holds; that is $\hat{U}$ is the 6-dimensional natural module for $\dot{K}_1 = F^*(\dot{H}) \cong A_8$. Let $D := Z_U^g$, $Z_D := Z \cap Z(I_2)$, and $V_D = \langle Z_D^{K_1} \rangle$. By 12.8.13.5, $K_1 \in \mathcal{L}_f(G, T)$, K_1 acts nontrivially on the submodule V_D of $Z_U \in \mathcal{R}_2(KT)$, and $K_1 = [K_1, D]$.

As $K_1 \in \mathcal{L}_f(G, T)$, $K_1 \leq K \in \mathcal{L}_f^*(G, T)$ by 1.2.9.2. Then either $K_1 = K$, or $K/O_2(K) \cong L_5(2)$, M_{24}, or J_4 by A.3.12. Thus $K_1 = K \in \mathcal{L}_f^*(G, T)$ by 12.9.2.

As $F^*(\dot{H}) \cong A_8$, $\dot{H} \cong A_8$ or S_8, so as T normalizes L_1 with $L_1/O_2(L_1) \cong L_3(2)$, we conclude $\dot{H} \cong A_8$. Thus $\dot{L}_1$ is a maximal parabolic of $\dot{H}$ corresponding to an end node. Next set $L_0 := O^2(C_L(V_2)) = O^2(C_{L_1}(V_2))$. Then $L_0^* T^*$ is the minimal parabolic of L_1^* centralizing $\tilde{V}_2$. As $\hat{V}_2$ is a singular 1-space in the orthogonal space $\hat{U}$, $\dot{L}_0 \dot{T}$ is one of the two permuting minimal parabolics in the maximal parabolic $\dot{P}_0 := C_{\dot{H}}(\hat{V}_2)$ corresponding to the middle node of the Dynkin diagram for $\dot{H}$; in particular P_0 normalizes L_0. Similarly $\dot{L}_1$ is the maximal parabolic of $\dot{H}$ normalizing the totally singular 3-subspace $\hat{V}$ of $\hat{U}$, and so corresponds to an end node of the diagram for $\dot{H}$, with $\dot{L}_0 = \dot{P}_0 \cap \dot{L}_1$. Finally $\bar{L}_0 \bar{T}$ is the minimal parabolic of $\bar{L}$ centralizing V_2, with $\bar{I}_2 \bar{T}$ the other minimal parabolic in the maximal parabolic $N_{\bar{L}}(V_2)$ for the middle node, so that I_2 normalizes L_0.

By 12.8.13.2, $D \leq C_T(V) = O_2(LT)$, and hence $\dot{D} \leq O_2(\dot{L}_1 \dot{T})$. By 12.8.12.2, $\dot{D} \trianglelefteq \dot{H}_2 := H \cap G_2$, so as $L_0 T \leq H_2$, $\dot{D} \trianglelefteq \dot{L}_0 \dot{T}$. By 12.8.10.4, $[\hat{V}_2^\perp, D] = [\hat{W}, D] \leq \hat{V}_2$, so $\dot{D} \leq O_2(\dot{P}_0)$. Therefore

$$1 \neq \dot{D} \leq \dot{D}_0 := O_2(\dot{L}_1 \dot{T}) \cap O_2(\dot{P}_0) \cong E_4.$$

Then as $\dot{L}_0$ is irreducible on $\dot{D}_0$, and $1 \neq \dot{D} \trianglelefteq \dot{L}_0 \dot{T}$, we conclude $\dot{D} = \dot{D}_0 \cong E_4$.

Next by 12.8.10.6, $C_D(\hat{U}) = (D \cap Z_U)V_1^g$, so using symmetry between U and U^g,

$$2 = m(\dot{D}) = m(D/C_D(\hat{U})) = m(D/(D \cap Z_U)V_1^g) = m(Z_U/(D \cap Z_U)V_1). \quad (*)$$

Recall $g \in I_2 \leq N_G(L_0)$. Further $\dot{D}$ centralizes $(D \cap Z_U)V_1$; but $\dot{D} \neq 1$ does not centralize Z_U since $K_1 = [K_1, D]$ and K_1 is nontrivial on Z_U. Therefore as $\dot{L}_0$ is irreducible on $\dot{D}$, and hence on its g^{-1}-conjugate $Z_U/(D \cap Z_U)V_1$, while $\dot{L}_0$ normalizes $C_{Z_U}(\dot{D})$, we conclude from $(*)$ that $C_{Z_U}(\dot{D}) = (D \cap Z_U)V_1$ is of index 4 in Z_U. Recall $K_1 \in \mathcal{L}_f^*(G,T)$, and $V_D \in \mathcal{R}_2(K_1T)$. By Theorem 12.6.34, each $I_D \in Irr_+(K_1, V_D)$ is a natural 4-dimensional module for K_1^*. As L_0 is irreducible on $Z_U/C_{Z_U}(\dot{D})$, $m(I_D/C_{I_D}(D)) = 2$, so $Z_U = C_{Z_U}(D)I_D$ and $C_{I_D}(D) = [I_D, D] =: D_I$ is of order 4. As $Z_U \leq C_G(V_1^g) \leq N_G(D)$, $D_I \leq D$. Further as $K_1 = [K_1, D]$, K_1 centralizes Z_U/I_D, so $I_D = [Z_U, K_1]$. Therefore $[V_2, O^2(P_0)] \leq [V_2 Z_U, O^2(P_0)] = V_2 I_D$, so P_0 acts on $C_{V_2 I_D}(L_0) = V_2 C_{I_D}(L_0)$. Therefore if $C_{I_D}(L_0) = 1$, then P_0 acts on V_2, contrary to 12.8.13.8.

Thus $C_{I_D}(L_0) \neq 1$, so since $Aut_{L_0}(I_D)$ is a minimal parabolic of $GL(I_D)$ and $\dot{L}_0$ normalizes $\dot{D}$, $C_{I_D}(L_0) = D_I$, and so P_0 acts on $V_2 D_I$ and on D_I. Finally I_2 acts on V_2 and centralizes D_I by 12.8.10.2, as $D_I \leq Z_U \cap Z_U^g$, so $Y := \langle P_0, I_2 \rangle$ acts on $V_2 D_I$ and D_I. Then $Aut_{I_2 T}(V_2 D_I)$ and $Aut_{P_0}(V_2 D_I)$ are the two minimal parabolics of $GL(V_2 D_I) \cong L_4(2)$ stabilizing the 2-subspace D_I; in particular, $I_2 C_Y(V_2 D_I)$ is normal in Y. But now as I_2 centralizes D_I, P_0 normalizes $[V_2 D_I, I_2 C_Y(V_2 D_I)] = V_2$, a case we eliminated in the previous paragraph. This contradiction completes the proof of 12.9.4. $\qquad \square$

By 12.9.4, $\langle V^{G_1} \rangle$ is abelian, and hence (cf. 12.8.6) so is $U_H = \langle V^H \rangle$ for each $H \in \mathcal{H}_z$.

LEMMA 12.9.5. *(1)* $C_{G_1}(K_1/O_2(K_1)) \leq M$, *so* $K_1 T \in \mathcal{H}_z$.
(2) $K_1/O_2(K_1) \cong A_7$, $L_4(2)$, *or* $L_5(2)$.
(3) If $n = 4$ *and* $K_1/O_2(K_1) \cong L_5(2)$, *then* $L_1 O_2(K_1)/O_2(K_1)$ *is the centralizer of a transvection in* $K_1/O_2(K_1)$.

PROOF. Observe first that $Out(K_1/O_2(K_1))$ is a 2-group for each possibility in 12.9.3, including $K_1 = L_1$, so that $G_1 = K_1 T C_{G_1}(K_1/O_2(K_1))$.

We will combine the proofs of the three parts of the lemma, but in proving (2) we will assume that (1) has already been proved. Thus when proving (2), $L_1 < K_1$ since $G_1 \nleq M$, so that K_1 is described in 12.9.3. We consider three cases:

Case I. If (1) fails, pick $H_1 \in \mathcal{H}_*(T, M)$ with $O^2(H_1) \leq C_{G_1}(K_1/O_2(K_1))$, and let $H := H_1 L_1$.

Case II. If (2) fails, then $L_1 < K_1$ but $K_1/O_2(K_1)$ is not A_7, $L_4(2)$, or $L_5(2)$, and we let $H := K_1 T$.

Case III. If (3) fails, then $L_1 O_2(K_1)/O_2(K_1)$ is a parabolic determined by an end node and the adjacent node in the Dynkin diagram for $L_5(2)$, and we pick $H_1 \in \mathcal{H}_*(T, M)$ to be the minimal parabolic of K_1 determined by the remaining end node, and let $H := H_1 L_1$.

In each case $H \in \mathcal{H}_z$. As 12.9.4 provides condition (2) of 12.8.6, the latter result says that H satisfies Hypotheses F.8.1 and F.9.8 with V in the role of "V_+". Thus we may apply the results in sections F.8 and F.9. In particular we adopt the notation of sections F.7 and F.8 (or F.9) for the amalgam generated by H and LT.

Suppose first that we are in Case II. Then by the choice made in the first paragraph of the proof, K_1^* is one of the groups listed in 12.9.3 other than A_7, $L_4(2)$ or $L_5(2)$, and $H = K_1T$. Then unless $K_1^* \cong SL_2(7)/E_{49}$, $K_1/O_2(K_1)$ is quasisimple, so the hypotheses of F.9.18 are satisfied. But then F.9.18.4 supplies a contradiction, as none of the groups other than A_7, $L_4(2)$, and $L_5(2)$ appear in both F.9.18.4 and 12.9.3. Thus we have reduced to $K_1^* \cong SL_2(7)/E_{49}$. This case is impossible, since by F.9.16.3, $q(H^*, \tilde{U}_H) \leq 2$, contrary to D.2.17 applied to $K_1^*T^*$ in the role of "G".

Thus Case I or III holds, and in either case, $H = H_1L_1$ with $[L_1, O^2(H_1)] \leq O_2(L_1)$, so $C_H(L_1/O_2(L_1)) \leq H_1$. As $H_1 \in \mathcal{H}_*(T, M)$, 3.3.2 says H_1 is a minimal parabolic in the sense of Definition B.6.1. By F.8.5, the parameter b is odd and $b \geq 3$. Then by F.7.3.2, there is $g \in \langle LT, H \rangle = G_0$ mapping the edge γ_{b-1}, γ to γ_0, γ_1, and $h \in H$ with $\gamma_2 h = \gamma_0$. Set $\beta := \gamma_1 g$, $\delta := \gamma h$, and let $\alpha \in \{\beta, \delta\}$; then $U_\alpha \leq O_2(G_{\gamma_0, \gamma_1})$ by F.8.7.2. Therefore as $L_1 \trianglelefteq H \geq G_{\gamma_0, \gamma_1}$, $[L_1, U_\alpha] \leq O_2(L_1)$, and hence $U_\alpha \leq C_H(L_1/O_2(L_1)) \leq H_1$. Further as H_1 is a minimal parabolic, for each nontrivial H_1-chief factor E_1 on $\tilde{U}$, $m(U_\alpha/C_{U_\alpha}(E_1)) \leq m(E_1/C_{E_1}(U_\alpha))$ by B.6.9.1. However by 12.8.5.1, each H-chief section E on $\tilde{U}$ is the sum of $n - 1$ chief sections under H_1, so that $(n-1)m(U_\alpha/C_{U_\alpha}(E)) \leq m(E/C_E(U_\alpha))$. Hence as $n \geq 4$, if $U_\alpha^* \neq 1$ then

$$2m(U_\alpha^*) < (n-1)m(U_\alpha^*) \leq m(\tilde{U}/C_{\tilde{U}}(U_\alpha^*)). \qquad (*)$$

Now take $\alpha = \beta$. Then $U_\alpha = U^g$ and $U = U_\gamma^g$, so $(*)$ shows that U_H does not induce transvections on U_γ. Therefore by F.9.16.1, $D_\gamma < U_\gamma$, so by F.9.16.4, we may choose γ so that $U_\gamma^* \in \mathcal{Q}(H^*, \tilde{U}_H)$. Then taking $\alpha = \delta$, we have a contradiction to $(*)$, completing the proof of (1) and (3), and hence of 12.9.5. $\qquad \square$

LEMMA 12.9.6. (1) $[V_2, O_2(K_1)] \neq 1$.
(2) $I_2 := \langle O_2(G_1)^{G_2} \rangle \trianglelefteq G_2$, $I_2/O_2(I_2) \cong S_3$, $O_2(I_2) = C_{I_2}(V_2)$, and I_2T is a minimal parabolic of LT.
(3) $m_3(C_G(V_2)) \leq 1$.

PROOF. By 12.9.5.2, $K_1/O_2(K_1) \cong A_7$, $L_4(2)$ or $L_5(2)$; and by 12.9.5.1, $H := K_1T \in \mathcal{H}_z$. Let $Q := O_2(LT)$, $Q_1 := O_2(K_1)$, and $H^* := H/Q_1$.

Assume that Q_1 centralizes V_2. Then Q_1 centralizes $\langle V_2^H \rangle$, and by 12.8.8.6, $U_H = \langle V_2^H \rangle$, so that Q_1 centralizes U_H. Thus as $K_1/O_2(K_1)$ is simple, $U_H \in \mathcal{R}_2(K_1Q)$. Next as Q_1 centralizes V, $Q_1 \leq Q < R_1$, with R_1/Q the natural module for $L_1/R_1 \cong L_{n-1}(2)$. As $H \not\leq M \geq N_G(Q)$, $Q_1 < Q$, so $Q^* \neq 1$. Therefore $1 \neq Q^* < R_1^* = O_2(L_1^*)$. As $O_2(L_1^*) \neq 1$, K_1^* is not A_7, so that $K_1^* \cong L_4(2)$ or $L_5(2)$. As $1 \neq Q^* < O_2(L_1^*)$, the parabolic $L_1^*T^*$ of K_1^* is not irreducible on $O_2(L_1^*)$, so we conclude that $n = 4$ and $K_1^* \cong L_5(2)$. Then using 12.9.5.3, $R_1^* \cong 2^{1+6}$ and $Q^* = O_2(P^*) \cong E_{16}$ for some end-node maximal parabolic P^* of K_1^*. But then $P \leq N_G(Q) \leq M$, contradicting $L_1 \trianglelefteq M_1$. This completes the proof of (1).

Let P_2 be the minimal parabolic of LT nontrivial on V_2, and $R := O_2(G_1)$. Now as $C_G(V_2) \leq G_1$ and P_2 induces $GL(V_2)$,

$$R^{G_2} = R^{C_G(V_2)P_2} = R^{P_2} \subseteq P_2,$$

so $\langle R^{G_2} \rangle = I_2 \leq P_2$. Further by (1), R does not centralize V_2, so $P_2 = I_2T$ and (2) follows. Finally $[I_2, C_G(V_2)] \leq C_{I_2}(V_2) = O_2(V_2)$ by (2), so

$$2 \geq m_3(G_2) = m_3(I_2) + m_3(C_G(V_2)) = 1 + m_3(C_G(V_2)),$$

establishing (3). $\qquad\qquad\qquad\qquad\qquad\qquad\qquad\qquad\qquad\qquad\qquad$ $\square$

LEMMA 12.9.7. $G_i \leq M_V$ for $1 < i < n$.

PROOF. Recall $M_i \leq M_V$ as V is a TI-set in M by 12.2.6, so it suffices to show $G_i \leq M$. Let $1 < i < n$. As $GL(V_i) = Aut_M(V_i)$, $G_i = M_i C_G(V_i)$, so it suffices to show $C_G(V_i) \leq M$. As $C_G(V_i) \leq C_G(V_2)$, it remains to show $C_G(V_2) \leq M$. Set $(K_1 T)^* := K_1 T / O_2(K_1 T)$. As $Out(K_1^*)$ is a 2-group, $G_1 = DK_1 T$, where $D := C_{G_1}(K_1^*)$ and $D \leq M_1$ by 12.9.5.1. As $[D, L_1^*] = 1$, $[D, \bar{L}_1] \leq O_2(\bar{L}_1)$, so D centralizes $\tilde{V}_2$. Thus $C_{G_1}(\tilde{V}_2) = DC_{K_1 T}(\tilde{V}_2)$, so it suffices to show that $Y := C_{K_1 T}(V_2)T = C_{K_1 T}(\tilde{V}_2) \leq L_1 T$.

Now Y is an overgroup of T in $K_1 T$ with $I := O^2(L_1 \cap G_2) \leq Y$, and IT is a parabolic of LT of Lie rank $n-3$ contained in $L_1 T$. By 12.9.5.2, $K_1^* \cong A_7$, $L_4(2)$, or $L_5(2)$.

We assume that $Y \not\leq L_1 T$ and derive a contradiction. Then $IT < Y$. If $K_1^* \cong L_4(2)$ or $L_5(2)$, then Y is a parabolic in $K_1 T$ of rank at least $n - 2 \geq 2$, so $(Y \cap K_1)O_2(Y)/O_2(Y) \cong S_3 \times S_3$, $L_3(2)$, $S_3 \times L_3(2)$, or $L_4(2)$. If $K_1^* \cong A_7$, then examining overgroups of $(T \cap K_1)^*$, we conclude that $(Y \cap K_1)O_2(Y)/O_2(Y)$ is $L_3(2)$, A_6, or a subgroup of index 2 in $S_4 \times S_3$. However by 12.9.6.3, $m_3(Y) \leq 1$, so $(Y \cap K_1)O_2(Y)/O_2(Y) \cong L_3(2)$. Then as Y has Lie rank at least $n-2$, we conclude that $n = 4$, so that $L_1/O_2(L_1) \cong L_3(2)$ and $IT/O_2(IT) \cong S_3$. As T induces inner automorphisms on $L_1/O_2(L_1)$, $T^* \leq K_1^*$, so $Y^* \leq K_1^*$ and $Y/O_2(Y) \cong L_3(2)$.

Set $H := \langle Y, L_1 \rangle$, so that $H \in \mathcal{H}_z$. Now Y and $L_1 T$ are of Lie rank 2 and intersect in IT of Lie rank 1, so we conclude from the lattice of overgroups of T in $K_1 T$ that $H/O_2(H)$ is A_7 or $L_4(2)$. In either case as L_1^* does not centralize $\tilde{V}_2$, $C_H(\tilde{V}_2) = Y$; so as $Y/O_2(Y) \cong L_3(2)$, we conclude from B.4.12 that $\tilde{U}_H = \langle \tilde{V}_2^H \rangle$ is a 4-dimensional module for $H/O_2(H) = A_7$ or $L_4(2)$. Thus $O^2(Y)$ is irreducible on U_H/V_2, so $U_H = \langle V^{O^2(Y)} \rangle$. Define I_2 as in 12.9.6.2. By that result, I_2 normalizes V, $I_2 \triangleleft G_2$, and $I_2/O_2(I_2) \cong S_3$; therefore as $Y \leq G_2$ with $Y/O_2(Y) \cong L_3(2)$, $O^2(Y)$ centralizes $I_2/O_2(I_2)$, and hence I_2 normalizes $O^2(O^2(Y)O_2(I)) = O^2(Y)$. Hence I_2 acts on $\langle V^{O^2(Y)} \rangle = U_H$. But then $LT = \langle I_2, L_1 \rangle \leq N_G(U_H)$, so that $N_G(U_H) \leq M = !\mathcal{M}(LT)$, contrary to $H \not\leq M$. This contradiction completes the proof of 12.9.7. $\qquad\qquad\qquad$ $\square$

LEMMA 12.9.8. (1) $m(V \cap V^g) \leq 1$ for $g \in G - M$.
(2) If $V \cap V^g \neq 1$, then $[V, V^g] = 1$.

PROOF. We may assume $V \cap V^g < V$ as $N_G(V) \leq M$. Then if $V \cap V^g \neq 1$, by 12.8.3.2 we may assume $V \cap V^g = V_i$ for some $1 \leq i < n$, and take $g \in G_i$ by 12.8.3.3. If $i > 1$, we have $G_i \leq M$ by 12.9.7, proving (1). Part (2) follows from 12.9.4 and 12.8.6. $\qquad\qquad\qquad\qquad\qquad$ $\square$

LEMMA 12.9.9. (1) $W_0 := W_0(T, V)$ centralizes V, so $N_G(W_0) \leq M$.
(2) If $A := V^g \cap M$ is a hyperplane of V^g contained in T, then $C_A(V) = 1$.

PROOF. Suppose $A := V^g \cap M \leq T$ with $[A, V] \neq 1$ and $m(V^g/A) \leq 1$. Let $I := N_V(A)$. By 12.9.8.2, $A \cap V = 1$, so $[A, I] \leq A \cap V = 1$ and hence $I < V$. By 12.9.7, for each noncyclic subgroup B of A, $C_V(B) \leq N_V(V^g) \leq N_V(A) = I = C_V(A) \leq C_V(B)$, so $C_V(B) = I$. Thus $m(C_A(W)) \leq 1$ for each A-submodule W of V not contained in I; in particular as $I < V$, $m(C_A(V)) \leq 1$.

Assume $C_A(V) \neq 1$. Conjugating in $N_G(V^g)$, we may assume that $C_A(V) = V_1^g$. Thus $V \leq G_1^g \leq N_G(U^g)$, where $U := \langle V^{G_1} \rangle$. Hence $[V, A] \leq U^g \leq C_G(V^g)$ as U is abelian by 12.9.4.

We now prove (2), so we may assume $A < V^g$. Then $[V, A]$ is cyclic: for otherwise $V^g \leq C_G([V, A]) \leq N_G(V) \leq M$ by 12.9.7, contrary to $A < V^g$. As $[V, A]$ is cyclic, A induces a group of transvections on V with center $[V, A]$; so as $C_V(B) = I = C_V(A)$ for each noncyclic subgroup B of A, $|\bar{A}| = 2$. But now $C_A(V)$ is noncyclic, contrary to paragraph one. This completes the proof of (2).

If $W_0 \leq C_T(V)$ then $N_G(W_0) \leq M$ by E.3.34.2. Thus we may assume $A = V^g$, and it remains to derive a contradiction. Suppose first that A acts nontrivially on V_{n-1}. Then $V_{n-1} \not\leq C_V(A) = I$ and hence $m(C_A(V_{n-1})) \leq 1$ by paragraph one. Let $M_{n-1}^* := M_{n-1}/C_M(V_{n-1}))$, and observe $M_{n-1}^* \cong L_{n-1}(2)$. Then

$$m_2(M_{n-1}^*) \geq m(A^*) \geq n - 1,$$

so we conclude $n = 5$ and $A^* = J(T^*)$. But now $C_{V_4}(A) = V_2 < C_{V_4}(B)$ for B a 4-subgroup of A with B^* inducing transvections on V_4 with a fixed axis, contrary to an observation in the first paragraph.

Therefore A centralizes V_{n-1}, so $A \leq R_{n-1}$. Then as $m(C_A(V)) \leq 1$,

$$m(\bar{A}) \geq m(A) - 1 = n - 1 = m(\bar{R}_{n-1}),$$

so that $AC_T(V) = R_{n-1}$. Thus $L_1 = [L_1, A]$ and $A_1 := C_A(V)$ is of order 2, so by paragraph two we may assume $A_1 = V_1^g$, $V \leq N_G(U^g)$, and $V_{n-1} = [A, V] \leq U^g$. Let $Q := O_2(G_1)$. For $y \in Q$, $[U, y] \leq V_1$ by 12.8.4.2, so $m(U/C_U(y)) \leq 1$ and hence as $n \geq 4$, $C_{V_{n-1}}(y^g)$ is noncyclic. Thus $y^g \in C_G(C_{V_{n-1}}(y^g)) \leq N_G(V)$ by 12.9.7, so $[Q^g, V] \leq Q^g \cap V = V_{n-1} \leq U^g$. If $[K_1^g, V] \leq O_2(K_1^g)$, then $V \leq N_G(A)$ by 12.9.5.1, contrary to $I < V$. Thus $K_1^g = [K_1^g, V]$, so K_1^g centralizes Q^g/U^g. Then $[K_1, Q] \leq U \leq C_Q(U)$ as U is abelian by 12.9.4. Therefore $[K_1, O_2(K_1)] \leq C_G(U) \leq C_G(V)$.

Let $P := O_2(K_1T)$ and choose X of order 7 or 5 in L_1 for $n = 4$ or 5, respectively. Recall $K_1T \in \mathcal{H}_z$ by 12.9.5.1, so that $[\tilde{V}, P] = 1$ by 12.8.4.2. Further $V = V_1 \times [V, X]$, and by Coprime Action, $P = C_P(X)[P, X] = C_P(X)[P, K_1] = C_P(X)[O_2(K_1), K_1]$. Now P acts on V, and $[O_2(K_1), K_1]$ centralizes V by the previous paragraph; then $C_P(X)$ acts on $[V, X]$, and hence P acts on $[V, X]$. Therefore as X is irreducible on $[V, X]$ and normalizes P, P centralizes $[V, X]$, so as $P \leq G_1$, P centralizes V. As $O_2(K_1) \leq P$ and $V_2 \leq V$, this is contrary to 12.9.6.1, so the proof of 12.9.9 is complete. $\qquad \square$

We are now in a position to complete the proof of Theorem 12.9.1.

By 12.9.5.2, $K_1^* \cong A_7$, $L_4(2)$, or $L_5(2)$. In particular, there is an overgroup H of T in K_1T not contained in M with $H/O_2(H) \cong S_3$. By 12.9.9.1, $N_G(W_0) \leq M$, so by E.3.15, $W_0 \not\leq O_2(H)$. Thus there is $A := V^g \leq T$ with $A \not\leq O_2(H)$. If $V_1 \leq A$, then by 12.8.3.3 and 12.9.4, $A \in V^{G_1} \cap H \subseteq O_2(H)$, contrary to our choice of A. Thus $V_1 \cap A = 1$.

Now $H \notin \mathcal{H}_z$ since H does not contain L_1, but we define some notation similar to that in Notation 12.8.2: Let $U_H := \langle V_2^H \rangle$ and $Q_H := O_2(H)$. Then $U_H \leq \langle V^{G_1} \rangle$, so U_H is abelian by 12.9.4. Indeed Hypothesis G.2.1 is satisfied with H, V_2, 1 in the roles of "G, V, L", so by G.2.2.1, $\tilde{U}_H \in Z(\tilde{Q}_H)$. By 12.9.7, $V_2 < U_H$. As $H/Q_H \cong S_3$ and $A \not\leq Q_H$, $B := A \cap Q_H$ is of index 2 in A. Then $[U_H, B] \leq V_1$, so for $u \in U_H$, $m(B/C_B(u)) \leq m(V_1) = 1$, so $C_B(u)$ is noncyclic, and hence by 12.9.7, $u \in N_G(A)$. Thus $U_H \leq N_G(A)$, and so $[U_H, B] \leq A \cap V_1 = 1$.

As $O^2(H) \leq \langle A^H \rangle$ and $V_2 < U_H$, there is $h \in H$ such that A^h does not act on V_2. But again using 12.9.7,

$$D := B^h \leq C_G(U_H) \leq C_G(V_2) \leq N_G(V).$$

If $[D, V] = 1$, then $V \leq C_G(D) \leq N_G(A^h)$ by 12.9.7, so $A^h \leq C_G(V) \leq C_G(V_2)$ by 12.9.9.1, contrary to our choice of h. Thus $\bar{D} \neq 1$, so $D = V^{gh} \cap M$ by 12.9.9.2 However

$$1 \neq [U_H, A^h] \leq U_H \cap A^h \leq C_D(V),$$

since U_H is abelian. As D is a hyperplane of A^h with $D = V^{gh} \cap M$, 12.9.9.2 supplies a contradiction.

This final contradiction completes the proof of Theorem 12.9.1.

CHAPTER 13

Mid-size groups over $\mathbf{F}_2$

In this chapter we consider the cases remaining in the Fundamental Setup (3.2.1) after the work of the previous chapter. We make more use of the generic methods for the $\mathbf{F}_2$ case, such as results from sections F.7, F.8, and F.9.

In Hypothesis 13.1.1, we essentially extend Hypothesis 12.2.3 which began the previous chapter, by adding the assumption that G is not one of the groups which arose in the course of that chapter. Then after some reductions in the initial sections 13.1 and 13.2, in the remainder of the chapter we assume an additional refinement in Hypothesis 13.3.1.

In particular in 13.1.2.3, we observe that the remaining possibilities for the section $L/O_2(L)$, with $L \in \mathcal{L}_f^*(G,T)$ in the FSU, are A_5, $L_3(2)$, A_6, $\hat{A}_6$, and $U_3(3) \cong G_2(2)'$. The main goal of the chapter is to treat the latter three groups, thus reducing the FSU to the case where $L/O_2(L)$ is $L_3(2)$ or A_5.

In the natural logical sequence, the smallest simple group A_5 is treated last; thus at that point, all other groups are eliminated, so that $K/O_2(K) \cong A_5$ for all $K \in \mathcal{L}_f^*(G,T)$. However, to avoid repeating arguments common to both A_5 and A_6, we prove such results simultaneously for both in sections 13.5 and 13.6. To do so, we *assume* in part (4) of Hypothesis 13.3.1 (and similarly in the hypothesis of 13.2.7) that $K/O_2(K) \cong A_5$ for all $K \in \mathcal{L}_f^*(G,T)$, when the subgroup $L \in \mathcal{L}_f^*(G,T)$ we've chosen satisfies $L/O_2(L) \cong A_5$. That is, we don't make this choice until we are forced to do so, after the treatment of the other groups.

13.1. Eliminating $\mathbf{L} \in \mathcal{L}_f^*(\mathbf{G}, \mathbf{T})$ with $\mathbf{L}/\mathbf{O}_2(\mathbf{L})$ not quasisimple

We now state the initial hypothesis for the chapter, which excludes the groups in the Main Theorem that have arisen so far under the FSU. Namely throughout this section, we assume:

HYPOTHESIS 13.1.1. *(1) G is a simple QTKE-group and $T \in Syl_2(G)$.*
(2) G is not a group of Lie type of Lie rank 2 over $\mathbf{F}_{2^n}$, $n > 1$.
(3) G is not $L_4(2)$, $L_5(2)$, A_9, M_{22}, M_{23}, M_{24}, He, or J_4.

As usual let $Z := \Omega_1(Z(T))$.

As mentioned earlier, Hypothesis 13.1.1 essentially contains Hypothesis 12.2.3, aside from the assumption in Hypothesis 12.2.1 that there is some $L \in \mathcal{L}_f^*(G,T)$ with $L/O_2(L)$ quasisimple. In Theorem 13.1.7, we show for each $K \in \mathcal{L}_f^*(G,T)$ that $K/O_2(K)$ is quasisimple.

We record some elementary consequences of Hypothesis 13.1.1.

LEMMA 13.1.2. *Assume there is $L \in \mathcal{L}_f^*(G,T)$ with $L/O_2(L)$ quasisimple and set $M := N_G(L)$. Then L is T-invariant, there exists a T-invariant member V of $Irr_+(L, R_2(LT))$, and:*

(1) *Hypothesis 12.2.3 holds.*
(2) $C_G(v) \not\leq M$ *for some* $v \in V^{\#}$.
(3) $L/O_2(L) \cong A_5$, A_6, $\hat{A}_6$, $L_3(2)$, *or* $G_2(2)'$.
(4) *If* $L/O_2(L) \cong \hat{A}_6$, *then* $V/C_V(L)$ *is the natural module for* A_6.
(5) *If* $L_1 \in \mathcal{L}(G, T)$ *and* $L_1 \leq L$, *then* $L_1 = L \in \mathcal{L}^*(G, T)$.

PROOF. As $L \in \mathcal{L}_f^*(G, T)$ with $L/O_2(L)$ quasisimple, part (1) of Hypothesis 13.1.1 implies that Hypothesis 12.2.1 holds, allowing us to apply Theorem 12.2.2. Parts (2) and (3) of Hypothesis 13.1.1 exclude the groups in conclusions (1) and (2) of Theorem 12.2.2, so that conclusion (3) of that result holds. Hence T normalizes L and Hypothesis 12.2.3 holds, establishing (1). Part (3) of Hypothesis 13.1.1 excludes the groups in conclusions (2)–(4) of Theorem 12.2.13, as well as the groups in the conclusions of Theorems 12.3.1, 12.7.1, and 12.9.1. Hence those results eliminate the corresponding cases from conclusion (3) in Theorem 12.2.2 and so establish (2)–(4). Finally as the groups in (3) are of Lie type and either of Lie rank 2 over $\mathbf{F}_2$, or A_5 of Lie rank 1, each proper T-invariant subgroup of L is solvable. Then (5) follows from 1.2.4. $\square$

Define
$$\mathcal{L}_+(G, T) := \{L \in \mathcal{L}_f(G, T) : L/O_2(L) \text{ is } \textit{not} \text{ quasisimple }\},$$
and suppose for the moment that $\mathcal{L}_+(G, T)$ is empty. If $K \in \mathcal{L}_f(G, T)$ then by 1.2.9, $K \leq L \in \mathcal{L}_f^*(G, T)$. As $\mathcal{L}_+(G, T) = \emptyset$, $L/O_2(L)$ is quasisimple, so $K = L \in \mathcal{L}_f^*(G, T)$ by 13.1.2.5. That is, once we show that $\mathcal{L}_+(G, T)$ is empty, we will be able to conclude that $\mathcal{L}_f(G, T) = \mathcal{L}_f^*(G, T)$.

REMARK 13.1.3. Recall that non-quasisimple $\mathcal{C}$-components are allowed by the general quasithin hypothesis: they appear as cases (3) and (4) of A.3.6, and cases (c) and (d) of 1.2.1.4. On the other hand, they do not actually arise in $\mathcal{L}_f(G, T)$ in any of the groups in our Main Theorem. Thus after Theorem 13.1.7, we will finally be rid of this nuisance. In particular, if $\mathcal{L}_f^*(G, T)$ is nonempty, then by 3.2.3 there will exist tuples in the Fundamental Setup. Furthermore, as we just observed, we will also have $\mathcal{L}_f(G, T) = \mathcal{L}_f^*(G, T)$.

If $L \in \mathcal{L}_+(G, T)$ then L appears in case (c) or (d) of 1.2.1.4, so $m_p(L) = 2$ for some odd prime p dividing the order of $O_{2,F}(L)$, and $T \leq N_G(L)$ by 1.2.1.3. Also in the notation of chapter 1, $1 \neq \Xi_p(L) \in \Xi(G, T)$ by 1.3.3.

Recall the basic facts about $\Xi(G, T)$ from that chapter. Recall also from Definition 3.2.12 that $\Xi_-(G, T)$ consists of those $X \in \Xi(G, T)$ such that either X is a $\{2, 3\}$-group, or $X/O_2(X)$ is a 5-group and $Aut_G(X/O_2(X))$ is a 2-group. Further $\Xi_+(G, T)$ is defined to be $\Xi(G, T) - \Xi_-(G, T)$. Set
$$\Xi_+^*(G, T) := \Xi_+(G, T) \cap \Xi^*(G, T).$$
We will make repeated use of results from section A.4 such as A.4.11.

We next collect some useful properties of the members L of $\mathcal{L}_+(G, T)$. Although the proof of the next lemma contains an appeal to 13.1.2.3, we could in fact have stated and proved 13.1.4 much earlier, after chapter 11. On the other hand, many arguments from now on (eg. the proof of 13.1.9.1) make strong use of 13.1.2.5— which does depend on work done in chapters after chapter 11.

LEMMA 13.1.4. *Assume* $L \in \mathcal{L}_+(G,T)$. *Then*

(1) *Either*

(i) $L \in \mathcal{L}^*(G,T)$, *or*

(ii) $L/O_{2,F}(L) \cong SL_2(5)$, *and* $L_+ \in \mathcal{L}_+(G,T)$ *for each* $L_+ \in \mathcal{L}(G,T)$ *with* $L < L_+$.

(2) *There is* $M_c \in \mathcal{M}(T)$ *with* $M_c = !\mathcal{M}(LT)$. *If* $L \in \mathcal{L}^*(G,T)$, *then* $M_c = N_G(L)$.

(3) *For some prime* $p > 3$, $X := \Xi_p(L) \neq 1$, *and for each such* X, $X \in \Xi(G,T)$ *and either*

(i) $X/O_2(X) \cong E_{p^2}$ *and* $L/X \cong SL_2(p)$, *or*

(ii) $L/O_{2,F}(L) \cong SL_2(5)$.

(4) $X \in \Xi_+^*(G,T)$, *and* $M_c = !\mathcal{M}(XT) = N_G(X)$.

(5) $C_L(R_2(M_c)) = O_\infty(L)$. *In particular,* X *centralizes* $R_2(M_c)$.

(6) $M_c = !\mathcal{M}(C_G(Z))$.

PROOF. Assume $L \leq L_+ \in \mathcal{L}(G,T)$. As $L \in \mathcal{L}_+(G,T)$, $L \in \mathcal{L}_f(G,T)$, so $L_+ \in \mathcal{L}_f(G,T)$ by 1.2.9.1. Recall T acts on L, so T acts on L_+ by 1.2.4.

As $L \in \mathcal{L}_+(G,T)$, $X := \Xi_p(L) \neq 1$ for some prime $p > 3$ by 1.2.1.4, and $X \in \Xi(G,T)$ by 1.3.3. Indeed (3) holds by 1.2.1.4.

Suppose that X is not normal in L_+. Then by 1.3.4, L_+ appears on the list of 1.3.4; in particular $L_+/O_2(L_+)$ is quasisimple in each case. As T acts on L_+, conclusion (1) of 1.3.4 does not hold, and as $p > 3$, conclusion (4) does not hold. Thus $L_+/O_2(L_+) \cong (S)L_3(p)$ or $Sp_4(2^n)$, with n even. Furthermore $L_+ \in \mathcal{L}_f^*(G,T)$ using 1.3.9.1. But this is contrary to the list of possibilities in 13.1.2.3.

This contradiction shows that $X \trianglelefteq L_+$, so $L_+/O_2(L_+)$ is not quasisimple and hence $L_+ \in \mathcal{L}_+(G,T)$ by definition. Further taking L_+ maximal, $L_+ \in \mathcal{L}^*(G,T)$. Therefore $X \in \Xi^*(G,T)$ by 1.3.8. If $L = L_+$, then (1i) holds. Otherwise by 1.2.4, the inclusion $L < L_+$ is described in A.3.12 (see A.3.13 for further detail in this case); so $1 \neq O_\infty(L) \leq O_\infty(L_+)$ and (1ii) holds. This completes the proof of (1).

As $X \in \Xi^*(G,T)$, $M_c := N_G(X) = !\mathcal{M}(XT)$ by 1.3.7. As $p > 3$ and $Aut_L(X)$ is not a 2-group, $X \in \Xi_+(G,T)$; thus $X \in \Xi_+^*(G,T)$, completing the proof of (4). Further as $X \leq L$, it follows that also $M_c = !\mathcal{M}(LT)$. If $L \in \mathcal{L}^*(G,T)$, then $L \in \mathcal{C}(M_c)$ by 1.2.7.1, and then $L \trianglelefteq M_c$ by 1.2.1.3, completing the proof of (2).

Recall $L_+ \in \mathcal{L}^*(G,T)$, so $L_+ \trianglelefteq M_c$ by (2). As $L_+ \in \mathcal{L}_f(G,T)$, $C_{L_+}(R_2(M_c)) < L_+$ by A.4.11. We also saw earlier that $O_\infty(L) \leq O_\infty(L_+)$. Let $Y := O^2(O_{2,F}(L_+))$. Then Y centralizes $R_2(L_+T)$ by 3.2.14, and $Y \trianglelefteq M_c$, so Y centralizes $R_2(M_c)$ by A.4.11. Then as $L_+ \trianglelefteq M_c$ and $R_2(M_c)$ is 2-reduced, $O_{2,F}(L_+) \leq C_{L_+}(R_2(M_c))$, and hence $O_\infty(L_+) = O_{2,F,2}(L_+) = C_{L_+}(R_2(M_c))$. So as $O_\infty(L) = O_\infty(L_+) \cap L$, we conclude that (5) holds.

Finally as $M_c \in \mathcal{H}^e$ by 1.1.4.6, $Z \leq R_2(M_c)$ by B.2.14, so (4) and (5) imply (6). $\square$

Let $\mathcal{L}_+^*(G,T)$ denote the maximal members of $\mathcal{L}_+(G,T)$; thus $\mathcal{L}_+^*(G,T)$ is nonempty whenever $\mathcal{L}_+(G,T)$ is nonempty. By 13.1.4.1,

$$\mathcal{L}_+^*(G,T) \subseteq \mathcal{L}_f^*(G,T).$$

LEMMA 13.1.5. *Assume* $\Xi_+^*(G,T) \neq \emptyset$. *Then*

(1) *There is* $M_c \in \mathcal{M}(T)$ *with* $M_c = !\mathcal{M}(C_G(Z))$.

(2) $\Xi_+^*(G,T) \subseteq M_c$, *so* $M_c = N_G(X) = !\mathcal{M}(XT)$ *for each* $X \in \Xi_+^*(G,T)$.

PROOF. Assume $X \in \Xi_+^*(G,T)$. Then $X \in \Xi^*(G,T)$, so $M_c := N_G(X) =$ $!\mathcal{M}(XT)$ by 1.3.7. Also $X \in \Xi_+(G,T)$, so by 3.2.13, $X \notin \Xi_f(G,T)$. Then by A.4.11, X centralizes $R_2(XT)$, so $R_2(XT)$ contains Z by B.2.14; hence $M_c = !\mathcal{M}(XT) = !\mathcal{M}(C_G(Z))$, so that (1) holds. This also establishes (2), as we may vary $X \in \Xi_+^*(G,T)$ independently of Z. □

LEMMA 13.1.6. *Assume* $X \in \Xi_+^*(G,T)$, *let* $M_c \in \mathcal{M}(XT)$, *and assume* $M \in \mathcal{M}(T) - \{M_c\}$. *Then either*

(1) There exists an odd prime d *and* $Y = O^2(Y) \trianglelefteq M$ *such that* $Y \not\leq M_c$, $[Z, Y] \neq 1$, *and* $Y/O_2(Y)$ *is a* d-*group of exponent* d *and class at most 2, or*

(2) There exists $Y \in \mathcal{C}(M)$ *with* $Y \not\leq M_c$. *For each such* Y, $Y/O_2(Y)$ *is quasisimple,* $Y \trianglelefteq M$, $[Z, Y] \neq 1$, *and* $Y \in \mathcal{L}_f^*(G,T)$.

PROOF. By 13.1.5, $N_G(X) = !\mathcal{M}(XT) = M_c = !\mathcal{M}(C_G(Z))$.

Suppose first that there is $Y \in \mathcal{C}(M)$ with $Y \not\leq M_c$. Then as $M_c = !\mathcal{M}(C_G(Z))$, $[Z, Y] \neq 1$, so that $Y \in \mathcal{L}_f(G,T)$. Let $Y \leq Y_1 \in \mathcal{L}^*(G,T)$; by 1.2.9.2, $Y_1 \in \mathcal{L}_f^*(G,T)$. If $Y_1 \in \mathcal{L}_+(G,T)$, then by (2) and (6) of 13.1.4, $N_G(Y_1) = !\mathcal{M}(Y_1T) = !\mathcal{M}(C_G(Z))$, contrary to our assumption that $Y \not\leq M_c$. Thus $Y_1/O_2(Y_1)$ is quasisimple, so that $Y = Y_1 \trianglelefteq M$ by 13.1.2. Therefore (2) holds in this case.

We may assume that (2) fails, so $\langle \mathcal{C}(M) \rangle \leq M_c$ by the previous paragraph. Let $M^* := M/O_2(M)$, and for d an odd prime, let $\theta_d(M)$ be the preimage of the group $\theta_d(M^*)$ defined in G.8.9; recall that $\theta_d(M^*)$ is of class at most 2 and of exponent d using A.1.24. Let $\theta(M)$ be the product of the groups $\theta_d(M)$, for $d \in \pi(F(M^*))$.

Suppose that $\theta(M) \leq M_c$. Then as $\langle \mathcal{C}(M) \rangle \leq M_c$, $\theta(M)O_{2,E}(M) =: Y \leq M_c$. with M, M_c in the roles of "H, K", $R := O_2(M_c \cap M) = O_2(M)$ and $C(M_c, R) \leq M_c \cap M$. Then M_c, R, $M_c \cap M$ satisfy Hypothesis C.2.3 in the roles of "H, R, M_H". As $X \in \Xi_+(G,T)$, X is a $\{2,p\}$-group for some prime $p > 3$, so X contains no A_3-blocks. Thus by C.2.6.2, $X \leq M_c \cap M$, contrary to $M \neq M_c = !\mathcal{M}(XT)$.

This contradiction shows that $\theta(M) \not\leq M_c$; hence there is some d with $Y := \theta_d(M) \not\leq C_G(Z)$ and $Y = O^2(Y) \trianglelefteq M$; so (1) holds. □

We are now prepared for the main result of the section:

THEOREM 13.1.7. *Assume Hypothesis 13.1.1. Then* $\mathcal{L}_+(G,T) = \emptyset$.

Until the proof of Theorem 13.1.7 is complete, assume G is a counterexample. As $\mathcal{L}_+(G,T)$ is nonempty, we may choose $L \in \mathcal{L}_+^*(G,T)$, so $L \in \mathcal{L}_f^*(G,T)$ by an earlier remark. Set $M_c := N_G(L)$; then $M_c = !\mathcal{M}(LT)$ by 13.1.4.2. By Theorem 2.1.1, $|\mathcal{M}(T)| > 1$, so $\mathcal{H}_*(T, M_c)$ is nonempty.

Let $\mathcal{X}$ consist of the groups $\Xi_p(L)$, $p \in \pi(F(L/O_2(L)))$. By 13.1.4, each $X \in \mathcal{X}$ is in $\Xi_+^*(G,T)$ and

$$M_c = N_G(X) = !\mathcal{M}(XT) = !\mathcal{M}(C_G(Z)). \qquad (+)$$

Set $V_c := R_2(M_c)$, $M_c^* := M_c/C_{M_c}(V_c)$, and

$$U := [V_c, L].$$

Define

$$\mathcal{L}_1 := \{L_1 \in \mathcal{L}(G,T) : L = O_\infty(L)L_1\}.$$

LEMMA 13.1.8. *(1)* $L^* \cong L_2(p)$ *for some prime* $p > 3$.
(2) $1 \neq [V_c, L] = [V_c, L_1]$ *for each* $L_1 \in \mathcal{L}_1$.

PROOF. By 13.1.4.5, $O_\infty(L) = C_L(V_c)$. Hence (1) and the statement in (2) that $[V_c, L] \neq 1$ follow from 13.1.4.3. If $L_1 \in \mathcal{L}_1$, then $L^* = L_1^*$ by (1), so (2) holds. $\square$

We next establish an important technical result:

LEMMA 13.1.9. (1) For each $L_1 \in \mathcal{L}_1$, $M_c = !\mathcal{M}(L_1 T)$.
(2) $L = [L, J(T)]$.

PROOF. We will show in the first few paragraphs that (1) implies (2).

Set $R := C_T(L/O_\infty(L))$ and let $L_1 \in \mathcal{L}_1$. Observe by 13.1.4.3 that R is Sylow in $O_\infty(LT)$, with $R/O_2(LT)$ cyclic and $\Omega_1(R/O_2(LT))$ inverts $O_{2,F}(L)/O_{2,\Phi(F)}(L)$. Also for $X \in \mathcal{X}$, $O_2(X) \leq R$, so that $R \in Syl_2(XR)$. Set $L_R := N_L(R)^\infty$; by a Frattini Argument, $L_1 = O_\infty(L_1)N_{L_1}(R) = O_\infty(L_1)N_{L_1}(R)^\infty$. As $R \trianglelefteq T$, T acts on L_R, so $L_R \in \mathcal{L}(G, T)$. As $\Omega_1(R/O_2(LT))$ inverts $O_{2,F}(L)/O_{2,\Phi(F)}(L)$, $O_\infty(LT) \cap L_R T = R$, so $R = O_2(L_R T)$ and $L_R/O_2(L_R) \cong L^* \cong L_2(p)$ for some $p > 3$ by 13.1.8.1. As $L \in \mathcal{L}_f(G, T)$, $L_R \in \mathcal{L}_f(G, T)$ by A.4.10.3. As $N_{L_1}(R)^\infty$ is an R-invariant subgroup of L_R and $L_R/O_2(L_R)$ is simple, $N_{L_1}(R)^\infty = L_R$.

Now we assume that (1) holds, but (2) fails. We saw at the outset of the proof of Theorem 13.1.7 that we may choose some $H \in \mathcal{H}_*(T, M_c)$. We will appeal to 3.1.7 with $M_0 := L_R T$, so we begin to verify the hypotheses of that result: We've seen that $R = O_2(M_0)$. As we are assuming that (2) fails, $J(T) \leq O_\infty(LT) \cap L_R T = R$. Thus it remains to verify Hypothesis 3.1.5.

Take $V := R_2(M_0)$. As $L_R \in \mathcal{L}_f(G, T)$, $[V, M_0] \neq 1$ by 1.2.10, so as M_0/R is simple, $R = C_T(V)$. Finally we verify condition (I) of Hypothesis 3.1.5: Let $B := O^2(H \cap M_c)$. As $H \not\leq M_c = !\mathcal{M}(XT)$ by (+), $X \not\leq H$, so as T is irreducible on $X/O_{2,\Phi}(X)$, $B \cap X \leq O_{2,\Phi}(X)$. As this holds for each $X \in \mathcal{X}$, $B \cap O_{2,F}(L) \leq O_{2,\Phi(F)}(L)$, so $H \cap O_\infty(L)$ is 2-closed, and hence $H \cap M_c$ acts on $R \cap L$. Thus $H \cap M_c$ acts on $L_R = N_L(R \cap L)$. This completes the verification of Hypothesis 3.1.5.

Applying our assumption that (1) holds to $L_R \in \mathcal{L}_1$, $M_c = !\mathcal{M}(L_R T)$. Then $O_2(\langle M_0, H \rangle) = 1$, which rules out conclusion (2) of 3.1.7. As $M_c = !\mathcal{M}(C_G(Z))$ by (+), $Z \not\leq Z(H)$, which rules out the remaining conclusion (1) of 3.1.7. This contradiction completes the proof that (1) implies (2).

So we may assume that $L_1 \in \mathcal{L}_1$ and $M \in \mathcal{M}(L_1 T) - \{M_c\}$, and it remains to derive a contradiction. By paragraph one, $L_1 = O_\infty(L_1)L_R$, so $L_R \in \mathcal{L}(M, T)$. Thus $L_R \leq L_M \in \mathcal{C}(M)$ by 1.2.4, and as T normalizes L_R, $L_M \trianglelefteq M$ by 1.2.1.3.

We apply 13.1.6 to M and choose Y as in case (1) or (2) of that result. In particular $Y \not\leq M_c$, $Y \trianglelefteq M$, and $[Z, Y] \neq 1$. We claim that $[Y, L_M] \leq O_2(Y)$: In case (2) of 13.1.6, $Y \in \mathcal{C}(M)$, so by 1.2.1.2, either $Y = L_M$ or $[Y, L_M] \leq O_2(Y)$. But by 13.1.6, $Y \in \mathcal{L}_f^*(G, T)$ with $Y/O_2(Y)$ quasisimple, so if $Y = L_M$ then $Y = L_R$ by 13.1.2.5, contradicting $Y \not\leq M_c$. Thus the claim holds in this case. Now assume that case (1) of 13.1.6 holds and let $\dot{M} := M/O_2(M)$. In this case $\dot{Y}$ is of class at most 2 and exponent d for an odd prime d, with $m_d(\dot{Y}) \leq 2$. Thus as both Y and L_M are normal in M, using 1.2.1.4, either $[Y, L_M] \leq O_2(Y)$ as claimed, or $1 \neq D := [O^{d'}(Y \cap L_M), L_M] \trianglelefteq L_M$, with $\dot{D} \cong E_{d^2}$ or d^{1+2} and L_M irreducible on $\dot{D}/\Phi(\dot{D})$. In the latter case $Y = D$ by A.1.32.2 applied with D, Y in the roles of "P, R". Then as $[Z, Y] \neq 1$, $L_M \in \mathcal{L}_+(G, T)$, so $Y \leq O_\infty(L_M) \leq C_G(Z)$ by 13.1.4.5, contrary to $Y \not\leq M_c = !\mathcal{M}(C_G(Z))$. This contradiction completes the proof of the claim.

In particular by the claim, L_R centralizes $Y/O_2(Y)$, and hence Y normalizes $(L_R O_2(Y))^\infty = L_R$. As $Y \trianglelefteq M$, $O_2(Y) \le O_2(L_R T) = R$ by a remark in the first paragraph, so $R \in Syl_2(YR)$. Now $L_R/O_2(L_R) \cong L_2(p)$ has 3-rank 1 and centralizes $Y/O_2(Y)$, so $m_3(Y) \le 1$ as $L_R Y$ is an SQTK-group. In case (2) of 13.1.6, $Y \in \mathcal{L}_f^*(G,T)$, so by 13.1.2.3, $Y/O_2(Y) \cong A_5$ or $L_3(2)$. In case (2) of 13.1.6, either $d = 3$ and $Y/O_2(Y) \cong \mathbf{Z}_3$, or $Y/O_2(Y)$ is a d-group for $d > 3$.

Recall X is a solvable $3'$-group and $R \in Syl_2(XR)$, so we may apply a standard Thompson factorization theorem 26.18 in [**GLS96**] to conclude that

$$XR = N_{XR}(J(R))N_{XR}(E_R), \text{ where } E_R := \Omega_1(Z(J_1(R))).$$

As $X \in \Xi(G,T)$, T is irreducible on the Frattini quotient of $X/O_2(X)$, so $J(R)$ or E_R is normal in XR; set $J := J_j(R)$ where $j := 0$ or 1 in the respective case, so in either case $J(R) \le J$ and $N_G(J) \le M_c$ since $M_c = !\mathcal{M}(XT)$ by (+).

Set $K := [Y, J]$. If $K \le O_2(Y)$ then Y normalizes $R_1 := JO_2(Y) \le R$; but $J = J_j(R_1)$ by B.2.3.3, and hence $Y \le N_G(J) \le M_c$, contrary to our choice of $Y \not\le M_c$. This contradiction shows that $K \not\le O_2(Y)$. Thus $K = Y$ in case (2) of 13.1.6, since there $Y \in \mathcal{C}(M)$ with $Y/O_2(Y)$ quasisimple. In case (1) of 13.1.6, $Y = KN_Y(J)$ by a Frattini Argument applied to KJ, and hence $Y = K(Y \cap M_c)$. Thus in either case $K = [K, J]$ and as $Y \not\le M_c$,

$$K \not\le M_c, \text{ and in particular } [Z, K] \ne 1. \tag{!}$$

As $L_R T$ normalizes Y and J, it also normalizes $[Y, J] = K$ and hence normalizers KR. Further $K \le Y \le C_G(L_R/O_2(L_R))$ as we saw after the claim, so $[L_R, KR] \le O_2(L_R) \cap KR \le O_2(KR)$.

As K is subnormal in M, $K \in \mathcal{H}^e$ by 1.1.3.1. As $R \in Syl_2(YR)$, $R \in Syl_2(KR)$. Thus if we set $D := \langle \Omega_1(Z(R))^{KR} \rangle$, then $D \in \mathcal{R}_2(KR)$ by B.2.14, and D is $L_R T$-invariant as R and KR are $L_R T$-invariant. Set $H := L_R KR$ and $\hat{H} := H/C_H(D)$. We saw that $L_R \in \mathcal{L}_f(G,T)$ and $R = O_2(L_R T)$, so $[\Omega_1(Z(R)), L_R] \ne 1$ by A.4.8.5 with L_R, $L_R T$, R, T in the roles of "X, M, R, S", and hence $[D, L_R] \ne 1$, so that $\hat{L}_R \ne 1$. As $[Z, K] \ne 1$ by (!), $[D, K] \ne 1$, so that $\hat{K} \ne 1$. We have seen that $[L_R, KR] \le O_2(L_R) \cap KR \le O_2(KR)$, while $O_2(\hat{K}\hat{R}) = 1$ as $D \in \mathcal{R}_2(KR)$, so $[\hat{L}_R, \hat{K}\hat{R}] = 1$. Further $O_2(L_R)$ centralizes $\Omega_1(Z(R))$, so $\hat{H} = \hat{L}_R \times \hat{K}\hat{R}$. In particular $F^*(\hat{H}) = \hat{L}_R \times \hat{K}$, so $\hat{R}$ is faithful on $\hat{K}$.

As T acts on D, $1 \ne [D, L_R] \cap Z =: Z_0 \le Z$. Then as $C_G(Z_0) \le M_c$ by (+), $K \not\le C_G(Z_0)$ by (!), and hence $[D, L_R, K] \ne 1$. As $K = [K, J]$ and $\hat{K} \ne 1$, $\hat{J} \ne 1$, so there is $A \in \mathcal{A}_j(R)$ with $\hat{A} \ne 1$. As $\hat{R}$ is faithful on $\hat{K}$, so is $\hat{A}$.

In a moment we will define a subgroup K_B of K, with $\hat{K}_B = [\hat{K}_B, \hat{A}]$ and $\hat{K}_B$ nontrivial on $[D, L_R]$. Set $H_1 := L_R K_B A$. As $\hat{H} = \hat{L}_R \times \hat{K}\hat{A}$, $\hat{H}_1 = \hat{L}_R \times \hat{K}_B \hat{A}$. As $\hat{L}_R$ is simple, we can choose an H_1-chief section W in $[D, L_R]$ with $\hat{L}_R$ faithful on W and $\hat{K}_B$ nontrivial on W. Set $\bar{H}_1 := H_1/C_{H_1}(W)$ and $\epsilon := m(\hat{A}) - m(\bar{A})$; then $\bar{H}_1 = \bar{L}_R \times \bar{K}_B \bar{A}$, and we will see that $\epsilon = 0$ or 1.

Assume $\hat{K}$ is simple. Then as $1 \ne \hat{A}$ is faithful on $\hat{K}$, $\hat{K} = [\hat{K}, \hat{A}]$; and as $[D, L_R, K] \ne 1$, $\hat{K}$ is faithful on $[D, L_R]$. In this case, we set $K_B := K$. As $\hat{A}$ is faithful on $\hat{K}$ and $\hat{K}$ is faithful on W, $\hat{A}$ is faithful on W, so that $m(\bar{A}) = m(\hat{A})$ and $\epsilon = 0$ in this case.

So assume $\hat{K}$ is not simple. Then $\hat{K}$ is a d-group for $d > 3$, and as $\hat{K}$ is of class at most 2, exponent d, and d-rank at most 2, we conclude from A.1.24 that $\hat{K} \cong E_{d^2}$ or d^{1+2}. Therefore $\hat{A} \le GL_2(d)$, so $m(\hat{A}) \le 2$. Now as $K = [K, J]$ and

K is nontrivial on $[D_L, R]$, we may choose A so that $K_A := [K, A]$ is nontrivial on $[D_L, R]$. Also by A.1.17, $\hat{K}_A$ is generated by the fixed points of hyperplanes of $\hat{A}$; so we may choose a hyperplane $\hat{B}$ of $\hat{A}$ and a subgroup $\hat{K}_B = [\hat{K}_B, \hat{A}]$ of $C_{\hat{K}_A}(\hat{B})$ of order d, such that $\hat{K}_B$ is nontrivial on $[D_L, R]$. By the Thompson $A \times B$-Lemma, $\hat{K}_B$ is nontrivial on $C_{[D, L_R]}(B)$. In this case as $m(\hat{A}) \leq 2$, $\epsilon = 0$ or 1.

Let I be an irreducible $K_B A$-submodule of W. Then as $\bar{H}_1 = \bar{L}_R \times \bar{K}_B \bar{A}$, Clifford's Theorem says that W is the direct sum of r copies of I for some r, and $C_{GL(W)}(\bar{K}_B \bar{A}) \cong GL_r(F)$, where $F := End_{\bar{K}_B}(I)$. As $\bar{L}_R \leq C_{GL(W)}(\bar{K}_B)$ with $\bar{L}_R$ nonabelian, $r > 1$.

As $A \in \mathcal{A}_j(R)$, by B.2.4.1,

$$m(D/C_D(A)) \leq j + m(\hat{A}).$$

Then

$$r \cdot m(I/C_I(\bar{A})) = m(W/C_W(\bar{A})) \leq m(D/C_D(A)) \leq j + m(\hat{A}) = j + \epsilon + m(\bar{A}). \quad (*)$$

Thus as $r > 1 \geq \epsilon$, it follows from $(*)$ that $m(I/C_I(\bar{A})) \leq m(\bar{A})$, so that I is an FF-module for $\bar{K}_B \bar{A}$. Therefore by Theorem B.5.6, $\bar{K}_B$ is not a d-group for a prime $d > 3$. Therefore $K_B = K$, $\epsilon = 0$, and $\bar{K} \cong \mathbf{Z}_3$, A_5, or $L_3(2)$. Further if I is the natural $L_2(4)$-module, then $m(\bar{A}) \leq m_2(Aut(L_2(4))) = 2 \leq m(I/C_I(\bar{A}))$, contrary to $(*)$. Thus if $\hat{K} \cong A_5$, then I is the A_5-module by B.4.2. It follows from B.4.2 that $F = \mathbf{F}_2$ for each of the possible irreducible FF-modules I for each $\bar{K}$, so $\bar{L}_R \leq GL_r(\mathbf{F}_2)$. Since $\bar{L}_R \cong L_2(p)$ for $p \geq 5$, we conclude $r \geq 3$, with $\bar{L}_R \cong L_2(7) \cong L_3(2)$ and each $I_R \in Irr_+(L_R, W)$ of rank 3 in case of equality. Now $m(\bar{A}) \leq m_2(Aut(\bar{K})) \leq 2$ and $m(W/C_W(\bar{A})) \geq r \geq 3$, so all inequalities in $(*)$ must be an equalities, and in particular $\bar{L}_R \cong L_3(2)$, $m(I_R) = r = 3$, $m(I/C_I(\bar{A})) = 1$, $j = 1$, and $m(\bar{A}) = 2$. As $r = 3$, $\bar{K} \cong L_3(2)$ and $m(I) = 3$, so $N_{GL(W)}(\bar{L}_R) \cong L_3(2) \times L_3(2)$; hence $\bar{A} \leq \bar{K}$, $\bar{L}_R \bar{K} \cong L_3(2) \times L_3(2)$, and W is the tensor product of natural modules for the two factors. Furthermore as $m(\bar{A}) = 2$ and $m(I/C_I(\bar{A})) = 1$, $\bar{A}$ is the group of transvections in $\bar{K}$ with a common axis on I. In particular $\bar{A}$ is the unique such subgroup of $\overline{T \cap K}$, so $\bar{A} = \bar{J} \cong E_4$, and hence $J = J_1(O_2(H_1)A)$. Then $N_K(\bar{A}) \leq N_G(J) \leq M_c$ by an earlier observation. Now $\hat{K}$ is simple, so $\hat{A}$ is faithful on each nontrivial $\hat{K}$-section of D. Since $(*)$ is an equality, we conclude that $[D, \hat{K}] = W \in Irr_+(L_R K T)$. Now $Z \cap W$ is a 1-subspace of W centralized by the parabolic of $\hat{K}$ stabilizing the group of transvections on I with a common center, and $C_G(Z \cap W) \leq M_c = !\mathcal{M}(C_G(Z))$ by $(+)$. Thus $K = \langle C_K(Z \cap W), N_K(\bar{A}) \rangle \leq M_c$, contrary to $(!)$. This contradiction completes the proof of (1), and hence of 13.1.9. $\qquad \square$

LEMMA 13.1.10. *One of the following holds:*

(1) $L^* \cong L_2(4)$ *and* $U/C_U(L)$ *is the* $L_2(4)$*-module.*

(2) $L^* T^* \cong S_5$ *and* U *is the* S_5*-module.*

(3) $L^* \cong L_3(2)$ *and* U *is the sum of at most two isomorphic natural modules.*

(4) $L^* \cong L_3(2)$, $m(U) = 4$, *and* $[Z, L] = 1$.

PROOF. By 13.1.9.2, V_c is an FF-module for M_c^* with $L^* \leq J(M_c^*, V_c)$. By 13.1.8.1, $L^* \cong L_2(p)$ for a prime $p \geq 5$. Thus L^* is isomorphic to $L_2(5)$ or $L_2(7)$ by B.5.5.1iv. Then cases (2) and (4) of Theorem B.5.6 do not hold; and in cases (3) and (5) of B.5.6, we see that $U = [V_c, L]$ satisfies one of conclusions (1)–(4). Finally if case (1) of B.5.6 holds, then $L^* = F^*(J(M_c^*, V_c))$ and U is described in Theorem

B.5.1. Now by Theorem B.5.1, either conclusion (3) holds or $U \in Irr_+(L, V_c)$, in which case B.4.2 says U satisfies one of conclusions (1), (2), or (4), using I.1.6.4 in the latter case. Furthermore in any case $Z \leq R_2(M_c) = V_c$, and $\langle Z^L \rangle = UC_Z(L)$ by B.2.14. Thus in conclusion (4), since $Z \cap U = C_U(L)$ by B.4.8.2, we conclude that $[Z, L] = 1$. □

LEMMA 13.1.11. *For $X \in \mathcal{X}$, if $X \leq X_1 \leq M_c$ and X_1 acts irreducibly on $X/O_{2,\Phi}(X)$, then X_1 is contained in a unique conjugate of M_c.*

PROOF. Suppose $X_1 \leq M_c^g$ and let p be the odd prime in $\pi(X)$, $P \in Syl_p(X)$, and $\hat{M}_c^g := M_c^g/O_2(M_c^g)$. We apply A.1.32 with $\hat{M}_c^g$, $\hat{X}^g$, $\hat{P}$, p, p in the roles of "G, R, P, r, p". The second case of A.1.32.2 does not arise as $\hat{X}^g$ is not of order p. Thus $\hat{X}^g = \hat{X}$, so $X = O^2(X) \leq O^2(X^g) = X^g$, and hence $g \in N_G(X) = M_c$, so $M_c = M_c^g$. □

LEMMA 13.1.12. *Let $u \in U^\#$, $G_u := C_G(u)$, $M_u := C_{M_c}(u)$, and pick u so that $T_u := C_T(u) \in Syl_2(M_u)$. Then*

(1) M_u is irreducible on $X/O_{2,\Phi}(X)$ for each $X \in \mathcal{X}$.
(2) $|T : T_u| \leq 2$.
(3) $G_u \leq M_c$.

PROOF. Let

$$\mathcal{U} := \{u \in U^\# : |T_u| < |T|\}.$$

If $u \in Z$ then $T_u = T$ is irreducible on $X/O_{2,\Phi}(X)$ for each $X \in \mathcal{X}$, while $G_u \leq M_c = !\mathcal{M}(C_G(Z))$ by $(+)$, so that (1)–(3) hold. Thus we may assume that $\mathcal{U}$ is nonempty, and it suffices to establish (1)–(3) for $u \in \mathcal{U}$. In particular M_c is not transitive on $U^\#$, so $C_U(L) \neq 1$ in case (1) of 13.1.10, and U is the sum of two irreducibles for L^* in case (3).

Let $u \in \mathcal{U}$, and recall from 13.1.4.5 that X centralizes $R_2(M_c) = V_c$ and $U = [V_c, L]$, so that $X \leq M_u$. As $X \in \Xi_+(G, T)$, $X \cong E_{p^2}$ or p^{1+2} for some prime $p > 3$. Set $M_u^+ := M_u/C_{M_u}(X/O_{2,\Phi}(X))$; thus $M_u^+ \leq GL_2(p)$. To prove (1), it will suffice to show that M_u^+ is nonabelian; as $C_G(X) \leq O_{2,F}(L) \leq C_G(U)$, it also suffices to show $C_L(u)^*$ is nonabelian. Indeed $L/O_{2,F}(L) \cong SL_2(q)$ for $q = 5$ or 7, so if $C_L(u)^*$ contains a 4-group, then the preimage of this 4-group in $C_L(u)/O_{2,F}(X)$ is the nonabelian group Q_8, which again suffices. To prove (2), it will suffice to show for each orbit $\mathcal{O}$ of LT on $\mathcal{U}$ that $|\mathcal{O}| \equiv 2 \mod 4$.

Assume case (1) of 13.1.10 holds. By I.2.3.1, U is a quotient of the rank-6 extension U_0 of $U/C_U(L)$, so $m(C_U(L)) = 1$ or 2. However if $m(C_U(L)) = 1$ or $T^* \leq L^*$, then all members of $U^\#$ are 2-central in M_c. Therefore $m(C_U(L)) = 2$, so $U = U_0$ and hence U admits an $\mathbf{F}_4$-structure by I.2.3.1. Further $T^* \nleq L^*$, so $L^*T^* \cong S_5$. Then LT has two orbits on $\mathcal{U}$: one of length 2 in $C_U(L)$, and one of length 30 on $U - C_U(L)$. In either case, $|u^{LT}| \equiv 2 \mod 4$, so that (2) holds by the previous paragraph. Also $(T_u \cap L)^* \in Syl_2(L^*)$, so (1) also holds by the previous paragraph.

Assume case (3) of 13.1.10 holds. We saw U is the sum of two natural modules, so L is transitive on

$$\mathcal{U}_1 := U - \bigcup_{I \in Irr_+(L,U)} I$$

of order 42 with $C_L(u_1)^* \cong E_4$ for $u_1 \in \mathcal{U}_1$. Further either $\mathcal{U} = \mathcal{U}_1$, or $Aut_{M_c}(U) \cong L_3(2) \times \mathbf{Z}_2$ and $\mathcal{U} = \mathcal{U}_1 \cup \mathcal{U}_2$, where

$$\mathcal{U}_2 := \bigcup_{I \in Irr_+(L,U)} I - U_1 \text{ is of order 14,}$$

where U_1 is the unique T-invariant member of $Irr_+(L,U)$, and $C_L(u)^* \cong S_4$ for $u \in \mathcal{U}_2$. In either case, (1) and (2) hold.

In case (2) of 13.1.10, $\mathcal{U}$ is the set of nonsingular vectors in the orthogonal space U, so $|\mathcal{U}| = 10$, and $C_{L^*}(u) \cong S_3$ is nonabelian; thus (1) and (2) hold in this case. Finally in case (4) of 13.1.10, L is transitive on $\mathcal{U} = U - C_U(L)$ of order 14, so $C_{L^*}(u) \cong A_4$ is nonabelian, completing the proof of (1) and (2). Observe we also showed in cases (1), (3), and (4) of 13.1.10 that $(T_u \cap L)^+$ contains a Q_8-subgroup.

It remains to prove (3), so we assume $G_u \not\le M_c$, and derive a contradiction.

We next claim that $T_u \in Syl_2(G_u)$: For if not, then by (2), $u^g \in Z$ for some $g \in G$. Then $G_u^g = C_G(u^g) \le M_c = !\mathcal{M}(C_G(Z))$ by (+), and hence $X \le M_u = M_c \cap G_u \le M_c^{g^{-1}}$. By (1), we may apply 13.1.11 to conclude that M_u is contained in a unique conjugate of M_c, so that $M_c = M_c^{g^{-1}}$. Then $g \in M_c$ as $M_c \in \mathcal{M}$, so u is centralized by an M_c-conjugate of T, whereas $u \in \mathcal{U}$ so $|M_c : M_u|$ is even. Hence the claim is established.

Set $R_X := O_2(XT)$ and $R := C_{R_X}(u)$. Observe that $[L, R_X] \le O_\infty(L) = C_L(U)$, so L^* centralizes R_X^*. We claim that $R = C_{R_X}(U)$. Since $C_{R_X}(U) \le R$, it suffices to show that R centralizes U. If $U \in Irr_+(L,U)$, then as L^* centralizes R_X^*, R_X centralizes U by A.1.41, so that the claim holds. Therefore we may assume that $U \notin Irr_+(L,U)$, so case (3) of 13.1.10 holds. Then $Aut_{M_c}(U) \le L_3(2) \times S_3$, so we may assume $Aut_R(U) = C_{Aut_{M_c}}(U) \cong \mathbf{Z}_2$. But then $C_U(R)$ is a T-invariant member of $Irr_+(L,U)$, so as $u \in C_U(R)$, u is 2-central in M_c, contrary to $u \in \mathcal{U}$. This completes the proof of the claim.

By the claim, $R \trianglelefteq XT$, so as $M_c = !\mathcal{M}(XT)$ by (+),

$$C(G_u, R) \le M_c \cap G_u = M_u.$$

Next, the hypotheses of A.4.2.7 are satisfied with G_u, M_u, T_u in the roles of "G, M, T": For $N_{G_u}(R) \le M_u$ and $X \trianglelefteq M_u$, with T_u Sylow in M_u and G_u, and $R = O_2(XT) = O_2(XT_u)$. Therefore $R \in \mathcal{B}_2(G_u)$ and $R \in Syl_2(\langle R^{M_u} \rangle)$ by A.4.2.7. Thus the pushing up Hypothesis C.2.3 is satisfied with G_u, M_u in the role of "H, M_H". However, before we apply pushing up results from section C.2, we will establish a number of further preliminary results.

We claim next that $O_{2,F}(G_u) \le M_u$: Set $\hat{G}_u := G_u/O_2(G_u)$ and recall p is the odd prime in $\pi(X)$. Let $\hat{R}_r$ denote a supercritical subgroup of $O_r(\hat{G}_u)$. As M_u is irreducible on $X/O_{2,\Phi}(X)$ by (2), we may apply A.1.32 with $\hat{G}_u$, $\hat{R}_r$, $\hat{X}$ in the roles of "G, R, P". If $r \ne p$, then by part (1) of that result, $\hat{X}$ centralizes $\hat{R}_r$, and hence $O^p(O_{2,F}(G_u))$ normalizes X. If $r = p$, then by part (2) of A.1.32, either $\hat{X} = \hat{R}_p$, or $\mathbf{Z}_p \cong \hat{R}_p = Z(\hat{X})$ and $\hat{X} \cong p^{1+2}$. In the former case, $O_p(\hat{G}_u)$ normalizes the characteristic subgroup $\hat{R}_p = \hat{X}$, so the claim holds. In the latter case, since the supercritical subgroup $\hat{R}_p$ contains all elements of order p in $C_{O_p(\hat{G}_u)}(\hat{R}_p)$, we conclude that $O_p(\hat{G}_u)$ is cyclic. Then as M_u is irreducible on $X/O_{2,\Phi}(X)$, $\hat{X}$ centralizes $O_p(\hat{G}_u)$, completing the proof of the claim. We have also shown that $\hat{X}$

centralizes $O^p(F(\hat{G}_u))$ and either $\hat{X} = \hat{R}_p \leq O_p(\hat{G}_u)$, or $\hat{X}$ centralizes $O_p(\hat{G}_u)$ and hence $F(\hat{G}_u)$.

Let $z \in Z^{\#}$. By 1.1.6, the hypotheses of 1.1.5 are satisfied with $\langle u \rangle$, G_u, T_u, M_c in the roles of "B, H, S, M". By the claim, $O(F(G_u)) \leq N_{G_u}(X) \leq M_u$, so $O(F(G_u)) \leq C_{G_u}(z)$ for $z \in Z \cap O_2(X)^{\#}$. But by 1.1.5.2, z inverts $O(G_u)$, so $O(G_u) = 1$.

Assume that $O_{2,F^*}(G_u) \leq N_{G_u}(X)$. Then $\hat{X}$ centralizes $E(\hat{G}_u)$. We saw that $\hat{X}$ centralizes $O^p(F(\hat{G}_u))$ and either $\hat{X} = \hat{R}_p$ or $\hat{X}$ centralizes $F(\hat{G}_u)$. In the latter case $\hat{X}$ centralizes $F^*(\hat{G}_u)$, so that $\hat{X} \leq O_p(\hat{G}_u)$, and then as $m_p(\hat{X}) = 2 = m_p(\Omega_1(O_p(\hat{G}_u)))$, $\hat{X} = \Omega_1(O_p(\hat{G}_u))$. Hence in either case $G_u \leq N_{G_u}(X) = M_c$, contrary to assumption.

Therefore there exists $K \in \mathcal{C}(G_u)$ with $K \nleq N_G(X)$ and $K/O_2(K)$ quasisimple. Then $X = O^2(X)$ normalizes K by 1.2.1.3, and $K = [K, X]$ by A.3.3.7. Set $K_0 := \langle K^{T_u} \rangle$.

Suppose $N_{M_u}(K)$ is irreducible on $X/O_{2,\Phi}(X)$. Then $C_X(\hat{K}) \leq O_{2,\Phi}(X)$ and as $\hat{K}$ is described in Theorem C (A.2.3), $m_p(Out(\hat{K})) \leq 1$ since $p > 3$. Therefore since $N_{M_u}(\hat{K})$ is irreducible on $X/O_{2,\Phi}(X)$, X induces inner automorphisms on $\hat{K}$. Then $m_p(K) > 1$, so $K = O^{p'}(G_u)$ by A.3.18. Thus $K_0 = K$ and $X \leq K$. Also $T_u X = X T_u$ with $T_u \in Syl_2(G_u)$, so $\hat{K}$ and the embedding of $\hat{X}$ in $\hat{K}$ are described in A.3.15. As $m_p(Aut_X(\hat{K})) > 1$, and $N_{M_u}(K)$ is irreducible on $X/O_{2,\Phi}(X)$, conclusion (3) of A.3.15 is eliminated, so conclusion (2) or (5) of A.3.15 holds.

Let $P \in Syl_p(X)$. During the proof of (1) and (2), we saw that T_u is reducible on $X/O_{2,\Phi}(X)$ in case (2) of 13.1.10, and in the remaining cases T_u is irreducible and $Aut_{T_u \cap L}(P)$ is noncyclic. Suppose T_u is irreducible on $X/O_{2,\Phi}(X)$. Then $X \in \Xi(G_u, T_u)$. We observe that the proof of 1.3.4 does not require the hypotheses that $H \in \mathcal{H}(XT)$, but only that $H \in \mathcal{H}(X)$, and $N_{T \cap H}(X)$ is irreducible on $X/O_{2,\Phi}(X)$. Thus we may apply the analogue of that result with G_u, T_u, K in the roles of "$H, T, \langle L^T \rangle$", to conclude that $\hat{K}$ is described in 1.3.4. Therefore if A.3.15.5 holds, then $\hat{K} \cong Sp_4(2^n)$ with $Aut_{T_u}(P)$ cyclic, contrary to a remark earlier in the paragraph. Thus T_u is reducible on $X/O_{2,\Phi}(X)$, so we are in case (2) of 13.1.10, where $C_L(u)^+$ contains an S_3-section. In case (2) of A.3.15, T_u is irreducible on $X/O_{2,\Phi}(X)$, so we are in case (5) of A.3.15. Then as $P \leq K$ with $PT_u = T_uP$ and $p > 3$, $\hat{K}$ is of Lie type over $\mathbf{F}_{2^n}$ with $2^n \equiv 1 \mod p$, and P is contained in the Borel subgroup $N_K(T_u \cap K)$. Hence the S_3-section is induced by outer automorphisms of $\hat{K}$, so from the structure of $Out(K_0/O_2(K_0))$, $\hat{K} \cong (S)L_3(2^n)$ with n even.

Having discussed the case where $N_{M_u}(K)$ is irreducible on $X/O_{2,\Phi}(X)$, we now turn to the remaining case where it is reducible. If T_u normalizes K, then so does M_u by 1.2.1.3, and then (1) contradicts the assumption in this case. Therefore $K < K_0$. However M_u acts on K_0 and is irreducible on $X/O_{2,\Phi}(X)$ by (1). Further by 1.2.1.3, $Out(\hat{K})$ is cyclic. so as $Out(\hat{K}_0)$ is $Out(\hat{K})$ wr $\mathbf{Z}_2$, again X induces inner automorphisms on $\hat{K}_0$. By 1.2.2.a, $K_0 = O^{p'}(G_u)$, so a Sylow p-subgroup P of X is containied in K_0 and $P = P_K \times P_K^t$, where $P_K := P \cap K \cong \mathbf{Z}_p$ and $t \in T_u - N_{T_u}(K)$. As $T_u P = PT_u$, we conclude from 1.2.1.3 and A.3.15 that $\hat{K}$ is isomorphic to $L_2(2^n)$ or $Sz(2^n)$ with $2^n \equiv 1 \mod p$, J_1 with $p = 7$, or $L_2(r)$ for a suitable odd prime r.

Summarizing our list of possibilities, $X \leq K_0$ and either

(i) $K = K_0$, with $\hat{K}$ isomorphic to $L_3(p)$ or $(S)L_3(2^n)$ where $2^n \equiv 1 \mod p$, or

(ii) $K_0 = KK^t$ for $t \in T_u - N_{T_u}(K)$, and $\hat{K}$ is isomorphic to $Sz(2^n)$ or $L_2(2^n)$ with $2^n \equiv 1 \mod p$, $L_2(r)$ for a suitable odd prime r, or J_1 with $p = 7$.

Recall we saw earlier that Hypothesis C.2.3 holds. We are now ready to apply pushing up results from section C.2.

Suppose first that $F^*(K) = O_2(K)$. If $R \not\leq N_{G_u}(K)$ then $K < K_0$, and by C.2.4.2, $R \cap K \in Syl_2(K)$, so K is a χ_0-block of G_u by C.2.4.1. Then from our list of possibilities for $\hat{K}$, $\hat{K} \cong L_2(2^n)$. On the other hand if $R \leq N_{G_u}(K)$, then K is described in C.2.7.3. We compare the list of C.2.7.3 with our list of possibilities for $\hat{K}$ in (i) and (ii): If $\hat{K} \cong (S)L_3(2^n)$, then case (g) of C.2.7.3 occurs, so $K \cap M_c$ is a maximal parabolic of $SL_3(2^n)$, impossible as $X \trianglelefteq K \cap M_c$. The only remaining possibility in both lists is case (a) of C.2.7.3 with K a χ-block, so again $\hat{K} \cong L_2(2^n)$ and $K < K_0$.

Thus in any case, $K_0 = KK^t$ for $t \in T_u - N_{T_u}(K)$, and $[K, K^t] = 1$ by C.1.9. Let $\mathcal{P}$ be the set of subgroups P_0 of P of order p such that $[C_{O_2(X)}(P_0), P] \neq 1$, and set $X_K := X \cap K$ and $P_K := P \cap K$. Then $X = X_K X_K^t$ and $\mathcal{P} = \{P_K, P_K^t\}$. But $M_c = N_G(X)$, so $N_{M_c}(P)$ permutes $\mathcal{P}$, contrary to the fact that $N_L(P)$ induces either $SL_2(p)$ or $SL_2(5)$ on P, and thus has no orbit of order 2 on the set of subgroups of P of order p.

Therefore $F^*(K) \neq O_2(K)$, so as $O(G_u) = 1$ and $K/O_2(K)$ is quasisimple, K is quasisimple. Then as $X \leq K_0$, and $F^*(X) = O_2(X)$, we conclude by comparing the list in 1.1.5.3 with our list of possibilities for $\hat{K}$ in (i) and (ii) that $K_0/Z(K_0) \cong (S)L_3(2^n)$, $L_2(2^n) \times L_2(2^n)$, or $Sz(2^n) \times Sz(2^n)$ for some n, or $L_2(r) \times L_2(r)$ for r a Fermat or Mersenne prime. In the latter three cases the components commute, so as in the previous paragraph we conclude that $N_{M_c}(P)$ permutes the subgroups $P \cap K$ and $P \cap K^t$, for the same contradiction. Furthermore a similar argument works in the first case: Namely X lies in a Borel subgroup of K, so that $O_2(X)$ is the full unipotent group A, which is special of order 2^{3n} with center of order 2^n. Therefore $N_{M_c}(P)$ acts on $C_P(Z(A)) \cong \mathbf{Z}_p$, for the same contradiction. This finally completes the proof of 13.1.12. $\qquad \square$

LEMMA 13.1.13. U is a TI-set in G.

PROOF. Suppose $1 \neq u \in U \cap U^g$ for some $g \in G$. Then by 13.1.12.3, $X^g \leq C_{M_c^g}(u) \leq M_c$ for $X \in \mathcal{X}$, and by 13.1.12.1, $C_{M_c^g}(u)$ is irreducible on $X^g/O_{2,\Phi}(X^g)$, so 13.1.11 says $M_c = M_c^g$. Thus $g \in N_G(M_c) = M_c$ as $M_c \in \mathcal{M}$, so $U = U^g$, completing the proof. $\qquad \square$

Recall the weak-closure parameters $w(G, U)$ and $r(G, U)$ from Definitions E.3.23 and E.3.3.

LEMMA 13.1.14. (1) $W_i(T, U)$ centralizes U for $i = 0, 1$, so $N_G(W_0(T, U)) \leq M_c$.

(2) $w(G, U) > 1 < r(G, U)$.

(3) If $H \in \mathcal{H}(T)$ with $n(H) = 1$, then $H \leq M_c$.

PROOF. As U is a TI-set in G by 13.1.13, if $N_{U^g}(U) \neq 1$ and $\langle U, U^g \rangle$ is a 2-group, then $[U, U^g] = 1$ by I.7.6. Therefore as $U \trianglelefteq T$, we conclude that $W_0 := W_0(T, U)$ centralizes U. Hence $W_0 \leq C_T(U) =: R$, so that $W_0 = W_0(R, U)$. Now by a Frattini Argument, $L = O_\infty(L)N_L(W_0)$, so $N_G(W_0) \leq M_c$ by 13.1.9.1.

Next assume $W_1(T, U)$ does not centralize U. Then by the previous paragraph, there is $g \in G$ with $A := U^g \cap T$ a hyperplane of U^g and $A^* \neq 1$. As $A^* \neq 1$, I.6.2.2a says that A^* is the full group of $\mathbf{F}_2$-transvections on U with axis $U \cap M_c^g$. Inspecting the cases listed in 13.1.10, we conclude $L^* \cong L_3(2)$, $m(U) = 3$, and $E_4 \cong A \leq L^*$. Hence A induces a faithful 4-group of inner automorphisms on $L/O_{infy}(L)$. This is impossible as $L/O_2(L) \cong SL_2(7)/E_{49}$, so $Aut(L/O_2(L)) \cong GL_2(7)/E_{49}$. This completes the proof of (1).

By (1), $w(G, U) > 1$, and by 13.1.13, $r(G, U) = m(U) > 1$. Thus (2) holds. Finally the hypotheses of E.3.35 are satisfied with U, R, M_c in the roles of "V, Q, M", so (2) and E.3.35.1 imply (3).   □

We are now in a position to complete the proof of Theorem 13.1.7.

We saw at that outset of the proof that $|\mathcal{M}(T)| > 1$, so that there is some $M \in \mathcal{M}(T)$ with $M \neq M_c$. Thus we may choose Y as in 13.1.6. Then $Y \not\leq M_c$, so $n(YT) > 1$ by 13.1.14.3. Thus Y is not solvable by E.1.13, so case (2) of 13.1.6 holds, and in particular $Y \in \mathcal{L}_f^*(G, T)$. Therefore $Y/O_2(Y)$ is described in 13.1.2.3, so as $n(YT) > 1$, $Y/O_2(Y) \cong A_5$ using E.1.14.

Next as $Y \not\leq M_c$, $Y \not\leq N_G(W_0(T, U))$ in view of 13.1.14.1. Thus by E.3.15, there is $A := U^g \leq T$ for some $g \in G$ with $A \not\leq Q := O_2(YT)$. Let $A_Q := A \cap Q$; then $m(A/A_Q) \leq m_2(YT/O_2(YT)) = 2$, so as $m(A) \geq 3$ by 13.1.10, it follows that $A_Q \neq 1$.

As $A \not\leq O_2(YT)$, $O^2(\langle A^Y \rangle) = Y$. As $Y \not\leq M_c$, there is $h \in Y$ with $A^h \not\leq M_c$. But $A^h \leq YT$, so if $U \leq O_2(YT) = Q$, then $\langle A^h, U \rangle$ is a 2-group with $1 \neq A_Q^h = A^h \cap Q \leq N_G(U)$; then by I.7.6, $A^h \leq C_G(U) \leq N_G(U) = M_c$, contrary to our choice of A^h. Thus $U \not\leq Q$, so we may take $A = U$.

Let $I := \langle U, U^h \rangle$. Then as $m_2(YT/Q) \leq 2$ and $Q = \ker_{YT}(N_{YT}(U))$, $V := U \cap M_c^h$ and $B := U^h \cap M_c$ are of codimension at most 2 in U and U^h, respectively. Therefore since $I/O_2(I)$ is a section of $Y/O_2(Y) \cong A_5$, (a) and (c) of I.6.2.2 say that $O_2(I) = V \times B$, and $I/O_2(I) \cong D_6$, D_{10}, or $L_2(4)$, and $O_2(I)$ is a direct sum of natural modules for $I/O_2(I)$. However in the first two cases, B is of index 2 in U^h, so by 13.1.14.1, $[B, U] = 1$, and then $[U^h, U] = 1$, a contradiction. Hence $I/O_2(I) \cong L_2(4)$. Let $D \in Syl_3(N_I(U))$; then $V = [V, D]$ since $O_2(I)$ is the sume of natural modules for $I/O_2(I)$, and so $U = [U, D]$; thus U is the natural module for $L^* \cong L_2(4)$ by 13.1.10. Hence $B \cong E_4$ and $V = C_U(b)$ for each $b \in B^{\#}$, so B induces a faithful 4-group of inner automorphisms on $L/O_{\infty}(L)$. As in the proof of 13.1.14, this is a contradiction. This completes the proof of Theorem 13.1.7.

13.2. Some preliminary results on $\mathbf{A}_5$ and $\mathbf{A}_6$

In this section we establish some technical results used in our treatment of the cases $L/O_{2,Z}(L) \cong A_5$ or A_6 in the FSU. Thus in section 13.2, we assume Hypothesis 12.2.3 from the previous chapter. In particular $M = N_G(L)$ for some $L \in \mathcal{L}_f^*(G, T)$ with $L/O_2(L)$ quasisimple and $V \in Irr_+(L, R_2(LT))$ is T-invariant and satisfies the Fundamental Setup (3.2.1).

As usual we adopt the conventions of Notation 12.2.5; e.g., $Z = \Omega_1(Z(T))$, $M_V = N_G(V)$, and $\bar{M}_V = M_V/C_G(V)$. We also set

$$Z_V := C_V(L) \quad \text{and} \quad \hat{V} := V/Z_V.$$

Throughout this short section we assume that $\bar{L} \cong A_n$ for $n = 5$ or 6. Then we are in case (d) of 12.2.2.3, with $\hat{V}$ the 4-dimensional chief factor in a rank-n

permutation module for $\bar{L}$. In particular if $L/O_2(L) = \hat{A}_6$, then $O_{2,Z}(L) \leq C_L(V)$. Therefore as $Out(\bar{L})$ is a 2-group and V is T-invariant, $\bar{M}_V = \bar{L}\bar{T} \cong A_n$ or S_n from the structure of $N_{Aut(\bar{L})}(V)$. We also adopt the notational conventions of section B.3; in particular, $\{1, 2, 3, 4\}$ is an orbit under T.

By B.3.3, if $Z_V \neq 1$ then $n = 6$, V is the core of the permutation module for $\bar{L}$ on $\Omega := \{1, \ldots, n\}$, and Z_V is generated by e_Ω. In any event $\hat{V}$ is the irreducible quotient of the core of the permutation module modulo $\langle e_\Omega \rangle$.

When $n = 6$ we can also view $\hat{V}$ as a 4-dimensional symplectic space over $\mathbf{F}_2$ for $\bar{L} \cong Sp_4(2)'$. When $n = 5$, $\hat{V} = V$ since $\hat{V}$ is projective for $\bar{L} \cong A_5$ (cf. I.1.6.1), and we can view $\hat{V}$ as a 4-dimensional orthogonal space for $\bar{L} \cong \Omega_4^-(2)$. Thus we can speak of isotropic or singular vectors in $\hat{V}$, nondegenerate subspaces of $\hat{V}$, etc. We also adopt the following notational conventions:

NOTATION 13.2.1. For $1 \leq i \leq 4$, let V_i be the preimage in V of an i-dimensional subspace of $\hat{V}$ stabilized by T. Set $M_i := N_{LT}(V_i)$, $L_i := O^2(M_i)$, and let R_i be the preimage in T of $O_2(\bar{M}_i)$. Notice for $i < 4$ that $|R_i : O_2(L_iT)| \leq 2$, with equality iff $L/O_2(L) \cong \hat{A}_6$ and $\bar{T} \not\leq \bar{L}$, in which case $O_2(L_iT) = O_2(L_i)O_2(LT)$. When $L/O_2(L) \cong \hat{A}_6$, define $L_0 := O^2(O_{2,Z}(L))$, and for $i = 1, 2$, set $L_{i,+} := O^2([L_i, T \cap L])$; observe $|L_0 : O_2(L_0)| = 3 = |L_{i,+} : O_2(L_{i,+})|$.

13.2.1. Results on $\mathbf{A_6}$. We collect a number of results on A_6 into a single lemma. The first few are easy calculations invoving only L and V, which do not require Hypothesis 12.2.3.

LEMMA 13.2.2. *Assume $n = 6$ and set $Q := O_2(LT)$. Then*

(1) *L is transitive on $\hat{V}^\#$.*

(2) *Each $v \in V^\#$ is in the center of a Sylow 2-subgroup of LT.*

(3) *If $L/O_2(L) \cong \hat{A}_6$, then $L_i = L_{i,+}L_0$ for $i = 1, 2$.*

(4) *If $L = [L, J(T)]$, then $\bar{L}_1 = [\bar{L}_1, J(T)]$.*

(5) *LT controls G-fusion in V.*

(6) *$m_2(R_1) = m_2(Q)$, so $V \leq J(R_1)$.*

(7) *Either there is a nontrivial characteristic subgroup of $Baum(R_1)$ normal in LT (and hence $N_G(Baum(R_1)) \leq M$), or L is an A_6-block.*

(8) *If $L/O_2(L) \cong \hat{A}_6$, then $J(O_2(L_1T)) = J(O_2(LT))$, so every nontrivial characteristic subgroup of $Baum(O_2(L_1T))$ is normal in LT.*

(9) *If $L/O_2(L) \cong \hat{A}_6$, then $N_G(L_1) \leq M \geq N_G(L_0)$.*

(10) *One of the following holds:*

    (I) Some nontrivial characteristic subgroup of $Baum(T)$ is normal in LT.

    (II) L is an A_6-block, and $\mathcal{A}(O_2(LT)) \subseteq \mathcal{A}(T)$.

    (III) $\bar{L}_2 = [\bar{L}_2, J_1(T)]$.

PROOF. Parts (1) and (3) are easy calculations, and (2) follows from (1) since the elements of V_1 are central in T. If $Z_V = 1$,, then (5) follows from (1). By Burnside's Fusion Lemma A.1.35, $N_G(T)$ controls fusion in Z, while $N_G(T) \leq M$ by Theorem 3.3.1. Therefore if $Z_V \neq 1$, then $M = M_V$ controls fusion in V_1, so as $\bar{M}_V = \bar{L}\bar{T}$, (5) follows from (1) in this case also.

Next we establish (4) and (6), which will follow fairly easily from B.3.4. First $\bar{R}_1$ contains no strong FF*-offenders by parts (1) and (2ii) of B.3.4, so by B.2.4.3, $m_2(R_1) = m_2(Q)$ and $\mathcal{A}(Q) \subseteq \mathcal{A}(R_1)$. Then as $V \leq \Omega_1(Z(Q))$, $V \leq J(Q) \leq J(R_1)$, completing the proof of (6).

If $J(T) \leq C_T(V) = O_2(LT)$, then (4) is vacuously true. Thus we may suppose that there is $A \in \mathcal{A}(T)$ with $\bar{A} \neq 1$; in particular $L = [L, J(T)]$. Now the hypotheses of B.2.10.2 (and hence of B.2.10.1) are satisfied with LT, T in the roles of "G, R", so $\mathcal{P} := \mathcal{P}_{T,LT}$ is a nonempty stable subset of $\mathcal{P}(\bar{L}\bar{T}, V)$ by B.2.10.1. Similarly using R_1 in the role of "R", $J(R_1) \not\leq Q$ iff $\mathcal{Q} := \mathcal{P}_{R_1,LT}$ is a nonempty stable subset of $\mathcal{P}(\bar{L}\bar{T}, V)$. Moreover by definition in B.2.5, $\overline{J(T)} = J_{\mathcal{P}}(\bar{T})$ and $\overline{J(R_1)} = J_{\mathcal{Q}}(\bar{R}_1)$.

If $\bar{M}_V \cong A_6$, then from B.3.4.1, $J_{\mathcal{P}}(\bar{T}) = \bar{R}_2$, so $\bar{L}_1 = [\bar{L}_1, J_{\mathcal{P}}(\bar{T})]$, and hence (4) holds. So assume instead that $\bar{M}_V \cong S_6$. If $J_{\mathcal{Q}}(\bar{R}_1) = 1$, then (4) follows from B.3.4.2iv, while if $J_{\mathcal{Q}}(\bar{R}_1) \neq 1$, then (4) follows from B.3.4.2v. This completes the proof of (4).

For part (7), we observe that the hypotheses of C.1.37 are satisfied with R_1 in the role of "R" and $P := L_1 T$, except when $\bar{M}_V \cong \hat{S}_6$, when we take $P := L_{1,+}T$. Thus conclusion (1) or (2) of C.1.37 holds, giving the alternatives of conclusion (7) of the present result. (Recall that L is not a $\hat{A}_6$-block as $O_{2,Z}(L) = C_L(V)$.)

Next we will prove (8) and (9), so we may assume $L/O_2(L) \cong \hat{A}_6$. Set $R := O_2(L_1 T)$. We claim first that $J(R) = J(Q)$: If $\bar{M}_V = A_6$ then $R = R_1$ by 13.2.1, and by B.3.4.1, $J(R_1) \leq Q \leq R_1$ so that $J(R) = J(R_1) = J(Q)$. If $\bar{M}_V = \hat{S}_6$, then by 13.2.1, $|R_1 : R| = 2$ and $R = O_2(L_1)Q \leq LQ$. But by B.3.4.2v, if $\overline{J(R)} \neq 1$ then $\overline{J(R)} \not\leq \bar{L}$, so again $J(R) \leq Q$ and $J(R) = J(Q)$, completing the proof of the claim. Then by B.2.3.5, $\mathrm{Baum}(R) = \mathrm{Baum}(Q)$, establishing (8).

Recall $L_0 = O^2(O_{2,Z}(L))$. Therefore $L_0 \trianglelefteq LT$, so that $N_G(L_0) \leq M = !\mathcal{M}(LT)$. Finally $N_G(L_1) = C_G(L_1/O_2(L_1))N_G(R)$ by A.4.2 and a Frattini Argument, so as $N_G(R) \leq N_G(J(R)) \leq M$ by (8), and $C_G(L_1/O_2(L_1))$ normalizes L_0 with $N_G(L_0) \leq M$, we conclude $N_G(L_1) \leq M$, completing the proof of (9).

Finally we prove (10). Let $S := \mathrm{Baum}(T)$. If $J(T) \leq C_T(V) = O_2(LT)$, then using B.2.3.3, $J(T)$ is a nontrivial characteristic subgroup of S normal in LT, so (I) holds. Thus we may assume there is $A \in \mathcal{A}(T)$ with $1 \neq \bar{A} \in \mathcal{P}$. If B is a hyperplane of A with $\bar{B}^L \cap \bar{T} \not\subseteq \bar{R}_2$, then as $B \in \mathcal{A}_1(T)$, (III) holds. Thus we may assume no such B exists. Therefore by B.3.4.2vi, $|\bar{A}| = 2$ for each such A. In particular, $\overline{J(T)}$ lies in the subgroup $\bar{R}_2$ of $\bar{T}$ generated by transvections, so $\mathrm{Baum}(T) = \mathrm{Baum}(R_2)$ by B.2.3.5. Observe that we now have the hypothesis of C.1.37 with R_2 in the role of "R" and $P := L_2 T$, unless $LT/O_2(LT) \cong \hat{S}_6$, when we take $P := L_{2,+}T$. Further conclusion (5) of C.1.37 does not hold, as there are no FF*-offenders with image of order greater than 2, so only conclusions (1) or (2) of that lemma can hold. In case (1), (I) holds, and in case (2), L is an A_6-block (Again L is not a $\hat{A}_6$-block as $C_L(V) = O_{2,Z}(L)$). Further FF*-offenders of order 2 are not strong by B.3.4.2i, so that $\mathcal{A}(Q) \subseteq \mathcal{A}(T)$ by B.2.4.3, and hence (II) holds. This completes the proof of (10), and of 13.2.2. $\qquad\square$

13.2.2. Results on $\mathbf{A}_5$. In this subsection we assume $n = 5$ and establish a series of results culminating in an important reduction: Theorem 13.2.7. Notice that as $n = 5$, we have Hypothesis 5.0.1, of section 5.1, so we can use results from that section and the subsequent sections of chapters 5 and 6.

LEMMA 13.2.3. *If $n = 5$ then*

(1) $O_2(LT) = C_{LT}(V) = C_{LT}(V_3)$.
(2) $N_G(V_3) \leq M_V$.

PROOF. Part (1) follows from the structure of the A_5-module. Then by (1), $R := O_2(LT) \in Syl_2(C_M(V_3))$, so as $C(G, R) \le M = !\mathcal{M}(LT)$, $N_G(R) \le M$ and $R \in Syl_2(C_G(V_3))$. Therefore by a Frattini Argument,

$$N_G(V_3) = C_G(V_3)(N_G(R) \cap N_G(V_3)),$$

so it remains to show that $C_G(V_3) \le M$—since then $N_G(V_3) \le M_V$ by 12.2.6. So assume $C_G(V_3) \not\le M$. Then there is $H \in \mathcal{H}_*(T, M)$ with $O^2(H) \le C_G(V_3)$, and hence $R \in Syl_2(O^2(H)R)$. Then by Theorem 3.1.1 there is $1 \ne R_0 \le R$ with $R_0 \trianglelefteq \langle LT, H \rangle$, and so $H \le N_G(R_0) \le M = !\mathcal{M}(LT)$, contrary to assumption. $\square$

LEMMA 13.2.4. *Assume* $n = 5$. *Then for any* $W \in \mathcal{R}_2(LT)$ *with* $[W, L] \ne 1$:

(1) $R_1 = (T \cap L)O_2(LT) = O_2(C_{LT}(Z \cap [W, L]))$. *Further* $J(R_1) = J(C_T(W))$ *and* $Baum(R_1) = Baum(C_T(W))$, *so that* $C(G, Baum(R_1)) \le M$.
(2) *Let* $S := Baum(T)$; *then either*:

   (a) $S \le C_T(W)$ *so that* $J(T) = J(C_T(W))$, $C(G, S) \le M$, *and* $\mathcal{H}_*(T, M) \subseteq C_G(Z)$, *or*

   (b) $\bar{L}\bar{T} \cong S_5$, $\bar{S} = \overline{J(T)} \cong E_4$ *is generated by the two transvections in* $\bar{T}$, $\langle Z^L \rangle = V \oplus C_Z(L)$, *and* $C_V(S) = V_2$.

PROOF. Recall that Hypothesis 12.2.3 excludes the groups in conclusions (2) and (3) of Theorem 6.2.20. Thus case (1) of Theorem 6.2.20 holds, so for any $W \in \mathcal{R}_2(LT)$ with $[W, L] \ne 1$, $[W, L]$ is a sum of at most two A_5-modules. Further $O_2(LT)$ is the kernel of the action of L on both W and V. Thus $N_{\bar{L}\bar{T}}(Z \cap [W, L])$ is the Borel subgroup over $\bar{T}$, so the first sentence in (1) holds. Next by B.3.2.4, each member of $\mathcal{P}(\bar{T}, V)$ is generated by transpositions, and hence none lie in $\bar{R}_1$. Thus $J(R_1) \le C_T(W) = O_2(LT)$, so that $J(R_1) = J(C_T(W))$ and $Baum(R_1) = Baum(C_T(W))$ by B.2.3.5; hence $C(G, Baum(R_1)) \le M = !\mathcal{M}(LT)$, so (1) holds.

Part (2) is essentially 5.1.2 applied to W in the role of "V". When $J(T) \le C_T(W)$, the final statement in (a) follows from Theorem 3.1.8.3. When $J(T) \not\le C_T(W)$, the statements about $\bar{S}$ and V_2 follow from E.2.3. $\square$

LEMMA 13.2.5. *If* $n = 5$ *then* $N_G(Baum(T)) \le M$.

PROOF. The lemma follows from 5.1.7. $\square$

LEMMA 13.2.6. *If* $n = 5$ *then*

(1) $C_T(v) \in Syl_2(C_G(v))$ *for* $v \in V_2 - V_1$.
(2) *Singular vectors of* V *are not fused in* G *to nonsingular vectors of* V, *so that* L *controls fusion of involutions in* V.

PROOF. Let $v \in V_2 - V_1$. By 13.2.4.2, $v \in V_2 \le C_V(J(T))$, so $S := Baum(T) \le T_v := C_T(v)$; then $S = Baum(T_v)$ by B.2.3.5. Let $T_v \le T_0 \in Syl_2(C_G(v))$. Then $N_{T_0}(T_v) \le N_{T_0}(S) \le M$ by 13.2.5, so as $T_v \in Syl_2(C_M(v))$, $T_v = T_0$ and hence (1) holds. Then (1) implies that $v \notin z^G$, where z is a singular vector in V, so that (2) holds. $\square$

Most of the remainder of the subsection is devoted to Theorem 13.2.7. This result assumes the hypothesis (*) below, which appears later as part (4) of Hypothesis 13.3.1.

THEOREM 13.2.7. *Assume* $n = 5$ *and*

$$L_+/O_2(L_+) \cong A_5 \text{ for each } L_+ \in \mathcal{L}_f(G, T). \tag{*}$$

Then $\mathcal{H}_*(T, M) \subseteq C_G(Z)$.

In the remainder of the section, we assume G is a counterexample to Theorem 13.2.7; thus there is $H \in \mathcal{H}_*(T, M)$ with $H \not\leq C_G(Z)$. Hence:

Conclusion (b) of 13.2.4.2 holds.

In particular, $\bar{L}\bar{T} \cong S_5$ rather than A_5. Let $U_H := \langle Z^H \rangle$, $V_H := [U_H, H]$, $L_H := O^2(H)$, and $H^* := H/C_H(U_H)$. As $H \not\leq C_G(Z)$, by 5.1.7.2:

$$L_H = [L_H, J(T)] \text{ and } L = [L, J(T)].$$

As $L_H = [L_H, J(T)]$, we conclude from B.6.8.6d that $[U_H, J(T)] \neq 1$ Therefore $S := \text{Baum}(T)$ does not centralize U_H, and U_H is an FF-module for H^*. Let $Q := O_2(LT)$.

LEMMA 13.2.8. *(1)* H *is solvable.*

(2) $U_H = V_H \oplus C_Z(H)$.

(3) Either

  (i) $H^* \cong S_3$ *and* $m(V_H) = 2$, *or*

  (ii) $H^* = (H_1^* \times H_2^*)\langle t^* \rangle \cong S_3 \text{ wr } \mathbf{Z}_2$ *and* $V_H = U_1 \oplus U_2$, *where* t^* *is an involution with* $H_1^{*t} = H_2^*$, $H_1^* \cong S_3$, *and* $U_1 := [U_H, H_1^*] \cong E_4$.

  (4) $S^* \in Syl_2(H^*)$ *in (3i), and* $J(T)^* = S^* \in Syl_2(H_1^* H_2^*)$ *in (3ii).*

  (5) $S \in Syl_2(L_H S)$.

  (6) Let $E := \Omega_1(Z(J(T)))$; *set* $s := 1$ *and* $U_1 := V_H$ *in case (i), and set* $s := 2$ *in case (ii). Then* $E = C_E(L_H) \oplus E_1 \oplus \cdots \oplus E_s$, *where*

$$E_i := \langle e_i \rangle = C_{U_i}(S) \cong \mathbf{Z}_2.$$

PROOF. Assume H is not solvable. Then L_H is the product of T-conjugates of members of $\mathcal{L}_f(G, T)$, so by hypothesis (*), $L_H^* \cong A_5$; indeed it follows from (*) that $L_H \in \mathcal{L}_f^*(G, T)$. But then $n(H) > 1$, so that the hypothesis of Theorem 5.2.3 is satisfied. Conclusions (2) and (3) of 5.2.3 are ruled out by Hypothesis 12.2.3, while conclusion (1) of 5.2.3 does not hold as $L_H \not\leq C_G(Z)$. This contradiction establishes part (1) of 13.2.8. Then as U_H is an FF-module for H^*, we conclude from Theorem B.5.6 and B.2.14 that (2) holds, and from E.2.3.2 that (3)–(6) hold. $\square$

We now adopt the notation of 13.2.8.6. Two cases appear in 13.2.8.3: $s = 1$ and $s = 2$. When $s = 2$, define H_i^* as in case (ii) of 13.2.8.3, and let H_i be the preimage in H of H_i^*.

LEMMA 13.2.9. $O_2(H) = C_H(U_H)$. *Thus* $L_H/O_2(L_H) \cong E_{3^s}$.

PROOF. Set $\dot{H} := H/O_2(H)$ and $J := \ker_{H \cap M}(H)$. By 13.2.8 and B.6.8.2, $\dot{L}_H$ is a 3-group, $\dot{J} = \Phi(\dot{L}_H)$, and T is irreducible on $\dot{L}_H/\dot{J}$. As $\Phi(L_H^*) = 1$ by 13.2.8.3, $J = C_H(U_H)$. Thus we may assume that $X := O^2(J) \neq 1$, and it remains to derive a contradiction.

First $X \leq M = N_G(L)$. If X centralizes $L/O_2(L)$, then L normalizes $X = O^2(XO_2(L))$, so $H \leq N_G(X) \leq M = !\mathcal{M}(LT)$, contrary to $H \not\leq M$. Therefore $L = [L, X]$. Let $R := O_2(XT)$. As $XT = TX$, X acts on $T \cap L$, so $R \in Syl_2(LR)$. As $L = [L, X]$, R induces inner automorphisms on $L/O_2(L)$, and $J(R) = J(O_2(LR)) \trianglelefteq$

LT by 13.2.4.1, so $N_G(J(R)) \leq M$. To complete the proof we show H acts on $J(R)$, contrary to $H \not\leq M$.

Assume H does not act on $J(R)$. Then $O_2(H) < R$, so by E.2.1, $L_H \cong \mathbf{Z}_3$, E_9, of 3^{1+2}, and as $\dot{X} \neq 1$, the last case must hold. Then byt 13.2.8.3, $H^* \cong S_3$ wr $\mathbf{Z}_2$, and as $\dot{R} = C_{\dot{T}}(\dot{X})$, $\dot{R} \cong \mathbf{Z}_4$. But then from the action of H^* on U_H, $J(R) = J(O_2(H))$, contrary to assumption. $\qquad\square$

LEMMA 13.2.10. *If $s = 2$, then $L_H \in \Xi_f^*(G,T)$.*

PROOF. Assume $s = 2$. Then by 13.2.8.3 and 13.2.9, T is irreducible on $L_H/O_2(L_H) \cong E_9$, so that $L_H \in \Xi(G,T)$. As $[Z,L_H] \neq 1$, $L_H \in \Xi_f(G,T)$. Further if $L_H \leq L_0$ for some $L_0 := \langle L_+^T \rangle$ with $L_+ \in \mathcal{L}(G,T)$, then $L_+ \in \mathcal{L}_f(G,T)$. Therefore by hypothesis (*), $L_+/O_2(L_+) \cong A_5$ and $L_+ \in \mathcal{L}_f^*(G,T)$, so $L_0 = L_+$ since conclusion (3) of Theorem 12.2.2 holds by Hypothesis 12.2.3; but this contradicts $m_3(L_H) = 2$. Thus no such L_0 exists, so by definition $L_H \in \Xi_f^*(G,T)$. $\qquad\square$

LEMMA 13.2.11. *Assume $Z(H) = 1$. Then*

(1) $C_T(L) = C_T(L_H) = C_E(L) = C_E(L_H) = 1$.

(2) $\overline{J(T)} = \bar{S} = \langle (1,2),(3,4) \rangle \cong E_4$.

(3) $s = 2$, $E = \langle e_{1,2}, e_{3,4} \rangle = \langle e_1, e_2 \rangle \cong E_4$, and $Z = \langle e_1 e_2 \rangle$ is of order 2, and (interchanging H_1 and H_2 if necessary) we may take $e_1 = e_{1,2}$ and $e_2 = e_{3,4}$.

(4) $T_0 := N_T(H_1) = N_T(H_2) = C_T(e_1) = C_T(e_2) = QS = O_2(H)S$.

(5) L is not an A_5-block.

(6) $O^2(H_2)$ is not an A_3-block.

PROOF. As T acts on $C_E(L_H)$ and $Z(H) = 1$, $C_E(L_H) = 1 = C_T(L_H)$, and hence $E \cong E_{2^s}$ by 13.2.8.

Next as we saw that $L = [L, J(T)]$, (2) follows from 13.2.4.2. Thus $V \cap E$ contains $\langle e_{1,2}, e_{3,4} \rangle \cong E_4$, so as $E \cong E_{2^s}$, we conclude $s = 2$ and $E = V \cap E \leq V$. As $Z \leq E \leq V$, $Z = C_V(T)$ has order 2 and is generated by $z := e_{1,2}e_{3,4}$. As $Z \leq V$, $C_T(L) = 1$, completing the proof of (1). Further $E = \langle e_{1,2}, e_{3,4} \rangle = E_1 E_2$. Then $Z = \langle z \rangle = \langle e_1 e_2 \rangle$, so (3) holds. Most of the equalities in (4) are clear; observe $T_0 = O_2(H)S$ by 13.2.8.4, and $T_0 = QS$ by (2) and (3).

If L is an A_5-block, then by C.1.13.c, $Q = O_2(LT) = V \times C_T(L)$. Thus $Q = V$ by (1). Now $\bar{T}_0 = \langle (1,2),(3,4) \rangle$ by (2) and (4), so as $Q = V$, $T_0 \cong D_8 \times D_8$. Thus $L_H T_0 \cong S_4 \times S_4$, so $O^2(H_2) =: K_2$ is an A_3-block.

Therefore if (5) fails, then so does (6); so to prove both parts of the lemma, we may assume that K_2 is an A_3-block. Thus $K_1 = K_2^t$ is also an A_3-block; and again by C.1.13.c, $K_i \cong A_4$ and $O_2(H) = C_T(L_H) \times V_H$. Thus $O_2(H) = V_H$ by (1), so $H \cong S_4$ wr $\mathbf{Z}_2$.

By 13.2.10, $L_H \in \Xi^*(G,T)$, so $M_1 := N_G(L_H) = !\mathcal{M}(H)$ by 1.3.7. As $O_2(H) = V_H = O_2(L_H)$, $O_2(H) = O_2(M_1)$ using A.1.6. Then as $F^*(M_1) = O_2(M_1)$, $C_{M_1}(V_H) = V_H$ so that $M_1/V_H \leq GL(V_H)$. Then as $H/V_H \cong O_4^+(2)$ is a maximal subgroup of $L_4(2)$ with Sylow group $T/V_H \cong D_8$, we conclude that $M_1 = H$. But now Theorem 13.9.1 contradicts the simplicity of G. $\qquad\square$

Set $H_0 := \langle H, L_1 \rangle$.

LEMMA 13.2.12. $O_2(H_0) \neq 1$.

PROOF. Assume that $O_2(H_0) = 1$. Since $L_1 = O^2(N_L(T \cap L))$, we conclude from 5.1.7.2iii that $Z(H) = 1$. Thus we can appeal to 13.2.11. In particular, by that lemma $s = 2$ and $E = \langle e_1, e_2 \rangle$, where $e_1 = e_{1,2}$ and $e_2 = e_{3,4}$ are nonsingular. Further $T_0 = C_T(V_2) \in Syl_2(C_G(e_2))$ by 13.2.6.1. Set $K_1 := O^2(H_1)$, $K_2 := O^2(C_L(e_2))$, $G_i := K_i T_0$, and $G_0 := \langle G_1, G_2 \rangle$. Then $G_0 \le C_G(e_2)$, so in particular $T_0 \in Syl_2(G_0)$ and G_0 is an SQTK-group. Therefore (G_0, G_1, G_2) is a Goldschmidt triple of Definition F.6.1 in section F.6, so we can appeal to results in that section.

Let $X := O_{3'}(G_0)$, $\dot{G}_0 := G_0/X$, $\alpha := (\dot{G}_1, \dot{T}_0, \dot{G}_2)$, and $Q_i := O_2(G_i)$. Observe that $\bar{K}_2 = \langle (1,2), (1,5) \rangle$ and $\bar{Q}_2 = \langle (3,4) \rangle$. Further X is 2-closed by F.6.11.1.

Suppose first that $Q_1 = Q_2$. By Theorem 4.3.2, $M = !\mathcal{M}(L)$, so as $K_1 \not\le M$, no nontrivial characteristic subgroup of Q_2 is normal in LQ_2. On the other hand the hypotheses of C.1.24 are satisfied with Q_2 in the role of "R", so L is an A_5-block by C.1.24, contrary to 13.2.11.5.

Therefore $Q_1 \ne Q_2$. In particular $\dot{\alpha}$ is a Goldschmidt amalgam by F.6.11, so as G_0 is an SQTK-group, $\dot{G}_0$ is described in Theorem F.6.18. Further by the previous paragraph, case (1) of F.6.18 does not arise.

Suppose next that $e_1 \in O_2(G_0)$. Then $W := \langle e_1^{G_0} \rangle \le O_2(G_0)$. As the generator $z := e_1 e_2$ of Z lies in $W \langle e_2 \rangle$, $N_G(W \langle e_2 \rangle) \in \mathcal{H}^e$ by 1.1.4.3, and hence $A := N_G(W) \cap C_G(e_2) \in \mathcal{H}^e$ by 1.1.3.2. Then as $T_0 \in Syl_2(A)$ since $T_0 \in Syl_2(C_G(e_2))$ and $T_0 \le G_0 \le A$, we conclude $G_0 \in \mathcal{H}^e$ by 1.1.4.4. As $[K_i, e_1] \ne 1$, $C_{G_i}(W) \le Q_i$ for $i = 1, 2$, so $C_{G_0}(W)$ is 2-closed and solvable by F.6.8. Further as $T_0 \in Syl_2(G_0)$ and $e_1 \in Z(T_0)$, $W \in \mathcal{R}_2(G_0)$ by B.2.13. As $\bar{K}_2 = \langle (1,2), (1,5) \rangle$, it follows from 13.2.11.2 that $K_2 = [K_2, J(T)]$ and $J(T) = J(T_0)$. By 13.2.8.4, $K_1 = [K_1, J(T)]$. Therefore W is an FF-module for $G_0^* := G_0/C_{G_0}(W)$ with $K_i^* = [K_i^*, J(T_0)^*] \ne 1$.

Assume first that $\dot{G}_0$ satisfies one of conclusions (3)–(13) of F.6.18, and let $L_0 := G_0^\infty$ and $W_0 := [W, L_0]$. Then from F.6.18, $\dot{L}_0$ is quasisimple, so as X is 2-closed, $L_0 \in \mathcal{C}(G_0)$ by A.3.3. As we saw $G_0 \in \mathcal{H}^e$, $L_0 \in \mathcal{H}^e$ by 1.1.3.1, so that $L_0 T_0 \in \mathcal{H}^e$. By F.6.18, $\dot{L}_0$ contains $\dot{K}_1$ or $\dot{K}_2$; so as $K_i = [K_i, J(T)]$, $L_0 = [L_0, J(T_0)]$. Hence $L_0^* J(T)^*$ is described in Theorem B.5.1. Comparing that list with the list in F.6.18, we conclude that $\dot{L}_0 \cong L_3(2)$, $Sp_4(2)'$, $G_2(2)'$, or A_7, and $W_0/C_{W_0}(L_0)$ is a natural module for L_0^*, a 4-dimensional module for $L_0^* \cong A_7$, or the sum of two isomorphic natural modules for $L_0^* \cong L_3(2)$. In each case F.6.18 says $L_0 = O^2(G_0)$, so $K_i \le L_0$. Then the condition that neither K_1 nor K_2 centralizes $e_1 \in C_W(T_0)$ eliminates all cases except the one where W_0 is the natural module for $G_0^* \cong S_7$ and (in the notation of section B.3) for $i := 1$ or 2, G_i^* is the stabilizer of a partition of type $2^2, 3$, while G_{3-i}^* is the stabilizer of a partition of type $2^3, 1$. This is impossible, as in that case $J(T)^* = O_2(G_{3-i}^*)$, contrary to $K_j^* = [K_j^*, J(T)]$ for each j.

This contradiction shows that $\dot{G}_0$ satisfies none of conclusions (3)–(13) of F.6.18; as case (1) of F.6.18 was eliminated earlier, we conclude that case (2) of F.6.18 holds. Therefore $\dot{G}_0 \cong S_3 \times S_3$ or $E_4/3^{1+2}$. As W is an FF-module for G_0^* and $K_i = [K_i, J(T)]$ for $i = 1$ and 2, it follows from Theorem B.5.6 that $K_i^* \trianglelefteq G_0^* \cong L_2(2) \times L_2(2)$, and $W = [W, K_1] \oplus [W, K_2]$, with $[W, K_i] \cong E_4$. Recall that $\bar{K}_2 = \langle (1,2), (1,5) \rangle$, so that as $e_1 = e_{1,2}$, $\langle e_1^{K_2} \rangle = \langle e_{1,2}, e_{1,5} \rangle = [W, K_2]$ is a proper G_0-invariant subgroup of W, whereas by definition $W = \langle e_1^{G_0} \rangle$. This contradiction finally eliminates the subcase $e_1 \in O_2(G_0)$.

So we turn to the remaining subcase $e_1 \notin O_2(G_0)$. First $C_G(z) \in \mathcal{H}^e$ by 1.1.4.3, so that $C := C_G(z) \cap C_G(e_2) \in \mathcal{H}^e$ by 1.1.3.2. Then as $T_0 \in Syl_2(C)$ since $T_0 \in Syl_2(C_G(e_2))$, we conclude from 1.1.4.4 that $C_{G_0}(z) \in \mathcal{H}^e$. Hence $C_{O(G_0)}(z) \le O(C_{G_0}(z)) = 1$.

Next $z = e_1 e_2 = e_{1,2} e_{3,4}$ generates Z, and as $\bar{K}_2 = \langle (1,2),(1,5)\rangle$, $\langle z^{K_2}\rangle =: F \cong E_8$, with each coset fE_2 of $E_2 = \langle e_2\rangle = \langle e_{3,4}\rangle$ in F distinct from E_2 containing a K_2-conjugate z_f of z. Therefore $C_{O(G_0)}(f) = C_{O(G_0)}(z_f) = 1$ using the previous paragraph. Thus no hyperplane of F centralizes an element of $O(G_0)$, so by Generation by Centralizers of Hyperplanes A.1.17, $O(G_0) = 1$.

Now $e_1 \in Z(T_0)$, so e_1 centralizes $O_2(G_0)$, but $e_1 \notin O_2(G_0)$ by assumption. Hence as $O(G_0) = 1$, $L_0 = [L_0, e_1]$ for some component L_0 of G_0. Thus $\dot{L}_0 = E(\dot{G}_0)$ is described in one of cases (3)–(13) of Theorem F.6.18. As $[K_i, e_1] \neq 1$ for $i = 1, 2$, and $\dot{e}_1 \in Z(\dot{Q}_i)$, $\dot{K}_i$ does not centralize $Z(\dot{Q}_i)$. Therefore $\dot{G}_0$ must satisfy conclusion (6) or (8) of F.6.18. But then $K_i \cong A_4$, contrary to 13.2.11.6.

This contradiction finally completes the proof of 13.2.12. $\square$

By 13.2.12, $H_0 \in \mathcal{H}(H)$. Let $U := \langle Z^{H_0}\rangle$, so that $\langle Z^H\rangle = U_H \leq U$, and let $H_0^* := H_0/C_{H_0}(U)$.

LEMMA 13.2.13. $O^2(H_0/O_{3'}(H_0))$ *is not a 3-group.*

PROOF. Assume that $O^2(H_0/O_{3'}(H_0))$ is a 3-group. Then

$$O_2(L_H) \leq O_{3',3}(H_0) \cap T \leq C_T(L_1/O_2(L_1)) = O_2(L_1T) = R_1,$$

so $R_1 \in Syl_2(R_1 L_H)$. By 13.2.4.1, $B := \mathrm{Baum}(R_1) = \mathrm{Baum}(Q)$. Thus as $L_H \not\leq M$, $J(R_1)$ is not normal in $R_1 L_H$, so as $[Z, L_H] \neq 1$, $B \in Syl_2(BL_H)$ by E.2.3.2. Thus $Q \in Syl_2(QL_H)$, so by Theorem 3.1.1 applied to LT, Q in the roles of "M_0", "R", $O_2(\langle LT, H\rangle) \neq 1$, and hence we obtain our usual contradiction to $H \not\leq M$. $\square$

LEMMA 13.2.14. $s = 1$.

PROOF. Assume that $s = 2$. By 13.2.10, $L_H \in \Xi_f^*(G, T)$, so $L_H \trianglelefteq H_0$ by 1.3.5. Therefore $H_0 = L_H L_1 T = L_1 H$. Recall we are in case (b) of 13.2.4.2, so that $[Z, L_1] = 1$, and hence $U = \langle Z^{H_0}\rangle = \langle Z^{L_1 H}\rangle = \langle Z^H\rangle = U_H$. By 13.2.8.6, $U_H = U_1 \oplus U_2 \oplus C_Z(H)$. Now $L_1 = O^2(L_1)$ fixes the two subgroups $O^2(H_i)$ with image of index 3 in $L_H/O_2(L_H)$ such that $C_{U_H}(L_H) < C_{U_H}(O^2(H_i)) < U_H$. Hence L_1 acts on U_1, U_2, and $C_Z(H)$. Therefore as $[Z, L_1] = 1$ and H_i induces $GL(U_i)$ on U_i, we conclude $[U_H, L_1] = 1$. Therefore $[L_1, L_H] \leq C_{L_H}(U_H) = O_2(L_H)$, so as $H_0 = L_1 H$, $H_0/O_{3'}(H_0)$ is a 3-group, contrary to 13.2.13. $\square$

We are now ready to complete the proof of Theorem 13.2.7.

As $s = 1$ by 13.2.14, $H/O_2(H) \cong S_3$ by 13.2.9. Hence $(H_0, L_1 T, H)$ is a Goldschmidt triple. As $O_2(H_0) \neq 1$ by 13.2.12, H_0 is an SQTK-group. Let $\dot{H}_0 := H_0/O_{3'}(H_0)$ and $\alpha := (\dot{L}_1 \dot{T}, \dot{T}, \dot{H})$. By 13.2.13 and F.6.11.2, α is a Goldschmidt amalgam; hence as H_0 is an SQTK-group, $\dot{H}_0$ is described in Theorem F.6.18.

Let $L_0 := H_0^\infty$. By 13.2.13, neither conclusion (1) nor (2) of F.6.18 holds, so $\dot{L}_0$ is quasisimple and described in one of cases (3)–(13) of F.6.18. By F.6.11.1, $O_{3'}(H_0)$ is 2-closed, so $L_0 \in \mathcal{C}(H_0)$ by A.3.3. Thus $L_0 \in \mathcal{L}(G, T)$; so if $[Z, L_0] \neq 1$, then $L_0/O_2(L_0) \cong A_5$ by hypothesis (*) of Theorem 13.2.7. As $\dot{L}_0$ is not A_5 in any of the conclusions of F.6.18, we conclude $[Z, L_0] = 1$. Thus $L_H \not\leq L_0$, so case (3) of F.6.18 holds; that is, $O^2(\dot{H}_0) = \dot{D} \times \dot{L}_0$, where $\dot{L}_0 \cong L_2(q)$, $q \equiv 11$ or 13 mod 24, and $\dot{D} \cong \mathbf{Z}_3$. Let D be a Sylow 3-subgroup of the preimage of $\dot{D}$ which permutes with T. Then D does not centralize Z as $O^2(H) = L_H$ does not. Further $\dot{L}_1 \leq C_{O^2(\dot{H})}(Z) = \dot{L}_0$, so $\dot{L}_1 \dot{T} \cong D_{24}$ and $L_1/O_2(L_1)$ is inverted in $C_T(D)$. Thus we may choose D to permute with L_1. Then $[D, O_2(DT)] \leq O_{3'}(H_0) \cap T \leq R_1$, so R_1 is Sylow in $R_1 D$.

We argue as in the proof of 13.2.13: Assume that $D \not\leq M$. Then as $R_1 \in Syl_2(R_1 D)$, $B := \mathrm{Baum}(R_1) \in Syl_2(BD)$ by E.2.3.2. But $B = \mathrm{Baum}(Q)$ by 13.2.4.1, so $Q \in Syl_2(QD)$. Then by Theorem 3.1.1 applied with Q, LT, DT in the roles of "R, M_0, H", $O_2(\langle LT, DT \rangle) \neq 1$, so that $D \leq M = !\mathcal{M}(LT)$, contrary to our assumption that $D \not\leq M$. Therefore $D \leq M = N_G(L)$. Now as $L_1/O_2(L_1)$ is inverted in $C_T(D)$, D centralizes $L/O_2(L)$, so L acts on $Y := O^2(DO_2(LT)) = \langle D^T \rangle$, and hence $N_G(Y) \leq M = !\mathcal{M}(L)$ by Theorem 4.3.2. Then $L_0 \leq N_{H_0}(Y) \leq H_0 \cap M$, so $H \leq H_0 = DL_0 T \leq M$, for our usual contradiction to $H \not\leq M$.

This contradiction completes the proof of Theorem 13.2.7.

13.3. Starting mid-sized groups over $\mathbf{F}_2$, and eliminating $U_3(3)$

In this section, with the preliminary results from sections 13.1 and 13.2 in hand, we begin to treat those pairs L, V in the Fundamental Setup (3.2.1) which constitute the main topic of the chapter: the pairs such that $L/O_2(L)$ is an intermediate-sized group A_5, A_6, $\hat{A}_6$, or $U_3(3)$ over $\mathbf{F}_2$. As in the previous chapter, we begin by stating our working hypothesis for this chapter, which excludes the groups in the Main Theorem which have arisen in previous sections. In particular, Hypothesis 13.3.1 extends Hypotheses 12.2.3 and 13.1.1. Each section treats one or more pairs L, V in the FSU; the treatment of a given case assumes the existence of $L \in \mathcal{L}_f(G,T)$ with $L/C_L(V)$ of the given type.

We also recall, as mentioned in the introduction to the chapter, that to avoid repetition of arguments, we treat the case $L/O_2(L) \cong A_5$ simultaneously with the other cases. However in the actual logical sequence, that case is the final one in our treatment of the FSU, so we actually consider it only when all other groups have been eliminated. This necessitates the assumption in part (4) of Hypothesis 13.3.1; the effect of this part of Hypothesis 13.3.1 is that we choose $L \in \mathcal{L}_f(G,T)$ with $L/O_2(L) \cong A_5$ only when we are forced to do so, because no other choice is possible. Thus for the purposes of the proof of the Main Theorem, Hypothesis 13.3.1.4 and the results in this chapter which depend on it, are actually invoked only when we reach that final case.

Thus in section 13.3 and indeed for the remainder of the chapter, we assume the following hypothesis:

HYPOTHESIS 13.3.1. *(1) G is a simple QTKE-group, $T \in Syl_2(G)$, and $L \in \mathcal{L}_f(G,T)$.*

(2) G is not a group of Lie type over $\mathbf{F}_{2^n}$, with $n > 1$.

(3) G is not $L_4(2)$, $L_5(2)$, A_9, M_{22}, M_{23}, M_{24}, He, or J_4.

(4) If $L/O_2(L) \cong A_5$, then $K/O_2(K) \cong A_5$ for each $K \in \mathcal{L}_f(G,T)$.

The next result describes the members K of $\mathcal{L}_f(G,T)$ which can arise under Hypothesis 13.3.1; as in Remark 12.2.4 of the previous chapter, we can usually replace our chosen pair L, V in the FSU by K, V_K for some suitable $V_K \in Irr_+(K, R_2(KT))$.

LEMMA 13.3.2. *If $K \in \mathcal{L}_f(G,T)$, then*
(1) $K/O_2(K) \cong A_5$, $L_3(2)$, A_6, $\hat{A}_6$, or $U_3(3)$.
(2) $K \trianglelefteq KT$ and $K \in \mathcal{L}_f^*(G,T)$. Hence $N_G(K) = !\mathcal{M}(KT)$.

(3) There is a T-invariant $V_K \in Irr_+(K, R_2(KT))$ and further each member of $Irr_+(K, R_2(KT), T)$ is T-invariant. The pair K, V_K satisfies the FSU and either V_K is the natural module for $K/C_K(V_K) \cong A_5$, A_6, $L_3(2)$, or $U_3(3)$, or V_K is the 5-dimensional core of a 6-dimensional permutation module for $K/C_K(V_K) \cong A_6$.

(4) Hypotheses 13.1.1, 12.2.1, and 12.2.3 are satisfied with K in the role of "L".

(5) Hypothesis 13.3.1 is satisfied with K in the role of "L" unless $K/O_2(K)$ is A_5 but $L/O_2(L)$ is not A_5.

PROOF. The initial argument is similar to that in 13.1.2: First $K \leq I \in \mathcal{L}^*(G, T)$, and by 1.2.9, $I \in \mathcal{L}_f^*(G, T)$. By Theorem 13.1.7, $I/O_2(I)$ is quasisimple, so $K = I$ by 13.1.2.5. Therefore Hypothesis 12.2.3 holds with K in the role of "L" by 13.1.2.1. Hence (4) is established. Furthermore parts (1)–(3) of Hypothesis 13.3.1 are satisfied by K in the role of "L', so (5) also follows as part (4) of Hypothesis 13.3.1.4 is satisfied by K unless $K/O_2(K)$ is A_5, but $L/O_2(L)$ is not.

Part (1) follows from 13.1.2.3. Further 13.1.2 says that K is T-invariant and the first sentence of (3) holds. Then $N_G(K) = !\mathcal{M}(KT)$ by 1.2.7.3, completing the proof of (2).

It remains to show V_K is one of the modules described in (3). Theorem 12.2.2.3 supplies an initial list of possibilities for V_K, and by Remark 12.2.4 the list of 12.2.2.3 can be refined using results from the previous chapter. If $C_{V_K}(K) \neq 1$, then Theorem 12.4.2 rules out the indecomposables in cases (b) and (f) of 12.2.2.3, leaving only case (d) with V_K the core of a 6-dimensional permutation module for $K/C_K(V_K) \cong A_6$. Otherwise $C_{V_K}(K) = 1$, so either V_K is one of the natural modules listed in 13.3.2.3, or V_K is the 6-dimensional faithful module for $\hat{A}_6$. The last case is out by Theorem 12.7.1 and the exclusions in Hypothesis 13.3.1.3. $\square$

Of course we may apply 13.3.2 to L in the role of "K", so $V \in Irr_+(L, R_2(LT))$ is T-invariant and V is one of the modules listed in 13.3.2.3. By 13.3.2, L satisfies Hypothesis 12.2.3, so we may appeal to the results from the previous chapter, and when $L/C_L(V) \cong A_5$ or A_6 we may appeal to results from section 13.2 of this chapter. We adopt the conventions in Notation 12.2.5 from the previous chapter.

We will refer to a module V which is the core of a 6-dimensional permutation module for $L/C_L(V) \cong A_6$ as a *5-dimensional module for A_6*. In addition we adopt:

NOTATION 13.3.3. If $\bar{L} \cong L_3(2)$, A_5, or A_6, define the T-invariant subspaces V_i of V for $1 \leq i \leq \dim(V/C_V(L))$ as in Notations 12.8.2 and 13.2.1. When $\bar{L}$ is $U_3(3)$, V is the 6-dimensional module for $\bar{L}$ regarded as $G_2(2)'$, which is the quotient of the Weyl module discussed in [Asc87]; see also B.4.6. In particular, V admits a symplectic form preserved by $\bar{M}_V$, so we can speak of nondegenerate and totally isotropic subspaces of V. In this case, define V_i to be the unique T-invariant subspace of V of dimension i. Notice that if $C_V(L) = 1$, then $m(V_i) = i$ in each case.

In each case define $G_i := N_G(V_i)$, $M_i := N_M(V_i)$, and $L_i := O^2(N_L(V_i))$. When $L/O_2(L)$ is not $\hat{A}_6$, define $R_i := O_2(L_iT)$. When $L/O_2(L) \cong \hat{A}_6$, define R_i L_0, and $L_{i,+}$ as in Notation 13.2.1.

LEMMA 13.3.4. *(1) $V_1 = Z \cap V$.*
(2) $V = \langle (Z \cap V)^L \rangle$.

(3) The proper overgroups of $\bar{T}$ in $\bar{L}\bar{T} = Aut_G(\bar{L})$ are $\bar{L}_1\bar{T}$ and $\bar{L}_2\bar{T}$—except when $\bar{L} \cong A_5$, when only $\bar{L}_1\bar{T}$ occurs. In particular, all proper overgroups of T in LT are $\{2,3\}$-groups.

(4) Statements analogous to (1)–(3) hold for any $K \in \mathcal{L}_f(G,T)$ and $V_K \in Irr_+(KT, R_2(KT), T)$ in the roles of "L, V".

PROOF. Part (1) follows from an inspection of the modules listed in 13.3.2.3. Then (2) follows since $V \in Irr_+(LT, R_2(LT))$. Part (3) follows from the well-known fact that the overgroups of T in an untwisted group of Lie type over $\mathbf{F}_2$ are parabolics, and as $Out(\bar{L})$ is a 2-group. Finally (4) follows since 13.3.2.3 also applies to each K and V_K. $\qquad\square$

As usual in the FSU, by 3.3.2.4, we may apply the results of section B.6 to members $H \in \mathcal{H}_*(T, M)$. Recall that for $v \in V^\#$, $G_v = C_G(v)$ in Notation 12.2.5.3.

LEMMA 13.3.5. *(1) If $\bar{L} \cong L_3(2)$ or $U_3(3)$ then $G_v \not\leq M$ for each $v \in V^\#$.*

(2) If $\bar{L} \cong A_5$ then $\mathcal{H}_*(T, M) \subseteq C_G(Z)$, so $G_z \not\leq M$ for z generating $Z \cap V = V_1$.

(3) If $\bar{L} \cong A_6$, then $G_v \not\leq M$ for some $v \in V_1 - C_V(L)$.

PROOF. As Hypothesis 13.3.1 excludes the groups in conclusions (2)–(4) of Theorem 12.2.13, conclusion (1) of that result holds: namely $G_v \not\leq M$ for some $v \in V^\#$. Next V is described in 13.3.2.3. In particular $C_V(L) = 1$ unless V is a 5-dimensional module for $\bar{L} \cong A_6$, and L is transitive on $(V/C_V(L))^\#$ unless $\bar{L} \cong A_5$. Therefore (1) holds, and if $\bar{L}$ is A_6, then $G_v \not\leq M$ for some $v \in V_1$. Further if $C_V(L) \neq 1$, then $C_V(L) \leq Z(LT)$, so as $M = !\mathcal{M}(LT)$, (3) holds. Finally when $\bar{L} \cong A_5$, $\mathcal{H}_*(T, M) \subseteq C_G(Z)$ by 13.2.7. Then as $Z \cap V = V_1$ by 13.3.4.1, $G_z \not\leq M$ for z generating $Z \cap V$, so (2) holds. $\qquad\square$

By 13.3.5:

LEMMA 13.3.6. *Either $G_1 \not\leq M$, or $C_V(L) \neq 1$ so that V is a 5-dimensional module for $\bar{L} \cong A_6$.*

As usual we let $\theta(X)$ denote the subgroup generated by all elements of order 3 in a group X.

LEMMA 13.3.7. *Assume $\bar{L} \cong A_6$. Then either*

(1) $C_G(V)$ is a $3'$-group, or
(2) $L/O_2(L) \cong \hat{A}_6$, $m_3(C_G(V)) = 1$, and $L_0 = \theta(C_G(V))$.

PROOF. Let $D := \theta(C_G(V))$ and $P \in Syl_3(C_G(V))$. Recall that we may apply 12.2.8; then $\theta(M) = L$ so that $D \leq L$, and hence either $D = 1$, or $L/O_2(L) \cong \hat{A}_6$ with $D = \theta(C_L(V)) = L_0$. In the first case, conclusion (1) holds. In the second, as $\Omega_1(P) \leq D = L_0$, $\Omega_1(P)$ is of order 3, so conclusion (2) holds. Thus the lemma is established. $\qquad\square$

LEMMA 13.3.8. *Assume $K \in \mathcal{L}_f(G,T)$, let $M_K := N_G(K)$, and assume $H \in \mathcal{H}(T, M_K)$ and $Y = O^2(Y) \trianglelefteq H$ with $Y \leq M_K$. Then*

(1) $K \not\leq Y C_{M_K}(K/O_2(K))$.
(2) Y is a $\{2,3\}$-group.

PROOF. As $K \in \mathcal{L}_f(G, T)$, $M_K = !\mathcal{M}(KT)$ by 13.3.2.2,; then as $H \not\leq M_K$,

$$O_2(\langle K, H \rangle) = 1. \tag{$*$}$$

Let $M_K^* := M_K / C_{M_K}(K/O_2(K))$. As $K/O_2(K)$ is quasisimple and $T \leq M_K$, $K = [K, T \cap K]$. Suppose (1) fails, so that $K^* \leq Y^*$. Then $K^* = [K^*, (T \cap K)^*] = [Y^*, (T \cap K)^*] = [Y, T \cap K]^*$, and as Y is T-invariant, $[Y, T \cap K] \leq Y$. Thus $K = (K \cap Y)O_2(K)$, so as T acts on Y, $K \leq Y \leq H$, contrary to $(*)$ as $O_2(H) \neq 1$. Thus (1) holds.

Let $Y_0 := O^{\{2,3\}}(Y)$; then $Y_0^* < M_K^*$ by (1). But by 13.3.4, the proper overgroups of T^* in $M_K^* = Aut_G(K^*)$ are $\{2,3\}$-groups, so we conclude that $Y_0^* = 1$. Then $Y_0 \leq C_G(K/O_2(K))$, so K normalizes $O^{\{2,3\}}(Y_0 O_2(K)) = Y_0$. However if $Y_0 \neq 1$, then $O_2(Y_0) \neq 1$ by 1.1.3.1, contrary to $(*)$. Thus (2) holds. $\square$

LEMMA 13.3.9. *Assume* $\bar{L} \cong A_6$, $H \in \mathcal{H}(T, M)$, *and* $Y = O^2(Y) \trianglelefteq H$ *with* $Y \leq C_M(v)$ *for some* $v \in V_1 - C_V(L)$. *Then either*

(1) $Y = 1$, *or*

(2) $\bar{Y} = \bar{L}_1$. *Further if* $L/O_2(L) \cong A_6$ *then* $Y = L_1$, *while if* $L/O_2(L) \cong \hat{A}_6$ *then* $Y = L_{1,+}$.

PROOF. As in the proof of the previous lemma, with L in the role of "K",

$$O_2(\langle L, H \rangle) = 1. \tag{$*$}$$

By hypothesis $Y = O^2(Y)$, and as Y centralizes v, $Y \leq M_V$ by 12.2.6. Therefore $\bar{Y} \leq O^2(\bar{M}_V) = \bar{L}$ by 12.2.10.2; and by 13.3.8.1, $\bar{Y} < \bar{L}$. By 13.3.8.2, $\bar{Y}$ is a $\{2,3\}$-group. By 1.1.3.1, $O_2(Y) \neq 1$.

If $\bar{Y} = 1$, then L normalizes $O^2(YO_2(L)) = Y$, and hence (1) holds by $(*)$. Thus we may assume that $\bar{Y} \neq 1$, so that $\bar{Y} = \bar{L}_i$ for $i = 1$ or 2 by 13.3.4.3. Then as Y centralizes v, $i = 1$. Further $Y_1 := \theta(Y) \leq \theta(M) = L$ by 12.2.8, so $Y_1 \leq L_1$.

Suppose first that $C_{Y_1}(V) \not\leq O_2(Y_1)$. Then by 13.3.7, $L/O_2(L) \cong \hat{A}_6$ and $L_0 \leq Y_1$. Now $L_0 \leq Y_1 \leq L_1$, so $Y_1 = L_0$ or L_1, and in either case $H \leq N_G(Y_1) \leq M$ by 13.2.2.9, contrary to $(*)$. Therefore $C_{Y_1}(V) \leq O_2(Y_1)$. So as Y is a $\{2,3\}$-group, $C_Y(V) \leq O_2(Y)$, and hence $\bar{Y} = \bar{Y}_1$ is of order 3. Therefore $Y = Y_1$ and $|Y : O_2(Y)| = 3$, so (2) holds. $\square$

LEMMA 13.3.10. *(1) If* $\bar{L} \cong A_5$ *then* $J(R_1) = J(O_2(LT))$, $B := Baum(R_1) = Baum(O_2(LT))$, *and* $C(G, B) \leq M$.

(2) If $\bar{L} \cong A_6$ *or* $U_3(3)$ *then either there is a nontrivial characteristic subgroup of* $B := Baum(R_1)$ *normal in* LT *(so that* $N_G(B) \leq M$*), or* L *is an* A_6-*block or a* $G_2(2)$-*block. Moreover if* L *is a* $G_2(2)$-*block, then* $N_G(B) \leq M$.

(3) If $\bar{L} \cong A_6$ *then either some nontrivial characteristic subgroup of* $B := Baum((T \cap L)O_2(LT))$ *is normal in* LT *(so that* $N_G(B) \leq M$*), or* L *is an* A_6-*block.*

(4) If $\bar{L} \cong L_3(2)$, *then either some nontrivial characteristic subgroup of* $B := Baum(R_1)$ *is normal in* LT *(so that* $N_G(B) \leq M$*), or* L *is an* $L_3(2)$-*block.*

PROOF. Part (1) follows from 13.2.4.1, and part (3) follows from case (b) of C.1.24; L is not a $\hat{A}_6$-block since $V/C_V(L)$ is the A_6-module by 13.3.2.3. Similarly the first sentence in (2) follows from 13.2.2 when $\bar{L} \cong A_6$, and from C.1.37 when $\bar{L} \cong U_3(3)$. When $\bar{L} \cong L_3(2)$, C.1.37 also establishes (4).

Thus it only remains to establish the final sentence of (2), so we assume that L is a $G_2(2)$-block, but that $N_G(B) \not\leq M$, and it remains to derive a contradiction.

We check that Hypothesis C.6.2 is satisfied, with L, B, T, LT, $N_G(B)$ in the roles of "L, R, T_H, H, Λ": For example C.6.2.3 holds, since $M = !\mathcal{M}(LT)$. The only part of Hypothesis C.6.2 which is not evident is that $Q := O_2(LB) \leq B$, and this was established during the proof of C.1.37 using Baumann's Argument B.2.18. Thus we may apply C.6.3.1 to conclude that there exists $x \in N_G(B)$ with $V^x \not\leq Q$. Now reversing the roles of V and V^x if necessary, we may assume that $m(\bar{V}^x) \geq m(V/C_V(V^x))$. Further since $\bar{T}$ contains no strong FF*-offenders on V by B.4.6.13, B.2.4.3 says $V \leq B$, so that $V^x \leq B \leq R_1$. Also by B.1.4.6, $\bar{V}^x \in \mathcal{P}(\bar{R}_1, V)$; then by parts (13) and (3) of B.4.6, we have the hypotheses of B.2.20, so $\bar{V}^x = \overline{J(R_1)} = \bar{B}$ is the unique member $\bar{B}$ of $\mathcal{P}(\bar{R}_1, V)$, and $m(V/C_V(V^x)) = m(\bar{B})$.

Next since $C_V(L) = 1$ by 13.3.2.3,

$$m(V/C_V(V^x)) = 3 = m(\bar{B}) = m(B/C_B(V)) = m(B/C_B(V^x)) = m(V)/2.$$

Therefore as $\bar{L} \leq \langle \bar{B}, \bar{B}^l \rangle$ for suitable $l \in L$, $L \leq \langle V^x, V^{xl}, V \rangle$, and

$$m(Q/(C_Q(V^x) \cap C_Q(V^{xl}))) \leq 2m(B/C_B(V^x)) = m(V).$$

Hence $Q = V \times C_B(\langle V^x, V^{xl} \rangle) = V \times C_B(L)$, and in particular V^x centralizes $C_B(L)$. Also $E_8 \cong [V, V^x] \leq V \cap V^x$, so as $m(\bar{V}^x) = 3$, $C_{V^x}(V) = V \cap V^x$. Then $|V^x C_B(L)| = |V C_B(L)| = |C_B(V)|$ and hence $C_B(V^x) = V^x \times C_B(L)$. Therefore as $x \in N_G(B)$,

$$\Phi(C_B(L)) = \Phi(C_B(V^x)) = \Phi(C_B(V))^x = \Phi(C_B(L))^x.$$

Thus if $\Phi(C_B(L)) \neq 1$, then $x \in N_G(\Phi(C_B(L)) \leq M = !\mathcal{M}(LT)$; but then as $M = N_G(L)$, $V^x \leq O_2(L) \leq O_2(LB) = Q$, contrary to the choice of x. Therefore $\Phi(C_B(L)) = 1$, so that $C_B(V) = Q$ is elementary abelian; and then $\mathcal{A}(B) = \{Q, Q^x\}$ is of order 2 by B.2.21 using B.4.6.6. Hence $O^2(N_G(B)) \leq N_G(Q) \leq M$, and then $N_G(B) = O^2(N_G(B))T \leq M$, contrary to our assumption. This contradiction completes the proof of (2), and hence of 13.3.10.   □

LEMMA 13.3.11. *Assume* $\bar{L} \cong A_5$. *Then*

(1) *For each* $v \in V^\#$, $\{U \in V^G : v \in U\} = V^{G_v}$.
(2) $V_2^L = V_2^G \cap V$ *and* $V_3^L = V_3^G \cap V$.
(3) V *is the unique member of* V^G *containing* V_3.
(4) $V^{G_2} = \{U \in V^G : V_2 \leq U\}$.
(5) *If* $g \in G$ *with* $[V_3, V_3^g] = 1$, *then* $[V, V^g] = 1$.

PROOF. Part (1) follows from 13.2.6.2 and A.1.7.1. As V_k^L, $k = 2, 3$, are the unique classes of subgroups of V of rank k containing a unique singular point, 13.2.6.2 also implies (2). Then (2) and A.1.7.1 imply (4), as well as the analogous statement for V_3 and G_3. Thus as $G_3 \leq M_V$ by 13.2.3.2, (3) holds. If $[V_3, V_3^g] = 1$, then by (3), V_3^g acts on V. Therefore as $C_{\bar{M}_V}(V_3) = 1$, $V \leq C_G(V_3^g) \leq N_G(V^g)$ again using (3), so that $V \leq C_G(V^g)$ again using $C_{\bar{M}_V}(V_3) = 1$. Thus (5) holds.   □

LEMMA 13.3.12. *Assume* $\bar{L} \cong U_3(3)$. *Then*

(1) $s(G, V) > 1$.
(2) *If* $U \leq V$ *with* $C_G(U) \not\leq M$, *then* U *is totally isotropic. Hence* $r(G, V) \geq 3$.
(3) *If* $r(G, V) = 3$, *then* $C_G(V_3) \not\leq M$.
(4) *If* $g \in G$ *with* $1 \neq [V, V^g] \leq V \cap V^g$, *then* $V \cap V^g = [V, V^g] = C_V(V^g) \in V_3^G$, *and we may take* $g \in C_G(V \cap V^g)$, *so that* $C_G(V_3) \not\leq M$.

PROOF. Recall V is a TI-set in M by 12.2.6, so Hypothesis E.6.1 is satisfied, and for $1 \neq U \leq V$, $C_M(U) \leq M_V$. By parts (4)–(6) of B.4.6, $m(\bar{M}_V, V) > 1$, so $C_M(W) = C_M(V)$ for each hyperplane W of V. Further the hyperplanes of V are of the form $v^\perp$ for $v \in V^\#$, so as L is transitive on $V^\#$, L is transitive on hyperplanes. Hence each hyperplane is invariant under a Sylow 2-subgroup of LT, so that $r(G, V) > 1$ by E.6.13. Hence (1) is established.

Next we establish some preliminary results, phrased in terms of the usual geometry of points and lines on V: From section 5 in [Asc87], we can identify the points and lines of the generalized hexagon of $\bar{G}_0 := N_{GL(V)}(\bar{L}) \cong G_2(2)$ with the points and *doubly singular* lines of V (i.e., totally isotropic as well as singular in the Dickson trilinear form; see p. 194 of [Asc87]). By 5.1 in [Asc87], $\bar{G}_0$ is transitive on nondegenerate lines of V, and each such line l is generated by a pair u, v of points opposite (i.e., at maximal distance) in the hexagon. Now $N_{\bar{G}_0}(l) = \bar{H}_1 \times \bar{H}_2$, where $\bar{H}_1 := C_{\bar{G}_0}(l) = C_{\bar{G}_0}(u) \cap C_{\bar{G}_0}(v) \cong S_3$ by F.4.5.5, and $\bar{H}_2 := C_{\bar{G}_0}(\bar{H}_1) \cong S_3$ acts faithfully on l. Further $\bar{H}_1\bar{H}_2$ acts faithfully on the 4-space $l^\perp$, with $l^\perp = [l^\perp, O^2(H_i)]$. Now $N_{\bar{M}_V}(l)$ is of index 1 or 2 in $N_{\bar{G}_0}(l)$ in the cases $\bar{M}_V = \bar{G}_0$ or $\bar{L}$, respectively. In particular $Q := O_2(LT)$ is of index at most 2 in $T_H := C_T(l)$, so $J(T_H) = J(Q)$ in view of B.4.6.13. Further $Q = O_2(K_1T_H)$, where $K_1 := O^2(H_1)$, and $K_2 := O^2(H_2)$ induces $\mathbf{Z}_3$ on l. Set $H := C_G(l)$, so that $T_H \in Syl_2(H \cap M)$. As $N_G(Q) \leq M = !\mathcal{M}(LT)$, $C(H, Q) \leq H \cap M =: M_H$. In particular as $J(T_H) = J(Q)$, $N_T(T_H) \leq M_H$, so that $T_H \in Syl_2(H)$. It also follows as $K_1 \leq M_H$ that $Q = O_2(M_H) = O_2(N_H(Q))$. Thus Hypothesis C.2.3 is satisfied with Q in the role of "R".

We are now ready to establish our main preliminary result: we claim that $H = C_G(l) \leq M$. So we assume that $H \nleq M$, and derive a contradiction. Observe first that as l contains 2-central involutions, $H \in \mathcal{H}^e$ by 1.1.4.3. Next Q is Sylow in $O_{2,F}(H)Q$ by C.2.6.1, and as $M = !\mathcal{M}(LT)$,

$$N_H(W_0(Q, V)) \leq M_H \geq C_H(C_1(Q, V)).$$

Hence as $n(O_{2,F}(H)) = 1$ by E.1.13, $O_{2,F}(H) \leq M$ by (1) and E.3.19. On the other hand, if $O_{2,F^*}(H) \leq M_H$, then $O_2(H) = Q$ by A.4.4.1; thus $H \leq N_G(Q) \leq M$, contrary to our assumption.

This contradiction shows that there is $K \in \mathcal{C}(H)$ with $K/O_2(K)$ quasisimple, and $K \nleq M$. By 1.1.3.1, $K \in \mathcal{H}^e$. Suppose first that $Q \nleq N_H(K)$. Then C.2.4.2 shows that $Q \cap K \in Syl_2(K)$, and as $K \nleq M$, C.2.4.1 then shows that K is a χ_0-block. In particular $m_3(K) = 1$ and hence $m_3(\langle K^Q \rangle) = 2$. But then $m_3(K_2\langle K^Q \rangle) \geq 3$, contrary to $N_G(l)$ an SQTK-group.

Therefore $Q \leq N_G(K)$, so that K is described in C.2.7.3. Notice if case (g) of C.2.7.3 occurs with n even, then we are in one of cases (1)–(4) of C.1.34, in which $Z(O_2(K))$ is the sum of at most two natural modules for $K/O_2(K) \cong SL_3(2^n)$; this case is ruled out by A.3.19 as $K_2 \nleq K$. The remaining cases of C.2.7.3 where $m_3(K) = 2$ are eliminated by A.3.18 as $K_2 \nleq K$. Thus $m_3(K) = 1$. Also K is not an A_5-block as M_H contains the Sylow 2-group T_H of H. Thus inspecting C.2.7.3, one of the following holds:

(i) K is an $L_2(2^m)$-block, Q is Sylow in KQ, and $M_K := M_H \cap K$ is a Borel subgroup of K.

(ii) $K/O_2(K) \cong L_3(2^n)$, n odd, M_K is a maximal parabolic of K, and K is described in C.1.34.

Furthermore $O^{3'}(H) = K$, again using the fact that $m_3(K_2 O^{3'}(H)) \leq m_3(N_G(l)) = 2$. Thus $K_1 \leq K$, so as $K_1 \trianglelefteq M_H$, we conclude $n = 1$ in case (ii). Next $K_1 = [K_1, t]$ for some $t \in N_{T \cap L}(l)$. Thus if K is an $L_2(2^m)$-block, t induces a field automorphism on $K/O_2(K)$ and $[M_K, t] \leq C_L(l)$, so $[M_K, t]$ is a $\{2, 3\}$-group; we conclude in case (i) that K is an $L_2(4)$-block.

Next

$$E_{16} \cong l^{\perp} = [l^{\perp}, K_1] \leq Z(Q) = Z(O_2(K_1 T_H)). \qquad (*)$$

Assume $K/O_2(K) \cong L_3(2)$. Then by $(*)$, case (2) of C.1.34 occurs, with $O_2(K)$ the direct sum of two isomorphic natural modules for $K/O_2(K)$. Hence K_1 has exactly three noncentral 2-chief factors, two of which are in $l^{\perp} \leq V$. Thus as $O_2(\bar{K}_1) = 1$, K_1 has one noncentral chief factor on Q/V, impossible as K_1 has more than one noncentral chief factor on each nontrivial irreducible on Q/V under $\bar{L} \cong U_3(3)$.

Therefore K is an $L_2(4)$-block, and Q is Sylow in QK. Now $Z(Q) \leq C_{KQ}(O_2(KQ)) = Z(O_2(KQ))$ as $KQ \in \mathcal{H}^e$. Hence

$$E_{16} \cong l^{\perp} = [l^{\perp}, K_1] \leq [Z(Q), K_1] \leq Z(Q) \cap U(K).$$

However this is impossible as K_1 has at most one noncentral chief factor on $Z(Q) \cap U(K)$, since $Q \in Syl_2(KQ)$ and K is an $L_2(4)$-block. This contradiction finally establishes the claim that $H = C_G(l) \leq M$.

Now (2) follows directly from the claim. Further if $r(G, V) = 3$, then there is a totally isotropic 3-subspace U of V with $C_G(U) \not\leq M$. By 7.3.3 in [**Asc87**], L has two orbits on such subspaces, represented by V_3 and Y where $N_{\bar{M}_V}(Y) \cong L_3(2)$ is faithful on Y. Thus $C_M(Y) = C_M(V)$, so as $r(G, V) > 1$ by (1), we conclude $C_G(Y) \leq M$ by E.6.12. So $U \in V_3^L$, and (3) follows.

Assume the hypotheses of (4). Then interchanging V and V^g if necessary, we may assume $m(\bar{V}^g) \geq m(V/C_V(V^g))$. We apply B.4.6.13 much as in the proof of 13.3.10: First $\bar{T}$ contains no strong FF*-offenders, so that $\bar{V}^g \in \mathcal{P}(\bar{T}, V)$ by B.1.4.6; then there is a unique conjugacy class of FF*-offenders in $\bar{L}\bar{T}$ represented by the subgroup "A_1" of that lemma, so we may assume that $\bar{V}^g = A_1$ and hence $C_V(V^g) = [V, V^g] \in V_3^G$. Thus we may take $V_3 = [V, V^g]$. By hypothesis, $[V, V^g] \leq V \cap V^g$, and $V \cap V^g \leq C_V(V^g) = V_3$, so $V_3 = V \cap V^g$. Also $m(V/(V \cap V^g)) = 3 = m(V^g/(V \cap V^g))$, so we have symmetry between V and V^g. We conclude $V_3 \in V_3^{gL^g}$, and hence we may take $g \in G_3$. Let U_5 be the preimage in V^g of $\bar{V}^g \cap \bar{L}$. By (3) and (4) of B.4.6,

$$U_5 = \{u \in V^g : C_V(u) > V_3\}$$

and similarly

$$V_5 = V_1^{\perp} = \{v \in V : C_{V^g}(v) > V_3\},$$

so $U_5 \in V_5^{g N_{L^g}(V_3)}$, and hence we may take $V_5^g = U_5$. Then as $V_1 = [U_5, V_5]$, $V_1^g = V_1$, so $g \in G_1 \cap G_3$. But $Aut_{LT}(V_3)$ is the stabilizer in $GL(V_3)$ of V_1, so $G_1 \cap G_3 = N_{LT}(V_3)C_G(V_3)$. Then as LT normalizes V, we may take $g \in C_G(V_3)$, and hence $C_G(V_3) \not\leq M$ as $C_M(V_3) \leq M_V$ by 12.2.6. This completes the proof of (4). $\qquad \square$

During the remainder of this subsection, we will assume the following hypothesis:

HYPOTHESIS 13.3.13. $C_V(L) = 1$, so that V is not a 5-dimensional module when $\hat{L} \cong A_6$. Set $Q_1 := O_2(G_1)$.

We recall from Notation 13.3.3 that since $C_V(L) = 1$ by Hypothesis 13.3.13, we have $m(V_i) = i$. In particular by 13.3.4.1, $V_1 = Z \cap V$ is of order 2, and from 13.3.2.3, $V_3 = [V_3, L_1] = \langle V_2^{L_1} \rangle$. Also $G_1 = N_G(V_1) = C_G(V_1)$ and $G_1 \in \mathcal{H}^e$ as G is of even characteristic.

LEMMA 13.3.14. *Assume Hypothesis 13.3.13. Then Q_1 does not centralize V_2.*

PROOF. We assume that $[V_2, Q_1] = 1$ and derive a contradiction. The bulk of the proof proceeds by a series of reductions labeled (a)–(g).

Set $U := \langle V_3^{G_1} \rangle$ and $G_1^* := G_1/Q_1$. Set $L_+ := L_{1,+}$ if $L/O_2(L) \cong \hat{A}_6$, and $L_+ := L_1$ otherwise. Then $O_2(L_+ T) = R_1$ is of index 2 in T, and as $V_3 = [V_3, L_1] = \langle V_2^{L_1} \rangle$, $V_3 = [V_3, L_+] = \langle V_2^{L_+} \rangle$.

Observe that Hypothesis G.2.1 is satisfied with V_3, G_1, L_+ in the roles of "V, H, L". Therefore by G.2.2.1, $U \leq Q_1$. As $V_3 = \langle V_2^{L_+} \rangle$, $U = \langle V_2^{G_1} \rangle$. Then as $[V_2, Q_1] = 1$:

(a) $U \leq \Omega_1(Z(Q_1))$.

Observe that as $C_V(L) = 1$ by Hypothesis 13.3.13, $G_1 \not\leq M$ by 13.3.6. Set $Y := O^2(C_G(U))$. Then $Y \leq C_G(V_3) \leq C_G(V_2) \leq G_1$. Also $Y \trianglelefteq G_1$, and $Y \cap M \leq M_V$ by 12.2.6.

(b) Y is solvable with $m_p(Y) \leq 1$ for each odd prime p.

For if $Y = 1$ then certainly (b) holds, so we may assume $Y \neq 1$. Consider any T-invariant subgroup $Y_0 = O^2(Y_0)$ of $Y \cap M$. As $C_{\bar{M}_V}(V_3)$ is a 2-group, Y_0 centralizes V. Then $[L, Y_0] \leq C_L(V) = O_{2,Z}(L)$, so $L = L^\infty$ centralizes $Y_0 O_2(L)/O_2(L)$. Hence as Y_0 is T-invariant, L acts on $O^2(Y_0 O_2(L)) = Y_0$. Therefore if $Y_0 \neq 1$, then $N_G(Y_0) \leq M = !\mathcal{M}(LT)$. In particular $Y \not\leq M$, as otherwise $G_1 \leq N_G(Y) \leq M$.

Now if $\bar{L} \cong L_3(2)$, then $V = V_3 \leq U$, so $Y \leq C_G(U) \leq C_G(V_3) = C_G(V) \leq M$, contrary to the previous paragraph. Similarly if $\bar{L} \cong A_5$, then $C_G(V_3) \leq M$ by 13.2.3.2, for the same contradiction. Therefore we may assume that $\bar{L}$ is A_6 or $U_3(3)$.

Let $g \in L_2 - G_1$ be of order 3. Then $m(V_3 V_3^g) = 4$, so $V_3 V_3^g = V$ if $\bar{L} \cong A_6$, and $V_3 V_3^g$ is not totally isotropic if $\bar{L}$ is $U_3(3)$. In the latter case $C_G(V_3 V_3^g) \leq M$ by 13.3.12.2, while if $\bar{L}$ is A_6, then $C_G(V_3 V_3^g) = C_G(V) \leq M$. So in either case, $C_G(V_3 V_3^g) \leq M$.

Set $Y_1 := O^2(Y \cap Y^g)$ and $Y_M := O^2(Y \cap M)$. Then $Y_1 \leq C_G(V_3 V_3^g) \leq M$, so $Y_1 \leq Y_M$. Further Y centralizes V_2, so $Y \leq G_1^g \leq N_G(Y^g)$, and hence $Y_1 \trianglelefteq Y$; then by symmetry, $Y_1 \trianglelefteq Y^g$. Next $T \leq M_1 \leq N_G(Y_M)$, so using Y_M in the role of "Y_0" above, L acts on Y_M and $N_G(Y_M) \leq M$; hence $Y_M = Y_M^g \leq Y_1$ as $g \in L_2$. We conclude $Y_M = Y_1$, so if $Y_1 \neq 1$, then $Y \leq N_G(Y_1) \leq M$, contrary to the first paragraph. Therefore $Y_1 = 1$, so that $Y \cap Y^g$ is a 2-group.

Set $\hat{G}_2 := G_2/O_2(G_2)$. As $Y \trianglelefteq G_1$ while $C_G(V_2) \leq G_1$, $Y \trianglelefteq C_G(V_2)$, so that $O_2(Y) \leq O_2(G_2)$, and hence $Y_+ := \langle Y, Y^g \rangle \trianglelefteq C_G(V_2)$. Then as Y and Y^g are normal in Y_+, and $Y \cap Y^g$ is a 2-group normal in Y_+, $\hat{Y}_+ = \hat{Y} \times \hat{Y}^g$.

Therefore since G_2 is an SQTK-group, $m_p(Y) = 1$ for each odd prime p. Further if Y is not solvable, then by 1.2.1.1, there is $K \in \mathcal{C}(Y)$, and as $Y \trianglelefteq G_2$, $K \in \mathcal{C}(G_2)$. Then as g is of order 3, g acts on K by 1.2.1.3, contradicting $Y \cap Y^g$ a 2-group. This contradiction completes the proof of (b).

(c) $O_2(L_+^*) \neq 1$.

If $O_2(L_+^*) = 1$, then $O_2(L_+) \leq Q_1 \leq C_G(V_3)$ by (a), impossible as L_+ induces A_4 on V_3/V_1.

(d) $O_2(L_+^*)$ centralizes $F(G_1^*)$.

Assume $O_2(L_+^*)$ is nontrivial on $O_p(G_1^*)$ for some odd prime p. Then as $L_+/O_2(L_+)$ has order 3, $Aut_{O_2(L_+)}(O_p(G_1^*))$ is noncyclic, so by A.1.21 and A.1.25, there is a noncyclic supercritical subgroup P^* of $O_p(G_1^*)$ such that $P^* \cong E_{p^2}$ or p^{1+2} and $Aut(P^*)/O_p(Aut(P^*))$ is a subgroup of $GL_2(p)$. Hence $Aut_{L_+}(P^*) \cong SL_2(3)$. Let $P := O^2(P_+)$, where P_+ is the preimage of P^* in G_1. Then $PL_+T =: H \in \mathcal{H}(T) \cap G_1$. Further as L_+ is normal in M_1 but L_+ is not P-invariant, $P \not\leq M$.

As $Aut_{L_+}(P^*) \cong SL_2(3)$, $P \in \Xi(G,T)$ with $Aut_{T \cap L_+}(P^*) \cong Q_8$. Also $[U,P] \neq 1$ as $m_p(Y) = 1$ by (b). Since $U \leq \Omega_1(Z(Q_1))$ by (a), $P \in \Xi_f(G,T)$ by an application of A.4.9 to P, G_1 in the roles of "X, M". Assume $P \leq \langle K^T \rangle$ for some $K \in \mathcal{C}(G,T)$ with $K/O_2(K)$ quasisimple. Then $\langle K^T \rangle$ is described in 1.3.4. Further $K \in \mathcal{L}_f(G,T)$ by 1.3.9.2, so $K = \langle K^T \rangle$ by 13.3.2.2, and $K/O_2(K)$ is described in 13.3.2.1. As the lists in 1.3.4 and 13.3.2.1 do not intersect, there is no such K, so $P \in \Xi_f^*(G,T)$. Then by 3.2.13, $P \in \Xi_-(G,T)$. Since $Aut_G(P/O_2(P))$ involves $SL_2(3)$ which is not a $\{2,5\}$-group, we conclude from Definition 3.2.12 that P is a $\{2,3\}$-group, so that $p = 3$. As $m_3(PL_+) \leq 2$ with $Aut_{L_+}(P^*) \cong SL_2(3)$, we conclude $P/O_2(P) \cong P^* \cong E_9$ rather than 3^{1+2}. Let $W := R_2(PT)$; as $Aut_T(P^*) \cong Q_8$, we conclude from D.2.17 that $\hat{q}(Aut_{PT}(W), W) > 2$. However $N_G(P) = !\mathcal{M}(PT)$ by Theorem 1.3.7, so that we may apply Theorem 3.1.8.1 to P, W in the roles of "L_0, V" to obtain $\hat{q}(Aut_{PT}(W), W) \leq 2$, contrary to the previous observation. This contradiction completes the proof of (d).

Since $O_2(G_1^*) = 1$, (c) and (d) say there is $K \in \mathcal{C}(G_1)$ with K^* a component of G_1^* and $[K^*, O_2(L_+^*)] \neq 1$. By 1.2.1.3, $L_+ = O^2(L_+)$ normalizes K. In particular, $K/O_2(K)$ is quasisimple and $K = [K, L_+]$.

(e) $K \in \mathcal{L}_f^*(G,T)$, $G_1 \leq N_G(K) = !\mathcal{M}(KT)$, $L \not\leq N_G(K)$, and $K \not\leq M$.

First $[U,K] \neq 1$ by (b), so using (a) and A.4.9 as in the proof of (d), $K \in \mathcal{L}_f(G,T)$. Then by 13.3.2.2, $K \in \mathcal{L}_f^*(G,T)$, $K \trianglelefteq KT$, and $G_1 \leq N_G(K) = !\mathcal{M}(KT)$. As $G_1 \not\leq M$, $N_G(K) \neq M$. So as $M = !\mathcal{M}(LT)$, $L \not\leq N_G(K)$, and as $N_G(K) = !\mathcal{M}(KT)$, $K \not\leq M$, completing the proof of (e).

If $L_2 \leq N_G(K)$, then $L = \langle L_1, L_2 \rangle \leq N_G(K)$, contrary to (e). So:

(f) $L_2 \not\leq N_G(K)$.

As $L_2 \not\leq N_G(K) \geq C_G(Z)$ by (e) and (f), $L_2 T$ contains some $H \in \mathcal{H}_*(T, N_G(K))$, and $H \not\leq C_G(Z)$. By (e), $K \in \mathcal{L}_f^*(G,T)$, and we saw $K/O_2(K)$ is quasisimple, so $O_2(KT) = C_T(R_2(KT))$ by 1.4.1.4b. Then applying 3.1.8.3 to K, $R_2(KT)$ in the roles of "L, V", $K = [K, J(T)]$. Set $J := KL_+T$, $W := R_2(J)$, $J^+ := J/C_J(W)$, and $W_K := [W, K]$. Then $R_2(KT) \leq W$ by A.1.11, so that $Irr_+(K, R_2(KT)) \subseteq Irr_+(K, W_K)$. Now $K/O_2(K)$ is described in 13.3.2.1, and the members of the set $Irr_+(KT, R_2(KT), T)$ are described in 13.3.2.3. Thus applying Theorem B.5.6 to the FF-module W_K for J, we conclude that either $W_K \in Irr_+(KT, R_2(KT), T)$ or $K/O_2(K) \cong L_3(2)$ and W_K is the sum of two isomorphic natural modules.

(g) $L_+ \not\leq K$.

Assume that that $L_+ \leq K$. Define $V(K)$ as in Definition A.4.7, and set $\hat{J} := J/C_J(V(K))$. By (a) and A.4.8.4, $V_3 = [V_3, L_+] \leq [\Omega_1(Z(Q_1)), K] \leq V(K)$. Let X be of order 3 in L_+ and set $Q_J := O_2(J)$. By A.4.8.1, $\hat{Q}_J$ centralizes $\hat{X}$. Thus by the Thompson $A \times B$-Lemma, X is faithful on $C_{V_3}(Q_J)$, so Q_J centralizes V_3. Now by A.4.8.4, $V_3 \leq W_K$, so $1 \neq V_1 \leq C_{W_K}(K)$. However we saw that either $W_K \in Irr_+(KT, R_2(KT), T)$ and so is described in 13.3.2.3, or W_K is the sum of natural modules for $K/O_2(K) \cong L_3(2)$. Thus as $C_{W_K}(K) \neq 1$, W_K is a 5-dimensional module for $K^+ \cong A_6$. Therefore $A_4 \cong L_+^+ \leq \hat{K}^+$ and V_3/V_1 is an L_+-invariant line in W_K/V_1, with $[V_3, O_2(L_+)] = V_1$, whereas in the 5-dimensional module W_K, the preimage of such a line is centralized by $O_2(L_+^+)$. This contradiction establishes (g).

Now as $L_+ \not\leq K$ by (g), but $m_3(KL_+) \leq 2$, A.3.18 eliminates the possibilities for $K/O_2(K)$ of 3-rank 2 in 13.3.2.1. Thus $m_3(K) = 1$, so that $K^+ \cong A_5$ or $L_3(2)$. As $L_+ \not\leq K = [K, L_+]$, and $Out(K^+)$ is a 2-group, L_+ is diagonally embedded in $L_K L_C$, where $L_C := C_{KL_+}(K/O_2(K))$ and $L_K = O^2(L_K)$ is the projection of L_+ on K. But if $K/O_2(K) \cong L_3(2)$, then $L_K = [L_K, T \cap K]$, contrary to the fact that L_+ is T-invariant. Thus $K/O_2(K) \cong A_5$, and from earlier discussion W_K is the A_5-module. Then as $L_K^+ \neq 1$, $L_K T = (T \cap K)O_2(KT)$, so as $R_1 = O_2(L_+T)$ is of index 2 in T, $R_1 = (T \cap K)O_2(KT)$. Since K satisfies Hypothesis 12.2.3 by 13.3.2.4, we may apply 13.2.4.1 with K in the role of "L to conclude that $C(G, \text{Baum}(R_1)) \leq N_G(K)$. It follows as $K \not\leq M$ by (e) that $N_G(\text{Baum}(R_1)) \not\leq M$. Therefore by 13.3.10, L is an $L_3(2)$-block or an A_6-block, and in either case L_+ has exactly two noncentral 2-chief factors.

As W_K is the A_5-module, $End_K(W_K) = \mathbf{F}_2$, so $[W_K, L_C] = 0$. Thus L_+ has at least one noncentral 2-chief factor on W_K, as well as one on $O_2(L_K^+)$; so as L_+ has just two noncentral 2-chief factors, $[O_2(J), L_+] \leq W_K$. Hence as $K = [K, L_+]$, K is an A_5-block. Then as L_C centralizes K^+ and W_K, $[K, L_C] = 1$ by Coprime Action. By C.1.13.c, $O_2(J) = C_{O_2(J)}(K) \times W_K$, so as $J \in \mathcal{H}^e$, L_C has a noncentral 2-chief factor in $C_{O_2(J)}(K)$, contradicting $[O_2(J), L_+] \leq W_K$. This contradiction finally completes the proof of 13.3.14. $\qquad \square$

LEMMA 13.3.15. *Assume Hypothesis 13.3.13. and that $\bar{L} \not\cong A_5$. Then*

(1) $I_2 := \langle Q_1^{G_2} \rangle \trianglelefteq G_2$.
(2) $I_2 = XQ_1$, where $X := L_{2,+}$ when $L/O_2(L) \cong \hat{A}_6$ and $X := L_2$ otherwise.
(3) $C_{I_2}(V_2) = O_2(I_2)$ and $I_2/O_2(I_2) \cong S_3$, with $O^2(I_2) = X$.
(4) $C_{Q_1}(V_2) \leq O_2(I_2) \leq O_2(G_2)$.
(5) $m_3(C_G(V_2)) \leq 1$.
(6) $C_G(V_3) \leq M_V$. Hence $[V, C_G(V_3)] \leq V_1$.

PROOF. Part (1) holds by construction. As $L \not\cong A_5$, $X/O_2(X)$ is of order 3. Recall one consequence of Hypothesis 13.3.13 is that V_2 is of rank 2. Then as $[Q_1, V_2] \neq 1$ by 13.3.14, XQ_1 induces $GL(V_2)$ on V_2, with X transitive on $V_2^\#$. Hence $G_2 = C_G(V_2)Q_1X$, with $C_G(V_2)Q_1 \leq G_1$, so

$$I_2 = \langle Q_1^{G_2} \rangle = \langle Q_1^X \rangle \leq XQ_1,$$

and $X = [X, Q_1] \leq I_2$, so $I_2 = XQ_1$ and (2)–(4) hold. As $X = O^2(I_2) \trianglelefteq G_2$ by (1) and (3),

$$[C_G(V_2), X] \leq C_X(V_2) = O_2(X);$$

so as $m_3(G_2) \leq 2$ and $X/O_2(X)$ is faithful on V_2, (5) holds.

Next $C_G(V_3) \le G_2 \le N_G(X)$, so as $N_L(V_3)$ normalizes $C_G(V_3)$, $C_G(V_3)$ acts on $X^{N_L(V_3)}$. Then as $L = \langle X^{N_L(V_3)} \rangle$, we conclude $C_G(V_3) \le N_G(L) = M$. Hence $C_G(V_3) \le N_G(V) = M_V$ as V is a TI-set in M by 12.2.6. This establishes (6). $\quad\square$

13.3.1. Eliminating $U_3(3)$. With the technical results from the earlier part of the section in hand, we are now ready to embark on the main project in this chapter: the treatment of the cases $\bar{L} \cong U_3(3)$, A_6, and A_5.

In this subsection, we handle the easiest of these cases:

THEOREM 13.3.16. *Assume Hypothesis 13.3.1. Then $\bar{L}$ is not $U_3(3)$.*

In the remainder of this section, assume G, L is a counterexample to Theorem 13.3.16. By 13.3.2.3, $C_V(L) = 1$ and $m(V) = 6$. In particular Hypothesis 13.3.13 is satisfied, so we can appeal to 13.3.14 and 13.3.15. Let z be a generator for V_1, so that $G_1 := C_G(z) = G_z$, and $\tilde{G}_1 = G_1/V_1$. As usual define

$$\mathcal{H}_z = \{ H \in \mathcal{H}(L_1 T) : H \le G_1 \text{ and } H \not\le M \}.$$

By 13.3.6, $G_1 \not\le M$, so $G_1 \in \mathcal{H}_z$ and hence $\mathcal{H}_z \ne \emptyset$.

We first observe:

LEMMA 13.3.17. *(1) $G_1 \cap G_3 \le M_V \ge C_G(V_3)$.*
(2) $r(G, V) > 3$.
(3) If $[V, V^g] \le V \cap V^g$, then $[V, V^g] = 1$.
(4) $[O_2(G_1), V_2] \ne 1$.

PROOF. Part (4) holds by 13.3.14, and $C_G(V_3) \le M_V$ by 13.3.15.6. Further $Aut_{M_1}(V_3)$ is the full stabilizer in $GL(V_3)$ of V_1, so $G_1 \cap G_3 = C_G(V_3) N_{M_1}(V_3)$. As V is a TI-set in M by 12.2.6, this completes the proof of (1). Then (1) together with parts (2) and (3) of 13.3.12 imply (2), while (1) and part (4) of 13.3.12 imply (3). $\quad\square$

LEMMA 13.3.18. *(1) For each $H \in \mathcal{H}_z$, Hypothesis F.9.1 is satisfied with V_3 in the roles of "V_+".*
(2) $\langle V^{G_1} \rangle$ is abelian.

PROOF. We check the various parts of Hypothesis F.9.1:
First hypothesis (c) of F.9.1 follows from 13.3.17.1, and by construction L_1 is irreducible on $\tilde{V}_3$, so hypothesis (b) holds. As $H \in \mathcal{H}(T)$, $H \in \mathcal{H}^e$ by 1.1.4.6. Also by Coprime Action and 13.3.17.1, $Y := O^2(C_H(\tilde{V}_3)) \le C_{M_V}(V_3))$, so as $O^2(C_{\bar{M}_V}(\tilde{V}_3)) = 1$, $Y \le C_M(V) \le C_M(L/O_2(L))$ and therefore L normalizes $Y = O^2(Y O_2(L))$. Thus if $Y \ne 1$, then $H \le N_G(Y) \le M = !\mathcal{M}(LT)$, contrary to the definition of $H \in \mathcal{H}_z$. Thus $C_H(\tilde{V}_3)$ is a 2-group, so hypothesis (a) follows. As $M = !\mathcal{M}(LT)$ and $H \not\le M$, hypothesis (d) holds. Finally 13.3.17.3 implies hypothesis (e), completing the proof of (1).

Now let $H := G_1$, and as in Hypothesis F.9.1, define $U_H := \langle V_3^H \rangle$, $V_H := \langle V^H \rangle$, $Q_H = O_2(H)$ and $H^* := H/C_H(\tilde{U}_H)$. It remains to prove (2), so we may assume V_H is nonabelian.

Observe that $O_2(\bar{L}_1) \cong \mathbf{Z}_4^2$ and $\bar{R}_1 = O_2(\bar{L}_1)$ in case $\bar{M}_V = \bar{L}$, while in case $\bar{M}_V \cong G_2(2)$, $\bar{R}_1$ is $O_2(\bar{L}_1)$ extended by an involution $\bar{r}$ inverting $O_2(\bar{L}_1)$ and centralizing a supplement to $O_2(\bar{L}_1)$ in $\bar{L}_1$. In particular, $\bar{A}_0 := \Omega_1(O_2(\bar{L}_1))$ is the unique nontrivial normal elementary abelian subgroup of $\bar{M}_1$ in case $\bar{M}_V = \bar{L}$, while

$\bar{A}_1 := \langle \bar{r} \rangle \bar{A}_0$ and $\bar{A}_0$ are the only such subgroups in case $\bar{M}_V \cong G_2(2)$. Further $C_{\bar{M}_V}(V_3)$ is $\bar{A}_0$ or $\bar{A}_1$ in the respective case.

By F.9.2, $\tilde{U}_H \le \Omega_1(Z(\tilde{Q}_H))$, $O_2(H^*) = 1$, $\Phi(U_H) \le V_1$, and $Q_H = C_H(\tilde{U}_H)$. Thus if V centralizes U_H, then $V \le Q_H \le \ker_H(N_H(V))$, and then for $h \in H$, $[V, V^h] \le V \cap V^h$, so that $[V, V^h] = 1$ by 13.3.17.3. But then V_H is abelian, contrary to our assumption. Therefore $[U_H, V] \neq 1$, so that $V^* \neq 1$, and as $\Phi(U_H) \le V_1$, $1 \neq \bar{U}_H$ is a normal elementary abelian subgroup of $\bar{M}_1$, so by the previous paragraph, $\bar{U}_H = \bar{A}_0$ or $\bar{A}_1$. In particular U_H centralizes V_3, so $V_3 \le Z(U_H)$ and thus $U_H = \langle V_3^H \rangle$ is elementary abelian.

Let $Z_H := Z(Q_H)$. By 13.3.17.4, $[Q_H, V_2] \neq 1$, so as L_1 is irreducible on $\tilde{V}_3$, $V_3 \cap Z_H = V_1$. Furthermore as $V_2 \le U_H$, $Q_C := C_{Q_H}(U_H)$ is properly contained in Q_H.

For $x \in Q_C$, define $\varphi(xQ_C) : U_H/Z_H \to V_1$ by $\varphi(xQ_C) : uZ_H \mapsto [x, u]$ for $u \in U_H$. By F.9.7, φ is an H-equivariant isomorphism between Q_H/Q_C and the $\mathbf{F}_2$-dual space of U_H/Z_H.

As $[V, \bar{A}_0] = [V, \bar{A}_1] = V_3$, $[V^*, \tilde{U}_H] = \tilde{V}_3$. Then as $V_3 \cap Z_H = V_1$, V^* is nontrivial on U_H/Z_H and hence also on Q_H/Q_C as φ is an equivariant isomorphism. But $Q_H \le T \le N_G(V)$, so $[Q_H/Q_C, V] \le Q_C(V \cap Q_H)/Q_C$, and hence $V \cap Q_H \not\le Q_C$.

Next $V_5 = V_1^\perp$ is a hyperplane of V, with $[v, R_1]V_3/V_3 = V_5/V_3$ for each $v \in V - V_5$. Thus as L_1 is irreducible on V_5/V_3 and $V \cap Q_H \not\le Q_C \ge V_3$, we conclude that $V_5 = V \cap Q_H$ and $V_3 = V \cap Q_C = V \cap U_H$. Also by 13.3.15.4,

$$V_5 \le C_{Q_H}(V_2) \le O_2(G_2),$$

so $V = \langle V_5^{L_2} \rangle \le O_2(G_2)$.

Now V^* is of order 2 and $O_2(H^*) = 1$, so by the Baer-Suzuki Theorem we can pick $h \in H$ so that for $I := \langle V, V^h \rangle$, $I^* \cong D_{2m}$ with $m > 1$ odd. Then V^* is conjugate to V^{*h} in I, so we may assume $h \in I$.

Suppose $[V_3, V_5^h] \neq 1$. Then as $C_{\bar{M}_V}(\tilde{V}_5) \le C_{\bar{M}_V}(V_3)$, $[V_5, V_5^h]$ contains a hyperplane of V_3 containing V_1. As all such hyperplanes are fused under L_1, we may take $V_2 \le [V_5, V_5^h]$. Now $V_5 = V \cap Q_H$ is normal in Q_H, and hence $V_5^h = V^h \cap Q_H$ is normal in Q_H, so that $V_2 \le V_5 \cap V_5^h \le V \cap V^h \le Z(I)$. Thus $I \le G_2$, impossible as I is not a 2-group, while $V \le O_2(G_2)$ and $h \in I$.

This contradiction shows that V_5^h centralizes V_3, and hence by symmetry, $V_5 V_5^h$ centralizes $V_3 V_3^h$. In particular $\bar{V}_5^h \le \bar{A}_1$. Therefore $[V, V_5^h] \le V_3$, and by symmetry I centralizes $V_5 V_5^h / V_3 V_3^h$. Hence as $h \in I$, $V_5 V_3^h = V_5(V_3 V_3^h) = V_5^h(V_3 V_3^h) = V_5^h V_3$. Therefore as $V_1 \le V_3^h$, $\bar{V}_5^h = \bar{V}_3^h$ is of rank $m(\bar{V}_3^h) \le m(V_3^h/V_1) \le 2$. Therefore as $r(G, V) > 3$ by 13.3.17.2, $V \le C_G(C_{V_5^h}(V)) \le N_G(V^h)$, once again contradicting I not a 2-group. This completes the proof of 13.3.18. $\qquad\square$

LEMMA 13.3.19. (1) If $g \in G$ with $V \cap V^g \neq 1$, then $[V, V^g] = 1$.
(2) $W_2(T, V)$ centralizes V.
(3) $H \le M$ for each $H \in \mathcal{H}(T)$ with $n(H) \le 2$.

PROOF. Suppose $1 \neq V \cap V^g$. As L is transitive on $V^\#$, we may take $z \in V^g$ and $g \in G_1$ by A.1.7.1. But then $[V, V^g] = 1$ by 13.3.18.2. Thus (1) is established.

Let $A := V^g \cap M \le T$ with $m(V^g/A) =: k \le 2$, and suppose $\bar{A} \neq 1$. Let $U := N_V(V^g)$. By (1), $V \cap V^g = 1$, so as $[U, A] \le V \cap V^g$, $U \le C_V(A)$. In particular $U < V$ as $\bar{A} \neq 1$. On the other hand, if $B \le A$ with $m(A/B) < 4 - k$, then as $r(G,V) > 3$ by 13.3.17.2, $C_G(B) \leq N_G(V^g)$, so that $C_V(B) = U$. Thus $\bar{A} \in \mathcal{A}_{4-k}(\bar{T}, V)$, so by B.4.6.9, $k = 2$ and $\bar{A} \in \bar{A}_1^L$. Then without loss, $\bar{A} = \bar{A}_1$, so from the action of $\bar{A}_1$ on V,

$$V_5 = \langle C_V(\bar{a}) : \bar{a} \in \bar{A}_1^{\#} \rangle \leq U,$$

and hence $V_5 = U$ as $U < V$. As $k = 2$, $W_1(T,V)$ centralizes V. Therefore as $m(V/U) = 1$ and $U = N_V(V^g)$, U centralizes V^g. Then since $r(G,V) > 3$, $V^g \leq C_G(U) \leq N_G(V)$, so that $V^g = A$, contrary to $k = 2$. This proves (2). Finally by (2) and 13.3.17.2, $\min\{w(G,V), r(G,V)\} \geq 3$, so E.3.35.1 implies (3). $\square$

LEMMA 13.3.20. $n(H) \leq 2$ for each $H \in \mathcal{H}_*(T,M) \cap G_1$.

PROOF. By 13.3.17.1, $G_1 \cap G_3 \leq M$, so hypothesis (c) of 12.2.11 is satisfied. Therefore as $H \leq G_1$, we may apply 12.2.11 with V_1 in the role of "U" to conclude that $n(H) \leq 2$. $\square$

We can now complete the proof of Theorem 13.3.16: Recall $T \leq G_1$, and $G_1 \not\leq M$ by 13.3.6, so there exists $H \in \mathcal{H}_*(T,M) \cap G_1$. Then $n(H) \leq 2$ by 13.3.20, so that $H \leq M$ by 13.3.19.3, for our final contradiction.

13.4. The treatment of the 5-dimensional module for $\mathbf{A}_6$

In section 13.4 we prove:

THEOREM 13.4.1. *Assume Hypothesis 13.3.1 with $C_V(L) \neq 1$. Then $G \cong Sp_6(2)$.*

Set $Z_V := C_V(L)$. By hypothesis, $Z_V \neq 1$, so by 13.3.2.3,

V is a 5-dimensional module for $L/C_L(V) \cong A_6$.

Recall this means that V is the core of the permutation module for A_6 acting on $\Omega := \{1, \ldots, 6\}$. Accordingly we adopt the notational conventions of section B.3. We also adopt the conventions of Notations 12.2.5 and 13.2.1.

Of course the parabolic of the target group $Sp_6(2)$ stabilizing a point in the natural module has this structure. Eventually we identify G with $Sp_6(2)$ during the proof of Proposition 13.4.9. We begin that process by setting up some notation to discuss $Sp_6(2)$.

Let $\dot{G} = Sp_6(2)$, $\dot{T} \in Syl_2(\dot{G})$, and $\dot{P}_i$, $1 \leq i \leq 3$, the maximal parabolics of $\dot{G}$ over $\dot{T}$ stabilizing an i-dimensional subspace of the natural module for $\dot{G}$. The pair $(\dot{G}, \{\dot{P}_1, \dot{P}_2, \dot{P}_3\})$ is a C_3-system in the sense of section I.5. Notice $\bar{L} \cong A_6 \cong \dot{P}_1'/O_2(\dot{P}_1)$. We will produce a corresponding C_3-system for G, and then use Theorem I.5.1 to conclude that $G \cong Sp_6(2)$. To do so, we will need to study the centralizer G_z of a suitable involution $z \in V_1 - Z_V$, and show $G_z/O_2(G_z) \cong S_3 \times S_3 \cong \dot{P}_2/O_2(\dot{P}_2)$. We must also construct a third 2-local H_0 and show $H_0/O_2(H_0) \cong L_3(2) \cong \dot{P}_3/O_2(\dot{P}_3)$. Then it is not difficult to construct our C_3-system.

13.4.1. Preliminary results on the structure of certain 2-local subgroups. As usual $Z = \Omega_1(Z(T))$ from Notation 12.2.5. Notice $Z_V \leq Z_L := C_Z(L)$. Recall that Z_V is of order 2 and is of index 2 in $V_1 = Z \cap V$ by 13.3.4.1.

As usual we let $\theta(X)$ denote the subgroup generated by all elements of order 3 in a group X.

LEMMA 13.4.2. *(1)* $M = N_G(L) = N_G(V) = C_G(Z_V)$.
(2) $C_G(Z) \le C_G(Z_L) \le C_G(Z_V) = M$, *and* $M = !\mathcal{M}(C_G(Z_L))$.
(3) For each $v \in V^\#$, $C_G(v)$ *is transitive on conjugates of* V *containing* v. *In particular,* V *is the unique member of* V^G *containing* Z_V.
(4) $C_M(V) = C_M(L/O_2(L)) = C_M(V/Z_V)$.
(5) $L = \theta(M)$, *and if* $L/O_2(L) \cong A_6$, *then* $L = O^{3'}(M)$.

PROOF. Theorem 12.2.2.3 shows that $M = N_G(L)$, and (since $Z_V \ne 1$) that $V \trianglelefteq M$. Hence $M \le C_G(Z_V)$, and (1) follows as $M \in \mathcal{M}$. As $Z_V \le Z_L \le Z$ and $M = !\mathcal{M}(LT)$, (2) holds. Since LT controls fusion in V by 13.2.2.5, (3) follows from A.1.7.1. Observe (5) follows from 12.2.8. Finally $C_M(L/O_2(L)) = C_M(V/Z_V) = C_M(V)$ by A.1.41, establishing (4). $\qquad\square$

As $M = N_G(V)$ by 13.4.2.1, $\bar{M} := M/C_M(V)$ from Notation 12.2.5.2. Recall $V_1 = V \cap Z$, and by 13.3.5.3 there is $z \in V_1 - Z_V$ with $L_1 T \le G_z := C_G(z) \not\le M$. Fix a choice of z and observe z has weight 2 or 4 in V. Eventually we will see that there is a unique $z \in V_1$ with $G_z \not\le M$. As usual define

$$\mathcal{H}_z := \{H \in \mathcal{H}(L_1 T) : H \le G_z \text{ and } H \not\le M\}.$$

In particular $G_z \in \mathcal{H}_z$ so $\mathcal{H}_z \ne \emptyset$. Recall that R_1 is defined in Notation 13.2.1.

LEMMA 13.4.3. *(1)* $L = [L, J(T)]$.
(2) $VZ = VZ_L$ *and* $|Z : Z_L| = 2$.
(3) $Z = V_1 Z_L$, *so* L_1 *centralizes* Z.

PROOF. As $1 \ne Z_V \le Z_L$, (1) follows from 3.1.8.3. Then by (1) and Theorem B.5.1, $[VZ, L] = V$, so $VZ = VZ_L$ by B.2.14. Then as $|Z \cap V : Z_V| = 2$ by 13.3.4.1, (2) and (3) hold. $\qquad\square$

LEMMA 13.4.4. *If* $H \in \mathcal{H}_z$ *and* $V_H \in \mathcal{R}_2(H)$ *with* $Z_L \cap V_H \langle z \rangle \ne 1$, *then*
(1) $C_H(V_H) \le M$.
(2) Set $L_+ := L_1$ *or* $L_{1,+}$, *for* $L/O_2(L) \cong A_6$ *or* $\hat{A}_6$, *respectively. Then either:*
   (i) $O^2(C_H(V_H)) = 1$ *and* $C_H(V_H) = O_2(H) \le O_2(L_1 T) \le R_1$, *or*
   (ii) $L_+ = O^2(C_H(V_H)) \trianglelefteq H$, *and* $H_1 := C_H(L_+/O_2(L_+))$ *is of index 2 in* H *with* $R_1 \in Syl_2(H_1)$.
(3) $L_2 \not\le H$, *and if* $L/O_2(L) \cong \hat{A}_6$, $L_{2,+} \not\le H$.

PROOF. As neither L_2 nor $L_{2,+}$ centralizes z, (3) holds. Next $C_H(V_H) = C_H(V_H \langle z \rangle) \le C_G(Z_L \cap V_H \langle z \rangle) \le M = !\mathcal{M}(LT)$ by 13.4.2.2, so (1) holds.

Set $Y := O^2(C_H(V_H))$. By (1), $Y \le M$, and as $H \in \mathcal{H}_z$, Y centralizes $z \in V_1 - Z_V$. Thus the hypotheses of 13.3.9 are satisfied, so we can appeal to that lemma. If $Y = 1$, then $C_H(V_H)$ is a 2-group; so as $V_H \in \mathcal{R}_2(H)$, $C_H(V_H) = O_2(H)$. Further $L_1 T \le H$, so $O_2(H) \le O_2(L_1 T)$ by A.1.6, and hence conclusion (i) of (2) holds. Thus we may assume conclusion (2) of 13.3.9 holds. In particular $L_+ = Y \trianglelefteq H$, so conclusion (ii) of (2) holds. $\qquad\square$

LEMMA 13.4.5. *Assume* $H \in \mathcal{H}(T, M)$ *is nonsolvable. Then*
(1) There exists $K \in \mathcal{C}(H)$, *and for each such* K, $K \in \mathcal{L}_f^*(G, T)$, $K \trianglelefteq H$, *and* $K/O_2(K) \cong A_5$, $L_3(2)$, A_6, *or* $\hat{A}_6$.
(2) $K \not\le M$, $L \not\le N_G(K)$, *and* $[Z_V, K] \ne 1$.

(3) $Irr_+(K, R_2(KT)) \subseteq Irr_+(KT, R_2(H))$, there is $V_K \in Irr_+(K, R_2(KT), T)$, and for each such V_K, $V_K \trianglelefteq T$, the pair K, V_K satisifies the FSU, and $V_K = \langle (Z \cap V_K)^K \rangle$ is either a natural module for $K/C_K(V_K) \cong A_5$, $L_3(2)$, or A_6, or a 5-dimensional module for $K/C_K(V_K) \cong A_6$.

(4) For V_K as in (3), $O_{2,Z}(K) = C_K(V_K) = C_K(R_2(H))$. In particular, $R_2(H)$ contains no faithful 6-dimensional modules when $K/O_2(K) \cong \hat{A}_6$.

PROOF. As H is nonsolvable, there exists $K \in \mathcal{C}(H)$ by 1.2.1.1. If $K \leq M$ then we obtain a contradiction by applying 13.3.8.2 with L, M, $\langle K^T \rangle$ in the roles of "K, M_K, Y". Thus $K \not\leq M$, so $[Z_V, K] \neq 1$ by 13.4.2.1. Thus as $Z_V \leq Z$, $[R_2(H), K] \neq 1$, so $K \in \mathcal{L}_f(G, T)$ by 1.2.10. Therefore $N_G(K) = !\mathcal{M}(KT)$, and (1) holds by parts (1) and (2) of 13.3.2, since Theorem 13.3.16 rules out $K/O_2(K) \cong U_3(3)$. As $K \not\leq M$ and $M = !\mathcal{M}(LT)$, $L \not\leq N_G(K)$ so (2) holds.

By 13.3.2.3, there is $V_K \in Irr_+(K, R_2(KT), T)$ and $V_K \trianglelefteq T$. As $R_2(KT) \leq R_2(H)$ by A.1.11, $V_K \in Irr_+(K, R_2(H))$. Further the action of K on V_K described in 13.3.2.3, and $V_K = \langle (Z \cap V_K)^K \rangle$ by 13.3.4.2, completing the proof of (3).

By (3), either $C_K(V_K) = O_2(K)$, or $K/O_2(K) \cong \hat{A}_6$ with $C_K(V_K) = O_{2,Z}(K)$. Therefore as $O_2(K) \leq C_K(R_2(H)) \leq C_K(V_K) = O_{2,Z}(K)$, either (4) holds, or $K/O_2(K) \cong \hat{A}_6$ with $C_K(R_2(H)) = O_2(K)$. However in the latter case by A.1.42, there is $I \in Irr_+(K, R_2(H), T)$ with $K/C_K(I) \cong \hat{A}_6$, and $I \in Irr_+(K, R_2(KT), T)$ by A.1.41, contrary to (3). Thus (4) holds. $\square$

When G is $Sp_6(2)$, G_z is solvable; thus we must eventually eliminate the case where G_z is nonsolvable. In that case by 13.4.5 there is $K \in \mathcal{C}(G_z)$ with $K \in \mathcal{L}_f^*(G, T)$, so that we can use our knowledge of groups in $\mathcal{L}_f^*(G, T)$ to restrict the structure of G_z. We begin with 13.4.6; notice in particular the very strong restrictions in part (4).

LEMMA 13.4.6. Assume $H \in \mathcal{H}_z$ is nonsolvable. Then

(1) There exists $K \in \mathcal{C}(H)$, and for each such K, $K \in \mathcal{L}_f^*(G, T)$ and $K \trianglelefteq H$. Further $K \not\leq M$, $L \not\leq N_G(K)$, and $[Z_V, K] \neq 1$.

(2) $K \leq G_z \leq N_G(K)$.

(3) $K = [K, J(T)]$.

(4) Either $M = LT$, or $L/O_2(L) \cong \hat{A}_6$ and $M = LXT$, where X is a cyclic Sylow 3-subgroup of $C_M(L/O_2(L)) = C_M(V)$.

PROOF. Part (1) is a restatement of parts (1) and (2) of 13.4.5. As $KT \leq H \leq G_z$ and $N := N_G(K) = !\mathcal{M}(KT)$ by (1), $G_z \leq N$, so (2) holds.

Let $L_- := L_2$ if $L/O_2(L) \cong A_6$, and $L_- := L_{2,+}$ if $L/O_2(L) \cong \hat{A}_6$. Then $L = \langle L_1, L_- \rangle$, and by (1), $L_1 \leq N$ but $L \not\leq N$; thus $L_-T \in \mathcal{H}_*(T, N)$. Now $[Z, L_-] \neq 1$ as L_- does not centralize $V_1 = Z \cap V$, so (3) follows by applying 3.1.8.3 with $N_G(K)$, $R_2(KT)$ in the roles of "M, V".

Let $Y := O^2(C_M(V))$. As $\bar{M} = \bar{L}\bar{T}$, $M = LTY$ and $Y \trianglelefteq M$. Further $Y \leq G_z \leq N$ by (2), so the hypotheses of 13.3.8 are satisfied with M, N in the roles of "H, M_K". Therefore Y is a $\{2,3\}$-group by 13.3.8.2. In particular if $m_3(C_M(V)) = 0$, then $C_M(V)$ is a 2-group, so that $M = LT$ and (4) holds. So assume $m_3(C_M(V)) \geq 1$. Then by 13.3.7, $L/O_2(L) \cong \hat{A}_6$ and $m_3(C_M(V)) = 1$ with $L_0 = \theta(C_M(V))$. Thus $C_M(V) = C_T(V)X$, where $X \in Syl_3(C_M(V))$ is cyclic, and once again (4) holds as $C_M(V) = C_M(L/O_2(L))$ by 13.4.2.4. $\square$

We will now begin to produce subgroups H_0 of G which are generated by subgroups H_1 and H_2 in $\mathcal{H}(T)$, such that (H_0, H_1, H_2) is a Goldschmidt triple in the sense of Definition F.6.1. Proposition 13.4.7.5 gives fairly strong information about those pairs which also satisfy conditions (a)–(c) of the Proposition. In particular subgroups satisfying (1i) and (1ii) of Proposition 13.4.7 will eventually be identified as the parabolics $\dot{P}_2$ and $\dot{P}_3$ of $Sp_6(2)$.

PROPOSITION 13.4.7. *Assume $H_i \in \mathcal{H}(T)$, $i = 1, 2$, are distinct with $H_i/O_2(H_i)$* $\cong S_3$. *Let $K_i := O^2(H_i)$, $H_0 := \langle H_1, H_2 \rangle$, and $V_0 := \langle Z^{H_0} \rangle$. Assume:*

(a) Either K_1 or K_2 has at least two noncentral 2-chief factors,
(b) $|Z : C_Z(H_i)| = 2$, for $i = 1$ and 2, and
(c) If $H_0 \in \mathcal{H}(T)$, then $K_i \trianglelefteq C_{H_0}(C_Z(K_i))$ for $i = 1$ and 2, and

$$K_j = O^{3'}(C_{H_0}(C_Z(K_j))) \text{ for } j := 1 \text{ or } 2.$$

Then

(1) $H_0 \in \mathcal{H}(T, M)$, $Z = V_1 = C_Z(H_1) \times C_Z(H_2)$ with $|C_Z(H_i)| = 2$, and one of the following holds:

(i) $H_0 = H_1 H_2$, $[K_1, K_2] \leq O_2(K_1) \cap O_2(K_2)$, $H_0/O_2(H_0) \cong S_3 \times S_3$ or $\mathbf{Z}_2/E_9$, and $V_0 = V_1 \oplus V_2$, where $V_i := [V_0, K_i] = C_{V_0}(K_{3-i})$ is of rank 2.

(ii) $H_0 = K_0 T$ where $K_0 \in \mathcal{C}(H_0)$ such that $K_0 \in \mathcal{L}_f^*(G, T)$, $K_0/O_2(K_0) \cong L_3(2)$, $J(T) \trianglelefteq H_0$, and V_0 is either the sum of two nonisomorphic natural modules for $K_0/O_2(K_0)$, or the core of the 7-dimensional permutation module.

(iii) $O_2(H_0) = C_{H_0}(V_0)$, $H_0/O_2(H_0) \cong E_4/3^{1+2}$, $m(V_0) = 6$, and $J(T) \trianglelefteq$ H_0.

(iv) $H_0 = K_0 T$, where $K_0 \in \mathcal{C}(H_0)$ with $K_0/O_2(K_0) \cong A_6$, and $J_i(T) \trianglelefteq$ H_0 for $i = 0, 1$.

(2) Assume conclusion (ii) holds, with $K_i = [K_i, J_1(T)]$ for some i, and $X \in$ $\mathcal{H}(H_0)$. Then $V_0 = \langle Z^X \rangle$ and $X = H_0 C_X(V_0)$; so if $C_X(Z) = T$, then $X = H_0$.

(3) If $K_2 = L_2$, then conclusion (iii) does not hold.

(4) Assume $K_2 = L_2$. Then conclusion (iv) does not hold, and if conclusion (ii) holds, then $K_2 = [K_2, J_1(T)]$.

PROOF. Let $Q_0 := O_2(H_0)$ and $H_0^* := H_0/C_{H_0}(V_0)$. Observe that the hypotheses say that (H_0, H_1, H_2) is a Goldschmidt triple in the sense of Definition F.6.1, so

$$(H_1/Q_0, T/Q_0, H_2/Q_0)$$

is a Goldschmidt amalgam by F.6.5.1, and hence is listed in F.6.5.2.

Assume that $Q_0 = 1$. By hypothesis (a), some K_i has at least two noncentral 2-chief factors, which eliminates cases (i)–(v) of F.6.5.2, and in case (vi) also eliminates cases (1) and (2) of F.1.12. In cases (3), (8), (12), and (13) of F.1.12, $Z \leq H_i$ for exactly one value of i, contrary to (b). Therefore $Q_0 \neq 1$, and hence $H_0 \in \mathcal{H}(T)$. Then $H_0 \in \mathcal{H}^e$ by 1.1.4.6, so $V_0 \in \mathcal{R}_2(H_0)$ by B.2.14, and hence $O_2(H_0^*) = 1$.

By (b), $[Z, K_i] \neq 1$, so $C_{K_i}(V_0) \leq O_2(K_i)$ and $K_1^* \neq 1 \neq K_2^*$. Therefore $C_T(V_0) \leq Q_0$ by F.6.8, so as $V_0 \in \mathcal{R}_2(H_0)$, $Q_0 = C_T(V_0)$. By (c),

$$C_{H_0}(V_0) \leq C_{H_0}(Z) \leq N_G(K_1) \cap N_G(K_2).$$

Also by (c), there exists an index j such that $K_j = O^{3'}(C_{H_0}(C_Z(K_j)))$. Then $O^{3'}(C_{H_0}(V_0)) \leq K_j$, so in fact $O^{3'}(C_{H_0}(V_0)) = 1$ since $C_{K_j}(V_0) \leq O_2(K_j)$. That is, $C_{H_0}(V_0)$ is a 3'-group.

Let $H_0^+ := H_0/O_{3'}(H_0)$. As $C_{H_0}(V_0)$ is a $3'$-group, H^+ is a quotient of H^*. Observe that H^+ is described in F.6.11. By F.6.6, $O^2(H_0) = \langle K_1, K_2 \rangle$, so $O^2(H_0) = \theta(H_0)$.

Suppose $H_0 \leq M$. Then $O^2(H_0) \leq \theta(M) = L$ by 13.4.2.5, so as each K_i is T-invariant, $K_i \leq L_{k(i)}$ for some $k(i) := 1$ or 2 by 13.3.4.3. By 13.4.3.3, L_1 centralizes Z; and if $L/O_2(L) \cong \hat{A}_6$ then $L_0 \leq L_1$, so L_0 centralizes Z. Therefore as $[Z, K_i] \neq 1$ for each i by (b), it follows that $K_1 = K_2 = L_2$ if $L/O_2(L) \cong A_6$, while $K_1 = K_2 = L_{2,+}$ if $L/O_2(L) \cong \hat{A}_6$. But $K_1 \neq K_2$, as otherwise $H_1 = TK_1 = TK_2 = H_2$, contrary to hypothesis. Hence $H_0 \not\leq M$.

As $H_0 \not\leq M$, $H_k \not\leq M$ for some $k := 1$ or 2. Therefore $C_{Z_L}(H_k) = 1$, as otherwise $H_k \leq C_G(C_{Z_L}(H_k) \leq M = !\mathcal{M}(LT)$. But $C_Z(H_k)$ is a hyperplane of Z by (b), while $1 \neq Z_V \leq Z_L$ and Z_L is also a hyperplane of Z by 13.4.3.2. We conclude that $m(Z) = 2$ and $Z_L = Z_V$ and $C_Z(H_k)$ are of rank 1. As $K_j = O^{3'}(C_{H_0}(C_Z(K_j)))$ by (c) and $K_1 \neq K_2$, $C_Z(H_1) = C_Z(K_1) \neq C_Z(K_2) = C_Z(H_2)$. Then we conclude from (b) that $Z = C_Z(H_1) \times C_Z(H_2)$ with $m(C_Z(H_i)) = 1$ for each i. As $V_1 = Z \cap V$ is of rank 2, $Z = V_1 \leq V$. Thus we have established the initial conclusions of (1), so it remains to show that one of conclusions (i)–(iv) of (1) holds.

Suppose first that $[K_1^*, K_2^*] = 1$. Then $O^2(H_0^*) = K_1^* K_2^*$, so $K_i C_{H_0}(V_0)$ is normal in H_0 for $i = 1, 2$. Furthermore $C_{H_0}(V_0)$ normalizes K_i by (c), and we saw $C_{H_0}(V_0)$ is a $3'$-group, so $K_i = O^{3'}(K_i C_{H_0}(V_0)) \trianglelefteq H_0$. It follows that $H_0 = H_1 H_2$ and $[K_1, K_2] \leq O_2(K_1) \cap O_2(K_2) \leq O_2(H_0)$; hence $H_0/O_2(H_0) \cong S_3 \times S_3$ or $\mathbf{Z}_2/E_9$. Set $V_i := [V_0, K_i]$. As $Z = C_Z(H_1) \times C_Z(H_2)$ with $|C_Z(H_i)| = 2$, $V_i = \langle C_Z(H_{3-i})^{H_i} \rangle \cong E_4$ is centralized by H_{3-i}, so we conclude that case (i) of (1) holds. Thus we may assume from now on that $[K_1^*, K_2^*] \neq 1$; under this assumption, we will show that one of (ii)–(iv) holds.

We first consider the case where H_0^* is not solvable, which will lead to (ii) or (iv). By 1.2.1.1, there is $K_0 \in \mathcal{C}(H_0)$ with $K_0^* \neq 1$. Then by 13.4.5.1, $K_0 \trianglelefteq H_0$, $K_0 \in \mathcal{L}_f^*(G, T)$, and $K_0/O_2(K_0)$ is listed in 13.4.5.1. In particular K_0 is not a $3'$-group, so that $K_0^+ \neq 1$; hence H_0^+ is described in F.6.18 by F.6.11.2. Also K_0^+ is a quotient of $K_0/O_2(K_0)$, so comparing the possibilities for $K_0/O_2(K_0)$ in 13.4.5 with the possible quotients K_0^+ in cases (3)–(13) of F.6.18, we conclude K_0^+ must be $L_3(2)$, A_6, or $\hat{A}_6$, with $H_0^+ = K_0^+ T^+$ appearing in case (6) or (8) of F.6.18. Furthermore if $K_0/O_2(K_0) \cong \hat{A}_6$, then as $C_{H_0}(V_0)$ is a $3'$-group, $K_0^* \cong K_0/O_2(K_0) \cong \hat{A}_6$, contrary to 13.4.5.4. Thus in any case $C_{K_0}(V_0) = O_2(K_0) = O_{3'}(K_0)$, so $K_0/O_2(K_0) \cong K_0^* \cong K_0^+$, and $K_0/O_2(K_0)$ is not $\hat{A}_6$.

As $H_0^+ = K_0^+ T^+$, $H_0 = K_0 T O_{3'}(H_0)$, so as $K_0 \trianglelefteq H_0$, $K_0 = O^{3'}(H_0)$. Thus $K_i \leq K_0$ for $i = 1, 2$, so $K_0 = O^2(H)$ by F.6.6, and hence $H_0 = K_0 T$. Since $K_1 \neq K_2$, H_1^* and H_2^* are the minimal parabolics of H^* over T^*.

By 13.4.5.3, we may choose a T-invariant $I \in Irr_+(K_0, V_0)$ in the FSU. By 13.4.5.4, $C_{K_0}(I) = C_{K_0}(V_0)$, so that $K_0^* = K_0/C_{K_0}(I)$. By 13.4.5.3, I is either a natural module for K_0^* or a 5-dimensional module for $K_0^* \cong A_6$. As H_1^* and H_2^* are the minimal parabolics of H^*, some K_i (say K_1) centralizes $Z \cap I$, so as $[Z, K_1] \neq 1$ by hypothesis (b), $IZ > IC_Z(K_0)$. Thus K_0 is nontrivial on V_0/I. Hence if $J(T) \not\leq O_2(H_0)$, then by Theorem B.5.1, $[V_0, K_0]$ is the sum of two isomorphic natural modules for $K_0^* \cong L_3(2)$; since $m(Z) = 2$, this contradicts $[Z, K_1] \neq 1$. Therefore $J(T) = J(O_2(H_0)) \trianglelefteq H_0$.

As $Z = C_Z(H_1) \times C_Z(H_2)$ with $C_Z(H_i) \cong \mathbf{Z}_2$, there is $v \in C_Z(H_2) - C_Z(H_1)$; set $V_v := \langle v^{K_0} \rangle$.

Assume first that $K_0/O_2(K_0) \cong L_3(2)$; we will show (ii) holds. As v centralizes H_2 but not H_1, V_v is a quotient of the 7-dimensional permutation module for K_0^* on the coset space K_0/H_2, with $m(V_v) > 1$. Thus by H.5.3, $[V_v, K_0]$ is either the 3-dimensional dual of I or the 6-dimensional core of the permutation module. Thus as $m(Z) = 2$, $\dim(V_0) = 6$, and so $V_0 = V_v \oplus I$ or V_v, respectively. This completes the verification of (ii).

So to complete the treatment of the case H_0^* not solvable, we may assume that $K_0^* \cong A_6$; then to establish (iv), it remains to show that $J_1(T) \trianglelefteq H_0$. As above, V_v is a quotient of the 15-dimensional permutation module on H_0/H_2, so by G.5.3, V_v has a K_0^*-irreducible quotient W_2 isomorphic to the conjugate of $W_1 := I/C_I(K_0)$ under a graph automorphism. Therefore K_0 has chief factors isomorphic to W_1 and W_2 on V_0. We may assume that $J_1(T) \not\leq O_2(H_0) = C_{H_0}(V_0)$, so that there is $A \in \mathcal{A}_1(T)$ with $A^* \neq 1$. By B.2.4.1, $m(V_0/C_{V_0}(A)) \leq m(A^*) + 1$, so either

(I) $m(A^*) = 1$ and $m(W_i/C_{W_i}(A)) = 1$ for $i = 1$ and 2, or

(II) $m(A^*) \geq 2$, so that $m(W_i/C_{W_i}(A)) \geq 2$ for $i = 1$ and 2, and hence $m(A^*) = 3$ and A^* is a strong FF*-offender on both W_1 and W_2.

These cases are impossible since by B.3.4, no involution induces a transvection on both W_1 and W_2, nor does there exists a subgroup which is a strong FF*-offender on both W_1 and W_2. This contradiction completes the verification of (iv), and establishes (1) when H_0^* is nonsolvable.

Thus we have reduced to the case where H_0^* is solvable, but $[K_1^*, K_2^*] \neq 1$. This time let $K_0 := O^2(H_0)$ and let $Q_i := O_2(H_i)$.

Suppose first that $Q_1 = Q_2$. Then $Q_0 = Q_1$, $T^* = \langle t^* \rangle$ is of order 2, and $K_i^* \cong S_3$. Thus $m(V_0) \leq 2m(C_{V_0}(t)) = 2m(Z) = 4$. Inspecting the solvable subgroups H_0^* of $GL_4(2)$ with $O_2(H_0^*) = 1$ and generated by a pair of distinct S_3-subgroups with a common Sylow 2-subgroup, we conclude H_0^* is E_9 extended by $\mathbf{Z}_2$. But this contradicts the fact that $[K_1^*, K_2^*] \neq 1$. This contradiction shows that $Q_1 \neq Q_2$.

Suppose next that K_i does not centralize $O_p(H_0/Q_0)$, for some prime $p > 3$ and $i = 1$ or 2, say $i = 1$. Then by A.1.21 there is a supercritical subgroup P of a Sylow p-subgroup of the preimage of $O_p(H_0/Q_0)$. By a Frattini Argument, $H_0 = N_{H_0}(P)Q_0$. Let $\dot{H}_0 := H_0/C_{H_0}(P/\Phi(P))Q_0$. By A.1.25, $\dot{H}_0 = \langle \dot{H}_1, \dot{H}_2 \rangle$ is a subgroup of $GL_2(p)$; of course $\dot{H}_0$ is solvable and $\dot{H}_1/O_2(\dot{H}_1) \cong S_3$. Thus P is noncyclic. Further if $\dot{K}_2 \neq 1$ then we conclude from Dickson's Theorem A.1.3 that $\dot{K}_0 \cong SL_2(3)$ or $\dot{K}_0$ is cyclic, and in either case $\dot{Q}_1 = \dot{Q}_2$, so that $Q_1 = Q_2$, contrary to the previous paragraph. Thus $K_P := \langle K_2^{H_0} \rangle$ centralizes PQ_0/Q_0. Let $1 \neq P_0 \leq P_1 \in Syl_p(K_P)$ with P_1 acting on P; as $m_p(H_0) = 2 = m_p(P)$, P contains all elements of order p in P_0, so $Aut_{K_P}(P_0)$ is a p-group by A.1.21, and hence K_P is p-nilpotent by the Frobenius Normal p-Complement Theorem 39.4 in [**Asc86a**]. As K_P is generated by 3-elements, K_P is a p'-group, so as K_1 is a $\{2,3\}$ group, we conclude $K_0 = \langle K_1, K_2 \rangle = K_1 K_P$ is a p'-group, contrary to $P \leq K_0$.

Therefore K_i centralizes $O^3(F(H_0/Q_0))$ for $i = 1$ and 2, so by F.6.9, H_0 is a $\{2,3\}$-group. Then as $C_{H_0}(V_0)$ is a $3'$-group, we conclude that $Q_0 = C_{H_0}(V_0)$. Therefore $H_0^* = H_0/Q_0 = H_0^+$ and $H_0 = K_0 T$ with K_0 a $\{2,3\}$-group. Further (H_1^*, T^*, H_2^*) is a Goldschmidt amalgam by F.6.5.1. Since $Q_1 \neq Q_2$ by an earlier reduction, $H_0^* = H_0^+$ is described in Theorem F.6.18 by F.6.11.2. As H is solvable and $Q_1 \neq Q_2$, conclusion (2) of F.6.18 holds, and by earlier reduction $[K_1^*, K_2^*] \neq 1$, so that $H_0^* \cong E_4/3^{1+2}$. Let $X_0^* := Z(K_0^*)$; then $V_0 = U_0 \oplus U_1$, where $U_1 := C_{V_0}(X_0)$ and $U_0 := [V_0, X_0]$ is of rank $6k$ for some $k \geq 1$. Now $Z = Z_0 \oplus Z_1$ is of rank 2 where $Z_i := Z \cap U_i$; so either $m(Z_i) = 1$ for $i = 0$ and 1, or $U_1 = 0$ and $Z = Z_0$. As T^* has order 4, $6 \leq 6k = m(U_0) \leq 4m(Z_0) \leq 8$; hence $k = 1$ and $Z = Z_0$ with $U_1 = 0$, so $V_0 = U_0$ is of rank 6. By Theorem B.5.6, $J(T) \trianglelefteq H_0$. Now conclusion (1iii) of the lemma holds. Hence the proof of (1) is at last complete.

We next prove (2). So assume that the hypotheses of (2) hold. As conclusion (ii) of (1) holds, $K_0 \trianglelefteq X$ by 13.4.5.1, so $O_2(K_0) \leq O_2(X)$. Set $V_X := \langle Z^X \rangle$, so that $V_0 \leq V_X \in \mathcal{R}_2(X)$ by B.2.14. We must show that $V_0 = V_X$ and $X = C_X(V_0)H_0$. Let $\hat{X} := X/C_X(V_X)$; then $C_X(V_X) \leq C_X(V_0)$, so H_0^* is a quotient of $\hat{H}_0$. Furthermore as conclusion (ii) holds, $C_{K_0}(V_0) = O_2(K_0)$, so that $C_{K_0}(V_0) = C_{K_0}(V_X)$ since $O_2(K_0) \leq O_2(X)$, and hence $\hat{K}_0 \cong K_0^*$.

From the structure of V_0 in either case of (ii), $O^2(N_{GL(V_0)}(K_0^*)) = K_0^*$. Suppose we have shown that $V_0 = V_X$. Then $O^2(\hat{X}) = \hat{K}_0$, so that $\hat{X} = \hat{K}_0\hat{T} = \hat{H}_0$. Hence $X = H_0 C_X(V_0)$, giving the remaining conclusion of (2). Thus it suffices to show $V_0 = V_X$, or equivalently that $V_X \leq V_0$.

By the hypotheses of (2), $J_1(T) \not\leq O_2(H_0)$, so as $H_0 = K_0T$, there is $A \in \mathcal{A}_1(T)$ with $K_0 = [K_0, A]$. Thus by B.2.4.1,

$$m(A^*) \geq m(V_0/C_{V_0}(A)) - 1 \text{ and } m(\hat{A}) \geq m(V_X/C_{V_X}(A)) - 1.$$

However V_0 is not an FF-module for $K_0^*T^*$ by Theorem B.5.1, so $m(A^*) < m(V_0/C_{V_0}(A))$, and hence $m(A^*) = m(V_0/C_{V_0}(A)) - 1$. Then by B.2.4.2, $B := V_0 C_A(V_0) \in \mathcal{A}(T)$; recall $C_A(V_0) = A \cap Q_0$, so as $V_0 \leq C_X(V_X)$,

$$\hat{B} \leq \hat{A} \cap \hat{Q}_0 \leq C_{\hat{X}}(\hat{K}_0).$$

Suppose first that $\hat{B} = 1$, so that $C_A(V_0) = C_A(V_X)$. Then

$$m(V_0/C_{V_0}(A)) - 1 = m(A^*) = m(\hat{A}) \geq m(V_X/C_{V_X}(A)) - 1,$$

so $V_X = V_0 C_{V_X}(A)$ and hence $[V_X, A] \leq V_0$. Therefore as $K_0 = [K_0, A]$, $V_0 = [V_0, K_0] = [V_X, K_0]$ is X-invariant; then as $Z \leq V_0$, $V_X = \langle Z^X \rangle = V_0$, so we are done.

Thus we may assume instead that $\hat{B} \neq 1$. Then as $B \in \mathcal{A}(T)$ and $\hat{B}$ centralizes $\hat{K}_0$, $J(C_X(\hat{K}_0)) =: Y \not\leq C_X(V_X)$. Hence as $m_3(X) \leq 2$ with $\hat{K}_0 \cong L_3(2)$, $m_3(\hat{Y}) \leq 1$, so we conclude from Theorem B.5.6 that either $\hat{Y} \cong S_3$ (in which case we set $Y_0 := Y$), or $\hat{Y} = \hat{Y}_0$ for some $Y_0 \in \mathcal{C}(X)$ with $m_3(Y_0) = 1$. In either case $\hat{Y}_0$ is normal in $\hat{X}$. In the latter case since $O_2(Y_0) \leq O_2(X) \leq C_X(V_X)$, we obtain $\hat{Y}_0 \cong L_3(2)$ or A_5 from 13.4.5.1; then by Theorem B.5.1 and 13.4.5.1, $[V_X, Y_0]$ is either the natural module for $\hat{Y}_0$ or the sum of two natural modules for $\hat{Y}_0 \cong L_3(2)$. Then $End_{\hat{Y}_0}([V_X, Y_0])$ is either a field or the ring of 2×2 matrices over $\mathbf{F}_2$, so that $[V_X, Y_0, K_0] = 1$. Hence $[Z, Y_0] \leq [K_0, V_X, Y_0] = 1$ using the Three-Subgroup Lemma. So as $\hat{Y}_0 \trianglelefteq \hat{X}$, Y_0 centralizes $V_X = \langle Z^X \rangle$, contrary to $\hat{Y}_0 \neq 1$. Thus the proof of (2) is complete.

We next prove (4), so we assume that $K_2 = L_2$, and that either conclusion (ii) or (iv) of (1) holds. Now $J(T) \trianglelefteq H_0$ in either of those cases, so that $S := \text{Baum}(T) = \text{Baum}(O_2(H_0))$ by B.2.3.4. Hence as $H_0 \not\leq M = !\mathcal{M}(LT)$, no nontrivial characteristic subgroup of S is normal in LT. Thus conclusion (I) of 13.2.2.10 does not hold. If conclusion (III) of 13.2.2.10 holds, then $K_2 = [K_2, J_1(T)]$, so we are not in case (iv), as there $J_1(T) \trianglelefteq H_0$. Hence we are in case (ii), so that (4) holds. Thus we may assume that conclusion (II) of 13.2.2 holds, so that L is an A_6-block with $\mathcal{A}(O_2(LT)) \subseteq \mathcal{A}(T)$. As L is an A_6-block, L_2 has exactly three noncentral 2-chief factors. Let $k := 2$ in case (ii), and $k := 3$ in case (iv). As $L_2 = K_2$, L_2 has at least k chief factors on V_0 and one on $O_2(L_2)^*$, so (ii) holds and $[O_2(K_0T), K_0] \leq V_0$. Thus as $J(T) \trianglelefteq H_0$, each $A \in \mathcal{A}(T)$ contains V_0, so $[A, K_0] \leq [O_2(K_0T), K_0] = V_0 \leq A$, and hence $A \trianglelefteq K_0 A$. Further as $\mathcal{A}(O_2(LT)) \subseteq \mathcal{A}(T)$, $J(O_2(LT)) = \langle A \in \mathcal{A}(T) : A \leq O_2(LT) \rangle$, so $K_0 \leq N_G(J(O_2(LT))) \leq M$, contrary to $H_0 \not\leq M$ by (1).

It remains to prove (3), so we may assume that $K_2 = L_2$ and conclusion (iii) holds, and we must produce a contradiction. As $m_3(L_2) = m_3(K_2) = 1$ by hypothesis, $L/O_2(L)$ is A_6 rather than $\hat{A}_6$. Let Y_Z be the preimage in H_0 of $Z(O_3(H_0^*))$, and set $Y_2 := O^2(Y_Z)$. Notice that as Y_2^* is fixed point free on V_0 of rank 6, while $Z = V_1$ is of rank 2, $[Z, Y_2]$ is of rank 4. In particular $Y_2 \not\leq C_G(Z_V) = M$ by 13.4.2.1.

Set $Y := \langle L_1T, Y_2T \rangle$, $Q_Y := O_2(Y)$, and $V_Y := \langle Z^Y \rangle$. Observe (Y, L_1T, Y_2T) is a Goldschmidt triple, so $(L_1T/Q_Y, T/Q_Y, K_2T/Q_Y)$ is a Goldschmidt amalgam by F.6.5.1, and hence is listed in F.6.5.2. Now $K_2 = L_2$ has at least three noncentral 2-chief factors in L; so as this does not hold in any case in F.6.5.2, we conclude $Q_Y \neq 1$, so that $Y \in \mathcal{H}(T)$. Hence $V_Y \in \mathcal{R}_2(Y)$ by B.2.14.

We saw $Y_2 \not\leq M$, so $Y \not\leq M$. On the other hand, for $z \in V_1 - Z_V$, $C_Y(V_Y) \leq C_Y(Z_V) \cap C_Y(z) \leq C_M(z)$, so applying 13.3.9 with Y, $O^2(C_Y(V_Y))$ in the roles of "H, Y", and recalling that $L/O_2(L) \cong A_6$, we conclude that $O^2(C_Y(V_Y)) = 1$ or L_1.

In the latter case, Y_2 acts on L_1, and hence centralizes $L_1/O_2(L_1)$ so that L_1 normalizes $O^2(Y_2O_2(L_1)) = Y_2$. Then as $L_2 = K_2$, $L = \langle L_1, L_2 \rangle \leq N_G(Y_2)$, contrary to $Y_2 \not\leq M = !\mathcal{M}(LT)$. Thus $O^2(C_Y(V_Y)) = 1$ so that $C_Y(V_Y) = Q_Y \leq O_{3'}(Y)$. In addition this argument shows that $[L_1, Y_2] \not\leq O_2(L_1)$.

Let $Y^* := Y/C_Y(V_Y)$ and $Y^+ := Y/O_{3'}(Y)$, so that Y^+ is a quotient of Y^*, and is described in F.6.11.2. Now $L = [L, J(T)]$ by 13.4.3.1, so that $L_1 = [L_1, J(T)]$ by 13.2.2.4; so as $J(T)^*$ centralizes $O^3(F^*(Y^*))$ by Theorem B.5.6, so does L_1. In particular L_1 centralizes $F^*(O_{3'}(Y^*))$, so as L_1 is generated by conjugates of an element of order 3, we conclude from A.1.9 that $L_1 \leq C_Y(O_{3'}(Y^*))$. Thus $L_1 = O^{3'}(L_1O_{3'}(Y))$, so if $[Y_2^+, L_1^+] \leq O_2(L_1^+)$, then $[Y_2, L_1] \leq O_2(L_1)$, which we showed earlier is not the case. We conclude $[Y_2^+, L_1^+] \not\leq O_2(L_1^+)$. Now as $J(T) \trianglelefteq Y_2T$ since we are in case (iii), but $L_1 = [L_1, J(T)]$, $O_2(Y_2T) \neq O_2(L_1T)$. Thus case (i) of F.6.11.2 holds, so Y^+ is described in F.6.18, where F.6.18.1 is similarly ruled out. As $L_1 = [L_1, J(T)]$, Y^+ is not $E_4/3^{1+2}$ by Theorem B.5.6, while the condition $[Y_2^+, L_1^+] \not\leq O_2(L_1^+)$ rules out the other possibility in F.6.18.2. In the remaining cases in Theorem F.6.18, Y is not solvable, so there is $K_Y \in \mathcal{C}(Y)$, and by 13.4.5, $K_Y/O_2(K_Y) \cong A_5$, $L_3(2)$, A_6, or $\hat{A}_6$. The A_5 case is ruled out, as A_5 does not appear as a composition factor in the groups listed in Theorem F.6.18. Similarly conclusion (3) of F.6.18 does not hold, so $L_1 \leq K_Y$.

As $L_1 = [L_1, J(T)]$, $K_Y = [K_Y, J(T)]$, so by Theorem B.5.1 and 13.4.5.3, $[V_Y, K_Y]$ is a natural module for $K_Y^* \cong L_3(2)$ or A_6, a 5-dimensional module for $K_Y^* \cong A_6$, or the sum of two natural modules for $K_Y^* \cong L_3(2)$. As $Z = V_1$ is of rank 2, with $\langle Z^{Y_2} \rangle \cong E_{16}$, V_Y is the sum of two natural modules for $K_Y^* \cong L_3(2)$. As $L_1^*T^*$ is the parabolic of K_Y^* centralizing Z, $J(R_1) \leq C_Y(V_Y) = Q_Y$, and hence $\mathrm{Baum}(R_1) = \mathrm{Baum}(Q_Y)$ by B.2.3.5. Then each nontrivial characteristic subgroup of Baum(R_1) is normal in Y, and hence not normal in LT as $Y \not\leq M = !\mathcal{M}(LT)$. Therefore L is an A_6-block by 13.2.2.7, and in particular L_1 has two noncentral chief factors. This is impossible, as L_1 has two noncentral chief factors on V_Y and one on $O_2(L_1^*)$. So the proof of (3), and hence of Proposition 13.4.7, is finally complete. $\qquad\square$

13.4.2. The case $\mathbf{G_z}$ solvable, leading to $\mathbf{Sp_6(2)}$. Recall the definitions of z and $\mathcal{H}_z$ given before 13.4.3, and recall that $G_z := C_G(z)$. In the next lemma, we begin to identify G_z and a suitable 2-local H_0 with the parabolics $\dot{P}_2$ and $\dot{P}_3$ of $\dot{G} = Sp_6(2)$.

LEMMA 13.4.8. *Assume $H \in \mathcal{H}_z$ is solvable, choose $V_H \in \mathcal{R}_2(H)$ with $Z_V \leq V_H$, and let $I := \langle J(R_1)^H \rangle$. Then*

(1) $L/O_2(L) \cong A_6$.
(2) $I/O_2(I) \cong S_3$ with $m([V_H, I]) = 2$.
(3) $H = L_1 IT$ and $H/O_2(H) \cong S_3 \times S_3$. In particular IT is the unique member of $\mathcal{H}_*(T, M)$ in H.

PROOF. Let $H^* := H/C_H(V_H)$. As usual, $O_2(H^*) = 1$ by B.2.14. As $Z_V \leq V_H$, we may apply 13.4.4.2, to conclude that

$$C_{R_1}(V_H) = O_2(H).$$

For in case (i), $C_H(V_H) = O_2(H) \leq R_1$; and in case (ii), $R_1 \in Syl_2(C_H(L_+/O_2(L_+))$, where $L_+ = O^2(C_H(V_H))$.

We claim that $[V_H, J(R_1)] \neq 1$, so we assume that $[V_H, J(R_1)] = 1$ and derive a contradiction. Then

$$B := \mathrm{Baum}(R_1) = \mathrm{Baum}(C_{R_1}(V_H)) = \mathrm{Baum}(O_2(H)), \qquad (*)$$

by B.2.3.5 and the previous paragraph. Hence as $H \not\leq M = !\mathcal{M}(LT)$, no nontrivial characteristic subgroup of B is normal in LT, so by 13.2.2.7, L is an A_6-block. In particular, $L/O_2(L) \cong A_6$ rather than $\hat{A}_6$.

Calculating in the core V of the permutation module:

$$V_3 = [V, L_1] = [V, T \cap L_1] = [V, O_2(L_1)] = \{e_J : J \subseteq \{1, 2, 3, 4\} \text{ and } |J| \text{ is even}\},$$

and $[V_3, O_2(L_1)] = \langle e_{1,2,3,4} \rangle$. Further if $\bar{M} \cong S_6$, then also $Z_V = \langle e_\Omega \rangle \leq [V, R_1]$.

By 13.2.2.6, $V \leq J(R_1)$, so by $(*)$, $V \leq O_2(H) \leq N_H(V)$ and V centralizes V_H. Hence $U := \langle V^H \rangle \leq O_2(H)$ and $[V, V^h] \leq V \cap V^h$ for each $h \in H$. If $U \leq C_T(V) =: Q$, then L normalizes U because $[Q, L] = V$ since L is a block. But then $H \leq N_G(U) \leq M = !\mathcal{M}(LT)$, contrary to $H \not\leq M$. So we conclude instead that $[V, V^h] \neq 1$ for some $h \in H$.

Suppose that $L_1 \trianglelefteq H$. Then as $V_3 = [V_3, L_1]$, $V_3^h = [V_3^h, L_1] \leq O_2(L_1)$. Thus either $V_3^h = [C_{O_2(L_1)}(V), L_1] = V_3$, or $V_3^h \cap Q = V_1$, $O_2(L_1) = V_3^h C_{O_2(L_1)}(V)$ and $\bar{V}^h = \bar{R}_1 \not\leq \bar{L}$, since V_3/V_1 is the unique minimal L_1-invariant subgroup of V/V_1. Assume the former case holds. Then V^h centralizes V_3, so $\bar{V}^h = \langle (5, 6) \rangle$ and hence $[V, V^h] = \langle e_{5,6} \rangle$. But then $Z_V \leq V_3 \langle e_{5,6} \rangle \leq V \cap V^h$, contrary to 13.4.2.3. In the latter case, $Z_V \leq [V, R_1] = [V, V^h]$, for the same contradiction.

This contradiction shows that L_1 is not normal in H. Hence $[V_H, L_1] \neq 1$ by 13.4.4.2. We saw $V_H \leq Q$, so $1 \neq [V_H, L_1] \leq [Q, L] = V$, and hence $[V_H, L_1] = [V, L_1] = V_3$. By C.1.13.d, $O_2(L) \leq C_T(Q) \leq C_H(V_H)$, so L_1^* is a quotient of $\bar{L}_1 \cong A_4$. Then by A.1.26, $O_2(L_1^*)$ centralizes $F^*(H^*)$ of odd order, so $O_2(L_1^*) = 1$, and hence $O_2(L_1) \leq C_H(V_H) \leq C_H(V_3)$, whereas we saw $[V_3, O_2(L_1)] \neq 1$.

This establishes the claim that $[V_H, J(R_1)] \neq 1$. By the first paragraph of the proof, $C_{R_1}(V_H) = O_2(H)$, so we may apply B.2.10.1 to conclude that

$$\mathcal{P}_{R_1, H} = \{A^{*h} \neq 1 : A \in \mathcal{A}(R_1), h \in H\}$$

is a nonempty stable subset of $\mathcal{P}(H^*, V_H)$. Hence by B.1.8.5, $I^* = \langle J(R_1)^{*H} \rangle = I_1^* \times \cdots \times I_s^*$ with $I_i^* \cong S_3$, and $[V_H, I]$ is the direct sum of the subgroups $U_i := [V_H, I_i] \cong E_4$. Further $s \leq 2$ by Theorem B.5.6.

Recall that $L = \theta(M)$, $L_1 = O^{3'}(C_L(z))$, and $H \leq G_z$. Thus $L_1 = \theta(C_M(z)) = \theta(H \cap M)$. Similarly if $L/O_2(L) \cong A_6$, then $L_1 = O^{3'}(H \cap M)$ using 13.4.2.5.

Next by B.1.8.5, $J(R_1)^* \in Syl_2(I^*)$; thus $J(R_1)^*$ is self-normalizing in I^*. We claim that $O^2(I^*) \cap L_1^* = 1$: If $L/O_2(L) \cong A_6$, then L_1 normalizes R_1, so this follows from the previous observation. So suppose $L/O_2(L) \cong \hat{A}_6$. Then $L_{1,+}$ normalizes R_1, so $O^2(I^*) \cap L_1^*$ is trivial or L_0^*. Assume the latter case holds. Then as L_0 is T-invariant, $L_0^* = O^2(I_i^*)$ for some i, and then $L_0^* \trianglelefteq H^*$ since $s \leq 2$. In case (i) of 13.4.4.2, $C_H(V_H) = O_2(H)$ acts on L_0, so $L_0 = O^2(L_0 C_H(V_H)) \trianglelefteq H$. In case (ii) of 13.4.4.2, $L_{1,+} = O^2(C_H(V_H))$, so $L_1 = L_{1,+} L_0 = O^2(L_0 C_H(V_H)) \trianglelefteq H$. In either case $H \leq M$ by 13.2.2.9, contrary to $H \not\leq M$. This contradiction completes the proof of the claim that $O^2(I^*) \cap L_1^* = 1$.

Since $L_1 = \theta(H \cap M)$, it follows from the claim that $I \not\leq M$. Furthermore $O^2(I^*) = O^{3'}(N_{GL([V_H, I])}(I^*))$, so the claim says $I^* L_1^* = I^* \times L_1^*$. Thus when $L_1^* \neq 1$, it follows from A.1.31.1 applied in the quotient $I^* L_1^* / O_2(L_1^*)$ that $s = 1$.

We first treat case (i) of 13.4.4.2, where $C_H(V_H) = O_2(H)$. Then $m_3(L_1) = m_3(L_1^*) = 1$, so $s = 1$ by the previous paragraph and $L/O_2(L) \cong A_6$. Thus (1) and (2) hold. By 13.4.3.2, $|Z : Z_L| = 2$, so as $z \in Z(H)$ does not lie in U_1,

$$1 \neq Z_L \cap \langle z \rangle (Z \cap U_1) =: Z_1$$

and $H = I C_H(U_1) = I C_H(Z_1) = I(H \cap M)$, where the final equality holds as $C_G(Z_1) \leq M = !\mathcal{M}(LT)$. As $C_H(V_H) \leq M$, $|H : H \cap M| = |I : O_2(I)| = 3$, so $O^{\{2,3\}}(H) \leq C_M(z)$. Then applying 13.3.9 to $O^{\{2,3\}}(H)$ in the role of "Y", we conclude that H is a $\{2,3\}$-group. So as $L_1 = O^{3'}(H \cap M)$, $H = I(H \cap M) = I L_1 T$, with $H/O_2(H) \cong S_3 \times S_3$, since $R_1 \leq C_H(L_1/O_2(L_1))$. Thus (3) holds.

We must treat case (ii) of 13.4.4.2, where $O_2(H) < C_H(V_H)$ with $O^2(C_H(V_H)) = L_+ = L_1$ or $L_{1,+}$, when $L/O_2(L) \cong A_6$ or $\hat{A}_6$, respectively, and R_1 is Sylow in the normal subgroup $H_1 := C_H(L_+/O_2(L_+))$ of H. Thus $I = \langle J(R_1)^H \rangle \leq H_1$, and hence $R_1 \in Syl_2(I R_1)$.

Assume that $L/O_2(L) \cong \hat{A}_6$. As $L_{1,+} = O^2(C_H(V_H))$, $L_1^* = L_0^*$ is of order 3, and hence $s = 1$ and $L_1/O_2(L_1) \cong E_9 \cong O^2(I^*) \times L_1^*$ by an earlier remark. Therefore as $m_3(H) \leq 2$, $O^2(I) L_1/O_2(O^2(I) L_1) \cong 3^{1+2}$. Then $O^2(I)$ normalizes $O^2(L_1 O_2(O^2(I) L_1)) = L_1$, so that $I \leq N_G(L_1) \leq M$ by 13.2.2.9, contrary to $I \not\leq M$.

Therefore $L/O_2(L) \cong A_6$, so (1) holds. If $s = 1$, then (2) holds, and an argument above shows that (3) holds. Thus we may assume that $s = 2$. Then as $L_1 = L_+ = O^2(C_H(V_H))$ and $m_3(H) \leq 2$, $I/O_2(I) \cong E_4/3^{1+2}$ with $L_1 = O^2(O_{2,\Phi}(I))$. This is impossible, since $R_1 \in Syl_2(I R_1)$, and $J(R_1)$ centralizes $L_1/O_2(L_1)$. This completes the proof of 13.4.8. $\square$

PROPOSITION 13.4.9. *If G_z is solvable then $G \cong Sp_6(2)$.*

PROOF. Assume G_z is solvable. Then using B.2.14 as usual, the pair $H := G_z$, $V_H := \langle Z^{G_z} \rangle$ satisfy the hypotheses of 13.4.8. Therefore by 13.4.8.3, $H = IL_1T$, where $I := \langle J(R_1)^H \rangle$ and $H/O_2(H) \cong S_3 \times S_3$. Also $M = LTC_M(V) = LC_M(z) = L(H \cap M)$ and $H \cap M = L_1T$, so $M = LT$.

We next check next that the hypotheses of Proposition 13.4.7 are satisfied with IT, L_2T in the roles of "H_1, H_2": For example, L_2 has at least three noncentral 2-chief factors, two on V and one on $O_2(\bar{L}_2)$, giving (a). Further $Z_L = C_Z(L_2T)$ is of index 2 in Z by 13.4.3.2; while $C_Z(IT)$ is of index 2 in Z as $m([V_H, I]) = 2$ by 13.4.8.2, and $V_H = [V_H, I]C_Z(I)$ by B.2.14, so that (b) holds. Suppose $H_0 := \langle IT, L_2T \rangle \in \mathcal{H}(T)$. As $H = IL_1T \not\leq M$, $I \not\leq M$, so $H_0 \not\leq M = !\mathcal{M}(LT)$ and hence $L \not\leq H_0$. But $L_2 \leq H_0$ and L_2T is maximal in $LT = M$, so $L_2 = O^{3'}(H_0 \cap L) = O^{3'}(H_0 \cap M)$ since $L/O_2(L) \cong A_6$ by 13.4.8.1. Hence $L_1 \cap H_0 = O_2(L_1)$. Further $Z_L = C_Z(L_2)$ and $C_G(Z_L) \leq M$, so $L_2 = O^{3'}(C_{H_0}(Z_L))) \unlhd C_{H_0}(Z_L)$. As $G_z = H = IL_1T$ with $L_1 \cap H_0 = O_2(L_1)$, $O^2(I) = O^{3'}(C_{H_0}(z)) \unlhd C_{H_0}(z)$, and $C_{H_0}(C_Z(I)) \leq C_{H_0}(z)$. Hence (c) holds. This completes the verification of the hypotheses of Proposition 13.4.7.

Now by 13.4.7.1, $H_0 \in \mathcal{H}(T)$ and $m(Z) = 2$. Therefore $m(V_H) = 3$ as $z \notin [V_H, I] \cong E_4$. Furthermore one of the cases (i)–(iv) holds. As $L_2 = O^2(H_2)$, conclusion (iii) is ruled out by 13.4.7.3, and conclusion (iv) is ruled out by 13.4.7.4. If $[O^2(I), L_2] \leq O_2(O^2(I))$, then $LT = \langle L_1T, L_2T \rangle \leq N_G(O^2(I))$, contrary to $I \not\leq M = !\mathcal{M}(LT)$; this rules out conclusion (i). Thus H_0 satisfies conclusion (ii), and so $H_0/O_2(H_0) \cong L_3(2)$.

Let $E_0 := M$, $E_1 := H$, $E_2 := H_0$, $\mathcal{F} := \{E_0, E_1, E_2\}$, and $E := \langle \mathcal{F} \rangle$. We show that $(E, \mathcal{F})$ is a C_3-system as defined in section I.5. First hypothesis (D5) holds as $Z_V \leq Z(E_0)$. By 13.4.8.1, $E_0/O_2(E_0) \cong A_6$ or S_6, verifying hypothesis (D1). We have already observed that hypothesis (D2) holds, and hypothesis (D3) holds by construction. Finally as $M \in \mathcal{M}$ and $H \not\leq M$, $\ker_T(E) = 1$, so hypothesis (D4) is satisfied.

As $(E, \mathcal{F})$ is a C_3-system, $E \cong Sp_6(2)$ by Theorem I.5.1. Thus it remains to show that $E = G$. To do so we appeal to a fairly deep result on groups disconnected at the prime 2, which we used earlier in our appeal to Goldschmidt's Theorem in chapter 2. Let $W := O_2(E_2)$; as $E \cong Sp_6(2)$, W is the core of the permutation module for E_2/W and $W = J(T)$. Thus H.5.3.4 tells us that E_2 has four orbits β_1, α_2, γ_2, β_3 on $W^\#$, consisting of vectors of weights 6, 4, 2, 4, and the orbits have length 7, 7, 21, 28, respectively. As $W = J(T)$, E_2 controls G-fusion in W by Burnside's Fusion Lemma A.1.35. As $E \cong Sp_6(2)$, it follows from [**AS76a**] that E has four classes of involutions, determined by the Suzuki type of each on the natural module—so these orbits contain representatives for the classes, namely the Suzuki types b_1, a_2, c_2, b_3 suggested by the notation above. Hence E controls G-fusion of its involutions. As $M = C_G(Z_V) \leq E$, it follows that E is the unique fixed point on G/E of a generator d of Z_V. For $j \notin \beta_3$, we may choose $T \in Syl_2(C_{E_2}(j))$, so that $F^*(C_G(j)) = O_2(C_G(j))$ by 1.1.4.6; hence $d \in O_2(C_G(j))$, so E is the unique fixed point of $O_2(C_G(j))$ on G/E, and hence $C_G(j) \leq E$.

Set $D := d^G$. We claim that D is product-disconnected in G with respect to E, in the sense of Definition ZD on page 20 of [**GLS99**]; cf. the proof of I.8.2. Condition (a) of that definition is trivial. Since $E \cong Sp_6(2)$ we check that $d^E \cap T = b_1 \cap T = \beta_1$. Since E controls G-fusion of its involutions, $D \cap E = d^E$, while by the previous paragraph, $C_G(d) \leq E$. Thus condition (b) of the definition holds by A.1.7.2. Finally consider any $e \in C_D(d) - \{d\}$. Then $e \in C_G(d) \leq E$, so since E controls fusion of its involutions, by conjugating in E we may assume that $e \in \beta_1$. Then $de \in \gamma_2$, so that $C_G(de) \leq E$ by the previous paragraph, verifying condition (c) of the definition and establishing the claim.

Therefore as G is simple, we may apply Corollary ZD on page 22 of [**GLS99**], to conclude that G is a simple Bender group, and E is a Borel subgroup, which is strongly embedded in G. This is impossible by 7.6 in [**Asc94**], as E has more than one class of involutions. $\qquad\square$

13.4.3. Eliminating the case $\mathbf{G_z}$ nonsolvable. If G_z is solvable then Theorem 13.4.1 holds by Proposition 13.4.9. Thus we may assume for the remainder of the proof of the Theorem that G_z is not solvable, and we will work to a contradiction.

In particular there exist nonsolvable members of $\mathcal{H}_z$. Our first result is a refinement of the information produced earlier in 13.4.6.

LEMMA 13.4.10. *Let $H \in \mathcal{H}_z$ be nonsolvable. Then*

(1) There exists $K \in \mathcal{C}(H)$, $K \not\leq M$, $K \in \mathcal{L}_f^*(G,T)$, and $K \unlhd H$. Set $V_H := \langle Z^K \rangle$ and $(KT)^* := KT/C_{KT}(V_H)$; then $K^* \cong L_3(2)$ or A_6.

(2) $L_1 \leq K$.

(3) $L/O_2(L) \cong A_6$ and if $K^* \cong A_6$, then $K/O_2(K) \cong A_6$.

(4) Let $V_K := [V_H, K]$. Then $V_K = [R_2(KT), K]$ and either V_K is the natural module for K^*, or V_K is a 5-dimensional module for $K^* \cong A_6$ with $\langle z \rangle = C_{V_K}(K)$.

(5) $|Z| = 4$, so $Z = Z \cap V = V_1$, $|Z_L| = 2$, and $C_Z(K) = \langle z \rangle$.

(6) $Z_V = Z_L$.

(7) $M = LT$ and $H = KT = G_z$. Thus $\mathcal{H}_z = \{G_z\}$ and $V_H = \langle Z^H \rangle$. Furthermore G_z contains a unique member of $\mathcal{H}_*(T,M)$: the minimal parabolic of H over T distinct from $L_1 T$.

(8) Let $H_2 \in \mathcal{H}(T)$ be the minimal parabolic of H distinct from $L_1 T$, and set $H_0 := \langle H_2, L_2 T \rangle$. Then $H_0 \in \mathcal{H}(T)$, H_2 is the unique member of $\mathcal{H}_*(T,M)$ in H_0, and either:

(i) Conclusion (i) of 13.4.7.1 holds, z is of weight 4 in V, and $Z_V \leq V_K$. Further if V_K is a 5-dimensional module for $K^* \cong A_6$, then Z_V is of weight 4 in V_K.

(ii) Conclusion (ii) of 13.4.7.1 holds and $H_0 = N_G(J(T)) \in \mathcal{M}(T)$.

PROOF. First by 13.4.6.1, there exists $K \in \mathcal{C}(H)$, $K \in \mathcal{L}_f^*(G,T)$, $K \unlhd H$, and $K \not\leq M$. By 13.4.5.1, $K/O_2(K)$ is A_5, $L_3(2)$, A_6, or $\hat{A}_6$.

Set $U := [R_2(KT), K]$. By 13.4.5.3 with KT in the role of "H", there is $W_K \in Irr_+(K, R_2(KT), T)$, and for each such W_K, $W_K = \langle (Z \cap W_K)^K \rangle \leq U$ and W_K is either a natural module for $K/O_{2,Z}(K)$ or a 5-dimensional module for $K/O_{2,Z}(K) \cong A_6$.

Now $K = [K, J(T)]$ by 13.4.6.3, so Theorem B.5.1 shows that either $U \in Irr_+(K, R_2(KT))$, or U is the sum of two isomorphic natural modules for $K^* \cong L_3(2)$, which are T-invariant since then $T^* \leq K^*$. In particular U is the A_5-module if $K^* \cong A_5$, and $U = \langle (Z \cap U)^K \rangle \leq V_H$, so $U = V_K$. By B.2.14, $V_H = UC_Z(K)$, so $C_{KT}(U) = C_{KT}(V_H)$.

As $V_H = UC_Z(K)$, $C_{KT}(Z) = C_K(Z \cap U)T$, so that $C_K(Z)^*T^*$ is a maximal parabolic of K^*T^* containing T^*. Now $C_K(Z)T = L_K T$, where $L_K :=$

$O^2(C_K(Z)) \leq M$ by 13.4.2.2. Then $L_K \leq \theta(M) = L$ by 13.4.2.5, so that $L_K \leq O^2(C_L(Z)) \leq O^2(C_L(z)) = L_1$. Let $L_C := O^2(C_{L_1}(K/O_2(K))$; as $L_K \leq L_1 \leq H \leq N_G(K)$ and $Out(K^*)$ is a 2-group, it follows that $L_1 = L_K L_C$. In each case $L_K/O_2(L_K)$ is an elementary abelian 3-group of rank 1 or 2; similarly $L_1/O_2(L_1)$ is of rank 1 or 2 for $L/O_2(L)$ isomorphic to A_6 or $\hat{A}_6$, respectively. In particular if $L/O_2(L) \cong A_6$, then $3 \leq |L_K : O_2(L_K)| \leq |L_1 : O_2(L_1)| = 3$, so equality holds and $L_K = L_1$.

Now set $R := O_2(L_1 T)$, $S := \text{Baum}(R)$, and $T_K := R(T \cap K)$. Then $T_K \in Syl_2(KT_K)$. Further $[T \cap K, L_C] \leq O_2(K) \leq R$, so as $L_1 = L_K L_C$, $O_2(L_K T_K) = O_2(L_1 T_K) = R$. Also $C_{KT_K}(Z) = L_K T_K$, so $R = O_2(L_K T_K) = O_2(C_{KT_K}(Z))$.

We are now in a position to complete the proof of (1). We showed that $K^* \cong A_5$, $L_3(2)$, or A_6; thus it remains to assume $K^* \cong A_5$, and derive a contradiction. In this case we saw that U is the A_5-module, and we also saw that $R^* = O_2(L_K^* T^*)$, so $R^* = T_K^*$. Then R^* contains no FF^*-offenders on U by B.3.2.4, so by B.2.10.1,

$$S = \text{Baum}(R) = \text{Baum}(O_2(KT_K)) \trianglelefteq KT_K.$$

If C is a nontrivial characteristic subgroup of S normal in LT, then $K \leq N_G(C) \leq M = !\mathcal{M}(LT)$, contrary to $K \not\leq M$; hence no such C exists. This eliminates the case $L/O_2(L) \cong \hat{A}_6$, since there 13.2.2.8 shows that each C is indeed normal in LT. Thus $L/O_2(L) \cong A_6$, so by an earlier remark $L_K = L_1$, and hence $R = R_1$. Now 13.2.2.7 shows that L is an A_6-block. Therefore L_1 has exactly two noncentral 2-chief factors; so also K is an A_5-block since $L_1 = L_K$. As $S = \text{Baum}(O_2(KT_K))$, S centralizes $O_2(K)$ by C.1.13.c; so by B.2.3.7, each $A \in \mathcal{A}(S)$ contains $O_2(K)$. Then $[A, K] \leq [O_2(KS), K] = O_2(K) \leq A$, so $A \trianglelefteq KA$. However $m_2(O_2(LT)) = m_2(S)$ by 13.2.2.6, so that $\mathcal{A}(O_2(LT)) \subseteq \mathcal{A}(S)$; hence $J(O_2(LT)) \trianglelefteq KT$, so that $K \leq M$ for our usual contradiction. Therefore K^* is not A_5, completing the proof of (1).

We next prove (2), so we suppose that $L_1 \not\leq K$, and derive a contradiction. If K^* is A_6, then $K = \theta(H)$ by 12.2.8, and hence $L_1 \leq K$, contrary to our assumption. Thus K^* is $L_3(2)$ by (1). If $L/O_2(L) \cong A_6$, then we saw earlier that $L_1 = L_K \leq K$, contrary to our assumption. Thus $L/O_2(L) \cong \hat{A}_6$. As L_0 and $L_{1,+}$ are the T-invariant subgroups with images of order 3 in $L_1/O_2(L_1)$, we conclude that $\{L_C, L_K\} = \{L_0, L_{1,+}\}$. Indeed as $K \not\leq M$, while K acts on $O^2(L_C O_2(K)) = L_C$ and $N_G(L_0) \leq M$ by 13.2.2.9, we conclude that $L_K = L_0$ and $L_{1,+} = L_C$.

As $K^* \cong L_3(2)$, $R^* = O_2(L_K^* T_K^*)$ is the unipotent radical of the maximal parabolic $L_K^* T_K^*$ of K^* stabilizing $Z \cap U$. As $L/O_2(L) \cong \hat{A}_6$, $S \trianglelefteq LT$ by 13.2.2.8, so no nontrivial characteristic subgroup of S is normal in KT, since $K \not\leq M$. Therefore we may apply C.1.37 to conclude that K is an $L_3(2)$-block. But then L_K has just two noncentral 2-chief factors, whereas we saw earlier that $L_K = L_0$, and L_0 has at least three noncentral chief factors on an L-chief section of $O_2(L)$ not centralized by L_0. This contradiction shows that $L_1 \leq K$, completing the proof of (2).

Recall that $L_K \leq L_1$, while by (2), $L_1 \leq L_K$, so $L_1 = L_K$. Thus $L/O_2(L) \cong \hat{A}_6$ iff $m_3(L_1) = 2$ iff $m_3(L_K) = 2$ iff $K/O_2(K) \cong \hat{A}_6$. But then by 13.2.2.8 applied to both LT and KT,

$$J(O_2(LT)) = J(O_2(L_1 T)) = J(O_2(L_K T)) = J(O_2(KT)),$$

so that $K \leq M$ for usual contradiction, establishing (3).

As $L/O_2(L) \cong A_6$ by (3), $R = R_1$. Also either $U \in Irr_+(K, R_2(KT))$, or $K^* \cong L_3(2)$ and U is a sum of two isomorphic natural modules. Suppose that the latter case holds. Again R^* is the unipotent radical of the parabolic $C_{K^*}(Z \cap U)T_K^*$ fixing a point in each summand of U, so we can finish much as in the proof of (1): R^* contains no FF*-offenders on U, so $S \trianglelefteq KT_K$ by B.2.10.1. Then no nontrivial characteristic subgroup of S is normal in LT, so L is an A_6-block by 13.2.2.7, and L_1 has exactly two noncentral 2-chief factors. This is a contradiction since $L_1 \leq K$ by (2), so that L_1 has a noncentral chief factor on each summand of U, plus one more on $O_2(L_1^*)$.

This contradiction shows that $U \in Irr_+(K, R_2(KT))$. Thus from earlier remarks, $V_H = UC_Z(K)$ and U is the natural module for K^* or a 5-dimensional module for $K^* \cong A_6$. In particular, $Z = (Z \cap U)C_Z(K)$. Next $C_{Z_L}(K) = 1$, as otherwise $K \leq C_G(C_{Z_L}(K) \leq M$ by 13.4.2.2. By 13.4.3.2, $|Z : Z_L| = 2$, so $|C_Z(K)| \leq 2$, and hence $C_Z(K) = \langle z \rangle$. In particular if $K^* \cong A_6$ and $m(U) = 5$, then $C_U(K) = \langle z \rangle$, establishing (4). Also $|Z \cap U : C_{Z \cap U}(K)| = 2$, so as $Z = (Z \cap U)C_Z(K)$ and $C_Z(K) = \langle z \rangle$, (5) and (6) hold.

Using (3) and 13.4.6.5, $M = LT$ so $C_G(Z) = L_1 T$. Let $W_H := \langle Z^H \rangle$ and $U_H := [W_H, K]$. By 13.4.5.4, $O_{2,Z}(K) = C_K(W_K) = C_K(U_H) = C_K(W_H)$. As $K = [K, J(T)]$, Theorems B.5.1 and B.5.6 say that either $U_H \in Irr_+(K, W_H)$, so that $U_H = U$, or $K^* \cong L_3(2)$ and U_H is the sum of two isomorphic natural modules for $K^* \cong L_3(2)$. Assume the latter holds. Then as K is irreducible on U and $O_2(H/C_H(V_H)) = 1$ by B.2.14, $Aut_H(U_H) = L_3(2) \times L_2(2)$ and U_H is the tensor product module. Then $Aut_R(U_H)$ contains no FF*-offenders, so as in earlier arguments we obtain a nontrivial characteristic subgroup of R normal in KT and M, a contradiction. Thus $U_H = U$.

By A.1.41, $C_H(K/O_2(K)) \leq C_H(U)$, so as $Z = (Z \cap U)\langle z \rangle$ and $Out(K/O_2(K))$ is a 2-group, $H = KTC_H(K/O_2(K)) = KC_H(U) = KC_H(Z) = KL_1T = KT$ since $L_1 \leq K$ by (2). Since G_z satisfies the hypotheses for H, we conclude $G_z = KT = H$. Thus (7) holds since $K \not\leq M$.

Define H_2 and H_0 as in (8), and let $H_1 := L_2 T$. Observe that the hypotheses of Proposition 13.4.7 are satisfied: For example (5) establishes part (b), with $Z_V = C_Z(L_2)$ and $\langle z \rangle = C_Z(H_2)$. Also if $X \in \mathcal{H}(H_0)$, then $L_1 \not\leq X$: as otherwise $M = LT = \langle L_1, L_2 T \rangle \leq X$ whereas $H_2 \leq X$ but $H_2 \not\leq M$ by (7). Therefore as $L_2 T$ is maximal in $M = C_G(Z_V)$ and H_2 is maximal in $H = G_z$, we conclude that $L_2 T = C_X(Z_V)$, so $L_2 = O^{3'}(C_{H_0}(Z_V))$; and $H_2 = C_X(z)$, so $O^2(H_2) = O^{3'}(C_{H_0}(z))$. Thus part (c) holds. Finally L_2 has at least three noncentral 2-chief factors, two on V and one on $O_2(\bar{L}_2)$, giving part (a). We conclude from 13.4.7.1 that $H_0 \in \mathcal{H}(T)$ and one of conclusions (i)—(iv) of that result holds. In applying 13.4.7, we interchange the roles of "H_1" and "H_2", so the hypothesis "$K_2 = L_2$" in parts (3) and (4) of 13.4.7 also holds; hence conclusions (iii) and (iv) do not hold here.

Suppose conclusion (ii) holds. Then $J(T) \trianglelefteq H_0$. Further for any $X \in \mathcal{H}(H_0)$,

$$C_X(Z) = C_X(Z_V) \cap C_X(z) = L_2 T \cap H_2 = T,$$

so we conclude from 13.4.7.2 that $X = H_0$. Thus $H_0 \in \mathcal{M}(T)$, and in particular $H_0 = N_G(J(T))$. That is, conclusion (ii) of (8) holds.

Finally suppose that conclusion (i) of 13.4.7.1 holds. Then $V_0 := \langle Z^{H_0} \rangle$ is of rank 4. Set $K_2 := O^2(H_2)$; then $[L_2, K_2] \leq O_2(L_2) \cap O_2(K_2)$, so L_2 and K_2

normalize each other. Thus L_2 centralizes $U_2 := \langle Z_V^{K_2} \rangle$, and K_2 centralizes $U_1 := \langle z^{L_2} \rangle$. But as conclusion (i) of 13.4.7.1 holds, $C_{V_0}(L_2) =: U_2' \cong E_4$, so that $U_2 = U_2' \cong E_4$; in particular, $Z_V \leq U_2 \leq V_K$. Similarly $U_1 \cong E_4$, so it follows that z is of weight 4 rather than 2 in V. Similarly Z_V is of weight 4 in V_K when V_K is the 5-dimensional module for $K^* \cong A_6$. Thus conclusion (i) of (8) holds, and the proof of 13.4.10 is complete. $\qquad \square$

LEMMA 13.4.11. *(1)* $L/O_2(L) \cong A_6$.
(2) $M = LT$.
(3) $\mathcal{H}_z = \{G_z\}$.

PROOF. Part (1) follows from 13.4.10.3, and (2) and (3) follow from 13.4.10.7. $\qquad \square$

LEMMA 13.4.12. *(1)* $Z = V_1$ *has rank* 2*, and there exists a unique* $z \in Z^{\#}$ *such that* $C_G(z) \not\leq M$.
(2) *There is a unique member* H_2 *of* $\mathcal{H}_*(T, M)$ *contained in* G_z.

PROOF. By 13.4.10.5, $Z = V_1$ is of rank 2. Recall $z \in V_1^{\#}$ with $G_z \not\leq M$, and z has weight 2 or 4 in V while a generator of Z_V is of weight 6. Let z_k denote the element of V_1 of weight k and choose m with $G_{z_m} \not\leq M$. In this subsection G_{z_m} is not solvable, so by parts (1) and (7) of 13.4.10, $G_{z_m} = K_{z_m} T$ for $K_{z_m} \in \mathcal{C}(G_{z_m})$, and there is a unique $H_{m,2} \in \mathcal{H}_*(T, M)$ contained in G_{z_m}. Set $K_{m,2} := O^2(H_{m,2})$.

As $H_{m,2}$ is the unique member of $\mathcal{H}_*(T, M)$ contained in G_{z_m}, (2) holds. Moreover by 13.4.10.5, $C_Z(H_{m,2}) = \langle z_m \rangle$. As $z_2 \neq z_4$, $H_{2,2} \neq H_{4,2}$.

It remains to prove the final statement in (1), so we assume that $G_{z_m} \not\leq M$ for both $m = 2$ and 4. Set $H_{m,0} := \langle L_2 T, H_{m,2} \rangle$. Then by 13.4.10.8, $H_{m,0} \in \mathcal{H}(T)$, and $H_{m,0}$ satisfies conclusion (i) or (ii) of both 13.4.7.1 and 13.4.10.8. As z_2 has weight 2 in V, $H_{2,0}$ satisfies conclusion (ii) rather than (i) of 13.4.10.8, and hence

$$H_{2,0} = N_G(J(T)) \in \mathcal{M}(T).$$

Suppose that $H_{4,0}$ also satisfies conclusion (ii) of both results. Then by 13.4.10.8, $H_{2,0} = N_G(J(T)) = H_{4,0}$ and $H_{m,0}$ contains a unique member $H_{m,2}$ of $\mathcal{H}_*(T, M)$. Therefore $H_{2,2} = H_{4,2}$, contrary to an earlier observation. Hence $H_{4,0}$ satisfies conclusion (i) of both results.

Next let $H_0 := \langle H_{2,2}, H_{4,2} \rangle$. We check that the hypotheses of Proposition 13.4.7 are satisfied: We already observed that $m(Z) = 2$ and $\langle z_k \rangle = C_Z(H_{k,2})$, establishing (b). We saw that $H_{k,2}$ does not centralize z_{6-k}, so $H_0 \not\leq G_{z_k}$ and hence $O^{3'}(C_{H_0}(z_k))T < G_{z_k}$. Now $G_{z_k} = K_{z_k} T$ for $k = 2, 4$, and in each case $K_{k,2} T$ is a maximal subgroup of G_{z_k}, so we conclude $K_{k,2} = O^{3'}(C_{H_0}(z_k))$, giving (c). Finally by 13.4.10.4, K_{z_k} has at least two noncentral 2-chief factors, one in $V_{G_{z_k}}$ and one in $K_{z_k}/C_{K_{z_k}}(V_{G_{z_k}})$, giving (a).

So we may apply 13.4.7. Assume first that H_0 satisfies one of conclusions (ii)–(iv) of 13.4.7.1. Then $H_0 \leq N_G(J(T)) = H_{2,0}$. Recall however that H_0 is generated by distinct members $H_{k,2}$ of $\mathcal{H}_*(T, M)$, whereas $H_{2,2}$ is the unique member of $\mathcal{H}_*(T, M)$ contained in $H_{2,0}$.

Therefore H_0 satisfies conclusion (i) of 13.4.7.1. Thus $K_{2,2} \leq N_G(K_{4,2})$ and hence $H_{2,0} = \langle K_{2,2}, L_2 T \rangle \leq N_G(K_{4,2})$, since $H_{4,0}$ also satisfies conclusion (i) of 13.4.7.1. Indeed we saw $H_{2,0} \in \mathcal{M}(T)$, so $H_{2,0} = N_G(K_{4,2})$, and hence $K_{4,2} \leq O_{2,3}(H_{2,0})$. However, we also saw that $H_{2,0}$ satisfies conclusion (ii) of 13.4.7.1, so that $H_{2,0}/O_2(H_{2,0}) \cong L_3(2)$, and hence $O_{2,3}(H_{2,0}) = O_2(H_{2,0})$ is a 2-group. This final contradiction completes the proof of 13.4.12. $\qquad\square$

By 13.4.12, there is a unique $z \in V_1^{\#} = Z^{\#}$ with $G_z \not\leq M$. For the remainder of the section, set $H := G_z$ and $V_H := [\langle Z^H \rangle, O^2(H)]$. By 13.4.10.7 there is a unique $H_2 \in \mathcal{H}_*(T, M)$ contained in H. Let $K_2 := O^2(H_2)$, $H_0 := \langle L_2 T, H_2 \rangle$, and $V_0 := \langle Z^{H_0} \rangle$.

LEMMA 13.4.13. *(1)* $H = KT$ *with* $K \in \mathcal{L}_f^*(G, T)$, $K/O_2(K) \cong L_3(2)$ *or* A_6, *and* V_H *is the natural module for* $K/O_2(K)$ *or a 5-dimensional module for* A_6.
(2) $\langle Z^H \rangle = V_H \langle z \rangle$, *so* $V_H \in \mathcal{R}_2(H)$.

PROOF. By 13.4.10, (1) holds and $Z = Z_V \langle z \rangle$ where $\langle z \rangle = C_Z(K)$. So by B.2.14, $\langle Z^H \rangle = [Z, K] C_Z(K) = V_H \langle z \rangle$ and $V_H \in \mathcal{R}_2(H)$. $\qquad\square$

LEMMA 13.4.14. H_0 *satisfies conclusion (i) or (ii) of 13.4.7.1, and:*

(1) If V_0 *is semisimple then* z *is of weight 4 in* V *and* $Z_V \leq [Z, K_2] \leq V_H$. *If further* V_H *is the 5-dimensional module for* A_6, *then* Z_V *is of weight 4 in* V_H.

(2) If $H_0/O_2(H_0) \cong L_3(2)$ *and* V_0 *is the core of the permutation module, then either:*

    (i) $Z_V \not\leq Soc(V_0)$, z *is of weight 4 in* V, *and either*

        (a) $Z_V \not\leq V_H$, *or*

        (b) V_H *is a 5-dimensional module for* A_6 *and* Z_V *is of weight 2 in* V_H;

or else

    (ii) $Z_V \leq Soc(V_0)$, z *is of weight 2 in* V, $Z_V \leq V_H$; *and if* V_H *is a 5-dimensional module for* A_6 *then* Z_V *is of weight 4 in* V_H.

PROOF. The initial statement follows from 13.4.10.8. In the remainder of the proof, we extend arguments used in the last few lines of the proof of that result: First z is of weight 4 in V iff $z \in [Z, L_2]$. Further $Z_V \leq [Z, K_2]$ iff $Z_V \leq V_H$ with Z_V of weight 4 in V_H when V_H is of dimension 5. Thus the subcase of conclusion (1) where H_0 is solvable can be treated exactly like the subcase in the earlier proof corresponding to 13.4.10.8i.

So assume $H_0/O_2(H_0) \cong L_3(2)$. Then $Z_V = C_{V_0}(L_2 T)$ and $\langle z \rangle = C_{V_0}(K_2 T)$. In case (1), where V_0 is semisimple, $C_{V_0}(L_2 T) \leq [Z, K_2]$ and $C_{V_0}(K_2 T) \leq [Z, L_2]$, completing the proof of (1) in view of the equivalences in the previous paragraph.

It remains to prove (2), so we assume V_0 is the core of the permutation module. Suppose first that $Z_V \not\leq Soc(V_0)$. Then L_2 centralizes the generator for Z_V, which lies in $V_0 - Soc(V_0)$. Thus we may apply section H.5 with L_2, K_2 in the roles of "L_p, L_l": Then $\langle z \rangle = C_{V_0}(K_2 T) \leq Soc(V_0)$ by H.5.2.5 and H.5.3.3, and $z \in [Z, L_2]$ by H.5.4.2, so that z is of weight 4 in V. Further $Z_V = C_{V_0}(L_2 T) \not\leq [Z, K_2]$ by H.5.4.1. Therefore if $Z_V \leq V_H$, then $z \in Z \leq Z_V[Z, K_2] \leq V_H$, so V_H is a 5-dimensional module for A_6, and hence Z_V is of weight 2 in V_H by our earlier equivalences. Thus either (a) or (b) of (2i) holds in this case.

On the other hand if $Z_V \leq Soc(V_0)$, then the roles of L_2 and K_2 are reversed in the application of section H.5. Thus $Z_V = C_{V_0}(L_2 T) \leq [Z, K_2]$ and $z \notin [Z, L_2]$, so that (2ii) holds. $\qquad\square$

Recall since $L/O_2(L) \cong A_6$ by 13.4.11 that $R_2 = O_2(L_2 T)$.

LEMMA 13.4.15. *Assume for some* $g \in G$ *that* $V_0^g \leq R_2$ *and* $V \leq R_2^g$. *Then* $1 = [V, V_0^g]$.

PROOF. Assume $[V, V_0^g] \neq 1$. By hypothesis $V_0^g \leq R_2$ and $V \leq R_2^g$, so V_0^g and V normalize each other. Let $H_0^{g*} := H_0^g/O_2(H_0^g)$. By 13.4.14, case (i) or (ii) of 13.4.7.1 holds. Now $1 \neq V^* \leq R_2^{g*}$. But in case (i), as R_2^{g*} centralizes L_2^{g*}, $H_0^{g*} \cong S_3 \times S_3$, V^* is of order 2, and $[V_0^g, V] \leq [V_0^g, K_2^g] = [Z, K_2]^g$. Similarly in case (ii), $V^* \leq R_2^{g*} \cong E_4$, so $V^* = R^{*g}$ if $|V^*| > 2$; further by 13.4.7.1, H_0^{g*} contains no transvections on V_0^g.

Suppose first that $|V^*| = 2$. Then $\bar{V}_0^g$ induces a nontrivial group of transvections on V with axis $C_V(V_0^g)$, so as V is a 5-dimensional module for $\bar{L} \cong A_6$, it follows that $[V, V_0^g] = \langle v \rangle$ with v of weight 2 in V. Conjugating in L, we may assume $v \in V_1$. Further $2 = |\bar{V}_0^g| = |V_0^g/C_{V_0^g}(V)|$. Hence V^* is a group of transvections on V_0^g with center $\langle v \rangle$, so $H_0^{g*} \cong S_3 \times S_3$ and $v \in [V_0^g, V] \leq [Z, K_2]^g$ by paragraph one. But by 13.4.14.1, $Z_V \leq [Z, K_2]$, so $\langle v \rangle$ is conjugate in G to Z_V of weight 6, contradicting 13.2.2.5 since v has weight 2.

Therefore $|V^*| > 2$, so by paragraph one, $H_0^{g*} \cong L_3(2)$ and $V^* = R^{g*}$ is of order 4. From the action of H_0 on V_0, $[V, V_0^g] = C_{V_0^g}(V)$ and $V_0^g/C_{V_0^g}(V)$ are of rank 3: this is clear if V_0 is semisimple, and it follows from H.5.2 if V_0 is the core of the permutation module. Hence $\bar{V}_0^g = \bar{R}_2$ and $m(\bar{R}_2) = 3$. As $m(\bar{R}_2) = 3$, $Z_V \leq [V, R_2] = [V, V_0^g] = C_V(V_0^g)$. Then Z_V is weakly closed in $[V, V_0^g]$ by 13.2.2.5. Also $(L_2 T)^g$ acts on $[R_2^g, V_0^g] = [V, V_0^g]$, and then also on the subgroup Z_V weakly closed in $[V, V_0^g]$, so $(L_2 T)^g \leq M$. Then T^g is conjugate to T in M, so as $N_G(T) \leq M$ by Theorem 3.3.1, $g \in M$. Now as $\bar{R}_2 = \bar{V}_0^g \leq \bar{R}_2^g = O_2(\bar{L}^g\bar{T}^g)$, $L_2 T = (L_2 T)^g$. As $M = LT$ by 13.4.11, $L_2 T$ is maximal in M, so $g \in L_2 T \leq H_0$, so $H_0 = H_0^g$. Thus $V_0^g = V_0 \trianglelefteq T$, so as $C_{V_0^g}(V) = [V, V_0^g] \leq V \cap V_0^g$,

$$[O_2(LT), V_0] \leq C_{V_0}(V) \leq V.$$

Therefore $[O_2(LT), L] = V$ and L is an A_6-block. Set $K_0 := O^2(H_0)$; similarly $[O_2(H_0), V] \leq V_0$ and then $[O_2(H_0), K_0] = V_0$.

If $V_0 = U_1 \oplus U_2$ is the sum of non-isomorphic 3-dimensional modules for K_0, we saw that $Z_V \leq U := U_i$ for $i := 1$ or 2 during the proof of 13.4.14.1. If instead V_0 is the core of the permutation module and $Z_V \leq Soc(V_0)$, set $U := Soc(V_0)$. In either of these two cases, since $V^* = R_2^*$ and L_2 centralizes Z_V, $[U, V] = C_U(L_2) = Z_V = C_U(V)$, so U induces a 4-group of transvections on V with center Z_V, impossible as $C_M(V/Z_V) = C_M(V)$ by 13.4.2.4. Therefore we are in case (i) of 13.4.14.2, where V_0 is the core of the permutation module and $Z_V \not\leq Soc(V_0)$; so by that result, z is of weight 4 in V.

As L is an A_6-block, L_1 has two noncentral 2-chief factors, so K is an $L_3(2)$-block or an A_6-block using 13.4.13.1. Further as z is of weight 4 in V, $\langle z \rangle = [V \cap O_2(L_1), O_2(L_1)]$, so that $z \in V_H$. Therefore since $z \in Z(H)$, V_H is the 5-dimensional module for the A_6-block K. By 13.4.12.1, $Z = Z_V\langle z \rangle$ is of order 4, and by symmetry between L and K, $Z \leq V_H$ and $Z \cap O_2(L_1) \not\leq Z(K) \cap V_H = \langle z \rangle$; so as $z \in L_1$, $Z \leq L_1$. Calculating in the A_6-block K, $|Z(K)| \leq 4$ and $O_2(L_1/Z(K)) \cong Q_8^2$, so $|Z(O_2(L_1)/\langle z \rangle)| \leq 4$. Therefore as $[V \cap O_2(L_1), O_2(L_1)] = \langle z \rangle$ and $|V \cap O_2(L_1)| = 8|Z_V \cap O_2(L_1)| = 16$, we have a contradiction. $\square$

LEMMA 13.4.16. *If $g \in G$ with $V_0^g \leq R_2$ and $V_0 \leq R_2^g$, then $[V_0, V_0^g] = 1$.*

PROOF. Assume V_0^g is a counterexample, and let $H_0^* := H_0/C_{H_0}(V_0)$. Interchanging V_0 and V^g if necessary, we may assume that $m(V_0^{g*}) \geq m(V_0/C_{V_0}(V_0^g))$, so V_0 is an FF-module for H_0^*. The modules V_0 in case (ii) of 13.4.7.1 are not

FF-modules by Theorem B.5.1, so by 13.4.14, we are in case (i) of 13.4.7.1. Arguing as in the proof of the previous lemma, $m(R_2^*) = 1$ and then $m(V_0^{g*}) = 1 = m(V_0/C_{V_0}(V_0^g))$ and $[V_0, V_0^g] = Z_V$. By symmetry between V_0 and V_0^g, $Z_V^g = [V_0, V_0^g] = Z_V$, so $g \in N_G(Z_V) = M = N_G(V)$ by 13.4.2.1. Then $V = V^g \leq R_2^g$, so by 13.4.15, $[V, V_0^g] = 1$. Thus as $V = V^g$, also $[V, V_0] = 1$.

Let $U_0 := [Z, K_2]$, so that U_0 is a 4-group as case (i) of 13.4.7.1 holds. As H_0 is solvable, z is of weight 4 in V and $Z_V \leq U_0 \leq V_H$ by 13.4.14.1. Therefore $U_0 = \langle Z_V^{K_2} \rangle$ and $C_{K_2 T}(U_0) = O_2(K_2 T)$. By 13.4.12.1, $C_G(v) \leq M = N_G(V)$ for each $v \in V^\#$ not of weight 4, so by 13.4.2.3, V is the unique member of V^G containing v. But up to conjugation under L, $\langle e_{1,2,3,4}, e_{1,2,5,6} \rangle$ is the unique maximal subspace U of V all of whose nontrivial vectors are of weight 4, so $r(G, V) \geq 3$.

Let $W_0 := W_0(T, V)$. We claim $[V, W_0] = 1$, so that $N_G(W_0) \leq M$ by E.3.34.2: For suppose $A := V^y \cap M$ with $\bar{A} \neq 1$. Assume for the moment that also $V \leq M^y$. Then $1 \neq [V, A] \leq V \cap V^y$, so by the previous paragraph, $[V, A]^\#$ contains only vectors of weight 4. We conclude that all involutions of $\bar{A}$ are of cycle type 2^3, and hence $|\bar{A}| = 2$ and $|V : C_V(V^y)| \geq |V : C_V(A)| = 4$. Therefore $A < V^y$ when $V \leq M^y$—since if $A = V^y$, then we have symmetry between V and V^y, so that $2 = |\bar{A}| = |V : C_V(V^y)| = 4$.

Now assume $A = V^y$; then $V \not\leq M^y$ by the previous paragraph. Therefore $m(\bar{A}) \geq r(G, V) \geq 3$, and hence $m(\bar{A}) = 3$ as $m_2(\bar{M}) = 3$. Further

$$U = \langle C_V(\bar{B}) : 1 \neq \bar{B} \leq \bar{A} \rangle = \langle C_V(D) : m(A/D) < 3 \rangle \leq M^y.$$

Thus $U < V$, which is not the case if $\bar{A}$ is conjugate to $\bar{R}_2$. Therefore $\bar{A}$ is conjugate to $\bar{R}_1$, and then $m(V/U) = 1$ so that $U = V \cap M^y$ and $m(U/C_U(A)) = 2$. But applying the previous paragraph with U, A in the roles of "A, V", we conclude that $m(U/C_U(A)) = 1$. This contradiction establishes the claim that $W_0 \leq C_T(V)$ and $N_G(W_0) \leq M$.

Thus W_0 is not normal in $K_2 T$, as $K_2 \not\leq M$, and hence $W_0 \not\leq O_2(K_2 T) = C_{K_2 T}(U_0)$ by E.3.15. Therefore there is $D := V^x \leq T$ for some $x \in G$ with $[U_0, D] \neq 1$. But $|D : C_D(U_0)| = 2$ as $|T : O_2(K_2 T)| = 2$, so $U_0 \leq C_G(C_D(U_0)) \leq N_G(V^x)$ as $r(G, V) \geq 3$. Then as D does not centralize U_0, $Z_V = [U_0, D] \leq D$. By 13.4.2.3, V is the unique member of V^G containing Z_V, so $D = V$. But now $[D, U_0] \neq 1$, whereas $U_0 \leq V_0$ and $[V, V_0] = 1$ by the first paragraph. This contradiction completes the proof of 13.4.16. $\qquad\square$

Our final lemma shows that the 2-locals M and H_0 resemble the parabolics $\dot{P}_1$ and $\dot{P}_3$ of $Sp_6(2)$, except that z is of weight 2 in V and $Z_V \leq Soc(V_0)$. Still with this information we will be able to obtain a contradiction to our assumption that G_z is not solvable, completing the proof of Theorem 13.4.1.

LEMMA 13.4.17. *(1) z is of weight 2 in V.*

(2) There exists $g \in H$ such that $[V, V^g] = \langle z \rangle \leq V^g$.

(3) $H_0/O_2(H_0) \cong L_3(2)$, V_0 is the core of the permutation module, $Z_V \leq Soc(V_0)$, $Z_V \leq V_H$, and V_H is not a 5-dimensional module if $H/O_2(H) \cong A_6$.

(4) H is transitive on $V_H^\#$, and each subgroup of V_H of order 2 is in Z_V^G.

(5) $r(G, V) \geq 3$.

(6) For each $g \in G - M$, $V^\# \cap V^g$ consists of elements of weight 2.

PROOF. Let $G_1 := LT = M$ and $G_2 := H_0$, and form the coset graph Γ with respect to these groups as in section F.7. Adopt the notational conventions that section including Definition F.7.2, where in particular γ_0, γ_1 are the cosets G_1, G_2. For $\gamma = \gamma_0 g$ set $V_\gamma := V^g$, while for $\gamma = \gamma_1 g$ set $V_\gamma := V_0^g$. Let

$$\alpha := \alpha_0, \ldots, \alpha_n =: \beta$$

be a geodesic in Γ, of minimal length n subject to $V_\alpha \not\leq G_\beta^{(1)}$; such an n exists by F.7.3.8. As $G_\beta^{(1)} = O_2(G_\beta) = C_{G_\beta}(V_\beta)$ for each $\beta \in \Gamma$, $[V_\alpha, V_\beta] \neq 1$, and so we have symmetry between α and β. This symmetry is fairly unusual among our applications of section F.7, as we almost always consider only geodesics whose orgin is conjugate to γ_0; however the approach in this lemma is the one most commonly used in the amalgam method in the literature. By minimality of n,

$$V_\alpha \leq G_{\alpha_{n-1}}^{(1)} \leq G_\beta,$$

so V_α acts on V_β, and by symmetry, V_β acts on V_α. By F.7.9.1, $V_\alpha \leq O_2(G_{\alpha,\alpha_{n-1}}^{(1)})$, and

$$O_2(G_{\alpha,\alpha_{n-1}}^{(1)}) = O_2(H_0 \cap M) = O_2(L_2 T)^g = R_2^g,$$

$$\text{for } g \in G \text{ with } \{\gamma_0 g, \gamma_1 g\} = \{\alpha_{n-1}, \beta\} \tag{*}$$

using transitivity of $\langle G_1, G_2 \rangle$ on the edges of the graph in F.7.3.2. Thus $V_\alpha \leq R_2^g$.

Suppose first that $\beta = \gamma_1 g$; then $V_\beta = V_0^g$. If α is conjugate to γ_1 then we may take $\alpha = \gamma_0$ and $\alpha_1 = \gamma_1$, so $V_0 \leq R_2^g$, and by (*) and symmetry between α and β, $V_0^g = V_\beta \leq R_2$. As $1 \neq [V_\alpha, V_\beta] = [V_0, V_0^g]$, we have a contradiction to 13.4.16. Thus at most one of α and β is conjugate to γ_1. Therefore if $\beta \notin \gamma_0 G$, then $\alpha \in \gamma_0 G$, so reversing the roles of α and β, we may assume $\beta \in \gamma_0 G$. Then conjugating in G, we may take $\beta := \gamma_0$ and $\alpha_{n-1} := \gamma_1$. Thus $V_\alpha \leq R_2$ by (*). Similarly $\alpha = \gamma_i g$ for $i = 0$ or 1, and by (*) we may take $V \leq R_2^g$.

If $\alpha = \gamma_1 g$ then $V_\alpha = V_0^g$, contrary to 13.4.15. Hence $\alpha = \gamma_0 g$ and $V_\alpha = V^g$. In particular $1 \neq [V, V^g] \leq V \cap V^g$.

Let $v \in [V, V^g]^\#$. If $C_G(v) \leq M$, then by 13.4.2.3, V is the unique member of V^G containing v; hence $C_G(v) \not\leq M$. However for any $t \in T - C_T(V)$, $[V, t]$ contains a vector of weight 2, so z is of weight 2 by the uniqueness of z in 13.4.12.1; thus (1) holds. Indeed by (1) and that uniqueness, $C_G(w) \leq M$ for $w \in V$ of weight 4, and hence V is the unique member of V^G containing w by 13.4.2.3. This establishes (6).

By (6), all vectors in $[V, V^g]^\#$ are of weight 2, so $[V, V^g]$ is of rank 1—since up to conjugacy, $E := \langle e_{5,6}, e_{4,6} \rangle$ is the unique maximal subspace of V with all nonzero vectors of weight 2, and $E \neq [V, A]$ for any elementary 2-subgroup of $\bar{M}$. Then conjugating in $L_2 \leq G_\beta$, we may assume $[V, V^g] = \langle z \rangle$. Now by 13.4.2.3, we may take $g \in H$, so (2) is established.

As z is of weight 2 in V, we are in case (ii) of 13.4.14.2. Hence either (3) holds, or else V_H is a 5-dimensional module for $H/O_2(H) \cong A_6$ and Z_V of weight 4 in V_H. But in the latter case we have symmetry between L, V and $K := O^2(H)$, V_H, so as Z_V is weight 4 in V_H, we have a contradiction to (1) applied to K, V_H. Hence (3) is established. By (3), V_H is not a 5-dimensional module for $K/C_K(V_H) \cong A_6$, and in the remaining two cases in 13.4.13, V_H is the natural module for $K/O_2(K)$, so H is transitive on the points of V_H; thus (4) is established as $Z_V \leq V_H$ by (3).

If $U \leq V$ with $C_G(U) \not\leq M$, then all vectors in $U^\#$ are of weight 2 by (6). But we saw that up to conjugation, the unique maximal subspace with this property is $\langle e_{5,6}, e_{4,6} \rangle$ of rank 2, so (5) holds. □

We will now obtain a contradiction to our assumption that H is not solvable. This contradiction will complete the proof of Theorem 13.4.1.

Pick $g \in H$ as in 13.4.17.2. Then $\bar{V}^g = \langle (5,6) \rangle$, so there is $l \in L$ with $\bar{V}^{gl} = \langle (3,4) \rangle$. Let $y := gl$. Then $A := V^y \leq T$ with $L_1 = [L_1, A]$. Let $K := O^2(H)$, so that $K \in \mathcal{C}(H)$ by 13.4.10 and our assumption that H is not solvable. As $L_1 = [L_1, A]$, $K = [K, A]$, and hence $[V_H, A] \neq 1$ as $[V_H, K] \neq 1$. Let $U := V_H \cap M^y$ so that $[U, A] \leq U \cap A$, and set $(KT)^* := KT/C_{KT}(V_H)$.

Suppose first that $[A, U] \neq 1$. By 13.4.17.4, H is transitive on $V_H^{\#}$ and $Z_V \leq V_H$, so $Z_V^h \leq [A, U] \leq A \cap U$ for some $h \in H$. Then as V^h is the unique member of V^G containing Z_V^h by 13.4.2.3, $V^h = A = V^y$, and hence $Z_V^h = Z_V^y$ as $N_G(V) = M = C_G(Z_V)$. Indeed this argument shows Z_V^y is the unique point of $V_H \cap A$, and hence of $[A, U]$; thus $[A, U] = Z_V^y$, and hence U induces transvections on A with center Z_V^y, whereas $\bar{M}$ contains no such transvection, since $C_M(V) = C_M(V/Z_V)$ by 13.4.2.4.

This contradiction shows that $[A, U] = 1$. In particular $V_H \not\leq M^y$, as $[V_H, A] \neq 1$; hence as $r(G, V) \geq 3$ by 13.4.17.5, $m_2(K^*T^*) \geq m(A^*) > 2$, and then examining the cases listed in 13.4.13, we conclude that $H^* \cong S_6$ and $m(A^*) = 3$. Hence for $1 \neq b^* \in A^*$, $\langle b^* \rangle = B^*$ for some $B \leq A$ with $m(A/B) \leq 2$, so $C_{V_H}(b^*) = C_{V_H}(B) \leq M^y$ as $r(G, V) \geq 3$, and therefore

$$\langle C_{V_H}(b^*) : 1 \neq b^* \in A^* \rangle \leq U \leq C_{V_H}(A^*),$$

so that $A^* \in \mathcal{A}_3(T^*, V_H)$. However H^* has no such rank-3 subgroup, since each such subgroup is the radical of some minimal parabolic and hence contains a transvection whose axis is centralized only by that transvection.

This contradiction establishes Theorem 13.4.1.

13.5. The treatment of A$_5$ and A$_6$ when $\langle \mathbf{V}_3^{\mathbf{G_1}} \rangle$ is nonabelian

In this section, we continue our treatment of the remaining alternating groups A_5 and A_6, postponing treatment of the final group $L_3(2)$ of $\mathbf{F}_2$-type until the following chapter. More specifically, this section begins the treatment of the case where $\langle V^{G_1} \rangle$ is nonabelian, by handling in Theorem 13.5.12 the subcase $\langle V_3^{G_1} \rangle$ nonabelian. In fact if $L/O_2(L)$ is A_5 and $\langle V_3^{G_1} \rangle$ is abelian, we will see that $\langle V^{G_1} \rangle$ is also abelian; thus in this section we also deal with the case where $L/O_2(L) \cong A_5$ and $\langle V^{G_1} \rangle$ is nonabelian.

In this section, with Theorem 13.4.1 now established, we assume the following hypothesis:

HYPOTHESIS 13.5.1. *Hypothesis 13.3.1 holds and G is not $Sp_6(2)$.*

In addition we continue the notation established earlier in the chapter, and the notational conventions of section B.3. In particular we adopt Notations 12.2.5 and 13.2.1.

LEMMA 13.5.2. *Assume Hypothesis 13.5.1. If $K \in \mathcal{L}_f(G, T)$, then*

(1) $K/O_2(K) \cong A_5$, $L_3(2)$, A_6, or $\hat{A}_6$.

(2) $K \trianglelefteq KT$ and $K \in \mathcal{L}^*(G, T)$.

(3) There is $V_K \in Irr_+(K, R_2(KT), T)$, and for each such V_K, $V_K \leq R_2(KT)$, $V_K \trianglelefteq T$, the pair K, V_K satisfies the FSU, $C_{V_K}(K) = 1$, and V_K is the natural module for $K/C_K(V_K) \cong A_5$, A_6, or $L_3(2)$.

PROOF. As in the proof of 13.4.5, this follows from 13.3.2, once we observe that by 13.3.2, we may apply various results to K in the role of "L": Theorem 13.3.16 says $K/O_2(K)$ is not $U_3(3)$. Then since Hypothesis 13.5.1 excludes $G \cong Sp_6(2)$, we conclude from Theorem 13.4.1 that $C_{V_K}(K) = 1$ when $K/O_{2,Z}(K) \cong A_6$. $\square$

13.5.1. Setting up the case division on $\langle \mathbf{V_3^{G_1}} \rangle$ for $\mathbf{A_5}$ and $\mathbf{A_6}$.

REMARK 13.5.3. In the remainder of this section (and indeed in the remainder of the chapter), in addition to assuming Hypothesis 13.5.1, we also assume $L/O_2(L)$ is not $L_3(2)$; that is, we restrict attention to the cases where $L/O_2(L) \cong A_5$, A_6, or $\hat{A}_6$.

Then by 13.5.2.3, $C_V(L) = 1$ and V is the natural module for $L/C_L(V) \cong A_n$, $n = 5$ or 6.

As usual we adopt the notational conventions of section B.3 and Notation 13.2.1. We view V as the quotient of the core of the permutation module for $L/C_L(V)$ on $\Omega := \{1, \ldots, n\}$, modulo $\langle e_\Omega \rangle$. Recall from Notation 12.2.5.2 that $M_V := N_M(V)$ and $\bar{M}_V := M_V/C_M(V)$. So there is an $\bar{M}_V$-invariant symplectic form on V, and when $n = 5$, an invariant quadratic form. Thus we use terminology (e.g., of isotropic or singular vectors) associated to those forms.

As in Notation 13.2.1, V_i is the T-invariant subspace of V of dimension i and $G_i := N_G(V_i)$.

LEMMA 13.5.4. When $L/O_2(L) \cong A_6$ or $\hat{A}_6$, set $I_2 := O_2(G_1)L_2$ or $O_2(G_1)L_{2,+}$, respectively. Then:

(1) $I_2 = \langle O_2(G_1)^{G_2} \rangle \trianglelefteq G_2$.
(2) $C_{I_2}(V_2) = O_2(I_2)$ and $I_2/O_2(I_2) \cong S_3$.
(3) $m_3(C_G(V_2)) \leq 1$.
(4) $C_G(V_3) \leq M_V$. Hence $[V, C_G(V_3)] \leq V_1$.
(5) $[O_2(G_1), V_2] \neq 1$.
(6) $O^2(I_2) = L_2$ or $O^2(L_{2,+})$, respectively, $O^2(\bar{I}_2) = \bar{L}_2$, and $O_2(O^2(I_2))$ is nonabelian.

PROOF. The equalities in (6) follow from the definition of I_2 and the fact that L_2 and $L_{2,+}$ are T-invariant. Then as $V = [V, \bar{L}_2]$ and $O_2(\bar{L}_2) \neq 1$, the remaining statement in (6) follows. Hypothesis 13.3.13 is satisfied by 13.5.2.3, so (5) follows from 13.3.14. Then parts (1)–(4) of the lemma follow from 13.3.15. $\square$

LEMMA 13.5.5. $G_1 \cap G_3 \leq M_V$.

PROOF. When $n = 5$, $G_3 \leq M_V$ by 13.2.3.2. When $n = 6$, $Aut_{L_1 T}(V_3) = C_{GL(V_3)}(V_1)$, so as $C_G(V_3) \leq M_V$ by 13.5.4.4, and as $L_1 T \leq M_V$, $G_1 \cap G_3 \leq M_V$. $\square$

As $C_V(L) = 1$:
$$V_1 = Z \cap V \text{ is of order } 2.$$
Let z be a generator for V_1. By 13.3.6,
$$G_1 = C_G(z) \not\leq M,$$
so $G_1 \in \mathcal{H}_z \neq \emptyset$, where as usual
$$\mathcal{H}_z := \{H \in \mathcal{H}(L_1 T) : H \leq G_1 \text{ and } H \not\leq M\}$$
and $\tilde{G}_1 := G_1/V_1$.

LEMMA 13.5.6. *Assume* $n = 6$; *then*

(1) $V \leq O_2(G_2)$.

(2) $V_2^G \cap V = V_2^L$.

(3) If $V_2 \leq V \cap V^g$ *then either* $[V, V^g] = 1$ *or* $[V, V^g] = V_2$, *and in the latter case* $V^g \leq M_V$ *and* $\bar{V}^g = O_2(\bar{L}_2)$.

PROOF. As L has two orbits on 4-subgroups of V, either (2) holds or for some $g \in G$, V_2^g is a nondegenerate 2-subspace of V.

Assume the latter holds. Then $Q_0 := C_T(V_2^g) \in Syl_2(C_M(V_2^g))$, and either $\bar{T} \leq \bar{L}$ and $Q_0 = Q := O_2(LT) = C_T(V)$, or Q_0 is the preimage in LT of the subgroup generated by a transposition. Now if $N_G(Q_0) \leq M$, then $Q_0 \in Syl_2(C_G(V_2^g))$, contradicting $|C_G(V_2^g)|_2 = |T|/2 > |Q_0|$.

Thus $N_G(Q_0) \not\leq M$, so $Q < Q_0$, and as $M = !\mathcal{M}(LT)$, no nontrivial characteristic subgroup of Q_0 is normal in LT. Therefore L is an A_6-block by case (c) of C.1.24. Now $Q = O_2(C_M(V_2^g))$, and $O^{3'}(N_M(V_2^g)) = X_1 \times X_2$ is the product of A_3-blocks, with $V = O^{2'}(X_1 X_2)$. Let $X := O^2(I_2)$, in the notation of 13.5.4. Now $X_1 X_2$ acts on X^g by 13.5.4.1, so $[X_1 X_2, X^g] \leq O_2(X^g)$ by 13.5.4.2. Therefore as $m_3(G_2) \leq 2$, $X^g \leq X_1 X_2$, impossible as $O^{2'}(X)$ is nonabelian by 13.5.4.6, while $V = O^{2'}(X_1 X_2)$ is abelian.

This contradiction establishes (2); thus it remains to prove (1) and (3). Observe that Hypothesis G.2.1 is satisfied with V_2, V in the roles of "V_1, V" as L_2 is irreducible on V/V_2. Let $U := \langle V^{G_2} \rangle$. By G.2.2, (1) holds and $V_2 \geq \Phi(U)$.

Finally suppose $V_2 \leq V \cap V^g$. By (2) and A.1.7.1 we may take $g \in G_2$, so $V^g \leq U \leq M_V$ and $[V, V^g] \leq V \cap V^g$ as $V \trianglelefteq U$. Further if U is abelian, then $[V, V^g] = 1$ and (3) holds. Thus we may take U nonabelian. As $X \trianglelefteq G_2$ by 13.5.4.1 while $V = [V, X]$, $U = [U, X] \leq O_2(X)$. Therefore using 13.5.4.6, $\bar{U} = O_2(\bar{L}_2)$ is of rank 2. Thus $\bar{V}^g \leq O_2(\bar{L}_2)$, so $[V, V^g] = V_2$ and $m(V/C_V(V^g)) = 2$, so by symmetry, $m(V^g/C_{V^g}(V)) = 2$ and hence $\bar{V}^g = O_2(\bar{L}_2)$, completing the proof of (3). $\square$

LEMMA 13.5.7. *For each* $H \in \mathcal{H}_z$, *Hypothesis F.9.1 is satisfied with* V_3 *in the role of "*V_+*".*

PROOF. Most of this proof is exactly parallel to that of 13.3.18.1: namely part (c) of F.9.1 follows, this time using 13.5.5 rather than 13.3.17.1 to obtain $G_1 \cap G_3 \leq M_V$; parts (b) and (d) follow just as before; and part (a) is proved as before. Thus it remains to verify F.9.1.e.

Assume $1 \neq [V, V^g] \leq V \cap V^g$; then $\bar{V}^g$ is quadratic on V. To verify hypothesis F.9.1.3, we may assume that $g \in G_1$ with $[V^g, \tilde{V}_3] = 1 = [V, \tilde{V}_3^g]$. Then $\bar{V}^g \leq O_2(\bar{L}_1 \bar{T}) = \bar{R}_1$, so as $\bar{V}^g$ is quadratic on V, either $m(\bar{V}^g) = 1$; or $n = 6$, $m(\bar{V}^g) = 2$, and conjugating in L_1, we may assume that $V_2 = [V, V^g]$. But in the latter case as $[V, V^g] \leq V \cap V^g$, $V_2 \leq V \cap V^g$; then by 13.5.6.3, $\bar{V}^g = O_2(\bar{L}_2)$, contradicting $\bar{V}^g \leq \bar{R}_1$.

Therefore $m(V^g/C_{V^g}(V)) = 1$, and hence also $1 = m(V/C_V(V^g))$ by symmetry. Suppose $[V_3, V^g] = 1$. Then as $[V, V^g] \neq 1$, $n \neq 5$, since in that case $C_M(V_3) = C_M(V)$. Thus $n = 6$ and $\bar{V}^g$ is generated by a transvection with center V_1, so $[V, V^g] = V_1$. Thus V induces a transvection on V^g with center V_1, so $C_{V^g}(V) = V_1^\perp = V_3^g$; hence $[V_3^g, V] = 1$, and F.9.1.e holds.

It remains to treat the case where $[V_3, V^g] = V_1 = [V_3^g, V]$. Here $m(V^g/C_{V^g}(V)) = 1$, so V induces a transvection with center V_1 on V^g, and so again $[V_3^g, V] = 1$, contrary to assumption. This contradiction completes the proof of 13.5.7. $\qquad\square$

NOTATION 13.5.8. Recall $\tilde{G}_1 = G_1/V_1$. By 13.5.7, we can appeal to the results of section F.9 with V_3 in the role of "V_+" in F.9.1. Recall from Hypothesis F.9.1 that for $H \in \mathcal{H}_z$, $U_H := \langle V_3^H \rangle$, $V_H := \langle V^H \rangle$, and $Q_H := O_2(H)$. By F.9.2.1, $U_H \leq Q_H$, and by F.9.2.2, $\Phi(U_H) \leq V_1$. By F.9.2.3, $Q_H = C_H(\tilde{U}_H)$; set $H^* := H/Q_H$.

Notice $G_1 \cap G_3 \leq M$ by 13.5.5, and $H \not\leq M$, so that:

LEMMA 13.5.9. $V_3 < U_H$

We begin our treatment of the case $\langle V^{G_1} \rangle$ nonabelian by considering the subcase where $\langle V_3^{G_1} \rangle$ is nonabelian; the next observation shows that if $n = 5$ and $\langle V^{G_1} \rangle$ is nonabelian, then $\langle V_3^{G_1} \rangle$ is also nonabelian:

LEMMA 13.5.10. *If $n = 5$ and $H \in \mathcal{H}_z$, then the following are equivalent:*

(1) U_H is abelian.
(2) V_H is abelian.
(3) $V \leq Q_H$.

PROOF. When $n = 5$, $C_M(V_3) = C_M(V)$, so the lemma follows from F.9.4.3. $\qquad\square$

LEMMA 13.5.11. *If $n = 5$ and $V_2 \leq V \cap V^g$, then $[V, V^g] = 1$ or V_2; in either case, $V^g \leq M_V$.*

PROOF. We may assume $[V, V^g] \neq 1$. By hypothesis $V_1 \leq V \cap V^g$, so by 13.3.11.1 we may take $g \in G_1$. By 13.3.11.5, $[V_3, V_3^g] \neq 1$, so by F.9.2.2, $[V_3, V_3^g] = V_1$. Thus $X := V_3 V_3^g \cong D_8 \times \mathbf{Z}_2$, and $\mathcal{A}(X) = \{V_3, V_3^g\}$. Now V_3^g acts on V_3, and also $V_3^g \leq U_{G_1} \leq M_V$, so $[V, V_3^g] \leq V_3 \leq X$, and hence V acts on X. Then as $\mathcal{A}(X) = \{V_3, V_3^g\}$, $V \leq N_{G_1}(V_3^g) \leq N_G(V^g)$ by 13.2.3.2. By symmetry V^g acts on V, so $[V, V^g] \leq V \cap V^g \leq C_V(V^g)$. As $[V_3, V_3^g] = V_1$ is singular, $\bar{V}^g$ does not induce a transvection on V, so $m(C_V(V^g)) \leq 2 \leq m([V, V^g])$, and hence $[V, V^g] = V \cap V^g$ is of rank 2. Then as $V_2 \leq V \cap V^g$ by hypothesis, we conclude $[V, V^g] = V_2$. This completes the proof. $\qquad\square$

13.5.2. The treatment of the subcase $\langle \mathbf{V}_3^{\mathbf{G}_1} \rangle$ nonabelian. We come to the main result of the section, which determines the groups where $\langle V_3^{G_1} \rangle$ is nonabelian:

THEOREM 13.5.12. *Assume Hypothesis 13.3.1 with $L/C_L(V) \cong A_n$ for $n = 5$ or 6, $G \not\cong Sp_6(2)$, and $\langle V_3^{G_1} \rangle$ nonabelian. Then either*

(1) $n = 5$ and $G \cong U_4(2)$ or $L_4(3)$.
(2) $n = 6$ and $G \cong U_4(3)$.

The remainder of this section is devoted to the proof of Theorem 13.5.12.

Observe since $G \not\cong Sp_6(2)$ that Hypothesis 13.5.1 holds. Thus we may apply results from earlier in the section; in particular by 13.5.7, we may apply results from section F.9, and continue to use the conventions of Notation 13.5.8.

In the remainder of the section we assume $\langle V_3^{G_1} \rangle$ is nonabelian. Thus as $G_1 \in \mathcal{H}_z$, there exists $H \in \mathcal{H}_z$ such that U_H is nonabelian.

In the remainder of the section, H will denote any member of $\mathcal{H}_z$ with U_H nonabelian.

Then $\Phi(U_H) = V_1$ by F.9.2.2. By F.9.4.1, $V \not\leq Q_H$, while by F.9.2.1, $V_3 \leq Q_H$. Thus as $|V : V_3| = 2$, $V_3 = V \cap Q_H$ and V^* is of order 2. By F.9.2.1, $\tilde{U}_H \in \mathcal{R}_2(\tilde{H})$.

LEMMA 13.5.13. *(1) If $g \in H$ with $V_1 < V \cap V^g$, then $\langle V, V^g \rangle$ is a 2-group.*
(2) The hypotheses of F.9.5.5 and F.9.5.6 hold.

PROOF. As $C_H(\tilde{U}_H) = Q_H$ is a 2-group, we may assume $V^* \neq V^{g*}$; so as V^* is of order 2, $V \cap V^g \leq V \cap Q_H = V_3$. Then as L_1 is transitive on $\tilde{V}_3^\#$, we may take $V_2 \leq V \cap V^g$. Now 13.5.11 and 13.5.6.3 show that V^g normalizes V, and so (1) follows.

By (1), we have the hypothesis of F.9.5.5. Further by 13.5.5, $C_H(V_3) \leq C_M(V_3)$. Now if $n = 5$ then $C_M(V_3) = C_M(V)$, while if $n = 6$ then $\overline{C_M(V_3)}$ is trivial or induces transvections with center V_1 on V. Thus we also have the hypotheses for F.9.5.6. $\qquad\square$

LEMMA 13.5.14. *If $n = 6$, assume $\tilde{U}_H = [\tilde{U}_H, L_1]$. Let $l \in L - L_1 T$, and if $n = 6$, choose $\bar{l}$ to fix a point $\omega \in \Omega$ fixed by $\bar{L}_1$. Set $K := \langle U_H, U_H^l \rangle$ and $L_- := O^2(O_2(LT)K)$. Then*

(1) If $\tilde{U}_H = [\tilde{U}_H, L_1]$ then $U_H = [U_H, L_1] \leq L_1 \leq L$.
(2) $\bar{U}_H = O_2(\bar{L}_1) \cong E_4$.
(3) If $n = 5$ then $\bar{K} = \bar{L}$ and $L = L_-$, while if $n = 6$, then $\bar{K} \cong A_5$ is the stabilizer in $\bar{L}$ of ω. Thus in any case $L_1 \leq K$.
(4) The hypotheses of G.2.4 are satisfied with V_1, V_3, V, L_-, U_H, K in the roles of "V_1, V, V_L, L, U, I", so $K = L_-$ and K is described in that lemma.

PROOF. Suppose first that $\tilde{U}_H = [\tilde{U}_H, L_1]$. Then as $V_1 \leq [V_3, L_1]$, $U_H = [U_H, L_1]$. Thus (1) holds. Moreover if $n = 6$, then $\tilde{U}_H = [\tilde{U}_H, L_1]$ by hypothesis, so $U_H \leq L$ by (1).

As $U_H = \langle V_3^H \rangle$ is nonabelian, $\bar{U}_H \neq 1$, and as $L_1 T \leq H \leq N_H(U_H)$, $\bar{U}_H \trianglelefteq \bar{L}_1 \bar{T}$. Thus (2) holds if $n = 5$. Similarly if $n = 6$, then $\bar{U}_H \leq \bar{L}$ by the first paragraph, so as $\bar{U}_H \trianglelefteq \bar{L}_1 \bar{T}$, (2) holds again. Part (3) is immediate from (2) and the choice of l. Then (3) implies the first statement in (4). Finally $L_1 \leq O^2(K) = L_-$ by (3) and $U_H \leq L_1$ by (1), so that $K = L_- U_H = L_-$ by G.2.4. $\qquad\square$

13.5.2.1. *Identifying the groups.* In the branch of the argument that will lead to the groups in Theorem 13.5.12, $L_1 \trianglelefteq G_1$ and G_1 is the unique member of $\mathcal{H}_z$. We begin by deriving some elementary consequences of the hypothesis that L_1 is normal in some member H of $\mathcal{H}_z$ with U_H nonabelian.

LEMMA 13.5.15. *Assume $L_1 \trianglelefteq H$, U_H is nonabelian, and $L/O_2(L)$ is not $\hat{A}_6$. Then*

(1) $U_H = [U_H, L_1] \leq L$ and $L_1^* \cong \mathbf{Z}_3$.
(2) $\bar{U}_H = O_2(\bar{L}_1) \cong E_4$.
Choose l and $K := \langle U_H, U_H^l \rangle$ as in 13.5.14. Then
(3) K is an A_5-block contained in L.
(4) If $n = 5$ then $L = K$, so that L is an A_5-block; if $n = 6$, then L is an A_6-block.
(5) $O_2(L_1) = U_H \cong Q_8^2$.

(6) $H = G_1$ and G_1 is the unique member of $\mathcal{H}_z$.

(7) $M = LT$ and $V = O_2(M)$.

(8) If $n = 5$ then $Q_H = U_H$, $H^* \leq \Omega_4^+(2)$, and either

    (i) $M = L$ with $H^* \cong S_3 \times \mathbf{Z}_3$, or

    (ii) $M/V \cong S_5$ with $H^* = \Omega_4^+(2) \cong S_3 \times S_3$.

(9) If $n = 6$ then $H^* = \Omega_4^+(2) \cong S_3 \times S_3$, and either

    (i) $M = L$ with $Q_H = U_H$, or

    (ii) $M/V \cong S_6$ with $Q_H = U_H C_T(L_1)$ and $|C_T(L_1)| = 4$.

PROOF. We saw $\tilde{U}_H \in \mathcal{R}_2(\tilde{H})$, so as $L_1 \trianglelefteq H$, $O_2(L_1^*) = 1$. Hence $L_1^* \cong \mathbf{Z}_3$. Also $\tilde{V}_3 = [\tilde{V}_3, L_1]$, so as $U_H = \langle V_3^H \rangle$, $\tilde{U}_H = [\tilde{U}_H, L_1]$. Then (1) follows from 13.5.14.1, and (2) from 13.5.14.2.

Choose l and K as in 13.5.14, and let $Q := O_2(LT)$ and $L_- := O^2(KQ)$. By 13.5.14.4, the hypotheses of G.2.4 are satisfied and $K = L_-$. By (1), the hypotheses of G.2.4.8 are satisfied. Therefore by G.2.4.8, K is an A_5-block with $V = O_2(K)$ and $U_H = O_2(L_1) \cong Q_8^2$. Now if $n = 5$, then $L = L_-$ by construction. If $n = 6$, then as Q acts on L_- and $L_- = K$, $[K, Q] \leq O_2(K) = V$, so L is an A_6-block. Thus (3), (4), and (5) are established.

We saw V^* is of order 2 and $O_2(H^*) = 1$, so by the Baer-Suzuki Theorem, there is $g \in H$ with I^* not a 2-group, where $I := \langle V, V^g \rangle$. Now by 13.5.13.2, we may apply F.9.5.6 to conclude that $O_2(I) = U_I := V_3 V_3^g \cong Q_8^2$ and $I/O_2(I) \cong I^* \cong S_3$. Therefore since $U_I \leq U_H$ and $U_H \cong Q_8^2$, we conclude $O_2(I) = U_H$, and hence $I^* = IQ_H/Q_H \cong I/U_H \cong S_3$.

Next as $C_H(\tilde{U}_H) = Q_H$, $H^* = H/Q_H \leq Out(U_H) \cong O_4^+(2)$. As $L_1^* \cong \mathbf{Z}_3$ is normal in H^*, and L_1^* centralizes V^*, L_1^* centralizes I^*. But the centralizer in $Out(U_H) \cong O_4^+(2)$ of L_1^* is isomorphic to S_3, so we conclude $I^* = C_{H^*}(L_1^*)$. Therefore either $H^* \cong S_3 \times S_3$, or $H^* = I^* \times L_1^* \cong S_3 \times \mathbf{Z}_3$ with $T = O_2(L_1 T)$, and the latter case can only occur when $n = 5$ and $\bar{M}_V = \bar{L} \cong A_5$. In either case $H = Q_H L_1 IT = L_1 IT$. Further if we establish (7), then $Q_H \cap Q = Q_H \cap V = V_3$, so $|Q_H| = 8|\bar{Q}_H|$, and then (8) and (9) follow using (5). So it remains to prove (6) and (7).

As Q_H acts on V and V^g, Q_H acts on I. Then as $I^* \trianglelefteq H^*$, H acts on $O^2(IQ_H) = O^2(I)$, and then $L_1 T$ acts on $O^2(I)V = I$. Thus $I \trianglelefteq L_1 IT = H$.

As K is an A_5-block by (3), $Q = V \times C_T(K)$ by C.1.13.c. Further $L_1 \leq K$ by 13.5.14.3, and $C_T(L_1) = V_1 C_T(K)$. Thus as $U_H = [U_H, L_1] \leq K$, $C_T(L_1) \leq C_T(U_H)$, so $[I, C_T(L_1)] \leq C_I(U_H) = V_1$, and therefore $[O^2(I), C_T(L_1)] = 1$ by Coprime Action. In particular $O^2(I)$ centralizes $C_T(L)$. As $O^2(I) = [O^2(I), V]$ and $V \leq O_2(M)$, $O^2(I) \nleq M$. Therefore as $M = !\mathcal{M}(LT)$, we conclude $C_T(L) = 1$.

We will show next that $D := C_T(K) = 1$. If $n = 5$, then $K = L$, so $D = C_T(L) = 1$; thus we may assume $n = 6$. Then as L is an A_6-block, we conclude from C.1.13.b and I.1.6.5 that either $D = C_T(L) = 1$ or $|D| = 2 = |Q : V|$. Assume that the latter case holds and set $G_D := C_G(D)$. We saw that $O^2(I)$ centralizes $C_T(L_1) \geq D$, so $\langle O^2(I), K \rangle \leq G_D$. Let $G_D^+ := G_D/D$ and $T_0 := C_T(D) \in Syl_2(C_M(D))$. Let $T_0 \leq T_D \in Syl_2(G_D)$; then $|T_D : T_0| \leq |T : T_0| = 2$, so as $K \in \mathcal{L}(G_D, T_0)$, there exists a unique $K_D \in \mathcal{C}(G_D)$ containing K by 1.2.5, and $O^2(I)$ normalizes K_D by 1.2.1.3. As $O^2(I) = [O^2(I), V]$ and K is irreducible on V, $V \cap O_2(K_D D) = 1$. Let $T_1 := T_0 \cap K_D D$; thus $T_1 \in Syl_2(M \cap K_D D)$. Then as $N_G(Q) \leq M = !\mathcal{M}(LT)$, $T_1 \in Syl_2(N_{K_D D}(Q))$. Therefore as $Q = O_2(KT_1) = $

$V \times D$, we conclude $Q \in \mathcal{B}_2(K_D D)$, and in particular Q contains $O_2(K_D D)$ by C.2.1.2. Then as $V \cap O_2(K_D D) = 1$, $D = O_2(K_D D)$. As $C_{C_M(D)}(Q) = Q$, $KT_0 = C_M(D)$. Hence $K^+ T_1^+ \cong KT_1/D$ is the 2-local $N_{K_D^+}(Q^+) = N_{K_D^+}(V^+)$ in the quasisimple group K_D^+. But inspecting the groups in Theorem C (A.2.3), we find no such 2-local. This contradiction establishes the claim that $C_T(K) = 1$.

As $C_T(K) = 1$, $V = Q = O_2(LT)$, so $O_2(M) = V$ and $M = LT$ by 3.2.11. Thus (7) holds, and hence also (8) and (9) by an earlier observation. Thus it remains to establish (6).

By A.1.6, $Q_1 := O_2(G_1) \leq Q_H$. Also $Q_1 \leq O_2(L_1 T)$, and by (8) and (9), either $O_2(L_1 T) = O_2(L_1) V = U_H V$, or $L/V \cong S_6$ and $O_2(L_1 T) = U_H V C_T(L_1)$, with $C_T(L_1)$ of order 4. Next $U_H V \cap Q_H = U_H$ and as $H \leq G_1$, $U_H \leq U_{G_1} \leq Q_1$. We conclude that $Q_1 = U_H$ or $U_H C_T(L_1)$. In either case, $Q_1 = U_H Z_1$ where $Z_1 := Z(Q_1)$ is of order at most 4, and $\Phi(Q_1) = V_1$. Thus G_1 preserves the usual symplectic form on $\hat{Q}_1 := Q_1/Z_1$. Now $m(\hat{Q}_1) = 4$ as $U_H \cong Q_8^2$, so $G_1/Q_1 \leq Sp(\hat{Q}_1) \cong S_6$. Then as $T \leq H$ and $H/Q_1 \cong S_3 \times S_3$ or $S_3 \times \mathbf{Z}_3$, it follows that $G_1 = H$. Thus $L_1 \trianglelefteq G_1$. Finally for any $H_1 \in \mathcal{H}_z$, as $L_1 \leq H_1 \leq G_1$ and $L_1 \trianglelefteq G_1$, $L_1 \trianglelefteq H_1$; so by symmetry between H and H_1, $H_1 = G_1$. This completes the proof of (6), and hence of the lemma. $\qquad \square$

We can now proceed to the identification of the groups in Theorem 13.5.12, under the assumption that L_1 is normal in H.

PROPOSITION 13.5.16. *If $L/O_2(L) \cong A_6$ and $L_1 \trianglelefteq H$, then $G \cong U_4(3)$.*

PROOF. By 13.5.15.6, $H := G_1$ is the unique member of $\mathcal{H}_z$. Let $U := U_H$ and $y \in L_2 - T$, so that $U \cong Q_8^2$ by 13.5.15.5. We consider the two cases of 13.5.15.9.

Suppose first that $M = L$. Then $O_2(G_1) = U$ by 13.5.15.9, and $U \cap U^y = V_2$. Hence G is of type $U_4(3)$ in the sense of section 45 (page 244) of [**Asc94**], so by 45.11 in [**Asc94**], $G \cong U_4(3)$.

Thus we may assume that $M/V \cong S_6$; in this case we will obtain a contradiction using transfer, eliminating shadows of extensions of $U_4(3)$. By 13.5.15.9, $Z(Q_H) = C_T(L_1)$ is of order 4.

Let $T_L := T \cap L \in Syl_2(L)$, and define I as in the proof of 13.5.15. From the proof of 13.5.15, $U = O_2(L_1) = O_2(I)$, $VU \in Syl_2(I)$, and $[L_1, I] \leq U$. Then $VU = V O_2(L_1) \leq T_L \in Syl_2(L)$, so $T_L \in Syl_2(T_L I L_1)$. Now L is transitive on $V^\#$, while $I L_1$ is transitive on the involutions in $U - V_1$, and all involutions in L are fused into U under L, so we conclude all involutions in T_L are in z^G.

Suppose that Q_H is not weakly closed in H with respect to G. Observe that $V_1 = \Phi(Q_H)$, so $N_G(Q_H) = H$. Then by A.1.13 there is $x \in G$ with $Q_H \neq Q_H^x$ and $[Q_H, Q_H^x] \leq Q_H \cap Q_H^x$. In particular $Q_H \leq N_G(Q_H^x) = C_G(z^x)$, so that $z^x \in C_H(Q_H) = Z(Q_H)$. As $Q_H \neq Q_H^x$, $x \notin H$, so $z^x \neq z$; thus $E_4 \cong \langle z, z^x \rangle = Z(Q_H)$, and then by symmetry between Q_H and Q_H^x, also $\langle z, z^x \rangle = Z(Q_H^x)$. Now 13.5.15 shows that H^* acts as $\Omega_4^+(2)$ on U/V_1, so that $Q_H/Z(Q_H) = J(T/Z(Q_H))$; hence $Q_H = Q_H^x$, contrary to the choice of Q_H^x.

This contradiction shows that Q_H is weakly closed in H. Hence H controls fusion in $Z(Q_H)$ by Burnside's Fusion Lemma A.1.35, so that z is weakly closed in $Z(Q_H)$ with respect to G. Now $Z(Q_H) = \langle z, j \rangle$ with $j \in T - T_L$. Therefore if j is an involution then $j \notin z^G$, so as all involutions in T_L are in z^G, Thompson Transfer gives $j \notin O^2(G)$, contrary to the simplicity of G. Hence $Z(Q_H) = \langle j \rangle \cong \mathbf{Z}_4$, so $Z(Q_H) \cap Z(Q_H)^y = 1$ as $z \neq z^y$.

Next as $y \in L_2 - H$, $z \neq z^y \in V_2 \leq Q_H \leq C_G(j)$ so that $j \in C_G(z^y) = H^y$, and similarly $j^y \in H$. Therefore $[j, j^y] \leq Z(Q_H) \cap Z(Q_H^y) = 1$. Thus $\langle j^{L_2} \rangle =: B$ is abelian with $\Phi(B) = V_2$; so as $\bar{B}$ is the E_8-subgroup generated by the transvections in $\bar{T}$, $B \cong \mathbf{Z}_4^2 \times \mathbf{Z}_2$. Let $A := \Omega_1(B)$. Then $\bar{A}$ is of order 2 and normal in $\bar{L}_2\bar{T}$, so $\bar{A} = \langle (1,2)(3,4)(5,6) \rangle$. Further for $a \in A - V_2$, $[a, V] = V_2$, so V is transitive on $A - V_2$, and hence $\overline{C_M(a)} = C_{\bar{M}}(\bar{a}) \cong \mathbf{Z}_2 \times S_4$ for each such a. Therefore $B = O_2(C_M(a))$ and $X := O_2(O^2(C_M(a))) \cong \mathbf{Z}_4^2$ with $V_2 = \Phi(X)$. As before by Thompson Transfer, there is $r \in G$ with $a^r = z$. Then $O^2(C_M(a))^r \leq O^2(H)$, so as U_H is Sylow in $O^2(H)$ by 13.5.15, $X^r \leq U_H$. Then $V_2^r = \Phi(X)^r \leq \Phi(U_H) = V_1$ of rank 1. This contradiction completes the proof. $\qquad\square$

PROPOSITION 13.5.17. *If $n = 5$ and $L_1 \trianglelefteq H$, then $G \cong U_4(2)$ or $L_4(3)$.*

PROOF. By 13.5.15.6, $H := G_1$ is the unique member of $\mathcal{H}_z$. By 13.5.15.7, $V = O_2(M)$. By 13.5.15.8, $Q_H = U_H =: U \cong Q_8^2$, and either $M = L$ with $H^* \cong S_3 \times \mathbf{Z}_3$, or $M/V \cong S_5$ and $H^* \cong S_3 \times S_3$. Let $T_L := T \cap L \in Syl_2(L)$, so that $|T : T_L| = 1$ or 2. Define I as in the proof of 13.5.15. Observe that Hypothesis F.1.1 is satisfied with I, L, T in the roles of "L_1, L_2, S": In particular, recall that during the proof of 13.5.15 we showed that $I \trianglelefteq H$ and $H = L_1IT$, so that $O_2(\langle I, L, T \rangle) = 1$ as $H \not\leq M$ and $M = !\mathcal{M}(LT)$.

Therefore $\gamma := (H, L_1T, M)$ is a weak BN-pair by F.1.9. As $T \cap I$ is self-normalizing in I, the hypotheses of F.1.12 are satisfied; so as $I/O_2(I) \cong S_3$, while $L/O_2(L) \cong A_5$ does not centralize Z, we conclude from F.1.12 that γ is of type $U_4(2)$ when $M = L$, and γ is of type $O_6^-(2)$ when $M/V \cong S_5$.

Next we verify the hypotheses of Theorem F.4.31: Let $G_0 := \langle M, H \rangle$. Then the inclusion $\gamma \to G_0$ is a faithful completion of γ. As $M \in \mathcal{M}$, $M = N_G(V)$. We saw $H = G_1 = C_G(z)$. Thus hypotheses (a) and (b) of F.4.31 hold. Hypotheses (c) and (d) are vacuously satisfied, and hypothesis (e) holds as G is simple.

We now appeal to Theorem F.4.31, and conclude as G is simple that either $M = L$ and $G \cong U_4(2)$ or $M/V \cong S_5$ and $G \cong L_4(3)$. $\qquad\square$

We mention that the shadows of extensions of $U_4(2)$ and $L_4(3)$ were essentially eliminated during the proof of Theorem F.4.31.

13.5.2.2. *Obtaining a contradiction in the remaining cases.* During the remainder of the section we assume that G is a counterexample to Theorem 13.5.12. Thus appealing to 13.5.16 and 13.5.17, it follows that:

LEMMA 13.5.18. *Assume $L/O_2(L)$ is not $\hat{A}_6$. Then L_1 is not normal in any $H \in \mathcal{H}_z$ such that U_H is nonabelian.*

LEMMA 13.5.19. *Let $Y := L_0$ if $L/O_2(L) \cong \hat{A}_6$, and $Y := L_1$ otherwise. Let $H \in \mathcal{H}_z$ and $H_1 := N_H(Y)$. Then*

(1) *$H_1^* = N_{H^*}(Y^*)$, and*

(2) *$V \leq O_2(H_1)$.*

PROOF. Recall that $Q_H = C_H(\tilde{U}_H)$; hence as $Y = O^2(YQ_H)$, (1) holds. Notice (2) holds when $L/O_2(L) \cong \hat{A}_6$, since $N_G(L_0) \leq M$ by 13.2.2.9. Therefore we may assume $L/O_2(L)$ is A_5 or A_6, and $V \not\leq O_2(H_1)$. Hence $H_1 \not\leq M$, so $H_1 \in \mathcal{H}_z$.

As $V \not\leq O_2(H_1)$, by the Baer-Suzuki Theorem there is $h \in H_1$ such that I^* is not a 2-group, where $I := \langle V, V^h \rangle$. By 13.5.13.2, we may apply F.9.5.6 to conclude that $\langle V_3^I \rangle$ is nonabelian. Thus U_{H_1} is nonabelian, contrary to 13.5.18. $\qquad\square$

LEMMA 13.5.20. V^* centralizes $F(H^*)$.

PROOF. If $[O_p(H^*), V^*] \neq 1$ for some prime p, then by 13.5.13.2, we may apply F.9.5.6 to conclude that $p = 3$. Let P^* be a supercritical subgroup of $O_3(H^*)$. Then $[P^*, V^*] \neq 1$, and $m(P^*) \leq 2$ since $H^* = H/Q_H$ is an SQTK-group. Define Y and H_1 as in 13.5.19. By definition Y normalizes V, so as V^* is of order 2, Y^* centralizes V^*. Suppose $O_2(Y^*)$ centralizes P^*. Then as $Y^*/O_2(Y^*)$ is of order 3, the Thompson $A \times B$ Lemma shows that $[C_{P^*}(Y^*), V^*] \neq 1$. This is a contradiction as $C_{P^*}(Y^*) \leq H_1^*$ by 13.5.19.1, and $V \leq O_2(H_1)$ by 13.5.19.2.

Therefore $O_2(Y^*)$ is nontrivial on P^*; then as $Y = O^{3'}(Y)$, P^* is not cyclic, so using A.1.25, $P^* \cong E_9$ or 3^{1+2} and $Y/C_Y(P^*/\Phi(P^*)) \cong SL_2(3)$. In particular Y is irreducible on $P^*/\Phi(P^*)$, so as $[Y^*, V^*] = 1$, $V^* = Z(Y^*)$ inverts $P^*/\Phi(P^*)$. However by F.9.5.2, $m([\tilde{U}_H, V^*]) = 2$; since a faithful irreducible for $SL_2(3)/E_9$ is of rank 8 and the commutator space of $Z(SL_2(3))$ on such a module is of rank 4, we conclude $P^* \cong 3^{1+2}$. But now X of order 3 in Y centralizes an E_9-subgroup of P, contradicting $m_3(H) \leq 2$. $\square$

By 13.5.20, $[K^*, V^*] \neq 1$ for some $K \in \mathcal{C}(H)$ with $K^* \cong K/O_2(K)$ quasisimple. *Let K have this meaning for the remainder of the section.*

LEMMA 13.5.21. (1) $K^* = [K^*, V^*]$ and $L_1 \leq K$.
(2) $K^*V^* \cong S_6$, A_8, or $G_2(2)'$.
(3) $\tilde{U}_H = [\tilde{U}_H, K]$, and $\tilde{U}_H/C_{\tilde{U}_H}(K^*)$ is the natural module for K^*.
(4) If $n = 6$, then $L/O_2(L) \cong A_6$ rather than $\hat{A}_6$.
(5) $K \trianglelefteq H$, $L_1^*T^*$ is the stabilizer in K^*T^* of the 2-subspace $\tilde{V}_3 = [\tilde{U}_H, V^*]$ of $\tilde{U}_H$, and $U_H = [Q_H, K]$.
(6) $\tilde{U}_H = [\tilde{U}_H, L_1]$.

PROOF. As V^* has order 2 and $V \trianglelefteq T$, $V^* \leq Z(T^*)$. Therefore V^* centralizes $(T \cap K)^* \in Syl_2(K^*)$, and hence normalizes K^* by 1.2.1.3, as does $L_1 = O^2(L_1)$ by that result. Then as $[K^*, V^*] \neq 1$ by choice of K, $K^* = [K^*, V^*]$, establishing the first part of (1).

Define $Y \leq L_1$ as in 13.5.19. As $K^* = [K^*, V^*]$, $K^* \not\leq H_1^*$ by 13.5.19.2, so Y^* does not centralize K^* by 13.5.19.1. Therefore $K^* = [K^*, Y^*]$ as K^* is quasisimple. Then since $Y \leq L_1$, $K^* = [K^*, L_1^*]$.

Let $T_X := N_T(K)$, $X := KL_1T_X$, and $\hat{X} := X/C_X(K^*)$. As $T_X \in Syl_2(N_H(K))$ by 1.2.1.3, $T_X \in Syl_2(X)$. As K^* is quasisimple, $F^*(\hat{X}) = \hat{K}$ is simple. We claim:

(i) $\hat{V}$ is generated by an involution in the center of the Sylow 2-subgroup $\hat{T}_X$ of $\hat{X}$.

(ii) $1 \neq \hat{Y} \leq \hat{L}_1 \leq O_{2,3}(C_{\hat{X}}(\hat{V}))$.

(iii) $[\tilde{U}_H, V] = \tilde{V}_3 = [\tilde{V}_3, L_1]$ is of rank 2.

(iv) If $\langle \hat{V}, \hat{V}^x \rangle$ is not a 2-group, then $\langle \hat{V}, \hat{V}^x \rangle \cong S_3$.

Part (i) follows as $V^* \leq Z(T^*)$ and V^* is of order 2 and faithful on K^*. Part (iii) follows from F.9.5.2, and (iv) is a consequence of F.9.5.6.2. As K^* is quasisimple with $O_2(K^*) = 1$, $C_{\hat{K}}(\hat{V}) = \widehat{C_{K^*}(V^*)}$. Further by F.9.5.3, $C_{K^*}(V^*) = N_K(V)^*$. Then as $L_1 \trianglelefteq H \cap M_V = N_H(V)$, $C_{\hat{K}}(\hat{V})$ acts on $\hat{L}_1$. Thus as L_1T acts on L_1 and $C_{\hat{X}}(\hat{V}) = C_{\hat{K}}(\hat{V})\hat{L}_1\hat{T}_X$, and as we saw earlier that $K^* = [K^*, Y^*]$, we conclude that (ii) holds.

By (iii), $m([\tilde{U}_H, V]) = 2$, so as $m(V^*) = 1$, $q(K^*V^*, \tilde{U}_H) \leq 2$. Therefore B.4.2 and B.4.5 describe K^* and the possible noncentral 2-chief factors W for KV on $\tilde{U}_H$. As $m(K^*V^*, \tilde{U}_H) = 2$, $\hat{K}$ is not one of the sporadic groups in B.4.5; cf. chapter H of Volume I, and recall that the 12-dimensional module for J_2 is the restriction of the natural module for $G_2(4)$. If $\hat{K} \cong A_7$, then by (iv), $\hat{V}$ induces a transposition on $\hat{K}$, contradicting (ii).

In the remaining cases in B.4.2 and B.4.5, $\hat{K}$ is of Lie type over $\mathbf{F}_{2^m}$ for some m. By (i), either $\hat{V}$ is generated by a long-root involution, or $\hat{K} \cong Sp_4(2^m)'$. Then by (iv), $m = 1$, and if $\hat{K}$ is A_6, then $\hat{V} \not\leq \hat{K}$. Furthermore if $K^* \cong \hat{A}_6$, then as $H \in \mathcal{H}^e$, for some choice of W, W is the faithful 6-dimensional module for K^*. But then as $\hat{V} \not\leq \hat{K}$, $m([\tilde{U}_H, V]) \geq 3$, contrary to (iii). Thus K^* is not $\hat{A}_6$, so in particular $Z(K^*) = 1$, and hence $K^* \cong \hat{K}$ is simple. Similarly by (iii), W is never the 10-dimensional module for $K^* \cong L_5(2)$.

The cases remaining appear in B.4.2. Further $\hat{K}\hat{V} \cong L_l(2)$, $3 \leq l \leq 5$, S_6, or $G_2(2)'$, (keeping in mind that $\hat{V} \not\leq \hat{K}$ iff $\hat{K} \cong A_6$) and W is either the natural module for K^*, or the 6-dimensional orthogonal module for $K^* \cong L_4(2)$. As $\hat{V}$ centralizes $\hat{Y} \neq 1$ by (ii), $\hat{K}$ is not $L_3(2)$. Therefore $m_3(K) = 2$, so by A.3.18, $L_1 \leq O^{3'}(H) = K$, completing the proof of (1). Further in each case $m_3(C_{\hat{K}}(\hat{V})) = 1$, so as $\hat{L}_1 \leq C_{\hat{K}}(\hat{V})$, we conclude that $m_3(L_1) = 1$, and hence (4) holds. Next $W = U_1/U_2$ for suitable submodules U_i of U_H, and by (iii),

$$[W, V] \leq V_W := (V_3 \cap U_1)U_2/U_2,$$

and L_1 is irreducible on $\tilde{V}_3$, so $V_W = [V_W, L_1]$ is of rank 2 and $V_W = [W, V]$. This eliminates the possibility that W is a natural module for $L_4(2)$ or $L_5(2)$, since there $\hat{V}$ is a long-root involution, so V induces a transvection on W. Hence W is the natural module for $K^*V^* \cong S_6$, A_8, or $G_2(2)'$, establishing (2). Furthermore by (iii), W is the unique noncentral chief factor for K on U_H. As $\tilde{V}_3 = [\tilde{V}_3, L_1] \leq [\tilde{U}_H, K]$, $\tilde{U}_H = \langle \tilde{V}_3^H \rangle = [\tilde{U}_H, K]$. This completes the proof of (3).

Finally we verify (5) and (6). As $L_1 \leq K$ and T acts on L_1, T acts on K, so $K \trianglelefteq H$ by 1.2.1.3. By (iii), $[\tilde{U}_H, V] = \tilde{V}_3$ is of rank 2 and is $L_1^*T^*$-invariant, and in each of the cases in (2), $P^* := N_{K^*T^*}([\tilde{U}_H, V])$ is a minimal parabolic of K^*T^*, so $|P^* : T^*| = 3$. Thus $P^* = L_1^*T^*$. Further $[V, Q_H] \leq V \cap Q_H = V_3$, so that $U_H = [Q_H, K]$. This completes the proof of (5). From the action of P^* on $\tilde{U}$, we determine that (6) holds in each case. Thus the proof of the lemma is complete. $\square$

By 13.5.21.6, the hypotheses of 13.5.14 are satisfied. Choose l as in 13.5.14, and set $L_- := \langle U_H, U_H^l \rangle$. By 13.5.14, $U_H \leq L_1$, and $L_- = O^2(L_-O_2(LT))$ is described in G.2.4. Further if $n = 5$ then $L_- = L$ by 13.5.14.3. As in G.2.4, let $S := O_2(L_-)$, $S_2 = V(U_H \cap U_H^l)$ and let s denote the number of chief factors for L_- on S/S_2, as in G.2.4.6. We maintain this notation throughout the remainder of the section.

LEMMA 13.5.22. *(1)* $|S| = 2^{4(s+1)}|S_2 : V|$.
(2) L_1 has $2s + 2$ noncentral 2-chief factors.
(3) $|U_H| = 2^{2s+5}|S_2 : V|$.
(4) $U_H \leq L_1$.

PROOF. By G.2.4, L_- has s natural chief factors on S/S_2 and one A_5-factor on V, so (1) and (2) hold. We already observed that (4) holds, and (3) follows from G.2.4.7. $\square$

LEMMA 13.5.23. K^* is not A_8.

PROOF. Assume $K^* \cong A_8$. Then by 13.5.21.3 and I.1.6.1, $\tilde{U}_H$ is either the 7-dimensional core of the permutation module for K^*, or its 6-dimensional irreducible quotient, which we regard as an orthogonal space for $K^* \cong \Omega_6^+(2)$. By 13.5.21.5, $[\tilde{U}_H, V^*] = \tilde{V}_3$ is of rank 2, while if $\tilde{U}_H$ were 7-dimensional, then $[\tilde{U}_H, V^*]$ would be of rank 3. Therefore $\tilde{U}_H$ is 6-dimensional orthogonal space. Moreover $C_K(\tilde{V}_2)^*$ is of 3-rank 2, so $n = 5$ by 13.5.4.3.

By 13.5.21.5, $U_H = [Q_H, K]$, and $L_1^* T^*$ is the minimal parabolic of $K^* T^*$ stabilizing the 2-space $\tilde{V}_2$. Thus L_1 has exactly four noncentral 2-chief factors, two on $\tilde{Q}_H$ and two on $O_2(L_1)^* = O_2(L_1^*) \cong Q_8^2$. Therefore by 13.5.22.2, the parameter s of 13.5.22 is equal to 1. Thus by 13.5.22.3,

$$|U_H| = 2^{2s+5}|S_2 : V| = 2^7 \cdot |S_2 : V|. \qquad (*)$$

Next as $\tilde{U}_H$ is the orthogonal module, $U_H \cong D_8^3$ is of order 2^7. Thus $S_2 = V$ by (*), and $|S| = 2^8$ by 13.5.22.1, so

$$|O_2(L_1)| \le |O_2(\bar{L}_1)| \cdot |S| = 2^2 \cdot 2^8 = 2^{10}. \qquad (**)$$

Further $|O_2(L_1)^*| = 2^5$ using 13.5.21.5, and $U_H \le O_2(L_1)$ by 13.5.22.4, so that $|O_2(L_1)| \ge 2^{12}$ by (*), contrary to (**). This contradiction completes the proof of 13.5.23. $\qquad \square$

LEMMA 13.5.24. K^* is not A_6.

PROOF. Assume $K^* \cong A_6$. Then by 13.5.21.3 and I.1.6.1, $\tilde{U}_H$ is either the 5-dimensional core of the permutation module for K^*, or its 4-dimensional irreducible quotient. In either case by 13.5.21.5, $U_H = [Q_H, K]$ and $L_1^* T^*$ is the maximal parabolic stabilizing the line $\tilde{V}_3$. Thus L_1 has two noncentral chief factors on $\tilde{U}_H$, and one on $O_2(L_1^*)$, so L_1 has exactly three noncentral 2-chief factors. This is a contradiction, since by 13.5.22.2, the number of noncentral 2-chief factors of L_1 is even. This completes the proof of 13.5.24. $\qquad \square$

LEMMA 13.5.25. K^* is not $G_2(2)'$.

PROOF. Assume $K^* \cong G_2(2)'$. Then by 13.5.21.3 and I.1.6.5, $\tilde{U}_H$ is either the 7-dimensional Weyl module for K^* or its 6-dimensional irreducible quotient. However $U_H = Z(U_H)U_0$ where U_0 is extraspecial, and if $\Phi(Z(U_H)) = 1$, then H preserves a quadratic from on $U_H/Z(U)$. Therefore as $G_2(2)'$ does not preserve a quadratic form on its 6-dimensional module, we conclude $m(\tilde{U}) = 7$ and $Z(U_H) = \langle j \rangle \cong \mathbf{Z}_4$.

Let $X \in Syl_3(L_1)$; then $\langle j \rangle = C_{U_H}(X)$. But by 13.5.22.4, $U_H \le O_2(L_1)$, and by G.2.4.6, S/S_2 is the sum of natural modules for L_-/S. So $j \in C_{O_2(L_1)}(X) \le S_2$. However $S_2 = V(U_H \cap U_H^l)$ with $\Phi(U_H \cap U_H^l) \le \Phi(U_H) \cap \Phi(U_H^l) = V_1 \cap V_1^l = 1$, so

$$V_1 = \Phi(\langle j \rangle) \le \Phi(S_2) = \Phi(V)\Phi(U_H \cap U_H^l) = 1,$$

a contradiction. $\qquad \square$

By 13.5.21.2, K^* is A_6, A_8, or $G_2(2)'$. But this contradicts 13.5.24, 13.5.23, and 13.5.25. This contradiction completes the proof of Theorem 13.5.12.

13.6. Finishing the treatment of $\mathbf{A}_5$

In this section, we complete the treatment of the case in the Fundamental Setup where $L/O_2(L) \cong A_5$, using assumption (4) in Hypothesis 13.3.1 as discussed earlier. To do so, we treat the only case remaining after the reduction in the previous section 13.5. We adopt the notational conventions of section B.3 and Notations 12.2.5 and 13.2.1.

We will prove a result summarizing the work of both this section and the previous section:

THEOREM 13.6.1. *Assume Hypothesis 13.3.1 with $L/O_2(L) \cong A_5$. Then G is isomorphic to $U_4(2)$ or $L_4(3)$.*

The groups in Theorem 13.6.1 have already appeared as conclusions in Theorem 13.5.12.1, under the hypothesis that $\langle V_3^{G_1} \rangle$ is nonabelian; we will prove that there are no examples in the remaining case. (Indeed, as far as we can tell, there are not even any shadows.) We assume throughout this section that G is a counterexample to Theorem 13.6.1, and work toward a contradiction. The contribution from the previous section 13.5 is:

LEMMA 13.6.2. $\langle V^{G_1} \rangle$ *is abelian, so V_H is abelian for each $H \in \mathcal{H}_z$.*

PROOF. By Theorem 13.5.12.1, $\langle V_3^{G_1} \rangle$ is abelian; hence $\langle V^{G_1} \rangle$ is abelian by 13.5.10.   □

LEMMA 13.6.3. *(1) $C_G(z) \nleq M$.*
(2) $C_Z(L) = 1$.

PROOF. This follows from 13.3.5.2 and the fact that $M = !\mathcal{M}(LT)$.   □

Set $Q := O_2(LT)$, $S := \mathrm{Baum}(T)$, let $v \in V_2 - V_1$, and let $G_v := C_G(v)$ and $M_v := C_M(v)$. In the notation of section B.3, the generator z of V_1 is $e_{1,2,3,4}$, and we may take $v = e_{1,2}$. Set $Q_v := O_2(G_v)$, $\check{G}_v := G_v/\langle v \rangle$, $L_v := O^2(C_L(v))$ and $V_v := \langle z^{L_v} \rangle$. Then $L_v/O_2(L_v)$ is of order 3, $V_v = \langle v \rangle \times [V, L_v]$, and $\check{V}_v = [\check{V}, L_v]$ is a natural module for $L_v/O_2(L_v)$, of rank 2. By 13.2.6.1,

$$T_v := C_T(v) \in Syl_2(G_v).$$

By 13.2.4.2, $S \le T_v$, so it follows from B.2.3 that:

LEMMA 13.6.4. $S = \mathrm{Baum}(T_v)$ *and* $J(T) = J(T_v)$.

Observe also that there is $a \in z^G \cap C_V(L_v)$ (e.g., $a = e_{1,3,4,5}$) and $\check{a} \in Z(\check{T}_v)$.

LEMMA 13.6.5. $N_{G_v}(S) \le M_v = C_{M_V}(v)$.

PROOF. First $N_G(S) \le M$ by 13.2.5, and then $M_v \le M_V$ by 12.2.6.   □

LEMMA 13.6.6. $F^*(G_v) = Q_v$.

PROOF. Assume that $Q_v < F^*(G_v)$. Then by 1.1.4.3, $z \notin Q_v$. By 1.1.6 we can appeal to lemma 1.1.5 with G_v, T_v, G_1, z in the roles of "H, S, M, z". In particular $F^*(C_{G_v}(z)) = O_2(C_{G_v}(z))$ and z inverts $O(G_v)$. On the other hand, $z \in V_v = \langle v \rangle \times [V, L_v]$, and $[V, L_v]$ centralizes $O(G_v)$ by A.1.26.1, so z centralizes $O(G_v)$, and hence $O(G_v) = 1$. Thus there is a component K of G_v. By 1.1.5.3, $K = [K, z]$ and K is described in 1.1.5.3. Also $L_v = O^2(L_v)$ normalizes K by 1.2.1.3, so as $z \in \langle v \rangle L_v$, also $K = [K, L_v]$.

Notice cases (c) and (d) of 1.1.5.3 cannot arise: for in those cases z induces an outer automorphism on K, whereas L_v induces inner automorphisms on K by A.3.18, so that $z \in \langle v \rangle L_v$ does too. In the remaining cases of 1.1.5.3, z induces an inner automorphism of K.

Recall there is $a \in z^G \cap C_V(L_v)$ with $\check{a} \in Z(\check{T}_v)$. As $a \in z^G$, $F^*(C_{G_v}(a)) = O_2(C_{G_v}(a))$ by 1.1.4.3 and 1.1.3.2. Therefore $[K, a] \neq 1$, and then as $\check{a} \in Z(\check{T}_v)$, a normalizes K and $K = [K, a]$. Set $X := N_{G_v}(K)$, $T_X := T_v \cap X$, and $X^* := X/C_X(K)$. Since L_v^* centralizes a^* but not z^*, $a^* \neq z^*$. Then

$$E_4 \cong E^* := \langle a^*, z^* \rangle \leq Z(T_X^*),$$

so neither case (e) or (f) of 1.1.5.3 holds. Thus we have reduced to cases (a) or (b) of 1.1.5.3, with K^* of Lie type over $\mathbf{F}_{2^m}$ for some m. Again as $E^* \leq Z(T_X^*)$, either $m > 1$ and $E^* \leq K^*$, or $X^* \cong S_6$. Suppose the latter case holds. Then $L_v \leq O^{3'}(G_v) = K$ by A.3.18, so $L_v \cong A_4$, and hence L is an A_5-block. By 13.6.3.2, $C_T(L) = 1$. Therefore $V = O_2(LT)$ by C.1.13.c, so $V = O_2(M) = C_G(V)$ by 3.2.11. Thus as $\langle V^{G_1} \rangle$ is abelian by 13.6.2, $G_1 \leq N_G(V) \leq M$, contrary to 13.6.3.1. This contradiction shows that $m > 1$ with $E^* \leq K^*$.

Next we conclude from A.3.18 that one of the following holds:

(i) $L_v \leq O^{3'}(G_v) = K$.

(ii) $m_3(K) \leq 1$.

(iii) $K^* L_v^* \cong PGL_3^\epsilon(2^m)$ or $L_3^{\epsilon,\circ}(2^m)$ and $K = O^{3'}(E(G_v))$.

We claim that T_v normalizes K; assume otherwise and let $K_0 := \langle K^{T_v} \rangle$ and $T_0 := T_v \cap K_0$. In (iii), $K \trianglelefteq G_v$; and in (i), T_v acts on K since T_v acts on L_v. Therefore we may assume that (ii) holds. Comparing the groups in (a) or (b) of 1.1.5.3 to those in 1.2.1.3, we conclude that $K^* \cong L_2(2^m)$ or $Sz(2^m)$. If $K^* \cong L_2(2^m)$ then by 1.2.2.a, $L_v \leq O^{3'}(G_v) = K_0$, so $L_v^* \leq K_0^* = K^*$ and L_v^* centralizes a^*, impossible as involutions of K^* are centralized only by a Sylow 2-subgroup. Therefore $K^* \cong Sz(2^m)$, so $K_0/Z(K_0) \cong Sz(2^m) \times Sz(2^m)$. Let B be the Borel subgroup of K_0 containing T_0, and set $V_0 := \Omega_1(T_0)$. Then (using I.2.2.4 when $Z(K_0) \neq 1$ and in particular $m = 3$) $J(T_v)$ centralizes V_0, so by B.2.3.5, $\mathrm{Baum}(T_v) \trianglelefteq B$. Then $B \leq C_{M_V}(v)$ by 13.6.4 and 13.6.5. Now $O^2(B) \leq O^{2,3}(C_{M_V}(v)) \leq C_M(V)$. Hence $[L_v, O^2(B)] \leq O_2(L_v)$, impossible as as subgroup of order 3 in L_v induces field automorphisms on K. This contradiction completes the proof of the claim that T_v normalizes K.

Hence $T_v = T_X$. Also $L_v^* \leq K^*$ in cases (i) and (ii); this is clear in (i), and it holds in (ii) as $L_v = [L_v, T_v]$ while $Out(K^*)$ is abelian when $m_3(K^*) \leq 1$.

Next as T_v acts on L_v, by inspection of the groups in (a) or (b) of 1.1.5.3, L_v^* is contained in a Borel subgroup B^* of $K^* L_v^*$. Hence as

$$[L_v^*, z^*] \neq 1 = [L_v^*, a^*]$$

and $E^* \leq Z(T_X^*)$, we conclude $K^* \cong Sp_4(2^m)$, as otherwise $Z(T_X^* \cap K^*) =: R^*$ is a root subgroup of K^*, so $C_{B^*}(R^*) = C_{B^*}(z^*)$.

As $m > 1$, K is simple by I.1.3, so $J(T_v) = J_K \times J(T_C)$, where $T_K := T_v \cap K$ and $T_C := C_{T_v}(K)$. Thus $Z_J := \Omega_1(Z(J(T_v))) = Z_K \times Z_C$, where $Z_K := Z(T_K)$ and $Z_C := \Omega_1(Z(J(T_C)))$. Then using 13.6.4,

$$S = \mathrm{Baum}(T) = \mathrm{Baum}(T_v) = T_K \times S_C,$$

where $S_C := \mathrm{Baum}(T_C)$. Therefore $B \leq N_G(S)$, so as before $B \leq C_{M_V}(v)$ by 13.6.5. Let $B_C := O^2(C_B(V))$. Since $|B : O_2(B)| > 3$ and $\bar{B} = \bar{\bar{L}}_v$ with $|\bar{L}_v :$

$O_2(\bar{L}_v)| = 3$, we conclude $|B : B_C O_2(B)| = 3$ and $B_C \neq 1$. Then $L_v/O_2(L_v)$ is inverted in $T_v \cap L \leq C_{T_v}(B_C/O_2(B_C))$. This is impossible since each element in $Aut(Sp_4(2^m))$ acting on B and inverting an element of order 3 in B^* induces a field automorphism on K^* inverting $\Omega_1(O_3(B^*/O_2(B^*))) \cong E_9$. This contradiction completes the proof of 13.6.6. $\qquad\square$

We come to the main technical result of the section, requiring the bulk of the argument; afterwards the remainder of the proof of Theorem 13.6.1 is fairly brief.

THEOREM 13.6.7. *(1)* $G_v \leq M_V$. *Hence* $G_v = C_{M_V}(v)$.
(2) $N_G(V_v) \leq M$.

Until the proof of Theorem 13.6.7 is complete, assume G is a counterexample. We begin a series of reductions.

Recall $V_v = \langle v \rangle \times [V, L_v]$ with $\{v\} \cup [V, L_v]^\#$ the set of nonsingular vectors in V_v. Therefore by 13.2.6.2, $N_G(V_v)$ acts on the three singular vectors of V_v, and hence preserves their product v—so that $N_G(V_v) \leq G_v$, and hence (1) implies (2). On the other hand, if $V_v \trianglelefteq G_v$, then G_v permutes

$$\mathcal{V} := \{V_u : u \in V_v \text{ and } u \text{ is nonsingular}\},$$

so G_v acts on $V = \langle \mathcal{V} \rangle$, and hence $G_v \leq N_G(V) = M_V$, so (1) holds. Thus we may assume:

LEMMA 13.6.8. $G_v > M_v$ *and* V_v *is not normal in* G_v.

Set $U_v := \langle z^{G_v} \rangle$ and $G_v^* := G_v/C_{G_v}(U_v)$. Regard G_v^* as a subgroup of $GL(U_v)$ and write u^{x^*} for the image of $u \in U_v$ under $x^* \in G_v^*$.

As $L_v \leq G_v$, $V_v \leq U_v$. As $z \in Z(T_v)$, by 13.6.6 we may apply B.2.14 to conclude $U_v \in \mathcal{R}_2(G_v)$. In particular $O_2(G_v^*) = 1$ and $U_v \leq Z(Q_v)$.

If $[U_v, J(T_v)] = 1$, then $S = \mathrm{Baum}(C_{T_v}(U_v))$ by B.2.3.5 and 13.6.4, so by a Frattini Argument and 13.6.5,

$$G_v = C_{G_v}(U_v)N_{G_v}(S) \leq C_G(U_v)C_{M_V}(v).$$

But then $V_v^{G_v} = V_v^{M_v} = \{V_v\}$, contrary to 13.6.8. Hence

LEMMA 13.6.9. $J(G_v)^* \neq 1$.

Thus U_v is an FF-module for G_v^*.

LEMMA 13.6.10. *If* $L_v^* = [L_v^*, J(T_v)^*]$, *then* L_v^* *is not subnormal in* G_v^*.

PROOF. Suppose otherwise. Then $O_2(L_v^*) \leq O_2(G_v^*) = 1$, so that L_v^* has order 3. Further $m([U_v, L_v^*]) = 2$ by Theorem B.5.6, so $[U_v, L_v] = [V, L_v]$, and hence $V_v = \langle v \rangle \times [V, L_v] = \langle v \rangle [U_v, L_v]$. Now Theorem B.5.6 also shows that $|L_v^{*G_v^*}| \leq 2$, so as L_v is T_v-invariant, L_v^* is normal in G_v^*, so that $\langle v \rangle [U_v, L_v] = V_v$ is G_v-invariant, contrary to 13.6.8. $\qquad\square$

Let $\mathcal{X}_0$ be the set of $L_v^* T_v^*$-invariant subgroups $X^* = O^2(X^*)$ of G_v^* such that $1 \neq X^* = [X^*, J(T_v)^*]$. Let $\mathcal{X}$ denote the set of all members of $\mathcal{X}_0$ normal in G_v^*, and $\mathcal{X}_z$ the set of those X^* in $\mathcal{X}_0$ with $[z, X^*] \neq 1$. For $X^* \in \mathcal{X}_0$, set $U_X := [\langle z^{X^* L_v^*} \rangle, X^*]$.

LEMMA 13.6.11. *For each* $X^* \in \mathcal{X}_z$, $[z, X^*, L_v^*] \neq 1$, *so* $[U_X, L_v^*] \neq 1$.

PROOF. Let $U := [z, X^*] \neq 1$ and suppose that $[U, L_v^*] = 1$. Then $[X^*, L_v^*] \leq C_{X^*}(U) \leq C_{X^*}(z)$ by Coprime Action, so

$$1 = [X^*, L_v^*, z] = [z, X^*, L_v^*],$$

and hence by the Three-Subgroup Lemma, $[L_v^*, z, X^*] = 1$. Then X^* centralizes $\langle v \rangle [z, L_v^*]$ which contains z, contrary to our choice of X^*. $\square$

LEMMA 13.6.12. $\mathcal{X} \subseteq \mathcal{X}_z$.

PROOF. If $X^* \in \mathcal{X}$ with $[z, X^*] = 1$, then as $X^* \trianglelefteq G_v^*$, X^* centralizes $\langle z^{G_v^*} \rangle = U_v$, contrary to $X^* \neq 1$. $\square$

LEMMA 13.6.13. No member of $\mathcal{X}$ is solvable.

PROOF. Suppose $X^* \in \mathcal{X}$ is solvable, and choose X^* minimal subject to this constraint. By Theorem B.5.6, $X^* = O^2(H^*)$ for some normal subgroup H^* of G_v^* with $H^* = J(H)^* = H_1^* \times \cdots \times H_s^*$ where $H_i^* \cong S_3$ and $s \leq 2$. Further $U_H := [U_v, H] = U_1 \oplus \cdots \oplus U_s$, where $U_i := [U_H, H_i]$ is of rank 2. By minimality of X^*, T_v^* is transitive on the H_i^*, so $X^* T_v^*$ is irreducible on U_H. Now $[z, X] \neq 1$ by 13.6.12, so $U_X = U_H$ as $X^* T_v^*$ is irreducible on U_H. By 13.6.11 $[U_H, L_v^*] \neq 1$, so the projection of L_v^* on H^* with respect to the decomposition $H^* \times C_{G_v^*}(U_X)$ is nontrivial. Then as L_v is T_v-invariant, it follows that $L_v^* = [L_v^*, J(T)^*] = O^2(H_i^*)$ for some i, contrary to 13.6.10. $\square$

We conclude from 13.6.13 that $F(J(G_v)^*) = Z(J(G_v)^*)$. Then by 13.6.9 and Theorem B.5.6, $J(G_v)^*$ is a product of components of G_v^*. By C.1.16, $J(T_v)^*$ normalizes the components of G_v^*. Thus there exists $K_+ \in \mathcal{C}(G_v)$ such that K_+^* is a component of G_v^* and $K_+^* = [K_+^*, J(T_v)]$. Thus $\langle K_+^{*T_v} \rangle$ is normal in G_v^* by 1.2.1.3, and so lies in $\mathcal{X}$. Hence $\langle K_+^{*T_v} \rangle \in \mathcal{X}_z$ by 13.6.12.

Let $\mathcal{Y}_z$ consist of those $K \in \mathcal{L}(G_v, T_v)$ such that $K^*/O_2(K^*)$ is quasisimple and $\langle K_-^{*T_v^*} \rangle \in \mathcal{X}_z$. By the previous paragraph, $K_+ \in \mathcal{Y}_z$, so $\mathcal{Y}_z$ is nonempty. Observe that if $K \in \mathcal{Y}_z$ and $K_0 \in \mathcal{L}(G_v, T_v)$ with $K_0^* = K^*$, then $K_0 \in \mathcal{Y}_z$.

For $K \in \mathcal{Y}_z$, let $K_- := \langle K^{T_v} \rangle$, $W_K := \langle z^{K_- L_v} \rangle$, and set $(K_- L_v T_v)^+ := K_- L_v T_v / C_{K_- L_v T_v}(W_K)$. Since $F^*(G_v) = O_2(G_v)$, $W_K \in \mathcal{R}_2(K_- L_v T_v)$ by B.2.14.

In the remainder of the proof of Theorem 13.6.7, let $K \in \mathcal{Y}_z$.

Then K^+ is a quotient of $K^*/O_2(K^*)$, so K^+ is also quasisimple, and W_K is an FF-module for $K_-^+ T_v^+$. Also the action of $K_- L_v^+ T_v^+$ on W_K is described in Theorem B.5.6.

LEMMA 13.6.14. K is T_v-invariant, $K^* \in \mathcal{X}_z$, and $U_K = [W_K, K]$.

PROOF. Assume 13.6.14 fails. By 1.2.1.3, $K_- = KK^t$ for some $t \in T_v - N_{T_v}(K)$, and comparing the list of groups in 1.2.1.3 to that in Theorem B.5.6, $K^+ \cong L_2(2^m)$ or $L_3(2)$. Then by 1.2.2, $L_v \leq K_-$. Since L_v is T_v-invariant with $L_v/O_2(L_v)$ of order 3, L_v^+ is diagonally embedded in K_-^+, K^+ is not $L_3(2)$, and $K_-^+ T_v^+$ is not $S_5 \text{ wr } \mathbf{Z}_2$. Therefore by Theorem B.5.6, $U_{K_-} = U_K U_K^t$, where $U_K := [W_K, K]$ and $U_K/C_{U_K}(K)$ is the natural module for $K^+ \cong L_2(2^m)$. Thus by E.2.3.2, $\text{Baum}(T_v)$ is normal in the preimage B of the Borel subgroup B^+ of K^+ normalizing $(T_v \cap K_-)^+$. But $S = \text{Baum}(T_v)$ by 13.6.4 and $N_B(S) \leq M_V$ by 13.6.5, so $B \leq M_V$. As $z \in Z(T_v)$, the projection of z on U_{K_-} is diagonally embedded in $U_K U_K^t$, so that $C_B(V) \leq C_B(\langle z^B \rangle) = O_2(B)$. This is a contradiction as $B/O_2(B)$ is noncyclic of odd order, while $O^2(\bar{M}_V) \cong A_5$ had cyclic Sylow groups for odd primes. This contradiction completes the proof. $\qquad\square$

LEMMA 13.6.15. *If $K^+ \cong A_5$, then no $I \in Irr_+(K, \langle z^K \rangle)$ is the A_5-module.*

PROOF. Assume $K^+/O_2(K^+) \cong A_5$ and some $I \in Irr_+(K, \langle z^K \rangle)$ is the A_5-module. Then as $K^+ = [K^+, J(T_v)^+]$, $U_K = I$ by Theorem B.5.6.

Further as there are no strong FF*-offenders on I by B.4.2.5, by that result and B.2.9.1, there is $A \in \mathcal{A}(T)$ with A^+ of order 2 inducing a transposition on K^+ and $K^+ = [K^+ L_v^+, A^+] = K^+$. By 13.6.11, $[U_K, L_v] \neq 1$. Then as T_v acts on L_v, the projection L_K^+ of L_v^+ on K^+ is a Borel subgroup of K^+ with $L_K^+ = [L_K^+, A]$. Therefore $L_v^+ = [L_v^+, A^+] = L_K^+$ as L_v is T_v-invariant and $K^+ = [K^+ L_v^+, A^+]$. As $L_v^+ = L_K^+$ and U_K is the A_5-module, the L_v-module $[W_K, L_v] = [U_K, L_K]$ is an indecomposable extension of the trivial module $C_{U_K}(L_v^+)$ by a natural module for $L_v^+/O_2(L_v^+) \cong \mathbf{Z}_3$. This is a contradiction, as $[V, L_v]$ is an L_v-submodule of $[W_K, L_v]$ of rank 2. $\qquad\square$

LEMMA 13.6.16. *Assume $K_0 \in \mathcal{L}(G_v, T_v)$ is T_v-invariant with $[z, K_0] \neq 1$. Then*

(1) $O_2(\langle K_0, T \rangle) = 1$.

(2) *If C is a nontrivial characteristic subgroup of T_v, then $O_\infty(K_0) N_{K_0}(C) < K_0$.*

(3) $K_0 = [K_0, J(T_v)]$.

PROOF. Let $H := \langle K_0, T \rangle$ and assume $O_2(H) \neq 1$; then $H \in \mathcal{H}(T)$. As $|T : T_v| = 2$, by 1.2.5 there is $K_2 \in \mathcal{C}(N_G(O_2(H)))$ containing K_0. As $[z, K_0] \neq 1$, $K_2 \in \mathcal{L}_f(G, T)$. By 13.3.2.2, $K_2 \in \mathcal{L}_f^*(G, T)$. We now make a particularly fundamental use of the special assumption in part (4) of Hypothesis 13.3.1 that we have chosen L with $L/O_2(L) \cong A_5$ only in the final case of the FSU when no other choice was possible: namely by Hypothesis 13.3.1.4, $K_2/O_2(K_2) \cong A_5$. Thus as A_5 is a minimal nonsolvable group, $K_2 = O_2(K_2) K_0$, and then as $|T : T_v| = 2$ and K_2 is perfect, $K_2 = K_0$. As v centralizes K_0, $1 \neq C_T(K_2)$, contrary to 13.6.3.2, since by 13.3.2 K_2 satisfies Hypothesis 13.3.1 in the role of "L". This completes the proof of (1).

Next assume thht C is a nontrivial characteristic subgroup of T_v with $K_0 = O_\infty(K_0) N_{K_0}(C)$. Then there is $K_1 \in \mathcal{L}(N_{K_0}(C), T_v)$ with $K_0 = O_\infty(K_0) K_1$. Replacing K_0 by K_1, we may assume K_0 acts on C. Since T_v is of index 2 in T, C is normal in T, and hence $1 \neq C \leq O_2(\langle K_0, T \rangle)$, contrary to (1). This establishes (2).

Finally if $J(T_v) \leq O_\infty(K_0)$, then $K_0 = O_\infty(K_0) N_{K_0}(J(T_v))$ by a Frattini Argument, contrary to (2). Thus (3) holds. $\qquad\square$

LEMMA 13.6.17. *Assume $a := z^g$ with $\tilde{a} \in Z(\tilde{T}_v)$. Then $[a, K] \neq 1$.*

PROOF. Assume $K \leq G_a := C_G(a)$. As a centralizes a subgroup of T_v of index 2, $|O_2(G_a) : (O_2(G_a) \cap N_G(K))| \leq 4$. Thus as $K = K^\infty$, K centralizes $O_2(G_a)/(O_2(G_a) \cap N_G(K))$, and hence $K \trianglelefteq K O_2(G_a)$. Since $[z, K] \neq 1$ with $z \in U_v \in \mathcal{R}_2(G_v)$, $V(K) \neq 1$ in the language of Definition A.4.7. Then $[Z(O_2(K O_2(G_a))), K] \neq 1$ by A.4.9 with K, $K O_2(G_a)$ in the roles of "X, M". Then as $G_a \in \mathcal{H}^e$, K does not centralize $Z_a := Z(O_2(G_a))$. By 1.2.1.1, $\langle K^{T^g} \rangle =: K_0 \leq \langle \mathcal{C}(G_a) \rangle$, so as $[Z_a, K] \neq 1$, some $K_1 \in \mathcal{C}(K_0)$ is in $\mathcal{L}_f(G, T^g)$ by A.4.9 with

K_1, G_a in the roles of "X, M". Thus as $a \in Z(T^g)$ centralizes K_1, 13.6.3.2 applied to K_1 in the role of "L" supplies a contradiction. □

We now begin to eliminate various cases for K^+ from the list of possible quasisimple groups in Theorem B.4.2.

LEMMA 13.6.18. K^+ is not $L_2(2^m)$.

PROOF. Assume otherwise. Then by Theorem B.5.6 and 13.6.15, $U_K/C_{U_K}(K)$ is the natural module for $K^+ \cong L_2(2^m)$. Let K_0 be a minimal member of $\mathcal{L}(KT_v, T_v)$. Then $K_0^* = K^*$, so $K_0 \in \mathcal{Y}_z$, and by minimality of K_0, K_0 is a minimal parabolic in the sense of Definition B.6.1. Then by 13.6.16.2 and C.1.26, K_0 is an $L_2(2^m)$-block. Hence replacing K by K_0, we may assume K is a block. Then by E.2.3.2, $J(T_v)$ is normal in the Borel subgroup B of KT_v over T_v, and $S = \mathrm{Baum}(T_v)$ by 13.6.4, so B acts on S in view of B.2.3.4. Hence $B \le C_{M_V}(v)$ by 13.6.5. Thus $O^2(B) \le O^2(C_{M_V}(v)) \le L_v C_M(V)$, so that $[V, O^{2,3}(B)] = 1$. But if $n > 2$, then $z \in C_{W_K}(O^{2,3}(B)) \le C_{W_K}(K)$ as $U_K/C_{U_K}(K)$ is the natural module for $L_2(2^m)$, contradicting $[z, K] \ne 1$. Thus $n = 2$ and $BC_M(V) = L_v C_M(V)$. Then B centralizes the element $a := z^g \in C_V(L_v)$ with $\breve{a} \in Z(\breve{T}_v)$ described before 13.6.5. Therefore as B contains a Borel subgroup of K, and K is an $L_2(4)$-block, $K \le C_G(a)$. This contradicts 13.6.17, completing the proof of the lemma. □

LEMMA 13.6.19. K^+ is not $SL_3(2^m)$, $Sp_4(2^m)$, or $G_2(2^m)$ with $m > 1$.

PROOF. If the lemma fails, then for some maximal parabolic P of K containing $T_v \cap K$, $K_1 := P^\infty$ does not centralize z. Then $K_1 \in \mathcal{L}(G_v, T_v)$ with $K_1^+/O_2(K_1^+) \cong L_2(2^m)$. As K_1 is not a block, this contradicts 13.6.16.2 in view of C.1.26. □

By Theorem B.5.6, K^+ is either a Chevalley group over a field of characteristic 2 in Theorem C (A.2.3), or $\hat{A}_6$ or A_7. Lemmas 13.6.18 and 13.6.19 say in the former case that K^+ is a group over $\mathbf{F}_2$. Therefore the list of B.5.6 is reduced to $K^+ \cong L_3(2)$, $Sp_4(2)'$, $G_2(2)'$, $\hat{A}_6$, A_7, $L_4(2)$, or $L_5(2)$. We next show:

LEMMA 13.6.20. $L_v \le K$.

PROOF. If $m_3(K) = 2$, then by A.3.18, $L_v \le \theta(KL_v) = K$. Thus we may assume $m_3(K) = 1$, so $K^+ \cong L_3(2)$. By Theorems B.5.1 and B.5.6, either $U_K/C_{U_K}(K)$ is a natural module for K^+, or U_K is the sum of two isomorphic natural modules for K^+. By 13.6.11, $[U_K, L_v] \ne 1$. So either $K^+ = [K^+, L_v^+]$, or $[K^+, L_v^+] = 1$ and U_K is the sum of two isomorphic natural modules for K^+, with $L_v^+ \le \mathrm{Aut}_{K^+}(U_K) \cong L_2(2)$.

Assume first that $[K^+, L_v^+] = 1$. Then $J(T_v)^+$ is the unipotent radical $O_2(P^+)$ of the maximal parabolic P^+ of K^+ stabilizing a line in each summand of U_K, and $S^+ = J(T_v)^+$ by B.2.20. Therefore since $S = \mathrm{Baum}(T_v)$ by 13.6.4, $P^+ = N_K(S)^+$ by a Frattini Argument, while $N_K(S) \le C_{M_V}(v)$ by 13.6.5. But as $[K^+, L_v^+] = 1$, $O^2(N_K(S))$ acts on $O^2(O_2(K)L_v) = L_v$, and hence as $N_K(S) \le M_v$, $O^2(N_K(S))$ also acts on $[V, L_v] = [z, L_v] \cong E_4$. But $[z, L_v]$ contains a point of each summand of U_K, and so is generated by those two points; whereas we saw that P is the stabilizer of a line in each summand, so that $P^+ = N_P(S)^+$ acts irreducibly on each such line.

Therefore $[K^+, L_v^+] \ne 1$. Hence the projection L_K^+ of L_v^+ on K^+ is nontrivial. So as L_v^+ is T_v-invariant, it is a maximal parabolic of $K^+ \cong L_3(2)$, and hence $L_K = [L_K, T_v \cap K]$. Then as L_v is T_v-invariant, $L_v = L_K \le K$, as desired. □

LEMMA 13.6.21. $K \not\le M$.

PROOF. By 13.6.20, K^* does not act on L_v^*, so as $L_v \trianglelefteq M_v$, 13.6.21 holds. $\square$

LEMMA 13.6.22. K^+ is not $L_3(2)$, $Sp_4(2)'$, $G_2(2)'$, $\hat{A}_6$, or A_7.

PROOF. Assume otherwise. Let K_0 be minimal subject to $K_0 \in \mathcal{L}(KT_v, T_v)$ and $K_0^* = K^*$. Then $K_0/O_2(K_0)$ is quasisimple by minimality of K_0, and $K_0 \in \mathcal{Y}_z$ as $K_0^* = K^*$, so replacing K by K_0, we may assume $K/O_2(K)$ is quasisimple.

If K^+ is not A_7, then using 13.6.6 and 13.6.16.2, (KT_v, T_v) is an MS-pair in the sense of Definition C.1.31. So by C.1.32, either K is a block of type A_6, $\hat{A}_6$, or $G_2(2)$, or K^+ is $L_3(2)$ and by C.1.32.5, K is described in C.1.34. Similarly if K^+ is A_7, C.1.24 says K is an A_7-block or exceptional A_7-block. Set $U := [Z(O_2(K)), K]$. When K is a block, $U_K = U \in Irr_+(K, W_K)$. If K is not a block, then $K/O_2(K) \cong L_3(2)$ and K is described in C.1.34, so $U = U_K$ is a sum of at most two isomorphic natural modules for $L_3(2)$.

As $L_v \le K$ by 13.6.20, L_v^+ is a T_v^+-invariant $\{2,3\}$-subgroup of K^+ with Sylow 3-group of order 3. Set $P := L_v(T_v \cap K)$. When K^+ is $L_3(2)$, $Sp_4(2)'$, or $G_2(2)'$, P^+ is a minimal parabolic.

Suppose first that K is an A_7-block. Then by B.3.2.4 and B.2.9.1, $J(T_v)^+$ is the subgroup of T_v^+ generated by its three transpositions, and $S^+ = J(T_v)^+$ by B.2.20. Further $N_{K^+}(S^+) = N_K(S)^+$ by a Frattini Argument, and $N_K(S) \le M_v$ by 13.6.5. From the structure of S_7, $N_{KT_v}(S)$ is maximal in KT_v subject to containing a normal subgroup $\{2,3\}$-subgroup which is not a 2-group, so it follows that $N_{KT_v}(S) = (T_v \cap K)L_v$ and $L_v = O^2(N_K(S))$. Now $L_v T_v \le K_1 T_v \le KT_v$ with $K_1/O_2(K_1) \cong A_6$, and as $[z, L_v] \ne 1$, $[z, K_1] \ne 1$. Further $K_1^+ = [K_1^+, J(T_v)^+]$, so $K_1 \in \mathcal{Y}_z$; thus replacing K by K_1, we may assume K is not an ordinary A_7-block.

Similarly if K is an $\hat{A}_6$-block, then U has the structure of a 3-dimensional $\mathbf{F}_4 K$-module and $J(T_v)^+$ is the 4-subgroup of $T_v^+ \cap K^+$ centralizing an $\mathbf{F}_4$-line U_2 of U, so $S^+ = J(T_v)^+$, $N_K(S)^+ = N_K(U_2)^+$, and hence $N_K(U_2) \le M_V$ by 13.6.5.

Let $l := [V, L_v]$. Then l is an $L_v T_v$-invariant line in $[z, L_v] \le [z, K] \le U$ with $l = [l, L_v]$. It follows that if $K/O_2(K) \cong L_3(2)$, then $m(U) \ne 4$, since in that case no minimal parabolic of K^+ acts on such a line (cf. B.4.8.2). If K is an $\hat{A}_6$-block, then from the previous paragraph, $N_K(U_2) \le M_V$, so $N_K(U_2)$ acts on l, a contradiction as $N_K(U_2)$ acts on no E_4-subgroup of U.

Let $\hat{U} := U/C_U(K)$. In the remaining cases, if K is irreducible on $\hat{U}$, then there is a unique T_v-invariant line in U, so $\hat{l}$ is that line. Then if K is not an exceptional A_7-block, P^+ is the parabolic stabilizing $\hat{l}$, while if K is an exceptional A_7-block, then L_v^+ is one of the three T_v-invariant subgroups $L_0 = O^2(L_0)$ of $N_K(\hat{l})$ with $L_0/O_2(L_0) \cong \mathbf{Z}_3$ and $\hat{l} = [\hat{l}, L_0]$. If K is not irreducible on $\hat{U}$, then U is the sum of two isomorphic modules for $K^+ \cong L_3(2)$, $\hat{l}$ is a T_v-invariant line in one of those irreducibles, and $P = N_K(l)$. For our purposes the important fact is that in each case $C_{\hat{U}}(L_v) = 0$, so $C_{\hat{U}}(L_v) = C_{\hat{U}}(K)$.

Let $Z_K := Z(O_2(K))$ and Z_0 the preimage in K of $Z(O_2(\check{K}))$. If K is a block, then by definition $U = [Z_K, K] = [O_2(K), K]$, so $[Z_0, K] = U$. If K is not a block, then from the description of K in C.1.34, again $[Z_0, K] = U$. So in any event $[Z_0, K] = U$.

Now recall there is $a \in z^G \cap C_V(L_v)$ with $\check{a} \in Z(\check{T}_v) \le Z(O_2(\check{K}\check{T}_v))$ since $F^*(\check{K}\check{T}_v) = O_2(\check{K}\check{T}_v)$. Then

$$[\check{a}, K] \le [Z(O_2(\check{K})), K] = \check{U},$$

by the previous paragraph, so K acts on $\check{F} := \langle \check{a} \rangle \check{U}$. Further by B.2.14, $\check{F} = \check{U} C_{\check{F}}(K)$. Then as we saw earlier that $C_{\check{U}}(L_v) = C_{\check{U}}(K)$, it follows that K centralizes $\check{a}$, and hence K centralizes a by Coprime Action, contrary to 13.6.17. $\qquad \square$

LEMMA 13.6.23. *(1)* $K = O^{3'}(G_v)$.
(2) $K^+ \cong L_4(2)$ *and* T_v^+ *is nontrivial on the Dynkin diagram of* K^+.
(3) $L_v^+ T_K^+$ *is the middle-node minimal parabolic of* $K^+ T_K^+$, *where* $T_K := T_v \cap K$.

PROOF. By 13.6.22 and the remarks before 13.6.20, we have reduced to the cases where $K^+ \cong L_m(2)$ for $m = 4$ or 5. Since $L_v \le K$ by 13.6.20 we conclude as in the proof of 13.6.22 that $L_v^+ T_K^+$ is a minimal parabolic of K^+.

If T_v is trivial on the Dynkin diagram of K^+, then T_v acts on a parabolic K_1^+ of K^+ containing L_v^+ with $K_1^+ / O_2(K_1^+) \cong L_3(2)$. However as $[z, L_v] \ne 1$, also $[z, K_1] \ne 1$. By 13.6.16.3, $K_1^+ = [K_1^+, J(T_v)]^+$ so that $K_1^\infty \in \mathcal{Y}_z$ and 13.6.22 supplies a contradiction.

Hence T_v is nontrivial on the Dynkin diagram of K^+. So as T_v acts on the minimal parabolic $L_v T_K$, $m = 4$ and $L_v T_K$ is the middle-node minimal parabolic of K. Thus (2) and (3) hold.

By 1.2.4, $K \le K_+ \in \mathcal{C}(G_v)$; then $K_+ \in \mathcal{Y}_z$, so by symmetry between K_+ and K, $K/O_2(K) \cong L_4(2) \cong K_+/O_2(K_+)$, and hence $K = K_+$. Then by A.3.18, $K = O^{3'}(G_v)$, so (1) is established. $\qquad \square$

By 13.6.23, $K^+ T_v^+ \cong S_8$ with $L_v^+ T_K^+$ the middle-node minimal parabolic of K^+. As $[z, L_v] \ne 1$ and U_K is an FF-module for K^+, we conclude from Theorem B.5.1 that $U_K/C_{U_K}(K)$ is the 6-dimensional orthogonal module. Thus $C_K(z)$ is the maximal parabolic determined by the end nodes, so using 13.6.23.1 we conclude that

$$X := O^{3'}(C_K(z)) = O^{3'}(C_{G_v}(z)) = O^{3'}(C_G(V_2)),$$

and hence that X is T-invariant and $XT_v/R_v \cong S_3 \text{ wr } \mathbf{Z}_2$, where $R_v := O_2(XT_v)$. As $U_K/C_{U_K}(K)$ is the orthogonal module, $J(R_v) = J(O_2(KT_v))$ by B.3.2.4. But as T acts on X and T_v, T acts on R_v, so that $J(R_v) \trianglelefteq \langle K, T \rangle$, contrary to 13.6.16.1. This contradiction finally completes the proof of Theorem 13.6.7.

With Theorem 13.6.7 now in hand, we can now use elementary techniques such as weak closure in a fairly short argument to complete the proof of Theorem 13.6.1.

LEMMA 13.6.24. *If* $g \in G - N_G(V)$ *with* $V \cap V^g \ne 1$, *then*

(1) $V \cap V^g$ *is a singular point of* V, *and*
(2) $[V, V^g] = 1$.

PROOF. By 13.3.11.1, G_v is transitive on $\{V^x : v \in V^x\}$, so as $G_v \le M_V$ by Theorem 13.6.7.1, V is the unique member of V^G containing v. Hence $V \cap V^g$ is totally singular, so that (1) holds. In particular conjugating in L we may assume $V \cap V^g = V_1$, and then take $g \in C_G(z) = G_1$ by 13.3.11.1. Hence (2) follows from 13.6.2. $\qquad \square$

LEMMA 13.6.25. *(1) $W_i(T,V) \trianglelefteq LT$ for $i = 0, 1$.*
(2) $n(H_1) > 1$ for each $H_1 \in \mathcal{H}(T, M)$.
(3) Each solvable member of $\mathcal{H}(T)$ is contained in M.
(4) $r(G,V) = 3$ and $w(G,V) > 1$.

PROOF. We first observe that by 13.6.3.1 and 13.6.24.1, $r(G,V) = 3$. Let $g \in G - M$ with $A := V^g \cap M \le T$ and B a hyperplane of A. Suppose $m(V^g/A) \le 1$ but $[V, A] \ne 1$. Then $m(V^g/B) \le 2 < r(G,V)$, so $C_V(B) \le N_G(V^g)$ and hence $[C_V(B), A] \le V \cap V^g$; therefore $[C_V(B), A] = 1$ by 13.6.24.2. Thus $\bar{A} \in \mathcal{A}_2(\bar{M}_V, V)$, whereas we compute directly that $a(\bar{M}_V, V) = 1$. This contradiction shows that $W_i(T,V) \le C_T(V) = O_2(LT)$ for $i = 0, 1$, establishing (1).

By (1), $w(G,V) > 1$, where $w(G,V)$ appears in Definition E.3.23; this completes the proof of (4). By (4), $\min\{r(G,V), w(G,V)\} > 1$, so (2) and (3) follow from E.3.35.1. $\square$

LEMMA 13.6.26. *For $H \in \mathcal{H}_z$:*
(1) No member of $\mathcal{C}(H)$ is contained in M.
(2) $O_{2,p}(H) = Q_H$ for each prime $p > 3$.

PROOF. Assume $K \in \mathcal{C}(H)$. Part (1) follows from 13.3.8.2 with L, M, $\langle K^T \rangle$ in the roles of "K, M_K, Y". Let $p > 3$ be prime and set $X := O^{p'}(O_{2,p}(H))$. By 13.6.25.3, $X \le M$, so as $p > 3$, $X = 1$ by 13.3.8.2. Hence (2) holds. $\square$

LEMMA 13.6.27. *There exists $K \in \mathcal{C}(G_1)$ such that one of the following holds:*
(1) $G_1 = KT$ and $K/O_2(K) \cong J_2$ or M_{23}.
(2) $K/O_2(K) \cong L_3(4)$, T is nontrivial on the Dynkin diagram of $K/O_2(K)$, $G_1 = O^{3'}(G_1)T$, and either $K = O^{3'}(G_1)$ or $O^{3'}(G_1/O_2(G_1)) \cong PGL_3(4)$.

PROOF. By 3.3.2 there exists $H_1 \in \mathcal{H}_*(T, M)$. By 13.3.5.2, $H_1 \le G_1$, and by 13.6.25.2, $n(H_1) > 1$. Now we apply Theorem 5.2.3: Hypothesis 13.3.1 rules out conclusions (2) and (3) of that Theorem, so we are left with conclusion (1) of 5.2.3. In particular $K_1 := O^2(H_1)$ lies in some $K \in \mathcal{C}(C_G(Z))$. As T normalizes K_1, it normalizes K, and as $K_1 \not\le M$, $KT \in \mathcal{H}(T, M)$. Thus $n(KT) > 1$ by 13.6.25.2, so in particular $K/O_2(K) \not\cong A_7$. So by Theorem 5.2.3.1, either

(a) $K_1/O_2(K_1) \cong L_2(4)$ and $K/O_2(K) \cong J_2$ or M_{23}, or

(b) $K_1 = K$ with $K/O_2(K) \cong L_3(4)$, and T nontrivial on the Dynkin diagram of $K/O_2(K)$ by E.2.2.

Next as $C_G(Z) \le G_1$, $K \in \mathcal{L}(G_1, T)$, so $K \le K_+ \in \mathcal{C}(G_1)$ by 1.2.4. If $K/O_2(K) \cong J_2$ or M_{23}, we conclude $K = K_+$ from 1.2.8.4. If $K/O_2(K) \cong L_3(4)$, then either $K = K_+$ or $K_+/O_2(K_+) \cong M_{23}$ by A.3.12, and the latter case is impossible as T is nontrivial on the Dynkin diagram of $K/O_2(K)$. Thus $K = K_+ \in \mathcal{C}(G_1)$.

By A.3.18, either $O^{3'}(G_1) = K$ or $O^{3'}(G_1/O_2(G_1)) \cong PGL_3(4)$. In particular K is the unique member of $\mathcal{C}(G_1)$ which is not a $3'$-group, and $O_{2,3}(G_1) = 1$ so that $O_{2,F}(G_1) = O_2(G_1)$ using 13.6.26.2.

Suppose $K_0 \in \mathcal{C}(G_1) - \{K\}$. By an earlier observation, K_0 is a $3'$-group, so $K_0/O_2(K_0)$ is $Sz(2^m)$. By 13.6.26.1, $K_0 \not\le M$, while by 13.6.25.3, a Borel subgroup of K_0 is contained in M. Therefore $\langle K_0, T \rangle \in \mathcal{H}_*(T, M)$, which is contrary to Theorem 5.2.3 as we saw above.

Let $\dot{G}_1 := G_1/O_2(G_1)$; we have shown that $\dot{K} = F^*(\dot{G}_1)$. But $Out(\dot{K})$ is a 2-group if $\dot{K}$ is J_2 or M_{23}, while $Out(L_3(4)) \cong \mathbf{Z}_2 \times S_3$. It follows that either

$G_1 = KT$, or $G_1 = O^{3'}(G_1)T$ with $O^{3'}(\dot{G}_1) \cong PGL_3(4)$. Hence the proof of the lemma is complete.   □

Let $H := G_1$ and $K := G_1^\infty$. Now $H \in \mathcal{H}_z$, so by 13.5.7, Hypothesis F.9.1 is satisfied with V_3 in the role of "V_+". Then by 13.6.24.2, Hypothesis F.9.8.f is satisfied, while case (i) of Hypothesis F.9.8.g holds in view of 13.2.3.2. We now adopt the standard conventions from section F.9 given in Notation 13.5.8, including $H^* := H/Q_H$, $U_H := \langle V_3^H \rangle$, and $\tilde{H} := H/V_1$. By 13.6.27, $F^*(H^*) = K^* \cong L_3(4)$, M_{23}, or J_2, and in the first case T^* is nontrivial on the Dynkin diagram of K^*. Therefore $q(H^*, \tilde{U}_H) > 2$ by B.4.2 and B.4.5, contrary to F.9.16.3. This contradiction completes the proof of Theorem 13.6.1.

13.7. Finishing the treatment of $\mathbf{A_6}$ when $\langle \mathbf{V^{G_1}} \rangle$ is nonabelian

In this section, and also in the final section 13.8 of the chapter, we adopt a hypothesis excluding the groups identified in previous sections:

HYPOTHESIS 13.7.1. *Hypothesis 13.3.1 holds, $L/C_L(V) \cong A_6$, and G is not $Sp_6(2)$ or $U_4(3)$.*

Thus since Hypothesis 13.7.1 includes Hypothesis 13.3.1 and Hypothesis 13.5.1, we may appeal to results in sections 13.4 and 13.5.

Set $Q := O_2(LT)$. We continue with the notation established in section 13.5: Namely we adopt the notational conventions of section B.3 and Notations 12.2.5 and 13.2.1.

By 13.5.2.3, $C_V(L) = 1$, so that V is the core of permutation module for $\bar{L} \cong A_6$, given by the vectors e_S for subsets S of even order in $\Omega := \{1, 2, 3, 4, 5, 6\}$, modulo e_Ω. In particular $V_1 = Z \cap V$ is generated by $z := e_{1,2,3,4} \equiv e_{5,6}$.

By 13.5.7, Hypothesis F.9.1 is satisfied with V_3 in the role of "V_+", so we may use results from section F.9. We also adopt the conventions from that section given in Notation 13.5.8, including $\tilde{G}_1 := G_1/V_1$. As usual define

$$\mathcal{H}_z := \{H \in \mathcal{H}(L_1T) : H \leq G_1 \text{ and } H \not\leq M\}.$$

By 13.3.6, $G_1 \in \mathcal{H}_z$, and so $\mathcal{H}_z$ is nonempty.

In the remainder of the section, H denotes a member of $\mathcal{H}_z$.

From Notation 13.5.8 $U_H := \langle V_3^H \rangle$, $V_H := \langle V^H \rangle$, $Q_H := O_2(H)$, and $H^* := H/Q_H$ so that $O_2(H^*) = 1$. By F.9.2.3, $Q_H = C_H(\tilde{U}_H)$. Set $H_C := C_H(U_H)$; then $H_C \leq Q_H$.

By Theorem 13.5.12:

LEMMA 13.7.2. $\langle V_3^{G_1} \rangle$ *is abelian, so U_H is abelian.*

There are no quasithin examples satisfying 13.7.2, so in the remainder of this section we will be working toward a contradiction. As far as we can tell, there are not even any shadows.

13.7.1. Preliminary results. We begin with several consequences of Hypothesis 13.7.1 and 13.7.2, which we can apply both in the next subsection where $\langle V^{G_1} \rangle$ is nonabelian, and in the final section 13.8 where $\langle V^{G_1} \rangle$ is abelian.

LEMMA 13.7.3. *(1)* $V_H \leq Q_H$.

(2) $U_H \leq Z(V_H)$, *so that* $V_H \leq H_C$.

(3) $\langle U_H^L \rangle \leq O_2(LT) = Q$.

(4) For $h \in H$, either $[V, V^h] = 1$, or $[V, V^h] = [V, V_H] = V_1$ with $\bar{V}_H = \bar{V}^h = \langle (5,6) \rangle$.

(5) Either V_H is abelian, or $\Phi(V_H) = V_1$.

(6) $O_2(\bar{L}_1) \leq \bar{Q}_H \leq \bar{R}_1$, $V_3 = [V, Q_H]$, $V_1 = [V_3, Q_H] = [U_H, Q_H] = C_{V_3}(Q_H)$, and $[V_H, Q_H] = U_H$.

(7) Either

(i) $H_C \leq Q$, so $H_C = C_H(V_H)$, or

(ii) $|H_C : Q \cap H_C| = 2$, so $\bar{H}_C = \langle (5,6) \rangle$, $[V_H, H_C] = V_1$, and $H_C \leq C_H(\tilde{V}_H)$.

(8) If $L/O_2(L) \cong \hat{A}_6$, then V_H is abelian.

(9) $H \cap M = N_H(V)$ and $L_1 = \theta(H \cap M)$.

PROOF. As U_H is abelian,

$$\bar{U}_H \leq C_{\bar{T}}(V_3) = C_{\bar{T}}(\tilde{V}),$$

so $V \leq C_H(\tilde{U}_H) = Q_H$ and hence (1) holds. By (1) and F.9.3, $V \leq C_{Q_H}(U_H)$, so (2) and (3) hold.

Let $h \in H$. Then by (2),

$$V^h \leq C_{Q_H}(V_3) = C_{Q_H}(\tilde{V}),$$

so (4) holds, since $(5,6)$ is the transvection in $\bar{T}$ with center V_1. Then (4) implies (5).

If $[L_1, Q_H] \leq Q = C_T(V)$, then $[L_1, Q_H] \leq C_{L_1}(V_3)$; so as $L_1/C_{L_1}(V_3) \cong A_4$ has trivial centralizer in $GL(V_3)$, $[V_3, Q_H] = 1$, contrary to 13.5.4.5 since $O_2(G_1) \leq Q_H$. Thus $[L_1, Q_H] \not\leq Q$, so $O_2(\bar{L}_1) = \overline{[Q_H, L_1]} \leq \bar{Q}_H \leq \bar{R}_1$, and hence $V_3 = [V, O_2(\bar{L}_1)] = [V, Q_H]$. Then as $U_H = \langle V_3^H \rangle$, (6) holds.

Observe $\bar{H}_C \leq C_{\bar{R}_1}(V_3) = 1$ or $\langle (5,6) \rangle$. If $\bar{H}_C = 1$, then (7i) holds. Otherwise $\bar{H}_C = \langle (5,6) \rangle$, and then as $[V, (5,6)] = V_1$, $[V, H_C] = V_1$, so that (7ii) holds.

If $L/O_2(L) \cong \hat{A}_6$, then each $t \in T$ inducing a transposition on $\bar{L}$ inverts $L_0/O_2(L_0)$ (see Notation 13.2.1), and hence $t \notin Q_H$ as $L_0 \leq L_1 \leq H$. We conclude $[V, V^h] = 1$ for all $h \in H$—since if not, some $t \in V^h$ induces a transposition on $\bar{L}$ by (4), whereas $V^h \leq Q_H$ by (1), contrary to $t \notin Q_H$. Thus (8) is established.

Finally as $H \leq G_z$, $H \cap M = N_H(V)$ by 12.2.6, so the remaining statement of (9) follows using 13.3.7. $\qquad\square$

By 13.7.2, $U_H \leq H_C$.

LEMMA 13.7.4. *(1)* If $L/O_2(L) \cong A_6$, then H^* is faithful on $U_H/C_{U_H}(Q_H)$.

(2) There is an H-isomorphism φ from Q_H/H_C to the dual of $U_H/C_{U_H}(Q_H)$, defined by $\varphi(xH_C) : uC_{U_H}(Q_H) \mapsto [x, u]$.

PROOF. Part (2) holds by F.9.7, so it remains to establish (1). Set $U_0 := C_{U_H}(Q_H)$ and observe $Q_H = C_H(\tilde{U}_H) \leq C := C_H(U_H/U_0)$. On the other hand $\tilde{V}_3 = [\tilde{V}_3, L_1]$ with $V_1 = V \cap U_0$ by 13.7.3.6, so that $L_1 \not\leq C$.

Assume that $L/O_2(L) \cong A_6$, but that (1) fails. Then $C^* \neq 1$, so as $O_2(H^*) = 1$, either $E(C^*) \neq 1$, or $O_p(C^*) \neq 1$ for some odd prime p. In the former case there is $K \in \mathcal{C}(C)$ with $K^* \cong K/O_2(K)$ quasisimple. In the latter case we take

$K := O^2(K_1)$, where K_1^* is a minimal normal subgroup of H^* contained in $O_p(C^*)$; thus $K^* \cong K/O_2(K) \cong E_{p^n}$, $n := 1$ or 2 since H is an SQTK-group. The subgroup K satisfies one of these two hypotheses throughout the rest of the proof.

In either case, $K = O^2(K)$ is subnormal in H, so $O_2(K) \leq Q_H$ and Q_H normalizes K. As $K \leq C$, $[U_H, K] \leq U_0$, so

$$1 \neq [U_H, K] = [U_0, K] \leq O_2(K) \leq Q_H \leq C_H(U_0). \tag{!}$$

Then $1 \neq [U_0, K] \leq [\Omega_1(Z(O_2(K))), K]$, so that $K \in \mathcal{X}_f$.

Consider for the moment the case where $K \in \mathcal{C}(H)$. Then $K \in \mathcal{L}_f(G, T)$, so that $K^* \cong K/O_2(K)$ is described in the list of 13.5.2.1, and $K \trianglelefteq H$ by 13.3.2.2. We saw $L_1 \not\leq C$, so $L_1 \not\leq K$. Thus $K/O_2(K)$ is not A_6 or $\hat{A}_6$ by A.3.18, so $K^* \cong L_3(2)$ or A_5 by 13.5.2.1. Then as $L_1 = [L_1, T]$, either $[K, L_1] \leq O_2(K)$, or $K^* \cong A_5$ and $K = [K, L_1]$. Further if K^* is A_5, then by 13.3.2.3, each $I \in Irr_+(K, R_2(KT), T)$ is a T-invariant A_5-module.

We now return to consideration of both cases. By the previous paragraph we have $K \trianglelefteq H$. Since K^* is either simple or a p-group for p odd, and $C_K(\tilde{U}_H) = O_2(K) \leq C_K(U_0)$ by (!), we conclude from Coprime Action that

$$C_K(U_0) = C_K(\tilde{U}_H) = O_2(K). \tag{*}$$

Observe that if $K \leq M$, then K normalizes V by 13.7.3.9, so $[V_3, K] \leq V \cap U_0 = V_1$, and hence $\bar{K} \neq \bar{L}_1$, contrary to 13.3.9 applied to K in the role of "Y". Thus $K \not\leq M$.

Suppose next that L is an A_6-block. Then L_1 has just two noncentral 2-chief factors, while L_1 is nontrivial on $V_3 U_0/U_0$ and hence also nontrivial on Q_H/H_C by (2). Therefore L_1 centralizes U_0, and hence $[K, L_1] \leq C_K(U_0) = O_2(K)$ using (*), so K acts on $O^2(L_1 O_2(K)) = L_1$. Then as $V_3 \leq L_1$ and $K \leq C$, $[V_3, K] \leq O_2(L_1) \cap U_0 \leq Z(L)V_1$, so K centralizes V_3 by Coprime Action as $|Z(L)| \leq 2$ by C.1.13.b. Thus $K \leq G_1 \cap G_3 \leq M_V$ by 13.5.5, whereas we saw $K \not\leq M$.

Therefore L is not an A_6-block. Then by 13.2.2.7, $N_G(B) \leq M$, where $B := Baum(R_1)$. In particular as $K \not\leq M$, B is not normalized by K.

Assume next that $K^* \cong A_5$ and $K = [K, L_1]$. Then $R_1 = (K \cap T)O_2(KR_1)$, and we saw earlier that each $I \in Irr_+(K, R_2(KT), T)$ is an A_5-module, so $J(R_1)$ centralizes I by B.4.2. Then $B \trianglelefteq KT$ by B.2.3.5, contrary to the previous paragraph. This contradiction shows that $[K, L_1] \leq O_2(K)$ in the case that $K \in \mathcal{C}(H)$.

Next consider the case where K^* is a p-group. As L_1 acts on $O_2(K)$, $O_2(K) \leq R_1$, so $R_1 \in Syl_2(KR_1)$. As $C_K(U_0) = O_2(K)$ by (*), $C_{KR_1}(R_2(KR_1)) = O_2(KR_1)$ by A.1.19. Thus if $J(R_1)$ centralizes $R_2(KR_1)$, then $B \trianglelefteq KR_1$ using B.2.3.5, whereas we saw B is not normalized by K. Thus KR_1 satisfies case (2) of Solvable Thompson Factorization B.2.16; so in particular $p = 3$. Since K^* is a minimal normal subgroup of H^*, T is irreducible on K^*, so by B.2.16.2, $J(KR_1)/O_2(J(KR_1)) \cong S_3$ or $S_3 \times S_3$. As $L_1 \not\leq C$ acts on $J(KR_1)$, the latter case is impossible as $m_3(H) \leq 2$. Thus if K^* is a p-group, we conclude $K^* \cong \mathbf{Z}_3$.

We have now shown that $K^* \cong \mathbf{Z}_3$, A_5, or $L_3(2)$, and that L_1 centralizes K^*. In particular, K acts on $O^2(O_2(K)L_1) = L_1$. Let $U \leq U_0$ be minimal subject to $U \trianglelefteq X := KL_1 T$ and $[U, K] \neq 1$. Set $X^+ := X/C_X(U)$; we claim that $O_2(X^+) = 1$: For $O_2(K^+) = 1$ as $O_2(K)$ centralizes U_0, so K^+ centralizes $O_2(X^+)$, and hence $O_2(X^+) = 1$ by the Thompson $A \times B$ Lemma and minimality of U, as claimed. As K^* is simple, $K^* \cong K^+$ and $C_{KR_1}(U) = O_2(KR_1)$.

As L_1 centralizes K^*, L_1 acts on $T \cap K$, so $R_1 \in Syl_2(KL_1R_1)$. Thus if $J(R_1) \leq C_{R_1}(U)$, then $B = \text{Baum}(O_2(KR_1))$ by B.2.3.5, whereas we saw K does not normalize B. Hence $K^+ = [K^+, J(R_1)^+]$. Also K acts on $[C_{\tilde{U}_H}(O_2(L_1)), L_1] =:$ $\tilde{U}_1$. If $[U_1, K] = 1$ then $K \leq C_G(V_3) \leq M_V$ using 13.5.4.4, again contrary to $K \not\leq M$. Thus $[U_1, K] \neq 1$, so as $[U_1, K] \leq U_0$, $U_2 := [U_0, L_1, K] \neq 1$. Thus we may take $U \leq U_2$, so $U = [U, L_1]$. Then as L_1 centralizes K^*, $L_1^+ \trianglelefteq X^+$, so that $L_1^+ \cong \mathbf{Z}_3$ as $O_2(X^+) = 1$. Then since $K^+ = [K^+, J(R_1)^+]$, it follows from Theorem B.5.6 that U is the sum of two isomorphic natural modules for $K^+ \cong L_3(2)$. Therefore by B.2.20, $B^+ = J(R_1)^+$ is the unipotent radical of a minimal parabolic $K_0^+ R_1^+$ of $K^+ R_1^+$. Then by B.2.3.4, $B = \text{Baum}(O_2(K_0R_1))$, so that $K_0 \leq N_G(B) \leq M$. Then $O^2(K_0) \leq O^2(H \cap M) = L_1$ by 13.7.3. But $|L_1|_3 = 3$ since we are assuming that $L/O_2(L)$ is A_6 rather than $\hat{A}_6$, so $O^2(K_0) = L_1$, contradicting $L_1 \not\leq K$. Thus the proof of (1) and hence of the lemma is at last complete. $\square$

LEMMA 13.7.5. *Let* $X := L_1$ *if* $L/O_2(L) \cong A_6$, *and* $X := L_{1,+}$ *if* $L/O_2(L) \cong \hat{A}_6$. *Assume* $K \in \mathcal{C}(H)$ *and* $X \leq K$. *Then*

(1) $K \trianglelefteq H$.

(2) $U_H = [U_H, K]$.

(3) $N_H(\tilde{V}_3) \leq M$.

(4) *If* $m_3(N_K(\tilde{V}_3)) = 2$, *then* $L/O_2(L) \cong \hat{A}_6$ *and* $L_1 = \theta(N_K(\tilde{V}_3))$.

(5) *If* $L_1 \leq K$ *and* $m_3(N_K(\tilde{V}_3)) = 1$, *then* $L/O_2(L) \cong A_6$.

(6) $O^{3'}(N_K(\tilde{V}_3))$ *is solvable.*

(7) *If* $L_1 \not\leq K$, *then* $Aut_{L_1}(K/O_2(K)) \neq Aut_X(K/O_2(K))$.

PROOF. Since T normalizes $X \leq K$, $K \trianglelefteq H$ by 1.2.1.3, proving (1). Further $V_3 = [V_3, X] \leq [U_H, K]$, so that $U_H = \langle V_3^H \rangle = [U_H, K]$, establishing (2). Part (3) follows from 13.5.5. Then by (3) and 13.7.3, either $\theta(N_K(V_3)) = L_1$ or $L/O_2(L) \cong \hat{A}_6$ with $\theta(N_K(V_3)) = X$. Now $m_3(X) = 1$, while $m_3(L_1) = 1$ when $L/O_2(L) \cong A_6$, and $m_3(L_1) = 2$ when $L/O_2(L) \cong \hat{A}_6$; so it follows that (4) and (5) hold. As $\theta(N_K(V_3)) \leq L_1$ which is solvable, (6) holds.

Finally suppose that $L_1 \not\leq K$, but the conclusion of (7) fails. Since $X \leq K$, $X < L_1$, so $L/O_2(L) \cong \hat{A}_6$; then as (7) fails, $L_1 = XL_C$, where $L_C = O^2(C_{L_1}(K/O_2(K)))$. As X and L_0 are the only proper nontrivial T-invariant subgroups Y of L_1 with $Y = O^2(Y)$, it follows that $L_C = L_0$. But then K normalizes $O^2(O_2(K)L_0) = L_0$ and so lies in M by 13.2.2.9, contrary to 13.3.9. $\square$

The next result eliminates various possibilities for H^* and its action on $\tilde{U}_H$. As usual Theorem C (A.2.3) determines the possibilities for n in (1) and (3). The lemma considers all cases where $[\tilde{U}_H, K] \in Irr_+(\tilde{U}_H, K)$ is an FF-module, except the cases where the noncentral chief factor for K on $\tilde{U}_H$ is the natural module for $K/O_2(K) \cong L_2(2^n)$ or $\hat{A}_6$.

LEMMA 13.7.6. *Assume* $K \in \mathcal{C}(H)$ *and let* $U_K := [U_H, K]$. *Then*

(1) *If* $K/O_2(K) \cong L_n(2)$ *and* $\tilde{U}_K/C_{\tilde{U}_K}(K)$ *is the natural* $K/O_2(K)$-module, *then* $n = 4$, $\tilde{U}_H$ *is the natural module for* $H^* \cong L_4(2)$, *and* $L/O_2(L) \cong \hat{A}_6$.

(2) *If* $K/O_2(K) \cong L_5(2)$, *then* $\tilde{U}_K/C_{\tilde{U}_K}(K)$ *is not a 10-dimensional irreducible for* $K/O_2(K)$.

(3) *If* $K/O_2(K) \cong A_n$ *and* $\tilde{U}_K/C_{\tilde{U}_K}(K)$ *is the natural module, then* $U_H = U_K$, $L_1 \leq K$, $H = KT$, *and applying the notation of section B.3 to* $\tilde{U}_H$, *either*

(a) $n = 6$, $L/O_2(L) \cong A_6$, $\tilde{V}_2 = \langle e_{1,2,3,4} \rangle$, and L_1 has two noncentral chief factors on U_H, or

(b) $n = 7$, $L/O_2(L) \cong \hat{A}_6$, $\tilde{V}_2 = \langle e_{5,6} \rangle$, and $\tilde{V}_3 = \langle e_{5,6}, e_{5,7} \rangle$.

(4) If $K/O_2(K) \cong A_7$ then $\tilde{U}_K$ is not a 4-dimensional A_7-module.

(5) If $K/O_2(K) \cong (S)L_3(2^n)$, $Sp_4(2^n)'$, or $G_2(2^n)'$ and $\tilde{U}_K/C_{\tilde{U}_K}(K)$ is a natural module for K^*, then $n = 1$.

PROOF. Assume $K/O_2(K)$, $\tilde{U}_K$ is one of the pairs considered in the lemma. We obtain a contradiction in (2) and (4), and in (5) under the assumption that $n > 1$. In (1) and (3), we establish the indicated restrictions. Observe that, except possibly in (5) when $K/O_2(K) \cong SL_3(2^n)$, $K/O_2(K)$ is simple so that $K^* \cong K/O_2(K)$. In that exceptional case $\tilde{U}_K$ is a natural module by hypothesis, so $C_K(\tilde{U}_K) = O_2(K)$ and thus again $K^* \cong K/O_2(K)$.

The first part of the proof treats the case where $L_1 \leq K$. Here $K \trianglelefteq H$ by 13.7.5.1, and $U_H = U_K$ by 13.7.5.2.

Next $\tilde{V}_3 = \langle \tilde{V}_2^{L_1} \rangle$ is a T-invariant line in $\tilde{U}_K$, so:

(i) If $K^* \cong L_3(2)$ then $C_{\tilde{U}_H}(K) = 0$ (cf. B.4.8.2).

(ii) Under the hypotheses of (3), $n > 5$.

Further

(iii) $\tilde{V}_2$ is a T-invariant $\mathbf{F}_2$-point of $\tilde{U}_H$. Set $K_0 := O^2(C_K(V_2))$, so that also $K_0 = O^2(C_K(\tilde{V}_2))$.

By 13.5.4.3, $m_3(K_0) \leq 1$, so we conclude from (iii) and the structure of $C_{K^*}(\tilde{V}_2)$ that: (2) holds; $n < 5$ in (1); in (3), $n \leq 7$ and in case of equality $\tilde{V}_2 = \langle e_{5,6} \rangle$, when $\tilde{U}_H$ is described in the notation of section B.3.

Assume the hypothesis of (3) with $n = 7$. As $\tilde{V}_2 = \langle e_{5,6} \rangle$ and $\tilde{V}_3 = \langle \tilde{V}_2^{L_1} \rangle$ is a T-invariant line, $\tilde{V}_3 = \langle e_{5,6}, e_{5,7} \rangle$. Hence $N_K(V_3)$ has 3-rank 2, so $L/O_2(L) \cong \hat{A}_6$ by 13.7.5.4. Since $N_{GL(\tilde{U}_K)}(K^*) \cong S_7$, $H = KT$. Hence conclusion (b) of (3) holds.

Thus under the hypotheses of (3), we have reduced to the case $n = 6$. Then as $\tilde{V}_3 = \langle \tilde{V}_2^{L_1} \rangle$ is a T-invariant line, $\tilde{V}_2 = \langle e_{1,2,3,4} \rangle$, L_1 has two noncentral chief factors on $\tilde{U}_H$, and $m_3(N_K(\tilde{V}_3)) = 1$, so that $L/O_2(L) \cong A_6$ by 13.7.5.5. Since $End_K(\tilde{U}_K)$ is of order 2, we conclude that $K^* = F^*(H^*)$. As T normalizes U_K, it is trivial on the Dynkin diagram of K^*, so as $Out(K^*)$ is a 2-group, we conclude that $H = KT$. This gives conclusion (a) of (3), and so completes the proof of (3).

Similarly when $n = 4$ in case (1), or in case (4), $N_K(V_3)$ has 3-rank 2, so that $L/O_2(L) \cong \hat{A}_6$ and $L_1 = O^2(N_K(V_3))$ by 13.7.5.4. Thus $L_1T/O_2(L_1T) \cong S_3 \times \mathbf{Z}_3$ or $S_3 \times S_3$ from the structure of L. Further as $U_H = U_K$ and $K^* \trianglelefteq H^*$, $H^* = N_{GL(\tilde{U}_H)}(K^*) = K^*$. Thus in case (4) where $K^* \cong A_7$, $L_1T/O_2(L_1T)$ is E_9 extended by an involution inverting the E_9, so this case is eliminated. When $K^* \cong L_4(2)$, the conclusions of (1) hold using I.1.6.6.

Assume the hypotheses of (5) with $n > 1$, and let $U_H^+ := U_H/C_{U_H}(K)$. By (iii), V_2^+ is contained in a T-invariant $\mathbf{F}_{2^n}$-point W of U_H^+. As $L_1 \leq K$ and L_1 is T-invariant, L_1 is contained in the Borel subgroup of K containing $T \cap K$. In particular, n is even. Thus L_1 acts on W, so $V_3^+ = [V_2^+, L_1] \leq W$. But now as $O^{3'}(C_K(W))$ is not solvable, 13.7.5.6 supplies a contradiction, establishing that $n = 1$ under the hypotheses of (5).

So to complete the treatment of the case $L_1 \leq K$, it remains only to eliminate case (1) with $n = 3$. So suppose $K^* \cong L_3(2)$. Since $L_1 \leq K$ which has 3-rank 1, $L/O_2(L) \cong A_6$ by 13.7.5.5, and $L_1^* T^*$ is a maximal parabolic of $K^* T^*$. Set $\mathcal{F} := (LT, K_0 L_2 T, KT)$ and $G_0 := \langle \mathcal{F} \rangle$. Here $K_0 T/O_2(K_0 T) \cong S_3$, and by 13.5.4.1, $[K_0, L_2] \leq O_2(L_2)$. As $M = !\mathcal{M}(LT)$, $O_2(G_0) = 1$. Thus if $|Q| > 2^5$, then $(G_0, \mathcal{F})$ is a C_3-system as defined in section I.5, so by Theorem I.5.1, $G_0 \cong Sp_6(2)$. Therefore $Z(KT) = 1$, whereas $z \in Z(KT)$ as $H \in \mathcal{H}_z$. Thus $|Q| \leq 2^5$, so L is an A_6-block. But then L_1 has just two noncentral 2-chief factors, whereas L_1 has noncentral chief factors on each of $O_2(L_1^*)$, $\tilde{U}_H$, and Q_H/H_C by 13.7.4.2. This contradiction completes the proof of (1) and of the lemma in the case $L_1 \leq K$.

It remains to treat the case $L_1 \not\leq K$. When $m_3(K) = 2$, $K = O^{3'}(H)$ by A.3.18 and A.3.19, so $K/O_2(K) \cong L_3(2^n))$, n odd, or A_5. Let $X := L_1$ if $L/O_2(L) \cong A_6$, and $X := L_{1,+}$ if $L/O_2(L) \cong \hat{A}_6$. Suppose $[K^*, X^*] = 1$. Then as $End_K(\tilde{U}_K/C_{\tilde{U}_K}(K)) = \mathbf{F}_{2^n}$ with n odd, X centralizes $\tilde{U}_K = [K, \tilde{U}_H]$. But then by the Three-Subgroup Lemma, $[\tilde{U}_H, X, K] = 1$; so as $\tilde{V}_3 = [\tilde{V}_3, X]$, $K = O^2(K) \leq C_G(V_3) \leq M_V$ by 13.5.4.4, and hence $\langle K^T \rangle \leq M$, contrary to 13.3.9. Therefore $K = [K, X]$. So as $X = [X, T]$, we conclude that X induces inner automorphisms on $K/O_2(K)$. Then again as $X = [X, T]$, $K/O_2(K)$ is not $L_3(2^m)$ for $m > 1$, and either:

(a) $X \leq K$, and hence $X < L_1$ as $L_1 \not\leq K$, so that $L/O_2(L) \cong \hat{A}_6$, or

(b) $K^* \cong A_5$ and X^* is diagonally embedded in $K^* C_{K^* X^*}(K^*)$, with $\tilde{V}_3$ projecting nontrivially on $\tilde{U}_K$.

But now $K^* \cong A_5$ is ruled out, since in both cases (a) and (b), $\tilde{V}_3 = [\tilde{V}_3, X]$, while either X^* or its projection on K^* is a Borel subgroup of K^*, which has no such T-invariant submodule of rank 2 on the A_5-module $\tilde{U}_K$. Therefore case (a) holds and $K^* \cong L_3(2)$. Now 13.7.5.7 supplies a contradiction, completing the proof. $\square$

LEMMA 13.7.7. $[V_H, H] \not\leq U_H$.

PROOF. If $[V_H, H] \leq U_H$, then $V_H = \langle V^H \rangle = V U_H$. Then by 13.7.3.6,

$$U_H = [V_H, Q_H] = [V, Q_H][U_H, Q_H] = V_3 V_1 = V_3,$$

contrary to 13.5.9. $\square$

13.7.2. The elimination of the case $\langle V^{G_1} \rangle$ nonabelian.

We come to the main result of this section, which reduces the treatment of Hypothesis 13.7.1 to the case where $\langle V^{G_1} \rangle$ is abelian. Then in the following section 13.8, that remaining case is also shown to lead to a contradiction.

THEOREM 13.7.8. *Assume Hypothesis 13.7.1. Then* $\langle V^{G_1} \rangle$ *is abelian.*

Until the proof of Theorem 13.7.8 is complete, assume G is a counterexample. Then the set $\mathcal{H}_1$ of those $H \in \mathcal{H}_z$ with V_H nonabelian is nonempty, since $G_1 \in \mathcal{H}_1$.

For the remainder of the section, let H denote a member of $\mathcal{H}_1$.

Then V_H is nonabelian, though U_H is abelian by 13.7.2.

LEMMA 13.7.9. *(1)* $\Phi(V_H) = V_1$, $\bar{V}_H = \langle (5,6) \rangle$, *and* $\bar{Q}_H = \bar{R}_1$.
(2) $L/O_2(L) \cong A_6$ *rather than* $\hat{A}_6$. *In particular* $|L_1|_3 = 3$.
(3) H^* *is faithful on* $U_H/C_H(Q_H)$.

PROOF. Observe (1) follows from parts (4) and (6) of 13.7.3, and (2) follows from part (8) of 13.7.3. Finally (3) follows from (2) and 13.7.4.1. □

Pick $g \in L$ with $\bar{g}^2 = 1$ and V_1^g not orthogonal to V_1. Set $I := \langle V_H, V_H^g \rangle$ and $Z_I := V_H \cap V_H^g$. Observe Q normalizes V_H, and also V_H^g since $g \in N_G(Q)$, so Q normalizes I.

Recall that we can appeal to results in section F.9. In particular, as in F.9.6 define $D_H := U_H \cap Q_H^g$, $D_{H^g} := U_H^g \cap Q_H$, $E_H := V_H \cap Q_H^g$ and $E_{H^g} := V_H^g \cap Q_H$. Since we chose $\bar{g}^2 = 1$, F.9.6.2 says that

$$(D_H)^g = D_{H^g} \quad \text{and} \quad (E_H)^g = E_{H^g}.$$

Let $A := V_H^g \cap Q$, $U_0 := C_{U_H}(Q_H)$, $U_H^+ := U_H/U_0$, and recall $H_C = C_H(U_H)$. By 13.7.9.3, H^* is faithful on U_H^+, and by 13.7.4.2, Q_H/H_C is dual to U_H^+ as an H-module.

Let $U_L := \langle U_H^L \rangle$. By 13.7.3.3, $U_L \leq Q$. In particular $U_H^g \leq Q$, so that $U_H^g \leq V_H^g \cap Q = A$.

LEMMA 13.7.10. *(1) $V_1^g \not\leq U_H$.*
(2) $O_2(I) = (V_H \cap Q)(V_H^g \cap Q)$ and $I/O_2(I) \cong S_3$.
(3) $O_2(I)/Z_I$ is elementary abelian and the sum of natural modules for $I/O_2(I)$, and $Z_I/V_1V_1^g$ is centralized by I.
(4) $\langle U_H^I \rangle Z_I = U_H U_H^g Z_I$ and $U_H^g Z_I = \{x \in V_H^g : [V_H, x] \leq U_H Z_I\}$.
(5) $\langle D_H^I \rangle = D_H D_{H^g} = V_1 V_1^g (D_H \cap D_{H^g}) \leq Z_I$.
(6) $[D_H, A] = 1$ and $[D_{H^g}, V_H] \leq V_1$.
(7) $E_H = E_{H^g} = Z_I \leq Q \cap H_C$, so $[E_{H^g}, V_H] \leq V_1$.
(8) L_1 has $m(A^*) + 2$ noncentral 2-chief factors.
(9) $U_H^g \cap V_3$ is a complement to V_1 in V_3, and $V_3 \leq Z_I$.
(10) $A \cap Q_H = E_{H^g}$.

PROOF. If $V_1^g \leq U_H$, then $V = V_3 V_1^g \leq U_H$, so $V_H \leq U_H$ is abelian, contrary to the choice of G as a counterexample to Theorem 13.7.8. Thus (1) holds.

By 13.7.9.1 and the choice of g, $\bar{I} \cong S_3$; e.g., if $\bar{g} = (4, 5)$ then $\bar{I} = \langle (5, 6), (4, 6) \rangle$. Let $P := (V_H \cap Q)(V_H^g \cap Q)$. By 13.7.9.1, $\Phi(V_H) = V_1$, so $\Phi(V_H) \leq V_1 V_1^g \trianglelefteq I$; e.g., $V_1 V_1^g = \langle e_{5,6}, e_{4,6} \rangle$. Arguing as in G.2.3 with I, $V_1 V_1^g$ in the roles of "L, V", (2) and (3) hold. In particular $Z_I \leq O_2(I) \leq Q$, so that $Z_I \leq A$.

Let $\hat{P} := P/Z_I$. For $v \in V_H \cap Q - Z_I$, $\hat{P}_v := \langle \hat{v}^I \rangle \cong E_4$ as $\hat{P}$ is the sum of natural modules for I/P. Thus if $v \in U_H$, then $\hat{P}_v \leq \hat{U}_H \hat{U}_H^g$ and hence $\langle U_H^I \rangle Z_I = U_H U_H^g Z_I$, proving (4).

By F.9.6.3, $[D_H, U_H^g] \leq V_1^g \cap U_H = 1$ using (1). Then by symmetry, $D_{H^g} \leq H_C$, so by 13.7.3.7, $[D_{H^g}, V_H] \leq V_1$ and $[D_H, A] \leq V_1^g \cap D_H = 1$. Hence (6) is established.

By 13.7.9.1 and (6), $[I, D_H D_{H_g}] \leq V_1 V_1^g$, so

$$\langle D_H^I \rangle = D_H V_1^g = D_{H^g} V_1$$

and hence (5) holds.

By 13.7.3.6, $[E_{H^g}, V_H] \leq U_H$, so for $v \in V_H - Q$, $[E_{H^g}, v] \leq U_H$, and hence $E_{H^g} \leq U_H^g Z_I$ by (4). Thus

$$E_{H^g} = E_{H^g} \cap U_H^g Z_I = (E_{H^g} \cap U_{H^g}) Z_I = D_{H^g} Z_I$$

and $D_{H^g} \leq Z_I$ by (5), so that $E_{H^g} \leq Z_I \leq V_H$, and hence $E_{H^g} \leq Q \cap H_C$ by (2) and 13.7.3.2. But $Z_I \leq E_{H^g}$, so $E_{H^g} = Z_I$, and then by symmetry $E_{H^g} = Z_I = E_H \leq V_H$. Then $[E_{H^g}, V_H] \leq V_1$ by 13.7.3.5, completing the proof of (7).

Next there exists $l \in L$ with $\bar{L}_1^l = O^2(\bar{I})O_2(\bar{L}_1^l) \cong A_4$. We saw Q acts on I, so L_1^l has $k+1$ noncentral 2-chief factors, where k is the number of noncentral 2-chief factors of I. One of those k factors is $V_1 V_1^g$, and by (2) and (3) there are $k-1 = m(O_2(I)/Z_I)/2 = m(A/Z_I)$ factors on $O_2(I)/Z_I$. Now

$$m(A/Z_I) = m(A/E_{H^g}) = m(A/(A \cap Q_H)) = m(A^*)$$

by (7), so that (8) holds. As $V_3 \cap V_3^g$ is a complement to V_1 in V_3, and $V_1 \not\leq U_H^g$ by (1), (9) holds.

Since $E_{H^g} \leq Q$ by (7), (10) is immediate from the definitions of A and E_{H^g}.   □

LEMMA 13.7.11. *(1)* $D_H < U_H$.
(2) $1 \neq U_H^{g*}$ *and* $U_H^g \leq A$.

PROOF. Recall $U_H^g \leq A$, $U_{H^g} = (U_H)^g$, and $D_{H^g} = (D_H)^g$. Therefore $D_H = U_H$ iff $D_H^g = U_{H^g}$ iff $U_{H^g} \leq Q_H = C_H(\tilde{U}_H)$; and hence (1) and (2) are equivalent. Thus we may assume that $U_H = D_H$, and it remains to derive a contradiction. By 13.7.10.6, $U_H = D_H$ centralizes A, so $A \leq Q_H$. Thus by 13.7.10, $A = A \cap Q_H = E_{H^g}$, while by 13.7.10.7, $E_{H^g} = E_H \leq V_H$; so using symmetry we conclude $V_H^g \cap Q = V_H \cap Q$. Let Λ be the graph on the points of V obtained by joining non-orthogonal points. Then Λ is connected, so $V_H \cap Q = V_H^x \cap Q$ for all $x \in L$. Therefore L acts on $V_H \cap Q$. Now $\Phi(V_H) = V_1$ by 13.7.9.1, so as L does not act on V_1, $\Phi(V_H \cap Q) = 1$. Also by 13.7.9.1, $V_H = Z(V_H)V_0$ with V_0 extraspecial and $|V_H : V_H \cap Q| = 2$; so we conclude $\Phi(Z(V_H)) = 1$ and $V_0 \cong D_8$. But now, V_H has just two maximal elementary abelian subgroups, one of which is $Z(V_H)V$; so both are normal in $O^2(H)T = H$, and hence $\langle V^H \rangle = V_H = Z(V_H)V$ is abelian, contrary to our choice of G as a counterexample to Theorem 13.7.8.   □

Recall that $U_0 = C_{U_H}(Q_H)$, and from 13.7.4 and 13.7.9.2 that $U_H^+ = U_H/U_0$ is H-dual to Q_H/H_C and H^* is faithful on U_H^+.

LEMMA 13.7.12. *(1)* A^* *centralizes* $(Q_H \cap Q)H_C/H_C$ *of corank 2 in* Q_H/H_C.
(2) $[U_H^+, A^*] \leq V_3^+$.
(3) For $F \in \{A^*, U_H^{g*}\}$, $r_{F,\tilde{U}_H} \leq 1 \geq r_{F,U_H^+}$, *so* F *contains* FF^*-*offenders on each of these* FF-*modules.*

PROOF. As $g \in N_G(Q)$, Q acts on A, so by 13.7.10, $[A, Q_H \cap Q] \leq A \cap Q_H = E_{H^g}$, and $E_{H^g} \leq H_C$ by 13.7.10.7. Further by 13.7.9.1, $\bar{Q}_H = \bar{R}_1 \cong E_8$ and $\bar{V}_H$ is of order 2, so $V_H = H_C$ by parts (2) and (7) of 13.7.3. Thus $|Q_H : (Q_H \cap Q)H_C| = 4$, completing the proof of (1). Then since V_3 centralizes Q and $C_{V_3}(Q_H) = V_1$ is of index 4 in V_3, V_3^+ corresponds to $(Q_H \cap Q)H_C/H_C$ under the duality between U_H^+ and Q_H/H_C, so part (2) is the dual of (1). By 13.7.10.6, $[D_H, A] = 1$, and

$$m(U_H^{g*}) = m(U_H^g/D_{H^g}) = m(U_H^g/D_H^g) = m(U_H/D_H),$$

so $r_{F,\tilde{U}_H} \leq 1$ for $F \in \{A^*, U_H^{g*}\}$, keeping in mind that $1 \neq U_H^{g*} \leq A^*$ by 13.7.11.2. Then $r_{F,U_H^+} \leq 1$ also holds using 13.7.4.1. Thus both modules are FF-modules for H^*, and B.1.4.4 shows that F contains FF^*-offenders on the modules.   □

LEMMA 13.7.13. *If* $m(\tilde{U}_H) = 4$ *then* $[\tilde{U}_H, L_1] < \tilde{U}_H$.

PROOF. Assume otherwise. Observe there is $X_1 \in Syl_3(L_1)$ with $V_1 V_1^g = C_V(X_1)$, and we can take $g \in N_L(X_1)$. Thus $X_1 \leq H \cap H^g$, so X_1 acts on D_H, and as $\tilde{U}_H = [\tilde{U}_H, L_1]$ is of rank 4, X_1 is irreducible on $\tilde{U}_H / \tilde{V}_3$. Thus X_1 has two nontrivial chief factors on $\tilde{U}_H = U_H^+$. By (7) and (9) of 13.7.10, $V_3 \leq Z_I = E_H$, so $V_3 \leq U_H \cap E_H = D_H$. Then as $D_H < U_H$ by 13.7.11.1, we conclude $D_H = V_3$, so that L_1 is irreducible on U_H / D_H. Then X_1 is also irreducible on $U_H^g / D_H^g \cong U_H^{g*} \leq O_2(L_1^*)$, so L_1 has a noncentral 2-chief factor not in Q_H. Also L_1 has two noncentral chief factors on each of U_H^+ and Q_H / H_C by 13.7.4.2, so L_1 has at least five noncentral 2-chief factors, with $[H_C, L_1] \leq U_H$ in case of equality. Therefore by 13.7.10.8, $m(A^*) \geq 3$, and $[H_C, L_1] \leq U_H$ in case of equality. Then as $m(U_H^+) = 4$, inspecting the subgroups H^* of $GL_4(2)$ of 2-rank at least 3 with $O_2(H^*) = 1$, we conclude $H^* \cong L_4(2)$ or S_6, with $m(A^*) = 3$ in the second case. The first case is impossible by 13.7.6.1 and 13.7.9.2. In the second, $[H_C, L_1] \leq U_H$, so $[V_H, H] \leq U_H$ in view of 13.7.3.2, contrary to 13.7.7. $\qquad \square$

LEMMA 13.7.14. $F(H^*)$ is centralized by each minimal FF*-offender B^* on U_H^+ contained in A^*.

PROOF. Set $\mathcal{P} := \{C^* \in \mathcal{P}^*(H^*, U_H^+) : [F(H^*), C^*] \neq 1\}$ and suppose $B^* \in \mathcal{P}$ with $B^* \leq A^*$. Then by B.1.9, there is a normal subgroup N^* of H^* such that $N^* = H_1^* \times \cdots \times H_s^*$, $H_i^* \cong L_2(2)$, with $s = 1$ or 2 since $m_3(H) \leq 2$, $U_N^+ := [U_H^+, N^*] = U_1^+ \oplus \cdots \oplus U_s^+$ with $U_i^+ := [U_H^+, H_i^*] \cong E_4$ affording the natural module for H_i^*, and

$$\mathcal{P} = \bigcup_i Syl_2(H_i^*).$$

In particular we may take $B^* \in Syl_2(H_1^*)$. Then $[U_1^+, B^*] = [U_H^+, B^*] \leq [U_H^+, A^*] \leq V_3^+$ by 13.7.12.2. As $s \leq 2$ and $L_1 = O^2(L_1)$, L_1 acts on H_i^* for each i, and hence also on U_i^+. Then $V_3^+ = [U_1^+, B^*, L_1^*] \leq U_1^+$, so $V_3^+ = U_1^+$ as both are of rank 2. However as $A \leq Q$, $B^* \leq A^* \leq R_1^* \trianglelefteq L_1^* T^*$, so L_1^* acts on $R_1^* \cap H_1^* = B^*$ and hence also on $[U_1^+, B^*]$ of rank 1. This is impossible as L_1^* is irreducible on $V_3^+ = U_1^+$. $\qquad \square$

Since $O_2(H^*) = 1$, by 13.7.14 some member B^* of $\mathcal{P}^*(H^*, U_H^+)$ contained in A^* acts nontrivially on $E(H^*)$. So there is $K \in \mathcal{C}(H)$ with K^* quasisimple and $[K^*, B^*] \neq 1$. Let $K_0 := \langle K^T \rangle$ and $U_K := [U_H, K]$. By B.1.5.4, B^* acts on K^*, so $K^* = [K^*, B^*]$.

LEMMA 13.7.15. (1) $K^* \cong L_n(2)$, A_n, $SL_3(4)$, or $Sp_4(4)$.
(2) $U_K^+ \in Irr_+(K^*, U_H^+)$.
(3) $K_0 = K$.
(4) $U_H^+ = U_K^+$.

PROOF. By B.1.5.1, $Aut_B(U_K^+)$ is an FF*-offender on U_K^+. Therefore by B.5.6 and B.5.1.1, either $U_K^+ \in Irr_+(K^*, U_H^+)$, or one of conclusions (ii)–(iv) of B.5.1.1 holds.

In the first case, (2) holds, with K^* and $\hat{U}_K := U_K^+ / C_{U_K^+}(K)$ described in B.4.2. However by 13.7.12.2, $m([\hat{U}_K, B^*]) \leq 2$, so we conclude K^* is one of the groups listed in (1) in this case: Recall B.4.6.13 eliminates $K^* \cong G_2(2)'$, and K^* is not $\hat{A}_6$ since $m([\hat{U}_K, B^*]) = 4$ for the unique FF*-offender B^* in B.4.2.8.

So assume that the second case holds. As $m([U_K^+, B^*]) \leq 2$, U_K^+ has exactly two chief factors U_1 and U_2, and B^* induces a group of transvections with fixed center on U_i; but this contradicts the structure of FF*-offenders in conclusions (ii)–(iv) of B.5.1.1. This completes the proof of (1) and (2).

Suppose $K < K_0$. Then by 1.2.1.3 and (1), $K^* \cong L_3(2)$ or A_5, and by 1.2.2.1, $K_0 = O^{3'}(H)$, so $L_1 \le K_0$. Then as T acts on L_1, we conclude $K^* \cong A_5$ and L_1^* is diagonally embedded in K_0^*. Since $B^* \le R_1^*$, B^* induces inner automorphisms on K^*, so by B.4.2, $\hat{U}_K$ is the natural module for K^* rather than the A_5-module, and $V_3^+ = [U_H^+, B^*] \le U_K^+$. Now as T acts on V_3, T acts on U_K and hence also on K, contrary to assumption. This establishes (3).

Next $1 \ne [U_K^+, B^*] \le V_3^+ \cap U_K^+$ by 13.7.12.2. Then as L_1 acts on K, and acts irreducibly on V_3^+,

$$V_3^+ = [V_3 \cap U_K^+, L_1] \le U_K^+,$$

so $U_H^+ = \langle V_3^{+H} \rangle = U_K^+$ since U_K is H-invariant by (3), and hence (4) holds. $\qquad\square$

LEMMA 13.7.16. (1) T^* is faithful on K^*.
(2) $[U_0, K] = 1$.
(3) $\tilde{U}_K \in Irr_+(K^*, \tilde{U}_H)$.

PROOF. By parts (2) and (4) of 13.7.15 and A.1.41, $C_{H^*}(K^*)$ is of odd order, so (1) holds. Also (3) follows from 13.7.15.2 if (2) holds, so it remains to prove (2).

Assume (2) fails. Then as $U_0 \le Z(Q_H)$, $K \in \mathcal{L}_f(G,T)$, so by 13.5.2.1, K^* is A_5, $L_3(2)$, A_6, or $\hat{A}_6$. Further K acts nontrivially on U_0 and U_H^+, so K has at least two noncentral chief factors on $\tilde{U}_H$. On the other hand by 13.7.12.3, U_H^{g*} contains an FF*-offender D^* on $\tilde{U}_H$, and by (1), D^* is faithful on K^*, so by B.1.5.1, $Aut_D(\tilde{U}_K)$ is an FF*-offender on the FF-module $\tilde{U}_K$. Then by Theorems B.5.6 and B.5.1.1, $K^* \cong L_3(2)$ and $\tilde{U}_K$ is the sum of two isomorphic natural modules for K^*. Then as $U_H^g \le Q \le O_2(L_1 T)$, $L_1^*(T \cap K)^*$ is the stabilizer of a line in each irreducible and $Aut_{U_H^g}(\tilde{U}_K) = O_2(Aut_{L_1}(\tilde{U}_K))$.

By 13.5.2.3, $U_1 := [U_0, K]$ is a natural module, so U_1 is isomorphic to $\tilde{U}_1$, and so $C_{U_1}(L_1) = 1$. But $Z_1 := Z \cap U_1$ is of order 2, and as $Aut_{U_H^g}(U_1) = O_2(Aut_{L_1}(U_1))$, $Z_1 \le [U_0, U_H^g]$. Then as $U_1 \le U_H \le N_Q(U_H^g)$ by 13.7.3.3, $Z_1 \le U_H^g$, so Z_1 is centralized by $\langle V_H, V_H^g \rangle = I$ and by T. Thus $L \le \langle I, T \rangle \le C_G(Z_1)$, contrary to $C_{U_1}(L_1) = 1$. $\qquad\square$

LEMMA 13.7.17. (1) $Z_I \cap U_0 = V_1$.
(2) If $U_0 \le [U_H, A]V_3$, then $U_0 = V_1$.

PROOF. First Q_H and $I = \langle V_H, V_H^g \rangle$ centralize $U_0 \cap U_H^g =: U_1$, so $L_0 := \langle I, Q_H \rangle \le C_G(U_1)$. However $\bar{Q}_H = \bar{R}_1$ by 13.7.9.1, and $\bar{L}\bar{T} \le \langle \bar{R}_1, \bar{I} \rangle$, so $LT = L_0 Q$. Further Q and L_0 act on U_1, so $LT = L_0 Q \le N_G(U_1)$.

If $U_1 \ne 1$ then $N_G(U_1) \le M = !\mathcal{M}(LT)$. But then by 13.7.16.2, $K \le C_G(U_1) \le M$, contrary to 13.3.9. Thus $U_1 = 1$.

Next by 13.7.10.5, $D_H = (D_H \cap D_{H^g})V_1$, and by 13.7.10.7, $E_H = Z_I$, so

$$Z_I \cap U_0 = E_H \cap U_0 = D_H \cap U_0 = (D_H \cap D_{H^g})V_1 \cap U_0$$

$$= (D_H \cap D_{H^g} \cap U_0)V_1 = (U_{H^g} \cap U_0)V_1 = U_1 V_1 = V_1,$$

establishing (1). By 13.7.3.2, $U_H \le Q \cap V_H$, so $[A, U_H] \le Z_I$ by 13.7.10.3. By 13.7.10.9, $V_3 \le Z_I$. Thus if $U_0 \le [U_H, A]V_3$ then $U_0 \le Z_I$, so (1) implies (2). $\qquad\square$

LEMMA 13.7.18. *Either*

(1) $K^* \cong L_2(4)$ *and* $\tilde{U}_K / C_{\tilde{U}_K}(K)$ *is the natural module, or*

(2) $K^* \cong A_6$ *and* $\tilde{U}_H / C_{\tilde{U}_H}(K)$ *is a natural module on which* L_1 *has two non-central chief factors.*

PROOF. By 13.7.16.3,

$$\tilde{U}_K / C_{\tilde{U}_K}(K) \cong U_K^+ / C_{U_K^+}(K)$$

is an irreducible K-module. Also $m([U_H^+, A]) \leq 2$ by 13.7.12.2, while B^* is an FF*-offender on the FF-module U_H^+. Further K^* appears in 13.7.15.1, so applying the remark before 13.7.6 to the restricted list in 13.7.15.1: either $U_K^+ / C_{U_K^+}(K)$ is the natural module for $K^* \cong L_2(4)$, or K^*, $U_K^+ / C_{U_K^+}(K)$ is one of the pairs considered in 13.7.6. In the former case (1) holds, so we may assume the latter. If $K^* \cong A_6$, then (2) holds by 13.7.6.3. Therefore we must eliminate the remaining cases in 13.7.15.1.

Observe that part (1) of 13.7.6 eliminates $L_3(2)$, parts (1) and (2) of that result eliminate $L_5(2)$, and $L_4(2) \cong A_8$ is eliminated by parts (1) and (3) of that result and 13.7.9.2. The natural module for A_5 is eliminated by part (3) of 13.7.6, and A_7 is eliminated by parts (3) and (4) and 13.7.9.2. Finally $SL_3(4)$ and $Sp_4(4)$ are eliminated by part (5) of 13.7.6. $\qquad\square$

LEMMA 13.7.19. $L_1 \leq K$.

PROOF. Assume $L_1 \not\leq K$. Then case (1) of 13.7.18 holds by 13.7.18 and A.3.18. Then as $L_1 = [L_1, T]$, while $|L_1|_3 = 3$ by 13.7.9.2, we conclude $H^* \cong \Gamma L_2(4)$ and either $L_1^* = O_3(H^*)$, or L_1^* is diagonally embedded in $O_3(H^*) \times K^*$. Hence $R_1 = (T \cap K)O_2(KR_1)$, so $m_3(N_H(R_1)) > 1$. Therefore as $L_1 = O^{3'}(H \cap M)$ by 13.7.3.9, and this group has 3-rank 1 by 13.7.9.2, $N_H(R_1) \not\leq M$. Thus L is an A_6-block by 13.2.2.7. Therefore L_1 has just two noncentral 2-chief factors. But if $L_1^* = O_3(H^*)$, then L_1 has two noncentral chief factors on U_H^+, and hence also two on Q_H/H_C by the duality 13.7.4.2. Therefore L_1^* is diagonally embedded, so L_1^* has one chief factor on $O_2(L_1^*)$, plus one each on U_H^+ and Q_H/H_C, again a contradiction. $\qquad\square$

LEMMA 13.7.20. $\tilde{U}_K = \tilde{U}_H$.

PROOF. This follows from 13.7.19 and 13.7.5.2. $\qquad\square$

LEMMA 13.7.21. K^* *is not* $L_2(4)$.

PROOF. Assume $K^* \cong L_2(4)$. By 13.7.18.1 and 13.7.20, $\tilde{U}_H / C_{\tilde{U}_H}(K)$ is the natural module, while by 13.7.19, $L_1 \leq K$, so $\tilde{V}_3 = [\tilde{V}_3, L_1]$ is a complement to $C_{\tilde{U}_H}(K)$ in $C_{\tilde{U}_H}(T \cap K) =: \tilde{W}$. If $C_{\tilde{U}_H}(K) = 1$, then $\tilde{U}_H = [\tilde{U}_H, L_1]$ is of rank 4, contrary to 13.7.13. Thus $C_{\tilde{U}_H}(K) \neq 1$.

By B.4.2.1, $(T \cap K)^*$ is the unique FF*-offender in T^*, so $A^* = (T \cap K)^*$ by 13.7.12.3. But for each $1 \neq a^* \in A^*$, $[U_H^+, a] = V_3^+$, so $C_{U_H^+}(K) = 1$ and hence $\tilde{U}_0 = C_{\tilde{U}_H}(K)$. Thus $V_1 < U_0$ and $U_0 \leq [U_H, A]V_3$. This contradicts 13.7.17.2. $\qquad\square$

By 13.7.18 and 13.7.21, $K^* \cong A_6$ and $\tilde{U}_H / C_{\tilde{U}_H}(K)$ is a natural module on which L_1 has two noncentral chief factors. Now L_1 has two noncentral chief factors on each of U_H and Q_H/H_C by 13.7.4.2, one on $O_2(L_1^*)$, and at least one on V_H/U_H by 13.7.7;

so L_1 has at least six noncentral 2-chief factors. Therefore $m(A^*) \geq 4$ by 13.7.10.8. On the other hand as $End_K(\tilde{U}_H/C_{\tilde{U}_H}(K)) \cong \mathbf{F}_2$, we conclude $K^* = F^*(H^*)$; so A^* acts faithfully on K^*, and hence $m(A^*) \leq m_2(Aut(K^*)) = 3$. This contradiction completes the proof of Theorem 13.7.8.

13.8. Finishing the treatment of $\mathbf{A}_6$

In this section, we complete the treatment of A_6. We prove:

THEOREM 13.8.1. *Assume Hypothesis 13.3.1 with $L/O_{2,Z}(L) \cong A_6$. Then G is isomorphic to $Sp_6(2)$ or $U_4(3)$.*

Throughout this section, we assume that G is a counterexample to Theorem 13.8.1.

Since $L/O_{2,Z}(L) \cong A_6$, we continue with the notation established in section 13.5: Namely we adopt the notational conventions of section B.3 and Notations 12.2.5 and 13.2.1.

As G is a counterexample to Theorem 13.8.1, G is not isomorphic to $U_4(3)$ or $Sp_6(2)$. Thus Hypotheses 13.5.1 and 13.7.1 hold, so we may apply results from sections 13.5 and 13.7. In particular recall from 13.5.2.3 that V is the 4-dimensional A_6-module. The main result Theorem 13.7.8 of section 13.7 has reduced us to the following situation (where $\mathcal{H}_z$ is defined below):

LEMMA 13.8.2. $\langle V^{G_1} \rangle$ *is abelian, so V_H is abelian for each $H \in \mathcal{H}_z$.*

As in the previous section, there are no quasithin examples under this restriction, so we are continuing to work toward a contradiction. Again as far as we can tell, there are not even any shadows.

LEMMA 13.8.3. *If $g \in G$ with $1 \neq V \cap V^g$, then $[V, V^g] = 1$.*

PROOF. As L is transitive on $V^{\#}$, G_1 is transitive on conjugates of V containing V_1 by A.1.7.1, so we may take $g \in G_1$. Then $\langle V, V^g \rangle \leq \langle V^{G_1} \rangle$, so the result follows from 13.8.2. $\square$

As usual z is a generator for V_1, and as in Notation 13.5.8, $\tilde{G}_1 := G_1/V_1$. By 13.3.6, $G_1 \not\leq M$, so $\mathcal{H}_z \neq \emptyset$, where

$$\mathcal{H}_z := \{H \in \mathcal{H}(L_1T) : H \leq G_1 \text{ and } H \not\leq M\}.$$

For the remainder of the section, let H denote some member of $\mathcal{H}_z$.

By 13.5.7, Hypothesis F.9.1 is satisfied with V_3 in the role of "V_+". From Notation 13.5.8, $U_H := \langle V_3^H \rangle$, $V_H := \langle V^H \rangle$, $Q_H := O_2(H) = C_H(\tilde{U}_H)$, and $H^* := H/Q_H$ so that $O_2(H^*) = 1$. Furthermore set $H_C := C_H(U_H)$; then $H_C \leq Q_H$.

Now condition (f) of Hypothesis F.9.8 is satisfied by 13.8.3, and condition (g.i) of Hypothesis F.9.8 is satisfied since $[V, C_H(V_3)] \leq V_1$ by 13.5.4.4; indeed $C_{\bar{M}_V}(V_3) \leq \langle (5,6) \rangle$, with $(5,6)$ inducing the transvection on V with center V_1.

Thus we can appeal to the results in sections F.7 and F.9. In particular, we form the coset geometry Γ of Definition F.7.2 on the pair of subgroups LT and H, and let $b := b(\Gamma, V)$. Choose $\gamma \in \Gamma$ with $d(\gamma_0, \gamma) = b$ and $V \not\leq G_{\gamma}^{(1)}$. By F.9.11.1, b is odd and $b \geq 3$. Without loss γ_1 is on the geodesic

$$\gamma_0, \gamma_1, \ldots, \gamma_b := \gamma$$

from γ_0 to γ.

Recall we may choose g_b with $(\gamma_0, \gamma_1)g_b = (\gamma_{b-1}, \gamma)$. Then $U_\gamma := U_H^{g_b}$, $V_\gamma := V_H^{g_b}$, $Q_\gamma := Q_H^{g_b}$, and $A_1 := V_1^{g_b}$. Further $D_H := C_{U_H}(U_\gamma/A_1)$, $E_H := C_{V_H}(U_\gamma/A_1)$, $D_\gamma := C_{U_\gamma}(\tilde{U}_H)$, and $E_\gamma := C_{V_\gamma}(\tilde{U}_H)$. We will appeal extensively to lemmas F.9.13 and F.9.16.

Set $U_L := \langle U_H^L \rangle$, and $Q := O_2(LT)$.

LEMMA 13.8.4. *(1) $b \geq 3$ is odd.*
(2) $U_L \leq Q = O_2(LT)$.
(3) If $b > 3$, then U_L is abelian.
(4) If $b = 3$, then $A_1 \leq V^h$ for some $h \in H$.
(5) $V_3 = V \cap U_H < U_H$, and V_H/U_H is a quotient of the $\mathbf{F}_2 H^*$-permutation module on $H^*/(H \cap M)^*$ with $[V_H/U_H, H] \neq 0$.
(6) V_γ^* is quadratic on V_H/U_H and $\tilde{U}_H$.
(7) $V < U_L$.

PROOF. We have already observed that (1) holds. Part (2) follows from 13.7.3.3. Parts (3) and (4) follow from parts (1) and (2) of F.9.14.

By 13.7.7, $[V_H, H] \not\leq U_H$, so $V \not\leq U_H$. Then as $V_3 \leq V \cap U_H$ with V_3 of index 2 in V, $V_3 = V \cap U_H$. By 13.7.3, $H \cap M$ acts on $VU_H/U_H \cong V/(V \cap U_H) = V/V_3 \cong \mathbf{Z}_2$, so as $V_H = \langle V^H \rangle$ and $[V_H, H] \not\leq U_H$, (5) holds.

As V_γ is abelian and V_H and V_γ normalize each other by F.9.13.2, (6) follows. As $V_3 \leq U_H$, $V = \langle V_3^L \rangle \leq U_L$, and as $V \not\leq U_H \leq U_L$, $V < U_L$. Thus (7) holds. $\square$

LEMMA 13.8.5. *If some element of H^* induces an $\mathbf{F}_2$-transvection on $\tilde{U}_H$, then*
(1) $H = KT$ with $K \in \mathcal{C}(H)$.
(2) Either

(a) $H^* \cong S_6$, $L/O_2(L) \cong A_6$, and L_1 has two noncentral chief factors on $\tilde{U}_H$, or

(b) $H^* \cong S_7$ or $L_4(2)$, and $L/O_2(L) \cong \hat{A}_6$.

(3) $\tilde{U}_H$ is a natural module for H^* or the 5-dimensional cover of such a module for $H^* \cong S_6$.

PROOF. Let $t^* \in T^*$ induce an $\mathbf{F}_2$-transvection on $\tilde{U}_H$. If $K^* = [K^*, t^*] \neq 1$ for some $K \in \mathcal{C}(H)$, then as t^* is an $\mathbf{F}_2$-transvection, we conclude from G.6.4 that K^* is $L_n(2)$ or A_n and $\tilde{U}_K/C_{\tilde{U}_K}(K)$ is a natural module, where $U_K := [U_H, K]$. Hence the lemma follows from parts (1) and (3) of 13.7.6, using I.1.6.1 in the latter case.

So we may assume instead that $K^* := \langle t^{*H} \rangle$ is solvable, and we derive a contradiction. By B.1.8, $K^* = K_1^* \times \cdots K_s^*$, $K_i^* \cong L_2(2)$, with $s \leq 2$ since $m_3(H) \leq 2$, and $\tilde{U}_K = [\tilde{U}_H, K^*] = \tilde{U}_1 \oplus \cdots \oplus \tilde{U}_s$, where $\tilde{U}_i := [\tilde{U}_H, K_i] \cong E_4$. Then L_1 acts on each K_i. Thus if $s = 2$, then as $m_3(H) \leq 2$, $L_1^* \leq K^*$. This is impossible as T normalizes a subgroup of order 3 of L_1^*, whereas T is irreducible on $K^* = \langle t^{*H} \rangle$ by construction.

Hence $s = 1$ and $K^* = K_1^* \trianglelefteq H^*$. This time we conclude from the T-invariance of L_1 that either

(a) $L/O_2(L) \cong A_6$ so that $|L_1|_3 = 3$, and either $L_1^* = O^2(K^*)$ or $[K^*, L_1^*] = 1$, or

(b) $L/O_2(L) \cong \hat{A}_6$ so that L_1 has 3-rank 2, and hence $O^2(K) = L_0$ or $L_{1,+}$.

If $O^2(K) = L_0$, then $H \leq N_G(O^2(K)) = N_G(L_0) \leq M$ by 13.2.2.9, contrary to $H \not\leq M$. If $O^2(K) = L_1$ or $L_{1,+}$, then $\tilde{U}_K = \tilde{V}_3$, so $H \leq G_1 \cap G_3 \leq M$ by 13.5.5 for the same contradiction. Thus $[K^*, L_1^*] = 1$, so

$$\tilde{V}_3 = [\tilde{V}_3, L_1] \leq [\tilde{U}_H, L_1] \leq C_{\tilde{U}_H}(K),$$

and then as $K^* \trianglelefteq H^*$, $\tilde{U}_H = \langle \tilde{V}_3^H \rangle \leq C_{\tilde{U}_H}(K)$, contrary to $K^* \neq 1$. $\qquad\square$

LEMMA 13.8.6. *Assume $H = KT$ with $K \in \mathcal{C}(H)$, $K^* \cong A_6$, and $\tilde{U}_H$ is a natural module for K^* or its 5-dimensional cover. Let $K_2 := O^2(C_H(V_2))$ and $U_2 := \langle V^{K_2} \rangle$. Then*

(1) $\tilde{V}_2$ is generated by a vector of weight 4 in $\tilde{U}_H$ and $K_2 T / O_2(K_2 T) \cong S_3$.

(2) $[K_2, L_2] \leq O_2(K_2) \cap O_2(L_2)$.

(3) $U_2 = [U_2, L_2] \leq U_L$.

(4) If $m(\tilde{U}_H) = 4$ and $\bar{L}\bar{T} \cong S_6$, then $m(U_2) = 6$ and U_L/V has a quotient isomorphic to the 16-dimensional Steinberg module for $\bar{L}\bar{T}$.

(5) If $m(\tilde{U}_H) = 5$ and $U_1 := C_{U_H}(K)$, then $m(U_2) = 8$, $U_0 := \langle U_1^L \rangle \leq U_L$, and U_0/V is a quotient of the 15-dimensional permutation module for $\bar{L}\bar{T}$ on $\bar{L}\bar{T}/\bar{L}_1\bar{T}$.

(6) $L/O_2(L) \cong A_6$.

PROOF. Observe that (1) and (6) hold by 13.7.6.3. In particular, $K_2^* T^*$ is the parabolic of H^* stabilizing the point $\tilde{V}_2$ generated by a vector of weight 4, and $\tilde{V}_3$ is a line with all vectors of weight 4.

By (6) and parts (1) and (6) of 13.5.4, $L_2 \trianglelefteq G_2$, so $[L_2, K_2] \leq C_{L_2}(V_2) = O_2(L_2)$, and hence (2) holds. Now

$$U_2 = \langle V^{K_2} \rangle = \langle V_3^{L_2 K_2} \rangle = \langle V_3^{K_2 L_2} \rangle \leq \langle U_H^{L_2} \rangle \leq U_L,$$

and as $L_2 \trianglelefteq L_2 K_2$ and $V = [V, L_2]$, $U_2 = [U_2, L_2]$, so (3) holds. Set $\hat{U}_2 := U_2/V_2$; it follows that $m(\hat{U}_2) = 2m(\langle \hat{V}_3^{K_2} \rangle)$. Thus $m(U_2)$ is 6 in case (4), and 8 in case (5).

Assume the hypotheses of (4), and recall $V < U_L$ by 13.8.4.7. Let $V \leq W < U_L$ with LT irreducible on U_L/W. By 13.8.5.2a, $\tilde{U}_H = [\tilde{U}_H, L_1]$, so that $U_H = [U_H, L_1]$ since $V_1 = [V_3, O_2(L_1)]$; hence $U_H V/V = [U_H V/V, L_1] \cong E_4$. As $W < U_L = \langle U_H^L \rangle$, $U_H \not\leq W$, so that $U_H W/W$ is $L_1 T$-isomorphic to $U_H V/V$. Similarly by (3), $U_2 = [U_2, L_2] \leq U_L$, and as we saw $m(U_2) = 6$ in this case, $U_2/V = [U_2/V, L_2] \cong E_4$, so $U_2 W/W$ is $L_2 T$-isomorphic to U_2/V. Hence (4) holds by G.5.2.

Finally $[U_1, L_1 T] \leq V_1 \leq V$, so (5) holds. $\qquad\square$

LEMMA 13.8.7. *Assume $H = G_1$. Then $D_H < U_H$ iff $D_\gamma < U_\gamma$.*

PROOF. Assume the lemma fails. If $D_H = U_H$ but $D_\gamma < U_\gamma$, then $U_\gamma \not\leq Q_H$, and in particular $V_\gamma \not\leq Q_H$. Thus there is some $\beta \in \Gamma(\gamma)$ with $V_\beta \not\leq Q_H$. By F.7.9.1, $d(\beta, \gamma_1) = b$. Thus we have symmetry (cf. the first part of Remark F.9.17) between the edges γ_0, γ_1 and β, γ, so we may assume that $D_H < U_H$ but $D_\gamma = U_\gamma$. Then case (i) of F.9.16.1 holds, so that U_H induces a nontrivial group of transvections on U_γ with center V_1. Recall there is $g \in G_0 := \langle LT, H \rangle$ with $\gamma g = \gamma_1$, and setting $\alpha := \gamma_1 g$ and $U_\alpha = U_H^g$, $U_\alpha^* \neq 1$ but $[U_H, U_\alpha] = V_1^g =: A_1$. Then U_α induces a group of transvections on $\tilde{U}_H$ with center $\tilde{A}_1$, so by 13.8.5, $H = KT$ for some $K \in \mathcal{C}(H)$, and $\tilde{U}_H$ is a natural module for $H^* \cong L_4(2)$, S_6, or S_7, or the 5-dimensional cover of a natural module for $H^* \cong S_6$.

Suppose one of the first three cases holds, namely $\tilde{U}_H$ is an irreducible module. To eliminate these cases, it will suffice to show:

$$V_1^{gh} \leq V_2 \quad \text{for some} \quad h \in H. \tag{*}$$

For if (*) holds, then $V_1^{gh} = V_1^l$ for $l \in L_2T$ with $l^2 \in H$. As $G_1 = H$, $U_H \trianglelefteq G_1$, so as $V_1^{gh} = V_1^l$, also $U_\alpha^h = U_H^{gh} = U_H^l$. Thus as $l^2 \in H$, l interchanges U_H and U_α^h, and also Q_H and Q_α^h, impossible as $U_\alpha \not\leq Q_H$ but $U_H \leq Q_\alpha$. This completes the proof of the sufficiency of (*). Now we establish (*) in each of the first three cases: If $\tilde{U}_H$ is the $L_4(2)$-module or S_6-module, then (*) holds as H is transitive on $\tilde{U}_H^\#$. If $\tilde{U}_H$ is the S_7-module, then (*) follows from 13.7.6.3b, which says $\tilde{V}_2$ is of weight 2, using the fact that the center $\tilde{V}_1^g$ of the transvection U_α^* is of weight 2.

Thus we may assume that $\tilde{U}_H$ is a 5-dimensional module for $H^* \cong S_6$. As U_α^* induces transvections on $\tilde{U}_H$ with center $\tilde{A}_1$, U_α^* has order 2, so $D_\alpha := U_\alpha \cap Q_H$ is a hyperplane of U_α; and as $D_\gamma = U_\gamma$, $[D_\alpha, U_H] = 1$ by F.9.13.7. As $U_\alpha^* \neq 1$, without loss $V_3^{g*} \neq 1$ and $[V_3^g, V_3] \neq 1$. Thus as we saw $[U_\alpha, U_H] = V_1^g$, $[V_3^g, V_3] = V_1^g \leq V_3^g$; so $V_3 \leq C_G(V_1^g) \cap N_G(V_3^g) \leq M_V^g$ by 13.5.5. Then V_3 lies in the unipotent radical of the stabilizer in M_V^g of V_1^g, and is nontrivial on the hyperplane V_3^g orthogonal to V_1^g, so $[V^g, V_3] > V_1^g$.

Define C to the preimage in U_H of $C_{\tilde{U}_H}(V_3^g)$; then $[U_\alpha, C] \leq V_1^g \cap V_1 = 1$, so $C \leq H_C^g$ and hence $[V^g, C] \leq V_1^g$ by 13.7.3.7. Thus from the action of S_6 on the core of the permutation module, $V_3^{g*} = V^{g*}$ is the group of transvections with center $\tilde{A}_1$, so $V^g = V_3^g(V^g \cap Q_H)$. Now $[V^g \cap Q_H, V_3] \leq V^g \cap V_1 = 1$ by 13.8.3. Thus $[V^g, V_3] = [V_3^g(V^g \cap Q_H), V_3] = [V_3^g, V_3] = V_1^g$, contrary to the previous paragraph. $\square$

LEMMA 13.8.8. *Either:*

(1) $D_\gamma = U_\gamma$ or $D_H = U_H$, and U_δ or V_δ induces a nontrivial group of transvections on $\tilde{U}_H$, for $\delta := \gamma g_b^{-1}$ or γ, respectively. Hence $H = KT$ for some $K \in \mathcal{C}(H)$, and H^* and its action on $\tilde{U}_H$ are described in 13.8.5.

(2) $D_\gamma < U_\gamma$, $D_H < U_H$, and we may choose γ so that $0 < m(U_\gamma^*) \geq m(U_H/D_H)$, and $U_\gamma^* \in \mathcal{Q}(H^*, \tilde{U}_H)$. Further there is $h \in H$ with $\gamma_2 h = \gamma_0$, and setting $\alpha := \gamma h$, $V_\alpha = V_\gamma^h \leq O_2(L_1T) \leq R_1$.

PROOF. If $D_\gamma = U_\gamma$, then (1) holds by F.9.16.1. Similarly as in the proof of the previous lemma, (1) holds if $D_H = U_H$. Thus we may assume $D_\gamma < U_\gamma$, so by F.9.16.4, we may choose γ as in conclusion (2); then the final statement of conclusion (2) follows from parts (1) and (2) of F.9.13. $\square$

LEMMA 13.8.9. *Assume some* $F \leq U_H$ *is* V_γ-*invariant and* $G_\gamma = \langle F^{G_\gamma} \rangle G_{\gamma, \gamma_{b-1}}$. *Then*

(1) $[F, V_\gamma] \not\leq U_\gamma$.

(2) *If* $[\tilde{F}, U_\gamma] = [\tilde{F}, V_\gamma]$, *then* $V_1 \not\leq U_\gamma$ *and* F *induces a group of transvections on* V_γ/U_γ *with center* $V_1 U_\gamma/U_\gamma$.

PROOF. Assume $[F, V_\gamma] \leq U_\gamma$. Then F centralizes V_γ/U_γ, so $X := \langle F^{G_\gamma} \rangle$ does also. But by 13.8.4.5, $V_{\gamma_{b-1}} U_\gamma/U_\gamma$ is of order 2, so the section is centralized by $G_{\gamma, \gamma_{b-1}}$, and hence also by $G_\gamma = XG_{\gamma, \gamma_{b-1}}$. But then as $V_\gamma = \langle V_{\gamma_{b-1}}^{G_\gamma} \rangle$, G_γ centralizes V_γ/U_γ, contrary to 13.7.7. Thus (1) is established.

So assume that $[\tilde{F}, U_\gamma] = [\tilde{F}, V_\gamma]$. Then $[F, V_\gamma] \le [F, U_\gamma]V_1 \le U_\gamma V_1$ as U_H acts on U_γ. If $V_1 \le U_\gamma$, then $[F, V_\gamma] \le U_\gamma$, contrary to (1), so (2) holds. $\qquad\square$

LEMMA 13.8.10. *If $m(U_\gamma^*) = 1$ and $U_H < D_H$, then*

(1) $m(U_H/D_H) = 1$, so we have symmetry between γ_1 and γ in the sense of Remark F.9.17.

(2) Either $V_1 \le U_\gamma$, or U_H induces transvections on U_γ with axis D_γ.

PROOF. Assume $m(U_\gamma^*) = 1$, so in particular case (2) of 13.8.8 holds. Then

$$1 = m(U_\gamma^*) = m(U_\gamma/D_\gamma) \ge m(U_H/D_H)$$

and $D_H < U_H$ by hypothesis, so we conclude that $m(U_H/D_H) = 1$, and we have symmetry between γ_1 and γ as discussed in Remark F.9.17. Now by F.9.13.6, $[D_\gamma, U_H] \le V_1 \cap U_\gamma$, so (2) follows. $\qquad\square$

LEMMA 13.8.11. *Assume $U_\gamma^* \ne 1$ and $G_\gamma = \langle U_H^{G_\gamma} \rangle G_{\gamma, \gamma_{b-1}}$. Then*

(1) If either $V_1 \le U_\gamma$, or no element of H induces a transvection on V_H/U_H, then $U_\gamma^* < V_\gamma^*$, so $m(V_\gamma^*) > 1$.

(2) If U_H does not induce a transvection on U_γ with axis D_γ, then $m(V_\gamma^*) > 1$.

PROOF. By hypothesis $U_\gamma^* \ne 1$ and $U_H \not\le U_\gamma$, so that case (2) of 13.8.8 holds and $D_H \ne U_H$. If $U_\gamma^* = V_\gamma^*$, then $[\tilde{U}_H, V_\gamma] = [\tilde{U}_H, U_\gamma]$, and so 13.8.9.2 supplies a contradiction with U_H in the role of "F"; hence $U_\gamma^* < V_\gamma^*$ so that (1) holds. Assume the hypotheses of (2) but with $m(V_\gamma^*) = 1$. Then as $1 \ne U_\gamma^* \le V_\gamma^*$, $U_\gamma^* = V_\gamma^*$ is of rank 1. Thus $V_1 \le U_\gamma$ by 13.8.10.2, contrary to (1); hence (2) holds. $\qquad\square$

LEMMA 13.8.12. *(1) If $K \in \mathcal{C}(H)$, then $K \not\le M$ and $\langle K, T \rangle L_1 \in \mathcal{H}_z$.*
(2) Let $X := O^2(O_{2,F}(H) \cap M)$. Then one of the following holds:

    (a) $X = 1$.
    (b) $L/O_2(L) \cong A_6$ and $X = L_1$.
    (c) $L/O_2(L) \cong \hat{A}_6$ and $X = L_{1,+}$.

PROOF. First $K \not\le M$ by 13.3.9 with $\langle K^T \rangle$ in the role of "Y", so (1) holds.

Now define X as in (2), and assume none of (a)–(c) holds. Then $O^2(O_{2,F}(H)) \le M$ by 13.3.9, so $O_{2,F}(H)T \in \mathcal{H}(T, M)$. Let F denote a T-invariant subgroup of $O_{2,F}(H)T$ minimal subject to $X \le F = O^2(F)$ and $FT \in \mathcal{H}(T, M)$. Then $XO_2(F) < F$ since $F \not\le M$, so as $F/O_2(F)$ is nilpotent, $X < O^2(N_F(XO_2(F)))$. But also $F \cap M = X$, so $O^2(N_F(XO_2(F))) = F$ by minimality of F. Then F normalizes $O^2(XO_2(F)) = X$, again contrary to 13.3.9, now with X, FT in the roles of "Y, H". $\qquad\square$

LEMMA 13.8.13. *Each solvable overgroup of L_1T in G_1 is contained in M.*

PROOF. If not, we may choose H solvable, and minimal subject to $H \in \mathcal{H}_z$. Then case (2) of 13.8.8 holds; in particular $1 \ne U_\gamma^* \in \mathcal{Q}(H^*, \tilde{U}_H)$ and $1 \ne V_\alpha^* \le R_1^*$. By 13.8.12, $O^2(O_{2,F}(H) \cap M) = 1$ or X, where $X := L_1$ if $L/O_2(L) \cong A_6$ and $X := L_{1,+}$ if $L/O_2(L) \cong \hat{A}_6$.

Now as $O_2(H^*) = 1$, there exists an odd prime p with $[O_p(H^*), V_\alpha^*] \ne 1$. So by the Supercritical Subgroups Lemma A.1.21 and A.1.24, there exists a subgroup $P \cong \mathbf{Z}_p$, E_{p^2}, or p^{1+2} such that $P^* \trianglelefteq H^*$, and V_α^* is nontrivial on P^*. If $P \le M$ then $P \le O^2(O_{2,F}(H) \cap M) \le X \le L_1$, so $[V_\alpha^*, P^*] \le O_2(L_1^*) \cap P^* = 1$, a contradiction. Thus $P \not\leq M$, so by minimality of H, $H = PL_1T$ and L_1T is irreducible on $P^*/\Phi(P^*)$. As $P \not\leq M$, $X^* \neq P^*$.

As $U_\gamma^* \in \mathcal{Q}(H^*, \tilde{U}_H)$, $p = 3$ or 5 by D.2.13.1.

Suppose first that $p = 3$. If P^* is of order 3, then as $m_3(H) \leq 2$ and $P \not\leq M$, $O_3(H^*) = P^* \times L_1^* \cong E_9$; hence $H^* \cong S_3 \times S_3$ as V_α^* is nontrivial on P^*. Therefore $V_\alpha^* = O_2(L_1^*T^*) \cong \mathbf{Z}_2$, so $m(V_\gamma^*) = m(U_\gamma^*) = 1$. Also $L_1 \trianglelefteq H$, so $\tilde{U}_H = [\tilde{U}_H, L_1]$, and hence $m(\tilde{U}_H) = 2m \geq 4$, where $m := m([\tilde{U}_H, U_\gamma^*])$. Now by 13.8.10.1, we have symmetry between γ and γ_1, so U_H does not induce transvections on U_γ/A_1. Hence $V_1 \leq U_\gamma$ by 13.8.10.2. Further $H = L_1T\langle U_\gamma^H\rangle$, so by symmetry, $G_\gamma = G_{\gamma,\gamma_{b-1}}\langle U_H^{G_\gamma}\rangle$, and hence $m(V_\gamma^*) > 1$ by 13.8.11.1, contradicting $|V_\gamma^*| = 2$.

Therefore $P^* \cong E_9$ or 3^{1+2}. Suppose $L_1^* \not\leq P^*$. As L_1T is irreducible on $P^*/\Phi(P^*)$, H induces $SL_2(3)$ or $GL_2(3)$ on $P^*/\Phi(P^*)$. So in particular if $P^* \cong E_9$, then $m([\tilde{U}_H, P]) \geq 8$; as $U_\gamma^* \in \mathcal{Q}(H^*, \tilde{U}_H)$, this contradicts D.2.17. Hence $P^* \cong 3^{1+2}$, so that $m_3(L_1P) > 2$, contradicting H an SQTK-group. Therefore $L_1^* \leq P^*$, so $L_1^* < P^*$ as $X^* \neq P^*$. Then as L_1T is irreducible on $P^*/\Phi(P^*)$, $P^* \cong 3^{1+2}$ and $L_1^* = Z(P^*)$, so that $L/O_2(L) \cong A_6$. Then $O_2(L_1^*T^*) = C_{T^*}(L_1^*)$ is of 2-rank at most 1, so $m(V_\gamma^*) = 1$. As $U_\gamma^* \in \mathcal{Q}(H^*, \tilde{U}_H)$, U_γ^* inverts $P^*/\Phi(P^*)$ by D.2.17.4. Now we obtain a contradiction as in the previous paragraph.

We have reduced to the case $p = 5$. As $U_\gamma^* \in \mathcal{Q}(H^*, \tilde{U}_H)$ and H is minimal, we conclude from D.2.17 that $P = P_1 \times \cdots \times P_s$ with $s \leq 2$, and $[P, \tilde{U}_H] = \tilde{U}_1 \oplus \cdots \oplus \tilde{U}_s$ with $P_i^* \cong \mathbf{Z}_5$, where $\tilde{U}_i := [P_i, \tilde{U}_H]$ is of rank 4. If $s = 1$ then $U_\gamma^* \cong \mathbf{Z}_2$, while if $s = 2$, then either $U_\gamma^* \cong \mathbf{Z}_2$ with $[U_\gamma^*, P_2^*] = 1$, or $U_\gamma^* = B_1^* \times B_2^*$ with $B_i^* \cong \mathbf{Z}_2$ centralizing P_{3-i}^*. However if U_γ^* is of order 4 then $m(\tilde{U}_H/C_{\tilde{U}_H}(U_\gamma)) = 4 = 2m(U_\gamma^*)$, and so F.9.16.2 shows that $m(U_H/D_H) = 2$ and U_γ^* acts faithfully on $\tilde{D}_H$ as a group of transvections with center $\tilde{A}_1$. Then $m([\tilde{U}_H, U_\gamma^*]) \leq 3$, whereas this commutator space has rank 4 since $s = 2$.

Therefore $m(U_\gamma^*) = 1$, so as before we have symmetry bewtween γ_1 and γ by 13.8.10.1; and as L_1T is irreducible on P^*, $G_\gamma = G_{\gamma,\gamma_{b-1}}\langle U_H^{G_\gamma}\rangle$. As $p = 5$, no element of H^* induces a transvection on $\tilde{U}_H$ by G.6.4; hence we conclude from 13.8.11.2 that $m(V_{\gamma^*}) > 1$. In particular as V_γ^* is faithful on P^*, P^* is not cyclic, so $s = 2$ and $V_\gamma^* = U_\gamma^* \times B_2^*$ with $B_2^* \cong \mathbf{Z}_2$ centralizing P_1^*.

Let $C_H := C_{U_H}(U_\gamma)$ and $\tilde{F}_H := C_{\tilde{U}_H}(U_\gamma)$. By 13.8.10.1, $m(U_H/D_H) = 1$, and by F.9.13.6, $[U_\gamma, D_H] \leq A_1$. Thus if $F_H \not\leq D_H$, then $U_H = D_H F_H$, so $[\tilde{U}_H, U_\gamma^*] \leq \tilde{A}_1$, contrary to $m([\tilde{U}_H, U_\gamma^*]) = 2$. Hence $F_H \leq D_H$. Then by F.9.13.6,

$$[F_H, U_\gamma] \leq V_1 \cap [D_H, U_\gamma] \leq V_1 \cap A_1 = 1$$

and hence $U_2 \leq F_H = C_H$. Then $[U_2, V_\gamma] \leq [C_H, V_\gamma] \leq A_1$ by 13.7.3.7, with $\tilde{A}_1 = [\tilde{D}_H, U_\gamma] \leq \tilde{U}_1$. On the other hand, $1 \neq [\tilde{U}_2, B_2] \leq [\tilde{U}_2, V_\gamma] \cap \tilde{U}_2 \leq \tilde{A}_1 \cap \tilde{U}_2$, contrary to $\tilde{U}_1 \cap \tilde{U}_2 = 0$. $\qquad\square$

By 13.8.13 and 13.8.12.2:

LEMMA 13.8.14. *Let* $X := O^2(O_{2,F}(H))$; *then one of the following holds:*

(a) $X = 1$.

(b) $L/O_2(L) \cong A_6$ *and* $X = L_1$.

(c) $L/O_2(L) \cong \hat{A}_6$ *and* $X = L_{1,+}$.

By 13.8.13, H is nonsolvable, so there exists $K \in \mathcal{C}(H)$. By 13.8.14, $F(H^*) = Z(O^2(H^*))$, so K^* is quasisimple. Then by 13.8.12.1:

LEMMA 13.8.15. $K \not\leq M$, so $\langle K^T \rangle L_1 T \in \mathcal{H}_z$. Further K^* is quasisimple.

LEMMA 13.8.16. (1) $K \trianglelefteq H$, so $KL_1T \in \mathcal{H}_z$. In particular, F.9.18.4 applies.
(2) $K/O_2(K)$ is not $Sz(2^n)$.

PROOF. Assume $K_0 = \langle K^T \rangle > K$. Then $K_0 = KK^t$ for $t \in T - N_T(K)$ by 1.2.1.3. Let $K_1 := K$ and $K_2 := K^t$. By 13.8.15, we may take $H = K_0L_1T$. By F.9.18.5, $K^* \cong L_2(2^n)$, $Sz(2^n)$, or $L_3(2)$. Further unless $K^* \cong Sz(2^n)$, $K_0 = O^{3'}(H)$ by 1.2.2.a so $L_1 \leq K_0$.

Suppose first that $K^* \cong L_3(2)$. Then $L_1 \leq H_1 \leq H$ where $H_1/O_2(H_1) \cong S_3$ wr $\mathbf{Z}_2$, so $L_1 = \theta(H \cap M) = O^2(H_1)$ using 13.7.3.9. As $m_3(O^2(H_1)) = 2$, $L/O_2(L) \cong \hat{A}_6$, so that $Aut_M(L_1/O_2(L_1)) \cong E_4$, whereas we have seen just above that $Aut_{H \cap M}(L_1/O_2(L_1)) \cong D_8$.

Therefore $K^* \cong L_2(2^n)$ or $Sz(2^n)$. Let B_0^* be a Borel subgroup of K_0^* containing $T_0^* := T^* \cap K_0^*$, and set $B := O^2(B_0)$. As $L_1T = TL_1$, L_1 acts on B_0. Therefore $B_0 \leq M$ by 13.8.13.

Let $\tilde{W}$ denote an H-submodule of $\tilde{U}_H$ maximal subject to $[\tilde{U}_H, K_0] \not\leq \tilde{W}$; thus $[\tilde{U}_H, K_0]\tilde{W}/\tilde{W}$ is an irreducible K_0-module. As $K_0^*T^*$ has no strong FF-modules by B.4.2, it follows from parts (5) and (6) of F.9.18 that either

(a) U_H/W and $\tilde{W}$ are FF-modules for $K_0^*T^*$, or
(b) $[\tilde{U}_H, K_0] = \tilde{I}_H = \langle \tilde{I}^H \rangle$ for some $\tilde{I} \in Irr_+(K_0, \tilde{U}_H, T)$, and $[\tilde{W}, K_0] = 0$.

Let $U := U_H/W$ or $\tilde{I}_H$ in case (a) or (b), respectively, and let V_U denote the projection of $\tilde{V}_3$ on U.

Suppose for the moment that case (a) holds. Then by Theorems B.5.1 and B.5.6, $K^* \cong L_2(2^n)$, and $U = U_1 \oplus U_2$, where U_i is the natural module or orthogonal module for K_i^*, and $[K_i, U_{3-i}] = 0$. Further as $U_H = \langle V_3^H \rangle$, $V_3 \not\leq W$, so as L_1 is irreducible on $\tilde{V}_3$, V_U is isomorphic to $\tilde{V}_3$.

Now suppose for the moment that case (b) holds. Then by F.9.18.5, either

(b1) $U = U_1 + U_2$ with $U_i := [U, K_i]$ and $U_i/C_{U_i}(K_i)$ the natural or A_5-module for K^*, or
(b2) U is the natural orthogonal module for $K_0^* \cong \Omega_4^+(2^n)$.

Here if $V_3 \leq U$, then $U = \langle V_3^H \rangle = U_H$. In particular this subcase holds when $K^* \cong L_2(2^n)$, since there we saw that $L_1 \leq K_0$, so that $V_3 \leq [U_H, L_1] \leq [U_H, K_0] = U$.

We first eliminate the case $K^* \cong L_2(2^n)$. Since $L_1 \leq K_0$, $L_1 \leq N_{K_0}(B_0) = B_0$, and hence n is even. Then $m_3(B_0) = 2$, so as $B_0 \leq M$, $L/O_2(L) \cong \hat{A}_6$ by 13.7.3.9. As $t \in T - N_T(K)$ acts on L_0 and $L_{1,+}$, these groups are diagonally embedded in K_0. Let $B := O^2(B_0)$. As $L_{1,+}/O_2(L_{1,+})$ is inverted by $s \in T \cap L$, and $[B, s] \leq L$, $[B, s]$ is a $\{2, 3\}$-group. We conclude that $n = 2$ and $L_1 = B$.

Assume that case (a) or (b1) holds. Then $V_U \not\leq U_i$ as V_U is T-invariant. Thus the projections V_U^i of V_U on U_i are nontrivial. As $V_U = [V_U, L_{1,+}]$, also $V_U^i = [V_U^i, L_{1,+}]$. Similarly L_0 centralizes V_U^i. This is impossible, as $L_0K_2 = L_{1,+}K_2$ since L_0 and $L_{1,+}$ are diagonally embedded in K_0, and $[U_1, K_2] = 0$.

Therefore case (b2) holds, so $U = \tilde{U}_H$ is the orthogonal module. In particular H^* contains no $\mathbf{F}_2$-transvections, so case (2) of 13.8.8 holds. Hence the K_0-conjugate V_α^* of V_γ^* defined in that case is contained in $O_2(L_1^*)$. Further

$U_\alpha^* \in \mathcal{Q}(H^*, \tilde{U}_H)$, so in particular U_α^* acts quadratically on $\tilde{U}_H$, and hence it follows from the facts that $n = 2$, U_γ^* is a 4-group, and $m(\tilde{U}_H/C_{\tilde{U}_H}(U_{\gamma^*})) = 4$. Now by F.9.16.2, $m(U_H/D_H) = 2$, which is impossible as $[\tilde{D}_H, U_\gamma^*] = \tilde{A}_1$ by F.9.13.6, whereas no 4-group in K_0^* induces a group of transvections on a subspace of codimension 2 in $\tilde{U}_H$ of dimension 8.

It remains to eliminate the case $K^* \cong Sz(2^n)$. Since $B \le M$, $[B, L_1] \le L_1 \cap B \le O_2(L_1)$, so L_1^* centralizes K_0^*. However case (a) or (b1) holds, so that $U = U_1 \oplus U_2$ with U_i the natural module for K^*; then $End_{K^*}(U_i) = \mathbf{F}_{2^n}$ with n odd, and hence $[U, L_1] = 1$. This is a contradiction, since $L_1 \trianglelefteq H$ and $\tilde{V}_3 = [\tilde{V}_3, L_1]$, so $\tilde{U}_H = [\tilde{U}_H, L_1]$.

Essentially the same argument establishes (2): We conclude from parts (4) and (7) of F.9.18 that $[\tilde{U}_H, K]/C_{[\tilde{U}_H, K]}(K)$ is the natural module for $K/O_2(K) \cong Sz(2^n)$. Again L_1^* centralizes K^* and then also $[\tilde{U}_H, K]$, for the same contradiction. $\qquad\square$

By 13.8.16 and F.9.18.4:

LEMMA 13.8.17. $K^* \cong L_2(2^n)$, $(S)L_3(2^n)^\epsilon$, $Sp_4(2^n)'$, $G_2(2^n)'$, $L_4(2)$, $L_5(2)$, A_7, $\hat{A}_6$, M_{22}, or $\hat{M}_{22}$.

In the remainder of the section, we successively eliminate the cases listed in 13.8.17.

Observe that the second case of 13.8.8 holds, unless K^* is one of the groups A_6, A_7, or $L_4(2)$ allowed by 13.8.5.2 in the first case.

LEMMA 13.8.18. If $H = KL_1T$, then

(1) $G_\gamma = \langle F^{G_\gamma}\rangle G_{\gamma, \gamma_{b-1}}$ for each $F \le U_H$ with $F \not\le D_H$.

(2) In case (2) of 13.8.8, the hypotheses of 13.8.11 are satisfied.

(3) If case (2) of 13.8.8 holds and U_H does not induce a transvection on U_γ, then $m(V_\gamma^*) > 1$.

(4) If no member of H^* induces a transvection on $\tilde{U}_H$, then $m(V_\gamma^*) > 1$.

PROOF. By F.9.13.2 $U_H \le O_2(G_{\gamma, \gamma_{b-1}})$, while as $H = KL_1T$, for g_b with $(\gamma_0, \gamma_1)g_b = (\gamma_{b-1}, \gamma)$ we have $G_\gamma = K^{g_b} G_{\gamma, \gamma_{b-1}}$. Thus if $F \not\le D_H$, then $K^{g_b} = [K^{g_b}, F]$, so (1) holds. In case (2) of 13.8.8, $D_H < U_H$, so (2) follows by an application of (1) with U_H in the role of "F". Finally 13.8.11.2 and (2) imply (3), and 13.8.8 and (3) imply (4). $\qquad\square$

LEMMA 13.8.19. H^* is not $L_3(2)$.

PROOF. Assume $H^* \cong L_3(2)$. Then $L_1^*T^*$ is a maximal parabolic of H^*; let P^* be the remaining maximal parabolic of H^* containing T^*. Since $\tilde{U}_H = \langle \tilde{V}_3^H\rangle$ with L_1T inducing S_3 on $\tilde{V}_3$, H.6.5 says $\tilde{U}_H$ is one of the following: the natural module W in which P^* stabilizes a point, the core U_2 of the permutation module on H^*/P^*, the Steinberg module S, $W \oplus S$, or $U_2 \oplus S$. By 13.7.6.1, $\tilde{U}_H$ is not natural. Then since $U_\alpha^* \in \mathcal{Q}(H^*, \tilde{U}_H)$ by 13.8.8, it follows using B.5.1 and B.4.5 that $\tilde{U}_H = U_2$. By 13.8.8, $V_\alpha^* \le R_1^*$, so as R_1^* is not quadratic on $\tilde{U}_H = U_2$, it follows that $m(V_\alpha^*) = 1$. This contradicts 13.8.18.4 in view of G.6.4. $\qquad\square$

LEMMA 13.8.20. K^* is not of Lie type over $\mathbf{F}_{2^n}$ for any $n > 1$.

PROOF. Assume otherwise. By 13.8.16, we may take $H = KL_1T$. By 13.8.17, $K^* \cong L_2(2^n)$, $(S)L_3^{\epsilon}(2^n)$, $Sp_4(2^n)$, or $G_2(2^n)$. By 13.8.5, case (2) of 13.8.8 holds, so in particular $U_{\gamma}^* \in \mathcal{Q}(H^*, \tilde{U}_H)$.

Let B_0^* be the Borel subgroup of K^* containing $T_0^* := T^* \cap K^*$, and let $B := O^2(B_0)$. As K is defined over $\mathbf{F}_{2^n}$ with $n > 1$, and $L_1T = TL_1$, L_1 acts on B; so by 13.8.13, $B \leq M$. Then using 13.7.3.9, $L_1 = \theta(BL_1)$, and $BL_1 = B_CL_1$, where

$$B_C := O^2(C_{BL_1}(L/O_2(L))) \leq C_M(V).$$

Let $X := L_1$ if $L/O_2(L) \cong A_6$, and $X := L_{1,+}$ if $L/O_2(L) \cong \hat{A}_6$. If $L/O_2(L) \cong A_6$ then $B_C = O^3(B)$, while if $L/O_2(L) \cong \hat{A}_6$, then $B_C = O^3(B)L_0$. In either case, $|BX : B_CO_2(B)| = 3$.

Next $X/O_2(X)$ is inverted by some $t \in T \cap L$, and $[B_C, t] \leq O_2(B_C)$. Now from the structure of $Aut(K^*)$, one of the following holds:

(i) $C_{T^*}(O^3(B^*)O_2(B^*)/O_2(B^*)) = O_2(B_0^*)$.
(ii) $n = 2$ or 6, and K^* is not $U_3(2^n)$.
(iii) $K^* \cong (S)U_3(8)$.

In case (i) as $[t^*, O^3(B^*)] \leq O_2(B^*)$, $t^* \in O_2(B^*)$, a contradiction as t^* inverts $X^*/O_2(X^*)$. In case (ii) if t^* induces an outer automorphism on K^*, then $|[B^*, t^*]/O_2([B^*, t^*])| > 3$ unless K^* is $(S)L_3(4)$ or $L_2(4)$. Therefore we conclude that either:

(a) $X \not\leq K$, $X^* \leq C_{H^*}(K^*)$ so that $X \trianglelefteq KL_1T = H$, and X^* is inverted in $C_{H^*}(K^*)$, or
(b) $K^* \cong L_2(4)$, $(S)U_3(8)$, or $(S)L_3(4)$, and t^* induces an outer automorphism on K^*.

Assume first that (a) holds. Then as H is an SQTK-group, $m_3(K) = 1$, so that $K^* \cong L_2(2^n)$, $L_3(2^n)$ for $n > 1$ odd, or $U_3(2^n)$ for n even. Further $X \trianglelefteq H$ and $\tilde{V}_3 = [\tilde{V}_3, X]$, so $\tilde{U}_H = [\tilde{U}_H, X]$. Hence as X^* is inverted in $C_{H^*}(K^*)$, each noncentral chief factor for H on $\tilde{U}_H$ is the sum of a pair of isomorphic K^*-modules. Then case (ii) of F.9.18.4 holds, so that each $\tilde{I} \in Irr_+(K, \tilde{U}_H, T)$ is a T-invariant FF-module for KT. Therefore $\tilde{I}_H := \langle \tilde{I}^H \rangle$ is the sum of two X-conjugates of $\tilde{I}$, and K^* is not $(S)U_3(2^n)$.

Suppose K^* is $L_2(2^n)$. Observe that if n is even, then $m_3(XB) > 1$, so we conclude from 13.7.3.9 that $L/O_2(L) \cong \hat{A}_6$. We saw earlier that $B_C = O^3(B)L_0$, with $|BX : O_2(B)B_C| = 3$; then since $X \not\leq K$ as case (a) holds, we conclude that $B = B_C$ centralizes V. On the other hand if n is odd, then B is a $3'$-group, so again $B = B_C$ centralizes V.

Next as K^*T^* has no strong FF-modules by B.4.2, applying F.9.18.6 to $\tilde{I}_H$ in the role of "$\tilde{W}$", we conclude $[\tilde{U}_H, K] = \tilde{I}_H$. As $\tilde{I}/C_{\tilde{I}}(K)$ is an FF-module, by B.4.2 it is either the natural $L_2(2^n)$-module or the A_5-module. In the first case as B centralizes V, $\tilde{V}_3 \leq C_{\tilde{U}_H}(BT_0^*) = C_{\tilde{U}_H}(K)$, a contradiction since $U_H = \langle V_3^H \rangle$ and $K^* \neq 1$. Thus $\tilde{I}$ is the A_5-module, so that $J(H^*) \cong S_5$ by B.4.2.5; hence $H^* \cong S_5 \times S_3$ and $\tilde{U}_H$ is the tensor product of the S_5-module and S_3-module. Since case (2) of 13.8.8 holds, there is an H-conjugate α of γ such that $V_{\alpha}^* \leq O_2(L_1^*T^*) = T_0^* \leq K^*$. Then as V_{γ}^* is quadratic on $\tilde{U}_H$, $|\tilde{V}_{\gamma}^*| = 2$, contrary to 13.8.18.4.

This leaves the case $K^* \cong L_3(2^n)$, $n > 1$ odd. This time the FF-module $\tilde{I}$ is natural by B.4.2, so $\tilde{I}_H$ is the tensor product of natural modules for K^* and S_3.

As n is odd, $B = B_C$, so B centralizes $\tilde{V}_3$. Therefore as $C_{\tilde{I}_H}(B) = 0$, we conclude $V_3 \not\leq I_H$. If $I_H = [U_H, K]$, then $V_3 I_H$ is invariant under $KL_1 T = H$, so $U_H = V_3 I_H$. Then as T_0 centralizes $\tilde{V}_3$, $\tilde{U}_H = \tilde{I}_H \oplus C_{\tilde{U}_H}(K)$ and $C_{\tilde{U}_H}(K) = C_{\tilde{U}_H}(B) = \tilde{V}_3$. But now $U_H = \langle V_3^H \rangle = V_3$, contrary to 13.5.9. Hence K^* is faithful on U_H/I_H, so case (b) or (c) of F.9.18.6 holds with $\tilde{I}_H$ in the role of "$\tilde{W}$". Therefore $[U_H, K]/I_H$ is an FF-module for $K^* T^*$, and hence this quotient is also the tensor product of natural modules for K^* and S_3. Then again B centralizes $\tilde{V}_3$, but is fixed-point-free on $[\tilde{U}_H, K]$, so that $V_3 \not\leq [U_H, K]$. Now we obtain a contradiction as in the earlier case, arguing on $[U_H, K]$ in place of I_H.

Therefore (b) holds. As $q(H^*, \tilde{U}_H) \leq 2$, K^* is not $L_3(4)$ by B.4.5. Thus $K^* \cong L_2(4)$, $SL_3(4)$, or $(S)U_3(8)$. We claim $L_1 \leq K$; so assume otherwise. As $1 \neq O^{3'}(B) \leq L_1$ but $L_1 \not\leq K$, it follows that $L/O_2(L) \cong \hat{A}_6$ and $|B|_3 = 3$. Hence $K^* \cong U_3(8)$ or $L_2(4)$. In the first case, A.3.18 supplies a contradiction as $L_1/O_2(L_1) \cong E_9$ and T acts on L_1 but does not permute with the subgroup generated by the element x^* in A.3.18.b. Thus $K^* \cong L_2(4)$, and as L_0 and $L_{1,+}$ are the only proper T-invariant subgroups of L_1 which are not 2-groups, $K^* L_1^* = K^* \times L_C^*$, where $L_C = L_0$ or $L_{1,+}$. In the former case, $K \leq N_G(L_0) = M$ by 13.2.2.9, a contradiction. In the latter case, as $[L_0, t] \leq O_2(L_0)$ and case (a) fails, we have a contradiction. Thus the claim is established.

By the claim and 13.7.5.2, $U_H = [U_H, K]$. We next observe that K^* is not $(S)U_3(8)$: For otherwise we may apply F.9.18.7 and B.4.5 to conclude that $\tilde{U}_H$ is the natural module for $K^* \cong SU_3(8)$, defined over $\mathbf{F}_8$. But then there is no B-invariant subspace $\tilde{V}_3 = [\tilde{V}_3, L_1]$ of 2-rank 2.

Suppose K^* is $SL_3(4)$. By B.4.5, any $I \in Irr_+(K, \tilde{U}_H, T)$ is the natural module. Further $B = \theta(B) \leq L_1$ by 13.7.3.9, and B is of 3-rank 2, so $L/O_2(L) \cong \hat{A}_6$. Then as $X = L_{1,+}$ is inverted by $t \in C_T(L_0/O_2(L_0))$, we conclude that either t induces a graph-field automorphism on K^* with $L_0^* = C_{L_1^*}(t^*) = Z(K^*)$, or t induces a graph automorphism on K^* and $X^* = [L_1^*, t^*] = Z(K^*)$. In the first case, $H \leq N_G(L_0) \leq M$ by 13.2.2.9, contrary to $H \not\leq M$; so the second case holds. Now case (iii) of F.9.18.4 holds, with $\tilde{I}_H = \tilde{I} \oplus \tilde{I}^t$, where $\tilde{I} \in Irr_+(\tilde{U}_H, K, T)$ is a natural module for K^* and $\tilde{I}^t$ is its dual. By F.9.18.7, $\tilde{I}_H = [\tilde{U}_H, K]$, so $I_H = U_H$ by the previous paragraph. Further $U_{\gamma^*} \in \mathcal{Q}(H^*, \tilde{U}_H)$ is either a root group of K^* of rank 2 with $m(\tilde{U}_H/C_{\tilde{U}_H}(U_\gamma)) = 4$, or $m(U_\gamma^*) \geq 3$ with $m(\tilde{U}_H/C_{\tilde{U}_H}(U_\gamma)) = 6$. In the first case by F.9.16.2, U_γ^* is faithful on $\tilde{D}_H$ of corank 2 in $\tilde{U}_H$; and in the second, at least $m(U_H/D_H) \leq m(U_\gamma^*) \leq m_2(H^*) = 4$. In either case, no subspace $\tilde{D}_H$ of this corank in $\tilde{U}_H$ satisfies the requirement $[U_\gamma^*, \tilde{D}_H] = \tilde{A}_1$ of F.9.13.6.

We are left with the case $L_1 \leq K^* \cong L_2(4)$. Thus $L/O_2(L) \cong A_6$ by 13.7.5.5. As $L_1 = [L_1, T]$ and $H = KL_1 T = KT$, $H^* \cong S_5$. Then as case (2) of 13.8.8 holds, $V_\alpha^* \leq R_1^* \in Syl_2(K^*)$, and $m_2(R_1^*) = 2$, so $V_\alpha^* = R_1^*$ by 13.8.18.4. Now by 13.8.4.5, V_H/U_H is a nontrivial quotient of the 5-dimensional permutation module for $H^* \cong S_5$. Then as $V_\alpha^* = R_1^*$, V_γ^* is not quadratic on V_H/U_H, contrary to 13.8.4.6. $\qquad\square$

LEMMA 13.8.21. *(1) $L_1 \leq K$.*
(2) $K/O_2(K)$ is $L_n(2)$, $3 \leq n \leq 5$, A_6, A_7, or $G_2(2)'$.
(3) $H = KT$. In particular if $K \cong L_3(2)$, then $H^* \cong Aut(L_3(2))$.
(4) $U_H = [U_H, K]$.

PROOF. We begin with the proof of (2); as usual, we may take $H = KL_1T$. By 13.8.17 and 13.8.20, $K^* \cong L_4(2)$, $L_5(2)$, A_6, A_7, $G_2(2)'$, $\hat{A}_6$, M_{22}, or $\hat{M}_{22}$. Thus to establish (2) we may assume $K/O_2(K) \cong \hat{A}_6$, M_{22}, or $\hat{M}_{22}$, and it remains to derive a contradiction.

By A.3.18, $L_1 \leq \theta(H) = K$. Then L_1 is solvable and normal in $J := K \cap M$. It follows when $K/O_2(K) \cong \hat{A}_6$ that $J/O_{2,Z}(K)$ is a maximal parabolic subgroup of $K/O_{2,Z}(K)$, and when $K/O_{2,Z}(K) \cong M_{22}$ that $J/O_{2,Z}(K)$ is a maximal parabolic of the subgroup $K_1/O_{2,Z}(K) \cong A_6/E_{2^4}$ of $K/O_{2,Z}(K)$.

Assume $K/O_2(K) \cong M_{22}$. By the previous paragraph, $|L_1|_3 = 3$, so $L/O_2(L) \cong A_6$ rather than $\hat{A}_6$. Further case (i) of F.9.18.4 holds with $\tilde{I} \in Irr(K, \tilde{U}_H)$, and $\tilde{I}$ is the code module in view of F.9.18.2 and B.4.5. As M_{22} has no FF-modules by B.4.2, $\tilde{I} = [\tilde{U}_H, K]$ by F.9.18.7, so that $\tilde{V}_3 = [\tilde{V}_3, L_1] \leq \tilde{I}$. By the previous paragraph, $C_{\tilde{I}}(O_2(L_1T)) \leq C_{\tilde{I}}(O_2(K_1T))$, while $m(C_{\tilde{I}}(O_2(K_1T))) = 1$ by H.16.2.1. This is a contradiction, since L_1T induces $GL(\tilde{V}_3)$ on $\tilde{V}_3$ of rank 2 in $\tilde{I}$, so that $O_2(L_1T)$ centralizes $\tilde{V}_3$.

Thus we may assume that $K/O_2(K) \cong \hat{A}_6$ or $\hat{M}_{22}$. Then $Y := O^2(O_{2,Z}(K)) \neq 1$; by 13.8.13, $Y \leq M$, so $Y \leq \theta(H \cap M) = L_1$ by 13.7.3.9. Then if $Y = L_1$, each solvable overgroup of YT in H is contained in M by 13.8.13. However there is $K_1 \in \mathcal{L}(KT, T)$ with $K_1/O_{2,Z}(K_1) \cong A_6$, so either $K_1 \in \mathcal{C}(H \cap M)$ or $K = K_1$ and T is nontrivial on the Dynkin diagram of K^*. In the former case $K_1 = L$, contradicting $M = !\mathcal{M}(LT)$. As $q(H^*, \tilde{U}_H) \leq 2$, the latter is impossible by B.4.5. Thus $Y < L_1$, so $L/O_2(L) \cong \hat{A}_6$. Then as $N_G(L_0) = M$ using 13.2.2.9, $Y \neq L_0$, so $Y = L_{1,+}$. Further if $K/O_2(K) \cong \hat{M}_{22}$, replacing K by K_1, we reduce to the case $K/O_2(K) \cong \hat{A}_6$.

Now $L_1 = \theta(H \cap M)$ by 13.7.3.9, and $H \cap M$ is a maximal parabolic of H. As $\tilde{V}_3 = [\tilde{V}, L_{1,+}]$ and $Y = L_{1,+} \trianglelefteq H$, $\tilde{U}_H = [\tilde{U}_H, Y]$.

Since K^* is $\hat{A}_6$, case (2) of 13.8.8 holds, so that $U_\gamma^* \in \mathcal{Q}(H^*, \tilde{U}_H)$. Let $\tilde{I} \in Irr_+(K, \tilde{U}_H, T)$; by B.4.5, $\tilde{I}$ is a 6-dimensional module for H^*. Further as H^* has no faithful strong FF-modules by B.4.2.8, F.9.18.6 says that either $\tilde{I} = \tilde{U}_H$ or $\tilde{U}_H/\tilde{I}$ is 6-dimensional. Set $W := \tilde{U}_H$ or $\tilde{U}_H/\tilde{I}$ in the respective cases. Now L_1T acts on $\tilde{V}_3$ and hence also on its image in W, so $L_1^*T^*$ is the stabilizer of an $\mathbf{F}_4$-point in W. Choose α as in case (2) of 13.8.8; then $V_\alpha^* \leq O_2(L_1^*T^*) \cong E_4$, so by 13.8.18.4, $V_\alpha^* = O_2(L_1^*T^*)$. This is a contradiction as V_α^* is quadratic on U_H by 13.8.4.6. Thus (2) is established.

We next prove (1); we may continue to assume $H = KL_1T$ but $L_1 \not\leq K$. Therefore $m_3(K) = 1$ by (2) and A.3.18, so $K^* \cong L_3(2)$. Let $X := L_{1,+}$ if $L/O_2(L) \cong \hat{A}_6$, and $X := L_1$ if $L/O_2(L) \cong A_6$. As $X = [X, T]$, either $X \leq K$ or $[X, K] \leq O_2(K)$.

Assume first that $L/O_2(L) \cong A_6$. Then $L_1^* = X^*$ centralizes K^* by the previous paragraph, so that $F^*(H^*) = K^* \times X^*$. As $\tilde{V}_3 = [\tilde{V}_3, X^*]$ and $X^* \trianglelefteq H^*$, $\tilde{U}_H = [\tilde{U}_H, X]$. If $H^*/C_{H^*}(K^*)$ is not $Aut(L_3(2))$, then KX is generated by a pair of solvable overgroups of X, so that $KX \leq M$ by 13.8.13, contrary to 13.8.12.1. On the other hand if $H^*/C_{H^*}(K^*) \cong Aut(L_3(2))$, then since $\tilde{U}_H = [\tilde{U}_H, X]$, for each chief factor W for H^* on $\tilde{U}_H$ with $[K^*, W] \neq 1$, W consists of either a pair of Steinberg modules, or a pair of natural modules and a pair of duals for K^*, contradicting $q(H^*, \tilde{U}_H) \leq 2$ by B.4.5.

Thus $L/O_2(L) \cong \hat{A}_6$, so that L_0 and $L_{1,+} = X$ are the two T-invariant subgroups of 3-rank 1 in L_1. As usual $K \not\leq M$ by 13.8.12.1, so that K does not act on L_0 in view of 13.2.2.9. Then $C_{L_1}(K^*) = X$ rather than L_0, so that $L_0 \leq K$. Now $X = [X, t]$ for $t \in T \cap L \leq C_T(L_0/O_2(L_0))$, so $[K^*, t^*] = 1$. Also T acts on L_0, and hence is trivial on the Dynkin diagram of K^*, so $H^* \cong L_3(2) \times S_3$. As earlier, $\tilde{U}_H = [\tilde{U}_H, X]$, so an H-chief factor W in $\tilde{U}_H$ is the tensor product of natural modules for the factors, as usual using B.4.5 and the fact that $U_\gamma^* \in \mathcal{Q}(H^*, \tilde{U}_H)$. As L_0 centralizes $\tilde{V}_3$, $L_0^* T^*$ is the stabilizer of a point in these natural modules. Then as $V_\alpha^* \leq O_2(L_1^* T^*)$ and V_α^* is quadratic on U_H, V_α^* is of order 2, contrary to 13.8.18.4. So (1) is established.

By (1), L_1 is contained in each $K \in \mathcal{C}(H)$, so there is a unique $K \in \mathcal{C}(H)$. Then by 13.8.14, $K^* = F^*(H^*)$, so (3) holds as $Out(K^*)$ is a 2-group for each of the groups listed in (2); if $K^* \cong L_3(2)$ that $H^* \cong Aut(L_3(2))$ by 13.8.19. Part (4) follows from (1) and 13.7.5.2. □

Let $\tilde{W}$ be a proper H-submodule of $\tilde{U}_H$ and set $\hat{U}_H := \tilde{U}_H/\tilde{W}$. As $\tilde{U}_H = \langle V_3^H \rangle$ and L_1 is irreducible on $\tilde{V}_3 \cong E_4$, it follows that $\hat{V}_3 \cong E_4$ is $L_1 T$-isomorphic to $\tilde{V}_3$. By 13.8.21, $\hat{U}_H = [\hat{U}_H, K]$ and $K^* = F^*(H^*)$ is simple, so that H^* is faithful on $\hat{U}_H$.

LEMMA 13.8.22. *Assume K is nontrivial on $\tilde{W}$. Then H^* is faithful on $\tilde{W}$, case (2) of 13.8.8 holds, and either*

(1) $A_1 \leq W$, U_γ^* *contains an FF^*-offender on the FF-module $\hat{U}_H$, and either U_γ^* contains a strong FF^*-offender on $\hat{U}_H$, or $W \leq D_H$ and $[\tilde{W}, U_\gamma^*] = \tilde{A}_1$.*

(2) $A_1 \not\leq W$, U_γ^* *contains an FF^*-offender on the FF-module $\tilde{W}$, and either U_γ^* contains a strong FF^*-offender on $\tilde{W}$, or $U_H = W D_H$ and $[\hat{U}_H, U_\gamma] = \hat{A}_1$, so that $A_1 \leq U_H$.*

PROOF. As K^* is nontrivial on $\tilde{W}$ and $K^* = F^*(H^*)$ is simple, H^* is faithful on $\tilde{W}$. As H^* is also faithful on $\hat{U}_H$, no member of H^* induces a transvection on $\hat{U}_H$, so case (2) of 13.8.8 holds.

Suppose $A_1 \leq W$. Then using F.9.13.6, $[\hat{D}_H, U_\gamma] \leq \hat{A}_1 = 1$, so $\hat{D}_H < \hat{U}_H$ and

$$m(\hat{U}_H/C_{\hat{U}_H}(U_\gamma)) \leq m(\hat{U}_H/\hat{D}_H) \leq m(U_H/D_H) \leq m(U_\gamma^*),$$

so by B.1.4.4, U_γ^* contains an FF^*-offender on the FF-module $\hat{U}_H$. Indeed either U_γ^* contains a strong FF^*-offender, or all inequalities are equalities, so that $m(\hat{U}_H/\hat{D}_H) = m(U_H/D_H)$, and hence $W \leq D_H$. In the latter case, $[\tilde{W}, U_\gamma] \leq [\tilde{D}_H, U_\gamma] = \tilde{A}_1$, so that (1) holds.

So assume $A_1 \not\leq W$. Then $[D_H \cap W, U_\gamma] \leq W \cap A_1 = 1$, so

$$m(W/C_W(U_\gamma)) \leq m(W/(D_H \cap W)) \leq m(U_H/D_H) \leq m(U_\gamma^*)$$

since case (2) of 13.8.8 holds. So by B.1.4.4, U_γ^* contains an FF^*-offender on the FF-module $\tilde{W}$. Further if U_γ^* does not contain a strong FF^*-offender, then all inequalities are equalities, so that $m(W/(D_H \cap W)) = m(U_H/D_H)$ and hence $U_H = W D_H$. But then

$$[\hat{U}_H, U_\gamma] \leq [\hat{D}_H, U_\gamma] \leq \hat{A}_1,$$

so that (2) holds. □

LEMMA 13.8.23. *Assume* $m(U_\gamma^*) = 1$, *and* K *is nontrivial on* $\tilde{W}$. *Then*

(1) U_γ *induces transvections on* $\tilde{W}$ *and* $\hat{U}_H$, D_H *is a hyperplane of* U_H, $C_H :=$
$C_{U_H}(U_\gamma)$ *is a hyperplane of* D_H, *and* $\hat{C}_H = C_{\hat{U}_H}(U_\gamma^*)$.
(2) $U_\gamma^* < V_\gamma^*$.
(3) Either $\tilde{A}_1 \nleq W$ *and* $[C_{\tilde{W}}(U_\gamma^*), V_\gamma^*] = 1$, *or* $A_1 \leq W$ *and* $[C_{\hat{U}_H}(U_\gamma^*), V_\gamma^*] = 1$.
(4) $U_\gamma^* < C_{H^*}(C_E(U_\gamma^*))$ *for at least one of* $E := \tilde{W}$ *or* $\hat{U}_H$.

PROOF. By 13.8.22, H^* is faithful on $\tilde{W}$ and case (2) of 13.8.8 holds, so $U_\gamma^* \in$
$\mathcal{Q}(H^*, \tilde{U}_H)$. Therefore as $m(U_\gamma^*) = 1$ it follows that $m(\tilde{U}_H/C_{\tilde{U}_H}(U_\gamma)) \leq 2$; then
since K is nontrivial on $\tilde{W}$, equality holds and U_γ^* induces transvections on both
$\tilde{W}$ and $\hat{U}_H$. Therefore by 13.8.10, D_H is a hyperplane of U_H.

Let $C_H := C_{U_H}(U_\gamma)$. By F.9.13.7, $[D_\gamma, D_H] = 1$, so as $m(U_\gamma^*) = 1$ and
$[D_H, U_\gamma] \leq A_1$ by F.9.13.6, C_H is a hyperplane of D_H. Therefore $\tilde{C}_H = C_{\tilde{U}_H}(U_\gamma)$
as both subgroups are of codimension 2 in $\tilde{U}_H$. Hence (1) holds.

Part (2) follows from 13.8.18.4. Next $[\tilde{C}_H, V_\gamma] \leq \tilde{A}_1$ by 13.7.3.7. Further by
(1), U_γ^* is not a strong FF*-offender on $\hat{U}_H$ or $\tilde{W}$. Assume $A_1 \leq W$. Then $W \leq D_H$
by 13.8.22.1, so $\hat{D}_H = C_{\hat{U}_H}(U_\gamma^*) = \hat{C}_H$ by (1). Thus if $[C_{\hat{U}_H}(U_\gamma^*), V_\gamma^*] \neq 1$, $D_H >$
WC_H, and hence $W \leq C_H$ as $|D_H : C_H| = 2$ by (1). However this contradicts
$[\tilde{W}, U_\gamma] \neq 1$. So suppose instead $A_1 \nleq W$. Then by F.9.13.6, $[D_H \cap W, U_\gamma] \leq$
$A_1 \cap W = 1$, so $D_H \cap W \leq C_H$, and hence

$$[\widetilde{D_H \cap W}, V_\gamma] \leq [\tilde{C}_H, V_\gamma] \cap \tilde{W} \leq \tilde{A}_1 \cap \tilde{W} = 1.$$

Since $C_{\tilde{W}}(U_\gamma^*) \leq \widetilde{D_H \cap W}$, this establishes (3).
Finally (2) and (3) imply (4). $\square$

LEMMA 13.8.24. K^* *is not isomorphic to* A_7.

PROOF. Assume $K^* \cong A_7$. We adopt the notational conventions of section
B.3, and represent H^* on $\Omega := \{1, \dots, 7\}$, so that T^* has orbits $\{1, 2, 3, 4\}$, $\{5, 6\}$,
and $\{7\}$. Let $\beta := \gamma g_b^{-1}$ for g_b as defined earlier, and let $\delta \in \{\beta, \gamma\}$. By (1) and (2)
of F.9.13, $V_\delta^{*y} \leq O_2(L_1^* T^*)$ for some $y \in H$.

Suppose first that case (1) of 13.8.8 holds, and pick δ as in that case. Then V_δ or
U_δ induces a nontrivial group of transvections on $\tilde{U}_H$, so in particular $K^* T^* \cong S_7$.
But as case (2b) of 13.8.5 holds, $L/O_2(L) \cong \hat{A}_6$ so $|L_1|_3 = 3^2$, and hence $L_1^* T^* \cong$
$S_4 \times S_3$ is the stabilizer of the partition $\{\{1, 2, 3, 4\}, \{5, 6, 7\}\}$ of Ω. Thus $O_2(L_1^* T^*)$
contains no transvections, whereas we showed $V_\delta^{*y} \leq O_2(L_1^* T^*)$ and $U_\delta \leq V_\delta$.

Therefore case (2) of 13.8.8 holds. Define α as in that case; thus $V_\alpha^* \leq O_2(L_1^* T^*)$
and $U_\gamma^* \in \mathcal{Q}(H^*, \tilde{U}_H)$.

Pick $\tilde{W}$ maximal in $\tilde{U}_H$, so that $\hat{U}_H$ is an irreducible H^*-module. It will suffice
to show $m(U_\gamma^*) = 1$ and $[\tilde{W}, K] \neq 1$: for then 13.8.23.4 supplies a contradiction,
since for each faithful $\mathbf{F}_2 H^*$-module E on which some $h^* \in H^*$ induces a transvection (that is, with $[E, H]$ the A_7-module), $\langle h^* \rangle = C_{H^*}(C_E(h^*))$.

As $L_1 T = T L_1$, $L_1 T$ stabilizes either $\{1, 2, 3, 4\}$ or a partition of type $2^3, 1$.
Assume the first case holds. Then the stabilizer S in H of $\{1, 2, 3, 4\}$ is solvable,
so $S \leq M$ by 13.8.13. Thus $S = H \cap M$; hence by 13.7.3.9, $L_1 = \theta(S)$ is of 3-rank
2, so that $L/O_2(L) \cong \hat{A}_6$. Next either $L_0^* = \langle (5, 6, 7) \rangle$ and $L_{1,+}^* = O^2(K_{5,6,7}^*)$, or
vice versa. As $L_{1,+}^*$ is inverted in $T \cap L \leq C_T(L_0^*)$, $H^* \cong S_7$. As $q(H^*, \tilde{U}_H) \leq 2$

and $H^* \cong S_7$, B.4.2 and B.4.5 say that $\hat{U}_H$ is either a natural module or the sum of a 4-dimensional module and its dual. As $\hat{V}_3$ is of rank 2 and T-invariant, with $\hat{V}_3 = [\hat{V}_3, L_{1,+}] \le C_{\hat{U}_H}(L_0)$, we conclude that $\hat{U}_H$ is natural, and $L_{1,+}^* = \langle (5,6,7) \rangle$. Recall $V_\alpha^* \le O_2(L_1^* T^*) = O_2(L_{1,+}^*)$. As V_α^* is quadratic on $\hat{U}_H$ by 13.8.4.6, it follows that $m(V_\alpha^*) = 1$, so U_H induces transvections on U_γ by 13.8.18.3. But then V_β^{*y} induces transvections on $\tilde{U}_H$, whereas $V_\beta^{*y} \le O_2(L_1^* T^*)$, which contains no transvections.

Thus $L_1 T$ is the stabilizer of a partition of type $2^3, 1$. In particular $m_3(L_1) = 1$, so $L/O_2(L) \cong A_6$ as $L_1 = \theta(H \cap M)$ by 13.7.3.9. As $U_\gamma^* \in \mathcal{Q}(H^*, \hat{U}_H)$, B.4.2 and B.4.5 say that $\hat{U}_H$ is either of dimension 4 or 6, or else the sum $4 + 4'$ of 4 and its dual $4'$. But L_1^* stabilizes the T^*-invariant line $\hat{V}_3 \le \hat{U}_H$, so as $L_1 T$ is the stabilizer of a partition of type $2^3, 1$, $\dim(\hat{U}_H) \ne 4$ or 8, and hence $\dim(\hat{U}_H) = 6$. If $[\tilde{W}, K] = 1$, then $[\tilde{U}_H, K] \cong \hat{U}_H$ is the natural module for K^*, so as $L/O_2(L) \cong A_6$, 13.7.6.3 supplies a contradiction. Thus $[\tilde{W}, K] \ne 1$, and so we may apply 13.8.22. As H^* has no strong FF-modules, we conclude from 13.8.22 that U_γ^* induces a group of transvections on $\hat{U}_H$ or $\tilde{W}$. Therefore $m(U_\gamma^*) = 1$, and we saw this suffices to complete the proof. $\square$

LEMMA 13.8.25. *If $K/O_2(K) \cong L_n(2)$ for $3 \le n \le 5$, then $n = 4$ and*

(1) $L/O_2(L) \cong \hat{A}_6$, and $\tilde{U}_H$ is a 4-dimensional natural module for $H^* \cong L_4(2)$.
(2) $H = G_1$.

PROOF. Assume otherwise. If case (1) of 13.8.8 holds, then conclusion (1) holds by 13.8.5. In particular $n = 4$, and we will see below that this implies conclusion (2); so we may assume that case (2) of 13.8.8 holds.

Then $U_\gamma^* \in \mathcal{Q}(H^*, \tilde{U}_H)$. Let $T_K^* := T^* \cap K^*$. As $L_1 \le K$ by 13.8.21.1, $L_1^* T_K^*$ is a T^*-invariant parabolic of K^*. Indeed $L_1^* T_K^*$ is a minimal parabolic when $L/O_2(L) \cong A_6$, since $|L_1|_3 = 3$ in that case, whereas $L_1^* T^* / O_2(L_1^* T^*) \cong S_3 \times S_3$ when $L/O_2(L) \cong \hat{A}_6$.

If $L_1^* T_K^*$ is a minimal parabolic, then as $L_1^* T_K^*$ is T^*-invariant, either $T^* = T_K^*$, or $n = 4$ and $L_1^* T_K^*$ is the middle-node parabolic of K^*. This allows us to eliminate the case $n = 3$: For if $n = 3$, then $m_3(K) = 1$, so $L_1^* T_K^*$ is a minimal parabolic and hence $T^* = T_K^*$. contrary to 13.8.21.3.

Further if $n = 5$, then $L/O_2(L) \cong \hat{A}_6$: For otherwise we have seen that $L_1^* T_K^*$ is a minimal parabolic and $T^* = T_K^*$. Therefore $L_1 T \le H_1 \le H$ with $H_1/O_2(H_1) \cong S_3 \times S_3$. But now $H_1 \le M$ by 13.8.13, so $L_1 = \theta(H \cap M)$ is of 3-rank 2 by 13.7.3.9, contradicting $L/O_2(L) \cong A_6$.

In the next few paragraphs, we assume $L/O_2(L) \cong A_6$ and derive a contradiction. Here the arguments above have reduced us to the case $n = 4$.

Suppose $T_K = T$. Then $L_1 \le K_1 \in \mathcal{L}(K, T)$ with $K_1/O_2(K_1) \cong L_3(2)$. But now $K_1 T \in \mathcal{H}_z$, a case already eliminated. Thus $T_K < T$, so we have seen that $L_1^* T_K^*$ is the middle-node parabolic.

Let $\tilde{W}$ be a maximal H^*-submodule of $\tilde{U}_H$, so that $\hat{U}_H$ is irreducible. As $U_\gamma^* \in \mathcal{Q}(H^*, \hat{U}_H)$ and $T_K^* < T^*$, B.4.2 and B.4.5 say that either $m(\hat{U}_H) = 6$, or $\hat{U}$ is the sum of a natural K^*-module and its dual. The latter is impossible, as $L_1^* T_K^*$ is a middle-node minimal parabolic and $\hat{V}_3 = [\hat{V}_3, L_1]$ is an $L_1 T$-invariant line in $\hat{U}_H$.

Thus $m(\hat{U}_H) = 6$. Since the case with a single nontrivial 2-chief factor which is an A_8-module is excluded by 13.7.6.3, $[\tilde{W}, K] \neq 1$, so we can appeal to 13.8.22.

If U_γ^* contains a strong FF*-offender on $\hat{U}_H$, then B.3.2.6 says that $U_\gamma^* \cong E_{16}$ is generated by the transpositions in T^* and $m(\hat{U}_H/C_{\hat{U}_H}(U_\gamma)) = 3$. Thus as $U_\gamma^* \in \mathcal{Q}(H^*, \tilde{U}_H)$,

$$m(\tilde{W}/C_{\tilde{W}}(U_\gamma)) \leq 2m(U_\gamma^*) - 3 = 5;$$

so as $\tilde{W}$ is a faithful module for $H^* \cong S_8$, we conclude $[\tilde{W}, H]$ is the 6-dimensional module or its 7-dimensional cover, and $m(\tilde{W}/C_{\tilde{W}}(U_\gamma)) = 3$. Thus

$$m(\tilde{U}_H/C_{\tilde{U}_H}(U_\gamma^*)) \geq m(\hat{U}_H/C_{\hat{U}_H}(U_\gamma^*)) + m(\tilde{W}/C_{\tilde{W}}(U_\gamma^*)) = 6.$$

Now by 13.8.8, $m(U_H/D_H) \leq m(U_\gamma^*) = 4$, so U_γ^* does not centralize D_H. Then by F.9.13.6, $A_1 = [D_H, U_\gamma] \leq U_H$. So as the transpositions in U_γ^* induce transvections on $\hat{U}_H$ with distinct centers, we conclude $C_{U_\gamma^*}(D_H)$ is a hyperplane of U_γ^*, so $m(U_H/C_{U_H}(U_\gamma)) \leq 5$, contrary to our previous calculation.

Therefore U_γ^* contains no strong FF*-offender on $\hat{U}_H$, so by 13.8.22 either

(i) $U_\gamma^* \cong \mathbf{Z}_2$ induces a transvection with center $\tilde{A}_1$ on $\tilde{W}$, or a transvection with center $\hat{A}_1$ on $\hat{U}_H$, or

(ii) $A_1 \not\leq W$, and U_γ^* contains a strong FF*-offender on $\tilde{W}$.

In case (ii), as in the previous paragraph, we conclude $U_\gamma^* \cong E_{16}$ and $\tilde{W}$ is the orthogonal module, leading to the same contradiction.

So case (i) holds. Then $m(U_\gamma^*) = 1$ and $[\tilde{W}, K] \neq 1$, so 13.8.23.4 supplies a contradiction, since $C_{H^*}(C_E(h^*)) = \langle h^* \rangle$ for each faithful $\mathbf{F}_2 H^*$-module E (namely with $[E, K]$ of dimension 6 or 7) on which some $h^* \in H^*$ induces a transvection. This contradiction completes the elimination of the case $L/O_2(L) \cong A_6$.

Therefore $L/O_2(L) \cong \hat{A}_6$. We eliminated $n = 3$ earlier, so $n = 4$ or 5. As T acts on the two minmal parabolics determined by L_0 and $L_{1,+}$, $T_K^* = T^*$. Further as $L_1 \trianglelefteq H \cap M$, $L_1 T = H \cap M$. Observe that $L_1^* T_K^*$ is a parabolic of rank 2 determined by two nodes not adjacent in the Dynkin diagram.

Suppose $n = 5$. By 13.2.2.9, $N_K(L_0) \leq K \cap M \leq N_K(L_{1,+})$, the node β determined by L_0 is an interior node, and the node δ determined by $L_{1,+}$ is the unique node not adjacent to β. Thus we may take δ and β to be the first and third nodes of the diagram for H^*. Then $L_1 T \leq H_2 \leq H$ with $H_2/O_2(H_2) \cong S_3 \times L_3(2)$. As $L_1 T = H \cap M$, $H_2 \not\leq M$, so $H_2 \in \mathcal{H}_z$, contrary to 13.8.21.1.

Therefore we have established that $n = 4$ in each case of 13.8.8. As mentioned earlier, we can now show that (2) holds: For $H \leq G_1$, so $K \leq K_1 \in \mathcal{C}(G_1)$ by 1.2.4. Now $K_1 T \in \mathcal{H}_z$ and we have shown $K_1/O_2(K_1)$ is not $L_5(2)$. Hence $K_1 = K \in \mathcal{C}(G_1)$ by 13.8.21.2 and A.3.12. As $G_1 \in \mathcal{H}_z$, we conclude from 13.8.21.3 that $H = KT = K_1 T = G_1$, so that (2) holds.

Thus (2) is established, and we've shown that $L/O_2(L) \cong \hat{A}_6$ and $H^* \cong L_4(2)$. We may assume (1) fails, so that $\hat{U}_H$ is not the natural module for H^*. Now $U_\gamma^* \in \mathcal{Q}(H^*, \hat{U}_H)$, so $\hat{U}_H$ is of dimension 4 or 6 by B.4.2 and B.4.5. Then as the maximal parabolic $L_1^* T^*$ determined by the end nodes stabilizes the line $\hat{V}_3$, $\dim(\hat{U}_H) = 4$. As (1) fails, $[\tilde{W}, K] \neq 1$; hence we can appeal to 13.8.22 and 13.8.23. Moreover $m(V_\gamma^*) > 1$ by 13.8.18.4.

We claim $\tilde{U}_H$ has a unique maximal submodule $\tilde{W}$. Assume not; then (writing $J(\tilde{U}_H)$ for the Jacobson radical of $\tilde{U}_H$)

$$\dot{U}_H := \tilde{U}_H/J(\tilde{U}_H) = \dot{U}_1 \oplus \cdots \oplus \dot{U}_s$$

is the sum of $s > 1$ four-dimensional irreducibles. Further the projection $\dot{V}_3^i$ of $\tilde{V}_3$ on $\dot{U}_i$ is faithful for each i and centralized by L_0, so the $\dot{U}_i$ are isomorphic natural modules. As $C_{\dot{U}_i}(T^*)$ is a point, each $L_1 T$-invariant line is contained in a member of $Irr_+(H, \dot{U}_H)$, so $\dot{V}_3 \leq \dot{U}_0$ for some irreducible H-submodule $\dot{U}_0$. But then $\dot{U}_H = \langle \dot{V}_3^H \rangle = \dot{U}_0$, contrary to $s > 1$. Thus the claim is established.

Define α as in case (2) of 13.8.8; thus $V_\alpha^* \leq O_2(L_1^*T^*) = O_2(L_1^*)$ and $m(V_\alpha^*) = m(V_\gamma^*) > 1$.

Let B be a noncentral chief factor for H on $\tilde{W}$. We claim $m(B) = 4$. For otherwise, as $q(H^*, \tilde{U}_H) \leq 2$, B is of rank 6 by B.4.2 and B.4.5. Thus as V_α^* is a noncyclic subgroup of the unipotent radical $O_2(L_1^*)$ of the parabolic $L_1^*T^*$ stabilizing a point of B and acting quadratically on B, it follows that $m(V_\alpha^*) = 2$ and V_α^* contains no FF*-offender on B by B.3.2.6. Therefore case (1) of 13.8.22 holds, so $A_1 \leq W$. As $T^* = T_K^*$, no member of H^* induces a transvection on B, so $[\tilde{W}, U_\gamma^*] > \tilde{A}_1$ and hence $W \not\leq D_H$ by F.9.13.6. Thus we conclude from 13.8.22 that U_γ^* contains a strong FF*-offender on $\hat{U}_H$. As $m(V_\gamma^*) = 2$ and U_γ^* contains a strong FF*-offender, we conclude $V_\gamma^* = U_\gamma^* \cong E_4$. Then $m(U_H/D_H) \leq m(U_\gamma^*) = 2$, so as $W \not\leq D_H$, $m(\hat{U}_H/\hat{D}_H) \leq 1$, with $m(\hat{U}_H/\hat{D}_H) = 2$ in case of equality. However U_γ^* centralizes $\hat{D}_H$ by F.9.13.6 as $A_1 \leq W$. Therefore $m(\hat{U}_H/\hat{D}_H) = 1$ and $m(U_H/D_H) = 2 = m(U_\gamma^*)$. Thus we have symmetry between γ_1 and γ. In particular as $A_1 \leq W \leq U_H$, $V_1 \leq U_\gamma$; further in view of 13.8.18.2, we may apply 13.8.11.1 to conclude $U_\gamma^* < V_\gamma^*$, contrary to an earlier remark. This establishes the latest claim that $m(B) = 4$.

Thus we have shown that all noncentral chief factors of $\tilde{U}_H$ are 4-dimensional. Then as the 1-cohomology of 4-dimensional modules is trivial by I.1.6.6, and $\tilde{W}$ is the unique maximal submodule of $\tilde{U}_H$, all chief factors are 4-dimensional.

Observe next that no noncyclic subgroup of V_γ^* centralizes a hyperplane of $\hat{U}_H$: For otherwise as V_γ^* is quadratic on $\tilde{U}_H$ by 13.8.4.6, the quotient module $\tilde{U}_H$ splits over the submodule $\tilde{W}$ by B.4.9.1, contradicting $\tilde{W}$ the unique maximal submodule of $\tilde{U}_H$. So as V_α^* lies in the unipotent radical $O_2(L_1^*)$ of the stabilizer of a line in the natural module for $L_4(2)$, it follows that $m(U_\gamma^*) \leq m(V_\gamma^*) \leq 3$.

Now let $\tilde{I}$ denote any member of $Irr_+(H^*, \tilde{W})$, so that in particular $\tilde{I}$ is 4-dimensional. Applying the dual of B.4.9.1, we conclude similarly that no noncyclic subgroup of V_γ^* acts as a group of transvections with a fixed center on $\tilde{I}$.

Suppose next that $A_1 \leq I$. Then $C_H(A_1)^*$ is the maximal parabolic fixing $\tilde{A}_1$. Then as $H = G_1$, $V_\gamma \trianglelefteq N_G(A_1)^*$, so $V_\gamma^* = O_2(C_H(A_1))^*$ as $N_G(A_1)^*$ is irreducible on $O_2(C_H(A_1))^*$. This is impossible as $V_\gamma^* \leq O_2(L_1)^*$ where $L_1^*T^*$ stabilizes a line of $\tilde{I}$.

Therefore $A_1 \not\leq I$, so $[I \cap D_H, U_\gamma] \leq I \cap A_1 = 1$; then by 13.7.3.7, $[I \cap D_H, V_\gamma] \leq A_1 \cap I = 1$. In particular $I \not\leq D_H$.

Suppose next that $m(U_\gamma^*) = 1$. We saw V_γ centralizes $D_H \cap I$, which is a hyperplane of I by 13.8.23.1. Then as $V_\alpha^* \leq O_2(L_1^*)$ and $L_1^*T^*$ is the parabolic stabilizing a line in I, we conclude $m(V_\alpha^*) \leq 2$, and hence $m(V_\alpha^*) = 2$ as $m(V_\gamma^*) > 1$

by 13.8.18.4. Also by 13.8.23.1, U_γ^* induces transvections on $\tilde{W}$ and $\hat{U}_H$, so H^* has a unique noncentral chief factor on W, and hence $W = I$. Again by 13.8.23.1, D_H is a hyperplane of U_H, so as $W = I \not\leq D_H$, $\hat{U}_H = \hat{D}_H$ and hence $\hat{A}_1 = [\hat{U}_H, U_\gamma]$ and $\tilde{A}_1[\tilde{W}, U_\gamma^*] = [\tilde{U}_H, U_\gamma^*]$ is of rank 2. Now $U_\alpha^* = Z(T^*)$, so T^* acts on $[\tilde{U}_H, U_\alpha^*]$ and centralizes $\tilde{W}_1 \tilde{V}_2$ where $\tilde{W}_1 := [\tilde{W}, U_\alpha^*]$. Thus

$$\tilde{W}_1 \tilde{A}_1^h = [\tilde{U}_H, U_\alpha^*] = \tilde{W}_1 \tilde{V}_2,$$

where $h \in H$ with $\gamma h = \alpha$. Thus the middle-node minimal parabolic H_0^* of H^* containing T^* centralizes $[\tilde{U}_H, U_\alpha^*]$, and in particular $\tilde{A}_1^h$, so H_0^* acts on V_α^* since $G_1 = H$ by (2). This is impossible as $H_0^* \cong S_3/D_8^2$ has no normal E_4-subgroup.

So $m(U_\gamma^*) > 1$. Now $V_\alpha^* \leq O_2(L_1^*)$, and we've seen that U_γ^* is noncyclic and U_α^* does not induce a group of transvections with fixed center on $\tilde{I}$; thus $[\tilde{I}, U_\alpha^*]$ is the line in $\tilde{I}$ fixed by L_1^*, and hence $[\tilde{I}, U_\alpha^*] = [\tilde{I}, V_\alpha^*]$. Therefore by 13.8.9.2 applied to I in the role of "F", $V_1 \not\leq U_\gamma$. Also we saw V_γ centralizes $D_H \cap I$, so $m(I/D_H \cap I) \geq 2$.

Suppose $m(U_\gamma^*) = m(U_H/D_H)$. Then we have symmetry between γ_1 and γ, so $A_1 \not\leq U_H$, and hence $[D_H, U_\gamma] \leq A_1 \cap U_H = 1$. Further as U_γ^* is noncyclic and we saw earlier that U_γ^* does not centralize any hyperplane of $\hat{U}_H$, $m(\hat{U}_H/\hat{D}_H) \geq 2$. Hence as $m(I/I \cap D_H) \geq 2$, $m(U_\gamma^*) = m(U_H/D_H) \geq 4$, contrary to our earlier observation that $m(U_\gamma^*) \leq 3$.

Therefore $m(U_\gamma^*) > m(U_H/D_H)$. So as $m(U_\gamma^*) \leq 3$, we conclude

$$3 \geq m(U_\gamma^*) > m(U_H/D_H) \geq 2,$$

where the final inequality holds since we saw $m(I/D_H \cap I) \geq 2$. Thus $m(U_\gamma^*) = 3$ and $m(U_H/D_H) = 2$. Hence $\hat{U}_H = \hat{D}_H$ since $m(I/D_H \cap I) \geq 2$, so $[\hat{U}_H, U_\gamma] = [\hat{D}_H, U_\gamma] = \hat{A}_1$ by F.9.13.6. This is impossible as $U_\alpha^* \leq O_2(L_1^*)$ with $m(U_\alpha) = 3$. Thus the proof of 13.8.25 is at last complete. $\square$

LEMMA 13.8.26. *If $K^* \cong A_6$, then $\tilde{U}_H$ is the natural module for K^* on which L_1 has two noncentral chief factors or its 5-dimensional cover.*

PROOF. In case (1) of 13.8.8, this holds by 13.8.5, so we may assume case (2) of 13.8.8 holds. Then $U_\gamma^* \in \mathcal{Q}(H^*, \tilde{U}_H)$, so each noncentral chief factor for K^* on $\tilde{U}_H$ is of rank 4 by B.4.2 and B.4.5. Suppose K has more than one such factor, and pick $\tilde{W}$ as in 13.8.22.

First assume U_γ^* contains a strong FF*-offender on $N := \hat{U}_H$ or $\tilde{W}$. Then by B.3.4.2i, $U_\alpha^* = R_1^* \cong E_8$ is generated by the transvections on N in T^*. But by 13.8.4.5, V_H/U_H has a quotient B which is the 4-dimensional H^*-module on which L_1T fixes a point. Then as $U_\alpha^* = R_1^*$, U_α^* is not quadratic on B, contrary to 13.8.4.6.

Thus U_γ^* contains no strong FF*-offender on either $\hat{U}_H$ or $\tilde{W}$, so by 13.8.22, U_γ^* induces transvections on $E =: \hat{U}_H$ or $\tilde{W}$, and hence $m(U_\gamma^*) = 1$. This is a contradiction to 13.8.23.4, as $U_\gamma^* = C_{H^*}(C_E(U_\gamma^*))$ for any transvection.

Thus $\tilde{U}_H$ has a unique noncentral chief factor. Since $U_H = \langle V_3^H \rangle$ with $\tilde{V}_3 = [\tilde{V}_3, L_1]$ a nontrivial irreducible for L_1, $\tilde{U}_H/C_{\tilde{U}_H}(K)$ is the 4-dimensional natural module on which L_1 has two noncentral chief factors. Then by I.1.6.1, $\tilde{U}_H$ is either natural or a 5-dimensional cover, completing the proof. $\square$

LEMMA 13.8.27. *(1) Either*

(a) $L/O_2(L) \cong A_6$, $H^* \cong A_6$ or S_6, and $\tilde{U}_H$ is the natural module for K^* on which L_1 has two noncentral chief factors or its 5-dimensional cover, or

(b) $L/O_2(L) \cong \hat{A}_6$, $H^* \cong L_4(2)$, and $\tilde{U}_H$ is a 4-dimensional natural module for H^*.

(2) $G_1 = H = KT$.

(3) If case (1) of 13.8.8 holds then $D_H = U_H$, $D_\gamma = U_\gamma$, V induces a group of transvections on U_γ with center V_1, and $V_1 \leq U_\gamma$. Further $V_\gamma \not\leq Q_H$, so we have symmetry between γ and γ_1.

PROOF. By 13.8.24 and 13.8.25, the list of 13.8.21.2 has been reduced to $K^* \cong A_6$, $L_4(2)$, or $G_2(2)'$. Further $K \leq K_1 \in \mathcal{C}(G_1)$ by 1.2.4, and as $G_1 \in \mathcal{H}_z$, $K_1/O_2(K_1) \cong A_6$, $L_4(2)$, or $G_2(2)'$. So as A.3.12 contains no inclusions between any pair on this list, we conclude that $K = K_1$. Thus $G_1 = K_1T = KT = H$ by 13.8.21.3, so (2) holds.

By (2) and 13.8.7, $D_H = U_H$ and $D_\gamma = U_\gamma$. Thus V induces a group of transvections on U_γ with center V_1 by F.9.16.1, so $V_1 \leq U_\gamma$. Thus to complete the proof of (3), we assume $V_\gamma \leq Q_H$ and derive a contradiction. Then $[U_H, V_\gamma] \leq V_1 \cap A_1 = 1$. Thus $V^g \leq C_G(V_3) \leq M_V$, so that $[V, V_3^g] = V_1 = [V, V^g]$. Then $C_{V^g}(V)$ is a hyperplane of V^g and hence conjugate to V_3^g, so $V \leq C_G(C_{V^g}(V)) \leq M_V^g$ by 13.5.4.4. Then $V_1 = [V, V^g] \leq V \cap V^g$, contrary to 13.8.3.

It remains to prove (1). However if $K^* \cong L_4(2)$ or A_6, then (1) holds by 13.8.25 or 13.8.26, so we may assume that $K^* \cong G_2(2)'$ and derive a contradiction. Thus case (2) of 13.8.8 holds as $G_2(2)'$ does not appear in 13.8.5. As H^* has no strong FF-modules and no transvection modules by B.4.2, 13.8.22 and 13.8.21.4 say $\tilde{U}_H \in Irr_+(K, \tilde{U}_H)$. So as $U_\gamma^* \in \mathcal{Q}(H^*, \tilde{U}_H)$ by 13.8.8, B.4.2, B.4.5, and I.1.6.5 say that $\tilde{U}_H$ is the 7-dimensional Weyl module or its 6-dimensional quotient module. Thus $m(U_H) \leq 8$.

By 13.8.18.2 and 13.8.11.1, $U_\gamma^* < V_\gamma^*$. Thus $m(U_\gamma^*) < m_2(H^*) = 3$, so by B.4.6.13, $r_{U_\gamma^*, \tilde{U}_H} > 1$. But by the choice of γ in case (2) of 13.8.8, $m(U_\gamma^*) \geq m(U_H/D_H)$, and $[U_\gamma, D_H] \leq A_1$ by F.9.13.6, so we conclude $A_1 \leq U_H$. Thus $H_1^* := C_{H^*}(\tilde{A}_1) = C_H(A_1)^*$ is a maximal parabolic of H^*, and U_γ^* is elementary abelian and normal in $C_H(A_1)^*$. Therefore as $m(U_\gamma^*) < 3$, $U_\gamma^* \cong E_4$ (cf. B.4.6.3). Thus $m(D_\gamma \cap U_H) \geq m([U_\gamma, U_H]) \geq 3$. Next by 13.7.4.2, Q_H/H_C is H^*-isomorphic to $U_H/C_{U_H}(Q_H)$, so $1 \neq [C_{Q_H}(A_1)/H_C, U_\gamma] \leq D_\gamma H_C/H_C$, so $D_\gamma \not\leq H_C$. Finally as V_H is abelian, $V_H \leq H_C$, and by (2) and 13.8.4.5, $H^* \cong G_2(2)'$ or $G_2(2)$ is faithful on V_H/U_H; so as $U_\gamma^* \cong E_4$, $m((D_\gamma \cap V_H)U_H/U_H) \geq m([V_H/U_H, U_\gamma]) \geq 3$. Thus as $r_{U_\gamma^*, \tilde{U}_H} > 1$,

$$m(U_\gamma) > m(U_\gamma^*) + m((D_\gamma \cap V_H)U_H/U_H) + m(D_\gamma \cap U_H) \geq 2 + 3 + 3 = 8,$$

contrary to the previous paragraph. $\square$

THEOREM 13.8.28. $K^* \cong A_6$.

Until the proof of Theorem 13.8.28 is complete, assume G is a counterexample. Then $H^* \cong L_4(2)$, $L/O_2(L) \cong \hat{A}_6$, and $m(U_H) = 5$ by 13.8.27.1. Recall $G_2 = N_G(V_2)$, and set $K_2 := O^2(N_H(V_2))$, $Q_2 := O_2(G_2)$, and $\dot{G}_2 := G_2/Q_2$. Set $U_0 := \langle U_H^{G_2} \rangle$ and $V_0 := \langle V_H^{G_2} \rangle$.

Since $L/O_2(L) \cong \hat{A}_6$, by 13.5.4, $I_2 = O_2(G_1)L_{2,+} \trianglelefteq G_2$ with $O_2(I_2) = C_{I_2}(V_2) = I_2 \cap Q_2$ and $\dot{I}_2 \cong S_3$. Let $g \in L_{2,+} - H$, so that $\widetilde{V_1^g} = \tilde{V}_2$.

LEMMA 13.8.29. (1) $K_2 \in \mathcal{C}(G_2)$ with $K_2/O_2(K_2) \cong L_3(2)$.

(2) $G_2 = K_2 L_{2,+} T$ and $\dot{G}_2 = \dot{K}_2 \times \dot{I}_2 \cong L_3(2) \times S_3$.

(3) $U_0 = V_H \cap V_H^g = U_H U_H^g$ and U_0/V_2 is the tensor product of the natural modules V/V_2 and U_H/V_2 for $\dot{I}_2$ and $\dot{K}_2$.

(4) $V_0 = V_H V_H^g V_H^{g^2}$, and V_0/U_0 is the tensor product of V/V_2 or $V/V_2 \oplus \mathbf{F}_2$ with the dual of U_H/V_2.

(5) V_H/U_H is the 6-dimensional orthogonal module for $H^* \cong L_4(2)$.

PROOF. As $\tilde{U}_H$ is the natural module for $H^* \cong L_4(2)$, $N_H(V_2)^* = C_{H^*}(\tilde{V}_2)$ is the parabolic subgroup $L_3(2)/E_8$ of H^* stabilizing the point $\tilde{V}_2$, so $K_2 \in \mathcal{C}(H \cap G_2)$ with $K_2/O_2(K_2) \cong L_3(2)$. As $I_2 \trianglelefteq G_2$, and I_2 acts transitively on $V_2^\#$, $G_2 = I_2(H \cap G_2)$ with $H \cap G_2 = K_2 T$ and $[\dot{K}_2, \dot{I}_2] = 1$. Thus (1) and (2) hold.

Next U_H/V_2 is the natural module for $\dot{K}_2 \cong L_3(2)$. Thus as $\dot{K}_2 \trianglelefteq \dot{G}_2$, U_0/V_2 is the direct sum of I_2-conjugates of U_H/V_2. Further $U_H = \langle V_3^{K_2} \rangle$ with V/V_2 the natural module for $\dot{I}_2$, so as $I_2 \trianglelefteq G_2$, U_0/V_2 is the direct sum of conjugates of V/V_2. Thus $U_0/V_2 = U_H U_H^g/V_2$ is the tensor product of V/V_2 and U_H/V_2. Further $U_H^g = \langle V_3^{gK_2} \rangle \leq V_H$, so $U_0 = U_H U_H^g \leq V_H$ and so $U_0 = U_0^g \leq V_H \cap V_H^g$.

Let $\hat{V}_H := V_H/U_H$. Then $\hat{V}_H = \langle \hat{V}^H \rangle$ with the maximal parabolic $L_1^* T^*$ of H^* centralizing the point $\hat{V}$, and $\langle \hat{V}^{K_2} \rangle \cong U_H^g/V_2 \cong U_H/V_2$ as a K_2-module, so we conclude from B.4.13 that (5) holds. In particular K_2 is irreducible on V_H/U_0, so either $U_0 = V_H \cap V_H^g$ or $V_H = V_H^g$. In the latter case, both $LT = \langle L_{2,+}, L_1 T \rangle$ and H act on V_H, contrary to $H \not\leq M = !\mathcal{M}(LT)$. This completes the proof of (3).

By (5), V_H/U_0 is isomorphic to the dual of U_0/U_H as a K_2-module, and by (3), $V_H < V_0$. Thus (4) holds. $\qquad \square$

LEMMA 13.8.30. L_0 has at least 9 noncentral 2-chief factors.

PROOF. Recall $V < U_L = \langle U_H^L \rangle \leq O_2(LT) = Q$ by (7) and (2) of 13.8.4. Let W be a normal subgroup of L maximal subject to being proper in U_L, and set $\hat{U}_L := U_L/W$. As U_H/V_3 is a 2-dimensional irreducible for $L_0 \trianglelefteq L$, and $U_L = \langle U_H^L \rangle$, $\hat{U}_L = \langle \hat{U}_H^L \rangle = [\hat{U}_L, L_0]$ is a faithful irreducible for $L^+ := L/O_2(L) \cong \hat{A}_6$, and so may be regarded as an $\mathbf{F}_4$-module on which $L_0^+ \cong \mathbf{Z}_3$ acts by scalar multiplication. In particular from the 2-modular character table for $\hat{A}_6$, $\dim_{\mathbf{F}_4}(\hat{U}_L) = 3$ or 9, so to complete the proof, it suffices to show $\dim_{\mathbf{F}_4}(\hat{U}_L) > 3$.

From 13.8.29.3, $\hat{S}_2 := \langle \hat{U}_H^{L_2} \rangle \cong \langle U_H^{L_2} \rangle/V$ is of $\mathbf{F}_4$-dimension 2. Let $\hat{S}_3 := \langle \hat{S}_2^{L_1} \rangle$; from 13.8.29.5, $\hat{S}_3/\hat{U}_L \cong \langle U_0^{L_1} \rangle/U_H V$ is of $\mathbf{F}_4$-dimension 2, so $\dim_{\mathbf{F}_4}(\hat{S}_3) = 3$. Finally by 13.8.29.4, $L_{2,+}$ does not act on $\langle U_0^{L_1} \rangle/U_0$, so $\hat{U}_L > \hat{S}_3$, completing the proof. $\qquad \square$

LEMMA 13.8.31. (1) $A_1 \not\leq U_H$.

(2) Case (2) of 13.8.8 holds.

(3) $[H_C, K] \not\leq V_H$; and if K has a unique noncentral chief factor on H_C/V_H, it is not a 4-dimensional module for $H^* \cong L_4(2)$.

PROOF. In case (1) of 13.8.8, $A_1 \leq U_H$ by 13.8.27.3, so to prove (2), it will suffice to establish (1).

Assume (1) fails, so that $A_1 \leq U_H$. Then as H is transitive on $\tilde{U}_H^\#$, there is $k \in H$ with $\tilde{A}_1^k = \tilde{V}_2 = \tilde{V}_1^g$; and since $[V_3, Q_H] = V_1$ by 13.7.3.6, we may assume $A_1^k = V_1^g$. Then as $G_1 = H$ by 13.8.27.2, $\gamma k = \gamma_1 g$, so that by 13.8.29.3,

$U_\gamma^k = U_H^g \leq V_H$. Thus as V_H is abelian, V_H centralizes U_γ^k, and hence also U_γ. Therefore $[U_H, U_\gamma] = 1$, and hence case (1) of 13.8.8 holds.

Next $V_\gamma^* \neq 1$, so $1 \neq V_\gamma^{*k} = V_H^{g*}$. However as $\widetilde{V_1^g} = \tilde{V}_2$, $V_H^{g*} \trianglelefteq C_{H^*}(\tilde{V}_2) = K_2^*$, so $V_H^g/E \cong V_H^{g*} = O_2(K_2^*) \cong E_8$, where $E := Q_H \cap V_H^g$. But $U_0 = V_H \cap V_H^g \leq E$, so $E = U_0$ as $m(V_H^g/E) = 3 = m(V_H^g/U_0)$ by parts (3) and (5) of 13.8.29. Also $H_C \leq C_G(A_1^k) \leq N_G(V_H^g)$, so $[H_C, V_H^g] \leq H_C \cap V_H^g \leq E = U_0 \leq V_H$, and hence $K = [K, V_H^g]$ centralizes H_C/V_H. Thus to complete the proof of (1) and hence of (2), it will suffice to establish (3).

Appealing to 13.8.29.5 and the duality in 13.7.4.2, K has the following noncentral 2-chief factors on Q_H/H_C and V_H: The natural module $\tilde{U}_H$, its dual Q_H/H_C, and the orthogonal module V_H/U_H. Therefore L_0 has six noncentral 2-chief factors not in H_C/V_H: two on $O_2(L_0^*)$, one each on Q_H/H_C and $\tilde{U}_H$, and two on V_H/U_H. Therefore by 13.8.30, L_0 has at least three noncentral chief factors on H_C/V_H, so (3) holds and the proof of the lemma is complete. $\square$

LEMMA 13.8.32. *(1)* $m(U_\gamma^*) = 1$.
(2) $m(U_\gamma \cap V_H) \geq 3$.
(3) $A_1 \leq V_H$.

PROOF. By 13.8.31.2, $U_\gamma^* \neq 1$; thus $m(U_H \cap U_\gamma) \geq m([U_H, U_\gamma]) > 0$. Further by 13.8.29.5, no member of H^* induces a transvection on V_H/U_H, so

$$m((U_\gamma \cap V_H)/(U_\gamma \cap U_H)) = m((U_\gamma \cap V_H)U_H/U_H) \geq m([V_H/U_H, U_\gamma] \geq 2, \quad (*)$$

with equality only if (1) holds. In particular this establishes (2), and moving on to the proof of (1), we may assume that $m(U_\gamma \cap V_H) \geq 4$. But then as $m(U_\gamma^*) = m(U_\gamma/(U_\gamma \cap Q_H)) \leq m(U_\gamma/(U_\gamma \cap V_H))$ and $m(U_\gamma) = 5$, it follows again that $m(U_\gamma^*) = 1$, completing the proof of (1). Thus (1) and (2) are established.

By (1) and 13.8.10, $m(U_H/D_H) = 1$, and we have symmetry between γ_1 and γ in the sense of Remark F.9.17. By 13.8.31.1, $A_1 \not\leq U_H$, so by symmetry and 13.8.10.2, $V_1 \not\leq U_\gamma$, and U_γ induces transvections on U_H with axis D_H.

Let $\beta \in \Gamma(\gamma)$; by F.7.3.2 there is $y \in G$ with $\gamma_1 y = \gamma$ and $V^y = V_\beta$. By 13.5.4.4, $[C_G(V_3^y), V^y] \leq A_1$, so as $A_1 \not\leq U_H$,

$$[C_{U_H}(V_3^y), V^y] \leq U_H \cap A_1 = 1. \quad (**)$$

But $V_3^y \leq U_\gamma \leq C_H(D_H)$, so $V_\beta = V^y$ centralizes D_H by $(**)$. As this holds for each $\beta \in \Gamma(\gamma)$, V_γ centralizes D_H. Therefore V_γ^* induces a group of transvections on $\tilde{U}_H$ with axis $\tilde{D}_H$. We saw that no member of H^* induces a transvection on V_H/U_H, so we conclude from 13.8.18.2 and 13.8.11.1 that $U_\gamma^* < V_\gamma^*$. By parts (1) and (2) of F.9.13, $V_\gamma^* \leq O_2(L_1^* T^*)^x$ for some $x \in H$. So as $L_1^* T^*$ is the parabolic subgroup of H^* stabilizing the 2-subspace $\tilde{V}_3$ of the 4-dimensional module $\tilde{U}_H$, while V_γ^* centralizes the hyperplane $\tilde{D}_H$ of $\tilde{U}_H$, we conclude that $m(V_\gamma^*) = 2$. By symmetry, $E_H = V_H \cap Q_H^y$ is of corank 2 in V_H, so as $|U_H : D_H| = 2$, $E_H U_H/U_H$ is a hyperplane of V_H/U_H. Thus $1 \neq [E_H U_H/U_H, U_\gamma] \leq A_1 U_H/U_H$ by F.9.13.6, establishing (3). $\square$

We are now in a position to obtain a contradiction, and hence establish Theorem 13.8.28. Recall $H^* \cong L_4(2)$, $m(U_H) = 5$, and $L/O_2(L) \cong \hat{A}_6$. Now $|Q_H : (Q \cap Q_H)| = |Q_H Q : Q| \leq |O_2(L_1 T) : Q|$, and as $L/O_2(L) \cong \hat{A}_6$, $|\overline{O_2(L_1 T)}| = 4$. Next by 13.7.4.2, $|Q_H/H_C : C_{Q_H}(V_3)/H_C| = |V_3/V_1| = 4$. So as $Q \cap Q_H \leq$

$C_{Q_H}(V_3)$, we conclude that $H_C \leq C_{Q_H}(V_3) = Q \cap Q_H \leq Q$. Thus H_C centralizes V, and hence H_C also centralizes $\langle V^H \rangle = V_H$. Therefore as $A_1 \leq V_H$ by 13.8.32.3, $H_C \leq C_G(A_1) = G_\gamma$, since $H = G_1$ by 13.8.27.2; thus $[H_C, U_\gamma] \leq H_C \cap U_\gamma$. But $m(U_\gamma \cap Q_H) = 4$ by 13.8.32.1, and by 13.8.32.2, $m(U_\gamma \cap V_H) \geq 3$. So $m((U_\gamma \cap H_C)V_H/V_H) \leq 1$. Thus as $[H_C, U_\gamma] \leq H_C \cap U_\gamma$, K has at most one noncentral chief factor on H_C/V_H, and by G.6.4, that factor is 4-dimensional if it exists. But this contradicts 13.8.31.3. This completes the proof of Theorem 13.8.28.

By Theorem 13.8.28, case (a) of 13.8.27.1 holds: Namely $L/O_2(L) \cong A_6$, $H^* \cong A_6$ or S_6, and $\tilde{U}_H$ is the natural module for K^* on which L_1 has two noncentral chief factors, or its 5-dimensional cover.

LEMMA 13.8.33. *Case (2) of 13.8.8 holds; that is, $D_\gamma < U_\gamma$.*

PROOF. Assume instead that case (1) of 13.8.8 holds. By 13.8.27.3, $D_H = U_H$, $D_\gamma = U_\gamma$, V induces a group of transvections with center V_1 on U_γ, $V_\gamma \not\leq Q_H$, and we have symmetry between γ_1 and γ, (cf. the first part of Remark F.9.17), so V_γ^* induces a transvection on $\tilde{U}_H$ with center $\tilde{A}_1$, and $A_1 \leq U_H$. As usual choose $g := g_b \in \langle LT, H \rangle$ with $\gamma_1 g = \gamma$. By F.9.13.7, $[U_H, U_\gamma] = 1$. Therefore $U_\gamma \leq C_G(V_3) \leq M_V$ by 13.5.4.4. By 13.8.5, $H = KT$ and $H^* \cong S_6$. In particular, we can appeal to 13.8.6 and adopt the notation of that lemma. As $[V, U_\gamma] \neq 1$, we may pick g so that $[V_3^g, V] \neq 1$. Thus as $[V_3, V_3^g] \leq [U_H, U_\gamma] = 1$, 13.5.4.4 says $V_1 = [V, V_3^g]$ and $\bar{V}_3^g = \langle (5,6) \rangle$, so that $\bar{L}\bar{T} \cong S_6$.

Notice that if $m(\tilde{U}_H) = 4$ then $\tilde{A}_1 \leq \tilde{V}_3^h$ for some $h \in H$. Assume instead for the moment that $m(\tilde{U}_H) = 5$. Then $\tilde{A}_1$ is of weight 2, while by 13.8.6.1, $\tilde{V}_3$ consists of vectors of weight 4, so A_1 is not contained in an H-conjugate of V_3. Thus as $V_3 = V \cap U_H$ by 13.8.4.5, we conclude from 13.8.4.4 that $b > 3$ when $m(\tilde{U}_H) = 5$.

We claim that U_L is abelian; the proof will require several paragraphs. Assume U_L is nonabelian. Then $b = 3$ by (1) and (3) of 13.8.4, so by the previous paragraph, $m(\tilde{U}_H) = 4$ and $A_1 \leq V_3^h$ for some $h \in H$. Thus $V_1 = V_1^h$ is orthogonal to A_1 in V^h, so V_1 is orthogonal to $A_1^{h^{-1}}$ in V, and $1 = [U_H, U_\gamma] = [U_H, U_H^g] = [U_H, U_H^{gh^{-1}}]$. Now $V_1^{gh^{-1}} = A_1^{h^{-1}} = V_1^y$ for some $y \in L$, so as $H = G_1$ by 13.8.27.2, we conclude that $U_H^{gh^{-1}} = U_H^y$. Finally $L_1 T$ is transitive on the points of V distinct from V_1 and orthogonal to V_1, and T is transitive on the points of V not orthogonal to V_1; so since we are assuming U_L is nonabelian, $[U_H, U_H^l] \neq 1$ for some $l \in L$ with V_1^l not orthogonal to V_1, and hence for all such V_1^l by transitivity of T on this set. Therefore $U_H^{l*} \neq 1$: for otherwise $U_L \leq Q_H$, and hence $U_L \leq Q_H^l$, so that by 13.7.3, $[U_H, U_H^l] \leq V_1 \cap V_1^l = 1$, contrary to our choice of l.

Choose l with $l^2 \in Q$. Then as $V_3 = V \cap U_H$, $W_2 := V \cap U_H \cap U_H^l$ is a complement to V_1 in V_3, and to $V_1 V_1^l$ in V. Further $X_1 := O^2(C_{L_1}(V_1 V_1^l))$ acts on U_H and U_H^l, with $W_2 = [W_2, X_1]$. By 13.8.27, L_1 has two nontrivial chief factors on U_H, so $[U_H V/V, X_1] = U_H V/V \cong E_4$, and hence $[U_H^l V/V, X_1] = U_H^l V/V \cong E_4$. Then X_1 is irreducible on $U_H^l V/V$, so as $U_H^{l*} \neq 1$, $U_H^{l*} = O_2(L_1^*) \cong E_4$ and $U_H^l \cap Q_H = V_3^l = U_H^l \cap V$.

Next by 13.7.3.7, $|H_C : (H_C \cap H^l)| \leq 2$, and as $U_L \leq Q \leq N_G(H_C)$ by 13.7.3,

$$[U_H^l, H_C \cap H^l] \leq U_H^l \cap H_C \leq U_H^l \cap Q_H = U_H^l \cap V \leq V_H.$$

Since $U_H^{l*} = O_2(L_1^*) \cong E_4$ does not centralize a hyperplane in any nontrivial irreducible for K^*, we conclude that $[H_C, K] \leq V_H$. Further $V_H \leq H_C$ by 13.7.3.2, so that $|V_H : V_H \cap H^l| \leq 2$, and $[U_H^l, V_H \cap H^l] \leq U_H^l \cap Q_H = U_H^l \cap V$ with VU_H/U_H of rank 1 by 13.8.4.5. Thus $O_2(L_1^*)$ induces transvections with a common center on $(V_H \cap H^l)U_H/U_H$ of index at most 2 in V_H/U_H. So we conclude that K has at most one nontrivial chief factor on V_H/U_H, and such a factor must be the natural module on which L_1 has one noncentral chief factor. So since L_1 has two noncentral chief factors on U_H, and Q_H/H_C is H-isomorphic to $\tilde{U}_H$ of rank 4 by 13.7.4.2, we conclude that L_1 has at most six noncentral 2-chief factors. However by 13.8.6.4, L_1 has at least six noncentral chief factors on U_L/V, and hence at least eight noncentral 2-chief factors including those on $O_2(\bar{L}_1)$ and V. This contradiction establishes the claim that U_L is abelian.

Since U_L is abelian, $U_L \leq H_C$. Also we saw $A_1 \leq U_H$, so $U_L \leq C_G(A_1) = H^g \leq N_G(U_\gamma)$ as $H = G_1$. But also $U_\gamma \leq M$, so U_L and U_γ act quadratically on each other. In particular, $U_L^{g^{-1}*} \leq Q_H^{g^{-1}*} \leq O_2(C_{H^*}(\tilde{V}_1^{g^{-1}}))$, so $m(U_L^{g^{-1}*}) \leq 2$, as $O_2(C_{H^*}(\tilde{V}_1^{g^{-1}})) \cong E_8$ is not quadratic on $\tilde{U}_H$. Indeed as $V \not\leq Q_H^g$, $1 \neq V^{g^{-1}*} \leq U_L^{g^{-1}*}$, so $|U_L : V(U_L \cap Q_H^g)| \leq 2$. Thus as $[U_L \cap Q_H^g, V_3^g] \leq A_1$, there is a subgroup B/V of index at most 2 in U_L/V such that $[V_3^g, B/V] \leq A_1 V/V$. In particular, $C_{U_L/V}(V_3^g)$ is of codimension at most 2 in U_L/V, so as $m(H^*, S) = 8$ for the Steinberg module S for H^*, we conclude from 13.8.6.4 that $m(\tilde{U}_H) = 5$.

Define U_1 and U_0 as in 13.8.6.5, and recall that $V \leq U_0$. Assume first that $U_0 < U_L$. Then as L_1 is irreducible on U_H/V_3U_1, $V_3U_1 = U_H \cap U_0$, so the image of U_H in U_L/U_0 is a T-invariant 4-group. Similarly define U_2 and K_2 as in 13.8.6.5, and set $U_{2,1} := \langle V_3^{K_2} \rangle$. Then $\tilde{U}_{2,1} = \tilde{V}_2^\perp$ in the 5-dimensional orthogonal space $\tilde{U}_H$, so $U_{2,1}/V_2 \cong E_8$ with $|U_{2,1} : U_1V_3| = 2$ and $U_H = \langle U_{2,1}^{L_1} \rangle$. By 13.8.6.3, $U_2 = [U_2, L_2]$, and by 13.8.6.5, $m(U_2) = 8$, so $U_2/V_2 = U_{2,1}/V_2 \oplus U_{2,1}^l/V_2$ for $l \in L_2 - H$ and $U_1U_1^lV \leq U_0 \cap U_2$ with L_2 irreducible on $U_2/U_1U_1^lV \cong E_4$. As $U_H = \langle U_{2,1}^{L_1} \rangle$ and $U_L > U_0$, $U_2 \not\leq U_0$; so $U_0 \cap U_2 = U_1U_1^lV$, and hence $U_2/(U_2 \cap U_0)$ is also a T-invariant 4-group. We conclude just as in 13.8.6.4 that U_L/U_0 has a Steinberg module as a quotient, and then obtain a contradiction as in the previous paragraph.

Therefore $U_0 = U_L$. As $U_H = [U_H, L_1]$, $U_L = [U_L, L]$. By 13.8.6.5, $U_L/V = U_0/V$ is a quotient of the 15-dimensional permutation module on L/L_1T; so as $U_L/V = [U_L/V, L]$, G.5.3.3 says that either U_L/V is L-isomorphic to V, or U_L/V has a quotient U_L/E isomorphic to the 5-dimensional cover of V. Indeed as V_3^g centralizes a subspace of U_L/V of codimension at most 2, in the latter case G.5.3 implies that $V = E$.

So $m(U_L/V) = 4$ or 5, and hence $m(U_L) = 8$ or 9. If $m(U_L) = 8$, then $U_L = U_2$ by 13.8.6.5, so $K_2 \leq N_G(U_L) \leq M = !\mathcal{M}(LT)$, and then $H = \langle K_2, L_1T \rangle \leq M$, contrary to $H \in \mathcal{H}_z$. Therefore $m(U_L/V) = 5$.

Let $u_1 \in U_1 - V$. Suppose $U_1 \leq Z(Q)$. Then $U_L = U_0 \leq Z(Q)$. Also $W_1 := \langle u_1^L \rangle$ is a quotient of the 15-dimensional permutation module on $L/L_1(T \cap L)$ with $W_1/(W_1 \cap V) \cong U_L/V$ of rank 5, so we conclude from G.5.3 that W_1 is the 5-dimensional cover of a copy of V. This is contrary to Theorem 13.4.1 and our choice of G as a counterexample to Theorem 13.8.1.

Thus $U_1 \not\leq Z(Q)$, so that $|Q : C_Q(u_1)| = 2$. Now U_L is generated by V and a set I of 5 conjugates of u_1, so

$$C_Q(U_L) = \bigcap_{i \in I} C_Q(i).$$

Therefore as $|Q : C_Q(u_1)| = 2$, $m(Q/C_Q(U_L)) \leq 5$. Also $C_L(u_1 V/V) = L_1 T$, so $C_{LT}(u_1)$ is of index 2 in $L_1 T$. We conclude from G.5.3.3 that $Q/C_Q(U_L)$ is a copy of V as a $\bar{L}$-module, or its 5-dimensional cover. Therefore L_1 has one noncentral 2-chief factor on each of $O_2(\bar{L}_1)$, $Q/C_Q(U_L)$, U_L/V, and V. Let k and j be the number of noncentral chief factors of L_1 on $C_Q(U_L)/U_L$ and H_C/U_H, respectively; thus L_1 has $n := 4 + k$ noncentral 2-chief factors. Next $C_Q(U_L) \leq H_C$, while the two noncentral chief factors for L_1 on U_L are the two contained in U_H, so $k \leq j$. On the other hand, L_1 has two noncentral 2-chief factors on U_H, and hence also two on Q_H/H_C by 13.7.4.2, and one on $O_2(L_1^*)$, so that $5 + j = n = 4 + k$. But now $j + 1 = k \leq j$, a contradiction. This contradiction completes the proof of 13.8.33. $\qquad\square$

LEMMA 13.8.34. *(1) $A_1 \leq U_H$ and $V_1 \leq U_\gamma$.*
(2) $m(U_\gamma^*) = 1$.
(3) $O_2(L_1^*) \not\leq V_\alpha^*$, so $m(V_\gamma^*) \leq 2$.

PROOF. By 13.8.33, case (2) of 13.8.8 holds. Then $1 \neq V_\alpha^* \leq R_1^* \cong E_4$ or E_8. By 13.8.4.5, V_H/U_H is a nontrivial quotient of the 15-dimensional $\mathbf{F}_2 H^*$-permutation module on $H^*/L_1^* T^*$, and by 13.8.4.6, V_α^* is quadratic on V_H/U_H. So $O_2(L_1^*) \not\leq V_\alpha^*$, and hence (3) holds. Further $R_1^* = C_{H^*}(\tilde{U}_1 \tilde{V}_3)$, where $\tilde{U}_1 := C_{\tilde{U}_H}(H)$ and $m(U_H/U_1 V_3) = 2$.

Suppose that $m(U_\gamma^*) > 1$. Then as $O_2(L_1^*) \not\leq V_\alpha^*$, $m(U_\alpha^*) = 2$ and $[\tilde{U}_H, U_\alpha] = \tilde{U}_1 \tilde{V}_3$, so as $[U_H, U_\alpha] \leq U_\alpha$, $U_1 V_3 \leq U_\alpha V_1$. Thus $U_H \cap U_\gamma$ is of codimension at most 3 in U_H, so

$$m(D_\gamma U_H/U_H) = m(D_\gamma/(D_\gamma \cap U_H)) = m(D_\gamma) - m(U_\gamma \cap U_H)$$

$$= m(U_H) - m(U_\gamma^*) - m(U_\gamma \cap U_H) \leq 1.$$

But $[V_H, U_\gamma] \leq V_H \cap U_\gamma \leq D_\gamma$, and hence $m([V_H/U_H, U_\gamma]) \leq 1$. However by 13.7.7, H^* is faithful on V_H/U_H, whereas by G.6.4, U_γ^* does not induce transvections with a common center on any faithful $\mathbf{F}_2 H^*$-module. This contradiction shows that $m(U_\gamma^*) = 1$, so (2) holds.

Assume that (1) fails. By 13.8.10.1, $m(U_H/D_H) = 1$, and we have symmetry between γ_1 and γ in the sense of Remark F.9.17. Therefore interchanging γ and γ_1 if necessary, we may assume that $A_1 \not\leq U_H$, and hence by 13.8.10.2 that U_γ induces transvections on U_H with axis D_H. Then as $A_1 \not\leq U_H$, 13.7.3.7 says $[V_\gamma, D_H] \leq A_1 \cap U_H = 1$. Thus V_γ^* induces transvections on $\tilde{U}_H$ with axis $\tilde{D}_H$, so V_γ^* is of rank 1, and hence $V_\gamma^* = U_\gamma^*$. Then by 13.8.9.2, $V_1 \not\leq U_\gamma$ and U_H induces transvections on V_γ/U_γ with center $V_1 U_\gamma/U_\gamma$. Since $V_1 \not\leq U_\gamma$, U_H induces transvections on U_γ/A_1 by 13.8.10.2. However by 13.8.4.5, V_γ/U_γ is a nontrivial quotient of the 15-dimensional $\mathbf{F}_2 H^g$-permutation module for $G_\gamma/Q_\gamma \cong A_6$ or S_6 on $H^g/L_1^g T^g$. Thus as $L_1^g T^g$ is not the stabilizer of a point in U_γ/A_1, V_γ/U_γ has a quotient which is the conjugate of U_γ/A_1 by an outer automorphism of G_γ/Q_γ. Therefore as U_H induces transvections on U_γ/A_1, it does not induce a transvection on V_γ/U_γ, a contradiction. This completes the proof of (1) and of the lemma. $\quad\square$

We are now ready to complete the proof of Theorem 13.8.1. As $A_1 \leq U_H$ by 13.8.34.1, $\tilde{A}_1 \leq Z(\tilde{T}^h)$ for some $h \in H$, and $C_{H^*}(\tilde{A}_1)$ is a maximal parabolic of H^* stabilizing a point of $\tilde{U}_H$. Next by 13.8.34.1 we may apply 13.8.11.1 to conclude that $U_\gamma^* < V_\gamma^*$. Thus as V_γ^* and U_γ^* are normal in $C_H(A_1)^* = C_{H^*}(\tilde{A}_1)$, it follows that $O_2(L_1^{h*}) \leq V_\gamma^*$. But this contradicts 13.8.34.3.

This final contradiction establishes Theorem 13.8.1.

Observe in fact that Theorems 13.3.16, 13.6.1, and 13.8.1 complete the treatment of Hypothesis 13.3.1 for all possibilities for $L/O_2(L)$ (cf. 13.3.2.1) except $L_3(2)$.

13.9. Chapter appendix: Eliminating the A_{10}-configuration

This section eliminates the shadow of the group A_{10}, by ruling out the existence of $M \in \mathcal{M}$ with $M \cong S_4$ wr $\mathbf{Z}_2$. We prove:

THEOREM 13.9.1. *There is no simple QTKE-group G such that there exists $T \in Syl(G)$ and $M \in \mathcal{M}(T)$ satisfying $M \cong S_4$ wr $\mathbf{Z}_2$.*

Throughout the section, we assume G, T, M is a counterexample to Theorem 13.9.1. As usual we will begin with a number of preliminary lemmas describing the structure of M.

Observe that $J(M) = M_1 \times M_2$ with $M_i \cong S_4$, and $M_1^s = M_2$ for s an involution in $M - J(T)$. Let $T_i := T \cap M_i$ and $\langle t \rangle := Z(T_1)$. Notice $Z(T) = \langle z \rangle$ is of order 2 where $z := tt^s$. Let $A := O_2(M)$, so that $A \cong E_{16}$. For $X \subseteq G$, let $G_X := C_G(X)$, and set $\tilde{G}_z := G_z/\langle z \rangle$.

Let $\hat{G} := A_{10}$ be the alternating group on $\Omega := \{1, \ldots, 10\}$, and $\hat{M}$ the subgroup of $\hat{G}$ permuting

$$\{\{1,2,3,4\}, \{5,6,7,8\}, \{9,10\}\}.$$

There is an isomorphism $\alpha : M \to \hat{M}$ such that $\hat{T} := \alpha(T) \in Syl_2(\hat{G})$ and $\hat{M} \in \mathcal{M}(\hat{T})$. Let $\hat{M}_i := \alpha(M_i)$, $\hat{z} := \alpha(z)$, etc. We may choose our isomorphism α so that $\hat{M}_1 = \hat{G}_{5,\ldots,10}$ and $\hat{z} = (1,2)(3,4)(5,6)(7,8)$.

We will show that the 2-local subgroups and 2-fusion in G are the same as that of $\hat{G}$; this is a contradiction since G is quasithin while $\hat{G}$ is not. From time to time, we use the identification α of M with $\hat{M}$ to compute facts about M and its subgroup T.

LEMMA 13.9.2. *(1) $\mathcal{A}(T) = \{A_i : 1 \leq i \leq 4\}$, with $A_1 := A$, and $B := A_2$ both normal in T, while $A_3^s = A_4$. Further $J(T) = T_1 \times T_2 = AB$ and $A \cap B = Z(J(T))$.*
(2) $N_G(J(T)) = T$.
(3) A and B are weakly closed in T with respect to G. Hence fusion in A is controlled by $M = N_G(A)$, and in B by $N_G(B)$.
(4) $M = N_G(A)$, $a^G \cap A = a^M$ for each $a \in A$, $|z^M| = 9$, and $|t^M| = 6$.
(5) $t \notin z^G$.
(6) $J(T) \in Syl_2(G_t)$.

PROOF. Part (1) is an easy calculation. As $M \in \mathcal{M}$, $M = N_G(A)$. Let $X := N_G(J(T))$ and $X^* := X/J(T)$. Then X acts on $\mathcal{A}(T)$, and as $M = N_G(A)$, $T = N_M(J(T)) = N_X(A)$, so $J(T)$ is the kernel of the action of X on $\mathcal{A}(T)$. Thus $X^* \leq Sym(\mathcal{A}(T)) \cong S_4$ with $\mathbf{Z}_2 \cong T^* \in Syl_2(X^*)$, so either $X = T$, or $X^* \cong S_3$. The latter is impossible, as $Aut(J(T))$ is a 2-group. Thus (2) holds.

As $J(T)$ is weakly closed in T, and each member of $\mathcal{A}(T)$ is normal in $J(T)$, we may apply the Burnside Fusion Lemma A.1.35 to these normal subsets to conclude for each $D \in \mathcal{A}(T)$ that $D^G \cap J(T) = D^{N_G(J(T))}$, and hence $D^G \cap J(T) = D^T$ by (2). In particular as A and B are normal in T, they are weakly closed in T.

Hence (3) holds by application of the Burnside Fusion Lemma to the elements of A and B. Next as $M = N_G(A)$, (4) follows from (3) and the identification α given after the statement of Theorem 13.9.1, which says A is the orthogonal module for $M/A \cong O_4^+(2)$ with z^M the singular points and t^M the nonsingular points. Now (4) implies (5), and then as $Z(T) = \langle z \rangle$ has order 2, t is not 2-central by (5), so (6) holds. $\qquad\qquad\qquad\qquad\qquad\qquad\qquad\qquad\qquad\qquad\qquad\qquad\qquad\qquad\quad$ $\square$

From now on let B be the group defined in 13.9.2.1, and set $K := N_G(B)$. As $B \trianglelefteq T$ by that result, $K \in \mathcal{H}(T) \subseteq \mathcal{H}^e$ by 1.1.4.6.

In the following lemma, "diagonal involutions" in $J(M)$ are those projecting nontrivially on both factors of the decomposition $J(M) = M_1 \times M_2$. The next two lemmas follow from straightforward calculations.

LEMMA 13.9.3. M *has 6 classes of involutions* Δ_i, $1 \le i \le 6$, *where*

(1) $\Delta_1 := z^M$ *consists of the diagonal involutions in* A.
(2) $\Delta_2 := t^M = (A \cap T_1)^\# \cup (A \cap T_2)^\#$.
(3) Δ_3 *consists of the involutions in* $M_1 - A$ *and* $M_2 - A$.
(4) Δ_4 *consists of the diagonal involutions* $i_1 i_2$ *with* $i_k \in M_k \cap \Delta_3$, $k = 1, 2$.
(5) Δ_5 *consists of the diagonal involutions* ij *with* $i \in M_k \cap \Delta_3$ *and* $j \in M_{3-k} \cap \Delta_2$, $k = 1, 2$.
(6) $\Delta_6 := s^M$ *consists of the involutions in* $M - J(M)$.

LEMMA 13.9.4. $B \cap \Delta_1 = \{z\}$, $|B \cap \Delta_2| = 2$, *and* $|B \cap \Delta_i| = 4$ *for* $i = 3, 4, 5$. *Further each set is an orbit under* T.

LEMMA 13.9.5. $G_z > T$, *so* $G_z \not\le M$.

PROOF. Assume that $G_z = T$. We will obtain a contradiction using Thompson Transfer A.1.36 on s, based on an analysis of fusion which will eventually include the explicit identification of $O^2(G_t)$.

First $C_M(s) \cong \mathbf{Z}_2 \times S_4$ is not a 2-group, so $s \notin z^G$. Suppose that z is weakly closed in B with respect to G. Then $z^G \cap M = \Delta_1 = z^M$ by 13.9.4 and 13.9.3, and $C_G(z) = T \le M$ by hypothesis. Therefore by 7.3.1 in [**Asc94**], M is the unique fixed point of z on G/M. Hence by 7.4.2 in [**Asc94**], $s^G \cap M = s^M$. Therefore as $s^M \subseteq M - J(M)$, $s \notin O^2(G)$ by Thompson Transfer, contradicting G simple.

Therefore z is not weakly closed in B with respect to G. On the other hand, $C_M(i)$ is not a 2-group for $i \in \Delta_2 \cup \Delta_3$, so

$$z^G \cap M \subseteq \Delta_1 \cup \Delta_4 \cup \Delta_5 \qquad\qquad (*)$$

by 13.9.3. Also by 13.9.2.3, $z^G \cap B = z^K$. Thus by $(*)$, and since the sets in 13.9.4 are T-orbits, z^K is of order 5 or 9, so in particular, $T < K$. Set $V := \langle z^K \rangle$ and $K^* := K/C_K(V)$. Then $O_2(K^*) = 1$ by B.2.14, so that $O_2(K) \le C_K(V)$. Also $C_K(V)$ is a 2-group as $T = G_z$, so $C_K(V) = O_2(K)$. On the other hand if $A \le O_2(K)$, then $J(T) = AB \le O_2(K)$, so $K \le N_G(J(T)) = T$ by 13.9.2.2, contrary to $T < K$. Thus $A \not\le C_K(V)$, and hence by B.2.5, V is an FF-module for K^*. Then $3 \in \pi(K^*)$ by Theorem B.5.6, while $C_K(z) = T$ is a 2-group, so

$$|K : T| = |z^K| = 9$$

rather than 5. Hence the inclusion in $(*)$ is an equality, and as $B = \langle \Delta_4 \cap B, \Delta_5 \cap B \rangle$, $V = B \cong E_{16}$. Then as $B = C_T(B)$, we conclude that $O_2(K) = B$. Inspecting the subgroups of $GL_4(2)$ with Sylow group D_8 and of order 72, we conclude B is the orthogonal module for $K/B \cong O_4^+(2)$ and $|t^K| = 6$. Therefore $t^G \cap J(M) = \Delta_2 \cup \Delta_3$ in view of 13.9.4. Thus all involutions in $J(T)$ are fused to t or z. So writing $I(S)$ for the set of involutions in a subgroup S of G:

(!) $t^G \cap J(T) = I(T_1) \cup I(T_2)$. So T_1 and T_2 are the subgroups S of $J(T)$ maximal subject to the property that $I(S) \subseteq t^G$. In particular each 4-subgroup of $J(T)$ consisting of members of t^G contains either t or t^s, and lies in either T_1 or T_2.

Set $X_1 := O^2(C_M(t)) = O^2(M_2)$, $X_2 := O^2(C_K(t))$, $X := O^2(G_t)$, and $\bar{G}_t := G_t/\langle t \rangle$. Thus $X_i \leq X$. We begin the explicit determination of X mentioned earlier. Recall $J(T) \in Syl_2(G_t)$ by 13.9.2.6.

For $i = 1, 2$, A_i is the orthogonal module for $N_G(A_i)/A_i$, with $t^{N_G(A_i)}$ the set of nonsingular vectors in A_i, so $X_i \cong A_4$ with $O_2(X_i) = A_i \cap X_i$ and $t \notin O_2(X_i)^\# \subseteq t^G$. Thus we conclude from (!) that $O_2(X_i) \leq T_2$, so that $O_2(X_i) = A_i \cap T_2 = X_i \cap T_2$, and hence $T_2 = O_2(X_1)O_2(X_2) \leq X$.

If U is a 4-subgroup of T_1 and $g \in G_t$ with $U^g \leq J(T)$, then $t \in U^g$, so $U^g \leq T_1$ by (!). Hence $I(T_1)$ is strongly closed in the Sylow group $J(T)$ with respect to G_t, so as $\bar{T}_1 \leq Z(\overline{J(T)})$, $N_{\bar{G}_t}(\bar{T}_1)$ controls fusion in $\overline{J(T)}$ by the Burnside Fusion Lemma A.1.35. Then as $Aut(T_1)$ is a 2-group and $\bar{T}_1$ is central in the Sylow group $\overline{J(T)}$, each element of $\bar{T}_1$ is strongly closed in $\overline{J(T)}$ with respect to $\bar{G}_t$; so by Thompson Transfer, $\bar{T}_1 \cap \bar{X} = 1$ as $X = O^2(X)$. Then $\langle t \rangle T_2 \in Syl_2(\langle t \rangle X)$, so by Thompson Transfer, $t \notin X$, and we conclude $T_1 \cap X = 1$. Hence $T_2 \in Syl_2(X)$ as $T_2 \leq X$ and $J(T) = T_1 T_2$ is Sylow in G_t. Further $C_X(tz) = C_X(z) = T \cap X = T_2 \cong D_8$, and from the action of the X_i on the $O_2(X_i)$, X has one class of involutions represented by tz. Thus by I.4.1, $X \cong L_3(2)$ or A_6. In particular the involutions in $X = O^2(G_t)$ are in t^G.

As G is simple, by Thompson Transfer, $s^G \cap J(T) \neq \emptyset$. We showed that z, t are representatives for the G-classes of involutions in $J(T)$, and that $s \notin z^G$. Thus $s \in t^G$. We also saw that $O^2(C_M(s)) \cong A_4$, with $I(O^2(C_M(s))) \subseteq z^G$. This is impossible, as we saw $I(O^2(G_t)) \subseteq t^G$. This contradiction completes the proof of 13.9.5. $\qquad\square$

Recall $\hat{G} := A_{10}$ and set $\hat{Q} := O_2(O^2(\hat{G}_{\hat{z}}))$. We may check directly from the structure of A_{10} that $\hat{Q} \cong Q_8^2$ and $J(\hat{Q}/\langle \hat{z} \rangle) = \hat{Q}/\langle \hat{z} \rangle \cong E_{16}$. Let $Q := \alpha^{-1}(\hat{Q})$. Since $\alpha : M \to \hat{M}$ is an isomorphism:

LEMMA 13.9.6. $Q \cong Q_8^2$ and $\tilde{Q} = J(\tilde{T}) \cong E_{16}$.

Furthermore from the structure of A_{10}, $\hat{G}_{\hat{z}}/\hat{Q} \cong S_3 \times \mathbf{Z}_2$. We wish to establish analogous statements in G, starting with:

LEMMA 13.9.7. $G_{z,t} = J(T)$.

PROOF. First by 13.9.2.6, $J(T) \in Syl_2(G_t)$. As $z \in Z(T)$, $F^*(G_z) = O_2(G_z)$ by 1.1.4.6, so $F^*(G_{t,z}) = O_2(G_{t,z})$ by 1.1.3.2; then setting $G_{t,z}^* := G_{t,z}/\langle t, z \rangle$, we obtain $F^*(G_{t,z}^*) = O_2(G_{t,z}^*)$ from A.1.8. As the Sylow group $J(T)^*$ of G_t^* is abelian,

$$J(T)^* \leq C_{G_{t,z}^*}(O_2(G_{t,z}^*)) \leq O_2(G_{t,z}^*),$$

so $J(T) = O_2(G_{t,z})$. Then the lemma follows from 13.9.2.2. $\qquad\square$

LEMMA 13.9.8. *(1)* $Q \trianglelefteq G_z$ and $G_z/Q \cong S_3 \times \mathbf{Z}_2$.
(2) $B \trianglelefteq G_z$, and hence $G_z \leq K$.

PROOF. We claim first that $Q \trianglelefteq G_z$. Let $Q_z := O_2(G_z)$. By G.2.2 with $\langle z \rangle$, $\langle t, z \rangle$, 1 in the roles of "V_1, V, L", $\tilde{U} := \langle \tilde{t}^{G_z} \rangle \leq \Omega_1(Z(\tilde{Q}_z))$ and $\tilde{U} \in \mathcal{R}_2(\tilde{G}_z)$. Now $C_{G_z}(\tilde{U})/C_{G_{z,t}}(\tilde{U})$ is of order at most 2, so by 13.9.7, $C_{\tilde{G}_z}(\tilde{U})$ is a 2-group, and hence $\tilde{Q}_z = C_{\tilde{G}_z}(\tilde{U})$. Let $G_z^* := G_z/Q_z$, so that $O_2(G_z^*) = 1$ and $G_z^* \leq GL(\tilde{U})$.

If $Q \leq Q_z$, then $\tilde{Q} = J(\tilde{Q}_z)$ by 13.9.6, so that $Q \trianglelefteq G_z$, as claimed. Thus we may assume $Q \not\leq Q_z$, so in particular $m(\tilde{U}) < m(J(\tilde{T})) = 4$ by 13.9.6. Further using the identification α, no element of $\tilde{T}$ induces a transvection on $\tilde{Q}$; so if $|Q : Q \cap Q_z| = 2$, then $\tilde{U} \leq C_{\tilde{T}}(\tilde{Q} \cap \tilde{Q}_z) \leq C_{\tilde{T}}(\tilde{Q})$, and then $\tilde{Q} \leq C_{\tilde{T}}(\tilde{U}) = \tilde{Q}_z$ by the first paragraph, contrary to assumption. Thus $|Q^*| > 2$, so $m(\tilde{U}) > 2$ and hence $m(\tilde{U}) = 3$. Then $G_z^* \leq GL(\tilde{U}) = L_3(2)$, with Sylow group T^* of order at least 4 and $O_2(G_z^*) = 1$, so we conclude $G_z^* = GL(\tilde{U})$. Then $G_{t,z}$ has order divisible by 3, contrary to 13.9.7, completing the proof of the claim.

By the claim, $Q \trianglelefteq G_z$. In particular as $t \in Q$, we have $U \leq Q \leq Q_z$. Now $C_{\tilde{G}_z}(\tilde{Q}) \leq C_{\tilde{G}_z}(\tilde{U}) = \tilde{Q}_z$, so as $\tilde{Q}$ is self-centralizing in $\tilde{T}$, we conclude $C_{\tilde{G}_z}(\tilde{Q}) = \tilde{Q}$. Hence $G_z' := G_z/Q$ lies in the orthogonal group $O(\tilde{Q}) \cong O_4^+(2)$ with Sylow group $T' \cong E_4$. As $T < G_z$ by 13.9.5, we conclude that either (1) holds, or $G_z^* \cong S_3 \times S_3$. In the latter case G_z is transitive on the involutions in $Q - \langle z \rangle$. This is impossible as $A \cap Q \cong E_8$ contains an element of Δ_1, and hence t is fused into z^G in G_z, contrary to 13.9.2.5. This completes the proof of (1).

Let $b \in B - Q$. As $B \trianglelefteq T$ and $m([\tilde{Q}, b]) = 2$, $[b, Q] = B \cap Q \cong E_8$. Similarly for $a \in A - Q$, $[Q, a] = A \cap Q \cong E_8$. Thus a and b interchange the two Q_8-subgroups Q_1 and Q_2 of Q. Now $T/Q \cong E_4$ has three subgroups E_i/Q, $1 \leq i \leq 3$, of order 2, with $E_1 := N_T(Q_1)$. Then $Q_z \neq E_1$, since (1) shows that Q_z/Q centralizes an a-invariant subgroup of G_z/Q of order 3, whereas E_1/Q does not. Furthermore E_1 is not AQ or BQ as we saw these subgroups interchange Q_1 and Q_2. Let $AQ =: E_2$; then $C_{AQ}([Q, a]) \cong E_{16}$ and $[Q, a] = [Q, i]$ for each $i \in AQ - Q$. Therefore $A = C_{AQ}([Q, a])$ and $BQ \neq AQ$. Thus $BQ = E_3$. As $M = N_G(A)$ and $T = C_M(z) < G_z$ by 13.9.5, A is not normal in G_z; therefore $A \not\leq Q_z$ as A is weakly closed in T by 13.9.2.3. Thus $BQ = E_3 = Q_z$, and then $B \trianglelefteq G_z$, as B is weakly closed in T by 13.9.2.3. This completes the proof of 13.9.8. $\square$

LEMMA 13.9.9. *(1) B is the natural module for $K/B \cong O_4^-(2)$.*

(2) z^K and t^K are of order 5 and 10, respectively, and afford the set of singular and nonsingular points in the orthogonal space B.

(3) $z^G \cap M = \Delta_1 \cup \Delta_3 \cup \Delta_6$.

(4) $t^G \cap M = \Delta_2 \cup \Delta_4 \cup \Delta_5$.

(5) G has two classes of involutions with representatives z and t.

PROOF. First $Q = \langle s \rangle [J(T), s]$ with $[J(T), s] = C_{J(T)}(s)\langle t \rangle$. This allows us to calculate that T has four orbits Γ_i, $1 \leq i \leq 4$, on the set Γ of 18 involutions in $Q - \langle z \rangle$: $\Gamma_1 := \{t, tz\} \subseteq \Delta_2$, $\Gamma_2 \subseteq \Delta_1$ of order 4, $\Gamma_3 := s^T$ of order 8 containing $s \in \Delta_6$, and $\Gamma_4 := B \cap \Delta_4$ of order 4. On the other hand, from 13.9.8.1, G_z has two orbits on Γ: Γ^1 of length 6, and Γ^2 of length 12, with $t \in \Gamma^1$ as $\langle t \rangle = Z(\tilde{T})$. As $t \notin z^G \supseteq \Delta_1 \supseteq \Gamma_2$, we conclude

$$\Gamma^1 = \Gamma_1 \cup \Gamma_4 = t^G \cap Q \quad \text{and} \quad \Gamma^2 = \Gamma_2 \cup \Gamma_3 = z^G \cap Q - \{z\}. \quad (*)$$

In particular $\Delta_4 \subseteq t^G$. Next $M/O^2(M) \cong D_8$; let M_0 be the subgroup of M of index 2 with $M_0/O^2(M) \cong \mathbf{Z}_4$. Then $\Delta_1 \cup \Delta_2 \cup \Delta_4$ is the set of involutions in M_0, so each is in $z^G \cup t^G$. Hence (5) follows from Thompson Transfer.

Next by 13.9.8 and (*), $G_z \leq K$ and $B \cap Q = \langle z, t^{G_z} \rangle$. We conclude using 13.9.8.1 that the orbits of G_z on $B^{\#}$ are $\Sigma_0 := \{z\}$, $\Sigma_1 := \Gamma^1$ of length 6, and two orbits Σ_i, $i = 2, 3$, on $B - Q$ of length 4. Then since $\Gamma_4 = B \cap \Delta_4 \subseteq Q$, appealing to 13.9.4, we may choose notation so that $\Sigma_2 = B \cap \Delta_3$ and $\Sigma_3 = B \cap \Delta_5$.

By 13.9.2.3, K controls fusion in B, so it follows from (5) that the G_z-orbit Σ_3 is fused to z or t under K. In particular, $G_z < K$. Thus there are three possibilities for z^K: $\{z\} \cup \Sigma_3$, $\{z\} \cup \Sigma_2$, or $\{z\} \cup \Sigma_2 \cup \Sigma_3$—of order 5, 5, or 9, respectively. Now by 13.9.7, $C_G(B) = C_{J(T)}(B) = B$, so $K/B \leq GL(B)$. As $|GL_4(2)|$ is not divisible by 27, while 3 divides the order of $C_K(z)$ by 13.9.8, we conclude $|z^K| = 5$ rather than 9. Set $K^* := K/B$; then $C_K(z)^* = G_z^* \cong S_4$ by 13.9.8. Further $B = \langle \Sigma_i \rangle$ for $i = 2, 3$, so B is the kernel of the action of $C_K(z)$ on Σ_2 and Σ_3. We conclude K^* acts faithfully as S_5 on z^K. As K has orbits of length 5 and 10 on $B^{\#}$, it follows that B is the natural module for $K^* \cong O_4^-(2)$, with z^K the singular points of the orthogonal space B and t^K the nonsingular points. This establishes (1) and (2). Also if $k \in K - G_z$ then $zz^k \in t^K$, while if $z^k \in \Delta_5$, then $zz^k \in \Delta_5$. Thus $\Sigma_3 = B \cap \Delta_5 \not\subseteq z^K$, so $z^K = \{z\} \cup \Sigma_2 = \{z\} \cup (B \cap \Delta_3)$, and $\Delta_5 \cap B = \Sigma_3 \subseteq t^K$. Now it follows using (*) that (3) and (4) hold. $\square$

LEMMA 13.9.10. Let $E := T_1 \cap A = O_2(M_1)$. Then E centralizes $O^2(C_K(t)) \cong A_4$.

PROOF. From the structure of K described in 13.9.9, and as $J(T) \in Syl_2(G_t)$ by 13.9.2.6, $C_K(t) = R_1 \times X$, where $t \in R_1 \cong D_8$, and $X \cong S_4$ with $O_2(X)^{\#} \subseteq t^K$. Let $R_2 := T \cap X$. Then $T_1 \times T_2 = J(T) = R_1 \times R_2$ with $\langle t \rangle = [R_1, R_1] = [T_1, T_1]$, $\langle t^s \rangle = [R_2, R_2] = [T_2, T_2]$, and $z = tt^s$. Now by the Krull-Schmidt Theorem A.1.15, $T_i \langle z \rangle = R_i \langle z \rangle$ for each $i = 1, 2$. Therefore $E \leq T_1 \leq R_1 \langle z \rangle$.

Suppose $E \not\leq R_1$. Then there is $e \in E - \langle t \rangle$ with $ez \in R_1$. As $E \cap B = \langle t \rangle$, $e \notin B$ and hence $ez \notin B$. Further $ez \in z^M$ by 13.9.3.1, so $ez = z^g \in (R_1 \cap z^G) - B$ for some $g \in G$. Then as X centralizes z^g, from the description of G_z in 13.9.8, $O_2(X) = [O_2(X), O^2(X)] \leq Q^g$. So as $O_2(X)^{\#} \subseteq t^K$, it follows from (*) in the proof of 13.9.9.1 that

$$U := \langle t^G \cap Q^g \rangle = \langle z^g \rangle \times O_2(X).$$

By 13.9.8, $\langle t^{G_z} \cap Q \rangle = B \cap Q$, while $C_G(B \cap Q) \leq G_{z,t} = J(T)$ by 13.9.7, so we conclude $U = B^g \cap Q^g$ and $C_G(U) \leq C_{J(T)^g}(U) = B^g$. Now $C_{R_1 O_2(X)}(U) = C_{R_1 O_2(X)}(z^g) \cong E_{16} \cong C_G(U)$, so $B^g = C_{R_1 O_2(X)}(U) \leq K = N_G(B)$. Hence $B^g = B$ as B is weakly closed in T by 13.9.2.3, contradicting $z^g \notin B$.

This contradiction shows that $E \leq R_1$. Hence E centralizes $O^2(X) = O^2(C_K(t)) \cong A_4$. $\square$

We are now in a position to prove Theorem 13.9.1. The argument will be much like that in the proof of 13.9.5.

Let E be as in 13.9.10, and recall $G_E = C_G(E)$. As $J(T) \in Syl_2(G_t)$ by 13.9.2.6 and $t \in E \trianglelefteq J(T)$, $E \times T_2 = C_{J(T)}(E) \in Syl_2(G_E)$. Let $H := O^2(G_E)$, $H_1 := O^2(M_2)$, and $H_2 := O^2(C_K(t))$. Thus $H_i \cong A_4$ centralizes E, using 13.9.10 in the case of H_2. Therefore $H_i \leq H$. Furthermore $O_2(H_1) = T_2 \cap A$, while $O_2(H_2) = H_2 \cap B$, with $O_2(H_1) \cap O_2(H_2) = \langle tz \rangle$. Therefore as $tz \in O_2(H_2)^{\#} \subseteq t^G$, $O_2(H_2) \leq T_2$, so as $H_2 \cap B \not\leq A$, $O_2(H_2)$ normalizes but does not centralize, $O_2(H_1)$, and then $T_H := O_2(H_1)O_2(H_2) \cong D_8$. As $H_i \leq H$, $T_H \leq H$. As $E \leq Z(H)$, by Thompson Transfer, $E \cap H = 1$, so that $T_H \in Syl_2(H)$. As all involutions in T_H are fused under H_1 and H_2, H has one class of involutions. Further $C_{G_t}(z) = J(T)$ by 13.9.7, so $C_H(tz) = C_H(z) = J(T) \cap H = T_H \cong D_8$. Therefore by I.4.1, $H \cong L_3(2)$ or A_6.

Now $M_1 \leq N_G(E) \leq N_G(H)$, and M_1 centralizes $O^2(M_2) = H_1 \cong A_4$, so from the structure of $Aut(H)$, $O^2(M_1) \leq C_G(H)$, and indeed M_1 centralizes H if $H \cong L_3(2)$. Therefore as $m_3(M_1 H) \leq 2$ since $N_G(H)$ is an SQTK-group, $H \cong L_3(2)$ and $M_1 H = M_1 \times H$. But by 13.9.9.3, there is $z^g \in M_1 - O^2(M_1)$, so $L_3(2) \cong H \leq G_z^g$, contrary to 13.9.8.1. This contradiction completes the proof of Theorem 13.9.1.

CHAPTER 14

$L_3(2)$ in the FSU, and $L_2(2)$ when $\mathcal{L}_f(G, T)$ is empty

The previous chapter reduced the treatment of the Fundamental Setup (3.2.1) to the case $\bar{L} \cong L_3(2)$—which we handle in this chapter. This in turn reduces the proof of the Main Theorem to the case $\mathcal{L}_f(G, T) = \emptyset$.

Recall that the case in the FSU where $\bar{L} \cong A_5$ is actually treated last in the natural logical order, but because of similarities with the case $\bar{L} \cong A_6$, those cases were treated together in the previous chapter; this was accomplished by introducing assumption (4) in Hypothesis 13.3.1.

In this chapter it will again be convenient to take advantage of some similarities in the treatment of two small linear groups: namely between the case $\bar{L} \cong L_3(2)$ for $L \in \mathcal{L}_f(G, T)$, and suitable $L \in \mathcal{M}(T)$ such that $LT/O_2(LT) \cong L_2(2)$ acts naturally on some 2-dimensional member of $\mathcal{R}_2(LT)$. The latter situation is the most difficult subcase of the case $\mathcal{L}_f(G, T) = \emptyset$, which of course remains after the Fundamental Setup is treated. As a result, we begin this chapter with several sections providing preliminary results on the case $\mathcal{L}_f(G, T) = \emptyset$, and in particular on the subcase with $L/C_L(V) \cong L_2(2)$.

14.1. Preliminary results for the case $\mathcal{L}_f(G, T)$ empty

As usual, $T \in Syl_2(G)$ and $Z := \Omega_1(Z(T))$.

This chapter includes the beginning of the treatment of the case $\mathcal{L}_f(G, T) = \emptyset$. The first few results below are based only on that assumption, but afterwards we will assume the stronger Hypothesis 14.1.5.

We use the following notation through the section:

NOTATION 14.1.1. Let $E := \Omega_1(Z(J(T)))$, $M \in \mathcal{M}(T)$, $V \in \mathcal{R}_2(M)$, and $\bar{M} := M/C_M(V)$.

Recall from section A.5 of Volume I that for $H \in \mathcal{H}(T)$, in this section we deviate from our usual meaning of $V(H)$ in definition A.4.7, instead using the meaning in notation A.5.1, namely

$$V(H) := \langle Z^H \rangle.$$

Recall the partial ordering on $\mathcal{M}(T)$ given by $M_1 \overset{<}{\sim} M_2$ whenever

$$M_1 = C_{M_1}(V(M_1))(M_1 \cap M_2).$$

Recall $V(H) \in \mathcal{R}_2(H)$ by B.2.14.

The first result below does not even require the hypothesis $\mathcal{L}_f(G, T) = \emptyset$:

LEMMA 14.1.2. Assume $J(T) \le C_M(V)$ and set $S := Baum(T)$. Then
(1) $V \le E$ and $S = Baum(C_T(V))$.
(2) Assume either

(a) M is maximal in $\mathcal{M}(T)$ under $\overset{\scriptscriptstyle\leq}{\sim}$ and $V = V(M)$, or

(b) $V = \langle (V \cap Z)^M \rangle$, and M is the unique maximal member of $\mathcal{M}(T)$ under $\overset{\scriptscriptstyle\leq}{\sim}$.

Then $M = !\mathcal{M}(N_M(S))$ and $C(G,S) \leq M$.

PROOF. Part (1) follows from B.2.3.5. Assume one of the hypotheses of (2). Then $M = !\mathcal{M}(N_M(C_T(V))) = !\mathcal{M}(N_M(S))$ by A.5.7.2, so that $C(G,S) \leq M$. □

The next two preliminary results do assume $\mathcal{L}_f(G,T) = \emptyset$:

LEMMA 14.1.3. Assume $\mathcal{L}_f(G,T) = \emptyset$. Then $H^\infty \leq C_H(U)$ for each $H \in \mathcal{H}(T)$ and $U \in \mathcal{R}_2(H)$.

PROOF. By 1.2.1.1, H^∞ is the product of groups $L \in \mathcal{C}(H)$. Then $L \in \mathcal{L}(G,T)$. By hypothesis, $L \notin \mathcal{L}_f(G,T)$, so by 1.2.10, $[U,L] = 1$. □

LEMMA 14.1.4. Assume $\mathcal{L}_f(G,T) = \emptyset$, M is maximal in $\mathcal{M}(T)$ under $\overset{\scriptscriptstyle\leq}{\sim}$, and $J(T) \leq C_M(V(M))$. Then M is the unique maximal member of $\mathcal{M}(T)$ under $\overset{\scriptscriptstyle\leq}{\sim}$.

PROOF. Let $S := \mathrm{Baum}(T)$. By 14.1.2.2, $C(G,S) \leq M$ and $M = !\mathcal{M}(N_M(S))$. In particular $N_G(J(T)) \leq M$.

Let $M_1 \in \mathcal{M}(T) - \{M\}$, and $V := V(M_1)$. If $[V, J(T)] = 1$, then by a Frattini Argument, $M_1 = C_{M_1}(V)N_{M_1}(J(T))$, so we conclude $M_1 \overset{\scriptscriptstyle\leq}{\sim} M$ as $N_G(J(T)) \leq M$.

Hence we may assume $[V, J(T)] \neq 1$. Set $M_1^* := M_1/C_{M_1}(V)$ and $I := J(M_1)$. By a Frattini Argument, $M_1 = IN_{M_1}(J(T)) = I(M_1 \cap M)$, so it will suffice to show that $I = C_{M_1}(V)(I \cap M)$. By 14.1.3, M_1^* is solvable, so by Solvable Thompson Factorization B.2.16, $I^* = I_1^* \times \cdots \times I_s^*$ with $I_i^* \cong S_3$, and $s \leq 2$ by A.1.31.1. Now $I^* \leq O^{2'}(I^*T^*)$, and by B.6.5, $O^{2'}(IT)$ is generated by minimal parabolics H above T, so it will suffice to show that $H \leq M$ for those H with $H^* \neq 1$. We apply Baumann's Lemma B.6.10 to H to conclude $S \in Syl_2(O^2(H)S)$. Then we apply Theorem 3.1.1 with $N_M(S)$, S in the roles of "M_0, R", and as $M = !\mathcal{M}(N_M(S))$, we conclude that $H \leq M$ as required. □

We next discuss the basic hypothesis which we will use during the bulk of our treatment of the case $\mathcal{L}_f(G,T)$ empty:

The final result in this chapter, Theorem D (14.8.2), determines the QTKE-groups G in which $\mathcal{L}_f(G,T) \neq \emptyset$. Then in the following chapter we determine those QTKE-groups G such that $\mathcal{L}_f(G,T) = \emptyset$. As in the previous chapters on the Fundamental Setup (3.2.1), we may also assume that $|\mathcal{M}(T)| > 1$, since Theorem 2.1.1 determined the groups for which that condition fails. Finally we divide our analysis of groups G with $\mathcal{L}_f(G,T) = \emptyset$ into two subcases: the subcase where $|\mathcal{M}(C_G(Z))| = 1$, and the subcase where $|\mathcal{M}(C_G(Z))| > 1$. The second subcase is comparatively easy to handle, perhaps because all the examples other than $L_3(2)$ and A_6 occur in the first subcase.

Thus in this section, and indeed in most of those sections in this and the following chapter which are devoted to the case $\mathcal{L}_f(G,T)$ empty, we assume the following hypothesis:

HYPOTHESIS 14.1.5. G is a simple QTKE-group, $T \in Syl_2(G)$, and

(1) $\mathcal{L}_f(G,T) = \emptyset$.

(2) There is $M_c \in \mathcal{M}(T)$ satisfying $M_c = !\mathcal{M}(C_G(Z))$.

(3) $|\mathcal{M}(T)| > 1$.

LEMMA 14.1.6. *(1)* $M^\infty \leq C_M(V)$.
(2) $\mathcal{L}^*(G, T) = \mathcal{C}(M_c)$, *so that* $M_c = !\mathcal{M}(\langle K, T \rangle)$ *for each* $K \in \mathcal{C}(M_c)$.
(3) For each $H \in \mathcal{H}(T)$, $H^\infty \leq C_G(Z) \leq M_c$.

PROOF. Part (1) follows from 14.1.3, and (3) follows from 14.1.3 applied to $V(H)$, using Hypothesis 14.1.5.2.

Let $L \in \mathcal{L}(G, T)$. Then $\langle L, T \rangle \in \mathcal{H}(T)$, so $L \leq M_c$ by (3). Therefore if $L \in \mathcal{L}^*(G, T)$, then by 1.2.7.3, $N_G(\langle L^T \rangle) = !\mathcal{M}(\langle L, T \rangle) = M_c$, and hence $L \in \mathcal{C}(M_c)$. Conversely let $L \in \mathcal{C}(M_c)$ and embed $L \leq K \in \mathcal{L}^*(G, T)$. We just showed $K \in \mathcal{C}(M_c)$, so $L = K$ and hence $L \in \mathcal{L}^*(G, T)$. Thus (2) is established. $\qquad\square$

LEMMA 14.1.7. *Assume* $J(T) \not\leq C_M(V)$, *and either* $M \neq M_c$ *or* $|Z| = 2$. *Then either*

(1) $m(V) = 2$, $\bar{M} = GL(V) \cong L_2(2)$, *and* $E \cap V = Z$ *is of order 2, or*
(2) $m(V) = 4$ *and* $\bar{M} \cong O_4^+(V)$. *Thus* $\bar{M} = (\bar{Y}_1 \times \bar{Y}_2)\langle \bar{t} \rangle$, *where* $\bar{Y}_i \cong L_2(2)$, $\bar{t}$ *is an involution interchanging* $\bar{Y}_1$ *and* $\bar{Y}_2$, *and* $V = V_1 \times V_2$, *where* $V_i := [V, Y_i] \cong E_4$, *and* $E \cap V$ *of order 4 contains* Z *of order 2.*

PROOF. Set $Y := J(M)$. By 14.1.6.1, $\bar{M}$ is solvable; so by Solvable Thompson Factorization B.2.16 $\bar{Y} = \bar{Y}_1 \times \cdots \times \bar{Y}_r$ with $\bar{Y}_i \cong L_2(2)$ and $V = V_1 \times \cdots \times V_r \times C_V(Y)$, where $V_i := [V, Y_i] \cong E_4$ for the preimage Y_i of $\bar{Y}_i$. As M is an SQTK-group, $r \leq 2$ by A.1.31.1. Thus either $\bar{M} = \bar{Y} \times C_{\bar{M}}(\bar{Y})$, or $r = 2$ and $\bar{M} = (\bar{Y} \times C_{\bar{M}}(\bar{Y}))\langle \bar{t} \rangle$, where t interchanges $\bar{Y}_1$ and $\bar{Y}_2$. Then as $\text{End}_{\bar{Y}}(V_i) = \mathbf{F}_2$, $C_M(\bar{Y})$ centralizes $[V, Y]$.

Next $Z \cap [V, Y] \neq 1$ and $[V, Y]$ is T-invariant, so by 14.1.5.2,

$$C_M(\bar{Y}) \leq C_M([V, Y]) \leq C_M(Z \cap [V, Y]) \leq M_c.$$

Suppose that $C_V(Y) \neq 1$. Then $C_Z(Y) \neq 1$, so $|Z| > 2$, and $Y \leq C_G(C_Z(Y)) \leq M_c$. Therefore $M = C_M(\bar{Y})YT \leq M_c$, and hence $M = M_c$, contrary to our hypothesis that $M \neq M_c$ when $|Z| > 2$.

Therefore $C_V(Y) = 1$ so that $[V, Y] = V$. Then $C_{\bar{M}}(\bar{Y})$ centralizes V, so that $C_{\bar{M}}(\bar{Y}) = 1$. Hence if $r = 1$, then (1) holds, so we may assume $r = 2$. If $\bar{Y} < \bar{M}$, then (2) holds, so we may assume $\bar{M} = \bar{Y} = \bar{Y}_1 \times \bar{Y}_2$. But then $Z_i := Z \cap V_i \neq 1$, so $|Z| = 4$ and $Y_{3-i} \leq C_G(Z_i) \leq M_c$, so that $M = C_M(V)Y_1 Y_2 \leq M_c$, and thus $M = M_c$, again contrary to our choice of M when $|Z| > 2$. $\qquad\square$

LEMMA 14.1.8. *Assume* $M \neq M_c$, $\bar{X} \leq \bar{M}$ *is* T-invariant of odd order, and $X \not\leq M_c$. *Then* $V = [V, X]$.

PROOF. As $\bar{X}$ is of odd order, $V = [V, X] \times C_V(X)$. Suppose $C_V(X) \neq 1$. As $\bar{X}$ is T-invariant, $C_Z(X) \neq 1$. But then by 14.1.5.2, $X \leq C_G(C_Z(X)) \leq M_c$, contrary to hypothesis. $\qquad\square$

LEMMA 14.1.9. *If* M *is the unique maximal member of* $\mathcal{M}(T)$ *under* $\stackrel{\sim}{\leq}$, *then* $M \neq M_c$.

PROOF. Assume $M = M_c$. By uniqueness of M and the definition of $\stackrel{\sim}{\leq}$, for each $M_1 \in \mathcal{M}(T)$ we have

$$M_1 = C_{M_1}(V(M_1))(M \cap M_1) \leq M$$

since $C_G(V(M_1)) \leq C_G(Z) \leq M_c = M$. This is impossible as $|\mathcal{M}(T)| > 1$ by 14.1.5.3. $\qquad\square$

LEMMA 14.1.10. *Assume M has a subnormal A_3-block X, and $O_2(M) \leq R \leq T$ such that $X = [X, J(R)]$. Then $M = M_c$ and $|Z| > 2$.*

PROOF. Let $X_0 := \langle X^M \rangle$. Thus as $m_3(M) \leq 2$, either $X_0 = X$, or $X_0 = X_1 \times X_2$ with $X = X_1$ while $X_2 = X^t$ for $t \in T - N_M(X)$. Set $K := C_M(X_0)$. As $X = [X, J(R)]$ and $O_2(M) \leq R$, $Aut_{X_0 T}(X_0) = Aut(X_0)$, so as $Z(X_0) = 1$ we conclude

$$M = (K \times X_0)T. \tag{$*$}$$

Since $Z_0 := Z \cap [O_2(X_0), X_0] \neq 1$, $K \leq C_G(Z_0) \leq M_c = !\mathcal{M}(C_G(Z))$ by 14.1.5.2. Now if $C_T(X_0) \neq 1$, then $C_Z(X_0) \neq 1$, so $|Z| > 2$ and $X_0 \leq C_G(C_Z(X_0)) \leq M_c = !\mathcal{M}(C_G(Z))$. Then $M = M_c$ by $(*)$, so the lemma holds.

Therefore we may assume that $C_T(X_0) = 1$. Then as $F^*(M) = O_2(M)$, we conclude from $(*)$ that $K = 1$. Thus as $Aut_{X_0 T}(X_0) = Aut(X_0)$, $M = X_0 T \cong S_4$ or S_4 wr $\mathbf{Z}_2$. In the first case, $T \cong D_8$, so $G \cong L_3(2)$ or A_6 by I.4.3. But then $T = C_G(Z)$, contrary to Hypothesis 14.1.5. In the second case, Theorem 13.9.1 supplies a contradiction. $\qquad\square$

LEMMA 14.1.11. *There exists a nontrivial characteristic subgroup $C_2 := C_2(T)$ of $Baum(T)$, such that for each $M \in \mathcal{M}(T)$, either*

(1) $M = C_M(V(M))N_M(C_2)$, or
(2) $M = M_c$ and $|Z| > 2$.

PROOF. Let $V := V(M)$. By 14.1.5.2, $M_c = !\mathcal{M}(C_G(Z))$, so $C_M(V) \leq C_M(Z) \leq M_c$. Let $S := Baum(T)$ and choose $C_i := C_i(T)$ for $i = 1, 2$ as in the Glauberman-Niles/Campbell Theorem C.1.18. Thus $1 \neq C_2$ char S and $1 \neq C_1 \leq Z$. In particular $C_G(C_1) \leq M_c = !\mathcal{M}(C_G(Z))$.

Suppose first that $[V, J(T)] = 1$. Then $S = Baum(C_T(V))$ by 14.1.2.1, so (1) holds by a Frattini Argument since C_2 is characteristic in S.

Thus we may assume that $[V, J(T)] \neq 1$, and that (2) fails, so that one of the conclusions of 14.1.7 holds. In either case $|Z| = 2$, so as $1 \neq C_1 \leq Z$ we conclude $C_1 = Z$. Further from the structure of $\bar{M}$, $C_M(C_1) = C_M(Z) = C_M(V)T \leq C_M(V)N_M(C_2)$. Therefore as we also may assume that conclusion (1) fails, $\langle C_M(C_1), N_M(C_2) \rangle < M$. Thus conclusion (2) of C.1.28 holds; in particular, there is a χ-block X of M with $X = [X, J(T)]$ such that X does not centralize V. Therefore as $\bar{M}$ is solvable by 14.1.6.1, we conclude that each such X is an A_3-block of M, and then 14.1.10 contradicts our assumption that (2) fails. $\qquad\square$

In the remainder of the section let $C_2 := C_2(T)$ be the subgroup defined as in 14.1.11 and its proof.

LEMMA 14.1.12. *Let $M_f \in \mathcal{M}(N_G(C_2))$ and $V(M_f) \leq V_f \in \mathcal{R}_2(M_f)$. Then*

(1) M_f is maximal in $\mathcal{M}(T)$ under $\stackrel{\sim}{<}$, M_f is the unique maximal member of $\mathcal{M}(T) - \{M_c\}$ under $\stackrel{\sim}{<}$, and if $|Z| = 2$ then M_f is the unique maximal member of $\mathcal{M}(T)$ under $\stackrel{\sim}{<}$.
(2) $M_f = !\mathcal{M}(N_{M_f}(C_T(V_f)))$.
(3) $C_{M_f}(V_f) \leq M$ for each $M \in \mathcal{M}(T)$.
(4) $M_f \neq M_c$.

PROOF. If $M \in \mathcal{M}(T)$ and either $M \neq M_c$ or $|Z| = 2$, then by 14.1.11, $M = C_M(V(M))N_M(C_2)$; so as $N_M(C_2) \leq M \cap M_f$, $M \stackrel{\sim}{<} M_f$. In particular if

$M_c = M_f$, then M_c is the unique maximal member of $\mathcal{M}(T)$ under $\stackrel{<}{\sim}$, contrary to 14.1.9. Thus $M_c \neq M_f$, proving (4). So as $C_{M_f}(V(M_f)) \leq M_c = !\mathcal{M}(C_G(Z))$, $M_f \stackrel{<}{\not\sim} M_c$, and then (1) follows from the first sentence of the proof.

As $V(M_f) \leq V_f$, $C_{M_f}(V_f) \leq C_{M_f}(V(M_f))$. By a Frattini Argument,

$$M_f = C_{M_f}(V_f)N_{M_f}(C_T(V_f)) = C_{M_f}(V(M_f))N_{M_f}(C_T(V_f)),$$

so (2) follows from A.5.7.1 and (1).

Next $C_{M_f}(V_f) \leq C_G(Z) \leq M_c$, while by (1) we may apply A.5.3.3 to each $M \in \mathcal{M}(T) - \{M_c\}$ to conclude that

$$C_{M_f}(V_f) \leq C_{M_f}(V(M_f)) \leq C_M(V(M)),$$

so (3) holds.   □

LEMMA 14.1.13. *Assume* $T \leq H \leq M$ *with* $R := O_2(H) \neq 1$ *and* $C(M, R) \leq H$. *Then either*

(1) $O_{2,F^*}(M) \leq H$ *and* $O_2(H) = O_2(M)$, *or*

(2) $|Z| > 2$, *and* $M = M_c = !\mathcal{M}(H)$.

PROOF. Observe that the triple R, H, M satisfies Hypothesis C.2.3 in the roles of "R, M_H, H". Thus we can appeal to results in section C.2, and in particular we conclude from C.2.1.2 that $O_2(M) \leq R$.

Suppose $L \in \mathcal{C}(M)$ with $L/O_2(L)$ quasisimple and $L \not\leq H$. By 14.1.5.1, $L \notin \mathcal{L}_f(G, T)$, so L is not a block. Thus $R \cap L \notin Syl_2(L)$ by C.2.4.1, so by C.2.4.2, $R \leq N_M(L)$. Then by C.2.2.3, $R \in \mathcal{B}_2(LR)$, so that $O_2(LR) \leq R$ by C.2.1.2. Further $Z(R) \leq O_2(LR)$ as $F^*(LR) = O_2(LR)$. Then as $L \notin \mathcal{L}_f(G, T)$, L centralizes $\Omega_1(Z(O_2(LR))) \geq \Omega_1(Z(R)) =: Z_R$, so

$$L \leq C_M(Z_R) \leq C(M, R) \leq H.$$

Thus we conclude $O_{2,E}(M) \leq H$.

Next set $F := O_{2,F}(M)$. By C.2.6, $R \in Syl_2(FR)$ and either

(i) $FR \leq H$, and hence also $O_{2,F^*}(M) \leq H$, or

(ii) $FR = (FR \cap H)X_0$, where X_0 is the product of A_3-blocks X subnormal in M with $X = [X, J(R)]$.

If (i) holds, then $O_2(M) = O_2(M \cap H) = O_2(H)$ by A.4.4.1, so that conclusion (1) holds. Thus we may assume (ii) holds. Therefore $M = M_c$ and $|Z| > 2$ by 14.1.10, so it remains to show that $M_c = !\mathcal{M}(H)$. Let $K := C_M(X_0)$. Then from (ii) and (*) in the proof of 14.1.10, we see that $KT = C_G(Z)$ and $O_{2,F^*}(K) = O_{2,F^*}(M) \cap H \leq H$, so that $O_{2,F^*}(KH) \leq H$.

We conclude from A.4.4.1 that $O_2(KH) = O_2(H) = R$. Thus $C_G(Z) = KT \leq C(M, R) \leq H$, so that $M_c = !\mathcal{M}(H)$ by 14.1.5.2. This completes the proof that (2) holds.   □

LEMMA 14.1.14. *If* $M_1 \in \mathcal{M}(T) - \{M\}$, *then* $O_2(M) < O_2(M_1 \cap M) > O_2(M_1)$.

PROOF. Let $H := M \cap M_1$, $R := O_2(H)$, $\{M_2, M_3\} = \{M, M_1\}$, and assume that $R = O_2(M_2)$. Then $C(G, R) \leq M_2$, so that $C(M_3, R) \leq H$. Thus by 14.1.13, either $R = O_2(M_3)$, or $M_3 = M_c = !\mathcal{M}(H)$. In the first case, $O_2(M_2) = R = O_2(M_3)$, so $M_1 = M$, contrary to the choice of M_1. In the second case as $H \leq M_2$, $M = M_1$ for the same contradiction.   □

LEMMA 14.1.15. $M = !\mathcal{M}(O_{2,F^*}(M)T)$.

PROOF. Suppose $M_1 \in \mathcal{M}(O_{2,F^*}(M)T)$ and let $H := M \cap M_1$. As $O_{2,F^*}(M) \leq H$, $O_2(M) = O_2(H)$ by A.4.4.1. Thus $M = M_1$ by 14.1.14. $\square$

LEMMA 14.1.16. *If $T \leq H \leq M$ with $1 \neq O_2(H)$ and $C(M, O_2(H)) = H$, then $M = !\mathcal{M}(H)$.*

PROOF. Suppose $M_1 \in \mathcal{M}(H) - \{M\}$. Then $|\mathcal{M}(H)| > 1$, so $O_{2,F^*}(M) \leq H$ by 14.1.13, contrary to 14.1.15. $\square$

LEMMA 14.1.17. *Let $M_1 \in \mathcal{M}(T) - \{M\}$ and assume either*

(a) $M_1 \overset{\leq}{\sim} M$ and $V = V(M)$, or

(b) $M_1 = M_c$.

Let $R := O_2(M_1 \cap M)$, assume there is T-invariant subgroup Y_0 of M with $\bar{Y}_0$ of odd order, and set $Y := O^2(\langle R^{Y_0 T}\rangle)$ and $M^* := M/O_2(M)$. Then

(1) $\bar{R} \neq 1$.

(2) $\bar{Y} = [\bar{Y}_0, \bar{R}]$.

(3) $[C_M(V)^*, Y^* R^*] = 1$ and $C_Y(V)^* \leq O(Z(C_M(V)^* R^*))$.

(4) If $1 \neq r^* \in R^*$ is faithful on $O_p(M^*)$ for some odd prime p, then $C_{M^*}(r^*)$ has cyclic Sylow p-groups, so $m_p(C_M(V)) \leq 1$.

(5) $R = O_2(C_M(V)R)$, so $N_{\bar{M}}(\bar{R}) = \overline{N_M(R)}$.

PROOF. In case (a), $M_1 \overset{\leq}{\sim} M$ and $V = V(M)$, so $C_M(V) \leq M_1$ by A.5.3.3. In case (b), $M_1 = M_c$ and $C_M(V) \leq C_M(Z \cap V) \leq M_c = !\mathcal{M}(C_G(Z))$. So in either case, $C_M(V) \leq M \cap M_1 \leq N_M(R)$. Since $V \in \mathcal{R}_2(M)$ by 14.1.1, it follows that $C_R(V) = O_2(C_M(V)) = O_2(M)$. Then $R = O_2(C_M(V)R)$, and hence (5) holds. Further if $\bar{R} = 1$, then $R = C_R(V) = O_2(M)$, contrary to 14.1.14. Hence (1) is established.

Next by Coprime Action, $\bar{Y}_0 = \bar{Y}_+ \bar{Y}_-$, where $\bar{Y}_+ := C_{\bar{Y}_0}(\bar{R})$ and $\bar{Y}_- := [\bar{Y}_0, \bar{R}]$ are T-invariant since Y_0 and R are T-invariant. By (5), $\bar{Y}_+ \leq \overline{N_M(R)}$, so $\bar{Y}_R := \langle \bar{R}^{Y_0 T}\rangle = \langle \bar{R}^{\bar{Y}_-}\rangle$ and $\bar{Y}_R = \bar{R}\bar{Y}_-$ with $\bar{Y}_- = \bar{Y}$. In particular (2) holds. Also $[C_M(V), R] \leq C_R(V) = O_2(M)$, so $[C_M(V)^*, R^*] = 1$, and hence $Y^* = [Y^*, R^*]$ centralizes $C_M(V)^*$, so that (3) holds. Part (4) follows from A.1.31.1 applied to the product of a Sylow p-subgroup of $O_p(M^*)C_{M^*}(r^*)$ with $\langle r^*\rangle$. $\square$

LEMMA 14.1.18. *Let $M := M_f$ as in 14.1.12, and assume $V := V(M)$ is of rank 2 with $\bar{M} \cong S_3$. Let $R_c := O_2(M \cap M_c)$, $Y := O^2(\langle R_c^M\rangle)$, $R := C_T(V)$, and $M^* := M/O_2(M)$. Then*

(1) M is the unique maximal member of $\mathcal{M}(T)$ under $\overset{\leq}{\sim}$.

(2) $R_c R = T$ and $M \cap M_c = C_M(V)R_c$.

(3) $\bar{Y} = O^2(\bar{M}) \cong \mathbf{Z}_3$ and $O_2(Y) = C_Y(V)$.

(4) $M = !\mathcal{M}(YT)$.

(5) $M^* = Y^* R_c^* \times C_M(V)^*$ with $Y^* R_c^* \cong S_3$ and $m_3(C_M(V)) \leq 1$.

(6) Z is of order 2 and $M_c = C_G(Z)$.

(7) $\mathcal{M}(T) = \{M, M_c\}$.

PROOF. As $V = V(M)$ is of rank 2 and $\bar{M} \cong S_3$, Z is of order 2, so (1) follows from part (1) of 14.1.12. Further $M_c \neq M$ by part (4) of that result, so case (b) of the hypothesis of 14.1.17 holds. By 14.1.17.1, $\bar{R}_c \neq 1$, so as $\bar{T}$ is of order 2, $\bar{R}_c = \bar{T}$ and hence $T = RR_c$. As $C_M(V) \leq C_G(Z) \leq M_c = !\mathcal{M}(C_G(Z))$, but $M \nleq M_c$, it follows that $M \cap M_c = C_M(V)R_c$, so that (2) holds. Further applying 14.1.17 to the preimage Y_0 in M of $O(\bar{M})$, we conclude $\bar{Y} = \bar{Y}_0$ and $C_Y(V)^* \leq O(Z(C_M(V)^* R_c^*))$. By the first remark, $\bar{M} = \bar{Y}\bar{T}$, so (4) holds by A.5.7.1. By (2) there is $r \in R_c$ inverting $\bar{Y}$, so as $[C_Y(V)^*, R_c^*] = 1$, r inverts y of order 3 in $Y - C_Y(V)$, and $Y^* = [Y^*, R_c^*] = \langle y^* \rangle$. Therefore $Y^* \cong \mathbf{Z}_3$, completing the proof of (3). As $[C_M(V), YR_c] \leq O_2(M)$, the first two statements in (5) hold, while the third follows from 14.1.17.4.

Let $K \in \mathcal{M}(T)$. By (1) and A.5.3.1, $V(K) \leq V$; so as $|V| = 4$, it follows that $V(K) = Z$ or V. In the latter case $K = M$ by A.5.4; in the former, $K \leq C_G(Z) \leq M_c$ so that $K = M_c$, completing the proof of (6) and (7). $\qquad\square$

14.2. Starting the $\mathbf{L_2}(2)$ case of $\mathcal{L}_f$ empty

We now state Hypothesis 14.2.1, which in effect is the special case of Hypothesis 14.1.5 where $V(M_f)$ is of of rank 2, for M_f the member of $\mathcal{M}(T)$ defined in 14.1.12. Namely Hypothesis 14.2.1 implies Hypothesis 14.1.5, and conversely when Hypothesis 14.1.5 holds and $V(M_f)$ is of rank 2, then Hypothesis 14.2.1 is satisfied with M_f in the role of "M" by 14.1.18. Indeed 14.1.18 supplies a normal subgroup Y of M with $YT/O_2(YT) \cong L_2(2)$ and $M =!\mathcal{M}(YT)$. Thus we view Y as a solvable analogue of $L \in \mathcal{L}_f^*(L, T)$, and then Hypothesis 14.2.1 allows us to treat the case $LT/O_2(LT) \cong L_2(2)$ in parallel with the final case in the Fundamental Setup where $L/O_2(L) \cong L_3(2)$.

Thus in this section, and as appropriate in the later sections of this chapter, we assume:

HYPOTHESIS 14.2.1. *G is a simple QTKE-group, $T \in Syl_2(G)$, $Z := \Omega_1(Z(T))$, and*

(1) $\mathcal{L}_f(G, T) = \emptyset$.
(2) $M_c := C_G(Z) \in \mathcal{M}(T)$.
(3) *There exists a unique maximal member M of $\mathcal{M}(T)$ under $\overset{\leq}{\sim}$.*
(4) *$V := V(M) = \langle Z^M \rangle$ is of rank 2, and $\bar{M} := M/C_M(V) \cong Aut(V) \cong L_2(2)$.*
(5) $|\mathcal{M}(T)| > 1$.

We observe that by parts (1), (2), and (5) of Hypothesis 14.2.1, Hypothesis 14.1.5 is satisfied. Indeed by 14.2.1.3 and 14.1.12.1, M is the maximal 2-local M_f containing $N_G(C_2(T))$ appearing in 14.1.12. Then by 14.2.1.4, the hypotheses of 14.1.18 are satisfied. As in 14.1.18, we set

$$R_c := O_2(M \cap M_c) \quad \text{and} \quad Y := O^2(\langle R_c^M \rangle).$$

Then applying 14.1.18 we conclude:

LEMMA 14.2.2. *(1) $T = C_T(V)R_c$, and $M \cap M_c = C_M(V)R_c$ so that $O^2(M \cap M_c) \leq C_M(V)$.*

(2) $\bar{Y} = O^2(\bar{M}) \cong \mathbf{Z}_3$ and $O_2(Y) = C_Y(V)$.

(3) $M = !\mathcal{M}(YT)$.

(4) $M/O_2(M) = YR_c/O_2(M) \times C_M(V)/O_2(M)$ with $YR_c/O_2(M) \cong L_2(2)$ and $m_3(C_M(V)) \leq 1$.

(5) $\mathcal{M}(T) = \{M, M_c\}$.

(6) $|Z| = 2$, and hence $C_T(Y) = 1$.

(7) $N_G(T) \leq M \cap M_c$.

(8) For each $H \in \mathcal{H}_*(T, M)$, $H \cap M$ is the unique maximal subgroup containing T, and H is a minimal parabolic described in B.6.8, and in E.2.2 when H is nonsolvable.

PROOF. Parts (1)–(6) follow from 14.1.18, so it remains to prove (7) and (8). As $Z = \Omega_1(Z(T))$ is of order 2, $N_G(T) \leq C_G(Z) = M_c$. As $N_G(T)$ preserves $\stackrel{<}{\sim}$, $N_G(T) \leq M$ by 14.2.1.3, completing the proof of (7). Then (8) follows from (7) just as in the proof of 3.3.2.4. $\square$

For the remainder of the section, H will denote a member of $\mathcal{H}(T, M)$.

Set $M_H := M \cap H$, $U_H := \langle V^H \rangle$, $Q_H := O_2(H)$, and $H^* := H/Q_H$. Let $\tilde{M}_c := M_c/Z$. Since $T \leq H \not\leq M$, we conclude from 14.2.2.5 that:

LEMMA 14.2.3. $C_G(Z) = M_c = !\mathcal{M}(H)$.

In particular $\tilde{H} := H/Z$ makes sense. Next observe using 14.2.2 that:

LEMMA 14.2.4. *Case (2) of Hypothesis 12.8.1 is satisfied with Y in the role of "L".*

In Notation 12.8.2, we have $V_2 = V$, $L_2 = Y$, $V_1 = Z$, and $L_1 = 1$. Defining $\mathcal{H}_z$ as in Notation 12.8.2, 14.2.3 says:

LEMMA 14.2.5. $\mathcal{H}_z = \mathcal{H}(T, M)$.

By 14.2.5, results from section 12.8 apply to H. In particular recall from 12.8.4 that:

LEMMA 14.2.6. *(1) Hypothesis G.2.1 is satisfied.*
(2) $\tilde{U}_H \leq \Omega_1(Z(\tilde{Q}_H))$ and $\tilde{U}_H \in \mathcal{R}_2(\tilde{H})$.
(3) $\Phi(U_H) \leq V_1$.
(4) $Q_H = C_H(\tilde{U}_H)$.

Part (2) of Hypothesis 14.2.1 excludes the quasithin examples $L_3(2)$ and A_6, which will be treated in the final section of the next chapter. In the remainder of this section, we will identify the other quasithin examples corresponding to $\bar{L} \cong L_2(2)$, which do satisfy Hypothesis 14.2.1. These examples arise in the cases where some $H \in \mathcal{H}_*(T, M)$ has one of three possible structures: $n(H) > 1$; $H/O_2(H) \cong D_{10}$ or $Sz(2) \cong F_{20}$; or $H/O_2(H) \cong L_2(2)$. In each case we will show that G possesses a weak BN-pair of rank 2, as discussed in section F.1; then we appeal to section F.1 and the subsequent sections in chapter F of Volume I, to identify G. Then in later sections we show that no further quasithin groups arise under Hypothesis 14.2.1, although certain shadows are eliminated in those sections.

14.2.1. The treatment of $\mathbf{n(H) > 1}$. The first major result of this section is:

THEOREM 14.2.7. *Either*
(1) $n(H) = 1$ for each $H \in \mathcal{H}_*(T, M)$, or
(2) G is $^3D_4(2)$, J_2, or J_3.

Until the proof of Theorem 14.2.7 is complete, we assume $H \in \mathcal{H}_*(T, M)$ with $n(H) > 1$. By 14.2.2.8 and E.1.13, the structure of H is described in E.2.2. As $n(H) > 1$, only cases (1a), (2a), or (2b) of E.2.2 can hold. Set $K := O^2(H)$. In each case, we next define a Bender subgroup K_1 of K which, together with Y, will be used to construct our weak BN-pair:

NOTATION 14.2.8. One of the following holds:

(1) $K/O_2(K)$ is $L_2(2^n)$ or $Sz(2^n)$, and we set $K_1 := K$.

(2) $K/O_2(K)$ is the product of two commuting Bender groups interchanged by T, and we choose $K_1 \in \mathcal{C}(H)$.

(3) $K/O_2(K)$ is $(S)L_3(2^n)$ or $Sp_4(2^n)$ for $n \geq 2$, with T inducing an automorphism nontrivial on the Dynkin diagram of $K/O_2(K)$, and we set $K_1 := P_1^\infty$, where $P_i/O_2(K)$, $i = 1, 2$, are the maximal parabolics of $K/O_2(K)$ with $T \cap K \leq P_i$.

Let $S := N_T(K_1)$. In each case in 14.2.8, $K_1/O_2(K_1)$ is a Bender group with $K_1 \in \mathcal{C}(K_1 S)$ and $K_1 \not\leq M$. In case (1), $K = K_1$ and $S = T$, while in cases (2) and (3), $K_1 < K$ and $|T : S| = 2$.

By 14.2.3, $H \leq M_c = C_G(Z)$.

LEMMA 14.2.9. Assume $S < T$. Then K_1 is contained in some $K_c \in \mathcal{C}(M_c)$, and one of the following holds:

(1) Case (3) of 14.2.8 holds, and $K = K_c$ is of 3-rank 2, with $K = \langle K_1^R \rangle$ for each $R \in Syl_2(M_c)$ with $S \leq R$.

(2) Case (2) of 14.2.8 holds, $K_1 = K_c$, $K = \langle K_1^T \rangle$, and either K has 3-rank 2, or $K_1/O_2(K_1) \cong Sz(2^n)$.

(3) Case (2) of 14.2.8 holds, $K_1/O_2(K_1) \cong L_2(4)$, $K_c/O_2(K_c) \cong J_1$ or $L_2(p)$ for p an odd prime with $p^2 \equiv 1 \mod 5$, $S = N_T(K_c)$, and $\langle K_1^R \rangle$ is of 3-rank 2 for each $R \in Syl_2(M_c)$ with $S \leq R$.

PROOF. The existence of K_c follows from 1.2.4. In case (3) of 14.2.8, $K/O_2(K) \cong (S)L_3(2^n)$ or $Sp_4(2^n)$, so that $K \in \mathcal{L}^*(G, T)$ by 1.2.8.4—except when $K/O_2(K) \cong L_3(4)$, where $K \in \mathcal{L}^*(G, T)$ by 1.2.8.3, since T is nontrivial on the Dynkin diagram of $K/O_2(K)$. Thus $K \in \mathcal{C}(M_c)$ by 14.1.6.2, so that $K = K_c$, and conclusion (1) holds in this case. In case (2) of 14.2.8, $K_1 \in \mathcal{L}(G, T)$, so by 1.2.8.2, either $K_1 \in \mathcal{L}^*(G, T)$ so that $K_1 = K_c$ and (2) holds; or else (3) holds. $\square$

LEMMA 14.2.10. If $S < T$, then $M_c = !\mathcal{M}(\langle K_1, T_1 \rangle)$ for each $T_1 \in Syl_2(M_c)$ containing S.

PROOF. By Sylow's Theorem, $T_1 = T^g$ for some $g \in M_c$. If $K_1 = K_c$, the result follows from 14.1.6.2 applied to T_1 in the role of "T". Thus we may assume $K_1 < K_c$, so that conclusion (1) or (3) of 14.2.9 holds.

Let $H_1 := \langle K_1, T_1 \rangle$ and $M_1 \in \mathcal{M}(H_1)$. By 14.2.2.5, $M_1 = M_c$ or M^g, and we may assume the latter. As case (1) or (3) of 14.2.9 holds, $\langle K_1^R \rangle$ is of 3-rank 2 for each $R \in Syl_2(M_c)$ containing S, so in particular $H_1 = \langle K_1, T_1 \rangle$ is of 3-rank 2. Then $O^2(H_1) \leq O^2(M_c \cap M^g) \leq C_{M^g}(V^g)$ by 14.2.2.1, contrary to 14.2.2.4. $\square$

Let B be a Hall $2'$-subgroup of $K \cap M$, and set $B_1 := B \cap K_1$.

LEMMA 14.2.11. B acts on K_1, $BT = TB$, $BS = SB$, and $B \leq C_M(V)$.

PROOF. As $M_H = BT$, $BT = TB$. Then as B acts on K_1, $BS = SB$. As $H \leq M_c$, $B \leq O^2(M \cap M_c) \leq C_G(V)$ by 14.2.2.1. $\square$

LEMMA 14.2.12. Either $O_2(M) \leq S$, so that $S \in Syl_2(YS)$, or the following hold:

(1) $K/O_2(K) \cong L_3(4)$, and some element of T induces a graph automorphism on $K/O_2(K)$.

(2) $B = B_1$ is of order 3 and $B \leq C_M(V)$.

(3) $K = O^{3'}(M_c^\infty)$.

PROOF. Assume $Q_M := O_2(M) \not\leq S$; in particular, $S < T$, so one of the cases of 14.2.9 holds. Now $Q_M = [Q_M, B]C_{Q_M}(B)$ by Coprime Action, and using A.1.6, $[Q_M, B] \leq [O_2(BT), B] \leq S$. Thus if $C_T(B) \leq S$, then $Q_M \leq S$, contrary to assumption; so $C_T(B) \not\leq S$, and then the cases in 14.2.9, only conclusion (1) of the present result can hold.

As $K/O_2(K) \cong L_3(4)$ by (1), $B = B_1$ is of order 3. By 14.2.11, $B \leq C_M(V)$, so (2) holds. By 14.2.9, $K = K_c \in \mathcal{C}(M_c)$. By A.3.18, $C_{M_c}(K/O_2(K))$ is a $3'$-group, so (3) holds. $\qquad \square$

LEMMA 14.2.13. $O_2(Y) \leq S$.

PROOF. Assume not. Then as $O_2(Y) \leq O_2(M)$, $O_2(M) \not\leq S$, so conclusions (1)–(3) of 14.2.12 are satisfied. In particular $K \in \mathcal{C}(M_c)$ and $K/O_2(K) \cong L_3(4)$. By 14.2.5, $M_c \in \mathcal{H}_z$. Let $U := \langle V^{M_c} \rangle$.

We first show that U is abelian. Suppose not. Let $y \in Y$ be of order 3 and set $I := \langle U^Y \rangle$. We appeal to 12.8.9; recall $V_2 = V$, and Y, I play the roles of "$O^2(P)$, I_2". Thus by 12.8.9.2, $O_2(I) = U_I U_I^y$, where $U_I := U \cap O_2(I)$. By 12.8.9.1, $Y = O^2(I)$ and T acts on I. Thus T acts on $O_2(I) = U_I U_I^y$, so that as $U \leq Q_H$ by 14.2.6.2, U_I^{y*} is a normal elementary abelian subgroup of T^*. Thus as $K^*T^* \leq Aut(L_3(4))$, we conclude $U_I^{y*} \leq K^*$. But then $O_2(Y) \leq O_2(I) = U_I U_I^y \leq S$, contrary to our hypothesis.

Therefore U is abelian. So by 12.8.6.5, Hypothesis F.9.8 is satisfied, for each $H \in \mathcal{H}_z$, with Z, V in the roles of "V_1, V_+". As $K^* \cong L_3(4)$ and T^* is nontrivial on the Dynkin diagram of K^*, H^* has no FF-modules by Theorem B.4.2, so we conclude from (3) and (4) of F.9.18 that there is $\tilde{I} \in Irr_+(K, \tilde{U}_H)$ with $I \trianglelefteq H$ and $q(Aut_H(\tilde{I}), \tilde{I}) \leq 2$. This contradicts B.4.2 and B.4.5. $\qquad \square$

Set $S_2 := O_2(Y)(T \cap K)$. We begin to verify the hypotheses of F.1.1 with K_1, YS_2, S in the roles of "L_1, L_2, S": By 14.2.13, $O_2(Y) \leq S$, while $T \cap K \leq S$ by definition, so that $S_2 \leq S$ and $S_1 := S \cap K_1 \in Syl_2(K_1)$. By construction $O_2(Y) \leq S_2$, so that $S \cap YS_2 = S_2 \in Syl_2(YS_2)$. Thus hypothesis (b) of F.1.1 holds. By definition, S acts on K_1. As S acts on K and Y, S acts on YS_2. Thus hypothesis (a) of F.1.1 holds. Next $K_1/O_2(K_1)$ is a Bender group by construction, and so satisfies (c) of F.1.1. Since $Y/O_2(Y) \cong \mathbf{Z}_3 \cong L_2(2)'$, to verify (c) for YS_2 we must show:

LEMMA 14.2.14. $Y = [Y, S_2]$.

PROOF. If not, then $S_2 \trianglelefteq YT$, so Theorem 3.1.1 applied to S_2, YT in the roles of "R, M_0" says $O_2(\langle YT, H \rangle) \neq 1$, contrary to 14.2.2.3. $\qquad \square$

Next $N_{K_1}(S_1) = S_1 B_1 =: C_1$ lies in M and so normalizes Y, and hence normalizes YS_2 by construction, and $C_2 := N_{YS_2}(S_2) = S_2$ normalizes K_1. Thus (d) of F.1.1 holds with C_1, C_2 in the roles of "B_1, B_2"; and (f) of F.1.1 also follows by construction. Therefore it remains to establish hypothesis (e) of F.1.1.

Let $G_1 := K_1 S$, $G_2 := B_1 Y S$, and $G_{1,2} := G_1 \cap G_2 = S B_1$. Consider the amalgam $\alpha := (G_1, G_{1,2}, G_2)$, and let $G_0 := \langle G_1, G_2 \rangle$. To establish hypothesis (e) of F.1.1, we need to show:

LEMMA 14.2.15. $O_2(G_0) = 1$.

PROOF. Assume $O_2(G_0) \neq 1$, and let $M_1 \in \mathcal{M}(G_0)$. Then $T \not\leq M_1$, since otherwise by 14.2.2.3, $M = !\mathcal{M}(YT) = M_1$, contrary to $\langle K_1, T \rangle = H \not\leq M$. Thus $S < T$, and hence one of the cases of 14.2.9 holds.

Let $Z_S := \Omega_1(Z(S))$. As T normalizes S, $Z \leq Z_S$. By 14.2.9, $K_1 \leq K_c \in \mathcal{C}(M_c)$. As $O_2(\langle K_c, T \rangle) \leq N_T(K_1) = S$, $Z_S \leq \Omega_1(Z(O_2(\langle K_c, T \rangle)))$. Thus as $\mathcal{L}_f(G, T) = \emptyset$ by 14.2.1.1,

$$K \leq \langle K_c^T \rangle \leq C_G(Z_S), \qquad (!)$$

so that $Z_S \trianglelefteq \langle K_c, T \rangle$. Hence $N_G(Z_S) \leq M_c = !\mathcal{M}(\langle K_1, T \rangle)$ by 14.2.10. As $|T : S| = 2$, S is normal in a Sylow 2-subgroup T_1 of M_1, and hence $T_1 \leq N_{M_1}(Z_S) \leq M_c$. If $S < T_1$, then $T_1 \in Syl_2(M_c)$, so $M_c = !\mathcal{M}(\langle K_1, T_1 \rangle) = M_1$ by 14.2.10, a contradiction as $M_c \neq M = !\mathcal{M}(YT)$.

So $S \in Syl_2(M_1)$. Therefore we can embed K_1 in some $L \in \mathcal{C}(G_0)$ by 1.2.4. Now $Y = O^2(Y)$ normalizes L by 1.2.1.3, and $S \leq N_G(K_1) \leq N_G(L)$, so $G_0 = \langle K_1 S, Y \rangle = LYS$.

Suppose that $L \leq C_G(Z)$. Then L centralizes $V = \langle Z^Y \rangle$, so $\langle L, T \rangle \leq N_G(V) = M$, a contradiction as $M \neq M_c = !\mathcal{M}(\langle K_1, T \rangle)$.

Therefore $[L, Z] \neq 1$. In particular, $K_1 < L$, so as $G_0 = LYS$, $L = [L, Y]$. Let $R := O_2(YS)$. Then $R \trianglelefteq YT$, so $C(G, R) \leq M = !\mathcal{M}(YT)$. Moreover if $Y \not\leq L$ then $YS \cap L = S \cap L$ is Y-invariant, so $S \cap L \leq R$ and hence $R \in Syl_2(LR)$.

Next $C_T(O_2(M_1)) \leq M_1$ as $M_1 \in \mathcal{M}$, and as $S \in Syl_2(M_1)$ and $S \leq M_c$, $O_2(M_1) \leq O_2(G_0) \leq O_2(M_c \cap G_0) \leq S$ by A.1.6, so that

$$C_{O_2(M_c)}(O_2(M_c \cap G_0)) \leq C_T(O_2(G_0)) \leq C_T(O_2(M_1)) \leq S \leq G_0, \qquad (*)$$

and hence G_0, M_c, S satisfy the hypotheses of 1.1.5 in the roles of "H, M, T_H". By (*), hypothesis (b) of 1.2.11 is satisfied, and since a generator z for Z is in $V = [V, Y]$, hypothesis (a) of 1.2.11 is also satisfied. Thus by 1.2.11, either $G_0 \in \mathcal{H}^e$, or L is quasisimple.

Assume first that L is quasisimple. Then L is described in 1.1.5.3, and $L = [L, z]$. As $K_1/O_2(K_1)$ is a Bender group over $\mathbf{F}_{2^n}$ with $n > 1$, and $K_1 \in \mathcal{L}(C_L(z), S)$, comparing the list of 1.1.5.3 with that of A.3.12, we conclude that either $L/Z(L)$ is $Sp_4(2^n)$, $G_2(2^n)$, $^2F_4(2^n)$, or $^3D_4(2^{n/3})$, or else $K_1/O_2(K_1) \cong L_2(4)$ and $L \cong J_2$, J_4, HS, or Ru. Then by A.3.18, $O^{3'}(G_0) = L$, so that $G_0 = LYS = LS$.

Suppose first that L is of Lie type. As $YS = SY$, either $L \cong {}^3D_4(2)$, or n is even and Y is contained in the Borel subgroup of L over S. But in the latter case, Y lies in the parabolic P of L with $K_1 = P^\infty$, so $G_0 = \langle K_1 S, Y \rangle \leq P < L$, contrary to $G_0 = LS$.

Therefore $L \cong {}^3D_4(2)$ with $K_1/O_2(K_1) \cong L_2(8)$, or J_2, J_4, HS, or Ru with $K_1/O_2(K_1) \cong L_2(4)$; note that case (2) of 14.2.8 holds. Now $1 \neq O_2(G_0) \leq C_S(L)$, so $1 \neq C_{Z_S}(L) =: Z_L$ is in the center of $G_0 = LS$. Thus we may assume $M_1 \in \mathcal{M}(C_G(Z_L))$. Then by (!), $K \leq C_G(Z_L) \leq M_1$. Further $K = O^{3'}(K)$ since $K_1/O_2(K_1) \cong L_2(2^m)$ for some m. By 1.2.4 and 1.2.8.4, $L \in \mathcal{C}(M_1)$, and by A.3.18, $L = O^{3'}(M_1)$. Thus $K \leq L$. However, when L is $^3D_4(2)$, J_2, J_4, HS, or Ru, $C_L(z)$ has no subgroup isomorphic to K satisfying 14.2.9—namely containing the product of two conjugates of K_1, since case (2) of 14.2.8 holds—see e.g. 16.1.4 and 16.1.5. This contradiction completes the treatment of the case that L is quasisimple.

Therefore $G_0 \in \mathcal{H}^e$, so $V_0 := \langle Z^{G_0} \rangle \in \mathcal{R}_2(G_0)$ by B.2.14. Then $[V_0, L] \neq 1$ since we saw $[L, Z] \neq 1$. If C is a nontrivial characteristic subgroup of S with $L \leq N_G(C)$, then $H = KT \leq \langle L, T \rangle \leq N_G(C)$, so $L \leq N_G(C) \leq M_c = {!}\mathcal{M}(H)$ by 14.2.3, contradicting $[L, Z] \neq 1$. Hence no such C exists, so as $L/O_{2,F}(L)$ is quasisimple by 1.2.1.4, $L = [L, J(S)]$. Then appealing to Thompson Factorization B.2.15, V_0 is an FF-module for $LS/C_{LS}(V_0)$, so by Theorems B.5.1 and B.5.6, $L/C_L(V_0) \cong L_2(2^n)$, $SL_3(2^n)$, $Sp_4(2^n)$, $G_2(2^n)$, $L_n(2)$, $\hat{A}_6$, or A_7. As $K_1 < L$ and S acts on K_1 with $n(K_1) > 1$, $L/C_L(V_0)$ is not $L_n(2)$ or a group over $\mathbf{F}_2$ or $\hat{A}_6$, and also L is not a χ_0-block. Further $L/O_2(L)$ is not A_7, since the FF-modules in Theorem B.5.1 do not satisfy the condition $[K_1, Z_S] = 1$ in (!). Therefore $L/C_L(V_0)$ is $SL_3(2^n)$, $Sp_4(2^n)$, or $G_2(2^n)$, and $K_1/O_2(K_1) \cong L_2(2^n)$ for $n > 1$. Recall $R = O_2(YS)$. If $Y \not\leq L$, then as we observed earlier, $R \in Syl_2(LR)$; while if $Y \leq L$ then Y is contained in a Borel subgroup of L, and then once again, R is Sylow in LR. We also saw $C(G, R) \leq M$, while $L \not\leq M$ as $K_1 \not\leq M$; thus L is a χ_0-block by C.1.29, contrary to an earlier observation. This contradiction completes the proof of 14.2.15. $\square$

LEMMA 14.2.16. α is a weak BN-pair of rank 2, $K = K_1$, $T = S$, $Q := O_2(K) = O_2(M_c)$ is extraspecial, and either

(1) α is isomorphic to the $^3D_4(2)$-amalgam, $|Q| = 2^{1+8}$, and $K/Q \cong L_2(8)$, or

(2) α is parabolic isomomorphic to the J_2-amalgam or $Aut(J_2)$-amalgam, $|Q| = 2^{1+4}$, and $K/Q \cong L_2(4)$.

PROOF. Recall 14.2.15 completed the verification of Hypothesis F.1.1 with K_1, YS_2, S in the roles of "L_1, L_2, S". Then by F.1.9, α is a weak BN-pair of rank 2. Furthermore we saw $B_2 = S_2$, so α appears in the list of F.1.12. Since $G_2/C_{G_2}(V) \cong S_3$, while K_1 is nonsolvable and centralizes Z, we conclude that α is either isomorphic to the $^3D_4(2)$-amalgam, or is parabolic-isomorphic to the J_2-amalgam or the $Aut(J_2)$-amalgam. In each case $Z_S \cong \mathbf{Z}_2$, $\langle Z_S^Y \rangle \cong E_4$, and $Q = O_2(K_1) = O_2(K_1S)$ is extraspecial of order 2^{1+8} or 2^{1+4}, while $K_1/Q \cong L_2(8)$ or $L_2(4)$.

As Z_S is of order 2, $Z = Z_S$. Also K_1 is irreducible on Q/Z, so $Q = O_2(M_c)$ using A.1.6. Further the action of K_1 on Q/Z does not extend to $(S)L_3(2^n)$, $Sp_4(2^n)$, or $L_2(2^n) \times L_2(2^n)$, so as $K = \langle K_1^T \rangle$, case (1) of 14.2.8 holds, so $K = K_1$ and $T = S$. $\square$

We say G is of *type* J_3 or J_2 if α is parabolic isomorphic to the J_2-amalgam, and G has 1 or 2 classes of involutions, respectively.

LEMMA 14.2.17. *Assume α is parabolic isomorphic to the J_2-amalgam or the $Aut(J_2)$-amalgam. Then*

(1) α *is parabolic-isomorphic to the J_2-amalgam, and G is of type J_2 or J_3.*

(2) *If G is of type J_2, then $G \cong J_2$.*

(3) *If G is of type J_3, then $G \cong J_3$.*

PROOF. By 14.2.16, $Q = O_2(M_c)$, so as $Out(Q) \cong S_5$, $KT = M_c = C_G(Z)$.

Assume first that α is parabolic isomorphic to the $Aut(J_2)$-amalgam. Then by 46.1 and 46.11 in [**Asc94**], K has three orbits on involutions in K, with representatives $z \in Z$, $s \in V - Z$, and $t \in K - Q$ with $C_T(t) \in Syl_2(C_{KT}(t))$. Then $s \in z^Y$, so as J_2 has two classes of involutions, $C_T(t)$ is isomorphic to a Sylow 2-subgroup of the centralizer in $Aut(J_2)$ of a non-2-central involution of J_2. Hence $C_T(t) = A\langle k\rangle$, where $A := C_{T \cap K}(t) \cong E_{16}$ and k is an involution acting freely on A. Next as α is parabolic isomorphic to the amalgam of $Aut(J_2)$, there is $j \in T - K$ with $\mathbf{Z}_2 \times D_{16} \cong C_T(j) \in Syl_2(C_{KT}(j))$. Now $N_G(C_T(j))$ normalizes $\Omega_1(\Phi(C_T(j))) = Z$ and hence lies in $C_G(z) = KT$; it follows that $C_T(j)$ is Sylow in $C_G(j)$—for otherwise $C_T(j) < X \in Syl_2(C_G(j))$ so that $C_T(j) < N_X(C_T(j)) \leq C_{KT}(j)$, contrary to $C_T(j) \in Syl_2(C_{KT}(j))$. But $C_T(j)$ does not contain a copy of $C_T(t)$ or T, so $j^G \cap K = \emptyset$. Therefore by Thompson Transfer, $j \notin O^2(G)$, contrary to the simplicity of G.

Therefore α is not parabolic isomorphic to the $Aut(J_2)$-amalgam, so α is parabolic isomorphic to the J_2-amalgam, and $K = C_G(z)$. G is of type J_2 or J_3, completing the proof of (1). Then (2) and (3) follow from existing classification theorems which we have stated in Volume I as I.4.7. $\qquad\square$

In view of 14.2.17, to complete the proof of Theorem 14.2.7, it remains to treat the ${}^3D_4(2)$-case. So assume α is the ${}^3D_4(2)$ amalgam. Let $Z = \langle z\rangle$, $\hat{G} := {}^3D_4(2)$, and $\dot{G} := Aut(\hat{G})$.

LEMMA 14.2.18. *Assume α is the ${}^3D_4(2)$-amalgam. Then $M_c = C_G(z)$ and either*

(1) $M_c = K$, or

(2) $M_c = KA$, where $A \leq M_c \cap M$ is of order 3 and induces field automorphisms on K/Q. Moreover $\dot\alpha := (M_c, M_c \cap M, M)$ is the $\dot G$-extension of α, in the sense of Definition F.4.3.

PROOF. By 14.2.1.2, $M_c = C_G(z)$. By 14.2.16, $Q = O_2(K) = O_2(M_c)$, so M_c/Q is faithful on $\tilde{Q}$ by A.1.8. Now $K \in \mathcal{L}(M_c, T)$ with $K/O_2(K) \cong L_2(8)$, and $T/Q \cong E_8$ is Sylow in M_c/Q, so we conclude from 1.2.4 and A.3.12 that $K \in \mathcal{C}(M_c)$. Then $K \trianglelefteq M_c$ by 1.2.1.3 since $T \leq K$. As the normalizer in $GL(\tilde{Q})$ of K/Q is isomorphic to $Aut(K/Q)$, either (1) holds or $M_c/Q \cong Aut(L_2(8))$, and we may assume the latter. Thus $M_c = KA$ where $A \leq N_G(T)$ is of order 3 and induces field automorphisms on K/Q. Then A acts on $C_{\tilde{Q}}(T) = \tilde{V}$, so $A \leq N_G(V) = M$. As $M_c = KA$, $M \cap M_c = B_1 A$, so $M = Y(M \cap M_c) = YTA$. Then $\dot\alpha := (M_c, M \cap M_c, M)$ satisfies Hypothesis F.1.1 just as α did, and hence by F.1.9, $\dot\alpha$ is a weak BN-pair of rank 2. Then $\dot\alpha$ is an extension of its sub-amalgam α, which we have already identified; so $\dot\alpha$ is the $\dot G$-extension of α. $\qquad\square$

LEMMA 14.2.19. *If α is the ${}^3D_4(2)$ amalgam, then $G \cong {}^3D_4(2)$.*

PROOF. Let $\gamma := \alpha$ in case (1) of 14.2.18, and $\gamma := \dot\alpha$ in case (2) of 14.2.18. In either case, by 14.2.18, γ is an extension of the ${}^3D_4(2)$-amalgam, with the role of "G_1" played by $M_c = C_G(z)$. Thus the hypotheses of Theorem F.4.31 are satisfied since $G = O^2(G)$, so by that Theorem, G is an extension of ${}^3D_4(2)$ of odd degree, and hence isomorphic to ${}^3D_4(2)$ since G is simple. $\qquad\square$

Observe that 14.2.17 and 14.2.19 establish Theorem 14.2.7.

14.2.2. The treatment of certain cases where H is solvable. We next analyze the case where for some $H \in \mathcal{H}_*(T, M)$, $H/O_2(H)$ is either a group of Lie rank 1 over $\mathbf{F}_2$ isomorphic to $L_2(2)$ or $Sz(2) \cong F_{20}$, or $H/O_2(H) \cong D_{10}$. We do not treat the case where $H/O_2(H)$ is $U_3(2)$.

We prove:

THEOREM 14.2.20. *Let $H \in \mathcal{H}_*(T, M)$. Then*

(1) If $H/O_2(H) \cong D_{10}$ or $Sz(2)$, then $H/O_2(H) \cong Sz(2)$ and $G \cong {}^2F_4(2)'$.

(2) If $H/O_2(H) \cong L_2(2)$ then $G \cong M_{12}$ or $G_2(2)'$.

Again we assume that H satisfies one of the hypotheses of Theorem 14.2.20, and we begin a series of reductions.

Let $G_1 := H$, $G_2 := YT$, and $G_0 := \langle G_1, G_2 \rangle$. Then $G_1 \cap G_2 = T$. We check easily that G_1, G_2, T satisfy Hypothesis F.1.1 in the roles of "L_1, L_2, S": For example since $H \not\leq M$, $O_2(G_0) = 1$ by 14.2.2.3. By F.1.9, $\alpha := (G_1, T, G_2)$ is a weak BN-pair of rank 2. Set $K := O^2(H)$.

LEMMA 14.2.21. *$H = M_c$, $M = YT$, and one of the following holds:*

(1) α is the amalgam of ${}^2F_4(2)$ or of the Tits group ${}^2F_4(2)'$.

(2) α is the amalgam of M_{12} or of $\mathrm{Aut}(M_{12})$.

(3) α is the amalgam of $G_2(2)'$ or of $G_2(2)$.

PROOF. Since $T = N_{G_i}(T)$, the hypothesis of F.1.12 holds. Since $G_2/C_{G_2}(V) \cong L_2(2)$, while $G_1/O_2(G_1)$ is D_{10}, $Sz(2)$, or $L_2(2)$ with G_1 centralizing Z, we conclude from the list of F.1.12 that either α appears in conclusions (1)–(3) of 14.2.21, or α is the amalgam of $Sp_4(2)$. However in the latter case, $|Z| = 4$, contrary to 14.2.2.6.

Thus it remains to show $M_c = H$ and $M = YT$. If $M_c = H$, then $M \cap M_c = T$, so $C_M(V) = C_T(V)$, and then $M = YT$ by 14.2.2.1. So it suffices to show $M_c = H$.

Let $K_c := O^2(M_c)$. If $K = K_c$, then $H = KT = K_cT = M_c$, so we may assume $K < K_c$, and it remains to derive a contradiction.

Let $Q := O_2(M_c)$. Then $Q \leq Q_H$ by A.1.6, and $F^*(\tilde{M}_c) = \tilde{Q}$ by A.1.8, so

$$Z(\tilde{Q}_H) \leq C_{\tilde{M}_c}(\tilde{Q}) \leq \tilde{Q} \leq \tilde{Q}_H. \qquad (*)$$

Suppose first that α is the amalgam of $G_2(2)'$, $G_2(2)$, or M_{12}. Then $\tilde{Q}_H$ is abelian, so $Q_H = Q$ by $(*)$. Hence if α is the $G_2(2)'$-amalgam, then $Q \cong Q_8 * \mathbf{Z}_4$ is the central product of Q_8 and $\mathbf{Z}_4$. Therefore $O^2(\mathrm{Aut}(Q)) \cong A_4 \cong \mathrm{Aut}_K(Q)$, so $K = K_c$, contrary to our assumption. Hence α is the amalgam of $G_2(2)$ or M_{12}, so $Q = Q_H \cong Q_8^2$, and hence $\mathrm{Out}(Q) \cong O_4^+(2)$. Then as $K < K_c$, $K_c \cong SL_2(3) * SL_2(3)$. Next $YT \cong D_{12}/\mathbf{Z}_4^2$, so $V \leq E \leq Q$, where $E_8 \cong E \trianglelefteq YT$. Hence $N_G(E) \leq M$ by 14.2.2.3. But $\tilde{E}$ is a maximal totally singular subspace of $\tilde{Q}$, so from the structure of K_c, $S_4 \cong \mathrm{Aut}_{M_c}(E)$ is the stabilizer in $GL(E)$ of z. Then since Y does not centralize Z, $\mathrm{Aut}_G(E) \cong L_3(2)$, contradicting $N_G(E) \leq M = N_G(Y)$.

Assume next that α is the $\mathrm{Aut}(M_{12})$-amalgam. Then $Q_K := [Q_H, K] \cong Q_8^2$ and $\tilde{Q}_H \cong E_4 \, \mathrm{wr} \, \mathbf{Z}_2$. Therefore we conclude from $(*)$ that either Q is Q_H or Q_K, or else $\tilde{Q} \cong E_8$ is the maximal abelian subgroup of $\tilde{Q}_H$ distinct from $\tilde{Q}_K$.

Assume this last case holds. Then $Q \cong E_{16}$ and $S_4 \cong H/Q \leq M_c/Q$, with M_c/Q contained in the stabilizer $L_3(2)/E_8$ in $GL(Q)$ of the point Z of Q. Further $T/Q \cong D_8$ is Sylow in M_c/Q, so as $K < K_c$, we conclude $M_c/Q \cong L_3(2)$ acts indecomposably on Q. But then $M_c \in \mathcal{L}_f(G, T)$, contrary to 14.2.1.1.

So $Q = Q_H$ or Q_K, and therefore $\tilde{Q}_K = J(\tilde{Q}) \trianglelefteq \tilde{M}_c$. In either case, $Q_8^2 \cong Q_K \trianglelefteq M_c$. Then as above, $K < K_c$ implies $K_c \cong SL_2(3) * SL_2(3)$. Now from the structure of $\mathrm{Aut}(M_{12})$, $J(T) \cong E_{16}$ is normal in YT, so $N_G(J(T)) \leq M = !\mathcal{M}(YT)$. But $N_{K_c}(J(T))$ does not act on V, a contradiction.

Thus it remains to deal with the case where α is the $^2F_4(2)$-amalgam or the Tits amalgam. The subgroups G_1 and G_2 are described in section 3 of [**Asc82b**]. In particular $E := [Q_H, Q_H] \cong E_{32}$, and $Z(\tilde{Q}_H) = \tilde{F}$, where $F := C_H(E)$. Further $F = E$ if α the Tits amalgam, while if α is the $^2F_4(2)$ amalgam, then $F = \langle v_5 \rangle E$ with $\langle v_5 \rangle := C_{Q_H}(K) \cong \mathbf{Z}_4$. In particular $F \leq Q$ by (*). Next H is irreducible on Q_H/F of rank 4, so $Q = F$ or Q_H. In the former case, F and $E = \Omega_1(F)$ are normal in M; in the latter, $E = [Q, Q]$ and $F = C_Q(E)$ are normal in M.

Now $H/F \leq M_c/F$, with M_c/F contained in the stabilizer $\Lambda \cong L_4(2)/E_{16}$ of Z in $GL(E)$, and $H/F \cong Sz(2)/E_{16}$ or D_{10}/E_{16} contains a Sylow 2-group T/F of M_c/F, with $Q_H/F = O_2(\Lambda)$. Thus $Q_H \trianglelefteq M_c$, so $Q = Q_H$ using (*). Further the Sylow 2-group T/Q of M_c/Q is cyclic, so by Cyclic Sylow-2 Subgroups A.1.38, M_c/Q is 2-nilpotent. Therefore $K_c/Q = O(M_c/Q)$ is of odd order and contains $K/Q \cong \mathbf{Z}_5$; then as $K < K_c$, $K_c/Q \cong \mathbf{Z}_{15}$ from the structure of $L_4(2)$. But by 3.2.11 in [**Asc82b**], H is transitive on the involutions in $Q - F$, so if j is such an involution, then $M_c = HC_{M_c}(j)$ by a Frattini Argument. In particular, j centralizes an element of order 3 in M_c, impossible as K_c/Q of order 15 is regular on $(Q/F)^\#$. This completes the proof of 14.2.21. $\qquad\square$

By 14.2.21, α is isomorphic to the amalgam of $\hat{G}$, where $\hat{G}$ is $^2F_4(2)$, the Tits group $^2F_4(2)'$, $G_2(2)$, $G_2(2)' \cong U_3(3)$, M_{12}, or $Aut(M_{12})$. As G and $\hat{G}$ are both faithful completions of the amalgam α, there exist injections $\beta_J : \hat{G}_J \to G_J$ of the parabolics $\hat{G}_J, G_J$ for each $\emptyset \neq J \subseteq \{1, 2\}$, such that $\beta_{1,2}$ is the restriction of β_i to $\hat{G}_{1,2}$ and $\beta_i(\hat{G}_i) = G_i$ for $i = 1, 2$. We abuse notation and write β for each of the maps β_J. Let $\hat{T} := \beta^{-1}(T)$.

LEMMA 14.2.22. *(1) α is not the amalgam of $^2F_4(2)$, $G_2(2)$, or $Aut(M_{12})$.*
(2) If α is the amalgam of $G_2(2)'$, M_{12}, or $^2F_4(2)'$, then $G \cong \hat{G}$.

PROOF. First if α is of type $^2F_4(2)'$, $^2F_4(2)$, or $G_2(2)$, then $G_1 = H = M_c = C_G(Z)$ by 14.2.21, so that the hypotheses of Theorem F.4.31 are satisfied. Then $G \cong \hat{G}$ by F.4.31, and hence as G is simple, α is the amalgam of $^2F_4(2)'$ and $G \cong {}^2F_4(2)'$, so that (2) holds.

Thus we may assume that α is of type $G_2(2)'$, M_{12}, or $Aut(M_{12})$.

Suppose first that α is of type $Aut(M_{12})$. Let $R := \beta(\hat{T} \cap O^2(\hat{G}))$. Then $J(T) \cong E_{16}$ is normal in YT and $M = YT$ by 14.2.21, so M controls fusion in $J(T)$ by Burnside's Fusion Lemma A.1.35. Thus for $j \in J(T) - R$, $j^G \cap J(T) \cap R = \emptyset$. But each involution in R is fused into $J(T) \cap R$ under $G_1 \cup G_2$, so $j^G \cap R = \emptyset$, and hence $j \notin O^2(G)$ by Thompson Transfer, contrary to the simplicity of G.

In the remaining cases we appeal to existing classification theorems stated in Volume I: If α is of type M_{12}, then $G \cong M_{12}$ by I.4.6, and if α is of type $G_2(2)'$, then $G \cong G_2(2)'$ by I.4.4. $\qquad\square$

Notice 14.2.21 and 14.2.22 establish Theorem 14.2.20.

14.3. First steps; reducing $\langle V^{G_1} \rangle$ nonabelian to extraspecial

As mentioned at the beginning of the chapter, the work of the previous two sections allows us to treat the most important subcase of the case $\mathcal{L}_f(G, T) = \emptyset$ where $M_f/C_{M_f}(V(M_f)) \cong L_2(2)$ in parallel with the final case $L/O_2(L) \cong L_3(2)$ in the Fundamental Setup (3.2.1). As usual we define an appropriate hypothesis, which excludes the quasithin examples characterized in earlier sections.

Thus in this section, and indeed for the remainder of the chapter, we assume:

HYPOTHESIS 14.3.1. *Either*

(1) Hypothesis 13.3.1 holds with $L/O_2(L) \cong L_3(2)$, and G is not $Sp_6(2)$ or $U_4(3)$; or

(2) Hypothesis 14.2.1 holds, and G is not J_2, J_3, $^3D_4(2)$, the Tits group $^2F_4(2)'$, $G_2(2)' \cong U_3(3)$, or M_{12}.

Observe that in case (1) of Hypothesis 14.3.1, parts (4) and (5) of 13.3.2 say that Hypotheses 13.1.1, 12.2.1, and 12.2.3 are satisfied, and 13.3.1 is satisfied for any $K \in \mathcal{L}_f(G, T)$ with $K/O_2(K) \cong L_3(2)$. Thus we may make use of appropriate results from the previous chapters 12 and 13, including (in view of the exclusions in 14.3.1.1) results depending on Hypotheses 13.5.1 and 13.7.1. Similarly the exclusions in case (2) allow us to make use of results from the previous section 14.2.

As usual, we let $Z := \Omega_1(Z(T))$, $M_V := N_M(V)$, and $\bar{M}_V := M_V/C_M(V)$.

NOTATION 14.3.2. In case (1) of 14.3.1, L is the member of $\mathcal{L}_f^*(G, T)$ appearing in Hypothesis 13.3.1, while in case (2), take $L := O^2(\langle O_2(M \cap M_c)^M \rangle)$. (Thus L plays the role of the group "Y" in section 14.2.)

Observe:

LEMMA 14.3.3. *(1) $L \trianglelefteq M$.*
(2) $M = !\mathcal{M}(LT)$.
(3) $N_G(T) \leq M$, and each $H \in \mathcal{H}_*(T, M)$ is a minimal parabolic described in B.6.8, and in E.2.2 when H is nonsolvable.
(4) V is a TI-set in M.
(5) $N_G(V) = M_V$.
(6) If $H \leq N_G(U)$ for some $1 \neq U \leq V$, then $H \cap M = N_H(V)$.

PROOF. Part (1) follows from 13.3.2.2 in case (1) of 14.3.1, and by construction in case (2). Part (2) follows from 1.2.7.3 or 14.2.2.3, and (3) follows either from Theorem 3.3.1 together with 3.3.2.4, or from parts (7) and (8) of 14.2.2. Further (5) follows from (2); and (4) follows by construction of $M = N_G(V)$ in case (2) of Hypothesis 14.3.1, and from 12.2.2.3 in case (1). Finally as in the proof of 12.2.6, (6) follows from (4) using 3.1.4.1. $\square$

We typically distinguish the two cases of Hypothesis 14.3.1 by writing $L/O_2(L) \cong L_3(2)$ or $L_2(2)'$.

14.3.1. Preliminary results under Hypothesis 14.3.1.

LEMMA 14.3.4. *If there exists $K \in \mathcal{L}_f(G, T)$, then*

(1) $K/O_2(K) \cong A_5$ or $L_3(2)$.
(2) $K \trianglelefteq KT$ and $K \in \mathcal{L}^*(G, T)$.
(3) Each $V_K \in Irr_+(K, R_2(KT), T)$ is T-invariant, K, V_K satisfies the FSU, and V_K is the natural module for $K/O_2(K) \cong A_5$ or $L_3(2)$.
(4) Case (1) of 14.3.1 holds, so that $L/O_2(L) \cong L_3(2)$.

PROOF. First case (1) of 14.3.1 must hold, since in case (2), $\mathcal{L}_f(G, T) = \emptyset$ by Hypothesis 14.2.1.1. In particular, (4) holds. Further 14.3.1.1 excludes $G \cong Sp_6(2)$ or $U_4(3)$, so $K/O_{2,Z}(K)$ is not A_6 by Theorem 13.8.1. Also we saw Hypothesis 13.5.1 holds, so (1)–(3) follow from 13.5.2. $\square$

LEMMA 14.3.5. *Assume* $L/O_2(L) \cong L_2(2)'$ *and* $H \in \mathcal{H}(T)$ *with* $|H : T| = 3$ *or* 5. *Then* $H \leq M$.

PROOF. Assume $H \not\leq M$. By 14.3.3.3, $H \not\leq N_G(T)$ so that $H/O_2(H) \cong S_3$, D_{10}, or $Sz(2)$. But the groups G appearing as conclusions in Theorem 14.2.20 are excluded by Hypothesis 14.3.1.2, so we conclude that the lemma holds. $\qquad\square$

LEMMA 14.3.6. *Assume* $L/O_2(L) \cong L_2(2)'$ *and* $H \in \mathcal{H}(T, M)$ *such that* $K := O^2(H) = \langle K_1^T \rangle$ *for some* $K_1 \in \mathcal{L}(G, T)$. *Then*

(1) *If* $K/O_2(K)$ *is of Lie type over* $\mathbf{F}_{2^n}$ *of Lie rank 1 or 2, then either*

(i) $n = 1$, $K/O_2(K) \cong L_3(2)$ *or* A_6, *and* T *is nontrivial on the Dynkin diagram of* $K/O_2(K)$, *or*

(ii) M *does not contain the Borel subgroup of* K *over* $T \cap K$.

(2) *If* $K/O_2(K)$ *is of Lie type over* $\mathbf{F}_2$ *of Lie rank 2, then* $K/O_2(K) \cong L_3(2)$ *or* A_6, *and* T *is nontrivial on the Dynkin diagram of* $K/O_2(K)$.

(3) *If* $K/O_2(K)$ *is of Lie type over* $\mathbf{F}_4$, *then* $KT/O_2(KT) \cong Aut(Sp_4(4))$ *or* S_5 *wr* $\mathbf{Z}_2$.

(4) *If* $K/O_2(K) \cong L_4(2)$ *or* $L_5(2)$, *then* T *is nontrivial on the Dynkin diagram of* $K/O_2(K)$.

(5) $K/O_2(K)$ *is not* A_7.

(6) $K/O_2(K)$ *is not* M_{12}, M_{22}, *or* $\hat{M}_{22}$.

PROOF. Assume that K either satisfies the hypotheses of one of (1)–(4) or is a counterexample to (5) or (6). Then $K/O_2(K)$ is either quasisimple, or else semisimple of Lie type in characteristic 2, and of Lie rank 1 or 2 using Theorem C (A.2.3). Thus as $K = \langle K_1^T \rangle$ with $K \in \mathcal{L}(G, T)$, using 1.2.1.3 we conclude that either $K/O_2(K)$ is quasisimple, or K is the product of two T-conjugates of $K_1 < K$ with $K_1/O_2(K_1) \cong L_2(2^n)$ or $Sz(2^n)$ and $n > 1$.

Assume the hypotheses of (1). We may assume that (ii) fails, so that $M \cap K$ contains the Borel subgroup B of K over $T \cap K$. Let $\mathcal{H}_0$ be the set of subgroups $\langle P, T \rangle$, such that P is a rank one parabolic of K over B. Then $H = \langle \mathcal{H}_0 \rangle$. So as $H \not\leq M$, there exists $H_0 \in \mathcal{H}_0$ with $H_0 \not\leq M$. Then $H_0 = H_2 B$ where $H_2 \in \mathcal{H}_*(T, M)$. Since Hypothesis 14.3.1 excludes the groups in Theorem 14.2.7, we conclude that $n(H_2) = 1$. Hence $K/O_2(K)$ is defined over $\mathbf{F}_2$. Then from the first paragraph, $K/O_2(K)$ is quasisimple. If T is trivial on the Dynkin diagram of K, then H_0 is a rank one parabolic, so as $K/O_2(K)$ is quasisimple and defined over $\mathbf{F}_2$, $|H_2 : T| = 3$ or 5 from the list of such groups $K/O_2(K)$ in Theorem C, contrary to 14.3.5. Thus T is nontrivial on the diagram, so again from that list, conclusion (i) of (1) holds. This completes the proof of (1).

If (2) fails, then conclusion (ii) of (1) must hold, so $B \not\leq M$. In particular, a Cartan subgroup of B is nontrivial, so as $K/O_2(K)$ is defined over $\mathbf{F}_2$, we conclude from the list of Theorem C that $K/O_2(K) \cong {}^3D_4(2)$ and $|B : T \cap K| = 7$. Now $B \leq N_G(T)$ since $Out(K/O_2(K))$ is of odd order, so $B \leq M$ by 14.3.3.3, contrary to the first sentence of this paragraph. Thus (2) is established.

Assume the hypotheses of (3); then by the first paragraph, $K/O_2(K)$ is either quasisimple of Lie rank at most 2, or $L_2(4) \times L_2(4)$. Let B be the T-invariant Borel subgroup of K. By (1), $B \not\leq M$, so there exists $H_2 \in \mathcal{H}_*(T, M)$ with $H_2 \leq BT$. Inspecting the groups in Theorem C defined over $\mathbf{F}_4$, either $B/O_2(B) \cong \mathbf{Z}_3$ or E_9; or $K/O_2(K) \cong U_3(4)$ with $B/O_2(B) \cong \mathbf{Z}_{15}$; or $K/O_2(K) \cong {}^3D_4(4)$ with $B/O_2(B) \cong$

$\mathbf{Z}_3 \times \mathbf{Z}_{63}$. By 14.3.5, any subgroup of order 3 or 5 permuting with T is contained in M, so as $H_2 \le BT$ but $B \not\le M$, we conclude that either $K/O_2(K) \cong {}^3D_4(4)$ with $(B \cap M)/O_2(B \cap M) \cong E_9$, or $B/O_2(B) \cong E_9$ and T is irreducible on $B/O_2(B)$. In the latter case, the irreducible action of T implies that (3) holds. In the former, $m_3(K \cap M) = 2$. However by 14.2.2.5, $K \le M_c$, so $O^2(K \cap M) \le C_M(V)$ by Coprime Action, whereas $m_3(C_M(V)) \le 1$ by 14.2.2.4. This completes the proof of (3).

Finally suppose $K/O_2(K)$ is one of the groups in (4)–(6), and T is trivial on the Dynkin diagram of $K/O_2(K)$ in (4). Then in each case H is generated by the set $\mathcal{H}_1$ of T-invariant subgroups H_2 with $H_2/O_2(H_2) \cong L_2(2)$. Thus $H \le M$ by 14.3.5, completing the proof of 14.3.6. $\qquad\square$

Next recall from our discussion at the beginning of the section that in case (1) of Hypothesis 14.3.1, Hypotheses 12.2.3 and 13.3.1 hold, so case (1) of Hypothesis 12.8.1 holds. Further by 14.2.4, case (2) of Hypothesis 12.8.1 holds in case (2) of Hypothesis 14.3.1. Thus we can appeal to the results in section 12.8, and we adopt Notation 12.8.2 from that section. In particular V_i is the T-invariant subspace of V of dimension i for $i \le \dim(V)$, $G_i := N_G(V_i)$, $L_i := O^2(N_L(V_i))$, $R_i := O_2(L_iT)$, etc.

Notice $V_1 = Z \cap V$, and indeed in case (2) of 14.3.1, $V_1 = Z$ by 14.2.1.4, and so $G_1 = M_c$ by 14.2.1.2. Recall $\tilde{G}_1 := G_1/V_1$, and by 12.8.3.4,

$$G_1 \not\le M, \text{ so } G_1 \in \mathcal{H}(T,M).$$

Observe since LT induces $GL(V)$ on V that:

LEMMA 14.3.7. $M_V = LC_M(V) = L(M \cap G_1)$. In particular if $M \cap G_1 = L_1T$ and $V \trianglelefteq M$, then $M = LT$.

LEMMA 14.3.8. Assume $L/O_2(L) \cong L_3(2)$. If $H \le G_1$ with $HL_i = L_iH$ for $i = 1, 2$, then $H \le M$.

PROOF. First $V_1^{L_2H} = V_1^{HL_2} = V_1^{L_2}$, so H acts on $\langle V_1^{L_2} \rangle = V_2$. Similarly $V_2^{L_1H} = V_2^{HL_1} = V_2^{L_1}$, so H acts on $\langle V_2^{L_1} \rangle = V$, so $H \le N_G(V) \le M$ by 14.3.3.5. $\qquad\square$

LEMMA 14.3.9. Assume $L/O_2(L) \cong L_3(2)$. Then

(1) If $J(R_1) \not\le O_2(LT)$ then there exists $A \in \mathcal{A}(R_1)$ and $g_i \in L$ with $A^{g_i} \le T$ but $A^{g_i} \not\le R_i$ for $i = 1, 2$.

(2) If $J(T) \le R_1$ then $J(T) \trianglelefteq LT$.

(3) If $J(T) \not\le O_2(LT)$ then $J_1(T) \not\le R_i$ for $i = 1, 2$.

PROOF. Notice (1) implies (2): For if $J(T) \le R_1$, then $J(T) = J(R_1)$ by B.2.3.3, so $J(R_1) \le O_2(LT)$ assuming (1), and hence $J(T) = J(R_1) = J(O_2(LT)) \trianglelefteq LT$.

Assume $J(R_1) \not\le O_2(LT)$. Then there is $A \in \mathcal{A}(R_1)$ with $\bar{A} \ne 1$, and either $\bar{A}$ has rank 1, or $\bar{A} = \bar{R}_1$ has rank 2. Since $\bar{R}_1$ is not a strong FF*-offender on V, in the latter case B.2.9.2 says we may make a new choice of A so that $\bar{A}$ has rank 1. Then there exists g_i as claimed. Thus (1) and hence (2) are established, so it remains to prove (3).

Assume the hypothesis of (3), so there is $D \in \mathcal{A}(T)$ with $\bar{D} \ne 1$. Now as $m(\bar{D}) \le 2$, we may choose B of index at most 2 in D, with $C_D(V) \le B$ and $\bar{B}$ of rank 1. Thus for either choice of $i = 1, 2$, there exists $g_i \in L$ with $B^{g_i} \leq T$ but $B^{g_i} \not\leq R_i$. Hence (3) holds. $\square$

14.3.2. Preliminary results for the case $\langle V^{G_1} \rangle$ is nonabelian. When $\langle V^{G_1} \rangle$ is nonabelian, we will concentrate on G_1, as opposed to an arbitrary member of $\mathcal{H}_z$; recall the latter set was defined in Notation 12.8.2.3. Thus in the remainder of this section, and indeed in the subsequent section 14.4, we assume:

HYPOTHESIS 14.3.10. *Assume Hypothesis 14.3.1 with $U := \langle V^{G_1} \rangle$ nonabelian. Take $H := G_1$.*

Observe that U plays the role of "U_H" in Notation 12.8.2; in particular by 12.8.4.2, $\tilde{U}$ is elementary abelian.

Since U is nonabelian, we also adopt the notation of the second subsection of section 12.8. Since $H \not\leq M$, $V < U$. Write $Q := O_2(H)$, rather than Q_H as in section 12.8, set $H^* := H/Q$, $Z_U := Z(U)$, $\hat{H} := H/Z_U$, $\dot{H} := H/C_H(\hat{U})$, pick $g \in N_L(V_2) - H$, let $I_2 := \langle U^{L_2} \rangle$, $W := C_U(V_2)$, and $E := W \cap W^g$. Let $d := m(\hat{U})$.

By 12.8.8.1, $U = U_0 Z_U$ with U_0 extraspecial and $\Phi(U_0) = V_1$, and $\dot{H}$ preserves a symplectic form on $\hat{U}$ of dimension d. By 12.8.12, this action satisfies Hypothesis G.10.1, with $\dot{H}$, $\hat{U}$, $\hat{V}_2$, $\hat{E}$, $\dot{W}^g$, $\dot{Z}_U^g$ in the roles of "G, V, V_1, W, X, X_0", and Hypothesis G.11.1 is also satisfied. Thus we may make use of results from sections G.10 and G.11. Recall also from G.10.2 that the bound (*) of sections G.7 and G.9 holds, so that we may apply the results of section G.9.

By 12.8.8.3:

LEMMA 14.3.11. $m(\hat{V}) = m(\tilde{V})$.

LEMMA 14.3.12. *Assume $m(\dot{W}^g) \leq d/2 - 1$. Then*

(1) $m(\dot{W}^g) = d/2 - 1$.

(2) $m(\hat{E}) = d/2$.

(3) $Z_U = V_1$, so U is extraspecial, $\hat{U} = \tilde{U}$, and $\dot{H} = H^*$.

(4) H preserves a quadratic form on $\tilde{U}$ of maximal Witt index in which $\hat{E}$ is totally singular.

PROOF. As $m(\dot{W}^g) \leq d/2 - 1$, the first inequality in G.10.2 is an equality with $\dot{Z}_U^g = 1$. Thus (1) holds. Then (1) and 12.8.11.5 imply (2). As $\dot{Z}_U^g = 1$, $Z_U = V_1$ by 12.8.13.4. Thus U is extraspecial by 12.8.8.1, so $\hat{U} = \tilde{U}$. By 12.8.4.4, $Q = C_H(\tilde{U})$, so $H^* = \dot{H}$. Thus (3) holds. Also $\Phi(Z_U) = \Phi(V_1) = 1$, so by 12.8.8.2, H preserves a quadratic form $q(\tilde{u}) := u^2$ on $\tilde{U}$. By 12.8.11.2, $\Phi(E) = 1$, so $\hat{E}$ is a totally singular subspace of the orthogonal space $\tilde{U}$, of rank $d/2$ by (2). Thus $\tilde{U}$ is of maximal Witt index, completing the proof of (4). $\square$

LEMMA 14.3.13. *$\dot{H}$ and its action on $\hat{U}$ satisfy one of the conclusions of Theorem G.11.2, but not conclusion (1), (4), (5), or (12).*

PROOF. By 12.8.12.4, $\dot{H}$ and its action on $\hat{U}$ satisfy one of the conclusions of G.11.2. By (6) and (7) of 12.8.13, conclusions (4) and (12) are not satisfied. If conclusion (5) is satisfied, then by 12.8.13.5, there is $K \in \mathcal{L}_f(G, T)$ with $K/O_2(K) \cong A_8$, contrary to 14.3.4.1.

Assume conclusion (1) is satisfied. Then $d = 2$ and $\dot{H} \cong S_3$. By 14.3.11, $m(\hat{V}) = m(\tilde{V})$, so if $L/O_2(L) \cong L_3(2)$, then $m(\hat{U}) = 2 = m(\hat{V})$, and hence

$U = V Z_U$, contradicting U nonabelian. Thus $L/O_2(L) \cong L_2(2)'$. Here by 12.8.13.3, $m(\tilde{Z}_U) = m(\dot{Z}_U^g) \le m_2(\dot{H}) = 1$, so by Coprime Action, $O^2(C_H(\hat{U}))$ centralizes $\tilde{U}$, and then by (2) and (4) of 12.8.4, $C_H(\hat{U}) = C_H(\tilde{U}) = Q$. Then $H^* \cong \dot{H} \cong S_3$, so that $|H : T| = 3$, and then 14.3.5 contradicts $H \nleq M$. $\qquad\square$

LEMMA 14.3.14. *One of the following holds:*

(1) $m(\dot{W}^g) = d/2 - 1$, *so that the conclusions of 14.3.12 hold.*

(2) $d = 4$ *and* $m(\dot{W}^g) \ge 2$. *Further $\dot{H}$ contains A_5 or $S_3 \times S_3$.*

(3) $d = 6$, $\dot{H} \cong G_2(2)$, *and* $m(\dot{W}^g) = 3$.

PROOF. If $m(\dot{W}^g) \le d/2 - 1$, then (1) holds by 14.3.12. Thus we may assume $m(\dot{W}^g) \ge d/2$. But by 14.3.13, $\dot{H}$ and $\hat{U}$ appear in one of the cases of G.11.2 other than (1), (4), (5), and (12). Thus as $m_2(\dot{H}) \ge d/2$, case (2), (6), or (7) of G.11.2 holds. Case (6) of G.11.2 gives conclusion (3), and case (2) gives conclusion (2) as $m_2(\dot{H}) \ge 2$ and $\dot{H}$ is a subgroup of $Sp_4(2)$ whose order is divisible by 10 or 18. Finally in case (7) of G.11.2, $\dot{W}^g \nleq E(\dot{H})$, so $m(\dot{W}^g) \le 3 < d/2$, contrary to assumption. $\qquad\square$

LEMMA 14.3.15. *Assume* $L/O_2(L) \cong L_2(2)'$. *Then* $U \cong Q_8^2$ *and* $H^* \cong O_4^+(2)$ *with $\tilde{E}$ totally singular.*

PROOF. Suppose first that $\dot{H}$ is not solvable. Then appealing to 14.3.13, and inspecting the list of G.11.2, there exists a component $\dot{K}_1$ of $\dot{H}$ isomorphic to $L_2(4)$, A_6, $G_2(2)'$, A_7, $L_2(8)$, or $\hat{M}_{22}$. By 1.2.1.4 we may choose $K \in \mathcal{L}(G, T)$ with $K/O_2(K)$ quasisimple and $\dot{K} = \dot{K}_1$, although K may not be in $\mathcal{C}(H)$; set $K_0 := \langle K^T \rangle$. From G.11.2, either $K = K_0$, or conclusion (7) of G.11.2 holds and $K_0/O_2(K_0) \cong \Omega_4^+(4)$. Further if $\dot{K} \cong A_6$, then from G.11.2, T is trivial on the Dynkin diagram of $K/O_2(K)$. Finally if $\dot{K}_0 \cong \Omega_4^+(4)$, then $K_0T/O_2(K_0T)$ is not S_5 wr $\mathbf{Z}_2$ since $N_{Sp(\hat{U})}(\dot{K}_0)$ is a proper subgroup of index 2 in S_5 wr $\mathbf{Z}_2$. We conclude using 14.3.6 that $K = K_0 \cong L_2(8)$, and $K \cap M = T$. However $Out(L_2(8))$ is of odd order, so $N_K(T)$ is a Borel subgroup of K. Then as $N_G(T) \le M$ by 14.3.3.3, $K \cap M > T$, contrary to the previous remark.

This contradiction shows that $\dot{H}$ is solvable. Thus in view of 14.3.13, $\dot{H}$ and its action on $\hat{U}$ are described in conclusion (2) or (3) of G.11.2. Indeed $\dot{H}$ and $\hat{U}$ are described in Theorem G.9.4 if H is irreducible on $\hat{U}$, and in G.10.5.2 if H is not irreducible on $\hat{U}$.

Suppose first that $V_1 = Z_U$. Then arguing as in the proof of (3) and (4) of 14.3.12, U is extraspecial with $\hat{U} = \tilde{U}$, and $\dot{H} = H^*$ preserves the quadratic form on $\tilde{U}$ in which $\tilde{E}$ is totally singular. In particular if $d = 4$ and $\dot{H}$ has order divisible by 9, then as 9 does not divide $|O_4^-(2)|$, $U \cong Q_8^2$ and so $\dot{H} = H^*$ lies in $O_4^+(2)$; further by 12.8.9.2, W^g/E is a natural $L_2(2)$-module, so that $\dot{W}^g = W^{*g}$ has rank 2. So since $H \nleq M$, 14.3.5 reduces cases (1)–(4) of G.9.4 and G.10.5.2 to $H^* \cong O_4^+(2)$, so that the lemma holds. Otherwise we have case (5) of G.9.4, with H^* a subgroup of $SD_{16}/3^{1+2}$ acting irreducibly on $O_3(H^*)/Z(O_3(H^*))$. Let X be the preimage in H of $Z(O_3(H^*))$; again $X \le M$ by 14.3.5. This is impossible, since $\tilde{U} = [\tilde{U}, X]$ in G.9.4.5, so that X does not act on the subspace $\tilde{V}$ of rank 1.

Thus $V_1 < Z_U$. Hence by 14.3.12.3, $m(\dot{W}^g) \ge d/2$, so case (2) of 14.3.14 holds as $\dot{H}$ is solvable; that is, $d = 4$, $m(\dot{W}^g) \ge 2$, and $\dot{H}$ contains $S_3 \times S_3$. It follows that $\dot{H} \cong S_3 \times S_3$ or $O_4^+(2)$, and $m(\dot{W}^g) = 2 = m_2(O_4^+(2))$.

Next by 12.8.13.3, $\dot{Z}_U^g \cong \tilde{Z}_U$, so $\dot{Z}_U^g \neq 1$. Let $K := O^2(\langle Z_U^{gH} \rangle)$. By 12.8.13.3, Z_U^g centralizes Z_U, so K centralizes Z_U. Thus using Coprime Action, $O^2(C_K(\hat{U}))$ centralizes $\tilde{U}$, and hence $1 \neq \dot{K} \cong K^*$ by parts (2) and (4) of 12.8.4. So if $\dot{H} \cong S_3 \times S_3$ then $K \leq M_V$ by 14.3.5 and 14.3.3.6, so K centralizes $\hat{V}$ of order 2. But then as $K \trianglelefteq H$, K centralizes $\hat{U} = \langle \hat{V}^H \rangle$, contrary to $\dot{K} \neq 1$.

Therefore $\dot{H} \cong O_4^+(2)$. By 12.8.12.2, $\dot{Z}_U^g \trianglelefteq \dot{T}$, so $O_3(\dot{H}) = [O_3(\dot{H}), \dot{Z}_U^g]$, and hence $O_3(\dot{H}) \leq \dot{K}$. Thus $K^* \cong E_9$, so $\hat{U} = [\hat{U}, K]$. Then as K centralizes Z_U, $\tilde{U} = [\tilde{U}, K] \oplus \tilde{Z}_U$ with $\tilde{Z}_U \neq 0$. Further as $\dot{Z}_U^g \trianglelefteq \dot{T}$ and $W \leq C_G(V) \leq C_G(Z_1^g)$, $[Z_U^g, W] \leq Z_U^g \cap W$. As $L/O_2(L) \cong L_2(2)'$, $Z(I_2) = 1$ by 12.8.13.3. Thus $Z_U^g \cap W = V_1^g$ by 12.8.10.3, so as $\dot{Z}_U^g$ acts nontrivially on the hyperplane $\hat{W}$ of $\hat{U}$ and centralizes $\tilde{Z}_U$,

$$\tilde{V} = \tilde{V}_1^g \leq [\tilde{W}, Z_U^g] \leq [\tilde{U}, K],$$

so $\tilde{U} = \langle \tilde{V}^H \rangle \leq [\tilde{U}, K]$, contradicting $0 \neq \tilde{Z}_U \not\leq [\tilde{U}, K]$. Thus the proof of 14.3.15 is complete. $\qquad \square$

14.3.3. Eliminating $\mathbf{L_2(2)}$ when $\langle \mathbf{V}^{\mathbf{G}_1} \rangle$ is nonabelian. Recall that $\langle V^{G_1} \rangle$ is nonabelian in the quasithin examples for $L/O_2(L) \cong L_2(2)'$ characterized in section 14.2; but of course those groups are now excluded in Hypothesis 14.3.1.

Thus in this subsection we prove:

THEOREM 14.3.16. *Assume Hypothesis 14.3.10. Then case (1) of Hypothesis 14.3.1 holds, namely $L/O_2(L) \cong L_3(2)$.*

REMARK 14.3.17. In proving Theorem 14.3.16, we will be dealing in effect only with the shadows of extensions of $U_4(3)$ which interchange the two classes of 2-locals isomorphic to A_6/E_{16}. These extensions satisfy our hypotheses except they are not simple, and sometimes not quasithin. Thus we construct 2-local subgroups which appear in those shadows, and eventually achieve a contradiction by showing $O^2(G) < G$ using transfer.

Until the proof of Theorem 14.3.16 is complete, assume G is a counterexample. Thus case (2) of 14.3.1 holds, so $V_1 = Z = \langle z \rangle$, $V = V_2$ and $G_2 = N_G(V) = M$. Recall $Q = O_2(H)$. By 14.3.15, $U \cong Q_8^2$, and $\tilde{U}$ has an orthogonal structure over $\mathbf{F}_2$ preserved by $H = G_1$, with $H^* = H/Q = O(\tilde{U}) \cong O_4^+(2)$ and $\tilde{E}$ totally singular. Thus H is a $\{2,3\}$-group, so in particular, H is solvable.

Recall $I_2 = \langle U^{L_2} \rangle = \langle U^L \rangle$, and by 12.8.9.1, $I_2 \trianglelefteq G_2 = M$ and $L = O^2(I_2)$.

LEMMA 14.3.18. *(1)* $V = Q \cap U^g$.

(2) $I_2 \trianglelefteq M$, $R := O_2(I_2) = O_2(L)$.

(3) R^* *is the 4-subgroup of T^* containing no transvections, and hence lying in $\Omega_4^+(\tilde{U})$; so $|T : RQ| = 2$.*

(4) $R = AA^t$, *where $A \cong E_{16}$ and A^t are the maximal elementary abelian subgroups of R, $|R| = 2^6$, $t \in T - RQ$, $V = A \cap A^t$, and $A \trianglelefteq I_2Q$.*

(5) $U = O_2(O^2(H))$.

(6) $N_H(A)^* = C_{H^*}(A^*) \cong \mathbf{Z}_2 \times S_3$.

PROOF. First I_2 plays the role of "I" in 12.8.8; then by 12.8.8.4 we may apply G.2.3.4 to conclude that $E = W \cap W^g$ is T-invariant. But we saw $\tilde{E}$ is totally singular and $H^* = O(\tilde{U}) \cong O_4^+(2)$, so T acts on no totally singular 2-subspace of

$\tilde{U}$; hence $\tilde{E}$ has rank 1, and so $V = E$. On the other hand, $U^g \cap Q \leq U^g \cap G_1 = W^g$, and $Q \cap W^g = E$ by 12.8.9.5. Thus (1) holds.

Recall $I_2 \trianglelefteq G_2 = M$ and $L = O^2(I_2)$. As $V = E$ by (1) and $|Q| = 2^5$, 12.8.9.2 says $R/V = W/V \oplus W^g/V$ is the sum of $m(W/V) = 2$ natural modules for $I_2/R \cong L_2(2)$. Therefore $R^* = W^{g*} \cong E_4$ and $R = [R, L] \leq L$, so that $R = O_2(L)$ as $L \trianglelefteq I_2$. Thus (2) holds. Recall Hypothesis G.10.1 is satisfied; then (3) follows from part (d) of G.10.1 and the fact that transvections in $O(\tilde{U})$ have nonsingular centers.

As R/V is the sum of two natural modules for I_2/R, R has order 2^6, and I_2 has three irreducibles $R(i)/V$, $1 \leq i \leq 3$, on R/V. As $W/V = C_{R/V}(U)$, each $R(i)$ contains some $r_i \in W - V$. Since $U \cong Q_8^2$, $W \cong \mathbf{Z}_2 \times D_8$. Thus we can choose notation so that $\langle r_i \rangle V \cong E_8$ for $i = 1$ and 2, and $\mathbf{Z}_4 \times \mathbf{Z}_2$ for $i = 3$. Then as I_2 is transitive on $(R(i)/V)^\#$, $R(1) \cong R(2) \cong E_{16}$ and $V = \Omega_1(R(3))$. It follows that $A := R(1) \cong E_{16}$ and $A' := R(2)$ are the maximal elementary abelian subgroups of R, $AA' = R$, and A^* is of order 2 in T^*, with $C_{\tilde{U}}(A^*) = \widetilde{A \cap U}$ a totally singular line. Thus $A^* \neq Z(T^*)$, so $A^t = A'$ for $t \in T - RQ$. Therefore (4) holds as A is I_2-invariant by construction, and $C_{H^*}(A^*) \cong \mathbf{Z}_2 \times S_3$.

Next $[W^g, Q] \leq W^g \cap Q = V$ using (1), so $O^2(H) = [O^2(H), W^g]$ centralizes Q/U, and hence (5) holds. Thus if H_A is the preimage of $C_{H^*}(A^*)$, $O^2(H_A)$ acts on AU and hence on $A = J(AU)$, completing the proof of (6). $\qquad \square$

From now on, let A be defined as in 14.3.18.4. We will show next that $A_6/E_{16} \leq N_G(A) \leq S_6/E_{32}$. Set $D := C_Q(U)$.

LEMMA 14.3.19. *Let $K := \langle O^2(N_H(A)), L \rangle$. Then*

(1) $Q = UD$.

(2) *Either*

   (i) $[A, D] = 1$ *with* $Aut_T(A) \cong D_8$, *or*

   (ii) D *induces the transvection on A with axis $A \cap U$ and center V_1.*

(3) $Aut_{RQ}(A) \in Syl_2(Aut_G(A))$, *and* $Aut_{RQ}(A) \cong D_8$ *or* $\mathbf{Z}_2 \times D_8$.

(4) K *is an A_6-block and $A = O_2(K)$.*

(5) $C_G(K) = 1$.

(6) $N_G(K) = KD$, *and D is a subgroup of D_8, with $D \cong D_8$ iff $|N_G(K) : K| = 4$ and $A < C_G(A)$.*

(7) $N_G(A) = N_G(K)$.

(8) $RU \in Syl_2(K)$.

(9) K *splits over A.*

(10) $Aut(K) = K\langle \alpha, \beta \rangle$, *with $A\langle \alpha \rangle = C_{Aut(K)}(A) \cong E_{32}$ the quotient of the permutation module for K/A modulo the fixed space of K/A, β is an involution inducing a transposition on a complement to A in K, and $D_8 \cong \langle \alpha, \beta \rangle = C_{Aut(K)}(U)$.*

PROOF. By 12.8.4.2, Q centralizes $\tilde{U}$, while as $U \cong Q_8^2$, $Inn(U) = C_{Aut(U)}(\tilde{U})$ by A.1.23, so (1) holds. Next D centralizes the hyperplane $A \cap U$ of A, and $[A, D] \leq C_A(U) = V_1$, so (2) holds.

As $A \cong E_{16}$, $Aut_K(A) \leq Aut_G(A) \leq GL(A) \cong L_4(2)$. As $Z = V_1 = C_A(U)$ and $RQ \in Syl_2(N_H(A))$, $Z = C_A(RQ)$, and hence $N_{N_G(A)}(RQ) \leq H$, so that $RQ = N_T(A) \in Syl_2(N_G(A))$. From 14.3.18.6, $Aut_{RU}(A) \cong D_8$ and $C_H(A)$ is a 2-group, so (2) implies (3), and as $Z \leq A$, $C_G(A) = C_H(A)$ is a 2-group.

By 14.3.18.4, $A \trianglelefteq I_2 Q$, so that as $L = O^2(I_2)$, $K \leq N_G(A)$. Indeed from 14.3.18, $Aut_{I_2}(A) \cong S_4$, $A = [A, L]$, and setting $Y_A := O^2(N_H(A))$, $Aut_{RY_A}(A) \cong S_4$ is of index at most 2 in the stabilizer in $Aut_G(A)$ of V_1. We conclude from the structure of $L_4(2)$ that $Aut_K(A)$ is A_6. Hence as $C_G(A)$ is a 2-group, and as $K = O^2(K)$ and K is $C_G(A)$-invariant by definition, it follows that $K = O^2(KC_G(A))$ and $K/O_2(K) \cong A_6$. Then as $[R, C_G(A)] \leq C_R(A) = A$, $K = [K, R]$ centralizes $C_G(A)/A$, so K is an A_6-block. Next $R = [R, L]$ and $U = [U, Y_A]$, so $RU \leq K$. Then as $R/A \cong E_4$ is elementary abelian, K/A does not involve the double cover of A_6, so $A = O_2(K)$, and $N_G(K) \leq N_G(A)$, completing the proof of (4). As $RU \leq K$ and $|RU| = 2^7 = |K|_2$, (8) is established. For $u \in U - R$ an involution, u acts on a complement B to V in A^t, so $B\langle u \rangle$ is a complement to A in RU, and hence (9) holds using Gaschütz's Theorem A.1.39. Let K_0 be a complement to A in K.

Let $J := Aut(K)$ and $A_0 := C_J(A)$. By (9), with 17.2 and 17.6 in [**Asc86a**], A_0 is elementary abelian with $A_0/A \cong H^1(K_0, A)$. Hence $A_0 \cong E_{32}$ by I.1.6.1. Further by 17.7 in [**Asc86a**] and a Frattini Argument, $J = A_0 J_0$, where $J_0 := N_J(K_0)$, and of course J_0 is the subgroup of $Aut(K_0)$ stabilizing the representation of K_0 on A, so $J_0 \cong S_6$. Thus (10) holds.

Recall $N_T(A) \in Syl_2(N_G(A))$, so $N_T(A) \in Syl_2(N_G(K))$. As $L = O^2(P)$ where P is the minimal parabolic of K with $A = [A, P]$, $C_T(K) = C_T(L)$ from the structure of $Aut(K)$ described in (10). Further $C_T(L) = 1$, since $Z \cong \mathbf{Z}_2$ by 14.2.2.6, while Z is not centralized by L. Thus $C_T(K) = 1$, so (5) holds since $C_G(K) \leq C_H(A)$ and we saw $C_H(A)$ is a 2-group.

As $N_G(K) \leq N_G(A)$ with K transitive on $A^\#$, by a Frattini Argument, $N_G(K) = KN_H(A) = KQR$. Thus $N_G(K) = KD$ by (1) and (8). Now (6) follows from (10).

As $N_T(A)$ acts on K and is Sylow in $N_G(A)$, $K \leq K_1 \in \mathcal{C}(N_G(A))$. Then we conclude from the structure of $GL_4(2)$ and A.3.12 that either $K = K_1$ or $Aut_{K_1}(A) \cong A_7$, and the latter case is impossible as we saw $Aut_H(A)$ is solvable. Thus $K \trianglelefteq N_G(A)$ by 1.2.1.3, so (7) holds as we saw $N_G(K) \leq N_G(A)$.   □

LEMMA 14.3.20. $A = C_G(A)$.

PROOF. Let $A_1 := C_G(A)$ and suppose $A < A_1$. Let $G_A := N_G(A)$. Then $G_A \leq Aut(K)$ by (5) and (7) of 14.3.19, so we conclude from the structure of $Aut(K)$ described in (10) of 14.3.19 that $E_{32} \cong A_1 = O_2(G_A)$. As the element t defined in 14.3.18.4 acts on $N_T(A)$, $A_1^t \leq G_A$. As $[A, A^t] \neq 1$, $[A, A_1^t] \neq 1$. By B.3.2.4, $K/O_2(K) \cong A_6$ contains no FF*-offenders on A_1, so $A_1 = J(KA_1)$. Thus $A_1^t \not\leq KA_1$, so $|G_A : K| = 4$, and hence $G_A = KA_1 A_1^t \cong Aut(K)$ and $D = C_Q(U) \cong D_8$ using 14.3.19.6.

Next by 14.3.19.10, the action of G_A/A_1 on A_1 is described in section B.3. In the notation of that section, $z = e_{5,6}$, so as $UA/A = O_2(C_K(z)/A)$, $D \cap A_1 = C_{A_1}(U) = \langle e_5, e_6 \rangle$. In particular $d := e_6 \in A_1 \cap D$ with $K_d := C_K(d)$ an A_5-block, and $C_{G_A}(d) \cong S_5/E_{32}$. Further $O^2(H)$ centralizes D as $D \cong D_8$. As $[d, RQ] = V_1$, $RQ = DC_{RQ}(d)$, so $C_{RQO^2(H)}(d) = C_{RQ}(d)O^2(H) \cong (S_3 \times S_3)/(Q_8^2 \times \mathbf{Z}_2)$. Let $T_d := C_T(d)$ and $S_d := RQ \cap T_d$. As $O^2(H)$ centralizes d, $T_d \in Syl_2(C_H(d))$. Further $Z(T_d) = Z(S_d) = \langle z, d \rangle$, and $|T_d : S_d| \leq |T : RQ| = 2$. As H is a 5'-group by 14.3.15, $d \notin z^G$. Thus as $dz \in d^K$, z is weakly closed in $Z(T_d)$, so that $N_G(T_d) \leq G_1 = H$, and hence T_d is Sylow in $G_d := C_G(d)$.

Now $z \notin O_2(G_d)$: for otherwise $A = \langle z^{K_d} \rangle \leq O_2(G_d)$, impossible as $A \not\leq O_2(C_H(d))$. Thus as K_d is irreducible on A, $A_1 \cap O_2(G_d) = \langle d \rangle$. Now $T_d \in Syl_2(G_d)$,

$O_2(G_d) \leq S_d$, and $A_1 = O_2(K_d S_d)$, so we conclude that $\langle d \rangle = O_2(G_d)$. Let $\check{G}_d := G_d/\langle d \rangle$. Next $K_d \in \mathcal{L}(G_d, S_d)$ and $|T_d : S_d| \leq 2$, so $K_d \leq L_d \in \mathcal{C}(G_d)$ by 1.2.5. As $O_2(G_d) = \langle d \rangle$, $G_d \notin \mathcal{H}^e$, so L_d is quasisimple by 1.2.11 applied with V, G_d in the roles of "U, H". As the hypotheses of 1.1.6 are satisfied with G_d in the role of "H", L_d is described in 1.1.5.3. As $C_{\check{G}_d}(\check{z})$ has a subgroup of index at most 2 isomorphic to $(S_3 \times S_3)/Q_8^2$, we have a contradiction to the 2-local structure of the groups on that list. $\qquad\square$

LEMMA 14.3.21. (1) If $|T : RU| = 2$, then there exist involutions in $T - RU$.
(2) No involution in $T - RQ$ is in z^G.
(3) All involutions in RU are in z^G.

PROOF. First $RU \in Syl_2(K)$ by 14.3.19.8, and K is transitive on $A^\#$, while all involutions in $K - A$ are fused into A^t, so (3) holds.

Assume $|T : RU| = 2$. As $I_2 = LU$ by G.2.3.2 and $I_2/R \cong S_3$, $LT/R \cong S_3 \times \mathbf{Z}_2$. Further for X of order 3 in I_2, $C_R(X) = 1$. Thus $C_{O_2(LT)}(X) = \langle t_X \rangle$ with t_X an involution in $T - RU$, proving (1).

It remains to prove (2). So suppose some $t \in T - RQ$ is of the form $t = z^y$ for some $y \in G$. Let $I_t := C_{I_2}(t)$, $R_t := R\langle t \rangle$, and $R_t^+ := R_t/V$. By 14.3.18, $A \cap A^t = V$, and A, A^t are the maximal elementary abelian subgroups of R, so that $V = \Omega_1([R, t]) \geq \Omega_1(C_R(t))$ and R is transitive on $[A^+, t^+]t^+$; hence

(*) Each coset of V in $[R, t]\langle t \rangle$ not contained in $[R, t]$ contains a conjugate of t.

We claim that $z \in Q^y$. First consider the case where $[V, t] = 1$. Here $R_t = C_{I_2 R_t}(V) \trianglelefteq I_2 R_t$. Further by (*), each element of $[R, t]t$ is an involution, so that t inverts $[R, t]$; hence $C_R(t) = V$ and R is transitive on $[R, t]t$. Thus R is transitive on the involutions in Rt, so that $I_t/C_R(t) \cong S_3$. As $C_R(t) = V$, we conclude $I_t \cong S_4$. Therefore $V = [V, O^2(I_t)] \leq U^y$. In particular, $z \in Q^y$, as claimed. Now consider the case where $[V, t] \neq 1$. Then by Exercise 2.8 in [**Asc94**], R is transitive on involutions in Rt and $|C_R(t)| = 8$, so since $\Omega_1(C_R(t)) \leq V$, we conclude $C_R(t) \cong Q_8$. Then as H/Q has no Q_8-subgroup, $z \in Q^y$, completing the proof of the claim.

By the claim, $z \in Q^y$. Thus $t \in \Phi(C_{U^y}(z))$. This is a contradiction as $t \notin RQO^2(H)$ which is of index 2 in H. $\qquad\square$

LEMMA 14.3.22. $D = Z$, so $U = Q$.

PROOF. Notice by 14.3.19.1 that $U = Q$ if $D = Z$. So we assume $Z < D$, and will derive a contradiction.

As $A = C_G(A)$ by 14.3.20, $|D| \leq 4$ by 14.3.19.6. So $|D| = 4$, and we take $d \in D - Z$. By 14.3.19.10, $C_{Aut(K)}(U) \leq \langle \alpha, \beta \rangle$ where $\langle \alpha \rangle A = C_{Aut(K)}(A)$ and $\langle \beta \rangle A^t = C_{Aut(K)}(A^t)$. Thus $d \neq \alpha$ or β as A is self-centralizing in G, so d induces $\alpha\beta$ on K, and hence $D = \langle d \rangle \cong \mathbf{Z}_4$.

As $D \trianglelefteq H = C_G(d^2)$, D is a TI-set in G. Then the standard result I.7.5 from the theory of TI-sets says that $X := \langle D^G \cap T \rangle$ is abelian. Now L is transitive on $V^\#$ and $D \leq C_T(V) = O_2(LT)$, so $V \leq \Omega_1(\langle D^L \cap T \rangle) \leq X$. Then $X \leq C_T(V)$ so X is weakly closed in $O_2(LT)$, and hence $X \trianglelefteq LT$. Then as $M = !\mathcal{M}(LT)$, $N_G(X) = N_M(X) = LN_{H \cap M}(X)$ using 14.3.7. As X is abelian and weakly closed, we may apply Burnside's Fusion Lemma A.1.35 to conclude $D^G \cap T = D^{N_G(X)} = D^L$ is of order 3.

Let $G_A^+ := G_A/A$. From the structure of $Aut(K)$ in 14.3.19.10, since $A = C_G(A)$, $G_A^+ \cong S_6$ with $d^+ = (5,6)$. Recall $g \in N_L(V_2) - H$, so that $w := dd^g d^{g^2}$ is an involution with $w^+ = (1,2)(3,4)(5,6)$, and hence $X \cong \mathbf{Z}_4^2 \times \mathbf{Z}_2$, with $\Omega_1(X \cap U) = V$. Now I_2 acts on $\Omega_1(X) = V \times \langle w \rangle$, so as $[A, w] = V$ and $A \leq R$, $[R, w] = V$. Therefore R is transitive on Vw, so by a Frattini Argument, $I_2 = RC_{I_2}(w)$, and hence $C_{I_2}(w)/C_R(w) \cong S_3$. Further $|C_R(w)| = |R|/4 = |X \cap R|$, so $C_R(w) = X \cap R \cong \mathbf{Z}_4^2$. Also for $u \in U - R$, $d^{+g}d^{+g^2} = [d^{+g}, u]$, so $d^g d^{g^2} \equiv [d^g, u] \mod V$ since $V = X \cap A$. Then $[d^g, u] \in (X \cap U) - V$, so $[d^g, u]$ is of order 4 as $\Omega_1(X \cap U) = V$. Thus $d^g d^{g^2} \in U$ has order 4 and hence as $O^2(H)$ centralizes d,

$$C_{O^2(H)}(w) = C_{O^2(H)}(d^g d^{g^2}) \cong \mathbf{Z}_4 * SL_2(3).$$

Further choosing T so that $T_w := C_T(w) \in Syl_2(C_H(w))$, $\Omega_1(Z(T_w)) = \langle w, z \rangle$ and $wz \in w^U$.

Set $G_w := C_G(w)$. As $C_R(w) \cong \mathbf{Z}_4^2$, $O^2(C_{I_2}(w)) \cong \mathbf{Z}_3/\mathbf{Z}_4^2$, while by (5) of 14.3.18, $O_2(O^2(H)) = U \cong Q_8^2$ has no $\mathbf{Z}_4^2$-subgroup, so we conclude $w \notin z^G$. Thus as $\Omega_1(Z(T_w)) = \langle w, z \rangle$ and $wz \in w^G$, z is weakly closed in $Z(T_w)$, so that $N_G(T_w) \leq H$ and hence $T_w \in Syl_2(G_w)$.

If $z \in O_2(G_w)$, then $V = \langle z^{C_{I_2}(w)} \rangle \leq Z(O_2(G_w))$, impossible since $V \nleq Z(\langle V^{C_H(w)} \rangle)$. Thus $z \notin O_2(G_w)$; now $T_w \in Syl_2(G_w)$, $O_2(C_H(w)) \leq G_A$, and z is contained in each nontrivial normal subgroup of $G_w \cap G_A$ other than $\langle w \rangle$, so we conclude that $O_2(G_2) = \langle w \rangle$. As in the the proof of 14.3.20, we appeal to 1.2.11, 1.1.6, and 1.1.5.3; this time from the structure of $C_H(w) = C_{G_w}(z)$ and $C_{I_2}(w)$, we conclude $G_w/\langle w \rangle \cong G_2(2)$, so $G_w \cong \mathbf{Z}_2 \times G_2(2)$ since $G_2(2)$ has trivial Schur multiplier by I.1.3. Set $L_w := G_w^\infty$, and observe that L_w has one class z^{L_w} of involutions, and so the set $\{w\} \cup (zw)^{L_w}$ of involutions in wL_w is contained in w^G since we saw w is conjugate to zw. Also $T_w \cap \langle w \rangle L_w = XC_U(t) \leq RQ$, so $T_w \cap \langle w \rangle L_w = T_w \cap RQ$. By 14.3.21.2, no involution in $T - RQ$ is in z^G, so $z^G \cap G_w = z^{G_w}$, and hence $w^G \cap H = w^H$ since G is transitive on commuting pairs from $z^G \times w^G$. But then as $H/O^2(H)R$ is of order 4 and $w \notin RU$, it follows that $w \notin O^2(G)$ from Generalized Thompson Transfer A.1.37.2, contrary to the simplicity of G. $\qquad \square$

We are now in a position to derive a contradiction, and hence establish Theorem 14.3.16. By 14.3.22, $Q = U$, so $|T : RU| = 2$. Thus by 14.3.21.1, there is an involution $t \in T - RU$. By 14.3.21.2, $t \notin z^G$, while by 14.3.21.3, all involutions in RU are in z^G. Thus $t^G \cap RU = \emptyset$, so $t \notin O^2(G)$ by Thompson Transfer, contrary to the simplicity of G.

14.3.4. Characterizing HS by $\langle V^{G_1} \rangle$ nonabelian but not extraspecial. In this subsection we continue to assume Hypothesis 14.3.10. By Theorem 14.3.16, case (1) of Hypothesis 14.3.1 holds. Thus in the remainder of our treatment of the case U nonabelian in this section and the next, we have $L/O_2(L) \cong L_3(2)$.

In this final subsection, we first prove several more preliminary results, and then reduce to the case where U is extraspecial, by showing HS is the only quasithin example with $V_1 < Z_U$. The treatment of the extraspecial case occupies the following section 14.4.

LEMMA 14.3.23. $d \geq 4$. If $d = 4$, then

(1) $\hat{V} = \hat{E} \cong E_4$.

(2) One of the following holds:

(i) $\dot{H} \cong S_3 \times S_3$, with $\dot{L}_1 \trianglelefteq \dot{H}$. Further if $\dot{Z}_U^g \neq 1$ then $\dot{Z}_U^g = C_{\dot{H}}(\hat{V})$ is of order 2, and setting $K := \langle Z_U^{gH} \rangle$, $\dot{K} \cong S_3$, $\dot{H} = \dot{K}\dot{L}_1\dot{T}$, and $K \not\leq M$.

(ii) $\dot{H} \cong S_5$, $\hat{U}$ is the $L_2(4)$-module, and $\mathbf{Z}_2 \cong \dot{Z}_U^g \leq E(\dot{H})$.

(iii) $\dot{H} \cong A_6$ or S_6, and $m(\dot{Z}_U^g) = 1$ or 2.

(iv) $\dot{H}$ is E_9 extended by $\mathbf{Z}_2$, $\dot{L}_1 \trianglelefteq \dot{H}$, and $U \cong Q_8^2$.

(3) $m(\dot{W}^g/\dot{Z}_U^g) = 1$ and Z_U^g centralizes $\hat{V}$.

(4) $H > (H \cap M)C_H(\hat{U})$.

PROOF. By 12.8.13.1, $V \leq E$. By 14.3.11, $m(\hat{V}) = m(\tilde{V}) = 2$. But by 12.8.11.2, $\hat{E}$ is totally isotropic in the symplectic space $\hat{V}$, so $2 = m(\hat{V}) \leq m(\hat{E}) \leq d/2$, and hence $d \geq 4$. Further if $d = 4$, these inequalities are equalities, so (1) holds.

Assume $d = 4$. By 12.8.11.5 and (1), $m(\dot{W}^g/\dot{Z}_U^g) = 1$, while by 12.8.13.2, Z_U^g centralizes V, and then by 12.8.11.3, Z_U^g is the kernel of the action of W^g on $\hat{V}$. Thus (3) is established. By 14.3.3.6, $H \cap M$ acts on V; so if (4) fails, then $\dot{H}$ acts on $\hat{V}$, contrary to $\hat{U} = \langle \hat{V}^H \rangle$ and $d = 4$. Thus (4) holds.

Observe that if $\dot{H} \leq O_4^+(2)$, then $O^2(\dot{H})$ is abelian, so $\dot{L}_1 \trianglelefteq O^2(\dot{H})$. Thus $\dot{L}_1 \trianglelefteq O^2(\dot{H})\dot{T} = \dot{H}$. If $O^2(\dot{H})$ is of order 3, then $\dot{H} = \dot{L}_1\dot{T}$, contrary to (4). Thus $O^2(\dot{H}) \cong E_9$, so as $\dot{L}_1 \trianglelefteq \dot{H}$, we conclude $\dot{H} < O_4^+(2)$ in this case.

Suppose first that $m_2(\dot{H}) = 1$. Then by (3) and (4) of 14.3.12, $U \cong Q_8^2$, so $\dot{H} \leq O_4^+(2)$. Then by the previous paragraph, $O^2(\dot{H}) \cong E_9$, so as $m_2(\dot{H}) = 1$, (2iv) holds.

Thus we may assume $m_2(\dot{H}) \geq 2$. Suppose first that $\dot{H} \leq O_4^+(2)$. Then $\dot{H} \cong S_3 \times S_3$ by remarks in paragraph three. Assume that $\dot{Z}_U^g \neq 1$. Then as Z_U^g centralizes $\hat{V}$ by (3), and as $\hat{V} = [\hat{V}, \dot{L}_1]$, $\dot{Z}_U^g$ is of order 2, $\dot{L}_1 = C_{O^2(\dot{H})}(\dot{Z}_U^g)$, $\dot{K} := \langle \dot{Z}_U^{gH} \rangle \cong S_3$, and $\dot{H} = \dot{L}_1\dot{K}\dot{T}$, and so $K \not\leq M$ by (4). This completes the proof that (2i) holds

Thus we may assume that $\dot{H}$ is not contained in $O_4^+(2)$. But by 14.3.14.2, $\dot{H}$ is a subgroup of $Sp_4(2)$ containing $S_3 \times S_3$ or A_5, so we conclude $F^*(\dot{H}) \cong L_2(4)$ or A_6.

Suppose $Z_U = V_1$. Then U is extraspecial, so $\dot{H} \leq O_4^\epsilon(2)$, and by the assumption in previous paragraph, $\epsilon = -1$. This is impossible, as $\tilde{U}$ contains the totally singular line $\tilde{V}$. We conclude $Z_U > V_1$, so $\dot{Z}_U^g \neq 1$ by 12.8.13.4.

Suppose $F^*(\dot{H}) \cong L_2(4)$. As $\dot{L}_1 \trianglelefteq \dot{L}_1\dot{T}$ and $\hat{V} = [\hat{V}, \dot{L}_1] \cong E_4$, it follows that $\hat{U}$ is the $L_2(4)$-module, and $\hat{V}$ is the $\mathbf{F}_4$-line invariant under T. As $\dot{W}^g$ is nontrivial on $\hat{V}$ by 12.8.11.3, $\dot{H} \cong S_5$. Then as $\dot{W}^g$ is elementary abelian and we saw that $1 \neq \dot{Z}_U^g < \dot{W}^g$ and $\dot{Z}_U^g$ centralizes $\hat{V}$, (2ii) holds. A similar argument shows (2iii) holds if $F^*(\dot{H}) \cong A_6$. $\square$

LEMMA 14.3.24. Assume $V_1 < Z_U$, so that U is not extraspecial. Then either:

(1) $d = 6$ and $\hat{U}$ is the natural module for $\dot{H} \cong G_2(2)$, or

(2) $d = 4$ and one of conclusions (i)—(iii) of 14.3.23.2 holds.

PROOF. By 14.3.12.3, $m(\dot{W}^g) \geq d/2$, so case (1) of 14.3.14 does not hold. Case (3) of 14.3.14 is conclusion (1), and in case (2) of 14.3.14, $d = 4$ so one of the conclusions of 14.3.23.2 holds, with conclusion (iv) ruled out as there U is extraspecial. $\square$

LEMMA 14.3.25. $Z(LT) \cap U = 1$.

PROOF. Assume $Z_L := Z(LT) \cap U \neq 1$. Set $V_H := \langle Z_L^H \rangle$; then $V_H \leq Z_U$, and as usual $V_H \in \mathcal{R}_2(H)$ by B.2.14. As $Z_L \trianglelefteq LT$ and $M = !\mathcal{M}(LT)$, $C_G(V_H) \leq C_G(Z_L) \leq M$. As $Z_L \neq 1$, $Z_U > V_1$, so by 14.3.24, either $d = 4$ and one of conclusions (i)–(iii) of 14.3.23.2 holds, or $d = 6$ and $\hat{U}$ is the natural module for $\dot{H} \cong G_2(2)$. In any case, $\dot{Z}_U^g \neq 1$ by 12.8.13.4.

Assume first that $\dot{H}$ is not solvable. Then from the previous paragraph, $F^*(\dot{H})$ is quasisimple, so there is $K \in \mathcal{C}(H)$ with $\dot{K} = F^*(\dot{H})$. As K is irreducible on $\hat{U}$ and $\hat{U} > \hat{V}$ in each case, K does not act on $\hat{V}$. Then as $K \cap M \leq M_V$ by 14.3.3.6, $K \not\leq M$. Thus as $C_G(V_H) \leq M$, $[V_H, K] \neq 1$. Therefore $K \in \mathcal{L}_f(G, T)$ by 1.2.10, and then as $\dot{K}$ is not $L_3(2)$ from 14.3.24, $K/O_2(K) \cong L_2(4)$ by 14.3.4.1. Thus case (ii) of 14.3.23.2 holds, so that $\mathbf{Z}_2 \cong \dot{Z}_U^g \leq \dot{K}$; in particular $K = [K, Z_U^g]$. As $m(\dot{Z}_U^g) = 1$, $C_{Z_U^g}(\hat{U})$ is a hyperplane of Z_U^g, so $Z_0 := (Z_U^g \cap Z_U)V_1$ is a hyperplane of Z_U by 12.8.10.6. Thus Z_U^g induces transvections on V_H with axis $Z_0 \cap V_H$. This is impossible, as Z_U^g induces inner automorphisms on $\dot{K}$ and we saw $K = [K, Z_U^g]$.

Therefore $\dot{H}$ is solvable. Hence by the first paragraph, case (i) of 14.3.23.2 holds, so $d = 4$, $\dot{H} \cong S_3 \times S_3$, $\dot{L}_1 \trianglelefteq \dot{H}$, and setting $K := \langle Z_U^{gH} \rangle$, $\dot{K} \cong S_3$, $\dot{H} = \dot{K}\dot{L}_1\dot{T}$, and $K \not\leq M$. Then $K \cap M \leq (K \cap T)C_K(\hat{U})$, since the latter group is maximal in $KC_H(\hat{U})$. Set $H^+ := H/C_H(V_H)$. As in the previous paragraph, Z_U^g induces transvections on V_H with axis $Z_0 \cap V_H$. By the first paragraph of the proof, $C_K(V_H) \leq M$, so that $C_K(V_H) \leq (K \cap T)C_K(\hat{U})$. Therefore K^+ has the quotient group $\dot{K} \cong S_3$ and $C_K(V_H) \leq C_K(\hat{U})$. Thus we conclude from the structure of SQTK-groups generated by transvections (e.g., G.6.4) that $K^+ \cong S_3$, and hence $C_K(V_H) = C_K(\hat{U})$ and $[V_H, K]$ is of rank 2. Indeed as Z_0 is a hyperplane of Z_U centralized by Z_U^g, $Z_U = [V_H, K] \times C_{Z_U}(K)$ and $C_{Z_U}(K) \trianglelefteq H$. Set $\check{H} := H/C_{Z_U}(K)$ and $H^! := H/C_H(\check{U})$; observe that $C_H([V_H, K]) \leq C_H(\check{Z}_U)$, and $\check{U}$ is a quotient of $\tilde{U}$ and so elementary abelian. As $\check{U} = \langle \check{V}_2^H \rangle$ and $\check{V}_2 \leq \Omega_1(Z(\check{T}))$, $O_2(H^!) = 1$ by B.2.13. As $C_K(\hat{U}) = C_K(V_H) \leq C_K([V_H, K]) \leq C_K(\check{Z}_U)$, $C_K(\hat{U})^! \leq O_2(K^!) = 1$. Therefore $C_K(\check{U}) = C_K(\hat{U})$, so $K^! \cong \dot{K} \cong S_3$. Next $[C_H(\hat{U}), K] \leq C_K(\hat{U}) = C_K(V_H)$, so that $[C_H(\hat{U})^+, K^+] = 1$. Then as $End_{K^+}([V_H, K]) \cong \mathbf{F}_2$, $C_H(\hat{U}) \leq C_H([V_H, K]) \leq C_H(\check{Z}_U)$, so $C_H(\hat{U})^! \leq O_2(H^!) = 1$. Therefore $C_H(\hat{U}) = C_H(\check{U})$, and hence $\dot{H} \cong H^!$. Next $\check{U} = \langle \check{V}^H \rangle$, while $\check{V} = [\check{V}, L_1]$ and $L_1^! \trianglelefteq H^!$ as $H^! \cong \dot{H}$, so we conclude $\check{U} = [\check{U}, L_1]$, contrary to $1 \neq [\check{V}_H, K] \leq C_{\check{U}}(L_1)$ since $\check{U}$ is elementary abelian. This contradiction completes the proof of 14.3.25. $\qquad \square$

THEOREM 14.3.26. *Assume Hypothesis 14.3.10. Then either $Z_U = V_1$ so that U is extraspecial, or $G \cong HS$.*

REMARK 14.3.27. If Hypothesis 14.3.1 did not exclude the possibility that $K/O_2(K) \cong A_6$ for some $K \in \mathcal{L}_f(G, T)$, then $Sp_6(2)$ would also appear as a conclusion in Theorem 14.3.26. Its shadow will be eliminated during the proof of lemma 14.3.31. Recall that the case leading to $Sp_6(2)$ was treated in Theorem 13.4.1.

Until the proof of Theorem 14.3.26 is complete, assume G is a counterexample. Thus $V_1 < Z_U$. Then by 12.8.13.4, $\dot{Z}_U^g \neq 1$.

Recall $V_2 = V_1 V_1^g$. As $L/O_2(L) \cong L_3(2)$ by Theorem 14.3.16, we may choose $l \in C_L(V_1^g)$ with $V = V_2 V_1^l$. In particular $V_2^l = V_1^g V_1^l$, so we may apply results from section 12.8 with V_2^l in the role of "V_2". Similarly $V_1 V_1^l$ can play the role of "V_2".

LEMMA 14.3.28. *(1)* $Z(I_2^l) = Z_U^g \cap Z_U^l$.
(2) $Z_U \cap Z(I_2^l) = 1$.

PROOF. As $(U, U^g)^l = (U^l, U^g)$, part (1) follows from 12.8.10.2. Then by (1) and 12.8.10.2,

$$Z_U \cap Z(I_2^l) = Z_U \cap Z_U^g \cap Z_U^l = Z(I_2) \cap Z(I_2^l) \le C_U(L) = 1,$$

since $L = \langle L_2, L_2^l \rangle$, and $C_U(L) = 1$ by 14.3.25. $\square$

LEMMA 14.3.29. *Assume there exists* $1 \ne e \in Z(I_2) \cap Z$, *and let* $V_e := \langle e^L \rangle$. *Then*

(1) V_e *is of dimension* 3, 4, 6, *or* 7, *and* V_e *has an quotient* L-*module isomorphic to the dual of* V.
(2) $J(T) \trianglelefteq LT$.

PROOF. By 12.8.10.2, $Z(I_2) \le Z_U$, so by choice of e and 14.3.25, $[L, e] \ne 1$. Thus $I_2 T = C_{LT}(e)$, so $|e^{LT}| = 7$. Thus (1) follows from H.5.3. As usual $VV_e \in \mathcal{R}_2(LT)$ by B.2.14, so as there is a quotient of V_e isomorphic to the dual of V as an LT-module, (2) follows from Theorem B.5.6. $\square$

LEMMA 14.3.30. *(1)* $|Z(I_2)| \le 2$.
(2) If $Z(I_2) \ne 1$, then the image of $Z(I_2^l)$ in $\dot{H}$ is the subgroup of order 2 generated by an involution of type a_2 in $Sp(\hat{U})$ with $[\hat{U}, Z(I_2^l)] = \hat{V}$.

PROOF. We may assume $Z(I_2) \ne 1$. By 14.3.28.1, $Z(I_2^l) = Z_U^g \cap Z_U^l$, so by 12.8.10.4,

$$[Z(I_2^l), W] \le [Z_U^g, W] \le Z_U V_2 = Z_U V_1^g \quad \text{and} \quad [Z(I_2^l), U \cap H^l] \le Z_U V_1^l.$$

By 12.8.4.1 and G.2.5.1, $\bar{U} = O_2(\bar{L}_1)$, so $U = C_U(V_1^g) C_U(V_1^l) = W(U \cap H^l)$ from the action of $\bar{L}$ on V, and hence $[U, Z(I_2^l)] \le Z_U V$, with $\hat{V} \cong V Z_U / Z_U$ of rank 2. Thus the image of $Z(I_2^l)$ in $\dot{H}$ is either trivial, or is $\langle \dot{a} \rangle$ of order 2, where $\dot{a}$ is the element of $Sp(\hat{U})$ of type a_2 with $[\hat{U}, \dot{a}] = \hat{V}$, and in the latter case (2) holds. We will show that $Z(I_2^l)$ is faithful on $\hat{U}$. This will prove (1), and complete the proof of (2).

So let $A := C_{Z(I_2^l)}(\hat{U})$; we must show $A = 1$. Applying 12.8.10.6 with $V_2 = V_1 V_1^g$ and $V_1 V_1^l$ in the role of "V_2", $A \le V_1^g Z_U \cap V_1^l Z_U = Z_U$, so $A \le Z_U \cap Z(I_2^l) = 1$ by 14.3.28.2, completing the proof. $\square$

LEMMA 14.3.31. $Z(I_2) = 1$.

PROOF. Assume $Z(I_2) \ne 1$. Then by 14.3.30.1, $Z(I_2) = \langle e \rangle$ is of order 2, and $e \in Z_U$ by 12.8.10.2. Further as T normalizes I_2, $e \in Z$. Let $a := e^l$. By 14.3.30.2, $\dot{a}$ is the involution in $Sp(\hat{U})$ of type a_2 with $[\hat{U}, \dot{a}] = \hat{V}$.

Let $K := \langle a^H \rangle$. Then $[a, Z_U] \le Z_U \cap Z(I_2^l) = 1$ by 14.3.28.2. Thus a centralizes Z_U, so K does too. In particular $\langle K, I_2 T \rangle \le C_G(e) =: G_e$. Also then $C_K(\hat{U}) = C_K(\tilde{U}) = O_2(K)$ using 12.8.4.4, so $K/O_2(K) \cong \dot{K} \cong K^*$.

By 14.3.24, either $\hat{U}$ is the natural module for $\dot{H} \cong G_2(2)$, or $d = 4$ and one of conclusions (i)–(iii) of 14.3.23.2 holds.

Assume first that one of the cases other than case (i) of 14.3.23.2 holds. Then $F^*(\dot{H})$ is simple, so $F^*(\dot{H}) \le \dot{K}$ and $K_1 := K^\infty \in \mathcal{C}(H)$ with $\dot{K}_1 = \dot{F}^*(\dot{H})$. If

$\dot{K}_1$ is $G_2(2)'$ or A_6, then K_1 contains all elements of order 3 in H by A.3.18, so $L = \langle L_1, L_2 \rangle \leq \langle K_1, I_2 \rangle \leq G_e$, contrary to 14.3.25. On the other hand if $\dot{H} \cong S_5$, then $\dot{H}$ contains no involution of type a_2, contrary to 14.3.30.2.

Therefore case (i) of 14.3.23.2 holds. so $\dot{H} \cong S_3 \times S_3$. Since $\dot{a}$ has type a_2, $\hat{U} = [\hat{U}, K]$, and since $[\hat{U}, \dot{a}] = \hat{V}$, $\dot{a}$ centralizes $\hat{V}$, so $\langle \dot{a} \rangle = \dot{Z}_U^g$, $\check{K} \cong S_3$, $\dot{H} = \dot{K}\dot{L}_1\dot{T}$, and $K \not\leq M$ by 14.3.23.2.

We saw earlier that $K_e := \langle KT, I_2 T \rangle \leq G_e$; set $U_e := \langle V_1^{K_e} \rangle$, $K_e^+ := K_e/C_{K_e}(U_e)$, and $\check{K}_e := K_e^+/O_{3'}(K_e^+)$. Then $O_2(K_e^+) = 1$ by B.2.14, so $\alpha := (I_2^+ T^+, T^+, K^+ T^+)$ is a Goldschmidt amalgam in the sense of Definition F.6.1. Observe that $V_2 = \langle V_1^{I_2} \rangle \leq U_e$, so $U_1 := \langle V_2^K \rangle \leq U_e$. Now $K/O_2(K) \cong S_3$, $\hat{U} = [\hat{U}, K]$, and $[V_2, U] = V_1$; so $F^*(K/C_K(U_1)) = O_2(K/C_K(U_1))$ and hence $F^*(K^+) = O_2(K^+)$.

By 14.3.29.2, $J(T) \trianglelefteq LT$. Hence $J(T) \leq O_2(I_2 T)$, and as $K \not\leq M = !\mathcal{M}(LT)$, $J(T) \not\leq O_2(KT)$ in view of B.2.3.3, so $O^2(K) = [O^2(K), J(T)]$. Thus $O_2(K^+ T^+) \neq O_2(I_2^+ T^+)$, and U_e is an FF-module for K_e^+. By F.6.11.1, $O_{3'}(K_e^+)$ is of odd order, so $K^+ T^+ \cong \check{K}\check{T}$ and $I_2^+ T^+ \cong \check{I}_2 \check{T}$, and hence $F^*(\check{K}) = O_2(\check{K})$. Then as $O_2(K^+ T^+) \neq O_2(I_2^+ T^+)$, F.6.11.2 says $K_e^+ \cong \check{K}_e$ is described in Theorem F.6.18. As $F^*(\check{K}) = O_2(\check{K})$, cases (1) and (2) of F.6.18 are ruled out. In the remaining cases, $K_e^+ \cong \check{K}_e$ is not solvable, so $K_0 := K_e^\infty \in \mathcal{L}_f(G, T)$ by 1.2.10. Then by 14.3.4.1, $K_0/O_2(K_0) \cong L_3(2)$ since A_5 is not a composition factor of any group in F.6.18. Then $\check{K}_e$ appears in case (6) of F.6.18, so $K_e = K_0 Y$ with Y the preimage in K_e of $O_{3'}(K_e^+)$. As $O^2(\check{K}) = [O^2(\check{K}), T \cap K_0]$ and $O^2(K)$ is T-invariant, $O^2(K) \leq K_0$. Similarly $O^2(I_2) \leq K_0$, so $K_0 = O^2(K_e)$ using F.6.6, and hence $K_e = K_0 T$. Also $V_2 = \langle V_1^{I_2} \rangle$ and KT centralizes V_1, so by H.5.5, $U_e = \langle V_1^{K_e} \rangle$ is a 3-dimensional natural module for $K_e^+ \cong L_3(2)$. Thus $U_e = \langle V_2^K \rangle$. We saw earlier that $\hat{U} = [\hat{U}, K]$, K centralizes Z_U, and $C_K(\hat{U}) = O_2(K)$. Therefore $\tilde{U} = [\tilde{U}, K] \oplus \tilde{Z}_U$. Now as $\dot{H} = \dot{K}\dot{L}_1\dot{T}$, $U = \langle V_2^{L_1 K} \rangle$, so $V_2 \not\leq [K, U]$ and $U_e = \langle V_2^K \rangle$ has rank greater than 3, contradicting $m(U_e) = 3$. $\qquad \square$

LEMMA 14.3.32. *(1)* $\hat{U}$ *is the* $L_2(4)$*-module for* $\dot{H} \cong S_5$.
(2) $U \cong Q_8^2 * \mathbf{Z}_4$.
(3) $Q = C_H(\hat{U})$, *so that* $\dot{H} \cong H^*$.
(4) $H = KT$ *with* $K \in \mathcal{C}(H)$, $U = [O_2(K), K]$, *and* K *acts indecomposably on* $\tilde{U}$.

PROOF. By 14.3.31, $Z(I_2) = 1$, so that by 12.8.10.6,

$$C_{Z_U^g}(\hat{U}) = V_1^g, \text{ so } \dot{Z}_U^g \cong \tilde{Z}_U \neq 1. \tag{*}$$

By 14.3.24, either case 14.3.14.3 holds with $\dot{H} \cong G_2(2)$, or $\dot{H}$ is described in one of cases (i)–(iii) of 14.3.23.2. Then $d = 6$ or 4, respectively. By 12.8.11.2, $m(\hat{E}) \leq d/2$. Then we can use 12.8.11.5 to show $m(\hat{E}) = d/2$ and $m(\dot{W}^g/\dot{Z}_U^g) = d/2-1$: For when $d = 4$, $\hat{E} = \hat{V} \cong E_4$ by 14.3.23.1, and when $d = 6$, $m(\dot{W}^g) = 3$ by 14.3.14.3 and $\dot{Z}_U^g \neq 1$ by (*). This also shows $m(\dot{Z}_U^g) = 1$ when $\dot{H} \cong G_2(2)$. When $d = 4$, $m(\dot{W}^g/\dot{Z}_U^g) = 1$ by 14.3.23.3, so as $\dot{Z}_U^g \neq 1$ by (*), either $m_2(\dot{H}) = 2$, $m(\dot{W}^g) = 2$, and $m(\dot{Z}_U^g) = 1$, or case (iii) of 14.3.23.2 holds with $\dot{H} \cong S_6$, $m(\dot{W}^g) = 3$, and $m(\dot{Z}_U^g) = 2$.

Thus in view of (*), we have shown that either $|\tilde{Z}_U| = 2$, or $\dot{H} \cong S_6$ and $|\tilde{Z}_U| = 4$. In either case, H^∞ centralizes Z_U by Coprime Action, and in the former H centralizes $\tilde{Z}_U$. Thus as $H = H^\infty T$ in the latter case, (3) holds by 12.8.4.4.

Suppose next that case (i) of 14.3.23.2 does not hold; we will eliminate that case at the end of the proof. Then there is $K \in \mathcal{C}(H)$ with $K^* = F^*(H^*)$. As K centralizes Z_U and T acts on V_2 with $[V_2, Q] = V_1$, $C_K(\hat{V}_2)^* = C_K(V_2)^*$ by Coprime Action. Then as $H^* \cong \dot{H}$ by (3), $C_K(\hat{V}_2)^*$ acts on Z_U^{g*} and W^{g*} by 12.8.12.2. But when $H^* \cong G_2(2)$, we saw $\dot{Z}_U^g$ has order 2, whereas $C_K(\hat{V}_2)^*$ is the stabilizer of a 4-subgroup of T^*, and in particular does not normalize $Z(T^*)$ of order 2.

Thus $d = 4$, so $\hat{V} = \hat{E}$ by 14.3.23.1. Further since $I_2 \trianglelefteq G_2$ by 12.8.9.1, $C_K(\hat{V}_2)^* = C_K(V_2)^*$ normalizes $E = U \cap U^g$. But in case (iii) of 14.3.23.2, the maximal parabolic $C_K(\hat{V}_2)^*$ does not normalize $\hat{V}$.

Thus we have reduced to case (ii) of 14.3.23.2, so that (1) holds, and also $|Z_U| = 4$. If $Z_U \cong E_4$ then $\dot{H}$ preserves a quadratic form on $\hat{U}$ by 12.8.8.2, which is not the case as here $\hat{U}$ is a natural $L_2(4)$-module. Thus (2) holds.

Next as Q normalizes V_2 with $[V_2, U] = V_1$, $Q = UC_Q(V_1^g)$. By 12.8.9.5, $W^g \cap Q = E$. Thus

$$[Q, W^g] = [U, W^g][C_Q(V_1^g), W^g] \leq U(W^g \cap Q) = U.$$

Then as $K = [K, W^g]$, $[Q, K] \leq U$. If $[U, K] < U$, then $[U, K]$ is extraspecial by (2), impossible as $\hat{U}$ is the $L_2(4)$-module for $\dot{K}$. Thus $U = [U, K] = [O_2(K), K]$, so K is indecomposable on $\tilde{U}$. By (1) and (3), $H = KT$, completing the proof of (4).

Finally we must eliminate case (i) of 14.3.23.2. Here $\dot{L}_1 \trianglelefteq \dot{H}$, so as $\dot{H} \cong H^*$ by (3), $L_1 \trianglelefteq H$, and hence $\tilde{U} = [\tilde{U}, L_1]$ by 12.8.5.1. This is a contradiction as we saw H centralizes $\tilde{Z}_U$. $\square$

LEMMA 14.3.33. (1) $P := O_2(L) = \langle Z_U^L \rangle \cong \mathbf{Z}_4^3$, with P/V isomorphic to V as an L-module.

(2) $U = O_2(K)$ and $PU \in Syl_2(K)$.

(3) $M = L$ and $H = KT$ with $U = O_2(H)$.

PROOF. By 14.3.32.2, $C_U(V) = VZ_U$, and $Z_U \cong \mathbf{Z}_4$ is centralized by L_1. By 14.3.23.1, $\hat{E} = \hat{V}$, so $V \leq U \cap U^g = E \leq VZ_U$ and hence $E = V(Z_U \cap U^g)$. By (*) in the proof of 14.3.32 and symmetry, $Z_U \cap U^g = V_1$, so $E = V$. By 12.8.8.4, $O_2(LU)/V$ is described in G.2.5; thus as $E = V$ and $m(W/V) = 1$, we conclude that $O_2(LU)/V$ is isomorphic to V as an L-module, and hence $O_2(LU) = \langle Z_U^L \rangle$ and $O_2(LU) = [O_2(LU), L] = O_2(L) = P$. As Z_U is a cyclic normal subgroup of $H = C_G(\Omega_1(Z_U))$, Z_U is a TI-set in G. Further $Z_U \leq C_T(V)$, so $[Z_U, Z_U^y] = 1$ for $y \in L$ by I.7.5, and hence (1) holds.

By (1), $V = \Omega_1(O_2(L)) \trianglelefteq M$. By 14.3.32, $H = KT$, with $KQ/Q \cong A_5$, so $H \cap M = L_1T$, and hence $M = LT$ by 14.3.7.

From the structure of L, $PU = O_2(L_1)$; so as $L_1 \leq O^2(H) = K$, $PU \leq K$. By 14.3.32.4, $U = [O_2(K), K]$, so if $U < O_2(K)$, then $K/U \cong SL_2(5)$; but this is impossible, as the central 2-chief factors of L_1 are in Z_U by (1). Thus $U = O_2(K)$. Then $|PU| = |K|_2$, so (2) holds.

Now $[K, C_T(U)] \leq C_K(U) = Z_U$ with Z_U centralized by K, so $K = O^2(K)$ centralizes $C_T(U)$ by Coprime Action. In particular $C_T(U) = C_T(K)$ since $U \leq K$. Then by (2), $C_T(L) \leq C_T(PU) \leq C_T(U) = C_T(K)$. But $K \not\leq LT = M$, while if $C_T(L) \neq 1$, then $N_G(C_T(L)) \leq M = !\mathcal{M}(LT)$; so we conclude $C_T(L) = 1$. By (2), $C_T(K)$ centralizes PU; so as $PU = O_2(L_1)$, from the structure of $Aut(L)$, $C_T(K) \leq C_T(PU) \leq C_T(L)Z_U = Z_U$. Thus $C_T(U) = C_T(K) = Z_U$.

Let X_1 be of order 3 in L_1. Then $Q = [Q, X_1]C_Q(X_1)$ with $[Q, X_1] = [U, X_1] \cong Q_8^2$ by 14.3.32. Now if Q_1 is the preimage of an irreducible X_1-submodule of $[\widetilde{Q, X_1}]$, then by 12.8.4.2, $C_Q(X_1)$ normalizes Q_1; further $C_{Q_1}(C_Q(X_1)) > V_1 = C_{Q_1}(X_1)$ by the Thompson $A \times B$-Lemma, so $C_Q(X_1)$ centralizes Q_1 as X_1 is irreducible on $\tilde{Q}_1$. Thus $C_Q(X_1)$ centralizes $[Q, X_1] = [U, X_1]$, so $Z_U = C_T(U) = C_Q(X_1) \cap C_Q(Z_U)$ is of index at most 2 in $C_Q(X_1)$ as $Z_U \cong \mathbf{Z}_4$. Thus either $C_Q(X_1) = Z_U$, or $C_Q(X_1)$ is dihedral or quaternion of order 8.

Suppose first that $C_Q(X_1) = Z_U$. Then $Q = U$, so as $H = KT$, $|H|_2 = 2^9$ by 14.3.32. Hence as we saw $M = LT$, $M = L$ using (1), so (3) holds.

So we assume $C_Q(X_1)$ is of order 8, and it remains to derive a contradiction.[1] Now $C_Q(X_1) \leq O_2(LT)$, so $O_2(LT) = PC_Q(X_1)$. Then as $M = LT$, $M = LC_Q(X_1)$.

For $r \in C_Q(X_1) - U$, r centralizes the supplement $[U, X_1]$ to P in $O_2(L_1)$, so from the structure of $Aut(L_3(2))$, r centralizes L/P. Then by Gaschütz's Theorem A.1.39, we may choose r so that $[r, L] \leq V$. Now as L is irreducible on V, r is an involution, and as $C_T(L) = 1$, P induces the full group of transvections on $V\langle r \rangle$ with axis V. So $L = PC_L(r)$ by a Frattini Argument, and r inverts P.

Let $T_L := T \cap L$, so that T_L is of index 2 in T. As G is simple, Thompson Transfer says there is $g \in G$ with $r^g \in T_L$. We show that any such r^g is not extremal in M; then the standard transfer result Exercise 13.1 in [**Asc86a**] contradicts $r \in O^2(G)$.

As H contains no $L_3(2)$-section, $r^G \cap V = \emptyset$. Thus $r^g \in T_L - P$ as $V = \Omega_1(P)$, and conjugating in L, we may take $r^g \in O_2(L_1) = PU$. By 14.3.32.2, each nontrivial coset of Z_U in U contains exactly two involutions fused under U, and by 14.3.32.1, K is transitive on $\hat{U}^\#$, so K is transitive on involutions in $U - V_1$. Thus as $r^G \cap V = \emptyset$, $r^g \notin U$. Then as $P^* \in Syl_2(K^*)$, $\hat{V} = C_{\hat{U}}(r^g)$; so as PU centralizes Z_U, $C_U(r^g) = Z_U C_V(r^g) = Z_U V_2$. Thus $|U : C_U(r^g)| = 2^3$, so $|C_T(r^g)| \leq 2^7$ as $|T| = 2^{10}$. Therefore if r^g is extremal in M, then $C_T(r^g) = C_T(r)^g$. As V is the natural module for $C_L(r)/V \cong L_3(2)$, $V_1 = Z(C_T(r)) \cap \Phi(C_T(r))$. Then as $V_1 \leq Z(C_T(r^g)) \cap \Phi(C_T(r^g))$, we conclude $g \in H$. This is impossible as $r \in Q = O_2(H)$, while $r^g \in PU$ but $r^g \notin U = Q \cap PU$. This contradiction completes the proof of (3), and hence of 14.3.33. $\qquad\square$

At this point, we can complete the identification of G as HS, and hence establish Theorem 14.3.26. Namely by 14.3.33 and 14.3.32, $U = Q = O_2(H) \cong \mathbf{Z}_4 * Q_8^2$ with $H/U \cong S_5$. By 14.3.32.4, $\tilde{U}$ is an indecomposable module under the action of H. Further by 14.3.33, $F^*(M) = P \cong \mathbf{Z}_4^3$, and $M/P \cong L_3(2)$. Thus G is of type HS in the sense of section I.4 of Volume I, so we quote the classification theorem stated there as I.4.8 to conclude that $G \cong HS$.

14.4. Finishing the treatment of $\langle \mathbf{V^{G_1}} \rangle$ nonabelian

In this section, we assume Hypothesis 14.3.1 holds, and continue the notation of section 14.3. In addition, we assume $U := \langle V^{G_1} \rangle$ is extraspecial. In particular, Hypothesis 14.3.10 holds, and we can appeal to results in the later subsections of section 14.3.

---

[1]Notice we are here eliminating the shadow of $Aut(HS)$.

Theorem 14.3.26 handled the case where U is nonabelian but not extraspecial, so this section will complete the treatment of the case U nonabelian. Recall also by Theorem 14.3.16 that $L/O_2(L) \cong L_3(2)$.

Recall that in Hypothesis 14.3.10, $H := G_1$, $U = \langle V^{G_1} \rangle$, and we can appeal to results in both subsections of section 12.8. Also $g \in N_L(V_2) - H$ and $W := C_U(V_2)$. Let s be the generator of V_1^g. As U is extraspecial, $Z_U = V_1$, so that $\hat{U} = \tilde{U}$, $\dot{Z}_U^g = 1$, $\dot{H} = H^*$, and $d := m(\hat{U}) = m(\tilde{U})$. By 12.8.4.4, $Q := O_2(H) = C_H(\tilde{U})$. Let $K := O^2(H)$. By 12.8.8.2, H^* preserves a quadratic form on $\tilde{U}$, so $H^* \le O(\tilde{U}) \cong O_d^\epsilon(2)$ for $\epsilon := \pm 1$. Notice $C_H(\tilde{V}_2) = N_H(V_2) \le G_2$, so since $I_2 \trianglelefteq G_2$ by 12.8.9.1, $C_{H^*}(\tilde{V}_2)$ acts on W^{g*} by 12.8.12.2, and on $\tilde{E}$ since $E = W \cap W^g = W \cap W^l$, where V_1^l is the point of V_2 distinct from V_1 and V_1^g.

As $Z_U^{g*} = 1$, 12.8.11.5 becomes:

LEMMA 14.4.1. $m(\tilde{E}) + m(W^{g*}) = m(\tilde{U}) - 1 = d - 1$.

We next obtain a list of possiblities for H^* and U from G.11.2; all but the second case will eventually be eliminated, although several correspond to shadows which are not quasithin.

LEMMA 14.4.2. $m(\tilde{E}) = d/2$, so $m(W^{g*}) = d/2 - 1$, $U \cong Q_8^{d/2}$, and one of the following holds:

(1) $d = 4$ and $H^* \cong S_3 \times S_3$.

(2) $d = 4$ and H^* is E_9 extended by $\mathbf{Z}_2$.

(3) $d = 8$, $\tilde{U}$ is the natural module for $K^* \cong \Omega_4^+(4)$, and $W^{g*} = C_{T^* \cap K^*}(x^*)\langle x^* \rangle$, where $x^* \in W^{g*} - K^*$ interchanges the two components of K^*, and $m([\tilde{U}, x^*]) = 4$.

(4) $d = 8$, $H^* \cong S_7$, $\tilde{U} = \tilde{U}_1 \oplus \tilde{U}_2$, where $\tilde{U}_i$ is a totally singular K-module of rank 4, and $U_1^x = U_2$ for $x \in W^g - N_H(U_1)$.

(5) $d = 8$, $H^* \cong S_3 \times S_5$ or $S_3 \times A_5$, and $\tilde{U}$ is the tensor product of the natural module for S_3 and the natural or A_5-module for $L_2(4)$.

(6) $d = 12$ and $H^* \cong \mathbf{Z}_2 / \hat{M}_{22}$.

PROOF. Notice the assertion that $m(W^{g*}) = d/2 - 1$ will follow from 14.4.1 once we show $m(\tilde{E}) = d/2$, as will the assertion that $U \cong Q_8^{d/2}$.

By 14.3.13, H^* and its action on $\tilde{U}$ satisfy one of the conclusions of G.11.2, but not conclusion (1), (4), (5), or (12). Further by 14.3.23: $d \ge 4$, and if $d = 4$ then $\tilde{E} = \tilde{V}$ is of rank $2 = d/2$, so that either (1) or (2) of 14.4.2 holds, since in conclusions (ii) and (iii) of 14.3.23.2, $1 \ne Z_U^{g*}$, contrary to an earlier remark.

Suppose $d = 6$. Then conclusion (3) or (6) of G.11.2 holds. In either case, 27 divides the order of H^*, so $\epsilon = -1$ as 27 does not divide the order of $O_6^+(2)$. Therefore $m(E) \le m_2(U) = 3$, so $\tilde{E} = \tilde{V}$ is of rank 2 as $V \le E$ by 12.8.13.1, and hence $m(W^{g*}) = 3$ by 14.4.1. Thus conclusion (3) of G.11.2 does not hold, as there $m_2(H^*) = 2$. In conclusion (6), $C_{H^*}(\tilde{V}_2)$ acts on $\tilde{E}$ of rank 2, impossible as $C_{H^*}(\tilde{V}_2)$ is the stabilizer in H^* of a point of $\tilde{U}$, and so acts on no line of $\tilde{U}$.

In the remaining cases of G.11.2, we have $d = 8$ or 12. So $m(W^{g*}) = d/2 - 1$ by 14.3.14, and thus $m(\tilde{E}) = d/2$ by 14.4.1, completing the proof of the initial conclusions of the lemma as mentioned earlier.

If $d = 12$, then conclusion (13) of G.11.2 holds, and hence conclusion (6) of 14.4.2 holds. Thus we may assume one of conclusions (7)–(11) of G.11.2 holds, where $d = 8$.

Conclusion (10) of G.11.2 is impossible, as $m_3(H^*) = m_3(H) \leq 2$ since H is an SQTK-group. As $L_1^* T^* \leq H^*$ with $L_1^* T^* / O_2(L_1^* T^*) \cong S_3$, conclusion (11) of G.11.2 does not hold. Conclusions (8) and (9) of G.11.2 appear as conclusion (4) and (5) of 14.4.2. So it remains to show that conclusion (7) of G.11.2 leads to conclusion (3) of 14.4.2. In case (7) of G.11.2, $W^{g^*} \not\leq K^*$. Then as we saw $W^{g^*} \trianglelefteq C_{H^*}(\tilde{V}_2)$, it follows that $W^{g*} = \langle x^* \rangle Y^*$, where x^* is an involution interchanging the two components of H^*, $m([\tilde{U}, x^*]) = 4$, and $Y^* = C_{T^* \cap K^*}(x^*)$, as desired. $\qquad \square$

14.4.1. Characterizing $G_2(3)$ when $d = 4$. The only quasithin example satisfying Hypotheses 14.3.1 with U extraspecial is $G_2(3)$, occurring when $d = 4$, so our first main result treats this case:

THEOREM 14.4.3. *Assume Hypothesis 14.3.10 with U extraspecial. If $d = 4$, then $G \cong G_2(3)$.*

Until the proof of Theorem 14.4.3 is complete, assume G is a counterexample.

LEMMA 14.4.4. *(1) $K^* \cong E_9$.*
(2) $V = E$ and W^{g*} is of order 2, inverts K^*, and is generated by an involution of type c_2 on $\tilde{U}$.
(3) Either $H^* = K^* W^{g*}$, or $H^* \cong S_3 \times S_3$.
(4) L is an $L_3(2)$-block with $V = O_2(L)$.
(5) $UW^g \in Syl_2(L)$ and $U = O_2(L_1)$.
(6) L does not split over V, and $m_2(UW^g) = 3$.
(7) $H = KT$ and $M = LT$.

PROOF. As $d = 4$, case (1) or (2) of 14.4.2 holds, establishing (1) and (3) since $m(W^{g*}) = 1$ by 14.4.2. By 14.3.23.1, $V = E$. Thus the first two statements in (2) are established. By (1), H is a $\{2,3\}$-group. As $L_1^* \trianglelefteq H^*$ by (1), $\tilde{U} = [\tilde{U}, L_1]$ by 12.8.5.1, so that $U = [U, L_1] \leq L$. By 12.8.8.4, $O_2(LU) = O_2(L)$ is described in G.2.5. Therefore since $E = V$ and $m(U/V) = 2 = m_2(O_2(\bar{L}_1))$, $V = O_2(L)$, giving (4); and $\bar{U} = O_2(\bar{L}_1)$ so $U = O_2(L_1)$, and hence $T_L := T \cap L = UW^g$, giving (5).

Let $a \in W^g - U$. Then a inverts L_1^* with $\tilde{U} = [\tilde{U}, L_1]$, so using the structure of $O_4^+(2)$, either the remaining two statements of (2) hold, or a^* is of type a_2, $A := \langle a, [U, a] \rangle \cong E_{16}$, and $H^* = N_H(A)^* L_1^*$. In the latter case, a acts on a complement to V in U, so that UW^g splits over V; then by Gaschütz's Theorem A.1.39, L splits over V. Conversely if L splits over V, then from the structure of the split extension of E_8 by $L_3(2)$, $J(T_L) \cong E_{16}$, so a^* is of type a_2 and $A = J(T_L)$. Thus to complete the proof of (2) and (6), it remains to assume L splits over V, and obtain a contradiction. Set $N^+ := N_G(A)/C_G(A)$; then $[O_2(L_2), L_2] = A$, so that $L_2^+ T_L^+ \cong \mathbf{Z}_2 \times S_3$ while $N_K(A)^+ \cong A_4$. Then from the structure of $Aut(A) \cong GL_4(2)$, the subgroup of N^+ generated by $L_2 T$ and $N_K(A)$ is isomorphic to A_7. But then the stabilizer of z in this subgroup is $L_3(2)$, contradicting H a $\{2,3\}$-group.

For (7), observe $V = O_2(L) \trianglelefteq M$ by (4). Then as $H = KT$ and $KQ/Q \cong E_9$ by (1), $H \cap M = L_1 T$, so that $M = LT$ by 14.3.7. $\qquad \square$

LEMMA 14.4.5. *(1) $L = M$.*
(2) $U = O_2(H)$ and $H^* \cong \mathbf{Z}_2 / E_9$.
(3) $T = UW^g$.

PROOF. Let $K_1 := \langle W^{gH} \rangle$. By (1) and (2) of 14.4.4, K_1^* is $K^* \cong E_9$ extended by $W^{g*} \cong \mathbf{Z}_2$; so as $V \leq W^g$, $U = \langle V^H \rangle \leq K_1$. Hence using 14.4.4.5, UW^g of order

2^6 is Sylow in both L and K_1. Then $[K_1, C_T(L)] \le [K_1, C_T(U)] \le C_{K_1}(U) = V_1$, so $K \le C_G(C_T(L))$ by Coprime Action; therefore $C_T(L) = 1$ as $K \not\le M = !\mathcal{M}(LT)$. Let $A := O_2(M)$. As $M = LT$ and L is an $L_3(2)$-block with $V = O_2(L)$ by parts (4) and (7) of 14.4.4, A is elementary abelian by C.1.13.1, while $m(A/V) \le \dim H^1(L/V, V) = 1$ by C.1.13.b and I.1.6.4. Thus either (1) holds, or $A \cong E_{16}$ and T/A is regular on $A - V$ from the structure in B.4.8.3 of the unique indecomposable A with $[A, L] = V$. But in the latter case, $A = J(T)$ using 14.4.4.6, and all involutions in $T - L$ are in A. However as $J(T) = A$, $N_G(A) = M$ controls fusion in A by Burnside's Fusion Lemma A.1.35, so $a^G \cap L = \emptyset$ for $a \in A - L$, and then Thompson Transfer contradicts the simplicity of G. [2]

Therefore (1) is established. Now (3) follows from (1) and 14.4.4.5. Then $H = KT = K_1$, and (2) holds. $\qquad \square$

We are now in a position to complete the proof of Theorem 14.4.3. We will show G is of $G_2(3)$-type in the sense of section I.4, and then conclude $G \cong G_2(3)$ by the classification theorem stated in Volume I as I.4.5.

First by 14.4.4.4 and 14.4.5.1, $F^*(M) = V \cong E_8$ and $M/V \cong L_3(2)$. Second $U = O_2(H)$ by 14.4.5.2, and as $d = 4$, $U \cong Q_8^2$ by 14.4.2. By 14.4.4.1, $K^* \cong E_9$, so $K = K_1 K_2$, where $K_i \cong SL_2(3)$, $[K_1, K_2] = 1$, and $K_1 \cap K_2 = V_1$. By 14.4.5.2, $|H : K| = 2$. Further by 14.4.4.2, W^{g*} inverts K^*; so W^g, and hence also H, acts on K_i. Thus G is of $G_2(3)$-type, completing the proof of Theorem 14.4.3.

14.4.2. Eliminating the case d > 4. Having established Theorem 14.4.3, we may assume for the remainder of this section that $d > 4$; as no quasithin examples arise, we are working toward a contradiction. In fact $d = 8$ or 12 since one of cases (3)–(6) of 14.4.2 holds.

LEMMA 14.4.6. *If a^* is an involution in H^* then either*

(1) $m([\tilde{U}, a^*]) > 2$, or

(2) $H^* \cong S_3 \times S_5$ or $F^*(H) \cong \Omega_4^+(4)$, and in either case $\tilde{V}_2 \not\le [\tilde{U}, a^*]$ and $m([\tilde{U}, a^*]) = 2$.

PROOF. Assume (1) fails. Then by inspection of cases (3)–(6) in 14.4.2, either:

(a) conclusion (5) of 14.4.2 holds, with $H^* = H_1^* \times H_2^*$ where $H_1^* \cong S_3$, $H_2^* \cong S_5$, $\tilde{U}$ is the tensor product of the natural module for H_1^* and the A_5-module for H_2^*, and a^* is a transposition in H_2^*, or

(b) conclusion (3) of 14.4.2 holds, with a^* inducing an $\mathbf{F}_4$-transvection on $\tilde{U}$.

In case (a), $\tilde{U} = \tilde{U}_1 \oplus \tilde{U}_2$ is the sum of two irreducible H_2^*-modules with $C_{\tilde{U}_i}(T^* \cap H_2^*) = \langle \tilde{u}_i \rangle$ and $\tilde{u}_i$ singular in the orthogonal space $\tilde{U}_i$. Therefore as the generator $\tilde{s}$ of $\tilde{V}_2$ centralizes T^*, $\tilde{s} = \tilde{u}_1 + \tilde{u}_2$. However $[\tilde{U}_i, a^*] = \langle \tilde{v}_i \rangle$ with $\tilde{v}_i$ nonsingular, so $\tilde{s} \notin [\tilde{U}, a^*]$, and hence (2) holds.

Similarly in case (b), $\tilde{V}_2$ is contained in a singular $\mathbf{F}_4$-point of $\tilde{U}$, while $[\tilde{U}, a^*]$ is a nonsingular $\mathbf{F}_4$-point, so again (2) holds. $\qquad \square$

LEMMA 14.4.7. $U = O_2(H) = Q$.

PROOF. As U is extraspecial, $O_2(H^g) = U^g D$, where $D := C_{H^g}(U^g)$. Now as $g \in N_L(V_2)$, $V_2 \le U^g$, so $[D, W] \le C_W(U^g)$. But $C_W(U^g) \le U^g$ by 12.8.9.5, so that $C_W(U^g) = V_1^g$. Therefore if $D^* \ne 1$, either D induces transvections on $\tilde{U}$ with

---

[2]Notice here we are eliminating the shadow of $Aut(G_2(3))$.

axis $\tilde{W}$, or $[\tilde{W}, D] = \tilde{V}_1^g = \tilde{V}_2$ with $m([\tilde{U}, d]) \le 2$ for each $d \in D$. This contradicts 14.4.6, so D centralizes $\tilde{U}$, and hence $[D, W] \le V_1 \cap V_1^g = 1$. Recall that W^{g*} is elementary abelian of rank $d/2 - 1 > 1$ by 14.4.2, and this forces $K^* = [K^*, W^{g*}]$ in each of cases (3)–(6) of 14.4.2. Thus by symmetry $K^g = [K^g, W] \le C_G(D)$. But $K^g \not\le M$ and $M = !\mathcal{M}(LT^g)$, so as T^g acts on $C_D(L)$, it follows that $C_D(L) = 1$.

Next we saw D centralizes $\tilde{U}$ so that $[D, U] \le V_1 \le V$, and hence by symmetry, $[D, U^x] \le V_1^x \le V$ for each $x \in L$. Thus $L \le \langle U^x : x \in L \rangle =: I$ centralizes DV/V. Further by 12.8.8.4, I is described by G.2.5, so $S := U^g W C_{U^l}(V)$ is Sylow in LS, for $l \in L - L_2 T$. As W centralizes D, so does $C_{U^l}(V)$ by symmetry, so that S centralizes D; then we conclude from Gaschütz's Theorem A.1.39 that $DV = V \times C_D(L)$ with $C_D(L)$ a complement to V_1^g in D. Then as $C_D(L) = 1$, $D = V_1^g$. Then $Q^g = U^g$, so $Q = U$, completing the proof of the lemma. $\qquad\square$

We now define certain $\{2, 3\}$-subgroups X of H, which are analogous to L_1: for example, 14.4.8 will show that $\langle X, L_2 \rangle =: L_X$ satisfies the hypotheses of L. Then 14.4.13 will show that $\langle LT, L_X \rangle \cong L_4(2)$, leading to our final contradiction.

So let $\mathcal{X}$ consist of the set of T-invariant subgroups $X = O^2(X)$ of H such that $|X : O_2(X)| = 3$. Let $\mathcal{Y}$ consist of those $X \in \mathcal{X}$ such that $V_X := [V_2, X]$ is of rank 3 and contained in E, and set $L_X := \langle L_2, X \rangle$.

LEMMA 14.4.8. *(1)* $L_1 \in \mathcal{Y}$, with $V_{L_1} = [V_2, L_1] = V$ and $L_{L_1} = L$.

(2) If $X \in \mathcal{Y}$ *then* $L_X \in \mathcal{L}_f^*(G, T)$, $L_X/O_2(L_X) \cong L_3(2)$, $L_X T$ *induces* $GL(V_X)$ *on* V_X *with kernel* $O_2(L_X T)$, *and* I_2 *and* XT *are the maximal parabolics of* $L_X T$ *over* T.

PROOF. By construction, $L_1 \in \mathcal{X}$ with $V = [V_2, L_1]$, and $V \le E$ by 12.8.13.1. Thus (1) holds.

Assume $X \in \mathcal{Y}$. Then $V_2 \le V_X \le E \le U \cap U^g$, so U and U^g act on V_X, and hence also $I_2 = \langle U, U^g \rangle$ acts on V_X. Then $Aut_{I_2}(V_X)$ is the maximal subgroup of $GL(V_X)$ stabilizing the hyperplane V_2 of V_X, and X does not act on that hyperplane as $V_X = [V_X, X]$, so $L_X/C_{L_X}(V_X) = GL(V_X)$. Thus there is $L_+ \in \mathcal{C}(L_X)$ with $L_+ C_{L_X}(V_X) = L_X$, so $L_+ \in \mathcal{L}_f(G, T)$. Then by 14.3.4.1, $L_+ \in \mathcal{L}_f^*(G, T)$ and $L_+/O_2(L_+) \cong L_3(2)$. The projection P of L_2 on L_+ satisfies $P = [P, T \cap L_+]$, so as T acts on L_2, $L_2 = [L_2, T \cap L_+] \le L_+$. Similarly $X \le L_+$, so $L_X = L_+$, and (2) holds. $\qquad\square$

The shadow of the Harada-Norton group F_5 is eliminated in the proof of the next lemma. We obtain a contradiction in the 2-local which would correspond to the local subgroup $\Omega_6^-(2)/E_{2^6}$ in F_5.

LEMMA 14.4.9. *Case (3) of 14.4.2 does not hold.*

PROOF. Assume case (3) of 14.4.2 holds. Then we can view $\tilde{U}$ as a 4-dimensional orthogonal space over $\mathbf{F}_4$ preserved by K^*. In particular $\tilde{V}_2 = C_{\tilde{U}}(W^{g*})$ lies in some totally singular $\mathbf{F}_4$-point $\tilde{U}_2$ of $\tilde{U}$. Further $\mathcal{X} = \{X_1, X_2\}$, where a subgroup of order 3 in each X_i^* is diagonally embedded in K^*, and we may choose notation so that $[X_2, \tilde{U}_2] = 1$ and $\tilde{U}_2 = [X_1, \tilde{U}_2]$. Thus $L_1 = X_1$ and $V = U_2$ by 14.4.8.1.

Therefore the subspace $\tilde{V}^{\perp_2}$ orthogonal to $\tilde{V}$ in the $\mathbf{F}_2$-orthogonal space $\tilde{U}$ is the same as the subspace $V^{\perp_4} =: \tilde{W}_1$ orthogonal to $\tilde{V}$ in $\mathbf{F}_4$-orthogonal space $\tilde{U}$. Choose $k \in K$ so that $\tilde{V}^k \not\le \tilde{W}_1$. As $\tilde{W}_1$ is an $\mathbf{F}_2$-hyperplane of $\tilde{W}$ and L_1 is transitive on $\tilde{V}^{\#}$, we can choose k so that $s^k \in W$ (recall s is the generator of V_1^g).

As the preimage W_1 of $\tilde{W}_1$ satisfies $W_1 = C_U(V)$, $W_1^{g*} = C_{W^{g*}}(V) = W^g \cap K^*$ since $C_{H^*}(V) \leq K^* \cong L_2(4) \times L_2(4)$ as case (3) of 14.4.2 holds. Therefore as $s^k \in W - W_1$, for some $x \in G$ there is $i := z^x \in W^g$ with $i^* \notin K^*$.

Then $K = K_1 K_1^i$, where $K_1 \in \mathcal{C}(H)$ and $K_1^* \cong L_2(4)$. As case (3) of 14.4.2 holds, $m([\tilde{U}, i^*]) = 4$. Thus by Exercise 2.8 in [**Asc94**], $C_{H^*}(\tilde{i}) = C_{H^*}(i^*)$. Let $K_0 := O^2(C_H(\tilde{i}))$. Then $K_0^* \cong L_2(4)$ is diagonally embedded in K^*, and K_0 centralizes $\langle i, z \rangle$, so $K_0 = O^2(C_H(i))$. Of course K_0 acts on $[\tilde{U}, i]$, and since diagonal subgroups of K^* of order 3 centralize a subspace of $\tilde{U}$ of rank exactly 4, it follows that $[\tilde{U}, i]$ is the A_5-module for K_0^*.

Let $D := [U, i]\langle i, z \rangle$. Then $D \cong E_{64}$ since K_0 is irreducible on $[\tilde{U}, i]$ of rank 4. Further as U is extraspecial, $K_0 U$ acts on D with $C_D(u) \leq [U, i]V_1$ for each $u \in U - [U, i]V_1$, and $U/[U, i]V_1$ induces the full group of transvections on $[U, i]V_1$ with center V_1. In particular $C_D(U) = \langle z \rangle$, $D = C_{UD}(D)$, and $U/[U, i]V_1$ is also the A_5-module for $K_0 U/U$. Further as $Q = U$ by 14.4.7, $D = O_2(C_H(i)) = O_2(C_G(\langle z, i \rangle))$ and $UD = O_2(K_0 UD)$.

Next $K_0 = O^2(C_{H^x}(z))$, so we conclude that z interchanges the two members of $\mathcal{C}(H^x)$. Thus we have symmetry between i and z, and so U^x acts on D with $C_D(U^x) = \langle i \rangle$. Therefore as D is an indecomposable $K_0 U$-module with chief series $1 < V_1 < [U, i]V_1 < D$, it follows that $Y := \langle K_0 U, U^x \rangle$ is irreducible on D.

Now let $T_D := N_T(D) \in Syl_2(N_H(D))$, and $G_D := N_G(D)$. As Y is irreducible on D, $D \leq Z(O_2(G_D))$, so as $D = C_{UD}(D)$, $D = UD \cap O_2(G_D)$. As $C_D(T_D) \leq C_D(U) = \langle z \rangle$, $N_G(T_D) \leq H$ so that $T_D \in Syl_2(G_D)$.

Next $K_0 \in \mathcal{L}(G_D, T_D)$, so $K_0 \leq K_+ \in \mathcal{C}(G_D)$ by 1.2.4, and as $D = DU \cap O_2(G_D)$, $K < K_+$. However A.3.14 contains no "B" with $O_2(B)$ the A_5-module UD/D for $K_0 D/UD$. This contradiction completes the proof of 14.4.9. $\qquad\square$

The elimination of the A_5-module in part (3) of the next lemma 14.4.10 rules out the shadow of the non-quasithin group $\Omega_8^-(2)$. Again we obtain a contradiction working in the 2-local corresponding to the local $\Omega_6^-(2)/E_{2^6}$ in the shadow.

LEMMA 14.4.10. *Assume case (5) of 14.4.2 holds. Then*

(1) $\mathcal{X} = \{X_1, X_2\}$ *where* $X_1 := O^2(O_{2,3}(H))$ *and* $X_2 := O^2(B)$ *for* B *a* T-*invariant Borel subgroup of* $K_0 := H^\infty$.

(2) There is a unique T-*invariant chief factor* $\tilde{U}_1$ *for* K_0, *and* $\tilde{V}_2 \leq C_{\tilde{U}}(T) \leq \tilde{U}_1$.

(3) $\tilde{U}_1$ *is the* $L_2(4)$-*module for* K_0^*.

(4) $\mathcal{X} = \mathcal{Y}$.

(5) $H^* \cong S_3 \times S_5$ *and* $X_1 X_2 T/O_2(X_1 X_2 T) \cong S_3 \times S_3$.

PROOF. Assume conclusion (5) of 14.4.2 holds. Let $K_0 := H^\infty$. It is easy to check that (1) and (2) hold, with $\tilde{U}_1 := [\tilde{U}, x]$ for $x^* \in C_{W^{g*}}(K_0^*)$; such an x^* exists since W^{g*} is of rank 3 by 14.4.2. Also $[\tilde{U}, X_1] = \tilde{U}$, so V_{X_1} is of rank 3, and there is a K_0-complement $\tilde{U}_2$ to $\tilde{U}_1$. Recall that $[W^g, W] \leq E$ by 12.8.11.1, and that $\tilde{W} = \tilde{V}_2^\perp$. Then computing in either module for A_5 in case (5) of 14.4.2, we obtain

$$\tilde{V}_{X_1} \cap \tilde{U}_2 \leq [\tilde{V}_2^\perp \cap \tilde{U}_2, W^{g*} \cap K_0^*] \leq \tilde{E}.$$

So as $V_{X_1} = \langle V_2, V_{X_1} \cap U_2 \rangle$, $X_1 \in \mathcal{Y}$.

Assume first that $\tilde{U}_1$ is the $L_2(4)$-module for K_0^*. Then (3) holds, and $\tilde{V}_{X_2} = [\tilde{V}_2, X_2]$ is the $\mathbf{F}_4$-point in $\tilde{U}_1$ containing $\tilde{V}_2$. Now $[\tilde{U}, x] \leq \tilde{W}$ as x acts on the hyperplane $\tilde{W}$ of $\tilde{U}$, so $\tilde{V}_{X_2} \leq [\tilde{U}_1, W^g] \leq \tilde{E}$. Thus $X_2 \in \mathcal{Y}$, and hence (4) holds. As $X_2 \in \mathcal{Y}$, $X_2 T / O_2(X_2 T) \cong S_3$ by 14.4.8.2, so (5) holds, completing the proof of the lemma for the $L_2(4)$-module.

Thus as conclusion (5) of 14.4.2 holds, we may assume instead that $\tilde{U}_1$ is the A_5-module, and it remains to derive a contradiction.

As $\tilde{U}_1$ is the A_5-module, X_2 centralizes $\tilde{V}_2$, so that $X_2 \notin \mathcal{Y}$. Hence we conclude from (1) and 14.4.8.1 that $L_1 = X_1$ and $V = V_{X_1}$. Then X_2 centralizes $\langle \tilde{V}_2^{X_1} \rangle = \tilde{V}$. Thus $X_2 \leq C_G(V) \leq M$ using Coprime Action, and then $[L, X_2] \leq C_L(V) = O_2(L)$, so that X_2 acts on L_2 and hence on $\langle U^{L_2} \rangle = I_2$. Let $G_0 := \langle I_2, K_0 \rangle$, $V_0 := \langle z^{G_0} \rangle$, and $G_0^+ := G_0 / C_{G_0}(V_0)$.

Suppose $O_2(G_0) = 1$. Then Hypothesis F.1.1 is satisfied with K_0, I_2, T in the roles of "L_1, L_2, S"; for example we just saw that $B_1 := N_{K_0}(T \cap K_0) = X_2(T \cap K_0)$ normalizes I_2. Thus $\alpha := (K_0 T, T X_2, I_2 X_2)$ is a weak BN-pair by F.1.9. Further $B_2 := N_{I_2}(K_0) = T \cap I_2$, so $T \trianglelefteq T B_2$, and hence the hypotheses of F.1.12 are satisfied. Therefore α is described in F.1.12. This is a contradiction as $U = O_2(K_0) \cong Q_8^4$ and $K_0/U \cong L_2(4)$, a configuration not appearing in F.1.12.

Thus $O_2(G_0) \neq 1$, so $G_0 \in \mathcal{H}(T)$, and $V_0 \in \mathcal{R}_2(G_0)$ by B.2.14. By 1.2.4, $K_0 \leq J \in \mathcal{C}(G_0)$. Then $1 \neq [V_2, K_0] \leq [V_0, J]$, so that $J \in \mathcal{L}_f(G, T)$ by 1.2.10. Then $J \in \mathcal{L}_f^*(G, T)$ by 14.3.4, so that $K_0 = J$ by 13.1.2.5. Now $I_2 = O^2(I_2)$ normalizes K_0 by 1.2.1.3, and hence acts on $Z(O_2(K_0)) = V_1$, contradicting $I_2 \nleq G_1$. $\qquad \square$

LEMMA 14.4.11. *Assume case (4) of 14.4.2 holds. Then*

(1) We can represent $H^* \cong S_7$ on $\Omega := \{1, \ldots, 7\}$ so that T preserves the partition $\{\{1, 2, 3, 4\}, \{5, 6\}, \{7\}\}$ of Ω.

(2) $\mathcal{Y} = \{X_1, X_2\}$, where $X_1 := O^2(H_{1,2,3,4})$ and $X_2 = O^2(H_{5,6,7})$. In particular, $X_1 X_2 T / O_2(X_1 X_2 T) \cong S_3 \times S_3$.

PROOF. Part (1) is trivial; cf. the convention in section B.3. Further $\mathcal{X} = \{X_1, X_2, X_3\}$, where X_1 and X_2 are defined in (2), $X_3 := O^2(P)$ for P the stabilizer of the partition $\{\{1, 2\}, \{3, 4\}, \{5, 6\}, \{7\}\}$, and $X_1 X_2 T / O_2(X_1 X_2 T) \cong S_3 \times S_3$.

Next $\tilde{U} = \tilde{U}_1 \oplus \tilde{U}_2$, where $\tilde{U}_1$ is a 4-dimensional irreducible for K, and $\tilde{U}_2 = \tilde{U}_1^x$ for $x^* \in W^{g*} - K^*$ is dual to $\tilde{U}_1$. Now $C_{\tilde{U}_i}(N_T(U_1)) = \langle \tilde{u}_i \rangle$ for suitable $\tilde{u}_i$, so $\tilde{s} = \tilde{u}_1 \tilde{u}_2$, with $C_{H^*}(\tilde{s}) = P^* = X_3^* T^*$ from the structure of the sum of $\tilde{U}_1$ and its dual. In particular, $X_3 \notin \mathcal{Y}$. Recall $W^{g*} \trianglelefteq C_{H^*}(\tilde{V}_2) = P^*$ and $m(W^{g*}) = 3$ by 14.4.2, so $W^{g*} = O_2(P^*) = \langle x_i^* : 1 \leq i \leq 3 \rangle$, where $x_i^* := (2i - 1, 2i)$ on Ω. Let $[U, x_i] =: D_i$. Then $D_i \leq W$ and $\tilde{D}_i$ is of rank 4, so $\tilde{D}_i$ is the $L_2(4)$-module for $Y_i^* := C_{K^*}(x_i^*) \cong S_5$ since elements of order 3 in Y_i^* are fixed-point-free on $\tilde{U}$. As such elements lie in X_i for $i = 1, 2$, V_{X_i} is of rank 3. Further $X_2^* \leq Y_3^*$ with $V_{X_2} \leq [D_3, x_1^* x_2^*] \leq [D_3, W^g] \leq E$ by 12.8.11.1, so $X_2 \in \mathcal{Y}$. Similarly a Sylow 3-group B^* of X_1^* is contained in Y_1^* with $\tilde{V}_{X_1} = [\tilde{V}_2, B] = [\tilde{V}_2, x_2^* x_3^*] \leq [\tilde{D}_1, W^g] \leq \tilde{E}$, so $X_1 \in \mathcal{Y}$, completing the proof of (2). $\qquad \square$

LEMMA 14.4.12. *Assume case (6) of 14.4.2 holds. Then $\mathcal{Y} = \{X_1, X_2\}$ where $X_1 := O^2(O_{2,3}(H))$ and $X_1 X_2 T / O_2(X_1 X_2 T) \cong S_3 \times S_3$.*

PROOF. First (cf. H.12.1.5) $C_{\tilde{U}}(T) = \tilde{V}_2$ and $C_{H^*}(\tilde{V}_2) \cong S_5 / E_{32}$. We have seen that $W^{g*} \trianglelefteq C_{H^*}(\tilde{V}_2)$, and by 12.8.1, $m(W^{g*}) = 5$, so $W^{g*} = O_2(C_{H^*}(\tilde{V}_2))$.

Next we calculate that $\mathcal{X} = \{X_1, X_2, X_3\}$, where $X_1 := O^2(O_{2,3}(H))$, $X_3 := O^2(B)$ where B^* is a Borel subgroup of $C_{H^*}(\tilde{V}_2)$, and $X_2 T$ is a minimal parabolic in the remaining maximal 2-local $D^* := \langle TX_2, X_3 \rangle^* \cong S_6/E_{16}/\mathbf{Z}_3$ of H^* over T^*, which does not centralize $\tilde{V}_2$. As X_3 centralizes $\tilde{V}_2$, $X_3 \notin \mathcal{Y}$.

Let $\tilde{U}_D := \langle \tilde{V}_2^D \rangle$; then $O_2(D^*)$ centralizes $\tilde{U}_D$, and $\tilde{U}_D$ is the 6-dimensional irreducible for $D^+ := D^*/O_2(D^*) \cong \hat{S}_6$. Now $\tilde{U}$ has the structure of a 6-dimensional $\mathbf{F}_4$-space preserved by K^*, with the $\mathbf{F}_4$-points the irreducibles for X_1^*. This $\mathbf{F}_4$-space structure restricts to $\tilde{U}_D$ of $\mathbf{F}_4$-dimension 3, and $\tilde{V}_{X_1}$ is the $\mathbf{F}_4$-point containing $\tilde{V}_2$, so that $V_{X_1} \cong E_8$. Further $T^*X_1^*X_2^*$ is the stabilizer of an $\mathbf{F}_4$-line $\tilde{U}_0$ of $\tilde{U}_D$ containing $\tilde{V}_{X_1}$, with $X_1X_2T/O_2(X_1X_2T) \cong S_3 \times S_3$. In particular X_2 is fixed-point-free on $\tilde{U}_0$, so $V_{X_2} \cong E_8$. Thus to complete the proof, it remains to show that $V_{X_i} \leq E$ for $i = 1, 2$. Now $\tilde{U}_D$ is totally singular, since $\tilde{U}_D$ is not self-dual as a D-module. Thus $\tilde{U}_D \leq \tilde{V}_2^\perp = \tilde{W}$, so by 12.8.11.1 it suffices to show $\tilde{V}_{X_i} \leq [\tilde{U}_D, W^g]$. But there is $x \in W^g$ inverting X_1^+ with x^+ centralizing X_2^+. As x^+ inverts X_1^+, $C_{\tilde{U}_D}(x) = [\tilde{U}_D, x]$ is of rank 3, and $\tilde{V}_2 \leq C_{\tilde{U}_D}(x)$. Then $\tilde{V}_{X_2} = [\tilde{V}_2, X_2] \leq [\tilde{U}_D, x] \leq [\tilde{U}_D, W^g]$, as required. As $W^{g*} \cap K^*$ induces a group of $\mathbf{F}_4$-transvections on $\tilde{U}_D$ with center $\tilde{V}_{X_1}$, $\tilde{V}_{X_1} \leq [\tilde{U}_D, W^{g*} \cap K^*] \leq \tilde{E}$. This completes the proof. $\qquad \square$

By 14.4.9–14.4.12, we have reduced to the situation where one of cases (4)–(6) of of 14.4.2 holds, and in case (5) the chief factors for H^∞ on $\tilde{U}$ are $L_2(4)$-modules. By 14.4.8.1, $L_1 \in \mathcal{Y}$; hence 14.4.10–14.4.12 show that in each case $\mathcal{Y} = \{L_1, X\}$ is of order 2, with $XL_1T/O_2(XL_1T) \cong S_3 \times S_3$.

Let $H_1 := LT$, $H_2 := L_1XT$, and $H_3 := L_XT$. Set $\mathcal{F} := \{H_1, H_2, H_3\}$ and $G_0 := \langle \mathcal{F} \rangle$.

LEMMA 14.4.13. $G_0 \cong L_4(2)$.

PROOF. We show that $(G_0, \mathcal{F})$ is an A_3-system as defined in section I.5. Then the lemma follows from Theorem I.5.1. We just observed that $H_2/O_2(H_2) \cong S_3 \times S_3$ and $H_i/O_2(H_i) \cong L_3(2)$ for $i = 1, 3$ by 14.4.8.2, so (D1) and (D2) hold. As L_2T is maximal in H_1 and H_3 but $X \neq L_1$, $L_2T = H_1 \cap H_3$, so $L_2 = L \cap H_3 \trianglelefteq M \cap H_3$ and hence $L_2T = M \cap H_3$. Thus as $M = !\mathcal{M}(LT)$, $O_2(G_0) = 1$, [3] so hypothesis (D4) holds. Similarly $L_1T = H_1 \cap H_2$ and $XT = H_2 \cap H_3$, so (D3) holds. Finally (D5) is vacuous for a system of type A_3. $\qquad \square$

We are now in a position to obtain a contradiction to our assumption that $d > 4$. Namely as $|T| \geq |U| > 2^9$, G_0 is not $L_4(2)$, contrary to 14.4.13. This contradiction shows:

THEOREM 14.4.14. Assume Hypothesis 14.3.1 holds with $\langle V^{G_1} \rangle$ nonabelian. Then $L/O_2(L) \cong L_3(2)$ and G is isomorphic to HS or $G_2(3)$.

PROOF. By assumption, Hypothesis 14.3.10 holds. Thus $L/O_2(L) \cong L_3(2)$ by Theorem 14.3.16. Then by Theorem 14.3.26, either $U = \langle V_1^{G_1} \rangle$ is extraspecial or $G \cong HS$, and we may assume the former. Hence if $d = m(\tilde{U}) = 4$, then $G \cong G_2(3)$ by Theorem 14.4.3. Finally we just obtained a contradiction under the assumption that U is extraspecial and $d > 4$, so the proof of Theorem 14.4.14 is complete. $\qquad \square$

---

[3]The group J_4 has the involution centralizer appearing in case (6) of 14.4.2, and there is $L \in \mathcal{L}_f(G, T)$ with $L/O_2(L) \cong L_3(2)$, but the condition $O_2(G_0) = 1$ fails as $L \notin \mathcal{L}_f^*(G, T)$.

14.5. Starting the case $\langle V^{G_1} \rangle$ abelian for $L_3(2)$ and $L_2(2)$

In this section, and indeed in the remainder of the chapter, we assume:

HYPOTHESIS 14.5.1. *Hypothesis 14.3.1 holds and* $U := \langle V^{G_1} \rangle$ *is abelian.*

As U is abelian and $\Phi(V) = 1$, U is elementary abelian. Recall from the discussion after 14.3.6 that Hypothesis 12.8.1 holds. In particular $G_1 \not\leq M$ by 12.8.3.4, so that $V < U$. Recall also the definitions of G_i, L_i, and V_i, for $i \leq \dim(V)$, from Notation 12.8.2.

LEMMA 14.5.2. *If* $g \in G$ *with* $1 \neq V \cap V^g$, *then* $[V, V^g] = 1$.

PROOF. As $\langle V^{G_1} \rangle$ is abelian by Hypothesis 14.5.1, the results follows from the equivalence of (2) and (3) in 12.8.6. $\square$

14.5.1. A result on $X \in \mathcal{H}(T)$ with $X/O_2(X) = L_2(2)$. Recall that under case (2) of Hypothesis 14.3.1 where $L/O_2(L) \cong L_2(2)'$, 14.3.5 says there exists no $X \in \mathcal{H}(T, M)$ such that $X/O_2(X) \cong L_2(2)$. In this subsection, we establish a result providing some restrictions on such subgroups in case (1) of Hypothesis 14.3.1, where $L/O_2(L) \cong L_3(2)$. Namely we prove:

THEOREM 14.5.3. *Suppose* $Y = O^2(Y) \leq G_1$ *is* T-*invariant with* $YT/O_2(YT) \cong L_2(2)$. *Then*

(1) Either $Y \leq M$, *or case (1) of Hypothesis 14.3.1 holds and* $[V_2, Y] = 1$.
(2) If $YL_1 = L_1Y$, *then* $Y \leq M$.
(3) $\langle \tilde{V}_2^Y \rangle$ *is not isomorphic to* E_8.

Until the proof of Theorem 14.5.3 is complete, assume Y is a counterexample.

LEMMA 14.5.4. *(1)* $Y \not\leq M$.
(2) Case (1) of Hypothesis 14.3.1 holds, namely $L/O_2(L) \cong L_3(2)$.

PROOF. Assume (1) fails, so that $Y \leq M$. Then conclusions (1) and (2) of Theorem 14.5.3 are satisfied. Further Y acts on V by 14.3.3.6. Thus as $V_2 \leq V$, $m(\langle \tilde{V}_2^Y \rangle) \leq m(\tilde{V}) \leq 2$, so that conclusion (3) of 14.5.3 holds. This contradicts our assumption that we are working in a counterexample.

Thus (1) is established. Then (1) and 14.3.5 imply (2). $\square$

Set $X := L_2$, and $H := \langle X, Y, T \rangle$. Notice that $H \not\leq G_1$ since $X \not\leq G_1$. Set $V_H := \langle V_1^H \rangle$, $Q_H := O_2(H)$, $\dot{H} := H/Q_H$, and $H^* := H/C_H(V_H)$. Observe that (H, XT, YT) is a Goldschmidt triple (in the language of Definition F.6.1), so by F.6.5.1, $\alpha := (\dot{X}\dot{T}, \dot{T}, \dot{Y}\dot{T})$ is a Goldschmidt amalgam, and so is described in F.6.5.2.

LEMMA 14.5.5. $Q_H \neq 1$.

PROOF. Assume $Q_H = 1$. By 1.1.4.6, XT and YT are in $\mathcal{H}^e$, and so satisfy Hypothesis F.1.1 in the roles of "L_1, L_2", with T in the role of "S". Then α is a weak BN-pair of rank 2 by F.1.9, and the hypothesis of F.1.12 is satisfied, so that α is described in case (vi) of F.6.5.2. Then as X has at least two noncentral 2-chief factors (from V and the image of $O_2(L_2)$ in $L/O_2(L) \cong L_3(2)$), by inspection of that list, α is isomorphic to the amalgam of $G_2(2)'$, $G_2(2)$, M_{12}, or $Aut(M_{12})$, and X has exactly two such factors. In each case, $Z = \Omega_1(Z(T))$ is of order 2, so $V_1 = Z$.

Next we saw $V < \langle V^{G_1} \rangle = U \trianglelefteq YT$, so $m_2(U) \geq 4$ since $m(V) = 3$ by 14.5.4.2. As the 2-rank of $G_2(2)'$, $G_2(2)$, and M_{12} is at most 3, it follows that α is the $Aut(M_{12})$-amalgam and $m(U) = 4$. Thus $A := Aut(M_{12})$ is a faithful completion of α, so identifying YT with its image under this completion, we may assume $YT \leq A$. As $m_2(M_{12}) = 3$, there is an involution $u \in U - M_{12}$. Thus $C_A(u) \cong \mathbf{Z}_2/(E_4 \times J)$ where $J := C_G(u)^\infty \cong A_5$. Therefore $U = J(C_T(u)) = O_2(C_A(u)) \times (U \cap J)$. Then from the structure of A, $N_A(U) = T(N_A(U) \cap C_A(u))$, and $V_1 = Z = C_U(T) \leq J$. Thus $|YT| = 2^7 \cdot 3 = |N_A(U)|$, so $N_A(U) = YT$ as $U \trianglelefteq YT$. This is a contradiction as YT centralizes V_1 but $Z(N_J(U)) = 1$. $\qquad \square$

By 14.5.5 and 1.1.4.6, $H \in \mathcal{H}(T) \subseteq \mathcal{H}^e$.

LEMMA 14.5.6. Y^* does not act on X^*.

PROOF. Assume otherwise. Then $V_1^{X^*Y^*} = V_1^{Y^*X^*} = V_1^{X^*}$ as $Y \leq G_1$. Therefore Y acts on $\langle V_1^X \rangle = V_2$, and hence $[Y, V_2] = 1$ by Coprime Action. Thus Y is not a counterexample to conclusion (1) or (3) of 14.5.3, so Y must be a counterexample to conclusion (2). Therefore $YL_1 = L_1Y$, and hence Y acts on $\langle V_2^{L_1} \rangle = V$, contradicting 14.5.4.1. $\qquad \square$

Set $H^+ := H/O_{3'}(H)$.

LEMMA 14.5.7. (1) $C_X(V_H) \leq O_2(X)$ and $C_Y(V_H) \leq O_2(Y)$.
(2) $Q_H = C_T(V_H)$.
(3) $V_H \leq Z(Q_H)$ and $O_2(H^*) = 1$.
(4) $Q_H \in Syl_2(O_{3'}(H))$, so $O_{3'}(H)$ is 2-closed and in particular solvable.
(5) Either

      (i) H^+ is described in Theorem F.6.18, or
      (ii) $O_2(XT) = O_2(YT) = Q_H$, and $H^+ \cong S_3$.

PROOF. We saw $H \in \mathcal{H}^e$, so as $V_1 \leq Z$, part (3) follows from B.2.14. Next $C_X(V_H) \leq O_2(X)$ as $X \nleq G_1$. Thus if $Y \leq C_H(V_H)$, then $Y^* = 1$, so Y^* acts on X^*, contrary to 14.5.6. Hence (1) holds. By (3), $Q_H \leq C_T(V_H)$, while by (1), we may apply F.6.8 to $C_H(V_H)$ in the role of "X" to conclude that $C_T(V_H) \leq Q_H$, so (2) holds. Similarly F.6.11.1 implies (4), and F.6.11.2 implies (5) as H is an SQTK-group. $\qquad \square$

LEMMA 14.5.8. H is solvable.

PROOF. Assume H is nonsolvable. Then by 1.2.1.1 there is $K \in \mathcal{C}(H)$, and by 14.5.7.2, $C_H(V_H)$ is 2-closed and hence solvable, so $K^* \neq 1$. Then $K \in \mathcal{L}_f(G, T)$ by 1.2.10, so by 14.3.4.1, $K/O_2(K) \cong A_5$ or $L_3(2)$. Now $O_2(K) = O_{3'}(K) = C_K(V_H)$, so $K^+ \cong K/O_2(K) \cong K^*$. By 14.5.7.5, H^+ is described in F.6.18, so we conclude that case (6) of F.6.18 holds, with $H^+ = K^+ \cong L_3(2)$. Hence $K = O^{3'}(H) = \langle X, Y \rangle$. Then $K = O^2(H)$ by F.6.6.3, so that $H = KT$. Now as $[V_1, Y] = 1$ and $\langle V_1^X \rangle = V_2 \cong E_4$, V_H is the natural module for K^* by H.5.5. In particular $V_2 = \langle V_1^X \rangle \leq V_H$ and $V_H = \langle V_2^Y \rangle$.

By 14.3.4.2, $K \in \mathcal{L}_f^*(G, T)$, so by our discussion after Hypothesis 14.3.1, part (1) of that Hypothesis holds with K in the role of "L". Thus by Theorem 14.4.14, either $\langle V_H^{G_1} \rangle$ is abelian, or $G \cong G_2(3)$ or HS. However in the latter two cases, L is the unique member of $\mathcal{L}_f^*(G, T)$, so $K = L \leq M$, contrary to 14.5.4.1. Therefore

$\langle V_H^{G_1} \rangle$ is abelian, so we have symmetry between LT, V and $H = KT$, V_H; that is, Hypothesis 14.5.1 holds with H, V_H in the roles of "LT, V".

Now $Y \not\leq M = !\mathcal{M}(LT)$, so that $O_2(\langle LT, H \rangle) = 1$. Hence Hypotheses F.7.1 and F.7.6 are satisfied with LT and H in the roles of "G_1" and "G_2", so we can form the coset geometry Γ of Definition F.7.2 with respect to this pair. Similarly we can form the dual geometry Γ' where the roles of LT and H are reversed. Let $\gamma_0 := LT$, $\gamma_1 := H$, and for $g, h \in \langle LT, H \rangle$ let $V_{\gamma_0 g} := V^g$ and $V_{\gamma_1 h} := V_H^h$. Also for $\sigma \in \Gamma$ let $Q_\sigma := O_2(G_\sigma)$. Observe $G_{\gamma_0, \gamma_1} = LT \cap H = XT$ and $\ker_{XT}(G_i) = O_2(G_i)$ for $i = 1, 2$, is the centralizer in G_i of V or V_H, respectively, so

$$Q_\sigma = G_\sigma^{(1)} = C_{G_\sigma}(V_\sigma).$$

Next as usual choose a geodesic

$$\alpha := \alpha_0, \ldots, \alpha_b =: \beta$$

in Γ of minimal length b, subject to $V_\alpha \not\leq Q_\beta$. Then $b = \min\{b(\Gamma, V), b(\Gamma', V_H)\}$, so by F.7.9.1, $V_\alpha \leq G_\beta$ and $V_\beta \leq G_\alpha$, and hence

$$1 \neq [V_\alpha, V_\beta] \leq V_\alpha \cap V_\beta. \tag{$*$}$$

Thus by 14.5.2 and the corresponding result for V_H, β is not conjugate to α, so b is odd. Replacing Γ by Γ' if necessary, we may assume $V_\alpha = V$, and we may assume $z \in V \cap V_\beta$ by transitivity of L on $V^\#$. As H is also transitive on $V_H^\#$, $V_\beta = V_H^g$ for some $g \in G_1$ by A.1.7.1, so

$$\langle V_\beta^{G_1} \rangle = \langle V_H^{G_1} \rangle = \langle \langle V_2^Y \rangle^{G_1} \rangle = \langle V_2^{G_1} \rangle = \langle V^{G_1} \rangle$$

since $V = \langle V_2^{L_1} \rangle$. Then as $\langle V^{G_1} \rangle$ is abelian, V_β centralizes $V = V_\alpha$, contrary to $(*)$. $\qquad \square$

LEMMA 14.5.9. $[V_H, J(T)] = 1$ and $J(T) \trianglelefteq H$.

PROOF. If $J(T)$ centralizes V_H, then $J(T) = J(Q_H)$ by 14.5.7.2 and B.2.3.5, so the lemma holds. Thus we assume $[V_H, J(T)] \neq 1$, and derive a contradiction. By 14.5.8, we may apply Solvable Thompson Factorization B.2.16 to conclude that $J(H)^* = K_1^* \times \cdots \times K_s^*$, with $K_i^* \cong S_3$ and $V_i := [V_H, K_i] \cong E_4$. Notice $s \leq 2$ by A.1.31.1. As $X = [X, T]$ either $X^* = O^2(K_i^*)$ for some i, or $[X^*, J(H)^*] = 1$. The same holds for Y as $Y = [Y, T]$. Thus if $X^* = O^2(K_i^*)$, then Y^* normalizes X^*, contrary to 14.5.6. Therefore X^* centralizes $J(H)^*$, so that $J(H) \cap X \leq O_2(X)$. Similarly $J(H) \cap Y \leq O_2(Y)$. Then we may apply F.6.8 to $J(H)$, to conclude that $J(T) \leq T \cap J(H) \leq Q_H \leq C_H(V_H)$, contrary to our assumption. $\qquad \square$

LEMMA 14.5.10. $J(T) = J(O_2(XT)) \not\leq O_2(LT)$ and $X = [X, J_1(T)]$.

PROOF. By 14.5.9 and 14.5.7.2, $J(T) \leq C_T(V_H) = Q_H \leq O_2(XT)$, so $J(T) = J(O_2(XT))$ by B.2.3.3. If $J(T) \not\leq O_2(LT)$, then $J_1(T) \not\leq R_2$ by 14.3.9.3, and hence the lemma holds. On the other hand if $J(T) = J(O_2(LT))$ then by 14.5.9, $H \leq N_G(J(T)) \leq M = !\mathcal{M}(LT)$, contradicting 14.5.4.1. $\qquad \square$

LEMMA 14.5.11. (1) H^* is a $\{2, 3\}$-group.
(2) $O_{3'}(H)) \leq C_H(V_H)$, so H^* is a quotient of H^+.

PROOF. Assume $[O_{3'}(H^*), X^*] \neq 1$. Then as $O_{3'}(H^*)$ is solvable of odd order by (2) and (4) of 14.5.7, $[R^*, X^*] \neq 1$ for some prime $p > 3$ and some supercritical subgroup R^* of $O_p(H^*)$ by A.1.21. As $X^* = [X^*, T^*]$, R^* is not cyclic, so $R^* \cong E_{p^2}$

or p^{1+2} by A.1.25. As $m_2(Aut(R^*)) \leq 2$ and $X^* = [X^*, J_1(T)]$ by 14.5.10, the hypothesis of D.2.17 is satisfied for each indecomposable pair in a decomposition of $(R^* X^* J_1(T)^*, V_H)$. So as $p > 3$ and R^* is not cyclic, we conclude from D.2.17 that $p = 5$, and that there are two indecomposable components: that is, $R^* = R_1^* \times R_2^*$ with $R_i^* \cong \mathbf{Z}_5$, $[V_H, R] = V_{H,1} \oplus V_{H,2}$, and $V_i := [V_H, R_i]$ is of rank 4. But by definition of the decomposition, X^* acts on each component, contradicting $[R^*, X^*] \neq 1$.

Therefore $[O_{3'}(H^*), X^*] = 1$, so (1) follows from F.6.9. Of course (1) implies (2). $\qquad\square$

LEMMA 14.5.12. *(1) H^+ is described in Theorem F.6.18.*
(2) $O_2(XT) \neq O_2(YT)$; in particular, case (2) of F.6.18 holds.

PROOF. By 14.5.11, H^* is a quotient of H^+, and by 14.5.7.1, $X^* \neq 1 \neq Y^*$. Thus if $H^+ \cong S_3$, then $H^* \cong S_3$, so that $Y^* = X^*$, contrary to 14.5.6. Thus (1) follows from 14.5.7.5. As H^+ is solvable by 14.5.8, case (1) or (2) of F.6.18 holds. As $O_{3'}(H^+) = 1$ by definition, H^+ is a $\{2,3\}$-group by F.6.9.

Assume (2) fails; then $O_2(XT) = O_2(YT) = Q_H$, so $X^+T^+ \cong Y^+T^+ \cong S_3$. As T^+ is of order 2, case (1) of F.6.18 holds, and we may apply Cyclic Sylow 2-Subgroups A.1.38 and F.6.6 to conclude that

$$\langle X^+, Y^+ \rangle = O^2(H^+) = O(H^+).$$

Then as H^+ is a $\{2,3\}$-group, $O^2(H^+) =: P^+$ is a 3-group. Furthermore P^+ is noncyclic in case (1) of F.6.18, so that $m_3(P^+) = 2$ as H is an SQTK-group.

We claim $P^+ \cong 3^{1+2}$; the proof will require several paragraphs. By 14.5.6, P^+ is nonabelian with X^+ and Y^+ of order 3, so $\Omega_1(P^+)$ is nonabelian. Thus as we saw $m_3(P^+) = 2$, if P^+ is of symplectic type (cf. p. 109 in [**Asc86a**]), then $\Omega_1(P^+) \cong 3^{1+2}$ and the claim holds.

So assume P^+ is not of symplectic type. Then P^+ has a characteristic subgroup $E^+ \cong E_9$. If X^+ or Y^+ is contained in E^+, say X^+, then $P^+ = \langle X^+, Y^+ \rangle = E^+ Y^+ \cong 3^{1+2}$, and again the claim holds, so we may assume neither X^+ nor Y^+ is contained in E^+. Now $F^+ := C_{P^+}(E^+)$ is of index 3 in P^+, and $E^+ = \Omega_1(F^+)$.

Let $T^+ = \langle t^+ \rangle$. Then t^+ inverts X^+, so as $X^+ E^+ \cong 3^{1+2}$, $B^+ := C_{E^+}(t^+) \cong \mathbf{Z}_3$, and hence $N_E(T)^+ \neq 1$. But $N_G(T) \leq M$ by 14.3.3.3, so either $E = \langle N_E(T)^X \rangle \leq M$, or $B^+ = \Omega_1(Z(P^+))$. The former case is impossible, as $X \trianglelefteq H \cap M$, whereas E^+ does not normalize X^+. Thus the latter case holds, and we let $B_0 \in Syl_3(B \cap M)$, where B is the preimage of B^+, and set $B_M := O^2(B_0 Q_H)$. Observe that $O_2(B_M) \neq 1$ since $H \in \mathcal{H}^e$. By a Frattini Argument, $H = O_{3'}(H) N_H(B_0)$, so $H^+ = N_H(B_M)^+$. As $X \not\leq B_M$, with $X/O_2(X)$ inverted in $T \cap L$ and $T B_M = B_M T$, we conclude $B_M \leq C_M(L/O_2(L))$, so L normalizes $O^2(B_M O_2(L)) = B_M$. Hence $N_G(B_M) \leq M = !\mathcal{M}(LT)$. As $H^+ = N_H(B_M)^+$ and $X \trianglelefteq H \cap M$ but $E^+ \not\leq N_{H^+}(X^+)$, this is a contradiction.

This establishes the claim that $P^+ \cong 3^{1+2}$. Thus t^+ inverts $P^+/Z(P^+)$ as t^+ inverts X^+ and Y^+. Hence t^+ centralizes $Z(P^+)$. Then we obtain a contradiction as in the previous paragraph. $\qquad\square$

LEMMA 14.5.13. *(1) $\langle X^*, Y^* \rangle = P^* = O_3(H^*) \cong 3^{1+2}$ and $H = PT$, where $P \in Syl_3(H)$.*
(2) $T^* \cong E_4$.
(3) $C_H(V_H) = O_{3'}(H)$.

PROOF. By 14.5.11.2, H^* is a quotient of H^+, while by 14.5.12.2, H^+ is described in case (2) of Theorem F.6.18. Thus $\langle X^+, Y^+ \rangle = O_3(H^+) \cong 3^{1+2}$ or E_9, and $T^+ \cong E_4$. Then 14.5.6 completes the proof. $\qquad \square$

We are now in a position to obtain a contradiction, and hence establish Theorem 14.5.3. Let $B^* := Z(P^*)$. By 14.5.10, $J_1(H)^* \neq 1$, so as $T^* \cong E_4$, the hypothesis of D.2.17 holds. Thus in view of 14.5.13, case (4) of D.2.17 holds, with $[V_H, P^*] = [V_H, B^*]$ of rank 6. Then $V_2 = [V_2, X] \leq [V_H, P^*]$, so $V_H = \langle V_1^H \rangle \leq [V_H, P^*]$ and hence $V_H = [V_H, P^*]$.

In particular, $V_H = V_X \oplus V_X^y \oplus V_X^{y^2}$, where $\langle y^* \rangle = Y^*$ and $V_X := C_{V_H}(X)$ is of rank 2. Further $C_{V_H}(T) = \langle w, z \rangle$ where $\langle w \rangle = C_{V_X}(T)$ and $z := w w^y w^{y^2}$. Thus $\langle z \rangle = C_{V_H}(YT)$, so $V_1 = \langle z \rangle$. On the other hand,

$$z \in V_2 = [V_2, X] \leq [V_H, X],$$

and X acts on $V_X^{y^i}$, since $V_X^{y^i} = C_{V_H}(X^{*y^i})$ and X^{*y^i} is contained in the abelian group $X^* B^*$. Therefore $[V_H, X] = V_X^y \oplus V_X^{y^2}$. This is a contradiction as $z \notin V_X^y \oplus V_X^{y^2}$ but we saw $z \in [V_H, X]$.

This contradiction completes the proof of Theorem 14.5.3.

14.5.2. Further preliminaries for the case U abelian. Recall we have adopted Notation 12.8.2, including: $V_1 = \langle z \rangle$, and

$$\mathcal{H}_z := \{H \leq G_1 : L_1 T \leq H \text{ and } H \not\leq M\}.$$

In the remainder of this section, H denotes a member of $\mathcal{H}_z$.

In contrast to the case where $\langle V^{G_1} \rangle$ was non-abelian, when $\langle V^{G_1} \rangle$ is abelian we work with members H of $\mathcal{H}_z$ possibly smaller than G_1.

Recall $U_H = \langle V^H \rangle$, $Q_H = O_2(H)$, and $\tilde{G}_1 = G_1/V_1$.

LEMMA 14.5.14. *(1) Hypothesis F.8.1 is satisfied in H.*
(2) Hypothesis F.9.8 is satisfied in H, with V in the role of "V_+".

PROOF. In view of Hypothesis 14.5.1, this follows from the list of equivalences in 12.8.6. $\qquad \square$

By 14.5.14, we may appeal to the results of sections F.8 and F.9.

LEMMA 14.5.15. *(1) $\tilde{U}_H \leq Z(\tilde{Q}_H))$, and $\tilde{U}_H \in \mathcal{R}_2(\tilde{H})$.*
(2) U_H is elementary abelian.
(3) Assume $L/O_2(L) \cong L_3(2)$, and $L_1 \trianglelefteq H$. Then $\tilde{U}_H$ is the direct sum of isomorphic natural modules for $L_1/O_2(L_1) = L_1/C_{L_1}(U_H) \cong \mathbf{Z}_3$.
(4) $Q_H = C_H(\tilde{U}_H)$.

PROOF. Parts (1) and (4) follow from 12.8.4, (2) follows from Hypothesis 14.5.1 and 12.8.6, and (3) follows from 12.8.5.1. $\qquad \square$

NOTATION 14.5.16. By 14.5.14, Hypotheses F.8.1 and F.9.8 are satisfied in H, so we can form the coset geometry Γ with respect to LT and H. Let $b := b(\Gamma, V)$, and choose a geodesic

$$\gamma_0, \gamma_1, \ldots, \gamma_b =: \gamma$$

as in section F.9. Define $U_H, U_\gamma, D_H, D_\gamma$, etc., as in section F.9; in particular set $A_1 := V_1^{g_b}$, recalling b is odd by F.9.11.1.

Since V plays the role of "V_+" in 14.5.14.2 in the notation of section F.9, $U_H = \langle V^H \rangle =: V_H$, and hence $D_H = E_H$. These identifications simplify the statements of various results in section F.9. In particular:

LEMMA 14.5.17. $D_H < U_H$.

PROOF. By F.9.13.5, $V \nleq D_H$, so the remark follows as $V \le V_H = U_H$. □

LEMMA 14.5.18. (1) If $U_\gamma = D_\gamma$, then U_H induces a nontrivial group of transvections with center V_1 on U_γ.

(2) If $m(U_\gamma^*) \ge m(U_H/D_H)$, then $U_\gamma^* \ne 1$ and $U_\gamma^* \in \mathcal{Q}(H^*, \tilde{U}_H)$. In case

$$2m(U_\gamma^*) = m(\tilde{U}_H/C_{\tilde{U}_H}(U_\gamma^*)),$$

then also $m(U_\gamma^*) = m(U_H/D_H)$, and U_γ^* acts faithfully on $\tilde{D}_H$ as a group of transvections with center $\tilde{A}_1$.

(3) $q(H^*, \tilde{U}_H) \le 2$.

(4) If we can choose γ with $D_\gamma < U_\gamma$, then we can choose γ with

$$0 < m(U_\gamma^*) \ge m(U_H/D_H),$$

in which case $U_\gamma^* \in \mathcal{Q}(H^*, \tilde{U}_H)$.

(5) Let $h \in H$ with $\gamma_0 = \gamma_2 h$ and set $\alpha := \gamma h$. Then $U_\alpha \le R_1$ and if $D_\gamma < U_\gamma$ then $U_\alpha^* \in \mathcal{Q}(H^*, \tilde{U}_H)$.

PROOF. Part (3) holds by F.9.16.3, while (1), (2), and (4) follow from 14.5.17 and the corresponding parts of F.9.16. Assume the hypotheses of (5). By parts (1) and (2) of F.9.13, $U_\alpha \le R_1$, and if $U_\gamma^* \ne 1$, then since we can choose γ so that $U_\gamma^* \in \mathcal{Q}(H^*, \tilde{U}_H)$ in (4), also $U_\alpha^* \in \mathcal{Q}(H^*, \tilde{U}_H)$, completing the proof of (5). □

LEMMA 14.5.19. If $K \in \mathcal{C}(H)$ then $K \nleq M$, so $K_0 L_1 T \in \mathcal{H}_z$, where $K_0 := \langle K^T \rangle$.

PROOF. This follows from 13.3.8.2 applied to L, K_0 in the roles of "K, Y". □

LEMMA 14.5.20. Assume $Y \trianglelefteq H$ with $Y/O_2(Y)$ a p-group of exponent p. Then either

(1) $Y \cap M = O_2(Y)$, or

(2) $p = 3$, $L/O_2(L) \cong L_3(2)$, $L_1 \le Y$, and one of the following holds:

(i) $L_1 = Y \trianglelefteq H$.

(ii) $Y/O_2(Y) \cong 3^{1+2}$, $L_1 = O^2(O_{2,Z}(Y)) = O^2(Y \cap M)$, and T is irreducible on $Y/L_1 O_2(Y)$.

(iii) $Y/O_2(Y) \cong E_9$ and there exists $Y_0 \le H$ such that $L_1 \le Y_0 \trianglelefteq Y_0 T$ with $Y_0/O_2(Y_0) \cong \mathbf{Z}_9$ and $Y_0 \nleq M$.

PROOF. We may assume that (1) fails, so that $Y_M := O^2(Y \cap M) \ne 1$.

Let $\mathcal{X}$ be the set of T-invariant subgroups X of H such that $1 \ne X = O^2(X) \le C_M(L/O_2(L))$. Then using the T-invariance of X, L normalizes $O^2(X O_2(L)) = X$, so $N_G(X) \le M = !\mathcal{M}(LT)$. In particular as $H \nleq M$:

For each $X \in \mathcal{X}$, $N_G(X) \le M$, so X is not normal in H.   (!)

Set $Y_Z := O^2(O_{2,Z}(Y))$, and $Y_C := O^2(C_{Y_M}(L/O_2(L))$. By (!), $Y_Z \notin \mathcal{X}$, so $Y_Z \not\leq Y_C$. On the other hand if $Y_C = Y_M$ then $Y_M \in \mathcal{X}$, so $Y_Z \leq N_G(Y_M) \leq M$ by (!); then $Y_Z \leq Y_M = Y_C$, contrary to the previous remark, so:

$$Y_C < Y_M. \qquad (*)$$

It follows that $p = 3$: For if $p > 3$ then T permutes with no p-subgroup of $L/O_2(L)$, so that $Y_M \leq Y_C$, contrary to (*). If $L/O_2(L) \cong L_2(2)'$, then Y_M centralizes V/V_1 and V_1 of order 2, and hence centralizes V by Coprime Action, so again Y_M centralizes $L/O_2(L)$, contrary to (*). Therefore $L/O_2(L) \cong L_3(2)$. Next we claim:

$$L_1 \leq Y. \qquad (!!)$$

For if $L_1 \not\leq Y$, then

$$[Y_M, T \cap L] \leq C_L(V_1) \cap Y_M = L_1 O_2(L) \cap Y_M \leq O_2(Y_M),$$

so Y_M centralizes $(T \cap L)/O_2(L)$ and hence also $L/O_2(L)$ by the structure of $Aut(L_3(2))$, again contrary to (*). We have established the first three statements in (2), so it remains to show that one of cases (i)–(iii) holds.

If $Y/O_2(Y)$ is cyclic then $Y = L_1$ by (!!) since $Y/O_2(Y)$ is of exponent 3, so conclusion (i) of (2) holds. Therefore by A.1.25.1, we may assume $Y/O_2(Y) \cong E_9$ or 3^{1+2}. In the latter case, Y_Z satisfies the hypotheses of "Y", so we conclude $L_1 = Y_Z$ from (!!). Thus in either case, L_1 is normal in Y.

Let $H^* := H/Q_H$. As $M = LC_M(L/O_2(L))$ and $L_1 \not\leq Y_C$:

$$Y_M^* = L_1^* \times Y_C^*. \qquad (**)$$

In particular if $Y^* \cong 3^{1+2}$ then $Y \not\leq M$ by (**).

Next we claim that if $Y_1 = O^2(Y_1) \leq Y$ is T-invariant with $Y_1/O_2(Y_1)$ of order 3, then $Y_1 \leq M$: For if $Y_1 \not\leq M$, then as $N_G(T) \leq M$ by 14.3.3.3, $Y_1 T/O_2(Y_1 T) \cong S_3$. Then as L_1 is normal in Y, the claim follows from 14.5.3.2. It then follows from the claim that if T acts reducibly on $Y/O_{2,\Phi}(Y)$, then $Y \leq M$. Now if $Y^* \cong 3^{1+2}$ we saw $Y \not\leq M$ and $L_1 = Y_Z$, so T acts irreducibly on Y^*/L_1^* and $L_1 = Y_M$, so that conclusion (ii) of (2) holds.

Thus we may assume that $Y^* \cong E_9$. Then $L_1 < Y$ so that T acts reducibly on Y^*, and hence $Y \leq M$ by an earlier remark. Then $Y^* = L_1^* \times Y_C^*$ by (**), with Y_C^* of order 3. Then $Y_C \in \mathcal{X}$, so Y_C is not normal in H by (!). Therefore as $Aut(Y^*) \cong GL_2(3)$ with $Aut_T(Y^*)$ normalizing Y_C, there is some 3-element $y \in H - Y$ inducing an automorphism of order 3 on Y^* centralizing L_1^*, with T acting on $Y_+ := Y\langle y \rangle$. As $M = LC_M(L/O_2(L))$, $Y_+ \not\leq M$, so $Y_+ T \in \mathcal{H}_z$, and then we may assume $H = Y_+ T$. If y^* has order 3, then $Y_+^* \cong 3^{1+2}$. As T is not irreducible on Y_+^*/L_1^*, this is contrary to an earlier reduction. Hence y has order 9, and we may choose y so that $Y_0 := \langle y, L_1 \rangle \trianglelefteq H$ with $Y_0/O_2(Y_0) \cong \mathbf{Z}_9$, and thus conclusion (iii) of (2) holds. $\qquad \square$

LEMMA 14.5.21. *(1) The map φ defined from $Q_H/C_{Q_H}(U_H)$ to the dual space of $U_H/C_{U_H}(Q_H)$ by $\varphi : xC_{Q_H}(U_H) \mapsto C_{U_H}(x)/C_{U_H}(Q_H)$ is an H-isomorphism.*

(2) $[U_H, Q_H] = V_1$.

(3) $C_H(V_2)$ acts on L_2, and $m_3(C_H(V_2)) \leq 1$.

PROOF. Part (1) is F.9.7.

Assume case (1) of Hypothesis 14.3.1 holds. Then (2) follows from 13.3.14 and 14.5.15.1, while (3) follows from parts (1), (2), and (5) of 13.3.15.

Assume case (2) of Hypothesis 14.3.1 holds. Then (3) follows from 14.2.2.4. Assume $[U_H, Q_H] = 1$. Then by 14.5.15.4, $Q_H = C_H(U_H)$. By 14.5.15.1, $O_2(H/Q_H) = 1$, so that $U_H \in \mathcal{R}_2(H)$. Suppose there exists $K \in \mathcal{C}(H)$. As $Q_H = C_H(U_H)$, $K \in \mathcal{L}_f(G, T)$, contradicting 14.3.4.4. So H is solvable by 1.2.1.1, and hence $O(H^*) \neq 1$. Then $U_H = [U_H, O(H^*)] \oplus C_{U_H}(O(H^*))$ by Coprime Action. As $H > Q_H = C_H(U_H)$, $[U_H, O(H^*)] \neq 0$. Then $Z \cap [U_H, O(H^*)] \neq 0$, contradicting $H \leq C_G(V_1)$ since $Z = V_1$ when $L/O_2(L) \cong L_2(2)'$. $\qquad \square$

14.6. Eliminating $\mathbf{L_2(2)}$ when $\langle \mathbf{V^{G_1}} \rangle$ is abelian

In this section we assume Hypothesis 14.5.1 holds with $L/O_2(L) \cong L_2(2)'$; in particular, $U := \langle V^{G_1} \rangle$ is abelian. Also Hypotheses 14.3.1.2 and 14.2.1 are satisfied, so we can appeal to results in sections 14.2 (with L in the role of "Y"), 14.3, and 14.5.

We will see in Theorem 14.6.25 that no further quasithin examples arise beyond those which we characterized earlier in Theorems 14.2.7 and 14.2.20, where U was nonabelian. Thus in this section we will be working toward a contradiction. Indeed as far as we can tell, there are no shadows.

As usual $Z := \Omega_1(Z(T))$ for $T \in Syl_2(G)$. Recall that by Hypothesis 14.2.1.4, V is of rank 2 with $V \trianglelefteq M$. Recall also that $C_T(L) = 1$ by 14.2.2.2.

We also adopt Notation 12.8.2: Thus $V_1 := Z \cap V = Z$ since Z is of order 2 by 14.2.2.6, and $G_1 = N_G(V_1) = C_G(Z) = M_c \in \mathcal{M}(T)$. Recall also that $L_1 := O^2(C_L(V_1)) = 1$; this simplifies the application of results from sections 14.3 and 14.5 involving L_1. For example as $L_1 = 1$, 14.2.5 says that:

$$\mathcal{H}(T, M) = \mathcal{H}_z.$$

For the remainder of this section, H denotes a member of $\mathcal{H}(T, M)$.

Recall $\tilde{G}_1 := G_1/V_1$ and notice $\tilde{H}$ makes sense as $H \leq G_1$ by definition of $\mathcal{H}_z$. As U is elementary abelian and $H \leq G_1$, $U_H := \langle V^H \rangle \leq U$ is also elementary abelian (cf. 14.5.15.2).

LEMMA 14.6.1. *(1)* $G_1 = !\mathcal{M}(H)$.
(2) $O_{2,p}(H) \cap M = O_2(H)$ *for each odd prime p.*
(3) If $K \in \mathcal{C}(H)$, then $K \not\leq M$.
(4) If $1 \neq X = O^2(X) \trianglelefteq H$, then $XT \in \mathcal{H}(T, M)$.
(5) $O_{2,F^*}(H)$ *centralizes $\Omega_1(Z(O_2(H)))$.*
(6) If $O_2(H) \leq T_1 \trianglelefteq T$, then $N_G(T_1) \leq N_G(\Omega_1(Z(T_1))) \leq G_1$.

PROOF. Part (1) is 14.2.3, part (3) is 14.5.19, and part (2) follows as case (1) of 14.5.20 holds because $L/O_2(L) \not\cong L_3(2)$. Under the hypotheses of (4), $O_2(X) < O_{2,F^*}(X)$, and $O_{2,F^*}(X) \not\leq M$ by (2) and (3), so (4) holds.

Let $R := O_2(H)$, $W := \Omega_1(Z(R))$, and $\hat{H} := H/C_H(W)$. Suppose there is $K \in \mathcal{C}(H)$ with $[W, K] \neq 1$. Then as R centralizes W, $K \in \mathcal{L}_f(G, T)$ by A.4.9, contrary to 14.3.4.4. This contradiction shows that $O_{2,E}(H)$ centralizes W.

Suppose (5) fails. Then by the previous remark, for some odd prime p, $X := O^2(O_{2,p}(H))$ is nontrivial on W. As $O_2(X) \leq R \leq C_H(W)$, $\hat{X}$ is of odd order, so $W = [W, X] \oplus C_W(X)$ by Coprime Action. Then as $[W, X] \neq 0$, $Z \leq [W, X]$ since Z has order 2. However $X \leq H \leq G_1$ by (1), so also $Z \leq C_W(X)$. This contradiction establishes (5).

Assume the hypotheses of (6), and let $Z_1 := \Omega_1(Z(T_1))$. As $R \leq T_1$ by hypothesis, and $H \in \mathcal{H}^e$, $Z_1 \leq W$, so that $[O_{2,F^*}(H), Z_1] = 1$ by (5). As $T_1 \trianglelefteq T$, T acts on Z_1, so $H_1 := O_{2,F^*}(H)T \leq N_G(Z_1)$. Now $H_1 \in \mathcal{H}(T, M)$ by (4), so $G_1 = !\mathcal{M}(H_1)$ by (1), and then (6) follows. $\qquad\square$

LEMMA 14.6.2. *If* $1 \neq X = O^2(X) \leq O_{2,F^*}(H)$ *with* $Q_H \leq N_H(X)$, *then* $Z \leq [U_H, X]$.

PROOF. As $C_H(\tilde{U}_H) = Q_H$ while $X = O^2(X) \neq 1$, $U_X := [U_H, X] \neq 1$. As Q_H acts on X, U_X is normal in Q_H, but U_X is not central in Q_H by 14.6.1.5. Then $1 \neq [U_X, Q_H] \leq U_X \cap Z$ using 14.5.15.1, so as $|Z| = 2$ we conclude that $Z \leq U_X$. $\qquad\square$

14.6.1. Preliminary results on suitable involutions in $\mathbf{U_H}$. In the proof of Theorem 14.6.18 and also at the end of the section, we will need to control the centralizers of involutions in U_H which satisfy certain special conditions (cf. 14.6.17.3 and 14.6.24.1). Thus we are led to define $\mathcal{U}(H)$ to consist of those u satisfying

(U0) $u \in U_H$,
(U1) $T_u := C_T(u) \in Syl_2(C_H(u))$, and $T_0 := C_T(\tilde{u})$ is of index 2 in T,
(U2) $[O_2(G_1), u] \neq 1 \neq [O_2(G_1), uu^t]$ for $t \in T - T_0$, and
(U3) $T = N_{G_1}(T_0)$.

LEMMA 14.6.3. *Assume* $u \in \mathcal{U}(H)$. *Then*

(1) $|T : T_u| = 4$, $|T_0 : T_u| = 2$, $T_0 = N_T(T_u)$, *and* $T_0 = O_2(G_1)T_u = Q_H T_u$.
(2) $N_G(T_0) = T$.
(3) $C_{Q_H}(u) \not\leq C_{Q_H}(V)$ *and* $L = [L, C_{O_2(G_1)}(u)] = [L, C_{Q_H}(u)]$.
(4) $N_G(T_u) = T_0$, $T_u \in Syl_2(C_G(u))$, *and* $u \notin z^G$.

PROOF. Set $Q_1 := O_2(G_1)$. First $[Q_1, u] \neq 1$ by (U2) and $u \in U_H \leq U$ by (U0), so $[Q_H, u] = [Q_1, u] = V_1$ is of order 2 by 14.5.15.1. Hence $C_{Q_1}(u) = Q_1 \cap T_u$ is of index 2 in Q_1, T_u is of index 2 in T_0, and $T_0 = Q_1 T_u = Q_H T_u$ as $C_T(\tilde{u}) = T_0$ and $C_T(u) = T_u$ by (U1).

Pick $t \in T - T_0$. If t normalizes $C_{Q_1}(u)$, then

$$C_{Q_1}(u) = C_{Q_1}(u)^t = C_{Q_1}(u^t).$$

Therefore for $x \in Q_1 - C_{Q_1}(u)$, $z = [x, u] = [x, u^t]$, and hence $Q_1 = \langle x, C_{Q_1}(u) \rangle$ centralizes uu^t, contrary to (U2). Thus t does not normalize $C_{Q_1}(u)$, so as $N_T(T_u)$ normalizes $T_u \cap Q_1 = C_{Q_1}(u)$, $t \notin N_T(T_u)$. As $|T_0 : T_u| = 2$ we conclude that $T_0 = N_T(T_u)$, and as $|T : T_0| = 2$ by (U1), $|T : T_u| = 4$, completing the proof of (1).

As $Q_H \leq T_0$ by (1), and $T_0 \trianglelefteq T$ by (U1), we may apply 14.6.1.6 to conclude that $N_G(T_0) \leq G_1$. Then as $N_{G_1}(T_0) = T$ by (U3), (2) holds. By (1), $T_0 = N_T(T_u)$, so $T_0 \in Syl_2(N_G(T_u))$ by (2).

As $[U, Q_1] = V_1$ by 14.5.21.2, and $U = \langle V^{G_1} \rangle$, also $[V, Q_1] = V_1$, so that $C_{Q_1}(V)$ is of index 2 in Q_1 since $m(V) = 2$. Suppose that $C_{Q_1}(u) \leq C_{Q_1}(V)$. Then $C_{Q_1}(u) = C_{Q_1}(V)$, as both are of index 2 in Q_1, so $\langle u \rangle C_U(Q_1) = V C_U(Q_1)$ by the duality in 14.5.21.1. Thus for $t \in T - T_0$, $\langle u^t \rangle C_U(Q_1) = V C_U(Q_1)$, so that $uu^t \in C_U(Q_1)$. This is impossible since Q_1 does not centralize uu^t by (U2), so

$C_{Q_1}(u) \not\leq C_{Q_1}(V)$. Since $Q_1 \leq Q_H$, $C_{Q_H}(u) \not\leq C_{Q_H}(V)$, and since $L = O^2(L)$ induces $\mathbf{Z}_3$ on V, also

$$L = [L, C_{Q_1}(u)] = [L, C_{Q_H}(u)],$$

completing the proof of (3).

Next $N_{G_1}(T_u)$ normalizes $T_u Q_1 = T_0$ using (1), so $N_{G_1}(T_u) \leq T$ by (2); hence again using (1), $N_{G_1}(T_u) = N_T(T_u) = T_0$.

We now show that to prove (4) it will suffice to establish that $I := N_G(T_u) \leq G_1$: For in that case $I = T_0$ by the previous paragraph, establishing the first assertion of (4). Next let $T_u \leq S \in Syl_2(C_G(u))$. Then $N_S(T_u) \leq I = T_0 \leq H$, so as $T_u \in Syl_2(C_H(u))$ by (U1), $S = T_u$. In particular $u \notin z^G$ as $|T_u| < |T|$. This completes the proof that (4) holds if $I \leq G_1$, so we may assume that $I \not\leq G_1$, and it remains to establish a contradiction. We saw earlier that $T_0 \in Syl_2(I)$, so in particular $T_0 < I$ as $T_0 \leq G_1$.

We claim that $N_I(C) = T_0$ for each $1 \neq C \leq T_0$ with $C \trianglelefteq T$, so we assume that $T_0 < N_I(C)$ and derive a contradiction. We saw that $T_0 = N_{G_1}(T_u)$, so $N_I(C) \not\leq G_1$. Hence as $\mathcal{M}(T) = \{M, G_1\}$ by 14.2.2.5, we must have $N_G(C) \leq M$. Therefore as $|M : M \cap G_1| = 3$ by 14.2.2.1, and $N_I(C) \not\leq G_1$, $M = (M \cap G_1)N_I(C)$; hence as I normalizes $Z(T_u)$,

$$V = \langle Z^M \rangle = \langle Z^{N_I(C)} \rangle \leq Z(T_u).$$

But then $C_{Q_H}(u) = T_u \cap Q_H \leq C_{Q_H}(V)$, contrary to (3), so the claim is established. In particular $C(I, T_0) = T_0$ as $T_0 \trianglelefteq T$.

We have seen that $T_0 \in Syl_2(I)$, with $|T_0 : T_u| = 2$, so that I/T_u and hence also I is solvable by Cyclic Sylow 2-Subgroups A.1.38. Also $F^*(I) = O_2(I)$ by 1.1.4.3 as $Z \leq T_u$. So since $C(I, T_0) = T_0 < I$, we may apply the Local $C(G, T)$-Theorem C.1.29 to conclude that $I = T_0 B$, where B is the product of $s := 1$ or 2 blocks of type A_3 which are not contained in G_1. Further $N_I(J(T_0)) = T_0$ as $C(I, T_0) = T_0$, so Solvable Thompson Factorization B.2.16 says that $I/O_2(I)$ contains the direct product of s copies of S_3. Therefore if $s = 2$, then $I/O_2(I)$ contains $S_3 \times S_3$, contradicting $T_u \trianglelefteq I$ and $|T_0 : T_u| = 2$. Thus $s = 1$, so $B \cong A_4$ by C.1.13.c.

Now the hypotheses of Theorem C.6.1 are satisfied with I, T, T_0 in the roles of "H, Λ, T_H"; for example, part (iv) of that hypothesis follows from the claim and the facts that $T_0 < I$ and $|T : T_0| = 2$. Therefore case (a) or (b) of Theorem C.6.1.6 holds since $s = 1$; thus $I \cong S_4$ or $\mathbf{Z}_2 \times S_4$, and in particular $T_u = O_2(I)$. By C.6.1.1, $T_0 = J(T_0) = O_2(I)O_2(I)^x$ for each $x \in T - T_0$, and hence $T_0 = T_u T_u^x$. However by (3), T_u is nontrivial on V, so that $T = T_0 C_T(V)$ since $|T : C_T(V)| = 2$; thus we may take $x \in C_T(V)$. Next as $T_0 = O_2(I)O_2(I)^x$, $C_{\tilde{T}_0}(x) = \tilde{Z}_0 \langle \tilde{b}\tilde{b}^x \rangle$, where $Z_0 := Z(T_0)$ and $O_2(O^2(I)) =: \langle b, z \rangle$. Further if $I \cong \mathbf{Z}_2 \times S_4$, then $Z_0 \cong E_4$, and hence $[Z_0, x] = Z$ as Z has order 2. However in either case, bb^x is of order 4, so that $\Omega_1(C_{T_0}(x)) = Z$; this is a contradiction, as $x \in C_T(V)$ and $V \leq T_u \leq T_0$. This contradiction completes the proof of (4), and hence of 14.6.3. $\qquad\square$

For the remainder of this subsection, u denotes a member of $\mathcal{U}(H)$.

Define $\mathcal{I} := \mathcal{I}(T, u)$ to be the set of $I \in \mathcal{H}(T_u)$ such that I is contained in neither G_1 nor M. We will see later (cf. 14.6.17.5 and 14.6.24.4) that for suitable $u \in \mathcal{U}(H)$, $C_G(u) \in \mathcal{I}$, so that $\mathcal{I}$ is nonempty.

Let $\mathcal{I}^*$ consist of those $I \in \mathcal{I}$ such that $T \cap I$ is not properly contained in $T \cap J$ for any $J \in \mathcal{I}$. Finally let $\mathcal{I}_*$ be the minimal members of $\mathcal{I}^*$ under inclusion.

For $I \in \mathcal{I}$, set $T_I := T \cap I$ and $I_z := I \cap G_1$.

The next two observations are straightforward from the definitions:

LEMMA 14.6.4. *If $I \in \mathcal{I}^*$ and $T_I \le J \in \mathcal{I}$, then $J \in \mathcal{I}^*$ and $T_J = T_I$.*

LEMMA 14.6.5. *If $I \in \mathcal{I}$ and $I \le J \in \mathcal{H}$, then $J \in \mathcal{I}$. If further $I \in \mathcal{I}^*$, then $J \in \mathcal{I}^*$ and $T_J = T_I$.*

Recall from Definition F.6.1 the discussion of Goldschmidt triples.

LEMMA 14.6.6. *Assume $I \in \mathcal{I}^*$, and let $L_I := O^2(L \cap I)$. Then*

(1) T_I *is either* T_u *or* T_0.
(2) $T_I \in Syl_2(I)$.
(3) *If* $I \cap M \not\le G_1$ *then* $L = L_I O_2(L)$ *and* $LT = L_I T_I O_2(LT)$.
(4) *Either* $C(I, T_I) \le I_z$, *or* $L = L_I O_2(L)$ *and* $LT = L_I T_I O_2(LT)$.
(5) *If* $I \in \mathcal{I}_*$ *then either* I_z *is the unique maximal subgroup of I containing T_I, or* $L = L_I O_2(L)$ *and* $LT = L_I T_I O_2(LT)$.
(6) *Assume* $|T| > 2^9$ *and* $L = L_I O_2(L)$. *Assume further that there exists H_2 with $T_0 \le H_2 \le C_H(\tilde{u})$, $H_2/O_2(H_2) \cong S_3$, $H_2 \not\le M$, and H_2 has at least two noncentral 2-chief factors. Then setting $I_2 := O^2(H_2)T_I$, $I_1 := L_I T_I$, and $I_0 := \langle I_1, I_2 \rangle$, we have $I_0 \in \mathcal{I}^*$ and (I_0, I_1, I_2) is a Goldschmidt triple.*
(7) T_I *is not normal in* I.

PROOF. We first establish (1) and (2). Let $T_I \le S \in Syl_2(I)$ and set $Q_1 := O_2(G_1)$. Now $N_G(T_u) = T_0$ by 14.6.3.4, so as T_u is of index 2 in T_0 by 14.6.3.1, $N_S(T_u) = T_u$ or T_0. In the first case, $S = T_u = T_I$, so that (1) and (2) hold. In the second case I is not contained in M or G_1, so that $T_I < T$ by 14.2.2.5, and hence $T_I = T_0$ since $|T : T_0| = 2$ by (U1). Then as $N_G(T_0) = T$ by 14.6.3.2, $N_S(T_I) \le N_{T \cap I}(T_0) = T_I$, so that $S = T_I = T_0$, and so (1) and (2) hold in this case also.

Next we prove (3), so assume $X := I \cap M \not\le G_1$. As L is transitive on $V^\#$, $M = L(M \cap G_1)$ and $|M : M \cap G_1| = 3$ is prime, so $M = X(M \cap G_1)$. Next $T_u \le T_I$, so by 14.6.3.3, $L = [L, a]$ for some $a \in Q_1 \cap T_I \le X$. As $LQ_1 \trianglelefteq L(M \cap G_1) = M$, $\langle a^X \rangle \le LQ_1 \cap X$. If $a^X \subseteq Q_1$, then as $M = X(G_1 \cap M)$ and $Q_1 \trianglelefteq G_1$, $a^M \subseteq Q_1$ so that $\langle a^M \rangle$ is a 2-group and hence $a \in O_2(M)$, contradicting $L = [L, a]$. Thus $a^X \not\subseteq Q_1$, so as Q_1 is of index 3 in LQ_1 and $L = O^2(LQ_1)$, $L \le \langle a^X \rangle O_2(LQ_1)$. Then as $L = O^2(LQ_1)$, $O^2(\langle a^X \rangle) \le L \cap X$, so that $L = (L \cap X)O_2(L) = L_I O_2(L)$, and as $L = [L, a]$, $LT = L_I T_I O_2(LT)$. Hence (3) holds.

Next suppose there is $1 \ne C$ char T_I with $N_I(C) \not\le I_z$. As $T_I < T$ by (1), T_I is proper in $N_T(T_I) \le N_G(C)$. Then as $I \in \mathcal{I}^*$, $N_G(C) \notin \mathcal{I}$ by 14.6.4, and hence $N_G(C) \le M$ since $N_I(C) \not\le I_z$. Therefore $I \cap M \not\le G_1$, so (4) follows from (3).

Next assume $I \in \mathcal{I}_*$ and let Y be a maximal subgroup of I containing T_I. Then by minimality of I, Y is contained in G_1 or M, so that Y is I_z or $I \cap M$ by maximality of Y. Thus (5) also follows from (3).

Assume the hypotheses of (6), and set $I_1 := L_I T_I$. By (2), $T_I \in Syl_2(I)$, so that $T_I \in Syl_2(I_1)$. As $L = L_I O_2(L)$, we conclude from 14.6.3.3 that $I_1/O_2(I_1) \cong S_3$.

Next since $T_I \le T_0 \le H_2$ using (1) and the hypothesis for (6), $I_2 := O^2(H_2)T_I$ is a subgroup of H_2 with $O^2(I_2) = O^2(H_2)$. Also $O^2(H_2)$ centralizes $\tilde{u}$ and hence also u, so as $T_u \in Syl_2(C_H(u))$ by 14.6.3.4, $T_u \in Syl_2(O^2(H_2)T_u)$. Thus as $T_u \le T_I$, $T_I \in Syl_2(I_2)$. By (U1), $T_0 \in Syl_2(C_H(\tilde{u}))$ so that $H_2 = O^2(H_2)T_0$, while

$H_2/O_2(H_2) \cong S_3$ by the hypothesis of (6). Then since $T_0 = Q_1 T_u$ by 14.6.3.1, and Q_1 is normal in H, we conclude $I_2/O_2(I_2) \cong S_3$.

Suppose first that $O_2(I_0) \neq 1$. Since $L = L_I O_2(L)$, we have $I_1 \not\leq G_1$, while $H_2 \not\leq M$ by hypothesis, so $I_2 \not\leq M$ since we saw $H_2 = O^2(H_2)T_0$. Thus $I_0 \in \mathcal{I}$, and indeed as $T_I \leq I_0$, $I_0 \in \mathcal{I}^*$ and $T_{I_0} = T_I$ by 14.6.4, so that $T_I \in Syl_2(I_0)$ by (2). We conclude that (I_0, I_1, I_2) is a Goldschmidt triple in the sense of Definition F.6.1, so that (6) holds in this case.

So we suppose instead that $O_2(I_0) = 1$, and it remains to derive a contradiction. By construction, Hypothesis F.1.1 is satisfied with I_1, I_2, T_I in the roles of "L_1, L_2, S". So by F.1.9, $\alpha := (I_1, T_I, I_2)$ is a weak BN-pair of rank 2, and as T_I plays the role of "B_j" for $j = 1, 2$, α appears on the list of F.1.12. Since $I_i/O_2(I_i) \cong S_3$, and $I_2/O_2(I_2)$ has at least two noncentral chief factors by hypothesis, it follows that α is of type $G_2(2)'$, $G_2(2)$, M_{12} or $Aut(M_{12})$. But then $|T_I| \leq 2^7$, so as $|T : T_I| \leq 4$ by (1) and 14.6.3.1, $|T| \leq 2^9$, contrary to the hypothesis for (6). This contradiction completes the proof of (6).

Finally observe that as $T_I = T_u$ or T_0 by (1), $N_G(T_I) \leq T$ by (2) or (4) of 14.6.3. Thus (7) holds since $I \not\leq M$. This completes the proof of (7), and hence of 14.6.6. $\square$

LEMMA 14.6.7. *Assume $I \in \mathcal{I}^*$. Then*

(1) The hypotheses of 1.1.5 are satisfied with I, G_1 in the roles of "H, M".
(2) $F^*(I_z) = O_2(I_z)$.
(3) $O(I) = 1$.
(4) If K is a component of I, then $K \not\leq I_z$ and $\langle K, T_I \rangle \in \mathcal{I}^*$.

PROOF. As $u \in U_H \leq U \leq O_2(G_1)$, $u \in O_2(I \cap G_1)$. Therefore

$$C_{O_2(G_1)}(O_2(I \cap G_1)) \leq C_{O_2(G_1)}(u) \leq T_u \leq I,$$

so (1) holds; hence we may apply 1.1.5. Then 1.1.5.1 implies (2). In view of (2), to prove (3) it suffices to show that $O(I) \leq G_1$. But as L is transitive on $V^\#$, $V \leq O_2(C_G(v))$ for each $v \in V^\#$ since $V \leq O_2(G_1)$ by 14.5.15.1. Therefore $[V, C_{O(I)}(v)] \leq O(I) \cap O_2(C_G(v)) = 1$. Then using Generation by Centralizers of Hyperplanes A.1.17, $O(I) \leq C_I(V) \leq G_1$, establishing (3).

Suppose K is a component of I. By 1.1.5.3, $K \not\leq I_z$. Further if $K \leq M$, then as $m(V) = 2$ by 14.2.1.4, $K \leq C_I(V) \leq I_z$, contrary to the previous remark; so also $K \not\leq M$. Thus $\langle K, T_I \rangle \in \mathcal{I}$, so that $\langle K, T_I \rangle \in \mathcal{I}^*$ by 14.6.4, completing the proof of (4). $\square$

LEMMA 14.6.8. *Assume $I \in \mathcal{I}^*$ and $F^*(I) \neq O_2(I)$. Then $m_2(I/O_2(I)) \geq m(U_H O_2(I)/O_2(I)) \geq m(U_H/C_{U_H}(Q_H))$.*

PROOF. By 14.6.7.3, $O(I) = 1$, so as $F^*(I) \neq O_2(I)$ by hypothesis, we conclude there is a component K of I. By 14.6.7.4, z does not centralize K, so that $Z \cap O_2(I) = 1$ as Z has order 2. Set $P := C_{Q_H}(u)$. By 14.5.15.1, $[U_H \cap O_2(I), P] \leq Z \cap O_2(I) = 1$. So since $Z \not\leq O_2(I)$, using the duality in 14.5.21.1 we obtain

$$U_H \cap O_2(I) < C_{U_H}(P) = \langle u \rangle C_{U_H}(Q_H).$$

Therefore

$$m_2(I/O_2(I)) \geq m(U_H O_2(I)/O_2(I)) = m(U_H/(U_H \cap O_2(I)))$$
$$> m(U_H/C_{U_H}(P)) = m(U_H/C_{U_H}(Q_H)) - 1,$$

so the lemma is established.   □

LEMMA 14.6.9. *Assume $I \in \mathcal{I}^*$, $|T : Q_H| > 4$, and $m(U_H/C_{U_H}(Q_H)) \geq 4$. Then $|T| > 2^{11}$, and*

$$LT = O^2(L \cap I)T_I O_2(LT).$$

PROOF. Observe first that by the duality in 14.5.21.1,

$$m(Q_H/C_{Q_H}(U_H)) = m(U_H/C_{U_H}(Q_H)) =: m,$$

with $Z \leq C_{U_H}(Q_H)$, so that $|Q_H| \geq 2^{2m+1} \geq 2^9$ since $m \geq 4$ by hypothesis. As we also assume $|T : Q_H| > 4$, $|T| > 2^{11}$, establishing the first conclusion of 14.6.9. By 14.6.6.1, $T_I = T_u$ or T_0, so $|T : T_I| \leq 4$ by 14.6.3.1, and hence $|T_I| > 2^9$.

Thus we may assume that $LT > O^2(L \cap I)T_I O_2(LT)$, and it remains to derive a contradiction. Then by 14.6.6.4, $C(I, T_I) \leq I_z$. As we are working toward a contradiction, we may also assume that I is minimal under inclusion; that is, $I \in \mathcal{I}_*$. Then by 14.6.6.5, I_z is the unique maximal subgroup of I containing T_I. Since T_I is not normal in I by 14.6.6.7, I is a minimal parabolic in the sense of Definition B.6.1.

We first treat the lengthier case where $F^*(I) = O_2(I)$. Here since $T_I \in Syl_2(I)$ by 14.6.6.2, and I is a minimal parabolic, we may apply C.1.26: Since $C(I, T_I) \leq I_z < I$, we conclude that $I = T_I K_1 \cdots K_s$, where K_i is a χ_0-block of I not contained in I_z, and T_I is transitive on the K_i. Further $s = 1$ or 2 as I is an SQTK-group, and the action of $J(T_I)$ on $O_2(K)$ is described in E.2.3. Also K_1 is not an $L_2(2^n)$-block for $n > 1$, as $I_z = C_I(z)$ is the unique maximal overgroup of T_I in I, whereas when K_1 is an $L_2(2^n)$-block, the center of that overgroup is $Z(I)$. Thus K_1 is a block of type A_3 or A_5.

Observe using 14.6.5 and 14.6.6.2 that:

(a) If $1 \neq S \leq T_I$ with $S \trianglelefteq I$, then $N_G(S) \in \mathcal{I}^*$ and $N_T(S) = T_I \in Syl_2(N_G(S))$.

Since $T_I < T$ by 14.6.6.1, we may choose $r \in N_T(T_I) - T_I$ with $r^2 \in T_I$. Then by (a),

(b) r acts on no nontrivial subgroup S of T_I normal in I.

Set $K := K_1 \cdots K_s$, so that $I = KT_I$. Assume first that K is not the product of two A_5-blocks. As $F^*(I) = O_2(I)$, this assumption establishes part (i) of the hypothesis of Theorem C.6.1, with I, $T_I\langle r \rangle$, T_I in the roles of "H, Λ, T_H", while (a) gives part (iv) of that hypothesis, and (ii) and (iii) are immediate. If K is an A_3-block then $|T_I| \leq 16$ since case (a) or (b) of C.6.1.6 must hold, contrary to $|T_I| > 2^9$ in the first paragraph of the proof. Therefore K is an A_5-block or a product of two A_3-blocks. In either case by C.1.13.c, $O_2(I) = D \times O_2(K)$, where $D := C_{T_I}(K)$, and by C.6.1.4, D is elementary abelian, so that $D \leq D_I := \Omega_1(Z(J(T_I)))$. Then we conclude from the action of $J(T_I)$ on $O_2(K)$ described in E.2.3, that $|D_I : D| = 4$. As $D \cap D^r$ is normalized by $KT_I = I$ and r, $D \cap D^r = 1$ by (b), so that $|D| \leq 4$. But now $|T_I| \leq 4|Aut(K)|_2 \leq 2^9$, again contrary to the first paragraph.

Therefore $K = K_1 \times K_2$ is the product of two A_5-blocks. Set $K_z := O^2(I_z)$ and $R_z := O_2(I_z)$. Then $K_z T_I/R_z \cong S_3$ wr $\mathbf{Z}_2$, and $J(R_z) = J(O_2(I))$ using E.2.3.3 and B.2.3.3. So applying (a) to $J(R_z)$ in the role of "S", we obtain $T_I \in Syl_2(N_G(J(R_z)))$; hence $T_I \in Syl_2(N_{G_1}(R_z))$. Thus as $R_z = O_2(I_z)$,

$R_z = O_2(N_{G_1}(R_z))$ by A.1.6—that is $R_z \in \mathcal{B}_2(G_1)$, so setting $Q_1 := O_2(G_1)$, we conclude from C.2.1.2 that

(c) $Q_1 \leq R_z$.

But $T_0 = T_u Q_1$ by 14.6.3.1, so as $R_z \leq T_I$, we conclude from (c) and 14.6.6.1 that

(d) $T_I = T_0$.

By (U1), $|T : T_0| = 2$, so $T = T_I \langle r \rangle$ by (d). Further as $G_1 \in \mathcal{H}^e$, (c) says

(e) $Z_z := \Omega_1(Z(R_z)) \leq \Omega_1(Z(Q_1)) =: Z_1$.

By 14.6.1.5 and (e):

(f) $Y := O^2(O_{2,F^*}(G_1))$ centralizes Z_1 and Z_z.

Next $K_z = X_1 \times X_2$ where $X_i := K_z \cap K_i$, $R_i := O_2(X_i) \cong Q_8^2$, and $|X_i : R_i| = 3$. Further as T_I is of index 2 in T, Hypothesis C.5.1 is satisfied with I, T_I, T_I, T in the roles of "H, T_H, R, M_0". Similarly Hypothesis C.5.2 is satisfied using (b), as is the hypothesis $|T : T_I| = 2$ in C.5.6.7. So by C.5.6.7, $D := C_{T_I}(K) \leq Z(\text{Baum}(T_I))$ is elementary abelian, and $O_2(I) = DO_2(K)$. Hence setting $Z_0 := Z(R_1 R_2)$, we have

(g) $O_2(I) = DO_2(K)$ and $Z_z = DZ_0$.

Observe since $|T : T_0| = 2 = |Z|$ that z is diagonally embedded in $Z(R_1) \times Z(R_2) = Z_0$.

We claim that $D = 1$. Suppose instead that $D \neq 1$. Then as D is normal in $KT_I = I$, (a) and (d) say that $I_D := N_G(D) \in \mathcal{I}^*$, and $T_I = T_0 \in Syl_2(I_D)$.

Assume first that $LT = L_D T_I O_2(LT)$, where $L_D := O^2(L \cap I_D)$. As $T_I = T_0$ and

$$|O_2(L) : O_2(L) \cap T_0| \leq |T : T_0| = 2,$$

L_D centralizes $O_2(L)/(O_2(L) \cap T_0)$, and hence $L = O^2(L) = L_D$. Next $K \leq C_G(D) \leq I_D$, and indeed $K_1 \in \mathcal{L}(I_D, T_0)$, so that $K_1 \leq K_1^+ \in \mathcal{C}(I_D)$ with K_1^+ described in 1.2.8.2. Then using 1.2.2.a, $L \leq O^{3'}(I_D) = \langle K_1^{+T_0} \rangle \leq C_G(D)$. Therefore $C_T(L) \neq 1$, contrary to 14.2.2.6 as we mentioned at the start of the section.

This contradiction shows that $LT > L_D T_I O_2(LT)$. Next assume $F^*(I_D) \neq O_2(I_D)$. Since $O(I_D) = 1$ by 14.6.7.3, I_D has a component K_D. By 14.6.7.1, K_D appears in the list of 1.1.5.3. As that list does not contain the possible proper overgroups of KT_I in 1.2.8.2, we conclude K centralizes K_D. But each component in that list has order divisible by 3 or 5, so $m_p(KK_D) > 2$ for $p = 3$ or 5, contrary to I_D an SQTK-group. Thus $F^*(I_D) = O_2(I_D)$.

Since $L_D T_I O_2(LT) < LT$, 14.6.6.4 says that $C(I_D, T_I) \leq I_{D,z} := I_D \cap G_1$. Thus as $F^*(I_D) = O_2(I_D)$, we may apply the local $C(G, T)$-Theorem C.1.29 to conclude that I_D is the product of $I_{D,z}$ with one or two χ_0-blocks. Since I_D contains $I = KT_I$, where K is the product of two A_5-blocks not in $I_{D,z}$, and no A_5-block is contained in a larger χ_0-block, we conclude that the blocks in K are the blocks in I_D, and $K \trianglelefteq I_D = KI_{D,z}$. By (e) and (f), YQ_1 centralizes Z_z, so $YQ_1 \leq I_{D,z}$ by (g). Then by A.4.4.1 applied with G_1, I_D, $I_{D,z}$, $Q_1 Y$ in the roles of "H, K, $H \cap K$, X", we conclude that $Q_1 = O_2(I_{D,z})$. Using A.1.6, $O_2(I_D) \leq O_2(I_{D,z}) = Q_1$ and $O_2(I_D) \leq O_2(I)$. Further $O_2(I) = O_2(K)D$ by (g), and $O_2(K) \leq O_2(I_D)$ as $K \trianglelefteq I_D$, so we conclude that $O_2(I_D) = O_2(I)$. Therefore $O_2(I) = O_2(I_D) \leq Q_1 \leq R_z$ by (c). As K_i is an A_5-block, $J(R_z) = J(O_2(I))$, so $J(O_2(I)) = J(Q_1)$ by B.2.3.3. Therefore $I \leq N_G(J(Q_1)) = G_1$ as $G_1 \in \mathcal{M}$ by 14.6.1.1, contrary to $I \in \mathcal{I}$.

This contradiction establishes the claim that $D = 1$. Now by (e) and (g):

(h) $O_2(I) = O_2(K)$ and $Z_0 = Z_z \le Z_1$.

Thus $I \cong (S_5/E_{16})$ wr $\mathbf{Z}_2$. It follows also that $I_z = C_I(z) \cong (S_4/E_{16})$ wr $\mathbf{Z}_2$, and:

(i) $K_z = O^2(I_z) \cong \mathbf{Z}_3/Q_8^2 \times \mathbf{Z}_3/Q_8^2$, $C_{R_z}(O_2(K_z)) = Z_0$, and $C_{R_z}(O_2(K_z)/Z_0) = O_2(K_z)$.

Set $G_1^+ := G_1/Z_1$ and $C_1 := C_G(Z_1)$. As $C_1 \trianglelefteq G_1 \in \mathcal{H}^e$, $C_1 \in \mathcal{H}^e$ by 1.1.3.1, so that $Q_1 = O_2(C_1) = F^*(C_1)$. Then $Q_1^+ = F^*(C_1^+)$ by A.1.8. Let X be the preimage in G_1 of $F^*(G_1^+)$; as Y centralizes Z_1 by (f), $O^2(X) \le Y \le C_1$; so as $Q_1^+ = F^*(C_1^+)$, $O^2(X) = 1$ and hence $F^*(G_1^+) = Q_1^+$. Thus using B.2.14:

(j) $E^+ := \Omega_1(Z(T^+)) \cap R_1^+ R_2^+ \le \Omega_1(Z(Q_1^+)) =: F^+$.

Next $O_2(K) = U_1 \times U_2$, where $U_i := O_2(K_i)$, and $E_i := U_i \cap R_i$ is a hyperplane of U_i. Let $E_0 := E_1 E_2$. Then as $I \cong S_5/E_{16}$ wr $\mathbf{Z}_2$, $Z(T_0/Z_0) \le E_0/Z_0$ and I_z is irreducible on E_0/Z_0; so as $Z_0 \le Z_1$ by (h), we conclude from (j) that

(k) $E_0 \le E \le F \le Q_1 \le O_2(K_z T_I)$,

where E and F are the preimages of E^+ and F^+ in G_1.

Recall $r \in T - T_I$, $T_I = T_0$, and case (iii) of C.5.6.7 holds. Hence by C.5.6.7, $A := O_2(K)$ and A^r are the two T_0-invariant members of $\mathcal{A}(T_0)$, and $A \cap A^r = [A, A^r]$ is of rank 4. Thus as E_0 is of rank 6, $E_0 \not\le A \cap A^r$, so as $E_0^r \le A^r$, $E_0^r \not\le A$. Now F is normal in G_1, so $E_0^r \le F \le O_2(K_z T_I)$ by (k). Then as $E_0^r \not\le A$ and $K_z T$ is irreducible on $O_2(K_z T_I)/A$:

(l) $O_2(K_z) = E_0[E_0^r, K_z] \le F \le Q_1 \le R_z$.

It follows from (l) that $Z_1 = \Omega_1(Z(Q_1)) \le C_{R_z}(O_2(K_z))$, so we conclude from (h) and (i) that:

(m) $Z_1 = Z_0$.

Then as $F^+ = \Omega_1(Z(Q_1^+))$, (l) says $Q_1 \le C_{R_z}(O_2(K_z)^+)$, while as $Z_1 = Z_0$ by (m), $C_{R_z}(O_2(K_z)^+) = O_2(K_z)$ by (i). So we conclude from (l) that:

(n) $Q_1 = O_2(K_z)$.

From (i), $\widetilde{O_2(K_z)} \cong Q_8^4$, so by 14.5.15.1 and (n), $\tilde{V} \le Z(\tilde{Q}_1) = Z(\widetilde{O_2(K_z)}) = \tilde{Z}_0$. Thus $V = Z_0 = Z_1 \trianglelefteq G_1$, contrary to $G_1 \not\le M = N_G(V)$. This contradiction finally completes the treatment of the case $F^*(I) = O_2(I)$.

Thus it remains to treat the case $F^*(I) \ne O_2(I)$. As $O(I) = 1$ by 14.6.7.3, there is a component K of I. As I_z is the unique maximal overgroup of T_I in the minimal parabolic I, I and I_z are described in E.2.2, and in particular $I = K_0 T_I$, where $K_0 := \langle K^{T_I} \rangle$. On the other hand by 14.6.7.1, K is described in 1.1.5.3; in particular $K = [K, z]$ with z 2-central in I.

We consider the possibilities from the intersection of the lists of E.2.2 and 1.1.5.3: First suppose $K/O_2(K)$ is a Bender group. Then by E.2.2, I_z is the normalizer of a Borel subgroup B of K_0, and centralizes no element of $(O_2(B)/O_2(K_0))^\#$, whereas I_z centralizes the projection of z on $O_2(B)/O_2(K_0)$. Similarly if $K/O_2(K) \cong Sp_4(2^n)'$ or $L_3(2^n)$, then $N_{T_I}(K)$ is nontrivial on the Dynkin diagram of $K/O_2(K)$ by E.2.2, so again I_z is the normalizer of a Borel subgroup B of K_0, and hence $n = 1$ since I_z centralizes the projection of z on $O_2(B)/O_2(K_0)$. Thus $K/O_2(K)$ is $L_2(p)$ with $p > 7$ a Fermat or Mersenne prime, or $K/O_2(K)$ is $L_3(2)$ or A_6 with $N_{T_I}(K)$ nontrivial on the Dynkin diagram of $K/O_2(K)$.

Set $I^* := I/O_2(I)$. Then $U_H^* \unlhd T_I^*$, while $m(U_H/C_{U_H}(Q_H)) \geq 4$ by hypothesis. Therefore by 14.6.8, $m_2(I^*) \geq m(U_H^*) \geq 4$, so from the previous paragraph, $K_0 > K$ and $K/O_2(K) \cong L_2(p)$ for $p \geq 7$ a Fermat or Mersenne prime, with $Aut_I(K^*) \cong PGL_2(7)$ if $p = 7$. But in these groups T_I^* has no normal elementary abelian subgroup of rank at least 4. This contradiction completes the proof of 14.6.9. $\qquad\square$

This subsection culminates in the technical lemma 14.6.10. In each of the subsequent two subsections, the final contradiction will be to part (5) of 14.6.10.

LEMMA 14.6.10. *Assume the hypotheses of 14.6.9 and let $L_I := O^2(L \cap I)$. Assume that $I = \langle I_1, I_2 \rangle$, where $I_1 := L_I T_I$ and $T_I \leq I_2 \leq H$ with $I_2/O_2(I_2) \cong S_3$. Set $R_i := O_2(I_i)$. Then*

(1) $C(G, R_1) \leq M$.
(2) $R_1 \neq R_2$.
(3) If P is an I_1-invariant subgroup of I, then either $L_I \leq P$ or $P \leq C_M(V)$.
(4) $F^*(I) = O_2(I)$.
(5) If $T_I = T_u$, assume further that $I \leq C_G(u)$. Then $m(\langle V^{I_2} \rangle) = 3$.

PROOF. As the hypotheses of 14.6.9 hold, by that result $LT = L_I T_I O_2(LT) = I_1 O_2(LT)$. In particular $L_I \not\leq G_1$, $I_1/O_2(I_1) \cong S_3$, $L_I/O_2(L_I) \cong \mathbf{Z}_3$, and $R_1 = O_2(LT) \cap T_I$. By 14.6.6.1, $T_I < T$, so $T_I < N_T(T_I) \leq N_T(R_1)$ since $R_1 = O_2(LT) \cap T_I$. Then as $I \in \mathcal{I}^*$ and $N_{LT}(R_1)$ contains $I_1 \not\leq G_1$, we conclude from 14.6.4 that $M = !\mathcal{M}(N_{LT}(R_1))$, so (1) holds. Since $I \not\leq M$ but $I_1 \leq M$, $I_2 \not\leq M$, so (1) implies (2).

Assume P is a counterexample to (3). If $P \leq G_1$, then as P is I_1-invariant, P centralizes $\langle Z^{I_1} \rangle = V$, so that $P \leq C_G(V) = C_M(V)$, contrary to the choice of P as a counterexample; thus $P \not\leq G_1$. Set $M^+ := M/O_2(M)$. By 14.2.2.4, $M^+ = L^+ R_c^+ \times C_M(V)^+$, where $R_c := O_2(M \cap G_1)$. As $L_I \neq 1$ while $L = [L, C_{O_2(G_1)}(u)]$ by 14.6.3.3, $L^+ R_c^+ = I_c^+$, where $I_c := I_1 \cap LR_c$. As we are assuming that P is I_1-invariant with $L_I \not\leq P$, $P \cap L_I \leq O_2(L_I)$, so as $O^2(L \cap P) \leq O^2(L \cap I) = L_I$, $O^2(L \cap P) = 1$. If $P \leq M$ then $[P, I_c] \leq P \cap LR_c \leq O_2(L \cap P)R_c = R_c$, so $P^+ \leq C_{M^+}(O^2(I_c^+)) \leq C_M(V)^+$, again contrary to the choice of P since $O_2(M) \leq C_M(V)$. Therefore P is contained in neither M nor G_1, and as $PT_I \leq I$, $PT_I \in \mathcal{H}(T_u)$. Hence $PT_I \in \mathcal{I}^*$ by 14.6.4. Thus we may apply 14.6.9 to PT_I in the role of "I", to conclude that $O^2(L \cap P) \neq 1$, contrary to an earlier observation. So (3) is established.

Set $I^* := I/O_{3'}(I)$. Observe as $T_I \in Syl_2(I)$ by 14.6.6.2, that (I, I_1, I_2) is a Goldschmidt triple in the sense of Definition F.6.1. In view of (2), case (i) of F.6.11.2 holds, so I^* is a Goldschmidt amalgam, and hence as I is an SQTK-group, I^* is described in Theorem F.6.18.

To prove (4), we assume $F^*(I) \neq O_2(I)$, and derive a contradiction. By hypothesis $m(U_H/C_{U_H}(Q_H)) \geq 4$, so since $O_2(I) \in Syl_2(O_{3'}(I))$ by F.6.11.1, $m(U_H^*) \geq 4$ by 14.6.8. Now the only case of Theorem F.6.18 in which $m_2(I^*) \geq 4$ is case (13), where $I^* \cong Aut(M_{12})$. Thus $|I_z^* : T_I^*| = 3 = |I_2^* : T_I^*|$, so $I_2^* = I_z^*$. Thus as $I_2 \leq H$, $U_H^* \unlhd I_z^*$ with $m(U_H^*) \geq 4$, whereas in $Aut(M_{12})$ (as we saw during the proof of 14.5.5), I_z^* has no such normal subgroup. This contradiction establishes (4).

Assume the hypotheses of (5). By (2), conclusion (1) of Theorem F.6.18 does not hold. If either case of conclusion (2) of F.6.18 holds, then there is a normal subgroup P of I with $I = PI_1$ and $P \cap L = O_2(L)$. But then by (3), $P \leq C_M(V)$, so $I = I_1 P \leq M$, contrary to $I \in \mathcal{I}$.

In the remaining conclusions of F.6.18, there is $K \in \mathcal{C}(I)$ with $K \trianglelefteq I$, and either $I = KT_I$, or case (3) of F.6.18 holds with KT_I of index 3 in I. Since $O_{3'}(I)$ is 2-closed by F.6.11.1, $K/O_2(K)$ is quasisimple by 1.2.1.4. Next by (1), $C(I, R_1) \leq M_I := I \cap M$. Further $L_I \trianglelefteq M_I$ and $R_1 = T_I \cap O_2(L_I T_I) \in Syl_2(C_{M_I}(L_I/O_2(L_I)))$, so $R_1 \in \mathcal{B}_2(M_I)$ by C.1.2.4; then as $N_I(R_1) \leq M_I$, $R_1 \in \mathcal{B}_2(I)$. Now Hypothesis C.2.3 is satisfied with I, R_1, M_I in the roles of "H, R, M_H". As K appears in F.6.18, $K/O_2(K)$ is not $L_2(2^n)$, so that K is not a χ_0-block. Now as K is T_I-invariant and $K/O_2(K)$ is quasisimple, we may apply C.2.7 to conclude that K is described in C.2.7.3. Comparing the lists of C.2.7.3 and F.6.18, we conclude that $O_2(I) = O_{3'}(I)$, $I^* = I/O_2(I) \cong L_3(2)$, $\hat{A}_6$, A_7, S_6, S_7, or $G_2(2)$, and except possibly in the first case, K is a block. In particular case (3) of F.6.18 is now ruled out, so $I = KT_I$. Then again using F.6.6, $K = O^2(I) = \langle K_1, K_2 \rangle$, where $K_i := O^2(I_i)$. Thus $L_I = K_1 \leq K$, so

$$V = [Z, L_I] \leq [\Omega_1(Z(O_2(K))), K] =: W.$$

To prove (5), we must show that $V_0 := \langle V^{I_2} \rangle$ is of rank 3, so we assume $m(V_0) \neq 3$, and it remains to derive a contradiction.

Suppose first that $K^* \cong L_3(2)$. Then case (g) of C.2.7.3 occurs, so we may apply C.1.34 to conclude that W is either a natural module, the sum of two isomorphic natural modules, or a 4-dimensional indecomposable module with a 1-dimensional submodule. As $V = [V, L_I]$ is a T_I-invariant projective line in W, it follows that $m(W) \neq 4$, and that $\langle V^K \rangle$ is an irreducible K-submodule of W of rank 3, so $V_0 = \langle V^{I_2} \rangle = \langle V^K \rangle$ is of rank 3, contrary to assumption. Therefore K is a block.

Suppose first that K is an $\hat{A}_6$-block. Then since $K = \langle K_1, K_2 \rangle$, $K_1 = L_I \not\leq X := O^2(O_{2,Z}(K))$, and of course X is normalized by $K_1 = I_1$. Thus $X \leq C_M(V)$ by (3), impossible as $C_W(X) = 1$ in an $\hat{A}_6$-block.

Next $V = [V, L_I]$ is a T_I-invariant line and I_2 stabilizes the point Z on that line. In particular if K is a $G_2(2)$-block then V is a doubly singular line in the language of [**Asc87**], and so V_0 is of rank 3, contrary to assumption. Similarly when $m(W) = 4$ and $K^* \cong A_6$ or A_7, we compute that Z and V_0 have ranks 1 and 3, respectively—again contrary to assumption.

Thus K is an A_n-block for $n := 6$ or 7, $I^* \cong A_n$ or S_n, with $m(W) = 5$ when $n = 6$, and we can represent I on $\Omega := \{1, \ldots, n\}$ as in section B.3, so that W is the core of the permutation module on Ω. Further M_I is the stabilizer in I of the T_I-invariant line V. So when $n = 6$, $M_I^* = I_1^*$ is the stabilizer of the partition $\Lambda := \{\{1,2\}, \{3,4\}, \{5,6\}\}$, $V = \langle e_{1,2,3,4}, e_{1,2,5,6} \rangle$, $z = e_{1,2,3,4}$, and $V_0 = \{e_J : |J \cap \{1,2,3,4\}| \equiv 0 \mod 2\}$, while $I_2^* = I_z^*$ is the stabilizer of the partition $\{\{1,2,3,4\}, \{5,6\}\}$. Next assume for the moment that $n = 7$. Then I_1^* and I_2^* are (in some order) the stabilizers of the partitions $\Lambda' := \Lambda \cup \{7\}$ and $\theta := \{\{1,2\}, \{3,4\}, \{5,6,7\}\}$. However if I_1^* is the stabilizer of θ then $V = \langle e_{5,6}, e_{5,7} \rangle$ and $z = e_{5,6}$, impossible as I_2 centralizes z but the stabilizer of Λ' does not. Thus $M_I^* = I_1^*$ is the stabilizer of Λ', I_2^* is the stabilizer of θ, and as before $V = \langle e_{1,2,3,4}, e_{1,2,5,6} \rangle$ and $z = e_{1,2,3,4}$, while now $V_0 = \langle V, e_{5,6}, e_{5,7} \rangle$. Observe in this case that I_2^* is a proper subgroup of the stabilizer I_z^* of the partition $\{\{1,2,3,4\}, \{5,6,7\}\}$.

Suppose first that $T_I = T_0$. We saw earlier that $|LT : L_I T_I| = |T : T_I|$, so as $|T : T_0| = 2$, $L_I = O^2(L_I T_I) = O^2(LT) = L$. Further $C_T(L) = 1$ by 14.2.2.6. However by the previous paragraph, $L = L_I$ centralizes $e_{1,2,3,4,5,6}$, contrary to $C_T(L) = 1$.

Thus $T_I = T_u$ by 14.6.6.1. Therefore by the hypothesis of part (5), $I \leq C_G(u)$. Further as $I_2 \leq H$, $V_+ := V_0\langle u \rangle \leq U_H$, and then from the discussion above, $V_- := C_{V_+}(T_u) = \langle z, e_{5,6}, u \rangle$.

Suppose that $u \notin W$. Then $W \cap V_- = \langle z, e_{5,6} \rangle$, so that $m(V_-) = 3$. Therefore as $[U_H, Q_H] \leq Z$ by 14.5.15.1, while $T_0 = Q_H T_u > T_u$ by 14.6.3.1, we conclude $V_\# := C_{V_-}(Q_H) = C_{V_-}(T_0)$ is a hyperplane of V_- with $u \notin V_\#$, so that $V_- = V_\#\langle u \rangle$. Let v_- be the projection on $V_\#$ of $e_{1,2,3,4,5,6}$, and set $J := C_K(v_-)$; then $v_- \neq 1$ as $u \notin W$. Now $J^* \cong A_6$, so J^* is contained in neither M_I^* nor I_z^* which are solvable from the discussion above, and hence J is contained in neither M nor G_1. But then $\langle T_0, J \rangle \leq C_G(v_-) \in \mathcal{I}$, contrary to 14.6.4 since $T_0 > T_u = T_I$.

Therefore $u \in W$. Since $I \leq C_G(u)$, $C_W(K) \neq 1$, and hence $n = 6$ and $\langle u \rangle = C_W(K)$, so that $u = e_{1,2,3,4,5,6}$. Let $Q_I := O_2(I)$. Since T_u is nontrivial on V by 14.6.3.3, and $|T : C_T(V)| = 2$, $T_0 = T_u C_{T_0}(V)$, so we may choose $t \in C_{T_0}(V) - T_u$. Since t normalizes T_u and $W \trianglelefteq T_u$, both $B := W^t$ and $WW^t = WB$ are normal in $T_u = T_I$. If $B \leq Q_I$, then as $[Q_I, K] = W$ since K is a block, $J := \langle K, T_I, t \rangle$ acts on WB, and J contains $KT_I = I$ and $T_0 \geq T_J > T_I$, contradicting 14.6.5. Thus $B \not\leq Q_I$, so that $B^* \neq 1$. By (U1), T_0 acts on $\langle z, u \rangle$, so as $V \trianglelefteq T$, T_0 acts on $V_u := V\langle u \rangle$. Therefore as $V \leq W$ and $\langle u \rangle = C_W(K)$, $V_u \leq W \cap B$. Similarly $u^t \in V_u \cap Z(T_u)$, and this latter group is generated by $z = e_{1,2,3,4}$ and $u = e_{1,2,3,4,5,6}$. Therefore as $u^t \notin z^K$ by 14.6.3.4, we conclude that $u^t = e_{5,6}$.

Notice for $v \in V^\#$ that $W_v := \langle V^{C_K(v)} \rangle$ is a hyperplane of W, and if $V = \langle v, w \rangle$, then $W = W_v W_w$. For example $W_z = V_0$. Thus $B^* = W^{t*} = W_v^{t*} W_w^{t*}$. Now $\langle V^{C_G(v)} \rangle$ is abelian by Hypothesis 14.5.1 and the transitivity of L on $V^\#$, and we chose t to centralize V, so $W_v^t \leq \langle V^{C_G(v)} \rangle \leq C_G(W_v)$. Therefore from the action of S_6 on its permutation module, $W_v^{t*} = \langle (i,j) \rangle$, where $v := e_{\Omega - \{i,j\}}$. Then as $W_v^{t*} W_w^{t*} = B^* \trianglelefteq T_I^*$, and the only normal subgroup of T_I^* containing $W_z^{t*} = \langle (5,6) \rangle$ generated by at most two transpositions is $\langle (5,6) \rangle$, we conclude that $B^* = W_z^{t*} = \langle (5,6) \rangle$. Thus $[W, W_z^t] = \langle e_{5,6} \rangle = \langle u^t \rangle$. This is impossible, as $C_I(W/\langle u \rangle) = C_I(W)$, so that $C_{I^t}(W_z^t/\langle u^t \rangle) = C_{I^t}(W_z^t)$.

This contradiction completes the proof of (5), and hence of 14.6.10. $\qquad\square$

14.6.2. Showing $\mathbf{O}(\mathbf{H}/\mathbf{O}_2(\mathbf{H})) = \mathbf{1}$.

Recall that H denotes a member of $\mathcal{H}(T, M) = \mathcal{H}_z$, and we have adopted Notation 12.8.2. In the remaining two subsections we adopt Notation 14.5.16 and use notation and results from section F.9. For example Γ is the coset geometry determined by LT and H as in section F.7, with the parameter b, the geodesic $\gamma_1, \dots, \gamma = \gamma_b$, the element g_b taking γ_1 to γ, and the subgroups U_H, U_γ, D_H, D_γ etc., as well as $Z_\gamma := Z^{g_b}$ defined in section F.9—where Z_γ was often denoted by A_1.

This second subsection is devoted to the proof of a key intermediate result:

THEOREM 14.6.11. *$O(H^*) = 1$ for each $H \in \mathcal{H}(T, M)$.*

Until Theorem 14.6.11 is established, assume H is a counterexample. Thus H is a member of $\mathcal{H}(T, M)$ with $O(H^*) \neq 1$, and we must derive a contradiction from the existence of such an H.

Let P_0^* be a minimal normal subgroup of H^* contained in $O(H^*)$; then P_0^* is an elementary abelian p-group and $P_0^* = P^*$ for $P \in Syl_p(P_0)$. Indeed $PT \in \mathcal{H}(T, M)$ by 14.6.1.4; so replacing H by PT, we may assume $H = PT$ with P^* a minimal normal subgroup of H^*. Thus T is maximal in $PT = H$, and $P \cong \mathbf{Z}_p$ or E_{p^2}, since H is an SQTK-group.

Set $K := O^2(H)$, so that $K^* = P^*$.

LEMMA 14.6.12. (1) $p = 3$ or 5.

(2) There is a subgroup H_0 of index 2 in H such that $H_0^* = H_1^* \times H_2^*$, $H_i^* \cong D_{2p}$, $H_2 = H_1^t$ for $t \in T - N_T(H_1)$ and H_i the preimage of H_i^* in H, and $[\tilde{U}_H, H] = \tilde{U}_{H,1} \oplus \tilde{U}_{H,2}$, where $\tilde{U}_{H,i} := [\tilde{U}_H, H_i]$ is of rank 4 when $p = 5$, and of rank 2 or 4 when $p = 3$.

(3) $Z \le [U_H, O^2(H_i)]$ for each i.

PROOF. By 14.5.18.3, $q(H^*, \tilde{U}_H) \le 2$. Let $H_0^* := \langle \mathcal{Q}_*(H^*, \tilde{U}_H) \rangle$; as T is maximal in H, $H = H_0 T$. By D.2.17, $H_0^* = H_1^* \times \cdots \times H_s^*$ and $[\tilde{U}_H, H_0] = \tilde{U}_{H,1} \oplus \cdots \oplus \tilde{U}_{H,s}$, where $(H_i^*, \tilde{U}_{H,i})$ are indecomposables in the sense of D.2.17. In particular $p = 3$ or 5 by D.2.17, so that (1) holds. Further $O_p(H_0)^*$ is not of order p by 14.3.5. Hence $P^* \cong E_{p^2}$, and as T is irreducible on P^*, our indecomposables appear only in conclusions (1) or (2) of D.2.17, so that (2) holds. Finally (3) follows from 14.6.2. $\square$

During the remainder of the proof of Theorem 14.6.11, we adopt the notation of 14.6.12.2, with $U_{H,i}$ the preimage in U_H of $\tilde{U}_{H,i}$. Also set $U_K := [U_H, H]$.

LEMMA 14.6.13. Either

(1) $p = 3$, $\tilde{U}_K$ is a 4-dimensional orthogonal space over $\mathbf{F}_2$ for

$$H^* = O(\tilde{U}_K) \cong O_4^+(2),$$

and $[\tilde{U}_{H,1}, U_H^g] \ne 1$ for some $g \in G - G_1$ such that $U_H^g \le N_H(U_{H,1})$ and $U_H \le H^g$, or

(2) $p = 3$ or 5, $m(\tilde{U}_K) = 8$, $D_\gamma < U_\gamma$, and we may choose γ so that $U_\gamma^* \le H_1^*$, $Z_\gamma \le U_{H,1}$, and $Z \le U_{H,1}^g$, for $g \in G - G_1$ with $\gamma_1 g = \gamma$.

PROOF. Suppose first that $D_\gamma = U_\gamma$. Then by 14.5.18.1, U_H induces a nontrivial group of transvections on U_γ with center Z, so by 14.6.12, $p = 3$, and H^* acts as $O_4^+(2)$ on $\tilde{U}_K$ of rank 4. Since $b \ge 3$ is odd by F.9.11.1, in this case there is $g \in \langle LT, H \rangle$ with $\gamma_1 = \gamma g$. Then U_H^g induces a nontrivial group of transvections on U_H with center Z^g, so $U_H^g \le N_H(U_{H,1})$, and we may choose notation so that $[\tilde{U}_{H,1}, U_H^g] \ne 1$. By F.9.13.2, $U_\gamma \le H$, so $U_H = U_\gamma^g \le H^g$. Thus (1) holds when $D_\gamma = U_\gamma$.

Hence we may suppose instead that $D_\gamma < U_\gamma$. So by 14.5.18.4, we may choose γ with $m(U_\gamma^*) \ge m(U_H/D_H) > 0$ and $U_\gamma^* \in \mathcal{Q}(H^*, \tilde{U}_H)$; in particular U_γ is quadratic on U_H, and hence either U_γ acts on $\tilde{U}_{H,1}$, or else the quadratic action forces $U_\gamma^* = \langle x^* \rangle$ to be of order 2 with $U_{H,1}^x = U_{H,2}$. Let $g \in \langle LT, H \rangle$ with $\gamma_1 g = \gamma$.

Suppose first that $U_\gamma^* = \langle x^* \rangle$ is of order 2 with $U_{H,1}^x = U_{H,2}$. As $U_\gamma^* \in \mathcal{Q}(H^*, \tilde{U}_H)$,

$$m(\tilde{U}_H / C_{\tilde{U}_H}(U_\gamma)) \le 2m(U_\gamma^*) = 2,$$

while $C_{\tilde{U}_H}(U_\gamma) = [\tilde{U}_H, U_\gamma]$ since x^* is an involution with $U_{H,1}^x = U_{H,2}$. Therefore $m(\tilde{U}_H) = 4$, and the inequality is an equality. Again by 14.6.12, $p = 3$, $\tilde{U}_K$ is a 4-dimensional orthogonal space over $\mathbf{F}_2$, and $H^* = O(\tilde{U}_K)$. Further $Z_\gamma = [U_\gamma, D_H]$ by F.9.13.6, so $\tilde{z}^g$ is a singular vector in $\tilde{U}_K$ since $\tilde{U}_{H,1}^\# \cup \tilde{U}_{H,2}^\#$ is the set of nonsingular vectors of $\tilde{U}_K$. Then $S^* := C_S(z^g)^*$ for some Sylow 2-subgroup S of H containing U_γ. Now $H \le G_1$ with $U_H = \langle V^H \rangle \le \langle V^{G_1} \rangle = U$; thus $U_\gamma = U_H^g \le U^g = \langle V^{gG_1^g} \rangle$, so since U is abelian by Hypothesis 14.5.1, $[U^g, C_S(z^g)] \le U^g \le C_G(U_\gamma)$. Now as $U_\gamma^* \le S^* \cong D_8$ with $U_{H,1}^x = U_{H,2}$, $[U_\gamma, C_S(z^g)]$ contains s with $s^* = Z(S^*)$. Then $s \in [U^g, C_S(z^g)] \le C_G(U_\gamma)$, whereas s^* does not centralize $[U_K, x] \le U_\gamma$.

Therefore U_γ acts on $U_{H,1}$. Suppose first that $m(\tilde{U}_K) = 4$. Then by 14.6.12.2, $p = 3$ and $\tilde{U}_K$ is a 4-dimensional orthogonal space for H^*. This time choose g so that $U_\gamma = U_H^g$, and choose notation so that $[\tilde{U}_{H,1}, U_H^g] \ne 1$; now $U_H \le H^g$ by F.9.13.2, completing the verification that (1) holds.

Thus it remains to treat the case in 14.6.12 with $m(\tilde{U}_K) = 8$. By the choice of γ:
$$0 < m(U_H/D_H) \le m(U_\gamma^*) \le m_2(H/C_H(\tilde{U}_K)) = 2. \qquad (*)$$
As $m(\tilde{U}_K) = 8$, if $U_\gamma^* \not\le H_i^*$ for $i = 1$ or 2, then $m(\tilde{U}_K/C_{\tilde{U}_K}(u_\gamma^*)) = 4$ for suitable $1 \ne u_\gamma^* \in U_\gamma^*$; this is a contradiction as $[D_H, U_\gamma] \le Z_\gamma$ by F.9.13.2, while $m(Z_\gamma) = 1$ and $m(U_H/D_H) \le 2$ by $(*)$. Therefore we may assume $U_\gamma \le H_1$, so that $m(U_\gamma^*) = 1$ from the structure of H^* in 14.6.12, and hence $m(U_H/D_H) = 1$ by $(*)$. Then since $[\tilde{U}_{H,1}, U_\gamma]$ has rank 2, $1 \ne [D_H \cap U_{H,1}, U_\gamma]$, so that $Z_\gamma \le U_{H,1}$ by F.9.13.6. Also $m(U_\gamma/D_\gamma) = 1 = m(U_H/D_D)$, so our hypotheses are symmetric in γ and γ_1, as discussed in Remark F.9.17. Hence we may choose notation so that $Z \le U_{H,1}^g$, so that (2) holds, completing the proof of 14.6.13. $\qquad \square$

Recall that G_1 is a member of $\mathcal{H}(T, M)$, so that the notational conventions of section 14.5 apply also to G_1 in the role of "H". Our convention in this subsection is to define $U := U_{G_1} = \langle V^{G_1} \rangle$, and set $Q_1 := O_2(G_1)$. Set $\hat{G}_1 := G_1/Q_1$ and $K_z := \langle K^{G_1} \rangle$.

Now we further specify our choice of $H \in \mathcal{H}(T, M)$, so that the odd prime $p \in \pi(H)$ is maximal over odd primes such that $O_p(H_0^*) \ne 1$ for some $H_0 \in \mathcal{H}(T, M)$; that is, in view of 14.6.12.1, we choose H with $p := 5$ if $O_5(H_0^*) \ne 1$ for some $H_0 \in \mathcal{H}(T, M)$, and otherwise $p := 3$.

LEMMA 14.6.14. *One of the following holds:*

(1) $K_z = K$, and if $p = 3$ then G_1 is a $\{2, 3\}$-group.

(2) $p = 3$, $K_z \in \mathcal{C}(G_1)$, and $\hat{G}_1 \cong Aut(L_n(2))$ for $n := 4$ or 5.

(3) $p = 3$, $K_z = K_1 K_1^s$ for $s \in T - N_T(K_1)$ with $K_1 \in \mathcal{C}(G_1)$, and $\hat{G}_1 \cong S_5 \ wr \ \mathbf{Z}_2$ or $L_3(2) \ wr \ \mathbf{Z}_2$.

(4) $p = 5$, $K_z = K_1 K_1^s$ for $s \in T - N_T(K_1)$ with $K_1 \in \mathcal{C}(G_1)$, and $\hat{G}_1 \cong Aut(L_2(16)) \ wr \ \mathbf{Z}_2$.

PROOF. First suppose H_+ is a solvable overgroup of H in G_1. If $X \trianglelefteq H_+$ with $X/O_2(X)$ a q-group for some odd prime q, then $XT \in \mathcal{H}(T, M)$ by 14.6.1.4, and so $q \le p \le 5$ by 14.6.12 and our maximal choice of p. Thus setting $\dot{H}_+ := H_+/O_2(H_+)$,
$$F^*(\dot{H}_+) = \prod_{q \le p} O_q(\dot{H}_+),$$
with $m_q(O_q(\dot{H}_+)) \le 2$ since H_+ is an SQTK-group. Therefore using A.1.25 and inspecting the order of $GL_2(q)$, we conclude H_+ is a $\{2, 3\}$-group if $p = 3$, and a $\{2, 3, 5\}$-group if $p = 5$.

We claim next that for $J \in \mathcal{C}(G_1)$, $\hat{J}$ is not a Suzuki group: For if $\hat{J} \cong Sz(2^m)$ for some odd $m > 1$, then the T-invariant Borel subgroup B of $J_0 := \langle \hat{J}^T \rangle$ has order divisible by each prime dividing $2^m - 1$, and one of these primes is larger than 5. On the other hand $H \cap J_0$ is a solvable overgroup of $T \cap J_0$ in J_0, and hence is 2-closed, so H acts on $N_{J_0}(T \cap J_0) = B$, contrary to the previous paragraph applied to HB in the role of "H_+".

Now $K \in \Xi(G, T)$ by 14.6.12, so by 1.3.4, either $K = K_z$, or $K_z = \langle K_1^T \rangle$ for some $K_1 \in \mathcal{C}(G_1)$ with $K_1/O_2(K_1)$ quasisimple, and K_z is described in 1.3.4.

Supose $K = K_z$. If $p = 5$ then (1) holds, so we may assume that $p = 3$. Now $\hat{K}$ contains all elements of order 3 in $C_{\hat{G}_1}(\hat{K})$ since $m_3(\hat{K}) = 2$ and G_1 is an SQTK-group. Thus if $J \in \mathcal{C}(G_1)$ then J is a $3'$-group, which is impossible by the claim, so we conclude from 1.2.1.1 that G_1 is solvable. Then G_1 is a $\{2, 3\}$-group by the first paragraph applied to G_1 in the role of "H". Therefore (1) holds when $K = K_z$.

Thus we may assume that $K < K_z = \langle K_1^T \rangle$ with $K_1 \in \mathcal{C}(G_1)$. As $K_1/O_2(K_1)$ is quasisimple, K_z is described in part (4) or (5) of F.9.18. Comparing the lists of 1.3.4 and F.9.18, we conclude that either:

(i) $K_z = K_1 K_2$ with $K_2 := K_1^s$ for $s \in T - N_T(K_1)$, and either $K_1^* \cong L_2(2^m)$ with $2^m \equiv 1 \mod p$, or $p = 3$ and $K_1^* \cong L_3(2)$.

(ii) $p = 3$ and $K_1 T/O_2(K_1 T) \cong Aut(L_n(2))$, $n = 4$ or 5.

Notice that the $Sp_4(2^n)$-case in 1.3.4.3 is excluded, as here $Aut_T(P)$ is noncyclic by 14.6.12.

Observe that $K_z = O^{p'}(G_1)$: in case (i), this follows from 1.2.2, and in case (ii) from A.3.18. Furthermore when $p = 5$ we have case (i) with $\hat{K}_1 \cong L_2(2^m)$ for m divisible by 4, so that $K_z = O^{3'}(G_1)$ by 1.2.2. Thus $C_{\hat{G}_1}(\hat{K}_z)$ is a $3'$-group, and if $p = 5$, then $C_{\hat{G}_1}(\hat{K}_z)$ is a $\{3, 5\}'$-group. Therefore applying the first paragraph to $HO_{2,F}(G_1)$ in the role of "H_+", we conclude $F(\hat{G}_1) = 1$, and by the second paragraph, $\hat{G}_1$ has no Suzuki components. Therefore as $C_{\hat{G}_1}(\hat{K}_z)$ is a $3'$-group, $\hat{K}_z = F^*(\hat{G}_1)$.

As $F^*(\hat{G}_1) = \hat{K}_z$, conclusion (2) of the lemma holds in case (ii), so we may assume case (i) holds. Similarly conclusion (3) holds if $\hat{K}_1 \cong L_3(2)$, since $N_T(K_1)$ is trivial on the Dynkin diagram of $\hat{K}_1$ because T acts on K. Thus we may assume that $\hat{K}_1 \cong L_2(2^m)$. Applying the first paragraph to BT in the role of "H_+", where B is a Borel subgroup of K_z over $T \cap K_z$, we conclude that $m = 2$ if $p = 3$, and that $m = 4$ if $p = 5$. If $p = 3$, then as T^* induces D_8 on K^*, $\hat{G}_1 \cong S_5$ wr $\mathbf{Z}_2$, so conclusion (3) holds. Finally if $p = 5$ we showed $BT \in \mathcal{H}(T, M)$, so for $B_3 \in Syl_3(B)$, $B_3 T \in \mathcal{H}(T, M)$ by 14.6.1.4. Applying 14.6.12 to $B_3 T$ in the role of "H", we conclude $B_3 T/O_2(B_3 T) \cong O_4^+(2)$. Therefore $\hat{G}_1 \cong Aut(L_2(16))$ wr $\mathbf{Z}_2$, so that conclusion (4) holds. This completes the proof of 14.6.14. $\qquad \square$

LEMMA 14.6.15. $p = 3$.

PROOF. Assume otherwise. Then by 14.6.12.1 we may assume $p = 5$, and it remains to derive a contradiction. As $p = 5$, conclusion (2) of 14.6.13 holds, so we may choose γ as in 14.6.13.2; in particular $Z_\gamma \leq U_{H,1}$. Also since $p = 5$, case (1) or (4) of 14.6.14 holds. Let $U_z := [U, K_z]$. As $U_H \leq U$ and $K \leq K_z$, $U_K = [U_H, K] \leq U_z$.

In the next several paragraphs we assume $K < K_z$ and establish some preliminary results in that case. First case (4) of 14.6.14 holds, so $G_1 = K_z T$ and $K_z = K_1 K_1^s$ for $K_1 \in \mathcal{C}(G_1)$ with $K_1/O_2(K_1) \cong L_2(16)$ and $s \in T - N_T(K_1)$. We now apply F.9.18 to K_1, G_1 in the roles of "K, H": As the $O_4^+(16)$-module in case (i) of F.9.18.5 does not extend to $\hat{G}_1 \cong Aut(L_2(16))$ wr $\mathbf{Z}_2$, case (iii) of F.9.18.5 holds. Indeed as $\hat{K}_1 \cong L_2(16)$, subcase (a) of case (iii) holds, so for $\tilde{I} \in Irr_+(K_z, \tilde{U}, T)$, $I_0 := \langle I^T \rangle = II^s$, and we may choose notation so that $\tilde{I}/C_{\tilde{I}}(K_1)$ is the natural or orthogonal module for $\hat{K}_1$ and $[\tilde{I}, K_1^s] = 1$.

We claim that $U_z = I_0$. For if not, case (a) of F.9.18.6 does not hold and G_1^* has no strong FF-modules, so that case (c) of F.9.18.6 does not hold. Thus case (b) of F.9.18.6 holds, so that $W_z := U_z/I_0$ and $\tilde{I}_0/C_{\tilde{I}_0}(K_z)$ are nontrivial FF-module for G_1, and hence $W_z/C_{W_z}(K_z)$ and $\tilde{I}_0/C_{\tilde{I}_0}(K_z)$ are both natural modules for $L_2(16)$ by Theorem B.4.2. Indeed since $D_\gamma < U_\gamma$ in case (2) of 14.6.13, we may choose α as in 14.5.18.5; then $\hat{U}_\alpha \in \mathcal{Q}(\hat{G}_1, \tilde{U})$, so since $\hat{G}_1$ has no strong FF-modules by Theorem B.4.2, $\hat{U}_\alpha^*$ is an FF*-offender on both $\tilde{I}_0$ and W_z. Therefore either $\hat{U}_\alpha$ is Sylow in $\hat{K}_z$, or interchanging $\hat{K}_1$ and $\hat{K}_1^s$ if necessary, we may assume that $\hat{U}_\alpha$ is Sylow in $\hat{K}_1$. In either case, $m(\tilde{U}/C_{\tilde{U}}(\hat{U}_\alpha)) = 2\,m(\hat{U}_\alpha)$, so we conclude from 14.5.18.2 that $m(U/D) = m(\hat{U}_\alpha)$ where $D := D_{G_1}$, and that $\hat{U}_\alpha$ acts faithfully on $\tilde{D}$ as a group of $\mathbf{F}_2$-transvections with center $\tilde{Z}_\alpha$. As $\hat{U}_\alpha$ is Sylow in $\hat{K}_1$ or $\hat{K}_z$, and $\tilde{I}/C_{\tilde{I}}(\hat{K}_1)$ is the natural $\hat{K}_1$-module, $\hat{U}_\alpha$ does not induce a nontrivial group of $\mathbf{F}_2$-transvections on any subspace of $\tilde{I}_0$, so $\tilde{D} \cap \tilde{I}_0 = C_{\tilde{I}_0}(\hat{U}_\alpha)$ is of codimension $m(\hat{U}_\alpha)$ in $\tilde{I}_0$, and hence $U = I_0 D$. But this is impossible as $\hat{U}_\alpha$ does not induce a nontrivial group of $\mathbf{F}_2$-transvections on W_z. Thus the claim is established.

Set $K_2 := K_1^s$. Since case (iii.a) of F.9.18.5 holds, with $I_0 = U_z$ by the claim, $\tilde{U}_z = \tilde{U}_1 + \tilde{U}_2$ with $U_i := [U, K_i]$, and $\tilde{U}_i/C_{\tilde{U}_i}(K_i)$ the 2-dimensional natural or 4-dimensional orthogonal module for $K_i/O_2(K_i)$. Then as $U_H \le U_z$, we can choose notation so that $O^2(H_i) \le K_i$, and hence $U_{H,i} \le U_i$.

This completes our preliminary treatment of the case $K < K_z$. In the case where $K = K_z$ we establish a similar setup: Namely in this case we set $K_i := O^2(H_i)$ and $U_i := U_{H,i}$.

Thus in any case $Z_\gamma \le U_{H,1} \le U_1$, so that Z_γ centralizes K_2. Choose g as in case (2) of 14.6.13, and for $X \le G$, let $\theta(X)$ be the subgroup generated by the elements of order 5 in X. Then $K_2 \le \theta(C_G(Z_\gamma)) = K_z^g$, and by 14.6.13.2, $Z \le U_{H,1}^g \le U_1^g$, so $K_2 \le \theta(C_{K_z^g}(Z)) = K_z^g$. Therefore $K_2 = K_z^g$, so $g \in N_G(K_2)$.

Set $G_2 := N_G(K_2)$; since $g \in G - G_1$ in 14.6.13.2, $G_2 \not\le G_1$. Set $T_2 := N_T(K_2)$ and $G_{1,2} := G_1 \cap G_2$, so that $|G_1 : G_{1,2}| = |T : T_2| = 2$, and in particular $G_{1,2} \trianglelefteq G_1$. As $Q_1 = O_2(K_z T_2)$, and $K_z T_2 \le G_{1,2}$, we conclude $Q_1 = O_2(G_{1,2})$. Then as $G_1 \in \mathcal{M}$ by 14.6.1.1, $C(G_2, Q_1) \le G_{1,2} = N_{G_2}(Q_1)$, so $Q_1 \in \mathcal{B}_2(G_2)$. Thus Hypothesis C.2.3 is satisfied with G_2, Q_1, $G_{1,2}$ in the roles of "H, R, M_H". As $Z \le [U, K_2] \le O_2(K_2)$ using 14.6.12.3, $F^*(G_2) = O_2(G_2)$ by 1.1.4.3.

Suppose $O_{2,F^*}(G_2) \le G_{1,2}$. Then $O_2(G_2) = O_2(G_{1,2})$ by A.4.4.1, and we saw $G_{1,2} \trianglelefteq G_1$, so $G_2 \le N_G(O_2(G_{1,2})) = G_1$ as $G_1 \in \mathcal{M}$, contrary to an earlier remark. Thus $O_{2,F^*}(G_2) \not\le G_{1,2}$.

Next $G_1 = N_G(K_z)$ as $G_1 \in \mathcal{M}$. If X is an A_3-block of G_2, then as G_2 is an SQTK-group, $|X^{G_2}| \le 2$; hence $K_z = O^{5'}(K_z)$ normalizes X, and then centralizes X as $Aut(X) \cong S_4$. Thus $X \le C_{G_2}(K_z) \le G_{1,2}$. Therefore $O_{2,F}(G_2) \le G_{1,2}$ by C.2.6, so there is $J \in \mathcal{C}(G_2)$ with $J/O_2(J)$ quasisimple and $J \not\le G_{1,2}$. If K_z centralizes $J/O_2(J)$, then J normalizes $O^2(K_z O_2(J)) = K_z$, contrary to $J \not\le G_1 = N_G(K_z)$, so we conclude $J = [J, K_z]$. Furthermore $[J, K_2] \le O_2(J)$ by 1.2.1.2, so as

$K_z = K_1 K_2$, we obtain $J = [J, K_1]$. As $m_5(G_2) \leq 2$ and $m_5(K_2) = 1$, $m_5(J) \leq 1$, with $J = O^{5'}(C_{G_2}(K_2)) \trianglelefteq G_2$ in case of equality.

Now either Q_1 does not normalize J, so that C.2.4.1 holds, or Q_1 normalizes J, so that C.2.7.3 holds. Set $J_+ := \langle J^{Q_1} \rangle$, and observe that $B := J_+ \cap G_{1,2}$ normalizes K_2 and hence also K_1.

Suppose first that $K_1 \not\leq J$. Then an element k of K_1 of order 5 induces an outer automorphism of order 5 on J, since we saw that $J = O^{5'}(C_{G_2}(K_2))$ when $5 \in \pi(J)$. Inspecting C.2.4.1 and C.2.7.3 for cases where $J/O_2(J)$ admits an outer automorphism of order 5, we conclude $J/O_2(J)$ is $L_2(2^n)$ or $SL_3(2^n)$ with 5 dividing n, k induces a field automorphism on $J/O_2(J)$, and B is a Borel subgroup or a minimal parabolic of J_+, respectively. But then B does not normalize K_1, contrary to the previous paragraph.

Thus $K_1 \leq J$, so that $m_5(J) = 1$ and $J \trianglelefteq G_2$. Now we examine the list of C.2.7.3 for those $\hat{J}$ of 5-rank 1 with a subgroup $\hat{K}_1$ normalized by $\hat{B}$, such that $\hat{K}_1/O_2(\hat{K}_1) \cong \mathbf{Z}_5$ or $L_2(16)$. We conclude that $K_1/O_2(K_1) \cong \mathbf{Z}_5$, J is an $L_2(2^m)$-block, and B is a Borel subgroup of J. But then as $Z \leq B \leq G_{1,2} \leq C_G(Z)$, $Z \leq Z(B) = Z(J)$ using the structure of an $L_2(2^m)$-block; so $J \leq C_G(Z) = G_1$, contrary to $J \not\leq G_{1,2}$. This contradiction completes the proof of 14.6.15. $\qquad\square$

We will see shortly in 14.6.17 that the group T_0 in the following result can play the role of "T_0" in (U1) in the first subsection.

LEMMA 14.6.16. *Let* $T_0 := N_T(H_1)$. *Then* $|T : T_0| = 2$ *and* $N_{G_1}(T_0) = T$.

PROOF. From 14.6.12, $|T : T_0| = 2$ and $T = N_H(T_0)$. Further $p = 3$ by 14.6.15, and in particular case (4) of 14.6.14 does not hold.

Suppose case (2) or (3) of 14.6.14 holds. Then $G_1 = K_z T$ and $\hat{B} := N_{\hat{K}_z}(\hat{Q}_H)$ is a parabolic subgroup of $\hat{K}_z$ with unipotent radical $\hat{Q}_H$ and $\hat{H} = \hat{B}\hat{T} = N_{\hat{G}_1}(\hat{Q}_H)$. Thus Q_H is weakly closed in T with respect to G_1 by I.2.5, so $N_{G_1}(T_0) \leq N_{G_1}(Q_H) = H$, and hence $N_{G_1}(T_0) = N_H(T_0) = T$.

Finally assume case (1) of 14.6.14 holds. Then $\hat{K} = \hat{K}_1 \times \hat{K}_2$ where $K_i := O^2(H_i)$ and $\hat{K}_1$ and $\hat{K}_2$ are the two T_0-invariant subgroups of $\hat{K}$ of order 3. Thus $X := O^2(N_{G_1}(T_0))$ acts on $\hat{K}_i$ and hence X centralizes $\hat{K}$. Then as $C_{\hat{K}}(T_0) = 1$ and $m_3(\hat{G}_1) = 2$, X is a 3'-group. However as case (1) of 14.6.14 holds, G_1 is a $\{2, 3\}$-group, so again we conclude that $N_{G_1}(T_0) = T$. $\qquad\square$

We can now determine H^*, and show that the set $\mathcal{U}(H)$ of involutions discussed in the first subsection is nonempty.

LEMMA 14.6.17. *(1)* $\tilde{U}_K$ *is a 4-dimensional orthogonal space over* $\mathbf{F}_2$ *for* $H^* = O(\tilde{U}_K) \cong O_4^+(2)$.
(2) $Z_\gamma \not\leq U_{H,1}$.
(3) Let $u \in U_K$ *with* $\tilde{u}$ *nonsingular in the orthogonal space* $\tilde{U}_K$ *and centralized by* $N_T(H_1)$. *Then* $u \in \mathcal{U}(H)$.
(4) $C_G(u) \in \mathcal{I}$, *so* $\mathcal{I}^*$ *is nonempty.*
(5) $m(\langle V^{O^2(H_2)} \rangle) = 4$.

PROOF. Set $T_0 := N_T(H_1)$ and let $u_1 \in U_{H,1} - Z$ with $\tilde{u}_1 \in Z(\tilde{T}_0)$. We first show that $u_1 \in \mathcal{U}(H)$ in the sense of Subsection 1. By choice of u_1, (U0) and (U1) are satisfied, and (U3) holds by 14.6.16. Next for $t \in T - T_0$, $u_1^t \in U_{H,2}$, so

$[K, u_1] \neq 1 \neq [K, u_1 u_1^t]$. Further in all cases of 14.6.14, $K \leq O^2(O_{2,F^*}(G_1)) =: X$, so $[X, u_1] \neq 1 \neq [X, u_1 u_1^t]$. Now by 14.6.1.5 applied to G_1 in the role of "H", X centralizes $\Omega_1(Z(Q_1))$, so as $F^*(G_1) = Q_1$, Q_1 does not centralize u_1 or $u_1 u_1^t$. Thus (U2) holds, completing the proof that $u_1 \in \mathcal{U}(H)$.

Assume for the moment that $Z_\gamma \leq U_{H,1}$, and if case (2) of 14.6.13 holds, assume further that $U_\gamma^* \leq H_1^*$; we will show that these assumptions lead to a contradiction. Let $Z_\gamma = \langle u_\gamma \rangle$. We claim first that $u_\gamma \in \mathcal{U}(H)$. Suppose that case (2) of 14.6.13 holds, so that $U_\gamma^* \leq H_1^*$ by assumption. By 14.6.15, $p = 3$, so $U_\gamma^*(T^* \cap H_2^*)$ and T_0^* are Sylow in H_0^*, and therefore conjugating in H_1, we may take $T_0^* = U_\gamma^*(T^* \cap H_2^*)$. Then $\tilde{u}_\gamma$ is centralized by $\tilde{T}_0$, so by the previous paragraph, $u_\gamma \in \mathcal{U}(H)$, establishing the claim in this case. Suppose instead that case (1) of 14.6.13 holds. Then each member of $U_{H,1} - Z$ is conjugate to an element in $Z(\tilde{T}_0)$, so as before $u_\gamma \in \mathcal{U}(H)$, completing the proof of the claim. But then by by the claim, we may apply 14.6.3.4 to conclude that $u_\gamma \notin z^G$, contrary to $\langle u_\gamma \rangle = Z_\gamma = Z^{gb}$. Thus the hypotheses of the first sentence of this paragraph lead to a contradiction.

If case (2) of 14.6.13 holds, that result shows we may choose γ so that $U_\gamma^* \leq H_1^*$ and $Z_\gamma \leq U_{H,1}$, contrary to the previous paragraph. Thus case (1) of 14.6.13 holds, establishing (1). Then (2) follows from the previous paragraph.

Next by (1), H has two orbits on $\tilde{U}_K$: the singular and nonsingular vectors, with $\tilde{U}_{H,1}^{\#} \cup \tilde{U}_{H,2}^{\#}$ the set of nonsingular vectors. Thus (3) follows from the first paragraph.

Choose u as in (3). By 14.6.3.4, $T_u \in Syl_2(C_G(u))$, and by 14.6.3.1, $|T : T_u| = 4$. But if $w \in U_K$ with $\tilde{w}$ singular, then $|C_H(w)| = |T|/2 > |T_u|$, so that $w \notin u^G$. Therefore $u^G \cap U_K = u^H$, so using A.1.7.1:

$$C_G(u) \text{ is transitive on the } G\text{-conjugates of } U_K \text{ containing } u. \qquad (*)$$

As case (1) of 14.6.13 holds, $[\tilde{U}_{H,1}, U_H^g] \neq 1$ for some $g \in G$ with $U_H^g \leq N_H(U_{H,1})$ and $U_H \leq H^g$. In particular by (1) we may take $u \in [U_{H,1}, U_H^g] \leq U_K^g$. By $(*)$, $U_K^g = U_K^h$ for some $h \in C_G(u)$. Therefore as $[U_{H,1}, U_K^h] \neq 1$, while $U_K \leq \langle V^{G_1} \rangle$ and $\langle V^{G_1} \rangle$ is abelian, $h \notin G_1$. Thus $C_G(u) \not\leq G_1$. Finally as $u \in U_{H,1}$, u is centralized by K_2, so $C_H(u) \not\leq M$. Thus $C_G(u)$ is in the set $\mathcal{I} = \mathcal{I}(T, u)$ defined in the first subsection, so (4) holds.

As $\tilde{V} =: \langle \tilde{v} \rangle \leq Z(\tilde{T})$, $\tilde{v} = \tilde{u}_1 \tilde{u}_2 \tilde{c}$, where $\langle \tilde{u}_i \rangle = C_{\tilde{U}_{H,i}}(T_0)$ and $\tilde{c} \in C_{\tilde{U}_H}(H)$. Therefore $\langle V^{O^2(H_2)} \rangle = \langle u_1 c, U_{H,2} \rangle$ is of rank 4 since $Z \leq U_{H,2}$ by 14.6.12.3, so (5) holds, completing the proof of 14.6.17. $\qquad \square$

We are now in a position to derive a contradiction, and hence establish Theorem 14.6.11.

Let T_0 and u be defined as in 14.6.17. By 14.6.17.4, $C_G(u) \in \mathcal{I}$, so $\mathcal{I}^*$ is nonempty, and if $T_u = T_I := T \cap I$ for some $I \in \mathcal{I}^*$, then also $C_G(u) \in \mathcal{I}^*$ by 14.6.4. By 14.6.17.1, $|T : Q_H| > 4$ and $m(U_H/C_{U_H}(Q_H)) = 4$. Thus the hypotheses of 14.6.9 are satisfied for any $I \in \mathcal{I}^*$, so by that result $|T| > 2^{11}$, and for any such I, setting $L_I := O^2(L \cap I)$ we have $LT = L_I T_I O_2(LT)$. Pick $I \in \mathcal{I}^*$, choosing $I := C_G(u)$ if $T_I = T_u$ for some $I \in \mathcal{I}^*$. Set $I_2 := O^2(H_2)T_I$, $I_1 := L_I T_I$, and $I_0 := \langle I_1, I_2 \rangle$. Observe H_2 has a noncentral 2-chief factor on U_H and on $Q_H/C_{Q_H}(U_H)$ by the duality in 14.5.21.1. Therefore $I_0 \in \mathcal{I}^*$ by 14.6.6.6. Further $O^2(H)$ centralizes u by Coprime Action; so if $T_I = T_u$, then $O^2(H_2)T_u = I_2$ centralizes u, while $I_1 \leq C_G(u)$ by our choice of I, so that $I_0 \leq C_G(u)$. Thus I_0

satisfies the hypotheses of 14.6.10.5, so $m(\langle V^{I_2} \rangle) = 3$ by that lemma, contrary to 14.6.17.5.

This contradiction completes the proof of Theorem 14.6.11.

As a corollaries to Theorem 14.6.11 we have:

THEOREM 14.6.18. *Each solvable member of $\mathcal{H}(T)$ is contained in M.*

LEMMA 14.6.19. *Let $H \in \mathcal{H}(T, M)$. Then*

(1) $O_{2,2'}(H) = O_2(H)$.
(2) If $K \in \mathcal{C}(H)$, then $K/O_2(K)$ is simple, and hence is described in F.9.18.

PROOF. Part (1) follows from 14.6.1.4 in view of 14.6.18. Then (1) implies (2). $\qquad \square$

14.6.3. The final elimination of U abelian when $L/O_2(L)$ is $L_2(2)$.

LEMMA 14.6.20. *If $H \in \mathcal{H}(T, M)$ and $K \in \mathcal{C}(H)$, then $K/O_2(K) \cong L_3(2)$ or A_6, and $N_T(K)$ is nontrivial on the Dynkin diagram of $K/O_2(K)$.*

PROOF. Let $K_0 := \langle K^T \rangle$. As $L_1 = 1$, $K_0 T \in \mathcal{H}(T, M)$ by 14.5.19, so without loss $H = K_0 T$. By 14.6.19.2, $K/O_2(K)$ is simple, and is described in (4) or (5) of F.9.18, so $K/O_2(K)$ is a group of Lie type and characteristic 2, A_7, or M_{22}. If $K/O_2(K) \cong A_7$, then KT is generated by solvable overgroups of T, which lie in M by 14.6.18, contrary to $H \not\le M$. If $K/O_2(K) \cong M_{22}$, solvable overgroups of T generate a subgroup J of KT with $O^2(J)/O_2(K) \cong A_6/E_{2^4}$, so that $J \le K \cap M$; then $J \le C_M(V)$, impossible as $m_3(C_M(V)) \le 1$ by 14.2.2.4. Thus $K/O_2(K)$ is of Lie type and characteristic 2. Set $B := N_K(T \cap K)$; then B is a Borel subgroup of K, so BT is solvable, and hence $BT \le M$ by Theorem 14.6.18.

Suppose first that $K = K_0$. If $K/O_2(K)$ is of Lie rank at most 2, then as $B \le M$ by the previous paragraph, the lemma follows from 14.3.6.1. Thus we may assume $K/O_2(K)$ is of higher Lie rank, and hence $K/O_2(K)$ is $L_4(2)$ or $L_5(2)$ by F.9.18. Let P be the product of the end-node minimal parabolics of K. Then $PT \le H \cap M$ by Theorem 14.6.18, so $P \le C_M(V)$ by 14.2.2.1, contrary to 14.2.2.4.

Therefore we may assume $K < K_0$. By F.9.18.5, $K/O_2(K)$ is either a Bender group or $L_3(2)$. In the former case, since $B \le M$, we contradict 14.3.6.1.ii; so $K/O_2(K) \cong L_3(2)$. Further by Theorem 14.6.18, K_0 is not generated by T-invariant solvable parabolics, so $N_T(K)$ is nontrivial on the Dynkin diagram of $K/O_2(K)$. This completes the proof of the lemma. $\qquad \square$

In the remainder of the section, we fix G_1 as our choice for $H \in \mathcal{H}(T, M)$, and use the symbol H to denote this group. As in the previous subsection, we adopt the setup of Notation 14.5.16, including the notation reviewed in that subsection involving the coset geometry Γ determined by LT and H, the vertex γ at distance b from γ_0, and the subgroups $U = U_H$, $D := D_H$, U_γ, etc. By Theorem 14.6.18, G_1 is not solvable, so there is $K \in \mathcal{C}(H)$. Then K is described in 14.6.20. Set $U_K := [U, K]$. Recall $H^* = H/Q_H$; as $H = G_1$ in this subsection, we do not require the convention $\hat{G}_1 = G_1/O_2(G_1)$ of the previous subsection.

LEMMA 14.6.21. *One of the following holds:*

(1) $H^\infty = K$, with $K/O_2(K) \cong L_3(2)$ or A_6.
(2) $H^\infty = KK^t$ for some $t \in T - N_G(K)$, with $K/O_2(K) \cong L_3(2)$.

(3) $H^\infty = KK_+$ with K, K_+ normal $\mathcal{C}$-components of H, and $K/O_2(K) \cong K_+/O_2(K_+) \cong L_3(2)$.

PROOF. If $H^\infty = K$ then (1) holds by 14.6.20, so assume $H^\infty > K$. Then by 1.2.1.1, there is $K_+ \in \mathcal{C}(H) - \{K\}$, and K_+ is also described in 14.6.20. As $m_3(H) \leq 2$, we conclude from 14.6.20 that $K/O_2(K) \cong K_+/O_2(K_+) \cong L_3(2)$ and $H^\infty = KK_+$. Then by 1.2.1.3, either (2) or (3) holds. $\square$

LEMMA 14.6.22. (1) $\tilde{U}_K = \tilde{U}_{K,1} + \tilde{U}_{K,2}$, where $\tilde{U}_{K,1}$ is a natural module for $K/O_2(K)$ or the 5-dimensional cover of a natural module for $K/O_2(K) \cong A_6$, and $U_{K,2} = U_{K,1}^s$ for $s \in N_T(K)$ acting nontrivially on the Dynkin diagram of $K/O_2(K)$.
(2) If there exists $K_+ \in \mathcal{C}(H) - \{K\}$, then $[U_K, K_+] = 1$.
(3) $Z \leq U_{K,i}$ for $i = 1, 2$.

PROOF. Let $K_0 := \langle K^T \rangle$ and $\tilde{I} \in Irr_+(K_0, \tilde{U})$. As $T_K := N_T(K)$ is nontrivial on the Dynkin diagram of $K/O_2(K)$ by 14.6.20, $KT_K/O_2(KT_K)$ has no FF-modules by Theorem B.5.1. By 14.6.19.2, we may apply F.9.18, so $[\tilde{U}, K_0] = \langle \tilde{I}^H \rangle$ by part (7) of that result. Next as T is nontrivial on the Dynkin diagram of $K/O_2(K)$, F.9.18 says $[\tilde{U}, K_0]$ is described in case (iii) of part (4) of F.9.18 if $K = K_0$, and in case (iii.b) of part (5) if $K < K_0$. Next if $C_{\tilde{I}}(K) \neq 1$, then $\tilde{I}$ is described in I.1.6.1; in particular $\tilde{I}$ is 5-dimensional when $K^* \cong A_6$. On the other hand if $K^* \cong L_3(2)$, then $\tilde{I}$ is the extension in B.4.8.2, and that result says $q(H^*, \tilde{U}_H) > 2$, contrary to part (2) of F.9.18. This completes the proof of (1). Also (2) follows, since $\tilde{U}_{K,1}$ is not K-isomorphic to $\tilde{U}_{K,2}$ and $End_K(U_{K,i}/C_{U_{K,i}}(K))$ is a field by A.1.41. Finally (3) follows from 14.6.2. $\square$

LEMMA 14.6.23. (1) $H^\infty = O^2(H)$, so $H = H^\infty T$.
(2) $M = LT$ and $T = M \cap H$.
(3) We have
$$\tilde{U} = \left(\bigoplus_{K \in \mathcal{C}(H)} \tilde{U}_K \right) + C_{\tilde{U}}(H).$$

PROOF. In view of 14.6.19.1, we obtain $F^*(H^*) = H^{\infty *}$ from 1.2.1.1. By 14.6.21, $Out(H^{\infty *})$ is a 2-group, so (1) holds.
Let $\tilde{U}_0 := [\tilde{U}, H^\infty]$. By 14.6.21 and 14.6.22, $\tilde{U}_0 = \bigoplus_{K \in \mathcal{C}(H)} \tilde{U}_K$. Now T centralizes $\tilde{V}$ of order 2, and $\tilde{U} = \langle \tilde{V}^H \rangle$, while $H = H^\infty T$ by (1), so (3) follows using Gaschütz's Theorem A.1.39. Further the projection $\tilde{V}_K$ of $\tilde{V}$ on $\tilde{U}_K$ is of order 2 and centralized by $N_T(K)$ for each $K \in \mathcal{C}(H)$. By 14.6.22.1, $C_{K^*}(\tilde{V}_K) = T^* \cap K^*$, so $T = C_H(\tilde{V})$ by (1). Therefore $T = M \cap H$, so $M = LT$ by 14.3.7. Thus (2) holds. $\square$

We next choose an element $u \in U$, which we will show lies in the set $\mathcal{U}(G_1)$ of the first subsection. In cases (1) and (3) of 14.6.21, pick $u \in U_{K,1}$ such that $[\tilde{u}, K] \neq 1$ and $N_T(U_{K,1})$ centralizes $\tilde{u}$. (This choice is possible when $U_{K,1}$ is the 5-dimensional cover of a natural module for $K/O_2(K) \cong A_6$ by I.2.3.1ia). In case (2) of 14.6.21, pick $u \in U_K - Z$ such that $N_T(K)$ centralizes $\tilde{u}$.

LEMMA 14.6.24. (1) $u \in \mathcal{U}(H)$.
(2) $K/O_2(K) \cong L_3(2)$.
(3) $K = H^\infty$.

(4) $C_G(u) \in \mathcal{I}$, so that $\mathcal{I}^*$ is nonempty.

PROOF. Set $T_0 := C_T(\tilde{u})$. To prove (1) we must verify (U1), (U2), and (U3). By construction T_0 is $N_T(U_{K,1})$ or $N_T(K)$, and so is of index 2 in T. Then as $T_0 \in Syl_2(C_H(\tilde{u}))$, (U1) holds. By 14.6.21, $N_{H^\infty T_0}(T_0) = T_0$, so as $G_1 = H = H^\infty T$ by 14.6.23.1, $N_{G_1}(T_0) = T$, establishing (U3). As $u \in U_K - Z$, $[K, u] \neq 1$, and for $t \in T - T_0$, u^t lies in either $U_{K,2}$ or U_{K^t}, so $1 \neq [K, uu^t]$. By 14.6.1.5, K centralizes $\Omega_1(Z(O_2(G_1)))$, so as $F^*(G_1) = Q_H$, neither u nor uu^t centralizes Q_H, establishing (U2). This completes the proof of (1).

In view of 14.6.22.1 and 14.6.23, we conclude from Theorem B.5.6 that for any $K \in \mathcal{C}(H)$,

$$\tilde{U} \text{ is not an FF-module for } H^*, \text{ and } \tilde{U}_K \text{ is not an FF-module for } Aut_H(\tilde{U}_K). \quad (a)$$

In particular, no member of H^* induces a transvection on $\tilde{U}$, so by 14.5.18.1, $D_\gamma < U_\gamma$. Therefore by 14.5.18.4, we can choose γ as in 14.5.18.4; in particular $U_\gamma^* \in \mathcal{Q}(H^*, \tilde{U})$, and from that choice and (a):

$$0 < m(U/D) \leq m(U_\gamma^*) < m(\tilde{U}/C_{\tilde{U}}(U_\gamma^*)). \quad (b)$$

In view of (b), $[\tilde{D}, U_\gamma^*] \neq 1$ by (a), so $\tilde{Z}_\gamma = [\tilde{D}, U_\gamma^*]$ by F.9.13.6. Then $\tilde{Z}_\gamma \leq [\tilde{U}, H^\infty]$ by 14.6.23.3. Set $g := g_b$, so that $\gamma_1 g = \gamma$ and $Z_\gamma := Z^g$ plays the role of "A_1" of section F.9.

Now we begin the proof of (3), which will be lengthier. Thus we assume that $K < H^\infty$ and derive a contradiction. Observe that case (2) or (3) of 14.6.21 holds, so that $\mathcal{C}(H) = \{K, K_+\}$ with $K/O_2(K) \cong K_+/O_2(K_+) \cong L_3(2)$. By 14.6.22.1 and 14.6.23.3, $\tilde{U} = \tilde{U}_K \oplus \tilde{U}_+ \oplus C_{\tilde{U}}(H)$, where $U_+ := [K_+, U]$. By 14.6.22.1, $\tilde{U}_K$ and $\tilde{U}_+$ have rank 6.

Assume that some $a^* \in U_\gamma^*$ does not normalize K^*. Then $C_{H^*}(a^*) \cong \mathbf{Z}_2 \times L_3(2)$, and

$$m(\tilde{U}/C_{\tilde{U}}(U_\gamma)) \geq m(\tilde{U}/C_{\tilde{U}}(a)) = m(\tilde{U}_K) = 6 = 2m_2(C_{H^*}(a^*)),$$

with $\langle a^* \rangle$ the kernel of the action of $C_{H^*}(a^*)$ on $C_{\tilde{U}}(a)$ of corank 6 in $\tilde{U}$. Thus $m(U_\gamma^*) \leq m_2(C_{H^*}(a^*)) = 3$, while $\tilde{U}/C_{\tilde{U}}(U_\gamma)$ is of rank 6 if $U_\gamma^* = \langle a^* \rangle$ is of rank 1, and rank greater than 6 if $m(U_\gamma) = 2$ or 3, contrary to $U_\gamma^* \in \mathcal{Q}(H^*, \tilde{U})$.

Thus U_γ normalizes K and K_+, and hence also U_K and U_+. So as U_γ^* is faithful on $F^*(H^*) = K^* K_+^*$ we may choose notation so that $K^* = [K^*, U_\gamma^*]$.

We claim that $\tilde{Z}_\gamma \leq \tilde{U}_K$ or $\tilde{U}_+$. Suppose otherwise. Then as $[D, U] \leq Z_\gamma$ by F.9.13.6, U_γ centralizes $\tilde{D} \cap \tilde{U}_K$ and $\tilde{D} \cap \tilde{U}_+$. Then by (a),

$$m(\tilde{U}_K/(\tilde{U}_K \cap \tilde{D})) \geq m(\tilde{U}_K/C_{\tilde{U}_K}(U_\gamma)) \geq m_2(Aut_{U_\gamma}(K^*)) + 1. \quad (c)$$

Set $U^+ := \tilde{U}/(\tilde{U}_K + C_{\tilde{U}}(H))$. As U_γ^* is faithful on $K^* K_+^*$ and normalizes both factors,

$$m(U_\gamma^*) \leq m(Aut_{U_\gamma}(K^*)) + m(Aut_{U_\gamma}(K_+^*)), \quad (d)$$

so using (b)–(d):

$$m := m(Aut_{U_\gamma}(K_+^*)) \geq m(U_\gamma^*) - m(Aut_{U_\gamma}(K^*)) \geq m(U/D) - (m(\tilde{U}_K/(\tilde{U}_K \cap \tilde{D})) - 1)$$

$$\geq m(U^+/D^+) + 1 \geq m(U^+/C_{U^+}(U_\gamma)) - m + 1, \quad (e)$$

where the last inequality follows from the fact that U_γ induces a group of transvections on D^+ with center Z_γ^+.

By (e),
$$2m \geq m(U^+/C_{U^+}(U_\gamma)) + 1. \tag{f}$$

In particular $m > 0$, so that U_γ^* is nontrivial on $\tilde{U}_+$, and hence also on K_+^*. Therefore $m(U^+/C_{U^+}(U_\gamma)) > 1$ by (a), so that (f) now gives $m > 1$. Hence $m = 2$ as $m_2(Aut_H(K_+^*)) = 2$. As $m = 2$, we conclude from (e) and the structure of $\tilde{U}_+$ that U_γ^* induces inner automorphisms on K_+^*, and
$$m(U^+/C_{U^+}(U_\gamma)) = m(\tilde{U}_+/C_{\tilde{U}_+}(U_\gamma)) = 3. \tag{g}$$

Therefore (f) is an equality, and hence all inequalities in (c)–(f) are equalities. Then (d) becomes:
$$m(U_\gamma^*) = m(Aut_{U_\gamma}(K^*)) + m(Aut_{U_\gamma}(K_+^*)). \tag{h}$$

As the inequalities in (c) are equalities,
$$\tilde{D} \cap \tilde{U}_K = C_{\tilde{U}_K}(U_\gamma) \text{ is of codimension } m_2(Aut_{U_\gamma}(K^*)) + 1 \text{ in } \tilde{U}_K, \tag{i}$$

and since $m = 2$ and the inequalities in (e) are equalities, we obtain $m(U^+/D^+) = m(U^+/C_{U^+}(U_\gamma)) - 2$. Thus by (g):
$$D^+ \text{ is a hyperplane of } U^+. \tag{j}$$

We had chosen notation so that $K^* = [K^*, U_\gamma^*]$, but we also saw after (f) that $K_+^* = [K_+^*, U_\gamma^*]$. Thus we have symmetry between K and K_+, so we conclude $m(Aut_{U_\gamma}(K^*)) = 2$ and U_γ^* induces inner automorphisms on K^*. Then by (h), $U_\gamma^* = A^* \times A_+^*$, where A^* and A_+^* are 4-subgroups of K^* and K_+^*, respectively. Since U_γ induces a group of transvections on D^+ with center Z_γ^+, and we are assuming that $\tilde{Z}_\gamma$ is not contained in $\tilde{U}_K$ or $\tilde{U}_+$, it follows from (j) that Z_γ is generated by $z^g = z_1 z_2$, where $1 \neq \tilde{z}_1 \in \tilde{U}_{K,1}$ and $1 \neq \tilde{z}_2 \in \tilde{U}_{K_+,1}$. Therefore from the structure of U_K in 14.6.22, $C_H(z^g) = C_G(ZZ_\gamma)$ has a Sylow 3-subgroup P isomorphic to E_9. However by 14.5.21.2 and 14.5.15.1, Q_H and Q_H^g induce transvections on ZZ_γ with centers Z and Z_γ, respectively, so that $m_3(C_G(V)) \leq 1$ by A.1.14.4. This contradiction finally completes the proof of the claim.

By the claim we may choose notation so that $\tilde{Z}_\gamma \leq \tilde{U}_K$, and hence $Z_\gamma = Z^g$ centralizes K_+. Set $\hat{H}^g := H^g/Q_H^g$; then $K_+ \leq C_G(Z^g) = H^g = N_G(U_\gamma)$ since $H = G_1 \in \mathcal{M}$ by 14.6.1.1. Therefore U_γ^* centralizes K_+^*, and hence $m(U_\gamma^*) \leq m_2(C_{H^*}(K_+^*)) = 2$. Also $H^g = K^g K_+^g T^g$ by 14.6.23.1, so either $\hat{K}_+$ is $\hat{K}^g$ or $\hat{K}_+^g$, or else $\hat{K}_+$ is a full diagonal subgroup of $\hat{K}^g \times \hat{K}_+^g$. Suppose this last case holds. Then as $\hat{K}_+$ also acts on $\hat{U}$, $\hat{U} = \langle \hat{w} \rangle$ for $\hat{w}$ an involution interchanging K^g and K_+^g, and hence $U_K^{gw} = U_+^g$. Then as $[D_\gamma, U] \leq Z$ by F.9.13.6 and $C_{U_\gamma}(w)$ is of codimension 6 in U_γ, $m(U_\gamma^*) = m(U_\gamma/D_\gamma) \geq m(\tilde{U}_K) - 1 = 5 > 2$, contradicting $m(U_\gamma^*) \leq 2$. Thus $\hat{K}_+ = \hat{J}$ where $J := K^g$ or K_+^g. Hence $K_+ \leq JQ_H^g$, so that $K_+ = K_+^\infty \leq (JQ_H^g)^\infty = J$.

Suppose $K_+ = J$. Then by 14.6.22.3, $z^g \in [U^g, J] \leq O_2(J)$. Then $U_K = \langle z^{gN_G(K)} \rangle \leq O_2(J) \leq Q_H^g$, so by 14.5.15.1, $[U_K, U_\gamma] \leq \langle z^g \rangle$, contrary to (a). Hence $K_+ < J$, so in particular $|K_+| < |J|$, and hence $J = K^g$. Further K and K_+ have different orders and so are normal in H, so that case (3) of 14.6.21 holds. As $K_+ < J = K^g$ with $J/O_2(J) \cong L_3(2) \cong K/O_2(K)$ by 14.6.21.3, K_+ has a noncentral chief factor on $O_2(J)$ not in $O_2(K_+)$, and this factor has dimension at least 3. However $K_+ = C_G(\langle z, z^g \rangle)^\infty$ is invariant under $S \in Syl_2(C_G(\langle z, z^g \rangle))$, so $S \cap J \notin Syl_2(J)$ and hence z does not centralize J, and $|Q_H^g S : S| \geq 8$, so that $|C_H(z^g)|_2 = |S| \leq |T|/8$.

Suppose first that $m(U/D) = m(U_\gamma^*)$. Then, as in Remark F.9.17, we have symmetry of hypotheses between γ_1 and γ, so there exists a unique $J_1 \in \mathcal{C}(H^g)$ such that $J_1 = C_{J_1}(z)O_2(J_1)$. Thus as z centralizes $K_+ \leq J$, $J = J_1$, whereas we saw $[J, z] \neq 1$. Therefore $m(U/D) < m(U_\gamma^*)$, and we saw earlier that $m(U_\gamma^*) \leq 2$, so we conclude from (b) that $m(U_\gamma^*) = 2$ and $m(U/D) = 1$. Thus $m(U_{K,i}/(D \cap U_{K,i})) \leq 1$, and as U_γ^* is of rank 2, U_γ^* does not centralize a hyperplane of both $U_{K,1}$ and its dual $U_{K,2}$. Therefore as $[D, U_\gamma] \leq Z_\gamma$ by F.9.13.6, we may take $z^g \in U_{K,1}$. But then as $K \trianglelefteq H$, $|C_H(z^g)|_2 = |T|/4$, contrary to the previous paragraph. This contradiction finally completes the proof of (3).

Suppose next that (2) fails. Then $K/O_2(K) \cong A_6$ by 14.6.21.1, and $m(U_{K,i}) = 4$ or 5 by 14.6.22.1. Then for i^* an involution in H^*, $m([\tilde{U}_{K,1}, i^*]) \geq 1$, and in case of equality, i^* induces a transposition on K^*. But also by 14.6.22.1, $\tilde{U}_{K,2} = U_{K,1}^s$ for $s \in N_T(K)$ nontrivial on the Dynkin diagram of K^*, so if i^* acts as a transvection on $\tilde{U}_{K,1}$, then $m([\tilde{U}_{K,2}, i^*]) = 2$. We conclude $m(U_\gamma^*) > 1$, since $U_\gamma^* \in \mathcal{Q}(H^*, \tilde{U})$. Next as U_γ^* is quadratic on $\tilde{U}_{K,1}$, from the action of $N_H(U_{K,1})$ on $U_{K,1}$, U_γ^* is contained in a 2-subgroup of H^* generated by transpositions. Then again as $\tilde{U}_{K,2} = U_{K,1}^s$ and U_γ^* is quadratic on $\tilde{U}_{K,2}$, U_γ^* is a 4-group F^* generated by a transposition and the product of three commuting transpositions. Then $m(\tilde{U}/C_{\tilde{U}}(U_\gamma^*)) = 4 = 2m(U_\gamma^*)$. This contradicts 14.5.18.2, as F^* does not induce a group of transvections on any subspace of $\tilde{U}_K$ of codimension 2. This establishes (2).

By (2) and (3), $H^\infty = K$ with $H^* \cong Aut(L_3(2))$, so by 14.6.22.1, $\tilde{U}_K = \tilde{U}_{K,1} \oplus \tilde{U}_{K,2}$ with $\tilde{U}_{K,1}$ natural and $\tilde{U}_{K,2}$ its dual. Thus H has three orbits on $U_K - Z$: $U_{K,1}^\# \cup U_{K,2}^\# = u^H$ plus two diagonal classes, one of which is 2-central in $\tilde{H}$. Denote this latter 2-central class by $\mathcal{C}$. Recall that $u \in \mathcal{U}(H)$ by (1), so that $u \notin z^G$ by 14.6.3.4. As $C_K(u)$ is not a 2-group, $C_K(u) \not\leq M$ since $H \cap M = T$ by 14.6.23.2. Thus to prove (4), we must also show that $C_G(u) \not\leq G_1 = H$.

Now U_γ^* is of rank 1 or 2 as $m_2(H^*) = 2$. Suppose first that $m(U_\gamma^*) = 1$. Then $[\tilde{U}_K, U_\gamma] = \langle \tilde{u}_1, \tilde{u}_2 \rangle$ with $u_i \in U_{K,i}$ and $u_1 u_2 \in \mathcal{C}$. Conjugating in H, we may take $u = u_1$. Recall $Z_\gamma \leq [U, U_\gamma] \leq \langle u_1, u_2, z \rangle$, with $u_i \notin z^G$ as $u \notin z^G$, so that Z_γ is generated by $u_1 u_2$ or $u_1 u_2 z$, and hence $\mathcal{C} \subseteq z^G$. Since $m(U_\gamma^*) = 1$, $m(U/D) = 1$ by (b), and so our hypotheses are symmetric between γ and γ_1. Thus $[U, U_\gamma] = \langle u_1', u_2', z^g \rangle$, with $u_i' \in U_{K,i}^g$ and $u_1' u_2' \in \mathcal{C}^g$. As $\mathcal{C} \subseteq z^G$, $\mathcal{C}^g \subseteq z^G$, so $u \notin \mathcal{C}^g$, and hence $u \in U_{K,1}^g$ or $U_{K,2}^g$. So since H is transitive on $U_{K,1}^\# \cup U_{K,2}^\#$, we may take $g \in C_G(u)$. Thus as $Z \neq Z^g$, $C_G(u) \not\leq H$. Hence $C_G(u) \in \mathcal{I}$, and so (4) holds.

So suppose instead that $m(U_\gamma^*) = 2$. Then $[\tilde{U}_K, U_\gamma] = \langle \tilde{U}_1, \tilde{u}_2 \rangle$ where $\tilde{U}_1$ is a hyperplane of $\tilde{U}_{K,1}$ and $u_2 \in U_{K,2}$, with $U_1^\# u_2 \subseteq \mathcal{C}$. If $m(U/D) = 2$, we again have symmetry between γ and γ_1, so the argument of the previous paragraph establishes (4) in this case also. Thus by (b) we may take $m(U/D) = 1$. As U_γ^* is of rank 2, U_γ^* does not centralize a hyperplane of both $U_{K,1}$ and its dual $U_{K,2}$, so $Z_\gamma = [U_\gamma, D \cap U_{K,i}] \leq U_{K,i}$ for $i = 1$ or 2, contrary to $u \notin z^G$. This contradiction completes the proof of (4), and of 14.6.24. $\qquad \square$

We now derive a contradiction, hence showing that no examples satisfy the hypotheses of this section.

By 14.6.24.4, $C_G(u) \in \mathcal{I}$, so that $\mathcal{I}^* \neq \emptyset$, and if $T_u = T_I := T \cap I$ for some $I \in \mathcal{I}^*$, then also $C_G(u) \in \mathcal{I}^*$ by 14.6.4. By 14.6.20, $|T : O_2(H)| > 4$, and by 14.6.22.1, $m(U/C_U(Q_H)) \geq 4$. Thus the hypotheses of 14.6.9 are satisfied for any $I \in \mathcal{I}^*$, and that result shows that $|T| > 2^{11}$ and $LT = L_I T_I O_2(LT)$, where $L_I := O^2(L \cap I)$. Set $H_2 := C_H(\tilde{u})$. By 14.6.24.2 and 14.6.22.1, $H_2/O_2(H_2) \cong S_3$, with $H_2 \not\leq M$ as $H \cap M = T$ by 14.6.23.2. By construction, $T_0 := C_T(\tilde{u}) = N_T(U_{K,1}) \in Syl_2(H_2)$, and H_2 has nontrivial chief factors on each $\tilde{U}_{K,i}$. Pick $I \in \mathcal{I}^*$, choosing $I := C_G(u)$ if $T_I = T_u$ for some $I \in \mathcal{I}^*$, and let $I_2 := O^2(H_2)T_I$, $I_1 := L_I T_I$, and $I_0 := \langle I_1, I_2 \rangle$. Then $I_0 \in \mathcal{I}^*$ by 14.6.6.6. Further $O^2(H_2)$ centralizes u by Coprime Action, so if $T_I = T_u$, then $I_2 = O^2(H_2)T_u$ centralizes u, while $I_1 \leq C_G(u)$ by our choice of I, so that $I_0 \leq C_G(u)$. Thus I_0 satisfies the hypotheses of 14.6.10.5, and hence $m(\langle V^{I_2} \rangle) = 3$ by that result. However as $\tilde{V} = \langle \tilde{v} \rangle \leq Z(\tilde{T})$, from the module structure in 14.6.22.1, $\tilde{v} = \tilde{u}\tilde{u}_2\tilde{c}$, where $\tilde{u}_2$ generates $C_{\tilde{U}_{K,2}}(T)$ and $\tilde{c} \in C_{\tilde{U}}(H)$. Therefore $\langle V^{I_2} \rangle = \langle uc \rangle [U_{K,2}, I_2]$ is of rank 4, since $Z \leq [U_{K,2}, O^2(H_2)]$ by 14.6.2.

This contradiction completes our analysis of the $L_2(2)$-case under Hypothesis 14.2.1; namely we have now proved:

THEOREM 14.6.25. *Assume Hypothesis 14.2.1. Then G is isomorphic to J_2, J_3, $^3D_4(2)$, the Tits group $^2F_4(2)'$, $G_2(2)' \cong U_3(3)$, or M_{12}.*

PROOF. We may assume that the Theorem fails, so that case (2) of Hypothesis 14.3.1.2 is satisfied. By Theorem 14.3.16, $U = \langle V^{G_1} \rangle$ is abelian, so that the hypotheses of this section are satisfied. Finally, as we just saw, those hypotheses lead to a contradiction, so the Theorem is established. $\square$

14.7. Finishing $\mathbf{L_3(2)}$ with $\langle \mathbf{V^{G_1}} \rangle$ abelian

In this section we continue to assume Hypothesis 14.5.1, but now assume that $L/O_2(L) \cong L_3(2)$; that is, we treat case (1) of Hypothesis 14.3.1, so in particular Hypothesis 13.3.1 holds, with $G \not\cong Sp_6(2)$ or $U_4(3)$, and $U := \langle V^{G_1} \rangle$ is abelian. Further by 13.3.2.4, Hypothesis 12.2.3 holds, and hence so does case (1) of Hypothesis 12.8.1. Thus we can appeal to results in sections 12.8, 13.3, 14.3, and 14.5.

We will see in Theorem 14.7.75 that the Rudvalis group Ru is the only quasithin example which arises under the hypothesis of this section; as far as we can tell, there are no shadows.

We adopt Notation 12.8.2, including the T-invariant subspaces V_i of V for $i = 1, 2$, and the subgroups $G_i := N_G(V_i)$, $M_i := N_M(V_i)$, $L_i := O^2(N_L(V_i))$, and $R_i := O_2(L_iT)$. In particular $V_1 = V \cap Z$ where as usual $Z := \Omega_1(Z(T))$, z is the generator for V_1, $\tilde{G}_1 := G_1/V_1$, and $\mathcal{H}_z$ consists of the members of $\mathcal{H}(L_1T, M)$ which lie in G_1.

In this section, H denotes a member of $\mathcal{H}_z$.

By 14.5.14 we may adopt Notation 14.5.16; in particular, form the coset geometry Γ of Hypothesis F.9.1 with respect to LT and H, set $b := b(\Gamma, V)$, choose a geodesic
$$\gamma_0, \gamma_1, \ldots, \gamma_b =: \gamma,$$
define U_H, U_γ, D_H, D_γ, etc. as in section F.9, and set $A_1 := V_1^{g_b}$ where $\gamma_1 g_b = \gamma$, using the fact from F.9.11 that b is odd.

Often we can show that $D_\gamma < U_\gamma$, and in those situations we also adopt:

NOTATION 14.7.1. If $D_\gamma < U_\gamma$, choose γ as in 14.5.18.4, so that
$$m(U_\gamma^*) \geq m(U_H/D_H) > 0,$$
and $U_\gamma^* \in \mathcal{Q}(H^*, \tilde{U}_H)$, and (as in 14.5.18.5) choose $h \in H$ with $\gamma_0 = \gamma_2 h$, and set $\alpha := \gamma h$ and $Q_\alpha := O_2(G_\alpha)$; then $U_\alpha \leq R_1$ and $U_\alpha^* \in \mathcal{Q}(H^*, \tilde{U}_H)$.

Set $Q := O_2(LT) = C_T(V)$ and
$$S := \langle U_H^L \rangle.$$
Since $Out(L_3(2))$ is a 2-group and T induces inner automorphisms on $L/O_2(L)$ (because T acts on V):
$$M = LC_M(L/O_2(L)).$$

14.7.1. Preliminary reductions.

LEMMA 14.7.2. Let $\tilde{I}$ be a proper H-submodule of $\tilde{U}_H$, and assume that $Y = O^2(Y) \leq H$ with $YT/O_2(YT) \cong S_3$. Set $\hat{U}_H := U_H/I$. Then

(1) $\hat{V}$ is isomorphic to $\tilde{V}$ as an $L_1 T$-module.

(2) $\langle \hat{V}_2^Y \rangle$ is of rank 1 or 2.

(3) If $[\hat{V}_2, Y] = 1$, then $[V_2, Y] = 1$.

PROOF. Observe as $I < U_H$ that $V \not\leq I$ since $U_H = \langle V^H \rangle$. Then as L_1 is irreducible on $\tilde{V}$, $V \cap I = V_1$, so part (1) follows. Next as $\tilde{V}_2$ is centralized by T of index 3 in YT, $\tilde{E} := \langle \tilde{V}_2^Y \rangle$ is of rank $\tilde{e} = 1, 2,$ or 3, with $\hat{E} := \langle \hat{V}_2^Y \rangle$ of rank $\hat{e} \leq \tilde{e}$. By Theorem 14.5.3.3, $\tilde{e} < 3$, so that (2) holds. If $[\hat{V}_2, Y] = 1$, then $\hat{e} = 1$ so $\tilde{E}$ has the 1-dimensional quotient $\hat{E}$, and therefore $\tilde{e} = 1$ or 3. But we just saw $\tilde{e} < 3$, so $\tilde{e} = 1$, and hence (3) holds. $\square$

LEMMA 14.7.3. (1) $b \geq 3$ is odd.

(2) $S \leq Q$.

(3) S is abelian iff $b > 3$.

(4) If $H = G_1$ and $A_1^h \leq V$ for some $h \in H$, then $b = 3$ and $U_{\gamma h} \in U_H^L$.

PROOF. Part (1) is F.9.11.1. As U_H is abelian, $U_H \leq C_{LT}(V) = Q$, so (2) holds. Part (3) is F.9.14.1, and part (4) follows from F.9.14.3 as L is transitive on $V^\#$ since $\bar{L} = GL(V)$. $\square$

LEMMA 14.7.4. (1) $[V_2, O_2(G_1)] = V_1$.

(2) $I_2 := \langle O_2(G_1)^{G_2} \rangle \trianglelefteq G_2$, $I_2 = L_2 O_2(G_1)$, $I_2/O_2(I_2) \cong S_3$, and $L_2 = O^2(I_2)$.

(3) $m_3(C_G(V_2)) \leq 1$.

(4) $QQ_H = R_1$, so $R_1^* = Q^*$.

PROOF. Part (1) follows from 14.5.21.2, and 13.3.15 implies (2) and (3). By (1), $1 \neq \bar{Q}_H \leq \bar{R}_1$, so as L_1 is irreducible on $\bar{R}_1$, (4) holds. $\square$

LEMMA 14.7.5. Assume $L_1^* \trianglelefteq H^*$. Then

(1) $D_\gamma < U_\gamma$, so we may adopt Notation 14.7.1.

(2) $QQ_H = R_1$.

(3) $L_1^* \cong \mathbf{Z}_3$, and $R_1^* = Q^* = C_{T^*}(L_1^*)$ is of index 2 in T^*.

(4) $[U_\gamma^*, L_1^*] = 1$.

(5) $\tilde{U}_H = [\tilde{U}_H, L_1]$.

PROOF. As $L_1^* \trianglelefteq H^*$, H normalizes $O^2(L_1 Q_H) = L_1$. Then 14.5.15.3 says that $L_1^* \cong \mathbf{Z}_3$ and (5) holds. As $L_1 T/R_1 \cong S_3$, it follows that $|T^* : C_{T^*}(L_1^*)| = 2$. Further $L_1 \trianglelefteq G_{\gamma_0,\gamma_1}$, so as γ_2 is conjugate to γ_0 in $H \le G_1$, $L_1 \trianglelefteq G_{\gamma_1,\gamma_2}$. Then as $L_1^* \cong \mathbf{Z}_3$, L_1^* centralizes $O_2(G_{\gamma_1,\gamma_2}^*)$. Thus (4) follows as $U_\gamma \le O_2(G_{\gamma_1,\gamma_2})$ by F.9.13.2, and similarly $[U_H, L_\gamma] \le O_2(G_\gamma)$, where $L_\gamma := L_1^{gb}$. Therefore as $U_\gamma/A_1 = [U_\gamma/A_1, L_\gamma]$ by (5) where L_γ has action of order 3 commuting with that of U_H, $m([U_\gamma/A_1, u])$ is even for each $u \in U_H$, so u does not induce a transvection on U_γ/A_1. Thus $D_\gamma < U_\gamma$ by 14.5.18.1, establishing (1).

Part (2) is contained in 14.7.4.4. Then by (2), $Q^* = C_{T^*}(L_1^*)$, completing the proof of (3).   □

LEMMA 14.7.6. $F(H^*)$ is a 3-group.

PROOF. Suppose H is a minimal counterexample, let $p > 3$ be prime with $H_1^* := \Omega_1(Z(O_p(H^*))) \ne 1$, and pick $P \in Syl_p(H_1)$ where H_1 is the preimage of H_1^*. Since $p > 3$, $H_1 \cap M = Q_H$ by 14.5.20, so $H = PL_1T$ by minimality of H. Similarly H^* is irreducible on P^*. By 14.5.18.3, $q(H^*, \tilde{U}_H) \le 2$, so by D.2.17, $p = 5$ and $P \le K \trianglelefteq H$ with $K^* = K_1^* \times \cdots \times K_s^*$, $K_i^* \cong D_{10}$, and $\tilde{U}_i := [\tilde{U}_H, K_i]$ of rank 4. As usual $s = m_5(H^*) \le 2$ as H is an SQTK-group, so $L_1^* = O^2(L_1^*)$ normalizes K_i^*. Then $[K_i^*, L_1^*] = 1$, so that $L_1 \trianglelefteq KL_1T = H$. Hence 14.7.5 says we may adopt Notation 14.7.1, $Q^* = C_{T^*}(L_1^*)$, U_γ^* centralizes L_1^* of order 3, and $\tilde{U}_H = [\tilde{U}_H, L_1]$. In particular, U_γ^* is faithful on P^* since $F^*(H^*) = L_1^* P^*$. Then since $U_\gamma^* \in \mathcal{Q}(H^*, \tilde{Q}_H)$, either $\mathbf{Z}_2 \cong U_\gamma^* \le K_i^*$ for some i, or $s = 2$ and $E_4 \cong U_\gamma^* \le K_1^* K_2^*$. In either case $2m(U_\gamma^*) = m(\tilde{U}_H/C_{\tilde{U}_H}(U_\gamma^*))$, so by 14.5.18.2, U_γ^* induces a faithful group of transvections with center $\tilde{A}_1$ on a subspace $\tilde{D}_H$ of $\tilde{U}_H$ of codimension $m(U_\gamma^*)$. But if U_γ^* is of rank 2, this is not the case, so we may choose notation so that $\mathbf{Z}_2 \cong U_\gamma^* \le K_1^*$. Therefore $A := [U_H, U_\gamma] \le U_1$. Since L_1^* centralizes U_γ^*, L_1^* normalizes $\tilde{A}$.

Now by the choice of γ in Notation 14.7.1, $1 = m(U_\gamma^*) \ge m(U_H/D_H) \ge 1$, so $m(U_\gamma^*) = 1 = m(U_H/D_H)$, and hence as discussed in Remark F.9.17, our hypotheses are symmetric between γ_1 and γ. As U_γ centralizes no hyperplane of $\tilde{U}_H$, $A_1 = [D_H, U_\gamma] \le A$ by F.9.13.6. Thus by the symmetry, $V_1 \le A$, so that $m(A) = 3$ as $m(\tilde{A}) = 2$, and L_1 acts on A as L_1^* acts on $\tilde{A}$.

Assume first that $s = 2$, and let $T_1 := N_T(K_1)$. Then T_1 is of index 2 in a Sylow 2-subgroup of G, $T_1 \in Syl_2(N_H(A))$, and by 14.5.21.1, $L_1 T_1$ induces A_4 or S_4 on A and centralizes V_1. Again by the symmetry between γ and γ_1, $N_{G_\gamma}(A)$ induces A_4 or S_4 on A and centralizes $A_1 \ne V_1$, so we conclude that $N_G(A)$ induces $GL(A) \cong L_3(2)$ on A. Therefore by 1.2.1.1, $N_G(A) = L_A C_G(A)$ for some $L_A \in \mathcal{C}(N_G(A))$ with $L_A/C_{L_A}(A) \cong L_3(2)$. By 1.2.1.4, $L_A/O_2(L_A) \cong L_3(2)$ or $SL_2(7)/E_{49}$, and in either case $Aut(L_A/O_2(L_A))$ is a $5'$-group. Thus as $K_0 := O^{5'}(K_2)$ acts on A, $[L_A, K_0] \le O_2(L_A)$. Also L_1 is nontrivial on A, so either $L_1 \le L_A$ or L_1 is diagonally embedded in $L_A C_G(A)$. As $[L_1^*, K_2^*] = 1$, L_1 acts on K_0.

Next $[K_0, T_1 \cap L_A] \le K_0 \cap O_2(L_A) \le O_2(K_0) \le Q_H$, so $(T_1 \cap L_A)^*$ centralizes K_0^*. Therefore as $C_{GL(\tilde{U}_2)}(K_2^*) \cong \mathbf{Z}_{15}$, $T_1 \cap L_A$ centralizes $\tilde{U}_2$, and hence also centralizes L_1^* and $L_1/O_2(L_1)$.

Let L_0 be the preimage in L_A of $Aut_{L_1}(A)$. As $Aut_{L_1}(A) \cong A_4$, $N_G(L_0) \cap N_G(A)$ contains a Sylow 2-group of $N_G(A)$. Thus $T_1 \le T_A \in Syl_2(N_G(A))$ with T_A acting on L_0. Further each $t \in T_A - O_2(L_0 T_A)$ inverts $L_0/O_2(L_0)$, so $t \notin T_1$ by the previous paragraph. Therefore as $|T_A : T_1| \leq |T : T_1| = 2 = |T_A : O_2(L_0T_A)|$, we conclude $T_A \in Syl_2(G)$. As $A \cap Z(T_A) \neq 1$, $L_A \in \mathcal{L}_f(G, T_A)$, so $L_A \in \mathcal{L}_f^*(G, T_A)$ by 14.3.4.2 and $T_1 \cap L_A = O_2(L_0T_A) \cap L_A$. In particular $O_2(L_A) \leq T_1 \leq N_G(K_0)$, so as $[L_A, K_0] \leq O_2(L_A)$, L_A normalizes $O^2(K_0O_2(L_A)) = K_0$, as does $T_1(T_A \cap L_A) = T_A$. Then as $N_G(L_A) = !\mathcal{M}(N_G(L_AT_A))$ by 1.2.7.3, $K_1 \leq N_G(K_0) \leq N_G(L_A)$. Therefore as $[V_1, K_1] = 1$, K_1 acts on $A = \langle V_1^{L_A} \rangle$, so as $m(A) = 3$, we conclude $K_1 = O^{5'}(K_1)$ centralizes A, whereas $K_1/O_2(K_1)$ is fixed-point-free on $\tilde{U}_1 \geq \tilde{A}$.

This contradiction shows that $s = 1$. Thus $H = K_1TL_1$, so $K_1 \not\leq M$. As $L_1/O_2(L_1)$ is inverted in T, and involutions in $GL(\tilde{U}_1)$ normalizing K_1^* centralize L_1^*, we conclude that $T^* \cong \mathbf{Z}_4$. Thus $\Omega_1(T) \leq Q_HQ$ using parts (2) and (3) of 14.7.5. Then as all involutions in LT/Q are fused to involutions in T/Q inverting $L_1/O_2(L_1)$, $\Omega_1(T) \leq Q$. Therefore $J(T) \leq J_1(T) \leq Q$, so using B.2.3.3 we conclude that $N_G(J(T))$ and $N_G(Z(J_1(T)))$ lie in $M = !\mathcal{M}(LT)$. Therefore as $K_1 \not\leq M$, $K_1 = [K_1, J(T)]$. Then as $p > 3$, a standard result of Thompson (see 26.18.a in [**GLS96**]) shows that $K_1 \leq N_G(J(T))N_G(Z(J_1(T))) \leq M$, a contradiction. $\quad\square$

LEMMA 14.7.7. *If $L_1^* \trianglelefteq H^*$, then $O_{3'}(H^*) = 1$.*

PROOF. Suppose H is a counterexample. Then $O_{3'}(E(H^*)) \neq 1$ by 14.7.6, so there is $K \in \mathcal{C}(H)$ with $K^* \cong Sz(2^n)$ for some odd $n \geq 3$. Let $K_1 := \langle K^T \rangle$; by 14.5.19, $K_1L_1T \in \mathcal{H}_z$, so without loss $H = K_1L_1T$.

As $L_1^* \trianglelefteq H^*$, 14.7.5 says that $L_1^* \cong \mathbf{Z}_3$ and $\tilde{U}_H = [\tilde{U}_H, L_1]$. As $Sz(2^n)$ has no FF-module by Theorem B.4.2, examining parts (4)–(6) of F.9.18 we conclude that $W := [\tilde{U}_H, K]/C_{[U_H, K]}(K)$ is the natural module for K^*. This is impossible as $[\tilde{U}_H, L_1] = \tilde{U}_H$ and $[L_1^*, K^*] = 1$, whereas $End_{\mathbf{F}_2K^*}(W) \cong \mathbf{F}_{2^n}$ has multiplicative group of order coprime to 3 since n is odd. $\quad\square$

LEMMA 14.7.8. *There is no $H \in \mathcal{H}_z$ with $O^2(H^*)$ a cyclic 3-group.*

PROOF. Assume $O^2(H^*)$ is a cyclic 3-group. Then as $L_1 \leq H$, $H = PT$ with $P \cong \mathbf{Z}_{3^n}$, and $L_1 = O^2(\Omega_1(P)O_2(H)) \trianglelefteq H$. Furthermore $n > 1$ since $H \not\leq M$. But then $Q_H = O_2(L_1T) = R_1$, so $U_\gamma \leq Q_H$ by 14.7.5, whereas $U_\gamma^* \neq 1$ by 14.7.5.1. $\quad\square$

Observe that 14.7.8 eliminates case (2.iii) of 14.5.20, so we may strengthen 14.5.20 to read:

LEMMA 14.7.9. *Assume $Y = O^2(Y) \trianglelefteq H$ with Y^* a p-group of exponent p, and $O_2(Y) < Y \cap M$. Then $p = 3$, and either*

(1) $Y = L_1$, or
(2) $Y^* \cong 3^{1+2}$, $L_1^* = Z(Y^*)$, T is irreducible on Y^*/L_1^*, and $L_1 = O^2(Y \cap M)$.

LEMMA 14.7.10. *Either*

(1) L_1 has at most three noncentral 2-chief factors, or
(2) $N_G(Baum(R_1)) \leq M$.

PROOF. Let $S_1 := Baum(R_1)$. We apply the Baumann Argument C.1.37 to the action of LT on V. If (1) fails, then by C.1.37 there is a nontrivial characteristic subgroup C of S_1 normal in LT. Thus as $M = !\mathcal{M}(LT)$, $N_G(S_1) \leq N_G(C) \leq M$, so (2) holds. $\quad\square$

LEMMA 14.7.11. *H^* is not $L_3(2)$, A_6, or S_6.*

PROOF. Assume otherwise and let $H_1 := L_1 T$ and H_2 the minimal parabolic of H over T distinct from H_1. Set $Y := O^2(H_2)$. Then $H = \langle L_1 T, Y \rangle \nleq M$, so $Y \nleq M$. Thus $[V_2, Y] = 1$ by 14.5.3.2, so $YL_2/O_2(YL_2) \cong E_9$ by 14.7.4.2.

Let $D_0 := H$, $D_1 := H_2 L_2$, $D_2 := LT$, $\mathcal{F} := (D_0, D_1, D_2)$, and $D := \langle \mathcal{F} \rangle$. We will show $(D, \mathcal{F})$ is an A_3-system or C_3-system in the sense of section I.5.

Set $P_i := O_2(D_i)$ and $\dot{D}_1 := D_1/P_1$. We saw $O^2(\dot{D}_1) = \dot{L}_2 \times \dot{Y} \cong E_9$. Further $O_2(G_1) \le P_0$ by A.1.6, so $\dot{L}_2 = [\dot{L}_2, \dot{P}_0]$ by 14.7.4.2. On the other hand, $Y \le H \le N_G(P_0)$, so $\dot{P}_0$ centralizes $\dot{Y}$, and hence $\dot{P}_0$ is of order 2. Next from the structure of H^* under our hypothesis, $Y^* = [Y^*, T^*]$, so $\dot{Y} = [\dot{Y}, \dot{T}]$, and hence $\dot{D}_1 \cong L_2(2) \times L_2(2)$. Of course $D_2/Q_2 = LT/O_2(LT) \cong L_3(2)$, so hypothesis (D2) of section I.5 holds. By the hypotheses of this lemma, hypothesis (D1) holds, and by construction hypothesis (D3) holds. By definition, $D = \langle \mathcal{F} \rangle$. As $H \nleq M = !\mathcal{M}(LT)$, $\ker_T(D) = 1$, so hypothesis (D4) is satisfied. As $V_1 \le Z(H)$, hypothesis (D5) holds. This completes the verification that $(D, \mathcal{F})$ is an A_3-system or C_3-system.

As $(D, \mathcal{F})$ is an A_3-system or C_3-system, $D \cong L_4(2)$ or $Sp_6(2)$ by Theorem I.5.1. But then $O_2(H)$ is abelian, contrary to 14.7.4.1 as $O_2(G_1) \le Q_H$. $\quad\square$

Recall that when $D_\gamma < U_\gamma$, we adopt Notation 14.7.1, and in particular we obtain α with $U_\alpha \le R_1$.

LEMMA 14.7.12. (1) L acts 2-transitively on the subgroups U_H^L generating S.
(2) Assume $D_\gamma < U_\gamma$ and $b = 3$. Then $U_H^L = \{U_H\} \cup U_\alpha^{L_1 T}$.

PROOF. As $N_L(U_H) = H \cap L$ is a maximal parabolic of L and $L/O_2(L) \cong L_3(2)$, L is 2-transitive on $L/N_L(U_H)$, so that (1) holds.

Assume the hypotheses of (2). As $b = 3$, $\gamma \in \Gamma(\gamma_2)$, so $\alpha = \gamma h \in \Gamma(\gamma_2 h) = \Gamma(\gamma_0)$, and hence $U_\alpha \in U_H^L$. Therefore (2) follows from (1). $\quad\square$

LEMMA 14.7.13. (1) Set $E := [U_H, Q]$ and $R := \langle E^L \rangle$. Then $[S, Q] = R$.
(2) Assume $[\tilde{E}, Q] = \tilde{V}$. Then $[R, Q] = V$.
Assume further that $D_\gamma < U_\gamma$ and $b = 3$.
(3) If $[E, U_\alpha] = 1$ then $R \le Z(S)$.
(4) Set $A := [U_H, U_\alpha]$ and $B := \langle A^L \rangle$. Then $\Phi(S) = [S, S] = B$.

PROOF. Observe that (1) and (2) follow directly from the definitions of $S = \langle U_H^L \rangle$ and $R = \langle E^L \rangle$. Now assume that $D_\gamma < U_\gamma$ and $b = 3$, so that in particular Notation 14.7.1 holds. Suppose E commutes with U_α. As E also commutes with U_H and $E \trianglelefteq L_1 T$, $E \le Z(S)$ by 14.7.12.2. Thus (3) holds. Similarly 14.7.12.1 implies (4). $\quad\square$

We close Subsection 1 with a brief overview of an argument used to analyze the most difficult configurations in Subsections 2 and 5:

(a) Begin with a particular structure for H^*, and possibly $\tilde{U}_H$.
(b) Determine the structure of Q_H, and hence of H—cf. 14.7.20 and 14.7.71.1.
(c) Determine the structure of S, and hence of LT—cf. 14.7.24, 14.7.25, and 14.7.71–14.7.72.

In Subsection 2 we will obtain a contradiction from this analysis, while in Subsection 5 we will determine G_1 and M, and this information is sufficient to identify G as Ru.

14.7.2. Eliminating solvable members of $\mathcal{H}_z$. As was the case in Theorem 14.6.18 where $LT/O_2(LT) \cong L_2(2)$, in Theorem 14.7.29 of this subsection we will be able to show that no member of $\mathcal{H}_z$ is solvable. The most complicated configuration we must treat is that of case (2) of 14.7.9. We eliminate that case in the following result:

THEOREM 14.7.14. *There exists no $H \in \mathcal{H}_z$ such that $O^2(H^*) \cong 3^{1+2}$.*

Until the proof of Theorem 14.7.14 is complete, assume H is a counterexample. Let $K := O^2(H)$ and $P := O_2(K)$. By hypothesis, $K^* \cong 3^{1+2}$.

LEMMA 14.7.15. *(1) $L_1 \trianglelefteq H$ with $L_1^* = Z(K^*)$.*
(2) $R_1^* = Q^* = C_{T^*}(L_1^*) \cong \mathbf{Z}_4$ or Q_8.
(3) $H = G_1 = KT$ is the unique member of $\mathcal{H}_z$.
(4) $N_G(K) = !\mathcal{M}(KT)$.
(5) $D_\gamma < U_\gamma$, so that $U_\gamma^* \neq 1$.

PROOF. Let K_Z be the preimage of $Z(K^*)$ in K, and $K_0 := O^2(K_Z)$. Then L_1 acts on K_0, so if $K_0 = [K_0, T]$ then $K_0 \leq M$ by 14.5.3.2. If $K_0 > [K_0, T]$, then $K_0 \leq N_G(T) \leq M$ by Theorem 3.3.1.

Thus in any case, $K_0 \leq M$, so we may apply 14.7.9 to K in the role of "Y" to conclude that $L_1 = K_0$, T is irreducible on K^*/L_1^*, and $L_1 = O^2(K \cap M)$. Thus (1) holds, and by (1) we can apply 14.7.5. By 14.7.5.1, (5) holds. By 14.7.5.3, $R_1^* = Q^* = C_{T^*}(L_1^*)$ is of index 2 in T^*, while as T is irreducible on K^*/L_1^*, the remainder of (2) follows from the structure of $Out(K^*) \cong GL_2(3)$; and we also conclude that $K \in \Xi(G, T)$ in the sense of chapter 1. Then by 1.3.6, $K \in \Xi^*(G, T)$, so (4) follows from 1.3.7. In particular $K \trianglelefteq G_1$, so also $L_1 \trianglelefteq G_1$.

Let $\dot{G}_1 := G_1/O_2(G_1)$, $\dot{C}_1 := C_{G_1}(\dot{K})$, and $Y_1 \in Syl_3(\dot{C}_1)$. As $m_3(G_1) \leq 2$, $\dot{Y}_1$ is cyclic. Thus $\Omega_1(\dot{Y}_1) = \dot{L}_1 \leq Z(\dot{C}_1)$, and hence $\dot{Y}_1 \leq Z(N_{\dot{C}_1}(\dot{Y}_1))$, so $\dot{C}_1$ is 3-nilpotent by Burnside's Normal p-complement Theorem 39.1 in [**Asc86a**]. As $L_1 \trianglelefteq G_1$, we may apply 14.7.7 with G_1 in the role of "H" to conclude that $O_{3'}(\dot{G}_1) = 1$, so that $\dot{C}_1 = \dot{Y}_1$ is a cyclic 3-group. Thus $Y_1 \leq M$ by 14.7.8. Then as $M = LC_M(L/O_2(L))$, $Y_1 = (Y_1 \cap L_1) \times C_{Y_1}(L/O_2(L))$, so we conclude $|Y_1| = 3$ as Y_1 is cyclic. Then as $\dot{Y}_1 = \dot{C}_1$, $\dot{K} = F^*(\dot{G}_1)$. Therefore as $O^2(\dot{G}_1) \leq GL_3(4)$, either $G_1 = KT$, or $O^2(\dot{G}_1)$ is the split extension of 3^{1+2} by $SL_2(3)$ in view of (2). In the latter case, $m_3(G_1) = 3$, contradicting G_1 an SQTK-group, so the former case holds with $H = KT = G_1$, completing the proof of (3). $\square$

By 14.7.15.3, $G_1 = H$ is the unique member of $\mathcal{H}_z$, so $U_H = \langle V^{G_1} \rangle = U$. Similarly set $D := D_H$. Also in view of 14.7.15.5 and 14.7.5.1:

During the remainder of the proof of Theorem 14.7.14, we adopt Notation 14.7.1.

LEMMA 14.7.16. *(1) $\tilde{U} = [\tilde{U}, L_1]$ is a 6-dimensional faithful irreducible module for K^*.*
(2) $U_\alpha^* = Z(T^*)$ is of order 2.
(3) $[U, U_\alpha] = V$ and $\tilde{V} = C_{\tilde{U}}(Q^*)$.
(4) $U^L = \{U\} \cup U_\alpha^{L_1 T}$.
(5) $b = 3$.
(6) $m(U/D) = 1 = m(U_\alpha^*)$.
(7) $N_G(Baum(R_1)) \leq M$.

REMARK 14.7.17. Notice (6) shows that our hypotheses are symmetric between γ_1 and γ, in the sense discussed in Remark F.9.17; therefore if a result $S(\gamma_1, \gamma)$ (proved under the choice $m(U_\gamma^*) \geq m(U_H/D_H) > 0$ made in Notation 14.7.1) holds, then $S(\gamma, \gamma_1)$ also holds. Similarly as α is an H-translate of γ, $S(\gamma_1, \alpha)$ and $S(\alpha, \gamma_1)$ hold too.

PROOF. (of 14.7.16) By 14.7.15.1, we may apply 14.7.5 to H, so 14.7.5.5 says that $\tilde{U} = [\tilde{U}, L_1]$.

From Notation 14.7.1, $U_\alpha \leq R_1$, so as U is elementary abelian, (2) follows from 14.7.15.2. Then from our choice of γ,

$$1 = m(U_\gamma^*) \geq m(U/D) > 0,$$

and hence $m(U/D) = 1$. Thus we have established (6), and hence also the symmetry between γ_1 and γ discussed in Remark 14.7.17. As $\mathbf{Z}_2 \cong U_\alpha^* \in \mathcal{Q}(H^*, \tilde{U})$, $m(\tilde{U}/C_{\tilde{U}}(U_\alpha)) \leq 2$. Then as $\tilde{U} = [\tilde{U}, L]$, (1) holds by D.2.17.

From 14.7.15.2 and the action of $Aut(K^*)$ on the module $\tilde{U}$ for K^* in (1), $[\tilde{U}, U_\alpha] = C_{\tilde{U}}(R_1)$ is of rank 2; then as $\tilde{V}$ is of rank 2 and centralizes $R_1^* = Q^*$, we conclude that $[\tilde{U}, U_\alpha] = \tilde{V} = C_{\tilde{U}}(Q^*)$. Therefore U_γ^* does not induce transvections on $\tilde{U}$, so U_γ^* does not centralize D, and hence $A_1 \leq [U, U_\gamma]$ by F.9.13.6. Thus by symmetry between γ_1 and α, $V_1 \leq [U, U_\alpha]$, so that $[U, U_\alpha] = V$, completing the proof of (3). In particular $A_1^h \leq V$, so as $H = G_1$, (5) follows from 14.7.3.4, and (4) follows from (5) and 14.7.12.2.

By (1) and 14.5.21.1, L_1 has at least six noncentral 2-chief factors, so (7) follows from 14.7.10. $\square$

Let $E := [U, Q]$ and $R := \langle E^L \rangle$. By 14.7.16.1, $\tilde{U}$ has the structure of a 3-dimensional $\mathbf{F}_4$-module preserved by K^*Q^*, with the 1-dimensional $\mathbf{F}_4$-subspaces the L_1-irreducibles since $L_1^* = Z(K^*)$. Thus $\tilde{V}$ is a 1-dimensional $\mathbf{F}_4$-subspace, and from the action of Q^* on $\tilde{U}$, $\tilde{E} = C_{\tilde{U}}(U_\alpha^*)$ is a 2-dimensional $\mathbf{F}_4$-subspace. Then as $V_1 \leq [U, U_\alpha]$ by 14.7.16.3, $m(E) = 5$. Set $E_H := E^{h^{-1}}$, so that $\tilde{E}_H = C_{\tilde{U}}(U_\gamma^*)$. Define E_γ by $E_\gamma/A_1 = C_{U_\gamma/A_1}(U)$, and set $D_\alpha := D_\gamma^h$ and $E_\alpha := E_\gamma^h$. Observe that these definitions of "E_H, E_γ" differ from those in section F.9, but the latter notation is unnecessary here, since U, D play the role of the groups "V_H, E_H" of section F.9.

LEMMA 14.7.18. $E_H = C_U(U_\gamma)$ is of index 2 in D, $E_\gamma = C_{U_\gamma}(U)$, $E = C_U(U_\alpha)$, and $E_\alpha = C_{U_\alpha}(U)$ is of rank 5.

PROOF. By F.9.13.7, $[D, D_\gamma] = 1$, while $m(U/D) = 1 = m(U_\gamma/D_\gamma)$ by 14.7.16.6. Also for $x \in U_\gamma - D_\gamma$, $[x, D] \leq A_1$ by F.9.13.6, so $m(U/C_U(U_\gamma)) \leq m(D/C_D(x)) + 1 \leq 2$. Thus as $C_U(U_\gamma) \leq E_H$ and $m(U/E_H) = m(U/E) = 2$, these inequalities are equalities, and so the first statement of the lemma follows. Then the second statement follows from the symmetry between γ_1 and γ in Remark 14.7.17, and then the third and fourth statements follow from the first and second via conjugation by h. $\square$

LEMMA 14.7.19. (1) $[S, Q] = R$.
(2) $[R, Q] = V$.
(3) $R \leq Z(S)$; in particular, R is abelian and $R \leq C_H(U)$.
(4) $\Phi(S) = [S, S] = V$.

PROOF. Recall $E = [U, Q]$ by definition, while $[\tilde{E}, Q] = \tilde{V}$ from the action of Q^* on the module $\tilde{U}$. Then (1) and (2) follow from the corresponding parts of 14.7.13. Furthermore $[E, U_\alpha] = 1$ by 14.7.18, and $[U, U_\alpha] = V$ by 14.7.16.3; then in view of 14.7.15.5 and 14.7.16.5, (3) and (4) follow from the corresponding parts of 14.7.13. $\qquad \square$

Recall that $K = O^2(H)$, $P = O_2(K)$, and $C_H(U) = C_{Q_H}(U)$ as $Q_H = C_H(\tilde{U})$.

LEMMA 14.7.20. (1) $Q_H/C_H(U)$ is isomorphic to $P/C_P(U)$ and to the dual of $\tilde{U}$ as an H-module.

(2) Either

(i) $[C_H(U), K] \le U$, or

(ii) H has a unique noncentral chief factor W on $C_H(U)/U$, W is of rank 6, and H^* is faithful on W.

(3) $O_2(L_1) = O_2(K) = P$.

(4) $|P : P \cap Q| = 4$ and $(P \cap Q)/C_P(U) = [P/C_P(U), Q]$.

PROOF. By 14.7.4.1, $[U, Q_H] \ne 1$, so as H is irreducible on $\tilde{U}$ by 14.7.16.1, $C_U(Q_H) = V_1$. Next $Q_H/C_H(U)$ is dual to $\tilde{U}$ as an H-module by 14.5.21.1, so as $\tilde{U} = [\tilde{U}, K]$, also $Q_H/C_H(U) = [Q_H/C_H(U), K]$. Thus $Q_H = PC_H(U)$, so that (1) holds. As $m(\tilde{V}) = 2$, the duality shows that $C_P(V) = P \cap Q$ is of corank 2 in P, and also that (4) holds, since $\tilde{V} = C_{\tilde{U}}(Q)$ by 14.7.16.3.

By 14.7.18, $E_\alpha = C_{U_\alpha}(U)$ is of rank 5, and $V \le U_\alpha \cap U \le E_\alpha$ using 14.7.16.3, so $m(E_\alpha U/U) \le 2$ as $m(V) = 3$. By 14.7.16.4, $U_\alpha = U^y$ for some $y \in L$; thus $V_1^y \le V \le U$. Then $C_H(U) \le C_H(V_1^y) \le N_H(U_\alpha)$ since $H = C_G(V_1)$, so $[C_H(U), U_\alpha] \le C_{U_\alpha}(U) = E_\alpha$. Hence if $W := W_1/W_2$ with $U \le W_1 \le W_2 \le C_H(U)$ is a noncentral chief factor for H on $C_H(U)/U$, then $m([W, U_\alpha]) \le 2$ as we saw $m(E_\alpha U/U) \le 2$. Therefore as U_α^* has rank 1 by 14.7.16.2, $\hat{Q}(H^*, W)$ is nonempty. Then by D.2.17, W is a 6-dimensional faithful module, and $[C_H(U), U_\alpha] \le W_2$, so W is the unique noncentral chief factor for K^* on $C_H(U)/U$. Therefore conclusion (ii) of (2) holds in this case, while conclusion (i) holds if no such chief factor exists; hence (2) is established.

By (1) and (2), all noncentral chief factors X for K on P satisfy $X = [X, L_1]$, so $O_2(L_1) = O_2(K) = P$, establishing (3). $\qquad \square$

LEMMA 14.7.21. (1) $H = C_G(z) \in \mathcal{M}$.

(2) $Z(P) \le Z(K)$.

PROOF. By 14.7.15.4, $H_K := N_G(K) = !\mathcal{M}(H)$. By 14.7.15.1, $L_1 \trianglelefteq H_K$. Set $C_K := C_{H_K}(K/O_2(K))$, and $Y_K := C_{H_K}(L_1/O_2(L_1))$, so that Y_K is of index 2 in $H_K = Y_K T$. Then as $R_1 = O_2(L_1 T)$, R_1 is Sylow in Y_K, and hence in $C_K R_1$. As

$$C_{\mathrm{Aut}(K^*))}(L_1^*))/\mathrm{Aut}_K(K^*) \cong SL_2(3) \text{ is 2-closed,}$$

$C_K R_1 \trianglelefteq H_K$. Let $B \in Syl_3(H_K)$, $B_K := B \cap K$, and $B_C := B \cap C_K$. As $m_3(H_K) \le 2 = m_3(K)$, B_C is cyclic with $B_1 := \Omega_1(B_C) = B_K \cap B_C$ Sylow in L_1. Then as $R_1 = O_2(L_1 T)$, $[B_1, R_1] \le O_2(L_1) \cap C_K \le O_2(C_K)$; so as R_1 is Sylow in $C_K R_1$, B_C is not inverted in its normalizer in $C_K R_1$. Therefore by Burnside's Normal p-complement Theorem 39.1 in [Asc86a], $C_K R_1$ has a normal 3-complement. Then by a Frattini Argument, we may take $B = B_K B_M$, where $B_C \le B_M := N_B(R_1)$, and $B_M \le M$ by 14.7.16.7. Therefore as $M = L C_M(L/O_2(L))$, $B_M = B_1 \times B_0$

where $B_0 := C_{B_M}(L/O_2(L))$. As B_1 is Sylow in L_1 and $B_1 \leq B_C \leq B_M$ with B_C cyclic, $B_C = B_1$. Further if $B_0 \neq 1$, then B_0 centralizes some a of order 3 in B_K which is inverted by $r \in R_1$ inverting B_K/B_1. But then $m_3(B_0 B_1 \langle a \rangle) = 3$, contradicting $m_3(H_K) = 2$.

Thus $B_0 = 1$, so $B = B_K \cong K^* \cong 3^{1+2}$ and $H_K = HO_{3'}(H_K)$ with $C_K = L_1 O_{3'}(H_K)$. Let $W := \langle z^{C_K} \rangle$, so that $W \in \mathcal{R}_2(C_K R_1)$ by B.2.14. As L_1 centralizes z and $L_1 \trianglelefteq H_K$, L_1 centralizes W, so that $C_K/C_{C_K}(W)$ is a $3'$-group. Since a $3'$-group has no FF-modules by Theorem B.4.2, and R_1 is Sylow in $C_K R_1$, $J(R_1)$ centralizes W by Thompson Factorization B.2.15. Also as $H = G_1 = C_G(z)$ by 14.7.15.3, $C_{C_K}(W) \leq C_K \cap H \leq L_1 O_2(H)$, so $C_{C_K}(W) = L_1 O_2(C_K)$. Then $\mathrm{Baum}(R_1) = \mathrm{Baum}(C_{R_1}(W)) = \mathrm{Baum}(O_2(C_K R_1))$ by B.2.3.5. Therefore $C_K \leq N_G(\mathrm{Baum} R_1) \leq M$ by 14.7.16.7. Then by 13.3.8 with L in the role of "K", $O^2(C_K)$ is a $\{2,3\}$-group; so as B_1 is Sylow in C_K, $O^2(C_K) = L_1$. Thus $H = HC_K = H_K \in \mathcal{M}$, and (1) is established.

Let $Z_0 := \Omega_1(Z(P))$ and assume (2) fails, so that $Z_P := [Z_0, K] \neq 1$. Now $Z_0 \in \mathcal{R}_2(K)$, so $Z_0 = Z_P \times C_{Z_0}(K)$ by Coprime Action. But $U \not\leq Z_0$ by 14.7.20.1, so Z_P is a faithful irreducible of rank 6 for K^* by 14.7.20.2. Hence $Z_P \in \mathcal{R}_2(KR_1)$ is not an FF-module for $K^* R_1^*$ by Theorem B.5.6; so as $R_1 \in Syl_2(KR_1)$, $\mathrm{Baum}(R_1) \trianglelefteq KR_1$ by Solvable Thompson Factorization B.2.16 and B.2.3.5. Thus $\mathrm{Baum}(R_1) \trianglelefteq KT = H$, contradicting 14.7.16.7. $\qquad\square$

LEMMA 14.7.22. *(1) R/V is isomorphic as an LT/Q-module to one of: the dual of V; the 6-dimensional core of the permutation module on $L/N_L(V_2)$, which we will denote by $Core$; the direct sum of the 8-dimensional Steinberg module with either $Core$ or the dual of V; or the Steinberg module.*

(2) L_1 has three noncentral chief factors on the Steinberg module, two on $Core$, and one on the dual of V.

PROOF. As E/V is the natural module for $L_1 T/O_2(L_1 T)$ and $R = \langle E^L \rangle$, (1) follows from H.6.5. Part (2) follows from H.6.3.3 and H.5.2. $\qquad\square$

LEMMA 14.7.23. $E = U \cap R \leq P$, and either

(1) Case (i) of 14.7.20.2 holds, R/V is isomorphic to the dual of V as an L-module, and E/V is the unique noncentral chief factor for L_1 on R/V; or

(2) Case (ii) of 14.7.20.2 holds, and $R/V \cong Core$.

PROOF. By (3) and (4) of 14.7.19, S is nonabelian while $R \leq Z(S)$; so $U \not\leq R$ as $S = \langle U^L \rangle$. Then as L_1 is irreducible on U/E and $R = \langle E^L \rangle$, $E = U \cap R$. Further $E = [E, L_1]$ in view of 14.7.16.1, so $E \leq P$. Thus the noncentral L_1-chief factors of R contained in U are the two in E, so E/V is the unique noncentral chief factor on R/V contained in U/V. Therefore if case (i) of 14.7.20.2 holds, then as $R \leq Z(S) \leq C_H(U)$, E/V is the unique noncentral L_1-chief factor on R/V, and hence R/V is dual to V by 14.7.22, so that (1) holds.

Thus we may assume instead that case (ii) of 14.7.20.2 holds. Then H has a unique noncentral chief factor W on $C_H(U)/U$, and H^* is faithful and irreducible on W of rank 6. Now $[R, Q] = V \leq U$ by 14.7.19.2, so that $[R \cap W, Q] = 1$, and hence $m(R \cap W) \leq 2$ from the action of Q^* on the 6-dimensional faithful irreducible W for K^*. As $U_\alpha \leq S$ by 14.7.16.4, $[Q, U_\alpha] \leq R$ by 14.7.19.1. Therefore as $C_H(U) \leq C_T(V) = Q$, $[W, U_\alpha] \leq R \cap W$. Thus as L_1 acts nontrivially on $[W, U_\alpha]$ in the 6-dimensional module W, we conclude that $R \cap W$ has rank 2, and is the unique noncentral L_1-chief factor on W contained in R/V. So as $R \leq C_H(U)$ by 14.7.19.3, and the first paragraph showed that E/V is the unique noncentral L_1-chief factor on R/V contained in U/V, we conclude there are exactly two noncentral L_1-chief factors on R/V. Then it follows from 14.7.22 that (2) holds.   □

Let $P_C := C_P(U)$ and $\hat{H} := H/U$.

LEMMA 14.7.24. *Assume case (1) of 14.7.23 holds. Then*

(1) $S/R \cong Core$.

(2) $V_1 < Z(K)$.

PROOF. As we are case in (1) of 14.7.23, case (i) of 14.7.20.2 holds, so that K centralizes $\hat{P}_C$. By 14.7.20.1, $P^+ := P/P_C$ is a 6-dimensional irreducible for H^*. Thus L_1 has exactly six nontrivial 2-chief factors, three each from $\tilde{U}$ and P^+. We next locate these factors relative to the series $Q > S > R > V$. By 14.7.20.4, one of the factors is $P^+/(P \cap Q)^+$, leaving five in $P \cap Q$. By 14.7.23, $E = U \cap R \leq P$, and the two factors in E are the factors appearing in V and R/V since case (1) of 14.7.23 holds; this leaves three factors to be located in Q/R. Further $UR/R \cong U/E$ is the natural module for $L_1T/O_2(L_1T)$. Therefore applying H.6.5 as in the proof of 14.7.22, S/R has one of the structures listed in 14.7.22.1. We will show that L_1 has exactly two noncentral chief factors in S/R, so that (1) will hold by applying 14.7.22.2 to the possibilities in 14.7.22.1.

First U/E is the only factor in S/R contained in UR/R. This leaves just the two factors from $(P \cap Q)^+$ to be located in Q/R. Now using (1) and (3) of 14.7.19, $[Q, S \cap P] \leq R \cap P \leq C_P(U) = P_C$, so that $(S \cap P)^+ \leq C_{P^+}(Q^*) =: A_0^+$. Observe $m(A_0^+) = 2$ by applying the duality in 14.7.19.1 to $C_{\tilde{U}}(Q^*) = \tilde{V}$ in view of 14.7.16.3. By 14.7.16.4, $U_\alpha \leq S$, and by 14.7.16.2, $U_\alpha^* = Z(T^*)$, so again applying 14.7.19.1, $[P^+, U_\alpha] = A_0^+ \leq (S \cap P)^+$. Hence $A_0^+ = (S \cap P)^+$ is of rank 2, so that L_1 has exactly two noncentral chief factors on S/R, given by A_0^+ and UR/R. As indicated earlier, this completes the proof of (1).

Define S_1 as the preimage in S of $Soc(S/R)$. Then $S_1/R \cong V$ as $S/R \cong Core$ by (1), so that $S_1/R = [S_1/R, L_1]$. Observe $U \not\leq S_1$ since S is generated by the L-conjugates of U, so we conclude from the proof of (1) that the noncentral L_1-chief factor in S_1/R comes from A_0^+ rather than from UR/R. Thus $S_1 = P_1R$, where $P_1 := P \cap S_1$ and $P_1^+ = A_0^+$ is of rank 2. So setting $C_1 := P_C \cap S_1$, $C_1R/R = C_{S_1/R}(L_1)$ has rank 1. But $C_{\tilde{U}}(L_1) = 1$, so $C_1 \not\leq U$, and hence $\hat{C}_1 \neq 1$.

Next as we are in case (i) of 14.7.20.2, $P \leq K \leq C_H(\hat{P}_C) \leq C_H(\hat{C}_1)$, so that $[C_1, P] \leq U$. Then as $C_1R/R = C_{S_1/R}(L_1)$ and $P \leq L_1$ by 14.7.20.3, $[C_1, P] \leq U \cap R = E$, so $[C_1U, P] \leq E$. Since K centralizes $\hat{C}_1$, K normalizes C_1U and hence also $[C_1U, P]$, so as K is irreducible on $\tilde{U}$, we conclude $[C_1, P] \leq V_1$—that is, P centralizes $\hat{C}_1$. Let D_1 be the preimage of $C_{\hat{C}_1\tilde{U}}(L_1)$. As P centralizes $\tilde{C}_1\tilde{U}$, by Coprime Action we have an L_1-module decomposition $\tilde{C}_1\tilde{U} = \tilde{D}_1 \times \tilde{U}$, and then $L_1 = O^2(L_1)$ centralizes D_1. In particular $D_1 \leq Z(P)$, and hence $D_1 \leq Z(K)$ by 14.7.21.2. As $\hat{C}_1 \neq 1$, $V_1 < D_1$, so (2) is established.   □

LEMMA 14.7.25. $V_1 < Z(K)$.

PROOF. In case (1) of 14.7.23 we obtained this result in 14.7.24, so we may assume we are in case (2) of 14.7.23. The proof proceeds much as did the proof of 14.7.24.2, except we analyze P_-/C_- rather than P/P_C, where $P_- := [P_C, L_1]$, and

C_- is the preimage in P_- of $C_{\hat{P}_-}(L_1)$. As case (ii) of 14.7.20.2 holds, P_-/C_- is a 6-dimensional faithful irreducible module for H^*. This time we work modulo V rather than modulo R, so we let R_0 denote the preimage in R of $Soc(R/V)$. Since we are in case (2) of 14.7.23, R_0/V is isomorphic to V as an L-module, so that $R_0/V = [R_0/V, L_1]$. Since R is generated by the L-conjugates of E, $E \not\leq R_0$, so from the analysis of case (2) in the proof of 14.7.23, the noncentral L_1-chief factor in R_0/V is $(R_0 \cap P_-)/(R_0 \cap C_-)$. Thus $R_0 = P_0 V$, where $P_0 := [P \cap R_0, L_1] \leq P_-$. This time we set $C_0 := P_0 \cap C_-$, so that $C_0 V/V = C_{R_0/V}(L_1)$ is of rank 1. Now $V \leq C_0$, but as L_1 is fixed-point-free on $\tilde{U}$, $C_0 \not\leq U$ and hence $\hat{C}_0 \neq 1$. As K is trivial on $\hat{C}_-$, K acts on $C_0 U$, and hence again K acts on $[C_0 U, P] = [C_0, P]$. Now $[C_0, P] \leq V$ as $C_0 V/V$ is T-invariant of rank 1, so as K is irreducible on $\tilde{U}$, we conclude P centralizes $\tilde{C}_0$. Let D_0 denote the preimage in $C_0 U$ of $C_{\tilde{C}_0 \tilde{V}}(L_1)$; just as at the end of the proof of 14.7.24.2, $D_0 \leq Z(K)$, so as $\hat{C}_0 \neq 1$, $V_1 < D_0$, completing the proof. $\qquad\square$

As $H = KT$, by 14.7.25 there is a subgroup D of order 4 in $Z(K)$ containing V_1 and normal in H.

LEMMA 14.7.26. *D is a TI-subgroup of G.*

PROOF. If D is cyclic, then $V_1 = \Omega_1(D)$, so as $D \trianglelefteq H = G_1$, the lemma holds. Thus we may assume $D \cong E_4$. If $D \leq Z(T)$ then $D \leq Z(H)$, so as $H \in \mathcal{M}$ by 14.7.21.1, the lemma follows from I.6.1.2. Therefore we may assume that $[D, T] = V_1$.

Let $d \in D - V_1$, and set $G_d := C_G(d)$, and $H_d := H \cap G_d$. Then $T_d := T \cap G_d$ is Sylow in H_d and of index 2 in T, so that $H_d := KT_d$ is of index 2 in $KT = H$; hence $H_d \trianglelefteq H$, and so $H_d \in \mathcal{H}^e$ by 1.1.3.1. Then $Z(T_d) \leq Z(O_2(H_d)) =: Z_d$. As $P = O_2(K)$ and $K \trianglelefteq H_d$, $P \leq O_2(H_d)$, so $[Z_d, K] \leq Z_d \cap P \leq Z(P) \leq Z(K)$ by 14.7.21.2. Therefore K centralizes Z_d by Coprime Action, and so $Z(T_d) \trianglelefteq KT = H$. Thus $N_G(T_d) \leq N_G(Z(T_d)) = H$ as $H \in \mathcal{M}$, so that $T_d \in Syl_2(G_d)$. In particular $d \notin z^G$, so that H controls fusion in D. So appealing to I.6.1.1, it suffices to show that $G_d \leq H$. Thus we assume $G_d \not\leq H$, and it remains to derive a contradiction. As $G_d \not\leq H$,

$$\mathcal{G}_0 := \{G_0 \leq G_d : H_d < G_0\}$$

is nonempty. The bulk of the proof consists of an analysis of $\mathcal{G}_0$.

Let $G_0 \in \mathcal{G}_0$; then $G_0 \in \mathcal{H}(T_d)$ as $d \in O_2(G_0)$. As $L_1 \trianglelefteq H \in \mathcal{M}$, $H = N_G(L_1)$, so $H_d = N_{G_d}(L_1)$ and in particular L_1 is not normal in G_0.

Suppose first that T_d is irreducible on K/L_1. Then $H_d \in \Xi(G_0, T_d)$, so the conclusions of 1.3.2 hold with T_d in the role of "T", and we may apply the proof of 1.3.4 to G_0 in the role of "H" (as that argument uses only 1.3.2 and the fact that G_0 is an SQTK-group, and does not actually require T to be Sylow in G_0) to conclude since $K/O_2(K)$ is not elementary abelian that $K \trianglelefteq G_0$, and hence $L_1 = O^2(O_{2,Z}(K)) \trianglelefteq G_0$, contrary to the previous paragraph.

Thus T_d is reducible on K/L_1, so as R_1 is irreducible on K/L_1 by 14.7.15.2, $R_1^* \not\leq T_d^*$ and hence $R_1 \not\leq T_d Q_H$. So $T_d Q_H < T$, and then as $|T : T_d| = 2$, $Q_H \leq T_d$. Therefore $Q_H = O_2(H_d)$. Further $H = N_G(Q_H)$ as $H \in \mathcal{M}$, so that $H_d = N_{G_0}(Q_H)$, and hence $C(G_0, Q_H) = H_d$. Therefore Hypothesis C.2.3 is satisfied with G_0, H_d, Q_H in the roles of "H, M_H, R".

Let $Y \in Syl_3(K)$, set $X := Y \cap L_1$, and let $\mathcal{I}$ consist of the Y-invariant subgroups I of G_0 with $3 \in \pi(I)$. Then for $I \in \mathcal{I}$, there is a Y-invariant Sylow 3-subgroup Y_I of I, and $X_I := \Omega_1(Z(YY_I) \cap I) = X$ since $m_3(Y) = 2$ and $m_3(G_0) \leq 2$. Thus $X \leq Z(Y_I)$ for each $I \in \mathcal{I}$.

Suppose next that $O_2(G_0) < O_{2,3}(G_0)$. Then $O_{2,3}(G_0) \in \mathcal{I}$, so X is in the center of a Sylow 3-subgroup of $O_{2,3}(G_0)$ by the previous paragraph. Then as $L_1 = X[O_2(G_0), X]$, $O_{2,F^*}(G_0) \leq N_{G_0}(L_1) = H_d$ using an earlier observation. Hence as H is a $\{2,3\}$-group by 14.7.15.3, $O_{2,F^*}(G_0)$ is a $\{2,3\}$-group. Then using A.1.25.3, G_0 is a $\{2,3\}$-group, so $G_0 \in \mathcal{I}$. Therefore X is in the center of a Sylow 3-group Y_I of G_0 containing Y, so that Y_I acts on $X[X, O_2(G_0)] = L_1$. Then $G_0 = Y_I T_d \leq N_{G_0}(L_1) = H_d$, contrary to $G_0 \in \mathcal{G}_0$. This contradiction shows that $O_{2,3}(G_0) = O_2(G_0)$, so that $O_3(G_0/O_2(G_0)) = 1$.

Now suppose J is a subnormal subgroup of G_0 contained in H_d. As H_d is a $\{2,3\}$-group, so is J, so as $O_{2,3}(J) \leq O_{2,3}(G_0)$, J is a 2-group by the previous paragraph. Hence $O_2(G_0)$ is the largest subnormal subgroup of G_0 contained in H_d.

Suppose that $L_0 \in \mathcal{C}(G_0)$ with $3 \in \pi(L_0)$. Then $Y = O^2(Y)$ acts on L_0 by 1.2.1.3, so $L_0 \in \mathcal{I}$, and hence $L_1 = X[X, O_2(G_0)] \leq L_0$. Therefore L_0 is the unique member of $\mathcal{C}(G_0)$ with $3 \in \pi(L_0)$.

Suppose next that $F^*(G_0) = O_2(G_0)$. Set $J := O_{2,3'}(G_0)$. Then $T_d \cap J \leq O_{2,3'}(H_d) = Q_H$, so Q_H is Sylow in JQ_H. Therefore as Hypothesis C.2.3 holds in G_0, we conclude from C.2.5 that $J \leq H_d$, so as J is normal in G_0, $J = O_2(G_0)$ by an earlier reduction. Thus $O_{3'}(G_0/O_2(G_0)) = 1 = O_3(G_0/O_2(G_0))$, so $O_{2,F^*}(G_0)$ is a product of $O_2(G_0)$ with members of $\mathcal{C}(G_0)$ whose order is divisible by 3. Then we conclude from the previous paragraph that $O^2(O_{2,F^*}(G_0)) =: L_0$ is the unique member of $\mathcal{C}(G_0)$ and $L_1 \leq L_0$. In particular $L_0 \trianglelefteq G_0$, so that L_0 is described in C.2.7.3. As $L_1 \leq L_0$ and $O_3(G_0/O_2(G_0)) = 1$, Y acts faithfully on $L_0/O_2(L_0)$. However no group K listed in C.2.7.3 has a group of automorphisms A containing $Inn(K)$ and a subgroup H_A of odd index in A with $O^2(H_A/O_2(H_A)) \cong 3^{1+2}$. Therefore $O_2(G_0) < F^*(G_0)$, so

$$H_d \text{ is maximal in } \{G_+ \leq G_d : F^*(G_+) = O_2(G_+)\}. \qquad (*)$$

Observe that by 1.1.6, Hypothesis 1.1.5 is satisfied with G_d, T_d, H in the roles of "H, S, M". However $U = [U, L_1]$ by 14.7.16.1, so U centralizes $O(G_d)$ by A.1.26. Then as $z \in U$, $O(G_d) = 1$ by 1.1.5.2. Thus there is a component L_d of G_d, and $L_d \not\leq H$ by 1.1.5.3.

Suppose first that L_d is a Suzuki group and set $L_0 := \langle L_d^{H_d} \rangle$. As H_d is a $\{2,3\}$-group, $H_d \cap L_0 = T_d \cap L_0$, so H_d acts on the Borel subgroup $B := N_{L_0}(T_d \cap L_0)$ of L_0. Therefore as $F^*(BH_d) = O_2(BH_d)$, $B \leq H_d$ by $(*)$, impossible as B is not a $\{2,3\}$-group.

Thus $3 \in \pi(L_d)$, so by an earlier reduction, L_d is the unique component of G_d and $L_1 \leq L_d$. Similarly $O_3(G_d/O_2(G_d)) = 1 = O_{3'}(G_d/O_2(G_d))$, so as before Y acts faithfully on L_d. This time L_d is described in 1.1.5.3, and again no subgroup A satisfying $Inn(L_d) \leq A \leq Aut(L_d)$ contains a subgroup H_A of odd index in A with $O^2(H_A/O_2(H_A)) \cong 3^{1+2}$. This contradiction finally completes the proof of 14.7.26. $\qquad \square$

We are now in a position to obtain a contradiction, and thus establish Theorem 14.7.14. To obtain our contradiction, we will show that the weak closure $X :=$

$W_0(R_1, D)$ of D in R_1 is normal in both LT and H, so that $H \leq N_G(X) \leq M = !\mathcal{M}(LT)$, contrary to $H \not\leq M$.

It suffices to show that X centralizes U: For then as $V \leq U$, $X \leq C_T(V) = Q$ and $X \leq C_T(U) \leq Q_H$, so $X = W_0(Q, D) = W_0(Q_H, D)$ is normal in LT and H using E.3.15. Thus we may assume there is $g \in G$ such that $A := D^g \leq R_1$, but A does not centralize U. By 14.7.26, A is a TI-subgroup of G, so:

(!) $[C_U(a), A] \leq A \cap U$ for each $a \in A^{\#}$.

Suppose first that $A \cap U = 1$. Then by (!),

(*) $C_U(a) = C_U(A)$ for each $a \in A^{\#}$.

In particular if $1 \neq a \in C_A(U)$, then A centralizes U, contrary to our assumption, so A is faithful on U. Thus A is not cyclic of order 4 by (*), so $A \cong E_4$. Now as $m_2(R_1^*) = 1$ by 14.7.15.2, $A \cap Q_H \neq 1$. Then as A is faithful on U, for each $b \in A \cap Q_H^{\#}$, $C_U(b)$ is a hyperplane of U in view of 14.7.4.1. However no element of $H - Q_H$ centralizes a hyperplane of $\tilde{U}$, and elements of $Q_H - bC_{Q_H}(U)$ centralize hyperplanes of U distinct from $C_U(b)$ by the duality in 14.5.21.1, so again using (*), we conclude $A^{\#} \subseteq bC_{Q_H}(U)$, a contradiction as A is faithful on U.

Therefore $A \cap U \neq 1$. Then as $|A| = 4$, $|A \cap U| = 2$, and hence A induces a group of transvections on U with center $A \cap U$ by (!). As no element of $H - Q_H$ centralizes a hyperplane of $\tilde{U}$, $A \leq Q_H$; hence $[A, U] = V_1$ by 14.7.4.1, so $A \cap U = V_1$. Therefore as D is a TI-subgroup of G by 14.7.26, $A = D \leq Z(K) \leq C_G(U)$ since $U = [U, K] \leq K$, contrary to our assumption that A does not centralize U.

Thus the proof of Theorem 14.7.14 is complete.

In the remainder of the subsection, H again denotes an arbitrary member of $\mathcal{H}_z$. We deduce various consequences of Theorem 14.7.14 for members of $\mathcal{H}_z$.

LEMMA 14.7.27. *For each $H \in \mathcal{H}_z$, either $O_3(H^*) = 1$ or $O_3(H^*) = L_1^*$.*

PROOF. Suppose H is a minimal counterexample, and let $P^* := O_3(H^*)$ with P a Sylow 3-group of the preimage of P^*. Let P_0 be a supercritical subgroup of P, so that $P_0 \cong \mathbf{Z}_3$, E_9, or 3^{1+2} by A.1.25.1. Further by definition, P_0 contains each subgroup of order 3 in $C_P(P_0)$, so if $|P_0| = 3$, then P is cyclic.

Suppose first that $P_0 \leq M$. Applying 14.7.9 with $O^2(P_0 Q_H)$ in the role of "Y" we conclude that $L_1^* = P_0^*$ is of order 3, so P is cyclic. But then $P \leq M$ by 14.7.8, so as $M = LC_M(L/O_2(L))$, $P = C_P(L/O_2(L)) \times (P \cap L_1)$; then as P is cyclic, $P^* = L_1^*$, contrary to the choice of H as a counterexample.

Thus $P_0 \not\leq M$, so by minimality of H, $H = P_0 L_1 T$. Let B be of order 3 in L_1; we may assume B acts on P.

Assume first that $B \not\leq P$. Then $L_1^* \not\leq O_3(H^*)$, so since L_1 is T-invariant in $H = P_0 L_1 T$, we conclude that $1 \neq O_2(L_1^*)$. Then by A.1.21.3, L_1^* is faithful on $P_0^*/\Phi(P_0^*)$, so H^* is the split extension of P_0^*, isomorphic to E_9 or 3^{1+2}, by $L_1^* T^* \cong GL_2(3)$. However if P_0^* is 3^{1+2}, then this split extension is of 3-rank 3, contradicting G quasithin. Therefore $P_0^* \cong E_9$. Now $q(H^*, \tilde{U}_H) \leq 2$ by 14.5.18.3, and the normal subgroup $J^* := \langle \mathcal{Q}(H^*, \tilde{U}_H) \rangle$ is either $H^* \cong GL_2(3)$ or $O_{3,Z}(H^*)$. But the first does not appear in D.2.17, and the second does not satisfy conclusion (3) of D.2.17, since irreducibles for H^* faithful on P_0^* have dimension 8 rather than 4.

Therefore $B \leq P$. If $P_0 \cap M \neq 1$, we may apply 14.7.9 to $O^2(P_0 Q_H)$ in the role of "Y" to conclude that $P_0^* \cong 3^{1+2}$. But now Theorem 14.7.14 supplies a contradiction. Hence $P_0 \cap M = 1$, so in particular $B \not\leq P_0$. Then as P_0 contains each subgroup of order 3 in $C_P(P_0)$, $P_1 := C_{P_0}(B) < P_0$. Now TL_1 acts on P_1^*, so as $P_1 < P_0$, $P_1 \leq M$ by minimality of H, contradicting $P_0 \cap M = 1$. This completes the proof of 14.7.27. $\qquad\square$

LEMMA 14.7.28. *For each $H \in \mathcal{H}_z$, either $O(H^*) = 1$ or $O(H^*) = L_1^*$.*

PROOF. Suppose H is a counterexample. Then by 14.7.27, $O_p(H^*) \neq 1$ for some prime $p > 3$. But this contradicts 14.7.6. $\qquad\square$

As a corollary to 14.7.28 we have

THEOREM 14.7.29. *Each solvable subgroup of G_1 containing L_1T is contained in M.*

PROOF. Assume $L_1T \leq H \not\leq M$ is solvable. Then $1 \neq O(H^*) = F^*(H^*) = L_1^* \cong \mathbf{Z}_3$ by 14.7.28, and hence $|H^* : C_{H^*}(F^*(H^*))| \leq 2$. Then $H = Q_H L_1 T \leq M$, contrary to assumption. $\qquad\square$

14.7.3. Reducing to $\mathbf{O^2(H^*)}$ isomorphic to $\mathbf{G_2(2)'}$ or $\mathbf{A_5}$. Let $H \in \mathcal{H}_z$. By Theorem 14.7.29 and 1.2.1.1, H contains $\mathcal{C}$-components. In this subsection, we establish restrictions on the $\mathcal{C}$-components of H: For example, 14.7.48 will show that H contains a unique $\mathcal{C}$-component K, and that $H = KT$. Then Theorem 14.7.52 will reduce our analysis to the cases where $K/O_2(K) \cong A_5$ or $G_2(2)'$.

Let $K \in \mathcal{C}(H)$. By 14.7.28, $|O(K^*)| \leq 3$, so $K/O_2(K)$ is quasisimple by 1.2.1.4. Also $K \not\leq M$ and $\langle K^T \rangle L_1 T \in \mathcal{H}_z$ by 14.5.19, and hence:

LEMMA 14.7.30. *For each $K \in \mathcal{C}(H)$, $K/O_2(K)$ is quasisimple, $K \not\leq M$, $\langle K^T \rangle L_1 T \in \mathcal{H}_z$, and $K/O_2(K)$ is described in F.9.18.*

LEMMA 14.7.31. *Suppose $C_G(V_2) \leq M$. Then*

(1) $W_0(R_1, V) \trianglelefteq LT$, so $N_G(W_0(R_1, V)) \leq M$.

(2) Let $U := \langle V^{G_1} \rangle$ and assume there is $Y \in \mathcal{H}^e$, $T_Y \in Syl_2(Y)$, and $V_Y \in \mathcal{R}_2(Y)$ with $Y/O_2(Y) \cong S_3$, $O_2(Y) = C_Y(V_Y)$, and $U^g \leq C_Y(V_Y)$ for each $V_1^g \leq V_Y$. Then $W_0(T_Y, V) \trianglelefteq Y$.

PROOF. Observe first that as $C_G(V_2) \leq M$ by hypothesis, $C_G(V_2) \leq M_V$ by 14.3.3. Thus as L is transitive on hyperplanes of V:

(*) $C_G(A) \leq N_G(V^g)$ for each $g \in G$ and each hyperplane A of V^g.

Suppose $V^g \leq R_1$ with $\bar{V}^g \neq 1$. Then

$$V = \langle C_V(A) : m(V^g/A) = 1 \rangle,$$

while for each hyperplane A of V^g, $[C_V(A), V^g] \leq V \cap V^g = 1$ by (*) and 14.5.2. Thus $[V, V^g] = 1$, contrary to assumption. We conclude $W_0(R_1, V) \leq C_T(V) = O_2(LT)$, so by E.3.15 and E.3.16, $W_0(R_1, V) = W_0(O_2(LT), V) \trianglelefteq LT$ and also $N_G(W_0(R_1, V)) \leq M = !\mathcal{M}(LT)$. Thus (1) holds.

Assume the hypotheses of (2), and suppose $V^g \leq T_Y$ with $[V_Y, V^g] \neq 1$. Then $A := V^g \cap O_2(Y)$ is a hyperplane of V^g, so by (*), $[V_Y, V^g] \leq V_Y \cap V^g$, and hence by transitivity of L on $V^\#$, we may take $V_1^g \leq V_Y$. Then $V^g \leq U^g \leq C_Y(V_Y)$ by hypothesis, contrary to assumption. Thus $W_0(T_Y, V) = W_0(O_2(Y), V) \trianglelefteq Y$ using E.3.15 just as in the proof of (1). $\qquad\square$

LEMMA 14.7.32. *T normalizes each $K \in \mathcal{C}(H)$, so $KL_1T \in \mathcal{H}_z$.*

PROOF. Let $K_0 := \langle K^T \rangle$. By 14.7.30, $K_0 L_1 T \in \mathcal{H}_z$, so without loss $H = K_0 L_1 T$. We assume T does not act on K and derive a contradiction. By 1.2.1.3, $K_0 = KK^t$ for $t \in T - N_T(K)$. Then by 14.7.30 we may apply F.9.18.5 to conclude that $K/O_2(K)$ is $L_2(2^n)$, $Sz(2^n)$, or $L_3(2)$.

Suppose first that $K^* \cong L_3(2)$. Then by 1.2.2, $K_0 = O^{3'}(H)$, and so $L_1 \leq K_0$. Therefore there is an overgroup H_1 of $L_1 T$ in H with $H_1/O_2(H_1) \cong S_3$ wr $\mathbf{Z}_2$, and hence by Theorem 14.7.29, $H_1 \leq M$. But then $O^2(H_1) = [L_1, H_1] \leq L$, so that $m_3(H_1 \cap L) = 2$, contrary to $m_3(L) = 1$.

So K^* is $L_2(2^n)$ or $Sz(2^n)$. Let B_0 be the preimage of the Borel subgroup of K_0^* containing $T_0^* := T^* \cap K_0^*$, and $B := O^2(B_0)$. Then B_0 is the unique maximal overgroup of $L_1 T \cap K_0$ in K_0, so $L_1 T$ normalizes B. Hence as B is solvable, $B \leq M$ by Theorem 14.7.29, so B acts on L_1. However if K^* is $Sz(2^n)$, then B^* acts on no subgroup L_1^* of $Aut(K^*)$ with $|L_1^* : O_2(L_1^*)| = 3$, so that $[K_0^*, L_1^*] = 1$. Hence $L_1^* \trianglelefteq H^*$, contrary to 14.7.7.

We now interrupt the proof of 14.7.32 briefly, to observe that we can use the previous argument to establish three further results:

LEMMA 14.7.33. *If* $H \in \mathcal{H}_z$ *and* $K \in \mathcal{C}(H)$, *then* $K/O_2(K)$ *is not* $Sz(2^n)$.

PROOF. By the reduction above, we may assume that T normalizes K, and take $H = KL_1 T$ using 14.7.30; then we repeat the argument for that reduction essentially verbatim. $\square$

Then using 14.7.33 and 1.2.1.4:

LEMMA 14.7.34. *If* $H \in \mathcal{H}_z$ *and* $K \in \mathcal{C}(H)$, *then* $m_3(K) = 1$ *or* 2.

By 14.7.28 and 14.7.34:

LEMMA 14.7.35. *For each* $H \in \mathcal{H}_z$, $O_{3'}(H) = Q_H$.

Now we return to the proof of 14.7.32. Recall we had reduced to the case where $K/O_2(K) \cong L_2(2^n)$ and $B_0 = K_0 \cap M$. Then 3 divides the order of K^*, so $L_1 \leq O^{3'}(H) = K_0$ by 1.2.2, and hence $L_1 \leq O^2(M \cap K_0) = B$, so n is even. As L_1 is T-invariant, L_1 is diagonally embedded in KK^t. Also $L_1/O_2(L_1)$ is inverted by a suitable $t_L \in T \cap L$, so either t_L induces a field automorphism on both K^* and K^{*t}, or t_L interchanges K^* and K^{*t}.

By 1.2.4, $K \leq K_1 \in \mathcal{C}(G_1)$; then as $K < K_0$, 1.2.8.2 says that K_1 is not T-invariant and either $K = K_1$, or $n = 2$ and $K_1/O_2(K_1) \cong J_1$ or $L_2(p)$ for $p^2 \equiv 1 \mod 5$. In the latter cases we replace H by $H_1 := \langle K_1, L_1 T \rangle$ and obtain a contradiction from the reductions above. Therefore $K \in \mathcal{C}(G_1)$ and $K^* \cong L_2(2^n)$ with n even. Again by 1.2.2, $K_0 = O^{3'}(G_1)$, so $C := C_{G_1}(K_0/O_2(K_0)) = O_2(G_1)$ by 14.7.35.

Next as $M = LC_M(L/O_2(L))$, $B = L_1 B_C$, where $B_C := O^2(C_B(L/O_2(L)))$ is of index 3 in B. Further $[B_C, t_L] \leq O_2(B_C)$, so that $n = 2$ and t_L does not induce a field automorphism on both K^* and K^{*t}; hence t_L interchanges K^* and K^{*t}.

As $n = 2$, $Out(K_0^*)$ is a 2-group, so as $C = O_2(G_1)$ we conclude

$$G_1 = K_0 T = H, \qquad\qquad (*)$$

and hence $U_H = \langle V^{G_1} \rangle = U$. As before, our convention will be to also abbreviate D_H by D, but we continue to write H for G_1.

As L_1 is T-invariant and diagonally embedded in K_0, no involution in H^* induces an outer automorphism on K^* centralizing K^{t*}. Thus H^* is A_5 wr $\mathbf{Z}_2$ or

A_5 wr $\mathbf{Z}_2$ extended by an involution inducing a field automorphism on both K^* and K^{*t}. In either case no element of H^* induces a transvection on $\tilde{U}$. Therefore $D_\gamma < U_\gamma$ by 14.5.18.1, so we may adopt Notation 14.7.1.

Let $\tilde{I}$ be a maximal H-submodule, and set $W := \tilde{U}/\tilde{I}$, so that W is H-irreducible. Let V_W denote the image of V in W. By 14.7.2.1 applied to L_1 in the role of "Y", V_W is isomorphic to $\tilde{V}$ as an L_1-module. Next as H is irreducible on W, either W is the tensor product $W_1 \otimes W_2$ of irreducibles W_i for $K_1 := K$ and $K_2 := K^t$, or $W = W_1 \oplus W_2$ with $W_i := [W, K_i]$ a K_i-irreducible. But in the latter case there is no BT-invariant line V_W of W with $V_W = [V_W, L_1]$. Thus $W = W_1 \otimes W_2$, and a similar argument shows that each W_i is the $L_2(4)$-module, so that W is the orthogonal module for $K_0^* \cong \Omega_4^+(4)$, and V_W is the T-invariant singular $\mathbf{F}_4$-point. Let $T_0 := T \cap K_0$. By (*), $H^* = K_0^* T^*$, so as K_0^* is faithful on W, so is H^*; then as $U_\gamma^* \in \mathcal{Q}(H^*, \tilde{U})$, $U_\gamma^* \in \mathcal{Q}(H^*, W)$. If a^* is an involution in H^* then either $C_{\tilde{U}}(a^*) = [\tilde{U}, a^*]$ or a^* induces an $\mathbf{F}_4$-transvection. Thus as U_γ^* is quadratic on $\tilde{U}$, $C_{\tilde{U}}(U_\gamma^*) = C_{\tilde{U}}(a^*)$ for each $a^* \in U_\gamma^{*\#}$ which is not an $\mathbf{F}_4$-transvection, and in particular for each $a^* \in K_0^*$. Then calculating in the orthogonal module, we conclude that one of the following holds:

(i) $U_\gamma^* = \langle t^* \rangle$, t^* an $\mathbf{F}_4$-transvection, and $[W, t]$ is a nonsingular $\mathbf{F}_4$-point of W.

(ii) U_γ^* is a 4-group with $[W, U_\gamma] = C_W(U_\gamma)$ of rank 4.

(iii) $U_\gamma^* = \langle t^* \rangle F^*$, where $F^* := C_{T_0^*}(t^*) \cong E_4$, and $[W, U_\gamma] = C_W(U_\gamma)$ is of rank 4.

Suppose case (iii) holds. Then $3 = m(U_\gamma^*) \geq m(U/D)$ by choice of γ in Notation 14.7.1, while by F.9.13.6, $[\tilde{D}, U_\gamma] \leq \tilde{A}_1$ with $m(\tilde{A}_1) = 1$. But then the image D_W of D has corank at most 3 in W, so D_W is not $C_W(U_\gamma)$, and we compute in the orthogonal module W that $[D_W, U_\gamma]$ has rank at least 2. This contradiction eliminates case (iii).

Thus case (i) or (ii) holds. Then as $U_\gamma^* \in \mathcal{Q}(H^*, \tilde{U})$ with $m(W/C_W(U_\gamma)) = 2m(U_\gamma^*)$, we conclude that W is the unique noncentral H-chief factor on $\tilde{U}$, and $W = [\tilde{U}, K_0]$. Further as $L_1 \leq K_0$ with $\tilde{V} = [\tilde{V}, L_1]$, $\tilde{U} = \langle \tilde{V}^H \rangle = W$. By 14.5.18.2, $m(U_\gamma^*) = m(U/D)$, so we have symmetry between γ_1 and γ (cf. Remark 14.7.17), and U_γ^* acts faithfully as a group of $\mathbf{F}_2$-transvections on $\tilde{D}$ with center $\tilde{A}_1$. This eliminates case (ii), for there D has corank 2 in $\tilde{U} = W$, while as U_γ^* contains a free involution, U_γ does not induce a 4-group of $\mathbf{F}_2$-transvections with fixed center on any subspace of corank 2. It also shows $A_1 \leq U$, and hence by symmetry, $V_1 \leq U_\gamma$.

Thus case (i) holds. Recall that under Notation 14.7.1, we choose α and h so that $U_\alpha \leq R_1$, and as in Remark 14.7.17, we also have symmetry between γ_1 and α. Then $U_\alpha^* = \langle t^* \rangle$, where $t^* \in T^*$ induces an $\mathbf{F}_4$-transvection on $\tilde{U} = W$, and $[\tilde{U}, t]$ is a nonsingular $\mathbf{F}_4$-point. We also saw that $m(U/D) = 1$ and that t^* induces an $\mathbf{F}_2$-transvection on the $\mathbf{F}_2$-hyperplane $\tilde{D}$ of $\tilde{U}$ with $[\tilde{D}, t] = \tilde{A}_1^h$.

To complete the proof, we will define subgroups Y, V_Y to which we apply 14.7.31.2, to construct a 2-local I, which we then use to derive a contradiction. We saw that $\tilde{V}$ is the T-invariant singular $\mathbf{F}_4$-point in $\tilde{U}$ containing $\tilde{V}_2$, and $H = G_1$, so $C_G(V_2) = C_H(V_2) \leq N_H(V) \leq M$.

Set $V_Y := V_1 A_1^h \cong E_4$. As H is irreducible on $\tilde{U}$, $[A_1^h, Q_H] = V_1$ by 14.5.21.1, and then by symmetry between γ_1 and α, also $[V_1, Q_\alpha] = A_1^h$. Thus Q_H and $O_2(G_\alpha)$ induce groups of transvections on V_Y with centers V_1 and A_1^h, so by A.1.14,

$Y_0 := \langle Q_H, O_2(G_\alpha) \rangle$ induces $GL(V_Y)$ with kernel $O_2(Y_0) = C_{Q_H}(V_Y)C_{O_2(G_\alpha)}(V_Y)$, and

$$N_G(V_Y) \leq N_G(Y_0). \tag{$**$}$$

Set $T_Y := U_\alpha Q_H C_{T \cap K_0 Q_H}(U_\alpha^*)$ and $Y := \langle T_Y, O_2(G_\alpha) \rangle$. Then T_Y centralizes U_α^* and preserves the $\mathbf{F}_4$-structure on $\tilde{U}$, so T_Y centralizes the $\mathbf{F}_4$-point $[\tilde{U}, U_\alpha^*]$ containing $\tilde{A}_1^h$, and hence acts on V_Y. Then by $(**)$, $Y_0 \trianglelefteq Y = Y_0 T_Y$, and Y acts on V_Y.

As $U_\alpha \leq R_1$, from the structure of H^*:

$$\langle t^{*L_1^* T^*} \rangle = \langle t^{*T^*} \rangle = \langle t^* \rangle C_{O_2(L_1^*)}(t^*) = T_Y^*;$$

that is, $T_Y \trianglelefteq L_1 T$.

Recall that $O_2(Y_0) = C_{Y_0}(V_Y)$, so $O_2(Y_0) \leq O_2(C_H(V_Y))$ by $(**)$, while $O_2(C_{H^*}(\tilde{V}_Y)) = U_\alpha^*$ from the action of H^* on the orthogonal module $\tilde{U}$, so $O_2(Y_0) \leq Q_H U_\alpha \leq T_Y$. Thus $T_Y \in Syl_2(Y)$, and as Y_0 induces $GL(V_Y)$ on V_Y, $C_Y(V_Y) = O_2(Y)$. Further $V_Y \leq U$, so $V_Y \leq U^y$ for each $y \in Y$. This completes the verification of the hypotheses for part (2) of 14.7.31, so we conclude from 14.7.31.2 that $W_0(T_Y, V) \trianglelefteq Y$.

Set $I := \langle L_1 T, Y \rangle$. We saw earlier that $L_1 T$ acts on T_Y, so I acts on $W_0(T_Y, V)$. Set $V_I := \langle V_1^I \rangle$ and $I^+ := I/C_I(V_I)$; as usual $V_I \in \mathcal{R}_2(I)$ by B.2.14. Also $V_Y \leq V_I$ as $Y \leq I$. We claim that L_1^+ is not subnormal in I^+: For otherwise $O_2(L_1)^+ = 1$, so that $O_2(L_1)$ centralizes V_Y. This is impossible, as $O_2(L_1)$ does not act on V_Y since $O_2(L_1^*) \in Syl_2(K_0^*)$ and $\tilde{A}_1^h$ is nonsingular. This completes the proof of the claim. By the claim, $L_1^+ \neq 1$ and also $Y_0 \not\leq N_G(L_1)$.

Now $L_0 := C_{L_1}(V_Y) \leq N_G(Y_0)$ by $(**)$, so $[Y_0, L_0] \leq C_{Y_0}(V_Y) = O_2(Y_0) \leq T_Y \leq N_G(L_0)$. Thus Y_0 acts on $O^2(L_0) =: L_Y$. Also $L_1^* = L_0^* O_2(L_1^*)$ from the action of H on $\tilde{U}$, so $L_1 = L_Y O_2(L_1)$.

Suppose next that $Y_0 \leq M$. Then as Y_0 normalizes L_Y, we conclude from the structure of $Aut(L_3(2))$ that $O^2(Y_0)$, and hence also $O^2(Y_0)T_Y = Y$, acts on L_1, whereas we saw that $Y_0 \not\leq N_G(L_1)$.

Therefore $Y_0 \not\leq M$. We claim next that $[V_I, J(T)] \neq 1$. For otherwise $J(T) \leq C_T(V_I) \leq C_T(V_Y) \leq R_1$ from the action of H^* on $\tilde{U}$. Then $J(T) = J(O_2(Y))$ by B.2.3.3, so that $Y_0 \leq N_G(J(T)) \leq M = !\mathcal{M}(LT)$ using 14.3.9.2, a contradiction establishing the claim.

By the claim, $J(I)^+ \neq 1$. If $J(I)^+$ is solvable, then by Solvable Thompson Factorization B.2.16, $J(I)^+$ has a direct factor $K_I^+ \cong S_3$, and there are at most two such factors by Theorem B.5.6, so that K_I^+ is normalized by $O^2(I^+)$ and L_1^+. If $J(I)^+$ is nonsolvable, then there is $K_I \in \mathcal{C}(J(I))$ with $K_I^+ \neq 1$, so $K_I \in \mathcal{L}_f(G, T)$ by 1.2.10—and then by parts (1) and (2) of 14.3.4, K_I^+ is A_5 or $L_3(2)$, and $K_I \trianglelefteq I$.

We saw that $L_Y = O^2(L_0)$ contains a Sylow 3-subgroup P_L of L_1, and that L_0 acts on Y_0. Since $V_Y = [V_Y, Y_0]$, $P_L \leq P \in Syl_3(Y_0 L_Y)$ with $P \cong E_9$. As $L_1^+ \neq 1$, $P_L^+ \neq 1$, and then as $P_L^+ = C_{P^+}(V_Y)$, $P^+ \cong E_9$. From the previous paragraph, P^+ normalizes K_I^+ and $Out(K_I^+)$ is a 2-group, so $P = P_K \times P_C$, where $P_K := P \cap K_I$ and $P_C := C_P(K_I^+)$. Now P_K has order at most 3 by the structure of K_I^+, and P_C has order at most 3 by A.1.31.1, so we conclude both P_K and P_C are of order 3. As $Y = (P \cap Y_0)T_Y$, $I = \langle L_1 T, Y \rangle = \langle L_1 T, P \rangle$, so as L_1^+ is not normal in I^+, P^+ does not act on L_1^+. Finally one of the following holds:

(a) $L_1^+ \leq K_I^+$.

(b) L_1^+ centralizes K_I^+.

(c) P_L^+ projects faithfully on both P_K^+ and P_C^+.

In case (a), $P_K^+ = P_L^+ \leq L_1^+ \leq K_I^+$, and hence P_C^+ centralizes L_1^+, so that $P^+ \leq L_1^+ P_C^+ \leq N_{I^+}(L_1^+)$, contrary to an earlier observation. In case (b), $P_C^+ = P_L^+ \leq L_1^+$, and P_K^+ centralizes L_1^+, so $P^+ \leq P_K^+ L_1^+ \leq N_{I^+}(L_1^+)$, for the same contradiction. Therefore case (c) holds. We saw that either T normalizes K_I^+ or $\langle K_I^{+T} \rangle \cong S_3 \times S_3$. However the latter case is impossible, as then by A.1.31.1, $P^+ \leq O^{3'}(I^+) = O(\langle K_I^{+T} \rangle)$, contradicting L_1^+ not subnormal in I^+. Thus T acts on K_I, so as T acts on L_1, it acts on the projections L_K^+ and L_C^+ of L_1^+ on K_I^+ and $C_{I^+}(K_I^+)$, respectively. Then as T acts on $L_1 = P_L O_2(L_1)$, and $P_L^+ \leq L_K^+ L_C^+ = O_2(L_K^+) O_2(L_C^+) P^+$, P^+ normalizes $O_2(L_K^+) O_2(L_C^+) P_L^+ = L_1^+$, for the same contradiction yet again. This finally completes the proof of 14.7.32. $\qquad \square$

LEMMA 14.7.36. *If* $K \in \mathcal{C}(H)$, *then* $K/O_{2,Z}(K)$ *is not sporadic.*

PROOF. Assume $K/O_{2,Z}(K)$ is sporadic. By 14.7.32, $KTL_1 \in \mathcal{H}_z$, so without loss $H = KTL_1$. We conclude from 14.7.30 and F.9.18.4 that $K^* \cong M_{22}$ or $\hat{M}_{22}$. As M_{22} and $\hat{M}_{22}$ have no FF-modules by B.4.2, $\tilde{I} := [\tilde{U}_H, K]$ is irreducible under K using F.9.18.7. As $q(H^*, \tilde{U}_H) \leq 2$ by 14.5.18.3, B.4.2 and B.4.5 say that $\tilde{I}$ is either the code module for M_{22} or the 12-dimensional irreducible for $\hat{M}_{22}$. In either case $\tilde{V}$ of rank 2 lies in $\tilde{I}$.

We first eliminate the case $K^* \cong M_{22}$, as in the proof of 13.8.21: First $L_1 \leq O^{3'}(H) = K$ by A.3.18. Since L_1 is solvable and normal in $J := K \cap M$, $J/O_2(K)$ is a maximal parabolic of $N/O_2(K) \cong A_6/E_{2^4}$. Then $C_V(O_2(L_1(T \cap K))) \leq C_V(O_2(NT))$, with $m(C_V(O_2(NT))) = 1$ by H.16.2.1. This is a contradiction, since $L_1 T$ induces $GL(\tilde{V})$ on $\tilde{V}$ of rank 2 in $\tilde{I}$, so that $O_2(L_1 T)$ centralizes $\tilde{V}$.

Thus we may assume that $K^* \cong \hat{M}_{22}$. By 14.7.28, $L_1^* = Z(K^*) \trianglelefteq H^*$, so $H = KT$, and $\tilde{I} = \tilde{U}_H = [\tilde{U}_H, L_1] = [\tilde{U}_H, K]$ by 14.7.5.5. As L_1^* is inverted in T^*, $H^* = K^* T^* \cong Aut(\hat{M}_{22})$. By 14.7.5.3, $Q^* \in Syl_2(K^*)$. By H.12.1.9, $m(C_{\tilde{U}_H}(T^*)) = 1$, so $\tilde{V}_2 = C_{\tilde{U}_H}(T^*)$, and then $\tilde{V} = [\tilde{V}_2, L_1]$. Now $H \cap M = N_H(V)$ by 14.3.3.6, so using H.12.1.7,

$$(H \cap M)^* = N_{H^*}(\tilde{V}) \cong S_5/E_{32}/\mathbf{Z}_3.$$

However from the structure of $Aut(\hat{M}_{22})$, there is an overgroup H_1 of $L_1 T$ in H (arising from the maximal parabolic of $A_6/E_{16}/\mathbf{Z}_3$ which is not contained in $S_5/E_{32}/\mathbf{Z}_3$) with $H_1/O_2(H_1) \cong S_3 \times S_3$ and $H_1^* \not\leq N_{H^*}(\tilde{V}) = (H \cap M)^*$, contrary to Theorem 14.7.29. $\qquad \square$

LEMMA 14.7.37. (1) $\tilde{U}_H > \langle \tilde{V}^{C_{H^*}(\tilde{V}_2)} \rangle$.

(2) $\tilde{U}_H$ *is not the natural module for* $O^2(H^*) \cong L_n(2)$, *with* $3 \leq n \leq 5$.

(3) $\tilde{U}_H$ *is not the natural module for* $H^* \cong S_7$.

PROOF. Set $H_0 := O^2(C_H(V_2))$; by Coprime Action, $H_0^* = O^2(C_{H^*}(\tilde{V}_2))$. Assume that (1) fails; then $U_H = \langle V^{H_0} \rangle$. By 14.7.4.2, H_0 acts on L_2, so $[L_2, H_0] \leq C_{L_2}(V_2) = O_2(L_2)$, and then L_2 acts on $O^2(H_0 O_2(L_2))) = H_0$. So as L_2 also acts on V, it acts on on $\langle V^{H_0} \rangle = U_H$. But then $LT = \langle L_1 T, L_2 \rangle$ acts on U_H, so as $M = !\mathcal{M}(LT)$, $H \leq N_G(U_H)) \leq M$, contrary to $H \in \mathcal{H}_z$. This contradiction establishes (1).

If (2) fails, then $C_{H^*}(\tilde{V}_2)$ is irreducible on U_H/V_2, contrary to (1); so (2) holds.

Assume (3) fails, and adopt the notation of section B.3 to describe $\tilde{U}_H$. Now L_1T induces $L_2(2)$ on $\tilde{V} \cong E_4$, so as we saw in the proof of 14.6.10, either

(i) $L_1^*T^*$ is the stabilizer in H^* of the partition $\Lambda := \{\{1,2\},\{3,4\},\{5,6\},\{7\}\}$, $\tilde{V}_2 = \langle e_{1,2,3,4} \rangle$, and $\tilde{V} = \langle e_{1,2,3,4}, e_{1,2,5,6} \rangle$, or

(ii) $L_1^*T^*$ is the stabilizer of the partition $\theta := \{\{1,2\},\{3,4\},\{5,6,7\}\}$, $\tilde{V}_2 = \langle e_{5,6} \rangle$, and $\tilde{V} = \langle e_{5,6}, e_{5,7} \rangle$.

However in case (i), $m_3(C_H(V_2)) = 2$, contrary to 14.7.4.3, so case (ii) must hold. Here $H_0^* \cong A_5$ stabilizes $\{5,6\}$, and $\tilde{U}_H = \langle \tilde{V}^{H_0} \rangle$, contrary to (1). $\square$

LEMMA 14.7.38. $U_\gamma > D_\gamma$.

PROOF. Assume $U_\gamma = D_\gamma$. By 14.5.18.1, U_H induces a nontrivial group of transvections on U_γ with center V_1. Recall that b is odd by 14.7.3.1, so by edge-transitivity in F.7.3.2, we may pick $g = g_b \in \langle LT, H \rangle$ such that $g : (\gamma_{b-1}, \gamma) \mapsto (\gamma_0, \gamma_1)$. Let $\beta := \gamma_1 g$, so that U_β induces a group of transvections with center $B_1 := V_1^g$ on U_H. By (1) and (2) of F.9.13, $U_\beta \leq O_2(G_{\gamma_0,\gamma_1}) = R_1$. Set $H_1 := \langle U_\beta^H \rangle$.

If H_1^* is solvable then by G.6.4, H_1^* is a product of copies of S_3, so by 14.7.28, $L_1^* = O^2(H_1^*)$ and hence $H_1^* = L_1^* U_\beta^* \cong S_3$, contradicting $U_\beta \leq R_1$. Therefore H_1^* is not solvable. Thus by 1.2.1.1, $K^* = [K^*, U_\beta^*]$ for some $K \in \mathcal{C}(H)$. Let $U_K := [U_H, K]$. As U_β^* induces transvections on $\tilde{U}_H$, G.6.4 says $\tilde{U}_K/C_{\tilde{U}_K}(K)$ is a natural module for $K^* U_\beta^*/C_{K^* U_\beta^*}(\tilde{U}_K) \cong S_n$ or $L_n(2)$.

Suppose first that $K^* \cong A_5$ or $L_3(2)$, and let L_K^* be the projection of L_1^* in K^* with respect to the decomposition $K^* \times C_{H^*}(K^*)$. As L_1 is T-invariant, L_K^* is T^*-invariant; so either $L_K^* \cong A_4$, or $L_K^* = 1$ so that $[L_1^*, K^*] = 1$. In case $K^* \cong A_5$, as U_β^* induces a transposition on K^* and $U_\beta \leq R_1$, $L_K^* = 1$, so $[L_1^*, K^*] = 1$. In case $K^* \cong L_3(2)$, as L_1 is T-invariant and $L_K^* = [L_K^*, T^* \cap K^*]$, either $L_1^* = L_K^* \leq K^*$ or $[L_1^*, K^*] = 1$. However if $[L_1^*, K^*] = 1$, then $[\tilde{U}_K, L_1] = 1$ since $End_{K^*}(\tilde{U}_K) \cong \mathbf{F}_2$, so $\tilde{V} = [\tilde{V}, L_1] \leq C_{\tilde{U}_H}(K)$, and then $\tilde{U}_H = \langle \tilde{V}^H \rangle \leq C_{\tilde{U}_H}(K)$, contradicting $K^* \neq 1$.

Therefore $L_1^* \leq K^* \cong L_3(2)$. Further $\tilde{V} = [\tilde{V}, L_1] \leq \tilde{U}_K$, so that $\tilde{U}_K = \tilde{U}_H$. Then as $End_{K^*}(\tilde{U}_H) \cong \mathbf{F}_2$, $C_{H^*}(K^*) = 1$ as H^* is faithful on $\tilde{U}_H$, so that $H^* = K^*T^*$. Then as the natural module $\tilde{U}_H$ is T-invariant, we conclude that $H^* \cong L_3(2)$, contrary to 14.7.11.

Therefore $K^* U_\beta^* \cong S_6$, S_7, S_8, $L_4(2)$, or $L_5(2)$. In particular by A.3.18, $K = O^{3'}(H)$, so $L_1 \leq K$, and then as above, $U_K = U_H$ and $H^* = K^* U_{\beta^*}$. By 14.7.11, H^* is not S_6, and by 14.7.37, H^* is not $L_n(2)$ or S_7.

Thus it remains to eliminate the case $H^* \cong S_8$. Here $\tilde{V}$ projects on a singular line in the orthogonal space $\tilde{U}_H/C_{\tilde{U}_H}(H)$, so $\tilde{V}_2$ projects on a singular point; hence

$$C_{H^*}(\tilde{V}_2)/O_2(C_{H^*}(\tilde{V}_2)) \cong S_3 \ \text{wr} \ \mathbf{Z}_2,$$

contrary to 14.7.4.3. $\square$

In view of 14.7.38, we establish the following convention:

In the remainder of the section, we adopt Notation 14.7.1.

REMARK 14.7.39. Whenever we can show that $m(U_\gamma^*) = m(U_H/D_H)$, our hypotheses are symmetric in γ and γ_1; see Remarks 14.7.17 and F.9.17 for a more extended discussion of this point.

THEOREM 14.7.40. *Assume* $H \in \mathcal{H}_z$ *such that* $H = KL_1T$ *for some* $K \in$
$\mathcal{C}(H)$ *with* $K/O_{2,Z}(K)$ *of Lie type over* $\mathbf{F}_{2^n}$ *for some* $n > 1$. *Then* $H^* \cong S_5$ *and*
$\tilde{U}_H/C_{\tilde{U}_H}(K)$ *is the* $L_2(4)$-*module.*

Until the proof of Theorem 14.7.40 is complete, assume the hypotheses of the
Theorem. By 14.7.30, we may apply F.9.18.4 to conclude that

(*) K^* is a Bender group, $(S)L_3(2^n)$, $Sp_4(2^n)$, or $G_2(2^n)$.

Let B_0^* be the Borel subgroup of K^* containing $T_0^* := T^* \cap K^*$ and let $B :=$
$O^2(B_0)$. As K is defined over $\mathbf{F}_{2^n}$ with $n > 1$, and $L_1T = TL_1$, L_1 acts on B,
so by Theorem 14.7.29, $B \leq H \cap M \leq N_H(L_1)$. Then as $M = LC_M(L/O_2(L))$,
$BL_1 = B_C L_1$, where $B_C := O^2(C_{BL_1}(L/O_2(L)))$. Also $L_1/O_2(L_1)$ is inverted by
some $t \in T \cap L$, and $[t, B_C] \leq O_2(L) \cap B_C \leq O_2(B_C)$, so $B_C O_2(L_1B)/O_2(L_1B)$
is the unique t-invariant complement to $L_1O_2(L_1B)/O_2(L_1B)$ in $L_1B/O_2(L_1B)$.
Choose $X_1 \in Syl_3(L_1)$ with X_1 inverted by t.

LEMMA 14.7.41. *Either*

(1) $L_1 \not\leq K$, $m_3(K) = 1$, $B_C = B$, $L_1 \trianglelefteq H$, *and* L_1^* *is inverted in* $C_{H^*}(K^*)$,

or

(2) $L_1^* \leq K^* \cong L_2(4)$, $U_3(8)$, *or* $(S)L_3(4)$.

PROOF. Suppose first that $L_1 \not\leq K$. Then $B^*/O_2(B^*)$ is a t-invariant complement to X_1^* in $X_1^* B^*/O_2(B^*)$, so as $B_C^* O_2(B^*)/O_2(B^*)$ is the unique such complement, $B_C = B$. Thus $X_1\langle t \rangle$ centralizes $B^*/O_2(B^*)$, so from the structure of
$Aut(K^*)$ for K^* on the list in (*), either X_1 induces inner automorphisms on K^*,
or $K^* X_1^* \cong PGL_3(4)$. As $L_1 \not\leq K$, K^* is not $GL_3(4)$ by 14.7.28. As $q(H^*, \tilde{U}_H) \leq 2$
by Notation 14.7.1, Theorems B.4.2 and B.4.5 eliminate the case $K^* X_1^* \cong PGL_3(4)$.
Thus $L_1^* \leq K^* C_{H^*}(K^*/O_2(K^*)) =: Y^*$, and as $L_1 \not\leq K$, $\theta(Y) \not\leq K$, where Y is
the preimage of Y^* in H. Therefore $m_3(K) < 2$ by A.3.18, so that $m_3(K) = 1$ by
14.7.34. Then as t centralizes $B^*/O_2(B^*)$, t also induces an inner automorphism
on K^*, from the structure of $Aut(K^*)$ for K^* in (*) of 3-rank 1. Indeed the projection of t on K^* then lies in $O_2(B^*) \leq R_1^*$, so we conclude $L_1 = [L_1, t_C]$ for some
$t_C \in C_T(K^*)$, and hence L_1 centralizes K^*. Therefore $L_1^* \trianglelefteq H^*$ as $H = KL_1T$,
so H normalizes $O^2(L_1Q_H) = L_1$, and hence (1) holds.
So assume instead that $L_1 \leq K$. As T acts on L_1, $L_1 \leq B$ and $B/O_2(B) =$
$L_1O_2(B)/O_2(B) \times B_C O_2(B)/O_2(B)$. Then as t inverts $L_1/O_2(L_1)$ but $[t, B_C] \leq$
$O_2(B_C)$ with B_C of index 3 in B, we conclude (2) holds from the structure of
$Aut(K^*)$ for K^* on the list of (*). $\qquad \square$

LEMMA 14.7.42. *If* $L_1 \not\leq K$ *then* $H^* \cong S_5 \times S_3$, $\tilde{U}_H = [\tilde{U}_H, K] \oplus C_{\tilde{U}_H}(K)$, *and*
$[\tilde{U}_H, K]$ *is the tensor product of the* S_3-*module and the* S_5-*module.*

PROOF. Assume $L_1 \not\leq K$. Then by 14.7.41, $L_1 \trianglelefteq H$, L_1^* is inverted in $C_{H^*}(K^*)$,
$m_3(K) = 1$, and $B = B_C$. Also $B \leq H \cap M = N_H(V)$ by 14.3.3.6, so B centralizes
V as $End_{L/O_2(L)}(V) \cong \mathbf{F}_2$.

As L_1^* is inverted in $C_{H^*}(K^*)$, each H-chief factor W on $\tilde{U}_H$ is the sum $W =$
$W_1 \oplus W_2$ of a pair of isomorphic K^*-modules W_i. Indeed since $U_\alpha^* \in \mathcal{Q}(H^*, \tilde{U}_H)$
by Notation 14.7.1, arguing as in the proof of F.9.18.6, U_α^* is an FF*-offender on
W_1 and W_2, so that K^* is $L_2(2^n)$, $SL_3(2^n)$, $Sp_4(2^n)$, or $G_2(2^n)$ by Theorem B.4.2.
As $m_3(K) = 1$, the last two cases are eliminated, and n is odd if $K^* \cong SL_3(2^n)$.

Pick $\tilde{I}$ to be an H-submodule of $\tilde{U}_H$ maximal subject to $[\tilde{U}_H, K] \not\le \tilde{I}$, and let $\hat{U}_H := U_H/I$; then we may take $W = [\hat{U}_H, K]$. As $L_1 \trianglelefteq H$, Q^* is Sylow in $C_{H^*}(L_1^*)$ by 14.7.5.3, so as Q centralizes V, and $\hat{U}_H = \langle \hat{V}^{C_{H^*}(L_1^*)} \rangle$, $\hat{U}_H = [\hat{U}_H, K] = W$ by Gaschütz's Theorem A.1.39. Now by B.4.2, W is either the sum of two natural modules for K^*, or the sum of two A_5-modules for $K^* \cong L_2(4)$. In the first case, as B centralizes V, $\hat{V} \le C_W(B^*) = 1$, contradicting $W = \langle \hat{V}^H \rangle$.

Thus the second case holds. As K^* has no strong FF-modules by B.4.2, $\tilde{I} = C_{\tilde{U}_H}(K)$ by 14.7.30 and F.9.18.6. Then $\tilde{U}_H = [\tilde{U}_H, K] \oplus C_{\tilde{U}_H}(K)$ as the A_5-module is K-projective, so the lemma holds. $\qquad\square$

LEMMA 14.7.43. K^* is not $U_3(8)$.

PROOF. Assume otherwise. By 14.7.42, $L_1 \le K$. By Theorems B.5.1 and B.4.2, H^* has no FF-modules, so by 14.7.30 we may apply parts (7) and (4) of F.9.18 to conclude that $\tilde{U}_H \in Irr_+(K, \tilde{U}_H)$. As $q(H^*, \tilde{U}_H) \le 2$ by Notation 14.7.1, we conclude from B.4.2 and B.4.5 that $\tilde{U}_H$ is the natural module for H^*. But then there is no B-invariant 2-subspace over $\mathbf{F}_2$ satisfying $\tilde{V} = [\tilde{V}, L_1]$. $\qquad\square$

LEMMA 14.7.44. K^* is not $(S)L_3(4)$.

PROOF. Assume otherwise. Again $L_1 \le K$ by 14.7.42.

Suppose first that $K^* \cong SL_3(4)$. By 14.7.28, $L_1^* = Z(K^*)$. Recall L_1^* is inverted in $C_{T \cap L}(B_C^*/O_2(B_C^*))$; thus from the structure of $Aut(SL_3(4))$, there is $t^* \in T^*$ inducing a graph automorphism on K^*. Choose I and I_H as in F.9.18.4; because t induces a graph automorphism, H^* has no FF-modules by Theorem B.5.1, so $U_H = I_H$ by F.9.18.7, and case (iii) of F.9.18.4 holds. Then as the 1-cohomology of the natural module is zero by I.1.6.4, $\tilde{U}_H = \tilde{I} \oplus \tilde{I}^t$, where $\tilde{I}$ is a natural module for K^* and $\tilde{I}^t$ is its dual. Further as $U_\alpha^* \in \mathcal{Q}(H^*, \tilde{U}_H)$, either U_α^* is a root group of K^* of rank 2 with $m(\tilde{U}_H/C_{\tilde{U}_H}(U_\alpha)) = 4$, or $m(U_\alpha^*) \ge 3$ and $m(\tilde{U}_H/C_{\tilde{U}_H}(U_\alpha)) = 6$. If $m(U_\alpha^*) = 2$ or 3, we get a contradiction from 14.5.18.2, since U_α^* does not induce $\mathbf{F}_2$-transvections on a subspace of $\tilde{U}_H$ of codimension $m(U_\alpha^*)$. If $m(U_\alpha^*) = 4$ at least $m(U_H/D_H) \le 4$ by Notation 14.7.1, whereas no subspace of $\tilde{U}_H$ of corank at most 4 satisfies the requirement $[U_\alpha^*, \tilde{D}_H] = \tilde{A}_1^h$ of F.9.13.6.

Thus $K^* \cong L_3(4)$, and hence H^* has no module $\tilde{U}_H$ with $q(H^*, \tilde{U}_H) \le 2$ by Theorems B.4.2 and B.4.5. This contradiction completes the proof. $\qquad\square$

LEMMA 14.7.45. (1) $K^* \cong A_5$.
(2) Either

    (a) $K \in \mathcal{C}(G_1)$, or
    (b) $L_1 \le K$ and $K \le K_1 \in \mathcal{C}(G_1)$ with $K_1/O_2(K_1) \cong A_7$.

PROOF. Conclusion (1) holds if $L_1 \not\le K$ by 14.7.42. If $L_1 \le K$, it holds since 14.7.43 and 14.7.44 eliminate the other possibilities in 14.7.41.2. Thus (1) is established.

Next as $K \in \mathcal{L}(G_1, T)$, $K \le K_1 \in \mathcal{C}(G_1)$ by 1.2.4, so $H_1 := K_1 L_1 T \in \mathcal{H}_z$ by 14.7.32. By 14.7.30, $K_1/O_2(K_1)$ is quasisimple. Applying 14.7.36 to G_1 in the role of "H", $K_1/O_{2,Z}(K_1)$ is not sporadic. Applying F.9.18.4 to H_1, either $K_1/O_{2,Z}(K_1)$ is of Lie type in characteristic 2 or $K_1/O_2(K_1) \cong A_7$. If $K = K_1$ then (2a) holds, so we may assume $K < K_1$. Then from the list of possible proper overgroups of A_5 in A.3.14 with $K_1/O_2(K_1)$ quasisimple, either $K_1/O_{2,Z}(K_1)$ is of Lie type over $\mathbf{F}_4$ of Lie rank 2, or $K_1/O_2(K_1) \cong A_7$. In the first case since $K_1/O_2(K_1)$ is defined over $\mathbf{F}_4$, we may apply (1) to H_1 to obtain a contradiction. In the second case $K_1 = O^{3'}(G_1)$ by A.3.18, so $L_1 \leq K_1$. Then as $K = O^2(N_{K_1}(K))$, $L_1 \leq K$, and (2b) holds. $\square$

LEMMA 14.7.46. $L_1 \leq K$.

PROOF. Assume $L_1 \not\leq K$; then H and its action on U_H are described in 14.7.42, and $K \in \mathcal{C}(G_1)$ by 14.7.45.2. Let B be the Borel subgroup of K containing $T \cap K$. Then $BT = C_K(\tilde{V}_2)$ from the module structure in 14.7.42, so B normalizes I_2 by 14.7.4.2. Further $L_1 \trianglelefteq H$ since case (1) of 14.7.41 holds, so B also centralizes $\tilde{V} = \langle \tilde{V}_2^{L_1} \rangle$. By 14.7.4.2, $I_2/O_2(I_2) \cong S_3$. Set $G_0 := \langle I_2, K, T \rangle$.

Suppose first that $O_2(G_0) = 1$. Then Hypothesis F.1.1 is satisfied with K, I_2, T in the roles of "L_1, L_2, S", so $\beta := (KT, BT, I_2BT)$ is a weak BN-pair of rank 2 by F.1.9. Further $T \trianglelefteq TN_{I_2}(T \cap I_2)$, so β is described in F.1.12. Then as KT centralizes V_1 with $KT/O_2(KT) \cong S_5$, and $I_2T/O_2(I_2T) \cong S_3$, it follows that β is parabolic isomorphic to the $Aut(J_2)$-amalgam. This is impossible, since in that amalgam, $O_2(KT) \cong Q_8 D_8$ while $U_H \leq O_2(KT)$ is of 2-rank 9 by 14.7.42.

Thus $G_0 \in \mathcal{H}(T)$, so $K \leq K_0 \in \mathcal{C}(G_0)$ by 1.2.4. If $K = K_0$, then $L_2 = O^2(I_2)$ acts on K by 1.2.1.3, so $LT = \langle L_1 T, L_2 \rangle$ acts on K; then as $M = !\mathcal{M}(LT)$, $K \leq N_G(K) \leq M$, contrary to 14.7.30. Thus $K < K_0$, so since $L_1 \not\leq K$, $K_0 \not\leq G_1$ by 14.7.45.2. Then $K_0 \in \mathcal{L}_f(G, T)$, so that $K_0/O_2(K_0) \cong A_5$ or $L_3(2)$ by 14.3.4.1, contrary to A.3.14. $\square$

We are now in a position to complete the proof of Theorem 14.7.40.

By 14.7.46, $L_1 \leq K$, so $H = KL_1T = KT$. Further $L_1T/O_2(L_1T) \cong S_3$. Therefore $H^* \cong S_5$ by 14.7.45.1.

As $L_1 \leq K$, $V = [V, L_1] \leq [U_H, K]$, so $U_H = [U_H, K]$. Suppose $\tilde{U}_H \in Irr_+(K, \tilde{U}_H)$. As $\tilde{V}$ is an L_1T-invariant line in $\tilde{U}_H$, $\tilde{U}_H$ is not the A_5-module. Then $\tilde{U}_H/C_{\tilde{U}_H}(K)$ is the $L_2(4)$-module, and hence Theorem 14.7.40 holds in this case.

Thus we may assume $\tilde{U}_H \notin Irr_+(K, \tilde{U}_H)$, and it remains to derive a contradiction. By Notation 14.7.1, $U_\alpha^* \leq R_1^*$ with R_1^* Sylow in K^*. Further $m(U_\alpha^*) =: k = 1$ or 2, $U_\alpha^* \in \mathcal{Q}(H^*, \tilde{U}_H)$, and $k \geq m(U_H/D_H)$ by choice of γ in 14.7.1. As $U_\alpha^* \leq R_1^* \leq K^*$, $m(W/C_W(U_\alpha^*)) \geq 2$ for each noncentral chief factor W for K on $\tilde{U}_H$, and as $\tilde{U}_H \notin Irr_+(K, \tilde{U}_H)$, there are at least two such chief factors. On the other hand, as $U_\alpha^* \in \mathcal{Q}(H^*, \tilde{U}_H)$, $2k \geq m(\tilde{U}_H/C_{\tilde{U}_H}(U_\alpha))$, so we conclude $k = 2$, and there are exactly two noncentral chief factors, both $L_2(4)$-modules. Further $2m(U_\gamma^*) = m(\tilde{U}/C_{\tilde{U}_H}(U_\alpha))$ so by 14.5.18.2, $m(U_H/D_H) = 2$, and U_γ^* acts as a group of transvections on $\tilde{D}_H$ with center $\tilde{A}_1$. This is impossible as $\tilde{U}_H$ has two $L_2(4)$-chief factors.

Thus Theorem 14.7.40 is at last established.

LEMMA 14.7.47. Let $K \in \mathcal{C}(H)$. Then

(1) $L_1 \leq K$, and

(2) $K/O_2(K) \cong L_n(2)$ or A_n for suitable n, or $G_2(2)'$.

PROOF. As $KTL_1 \in \mathcal{H}_z$ by 14.7.32, we may take $H = KTL_1$. By 14.7.36, $K/O_{2,Z}(K)$ is not sporadic, so by 14.7.30 we may apply F.9.18.4 to conclude that either $K/O_{2,Z}(K)$ is of Lie type in characteristic 2, or $K/O_2(K) \cong A_7$. Assume the first case holds. If $K/O_{2,Z}(K)$ is not defined over $\mathbf{F}_2$, then $H^* \cong S_5$ by Theorem

14.7.40, so the lemma holds. On the other hand, if $K/O_{2,Z}(K)$ is defined over $\mathbf{F}_2$, then from F.9.18.4, either (2) holds, or $K/O_2(K) \cong \hat{A}_6$, and T is trivial on the Dynkin diagram of $K/O_{2,Z}(K)$ from the possible modules listed in that result. However in the latter case, $L_1^* = Z(K^*)$ by 14.7.28, so as T is trivial on the Dynkin diagram of $K/O_{2,Z}(K)$, KT is generated by solvable overgroups of L_1T, which lie in M by Theorem 14.7.29, contrary to $H \not\leq M$. Thus (2) is established, and it remains to establish (1) when $K/O_2(K)$ is not A_5.

If $m_3(K) > 1$, then $K = O^{3'}(H)$ by A.3.18, so (1) holds. Thus we may assume $m_3(K) = 1$, so as K^* is not A_5, $K^* \cong L_3(2)$ by (2). Assume $L_1 \not\leq K$. Then L_1^* centralizes K^* as $Out(L_3(2))$ is of order 2 and $L_1 = [L_1, T]$. Now if $H^*/C_{H^*}(K^*) \not\cong Aut(L_3(2))$, then H is generated by a pair of solvable subgroups containing L_1T which lie in M by Theorem 14.7.29, contrary to $H \not\leq M$. Therefore $H^*/C_{H^*}(K^*) \cong Aut(L_3(2))$, so K^*T^* has no FF-modules by Theorem B.4.2. Therefore by parts (7) and (4) of F.9.18, either $\tilde{U}_H \in Irr_+(K, \tilde{U}_H)$ or $\tilde{U}_H = \tilde{I} + \tilde{I}^t$ with $\tilde{I}$ a natural K^*-module and t inducing an outer automorphism of K^*. In either case, $C_{GL(\tilde{U}_H)}(K^*) = 1$, impossible as L_1^* centralizes K^*. $\square$

LEMMA 14.7.48. *(1) There is a unique $K \in \mathcal{C}(H)$, and $H = KT$.*
(2) $U_H = [U_H, K]$.

PROOF. By Theorem 14.7.29, H is not solvable, so there exists $K \in \mathcal{C}(H)$. By 14.7.47.1, L_1 is contained in each $K \in \mathcal{C}(H)$, so K is unique. Then $C_{H^*}(K^*)$ is solvable by 1.2.1.1, and hence $C_{H^*}(K^*) = 1$ by 14.7.28, since $L_1 \leq K$ but $L_1^* \not\leq Z(K^*)$ by 14.7.47.2. So (1) holds as $Out(K^*)$ is a 2-group in each case listed in 14.7.47.2.

As $L_1 \leq K$, $V = [V, L_1] \leq [U_H, K]$, so $U_H = \langle V^H \rangle = [U_H, K]$, and (2) holds. $\square$

LEMMA 14.7.49. *K^* is not $L_3(2)$ or A_6.*

PROOF. Assume otherwise. First $H = KT$ by 14.7.48.1. By 14.7.11, H^* is not $L_3(2)$, A_6, or S_6. Thus T is nontrivial on the Dynkin diagram of K^*, a contradiction as $H = KT$ and T acts on L_1. $\square$

LEMMA 14.7.50. *K^* is not A_7.*

PROOF. Let $\tilde{I}$ be a maximal submodule of $\tilde{U}_H$, and $\hat{U}_H := \tilde{U}_H/\tilde{I}$. As $U_H = [U_H, K]$ by 14.7.48.2, $\hat{U}_H$ is a nontrivial irreducible for K. As $U_\alpha^* \in \mathcal{Q}(H^*, \tilde{U}_H)$ by Notation 14.7.1, $\hat{U}_H$ is of rank 4 or 6 by Theorems B.4.2 and B.4.5.

We first eliminate the case $\dim(\hat{U}_H) = 4$. Notice $H^* \cong A_7$ since $\hat{U}_H$ is not invariant under S_7. By 14.7.2.1, $\hat{V}$ is isomorphic to $\tilde{V}$ as an L_1T-module, so from the action of H^* on $\hat{U}_H$, $N_{H^*}(\hat{V})$ is the stabilizer $H_{4,3}^*$ in $H^* \cong A_7$ of a partition of type 4,3. Set $H_M := H \cap M$; by 14.3.3.6, $H_M = N_H(V)$. As $H_{4,3}^*$ is solvable and maximal in H^*, we conclude from Theorem 14.7.29 that $H_M^* = H_{4,3}^*$. Since $M = LC_M(L/O_2(L))$, an element $t \in T \cap L$ inverts $L_1/O_2(L_1)$ and centralizes $O^2(C_{H_M}(L/O_2(L)))$ modulo $O_2(M)$. This is a contradiction as $H^* \cong A_7$ rather than S_7, so elements of $H_{4,3}^* - O^2(H_{4,3}^*)$ invert $O^2(H_{4,3}^*)/O^2(O_2(H_{4,3}^*))$.

Thus $\dim(\hat{U}_H) = 6$, and if $H_M^* = H_{4,3}^*$ then $H^* \cong S_7$. As usual, we use the notation of section B.3 for the module $\hat{U}_H$. Since H_M normalizes L_1, H_M^* is a solvable overgroup of $L_1^*T^*$ in S_7, rather than one of the overgroups of T^*

containing a subgroup isomorphic to A_6 or $L_3(2)$. Thus either $H_M^* = H_{4,3}^*$, or H_M^* is the stabilizer $H_{2^3,1}^*$ of a partition of type $2^3, 1$.

Assume first that $H_M^* = H_{4,3}^*$, so that $H^* \cong S_7$. As $\hat{V} = [\hat{V}, L_1]$ is a T-invariant line, $L_1^* \cong \mathbf{Z}_3$ fixes 4 points, and $\hat{V}_2 = \langle e_{5,6} \rangle$. Set $Y := O^2(H_{2^3,1})$; then $\langle \hat{V}_2^Y \rangle$ is of rank 3, contrary to 14.7.2.2.

Finally assume that $H_M^* = H_{2^3,1}^*$. This time as $\hat{V}$ is a line, $\hat{V}_2 = \langle e_{1,2,3,4} \rangle$, so that $[\hat{V}_2, O^2(H_{4,3})] = 1$, and then $[V_2, O^2(H_{4,3})] = 1$ by 14.7.2.3. But then $m_3(C_G(V_2)) > 1$, contrary to 14.7.4.3. $\qquad \square$

LEMMA 14.7.51. K^* is not $L_n(2)$.

PROOF. Assume otherwise. In view of 14.7.49 and Theorem C (A.2.3), $n = 4$ or 5. Observe $P^* := L_1^*(T^* \cap K^*)$ is a T-invariant minimal parabolic of K^*.

Assume first that T^* is nontrivial on the Dynkin diagram of K^*. Then $n = 4$, P^* is the middle-node parabolic, and $H^* \cong S_8$. Define $\tilde{I}$ and $\hat{U}_H$ as in the proof of 14.7.50. Again using Theorems B.4.2 and B.4.5, we conclude that $m(U_H) = 4$ or 6, and since P^* acts on the T-invariant line $\hat{V}$, that $\hat{U}_H$ is the 6-dimensional orthogonal module for H^*, and $\hat{V}$ is a totally singular line. Thus $m_3(C_{H^*}(\hat{V}_2)) = 2$, so $m_3(C_H(V_2)) = 2$ by 14.7.2.3, again contrary to 14.7.4.3.

Thus T is trivial on the Dynkin diagram of K^*, so $K^* = H^*$. Thus H^* is generated by rank-2 parabolics H_1^* containing P^* which satisfy $H_1/O_2(H_1) \cong L_3(2)$ or $S_3 \times S_3$. Therefore $H_1 \le M$ by 14.7.49 or Theorem 14.7.29, contrary to $H \not\le M$. $\qquad \square$

THEOREM 14.7.52. (1) $H = KT = G_1$ is the unique member of $\mathcal{H}_z$, and $U_H = U$.

(2) $K^* \cong A_5$ or $G_2(2)'$.

PROOF. Part (2) follows since 14.7.49–14.7.51 eliminate all other possibilities from 14.7.47.2. As $K \in \mathcal{L}(G_1, T)$, $K \le K_1 \in \mathcal{C}(G_1, T)$ by 1.2.4. But $G_1 \in \mathcal{H}_z$, so since $K_1/O_2(K_1)$ is quasisimple by 14.7.30, (2) shows there is no proper containment $K < K_1$ in A.3.12, and hence $K = K_1 \in \mathcal{C}(G_1)$. Then by 14.7.48.1 applied to both H and G_1, $H = KT = G_1$, so (1) holds. $\qquad \square$

14.7.4. Eliminating the case $O^2(H^*)$ isomorphic to $G_2(2)'$. In the remainder of this section, set $M_1 := H \cap M$. Thus $M_1 = C_M(z)$ as $H = G_1$ by Theorem 14.7.52. Further $M_1 = N_H(V)$ by 14.3.3.6. Abbreviate U_H by U. Since in this subsection we use α in preference to γ, we will reserve the abbreviation D not for $D_H = U \cap Q_\gamma$ but instead for $U \cap Q_\alpha$.

In this subsection we show $K^* \cong A_5$ by proving:

THEOREM 14.7.53. $K/O_2(K)$ is not $G_2(2)'$.

Until the proof of Theorem 14.7.53 is complete, assume H is a counterexample. Recall we are operating under Notation 14.7.1, so we choose γ as in 14.5.18.4 and α as in 14.5.18.5, and in particular $U_\alpha^* \in \mathcal{Q}(H^*, \tilde{U})$.

LEMMA 14.7.54. $\tilde{U}$ is either the 7-dimensional indecomposable Weyl module for $K^* \cong G_2(2)'$, or its 6-dimensional irreducible quotient.

PROOF. By 14.7.48.2, $U = [U, K]$. By Theorems B.4.2 and B.4.5, the 6-dimensional module for K^* is the unique irreducible $\mathbf{F}_2 H^*$-module W satisfying

$q(H^*, W) \le 2$, and that module is not a strong FF-module. By B.4.6.1, the Weyl module is the unique indecomposable extension of that irreducible by a module centralized by K^*. By 14.7.30, we may apply parts (6) and (4) of F.9.18, so if the lemma does not hold there are exactly two noncentral chief factors W_1 and W_2 for H^* on U, and each is of dimension 6. Indeed as in the proof of F.9.18.6, $m(\tilde{U}/C_{\tilde{U}}(U_\gamma) = 2m(U_\gamma^*) = 6$ and U_γ^* is an FF^*-offender on both W_1 and W_2. Then 14.5.18.2 supplies a contradiction, as U_γ^* does not act as a group of transvections on any subspace of corank 3 in $\tilde{U}$. $\qquad\square$

In view of 14.7.54, we now appeal to B.4.6 and [Asc87] for the structure of $\tilde{U}$, and we use the terminology in [Asc87], such as "doubly singular line". As $\tilde{V} = [\tilde{V}, L_1]$ is T-invariant, we have:

LEMMA 14.7.55. *(1) $\tilde{V}$ is a doubly singular line of $\tilde{U}$.*

(2) $\tilde{V}_2$ is a singular point of $\tilde{U}$.

(3) The set $\mathcal{V}(V_1, V_2)$ of doubly singular lines in $\tilde{U}$ through $\tilde{V}_2$ is of order 3, and generates a subspace $\tilde{U}(V_1, V_2)$ of rank 3.

(4) $C_{K^*}(\tilde{U}(V_1, V_2)) =: B^* = B^*(V_1, V_2) \cong E_4$, $[\tilde{U}, b] \in \mathcal{V}(V_1, V_2)$ for each $1 \ne b^* \in B^*$, and

$$\tilde{W} := \tilde{W}(V_1, V_2) := \langle C_{\tilde{U}}(b^*) : 1 \ne b^* \in B^* \rangle = \tilde{V}_2^\perp$$

is a hyperplane of $\tilde{U}$. If $H^* \cong G_2(2)$, then $C_{H^*}(\tilde{U}(V_1, V_2)) =: A^* = A^*(V_1, V_2) \cong E_8$, and $\tilde{U}(V_1, V_2)C_{\tilde{U}}(H) = C_{\tilde{U}}(a^*) = C_{\tilde{U}}(B^*) = [\tilde{U}, A^*]$ for each $a^* \in A^* - B^*$.

(5) If $\tilde{U}$ is an FF-module for H^* then $H^* \cong G_2(2)$ and $A^*(V_1, V_2)^H$ is the set of FF^*-offenders in H^*.

(6) Let $Y := O^2(C_H(V_2))$. Then $YT/O_2(YT) \cong S_3$, $\tilde{U}(V_1, V_2) = [\tilde{U}(V_1, V_2), Y]$, and Y is transitive on $\mathcal{V}(V_1, V_2)$.

(7) The geometry $\mathcal{G}(\tilde{U})$ of singular points and doubly singular lines in $\tilde{U}$ is the generalized hexagon for $G_2(2)$. In particular, there is no cycle of length 4 in the collinearity graph of $\mathcal{G}(\tilde{U})$.

(8) $\{[\tilde{U}, b^*] : b^* \in B^*\} = \{[\tilde{w}, A^*] : \tilde{w} \in \tilde{W}\}$.

In the remainder of this subsection, we adopt the notation in 14.7.55.

LEMMA 14.7.56. *Let $\mathcal{V}(V_2)$ be the set of preimages in U of members of $\mathcal{V}(V_1, V_2)$, and $U(V_2)$ the preimage of $\tilde{U}(V_1, V_2)$. Then*

(1) $\mathcal{V}(V_2) = V^Y$ is the set of G-conjugates of V containing V_2.

(2) Y centralizes $L_2/O_2(L_2)$, and $G_2 = L_2YT$ acts on $U(V_2)$, with L_2 fixing $\mathcal{V}(V_2)$ pointwise and $G_2/C_G(U(V_2))$ the stabilizer in $GL(U(V_2))$ of V_2.

PROOF. By parts (1) and (6) of 14.7.55, $V^Y = \mathcal{V}(V_2)$. Then $U(V_2) = \langle V^Y \rangle$, so as $[V, L_2] = V_2$, while $[L_2, Y] \le C_{L_2}(V_2) = O_2(L_2)$ by 14.7.4.2, we have $[U(V_2), L_2] = V_2$, and hence L_2 fixes V^Y pointwise. Further as $H = G_1$, $C_G(V_2) = C_H(V_2) = YC_T(V_2)$, so since L_2T induces $GL(V_2)$ on V_2, $G_2 = L_2YT$. Then $V^{G_2} = V^Y$, so as L is transitive on the hyperplanes of V, (1) follows from A.1.7.1. Finally $P_0 := Aut_{G_2}(U(V_2)) \le N_{GL(U(V_2))}(V_2) =: P$ with $P = P_0O_2(P)$ and $1 \ne O_2(Aut_Y(U(V_2))) \le O_2(P_0)$, so $P = P_0$ as P is irreducible on $O_2(P)$. This completes the proof of (2). $\qquad\square$

LEMMA 14.7.57. *(1) $M_1 = N_H(V) = L_1T$ and $L_1^*T^*$ is the minimal parabolic of H^* over T^* other than Y^*T^*.*

(2) $QQ_H = R_1$. Thus $Q^* = R_1^* = O_2(M_1^*)$ is the unipotent radical of M_1^*.

(3) If $H^* \cong G_2(2)$ then M_1 is transitive on the three conjugates in R_1^* of $A^* := A^*(V_1, V_2)$ in 14.7.55.4.

PROOF. Recall $M_1 = N_H(V)$, so M_1^* is the minimal parabolic $N_{H^*}(\tilde{V}) = L_1^* T^*$ of H^*, and hence (1) holds. Next $R_1 = QQ_H$ by 14.7.4, so (2) follows from (1). Finally (3) follows from (1) and B.4.6.13. $\square$

Recall from Notation 14.7.1 that $h \in H$ with $\gamma_0 = \gamma_2 h$, $\alpha := \gamma h$, and $U_\alpha \leq R_1$. Set $Z_\alpha := A_1^h$, and let U_0 denote the preimage in U of $C_{\tilde{U}}(H)$. Let $D := U \cap Q_\alpha$.

LEMMA 14.7.58. *Assume $Z_\alpha \leq V$. Then*

(1) There exists $g \in G$ interchanging γ_1 and α.

(2) $V_1 \leq U_\alpha$ and $m(U_\alpha^*) = m(U/D)$.

PROOF. Part (1) follows as L is 2-transitive on $V^\#$. Then (1) implies (2). $\square$

Set $H_\alpha := C_H(Z_\alpha)$ and $U_- := U(V_1, V_2)U_0$. As $H = G_1$ by Theorem 14.7.52, H_α acts on U_α and hence on U_α^*, so that:

LEMMA 14.7.59. $O_2(H_\alpha^*) \neq 1$.

LEMMA 14.7.60. *Assume $Z_\alpha \leq U$. Then*

(1) Replacing α by a suitable conjugate under M_1, we may assume $Z_\alpha U_0 = V_2 U_0$.

(2) $H_\alpha^* = C_{H^*}(\tilde{V_2})$.

(3) Either

(a) $U_\alpha^* = A^* := A^*(V_1, V_2)$, $D \leq U_-$, and $H^* \cong G_2(2)$), or

(b) $U_\alpha^* = B^* := B^*(V_1, V_2)$, and either $D \leq U_-$ or $\tilde{D}\tilde{U}_- = \tilde{V}_2^\perp$.

PROOF. As $Z_\alpha \leq U$, H_α is a subgroup of index at most 2 in $C_H(\tilde{Z}_\alpha)$, so that $O_2(C_{H^*}(\tilde{Z}_\alpha)) \neq 1$ by 14.7.59. Therefore as $O_2(H^*) = 1$, $Z_\alpha \not\leq U_0$. It follows that there is $a \in H$ with $Z_\alpha U_0 = V_2^a U_0$. Indeed by 14.5.21.2, $[Q_H, Z_\alpha] = V_1$, so $H_\alpha^* = C_{H^*}(\tilde{Z}_\alpha)$ is the parabolic subgroup of H^* centralizing $\tilde{Z}_\alpha$. Thus $U_\alpha^* \trianglelefteq H_\alpha^*$ with $\Phi(U_\alpha^*) = 1$, so it follows from the structure of the parabolic H_α^* that U_α^* is one of the two subgroups $B^*(V_1, V_2^a)$ or $A^*(V_1, V_2^a)$ described in 14.7.55. In either case, 14.7.55.4 says that $C_{\tilde{U}}(U_\alpha^*) = \tilde{U}(V_1, V_2^a)\tilde{U}_0 =: \tilde{U}_-^a$. But $U_\alpha^* \leq R_1^* = C_{H^*}(\tilde{V})$ using 14.7.57, so the doubly singular line $\tilde{V}$ is contained in $\tilde{U}_-^a$. Therefore $V_2^a \leq V$ by 14.7.56.1 and the fact that the generalized hexagon $\mathcal{G}(\tilde{U})$ contains no cycle of length 3. Then as $M_1 = N_H(V)$ is transitive on $\tilde{V}^\#$, conjugating in M_1, we may take $V_2^a = V_2$, and maintain the constraint $U_\alpha \leq R_1$. Hence (1) holds. We saw $[Q_H, Z_\alpha] = V_1$, so (2) holds. Further H_α acts on $U \cap Q_\alpha = D$, so from the action of H_α^* on $\tilde{U}$, $\tilde{D}\tilde{U}_-$ is U_-, $\tilde{V}_2^\perp$, or $\tilde{U}$. As $[\tilde{D}, U_\alpha] \leq \tilde{Z}_\alpha$ by F.9.13.6, the third case is impossible as H^* induces no transvections on the module $\tilde{U}_H$ in 14.7.54. In the second $U_\alpha^* = B^*(V_1, V_2)$ by 14.7.55.4. Thus (3) is established. $\square$

LEMMA 14.7.61. *(1) $H^* \cong G_2(2)$, and replacing α by a suitable M_1-conjugate, we may assume $U_\alpha^* = A^* := A^*(V_1, V_2)$.*

(2) $[\tilde{U}, U_\alpha] = \widetilde{U \cap U_\alpha} = C_{\tilde{U}}(A^*) = \tilde{U}_-$.

(3) $D = U_-$.

(4) $m(U_\alpha^*) = 3 = m(U/D)$.

(5) We have symmetry between γ_1 and α, as discussed in Remark 14.7.39.

PROOF. Suppose first that $\tilde{U}_\alpha^*$ is not an FF*-offender on $\tilde{U}_H$. As $m(U_\alpha^*) \geq m(U/D)$, U_α does not centralize D, so that $Z_\alpha = [D, U_\alpha] \leq U$ using F.9.13.6. Therefore by 14.7.60.1, we may take $Z_\alpha U_0 = V_2 U_0$, and by 14.7.60.3, U_α^* is $B^* := B^*(V_1, V_2)$ or $A^* := A^*(V_1, V_2)$. As U_α^* is not an FF*-offender on $\tilde{U}$, $U_\alpha^* = B^*$, so $[U_\alpha, \tilde{U}] = \tilde{U}(V_1, V_2)$ by 14.7.55.4. Also by 14.7.60.3, either $\tilde{D} \leq \tilde{U}_-$ or $\tilde{D}\tilde{U}_- = \tilde{V}_2^\perp$. The first case is impossible, as $m(U/D) \leq m(U_\alpha^*) = 2$, whereas $m(\tilde{U}/\tilde{U}_-) = 3$. Thus $\tilde{D}\tilde{U}_- = \tilde{V}_2^\perp$, and hence $\tilde{Z}_\alpha = [\tilde{D}, U_\alpha] = [\tilde{V}_2^\perp, B^*] = \tilde{V}_2$, so that $Z_\alpha \leq V_2 \leq V$. Therefore $V_1 \leq U_\alpha$ and $m(U_\alpha^*) = 2 = m(U/D)$ by 14.7.58.2. Then as $\tilde{D}$ lies in the hyperplane $\tilde{D}\tilde{U}_- = \tilde{V}_2^\perp$ of $\tilde{U}$, while $m(\tilde{U}/\tilde{U}_-) = 3$, we obtain $m(\tilde{D}\tilde{U}_-/\tilde{D}) = 1$, and in particular $\tilde{U}_- \not\leq \tilde{D}$. Since $U_- = [U, U_\alpha]U_0$ and $[U, U_\alpha] \leq U \cap Q_\alpha = D$, we conclude there is $u_0 \in U_0 - D$. But this is impossible, as then $[U_\alpha, u_0] \leq V_1$, whereas no nontrivial element of G_α/Q_α induces a transvection on U_α/Z_α.

Thus U_α^* is an FF*-offender on $\tilde{U}$, so $H^* \cong G_2(2)$ by 14.7.55.5. As $U_\alpha^* \leq R_1^*$ by Notation 14.7.1, (1) follows from 14.7.57.3. Then (2) follows from 14.7.55.4.

Suppose $D \not\leq U_-$. Then as $\widetilde{C_U(U_\alpha)} = C_{\tilde{U}}(A^*) = \tilde{U}_-$, $1 \neq [D, U_\alpha]$, so as in the previous paragraph, $Z_\alpha \leq U$ by F.9.13.6. However this contradicts 14.7.60.3a since $U_\alpha^* = A^*$. Therefore $D \leq U_-$, so as $3 = m(U_\alpha^*) \geq m(U/D) \geq m(U/U_-) = 3$, we conclude (3) and (4) hold. Then (4) implies (5), completing the proof. $\qquad\square$

Set $H_\alpha^+ := H_\alpha/Q_\alpha$ and let W denote the preimage of $\tilde{W}(V_1, V_2) = \tilde{V}_2^\perp$ in U.

LEMMA 14.7.62. *(1)* $V_1 \not\leq U_\alpha$ *and* $Z_\alpha \not\leq U$.
(2) $U^+ = A^*(Z_\alpha, V_{2,\alpha})$ *for a suitable conjugate* $V_{2,\alpha}$ *of* V_2 *in* U_α *containing* Z_α.
(3) $W^+ = B^*(Z_\alpha, V_{2,\alpha})$.

PROOF. Recall that $U \leq G_\gamma \leq C_G(A_1)$, so that $A_1 \leq C_{U_\gamma}(U) \leq Q_H$ and hence also $Z_\alpha \leq Q_H$.

Suppose first that $V_1 \leq U_\alpha$. Then by 14.7.61.2, $U \cap U_\alpha = U_-$ is of codimension 3 in U_α, so as $m(U_\alpha^*) = 3$, $Q_H \cap U_\alpha = U \cap U_\alpha$. Thus $Z_\alpha \leq Q_H \cap U_\alpha \leq U$. Then $C_{Q_H}(Z_\alpha)$ is of index 2 in Q_H by 14.5.21.1, with $[C_{Q_H}(Z_\alpha), U_\alpha] \leq Q_H \cap U_\alpha \leq U$, so U_α^* centralizes a hyperplane of $Q_H/C_H(U)$. But this is impossible since by 14.5.21.1, $Q_H/C_H(U)$ is H^*-dual to $\tilde{U}$, and no member of H^* acts as a transvection on $\tilde{U}$.

Therefore $V_1 \not\leq U_\alpha$. Then by the symmetry in 14.7.61.5, $Z_\alpha \not\leq U$, so (1) holds.

By 14.7.61.1, U_α^* is an FF*-offender on $\tilde{U}$, so by symmetry U/D is also an FF*-offender on U_α/Z_α. In particular (2) holds.

As $V_1 \not\leq U_\alpha$, $C_{\tilde{U}}(a) = \widetilde{C_U(a)}$ for each $a \in U_\alpha$, so as each $w \in W$ is centralized by some $1 \neq b^* \in B^*$ by 14.7.55.4, $m(U_\alpha/C_{U_\alpha}(w)) \leq 2$. Thus W^+ is a hyperplane of U^+ such that $m(U_\alpha/C_{U_\alpha}(w^+)) \leq 2$ for each $w \in W$, so (3) follows from 14.7.55.4. $\qquad\square$

We now enter the last stages of our proof of Theorem 14.7.53.

From 14.7.62, in the symmetry between γ_1 and α appearing in 14.7.61.5, the tuple H, U, V_1, V_2, W, U_α^*, B^*, γ_1 corresponds to the tuple G_α, U_α, Z_α, $V_{2,\alpha}$, B, U^+, W^+, α, where B is the preimage in U_α of B^*.

Now since $V_1 \not\leq U_\alpha$, using 14.7.55.4 we see that

$$\mathcal{F} := \{[U, b] : 1 \neq b^* \in B^*\}$$

consists of three 4-subgroups, with

(a) $\mathcal{V}(V_2) = \{FV_1 : F \in \mathcal{F}\}$.

Pick $F_0 \in \mathcal{F}$. As V_2 and F_0 are distinct hyperplanes of $F_0 V_1 \cong E_8$, $V_2 \cap F_0 = V_1^l$ for a suitable $l \in L_2 T$ interchanging V_1 and V_1^l. Set $\beta := \gamma_0 l$. Then $Z_\beta \leq F_0 \leq [U, U_\alpha] \leq U_\alpha$, and as $V_1 \nleq U_\alpha$:

(b) For each $F \in \mathcal{F}$, $Z_\beta = V_2 \cap U_\alpha = V_2 \cap F$ is a complement to V_1 in V_2.

Set $\hat{U}_\beta := U_\beta / Z_\beta$ and consider the generalized hexagon $\mathcal{G}(\hat{U}_\beta)$. Since $\tilde{V}_2 = \tilde{Z}_\beta$ is a singular point of $\tilde{U}$, conjugating by l it follows that $\hat{V}_1$ is a singular point of $\hat{U}_\beta$.

Next $\tilde{F}$ is the set of lines in $\mathcal{G}(\tilde{U})$ through $\tilde{V}_2 = \tilde{Z}_\beta$, while by 14.7.56.2, L_2 fixes $\mathcal{V}(V_2) = \{FV_1 : F \in \mathcal{F}\}$ pointwise. Therefore conjugating by l, we conclude:

(c) $\{\hat{F}\hat{V}_1 : F \in \mathcal{F}\}$ is the set of lines through $\hat{V}_1$ in $\mathcal{G}(\hat{U}_\beta)$.

Now by 14.7.55.8,

$$\tilde{\mathcal{F}} = \{[\tilde{w}, U_\alpha] : 1 \neq \tilde{w} \in \tilde{W}\}.$$

Therefore as $[U_\alpha, w] \leq U_\alpha$ and $F = U_\alpha \cap FV_1$ for each $F \in \mathcal{F}$,

(d) $\{[U_\alpha, w] : w \in W\} = \mathcal{F} = \{[U, b] : b \in B\}$.

Applying symmetry to (a), and using (d) to conclude that $\mathcal{F}$ is invariant when interchanging γ_1 and α, it follows that

(a') $\mathcal{V}(Z_\alpha, V_{2,\alpha}) = \{FZ_\alpha : F \in \mathcal{F}\}$,

and then from (a') and (c) that:

(c') $\{\hat{F}\hat{Z}_\alpha : F \in \mathcal{F}\}$ is the set of lines through $\hat{Z}_\alpha$ in $\mathcal{G}(\hat{U}_\beta)$.

But now choosing F_1 and F_2 to be distinct members of $\mathcal{F}$, it follows from (c) and (c') that $\hat{Z}_\alpha$, $\hat{F}_1$, $\hat{V}_1$, $\hat{F}_2, \hat{Z}_\alpha$ is a 4-cycle in the collinearity graph of $\mathcal{G}(\hat{U}_\beta)$, contrary to 14.7.55.7.

This contradiction completes the proof of Theorem 14.7.53.

14.7.5. Identifying Ru when $\mathbf{O^2(H^*) = A_5}$. We summarize the major reductions achieved so far in this section:

THEOREM 14.7.63. *$H = C_G(z)$ is the unique member of $\mathcal{H}_z$, $H = KT$ where $K := O^2(H) \in \mathcal{C}(H)$, $H/O_2(H) \cong S_5$, $\tilde{U}$ is an indecomposable K-module, and $\tilde{U}/C_{\tilde{U}}(K)$ is the $L_2(4)$-module for $K/O_2(K)$.*

PROOF. By Theorem 14.7.52.1, $C_G(z) = H$ is the unique member H of $\mathcal{H}_z$. By 14.7.48.1, $H = KT$ for some $K \in \mathcal{C}(H)$; thus $K = O^2(H)$. By Theorem 14.7.52.2 and Theorem 14.7.53, $K/O_2(K)$ is A_5. Then Theorem 14.7.40 says $H^* \cong S_5$ and U/U_0 is the $L_2(4)$-module. Thus $\tilde{U}$ is indecomposable as $U_H = [U_H, K]$ by 14.7.48.2. $\square$

REMARK 14.7.64. We will be working with the following special case of I.1.6.1: Let $\check{U}$ be the largest $\mathbf{F}_2 H^*$-module such that $\check{U} = [\check{U}, H^*]$ and $\check{U}/C_{\check{U}}(H^*) \cong N := \check{U}/C_{\check{U}}(K^*)$. (cf. 17.12 in [**Asc86a**]) As N is the natural module for $K^* \cong L_2(4)$, $\check{U}$ has the structure of an $\mathbf{F}_4 K^*$-module, and as $\dim_{\mathbf{F}_4}(H^1(K^*, N)) = 1$, $\dim_{\mathbf{F}_4}(\check{U}) = 3$. Set $\check{U}_0 := C_{\check{U}}(K^*)$. There exists a 4-dimensional orthogonal space $\check{U}_1$ over $\mathbf{F}_4$ with $H^* \leq \Gamma O(\check{U}_1)$ such that $\check{U}_0$ is a nonsingular point of $\check{U}_1$ and $\check{U} = \check{U}_0^\perp$. This facilitates later calculations in the image $\tilde{U}$ of $\check{U}$.

Observe that by Theorem 14.7.63 and 14.3.3.6, $M_1 = H \cap M = N_H(V) = L_1 T$, and $R_1^* = O_2(L_1^*) \in Syl_2(K^*)$. Let U_0 be the preimage in U of $C_{\tilde{U}}(K)$. As $\tilde{V} = [\tilde{V}, L_1] \cong E_4$ and $\tilde{U}$ is a quotient of the module $\check{U}$ in Remark 14.7.64:

LEMMA 14.7.65. $\tilde{V}\tilde{U}_0 = C_{\tilde{U}}(R_1)$ and $\tilde{V} = [C_{\tilde{U}}(R_1), L_1]$.

In Notation 14.7.1 we chose $h \in H$ with $\gamma_0 = \gamma_2 h$, $\alpha := \gamma h$, and $U_\alpha \leq R_1$. Let $Z_\alpha := A_1^h$.

LEMMA 14.7.66. $U_\alpha^* \leq Q^* \in Syl_2(K^*)$.

PROOF. By 14.7.4.4, $Q^* = R_1^*$, so $Q^* \in Syl_2(K^*)$, and the lemma follows as $U_\alpha \leq R_1$. $\qquad\square$

LEMMA 14.7.67. (1) $G_2 \leq M$.

(2) If F is a hyperplane of V, then V is the unique member of V^G containing F.

(3) $K \in \mathcal{L}^*(G, T)$.

(4) $N_G(K) = H \in \mathcal{M}$.

(5) $LT = N_G(V)$.

PROOF. First as $H = C_G(z)$, $C_G(V_2) = C_H(V_2) \leq T$ from the action of H on U, so (1) holds since $L_2 T$ induces $GL(V_2)$ on V_2. Then as L is transitive on hyperplanes of V, (1) and A.1.7.1 imply (2). Similarly $Aut_{LT}(V) = GL(V)$, so $N_G(V) = LTC_G(V)$ with $C_G(V) = C_H(V) \leq C_H(V_2) \leq T$, so (5) holds.

Suppose $K < I \in \mathcal{L}(G, T)$. As $K = O^2(C_G(z))$, $[z, I] \neq 1$, so $I \in \mathcal{L}_f(G, T)$, and hence $I/O_2(I)$ is A_5 or $L_3(2)$ by 14.3.4.1. But then A.3.14 supplies a contradiction, establishing (3).

Let $M_K := N_G(K)$; by (3) and 1.2.7.3, $M_K = !\mathcal{M}(H)$. It remains only to prove (4), so we may assume $H < M_K$, and we must derive a contradiction. Let $D := C_{M_K}(K/O_2(K))$; then $M_K = KDT$ so $O^2(D) \neq 1$.

Set $D_1 := O^2(D \cap M)$. Then KT normalizes $O^2(D_1 O_2(K)) = D_1$, and D_1 normalizes $O^2(L_1 O_2(K)) = L_1$. Thus D_1 centralizes $L_1/O_2(L_1)$, and $D \cap L_1 \leq O_2(L_1)$ as $L_1 \leq K$, so as D_1 is T-invariant and $L_1 = [L_1, T \cap L]$, we conclude that D_1 centralizes $L/O_2(L)$. Thus LT normalizes $O^2(D_1 O_2(L)) = D_1$, so if $D_1 \neq 1$ then $K \leq N_G(D_1) \leq M = !\mathcal{M}(LT)$, a contradiction.

Therefore $D_1 = 1$, so that $D \cap M \leq T$. Also $D \cap H \leq C_H(K/O_2(K)) \leq T$ as $H = KT$. As $[D, L_1] \leq O_2(L_1)$, $D \cap T \leq R_1$, and hence $R_1 \in Syl_2(DR_1)$. Let $S_1 := \mathrm{Baum}(R_1)$. Now L_1 has two noncentral chief factors on $\tilde{U}$, and hence also two on $Q_H/C_H(U)$ by the duality in 14.5.21.1. Thus L_1 has at least four noncentral 2-chief factors, so $N_G(S_1) \leq M$ by 14.7.10.

Let $E := \langle V_1^D \rangle$; then $E \in \mathcal{R}_2(DR_1)$ by B.2.14, since $D \in \mathcal{H}^e$ by 1.1.3.1. Further $C_D(E) \leq D \cap H \leq T$, so $C_{DR_1}(E) = O_2(DR_1) = C_{R_1}(E)$. Thus if $J(R_1)$ centralizes E, then $S_1 = \mathrm{Baum}(O_2(DR_1))$ by B.2.3.5, and then $1 \neq O^2(D) \leq N_G(S_1) \leq M$, contrary to $D \cap M \leq T$. Therefore $J(R_1)$ does not centralize E, so by Thompson Factorization B.2.15, E is an FF-module for $(DR_1)^+ := DR_1/O_2(DR_1)$.

Suppose there exists $K_D \in \mathcal{C}(J(DR_1))$. Then as $[E, K_D] \neq 1$, $K_D \in \mathcal{L}_f(G, T)$ by 1.2.10, so we conclude from 14.3.4.1 that $K_D/O_2(K_D) \cong L_3(2)$ or A_5, $K_D \trianglelefteq M_K$, and for each $V_K \in Irr_+(K_D, E, T)$, V_K is the $L_3(2)$-module or A_5-module and is T-invariant. As $K_D = [K_D, J(R_1)]$, we conclude using Theorem B.5.1 and B.2.14 that $E = [E, K_D] \oplus C_E(K_D R_1)$, and $[E, K_D]$ is the A_5-module or the sum of at most two isomorphic $L_3(2)$-modules. Thus $O^2(C_{K_D}(V_1)) \neq 1$, impossible as $O^2(C_{K_D}(V_1)) \leq D \cap H \leq T$.

Thus $J := J(DR_1)$ is solvable by 1.2.1.1. As D centralizes $K/O_2(K)$ and $m_3(M_K) \leq 2$, $m_3(J) = 1$ and hence $J/O_2(J) \cong S_3$ by Solvable Thompson Factorization B.2.16. Let $W_0 := W_0(R_1, V)$. By (1) and 14.7.31.1, $N_G(W_0) \leq M$.

Next suppose $g \in G$ with $V_1^g \le E$. As K centralizes V_1 and D normalizes K, K centralizes $\langle V_1^D \rangle = E$, so $K \le O^2(C_G(V_1^g)) = K^g$, and hence $g \in N_G(K) = M_K$. Thus as $U \le O_2(K)$, $U^g \le O_2(K) \le O_2(JR_1)$. Also using an earlier remark, $C_{JR_1}(E) = JR_1 \cap C_{DR_1}(E) = C_{R_1}(E) = O_2(JR_1)$. Therefore we may apply 14.7.31.2 with JR_1, E in the roles of "Y, V_Y", to conclude that $W_0 \trianglelefteq JR_1$. But then $O^2(J) \le N_D(W_0) \le D \cap M \le T$, a contradiction which completes the proof of (4), and hence of 14.7.67. □

LEMMA 14.7.68. *(1)* $z^G \cap U_0 = \{z\}$.

(2) If $u \in U_0^\#$ with $[\tilde{u}, T] = 1$, then $C_G(u) \le H$, and U_0 is the unique member of U_0^G containing u.

PROOF. Assume u satisfies the hypotheses of (2) and set $G_u := C_G(u)$. Notice $T_u := C_T(u)$ is of index at most 2 in T and $K \le G_u$ by Coprime Action.

Suppose first that $G_u \le H$ holds; we will show that (1) and the remaining statement in (2) follow. Assume that u lies in some conjugate U_0^g. Then $K^g \le O^2(G_u) \le O^2(H) = K$, so that $K^g = K$. Thus $g \in N_G(K) = H$ by 14.7.67.4, so in particular g normalizes U_0, completing the proof of (2) in this case. Further as z satisfies these hypotheses in the role of "u", z is in a unique G-conjugate of U_0, so $z^G \cap U_0 = z^{N_G(U_0)}$ by A.1.7.1. But then as $H \in \mathcal{M}$ by 14.7.67.4, $N_G(U_0) = H = C_G(z)$ so that (1) also holds.

So to complete the proof of the lemma, we assume $G_u \not\le H$, and it remains to derive a contradiction. As K has more than one noncentral 2-chief factor by 14.5.21.1, KT_u is not a block, so by C.1.26 there is $1 \ne C$ char T_u with $C \trianglelefteq KT_u$. But then as T_u is of index at most 2 in T, $H = KT \le N_G(C)$ so that $N_G(C) = H$ since $H \in \mathcal{M}$. Thus if $T_u \le T_0 \in Syl_2(G_u)$, then $N_{T_0}(T_u) \le N_G(C) = H$, so that $T_u = N_{T_0}(T_u)$ and hence $T_u = T_0$. Therefore $K \le L_u \in \mathcal{C}(G_u)$ by 1.2.4, and $L_u \trianglelefteq G_u$ by 1.2.1.3 since T normalizes K. Thus $K < L_u$ as $G_u \not\le H = N_G(K)$. In particular $L_u \not\le H$ since $K \trianglelefteq H$, so as $H = C_G(z)$, $[z, L_u] \ne 1$. Observe further as U_α is elementary abelian and contained in R_1 with $R_1^* = Q^* \in Syl_2(K^*)$ that $L_u/O_2(L_u)$ does not involve $SL_2(5)$ on a group of odd order, and so is quasisimple by 1.2.1.4.

Observe that the hypotheses of 1.1.6 are satisfied with G_u, H in the roles of "H, M", so that we may apply 1.1.5. Suppose first that L_u is quasisimple, and hence a component of G_u. As $u \in [U, K] \le L_u \le G_u$, $Z(L_u)$ is of even order. On the other hand z is in the center of the Sylow 2-subgroup T_u of G_u, and $KT_u = C_{G_u}(z)$. Inspecting the list of possiblities for L_u in 1.1.5.3, we conclude from this structure of KT_u (in particular from the two noncentral 2-chief factors) that L_u is the covering group of Ru. Next V is the unique L_1-invariant complement to $\langle u \rangle$ in $\langle u \rangle V$, so as $L_u/\langle u \rangle \cong Ru$, $N_{L_u}(V) =: L_0$ satisfies $L_0/O_2(L_0) \cong L_3(2)$. Thus $L_0 \le O^2(N_G(V)) = L$ by 14.7.67.5, so as $|T : T_u| = 2$ and $L = O^2(L)$, we conclude $L = L_0$. Then as $Z(L_0)$ is of order 2 by I.1.3, $\langle u \rangle = Z(L) \cap T_u$, so $\langle u \rangle$ is T-invariant, contrary to $T_u \in Syl_2(G_u)$.

Therefore L_u is not quasisimple, so $F^*(L_u) = O_2(L_u)$ by 1.2.11. Let $R_u := O_2(KT_u)$. As $KT_u \trianglelefteq H$, $R_u = O_2(H) \cap KT_u \trianglelefteq H$, so since $H \in \mathcal{M}$, $C(G_u, R_u) \le H_u := H \cap G_u$ and $R_u = O_2(H_u)$. Thus Hypothesis C.2.3 is satisfied with G_u, R_u, H_u in the roles of "H, R, M_H". Then as $L_u \trianglelefteq G_u$, while $L_u \not\le H$ and $L_u/O_2(L_u)$ is quasisimple, L_u is described in C.2.7.3; and comparing the list in C.2.7.3 to the embeddings in A.3.14, we conclude that either L_u is a block with $L_u/O_2(L_u) \cong A_7$

or $Sp_4(4)$, or else $L_u/O_2(L_u) \cong SL_3(4)$. The first case is impossible as K has two noncentral 2-chief factors. In the remaining two cases, there is Y of order 3 in $C_{L_u}(K/O_2(K))$, so $Y \leq N_G(K) = H$, a contradiction as $C_H(K/O_2(K)) = Q_H$ by Theorem 14.7.63. $\square$

LEMMA 14.7.69. U_α^* is of order 2.

PROOF. Assume otherwise. Then as $U_\alpha^* \leq Q^* \cong E_4$ by 14.7.66, $U_\alpha^* = Q^*$. Therefore using Remark 14.7.64,

$$C_{\tilde{U}}(U_\alpha) = [\tilde{U}, U_\alpha] = \tilde{U}_0 \tilde{V}, \qquad (a)$$

so as

$$[U, U_\alpha] \leq U \cap U_\alpha =: F \leq C_U(U_\alpha), \qquad (b)$$

we conclude

$$[U, U_\alpha]V_1 = (U \cap U_\alpha)V_1 = FV_1 = C_U(U_\alpha) = U_0 V. \qquad (c)$$

From the action of H^* on $\tilde{U}$, for $u \in U - U_0 V$ we have $m([u, U_\alpha]) \geq 2$, so $[u, U_\alpha] \not\leq Z_\alpha$. Thus we conclude from 14.7.4.1 and (c) that

$$U \cap Q_\alpha = U_0 V = C_U(U_\alpha). \qquad (d)$$

Then $m(U_\alpha^*) = 2 = m(U/U_0 V) = m(U/U \cap Q_\alpha)$, so that we have symmetry between γ_1 and α as discussed in Remark 14.7.39. As $U \leq G_\alpha = C_G(Z_\alpha)$ and $C_G(U) \leq Q_H$:

$$Z_\alpha \leq Q_H \cap U_\alpha. \qquad (e)$$

Suppose first that $V_1 \leq U_\alpha$. Then $V_1 \leq U \cap U_\alpha = F$, so $F = U_0 V$ by (c). Hence $m(U_\alpha^*) = 2 = m(U/F) = m(U_\alpha/F)$, so

$$Q_H \cap U_\alpha = F \leq U.$$

Then using (e), $Z_\alpha \leq U$. It now follows from 14.7.4.1 that $m(Q_H/C_{Q_H}(Z_\alpha)) \leq 1$. But $C_{Q_H}(Z_\alpha) \leq N_G(U_\alpha)$ since $H = C_G(z)$, so $[C_{Q_H}(Z_\alpha), U_\alpha] \leq Q_H \cap U_\alpha \leq U$. This is impossible, since by 14.5.21.1, $Q_H/C_H(U)$ is dual to $U/C_U(Q_H)$ as an H-module, so U_α centralizes no hyperplane of $Q_H/C_H(U)$.

Therefore $V_1 \not\leq U_\alpha$. Hence $V_1 \not\leq F$, so we can now refine (b)–(d) to:

$$[U, U_\alpha] = U \cap U_\alpha = F \quad \text{and} \quad F \times V_1 = U_0 V = C_U(U_\alpha) = U \cap Q_\alpha. \qquad (f)$$

Suppose that $U_0 = V_1$. Then by (f), F is a hyperplane of $V = C_U(U_\alpha)$, and by symmetry between γ_1 and α, F is a hyperplane of $C_{U_\alpha}(U)$ and $C_{U_\alpha}(U) \in V^G$. Hence by 14.7.67.2, $C_U(U_\alpha) = V = C_{U_\alpha}(U)$, so that $V_1 \leq U_\alpha$, contrary to our assumption.

Therefore $U_0 > V_1$. By I.1.6.2, $m(\tilde{U}_0) \leq 2$, so that $m(U_0) = 2$ or 3.

Suppose first that $m(U_0) = 3$. Then $\tilde{U}$ is the module $\check{U}$ discussed in Remark 14.7.64. In particular the 2-dimensional $\mathbf{F}_4$-subspace $\tilde{F} = \widetilde{C_U(U_\alpha)}$ is partitioned [4] by $\tilde{V}$, $\tilde{U}_0$, and the three 1-dimensional $\mathbf{F}_4$-spaces spanned by the various $[\tilde{u}, s^*]$ for $s^* \in U_\alpha^\#$ and $\tilde{u} \in \tilde{U} - \tilde{U}_0 \tilde{V}$. So as $C_U(U_\alpha) = F \times V_1$ by (f), F has the partition

$$F = F_0 \cup F_1 \cup F_V,$$

where $F_V := F \cap V$, $F_0 := F \cap U_0$, and

$$F_1 := \{[x, y] : x \in U_\alpha - Q_H, y \in U - U_0 V\}.$$

---

[4]Following Suzuki, a *partition of a vector space* is a collection of subspaces such that each nonzero element is contained in a unique subspace.

Now F_1 is invariant under the symmetry interchanging γ_1 and α, so by this symmetry there is a similar partition of F given by

$$F = (F \cap V^g) \cup (F \cap U_0^g) \cup F_1,$$

for $g \in \langle LT, H \rangle$ with $V_1^g = Z_\alpha$. By 14.7.68.1, $z^G \cap U_0^g = \{z^g\}$, so as $z^g \notin F$ and $F_V^\# \subseteq z^G$, $F_V = F \cap V^g \leq V^g$. Then as F_V is a hyperplane of V, $V = V^g$ by 14.7.67.2, contrary to our earlier reduction $V_1 \not\leq U_\alpha$.

Therefore $m(U_0) = 2$. This time $\tilde{F}$ is partitioned by $\tilde{U}_0$ and $\tilde{F}_1$, so F has the partition $F = F_0 \cup F_1$, and again using the symmetry between γ_1 and α as above, we conclude that $F = (F \cap U_0^g) \cup F_1$ is also a partition, and then that $F_0 \leq U_0^g$. Further $\tilde{F}_0 = \tilde{U}_0 \leq Z(\tilde{H})$, so $U_0 = U_0^g$ by 14.7.68.2, and hence $g \in N_G(U_0) = H$ as $H \in \mathcal{M}$ by 14.7.67.4, contrary to $V_1^g = Z_\alpha \neq V_1$. This contradiction completes the proof of 14.7.69. $\square$

By choice of γ in Notation 14.7.1, $m(U_\alpha^*) \geq m(U/D) > 0$, where $D := U \cap Q_\alpha$; so as $m(U_\alpha^*) = 1$ by 14.7.69, also $m(U/D) = 1$. Thus again we have symmetry between α and γ_1, as discussed in Remark 14.7.39.

LEMMA 14.7.70. *(1) We may choose α so that $Z_\alpha \leq V_2$.*
(2) $m(U_0) \leq 2$.
(3) $U \cap U_\alpha = U_0 V = [U, U_\alpha]V = [U, U_\alpha]U_0$.
(4) $b = 3$ and $U_\alpha \in U^L$.

PROOF. Observe that if (1) holds, then so does (4) by 14.7.3.4. Thus it suffices to establish (1)–(3).

Let $F := [U, U_\alpha]$. By 14.7.66 and 14.7.69, U_α^* is a subgroup of $Q^* \in Syl_2(K^*)$ of order 2. Then using Remark 14.7.64, $FU_0 = VU_0$, $\tilde{V}\tilde{U}_0 = \tilde{F} \times \tilde{U}_0$, and U_α centralizes no $\mathbf{F}_2$-hyperplane of $\tilde{U}$; so $1 \neq [D, U_\alpha]$, and hence $Z_\alpha = [D, U_\alpha] \leq F \leq U$ using F.9.13.6. By the symmetry between γ_1 and α discussed above, also $V_1 = [D_\alpha, U] \leq F$. By 14.7.68.1, $Z_\alpha \not\leq U_0$.

By Remark 14.7.64, $m(\tilde{U}_0) \leq 2$, so that $m(U_0) \leq 3$. We now make some choices: We may conjugate in $N_H(R_1) = L_1 T$ and preserve the condition $U_\alpha \leq R_1$. As U_α^* is of order 2 in Q^*, conjugating in L_1, we may assume that $U_\alpha^* \leq Z(T^*)$; when $m(U_0) = 3$, we make this choice. When $m(U_0) \leq 2$, we make a more careful choice: As $Z_\alpha \leq F \leq U_0 V$, conjugating in L_1 we may assume that $Z_\alpha U_0 = V_2 U_0$. As $m(U_0) \leq 2$, T centralizes $\tilde{V}_2 \tilde{U}_0$ and hence also $\tilde{Z}_\alpha$. Further $[Z_\alpha, Q_H] = V_1$ by 14.7.4.1, so by a Frattini Argument, $T^* = C_T(Z_\alpha)^*$. Now as $H = C_G(z)$, $C_T(Z_\alpha) \leq N_G(U_\alpha)$, so again $U_\alpha^* \leq Z(T^*)$. Thus in either case our choice implies $U_\alpha^* \leq Z(T^*)$.

As $U_\alpha^* \leq Z(T^*)$, T acts on $[\tilde{U}, U_\alpha^*] = \tilde{F}$; hence as $V_1 = [D_\alpha, U] \leq F$, T also acts on F. Recall also that $Z_\alpha \leq F$, so

$$V_1 Z_\alpha \leq F \leq U \cap U_\alpha \leq C_U(U_\alpha) \leq FU_0 = VU_0 = FV. \qquad (*)$$

Suppose first that $V_1 = U_0$. Then (2) holds, and by our choice under this assumption, $Z_\alpha \leq V_2 U_0 = V_2$, so that (1) holds. Further (3) follows from $(*)$, completing the proof of the lemma in this case.

Thus we may suppose that $V_1 < U_0$. Recall $\tilde{F}$ is a complement to $\tilde{U}_0$ in $\tilde{V}\tilde{U}_0$. Further if $m(U_0) = 3$, then from Remark 14.7.64, $\tilde{F} \cap \tilde{V} = 1$, while if $m(U_0) = 2$ then $m(\tilde{F} \cap \tilde{V}) = 1$.

Suppose first that $m(U_0) = 2$. Again (2) holds. Also T acts on V, V_2, and F, and T centralizes $\tilde{Z}_\alpha$ by our choice when $m(U_0) \leq 2$; in particular, T^* centralizes $\tilde{F} \cap \tilde{V}$ of rank 1. As $H^* \cong S_5$ by Theorem 14.7.63, $m(C_{\tilde{V}}(T)) = 1 = m(C_{\tilde{F}}(T))$, so as $Z_\alpha \leq F$, we conclude that $\tilde{V}_2 = C_{\tilde{V}}(T) = \tilde{V} \cap \tilde{F} = C_{\tilde{F}}(T) = \tilde{Z}_\alpha$, since all these subspaces are of rank 1, and each successive pair is related by inclusion. Thus (1) holds. Then we saw that (4) also holds, so that $U_\alpha \in U^L$, and hence as $V \leq U$, also $V \leq U_\alpha$, so that $FV \leq U \cap U_\alpha$. Then (3) follows from (*), completing the proof of the lemma in this case.

Therefore we may assume $m(U_0) = 3$, and it remains to derive a contradiction. This time as $\tilde{F} \cap \tilde{V} = 1$ and $Z_\alpha \leq F$, we have $Z_\alpha \not\leq V$. Let $E := V_1 Z_\alpha$, $Y_E := \langle Q_H, Q_\alpha \rangle$, and $Y := O^2(Y_E)$. As H is irreducible on $\tilde{U}/\tilde{U}_0$ and $\tilde{Z}_\alpha \not\leq \tilde{U}_0$, it follows from 14.5.15.1 that $[Z_\alpha, Q_H] = V_1$. By the symmetry between γ_1 and α, $[V_1, Q_\alpha] = Z_\alpha$. Then by A.1.14, Y_E induces $GL(E)$ on E, $N_G(E) = Y_E C_G(E)$, and $Y_E \trianglelefteq N_G(E)$. As $Z_\alpha \not\leq U_0$ and $H = C_G(z)$, $C_G(z) = C_H(Z_\alpha)$ is a 2-group.

As R_1^* centralizes $\tilde{U}_0 \tilde{V}$ by 14.7.65, R_1 acts on E. We claim $T \leq N_G(E)$, so suppose otherwise. Then for $t \in T - R_1$, $F_0 := V_1 Z_\alpha Z_\alpha^t$ is of rank 3, so as T acts on F with $E = V_1 Z_\alpha \leq F \leq U \cap U_\alpha$, F_0 is contained in $U \cap U_\alpha \cap U_\alpha^t$. Therefore $Y_0 := \langle Y_E, Y_E^t \rangle$ induces $GL(F_0)$ on F_0, since $Aut_{Y_E}(F_0)$ is the stabilizer of E in $GL(F_0)$. But then there is an element of order 3 in $C_{Y_0}(z)$, impossible as $N_H(F_0) \leq T$.

Thus $T \leq N_G(E)$ as claimed, so T acts on $O^2(Y_E) = Y$, and further $\tilde{Z}_\alpha C_{\tilde{U}_0}(T) = C_{\tilde{U}}(T) = \tilde{V}_2 C_{\tilde{U}_0}(T)$. Therefore $\langle Z_\alpha^{L_1} \rangle = V Z_\alpha$ is of rank 4, as we saw $Z_\alpha \not\leq V$.

Let $I := \langle L_1 T, Y \rangle$, $V_I := \langle V_1^I \rangle$, $Q_I := O_2(I)$, and $I^+ := I/Q_I$. Then $(I, L_1 T, YT)$ is a Goldschmidt triple in the sense of Definition F.6.1, so $\alpha := (L_1^+ T^+, T^+, Y^+ T^+)$ is a Goldschmidt amalgam by F.6.5.1, and hence is described in F.6.5.2. Next L_1 has at least five noncentral 2-chief factors, one on $O_2(L_1^*)$ and two each on $\tilde{U}$ and $Q_H/C_H(U)$ using 14.5.21.1. Thus we conclude from F.6.5.2 that $Q_I \neq 1$. In particular I is an SQTK-group and $I \in \mathcal{H}(T) \subseteq \mathcal{H}^e$ by 1.1.4.6, so that $V_I \in \mathcal{R}_2(I)$ by B.2.14. As $E \leq V_I$ and $C_G(E)$ is a 2-group, $Q_I = C_I(V_I)$.

We finish much as at the end of the proof of 14.7.32: If Y^+ acts on L_1^+, then as T acts on Y, $I^+ = L_1^+ T^+ Y^+$, so $V_I = \langle V_1^{L_1^+ T^+ Y^+} \rangle = \langle V_1^{Y^+} \rangle = E$, impossible as L_1 does not act on E. Therefore Y^+ does not act on L_1^+, so in particular L_1^+ is not normal in I^+, and so $L_1^+ \neq 1$.

Suppose $Y \leq M$. As Y does not act on L_1 but T acts on Y, the projection of Y on $L/O_2(L)$ in $M/O_2(L) = L/O_2(L) \times C_M(L/O_2(L))/O_2(L)$ is the maximal parabolic $L_2 O_2(L)/O_2(L)$. Then $Y = [Y, T \cap L] \leq L$, so $E = \langle V_1^Y \rangle \leq V$, whereas we saw earlier that $Z_\alpha \not\leq V$. Thus $Y \not\leq M$.

Assume next that $J(T) \leq Q_I$. Then $J(T) = J(Q_I)$ by B.2.3.3, so that $I \leq N_G(J(T))$. Then since $M = !\mathcal{M}(LT)$ and $Y \not\leq M$, we conclude again using B.2.3.3 that $J(T) \not\leq O_2(LT)$. Thus $L_1 = [L_1, J(T)]$ by 14.3.9.2, contradicting $J(T) \leq Q_I$ and $L_1^+ \neq 1$.

Therefore $J(I)^+ \neq 1$, so by Theorem B.5.6, either $J(I)^+$ is solvable and the direct product of copies of S_3, or there is $K_I \in \mathcal{C}(J(I))$ with $K_I^+ \neq 1$. In the latter case, $K_I \in \mathcal{L}_f(G, T)$, so by 14.3.4.1, K_I^+ is $L_3(2)$ or A_5.

Let $I^! := I/O_{3'}(I)$. By F.6.11.2, either $I^!$ is described in Theorem F.6.18, or $I^! \cong S_3$. But in the latter case, and in case (1) of F.6.18, T^+ is of order 2, so that $T^+ = J(T)^+$, and $I^+ = \langle T^{+I^+} \rangle = J(I^+) \cong S_3$, contrary to L_1^+ not normal in I^+.

Therefore $I^!$ appears in one of the cases (2)–(13) of F.6.18. Further the subcase of case (2) of F.6.18 with $O^2(I^+) \cong 3^{1+2}$ is eliminated, since in that case there is no subnormal subgroup of I^+ isomorphic to S_3. Thus if I^+ is solvable, then by F.6.18, $I^! \cong S_3 \times S_3$, so there is a normal subgroup K_I^+ of I^+ contained in $J(I)^+$ isomorphic to S_3. Then as $Y = [Y, T]$, either $Y^+ = O^2(K_I)^+$ or Y^+ centralizes K_I^+. Similarly either $L_1^+ = O^2(K_I)^+$ or L_1^+ centralizes K_I^+. Therefore as Y^+ does not act on L_1^+, we conclude using F.6.6 that $O^2(I) = \langle Y, L_1 \rangle$ centralizes K_I^+, impossible as $O^2(K_I^+) \not\leq Z(K_I^+)$.

Therefore $I^!$ is nonsolvable, so as I^+ has a subnormal subgroup isomorphic to S_3, $L_3(2)$ or A_5, it follows from F.6.18 that $I^! \cong L_3(2)$. Thus $K_I = O^2(I) = \langle Y, L_1 \rangle$ and $I = K_I T$. But now $E_4 \cong E = [E, Y] \leq [V_I, K_I]$, so as $L_1 T$ centralizes V_1 and $K_I = O^2(I)$, $V_I = [V_I, K_I] = \langle V_1^{K_I} \rangle$ is of rank 3 by H.5.5. This is impossible, since we saw earlier that $\langle Z_\alpha^{L_1} \rangle$ is of rank 4. $\qquad \square$

LEMMA 14.7.71. *(1) H has two noncentral 2-chief factors, both isomorphic to $\tilde{U}$, one on U and one on $Q_H/C_H(U)$.*
(2) L_1 has five noncentral 2-chief factors, one in $O_2(\bar{L}_1)$, and four in S.
(3) $[Q, L] \leq S$.

PROOF. The proof is similar to some of the analysis in the second subsection, but is substantially easier. First L_1 has one noncentral chief factor on $O_2(L_1^*)$, two on $\tilde{U}$, and hence also two on $Q_H/C_H(U)$ by the duality in 14.5.21.1. Thus L_1 has at least five noncentral 2-chief factors.

Next as $U_\alpha \leq S$ by 14.7.70.4, using 14.7.66 we have

$$O_2(L_1^*) = \langle U_\alpha^{*L_1} \rangle \leq S^*. \qquad (*)$$

Set $Q_K := [Q_H, K]C_H(U)$. As $[Q_K/C_H(U), S] = [Q_K/C_K(U), O_2(L_1^*)]$ is of corank 2 in Q_K, with $[S, Q_K] \leq S$, and as $m(Q_K/C_{Q_K}(V)) = 2$ by the duality in 14.5.21.1, we conclude

$$C_{Q_K}(V) = Q_K \cap Q = (S \cap Q_K)C_H(U). \qquad (**)$$

Thus one noncentral 2-chief factor for L_1 in Q lies in S^*, two lie in $U \leq S$, and by $(**)$ a fourth factor also lies in S. Now if (1) holds, then L_1 has four noncentral 2-chief factors in Q_H, so L_1 has exactly five noncentral 2-chief factors by $(*)$. Then as L_1 has at least four noncentral chief factors on S, (2) holds, and of course (3) follows from (2).

Thus it remains to prove (1), so we must show that $[C_H(U), K] \leq U$. But $K = [K, U_\alpha]$, so it suffices to show $[C_H(U_\alpha), U_\alpha] \leq U$.

Now $C_H(U) \leq C_H(Z_\alpha) \leq N_G(U_\alpha)$, so $[U_\alpha, C_H(U)] \leq C_{U_\alpha}(U)$. We will show that $m(C_{U_\alpha}(U)/U \cap U_\alpha) \leq 1$; then as $m([W, U_\alpha]) \geq 2$ for any nontrivial H-chief factor W on $C_H(U)/U$ since $U_\alpha^* \leq K^*$ by 14.7.66, our proof will be complete.

By 14.7.69, $m(U_\alpha^*) = 1$, and by 14.7.70.3, $m(U_\alpha/U \cap U_\alpha) = 2$. So indeed $m(C_{U_\alpha}(U)/U \cap U_\alpha) \leq 1$, as desired. $\qquad \square$

LEMMA 14.7.72. *(1) $S = O_2(L) = [O_2(L), L]$.*
(2) S/V is the Steinberg module for L/S.
(3) $U_0 = V_1$.
(4) $V = Z(S) = \Phi(S) = [S, S]$.

PROOF. By 14.7.70.4, $b = 3$. Set $R := \langle U_0^L \rangle$, so that $V \leq R \leq S$, and $\langle (U_0 V)^L \rangle = RV = R$. From Theorem 14.7.63 and 14.7.66, $[U, Q] = VU_0$, and from

14.7.70.3, $VU_0 = U \cap U_\alpha = [U, U_\alpha]V$. Thus U_α centralizes $[U,Q]$, so $R \leq Z(S)$ by 14.7.13.3. Also $\Phi(S) = [S,S] = R$ by 14.7.13.4. In particular $U \not\leq R$ as $S = \langle U^L \rangle$, so as L_1 is irreducible on U/U_0V, $U_0V = R \cap U$. Therefore as $R \leq Z(S) \leq C_H(U)$, we conclude from 14.7.71.1 that $[R, L_1] \leq R \cap U = U_0V$, so $[R, L_1] = [U_0V, L_1] = V$ in view of 14.7.65. Thus $[R, L] \leq V$, so $R = U_0V$. Further $UR/R = [UR/R, L_1] \cong E_4$, so by H.6.5:

(*) S/R is one of: the Steinberg module, the dual of V, the core (denoted $Core$) of the permutation module for LT on LT/L_2T, or the sum of the Steinberg module with either the dual of V or $Core$.

Suppose first that $U_0 = V_1$, so that $R = V$. Then by 14.7.71.2, L_1 has three noncentral chief factors on S/V, so that $S/R = S/V$ must be the Steinberg module, since by 14.7.22.2, the Steinberg module is the only module listed in (*) with this property. It follows that $V = Z(S)$, and then the rest of the lemma holds: For example, $S = [Q, L]$ by 14.7.71.3, and then as $Q_H \cap O_2(L_1) - Q$ contains an involution H-conjugate to an involution in U_α, the double cover of $L_3(2)$ is not involved in L/S, so that $S = [O_2(L), L] = O_2(L)$.

Thus we assume that $V_1 < U_0$, and it remains to derive a contradiction. By 14.7.70.2, $m(U_0) = 2$, so as $R = U_0V$, $m(R/V) = 1$. Therefore as L is irreducible on V, either $R \leq Z(Q)$ or $[R, Q] = V$, and the latter is impossible as $|T : C_T(U_0)| \leq 2$. Thus $R \leq Z(Q)$ and $m(R/V) = 1$, but $|T : C_T(U_0)| \leq 2$ so R is not the extension in B.4.8.3; thus $R = V \oplus C_R(L)$ where $C_R(L) = R \cap Z(L)$ is of rank 1. Hence $C_R(L)V_1 = C_R(T) = C_R(L_1)$, so as $U_0 \leq C_R(L_1)$, there exists $u \in C_{U_0}(LT) - V_1$. But now by 14.7.68.2, $L \leq C_G(u) \leq H$, contrary to $H \not\leq M = !\mathcal{M}(LT)$. $\square$

Recall $M_1 = H \cap M = L_1T = N_H(V)$.

LEMMA 14.7.73. (1) $SO_2(K) = O_2(L_1) \in Syl_2(K)$.

(2) $|T \cap L : T \cap K| = 2$.

(3) Let $k \in K - M_1$. Then $K = \langle S, S^k \rangle$, $O_2(K) = (S \cap O_2(K))(S^k \cap O_2(K))$ is of order 2^{11}, $S \cap S^k = C_{O_2(K)}(U)$, and $O_2(K)/U$ is the 6-dimensional indecomposable for $K/O_2(K)$ with $C_{O_2(K)/U}(K) = (S \cap S^k)/U \cong E_4$ and $O_2(K)/(S \cap S^k)$ the $L_2(4)$-module.

PROOF. By 14.7.72.2 and H.6.3.5, $S/V = [S/V, L_1]$. Then as $V = [V, L_1]$, $S = [S, L_1] \leq O_2(L_1) \leq K$. We saw in (*) in the proof of 14.7.71 that $O_2(L_1^*) \leq S^*$, so we conclude that $O_2(L_1^*) = S^* \in Syl_2(K^*)$.

We can now argue much as in the proof of G.2.3: Let $k \in K - M_1$ and set $K_0 := \langle S, S^k \rangle$. Now $K^* = K_0^*$, so $K \leq K_0Q_H$; therefore as $Q_H \leq N_G(S)$, $S^K = S^{K_0}$, so $K \leq \langle S^{K_0} \rangle = K_0$. Then as $S \leq K \unlhd H$, $K = K_0$. Let $P := (S \cap Q_H)(S^k \cap Q_H)$. Then $[P, S] \leq S \cap Q_H \leq P$ and similarly $[P, S^k] \leq P$, so $P \unlhd K$; then as $PS/P \cong S/S \cap P \cong S^* \in Syl_2(K^*)$, $P = O_2(K)$.

Next $U \leq S \cap S^k$, and $[S, S] = \Phi(S) = V \leq U$ by 14.7.72.4, so $(S \cap S^k)/U \leq Z(K/U)$. Further setting $P^+ := P/S \cap S^k$,

$$P^+ = (S \cap P)^+ \oplus (S^k \cap P)^+.$$

For each $s \in S - P$, $[P^+, s] \leq (S \cap P)^+ \leq C_{P^+}(s)$ again since $[S, S] = V \leq S \cap S^k$ using 14.7.72.4. So by G.1.5.3 and Theorem G.1.3, P^+ is the sum of natural modules for $K/O_2(K)$. Hence as $U \leq S \cap S^k$, we conclude from 14.7.71.1 that P^+ is a natural module and $S \cap S^k = C_P(U)$. Therefore $P/(S \cap P) = [P/(S \cap P), L_1]$. Thus as $S \leq O_2(L_1) \leq SP$, $P = [P, L_1](S \cap P) \leq O_2(L_1)$ and $SO_2(K) = SP = O_2(L_1) \in$

$Syl_2(K)$. That is, (1) holds. Further as $S \leq O_2(L_1)$ and $|\bar{T} : O_2(\bar{L}_1)| = 2$, (2) holds.

Let $B \in Syl_3(L_1)$. By 14.7.72.2 and H.6.3.3, $|C_S(B)| = 8$, so as $P/C_P(U) = [P/C_P(U), B]$ and $C_U(B) = V_1$ using 14.7.72.3, $(S \cap S^k)/U = C_S(B)U/U$ is of order 4. As $S = [S, L_1]$, L_1 is indecomposable on P/U, so we conclude (3) holds. $\quad\square$

LEMMA 14.7.74. *(1)* $M = L$ and $S = O_2(M)$.
(2) $H = KT$ and $O_2(H) = O_2(K)$.

PROOF. By 14.7.72.2, S/V is the Steinberg module which is a projective L-module, so $Q/V = Q_C/V \oplus S/V$, where $Q_C/V = C_{Q/V}(L)$. Now $[Q_C, L_1] \leq V$ and L_1^* contains the Sylow 2-subgroup $O_2(L_1^*)$ of K^*, so by Gaschütz's Theorem A.1.39, $[Q_C, K] \leq U$. Then as $Q_C U$ centralizes V, $Q_C U$ centralizes $\langle V^K \rangle = U$, so Q_C centralizes $\langle U^L \rangle = S$.

Let $Q_L := C_Q(L)$, so that $Q_L \leq Q_C$. Then $[Q_L, K] \leq [Q_C, K] \leq U$. Further in the unique nonsplit L-module extension W of V in I.1.6 whose quotient is a trivial L-module, $O_2(L_1)$ does not centralize a vector in $W - V$ (cf. B.4.8.3), so $Q_L V_1 = C_{Q_C U}(L_1)$. Therefore since L_1 contains a Sylow 2-subgroup of K, $\tilde{Q}_L \tilde{U} = \tilde{Q}_L \times \tilde{U}$ with K centralizing $\tilde{Q}_L$ again using Gaschütz's Theorem A.1.39. Then K centralizes Q_L by Coprime Action. So since $K \not\leq M = \,!\mathcal{M}(LT)$, we conclude $Q_L = 1$.

Let $B \in Syl_3(L_1)$, and set $Q_B := C_{Q_C}(B)$. Then $Q_C = V Q_B$, so $\Phi(Q_B) = \Phi(Q_C) \trianglelefteq LT$. But L is irreducible on V, and $Q_B \cap V = V_1$, so $\Phi(Q_C) \cap V = 1$. Then $[\Phi(Q_C), L] \leq \Phi(Q_C) \cap V = 1$, so that $\Phi(Q_C) \leq C_Q(L) = 1$ by the previous paragraph. Since also $C_{Q_C}(L) = 1$, $m(Q_C/V) \leq \dim H^1(\bar{L}, V) = 1$, with $[Q_C, O_2(L_1)] = V$ in case of equality (again cf. B.4.8.3).

Suppose $V < Q_C$. By 14.7.73.1, $O_2(L_1) = SO_2(K)$, so as we saw that S centralizes Q_C, $[Q_C, O_2(K)] = [Q_C, O_2(L_1)] = V$. Then as $V_1 = [U, O_2(K)]$, $[Q_C U, O_2(K)] = [Q_C, O_2(K)]V_1 = V$. However, K normalizes $Q_C U$, and hence also normalizes $[Q_C U, O_2(K)] = V$, so $H = KT \leq N_G(V) \leq M$, contrary to $H \in \mathcal{H}_z$.

This contradiction shows that $Q_C = V$. Hence $Q = S = O_2(L) \leq O_2(M)$ by 14.7.72.1, so $O_2(L) = O_2(M)$ by A.1.6. By 14.7.72.4, V char S, so that $V \trianglelefteq M$. Thus $M = LT$ by 14.7.67.5, so as $LT = LO_2(LT)$, $O_2(LT) = O_2(M) = O_2(L)$, and hence (1) holds.

Finally using (1) and 14.7.72.2, $4|Q_H| = |R_1| = 4|S| = 2^{13}$, so $|Q_H| = 2^{11} = |O_2(K)|$ by 14.7.73.3, and hence (2) holds. $\quad\square$

Under the hypotheses of this section, we can now identify G as Ru.

THEOREM 14.7.75. *Assume Hypothesis 14.3.1 holds with* $L/O_2(L) \cong L_3(2)$ *and* $\langle V^{G_1} \rangle$ *abelian. Then* $G \cong Ru$.

PROOF. We verify that G is of type Ru as defined in section J.1. Then the Theorem follows from Theorem J.1.1.

By 14.7.74.1, $M = L$ and $S = O_2(L)$. Thus as L acts on V and $M \in \mathcal{M}$, $L = N_G(V)$ with $L/S \cong L_3(2)$. By 14.7.72.4, S is special with center V. Of course V is the natural module for L/S, and by 14.7.72.2, S/V is the Steinberg module. Thus hypothesis (Ru1) is satisfied.

As $F^*(L) = O_2(L) = S$ and $V = Z(S)$ by 14.7.72.4, $Z = C_V(T) = V_1$. By Theorem 14.7.63, $H = C_G(Z)$ with $H^* \cong S_5$. By 14.7.72.3, $C_{\tilde{U}}(H) = 1$, so by

Theorem 14.7.63, $\tilde{U}$ is the $L_2(4)$-module for H^*. By 14.7.74.2, $Q_H = O_2(K)$, so by 14.7.73.3, Q_H/U is a 6-dimensional indecomposable for H^*. Thus hypothesis (Ru2) is satisfied. Therefore G is of type Ru, completing the proof of the Theorem. $\square$

14.8. The QTKE-groups with $\mathcal{L}_f(G, T) \neq \emptyset$

We now come to a major watershed in this work: We complete the treatment of the case where $\mathcal{L}_f(G, T)$ is nonempty. We begin with the following preliminary result:

THEOREM 14.8.1. *Assume Hypothesis 13.3.1. Then one of the following holds:*

(1) $L/O_2(L) \cong A_6$ and $G \cong Sp_6(2)$ or $U_4(3)$.
(2) $L/O_2(L) \cong A_5$ and $G \cong U_4(2)$ or $L_4(3)$.
(3) $L/O_2(L) \cong L_3(2)$ and $G \cong Sp_6(2)$, $G_2(3)$, HS, or Ru.

PROOF. First by 13.3.2.1, $L/O_2(L) \cong A_5$, $L_3(2)$, A_6, $\hat{A}_6$, or $G_2(2)'$. By Theorem 13.3.16, $L/O_2(L)$ is not $G_2(2)'$. If $L/O_2(L) \cong A_5$, then (2) holds by Theorem 13.6.1. If $L/O_2(L)$ is A_6 or $\hat{A}_6$, then G is $Sp_6(2)$ or $U_4(3)$ by Theorem 13.8.1, so (1) holds. This leaves the case where $L/O_2(L) \cong L_3(2)$. Then G is not $U_4(3)$, as in that case there is no $L \in \mathcal{L}(G, T)$ with $L/O_2(L) \cong L_3(2)$. Further if $G \cong Sp_6(2)$, then (3) holds, so we may assume G is not $Sp_6(2)$. Therefore Hypothesis 14.3.1.1 is satisfied. Let $U := \langle V_1^{G_1} \rangle$. If U is nonabelian then G is $G_2(3)$ or HS by Theorem 14.4.14, so that (3) holds. Thus we may assume U is abelian. Then Theorem 14.7.75 shows that $G \cong Ru$, so that (3) holds, completing the proof. $\square$

We can now easily deduce our main result Theorem D (14.8.2) below from Theorem 14.8.1. Theorem 14.8.1 assumes Hypothesis 13.3.1, and some major reductions are concealed in Hypothesis 13.3.1, so we briefly recapitulate those reductions; they take place in the proof of 13.3.2. In Hypothesis 13.3.1 we assume that $\mathcal{L}_f(G, T) \neq \emptyset$. This rules out the groups in Theorem 2.1.1, so that $|\mathcal{M}(T)| > 1$, and allows us to appeal to the theory based on Theorem 2.1.1. The groups excluded in Hypothesis 13.1.1 are also excluded in Hypothesis 13.3.1, so we are able to apply Theorem 13.1.7 to conclude that $K/O_2(K)$ is quasisimple for each $K \in \mathcal{L}_f(G, T)$. By 1.2.9, $\mathcal{L}_f^*(G, T) \neq \emptyset$, and we pick $L \in \mathcal{L}_f^*(G, T)$. In particular, Hypothesis 12.2.1 is satisfied. The proof of Theorem 12.2.2 discusses how previous work leads to the groups in conclusions (1) and (2) of 12.2.2; Hypothesis 12.2.3 excludes these groups, but they are also excluded in Hypothesis 13.3.1, so Hypothesis 12.2.3 is also satisfied. This allows us to appeal to the work in chapter 12 which restricts the choice for the pair L, V in the Fundamental Setup to those listed in 13.3.2.

THEOREM 14.8.2 (Theorem D). *Assume that G is a simple QTKE-group, with $T \in Syl_2(G)$, and $\mathcal{L}_f(G, T) \neq \emptyset$. Then one of the following holds:*

(1) G is a group of Lie type over $\mathbf{F}_{2^n}$, $n > 1$, of Lie rank 2, but $G \cong U_5(2^n)$ only for $n = 2$.
(2) G is $L_4(2)$, $L_5(2)$, A_9, M_{22}, M_{23}, M_{24}, He, or J_4.
(3) G is $Sp_6(2)$, $U_4(2)$, $L_4^\epsilon(3)$, $G_2(3)$, HS, or Ru.

PROOF. Since the groups excluded in parts (2) and (3) of Hypothesis 13.3.1 appear as conclusions in Theorem D, we may assume that parts (1)–(3) of Hypothesis 13.3.1 are satisfied. Now choose $L \in \mathcal{L}_f^*(G, T)$ with $L/O_2(L)$ not A_5 if possible.

Suppose that $L/O_2(L) \cong A_5$. Then the choice of L above was forced, so that $K/O_2(K) \cong A_5$ for all $K \in \mathcal{L}_f^*(G, T)$. Therefore $J/O_2(J) \cong A_5$ for all $J \in \mathcal{L}_f(G, T)$ by 1.2.4 and A.3.12. Thus part (4) of Hypothesis 13.3.1 is satisfied, completing the verification of Hypothesis 13.3.1.

Now Theorem 14.8.1 completes the proof of Theorem D. $\qquad\square$

Part 6

The case $\mathcal{L}_f(G, T)$ empty

CHAPTER 15

The case $\mathcal{L}_{\mathbf{f}}(\mathbf{G}, \mathbf{T}) = \emptyset$

In this chapter, we complete the treatment of the case $\mathcal{L}_f(G,T)$ empty. Since the previous chapter completed the analysis of the case $\mathcal{L}_f(G,T)$ nonempty, this chapter will complete the proof of our Main Theorem.

Initially we assume Hypothesis 14.1.5, introduced at the start of the previous chapter, with $M := M_f$. Recall that $V(M)$ is defined just before 14.1.2: as mentioned in section A.5, in this chapter we are deviating from our usual meaning of $V(M)$ in definition A.4.7, instead using the meaning in notation A.5.1, namely $V(M) := \langle Z^M \rangle$. In the first two sections of this chapter, we reduce to the case where M and $V := V(M)$ satisfy $m(V) = 4$ and $M/O_2(M) \cong O_4^+(V)$. We treat that final difficult case in the third section. The fourth section then treats the remaining subcase of the case $\mathcal{L}_f(G,T)$ empty when Hypothesis 14.1.5 is not satisfied; this subcase quickly reduces to the situation $\mathcal{L}(G,T)$ empty, or equivalently each member of $\mathcal{H}(T)$ is solvable.

15.1. Initial reductions when $\mathcal{L}_{\mathbf{f}}(\mathbf{G}, \mathbf{T})$ is empty

In this section, and indeed until the final section of this chapter, we assume Hypothesis 14.1.5. This Hypothesis isolates the most important subcase of the case $\mathcal{L}_f(G,T)$ empty, and was already introduced at the beginning of the previous chapter. Recall Hypothesis 14.1.5 includes the assumption that $|\mathcal{M}(T)| > 1$, which is appropriate in view of Theorem 2.1.1. Hypothesis 14.1.5 also includes the assumption that there is a unique maximal 2-local M_c containing the centralizer in G of $Z := \Omega_1(Z(T))$; that is,

$$M_c = !\mathcal{M}(C_G(Z)).$$

The case where this condition fails will be treated in the final section of the chapter; in that case Hypothesis 15.4.1.2 of the final section is satisfied.

By 14.1.12.1, there is $M := M_f \in \mathcal{M}(T) - \{M_c\}$, which is maximal under the partial order $\overset{<}{\sim}$ on $\mathcal{M}(T)$ of Definition A.5.2, and M is the unique maximal member of $\mathcal{M}(T) - \{M_c\}$ under $\overset{<}{\sim}$. As in Definition A.5.8, set $V(M) := \langle Z^M \rangle$; as usual $V(M) \in \mathcal{R}_2(M)$ by B.2.14.

The uniqueness theorems in A.5.7, for overgroups of T in M which cover $M/C_M(V(M))$, replace the uniqueness theorems for members of $\mathcal{L}_f^*(G,T)$, used in the treatement of the Fundamental Setup (3.2.1), which are no longer available as $\mathcal{L}_f(G,T)$ is empty.

LEMMA 15.1.1. *Set* $V := V(M)$, $R := C_T(V)$, *and* $\bar{M} := Aut_M(V)$. *Then*

(1) Case (II) of Hypothesis 3.1.5 is satisfied with $N_M(R)$ *in the role of "M_0", and any* $H \in \mathcal{H}_*(T, M)$.

(2) $\hat{q}(\bar{M}, V) \leq 2$, *and if* $q(\bar{M}, V) > 2$ *then* $\hat{q}(\bar{M}, V) < 2$.

PROOF. As M is maximal in $\mathcal{M}(T)$ under $\stackrel{<}{\sim}$ and $V = V(M)$, we conclude from A.5.7.2 that $R = O_2(N_M(R))$, $V \in \mathcal{R}_2(N_M(R))$, $\bar{M} = Aut_{N_M(R)}(V) = \overline{N_M(R)}$, and $M = !\mathcal{M}(N_M(R))$. So as $V \trianglelefteq M$, (1) holds. Further for $H \in \mathcal{H}_*(T, M)$, $O_2(\langle H, N_M(R) \rangle) = 1$ as $M = !\mathcal{M}(N_M(R))$, so conclusion (1) of Theorem 3.1.6 does not hold. Then conclusion (2) or (3) of 3.1.6 holds, establishing (2) since $\overline{N_M(R)} = \bar{M}$. $\qquad\square$

By 15.1.1, $Aut_M(V(M))$ and its action on $V(M)$ are described in section D.2. Using the fact that $M_c = !\mathcal{M}(C_G(Z))$, we refine that description in the first lemma in this section, which provides the basic list of cases to be treated in the first three sections of this chapter. Recall $\hat{\mathcal{Q}}_*(Aut_M(V(M)), V(M))$ from Definition D.2.1, and set $\hat{J}(Aut_M(V(M)), V(M)) := \langle \hat{\mathcal{Q}}_*(Aut_M(V(M)), V(M)) \rangle$.

LEMMA 15.1.2. Let $V := V(M)$, and set $\bar{M} := M/C_M(V)$ and $\bar{M}_J := \hat{J}(\bar{M}, V)$. Then one of the following holds:

(1) $\bar{M}_J \cong D_{2p}$ and $m(V) = 2m$, where $(p, m) = (3, 1)$, $(3, 2)$, or $(5, 2)$.

(2) $m(V) = 4$ and $\bar{M} = \bar{M}_J = \Omega_4^+(V) \cong S_3 \times S_3$.

(3) $\bar{M}_J = \bar{M}_1 \times \bar{M}_2$ and $V = V_1 \oplus V_2$, with $\bar{M}_i \cong D_{2p}$, $V_i := [V, M_i]$ of rank $2m$, (p, m) as in (1), and $\bar{M}_1$ and $\bar{M}_2$ interchanged in $\bar{M}$.

(4) $\bar{M}_J = \bar{P}\langle \bar{t} \rangle$ where $\bar{P} := O^2(\bar{M}) \cong 3^{1+2}$, and $\bar{t}$ is an involution inverting $\bar{P}/\Phi(\bar{P})$. Further $m(V) = 6$, and T acts irreducibly on $\bar{P}/\Phi(\bar{P})$.

(5) $\bar{M}_J = \bar{P}\langle \bar{t} \rangle$ where $\bar{P} := O^2(\bar{M}) \cong E_9$ and $\bar{t}$ is an involution inverting $\bar{P}$. Further $m(V) = 4$, and $\bar{T} \cong \mathbf{Z}_4$.

(6) $\bar{M}_J \cong S_3$, $V = [V, M_J] \times C_V(M_J)$ with $m([V, M_J]) = 4$ and $C_V(M_J) \neq 1$, $M/C_M([V, M_J]) = \Omega_4^+([V, M_J])$, and $M \cap M_c = C_M([V, M_J])C_M(C_V(M_J))T$ is of index 3 in M.

PROOF. By 15.1.1.1, $\hat{q}(\bar{M}, V) \leq 2$, while by 14.1.6.1, $\bar{M}$ is solvable. Hence, in the language of the third subsection of section D.2, $(\bar{M}_J, [V, F(\bar{M}_J)])$ is a sum of indecomposables, so there is a partition

$$\hat{\mathcal{Q}}_*(\bar{M}, V) = \mathcal{Q}_1 \cup \cdots \cup \mathcal{Q}_s$$

such that $\bar{M}_J = \bar{M}_1 \times \cdots \times \bar{M}_s$ and $V_0 := [V, F(\bar{M}_J)] = V_1 \oplus \cdots \oplus V_s$, where $\bar{M}_i := \langle \mathcal{Q}_i \rangle$, $V_i := [V, M_i]$, and $(\bar{M}_i, V_i)$ is indecomposable as defined in section D.2. Further by D.2.17, each indecomposable $(\bar{M}_i, V_i)$ satisfies one of the conclusions of D.2.17. Let M_i denote the preimage in M of $\bar{M}_i$. As M permutes the set $\{\mathcal{Q}_i : 1 \leq i \leq s\}$ of orbits of M_J on $\hat{\mathcal{Q}}_*(\bar{M}, V)$, M permutes $\{M_i : 1 \leq i \leq s\}$.

Observe that $F^*(\bar{M}_i) = O_p(\bar{M})$ for some odd prime p (depending on i), so for each nontrivial 2-element $\bar{t}$ in $\bar{M}_i$, $C_{O_p(\bar{M})}(\bar{t})$ is cyclic by A.1.31.1. Thus if $\bar{M}_i$ is not normal in $\bar{M}$, then as the product $\bar{M}_J$ of the $\bar{M}_j$ is direct, $\bar{M}_i^{\bar{M}} = \bar{M}_i^T$ is of order 2, and $m_p(\bar{M}_i) = 1$, so that $\bar{M}_i$ falls into case (1) or (2) of D.2.17. In particular, if $m_p(\bar{M}_i) > 1$, then $\bar{M}_i \trianglelefteq \bar{M}$.

Let $K_1, \ldots, K_a$ be the groups $\langle M_i^M \rangle$, and set $W_i := [V, K_i]$; then $\bar{M}_J = \bar{K}_1 \times \cdots \times \bar{K}_a$ and $V_0 = W_1 \oplus \cdots \oplus W_a$. Further $V = V_0 \oplus C_V(F(\bar{M}_J))$ by Coprime Action.

Assume first that $J(T) \not\leq C_M(V)$. Then as $M \neq M_c$, we conclude from 14.1.7 that either (1) or (3) holds, with $(p, m) = (3, 1)$.

Thus in the remainder of the proof, we may assume that $J(T) \leq C_M(V)$. Therefore since M is maximal in $\mathcal{M}(T)$ under $\stackrel{<}{\sim}$, we may apply 14.1.4 to conclude that $M_c \lesssim M$. Then since $M_c \not\leq M$, A.5.6 gives

$$\text{There is no } 1 \neq X \leq V \text{ with } X = \langle (Z \cap X)^{M \cap M_c} \rangle \trianglelefteq M. \qquad (*)$$

Suppose first that $(\bar{M}_1, V_1)$ satisfies case (6) of D.2.17; we will derive a contradiction. For then $\bar{P} := O^2(\bar{M}_1) = \bar{P}_1 \times \bar{P}_2 \times \bar{P}_3$, with $\bar{P}_j \cong \mathbf{Z}_3$ for each j, and $V_1 = U_1 \oplus U_2 \oplus U_3$, where $U_j := [V, P_j]$ is of rank 2 for P_j the preimage of $\bar{P}_j$. As $m_3(\bar{M}_1) > 1$, M_1 and V_1 are normal in M by the second paragraph of the proof, so M permutes $\mathcal{X} := \{P_1, P_2, P_3\}$. Then T fixes some member of $\mathcal{X}$, say P_1, so that T acts on $[V, P_1] = U_1$, and on $U := U_2 \oplus U_3$. Thus $1 \neq Z_1 := Z \cap U_1$ and $1 \neq Z_U := Z \cap U$. So $P_1 \leq C_G(Z_1) \leq M_c = !\mathcal{M}(C_G(Z))$, and similarly $P_2 P_3 \leq C_M(Z_U) \leq M_c$. Then $V_1 = \langle Z_1^{P_1}, Z_U^{P_2 P_3} \rangle = \langle (Z \cap V_1)^{M_c \cap M} \rangle$, contrary to $(*)$. This completes the proof that no $(\bar{M}_i, V_i)$ satisfies conclusion (6) of D.2.17.

Next suppose for the moment that $(\bar{M}_1, V_1)$ satisfies conclusion (3) of D.2.17; in this case, we show that $M_1 T$ acts irreducibly on V_1. Again by the second paragraph, M_1 and V_1 are normal in M. As case (3) of D.2.17 holds, $\bar{M}_1 \cong \mathbf{Z}_2/E_9$, with $m(V_1) = 4$ and $O(\bar{M}_1)$ inverted in $\bar{M}_1$. Thus $Aut_M(V_1) \leq N_{GL(V_1)}(\bar{M}_1) \cong O_4^+(V_1)$. Assume now that $M_1 T$ acts reducibly on V_1. Then $Aut_T(V_1) \cong \mathbf{Z}_2$ or E_4, and in either case $Z \cap V_1 \cong E_4$ and $M = \langle C_M(z) : z \in Z^\# \cap V_1 \rangle$, so $M \leq M_c$ as $M_c = !\mathcal{M}(C_G(Z))$. This contradiction completes the proof that if $(\bar{M}_i, V_i)$ satisfies case (3) of D.2.17, then $M_i T$ acts irreducibly on V_i.

Next we introduce a basic case division for the proof of the lemma: Let $Z_i := Z \cap W_i$, and suppose that $W_i = \langle Z_i^{K_i} \rangle$ for some i, and that either $C_V(M_J) \neq 1$, or $a > 1$. Then $C_Z(K_i) \neq 1$, so that $K_i \leq C_G(C_Z(K_i)) \leq M_c = !\mathcal{M}(C_G(Z))$. Then W_i is generated by $Z_i^{K_i} \subseteq Z_i^{M \cap M_c}$, so as $W_i \trianglelefteq M$ since $K_i = \langle M_i^M \rangle$, we have a contradiction to $(*)$. Thus we conclude that either

(i) $a = 1$ and $V = V_0$, or

(ii) For each i, $\langle Z_i^{K_i} \rangle < W_i$.

We first assume that case (i) does not hold; then case (ii) holds, and we will show that conclusion (6) is satisfied in this case. Choose notation so that $W_1 = V_1 \oplus \cdots \oplus V_b$; then $b \leq 2$ by paragraph two. As (ii) holds, $\langle Z_1^{K_1} \rangle < W_1$, so $K_1 T$ acts reducibly on W_1, and hence $M_1 N_T(V_1)$ acts reducibly on V_1. Now in D.2.17, $\bar{M}_1$ acts reducibly only in case (3), and in case (1) when $m(V_1) = 4$. But earlier we showed $M_1 N_T(V_1)$ is irreducible on V_1 in case (3), so $\bar{M}_1 \cong S_3$ with $m(V_1) = 4$, and in particular $\hat{q}(\bar{M}_1, V_1) = 2$ and $Aut_M(V_1) \leq N_{GL(V_1)}(\bar{M}_1) = \Omega_4^+(V_1)$.

Since $\langle C_{V_1}(N_T(V_1))^{M_1} \rangle < V_1$, $m(C_{V_1}(N_T(V_1))) = 1$, so $|Aut_T(V_1)| > 2$. Then as $Aut_M(V_1) \leq \Omega_4^+(V_1)$, $Aut_T(V_1) \cong E_4$, so $Aut_M(V_1))$ is either $\Omega_4^+(V_1)$ or $S_3 \times \mathbf{Z}_2$. However in the latter case, $M = K_1 C_M(Z_1)$, so as $W_1 > \langle Z_1^{K_1} \rangle$, also $W_1 > \langle Z_1^M \rangle$, contrary to $V = \langle Z^M \rangle$.

Therefore $Aut_M(V_1) = \Omega_4^+(V_1)$. Now as $\bar{M}_1 = Aut(\bar{M}_1)$, $\bar{M}_0 := N_{\bar{M}}(\bar{M}_1) = \bar{M}_1 \times C_{\bar{M}_0}(\bar{M}_1)$, with $m_3(C_{\bar{M}_0}(\bar{M}_1)) \leq 1$ using A.1.31.1. Suppose $s > 1$. Then by symmetry, $\bar{M}_2 \cong S_3$, so $\bar{M}_0 = \bar{M}_1 \times \bar{M}_2 \times C_{\bar{M}_0}(\bar{M}_1 \bar{M}_2)$, and then $O(\bar{M}_1)O(\bar{M}_2) = O^{3'}(\bar{M}_0)$ by A.1.31.1. This is a contradiction as $Aut_M(V_1) = \Omega_4^+(V_1)$ and $[V_1, M_2] = 1$. Therefore $s = a = 1$, and hence $K_1 = M_1 = M_J$ and $W_1 = V_1 = V_0$. If $V = V_0$, then case (i) holds, contrary to our assumption, so we may assume that $C_V(M_J) \neq 0$. Now $C_M(V_0)$ and $C_M(C_V(M_J))$ lie in $M_c = !\mathcal{M}(C_G(Z))$, and $|M : C_M(V_0)C_M(C_V(M_J))T|$ divides $|Aut_M(V_0) : \bar{K}_1 \bar{T}| = 3$; then as $M \not\leq M_c$, we conclude that $M \cap M_c = C_M(V_0)C_M(C_V(M_J))T$ is of index 3 in M. This completes the proof that conclusion (6) of 15.1.2 holds if case (i) does not hold.

Thus we may assume that case (i) holds, so $\bar{M}_J = \bar{K}_1$ and $V = V_0 = W_1$. Thus $\bar{M}$ is faithful on W_1 and $\bar{M} \leq N_{GL(W_1)}(\bar{K}_1)$.

Assume first that $(\bar{M}_J, V)$ is indecomposable. Then $s = 1$, so $\bar{M}_J = \bar{M}_1$ and $V = V_1$. Recall we showed that conclusion (6) of D.2.17 does not hold for $(\bar{M}_J, V)$. Conclusions (1) and (2) of D.2.17 give conclusion (1) of 15.1.2.

Suppose conclusion (5) of D.2.17 holds. Then $\bar{M}_J = \Omega_4^+(2)$, so $\bar{M} = \bar{M}_J$, for otherwise $\bar{M} = N_{GL(V)}(\bar{M}_J) = O_4^+(V)$ contains transvections, whereas $\hat{q}(\bar{M}, V) = 3/2$ in case (5) of D.2.17. Thus conclusion (2) of 15.1.2 holds in this case.

Suppose conclusion (3) of D.2.17 holds. We showed earlier that $\bar{M} \leq O_4^+(V)$ and M acts irreducibly on V. As conclusion (3) of D.2.17 holds, $\hat{q}(\bar{M}, V) = 2$, so $\bar{T}$ contains no transvections on V. Hence as M is irreducible on V, $\bar{T} \cong \mathbf{Z}_4$, so conclusion (5) of 15.1.2 holds.

Suppose that case (4) of D.2.17 holds. Then $\bar{M}_J = \bar{P}\langle \bar{t} \rangle$, where $\bar{P} = F^*(\bar{M}_J) \cong 3^{1+2}$, $\bar{t}$ inverts $\bar{P}/\Phi(\bar{P})$, and $m(V) = 6$. Hence $\bar{M} \leq N_{GL(V)}(\bar{M}_J) \cong GL_2(3)/3^{1+2}$. If $O^2(\bar{M}) > \bar{P}$, then $m_3(C_{\bar{M}}(\bar{t})) > 1$, contrary to A.1.31.1; thus $O^2(\bar{M}) = \bar{P}$. Therefore if T is irreducible on $\bar{P}/\Phi(\bar{P})$, then conclusion (4) of 15.1.2 holds, so we may assume that T is reducible on $\bar{P}/\Phi(\bar{P})$, and it remains to derive a contradiction. Then $\bar{T} \cong \mathbf{Z}_2$ or E_4, and in either case T acts on subgroups $\bar{P}_1$ and $\bar{P}_2$ of order 3 generating $\bar{P}$. Thus $Z = E_1 E_2$, where $1 \neq E_i := C_V(\bar{P}_i \bar{T})$. Therefore the preimages P_i satisfy $P_i T \leq C_G(E_i) \leq M_c = !\mathcal{M}(C_G(Z))$, and hence $M = \langle P_1, P_2 \rangle T \leq M_c$, a contradiction.

Finally assume that $(\bar{M}_J, V)$ is decomposable. As (i) holds, $a = 1$. Then from the second paragraph of the proof, $s = 2$ and $(\bar{M}_i, V_i)$ satisfies case (1) or (2) of D.2.17. As $a = 1$, $\bar{M}_1$ and $\bar{M}_2$ are interchanged in M, so that conclusion (3) of 15.1.2 holds.

This completes the proof of 15.1.2. $\qquad\qquad\qquad\qquad\qquad\qquad\square$

15.1.1. Statement of the main theorem, and some preliminaries.

Our first goal is to show that case (1) or (3) of 15.1.2 holds, and $V(M)$ is an FF-module for $\bar{M}$. That is, we will prove that either $m(V(M)) = 2$ with $M/C_M(V(M)) = GL(V(M)) \cong L_2(2)$, or $m(V(M)) = 4$ with $M/C_M(V(M)) = O_4^+(V)$.

In the remaining cases there are no quasithin examples; indeed as far as we can tell, there are not even any shadows. But we saw in Theorem 14.6.25 of the previous chapter that quasithin examples do arise in the first case, and many shadows complicate our analysis of the second case, in the third section 15.3 of this chapter.

Thus the remainder of this section is devoted to the first steps in a proof of the following main result:

THEOREM 15.1.3. *Assume Hypothesis 14.1.5, and let $M := M_f$ as in 14.1.12. Then either*

(1) $m(V(M)) = 2$, $M/C_M(V(M)) \cong L_2(2)$, *and G is isomorphic to J_2, J_3, $^3D_4(2)$, the Tits group $^2F_4(2)'$, $G_2(2)' \cong U_3(3)$, or M_{12}; or*

(2) $m(V(M)) = 4$, *and* $M/C_M(V(M)) = O_4^+(V(M))$.

The proof of Theorem 15.1.3 involves a series of reductions, which will not be completed until the end of section 15.2. Thus in the remainder of this section, and throughout section 15.2, we assume G is a counterexample to Theorem 15.1.3. We also adopt the following convention:

NOTATION 15.1.4. Set $M := M_f$. We choose $V := V(M)$ in cases (1)–(5) of 15.1.2, but in case (6) of 15.1.2 we choose $V := [V(M), M_J]$, where M_J is the preimage in M of $\hat{J}(M/C_M(V(M)))$. Set $\bar{M} := M/C_M(V)$ and $\bar{M}_0 := \hat{J}(\bar{M}, V)$.

Observe that except in case (6) of 15.1.2, $\bar{M}_0$ coincides with $\bar{M}_J$. We review some elementary but fundamental properties of V:

LEMMA 15.1.5. *(1)* $V = \langle (Z \cap V)^{M_0} \rangle$, so $V \in \mathcal{R}_2(M)$.
(2) $C_G(V) \leq M_c$.

PROOF. From the description of V in 15.1.2, $V = \langle (Z \cap V)^{M_0} \rangle$. Then B.2.14 completes the proof of (1). Part (2) follows since $M_c = !\mathcal{M}(C_G(Z))$. $\qquad\square$

LEMMA 15.1.6. $O^2(C_M(Z)) \leq C_M(V)$.

PROOF. By 14.1.6.1, $M^\infty \leq C_M(V)$. Let $S := T \cap M^\infty$. By a Frattini Argument, $C_M(Z) = M^\infty K$, where $K := C_M(Z) \cap N_M(S)$. Since $K^\infty \leq N_{M^\infty}(S)$ and $N_{M^\infty}(S)$ is 2-closed, K is solvable. Thus $K = XT$, where X is a Hall $2'$-subgroup of K. Therefore it remains to show $X \leq C_M(V)$, so we may assume $\bar{X} \neq 1$. From the structure of $\bar{M}$ described in 15.1.2, $\bar{M}$ is 2-nilpotent, and hence so is $\bar{X}\bar{T}$. Therefore $\bar{X} = O(\bar{X}\bar{T}) \trianglelefteq \bar{X}\bar{T}$. Then as $\bar{X} \neq 1$, $Z \cap [V, X] = C_{[V,X]}(T) \neq 1$, and $C_{[V,X]}(\bar{X}) = 1$ by Coprime Action, whereas $X \leq K \leq C_M(Z)$. This contradiction completes the proof. $\qquad\square$

In the next result we review the cases from 15.1.2 which can occur in our counterexample, except that we reorder them according to the value of $m(V)$:

LEMMA 15.1.7. *One of the following holds:*
(1) $m(V) = 4$, *and* $\bar{M} = \bar{M}_0 \cong S_3$.
(2) $m(V) = 4$, $\bar{M}_0 \cong S_3$, *and* $\bar{M} \cong S_3 \times \mathbf{Z}_3$.
(3) $m(V) = 4$, *and* $\bar{M} = \bar{M}_0 = \Omega_4^+(V)$.
(4) $m(V) = 4$, $\bar{M}_0 = \bar{P}\langle \bar{t} \rangle$ *where* $\bar{P} := O^2(\bar{M}) \cong E_9$ *and* $\bar{t}$ *is an involution inverting* $\bar{P}$, *and* $\bar{T} \cong \mathbf{Z}_4$.
(5) $m(V) = 4$, $\bar{M}_0 \cong D_{10}$, $\bar{T} \cong \mathbf{Z}_2$ *or* $\mathbf{Z}_4$, *and either* $F(\bar{M}) = F(\bar{M}_0)$ *or* $F(\bar{M}) \cong \mathbf{Z}_{15}$.
(6) $m(V) = 8$, $\bar{M}_0 = \bar{M}_1 \times \bar{M}_2$ *where* $\bar{M}_i \cong D_{2p}$ *with* $p = 3$ *or* 5, $M_1^t = M_2$ *for some* $t \in T$, *and* $V = V_1 \oplus V_2$, *where* $V_i := [V, M_i]$.
(7) $m(V) = 6$, $\bar{M}_0 = \bar{P}\langle \bar{t} \rangle$ *where* $P := O^2(\bar{M}) \cong 3^{1+2}$, $\bar{t}$ *is an involution inverting* $\bar{P}/\Phi(\bar{P})$, *and* T *acts irreducibly on* $\bar{P}/\Phi(\bar{P})$.
Furthermore if $V < V(M)$, *then case (3) holds.*

PROOF. Suppose first that $V < V(M)$. Then by definition of V in 15.1.4, case (6) of 15.1.2 holds and $V = [V, M_J]$; it follows that conclusion (3) holds. Thus in the remainder of the proof we may assume that $V = V(M)$, and hence that one of cases (1)–(5) of 15.1.2 holds.

Assume first that $m(V) = 2$. Then case (1) of 15.1.2 holds with $(p, m) = (3, 1)$ and $\bar{M} \cong S_3$. Then as we observed at the start of section 14.2, 14.1.18 shows that Hypothesis 14.2.1 is satisfied. Therefore we may apply Theorem 14.6.25 to conclude that G is one of the groups listed in conclusion (1) of Theorem 15.1.3, contrary to the choice of G as a counterexample.

Thus $m(V) > 2$. Also since G is a counterexample, conclusion (2) of Theorem 15.1.3 does not hold. Thus if case (3) of 15.1.2 holds, then $m(V_i) = 4$ for each i, so that conclusion (6) holds. In case (2) of 15.1.2, conclusion (3) holds. Cases (4) and (5) of 15.1.2 are conclusions (7) and (4).

It remains to treat case (1) of 15.1.2. In this case, $m(V) = 4$ as $m(V) > 2$, so $\bar{M}_0$ is S_3 or D_{10}. If $\bar{P} := O^2(\bar{M}_0) = F^*(\bar{M})$, then $\bar{M} \leq Aut(\bar{P}) \cong S_3$ or $Sz(2)$, respectively, so that conclusion (1) or (5) holds. Thus we may assume that $\bar{P} < F^*(\bar{M})$. Now $\bar{M} \leq N_{GL(V)}(\bar{M}_0)$, and $N_{GL(V)}(\bar{M}_0)$ is $\Omega_4^+(V)$ or $\mathbf{Z}_4/\mathbf{Z}_{15}$, respectively. Since $\bar{P} < F^*(\bar{M})$, and $O_2(\bar{M}) = 1$ by 15.1.5.1, one of conclusions (2)–(5) of the lemma holds. This completes the proof. $\square$

Our assumption that G is a counterexample to Theorem 15.1.3 has ruled out the subcases of 15.1.2 in which $\bar{M}$ contains an FF*-offender on V; that is we are left with thoses cases where $q(\bar{M}, V) > 1$. Indeed:

LEMMA 15.1.8. *One of the following holds:*

(1) $\hat{q}(\bar{M}, V) = q(\bar{M}, V) = 2$.
(2) Case (3) of 15.1.7 holds, where $\hat{q}(\bar{M}, V) = 3/2$ and $q(\bar{M}, V) = 2$.

PROOF. The proof of 15.1.2 showed that one of the following holds:

(i) $V = V(M)$, and $(\bar{M}_J, V)$ is an indecomposable appearing in one of cases (1)–(5) of D.2.17.

(ii) $V = V(M)$, and the $(\bar{M}_i, V_i)$ are indecomposable and appear in case (1) or (2) of D.2.17; hence case (6) of 15.1.7 holds.

(iii) $V < V(M)$, and hence case (3) of 15.1.7 holds.

In cases (i) and (ii), $\bar{M}_J = \bar{M}_0$ by Notation 15.1.4, so we conclude from the values listed in the corresponding cases of D.2.17 that $\hat{q}(\bar{M}_0, V) = q(\bar{M}_0, V) = 2$— unless case (3) of 15.1.7 holds, where (M, V) appears in case (5) of D.2.17, $\bar{M}_0 = \bar{M}$, $\hat{q}(\bar{M}_0, V) = 3/2$, and $q(\bar{M}_0, V) = 2$. However by definition of $\hat{\mathcal{Q}}_*(\bar{M}, V)$, if $\hat{q}(\bar{M}, V) \leq 2$, then $\hat{q}(\bar{M}, V) = \hat{q}(\bar{M}_0, V)$ and $\hat{q}(\bar{M}, V) \leq q(\bar{M}, V)$. Thus the lemma holds in cases (i) and (ii). In case (iii), conclusion (2) of the lemma holds, again as $(\bar{M}, V)$ appears in case (5) of D.2.17. $\square$

Recall that $|\mathcal{M}(T)| > 1$ by Hypothesis 14.1.5.3, so that $\mathcal{H}_*(T, M)$ is nonempty. The next few results study properties of members of $\mathcal{H}_*(T, M)$.

LEMMA 15.1.9. *Set $R := C_T(V)$. Then*

(1) $[V, J(T)] = 1 = [V(M), J(T)]$, so that $Baum(T) = Baum(R)$ and further $C(G, Baum(T)) \leq M$.

(2) M is the unique maximal member of $\mathcal{M}(T)$ under $\overset{<}{\sim}$.

(3) $\mathcal{H}_*(T, M) \subseteq C_G(Z) \leq M_c$.

(4) For each $H \in \mathcal{H}_*(T, M)$, $O^2(H \cap M) \leq C_M(V)$.

(5) $M = !\mathcal{M}(N_M(R))$.

(6) $N_M(R) \in \mathcal{H}^e$, $V \in \mathcal{R}_2(N_M(R))$, $R = O_2(N_M(R))$, and $\overline{N_M(R)} = \bar{M}$; and case (II) of Hypothesis 3.1.5 is satisfied with $N_M(R)$ in the role of "M_0" for any $H \in \mathcal{H}_*(T, M)$.

(7) $N_G(T) \leq M$, and each $H \in \mathcal{H}_*(T, M)$ is a minimal parabolic described in B.6.8, and in E.2.2 if H is nonsolvable.

PROOF. If $J(T)$ does not centralize $V(M)$, then as $m(V) > 2$ by 15.1.7, 14.1.7 shows that conclusion (2) of Theorem 15.1.3 holds, contrary to the choice of G as a counterexample. Therefore $J(T)$ centralizes $V(M)$, and hence also centralizes V.

Since M is maximal in $\mathcal{M}(T)$ under $\overset{\sim}{<}$, we may now apply 14.1.4 to conclude that (2) holds; and apply 15.1.5.1, (2), and 14.1.2 to complete the proof of (1). Observe that $N_M(R) \in \mathcal{H}^e$ by 1.1.3.2. Using case (b) of the hypothesis of A.5.7.2 rather than case (a), the proof of 15.1.1.1 shows that (5) and (6) hold, and case (II) of Hypothesis 3.1.5 is satisfied with $N_M(R)$ in the role of "M_0" for any $H \in \mathcal{H}_*(T, M)$. Further $M = !\mathcal{M}(N_M(R))$ by (5), so (3) follows from 3.1.7. Then (4) follows from (3) and 15.1.6. Finally $N_G(T) \le M$ by (1), so (7) follows from 3.1.3.2. $\qquad \square$

LEMMA 15.1.10. *If case (6) of 15.1.7 holds with $p = 3$, then $\bar{M} \cong S_3$ wr $\mathbf{Z}_2$.*

PROOF. Since $C_{O_3(\bar{M})}(\bar{T} \cap \bar{M}_i)$ is cyclic by A.1.31.1, $O^2(\bar{M}_0) = O^2(\bar{M})$. Then as $O^2(\bar{M})$ acts on $\bar{M}_i$ and V_i for $i = 1, 2$, $C_{GL(V_i)}(\bar{M}_i) \cong L_2(2)$, and $O_2(\bar{M}) = 1$ by 15.1.5.1, the result follows. $\qquad \square$

LEMMA 15.1.11. *For $H \in \mathcal{H}_*(T, M)$:*

(1) $V \le O_2(H)$.
(2) $U_H := \langle V^H \rangle$ is elementary abelian.

PROOF. Set $R := C_T(V)$. By 15.1.9.5, $O_2(\langle N_M(R), H \rangle) = 1$, so Hypothesis F.7.1 is satisfied with $N_M(R)$, H in the roles of "G_1, G_2". Further as $R = O_2(N_M(R))$ by 15.1.9.6, $R = O_2(C_{N_M(R)}(V))$, so that Hypothesis F.7.6 is also satisfied. Now V is not an FF-module for $Aut_{N_M(R)}(V)$ by 15.1.8, so if (1) holds, we may apply F.7.11.8 to obtain (2).

So we may assume that $V \not\le O_2(H)$, and it remains to derive a contradiction. By 3.1.3.1, $H \cap M$ is the unique maximal subgroup of H containing T, and by 15.1.9.7, H is described in B.6.8. Then our assumption $V \not\le O_2(H)$ implies $V \not\le \ker_{H \cap M}(H)$ by B.6.8.5. Thus Hypothesis E.2.8 is satisfied with $H \cap M$ in the role of "M". Then by E.2.15, $r := \hat{q}(\bar{M}, V) < 2$, so that by 15.1.8, $m(V) = 4$, $\bar{M} = \Omega_4^+(2)$, and $r = 3/2$. Also by 15.1.9.4, $O^2(H \cap M) \le C_H(V)$. Hence by E.2.17, $Y = \langle V^H \rangle$ is isomorphic to S_3/Q_8^2, $L_3(2)/D_8^3$, or $(\mathbf{Z}_2 \times L_3(2))/D_8^3$. However in the last two cases, $|Aut_T(V)| \ge 8$ by E.2.17, contrary to $|\bar{M}|_2 = 4$. Therefore $Y \cong S_3/Q_8^2$. Set $P := O_2(Y)$, $X_0 := O^2(N_M(R))$, and $X := O^2([X_0, P])$. As $Aut_P(V) \cong E_4 \cong \bar{T}$ and $R = C_T(V)$, $T = PR$, so $\bar{X} = \bar{X}_0 \cong E_9$. Next

$$[R, P] \le C_P(V) = V \cap P, \qquad (*)$$

so P centralizes R/V, and hence $X \le [X_0, P]$ centralizes R/V. Then $V = [R, X]$ so as $F^*(M) = O_2(M) \le R$, $C_X(V)$ is a 2-group by Coprime Action. Then as $\bar{X} \cong E_9$ and $X = O^2(X)$, it follows that $X \cong A_4 \times A_4$ and $R = V \times C_R(X)$. By $(*)$, $[C_R(X), P] \le C_V(X) = 1$, so $TX = PRX = PX \times C_R(X)$ with $PX \cong S_4 \times S_4$. But now $[V, J(T)] \ne 1$, contrary to 15.1.9.1. $\qquad \square$

LEMMA 15.1.12. *Let $H \in \mathcal{H}_*(T, M)$ and $U_H := \langle V^H \rangle$. Then*

(1) H has exactly two noncentral chief factors U_1 and U_2 on U_H.
(2) There exists $A \in \mathcal{A}(T) - \mathcal{A}(O_2(H))$, and for each such A chosen with $AO_2(H)/O_2(H)$ minimal, A is quadratic on U_H, and setting $B := A \cap O_2(H)$, we have:

$$2m(A/B) = m(U_H/C_{U_H}(A)) = 2m(B/C_B(U_H));$$

$$2m(B/C_B(V^h)) = m(V^h/C_{V^h}(B))$$

for each $h \in H$ *with* $[B, V^h] \neq 1$; $m(A/B) = m(U_i/C_{U_i}(A))$; *and* $C_{U_H}(A) = C_{U_H}(B)$.

(3) $H/C_H(U_i) \cong S_3, S_5, S_3 \ wr \ \mathbf{Z}_2,$ *or* $S_5 \ wr \ \mathbf{Z}_2,$ *with* U_i *the direct sum of the natural modules* $[U_i, F]$, *as* F *varies over the* S_3-*factors or* S_5-*factors of* $H/C_H(U_i)$. *Further* $J(H)C_H(U_i)/C_H(U_i) \cong S_3, S_5, S_3 \times S_3,$ *or* $S_5 \times S_5,$ *respectively*.

(4) $[\Omega_1(Z(J_1(T))), O^2(H)] = 1$.

PROOF. Let $R := C_T(V)$. By 15.1.9.6, $\bar{M} = \overline{N_M(R)}$. We check that the hypothesis of 3.1.9 holds, with $N_M(R)$ in the role of "M_0": First case (II) of Hypothesis 3.1.5 is satisfied by 15.1.9.6. By 15.1.11, $V \leq O_2(H)$, giving (c). By 15.1.7 and B.1.8, V is not a dual FF-module for $\bar{M} = \overline{N_M(R)}$, giving (d). By 15.1.8, $q(\bar{M}, V) = 2$, giving (a). By 15.1.9.5, $M = !\mathcal{M}(N_M(R))$, giving (b). Finally by 15.1.9.4, $O^2(H \cap M) \leq C_G(V)$, so the hypotheses of part (5) of 3.1.9 are satisfied. Therefore by 3.1.9, (1)–(3) hold.

As A^* is an FF*-offender on U_i, it follows from (3) that there is a subgroup X of Y with $A^* \in Syl_2(X^*)$, $O_2(H) \leq X$, $X/O_2(X) \cong S_3$, and $H = \langle O^2(X), T \rangle$. Now we chose $A \in \mathcal{A}(T)$, and U_H is elementary abelian by 15.1.11.2, with $A \cap U_H \leq A \cap O_2(H) = B$, so $C_{U_H}(A) = A \cap U_H = B \cap U_H \leq C_B(U_H)$. Next by (2),

$$m(A/C_B(U_H)) = m(A/B) + m(B/C_B(U_H)) = 2m(A/B) = m(U_H/C_{U_H}(A)),$$

so $m(U_H C_B(U_H)) \geq m(A)$. Hence $U_H C_B(U_H) \in \mathcal{A}(T)$, so as $U_H C_B(U_H) \leq O_2(H) \leq O_2(X)$, also $U_H C_B(U_H) \in \mathcal{A}(O_2(X))$. Therefore by B.2.3.7, $\Omega_1(Z(J(T)))$ and $\Omega_1(Z(J(O_2(X))))$ are contained in $U_H C_B(U_H)$, so by B.2.3.2, $\Omega_1(Z(J_1(T))) =:$ E and $\Omega_1(Z(J_1(O_2(X)))) =: D$ are also contained in $U_H C_B(U_H)$. In particular, $E \leq O_2(X)$, so $E \leq D$.

If $[E, O^2(X)] = 1$, then $H = \langle O^2(X), T \rangle \leq N_G(E)$, and hence $K \leq \langle O^2(X)^H \rangle \leq C_G(E)$, so that (4) holds. Thus we may assume that $[E, O^2(X)] \neq 1$, and it remains to derive a contradiction. We saw $E \leq D$, so also $[D, O^2(X)] \neq 1$. Then as $O^2(X) = [O^2(X), A]$ by construction, $[D, A] \neq 1$, so in particular $D \not\leq A \cap O_2(X) =: B_X$. Observe that $B_X \geq A \cap O_2(H) = B$. On the other hand, $B_X \in \mathcal{A}_1(O_2(X))$ as $X/O_2(X) \cong S_3$, so D centralizes B_X, and then as $D \not\leq B_X$, $m(DB_X) > m(B_X) = m(A) - 1$, so $DB_X \in \mathcal{A}(T)$. Then as $D \leq U_H C_B(U_H) \leq O_2(H)$, by minimality of $AO_2(H)/O_2(H)$, $B_X \leq O_2(H) \cap A = B$, so that $B_X = B$. But by (2), $C_{U_H}(B) = C_{U_H}(A)$, so

$$D \leq C_B(U_H)U_H \cap C_G(B) = C_B(U_H)C_{U_H}(B) = C_B(U_H)C_{U_H}(A) \leq C_G(A),$$

contrary to an earlier observation. This contradiction completes the proof of 15.1.12. $\qquad\square$

LEMMA 15.1.13. *Let* $E_1 := \Omega_1(Z(J_1(T)))$. *Then*

(1) $C_G(E_1) \not\leq M$.

(2) $[V, J_1(T)] \neq 1$.

(3) *Either*

(i) *for all* $A \in \mathcal{A}_1(T)$ *with* $\bar{A} \neq 1$, $|\bar{A}| = 2$ *and* $\bar{A} \in \hat{\mathcal{Q}}_*(\bar{M}, V)$, *or*

(ii) *case (3) of 15.1.7 holds.*

(4) *Either*

(a) $\overline{J_1(T)} = \bar{T} \cap \bar{M}_0$ *and* $\overline{J_1(M)} = \bar{M}_0$, *or*

(b) *Case (3) of 15.1.7 holds, and* $\overline{J_1(T)}$ *is of order 2.*

PROOF. As $\mathcal{H}_*(T, M) \neq \emptyset$, (1) follows from 15.1.12.4. Next if $[V, J_1(T)] = 1$, then by B.2.3.5, $N_M(C_T(V))$ normalizes $J_1(T)$ and hence also normalizes E_1, so that $N_G(E_1) \leq M$ by 15.1.9.5, contrary to (1). This establishes (2).

By (2), there is $A \in \mathcal{A}_1(T)$ with $\bar{A} \neq 1$. Now $m(\bar{A}) \leq m_2(\bar{M})$ and $m_2(\bar{M}) \leq 2$ from 15.1.7. As $A \in \mathcal{A}_1(T)$, $m(V/C_V(A)) \leq m(\bar{A}) + 1$, while $q(\bar{M}, V) > 1$ by 15.1.8, so that $m(V/C_V(A)) = m(\bar{A}) + 1$. Hence for $m(\bar{A}) = 1$ or 2,

$$r_{\bar{A}, V} = \frac{m(V/C_V(A))}{m(\bar{A})} = 2 \text{ or } 3/2,$$

respectively. Assume that case (3) of 15.1.7 does not hold. Then by 15.1.8, $\hat{q}(\bar{M}, V) = q(\bar{M}, V) = 2$, and the calculation above shows that $m(\bar{A}) = 1$ and $r_{\bar{A}, V} = 2$ for each $A \in \mathcal{A}_1(T)$ with $\bar{A} \neq 1$, so that $\bar{A} \in \hat{\mathcal{Q}}_*(\bar{M}, V)$. This establishes (3).

It remains to prove (4). Suppose first that case (3) of 15.1.7 holds. Then $\bar{M} = \bar{M}_0$ and $\bar{T} \cong E_4$, and $\overline{J_1(T)} \neq 1$ by (2). Therefore either $\overline{J_1(T)} = \bar{T}$, and hence conclusion (a) of (4) holds, or $\overline{J_1(T)}$ is of order 2, and conclusion (b) holds. Thus we may assume that case (3) of 15.1.7 does not hold. Then by (3), $\overline{J_1(T)} \leq \bar{T} \cap \bar{M}_0 =: \bar{T}_0$. As case (3) of 15.1.7 does not hold, either case (6) of 15.1.7 holds or $|\bar{T}_0| = 2$. In the latter case, $\overline{J_1(T)} = \bar{T}_0$ so that $\overline{J_1(M)} = \bar{M}_0$, giving conclusion (a). In the former case, $\bar{A} \leq \bar{M}_i$ for $i = 1$ or 2 since $r_{\bar{A}, V} = 2$, and then $\overline{J_1(T)} = \langle \bar{A}, \bar{A}^t \rangle = \bar{T}_0$, so again conclusion (a) holds. This completes the proof of (4). $\quad\square$

LEMMA 15.1.14. *Let* $V_E := C_V(J_1(T))$. *Then*

(1) $O^2(C_G(Z)) \leq C_G(V_E)$.
(2) $N_M(J_1(T)) \leq N_G(V_E) \leq M_c$.
(3) $N_G(J_1(T)) \leq M \cap M_c$.

PROOF. By 15.1.6, $O^2(C_M(Z)) \leq C_M(V) \leq C_M(V_E)$. Thus if (1) fails, then

$$O^2(C_G(Z)) \not\leq \langle M \cap O^2(C_G(Z))T, O^2(C_G(Z)) \cap C_G(V_E) \rangle,$$

so there exists $H \in \mathcal{H}_*(T, M)$ with $H \leq C_G(Z)$ but $O^2(H) \not\leq O^2(C_G(V_E))$. However since $V_E \leq \Omega_1(Z(J_1(T)))$, this contradicts 15.1.12.4, so (1) is established. Then (1) implies (2) since $M_c = !\mathcal{M}(C_G(Z))$. Finally as $J(J_1(T)) = J(T)$ by (1) and (3) of B.2.3,

$$N_G(J_1(T)) = N_G(J(T)) \cap N_G(J_1(T)) \leq N_M(J_1(T))$$

by 15.1.9.1, so (2) implies (3). $\quad\square$

15.1.2. Eliminating some larger possibilities from 15.1.7. Our proof of Theorem 15.1.3 now divides into two cases:

Case I. $M = \langle C_M(Z_1), T \rangle$ for some nontrivial subgroup Z_1 of $C_V(J_1(T))$.
Case II. There exists a subgroup X of M containing T with $M = !\mathcal{M}(X)$ and $X/O_2(X) \cong S_3$, D_{10}, or $Sz(2)$.

Case II will be treated in the following section. Cases (1)–(3) and (5) of 15.1.7 appear in Case II, although this fact is not established until lemma 15.2.6 in that section. In the remainder of this section, we treat the three cases from 15.1.7 which appear in Case I. Namely we prove the following theorem:

THEOREM 15.1.15. *None of cases (4), (6) or (7) of 15.1.7 can hold.*

Until the proof of Theorem 15.1.15 is complete, assume G is a counterexample. Thus we are in case (4), (6), or (7) of 15.1.7. As in 15.1.13, let $E_1 := \Omega_1(Z(J_1(T)))$. By 15.1.13.1, $C_G(E_1) \not\le M$.

As case (3) of 15.1.7 does not hold, $\overline{J_1(T)} = \bar{T} \cap \bar{M}_0$ and $\overline{J_1(M)} = \bar{M}_0$ by 15.1.13.4. Also $V = V(M)$ by 15.1.7.

We begin by determining $V_E := C_V(J_1(T))$ in each of our three cases, and defining some notation:

NOTATION 15.1.16. (a) In case (6) of 15.1.7, $V = V_1 \oplus V_2$ for V_i defined there, and $V_E = Z_1 \oplus Z_2$, where $Z_i := C_{V_i}(T \cap M_0) \cong E_4$.

(b) In case (4) of 15.1.7, $V = V_1 \oplus V_2$, where V_1 and V_2 are the two 4-subgroups of V such that $\bar{M}_i := N_{\bar{M}}(V_i)$ is not a 2-group; in this case $V_E = Z_1 \oplus Z_2$ where $Z_i := V_E \cap V_i$ is of order 2.

(c) Finally in case (7) of 15.1.7, $V_E \cong E_{16}$. In this last case, $\bar{P} := O_3(\bar{M}) \cong 3^{1+2}$. Let $\bar{P}_Z := Z(\bar{P})$, pick $\bar{P}_i$ of order 3 in $\bar{M}_0$ inverted by $T \cap M_0$ for $i = 1, 2$ with $\bar{P} = \bar{P}_Z \bar{P}_1 \bar{P}_2$, and set $Z_i := C_V(\bar{P}_i)$ and $V_2 := [V, P_1]$, so that $V = Z_1 \oplus V_2$, $Z_i \cong E_4$, and $V_2 \cong E_{16}$. In this case $V_E = Z_1 \oplus Z_2$.

In each case, set $S := C_T(Z_1)$, $G_1 := C_G(Z_1)$, $M_Z := G_1 \cap M_c$, and $Q_1 := O_2(M_Z)$.

Observe that in each of the cases in Notation 15.1.16, Case I holds by construction: Namely $Z_1 \le V_E = C_V(J_1(T))$, and $M = \langle C_M(Z_1), T \rangle$. Also:

LEMMA 15.1.17. $S \in Syl_2(G_1 \cap M)$, $J(S) = J(T)$, $Baum(S) = Baum(T)$, and $C(G, Baum(S)) \le M$.

PROOF. By construction in Notation 15.1.16, S is Sylow in $G_1 \cap M$ and $C_T(V) \le C_T(Z_1) = S$. So as $J(T) \le C_T(V)$ by 15.1.9.1, $J(S) = J(T)$ and $Baum(S) = Baum(T)$ by (3) and (5) of B.2.3. Then 15.1.9.1 completes the proof. $\square$

LEMMA 15.1.18. (1) $Z_1 \le V_E$, and $O^2(C_G(Z)) \le M_Z$.
(2) $O^2(M \cap M_c \cap G_1) = O^2(C_M(V))$ and $C_M(Z_1) \not\le M_Z$.
(3) S, M_Z, and Q_1 are T-invariant.
(4) $S \in Syl_2(G_1)$.
(5) $C(G_1, Q_1) = M_Z = N_{G_1}(Q_1)$, so Hypothesis C.2.3 is satisfied with G_1, Q_1, M_Z in the roles of "H, R, M_H".
(6) Hypothesis 1.1.5 is satisfied with G_1, M_c in the roles of "H, M", for any $1 \ne z \in Z$.
(7) $M_c = !\mathcal{M}(M_Z T)$ and $C(G, Q_1) \le M_c$.

PROOF. We observed earlier that Case I holds, so in particular, $Z_1 \le V_E$ and $M = \langle C_M(Z_1), T \rangle$. Then as $O^2(C_G(Z)) \le C_G(V_E)$ by 15.1.14.1, and $M_c = !\mathcal{M}(C_G(Z))$, (1) follows; and as $M \not\le M_c$, $C_M(Z_1) \not\le M_Z$. By 15.1.17, $S \in Syl_2(G_1 \cap M)$ and $N_G(S) \le M$, so (4) holds. Hence S is also Sylow in M_Z and in $M \cap M_Z = G_1 \cap M \cap M_c$. Since $Z_1 \le V$, 15.1.5.2 says that $C_M(V) \le G_1 \cap M \cap M_c$. By construction in 15.1.16, $O^2(\overline{C_M(Z_1)})$ is of prime order, so as $C_M(Z_1) \not\le M_Z$ and $S \in Syl_2(M_Z)$, it follows that $O^2(M \cap M_c \cap G_1) = O^2(C_M(V))$, completing the proof of (2). We check in each case in 15.1.16 that $\bar{S} = C_{\bar{T}}(V_E)$, so that $S \trianglelefteq T$. By 15.1.9.2, $M_c \overset{<}{\sim} M$, so

$$M_c = N_M(V(M_c)) C_{M_c}(V(M_c)). \qquad (*)$$

By (1),

$$O^2(C_G(V(M_c))) \le O^2(C_G(Z)) \le O^2(M_Z), \qquad (**)$$

and then by (*) and (**),

$$O^2(M_Z) = O^2(N_{M \cap M_Z}(V(M_c)))O^2(C_{M_c}(V(M_c))).$$

Next from (2), $O^2(M \cap M_Z) = O^2(C_M(V))$, so

$$O^2(M_Z) = O^2(C_M(V))O^2(C_{M_c}(V(M_c))),$$

and hence $O^2(M_Z)$ is T-invariant. Therefore as S is Sylow in M_Z and normal in T, both $O^2(M_Z)S = M_Z$ and $O_2(M_Z) = Q_1$ are also T-invariant, completing the proof of (3). Then as $O^2(C_G(Z)) \le M_Z$ and $M_c = !\mathcal{M}(C_G(Z))$, (7) holds. Since $C(G, Q_1) \le M_c$, $C(G_1, Q_1) = M_Z = N_{G_1}(Q_1)$ and $Q_1 \in \mathcal{B}_2(G_1)$. Then we easily verify Hypothesis C.2.3 with G_1, Q_1, M_Z in the roles of "H, R, M_H", so that (5) holds. Finally for any $1 \ne z \in Z$, $M_c \in \mathcal{M}(C_G(z))$, so (6) follows from 1.1.6 applied to G_1, M_c in the roles of "H, M". $\qquad\square$

LEMMA 15.1.19. *(1)* $O(G_1) = 1$.
(2) If $K = O_{2,2'}(K)$ is an M_Z-invariant subgroup of G_1 with $F^*(K) = O_2(K)$, then $K \le M_Z$.
(3) $O_{2,F}(G_1) \le M_Z$.
(4) If $M_Z \le H \le G_1$ with $O_{2,F^*}(H) \le M_Z$, then $H = M_Z$.
(5) $O_\infty(G_1) \le M_Z$.
(6) There exists $L \in \mathcal{C}(G_1)$ with $L/O_2(L)$ quasisimple and $L \not\le M_Z$.
(7) For L as in (6), $O^2(N_M(Z_1))V_2$ acts on L and $[L, V_2] \ne 1$.

PROOF. Observe that $V_2 = [V_2, O^2(C_M(Z_1))]$ by construction in Notation 15.1.16, so V_2 centralizes $O(G_1)$ by A.1.26.1. Also by construction, $1 \ne Z \cap Z_1 Z_2 = Z \cap Z_1 V_2 =: Z_+$, so that Z_+ centralizes $O(G_1)$. Now by 15.1.18.6, we may apply 1.1.5.2 with any involution of $Z_+^\#$ in the role of "z", so (1) follows.

Assume K_Z is a counterexample to (2). Then K is M_Z-invariant and $S \in Syl_2(G_1)$ by 15.1.18.4, so $O_2(K) \le O_2(KM_Z) \le S \le M_Z$, and hence $O_2(K) \le O_2(M_Z) = Q_1$, so that $Q_1 \in Syl_2(KQ_1)$. Then by 15.1.18.5 and C.2.5, there is an A_3-block X of K with $X \not\le M_Z$. Let $Y := O^2(M_Z)$; then $[Y, X] \le O_2(K) \le N_G(Y)$, so X normalizes $O^2(YO_2(K)) = Y$. However as $M_c = !\mathcal{M}(M_Z T)$ by 15.1.18.7, $X \le N_G(Y) \le M_c$, contrary to the choice of X. This contradiction establishes (2). By (1), $F^*(O_{2,F}(G_1)) = O_2(O_{2,F}(G_1))$, so (2) implies (3).

Assume the hypotheses of (4). Then $Q_1 = O_2(H)$ by A.4.4.1 with M_Z in the role of "K", so $H \le N_G(Q_1) \le M_c$ by 15.1.18.5, establishing (4). By (3), we may apply (4) with $O_\infty(G_1)M_Z$ in the role of "H", to obtain (5). Similarly if (6) fails, then by (3), $O_{2,F^*}(G_1) \le M_Z$, so $G_1 = M_Z$ by (4), contrary to 15.1.18.2.

Finally by 1.2.1.3, $O^2(C_M(Z_1))$ acts on each L satisfying (6), and hence so does $V_2 = [V_2, O^2(C_M(Z_1))]$. Further if V_2 centralizes L, then so does $Z \cap Z_1 V_2 \ne 1$, so that $L \le M_c = !\mathcal{M}(C_G(Z))$, contrary to $L \not\le M_Z$. So V_2 is nontrivial on L, establishing (7). $\qquad\square$

Recall $J(S) = J(T)$ by 15.1.17, and $S \in Syl_2(G_1)$ by 15.1.18.4. Further 15.1.19.6 shows that there is $L \in \mathcal{C}(G_1)$ with $L/O_2(L)$ quasisimple and $L \not\le M_Z$, so the collection of subgroups studied in the following result is nonempty:

LEMMA 15.1.20. *Let* $L \in \mathcal{L}(G_1, S)$ *with* $L/O_2(L)$ *quasisimple and* $L \not\leq M_Z$. *Set* $S_L := S \cap L$ *and* $M_L := M_Z \cap L$. *Then* $S_L \in Syl_2(L)$ *and*

(1) $L \not\leq M$.

(2) Assume $F^*(L) = O_2(L)$. *Then* $L = [L, J(S)]$, *and one of the following holds:*

    (a) L *is a block of type* A_5 *or* $L_2(2^n)$, *and* M_L *is a Borel subgroup of* L.

    (b) $L/O_{2,Z}(L) \cong A_7$, $L_3(2)$, A_6, *or* $G_2(2)'$. *Further if* $L \in \mathcal{C}(G_1)$ *then* L *is a block of type* A_7, $L_3(2)$, A_6, *or* $G_2(2)$. *In the last three cases,* $M_L = C_L(Z)$ *is the maximal parabolic subgroup of* L *centralizing* $C_{U(L)}(S_L)$, *and in the first case* M_L *is the stabilizer of the vector of* $U(L)$ *of weight* 2 *centralized by* S_L.

    (c) $L/O_2(L) \cong L_4(2)$ *or* $L_5(2)$, *and* M_L *is a proper parabolic subgroup of* L.

(3) Assume L *is a component of* G_1. *Then* $Z(L)$ *is a* 2*-group, and one of the following holds:*

    (a) L *is a Bender group or* $L/O_2(L) \cong Sz(8)$, *and* M_L *is a Borel subgroup of* L.

    (b) $L \cong L_3(2^n)$ *or* $Sp_4(2^n)$, $n > 1$, *or* $L/O_2(L) \cong L_3(4)$, *and* M_L *is a Borel subgroup or a maximal parabolic of* L.

    (c) $L \cong G_2(2)'$, ${}^2F_4(2)'$, *or* ${}^3D_4(2)$, *and* $M_L = C_L(Z(S_L))$.

    (d) $L/O_2(L)$ *is a Mathieu group,* J_2, HS, He, *or* Ru, *and* $M_L = C_L(Z(S_L))$.

    (e) $L \cong L_4(2)$ *or* $L_5(2)$, *and* M_L *is a proper parabolic subgroup containing* $C_L(Z(S_L))$.

PROOF. If $L \leq M$, then by 14.1.6.1 and 15.1.5.2, $L \leq C_M(V) \leq M_c$, contrary to the choice of L; so (1) holds.

Since $S \in Syl_2(G_1)$ by 15.1.18.4, and $L \in \mathcal{L}(G_1, S)$ by hypothesis, $S_L = S \cap L$ is Sylow in the subnormal subgroup L of $\langle L, S \rangle$.

Suppose that $L/O_2(L)$ is a simple Bender group; we claim that M_L is the Borel subgroup B_L of L over S_L. For B_L is the unique maximal overgroup of S_L in L, so as $S_L \leq M_L < L$, it follows that $M_L \leq B_L$; then $N_{M_Z}(L)$ acts on $N_L(O_2(M_L)) = B_L$, and hence M_Z normalizes the Borel subgroup $B_0 := \langle B^S \rangle$ of $L_0 := \langle L^S \rangle$. As $L/O_2(L)$ is a Bender group, $F^*(B_0) = O_2(B_0)$, so $B_0 \leq M_Z$ by 15.1.19.2, and hence $M_L = B_L$, as claimed.

We now begin the proof of (2), so assume that $F^*(L) = O_2(L)$. Set $H := LS_H$, where $S_H := N_S(L)$. Then $S_H \in Syl_2(H)$ and as $F^*(L) = O_2(L)$, also $F^*(H) = O_2(H)$. Let $U := \langle Z^H \rangle$ and $H^* := H/C_H(U)$, so that $O_2(H^*) = 1$ by B.2.14. As $C_H(U) \leq C_H(Z) \leq M_Z$ and $L \not\leq M_Z$, $L^* \neq 1$; thus as $L/O_2(L)$ is quasisimple, L^* is quasisimple. As $N_G(J(S)) \leq M$ by 15.1.17, and $L \not\leq M$ by (1), $J(S) \not\leq O_2(\langle L, S \rangle)$ using B.2.3.3. Now by B.2.5, we may apply B.1.5.4 to conclude that $J(S)^*$ normalizes L^*, so that $L = [L, J(S)]$. Thus U is an FF-module for H^* by B.2.7. Therefore by Theorem B.4.2, L^* is one of $L_2(2^n)$, $SL_3(2^n)$, $Sp_4(2^n)'$, $G_2(2^n)'$, $L_n(2)$, for suitable n, or $\hat{A}_6$, or A_7.

Suppose $L^* \cong L_2(2^n)$. If L is a block then $L/O_2(L) \cong L_2(2^n)$, so M_L^* is a Borel subgroup of L^* by paragraph three, and hence conclusion (a) of (2) holds. So we assume L is not a block, and it remains to derive a contradiction. Now as $L/O_2(L)$ is quasisimple, H is a minimal parabolic, so we may apply C.1.26 to conclude that either $C_1(S_H)$ centralizes L, or $C_2(S_H) \trianglelefteq H$. By 15.1.17, $\mathrm{Baum}(T) = \mathrm{Baum}(S)$, and by C.1.16.3, $\mathrm{Baum}(S)$ acts on L, and hence $\mathrm{Baum}(S) = \mathrm{Baum}(S_H)$ by B.2.3.4.

Now by Remark C.1.19, we may take $C_2(S_H) = C_2(T)$ and $C_1(T) \leq C_1(S_H)$. Then $N_G(C_2(S_H)) \leq M$ by 15.1.17, so $C_2(S_H)$ is not normal in H by (1). Therefore $L \leq C_G(C_1(S_H)) \leq C_G(C_1(T)) \leq M_c$ since $C_1(T) \leq Z$ and $M_c = !\mathcal{M}(C_G(Z))$. However this contradicts our hypothesis that $L \not\leq M_Z$.

Suppose next that $L^* \cong SL_3(2^n)$, $Sp_4(2^n)$, or $G_2(2^n)$ with $n > 1$. Let P_i, $i = 1, 2$, be the maximal parabolics of H over S_H, and $L_i := O^2(P_i)$. Then $L_i \in \mathcal{L}(G_1, S)$ with $L_i/O_2(L_i) \cong L_2(2^n)$, but L_i is not a block since $O_2(L_i^*) \neq 1$ and $[U, L_i] \neq 1$, so by the previous paragraph, $L_i \leq M_Z$. Thus $L = \langle L_1, L_2 \rangle \leq M_Z$, contrary to hypothesis.

If $L^* \cong L_4(2)$ or $L_5(2)$, then as $S_L \leq M_L < L$ and $S_L \in Syl_2(L)$, M_L is a proper parabolic subgroup, so conclusion (c) of (2) holds. If L^* is one of the remaining possibilities, then $L/O_{2,Z}(L)$ is listed in conclusion (b) of (2). Thus to complete the proof of (2), it remains to assume that $L \in \mathcal{C}(G_1)$ with $L^* \cong L_3(2)$, A_6, A_7, $\hat{A}_6$, or $G_2(2)'$, and to verify the final two sentences of (2b).

As L is not a χ_0-block, by 15.1.18.5 we may apply C.2.4 to conclude that Q_1 acts on L. So since $L \in \mathcal{C}(G_1)$, the hypotheses of C.2.7 are satisfied, and hence L is listed in C.2.7.3. Set $Z_S := C_U(S_H)$ and $Z_U := C_{[U,L]}(S_H)$; then $Z \leq U \cap Z(S_H) = Z_S$. By B.2.14, $U = C_{Z_S}(L)[U, L]$, so $Z_S = C_{Z_S}(L)Z_U$. Since $M_c = !\mathcal{M}(C_G(Z))$,

$$C_L(Z_U) = C_L(Z_S) \leq C_L(Z) \leq M_L. \qquad (*)$$

Suppose either that $L/O_{2,Z}(L)$ is of rank 2 over $\mathbf{F}_2$, or that L is an exceptional A_7-block. The module $[U, L]/C_{U,L}(L)$ is described in case (i) or (ii) of Theorem B.5.1.1, and in each module U, $C_L(Z_U)$ is a maximal subgroup of L. Therefore as $M_L < L$, the inequalities in $(*)$ are equalities, so that $M_L = C_L(Z_U) = C_L(Z(S_L))$. Suppose instead that L is an ordinary A_7-block; then Z_U contains vectors z_w of weights $w = 2, 4, 6$, and there is $z \in Z \cap C_{Z_S}(L)z_w$ for some w. Then $C_L(z_w) \leq C_L(z) \leq M_L$. But unless $w = 2$, $Aut_{O_2(C_{LS_H}(z_w))}([U, L])$ contains no FF*-offenders by B.3.2.4, contrary to C.2.7.2. Thus $w = 2$ and as $C_L(z_2)$ is maximal in L, $C_L(z_2) = M_L$.

We have shown that if L is one of the four blocks listed in (2b), then (2b) holds, so we may assume L is not one of these blocks. If L^* is A_6 or $G_2(2)'$, then by C.2.7.3, L is an A_6-block or $G_2(2)$-block, contrary to this assumption. If $L^* \cong L_3(2)$, then by C.2.7.3, L is described in C.1.34. By the previous paragraph, M_Z is the parabolic centralizing Z_S, so case (1) or (5) of C.1.34 holds as the other cases exclude Q_1 normal in that parabolic; thus L is an $L_3(2)$-block, again contrary to assumption. If $L/O_{2,Z}(L) \cong A_7$, then by C.2.7.3, L is either an A_7-block or an exceptional A_7-block. Again the first case contradicts our assumption, and in the second case, we showed in the previous paragraph that M_L^* is the maximal subgroup of index 15 fixing Z_U, rather than the subgroup of index 35 in L^* appearing in case (d) of C.2.7.3.

Thus it only remains to eliminate case (c) of C.2.7.3, where L is a block of type $\hat{A}_6$: Here by B.4.2, the only parabolic P of L such that $O_2(P)$ contains an FF-offender is *not* the parabolic $C_L(Z) = M_L$, contrary to C.2.7.2. This completes the proof of (2).

Finally we prove (3), so assume L is a component of G_1. By 15.1.18.6 we can apply 1.1.5, and in particular L is described in 1.1.5.3. Since $O(G_1) = 1$ by 15.1.19.1, $Z(L) = O_2(L)$ is a 2-group. By 1.1.5.1, $M_Z \in \mathcal{H}^e$, so $M_L \in \mathcal{H}^e$ by 1.1.3.1. By 1.1.5.3, Z is faithfully represented on L with $Aut_Z(L) \leq Z(Aut_S(L))$. Thus if some $z \in Z^{\#}$ induces an inner automorphism on L, then as $M_c = !\mathcal{M}(C_G(Z))$, we have $C_L(Z(S_L)) \le C_L(z) \le M_L$.

We first treat cases (a)–(c) of 1.1.5.3, where $L/O_2(L)$ is of Lie type in characteristic 2, and hence described in case (3) or (4) of Theorem C (A.2.3). As M_L is contained in a proper overgroup of $S_L \in Syl_2(L)$, M_L is contained in a proper parabolic subgroup P_L of L (cf. 47.7 in [**Asc86a**]). In cases (a)–(c) of 1.1.5.3, either $L \cong A_6$, or Z induces inner automorphisms on L. In the latter case, $C_L(Z(S_L)) \le P_L$ by the previous paragraph, and in the former $C_L(S_L) = S_L \le P_L$ trivially.

Next if $L/O_2(L)$ is of Lie rank 1, then by 1.1.5.3, $L/O_2(L)$ is a simple Bender group, and conclusion (a) of (3) holds by paragraph three. Thus we may assume that $L/O_2(L)$ is of Lie rank at least 2. Then by 1.1.5.3, either L is simple, or $L/O_2(L) \cong L_3(4)$ or $G_2(4)$.

Assume first that $L/O_2(L)$ is defined over $\mathbf{F}_{2^n}$ with $n > 1$. This rules out case (4) of Theorem C, so that $L/O_2(L)$ is in one of the five families of groups of Lie rank 2 in case (3) of Theorem C. Further $L \trianglelefteq G_1$ by 1.2.1.3. If S is nontrivial on the Dynkin diagram of L, then either $L \cong L_3(2^n)$ or $Sp_4(2^n)$ with $n > 1$, or $L/O_2(L) \cong L_3(4)$; further the Borel subgroup B of L over S_L is the unique S-invariant proper parabolic subgroup of L containing S_L, so arguing as in paragraph three, $M_L = B$, and then conclusion (b) of (3) holds.

Thus we may assume that S normalizes both maximal parabolics P_i, $i = 1, 2$, of L over S_L. Then $L_i := P_i^{\infty} \in \mathcal{L}(G_1, S)$ with $F^*(L_i) = O_2(L_i)$, and $L_i/O_2(L_i)$ is either $L_2(2^m)$ (with m a multiple of n) or $Sz(2^n)$. By a Frattini Argument, $M_Z = M_L N_{M_Z}(S_L)$, and $N_{M_Z}(S_L) = O^2(N_{M_Z}(S_L))S$ acts on the two maximal overgroups P_1 and P_2 of S_L in L. Thus M_Z acts on each parabolic P containing M_L, so $O_{2,2'}(P) \le M_Z$ by 15.1.19.2. Then if $L_i \le M_Z$, $P_i \le M_Z$ by 15.1.19.4, so $P_i = M_L$ by maximality of P_i.

If L_i is not a block, then $L_i \le M_Z$ by (2). Thus if neither L_1 nor L_2 is a block, then $L = \langle L_1, L_2 \rangle \le M_Z$, contrary to hypothesis. Therefore L_i is a block for $i = 1$ or 2, so either L is $L_3(2^n)$ or $Sp_4(2^n)$, or $L/O_2(L) \cong L_3(4)$. So if $L_1 \le M_L$, then $M_Z = P_1$ by the previous paragraph, so that conclusion (b) of (3) holds. Thus we may assume that neither L_1 nor L_2 is contained in M_L. Then (cf. 47.7 in [**Asc86a**]), M_L is contained in the Borel subgroup $P_1 \cap P_2 = P$ over S_L, so $M_L = P$ by the previous paragraph, and again conclusion (b) of (3) holds.

Thus we may assume that $L/O_2(L)$ is defined over $\mathbf{F}_2$. Then from 1.1.5.3, and recalling $Z(L)$ is a 2-group, L is simple. So from Theorem C, L is $G_2(2)'$, ${}^2F_4(2)'$, ${}^3D_4(2)$, $Sp_4(2)'$, $L_3(2)$, $L_4(2)$ or $L_5(2)$. Recall from earlier discussion that $P_c := C_L(Z(S_L)) \le M_L \le P_L$ for some proper parabolic P_L of L. However in the first three cases, P_c is a maximal parabolic, so $M_L = P_c$, and hence conclusion (c) of (3) holds. Thus we may assume one of the remaining four cases holds. In those cases, *all* overgroups of S_L are parabolics, so M_L is a parabolic. Thus conclusion (e) of (3) holds if $L \cong L_4(2)$ or $L_5(2)$.

In cases (a)–(c) of 1.1.5.3, we have reduced to $L \cong L_3(2)$ or A_6. We now eliminate these cases, along with case (d) of 1.1.5.3. Since $Z(L) = O_2(L)$, $L \cong A_7$ in the last case. In each case $S_L \cong D_8$, and $Z(S_L)$ is of order 2.

We claim that Z contains a nontrivial subgroup Z_L inducing inner automorphisms on L. If $L \cong L_3(2)$, this follows from earlier discussion. In the other two cases, L is normal in G_1 by 1.2.1.3, so as $Out(L/O_2(L))$ is an elementary abelian

2-group, $\Phi(S)$ induces inner automorphisms on L. Then as $S \trianglelefteq T$ by 15.1.18.3, $1 \neq Z \cap \Phi(S)$ induces inner automorphisms, completing the proof of the claim.

Notice the claim eliminates case (d) of 1.1.5.3 where $L \cong A_7$, as there each $z \in Z^\#$ induces outer automorphisms on L. Thus $L \cong L_3(2)$ or A_6.

Next $Y := O^2(C_G(Z)) \leq M_Z$ by 15.1.18.1, and so Y acts on L by 1.2.1.3. However as L is $L_3(2)$ or A_6, $C_{Aut(L)}(Z(S_L))$ is a 2-group, so as Z_L induces $Z(S_L)$ on L, $O^2(C_{Aut(L)}(Z)) \leq O^2(C_{Aut(L)}(Z(S_L))) = 1$, and hence $Y \leq C_{G_1}(L)$. Then as $M_c = !\mathcal{M}(C_G(Z))$, $L \leq N_G(Y) \leq M_c$, contrary to our choice of L. This completes the treatment of cases (a)–(d) of 1.1.5.3.

Next suppose case (f) of 1.1.5.3 holds. Then $L/O_2(L)$ is sporadic, Z induces inner automorphisms on L, and $Z(S_L O_2(L)/O_2(L))$ is of order 2. Thus by paragraph one, Z induces $Z(S_L)$ on L and $C_L(Z(S_L)) \leq M_L$. Indeed if $L/Z(L)$ is not M_{22}, M_{23}, or M_{24}, then $C_L(Z(S_L))$ is a maximal subgroup, so that $C_L(Z(S_L)) = M_L$. Hence in these cases either conclusion (d) of (3) holds, or $L \cong J_4$, a case we postpone temporarily. Next assume $L/Z(L) \cong M_{22}$, M_{23}, or M_{24}. To complete our treatment of case (f) in these cases, we assume that $C_L(Z(S_L)) < M_L$, and derive a contradiction. Here since $F^*(M_L) = O_2(M_L)$ from paragraph one, the subgroup structure of L determines M_L uniquely as a block of type A_6, exceptional A_7, or $L_4(2)$, respectively. Therefore $[M_L, Z] \neq 1$ as $C_L(Z(S_L)) = C_L(Z_L) = C_L(Z)$. Then $M_L \leq M_c^\infty$, so 1.2.1.1 says M_c^∞ contains a member of $\mathcal{L}_f(G, T)$, contrary to Hypothesis 14.1.5.1.

To complete our treatment of case (f), we may assume $L \cong J_4$. Then there is $K \in \mathcal{L}(G_1, S)$ with $K \leq L$, $F^*(K) = O_2(K)$, and $K \cong M_{24}/E_{2^{11}}$. Now $K \leq M_L$ by (2). But then $L = \langle K, C_L(Z(S_L)) \rangle \leq M_L$, contrary to our choice of $L \not\leq M_Z$.

Finally suppose case (e) of 1.1.5.3 holds. We have already treated the cases where $L \cong L_2(4) \cong L_2(5)$ and $L \cong L_3(2) \cong L_2(7)$. Thus L is either $L_3(3)$, or $L_2(p)$ for $p > 7$ a Fermat or Mersenne prime. By 15.1.19.7, $X := O^2(C_M(Z_1))$ acts on L, and V_2 acts nontrivially on L. Thus X is nontrivial on L since $V_2 = [V_2, X]$ by construction of V_2 in 15.1.16. This is impossible since S_L acts on X, whereas neither $L_3(3)$ nor $L_2(p)$ has a subgroup of odd index in which an element of odd order acts nontrivially on a normal elementary abelian 2-subgroup. (Cf. Dickson's Theorem A.1.3 in the case of $L_2(p)$). $\qquad\square$

In the remainder of this section, we will eliminate cases from 15.1.20 until we have reduced to case (2c), at which point we will derive our final contradiction.

We begin by eliminating cases (2a) and (3ab) of 15.1.20:

LEMMA 15.1.21. *Assume $L \in \mathcal{C}(G_1)$. Assume further that either $F^*(L) = O_2(L)$ with L an $L_2(2^n)$-block or an A_5-block, or L is a component of G_1 with $L/O_2(L)$ a Bender group, $L_3(2^n)$ or $Sp_4(2^n)$. Then $L \leq M_Z$.*

PROOF. Set $S_L := S \cap L$, $M_L := L \cap M_Z$, and assume that $L \not\leq M_Z$. By 15.1.20, M_L is a either the Borel subgroup B_L of L over S_L, or a maximal parabolic of L. Set $L_0 := \langle L^S \rangle$; then $M_Z \cap L_0$ is either the Borel subgroup $B := \langle B_L^S \rangle$ of L_0 over $S \cap L_0$, or a maximal parabolic of L_0. So in any case, $B \leq M_Z$, and S normalizes B.

When L is a block, $L = [L, J(S)]$ by 15.1.20.2, so the action of $J(S)$ on L is described in Theorem B.4.2. When L is a component, $n > 1$ by 15.1.20.3. Then we conclude that one of the following holds:

(i) $J(S) \trianglelefteq SB$, and L is not an A_5-block.

(ii) $O^2(B) = [O^2(B), J(S)]$, and either $L \cong L_2(4)$ or L is an A_5-block.

(iii) $L \cong U_3(2^n)$, some $A \in \mathcal{A}(S)$ does not induce inner automorphisms on L, and $J(S) \trianglelefteq DS$, where D is the subgroup of B generated by all elements of order dividing $2^n - 1$.

Suppose that case (i) or (iii) holds. Let $B_0 := B$ in case (i), and $B_0 := D$ in case (iii). By 15.1.17, $J(T) = J(S)$ and $B_0 \leq N_G(J(S)) \leq M$. As $B \leq M_Z$, $B_0 \leq M \cap M_Z = M \cap M_c \cap G_1$, so by 15.1.18.2, $B_0 = O^2(B_0)$ centralizes V. Thus $V_2 \leq C_S(B_0) \leq C_S(L)$ from the structure of $Aut(L)$ in (i) or (iii), contrary to 15.1.19.7.

Therefore case (ii) holds. Now $J(T) = J(C_T(V))$ by 15.1.9.1, so that $M = C_M(V)N_M(J(T))$ by a Frattini Argument. Hence by construction of Z_1 and V_2 in 15.1.16, there is a p-subgroup Y of $N_M(J(T)) \cap G_1$ where $p := 3$ or 5, satisfying $SY = YS$ and $V_2 = [V_2, Y]$. Now $YS = SY$, Y acts on $J(T) = J(S)$, and $O^2(B) = [O^2(B), J(S)]$, so it follows from the structure of $Aut(L_0)$ that $[L, Y] = 1$. But then $V_2 = [V_2, Y]$ centralizes L_0, contrary to 15.1.19.7. This contradiction completes the proof. $\qquad\square$

We now define notation in force for the remainder of this section: By 15.1.19.6, we may choose $L \in \mathcal{C}(G_1)$ with $L/O_2(L)$ quasisimple and $L \not\leq M_Z$. Set $M_L := M_Z \cap L$ and $S_L := S \cap L$. Then L is described in 15.1.20. In the next lemma, we refine that description.

LEMMA 15.1.22. *One of the following holds:*

(1) L *is an* A_7*-block. Further* $L = \langle M \cap L, M_L \rangle$, $M \cap L$ *is a proper subgroup of* L *containing the stabilizer of the partition of type* $2^3, 1$ *stabilized by* S_L, *and* M_L *is the stabilizer of the vector of weight* 2 *in* $C_{U(L)}(S_L)$.

(2) L *is a block of type* $L_3(2)$, A_6, *or* $G_2(2)$, M_L *is the maximal parabolic subgroup of* L *centralizing* Z, $M \cap L$ *is the remaining maximal parabolic over* S_L, *and* $C_S(O^{3'}(M \cap L)) = C_S(L)$.

(3) $F^*(L) = O_2(L)$ *with* $L/O_2(L) \cong L_4(2)$ *or* $L_5(2)$, *and* M_L *and* $M \cap L$ *are proper parabolic subgroups of* L *which generate* L.

(4) $L \cong G_2(2)'$, $^2F_4(2)'$, *or* $^3D_4(2)$, $M_L = C_L(Z(S_L))$, $M \cap L$ *is the remaining maximal parabolic over* S_L, *and* $C_S(O^{3'}(M \cap L)) = C_S(L)$.

(5) $L/Z(L)$ *is* J_2, He, *or a Mathieu group other than* M_{11}, $M_L = C_L(Z(S_L))$, *and* $C_S(O^{3'}(M \cap L)) = C_S(L)$.

(6) $L \cong M_{11}$, $M_L = C_L(Z(S_L))$, *and* $O^2(N_{G_1}(J(S))) \leq C_{G_1}(L)$.

(7) $L \cong L_4(2)$ *or* $L_5(2)$, $L = \langle M_L, M \cap L \rangle$, *where* M_L *is a proper parabolic containing* $C_L(Z(S_L))$, $M \cap L$ *is a proper parabolic, and* $C_S(O^{3'}(M \cap L)) = C_S(L)$.

PROOF. Observe that $M \cap L < L$ by 15.1.20.1, and $M_L < L$ since $L \not\leq M_Z$ by the choice of L. By 15.1.19.1, $Z(L)$ is a 2-group.

We first establish some preliminary technical results. The first is on overgroups of S_L in L. Let $\mathcal{P}$ be the set of $N_S(L)$-invariant subgroups P of L such that $F^*(P) = O_2(P)$, $PN_S(L)/O_2(PN_S(L)) \cong S_3$ or S_3 wr $\mathbf{Z}_2$, and $O^2(P)$ is not a product of A_3-blocks. We show:

(*) For $P \in \mathcal{P}$, either $P \leq M \cap L$ or $P \leq M_L$.

For let $P_0 := \langle P, S \rangle$; then P_0 is a minimal parabolic in the sense of Definition B.6.1, so as $O^2(P)$ is not a product of χ_0-blocks, we conclude from C.1.26 that either $C_1(S)$ centralizes P, or $C_2(S)$ is normalized by P. But $\text{Baum}(T) = \text{Baum}(S)$ by

15.1.17, so by Remark C.1.19 we may take $C_2(S) = C_2(T)$ and $C_1(T) \leq C_1(S)$. Thus $N_G(C_2(S)) \leq M$ by 15.1.17, while $C_G(C_1(S)) \leq M_c$ since $C_1(T) \leq Z$ and $M_c = !\mathcal{M}(C_G(Z))$. This establishes (*).

Next let $G_L := N_{G_1}(L)$, $G_L^+ := G_L/C_{G_1}(L)$, and $Z_L^+ := \Omega_1(Z(N_S(L)^+)))$. We establish:

(**) If $|Z_L^+| = 2$, then $Z^+ = Z_L^+$ and $C_{L^+}(Z^+) = C_L(Z)^+ \leq (M_Z \cap L)^+$. Further if $O^{3'}(M \cap L) \not\leq M_L$, then $C_S(O^{3'}(M \cap L)) = C_S(L)$.

For assume the hypotheses of (**). As $L \not\leq M_Z$ but $M_c = !\mathcal{M}(C_G(Z))$, L does not centralize Z, and hence $Z^+ \neq 1$. Therefore since $|Z_L^+| = 2$, $Z^+ = Z_L^+$, so $O^2(C_{L^+}(Z_L^+)) = O^2(C_L(Z)^+)$ by Coprime Action. Then as the Sylow 2-group S of G_1 centralizes Z, $C_{L^+}(Z_L^+) = C_L(Z)^+ \leq (M_Z \cap G_L)^+$. Therefore if $O^{3'}(M \cap L) \not\leq M_L$, then $O^{3'}(M \cap L)^+$ does not centralize Z_L^+. However, $N_S(L)$ acts on $D := C_S(O^{3'}(M \cap L))$, so if $D^+ \neq 1$, then $1 \neq Z_L^+ \cap D^+$. Then as $|Z_L^+| = 2$, Z_L^+ lies in D^+, and so centralizes $O^{3'}(M \cap L)^+$, contrary to the previous remark. This completes the proof of (**).

Our final preliminary result says:

(!) If $\mathcal{P}_0 \subseteq \mathcal{P}$ with $\langle \mathcal{P}_0 \rangle \not\leq M_L$, then $P \leq M$ for each $P \in \mathcal{P}_0$ with $P \not\leq M_L$. If in addition $|Z_L^+| = 2$, then $C_S(O^{3'}(M \cap L)) = C_S(L)$.

Under the hypothesis of (!), the first statement follows from (*), and so in particular $O^{3'}(M \cap L) \not\leq M_L$. Then the second statement follows from (**).

We now begin to show that one of the conclusions of the lemma must hold. By 15.1.21, cases (2a), (3a), and (3b) of 15.1.20 do not hold.

Suppose that case (2b) or (3c) of 15.1.20 holds, but L is not an A_7-block. Therefore either L is a block of type $L_3(2)$, A_6, or $G_2(2)$, or $L \cong G_2(2)'$, $^2F_4(2)$, or $^3D_4(2)$. In each case $N_S(L)$ is trivial on the Dynkin diagram of $L/O_2(L)$; when L is a block, this follows since $U(L)/C_{U(L)}(L)$ is the natural module. Thus each minimal parabolic over S_L is $N_S(L)$-invariant. Further in each case, $C_L(Z_L^+)$ is one of these minimal parabolics, with $M_L = C_L(Z_L^+)$ by 15.1.20. Let P denote the other minimal parabolic over S_L, and set $\mathcal{P}_0 := \{P\}$. As $O^2(P)$ is not an A_3-block, $\mathcal{P}_0 \subseteq \mathcal{P}$. Thus if we can show that $|Z_L^+| = 2$, then conclusion (2) or (4) of 15.1.22 will hold by (!). When L^+ is simple, this is a well-known fact (cf. 16.1.4 and 16.1.5) about the structure of $Aut(L)$, so we may assume L is a block. Here $F^*(L^+) = O_2(L)^+ = O_2(L)^+$ by A.1.8, so also $F^*(L^+ N_S(L)^+) = O_2(L^+ N_S(L)^+) =: Q_L^+$, and hence $Z_L^+ \leq Q_L^+$. But then $Z_L^+ = C_{U(L)^+}(N_S(L)^+)$ by Gaschütz's Theorem A.1.39. Thus $|Z_L^+| = 2$ from the action of L on $U(L)$, completing the proof that the lemma holds in this case.

Next we consider the remaining case in (2b) of 15.1.20, where L is an A_7-block. Here we adopt the notation of section B.3, let P denote the preimage of the stabilizer of the partition $\{\{1,2\}, \{3,4\}, \{5,6\}, \{7\}\}$, and set $\mathcal{P}_0 := \{P\}$. Again $\mathcal{P}_0 \subseteq \mathcal{P}$. Further by 15.1.20, M_L is the stabilizer of the vector $e_{5,6}$ of $U(L)$, and hence $P \not\leq M_L$, so $P \leq M$ by (!), completing the proof that conclusion (1) holds in this case.

Now assume that case (2c) or (3e) of 15.1.20 holds, so that $L/O_2(L) \cong L_4(2)$ or $L_5(2)$. Then $S = N_S(L)$ by 1.2.1.3. Let P_c denote the parabolic generated by the minimal parabolics for the interior nodes in the diagram for $L/O_2(L)$. In case (3e), $|Z_L^+| = 2$, and by 15.1.20, M_L is a proper parabolic containing $P_c = C_L(Z_L^+)$.

In case (2c), M_L is some proper parabolic. In any case, let

$$\mathcal{P}_0 := \{\langle P^S \rangle : \quad P \text{ is a minimal parabolic and } P \not\leq M_L\}.$$

Now either $L^+ S^+ \cong Aut(L_5(2))$ and $F^*(L) = O_2(L)$, or $\mathcal{P}_0 \subseteq \mathcal{P}$. In the latter case, conclusion (3) or (7) of 15.1.22 holds by (*) and (!), so we may assume the former case holds. Here $L = \langle P_c, P_e \rangle$, where P_e is the parabolic generated by the two end-node minimal parabolics, $P_e \in \mathcal{P}$, and $P_c S / O_2(P_c S) \cong Aut(L_3(2))$. As $P_e \in \mathcal{P}$, P_e is contained in $M_e \in \{M_L, M \cap L\}$ by (*). Then as $P_e S$ is a maximal subgroup of LS, $M_e = P_e$.

If P_c centralizes Z, then $P_c \leq M_c$ as $C_G(Z) \leq M_c$, so $P_c = M_L$ by maximality of $P_c S$ in LS. Then $P_e = M \cap L$ by the previous paragraph, so that conclusion (3) of 15.1.22 holds. Thus we may assume that $[Z, P_c] \neq 1$. Now $W := \langle Z^{P_c} \rangle \in \mathcal{R}_2(P_c)$ by B.2.14, so as $P_c S / O_2(P_c S) \cong Aut(L_3(2))$, and the latter group has no FF-module by Theorem B.5.1, we conclude that $J(S) \leq C_{P_c S}(W) = O_2(P_c S)$. Therefore $J(S) = J(O_2(P_c S))$ by B.2.3.3, and hence $P_c \leq N_G(J(S)) \leq M$ by 15.1.17. As $M \cap L < L$ and P_c is a maximal S-invariant subgroup of L, we conclude that $P_c = M \cap L$, and then $P_e = M_L$ by the previous paragraph, so again conclusion (3) of 15.1.22 holds.

Finally we assume that case (d) of 15.1.20.3 holds, so that L is a component of G_1 with $L/Z(L)$ sporadic, and $Z(L) = O_2(L)$.

Suppose first that $L/Z(L)$ is HS or Ru. Then there is $K \in \mathcal{L}(LS, S) \cap L$ with $F^*(K) = O_2(K)$ and $K/O_2(K) \cong L_3(2)$. Further $O_2(K) \cong \mathbf{Z}_4^3$ or 2^{3+8}, respectively, so KS is not among the conclusions of C.1.34. Hence by C.1.34, there is a nontrivial characteristic subgroup C of S normal in K. Then as $S \trianglelefteq T$ by 15.1.18.3, $\langle K, T \rangle \leq N := N_G(C)$. Then $K \leq N^\infty \leq M_c$ by 14.1.6.3; but this is impossible, as $K \not\leq C_L(Z(S_L))$, whereas $C_L(Z(S_L)) = M_L$ by 15.1.20.3.

Therefore $L/Z(L)$ is a Mathieu group, J_2, or He, and $M_L = C_L(Z(S_L))$ by 15.1.20.3. Assume first that $L/Z(L)$ is not M_{11}, and set $K := \langle M_L, \mathcal{P} \rangle$. Then from the structure of $Aut(L)$, either $K = L$, or $L/Z(L) \cong M_{22}$ and $K/Z(L) \cong A_6/E_{16}$. Moreover in the latter case, $K > M_L$ as we saw in our treatment of M_{22} during the proof of 15.1.20. Thus in any case there is $P \in \mathcal{P}$ with $P \not\leq M_L$, and as $|Z_L^+| = 2$ in these groups, (!) completes the proof that conclusion (5) holds.

It remains to treat the case $L/Z(L) \cong M_{11}$, where $L \cong M_{11}$ by I.1.3, and $L \trianglelefteq G_1$ by 1.2.1.3. Then $Out(L) = 1$, so that $G_1 = L \times C_{G_1}(L)$; in particular $J(S) = J(C_S(L)) \times J(S_L)$, and hence $N_{G_1}(J(S)) \leq N_{G_1}(J(S_L))$. Further $J(S_L) \cong D_8$, so that $N_L(J(S_L)) = S_L$, and hence $O^2(N_{G_1}(J(S)))$ centralizes L, so that conclusion (6) holds.

This completes the proof of 15.1.22. $\square$

LEMMA 15.1.23. *If case (6) of 15.1.7 holds, then $p = 3$, so $\bar{M} \cong S_3$ wr $\mathbf{Z}_2$.*

PROOF. Assume case (6) of 15.1.7 holds. If $p = 3$, then $\bar{M} \cong S_3$ wr $\mathbf{Z}_2$ by 15.1.10. So we may assume that $p = 5$, and it remains to derive a contradiction. Then $\bar{M}_0 \cong D_{10} \times D_{10}$. Hence there is a 5-group $Y \leq C_M(Z_1)$ with $SY = YS$ and $V_2 = [V_2, Y]$. Set $G_0 := N_{G_1}(L)$ and $G_0^* := G_0/C_{G_0}(L)$. By 15.1.19.7, YV_2 acts on L, and V_2 acts nontrivially on L. Thus as Y is faithful and irreducible on V_2, YV_2 is faithful on L. Thus comparing the list in 15.1.22 to the possibilities for $L/O_2(L)$ in A.3.15, we conclude $L \cong {}^2F_4(2)'$, and $Aut_Y(L) \leq Aut_P(L)$, where $P := C_L(Z(S_L))$ with $P/O_2(P) \cong Sz(2)$. This is impossible, as $Aut_{YS}(L)$ does not act irreducibly on an E_{16}-subgroup $Aut_{V_2}(L)$ of P. $\square$

In the remainder of the section, let $Y := O^{3'}(G_1 \cap M)$. As G is a counterexample to Theorem 15.1.15, 15.1.23 says we are in case (4), case (6) with $p = 3$, or case (7) of 15.1.7, so that $\bar{M}_0$ is a $\{2, 3\}$-group. In particular $Y \not\le M_Z$ by 15.1.18.2.

LEMMA 15.1.24. (1) *Either case (4) of 15.1.7 holds, or case (6) of 15.1.7 holds with $p = 3$. In particular, $Z_1 \le V_1$.*

(2) *L is not an $L_3(2)$-block.*

(3) *For each $Y_0 = O^2(Y_0) \le Y$ with $Y_0 \not\le M_Z$, $V_2 = [V_2, Y_0]$ and $|Y_0 : C_{Y_0}(V_2)| = 3$. In particular $V_2 = [V_2, Y]$.*

(4) *$V_1 \le C_S(Y)$.*

PROOF. By construction of V_2 in 15.1.16, and since $p = 3$ when case (6) of 15.1.7 occurs, $\bar{Y}$ is of order 3 and $V_2 = [V_2, Y]$. Thus (3) follows from 15.1.18.2. Further from 15.1.16, in cases (4) and (6) of 15.1.7, $V_1 \ge Z_1$ and $\bar{Y}$ centralizes V_1, so (4) will follow once we prove (1).

To establish (1), we may assume that case (7) of 15.1.7 holds, and we must produce a contradiction. Let X_0 be the preimage in M of $Z(O^2(\bar{M}))$, $R := O_2(M \cap M_c)$, and $X_1 := O^2(\langle R^{X_0} \rangle)$. We apply 14.1.17 to M_c, X_0 in the roles of "M_1, Y_0" to conclude $\bar{X}_1 = \bar{X}_0 = [\bar{X}_0, R]$ and $[R, C_{X_1}(V)] \le O_2(M)$. As $\bar{X}_1 = \bar{X}_0$, $Z_1 = [Z_1, X_0]$. Then as $\bar{X}_1 = [\bar{X}_1, R]$ and $[R, C_{X_1}(V)] \le O_2(M)$, there is a subgroup X_2 of X_1 of order 3 with $Z_1 = [Z_1, X_2]$. So as $m_3(N_G(Z_1)) \le 2$, A.3.18 eliminates the possibilities in 15.1.22 of 3-rank 2, leaving the case where L is an $L_3(2)$-block. Then by 1.2.1.3, $X_2 = O^2(X_2)$ normalizes L and $O^{3'}(G_1) = LO^{3'}(C_{G_1}(L/O_2(L)))$. Therefore as $X_2 \not\le G_1$ and $m_3(N_{G_1}(Z_1)) \le 2$, $L = O^{3'}(G_1)$.

Thus to establish both (1) and (2), it suffices to assume L is an $L_3(2)$-block. As usual let $U(L) := [O_2(L), L]$. By 15.1.22, M_L is the parabolic of L centralizing $Z_S := \Omega_1(Z(S_L))$, and $M \cap L$ is the remaining maximal parabolic of L over S_L. Let $Y_0 := O^2(M \cap L)$, so that as $L \le G_1$, $Y_0 \le Y$ but $Y_0 \not\le M_L$; hence $V_2 = [V_2, Y_0]$ and $C_{Y_0}(V_2) = O_2(Y_0)$ by (3). Thus $V_2 \le Z(O_2(Y_0)) \le U(L)$, so $V_2 \le [C_{U(L)}(O_2(Y_0)), Y_0] =: U_2$. As L is an $L_3(2)$-block, U_2 is of rank 2, so $V_2 = U_2$ is of rank 2. This eliminates cases (6) and (7) of 15.1.7, and in particular completes the proof of conclusions (1) and (4) as mentioned earlier, though not yet of (2). Thus case (4) of 15.1.7 holds. Further $V_1 \le C_S(Y_0)$ by (4), and $C_S(Y_0) = C_S(L)$ as L is an $L_3(2)$-block, so L centralizes V_1. Since $Z_1 \le V_1$ by (1), $C_G(V_1) \le C_G(Z_1) = G_1$, and hence $L \in \mathcal{C}(C_G(V_1))$. Let $t \in T - S$ and $X \in Syl_3(Y_0^t)$; then X is of order 3, and by our construction of V_1 and V_2 in 15.1.16, $V_1 = [V_1, X]$ and $[V_2, X] = 1$. As $L \in \mathcal{C}(C_G(V_1))$, $Y_0^t = O^2(Y_0^t)$ acts on L by 1.2.1.3, and as X centralizes $V_2 = [U(L), Y_0]$, X centralizes L, since L is an $L_3(2)$-block. Then as $m_3(N_G(V_1)) \le 2$ and $X \not\le C_G(V_1)$, arguing as above, we conclude that $L = O^{3'}(C_G(V_1))$. Indeed as X centralizes L, so does $\langle X^{Y_0^t} \rangle = Y_0^t$. Then $U(L) \le C_S(Y_0^t) = C_S(L^t)$, and by symmetry, $U(L)^t$ centralizes L. Thus $\langle L, T \rangle$ acts on $W := U(L)U(L)^t$. Then setting $N := N_G(W)$, $L \le N^\infty \le M_c$ by 14.1.6.3, contrary to the choice of L. This contradiction completes the proof of (2), and hence of the lemma. $\square$

LEMMA 15.1.25. (1) *$L = O^{3'}(G_1)$.*

(2) *L is not M_{11}.*

(3) *V_1 does not centralize L.*

(4) *$|S : C_S(V_1)| = 2$.*

(5) *L is not an A_7-block.*

(6) *$|T : S| = 2$.*

PROOF. By 15.1.24.2, L is not an $L_3(2)$-block, while in all other cases of 15.1.22, $m_3(L) = 2$; then we obtain (1) from A.3.18.

Let $G_1^* := G_1/C_{G_1}(L/O_2(L))$. By (1), $Y \leq L$, so $Y = O^{3'}(L \cap M)$. By 15.1.17, $J(S) = J(T)$ and $N_G(J(S)) \leq M$. By 15.1.9.1, $J(T)$ centralizes V, so by a Frattini Argument, $Y = C_Y(V)Y_0$, where $Y_0 := O^2(N_Y(J(S)))$. Now $C_Y(V) \leq M_c$ by 15.1.5.2, but we saw $Y \not\leq M_Z$, so $Y_0 \neq 1$. However if $L \cong M_{11}$, then as $Y_0 \leq O^2(N_{G_1}(J(S)))$, Y_0 centralizes L as case (6) of 15.1.22 holds, contradicting $1 \neq Y_0 \leq L$. Hence (2) is established.

Next recall by 15.1.24.1, that we are in case (4) or (6) of 15.1.7, so that $|T : S| = 2$ from our construction in 15.1.16; thus (6) holds.

We turn to the proof of (3). Let $t \in T - S$; then $V_2^t = V_1$ since we are in case (4) or (6) of 15.1.7. By 15.1.24.3, $V_2 = [V_2, Y]$, and so $V_1 = [V_1, Y^t]$. Further a Sylow 3-group of L, and hence also of Y, is of exponent 3, so there is X of order 3 in Y^t faithful on V_1. However if V_1 centralizes L, then as $Z_1 \leq V_1$, $L = O^{3'}(C_G(V_1))$, while $X \not\leq L$ as X is faithful on V_1. This is a contradiction as $L = O^{3'}(N_G(V_1))$ by A.3.18, so (3) is established.

Part (4) follows from the action of $\bar{M}$ on V in cases (4) and (6) of 15.1.7; use 15.1.23 in case (6).

Finally suppose L is an A_7-block. Represent LS on $\Omega := \{1, \ldots, 7\}$, and adopt the notation of section B.3. By 15.1.22, Y^*S^* contains the stabilizer P^* of the partition $\{\{1,2\}, \{3,4\}, \{5,6\}, \{7\}\}$. Let P be the preimage of P^* and $Y_1 := O^2(P)$. By 15.1.24.4, $V_1 \leq C_S(Y_1)$, and from the representation of LS on $U(L)$, $C_S(Y_1) \leq C_S(L)\langle u, s\rangle$, where $u := e_\theta$, $\theta := \{1, \ldots, 6\}$, and $s^* := (1,2)(3,4)(5,6)$. Therefore as s^* does not induce a transvection on $U(L)$, we conclude from (4) that $V_1 \leq C_{LS}(U(L)) = O_2(LS)$. So as $V_1 \leq C_S(L)\langle u, s\rangle$, $V_1 \leq C_S(L)\langle u\rangle$, so V_1 centralizes $K := L_7^\infty$ with $K/O_2(K) \cong A_6$. As $Z_1 \leq V_1$, $K = O^{3'}(C_G(V_1))$ by (1), and then it follows from A.3.18 that $O^{3'}(N_G(V_1)) = K \leq C_G(V_1)$. This is a contradiction, as the subgroup X of order 3 defined earlier acts nontrivially on V_1. Hence the proof of (5) and of 15.1.25 is complete. $\qquad\square$

LEMMA 15.1.26. $F^*(L) = O_2(L)$ and $L/O_2(L) \cong L_4(2)$ or $L_5(2)$.

PROOF. Assume otherwise. In cases (2), (4), (5), and (7) of 15.1.22, $C_S(O^{3'}(M \cap L)) = C_S(L)$. Thus by 15.1.24.4, $V_1 \leq C_S(Y) \leq C_S(L)$, contrary to 15.1.25.3. Further cases (1) and (6) of 15.1.22 were eliminated in parts (5) and (2) of 15.1.25, leaving only case (3) of 15.1.22, where the lemma holds. $\qquad\square$

By 15.1.26, $F^*(LS) = O_2(LS)$. Let $U := \langle Z_2^L\rangle$ and $U_L := [U, L]$. By 15.1.24.1, case (4) or (6) of 15.1.7 holds, where by construction in 15.1.16, Z is a full diagonal subgroup of $Z_1 \oplus Z_2$, so $C_{G_1}(Z) = C_{G_1}(Z_2)$ and $S = C_T(Z_1) = C_T(Z_2)$. Thus $U \in \mathcal{R}_2(LS)$ by B.2.14. Set $(LS)^* := LS/C_{LS}(U)$, and recall $C_{LS}(U) = O_2(LS)$ since $L/O_2(L)$ is simple and $U \in \mathcal{R}_2(LS)$. From 15.1.20.2, $L = [L, J(S)]$, so that U is an FF-module for L^*S^*; then we conclude from Theorem B.5.1 and B.4.2 using I.1.6 that:

LEMMA 15.1.27. One of the following holds:

(1) U_L is the orthogonal module or its 7-dimensional cover for $L^* \cong L_4(2)$.

(2) U_L is a 10-dimensional irreducible for $L^* \cong L_5(2)$.

(3) U_L is the sum of the natural module and its dual.

(4) U_L *is a sum of at most* $n - 1$ *isomorphic natural modules for* $L^* \cong L_n(2)$, *where* $n = 4$ *or* 5.

By B.2.14, $U = U_L C_U(L)$. Let Z_U be the projection of Z_2 on U_L with respect to this decomposition.

LEMMA 15.1.28. U_L *is a natural module and* $M_L = C_L(Z)$ *is the stabilizer of the point* Z_U *in* U_L.

PROOF. First by 15.1.18.5, $C(G_1, Q_1) = M_Z$, and Hypothesis C.2.3 is satisfied with G_1, Q_1, M_Z in the roles of "H, R, M_H". Thus by C.2.1.2, $O_2(LS) \leq Q_1$. Further L is normal in G_1 by 15.1.25.1, so we may apply C.2.7.2 to conclude that Q_1 contains an FF-offender on U.

As $C_{G_1}(Z) = C_{G_1}(Z_2)$, $C_{LS}(Z_U) = C_{LS}(Z) \leq M_Z$. That is, M_L is an S-invariant proper parabolic containing $C_L(Z_U)$.

Suppose case (1), (2), or (4) of 15.1.27 holds. Then $C_L(Z_U)$ is a maximal parabolic, acting irreducibly on $O_2(C_L(Z_U)^*)$, so by the previous paragraph $M_L = C_L(Z_U)$ and $Q_1^* = O_2(M_L^*)$. Therefore as Q_1^* contains an FF*-offender, we conclude from B.3.2 or B.4.2 that case (1) holds with U_L the natural module for L, so that the lemma holds in this case.

Thus we may assume case that (3) of 15.1.27 holds. Therefore $U = U_1 \oplus U_2$, where U_1 is a natural module for L^*, and U_2 is its dual. Let $Z_0 := C_{U_L}(S)$, so that $Z_U \leq Z_0$. Then either S acts on U_1 and U_2 with $Z_0 = Z_{U,1} \oplus Z_{U,2}$, where $Z_{U,i}$ is the point of U_i fixed by S_L, or else S is nontrivial on the Dynkin diagram of L^*, with $Z_0 = \langle z_1 z_2 \rangle$ where $Z_{U,i} := \langle z_i \rangle$. In either case, $C_L(Z_0)$ contains the parabolic P determined by the interior node(s) of the diagram for L^*. Thus as $C_L(Z_0) \leq C_L(Z_U) \leq M_L$, $Q_1^* \leq O_2(P^*)$. But then by B.4.9.2, Q_1^* contains no FF*-offenders, contrary to an earlier remark. $\qquad\square$

LEMMA 15.1.29. U_L *is not a natural module.*

PROOF. By 15.1.24.1, we are in case (4) or (6) of 15.1.7. Adopt the notation of the proof of 15.1.28. By 15.1.28, the projection Z_U of Z_2 on U_L is of order 2. We saw $U = U_L C_U(L)$ and Z is a full diagonal subgroup of $Z_1 Z_2$ with $[Z_1, L] = 1$, $L \nleq M_c$, and $C_G(z) \leq M_c$ for each $z \in Z^\#$. Thus Z projects faithfully on U_L, so $|Z| = 2$. Therefore case (4) of 15.1.7 holds, rather than case (6) with $p = 3$, so V_2 is of rank 2. As S acts on V_2, and $V_2 = [V_2, Y] \leq [U, L] \leq U_L$, it follows that V_2 is the line in U_L stabilized by S. Thus $N_L(V_2)$ contains the minimal parabolic P_0 of LS over S which is not contained in the maximal parabolic $M_L S = C_L(Z)S$. By 15.1.9.1, $J(T) \leq C_T(V_2)$, and by 15.1.17, $J(T) = J(S)$ and $N_G(J(S)) \leq M$. Hence $J(S) = J(C_S(V_2)) = J(O_2(P_0))$ using B.2.3.3, so that $P_0 \leq M$, and hence $Y_0 := O^2(P_0) \leq Y$ with $V_2 = [V_2, Y_0]$. Let P_2 be the minimal parabolic adjacent to P_0 with respect to the Dynkin diagram of L, let $Y_2 := O^2(P_2)$, and let $K := \langle Y_0, Y_2 \rangle$. Thus $KS/O_2(KS) \cong L_3(2)$ with KS the rank-2 parabolic corresponding to an end node and its neighbor, and $KS \cap M_L S = P_2$. Let $Q := C_T(V)$ and $t \in T - S$. Let P_3 be the remaining end-node minimal parabolic of L, and set $L_0 := O^{3'}(M_L)$. Thus $L_0 S / O_2(L_0 S) \cong L_{n-1}(2)$, and P_2 and P_3 are the end-node minimal parabolics of $L_0 S$. By 15.1.25.1 and 15.1.28, $L_0 = O^{3'}(C_{G_1}(Z))$, so as $O^2(C_G(Z)) \leq C_G(V_E)$ by 15.1.14.1, and $Z_1 \leq V_E$ by 15.1.18.1, $L_0 = O^{3'}(C_G(Z))$. Thus T acts on L_0. Hence as P_2 and P_3 are the end-node minimal parabolics of L_0, Y_2^t is either Y_2 or $Y_3 := O^2(P_3)$.

Next $O_2(P_0) = C_S(V_2)$, and as case (4) of 15.1.7 holds, $C_S(V_2) = C_T(V)$, so $Q = O_2(P_0)$. Thus from the structure of the rank-2 parabolics of LS: $O_2(Y_3) \leq Q$ so that $Q \in Syl_2(QY_3)$, as the nodes determining P_0 and P_3 are not adjacent in the diagram of L; but $Q \notin Syl_2(P_2)$, as the nodes determining P_0 and P_2 are adjacent. Therefore as t acts on Q, $Y_2^t \neq Y_3$, so $Y_2^t = Y_2$.

Let $Q_2 := O_2(P_2)$; as T acts on Y_2 and on S, T acts on Q_2. But $P_2 = C_{KS}(Z)$, and K has noncentral 2-chief factors on both $O_2(L)$ and $O_2(K)O_2(L)/O_2(L)$, so that K is not an $L_3(2)$-block. We conclude from C.1.34 that there is a nontrivial characteristic subgroup C of Q_2 with $C \trianglelefteq KQ_2$, and hence $C \trianglelefteq KT$. Then $H := \langle T, K \rangle \leq N_G(C)$. On the other hand, Y_0 centralizes Z_1 but $V_2 = [V_2, Y_0]$, so from our construction in 15.1.16, $M = \langle T, Y_0 \rangle C_M(V)$. Then by A.5.7.1, $M = !\mathcal{M}(\langle T, Y_0 \rangle)$. Since $Y_0 \leq K$, we conclude $H \leq N_G(C) \leq M$. But then $K = K^\infty \leq C_M(V)$ by 14.1.6.1, contrary to $V_2 = [V_2, Y_0]$. This contradiction completes the proof of 15.1.29. $\qquad\square$

Observe that 15.1.28 and 15.1.29 supply a contradiction which establishes Theorem 15.1.15.

15.2. Finishing the reduction to $\mathbf{M_f}/\mathbf{C_{M_f}}(\mathbf{V}(\mathbf{M_f})) \simeq \mathbf{O_4^+(2)}$

In this section, we complete the proof of Theorem 15.1.3, begun in section 15.1. Thus we assume G is a counterexample to Theorem 15.1.3.

15.2.1. Preliminary reductions. Recall we are assuming Hypothesis 14.1.5; in particular by 14.1.5.2,

$$M_c = !\mathcal{M}(C_G(Z)).$$

We continue Notation 15.1.4: namely we set $M := M_f$, and set $V := V(M)$ unless case (6) of 15.1.2 holds, where we set $V := [V(M), M_J]$. Also M_0 is the preimage in M of $\hat{J}(\bar{M}, V)$.

Since Theorem 15.1.15 eliminated cases (4), (6), and (7) of 15.1.7, we have reduced to the remaining cases in 15.1.7, which we summarize below for convenience:

LEMMA 15.2.1. $m(V) = 4$, and one of the following holds:

(1) $\bar{M} = \bar{M}_0 \cong S_3$.
(2) $\bar{M}_0 \cong S_3$ and $\bar{M} \cong S_3 \times \mathbf{Z}_3$.
(3) $\bar{M} = \bar{M}_0 = \Omega_4^+(V)$.
(4) $\bar{M}_0 \cong D_{10}$, $\bar{T} \cong \mathbf{Z}_2$ or $\mathbf{Z}_4$, and either $F(\bar{M}) = F(\bar{M}_0)$ or $F(\bar{M}) \cong \mathbf{Z}_{15}$.
Furthermore if $V < V(M)$, then case (3) holds.

LEMMA 15.2.2. If $T \leq X \leq M$ with $M_0 \leq X C_M(V)$ or $M_0 \leq X N_M(Z \cap V)$, then $M = !\mathcal{M}(X)$.

PROOF. Let $M_1 \in \mathcal{M}(X)$. By 15.1.5.1, $V = \langle (Z \cap V)^X \rangle$, and by 15.1.9.2, $M_1 \overset{\sim}{\leq} M$, so $M_1 \leq M$ by A.5.6. $\qquad\square$

LEMMA 15.2.3. Let $R_c := O_2(M \cap M_c)$, $Y := O^2(\langle R_c^{O^2(M_0)T} \rangle)$, and $M^* := M/O_2(M)$. Then

(1) $1 \neq \bar{Y} = [\bar{Y}, \bar{R}_c]$, and one of the following holds:

    (i) $\bar{Y} = O^2(\bar{M}_0) \cong Y^*$. Further if case (3) of 15.2.1 holds, then $C_M(V)$ is a $3'$-group.

(ii) Case (3) of 15.2.1 holds, $\bar{R}_c \cong \mathbf{Z}_2$ inverts $O^2(\bar{M}_0) = O^2(\bar{M}) = \bar{Y}$, $Y^* \cong 3^{1+2}$, and $O^2(O_{2,Z}(Y))$ is the subgroup $\theta(C_M(V))$ generated by all elements of $C_M(V)$ of order 3.

(iii) Case (3) of 15.2.1 holds, $\bar{R}_c \cong \mathbf{Z}_2$ inverts $\bar{Y} \cong Y^* \cong \mathbf{Z}_3$, and a Sylow 3-subgroup of M is isomorphic to $\mathbf{Z}_3 \times \mathbf{Z}_{3^n}$ for some $n \geq 1$.

(2) $Y \trianglelefteq M$.

(3) If $\bar{Y} = O^2(\bar{M}_0)$, then $M = !\mathcal{M}(YT)$.

(4) $M = (M \cap M_c)O^2(M_0)$.

(5) $R_c^* Y^*$ centralizes $C_M(V)^*$.

PROOF. Part (3) follows from 15.2.2. Set $Y_0 := O^2(M_0)$. To establish the remaining parts, we apply case (b) of 14.1.17 with M_c in the role of "M_1". By 14.1.17: $\bar{R}_c \neq 1$, $\bar{Y} = [\bar{Y}_0, \bar{R}_c]$, and $R_c^* Y^*$ centralizes $C_M(V)^*$, so that (5) holds. As $\bar{R}_c \neq 1 = O_2(\bar{M})$ using 15.1.5.1, $\bar{Y} \neq 1$.

Assume for the moment that $\bar{Y}$ is cyclic. Then $\bar{Y}$ is of prime order from 15.2.1, and $\bar{Y}$ is inverted in $\bar{R}_c$. Hence as R_c^* centralizes $C_Y(V)^*$ by (5), we conclude $C_Y(V)^* = 1$, so that $\bar{Y} \cong Y^*$.

We next prove (1). If case (3) of 15.2.1 does not hold, then $\bar{Y}_0$ is of prime order, so $\bar{Y} = \bar{Y}_0$, and then conclusion (i) of (1) holds by the previous paragraph. Thus we may assume that case (3) of 15.2.1 holds. Since R_c^* is faithful on $F^*(M^*) \leq Y_0^* C_M(V)^*$, but R_c^* centralizes $C_M(V)^*$, R_c^* is faithful on $O_3(M^*)$, so that $m_3(C_M(V)) \leq 1$ by 14.1.17.4.

Suppose first that $\bar{Y} = \bar{Y}_0$, so that $\bar{Y} = O^2(\bar{M})$. Since $C_Y(V)^* \leq Z(Y^*)$ by (5), we conclude from A.1.21 and A.1.24 that either $Y^* \cong \bar{Y} \cong E_9$ or $Y^* \cong 3^{1+2}$. In the former case, conclusion (i) of (1) holds: for $C_M(V)^*$ is centralized by $Y^* \cong E_9$ by (5), so that $C_M(V)$ is a $3'$-group since $m_3(M) = 2$. In the latter case as $\bar{T} \cong E_4$, we conclude from (5) that $\bar{R}_c$ is the subgroup of order 2 in $\bar{T}$ which centralizes $C_Y(V)^*$, and hence inverts $\bar{Y}$; then since $m_3(C_M(V)) \leq 1$, conclusion (ii) of (1) holds.

Thus we may suppose that $\bar{Y} < \bar{Y}_0$. Therefore $\bar{Y}$ is of order 3, so $\bar{R}_c$ is of order 2, and $\bar{Y} \cong Y^*$ by the second paragraph of the proof. Since Y^* centralizes $C_M(V)^*$ and $m_3(C_M(V)) \leq 1$, conclusion (iii) of (1) holds, completing the proof of (1).

We next prove (4). First assume the subcase of case (4) of 15.2.1 where $\bar{T} \cong \mathbf{Z}_4$ and $O(\bar{M}) \cong \mathbf{Z}_{15}$ does not hold. In the remaining cases, $\bar{M} = \bar{Y}_0 N_{\bar{M}}(\bar{T})$, and $N_{\bar{M}}(\bar{T}) = \overline{N_M(T)}$ by a Frattini Argument, so as $N_M(T) \leq N_M(Z) \leq M_c = !\mathcal{M}(C_G(Z))$, (4) holds. Now consider the excluded subcase. By 15.1.13.4, $\overline{J_1(T)} = \bar{T} \cap \bar{M}_0$, so $\overline{J_1(T)}$ centralizes $O_3(\bar{M})$. Then $O^{3'}(M)$ acts on $V_E := C_V(J_1(T))$, so that $O^{3'}(M) \leq M_c$ by 15.1.14.2. Hence $M = YTO^{3'}(M) = Y_0(M \cap M_c)$, completing the proof of (4).

As $M \cap M_c$ acts on R_c and Y_0, $M \cap M_c$ and Y_0 act on $[Y_0, R_c] = Y$. Then as $M = (M \cap M_c)Y_0$ by (4), (2) holds. $\qquad\square$

Recall that 15.1.12 describes the possible structures for $H \in \mathcal{H}_*(T, M)$. We next eliminate one subcase of 15.1.12.3:

LEMMA 15.2.4. If $H \in \mathcal{H}_*(T, M)$, then $H/O_2(H)$ is not $S_5 \ wr \ \mathbf{Z}_2$.

PROOF. Assume otherwise, define R_c and Y as in 15.2.3, set $M^* := M/O_2(M)$ and $X := O^2(H \cap M)$. By 15.1.9.7 we may apply E.2.2 to conclude that $X^* \cong E_9$. Further $X \leq C_M(V) \leq M \cap M_c$ by 15.1.9.4 and 15.1.5.2. Therefore case (3) of 15.2.1

does not hold, as in that case $m_3(C_M(V)) \leq 1$ by 15.2.3.1. Thus $\bar{Y} = O^2(\bar{M}_0) \cong Y^*$ by 15.2.3.1, so $Y^* X^* = Y^* \times X^*$ as Y^* centralizes $C_M(V)^*$ by 15.2.3.5. Then as M is an SQTK-group, $O^2(\bar{M}_0) \cong Y^*$ is a $3'$-group, so case (4) of 15.2.1 holds.

Next $K := O^2(H) = K_1 K_1^t$ for $t \in T - N_T(K_1)$ and $K_1 \in \mathcal{C}(H)$ with $K_1/O_2(K_1) \cong A_5$ and $K_1 \not\leq M$. Observe that $F^*(K_1) = O_2(K_1)$ by 1.1.3.1. Let $X_i := X \cap K_i$ and $S := N_T(K_1)$, and observe that $O_2(XT) \leq S$, while $J(T) \leq S$ by 15.1.12.3. By 15.2.3.2, $O_2(Y) \leq O_2(M) \leq R_c$, and hence $R_c \in Syl_2(YR_c)$. Then as $X \leq M \cap M_c$, $R_c \leq O_2(XT) \leq S$, so $S \in Syl_2(G_1)$, where $G_1 := YX_1 S$. Also $S \in Syl_2(G_2)$, where $G_2 := K_1 S$. Let $G_0 := \langle G_1, G_2 \rangle$.

Suppose first that $O_2(G_0) = 1$. This assumption gives part (e) of Hypothesis F.1.1 with YR_c, K_1 in the roles of "L_1, L_2"; most other parts are straightforward, but we mention: As X^* centralizes $Y^* R_c^*$, $N_{K_1}(S \cap K_1) = X_1(S \cap K_1) \leq XS \leq N_G(YR_c)$. Recall that $Y = [Y, R_c]$ by construction in 15.2.3, so that $YR_c/O_2(YR_c) \cong D_{10}$ or $Sz(2)$. Thus the amalgam $\alpha := (G_1, X_1 S, G_2)$ is a weak BN-pair of rank 2 by F.1.9. Furthermore as $S = N_{YS}(R_c)$, α is described in F.1.12. This is a contradiction, as $YR_c/O_2(YR_c) \cong D_{10}$ or $Sz(2)$ while $K_1/O_2(K_1)) \cong A_5$, and no such pair appears in F.1.12.

Therefore $O_2(G_0) \neq 1$. Let $S \leq T_0 \in Syl_2(G_0)$. We saw $J(T) \leq S$, so that $J(T) = J(S)$ by B.2.3.3. As $|T : S| = 2$, $|T_0 : S| \leq 2$, so that T_0 normalizes S. We conclude from 15.1.9.1 that $T_0 \leq N_G(J(T)) \leq M$, so that either $T_0 = S$ or $T_0 \in Syl_2(M)$. But in the latter case, $M = !\mathcal{M}(YT_0)$ by 15.2.3.3, so $K_1 \leq G_0 \leq M$, whereas we saw $K_1 \not\leq M$. Thus $S \in Syl_2(G_0)$. Let $\hat{G}_0 := G_0/O_2(G_0)$ and $M_1 := G_0 \cap M$. If $F^*(\hat{G}_i) = O_2(\hat{G}_i)$ for $i = 1$ and 2, then as above $(\hat{G}_1, \hat{X}_1 \hat{S}, \hat{G}_2)$ is a weak BN-pair described in F.1.12, for the same contradiction as before. Thus either $\hat{Y} \cong \mathbf{Z}_5$ or $\hat{K}_1 \cong A_5$.

As $K_1 \in \mathcal{L}(G_0, S)$ and $S \in Syl_2(G_0)$, $K_1 \leq L_1 \in \mathcal{C}(G_0)$ by 1.2.4. As $K_1 \not\leq M$, $L_1 \not\leq M$. As S normalizes K_1, $L_1 \trianglelefteq G_0$ by 1.2.1.3. Indeed since $K_1 \leq L_1$ and $G_1 \leq M_1$, $G_0 = \langle G_1, G_2 \rangle = L_1 M_1$. As $|T : S| = 2$ and $S \leq M_1$, $F^*(M_1) = O_2(M_1)$ by 1.1.4.7.

We claim that $G_0 \in \mathcal{H}^e$: If $\hat{Y} \cong \mathbf{Z}_5$, then $V = [V, Y] \leq O_2(Y) \leq O_2(G_0)$, so that $G_0 \in \mathcal{H}^e$ by 1.1.4.3. Suppose on the other hand that $\hat{K} \cong A_5$. We have seen that $G_0 = L_1 M_1$ and $F^*(M_1) = O_2(M_1)$. Further $N_G(O_2(Y)) = M$ since $Y \trianglelefteq M$ and $M \in \mathcal{M}$. So it suffices by A.1.10 to show that $F^*(L_1) = O_2(L_1)$. Now as $K_1 \leq L_1$ and $\hat{K}_1 \cong A_5$, $O_2(K_1) \leq \leq O_2(L_1)$. Therefore $\hat{L}_1 \cong L_1/O_2(L_1)$ is quasisimple by 1.2.1.4. Then as $F^*(K_1) = O_2(K_1)$, L_1 does not centralize $O_2(L_1)$, so that $F^*(L_1) = O_2(L_1)$. This completes the proof of the claim that $G_0 \in \mathcal{H}^e$.

Let $R := O_2(YS)$. As Y and S are T-invariant, so is R; so as $M = !\mathcal{M}(YT)$ by 15.2.3.3, $C(G, R) \leq M$, and hence $C(G_0, R) \leq M_1$. Further as $Y \trianglelefteq M_1$, C.1.2.4 says $R \in \mathcal{B}_2(M_1)$ and $R \in Syl_2(\langle R^{M_1} \rangle)$, so $R \in \mathcal{B}_2(G_0)$. Thus Hypothesis C.2.3 is satisfied with G_0, M_1 in the roles of "H, M_H".

Suppose first that $R \in Syl_2(RL_1)$. Then L_1 is a χ_0-block by C.2.5, so as $K_1 \in \mathcal{L}(L_1, S)$, we conclude from A.3.14 that $L_1 = K_1$. As $\hat{Y} = O^{5'}(\hat{Y})$ normalizes $\hat{R}$, it centralizes the Sylow group $\hat{R} \cap \hat{K}_1$ of $\hat{K}_1 = \hat{L}_1$, and hence centralizes $\hat{L}_1$. Therefore K_1 normalizes $O^2(YO_2(G_0)) = Y$, and hence $K_1 \leq N_G(Y) \leq M = !\mathcal{M}(YT)$, a contradiction. Thus R is not Sylow in RL_1. However if $Y \not\leq L_1$, then Y normalizes $YS \cap L_1 = S \cap L_1$, so $S \cap L_1 \leq O_2(YS) = R$, contradicting $R \notin Syl_2(RL_1)$. Therefore $Y \leq L_1$.

Suppose $\hat{L}_1$ is not quasisimple. Then by 1.2.1.4, $\hat{F}_1 := F(\hat{L}_1) = F^*(\hat{L}_1)$ is a $3'$-group, so the preimage F_1 of $\hat{F}_1$ lies in M_1 by C.2.6.2. Then $Y \leq F_1$: for otherwise $[F_1, Y] \leq F_1 \cap Y \leq F_1 \cap O_2(Y) \leq O_2(F_1)$, and then $L_1 = [L_1, Y]$ centralizes $\hat{F}_1$, contradicting $F^*(\hat{L}_1) = \hat{F}_1$. Therefore $\hat{Y} \leq O_5(\hat{L}_1)$, so $\hat{L}_1 \cong SL_2(5)/E_{25}$ by 1.2.1.4. In particular S is irreducible on $\hat{F}_1$, impossible as S acts on Y and $Y < F_1$.

Therefore $\hat{L}_1$ is quasisimple, so as $L_1 \trianglelefteq G_0$, L_1 is described in C.2.7.3. Further $\hat{Y}$ is an $\hat{S}$-invariant subgroup of $\hat{L}_1$ with $|\hat{Y} : O_2(\hat{Y})| = 5$, so we may apply A.3.15; comparing the list of A.3.15 with the list of C.2.7.3, we conclude $L_1/O_2(L_1) \cong L_2(2^n)$ or $SL_3(2^n)$ with $n \equiv 0 \mod 4$. This is impossible, as $K_1 \in \mathcal{L}(L_1S, S)$ with $K_1/O_2(K_1) \cong A_5$. This contradiction completes the proof of 15.2.4. $\qquad\square$

We are now able to obtain the analogue of 14.2.2.5:

LEMMA 15.2.5. $\mathcal{M}(T) = \{M, M_c\}$.

PROOF. We assume $M_1 \in \mathcal{M}(T) - \{M, M_c\}$, and derive a contradiction. Set $H := M \cap M_1$ and $Z_V := Z \cap V$. As $M_c = !\mathcal{M}(C_G(Z))$, $C_{M_1}(V(M_1)) \leq C_G(Z) \leq M_c \geq N_G(Z_V)$. By 15.1.9.2, $M_1 \overset{\sim}{\sim} M$, so that $M_1 = HC_{M_1}(V(M_1))$, and hence as $C_{M_1}(V(M_1)) \leq M_c$ but $M_1 \not\leq M_c$, also $H \not\leq M_c$. Thus $H \not\leq N_M(Z_V)$, so as $C_M(V) \leq M_c$ by 15.1.5.2, it follows that $\bar{H} \not\leq \bar{M}_c$, so that $\bar{T} < \bar{H}$. On the other hand if $O^2(M_0) \leq HN_M(Z_V)$, then $M = !\mathcal{M}(H)$ by 15.2.2, contrary to $H \leq M_1 \neq M$. Thus $O^2(M_0) \not\leq HC_M(V)$, so that $\bar{H} < \bar{M}$.

Now if either case (1) or (2) of 15.2.1 holds, then $|M : N_M(Z_V)| = 3$ is prime, so as $H \not\leq N_M(Z_V)$, $M = N_M(Z_V)H$, which is contrary to the previous paragraph. Similarly in case (4) of 15.2.1, as $O^2(\bar{M}_0) \not\leq \bar{H}$ and $\bar{M} > \bar{H} > \bar{T}$, $F^*(\bar{M})$ has order 15, and $\bar{H} = O_3(\bar{M})\bar{T}$. But $\bar{M} = \overline{M \cap M_c}O^2(\bar{M}_0)$ by 15.2.3.4, so $\bar{H} = O_3(\bar{M})\bar{T} = \overline{M \cap M_c}$, contrary to the previous paragraph.

Thus case (3) of 15.2.1 holds, so as $\bar{M} > \bar{H} > \bar{T}$, $\bar{H} \cong \mathbf{Z}_2 \times S_3$. Set $M^* := M/O_2(M)$ and $R := O_2(H)$.

Suppose for the moment that $V < V(M)$. Then from Notation 15.1.4, case (6) of 15.1.2 holds. Let M_J denote the preimage in M of $\hat{J}(Aut_M(V(M)), V(M))$, and $V_J := C_{V(M)}(M_J)$; by 15.1.4, $V = [V(M), M_J]$, and by 15.1.2.6, $V(M) = V \times V_J$ with $V_J \neq 1$, $C_M(V)C_M(V_J)T = M_c$, and $|M : M \cap M_c| = 3$ is prime. So as $H \not\leq M_c$, $M = H(M \cap M_c) = HC_M(V)C_M(V_J)$ in this case. We now drop the assumption that $V < V(M)$.

Suppose that $\bar{R} = 1$. Observe that hypothesis (a) of 14.1.17 is satisfied with $V(M)$, $O^2(M)$ in the roles of "V, Y_0" so as $\bar{R} = 1$, we conclude $V < V(M)$ from 14.1.17.1, and we adopt the notation of the previous paragraph. As $\bar{R} = 1$, $R \leq C_M(V)$, so $[C_M(V_J), R] \leq C_M(V(M))$. Also $R = O_2(RC_M(V(M)))$ by 14.1.17.5 applied to $V(M)$ in the role of "V", so $C_M(V_J) \leq N_M(R)$. From the previous paragraph, $M = HC_M(V)C_M(V_J)$, so $M = C_M(V)N_M(R)$, and hence $M = !\mathcal{M}(N_M(R))$ by 15.2.2. Therefore $C(G, R) \leq M$, so $C(M_1, R) = H$, and hence $M_1 = !\mathcal{M}(H)$ by 14.1.16, contrary to $H \leq M \neq M_1$.

Therefore $\bar{R} \neq 1$. Since $\bar{R} \leq O_2(\bar{H})$ with $\bar{H} \cong S_3 \times \mathbf{Z}_2$, $\bar{R} = O_2(\bar{H})$ is of order 2. Let Y_0 denote the preimage in M of $O(\bar{H})$; then $\bar{H} = \bar{Y}_0\bar{T}$. As $O^2(\bar{M})$ is abelian and $T \leq H$, $\bar{Y}_0$ of order 3 is normal in $\bar{M}$.

Set $R_1 := O_2(M_1 \cap M_c)$, $V_1 := V(M_1)$, $\hat{M}_1 := M_1/C_{M_1}(V_1)$, and $M_1^+ := M_1/O_2(M_1)$. Recall $\hat{M}_1 = \hat{H}$ and $C_{M_1}(V_1) \leq M_c$, so $M_1 = (M_1 \cap M_c)H$, and $O^2(\hat{H}) \neq 1$ as $M_1 \not\leq M_c$.

We next construct a subgroup Y_1 of H with $\hat{Y}_1$ of order 3, $Y_1 \trianglelefteq M_1$, and $M_1 = (M_1 \cap M_c)Y_1$.

Suppose first that $V = V(M)$. Then by A.5.3.3, $C_M(V) \leq C_{M_1}(V_1) \cap M \leq H$, so as $\bar{H} = \bar{Y}_0 \bar{T}$, $Y_0 \leq H$ and $H = Y_0 T$. Therefore $\hat{Y}_0 \trianglelefteq \hat{Y}_0 \hat{T} = \hat{H} = \hat{M}_1$. Further $\hat{Y}_0 \neq 1$ as $O^2(\hat{H}) \neq 1$, so as $\bar{Y}_0$ has order 3 and $C_M(V) \leq C_{M_1}(V_1)$, we conclude that $\hat{Y}_0$ has order 3. In this case let Y_1 be the preimage in M_1 of $\hat{Y}_0$, so that $\hat{Y}_1 = \hat{Y}_0$ has order 3 and $\hat{M}_1 = \hat{Y}_1 \hat{T}$, so that $M_1 = Y_1(M_1 \cap M_c)$ and $Y_1 \trianglelefteq M_1$.

Suppose instead that $V < V(M)$. Then by our earlier discussion, $M = H(M \cap M_c)$ with $|M : M \cap M_c| = 3$. Thus $M = Y_0(M \cap M_c)$, and $\bar{Y}_0$ is the unique $\bar{T}$-invariant subgroup of $\bar{M}$ of order 3 not contained in $\overline{M \cap M_c}$. Let $R_c := O_2(M \cap M_c)$, and define Y as in 15.2.3. As $|M : M \cap M_c| = 3$, $\bar{Y} = [O^2(\bar{M}_0), \bar{R}_c] < O^2(\bar{M}_0) \cong E_9$, so case (iii) of 15.2.3.1 holds, and $Y^* \cong \bar{Y} = [O^2(\bar{M}_0), \bar{R}_c]$. By the uniqueness of $\bar{Y}_0$ mentioned above, $\bar{Y} = \bar{Y}_0$. Therefore $Y C_M(V) = (Y_0 \cap H)C_M(V)$. Now R_c acts on $Y_0 \cap H$, $Y^* R_c^* C_M(V)^* = Y^* R_c^* \times C_M(V)^*$ by 15.2.3.5, and $Y^* = [Y^*, R_c^*]$ since $Y^* \cong \bar{Y} = [\bar{Y}, \bar{R}_c]$. Thus $Y \leq H$, so as $|M : M \cap M_c| = 3$ is prime, $H = Y(H \cap M_c)$. Then as we saw $M_1 = H(M_1 \cap M_c)$, $M_1 = Y(M_1 \cap M_c)$. As $Y \trianglelefteq M$ and $\hat{M}_1 = \hat{H}$, $\hat{Y} \trianglelefteq \hat{M}_1$. As $Y \not\leq M_c \geq C_{M_1}(V_1) \geq C_M(V_1)$, $\hat{Y} \neq 1$, so as Y^* has order 3 and $C_Y(V) \leq C_{M_1}(V_1)$, we conclude $\hat{Y}$ has order 3. In this case, let Y_1 be the preimage in M_1 of $\hat{Y}$, so that $\hat{Y}_1 = \hat{Y}$ has order 3, and $M_1 = Y_1(M_1 \cap M_c)$. This completes the definition of Y_1 in our second case.

Now in either case we have the hypotheses of case (b) of 14.1.17, with M_1, M_c, R_1, Y_1 in the roles of "M, M_1, R, Y_0". We claim $\hat{Y}_1 = [\hat{Y}_1, R_1]$: For otherwise $\hat{R}_1$ is normal in $\hat{Y}_1 \widehat{M_1 \cap M_c} = \hat{M}_1$, whereas $O_2(\hat{M}_1) = 1$ by B.2.14. Set $Y_2 := O^2(\langle R_1^{Y_1} \rangle) = O^2(\langle R_1^{Y_1 T} \rangle)$, so that Y_2 plays the role of "Y" in 14.1.17. Since R_1 is normal in $M_1 \cap M_c$ and $M_1 = Y_1(M_1 \cap M_c)$, we have $Y_2 = O^2(\langle R_1^{M_1} \rangle)$ normal in M_1, so that $M_1 = N_G(Y_2)$ as $M_1 \in \mathcal{M}$. To complete the proof, we will show that $Y_2 \trianglelefteq M$, so that $M = N_G(Y_2) = M_1$, contrary to our choice of $M_1 \neq M$.

Since $\hat{Y}_1 = [\hat{Y}_1, \hat{R}_1]$ is of order 3, $\hat{Y}_2 = \hat{Y}_1 \cong \mathbf{Z}_3$. But by 14.1.17.3, $C_{Y_2}(V_1)^+$ centralizes R_1^+, so $Y_2^+ \cong \hat{Y}_2 \cong \mathbf{Z}_3$. Moreover $Y_2 C_{M_1}(V_1) = Y_1 C_{M_1}(V_1)$, so arguing as above when we showed $Y \leq H$, we conclude $Y_2 \leq H$. Further $Y_2^+ = [Y_1^+, R_1^+]$.

Suppose first that $V < V(M)$. Then by construction $Y \leq Y_1$ and $\hat{Y} = \hat{Y}_1 = [\hat{Y}_1, R_1]$, so that $O_2(Y) = C_Y(V_1)$ as $Y^* \cong \mathbf{Z}_3$. Then as R_1 acts on Y, $Y = [Y, R_1]$. Thus $\mathbf{Z}_3 \cong Y_2^+ = [Y_1^+, R_1] \geq [Y^+, R_1] = Y^+ \cong \mathbf{Z}_3$, so $Y_2^+ = Y^+$. Then $Y_2 = O^2(Y_2 O_2(M_1)) = O^2(Y O_2(M_1)) = Y \trianglelefteq M$ by 15.2.3, completing the proof in this case.

Thus we may assume that $V = V(M)$. Here we saw that $C_M(V) \leq H$, so $C_M(V) \leq M_1 = N_G(Y_2)$, and $[R, C_M(V)] \leq O_2(H) \cap C_M(V) \leq O_2(C_M(V)) \leq O_2(M)$, so that R^* centralizes $C_M(V)^*$. Further $\bar{R}$ centralizes $O(\bar{H})$, so case (ii) of 15.2.3.1 does not hold, since there $Y^* \cong 3^{1+2}$, in which case involutions not inverting $O^2(\bar{M})$ do not centralize $C_Y(V)^*$. Therefore a Sylow 3-subgroup P of M is abelian by 15.2.3.1. Choose P with $X := P \cap Y_2 \in Syl_3(Y_2)$. Then P centralizes X and normalizes $C_M(V)$. Now we saw $C_M(V) \leq C_{M_1}(V_1)$ and $Y_2 \trianglelefteq M_1$, so $\langle X^{C_M(V)} \rangle = Y_2$. Hence P acts on Y_2 so $HP = M$ acts on Y_2, completing the proof. $\square$

LEMMA 15.2.6. *Define R_c and Y as in 15.2.3. Then there exists a T-invariant subgroup $Y_1 := O^2(Y_1)$ of Y such that*

(1) $Y_1 R_c / O_2(Y_1 R_c) \cong S_3$, D_{10}, *or* $Sz(2)$.
(2) $C_{Y_1}(V) = O_2(Y_1)$.
(3) $M = !\mathcal{M}(Y_1 T)$.

PROOF. Assume case (3) of 15.2.1 does not hold. Here we take $Y_1 := Y$. Then by 15.2.3.1, $\bar{Y} = [\bar{Y}, \bar{R}_c] = O^2(\bar{M}_0)$ and $C_Y(V) = O_2(Y)$, so (2) holds, and (3) follows from 15.2.3.3. Conclusion (1) follows from the structure of $\bar{M}_0$ described in 15.2.1.

So assume that case (3) of 15.2.1 holds. In case (iii) of 15.2.3.1, we again choose $Y_1 := Y$, so that $\bar{Y} = [\bar{Y}, \bar{R}_c]$ is of order 3. In cases (i) and (ii) of 15.2.3.1, we choose Y_0 to be the preimage of a T-invariant subgroup Y_0^* of Y^* of order 3 with $\bar{Y}_0 = [\bar{Y}_0, \bar{R}_c]$ of order 3, and set $Y_1 := O^2(Y_0)$. In each case Y_1 satisfies (1) by construction. In case (iii) of 15.2.3.1, $\bar{Y}_1 = \bar{Y} \cong Y^*$, so (2) holds; in the remaining cases we chose Y_1 with $\bar{Y}_1 \cong Y^*$, so again (2) holds. Finally as $Y_1 = [Y_1, R_c]$, $Y_1 \not\le M_c$, so (3) follows from 15.2.5. $\qquad\square$

LEMMA 15.2.7. *Define Y as in 15.2.3 and Y_1 as in 15.2.6. Then*
(1) $M = !\mathcal{M}(YT)$.
(2) If $1 \ne X = O^2(X) \le C_M(V)$ is T-invariant, then
    (i) $N_G(X) \le M$, *and*
    (ii) if $|X : O_2(X)| = 3$, then X acts on Y_1.

PROOF. Part (1) follows from 15.2.6.3 as $Y_1 \le Y$. Assume X satisfies the hypotheses of (2); to prove (2), it suffices by (1) to show that Y acts on X, and that X acts on Y_1 if $|X : O_2(X)| = 3$. Let $M^* := M/O_2(M)$. As T acts on $X = O^2(X)$ and $Y_1 = O^2(Y_1)$, it suffices to show that Y^* acts on X^*, and that X^* acts on Y_1^* if $|X : O_2(X)| = 3$. But as $Y \trianglelefteq M$ by 15.2.3.2, $[X, Y] \le C_Y(V)$, so if $C_Y(V) = O_2(Y)$, then $[X^*, Y^*] = 1$, and the lemma holds. Thus by 15.2.3.1, we may assume that case (ii) of 15.2.3.1 holds. Then $[X^*, Y^*] \le Z(Y^*)$ with $Z(Y^*)$ of order 3. Thus if X^* is a $3'$-group, then $X^* = O^2(X^*)$ centralizes Y^* by Coprime Action, and as before the lemma holds. Finally if X^* is not a $3'$-group, then $Z(Y^*) \le X^*$ as case (ii) of 15.2.3.1 holds, so $[X^*, Y^*] \le Z(Y^*) \le X^*$, and once again Y^* acts on X^*. Also if $|X : O_2(X)| = 3$, then $X^* = Z(Y^*)$ acts on Y_1^*, completing the proof. $\qquad\square$

LEMMA 15.2.8. *If $H \in \mathcal{H}_*(T, M)$, then $H/O_2(H) \cong S_3$ wr $\mathbf{Z}_2$.*

PROOF. First $H/C_H(U_1)$ is described in 15.1.12.3, where U_1 is a noncentral chief factor for H on $U_H := \langle V^H \rangle$. In particular $O_2(H/C_H(U_1)) = 1$, so $O_2(H) \le C_H(U_1)$. Recall by 15.1.9.7 that H is a minimal parabolic described by B.6.8; thus $H \cap M = N_H(T \cap H)$ by 3.1.3.1, and $C_H(U_1) \le H \cap M$ by B.6.8.6a. If $C_H(U_1) > O_2(H)$, then $X := O^2(C_H(U_1)) \ne 1$, so that $X \le C_M(V)$ by 15.1.9.4; hence by 15.2.7.2, $H \le N_G(X) \le M$, contrary to $H \in \mathcal{H}_*(T, M)$. Therefore $O_2(H) = C_H(U_1)$.

Thus $H/C_H(U_1) = H/O_2(H)$ is described in 15.1.12.3. By 15.2.4, $H/O_2(H)$ is not S_5 wr $\mathbf{Z}_2$, and the lemma holds if $H/O_2(H)$ is S_3 wr $\mathbf{Z}_2$, so we may assume that $H/O_2(H) \cong S_3$ or S_5, and it remains to derive a contradiction.

Define R_c and Y as in 15.2.3, and Y_1 as in 15.2.6. We will verify that Hypothesis F.1.1 is satisfied with $Y_1 R_c$, H, T in the roles of "L_1, L_2, S". Most parts are straightforward, but we give a few details: First $Y_1 R_c / O_2(Y_1 R_c) \cong S_3$, D_{10}, or $Sz(2)$ by 15.2.6.1, while we saw $H/O_2(H) \cong S_3$ or S_5, so that part (c) holds. Next

$M = !\mathcal{M}(Y_1 T)$ by 15.2.6.3, so that $O_2(\langle Y_1 T, H\rangle) = 1$, and hence part (e) holds. To verify part (d), we must show that $H \cap M$ normalizes $Y_1 R_c$. By 15.1.9.3, $H \leq M_c$, so $H \cap M$ acts on R_c. By 15.1.9.4, $O^2(H \cap M) \leq C_M(V)$, so $O^2(H \cap M)$ acts on Y_1 by 15.2.7.2.

Hence $\alpha := (Y_1(H \cap M), H \cap M, H)$ is a weak BN-pair of rank 2 by F.1.9, and since $N_{Y_1 R_c}(T) = T$, α appears in the list of F.1.12. Now U_H is abelian by 15.1.11.2, and H has two noncentral 2-chief factors on U_H by 15.1.12.1, which are natural modules for $H/O_2(H) \cong S_n$ for $n = 3$ or 5 by 15.1.12.3. But these conditions are not satisfied by any member of F.1.12. $\qquad\square$

We now define notation which will be in force for the remainder of the section:

NOTATION 15.2.9. Pick $H \in \mathcal{H}_*(T, M)$, and let $Q_H := O_2(H)$, $U_H := \langle V^H\rangle$, and $H^* := H/O_2(H)$. Recall in particular by 15.1.9.3 that

$$H \leq M_c.$$

By 15.1.12.1, H has exactly two noncentral chief factors U_1 and U_2 on U_H. By 15.2.8, $H^* \cong S_3$ wr $\mathbf{Z}_2$. Thus by 15.1.12.4, $m(U_i) = 4$ and $H^* = O_4^+(U_i)$, so $U_i = U_{i,1} \oplus U_{i,2}$ with $U_{i,j} \cong E_4$, $j = 1, 2$, the two definite 2-dimensional subspaces of the othogonal space U_i. Also $H^* = (H_1^* \times H_2^*)\langle t^*\rangle$, where t^* is an involution with $H_1^t = H_2$ and $H_i^* \cong S_3$. This choice for H_1 and H_2 is not unique, but 15.1.12 supplies us with a distinguished choice: Pick $H_i := C_H(U_{1,3-i})$. In particular the subgroups H_i^*, $i = 1, 2$ contain the transvections in H^* on U_1. Let $K_i := O_2(H_i)$ and $K := O^2(H)$.

Next let Δ consist of those $A \in \mathcal{A}(H)$ such that A^* is minimal subject to $A \not\leq Q_H$. By 15.1.12.2, for each $A \in \Delta$, A^* is an FF*-offender on U_1 and U_2. From B.2.9.1 and the description of FF*-offenders in B.1.8.4, A^* is of order 2 by minimality of A^*, so A^* induces transvections on both U_1 and U_2. Thus A lies in either H_1 or H_2, and we can choose notation so that also $H_i = C_H(U_{2,3-i})$. Then $U_{j,i} = [U_j, H_i]$ and $U_{j,1}^t = U_{j,2}$.

For $A \in \Delta$, let $B(A) := A \cap Q_H$; thus $|A : B(A)| = |A^*| = 2$. Let $\Sigma := \{B(A) : A \in \Delta\}$.

Observe by 15.2.8 that $T = M \cap H = N_H(V)$, so $|V^H| = 9$. For $h \in H$, let $\Delta(V^h) := \Delta \cap T^h$, $\Delta'(V^h) := \Delta - \Delta(V^h)$; $\Sigma(V^h) := \{B(A) : A \in \Delta(V^h)\}$, and $\Sigma'(V^h) := \Sigma - \Sigma(V^h)$.

LEMMA 15.2.10. Let $\bar{D} \in \mathcal{Q}(\bar{T}, V)$. Then

(1) $\bar{D} \leq \bar{M}_0$ and $m(\bar{D}) = 1$.
(2) $m([V, D]) = 2$.
(3) $[V, D] \trianglelefteq T$.

PROOF. By 15.2.1, $\mathcal{Q}(\bar{T}, V) \subseteq \Omega_1(\bar{T}) \leq \bar{M}_0$. Then the lemma follows easily from 15.2.1: For example (3) follows as $\bar{T}$ is abelian, and in case (3) of 15.2.1, $m(\bar{D}) = 1$ since $\bar{D}$ acts quadratically on V. $\qquad\square$

LEMMA 15.2.11. Let $B \in \Sigma'(V)$ and $Z_S := [V, B]$. Then

(1) $\bar{B} \in \mathcal{Q}(\bar{M}, V)$.
(2) $Z_S \leq Z(K)$.
(3) $E_4 \cong Z_S \trianglelefteq T$.

PROOF. Recall $B = B(A)$ for some $A \in \Delta'(V)$, and we may assume without loss that $A \le H_1$. Let $X := \langle \Delta \cap T \cap H_1 \rangle$ and $I_1 := \langle X^{H_1} \rangle$; then $X \trianglelefteq N_T(H_1)$. As $A \le H_1$, $K_1 = [K_1, A] \le I_1$. Also $H_1 = Q_H \langle X, A \rangle$, so $I_1 = \langle X, A \rangle$.

We saw $T = N_H(V)$, so A does not normalize V, and hence $[V, A] \neq 1$. But $B \le O_2(H)$, so B does normalize V, and by 15.1.12.2, $C_{U_H}(A) = C_{U_H}(B)$; so $[V, B] \neq 1$. Therefore (1) holds by 15.1.12.2, and then (3) follows from 15.2.10. Recall A^* is of order 2; thus $1 = m(A/B) = m(B/C_B(U_H))$ by 15.1.12.2. Also 15.1.12.2 shows A is quadratic on U_H, so $Z_S = [V, B]$ is centralized by A. Further as $\Delta \subseteq \mathcal{A}(H)$, $X \le J(T) \le C_G(V)$ by 15.1.9.1, so Z_S is centralized by $\langle X, A \rangle = I_1$. Thus by (3), $Z_S \trianglelefteq \langle I_1, T \rangle = H$, so $K = \langle K_1^H \rangle \le C_G(Z_S)$, and hence (2) holds. $\square$

Recall from Notation 15.2.9 that $H \le M_c$. Set $K_c := \langle K^{M_c} \rangle$.

LEMMA 15.2.12. Let $B \in \Sigma'(V)$ and $Z_S := [V, B]$. Then

(1) $K \in \Xi(G, T)$.

(2) One of the following holds:

(i) $K_c = K$.

(ii) $K_c \in \mathcal{L}^*(G, T) = \mathcal{C}(M_c)$ and $K_c T/O_2(K_c T) \cong Aut(L_n(2))$, $n = 4$ or 5.

(iii) $K_c = LL^t$ with $L \in \mathcal{L}^*(G, T) = \mathcal{C}(M_c)$ and $L/O_2(L) \cong L_2(2^n)$, n even, or $L_2(p)$ for some odd prime p.

(3) $M_c = !\mathcal{M}(H)$ and $Z_S \trianglelefteq H$, so $N_G(Z_S) \le M_c$.

(4) $K_c = O^{3'}(M_c) \le C_G(Z)$.

(5) Case (6) of 15.1.2 does not hold, so $V = V(M)$.

PROOF. Part (1) is immediate from 15.2.8. By 1.3.4, either $K = K_c$, or $K_c = \langle L^T \rangle$ for some $L \in \mathcal{C}(M_c)$ described in (1)–(4) of 1.3.4. Suppose the latter holds. By 14.1.6.2, $L \in \mathcal{L}^*(G, T)$. As $Aut_T(K/O_2(K)) \cong D_8$, we conclude from 1.3.4 that either $LT/O_2(LT) \cong Aut(L_n(2))$, $n = 4$ or 5, or $L < K_c$ and $L/O_2(L) \cong L_2(2^n)$ or $L_2(p)$. Therefore (2) is established.

By 15.2.5, $M_c = !\mathcal{M}(H)$, while by (2) and (3) of 15.2.11, H normalizes Z_S, so (3) holds. In case (i) of (2), as $Aut_T(K/O_2(K)) \cong D_8$ and $Aut(K/O_2(K)) = GL_2(3)$, it follows as $T \in Syl_2(N_G(K))$ that $Aut_G(K/O_2(K)) = Aut_T(K/O_2(K))$, so $O^{3'}(M_c) = K O^{3'}(C_{M_c}(K/O_2(K)))$. Therefore as $K/O_2(K) \cong E_9$ and $m_3(M_c) \le 2$, we conclude $K_c = K = O^{3'}(M_c)$. In cases (ii) and (iii) of (2), we obtain $K_c = O^{3'}(M_c)$ using A.3.18 and 1.2.2.a. If $K < K_c$, then $K_c = K_c^\infty$ centralizes Z by 14.1.6.3. If $K = K_c$, this follows from 15.1.9.3. This completes the proof of (4).

By (4), $O^{3'}(M \cap M_c) \le C_M(Z)$. However in case (6) of 15.1.2, $|M : M \cap M_c| = 3$ and $O^2(\bar{M}) \cong E_9$ with $\bar{T} = C_{\bar{M}}(Z \cap V)$. This contradiction establishes (5). $\square$

LEMMA 15.2.13. (1) Either

(i) case (1) or (4) of 15.2.1 holds, with $\bar{M} \cong S_3$, D_{10}, or $Sz(2)$, or

(ii) case (3) of 15.2.1 holds, and $O(\bar{M}) = [O(\bar{M}), B]$ for each $B \in \Sigma'(V)$.

(2) Let $\bar{B}_0$ be the unique subgroup of $\bar{T}$ of order 2 with $O(\bar{M}) = [O(\bar{M}), \bar{B}_0]$. Then for each $B \in \Sigma'(V)$, $\bar{B} = \bar{B}_0$, and $C_V(B) = [V, B] = [V, \bar{B}_0] = C_V(\bar{B}_0)$.

PROOF. Assume conclusion (i) of (1) does not hold. Then either one of cases (2) or (3) of 15.2.1 holds, or case (4) of 15.2.1 holds with $F^*(\bar{M}) \cong \mathbf{Z}_{15}$. Pick $B \in \Sigma'(V)$. By 15.2.11.1 and 15.2.10.1, $\bar{B}$ is of order 2. Then either $\bar{X} := C_{O(\bar{M})}(\bar{B}) \cong \mathbf{Z}_3$, or case (3) of 15.2.1 holds with $O(\bar{M}) = [O(\bar{M}), \bar{B}]$. Assume the former case holds. Then we compute that $\bar{X}$ acts faithfully on $[V, B] =: Z_S$, so $X \leq N_G(Z_S) \leq M_c$ by 15.2.12.3. Hence $X \leq C_G(Z)$ by 15.2.12.4, impossible as X is nontrivial on Z_S, and $1 \neq Z \cap Z_S$. Therefore the latter case holds for each $B \in \Sigma'(V)$, and hence conclusion (ii) of (1) holds. This completes the proof of (1). Then (1) implies (2). $\qquad\square$

For the remainder of the section, we define $\bar{B}_0$ and $Z_S := [V, \bar{B}_0]$ as in 15.2.13.2. Thus $Z_S = [V, B]$ for each $B \in \Sigma'(V)$ by 15.2.13.2. Let $S := C_T(Z_S)$.

LEMMA 15.2.14. *(1) $M_c = C_G(Z)$.*
(2) $Z \leq Z_S$, and either
    (a) $S = T$ and $Z = Z_S \cong E_4$, or
    (b) $\bar{M} \cong Sz(2)$ or $\Omega_4^+(V)$, Z is of order 2, and $|T : S| = 2$.
(3) $\mathrm{Baum}(T) \leq S$.
(4) $Z_S = \Omega_1(Z(S))$.
(5) $M \cap M_c = C_M(Z) = C_M(V)T$.
(6) $m(\bar{M}, V) = 2$ and $a(\bar{M}, V) = 1$.

PROOF. As $\bar{M}$ is solvable, $a(\bar{M}, V) = 1$ by E.4.1. Then by inspection of the cases in 15.2.13.1, and recalling $V = V(M)$ by 15.2.12.5 so that $Z \leq V$, (2) and (6) hold, and $C_{\bar{M}}(Z) = \bar{T}$. Recall $C_M(V) \leq M_c$ by 15.1.5.2. In case (i) of 15.2.13.1, $\bar{T}$ is maximal in $\bar{M}$, so $C_M(V)T = M \cap M_c$ as $M_c \not\leq M$. In case (ii) of 15.2.13.1, this holds as $O^{3'}(M \cap M_c) \leq C_M(V)$ by 15.2.12.4. Thus (5) is established. Further $M_c = (M \cap M_c)C_{M_c}(V(M_c))$ since $M_c \overset{\sim}{<} M$ by 15.1.9.2, so M_c centralizes Z by (5), and hence (1) holds.

If $Z_S = Z$, then $S = T$ so (3) and (4) are trivial; thus we may assume that $\bar{M} \cong Sz(2)$ or $\Omega_4^+(2)$ with Z of order 2. Then $\mathrm{Baum}(T) \leq C_T(V) \leq S$ by 15.1.9.1, completing the proof of (3). Finally $Z_S \leq \Omega_1(Z(S)) =: Z_0$ and $\mathbf{Z}_2 \cong Z = C_{Z_0}(T)$; so as T/S is of order 2, $m(Z_0) \leq 2m(Z) = 2 = m(Z_S)$ using 15.2.11.3, and hence $Z_0 = Z_S$, establishing (4). $\qquad\square$

15.2.2. A uniqueness theorem. This subsection is devoted to establishing the following uniqueness theorem:

THEOREM 15.2.15. $M_c = !\mathcal{M}(C_{M_c}(Z_S))$.

The proof of Theorem 15.2.15 involves a series of reductions. Until it is complete, we assume $I \in \mathcal{H}(C_{M_c}(Z_S))$ with $I \not\leq M_c$, and work toward a contradiction. Set $M_I := M_c \cap I$ and $N_I := M \cap I$. In particular

$$M_I < I.$$

Since $Z \leq Z_S$ by 15.2.14.2, and $M_c = C_G(Z)$ by 15.2.14.1, while we chose $C_{M_c}(Z_S) \leq I$:

LEMMA 15.2.16. $C_G(Z_S) = C_{M_c}(Z_S) \leq M_I$.

Recall $H \leq M_c$ by Notation 15.2.9; so as $Z_S \leq Z(K)$ by 15.2.11.2, $KS \leq M_I$. As $C_{M_c}(Z_S) \leq I \not\leq M_c$ and $M_c = C_G(Z)$, $Z < Z_S$. Hence case (b) of 15.2.14.2 holds, so by that result:

LEMMA 15.2.17. *(1) $\bar{M} \cong Sz(2)$ or $\Omega_4^+(2)$.*
(2) $|Z| = 2$.
(3) $|T : S| = 2$.

LEMMA 15.2.18. *(1)* $S \in Syl_2(I)$.

(2) $B := Baum(T) = Baum(S)$ *and* $C(I, B) \leq N_I$.

(3) Either

   (i) $N_I \leq M_I$, *or*

   (ii) $N_I \not\leq M_I$, *case (ii) of 15.2.13.1 holds, with* $\bar{M} = \Omega_4^+(V)$, *and* $N_I = C_M(Z_1)$ *is of index 6 in* M, *for some complement* Z_1 *to* Z *in* Z_S.

PROOF. Recall Z_S is of order 4 by 15.2.11.3. Let $S \leq T_I \in Syl_2(I)$. By 15.2.14.3 and B.2.3.5, $B = Baum(S) = Baum(T_I)$, and $C(G, B) \leq M$ by 15.1.9.1. Thus (2) holds and also $T_I \leq N_I(B) \leq M$, so as $N_{\bar{M}}(\bar{S}) = \bar{T}$ and $|T : S| = 2$ by 15.2.17, T_I is either T or S. But if $T_I = T$, then $I \leq M$ by 15.2.5 since $I \not\leq M_c$, and we saw $K \leq M_I$; but this is contrary to $H \not\leq M$. Thus (1) is established.

Next using 15.2.16, $C_M(V) \leq C_M(Z_S) \leq M_I$, so if $O^2(N_I) \leq C_M(V)$, then conclusion (i) of (3) holds. Thus we may assume $X := O^2(N_I) \not\leq C_M(V)$.

Suppose $\bar{X} \trianglelefteq \bar{M}$. Then T acts on $SXC_M(V)$ and K, so T acts on $G_0 := \langle SXC_M(V), K \rangle$. Now $O_2(I) \leq S \leq G_0 \leq I$, so $TG_0 \in \mathcal{H}(T)$. But by 15.2.14.5, $M \cap M_c = C_M(V)T$, so as $X \not\leq C_M(V)$, $TG_0 \not\leq M_c$. Hence $M = !\mathcal{M}(TG_0)$ by 15.2.5, so $K \leq G_0 \leq M$, contrary to $H \not\leq M$.

Therefore $\bar{X}$ is not normal in $\bar{M}$, so case (ii) of 15.2.13.1 holds, and $\bar{X}$ is one of the two subgroups of $O(\bar{M})$ of order 3 not normal in $\bar{M}$. Thus conclusion (ii) of (3) holds, completing the proof. $\square$

Recall $C_1(S)$ from Definition C.1.18.

LEMMA 15.2.19. *Define* $B := Baum(S)$.

(1) If conclusion (i) of 15.2.18.3 holds, then $C(I, B) \leq M_I \geq C_I(C_1(S))$.

(2) If conclusion (ii) of 15.2.18.3 holds, then Hypothesis C.2.3 is satisfied with I, $O_2(N_I)$, N_I *in the roles of "H, R, M_H".*

PROOF. Assume first that conclusion (i) of 15.2.18.3 holds. Then $N_I \leq M_I$, so $C(I, B) \leq N_I \leq M_I$ by 15.2.18.2. Further $C_1(S) \trianglelefteq T$ as $S \trianglelefteq T$ by 15.2.17.3, so $1 \neq Z \cap C_1(S)$ and hence $C_I(C_1(S)) \leq C_G(Z \cap C_1(S)) \leq M_c = !\mathcal{M}(C_G(Z))$, completing the proof of (1).

Next assume conclusion (ii) of 15.2.18.3 holds, and let $R := O_2(N_I)$. Then $\bar{N}_I \cong S_3$, so $R \leq C_M(V)$. But $C_M(V) \leq C_M(Z_S) \leq N_I$, so $R = O_2(C_M(V)) = O_2(M)$, and hence $C(I, R) \leq N_I$. Then as $R = O_2(N_I)$, the remaining two conditions of Hypothesis C.2.3 are trivially satisfied, so (2) holds. $\square$

LEMMA 15.2.20. *(1) The hypotheses of 1.1.5 are satisfied with* I, M_c *in the roles of "H, M" for each* $z \in Z^\#$.

(2) $F^*(M_I) = O_2(M_I)$.

(3) $O(I) = 1$.

PROOF. Let $I_0 \in \mathcal{M}(I)$; then part (1) holds for I_0 in the role of "I" by 1.1.6. Then by 15.2.18.1, S is Sylow in I and I_0. In particular, $O_2(I_0 \cap M_c) \leq O_2(I \cap M_c) \leq S$ by A.1.6. Hence as I_0 satisfies the hypotheses of 1.1.5,

$$C_{O_2(M_c)}(O_2(I \cap M_c)) \leq C_{O_2(M_c)}(O_2(I_0 \cap M_c)) \leq T \cap I_0 = S \leq I,$$

and so (1) holds. Then (2) follows from (1) and 1.1.5.1. As usual $1 \neq U_K := [U_H, K]$ centralizes $O(I)$ by A.1.26.1, and $Z \leq U_K$ since $Z = \Omega_1(Z(T))$ has order 2 by 15.2.17.2, so (3) follows from 1.1.5.2. $\square$

LEMMA 15.2.21. *(1) If S is not irreducible on $K/O_2(K)$ then $KS = H_1 H_2$, $K_1^t = K_2$ for $t \in T - S$, and K_c centralizes Z_S, so that $K_c = O^{3'}(M_I)$.*

(2) If $K = K_c$ then $C_M(V)$ is a $3'$-group.

(3) Assume $K_c/O_2(K_c) \cong L_4(2)$, conclusion (ii) of 15.2.18.3 holds, and there is an S-invariant subgroup $Y_1 = O^2(Y_1)$ of N_I with $Y_1 S/O_2(Y_1 S) \cong S_3$ and $Y_1 \not\leq M_I$. Then S is irreducible on $K/O_2(K)$.

PROOF. Assume S is not irreducible on $K/O_2(K)$. From Notation 15.2.9, $K^* T^* \cong O_4^+(2)$, so T^* is irreducible on K^* and hence $S^* < T^*$. But $|T : S| = 2$ by 15.2.17.3, so $O_2(KT) \leq S$. As $J(T) \leq S$ by 15.2.18.2, and $|T^* : J(T)^*| = 2$ by 15.1.12.3, $J(T)^* = S^*$. Further $J(T)^* \in Syl_2(H_1^* H_2^*)$ by 15.1.12.3, so $KS = H_1 H_2$, and hence $K_1^t = K_2$ for $t \in T - S$.

Recall $K_c = \langle K^{M_c} \rangle = O^{3'}(M_c)$ is described in 15.2.12.2. If $K = K_c$, then (1) holds by 15.2.11.2, so we may assume $K < K_c$. Then $K_c = \langle L^T \rangle$ for some $L \in \mathcal{C}(M_c)$ described in 15.2.12.2. Now using A.1.6, $O_2(K_c T) \leq O_2(KT) \leq S = C_T(Z_S)$, so as $L \notin \mathcal{L}_f(G, T)$ by Hypothesis 14.1.5.1, $K_c \leq C_G(Z_S) \leq M_I$ by 1.2.10 and 15.2.16, completing the proof of (1).

Assume in addition the hypotheses of (3); we will obtain a contradiction to our assumption that S is not irreducible on $K/O_2(K)$, and hence establish (3). As $|Y_1 : O_2(Y_1)| = 3$ and $Y_1 \leq N_I$ but $Y_1 \not\leq M_I$, $\bar{Y}_1 \cong \mathbf{Z}_3$. By hypothesis, case (ii) of 15.2.18.3 holds, so $N_I = C_M(Z_1)$ is of index 6 in M, for some complement Z_1 to Z in Z_S; in particular, $\bar{Y}_1 = O(\bar{N}_I)$ is not T-invariant. Define Y as in 15.2.3, and set $\hat{M} := M/O_2(M)$. If case (iii) of 15.2.3.1 holds, then $\hat{Y} \cong \mathbf{Z}_3$ is T-invariant, so $YY_1/O_2(YY_1) \cong \bar{Y}\bar{Y}_1 \cong E_9$, and hence $C_M(V)$ is a $3'$-group as $m_3(M) \leq 2$.

Next KS is the maximal parabolic subgroup of $K_c S$ determined by the end nodes of the Dynkin diagram for $K_c/O_2(K_c)$. Let $Y_0 := O^2(P)$, where P is the minimal parabolic determined by the middle node. If $Y_0 T \not\leq M$, then $Y_0 T \in \mathcal{H}_*(T, M)$, contrary to 15.2.8; hence $Y_0 T \leq M$, so $Y_0 \leq C_M(V)$ by 15.2.14.5. Thus $C_M(V)$ is not a $3'$-group, so case (ii) of 15.2.3.1 holds by the previous paragraph. Therefore $\hat{Y} \cong 3^{1+2}$ and $\hat{Y}_0 = Z(\hat{Y})$. Now $\bar{S}$ inverts $\bar{Y}_1$ which is not $\bar{T}$-invariant, so as $|T : S| = 2$, $\bar{S}$ is the subgroup of order 2 of $\bar{T}$ inverting $\bar{Y}$, and so $\hat{S}$ centralizes $\hat{Y}_0$, This is impossible, as $K_c \leq M_I$ by (1), so $S \in Syl_2(K_c S)$ by 15.2.18.2, and then $SY_0/O_2(SY_0) \cong S_3$. So (3) is established.

Finally assume $K = K_c$. Then as $M \cap K = O_2(K)$ by 15.2.8, and $K = O^{3'}(M_c)$ by 15.2.12.4, we conclude $O^{3'}(M \cap M_c) = 1$, so (2) holds. $\qquad \square$

LEMMA 15.2.22. *Assume conclusion (ii) of 15.2.18.3 holds, $F^*(I) = O_2(I)$, and $C_M(V)$ is a $3'$-group. Assume S is not irreducible on $K/O_2(K)$, and there is $Y_1 = O^2(Y_1) \leq N_I$ which is S-invariant with $Y_1 S/O_2(Y_1 S) \cong S_3$. Then $[Y_1, K_2] \not\leq Y_1 \cap K_2$.*

PROOF. Assume $[Y_1, K_2] \leq Y_1 \cap K_2$. Then for $t \in T - S$, $[Y_1^t, K_1] \leq Y_1^t \cap K_1$ as $K_2^t = K_1$ by 15.2.21.1. Next as conclusion (ii) of 15.2.18.3 holds, $\bar{N}_I = C_{\bar{M}}(Z_1)$ is of index 6 in $\bar{M} \cong \Omega_4^+(V)$ for some complement Z_1 to Z in Z_S. Then as $Y_1 \leq N_I$ and $C_M(V)$ is a $3'$-group, $\bar{Y}_1 = O_3(\bar{N}_I)$, $\bar{Y}_1 \bar{Y}_1^t = O(\bar{M}) \cong E_9$, and M has Sylow 3-subgroups isomorphic to E_9. Define Y as in 15.2.3; as $Y \trianglelefteq M$ but Y_1 is not T-invariant, YY_1 contains a Sylow 3-subgroup of M, so by a Frattini Argument, we may take t to act on YY_1, and thus $YY_1 = Y_1 Y_1^t$ with $[Y_1, Y_1^t] \leq Y_1 \cap Y_1^t$.

Let $X := \langle Y_1, K_1 \rangle$; then $[X, X^t] \leq X \cap X^t$, in view of the commutator relations established in first sentence of the proof, along with the relations $[Y_1, Y_1^t] \leq Y_1 \cap Y_1^t$

and $[K_1, K_2] \leq K_1 \cap K_2$. Now S acts on X, $F^*(I) = O_2(I)$ by hypothesis, and $S \in Syl_2(I)$ by 15.2.18.1; thus $F^*(XS) = O_2(XS)$ by 1.1.4.4. Then $F^*(X) = O_2(X)$ by 1.1.3.1, so that $O_2(X) \neq 1$. It follows that $O_2(XX^t) \neq 1$. Then as T acts on XX^t, $XX^tT \in \mathcal{H}(T)$. This is a contradiction, as $Y_1Y_1^tT \leq XX^tT$ and $M = !\mathcal{M}(Y_1Y_1^tT)$ by 15.2.2, so that $K = K_1K_2 \leq M$, contrary to $H \not\leq M$. $\qquad\square$

Recall $C_{M_c}(Z_S) = C_G(Z_S)$ by 15.2.16. During the remainder of the proof of Theorem 15.2.15, take I minimal subject to $I \in \mathcal{H}(C_G(Z_S))$ and $I \not\leq M_c$. Recall $M_I < I$, so as $M_I = I \cap M_c$, M_I is a maximal subgroup of I.

For $X \leq G$, let $\theta(X)$ be the subgroup generated by all elements of X of order 3.

LEMMA 15.2.23. $F^*(I) \neq O_2(I)$.

PROOF. We assume $F^*(I) = O_2(I)$ and derive a contradiction. We begin with some preliminary reductions.

Suppose first that there is $X_0 = O^{3'}(X_0) = X_0^\infty \trianglelefteq I$ with X_0 nontrivial on $Z_0 := \Omega_1(Z(O_2(X_0)))$ and $X_0 \leq M_c$. As $X_0 = O^{3'}(X_0)$, $X_0 \leq K_c$ by 15.2.12.4. Since $S \in Syl_2(I)$ by 15.2.18.1, $S \cap O_2(K_c) = I \cap O_2(K_c)$, so X_0 acts on $S \cap O_2(K_c)$. Therefore as $|T : S| = 2$ by 15.2.17.3, $|O_2(K_c) : S \cap O_2(K_c)| \leq 2$, so $[O_2(K_c), X_0] \leq S \cap O_2(K_c) \leq N_G(X_0)$, Thus $X_0 = (X_0O_2(K_c))^\infty$ is $O_2(K_c)$-invariant. Hence $[X_0, O_2(K_c)] \leq O_2(X_0) \leq C_G(Z_0)$, so by the Thompson $A \times B$-lemma, X_0 is nontrivial on $C_{Z_0}(O_2(K_c))$. But since $K_c \in \mathcal{H}^e$ by 1.1.3.1, $C_{Z_0}(O_2(K_c) \leq \Omega_1(Z(O_2(K_c))) =: Z_c$. Hence K_c is nontrivial on Z_c, so that $K_c \in \mathcal{L}_f(G,T)$ by 1.2.10, contradicting part (1) of Hypothesis 14.1.5. Thus no such X_0 exists.

Next assume that either X is a χ-block of I, or $X \in \mathcal{C}(I)$ with $X/O_2(X) \cong L_3(2)$ and X is described in C.1.34. Set $X_0 := \langle X^S \rangle$ and $U_0 := \langle Z_S^{X_0} \rangle$. Observe that $X_0 \trianglelefteq I$ either by 1.2.1.3, or when X is an A_3-block since $|X^I| \leq m_3(I) \leq 2$. If X is an A_7-block, then $X = O^{3'}(I)$ by A.3.18, so $K \leq X$. In the remaining cases, $m_3(X) = 1$ and K acts on X, so as $m_3(KX) \leq 2$, $K_0 := O^2(K \cap X) \neq 1$.

Consider first the subcase where X is an A_3-block or an $L_2(2^n)$-block. By 15.2.11.2, Z_S centralizes K, so that $Z_S \cap U_0$ centralizes $\langle K_0^S \rangle = X_0$; this is impossible, as $Z_S \cap U_0 \not\leq C_{U_0}(X_0)$ in these blocks (cf I.2.3.1).

This leaves the subcases where either X is an A_5-block or an A_7-block, or $X/O_2(X) \cong L_3(2)$. Then $X_0 \not\leq M_c$ by paragraph two, so $M_0 := X_0 \cap M_I < X_0$. Notice $C_{X_0}(Z_S) \leq M_0$ by 15.2.12.3. If X is an A_5-block, then $C_{X_0}(Z_S)$ is a Borel subgroup of X_0, so M_0 is that Borel subgroup. If X is an A_7-block, then we saw $K \leq X$, so $K \leq M_0$ by 15.2.11.2, and hence $M_0 = K(X \cap S)$ is the maximal subgroup stabilizing the partition $\{\{1,2,3,4\}, \{5,6,7\}\}$, using the notation of section B.3. Finally if $X/O_2(X) \cong L_3(2)$, then since $C_{X_0}(Z_S) \leq M_0 < X_0$, case (5) of C.1.34 is eliminated by B.4.8.2, as in that case Z_S centralizes X_0. Thus $C_{X_0}(Z_S)$ is a maximal parabolic of X_0, so M_0 is that maximal parabolic.

Let $Q_I := O_2(KS)$; in each case $M_0 \trianglelefteq KS$, so $Q_I = O_2(M_0Q_I)$. Further M_0 contains a Sylow 2-group of X_0, so $O_2(X_0Q_I) \leq Q_I$ by A.1.6. Next $Q_I \trianglelefteq H$ as $|T : S| = 2$, so $C(I, Q_I) \leq M_I$ by 15.2.12.3. Then as $M_0 < X_0$, $J(Q_I) \not\leq O_2(X_0Q_I))$, so there is an FF*-offender in $Aut_{Q_I}(U_0)$ by B.2.10. Hence by B.3.2.4, X is not an A_5-block or an A_7-block. Further cases (2)–(4) of C.1.34 are eliminated since $C_{X_0}(Z_S) \leq M_0$, leaving case (1) of C.1.34 where X is an $L_3(2)$-block. Thus we have shown that I possesses no χ-blocks, and if $X \in \mathcal{C}(I)$ with $X/O_2(X) \cong L_3(2)$ and

X is described in C.1.34, then X is an $L_3(2)$-block. This completes our preliminary reductions.

Suppose first that conclusion (i) of 15.2.18.3 holds. Then by 15.2.19.1 and C.1.28, $I = M_I L_1 \cdots L_s$, where L_i is a χ-block. But then $M_I = I$ by the previous paragraph, a contradiction.

Therefore conclusion (ii) of 15.2.18.3 holds. Set $R := O_2(N_I)$. Then case (2) of 15.2.19 holds, so Hypothesis C.2.3 is satisfied by I, R, N_I in the roles of "H, R, M_H". If $O_{2,F}(I) \not\leq N_I$, then by C.2.6, there is an A_3-block X of I with $X \not\leq N_I$, contrary to an earlier reduction. Thus $O_{2,F}(I) \leq N_I$. On the other hand, if $O_{2,F^*}(I) \leq N_I$, then $R = O_2(I)$ by A.4.4.1, contradicting $N_I = N_I(R)$ and $K \not\leq N_I$. Thus there is $X \in \mathcal{C}(I)$ with $X/O_2(X)$ quasisimple and $X \not\leq N_I$. Further $K = O^2(K)$ normalizes X by 1.2.1.3.

If R does not act on X, then X is a χ-block by C.2.4, contrary to an earlier reduction. Thus R acts on X, so X is described in C.2.7.3. Let $M_X := M \cap X$, and this time set $M_0 := M_I \cap X$, so that $M_1 := C_X(Z_S) \leq M_0$ by 15.2.12.3.

Suppose that case (g) of C.2.7.3 holds. Then $X/O_2(X) \cong SL_3(2^n)$, M_X is a maximal parabolic of X, and (XR, R) is an MS-pair described in C.1.34. Assume first that $n > 1$. Then $M_1 = P^\infty$, where P is the maximal parabolic of XS over S other than M_X. As $m_3(KC_X(Z_S)) = 2$, $K_0 := O^2(K \cap M_1) \neq 1$; then as S acts on K_0, n is even. Then $O_{2,Z}(X) > O_2(X)$, so as $m_3(KO_{2,Z}(X)) = 2$, $O_{2,Z}(X) \leq K$. This is impossible, as $O_{2,Z}(X) \leq M_X$ while $K \cap M = O_2(K)$. Therefore $n = 1$, so by an earlier reduction, X is an $L_3(2)$-block and $M_0 = M_1$ is the maximal parabolic of X over $S \cap X$ other than M_X. If $X < \langle X^S \rangle =: X_0$, then $M_X S/O_2(M_X S) \cong O_4^+(2)$. This is impossible, as $\bar{M}_X \bar{S} \cong S_3$ since conclusion (ii) of 15.2.18.3 holds, while $O_4^+(2)$ has no such quotient group. Thus $X \trianglelefteq I$ by 1.2.1.3, so by minimality of I, $I = M_I X$. Then S is not irreducible on $K/O_2(K)$ since $m_3(K \cap X) = 1$, so $K_c \leq M_I$ by 15.2.21.1. Further $KS = H_1 H_2$ by 15.2.21.1, so

(*) K_1 and K_2 are the S-invariant subgroups K_+ of K with $|K_+ : O_2(K_+)| = 3$.

As $m_3(KX) = 2 = m_3(K)$, by (*) we may assume $K_1 = O^2(K \cap X)$. Then by another application of (*), $[X, K_2] \leq O_2(X)$. Further $K_1 = O^2(M_0) \trianglelefteq M_I$, so that $K = K_c$ by 15.2.12.2. Thus $C_M(V)$ is a $3'$-group by 15.2.21.2, so $O^{3'}(N_I) =: Y_1 = O^{3'}(M_X)$, and Y_1 is S-invariant with $Y_1 S/O_2(Y_1 S) \cong S_3$. As $[X, K_2] \leq O_2(X)$, $[Y_1, K_2] \leq K_2 \cap Y_1$, contrary to 15.2.22.

Thus case (g) of C.2.7.3 is eliminated. An earlier reduction showed that X is not a χ-block; this eliminates case (a) of C.2.7.3, and the subcases of (b) where X is a χ-block. In the remaining cases, $m_3(X) = 2$, and then $X = \theta(I)$ by A.3.18; so as $KS \leq C_G(Z_S)$ by 15.2.11.2, $KS \leq M_1 \leq M_0$. In particular $m_3(M_0) = 2$, with $KS/O_2(KS) \cong S_3 \times S_3$; so by inspection of the list in C.2.7.3 (recalling that $Out(Sp_4(4))$ is cyclic; cf. 16.1.4 and its underlying reference), either X is an A_7-block, or $X/O_2(X) \cong L_4(2)$ or $L_5(2)$. The former case was eliminated earlier, so the latter holds. Now M_1 is a proper parabolic of X containing K, so either $M_1 S = KS$ is determined by a pair of non-adjacent nodes, or $X/O_2(X) \cong L_5(2)$ and $M_1 S$ is a maximal parabolic determined by all the nodes except one interior node. Let $U := [\langle Z_S^X \rangle, X]$. By B.2.14, $\langle Z_S^X \rangle = UC_{\langle Z_S^X \rangle}(X)$, so that $C_X(U \cap Z_S) = C_X(Z_S)$. Now by C.2.7.2, U is an FF-module for $(XS)^+ := XS/O_2(XS)$, and hence is described in Theorem B.5.1. In particular one of the following holds:

(a) U is the sum of isomorphic natural modules, and M_1 is an end-node maximal parabolic.

(b) U is the sum of a natural module and its dual, and M_1 is the parabolic determined by the interior nodes.

(c) $U/C_U(X)$ is the 6-dimensional orthogonal module for $X^+ \cong L_4(2)$.

(d) U is a 10-dimensional module for $X^+ \cong L_5(2)$.

As $K \le M_1$, case (b) is eliminated. Let $K_I := O^{3'}(M_1)$. Assume case (d) occurs. Then $K_I/O_2(K_I) \cong \mathbf{Z}_3 \times L_3(2)$. Now $X = O^{3'}(I)$ by A.3.18; so as $C_G(Z_S) \le I$ and $M_1 = C_X(Z_S)$, $K_I = O^{3'}(C_G(Z_S))$ is T-invariant, and so $O^{3'}(O_{2,3}(K_I))$ is a T-invariant subgroup of 3-rank 1. But this is impossible as T is irreducible on $K/O_2(K)$. Assume case (a) occurs. Then as $K \le M_1$, $X/O_2(X) \cong L_5(2)$, and $K_I/O_2(K_I) \cong L_4(2)$. In particular as S acts on M_1, S is trivial on the Dynkin diagram of $X/O_2(X)$, and so S is not irreducible on $K/O_2(K)$. Then by 15.2.21.2, $K_c = O^{3'}(M_I)$, so $K_I = K_c$. As conclusion (ii) of 15.2.18.3 holds, $O^{3'}(N_I) \not\le M_I$, so the minimal parabolic P of X not contained in K_I is contained in N_I. Thus $Y_1 := O^2(P) \le N_I$ with $Y_1 S/O_2(Y_1 S) \cong S_3$, but $Y_1 \not\le M_I$. This contradicts 15.2.21.3.

Thus case (c) holds. In this case, $M_1 S = KS$ is the maximal parabolic determined by the end nodes. We apply an argument made in an earlier reduction, with M_1, U in the roles of "M_0, U_0", to conclude that for $Q_I = O_2(KS)$, $Aut_{Q_I}(U)$ contains an FF*-offender. But this is not the case for this parabolic and representation by B.3.2.6.

This contradiction finally completes the proof of 15.2.23. $\square$

By 15.2.20.3, $O(I) = 1$, so as $F^*(I) \ne O_2(I)$ by 15.2.23, $E(I) \ne 1$. By 15.2.20.2, $F^*(M_I) = O_2(M_I)$, so there is a component L of I with $L \not\le M_I$, and by 15.2.20.1, L is described in 1.1.5.3. Further as $O(I) = 1$, $Z(L)$ is a 2-group. Let $L_0 := \langle L^S \rangle$, $S_L := S \cap L_0$, and $M_L := L_0 \cap M_I$. As usual $L_0 \trianglelefteq I$ by 1.2.1.3. Recall by our minimal choice of I that M_I is a maximal subgroup of I; hence $I = L_0 M_I$. By 1.1.5.3, Z is faithful on L.

LEMMA 15.2.24. $L/Z(L)$ is not of Lie type and characteristic 2.

PROOF. Suppose otherwise. Then we are in one of cases (a)–(c) of 1.1.5.3, and $L \cong A_6$ in case (c), since $Z(L)$ is a 2-group.

Now $S_L \in Syl_2(L_0)$ and $S_L \le M_L$. Further $M_L < L_0$ since M_I is a maximal subgroup of $I = L_0 M_I$. So since the maximal S-invariant overgroups of S_L in L_0 are parabolics over S_L, M_L is such a parabolic. Also $Z \le C_I(M_L)$ by 15.2.14.1, and Z is faithful on L. We conclude from the list in (a)–(c) of 1.1.5.3 that L is defined over $\mathbf{F}_2$, and if $L \cong L_3(2)$, then $M_L = S_L$, so that $N_S(L)$ is nontrival on the Dynkin diagram of L. Now if $m_3(L_0) = 2$ then $L_0 = O^{3'}(I)$ by A.3.18 or 1.2.2.a, so $K \le C_L(Z) \le M_L$, and hence $m_3(C_L(Z)) \ge 2$; but this is not the case for the groups of 3-rank 2 defined over $\mathbf{F}_2$ in Theorem C (A.2.3). Therefore $m_3(L_0) = 1$, so $L_0 = L \cong L_3(2)$. But now $Aut_{M_I}(L)$ is a 2-group, so K centralizes L and hence $m_3(KL) = 3$, contrary to I an SQTK-group. $\square$

We are now in a position to complete the proof of Theorem 15.2.15. By 15.2.20.3, L is described 1.1.5.3, and indeed appears in one cases (d)–(f) by 15.2.24, and Z is faithful on L. Further in case (d), $L \cong A_7$ since $Z(L)$ is a 2-group.

If $L \cong L_2(p)$ for p a Mersenne or Fermat prime, then $p > 7$ by 15.2.24, and $C_{L_0}(Z) = S_L$. Then as K centralizes Z, and $Aut_{M_I}(L)$ is a 2-group, $[K, L_0] = 1$, so $L \leq N_I(K) \leq M_c = !\mathcal{M}(H)$ by 15.2.12.3, contrary to the choice of L.

Thus L is not $L_2(p)$. In the remaining cases $m_3(L) = 2$, so $L = O^{3'}(I)$ by A.3.18. Thus $K \leq C_L(Z)$, so $m_3(C_L(Z)) = 2$. Inspecting the list of groups remaining in 1.1.5.3, we conclude $L \cong J_4$. But then $O^2(C_L(Z))/O_2(O^2(C_L(Z)) \cong \hat{M}_{22}$, whereas $C_G(Z) = M_c$ by 15.2.14.1, $K_c = O^{3'}(M_c)$ by part (4) of 15.2.12, and K_c contains no such section by part (2) of the latter result. This contradiction completes the proof of Theorem 15.2.15.

15.2.3. The final contradiction. For $X \leq G$, write $\Lambda(X)$ for the subgroup generated by all involutions of X.

LEMMA 15.2.25. *(1) Case (1) or (4) of 15.2.1 holds, with $\bar{M} \cong S_3$, D_{10}, or $Sz(2)$.*

(2) $\Lambda(T) \leq S$.

PROOF. If (1) fails, then conclusion (ii) of 15.2.13.1 holds. In this case there is Z_1 of order 2 in Z_S with $C_{\bar{M}}(Z_1) \cong S_3$. In particular as $M \cap M_c = C_M(V)T$ by 15.2.14.5, $C_M(Z_1) \not\leq M_c$, contrary to Theorem 15.2.15. Therefore (1) holds. By (1), $\bar{S} = \Omega_1(\bar{T})$, so (2) holds. $\qquad\square$

Recall $U_H = \langle V^H \rangle$ from Notation 15.2.9.

LEMMA 15.2.26. *(1) $r(G, V) > 1$ and hence $s(G, V) > 1$.*

(2) $W_0(T, V) \leq C_T(V)$.

(3) $W_0(Q_H, V) \leq C_H(U_H)$.

PROOF. Let U be a hyperplane of V. Suppose first that $\bar{M}$ is not $Sz(2)$. Then $Z = Z_S$ is of rank 2 by examination of the cases in 15.2.25.1, and hence $Z \cap U \neq 1$. Then as $C_G(Z) = M_c$ by 15.2.14.1, $C_G(Z) = C_G(Z \cap U)$. Similarly for $g \in M - M_c$, $C_G(Z^g) = C_G(Z^g \cap U)$ and $ZZ^g = V$, so that

$$C_G(U) \leq C_G(Z \cap U) \cap C_G(Z^g \cap U)) = C_G(Z) \cap C_G(Z^g) = C_G(V).$$

Thus $r(G, V) > 1$ in this case.

So assume $\bar{M} \cong Sz(2)$. In this case $m(Z_S) = 2$ and $m(Z) = 1$. Here M has two orbits on nonzero vectors of V of lengths 5 and 10, and hence two orbits on hyperplanes of V, which are also of lengths 5 and 10. Notice by 15.1.9.5 that Hypothesis E.6.1 is satisfied, so if U is T-invariant then E.6.13 says $C_G(U) \leq N_G(V)$. If U is not T-invariant, then $|U^M| = 10$, so as $\bar{T}$ is cyclic, we may assume that s normalizes U and hence centralizes a nontrivial 2-subspace of U, so that $Z_S = C_V(s) \leq U$. As $V = \langle Z^M \rangle$, there exists $g \in M - M_c$ with $Z^g \not\leq U$. By Theorem 15.2.15, $C_G(Z_S^g \cap U) \leq M_c^g$, so as $M_c^g = C_G(Z^g)$ by 15.2.14.1, with $Z^g \not\leq Z_S^g \cap U \neq 1$ and Z_S^g of rank 2, we conclude that $C_G(Z_S^g \cap U) = C_G(Z_S^g)$. Thus $C_G(U) = C_G(V)$ as in the previous paragraph.

Therefore $r(G, V) > 1$ in either case. Since $m(\bar{M}, V) > 1$ by 15.2.14.6, also $s(G, V) > 1$, so that (1) holds. Furthermore $a(\bar{M}, V) = 1$ by 15.2.14.6. Now E.3.21.1 implies (2), and (2) implies (3). $\qquad\square$

LEMMA 15.2.27. *(1) $O_{2,F^*}(M_c) \leq K_c(M \cap M_c)$.*

(2) $V \leq O_2(M_c)$.

(3) If $Z_S \cap V^g \neq 1$, then $[V, V^g] = 1$.

PROOF. We claim that $O_{3'}(M_c) \leq M \cap M_c$: for otherwise we may choose a T-invariant $3'$-subgroup K of M_c minimal with respect to $J := KT \not\leq M$; then $J \in \mathcal{H}_*(T, M)$, whereas members of $\mathcal{H}_*(T, M)$ are not $3'$-groups by 15.2.8. So the claim holds, and hence as $O^{3'}(M_c) = K_c$ by 15.2.12.4, (1) holds.

If (2) fails, then $[K_c, V] \neq 1$ by (1), so $K < K_c$ as $V \leq O_2(H)$ by 15.1.11.1. Thus case (ii) or (iii) of 15.2.12.2 holds. and then by $K_c = [K_c, V]$ by (1). Let $Q_c := O_2(M_c)$, $V_c := V \cap Q_c$, $K_c^* T^* := K_c T/O_2(K_c T)$, and $\tilde{M}_c := M_c/Z$. Then $C_{M_c}(\tilde{Q}_c) \leq Q_c$ by A.1.8. Thus as $V^* \neq 1$, V does not centralize $\tilde{Q}_c$, so as $[\tilde{Q}_c, V] \leq \tilde{V}_c$, $m(\tilde{V}_c) \geq 1$. On the other hand since $m(Z) \geq 1 \leq m(V^*)$ and V has rank 4, $m(\tilde{V}_c) \leq 2$ with equality holding only if V^* and Z are of rank 1. Next in the groups in (ii) and (iii) of 15.2.12.2, no normal subgroup of T^* induces a group of $\mathbf{F}_2$-transvections with fixed center on a chief section of Q_c by B.4.2, keeping in mind in (ii) that T^* is nontrivial on the Dynkin diagram of K_c^*. Therefore we conclude that $[V, \tilde{Q}_c] = \tilde{V}_c$ is of rank 2, so that V^* and Z are indeed of order 2. It follows that $V^* = Z(T^*)$, $K_c T$ has just one noncentral 2-chief factor W, and (e.g., by D.3.10, B.4.2, and B.4.5) either

(i) $K_c^* T^* \cong S_8$ or $Aut(L_5(2))$, and either W is the 6-dimensional orthogonal module for S_8, or W is the sum of the natural module for $K_c^* \cong L_n(2)$ ($n = 4$ or 5) and its dual; or

(ii) $K_c^* T^* \cong L_3(2)$ wr $\mathbf{Z}_2$ and $W = W_1 \oplus W_2$, where $W_i := [W, K_{c,i}]$ is the natural module for the direct factor $K_{c,i}^* \cong L_3(2)$ of K_c^*.

Next as Z is of order 2 and Z_S is of order 4, $1 \neq \tilde{Z}_S$. As $\tilde{Z}_S \leq Z(\tilde{T})$, we conclude $\tilde{Z}_S \leq \widetilde{Q_c \cap V} = \tilde{V}_c$ using B.2.14. Thus the projection W_Z of $\tilde{Z}_S$ on W is nontrivial and centralized by H^* by 15.2.11.2. As $H^* \cong S_3$ wr $\mathbf{Z}_2$, it follows that in (i), $K_c^* T^* \cong S_8$ and W is the orthogonal module; and in (ii), H^* is the parabolic of $K_c^* T^*$ over T^* stabilizing a point of W. In either case, there is a parabolic P^* of $K_c^* T^*$ not contained in H^*, minimal subject to being T^*-invariant; further $P^*/O_2(P^*) \cong S_3$ in (i), and $P^* \cong S_3$ wr $\mathbf{Z}_2$ in (ii). By minimality of P^*, if the preimage P is not contained in M, then $P \in \mathcal{H}_*(T, M)$. We conclude from 15.2.8 that $P \leq M$ in (i), while in (ii) we get $P \leq M$ from 15.2.11.2 since $[W_Z, P] \neq 1$ by construction. Then by 15.2.14.5, $O^2(P) \leq C_P(V) \leq C_P(Z_S)$, again contrary to $[W_Z, P] \neq 1$. This contradiction completes the proof of (2).

Now by (2), $V^x \leq Q_c \leq T$ for each $x \in M_c$, so $[V, V^x] = 1$ by 15.2.26.2. Finally assume that $1 \neq Z_S \cap V^g$ for some $g \in G$. As $V \leq Z(J(T))$ and $N_G(J(T)) \leq M$ by 15.1.9.1, we may apply Burnside's Fusion Lemma A.1.35 to conclude that M controls fusion in V. Therefore we may take $g \in N_G(Z_S \cap V^g)$ by A.1.7.1, and hence $g \in M_c$ by Theorem 15.2.15. Then $[V, V^g] = 1$ by the initial remark of the paragraph, so (3) holds. $\qquad\square$

LEMMA 15.2.28. $[U_H, \Lambda(Q_H)] \leq Z_S$.

PROOF. Observe that $\Lambda(Q_H) \leq \Lambda(T) \leq S$ by 15.2.25.2, and S centralizes V/Z_S in each case of 15.2.25.1. Thus as $\Lambda(Q_H) \trianglelefteq H$ and $U_H = \langle V^H \rangle$, the lemma holds. $\qquad\square$

We are now in a position to complete the proof of Theorem 15.1.3.

Observe that the pair M, H satisfies Hypotheses F.7.1 and F.7.6 in the roles of "G_1, G_2". Form the coset graph Γ on the pair M, H, and adopt the notation of section F.7. In particular γ_0 and γ_1 are the points of Γ stabilized by M and H, respectively. For $\alpha := \gamma_1 g$ and $\beta := \gamma_0 y$, set $U_\alpha := U_H^g$, $Z_\alpha := Z_S^g$, and $V_\beta := V^y$. Let $b := b(\Gamma, V)$. Also we choose a geodesic

$$\gamma_0, \gamma_1, \ldots, \gamma_b =: \gamma$$

with $V \not\leq G_\gamma^{(1)}$. As V is not an FF-module for $\bar{M}$ by 15.1.8, b is odd by F.7.11.7. From 15.2.8, $Q_H = G_{\gamma_1}^{(1)}$. Thus as $V \leq Q_H$ by 15.1.11.2, $b > 1$. As b is odd, G_γ is a conjugate of H, so $G_\gamma^{(1)} = O_2(G_\gamma) =: Q_\gamma$.

While Hypothesis F.8.1 does not hold, we can still make use of arguments in section F.8. As in section F.8, define $D_\gamma := U_\gamma \cap Q_H$ and $D_H := U_H \cap Q_\gamma$. By choice of γ, $V \not\leq Q_\gamma$, so $D_H < U_H$. Indeed V does not centralize U_γ, so there is $g \in G$ with $\gamma_1 g = \gamma$ and $[V, V^g] \neq 1$. But if $D_\gamma = U_\gamma$, then $V^g \leq W_0(Q_H, V) \leq C_H(U_H) \leq C_H(V)$ by 15.2.26.3, a contradiction. Therefore $D_\gamma < U_\gamma$, so we have symmetry between γ_1 and γ (cf. Remark F.9.17).

Next

$$m(U_\gamma/D_\gamma) = m(U_\gamma^*) \leq m_2(H^*) = 2,$$

so by symmetry, $m(U_H/D_H) \leq 2$. Now $[U_H, D_\gamma] \leq Z_S$ by 15.2.28, so by symmetry $[U_\gamma, D_H] \leq Z_\gamma$. Then $[D_H, D_\gamma] \leq Z_S \cap Z_\gamma$, while as $[V, V_\gamma] \neq 1$, we conclude from 15.2.27.3 that $Z_S \cap Z_\gamma = 1$. Thus $[D_H, D_\gamma] = 1$. Next $m(D_\gamma/C_{D_\gamma}(V)) \leq m_2(\bar{M}) = 1$ by 15.2.25.1, and by symmetry $m(D_H/C_{D_H}(V^g)) \leq 1$, so

$$m(U_H/C_{U_H}(V^g)) \leq m(U_H/D_H) + 1 \leq 3.$$

Therefore V^{g*} induces a transvection on each of the 2-chief factors of H on U_H appearing in 15.1.12, so $V^{g^*} \leq H_i^*$ for $i = 1$ or 2. Hence $m(V^g/(V^g \cap Q_H)) = 1$. By symmetry, $m(V/(V \cap Q_\gamma)) = 1$, and

$$[V \cap Q_\gamma, V^g \cap Q_H] \leq [D_\gamma, D_H] = 1.$$

Therefore as $s(G, V) > 1$ by 15.2.26.1, E.3.6 says $V \cap Q_\gamma \leq C_G(V^g)$, and then by another application of those lemmas, $V^g \leq C_G(V \cap Q_\gamma) \leq C_G(V)$, contrary to the choice of V^g.

This contradiction completes the proof of Theorem 15.1.3.

15.3. The elimination of $\mathbf{M_f}/\mathbf{C_{M_f}}(\mathbf{V(M_f)}) = \mathbf{S_3}$ wr $\mathbf{Z_2}$

In this section, we complete our treatment of the groups satisfying Hypothesis 14.1.5, by proving:

THEOREM 15.3.1. *Assume Hypothesis 14.1.5. Then G is isomorphic to J_2, J_3, $^3D_4(2)$, the Tits group $^2F_4(2)'$, $G_2(2)'$, or M_{12}.*

Observe that the groups in Theorem 15.3.1 have already appeared in Theorem 15.1.3, so that we will be working toward a contradiction. On the other hand, the shadows of the groups $Aut(L_n(2))$, $n = 4$, 5, S_9, A_{10}, $Aut(He)$, and L wr $\mathbf{Z}_2$ for $L \cong S_5$ or L of rank 2 over $\mathbf{F}_2$ arise,   and cause difficulties: Each of these groups possesses $M \in \mathcal{M}(T)$ such that $V(M)$ is of rank 4 and $Aut_M(V(M)) = O^+(V(M))$.

For $X \leq G$, we let $\theta(X)$ denote the subgroup generated by all elements of X of order 3.

15.3.1. Preliminary results. Recall by Hypothesis 14.1.5.2 that

$$M_c = !\mathcal{M}(C_G(Z)).$$

Throughout section 15.3 we assume G is a counterexample to Theorem 15.3.1. Let $M := M_f$ be the unique maximal member of $\mathcal{M}(T) - \{M_c\}$ under the partial order $\overset{\sim}{\lesssim}$ of Definition A.5.2, supplied by 14.1.12. Recall in particular that $M \in \mathcal{M}(N_G(C_2))$, where $C_2 := C_2(\mathrm{Baum}(T))$ is the characteristic subgroup of $\mathrm{Baum}(T)$ from C.1.18. Let $V := V(M)$.

We summarize what has been established in this chapter so far:

LEMMA 15.3.2. *(1)* $m(V) = 4$ *and* $\bar{M} = O_4^+(V)$.
(2) $Z = C_V(T)$ *is of order 2.*
(3) $\mathcal{M}(T) = \{M, M_c\}$.
(4) If $T \leq X \leq M$, *then either*
  (i) $O^2(X) \leq C_M(V)$, *or*
  (ii) $\bar{X} = \bar{M}$ *and* $M = !\mathcal{M}(X)$.

(5) M *is the unique maximal member of* $\mathcal{M}(T)$ *under* $\overset{\sim}{\lesssim}$.
(6) $N_G(T) \leq M$. *In particular, members of* $\mathcal{H}_*(T, M)$ *are minimal parabolics described in B.6.8, and in E.2.2 when nonsolvable.*

PROOF. As G is a counterexample to Theorem 15.3.1, and the groups in 15.3.1 appear as conclusions in Theorem 15.1.3, conclusion (2) of 15.1.3 holds, giving (1). Then (1) implies (2) as $V = V(M) = \langle Z^M \rangle$, and (2) and 14.1.12.1 imply (5). Since $N_G(T)$ preserves the relation $\overset{\sim}{\lesssim}$ on $\mathcal{M}(T)$, $N_G(T) \leq M$ by (5); then 3.1.3.2 completes the proof of (6).

Assume the hypotheses of (4). By (1), $\bar{T}$ is maximal in $\bar{M}$, so either $O^2(X) \leq C_M(V)$ so that (i) holds, or $\bar{X} = \bar{M}$. In the latter case, (ii) holds by A.5.7.1. Thus (4) is established.

Finally suppose $M_1 \in \mathcal{M}(T)$. By (5), $M_1 = (M \cap M_1)C_{M_1}(V(M_1))$. As usual $C_{M_1}(V(M_1)) \leq M_c$ since $M_c = !\mathcal{M}(C_G(Z))$. We apply (4) to $M \cap M_1$ in the role of "X": in case (i) of (4), $M \cap M_1 \leq C_M(V) \leq C_G(Z) \leq M_c$, so that $M_1 \leq M_c$; in case (ii) of (4), $M = !\mathcal{M}(M \cap M_1)$ so that $M_1 = M$. Thus (3) holds. $\square$

LEMMA 15.3.3. *(1)* $M = !\mathcal{M}(N_M(C_2(\mathrm{Baum}(T))))$.
(2) If $\mathrm{Baum}(T) \leq S \leq T$, *then* $\mathrm{Baum}(T) = \mathrm{Baum}(S)$, *and further* $N_G(S) \leq N_G(\mathrm{Baum}(S)) \leq M$.

PROOF. Set $C_2 := C_2(\mathrm{Baum}(T))$, and recall $M \in \mathcal{M}(N_G(C_2))$. By 15.3.2.2 and 14.1.11, $M = C_M(V)N_M(C_2)$, so that (1) follows from 15.3.2.4. Choose S as in (2). Then $\mathrm{Baum}(T) = \mathrm{Baum}(S)$ by B.2.3.4, and $N_G(S) \leq N_G(\mathrm{Baum}(S)) \leq N_G(C_2)$, so that (2) follows from (1). $\square$

LEMMA 15.3.4. $M_c = C_G(Z)$.

PROOF. Let $U := V(M_c)$, and assume the lemma fails; then $U > Z$. Next $M_c \overset{\sim}{\lesssim} M$ by 15.3.2.5, so $M_c = C_M(U)X$ where $X := M \cap M_c$, and hence $U = \langle Z^X \rangle$. Thus as $Z < U$, $O^2(X) \not\leq C_G(V)$, so $M = !\mathcal{M}(X)$ by 15.3.2.4, contrary to $M_c \neq M$. $\square$

By 15.3.2.1, $\bar{M} = O_4^+(V)$ preserves an orthogonal-space structure on V, so $V = V_1 \oplus V_2$, where V_1 and V_2 are the two definite 2-dimensional subspaces of V.

Thus $\bar{M} = (\bar{M}_1 \times \bar{M}_2)\langle \bar{t} \rangle$, where $M_i := C_M(V_{3-i})$, $V_i = [V, M_i]$, $\bar{M}_i \cong O_2^-(V_i) \cong S_3$, and $\bar{t}$ is an involution with $M_1^t = M_2$. Set $S := N_T(V_1)$ and $Z_S := C_V(S)$.

LEMMA 15.3.5. *Let* $E := \Omega_1(Z(J(T)))$ *and* $B := Baum(T)$. *Then either*

(1) $[V, J(T)] = 1$, $B = Baum(C_T(V))$, $V \leq E$, *and* $C(G, B) \leq M$, *or*

(2) $\bar{B} = \overline{J(T)} = \bar{S} \cong E_4$ *and* $E \cap V = Z_S \cong E_4$.

PROOF. If $J(T) \leq C_T(V)$ then $V \leq E$ and $B = \text{Baum}(C_T(V))$ by B.2.3.5; thus $M = C_M(V)N_M(B)$ by a Frattini Argument, and hence $C(G, B) \leq M$ by 15.3.2.4. Otherwise $\overline{J(T)} \neq 1$, and then (2) follows from B.1.8. □

LEMMA 15.3.6. *(1)* $Baum(T) = Baum(S)$.

(2) $N_G(S) \leq N_G(Baum(S)) \leq M$.

(3) $S \in Syl_2(C_G(C_{V_1}(S)))$.

(4) $S \in Syl_2(N_G(V_1))$.

PROOF. By 15.3.5, $\text{Baum}(T) \leq S$, so that (1) and (2) follow from 15.3.3.2. As $S \in Syl_2(C_M(C_{V_1}(S)))$, (2) implies (3), and similarly (2) implies (4). □

LEMMA 15.3.7. *Let* $R_c := O_2(M \cap M_c)$ *and* $Y := O^2(\langle R_c^M \rangle)$. *Then* $\bar{Y} = O^2(\bar{M})$, $M \cap M_c = C_M(V)T$,

$$M = N_G(Y) = !\mathcal{M}(YT),$$

$O_2(YT) = C_T(V)$, *and either*

(1) $O_2(Y) = C_Y(V)$ *with* $Y/O_2(Y) \cong E_9$, *and* $Y = O^{3'}(M)$; *or*

(2) $Y/O_2(Y) \cong 3^{1+2}$, $O_{2,Z}(Y) = C_Y(V)$, $\bar{R}_c$ *is cyclic,* $Y = \theta(M)$, *and* $M \cap M_c$ *has cyclic Sylow 3-subgroups.*

PROOF. We apply case (b) of 14.1.17 with M_c in the role of "M_1", and $Y_0 := O^2(M)$. By 14.1.17.1, $\bar{R}_c \neq 1$. As $\bar{R}_c \trianglelefteq \bar{T}$, $\bar{R}_c$ contains $Z(\bar{T}) =: \langle \bar{r} \rangle$, and $\bar{r}$ inverts $\bar{Y}_0$ by 15.3.2.1, so $\bar{Y}_0 = [\bar{Y}_0, \bar{R}_c]$ and $M \cap M_c = C_M(V)T$. Further applying parts (2) and (3) of 14.1.17, we conclude $\bar{Y} = \bar{Y}_0$ and $Y^* R_c^*$ centralizes $C_M(V)^*$, where $M^* := M/O_2(M)$. In particular, $M = !\mathcal{M}(YT)$ by 15.3.2.4, and of course $M = N_G(Y)$ as $Y \trianglelefteq M \in \mathcal{M}$. Also $V = \langle Z^Y \rangle$, so that $V \in \mathcal{R}_2(YT)$ by B.2.14.

As $\bar{r}$ inverts $\bar{Y}_0 = O^2(\bar{M})$ and $[r^*, C_Y(V)^*] = 1$, r inverts y of order 3 in each coset of $C_Y(V)$ in Y. Therefore since Y^* centralizes $C_M(V)^*$, $Y^* = \Omega_1(Y^*)$ is a 3-group. As $\Phi(Y^*) \leq C_Y(V)^* \leq C_{Y^*}(r^*)$, it follows that $\Phi(Y^*) \leq Z(Y^*)$, and hence $Y^* \cong Y/O_2(Y) \cong E_9$ or 3^{1+2} by A.1.24. Further $O_2(YT) = C_T(V)$.

If $Y^* \cong E_9$, then as $M = YC_M(Y/O_2(Y))T$ and $m_3(M) \leq 2$, $Y = O^{3'}(M)$, so that (1) holds. If $Y^* \cong 3^{1+2}$, $m_3(C_M(V)) = 1$ by 14.1.17.4, so that $Y = \theta(M)$, and $C_M(V)T$ has cyclic Sylow 3-subgroups. Further $\bar{R}_c$ is embedded in $C_{Aut(Y^*)}(Z(Y^*)) \cong Q_8$, while $\bar{T} \cong D_8$, so $\bar{T}$ contains no Q_8-subgroup. Thus $\bar{R}_c$ is cyclic, completing the proof of (2). □

In the remainder of this section, define Y as in lemma 15.3.7.

LEMMA 15.3.8. $Z_S = \Omega_1(Z(S)) \cong E_4$.

PROOF. Let $Z_0 := \Omega_1(Z(S))$. As T/S is of order 2, $Z = C_{Z_0}(T)$ is of rank at least $m(Z_0)/2$, so $m(Z_0) \leq 2$ as Z is of order 2. As $Z_S = C_V(S)$ is of rank 2, the lemma follows. □

Finally we eliminate a configuration which appears at various points later, the case $V = O_2(Y)$:

LEMMA 15.3.9. $V < O_2(Y)$, so that $Y \not\cong A_4 \times A_4$.

PROOF. Assume $V = O_2(Y)$, or equivalently that $Y \cong A_4 \times A_4$. By 15.3.7, $Y \trianglelefteq M$. As $Aut(A_4) \cong S_4$, $Aut(Y) \cong S_4$ wr Z_2 with $C_{Aut(Y)}(V) = Aut_V(Y)$. Thus $YC_M(V) = Y \times C_M(Y)$. Since Z has order 2 and $C_T(Y) \trianglelefteq T$, $C_M(Y)$ has odd order; thus $C_M(Y) = 1$ as $F^*(M) = O_2(M)$. Therefore $M \leq Aut(Y)$, so as $\bar{M} \cong O_4^+(2)$ by 15.3.2.1, we conclude $M \cong S_4$ wr Z_2. But this is ruled out by Theorem 13.9.1. $\qquad\square$

15.3.2. Uniqueness theorems for Y and $O^2(C_Y(V_i))$. Our first main goal, in Theorem 15.3.45 in this subsection, is to show that $M = !\mathcal{M}(C_Y(V_i)S)$; to do so, we first show that $M = !\mathcal{M}(YS)$. We prove the two results simultaneously, adopting a suitable hypothesis to cover both cases, and eventually establish the common uniqueness result in 15.3.44.

Thus in this subsection, we assume:

HYPOTHESIS 15.3.10. *Either*

(1) $Y_+ := Y$, *or*
(2) $M = !\mathcal{M}(YS)$ *and* $Y_+ := O^2(C_Y(V_1))$.

Let
$$\mathcal{H}_+ := \mathcal{H}(Y_+S, M) = \{I \in \mathcal{H}(Y_+S) : I \not\leq M\},$$

and write $\mathcal{H}_{+,*}$ for the mimimal members of $\mathcal{H}_+$ under inclusion. As our goal is to show that $M = !\mathcal{M}(Y_+S)$, we will assume $\mathcal{H}_+$ is nonempty, and derive a contradiction. Given $I \in \mathcal{H}_+$, define $M_I := M \cap I$, $U_I := \langle Z^I \rangle$, $I^* := I/C_I(U_I)$, and $R := O_2(Y_+S)$.

LEMMA 15.3.11. *Assume* $I \in \mathcal{H}_+$. *Then*

(1) $S \in Syl_2(I)$.
(2) $C_I(U_I) \leq M_I$.
(3) If case (2) of Hypothesis 15.3.10 holds, then $Y_+ = O^2(Y \cap I)$ *and* $N_G(V_i) \leq M \geq N_G(Y_+)$ *for* $i = 1, 2$.
(4) $N_I(Y_+) = M_I$.
(5) Either:

(i) $Y_+/O_2(Y_+)$ *is* E_9 *or* 3^{1+2}, $Y_+S/O_{2,\Phi}(Y_+S) \cong S_3 \times S_3$, $R = O_2(YS) \trianglelefteq YT$, *and* $R = C_T(V)$. *Further* $R = O_2(N_I(R))$, $R \in Syl_2(\langle R^{M_I} \rangle)$, *and* $C(G, R) \leq M$. *Or:*

(ii) Case (2) of Hypothesis 15.3.10 holds, $Y_+S/R \cong S_3$, $Y_+ = O^{3'}(M_I)$, *and* $R = C_S(V_2)$.

(6) $Y_+ = \theta(M_I)$.
(7) $F^*(M_I) = O_2(M_I)$.
(8) $O(I) = 1$.
(9) $N_{I^*}(Y_+^*) = M_I^* < I^*$, *and* $Y_+^* \neq 1$.
(10) Either

(i) $Baum(R) = Baum(S)$ *and* $C(I, Baum(S)) \leq M_I$, *or*
(ii) $Y_+ = [Y_+, J(S)]$, *so* $Y_+^* \leq J(I)^*$.

(11) $J(I)^* \not\leq M_I^*$.
(12) If $L \leq I$ *with* $[V, O^2(Y \cap L)] \neq 1$ *and* $L \not\leq M$, *then no nontrivial characteristic subgroup of* S *is normal in* $\langle L, S \rangle$.

PROOF. Let $S \leq T_I \in Syl_2(I)$. By 15.3.6.2, $N_G(S) \leq M$, so as $|T : S| = 2$, $T_I \leq M$. Thus either (1) holds, or $T_I \in Syl_2(M)$, and in the latter case, 15.3.2.4 supplies a contradiction as $\bar{Y}_+ \neq 1$. Thus (1) is established.

We next prove (2)–(6).

Suppose first that case (1) of Hypothesis 15.3.10 holds. Then (3) holds vacuously, and (4) holds as $M = N_G(Y)$ by 15.3.7. Also $V = \langle Z^Y \rangle \leq U_I$, so (2) holds as $M = N_G(V)$. Further from 15.3.7 and the structure of $\bar{M}$ in 15.3.2.1, (6) and the first three statements of (5i) hold. As $R \trianglelefteq YT$, $M = !\mathcal{M}(N_G(R))$ by 15.3.7, so that $C(G, R) \leq M$. Also $R = C_T(V)$ as $O_2(\bar{Y}\bar{S}) = 1$. As $Y \trianglelefteq M_I$ and $R = O_2(YS)$ with $S \in Syl_2(I)$, $R \in \mathcal{B}_2(M_I)$ and $R \in Syl_2(\langle R^{M_I} \rangle)$ by C.1.2.4. Then as $N_I(R) \leq M_I$, $R = O_2(N_I(R))$, completing the proof that conclusion (i) of (5) holds in this case.

Now suppose that case (2) of Hypothesis 15.3.10 holds, so that $Y_+ = O^2(C_Y(V_1))$. Then $M = !\mathcal{M}(YS)$ by Hypothesis 15.3.10, so that $Y \nleq I$. Then as $|Y : Y_+ O_2(Y)| = 3$, $Y_+ = O^2(Y \cap I)$. Further as V_1, V_2, and Y_+ are normal in YS, the remainder of (3) and also (4) follow as $M = !\mathcal{M}(YS)$. Then as $V_2 \leq \langle Z^{Y_+} \rangle \leq U_I$, (2) follows from (3). We next prove (5) and (6). First suppose that conclusion (1) of 15.3.7 holds. Then $Y = O^{3'}(M)$, so $Y_+ = O^{3'}(M_I)$ as $Y_+ = O^2(Y \cap I)$. Thus (6) holds, and visibly conclusion (ii) of (5) holds. Therefore we may assume that conclusion (2) of 15.3.7 holds. Then $Y^* \cong 3^{1+2}$ and $Y_+/O_2(Y_+) \cong E_9$, with $Y_+ S/R \cong S_3 \times S_3$ using the structure of $\bar{M}$ in 15.3.2.1. This time $Y = \theta(M)$, so (6) holds. As $C_T(V) = C_S(Y_+^*)$, $R = C_T(V) = O_2(YS)$, so $C(G, R) \leq M$ as $M = !\mathcal{M}(YT)$. As $Y_+ \trianglelefteq M_I$ by (6) and $R = O_2(Y_+ S)$ with $S \in Syl_2(I)$, $R \in \mathcal{B}_2(M_I)$ and $R \in Syl_2(\langle R^{M_I} \rangle)$ by C.1.2.4. Then as $N_I(R) \leq M_I$, $R = O_2(N_I(R))$, so that conclusion (i) of (5) holds.

It remains to prove (7)–(12).

As $|T : S| = 2$ and $F^*(M) = O_2(M)$, 1.1.4.7 implies (7). In case (1) of Hypothesis 15.3.10, $V = [V, Y]$ so that $O(I) \leq C_I(V) \leq M_I$ by A.1.26.1, and hence (8) follows from (7). In case (2) of Hypothesis 15.3.10, $V_2 = [V_2, Y_+]$, and (8) follows similarly from (7) as $C_I(V_2) \leq M$ by (3).

Next $X := Y_+ C_I(U_I) \leq M_I$ by (2), so $Y_+ = \theta(Y_+ C_I(U_I))$ by (6); then (9) follows from (4) as Y_+ is nontrivial on $1 \neq [V_2, Y_+] \leq U_I$ by construction. By (5), $C_T(V) \leq R \leq S$. So if $[V, J(T)] = 1$, then $Baum(R) = Baum(S) = Baum(C_T(V))$ and $C(I, Baum(S)) \leq M_I$ by 15.3.5 and B.2.3.5, so that conclusion (i) of (10) holds. Then $I = J(I)N_I(J(S)) = J(I)M_I$ by a Frattini Argument, so as $I \nleq M$, $J(I) \nleq M$, and hence $J(I)^* \nleq M_I^*$ by (2), establishing (11). So suppose instead that $[V, J(T)] \neq 1$. Then case (2) of 15.3.5 holds, so that $\overline{J(T)} = \bar{S}$. But by (5), $Y_+ = [Y_+, S]$ and $C_T(V) \leq R \leq S$, so $Y_+ = [Y_+, J(S)]$, and hence conclusion (ii) of (10) holds. Thus if $J(I)^* \leq M_I^*$, then $Y_+^* = \theta(J(I)^*) \trianglelefteq I^*$ by (6), contrary to (9). This completes the proof of (10) and (11).

Assume the hypotheses of (12), with $1 \neq C$ char S and $C \trianglelefteq \langle L, S \rangle$. Then $\langle O^2(Y \cap L), T \rangle \leq N_G(C)$, so as $O^2(Y \cap L) \nleq C_M(V)$ by hypothesis, $N_G(C) \leq M$ by 15.3.2.4, a contradiction. $\square$

LEMMA 15.3.12. *Assume $I \in \mathcal{H}_+$ and there is $L \in \mathcal{C}(I)$ with $m_3(L) \geq 1$. Also assume $m_3(Y_+) = 2$. Then*

(1) $Y_L := O^2(Y_+ \cap L) \neq 1$. Further R normalizes L and $O_2(L)O_2(Y_L) \leq R$.

(2) Assume that $m_3(L) = 1$ and Y_+ induces inner automorphisms of $L/O_2(L)$. Then $Y_+ = Y_L Y_C$ where $Y_C := O^2(C_{Y_+}(L/O_2(L)))$, $|Y_L|_3 = 3 = |Y_C|$, and $Y_+/O_2(Y_+) \cong E_9$.

PROOF. First $Y_+ = O^2(Y_+)$ normalizes L by 1.2.1.3. Then $m_3(LY_+) \le 2$ since I is an SQTK-group, so as $m_3(Y_+) = 2$ and $m_3(L) \ge 1$ by hypothesis, $Y_L \ne 1$. As $[Y_L, R]$ is a 2-group, R normalizes L by 1.2.1.3. Further $O_2(L)O_2(Y_L)$ is normalized by Y_+, and so lies in R, completing the proof of (1).

Assume that $m_3(L) = 1$. As $Y_L \ne 1$ by (1) while Y_+ is of exponent 3, $|Y_L|_3 = 3$. Assume also that Y_+ induces inner automorphisms on $L/O_2(L)$; then as $m_3(L) = 1$ and Y_+ is of exponent 3, $\text{Aut}_{Y_+}(L) = \text{Aut}_{Y_L}(L)$. Hence $Y_+ = Y_L Y_C$, with $Y_L \cap Y_C$ a 2-group as $Z(L/O_2(L)) = 1$. In particular, $Y_+/O_2(Y_+) \cong E_9$ rather than 3^{1+2}, and $|Y_C|_3 = 3$, completing the proof of (2). $\qquad\square$

We now begin our analysis of the case $F^*(I) = O_2(I)$. Observe then that $U_I = \langle Z^I \rangle \in \mathcal{R}_2(I)$ by B.2.14.

We partition the problem into the subcases $m_3(Y_+) = 2$ and $m_3(Y_+) = 1$.

THEOREM 15.3.13. *Assume $I \in \mathcal{H}_+$ and $m_3(Y_+) = 2$. Then $F^*(I) \ne O_2(I)$.*

Until the proof of Theorem 15.3.13 is complete, assume I is a counterexample. Then $F^*(I) = O_2(I)$, so that $U_I \in \mathcal{R}_2(I)$ by B.2.14. As $m_3(Y^+) = 2$, case (i) of 15.3.11.5 holds, so $Y/O_2(Y) \cong E_9$ or 3^{1+2}, and $R = C_T(V)$. If case (2) of Hypothesis 15.3.10 holds, then as $Y_+ < Y$, $Y/O_2(Y) \cong 3^{1+2}$ and $C_Y(V)/O_2(C_Y(V)) \cong Z_3$. Thus:

LEMMA 15.3.14. *(1) $V \le Z(R)$.*

(2) If case (2) of Hypothesis 15.3.10 holds, then $Y/O_2(Y) \cong 3^{1+2}$ and $|C_Y(V) : O_2(C_Y(V))| = 3$.

LEMMA 15.3.15. *(1) Hypothesis C.2.3 is satisfied with I, M_I in the roles of "H, M_H".*

(2) There exists $L \in \mathcal{C}(J(I))$ with $L \not\le M_I$, $m_3(L) \ge 1$, $L = [L, Y_+]$, and L^* and $L/O_2(L)$ quasisimple.

(3) Each solvable Y_+S-invariant subgroup of I is contained in M_I.

PROOF. As case (i) of 15.3.11.5 holds with $R = C_T(V)$, (1) follows. By 15.3.11.11, $J(I)^* \not\le M_I^*$. In particular $J(I)^* \ne 1$, so that U_I is an FF-module for I by B.2.7, and hence $J(I)^*$ is described in Theorem B.5.6. If L^* is a direct factor of $J(I)^*$ isomorphic to S_3, then there are at most two such factors by Theorem B.5.6, so $Y_+^* = O^2(Y_+^*)$ normalizes and hence centralizes L^*, and then $L^* \le N_{I^*}(Y_+^*) = M_I^*$ by 15.3.11.9. Thus as $J(I)^* \not\le M_I^*$, Theorem B.5.6 says there exists $L \in \mathcal{C}(J(I))$ with $L \not\le M_I$, $m_3(L) \ge 1$, and L^* quasisimple. By 1.2.1.3, $Y_+ = O^2(Y_+) \le N_I(L)$. By 15.3.11.2, $C_I(U_I) \le M_I$, so as $L \not\le M_I$, $L = [L, Y_+]$ by 15.3.11.9. Further as L^* is quasisimple and not isomorphic to $Sz(2^m)$ by Theorem B.5.6, $O_{3'}(L) \le C_I(U_I) \le M_I$, so by 15.3.11.6, $[O_{3'}(L), Y_+] \le Y_+ \cap O_{3'}(L) \le O_2(L)$, and hence $L/O_2(L)$ is quasisimple by 1.2.1.4. Thus (2) holds. Then by 1.2.1.1, each member of $\mathcal{H}_+$ is nonsolvable, so (3) follows. $\qquad\square$

During the remainder of the proof of Theorem 15.3.13, pick L as in 15.3.15.2. Set $Y_L := O^2(Y_+ \cap L)$, $Y_C := O^2(C_{Y_+}(L/O_2(L)))$, $S_L := S \cap L$, $R_L := R \cap L$, and $M_L := M \cap L$. Let $W_L := \langle V^L \rangle$ and $(LRY_+)^+ := LRY_+/C_{LRY_+}(W_L)$.

LEMMA 15.3.16. *(1)* $W_L \in \mathcal{R}_2(LR) \cap \mathcal{R}_2(LRY_+)$ *and* L^+ *is quasisimple.*

(2) $m_3(L) \geq 1$, $Y_L \neq 1$, R *acts on* L, $L = [L, J(R)]$, *and* L *is described in* C.2.7.3.

PROOF. By 15.3.15, $m_3(L) \geq 1$ and we can appeal to the results in section C.2. As $m_3(L) \geq 1$, $Y_L \neq 1$, R acts on L, and $O_2(LR) \leq R$ by 15.3.12.1.

As $L/O_2(L)$ is quasisimple and $O_2(LR) \leq R = C_T(V)$, $W_L \in \mathcal{R}_2(LR) \cap \mathcal{R}_2(LRY_+)$. As R acts on L and $L \not\leq M_I$, L is described in C.2.7.3. By C.2.7.2, $L = [L, J(R)]$. As $C_G(V) \leq M$ but $L \not\leq M_I$, $L^+ \neq 1$, so as $L/O_2(L)$ is quasisimple, so is L^+. $\qquad\square$

LEMMA 15.3.17. *One of the following holds:*

(1) $m_3(L) = 1$ *and* $L/O_2(L) \cong L_2(2^n)$, n *even, or* $L_3(2)$.

(2) $m_3(L) = 2$ *and* $Y_+ \leq \theta(I) = L$.

(3) $m_3(L) = 2$ *and* $L^* \cong SL_3(2^n)$ *with* n *even.*

PROOF. By 15.3.16, $m_3(L) \geq 1$, $Y_L \neq 1$, and L is described in C.2.7.3.

Suppose first that $m_3(L) = 1$. Then from the list in C.2.7.3, $L/O_2(L) \cong L_2(2^n)$ or $L_3(2^m)$, with m odd. Now as $S \in Syl_2(I)$, $S_L = S \cap L \in Syl_2(L)$ and S_L acts on Y_L since Y_+ is S-invariant. It follows that n is even if $L/O_2(L) \cong L_2(2^n)$, and that $m = 1$ if $L/O_2(L) \cong L_3(2^m)$. That is, (1) holds in this case.

So assume $m_3(L) = 2$. Then (3) holds if $L^* \cong SL_3(2^n)$ with n even; otherwise $\theta(I) = L$ by A.3.18, so that (2) holds. $\qquad\square$

In the remainder of the proof of Theorem 15.3.13, we successively eliminate the various possibilities in C.2.7.3 given by 15.3.16.

LEMMA 15.3.18. L *is not an* $L_2(2^n)$*-block.*

PROOF. Assume otherwise. Then n is even by 15.3.17, while by 15.3.16, $Y_L \neq 1$ and R normalizes L.

Let $L_0 := \langle L^S \rangle$, so that $S_0 := S \cap L_0 \in Syl_2(L_0)$ and $M_0 := M \cap L_0 \geq Y_L$. Then M_0 is an overgroup of S_0 in L_0, so M_0 is contained in a unique Borel subgroup B_0 of L_0, and hence B_0 is M_I-invariant. Therefore as B_0 is solvable, $B_0 = M_0$ by 15.3.15.3. Then as $Y_+ \trianglelefteq M_I$ by 15.3.11.4, we conclude that Y_+ induces inner automorphisms on $L_0/O_2(L_0)$. By 15.3.12.2, $Y_+ = Y_L Y_C$ with $|Y_L|_3 = 3 = |Y_C|_3$, and $Y_+/O_2(Y_+) \cong E_9$. As S_0 is M_I-invariant, $S_0 \leq O_2(Y_+S) = R$; hence $R_L = R \cap L \in Syl_2(L)$, and $R \in Syl_2(LR)$. Thus as $V \leq Z(R)$ and $U(L) = [W_L, L]$ since L is a block, $W_L = C_{W_L}(R_L)U(L)$ by B.2.14, so that

$$V \leq C_{U(L)}(R)C_{W_L}(L). \qquad (*)$$

Now if $[V, Y_L] = 1$, then by $(*)$, we have $V \leq C_{U(L)C_R(L)}(Y_L) = C_R(L)$, since $U(L)/C_{U(L)}(L)$ is the natural module for $L_2(2^n)$. But then $L \leq C_G(V) \leq M$, contrary to $L \not\leq M$. Therefore $1 \neq [V, Y_L]$ is a B_0-invariant subgroup of $[C_{U(L)}(R_L), Y_L]$, so $[V, Y_L] = C_{U(L)}(R_L)$ by $(*)$ and the structure of coverings of the natural module. in I.2.3. But for $b \in B_0 - R$, $C_{U(L)}(b) = C_{U(L)}(L)$, while $O^{2,3}(M_L) \leq C_G(V) \leq C_G([V, Y_L])$ by 15.3.2.1, so we conclude that $(B_0 \cap L)/R_L$ is a 3-group, and hence $n = 2$. Thus as $[V, Y_L] = C_{U(L)}(R_L)$, $[V, Y_L]/C_{[V,Y_L]}(Y_L) \cong E_4$. As $V = V_1 \oplus V_2$ where V_1 and V_2 are the only Y_+-invariant 4-subgroups of V, $C_{U(L)}(L) = 1$ and we may take $V_2 = [V, Y_L]$, and hence $Y_L \leq C_M(V_1) = M_2$. Then by $(*)$, $V_1 \leq C_{W_L}(Y_L)$, so as $L \not\leq M$, also $N_G(V_1) \not\leq M$. Hence by 15.3.11.3, we are in case (1) of Hypothesis 15.3.10, where $Y_+ = Y$. Then $Y/O_2(Y) \cong E_9$, so as $Y_L \leq M_2$, $O^2(M_2) = Y_L$ is S-invariant. Hence S also acts on L and Y_C, so $Y_C = O^2(M_1)$. Let $t \in T - S$, and recall $|T : S| = 2$ so that t normalizes S. As $M_1^t = M_2$, $Y_L^t = Y_C$. Further Y_C centralizes L by C.1.10, and as Y_L contains a Sylow 2-group of L, $U(L) \leq O_2(Y_L)$. Then $U(L)^t \leq O_2(Y_L)^t = O_2(Y_C) \trianglelefteq LS$. Hence $\langle LS, t \rangle = \langle L, T \rangle$ acts on $U(L)U(L)^t$, so that $\langle L, T \rangle \leq M = !\mathcal{M}(YT)$ by 15.3.7, contrary to $L \not\leq M$. $\square$

LEMMA 15.3.19. $L/O_2(L)$ is not $L_2(2^n)$.

PROOF. Assume otherwise. Then by 15.3.16 and C.2.7.3, L is a block of type $L_2(2^n)$ or type A_5, so by 15.3.18, L is an A_5-block. Let $L_0 := \langle L^S \rangle$, $S_0 := S \cap L_0$, and $M_0 := M \cap L_0$. Arguing as in the proof of the previous lemma, we conclude that M_0 is the Borel subgroup of L_0 containing S_0, $Y_+ = Y_L Y_C$ with $|Y_L|_3 = 3 = |Y_C|_3$, $Y_+/O_2(Y_+) \cong E_9$, $R_L := R \cap L \in Syl_2(L)$, and $W_L = U(L) \times C_{W_L}(L)$, so

$$V \leq C_{U(L)}(R_L) \times C_{W_L}(L). \qquad (*)$$

Since $U(L)$ is the A_5-module, Y_L centralizes V by $(*)$, so as $C_Y(V) \neq 1$, case (2) of 15.3.7 holds, with $Y/O_2(Y) \cong 3^{1+2}$. In particular $|C_Y(V) : O_2(C_Y(V))| = 3$, so $Y_L = O^2(C_Y(V))$. Further as $Y_+/O_2(Y_+) \cong E_9$, $Y_+ < Y$, so that case (2) of Hypothesis 15.3.10 holds, with $V_2 = [V, Y_+]$, and $N_G(V_2) \leq M$ by 15.3.11.3. Now $\operatorname{End}_{L/O_2(L)}(U(L)) \cong \mathbf{F}_2$ so that Y_C centralizes $U(L)$. Thus $V_2 = [V_2, Y_+] \leq [U(L)C_R(L), Y_C] \leq C_R(L)$. Then $L \leq N_G(V_2) \leq M$, contrary to the choice of L. $\square$

LEMMA 15.3.20. $L/O_2(L)$ is not $SL_3(2^n)$ with $n > 1$ or $Sp_4(4)$.

PROOF. Assume otherwise. By 15.3.16, L is described in C.2.7.3, and in particular as W_L is an FF-module for $L^+ R^+$, S is trivial on the Dynkin diagram of L^+ by Theorem B.4.2. Further as S normalizes Y, $S_L Y_+ = Y_+ S_L$, so as each solvable overgroup of S_L in $L Y_+$ is 2-closed, Y_+ acts on S_L. Thus $Y_+ S$ acts on both maximal parabolics P_i of L. If $X_i := P_i Y_+ S \not\leq M$, then $X_i \in \mathcal{H}_+$, contrary to 15.3.19. Thus $L = \langle P_1, P_2 \rangle \leq M$, contrary to the choice of L. $\square$

LEMMA 15.3.21. L is not a block of type A_6, $G_2(2)$, $\hat{A}_6$, or A_7, and L is not an exceptional A_7-block.

PROOF. Assume L is one of the blocks appearing in 15.3.21. By 15.3.17, $Y_+ \leq L$, so that $Y_+ = Y_L$. As $Y_+ S_L = S_L Y_+$, L is not of type A_6 or $G_2(2)$, since no proper parabolic in these groups has 3-rank 2. Similarly if $L/O_2(L) \cong \hat{A}_6$, the preimage of a proper parabolic does not contain 3^{1+2}, and if $L/O_2(L) \cong A_7$, then L has abelian Sylow 3-groups; thus $Y_+ S/R \cong S_3 \times S_3$ in these two cases. Hence L is not an an exceptional A_7-block, since in that case $LS/O_2(LS) \cong A_7$ rather than S_7. Further L is not an ordinary A_7-block, since in that case M_L^+ has no normal E_9-subgroup by C.2.7.3. This leaves the case where L is an $\hat{A}_6$-block, where by C.2.7.3, $S^+ Y_+^+$ is the stabilizer of a 2-dimensional $\mathbf{F}_4$-subspace U of $[W_L, L]$. Now $[W_L, L]$ has the structure of an $\mathbf{F}_4 L$-module on which $Z(L^+)$ induces scalars in $\mathbf{F}_4$, and $U = U_1 \oplus U_2$ is the sum of two Y_+-invariant $\mathbf{F}_4$-points, so $U_i = V_i$. Thus is impossible as S interchanges the two $\mathbf{F}_4$-points in an $\hat{A}_6$-block, but S acts on V_1 and V_2 by definition. $\square$

LEMMA 15.3.22. $L/O_2(L) \cong L_n(2)$ for $n = 3$, 4, or 5.

PROOF. Observe that 15.3.19, 15.3.20, and 15.3.21 have eliminated all other possibilities for L^* from the list of C.2.7.3 provided by 15.3.16. Then as $L/O_2(L)$ is quasisimple by 15.3.15.2, and the Schur multiplier of L^* is a 2-group by I.1.3, $O_{2,Z}(L) = O_2(L)$ so that $L/O_2(L) \cong L^*$ is simple.   □

LEMMA 15.3.23. $L/O_2(L)$ is not $L_3(2)$.

PROOF. Assume $L/O_2(L) \cong L_3(2)$, and set $U_L := [W_L, L]$. By 15.3.12.2, $Y_+ = Y_L Y_C$ with $|Y_L|_3 = 3 = |Y_C|$ and $Y_+/O_2(Y_+) \cong E_9$. By 15.3.16, R acts on L, and by C.2.7.3, $M_L = S_L Y_L$ is a minimal parabolic of L and $R = O_2(Y_+ S) = O_2(LR)O_2(Y_L)$.

By C.2.7.3, (LR, R) is described in Theorem C.1.34. In particular L has $k := 1$, 2, 3, or 6 noncentral 2-chief factors. If $k = 6$, then case (4) of C.1.34 holds so that $m(\Omega_1(Z(S))) = 3$, contrary to 15.3.8. Thus $1 \le k \le 3$.

From C.1.34, either U_L is the direct sum of $s \le 2$ isomorphic natural modules, or U_L is a 4-dimensional indecomposable. Thus either $[U_L, Y_C] = 1$, or $s = 2$ and $U_L = [U_L, Y_C]$.

Assume first that $[V, Y_L] = 1$. Then as $Y_L \ne 1$ and Y is faithful on V in case (1) of 15.3.7, case (2) of 15.3.7 holds with $Y/O_2(Y) \cong 3^{1+2}$, and $Y_L = O^2(C_Y(V))$. We saw $Y_+/O_2(Y_+) \cong E_9$, so $Y_+ < Y$, and hence case (2) of Hypothesis 15.3.10 holds. Then $N_G(V_i) \le M$ by 15.3.11.3, Y_+ centralizes V_1, and $V_2 = [V_2, Y_+] = [V_2, Y_C] \le C_{W_L}(Y_L)$. As $N_G(V_i) \le M$ but $L \not\le M$, L centralizes neither V_1 nor V_2. From the previous paragraph, U_L is either a sum of isomorphic natural modules, or a 4-dimensional indecomposable with a natural quotient, while M_L is a minimal parabolic of L with $R = O_2(M_L R)$. Thus $V \le Z(R)$ while the fixed points of the unipotent radical R on any extension in B.4.8 of U_L with trivial quotient lie in U_L, so we conclude that

$$V \le U_L C_{W_L}(L).$$

By the previous paragraph, either $[U_L, Y_C] = 1$, or $s = 2$ and $U_L = [U_L, Y_C]$. In the first case,

$$V_2 = [V_2, Y_C] \le [U_L C_{W_L}(L), Y_C] = [C_{W_L}(L), Y_C] \le C_{W_L}(L),$$

contrary to an earlier remark. In the second case,

$$V_1 = C_V(Y_C) \le C_{U_L}(Y_C) C_{W_L}(L Y_C) = C_{W_L}(L Y_C),$$

contrary to the same remark.

Therefore $[V, Y_L] \ne 1$. Now $Y_L = [Y_L, S_L]$ while $[Y_C, S_L] \le O_2(Y_C)$, so from the action of S on Y_+, Y_L and Y_C are normal in $Y_+ S$, and $\{Y_L, Y_C\}$ is the set $\mathcal{Y}$ of S-invariant subgroups of Y_+ with Sylow 3-group of order 3. In particular, S acts on Y_L and hence on L.

Suppose that case (2) of Hypothesis 15.3.10 holds. Then by 15.3.14.2, $Y/O_2(Y) \cong 3^{1+2}$ with $C_Y(V) > O_2(Y)$. As $\mathcal{Y} = \{Y_L, Y_C\}$ while $[V, Y_L] \ne 1$, it follows that $Y_C = O^2(C_Y(V))$, so $N_G(Y_C) \le M$ as $M = !\mathcal{M}(N_G(Y_C))$ by 15.3.7.2. But L normalizes $O^2(Y_C O_2(L)) = Y_C$, contradicting $L \not\le M$.

Thus we are in case (1) of Hypothesis 15.3.10, so $Y = Y_+$, and hence from earlier discussion, $Y = Y_C Y_L$ and $Y/O_2(Y) \cong E_9$. Therefore we may take $Y_C = O^2(M_1)$ and $Y_L = O^2(M_2)$, since $\{Y_L, Y_C\} = \mathcal{Y}$. Thus $Y_L^t = Y_C$, and $V_2 = [V, Y_L]$. As $V_2 = [V, Y_L]$ is S-invariant, $Y_L S_L$ is the parabolic of L stabilizing the line V_2 in $Z(O_2(L))$. Hence case (5) of C.1.34 does not hold, as no such line exists in that case.

Set $Q := [O_2(L), L]$ as in C.1.34, and observe that $[Z, L] \leq U_L \leq Q$. We will complete the proof by showing that for $t \in T - S$, $W := QQ^t$ is normalized by LS, and hence also by T as $|T : S| = 2$. Then as $Y = Y_L Y_C \leq \langle Y_L, t \rangle$, $\langle L, T \rangle \leq N_G(W) \leq M = !\mathcal{M}(YT)$ by 15.3.7, contrary to the choice of L.

Assume that $k = 3$, so that case (3) of C.1.34 holds. Then as Y_L stabilizes the line V_2 in the natural module $Z(Q)$, $Q = [Q, Y_L]$, so $Q \leq O_2(Y_L)$. Further Y_C centralizes the natural module $Z(Q)$ since $\mathrm{End}_{L/O_2(L)}(Z(Q) = \mathbf{F}_2$. As $Q/Z(Q)$ is the direct sum of two natural modules, either Y_C centralizes Q, or $Q = [Q, Y_C]$. In the latter case $Q = O_2(Y_C) \cap O_2(Y_L)$, so Q is t-invariant, whereas Y_C and Y_L have three and two nontrivial 2-chief factors, on $Q/Z(Q)$, respectively. Therefore $[Q, Y_C] = 1$, so $Y_C = O^2(Y_C)$ centralizes L by Coprime Action. Then $Q^t \leq O_2(Y_L)^t = O_2(Y_C) \leq C_S(L)$, so that $W = QQ^t \trianglelefteq LS$, which suffices as mentioned above.

Suppose finally that $k = 1$ or 2, so that case (1) or (2) of C.1.34 holds. In each case $Q = [Q, Y_L]C_Q(P_L)$, for $P_L \in Syl_3(Y_L)$, and as Y_C centralizes the line V_2 stabilized by Y_L in a natural submodule in Q, Y_C centralizes L from the structure of $Aut(L)$. Thus $Q = O_2(Y_L)C_Q(P) \leq O_2(Y_L)C_S(P)$, for $P \in Syl_3(Y)$, and by a Frattini Argument we may assume $t \in T - S$ normalizes P, and hence also $C_S(P)$. Therefore $Q^t \leq O_2(Y_C)C_S(P) \leq O_2(LS)$, so $[Q^t, LS] \leq [O_2(LS), L] = Q$, and hence $W = QQ^t \trianglelefteq LS$, which again suffices as mentioned earlier. $\qquad\square$

LEMMA 15.3.24. $L/O_2(L)$ is not $L_4(2)$ or $L_5(2)$.

PROOF. Assume $L/O_2(L) \cong L_n(2)$ for $n := 4$ or 5. Then Y_+ is solvable and S-invariant of 3-rank 2, $Y_+ \leq L$ by 15.3.17, and $S \cap L \in Syl_2(L)$ as $S \in Syl_2(I)$. Thus $LS \in \mathcal{H}_+$, so we may take $I = LS$, and hence $U_I = \langle Z^L \rangle$. As Sylow 3-subgroups of L are isomorphic to E_9, $Y_+/O_2(Y_+) \cong E_9$ rather than 3^{1+2}. Then $Y_+S/R \cong S_3 \times S_3$ from the action of S on Y, so S is trivial on the Dynkin diagram of $L/O_2(L)$.

Suppose first that $n = 4$. Then Y_+S is the maximal parabolic of LS over S determined by the end nodes, so $Y_+S_L = L \cap M$ as $L \not\leq M$. This parabolic has unipotent radical $R_L/O_2(L) \cong E_{2^4}$.

Set $U_L := [W_L, L]$. By 15.3.16, $L = [L, J(R)]$, so there are FF-offenders on U_L with respect to R, and in particular U_L is an FF-module for $L/O_2(L)$. As $1 \neq [Z, Y_+] \leq U_L$, $U_L/C_{U_L}(L)$ is not the orthogonal module, so we conclude from Theorem B.5.1 that U_L is either the sum of a natural module and its dual, or the sum of at most $n - 1$ isomorphic natural modules. Now by B.2.14, $U_L Z = U_L C_{U_L Z}(L)$, and we let Z_L denote the projection of Z on U_L with respect to this decomposition. Then $Z \leq Z_L C_{W_L}(L)$, so that $C_L(Z_L) = C_L(Z)$.

Assume first that U_L is a sum of isomorphic natural modules. Then $W_L = U_L C_{W_L}(L)$ by I.1.6.6. Also $1 \neq O^2(C_{Y_+}(Z_L)) = O^2(C_{Y_+}(Z)) \leq O^2(C_{Y_+}(V))$, so case (2) of 15.3.7 holds. Hence $Y_+ < Y$, so that case (2) of Hypothesis 15.3.10 holds, and thus $N_G(V_1) \leq M$ by 15.3.11.3. But now $V_1 \leq C_{W_L}(Y_+) = C_{W_L}(L)$, so $L \leq C_G(V_1) \leq M$, contrary to the choice of L.

Therefore U_L is the sum of a natural module and its dual. Since $Y_+S_L = L \cap M$ is the maximal parabolic over S_L determined by the end nodes, each $L_3(2)$-parabolic P over S_L satisfies $P \not\leq M$ and $[Z_L, O^2(Y_+ \cap P)] \neq 1$. Then applying 15.3.11.12 to P in the role of "L", no nontrivial characteristic subgroup of S is normal in PS. Thus $(O^2(P)S, S)$ is an MS-pair, and hence is described in C.1.34. But P has two noncentral chief factors on U_L and one on $O_2(P)/O_2(L)$, so case (3) or (4) of C.1.34

holds, since in the other cases in C.1.34 there are at most two such factors. Case (4) is eliminated as $m(\Omega_1(Z(S))) \geq 3$, contrary to 15.3.8. Suppose case (3) holds, set $Q_P := [O_2(P), P]$, and let $W_L = W_1 \oplus W_2$ with $W_i \in Irr_+(L, W_L)$; we may choose notation so that $[W_1, P]$ is of rank 3 and $W_2 = [W_2, P]$. Then $Z(Q_P)$ is a natural module for P/Q_P and $Q_P/Z(Q_P)$ is the sum of two copies of the dual of $Z(Q_P)$, impossible as $[W_1, P]C_{W_2}(P) \leq Z(Q_P)$.

So $n = 5$. Let P be a maximal parabolic of L over S_L containing Y_+. Since $Y_+S/R \cong S_3 \times S_3$, L is generated by such parabolics, so we may assume $P \nleq M$. Thus $PS \in \mathcal{H}_+$, and we obtain a contradiction from earlier reductions as $P/O_2(P) \cong S_3 \times L_3(2)$ or $L_4(2)$. $\square$

Observe that 15.3.22, 15.3.23, and 15.3.24 establish Theorem 15.3.13.

We now complete the elimination of the case $F^*(I) = O_2(I)$ under Hypothesis 15.3.10, by treating the remaining subcase where $m_3(Y_+) = 1$ in the following result:

THEOREM 15.3.25. *If* $I \in \mathcal{H}_+$ *then* $F^*(I) \neq O_2(I)$.

Until the proof of Theorem 15.3.25 is complete, assume I is a counterexample. The proof will be largely parallel to that of Theorem 15.3.13, except this time our list of possibilities for L^* will come from Theorem B.5.6 rather than C.2.7.3, and the elimination of those cases will be somewhat simpler. As $F^*(I) = O_2(I)$, $U_I = \langle Z^I \rangle \in \mathcal{R}_2(I)$ by B.2.14.

LEMMA 15.3.26. *(1) Case (2) of Hypothesis 15.3.10 holds,* $Y_+S/R \cong S_3$, $Y_+ = O^{3'}(M_I)$, *and* $R = C_S(V_2)$.
(2) There is $L \in \mathcal{C}(J(I))$ *with* $L \nleq M_I$, $L = [L, Y_+]$, *and* L^* *and* $L/O_2(L)$ *quasisimple.*
(3) Each solvable Y_+S-*invariant subgroup of* I *is contained in* M_I.

PROOF. By Theorem 15.3.13, $m_3(Y_+) = 1$. Then (1) follows from 15.3.11.5. The proofs of (2) and (3) are the same as those in 15.3.15. $\square$

During the remainder of the proof of Theorem 15.3.25, pick L as in 15.3.26.2. As L^* is a component of $J(I)^*$, $L^* = [L^*, J(S)^*]$ and $U_L := [U_I, L]$ is an FF-module for $Aut_{LJ(S)}(U_L)$ by B.2.7, so L^* is described in Theorem B.5.6.

Recall $\mathcal{H}_{+,*}$ denotes the set of members of of $\mathcal{H}_+$ minimal under inclusion.

LEMMA 15.3.27. *(1)* $L \trianglelefteq I$.
(2) $Y_+ \leq L$.
(3) L^* *is not* $SL_3(2^n)$, n *even, or* $\hat{A}_6$.
(4) If $I \in \mathcal{H}_{+,*}$, *then* $I = LS$, *and* M_I *is the unique maximal subgroup of* I *containing* Y_+S.

PROOF. Suppose first that L is not normal in I, and let $L_0 := \langle L^S \rangle$. By 1.2.1.3 and Theorem B.5.6, $L^* \cong L_2(2^n)$ or $L_3(2)$. By 1.2.2, $L_0 = O^{3'}(I)$, so $Y_+ \leq L_0$. Then as $Y_+S/R \cong S_3$ by 15.3.26.1, and $S \in Syl_2(I)$ by 15.3.11.1, $L^* \cong L_2(2^n)$—since when $L^* \cong L_3(2)$, there is no S-invariant subgroup of L_0 with Sylow 3-group of order 3. Let B_0 be the Borel subgroup of L_0 containing $S \cap L_0$; then $M_0 := M \cap L_0 \leq B_0$, and B_0 is M_I-invariant and solvable, so $M_0 = B_0$ by 15.3.26.3. Then as $Y_+ \leq B_0$, n is even. But now $m_3(M_I) \geq m_3(B_0) > 1$, contrary to 15.3.26.1. This contradiction establishes (1).

By 15.3.26.2, $L^* = [L^*, Y_+^*]$. Comparing the list of Theorem B.5.6 with that in A.3.18, we conclude that one of the following holds:

(i) $m_3(L) = 1$ and L^* is $L_2(2^n)$ or $L_3(2^m)$, m odd.

(ii) $L = \theta(I)$.

(iii) $L^* \cong SL_3(2^n)$, n even.

In case (ii), (2) holds. In case (i), as $Y_+S/R \cong S_3$ and $Out(L/O_2(L))$ is abelian, Y_+^* induces inner automorphisms on L^*. Then the projection Y_L^* of Y_+^* on L^* is contained in M_I^* by 15.3.26.3, so the preimage Y_L is contained in $O^{3'}(M_I) = Y_+$ by 15.3.26.1, so that (2) holds again. Finally if (iii) holds, then $Z(L)^* \leq M_I^*$ by 15.3.26.3, so $O^{3'}(M_I)^* = Y_+^* = Z(L^*)$ by 15.3.26.1, contradicting $L^* = [L^*, Y_+^*]$. This completes the proof of (2), and the same argument shows that L^* is not $\hat{A}_6$, completing the proof of (3) also.

Finally assume that $I \in \mathcal{H}_{+,*}$. By (1), S acts on L, and by (2), $Y_+ \leq L$. Thus as $L \not\leq M$, $LS \in \mathcal{H}_+$, so $I = LS$ by minimality of I. Similarly M_I is the unique maximal subgroup of I containing Y_+S, so (4) holds. $\square$

Until the proof of Theorem 15.3.25 is complete, assume $I \in \mathcal{H}_{+,*}$. Thus $I = LS$ by 15.3.27.4.

LEMMA 15.3.28. *Let $M_L := M \cap L$; then one of the following holds:*

(1) $L^* \cong L_2(2^n)$, n even, and M_L^* is a Borel subgroup of L^*.

(2) $L^* \cong L_3(2)$, $Sp_4(2)'$, or $G_2(2)'$, and M_L^* is a minimal parabolic of L^*, so that S is trivial on the Dynkin diagram of L^*.

(3) $I^* \cong S_8$ and M_L^* is the middle-node minimal parabolic of L^*S^*.

PROOF. Suppose first that $L^* \cong A_7$. Then as $Y_+S/R \cong S_3$ by 15.3.26.1, $Y_+^*S^*$ is either the stabilizer of a partition of type $2^3, 1$, or is contained in the stabilizer $I_{4,3}^*$ of a partition of type $4, 3$. In the latter case, the preimage $I_{4,3}$ is contained in M_I by 15.3.27.4, whereas $m_3(M_I) = 1$ by 15.3.26.1. In the former, $Y_+^*S^* \leq I_1^* \cong A_6$ or S_6, and this time $I_1 \leq M_I$ for the same contradiction.

Then by 15.3.27.3 and Theorem B.5.6, L^* is of Lie type and characteristic 2, so as $S \in Syl_2(I)$, M_L^* is a maximal S-invariant parabolic of L^* by 15.3.27.4. As $O^{3'}(M_L^*) = Y_+^*$ with $Y_+S/R \cong S_3$ by 15.3.26.1, we conclude by inspection of the list of Theorem B.5.6 and appeals to parts (3) and (4) of 15.3.27 that one of cases (1)–(3) of the lemma holds. $\square$

LEMMA 15.3.29. *(1) No nontrivial characteristic subgroup of S is normal in I.*
(2) $N_G(V_1) \leq M$ and V_1 centralizes Y_+.

PROOF. As $Y_+ \leq L$ by 15.3.27.2 and $I = LS$, we may apply 15.3.11.12 to obtain (1). By 15.3.26.1, case (2) of Hypothesis 15.3.10 holds, so that V_1 centralizes Y_+, and $N_G(V_1) \leq M$ by 15.3.11.3, so (2) holds. $\square$

LEMMA 15.3.30. *L^* is not $L_2(2^n)$.*

PROOF. Assume L^* is $L_2(2^n)$. By 15.3.28, M_L^* is a Borel subgroup of L^*, so as $R = O_2(Y_+S)$, $R \in Syl_2(LR)$. Then by 15.3.27.4, LR is a minimal parabolic in the sense of Definition B.6.1, so we conclude from 15.3.29.1 and C.1.26 that L is a block of type $L_2(2^n)$ or A_5. Next M_I acts on $[V, Y_+] = V_2 \cong E_4$, so $V_2 \leq U(L)$ is an M_L-invariant line. Thus L is not an A_5-block, so L is an $L_2(2^n)$-block and in particular

$C_{Aut(L)}(Y_+) = 1$. Then by 15.3.29.2, $V_1 \leq C_S(Y_+) \leq C_S(L)$, so $L \leq C_G(V_1) \leq M$ by 15.3.29.2, contrary to the choice of L. $\square$

LEMMA 15.3.31. $I^* \cong S_8$.

PROOF. Assume otherwise; by 15.3.30 and 15.3.28, we may assume that case (2) of 15.3.28 holds; that is $L^* \cong L_3(2)$, $Sp_4(2)'$, or $G_2(2)$. As $V_2 = [V_2, Y_+] \cong E_4$ is a Y_+S-invariant line in U_L, it follows from Theorem B.5.6 that M_L^* is the parabolic stabilizing the line V_2 in some $W \in Irr_+(L^*, U_L)$, and hence for each such W when $[\Omega_1(Z(S)), L]$ is a sum of two isomorphic natural modules for $L^* \cong L_3(2)$. By 15.3.29.1, (LS, S) is an MS-pair in the sense of Definition C.1.31, and by 15.3.27.3, L is not a $\hat{A}_6$-block. Therefore by C.1.32, either L is a block of type A_6 or $G_2(2)$, or $L^* \cong L_3(2)$ and L is described in C.1.34. In particular, $L/O_2(L)$ is simple in each case, so that $L/O_2(L) = L^*$. Set $Q := [O_2(L), L]$. As $L \trianglelefteq I = LS$, $Q = [O_2(I), L]$.

In case (4) of C.1.34, $m(\Omega_1(Z(S))) = 3$, contrary to 15.3.8, and case (5) of C.1.34 does not hold, as M_L^* stabilizes the line V_2. Thus only cases (1)–(3) of C.1.34 can arise when $L^* \cong L_3(2)$.

Next by 15.3.29.2, $V_1 \leq C_I(Y_+) =: D$. It will suffice to show that $D = C_I(L)$: for then $L \leq C_G(V_1) \leq M$ by 15.3.29.2, contrary to the choice of L. Set $I^+ := I/C_I(L)$; it remains to show that $D^+ = 1$.

Suppose that D centralizes Q. Then $[D, Q] \leq C_L(Q) \leq O_2(L)$, so as L/Q is quasisimple, $[D, L] \leq Q$. Thus $[D, L] \leq C_Q(Q) = Z(Q)$. Therefore $O^2(D^+) = 1$ by Coprime Action. Further as $\Phi(Z(Q)) = 1$, $\Phi(D^+) = 1$ (cf. the argument in the proof of C.1.13); so as Y_+ centralizes D, $D^+ = 1$ from the structure of the the the covering of the L^*-module $Z(Q)/C_{Z(Q)}(L)$ in I.2.3.

Therefore we may assume $[D, Q] \neq 1$. Thus $Q \not\leq Y_+$. In case (3) of C.1.34, $Z(Q)$ is a natural module for L^* and $Q/Z(Q)$ is a sum of two modules dual to $Z(Q)$. In this case, and when L is a block of type A_6 or $G_2(2)$, since M_L^* is the parabolic stabilizing the line V_2 in $Z(Q)$, $Q = [Q, Y_+] \leq O_2(Y_+)$. Therefore case (1) or (2) of C.1.34 holds. Then $C_{I^*}(Y_+^*) = 1$, so $[D, L] \leq Q$. Further the intersection of Y_+ with each $W \in Irr_+(L, Q)$ is a hyperplane W_0 of W, so as $DQ \trianglelefteq DL$ and Q is abelian, DQ centralizes $\langle W_0^L \rangle = W$. Therefore DQ centralizes Q, a contradiction completing the proof. $\square$

Now $L/O_2(L)$ is quasisimple by 15.3.26.2, $L^* \cong A_8$ by 15.3.31, and the Schur multiplier of A_8 is a 2-group by I.1.3. Then as $I = LS$, $O_2(I) = C_I(U_L) = C_S(U_L)$.

LEMMA 15.3.32. (1) U_L is the 6-dimensional orthogonal module for I^*.

(2) $C_{Z_S}(L) = Z_S \cap V_1 =: Z_1$ is of order 2.

(3) $L = O^{3'}(C_G(Z_1))$.

(4) $X := O^{3'}(C_G(Z_S)) = O^{3'}(K)$, where K is the maximal parabolic of L over $S \cap L$ determined by the end nodes of the diagram of L^*.

(5) $W := \langle U_L, U_L^t \rangle = U_L \times U_L^t$ for $t \in T - S$, and XS normalizes W.

(6) Let $(XS)^+ := XS/C_S(W)$ and $P := O_2(XS)$. Then $P^+ = C_S(U_L)^+ \times C_S(U_L^t)^+$.

PROOF. By 15.3.31, $I^* \cong S_8$, and then by 15.3.28.3, M_L^* is the middle-node minimal parabolic. Therefore as U_L is an FF-module for I^*, B.5.1 says that either $U_L/C_{U_L}(L)$ is the orthogonal module, or U_L is the sum of a natural module and its dual. Then as $V_2 = [V_2, Y_+]$ is an S-invariant line of U_L, the former case holds with $C_{U_L}(L) = 1$, giving (1). Recall that $Z_S \cong E_4$ by 15.3.8, and that $Z_i := Z_S \cap V_i$ is of order 2 for $i = 1, 2$. Further $Z_1 V_2 = Z V_2 = \langle Z^{Y_+} \rangle \leq \langle Z^L \rangle = U_I$, so $U_I = Z_1 U_L$. Then as $Z_1 \leq Z(S)$, $U_I = U_L C_{U_I}(LS)$ by B.2.14, and $C_{U_I}(LS) = C_{Z_S}(L) = C_{Z_S}(Y_+) = Z_1$, so (2) holds.

Next $I_1 := C_G(Z_1) \in \mathcal{H}_+$, so $S \in Syl_2(I_1)$ by 15.3.11.1. Thus $L \leq L_1 \in \mathcal{C}(I_1)$ by 1.2.4, and A.3.12 says that either $L = L_1$ or $L_1/O_2(L_1) \cong L_5(2)$, M_{24}, or J_4. As S is nontrivial on the Dynkin diagram of L^*, it follows that $L = L_1$, and then (3) folllows from A.3.18.

Let K be the S-invariant maximal parabolic of L, and set $X := O^2(K)$ and $P := O_2(XS)$. Thus $XS/P \cong S_3$ wr $\mathbf{Z}_2$ is determined by the end nodes of the Dynkin diagram of L^*. By (1), $XS = C_I(Z_S)$, so (3) implies (4). Then as $Z_S \trianglelefteq T$, T acts on $O^2(C_I(Z_S)) = X$ and P. Let $t \in T - S$. Then T acts on $U_L \cap U_L^t$, so if $U_L \cap U_L^t \neq 1$, then $Z \leq U_L \cap U_L^t$, whereas $Z_S \cap U_L = Z_S \cap V_2 = Z_2$. Thus $U_L \cap U_L^t = 1$, so as $U_L \trianglelefteq XS$ and T acts on XS, (5) holds.

Adopt the notation of (6) and let $P_I := O_2(I)$. As XS is irreducible on P^*, either $P_I^t \leq P_I$, or $P = P_I P_I^t$ and (6) follows from (5). But in the former case $P_I^t = P_I$, so that $\langle T, L \rangle$ acts on P_I; then as $Y_+ \not\leq C_M(V)$, $L \leq M$ by 15.3.2.4, contrary to the choice of L. Thus (6) holds. $\square$

We can now complete the proof of Theorem 15.3.25. Let X, W, and P be defined as in 15.3.32.

Let $\mathcal{B}$ be the set of $A \in \mathcal{A}(S)$ such that $A^* \neq 1$, and A^* is minimal subject to this property. Choose some $A \in \mathcal{B}$. By B.2.5, $A^* \in \mathcal{P}^*(I^*, U_L)$. Now B.3.2.6 describes the possible FF-offenders, and the only strong FF-offender is generated by four transvections; so from B.2.9 we conclude that one of the following holds:

(i) $A^* \cong E_8$ is regular on $\Omega := \{1, \ldots, 8\}$.
(ii) A^* is generated by a transposition.
(iii) $A^* = D := \langle (1,2), (3,4), (5,6), (7,8) \rangle$.
(iv) $A^* = D \cap L^*$.

In particular in each case, $A \not\leq P$. Further $m(A^*) = m(U_L/C_{U_L}(A))$ except in case (iii), where $m(A^*) = 4$ and $m(U_L/C_{U_L}(A)) = 3$.

Let $\mathcal{C} := \mathcal{B} \cap \mathcal{B}^t$ for $t \in T - S$. As $A \not\leq P$, $Aut_A(U_L^t) \neq 1$ by 15.3.32.6, so there is $A_+ \leq A$ such that $Aut_{A_+}(U_L^t) \in \mathcal{P}^*(Aut_{I^t}(U_L^t), U_L^t)$ by B.1.4.4. Then $A_+ \not\leq P$ by the previous paragraph applied to U_L^t in place of U_L, so $A_+^* \neq 1$ again using 15.3.32.6. Hence by minimality of A^*, $A_+^* = A^*$. Thus $A \in \mathcal{C}$.

Let $\widehat{XT} := XT/O_2(XT)$, so that $\hat{S} \cong D_8$, and set $S_0 := S \cap LO_2(LS)$. Observe:

(I) In (i), $|\hat{A}| = 2$ and $\hat{S}_0 = \hat{A} \times Z(\hat{S})$.
(II) In (ii), $|\hat{A}| = 2$ and $\hat{A} \not\leq \hat{S}_0$.
(III) In (iii), $\hat{A}$ is the 4-subgroup of $\hat{S}$ distinct from $\hat{S}_0$.
(IV) In (iv), $\hat{A} = Z(\hat{S})$.

Let $B := C_A(W)$, $m_0 := m(\hat{A})$, $m_1 := m(A^* \cap P^*)$, $m_2 := m(Aut_{A \cap P}(U_L^t))$, $m_3 := m(U_L/C_{U_L}(A))$, and $m_4 := m(U_L^t/C_{U_L^t}(A))$. Then $m(A) \leq m_0 + m_1 + m_2 + m(B)$. Also $m(BW) = m(B) + m_3 + m_4$. Therefore as $m(A) \geq m(BW)$ since $A \in \mathcal{A}(S)$,

$$m_0 + m_1 + m_2 \geq m_3 + m_4. \qquad (!)$$

Suppose first that $\hat{S} < \hat{T}$. Then $\hat{T}$ is Sylow in $GL_2(3)$, so $\hat{T} \cong SD_{16}$, and hence $\hat{S}_0^t$ is the 4-subgroup in $\hat{S}$ distinct from $\hat{S}_0$. As $A \in \mathcal{C}$, it satisfies one of conclusions

(I)–(IV), and also one of the analogous conclusions on U_L^t. Then inspecting (I)–(IV), we conclude that either:

(a) A^* is in case (ii) and $Aut_A(U_L^t)$ is in case (i), or vice versa; or:
(b) A^* and $Aut_A(U_L^t)$ are in case (iv).

However in case (a), the tuple of parameters $(m_0, m_1, m_2, m_3, m_4)$ is $(1, 0, 2, 1, 3)$, contrary to (!). Similarly in case (b), the tuple is $(1, 2, 2, 3, 3)$, again contrary to (!).

Thus $\hat{T} = \hat{S}$. So as $\hat{S}_0$ and $Z(\hat{S})$ are normal in $\hat{S}$, their preimages are normal in T. Then inspecting (I)–(IV), we conclude that A^* and $Aut_A(U_L^t)$ always appear in the same case of (i)–(iv). In cases (i), (ii), and (iv), we calculate the tuple of parameters to be $(1, 2, 2, 3, 3)$, $(1, 0, 0, 1, 1)$, and $(1, 2, 2, 3, 3)$, again contrary to (!). We conclude A^* is in case (iii). In particular, $A \not\leq S_0$; so since A is an arbitrary member of $\mathcal{B}$, it follows that $J(S_0) \leq C_I(U_L)$. Thus $J(S_0) = J(C_I(U_L)) \trianglelefteq \langle T, L \rangle$, again contrary to $L \not\leq M = !\mathcal{M}(YT)$ by 15.3.2.4 since $Y_+ \not\leq C_M(V)$.

This completes the proof of Theorem 15.3.25.

Theorem 15.3.25 has reduced the treatment of Hypothesis 15.3.10 to the case $F^*(I) \neq O_2(I)$. As $O(I) = 1$ by 15.3.11.8, there is a component L of I, and $Z(L) = O_2(L)$. As $F^*(M_I) = O_2(M_I)$ by 15.3.11.7, $L \not\leq M$. Thus to complete our proof that $M = !\mathcal{M}(Y_+S)$, it remains to determine the possibilities for L, and then to eliminate each possibility.

Set $L_0 := \langle L^S \rangle$, $S_L := S \cap L$, and $M_L := M \cap L$. Let z denote a generator of Z.

LEMMA 15.3.33. *(1) If L_1 is a Y_+S-invariant subgroup of L_0 with $F^*(L_1) = O_2(L_1)$, then $L_1 \leq M_I$.*

(2) The hypotheses of 1.1.5 are satisfied with I, M_c is the roles of "H, M".

(3) If $K \in \mathcal{C}(I)$ then $K \not\leq M$, $K = [K, z]$ is described in 1.1.5.3, $O(K) = 1$, and

$$F^*(C_K(z)) = O_2(C_K(z)).$$

PROOF. Choose L_1 as in (1); if $L_1 \not\leq M$, then $L_1 Y_+ S \in \mathcal{H}_+$, contrary to Theorem 15.3.25, so (1) holds. Next let $H \in \mathcal{M}(I)$, so in particular $H = N_G(O_2(H))$. Then $H \in \mathcal{H}_+$, so $S \in Syl_2(H)$ by 15.3.11.1. Thus $S = T \cap H$ and $O_2(H) \leq O_2(I \cap M_c)$ by A.1.6, so

$$C_{O_2(M_c)}(O_2(I \cap M_c)) \leq C_T(O_2(H)) \leq T \cap H = S \leq I,$$

and hence (2) follows since $C_G(z) = M_c$ by 15.3.4. Then (2) and 1.1.5 imply (3), since we saw $O(I) = 1$. $\qquad\square$

Observe that 15.3.33.3 applies to L in the role of "K". We now begin to eliminate the various possibilities for L in 1.1.5.3.

LEMMA 15.3.34. *$L/Z(L)$ is not $Sz(2^n)$. Hence $m_3(L) \geq 1$.*

PROOF. If $L/Z(L) \cong Sz(2^n)$, then Y_+S acts on a Borel subgroup B of L_0, so $B = M_I \cap L_0$ by 15.3.33.1, since B is a maximal S-invariant subgroup of L_0. By 15.3.11.4, $M_I = N_I(Y_+)$, so as automorphisms of $L_0/O_2(L_0)$ of order 3 acting on B are nontrivial on $B/O_2(L_0)$, we conclude $Y_+ \leq C_I(B) = C_I(L_0)$. Thus $L_0 \leq N_I(Y_+) = M_I$ by 15.3.11.4, contrary to our choice of L. $\qquad\square$

Again we will divide the proof into two cases: $m_3(Y_+) = 2$ and $m_3(Y_+) = 1$. We eliminate the first case in the next theorem:

THEOREM 15.3.35. *Case (2) of Hypothesis 15.3.10 holds, $Y_+S/R \cong S_3$, $Y_+ = O^{3'}(M_I)$, $Y_+ < Y$, and $R = C_S(V_2)$.*

Until the proof of Theorem 15.3.35 is complete, assume I is a counterexample. Therefore case (i) of 15.3.11.5 holds, so:

LEMMA 15.3.36. *(1) $Y_+S/O_{2,\Phi}(Y_+S) \cong S_3 \times S_3$.*
(2) $R = C_T(V)$.

The next lemma eliminates the shadow of $G \cong A_{10}$, where $L \cong A_6$. It also eliminates the shadows of $G \cong L$ wr Z_2, for various groups L of Lie rank 2 over F_2.

LEMMA 15.3.37. *One of the following holds:*
(1) $Y_+ \leq L_0$.
(2) $L = L_0 \cong L_2(2^n)$ or $U_3(2^n)$ with n even, or $L_3(2)$. Further $Y_+ = Y_L Y_C$ where $Y_L := O^2(Y_+ \cap L)$, $Y_C := O^2(C_{Y_+}(L))$, $|Y_L|_3 = 3 = |Y_C|$, and $Y_+/O_2(Y_+) \cong E_9$.
(3) $L = L_0$, with $L \cong L_3(2^n)$, n even, or $L/O_2(L) \cong L_3(4)$. Further $Y_L := O^2(Y_+ \cap L) \neq 1$ and $Y_+ = Y_L\langle y \rangle$ with y of order 3 inducing a diagonal outer automorphism on L.

PROOF. By 15.3.34, $m_3(L) \geq 1$. Thus if $L < L_0$, then $L_0 = O^{3'}(I)$ by 1.2.2, so (1) holds. Therefore we may assume $L = L_0$. By 15.3.12.1, $Y_L \neq 1$.

Suppose first that $m_3(L) = 1$. Then $|Y_L|_3 = 3$. Further by 15.3.33.3 and 1.1.5.3, one of the following holds: L is $L_2(2^n)$, L is $L_3^\delta(2^m)$ with $2^m \equiv -\delta$ mod 3, or L is $L_2(p)$ for some Fermat or Mersenne prime p. Then as $Y_L \neq 1$ and S_L acts on Y_L, n is even in the first case; in the second case, either $L \cong L_3(2)$, or m is even and $L \cong U_3(2^m)$; and in the third case, $p = 5$ or 7, so that L is $L_2(4)$ or $L_3(2)$ and so appears in previous cases. Now if L is $L_2(2^n)$ or $U_3(2^m)$, then M_I acts on the Borel subgroup B of L containing S_L, so $B = M_L$ by 15.3.33.1 and maximality of B in L. Thus B acts on Y_+ by 15.3.11.4. Hence Y_+ induces inner automorphisms on L. This also holds if L is $L_3(2)$ since there $Out(L)$ is a 2-group. Then by 15.3.12.2, $Y_+ = Y_L Y_C$ with $|Y_L|_3 = 3 = |Y_C|_3$ and $Y_+/O_2(Y_+) \cong E_9$. Then (2) holds.

Finally suppose $m_3(L) = 2$. Again by 15.3.33.3, L is described in 1.1.5.3 with $O(L) = 1$, and then by A.3.18, either

(i) $L = \theta(I)$, or
(ii) $L \cong L_3^\epsilon(2^n)$ with $2^n \equiv \epsilon$ mod 3, or $L/O_2(L) \cong L_3(4)$. Further some y of order 3 in Y_+ induces a diagonal outer automorphism on L.

In case (i), $Y_+ \leq L$, so that (1) holds. In case (ii), $Y_+ = Y_L\langle y \rangle O_2(Y)$ is S-invariant of 3-rank 2, so $\epsilon = +1$ and hence (3) holds, completing the proof of the lemma. $\square$

The next lemma rules out conclusion (3) of 15.3.37, and eliminates the first appearance of a shadow of $Aut(He)$, where $L/Z(L) \cong L_3(4)$.

LEMMA 15.3.38. *If $m_3(L) = 2$, then $Y_+ \leq L$.*

PROOF. Assume otherwise, and let $I^+ := I/C_I(L)$. Then case (3) of 15.3.37 holds, and in particular some element y of order 3 in Y_+ induces a diagonal outer automorphism on L. If $L/O_2(L) \cong L_3(4)$, let $n := 2$; in the remaining cases $L \cong$

$L_3(2^n)$ with n even. As $S \in Syl_2(I)$ acts on Y_+, Y_+S acts on the Borel subgroup B of L over $S \cap L = S_L$, and then $Y_L := O^2(Y_+ \cap L) \leq B \leq M_I$ by 15.3.33.1. Hence by 15.3.11.6, $M_I = N_I(B)$, n is coprime to 3, and $Y_L/O_2(Y_L) \cong \mathbf{Z}_3$. Also $[V, Y_L] \neq 1$: for otherwise $V \leq C_S(Y_L) = C_S(L)$, so $L \leq N_G(V) = M$, contrary to the choice of L. Since Y_L is S-invariant with $Y_L/O_2(Y_L)$ of order 3, $Y_L \leq M_i$ for $i = 1$ or 2, and we may choose notation so that $i = 2$. Thus $E_4 \cong V_2 = [V, Y_L] \trianglelefteq B$, so $n = 2$, and V_2^+ is the root group $Z(S_L^+)$ of L^+. Further $V_1 = C_V(Y_L) \leq C_G(L)$, so $N_G(V_1) \nleq M$, and hence by 15.3.11.3, case (1) of Hypothesis 15.3.10 holds, so $Y = Y_+$ and $Y_L = O^2(Y \cap M_2)$. Hence $Y/O_2(Y) \cong E_9$. Let $Y_D := O^2(Y \cap M_1)$ and $t \in T - S$. Then $Y_L^t = O^2(Y \cap M_2)^t = O^2(Y \cap M_1) = Y_D$, and $Y = Y_LY_D$. As case (3) of 15.3.37 holds, an element of order 3 in Y_D induces a diagonal outer automorphism on L^+. Also $Y_D \leq M_1 \leq C_G(V_2)$, so from the structure of $PGL_3(4)$, $V_2^+ = C_{S_L^+}(Y_D)$ and $S_L = [S_L, Y_D]$. Of course $[S_L, Y_D] \leq O_2(Y_D)$, so as $O_2(Y_L) = S_L$ and $Y_L^t = Y_D$, $S_L = O_2(Y_D)$ by an order argument. Thus t acts on S_L, and hence also on $Z(S_L^+) = V_2^+$. Since t interchanges V_1 and V_2, $V_1 \leq V_2Z(L)$.

Now $C_{Z(L)}(Y_D) = 1$, and we showed $V_2^+ = C_{S_L^+}(Y_D)$. Further $Z(L) = C_{S_L}(Y_L)$. Thus $V_2 = C_{S_L}(Y_D)$ and $Z(L)^t = C_{S_L}(Y_L^t) = C_{S_L}(Y_D) = V_2 \cong E_4$, so $E_4 \cong V_1 = V_2^t = Z(L)$.

Next $C_S(Y_L) = C_S(L)$, so conjugating by t, $|C_S(Y_D)| = |C_S(L)|$, and as $Y = Y_LY_D$, $C_S(Y) = C_S(L) \cap C_S(Y_D)$. Then as $C_S(Y_D) \leq V_2C_S(L)$ and $|V_2C_S(L) : C_S(L)| = |V_2| = 4$, $|V_2C_S(L) : C_S(Y_D)| = 4$. Therefore $|C_S(L) : C_S(Y)| = |C_S(L) : C_S(L) \cap C_S(Y_D)| \leq 4$, so as $C_{V_1}(Y_D) = 1$, $C_S(L) = V_1C_S(Y)$. But by 15.3.8, $\Omega_1(Z(S)) = Z_S \leq V$, and $C_V(Y) = 1$, so we conclude that $C_S(Y) = 1$; hence $V_1 = C_S(L)$. Thus $C_T(V) = O_2(YS) = O_2(Y) = S_L$. As $YS/C_T(V) = \bar{Y}\bar{S} \cong S_3 \times S_3$, we conclude that $L^+Y_D^+S^+ = Aut(L^+)$. As $O(I) = 1$, $V_1 = C_S(L) = C_I(L)$, so $L = F^*(I)$ and hence $I = LY_DS$.

Let $X \in Syl_3(Y)$; then $C_{S_L}(X) = 1$, and by a Frattini Argument, $N_{YS}(X) \cong S_3 \times S_3$. Let $E := N_S(X)$; then $E = \langle \tau, f \rangle$, where τ and f are involutions inducing a graph and a field automorphism on L^+, respectively. Further $X = X_L \times X_D$, where $X_A := X \cap Y_A$ for $A := L, D$, f inverts X, and τf centralizes V_1. By a Frattini Argument, we may choose $t \in N_T(X)$, so $\langle t, \tau \rangle \cong \bar{T} \cong D_8$. Thus we may choose t to be an involution and $\tau^t = \tau f$.

Let w be of order 4 in $\langle \tau, t \rangle$, and set $W := \langle w, S_L \rangle$; then $|T : W| = 2$, so as $G = O^2(G)$, $\tau^G \cap W \neq \emptyset$ by Thompson Transfer. As $W/S_L = \bar{W} \cong \mathbf{Z}_4$ and $w^2 = f$, τ is fused to a member of $W_0 := \langle f \rangle S_L$.

Next $V_1 = Z(L) \trianglelefteq I$. Then $I_1 := N_G(V_1) \in \mathcal{H}_+$, so $S \in Syl_2(I_1)$ by 15.3.11.1, and thus $L \leq L_1 \in \mathcal{C}(I_1)$ by 1.2.4. Then $m_3(L_1) \geq m_3(L) = 2$, so applying our reductions so far to I_1, L_1 in the roles of "I, L", we conclude that $I_1/O_2(I_1) \cong Aut(L_3(4))$, and hence $L = L_1$ and $I = I_1$. That is, $I = N_G(V_1)$.

Let $\langle v \rangle = Z_S \cap V_1$, $G_v := C_G(v)$, and $\dot{G}_v := G_v/\langle v \rangle$. Then $\mathbf{Z}_2 \cong \dot{V}_1$ and $\dot{L}\dot{S} = C_{\dot{G}_v}(\dot{V}_1)$ as $I = N_G(V_1)$. By I.3.2, $L \leq O_{2',E}(G_v)$, so $F^*(G_v) \neq O_2(G_v)$, and hence $|G_v|_2 < |T|$ as G is of even characteristic. Thus as $S \leq G_v$ and $|T : S| = 2$, $S \in Syl_2(G_v)$. Therefore $L \leq L_v \in \mathcal{C}(G_v)$ by 1.2.4, with the embedding described in A.3.12. As $L \leq O_{2',E}(G_v)$, L_v is quasisimple. Indeed as $\dot{L}$ is a component of $C_{\dot{L}_v}(\dot{V}_1)$, while the only embedding of $L_3(4)$ appearing in A.3.12 is in M_{23}, and M_{23} has trivial Schur multiplier by I.1.3, we conclude $L_v = L$. Thus $G_v \leq N_G(L) \leq$

$N_G(V_1) = I$, so as M_1 is transitive on $V_1^{\#}$, V_1 is a TI-subgroup of G by I.6.1.1. Then as $V_1 \cong E_4$:

(*) V_1 is faithful on any subgroup $F = O^2(F)$ on which it acts nontrivially.

Recall τ is fused to an element of W_0. But L is transitive on the involutions in fL, and each involution in L fused into V under L and hence is in $z^G \cup v^G$. Thus to obtain a contradiction and complete the proof, it remains to show that τ is not fused to f, v, or z.

Recall that τf centralizes V_1. Suppose $\tau f = v^g$ for some $g \in G$. Then V_1 normalizes V_1^g since V_1 is a TI-subgroup of G, and hence by I.6.2.1, $[V_1, V_1^g] = 1$. Thus $V_1^g \le C_G(V_1) = C_I(V_1) = L\langle \tau f \rangle$, so V_1^g centralizes $F := O^2(C_L(\tau f)) \cong E_9$ by (*). Then as $|C_{Aut(L)}(F)| = 2$, $1 \ne V_1^g \cap V_1$, so $V_1^g = V_1$ as V_1 is a TI-subgroup, contrary to $v^g = \tau f \notin V_1$. Thus τf is not fused to v, so as $\tau^t = \tau f$, τ is not fused to v. Similarly using (*) and Generation by Centralizers of Hyperplanes A.1.17, $O(C_G(\tau f))$ centralizes V_1, so $O(C_G(\tau f)) \le O(C_I(\tau f)) \cong E_9$.

Next $O^2(C_L(\tau)) \cong L_2(4)$, so applying I.3.2 as above, we conclude $F^*(C_G(\tau)) \ne O_2(C_G(\tau))$, and hence τ is not fused to z in G. Therefore τ is fused to f in G, so τf is also.

Let $L_f := O^2(C_L(f))$ and $G_f := C_G(f)$; then $L_f \cong L_3(2)$, and again using I.3.2, $L_f \le O_{2',E}(G_f)$. We saw earlier that $O(C_G(\tau f))$ is an elementary abelian 3-group of rank at most 2, and f is fused to τf; so $O_{2',E}(G_f)^\infty = E(G_f)$, and hence $L_f \le E(G_f)$.

Suppose that there is a component L_1 of G_f of 3-rank 1. As f is fused to τf, and a Sylow 3-subgroup of $C_I(\tau f)$ is isomorphic to 3^{1+2}, there is $3^{1+2} \cong B \le G_f$. But now $m_3(L_1 B) = 3$ from the structure of $Aut(L_1)$ with L_1 of 3-rank 1 in Theorem C (A.2.3), contrary to G_f an SQTK-group. Thus no such component exists. But $L_f \le E(G_f)$ so there is a component L_2 of G_f which is not a $3'$-group, and then as $m_3(G_f) \le 2$, L_2 is of 3-rank 2 and $L_2 = O^{3'}(E(G_f))$, so $L_f \le L_2$. Indeed $L_3(2) \cong L_f$ is a component of $C_{L_2}(v)$, so we conclude using Theorem C that $L_2/Z(L_2) \cong L_3(4)$, J_2, or $L_3(7)$. By A.3.18, either $C_{G_f}(L_2)$ is a $3'$-group or $L_2/O_2(L_2)$ is $SL_3(4)$ or $SL_3(7)$, with $O(Z(L_2))$ the unique subgroup of order 3 in $C_{G_f}(L_2)$.

As τf is fused to f, there is a conjugate U_1 of V_1 with $L_3 := O^2(C_G(\langle U_1, f \rangle)) = O(C_{G_f}(U_1)) \cong 3^{1+2}$. In particular, U_1 acts nontrivially on L_2, and hence U_1 acts faithfully on L_2 by (*). By the previous paragraph, either L_3 is faithful on L_2 or $Z(L_2) = Z(L_3)$. We conclude from the structure of centralizers of involutions in $Aut(L_2)$ in our three cases that $L_3 = O(C_{G_f}(U_1)) \le C_{G_f}(L_2)$, contrary to the previous paragraph. $\qquad\square$

LEMMA 15.3.39. $L = L_0$.

PROOF. Assume $L < L_0$. Then $L_0 := LL^s$ for some $s \in S - N_S(L)$, with L described in 1.2.1.3. Then $m_3(L) = 1$ by 15.3.34, and by 15.3.33.3, L is also described in 1.1.5.3 with $O(L) = 1$, so L is simple and $L_0 = L \times L^s$. Therefore as $m_3(Y_+) = 2$ and $Y_+ \le L_0$ by 15.3.37, $Y_+ = XX^s$ where $X := O^2(Y_+ \cap L)$. Next there is $Y_2 \le Y_+ \cap M_2$, with $V_2 = [V, Y_2]$ and Y_2 is S-invariant. As Y_2 and V_2 are s-invariant, they are diagonally embedded in L_0, so X is the projection of Y_2 on L, and $V_L = [V_L, X]$, where V_L is the projection of V_2 on L. Similarly s acts on $O^2(C_{Y_+}(V_2)) =: Y_1$, so Y_1 is also diagonally embedded in L_0 with projection X on

L. Now Y_1 centralizes V_2 and hence also V_L, whereas $[V_L, Y_1] = [V_L, X] \neq 1$, a contradiction. $\square$

The next lemma rules out conclusion (1) of 15.3.37, and eliminates the shadow of $G \cong S_9$ where $L \cong A_5$, and also those of $G \cong L$ wr $\mathbf{Z}_2$ for $L \cong L_3(2)$ and A_5.

LEMMA 15.3.40. $Y_+ \leq L$.

PROOF. Assume $Y_+ \not\leq L$. Then $m_3(L) = 1$ by 15.3.38, and $L = L_0$ by 15.3.39, so that case (2) of 15.3.37 holds, and in that notation of the lemma, $Y_+ = Y_L Y_C$ with $|Y_L|_3 = 3 = |Y_C|_3$, and $Y_+/O_2(Y_+) \cong E_9$. As L is S-invariant, the subgroups Y_L and Y_C are S-invariant. By 15.3.36.1, $Y_+ S/R \cong S_3 \times S_3$, so from the structure of $\bar{M}$ in 15.3.2.1, $\{Y_C, Y_L\} = \{Y_1, Y_2\}$, where $Y_2 \leq Y_+ \cap M_2$ with $V_2 = [V, Y_2]$, and $Y_1 \leq Y_+ \cap M_1$.

If $Y_C = Y_2$ then $V_2 = [V_2, Y_C] \leq C_G(L)$, so as $L \not\leq M$, $C_G(V_2) \not\leq M$, and hence $Y = Y_+$ by 15.3.11.3. Then since $V_1 = [V_1, Y_1]$, interchanging the roles of V_1 and V_2 if necessary, we may assume instead that $Y_L = Y_2$. As $Y_L = Y_2$, $V_2 = [V_2, Y_L] \leq L$.

Suppose first that $L \cong L_2(2^n)$ or $U_3(2^n)$. Then M_I acts on the Borel subgroup over S_L, so M_L is that Borel subgroup by 15.3.33.1. In particular M_L acts on $V_2 \cong E_4$, so we conclude $n = 2$. Then as $Aut_{M_I}(V_2) \cong S_3$, $L \cong L_2(4)$.

Therefore $L \cong L_2(4)$ or $L_3(2)$ as case (2) of 15.3.37 holds, so $Y_2 = Y_L \cong A_4$. In particular $Y/O_2(Y)$ is E_9 rather than 3^{1+2} as Y_2 has one noncentral 2-chief factor, so $Y = Y_+ = Y_2 \times Y_2^t \cong A_4 \times A_4$ for $t \in T - S$, contrary to 15.3.9. $\square$

Assume for the remainder of the proof of Theorem 15.3.35 that $I \in \mathcal{H}_{+,*}$. By 15.3.39, $L \trianglelefteq I$, by 15.3.40, $Y_+ \leq L$, and by 15.3.33.3, $L \not\leq M$. Thus $LS \in \mathcal{H}_+$, so $I = LS$ by minimality of I. Let $I^+ := I/C_I(L)$.

LEMMA 15.3.41. (1) $F^*(I^+) = L^+$ is simple and described in 1.1.5.3.
(2) M_I^+ is a 2-local of I^+ containing a Sylow 2-subgroup S^+ of I^+ with $Y_+^+ \trianglelefteq M_L^+$, $Y_+^+ S^+/O_{2,\Phi}(Y_+^+ S^+) \cong S_3 \times S_3$, and M_I^+ is maximal in I^+ subject to $F^*(M_I^+) = O_2(M_I^+)$.

PROOF. Part (1) follows from 15.3.33.3. By 15.3.11.4 and Coprime Action, $M_I^+ = N_{I^+}(Y^+)$. The remaining two assertions follow from 15.3.36.1 and Theorem 15.3.25. $\square$

LEMMA 15.3.42. L^+ is of Lie type and characteristic 2.

PROOF. Assume otherwise. If $L^+ \cong A_7$, then as $Y_+ \leq L$ and Y_+ is S-invariant, $Y_+ \cong A_4 \times \mathbf{Z}_3$. Thus M_I is the stabilizer in I of a partition of type $4, 3$, as that stabilizer is the unique maximal subgroup of I containing $Y_+ S$. This contradicts $F^*(M_I) = O_2(M_I)$ in 15.3.11.7.

By 15.3.11.8, $O(L) = 1$. Thus by the previous paragraph, L must appear in case (e) or (f) of 1.1.5.3. Inspecting the 2-local subgroups of the groups in those cases for subgroups satisfying the conclusions of 15.3.41.2, we conclude that $I^+ \cong Aut(J_2)$. Then as $V \leq Z(R)$ by 15.3.36.2, and $[V, Y_+]$ is normal in M_I, we conclude $[V, Y_+] \cong E_4$, and hence case (2) of Hypothesis 15.3.10 holds, so $V_2 = [V, Y_+]$. But now $V_1 \leq C_I(Y_+) \leq C_I(L)$ from the structure of $Aut(J_2)$, so $L \leq C_G(V_1) \leq M$ by 15.3.11.3, contrary to the choice of L. $\square$

We are now in a position to complete the proof of Theorem 15.3.35.

By 15.3.42, L^+ is of Lie type and characteristic 2, and hence is described in cases (a)–(c) of 1.1.5.3. By 15.3.41.2, M_L^+ is a maximal S-invariant parabolic of I^+ and S^+ acts on Y_+^+ with $Y_+^+S^+/O_{2,\Phi}(Y_+^+S^+) \cong S_3 \times S_3$. Therefore either $L^+ \cong L_4(2)$ or $L_5(2)$, or L^+ is of Lie rank 2 and defined over $\mathbf{F}_{2^n}$ for $n > 1$ with Y_+ contained in the Borel subgroup B of L over S_L.

Assume the latter case holds. Then as $Y_+S/O_{2,\Phi}(Y_+S) \cong S_3 \times S_3$, we conclude from the structure of $Aut(L^+)$ that $L^+ \cong L_3(2^n)$ with n even. As $O^{2,3}(B) \leq C_M(V)$ by 15.3.2.1, and $[V, Y_+] \trianglelefteq B$, we conclude $n = 2$. Since $O(L) = 1$ by 15.3.33.3, $L \cong L_3(4)$, so that $m_3(B) = 1$, a contradiction.

Therefore $L \cong L_4(2)$ or $L_5(2)$. As $Y_+S/R \cong S_3 \times S_3$, we conclude that S is trivial on the Dynkin diagram of L. By 15.3.41.2, M_I^+ is maximal in I^+ subject to $F^*(M_I^+) = O_2(M_I^+)$, so $L^+ \cong L_4(2)$ and $M_I^+ = Y_+^+S^+$ is the maximal parabolic determined by the two end nodes. Therefore $Y_+^+S^+$ is irreducible on $O_2(Y_+^+S^+)$ of order 16, impossible as $E_4 \cong V_2 = [V_2, Y_+] \trianglelefteq Y_+S$.

This contradiction completes the proof of Theorem 15.3.35.

Finally we complete our analysis of Hypothesis 15.3.10 by eliminating the only possiblity left in Theorem 15.3.35:

THEOREM 15.3.43. Y_+S/R is not S_3.

PROOF. Assume otherwise; then case (ii) of 15.3.11.5 holds, so

(a) $Y_+ = O^{3'}(M_I)$ with $|Y_+ : O_2(Y_+)| = 3$.

In particular, case (2) of Hypothesis 15.3.10 holds, so that

(b) $[V_1, Y_+] = 1$ and $V_2 = [V_2, Y_+] \cong E_4$.

Further $N_G(V_i) \leq M$ by 15.3.11.3, so as $L \not\leq M$, $[V_1, L] \neq 1$. Therefore as $V_1 \leq C_S(Y_+)$ by (b):

(c) $[V_1, L] \neq 1$ and $C_S(Y_+) \neq C_S(L)$.

Suppose $Y_+ \cong A_4$. As $|Y_+|_3 = 3$, case (1) of 15.3.7 holds and $Y_+ = O^{3'}(M_2)$. Thus $Y = Y_+ \times Y_+^t \cong A_4 \times A_4$ for $t \in T - S$, contrary to 15.3.9. Therefore:

(d) Y_+ is not isomorphic to A_4.

Arguing as in the the proof of 15.3.37, one of the following holds:

(i) $Y_+ \leq \theta(I) = L_0$.

(ii) $L = L_0$ is of 3-rank 1 with $L/O_2(L) \cong L_2(2^n)$, $L_3^\delta(2^m)$ with $2^m \equiv -\delta$ mod 3, or $L_2(p)$ for some Fermat or Mersenne prime p.

(iii) $L = L_0 \cong L_3^\epsilon(2^n)$, $2^n \equiv \epsilon$ mod 3, or $L/O_2(L) \cong L_3(4)$. Further some y of order 3 in Y_+ induces a diagonal outer automorphism on L.

Suppose that Y_+ does not induce inner automorphisms on L. Then as $Y_+S/R \cong S_3$, conclusion (iii) holds. As $Y_+S = SY_+$, Y_+S acts on a Borel subgroup B of L over $S \cap L$, so by 15.3.33.1, $B \leq M_L$. But then $m_3(M_I) > 1$, contrary to (a).

Therefore Y_+ induces inner automorphisms on L, and hence case (i) or (ii) holds. As $L \not\leq M$, $[L, Y_+] \neq 1$ by 15.3.11.4, so the projection Y_L of Y_+ on L is nontrivial. Now $N_S(L)$ acts on Y_L, so $S \cap Y_L \in Syl_2(Y_L)$ and hence Y_L normalizes $O^2(Y_+O_2(S \cap Y_L)) = Y_+$, so that $Y_L \leq M_I$ by 15.3.11.4. Thus $Y_L \leq O^{3'}(M_I) = Y_+$, so $Y_L = Y_+$ since $|Y_+|_3 = 3$.

As S normalizes Y_+ and $Y_+ \leq L$, S normalizes L, and hence $L \trianglelefteq I$. As $Y_+ \leq L$ and $I \in \mathcal{H}_{+,*}$, $I = LS$. Let $I^+ := I/C_I(L)$. Now we obtain the following analogue of 15.3.41, using the same proof, but replacing the appeal to 15.3.36.1 by an appeal to (a):

(e) $F^*(I^+) = L^+$ is simple, and is described in 1.1.5.3. Further M_I^+ is a 2-local of I^+ containing a Sylow 2-subgroup S^+ of I^+ with $Y_+^+ = O^{3'}(M_I^+)$, $S^+ Y_+^+ / R^+ \cong S_3$, and M_I^+ is maximal subject to $F^*(M_I^+) = O_2(M_I^+)$.

We now eliminate the various possibilities for L^+ arising in 1.1.5.3 and satisfying condition (e).

Suppose first that L^+ is of Lie type over $\mathbf{F}_{2^n}$, and hence is described in cases (a)–(c) of 1.1.5.3. Then M_L^+ is a maximal S-invariant parabolic by (e).

Assume that $n > 1$. Then as $Y_+ = O^{3'}(M_L)$, we conclude $L^+ \cong L_3(2^n)$ or $L_2(2^n)$ with n even, or $U_3(2^n)$, and M_L^+ is a Borel subgroup of L^+. Then $E_4 \cong V_2 = [V, Y_+] \trianglelefteq M_L$ by (b), so we conclude that $n = 2$. But from the structure of $Aut(L)$, $C_S(Y_+) = C_S(L)$, contrary to (c).

So $n = 1$. As $Y_+ = O^{3'}(M_L)$ and M_L is a maximal S-invariant parabolic, either L^+ is of Lie rank 2, or $I^+ \cong S_8$ and M_L^+ is the middle-node minimal parabolic isomorphic to S_3/Q_8^2. As $E_4 \cong V_2 \trianglelefteq M_I$, the last case is eliminated. Now by (b), $V_1 \leq C_S(Y_+)$, but $[V_1, L] \neq 1$ by (c), and again $V_2 \trianglelefteq M_I$, so we conclude that $I^+ \cong S_6$. However in this case $Y_+ \cong A_4$, contrary to (d).

Suppose next that L^+ is sporadic. We inspect the list of possible sporadics in case (f) of 1.1.5.3 for subgroups I^+ of $Aut(L)$ such that there is a 2-local M_I^+ satisfying (e) and with $E_4 \cong V_2 = [V_2, Y_+] \trianglelefteq M_I$. We conclude $L^+ \cong M_{12}$. But then $C_S(Y_+) = C_S(L)$, contrary to (c).

If $L^+ \cong A_7$, then arguing as in the sporadic case, M_I is the stabilizer of a partition of type $2^3, 1$, so $Y_+ \cong A_4$, again contrary to (d).

From the list of 1.1.5.3, this leaves the case where L is $L_3(3)$ or $L_2(p)$, p a Fermat or Mersenne prime; we may take $p > 7$ as $L_2(5) \cong L_2(4)$ and $L_2(7) \cong L_3(2)$ were eliminated earlier. However in each case, there is no candidate for M_I satisfying (e). This completes the proof of 15.3.43. $\square$

By Theorems 15.3.35 and 15.3.43:

THEOREM 15.3.44. *Assume Hypothesis 15.3.10. Then $M = !\mathcal{M}(Y_+ S)$.*

In the remainder of this subsection we deduce information about the structure of M and of members of $\mathcal{H}(T, M)$ from these uniqueness results.

THEOREM 15.3.45. *For $i = 1, 2$:*

(1) $M = !\mathcal{M}(C_Y(V_i)S)$.
(2) $N_G(V_i) \leq M$.
(3) $C_G(C_{V_i}(S)) \leq M$.

PROOF. First Hypothesis 15.3.10.1 is satisfied with $Y_+ := Y$, so by 15.3.44, $M = !\mathcal{M}(YS)$. Therefore Hypothesis 15.3.10.2 also holds with $Y_+ := O^2(C_Y(V_1))$, so (1) holds since V_1 and V_2 are conjugate in M. Then as $V_i \trianglelefteq YS$ and $C_Y(V_i)S \leq C_G(C_{V_i}(S))$, (2) and (3) follow from (1). $\square$

Recall that we view V as a 4-dimensional orthogonal space of sign $+1$ over $\mathbf{F}_2$, and $\bar{M}$ as the isometry group of this space. In particular, there are two M-classes of involutions in V: the 9 singular involutions fused to z under M, and the 6 nonsingular involutions in $V_1^\# \cup V_2^\#$. We will show next that these classes are not fused in G. Recall weak closure parameters $r(G, V)$ and $w(G, V)$ from Definitions E.3.3 and E.3.23.

LEMMA 15.3.46. *(1) M controls G-fusion of involutions in V.*

(2) For $g \in G - M$, $V \cap V^g$ is totally singular. In particular if $1 < U < V$ with $N_G(U) \not\leq M$, then U is totally singular.

(3) $r(G, V) > 1$.

(4) $W_0(T, V)$ centralizes V, so that $w(G, V) > 0$, $V^G \cap M \subseteq C_G(V)$, and $N_G(W_0(T, V)) \leq M$.

(5) $N_G(Z_S) \leq M \geq C_G(v)$ for $v \in V$ nonsingular.

PROOF. Let $\langle v \rangle = Z_S \cap V_1$; as we just observed, v and z are representatives for the orbits of M on $V^{\#}$. Now $S \in Syl_2(C_M(v))$, and $C_G(v) \leq M$ by 15.3.45.3, so v is not 2-central in G, and hence there is not fused to the 2-central involution z. Thus (1) holds. As $C_G(v) \leq M = N_G(V)$, (1) and A.1.7.2 say that V is the unique member of V^G containing v, so (2) holds. As no hyperplane of V is totally singular, (2) implies (3). Similarly Z_S is not totally singular, so (5) holds.

It remains to prove (4), so suppose $g \in G - M$ and $A := V^g \leq T$. We must show $[V, A] = 1$, so assume otherwise.

Assume first that $V \leq N_G(A)$. Then we have symmetry between V and A, $1 \neq [V, A] \leq V \cap A$, and $[V, A]$ is totally singular by (2). As $[V, A]$ is totally singular, $m(\bar{A}) = 1$ and $\bar{A} = \langle \bar{a} \rangle$ with $V_1^a = V_2$. But as $m(\bar{A}) = 1$, V centralizes the hyperplane $C_A(V)$ of A, so that V induces a group of transvections on A, contrary to $V_1^a = V_2$ and symmetry.

Therefore $V \not\leq N_G(A)$. In the notation of Definition F.4.41, by (3), $U := \Gamma_{1,\bar{A}}(V) \leq N_V(A)$, so $U < V$. Hence $m(\bar{A}) > 1$ so $m(\bar{A}) = m_2(\bar{M}) = 2$, and so $\bar{A}$ is one of the two 4-subgroups of $\bar{T}$. As $V = \Gamma_{1,\bar{S}}(V)$, $\bar{A}$ is the 4-subgroup distinct from $\bar{S}$, so $U = Z^{\perp}$ and $C_U(A) = Z$. Let $B := C_A(V)$; then $m(B) = 2 = m(Aut_U(A))$. As $V \not\leq N_G(A)$, $C_G(B) \not\leq N_G(A)$, so B is totally singular in A by (2). This is impossible, as U centralizes B and $m(Aut_U(A)) = 2$, whereas the centralizer of a totally singular line is of 2-rank 1.

Therefore $W_0(T, V)$ centralizes V, and hence $w(G, V) > 0$ and $V^G \cap N_G(V) \subseteq C_G(V)$. Then by a Frattini Argument, $M = C_M(V)N_M(W_0(T, V))$, and it follows that $N_G(W_0(T, V)) \leq M$ by 15.3.2.4. $\qquad\square$

LEMMA 15.3.47. *If $x \in C_G(Z_S)$, then either $[V, x] = 1$ or $V^G \cap N_G([V, x]) \subseteq C_G(V)$.*

PROOF. Assume $[V, x] \neq 1$. As $x \in C_G(Z_S)$, $x \in M$ by 15.3.46.5. Therefore as $[V, x] \neq 1$ and x centralizes Z_S, $[V, x]$ is not totally singular, so $N_G([V, x]) \leq M$ by 15.3.46.2. But then $V^G \cap N_G([V, x]) \subseteq C_G(V)$ by part (4) of 15.3.46. $\qquad\square$

Observe that $M_c \in \mathcal{H}(T, M)$, and in particular $\mathcal{H}(T, M)$ is nonempty.

In the remainder of the section, H denotes a member of $\mathcal{H}(T, M)$.

Let $M_H := M \cap H$, $V_H := \langle V^H \rangle$, $U_H = \langle Z_S^H \rangle$, $Q_H := O_2(H)$, and $H^* := H/Q_H$. By 15.3.2.3 and 15.3.4, $H \leq M_c = C_G(Z)$, so we can form $\tilde{H} := H/Z$.

LEMMA 15.3.48. *If case (2) of 15.3.7 holds, assume that $C_Y(V) \leq H$. Then*

(1) Hypothesis F.9.1 is satisfied with Y, Z_S, Z in the roles of "L, V_+, V_1".

(2) $\tilde{Z}_S \leq Z(\tilde{T})$, $\tilde{U}_H \leq \Omega_1(Z(\tilde{Q}_H))$, and $\Phi(U_H) \leq Z$.

(3) $Q_H = C_H(\tilde{U}_H)$.

(4) $O_2(H^*) = 1$.

PROOF. By 15.3.46.5, $N_G(Z_S) \leq M = N_G(V)$, so that part (c) of Hypothesis F.9.1 holds. Let $L_1 := O^2(C_Y(Z))$. By 15.3.7, $C_M(Z) = TC_M(V)$, so $L_1 \leq C_M(V)$. Now part (b) of Hypothesis F.9.1 holds as $Z_S \trianglelefteq T$ and $\tilde{Z}_S$ is of order 2. Further $\tilde{Z}_S \leq Z(\tilde{T})$. Part (d) holds as $M = !\mathcal{M}(YT)$ by 15.3.7.

We next establish part (a) of F.9.1. As $C_G(Z_S) \leq M$ by 15.3.46.5, and $C_M(Z) = TC_M(V)$, $C_G(Z_S) = C_M(V)S$, so that using Coprime Action,

$$X := O^2(\ker_{C_H(\tilde{Z}_S)}(H)) \leq C_M(V),$$

and hence $[X, Y] \leq C_Y(V)$. In case (1) of 15.3.7, $C_Y(V) = O_2(Y)$; thus $[Y, X] \leq O_2(Y)$ and $L_1 = 1$ so $L_1T \leq H$. In case (2) of 15.3.7, $C_Y(V) = O_{2,Z}(Y)$, and $L_1 \leq C_Y(V) \leq H$ by hypothesis. If X is a 3′-group then again $[Y, X] \leq O_2(Y)$ as $Aut(Y/O_2(Y))$ is a $\{2, 3\}$-group. If X is not a 3′-group then as $O^2(C_Y(V)) = \theta(C_M(V))$ by 15.3.7, $[Y, X] \leq C_Y(V) \leq XO_2(Y)$. Thus in any case $[Y, X] \leq XO_2(Y)$, so as $X \trianglelefteq XT$, YT normalizes $O^2(XO_2(Y)) = X$. It follows that $X = 1$, as otherwise $H \leq N_G(X) \leq M = !\mathcal{M}(YT)$ by 15.3.7, contrary to $H \in \mathcal{H}(T, M)$. Thus $\ker_{C_H(\tilde{Z}_S)}(H)$ is a 2-group, and hence lies in Q_H. This completes the verification of part (a) of F.9.1.

Finally under the hypothesis of part (e) of F.9.1, $V^g \leq W_0(N_G(V)) \leq C_G(V)$ by 15.3.46.4, so part (e) holds. This completes the verification of (1). By (1), we may apply F.9.2 to obtain the remaining conclusions of 15.3.48. $\square$

The next result eliminates case (2) of 15.3.7; in particular Lemma 15.3.48 applies thereafter to all members of $\mathcal{H}(T, M)$.

LEMMA 15.3.49. (1) $O^2(M_H) \leq C_{M_H}(V)$.

(2) $Z = [Z_S, O_2(M_c)]$.

(3) Case (1) of 15.3.7 holds, so $Y/O_2(Y) \cong E_9$, $O_2(Y) = C_Y(V) = C_Y(Z)$, and $Y = O^{3'}(M)$.

(4) $C_M(V)$ and M_H are 3′-groups.

(5) $N_G(Z_S) = N_M(Z_S)$ is a 3′-group.

PROOF. Part (1) follows since $M \cap M_c = C_M(V)T$ by 15.3.7.

Since $C_Y(V) \leq C_G(Z) = M_c \in \mathcal{H}(T, M)$, enlarging H ir necessary, we may assume when case (2) of 15.3.7 holds that H contains $C_Y(V)$, so that 15.3.48 applies to H.

Let $U_C := C_{U_H}(Q_H)$; we claim:

(a) $O_{2,F^*}(H)$ centralizes U_C.

For $U_C \leq Z(Q_H)$, so as $\mathcal{L}_f(G, T) = \emptyset$ by Hypothesis 14.1.5, each member of $\mathcal{C}(H)$ centralizes U_C by A.4.11, and hence $O_{2,E}(H)$ centralizes U_C. Also by Coprime Action, $U_C = C_{U_C}(O_{2,F}(H)) \oplus [U_C, O_{2,F}(H)]$, so as $Z \leq C_{U_C}(M_c)$ by 15.3.4, and $H \leq M_c$, it follows that $[U_C, O_{2,F}(H)] = 1$, completing the proof of (a).

Set $\hat{U}_H := U_H/U_C$. By 15.3.48, H^* is faithful on $\tilde{U}_H$ and $O_2(H^*) = 1$, while $F^*(H^*)$ centralizes $\tilde{U}_C$ by (a); thus $F^*(H^*)$ is faithful on $\hat{U}_H$, and then also H^* is faithful on $\hat{U}_H$. In particular $U_H \not\leq Z(Q_H)$, so $[Z_S, Q_H] = Z$ by 15.3.48.2, and hence (2) holds since we may take M_c in the role of "H".

By 15.3.7, (3) holds iff $C_M(V)$ is a 3′-group, in which case $M \cap M_c = C_M(V)T$ is a 3′-group, and hence M_H is also a 3′-group as $H \leq M_c$. That is, (3) and (4) are equivalent. Further $N_G(Z_S) = N_M(Z_S)$ by 15.3.46.5, so as $N_{\bar{M}}(Z_S) = \bar{T}$ is a 2-group, (4) implies (5).

Thus we may assume that (3) fails, and it remains to derive a contradiction. Hence case (2) of 15.3.7 holds, so that $Y/O_2(Y) \cong 3^{1+2}$, $Y = \theta(M)$, $C_Y(V) = O_{2,Z}(Y)$, and $Y_0 := O^2(C_Y(V)) \neq 1$. Hence:

(b) $Y_0 = \theta(M_H)$.

Let $\Omega := \Omega_1(Q_H)$. Now $[Q_H, Y_0] \leq Q_H \cap Y_0 \leq O_2(Y_0)$, so $\bar{\Omega} \leq \Omega_1(\overline{C_T(Y_0/O_2(Y_0))}) = Z(\bar{T})$. Thus $\Omega \leq S$, so as $U_H = \langle Z_S^H \rangle$:

(c) $U_H \leq Z(\Omega)$.

(d) V_H is elementary abelian.

For $[V, Q_H] \leq V \cap Q_H \leq \Omega \leq C_H(U_H)$ by (c), so V_H centralizes $Q_H/C_H(U_H)$. Now Hypothesis F.9.1 holds by 15.3.48.1, and U_H is abelian by (c), so we may apply F.9.7 to conclude that $Q_H/C_H(U_H)$ is H-isomorphic to the dual of $\hat{U}_H$. So as H^* is faithful on $\hat{U}_H$, $V_H \leq Q_H$. In particular V_H normalizes V, so V commutes with each H-conjugate of V by 15.3.46.4, and hence V_H is abelian, establishing (d).

We next extend Hypothesis F.9.1 to:

(e) Hypothesis F.9.8 holds.

For suppose $Z \leq V \cap V^g$ for some $g \in G$. As $M_c = C_G(Z)$, and M controls G-fusion in V by 15.3.46.1, we conclude from A.1.7.1 that M_c is transitive on $\{U \in V^G : Z \leq U\}$. Thus we may take $g \in M_c$, and then $[V, V^g] = 1$ by (d) applied to M_c in the role of "H". Thus condition (f) of Hypothesis F.9.8 holds. Further case (ii) of condition (g) of Hypothesis F.9.8 holds by 15.3.47.

We now adopt the notation of the latter part of section F.9 and obtain:

(f) $[E_H, V_\gamma] = 1 = [E_\gamma, V_H]$. In particular, $C_{V_H}(U_\gamma/Z_\gamma) = C_{V_H}(U_\gamma)$.

For as V_H is elementary abelian by (d), $E_\gamma = V_\gamma \cap Q_H \leq \Omega$, and so $[E_\gamma, U_H] = 1$ by (c). Thus as $Z_S \leq U_H$, $E_\gamma \leq C_T(Z_S)$. Suppose E_γ does not centralize V. Then $1 \neq [V, E_\gamma] \leq V_\gamma$, so as V_γ is abelian, $V_\gamma \in V^G \cap C_G([V, E_\gamma]) \subseteq C_G(V)$ by 15.3.47, contradicting our assumption that $[V, E_\gamma] \neq 1$. Thus E_γ centralizes V. But as $E_\gamma \leq Q_H$, E_γ normalizes each H-conjugate of V, so this argument gives the second equality in (f). Before completing the proof of (f), we recall $[V, U_\gamma] \neq 1$ since $V \not\leq G_\gamma^{(1)}$, so as $[D_\gamma, V] \leq [E_\gamma, V] = 1$:

(g) $D_\gamma < U_\gamma$.

By (g) we have symmetry between γ_1 and γ as discussed in the first paragraph of Remark F.9.17, so that the remaining equality in (f) follows from that symmetry. Further by F.9.16.4, we can choose γ so that $0 < m(U_\gamma^*) \geq m(U_H/D_H)$, and hence by (f):

(h) U_γ^* and V_γ^* are quadratic FF*-offenders on $\tilde{U}_H$.

Choose $h \in H$ with $\gamma_0 = \gamma_2 h$, set $\alpha := \gamma h$, and observe $V_\alpha^* \leq O_2(Y_0^* T^*)$—since from the proof of 15.3.48, Y_0 plays the role of "L_1". Let $J_H := \langle V_\alpha^H \rangle$. We show:

(i) J_H is the product of $\mathcal{C}$-components L of J_H with $L = [L, Y_0]$.

For if J_H is not the product of members of $\mathcal{C}(J_H)$, then by (h) and Theorem B.5.6, there is L^* subnormal in J_H^* with $L^* \cong S_3$, $O_3(L^*) = [O_3(L^*), V_\alpha^*]$, and $[\tilde{U}_H, L]$ of rank 2. Further Y_0 acts on L^* as there are at most two H-conjugates of L^* in Theorem B.5.6 and $Y_0 = O^2(Y_0)$. As $O_3(L^*) = [O_3(L^*), V_\alpha^*]$ and $V_\alpha^* \leq O_2(Y_0^* T^*)$, $O_3(L^*) \neq Y_0^*$. Hence Y_0 centralizes $L/O_2(L)$ so that L normalizes $O^2(Y_0 O_2(L)) =$

Y_0. Thus $L \leq N_G(Y_0) = M$, contrary to (b) since we just saw $O_3(L^*) \neq Y_0^*$. This contradiction shows that J_H is the product of members of $\mathcal{C}(J_H)$. Similarly $L = [L, Y_0]$ for each $L \in \mathcal{C}(J_H)$, completing the proof of (i).

Applying (i) to any overgroup of Y_0T in H we conclude

(j) Each solvable overgroup of Y_0T in H is contained in M_H.

Pick $L \in \mathcal{C}(J_H)$ and let $L_0 := \langle L^T \rangle$ and $U_0 := [U_H, L_0]$. Then $L_0 Y_0 T \in \mathcal{H}(Y_0 T, M)$, so replacing H by $L_0 Y_0 T$, we may assume $H = L_0 Y_0 T$. As $\tilde{Z}_S \leq Z(\tilde{T})$ by 15.3.48, $\tilde{U}_H = \tilde{U}_0 C_{\tilde{U}_H}(H)$ by B.2.14. Let $\tilde{Z}_0$ be the projection of $\tilde{Z}_S$ on $\tilde{U}_0$ with respect to this decomposition; thus $C_{H^*}(\tilde{Z}_0) \leq N_H(Z_S)^* \leq M_H^*$ by 15.3.46.5.

By Theorems B.5.1 and B.5.6, L^* is A_7, $\hat{A}_6$, $L_n(2)$ for $n := 4$ or 5, or a group of Lie type of Lie rank 1 or 2 over some $\mathbf{F}_{2^e}$. Set $T_0 := T \cap L_0$. When L^* is of Lie type, let B denote the Borel subgroup of L_0 containing T_0.

Assume first that $Y_0 \not\leq L_0$. Then L^* is not A_7 or $\hat{A}_6$ by A.3.18. Further $T_0 = Y_0 T \cap L_0$ is $Y_0 T$-invariant, so $Y_0 T$ acts on B. Then $B \leq M_H$ by (j), so as we are assuming $Y_0 \not\leq L_0$, we conclude from (b) that B is a $3'$-group acting on Y_0. As $L_0 = [L_0, Y_0]$ by (i), we conclude from the structure of $Aut(L^*)$ for L^* of Lie type that $B = T_0$, and so L is defined over $\mathbf{F}_2$. Then $Out(L_0^*)$ is a $3'$-group from the list of possibilities in Theorem B.4.2, so Y_0 induces inner automorphisms on L_0^*, and this time we obtain a contradiction from (j) and (b) since the projection of Y_0^* on L_0^* is $Y_0 T$-invariant and nontrivial. Thus we have shown:

(k) $Y_0 \leq L_0$.

Suppose that L^* is of Lie type, and defined over $\mathbf{F}_{2^e}$ with $e > 1$; then from Theorem B.4.2, L^* is $L_2(2^e)$, $SL_3(2^e)$, $Sp_4(2^e)$, or $G_2(2^e)$. Further as T_0 acts on Y_0, Y_0 is contained in B, and e is even. Then by (b), $\theta(N_{L_0^*}(Y_0)) = Y_0^*$ is of 3-rank 1, so we conclude $L^* = L_0^* \cong L_2(2^e)$. As V_α^* is an FF*-offender contained in $O_2(Y_0^* T^*)$, we conclude from Theorem B.4.2 that $\tilde{U}_0 / C_{\tilde{U}_0}(L)$ is the natural module for L^*. But then $\tilde{Z}_0 \leq C_{\tilde{U}_0}(Y_0) \leq C_{\tilde{U}_0}(L)$, contrary to $U_H = \langle Z_S^H \rangle$. Therefore L^* is not of Lie type of $\mathbf{F}_{2^e}$ with $e > 1$.

Applying (j) and (b) as at the end of the proof of (k), we conclude that $L = L_0$ if $L^* \cong L_3(2)$; so using 1.2.1.3, we have reduced to:

(l) $L_0 = L$, L^* is $L_n(2)$, $3 \leq n \leq 5$, A_6, $\hat{A}_6$, A_7, or $G_2(2)'$, and either $Y_0^* T_0^*$ is a minimal parabolic of L^* of Lie type, or L^* is A_7 or $\hat{A}_6$.

(m) $V_H > U_H V$.

For suppose that $V_H = U_H V$; then because $[U_H, Q_H] \leq V_1$ by 15.3.48.2, $[V_H, Q_H] = [U_H, Q_H][V, Q_H] \leq V_1 V = V$. Further $[V, Q_H] \neq 1$ by (2), so $Z(\tilde{T}) \leq \tilde{Q}_H$ as $Z(\tilde{T})$ is of order 2. Thus $Z_S \leq [V, Q_H]$, and hence $Z_S \leq [V_H, Q_H] \leq V$. Therefore $[V_H, Q_H]$ is not totally singular in V, so $H \leq N_G([V_H, Q_H]) \leq M$ by 15.3.46.2, contrary to $H \in \mathcal{H}(T, M)$.

(n) V_γ^* is a strong FF*-offender on $\tilde{U}_H$.

Suppose otherwise. By the choice of γ, $m(U_\gamma^*) \geq m(U_H/D_H)$, and $U_\gamma \leq V_\gamma \leq C_H(D_H)$ by (f), so as V_γ^* is not a strong offender, we conclude that $\tilde{D}_H = C_{\tilde{U}_H}(V_\gamma)$, $U_\gamma^* = V_\gamma^*$, and $m(U_\gamma^*) = m(U_H/D_H)$. By the last equality we have symmetry between γ and γ_1 (as discussed in the second paragraph of Remark F.9.17) so also $V_H = U_H C_{V_H}(U_\gamma/Z_\gamma)$ by that symmetry. Further $C_{V_H}(U_\gamma/Z_\gamma) = C_{V_H}(U_\gamma)$ by (f), so U_γ centralizes V_H/U_H. Hence as $L = [L, U_\gamma]$, L centralizes V_H/U_H, so as $H = LY_0T$, $V_H = \langle V^H \rangle = U_H V$, contrary to (m). Thus (n) holds.

Observe that L^* is A_6 or $L_n(2)$, $3 \leq n \leq 5$, since in the remaining cases in (1), L^*T^* has no strong FF*-offenders by Theorem B.4.2, contrary to (n).

Suppose that $L^* \cong L_3(2)$. As $V_\alpha^* \leq O_2(Y_0^*T^*)$ and $Y_0^*T^*$ is the stabilizer of the point $\tilde{Z}_0$ in $\tilde{U}_0$, V_γ^* is not a strong offender on $\tilde{U}_H$ by Theorem B.5.1, contrary to (n). Thus L^* is not $L_3(2)$.

Suppose next that $L^* \cong L_n(2)$ for $n = 4$ or 5. As $Y_0^*T_0^*$ is a T-invariant minimal parabolic by (1), either LT is generated by overgroups H_1 of Y_0T with $H_1/O_2(H_1) \cong S_3 \times S_3$ or $L_3(2)$, or $H^* \cong S_8$ with $Y_0^*T_0^*$ the middle-node minimal parabolic of L^*. In the first case, $L \leq M$ by our previous reductions, contrary to $H \not\leq M$. In the second case, $Y_0T = C_H(\tilde{Z}_0)$, so by Theorem B.5.1, $\tilde{U}_0$ is the sum of the natural module and its dual; hence $O_2(Y_0^*T^*)$ contains no FF*-offender by B.4.9.2iii, whereas V_α^* is such an offender by (h).

Thus $L^* \cong A_6$. But then as $V_\alpha^* \leq O_2(Y_0^*T^*)$ with $Y_0^*T^*$ the stabilizer of the point $\tilde{Z}_0$, V_γ^* is not a strong FF*-offender on $\tilde{U}_H$ by B.3.2, contrary to (n). This contradiction completes the proof of 15.3.49. $\square$

15.3.3. The case $\langle \mathbf{V^{M_c}} \rangle$ nonabelian. Recall from 15.3.49.4 that case (1) of 15.3.7 holds, and in particular 15.3.48 applies to all $H \in \mathcal{H}(T, M)$.

In this subsection, we will assume that $\langle V^{M_c} \rangle$ is nonabelian, and derive a contradiction via an application of the methods in section 12.8; in particular we will use Theorem G.9.3. Thus we will reduce to the following situation, to be treated in the final subsection:

THEOREM 15.3.50. V_H is abelian for each $H \in \mathcal{H}(T, M)$.

Until the proof of Theorem 15.3.50 is complete, assume H is a counterexample. Then $\langle V^{M_c} \rangle$ is also nonabelian, so as usual in the nonabelian case of section F.9, we take $H := M_c$. Recall $M_c = C_G(Z)$ by 15.3.4, so $V_H = \langle V^{C_G(Z)} \rangle$. Set $U := U_H = \langle Z_S^{C_G(Z)} \rangle$.

LEMMA 15.3.51. *(1) $V^* \neq 1$.*

(2) Either

  (a) U is nonabelian, $\bar{U}$ is the 4-subgroup of $\bar{T}$ distinct from $\bar{S}$, and $\bar{U}$ is a Sylow group of $\Omega_4^+(V)$, or

  (b) U is elementary abelian, $U \leq S$, $Z(\bar{T}) \leq \bar{U}$, and $Z_S = V \cap U$.

(3) $Y = [Y, U]$.

(4) $[V, Q_H] \leq V \cap Q_H$ and $[V, U] \leq V \cap U$.

PROOF. If $V \leq Q_H$ then the members of V^H normalize V, so that V_H is abelian by 15.3.46.4, contrary to our choice of H as a counterexample. Thus (1) holds, so $[\tilde{U}, V] \neq 1$ by 15.3.48.3, and hence $\bar{U} \neq 1$. By 15.3.48.2, $\Phi(U) \leq Z$, so $\bar{U}$ is elementary abelian, and as $T \leq H$, $\bar{U} \trianglelefteq \bar{T}$, so $Z(\bar{T}) \leq \bar{U}$ as $Z(\bar{T})$ is of order 2. As $U = \langle Z_S^H \rangle$, U is nonabelian iff $U \not\leq C_T(Z_S) = S$ iff conclusion (a) of (2) holds. Thus if U is abelian then $U \leq S$, so as $Z(\bar{T}) \leq \bar{U}$, $Z_S \leq V \cap U \leq C_V(U) \leq C_V(Z(\bar{T})) = Z_S$, and hence conclusion (b) of (2) holds. As $Z(\bar{T}) \leq \bar{U}$, $\bar{Y} = [\bar{Y}, \bar{U}]$, so (3) holds. Part (4) follows as V normalizes Q_H and U, and vice versa. $\square$

LEMMA 15.3.52. *U is nonabelian.*

PROOF. Assume U is abelian; then case (b) of 15.3.51.2 holds. Thus $Z_S = V \cap U$ with $[U, V] \leq U \cap V$ by 15.3.51.4, so V^* induces a group of transvections on $\tilde{U}$ with center $\tilde{Z}_S$. Then $V_H^* = \langle V^{*H^*} \rangle$ is generated by transvections, $\tilde{U} = \langle \tilde{Z}_S^H \rangle$, and $V^* \trianglelefteq T^*$, so by G.6.4.4, $V_H^* = H^* \cong L_n(2)$, $2 \le n \le 5$, S_6, or S_7, and $\tilde{U}/C_{\tilde{U}}(H^*)$ is the natural module for H^*. As $C_{H^*}(\tilde{Z}_S)$ is a $3'$-group by 15.3.49.5, we conclude that $H^* \cong S_3$. Then $m(U) = 3$ and $Z_S = C_U(V) = U \cap V$. Now as $O_2(Y) = C_Y(V)$ by 15.3.49.3, $[O_2(Y), U] \le C_U(V) \le V$; then in view of 15.3.51.3, Y centralizes $O_2(Y)/V$, so that $V = O_2(Y)$. Thus $Y \cong A_4 \times A_4$, contrary to 15.3.9. $\qquad\square$

LEMMA 15.3.53. *(1)* $\bar{U}\bar{Y} = \Omega_4^+(V) = \bar{N}_1 \times \bar{N}_2$ *with* $\bar{N}_i \cong S_3$ *and* $V = [V, N_i]$.
(2) $V \cap Q_H = V \cap U = [U, V] = Z^\perp$ *is the hyperplane of* V *orthogonal to* Z. *Thus* V^* *is of order 2.*

PROOF. By 15.3.52, conclusion (a) of 15.3.51.2 holds, giving (1). Next by (1), $[U, V] = Z^\perp$, so as $V^* \ne 1$ by 15.3.51.1, and $[U, V] \le U \cap V$ by 15.3.51.4, (2) follows. $\qquad\square$

For the remainder of this subsection, define N_i as in 15.3.53, and set $Y_i := O^2(Y \cap N_i)$.

LEMMA 15.3.54. *Let* $g \in Y$ *with* Z^g *not orthogonal to* Z *in* V, *and set* $I := \langle U, U^g \rangle$, $P := O_2(I)$, *and* $W := U \cap P$. *Then*
(1) $I = YU$.
(2) $P = WW^g$ *and* $V \le Z(P)$.
(3) $U \cap U^g = W \cap W^g = Z^\perp \cap Z^{g\perp} \cong E_4$.
(4) $P/V = P_1/V \oplus P_2/V$, *where* $P_i/V := [P/V, Y_i] = C_{P/V}(N_{3-i})$, *and* P_i/V *is the sum of* s *natural modules for* $\bar{N}_i$.
(5) $[W^g, U] \le W$ *and* W^g *normalizes* U.

PROOF. We verify the hypotheses of G.2.6, with V, Y, Z, U in the roles of "V_L, L, V_1, U" By 15.3.481, G.2.2 is satisfied by the tuple of groups, and the remaining hypotheses of G.2.6 hold by 15.3.53. Hence the conclusions of G.2.6 hold with $V(U \cap U^g)$ in the role of "S_2". Thus conclusions (1) and (2) of 15.3.54 follow from G.2.6, and conclusion (4) will follow from G.2.6.5 once we show that $U \cap U^g \le V$.

As $W^g \le T$, W^g normalizes U, so $[W^g, U] \le P \cap U = W$, and hence (5) holds. Further $\Phi(U^g) \le Z^g$ by 15.3.48.2, so $[U \cap U^g, W^g] \le W \cap Z^g$. But $Z^g \le V$, so $W \cap Z^g \le U \cap V$, and hence $W \cap Z^g = 1$, since we chose $Z^g \not\le Z^\perp$, and $Z^\perp = U \cap V$ by 15.3.53.2. Thus W^g centralizes $U \cap U^g$, and by symmetry, W centralizes $U \cap U^g$, so using (2), $P_0 := (U \cap U^g)V \le Z(P)$. Further by G.2.6.4, I centralizes P_0/V, so since $P_0 \le Z(P)$, we may apply Coprime Action to conclude $P_0 = V \times C_{P_0}(Y)$. Now T normalizes $YU = I$, and hence normalizes the preimage P_0 of $C_{O_2(I)/V}(Y)$ in I, and then also normalizes $C_{P_0}(Y)$. Therefore as $\Omega_1(Z(T)) = Z \le V$, we conclude $C_{P_0}(Y) = 1$ so that $P_0 = V$, and hence $U \cap U^g \le V$. As mentioned earlier, this completes the proof of (4), and we established (5) earlier, so it remains to complete the proof of (3). But by 15.3.53.2, $U \cap V = Z^\perp$, so as $U \cap U^g \le V$,

$$U \cap U^g = (U \cap V) \cap (U^g \cap V) = Z^\perp \cap Z^{g\perp} = W \cap W^g \cong E_4.$$

$\qquad\square$

In the next few lemmas, we use techniques similar to those in section 12.8 to study the action of H on U.

For the remainder of the subsection, define g, W, P, P_i, and s as in 15.3.54.

LEMMA 15.3.55. *U is extraspecial, and* $V = Z(P)$.

PROOF. First U is nonabelian by 15.3.52, so that $Z = \Phi(U)$ by 15.3.48.2; hence $U = U_0 Z_U$, where U_0 is extraspecial and $Z_U := Z(U)$. Thus we must show that $Z_U = Z$. As $U = \langle Z_S^H \rangle$ is nonabelian and Z_S is of order 4, $Z_S \cap Z_U = Z$; then as $C_{\tilde{V}}(T) = \tilde{Z}_S$, $V \cap Z_U = Z$. Therefore $[V, Z_U] \leq V \cap Z_U = Z$, but no member of M induces a transvection on V with singular center, so $Z_U \leq C_U(V) = W$. Hence also $Z_U^g \leq W^g$.

As $Z_U \cap V = Z$, $Z_U^g \cap V = Z^g$, so by 15.3.54.3,

$$Z_U^g \cap U = Z_U^g \cap (U \cap U^g) = Z_U^g \cap (Z^\perp \cap Z^{g\perp}) \leq Z^g \cap Z^\perp = 1.$$

Then as W^g normalizes U and W normalizes U^g by 15.3.54.5, $[Z_U^g, W] \leq Z_U^g \cap U = 1$, so as $P = WW^g$ by 15.3.54.2, $Z_U^g \leq Z_P := Z(P)$. Therefore also $Z_U \leq Z_P$. By 15.3.54.4, the irreducibles for Y on P/V are not isomorphic to those on V, so $Z_P = V \oplus Z_0$, where Z_0 is the sum of the Y-irreducibles on Z_P not isomorphic to those on V. Thus T acts on Z_0, so as $Z \leq V$, $Z_0 = 1$. Thus $Z_P = V$, so as $Z_U \leq Z_P$, $Z_U = V \cap Z_U = Z$, completing the proof. $\square$

LEMMA 15.3.56. Let $y \in Y_1 - O_2(Y_1)$, $V_0 := \langle Z^{Y_1} \rangle$, $F := U \cap H^y$, $X := F^y$, $E := F \cap F^y$, and $t \in T - UC_T(V)$. Then

(1) The power map and commutator map make $\tilde{U}$ into an orthogonal space with $H^* \leq O(\tilde{U})$.

(2) $m(\tilde{U}) = 2(s+2)$.

(3) $X \cap Q_H = E$, $[X, F] \leq E$, $V_0 = ZZ^y$, and $\tilde{E}$ is totally singular of rank $s + 2$ in the orthogonal space $\tilde{U}$.

(4) $X^* \cong E_{2^{s+1}}$ induces the full group of tranvections on $\tilde{E}$ with center $\tilde{V}_0$.

(5) $\tilde{U} = \tilde{E} \oplus \tilde{E}^t$ and X^* induces the full group of transvections on $\tilde{E}^t$ with axis $\widetilde{C_{E^t}(V_0)}$.

(6) $X^* \cap X^{*t} = V^*$ is of order 2, and $X^* X^{*t} \cong D_8^s$.

(7) $\tilde{Z}_S = C_{\tilde{U}}(\langle X^*, t^* \rangle)$.

PROOF. Part (1) follows from 15.3.55. By 15.3.54.4, $|P| = 2^{4s+4}$, while by parts (2) and (3) of 15.3.54, $|P| = 2^{2(m(W)-1)}$. Thus $m(W) = 2s + 3$. By 15.3.53.1 and 15.3.54.1, $m(U/W) = 2$, so (2) follows.

As $y \in Y_1 - O_2(Y_1)$, $z^y \in Z^\perp - Z$, so $z^y \in U - Z$ by 15.3.53.2. Thus as U is extraspecial, $|U : F| = 2$; and the argument in 8.14 of [**Asc94**], which is essentially repeated in the proof of G.2.3, gives us the structure of $J := \langle U, U^y \rangle$: $J/O_2(J) \cong S_3$, $ZZ^y = V_0 \cong E_4$, $O_2(J) = FF^y = FX = C_J(V_0)$, $[E, J] \leq V_0$, and for some r, $O_2(J)/E$ is the direct sum of r natural modules for $J/O_2(J)$ with $[O_2(J)/E, U] = F/E$. Thus

$$J \text{ has } r + 1 \text{ noncentral 2-chief factors.} \tag{$*$}$$

Moreover J and E are normal in $N_G(V_0)$.

As $O_2(J)/E$ is abelian and $O_2(J) = XF$, $[X, F] \leq E$. Similarly as $[XF/E, U] = F/E$ and $|U : F| = 2$, for $u \in U - F$ the map $\varphi : X/E \to F/E$ defined by $\varphi(xE) := [u, x]E$ is a bijection. Therefore as $[U, Q_H] = Z \leq E$ by 15.3.52 and 15.3.48.2, $X \cap Q_H = E$. Finally $\Phi(E) \leq \Phi(U) \cap \Phi(U^y) = Z \cap Z^y = 1$, so by (1), $\tilde{E}$ is totally singular in the orthogonal space $\tilde{U}$.

Next $\bar{J} = \bar{Y}_1 \bar{U} \cong S_3 \times \mathbf{Z}_2$, with $\bar{F} = \bar{X} = Z(\bar{J}) = \bar{U} \cap \bar{N}_2$; in particular $[Z^\perp, X] = Z$. By 15.3.54.4, Y_1 has s noncentral chief factors on P/V, and by 15.3.53.1, Y_1 has two noncentral chief factors on V. Thus J has $s + 2$ noncentral

2-chief factors, so $s + 2 = r + 1$ by (*). Further $m(U/X) = 1$ and $m(X/E) = r$, so using (2),

$$m(\tilde{E}) = m(\tilde{U}) - (r + 1) = 2(s + 2) - (s + 2) = s + 2,$$

completing the proof of (3). Further $\bar{F} \neq \bar{F}^t$, so $F \cap F^t \leq P$. Also $[E \cap P, Y_1] \leq [E, J] \leq V_0 \leq V$, so $(E \cap P)/V \leq C_{P/V}(Y_1) = P_2/V$ by definition of P_2, and then $E \cap E^t \leq P_2 \cap P_2^t = V$.

Now (4) holds, using an argument in the proof of 12.8.11.3; indeed the argument is easier here since U is extraspecial. Next using 15.3.53.2, $V \cap U = Z^\perp = V_0 V_0^t$, and we saw earlier that $[Z^\perp, X] = Z$, so X centralizes $\widetilde{V \cap U}$. Therefore X acts on V_0 and V_0^t, so since E and E^t are normal in $N_G(V_0)$ and $N_G(V_0^t)$, respectively, X acts on E and E^t. We saw earlier that $E \cap E^t \leq V$, so

$$E \cap E^t \leq U \cap U^y \cap U^{yt} \cap V = Z^\perp \cap Z^{y\perp} \cap Z^{yt\perp} = Z.$$

Then as $m(\tilde{U}) = 2m(\tilde{E})$, $\tilde{U} = \tilde{E} \oplus \tilde{E}^t$. Since the action of H^* on $\tilde{U}$ is self-dual, the action of X^* on $\tilde{E}^t$ is dual to its action on $\tilde{E}$, so (5) holds. By 15.3.53.2, V^* is of order 2, so (4) and (5) imply (6) and (7). $\square$

In the remainder of the proof of Theorem 15.3.50, define V_0, X, E, and y as in lemma 15.3.56.

LEMMA 15.3.57. *(1) H is irreducible on $\tilde{U}$.*

(2) If $1 \neq K^* = O^2(K^*) \trianglelefteq H^*$ and the irreducibles of K^* on $\tilde{U}$ are of rank at least 3, then $[K^*, V^*] \neq 1$, and either

(a) K^* is irreducible on $\tilde{U}$, or

(b) $\tilde{U} = \tilde{U}_1 \oplus \tilde{U}_2$, where the $\tilde{U}_i$ are irreducible K^*-modules of rank $s + 2$, and V^* induces a transvection on each $\tilde{U}_i$.

PROOF. Let $\tilde{U}_0$ be a nonzero H-submodule of $\tilde{U}$. Then $C_{\tilde{U}_0}(T) \neq 0$, so by 15.3.56.7, $\tilde{Z}_S \leq \tilde{U}_0$. Thus $\tilde{U} = \langle \tilde{Z}_S^H \rangle \leq \tilde{U}_0$, so (1) holds.

Assume the hypothesis of (2). By (1) and Clifford's Theorem, $\tilde{U}$ is a semisimple K^*-module, and by hypothesis, each $\tilde{J} \in Irr_+(K^*, \tilde{U})$ is of rank at least 3. If $[K^*, V^*] = 1$ then K^* acts on $[\tilde{U}, V^*]$; this is impossible as $[\tilde{U}, V^*] = \widetilde{V \cap U}$ is of rank 2 by 15.3.53.2, contradicting $m(\tilde{J}) > 2$ for $\tilde{J} \in Irr_+(K^*, [\tilde{U}, V^*])$. Thus $[K^*, V^*] \neq 1$.

Similarly if V^* does not normalize some $\tilde{J}$, then $m([\tilde{U}, V^*]) \geq m(\tilde{J}) > 2$ by hypothesis, again contrary to 15.3.53.2. Thus we can write $\tilde{U} = \tilde{J}_1 \oplus \cdots \oplus \tilde{J}_k$ where $\tilde{J}_i \in Irr_+(K^*, \tilde{U})$ and $\tilde{J}_i$ is V^*-invariant. Again using 15.3.53.2,

$$2 = m([\tilde{U}, V^*]) = \sum_{i=1}^{k} m([\tilde{J}_i, V^*]) \geq k,$$

so that (2) holds. $\square$

The next lemma eliminates the shadow of $Aut(L_4(2))$, and begins to zero in on the shadows of $Aut(L_5(2))$ and $Aut(He)$.

LEMMA 15.3.58. *(1) $H^* \cong Aut(L_3(2))$.*

(2) $s = 1$ and $U \cong D_8^3$.

(3) $\tilde{U} = \tilde{U}_1 \oplus \tilde{U}_1^t$, for $t \in T - UC_T(V)$, and some natural submodule $\tilde{U}_1$ for $O^2(H^*) \cong L_3(2)$, such that $\tilde{U}_1^t$ is dual to $\tilde{U}_1$.

PROOF. Observe first that $s > 0$: For if $s = 0$, then by 15.3.54.4, $Y \cong A_4 \times A_4$, contrary to 15.3.9.

Next let $V_U := V \cap U$; as $V_U = Z^\perp$ by 15.3.53.2, $N_G(V_U) \le M$ by 15.3.46.2; so as $\tilde{V}_U = [\tilde{U}, V^*]$ by 15.3.53.2, $C_{H^*}(V^*) \le M_H^*$. Now using (4) and (5) of 15.3.49, $N_{H^*}(\tilde{V}_U)$, $C_{H^*}(\tilde{Z}_S)$, and $C_{H^*}(V^*)$ are $3'$-groups.

Let K^* be a minimal normal subgroup of H^*. As H^* is faithful and irreducible on $\tilde{U}$ using 15.3.57.1, $\tilde{U} = [\tilde{U}, K^*]$. If K^* is a 3-group, then as $C_{H^*}(V^*)$ is a $3'$-group, V^* inverts K^*; so by 15.3.56.2 and 15.3.53.2,

$$2(s + 2) = m(\tilde{U}) = 2m([\tilde{U}, V^*]) = 4,$$

contradicting $s > 0$.

Therefore K^* is not a 3-group, so each irreducible for K^* on $\tilde{U}$ has rank at least 3. Thus by 15.3.57.2, $[K^*, V^*] \ne 1$, and K^* satisfies one of the two conclusions of 15.3.57.2. Suppose K^* is solvable. Then K^* is a p-group for some prime $p > 3$. As $m([\tilde{U}, V^*]) = 2$, it follows (cf. D.2.13.2) that $[K^*, V^*] \cong \mathbf{Z}_5$. However as $s > 0$, there is a D_8-subgroup D^* of H^* with center V^* by 15.3.56.5. As V^*, and hence also D^*, is faithful on $[K^*, V^*]$, this is a contradiction.

Therefore K^* is not solvable, so as K is an SQTK-group, K^* is the direct product of at most two isomorphic nonabelian simple groups.

Suppose first that conclusion (b) of 15.3.57.2 holds. Then $\tilde{U} = \tilde{U}_1 \oplus \tilde{U}_2$ is the sum of two K^*-irreducibles $\tilde{U}_i$ of rank $s+2$ with V^* inducing a transvection on each $\tilde{U}_i$. By G.6.4.4, $K^*V^* \cong L_n(2)$, $3 \le n \le 5$, S_6, or S_7, and $\tilde{U}_i$ is a natural module for K^*. Let $\langle \tilde{u}_i \rangle = [\tilde{U}_i, V^*]$; then $\tilde{V}_U = \langle \tilde{u}_1, \tilde{u}_2 \rangle$, so as $N_{K^*}(\tilde{V}_U)$ is a $3'$-group, we conclude $K^* \cong L_3(2)$, and so $s = 1$.

Next as K^* is irreducible on $\tilde{U}_i$, and $\tilde{U}_i$ is not self-dual, $\tilde{U}_i$ is totally singular, and $\tilde{U}_2$ is dual to $\tilde{U}_1$. Thus $Irr_+(K^*, \tilde{U}) = \{\tilde{U}_1, \tilde{U}_2\}$ is permuted by H^*, and as H^* is irreducible on $\tilde{U}$ by 15.3.57.1, H^* is transitive on $\{\tilde{U}_1, \tilde{U}_2\}$. Further as $End_{K^*}(\tilde{U}_i) \cong \mathbf{F}_2$, $C_{H^*}(K^*) = 1$, so $H^* \cong Aut(L_3(2))$. completing the proof of the lemma in this case.

Thus we may assume that K^* is irreducible on $\tilde{U}$. By 15.3.56.6, $m_2(H^*) \ge s+1$, so using 15.3.56.2, $m(\tilde{U}) = 2(s+2) \le 2(m_2(H^*)+1)$. As $[K^*, V^*] \ne 1$ by 15.3.57.2, the hypotheses of Theorem G.9.3 are satisfied with K^*, $\tilde{U}$, X^* in the roles of "H, V, A", so H^* and its action on $\tilde{U}$ are described in Theorem G.9.3. As $m([\tilde{U}, V^*]) = 2$ with $V^* \le Z(T^*)$, we conclude: cases (0)–(2) and (15)–(17) do not hold (see e.g. chapter H of Volume I for the Mathieu groups); in cases (6)–(10), $n \le 2$; and in case (13), $\tilde{U}$ is a natural module rather than a 10-dimensional module. As $s > 0$, $m(\tilde{U}) \ge 6$; therefore case (3) does not hold, nor does (6) or (7) when $n \le 2$, completing the elimination of those cases. As $\tilde{Z}_S = C_{\tilde{U}}(T)$ is of order 2, and $C_{H^*}(\tilde{Z}_S)$ is a $3'$-group, the remaining cases are eliminated. $\qquad\square$

Let $K := O^2(H)$ and $T_K := T \cap K$, so that $K^* \cong L_3(2)$ by 15.3.58.1.

LEMMA 15.3.59. *(1)* $U = Q_H$.
(2) $T_K \in Syl_2(YU)$.
(3) $|T : T_K| = 2$.
(4) $M = YT$.

PROOF. Let $Q_C := C_T(U)$. As $\tilde{Q}_H = C_{\tilde{T}}(\tilde{U})$ by 15.3.48, and U is extraspecial, $Q_H = UQ_C$. Now $[Q_C, V] \le C_V(U) = Z$, so $[\tilde{Q}_C, V^*] = 1$. Then as $K = [K, V]$

and $K = O^2(K)$, we conclude using Coprime Action that K centralizes Q_C. Thus $Q_C = C_T(T_K)$. By 15.3.58, $U \leq K$ and $K^* \cong L_3(2)$, so $\hat{K} := K/U \cong L_3(2)$ or $SL_2(7)$ and hence $|T_K| \geq 2^{10}$ and $\hat{Q}_C \leq \Phi(\hat{T}_K)$.

Next $X^* = [X^*, O^2(N_K(V_0))]$, so as $X \cap Q_H = E \leq U$ by 15.3.56.3, $X \leq K$. Thus as $\hat{T}_K = \hat{X}\hat{X}^t\hat{Q}_C$ and $\hat{Q}_C \leq \Phi(\hat{T}_K)$, $T_K = \langle X, X^t \rangle U$. Further $X \leq \langle U^Y \rangle = YU$, so $T_K \leq YU$. Then (2) holds as $|T_K| \geq 2^{10}$ and $|YU|_2 = 2^{10}$ by 15.3.54.4 since $s = 1$.

Since $F^*(YU) = O_2(YU)$, since $V = Z(P)$ by 15.3.55, and since Q_C centralizes T_K, $Q_C V = C_{YUQ_C}(P)$. Thus by Coprime Action, $Q_C V = Q_Y \times V$, where $Q_Y := C_{Q_C V}(Y)$. Then as T acts on Q_Y, and $\Omega_1(Z(T)) = Z \leq V$, $Q_Y = 1$, so $Q_C \leq Z$, establishing (1) and (3). Then $O_2(YT) = O_2(Y)$, so as $Y \trianglelefteq M$, $F^*(M) = O_2(M) = O_2(Y)$ using A.1.6. By 15.3.58.2, $s = 1$, so from 15.3.53.1 and 15.3.54.4, $Aut_Y(B) = O^2(N_{GL(B)}(Aut_Y(B)))$ for $B \in \{V, O_2(Y)/V\}$. Therefore $Y = O^2(M)$ by Coprime Action, so (4) holds. $\qquad\square$

Let $D_M \in Syl_3(C_M(V_1))$ and $D_H \in Syl_3(H)$; observe D_M and D_H both have order 3. Let $\langle v \rangle = Z_S \cap V_1$ and $Z = \langle z \rangle$. By 15.3.46.5, $C_G(v) \leq M$.

By 15.3.59.4, $M = YT$, and $s = 1$ by 15.3.58.2, so by 15.3.54.4, $C_M(D_M) = D_M \times J_M$, where $J_M \cong S_4$ and $V_1 = O_2(J_M)$. By construction, an involution $t \in J_M - V_1$ induces a transvection on V, and hence $t \notin UC_T(V)$.

Next a Sylow 2-group of $C_M(D_M)$ is dihedral of order 8 with center $\langle v \rangle$, and as $C_G(v) \leq M$, $|C_G(D_M)|_2 = 8$. On the other hand, from the structure of H described in 15.3.58 and 15.3.59, $|C_H(D_H)|_2 = 2^4$, so $D_M \notin D_H^G$. Thus as $D_H \in Syl_3(C_G(z))$, $t \notin z^G$. Summarizing:

LEMMA 15.3.60. *(1) $D_M \notin D_H^G$.*
(2) An involution t in $T \cap J_M - V_1$ is not in $UC_T(V)$, and $t \notin z^G$.

LEMMA 15.3.61. *(1) $t \notin v^G$.*
(2) All involutions in K are in z^G or v^G.

PROOF. As $\tilde{U} = \tilde{U}_1 \oplus \tilde{U}_1^t$ by 15.3.58.3, $m(\tilde{U}) = 2m([\tilde{U}, t])$, so $\tilde{U}$ is transitive on involutions in $t\tilde{U}$. Thus $O^2(C_{H^*}(t^*)) = O^2(C_H(t)^*)$, and hence t centralizes a conjugate of D_H. But by 15.3.46.5, $C_M(v) = C_G(v)$, so D_M is Sylow in $C_G(v)$ by construction. Thus (1) follows from 15.3.60.1.

From the action of H on U described in 15.3.58, H has two orbits on involutions in $U - Z$: $(U_1 - Z) \cup (U_1^t - Z) \subseteq z^G$ and v^H. Let $a \in V - U$ with U_a the preimage in U of $C_{\tilde{U}}(a)$. Then all involutions in $K - U$ are fused into aU_a under H, so it remains to show that each such involution is in $z^G \cup v^G$. Now $|U : U_a| = 4 = |\bar{U}|$ by 15.3.53, so $U_a = C_U(V) = C_U(U \cap V)$. Thus $U_a \cong E_4 \times D_8$, and all involutions in $U_a V$ are in the two E_{16} subgroups A_1 and A_2 of $U_a V$.

Next $VU_a \leq P$; let $P^+ := P/V$. From the description of I in 15.3.54, $U_a^+ = [P^+, U]$ is an isotropic line in the orthogonal space P^+ with one singular point, and I is transitive on singular and nonsingular points of P^+. Thus A_i^+, $i = 1, 2$, are the nonsingular points in U_a^+. Therefore there is D_i of order 3 in I centralizing A_i^+ and $[Z, D_i]$ is a singular line in the orthogonal space V, so $[D_i, Z] \leq V^\perp = U \cap V$. Let a_i generate $C_{A_i}(D_i)$. If $a_i \in U$, then each member of A_i is fused into U under D_i, so that (2) holds. Thus we may assume $a_i \notin U$. Here each member of $A_i - \langle a_i \rangle [a_i, P]$ is fused into U, and P is transitive on $a_i[a_i, P]$, so it remains to show the a_i is fused to z or v.

Let $B_i := C_P(D_i)V$. Then $B_i \cong E_{64}$ and I has four orbits on $B_i^{\#}$: z^I and v^I, and orbits of length 12 and 36 on $B_i - V$. Further $B_i \leq T_K$, and $E_i := B_i \cap U$ is of rank at most $m(U) = 4$, while $m(B^*) \leq m(H^*) = 2$, so B_i^* is a 4-group in T_K^* and $\tilde{E}_i = C_{\tilde{U}}(B_i^*)$. Then $B_i = C_{T_K}(E_i)$ is invariant under $N_H(E_i) = N_H(B_i^*) \cong S_4$, so as $N_H(B_i^*)$ does not act on V^*, $N_G(B_i)$ does not act on V. Thus as $v \notin z^G$, the two orbits of I on $V^{\#}$ are fused to its two orbits on $B_i - V$, so all involutions in B_i are fused to z or v, completing the proof of (2). □

We now eliminate the shadows of $Aut(L_5(2))$ and $Aut(He)$, and establish Theorem 15.3.50.

First the involution t of 15.3.60.2 is in $T - T_K$, since $T_K = UC_T(V)$ by 15.3.59.2. By 15.3.59.3, $|T : T_K| = 2$, so as G is simple, $t^G \cap T_K \neq \emptyset$ by Thompson Transfer. Thus $t \in z^G \cup v^G$ by 15.3.61.2. However this contradicts 15.3.60.2 and 15.3.61.1. This contradiction completes the proof of Theorem 15.3.50.

15.3.4. The case $\langle \mathbf{V^{M_c}} \rangle$ abelian. By Theorem 15.3.50, V_H is abelian for each $H \in \mathcal{H}(T, M)$. This will allow us to use weak closure in 15.3.63, and to verify Hypothesis F.9.8. Then Hypothesis F.9.8 eventually leads to a contradiction.

LEMMA 15.3.62. *(1) M_c is transitive on $\{V^g : g \in G$ and $Z \leq V^g\}$.*
(2) If $V \cap V^g \neq 1$, then $[V, V^g] = 1$.

PROOF. Part (1) folllows from 15.3.46.1 using A.1.7.1, since $M_c = C_G(Z)$ by 15.3.4. If $g \in G - M$ and $V \cap V^g \neq 1$, then as $V \cap V^g$ is totally singular by 15.3.46.2 and M is transitive on singular vectors, we may take $Z \leq V \cap V^g$. Therefore $V^g \leq V_{M_c} \leq C_G(V)$ by (1) since V_{M_c} is abelian by Theorem 15.3.50. □

LEMMA 15.3.63. *Assume $r(G, V) \geq 3$. Then*
(1) $W_1(T, V) \leq C_T(V)$, so $w(G, V) > 1$.
(2) $n(H) > 1$ for each $H \in \mathcal{H}(T, M)$.

PROOF. Assume $W_1(T, V)$ does not centralize V, and let A be a hyperplane of V^g with $A \leq T$ and $\bar{A} \neq 1$. In particular $V \nleq M^g$ by 15.3.46.4, so as $r(G, V) \geq 3$ by hypothesis, $m(V^g/C_{V^g}(V)) = m(\bar{A}) + 1 > 2$, and hence $m(\bar{A}) = 2 = m_2(\bar{M})$. As $\bar{M}$ is solvable, $a(\bar{M}, V) = 1$ by E.4.1, so there is a hyperplane B of A with $C_A(V) \leq B$ such that $1 \neq [C_V(B), A] =: V_B$. As $r(G, V) \geq 3$ and $m(V^g/B) = 2$, $C_V(B) \leq M^g$, so $1 \neq V_B \leq V \cap V^g$, contrary to 15.3.62.2. Thus $[V, W_1(T, V)] = 1$, establishing (1).

By A.5.7.2, $M = !\mathcal{M}(N_M(C_T(V)))$, while $r(G, V) > 1 < w(G, V)$ by our hypotheses and (1). Thus (2) follows from E.3.35.1. □

Recall that Hypothesis F.9.1 holds by 15.3.48.1 and 15.3.49.3.. Further 15.3.62.2 gives part (f) of Hypothesis F.9.8, while case (ii) of part (g) of Hypothesis F.9.8 holds by 15.3.47. Thus Hypothesis F.9.8 holds, so we conclude from F.9.16.3 that:

LEMMA 15.3.64. $q(H^*, \tilde{U}_H) \leq 2$.

LEMMA 15.3.65. *(1) If $H \in \mathcal{H}_*(T, M)$, then $n(H) = 1$.*
(2) $r(G, V) = 2$.

PROOF. By 15.3.46.3, $r(G, V) \geq 2$, so if (2) fails then $r(G, V) \geq 3$, and hence $n(H) > 1$ for $H \in \mathcal{H}(T, M)$ by 15.3.63.2. Thus (1) implies (2), so it remains to establish (1).

Assume $H \in \mathcal{H}_*(T, M)$ with $n(H) > 1$. Then in view of 15.3.2.6, H is described in E.2.2. In particular $K_0 := O^2(H) = \langle K^T \rangle$ for some $K \in \mathcal{C}(H)$, $K_0/O_2(K_0)$ is of Lie type over $\mathbf{F}_{2^n}$ for some $n > 1$, and setting $M_0 := M \cap K_0$, M_0 is a Borel subgroup of K_0. As M_H is a $3'$-group by 15.3.49.4, n is odd.

By A.1.42.2, we may pick $\tilde{I} \in Irr_+(K_0, \tilde{U}_H, T)$; set $\tilde{I}_T := \langle \tilde{I}^T \rangle$. We apply parts (4) and (5) of F.9.18 to the list of possibilities in E.2.2 defined over $\mathbf{F}_{2^n}$ with n odd. In view of 15.3.64, we may also appeal to Theorems B.4.2 and B.4.5; this determines the modules from the restrictions given in F.9.18. In particular as n is odd, there is no orthogonal module for $L_2(2^n)$. We conclude that one of the following holds:

(i) $K/O_2(K)$ is a Bender group, and $\tilde{I}/C_{\tilde{I}}(K)$ is the natural module for $K/O_2(K)$. Further either $K = K_0$ and $I = I_T$; or $K < K_0$, $K/O_2(K) \cong L_2(2^n)$ or $Sz(2^n)$, and for $t \in T - N_T(K)$, $I_T = I + I^t$ and $[I, K^t] = 0$.

(ii) $K/O_2(K) \cong SL_3(2^n)$ or $Sp_4(2^n)$, T is nontrivial on the Dynkin diagram of $K/O_2(K)$, and $\tilde{I}_T/C_{\tilde{I}_T}(K)$ is the sum of a natural module and its conjugate by an outer automorphism nontrivial on the diagram.

(iii) $K_0/O_2(K_0) \cong \Omega_4^+(2^n)$, and $\tilde{I}_T$ is the orthogonal module.

Now by Theorems B.5.1 and B.4.2, $K_0T/O_2(K_0T)$ has no FF-modules, except in (i) with $K/O_2(K) \cong L_2(2^n)$, where $K_0T/O_2(K_0T)$ has no strong FF-modules. We conclude from F.9.18.6 that either $I_T = [U_H, K_0]$, or case (i) holds with $K/O_2(K) \cong L_2(2^n)$, and $[U_H, K]/I$ is an extension of the natural module for $K/O_2(K)$ over a submodule centralized by K. (Recall that $n > 1$ is odd).

As T centralizes $\tilde{Z}_S$ and $H = K_0T$, $\tilde{U}_H = [\tilde{U}_H, K_0]C_{\tilde{U}_H}(H)$ by B.2.14. By 15.3.49.1, $O^2(M_0)$ centralizes V, and hence M_0 centralizes $\tilde{Z}_S$. It follows from the structure of the modules described in (i)–(iii), that H centralizes $\tilde{Z}_S$. But then K centralizes Z_S by Coprime Action, and so K centralizes $\tilde{U}_H$, contrary to $K^* \neq 1$. This contradiction completes the proof of 15.3.65. $\qquad\square$

As $r(G, V) = 2$ by 15.3.65.2, there is $E_4 \cong E \leq V$ with and $G_E := N_G(E) \not\leq M$. Further E is totally singular by 15.3.46.2. Pick E so that $T_E := N_T(E) \in Syl_2(M_E)$, where $M_E := N_M(E)$. Let $Y_E := O^2(N_Y(E))$, $Q_E := O_2(G_E)$, and $V_E := \langle V^{G_E} \rangle$.

LEMMA 15.3.66. *(1)* $\bar{T}_E = \bar{T} \cap \Omega_4^+(V)$.

(2) $|T : T_E| = 2$.

(3) $\bar{T}_E$ *is the 4-subgroup of* $\bar{T}$ *distinct from* $\bar{S}$.

(4) $Z \leq E$.

(5) $Y_ET_E/O_2(Y_ET_E) \cong S_3$, $V = [V, Y_E]$, $Q_E \leq C_G(E)$, *and* $O_2(Y_ET_E) = C_{Y_ET_E}(E)$.

(6) $G_E = Y_ET_EC_G(E)$.

(7) $C_G(E) \leq M_c$.

PROOF. As E is a totally singular line in V, $Aut_M(E) = GL(E)$, so that $Q_E \leq C_G(E)$ and (1) and (5) hold. Then (1) implies (2)–(4), and as Y_ET_E induces $GL(E)$ on E, (6) holds. Finally $C_G(E) \leq C_G(Z) = M_c$ by (4) and 15.3.4. $\qquad\square$

LEMMA 15.3.67. *(1)* $R := C_T(V) = C_G(V)$ *and* $M = YT$.

(2) $T_E \in Syl_2(G_E)$ *and* $B_E := Baum(T_E) \leq R$, *so that* $C(G, B_E) \leq M$.

(3) $G_E = Y_EX_ET_E$, *where* $X_E := O^2(C_G(E)) \not\leq M$, *with* $X_E/O_2(X_E) \cong Y_E/O_2(Y_E) \cong \mathbf{Z}_3$, *and* Y_E *and* X_E *are normal in* G_E.

(4) $G_E/Q_E \cong S_3 \times S_3$ and $X_E = [X_E, J(R)]$.

PROOF. By 15.3.5.2, if $A \in \mathcal{A}(T)$ with $A \not\leq C_T(V) = R$, then either $\bar{A} \leq \bar{M}_i$ or $\bar{A} = \bar{S}$. Therefore by 15.3.66.3, $J(T_E) = J(R)$, so that $B_E = \text{Baum}(R)$ by B.2.3.5. Hence $C(G, B_E) \leq M$ as $M = !\mathcal{M}(N_M(R))$ by A.5.7.2. In particular $N_G(T_E) \leq M$, so as $T_E \in Syl_2(M_E)$, $T_E \in Syl_2(G_E)$, and hence (2) holds.

Let $Q_c := O_2(M_c)$ and $P_c := Q_c \cap G_E$. By 15.3.49.2, $\bar{Q}_c \neq 1$, so as $\bar{Q}_c \trianglelefteq \bar{T}$ while $Z(\bar{T})$ is of order 2 and lies in $\bar{T}_E$ by 15.3.66.3, $Z(\bar{T}) \leq \bar{P}_c$ and $Y_E = [Y_E, P_c]$. As P_c is $(M_c \cap G_E)$-invariant, $P_c^{G_E} = P_c^{Y_E}$ since $G_E = Y_E T_E C_G(E)$ by 15.3.66.6; thus as $Y_E = [Y_E, P_c]$, $Y_E = O^2(\langle P_c^{G_E} \rangle) \trianglelefteq G_E$. So as $V = [V, Y_E]$ by 15.3.66.5, $V_E = [V_E, Y_E]$. Further $V^{G_E} = V^{Y_E T_E C_G(E)} = V^{C_G(E)} \subseteq V^{M_c}$, so that $V_E \leq V_{M_c}$. Thus V_E is abelian since V_{M_c} is abelian by Theorem 15.3.50.

Let $S_E := O_2(Y_E T_E) = C_{Y_E T_E}(E)$ and $C_E := C_G(E)$. Then $S_E = C_T(E) \in Syl_2(C_E)$ by (2). Let $\dot{G}_E := G_E/Q_E$. Then $\dot{G}_E = \dot{Y}_E \langle \dot{\tau} \rangle \times \dot{C}_E$, where $\tau \in P_c - S_E$ and $\dot{Y}_E \langle \dot{\tau} \rangle \cong S_3$. As $\tau \notin S_E$ and $E \cong E_4$, $C_E(\tau) = Z$. As $G_E = Y_E \langle \tau \rangle C_E$ and $Y_E \langle \tau \rangle$ acts on V, $V_E = \langle V^{G_E} \rangle = \langle V^{C_E} \rangle$.

Let $\check{G}_E := G_E/E$. Now $[V, S_E] = E$ from 15.3.66.1, so Q_E centralizes $\check{V}_E$ as $Q_E \leq S_E$. Then as we saw $\dot{G}_E = \dot{Y}_E \langle \dot{\tau} \rangle \times \dot{C}_E$ and $\dot{Y}_E \langle \dot{\tau} \rangle \cong S_3$,

$$\check{V}_E = \check{V}_{E,1} \oplus \check{V}_{E,2}$$

is a C_E-invariant decomposition, where $V_{E,1} := C_{V_E}(\tau)$, and $V_{E,2} := C_{V_E}(\tau^y)$ for $1 \neq \dot{y} \in \dot{Y}_E$. Thus $V_{E,1}^y = V_{E,2}$.

Let $I := J(C_E)$. By (2), $J(T_E) \leq C_T(V) \leq S_E$, so that $J(T_E) = J(S_E)$ by B.2.3.3. Then $G_E := I N_{G_E}(J(T_E)) = I M_E$ by a Frattini Argument and (2), so as $G_E \not\leq M$, we conclude $I \not\leq M$. Thus $\dot{I} \neq 1$.

Let $I_0 := N_I(C_T(\check{V}_E))$. In order to determine the structure of I_0, temporarily replacing G_E by $N_{G_E}(C_T(\check{V}_E))$ if necessary, we may assume that $Q_E = C_T(\check{V}_E)$. We will drop this assumption later, once we have determined I_0, and then complete the proof of (1) and (3).

Let $\hat{G}_E := G_E/C_{G_E}(V_E)$. Now $\langle \tau, Q_E \rangle \leq T_E$, so $\Phi(\langle \bar{\tau}, \bar{Q}_E \rangle) \leq \Phi(\bar{T}_E) = 1$, and hence $\Phi(\langle \tau, Q_E \rangle \leq C_G(V)$. We saw earlier that $\dot{\tau}$ centralizes $\dot{C}_E$, so C_E acts on $\langle \tau, Q_E \rangle$. Then as $V_E = \langle V^{C_E} \rangle$, we conclude that $\Phi(\langle \tau, Q_E \rangle) \leq C_G(V_E)$, and hence $\Phi(\langle \hat{\tau}, \hat{Q}_E \rangle) = 1$. Therefore as Q_E centralizes $\check{V}_E$, $\hat{Q}_E$ induces a group of transvections on $V_{E,1}$ with center $C_E(\tau) = Z$. Next $[Y_E, Q_E] \leq O_2(Y_E) = C_{Y_E}(V)$, so as $[Y_E, Q_E] \trianglelefteq G_E$, $[Y_E, Q_E] \leq C_{G_E}(V_E)$, and hence $[\hat{Y}_E, \hat{Q}_E] = 1$. Then as $Y_{E,1}^y = Y_{E,2}$, $C_{Q_E}(V_{E_1}) = C_{Q_E}(V_E)$. Hence as Q_E induces a group of transvections on $V_{E,1}$ with center Z, we conclude $m(V_E/C_{V_E}(\hat{W})) = 2m(\hat{W})$ for each $\hat{W} \leq \hat{Q}_E$.

As $\dot{I} \neq 1$, there is $A \in \mathcal{A}(T_E)$ with $\dot{A} \neq 1$. Let $B := A \cap Q_E$ and $D := C_A(V_E)$. Then since $C_{V_E}(A) = A \cap V_E$ as $A \in \mathcal{A}(T_E)$,

$$m(\dot{A}) + m(\hat{B}) + m(D) = m(A) \geq m(DV_E)$$

$$\geq m(D) + m(V_E/(A \cap V_E)) = m(D) + m(V_E/C_{V_E}(A))$$

so that

$$m(\dot{A}) + m(\hat{B}) \geq m(V_E/C_{V_E}(A)).$$

Further using an earlier remark with $\hat{B}$ in the role of "$\hat{W}$",

$$m(V_E/C_{V_E}(A)) = m(V_E/C_{V_E}(B)) + N = 2m(\hat{B}) + N,$$

where $N := m(C_{V_E}(B)/C_{V_E}(A))$. Therefore $m(\dot{A}) \geq m(\hat{B}) + N$ and hence

$$2m(\dot{A}) \geq 2m(\hat{B}) + 2N = m(V_E/C_{V_E}(A)) + N.$$

Further

$$m(V_E/C_{V_E}(A)) \geq m(\check{V}_E/C_{\check{V}_E}(A)) = 2m(\check{V}_{E,1}/C_{\check{V}_{E,1}}(A)),$$

so $m(\dot{A}) \geq m(\check{V}_{E,1}/C_{\check{V}_{E,1}}(A)) + N/2$ with $N \geq 0$, and hence $\dot{A}$ is an FF*-offender for the FF-module $\check{V}_{E,1}$.

Next $m_3(G_E) \leq 2$ since G_E is an SQTK-group, and $\dot{G}_E \cong S_3 \times \dot{C}_E$, so $m_3(C_E) \leq 1$. Therefore by Theorem B.5.1, $\dot{I} \cong L_2(2^n)$, $L_3(2^m)$, m odd, or S_5, with $\check{V}_{E,1}/C_{\check{V}_{E,1}}(I)$ the natural module or the sum of two natural modules for $L_3(2^m)$. As $G_E = IM_E$, $V_E = \langle V^I \rangle$, and as $\check{V} \leq Z(\check{S}_E)$, $\check{V}_E = [\check{V}_E, I]\check{V}$ and $C_I(\check{V}) = C_I(\check{V}_\#)$, where $\check{V}_\# \neq 1$ is the projection of $\check{V}$ on $\check{V}_{E,1}$ in the decomposition of $\check{V}_E$. Also $\check{V}_\# \leq C_{\check{V}_{E,1}}(\check{S}_E)$, and by (2), $N_{\dot{I}}(J(S_E)) \leq \dot{C}_E \cap \dot{M}_E \leq C_{\dot{G}_E}(\check{V})$, so $N_{\dot{I}}(J(S_E)) \leq C_{\dot{I}}(\check{V}_\#)$. Using the structure of $J(S_E)$ from Theorem B.4.2 we conclude that $\dot{I} \cong S_3$, S_5, or $L_3(2)$. But in the last two cases, as $\check{V}_\# \leq Z(\check{S}_E)$, $O^{3'}(C_I(\check{V})) \neq 1$, contradicting 15.3.49.5.

Therefore $\dot{I}_0 \cong S_3$ and $m([\check{V}_{E,1}, I]) = 2$. At this point, we drop the temporary assumption that $Q_E = C_T(\check{V}_E)$.

By a Frattini Argument, $I = I_0 C_I(\check{V}_E)$, while $C_I(\check{V}_E) \leq N_I(V) \leq M_E$. Thus as $G_E = M_E I$, $G_E = M_E I_0$, so $|G_E : M_E| = 3$. Therefore $O^{2,3}(G_E) = O^{2,3}(C_G(V))$ is normal in M and G_E, so $O^{2,3}(G_E) = O^{2,3}(C_M(V)) = 1$ as $G_E \not\leq M \in \mathcal{M}$. Then as $C_M(V)$ is a $3'$-group by 15.3.49.4, (1) holds. Hence $M_E = Y_E T_E$, so as $|G_E : M_E| = 3$, (3) holds. Further $X_E = [X_E, J(R)]$ since $J(T_E) = J(R)$ by (2), so $G_E/Q_E \cong S_3 \times S_3$ since $Y_E T_E/O_2(Y_E T_E) \cong S_3$ and $R \leq O_2(M_E)$. This completes the proof of 15.3.67. $\square$

Next Z is contained in exactly two totally singular 4-subgroups E and $F := E^s$ of V, where $s \in S - T_E$. Observe $T_E = T_E^s = T_F$ acts on $Y_F := Y_E^s$ with $T_E \in Syl_2(Y_F T_E)$, and $Y = Y_E Y_F O_2(Y)$. Let $G_1 := Y_F T_E$, $G_2 := X_E T_E$, and $G_0 := \langle G_1, G_2 \rangle$. Set $L_i := O^2(G_i)$ and $Q_i := O_2(G_i)$ for $i = 1, 2$. Thus $G_i/Q_i \cong S_3$ and $T_E = G_1 \cap G_2 \in Syl_2(G_i)$.

LEMMA 15.3.68. *(1) $G_0 \leq N_G(Y_E)$.*
(2) $V \leq Z(Q_0)$, where $Q_0 := O_2(G_0)$.
(3) $T_E \in Syl_2(M_0)$ for each $M_0 \in \mathcal{M}(G_0)$.
(4) (G_0, G_1, G_2) is a Goldschmidt triple.
(5) $Q_0 = O_{3'}(G_0)$.

PROOF. By construction, $Y_E \trianglelefteq YT_E$, so G_1 acts on Y_E. By 15.3.67.3, G_2 acts on Y_E. Thus $G_0 = \langle G_1, G_2 \rangle$ acts on Y_E, establishing (1). By 15.3.66.5, $V = [V, Y_E]$, so $V \leq O_2(Y_E)$ and hence $V \leq Q_0$ by (1). Set $R := C_T(V)$ as in 15.3.67.1. Then $Q_0 \leq Q_1 \cap Q_2 = R = C_{T_E}(V)$, so $V \leq Z(Q_0)$. Hence (2) holds. Next let $M_0 \in \mathcal{M}(G_0)$. As $N_G(T_E) \leq M$ by 15.3.67.2, if $T_E \not\in Syl_2(M_0)$ then we may take $T \leq M_0$. But then $YT = \langle Y_F, T \rangle \leq M_0$, so $X_E \leq M_0 = M = !\mathcal{M}(YT)$ by 15.3.7, contrary to 15.3.67.3. Hence (3) holds and $T_E \in Syl_2(G_0)$, so (4) holds.

Let $P := O_{3'}(G_0))$. By F.6.11.1, P is 2-closed with $T_E \cap P = Q_0$, so P is solvable. By (2), $V \leq O_2(P)$, so $O(P) \leq C_G(V)$, and hence $O(P) = 1$ as $C_G(V) = R$ by 15.3.67.1. Thus $F^*(P) = Q_0$. Let $X := J(T_E)P$, $T_0 := T_E \cap X$, and $Z_0 := R_2(X)$. Then $T_0 \in Syl_2(X)$ as Q_0 is Sylow in P, and $F^*(X) = O_2(X)$.

By 15.3.67.2, $J(T_E) \leq R$, so that $J(T_E) \trianglelefteq Y_F T_E$, and so Y_F acts on X. As X is a $3'$-group with $F^*(X) = O_2(X)$, $X = C_X(Z_0) N_X(J(T_E))$ by Solvable Thompson Factorization B.2.16. As Y_F acts on X, $F = \langle Z^{Y_F} \rangle \leq Z_0$ by B.2.14, so $P = C_P(F) N_P(J(T_E))$. But by 15.3.67, $C_G(F)$ and $M = YT$ are $\{2,3\}$-groups, so $P = Q_0$ as $N_G(J(T_E)) \leq M$ by 15.3.67.2. Thus (5) holds, completing the proof of 15.3.68. $\qquad \square$

We are now in a position to complete the proof of Theorem 15.3.1.

Let $Q_0 := O_2(G_0)$ and $\dot{G}_0 := G_0/Q_0$. By 15.3.68.3 and F.6.5.1, $(\dot{G}_1, \dot{T}_E, \dot{G}_2)$ is a Goldschmidt amalgam. Since $G_1 \cap G_2 = T_E$, and $O_{3'}(G_0) = Q_0 \leq T_E$ by 15.3.68.5 case (i) of F.6.11.2 holds, so $\dot{G}_0$ is described in Theorem F.6.18.

Let $V_0 := \langle V^{G_0} \rangle$. By 15.3.68.2, $V_0 \leq \Omega_1(Z(Q_0))$. Also $C_{G_0}(V_0) \leq C_{G_0}(V) = R$ is a 2-group by 15.3.67.1, so $Q_0 = C_{G_0}(V_0)$. By 15.3.67.4, $X_E = [X_E, J(T_E)] \leq J(G_0) =: X$, so V_0 is an FF-module for $\dot{G}_0$. Thus examining the list of Theorem F.6.18 for groups appearing in Theorem B.5.6, and recalling that $J(T_E) \trianglelefteq G_1$ by 15.3.67.2, we conclude that $\dot{X} \cong S_3$, $L_3(2)$, A_6, S_6, A_7, S_7, $\hat{A}_6$, or $G_2(2)$.

Assume first that $\dot{X} \cong S_3$. Then $X_E = O^2(X)$, so $Z \leq C_{G_0}(X)$, and hence $F = \langle Z^{Y_F} \rangle \leq C_{G_0}(X)$. But then X_E acts on $EF = Z^\perp$, so $X_E \leq M$ by 15.3.46.2, contrary to 15.3.67.3.

In the remaining cases, $O^2(X) = O^2(G_0)$ by Theorem F.6.18, so $Y_F \leq X$. However $Q_0 O_2(Y_F) \leq C_{T_E}(V) = R$, so $O_2(\dot{Y}_F)$ centralizes V, while $[V, Q_1] = F$, so $\dot{Q}_1 > O_2(\dot{Y}_F)$. This eliminates the cases $\dot{G}_0 \cong L_3(2)$, A_6, A_7, or $\hat{A}_6$, so that $\dot{G}_0$ is S_6, $\hat{S}_6$, S_7, or $G_2(2)$. As $V = [V, Y_F] \leq [V_0, X]$, $V_0 = [V_0, X]$. Thus $O_2(\dot{Y}_F)$ centralizes the 4-dimensional subspace V of the FF-module $V_0 = [V_0, X]$ for $\dot{X}$, so we conclude using Theorem B.5.1 that $\dot{G}_0$ is $\hat{S}_6$ and $m(V_0) = 6$. But now $N_{\dot{X}}(V_1)$ has a quotient A_5, whereas $N_G(V_1) \leq M$ by 15.3.45.2, and M is solvable by 15.3.67.1.

This contradiction completes the proof of Theorem 15.3.1.

15.4. Completing the proof of the Main Theorem

In this section, we complete the treatment of the case $\mathcal{L}_f(G, T)$ empty, and hence also the proof of the Main Theorem. Our efforts so far have in effect reduced us to the case $\mathcal{L}(G, T)$ empty (cf. 15.4.2.1 below).

More precisely, since we have been assuming that $|\mathcal{M}(T)| > 1$, and since Theorem 15.3.1 completed the treatment of groups satisfying Hypothesis 14.1.5, we may assume that condition (2) of Hypothesis 14.1.5 fails. Thus in this section, we assume instead:

HYPOTHESIS 15.4.1. G is a simple QTKE-group, $T \in Syl_2(G)$, and

(1) $\mathcal{L}_f(G, T) = \emptyset$.
(2) Let $Z := \Omega_1(Z(T))$. Then $|\mathcal{M}(C_G(Z))| > 1$.

The section culminates in Theorem 15.4.24, where we see that $L_3(2)$ and A_6 are the only groups which satisfy Hypothesis 15.4.1.

We now define a collection of subgroups similar to the set $\Xi(G, T)$ of chapter 1: Let $\xi(G, T)$ consist of those T-invariant subgroups $X = O^2(X)$ of G such that $XT \in \mathcal{H}(T)$ and $|X : O_2(X)|$ is an odd prime. Let $\xi^*(G, T)$ consist of those $X \in \xi(G, T)$ such that $\exists! \mathcal{M}(XT)$.

Because of Hypothesis 15.4.1, members of $\xi(G,T)$ have few overgroups, so that $\xi^*(G,T)$ is nonempty in many situations—cf. 15.4.3.1 and 15.4.12.

15.4.1. Preliminary results, and the reduction to $\mathbf{C_G(Z)} = \mathbf{T}$. Recall $\Xi_+(G,T)$ is defined before 3.2.13.

LEMMA 15.4.2. *(1) $\mathcal{L}(G,T) = \emptyset$.*

(2) Each member of $\mathcal{M}(T)$ is solvable.

(3) If $X \in \Xi(G,T)$ then $[Z,X] \neq 1$ and $X \in \Xi_f^*(G,T)$ but $X \notin \Xi_+(G,T)$. In particular $X/O_2(X)$ is a 3-group or a 5-group.

(4) Assume $M_0 \in \mathcal{H}(T)$ with $M = !\mathcal{M}(M_0)$, and $V \in \mathcal{R}_2(M_0)$ with $R := C_T(V) = O_2(C_{M_0}(V))$ and $V \leq O_2(M)$. Then $\hat{q}(M_0/C_{M_0}(V), V) \leq 2$.

PROOF. Assume $\mathcal{L}(G,T) \neq \emptyset$. Then there is $L \in \mathcal{L}^*(G,T)$. Setting $L_0 := \langle L^T \rangle$, $N_G(L_0) = !\mathcal{M}(\langle L,T \rangle)$ by 1.2.7.3. But by Hypothesis 15.4.1.1, $\mathcal{L}_f(G,T) = \emptyset$, so $\langle L,T \rangle \leq C_G(Z)$ and hence $N_G(L_0) = !\mathcal{M}(C_G(Z))$, contrary to Hypothesis 15.4.1.2. Thus (1) holds, and since $\mathcal{C}(M) \subseteq \mathcal{L}(G,T)$ for $M \in \mathcal{M}(T)$, (1) and 1.2.1.1 imply (2).

Supppose $X \in \Xi(G,T)$. By (1), $X \in \Xi^*(G,T)$, so by 1.3.7, $N_G(X) = !\mathcal{M}(XT)$. Thus $[Z,X] \neq 1$ by Hypothesis 15.4.1.2, so $X \in \Xi_f^*(G,T)$. Hence $X \notin \Xi_+(G,T)$ by 3.2.13, completing the proof of (3).

Assume the hypotheses of (4), and let $\hat{q} := \hat{q}(M_0/C_{M_0}(V), V)$. Pick $H \in \mathcal{H}_*(T,M)$, and let $Q_H := O_2(H)$. Observe that Hypothesis D.1.1 is satisfied with M_0, H in the roles of "G_1, G_2": First, by hypothesis $M = !\mathcal{M}(M_0)$, so that $O_2(\langle M_0,H \rangle) = 1$, and hence part (3) of Hypothesis D.1.1 holds. Second, $C_T(V) = O_2(C_{M_0}(V))$, with $O_2(C_{M_0}(V)) = O_2(M_0)$ since $V \in \mathcal{R}_2(M_0)$, so that part (2) of D.1.1 holds. Finally by 3.1.3.1, $H \cap M$ is the unique maximal subgroup of H containing T, so that part (1) of D.1.1 holds. Now recall by B.5.13 that if the dual V^* is an FF-module for $M_0/C_{M_0}(V)$, then $q(M_0/C_{M_0}(V), V) \leq 2$. Thus combining conclusions (2), (3), and (4) of the qrc-lemma D.1.5 into case (ii) below, one of the following holds:

(i) $V \not\leq Q_H$.

(ii) $q(M_0/C_{M_0}(V), V) \leq 2$.

(iii) $V \leq R \cap Q_H \trianglelefteq H$, the dual V^* is not an FF-module for $M_0/C_{M_0}(V)$, $U := \langle V^H \rangle$ is abelian, and H has a unique noncentral chief factor on U.

If (ii) holds, then the conclusion of (4) holds and we are done.

Assume that (i) holds. We verify Hypothesis E.2.8 with $H \cap M$ in the role of "M": As $V \not\leq Q_H$, $T \not\leq Q_H$, so $H \not\leq N_G(T)$; hence by 3.1.3.2, H is a minimal parabolic in the sense of Definition B.6.1, and so is described in B.6.8. Therefore by B.6.8.5, $\ker_{H \cap M}(H)$ is 2-closed with Sylow group Q_H, so $V \not\leq \ker_{H \cap M}(H)$. Finally $V \leq O_2(M)$ by hypothesis, so that $V \leq O_2(H \cap M)$. Now $\hat{q}(Aut_H(V), V) \leq 2$ by E.2.13.2, and again (4) holds.

This leaves case (iii), so as H is solvable by (2), $R \in Syl_2(O^2(H)R)$ by D.1.4.4.2. But now $O_2(\langle M_0,H \rangle) \neq 1$ by Theorem 3.1.1, contrary to an earlier observation. This completes the proof of (4), and hence of 15.4.2. $\qquad\square$

Following the notational convention of chapter 1, set $\xi_f(G,T) := \xi(G,T) \cap \mathcal{X}_f$.

LEMMA 15.4.3. *Let $X \in \xi(G,T)$, with $|X : O_2(X)| = p$ where p is chosen maximal among such X, and suppose $p > 3$. Then*

(1) $\exists! \mathcal{M}(XT)$, so that $X \in \xi^*(G, T)$.
(2) $X \leq O_{2,p}(M)$ for $M \in \mathcal{M}(XT)$. In particular, $O_2(X) \leq O_2(M)$.
(3) $[Z, X] \neq 1$, so $X \in \xi_f(G, T)$.
(4) $p = 5$.

PROOF. Let $M \in \mathcal{M}(XT)$, and set $M^* := M/O_2(M)$. If $q \neq p$ is an odd prime such that $[O_q(M^*), X^*] \neq 1$, then $Aut_X(O_q(M^*))$ is embedded in $GL_2(q)$ by A.1.25.2, so p divides $q - \epsilon$ for $\epsilon := \pm 1$. This is impossible as $p \geq q$ by maximality of p, with p and q odd. Therefore $X^* \leq C_{M^*}(O^p(F(M^*))) =: H^*$, and as M is solvable by 15.4.2.2, $F^*(H^*) = O_p(H^*) \times O^p(Z(H^*))$. As $p > 3$, the solvable subgroup $Aut_H(O_p(H^*))$ of $GL_2(p)$ is p-closed using Dickson's Theorem A.1.3, so that $X^* = O^{p'}(X^*) \leq O_p(H^*) \leq O_p(M^*)$, establishing (2) for each $M \in \mathcal{M}(XT)$.

Let $N_G(X) \leq M_X \in \mathcal{M}$. We will show that $M = M_X$; then since M is an arbitrary member of $\mathcal{M}(XT)$, (1) will be established. Let $K := O_{2,F}(M)$ and $Y := O^2(N_K(X))$; by (2), $X \leq Y \leq M \cap M_X =: I$. Let X_0 be the preimage in M of X^*; as X is T-invariant, $X = O^2(X_0)$, so $N_M(X^*) = N_M(X)$. Thus $Y^* = O^2(N_{K^*}(X^*))$. Next $C_{K^*}(Y^*) \leq C_{K^*}(X^*) \leq Y^*$ as K^* is of odd order. Thus the hypotheses of case (b) of A.4.4 are satisfied with M, M_X, $YO_2(M)$, in the roles of "H, K, X". Therefore $R := O_2(I) = O_2(M)$ by A.4.4.1, and $C(M_X, R) \leq I$ by A.4.4.2, so Hypothesis C.2.3 is satisfied by M_X, I in the roles of "H, M_H". Further C.2.6.2 says that either $O_{2,F}(M_X) \leq I$, or M_X has an A_3-block L with $L \not\leq I$. In the first case $M = M_X$ by A.4.4.3, as desired. In the second $[L, Y] \leq O_2(L)$, so taking Y_Z to be the preimage in M of $Z(O_p(M^*)) \leq Y^*$ and $Y_0 := O^2(Y_Z)$, L normalizes $O^2(Y_0 O_2(L)) = Y_0$. Then $L \leq N_G(Y_0) = M$ as $M \in \mathcal{M}$, contrary to $L \not\leq I$. This contradiction completes the proof of (1).

If $[Z, X] = 1$ then $XT \leq C_G(Z)$ and $M = !\mathcal{M}(XT)$ by (1), so that also $M = !\mathcal{M}(C_G(Z))$, contrary to Hypothesis 15.4.1.2. Thus (3) holds.

Next $V := \langle Z^X \rangle \in \mathcal{R}_2(XT)$ and $V \leq O_2(M)$ by B.2.14, and as $[Z, X] \neq 1$ and $X/O_2(X)$ has prime order, $C_{XT}(V) = O_2(C_{XT}(V)) = C_T(V)$. Then by (1) we may apply 15.4.2.4 with XT in the role of "M_0", to conclude $\hat{q}(XT/C_{XT}(V), V) \leq 2$. Then as $p > 3$ by hypothesis, D.2.13.1 shows $p = 5$, so (4) holds. $\qquad\square$

LEMMA 15.4.4. Each member of $\mathcal{M}(T)$ is a $\{2, 3, 5\}$-group.

PROOF. Suppose some $M \in \mathcal{M}(T)$ has order divisible by $p > 5$, and choose p maximal subject to this constraint. As M is solvable by 15.4.2.2, there is a Hall $\{2, p\}$-subgroup H of M containing T. Let P denote a Sylow p-subgroup of the preimage in H of $\Omega_1(Z(H/O_2(H)))$; then $TP \in \mathcal{H}(T)$ with P elementary abelian, and $m_p(P) \leq 2$ since H is an SQTK-group. If $m_p(P) = 2$ and T is irreducible on P, then $H \in \Xi(G, T)$, contrary to 15.4.2.3. Thus there is $X \leq TP$ with $X \in \xi(G, T)$, contrary to 15.4.3.4. $\qquad\square$

LEMMA 15.4.5. $C_G(Z)$ is a $\{2, 3\}$-group.

PROOF. If not, arguing as in the proof of 15.4.4, there is a nontrivial elementary abelian 5-subgroup P of $C_G(Z)$ with $PT \in \mathcal{H}(T)$, and there is $X \leq PT$ with $X \in \Xi(G, T)$ or $\xi(G, T)$. Since $X \leq C_G(Z)$, the former is impossible by 15.4.2.3, and the latter by 15.4.3.3. $\qquad\square$

Recall from 14.1.4 that for $V(M) := \langle Z^M \rangle$:

LEMMA 15.4.6. *If M is maximal in $\mathcal{M}(T)$ with respect to $\overset{\scriptstyle <}{\sim}$ and $[V(M), J(T)] = 1$, then M is the unique maximal member of $\mathcal{M}(T)$ under $\overset{\scriptstyle <}{\sim}$.*

The two groups which satisfy Hypothesis 15.4.1 appear in conclusion (2) of the next result. The second subsection completes the treatment of Hypothesis 15.4.1, by eliminating the case where $|Z| > 2$, and the case where $|Z| = 2$ but conclusion (1) of the result fails.

LEMMA 15.4.7. *Assume Z is of order 2. Then either*

(1) There exists a nontrivial characteristic subgroup $C_2 := C_2(T)$ of $Baum(T)$ such that

$$M_f = !\mathcal{M}(N_G(C_2)),$$

and M_f is the unique maximal member of $\mathcal{M}(T)$ under $\overset{\scriptstyle <}{\sim}$, or

(2) $G \cong L_3(2)$ or A_6.

PROOF. Let $S := Baum(T)$ and choose $C_i := C_i(T)$ for $i = 1, 2$ as in the Glauberman-Niles/Campbell Theorem C.1.18. Thus $1 \neq C_2$ char S, and $1 \neq C_1 \leq Z$, so as Z is of order 2 by hypothesis, $C_1 = Z$.

Assume (2) fails; we claim that:

(*) For each $M \in \mathcal{M}(T)$, $M = N_M(C_2)C_M(V(M))$.

Assume (*) fails and set $V := V(M)$. If $[V, J(T)] = 1$, then $S = Baum(C_T(V))$ by B.2.3.5, so (*) holds by a Frattini Argument, contrary to our assumption. Thus $[V, J(T)] \neq 1$.

By 15.4.2.2, M is solvable, so by Solvable Thompson Factorization B.2.16, $\overline{J(M)} = \bar{Y} = \bar{Y}_1 \times \cdots \times \bar{Y}_r$ with $\bar{Y}_i \cong S_3$ and $V = V_1 \times \cdots \times V_r \times C_V(J(M))$, where $V_i := [V, Y_i] \cong E_4$, and Y and Y_i denote the preimages in M of $\bar{Y}$ and $\bar{Y}_i$. As M is an SQTK-group, $r \leq 2$ by A.1.31.1. As $|Z| = 2$, $C_V(J(M)) = 1$ and T is transitive on $\{Y_1, \ldots, Y_r\}$. Thus if $r = 1$, then $m(V) = 2$ and $\bar{M} = \bar{Y} = GL(V)$, while if $r = 2$ then $m(V) = 4$ and $\bar{M}$ is the normalizer $O_4^+(V)$ of $\bar{Y}$ in $GL(V)$. In either case as $C_1 = Z$, $C_M(C_1) = C_M(Z) = C_M(V)T$. Thus as (*) fails, $M > N_M(C_2)C_M(V) = N_M(C_2)C_M(C_1) = \langle N_M(C_2), C_M(C_1) \rangle$. Therefore as M is solvable, we conclude from C.1.28 that there is an A_3-block $A_4 \cong X \unlhd \unlhd M$ such that $X = [X, J(T)]$.

Let $X_0 := \langle X^M \rangle$. By the previous paragraph, either $r = 1$ and $X_0 = X$, or $r = 2$ and $X_0 = X_1 \times X_2$ with $X = X_1$ and $X_2 = X^t$ for suitable $t \in T - N_M(X)$. Let $H \in \mathcal{M}(X_0T)$. As H is solvable by 15.4.2.2, applying C.1.27 to H, X in the roles of "G, K", we conclude that $X_0 \unlhd H$, so $H = N_G(X_0)$ as $H \in \mathcal{M}$. Thus $H = !\mathcal{M}(X_0T)$, so $M = H = N_G(X_0) = !\mathcal{M}(X_0T)$ as $X_0T \leq M \in \mathcal{M}$. Next $X_0T/C_T(X_0) \cong S_4$ or S_4 wr $\mathbf{Z}_2$. As $|Z| = 2$, $Z \leq O_2(X_0)$, so $C_T(X_0) = 1$. Then as $M \in \mathcal{H}^e$, $C_M(X_0) = 1$, so $M = X_0T \cong S_4$ or S_4 wr $\mathbf{Z}_2$. In the second case, Theorem 13.9.1 supplies a contradiction, so suppose the first case holds. Then $T \cong D_8$, so as $F^*(C_G(Z)) = O_2(C_G(Z))$ by 1.1.3.2, we conclude $T = C_G(Z)$. Further Thompson Transfer shows that each noncentral involution of T is fused into $Z(T)$, so that G has one conjugacy class of involutions. Thus (2) holds by I.4.1.2, contrary to our assumption; this completes the proof of the claim (*).

Pick $M_f \in \mathcal{M}(N_G(C_2))$. By (*), for each $M \in \mathcal{M}(T)$, $M \overset{\scriptstyle <}{\sim} M_f$. Thus M_f is the unique maximal member of $\mathcal{M}(T)$ under $\overset{\scriptstyle <}{\sim}$—in particular M_f is uniquely determined since $\overset{\scriptstyle <}{\sim}$ is a partial order (cf. A.5.5, and in particular A.5.4). Thus (1) holds. $\square$

We come to the main result of this subsection:

THEOREM 15.4.8. $C_G(Z) = T$.

The proof of Theorem 15.4.8 involves a short series of reductions. Until the proof is complete, assume G is a counterexample. Let $\mathcal{X}$ consist of those $X \in \xi(G, T)$ such that $X \trianglelefteq C_G(Z)$. Recall by 15.4.5 that $X/O_2(X) \cong \mathbf{Z}_3$.

LEMMA 15.4.9. $\mathcal{X} \neq \emptyset$.

PROOF. Set $H := C_G(Z)$ and $\hat{H} := H/O_2(H)$. By 15.4.5, H is a $\{2,3\}$-group, so as G is a counterexample to Theorem 15.4.8, $O_3(\hat{H}) \neq 1$. Let $\hat{P} := \Omega_1(Z(O_3(\hat{H})))$. If $\hat{P}$ is of order 3, then $X := O^2(P) \in \mathcal{X}$, so we may assume that $|\hat{P}| > 3$. Then as $m_3(H) \leq 2$, $E_9 \cong \hat{P} = \Omega_1(O_3(\hat{H}))$, so that $C_{\hat{H}}(\hat{P}) = O_3(\hat{H})$ by Coprime Action, and $\hat{H}/O_3(\hat{H})$ is a subgroup of $GL_2(3)$. As H centralizes Z, $PT \notin \Xi(G, T)$ by 15.4.2.3, so $\hat{T}$ is not irreducible on $\hat{P}$. Therefore there is a normal subgroup $\hat{P}_1$ of $\hat{H}$ of order 3, and so $X := O^2(P_1) \in \mathcal{X}$. $\qquad \square$

LEMMA 15.4.10. For each $X \in \mathcal{X}$ and each $M \in \mathcal{M}(XT)$, $X \leq C_M(V(M))$.

PROOF. Assume X, M is a counterexample, and let $V := V(M)$ and $\bar{M} := M/C_M(V)$. In particular $\bar{X} \neq 1$. If $O_2(\bar{X}) = 1$, then $V = [V, X] \oplus C_V(X)$ by Coprime Action, and $Z \cap [V, X] \neq 1$ as X is T-invariant, contrary to $X \leq C_G(Z)$. Therefore $O_2(\bar{X}) \neq 1$, so $O_2(X) \not\leq O_2(M)$.

Let $M^* := M/O_2(M)$. Thus $1 \neq O_2(X)^* \leq O_2(X^*)$. We claim next that $O_2(X^*)$ centralizes $O_5(M^*)$, so suppose not. Then by A.1.25, $O_2(X^*)$ acts non-trivially on a supercritical subgroup P^* of $O_5(M^*)$, $P^* \cong \mathbf{Z}_5$, E_{25} or 5^{1+2}, and $Aut_X(P^*)$ is a subgroup of $Aut(P^*)/O_5(Aut(P^*)) \cong GL_2(5)$. As $O_2(X^*)$ does not centralize P^* and $X^* = O^2(X^*)$, we conclude that P^* is not of order 5 and $Aut_{X^*}(P^*) \cong \mathbf{Z}_3/Q_8$. Thus $O_2(X^*)$ is irreducible on $P^*/\Phi(P^*)$, and so the preimage P contains a member of $\Xi_+(G, T)$, contrary to 15.4.2.3. This establishes the claim.

As M is a solvable $\{2,3,5\}$-group by 15.4.4, $F^*(M^*) = O_3(M^*)O_5(M^*)$, so $O_2(X^*)$ is faithful on $O_3(M^*)$ by the claim. Again by A.1.25, $O_2(X^*)$ acts nontrivially on a supercritical subgroup P^* of $O_3(M^*)$, $P^* \cong E_9$ or 3^{1+2}, and $O_2(X^*) \cong Q_8$ is irreducible on $P^*/\Phi(P^*)$. Let $Y := O^2(P)$, so that $Y \in \Xi(G, T)$ and $Aut_X(P^*) \cong SL_2(3)$. If $P^* \cong 3^{1+2}$, then as $Aut_{X^*}(P^*) \cong SL_2(3)$, $m_3(XP) = 3$, contrary to M an SQTK-group; thus $P^* \cong E_9$.

Let $H := YXT$, $W := \langle Z^H \rangle$, and $H^+ := H/C_H(W)$; then $W = \langle Z^H \rangle \in \mathcal{R}_2(H)$ and $W \leq O_2(M)$ by B.2.14. As $[Z, Y] \neq 1$ by 15.4.2.3, and $O_2(X^*)$ is irreducible on P^*, $C_Y(W) = O_2(Y)$. Therefore H^+ is the split extension of $P^+ \cong E_9$ by either $SL_2(3)$ or $GL_2(3)$, so W contains an 8-dimensional faithful irreducible H-submodule I. Thus $\hat{q}(H^+, W) \geq \hat{q}(H^+, I) > 2$.

By 15.4.2.3, $Y \in \Xi^*(G, T)$, so that $N := N_G(Y) = !\mathcal{M}(YT)$ by 1.3.7, and as $YT \leq M$, $M = N$. Of course $YT \leq H$, so $M = !\mathcal{M}(H)$. Further $O_2(C_H(W)) = C_T(W)$ since $C_Y(V) = O_2(Y)$. Thus we may apply 15.4.2.4 to conclude $\hat{q}(H^+, W) \leq 2$, contrary to the previous paragraph. $\qquad \square$

We are now ready to complete the proof of Theorem 15.4.8. By Hypothesis 15.4.1.2, there exist distinct members M_1 and M_2 of $\mathcal{M}(C_G(Z))$. By 15.4.9, there is $X \in \mathcal{X}$. Now X is not normal in both M_1 and M_2, so we may assume X is not normal in M_1. Let $Y_1 := \langle X^{M_1} \rangle$, and set $M_i^i := M_i/O_2(M_i)$ for $i = 1, 2$. By

15.4.10, $X \leq C_{M_i}(V(M_i)) =: H_i$, so as $H_i \leq C_G(Z)$ and $X \trianglelefteq C_G(Z)$, $X \trianglelefteq H_i$. Thus $O_2(X) \leq O_2(H_i)$, so X^i is of order 3 for each i, and $Y_1^1 \cong E_{3^e}$ for some $e \geq 1$. Notice $e \leq m_3(M_1) \leq 2$, so as X is not normal in M_1, $e = 2$. Now $T \leq C_G(Z) \leq N_{M_1}(X)$, so as $Aut_{M_1}(Y_1^1) \leq GL_2(3)$ and $Y_1 = \langle X^{M_1} \rangle$, it follows that $O^2(Aut_{M_1}(Y_1^1)) \cong \mathbf{Z}_3$. Therefore $Y_1 = XX_1$, where $X_1 := O^2(X_0)$ and X_0 is the preimage in Y_1 of $C_{Y_1^1}(P^1)$ for $P \in Syl_3(M_1)$. Thus $X_1 \in \mathcal{X}$, so by 15.4.10, $X_1 \leq H_2$. As $H_2 \leq C_G(Z)$, X^2 and X_1^2 are normal in H_2^2. Therefore $O^2(H_2^2) \leq C_{H_2^2}(Y_1^2)$, so as $m_3(H_2) \leq 2$, $Y_1^2 = \Omega_1(O_3(H_2^2))$. Hence Y_1 is normal in M_2, and Y_1 is normal in M_1 by definition, contrary to the simplicity of G. This contradiction completes the proof of Theorem 15.4.8.

In the remainder of the subsection, we collect some useful consequences of Theorem 15.4.8.

LEMMA 15.4.11. *For each $M \in \mathcal{M}(T)$:*

(1) $O_2(M) = C_M(V(M))$.

(2) M is maximal with respect to $\stackrel{<}{\sim}$. In particular, there is no unique maximal member of $\mathcal{M}(T)$ under $\stackrel{<}{\sim}$.

(3) $[V(M), J(T)] \neq 1$.

PROOF. First $C_M(V(M)) \leq C_G(Z) = T$ by Theorem 15.4.8, so as $V(M)$ is 2-reduced, (1) holds. Now if $M \stackrel{<}{\sim} M_1 \in \mathcal{M}(T)$, then

$$M = C_M(V(M))(M \cap M_1) = O_2(M)(M \cap M_1) \leq M_1$$

by (1), so $M = M_1$. Thus M is maximal in $\mathcal{M}(T)$ under $\stackrel{<}{\sim}$, so since $|\mathcal{M}(T)| > 1$ by Hypothesis 15.4.1.2, (2) holds. Then (3) follows from (2) and 15.4.6. $\square$

Define $\mathcal{Y}$ to consist of those groups Y in $\Xi(G, T) \cup \xi(G, T)$ such that $Y = [Y, J(T)]$; we show $\mathcal{Y}$ is nonempty in the next lemma. Set $S := \text{Baum}(T)$ and $E := \Omega_1(Z(J(T)))$.

LEMMA 15.4.12. *Let $M \in \mathcal{M}(T)$. Then $\mathcal{Y} \cap M \neq \emptyset$. Further for each $Y \in \mathcal{Y} \cap M$:*

(1) $Y \trianglelefteq M$.

(2) $M = N_G(Y) = !\mathcal{M}(YT)$.

(3) For suitable $s(Y) = 1$ or 2, $Y/O_2(Y) \cong E_{3^{s(Y)}}$ and $m([V(M), Y]) = 2s(Y)$.

(4) $S \in Syl_2(YS)$.

(5) $R_2(YT) = [V(M), Y] \oplus E_Y$, where $E_Y := C_{\Omega_1(Z(O_2(YT)))}(Y)$ and $E = E_Y \oplus C_{[V(M),Y]}(S)$.

(6) Either

    (i) $s(Y) = 1$, $YT/O_2(YT) \cong S_3$, and $|E : E_Y| = 2$, or

    (ii) $s(Y) = 2$, $YT/O_2(YT) \cong O_4^+(2)$, $Y = Y_1 Y_2$ with $Y_i/O_2(Y_i) \cong \mathbf{Z}_3$, $[V(M), Y] = V_1 \oplus V_2$, where $V_i := [V(M), Y_i] \cong E_4$ for $i = 1, 2$, $Y_i = [Y_i, J(T)]$ is S-invariant, and $|E : E_Y| = 4$.

PROOF. Set $\bar{M} := M/C_M(V(M))$. By 15.4.11.3, $[V(M), J(T)] \neq 1$, so as M is solvable, we conclude from Solvable Thompson Factorization B.2.16 and A.1.31.1 that $\bar{X} := [O_3(\bar{M}), J(T)]$ and its action on $V(M)$ are described in (3). Let X be the preimage in M of $\bar{X}$ and $Y_0 := O^2(X)$.

By 15.4.11.1, $O_2(Y_0) = C_{Y_0}(V(M))$. If T acts irreducibly on $\bar{X}$, then Y_0 lies in $\Xi(G, T)$ or $\xi(G, T)$, and hence also in $\mathcal{Y} \cap M$, and Y_0 satisfies (1) and (3). Otherwise, $\bar{X} = \bar{X}_1 \times \bar{X}_2$ where $\bar{X}_i$ is T-invariant; setting $Y_i := O^2(X_i)$ for X_i the preimage in M of $\bar{X}_i$, Y_i lies in $\xi(G, T)$ and hence also in $\mathcal{Y} \cap M$, and Y_i satisfies (1) and (3). In particular $\mathcal{Y} \cap M \neq \emptyset$. By E.2.3, Y_i satisfies (4)–(6) for $i = 0$ or $i = 1, 2$ in our two cases.

Now consider any $Y \in \mathcal{Y} \cap M$. By 15.4.11.1, $O_2(Y) = C_Y(V(M))$. Since $Y = [Y, J(T)]$, $\bar{Y} \leq [O_3(\bar{M}), J(T)]$ by Solvable Thompson Factorization B.2.16. Hence as $Y = O^2(Y)$ is T-invariant, either $Y = Y_0$, or $Y = Y_1$ or Y_2, in our two cases. In particular $Y \trianglelefteq M$, so $M = N_G(Y)$ as $M \in \mathcal{M}$; similarly Y is normal in each member of $\mathcal{M}(YT)$, so (2) holds.   □

LEMMA 15.4.13. *For $Y \in \mathcal{Y}$, YT is not isomorphic to $\mathbf{Z}_2 \times S_4$.*

PROOF. Assume otherwise. Then $Z \cong E_4$ and $T \cong \mathbf{Z}_2 \times D_8$ with $\Phi(T)$ of order 2, so $O^2(Aut(T))$ centralizes Z. Thus as $N_G(T)$ controls fusion in Z by Burnside's Fusion Lemma A.1.35, the three involutions in Z are in distinct G-conjugacy classes. Pick an involution $t \in T - O_2(YT)$, and let $R := \langle t \rangle O_2(Y)$. Then $R \cong D_8$ has two YT-classes of involutions t^R and z^Y for $1 \neq z \in Z \cap R = \Phi(T)$. As the three involutions in Z are in distinct G-classes, at most one of the two involutions in $Z - R$ can be G-conjugate to t, so that some $i \in Z^\#$ satisfies $i \notin t^G \cup z^G$. Thus $i^G \cap R = \emptyset$, so by Thompson Transfer, $i \notin O^2(G)$, contrary to the simplicity of G.   □

15.4.2. The final contradiction. *In this subsection, we assume that G is not $L_3(2)$ or A_6.*

LEMMA 15.4.14. *(1) For each $Y \in \mathcal{Y}$, $C_Z(Y)$ is a hyperplane of Z.*
(2) $m(Z) = 2$. Thus $C_Z(Y) \neq 1$.

PROOF. Part (1) follows from 15.4.12.6. By the hypothesis of this subsection, conclusion (2) of 15.4.7 does not hold, and by 15.4.11.2, conclusion (1) of 15.4.7 does not hold. Thus $m(Z) > 1$ by 15.4.7. Indeed $|M(T)| > 1$ by Hypothesis 15.4.1.2, so by 15.4.12 there exist distinct $Y, X \in \mathcal{Y}$ with $!\mathcal{M}(XT) \neq !\mathcal{M}(YT)$, and hence $C_Z(Y) \cap C_Z(X) = 1$. Thus $m(Z) \leq 2$ by (1), so (2) holds.   □

LEMMA 15.4.15. *There exists at most one $M \in \mathcal{M}(T)$ such that $s(Y) = 1$ for some $Y \in \mathcal{Y} \cap M$.*

PROOF. Assume $M_i \in \mathcal{M}(T)$, $i = 1, 2$, are distinct, with $Y_i \in \mathcal{Y} \cap M_i$ such that $s(Y_i) = 1$. Let $G_i := Y_i T$ for $i = 1, 2$, and $G_0 := \langle G_1, G_2 \rangle$; then (G_0, G_1, G_2) is a Goldschmidt triple. By 15.4.12.2, $O_2(G_0) = 1$, so by F.6.5.1, $\alpha := (G_1, T, G_2)$ is a Goldschmidt amalgam, and hence α is described in F.6.5.2. As $G_i \in \mathcal{H}(T)$, $G_i \in \mathcal{H}^e$, so α appears in case (vi) of F.6.5.2—namely in one of cases (1), (2), (3), (8), (12), or (13) of F.1.12. By 15.4.14, $Z \not\leq Z(G_i)$ for $i = 1$ and 2, and $m(Z) = 2$. Thus by inspection of the possibilities for α, case (2) of F.1.12 holds; that is $G_1 \cong G_2 \cong \mathbf{Z}_2 \times S_4$. However this contradicts 15.4.13.   □

Recall that $S = \mathrm{Baum}(T)$.

LEMMA 15.4.16. *$N_G(S) \leq M$ for each $M \in \mathcal{M}(T)$.*

PROOF. Pick $Y \in \mathcal{Y} \cap M$, and apply Theorem 3.1.1 to S, $N_G(S)$, YT in the roles of "R, M_0, H". Note that by 15.4.12.4, $S \in Syl_2(YS)$, and by 15.4.12.6, YS is a minimal parabolic in the sense of Definition B.6.1, so the hypotheses of 3.1.1 are indeed satisfied. Therefore by Theorem 3.1.1, $O_2(\langle N_G(S), YT \rangle) \neq 1$, so $N_G(S) \leq M$ by 15.4.12.2. $\qquad\qquad\square$

LEMMA 15.4.17. *Assume* $BC_G(B) \leq H \leq N_G(B)$ *for some* $1 \neq B \leq T$ *such that* $T_H := T \cap H \in Syl_2(H)$. *Assume* $T_H \leq I \leq H$, *and for* $z \in Z^{\#}$, *let* $M_z \in \mathcal{M}(C_G(z))$. *Then*

(1) *The hypotheses of 1.1.5 are satisfied with* I, M_z, z *in the roles of "H, M, z".*

(2) *If* $Z \cap O_2(I) \neq 1$, *then* $F^*(I) = O_2(I)$.

PROOF. As $T_H \in Syl_2(H)$, and $Z \leq C_T(B) \leq T_H$ by hypothesis, 1.1.6 says that the hypotheses of 1.1.5 are satisfied with H, M_z, z in the roles of "H, M, z". Then as T_H is Sylow in H and I, $O_2(H \cap C_G(z)) \leq O_2(I \cap C_G(z))$ by A.1.6, so that (1) holds. Assume $Z \cap O_2(I) \neq 1$. Then $N := N_G(O_2(I)) \in \mathcal{H}^e$ by 1.1.4.3, so as $BC_N(B) \leq H \cap N \leq N_N(B)$, $H \cap N \in \mathcal{H}^e$ by 1.1.3.2. Hence as $T_H \leq I \leq H \cap N$, and T_H is Sylow in H, we conclude $I \in \mathcal{H}^e$ from 1.1.4.4. $\qquad\square$

LEMMA 15.4.18. $s(Y) = 2$ *for each* $Y \in \mathcal{Y}$.

PROOF. Assume $Y \in \mathcal{Y}$ with $s(Y) = 1$, and let $M_1 := N_G(Y)$; then $M_1 = !\mathcal{M}(YT)$ by 15.4.12.2. Pick $M_2 \in \mathcal{M}(T) - \{M_1\}$; by 15.4.12, we may choose $X \in \mathcal{Y} \cap M_2$, and again $M_2 = N_G(X) = !\mathcal{M}(XT)$. By 15.4.15, $s(X) = 2$, so by 15.4.12, $X = Y_2 Y_2^t$ where $Y_2 = O^2(Y_2) = [Y_2, J(T)]$ is S-invariant with $Y_2/O_2(Y_2) \cong \mathbf{Z}_3$ and $t \in T - N_T(Y_2)$. Set $T_I := N_T(Y_2)$, $Y_1 := Y$, $G_i := Y_i T_I$, and $I := \langle G_1, G_2 \rangle$. By 15.4.12.4, $S \in Syl_2(Y_i S)$, so as $S \leq T_I$, $T_I \in Syl_2(G_i)$. Notice $|T : T_I| = 2$.

As $S \in Syl_2(Y_i S)$, $E = \Omega_1(Z(J(T))) = E_i \times F_i$, where $E_i := C_E(Y_i)$, and $F_i := [E, Y_i] \cap E$ is of order 2. In particular E_i is a hyperplane of E. Similarly $E = E_X \times F_X$, where $E_X := C_E(X)$ and $F_X := [E, X] \cap E \cong E_4$. As T acts on $E_X \cap E_1$, if $E_X \cap E_1 \neq 1$, then $C_Z(X) \cap C_Z(Y) \neq 1$, then $M_1 = M_2$ by 15.4.12.2, contrary to the choice of M_2. Thus $E_X \cap E_1 = 1$, so as E_1 is a hyperplane of E, $m(E_X) \leq 1$. By 15.4.14, $1 \neq C_Z(X) \leq E_X$, so $C_Z(X) = E_X$ is of rank 1, and $m(E) = 3$. Thus as E_i is a hyperplane of E, $E_1 \cap E_2 =: E_0 \neq 1$; and G_i centralizes E_0, so $I \leq C_G(E_0)$. In particular, $IE_0 \in \mathcal{H}$, so that I is an SQTK-group.

Next $S = \text{Baum}(T) \leq T_I$, so that $S = \text{Baum}(T_I)$ by B.2.3.5. Thus $N_G(T_I) \leq N_G(S) \leq M_i$ by 15.4.16, so as $T_I = N_T(Y_2) \in Syl_2(C_{M_2}(E_0))$, we conclude that T_I is Sylow in $G_E := C_G(E_0)$, and hence also $T_I \in Syl_2(I)$. Thus (I, G_1, G_2) is a Goldschmidt triple. Let $I^* := I/O_{3'}(I)$. As $T \leq M_1$, $Y_2 \not\leq M_1$ since $M_2 = !\mathcal{M}(XT)$. Then as $M_1 = !\mathcal{M}(YT)$ and $O_2(G_1) \trianglelefteq YT$, we conclude $O_2(G_1) \neq O_2(G_2)$. Thus $\alpha := (G_1^*, T_I^*, G_2^*)$ is a Goldschmidt amalgam by F.6.11.2, so α and I^* are described in Theorem F.6.18.

As $Z \leq T_I \in Syl_2(G_E)$, we may apply 15.4.17 with I, G_E, E_0 in the roles of "I, H, B". We conclude that for $z \in Z^{\#}$ and $M_z \in \mathcal{M}(C_G(z))$, the hypotheses of 1.1.5 are satisfied with I, M_z, z in the roles of "H, M, z", and $F^*(I) = O_2(I)$ if $Z \cap O_2(I) \neq 1$. By 1.1.5.1, $F^*(I \cap M_z) = O_2(I \cap M_z)$, so by 1.1.3.2, $F^*(C_I(z)) = O_2(C_I(z))$. As $[Z, Y_i] \leq C_I(O(I))$ by A.1.26.1, and $1 \neq Z \cap [Z, Y_1]$, $O(I) = 1$ by 1.1.5.2.

Suppose first that $F^*(I) \neq O_2(I)$. Then there is a component L of I, and Z is faithful on L by 1.1.5.3. Now L is described in one of cases (3)–(13) of F.6.18, so as

$F^*(C_I(z)) = O_2(C_I(z))$ for each $z \in Z^\#$, and $m(Z) = 2$ by 15.4.14.2, we conclude $L \cong A_6$ and $L^*Z^* \cong S_6$. In particular $Y_i \le O^{3'}(I) = L$ for $i = 1, 2$, so $L = O^2(I)$ using F.6.6. Now T acts on $C_{T_I}(Y) = C_{T_I}(L) \times (Z \cap L)$, but T acts on no nontrivial subgroup of $C_{T_I}(L)$ as $C_Z(X) \cap C_Z(Y) = 1$. Therefore as $|T : T_I| = 2$, $C_{T_I}(L)$ is of order 2, and hence $C_{T_I}(L) = E_0$, so $G_i \cong E_4 \times S_4$. Thus $Y_2 \cong A_4$, so $X \cong A_4 \times A_4$. Then as $C_Z(X) = E_X$ is of order 2, $m_2(T_I) \ge 5$, contrary to $G_i \cong E_4 \times S_4$.

Therefore $F^*(I) = O_2(I)$. Let $V_I := \langle Z^I \rangle$, so that $V_I \in \mathcal{R}_2(I)$ by B.2.14. But $C_I(V_I) \le C_I(Z) = T_I$ using Theorem 15.4.8, so $C_I(V_I) = O_2(I)$. Let $\hat{I} := I/O_2(I)$. Now $Y_i = [Y_i, J(T)]$ as $Y_i \in \mathcal{Y}$, and $[Z, Y_i] \ne 1$ by 15.4.12.6, so V_I is an FF-module for $\hat{I}$. Then using Theorem B.5.6 to determine the FF-modules for the possible groups in Theorem F.6.18, it follows as $C_I(Z) = T_I$, that $\hat{I} \cong S_3 \times S_3$. But then Y_2 normalizes $O^2(Y_1 O_2(I)) = Y_1$, so that $Y_2 \le N_G(Y) = M_1$, contrary to an earlier remark. $\square$

We now define notation in force for the remainder of the subsection. By Hypothesis 15.4.1.2, we can pick distinct members M_1 and M_2 of $\mathcal{M}(T)$, and by 15.4.12, we can choose $X \in \mathcal{Y} \cap M_1$ and $Y \in \mathcal{Y} \cap M_2$. Thus $M_1 = N_G(X) = !\mathcal{M}(XT)$ and $M_2 = N_G(Y) = !\mathcal{M}(YT)$ by 15.4.12.2. Further $s(X) = s(Y) = 2$ by 15.4.18, so that $Y = Y_1 Y_2$ and $X = X_1 X_2$ as in 15.4.12.6. Let $T_0 := N_T(Y_1) \cap N_T(X_1)$. By 15.4.12.4, S is Sylow in XS and YS, so as $S \le T_0$ by 15.4.12.6, T_0 is Sylow in XT_0 and YT_0. Let $L_1 := X_1$ or X_2, and $L_2 := Y_1$ or Y_2. Set $G_i := L_i T_0$, and $I := \langle G_1, G_2 \rangle$. Let $V_i := [V(M_i), L_i]$, so that $V_i \cong E_4$ by 15.4.12.6. Observe $|T : T_0| \le 4$ since $|T : N_T(Y_i)| = 2 = |T : N_T(X_i)|$.

LEMMA 15.4.19. *(1)* $1 \ne C_E(I) \le Z(I)$. *In particular* $I \in \mathcal{H}$.
(2) $L_{3-i} \not\le M_i$.
(3) $Z \cap Z(I) V_1 \ne 1$.

PROOF. If $L_2 \le M_1$ then $YT = \langle L_2, T \rangle \le M_1$, contrary to $M_2 = !\mathcal{M}(YT)$ and the choice of $M_1 \ne M_2$. Thus (2) holds. Similarly $Z \cap Z(I) = 1$.

Let $E_I := C_E(T_0)$. Arguing as in the second paragraph of the proof of 15.4.18, $E_I = E_i \times F_i$, where $E_i := C_E(G_i)$ and $F_i := C_{V_i}(T_0) \cong \mathbf{Z}_2$. Thus $E_0 := C_E(I)$ is of corank at most 2 in E_I. As $C_Z(X) \ne 1$ by 15.4.14, and $C_{E \cap [Z, X]}(T_0) \cong E_4$ by 15.4.12.6, $m(E_I) \ge 3$, so (1) holds. Further as $Z \cap Z(I) = 1$ and $m(Z) = 2$ by 15.4.14.2, $E_I = E_0 \times Z$, so $1 \ne Z \cap E_0 F_1 \le Z(I) V_1$, and hence (3) holds. $\square$

By 15.4.19, $I \in \mathcal{H}$, so that $\mathcal{H}(I)$ is nonempty.

LEMMA 15.4.20. $T_0 \in Syl_2(I_0)$ *for each* $I_0 \in \mathcal{H}(I)$.

PROOF. Assume otherwise, and let $T_0 < T_I \in Syl_2(I_0)$, and $T_1 := T_I \cap M_1 \cap M_2$. As $S \le T_0 \le T_1$, $S = \mathrm{Baum}(T_1)$ by B.2.3.4, and hence $N_{T_I}(T_1) \le N_{T_I}(S) \le T_I \cap M_1 \cap M_2 = T_1$ by 15.4.16. Thus $T_I = T_1$, so we may take $T_I \le T$. Of course $T_I < T$, as otherwise I contains XT and YT, contrary to $!\mathcal{M}(XT) = M_1 \ne M_2 = !\mathcal{M}(YT)$. Therefore $|T : T_I| = 2$ since $|T : T_0| \le 4$. Also $T_I \in Syl_2(J)$ for any $J \in \mathcal{H}(I_0)$, and in particular, $T_0 \in Syl_2(N_G(O_2(I_0)))$.

As $T_I > T_0$, T_I does not normalize at least one of L_1 or L_2, so we may assume T_I does not normalize L_1. Then $X = \langle L_1^{T_I} \rangle \le I \le I_0$ and $R := O_2(XT_I) \trianglelefteq XT$, so as $M_1 = !\mathcal{M}(XT)$,

$$C(G, R) \le M_1.$$

Set $K := \langle X^{I_0} \rangle$ and recall $K \in \Xi(G, T)$. As $M_1 = N_G(X)$, X is not normal in I_0 since $I_0 \not\leq M_1$ by 15.4.19.2.

Observe that either:

(i) $K \in \mathcal{C}(I_0)$ with $KT_I/O_2(KT_I) \cong Aut(L_n(2))$, $n = 4$ or 5, or

(ii) $K = K_1 K_1^r$ for some $K_1 \in \mathcal{C}(I_0)$ and $r \in T_I - N_{T_I}(K_1)$, with $K_1/O_2(K_1) \cong L_2(2^n)$, n even, or $L_2(p)$ for some odd prime p.

This follows from 1.3.4, since $K/O_2(K)$ is not $L_3(3)$, M_{11}, or $Sp_4(2^n)$ because T_I induces D_8 on $X/O_2(X)$ since $XS/O_2(XS) \cong S_3 \times S_3$ by 15.4.12.6, while T_I does not act on L_1. Also $K = O^{3'}(I_0)$ by A.3.18 or 1.2.2, so $I \leq KT_I$, and hence without loss $I_0 = KT_I$.

Suppose first that $F^*(K) \neq O_2(K)$, so that K is a product of components of I_0. By an earlier remark, T_0 is also Sylow in $N_G(O_2(I_0))$, so we may apply 15.4.17 with I_0, $N_G(O_2(I_0))$, $O_2(I_0)$ in the roles of "I, H, B" to conclude that for $z \in Z^\#$ and $M_z \in \mathcal{M}(C_G(z))$, the hypotheses of 1.1.5 are satisfied with I_0, M_z, z in the roles of "H, M, z". Thus K is described in 1.1.5.3, and Z is faithful on K. Suppose first that case (ii) holds. As Z is noncyclic and in the center of T_I, while T_I induces D_8 on $X/O_2(X)$, K_1 is not $L_2(p)$ for p odd. Thus $K_1 \cong L_2(2^n)$, so as $L_2(4) \cong L_2(5)$ and n is even, $n \geq 4$. Further as $C(G, R) \leq M_1$, a Borel subgroup B of K is contained in M_1, and hence $B = O^2(B)$ acts on the 4-subgroup V_1 of 15.4.12.6; this is impossible, as when $n \geq 4$, B does not act on a 4-subgroup of $O_2(B)$. Thus (i) holds, in which case we again have a contradiction to Z 2-central, noncyclic, and faithful on K.

Therefore $F^*(K) = O_2(K)$. Let $I_X := I_0 \cap M_1$. Recall $C(G, R) \leq M_1$, so that Hypothesis C.2.3 is satisfied with I_X, I_0 in the roles of "M_H, H". If case (ii) holds, then as R centralizes $X/O_2(X)$, R normalizes K_1, so it follows from C.2.7.3 that either K_1 is a block of type $L_2(2^n)$ or A_5, or $K_1/O_2(K_1) \cong L_3(2)$.

Let $V_I := \langle Z^K \rangle$ so that $V_I \in \mathcal{R}_2(I_0)$ by B.2.14, and set $I_0^* := I_0/O_2(I_0)$. As $K/O_2(K)$ is semisimple, $O_2(I_0) = C_{I_0}(V_I)$. As $L_i = [L_i, J(T)]$, V_I is an FF-module for I_0^*. Then as $C_{I_0}(Z)$ is a 2-group by Theorem 15.4.8, it follows from the description of the modules in C.2.7.3 and C.1.34 for the groups in cases (i) and (ii), that K_1 is an $L_2(2^n)$-block. But once again a Borel subgroup B of K is contained in M_1, and hence $B = O^2(B)$ acts on the 4-subgroup V_1 of case (ii) of 15.4.12.6, so we conclude that K_1 is an $L_2(4)$-block. Finally by 15.4.14, $C_Z(X) \neq 1$, so $C_Z(X) \leq Z(I_0)$ from the structure of K. Thus $I_0 \leq C_G(C_Z(X)) \leq M_1 = !\mathcal{M}(XT)$, contrary to 15.4.19.2. $\qquad\square$

Recall $I \in \mathcal{H}$ by 15.4.19.1, so $T_0 \in Syl_2(I)$ by 15.4.20. Then (I, G_1, G_2) is a Goldschmidt triple. Set $Q_i := O_2(G_i)$ and $I^+ := I/O_{3'}(I)$.

LEMMA 15.4.21. *If $F^*(I) = O_2(I)$, then $I = L_1 L_2 T_0$ with $L_i \trianglelefteq I$.*

PROOF. Assume $F^*(I) = O_2(I)$ and let $V_I := \langle Z^I \rangle$. As $C_I(V_I) \leq C_I(Z) = T_0$ by Theorem 15.4.8, and $V_I \in \mathcal{R}_2(I)$ by B.2.14, we conclude that $C_I(V_I) = O_2(I)$. Let $I^* := I/O_2(I)$, so that I^+ is a quotient of I^*. As $L_i = [L_i, J(T)]$, V_I is an FF-module for I^* and L_i centralizes $O^3(F(I^*))$ by B.1.9.

We claim $\alpha := (G_1^+, T_0^+, G_2^+)$ is a Goldschmidt amalgam. For if not, by F.6.11.2, $Q_1 = Q_2$ and $I^+ \cong S_3$ with $O_{3'}(I)^* \neq 1$. Then $Q_1 = O_2(I)$ and I is solvable by F.6.11.1, so as L_i^* centralizes $O^3(F(I^*))$, I is a $\{2, 3\}$-group by F.6.9, contradicting $O_{3'}(I^*) \neq 1$. Thus the claim is established.

By the claim, I^+ is described in Theorem F.6.18. Therefore as V_I is an FF-module for I^*, either the lemma holds, or comparing the list of Theorem F.6.18 with that of Theorem B.5.1, we conclude that I^* is an extension of $L_3(2)$, A_6, A_7, $\hat{A}_6$, or $G_2(2)$, so that $C_I(Z)$ is not a 2-group. The latter case contradicts Theorem 15.4.8. $\qquad\square$

LEMMA 15.4.22. *If $F^*(I) \neq O_2(I)$ then $I = KT_0$, $K \cong A_6$, and $I/C_T(K) \cong S_6$.*

PROOF. Assume $F^*(I) \neq O_2(I)$. By 15.4.20, $T_0 \in Syl_2(N_G(O_2(I)))$, so we may apply 15.4.17 to conclude that the hypotheses of 1.1.5 are satisfied and $Z \cap O_2(I) = 1$. Since $V_1 = [V_1, L_1] \cong E_4$, V_1 centralizes $O(I)$ by A.1.26.1, so $1 \neq Z \cap V_1 Z(I)$ centralizes $O(I)$ by 15.4.19.3. Thus $O(I) = 1$ by 1.1.5.2.

If $Q_1 = Q_2$ then $Q_1 = O_2(I)$; but $Z \leq Q_1$ by B.2.14, contradicting $Z \cap O_2(I) = 1$. Thus $Q_1 \neq Q_2$, so (G_1^+, T_0^+, G_2^+) is a Goldschmidt amalgam by F.6.11.2, and I^+ is described in Theorem F.6.18. By F.6.11.1, $O_{3'}(I)$ is 2-closed, so as $Z \cap O_2(I) = 1$, $Z \cap O_{3'}(I) = 1$ and hence $Z \cong Z^+$ is noncyclic; then we conclude from Theorem F.6.18 that I^+ is either $L_2(p^2)$ extended by a field automorphism, or S_7. However $F^*(I \cap M_z) = O_2(I \cap M_z)$ for each $z \in Z^\#$ by 1.1.5.1, so that $F^*(C_I(C_Z(W))) = O_2(C_I(C_Z(W)))$ for $W := X, Y$ by 1.1.3.2. We conclude that $I^+ \cong S_6$. As $O_{3'}(I)$ is 2-closed and $F^*(I) \neq O_2(I)$, it follows that I has a component K with $K/O_2(K) \cong A_6$ and then that $K = O^{3'}(I)$ by A.3.18. Thus $K = O^2(I)$ by F.6.6, so $I = KT_0$. As $E_4 \cong Z$ is faithful on K, $Z(K) = 1$, so the lemma holds. $\qquad\square$

LEMMA 15.4.23. $F^*(I) \neq O_2(I)$.

PROOF. Assume $F^*(I) = O_2(I)$. Then by 15.4.21, $[L_2, L_1] \leq O_2(L_1)$. We may choose notation so that $L_1 := X_1$, and set $L_1' := X_2$ and $I' := \langle L_1' T_0, L_2 \rangle$. As $[L_1, L_1'] \leq O_2(L_1)$ and $[L_1, L_2] \leq O_2(L_1)$, we conclude $[O^2(I'), L_1] \leq O_2(L_1)$ from F.6.6. However by 15.4.21 and 15.4.22, I' contains an E_9-subgroup P, with $P \cap L_1 = 1$, since $I' \not\leq M_1$ by 15.4.19.2. Therefore as $[O^2(I'), L_1] \leq O_2(L_1)$, $m_3(L_1 P) = 3$, contrary to $N_G(I)$ an SQTK-group. $\qquad\square$

We are now ready to establish the main result of this section. By 15.4.23, we may apply 15.4.22, to conclude that $L_i \cong A_4$, $O_2(L_1)O_2(L_2) \cong D_8$, and $O_2(L_1) \cap O_2(L_2) \neq 1$. We may choose $L_1 := X_1$, and set $L_1' := X_2$ and $I' := \langle L_1', L_2 \rangle$. By symmetry, $O_2(L_1') \cap O_2(L_2) \neq 1$, so as $O_2(L_1) \cap O_2(L_1') = 1$,

$$O_2(L_2) = (O_2(L_1) \cap O_2(L_2))(O_2(L_1') \cap O_2(L_2)) \leq O_2(X) \cong E_{16}.$$

This is impossible as $O_2(L_1)O_2(L_2) \cong D_8$ and $O_2(L_1) \leq O_2(X)$.

Since we assume in this subsection that G is not $L_3(2)$ or A_6, this contradiction establishes:

THEOREM 15.4.24. *Assume Hypothesis 15.4.1. Then $G \cong L_3(2)$ or A_6.*

Then combining the main results of this chapter:

THEOREM 15.4.25 (Theorem E). *Assume G is a simple QTKE-group, $T \in Syl_2(G)$, $|\mathcal{M}(T)| > 1$, and $\mathcal{L}_f(G,T) = \emptyset$. Then G is isomorphic to J_2, J_3, $^3D_4(2)$, the Tits group $^2F_4(2)'$, $G_2(2)'$, M_{12}, $L_3(2)$, or A_6.*

PROOF. If Hypothesis 15.4.1 holds, the groups in Theorem 15.4.24 appear in the list of Theorem E. On the other hand if Hypothesis 15.4.1 fails, then there is a unique member M_c of $\mathcal{M}(C_G(Z))$, so that Hypothesis 14.1.5 holds, and the groups in Theorem 15.3.1 appear as conclusions in Theorem E. $\qquad\square$

Although by this point it may feel like something of an anticlimax, we have also completed the proof of the Main Theorem: For suppose G is a simple QTKE-group, with $T \in Syl_2(G)$. If $|\mathcal{M}(T)| = 1$, the groups appearing as conclusions in Theorem 2.1.1 of chapter 2 appear as conclusions in the Main Theorem. So assume that $|\mathcal{M}(T)| > 1$. If $\mathcal{L}_f(G, T) = \emptyset$, then the groups in the conclusion of Theorem E are among those in the conclusion of the Main Theorem. Finally if $\mathcal{L}_f(G, T) \neq \emptyset$, then the groups in the conclusion of Theorem D (14.8.2) appear as conclusions in the Main Theorem.

Thus, after a hiatus of roughly twenty years, there is at last a classification of the quasithin groups of even characteristic. In particular, this result fills that gap in the literature classifying the finite simple groups.

Part 7

The Even Type Theorem

CHAPTER 16

Quasithin groups of even type but not even characteristic

The original proof of the classification of the finite simple groups (CFSG) requires the classification of simple QTK-groups G of characteristic 2-type. (Recall G is of *characteristic 2-type* if $F^*(M) = O_2(M)$ for all 2-local subgroups M of G.) Mason produced a preprint [**Mas**] which goes a long way toward such a classification, but that preprint is incomplete. Our Main Theorem fills this gap in the "first generation" proof of CFSG, since we determine all simple groups in the larger class of QTK-groups of even characteristic. (Recall G is of *even characteristic* if $F^*(M) = O_2(M)$ only for those 2-locals M containing a Sylow 2-subgroup T of G.)

The "revisionism" project (see [**GLS94**]) of Gorenstein-Lyons-Solomon (GLS) aims to produce a "second-generation" proof of CFSG. In GLS, the notion of characteristic 2-type from the first-generation proof is replaced by the notion of *even type* (see p. 55 in [**GLS94**]). In a group of even type, centralizers of involutions are allowed to contain certain components (primarily of Lie type in characteristic 2). In particular, if the centralizer of a 2-central involution has a component, then G is not of even characteristic, and so does not satisfy the hypothesis of our Main Theorem.

To bridge the gap between these two notions of "characteristic 2", this final chapter of our work classifies the simple QTK-groups of even type. More precisely, our main result Theorem 16.5.14 (the Even Type Theorem) shows that J_1 is the only simple QTK-group which is of even type but not of even characteristic. Thus the simple QTK-groups of even type are the groups in our Main Theorem, of even type, along with J_1.

To prove Theorem 16.5.14, we will utilize a small subset of the machinery on standard components from the first generation proof of CFSG. In sections I.7 and I.8 of Volume I, we give proofs of all but one of the results we use; that result is Lemma 3.4 from [**Asc75**], which is a fairly easy consequence of Theorem ZD on page 21 in [**GLS99**].

We are grateful to Richard Lyons and Ronald Solomon for their careful reading of this chapter, and suggestions resulting in a number of improvements.

16.1. Even type groups, and components in centralizers

In this chapter, we assume the following hypothesis:

HYPOTHESIS 16.1.1. *G is a quasithin simple group, all of whose proper subgroups are $\mathcal{K}$-groups, but G is not of even characteristic. On the other hand, G is of even type in the sense of GLS.*

The definition of even type is given on p.55 of [**GLS94**]. We will not need to assume that $m_2(G) \geq 3$ as in part (3) of that definition. Part (2) of that definition says that if K is a component of the centralizer of an involution, then $K/Z(K)$ is in the set $\mathcal{C}_2$ of simple groups listed in Definition 12.1 on page 100 in [**GLS94**]; and also that $Z(K)$ satisfies further restrictions given in the final sentences of that definition. We will not reproduce that full list here, since as G is a QTK-group, $K/O_2(K)$ also appears in Theorem C (A.2.3). Instead, intersecting the list of possibilities from Theorem C (A.2.3) with the list of possibilities in Definition 12.1 on page 100 of [**GLS94**], it follows that when G is a QTK-group, G is of even type if and only if:

(E1) $O(C_G(t)) = 1$ for each involution $t \in G$.

(E2) If L is a component of $C_G(t)$ for some involution $t \in G$, then one of the following holds:

(i) $L/O_2(L)$ is of Lie type and characteristic 2 appearing in case (3) or (4) of Theorem C; but L is not $SL_2(q)$, $q = 5, 7, 9$ or $A_8/\mathbf{Z}_2$. Further if $L/O_2(L) \cong L_3(4)$, then $\Phi(O_2(L)) = 1$.

(ii) $L \cong L_3(3)$ or $L_2(p)$, p a Fermat or Mersenne prime.

(iii) $L/O_2(L)$ is M_{11}, M_{12}, M_{22}, M_{23}, M_{24}, J_2, J_4, HS, or Ru.

Observe that from Theorem C, in case (i) either $L/O_2(L)$ is of Lie rank 1, and so of Lie type $A_1 = L_2$, ${}^2B_2 = Sz$, or ${}^2A_2 = U_3$; or $L/O_2(L)$ is of Lie rank 2 and of Lie type A_2, B_2, G_2, ${}^2F_4$, or ${}^3D_4$; or L is $L_4(2)$ or $L_5(2)$.

In the remainder of this introductory section assume that L is a component of the centralizer of some involution of G, and set $\bar{L} := L/Z(L)$. Thus by Hypothesis 16.1.1, L is one of the quasisimple groups listed above. To provide a more self-contained treatment, in this introductory section we collect some facts about L which we use frequently.

First, inspecting the list of Schur multipliers in I.1.3 for the groups L in (E2), and recalling that $O(L) = 1$ by (E1), we conclude:

LEMMA 16.1.2. *(1) If L is not simple, then $Z(L) = O_2(L)$ and $\bar{L} \cong Sz(8)$, $L_3(4)$, $G_2(4)$, M_{12}, M_{22}, J_2, HS, or Ru.*

(2) Either $|Z(L)| \leq 2$, or $\bar{L} \cong Sz(8)$ or $L_3(4)$ with $Z(L) \cong E_4$, or $\bar{L} \cong M_{22}$ with $Z(L) \cong \mathbf{Z}_4$.

Occasionally we need more specialized information about the quasithin groups appearing in 16.1.2, which can be obtained from knowledge of the covering groups L of $\bar{L}$. Such facts are collected in I.2.2.

In the next two lemmas, we list the involutory automorphisms of L and their centralizers in $\bar{L}$. Notice that we write L rather than $\bar{L}$ in those cases where $Z(L) = 1$ by 16.1.2.1.

We begin with the groups of Lie type and characteristic 2 in case (i) of (E2), that is, in case (3) or (4) of Theorem C. Recall that the involutions in classical groups of characteristic 2 are determined up to conjugacy by their Suzuki type: In orthogonal and symplectic groups, the types are denoted a_k, b_k, c_k, as discussed in Definition E.2.6; in linear and unitary groups, the types are denoted j_k, as discussed in Aschbacher-Seitz [**AS76a**]. In each case, k is the dimension of the commutator space for the involution on the natural module for the classical group.

NOTATION 16.1.3. Recall that the types of twisted groups in Theorem C are $^2A_2 = U_3$, $^2B_2 = Sz$, $^3D_4$, and $^2F_4$. We adopt the convention of [**GLS98**] for labeling involutory outer automorphisms of these groups. We emphasize that this convention differs from that of Steinberg which is widely used in the literature (eg. in the Atlas [**C$^+$85**]) in which there are no graph automorphisms of twisted groups. Instead the convention in Definition 2.5.3 in [**GLS98**] is that all involutory automorphisms of groups of type $^2A_2 = U_3$ which are not inner-diagonal are called graph automorphisms, but involutory outer automorphisms of groups of type $^3D_4$ are called field automorphisms. All involutory automorphisms of groups of types $^2B_2 = Sz$ and $^2F_4$ are inner.

LEMMA 16.1.4. *Assume that $\bar{L} \cong X(2^n)$ is of Lie type X and characteristic 2. Let r be an involution in $Aut(L)$, and set $L_r := O^2(C_L(r))$. Then one of the following holds:*

(1) r induces an automorphism on L corresponding to a root involution of $\bar{L}$ (or in $Sp_4(2)$ or $G_2(2)$, if $L = Sp_4(2)'$ or $G_2(2)'$), and $L_r = O^2(C_P(r))$ for the proper parabolic P containing $C_L(r)$.

(2) $L \cong Sp_4(2^n)$, r induces an automorphism of type c_2, and $L_r = 1$.

(3) $L \cong L_4(2)$ or $L_5(2)$, r induces an automorphism of type j_2, and $L_r \cong A_4$ or $\mathbf{Z}_3/(E_4 \times E_4)$, respectively.

(4) r induces a field automorphism on $\bar{L}$ and $L_r \cong X(2^{n/2})'$. Further $Sz(2^n)$ and $^2F_4(2^n)$ have no involutory non-inner automorphisms, and $U_3(2^n)$ has no involutory field automorphisms.

(5) $\bar{L} \cong L_3(2^n)$, n even, r induces a graph-field automorphism on L, and $L_r \cong U_3(2^{n/2})$—unless $n = 2$, where $L_r \cong E_9$.

(6) $\bar{L} \cong L_3^\epsilon(2^n)$, r induces a graph automorphism on $\bar{L}$, and $L_r \cong L_2(2^n)'$.

(7) $L \cong Sp_4(2^n)$, n odd, r induces a graph-field automorphism on L, and $L_r \cong Sz(2^n)'$.

(8) $L \cong L_4(2)$ or $L_5(2)$, r induces a graph automorphism on L, and $L_r \cong A_6$.

(9) $L\langle r \rangle \cong S_8$, r is of type $2^3, 1^2$, and $L_r \cong A_4$.

PROOF. This follows from list of possibilities for L in (E2), and the 2-local structure of $Aut(L)$ (cf. Aschbacher-Seitz [**AS76a**]). $\qquad\square$

We turn to the cases in parts (ii) and (iii) of (E2):

LEMMA 16.1.5. *Assume that $\bar{L}$ is not of Lie type and characteristic 2, and r is an involution in $Aut(L)$. Then one of the following holds:*

(1) $L \cong L_3(3)$ and either r is inner with $C_L(r) \cong GL_2(3)$, or r is outer with $C_L(r) \cong S_4$.

(2) $L \cong L_2(q)$ for $q > 7$ a Fermat or Mersenne prime, and either r is inner with $C_L(r) \in Syl_2(L)$, or r is an outer automorphism in $PGL_2(q)$ with $C_L(r) \cong D_{q+\epsilon}$, where $q \equiv \epsilon \mod 4$.

(3) $L \cong M_{11}$, r is inner, and $C_L(r) \cong GL_2(3)$.

(4) $\bar{L} \cong M_{12}$ and either r is inner with $C_{\bar{L}}(r) \cong S_3/Q_8^2$ or $\mathbf{Z}_2 \times S_5$, or r is outer and $C_{\bar{L}}(r) \cong \mathbf{Z}_2 \times A_5$.

(5) $\bar{L} \cong M_{22}$ and either r in inner with $C_{\bar{L}}(r) \cong S_4/E_{16}$, or r is outer with $C_{\bar{L}}(r) \cong L_3(2)/E_8$ or $Sz(2)/E_{16}$.

(6) $L \cong M_{23}$, r is inner, and $C_L(r) \cong L_3(2)/E_{16}$.

(7) $L \cong M_{24}$, r is inner, and $C_L(r) \cong L_3(2)/D_8^3$ or S_5/E_{64}.

(8) $\bar{L} \cong J_2$ *and either* r *is inner with* $C_{\bar{L}}(r) \cong A_5/Q_8 * D_8$ *or* $E_4 \times A_5$, *or* r *is outer and* $C_{\bar{L}}(r) \cong Aut(L_3(2))$.

(9) $L \cong J_4$, r *is inner, and* $C_L(r) \cong Aut(\hat{M}_{22})/D_8^6$ *or* $Aut(M_{22})/E_{2^{11}}$.

(10) $\bar{L} \cong HS$, *and either* r *is inner with* $C_{\bar{L}}(r) \cong S_5/(Q_8^2*\mathbf{Z}_4)$ *or* $\mathbf{Z}_2 \times Aut(A_6)$, *or* r *is outer with* $C_{\bar{L}}(r) \cong S_8$ *or* S_5/E_{16}.

(11) $\bar{L} \cong Ru$, r *is inner, and* $C_{\bar{L}}(r) \cong S_5/E_{64}/E_{32}$ *or* $E_4 \times Sz(8)$.

PROOF. The Atlas [**C**$^+$**85**] contains a list of centralizers, as does [**GLS98**]. Neither reference includes proofs for the sporadic groups, but there are proofs in section 5 of chapter 4 of [**GLS98**] when $\bar{L}$ is of Lie type and odd characteristic. Proofs for M_{24}, He, and J_2 appear in [**Asc94**], for M_{11} and M_{12} in [**Asc03b**], and for HS in [**Asc03a**]. Proofs or references to proofs for the remaining groups can be found in [**AS76b**]. $\square$

LEMMA 16.1.6. *Assume* r *is a 2-element of* $Aut(L)$ *centralizing a Sylow 2-subgroup of* L. *Then either*

(1) $r \in Inn(L)$, *and if* L *appears in case (i) of (E2) and* r *is an involution, then either* $\bar{r}$ *is a long-root involution or* $\bar{L} \cong Sp_4(2^n)$; *or*

(2) $L \cong A_6$ *and* r *induces an automorphism in* S_6.

PROOF. This is well known; it follows from 16.1.4 and 16.1.5 when r is of order 2. $\square$

Our final preliminary results describe the possible embeddings among components of involution centralizers.

LEMMA 16.1.7. *Assume* t *is an involution in* G, L *is a component of* $C_G(t)$, *and* i *is an involution in* $C_G(\langle t, L \rangle)$. *Set* $K := \langle L^{E(C_G(i))} \rangle$. *Then one of the following holds:*

(1) $K = L$.

(2) $L/O_2(L) \cong L_2(2^n)$, $Sz(2^n)$, *or* $L_2(p)$, p *prime*, $K = K_1 K_1^t$ *with* K_1 *a component of* $C_G(i)$ *and* $K_1 \neq K_1^t$, $K_1/O_2(K_1) \cong L/O_2(L)$, *and* $L = C_K(t)^\infty$.

(3) K *is a component of* $C_G(i)$, $K = [K, t]$, L *is a component of* $C_K(t)$, *and one of the following holds:*

(a) $K/O_2(K) \cong X(2^{2n})$, *where* X *is a Lie type of Lie rank at most 2, but not* $Sz(2^n)$, $U_3(2^n)$, *or* $^2F_4(2^n)$, *and* t *induces a field automorphism on* $K/O_2(K)$ *with* $L/O_2(L) \cong X(2^n)'$.

(b) $K \cong L_3(2^{2n})$ *for* $n > 1$, t *induces a graph-field automorphism on* K, *and* $L \cong U_3(2^n)$.

(c) $K/O_2(K) \cong L_3^\epsilon(2^n)$ *for* $n > 1$, t *induces a graph automorphism on* $K/O_2(K)$, *and* $L \cong L_2(2^n)$.

(d) $K \cong Sp_4(2^n)$, $n > 1$ *odd*, t *induces a graph-field automorphism on* K, *and* $L \cong Sz(2^n)$.

(e) $K \cong L_4(2)$ *or* $L_5(2)$, t *induces a graph automorphism on* K, *and* $L \cong A_6$.

(f) $K/O_2(K) \cong M_{12}$ *or* J_2 *and* $L \cong A_5$.

(g) $K/O_2(K) \cong J_2$ *and* $L \cong L_3(2)$.

(h) $K/O_2(K) \cong HS$ *and* $L \cong A_6$ *or* A_8.

(i) $K/O_2(K) \cong Ru$ *and* $L \cong Sz(8)$.

PROOF. Let $\bar{K} := K/O_2(K)$. Since L is a component of $C_G(t)$ and i centralizes L by hypothesis, L is also a component of $C_{C_G(i)}(t)$. We apply I.3.2 with $C_G(i)$ in the role of "H": As $O(C_G(i)) = 1$ by (E1), $O_{2',E}(C_G(i)) = E(C_G(i))$ and the 2-components in that result are components. Then either

(i) $K = K_1 K_1^t$ for some component $K_1 \neq K_1^t$ of $C_G(i)$, $L/O_2(L) \cong \bar{K}_1$, and $L = C_K(t)^\infty$, or

(ii) K is a t-invariant component of $C_G(i)$, and L is a component of $C_K(t)$.

In case (i), since G is quasithin, the possibilities in (2) are obtained by intersecting the list of A.3.8.3 with that of (E2). Therefore we may assume that case (ii) holds, and $K > L$ since otherwise conclusion (1) holds. The simple group $\bar{K}$ is described in (E2). The cases in (3) arise by inspecting 16.1.4 and 16.1.5 for involutions $i \in Aut(K)$ such that $C_K(i)$ has a component. We use 16.1.2.1 to conclude that $O_2(K) = 1$ or $O_2(L) = 1$ when appropriate; in case (i) of (3), $Z(L) = 1$ from the structure of the covering group K of $\bar{K} \cong Ru$ in I.2.2.7a. $\qquad\square$

LEMMA 16.1.8. *Assume t is an involution in G, L is a component of $C_G(t)$, i is an involution in $C_G(\langle t, L \rangle)$, and $S \in Syl_2(C_G(i))$ with $|S : C_S(t)| \leq 2$. Then L is a component of $C_G(i)$.*

PROOF. Assume otherwise, and set $K := \langle L^{E(C_G(i))} \rangle$; then K is described in case (2) or (3) of 16.1.7, and it remains to derive a contradiction. As $S \in Syl_2(C_G(i))$ and K is subnormal in $C_G(i)$, $S_K := S \cap K \in Syl_2(K)$. Further $|S_K : C_{S_K}(t)| \leq |S : C_S(t)| \leq 2$. However in case (2) of 16.1.7,

$$|S_K : C_{S_K}(t)| \geq |K_1/O_2(K_1)|_2 > 2,$$

so case (3) must hold. But in each subcase of (3), $|S_K : C_{S_K}(t)| > 2$, a contradiction establishing the lemma. $\qquad\square$

16.2. Normality and other properties of components

Let $\mathcal{P}$ denote the set of pairs (z, L) such that z is a 2-central involution in G and L is a component of $C_G(z)$.

LEMMA 16.2.1. $\mathcal{P} \neq \emptyset$.

PROOF. Let $T \in Syl_2(G)$. By Hypothesis 16.1.1, G is not of even characteristic, so there is $M \in \mathcal{M}(T)$ such that $O^2(F^*(M)) \neq 1$ and there is $1 \neq z \in \Omega_1(Z(T)) \cap O_2(M)$. Then $O^2(F^*(M))) \leq O^2(F^*(C_M(z)))$, so that $F^*(C_M(z)) \neq O_2(C_M(z))$. Then as $M = N_G(O_2(M))$ since $M \in \mathcal{M}$, $F^*(C_G(z)) \neq O_2(C_G(z))$ by 1.1.3.2. On the other hand by Hypothesis 16.1.1, G is of even type, so by (E1), $O(C_G(z)) = 1$. Therefore $E(C_G(z)) \neq 1$, so there is a component L of $C_G(z)$, and then $(z, L) \in \mathcal{P}$. $\qquad\square$

In view of 16.2.1, we assume for the remainder of the chapter:

NOTATION 16.2.2. $T \in Syl_2(G)$, z is an involution in $Z(T)$, $(z, L) \in \mathcal{P}$, $G_z := C_G(z)$, $T_L := T \cap L$, and $T_C := C_T(L)$.

LEMMA 16.2.3. *If t is an involution in T_C with $|T : C_T(t)| \leq 2$, then L is a component of $C_G(t)$.*

PROOF. Let $C_T(t) \leq S \in Syl_2(C_G(t))$, so that $C_T(t) \leq C_S(z)$. Since $T \in Syl_2(G)$,

$$|S : C_S(z)| \leq |S : C_T(t)| \leq |T : C_T(t)| \leq 2,$$

and hence the lemma follows from 16.1.8 with z, t in the roles of "t, i". $\qquad \square$

Of course the component L is subnormal in G_z; the main result in this section is 16.2.4 below, showing that in fact L is normal in G_z.

Our eventual goal will be to show that L is *standard* in G, as defined in the next section. As Ronald Solomon has observed, rather than proving that L is normal in G_z, we might instead prove that L is "terminal" in the sense of [**GLS99**] (ie. for each $t \in C_G(L)$, L is a component of $C_G(t)$), and then appeal to Corollary PU_4 in chapter 3 of [**GLS99**] to prove that L is standard. Instead we show directly that L is standard, later in 16.3.2. This allows us to keep our treatment self-contained, and avoid an appeal to a fairly deep result such as Corollary PU_4 of [**GLS99**], with a minimal amount of extra effort.

THEOREM 16.2.4. $L \trianglelefteq G_z$.

Until the proof of Theorem 16.2.4 is complete, assume (z, L) is a counterexample. Set $L_0 := \langle L^{G_z} \rangle$ and $H := N_G(L_0)$. By A.3.8, $|T : N_T(L)| = 2$ and $L_0 = LL^u$ for $u \in T - N_T(L)$, so that $T \leq H$ and $[L, L^u] = 1$. The possibilities for L are obtained by intersecting the lists of A.3.8.3 and (E2); 16.1.2.1 allows only one case with $O_2(L) \neq 1$:

LEMMA 16.2.5. *Either* $L \cong L_2(2^n)$, $Sz(2^n)$, *or* $L_2(p)$ *with p odd, or* $L/O_2(L) \cong Sz(8)$ *with* $O_2(L) \neq 1$.

In the remainder of this section we will eliminate the possibilities in the list of 16.2.5.

LEMMA 16.2.6. *(1) L is a component of $C_G(t)$ for each involution $t \in \langle z \rangle L^u$.*
(2) If $L/O_2(L) \cong Sz(8)$ and $O_2(L) \neq 1$, then L is a component of $C_G(s)$ for each involution $s \in O_2(L)$.

PROOF. Let t be an involution in $\langle z \rangle L^u$. From our list in 16.2.5, either

(I) L is simple and has one conjugacy class of involutions, or
(II) $L/O_2(L) \cong Sz(8)$, and $O_2(L) \neq 1$.

If (I) holds, then conjugating in L, we may take $t \in Z(N_T(L))$; then as $|T : N_T(L)| = 2$, L is a component of $C_G(t)$ by 16.2.3, establishing (1) in this case.

Therefore we may assume that (II) holds. Let s be an involution in $O_2(L)$; then the same argument also establishes (2), since $O_2(L) = Z(L) \leq Z(N_T(L))$ as $Out(L/Z(L))$ is of odd order. Thus it remains to establish (1) in case (ii).

By (2), L is a component of $C_G(s)$. Thus we can apply 16.1.7 to s, t in the roles of "t, i". As $s \in L \leq O^2(C_G(t))$, s acts on each component of $C_G(t)$ by A.3.8.1, so that case (2) of 16.1.7 does not occur. Also the only subcase of case (3) of 16.1.7 in which $L/Z(L) \cong Sz(8)$ is subcase (i), and in that subcase L is simple, whereas here $O_2(L) \neq 1$. Thus case (1) of 16.1.7 holds, completing the proof that conclusion (1) holds. $\qquad \square$

LEMMA 16.2.7. $\langle N_G(L), N_G(L^u) \rangle \leq H$.

PROOF. Let $g \in N_G(L^u)$, and let t be an involution in L^u. By 16.2.6.1, L is a component of $C_G(t)$, so as $C_G(L^u) \leq C_G(t)$, L is a component of $C_G(L^u)$. Since $g \in N_G(L^u)$, L^g is also a component of $C_G(L^u)$. If $L \neq L^g$, then $C_G(L^u)$ contains three isomorphic components L^u, L, and L^g, contrary to A.1.34.2. Thus $L = L^g$, so g normalizes $LL^u = L_0$. Therefore $N_G(L^u) \leq N_G(L_0) = H$, so the lemma holds as $u \in H$. $\qquad \square$

In the next few lemmas, we will show that L is tightly embedded in G. Recall that a subgroup K of a finite group G is *tightly embedded* in G if K has even order, but $K \cap K^g$ is of odd order whenever $g \in G - N_G(K)$.

LEMMA 16.2.8. (1) If $g \in G$ such that $\langle z \rangle L^u \cap (\langle z \rangle L^u)^g$ has even order, then $g \in H$.
(2) Either L is tightly embedded in G, or $L/O_2(L) \cong Sz(8)$ and $O_2(L) \neq 1$.
(3) If X is a nontrivial 2-subgroup of $\langle z \rangle L^u$, then $N_G(X) \leq H$.

PROOF. Observe that (3) is a special case of (1). Assume the hypotheses of (1). Then there is an involution $t \in \langle z \rangle L^u \cap (\langle z \rangle L^u)^g$, so by 16.2.6, L and L^g are both components of $C_G(t)$. Then L^g normalizes L so that $L^g \leq H$ by 16.2.7, and hence L^g is a component of $C_H(t)$. Applying I.3.2, L^g lies in a 2-component of H, which is a member of $\mathcal{C}(H)$, so that by A.3.7, either $L^g \in \{L, L^u\}$ or $[LL^u, L^g] = 1$. The latter case is impossible, for since $L/O_2(L)$ is not $U_3(8)$, case (1.a) of A.1.34 holds, so that $O^{r'}(H) = LL^u$ for a suitable odd prime r; while in the former, either g or gu^{-1} lies in $N_G(L)$, so $g \in H$ by 16.2.7. Thus (1) holds.

Now if $L^u \cap L^{ug}$ has even order, then $g \in H$ by (1). Hence if $g \notin N_G(L)$, then $L^{ug} = L$, so that $1 \neq L^u \cap L \leq O_2(L)$. Then we conclude from 16.2.5 that $L/O_2(L) \cong Sz(8)$, so that (2) holds. $\qquad \square$

LEMMA 16.2.9. (1) Let p be a prime divisor of $2^n - 1$ if $L/O_2(L) \cong Sz(2^n)$ or $L_2(2^n)$, and let $p := 3$ if $L \cong L_2(r)$ for odd r. Then $L_0 = O^{p'}(H)$.
(2) $L_0 \nleq H^g$ for $g \in G - H$.

PROOF. Observe if $L \cong L_2(r)$ for r odd that 3 divides the order of some 2-local subgroup of L. Then part (1) follows as case (a) of A.1.34.1 holds. If $L_0 \leq H^g$ then $L_0 = O^{p'}(L_0) \leq O^{p'}(H^g) = L_0^g$ by (1), so that $g \in N_G(L_0) = H$, and (2) holds. $\qquad \square$

When analyzing a tightly embedded subgroup K of a group G, one focuses on the conjugates K^g such that $N_{K^g}(K)$ is of even order. (See e.g. the definition of $\Delta(K)$ in Section 4.) In our present setup, we need a slightly stronger condition, which we establish in the next lemma:

LEMMA 16.2.10. (1) The strong closure of T_L in $N_T(L)$ with respect to G properly contains $T_L \cup T_L^u$.
(2) There is $g \in G - H$ such that $|L^g \cap N_H(L)|_2 > 1$.

PROOF. Set $A_1 := T_L$, $A_2 := T_L^u$, and assume $A_1 \cup A_2$ is strongly closed in $N_T(L)$ with respect to G; we check that the hypotheses of Lemma 3.4 of [**Asc75**] are satisfied. First if $A_i^g \cap A_j \neq 1$ for some i, j, then $A_2^{wg} \cap A_2^v \neq 1$ for some choice of $v, w \in \{1, u^{-1}\}$; therefore $wgv^{-1} \in N_G(A_2) \leq H$ by 16.2.8.1, and hence also $g \in H$ as $u \in H$. Thus the subgroup H_0 of H generated by all such elements g plays the role of the group "H" in 3.4 of [**Asc75**]. Next as H permutes $\{L, L^u\}$, $A_1 \cup A_2$ is strongly closed in T with respect to H. Of course $N_T(A_i) = N_T(L)$, so hypothesis (*) of 3.4 of [**Asc75**] is satisfied.

Then since $H_0 \leq H < G$ and $T_L \not\leq T_L^u$, conclusion (3) of 3.4 in [**Asc75**] holds: namely $A_1 \cap A_2 \neq 1$, and A_1 is dihedral or semidihedral. But as $1 \neq A_1 \cap A_2 \leq L \cap L^u \leq O_2(L)$, $L/O_2(L) \cong Sz(8)$ by 16.2.5, so that A_1 is of 2-rank at least 3 and hence not dihedral or semidihedral. This contradiction completes the proof of (1).

As $T_L \in Syl_2(L)$ and H permutes $\{L, L^u\}$, (1) implies (2). $\qquad\square$

LEMMA 16.2.11. $O_2(L) = 1$, so L is tightly embedded in G.

PROOF. If L is not tightly embedded in G, then $1 \neq O_2(L) = Z(L)$ by 16.2.8.2, so to prove both assertions we may assume $Z(L) \neq 1$, and it remains to derive a contradiction. By 16.2.5, $L/Z(L) \cong Sz(8)$.

Set $Z_L := \Omega_1(T_L)$. From I.2.2.4, involutions of $T_L Z(L)/Z(L)$ lift to involutions of T_L, and these involutions are the nontrivial elements of Z_L, so Z_L is elementary abelian. Further $Out(L)$ is of odd order. From these remarks we deduce:

(*) If A is an elementary abelian 2-subgroup of $N_T(L)$ then $A \leq Z_L T_C$ and A centralizes Z_L.

Further as $Z(L) \neq 1$, $m(Z_L) > m(Z_L/Z(L)) = 3$.

By 16.2.10.2, there is $g \in G - H$ such that $L^g \cap N_H(L)$ contains an involution i, and as $T \in Syl_2(H)$ we may take $i \in T$. Then by (*), i centralizes Z_L, so as $C_G(i) \leq H^g$ by 16.2.8.3, $Z_L \leq H^g$; then conjugating in H^g, we may take $Z_L \leq T^g$. Hence by (*), $X := N_{Z_L}(L^g)$ centralizes $V := Z_L^g$. Further $|Z_L : X| \leq |H : N_H(L)| = 2 < |Z_L : Z(L)|$, and hence $X \not\leq Z(L)$. In particular $1 \neq X$, so $V \leq C_G(X) \leq H$ by 16.2.8.3. As $X \leq Z_L$ but $X \not\leq Z(L)$, $N_H(X) \leq N_H(L)$, so $V \leq N_H(L)$. Then as $m(V) > 3 = m_2(Aut(L))$, $C_V(L) \neq 1$, so $L \leq C_G(C_V(L)) \leq H^g$ by 16.2.8.3. Similarly $C_V(L^u) \neq 1$ so that $L^u \leq H^g$, and then $L_0 \leq H^g$, contrary to 16.2.9.2. $\qquad\square$

LEMMA 16.2.12. $T_L^G \cap T = \{T_L, T_L^u\}$.

PROOF. Assume otherwise. Then there is $g \in G - H$ with $S := T_L^g \leq T$ but S is not equal to T_L or T_L^u. Now as $|T_L| > 2 = |T : N_T(L)|$, $1 \neq N_S(L) \leq N_S(T_L)$; so as L is tightly embedded in G by 16.2.11, S centralizes T_L (and similarly T_L^u) by I.7.6 with G, L, T_L, T in the roles of "H, K, Q, S". Then $R := T_L \langle z \rangle \leq C_G(S) \leq H^g$ using 16.2.8.3. As R centralizes $S \in Syl_2(L^g)$, we conclude from 16.1.6 that R induces inner automorphisms on L^g. Then as $|R| = 2|S|$, $1 \neq C_R(L^g)$, so $L^g \leq C_G(C_R(L^g)) \leq H$ by 16.2.8.3. Similarly $L^{ug} \leq H$, so $L_0^g \leq H$, contrary to 16.2.9.2. $\qquad\square$

We are now in a position to complete the proof of Theorem 16.2.4.

By 16.2.10.2, there is $g \in G - H$ such that $L^g \cap N_T(L)$ contains an involution i. If i centralizes a Sylow 2-subgroup of L, we may assume by conjugating in L that i centralizes T_L. Then by 16.2.8.3, $T_L \leq C_G(i) \leq H^g$, and conjugating in H^g we may assume $T_L \leq T^g$. But now by 16.2.12, $T_L \in \{T_L^g, T_L^{ug}\}$, contrary to L tightly embedded in G by 16.2.11 since $g \notin H$. Thus i does not centralize any Sylow 2-subgroup of L. But as $O_2(L) = 1$ by 16.2.11, 16.2.5 says that L has one conjugacy class of involutions, so we conclude i induces an outer automorphism on L. Therefore by 16.1.4 and 16.1.5 applied to the list in 16.2.5, either

(i) $L \cong L_2(2^{2n})$, and i induces a field automorphism on L, or
(ii) $L\langle i \rangle \cong PGL_2(p)$.

In case (ii), $C_{L_0}(i) \cong D_{p+\epsilon} \times D_{p+\epsilon}$, where $p \equiv \epsilon \mod 4$ and $\epsilon = \pm 1$. But 3 divides $p + \epsilon$ as p is a Fermat or Mersenne prime, so $C_{L_0}(i)$ contains $E \cong E_9$. This is impossible, since $i \in T_L^g$, so using 16.2.9.1,

$$E \le O^{3'}(C_{H^g}(i)) \le C_{L_0^g}(i) \le C_{L^g}(i) \times L^{ug} \cong D_{p-\epsilon} \times L_2(p).$$

Similarly in case (i), $C_{L_0}(i) \cong L_2(2^n) \times L_2(2^n)$, and by 16.2.8.3 and 16.2.9.1,

$$1 \ne O^2(C_{L_0}(i)) \le O^{q'}(H^g) = L_0^g,$$

for any prime divisor q of $2^n - 1$. Since $L_2(4) \cong L_2(5)$, we may assume $n > 1$, so such primes q exist and $m_q(C_{L_0}(i)) = 2$ while $m_q(C_{L_0^g}(i)) = 1$. This contradiction completes the proof of Theorem 16.2.4.

16.3. Showing L is standard in G

In Theorem 16.3.7 of this section, we will show that the component L is in *standard form* in G, in the sense of [**Asc75**]: that is $C_G(L)$ is tightly embedded in G, $N_G(L) = N_G(C_G(L))$, and L commutes with none of its conjugates.

To show that L is standard, we show that L is "terminal" in the sense of [**GLS99**], as defined earlier. The next two lemmas show that if L is terminal then L is standard. The proof of the first lemma makes use of the normality of L in G_z which we established in Theorem 16.2.4.

LEMMA 16.3.1. *$C_G(L)$ contains at most one component isomorphic to L, and no component G-conjugate to L.*

PROOF. The first assertion follows from A.1.34.2 with $N_G(L)$ in the role of "H". Assume that L^g is a component of $C_G(L)$. Then by the first assertion,

$$\Theta(L) := \{L^x : x \in G \text{ and } L^x \text{ is a component of } N_G(L)\} = \{L, L^g\}.$$

Since L is not $SU_3(8)$, case (1.a) of A.1.34 holds, so that $LL^g = O^{r'}(N_G(L))$ for a suitable odd prime r, and hence $L^g = O^{r'}(C_G(L))$. It follows that $L = O^{r'}(C_G(L^g))$, so that $\Theta(L) = \Theta(L^g) = \Theta(L)^g$. Thus $g \in N_G(\Theta(L)) =: N$, and hence a Sylow 2-subgroup of N is transitive on $\Theta(L)$ of order 2. This is impossible, as by Theorem 16.2.4, $T \le N_G(L) \le N_G(\Theta(L)) = N$ so that T fixes $\Theta(L)$ pointwise but is also Sylow in N. $\qquad\square$

LEMMA 16.3.2. *Assume that L is a component of $C_G(t)$ for each involution $t \in C_G(L)$. Then L is standard in G.*

PROOF. Assume the hypothesis of the lemma. We first observe that $C_G(L)$ contains no conjugate of L, verifying the third condition in the definition of "standard form". For if $L^g \le C_G(L)$, then L^g is a component of $C_G(i)$ for each involution $i \in L$ by hypothesis, so as $C_G(L) \le C_G(i)$, L^g is a component of $C_G(i)$, contrary to 16.3.1.

Set $H := N_G(L)$, $X := C_G(L)$, and assume that $X \cap X^g$ is of even order for some $g \in G$. We will show that $g \in H$, which will suffice: For then since $C_G(L)$ is of even order and $H \le N_G(C_G(L))$, $C_G(L)$ is tightly embedded in G and $N_G(C_G(L)) = H$, verifying the remaining conditions for L to be in standard form.

Finally assume that $g \notin H$. Thus $L^g \ne L$, while as $X \cap X^g$ is of even order, there is an involution $t \in X \cap X^g$. By hypothesis, L and L^g are components of $C_G(t)$, so as $L^g \ne L$, $L^g \le C_G(L)$, contrary to our first observation. $\qquad\square$

Observe also (cf. I.7.2.5):

REMARK 16.3.3. If L is standard in G, then for each nontrivial 2-subgroup X of $C_G(L)$,

$$N_G(X) \le N_G(C_G(L)) = N_G(L).$$

To show that L is terminal, we need to eliminate the proper inclusions of L in K in parts (2) and (3) of 16.1.7. The first elimination makes use of an approach suggested by Richard Lyons. Although the method could be applied without appeal to Theorem 16.2.4, it goes more smoothly with such appeal. The method could also be used to eliminate other proper containments in 16.1.7, but it is easier to use other arguments like those in 16.3.9.

LEMMA 16.3.4. Assume t is an involution in T_C, and L is a component of $C_K(i)$ for some component K of $C_G(t)$ and some involution $i \in C_G(L\langle t \rangle)$ with $K = [K, i]$. Then K is not $U_3(2^n)$, M_{12}, J_2, HS, or Ru.

PROOF. Assume K is one of the components we wish to eliminate. Inspecting the list of possibilities for the pair L, K in 16.1.7, we find that L is simple, so $T_L T_C = T_L \times T_C$. By Theorem 16.2.4, T acts on L, so $|T : T_C| \le |Aut(L)|_2$, while by inspection of the pairs on our list, $|K|_2 > |Aut(L)|_2$, so $|K|_2 |T_C| > |T|$.

When $K \cong U_3(2^n)$ set $V := T_L$, and in the remaining cases choose V of order 2 in $T_L \cap Z(T)$. Thus in any case $T_C \cap V = 1$ and $T \le N_G(V)$, so as $T \in Syl_2(G)$, $T \in Syl_2(N_G(V))$. Thus for $S \in Syl_2(N_G(V))$, $S = T^g$ for some $g \in N_G(V)$, and setting $S_A := T_A^g$ for $A \in \{C, L\}$, $S_C \trianglelefteq S$, $|K|_2 |S_C| > |S|$ and $S_C \cap V = 1$.

Next we claim that there exists $S_K \in Syl_2(K)$ such that $V = Z(S_K)$: This is clear from the embedding of L in K when K is $U_3(2^n)$, while in the remaining cases we will show that 2-central involutions in L are 2-central in K, so the claim holds there too. Choose S_K so that $S_i := C_{S_K}(i) \in Syl_2(C_K(i))$. Then S_i contains an involution z in $Z(S_K)$, and by inspection of $C_K(i)$ for the pairs on our list, this forces $z \in L$: This is evident if $Z(S_i) \le L$, while in the remaining cases all involutions in $C_{S_i}(L)$ are not 2-central and all involutions in $Z(S_i) - L$ are fused into $C_{S_i}(L)$ under $N_K(S_i)$.

By the claim, $S_K \le N_K(V)$; let $S_K \le S \in Syl_2(N_G(V))$. Then $|S_K||S_C| = |K|_2 |S_C| > |S|$, so $S_K \cap S_C \ne 1$. By the claim, $Z(S_K) = V$, so as $1 \ne S_K \cap S_C \trianglelefteq S_K$, $V \cap (S_K \cap S_C) \ne 1$, contrary to $S_C \cap V = 1$. $\square$

LEMMA 16.3.5. Assume E is a subgroup of G of order 4, and K is a component of $C_G(e)$ for each $e \in E^{\#}$. Let i be an involution in $C_G(EK)$. Then one of the following holds:

(1) K is a component of $C_G(i)$.

(2) $K < I$, where I is a component of $C_G(i)$ such that E is faithful on I, $O_2(I) \ne 1$, $E \cong C_{Aut(I)}(Aut_K(I)) \cong E_4$, and either

    (a) $K \cong A_5$ and $I/O_2(I) \cong M_{12}$ or J_2, or

    (b) $K \cong Sz(8)$ and $I/O_2(I) \cong Ru$.

PROOF. Assume that conclusion (1) fails, and set $I := \langle K^{E(C_G(i))} \rangle$. Then I and the action of an involution $t \in E^{\#}$ on I are described in conclusion (2) or (3) of 16.1.7, with K, I in the roles of "L, K". Observe that $C_E(I) = 1$ since $K < I$ and K is a component of $C_G(e)$ for each $e \in E^{\#}$, so E is faithful on I.

Suppose the pair (t, I) satisfies case (2) of 16.1.7. Then $I = I_1 I_1^t$ with $I_1 \ne I_1^t$, for some component I_1 of $C_G(i)$ such that $I_1/Z(I_1) \cong K/Z(K)$. Thus $N_E(I_1) \ne 1$

as E is of order 4, so by hypothesis K is a component of the centralizer of an involution $e \in N_E(I_1)$. Thus as e centralizes K, which is a full diagonal subgroup of $I_1 I_1^t = I$, e centralizes I, contrary to E faithful on I.

Therefore (t, I) is described in case (3) of 16.1.7. Then as K centralizes the subgroup E of order 4 faithful on I, we conclude from 16.1.4 and 16.1.5 (applied to I described in 16.1.7.3) that $E \cong E_4$, and $I/O_2(I) \cong M_{12}, J_2, HS$, or Ru. By 16.3.4, $O_2(I) \neq 1$. We may assume that conclusion (2) fails, so $I/O_2(I) \cong HS$, with $E = \langle s, t \rangle$ a 4-group such that $E(C_I(s)) \cong A_8$ and $E(C_I(t)) \cong A_6$. But this contradicts the hypothesis that K is a component of $C_I(e)$ for each $e \in E^{\#}$. $\quad \square$

LEMMA 16.3.6. *Assume $E \leq T_C$ is of order 4, with L a component of $C_G(e)$ for each $e \in E^{\#}$. Then L is a component of $C_G(i)$ for each involution $i \in C_G(EL)$.*

PROOF. Assume otherwise. Let i be a counterexample to the lemma, set $G_i := C_G(i)$, and take $E \langle i \rangle \leq T_i \in Syl_2(N_{G_i}(L))$. As $T \in Syl_2(N_G(L))$ by Theorem 16.2.4, we may assume $T_i \leq T$, so that $T_i = C_T(i)$. Then $i \in C_T(L) = T_C$.

As L is a component of $C_G(e)$ for each $e \in E^{\#}$, and we are assuming the lemma fails, $L < I := \langle L^{E(G_i)} \rangle$, where I, E, and L are described in 16.3.5.2 with L, I in the roles of "K, I". In particular $O_2(I) \neq 1$. Set $R := C_{T_C}(i)$; as $T_i = C_T(i)$, $R = C_{T_i}(L)$, so $z \in R$. Also $T_L \leq T_i$ since $i \in C_G(L)$, so $C_{T_L T_C}(i) = T_L R$.

Let $R_0 := C_R(I)$. By 16.3.5.2, $E \cong E_4$ is faithful on I and $Aut_E(I) = C_{Aut(I)}(Aut_L(I))$, so $R = R_0 E$ with $E \cap R_0 = 1$. Next $R < T_C$: for otherwise $T_L T_C \leq T_i$, so that $|T : T_i| \leq |T : T_L T_C| \leq |Out(L)|_2 \leq 2$ by inspection of the cases in 16.3.5.2, contrary to 16.1.8 with z in the role of "t".

Now pick the counterexample i so that R is maximal. As $R < T_C$, there is $y \in N_{T_C}(R) - R$ with $y^2 \in R$. Suppose $X := R_0 \cap R_0^y \neq 1$. Then as R normalizes R_0 and y normalizes R, R also normalizes R_0^y, and hence normalizes X. Therefore there is an involution i_1 in X central in $R\langle y \rangle$, contrary to the maximality of R.

Therefore $R_0 \cap R_0^y = 1$, so R_0 is isomorphic to a subgroup of $R/R_0 = R_0 E/R_0 \cong E_4$, and in particular $\Phi(R_0) = 1$. As $O_2(I) \neq 1$, from (5b) and (7b) of I.2.2, non-2-central involutions of $I/Z(I)$ lift to 4-elements of I, so either $C_I(L) \cong Q_8$, or $I/O_2(I) \cong M_{12}$ and $C_I(L) \cong \mathbf{Z}_4$. In either case there is $e \in E^{\#}$ inducing an inner automorphism on I, so that $e = r_0 f$ with $r_0 \in R_0$ and $f \in C_I(L)$; then f is of order 4 with $f^2 \in Z(I)$, so r_0 is also of order 4, contradicting $\Phi(R_0) = 1$. $\quad \square$

We are now ready to state the main result of this section:

THEOREM 16.3.7. *L is standard in G.*

Until the proof of Theorem 16.3.7 is complete, assume L is a counterexample. Thus by 16.3.2, there is an involution $t \in C_G(L)$ such that L is not a component of $G_t := C_G(t)$. Recall $T_C = C_T(L) \in Syl_2(C_G(L))$, so we may assume $t \in T_C$.

LEMMA 16.3.8. *T_C is dihedral or semidihedral of order at least 8. In particular $C_{T_C}(t) = \langle z, t \rangle$.*

PROOF. As L is not a component of G_t, $t \neq z$, so $|T_C| > 2$. Since T normalizes T_C by Theorem 16.2.4, we may choose $E \leq T_C$ of order 4 with $E \trianglelefteq T$, and set $S := C_{T_C}(E)$. Then $|T : C_T(e)| \leq 2$ for each $e \in E^{\#}$, and hence L is a component of $C_G(e)$ by 16.1.8. Hence L is a component of $C_G(s)$ for each $s \in S^{\#}$ by 16.3.6. Therefore $t \in T_C - S$, and if $C_S(t) > \langle z \rangle$, then applying 16.3.6 to a subgroup of $C_S(t)$ of order 4 in the role of "E", we contradict our assumption that

L is not a component of G_t. Therefore $C_S(t) = \langle z \rangle$, so as S is of index 2 in T_C, $C_{T_C}(t) = \langle t, z \rangle \cong E_4$. Then by a lemma of Suzuki (cf. Exercise 8.6 in [**Asc86a**]), T_C is dihedral or semidihedral. As $T_C > S \geq E$ and $|E| = 4$, $|T_C| > 4$. $\square$

Let $K := \langle L^{E(G_t)} \rangle$. By assumption, $K > L$, so K, L, and the action of z on K are described in case (2) or (3) of 16.1.7. To prove Theorem 16.3.7, we will successively eliminate those possibilities, beginning with the reduction:

LEMMA 16.3.9. *One of the following holds, with $O_2(K) \neq 1$ in cases (2)–(4):*

(1) $K \cong L_4(2)$, $L \cong A_6$, and z induces a graph automorphism on K.

(2) $K/O_2(K) \cong M_{12}$ and $L \cong A_5$.

(3) $K/O_2(K) \cong J_2$ and $L \cong A_5$ or $L_3(2)$.

(4) $K/O_2(K) \cong HS$ and $L \cong A_6$ or A_8.

PROOF. We observe first that either

(i) $t \notin K$, so that $m_2(K\langle t \rangle) \geq m_2(K) + 1$, or

(ii) $t \in K$, so that $t \in Z(K)$ and hence $Z(K) \neq 1$.

On the other hand, T normalizes L by Theorem 16.2.4, and $m_2(T_C) = 2$ by 16.3.8, so

$$m_2(K\langle t \rangle) \leq m_2(T) = m_2(LT) \leq m_2(Aut(L)) + m_2(T_C) = m_2(Aut(L)) + 2. \quad (!)$$

We will eliminate those cases in 16.1.7 not appearing in the lemma, primarily by appeals to (!). Set $\bar{K} := K/Z(K)$, $L^* := L/Z(L)$, and $m := m_2(L^*)$.

Suppose first that K is not quasisimple, so that case (2) of 16.1.7 holds. By 16.1.2, either L is simple, or $L^* \cong Sz(8)$ with $m(Z(L)) \leq 2$. Furthermore as $Inn(L^*) \leq Aut(L) \leq Aut(L^*)$, using 16.1.4 and 16.1.5, $m_2(Aut(L)) = m$; Thus (!) says that

$$m_2(K\langle t \rangle) \leq m + 2. \quad (*)$$

Further $m_2(K) \geq 2m$. Thus if $t \notin K$, then $2m + 1 \leq m + 2$ by paragraph one and (*), so that $m \leq 1$, contrary to L^* simple. On the other hand if $t \in K$, then $Z(K) \neq 1$ and hence $L^* \cong \bar{K} \cong Sz(8)$. Thus $2m \leq m + 2$ by (*), contrary to $m = m_2(Sz(8)) = 3$.

Therefore K is quasisimple, and so K is described in one of the subcases of part (3) of 16.1.7.

Suppose first that one of subcases (a)–(d) holds, but that $\bar{K}$ is neither $Sp_4(4)$ nor $G_2(4)$. By 16.3.4, K is not $U_3(2^m)$. Further either K is simple, and then L is also simple in each case; or $Z(K) \neq 1$, and then by 16.1.2, $\bar{K} \cong L_3(4)$ and $\Phi(Z(K)) = 1$. Hence when $Z(K) \neq 1$, involutions in $\bar{K}$ lift to involutions in K, and so as $\Phi(Z(K)) = 1$, $m_2(K) = m_2(\bar{K}) + m(Z(K))$. Therefore in any case, $m_2(T) \geq m_2(\bar{K}) + 1$ from paragraph one. By inspection, $m_2(\bar{K}) \geq 2m$, and $m_2(Aut(L)) = m$. Thus from (!), $2m + 1 \leq m + 2$, contradicting $m > 1$.

The lemma holds if K is $L_4(2)$, or if K appears in case (f)–(h) of 16.1.7, using 16.3.4 to conclude that $O_2(K) \neq 1$ in the latter cases. So we may assume that $\bar{K}$ is $Sp_4(4)$, $G_2(4)$, $L_5(2)$, or Ru. Now by 16.1.2, either K is simple, or $\bar{K} \cong G_2(4)$ or Ru with $Z(K)$ of order 2. By inspection, $m_3(Aut(L)) = 3$ in each case, so (!) says that

$$m_2(K\langle t \rangle) \leq 5. \quad (!!)$$

However by inspection, $m_2(\bar{K}) \geq 6$, so if K is simple, then (!!) supplies a contradiction. Thus $\bar{K} \cong G_2(4)$ or Ru and $|Z(K)| = 2$. In the latter case $m_2(K) \geq 6$

by I.2.2.7b, contrary to (!!). Thus $\bar{K} \cong G_2(4)$. By I.2.2.5a, 2-central involutions of $\bar{K}$ lift to involutions of K; so since the unipotent radical of the stabilizer in $\bar{K}$ of a point in the natural representation contains a product of two long roots groups with elements permuted transtivitely by a subgroup $L_2(4)$ of a Levi complement, we conclude that $m_2(K) \geq 5$. Therefore $Z(K) = \langle t \rangle$ by (!!).

Next from I.2.2.5b, short-root involutions in $\bar{K}$ lift to elements of order 4 in K squaring to t. We may choose such a u of order 4 to normalize L. But now as T is Sylow in $N_G(L)$, if necessary replacing T by a Sylow 2-subgroup of $N_G(L)$ containing $\langle u, z \rangle$, we may assume that u lies in T and so normalizes T_C. But by 16.3.8, T_C is dihedral or semidihedral of order at least 8 and $C_{T_C}(t) = \langle z, t \rangle$, while as $u^2 = t$, t centralizes the characteristic subgroup of T_C isomorphic to $\mathbf{Z}_4$. $\qquad\square$

LEMMA 16.3.10. *Neither t nor tz is in z^G.*

PROOF. Suppose $z^g = t$. Then as L^g and K are distinct components of G_t described in 16.3.9, $m_3(KL^g) > 2$, contrary to G quasithin.

Therefore $t \notin z^G$. But by 16.3.8, $tz \in t^{T_C}$, so also $tz \notin z^G$. $\qquad\square$

LEMMA 16.3.11. *(1) $\langle t, z \rangle \in Syl_2(C_{G_t}(L))$.*
(2) $\langle t \rangle \in Syl_2(C_{G_t}(K))$.
(3) $K \cong A_8$, $L \cong A_6$, and z induces a transposition on K.

PROOF. By 16.3.8, $\langle t, z \rangle =: E \in Syl_2(C_{G_t}(L\langle z \rangle))$, and by 16.3.10, $tz \notin z^G$, so z is weakly closed in E with respect to G_t. Hence (1) holds, and of course (1) implies (2) since z does not centralize K.

Assume (3) fails. Then K appears in one of cases (2)–(4) of 16.3.9. Thus $1 \neq O_2(K)$, so by (2), $O_2(K) = \langle t \rangle$, and if z induces an inner automorphism on K, then $z \in K$. Let $\bar{K} := K/\langle t \rangle$.

Suppose z induces an inner automorphism on K. Then $z \in K$ by the previous paragraph, so as z centralizes L, we conclude from 16.1.5 that $\bar{z}$ is a non-2-central involution of $\bar{K}$. Then from I.2.2.5b, the lift in K of $\bar{z}$ is of order 4, a contradiction.

Thus z induces an outer automorphism on K. Again using 16.1.5, z centralizes a non-2-central involution $\bar{u}$ in $\bar{K}$. Thus a preimage u of $\bar{u}$ in K is of order 4, and $\bar{u}$ acts on $\bar{L} = O^2(C_{\bar{K}}(z))$, so u acts on L. Now the argument in the last paragraph of the proof of 16.3.9 supplies a contradiction. $\qquad\square$

Let $T_t := C_T(t)$; as $T \in Syl_2(G_z)$ and $L \trianglelefteq G_z$, we may choose t so that $T_t \in Syl_2(C_{G_t}(z))$. Let $T_t \leq P \in Syl_2(G_t)$.

As $K\langle z \rangle \cong S_8$ by 16.3.11.3, we can represent $K\langle z \rangle$ as the symmetric group on $\Omega := \{1, \ldots, 8\}$ with $z := (1, 2)$. Then there is an involution $u \in C_K(z)$ acting as $(1, 2)(3, 4)$ on Ω and inducing a transposition on $L \cong A_6$. Let y denote a generator for the characteristic cyclic subgroup Y of index 2 in T_C provided by 16.3.8. Choose $w \in \{u, tu\}$ with $|C_Y(w)|$ maximal.

LEMMA 16.3.12. $J(T) = R \times T_L \times \langle w \rangle$, *where either*

(1) w centralizes T_C, $R := T_C$ if T_C is dihedral, and R is the dihedral subgroup of T_C of index 2 if T_C is semidihedral; or

(2) $|Y| > 4$, $y^w = yz$, and $R = \langle y^2, t \rangle$.

PROOF. First $\langle u, t \rangle$ acts on Y with $y^t = y^{-1}$ or $y^{-1}z$ for T_C dihedral or semidihedral, respectively. Further $L\langle u \rangle \cong S_6$ by construction, so that $m_2(T/T_C) = 3$, while $m_2(T_C) = 2$ by 16.3.8; hence

$$m_2(T) \leq m_2(T/T_C) + m_2(T_C) = 3 + 2 = 5,$$

so as $m_2(T) \geq m_2(P) \geq m_2(S_8) + 1 = 5$, all inequalities are equalities. Hence $m_2(T) = 5$, and for each $A \in \mathcal{A}(T)$, $m(A/A \cap T_C) = 3$ and $m(A \cap T_C) = 2$. Thus $AT_C/T_C \cong S_6$, so $A \leq T_C L\langle u \rangle$ and hence $J(T) = J(T_0)$, where $T_0 := T_C T_L \langle u \rangle$.

If $\langle u, t \rangle$ is not faithful on Y, then w centralizes T_C since we chose $|C_Y(w)|$ maximal; therefore $T_0 = T_C \times T_L \langle w \rangle$, and (1) follows. Thus we may assume that $\langle u, t \rangle$ is faithful on Y, so $|Y| > 4$ and $y^w = yz$. Then we calculate that $\Omega_1(T_0) = \langle y^2, t \rangle \times T_L \langle w \rangle$, and then that (2) holds. □

We now complete the proof of Theorem 16.3.7.

As $L \cong A_6$, $T_L \cong D_8$. It follows from 16.3.12 and 16.3.8 that $\Omega_1(\Phi(J(T))) = \langle z, v \rangle$, where $\langle v \rangle = Z(T_L)$, and $\langle z, v \rangle \leq Z(T)$. On the other hand, by 16.3.11.2, $\langle t \rangle = C_P(K)$, so $P = \langle t \rangle \times Q$, where $Q := (P \cap K)\langle z \rangle \cong D_8$ wr $\mathbf{Z}_2$. Thus $J(P) = \langle t \rangle \times S_1 \times S_2$, with $S_i \cong D_8$, and $\Omega_1(\Phi(J(P))) = \langle s_1, s_2 \rangle$, where $\langle s_i \rangle = Z(S_i)$, s_i has cycle structure 2^2 on Ω, and $s_1 s_2$ has cycle structure 2^4.

Now by 16.3.12, $J(T) = R \times T_L \times \langle w \rangle$, where R and T_L are dihedral of order at least 8. Then by the Krull-Schmidt Theorem A.1.15, either $|R| > |T_L|$ and $N_G(J(T))$ normalizes $RZ(J(T))$ and $T_L Z(J(T))$, or $|R| = |T_L| = 8$ and $N_G(J(T))$ permutes the pair. Thus $N_G(J(T))$ permutes $\{z, v\}$ since $\langle z \rangle = \Omega_1(\Phi(RZ(J(T))))$ and $\langle v \rangle = \Omega_1(\Phi(T_L Z(J(T))))$. Next $J(P) \leq T^g$ for some $g \in G$, and $m_2(P) = 5 = m_2(T)$, so $J(P) \leq J(T^g)$. Then $\langle s_1, s_2 \rangle = \Phi(J(P)) \leq \Phi(J(T^g)) = \langle z, v \rangle^g$. This is impossible as $\langle z, v \rangle \leq Z(T)$, whereas $\langle s_1, s_2 \rangle \not\leq Z(P)$.

This contradiction completes the proof of Theorem 16.3.7.

LEMMA 16.3.13. *(1)* $L^G \cap C_G(L) = \emptyset$.

(2) L *is standard in* G.

(3) If $C_G(L) \cap N_G(L^g)$ *is of even order for some* $g \in G - N_G(L)$, *then* $L \not\leq N_G(L^g)$.

PROOF. Observe that (2) is just a restatement of Theorem 16.3.7, and (1) is a restatement of the condition in the definition of standard form that L commutes with none of its conjugates.

Assume the hypothesis of (3) and $L \leq N := N_G(L^g)$. Thus $L^g \neq L$, and there is an involution $i \in C_N(L)$. By Remark 16.3.3, L is a component of $C_N(i)$, so we may apply I.3.1 with N, $\langle i \rangle$ in the roles of "H, P", to conclude that $L \leq KK^i$, where K and K^i are (not necessarily distinct) 2-components of N. If $L^g \leq KK^i$, then $L^g \in \{K, K^i\}$, so as $i \in N = N_G(L^g)$, $L \leq KK^i = L^g$, contrary to $L \neq L^g$. Therefore $[L^g, KK^i] = 1$ by 31.4 in [**Asc86a**], so $L \leq C_G(L^g)$, contrary to (1). □

16.4. Intersections of $\mathbf{N_G(L)}$ with conjugates of $\mathbf{C_G(L)}$

Recall that in Notation 16.2.2, z is an involution in the center of T, and L is a component of $G_z = C_G(z)$. By Theorem 16.3.7, L is standard in G.

With this setup, we could now finish quickly by quoting some of the machinery on standard subgroups and tightly embedded subgroups in the Component Paper [**Asc75**] and the Tightly Embedded Subgroup Paper [**Asc76**], and some of the classification theorems in the literature based on that theory. But since GLS do not use this machinery, we will only use some comparatively elementary results from that theory, which we have reproduced in section I.7.

In this section we develop some technical tools, which we apply in the final section to show that J_1 is the only group satisfying Hypothesis 16.1.1.

Set $K := C_G(L)$, $H := N_G(L)$, and $H^* := H/K$. As L is standard in G, $H = N_G(K)$. Thus for each nontrivial 2-subgroup X of K, $N_G(X) \leq H$ by Remark 16.3.3. In particular,

$$G_z \leq H.$$

For $K' \in K^G$, define $L(K') := L^g$, where $g \in G$ with $K^g = K'$; as $N_G(K) = N_G(L) = H$, this definition is independent of the choice of g, and the set of such elements is a coset of H in G.

Recall $T_L = T \cap L$ and $T_C = T \cap K$.

Our discussion in this section will be based on an analysis of the set

$$\Delta = \Delta(K) := \{K' \in K^G - \{K\} : |N_{K'}(K)|_2 > 1\}.$$

To see that Δ is nonempty under our hypotheses, we appeal to I.8.2: Since G is simple, K is not normal in $G = \langle K^G \rangle$. Therefore if Δ is empty, then H is strongly embedded in G by I.8.2. Then by the Bender-Suzuki classification (see Theorem SE on p. 20 of [**GLS99**]) of simple groups with strongly embedded subgroups, $G = O^{2'}(G)$ is a Bender group, contrary to our assumption in Hypothesis 16.1.1 that G is not of even characteristic. Thus we conclude that

$$\Delta \text{ is nonempty.}$$

Recall that in Notation 16.2.2, $T \in Syl_2(G)$, and then $T \in Syl_2(H)$ using Theorem 16.2.4. Then as Δ is nonempty we can extend that earlier Notation by adopting:

NOTATION 16.4.1. $K' \in \Delta$, $L' := L(K')$, $H' := N_G(K')$, and $R \in Syl_2(N_{K'}(K))$ with $R \leq T$. For each involution r in R, set $L_r := O^2(C_L(r))$. Also set $H^* := H/K$.

Since $R \leq T$ in Notation 16.4.1, R normalizes T_L and T_C by Theorem 16.2.4. Our next result lists elementary properties of the members of $\Delta(K)$:

LEMMA 16.4.2. *(1)* $R \cong N_{T_C}(R) = C_{T_C}(R) = N_{T_C}(K') \in Syl_2(N_K(K'))$, *with* $R \cap K = 1$. *In particular R is faithful on L and* $|N_{K'}(K)|_2 = |N_K(K')|_2$.

(2) $L \not\leq H'$.

(3) $R = K' \cap T$.

(4) *There exists* $g \in G$ *with* $K' = K^g$ *and* $N_T(K') \leq T^g$.

(5) *For each* $1 \neq X \leq R$, $N_G(X) \leq H'$.

(6) *If* $N_T(R) \in Syl_2(N_H(R))$, *then* $C_T(R\langle z \rangle) \in Syl_2(C_G(R\langle z \rangle))$.

PROOF. Part (5) is a restatement of Remark 16.3.3. We apply parts (1) and (2) of I.7.7 with K', K, $T_C R$ in the roles of "K, K^g, S" to obtain $N_{T_C}(R) = C_{T_C}(R) \cong R$. By I.7.7.3, $N_{T_C}(R)$ is Sylow in $N_K(K')$, completing the proof of (1).

By (1), $|N_K(K')|$ is even, so (2) follows from 16.3.13.3. Part (3) holds since $R \in Syl_2(N_{K'}(K))$ and $R \leq T$. Let $g \in G$ with $K' = K^g$ and $N_T(K') \leq T' \in Syl_2(H')$. As $T^g \in Syl_2(H')$ there is $y \in H'$ with $T^{gy} = T'$, so replacing g by gy, we may take $N_T(K') \leq T^g$. Thus (4) holds.

If $N_T(R) \in Syl_2(N_H(R))$ then $C_T(R) \in Syl_2(C_H(R))$, so as $z \in Z(T)$, $C_T(R\langle z \rangle) \in Syl_2(C_H(R\langle z \rangle))$. Thus (6) follows as $G_z \leq H$. □

The next result says that Δ defines a symmetric relation on K^G.

LEMMA 16.4.3. *(1)* $K \in \Delta(K')$.
(2) $L' = [L', z]$.

PROOF. Part (1) is a consequence of 16.4.2.1. By 16.4.2.1, $z \in Z(T) \cap T_C \leq N_{T_C}(K')$. Thus (2) follows from 16.4.2.1 and the fact that Δ is symmetric. $\square$

LEMMA 16.4.4. *(1) Assume R is of order 2. Then $C_{T_C}(R) = \langle z \rangle$ is also of order 2, $m_2(RT_C) = 2$, and RT_C is dihedral or semidihedral. If furthermore $Z(L) \neq 1$, then $z \in Z(L)$.*
(2) If T_C is cyclic, then $K = T_C O(K)$ and $C_K(z) = T_C$.

PROOF. Assume R is of order 2. Then by 16.4.2.1, $C_{T_C}(R)$ is also of order 2, so that $C_{T_C}(R) = \langle z \rangle$. Hence by Suzuki's lemma (cf. Exercise 8.6 in [**Asc86a**]), RT_C is dihedral or semidihedral, so that (1) holds.

Next assume T_C is cyclic. Then by Cyclic Sylow 2-Subgroups A.1.38, $K = T_C O(K)$. Further as $G_z \leq H = N_G(K)$, $C_{O(K)}(z) \leq O(G_z) = 1$ by (E1), so that (2) holds. $\square$

Now we begin the process of obtaining restrictions on H, and in particular on the Sylow 2-subgroup R of $N_{K'}(K)$.

LEMMA 16.4.5. *Assume p is an odd prime with $m_p(L) > 1$, and i is an involution in K. Then either*
(1) $L = O^{p'}(C_G(i))$, or
(2) $p = 3$, $O^{3'}(C_G(i)^*) \cong PGL_3^\epsilon(2^n)$ or $L_3^{\epsilon, \circ}(2^n)$, with $2^n \equiv \epsilon$ mod 3, and $L = O^{3'}(LC_K(i))$. In particular, $O^{3'}(H^*) \cong PGL_3^\epsilon(2^n)$ or $L_3^{\epsilon, \circ}(2^n)$.

PROOF. Recall $L \trianglelefteq C_G(i)$ by Remark 16.3.3, and $O(L) = 1$ by (E1). Thus as L is in the list of (E2), $C_G(i)$ satisfies conclusion (1) or (2) of A.3.18, so the lemma holds. $\square$

The next lemma eliminates the shadow of $L_2(p^2)$ (p a Fermat or Mersenne prime) extended by a field automorphism, and the shadow of S_7. These groups are quasithin, and have a 2-central involution with centralizer $\mathbf{Z}_2 \times PGL_2(p)$, but the groups are neither simple nor of even type.

LEMMA 16.4.6. *If $L \cong L_2(q)$, q odd, then no involution in R induces an outer automorphism in $PGL_2(q)$ on L.*

PROOF. Let r denote an involution in R with $L\langle r \rangle \cong PGL_2(q)$. Recall q is a Fermat or Mersenne prime or 9 by (E2). Further if $q \neq 9$, $Aut(L) \cong PGL_2(q)$, so either $H^* \cong PGL_2(q)$, or $q = 9$ and $H^* \cong Aut(PGL_2(9)) \cong Aut(A_6)$.

If $q \neq 9$, let $R_0 := R \cap LK$; while if $q = 9$, let R_0 be the subgroup of R inducing automorphisms in S_6. Then $R = R_0\langle r \rangle$. If $R_0 \neq 1$ there is an involution r_0 in R_0, and $L = \langle C_L(r), C_L(r_0) \rangle \leq H'$ by 16.4.2.5, contrary to 16.4.2.2. Thus $R_0 = 1$, so $R = \langle r \rangle$ is of order 2. Hence by 16.4.4.1, $\langle z \rangle = C_{T_C}(R)$ is also of order 2. Choose T so that $N_T(R) \in Syl_2(N_H(R))$.

Let $E := \langle r, z \rangle$ and $T_E := C_T(E)$. As RT_L is dihedral, $C_{T_L}(r) =: \langle v \rangle$ is of order 2. Therefore as $C_{T_C}(R) = \langle z \rangle$, $T_E \cap LKR =: V = \langle v, z, r \rangle \cong E_8$, and either $H^* \cong PGL_2(q)$ and $T_E = V$, or $H^* \cong Aut(A_6)$ and $T_E \cong E_4 \times \mathbf{Z}_4$. Further $T_E \in Syl_2(C_G(E))$ by 16.4.2.6. As RT_L is dihedral, $rv \in r^{N_{T_L}(T_E)}$. From the structure of $Aut(L)$, $Z(T^*) = \langle v^* \rangle$, so $Z(T) \leq \langle v \rangle T_C \cap T_E = \langle v, z \rangle$, and hence $Z(T) = \langle z, v \rangle$.

We claim that z is weakly closed in $Z(T)$ with respect to G; the proof will require several paragraphs. Suppose the claim fails. Then using Burnside's Fusion Lemma A.1.35, $N_G(T)$ induces $\mathbf{Z}_3$ on $Z(T)$, and in particular is transitive on $Z(T)^\#$. Thus there are $h, k \in N_G(T)$ such that $v = z^h$ and $vz = z^k$; and in particular, $N_G(T)$ transitively permutes $\{T_C, T_C^h, T_C^k\} =: \mathcal{T}$. As K is tightly embedded in G, distinct members of $\mathcal{T}$ intersect trivially.

Since T_L and T_C^k are normal in T, $T_L \cap T_C^k$ is normal in T. Then as $Z(T) \cap T_L = \langle v \rangle$ is of order 2, v lies in $T_L \cap T_C^k$ if this group is nontrivial; but this is impossible as $v = z^h \in T_C^h$ and $T_C^h \cap T_C^k = 1$ by the previous paragraph. Therefore $[T_L, T_C^k] \le T_L \cap T_C^k = 1$, so that $T_C^k \cong T_C^{k*} \le C_{T^*}(T_L^*)$. Now by 16.1.6, either $C_{T^*}(T_L^*) = \langle v^* \rangle$ or $H^* \cong Aut(A_6)$ and $C_{T^*}(T_L^*) = \langle v^*, x^* \rangle$, where x induces a transposition on L and $L = \langle C_L(v), C_L(x) \rangle$. Thus by 16.4.2.2, $T_C^{k*} \ne \langle v^*, x^* \rangle$ when $H^* \cong Aut(A_6)$, so $|T_C^k| = 2$. Therefore $T_C = \langle z \rangle$ is of order 2. Then $z^h = v \in [T, T]$ since $L\langle r \rangle \cong PGL_2(q)$, so as $h \in N_G(T)$, $z \in [T, T]$. Thus $|T : T_L| > 4$, so that $q = 9$ and $H^* \cong Aut(A_6)$. Then $[T^*, T^*] = Y^*$, where $\mathbf{Z}_4 \cong Y \le T_L$, so $[T, T] \le T_C Y$ and hence $\langle v \rangle = \Phi([T, T]) \trianglelefteq N_G(T)$. This contradiction completes the proof of the claim that z is weakly closed in $Z(T)$ with respect to G. In particular, $z \notin v^G$.

By 16.4.2.4 there exists $g \in G$ with $K' = K^g$ and $N_T(K') \le T^g$. By 16.4.3.2, $L' = [L', z]$.

We next establish symmetry between L, r and L', z by showing that $L'\langle z \rangle \cong PGL_2(q)$. Assume otherwise and recall $E = \langle r, z \rangle$ and $T_E \in Syl_2(C_G(E))$ is abelian. Thus if $q = 9$, z does not induce an automorphism of L' contained in S_6. Therefore we may assume that z induces an inner automorphism on L'. By 16.4.2.1, R is Sylow in $C_{K'}(z)$, so as $z \in K'L'$ and R is of order 2, R is Sylow in K'. Hence $z^{L'} \cap Z(T^g) \ne \emptyset$, impossible as r is weakly closed in $Z(T^g)$ by the claim. Therefore $L'\langle z \rangle \cong PGL_2(q)$.

Recall that $rv \in r^L$, so if $v \in L'$, then by symmetry $zv \in z^{L'}$, contrary to the claim. Hence $v \notin L'$. Recall $\langle z, r, v \rangle = V = \Omega_1(T_E) \cong E_8$, and there is $t \in N_{T_L}(T_E)$ with $[t, r] = v$. As $v \notin L'$,

$$V \cap L' =: \langle u \rangle \ne \langle v \rangle,$$

and by symmetry there is $s \in N_{T^g}(T_E)$ with $[s, z] = u$. Set $X := N_G(V)$ and $X^+ := X/C_X(V)$, so that $X^+ \le GL(V) \cong L_3(2)$, and t^+ and s^+ are transvections in X^+ on V, with centers v, u, and axes $Z(T) = \langle z, v \rangle$, $Z(T^g) = \langle r, u \rangle$, respectively. As $Z(T) \ne Z(T^g)$ and $v \ne u$, $\langle t^+, s^+ \rangle$ is either D_8 or S_3 from the structure of $L_3(2)$. In the first case the unique hyperplane W of V normalized by $\langle t^+, s^+ \rangle \cong D_8$ is centralized by either t or s, say t; but then $W = Z(T)$ is not centralized by s, so that $z \ne z^s \in Z(T)$, contrary to the claim. Hence $\langle t^+, s^+ \rangle \cong S_3$, and so $V = V_1 \oplus V_2$, where

$$V_1 := \langle u, v \rangle = [V, \langle s, t \rangle] \cong E_4,$$

and

$$V_2 := C_V(\langle s, t \rangle) = Z(T) \cap Z(T^g) = \langle vz \rangle = \langle ur \rangle.$$

In particular as u is fused to v and all involutions in $D := T_L\langle u \rangle$ are in v^G. As $z \notin v^G$ by the claim, $z^G \cap D = \emptyset$.

Next if $T_C > \langle z \rangle$, then $N_{T_C}(V)^+$ is a transvection on V with axis $Z(T)$ and center $\langle z \rangle$, so $\langle N_{T_C}(V)^+, t^+, s^+ \rangle \cong S_4$ is the stabilizer in $GL(V)$ of vz, and hence is transitive on $V - \langle vz \rangle$, which is impossible since $v \notin z^G$ by the claim. Therefore $T_C = \langle z \rangle$, so $T_C T_L R = T_C \times D$ as $vz = ur$.

Suppose that $H^* \cong PGL_2(q)$. Then $|T : D| = 2$, so since we saw that $z^G \cap D = \emptyset$, $z \notin O^2(G)$ by Thompson Transfer, contrary to the simplicity of G.

Therefore $H^* \cong Aut(A_6)$, so $T_E \cong E_4 \times \mathbf{Z}_4$. Now $\langle s, t \rangle$ acts on T_E and $C_V(\langle t, s \rangle) = \langle vz \rangle$, so we conclude that $\langle vz \rangle = \Phi(T_E)$. Next $D = T_L \langle u \rangle$ and $T = T_E T_L$ with T_E abelian, so that $D \trianglelefteq T$. As $vz \notin D$ and $vz \in \Phi(T_E)$, $T/D \cong \mathbf{Z}_4$. Then since we saw that $z^G \cap D = \emptyset$, $z \notin O^2(G)$ by Generalized Thompson Transfer A.1.37.2, contrary to the simplicity of G, completing the proof of 16.4.6. $\qquad\square$

In the next lemma, we deal with the only case (other than those eliminated in 16.4.6 and the case $L \cong L_2(2^n)$) where for some involution r in R, L_r is not generated by its p-elements as p varies over those odd primes such that $m_p(L) > 1$.

LEMMA 16.4.7. *If* $L^* \cong M_{22}$, *then no involution in R induces an outer automorphism on L with $C_{L^*}(r) \cong Sz(2)/E_{16}$.*

PROOF. Assume r is a counterexample, and set $L_R := O^2(N_L(O_2(L_r)))$. Recall $L_r = O^2(C_L(r))$. Then $R = R_0 \langle r \rangle$, where $R_0 := R \cap LK$. From the structure of the extension of L^* by a 2-group in I.2.2.6a, L_r is isomorphic to $\mathbf{Z}_5/E_{16}$ if $|Z(L)| < 4$, but isomorphic to $\mathbf{Z}_5/Q_8 D_8$ if $Z(L) \cong \mathbf{Z}_4$.

We first show that $R_0 = 1$, so we assume that $R_0 \neq 1$ and derive a contradiction. Then as $L_r \leq H'$, $[L_r, C_{R_0}(r)] \leq L_r \cap K' =: Y \trianglelefteq L_r$, and $Y^* \neq 1$ as $C_{H^*}(L_r^*) = 1$. Thus Y^* contains the unique minimal normal subgroup $O_2(L_r^*)$ of L_r^*, so $O_2(L_r) \leq Y \leq K'$. It follows from 16.4.2.5 that $L_R \leq H'$. Further as $O(L') = 1$ by (E1), $L' = O^{3'}(H')$ by A.3.18, and hence $L_R \leq L'$. But then $O_2(L_r) \leq K' \cap L' \leq Z(L')$, impossible as $m_2(O_2(L_r)) \geq 3$ by the first paragraph, while $Z(L')$ is cyclic by 16.1.2.2.

Thus $R_0 = 1$, so $R = \langle r \rangle$ is of order 2. So by 16.4.4.1, $C_{T_C}(R) = \langle z \rangle$ is of order 2, and $T_C R$ dihedral or semidihedral. Thus if $Z(L) \neq 1$, $\langle z \rangle = \Omega_1(Z(L))$. Conversely if $z \in L$, then $Z(L) \neq 1$ and $\langle z \rangle = \Omega_1(Z(L))$. By 16.4.3.2, $L' = [L', z]$.

Again we establish symmetry between L, z, r and L', r, z, by showing that the action of z on L' is the same as that of r on L: First RT_C is dihedral or semidihedral and isomorphic to a Sylow group of H/L, so as $L_r \leq H'$ and L_r is irreducible on $O_2(L_r)/\Phi(O_2(L_r)) \cong E_{16}$, we conclude that $O_2(L_r) \leq L'$ and $O_2(L_r) \cap K' \leq \Phi(O_2(L_r))$. As z centralizes L_r, we conclude from 16.1.5.5 that z induces an outer automorphism of L' with $C_{L'/O_2(L')}(z) \cong Sz(2)/E_{16}$, establishing the symmetry.

In particular as $|Out(L')| = 2$, $z \notin [H', H']$. But if $Z(L) \cong \mathbf{Z}_4$, then we saw that $z \in [O_2(L_r)), O_2(L_r)$; so $|Z(L)| \leq 2$, and hence $O_2(L_r) \cong E_{16}$ from our earlier discusion. Further we saw $O_2(L_r) \leq L'$.

Set $E := \langle r, z \rangle$, $V := O_2(L_r)E$, and choose $g \in G$ with $K^g = K'$ and $N_T(K') \leq T^g$. Set $M := \langle N_H(V), N_{H'}(V) \rangle$ and $M^+ := M/C_M(V)$; then $N_L(V)^+ \cong S_5$.

Assume for the moment that $Z(L) \neq 1$, so that $Z(L) = \langle z \rangle$ is of order 2. Then V is a 6-dimensional indecomposable for $N_L(V)^+$, so by symmetry between r and z, also $N_{L'}(V)^+$ is indecomposable on V, and hence M is irreducible on V when $Z(L) \neq 1$. Now assume that $Z(L) = 1$. Then $V = \langle z \rangle \oplus O_2(L_r)\langle r \rangle$ as an $N_L(V)^+$-module, and $O_2(L_r)\langle r \rangle$ is a 5-dimensional indecomposable with trivial quotient.

Since we saw $O_2(L_r) \leq L'$, we conclude by symmetry that $O_2(L_r) \trianglelefteq M$ when $Z(L) = 1$.

Suppose that $T_C > \langle z \rangle$ and set $S_C := N_{T_C}(V)$. Then as $[V, S_C] = [R, S_C] \leq V \cap T_C = \langle z \rangle = C_{T_C}(R)$, S_C is of order 4 and $S_C^+ = \langle t^+ \rangle$, where t^+ induces a transvection on V with center $\langle z \rangle$ and axis $O_2(L_r)\langle z \rangle$. By symmetry there is $t_0 \in M \cap T_C^g - R$ which induces a transvection on V with center $\langle r \rangle$ and axis not containing z. As $\langle z \rangle$ is the center of t^+, $C_{M^+}(t^+) \leq C_M(z)^+ = S_C^+ \times N_L(V)^+$. In particular $S_C^+ = O_2(C_{M^+}(t^+))$, so either $O_2(M^+) = 1$ or $O_2(M^+) = S_C^+$. But in the latter case, M^+ acts on z, whereas $[t_0, z] \neq 1$. Thus $O_2(M^+) = 1$. Let $X^+ := \langle t^+, t_0^+ \rangle$; then $X^+ \cong S_3$ centralizes $O_2(L_r)$ from the structure of subgroups of the general linear group generated by a pair of transvections.

Assume first that $Z(L) = 1$. Then $O_2(L_r) \trianglelefteq M$ and $N_L(V)$ centralizes $V/O_2(L_r)$, so

$$[X^+, N_L(V)^+] \leq C_{M^+}(O_2(L_r)) \cap C_{M^+}(V/O_2(L_r)) \leq O_2(M^+) = 1$$

by Coprime Action, and hence $N_L(V)$ normalizes $[V, X] = \langle z, r \rangle$. This is a contradiction as r does not centralize L_V. Therefore $Z(L) \neq 1$, so M is irreducible on V by earlier remarks. As M^+ contains the transvection t^+, with $C_{M^+}(t^+) \leq C_M(z)^+ \cong \mathbf{Z}_2 \times S_5$, it follows from G.6.4 that $M^+ \cong S_7$ and V is the natural module. This is a contradiction as the noncentral chief factor for $N_L(V)^+$ on V is the natural module for $L_2(4)$, whereas the centralizer in A_7 of a transvection has the A_5-module as its noncentral chief factor. This contradiction completes the elimination of the case $T_C > \langle z \rangle$.

So $T_C = \langle z \rangle$. Then $T = T_L T_C R$ acts on V, so $T^+ = T_L^+ \cong D_8$ is Sylow in M^+. We may apply A.3.12 to conclude that $L_V \leq N \in \mathcal{C}(M)$, and the embedding of L_V^+ in N^+ is described in A.3.14. As D_8 is Sylow in N^+T^+ and there is a nontrivial $\mathbf{F}_2 N^+T^+$-representation of dimension 6, we conclude that either $N^+ = L_V^+$ or $N^+ \cong A_7$. In the first case, M acts on $C_V(L_V) = \langle z \rangle$, and then by symmetry, M also centralizes r, a contradiction as L_V does not centralize r. In the second case, as $m(V) = 6$ and $S_5 \cong N_L(V)^+ = C_{M^+}(z)$, we conclude that V is the core of the 7-dimensional permutation module for M^+, and that z is of weight 2 in that module. This is impossible, as we saw at the end of the previous paragraph. This contradiction completes the proof of 16.4.7. $\square$

LEMMA 16.4.8. *Let r be an involution of R. Then either*

(1) $L_r \leq L'$, *or*

(2) $O^{3'}(H^*) \cong PGL_3^\epsilon(2^n)$ *or* $L_3^{\epsilon, \circ}(2^n)$, $2^n \equiv \epsilon \bmod 3$, r *induces an inner automorphism on L, $n \neq 3$, and $O^3(L_r) \leq L'$.*

PROOF. As usual recall that $L_r \leq C_G(r) \leq H'$. Let $\bar{H}' := H'/K'$, and define

$$\Lambda(L_r) := \langle O^{p'}(L_r) : p \text{ is an odd prime such that } m_p(L) > 1 \rangle.$$

Applying 16.4.5 with L', r in the roles of "L, i", either $\Lambda(L_r) \leq L^g$ or $O^{3'}(\bar{H}') \cong PGL_3^\epsilon(2^n)$ or $L_3^{\epsilon, \circ}(2^n)$.

Assume first that the latter case holds. In particular, $n \geq 2$.

We will first treat the subcase where r induces an outer automorphism on L. Then by 16.1.4, either L_r is $PSL_3(2^{n/2})$, $U_3(2^{n/2})$ with $n > 2$, or $L_2(2^n)$; or r induces a graph-field automorphism on $L^* \cong L_3(4)$ and $L_r \cong E_9$. As $m_3(L') = 2$, $Z(L')$ is a 2-group, and $L'K'$ is an SQTK-group, $C_{K'}(r)$ is a $3'$-group; so as

$O_{3'}(L_r) = 1$, L_r is faithful on L' and $\bar{L}_r \trianglelefteq O^2(C_{\bar{H}'}(\bar{z}))$. We conclude from 16.1.4 first that z induces an outer automorphism on L', and second as $O^{3'}(\bar{H}')$ is $PGL_3^\epsilon(2^n)$ or $L_3^{\epsilon,\circ}(2^n)$ that either $\bar{L}_r \leq \bar{L}'$, or $O^2(\bar{H}') \cong PGL_3(4)$ and z induces a graph-field automorphism on L' with $O^2(C_{L'}(z)) \cong E_9 \cong L_r$. In the former case conclusion (1) holds, so we may assume the latter. Then $|C_{L^*}(r^*) : C_L(r)^*| \leq |Z(L)| =: m$, and by 16.1.2.2, $m \leq 4$. On the other hand, $C_{L^*}(r^*)$ contains a Q_8-subgroup faithful on L_r^*, so if $m < 4$, then r centralizes $x \in L$ with x^2 inverting L_r; but then $\bar{x}^2 \in \bar{L}'$ by 16.4.2.5, so that $L_r = [L_r, x^2] \leq L'$, and conclusion (1) holds again. Finally if $m = 4$, then r centralizes $Z(L)$ by I.2.2.3b, so that $Z(L) \leq C_G(r) \leq H'$ using 16.4.2.5. Then as $m_2(C_{Aut(L')}(L_r)) = 1$, $Z(L) \cap K' \neq 1$, contradicting K tightly embedded in G.

Next we treat the subcase where r induces an inner automorphism on L. Recall here that $|C_L(r)/O_2(C_L(r))| = (2^n - \epsilon)/3$. Assume that $n > 3$. Then there are prime divisors $p > 3$ of $|C_L(r)|$, and $m_p(L) > 1$ for each such p; hence $L = O^{p'}(H)$ by 16.4.5, so that $O^{p'}(L_r) \leq O^{p'}(H') = L'$. Thus $O^3(L_r) \leq L'$, so that conclusion (2) holds. Finally suppose that $n = 3$, so that $L \cong U_3(8)$, $|L_r : T_L| = 3$, and $O^2(L_r)$ centralizes $Z(T_L)$. Let x be of order 3 in L_r; we may assume (1) fails, so $x \notin L'$. By 16.4.5, $O^{3'}(\bar{H}') \cong PGU_3(8)$ or $U_3^\circ(8)$, and $C_{K'}(r)$ is a $3'$-group, so as $x \notin L'$, $\bar{x} \notin \bar{L}'$. As $1 \neq \bar{z} \in C_{\bar{H}'}(x)$, $C_{\bar{H}'}(x)$ is of even order, and $\bar{x}$ is of order 3; so from the structure of $Aut(U_3(8))$, and as $O^{3'}(\bar{H}') \cong PGU_3(8)$ or $U_3^\circ(8)$, $\bar{x} \in \bar{L}'$, a contradiction.

This completes the treatment of the case where $O^{3'}(\bar{H}')$ is described in case (2) of 16.4.5, so we may assume that $\Lambda(L_r) \leq L^g$. In particular if $L_r = \Lambda(L_r)$, then conclusion (1) holds, so we may assume that $\Lambda(L_r) < L_r$; thus $L_r \neq 1$.

Suppose first that $L \cong L_2(q)$, q odd. Then q is a Fermat or Mersenne prime or 9 by (E2), and r does not induce an outer automorphism in $PGL_2(q)$ on L by 16.4.6. If $q = 9$ and r induces an outer automorphism in S_6, then $L_r \cong A_4$, contrary to our assumption that $\Lambda(L_r) < L_r$. Thus we conclude from 16.1.4 that r induces an inner automorphism of L. But then $C_L(r)$ is a 2-group, so $L_r = 1$, again contrary to assumption.

Next suppose $L^* = X(2^n)$ is of Lie type. We can assume by the previous paragraph that L is not $L_2(4) \cong L_2(5)$, $L_3(2) \cong L_2(7)$, or $Sp_4(2)' \cong L_2(9)$. Hence as $\Lambda(L_r) < L_r$, we conclude from 16.1.4 and 16.1.2.1 that $L \cong L_2(2^n)$ for $n > 2$ even, and r induces a field automorphism on L. For each involution $i \in C_{T_C}(r)$, $L_r \cong L_2(2^{n/2})$ is a component of $C_{H'}(\langle i, r \rangle)$ using 16.4.2.5. Therefore either $L_r \leq L'$ so that conclusion (1) holds, or $L_r \leq K'$, and we may assume the latter. Then using 16.4.2.5,

$$L = \langle C_L(j) : \ j \text{ is an involution in } L_r \rangle \leq H',$$

contrary to 16.4.2.2.

It remains only to consider the cases in (E2) where $L \cong L_3(3)$ or L^* is sporadic. Inspecting centralizers of involutory automorphisms of L using 16.1.5, we conclude that $L_r = \Lambda(L_r)$, except in the situation which we already eliminated in 16.4.7. This completes the proof of 16.4.8. $\qquad\square$

We now focus on those members of Δ with the property that involutions of R induce inner automorphisms on L. Set

$$\Delta_0 = \Delta_0(K) := \{K' \in \Delta : \Omega_1(R) \leq KL \text{ for } R \in Syl_2(N_{K'}(K))\}.$$

We will first show in 16.4.9.2 that Δ_0 is nonempty. The shadow of S_{10} is eliminated toward the end of the proof: that is, a transposition in S_{10} is a 2-central involution with centralizer $\mathbf{Z}_2 \times S_8$, such that $\Delta_0 = \emptyset$; of course S_{10} is neither simple nor quasithin.

LEMMA 16.4.9. *(1) If $K' \notin \Delta_0$, then $|R| = 2$.*

(2) $\Delta_0 \neq \emptyset$.

(3) Assume $J \in K^G$, and some involution i in J induces a nontrivial inner automorphism on L. Then $J \in \Delta_0$.

PROOF. First assume (1) and the hypothesis of (3). Then $i \in LK - K$, so $J \neq K$ and $|N_J(K)|_2 > 1$, and hence $J \in \Delta$. Now if $|N_J(K)|_2 > 2$ then $J \in \Delta_0$ since we are assuming (1), while if $|N_J(K)|_2 = 2$, then $\langle i \rangle \in Syl_2(N_J(K))$ with $i \in LK$, so that $J \in \Delta_0$ by definition. Thus (1) implies (3), so it remains to establish (1) and (2).

If $\Delta = \Delta_0$ then (1) is vacuous and (2) holds as Δ is nonempty, so we may assume $K' \in \Delta - \Delta_0$ and pick some involution $r \in R$ inducing an outer automorphism on L. Then by 16.4.8, $L_r \leq L' \leq C_G(R)$.

We now prove (1). By inspection of the centralizers of involutory outer automorphisms of L^* listed in 16.1.4 and 16.1.5, one of the following holds:

(I) $C_{H^*}(L_r^*) = \langle r^* \rangle$.

(II) $L^* T^* \cong S_8$ and r^* is of type $2^3, 1^2$.

(III) $L^* \cong M_{12}$ and $L_r^* \cong A_5$.

In case (I), as R^* centralizes L_r^* and $R \cong R^*$, $R = \langle r \rangle$ is of order 2, and hence (1) holds in this case. Thus we may assume that (II) or (III) holds. In either case, $C_{H^*}(L_r^*) \cong E_4$, so either R is of order 2 and (1) holds, or $R^* = C_{H^*}(L_r^*) \cong E_4$, and we may assume the latter.

Suppose case (II) holds. Then there is $s \in R^\#$ with s^* of type $2, 1^6$. But then $[R^*, L_s^*] \neq 1$, a contradiction since $L_s \leq L' \leq C_G(R)$ by 16.4.8.

Therefore case (III) holds. Then there is $s \in R^\#$ with $s^* \in L^*$ but s^* not 2-central in L^*. Let s_L denote the projection of s on L. If $O_2(L) \neq 1$, then from I.2.2.5b, s_L is of order 4, so $s = s_L s_C$ with $s_C \in N_{T_C}(K')$ of order 4, impossible as $N_{T_C}(K') \cong R \cong R^* \cong E_4$ by 16.4.2.1. Therefore $O_2(L) = 1$, and hence $C_{L^*}(s^*) = C_L(s)^*$, with $C_L(s) \leq N_G(K')$ by 16.4.2.5. Thus $\langle s_L \rangle = [C_{T_L}(s), r] \leq T \cap K' = R$ by 16.4.2.3, and hence $s = s_L \in L$. Then

$$T_C = C_{T_C}(s) \leq N_{T_C}(K') = C_{T_C}(R) \cong R \cong E_4,$$

so $T_C \cong E_4$ centralizes R. Then as $L^* \langle r^* \rangle = Aut(L^*)$, $T = T_C T_L R \leq C_G(T_C)$ so $T_C \leq Z(T)$. Hence R is in the center of each Sylow 2-subgroup of H' containing R. As $C_T(s) \leq H'$ and $[R, C_{T_L}(s)] \neq 1$, this is a contradiction. Thus (1) is established.

We may assume that (2) fails, and it remains to derive a contradiction. Now $R = \langle r \rangle$ is of order 2 by (1). By 16.4.4.1, $C_{T_C}(r) = \langle z \rangle$ is of order 2, and $T_C R$ is dihedral or semidihedral. Set $E := \langle r, z \rangle$. By (1) and (3):

$$\text{If } J \in \Delta, \text{ then } |N_J(K)|_2 = 2 \text{ and } J \cap KL \text{ is of odd order.} \qquad (*)$$

By 16.4.3.1, $K \in \Delta(K')$, so we have symmetry between K and K'. Thus applying (*) with the roles of K and K' reversed, we conclude that z induces an outer automorphism on L'.

Next we show:

(+) If $r^l = rv$ for some $l \in L$ and $1 \neq v \in C_L(r)$, then r^l acts on K' with $\langle r \rangle \in Syl_2(C_{K'}(r^l))$ and $v \notin L'$.

For assume the hypotheses of (+), and let $J := (K')^l$. Then $r^l \in C_J(r) \leq N_J(K')$, and as R has order 2, $r^l \notin K'$, so that $J \neq K'$. Thus $J \in \Delta(K')$, and then by (*), $R \in Syl_2(N_{K'}(J))$, and also $J \cap K'L'$ is of odd order so that $v \notin L'$. As $\langle r \rangle = R \in Syl_2(N_{K'}(J))$ and $C_G(r^l) \leq N_G(J)$, also $\langle r \rangle \in Syl_2(C_{K'}(r^l))$, completing the proof of (+).

As r induces an outer automorphism on L, by 16.1.6, r does not centralize T_L unless L is A_6. In the first case, there is $l \in T_L$ with $r \neq r^l \in C_G(r)$, and in the second by inspection there is $l \in L$ with this property; thus in any case there is $l \in L$ with $r \neq r^l \in C_G(r)$. Let $J := (K')^l$.

Suppose that $T_C > \langle z \rangle$. Then r^l normalizes some subgroup $X = \langle x, r \rangle$ of order 4 in K', and $\langle r \rangle \in Syl_2(C_{K'}(r^l))$ by (+), so that $\langle r^l, X \rangle \cong D_8$, and hence $v := rr^l = r^{lx} \in J^x \cap L$. Thus $J^x \notin \Delta$ by (*), so $J^x = K$ and hence $v \in K \cap L = Z(L)$, so $z = v \in L$ as $\langle z \rangle = C_{T_C}(r)$. Further $r^L \cap C_G(r) = \{r, rz\}$, so $|C_{L^*}(r^*) : C_L(r)^*| = 2$ and $r^*C_L(r)^* \cap r^{*L^*} = \{r^*\}$, so there are involutions in $C_{L^*}(r^*) - C_L(r)^*$. Suppose $L^* \cong L_3(4)$ or M_{22}. Examining 16.1.4 and 16.1.5 for outer automorphisms r with $C_{L^*}(r)$ not perfect, we conclude that either $L^* \cong L_3(4)$, r is a graph-field automorphism, and $C_{L^*}(r^*) \cong Q_8/E_9$; or $L^* \cong M_{22}$ and $C_{L^*}(r^*) \cong \mathbf{Z}_4/\mathbf{Z}_5/E_{2^4}$. However in both cases each subgroup of $C_{L^*}(r^*)$ of index 2 contains all involutions in $C_{L^*}(r^*)$, contrary to an earlier remark.

We have shown that either $\langle z \rangle = T_C \in Syl_2(K)$, or $z \in L$ and L^* is not $L_3(4)$ or M_{22}.

As r induces an outer automorphism on L, $L_r \leq L'$ by 16.4.8. So if $rv = r^l$ for some $l \in L$ and $1 \neq v \in L_r$, then $v \in L'$ which is contrary to (+). Thus:

$$r^L \cap rL_r = \{r\}. \tag{!}$$

It is now fairly easy to eliminate most possibilities for the involutory outer automorphism r on L in 16.1.4 and 16.1.5; indeed the next few paragraphs will be devoted to the reduction to the following cases:

(i) $L \cong A_6$ or A_8.

(ii) $L^* \cong L_3(4)$, and r^* induces a graph-field automorphism on L^*.

We may assume that neither (i) nor (ii) holds, and will derive a contradiction. Since r induces an outer automorphism on L, L is not $L_3(2)$ by 16.4.6. Then as (ii) does not hold, by inspection of the outer automorphisms in the remaining cases in 16.1.4 and 16.1.5, L_r is of even order. Choose notation so that $C_T(r) \in Syl_2(C_H(r))$.

Suppose first that $L \cong M_{22}$ or HS, and let $T_r := T \cap RC_L(r))$; then $Z(T_r) = R \times Z_r$, where $\mathbf{Z}_2 \cong Z_r = \langle v \rangle \leq L_r$. So as $T_r < RT_L \in Syl_2(RL)$, $rv \in r^{N_{T_L}(T_r)}$, contrary to (!). Thus if $L^* \cong M_{22}$ or HS, we may assume that $O_2(L) \neq 1$.

Now assume that $O_2(L) = 1$, and recall L is not A_6, A_8, M_{22}, or HS by assumption. Therefore by inspection of the outer automorphisms in the possibilities remaining in 16.1.4 and 16.1.5, L is transitive on the involutions in rL. Then since L_r is of even order (recalling $L^* \not\cong L_3(4)$ as we are assuming that (ii) fails), (!) supplies a contradiction.

Thus to establish our reduction, it remains to treat the case $1 \neq O_2(L) = Z(L)$. Since L^* admits the outer automorphism r^* of order 2, we conclude from 16.1.2.1 that L^* is $L_3(4)$, $G_2(4)$, M_{12}, M_{22}, J_2, or HS. If $|T_C| = 2$, then $\langle z \rangle = Z(L)$. If

$|T_C| > 2$, we showed $z \in L$ and L^* is not $L_3(4)$ or M_{22}. So in any case $z \in Z(L)$; it then follows by 16.1.2.2 that $Z(L) = \langle z \rangle$ is of order 2.

Suppose first that $L^* \cong L_3(4)$, and recall we are assuming that r^* does not induce a graph-field automorphism on L^*. Assume that r^* induces a field automorphism on L^*, so that $L_r \cong L_3(2)$ by 16.1.4.4. Let $L_{r,1}$ be a maximal parabolic of L_r; then there is an r-invariant maximal parabolic L_1 of L with $L_{r,1} \leq L_1$ and $O_2(L_{r,1}) \leq O_2(L_1)$. Then $O_2(L_1)$ is transitive on $rO_2(L_{r,1})$, contrary to (!). Therefore by 16.1.4, r^* induces a graph automorphism on L^*, so $L_r \cong L_2(4)$ by 16.1.4.6. Let $U := T \cap L_r \cong E_4$ and $V := EU$; thus $V \cong E_{16}$ and as $Z(L) = \langle z \rangle$, $N_{T_L}(V)/V \cong E_4$ induces a group of $\mathbf{F}_2$-tranvections on V with axis $V_0 := \langle z, U \rangle \cong E_8$. Now $R \leq T \leq N_G(T_L)$, so that $[N_{T_L}(V), r] =: W \leq U$ is a hyperplane of V_0, and hence $W = U$. Then for $1 \neq v \in W$, $rv \in r^L$, contrary to (!). This establishes the reduction to (ii) when $L^* \cong L_3(4)$.

Next suppose that $L^* \cong G_2(4)$, M_{12}, J_2, or HS. Then $|Z(L)| = 2$, so $Z(L) = \langle z \rangle$; and from I.2.2.5b, r inverts y of order 4 in L with $y^2 = z$, so $r^y = rz$. Hence the involutions in $rZ(L)$ are fused under L: for example, from the description of $C_{L^*}(r^*)$ in 16.1.4 or 16.1.5, and observing that $C_{L^*}(r^*) \cong G_2(2)$ when $L^* \cong G_2(4)$, choose $y^* \in C_{L^*}(r^*) - L_r^*$, and observe $y^*r^* \in r^{*L}$, so r inverts y as required. But we showed earlier that $r^*v^* = r^{*l}$ for some involution $v^* \in L_r^*$ and $l \in L$, and now from I.2.2.5a, we may choose a preimage v to be an involution. So as $r^l \in rvZ(L)$ and the involutions in $rZ(L)$ are fused, $rv \in r^L$, again contrary to (!).

This leaves the case $L^* \cong M_{22}$, where in view of 16.1.5 and 16.4.7, $C_{L^*}(r^*) = L_r^* \cong L_3(2)/E_8$. Choose an involution s^* inducing an outer automorphism of the type in 16.4.7, with $r^* \in s^*O_2(L_s)$; then as we saw in the proof of 16.4.7, since $Z(L) = \langle z \rangle$ is of order 2, the group $O_2(L_s^*) \cong E_{16}$ splits over $Z(L)$, so $J(C_T(r))$ is the group V of rank 6 in the proof of 16.4.7, now with s in the role of "r". Now the argument in the final paragraph of that proof again supplies a contradiction. This finally completes the proof of the reduction to (i) or (ii).

Thus to prove (2), it remains to eliminate the groups in parts (i) and (ii) of the claim. Choose T so that $N_T(R) \in Syl_2(N_H(R))$. As in 16.4.2.4, choose $g \in G$ with $K^g = K'$ and $N_T(K') \leq T^g$.

Suppose first that (ii) holds. Then $L_r \cong E_9$ by 16.1.4.5, and $N_{LE}(E) = L_rQ$, where $Q/E \cong Q_8$ and $\langle z, r \rangle = E = C_Q(L_r)$.

Assume that $O_2(L) \neq 1$. Then since $T_C = \langle z \rangle$ is of order 2, $\langle z \rangle = Z(L)$. Also $C_Q(E)/E \cong \mathbf{Z}_4$ is irreducible on L_r, and Q induces a transvection on E with center $\langle z \rangle$. By symmetry, Q^g induces a transvection on E with center $\langle r \rangle$, so $X := \langle Q, Q^g \rangle$ induces S_3 on E. Since $N_T(R) \in Syl_2(N_H(R))$, $C_T(E) \in Syl_2(C_G(E))$ by 16.4.2.6, so by a Frattini Argument, $Y := O^2(N_G(E) \cap N_G(C_T(E)))$ is transitive on $E^\#$. Hence as $L_r \leq L'$,

$$L_r = O^2\left(\bigcap_{y \in N_G(E)} L^y \right) \trianglelefteq N_G(E),$$

and then as $C_T(E)$ is irreducible on L_r, either $[L_r, Y] = 1$ or Y induces $SL_2(3)$ on L_r. In the former case as $C_T(E)$ is irreducible on L_r, there is an element of order 3 in $Y - L_r$ centralizing L_r, contradicting G quasithin. In the latter, $N_T(E)Y/E \cong GL_2(3)$ has a unique Q_8-subgroup, so that $Q \trianglelefteq N_T(E)Y$, impossible as $Aut_Q(E)$ is not normal in $Aut_{QY}(E)$.

Therefore $O_2(L) = 1$, so $L \cong L_3(4)$. Then Q centralizes E, and there is a complement P to E in Q. Let $\bar{H}' := H'/K'$. Each Q_8-subgroup of $C_{\bar{H}'}(\bar{z})$ is contained in $\bar{L}'\langle \bar{z} \rangle$, so that $P \leq T^g \cap K'L'E \leq L'E$. Hence $1 \neq \Phi(P) \leq L'$. This contradicts (+) as $\Phi(P) = \langle v \rangle$ where v is an involution in $C_L(r)$ with $rv \in r^L$.

Thus case (i) holds, so $L \cong A_6$ or A_8. Since $R = \langle r \rangle$, $T_C = \langle z \rangle$, and $N_T(K') \leq T^g$, $r = z^g$. Further r induces an outer automorphism on L, and RL is not $PGL_2(9)$ by 16.4.6, so we conclude that $z^G \cap LT \subseteq LE$ and $LE = LR \times \langle z \rangle \cong S_n \times \mathbf{Z}_2$, with $n := 6$ or 8. Represent LE on $\Omega := \{1, \ldots, n\}$ with kernel $T_C = \langle z \rangle$. Observe that either $LT = LE$ and $T = T_LE$, or $H^* \cong Aut(A_6)$ and $|LT : LE| = |T : T_LE| = 2$.

By (*), $z^G \cap L\langle z \rangle = \{z\}$, so since $z^G \cap LT \subseteq LE$, we conclude $z^G \cap LT \subseteq \{z\} \cup rL \cup rzL$. If $n = 6$, we may choose notation so that r induces a transposition on Ω, and if $n = 8$, r is either a transposition, or of type $2^3, 1^2$. In any case, setting $m := n/2 + 1$, there is a subgroup A_0 of T_LR generated by a set of $m - 1$ commuting transpositions, and by the $m - 1$ conjugates of r under L in A_0. Thus there is $E_{2^m} \cong A \leq LE$ with $\alpha := \{z\} \cup (r^L \cap A)$ of order m.

Let a, b, c be a triple from α. Then $c := z^x$ for suitable $x \in G$; set $X := L^x\langle z^x \rangle$. By the previous paragraph,
$$z^G \cap X = \{z^x\}. \qquad (**).$$
Then $a, b \in L^xA - X$, so as $z^G \cap L^xT^x \subseteq L^xE^x$ and $|(LE)^x : X| = 2$, $ab \in X$. Hence by (**), neither ab nor abz^x is in z^G. Thus no product of two or three members of α is in z^G.

Now assume $n = 8$, let $\alpha = \{a_1, \ldots, a_5\}$, and take $a_5 = z^x$. By the previous paragraph, a_1a_2 and a_3a_4 are in X, so $a_1a_2a_3a_4 \in X$. Hence by (**), neither $a_1a_2a_3a_4$ nor $a_1a_2a_3a_4z^x = a_1a_2a_3a_4a_5$ is in z^G. Thus no product of four or five members of α is in z^G, so $z^G \cap A = \alpha$. But each involution in T is fused into A under L, so we conclude that $z^G \cap \langle rz \rangle L = \emptyset$, and hence $z \notin O^2(G)$ by Thompson Transfer, [1] contrary to the simplicity of G.

Therefore $n = 6$. Set $\alpha = \{a_1, \ldots, a_4\}$ and $z^G \cap T =: \beta$. First assume $H^* \not\cong Aut(A_6)$. Then $LT = LE = LA$ and $T = T_LR \times \langle z \rangle$. If $t := a_1a_2a_3a_4 \notin z^G$, then the argument of the previous paragraph supplies a contradiction, so $t \in z^G$. Hence $\beta = \alpha \cup \{t\}$ is of order 5. But $\mathcal{A}(T)$ is of order 2, while the member A of $\mathcal{A}(T)$ is normal in T, so A is weakly closed in T with respect to G. Hence by Burnside's Fusion Lemma A.1.35, $N_G(A)$ is transitive on β. Then as $N_{G_z}(A)/A$ induces S_3 on β, $Aut_G(A)$ is a subgroup of S_5 of order 30, a contradiction.

Therefore $H^* \cong Aut(A_6)$. Thus $\mathcal{A}(T) = \{A, A^s\}$ for $s \in T - AT_L$, so $C_G(z)$ is transitive on $\mathcal{A}(C_G(z))$; hence by A.1.7.1, $N_G(A)$ is transitive on β. In particular as $|N_T(A) : A| = 2$, $|\beta| \neq 4$, so $t \in \beta$ and $|\beta| = 5$, for the same contradiction as at the end of the previous paragraph.

This finally completes the proof of 16.4.9. $\qquad \square$

In view of 16.4.9:

In the remainder of this chapter, we choose $K' \in \Delta_0$. Thus $\Omega_1(R) \leq KL$, where $R \in Syl_2(N_{K'}(K))$.

LEMMA 16.4.10. $R \in Syl_2(K')$.

PROOF. This is more or less the argument on page 101 of [**Asc76**]: By 16.4.2.1, $R \cong N_{T_C}(K') = C_{T_C}(R) =: S$. Assume $R \notin Syl_2(K')$; then also $S < T_C$, so in

---

[1]Here we are in particular eliminating the shadow of S_{10}.

particular T_C does not centralize R. For $1 \neq r \in R$, observe by 16.4.2.5 that $C_{T_C}(r) \leq N_{T_C}(K') = C_{T_C}(R) = S$, so that $C_{T_C}(r) = S$. By parts (4) and (6) of I.7.7, $S \trianglelefteq T_C$ and R is abelian.

Set $R_0 := R \cap LK$, $U := \Omega_1(R)$, and recall $K' \in \Delta_0$ so that $U \leq R_0$. In particular $R_0 \neq 1$. For $r \in R_0$, we can write $r = s(r)l(r)$ with $s(r) \in T_C$ and $l(r) \in L$. Observe that $l(r)$ is determined up to an element of $L \cap K = Z(L)$, and then $s(r)$ is determined by $l(r)$. Also $s(r)Z(L) \subseteq C_{T_C}(s(r)) \leq C_{T_C}(r)$. Then as $C_{T_C}(r) = S < T_C$, $s(r)Z(L) \subseteq S$, but $s(r) \notin Z(T_C)$. Indeed $\hat{s}(r) := s(r)Z(L)$ is a uniquely determined element of $\hat{S} := S/Z(L)$, and $\hat{s} : R_0 \to \hat{S}$ is a group homomorphism. Also $R \cap L = 1$ as $C_{T_C}(r) = S < T_C$ for $r \in R^{\#}$, so $\ker(\hat{s}) = R_0 \cap Z(L) = 1$ and hence $\hat{s}$ is injective. For $R_1 \leq R_0$, let $s(R_1)$ be the preimage in S of $\hat{s}(R_1)$. As $s(r) \notin Z(T_C)$ for each $r \in R^{\#}$, $s(U) \cap Z(T_C) = 1$.

As R is abelian, U is elementary abelian. Suppose $s(U)$ is elementary abelian. Then since $s(U)/Z(L) \cong \hat{s}(U) \cong U \cong \Omega_1(S) \geq s(U)$, we conclude that $Z(L) = 1$ and the map $s : U \to \Omega_1(S)$ is an isomorphism. So as $S \trianglelefteq T_C$, we have a contradiction to $s(U) \cap Z(T_C) = 1$.

Thus for some $u \in U^{\#}$, we may choose $s(u)$, and hence also $l(u)$, of order at least 4. Thus $1 \neq l(u)^2 \in Z(L)$, so $Z(L) \neq 1$. Therefore L^* is not $L_3(4)$, since $Z(L)$ is elementary abelian, and from I.2.2.3b, involutions in L^* lift to involutions in L. Hence by inspection of the remaining groups in 16.1.2.1, $Out(L)$ has Sylow 2-groups of order at most 2, so $|R : R_0| \leq 2$. Therefore

$$|R| = |S| \geq |\hat{s}(R_0)||Z(L)| = |R_0||Z(L)| \geq |R||Z(L)|/2.$$

So as $|Z(L)| \geq 2$, all inequalities are equalities, and hence $S = s(R_0)$ with $Z(L) = \langle l(u)^2 \rangle$ of order 2. Thus there is $1 \neq v \in U$ with $E := \langle s(v) \rangle Z(L)$ a normal subgroup of T_C of order 4. Then $|T_C : C_{T_C}(s(v))| = |T_C : C_{T_C}(v)| = 2$, so $S = C_{T_C}(s(v))$ is of index 2 in T_C. Thus $RS \trianglelefteq RT_C$, so as $N_{T_C}(R) = S \cong R$ is abelian, while R is a TI-set in $T_C R$ under $N_G(T_C R)$ by I.7.2.3, $SR = R \times R^t$ for $t \in T_C - S$. Therefore $\Omega_1(S) = C_{UU^t}(t) = [U, t]$, and hence $s(w) \in Z(T_C)$ for each $w \in U$, contrary to $s(U) \cap Z(T_C) = 1$. This contradiction completes the proof. $\square$

The next lemma summarizes some fundamental properties of members of Δ_0; in particular it shows that Δ_0 defines a symmetric relation on K^G.

LEMMA 16.4.11. *(1)* R *centralizes* $T_C \cong R$. *In particular,* $T_C \leq H'$.
(2) $K \in \Delta_0(K')$.
(3) *If we choose* g *as in 16.4.2.4, then* $R = T_C^g$.

PROOF. By 16.4.10, $R \in Syl_2(K')$, so that $R \cong T_C$. Then since $R \cong C_{T_C}(R)$ by 16.4.2.1, T_C centralizes R, and so (1) holds using 16.4.2.5.

By 16.4.3.1, $K \in \Delta(K')$. Suppose that $K \notin \Delta_0(K')$. Then by 16.4.9.1, $|T_C| = 2$, and by 16.4.9.1, z induces an outer automorphism on L'. Applying 16.4.8 with the roles of K and K' reversed, we conclude that $L_z := O^2(C_{L'}(z)) \leq L$. As $C_G(r) \leq N_G(L')$, we conclude that $L_z \trianglelefteq L_r$. Comparing the fixed points of outer and inner automorphisms of order 2 in 16.1.4 and 16.1.5, we conclude $L^* \cong M_{12}$ and $L_r^* = L_z^* \cong A_5$. As r induces an inner automorphism on L, if $Z(L) \neq 1$, then I.2.2.5b says that the projection r_L of r on L is of order 4, so $r = r_C r_L$ with $r_C \in T_C$ of order 4, contradicting $|T_C| = 2$. Thus $Z(L) = 1$, so $C_L(r) \cong \mathbf{Z}_2 \times S_5$. However $C_L(r) \leq C_{H'}(z)$ by 16.4.2.5, and as z induces an outer automorphism on L', no element of $C_{H'}(z)$ induces an outer automorphism on L_z. Therefore (2) holds.

Choose g as in 16.4.2.4; that is, so that $N_T(K') \leq T^g$. Then $R \leq T^g_C \leq K'$ and $R \in Syl_2(K')$, so $R = T^g_C$, establishing (3). $\square$

16.5. Identifying J_1, and obtaining the final contradiction

In this final section, we first see that $G \cong J_1$ when $L/Z(L)$ is a Bender group. Then we eliminate all other possiblities for L appearing in (E2), to establish our main result Theorem 16.5.14.

Recall R is faithful on L by 16.4.2.1 and $H^* = H/K$. Set $U := \Omega_1(R)$; as $K' \in \Delta_0$,

$$U \leq KL.$$

Let u denote an element of $U^\#$, and set $U_C := \Omega_1(T_C)$. By 16.4.11.3, if we choose g as in 16.4.2.4, then $R = T^g_C$ and hence $U = U^g_C$; in particular, $U_C \in U^G$.

PROPOSITION 16.5.1. *If L^* is a Bender group, then $G \cong J_1$.*

PROOF. By hypothesis, $L^* \cong L_2(2^n)$, $U_3(2^n)$, or $Sz(2^n)$. Set $U_L := \Omega_1(T_L)$. Then there is $X \leq N_L(T_L)$ with $X \cong \mathbf{Z}_{2^n-1}$ and X regular on $U_L^{*\#}$. Now either $Z(L) = 1$, or from 16.1.2, $L^* \cong Sz(8)$, with $U_L^* = \Omega_1(T_L^*)$ from I.2.2.4. Thus as $U \leq KL$,

$$U \leq U_C U_L =: V = U_C \times [V, X],$$

with X regular on $[V, X]^\#$ and $U_C = C_V(X)$.

We claim that there is $g \in M := N_G(V)$ with $K^g = K'$. Suppose first that $L^* \cong L_2(2^n)$ or $Sz(2^n)$. Recall by 16.4.6 that if $L \cong L_2(4) \cong L_2(5)$, then no involution in U induces an outer automorphism on L, so that $V = \langle U^G \cap T \rangle$. In the remaining cases, no outer automorphism of L is an FF^*-offender on U_L, so that $V = J(T)$. Thus in each case, V is weakly closed in T with respect to G, so by Burnside's Fusion Lemma A.1.35, M controls fusion in V, and hence there is $g \in M$ with $U^g_C = U$ and hence $K^g = K'$, as claimed. So we may assume $L \cong U_3(2^n)$. Choose g as in 16.4.2.4, so that $N_T(K') \leq T^g$, and as observed earlier, $R = T^g_C$ and $U = U^g_C$. Fix $u \in U^\#$. Set $Y_u := O^3(L_u)$ if $n \neq 3$ and $Y_u := L_u$ if $n = 3$; then $Y_u \leq L'$ by 16.4.8. But in either case $T_L \leq Y_u$, so $T_L \leq L'$ and hence $T_L = T^g_L$ as $N_T(K') \leq T^g$. Thus $U_L = U^g_L$, so as $U = U^g_C$,

$$V^g = (U_L U_C)^g = U_L U^g_C = U_L U = V,$$

completing the proof of the claim.

Pick g as in the claim, and set $M^+ := M/C_G(V)$. As $U_C = V \cap K$ and K is tightly embedded in G, U_C is a TI-set in V under the action of M by I.7.2.3. Further $X^+ = N_L(V)^+ \trianglelefteq (M \cap H)^+ = N_{M^+}(U_C)$ since $N_G(T_C) \leq H$ by 16.4.2.5, and X^+ is regular on $[V, X]^\#$, while $T^+ \in Syl_2(M^+)$ acts on X^+. So we have a Goldschmidt-O'Nan pair in the sense of Definition 14.1 of [GLS96]; hence we may apply O'Nan's lemma Proposition 14.2 in [GLS96] with M^+, X^+, V in the roles of "X, Y, V": Observe that conclusion (i) of that result does not hold, since there M normalizes U_C—whereas here $g \in M - N_G(U_C)$. Similarly conclusions (ii) and (iii) of that result do not hold, since here T normalizes U_C, but does not normalize U_C there. Thus conclusion (iv) of that result holds: $m(U_C) = 1$, $m(V) = 3$, and $M^+ \cong Frob_{21}$. In particular, $n = 2$, so that $L \cong L_2(4)$ or $U_3(4)$. But the latter case is impossible, since we saw in the unitary case that g normalizes T_L, so that $M^+ = \langle X^+, X^{g+} \rangle$ acts on $\Phi(T_L) = U_L$, whereas M^+ is irreducible on V.

Thus $L \cong L_2(4)$. Then $H = LK$ by 16.4.6, so R induces inner automorphisms on L. Recall R is faithful on L, so R is elementary abelian; hence as $|U_C| = 2$, $U_C = T_C$ is of order 2. Therefore $C_K(z) = T_C$ by 16.4.4.2, so that $G_z = L \times T_C \cong L_2(4) \times \mathbf{Z}_2$, and hence G is of type J_1 in the sense of I.4.9. Then we conclude from that result that $G \cong J_1$. $\square$

In the remainder of the chapter, assume G is not J_1; therefore by Proposition 16.5.1, L^* is not a Bender group. To complete the proof of our main result Theorem 16.5.14, we must eliminate each remaining possibility for L in (E2).

Recall from 16.4.3.2 that $L' = [L', z]$, and that z induces an inner automorphism on L' since $K \in \Delta_0(K')$ by the symmetry in 16.4.11.2.

LEMMA 16.5.2. *(1) If $R \cap Z(T) \neq 1$ and g is chosen as in 16.4.2.4, then $g \in N_G(T)$.*

(2) Assume u, z are involutions in R, T_C whose projections on L, L^g are 2-central, and that $|T : T_C T_L| \leq 2$. Then either

(a) $R \cap Z(T_1) \neq 1$ for some $T_1 \in Syl_2(H)$, and we may choose T_1 so that $R \trianglelefteq T_1 \in Syl_2(H \cap H')$, or

(b) $T_C T_L^l =: T_0 \in Syl_2(H \cap H')$ for some $l \in L$, with $R \trianglelefteq T_0$, $|T_0| = |T|/2$, and there exists $g \in N_G(T_0)$ with $K^g = K'$.

PROOF. Assume that $R \cap Z(T) \neq 1$, and that g is chosen as in 16.4.2.4. Now $T \leq C_G(R \cap Z(T)) \leq N_T(K')$ using 16.4.2.5, so $T^g = T$ as $N_T(K') \leq T^g$ by the choice of g. Thus (1) holds.

Assume the hypotheses of (2). Then u centralizes T_L^l for some $l \in L$, so $T_L^l \leq H'$ by 16.4.2.5. Also $T_C \leq H'$ by 16.4.11.1, so $T_0 := T_C T_L^l$ acts on some $R_1 \in Syl_2(K' \cap H)$. But by 16.4.10, $R \in Syl_2(K')$, so by Sylow's Theorem there is $x \in K' \cap H$ with $R_1^x = R$, and thus $T_2 := T_0^x$ acts on R. Let $T_2 \leq T_1 \in Syl_2(H)$. By hypothesis $|T_1 : T_2| \leq 2$, so either $R \trianglelefteq T_1$, or $T_2 = N_{T_1}(R) \in Syl_2(H \cap H')$. In the former case, $R \cap Z(T_1) \neq 1$ and conclusion (a) of (2) holds, so we may assume the latter. Thus $T_0 = T_2^{x^{-1}} \in Syl_2(H \cap H')$. By 16.4.11.2 we have symmetry between K and K', so there is $S \in Syl_2(K'L')$ with S Sylow in $H \cap H'$. Thus there is $h \in H \cap H'$ with $T_0^h = S$, so as h acts on $K'L'$, T_0 is Sylow in $K'L'$. Let $y \in G$ with $K^y = K'$; then T_0 and T_0^y are Sylow in $K'L'$, so there is $w \in K'L'$ with $T_0^{yw} = T_0$, and hence conclusion (b) of (2) holds with $g := yw$. $\square$

We now begin the process of eliminating the possibilities for L remaining in (E2). Let u denote an involution in U, and recall $z \in T_C \cap Z(T)$. Also $R \cap K = 1$ by 16.4.2.1, so that by 16.4.11.1,

$$R^* \cong R \cong T_C.$$

In particular as $U \leq LK$,

$$U \cong U^* \leq T_L^*.$$

LEMMA 16.5.3. *L is not A_6.*

PROOF. Assume otherwise. Then $U \cong U^* \leq T_L^* \cong D_8$, and hence $U \cong \mathbf{Z}_2$, E_4 or D_8. Now in the notation of Definition F.4.41, $X := \Gamma_{1,U}(L) \leq H'$ using 16.4.2.5. But if $U^* \cong D_8$, then $X = L$, contrary to 16.4.2.2. Assume $U^* \cong E_4$. Then $X \cong S_4$ with $O_2(X^*) = U^*$; so as X acts on K' while $U = \Omega_1(R)$, X acts on $K' \cap O_2(XU) = U$, and hence $3 \in \pi(Aut_{H'}(U))$. Then as $Out(L')$ is a 3'-group, $3 \in \pi(N_{K'}(U))$, so that $m_{2,3}(H) > 2$, contradicting G quasithin. Hence

$U = \Omega_1(R)$ is of order 2, so as $R \cong T_C$, $m_2(R) = 1 = m_2(T_C)$. Recall u denotes the involution in U and z the involution in T_C. The projection v of u on L is 2-central in LT, so conjugating in L if necessary, without loss $v \in Z(T)$, and then $u \in \langle z, v \rangle =: E \leq Z(T)$. Now we may choose g as in 16.4.2.4, so that $u = z^g$ by 16.4.11.3, and $g \in N_G(T)$ by 16.5.2.1. Now

$$[T, T] = Y_L \times Y_C, \qquad (*)$$

where Y_C is the preimage in T_C of $[T/T_L, T/T_L]$, and Y_L is of index at most 2 in the cyclic subgroup Y of order 4 in T_L.

Assume that $Y_C = 1$; that is, that T/T_L is abelian. Set $Z := \Omega_1(Z(T))$. Then either $Z = E$, or $Z = E\langle t \rangle$ where t induces a transposition on L. Further $N_G(T)$ centralizes $[T, T] \cap Z = \langle v \rangle$ by $(*)$. However g does not centralize Z since $u = z^g$, so $g \notin T$, and hence we may assume g has odd order. As $g \in N_G(T)$, g centralizes v, so $Z = E\langle t \rangle = \langle v \rangle \times [Z, N_G(T)]$ where $[Z, N_G(T)]$ is of rank 2, and g induces an element of order 3 on Z. But then either z or $u = zv$ lies in $[Z, N_G(T)]$, so as $u = z^g$, we conclude $z \in [Z, N_G(T)]$, and then $zz^g = z(zv) = v \in [Z, N_G(T)]$, whereas we saw $N_G(T)$ centralizes v.

Therefore $Y_C \neq 1$. Then as $m_2(T_C) = 1$, $\langle z \rangle = \Omega_1(Y_C)$, so by $(*)$, $\Omega_1([T, T]) = E \leq \Omega_1(Z(T))$. Therefore as $g \in N_G(T)$, g induces an element of order 3 on E, and $N_G(T)$ is transitive on $E^\#$. Thus as Y_L is cyclic, so is Y_C by the Krull-Schmidt Theorem A.1.15, and then $Y_C \cong Y_L$, so that Y_C is cyclic of order at most 4.

Next $Y_C \trianglelefteq T$, so $Y_C^g \trianglelefteq T$. Now $s \in T_L - Y$ inverts Y and centralizes Y_C, so if $|Y_C| = 4$, then s does not act on Y_C^g, a contradiction.

Thus $Y_C = \langle z \rangle$, so $Y_L = \langle v \rangle$ as $Y_L \cong Y_C$. As $Y_L = \langle v \rangle$, L^*T^* is A_6 or S_6. Assume $L^* \cong A_6$. Then $T = T_L \times T_C$, so $T_C \cong Q_8$ since $m_3(T_C) = 1$ and $\langle z \rangle = Y_C \cong [T/T_L, T/T_L] \cong [T_C, T_C]$. But $R = T_C^g$, so $Q_8 \cong T_C \cong R \cong R^* \leq T^* = T_L^*$, impossible as $T_L^* \cong D_8$. Therefore $L^*T^* \cong S_6$, so $T^* \cong \mathbf{Z}_2 \times D_8$. Then as $T_C \cong R \cong R^* \leq T^*$, while $m_2(T_C) = 1$ and $[T, T_C] \neq 1$, we conclude that $T_C \cong \mathbf{Z}_4$ and $t \in T - T_C T_L$ inverts T_C. As $T^* \cong \mathbf{Z}_2 \times D_8$, we may pick t so that t centralizes T_L and $t^2 \in T_C$. Let $T_1 := T_C\langle t \rangle$; then $T = T_1 \times T_L$, with $T_1 \cong D_8$ or Q_8, and $T_L \cong D_8$. Now $g \in N_G(T)$ is transitive on $E^\#$; but this is impossible, as by the Krull-Schmidt Theorem A.1.15, $N_G(T)$ permutes $\{\Phi(T_1 Z(T)), \Phi(T_L Z(T))\}$. $\square$

LEMMA 16.5.4. L^* is not $L_3(4)$.

PROOF. Assume otherwise. As $U^* \leq T_L^*$ and all involutions in L are 2-central in L from I.2.2.3b, u centralizes a Sylow 2-group of L. Then as R centralizes T_C, we may take $u \in Z(T_L T_C)$.

Suppose first that $U^* \not\leq Z(T_L^*)$. Then $Y := \Gamma_{1,U}(L)$ contains a maximal parabolic P of L, and $Y \leq H'$ by 16.4.2.5. If $P \leq L'$, then $P \leq C_G(U)$, so $U^* \leq C_{L^*}(P^*) = 1$, a contradiction. Thus $P \not\leq L'$, so K' has an $L_2(4)$-section, and hence $m_{2,3}(H') > 2$, contradicting G quasithin.

Therefore $U^* \leq Z(T_L^*)$, so $U \cong \mathbf{Z}_2$ or E_4. Now $J(T) = J(T_L T_C) = T_L J(T_C) = T_L U_C$, where $U_C = \Omega_1(T_C) \cong U$. As $u \in Z(T_C T_L)$, $u \in Z(J(T))$. Recall $U_C \in U^G$ so there is $g \in G$ with $U^g = U_C$. Thus $u^g \in U_C \leq Z(J(T))$, and by Burnside's Fusion Lemma A.1.35, we may take $g \in M := N_G(J(T))$. Hence g acts on $Z(J(T)) =: V$, where $V = U_C \times V_L$, with $V_L := [V, X] \cong E_4$ for X of order 3 in $N_L(T_L)$. Now we argue, just as in the proof of Proposition 16.5.1, that X is regular in $V_L^\#$, and U_C is a TI-set under M, so again by Proposition 14.2 in [GLS96], conclusion (iv) of that result holds: namely, $m(V) = 3$ and $Aut_M(V) \cong Frob_{21}$. But then M acts irreducibly on V, impossible as M acts on $\Phi(J(T)) = V_L$. $\square$

NOTATION 16.5.5. For $u \in U^{\#}$, define $X_u := O^3(L_u)$ if $L \cong L_3(2^n)$, n even, and $X_u := L_u$ otherwise. In the former case, $n > 2$ by 16.5.4. Thus in any event, $X_u \leq L^g \leq C_G(R)$ by 16.4.8, so $R^* \leq C_{H^*}(X_u^*)$. Further for i an involution in T_C, we can define $X_i \leq L^g$ analogously. Then $X_u \leq X_i$ as $X_u \leq C_{L^g}(i)$, so by symmetry between L, u and L^g, i, $X_i \leq X_u$. Thus we may define $X := X_u = X_i$.

Inspecting the possibilities for L^* remaining in (E2) after 16.5.1, 16.5.3, and 16.5.4, we conclude from 16.1.4 and 16.1.5 that for each involution j^* in L^*, $X_j \neq 1$ except when $L \cong Sp_4(2^n)$ with $n > 1$, and j^* is of type c_2.

Observe that the fourth part of the next lemma supplies another assertion about the symmetry between K and K'.

LEMMA 16.5.6. Let $\bar{H}' := H'/K'$.

(1) $X = X_u = X_i$ for all involutions $i \in T_C$ and $u \in R$.

(2) $R^* \leq C_{L^*T^*}(X^*)$.

(3) If we choose choose g as in 16.4.2.4, then $g \in N_G(X)$.

(4) The following are equivalent:

   (a) Some involution in R^* is 2-central in L^*.
   (b) Each involution in $\bar{T}_C$ is 2-central in $\bar{L}'$.
   (c) Some involution in $\bar{T}_C$ is 2-central in $\bar{L}'$.
   (d) Each involution in R^* is 2-central in L^*.

(5) Assume that $Z(L) = 1$, and for each $J \in \Delta_0$ and each involution i in $N_J(K)$, that i^* is not 2-central in L^*. Let v be the projection of u on L, and suppose there is $l \in L$ with vv^l an involution of X. Then $\bar{v}\bar{v}^l$ is not 2-central in $\bar{L}'$.

PROOF. We already observed that (1) and (2) hold. We saw in 16.4.11.3 that if we choose g as in 16.4.2.4, then $T_C^g = R$, so (1) implies (3).

Suppose u^* is 2-central in L^*. By (1), for each involution $i \in T_C$, $X_u = X_i$, so by inspection of the centralizers of involutions of H^* listed in 16.1.4 and 16.1.5 remaining after 16.5.1, 16.5.3, and 16.5.4, we conclude that $\bar{i}$ is also 2-central in $\bar{L}'$. Thus (4a) implies (4b). Then as $K \in \Delta_0(K')$ by 16.4.11.2, by symmetry (4c) implies (4d). Of course (4b) implies (4c), and (4d) implies (4a), so (4) holds.

Assume the hypotheses of (5). As $Z(L) = 1$, $u = jv$ for some $j \in T_C$ with $j^2 = 1$, so $uu^l = (jv)(jv^l) = vv^l \in X \leq L'$ by hypothesis and (1), and $\bar{u}^l = \bar{v}\bar{v}^l$. Let $i := u^{lg^{-1}}$ and $J := (K')^{lg^{-1}}$. As $uu^l \in X$ while $g \in N_G(X)$ by (3), $u^{g^{-1}}u^{lg^{-1}} \in X$; thus as $u^{g^{-1}} \in K$, $i = u^{lg^{-1}} \in J \cap XK$, so $J \in \Delta_0$ by 16.4.9.3. By hypothesis i^* is not 2-central in L^*, so conjugating by g, $\bar{v}\bar{v}^l = \bar{u}^l$ is not 2-central in $\bar{L}'$, establishing (5). $\square$

LEMMA 16.5.7. Assume $Z(L) = 1$, $|C_{T^*}(T_L^*)| = 2$, and $|T^* : T_L^*| \leq 2$. Then

(1) R^* contains no 2-central involution of L^*.

(2) L has more than one class of involutions.

PROOF. As $U \leq LK$ while $Z(L) = 1$ by hypothesis, (1) implies (2). Hence we may assume that (1) fails, and it remains to derive a contradiction.

By 16.5.6.4, the hypotheses of 16.5.2.2 are satisfied. If case (a) of 16.5.2.2 holds, then replacing T by the subgroup "T_1" defined there, we may assume $R \trianglelefteq T$; further by 16.5.2.1, we may choose $g \in N_G(T)$ with $K^g = K'$. Otherwise case

(b) of 16.5.2.2 holds, and replacing T by the subgroup T^l defined there, we may assume $T_0 := T_C T_L = N_T(R)$ is of index 2 in T, and take $g \in N_G(T_0)$ with $K^g = K'$. Set $T_1 := T$ or T_0 in the respective cases, and set $Z_C := Z(T_1) \cap T_C$ and $Z_L := Z(T_1) \cap T_L$. Thus $g \in N_G(T_1)$, so g acts on $Z(T_1)$ and $T_1 = N_T(R)$. By hypothesis, $T_L T_C = T_L \times T_C$ and $C_T(T_L) = T_C Z(T_L)$ with $|Z(T_L)| = 2$, so $Z_L = Z(T_L) = C_{T_L}(T)$, and as $Z(T_1) \le C_T(T_L)$,

$$Z(T_1) = Z_L \times Z_C. \tag{$*$}$$

By 16.4.2.1, $Z_C \cap Z_C^g = 1$, so as g acts on $Z(T_1)$ and $|Z(T_1) : Z_C| = |Z_L| = 2$ by $(*)$, we conclude Z_C is also of order 2. Hence T centralizes $Z_L \times Z_C$. Then since R is normal in T_1, $1 \ne R \cap Z(T_1)$ is central in T by $(*)$, so that $R \cap Z(T) \ne 1$; thus case (a) of 16.5.2.2 holds, and hence $T_1 = T$.

As $T_1 = T$, $Z(T) = Z_L \times Z_C$ is of rank 2 and $g \in N_G(T)$. In particular $R = T \cap K^g = (T \cap K)^g = T_C^g$. Also as $Z_C^g \cap Z_C = 1$, g induces an element of order 3 on $Z(T)$ so either Z_C^g or $Z_C^{g^2}$ is not equal to Z_L, and replacing g^2 by g of necessary, we may assume $Z_C^g \ne Z_L$. As $T_C \trianglelefteq T$, also $R = T_C^g \trianglelefteq T$, so $R \cap L \trianglelefteq T$. Hence as $Z(T) \cap R = Z_C^g$ is of order 2 and does not lie in L, $R \cap L = 1$. Thus $[T_L, R] \le T_L \cap R = 1$, so $R^* \le C_{T^*}(T_L^*) = Z_L^*$. Therefore as $|Z_L^*| = 2$ and $R \cong R^*$, R is of order 2, and hence $T_C = Z_C$. Then as $|T^* : T_L^*| \le 2$, $|T : T_L| \le 4$, so that $[T, T] \le T_L$. By Proposition 16.5.1 and our assumption that G is not J_1, L is not $L_2(2^n)$, so by inspection of the groups in (E2), T_L is nonabelian. Therefore as Z_L is of order 2, $Z_L = Z(T) \cap [T, T] \trianglelefteq N_G(T)$. This is impossible, as $\langle g \rangle$ is irreducible on $Z(T) \cong E_4$. $\qquad \square$

LEMMA 16.5.8. *(1) L is not $L_2(p)$, p an odd prime, or $L_3(3)$.*
(2) L is not M_{11}, M_{22}, M_{23}.

PROOF. If L is a counterexample to (1) or (2), then L has one class of involutions, $Z(L) = 1$, $Out(L)$ is of order at most 2, and $C_{T^*}(T_L^*)$ is of order 2; hence the lemma follows from 16.5.7. $\qquad \square$

LEMMA 16.5.9. *L is not $U_3(3)$.*

PROOF. Assume otherwise. Then L has one class of involutions, $X \cong SL_2(3)$, and $C_{H^*}(X^*) \cong \mathbf{Z}_4$ or Q_8. Thus $R^* \cong R$ is of 2-rank 1 by 16.5.6.2, and $T_C \cong R$ by 16.4.11.1. Then $Z := \Omega_1(Z(T)) \cong E_4$, with $u \in Z := \langle z, v \rangle$, where $z \in T_C$ and $v \in T_L$. Choose g as in 16.4.2.4; then $T_C^g = R$ by 16.4.11.3. As $R \cap Z(T) \ne 1$, $g \in N_G(T)$ by 16.5.2.1, and as $T_C^g = R$, g is nontrivial on Z. Then as Z is of rank 2 we conclude that g induces an element of order 3 on Z. This is impossible, as $g \in N_G(X)$ by 16.5.6.3, so g acts on $Z(X) = \langle v \rangle$. $\qquad \square$

LEMMA 16.5.10. *Assume L^* is not of Lie type of Lie rank 2 over $\mathbf{F}_{2^n}$ for some $n > 1$. Then*

(1) If L^* is of Lie type in characteristic 2, then L is $^2F_4(2)'$, $^3D_4(2)$, $L_4(2)$, or $L_5(2)$.

(2) If L^* is not of Lie type and characteristic 2, then $L^* \cong M_{12}$, M_{22}, M_{24}, J_2, J_4, HS, or Ru; and if $L^* \cong M_{22}$, then $Z(L) \ne 1$.

(3) $|T^* : T_L^*| \le 2$.

(4) $|C_{T^*}(T_L^*)| \le 2$.

(5) Either $Z(L) \ne 1$, or R^* contains no 2-central involution of L^*.

(6) Assume that $Z(L) = 1$, v is the projection of u on L, and there is $l \in L$ with vv^l an involution in X. Then vv^l is not 2-central in L'.

PROOF. Part (2) follows from 16.5.8 and inspection of the groups in (E2).

Suppose that L^* is of Lie type in characteristic 2. By Proposition 16.5.1 and our assumption that G is not J_1, L is not of Lie rank 1, and by hypothesis L^* is not of Lie rank 2 over $\mathbf{F}_{2^n}$ for some $n > 1$. Thus from Theorem C, either L^* is of Lie rank 2 over $\mathbf{F}_2$, or $L^* \cong L_4(2)$ or $L_5(2)$. By 16.5.8 L is not $L_3(2) \cong L_2(7)$, by 16.5.9 L is not $U_3(3) \cong G_2(2)'$, and by 16.5.3 L is not $A_6 \cong Sp_4(2)'$. Thus (1) holds since L is simple by 16.1.2.1.

Next by inspection of $Aut(L^*)$ for L^* listed in (1) and (2), $|Out(L^*)|_2 \leq 2$, so (3) holds. Similarly (4) follows from inspection of $Aut(L^*)$. Then (5) follows from (3), (4), and 16.5.7.1. Finally assume the hypotheses of (6). Then the hypotheses of 16.5.6.5 are satisfied using (5), so (6) follows from that result. $\square$

LEMMA 16.5.11. *L^* is of Lie type in characteristic 2.*

PROOF. Assume otherwise; then L^* is in the list of 16.5.10.2.

Suppose first that u^* is 2-central in L^*. Then by 16.5.10.5, $Z(L) \neq 1$, so applying 16.1.2.1 to the list in 16.5.10.2, L^* is M_{12}, M_{22}, J_2, HS, or Ru. Furthermore using 16.1.5, we find that either $C_{H^*}(X^*)$ is of order 2, or L^* is HS and $C_{H^*}(X^*) \cong \mathbf{Z}_4$. Thus by 16.5.6.2, either $|R| = 2$, or L^* is HS and $R \cong \mathbf{Z}_4$. In any case $\langle u \rangle = \Omega_1(R)$ and $\langle z \rangle = \Omega_1(T_C)$. Further if L^* is M_{22}, then as $T_C \cong R \cong \mathbf{Z}_2$, $T_C = \langle z \rangle$ and $Z(L) \cong \mathbf{Z}_2$. Hence we conclude from 16.1.2.2 that in each case $\langle z \rangle = \Omega_1(T_C) = Z(L) \cong \mathbf{Z}_2$. Then from the structure of the covering group L of L^* in parts (5)–(7) of I.2.2, either:

(a) There is a unique $v \in uZ(L)$ such that there exists $x \in O_2(X)$ with $x^2 = v$.
(b) L^* is Ru, and setting $Y_1 := C_{O_2(X)}(\Phi(O_2(X)))$, $Y := [Y_1, Y_1]$ is of of order 2, and $Y^* = \langle u^* \rangle$.

In case (a) set $Y := \langle u \rangle$. Thus in any case $Y^* = \langle u^* \rangle$. Further T normalizes X, and hence centralizes Y, so T centralizes $Z(L)Y = \langle z, u \rangle$. Therefore $1 \neq u \in R \cap Z(T)$. Choose g as in 16.4.2.4; then $g \in N_G(T)$ by 16.5.2.1, and $g \in N_G(X)$ by 16.5.6.3. Next $|C_{T^*}(T_L^*)| = 2$ in each case, so $Z(T_L^*) = \langle u^* \rangle$ and hence $Z := \Omega_1(Z(T)) = \langle z, u \rangle \cong E_4$. By our choice of g and 16.4.11.3, $R = T_C^g$ and hence $u = z^g$. Then as g acts on T, g induces an element of order 3 on Z, and in particular $\langle g \rangle$ acts irreducibly on Z. This is impossible since g acts on X and hence on $Y < Z$.

Therefore u^* is not 2-central in L^*. Thus as M_{22} has one class of involutions, L^* is not M_{22}.

Inspecting the list of centralizers of non-2-central involutions in 16.1.5 for the remaining groups in 16.5.10.2, either $C_{H^*}(X^*)$ is of order 2, or L^* is M_{12}, M_{24}, J_2, HS, or Ru and $C_{H^*}(X^*) \cong E_4$. Arguing as in the previous paragraph, either $|R| = 2 = |T_C|$, or one of the exceptional cases holds with $R \cong E_4 \cong T_C$. In any case, $\Phi(R) = 1 = \Phi(T_C)$, so $R = U$.

Assume $L \cong J_4$ or M_{24}. Then $Out(L) = 1$, so $T = T_L \times T_C$ with $\Phi(T_C) = 1$, and hence $z \notin \Phi(T)$ for $T \in Syl_2(C_G(z))$, and T_C is in the center of $O^{2'}(C_G(z))$. Thus if $|T_C| = 2$, then $z^G \cap T_L \neq \emptyset$ by Thompson Transfer, whereas for each involution $a \in T_L$, $a \in \Phi(C_{T_L}(a))$ by 16.1.5.9. Thus $|T_C| > 2$, so $L \cong M_{24}$, and then U is not centralized by $O^{2'}(C_L(u))$; so as $U = R = T_C^g$, this is contrary to T_C in the center of $O^{2'}(C_G(z))$.

Therefore L^* is M_{12}, J_2, HS, or Ru. If $Z(L) \neq 1$, then from (5b) and (7b) of I.2.2, the projection v of u on L is of order 4, so $u = vt$ with $t \in T_C$ of order 4, contrary to $\Phi(T_C) = 1$. Hence $Z(L) = 1$, so if v is the projection of u on L and there is $l \in L$ with vv^l an involution of X, then by 16.5.10.6, vv^l is not a 2-central involution of L'. However $X \cong A_5$, A_5, A_6, or $Sz(8)$, respectively, with all involutions in X 2-central in L, and the involutions in vX are in v^L, so we have a contradiction which completes the proof of 16.5.11. $\qquad\square$

LEMMA 16.5.12. *L^* is of Lie type in characteristic 2 of Lie rank 2.*

PROOF. Assume otherwise. By 16.5.11 and 16.5.10.1, L is $L_n(2)$ for $n := 4$ or 5. Thus H^* is either L^* or $Aut(L^*)$, so $H = LKT$. Recall z is an involution in $T_C \cap Z(T)$. If T_C is cyclic, then by 16.4.4.2, $T_C = C_K(z)$, so that $G_z = LT$.

Next L has two classes j_1 and j_2 of involutions, where j_1 is the class of transvections and the 2-central class. Hence $u^* \in j_2$ by 16.5.10.5, so by 16.1.4.3, $X \cong A_4$ or $\mathbf{Z}_3/2^{4+4}$, for $n = 4$ or 5, respectively. Also $C_{T^*}(X^*) \cong E_4$, unless $L^*T^* \cong S_8$, in which case $C_{T^*}(X^*) \cong D_8$. Thus by 16.5.6.2, either R is a subgroup of E_4, or $H^* = L^*R^* \cong S_8$ and $R \cong \mathbf{Z}_4$; R is not D_8 as $\Omega_1(R^*) = U^* \leq T_L^*$.

Suppose $H^* \cong S_8$ with $R \cong \mathbf{Z}_4$. Then $T_C \cong R \cong \mathbf{Z}_4$, so $G_z = LRT_C$ by paragraph one. Hence $T = T_L T_C R$ centralizes T_C, so $T_C \leq Z(G_z)$. Then $R \leq Z(C_G(u))$, whereas $R^* \cong \mathbf{Z}_4$ is not central in a subgroup D_8 of $C_{L^*}(u^*)$.

Therefore $R \cong \mathbf{Z}_2$ or E_4, so that $R = U$ and hence $R^* \leq T_L^*$. Let v denote the projection of u on L.

Assume first that $n = 5$. Then

$$\Phi(O_2(X)) =: V = [V, X] \oplus C_V(X),$$

with the involutions in the 4-groups $[V, X]$ and $C_V(X)$ of type j_2, and the diagonal involutions of type j_1. Then $v \in C_V(X)$, and there is $l \in L$ with $v^l \in [V, X]$, so that vv^l is 2-central in L', contrary to 16.5.10.6.

Hence $n = 4$, so that $L \cong L_4(2)$. Assume $R \cong E_4$. Then for $r \in R - \langle u \rangle$, the projection of r on L is in $O_2(C_L(X)) - \langle v \rangle$, so as $C_G(u) \leq H'$ by 16.4.2.5, $v \in [r, C_{T_L}(u)] \leq K'$, so $u = v \in L$. Similarly $r \in L$, so that $R = O_2(C_L(X))$. Then there is $y \in L$ of order 3 faithful on R. But as $m_3(N_L(O_2(X))) = 2$ and G is quasithin, K is a 3'-group, so $y \in O^{3'}(H') = L' \leq C_G(R)$, a contradiction.

Therefore $R = \langle u \rangle$ is of order 2, so as $T_C \cong R$, $T_C = \langle z \rangle$ is of order 2, and $G_z = LT$ by paragraph one. Set $A := O_2(C_L(u))T_C$. Then $A \cong E_{32}$ and $A = J(C_{T_L T_C}(u))$. If $H^* \cong L_4(2)$ then $T = T_L \times \langle z \rangle$, so that $z \notin \Phi(T)$ with $T \in Syl_2(C_G(z))$. However $z^G \cap T_L \neq \emptyset$ by Thompson Transfer, and each involution $a \in L$ satisfies $a \in \Phi(C_L(a))$ (cf. parts (1) and (3) of 16.1.4). Therefore $LT \cong S_8 \times \mathbf{Z}_2$ with $D_8 \times E_8 \cong O_2(C_{LT}(u)) = J(O_2(C_{LT}(u)))$. Choose g with $T \cap H' = N_T(K') \leq T^g$ as in 16.4.2.4, so that $R = T_C^g$ by 16.4.11.3, and hence $u = z^g \in z^G \cap A - \{z\}$ and $A \leq T^g$.

Suppose first that $A \leq L'K'$. Then $A \leq T_L^g T_C^g$, so $A = J(C_{(T_L T_C)^g}(z)) = O_2(C_{L'}(z)) \times R$, and hence A plays the same role for the pairs L', T_C and L, R. Next $A \cap j_2$ is an orbit of length 6 on $A^\# \cap L$ under $N_H(A)$, with $Aut_H(A) \cong O_4^+(2)$. Further if $y \in G$ such that $z^y \in LK$ projects on a member of the class j_1, then $K^y \in \Delta_0$ by 16.4.9.3, contrary to 16.5.10.5. Thus no member of $z^G \cap LK$ projects on j_1, so as $N_H(A)$ has two orbits of length 6 on the elements of A projecting on members of j_1 and u is such a member, we conclude $z^G \cap A =: \alpha$ is of order 7 or 13. Set $M := N_G(A)$ and $M^+ := M/C_M(A)$. Since A plays the same role for both pairs

L', T_C and $L, R,$ $C_M(u)$ moves z, so z^M is of order 7 or 13. Further $A = \langle z^M \rangle$, so M^+ acts faithfully on z^M. Since $|L_5(2)|$ is not divisible by 13, $|z^M| = 7$, so $M^+ \le S_7$. As $C_{M^+}(z) \cong O_4^+(2)$, $|M^+| = 2^3 \cdot 3^2 \cdot 7$. But S_7 has no subgroup of index 10.

Therefore $A \not\le L'K'$. Hence $LT \cong S_8 \times \langle z \rangle$, so from the structure of S_8, $\mathcal{A}(T) = \{A, A_1, A_1^t, B\}$ for suitable B and $t \in T_L$, where $J(O_2(C_{LT}(u))) = \{A, A_1\}$. As $A \not\le L'K'$, $A_1 = O_2(C_{L'}(z))R = J(C_{T_L^g T_C^g}(u))$, so $A_1 \in A^G$. Observe that each member of $\mathcal{A}(T)$ is normal in $J(T)$, so by Burnside's Fusion Lemma, $I := N_G(J(T))$ is transitive on $A^G \cap J(T)$, and hence $A_1 \in A^I$. As $|T : J(T)| = 2$, I is not transitive on $\mathcal{A}(T)$, so $A^I = \{A, A_1, A_1^t\}$ and I induces S_3 on $\mathcal{A}(T)$. But $J(T) \cong \mathbf{Z}_2 \times D_8 \times D_8$, so by the Krull-Schmidt Theorem A.1.15, I permutes the two involutions generating the Frattini subgroups of the D_8-subgroups, so that $O^2(I)$ centralizes $\Phi(J(T))$. Then as $Z(J(T)) \cong E_8$, by Coprime Action, $O^2(I) \le O^2(C_G(Z(J(T)))) \le O^2(G_z) = O^2(LT) = L$; then $I = N_L(J(T))T \le N_G(A)$, contrary to $|A^I| = 3$. This completes the proof of 16.5.12. $\qquad \square$

LEMMA 16.5.13. *(1) u^* is not in the center of T^*.*
(2) L^* is not $L_3(2^n)$ or $Sp_4(2^n)$.

PROOF. By 16.5.12, $L^* \cong Y(2^n)'$, where Y is one of the Lie types A_2, C_2, G_2, $^2F_4$, or $^3D_4$. Further if $n = 1$, then L is the Tits group $^2F_4(2)'$ or $^3D_4(2)$ by 16.5.10.1, and so (1) holds by 16.5.10.5. Thus we may assume that $n > 1$. Further L^* is not $L_3(4)$ by 16.5.4, so by 16.1.2.1, either $Z(L) = 1$ or L^* is $G_2(4)$.

We first treat the case where u^* is a long-root involution. (When $L \cong Sp_4(2^n)$, either class of root involutions can be regarded as "long", as the classes are interchanged in $Aut(L)$.) Thus u^* is 2-central in L^*. Let Z denote the root group of the projection v of u on L—unless L^* is $G_2(4)$ with $Z(L) \ne 1$, where we let $Z := [N_L(Z_1), Z_1]$ where Z_1 is the preimage in L of the root group of u^*. Set $P := N_L(Z)$ and recall the definition of $X := X_u$ from Notation 16.5.5. As u^* is a long-root involution, either

(a) P is a maximal parabolic of L, and we check (cf. 16.1.4.1) that $X = P^\infty$, or

(b) $L \cong L_3(2^n)$ and $X = C_P(Z)$ if n is odd, while $X = O^3(C_P(Z))$ if n is even.

In case (b), $n > 2$ by 16.5.4. Thus in any case $X \ne 1$ and $Z^* = C_{L^*T^*}(X^*)$, so that $V := ZT_C = C_T(X)$, and $T_C \cong R \cong R^* \le Z^*$ by 16.5.6.2. In particular $\Phi(R) = \Phi(T_C) = 1$. Choose g as in 16.4.2.4; thus $V = ZT_C \le T \cap H' = N_T(K') \le T^g$. Further $X^g = X$ by 16.5.6.3, so

$$V = C_T(X) \le C_{T^g}(X) = C_{T^g}(X^g) = C_T(X)^g = V^g,$$

and hence $g \in N_G(V) \cap N_G(X) =: M$. Let $T_0 := N_T(X)$ and notice that either $T_0 = T$, or T^* is nontrivial on the Dynkin diagram of $L^* \cong Sp_4(2^n)$ with T_0 of index 2 in T. Let $\bar{M} := M/C_M(V)$. We can finish much as in the proof of Proposition 16.5.1: For $\bar{P} \cong \mathbf{Z}_{2^n-1}$ is regular on $Z^\# = [V, P]^\#$, $T_C = C_V(P)$ is a TI-set in V under the action of M by I.7.2.3, $N_M(T_C) \le N_M(P)$ by 16.4.2.5, and $\bar{T} \in Syl_2(\bar{M})$ acts on $\bar{P}$. Thus again we have the hypotheses for a Goldschmidt-O'Nan pair in Definition 14.1 of [GLS96], so we may apply O'Nan's lemma Proposition 14.2 in [GLS96]. Conclusion (i) of that result is eliminated since $g \in M - N_G(T_C)$, so either $m(V) = 3$ and $\bar{M} \cong Frob_{21}$ is irreducible on V, or T_C^M is of order 2^n where $n = m(Z^*) > 1$. The latter case is impossible as $|T : T_0| \le 2$. In the former as M

is irreducible on V and M normalizes X containing Z, $V \leq X \leq L$. Thus $T_C \leq L$ so $Z(L) \neq 1$, and hence L^* is $G_2(4)$ and $Z(L) = T_C$ is of order 2 by 16.1.2.2. Then $X/V \cong L_2(4)/E_{2^8}$ and the chief factors for X/V on $O_2(X)/V$ are natural modules. However $Y := O^{7'}(M)$ centralizes $X/O_2(X)$ as $Aut(A_5) = S_5$, so since the group of units of $End_{X/V}(O_2(X)/V)$ is $GL_2(4)$, Y centralizes $O_2(X)/V$. Then as $V = \Phi(O_2(X))$, Y centralizes $O_2(X)$ by Coprime Action, a contradiction as Y induces $\mathbf{Z}_7$ on V. Therefore u^* is not a long-root involution.

If $L^* \cong L_3(2^n)$ then all involutions of L^* are long-root involutions, so the lemma is established in this case. Further when L^* is of type G_2, $^2F_4$, or $^3D_4$, the 2-central involutions are the long-root involutions, so the lemma holds in these cases too. Thus we may assume $L^* \cong Sp_4(2^n)$, so that $Z(L) = 1$ by 16.1.2.1. As u^* is not a root involution, u^* is 2-central of type c_2 in L^* by 16.1.4.2, so we may take $u^* \in Z(T^*)$; thus the projection v of u is in $Z(T)$, so $u \in Z(T_L T_C)$ since R centralizes T_C. Proceeding as in the proof of 16.5.4, $U^* \leq Z(T_L^*)$. As $U^* \cong U$ and U^* contains no root elements, $m(U) \leq n$. Also as in 16.5.4, $J(T) = T_L J(T_C) = T_L U_C$, where $U_C = \Omega_1(T_C) \cong U$ is elementary abelian. Let $M := N_G(J(T))$. Recall that $U_C \in U^G$, so by Burnside's Fusion Lemma A.1.35, $U_C \in U^M$. Now $V := Z(J(T)) = U_C V_L$ is elementary abelian of order rq^2, where $V_L := Z(T_L) \cong E_{q^2}$, $q := 2^n$, and $r := |U_C|$. Further M acts on $\Phi(J(T)) = V_L$ and on V, so as $U_C \cap V_L = 1$, also $U_C^m \cap V_L = 1$ for each $m \in M$. Let β be the set of involutions in $V - V_L$ either contained in U_C, or projecting on a member of $V_L - (Z_1 \cup Z_2)$, where Z_1 and Z_2 are the two root groups in V_L. Then

$$|\beta| = (q^2 - 1 - 2(q-1) + 1)(r - 1) = ((q-1)^2 + 1)(r - 1).$$

Let γ be the set of involutions contained in a member of U_C^M. If $y \in G$ such that $z^y \in H$ and z^{y*} is a root involution of L^*, then $K^y \in \Delta_0$ by 16.4.9.3, contrary to 16.5.10.5. It follows that $\gamma \subseteq \beta$. Also $L \cap M$ contains a Cartan subgroup Y of $N_L(T_L)$ of order $(q-1)^2$, and Y acts regularly on U^{*Y} and hence also on U^Y. Therefore as K is tightly embedded in G, and $N_M(K)$ normalizes $V \cap K = U_C$, U_C is a TI-subset of V under the action of M by I.7.2.3, so

$$|\gamma| \geq ((q-1)^2 + 1)(r - 1) = |\beta|,$$

and hence as $\gamma \subseteq \beta$, we conclude that $\gamma = \beta$ and $U_C^M = \gamma$ is of order $1 + (q-1)^2$. This is impossible since $1 + (q-1)^2$ is even, while $T \leq N_M(U_C)$ and T is Sylow in G. This contradiction completes the proof of 16.5.13. $\qquad\square$

We are now in a position to establish our main result Theorem 16.5.14.

By 16.5.12 and 16.5.13.2, we have reduced the possibilities for L in (E2) to the case where $L^* \cong G_2(2^n)'$, $^2F_4(2^n)'$, or $^3D_4(2^n)$. By 16.5.10.1, $n > 1$ if $L^* \cong G_2(2^n)'$, and by 16.5.13.1, u^* is a short-root involution in L^*. By 16.1.2, either $Z(L) = 1$, or L^* is $G_2(4)$ and $Z(L)$ is of order 2. However in the latter case, from I.2.2.5b, u^* lifts to an v element of order 4, so $u = cv$ with $c \in T_C$ of order 4. This is impossible, as $C_{H^*}(X^*) \cong E_4$, so $\Phi(R^*) = 1$ by 16.5.6.2, and hence $R^* \cong R \cong T_C$ is elementary abelian. Thus $Z(L) = 1$.

Let V be the root group of the projection v of u on L—except when L is $^3D_4(2^n)$, where we set $V := Z(X)$. Then (cf. 16.1.4 and [AS76a] for further details) one of the following holds:

(a) $L \cong G_2(2^n)$, $X \cong L_2(2^n)/E_{2^{2n}}$ is an $L_2(2^n)$-block, and $E_{2^n} \cong V^* = C_{H^*}(X^*)$.

(b) $L \cong {}^2F_4(2^n)'$, $X/O_2(X) \cong L_2(2^n)'$, $Z(O_2(X)) = V \oplus W$, where $W :=$ $[Z(O_2(X)), X]$ is the natural module for $X/O_2(X)$, and $V = Z(X)$.

(c) $L \cong {}^3D_4(2^n)$, $X/O_2(X) \cong L_2(2^n)'$, $Z(O_2(X)) = V \oplus W$, where $W :=$ $[Z(O_2(X)), X]$ is the natural module for $X/O_2(X)$, and $V = Z(X) \cong E_{2^n}$.

In case (a), set $W := O_2(X)$. We finish as in several earlier arguments: In each case, $W^\#$ is the set of long root involutions in $Z(O_2(X))$, and $vv^l \in W^\#$ for suitable $l \in L$, contrary to 16.5.10.6.

This final contradiction establishes:

THEOREM 16.5.14 (Even Type Theorem). *Assume G is a quasithin simple group, all of whose proper subgroups are $\mathcal{K}$-groups. Assume in addition that G is of even type, but not of even characteristic. Then $G \cong J_1$.*

Bibliography and Index

Background References Quoted
(Part 1: also used by GLS)

[Asc86a] M. Aschbacher, *Finite group theory*, Cambridge Studies in Advanced Math., vol. 10, Cambridge U. Press, Cambridge, 1986, 274 pages.

[Asc94] _____, *Sporadic groups*, Cambridge Tracts in Math., vol. 104, Cambridge U. Press, Cambridge, 1994, 314 pages.

[BG94] H. Bender and G. Glauberman, *Local analysis for the odd order theorem*, London Math. Soc. Lecture Notes, vol. 188, Cambridge U. Press, Cambridge, 1994, 174 pages.

[Car72] R. Carter, *Simple groups of Lie type*, Wiley, London, 1972, 331 pages.

[Fei82] W. Feit, *The representation theory of finite groups*, North Holland, Amsterdam, 1982, 502 pages.

[FT63] W. Feit and J. G. Thompson, *Solvability of groups of odd order*, Pacific J. Math. **13** (1963), 775–1029, (Chapter V, and the supporting material from Chapters II and III only).

[GLS94] D. Gorenstein, R. Lyons, and R. M. Solomon, *The classification of the finite simple groups*, Math. Surveys and Monographs, vol. 40.1, Amer. Math. Soc., Providence RI, 1994, 165 pages.

[GLS96] _____, *The classification of the finite simple groups. Number 2. Part I. Chapter G. General group theory*, Surveys and Monographs, vol. 40.2, Amer. Math. Soc., Providence RI, 1996, 218 pages.

[GLS98] _____, *The classification of the finite simple groups. Number 3. Part I. Chapter A. Almost simple K-groups*, Surveys and Monographs, vol. 40.3, Amer. Math. Soc., Providence RI, 1998, 419 pages.

[GLS99] _____, *The classification of the finite simple groups. Number 4. Part II. Chapters 1–4. Uniqueness theorems*, Surveys and Monographs, vol. 40.4, Amer. Math. Soc., Providence RI, 1999, 341 pages.

[GLS02] _____, *The classification of the finite simple groups. Number 5. Part III, Chapters 1 - 6: The Generic Case, Stages 1–3a*, Surveys and Monographs, vol. 40.5, Amer. Math. Soc., Providence RI, 2002, 467 pages.

[Gor80] D. Gorenstein, *Finite groups (2nd ed.)*, Chelsea, New York, 1980, 527 pages.

[HB85] B. Huppert and N. Blackburn, *Finite Groups III*, Springer-Verlag, Berlin, 1985, 454 pages.

[Hup67] B. Huppert, *Endliche Gruppen I*, Springer-Verlag, Berlin, 1967, 793 pages.

[Suz86] M. Suzuki, *Group theory I, II*, Grundlehren der Math. Wissenschaften, vol. 247, Springer-Verlag, Berlin, 1982; 1986, (434 and 621 pages).

Background References Quoted
(Part 2: used by us but not by GLS)

[AMS01] M. Aschbacher, U. Meierfrankenfeld, and B. Stellmacher, *Counting involutions*, Illinois J. Math. **45** (2001), 1051–1060.

[AS76a] M. Aschbacher and G. Seitz, *Involutions in Chevalley groups over fields of even order*, Nagoya Math. J. **63** (1976), 1–91.

[AS76b] ———, *On groups with a standard component of known type*, Osaka J. Math. **13** (1976), 439–482.

[AS85] M. Aschbacher and L. Scott, *Maximal subgroups of finite groups*, J. Algebra **92** (1985), 44–80.

[AS91] M. Aschbacher and Y. Segev, *The uniqueness of groups of type J_4*, Inv. Math. **105** (1991), 589–607.

[Asc75] M. Aschbacher, *On finite groups of component type*, Ill. J. Math. **19** (1975), 87–115.

[Asc80] ———, *On finite groups of Lie type and odd characteristic*, J. Alg. **66** (1980), 400–424.

[Asc81a] ———, *A factorization theorem for 2-constrained groups*, Proc. London Math. Soc. **(3)43** (1981), 450–477.

[Asc81b] ———, *Some results on pushing up in finite groups*, Math. Z. **177** (1981), 61–80.

[Asc81c] ———, *Weak closure in finite groups of even characteristic*, J. Algebra **70** (1981), 561–627.

[Asc82a] ———, *GF(2)-representations of finite groups*, Amer. J. Math. **104** (1982), 683–771.

[Asc82b] ———, *The Tits groups as a standard subgroup*, Math. Z. **181** (1982), 229–252.

[Asc86b] ———, *Overgroups of Sylow subgroups in sporadic groups*, vol. 60, Memoirs A. M. S., no. 343, Amer. Math. Soc., Providence, 1986, 235 pages.

[Asc87] ———, *Chevalley groups of type G_2 as the group of a trilinear form*, J. Alg. **109** (1987), 193–259.

[Asc88] ———, *Some multilinear forms with large isometry groups*, Geometriae Dedicata **25** (1988), 417–465.

[Asc97] ———, *3-transposition groups*, Cambridge Tracts in Math., vol. 124, Cambridge U. Press, Cambridge, 1997, 260 pages.

[Asc02a] ———, *Characterizing $U_3(3)$ by counting involutions*, J. Ramanujan Math. Soc. **17** (2002), 35–49.

[Asc02b] ———, *Finite groups of $G_2(3)$-type*, J. Algebra **257** (2002), 197–214.

[Asc03a] ———, *A 2-local characterization of HS*, J. Algebra **260** (2003), 16–32.

[Asc03b] ———, *A 2-local characterization of M_{12}*, Illinois J. Math. **47** (2003), 31–47.

[Ben74a] H. Bender, *The Brauer-Suzuki-Wall theorem*, Ill. J. Math. **18** (1974), 229–235.

[DGS85] A. Delgado, D. Goldschmidt, and B. Stellmacher, *Groups and graphs: New results and methods*, DMV Seminar, vol. 6, Birkhaeuser, Basel, 1985, 244 pages.

[Fan86] P. Fan, *Amalgams of prime index*, J. Alg. **98** (1986), 375–421.

[Fro83] D. Frohardt, *A trilinear form for the third Janko group*, J. Algebra **83** (1983), 343–379.

[GM02] R. Guralnick and G. Malle, *Classification of 2F-modules, I*, J. Algebra **257** (2002), 348–372.

[GM04] ———, *Classification of 2F-modules, II*, "Finite Groups, 2003; Proceedings of the Gainesville Conference on Finite Groups, March 6 – 12, 2003" (Berlin-New York) (C.-Y. Ho, P. Sin, P. Tiep, and A. Turull, eds.), de Gruyter, 2004?, pp. Preprint 2003, 67 pages, http://math.usc.edu/ guralnic/Preprints.

[GN83] G. Glauberman and R. Niles, *A pair of characteristic subgroups for pushing-up in finite groups*, Proc. London Math. Soc.(3) **46** (1983), 411–453.

[Gol80]   D. Goldschmidt, *Automorphisms of trivalent graphs*, Ann. of Math.(2) **111** (1980), 377–406.

[Jam73]   G. D. James, *The modular characters of the Mathieu groups*, J. of Algebra **27** (1973), 57–111.

[Jan66]   Z. Janko, *A new finite simple group with abelian Sylow 2-subgroups and its characterization*, J. Algebra **3** (1966), 147–186.

[JP76]   W. Jones and B. Parshall, *On the 1-cohomology of finite groups of Lie type*, Proceedings of the Conference on Finite Groups (Univ. Utah, Park City, Utah, 1975) (New York) (W. Scott et al., ed.), Academic Press, 1976, pp. 313–328.

[Ku97]   C. Ku, *A characterization of M_{22}*, J. Algebra **187** (1997), 295–303.

[MS93]   U. Meierfrankenfeld and B. Stellmacher, *Pushing up weak BN-pairs of rank two*, Comm. Algebra **21** (1993), 825–934.

[SW91]   R. Solomon and S. K. Wong, *Characterizations without characters*, Illinois J. Math. **35** (1991), 68–84.

[TW02]   J. Tits and R. Weiss, *Moufang polygons)*, Monographs in Mathematics, Spring-Verlag, Berlin-New York, 2002, 535 pages.

[Zsi92]   K. Zsigmondy, *Zur Theorie der Potenzreste*, Monatshefte Math. Phy. **3** (1892), 265–284.

Expository References Mentioned

[ABG70]   J. Alperin, R. Brauer, and D. Gorenstein, *Finite groups with quasi-dihedral and wreathed Sylow 2-subgroups*, Trans. Amer. Math. Soc. **151** (1970), 1–261.

[AJL83]   H. Andersen, J. Jorgensen, and P. Landrock, *The projective indecomposable modules of $SL(2, p^n)$*, Proc. London Math. Soc. (3) **46** (1983), 38–52.

[Asc73]   M. Aschbacher, *A condition for the existence of a strongly embedded subgroup*, Proc. Amer. Math. Soc. **38** (1973), 509–511.

[Asc76]   _____, *Tightly embedded subgroups of finite groups*, J. Algebra **42** (1976), 85–101.

[Asc78a]   _____, *A pushing up theorem for characteristic 2 type groups*, Ill. J. Math. **22** (1978), 108–125.

[Asc78b]   _____, *Thin finite simple groups*, J. Alg. **54** (1978), 50–152.

[Asc81d]   _____, *Finite groups of rank 3*, Inv. Math. **63** (1981), 51–163.

[Asc81e]   _____, *On the failure of the Thompson factorization in 2-constrained groups*, Proc. London Math. Soc. (3)**43** (1981), 425–449.

[Asc84]   _____, *Finite geometries of type C_3 with flag transitive groups*, Geom. Dedicata **16** (1984), 195–200.

[Asc93]   _____, *Simple connectivity of p-group complexes*, Israel J. Math. **82** (1993), 1–43.

[Bau76]   B. Baumann, *Endliche nichtaufloesbare Gruppen mit einer nilpotenten maximalen Untergruppe*, J. Algebra **38** (1976), 119–135.

[Ben71]   H. Bender, *Transitive Gruppen gerader Ordnung, in denen jede Involution genau einen Punkt festlaesst*, J. Algebra **17** (1971), 527–554.

[Ben74b]   _____, *Finite groups with large subgroups*, Ill. J. Math. **18** (1974), 223–228.

[Bra57]   R. Brauer, *On the structure of groups of finite order*, Proceedings of the Inter. Congr. Math., Amsterdam, 1954 (Amsterdam), North-Holland, 1957, pp. 209–217.

[BS01]   C. Bennett and S. Shpectorov, *A remark on the theorem of J. Tits*, Proc. Amer. Math. Soc. **129** (2001), 2571–2579.

[C$^+$85]   J. H. Conway et al., *Atlas of finite groups*, Clarendon Press, Oxford, 1985, 252 pages.

[Cam79]   N. Campbell, *Pushing up in finite groups*, Ph.D. thesis, Caltech, 1979.

[CM]   B. Cooperstein and G. Mason, *Some questions concerning the representations of Chevalley groups in characteristic two*, unpublished preprint, U. Cal. Santa Cruz, about 1980, 93 pages.

[Con71]   J. H. Conway, *Three lectures on exceptional groups*, Finite Simple Groups (New York) (M. B. Powell and G. Higman, eds.), Academic Press, 1971, Oxford 1969, pp. 215–247.

[Coo78]   B. Cooperstein, *An enemies list for factorization theorem*, Commun. Algebra **6** (1978), 1239–1288.

[CW73]   J. H. Conway and D. B. Wales, *Construction of the Rudvalis group of order 145,926,144,000*, J. Algebra **27** (1973), 538–548.

[Fon67]   P. Fong, *Some Sylow groups of order 32 and a characterization of $U_3(3)$*, J. Algebra **6** (1967), 65–76.

[Fon70]   _____, *A characterization of the finite simple groups $PSp_4(q)$*, Nagoya Math. J. **39** (1970), 39–79.

[FS73]   P. Fong and G. Seitz, *Groups with a (B,N)-pair of rank 2, I*, Inv. Math. **21** (1973), 1–57.

[FW69]   P. Fong and W. Wong, *A characterization of the finite simple groups $PSp_{(}q)$, $G_2(q)$, $^2D_4(q)$*, Nagoya Math. J. **36** (1969), 143–184.

[Gla66]   G. Glauberman, *Central elements in core-free groups*, J. Algebra **4** (1966), 403–420.

[Gla73]   _____, *Failure of factorization in p-solvable groups*, Quart. J. Math. Oxford **24** (1973), 71–77.

[Gol74]   D. Goldschmidt, *2-fusion in finite groups*, Ann. of Math. **99** (1974), 70–117.

[Gro02]   J. Grodal, *Higher limits via subgroup complexes*, Ann. of Math.(2) **155** (2002), 405–457.

[GT85]   K. Gomi and Y. Tanaka, *On pairs of groups having a common 2-subgroup of odd indices*, Sci. Papers Coll. Arts Sci. U. Tokyo **35** (1985), 11–30.

[GW64]   D. Gorenstein and J. Walter, *The characterization of finite simple groups with dihedral Sylow 2-subgroups*, J. Algebra **2** (1964), 85–151,218–270,354–393.

[Hig68]   G. Higman, *Odd characterizations of finite simple groups*, U. Michigan lecture notes, 77 pages, 1968.

[HM69]   G. Higman and J. McKay, *On Janko's simple group of order* 50, 232, 960, Bull. London Math. Soc. **1** (1969), 89–94.

[HW68]   M. Hall and D. Wales, *The simple group of order* 604, 800, J. Algebra **9** (1968), 417–450.

[Jam78]   G. D. James, *The representation theory of the symmetric groups*, Lecture Notes in Math., vol. 682, Springer-Verlag, Berlin, 1978, 156 pages.

[Jan68]   Z. Janko, *Some new simple groups of finite order, I*, Symposia Math. **1** (1968), 25–64.

[Jan69]   ———, *A characterization of the simple group* $G_2(3)$, J. Algebra **12** (1969), 360–371.

[Jan72]   ———, *Nonsolvable finite groups all of whose 2-local subgroups are solvable, i*, J. Algebra **21** (1972), 458–517.

[JLPW95]   C. Jansen, K. Lux, R. Parker, and R. Wilson, *An Atlas of Brauer characters*, London Math. Soc. Monographs (new series), vol. 11, Oxford U. Press, Oxford, 1995, 327 pages.

[JW69]   Z. Janko and S. K. Wong, *A characterization of the Higman-Sims simple group*, J. Algebra **13** (1969), 517–534.

[Kan79]   W. Kantor, *Subgroups of classical groups generated by long root elements*, Trans. Amer. Math. Soc. **248** (1979), 347–379.

[Mas]   G. Mason, *The classification of finite quasithin groups*, typescript, U. California Santa Cruz, about 1981 about 800 pages,.

[Mas80]   ———, *Quasithin groups*, Finite Simple Groups II (Durham 1978) (New York) (M. J. Collins, ed.), Academic Press, 1980, pp. 181–197.

[McC82]   P. McClurg, *(roughly, on FF-modules for almost-simple groups)*, Ph.D. thesis, U. Cal. Santa Cruz, about 1982.

[McL67]   J. McLaughlin, *Some groups generated by transvections*, Arch. Math. **18** (1967), 364–368.

[McL69]   ———, *Some subgroups of* $SL_n(\mathbf{F}_2)$, Illinois J. Math. **13** (1969), 108–115.

[Mit11]   H. Mitchell, *Determination of the ordinary and modular ternary linear group*, Trans. Amer. Math. Soc. **12** (1911), 207–242.

[MS]   U. Meierfrankenfeld and B. Stellmacher, *The generic groups of p-type*, preprint, Mich. St. U., about 1997; see http://www.math.msu.edu/~meier/CGP/QT/qt.dvi, 89 pages.

[MS90]   U. Meierfrankenfeld and G. Stroth, *Quadratic GF(2)-modules for sporadic simple groups and alternating groups*, Commun. in Alg. **18** (1990), 2099–2139.

[Par72]   David Parrott, *A characterization of the Tits' simple group*, Canadian J. Math. **24** (1972), 672–685.

[Par76]   ———, *A characterization of the Rudvalis simple group*, Proc. London Math. Soc. (3) **32** (1976), 25–51.

[PW70]   D. Parrott and S. K. Wong, *On the Higman-Sims simple group of order* 44, 352, 000, Pacific J. Math. **32** (1970), 501–516.

[RS80]   M. A. Ronan and S. D. Smith, *2-Local geometries for some sporadic groups*, The Santa Cruz Conference on Finite Groups (Providence RI) (B. Cooperstein and G. Mason, eds.), Proc. Symp. Pure Math., vol. 37, Amer.Math.Soc., 1980, pp. 283–289.

[Rud84]   A. Rudvalis, *A rank-3 simple group of order* $2^{14} \cdot 3^3 \cdot 5^3 \cdot 7 \cdot 13 \cdot 29, I$, J. Algebra **86** (1984), 181–218.

[Ser80]   J.-P. Serre, *Trees*, Springer-Verlag, Berlin, 1980, 142 pages.

[Shu]   E. Shult, *On the fusion of an involution in its centralizer*, unpublished manuscript, early 1970s.

[Sim67]   C. C. Sims, *Graphs and finite permutation groups*, Math. Z. **95** (1967), 76–86.

[Smi75]   F. L. Smith, *Finite simple groups all of whose 2-local subgroups are solvable*, J. Algebra **34** (1975), 481–520.

[Smi80]   Stephen D. Smith, *The classification of finite groups with large extra-special 2-subgroups*, The Santa Cruz Conference on Finite Groups (Providence RI) (B. Cooperstein and G. Mason, eds.), vol. 37, Amer. Math. Soc., 1980, (Proc. Symp. Pure Math.), pp. 111–120.

[Smi81]   _____, *A characterization of some Chevalley groups in characteristic two*, J. of Algebra **68** (1981), 390–425.

[Ste86]   B. Stellmacher, *Pushing up*, Arch. Math. **46** (1986), 8–17.

[Ste92]   _____, *On the 2-local structure of finite groups*, Groups, Combinatorics, and Geometry (Cambridge) (M. Liebeck and J. Saxl, eds.), Durham 1990 Proceedings, Cambridge U. Press, 1992, pp. 159–182.

[Ste97]   _____, *An application of the amalgam method: the 2-local structure of N-groups of characteristic 2-type*, J. Algebra **190** (1997), 11–67.

[Str]   G. Stroth, *The uniqueness case*, unpublished MS, mid 1990s, 246 pages.

[Suz62]   M. Suzuki, *On a class of doubly transitive groups*, Ann. of Math. (2) **75** (1962), 105–145.

[Suz64]   _____, *On a class of doubly transitive groups II*, Ann. of Math. (2) **79** (1964), 514–589.

[Suz65]   _____, *Finite groups in which the centralizer of any element of order 2 is 2-closed*, Ann. of Math. **82** (1965), 191–212.

[Tho68]   John G. Thompson, *Nonsolvable finite groups all of whose local subgroups are solvable*, Bull. Amer. Math. Soc. **74** (1968), 383–437.

[Tim75]   F. G. Timmesfeld, *Groups with weakly closed TI-subgroups*, Math. Z. **143** (1975), 243–278.

[Tim78]   _____, *Finite simple groups in which the generalized Fitting group of the centralizer of some involution is extraspecial*, Ann. of Math. **107** (1978), 297–369.

[Tit74]   J. Tits, *Buildings of spherical type and finite BN-pairs*, Lecture Notes in Math., vol. 386, Springer-Verlag, Berlin, 1974, 299 pages.

[Tod66]   J. Todd, *A representation of the Mathieu group m_{24} as a collineation group*, Annali di Math. Pure et Appl. **71** (1966), 199–238.

[Tut47]   W. T. Tutte, *A family of cubical graphs*, Proc. Cambridge Philos. Soc. **43** (1947), 459–474.

[Wal69]   D. B. Wales, *Uniqueness of the graph of a rank-3 group*, Pacific J. Math. **30** (1969), 271–276.

[Wei90]   R. Weiss, *Generalized polygons and s-transitive graphs*, Finite Geometries, Buildings, and Related Topics (New York) (W. Kantor, R. Liebler, S. Payne, and E. Shult, eds.), Pingree Park CO 1988 Proceedings, Oxford U. Press, 1990, pp. 95–103.

[Won64a] W. Wong, *A characterization of the Mathieu group M_{12}*, Math. Z. **84** (1964), 378–388.

[Won64b] _____, *On finite groups whose 2-Sylow subgroups have cyclic subgroups of index 2*, J. Austral. Math. Soc. **4** (1964), 90–112.

[Yos]   S. Yoshiara, *Radical 2-subgroups of the Monster and Baby Monster*, ?? ?? (??), preprint, Tokyo Women's Christian U., 2003.

Index

$!\mathcal{M}(U)$, 499

$!\mathcal{N}(T)$, 113

A_7-block, exceptional, 129

A_n-block, 124

$Aut(\alpha)$ (for an amalgam α), 260

$C(G, S)$, 122

$C_1(S)$ (for pushing up), 126

$C_2(S)$ (for pushing up), 126

$C_i(X, V)$ (for higher weak closure), 233

$D_8^n Q_8^m$ (extraspecial group), 228

D_8^n (extraspecial group), 228

$E_i(G, T, \Omega)$, 210

$F(G)$ (Fitting subgroup), 20

$F^*(G)$ (generalized Fitting subgroup), 19

$Frob_{21}$ (Frobenius group of order 21), 255

$G(\{\alpha, \beta\})$ (setwise stabilizer), 445

$G_2(2)$-block, 135

$G_2(3)$-type, 417

G_A (pointwise stabilizer of A), 311

$G_\gamma^{(n)}$, 311

$H^1(G, V)$ (for module extensions), 408

$I(X)$ (involutions), 431

$Irr(X, V)$ (irreducible submodules), 31

$Irr_+(X, V)$ (irreducible modulo trivial), 31

$Irr_+(X, V, Y)$, 31

$J(G, V)$ (generated by offenders), 68

$J(M)$ (radical of a module), 343

$J(X)$ (Thompson subgroup), 74

$J^i(M)$ (iterated radical of a module), 343

$J_j(X)$ (higher Thompson subgroup), 74

$J_{\mathcal{P}}(H)$, 70

L-balance, 414

$L_2(2^n)$-block, 124

$L_n(2)$-block, 135

$L_{2'}(H)$ (2-layer), 414

$M(\lambda_i)$ (basic irreducible for Lie type group), 386

$O_{2,E}(U)$, 491

$O_{2,F}(U)$, 491

$O_{2,\Phi}(U)$, 491

$PG(V)$ (projective space of V), 327

$Q_8^m D_8^n$ (extraspecial group), 228

Q_8^m (extraspecial group), 228

$R_2(G)$ (maximal 2-reduced), 78

$S(\gamma_1, \gamma)$, 324

$SL_3(2^n)$-block, 135

$Sp_4(4)$-block, 135

TI-set, 21

$U(L)$ (internal module in a block), 123

$U_3(3)$-type, 417

$V(H)$ (for use with $\overset{<}{\sim}$ in final chapter), 61

$V(Y)$, 59

$V(Y, R)$, 59

$W(X)$, 210

$W(X, \Omega)$, 210

$W_i(X, V)$ (higher weak closure), 233

Y-amalgam (amalgam of subgroups), 267

$\Gamma(\gamma)$ (i.e., $\Gamma^1(\gamma)$), 311

$\Gamma^i(\gamma)$, 311

$\Gamma^{<i}(\gamma)$, 311

$\Gamma^{\leq i}(\gamma)$, 311

$\Gamma_{k,P}(H)$, 294

$\Omega_m^\epsilon(2^n)$-block, 135

$\Theta(H)$, 350

$\Xi(G, T)$, 508

$\Xi^*(G, T)$ (solvable uniqueness groups), 509

$\Xi_+(G, T)$, 584

$\Xi_+^*(G, T)$, 866

$\Xi_-(G, T)$, 584

$\Xi_f(G, T)$, 514

$\Xi_f^*(G, T)$, 514

$\Xi_p(L)$, 509

$\Xi_{rad}(G, T)$, 509

$\Xi_{rad}^*(G, T)$, 512

α (module parameter for $(F-1)$-offenders), 698

β (module parameter for odd action), 698

$\tilde{\Gamma}_{k,P}(H)$ (corank-k generated core), 237

χ (block types for pushing up), 124

χ-block, 124

χ_0 (block types for SQT pushing up), 124

χ_0-block, 124

$\hat{A}_6$ (triple cover), 43

$\hat{A}_6$-block, 129

$\hat{A}_7$ (triple cover), 43

$\hat{M}_{22}$ (triple cover), 43

$\hat{q}(G, V)$ (cubic module parameter), 85
$\hat{\mathcal{Q}}(G, V)$ ($\hat{q} \le 2$ offenders), 178
$\hat{\mathcal{Q}}_*(G, V)$, 178
$\hat{\mathcal{U}}(X)$, 112
$\ker_A(B)$, 113
λ_i (fundamental weight of Lie type group), 386
$\langle \ldots \rangle$ (subspace spanned by), 197
$\overset{\sim}{\lesssim}$ (ordering on $\mathcal{H}(T)$ and $\mathcal{M}(T)$), 61
$\overset{\sim}{\lesssim}$ (ordering on FF*-offenders), 68
$\pi(X)$ (primes dividing), 20
$\mathrm{Baum}(H)$ (Baumann subgroup), 75
$\tau(G, T, A)$, 209
$\theta(G)$ (for any odd prime p), 556
$\theta(H)$ (generated by elements of order 3), 51
$a(X, W)$ (module parameter), 232
$b(\Gamma, V)$, 313
$d(p)$, 347
$e(G)$ (maximum 2-local p-rank), 4, 482
e_S (set as vector in permutation module), 83
$gp(\alpha)$ (universal completion), 265
$m(X, W)$ (module parameter), 231
$m_p(M)$, 482
$m_p(M)$ (p-rank), 4
$m_{2,p}(G)$, 26
$n'(X)$, 239
$n(G)$, 210
p-rank, 4, 482
p^{1+2} (extraspecial of exponent p), 25
$q(G, V)$ (quadratic module parameter), 67
q^{1+2w} (special group of this order), 262
qrc-lemma, 177
$r(G, V)$ (weak closure parameter), 231
$r_{A,V}$ (action ratio parameter), 67
$s(G, V)$ (weak closure parameter), 232
w-offender (weak closure), 236
$w(G, V)$, 236
$\mathcal{A}(X)$ (maximal rank elementary), 74
$\mathcal{A}^2(G)$ (elementary 2-subgroups), 67
$\mathcal{A}_j(H)$ (corank j in maximal), 74
$\mathcal{A}_k(X, W)$ (for a-parameter), 232
$\mathcal{B}_2(G)$ (2-radical subgroups), 121
$\mathcal{C}$-component, 8, 41
$\mathcal{C}(G)$ ($\mathcal{C}$-components), 41
$\mathcal{E}_i$, 210
$\mathcal{E}(G, T, A)$, 209
$\mathcal{E}_i(G, T, A)$, 210
$\mathcal{G}(S)$, 126
$\mathcal{H}$ ("partial" 2-locals), 500
$\mathcal{H}(X)$, 500
$\mathcal{H}(X, Y)$, 500
$\mathcal{H}^e$, 499
$\mathcal{H}^e(X)$, 500
$\mathcal{H}^e(X, Y)$, 500
$\mathcal{H}_*(T, M)$, 571
$\mathcal{H}_G^e$, 499
$\mathcal{H}_v(T)$, 61

$\mathcal{K}$ (known simple groups), 4, 482
$\mathcal{K}$-group, 4, 482
$\mathcal{L}(G, T)$, 507
$\mathcal{L}(H, S)$, 505
$\mathcal{L}^*(H, S)$, 505
$\mathcal{L}_f(G, T)$, 507
$\mathcal{L}_f^*(G, T)$ (nonsolvable uniqueness groups), 507
$\mathcal{M}$ (maximal 2-locals), 499
$\mathcal{M}(X)$, 499
$\mathcal{P}(G, V)$ (FF*-offenders), 68
$\mathcal{P}^*(G, V)$, 68
$\mathcal{P}_G$ (FF-offenders), 76
$\mathcal{P}_{R,G}$, 77
$\mathcal{Q}(G, V)$ ($q \le 2$ offenders), 178
$\mathcal{R}_2(G)$ (2-reduced subgroups), 77
$\mathcal{S}_2(G)$ (2-subgroups), 121
$\mathcal{S}_2^e(G)$, 500
$\mathcal{U}(X)$, 112
$\mathcal{X}$ (set including $\mathcal{L}$ and Ξ), 58
$\mathcal{X}_f$, 59
(CPU) pushing up hypothesis, 122
(E) even characteristic hypothesis, 4, 482
(F-1)-module, 68
(F-1)-offender, 68
(F-j)-module, 256
(K) inductive "known" hypothesis, 4, 482
(PU) pushing up hypothesis, 122
(QT) quasithin hypothesis, 4, 482
(SQT) strongly quasithin, 32
(SQTK) strongly quasithin $\mathcal{K}$-group, 33
$\mathcal{U}_M(X, 2)$ (invariant 2-subgroups), 56
2-component, 414
2-layer, 414
2-local p-rank, 26
2-radical, 121
2-reduced, 77, 491
2-signalizers, 56
2-stubborn, 121
2F-modules, 10
5-dimensional module for A_6, 885

A×B Lemma (Thompson), 24
almost special (group), 333
almost-extraspecial 2-group, 357
Alperin, J., 416, 518
Alperin-Brauer-Gorenstein Theorem (semidihedral and wreathed Sylow 2-subgroups), 416
Alperin-Goldschmidt conjugation family, 518
Alperin-Goldschmidt Fusion Theorem, 519
amalgam, 14
amalgam (rank-2), 260
amalgam method, 6, 311
amalgam, subgroup, 260
Andersen, H., 329
apartment (of a rank-2 amalgam), 274
Aschbacher block, 123

Aschbacher Local $C(G,T)$-Theorem, 121
Aschbacher, M., 10, 209, 231, 429, 486
automorphism group of an amalgam, 260
axis (of a transvection), 23

b (amalgam parameter), 313
Background References, 3
backtracks, path without, 270
Baer-Suzuki Theorem, 20
balance, 414
base (for uniqueness system), 658
basic irreducible module $M(\lambda_i)$ (Lie type
    group), 386
Baumann subgroup (Baum(H)), 75
Baumann's Argument, 7, 9, 80
Baumann's Lemma, 9, 117
Baumann, B., 9, 80, 117, 176
Baumeister, B., 418
Bender groups (rank-1 Lie type), 487
Bender, H., 415, 429, 484, 518
Bender-Glauberman revision of Feit-Thompson,
    15
Bender-Suzuki Theorem (strongly embedded
    subgroups), 15
Bennett, C., 279
Blackburn, N., 15
block, 10, 123
BN-pair, 283
Borel-Tits Theorem, 414
Brauer trick, 431
Brauer, R., 416, 417
Brauer-Suzuki Theorem, 429, 435
Brauer-Suzuki-Wall Theorem, 415
building, Tits, 283
Burnside's Fusion Lemma, 30
Burnside's Lemma, 30

C(G,T)-Theorem, 134
Campbell, N., 10, 126
Cartan subgroup of L_0 or of $H \in \mathcal{H}_*(T,M)$,
    740
Carter, R., 48
center (of a transvection), 23
central extension (of a group), 407
CFSG (Classification), 3, 483
characteristic p-type, 484
characteristic 2-type, 4
characteristic of a group (abstract notions
    of), 3, 481
Classification (of the Finite Simple Groups),
    3, 483
Clifford's Theorem, 31
cocode module (for M_{22}), 395
cocode module (for M_{24}, M_{23}), 395
code module (for M_{22}), 395
code module (for M_{24}, M_{23}), 395
cohomology of small modules for SQTK-groups,
    408
commuting graph, 487

completion (of an amalgam), 14, 261
component, 483
conjugation family, 518
Conway, J., 396, 431, 712
Cooperstein, B., 86, 451
Coprime Action (various results), 24
core (of a permutation module), 83
coset complex, 14, 269
coset geometry, 269
covering (of a group), 407
covering (of a module), 408
covering group, 407
covering, dual (of a module), 408
CPU (pushing up hypothesis), 122
critical subgroup, 24
cubic (action on a module), 85
Cyclic Sylow 2-Subgroups (transfer), 31

Dedekind Modular Law, 19
defined over (Lie type group), 38
Delgado, A., 6, 487
Dickson's Theorem (on subgroups of $L_2(q)$),
    20
Dickson, L., 20
disconnected (at the prime 2), 487
doubly singular (line in G_2-geometry), 889
dual covering (of a module), 408

E (even characteristic hypothesis), 4, 482
elation (of a point-line geometry) , 283
equivalence (of completions), 265
even characteristic, 3, 499
even type, 1170
Even Type Theorem, 1203
example, 484
    A_5, 518, 520
    A_6, 1165
    A_8, 799, 807, 1078
    A_9, 799, 807, 1078
    $G_2(2)'$, 988, 989, 1042, 1086, 1120, 1165
    $G_2(2^n)$, $n > 1$, 649, 653, 659, 691, 1078
    $G_2(3)$, 1007, 1008, 1012, 1078
    HS, 1001, 1005, 1012, 1078
    He, 838, 842, 1078
    J_2, 982, 986, 1042, 1086, 1120, 1165
    J_3, 982, 986, 1042, 1086, 1120, 1165
    J_4, 695, 723, 1078
    $L_2(2^n)$, 518, 520
    $L_2(p)$, p Fermat or Mersenne, 518, 527,
        543
    $L_3(2)$, 1165
    $L_3(2^n)$, $n > 1$, 649, 653, 659, 691, 1078
    $L_3(3)$, 518, 527, 543
    $L_4(2)$, 799, 807, 1078
    $L_4(3)$, 918, 922, 926, 1078
    $L_5(2)$, 799, 807, 857, 1078
    M_{11}, 518, 527, 543
    M_{12}, 988, 989, 1042, 1086, 1120, 1165
    M_{22}, 686, 688, 691, 1078

M_{23}, 649, 653, 657, 691, 1078

M_{24}, 799, 807, 838, 843, 844, 857, 1078

Ru, 1077, 1078

$Sp_4(2)'$, 1165

$Sp_4(2^n)$, $n > 1$, 649, 653, 659, 691, 1078

$Sp_6(2)$, 896, 905, 946, 1078

$Sz(2^n)$, 518, 520

$U_3(2^n)$, 518, 520

$U_3(3)$, 988, 989, 1042, 1086, 1120, 1165

$U_4(2)$, 918, 922, 926, 1078

$U_4(2^n)$, $n > 1$, 649, 653, 659, 691, 1078

$U_4(3)$, 918, 921, 946, 1078

$U_5(4)$, 649, 653, 659, 691, 1078

$^2F_4(2)'$, 988, 989, 1042, 1086, 1120, 1165

$^2F_4(2^n)$, $n > 1$, 649, 653, 659, 691, 1078

$^3D_4(2)$, 982, 987, 1042, 1086, 1120, 1165

$^3D_4(2^n)$, $n > 1$, 649, 653, 659, 691, 1078

example (QTKE-group in Main Theorem), 484

exceptional A_7-block, 129

Expository References, 15

extension (of a generalized Lie amalgam), 274

extension of an amalgam, 262

extension, central (of a group), 407

extraspecial p-group p^{1+2} of exponent p, 25

extraspecial 2-group (notation), 228

extremal, 294

extremal (conjugate), 548

F-1-offender, 68

failure of factorization module, 68

failure of Thompson factorization, 78

faithful (completion of an amalgam), 14

faithful completion (of an amalgam), 261

Fan, P., 16, 262, 263, 494

Feit, W., 428

Feit-Thompson Theorem (Odd Order Paper), 15

FF*-offender, 68
  strong, 68

FF-module, 8, 68

FF-module, strong, 68

FF-offender, 76
  strong, 76

field of definition of Lie type group, 38

First Main Problem, 492

Fitting subgroup $F(G)$, 20

Fong, P., 16, 417

Fong-Seitz Theorem (split BN-pairs of rank 2), 16, 282, 629

Frattini Argument, 19

Frohardt, D., 418

FSU, 579

Fundamental Setup (FSU), 579

Fundamental Weak Closure Inequality (FWCI), 237

fundamental weight λ_i (of Lie type group), 386

FWCI (Fundamental Weak Closure Inequality), 237

Gaschütz's Theorem, 31

generalized m-gon, 270

generalized Fitting subgroup $F^*(G)$, 19

generalized Lie amalgam, 273

generalized Lie group, 273

generalized polygon, 270

Generalized Thompson Transfer, 31

Generation by Centralizers of Hyperplanes, 24

Generic Case, 629

geodesic (in a coset geometry), 270

Glauberman's Argument, 7, 10, 127

Glauberman's form of Solvable Thompson Factorization, 79

Glauberman, G., 10, 70, 78, 126, 127, 487

Glauberman-Niles/Campbell Theorem, 10, 16, 126, 487, 1158

GLS (Gorenstein, Lyons, and Solomon), 4

Goldschmidt amalgam, 304

Goldschmidt triple, 304

Goldschmidt's Fusion Theorem, 518, 519

Goldschmidt, D., 6, 16, 262, 263, 304, 487, 494, 518, 520

Goldschmidt-O'Nan pair, 520

Gomi, K., 487

Gorenstein, D., xiii, 4, 416, 482

Gorenstein-Walter Theorem (dihedral Sylow 2-subgroups), 415

Green Book (Delgado-Goldschmidt-Stellmacher), 259

Grodal, J., 414

Guralnick, R., xiii, 12, 16, 86, 451

Hall subgroups, 19

Hall's Theorem (on solvable groups), 19

Hall, M., 418

Hall, P., 19

hexad (of Steiner system), 396

higher Thompson subgroup $(J_j(H))$, 9, 74

Higman, G., 327, 418

Ho, C.-Y., xiv

Huppert, B., 15

indecomposable (for discussion of $\hat{q}$), 187

internal module, 74

intersections, notational convention for, 41

James, G., 16, 348, 393

Janko, Z., 4, 417, 418, 482, 486

Jansen, C., 346

Jones, W., 16, 91, 408

Jorgensen, J., 329

K (inductive "known" hypothesis), 4, 482

Kantor, W., 339
$\ker_A(B)$, 113
Krull-Schmidt Theorem, 23
Ku, C., 686

L-balance, 414
Landrock, P., 329
large (faithful completion of an amalgam), 278
large extraspecial 2-subgroup, 13
layer, 414
Lie amalgam, 263, 273
Lie-type groups, representation theory of, 386
Lie-type groups, structure of, 38
LIST (of Schur multipliers), 38
Local $C(G, T)$-Theorem, 134
locally isomorphic (amalgams), 262
Lux, K., 346
Lyons, R., xiii, 4, 482

Main Hypothesis., 4, 482
Main Theorem, 4, 482
Malle, G., xiii, 12, 16, 86, 451
Maschke's Theorem, 31
Mason, G., 86, 231, 242, 257, 451, 486
Mason, G. (preprint on quasithin groups), 3, 481
McClurg, P., 86
McKay, J., 418
McLaughlin, J., 338, 362, 364
Meierfrankenfeld, U., xiii, 6, 9, 171, 311, 630
Meierfrankenfeld-Stellmacher Theorem (rank-2 pushing up), 11, 16, 121, 135
minimal parabolic (abstract), 5, 112
minimal representation dimensions, 348, 356
Mitchell, H., 55
Modular Law, 19
morphism (of amalgams), 260
Moufang condition (for generalized m-gon), 283
Moufang generalized polygon, 14
MS-group (rank 2 pushing up), 135
MS-pair, 135
multipliers of quasithin $\mathcal{K}$-groups, 407
multipliers of SQTK-groups, 407

natural module for S_n, 83
natural module for dihedral group, 424
Niles, R., 10, 126

O'Nan, M., 520
obstructions (to pushing up), 122
octad (of Steiner system), 396
Odd Order Theorem (Feit-Thompson), 15
odd transpositions, 339
offender, 68
    $(F - 1)$, 68
    w (weak closure), 236
    FF, 76

strong, 76
FF*, 68
    strong, 68
    for q or $\hat{q} \leq 2$, 12
opposite (at maximal distance in building), 446

parabolic isomorphic (amalgams), 262
parabolic, minimal (abstract), 112
parabolics in Lie-type groups, structure of, 38
Parker, R., 346
Parrott, D., 418, 431
Parshall, B., 16, 91, 408
partition (of a vector space), 1072
path (in a coset geometry), 270
path without backtracks, 270
pointwise stabilizer, 311
polar space (for $Sp_4(2)$), 420
product-disconnected, 429
projective module, 408
PU (pushing up hypothesis), 122
pushing up, 5, 122
    rank 2 (Meierfrankenfeld-Stellmacher), 135

qrc-lemma, 177
QT (quasithin hypothesis), 4, 482
QTKE-group, 4, 482
quadratic (action on a module), 67
quasiequivalence (of completions), 266
quasiequivalence of modules (conjugacy in $Out(G)$), 111
quasithin, 4, 32
quasithin, strongly, 32
quintet (of Steiner system), 396

radical ($J(M)$, of a module M), 343
radical (2-radical) subgroup, 11, 121
rank 2 amalgam, 260
rank-2 pushing up (Meierfrankenfeld-Stellmacher), 135
recognition theorems, 6
reduced (2-reduced), 77
regular (transitive) permutation action, 22
representation (of a group or amalgam, in a category), 265
residually connected (geometry), 419
residue (of a vertex), 419
Ronan, M., 818
root group (in Lie type group), 88
Rudvalis rank 3 group, 445
Rudvalis, A., 431

Schreier property, 46
Schur multiplier, 407
Schur multipliers of quasithin $\mathcal{K}$-groups, 407
Schur multipliers of SQTK-groups, 407
Schur's Lemma, 31
Scott, L., 17, 197

Segev, Y., 723
Seitz, G., 16, 282
Serre, J.-P., 273
sextet (of Steiner system), 396
shadow, 484
  A_5 wr $\mathbf{Z}_2$, 1138
  A_{10}, 969, 1120, 1135
  $Aut(G_2(3))$, 1008
  $Aut(HS)$, 1005
  $Aut(He)$, 1120, 1135, 1148, 1151
  $Aut(L)$ for L a Bender group, 521
  $Aut(L_2(p^2))$, 1184
  $Aut(L_4(2))$, 1120, 1148
  $Aut(L_4(2^n))$, $n > 1$, 731
  $Aut(L_4(3))$, 922
  $Aut(L_5(2))$, 1120, 1148, 1151
  $Aut(L_5(2^n))$, $n > 1$, 731
  $Aut(L_6(2))$, 724, 725
  $Aut(L_7(2))$, 724, 725
  $Aut(U_4(2))$, 922
  $Aut(U_4(3))$, 921, 995
  Co_1, 719
  Co_2, 721, 723
  Co_3, 857
  F_3 (Thompson group), 857
  F_5 (Harada-Norton group), 495, 1009
  F'_{24}, 822
  G with V not an FF-module, 696
  L wr $\mathbf{Z}_2$ for L of rank 2 over $\mathbf{F}_2$, 1120, 1135
  $L_2(2^n)$ wr $\mathbf{Z}_2$, 521
  $L_2(p)$ wr $\mathbf{Z}_2$, 560
  $L_3(2)$ wr $\mathbf{Z}_2$, 1138
  $L_4(2^n)$, $n > 1$, 489, 759, 760, 770, 773–775
  McL, 807, 808
  $O_{10}^+(2)$, 822
  $O_{12}^+(2)$, 817
  $P\Omega_8^+(3)$, 822
  $PSL_2(p)$ wr $\mathbf{Z}_2$, 566
  S_5 wr $\mathbf{Z}_2$, 1120
  S_7, 1184
  S_9, 1120, 1138
  S_{10}, 1189, 1192
  $Sp_6(2^n)$, $n > 1$, 759, 760, 770, 775, 778
  $Sp_8(2)$, 822
  $Sp_{10}(2)$, 817
  $Sz(2^n)$ wr $\mathbf{Z}_2$, 521
  $U_6(2)$, 721, 723
  $U_6(2^n)$, $n > 1$, 716
  $U_7(2)$, 720
  $U_7(2^n)$, $n > 1$, 716
  $\Omega_7(3)$, 809
  $\Omega_8^-(2^n)$, $n > 1$, 759, 760
  $\Omega_7(3)$, 807
  $\Omega_8^+(2)$, 822, 823, 831
  $\Omega_8^-(2)$, 1010

  $\Omega_{10}^+(2)$, 817
  $\Omega_{12}^-(2)$, 817
  $\mathbf{Z}_2/L_3(2^n)$, 529, 532
  $\mathbf{Z}_2/Sp_4(2^n)$, 529, 532
  $\mathbf{Z}_2/\Omega_8^+(2^n)$, $n > 1$, 759, 760, 775, 778
  $\mathbf{Z}_3/\Omega_8^+(2)$, 586, 595, 599
  Conway groups, 494
  extensions of $L_4(3)$, 544, 547, 550, 558, 560, 565
  Fischer groups, 494, 711–713
  rank 2 groups, certain, 517
shadow (configuration "close" to Main Theorem), 484
short (group), 123
Shpectorov, S., 279
Shult's Fusion Theorem, 430, 518
Shult, E., 429, 518
Sims, C., 6, 10, 486
Sin, P., xiv
small (faithful completion of an amalgam), 278
small dimensional representations, 348, 356
Smith, F., 486
Smith, S., 13, 818
Solomon, R., xiii, 4, 416, 482
Solvable Thompson Factorization, 70, 79
split BN-pair of rank 2, 283
split BN-pairs of rank 2, 16
SQT, 32
SQTK, 33
SQTK-group, 7, 33
stable (subset of offenders $\mathcal{P}(G, V)$), 70
standard form, 1177
Standard Notation (for G, T), 499
standard subgroup of G, 1177
Steinberg module, 453
Steinberg relations (for Lie-type group), 356
Steiner system (for Mathieu groups), 396
Stellmacher, B., 6, 171, 311, 487
Stellmacher-Meierfrankenfeld qrc-lemma, 177
Stelmmacher, B., 9
strong FF*-offender, 68
strong FF-module, 68
strong FF-offender, 76
strongly closed, 518
strongly embedded (subgroup), 428
strongly quasithin, 4, 32
Stroth, G., 87, 311
structure of Lie-type groups, 38
structure of parabolics in Lie-type groups, 38
stubborn (2-stubborn), 121
subgroup amalgam, 14, 260
supercritical subgroup, 24
Supercritical Subgroups Lemma, 24
Suzuki 2-group, 532
Suzuki type a_i, 218
Suzuki type b_i, 218

Suzuki type c_i, 218
Suzuki type (of suitable involutions), 218
Suzuki, M., 164, 291, 296, 298, 415, 417, 427, 429, 569, 1072, 1180, 1184
symmetry (between γ_1 and γ), 317, 322, 324
symplectic type (p-group), 1016

Tanaka, Y., 487
tetrad (of Steiner system), 396
Theorem A, 33
Theorem B, 33
Theorem C, 33
Theorem D, 1078
Theorem E, 1165
thin (group), 4
Thompson A×B Lemma, 24
Thompson amalgam strategy, 487
Thompson factorization, 5, 78
Thompson Factorization for Solvable Groups, 79
Thompson Factorization Lemma, 78
Thompson Order Formula, 64
Thompson Replacement Lemma, 68
Thompson strategy, 487
Thompson subgroup, 74
    higher ($J_j(H)$), 74
    usual ($J(H)$), 74
Thompson subgroup ($J(X)$), 8
Thompson Transfer (Lemma), 30
Thompson's Dihedral Lemma, 20
Thompson, J., xiv, 231, 428
Three-Subgroup Lemma, 20
TI-set, 21
tightly embedded (subgroup), 425
Timmesfeld, F., 13, 364, 425, 494, 852
Tits amalgam, 281
Tits building, 283
Tits group $^2F_4(2)'$, 495
Tits sytem, 283
Tits, J., 16, 273, 274, 419
Tits-Weiss Theorem (Moufang buildings), 14, 16, 275, 282, 629
Todd module (for M_{22}), 395
Todd module (for M_{24}, M_{23}), 395
Todd, J., 396, 712
transvection, 23
triangulable (uniqueness system), 658
trio (of Steiner system), 396
Tutte, W., 6, 486
Tutte-Sims graph methods, 6, 304, 311, 486, 487
type $\mathcal{H}(2, \Omega_4^-(2))$, 418
type a_i of involution, 218
type b_i of involution, 218
type c_i of involution, 218
type $G_2(3)$, 417
type HS, 418
type J_1, 418

type J_2, 418
type J_3, 418
type j_k (of involution in linear/unitary group), 1170
type Ru, 431
type $U_3(3)$, 417
type (of a block), 124

uniqueness subgroup, 499
uniqueness system, 657
universal completion, 14
universal completion (of an amalgam), 261, 265
universal covering (of a group), 407
universal covering (of a module), 408
universal covering group, 407
universal dual covering (of a module), 408

w-offender (weak closure), 236
Wales, D., 418, 431
Wall, G. E., 415
Walter, J., 415
weak BN-pair of rank 2, 261
weak closure $W(X, \Omega)$ of Ω in X, 210
weak closure methods, 6
weak closure methods, basics of, 232
weakly closed, 22
weight λ_i, fundamental (of Lie type group), 386
weight (of vector in permutation module), 83, 395
weight theory for Lie-type representations, 386
Weiss, R., 16, 282
Weyl group (of an amalgam), 274
Wilson, R., 346
Wong, S. K., 416, 418
Wong, W., 416, 417

Yoshiara, S., 38

Zassenhaus groups, 415
Zsigmondy prime divisor, 22
Zsigmondy's Theorem, 22
Zsigmondy, K., 16, 22, 640, 731

Titles in This Series

112 **Michael Aschbacher and Stephen D. Smith,** The classification of quasithin groups II. Main theorems: The classification of simple QTKE-groups, 2004

111 **Michael Aschbacher and Stephen D. Smith,** The classification of quasithin groups I. Structure of strongly quasithin K-groups, 2004

110 **Bennett Chow and Dan Knopf,** The Ricci flow: An introduction, 2004

109 **Goro Shimura,** Arithmetic and analytic theories of quadratic forms and Clifford groups, 2004

108 **Michael Farber,** Topology of closed one-forms, 2004

107 **Jens Carsten Jantzen,** Representations of algebraic groups, 2003

106 **Hiroyuki Yoshida,** Absolute CM-periods, 2003

105 **Charalambos D. Aliprantis and Owen Burkinshaw,** Locally solid Riesz spaces with applications to economics, second edition, 2003

104 **Graham Everest, Alf van der Poorten, Igor Shparlinski, and Thomas Ward,** Recurrence sequences, 2003

103 **Octav Cornea, Gregory Lupton, John Oprea, and Daniel Tanré,** Lusternik-Schnirelmann category, 2003

102 **Linda Rass and John Radcliffe,** Spatial deterministic epidemics, 2003

101 **Eli Glasner,** Ergodic theory via joinings, 2003

100 **Peter Duren and Alexander Schuster,** Bergman spaces, 2004

99 **Philip S. Hirschhorn,** Model categories and their localizations, 2003

98 **Victor Guillemin, Viktor Ginzburg, and Yael Karshon,** Moment maps, cobordisms, and Hamiltonian group actions, 2002

97 **V. A. Vassiliev,** Applied Picard-Lefschetz theory, 2002

96 **Martin Markl, Steve Shnider, and Jim Stasheff,** Operads in algebra, topology and physics, 2002

95 **Seiichi Kamada,** Braid and knot theory in dimension four, 2002

94 **Mara D. Neusel and Larry Smith,** Invariant theory of finite groups, 2002

93 **Nikolai K. Nikolski,** Operators, functions, and systems: An easy reading. Volume 2: Model operators and systems, 2002

92 **Nikolai K. Nikolski,** Operators, functions, and systems: An easy reading. Volume 1: Hardy, Hankel, and Toeplitz, 2002

91 **Richard Montgomery,** A tour of subriemannian geometries, their geodesics and applications, 2002

90 **Christian Gérard and Izabella Łaba,** Multiparticle quantum scattering in constant magnetic fields, 2002

89 **Michel Ledoux,** The concentration of measure phenomenon, 2001

88 **Edward Frenkel and David Ben-Zvi,** Vertex algebras and algebraic curves, second edition, 2004

87 **Bruno Poizat,** Stable groups, 2001

86 **Stanley N. Burris,** Number theoretic density and logical limit laws, 2001

85 **V. A. Kozlov, V. G. Maz'ya, and J. Rossmann,** Spectral problems associated with corner singularities of solutions to elliptic equations, 2001

84 **László Fuchs and Luigi Salce,** Modules over non-Noetherian domains, 2001

83 **Sigurdur Helgason,** Groups and geometric analysis: Integral geometry, invariant differential operators, and spherical functions, 2000

82 **Goro Shimura,** Arithmeticity in the theory of automorphic forms, 2000

81 **Michael E. Taylor,** Tools for PDE: Pseudodifferential operators, paradifferential operators, and layer potentials, 2000

TITLES IN THIS SERIES

80 **Lindsay N. Childs,** Taming wild extensions: Hopf algebras and local Galois module theory, 2000

79 **Joseph A. Cima and William T. Ross,** The backward shift on the Hardy space, 2000

78 **Boris A. Kupershmidt,** KP or mKP: Noncommutative mathematics of Lagrangian, Hamiltonian, and integrable systems, 2000

77 **Fumio Hiai and Dénes Petz,** The semicircle law, free random variables and entropy, 2000

76 **Frederick P. Gardiner and Nikola Lakic,** Quasiconformal Teichmüller theory, 2000

75 **Greg Hjorth,** Classification and orbit equivalence relations, 2000

74 **Daniel W. Stroock,** An introduction to the analysis of paths on a Riemannian manifold, 2000

73 **John Locker,** Spectral theory of non-self-adjoint two-point differential operators, 2000

72 **Gerald Teschl,** Jacobi operators and completely integrable nonlinear lattices, 1999

71 **Lajos Pukánszky,** Characters of connected Lie groups, 1999

70 **Carmen Chicone and Yuri Latushkin,** Evolution semigroups in dynamical systems and differential equations, 1999

69 **C. T. C. Wall (A. A. Ranicki, Editor),** Surgery on compact manifolds, second edition, 1999

68 **David A. Cox and Sheldon Katz,** Mirror symmetry and algebraic geometry, 1999

67 **A. Borel and N. Wallach,** Continuous cohomology, discrete subgroups, and representations of reductive groups, second edition, 2000

66 **Yu. Ilyashenko and Weigu Li,** Nonlocal bifurcations, 1999

65 **Carl Faith,** Rings and things and a fine array of twentieth century associative algebra, 1999

64 **Rene A. Carmona and Boris Rozovskii, Editors,** Stochastic partial differential equations: Six perspectives, 1999

63 **Mark Hovey,** Model categories, 1999

62 **Vladimir I. Bogachev,** Gaussian measures, 1998

61 **W. Norrie Everitt and Lawrence Markus,** Boundary value problems and symplectic algebra for ordinary differential and quasi-differential operators, 1999

60 **Iain Raeburn and Dana P. Williams,** Morita equivalence and continuous-trace C^*-algebras, 1998

59 **Paul Howard and Jean E. Rubin,** Consequences of the axiom of choice, 1998

58 **Pavel I. Etingof, Igor B. Frenkel, and Alexander A. Kirillov, Jr.,** Lectures on representation theory and Knizhnik-Zamolodchikov equations, 1998

57 **Marc Levine,** Mixed motives, 1998

56 **Leonid I. Korogodski and Yan S. Soibelman,** Algebras of functions on quantum groups: Part I, 1998

55 **J. Scott Carter and Masahico Saito,** Knotted surfaces and their diagrams, 1998

54 **Casper Goffman, Togo Nishiura, and Daniel Waterman,** Homeomorphisms in analysis, 1997

53 **Andreas Kriegl and Peter W. Michor,** The convenient setting of global analysis, 1997

52 **V. A. Kozlov, V. G. Maz'ya, and J. Rossmann,** Elliptic boundary value problems in domains with point singularities, 1997

For a complete list of titles in this series, visit the
AMS Bookstore at **www.ams.org/bookstore/.**

DATE DUE

GAYLORD   #3522PI   Printed in USA